S0-ASF-468

POLYMERIC BIOMATERIALS

POLYMERIC BIOMATERIALS

Second Edition, Revised and Expanded

Severian Dumitriu
University of Sherbrooke
Sherbrooke, Quebec, Canada

MARCEL DEKKER

MARCEL DEKKER, INC. NEW YORK • BASEL

ISBN: 0-8247-0569-6

This book is printed on acid-free paper.

Headquarters
Marcel Dekker, Inc.
270 Madison Avenue, New York, NY 10016
tel: 212-696-9000; fax: 212-685-4540

Eastern Hemisphere Distribution
Marcel Dekker AG
Hutgasse 4, Postfach 812, CH-4001 Basel, Switzerland
tel: 41-61-261-8482; fax: 41-61-261-8896

World Wide Web
http://www.dekker.com

The publisher offers discounts on this book when ordered in bulk quantities. For more information, write to Special Sales/Professional Marketing at the headquarters address above.

Current printing (last digit):
10 9 8 7 6 5 4 3 2 1

PRINTED IN THE UNITED STATES OF AMERICA

Dedicated with affection

In memory of my parents

To my wife, Maria,
and my children, Daniela and Cristina,
who make everything worthwhile

Preface

The need for spare parts for the human body has never been greater than at present. More people than ever are living longer, but the human body, shaped by millions of years of evolution, isn't keeping up. Spare parts must be available to replace organs and tissues that are worn out from operating beyond their expiration dates, as well as those damaged by injury, disease, or developmental mishaps.

The biomaterials field has come a long way from its empirical beginnings, with researchers taking whatever materials were available and attempting to integrate them into the human body, sometimes with disastrous results. Now, the body's response to foreign materials is better understood than ever before. Furthermore, the past 15 years have seen great strides in tissue engineering. Spare parts consisting of living tissues are poised for significant clinical application. Tissue engineering, especially of tissues derived from the patient's own cells, offers total acceptance by and integration with the patient's body, unlike nonliving materials or living tissues from other humans or species.

The active ingredient in a medicine is only part of the arsenal against disease. The drug must somehow get to the right place at the right time. That's where drug delivery comes in. Drug delivery companies work to devise new dosage forms for medications. The main challenge is to create the technologies for the easier and most convenient systematic delivery systems. For proteins and other macromolecules, however, the oral route is by far the hardest to accomplish.

From a practical perspective, medical applications of polymers fall into three broad categories: (1) extracorporeal uses (catheters, tubing, and fluid lines; dialysis membranes/artificial kidney; ocular devices; wound dressings and artificial skin), (2) permanently implanted devices (sensory devices; cardiovascular devices; orthopedic devices; dental devices), and (3) temporary implants (degradable sutures; implantable drug delivery systems; polymeric scaffolds for cell or tissue transplants; temporary vascular grafts and arterial stents; temporary small bone fixation devices).

Today, biomaterial research has developed into a major interdisciplinary effort involving chemists, biologists, engineers, and physicians. Biomaterials research has provided the clinician with a large number of new materials and new medical devices. As the biomaterials device industry continues to grow, degradable polymers will increase at the expense of traditional biomaterials such as metals and conventional, biostable polymers.

This volume consists of two parts: Part I: Polymers as Biomaterials and Part II: Medical and Pharmaceutical Applications of Polymers. The fundamental questions of polymer synthesis, the types of polymers used for medical purposes, and modification of polymers to increase their biocompatibility, are presented in the first part. The applications of the two major groups of biomaterials—natural biomaterials (polysaccharides, cellulose, chitosan, proteins, etc.) and synthetic biomaterials (polyesters, silicones, elastomers, etc.)—are also reviewed. Part II deals with concrete utilization of polymeric biomaterials in the domains of tissue engineering, ophthalmic delivery, vascular prostheses (grafts), dental and maxillofacial surgery, blood contacting, skin graft polymers, sensors in biomedical applications, medical adhesives, medical textiles, and topical hemostat biomaterials.

The uses of polymers in the pharmaceutical domain fall into two areas: drug polymers and drug carrier polymers for controlled release. Part II provides the groundwork for understanding the fundamentals of drug delivery including conventional, nonconventional, and modulated systems; structure–property relations of selected supports and their role in drug delivery; delivery of drugs to sites such as the gastrointestinal tract, lung, skin, tumors, and blood vessels; and marketing considerations in new drug delivery systems.

This book is truly international, with authors from Austria, Canada, Finland, France, Germany, India, Israel, Italy, Japan, the Netherlands, Portugal, Slovenia, Spain, Switzerland, Thailand, the United Kingdom, and the United States. I am grateful to all the contributors.

Severian Dumitriu

Contents

Contributors

Christine Allen McGill University, Montreal, Quebec, Canada

Mansoor Amiji Northeastern University, Boston, Massachusetts

Maureen Andrew Hamilton Civic Hospitals Research Centre, Hamilton, Ontario, Canada

P. Auroy Université de Rennes, Rennes, France

T. Avramoglou Université de Paris, Villetaneuse, France

V. Baeyens Centre Interuniversitaire de Recherche et d'Enseignement, Archamps, France

Marcel B. Bally BC Cancer Agency, Vancouver, British Columbia, Canada, and University of British Columbia, Vancouver, British Columbia, Canada

P. G. Baraldi Ferrara University, Ferrara, Italy

G. Barratt UMR CNRS, Chatenay-Malabry, France

A. Bascones Universidad Complutense, Madrid, Spain

Alfonso Bentolila The Hebrew University of Jerusalem, Jerusalem, Israel

Andreas Bernkop-Schnürch University of Vienna, Vienna, Austria

Leslie R. Berry Hamilton Civic Hospitals Research Centre, Hamilton, Ontario, Canada

G. S. Bhuvaneshwar Sree Chitra Tirunal Institute of Medical Sciences and Technology, Thiruvananthapuram, India

Martin J. Bide University of Rhode Island, Kingston, Rhode Island

M. Bonnaure-Mallet Université de Rennes, Rennes, France

Larry R. Brown Harvard–Massachusetts Institute of Technology, Cambridge, Massachusetts

Riccardo Calafiore University of Perugia School of Medicine, Perugia, Italy

Boris Čeh University of Ljubljana, Ljubljana, Slovenia

Anthony K. C. Chan Hamilton Civic Hospitals Research Centre, Hamilton, Ontario, Canada

K. B. Chandran University of Iowa, Iowa City, Iowa

D. J. Chauvel-Lebret Université de Rennes, Rennes, France

C. C. Chu Cornell University, Ithaca, New York

Bogdan Comanita Institute for Chemical Process and Environmental Technology, Ottawa, Ontario, Canada

P. Corvi Mora EUPHAR Group, Piacenza, Italy

G. Couarraze UMR CNRS, Chatenay-Malabry, France

P. Couvreur UMR CNRS, Chatenay-Malabry, France

Kay C Dee Tulane University, New Orleans, Louisiana

Ferencz Denes University of Wisconsin, Madison, Wisconsin

D. de Zeeuw University of Groningen, Groningen, The Netherlands

Alain Domard University Claude Bernard Lyon I, Villeurbanne, France

Monique Domard University Claude Bernard Lyon I, Villeurbanne, France

Abraham J. Domb The Hebrew University of Jerusalem, Jerusalem, Israel

C. Dubernet UMR CNRS, Chatenay-Malabry, France

Severian Dumitriu University of Sherbrooke, Sherbrooke, Quebec, Canada

Nejät Düzgüneş University of the Pacific, San Francisco, California

S. Einmahl University of Geneva, Geneva, Switzerland

Adi Eisenberg McGill University, Montreal, Quebec, Canada

A. El-Kattan Pfizer Inc., Ann Arbor, Michigan

Haissam S. El-Zaim University of Texas Medical Branch, Galveston, Texas

E. Fattal UMR CNRS, Chatenay-Malabry, France

O. Felt University of Geneva, Geneva, Switzerland

Hiroshi Fujita Kyoto University, Kyoto, Japan

P. Furrer University of Lausanne, Lausanne, Switzerland

J. G. Gavilanes Universidad Complutense, Madrid, Spain

Danielle C. Giliberti Tulane University, New Orleans, Louisiana

R. Gref UMR CNRS, Chatenay-Malabry, France

R. Gurny University of Geneva, Geneva, Switzerland

M. Haas University of Groningen, Groningen, The Netherlands

John P. Heggers University of Texas Medical Branch, Galveston, Texas

R. F. G. Haverdings University of Groningen, Groningen, The Netherlands

Radi Hejazi Northeastern University, Boston, Massachusetts

Leaf Huang University of Pittsburgh, Pittsburgh, Pennsylvania

Yoshito Ikada Suzuka University of Medical Science, Mie, Japan

David S. Jones The Queen's University of Belfast, Belfast, Northern Ireland

J. Jozefonvicz Université de Paris, Villetaneuse, France

M. Jozefowicz Université de Paris, Villetaneuse, France

Weiyuan John Kao University of Wisconsin, Madison, Wisconsin

Jung Ju Kim Purdue University, West Lafayette, Indiana

Akio Kishida National Cardiovascular Center Research Institute, Osaka, Japan

Richard J. Knox Gnact Pharma Plc, Salisbury, England

R. J. Kok University of Groningen, Groningen, The Netherlands

F. Kratz Tumor Biology Center, Freiburg, Germany

Rajesh Krishna* BC Cancer Agency, Vancouver, British Columbia, Canada, and University of British Columbia, Vancouver, British Columbia, Canada

Neeraj Kumar The Hebrew University of Jerusalem, Jerusalem, Israel

D. Labarre UMR CNRS, Chatenay-Malabry, France

Danilo D. Lasic Liposome Consultations, Newark, California

Jin Ho Lee Purdue University, West Lafayette, Indiana

P. Legrand UMR CNRS, Chatenay-Malabry, France

Kenneth W. Liang University of Pittsburgh, Pittsburgh, Pennsylvania

M. A. Lizarbe Universidad Complutense, Madrid, Spain

Frank W. LoGerfo Beth Israel Deaconess Medical Center, Boston, Massachusetts

R. Karl Malcolm The Queen's University of Belfast, Belfast, Northern Ireland

Sorin Manolache University of Wisconsin, Madison, Wisconsin

Howard W. T. Matthew Wayne State University, Detroit, Michigan

Lawrence D. Mayer BC Cancer Agency, Vancouver, British Columbia, Canada, and University of British Columbia, Vancouver, British Columbia, Canada

Dusica Maysinger McGill University, Montreal, Quebec, Canada

Paul A. McCarron The Queen's University of Belfast, Belfast, Northern Ireland

D. K. F. Meijer University of Groningen, Groningen, The Netherlands

Roger G. Melton Gnact Pharma Plc, Salisbury, England

D. D. Michniak University of Medicine and Dentistry of New Jersey, Newark, New Jersey

Takashi Miyata Kansai University, Osaka, Japan

Rashid Mohammad Shriners Burns Hospital, Massachusetts General Hospital, Boston, Massachusetts, and Harvard Medical School, Cambridge, Massachusetts

Jeffrey R. Morgan Shriners Burns Hospital, Massachusetts General Hospital, Boston, Massachusetts, and Harvard Medical School, Cambridge, Massachusetts

F. Moolenaar University of Groningen, Groningen, The Netherlands

Kazuya Nagata Industrial Technology Center of Okayama Prefecture, Okayama, Japan

N. Olmo Universidad Complutense, Madrid, Spain

Edwin C. Ouyang University of Connecticut Health Center, Farmington, Connecticut

Kinam Park Purdue University, West Lafayette, Indiana

Maria C. Pedroso de Lima University of Coimbra, Coimbra, Portugal

Matthew D. Phaneuf Beth Israel Deaconess Medical Center, Boston, Massachusetts

Pedro Pires University of Coimbra, Coimbra, Portugal

G. Ponchel UMR CNRS, Chatenay-Malabry, France

William C. Quist Beth Israel Deaconess Medical Center, Boston, Massachusetts

A. V. Ramani T.T.K. Pharma, Ltd., Bangalore, India

* *Current affiliation*: Bristol-Meyers Squibb, Princeton, New Jersey

K. Riebeseel Tumor Biology Center, Freiburg, Germany

Joseph R. Robinson University of Wisconsin, Madison, Wisconsin

P. C. A. Rodrigues Tumor Biology Center, Freiburg, Germany

Pentti U. Rokkanen Helsinki University Central Hospital and University of Helsinki, Helsinki, Finland

Ronit Satchi Center for Polymer Therapeutics, London, England

Radoslav Savic McGill University, Montreal, Quebec, Canada

Leonard W. Seymour University of Birmingham, Birmingham, England

Robert L. Sheridan Shriners Burns Hospital, Massachusetts General Hospital, Boston, Massachusetts, and Harvard Medical School, Cambridge, Massachusetts

Tzviel Sheskin The Hebrew University of Jerusalem, Jerusalem, Israel

Sérgio Simões University of Coimbra, Coimbra, Portugal

Joram Slager The Hebrew University of Jerusalem, Jerusalem, Israel

Joseph Tam McGill University, Montreal, Quebec, Canada

Doron Teomim The Hebrew University of Jerusalem, Jerusalem, Israel

Naohide Tomita Kyoto University, Kyoto, Japan

J. Turnay Universidad Complutense, Madrid, Spain

Tadashi Uragami Kansai University, Osaka, Japan

C. Vauthier UMR CNRS, Chatenay-Malabry, France

J. M. Vega Universidad Complutense, Madrid, Spain

A. Warnecke Tumor Biology Center, Freiburg, Germany

Iain Webster Smith & Nephew Group Research Centre, York, England

Peter J. West Smith & Nephew Group Research Centre, York, England

Kyle K. White Tulane University, New Orleans, Louisiana

Nimit Worakul University of Wisconsin, Madison, Wisconsin, and Prince of Songkla University, Haad Yai, Thailand

A. David Woolfson The Queen's University of Belfast, Belfast, Northern Ireland

Catherine H. Wu University of Connecticut Health Center, Farmington, Connecticut

George Y. Wu University of Connecticut Health Center, Farmington, Connecticut

Yuan-Peng Zhang ALZA Corp., Mountain View, California

1

Polysaccharides as Biomaterials

Severian Dumitriu
University of Sherbrooke, Sherbrooke, Quebec, Canada

I. INTRODUCTION

Polymers have found applications in virtually every discipline of medicine, ranging from simple extracorporeal devices to intricately designed implants. Since each medical application has its own highly specialized requirements, a range of diverse materials with good biocompatibility but different chemical and physicomechanical properties must be available.

Polysaccharides constitute an important component of life matter. Polysaccharides display a perfect biocompatibility and biodegradability, which are the basic characteristics for polymers used as biomaterials. They have several characteristics not found in other natural polymers. Recently, specific properties of antivirals, antitumorals, gene modulators, etc., have been discovered for various classes of polysaccharides.

This chapter presents all polysaccharide classes with applications as biomaterials and as pharmaceutical systems. For the first group—polysaccharides used as biomaterials—a special emphasis is given to the presentation of membranes, implants, skin replacements, matrices for cellular engineering (scaffolds), and hemostatic agents. The second group—polysaccharides used as pharmaceutical systems—describes mainly the use of polysaccharides as components for drug conditioning and drug delivery systems fabrication. This chapter presents a review of the literature based on the three books previously published: *Polymeric Biomaterials*, *Polysaccharides in Medicinal Applications*, and *Polysaccharides: Structural Diversity and Functional Versatility*, all edited by S. Dumitriu and published by Marcel Dekker, Inc.

II. BACTERIAL POLYSACCHARIDES

Encapsulated bacterial diseases are still the most prevalent and serious infections in humans. There is increasing research of the chemical basis of immunogenicity to capsular polysaccharides and prevention of bacterial infections through immunization.

The capsular polysaccharides of bacteria are attractive vaccine candidates because they constitute the most highly conserved and most exposed bacterial surface antigens. They can be readily isolated and purified; they are nontoxic; and they are immunogenic in humans (with few exceptions) (1). Another important feature is that capsular polysaccharides can be chemically and physically defined, which is necessary for control over their efficacy as human biologicals. In fact, the use of capsular polysaccharides as immunoprophylactic agents against human disease caused by encapsulated bacteria is now firmly established (2,3). Efficacious vaccines composed of the capsular polysaccharides of *Nisseria meningitidis*, *Streptococcus pneumoniae*, *Haemophilus influenzae*, and *Salmonella typhi* have been developed as commercial products (Table 1).

A new generation of highly successful semisynthetic glycoconjugate vaccines in which the polysaccharides were

Table 1 Bacterial Polysaccharides

Pathogenic organism	Reference
1. *Neisseria meningitidis*	
Group A	4
Group B	5
Group C	5
2. *Streptococcus pneumoniae*	
Type 19F	6
Type 14	7
3. *Hemophilus influenzae*	
Type b	8

conjugated (covalently linked) to protein carriers were developed (1). During the last 15 years, comprehensive studies specially directed to the development of glycoconjugate vaccines as potential human vaccines have been reported (Table 2) (9–11).

Complex carbohydrates are essential cell surface components. Generally, the surface carbohydrates of bacteria are polysaccharides, either capsular polysaccharides or a part of cell wall lipopolysaccharides (LPS) (22–24). Bacterial cells are protected by capsules in inadvertent surroundings (25–29).

Capsular polysaccharides (CPS) are generally negatively charged (30). All structures of bacterial polysaccharides published until early 1982 have been recorded in a review (31).

Generally, bacterial polysaccharides have repetitive units that may consist of monosaccharides and octasaccharides and they may be linear or branched.

Cell wall lipopolysaccharides of gram-negative bacteria consist of a lipid moiety, a core oligosaccharide, and a polysaccharide moiety which is made up from repeating units (32). Mammalian glycoproteins contain oligosaccharides with different complexities, which are bound to hydroxyamino acid units or asparagine units of proteins. In proteoglycans, polysaccharides are bound to polypeptides.

Table 2 Synthetic Glycoconjugate Vaccines

Polysaccharide source	Protein carrier	Reference
Haemophilus influenzae	Diphtheria toxoid	12,13
	Tetanus toxoid	14
Streptococcus pneumonia	Tetanus toxoid	15
	Cholera toxin	16,17
	Pertussis toxin	16
Neisseria meningitidis	Tetanus toxoid	18
	Bovine serum albumin	19
Salmonella typhi	Bovine serum albumin	20
	Cholera toxin	20
	Tetanus toxoid	21

The crystal structure of the complex of an anti–Id Fab with a Fab specific for a *Brucella* polysaccharide antigen has previously been reported (33). To complement this study, the binding characteristics and immunological properties of this Ab2 and two others raised with a second anti-*Brucella* antibody were investigated, including quantitative kinetic measurements by surface plasmon resonance (34). Protective immunity to encapsulated bacteria is related to antibody response to polysaccharide (PS) antigen, interactions with T and B cells, and host defense mechanisms (35).

Klebsiella pneumoniae is one of the most frequently isolated gram-negative pathogens in severe nosocomial infections. Alvarez et al. (36) have demonstrated the existence of *K. pneumoniae* clinical isolates deficient in the lipopolysaccharide O side chain, the major factor for resistance to complement-mediated killing in this bacterial species. These isolates are complement resistant, and their mechanisms to resist complement were investigated by selecting transposon-generated complement-sensitive mutants. One mutant with a drastically reduced capacity to grow in nonimmune human serum carried the transposon inserted in an open reading frame of a gene cluster involved in capsule synthesis. Four additional clinical isolates representing four different K serotypes were studied, and results showed that capsular polysaccharide is a major complement resistance factor in these O side chain–deficient isolates.

Barnett et al. (37) compared responses to pneumococcal conjugate and polysaccharide vaccines in 48 otitis-free and 64 otitis-prone children. Pre- and postimmunization concentrations of antibodies to pneumococcal serotypes 6B, 14, 19F, and 23F were measured by enzyme-linked immunosorbent assay. Postimmunization mean concentrations of antibodies to all four serotypes were significantly higher for children receiving conjugate vaccine than for those receiving polysaccharide vaccine; the difference in responses was primarily due to a better response to conjugate vaccine in the otitis-prone group. Significantly higher postimmunization concentrations of antibodies to all four serotypes and to one of the four serotypes were found in otitis-prone children and otitis-free children who received conjugate vaccine, respectively. Pneumococcal conjugate vaccine has the potential to reduce the incidence of disease due to vaccine serotypes, even among children with recurrent otitis media.

Protein antigens induce significant mucosal immunoglobulin A (IgA) and systemic IgG responses when administered intranasally (i.n.) with the glyceride-polysorbate–based adjuvant RhinoVax (RV) both in experimental animals and humans (38). The pneumococcal polysaccharide conjugates (PNCs) induced significant systemic IgG and IgA antibodies after i.n. immunization only when given with RV and, for serotype 1, serum IgG titers were comparable to titers induced by subcutaneous immunization. In addition, i.n. immunization with PNC-1 in RV elicited detectable mucosal IgA. These results demonstrate that RV is a potent mucosal adjuvant.

Zhong et al. (39) describe the generation, molecular structure, and protective efficacy of a human monoclonal antibody (MAb) reactive with the capsular polysaccharide of serotype 8 *Streptococcus pneumoniae*. In vitro studies revealed MAb D11–dependent complement deposition on the capsule of serotype 8 organisms via either the classical or the alternative complement pathway. In vivo, MAb D11 prolonged the survival of both normal and C4-deficient mice with lethal serotype 8 *S. pneumoniae* infection.

III. CELLULOSE AND ITS DERIVATIVES

Polysaccharides currently used in medicine on a large scale include cellulose, cellulose derivatives, heparin, dextrin, pullulan, hyaluronic acid, and alginate.

Cellulosics, regenerated cellulose and its methyl, ethyl, amino ethyl, and acetate-phthalate derivatives (40) have been widely investigated as membranes in artificial kidneys, encapsulating materials for controlled drug delivery (41), sutures and bandages (42), and blood-compatible materials (43), and are used mostly for blood purification, as anticoagulant, and as plasma expander in aqueous solution.

A. Blood Purification

1. Hemodialysis

Hemodialysis is a therapeutic treatment of chronic renal disease to remove an excess of water and uremic toxins from the patient blood by dialysis under the extracorporeal blood circulation. The most important part of the hemodialysis system is the semipermeable membrane installed in the dialyzer. The first semipermeable membrane was reported in 1944 using cellophane. Cellulosic membranes are still used for the hemodialyzer, largely synthetic semipermeable membranes made of polysulfone (PS), polyacrylonitrile (PAN), poly(methyl methacrylate) (PMMA), and ethylene-vinyl alcohol copolymer (EVAL) (44).

The cellulosic hemodialysis membrane can be divided into regenerated cellulose and cellulose acetate (di- and triacetate). In the early years of hemodialysis therapy, flat-, tube-, hollow fiber–type membranes were employed. At present almost 100% of hemodialyzers appear to be made from hollow fibers. The cellulose membrane for hemodialysis is regenerated from cuproammonium solution of refined cotton linters. Cellulose hollow fibers are regenerated from cellulose acetate by deacetylation (45). Current extensive efforts to improve hemodialysis membranes are focusing on removal of β-microglobulin and improving blood compatibility. The improvement of hemocompatibility of cellulose membranes is achievement by surface modification through uretanation, acylation/sulfation, chloroacetylation/sulfonation, and polyacylation (46–48).

Ishihara et al. (49) have synthesized a water-soluble graft polymer composed of a cellulose and polymer having a phospholipid polar group, poly(2-methacryloyloxyethyl) phosphoacrylcholine (MGC) as a coating material on the cellulose hemodialysis membrane. The MGC could be coated on the hollow fiber probe and the performance of the probe did not decrease even after the coating. By recording rapid changes in the glucose concentration from 100 to 200 mg/dL, the time to reach 90% of the maximum value was within 7 min. To determine the glucose concentration in subcutaneous tissue, the hollow fiber probe modified with MGC was applied to human volunteers.

Both cellulose diacetate and triacetate are used for hemodialysis. The blood compatibility of the cellulose acetate membrane seems to be excellent if monolayer coverage of the surface with serum albumin is employed (50).

Collodion, a cellulose trinitrate derivative, was the first polymer to be used as an artificial membrane, and it played a central role in further investigations and applications (51,52). Cellophane and Cuprophan (53–57) membranes replaced collodion later because of their better performance and mechanical stability. However, due to their alleged lack of hemocompatibility, membranes made from unmodified cellulose lost their market share. They have been replaced by modified cellulosic and synthetic dialysis membranes which show a better hemocompatibility than unmodified cellulose membranes. Most of the new membrane materials are also available in high-flux modifications and for this reason are suitable as well for more effective therapy modes, such as hemodiafiltration and hemofiltration. The success of hemodialysis as a routine therapy is also the success of membrane development, because both a reproducible membrane production and an unlimited availability of dialysis membranes have increased the number of dialyzed patients to about 1 million worldwide in 1999 (58).

Dialysis membranes made from regenerated cellulose are under dispute because of their alleged lack of hemocompatibility. The introduction of membranes from synthetically modified cellulose, like cellulose acetate or Hemophan, has proven, however, that hemocompatible membranes can be fabricated from cellulose by means of chemical surface modifications (59,60). Diamantoglou et al. (59) have synthesized a series of cellulose carbamate derivatives to profit from the excellent hemocompatibility pattern of the urethane family. In vitro investigations on membranes made from these cellulose modifications proved a direct relationship between the degree of modification and hemocompatibility. This was proven for the following three representative hemocompatibility parameters: complement C5a generation, thrombin–antithrombin (TAT) III formation, and platelet count (PC). As already shown for modifications made from cellulose esters, a direct dependency between improved hemocompatibility and the degree of substitution (DS) in the cellulose molecule could be found.

2. Plasmapheresis

Plasmapheresis is a therapeutic treatment of blood to separate only plasma from the whole blood of a patient through extracorporeal circulation when the plasma contains pathogenic substances which are mostly high molecular weight proteins. The separated plasma is discarded and substituted with fresh plasma. The plasma separation is based on ultrafiltration by hollow fiber membrane with pore size from 0.1 to 0.5 μm, prepared from cellulose acetate, polyvinyl alcohol, polyethylene, and polypropylene.

3. Virus Removal

Cellulose is a very good candidate as the filter material since the cellulose surface exhibits the least protein adsorption of the conventional polymer surfaces. The membrane is prepared by coagulation of the cuproammonium solution. It was proved that this filter with an average pore size of 35 nm could completely remove hepatitis C virus (61,62).

4. Hemostasis

Oxidized cellulose has been used as hemostat. This is an acidic polysaccharide produced by oxidation of cellulose with periodic acid or nitrogen oxide. A more effective method for homeostasis is to use oxidized cellulose gauze together with a liquid-type hemostat. Wiseman et al. (63) evaluated two adhesion barriers composed of oxidized regenerated cellulose (ORC) in a model of bowel surgery, with and without bleeding. They concluded that with hemostasis both absorbable fabrics of ORC reduced adhesion formation between the injured cecum and abdominal side wall. The effectiveness of INTERCEED Barrier, but not nTC7, was reduced but not eliminated in the presence of bleeding. This confirms similar observations in models of gynecologic surgery.

A particularly interesting application of oxidized cellulose is as a barrier for the prevention of surgical adhesions (64). The use of oxidized cellulose (Surgicel®) was largely unsuccessful in animal models in reducing surgical adhesions in a reproducible fashion.

B. Drug Delivery Systems

Hydrophilic matrices based on polysaccharide carriers remain a highly popular design of sustained release dosage form. Control of drug release from tablets containing polysaccharide excipient system was found to be capable using a variety of different formulations and process methods to provide a variety of different release modalities which were capable of matrix-dimension independence. Controllability of drug release was achieved by manipulation of the synergistic interactions of the heterodisperse polysaccharides (65).

Release data from ethylcellulose (EC) matrix tablets were analyzed to determine which release equation provided the best fit to the data and to observe the effect of drug solubility on the release mechanism(s) (66). Theophylline, caffeine, and dyphylline were selected as nonelectrolyte xanthine derivatives, with solubilities from 8.3 to 330 mg/mL at 25°C. At high drug loading, drug was released by a diffusion mechanism with a rate constant that increased with an increase in aqueous solubility. At low drug loading, polymer relaxation also became a component of the release mechanism. However, its contribution to drug release was less pronounced as solubility decreased, becoming negligible in the case of theophylline.

Microcrystalline cellulose (MCC), sodium carboxymethylcellulose (NaCMC), hydroxypropylmethylcellulose (HPMC), hydroxyethylcellulose (HEC), hydroxypropylcellulose (HPC), and ethylcellulose were used for the production of time-controlled acetaminophen delivery systems using a spray-drying technique. The influence of factors such as polymer concentration, inlet temperature, and drug/polymer ratio were investigated (67). Dissolution studies in pH 1 dilute HCl and pH 6.8 phosphate buffer dissolution media showed that formulations consisting of 1% polymer with a drug/polymer ratio of 1:1 exhibited the slowest drug release, with the spheroids coated with NaCMC and HEC showing the longest T50% values (45 and 53 min at pH 1 and 49 and 55 min at pH 6.8, respectively). Slightly better sustained drug release in pH 6.8 dissolution medium was reached, showing the following trend: HEC > NaCMC > MCC > EC > HPMC. Concerning the additives, the trends in dissolution T50% of drug revealed TA > SA > CA > OA > PVP > PA > DBS in acidic pH 1 dissolution medium and PVP > OA > TA > SA > PA > CA > DBS in phosphate buffer at pH 6.8 (67).

Hydroxypropylmethylcellulose has been the most widely used hydrophilic drug carrier. Matrix tablets were prepared by wet granulation (68), slugging (69), or direct compression (70). The drug release was controlled through variations in the molecular weight of the polymer, drug solubility, the drug/polymer ratio, the particle size of the drug and polymer, and the addition of different additives (71,78).

The available grades of HPMC have varied molecular weight, degree of substitution, and particle size. The drug and HPMC particles sizes also influence the drug release parameters, although to a lesser extent. The major disadvantage of single-unit hydrophilic matrix systems such as tablets or capsules has been the possibility of uncontrolled erosion as a result of mechanical stress during passage through the gastrointestinal tract, causing erratic drug release and absorption.

A multiple-unit indomethacin delivery system based on HPMC as the hydrophilic carrier material was developed by a novel technique using the insolubility of cellulose ether at elevated temperatures and the ionotropic gelation of the polysaccharide, sodium alginate with calcium ions (79).

In order to develop nasal powder preparations with higher bioavailability for peptide delivery, the effect of a combination of hydroxypropylcellulose and microcrystalline cellulose used as base materials and microenvironment for the drugs in the preparations was examined (80). Significant enhanced absorption of leuprolide, calcitonin, and FITC-dextran was attained by the addition of a small amount of HPC to MCC. It is suggested that MCC works as an absorption enhancer by causing a locally high concentration of drugs in the vicinity of the mucus surface. On the other hand, HPC works to increase retention of drugs on the nasal mucus due to its gel-forming property (80).

The aqueous interaction of the sodium salt of ibuprofen with the cellulose ethers ethyl hydroxyethylcellulose (EHEC) and HPMC has been investigated in the concentration range 0–500 mM ibuprofen and 0.1–1% (w/w) polymer by cloud point, capillary viscometry, equilibrium dialysis, and fluorescence probe techniques (81). A combination of time-resolved and static fluorescence quenching shows that micellelike ibuprofen aggregates are formed in the solution. The average aggregation number of pure ibuprofen micelles in water is about 40.

IV. CURDLAN

Harada et al. (82) and Saito et al. (83) showed that more than 99% of the linkages in curdlan are $(1 \rightarrow 3)$-β-D-glycoside linkages. Very few $(1 \rightarrow 6)$-β-D-glycoside linkages seem to occur in this polysaccharide by the chemical (83) and enzymatic (84) methods. Curdlan is insoluble but swells in water.

A. Clinical Application of Curdlan

Maeda and Chihara (85) and Sasaki et al. (86) reported that curdlan, lentinan, and other similar polysaccharides inhibited the growth of Sarcoma-180. Moreover, Sasaki et al. (87) found that a serum factor is involved in the tumor inhibition by curdlan. When peritoneal macrophages from normal untreated mice were incubated in vitro with serum of mice treated with curdlan, they become cytotoxic to tumor cells. This was attributed to two serum factors, a peptide (mw = 4500) and probably a peptidoglycan (mw = 9000) found in serum after administration of the polysaccharide.

Morikawa et al. (88) showed that when curdlan is given i.p. to mice, it produces a high and persistent level of both polymorphonuclear (PMN) leukocytes and macrophages. The activated PMN leukocytes were spontaneously cytotoxic to mammary carcinoma cells in vitro.

Cellulose sulfate, curdlan sulphate, and sulfopropyl curdlan have been found to exhibit strong anti–human immunodeficiency virus (anti-HIV) activity (89,90). Yoshida et al. (91) showed that curdlan sulfate with a sulfur content of 14.4% at concentrations as low as 3.3 μg/mL completely inhibited infection by HIV.

Antitumor active polysaccharide against Sarcoma 180 was isolated by DEAE-Sepharose CL-6B and Sepharose 4B column chromatography from the hot water–soluble fraction of the mycelium of liquid-cultured *Agaricus blazei* mill (92). This polysaccharide did not react with antibodies of antitumor polysaccharides such as lentinan, gliforan, and FIII-2-b, which is one of the antitumor polysaccharides from *A. blazei*. Moreover, the analyses of ^{13}C-NMR and GC-MS suggested that this polysaccharide was preliminary glucomannan with a main chain of β-1,2–linked D-mannopyranosyl residues and β-D-glucopyranosyl-3-O-β-D-glucopyranosyl residues as a side chain. This polysaccharide was completely different from the antitumor polysaccharide obtained from fruiting body of *A. blazei*, β-1,6-glucan.

V. GUAR GUM

Targeting of drugs to the colon, following oral administration, can be accomplished by the use of modified biodegradable polysaccharides as vehicles. Guar gum (GG) was crosslinked with increasing amounts of trisodium trimetaphosphate (STMP) to reduce its swelling properties for use as a vehicle in oral delivery formulations, especially drug delivery systems aimed at localizing drugs in the distal portions of the small bowel (93). Swelling of GG in artificial gastrointestinal fluids was reduced from 100- to 120-fold (native GG) to 10- to 35-fold depending on the amount of crosslinker used, showing a bell shape dependency. As a result of the crosslinking procedure GG lost its nonionic nature and became negatively charged.

Functionalizing of GG crosslinked products (GGP) as possible colon-specific drug carriers was analyzed by studying (1) the release kinetics of preloaded hydrocortisone from GGP hydrogels into buffer solutions with or without GG-degrading enzymes (α-galactosidase and β-mannanase) and (2) direct measurements of the polymer's degradation in the cecum of conscious rates (94). The effect of a GG diet on α-galactosidase and β-mannanase activity in the cecum of the rat and GGP degradation was also measured. It was found that the product GGP-0.1 (loosely crosslinked with 0.1 equivalents of STMP) was able to prevent the release of 80% of its hydrocortisone load for at least 6 h in PBS, pH 6.4. In vivo degradation studies in the rat cecum showed that, despite the chemical modification of GG, it retained its enzyme-degrading properties in a crosslinker concentration–dependent manner. Eight days of GG diet prior to the study increased α-galactosidase activity in the cecum of the rat threefold compared to its activity without the diet.

A novel tablet formulation for oral administration using guar gum as the carrier and indomethacin as a model drug has been investigated for colon-specific drug delivery using in vitro methods (95). Drug release studies under conditions mimicking mouth-to-colon transit have shown that guar gum protects the drug from being released completely in the physiological environment of stomach and small intestine (95). Studies in pH 6.8 phosphate-buffered saline (PBS) containing rat cecal contents have demonstrated the susceptibility of guar gum to the colonic bacterial enzyme action with consequent drug release.

Also, it has been shown that the swelling of guar gum is affected by concentration of the drug and viscosity grade of the polymer. This study examines the mechanism of behavior of guar gum in a polymer–drug matrix (96). The swelling action of guar gum, in turn, is controlled by the rate of water uptake into the matrices. An inverse relationship exists between the drug concentration in the gel and matrix swelling. This implies that guar gum swelling is one of the factors affecting drug release. The swelling behavior of guar gum is therefore useful in predicting drug release.

Poly(vinyl alcohol)–guar gum interpenetrating network microspheres were prepared by crosslinking with glutaraldehyde. Nifedipine was loaded into these matrices before and after crosslinking in order to study its release patterns (97). The mean particle size of the microspheres was found to be around 300 μm. The molecular transport phenomenon, as studied by the dynamic swelling experiments, indicated that an increase in crosslinking affected the transport mechanism from Fickian to non-Fickian. The in vitro release study indicated that the release from these microspheres is not only dependent upon the extent of crosslinking, but also on the amount of the drug loaded as well as the method of drug loading (97).

Guar gum tablet formulations were prepared and evaluated under a variety of in vitro dissolution conditions. The formulations, along with Dilacor XR (diltiazem), were administered to a group of eight fasted, healthy volunteers in a four period crossover study (98). Dissolution of diltiazem from guar gum tablets was essentially independent of stir speed under normal conditions (USP Apparatus II).

VI. PULLULAN

Although many fungi are known to produce exopolysaccharides with an interesting range of chemical and physical properties (99), most of the research effort has been directed to the α-glucan pullulan, produced by *Aureobasidium pullulans* (100). Bernier (101) first described the production of two exopolysaccharides by Pullularia (*A. pullulans*), a heteropolysaccharide and a neutral glucan. These contain α-(1-4) and α-(1-6) glucosidic linkages.

Pullulan is a polysaccharide with high liver affinity. Considering this property, interferon–pullulan conjugation was promising for interferon (IFN) targeting to the liver with efficient exertion of its antiviral activity therein (102). The cyanuric chloride method enabled the preparation of an IFN–pullulan conjugate that retained approximately 7–9% of the biological activity of IFN. Pullulan conjugation enhanced the liver accumulation of IFN and the retention period with the results being reproducible. When injected intravenously to mice, the IFN–pullulan conjugate enhanced the activity of 2-5A synthetase in the liver. The activity could be induced at IFN doses much lower than those of free IFN injection. In addition, the liver 2-5A synthetase induced by conjugate injection was retained for 3 days, whereas it was lost within the first day for the free IFN–injected mice.

A chelating residue [diethylenetriamine pentaacetic acid (DTPA)] was introduced to pullulan (103). This DTPA–pullulan could conjugate with IFN through Zn^{2+}

coordination on mixing these three components. Intravenous injection of the IFN–DTPA–pullulan conjugate with Zn^{2+} coordination induced activity in the liver of an antiviral enzyme (103). Liver targeting of IFN by this conjugation technique based on Zn^{2+} coordination opens a new method of IFN therapy.

Gu et al. (104) and Wang et al. (105) reported a novel formula of hydrophobized polysaccharide nanoparticles which can deliver a HER2 oncoprotein containing an epitope peptide to the MHC class I pathway. A protein consisting of the 147 amino-terminal amino acids of oncogene erbB-2/neu/HER2 (HER2) was complexed with two kinds of hydrophobized polysaccharides, cholesteryl group–bearing mannan (CHM) and cholesteryl group–bearing pullulan (CHP), to form nanoparticles (CHM-HER2 and CHP-HER2). CHM-HER2 and CHP-HER2 were able to induce CD3+/CD8+ CTLs against HER2-transfected syngeneic fibrosarcoma cell lines. In addition, vaccination by CHM-HER2 complexes led to a strongly enhanced production of IgG antibodies against HER2, whereas vaccination with HER2 proteins alone resulted in a production of antibodies at a marginal level. Mice immunized with CHM-HER2 or CHP-HER2 before tumor challenge successfully rejected HER2-transfected tumors. The complete rejection of tumors also occurred when CHM-HER2 was applied not later than 3 days after tumor implantation.

The effect of liposomal adriamycin with tumor recognition molecule, 1-aminolactose (1-AL), on AH66 hepatoma transplanted into nude mice was investigated by Ichinose et al. (106). Adriamycin (ADM) was encapsulated in liposome coating with CHP to increase the stability in the blood stream. 1-Aminolactose (1-AL) was assembled to the outer layer of CHP-coated liposomal ADM as a tumor recognition molecule. In an in vivo therapeutic study, 1-AL/CHP-coated liposomal ADM restrained tumor growth more when compared with CHP-coated liposomal ADM. Thus, 1-AL/CHP-coated liposome seems to be a carrier of ADM to tumor cells.

Yamamoto et al. (107) synthesized CHP bearing 1-aminolactose and introduced a saccharide, cholesteryl pullulan bearing 1-aminolactose (1-AL/CHP), to an outer layer of the conventional liposome as a cell recognition element. Lectin recognized the β-galactose by aggregation of 1-AL/CHP–coated liposome (1-AL/CHP liposome). The uptake of this liposome to AH66 rat hepatoma cells was greater than in liposomes without 1-aminolactose in vitro. Furthermore, 1-AL/CHP liposomal adriamycin showed a stronger antitumor effect in comparison with other types of liposomal adriamycin in vitro. When in vivo tumor-targeting efficacy was investigated in AH66 tumor–transplanted mice using 3H-liposome, the tumor/serum radioactivity ratio in mice injected with 1-AL/CHP liposome was higher than that of mice injected with other liposomes. These observations suggest that 1-AL is effective as a cell recognition element. As a result, 1-AL/CHP liposome is considered to be a good carrier of anticancer drugs for the active targeting of tumor cells (107).

Insulin spontaneously and easily complexed with the hydrogel nanoparticle of CHP in water (108). The complexed nanoparticles (diameter 20–30 nm) thus obtained formed a very stable colloid (108). The original physiological activity of complexed insulin was preserved in vivo after i.v. injection.

Suzuki and Sunada (109) have investigated the influence of water-soluble polymers on the dissolution behavior of nifedipine from solid dispersions with combined carriers. All the solid dispersions of nifedipine were prepared by the fusion method using nicotinamide and four different water-soluble polymers: hydroxypropylmethylcellulose (HPMC), polyvinylpyrrolidone (PVP), partially hydrolyzed polyvinyl alcohol (PVA), and pullulan. HPMC, PVP, and PVA dissolved in the fused liquid of nicotinamide and operated efficiently on the amorphous formation of nifedipine in solid dispersions. In dissolution studies, the drug concentration for these dispersions increased to more than twice the intrinsic drug solubility. The rank order of the drug concentration was HPMC > PVP > PVA. However, since pullulan did not dissolve in the fused nicotinamide, nifedipine was present as a crystalline state in the solid dispersion; the supersaturation behavior of the drug was scarcely observed (109).

VII. DEXTRIN AND CYCLODEXTRIN

Dextrin consist of an α-(1→6)-linked glucan with branches attached to O-3 of the backbone chain units. The degree of branching is approximately 5%. Recently, enzymatic hydrolysis combined with chemical and nuclear magnetic resonance studies have enabled the ratio of single to multiple branches to be elegantly elucidated (110).

The molecular size is a pivotal importance for each of the pharmacological properties of dextrin, for example, colloid osmotic pressure, viscosity, cell surface adsorption, and steric exclusion principles (111,112).

The effects of dextrin on hemostasis have been reviewed (113,114). Although it is generally agreed that the coagulation mechanism remains normal after infusion of a standard clinical dose of dextrin (115,116). Dextran appears to be adsorbed to the surfaces of the vascular endothelium and various cells (117,118).

Whereas native dextrin is immunogenic in humans (119), lower molecular weight fractions in the clinical range were not found to be immunogenic (120,121). The antigenic determinants on the dextrin chain appear to correspond to segments of two to seven glucose units (121,122).

pH-sensitive dextran hydrogels were prepared by activation of dextrin (T-70) with 4-nitrophenyl chloroformate, followed by conjugation of the activated dextran with 4-aminobutyric acid and crosslinking with 1,10-diaminodecane (123). The crosslinking efficiencies determined by mechanical measurements were in the range of 52–63%. Incorporation of carboxylpropyl groups in dextran hydrogels led to a higher equilibrium and faster swelling under high pH conditions. The swelling reversibility of hydrogels was also observed after repeated changes in buffers between pH 2.0 and 7.4 (123).

Cyclodextrin (CDs) are naturally occurring homochiral oligosaccharides composed of from 6 to 13 α-1,4-linked D-glucopyranose units. They possess annular structures whose wide and narrow hydrophilic ends are delineated by OH (2) and OH (3) secondary and OH (6) primary hydroxyl groups, respectively, whereas their hydrophobic annular interiors are lined with methyl and methylene groups and ether oxygens. Interest in CDs stems from their ability to partially or completely include a wide range of guest species within their annuli to form inclusion complexes, also referred to a host–guest complexes (124–130). The bonding between the CDs and guests is solely secondary in nature; nevertheless, the inclusion complexes can exhibit considerable thermodynamic stability (131).

Cyclodextrins are potentially very interesting as the formulator in pharmaceutical technology. These cyclic oligosaccharides have the ability to form noncovalent complexes with a number of drugs and in so doing alter their physicochemical properties. In addition, the primary and secondary hydroxyl group of the native (α, β, γ) cyclodextrins are potential sites for chemical modifications (132). Cyclodextrins have remarkable properties in improving stability, solubility, and bioavailability of drugs after oral administration (133,134). Natural CDs undergo enzymatic degradation along the gastrointestinal tract, which probably occurs mainly in the colon (135). No definitive results of acute toxicity have been published because the highest doses administered to animals do not result in any mortality (133). Because CDs are most often used to enhance the solubility and consequently the bioavailability of poorly water-soluble active ingredients, intravenous administration is among the most interesting parenteral routes. Inclusion of a guest molecule in the cavity of CDs constitutes a protected state of the included molecule. This protection is especially effective in the solid state, with respect to the oxygen from ambient air (136).

Cyclodextrins have the potential to enhance drug release by increasing the concentration of diffusible species within matrix. Guo and Cooklock (137) used a range of additives including CDs to increase the solubility of the poorly water-soluble opoid analgesic, buprenorphine, and modify its release from buccal patches composed of poly(acrylic acid), poly(isobutylene), and poly(isoprene).

By inclusion of guest molecule inside the cavity of CDs, side effects decreased. Inclusion in CDs can reduce the bitterness of femoxetine (138), reduce the local irritation of pirprofen (139), and decrease the ulcerous effect of phenylbutazone (140) or indomethacin (141). In the case of active ingredients exhibiting a poor bioavailability due to water nonsolubility or low solubility, but without absorption problems, the improvement in apparent solubility can improve the bioavailability (142,143). Other examples of using CDs to promote drug release through dissolution-erosion mechanisms are given by Giunchedi et al. (144) and Song et al. (145).

Because of their ability to enhance the stability, solubility, or bioavailability of drugs, CDs have been the subject of studies concerning every administration route: oral (146–158); rectal (159–163); dermal (164–178); ocular (179–182); nasal (183–188); pulmonary (189–191); parenteral (192–196); intracerebral (197); intrathecal (198), and epidural (199,200) administration. Recently, modified CDs (201–212) were prepared either to allow the direct formation of targeting agents (213–217) or to enable them to be targeting agents.

VIII. STARCH

A. Microspheres and Microcapsules

Starch, in its native and modified form, has been subjected to extensive study over the past 40 years. Early interest in starch was associated with the food and paper industry, textile manufacture, and medicine. Crosslinked starch and starch networks have both the required biodegradability and a relatively high mechanical and chemical stability. Starch microspheres were prepared using epichlorohydrin as a crosslinking agent. Recently the need for three-dimensional matrices with controlled release properties for pharmaceutical and agrochemical uses has increased interest in starch gelation (218–223).

The condensation mechanism of epichlorohydrin with amylose involves epoxide ring opening, mediated by the nucleophilic attack of the alkali amylose, and subsequent chlorine displacement and epoxidation (224,225). The starch–epichlorohydrin reaction follows second order kinetics, first order with respect to epichlorohydrin as well as starch. The activation energy for the reaction is 38 kJ/mol and the temperature coefficient K_{323}/K_{303} is 2 (226).

The crosslinking of starch with epichlorohydrin under homogeneous and heterogeneous condition was studied with a particular view of measuring the extent of the side reaction, where starch was substituted to give a monoether derivative (227). The extent of this reaction was strongly dependent on the reaction conditions (temperature, time, molar ratio of all reactants). Depending on these conditions, 5–25% epichlorohydrin was bound as glycerol monoether substituent by starch. The degree of swelling of the crosslinked starch was linearly dependent on the water/starch molar ratio in the reaction mixture (227). The crosslinking of starch proceeded with remarkable efficiency when epichlorohydrin was applied in the vapor phase.

An interfacial crosslinking process was applied to hydrosoluble starch derivatives: hydroxyethyl starch and carboxymethyl starch (223). All crosslinked polysaccharide microcapsules were characterized by a total resistance to digestive media.

Shefer et al. (224) characterized the structure and morphology of starch networks formed by two distinct methods using cross-polarization magic angle–spinning ^{13}C-NMR spectroscopy (CP-MAS ^{13}C-NMR) combined with wide angle x-ray diffraction measurements. The first step in the process involves the gelatinization of amylose using sodium hydroxide. After the addition of NaOH solution (6.6% w/w) the resonance lines of the amylose were broadened. Broadening can be caused by distribution of isotropic chemical shifts due to the loss of crystallinity (224). The loss of crystallinity during this process is also observed in the wide angle x-ray diffractogram of a sample following treatment with NaOH.

The amorphous nature of the networks formed following the reaction of amylose with epichlorohydrin is also supported by x-ray diffraction analysis. The resonance of the C6 carbon in the network formed is of weaker intensity and is shifted downfield by about 1 ppm from about 61.4 ppm in the native amylose molecule to 60.4 ppm in the crosslinked network. The secondary OH(2) and OH(3) of the native amylose at 5.0–5.4 ppm appear in the crosslinked network spectrum. This indicates that both crosslink points as well as glycolic functional group (4.7 ppm) are formed (224). These observations suggest that epichlorohydrin crosslinks and reacts monofunctionally at C2, C3, and C6 (224). Their swelling degree, reflecting the number of glycerol dieter bridges in the polymeric network, and the number of noncrosslinking monoglycerol ether groups corresponding to a side reaction of epichlorohydrin with starch were determined (228). Degradation by α-amylase was surface controlled and could be modulated by the introduction in the polymeric network of (1) nonhydrolyzable α-1,6 bonds related to the presence of amylopectin in the raw starch, (2) glycerol diether, and (3) monoether groups, all of these being likely to block the activity of α-amylase.

A novel silicone polymer-grafted starch microparticle—starch microparticles (MPs) grafted with 3-(triethoxysilyl)-propyl–terminated polydimethylsiloxane (TS-PDMS)—was developed that is efficacious both orally and intranasally (229,230). Unlike most other microparticle systems, this novel system does not appear to retard the release of antigen or to protect antigen from degradation. The results indicate that a unique physiochemical relationship occurs between protein antigen and silicone in a starch matrix that facilitates the mucosal immunogenicity of antigen. This leads to predominance of Th2 antibody response (229,230).

The efficacy of temporary arterial emmobilization using degradable starch microspheres combined with hyperthermia was investigated in rabbits bearing VX2 tumors (231). Microsphere injection caused a marked decrease of tumor blood flow and pH. During heating, there was a marked increase of the maximum temperature in tumor tissue compared with normal muscle. Tumor growth was suppressed 330% times at 3 weeks after hyperthermia alone and 270% times following combined treatment with microspheres and hyperthermia. Damage to normal muscle tissue was mild (231).

Evaluation of reticular endothelial system specific magnetic starch microspheres (MSMs) as an i.v. contrast agent for MR imaging in a model of experimental liver metastases has been studied. A loss of liver signal intensity was obtained at all MSM dose levels. No metastases were detected in the precontrast images. The optimal detection rate of hepatic metastases was reached with the T1-weighted spin-echo (SE) sequence at a dose of 1.0 mg Fe/kg b.w. MSM and the diameters of the smallest lesions depicted were 1 mm (232). The use of MSM dramatically increased the detection of experimental hepatic metastases.

B. Starch Derivatives

1. Hydroxyethyl Starch

Hydroxyethyl starches (HES) are high-polymeric compounds obtained via hydrolysis and subsequent hydroxyethylation from the highly branched amylopectin contained in maize. The glucose units can be substituted at carbon 2, 3, and 6 leading to various substitution patterns. This pattern is described by the C2/C6 hydroxyethylation ratio. The higher the degree of substitution and the C2/C6 ratio, the less the starch is metabolized. The in vitro molecular weight, the degree of substitution, and the C2/C6 ratio are the main determinants of the in vivo molecular weight, which is clinically relevant. Hemorrhagic complications that occur after infusing larger volumes of HES can be avoided with a starch of low in vivo molecular weight (233). Furthermore high molecular weight HES macromolecules lead to a distinctive decrease in fibronectin concentration that reflects saturation of the reticuloendothelial system (RES). Another advantage of low in vivo molecular weight HES is its rather short half-life. Patients with an increased bleeding risk, microcirculatory disturbance, or affected RES should receive HES with low in vivo molecular weight. In the future, HES should be mainly characterized by the in vivo and not the in vitro molecular weight.

Artificial colloids affect hemostasis. Particularly, HES solutions may have detrimental effects on hemostatic mechanisms (234).

Hydroxyethyl starch is frequently used as a volume expander in critically ill patients. Hofbauer et al. (235) investigated whether HES influences the chemotaxis of polymorphonuclear leukocytes (PMNs) through endothelial cell monolayers by using a test system that allows the simultaneous treatment of both cell types; HES was shown to significantly reduce the chemotaxis of PMNs through endothelial cell monolayers.

The effects of HES on blood coagulation were investigated in 20 patients undergoing surgery to determine whether its use places recipients at risk of hemorrhage or thrombosis (236). The partial thromboplastin times are significantly prolonged; factor VIII activities and fibrinogen levels are decreased. After infusion of HES, no significant differences were detected in platelet count or prothrombin time. A decreased platelet aggregation was also found after the infusion of HES (236).

Jamnicki et al. (237,238) compared the effects of progressive in vitro hemodilution (30 and 60%) on blood coagulation in 80 patients receiving one of two different 6% HES solutions using thrombelastography (TEG). The newly developed solution has a mean molecular weight of 130 kD and a degree of substitution, defined as the average number of hydroxyethyl groups per glucose moiety, of 0.4 (HES 130/0.4); the conventional solution has a mean molecular weight of 200 kD and a degree of substitution of 0.5 (HES 200/0.5). Both HES solutions significantly compromised blood coagulation, as seen by an increase in reaction time and coagulation time and a decrease in angle alpha, maximal amplitude, and coagulation index (all $p < 0.05$).

2. Acetyl Starch

Acetyl starch (ACS) is a new synthetic colloid solution for plasma volume expansion and is now undergoing phase 2 clinical trials. Behne et al. (239) compared the pharmacokinetics of ACS with those of HES in 32 patients (ASA physical status I and II) undergoing elective surgery. In contrast to hydroxyethyl starch, this new agent undergoes rapid and nearly complete enzymatic degradation.

3. Carboxymethyl Starch

Claudius et al. (240) determined effects of sodium carboxymethyl starch (CMS) on the antimicrobial activity of vancomycin. In particular, the in vitro activity of vancomycin against two clinically relevant bacteria, *Staphylococcus aureus* and *Enterococcus faecalis*, was studied in the presence of varying concentrations of sodium CMS. From two independent studies conducted using an agar dilution method, it appeared that the binding of vancomycin to sodium carboxymethyl starch had no effect on the in vitro antimicrobial activity of vancomycin.

IX. FUCAN SULFATES

Marine algal sulfated polysaccharides have been found to possess various pharmacological activities, i.e., antibacterial, antiviral, antitumor (241,242), immunosuppressive, antilipemic, antihemostatic, and anticoagulant (243). Fucan sulfates are a type of sulfated polysaccharide occurring in brown marine algae.

The anticoagulant activity of fucan sulfates (244) was mainly assayed by activated partial thromboplastone time, which expresses the intrinsic pathway of blood anticoagulation; prothrombin time, which explores the extrinsic pathway; thrombin clotting time; and repilase clotting time methods (245). The anticoagulant components of brown, red, and green algae are found in fucan sulfates. They all comprise a family of polydisperse heteromolecules based on fucose, xylose, glucuronic acid, galactose, mannose, and half ester sulfate. They differ in sugar composition and sulfate content, and thus in structure (246–249).

The correlation between the sulfate and uronic acid contents and the anticoagulant activity of fucan sulfates was confirmed by a study on fucan sulfate from *Ecklenia kurome* (250). It has also been reported that fucoidans (pure fucans) from *F. vesiculosus* (251), *Eisenia bicyclis* (252), *Hizikia fusiforme* (253), *Laminaria angustata* (254), and *P. canaliculata* (255) showed antithrombin activity.

X. LECTINS

Lectins are proteins or glycoproteins of nonimmune origin capable of binding to one or more specific sugar residues and mediating a variety of biological processes, such as cell–cell and host–pathogen interactions, serum glycoprotein, and innate immune responses. Currently, over 200 three-dimensional structures of lectins from plants, animals, bacteria, and viruses and their complexes are available (256–262).

Excellent recent reviews on the structure and interactions of lectins are available (263–272). Thus, lectins possess various specificities that are associated with their ability to interact with acetylaminocarbohydrates, aminocarbohydrates, sialic acid, hexoses, pentose, and many other carbohydrates (258,273). Lectins from plant sources were the first proteins of this class to be studied (274–277). Human foods of both plant and animal origin contain a variety of simple and complex carbohydrates as well as lectins. Both saccharides and lectins have the capacity to interfere with bacterial and viral attachment to epithelial cell surfaces within the alimentary canal, as has the major mucosal immunoglobulin, secretory IgA. It is well known that the lectin from jack fruits can bind to serum IgA1, but secretory IgA also possesses oligosaccharide receptors for bacterial lectins in fimbriae and can agglutinate *E. coli* by this antigen nonspecific mechanism (264).

Membrane lectins of certain cells are capable of internalization of their ligands, and hence glycoconjugates specifically recognized by these lectins can be used as carriers of metabolite inhibitors and drugs (278). Galactose-terminated glycoproteins and neoglycoproteins have been used to carry antiparasitic (279) and antiviral drugs (280,281).

The potential for using lectins as a means of ''anchoring'' a drug delivery system to the mucosal surfaces of the eye has been investigated (282). In this study the acute local dermal irritancy of these lectins, in terms of their potential to cause inflammation and tissue necrosis, was investigated. There was no evidence of tissue necrosis, edema, or Evans blue infiltration with any of the lectin solutions administered. The rabbits did not display any signs of discomfort such as scratching or continued grooming throughout the experiment. Histological examination of the injection sites revealed little sign of any inflammation, such as heterophil migration, edema, or tissue damage. It was concluded that these lectins demonstrate minimal acute irritancy, and will therefore be taken forward for formulation and in vivo studies.

XI. HYALURONIC ACID, HYALURONAN, AND HYALURONAN DERIVATIVES

Hyaluronic acid (HA) is a natural mucopolysaccharide which consists of alternating residues of D-glucuronic acid and N-acetyl-D-glucosamine. Hyaluronic acid functions as the backbone of the proteoglycan aggregates necessary for the functional integrity of articulate cartilage of the knee.

Hyaluronan (sodium hyaluronate) and hyaluronan derivatives (hylans) have been developed as topical, injectable, and implantable vehicles for the controlled and localized delivery of biologically active molecules (283). Hyaluronan is the original lastoviscous, biocompatible polysaccharide developed for use in eye surgery and viscosurgery, orthopedic surgery, rheumatology, otology, plastic surgery, and veterinary medicine (284,285). Hyaluronic acid, either by itself or mixed with fibronectin, may be a potentially optimal bioimplant for the surgical management of vocal fold mucosal defects and lamina propria deficiencies (e.g., scarring) from a biomechanical standpoint (286).

Hyaluronic acid in the range of Mw 1300 kD may prove beneficial in minimizing bacterial contamination of surgical wounds when used in guided tissue regeneration surgery (287). The 1.0 mg/mL concentration of high molecular weight HA had the greatest overall bacteriostatic effect, inhibiting the growth of all six bacterial strains tested. Among the bacterial strains studied, HA was found to have no bactericidal effects, regardless of concentration or molecular weight.

An animal model study was conducted to compare the efficacy of recurrent topical applications of hyaluronic acid and gentamicin ointment for the treatment of noninfected, mechanical corneal erosions (288). Rabbit eyes treated with hyaluronic acid showed a significantly enhanced rate of epithelial defect closure compared with untreated eyes and a similar rate to that achieved with gentamicin ointment. In the eyes treated with hyaluronic acid a normal, multilayered epithelium was observed 48 h after complete healing, whereas the gentamicin-treated eyes showed an imperfectly layered epithelium, with irregularity of the cuboidal cells.

Through the esterification of the carboxyl group of the glucuronic acid moiety, polymeric prodrugs of hyaluronic acid have been prepared by several groups (289–291). Two drugs made up of HA derivatives have recently become available for patients in whom simple analgesics and conservative nonpharmacological therapy have failed. Leslie (292) reviews the epidemiology, pathogenesis, diagnosis, and medical management of osteoarthritis of the knee, with an emphasis on the physiologic and pharmacological mechanisms of HA. Health care providers may administer HA via intra-articular injection in primary care and rheumatologic or orthopedic settings or they may refer their patients to specialists for consultation.

Hyaluronic acid grafted with poly(ethylene glycol) (PEG) (PEG-g-HA) were synthesized. The materials characterization, enzymatic degradability, and peptide (insulin) release from solutions of the copolymers were examined (293). Insulin was preferentially partitioned into the PEG phase in a PEG/HA solution system. Leakage of insulin from the copolymers was dependent upon the PEG content. Leakage rate of insulin from copolymer containing between 7 and 39% by weight of PEG were similar. The conformational change of insulin was effectively prevented in PEG-g-HA solutions, although insulin was denatured in storage of both phosphate buffered solution and HA solution. Such a heterogeneous-structured polymeric solution may be advantageous as an injectable therapeutic formulation for ophthalmic or arthritis treatment.

To increase the availability of sodium butyrate over a longer period of time. Coradini et al. (294) covalently linked isodium butyrate to HA (a component of the extracellular matrix). Its major advantages as a drug carrier consist of its high biocompatibility and its ability to bind CD44, a specific membrane receptor frequently overexpressed on the tumor cell surface (294). The biological activity of hyaluronic acid–butyric ester derivatives was evaluated in terms of the inhibition of the growth of the MCF7 cell line and compared with that of sodium butyrate. After 6 days of treatment, we observed a progressive improvement of the antiproliferative activity up to DS = 0.20; thereafter, the antiproliferative effect of the ester derivatives decreased (294). Fluorescence microscopy showed that after 2 h of treatment fluorescein-labelled compounds appeared to be almost completely internalized into MCF7 cells, expressing CD44 standard and variant isoforms.

Two biomaterials based on hyaluronic acid modified by esterification of the carboxyl groups of the glucuronic acid (HYAFF 11 and ACP sponges) were tested as osteogenic or chondrogenic delivery vehicles for rabbit mesenchymal progenitor cells and compared with a well-characterized porous calcium phosphate ceramic delivery vehicle (295). The hyaluronic acid–based delivery vehicles are superior to porous calcium phosphate ceramic with respect to the number of cells loaded per unit volume of implant, and HYAFF 11 sponges are superior to the ceramics with regard to the amount of bone and cartilage formed. Additionally, hyaluronic acid–based vehicles have the advantage of degradation/resorption characteristics that allow complete replacement of the implant with newly formed tissue.

The tolerability and safety of hyaluronan-based three-dimensional scaffolds as a culture vehicle for mesenchymal progenitor cells was investigated (296). The proliferation patterns and extracellular matrix production of rabbit and human mesenchymal, bone marrow–derived progenitors first were characterized in vitro. Subsequently rabbit autologous cells were cultured in this hyaluronan-based scaffold and implanted in a full thickness osteochondral lesion. In vitro histologic findings showed that mesenchymal progenitor cells adhered and proliferated onto the hyaluronan-derived scaffold. In vivo data demonstrated that the biomaterial, with or without mesenchymal progenitors, did not elicit any inflammatory response and was completely degraded within 4 months after implantation (296).

The new composite biomaterial made from hydroxyapatite and collagen conjugated with hyaluronic acid has been studied (297). The structure evaluation of the composite showed more dense arrangement due to the formation of collagen hyaluronic acid conjugate, and particles of inorganic component are closely anchored in the structure (297). The test of contact cytotoxicity showed a very good biocompatibility of the biomaterial.

The adsorption of glycosaminoglycans (heparin, heparan sulfate, dermatan sulfate, highly sulfated chondroitin sulfate, chondroitin sulfate, and hyaluronan) onto coral has been investigated (298). Granules of natural coral of specific diameter, between 100 and 500 μm, having high content of calcium (>98%) and a homogeneous surface adsorb glycosaminoglycans with different capacity. Heparin (maximum adsorption 1.29 ± 0.10 mg/20 mg of coral, 6.45% w/w) is adsorbed more than highly sulfated chondroitin sulfate species (maximum adsorption 0.90 ± 0.06 mg/20 mg of coral, 4.50% w/w), chondroitin sulfate (maximum adsorption of 0.72 ± 0.06 mg/20 mg of coral, 3.60% w/w), dermatan sulfate (maximum adsorption of 0.70 ± 0.06 mg/20 mg of coral, 3.50% w/w), and heparan sulfate (maximum adsorption of 0.72 ± 0.07 mg/20 mg of coral, 3.60% w/w) (298). Hyaluronan is not adsorbed onto granules of coral. The percentage adsorption of polyanions onto coral depends mainly on their charge density, with sulfate groups being more important than carboxyl groups. The adsorption of glycosaminoglycans is driven by electrostatic interactions with calcium sites of coral that are dependent on pH and blocked in the presence of large amounts of salt. Due to these peculiar properties, the combination of granules of natural coral with glycosaminoglycans makes this material potentially useful in osseointegration in bone metabolism or periodontal therapy (298).

Several biomaterials are available for the purpose of soft tissue augmentation, but none of them has all the properties of the ideal filler material. The recent development of HA gels for dermal implantation give the physician new possibilities of effective treatment in this field (299). Stabilized, nonanimal hyaluronic acid gel is well tolerated and effective in augmentation therapy of soft tissues of the face. This material presents several advantages in comparison to previously used injectable biomaterials and expands the arsenal of therapeutic tools in the field of soft tissue augmentation.

With the aim of producing a biomaterial for surgical applications, the alginate–hyaluronate association has been investigated (300). Crossed techniques were used to assess the existence of polymer interactions in aqueous solutions up to 20 mg/mL.

XII. ALGINATE

The alginates is a copolymer composed of D-mannuronic acid (M) and L-guluronic acid (G) arranged in MM and GG blocks interrupted by regions of more random distribution of M and G units. Due to the presence of carboxylate groups, alginate is a polyelectrolyte at neutral pH, with one charge per repeating unit in the coil conformation (301).

A. Alginate Hydrogel

Interactions of alginate with univalent cations in solution have been investigated by circular dichroism (c.d.) (302) and rheological measurements. Poly-L-guluronate chain segments show substantial enhancement (approximately 50%) of c.d. ellipticity in the presence of excess of K^+, with smaller changes for other univalent cations: $Li^+ < Na^+ < K^+ > Rb^+ > Cs^+ > NH_4^+$ (302). Pass and Hales (303) have investigated the effect of the cation on the enthalpy of dilution of alkali metal salts of alginate.

Calcium alginate has been one of the most extensively investigated biopolymers for binding heavy metals from dilute aqueous solutions (304–307). Alginate forms gels in the presence of divalent ions at concentration of $>0.1\%$ w/w (308). The gels are not thermoreversible. The ratio of calcium to alginate over which thixotropic gels are formed depends on the alginate type, the pH, and the solids content of the system (309). Previous studies of alginate gelation by c.d. and other techniques (310–312) have shown that the primary event in network formation is dimerization of poly-L-guluronate chain sequences, in a regular 2(1) conformation (313,314) with specific chelation of Ca^{2+} ions between the participating chains (315).

For high M alginates (high D-mannuronic acid alginate), thixotropic gels exist at calcium levels that, in an alginate with a high proportion of L-guluronic acid blokcs, would be holding the alginate chains in a permanent gel structure. Specific intermolecular cooperative interactions occur between calcium and glucuronate blocks owing to the buckled ribbon structure of the polyguluronic acid. This observation has led to the well-known proposal of ''egg-box'' junction zones (308,316). ^{13}C-NMR spectroscopic studies have been made on alginate solutions undergoing sol–gel transition induced by four different divalent cations: Ca, Cu, Co, and Mn (317).

The rigid structure and large pore size of these gels are useful for the encapsulation of enzymes, proteins, drugs, liposomes, and living cells. Oral administration of calcium alginate hydrogels without buffering of stomach acid may result in desegregation of the calcium alginate complex. Several antacids can buffer gastric fluid at pH 4.5, which is an appropriate pH for pancreatic buffering in the duodenum. An increase in enteric pH to 6.5 will occur in the ileum. Buffering of gastric fluid with antacids would be necessary in order to facilitate release of drugs from calcium alginate into small intestine—the most appropriate application of this system.

The sodium alginate from *Sarassum fulvellum* showed a considerable antitumor activity against various murine tumors, such as Sarcoma-180, Ehrlich ascites carcinoma and IMC carcinoma (318).

With the aim of producing a biomaterial for surgical applications, the alginate–hyaluronate association has been investigated (319). Crossed techniques were used to assess the existence of polymer interactions in aqueous solutions up to 20 mg/mL. Viscometry measurements using the capillary technique or the Couette flow, together with circular dichroism investigations, evidenced the moderate significance of interactions between the two polysaccharides in dilute solutions. In addition, the case of more concentrated solutions and containing 20 mg/mL alginate was approached by rheological measurements in the flow mode; the behavior of the polymer associations appeared as a compromise between those of individual polysaccharides (319).

B. Alginate–Polyelectrolyte Complexation

Irreversible hydrogels are insoluble in water, as well as in other solvents, in a wide range of temperatures and dilutions. Due to their solubility profiles, irreversible hydrogels have found multiple applications as food additives, in cosmetics, in medicine, and in biotechnology (320–326).

The complexation of complementary polymers has been modeled theoretically (327,328). In this analysis the total free energy of complexation was divided into two contributions: one arising from the specific interactions between complexing functional groups and a second arising from configurational changes of the system upon complexation. Several authors have modeled the association of biological polymers (329–332).

The interactions of alginates of various compositions with basic polypeptides, namely, poly(L-lysine) and poly (Lys-Ala-Ala), have been studied by means of circular dichroism (333). The alginates used differ from each other in the content of L-guluronate and mixed sequences. The content of D-mannuronate sequences in all alginates is almost identical. The lower complexation efficiency with alginate II (MM 33%, GG 30%, MG 37%) than with alginate I (MM 30%, GG 20%, MG 50%) and the nearly zero efficiency with alginate III (MM 33%, GG 47%, MG 20%) are due to the presence of considerable amounts of noncomplexing L-guluronate sequences in the alginate structure (333). On the basis of the results of complexation achieved in this study (333), it may be suggested that the L-guluronan chain is more rigid, i.e., less adaptable to the changes in the surrounding medium than the D-galacturonan chain.

The influence of charge density of a polycation on complexation was studied with the sequentially regular poly (Lys-Ala-Ala), which is characterized by a charge density one-third that of poly(L-lysine) (333). Alginates interacted with poly(Lys-Ala-Ala) rather intensively. The difference in efficiency of interaction of L-guluronan and D-galacturonan with poly(L-lysine) results from the difference in the conformational flexibility of their polyanionic chains in solution. L-Guluronan maintains the rigid twofold symmetry in solution, and D-galacturonan is conformationally adaptable in the course of interaction.

Many of the present controlled-release devices for in vivo delivery of drugs involve elaborate preparations, often employing either harsh chemicals, such as organic solvent, or extreme conditions, such as elevated temperatures. These conditions have the potential to destroy the activity of sensitive macromolecule drugs, such as proteins or oligopeptides.

Drug delivery particulates were prepared using alginate, polylysine, and pectin. Theophylline, chlorothiazide, and indomethacin were used as the model drugs for in vitro assessments, and mannitol was the model for assessing paracellular drug absorption across Caco-2 cell monolayers. Alginate and pectin served as the core polymers, and polylysine helped to strengthen the particulates. Use of pectin specially helped in forming a more robust particulate that was more resistant in acidic pH and modulated the release profiles of the encapsulated model drugs in the alkaline pH. Alginate and pectin were also found to enhance the paracellular absorption of mannitol across Caco-2 cell monolayers by about three times. The release rate could be described as a first-order or square-root time process depending on the drug load. Use of alginate–polylysine–pectin particulates is expected to combine the advantages of bioadhesion, absorption enhancement, and sustained release. This particulate system may have potential use as a carrier for drugs that are poorly absorbed after oral administration (334).

The polyionic complex based on alginate were used to included the animal cells (335). Microencapsulation of islets of Langherhans in alginate–poly(L-lysine) capsules provides an effective protection against cell-mediated immune destruction, and ideally should allow the transplantation of islets in the absence of immunosuppression. It has previously been suggested that alginate rich in mannuronic acid (high M) is more immunogenic than alginate rich in guluronic acid (high G). The ability of these alginates to induce an antibody response in the recipient or act as an adjuvant to antibody responses against antigens leaked from the capsule was investigated (335). High G–alginate capsules are less immunogenic than high M capsules. Because encapsulation did not protect against the generation of antibodies against islet-like cell clusters (ICC), it can be assumed that antigen leakage from the capsules occurs, as no evidence was found for capsules breaking in vivo.

Alginate and proteins were also used in polyionic complexation (336). Freshly prepared gels of gelatine with alginate or pectate below the isoelectric point of gelatine (e.g., at pH 3.9) melt over the same temperature range as gelatine (30–40°C), but on aging they become thermostable (336). The gels are also stable in 7M urea, arguing against hydrogen bonding or hydrophobic interactions, but the enhanced thermal stability can be eliminated by high salt (e.g., 0.3M NaCl) or by raising the pH to above the isoelectric point of gelatine, as expected for an ionic network (336).

A systematic study of the alginate–polycation microencapsulation process, as applied to encapsulation of bioactive macromolecules such as protein, was conducted by Wheatly et al. (337). When protein drugs (myoglobin) were suspended in sodium alginate solution and sprayed into buffered calcium chloride solution to form crosslinked microcapsules. The drug-loaded capsules were coated with a final layer of poly(L-lysine).

C. Calcium Alginate as a Matrix for Delivery of Nucleic Acids

Advances in the design of genetically targeted agents offer new opportunities for drug therapy (338–340). Accordingly, the encapsulation of DNA and its derivatives may be useful for enteric targeting of nucleic acids as gene transfer agents, carriers for DNA intercalaters, and modified oligonucleotides (341).

The biodegradable microspheres based on sodium alginate were used to encapsulate plasmid DNA containing the bacterial β-galactosidase (LacZ) gene under the control of either the cytomegalovirus (CMV) immediate-early promoter or the Rous sarcoma virus (RSV) early promoter. Mice inoculated orally with microspheres containing plasmid DNA expressed LacZ in the intestine, spleen, and liver. Inoculation of mice with microspheres containing both the plasmid DNA and bovine adenovirus type 3 (BAd3) resulted in a significant increase in LacZ expression compared to those inoculated with microspheres containing only the plasmid DNA. Our results suggest that adenoviruses are capable of augmenting transgene expression by plasmid DNA both in vitro and in vivo (342).

Chitosan and poly(L-lysine) membranes, coating alginate beads, were almost totally inert to the enzymatic hydrolysis by lysozyme; chitosanase; and trypsin, chymotrypsin, or proteinase (343). Less than 2% of the membrane weight was hydrolyzed. It appears that either membrane material would be stable for in vivo application, and in particular in the protection of DNA during gastrointestinal transit. At chitosanase concentrations of 1.4 mg/mL and in the presence of sodium ions, 20% of the total double-stranded DNA was released from chitosan coated beads. An exchange of calcium for sodium within the bead liquefied the alginate core releasing DNA. The presence of calcium stabilized the alginate bead, retaining all the DNA. Highly pure DNA was recovered from beads through mechanical membrane disruption, core liquefaction in citrate, and use of DNA spin columns to separate DNA/alginate mixtures in a citrate buffer. DNA recovery efficiencies as high as 94% were achieved when the initial alginate/DNA weight ratio was 1000 (343).

Alginate gels produced by an external or internal gelation technique were studied so as to determine the optimal bead matrix within which DNA can be immobilized for in vivo application. The encapsulation yield of double-stranded DNA was over 97 and 80%, respectively, for beads formed using external and internal calcium gelation methods, regardless of the composition of alginate. Homogeneous gels formed by internal gelation absorbed half as much DNAse as compared with heterogeneous gels formed by external gelation. Testing of bead weight changes during formation, storage, and simulated gastrointestinal (GI) conditions (pH 1.2 and 7.0) showed that high alginate concentration, high G content, and homogeneous gels (internal gelation) result in the lowest bead shrinkage and alginate leakage. These characteristics appear best suited for stabilizing DNA during GI transit (344).

Co-guanidine membranes were shown to form intact, ionically complexed membranes on alginate beads, serving as an alternative to the commonly used polymers, poly(L-lysine) and chitosan. DNA was encapsulated (345). The level of DNA protection from nuclease diffusion and the degree of DNA complexation with co-guanidine membranes were all shown to be dependent on both polymer concentration and coating time. The highest level of DNAse exclusion was possible within beads coated with a polymer concentration of 5 mg/mL. Recovery of double-stranded DNA after nuclease exposure for 60 min reached 90% of that initially encapsulated. The level of DNA protection was found to be comparable to high molecular weight poly(L-lysine) membranes (197.1 kDa) (345). Co-guanidine membranes coating alginate result in a molecular weight cutoff sufficient to retain DNA and exclude 31-kDa DNAse, while providing access to the low molecular weight carcinogen, ethidium bromide.

Somatic gene therapy using nonautologous recombinant cells immunologically protected with alginate microcapsules has been successfully used to treat rodent genetic diseases. Ross et al. (346) have reported the delivery of recombinant gene products to the brain in rodents by implanting microencapsulated cells for the purpose of eventually treating neurodegenerative diseases with this technology. Alginate–poly(L-lysine)–alginate microcapsules enclosing mouse C2C12 myoblasts expressing the marker gene human growth hormone (hGH) at 95 ± 20 ng per million cells per hour were implanted into the right lateral ventricles of mice under stereotaxic guidance. Control mice were implanted similarly with nontransfected but encapsulated cells. Delivery of hGH to the different regions of the brain at various times postimplantation was examined. At 7, 28, 56, and 112 days postimplantation, hGH was detected at high levels around the implantation site and also at lower levels in the surrounding regions, while control mice showed no signal. Immunohistochemical staining of the implanted brains showed that on days 7, 56, and 112 postimplantation, hGH was localized in the tissues around the implantation site.

D. Calcium Alginate as Microparticles for Drug and Drug Proteins Delivery Systems

Present and future applications of alginates are mainly linked to the most striking feature of the alginate molecule, i.e., a sol–gel transition in the presence of multivalent cations, e.g., Ca^{2+}. The properties of alginate gels suggest biomedical and pharmaceutical uses. Calcium alginate has been extensively studied and employed in a number of pharmaceutical applications (347,348), especially in controlled drug release (349). For many drug candidates a modified in vivo drug release is desired to improve efficacy, sustain effect, or minimize toxicity. Polymeric delivery systems, such as microspheres, nanospheres, and polymeric films, have been extensively researched in an attempt to achieve modified drug release (350). Calcium alginate offer an alternative approach.

The release rate of nicardipine HCl from various alginate microparticles was investigated (351,352). The effect of drug/polymer weight ratio, $CaCl_2$ concentration, and curing time on parameters such as the time for 50% of the drug to be released (t50%) and the drug entrapment efficiency were evaluated with analysis of variance. The release of drug from alginate microparticles took place by both diffusion through the swollen matrix and relaxation of the polymer at pH 1.2–4.5. However, the release was due to diffusion and erosion mechanisms at pH 7–7.5.

Pellets of calcium–alginate, calcium–pectinate, and calcium–alginate–pectinate were produced via crosslinking in an aqueous medium for site-specific drug delivery in the gastrointestinal tract (353). In general, texture analysis of various pellets indicated that both strength and resilience profiles were in the order of calcium–alginate ≥ calcium–alginate–pectinate > calcium–pectinate. Calcium–alginate pellets were found to be viscoelastic, while calcium–pectinate was highly brittle.

Alginate gel beads containing tiaramide (TAM) were prepared (354) using a gelation of alginate with calcium cations. Bead performance was evaluated in vitro for different dissolution media, and beads were also subjected to coating. Tiaramide release was dependent both on its solubility in dissolution medium and the guluronate residue content of the alginate used. The release rate was in the following order: in pH 1.2 > pH 6.8 > water. The fast release rate in pH 1.2 is the result of the high solubility of TAM in acidic medium (354). Beads with high guluronate content gave the best controlled results.

Calcium-induced alginate gel beads containing chitosan salt (Alg-CS) were prepared using nicotinic acid (NA), a drug for hyperlipidemia, and investigated its two functions in gastrointestinal tract: (1) NA release from Alg-CS and (2) uptake of bile acids into Alg-CS. The amount of NA incorporated in Alg-CS increased according to increment of CS content. Nicotinic acid was rapidly released from Alg-CS in diluted HCl solution (pH 1.2) or physiological saline without disintegration of the beads. When Alg-CS was placed in bile acid solution it took bile acid into itself. About 80% of taurocholic acid dissolved in the medium was taken into Alg-CS. According to increment of bile acid concentration the uptake amount increased, and an approximately linear relationship existed among them (355).

Calcium alginate beads containing ampicillin were prepared (356). Morphological studies and drug contents, in vitro release, and erosion tests were carried out for the characterization of the prepared beads. The dried particles were characterized by irregular shape and a smooth or rough surface, depending on the viscosity grade of the alginate used. The control of the drug for different time intervals depended on the molecular weight of the polymer used; however, the pH change test showed that this capacity was much lower in the case of acid-treated particles. The results obtained show that the ampicillin beads prepared are suitable for intramammary therapy.

Ionotropic gelation by divalent metal interaction was employed of indomethacin–sodium alginate dispersion with calcium ions to induce the spontaneous formation of indomethacin–calcium alginate gel discs (357). A three-phase approach was developed to establish the critical curing parameters. Since curing involved crosslinking of the sodium alginate with calcium ions, an optimal concentration of calcium chloride (phase one) and crosslinking reaction time (phase two) had to be determined. Furthermore, the third phase involved the optimization of the air drying time of the gel discs. In phases one and two, stabilization of in vitro drug release characteristics was used as the marker of optimal crosslinking efficiency. Phase three was based on achieving fully dried gel discs by drying to constant weight at 21°C under an extractor. The study revealed that optimal crosslinking efficiency was achieved in 1% w/v calcium chloride solution for 24 h and air dried at 21°C under an extractor for 48 h (357).

Growth and progression of malignant brain tumors occurs in a micromilieu consisting of both tumor and normal cells. Several proteins have been identified with the potential of interfering directly with tumor cells or with the neovascularization process, thereby inhibiting tumor growth. A continuous delivery of such inhibitory proteins to the tumor microenvironment by genetically engineered cells could theoretically be of considerable therapeutic importance. Read et al. (358) have investigated the growth characteristics of cells encapsulated in alginate, which represents a potential delivery system for recombinant proteins that may have antitumor effects. Three different cell lines, NHI 3T3, 293, and BT4C, were encapsulated in alginate, which is an immunoisolating substance extracted from brown seaweed. Morphological studies showed that encapsulated cells proliferated and formed spheroids within the alginate in the in vitro cultures and after implantation into the brain. Even after 4 months in vivo a substantial amount of living cells were observed within the alginate beads. A vigorous infiltration of mononuclear cells was observed in the brain bordering the alginate beads 1 week after implantation (358).

The insecticide/nematicide carbofuran was incorporated in alginate-based granules to obtain controlled release (CR) properties (359). The basic formulation (sodium alginate 1.61%, carbofuran 0.59%, water) was modified by addition of sorbents. The effect on carbofuran release rate, caused by the incorporation of natural and acid-treated bentonite in alginate formulation, was studied by immersion of the granules in water under shaking (359). The use of alginate-based CR formulations resulted in a reduction of the leached amount of carbofuran compared with the total amount of pesticide leached using the technical product (50 and 75% for CR granules containing natural and acid-treated bentonite, respectively). Alginate–bentonite CR formulations might be efficient systems for reducing carbofuran leaching in clay soils, which would reduce the risk of groundwater pollution.

E. Calcium Alginate Wound Dressing

It is commonly accepted that the ideal wound covering should mimic many properties of human skin. It should be adhesive, elastic, durable, impermeable to bacteria, and occlusive. Alginates are highly absorbent, gel-forming materials with hemostatic properties, and it has long been known that more rapid wound healing occurs when a gel is formed at the wound surface and dehydration is prevented (306). Because of the biocompatibility, exudate absorbability, and film-forming properties of calcium alginate product, they are good candidates for burn and wound management uses.

Ueng et al. (360) investigated the calcium alginate dressing as a drug delivery system for the treatment of various surgical infections. Cytotoxicity of the calcium alginate dressing to fibroblasts and HeLa cells was evaluated by the 3-(4,5-dimethyl-2-thiazolyl)-2,5-diphenyl-2H tetrazolium bromide colorimetric assay. The calcium alginate dressing was mixed with vancomycin, and lyophilized or not lyophilized to form two types of antibiotic dressings. The results suggested that the antibiotic dressings present no obvious toxic risk to their use as a drug delivery system. All antibiotic dressings released bactericidal concentrations of the antibiotics in vitro for the period of time needed to treat surgical infections. This study offers a convenient method to meet the specific antibiotic requirement for different infections.

The healing of cutaneous ulcers requires the development of a vascularized granular tissue bed, filling of large tissue defects by dermal regeneration, and the restoration of a continuous epidermal keratinocyte layer. These processes were modelled in vitro in the present study, utilizing human dermal fibroblast, microvascular endothelial cell (HMEC), and keratinocyte cultures to examine the effect of calcium alginate on the proliferation and motility of these cultures, and the formation of capillarylike structures by HMEC (361). This study demonstrates that the calcium alginate tested increased the proliferation of fibroblasts but decreased the proliferation of HMEC and keratinocytes. In contrast, the calcium alginate decreased fibroblast motility but had no effect on keratinocyte motility. There was no significant effect of calcium alginate on the formation of capillarylike structures by HMEC. The effects of calcium alginate on cell proliferation and migration may have been mediated by released calcium ions. These results suggest that the calcium alginate tested may improve some cellular aspects of normal wound healing but not others (361).

Drug-impregnated polyelectrolyte complex (PEC) sponge composed of chitosan and sodium alginate was prepared for wound dressing application (362). Equilibrium water content and release of silver sulfadiazine (AgSD) could be controlled by the number of repeated in situ PEC reactions between chitosan and sodium alginate. The release of AgSD from AgSD-impregnated PEC wound dressing in PBS buffer (pH = 7.4) was dependent on the number of repeated in situ complex formations for the wound dressing (362). In vivo tests showed that granulation tissue formation and wound contraction for the AgSD plus dihydroepiandrosterone (DHEA)–impregnated PEC wound dressing were faster than any other groups.

Control of hemorrhage during excision and grafting is difficult and postoperative hematoma may reduce graft take. Kneafsey et al. (363) have found calcium alginate dressings can be of immense help in minimizing these technical problems. Calcium alginate dressings following hemorrhoidectomy effectively reduce postoperative pain compared to more bulky anal packs (364).

A prospective controlled trial was carried out to assess the healing efficacy of calcium alginate and paraffin gauze on split skin graft donor sites (365). Calcium alginate dressings provide a significant improvement in healing split skin graft donor sites.

F. Calcium Alginate as Tissue Engineering

New cartilage formation has been successfully achieved by technology referred to as tissue engineering. Recent advances in tissue engineering permit us to focus on production of larger amounts of cartilaginous tissue, such as might be needed for reconstructive surgery of the entire auricle.

Polymers and hydrogels such as poly(glycolic acid), calcium alginate, and poly(ethylene) and poly(propylene) hydrogels have been used as cell carriers to regenerate cartilage in the nude mouse model. This study compared the suitability of three polymers for generating tissue engineered elastic cartilage using autologous cells in an immunocompetent porcine animal model (366). When using pluronics as scaffold, histologic features resemble those of native elastic cartilage, showing a more organized arrangement of the cells, which seems to correlate to functional properties as elastin presence in the tissue.

Transplantation of isolated chondrocytes has long been acknowledged as a potential method for rebuilding small defects in damaged or deformed cartilages. Chalain et al. (367) describe modification of the basic techniques that lead to production of a large amount of elastic cartilage originated from porcine and human isolated chondrocytes. Small fragments of auricular cartilage were harvested from children undergoing ear reconstruction for microtia or extirpation of preauricular tags and from ears of juvenile pigs. Enzymatically isolated elastic chondrocytes were then agitated in suspension to form the chondronlike aggregates, which were further embedded in molded hydrogel constructs made of alginate and type I collagen augmented with κ-elastin. The constructs were then implanted in nude mice and harvested 4 and 12 weeks after heterotransplantation. The resulting neocartilage closely resembled native auricular cartilage at the gross, microscopic, and ultrastructural levels (367).

In vitro multiplication of isolated autologous chondrocytes is required to obtain an adequate number of cells to generate neocartilage, but is known to induce cell dedifferentiation. Marijnissen et al. (368) and Demoor-Fossard et al. (369) investigated whether multiplied chondrocytes can be used to generate neocartilage in vivo. Adult bovine articular chondrocytes were suspended in alginate at densities of 10 or 50 million/mL, or after multiplication in monolayer for one (P1) or three passages (P3). Alginate with cells was seeded in demineralized bovine bone matrix (DBM) or a fleece of polylactic/polyglycolic acid (E210) and implanted in nude mice for 8 weeks. The newly formed tissue was evaluated. Structural homogeneity of the tissue, composed of freshly isolated as well as serially passaged cells, was found to be enhanced by high-density seeding (50 million/mL) and the use of E210 as a carrier. The percentage of collagen type II, positive-staining P3 cells was generally higher when E210 was used as a carrier. Furthermore, seeding P3 chondrocytes at the higher density (50 million/mL) enhanced collagen type II expression.

Chondrocytes from 21-day-old rat fetal nasal cartilage were cultured in alginate beads for up to 20 days (370). It was found that chondrocytes retained their spherical shape and typical chondrocytic appearance. During the culture time, chondrocytes underwent differentiation, as demonstrated by the alkaline phosphatase–specific activity and rate of proteoglycan synthesis. Morphological data confirmed chondrocyte differentiation with the appearance of hypertrophic chondrocytes scattered in the alginate gel and a dense extracellular matrix containing filamentous structures and matrix vesicles (370). These results indicate that the alginate system represents a relevant model for studies of chondrogenesis and endochondral ossification. Furthermore, the encapsulation method could prove useful for studies of tissue–biomaterial interactions in an in vitro environment which more closely mirrors the cartilage matrix than other culture methods.

A potential approach to facilitate the performance of implanted hepatocytes is to enable their aggregation and re-expression of their differentiated function prior to implantation. Glicklis et al. (371) have examined the behavior of freshly isolated rat adult hepatocytes seeded within a novel three-dimensional (3-D) scaffold based on alginate. The attractive features of this scaffold include a highly porous structure (spongelike) with interconnecting pores, and pore sizes with diameters of 100–150 μm. Due to their hydrophilic nature, seeding hepatocytes onto the alginate sponges was efficient. DNA measurements showed that the total cell number within the sponges did not change over 2 weeks, indicating that hepatocytes do not proliferate under these culture conditions. More than 90% of the seeded cells participated in the aggregation; the high efficiency is attributed to the nonadherent nature of alginate. The 3-D arrangement of hepatocytes within the alginate sponges promoted their functional expression; within a week the cells secreted the maximal albumin secretion rate of 60 microg albumin/10(6) cells/day (371).

Tissue engineering, a field that combines polymer scaffolds with isolated cell populations to create new tissue, may be applied to soft tissue augmentation—an area in which polymers and cell populations have been injected independently. Marler et al. (372) have developed an inbred rat model in which the subcutaneous injection of a hydrogel, a form of polymer, under vacuum permits direct comparison of different materials in terms of both histologic behavior and their ability to maintain the specific shape and volume of a construct. Using this model, three forms of calcium alginate, (a synthetic hydrogel) were compared over an 8-week period: a standard alginate that was gelled following injection into animals (alginate postgel); a standard alginate that was gelled before injection into animals (alginate pregel); and alginate-RGD, to which the cell adhesion tripeptide RGD was linked covalently (RGD postgel). Parallel groups that included cultured syngeneic fibroblasts suspended within each of these three gels were also evaluated (alginate postgel plus cells, alginate pregel plus cells, and RGD postgel plus cells). Histologically, the gel remained a uniform sheet surrounded by a fibrous capsule in the alginate postgel groups. In the alginate pregel and RGD postgel groups, there was significant ingrowth of a fibrovascular stroma into the gel with fragmentation of the construct. In constructs in which syngeneic fibroblasts were included, cells were visualized throughout the gel but did not extend processes or appear to contribute to new tissue formation. Material compression testing indicated that the alginate and RGD postgel constructs became stiffer over a 12-week period, particularly in the cell-containing groups (372).

Bone morphogenetic proteins (BMPs) are unique molecules with a specific biological activity for inducing ectopic bone formation when implanted with a suitable carrier matrix. A novel BMP-2–derived oligopeptide,

NSVNSKIPKACCVPTELSAI, was coupled covalently to alginate. Then NSVNSKIPKACCVPTELSAI-linked alginate hydrogel composites were implanted into the calf muscle of rats and harvested 3 or 8 weeks after surgery. Ectopic bone formation was observed in alginate hydrogel linked with BMP-2–derived peptide. It is suggested that alginate hydrogel linked with an oligopeptide derived from BMP-2 might provide an alternative system for topical delivery of the morphogenetic signal of BMP-2 (373).

Alginate membrane was proposed as a self-setting barrier membrane that can be used for guided tissue regeneration. The alginate membrane can be prepared and placed at the bone defect during the surgical procedure. The procedure consists of two simple steps. First, the bone defect is filled with sodium alginate (Na-Alg) aqueous solution. Then calcium chloride aqueous solution is dropped on the surface of the Na-Alg aqueous solution. An alginate membrane is formed on the bone defect, keeping the inside of the bone defect filled with unreacted Na-Alg aqueous solution (374).

Many materials have been used for artificial tubular prostheses to assist peripheral nerve gap reconstruction. However, the clinical use of these devices has been restricted because a microsurgical procedure requires specialized techniques and expensive equipment, such as operating microscope systems. Therefore Suzuki et al., (375) developed a new gluing method, without sutures, that uses freeze-dried alginate gel. A 7 mm gap in the sciatic nerve of rats was bridged with freeze-dried alginate gel. Regeneration was evaluated by electrophysiologic testing and histologic study. Eighteen weeks after surgery, functional reinnervation of motor and sensory nerves had occurred, as demonstrated by recovery of compound muscle action potentials (CMAPs), compound nerve action potentials (CNAPs), and somatosensory-evoked potentials (SEPs). Histologically, many regenerated nerve fasciculi, including myelinated and unmyelinated fibers, were observed and the implanted alginate gel had disappeared. In conclusion, a gluing technique using alginate gel is a potential alternative to the conventional nerve autograph technique. Advantages include simple application and rapid repair. Freeze-dried alginate gel is a promising material for artificial nerve guides for peripheral nerves and also could be used for repair of disrupted pathways in central nervous tissue that is amorphous and cannot be sutured (375).

XIII. XANTHAN GUM

Xanthan gum is a widely used thickening agent in foods and is a recent addition to the hydrophilic matrix carrier list (376). The primary structure is a $(1 \rightarrow 4)$-linked β-D-glucan backbone (cellulose) substituted through position 3 on alternate glucose residues with a charged trisaccharide side chain (377). Xanthans secondary structure has been studied by x-ray fiber diffraction (378) and analyzed be molecular modeling (379). Addition of the side chain causes the backbone to change from a twofold ribbonlike cellulose conformation to a fivefold helix (380–383). A single helix stabilized by backbone–sidechain bonding has been proposed, but double helical models cannot be excluded (378,379).

The majority of studies of xanthan in the literature have been ''molecular,'' including optical rotation, NMR, DSC, and circular dichroism (381–385), but the interpretation of some of these have been complicated by the nature of the macromolecule. The order–disorder transition has first order kinetics, is fully reversible, and shows no thermal hysteresis. From this evidence Morris et al. (380) suggested a single helix stabilized intramolecularly by ordered packing of side chains along the polymer backbone.

A. Xanthan Gels

A number of polysaccharides interact with galactomannans resulting in synergistic viscosity increases or gel formation, which have been extensively investigated be Dea et al. (386,387) and other groups (388–398). Synergistic interactions of polysaccharides in binary mixtures have often been considered to be synonymous with intermolecular binding of the two polysaccharides. The synergisms between plant galactomannans (carob, tara, or enzymatically modified guar gum) and xanthan or certain algal polysaccharides (κ-carrageenan, furcelaran, or agarose) have been attributed to intermolecular binding of the backbone of the galactomannan and the helix of the order polysaccharide (386,390). The evidence for this and for similar (388,389) well-accepted such intermolecular models. The structural similarity of galacturonic acid and glucuronic acid blocks favors the formation networks between pectin and alginate (387). The crosslinking of cellulose fibrils by galactomannans has indeed been demonstrated by electron microscopy (391). Cairns et al. (392,393) have proposed a different model for the gelation of xanthan and galactomannans, based on x-ray fiber diffraction studies on stretched gels. Mixed junction zones were formed by interaction between the xanthan and galactomannan backbones, with the relative positions of xanthan side chains on either side of a sandwiched galactomannan molecule being staggered. The x-ray results indicate a repeat distance of 0.52 nm (393). Xanthan is widely used in foodstuffs in the form of synergistic gels with gluco- and galactomannans (394); these two types of polysaccharide per se and xanthan alone will not form gels, whereas mixtures of xanthan with either of the plant polysaccharides, when heated and allowed to cool, form thermoreversible gels (395,396).

The influence of the galactomannan characteristic ratios (M/G) on the temperature of gelation (Tg) and the gel strength of mixtures of galactomannan with xanthan is reported (397). Two galactomannans were investigated: one highly substituted from the seeds of *Mimosa scabrella* (M/G = 11) and the other, less substituted, from the endosperm of *Schizolobium parahybae* (M/G = 30). The xanthan/galactomannan systems (4:2 g l(−1), in 5 mM NaCl) showed a Tg of 24°C for that of *S. parahybae* (398) and 20°C for the galactomannan of *M. scabrella*, determined by viscoelastic measurements and microcalorimetry. A Tg of 40–50°C was found by Shatwell et al. (399) for locust bean gum (LBG) (M/G = 43). Lundin and Hermansson (400) reported a difference of 13°C Tg of two LBG samples, with M/G = 3 (40°C) and 5 (53°C), in mixtures with xanthan. It appears that the more substituted galactomannans have lower temperatures of gelation in the presence of xanthan. The mechanism of gelation depends also on the M/G ratio. For the lower values it involves only disordered xanthan chains in contrast to M/G ratios higher than 3. In addition, the presence of the galactomannan from *M. scabrella* increased slightly the temperature of the conformational change (Tm) of xanthan, probably due to the ionic strength contribution of proteins (3.9%) present in the galactomannan. On the other hand, the galactomannans from *S. parahybae*, with 1.5% of proteins, and *M. scabrella*, with 2.4% of protein, did not show this effect, the Tm of xanthan alone or in a mixture being practically unchanged (397).

Examples of current or potential applications of xanthan in pharmaceutics (401,402) or biomedical uses are as excipients in tablets or clear blood fluid substitutes. The function of xanthan in tablets is similar to other polysaccharides used in the same way, namely, to erode or dissolve slowly and thereby yield a delayed release of active ingredients compared to formulations not containing hydrocolloids. Recently reported is that xanthan used in tablets yields a comparable kinetics in the release of drugs to those of formulations containing N-CMC or carrageenan (403).

Oral candidiasis frequently occurs in individuals with dry mouth syndrome (xerostomia), in immunocompromised patients, and in denture wearers. Russien et al. (401) developed a formulation which will prolong the retention time of antimicrobial agents at the site of application. The activity against *Candida albicans* of a synthetic cationic peptide dhvar 1, based on the human fungicidal salivary peptide histatin 5, was tested in a mixture with the bioadhesive. Coupling caused a reduction of the viscosity and elasticity of the xanthan solution related to the applied concentration of the coupling agent. Incubation of the peptide with clarified human whole saliva resulted in proteolytic degradation of the peptide. In the presence of xanthan the degradation occurred more slowly. It was concluded that xanthan is an appropriate vehicle for antimicrobial peptides in a retention-increasing formulation.

XIV. PECTIN

Pectin is a general term for a group of natural polymers that occur as structural materials in all land-growing plants. Polymerized galacturonic acid partially esterified with methanol accounts for the major part of any commercial pectin. Commercial pectins are divided into low-ester pectin and high-ester pectins. Pectin is not degraded by enzymes secreted in the upper gastrointestinal tract and passes intact through the small intestine. Pectin has been used as fiber source in numerous studies of the effects on gastric emptying, gastric ulcer, glucose and insulin level, bile acid binding, gallstone, binding of divalent and trivalent cations, influence on the enzymatic activity of the upper digestive tract, lowering serum cholesterol level, healing intestinal wounds, colon cancer, colonic cell proliferation, and as source of volatile fatty acids in the colon (404–409). Pectin influences the viscosity of the meal, and its short-term effect on the rate of gastric emptying may be explained in this way. A sustained effect of pectin consumption in the studies by Schwartz et al. (410), however, cannot be explained by increased viscosity of the stomach content. Apparently, pectin delays absorption of glucose in two ways: by delaying gastric emptying and by increasing the intestinal barrier layer (407).

Kohen et al. (411) demonstrated a protective effect of pectin against oxidative damages of the jejunal mucosa in rats. Protection against radicals is important because the mucosa is exposed to oxidative stress from the diet.

The antidiarrhea effect of pectin (in the form of dried citrus peels or waste from potato starch manufacturing) in cattle is enhanced by mixing with lecitin from soy oil refining. The positive effect of lecithin supplementation is explained by increased adhesion of pathogenic bacteria to the pectin–lecithin complex (412).

The pectin-based raft-forming antireflux agent Aflurax (Idoflux) was examined, first regarding reduction of esophageal acid exposure and also as to its efficacy as maintenance treatment in patients with healed esophagitis (413). The median (interquartile range) acid exposure times in the upright position were 3.1% (1.6–13.0%) on Aflurax versus 6.7% (2.5–14.9%) on placebo ($p = 0.10$). In the supine position no difference was found (Aflurax 13.7%, placebo 13.2%). The time to recurrence of heartburn with Aflurax treatment was prolonged significantly; after 6 months the life table estimates were 48% of patients in remission on Aflurax versus 8% on placebo ($p = 0.01$).

Following treatment, erosive esophagitis was found in 17/34 on Aflurax versus 28/38 on placebo ($p < 0.05$). Aflurax significantly delays recurrence of moderate or severe heartburn and erosive esophagitis, when used as maintenance treatment. The acid exposure was not significantly reduced with pH monitoring (413).

Lowering the serum cholesterol level is, without doubt, the most studied physiological effect of pectin and other soluble dietary fibbers. Numerous studies in hypercholesterolemic and normolipidemic humans, rats, and other experimental animals have proven the serum cholesterol lowering effect of pectin (404–408,414–417). The conclusion from the last 25 years' studies is that a pectin dosage of 10–15 g/day leads to a decrease in the serum cholesterol level of 10%.

The dietary effect of the water-soluble dietary fibers (WSDF) guar gum, partially hydrolyzed guar gum (PHGG), glucomannan, and highly methoxylated (HM) pectin on the serum lipid level and immunoglobulin (Ig) production of Sprague–Dawley rats was compared with that of water-insoluble cellulose (418). Although serum total cholesterol and triglyceride levels were significantly lower in the rats fed with WSDF than in those fed with cellulose, a decrease in the level of phospholipids was only observed in the rats that had been fed on guar gum or glucomannan. In addition, all WSDF feeding enhanced IgA productivity in the spleen and mesenteric lymph node lymphocytes, although the increase in serum IgA level was only observed in the rats fed on WSDF, and not on PHGG. When mesenteric lymph node lymphocytes were cultured in the presence of various concentrations of guar gum or glucomannan, no significant increase in Ig production was apparent. These data suggest that WSDF indirectly enhanced the Ig production of lymphocytes, and that serum lipid reduction and IgA production–enhancing activities of WSDF were dependent on their molecular sizes (418).

Experimental hypercholesterolemia and its modulation by some natural dietary supplements (pectin, garlic, and ginseng) and by the drug gemfibrozil were studied (419). Results of the study demonstrated that feeding the cholesterol-enriched diet caused a significant increase in total, LDL, and HDL cholesterol; plasma MDA and post-heparin total; and hepatic lipase activities. On the other hand, serum Tg and erythrocyte superoxide dismutase were not changed. Histopathological examination revealed marked alteration in the aortic wall with the appearance of large multiple atheromatous plaques. Both garlic and pectin were successful in a significant reduction of the hypercholesterolemia in a way comparable to gemfibrozil. Garlic was the only treatment that has antilipid peroxidative property (419).

To investigate the effects of pectin on cholesterol metabolism, normal rats were fed for 3 weeks a diet containing 2.5 or 5% apple or orange pectin or without pectin (control) (420). Cholesterol concentrations were determined in feces after 1, 2, and 3 weeks of treatment, and in liver and serum at the end of the experimental trials. Cholesterol concentration in feces showed a significant increase by week 3 in rats fed 5% orange or apple pectin. Hepatic cholesterol concentration declined significantly in all pectin-fed groups. Serum cholesterol only declined significantly in apple-fed groups. The decrease of cholesterol levels in liver and serum and its increase in feces could explain the beneficial effect of including these fibbers in the diet to prevent some currently very frequent diseases (420).

Because a high daily consumption of polysaccharide-containing food is assessed to decrease the risk of cancer of the gastrointestinal system, different types of carbohydrates were investigated for their antimutagenic activity against different standard mutagens. Within the screening pronounced antimutagenic effects were found for xyloglucan and different pectins and pectinlike rhamnogalacturonans against 1-nitropyrene–induced mutagenicity. Inhibition rates were dose-dependent and varied between 20 and 50%. Concerning the mode of action, a direct interaction of the polymers with the cells is claimed, protecting the organisms from the mutagenic attack (421).

Anticarcinogenic and tumor growth–inhibiting effects of nonsoluble fibers have been described (422). In a preliminary study on methylnitrosourea-induced mammary carcinogenesis in female Sprague–Dawley rats, 15% oligofructose added to the basal diet modulated this carcinogenesis in a negative manner (423). There was a lower number of tumor-bearing rats and a lower total number of mammary tumors in oligofructose-fed rats than in the group fed the basal diet alone.

The effect of dietary nondigestible carbohydrates (15% oligofructose, inulin, or pectin incorporated into the basal diet) on the growth of intramuscularly transplanted mouse tumors, belonging to two tumor lines (TLT and EMT6), was also investigated (424). The results were evaluated by regular tumor measurements with a vernier caliper. The mean tumor surface in the experimental groups was compared with that in animals of the control group fed the basal diet containing starch as the only carbohydrate. The growth of both tumor lines was significantly inhibited by supplementing the diet with nondigestible carbohydrates. Such nontoxic dietary treatment appears to be easy and risk free for patients, applicable as an adjuvant factor in the classical protocols of human cancer therapy (424).

Among pectin, apple pectin exerts stronger bacteriostatical action on *Staphylococcus aureus*, *Streptococcus faecalis*, *Pseudomonas aeruginosa*, and *Escherichia coli* in comparison with citrus pectin. In this study, we used water-soluble methoxylated pectin from apple. The diet, supplemented by 20% apple pectin, significantly decreased the number of tumors and the incidence of colon tumors. PGE2 level in distal colonic mucus in 20% apple pectin–fed rats were lower than those in basal diet fed rats. Fecal β-glucuronidase activities in the apple pectin–fed group, which has been considered a key enzyme for the final activation of dimethylhydrazine metabolism to carcinogens in the colonic lumen, were significantly lower than those in control group at initiation stage of carcinogenesis. The concentrations of β-glucosidase and azoreductase were also decreased. The effect of apple pectin on the colon carcinogenesis may partially depend on PGE, concentration decrease in colonic mucus, and on the type of pectin, also related to fecal enzyme activities.

The resistance of pectin to degradation in the upper GI tract and its complete dissolution in the colon makes pectin an ideal ingredient for colon-specific delivery. The solubility in the GI fluids is suppressed by crosslinking with calcium or by chemical means. Coacervate with gelatine permits the formation of microglobules suitable for controlled-release products (425–427).

To develop an enzymatically controlled pulsatile drug release system based on an impermeable capsule body, which contains the drug and is enclosed by an erodible pectin/pectinase plug have been studied (428). The plug was prepared by direct compression of pectin and pectinase in different ratios. In addition to the disintegration times of the plugs, the lag times and the release profiles of the pulsatile system were determined as a function of pectin/enzyme ratio, the pH of the surrounding medium, and the addition of buffering or chelating agents. The drug release was controlled by the enzymatic degradation and dissolution of pectin (429).

The use of pectin in tablet formulations have been studied by Meshali and Gabr (430) using a blend of pectin and chitosan. Ashford et al. (431,432) protected a core tablet with a coat of HM pectin. It was shown that the tablets disintegrated in the colon. Coprecipitation of cationic drugs with pectin was shown to be applicable for sustained release of water-soluble drugs (433). Encapsulation of liver cells by coacervation of carboxymethylcellulose, chondroitin sulfate, chitosan, and polygalacturonic acid gave viable cells that could be used for extracorporeal liver support (434).

XV. HEPARIN

Heparin is the most biologically reactive member of the family of sulfated glycosaminoglycans (GAGs), which are widespread in animal tissue (435). Heparin and chondroitin sulfates, which are the most abounded sulfated glycosaminoglycans, contribute an enormously important role in the successful development of open heart surgery. Due to its antithrombotic and anticoagulant properties, heparin is extensively used in the management of cardiovascular diseases (436–438). Long-term treatments with heparin, as well as short-term ones in the case of patients susceptible to hemorrhage, involve risk of bleeding. Another unwanted side effect of heparin is its interaction with platelet function, which may lead to depletion of these blood components (thrombocytopenia, HIT). During the last decade, fundamental aspects of the pathogenesis of HIT have been resolved. The understanding of some the mechanisms underlying the development of new, paradox thromboembolic complications in HIT led to the concept that thrombin generation plays a key role in clinically manifest HIT. Consequently new therapeutic concepts imply the use of drugs with either indirect or direct antithrombin activity, such as donaparoid sodium and the recombinant hirudin lepirudin. In recent years results of the first prospective studies assessing various treatment regimens in HIT became available. Although data of randomized trials are still missing, some treatment recommendations can already be drawn from these studies (439).

The discovery of new activities of heparin, relating to inhibition of growth of smooth muscle cells (440), angiogenesis (441), the human immunodeficiency virus (HIV) (442), and tissue engineering (443), has widened the scope of potential uses of this GAG. Incorporation of heparin-binding peptides into fibrin gels enhances neurite extension, an example of designer matrices in tissue engineering.

Kratz et al. (444) have recently shown that heparin in combination with chitosan stimulates re-epithelialization in an in vitro model of human wound healing. The chitosan–heparin membrane stimulated the increased stabilization and concentration of growth factors in the wound area, which stimulated healing.

Some of the biological properties of heparin can be simulated by other GAGs or by chemically sulfated vegetal, algal, or microbial polysaccharides. However, the type of carbohydrate backbone and the sulfation pattern of the polysaccharide are important for the emergence and level of specific activities (445). As the prophylactic use of heparin continues to increase, nurses must be aware that heparin use may cause heparin-induced skin necrosis—a rare but serious complication. Although even more severe complications may occur from heparin use, this discussion will focus on skin necrosis caused by subcutaneous heparin. Should heparin-induced skin necrosis develop, heparin therapy must be discontinued immediately. Reports of one patient's reaction to this complication have been presented (446,447).

Angiogenesis is a prerequisite for tumor expansion and metastasis. The angiogenic potential of the heparin-binding growth factors acidic fibroblast growth factor (FGF) and basic FGF has been demonstrated in various publications. Zugmaier et al. (448) have studied the inhibitory effects of suramin and the polysulfated heparinoids pentosan polysulfate, dextran sulfate, and fucoidan on the action of FGF. Polysulfated heparinoids exert a selective inhibitory effect on heparin-binding angiogenesis factors at an IC50, which is 100 times below the IC50 of suramin. Therefore, the administration of polysulfated heparinoids might become a novel approach to tumor therapy based on blocking angiogenesis.

Aspirin, a potent antiplatelet drug, and heparin, an anticoagulant, are commonly used for postimplant complications such as thrombosis and thromboembolism. Aspirin and heparin were embedded in chitosan–polyethylene vinylacetate comatrix to develop a prolonged release form (449). The effect of these drugs toward the bioprosthetic calcification was investigated by in vitro and in vivo models (449). In vitro and in vivo evaluation suggest that the released aspirin–heparin from the comatrix had a synergistic effect in inhibiting glutaraldehyde pretreated bovine pericardium calcification. Biochemical, histological, and scanning electron microscopic evaluation of retrieved samples demonstrated a significant reduction in calcium deposition.

Many of the acute coronary ischemic syndromes are triggered by spontaneous or mechanical disruption of atherosclerotic plaques with resultant activation of platelets and coagulation. Given the central role of platelets and thrombin in arterial thrombosis, current strategies for its prevention and treatment focus on both inhibition of platelet aggregation and control of thrombin generation and activity. Although aspirin and unfractionated heparin are the cornerstones of current treatment strategies, both have limitations (450).

Among the biomedical applications of synthetic polymers, the development of blood-handling equipment (e.g., the hemodialysis and cardiopulmonary bypass, angiographic catheters, intra-aortic balloon pumps, cardiovascular prostheses, artificial heart devices) has been one of the most investigated areas in the past two decades. Biomaterial thrombogenicity remains the single most important concern preventing even more widespread application.

In recent years, low molecular weight heparins (LMWH) have been tested in the prevention and treatment of deep vein thrombosis and pulmonary embolism, as well as in the treatment of stroke and unstable angina. In all these situations, LMWH had a similar or even a better risk/benefit ratio than unfractionated heparin (UH) (451). Acute myocardial infarction and coronary angioplasty are among the new targets presently under investigation with various LMWH. Patients with mechanical heart valves require life-long anticoagulation. Therefore, the anticoagulation with LMWH after mechanical heart valve replacement were studied (451). Anticoagulation activity with LMWH after mechanical heart valve replacement appears feasible, provides adequate biological anticoagulation, and compares favorably with UH anticoagulation. Randomized studies are now needed to further evaluate this new therapeutic approach.

Albumin has also been utilized with heparin preadsorption of both molecules or by their covalent coupling, which results in heparin–albumin conjugates. Albumin preadsorbed onto surface reduce platelet adhesion, while heparin is able to interact with antithrombin III, preventing thrombus formation (452–456). Crosslinked gels of albumin as well as heparinized albumin gels, potential sealant of prosthetic vascular grafts, were studied with regard to in vitro stability, binding of basic fibroblast growth factor, and cellular interactions (457). It can be concluded that crosslinked gels of albumin to which heparin is immobilized, are candidate sealants for prosthetic vascular grafts and suitable substrates for endothelial cell seeding.

Inhibitory effects of heparin coupling on calcification of bioprosthetic vascular grafts of different origin were studied. Heparin-bonded and 0.625% glutaraldehyde crosslinked (GA) segments of porcine thoracic aorta (AO), pulmonary artery (PA), jugular vein (JV) and rabbit aorta (RA) were implanted subcutaneously in weanling rats for 5 months (458). Heparin bonding was ineffective in prevention of calcification of JV and RA. Calcium content of heparin-coupled PA and AO was significantly less when compared with their GA-treated counterparts. Calcification inhibition was achieved to a greater extent in heparin-bonded PA than in the AO coupled to heparin. Heparin-bonded porcine pulmonary artery seemed to be the best among all vascular bioprostheses in this study.

Clinical procedures involving extracorporeal blood circulation are potentially complicated by the interaction of various blood systems with foreign surfaces. In cardiopulmonary bypass, exposure of blood to synthetic surfaces generally leads to activation of cellular and humoral blood systems with activation of complement cascade. This reaction can be associated with a variety of postoperation clinical complications, such as increased pulmonary capillary permeability, anaphylactic reactions, and various degrees of organ failure which contribute to mortality in routine cardiac operations. Application of biocompatible materials in an extracorporeal circuit modifies the normal pattern of blood activation, and therefore may potentially reduce clinical complications in routine cardiac surgery (459). The use of heparin-coated circuits resulted in reduction of systemic leukocyte activation of cardiopulmonary bypass reflected by reduced leukocyte and neutrophil counts 24 h after operation ($p < 0.05$).

Various sugar-carrying polystyrenes (PS), which consist of synthetic styrene and sugar moieties, are glycoconjugates that are able to attach to polymeric surfaces. Heparin-carrying PS (HCPS) is especially able to retain the binding of heparin-binding growth factors (GFs) such as vascular endothelial GF 165 or fibroblast GF 2. Human skin fibroblast cells, human coronary smooth muscle cells, and human coronary endothelial cells have good adherence to the HCPS-coated plate. These results indicate that growth of various cells can be controlled by the HCPS coating, thereby retaining the bioactivity of molecules such as heparin-binding GFs. Thus, HCPS-coated surfaces control selective growth of various cells (460).

Using polystyrene microspheres coated with heparin or heparan sulfate, it was shown that coated microspheres specifically bound eukaryotic cells and were endocytosed by nonprofessional phagocytic cells. Coated microspheres displayed properties of binding to eukaryotic cells that were similar to those of Chlamydiae, and the microspheres were competitively inhibited by chlamydial organisms. Endocytosis of heparin-coated beads resulted in the tyrosine phosphorylation of a similar set of host proteins, as did endocytosis of Chlamydiae; however, unlike viable chlamydial organisms, which prevent phagolysosomal fusion, endocytosed beads were trafficked to a lysosomal compartment. These findings suggest that heparin-coated beads and *Chlamydia trachomatis* enter eukaryotic cells by similar pathways (461).

Poly(ethylene terephthalate) (PET) film was exposed to oxygen plasma glow discharge to produce peroxides on its surfaces. These peroxides were then used as catalysts for the polymerization of acrylic acid (AA) in order to prepare a carboxylic acid group–introduced PET (PET-AA). Insulin and heparin coimmobilized PET (PET-I-H) was prepared by the grafting of poly(ethylene oxide) to PET-AA, followed by reaction first with insulin and then heparin (462). The blood compatibility of the surface-modified PETs were examined using in vitro thrombus formation, plasma recalcification time, activated partial thromboplastin time, and platelet adhesion and activation.

The growing importance of polymer membrane–based potentiometric polyion sensors in biomedical research and clinical measurements has brought up the question of how accurate and reproducible these sensors are. Polymer membrane–based potentiometric sensors were developed earlier to provide a rapid and direct method of analysis for polyions such as heparin. These heparin sensors are irreversible, requiring a membrane renewal procedure between measurements which currently prevents the sensors from being used for continuous monitoring of blood heparin. Indeed, recent research has revealed that these sensors behave quite differently than classical ion-selective electrodes. Mathison and Bakker (463,464) explored ways to improve measurement reproducibility and long-term potential stability by considering the unique pseudo–steady state response mechanism of the polyion sensors developed so far. Heparin may be stripped out of the phase boundary membrane surface with a high sample NaCl concentration, and this characteristic is used to modify the calibration procedure in order to avoid memory effects. It is also attempted to reduce long-term potential drifts by continuously stripping heparin out of the membrane at the membrane inner filling solution side. A theoretical model is presented to explain the experimental results (464).

XVI. CHITOSAN

Chitosan is a polysaccharide comprising copolymers of glucosamine and N-acetylglucosamine. Chitosan is usually prepared from chitin (2-acetamido-2-deoxy β-1,4-D-glucan), and chitin has been found in a wide range of natural sources (crustaceans, fungi, insects, annelids, mollusks, coelenterate, etc.). However chitosan is only manufactured from crustaceans (crab and crayfish), primarily because a large amount of the crustacean exoskeleton is available as a byproduct of food processing. It is a natural, nontoxic, biodegradable polysaccharide available as solution, flake, fine powder, bead, and fiber. Due to the fact that chitosan has a large molecular weight, exhibits a positive charge, and demonstrates film-forming ability and gelation characteristics, the material has been extensively used in the industry, foremost as a flocculant in the clarification of wastewater (465,466), as a chelating agent for harmful metals for the detoxification of hazardous waste (467,468), for the clarification of beverages, such as fruit juices and beers (469), and for agricultural purposes such as fungicides (470). In addition chitosan has been exploited in the cosmetic industry, in the dental industry, for hair care products, and for ophthalmic applications, such as for contact lens coatings or the contact lens material itself (471–473). It has been extensively used as a biomaterial (474) due to its immunostimulatory activities, anticoagulant properties (475), antibacterial and antifungal action (476–481), and also as a promoter of wound healing in the field of surgery (482–484). Chitosan caused leakage of glucose and lactate dehydrogenase from *E. coli* cells. These data support the hypothesis that the mechanism of chitosan antibacterial action involves a crosslinkage between the polycations of chitosan and the anions on the bacterial surface that changes the membrane permeability.

Recently, Artursson et al. (485) and Schipper et al. (486) reported that the structural properties of chitosans, such as degree of acetylation and molecular weight, are very important for its absorption enhancement of hydrophilic drugs. They found that a low degree of acetylation and/or a high molecular weight appear to be necessary for chitosans to increase the epithelial permeability.

Chitosan ingestion effectively lowers serum cholesterol (487–489). Oligosaccharides of chain length less than five residues are ineffective. Orally administered chitosan binds fat in the intestine, blocking absorption, and has been shown to lower blood cholesterol in animals and humans. As a result it has been proposed that dietary supplementation with chitosan may inhibit the formation of atherosclerotic plaque (490,491). Animals were fed for 20 weeks on a diet containing 5% chitosan or on a control diet. Blood cholesterol levels were significantly lower in the chitosan-fed animals throughout the study, and at 20 weeks were 64% of control levels. When the area of aortic plaque in the two groups was compared, a highly significant inhibition of atherogenesis in both the whole aorta and the aortic arch was observed in the chitosan-fed animals—42 and 50%, respectively. Body growth was significantly greater in the chitosan-fed animals. This study shows a direct correlation between lowering of serum cholesterol with chitosan and inhibition of atherogenesis, and suggests that the agent could be used to inhibit the development of atherosclerosis in individuals with hypercholesterolemia.

The effects of chitosan have been investigated on 80 patients with renal failure undergoing long-term stable hemodialysis treatment (492). The patients were tested after a control treatment period of 1 week. Half were fed 30 chitosan tablets (45 mg chitosan per tablet) three times a day. Ingestion of chitosan effectively reduced total serum cholesterol levels (from 10.14 ± 4.40 to 5.82 ± 2.19 mM) and increased serum hemoglobin levels (from 58.2 ± 12.1 to 68 ± 9.0 g L-1). Significant reductions in urea and creatinine levels in serum were observed after 4 weeks of chitosan ingestion (492). During the treatment period, no clinically problematic symptoms were observed. These data suggest that chitosan might be uneffective treatment for renal failure patients, although the mechanism of the effect should be investigated further.

New silica–chitosan hybrid biomaterials were produced using biopolymer chitosan and its heparinlike derivative as the organic species to be incorporated into the silicon alkoxide–based network (493). These hybrid materials displayed good blood compatibility in comparison with their single component systems.

A. Biocompatibility and Bioadhesivity of Chitosan

Chitin and chitosan are substrate for lysozyme, an enzyme found in various mammalian tissues (494,495). The enzymatic hydrolysis of chitin and chitosan leads to the production of N-acetyl-D-glucosamine and D-glucosamine. It is believed that the glucosamine plays an important physiological role in in vivo biochemical processes. It is known that glucosamine takes part in detoxification functions of liver and kidney and possesses anti-inflamatory, hepatoprotective, antireactive, and antihypoxic qualities (496–498).

Chitosan lacks irritant or allergic effect and is biocompatible with both healthy and infected human skin (499). When chitosan was administered orally in mice, the LD_{50} was found to be in excess of 126 g/kg, which is higher than that of sucrose (500).

Chitosan has been proposed for the development of membranes and fibers for hemodialysis and blood oxygenators, skin substitute, and wound dressing materials (501–503). Chitosan as hemodialysis membranes promotes surface-induced thrombosis and embolization (501). When blood comes in contact with biomaterial surfaces like chitosan, there is an initial adsorption of plasma proteins, followed by adhesion and activation of platelets (504).

To improve blood compatibility, chitosan surface was modified by the complexation–interpenetration method using an anionic derivative of poly(ethylene glycol), methoxypropyl(ethylene glycol) sulfonate (505). The result of this study shows that chitosan surface can be permanently modified by complexation–interpenetration of anionic poly(ethylene glycol) derivative to improve blood compatibility of chitosan. When in contact with blood, the modified surface can resist plasma protein adsorption and cell adhesion by the steric repulsion mechanism (505).

Mucoadhesive properties of three viscosity grades of chitosan were investigated to assess the suitability of this polymer for gastroadhesive formulations. The influence of the pH, ionic strength, and temperature of the hydration medium on the rheological properties were studied. The mucoadhesive performance was assessed in vitro by means of rheological synergism and tensile stress testing (506). It was found that the interaction of chitosan with mucin decreases on increasing molecular weight.

Toxicity of chitosan also depends on its high charge density but appears to be less affected by the molecular weight (507,508).

B. Chitosan and Its Derivatives as Biomaterials for Wound Healing

It is commonly accepted that the ideal wound covering should mimic many properties of human skin. Because of the biocompatibility, exudate absorbability, and film-forming properties of chitosan products, they are good candidates for burn and wound management uses (509). Chitosan may be used to inhibit fibroplastia in wound healing and promote tissue growth and differentiation in tissue culture. Chitosan also shows a biological aptitude to stimulate cell proliferation and histoarchitectural tissue organization (510).

Earlier, Prudden (511) and Allen and Prudden (512) had investigated the tricking effect of cartilage preparations in accelerating wound healing. The phenomenon was later attributed to the presence of N-acetyl-D-glucosamine.

Chitosan has been reported as a wound healing accelerator (513). Cotton fiber type chitosan (DA = 18%) was applied on open skin wounds for 15 days, and the process of wound healing was evaluated histologically and immunohistochemically. On day 3 postwounding, the chitosan-treated wounds showed histologically severe infiltration of polymorphonuclear cells. Granulation was more pronounced on day 9 and 15. Immunohistochemical typing of collagen I, III, and IV showed increase of the production of type III collagen in the chitosan group. The appearance of mitotic cells occurred numerously in the control on postwounding day 3 and in the chitosan group on postwounding day 6. These results suggest chitosan has a function in the acceleration of infiltration of polymorphonuclear cells at the early stage of wound healing, followed by the production of collagen by fibroblasts (513).

Chitosan acetate films, which were tough and protective, had the advantages of good oxygen permeability, high water absorptivity, and slow lysozyme degradation (514).

Loke et al. (515) have prepared wound dressing consisting of two layers: the upper layer a carboxymethyl–chitin hydrogel material, while the lower layer is an antimicrobial impregnated biomaterial. The hydrogel layer acts as a mechanical and microbial barrier and is capable of absorbing wound exudate. From the in vitro release studies, the loading concentration was optimized to deliver sufficient antimicrobial drug into the wound area to sustain the antimicrobial activity for 24 h.

Chitosan may facilitate wound healing by stimulating granulation tissue formation and re-epithelialization (516,517). The 3M company has marketed TEGASORB, a wound healing product for human use containing chitosan as an excipient.

A photocrosslinkable chitosan to which both azide and lactose moieties were introduced (Az-CH-LA) was prepared as a biological adhesive for soft tissues, and its effectiveness was compared with that of fibrin glue. Introduction of the lactose moieties resulted in a much more water-soluble chitosan at neutral pH. Application of ultraviolet light irradiation to photocrosslinkable Az-CH-LA produced an insoluble hydrogel within 60 s (518). The binding strength of the chitosan hydrogel prepared from 30–50 mg/mL of Az-CH-LA was similar to that of fibrin glue. Compared to the fibrin glue, the chitosan hydrogel more effectively sealed air leakage from pinholes on isolated small intestine and aorta and from incisions on isolated trachea. Neither Az-CH-LA nor its hydrogel showed any cytotoxicity in cell culture tests of human skin fibroblasts, coronar endothelial cells, and smooth muscle cells. These results suggest that the photocrosslinkable chitosan developed here has the potential of serving as a new tissue adhesive in medical use.

Water-soluble chitin was prepared by controlling degree of deacetylation and molecular weight of chitin through alkaline and ultrasonic treatment (519). Its accelerating effect on wound healing in rats was compared with those of chitin and chitosan. The water-soluble chitin was found to be more efficient than chitin or chitosan as a wound-healing accelerator. The wound treated with water-soluble chitin solution was completely re-epithelialized, granulation tissues in the wound were nearly replaced by fibrosis, and hair follicles were almost healed at 7 days after initial wounding.

C. Chitosan Use as a Matrix for Drug Delivery Systems

As a pharmaceutical excipient, chitosan has been added for sustained release (520–522) and to improve dissolution of poorly soluble drugs (523). The applications of chitosan as an excipient in pharmaceutical products are listed in Table 3.

Kotze et al. (524) showed that chitosan hydrochloride and chitosan glutamate are potent absorption enhances in acidic environments. It was concluded that there is a need for chitosan derivatives with increased solubility, especially at neutral and basic pH values, for use as absorption enhancers aimed at the delivery of therapeutic compounds in the more basic environment of the large intestine and colon.

The efficiency of chitosan in tablets was dependent on chitosan cristallinity, degree of deacetylation, molecular weight, and particle size (525,526).

Table 3 Chitosan as a Pharmaceutical Excipient

Conventional formulations:
Gels
Films
Emulsions
Direct compression tablets
Wetting agent
Coating agent
Controlled release matrix tablets:
Microspheres and microcapsules
Transmucosal drug delivery
Vaccine delivery
DNA delivery

Sabnis et al. (527) studied the potential utility of chitosan (I) in inhibiting diclofenac sodium (II) release in the gastric environment from a directly compressible tablet formulation. Analyses of variance model was tested using SAS® (SAS Institute, Inc., NC) indicated that the degree of N-deacetylation of chitosan significantly affected drug release at pH 1.2 and 6.8 ($p < 0.0001$). An increase in the pH of the dissolution medium resulted in an increase in drug release ($p < 0.0001$). The ionic strength of the dissolution medium did not significantly affect drug release at any of the pHs studied ($p > 0.198$).

Using chitosan solution (pH < 6), gel formation will occur with a number of different multivalent anionic counterions. By adding a chitosan solution dropwise into a crosslinking solution having pH < 6, a real ionotropic gel is formed (528). The NH_2 groups of chitosan are protonated and an ionic crosslinking occurs.

Chitosan crosslinked with high molecular weight counterions (529–531) results in capsules (Table 4), while crosslinking with low molecular weight counterions results in globules (Table 4) (528). Using more hydrophobic counterions (Table 5) it is possible to prepare hydrophobic carriers (532).

Table 4 Possible Counterions for the Ionotropic Gelation of Chitosan

Polycation	Counterions
Chitosan–NH_2	Low molecular weight
	Pyrophosphate
	Tripolyphosphate
	Tetrapolyphosphate
	Octapolyphosphate
	Hexametaphosphate
	$[Fe(CN)_6]^{4-}$, $[Fe(CN)_6]^{3-}$
	High molecular weight
	Poly(1-hydroxy-1-sulfonate-2-propene)
	Poly(aldehydocarbonic acid)
	Xanthane
	Pectin

Table 5 Hydrophobic Counterions for the Ionotropic Gelation of Chitosan

Polycation	Hydrophobic counterions
Chitosan–NH_2	Octyl sulfate
	Lauryl sulfate
	Hexadecyl sulfate
	Cetylstearyl sulfate

D. Microspheres and Microcapsules

Over the past decade, several microsphere drug formulations have matured beyond the stage of promising investigational agents in preclinical and clinical trials to become viable pharmaceutical products approved for widespread human use. Chitosan microspheres are produced, either by an emulsification–crosslinking process or by use of complexation between oppositely charged macromolecules. The resulting crosslinked chitosan microspheres were characterized by being smooth, spherical, and in the size range of 45–300 μm.

Influence of chitosan molecular weight on drug loading and drug release of drug loaded chitosan microspheres was studied (533). Chitosans of 70,000, 750,000, and 2,000,000 molecular weight were employed. Ketoprofen (ket) was chosen as the model drug to be encapsulated. Prepared chitosan microparticle delivery systems can modulate ket release within 48 h.

A new microparticulate of chitosan controlled-release system, consisting of hydrophilic chitosan microcores entrapped in a hydrophobic cellulosic polymer, such as cellulose acetate butyrate (CAB) or ethyl cellulose (EC), was proposed (534). These microparticles were obtained with different types of chitosan and various core/coat ratios, with the particle size in all cases being smaller that 70 μm. Using sodium diclofenac (SD) and fluorescein isothiocyanate-labeled bovine serum albumin (FITC-BSA) as model compounds, the properties of these new microparticles for the entrapment and controlled release of drugs and proteins were investigated. The microparticles were stable at low pH and thus suitable for oral delivery without requiring any harmful crosslinkage treatment.

A new system which combines specific biodegradability and pH-dependent release is presented. The system consists of chitosan (CS) microcores entrapped within acrylic microspheres. Sodium diclofenac (SD), used as a model drug, was efficiently entrapped within CS microcores using spray-drying and then microencapsulated into Eudragit L-100 and Eudragit S-100 using an oil-in-oil solvent evaporation method (535). The size of the CS microcores was small (1.8–2.9 μm) and they were encapsulated within Eudragit microspheres (size between 152 and 233 μm) forming a multireservoir system. A combined mechanism of release is proposed, which considers the dissolution of the Eudragit coating, the swelling of the CS microcores and the dissolution of SD and its further diffusion through the CS gel cores.

Novel CS and CS/ethylene oxide–propylene oxide block copolymer (PEO-PPO) nanoparticles and their potential for the association and controlled release of proteins and vaccines were studied by Calvo et al. (536). The mechanism of protein association was elucidated using several proteins, bovine serum albumin, and tetanus and diphtheria toxoids, and varying the formulation conditions (different pH values and concentrations of PEO-PPO) and the stage of protein incorporation into the nanoparticle formation medium.

1. Microspheres and Microcapsules with Insulin

As a consequence of poor oral bioavailability and the current lack of alternative delivery routes, insulin is presently administered parenterally. The difficulties in achieving a normal physiological profile of insulin by injectable therapy has led to the investigation of alternative, nonparenteral routes for the delivery of insulin in an attempt to improve glycemic control. An insulin delivery system has been widely investigated as an alternative to subcutaneous injection for the treatment of diabetes.

Bugamelli et al. (537) investigated the production of chitosan microparticles containing insulin by interfacial crosslinkage of chitosan solution in the aqueous phase of a water/oil dispersion in the presence of ascorbil palmitate, thus permitting covalent bond formation with the amino group of chitosan when its oxidation to dehydroascorbyl palmitate takes place during microparticle preparation. This preparation method produced microparticles characterized by high loading levels of insulin, completely releasing the drug in about 80 h at an almost constant release rate.

Tozaki et al. (538) studied the colon-specific insulin delivery with chitosan capsules. A marked absorption of insulin and a corresponding decrease in plasma glucose levels was observed following the oral administration of these capsules that contain 20 IU of insulin and sodium glycocholate (PA% = 3.49%), as compared with the capsules containing only lactose or only 20 IU of insulin (PA% = 1.62%). The hypoglycemic effect started from 8 h after the administration of chitosan capsules when the capsules entered the colon, as evaluated by the transit time experiments with chitosan capsules. These findings suggest that chitosan capsules may be useful carriers for the colon-specific delivery of peptides including insulin.

2. Microspheres and Microcapsules for Cancer Therapy

Intravenously administered anticancer drugs are distributed throughout the body as a function of the physicochemical properties of the molecule. A pharmacologically active concentration is reached in the tumor tissue at the expense of massive contamination of the rest of the body. The use of microsphere drug carriers could represent a more rational approach to specific cancer therapy.

Kawashima et al. (539) have developed a novel mucoadhesive DL-lactide/glycolide copolymer (PLGA) nanosphere system to improve peptide absorption and prolong the physiological activity following oral administration. The desired PLGA nanospheres with elcatonin were prepared by the emulsion solvent diffusion method to coat the surface of the resultant nanospheres with a mucoadhesive polymer such as chitosan, poly (acrylic acid), and sodium alginate. The chitosan-coated nanospheres showed higher mucoadhesion to the everted intestinal tract in saline than the other polymer-coated nanospheres.

Nishioka et al. (540) produced glutaraldehyde crosslinked chitosan microspheres that contained cisplatin. Similar microspheres were prepared by Jameela and Jayakrishnan (541) containing mitoxantrone. Drug release was found to be effectively controlled by the degree of crosslinking. Wang et al. (542) described an orthogonal experimental design to optimize the formulation of cisplatin (CDDP)-loaded chitosan microspheres (namely, CDDP-DAC-MS), which were produced by an emulsion-chemical crosslinking technique. Seven factors and three levels for each factor that might affect the formulation of microspheres were selected and arranged in an L27(3(13)) orthogonal experimental table.

Chitosan nanoparticles, readily prepared without the use of organic solvents, are a suitable vehicle for the delivery of these immunostimulants from *Mycobacterium vaccae* (PS4A); the formulations might find application as antitumor agents (543).

The fate of 166Ho–chitosan complex, a radiopharmaceutical drug for cancer therapy, was determined by studying its absorption, distribution, and excretion in rats and mice (544). To determine the effects of chitosan in 166Ho–chitosan complex, 166Ho alone (0.75 mg of $Ho(NO_3)_3 \times 5H_2O$/head) was intrahepatically administered to male rats; and radioactive concentrations in blood, urinary and fecal excretion, and radioactive distribution were examined. The radioactive concentrations in tissues and the whole-body autoradiography images showed that most of the administered radioactivity was localized at the administration site, and only slight radioactivity was detected from the liver, spleen, lungs, and bones. These results strongly suggest that 166Ho is retained at the administration site only when it forms a chelate complex with chitosan. Autoradiographs after intratumoral administration of 166Ho–chitosan complex showed that radioactivity was localized at the site of administration without distribution to the other organs and tissues. These results strongly suggest that 166Ho is retained at the administration site only when it forms a chelate complex with chitosan. Autoradiographs after intratumoral administration of 166Ho–chitosan complex showed that radioactivity was localized at the site of administration without distribution to the other organs and tissues (544).

The potential of gadolinium neutron–capture therapy (Gd-NCT) for cancer was evaluated using chitosan nanoparticles as a novel gadolinium device (545). The nanoparticles, incorporating 1200 μg of natural gadolinium, were administered intratumorally twice in mice bearing subcutaneous B16F10 melanoma. The thermal neutron irradiation was performed for the tumor site, with the fluence of 6.32×10^{12} neutrons/cm^2 8 h after the second gadolinium administration. After the irradiation, the tumor growth in the nanoparticle administered group was significantly suppressed compared to that in the gadopentetate solution–administered group, despite radioresistance of melanoma and the smaller Gd dose than that administered in past Gd-NCT trials. This study demonstrated the potential usefulness of Gd-NCT using gadolinium-loaded nanoparticles (545).

Hojo et al. (546) have prepared a hybrid of a water-soluble chitosan and a laminin-related peptide, and have examined its inhibitory effect on experimental metastasis in mice. Four methods were tried to achieve a coupling reaction, the diphenylphosphoryl azide method, the diisopropylcarbodiimide/1-hydroxybenzotriazole method, the water-soluble carbodiimide, and the 2-(1H-benzotriazole-1-yl)-1,1,3,3-tetramethyluronium tetrafluoroborate (TBTU) method, but all four methods were unsuccessful. Therefore, a small spacer, tert-butyloxycarbonyl-Gly, was intercalated in chitosan by the TBTU method to facilitate its coupling with the peptide. Conjugation of the peptide with the larger chitosan molecule did not reduce the inhibitory effect of the peptide on experimental metastasis in mice, it actually potentiated the antimetastatic effect, demonstrating that chitosan may be effective as a drug carrier for peptides.

DNA was immobilized within alginate matrix using an external or an internal calcium source, and then membrane coated with chitosan or poly-L-lysine. Membrane thickness increased with decreasing polymer molecular weight and increasing degree of deacetylation of chitosan (547,548). Less than 1% of total double-stranded DNA remained unhydrolyzed within chitosan- or poly-L-lysine–coated beads, corresponding with an increase in DNA residuals (i.e., double- and single-stranded DNA, polynucleotides, bases). Chitosan membranes did not offer sufficient DNA protection from DNase diffusion since all of the double-stranded DNA was hydrolyzed after 40 min of exposure.

3. Microspheres and Microcapsules with Antiinflammatory Drugs

An inclusion complex composed of hydrocortisone acetate (HC) and hydroxypropyl-β-cyclodextrin (HPβCD) was prepared by the spray-drying method (549). Hydrocortisone acetate alone, HC inclusion complex, or HC with HPβCD as a physical mixture were incorporated into chitosan microspheres by spray-drying (549). The HC release rates from chitosan microspheres were influenced by the drug/polymer ratio in the manner that an increase in the release rate was observed when the drug loading was decreased. However, release data from all samples showed significant improvement of the dissolution rate for HC, with 25–40% of the drug being released in the first hour compared with about 5% for pure HC.

Mooren et al. (550) investigated the influence of chitosan microspheres on transport of the hydrophilic, antiinflammatory drug prednisolone sodium phosphate (PSP) across the epithelial barrier. The effect is dependent on the integrity of the intercellular cell contact zones and the microparticles are able to pass the epithelial layer. Their potential benefit under inflammatory conditions like in inflammatory bowel disease, in order to establish high drug doses at the region of interest, remains to be shown.

Chitosan microspheres were also used as support for the antiinflammatory drugs in order to assure the gastrointestinal tract protection. In this way the chitosan microspheres covalently linked with citric acid and loaded with indomethacin are prepared (551). Also, a pharmacokinetic model of colon-specific drug has been validated by use of 5-aminosalicylic acid (5-ASA) as a model antiinflammatory drug (552). The simulation curves obtained from the pharmacokinetic model were in good agreement with experimental data obtained after oral administration of 5-ASA–containing chitosan capsules. The concentrations of 5-ASA in the large intestinal mucus after drug administration were higher than after administration of the drug in carmellose suspension. These findings suggest that our pharmacokinetic model of colon-specific drug delivery can accurately evaluate this colon-specific delivery system and that 5-ASA–containing chitosan capsules are more effective than other 5-ASA formulations for treatment of colitis in rats.

4. Microspheres and Microcapsules as Local Injection

Local implantation or injection of microspheres containing bisphosphonates for site-specific therapy may aid in treating several pathological conditions associated with bone destruction. Chitosan microspheres containing two antiresorption and anticalcification agents, pamidronate and suberoylbisphosphonate, were prepared from a water-in-oil emulsion. Various formulation variables were studied for their effect on the release rate profile of these bone-seeking agents (553). Polymer coating of micromatrices yielded microspheres with the most retarded release rate, and the drug delivery system was found biocompatible in endothelial cell culture. Bisphosphonate released from chitosan microspheres effectively inhibited bioprosthetic tissue calcification in the rat subdermal model.

The formation of microcapsules which contain rosemary oil is herewith described (554). The process is based on two steps: (1) formation of oil-in-water emulsions using lecithin as emulsifier, thus imparting negative charges on the oil droplets and (2) addition of a cationic biopolymer, chitosan, in conditions that favor the formation of an insoluble chitosan–lecithin complex.

E. Gels

Hydrogels formed from chitosan are usually covalently crosslinked. Crosslinking agents such as glutaraldehyde (555–557), glyoxal (558), and diethyl squarate (559) have been employed in the fabrication of chitosan hydrogels. Chitosan has also been covalently crosslinked with the carbohydrate sceleroglucandialdehyde (560). In addition semi-interpenetring polymer network hydrogels have been produced between chitosan–polyethylene oxide diacrylate crosslinked by UV irradiation (561), glyoxal crosslinked chitosan–polyethylene oxide (562), and glutaraldehyde crosslinked chitosan combined with silk fibroin (563) or polyacrylic acid (564).

Noncovalent crosslinking has been used to prepare chitosan-based gels by the O- and N-acetylation of chitosan (565) and the attachment of C10-alkyl glycosides to chitosan (566). By the attachment of hydrophobic palmitoyl groups to glycol chitosan, a water-soluble chitosan derivative has been produced (567).

Thin films of a biopolymer chitosan (CHIT) were cast on glassy carbon electrodes, modified by grafting Lucifer Yellow VS dye (LYVS) onto chitosan chains, and crosslinked with glutaric dialdehyde (GDI) (568). The ion transport and ion exchange properties of such polymeric structures (CHIT, CHIT-LYVS, CHIT-LYVS-GDI) were studied using cyclic voltammetry, rotating disk electrode, and flow injection analysis. The results showed that the chitosan matrix supported a fast ion transport as demonstrated by aqueous-like values of the apparent diffusion coefficients of $Ru(NH_3)_6(3)^+$ and dopamine in the films (568). The results indicate that the chemically modified chitosan is an attractive new coating for the development of fast, selective, and reversible sensors.

Miyazaki et al. (569) and Kristl et al. (565) investigated the suitability of dried chitosan gels as vehicles for the sustained release of the poorly soluble drugs such as indomethacin, papaverine hydrochloride, and lidocaine. The gel showed zero order release into pH 7.4 buffer at 24 h. It was found that the degree of deacetylation and the chitosan content were important for the release properties.

The influence of the acid type used to dissolve chitosan on the resulting sponge, physical properties, and the consequent effect on the drug liberation were investigated (570). Chitosan was dissolved in different acid solutions and chitosan–gelatine sponges were produced by frothing up the polymer solution and then freeze-drying the foam. Prednisolone was used as a model drug. Using tartaric or citric acid resulted in unstable, soft, elastic, and disintegrating sponges with fast drug release. Elastic but harder sponges from stable foams were obtained when hydrochloric or lactic acid were used. The use of acetic or formic acid enabled the production of stable foams and soft and elastic sponges, allowing low drug release. The rate of drug release was decreased by crosslinking the polymers with glutaraldehyde. Therefore, it is possible to manipulate the mechanical properties and the drug liberation rate by using different acids to dissolve chitosan (570).

A biodegradable drug delivery system of a gentamicin-loaded chitosan bar with sustained antibiotic effect is described (571). Combined crosslinking, solvent evaporation, and a cylinder model cutting technique was used to prepare the chitosan bar. Sustained diffusion of gentamicin into the surrounding medium was seen using a release test in vitro. Approximately 11% gentamicin was released from the bar in the first 24 h. The gentamicin released from the bar showed significant antibacterial activity. The bar implanted in the proximal portion of the rabbit tibia produced a low blood concentration of gentamicin, but a much higher concentration was produced in local bone and in the hematoma. In all bone tissue around the bar, the gentamicin concentration exceeded the minimum inhibitory concentration for the common causative organisms of osteomyelitis for approximately 8 weeks (571). Based on these test results together with the chitosan characteristics of biodegradable, antibiotic, and immunologic activity, the gentamicin-loaded chitosan bar seems to be a clinically useful method for the treatment of bone infection. This system has an advantage over other systems in that it avoids a second operation for removal of the carrier.

Chitosan–ethylene diamine tetracetic acid (EDTA) conjugate gels have also been reported in which the carboxylic acid group of EDTA are covalently linked to the amino groups of chitosan (572,573). A recently developed chitosan–EDTA conjugate, neutralized with sodium hydroxide (NaChito-EDTA), has been tested for possible topical use (571). This antimicrobial activity of NaChito-EDTA can be explained by its highest binding affinity toward magnesium, which stabilizes the outer membrane of gram-negative bacteria.

New mucoadhesive polymers, exhibiting a high capacity to bind bivalent cations that are essential cofactors for intestinal proteolytic enzymes, were synthesized (574). Under the formation of amide bonds, the complexing agent EDTA was covalently bound to the primary amino groups of chitosan (574,575). Whereas proteolytic activity of the serine proteases trypsin, α-chymotrypsin, and elastase could not be inhibited, proteolytic activity of the zinc proteases carboxypeptidase A and aminopeptidase N was strongly inhibited by the chitosan–EDTA conjugate. The adhesive force of the conjugate was even higher than of chitosan HCl.

Berberine (the main ingredient of Coptis spp.) was incorporated into chitosan hydrogel to prepare ointments. The physicochemical properties of the ointments and the release profile of berberine were investigated. The results indicated that the viscosity of chitosan hydrogel increased with an increasing amount of lactic acid or EDTA (576). The release rate of berberine was inversely proportional to ointment viscosity.

The effect of heparin ionically linked to chitosan on the stimulation of re-epithelialization of full thickness wounds in human skin was investigated in an in vitro model (577). After 7 days of incubation, heparin–chitosan gel stimulated 9/10 of the full thickness wounds to re-epithelialize compared with only 3/10 of the wounds that were covered with chitosan gel or membrane; and none of the wounds incubated without gel or membrane or with heparin solution alone.

One approach taken by Hoffman et al. (578) was to synthesize hybrid copolymers by grafting temperature-responsive polymers (Pluronic) to chitosan backbones. The polymers gelled as the temperature was raised from 4 to 37°C and provided significant extension of drug release.

F. Chitosan–Polyelectrolyte Complexation

Polymer complexes are formed by the association of two or more complementary polymers and may arise from electrostatic forces, hydrophobic interactions, hydrogen bonding, van der Waals forces, or combinations of these interactions (579–583). The formation of complexes may strongly affect the polymer solubility, rheology, conductivity, and turbidity of polymer solutions (583). Hydrogen bonds are distinctly directional and specific and are more localized than any other type of weak intermolecular interaction (584). Many polymer complexes form as a result of electrostatic forces. Polymeric acids may form complexes by proton transfer to complementary polymeric bases, resulting in poly(cation)/poly(anion) pairs. In addition to the interpolymer complexes, intrapolymer complexes of polyampholytes have been studied (585–587). In these systems complexes may be formed between oppositely charged moieties on the same polymer chain, or even on the same monomer units. Finally, simple polyelectrolytes in solution may be linked together by multivalent ions to form gels or coacervates (588,589). The complex stability is dependent upon such variables as charge density, solvent, ionic strength, pH, and temperature. Polyelectrolytes of high charge density may form relatively insoluble complexes. Characteristics of the macromolecule complexes are largely based on their higher molecular weight, which are called polymer effects, such as cooperative interactions, concerted interactions, steric compatibility, and microenvironmental effects (590).

Gel technology based on ionic crosslinked polysaccharides is currently in use and of considerable interest for potential application in the extraction with solvents (591), as controlled release systems (592,593), as mechanochemical devices (594), in the oil industry (595–597), and also in several other biomedical applications (598,599).

The binding of chitosan to alginate beads was studied quantitatively by using radioactive-labelled fractions of chitosan (600). The binding of chitosan was markedly increased by reducing the number average molecular weight of chitosan below 20,000 Da and by increasing the porosity of the alginate gel. The porosity was increased by producing homogeneous gels, and by adding calcium chloride to the chitosan solution during the membrane-forming stage. The binding of chitosan was also found to increase with decreasing fraction of N-acetylations (FA) on chitosan, in the range of FA = 0.3 to FA = 0, and with increasing pH, in the range of pH 4 to 6. Capsules with a diameter of 500 μm had a higher weight ratio of chitosan to alginate after 24 h of binding than the capsules with the larger diameter of 1500 μm.

To prepare spray-dried composite particles containing alginate–chitosan complex [SD(L/AL-CS)], an aqueous solution of lactose and sodium alginate and the acetic acid solution of chitosan were concomitantly fed into the rotary atomizer of a spray-dryer (601). The drug release profiles of dry-coated tablet with SD(L/AL-CS) indicated a long induction period followed by a rapid drug release phase in the artificial intestinal fluid. The induction period for drug release to occur was increased with an increase in the degree of deacetylation of chitosan and in the amount of chitosan in the formulation. The prolongation of induction period was attributed to the formation of an insoluble ion complex between sodium alginate and chitosan in the composite particles, which could form a rigid gel structure on the tablet surface.

The stability of alginate–chitosan capsules was shown to depend strongly on the amount of chitosan bound to the capsules. When the capsules were made by dropping a solution of sodium alginate into a chitosan solution (one-stage procedure), all the chitosan was located in a thin alginate–chitosan membrane on the surface. These capsules were much weaker than the capsules made by reacting calcium alginate beads in an aqueous solution of chitosan and calcium chloride (two-stage procedure). Capsules with high mechanical strength were obtained after shorter reaction times when the number average molecular weight of the chitosan was reduced to around 15,000 Da, when the capsules were made more homogeneous, and when the capsule diameter was reduced to around 300 μm (602).

Noncovalent crosslinked chitosan gel mixtures can be prepared by the use of polyelectrolyte complexes of chitosan and polyanions such as carboxymethylcellulose (603) and xanthan (604,605).

Gelatine forms polyionic complexes with chitosan at the suitable pH value (606). These complexes show a slower rate of dissolution than chitosan at the corresponding pH. Therefore a combination of chitosan and gelatine in a sponge will first lead to an absorption of the wound fluid and second slowly release a wound-treating drug (607,608). The release profile and biodegradability of both polymers can be effectively controlled by glutaraldehyde crosslinking (609,610).

The complexation reaction between the polyions leads to structural changes in both polymers, xanthan and chitosan. A water-insoluble hydrogel is formed by blocking the hydrophilic functions (R—COOH in case of polyanions and R′—NH_2 for the polycations) and by the interaction between the macromolecular chains (Fig. 1). The latter effect created an ''ionic'' reticular network that immobilized the polymers. The hydrogel has the property of incorporat-

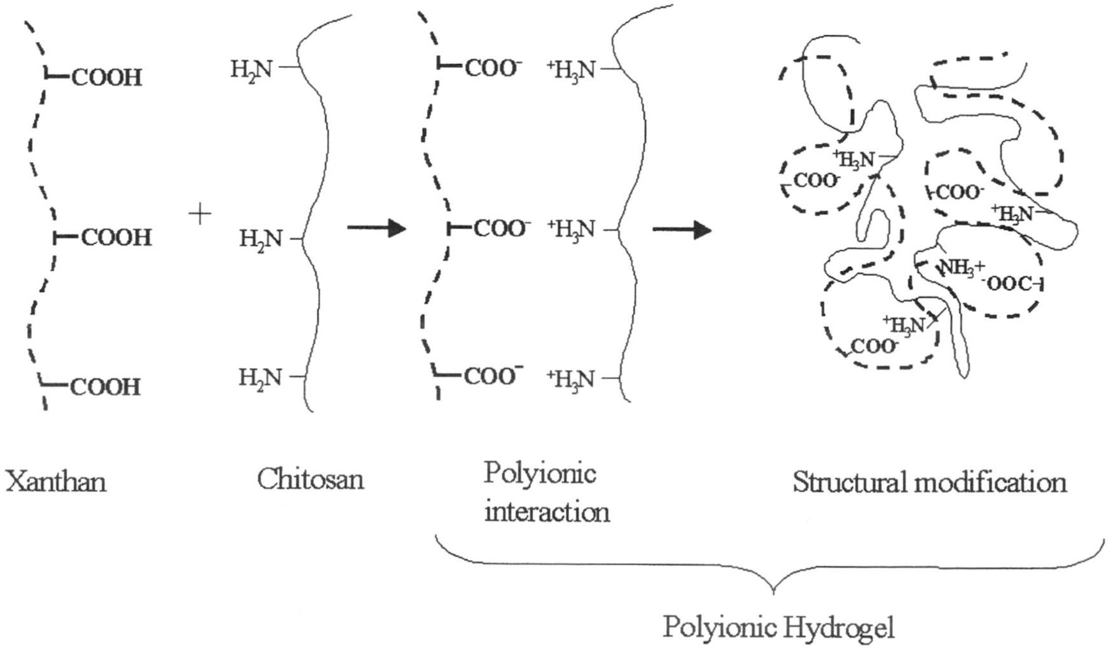

Figure 1.1 The complexation reaction between xanthan and chitosan.

ing a significant amount of structural water as well as the capacity of stabilizing, in its matrix, biological products (611–615).

Recently, we have studied the complexation reaction between xanthan and chitosan by determining the yields and the structure of the complex as well as its swelling capacity. The complexation efficiency readily reaches 90% of the available chitosan and the less deacetylated chitosan (DA = 34%) forms a stable hydrogel having the highest chitosan/xanthan ratio. In fact, the lower the DA, the higher the amount of xanthan needed to form a stable hydrogel. Additionally, we have studied the influence of the pH of the chitosan solution on the yield of complexation with xanthan. The results are shown in Table 6. We notice a poor complexation at low pH due to strong ionization of the amine groups, which impedes its reaction with the carboxylic groups present in the xanthan. Similar results have been reported for the complexation of chitosan with carboxymethyl cellulose (CMC) and alginic acid (616,617).

The decomposition of the chitosan–xanthan hydrogel depends on the pH and the ionic concentration of the buffer solution. In the pH range between 1.2 and 9.2 the hydrogel is stable no matter the ionic concentration present (Fig. 2).

The swelling characteristics and the water content of the hydrogel are shown in Table 7, from which we can notice that water constitutes, in all cases, the largest weight concentration of the hydrogel. The large swelling capacity of these hydrogels does not lead to their disintegration when prepared as microspheres, as in the case for the xanthan–CMC complex (616).

Scanning electron microscope (SEM) images of the hydrogels are shown in Fig. 3 for typical gel samples (24.25% chitosan, 75.75% xanthan). The images convey the message that the gels are quite porous and that formation of fibrilar structures takes place. The channels present in the gels have a pore size between 10^{-7} and 10^{-6} m (0.1–1 μm), whereas the fibrils have a typical dimension of 10^{-7} m (100 nm). The surface of the microsphere has a homogeneous porous structure (Fig. 3b), which allows the passage of polymeric substrates to the regions where the immobilized enzymes are lodged.

Transmission electron microscope (TEM) images of the hydrogels are shown in Fig. 4. Since the preparation of the samples for TEM inevitably leads to modifications of the sample, care has to be exercised in the interpretation of the micrographs. Figure 4 shows the presence of loose aggregates of fibrils probably resulting from the rupture of the elaborate network, observed via scanning electron microscopy (SEM), due to the drying and cutting procedures used. The aggregates are formed of fiber fragments having typical diameters of 50 to 100 nm in agreement

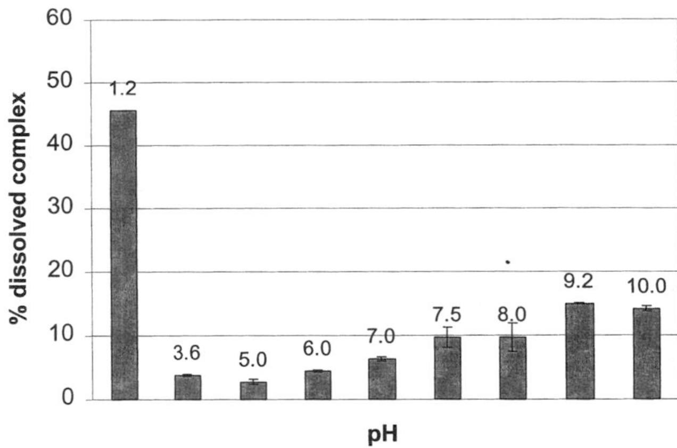

Figure 1.2 Evaluation of chitosan–xanthan complex stability at different pH conditions. Time: 48 h; buffer solution concentration: 0.2M.

Table 6 Influence of the pH of the Chitosan Solution on the Yield of Its Complexation with Xanthan

pH	Xanthan complexation yield (%)
1.5	21
2.5	36
3.5	56
4.5	60
5.5	98
6.3	82

Note: Ratio CH/X = 0.65 g/g. Mn chitosan = 691,390, determined by measuring the intrinsic viscosity. $[\eta] = 1.81 \times 10^{-3} M^{0.93}$, determined in 0.1 M acetic acid/0.2 M NaCl solution at 25°C.

Table 7 Influence of the Molecular Weight of Chitosan on the Absorption of Water by the Chitosan–Xanthan Hydrogel

Chitosan Mn[a]	Degree of acetylation (%)	Complexation yield (%)	α_{max} (%)
691,930	28.0	98.9	2560
452,875	27.9	97.2	1805
191,325	28.2	84.5	992
122,350	28.8	81.3	457

Note: pH = 5.0; buffer acetate 0.2 M.
[a] Determined by intrinsic viscosity $[\eta] = 1.81 \times 10^{-3} M^{0.93}$, in 0.1 M acetic acid/0.2 M NaCl solution at 25°C.

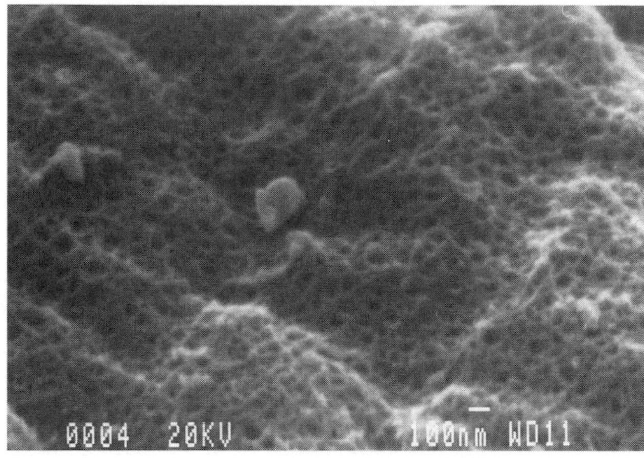

(a)

(b)

Figure 1.3 Scanning electron microscope images of typical gels (24.25% chitosan, 75.75% xanthan). (a) Image of external surface, 30,000×; (b) image of internal section, 60,000×.

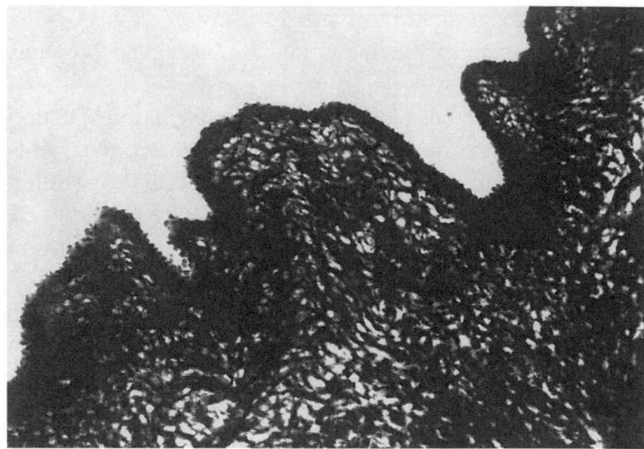

Figure 1.4 Transmission electron microscope images of typical gels (24.25% chitosan, 75.75% xanthan) 48,000×.

with the SEM observations. A concentration of material near the external surface of the sample is also evident. The ''open spaces'' (pores) are quite developed toward the center of the sample.

Evaluation of their bicompatibility was carried out with L929 fibroblasts cell line to determine the cytotoxicity and with J774 macrophages cell line to assess the inflammatory response. Cytokines and nitric oxide (NO) are also secreted by the macrophages as indicators of the inflammatory response, and they are known to induce other cells such as T lymphocytes to proliferate and synthesize proteins and additional factors which, in turn, enhance macrophage activation (618–621). Nitric oxide production and efficacy as an immune effector molecule has been also demonstrated in murine models (622). The TNF-α, IL-β, and NO secretion as markers for macrophages activation were assessed.

The CH-X complex did not show a cytotoxic effect regardless of degradation time periods and concentration used (Fig. 5) (623).

Production of TNF-α by macrophages was stimulated by CH-X particles for the concentration of 1 mg/mL. This effect was more evident at higher concentration (10 mg/mL). Production of IL-1β was enhanced after macrophage incubation with CH-X particles, while the concentration did not affect the secretion level (Fig. 6). Our study demonstrated clearly that macrophages incubated with

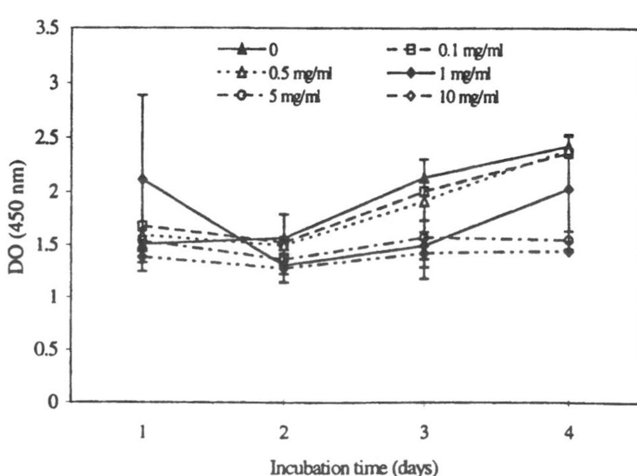

Figure 1.5 Effect of CH-X particles on L-929 cell viability particles (direct contact).

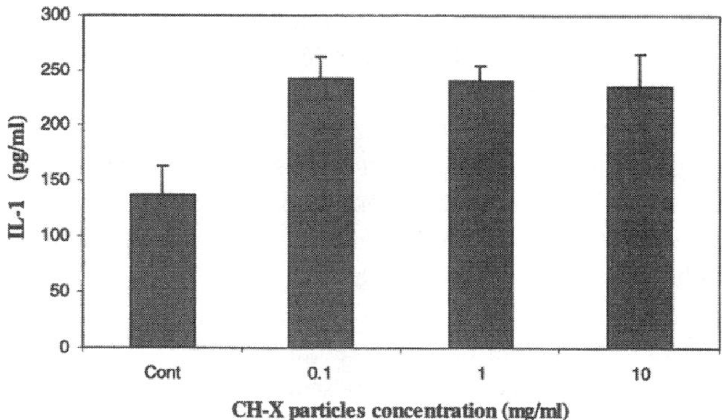

Figure 1.6 Effect of CH-X particles on IL-1β secretion.

CH-X particles were activated to produce IL-1β. The size of the particles could be not phagocytosable.

Secretion of TNF-α and NO (Fig. 7) is influenced in the same way by CH-X extract products with a slight sensitivity for TNF-α secretion.

Macrophages also play an important role in the phagocytosis of the material and its degradation in vivo. The inflammatory reaction is essential for preparing the wound for the production of a new extracellular matrix. Cytokines and growth factors released by the inflammatory cells attract the fibroblasts into the wound to initiate the reconstruction process. In vivo studies based on light microscopy observations showed a recruitment of specific cell populations responsible for the preparation of wound repair and deposition of new matrices. Their presence is shown in Fig. 8, and the presence of collagen attests to the healing process (Fig. 9). Degradation study on the animal

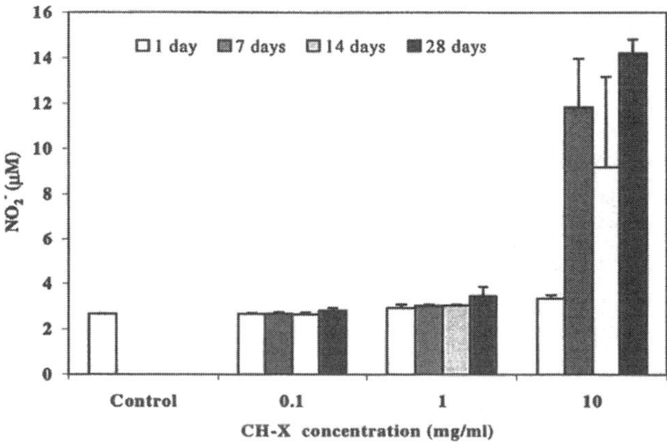

Figure 1.7 Effects of CH-X extract products on NO_2^- secretion by J-774 macrophages.

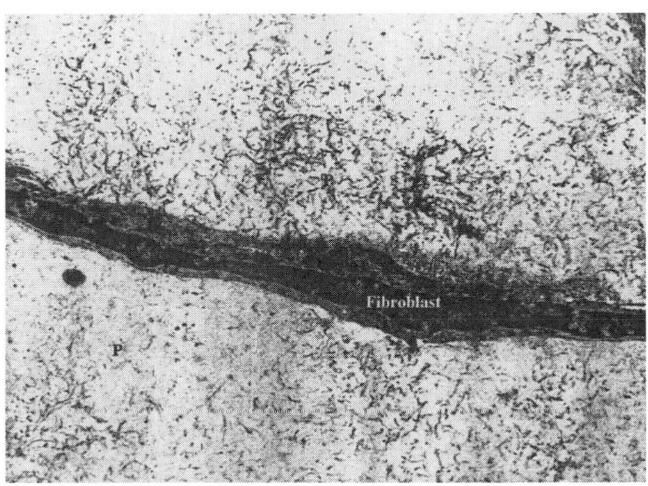

Figure 1.8 TEM micrograph showing fibroblast adhering along CH-X particles (P) after 8 weeks of implantation in rat subcutaneous tissue.

model used indicated a phagocytosis of the CH-X particles by macrophages, which can lead to the entire resorption of the hydrogel.

Shioya et al. (616) developed the encapsulation method using chitosan. The encapsulation is based on the electrostatic interactions of chitosan as a polycation with sodium carboxymethyl cellulose as a polyanion. The objective of this investigation was to identify the factors which control the transmembrane permeability of the capsules prepared with chitosan and CMC (616). The permeability slightly decreases with time. Longer reaction times did not improve

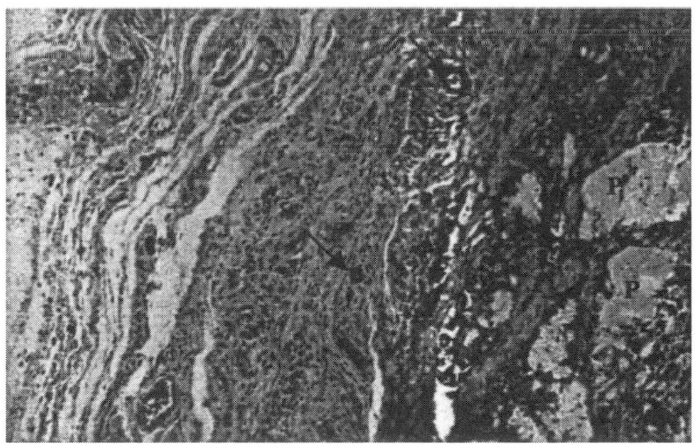

Figure 1.9 LM micrograph (156 ×) showing dense fibrosis formation with blood vessels (arrows) and inflammatory reaction around CH-X particles (P) after 2 weeks implantation in rat subcutaneous tissue.

the capsule membrane (616). The permeability increases with an increase in the salt concentration. Adding salt may cause a reduction in the effective charge by a shielding effect (623) and then prevent the chitosan from reacting with CMC. In parallel with permeability, the capsule membranes formed by chitosan with salt also have lower mechanical strengths because of a lower density of interpolymer bridge. The lower molecular weight of chitosan may give a more compact membrane structure, thus lower permeability.

The equilibrium adsorption rates of chitosan on oxidized cellulose substrates were determined and are plotted as a function of carboxylic acid group content (624). The increase in absorption with increase in carboxyl group content may arise from two effects (624):

1. Increased electrostatic interaction between polyanion (oxidized cellulose) and polycation (chitosan)
2. Increased substrate surface area due to increased swelling on immersion in water because of increase in the hydrophilic character of the substrate (oxidized cellulose) with increase in the extent of modification

The addition of NaCl has two opposing effects on the absorption of chitosan on oxidized cellulose (624). At low sodium chloride concentration, the dominant effect is reduction of the hydrodynamic volume of the chitosan molecules leading to an increase in the equilibrium adsorption. At higher sodium chloride concentration, competition between Na^+ ions and cationic groups on the chitosan for the carboxylic acid groups on the cellulose predominates, leading to a gradual decrease in the equilibrium adsorption of chitosan (624).

Remunan-Lopez and Bodmeier (625) investigated optimal conditions for the complex between chitosan and gelatine. All of the optimal conditions—pH, polymer ratio, polymer concentration, temperature, reaction time, and ionic strength—were essentially coincident with chitosan–polysaccharide complexes.

Glycol chitosan (GC) and methyl glycol chitosan (MGC) were, individually, reacted with potassium metaphosphate (MPK) to form a series of water-insoluble macromolecular complexes (MC) in aqueous solution at different hydrogen ion concentration (GC-MPK and MGC-MPK) systems (626). These coagulation curves indicate the dependence of hydrogen ion concentration on the mole ratio of the reacting group of MPK to that of GC, GC + $CaCl_2$, or MGC in the reaction mixture at the beginning and end of the coagulation.

Enzymatic-hydrolyzed chitosan was employed to prepare chitosan–tripolyphosphate and chitosan–polyphosphoric acid gel beads using a polyelectrolyte complexation method for the sustained release of anticancer agent, 6-mercaptopurine (627). The complexation mechanism of chitosan beads gelled in pentasodium tripolyphosphate or polyphosphoric acid solution was ionotropic crosslinking or interpolymer complexation, respectively. The drug release patterns of all chitosan gel beads in pH 6.8 seemed to be diffusionally based, whereas release profiles in pH 1.2 medium seemed to be non-Fickian diffusion controlled due to the swelling or matrix erosion.

Insulin-loaded chitosan nanoparticles were prepared by ionotropic gelation of chitosan with tripolyphosphate anions (628). The ability of chitosan nanoparticles to enhance the nasal absorption of insulin was investigated in a conscious rabbit model by monitoring the plasma glucose levels. Chitosan nanoparticles enhanced the nasal absorption of insulin to a greater extent than an aqueous solution of chitosan. The amount and molecular weight of chitosan did not have a significant effect on insulin response. In vivo absorption studies in animals have been carried out to evaluate the nasal absorption–promoting activity of chitosan using insulin as a model peptide (629–630).

F. Chitosan as Scaffolds

It is known that glycosaminoglycans are involved in cell–cell and cell–matrix interactions, and probably act as modulators of cell morphology, differentiation, movement, synthesis, and function (631–633). Chitosan, having structural similarity to glucosaminoglycans, was modified using several proteins (collagen, albumin, and gelation) to increase surface area and improve biocompatibility. In vitro, collagen-blended chitosan matrices were found to attach more readily to chromaffin cells than gelatine- or albumin-blended matrices (634). Morphological evidence showed that the chromaffin cells attached to collagen-blended chitosan substrate integrated well with the hydrogel matrix and survived for at least 2 weeks under in vivo conditions. Collagen-blended chitosan substrates produce significantly improved bovine adrenal medullary chromaffin cell attachment characteristics, which can be considered for neural tissue engineering studies. Morphological examinations revealed that the chromaffin cells survived for at least 2 weeks with collagen–chitosan scaffolds.

The effectiveness of chitosan as a scaffold of hepatocyte attachment was examined (635). Since chitosan gel was too fragile to use for cell culture, its free amino groups were crosslinked by glutaraldehyde to increase its strength. Rat hepatocytes seeded onto glutaraldehyde-crosslinked chitosan (GA–chitosan) gel could attach with stability to the surface, retaining its spherical form, the same as in vivo, and then release a very small amount of lactate dehydrogenase during the 5-day culture period. By contrast, hepatocytes on a collagen-coated surface spread flat, and they released much more lactate dehydrogenase than those on the GA–chitosan gel. Hepatocytes on GA–chitosan also retained higher urea synthesis activity, a liver-specific function, than those on the collagen-coated surface. These results indicate that chitosan is a promising biopolymer as a scaffold of hepatocyte attachment which can be applied to an effective bioartificial liver support system.

The quality of articulate cartilage engineered using a cell–polymer construct depends, in part, on the chemical composition of the biomaterial and whether that biomaterial can support the chondrocytic phenotype. Acknowledging the supportive influence of tissue-specific matrix molecules on the chondrocytic phenotype, Sechriest et al. (636) combined chondroitin sulphate-A (CSA) and chitosan, a glycosaminoglycan (GAG) analog, to develop a novel biomaterial to support chondrogenesis. Chitosan may be combined with the polyanionic CSA such that ionic crosslinking results in hydrogel formation. Bovine primary articular chondrocytes, when seeded onto a thin layer of CSA–chitosan, form discrete, focal adhesions to the material and maintain many characteristics of the differentiated chondrocytic phenotype, including round morphology, limited mitosis, and collagen type II and proteoglycan production.

With an aim of improving bone regeneration, chitosan sponge containing platelet-derived growth factor-BB (PDGF-BB) was developed. For fabrication of chitosan sponge, chitosan solution was freeze-dried, crosslinked, and freeze-dried again. PDGF-BB was incorporated into the chitosan sponge by soaking chitosan sponge into the PDGF-BB solution. Release kinetics of PDGF-BB, cell attachment, proliferation capacity, and bony regenerative potentials of PDGF-BB–loaded chitosan sponge were investigated (637). Release rate of PDGF-BB could be controlled by varying initial loading content of PDGF-BB to obtain optimal therapeutic efficacy. PDGF-BB–loaded chitosan sponge induced significantly high cell attachment and proliferation levels, which indicated good cellular adaptability. PDGF-BB–loaded chitosan sponge demonstrated marked increase in new bone formation and rapid calcification. Degradation of the chitosan sponge was proceeded at defect site and subsequently replaced with new bone. Histomorphometric analysis confirmed that PDGF-BB–loaded chitosan sponge significantly induced new bone formation. These results suggested that chitosan sponge and PDGF-BB–loaded chitosan sponge may be beneficial to enhance periodontal bone regeneration (637).

The ability of cultured human foreskin fibroblasts to bind and to contract lattices of collagen, collagen–chitosan, and collagen–chitosan sulfate was determined (638). Scanning electron microscopy of the cellular lattices showed fibers of the collagen–chitosan mixture to be the thickest and with altered organization. These results show that chitosan sulfation markedly enhances fibroblast adhesion and promotes contraction of a collagen lattice compared to the unsulfated material. By analogy to the in vivo sequence of hyaluronan followed by sulfated glycosaminoglycans in wounds, the results suggest that glycosaminoglycan sulfation may be a contributing signal for phenotypic transformation during wound healing (638).

G. Chitosan-Based Vector–DNA Complexes for Gene Delivery

Chitosan is a candidate nonviral vector for gene delivery. Therefore, formation of DNA particles with chitosan (639) or chemically modified chitosan (640), such as N,N,N-trimethylchitosan, has been investigated for gene transfer. Successful chitosan-mediated transfection has been reported for a few cell types, such as HEK 293 cells (639). Erbacher et al. (641) have found chitosan presents some characteristics favorable for gene delivery, such as the ability to condense DNA and to form homogeneous population of complexes, smaller than 100 nm, in defined conditions. Moreover, chitosan could easily be chemically modified by coupling ligands, such as lactose, in order to target cells expressing a galactose-binding membrane lectin. With the aim of developing a chitosan based vector system, various biophysical characteristics of chitosan-condensed DNA complexes were measured, and several transfection experiments mediated by chitosan and lactosylated chitosan were performed (642).

In order to achieve an efficient gene delivery via receptor-mediated endocytosis, the synthesis of a novel chitosan derivative having recognizable saccharide residues, N,N,N-trimethylchitosan–galactose conjugate, was performed (643). The formation of a polyelectrolyte complex with DNA and the cellular recognition of N,N,N-trimethylchitosan–galactose conjugate were tested, and then the possibility of its application as a gene delivery tool was investigated.

Chitosan is a polysaccharide that demonstrates much potential as a gene delivery system. The ability of a commercially available chitosan and depolymerized chitosan oligomers to condense plasmid was determined using TEM and microtitration calorimetry, while the diameter and stability of the resultant complexes were measured using laser light scattering (644). A selected complex containing a lytic peptide was administered in the upper small intestine and colon of rabbits, and reporter gene expression was measured in defined intestinal tissues. Reporter gene expression was enhanced in defined intestinal tissues, although levels of expression remained low. The combination of strong complex stability and low in vivo expression levels suggest that uptake and/or decomplexation, but not endosomal release, may be the critical rate-limiting steps in the uptake process.

Many cationic polymers (645–649) have been explored as nonviral vectors. Highly purified chitosan fractions of ≤5000 Da (N1), 5000–10,000 Da (N2), and ≥ 10,000 Da (N3) were prepared and characterized in respect of their cytotoxicity, ability to cause hemolysis, ability to complex DNA, as well as ability to protect DNA from nuclease degradation (650). Chitosan–DNA interaction at a charge of 1:1 was much greater than seen for poly(L-lysine), and complexation resulted of DNA degradation by DNase II: 99.9 ± 0.1, 99.1 ± 1.5, and 98.5 ± 2.0% for N1, N2, and N3, respectively. After intravenous injection, all the chitosans showed rapid blood clearance. The observations that the highly purified chitosan fractions used were neither toxic nor hemolytic, that they have ability to complex DNA and protect against nuclease degradation and that low molecular weight chitosan can be administered intravenously without liver accumulation suggest there is potential to investigate further low molecular weight chitosans as components of a synthetic gene delivery system (650).

Hydrophobically modified chitosan containing 5.1 deoxycholic acid groups per 100 anhydroglucose units was synthesized by an EDC-mediated coupling reaction. Formation and characteristics of self-aggregates of hydrophobically modified chitosan were studied by fluorescence spectroscopy and dynamic light scattering method (651). Charge complex formation between self-aggregates and plasmid DNA was confirmed by electrophoresis on an agarose gel.

Mao et al. (652) describe a new immunoprophylactic strategy using oral allergen gene immunization to modulate peanut antigen–induced murine anaphylactic responses. Oral administration of DNA nanoparticles synthesized by complexing plasmid DNA with chitosan resulted in transduced gene expression in the intestinal epithelium. Oral allergen gene immunization with chitosan–DNA nanoparticles is effective in modulating murine anaphylactic responses, and indicates its prophylactic utility in treating food allergy.

H. Chitosan Derivatives

In the 1990s chitosan turned out to be a useful excipient in various pharmaceutical formulations. By modifications of the primary amino group at the 2 position of this poly (β1 $\rightarrow$ 4-D-glucosamine), the features of chitosan can even be optimized according to a given task in drug delivery systems. For peroral peptide delivery these tasks focus on overcoming the absorption (I) and enzymatic barrier (II) of the gut. On the one hand, even unmodified chitosan proved to display a permeation-enhancing effect for peptide drugs. On the other hand, a protective effect for polymer embedded peptides toward degradation by intestinal peptidases can be achieved by the immobilization of enzyme inhibitors on the polymer. Whereas serine proteases are inhibited by the covalent attachment of competitive inhibitors such as the Bowman-Birk inhibitor, metallopeptidases are inhibited by chitosan derivatives displaying complexing properties such as chitosan–EDTA conjugates. In addition, because of the mucoadhesive properties of chitosan and most of its derivatives, a presystemic metabolism of peptides on the path between the dosage form and the absorption membrane can be strongly reduced. Based on these unique features, the co-administration of chitosan and its derivatives leads to a strongly improved bioavailability of many perorally given peptide drugs such as insulin, calcitonin, and buserelin. These polymers are therefore useful excipients for the peroral administration of peptide drugs (653).

To increase chitosan's utility, it has been N-derivatized by phosphonomethylation (PM-chitosan) or by carboxymethylation (CM-chitosan). Both derivatives disolve easily in water at neutral conditions. Moreover, these derivatives maintain the beneficial cosmetic properties of chitosan (654).

Chitosan succinate and phthalate matrices showed pH-dependent release profiles of the entrapped diclofenac sodium (655). Maximum drug release was observed under pH 7.4, in contrast to pH 2 at which the matrices resisted dissolution.

A simple carbohydrate polymer glycol chitosan (degree of polymerization approximately 800) has been investigated for its ability to form polymeric vesicle drug carriers. The attachment of hydrophobic groups to glycol chitosan should yield an amphiphilic polymer capable of self-assembly into vesicles. Chitosan is used because the membrane-penetration enhancement of chitosan polymers offers the possibility of fabricating a drug delivery system suitable for the oral and intranasal administration of gut-labile molecules. Glycol chitosan modified by attachment of a strategic number of fatty acid pendant groups (11–16 mol%) assembles into unilamellar polymeric vesicles in the presence of cholesterol. These polymeric vesicles are found to be biocompatible and hemocompatible and capable of entrapping water-soluble drugs (656). These polymeric vesicles efficiently entrap water-soluble drugs.

Investigations of chitosan citrate as a hydrocolloidal matrix material (657–659) and as the wall of a mcrocapsule (660) have been reported.

A chitosan with lauryl groups attached to amino groups to provide the hydrophobic moieties and carboxymethyl groups attached to hydroxy groups to provide the hydrophilic moieties [N-lauryl-carboxymethylchitosan (LCC)] was newly synthesized. The solubility of taxol in LCC micelles in aqueous solution was examined (661). It was found that LCC solubilized taxol by forming micelles with particle sizes less than 100 nm. This particle size was considered effective for passive targeting for tumors. The concentration of taxol in the micellar solution was very high, with a maximum of 2.37 mg/mL.

I. Macromolecular Produgs Based on Chitosan

5-Fluorouracil (5FU) has remarkable antitumor activity. In order to provide a polymeric prodrug of 5FU with reduced side effects, having affinity for tumor cells, and exhibiting high antitumor activity, four kinds of chitosan derivatives were synthesized carrying 5FU through some kinds of spacers via ether, amide, ester, or carbamoyl bonds (662,663). Chitosan-fixing 5FU at the 2 position through a hexamethylene spacer and carbamoyl linkages exhibits a higher effect with respect to prolongation of life than free 5FU. Small-sized chitosan gel nanospheres, containing 5FU or immobilizing 5FU derivatives (aminophenyl-carbamoyl-5FU or aminopentyl-ester-methylene-5FU), were prepared by the glutaraldehyde crosslinking technique and the emulsion method (663). When chitosan was crosslinked with glutaraldehyde, these 5FU derivatives were simultaneously immobilized to chitosan by means of Schiff's base formation. The chitosan gel nanospheres were coated with anionic polysaccharides.

Covalently drug-pendanted 6-O-carboxymethyl-chitin (CM-chitin) was synthesized, in which process the drug was coupled through an enzyme-susceptible bond (664). Chromofore-terminating peptides were coupled with CM-chitin as a model for a CM-chitin–drug conjugate. The amino acid composition of the spacer and the spacer length were found to be preliminary regulation factors for two-step hydrolysis of the polymeric drug.

The water-soluble conjugates of mitomycin C (MMC) with N-succinyl-chitosan (N-suc-chitosan) and glycol-chitosan (gly-chitosan), named N-suc-chitosan-glu-MMC and gly-chitosan-glu-MMC, respectively, were characterized mainly by the plasma concentration time profiles of MMC after intraperitoneal administration and their in vivo antitumor effect against P388 leukemia and Sarcoma-180. Before in vivo evaluation, polymer–drug binding characteristics were checked by gel chromatography (665). For the Sarcoma-180 solid tumor inoculated subcutaneously, the polymer characteristics affected the antitumor effect. That is, with the intravenous injection, gly-chitosan-glu-MMC hardly exhibited any tumor growth inhibition, but N-suc-chitosan-glu-MMC showed significant tumor growth suppression. As to the intratumoral administration, the tendency to suppress tumor growth was observed in MMC and both the conjugates.

J. Sterilization of Chitosan

Chitosan has potential biomedical applications that may require the final products to be sterilized before use. The gamma irradiation of purified and highly deacetylated chitosan fibers and films at sterilizing doses (up to 25 KGB) caused main chain scissions. The viscosity average molecular weight of the polymer decreased with increasing irradiation dose (666). Irradiation in anoxia did not affect film properties significantly.

REFERENCES

1. H. J. Jennings, R. A. Pon. Polysaccharides and glycoconjugates as human vaccines. In: *Polysaccharides in Medicinal Applications*, S. Dumitriu (Ed.). Marcel Dekker, New York, 1996, pp. 443–479.
2. H. J. Jennings. Capsular polysaccharides as human vaccines. *Adv. Carbohydr. Chem. Biochem*. 41:155 (1983).
3. H. J. Jennings. Capsular polysaccharides as vaccine candidates. *Curr. Top. Microbiol. Immunol*. 150:97 (1990).
4. D. R. Bundle, I. C. P. Smith, H. J. Jennings. Determination of the structure and conformation of bacterial polysaccharides by carbon 13 nuclear magnetic resonance: studies on the group-specific antigens of *Neisseria meningitidis* serogroups A and X. *J. Biol. Chem*. 249:2275 (1974).
5. A. K. Bhattacharjee, H. J. Jennings, C. P. Kenny, A. Martin, I. C. P. Smith. Structural determination of the sialic acid polysaccharide antigens of *Neisseria meningitidis* serogroups B and C with carbon 13 nuclear magnetic resonance. *J. Biol. Chem*. 250:1962 (1975).
6. C. J. Lee, B. A. Fraser. The structure of the cross-reactive types 19 (19F) and 57 (19A) pneumococcal capsular polysaccharides. *J. Biol. Chem*. 255:6847 (1980).
7. C. J. Lee, K. Koizumi, J. Henrichsen, B. Perch, C. S. Lin, W. Egan. Capsular polysaccharides of non-groupable streptococci that cross-react with pneumococcal group 19. *J. Immunol*. 133:2706 (1984).
8. R. M. Cristel, R. S. Baker, D. E. Dorman. Capsular polymer of *Haemophilus influenzae* type b. I. Structural characterization of capsular polymer of strain Eagan. *J. Biol. Chem*. 250:4926 (1975).
9. H. J. Jennings, R. K. Sood. Synthetic glycoconjugates as human vaccines. In: *Neoglycoconjugates: Preparation and Applications*, Y. C. Lee, R. Thee (Eds.). Academic Press, New York, 1994, pp. 325–371.
10. W. E. Dick, Jr., M. Beurret. Glycoconjugates of bacterial

carbohydrate antigens. In: *Contributions to Microbiology and Immunology*, Vol. 10, J. M. Cruse, R. E. Lewis (Eds.). S. Karger, Basel, 1989, pp. 48–114.
11. J. B. Robbins, R. Schneerson. Polysaccharide–protein conjugates: a new generation of vaccines. *J. Infect. Dis.* 161:821 (1990).
12. P. W. Anderson, M. E. Pichichero, R. A. Insel, R. Betts, R. Eby, D. H. Smits. Vaccines consisting of periodate-cleaved oligosaccharides from the capsule of *Haemophilus influenzae* type b coupled to a protein carrier: structural and temporal requirements for priming of the human infants. *J. Immunol.* 137:1181 (1986).
13. R. Schneerson, O. Barrera, A. Sutton, J. B. Robbins, Preparation, characterization and immunogenicity of *Haemophilus influenzae* type b polysaccharide–protein conjugates. *J. Exp. Med.* 152:361 (1980).
14. C. Chu, R. Schneerson, J. B. Robbins, S. C. Rastagi. Further studies on the immunogenicity of *Haemophilus influenzae* type b and pneumococcal type 6A polysaccharide–protein conjugates. *Infect. Immun.* 40:245 (1983).
15. R. K. Sood, F. Michon, J. M. de Muys, H. J. Jennings. Synthesis and immunogenicity in rabbits of *Streprococcus pneumoniae* type 14 polysaccharide–tetanus toxoid conjugates. Proceedings of the XIIth International Carbohydrate Symposium, Paris, 1992, p. 370.
16. R. Schneerson, J. B. Chu, A. Sutton, W. Vann, J. C. Vickers, W. T. London, B. Curfman, M. C. Hardegree, J. Shiloach, S. C. Rastogi. Serum antibody response of juvenile and infant Rhesus monkeys injected with *Haemophilus influenzae* type b and pneumococcus type 6A capsular polysaccharide–protein conjugates. *Infect. Immun.* 45:582 (1984).
17. R. Schneerson, J. B. Robbins, J. C. Parke, C. Bell, J. J. Schlessman, A. Sutton, Z. Wang, G. Schiffman, G. Karpas, J. Shiloach. Quantitative and qualitative analyses of serum antibodies elicited in adults by *Haemophilus influenzae* type b and pneumococcus type 6A capsular polysaccharide–tetanus toxoid conjugates. *Infect. Immun.* 52:519 (1986).
18. H. J. Jennings, R. Roy. Enhancement of the immune response to the group B polysaccharide of *Neisseria meningitidis* by means of its chemical modification. In: *The Pathogenic Neisseriae*, Schoolnik (Ed.). American Society for Microbiology, Washington, D.C., 1985, p. 628.
19. H. J. Jennings, C. Lugowski, Immunochemistry of group A, B and C meningococcal polysaccharide–tetanus toxoid conjugates. *J. Immunol.* 127:1011 (1981).
20. S. C. Szu, A. L. Stone, J. D. Robbins, R. Schneerson, J. B. Robbins. Vi capsular polysaccharide–protein conjugates for prevention of typhoid fever. *J. Exp. Med.* 166: 1510 (1987).
21. S. C. Szu, X. Li, R. Schneerson, J. H. Vickers, D. Bryla, J. B. Robbins. Comaparative immunogenicities of Vi polysaccharide–protein conjugates composed of cholera toxin or its B subunit as a carrier bound to higher or lower molecular weight Vi. *Infect. Immun.* 57:3823 (1989).
22. K. Jann, B. Jann (Eds.). *Current Topics in Microbiology and Immunology*, Vol. 150, *Bacterial Capsules*. Springer-Verlag, Heidelberg, Germany, 1990.
23. C. Whitfield, M. A. Valvano. Biosynthesis and expression of cell-surface polysaccharides. *Adv. Microb. Physiol.* 35: 135 (1993).
24. K. Jann, B. Jann. Structure and biosynthesis of O-antigens. In: *Handbook of Endotoxin*, Vol. 1, *Chemistry of Endotoxin*, E. Th. Rietschel (Ed.). Elsevier Science, Amsterdam, 1984, p. 138.
25. A. S. Cross. The biologic significance of bacterial encapsulation. *Cur. Top. Microbiol. Immunol.* 150:87 (1990).
26. K. Jann, B. Jann, Capsules of *Escherichia coli*, expression and biological significance, *Can. J. Microbiol.* 38:705 (1991).
27. D. L. Kasper. Bacterial capsules—old dogmas and new tricks. *J. Infect. Dis.* 3:407 (1986).
28. K. S. Kim, J. H. Kang, A. S. Cross. The role of capsular antigens in serum resistance and in vivo virulence of *Escherichia coli. FEMS Microbiol. Lett.* 35:275 (1986).
29. I. S. Roberts, F. K. Saunders, G. J. Boulnois. Bacterial capsules and interactions with complements and phagocytes. *Biochem. Soc. Trans.* 17:462 (1989).
30. K. Jann, B. Jann. Structure and biosynthesis of the capsular antigens of *Escherichia coli* I. In: *Current Topics in Microbiology and Immunology*, Vol. 150, K. Jann, B. Jann (Eds.). Springer-Verlag, Heidelberg, Germany, 1990, p. 19.
31. L. Kenne, B. Lindberg. Bacterial polysaccharides. In: *The Polysaccharide*, Vol. 2, G. O. Aspinal (Ed.). Academic Press, New York, 1983, p. 287.
32. K. Jann, B. Jann. Structure and biosynthesis of O-antigens. In: *Handbook of Endotoxin*, Vol. 1, *Chemistry of Endotoxin*, E. Th. Rietschel (Ed.). Elsevier Science, Amsterdam, 1984, p. 138.
33. S. V. Evans, D. R. Rose, R. To, N. M. Young, D. R. Bundle. Exploring the mimicry of polysaccharide antigens by anti-idiotypic antibodies. The crystallization, molecular replacement, and refinement to 2.8 Å resolution of an idiotope-anti-idiotope Fab complex and of the unliganded anti-idiotope Fab. *J. Mol. Biol.* 241:691–705 (1994).
34. N. M. Young, M. A. Gidney, B. M. Gudmundsson, C. R. MacKenzie, R. To, D. C. Watson, D. R. Bundle. Molecular basis for the lack of mimicry of Brucella polysaccharide antigens by Ab2gamma antibodies. *Molec. Immunol.* 36:339–347 (1999).
35. C. J. Lee. Bacterial capsular polysaccharides: immunogenicity and vaccines. In: *Polysaccharides in Medicinal Applications*, S. Dumitriu (Ed.). Marcel Dekker, New York, pp. 411–442 (1996).
36. D. Alvarez, S. Merino, J. M. Tomas, V. J. Benedi, S. Alberti. Capsular polysaccharide is a major complement resistance factor in lipopolysaccharide O side chain–deficient *Klebsiella pneumoniae* clinical isolates. *Infect. Immun.* 68:953–955 (2000).
37. E. D. Barnett, S. I. Pelton, H. J. Cabral, R. D. Eavey, C. Allen, M. J. Cunningham, E. R. McNamara, J. O. Klein. Immune response to pneumococcal conjugate and poly-

saccharide vaccines in otitis-prone and otitis-free children. *Clin. Infect. Dis.* 29:191–192 (1999).
38. H. Jakobsen, E. Saeland, S. Gizurarson, D. Schulz, I. Jonsdottir. Intranasal immunization with pneumococcal polysaccharide conjugate vaccines protects mice against invasive pneumococcal infections. *Infect. Immun.* 67:4128–4133 (1999).
39. Z. Zhong, T. Burns, Q. Chang, M. Carroll, L. Pirofski. Molecular and functional characteristics of a protective human monoclonal antibody to serotype 8 *Streptococcus pneumoniae* capsular polysaccharide. *Infect. Immun.* 67: 4119–4127 (1999).
40. T. Miyamoto, S. Takahashi, H. Ito, H. Inagaki, Y. Noishiki. Tissue biocompatibility of cellulose and its derivatives. *J. Biomed. Mater. Res.* 23:125–133 (1989).
41. J. W. Beyger, J. G. Nairn. Some factors affecting the microencapsulation of pharmaceuticals with cellulose acetate phthalate. *J. Pharm. Sci.* 75:573–578 (1986).
42. K. S. Devi, T. J. M. Sinha, P. Vasudeven. Biosoluble surgical material from 2,3-dialdehyde cellulose. *Biomaterials* 7:193–196 (1986).
43. T. Miyamoto, S. Takahashi, H. Ito, H. Inagaki, Y. Noishiki. Tissue biocompatibility of cellulose and its derivatives. *J. Biomed. Mater. Res.* 23:125–133 (1989).
44. Y. Ikada. Membranes as biomaterials. In: *Polysaccharides in Medicinal Applications*, S. Dumitriu (Ed.). Marcel Dekker, New York, 1996, pp. 663–688.
45. K. Kamide, H. Iijima. Recent advances in cellulose membranes. In: *Cellulosic Polymers*, R. D. Gilbert (Ed.). Hanser, Munich, 1994, p. 189.
46. T. Akizawa, K. Kino, E. Kinugasa, S. Koshikawa, Y. Ikada, A. Kishida, Y. Hatanaka, K. Imamura. Clinical effects of a polyethylene glycol grafted cellulose membrane on thrombogenicity and biocompatibility during hemodialysis. *ASAIO Trans.* 86:M640 (1990).
47. I. V. Yannas, J. F. Burke. Design of an artificial skin. I. Basic design principles. *J. Biomed. Mater. Res.* 14:65 (1980).
48. N. L. Sussman, M. G. Chong, T. Koussayer, D. R. He, T. A. Shang, H. H. Whisennand, J. H. Kelly. Reversal of fulminant hepatic failure using an extracorporeal liver assist device. *Hepatology* 16:60 (1992).
49. K. Ishihara, N. Nakabayashi, M. Sakakida, K. Nishida, M. Shichiri. Biocompatible microdialysis hollow-fiber probes for long-term in vivo glucose monitoring. In: *Polymers in Sensors*, N. Akmal, A. M. Usmani (Eds.). ACS Symposium Series 690. American Chemical Society, Washington, DC, 1998, pp. 25–33.
50. M. Ohno, M. Suzuki, M. Miyagi, T. Yagi, H. Sakurai, T. U. Kai. CTA hemodialysis membrane design for β2-microglobulin removal. In: *High-Performance Cellulosic Materials: Cellulosics: Chemical, Biochemical and Material Aspects*, J. F. Kennedy, G. O. Phillips, P. A. Williams (Eds.). Ellis Horwood, p. 418.
51. T. Urgami, M. Tamura, M. Sugihara. Characteristics of cellulose nitrate membranes, *Angew. Makromol. Chem.* 55:59 (1976).
52. T. Urgami, Y. Sugitani, M. Sugihara. Preparation and permeation characteristics of cellulose membrane. *Sep. Sci. Technol.* 17:307 (1982).
53. M. V. Mac-Kay, I. P. Fernandez, J. L. Sanchez Burson, G. F. Herrero. Experimental equations to calculate aminoglycoside clearances through cuprophan dialyzers. *Biopharm. Drug Disposition* 19:555–561 (1998).
54. V. Stefanovic, P. Vlahovic, S. Kostic, M. Mitic-Zlatkovic. *In vitro* blood compatibility evaluation of cuprophan and polyacrylonitrile membranes. *Nephron.* 79:350–351 (1998).
55. N. Faucheux, R. Warocquier-Clerout, B. Haye, M. D. Nagel. Cyclic AMP in cells adhering to bioincompatible (Cuprophan) and biocompatible (AN69) substrates. *J. Biomed. Mater. Res.* 39:506–510 (1998).
56. V. Stefanovic, V. Savic, M. Rajic, M. Mitic-Zlatkovic. β 2-microglobulin and IgE release during hemodialysis with cuprophan membrane. *Int. J. Artif. Organs* 20:713–714 (1997).
57. W. L. Hinrichs, H. W. ten Hoopen, G. H. Engbers, J. Feijen. *In vitro* evaluation of heparinized Cuprophan hemodialysis membranes. *J. Biomed. Mater. Res.* 35:443–450 (1997).
58. J. Vienken, M. Diamantoglou, W. Henne, B. Nederlof. Artificial dialysis membranes: from concept to large scale production. *Am. J. Nephrology* 19:355–362 (1999).
59. M. Diamantoglou, J. Platz, J. Vienken. Cellulose carbamates and derivatives as hemocompatible membrane materials for hemodialysis. *Artif. Organs* 23:15–22 (1999).
60. S. Daimon, T. Saga M. Nakayama, Y. Nomura, H. Chikaki, K. Dan, I. Koni. Dextran sulphate cellulose columns for the treatment of nephritic syndrome due to inactive lupus nephritis. *Nephrology, Dialysis, Transplantation* 15: 235–238 (2000).
61. E. Muchmore. International Symposium on Viral Hepatitis Liver Disease. The 8th Triennial Congress, Tokyo, May 10–14, no. 3196 (1993).
62. D. L. Prince, N. H. Prince, O. Thraenhart, E. Muchmore, E. Bonder, J. Pugh. Methodological approaches to disinfection of human hepatitis B virus. *J. Clin. Microbiol.* 31: 3296–3304 (1993).
63. D. M. Wiseman, L. Gottlick-Iarkowski, L. Kamp. Effect of different barriers of oxidized regenerated cellulose (ORC) on cecal and sidewall adhesions in the presence and absence of bleeding. *J. Invest. Surg.* 12(3):141–146 (1999).
64. S. Pulapura, J. Kohn, Trends in the development of bioresorbable polymers for medical applications. *J. Biomater. Appl.* 6:216–250 (1992).
65. J. N. Staniforth, A. R. Baichwal. Synergistically interacting heterodisperse polysaccharides. Function in achieving controllable drug delivery. *ACS Symp. Ser.* 520:327–343 (1993).
66. S. H. Neau, M. A. Howard, J. S. Claudius, D. R. Howard. The effect of the aqueous solubility of xanthine derivatives on the release mechanism from ethylcellulose matrix tablets. *Int. J. Pharm.* 179:97–105 (1999).

67. A. Billon, M. Petit, M. B. Doko, B. Bataille, M. Jacob. Effects of cellulose derivatives and additives in the spray-drying preparation of acetaminophen delivery systems. *Drug Dev. Pharm.* 25:1149–1156 (1999).
68. H. Lapidus, N. G. Lordi. Some factors affecting the release of a water-soluble drug from a compressed hydrophilic matrix. *J. Pharm. Sci.* 55:840–843 (1966).
69. H. E. Huber, G. L. Christenson. Utilization of hydrophilic gums for the control of drug substance release from tablet formulations. II. Influence of tablet hardness and demsity on dissolution behavior. *J. Pharm. Sci.* 57:164–167 (1968).
70. U. Conte, L. Maggi, M. L. Torre, P. Giunchedi, A. La Manna. Press-coated tablets for time-programmed release of drugs. *Biomaterials* 14:1017–1023 (1993).
71. Y. Kawashima, T. Serigano, T. Hino, H. Yamamoto, H. Takeuchi. A new powder design method to improve inhalation efficiency of pranlukast hydrate dry powder aerosols by surface modification with hydroxypropylmethylcellulose phthalate nanospheres. *Pharm. Res.* 15(11):1748–1752, (1998).
72. A. Nokhodchi, J. L. Ford, P. H. Rowe, M. H. Rubinstein. The effect of moisture on the heckel and energy analysis of hydroxypropylmethylcellulose 2208 (HPMC K4M). *J. Pharm. Pharmacol.* 48:1122–1127 (1996).
73. A. Nokhodchi, J. L. Ford, P. H. Rowe, M. H. Rubinstein. The influence of moisture content on the consolidation properties of hydroxypropylmethylcellulose K4M (HPMC 2208). *J. Pharm. Pharmacol.* 48:1116–1121, (1996).
74. B. Perez-Marcos, J. L. Ford, D. J. Armstrong, P. N. Elliott, C. Rostron, J. E. Hogan. Influence of pH on the release of propranolol hydrochloride from matrices containing hydroxypropylmethylcellulose K4M and carbopol 974. *J. Pharm. Sci.* 85:330–334 (1996).
75. G. Xu, H. Sunada. Influence of formulation change on drug release kinetics from hydroxypropylmethylcellulose matrix tablets. *Chem. Pharm. Bull.* 43:483–487 (1995).
76. A. Cernak. Coatel—2% hydroxypropylmethylcellulose (OPSIA). *Ceskoslovenska Oftalmologie* 50:380–381 (1994).
77. C. Alvarez-Lorenzo, J. L. Gomez-Amoza, R. Martinez-Pacheco, C. Sauto, A. Concheiro. Evaluation of low-substituted hydroxypropylcellulose (L-HPCs) as filler-binders for direct compression. *Int. J. Pharm.* 197:107–116 (2000).
78. V. Pillay, R. Fassihi. Electrolyte-induced compositional heterogeneity: a novel approach for rate-controlled oral drug delivery. *J. Pharm. Sci.* 88:1140–1148 (1999).
79. R. Bodmeier, O. Paeratakul. A novel multiple-unit sustained release indomethacin–hydroxypropyl methylcellulose delivery system prepared by ionotropic gelation of sodium at elevated temperatures. *Carbohydr. Polym.* 16: 399–408 (1991).
80. Y. Suzuki, Y. Makino. Mucosal drug delivery using cellulose derivatives as a functional polymer. *J. Controlled Release* 62:101–107 (1999).
81. A. Ridell, H. Evertsson, S. Nilsson, L. O. Sundelof. Amphiphilic association of ibuprofen and two nonionic cellulose derivatives in aqueous solution. *J. Pharm. Sci.* 88: 1175–1181 (1999).
82. T. Harada, A. Misaki, S. Saito. Curdlan: a bacterial gel-forming β-1,3-glucan. *Arch. Biochem. Biophys.* 124:292 (1968).
83. S. Saito, A. Misaki, T. Harada. A comparison of the structure of curdlan and pachyman. *Agric. Biol. Chem.* 32:1261 (1968).
84. J. Ebata. The application of an exo-β-D-(1 → 3)-glucanase on structural analysis of β-glucan. Abstract of the Eighth International Symposium on Carbohydrate Chemistry, 1976, p. 112.
85. Y. Y. Maeda, G. Chihara. Lentinan, a new immuno-accelerator of cell-mediated responses. *Nature* 229:634 (1971).
86. T. Sasaki, N. Abiko, Y. Sugino, K. Nitta. Dependence on chain length of antitumor activity of (1 → 3)-β-D-glucan from *Alcaligenes faecalis* var. myxogenes, IFO 13140, and its acid degraded products. *Cancer Res.* 38:379 (1978).
87. T. Sasaki, M. Tanaka, H. Uchida. Effect of serum from mice treated with antitumor polysaccharide on expression of cytotoxicity by mouse pertoneal macrophages. *J. Pharm. Dyn.* 5:1013 (1982).
88. K. Morikawa, R. Takeda, M. Yamazaki, D. Mizuno. Induction of tumoricidal activity of polymorphonuclear leukocytes by a linear β-1,3-D-glucan and other immunomodulators in murine cells. *Cancer Res.* 45:1496 (1985).
89. I. Yamamoto, K. Takayama, K. Komma, T. Gonda, K. Yamazaki, K. Matsuzaki, K. Hatanaka, T. Uryu, O. Yoshida, H. Nakashima, N. Yamamoto, Y. Kaneko, T. Mimura. Synthesis, structure and antiviral activity of sulfates of cellulose and its branched derivatives. *Carbohydr. Polym.* 14:53 (1991).
90. Z. Osawa, T. Morota, K. Hatanaka, T. Akaike, K. Matsuzuki, H. Nakashima, N. Yamamoto, E. Suzuki, H. Miyano, T. Mimura, Y. Kaneko. Synthesis of sulfated derivatives of curdlan and their anti-HIV activity. *Carbohydr. Polym.* 21:283 (1993).
91. T. Yoshida, K. Hatanaka, T. Uryu, Y. Kaneko, E. Suzuki, H. Miyano, T. Mimura, O. Yoshida, N. Yamamoto. Synthesis and structural analysis of curdlan sulfate with a potent inhibitory effect in vitro of AIDS virus infection. *Macromolecules* 23:3717 (1990).
92. M. Mizuno, K. Minato, H. Ito, M. Kawade, H. Terai, H. Tsuchida. Anti-tumour polysaccharide from the mycelium of liquid-cultured *Agaricus blazei* mill. *Biochem. Molec. Biol. Int.* 47:707–714 (1999).
93. I. Gliko-Kabir, B. Yagen, A. Penhasi, A. Rubinstein. Phosphated crosslinked guar for colon-specific drug delivery. I. Preparation and physicochemical characterization. *J. Controlled Release* 63:121–127 (2000).
94. I. Gliko-Kabir, B. Yagen, B. Baluom, A. Rubinstein. Phosphated crosslinked guar for colon-specific drug delivery. II. In vitro and in vivo evaluation in the rat. *J. Controlled Release* 63:129–134 (2000).

95. Y. V. Prasad, Y. S. Krishnaiah, S. Satyanarayana. *In vitro* evaluation of guar gum as a carrier for colon-specific drug delivery. *J. Controlled Release* 51:281–287 (1998).
96. P. Khullar, R. K. Khar, S. P. Agarwal. Evaluation of guar gum in the preparation of sustained-release matrix tablets. *Drug Dev. Ind. Pharm*. 24:1095–1099 (1998).
97. K. S. Soppimath, A. R. Kulkarni, T. M. Aminabhavi. Controlled release of antihypertensive drug from the interpenetrating network poly(vinyl alcohol)–guar gum hydrogel microspheres. *J. Biomater. Sci. Polym. Ed*. 11:27–43 (2000).
98. S. A. Altaf, K. Yu, J. Parasrampuria, D. R. Friend. Guar gum–based sustained release diltiazem. *Pharm. Res*. 15: 1196–1201, (1998).
99. R. J. Seviour, S. J. Stasinopoulos, D. P. F. Auer, P. A. Gibs. Production of pullulan and other exopolysaccharides by filamentous fungi. *Crit. Rev. Biotechnol*. 12:279 (1992).
100. M. S. Deshpande, V. B. Rale, J. M. Lynch. *Aureobasidium pullulans* in applied microbiology: a status report. *Enz. Microb. Technol*. 14:514 (1992).
101. B. Bernier. The production of polysaccharides by fungi active in the decomposition of wood and forest litter. *Can. J. Microbiol*. 4:195 (1958).
102. K. Xi, Y. Tabata, K. Uno, M. Yoshimoto, T. Kishida, Y. Sokawa, Y. Ikada. Liver targeting of interferon through pullulan conjugation. *Pharm. Res*. 13:1846–1850 (1996).
103. Y. Tabata, Y. Matsui, K. Uno, Y. Sokawa, Y. Ikada. Simple mixing of IFN with a polysaccharide having high liver affinity enables IFN to target to the liver. *J. Interferon Cytokine Res*. 19:287–292 (1999).
104. X. G. Gu, M. Schmitt, A. Hiasa, Y. Nagata, H. Ikeda, Y. Sasaki, K. Akiyoshi, J. Sunamoto, H. Nakamura, K. Kuribayashi, H. Shiku. A novel hydrophobized polysaccharide/oncoprotein complex vaccine induces *in vitro* and *in vivo* cellular and humoral immune responses against HER2-expressing murine sarcomas. *Cancer Res*. 58:3385–3390 (1998).
105. L. Wang, H. Ikeda, Y. Ikuta, M. Schmitt, Y. Miyahara, Y. Takahashi, X. Gu, Y. Nagata, Y. Sasaki, K. Akiyoshi, J. Sunamoto, H. Nakamura, K. Kuribayashi, H. Shiku. Bone marrow–derived dendritic cells incorporate and process hydrophobized polysaccharide/oncoprotein complex as antigen presenting cells. *Int. J. Oncology* 14:695–701 (1999).
106. K. Ichinose, M. Yamamoto, T. Khoji, N. Ishii, J. Sunamoto, T. Kanematsu. Antitumor effect of polysaccharide coated liposomal adriamycin on AH66 hepatoma in nude mice. *Anticancer Res*. 18:401–404, (1998).
107. M. Yamamoto, K. Ichinose, N. Ishii, T. Khoji, K. Akiyoshi, N. Moriguchi, J. Sunamoto, T. Kanematsu. Utility of liposomes coated with polysaccharide bearing 1-aminolactose as targeting chemotherapy for AH66 hepatoma cells. *Oncology Reports* 7:107–111 (2000).
108. K. Akiyoshi, S. Kobayashi, S. Shichibe, D. Mix, M. Baudys, S. W. Kim, J. Sunamoto. Self-assembled hydrogel nanoparticle of cholesterol-bearing pullulan as a carrier of protein drugs: complexation and stabilization of insulin. *J. Controlled Release* 54:313–320 (1998).
109. H. Suzuki, H. Sunada. Influence of water-soluble polymers on the dissolution of nifedipine solid dispersions with combined carriers. *Chem. Pharm. Bull*. 46:482–487 (1998).
110. N. W. H. Cheetham, E. Fiala-Beer, G. W. Walker. Dextrin, structural details from high-field proton NMR spectroscopy. *Carbohydr. Res*. 14:149 (1991).
111. M. Q. Zhang, C. D. Rees. A review of recent applications of cyclodextrins for drug discovery. *Expert Opinion on Therapeutic Patents* 9:1697–1719, (1999).
112. A. N. Belder. Medical Applications of dextrin and its derivatives. In: *Polysaccharides in Medicinal Applications*, S. Dumitriu (Ed.). Marcel Dekker, New York, 1996, pp. 505–524.
113. M. Aberg. On effect of dextrin on lysability and structure of ex vivo thrombi, platelet function and factor VIII. Dissertation, Malmo General Hospital, University of Lund, Malmo, 1978.
114. B. Alexander, K. Odake, D. Lawlor, M. Swanger. Coagulation, hemostasis and plasma volume expanders: a quarter century enigma. *Fed. Proc*. 34:1429 (1975).
115. U. F. Gruber. *Blood Replacement*. Springer-Verlag, Berlin, 1969, p. 55.
116. S. E. Bergentz, O. Eiken, I. M. Nilsson. The effect of dextrin of various molecular weights on the coagulation in dogs. *Thromb. Diath. Hemorrh*. 11:15 (1961).
117. W. L. Bloom, D. S. Harmer, M. F. Bryant, S. S. Brewer. Coating of vascular surfaces and cells: a new concept in prevention of intravascular thrombosis. *Proc. Soc. Exp. Biol. NY* 115:384 (1964).
118. S. Rothman, E. Adelson, A. Schwebel, R. D. Langdell. Adsorption of C-14 dextrin to human blood platelets and red blood cells in vitro. *Vox Sang* 2:104 (1957).
119. E. A. Kabat, A. E. Bezer. The effect of variation in molecular weight on the antigenicity of dextrin in man. *Arch. Biochem. Biophys*. 78:306 (1958).
120. A. Gronwall. *Dextrin and Its Colloidal Infusion Solutions*. Almqvist and Wiksell, Stockholm, 1957, p. 73.
121. W. Richter, L. Kagedal. Preparation of dextran–protein conjugates and studies of their immunogenicity. *Int. Arch. Allergy* 42:887 (1972).
122. E. A. Kabat. The upper limit for the size of the human anti-dextran combining site. *J. Immunol*. 84:82 (1960).
123. H. C. Chiu, G. H. Hsiue, Y. P. Lee, L. W. Huang. Synthesis and characterization of pH-sensitive dextrin hydrogels as a potential colon-specific drug delivery system. *J. Biomater. Sci. Polym. Ed*. 10:591–608 (1999).
124. M. L. Bender, M. Komiyama. *Cyclodextrin Chemistry*. Springer-Verlag, New York, 1977.
125. W. Saenger. Cyclodextrin inclusion compounds in research and industry. *Angew. Chem. Int. Ed. Engl*. 19:344 (1980).
126. T. Loftsson. Pharmaceutical applications of β-cyclodextrin. *Pharm. Technol*. 23:40–51, (1999).
127. J. Szejtli. *Cyclodextrin Technology*. Kluwer, Dordrecht, The Netherlands, 1988.

128. R. J. Clarke, J. H. Coates, S. F. Lincol. Inclusion complexes of the cyclomaltooligosaccharides (cyclodextrins). *Adv. Carbohydr. Chem. Biochem.* 46:205 (1989).
129. S. Li, W. C. Purdy. Cyclodextrins and their applications in analytical chemistry. *Chem. Rev.* 92:1457 (1992).
130. B. Chankvedze, G. Endresz, G. Blaschke. Charged cyclodextrin derivatives as chiral selectors in capillary electrophoresis. *Chem. Soc. Rev.* 25:141 (1996).
131. S. F. Lincoln, C. J. Easton. Inclusion complexes of the cyclodextrins. In: *Polysaccharides*, S. Dumitriu (Ed.). Marcel Dekker, New York, 1998, pp. 473–522.
132. D. C. Bibby, N. M. Davies, I. G. Tucker. Mechanisms by which cyclodextrins modify drug release from polymeric drug delivery systems. *Int. J. Pharm.* 197:1–11 (2000).
133. J. Szeijtli. The mechanism, toxicity and biological effects of cyclodextrins. In: *Cyclodextrins and Their Industrial Uses*, D. Ducheme (Ed.). Editions de Santé, Paris, p. 173.
134. K. Uekama, F. Hirayama, T. Irie. Pharmaceutical uses of cyclodextrin derivatives. In: *High Performance Biomaterials: A Comprehensive Guide to Medical and Pharmaceutical Applications*, M. Szycher (Ed.). Technomic, Lancaster, PA, 1991, p. 789.
135. R. N. Antenucci, J. K. Palmer. Enzymatic degradation of alpha- and beta-cyclodextrins by *Bacteroides* of the human colon. *J. Agric. Food Chem.* 32:1316 (1984).
136. D. Duchêne, D. Wouessidjewe. Pharmaceutical and medical applications of cyclodextrins. In: *Polysaccharides in Medicinal Applications*, S. Dumitriu (Ed.). Marcel Dekker, New York, 1996, pp. 575–603.
137. J. H. Guo, K. M. Cooklock. Bioadhesive polymer buccal patches for buprenorphine controlled delivery: solubility consideration. *Drug. Dev. Ind. Pharm.* 21: 2013–2019 (1995).
138. F. A. Mollgaard, H. Bundgaard. The influence of cyclodextrin on the stability of betamethasone-17-valerate. *Int. J. Pharm.* 20:155 (1984).
139. T. Hibi, M. Tatsumi, M. Hanabusa, T. Imai, M. Otagiri, K. Uekama. Stabilization and reduction of irritant taste of anti-inflammatory drug pirprofen by β-cyclodextrin complexation. *Yakugaku Zasshi* 104:990 (1985).
140. N. Nambu, K. Kikuchi, T. Kikuchi, Y. Takahashi, H. Ueda, T. Nagai. Influence of inclusion of nonsteroidal anti-inflammatory drugs with β-cyclodextrin on the irritation to stomach of rats upon oral administration. *Chem. Pharm. Bull.* 26:3609 (1978).
141. S. Z. Lin, D. Wouessidjewe, M. C. Poelman, D. Duchêne. In vivo evaluation of indomethacin/cyclodextrin complexes: gastrointestinal tolerance and dermal anti-inflammatory activity. *Int. J. Pharm.* 106:63 (1994).
142. D. Duchêne, F. Glomot, C. Vaution. Pharmaceutical application of cyclodextrins. In: *Cyclodextrins and Their Industrial Uses*, D. Duchêne (Ed.). Editions de Santé, Paris, 1987, p. 211.
143. K. Uekama, F. Hirayama, T. Irie. Modifications of drug release by cyclodextrin derivatives. In: *New trends in Cyclodextrins and Derivatives*, D. Duchêne (Ed.). Editions de Santé, Paris, 1991, p. 409.
144. P. Giunchedi, L. Maggi, A. La Manna, U. Conti. Modification of the dissolution behavior of a water-insoluble drug, naftazone, for zero-order release matrix preparation. *J. Pharm. Pharmacol.* 46:476–480 (1994).
145. C. X. Song, V. Labhasetwar, R. J. Levy. Controlled release of U-86983 from double-layer biodegradable matrices: effect of additives on release mechanism and kinetics. *J. Controlled Release* 45:177–192 (1997).
146. D. C. Bibby, N. M. Davies, I. G. Tucker. Mechanisms by which cyclodextrins modify drug release from polymeric drug delivery systems. *Int. J. Pharm.* 197:1–13 (2000).
147. K. Uekama, T. Fujinaga, F. Hirayama, M. Otagiri, M. Yamasaki, H. Seo, T. Hashimoto, M. Tsuruka. Improvement of the oral bioavailability of digitalis glycosides by cyclodextrin complexation. *J. Pharm. Sci.* 72:1338 (1983).
148. H. Seo, M. Tsuruka, T. Hashimoto, T. Fujinaga, M. Otagiri, K. Uekama. Enhancement of oral bioavailability of spironolactone by β- and γ-cyclodextrin complexations. *Chem. Pharm. Bull.* 31:286 (1983).
149. D. Duchene, D. Wouessidjewe, G. Ponchel. Cyclodextrins and carrier systems. *J. Controlled Release* 62:263–269 (2000).
150. M. Otagiri, T. Imai, K. Uekama. Enhancement oral bioavailability of the anti-inflammatory drug flurbiprofen in rabbits by tri-O-methyl β-cyclodextrin complexation. *J. Pharmacobio-Dyn.* 5:1027 (1982).
151. T. Imai, M. Otagiri, H. Saito, K. Uekama. Inclusion mode of flubiprofen with β-cyclodextrin and heptakis (2,3,6-tri-O-methyl) β-cyclodextrin, and improvements of some pharmaceutical properties of flurbiprofen by complexation. *Chem. Pharm. Bull.* 36:354 (1988).
152. G. Piel, B. Evrard, T. Hees, Van, G. Llabres, L. Delattre. Development of a parenteral and of an oral formulation of albendazole with cyclodextrins. *STP Pharma Sci.* 9: 257–261 (1999).
153. Y. Nakai, K. Yamamoto, K. Terada, H. Horibe, K. Ozawa. Interaction of tri-O-methyl β-cyclodextrin with drugs. II. Enhanced bioavailability of ketoprofen in rats when administered with tri-O-methyl β-cyclodextrin. *Chem. Pharm. Bull.* 31:3745 (1983).
154. T. Miyaji, Y. Inoue, F. Acartuk, T. Imai, M. Otagiri, K. Uekama. Improvement of oral bioavailability of fenbufen by cyclodextrin complexations. *Acta Pharm. Nord.* 4:17 (1992).
155. N. Bodor, J. Drustrup, W. Wu. Effect of cyclodextrins on the solubility and stability of a novel soft corticosteroid, loteprednol etabonate. *Pharmazie* 55:206–210 (2000).
156. K. Uekama, M. Otagiri, Y. Uemura, T. Fujinaga, K. Arimori, N. Matsuo, K. Tasaki, A. Sugii. Improvement of oral bioavailability of prednisolone by β-cyclodextrin complexation in humans. *J. Pharmacobio-Dyn.* 6:124 (1983).
157. M. Kikuchi, F. Hirayama, K. Uekama. Improvement of oral and rectal bioavailability of carmofur by methylated β-cyclodextrin complexations. *Int. J. Pharm.* 38:191 (1987).
158. R. Ficarra, P. Ficarra, M. R. Di Bella, D. Raneri, S. Tom-

masini, M. L. Calabro, A. Villari, S. Coppolino. Study of the inclusion complex of atenolol with β-cyclodextrins. *J. Pharmaceut. Biomed. Anal.* 23:231–237 (2000).
159. S. Emara, I. Morita, K. Tamura, S. Razee, T. Masujima, H. A. Mohamed, S. M. El-Gizawy, N. A. El-Rabbat. Effect of cyclodextrins on the stability of adriamycin, adriamycinol, adriamycinone and daunomycin. *Talanta* (Oxford) 51:359–365 (2000).
160. H. Matsuda, H. Arima. Cyclodextrins in transdermal and rectal delivery. *Adv. Drug Delivery Rev.* 36:81–101 (1999).
161. H. Arima, T. Kondo, T. Irie, F. Hirayama, K. Uekama, T. Miyaji, Y. Inoue. Use of water-soluble β-cyclodextrin derivatives as carriers of antiinflammatory drug biphenylacetic acid in rectal delivery. *Yakugaku Zassi* 112:65 (1992).
162. H. Arima, T. Kondo, T. Irie, K. Uekama. Enhanced rectal adsorption and reduced local irritation of the antiinflammatory drug ethyl 4-biphenylacetate in rats by complexation with water-soluble β-cyclodextrin derivatives and formulation as oleaginous suppository. *J. Pharm. Sci.* 81:1119 (1992).
163. H. W. Frijlink, S. Paiotti, A. C. Eissens, C. F. Lerk. The effects of cyclodextrins on drug release from fatty suppository bases. III. Application of cyclodextrin derivatives. *Eur. J. Pharm. Biopharm.* 38:174 (1992).
164. D. Duchêne, D. Wouessidjewe, M. C. Poelman. Dermal use of cyclodextrins and derivatives. In: *New Trends in Cyclodextrins and Derivatives*, D. Duchêne (Ed.). Editions de Santé, Paris, 1991, p. 447.
165. F. Glomont, L. Benkerrour, D. Duchêne, M. C. Poelman. Improvement in availability and stability of a dermocorticoid by inclusion in β-cyclodextrin. *Int. J. Pharm.* 46:49 (1988).
166. T. Loftsson, N. Leeves, B. Bjornsdottir, L. Duffy, M. Masson. Effect of cyclodextrins and polymers on triclosan availability and substantivity in toothpastes in vivo. *J. Pharm. Sci.* 88:1254–1259 (1999).
167. K. Uekama, H. Adachi, T. Irie, T. Yano, M. Suita, K. Noda. Improved transdermal delivery of prostaglandin El through hairless mouse skin: combined use of carboxymethyl-ethyl-β-cyclodextrin and penetration enhancers. *J. Pharm. Pharmacol.* 44:119 (1992).
168. H. Arima, H. Adachi, T. Irie, K. Uekama. Improved drug delivery through the skin by hydrophilic β-cyclodextrins: enhancement of antiinflammatory effect of 4-diphenylacetic acid in rats. *Drug Invest.* 2:155 (1990).
169. H. Okamoto, H. Komatsu, M. Hashida, M. Sezaki. Effects of β-cyclodextrin and di-O-methyl β-cyclodextrin on the percutaneous absorption of butyl paraben, indomethacin and sulfamic acid. *Int. J. Pharm.* 30:35 (1986).
170. T. Loftsson, N. Bodor. Effects of 2-hydroxypropyl β-cyclodextrin on the aqueous solubility of drugs and transdermal delivery of 17-β-estradiol. *Acta Pharm. Nord.* 1: 185 (1989).
171. M. O. Ahmed, I. El-Gibaly, S. M. Ahmed. Effect of cyclodextrins on the physicochemical properties and antimycotic activity of clotrimazole. *Int. J. Pharm.* 171:111–123 (1998).
172. A. C. Williams, S. R. S. Shatri, B. W. Barry. Transdermal permeation modulation by cyclodextrins: a mechanistic study. *Pharm. Dev. Technol.* 3:283–297 (1998).
173. A. R. Hedges. Industrial Applications of Cyclodextrins. *Chem. Rev.* 98:2035–2045 (1998).
174. T. Loftsson, O. J. Hjaltalin. Cyclodextrins: new drug delivery systems in dermatology. *Int. J. Dermatol.* 37:241–247 (1998).
175. T. Loftsson, H. Fridriksdottir, G. Ingvasdottir, B. Jonsdottir, H. Sigurdardottir. Influence of 2-hydroxypropyl-β-cyclodextrin on diffusion rates and transdermal delivery of hydrocortisone. *Drug Dev. Ind. Pharm.* 20:1699 (1994).
176. A. Rolland, B. Shroot. Preparation and evaluation of dermatological formulations based on the inclusion of tretinoin in β-cyclodextrin and hydroxypropyl β-cyclodextrin. Minutes of the 6th International Symposium on Cyclodextrin, A. R. Hedges (Ed.). Editions de Santé, Paris, 1992, p. 529.
177. N. Shankland, A. Pearson, J. R. Johnson, E. G. Salole. The influence of β-cyclodextrin on the release of hydrocortisone from topical cream base. *J. Pharm. Pharmacol.* 37: 107P (1985).
178. H. van Doorne, E. H. Bosch, C. F. Lerk. Formation and antimicrobial activity of β-cyclodextrin and some antimycotic imidazole derivatives. *Pharm. Weekbl.* 10:80 (1988).
179. B. Siefert, U. Pleyer, M. Muller, C. Hartmann, S. Keipert. Influence of cyclodextrins on the in vitro corneal permeability and in vivo ocular distribution of thalidomide. *J. Ocular Pharmacol. Therap.* 15:429–439 (1999).
180. T. Loftssona, T. Jarvinen. Cyclodextrins in ophthalmic drug delivery. *Adv. Drug Delivery Rev.* 36:59–81 (1999).
181. P. Saarinen-Savolainen, T. Jarvinen, K. Araki-Sasaki, H. Watanabe, A. Urtti. Evaluation of cytotoxicity of various ophthalmic drugs, eye drop excipients and cyclodextrins in an immortalized human corneal epithelial cell line. *Pharm. Res.* 15:1275–1281 (1998).
182. T. Jansen, B. Xhonneux, J. Mesens, M. Borgers. β-Cyclodextrins as vehicles in eye-drop formulations: an evaluation of their effects on rabbit corneal epithelium. *Lens Eye Tox. Res.* 7:459 (1990).
183. F. W. H. M. Merkus, J. C. Verhoef, E. Marttin, S. G. Romeijn, P. H. M. van der Kuy, W. A. J. J. Hermens, N. G. M. Schipper. Cyclodextrins in nasal drug delivery. *Adv. Drug Delivery Rev.* 36:41–59 (1999).
184. N. G. M. Schipper, M. J. M. Deurloo, S. G. Romeijn, J. C. Verhoef, F. W. M. Merkus. Nasal absorption enhancement of 17β-estradiol by dimethyl-β-cyclodextrin in rabbits and rats. *Pharm. Res.* 7:500 (1990).
185. F. W. M. Merkus, J. C. Verhoef, S. G. Romeijn, N. G. M. Schipper. Absorption enhancing effect of cyclodextrins on intranasally administered insulin in rats. *Pharm. Res.* 8: 588 (1991).
186. Z. Sho, R. Krishnamoorthy, K. Mitra. Cyclodextrins as nasal absorption promoters of insulin: mechanistic evaluations. *Pharm. Res.* 9:1157 (1992).

187. T. Irie, K. Wakamatsu, H. Arima, H. Arimoto, K. Uekama. Enhancing effects of cyclodextrins on nasal absorption of insulin in rats. *Int. J. Pharm.* 84:129 (1992).
188. N. G. M. Schipper, S. G. Romeijn, J. C. Verhoef, F. W. M. Merkus. Nasal insulin delivery with dimethyl-β-cyclodextrin as an absorption enhancer in rabbits: powder more effective than liquid formulations. *Pharm. Res.* 10: 682 (1993).
189. D. Duchene, G. Ponchel, D. Wouessidjewe. Cyclodextrins in targeting—application to nanoparticles. *Adv. Drug Delivery Rev.* 36:29–41 (1999).
190. Z. Shao, K. Mitra. Transport enhancing effects of cyclodextrin derivatives relative to pulmonary, nasal, and enteral absorption of insulin. *Pharm. Res.* 10:S7 (1993).
191. D. A. Wall, J. Marcello, D. Pierdomenico, A. Farid. Administration of hydroxypropyl β-cyclodextrin complexes does not slow rates of pulmonary drug absorption in rat. *STP Pharma Sci.* 4:63 (1994).
192. G. Piel, B. Evrard, T. Van Hees, G. Llabres, L. Delattre. Development of a parenteral and of an oral formulation of albendazole with cyclodextrins. *STP Pharma Sci.* 9: 257–261 (1999).
193. H. W. Frijlink, E. J. F. Franssen, A. C. Eissens, R. Oosting, C. F. Lerk, D. K. F. Meijer. The effects of cyclodextrins on the disposition of intravenously injected drugs in the rats. *Pharm. Res.* 8:380 (1991).
194. M. E. Brewster, W. R. Anderson, K. S. Estes, N. Bodor. Development of aqueous parenteral formulations for carbamazepine through the use of modified cyclodextrins. *J. Pharm. Sci.* 80:380 (1991).
195. M. Krenn, M. P. Gamcsik, G. B. Vogelsang, O. M. Colvin, K. W. Leong. Improvements in solubility and stability of thalidomide upon complexation with hydroxypropyl-β-cyclodextrin. *J. Pharm. Sci.* 81:685 (1992).
196. M. Vikmon, A. Stadler-Szoke, J. Szejtli. Solubilization of amphotericin B with γ-cyclodextrin. *J. Antibiotics* 38:1822 (1985).
197. T. L. Yaksh, J. Jang, Y. Nishiuchi, K. P. Braun, S. Ro, M. Goodman. The utility of 2-hydroxypropyl-β-cyclodextrin as a vehicle for the intrathecal and intracerebral administration of drugs. *Life Sci.* 48:623 (1991).
198. J. Jang, T. L. Yaksh, H. F. Hill. Use of 2-hydroxypropyl-β-cyclodextrin as an intrathecal drug vehicle with opioids. *J. Pharmacol. Exp. Ther.* 261:592 (1992).
199. T. F. Meert, W. Melis. Interactions between epidurally and intrathecally administered sufentanil and bupivacaine in hydroxypropyl-β-cyclodextrin in the rat. *Acta Anaesth. Belg.* 43:79 (1992).
200. F. Hirayama, M. Kurihara, K. Uekama. Mechanisms of deceleration by methylated cyclodextrins in the dehydration oristaglandin A2 in aqueous solution. *Chem. Pharm. Bull.* 34:5093 (1986).
201. P. Zhang, C. C. Ling, H. Parrot-Lopez, H. Galons. Formation of amphiphilic cyclodextrins via hydrophobic esterification in the secondary hydroxyl face. *Tetrahedron Lett.* 32:2769 (1991).
202. P. Zhang, H. Parrot-Lopez, P. Tchoreloff, C. C. Ling, C. De Rango, A. Coleman. Self-organizing systems based on amphiphilic cyclodextrins. *J. Phys. Org. Chem.* 5:518 (1992).
203. M. Skiba, F. Puisieux, D. Duchene, D. Wouessidjewe. Direct imaging of modified β-cyclodextrin nanospheres by photon scanning tunnelling and scanning force microscopy. *Int. J. Pharm.* 120:1–9 (1995).
204. P. L. Thuaut, B. Martel, G. Crini, U. Maschke, X. Coqueret, M. Morcellet. Grafting of cyclodextrins onto polypropylene nonwoven fabrics for the manufacture of reactive filters. I. Synthesis parameters. *J. Appl. Polym. Sci.* 77:2118–2126 (2000).
205. H. Okumura, M. Okada, Y. Kawaguchi, A. Harada. Complex formation between poly(dimethylsiloxane) and cyclodextrins: new pseudo-polyrotaxanes containing inorganic polymers. *Macromolecules* 33:4297–4299 (2000).
206. Y. Liu, B. Li, T. Wada, Y. Inoue. Enantioselective recognition of aliphatic amino acids by organoselenium modified β-cyclodextrins. *Supramolec. Chem.* 10:173–185 (1999).
207. P. Glockner, N. Metz, H. Ritter. Cyclodextrins in polymer synthesis: free-radical polymerization of methylated β-cyclodextrin complexes of methyl methacrylate and styrene controlled by N-Acetyl-L-cysteine as a chain-transfer agent in aqueous medium. *Macromolecules* 33:4288–4291 (2000).
208. S. Tian, H. Zhu, P. Forgo, V. T. D'Souza. Selectively monomodified cyclodextrins. Synthetic strategies. *J. Org. Chem.* 65:2624–2631 (2000).
209. C. Pean, C. Creminon, A. Wijkhuisen, J. Grassi, P. Guenot, P. Jehan, J. P. Dalbiez, B. Perly, F. Djedaini-Pilard. Synthesis and characterization of peptidyl-cyclodextrins dedicated to drug targeting. *J. Chem. Soc. Perkin Transactions II* 4:853–865 (2000).
210. Y. Liu, C. C. You, T. Wada, Y. Inoue, Yoshihisa. Effect of host substituent upon inclusion complexation of aliphatic alcohols with organoseleno P-cyclodextrins. *J. Chem. Res.* (Part S Synopsis) 2:90–93 (2000).
211. M. Sato, M Narita, N. Ogawa, F. Hamada. Fluorescent chemo-sensor for organic guests based on regioselectively modified 6A 6B-, 6A 6C-, and 6A 6D-bis-dansylglycine-modified b-cyclodextrins. *Analyt. Sci.* 15:1199–1207 (1999).
212. S. Kamitori, O. Matsuzaka, S. Kondo, S. Muraoka, K. Okuyama, K. Noguchi, M. Okada, A. Harada. A novel pseudo-polyrotaxane structure composed of cyclodextrins and a straight-chain polymer: crystal structures of inclusion complexes of b-cyclodextrin with poly(trimethylene oxide) and poly(propylene glycol). *Macromolecules* 33: 1500–1503 (2000).
213. K. S. Prabhu, C. S. Ramadoss. Penicillin acylase catalyzed synthesis of penicillin-G from substrates anchored in cyclodextrins. *Indian J. Biochem. Biophys.* 37:6–13 (2000).
214. J. B. Harper, C. J. Easton, S. F. Lincoln. Cyclodextrins to increase the utility of enzymes in organic synthesis. *Cur. Org. Chem.* 4:429 (2000).

215. M. Usuda, T. Endo, H. Nagase, K. Tomono, H. Ueda. Interaction of antimalarial agent Artemisinin with cyclodextrins. *Drug Dev. Ind. Pharm.* 26:613–621 (2000).
216. L. M. Hamilton, C. T. Kelly, M. W. Fogarty. Cyclodextrins and their interaction with amylolytic enzymes. *Enz. Microb. Technol.* 26:561–568 (2000).
217. W. S. Lee, A. Ueno. A new feature of bifunctional catalysis. Cyclodextrins bearing two imidazole moieties as hydrolysis enzyme model. *Chem. Lett.* 3:258–260 (2000).
218. R. E. Wing, S. Maiti, W. M. Doane. Effectiveness of jet-cooked pearl cornstarch as a controlled release matrix. *Starch/Staerke* 39:422 (1987).
219. D. Trimnell, B. S. Shasha. Autoencapsulation: a new method for entrapping pesticides within starch. *J. Controlled Release* 7:25 (1988).
220. R. E. Wing, S. Maiti, W. M. Doane. Amylose content of starch controls the release of encapsulated bioactive agents. *J. Controlled Release* 7:33 (1988).
221. J. P. Reed, F. R. Hall, D. Trimnell. Effect of encapsulating thiocarbamate herbicides within starch for overcoming enhanced degradation in soils. *Starch/Staerke* 41:184 (1989).
222. J. Heller, S. H. Pangburn, K. V. Roskos. Development of enzymatically degradable protective coating for use in triggered drug delivery systems: derivatized starch hydrogels. *Biomaterials* 11:345 (1990).
223. M. C. Levy, M. C. Andry. Microcapsules prepared through interfacial crosslinking of starch derivatives. *Int. J. Pharm.* 62:27 (1990).
224. A. Shefer, S. Shefer, J. Kost, R. Langer. Structural characterization of starch networks in the solid state by cross-polarization magic-angle-spinning ^{13}C NMR spectroscopy and wide angle X-ray diffraction. *Macromolecules* 25: 6756 (1992).
225. G. O. Apinall. *The Polysaccharides*, Vol. 3. Academic Press, New York, 1985.
226. K. P. R. Kartha, H. Srivastava. Reaction of epichlorohydrin with carbohydrate polymers. Part I. Starch reaction kinetics. *Starch/Staerke* 37:270 (1985).
227. L. Kuniak, R. H. Marchessault. Crosslinking reaction between epichlorohydrin and starch. *Starch/Staerke* 24:110 (1972).
228. G. Hamdi, G. Ponchel. Enzymatic degradation of epichlorohydrin crosslinked starch micropheres by alpha-amylase. *Pharm. Res.* 16:867–875 (1999).
229. M. R. McDermott, P. L. Heritage, V. Bartzoka, M. A. Brook. Polymer-grafted starch microparticles for oral and nasal immunization. *Immunol. Cell Biology* 76:256–262 (1998).
230. P. L. Heritage, B. J. Underdown, M. A. Brook, M. R. McDermott. Oral administration of polymer-grafted starch microparticles activates gut-associated lymphocytes and primes mice for a subsequent systemic antigen challenge. *Vaccine* 16:2010–2017 (1998).
231. T. Murata, K. Akagi, M. Imamura, R. Nasu, H. Kimura, K. Nagata, Y. Tanaka. Studies on hyperthermia combined with arterial blockade for treatment of tumors. Part II: effectiveness of hyperthermia combined with arterial embolization using degradable starch microspheres. *Oncology Reports* 5:705–708 (1998).
232. A. Sundin, C. Wang, A. Ericsson, A. K. Fahlvik. Magnetic starch microspheres in the MR imaging of hepatic metastases. A preclinical study in the nude rat. *Acta Radiologica* 39:161–166 (1998).
233. J. Treib, J. F. Baron. Hydroxethyl starch: effects on hemostasis. *Annales Francaises d'Anesthesie et de Reanimation* 17:72–81 (1998).
234. T. T. Niemi, A. H. Kuitunen. Hydroxyethyl starch impairs in vitro coagulation. *Acta Anaest. Scandinavica*. 42:1104–1109 (1998).
235. R. Hofbauer, D. Moser, S. Hornykewycz, M. Frass, S. Kapiotis. Hydroxyethyl starch reduces the chemotaxis of white cells through endothelial cell monolayers. *Transfusion* 39:289–294 (1999).
236. H. Turkan, A. U. Ural, C. Beyan, A. Yalcin. Effects of hydroxyethyl starch on blood coagulation profile. *Eur. J. Anaesthesiology* 16:156–159 (1999).
237. M. Jamnicki, A. Zollinger, B. Seifert, D. Popovic, T. Pasch, D. R. Spahn. Compromised blood coagulation: an in vitro comparison of hydroxyethyl starch 130/0.4 and hydroxyethyl starch 200/0.5 using thrombelastography. *Anesthesia & Analgesia* 87:989–993 (1998).
238. M. Jamnicki, A. Zollinger, B. Seifert, D. Popovic, T. Pasch, D. R. Spahn. The effect of potato starch derived and corn starch derived hydroxyethyl starch on in vitro blood coagulation. *Anaesthesia* 53:638–644 (1998).
239. M. Behne, H. Thomas, D. H. Bremerich, V. Lischke, F. Asskali, H. Forster. The pharmacokinetics of acetyl starch as a plasma volume expander in patients undergoing elective surgery. *Anesthesia & Analgesia* 86:856–860 (1998).
240. J. S. Claudius, S. H. Neau, M. T. Kenny, J. K. Dulworth. The antimicrobial activity of vancomycin in the presence and absence of sodium carboxymethyl starch. *J. Pharm. Pharmacol.* 51:1333–1337 (1999).
241. I. L. Valuev, S. N. Popovich, Yu. A. Talyzenkov, G. A. Sytov, V. V. Chupov, L. I. Valuev, N. A. Plate, J. Jozefonvicz. Complexation of polysaccharides of the fucan series with lectins and polypeptides. *Doklady Phys. Chem.* 357: 386–389 (1997).
242. D. Riou, S. Colliec-Jouault, D. Pinczon Du Sel, S. Bosch, S. Siavoshian, V. Lebert, C. Tomasoni, C. Sinquin, P. Durand, C. Roussakis. Antitumor and antiproliferative effects of a fucan extracted from *Ascophyllum nodosum* against a non–small-cell bronchopulmonary carcinoma line. *Anticancer Res.* 16:1213–1219 (1996).
243. T. Nomura. *Biologically Active Substances Produced by Marine Organisms*. Nanko-Do, Tokyo, 1978.
244. D. S. McLellan, K. M. Jurd. Anticoagulants from marine algae. *Blood Coagul. Fibrinolysis*. 3:69 (1992).
245. T. Naguni, T. Nishino. Fucan sulfates and their anticoagulant activities. In: *Polysaccharides in Medicinal Applications*, S. Dumitriu (Ed.). Marcel Dekker, New York, 1996, pp. 545–574.
246. E. Percival, R. H. McDowell. Algal walls: composition and biosynthesis. In: *Encyclopedia of Plant Physiology*,

New Series Vol. 13B, *Plant Carbohydrates*. II. *Extracellular Carbohydrates*, W. Tanner, F. A. Loewus (Eds.). Springer-Verlag, Berlin, 1981, pp. 276–316.
247. M. Tomoda. *The Structure of Polysaccharides and Their Physiological Activities*, T. Miyazaki (Ed.). Asakura Shoten, Tokyo, 1990, pp. 97–103.
248. A. C. Ribeiro, R. P. Vieira, P. A. S. Mourao, B. Mulloy. A sulfated a-L-fucan from sea cucumber. *Carbohydr. Res.* 255:225–241 (1994).
249. T. Nishino, C. Nishioka, H. Ura, T. Nagumo. Isolation and partial characterization of a novel amino sugar-containing fucan sulfate from commercial *Fucus vesiculosus* fucoidan. *Carbohydr. Res.* 255:213–225 (1994).
250. T. Nishino, G. Yokoyama, K. Dobashi, M. Fujihara, T. Nagumo. Isolation, purification, and characterization of fucose-containing sulfated polysaccharides from the brown seaweed *Ecklonia kurome* and their blood-anticoagulant activities. *Carbohydr. Res.* 186:119 (1989).
251. G. Gernardi, G. F. Springer. Properties of highly purified fucan. *J. Biol. Chem.* 237:75 (1962).
252. T. Usui, K. Asari, T. Mizuno. Isolated of highly purified ''fucoidan'' from *Eisenia bicyclis* and its anticoagulant and antitumor activities. *Agric. Biol. Chem.* 44:1965 (1980).
253. K. Dobashi, T. Nishino, M. Fujihara, T. Nagumo. Isolation and preliminary characterization of fucose-containing sulfated polysaccharides with blood-anticoagulant activity from the brown seaweed *Hizikia fusiforme. Carbohydr. Res.* 194:315 (1989).
254. K. Kitamura, M. Matsuo, T. Yasui. Fucoidan from brown seaweed *Laminaria angustata* var. *longissima. Agric. Biol. Chem.* 55:615 (1991).
255. V. Grauffel, B. Kloareg, S. Mabeau, P. Durand, J. Jozefonvicz. New natural polysaccharides with potent antithrombic activity: fucans from brown algae. *Biomaterials* 10: 363 (1989).
256. H. Lis, N. Sharon. Lectins: carbohydrate-specific proteins that mediate cellular recognition. *Chem. Rev*. 98:637–674 (1998).
257. M. Vijayan, N. Chandra. Lectins. *Current Opinion in Structural Biology* 9:707–715 (1999).
258. R. S. Singh, A. K. Tiwary, J. F. Kennedy. Lectins: sources, activities, and applications. *Crit. Rev. Biotechnol*. 19:145–178 (1999).
259. G. A. Fragkiadakis. Isolation of lectins from hemolymph of decapod crustaceans by adsorption on formalinized erythrocytes. *J. Biochem. Biophys. Methods* 44:109–115 (2000).
260. T. C. Elden. Effects of proteinase inhibitors and plant lectins on the adult alfalfa weevil (Coleoptera: Curculionidae). *J. Entomol. Sci*. 35:62–70 (2000).
261. D. C. Kilpatrick. Immunological aspects of the potential role of dietary carbohydrates and lectins in human health. *Eur. J. Nutr*. 38:107–118 (1999).
262. M. Ogino, K. Yoshimatsu, H. Ebihara, J. Arikawa. N-acetylgalactosamine (GalNAc)-specific lectins mediate enhancement of Hantaan virus infection. *Arch. Virology* 144:1765–1779 (1999).
263. H. Lis, N. Sharon. Lectins: carbohydrate-specific proteins that mediate cellular recognition. *Chem. Rev*. 98:637–674 (1998).
264. C. D. Kilpatrick. Immunological aspects of the potential role of dietary carbohydrates and lectins in human health. *Eur. J. Nutr*. 38:107–117 (1999).
265. J. M. Rhodes. Lectins, colitis and colon cancer. *J. Roy. Coll. Physicians Lond*. 34:191–197 (2000).
266. H. Rudiger, H.-C. Siebert, D. Solis, J. Jimenez-Barbero, A. Romero, C.-W. Lieth, T. Diaz-Maurino, H.-J. Gabius. Medicinal chemistry based on the sugar code: fundamentals of lectinology and experimental strategies with lectins as targets. *Curr. Med. Chem*. 7:389–417 (2000).
267. H. Wang, T. B. Ng, V. E. C. Ooi, W. K. Liu. Effects of lectins with different carbohydrate-binding specificities on hepatoma, choriocarcinoma, melanoma and osteosar-6ycoma cell lines. *Int. J. Biochem. Cell Biol*. 32:365–373 (2000).
268. M. Vijayan, N. Chandra. Proteins—Lectins. *Current Opinion in Structural Biology* 9:707–716 (1999).
269. J. Bouckaert, T. Hamelryck, L. Wyns, R. Loris. Novel structures of plant lectins and their complexes with carbohydrates. *Current Opinion in Structural Biology* 9:572–578 (1999).
270. R. Samtleben, T. Hajto, K. Hostanska, H. Wagner. Mistletoe lectins as immunostimulants (chemistry, pharmacology and clinic). *Prog. Inflammation Res*. 8:223–243 (1999).
271. R. Loris, T. Hamelryck, J. Bouckaert, J. Wyns. Legume lectin structure. *Biochim. Biophys. Acta* 1383:9–36 (1998).
272. W. I. Weiss, M. E. Taylor, K. Drickamer. The C-type lectin superfamily in the immune system. *Immunol. Rev*. 163: 19–34 (1998).
273. J. Jimenez-Barbero, J. Canada, J. L. Asensio, J. F. Espinosa, M. Martin-Pastor, E. Montero, A. Poveda. Applications of NMR spectroscopy to the study of the bound conformation of O- and C-glycosides to lectins and enzymes. *NATO ASI Series C Mathematical and Physical Sciences* 526:99–117 (1999).
274. N. M. N. Alencar, E. H. Teixeira, A. M. S. Assreuy, B. S. Cavada, C. A. Flores, R. A. Ribeiro. Leguminous lectins as tools for studying the role of sugar residues in leukocyte recruitment. *Mediators of Inflammation* 8:107–115 (1999).
275. C. J. Thomas, A. Surolia. Mode of molecular recognition of L-fucose by fucose-binding legume lectins. *Biochem. Biophys. Res. Commun*. 268:262–268 (2000).
276. E. C. Lavelle, G. Grant, A. Pusztai, U. Pfuller, D. T. O'Hagan. Mucosal immunogenicity of plant lectins in mice. *Immunology* 99:30–38 (2000).
277. K. Yamamoto, I. N. Maruyama, T. Osawa. Glycobiology and carbohydrate biochemistry—cyborg lectins: novel leguminous lectins with unique specificities. *J. Biochem.* (Tokyo) 127:137–143 (2000).
278. J. D. Smart, T. J. Nicholls, K. L. Green, D. J. Rogers, J. D. Cook. Lectins in drug delivery: a study of the acute

local irritancy of the lectins from *Solanum tuberosum* and *Helix pomatia. Eur. J. Pharm. Sci*. 9:93–99 (2000).
279. A. Trouet, M. Masquelier, R. Baurin, D. D.-D. Campeneere. A covalent linkage between daunorubicin and proteins that is stable in serum and reversible by hydrolases, as required for a lusosonotropic drug-carrier conjugate: in vitro and in vivo studies. *Proc. Natl. Acad. Sci. U.S.A*. 79: 626–629 (1982).
280. L. Fiume, B. Bassi, C. Busi, A. Mattioli, G. Spinosa. Drug targeting in antiviral chemotherapy: a chemically stable conjugate of 9-β-D-arabinofurasonyl adenine 5′ monophosphate with lactosaminated albumen accomplishes a selective delivery of the drug to liver cells. *Biochem. Pharmacol*. 35:967–972 (1986).
281. L. Fiume, C. Busi, A. Mattioli, P. G. Balboni, G. Barbanti-Brodano. Hepatocyte targeting of adenine- 9-β-D-arabinofurasonyl adenine 5′ monophosphate (ara-AMP) coupled to lactosaminated albumin. *FEBS Lett*. 129:261–264 (1981).
282. J. D. Smart, T. J. Nicholls, K. L. Green, D. J. Rogers, J. D. Cook. Lectins in drug delivery: a study of the acute local irritancy of the lectins from *Solanum tuberosum* and *Helix pomatia. Eur. J. Pharm. Sci*. 9:93–98 (1999).
283. N. E. Larsen, E. A. Balazs. Drug delivery systems using hyaluronan and its derivatives. *Adv. Drug Delivery Rev*. 7:279–293 (1991).
284. E. A. Balazs. Viscosurgery in the eye. *Ocular Inflam. Ther*. 1:91–92 (1983).
285. B. B. Sand, K. Marner, M. S. Norn. Sodium hyaluronate in the treatment of *keratoconjuctivitis sicca*. A double masked clinical trial. *Acta Ophthalmol*. 67:181–183 (1989).
286. R. W. Chan, I. R. Titze. Hyaluronic acid (with fibronectin) as a bioimplant for the vocal fold mucosa. *Laryngoscope* 109:1142–1149, (1999).
287. P. Pirnazar, L. Wolinsky, S. Nachnani, S. Haake, A. Pilloni, G. W. Bernard. Bacteriostatic effects of hyaluronic acid. *J. Periodontology* 70:370–374 (1999).
288. H. Stiebel-Kalish, D. D. Gaton, D. Weinberger, N. Loya, M. Schwartz-Ventik, A. Solomon. A comparison of the effect of hyaluronic acid versus gentamicin on corneal epithelial healing. *Eye* 12:829–833 (1998).
289. R. Varma, R. S. Varma. *Mucopolysaccharides—Glycosaminoglycans of Body Fluids in Heals and Diseases*, Walter de Gruyter & Co., New York, 1983, pp. 12, 107–109.
290. J. E. F. Reynolds (Ed.). *Martindale: The Extra Pharmacopoeia*, 28th ed., The Pharmaceutical Press, London, 1982, p. 1755.
291. L. Goei, E. Topp, V. Stella, L. Benedetti, F. Biviano, L. Callegaro. Drug release from hydrocortisone esters of hyaluronic acid: influence of ester hydrolysis rate and release rate. In: *Polymers in Medicine* (R. M. Ottenbrite, E. Chiellini, Eds.). Technomic, Lancaster, PA, 1992, pp. 85–101.
292. M. Leslie. Hyaluronic acid treatment for osteoarthritis of the knee. *Nurse Practitioner* 24:38, 41–48 (1999).
293. K. Moriyama, T. Ooya, N. Yui. Hyaluronic acid grafted with poly(ethylene glycol) as a novel peptide formulation. *J. Controlled Release* 59:77–86 (1999).
294. D. Coradini, C. Pellizzaro, G. Miglierini, M. G. Daidone, A. Perbellini. Hyaluronic acid as drug delivery for sodium butyrate: improvement of the anti-proliferative activity on a breast-cancer cell line. *Int. J Cancer* 81:411–416 (1999).
295. L. A. Solchaga, J. E. Dennis, Y. M. Goldberg, A. I. Caplan. Hyaluronic acid-based polymers as cell carriers for tissue-engineered repair of bone and cartilage. *J. Orthopaedic Res*. 17:205–213 (1999).
296. M. Radice, P. Brun, R. Cortivo R. Scapinelli, C. Battaliard, G. Abatangelo. Hyaluronan-based biopolymers as delivery vehicles for bone-marrow-derived mesenchymal progenitors. *J. Biomed. Mater. Res*. 50:101–109 (2000).
297. D. Bakos, M. Soldan, I. Hernandez-Fuentes. Hydroxyapatite–collagen–hyaluronic acid composite. *Biomaterials* 20:191–195 (1999).
298. N. Volpi. Adsorption of glycosaminoglycans onto coral—a new possible implant biomaterials for regeneration therapy. *Biomaterials* 20:1359–1363 (1999).
299. F. Duranti, G. Salti, B. Bovani, M. Calandra, M. L. Rosati. Injectable hyaluronic acid gel for soft tissue augmentation. A clinical and histological study. *Dermatologic Surg*. 24: 1317–1325 (1998).
300. S. Oerther, E. Payan, F. Lapicque, N. Presle, P. Hubert, S. Muller, P. Netter. Hyaluronate–alginate combination for the preparation of new biomaterials: investigation of the behaviour in aqueous solutions. *Biochim. Biophys. Acta* 1426:185–194 (1999).
301. S. Nilsson. A thermodynamic analysis of calcium–alginate gel formation in the presence of inert electrolyte. *Biopolymers* 32:1311–1315 (1992).
302. R. Seale, E. R. Morris, D. A. Rees. Interactions of alginates with univalent cations. *Carbohydr. Res*. 110:101 (1982).
303. G. Pass, P. W. Hales. Interaction between metal cations and anionic polysaccharides. In: *Solution Properties of Polysaccharides. ACS Symposium Series* 150:349 (1981).
304. K. L. Jang, W. Brand, M. Resong, W. Mainieri, G. G. Geesey. Feasibility of using alginate to absorb dissolved copper from aqueous media. *Environ. Prog*. 9:269 (1990).
305. L. K. Jang, S. L. Lopez, S. L. Eastman, P. Pryfogle. Recovery of copper and cobalt by biopolymer gels. *Biotechnol. Bioeng*. 37:266 (1991).
306. S. E. Barnett, S. J. Varley. The effects of calcium alginate on wound healing. *Ann. Roy. Coll. Surg. Engl*. 69:153–155 (1987).
307. O. Smidsrod. Molecular basis for some physical properties of alginates in the gel state. *Farad. Discuss. Chem. Soc*. 57:263 (1974).
308. M. Yalpani. *Polysaccharides: Syntheses, Modifications and Structure/Property Relations*. Elsevier, Amsterdam 1988, p. 405.
309. W. J. Sime. Alginates. In: *Food Gels* (P. Harris, Ed.). Elsevier, London, 1989, p. 53.
310. E. R. Morris, D. A. Ress, D. Thom. Characterization of

polysaccharide structure and interactions by circular dichroism: order–disorder transition in the calcium alginate system, *J.C.S. Chem. Commun*. 245 (1973).
311. E. R. Morris, D. A. Ress, D. Thom, J. Boyd. Chiroptical and stoichiometric evidence of a specific, primary dimerization process in alginate gelation. *Carbohydr. Res*. 66: 145 (1978).
312. O. Smidsrod. Molecular basis for some physical properties of alginates in the gel state. *Faraday Discuss. Chem. Soc*. 57:263 (1974).
313. W. Mackie. Degree of polymerization and polydispersity of mannan from the cell wall of the green seaweed *Codium fragile*. *Biochem. J*. 125:89 (1971).
314. E. D. T. Atkins, I. A. Nieduszynski, W. Mackie, K. D. Parker, E. E. Smolko. Structural components of alginic acid. II. The crystalline structure of poly-α-L-guluronic acid. Results of X-ray diffraction and polarized infrared studies. *Biopolymers* 12:1879 (1973).
315. G. T. Grant, E. R. Morris, D. A. Rees, P. J. C. Smith, D. Thom. Biological interactions between polysaccharides and divalent cations: egg-box model. *FEBS Lett*. 32:195 (1973).
316. N. Nesle, R. Kimmich. Susceptibility NMR microimaging of heavy metal uptake in alginate biosorbents. *Magnetic Resonance Imaging* 14:905–906 (1996).
317. Z. Y. Wang, Q. Z. Zhang, M. Konno, S. Saito. Sol–gel transition of alginate solution by the addition of various divalent cations: ^{13}C-NMR spectroscopy study. *Biopolymers* 33:703 (1993).
318. M. Fujihara, T. Nagumo. An influence of the structure of alginate on the chemotactic activity of macrophages and the antitumor activity. *Carbohydr. Res*. 243:211–216 (1993).
319. S. Oerther, E. Payan, F. Lapicque, N. Presle, P. Hubert, S. Muller, P. Netter. Hyaluronate–alginate combination for the preparation of new biomaterials: investigation of the behaviour in aqueous solutions. *Biochim. Biophys. Acta* 1426:185–194 (1999).
320. S. E. Barnett, S. J. Varley. The effects of calcium alginate on wound healing. *Ann. Roy. Coll. Surg. Eng*. 69:153 (1987).
321. R. Robitaille, F. A. Leblond, N. Henley, G. J. Prud' homme, E. Drobetsky, J. P. Hall. Alginate–poly-L-lysine microcapsule biocompatibility: a novel RT-PCR method for cytokine gene expression analysis in pericapsular infiltrates. *J. Biomed. Mater. Res*. 45:223–230 (1999).
322. S. Dumitriu. Drug-loaded ophthalmic prostheses. In: *High Performance Biomaterials*, M. Szycher (Ed.). Technomic, Lancaster, PA, 1992, p. 669.
323. S. Dumitriu, M. Dumitriu, G. Teaca. Bioactive polymers. 65. Studies of cross-linked xanthan hydrogels as supports in drug retardation. *Clin. Mater*. 6:265 (1990).
324. R. Robitaille, F. A. Leblond, Y. Bourgeois, N. Henley, M. Loignon, J. P. Halle. Studies on small (<350 microm) alginate–poly-L-lysine microcapsules. V. Determination of carbohydrate and protein permeation through microcapsules by reverse-size exclusion chromatography. *J. Biomed. Mater. Res*. 50:420–427 (2000).
325. S. Dumitriu, E. Chornet. Inclusion and release of proteins from polysaccharide-based polyion complex. *Adv. Drug Delivery Rev*. 31:223–246 (1998).
326. Y. S. Choi, S. R. Hong, Y. M. Lee, K. W. Song, M. H. Park, Y. S. Nam. Study on gelatin-containing artificial skin I. Preparation and characteristics of novel gelatin–alginate sponge. *Biomaterials* 20(5):409–417 (1999).
327. E. A. Bekturov, L. A. Bimendina. Interaction of heterogeneous macromolecules. *Vestn. Akad. Nauk Kaz. SSR* 9:7 (1980).
328. I. M. Papisov, A. A. Litmanovich. Specificity of co-operative interactions between simple synthetic macromolecules and its relation to chain length. *Vysocomol. Soyed*. A19:716 (1977).
329. W. S. Magee, J. H. Gibbs, B. H. Zimm. Theory of helix-coil transitions involving complementary poly- and oligonucleotides. I. The complete binding case. *Biopolymers* 1: 133 (1963).
330. V. N. Damle. On the helix-coil equilibrium in two- and three-stranded complexes involving complementary poly- and oligonucleotides. *Biopolymers* 9:353 (1970).
331. D. Poland. *Cooperative Equilibria in Physical Biochemistry*. Oxford University Press, Oxford, 1978.
332. D. Poland, H. A. Sheraga. *Theory of Helix-Coil Transitions in Biopolymers*. Academic Press, New York, 1970.
333. S. Bystricky, A. Malovikova, T. Sticzay. Interaction of alginates and pectins with cationic polypeptides. *Carbohydr. Polym*. 13:283 (1990).
334. P. Liu, T. R. Krishnan. Alginate–pectin–poly-L-lysine particulate as a potential controlled release formulation. *J. Pharm. Pharmacol*. 51:141–149 (1999).
335. B. Kulseng, G. Skjak-Braek, L. Ryan, A. Andersson, A. King, A. Faxvaag, T. Espevik. Transplantation of alginate microcapsules: generation of antibodies against alginates and encapsulated porcine islet-like cell clusters. *Transplantation* 67:978–984 (1999).
336. E. R. Morris. Mixed polymer gels. In: *Food Gels*, P. Harris (Ed.). Elsevier Applied Science, London, 1990, p. 291.
337. M. A. Wheatley, M. Chang, E. Park, R. Langer. Coated alginate microspheres: factors influencing the controlled delivery of macromolecules. *J. Appl. Polym. Sci*. 43: 2123–2135 (1991).
338. S. R. Byrn et al. Drug–oligonucleotide conjugates. *Adv. Drug Delivery Rev*. 6:287–308 (1991).
339. T. J. Smith. Calcium alginate hydrogel as a matrix for enteric delivery of nucleic acids. *Pharm. Techn*. April: 26–29 (1994).
340. T. L. Bowersock, H. HogenEsch, M. Suckow, P. Guimond, S. Martin, D. Borie, S. Torregrosa, H. Park, K. Park. Oral vaccination of animals with antigens encapsulated in alginate microspheres. *Vaccine* 17:1804–1811 (1999).
341. P. L. Gelgner et al. Lipofection: a highly efficient, lipid-mediated DNA-transfection procedure. *Proc. Natl. Acad. Sci. USA* 84:7413–7417 (1987).

342. N. Aggarwal, H. HogenEsch, P. Guo, A. North, M. Suckow, S. K. Mittal. Biodegradable alginate microspheres as a delivery system for naked DNA. *Can. J. Veterinary Res.* 63:148–152 (1999).
343. D. Quong, J. N. Yeo, R. J. Neufeld. Stability of chitosan and poly-L-lysine membranes coating DNA-alginate beads when exposed to hydrolytic enzymes. *J. Microencapsulation* 16:73–82 (1999).
344. D. Quong, R. J. Neufeld, G. Skjak-Braek, D. Poncelet. External versus internal source of calcium during the gelation of alginate beads for DNA encapsulation. *Biotechnol. Bioeng.* 57:438–446 (1998).
345. D. Quong, R. J. Neufeld. DNA encapsulation within co-guanidine membrane coated alginate beads and protection from extracapsular nuclease. *J. Microencapsulation* 16:573–585 (1999).
346. C. J. Ross, M. Ralph, P. L. Chang. Delivery of recombinant gene products to the central nervous system with nonautologous cells in alginate microcapsules. *Hum. Gene Ther.* 10:49–59, (1999).
347. F. Edwards-Levy, M. C. Levy. Serum albumin–alginate coated beads: mechanical properties and stability. *Biomaterials* 20:2069–2084 (1999).
348. G. Fundueanu, C. Nastruzzi, A. Carpov, J. Desbrieres, M. Rinaudo. Physico-chemical characterization of Ca-alginate microparticles produced with different methods. *Biomaterials* 20:1427–1435 (1999).
349. I. Aynie, C. Vauthier, H. Chacun, E. Fattal, P. Couvreur. Spongelike alginate nanoparticles as a new potential system for the delivery of antisense oligonucleotides. *Antisense & Nucleic Acid Drug Dev.* 9:301–312, (1999).
350. T. Imai, C. Kawasaki, T. Nishiyama, M. Otagiri. Comparison of the pharmaceutical properties of sustained-release gel beads prepared by alginate having different molecular size with commercial sustained-release tablet. *Pharmazie* 55:218–222 (2000).
351. F. Acarturk, S. Takka. Calcium alginate microparticles for oral administration. II. Effect of formulation factors on drug release and drug entrapment efficiency. *J. Microencapsulation* 16:291–301 (1999).
352. S. Takka, F. Acarturk. Calcium alginate microparticles for oral administration. I: Effect of sodium alginate type on drug release and drug entrapment efficiency. *J. Microencapsulation* 16:275–290 (1999).
353. V. Pillay, R. Fassihi. *In vitro* release modulation from crosslinked pellets for site-specific drug delivery to the gastrointestinal tract. II. Physicochemical characterization of calcium-alginate, calcium-pectinate and calcium-alginate-pectinate pellets. *J. Controlled Release* 59:243–256 (1999).
354. M. Fathy, S. M. Safwat, S. M. El-Shanawany, S. Shawky Tous, M. Otagiri. Preparation and evaluation of beads made of different calcium alginate compositions for oral sustained release of tiaramide. *Pharm. Dev. Techn.* 3:355–364 (1998).
355. Y. Murata, S. Toniwa, E. Miyamoto, S. Kawashima. Preparation of alginate gel beads containing chitosan nicotinic acid salt and the functions. *Eur. J. Pharm. Biopharm.* 48:49–52 (1999).
356. M. L. Torre, P. Giunchedi, L. Maggi, R. Stefli, E. O. Machiste, U. Conte. Formulation and characterization of calcium alginate beads containing ampicillin. *Pharm. Dev. Techn.* 3:193–198 (1998).
357. V. Pillay, C. M. Dangor, T. Govender, K. R. Moopanar, N. Hurbans. Ionotropic gelation: encapsulation of indomethacin in calcium alginate gel discs. *J. Microencapsulation* 15:215–226 (1998).
358. T. A. Read, V. Stensvaag, H. Vindenes, E. R. Ulvestad, F. Thorsen. Cells encapsulated in alginate: a potential system for delivery of recombinant proteins to malignant brain tumours. *Int. J. Dev. Neurosci.* 17:653–663 (1999).
359. M. Fernandez-Perez, M. Villafranca-Sanchez, E. Gonzalez-Pradas, F. Martinez-Lopez, F. Flores-Cespedes. Controlled release of carbofuran from an alginate-bentonite formulation: water release kinetics and soil mobility. *J. Agric. Food Chem.* 48:938–943 (2000).
360. S. W. Ueng, S. S. Lee, E. C. Chan, K. T. Chen, C. Y. Yang, C. Y. Chen, Y. S. Chan. In vitro elution of antibiotic from antibiotic-impregnated biodegradable calcium alginate wound dressing. *J. Trauma-Inj. Infect. Crit. Care* 47:136–141 (1999).
361. J. W. Doyle, T. P. Roth, R. M. Smith, Y. Q. Li, R. M. Dunn. Effects of calcium alginate on cellular wound healing processes modeled in vitro. *J. Biomed. Mater. Res.* 32:561–568 (1996).
362. H. J. Kim, H. C. Lee, J. S. Oh, B. A. Shin, C. S. Oh, R. D. Park, K. S. Yang, C. S. Cho. Polyelectroylte complex composed of chitosan and sodium alginate for wound dressing application. *J. Biomater. Sci. Polym. Ed.* 10:543–556 (1999).
363. B. Kneafsey, M. O'Shaughnessy, K. C. Condon. The use of calcium alginate dressings in deep hand burns. *Burns* 22:40–43 (1996).
364. M. Ingram, T. A. Wright, C. J. Ingoldby. A prospective randomized study of calcium alginate (Sorbsan) versus standard gauze packing following haemorrhoidectomy. *J. Roy. Coll. Surg. Edinburgh* 43:308–309 (1998).
365. J. M. O'Donoghue, S. T. O'Sullivan, E. S. Beausang, J. I. Panchal, M. O'Shaughnessy, T. P. O'Connor. Calcium alginate dressings promote healing of split skin graft donor sites. *Acta Chirurgiae Plasticae* 39:53–55 (1997).
366. Y. Cao, A. Rodriguez, M. Vacanti, C. Ibarra, C. Arevalo, C. A. Vacanti. Comparative study of the use of poly(glycolic acid), calcium alginate and pluronics in the engineering of autologous porcine cartilage. *J. Biomater Sci. Polym. Ed.* 9:475–487 (1998).
367. T. Chalain, J. H. Phillips, A. Hinek. Bioengineering of elastic cartilage with aggregated porcine and human auricular chondrocytes and hydrogels containing alginate, collagen, and kappa-elastin. *J. Biomed. Mater. Res.* 44:280–288 (1999).
368. W. J. Marijnissen, G. J. van Osch, J. Aigner, H. L. Verwoerd-Verhoef, J. A. Verhaar. Tissue-engineered cartilage using serially passaged articular chondrocytes. Chon-

drocytes in alginate, combined in vivo with a synthetic (E210) or biologic biodegradable carrier (DBM). *Biomaterials* 21:571–580 (2000).

369. M. Demoor-Fossard, M. Boittin, F. Redini, J. P. Pujol. Differential effects of interleukin-1 and transforming growth factor beta on the synthesis of small proteoglycans by rabbit articular chondrocytes cultured in alginate beads as compared to monolayers. *Molec. Cell. Biochem.* 199: 69–80 (1999).
370. S. Loty, J. M. Sautier, C. Loty, H. Boulekbache, T. Kokubo, N. Forest. Cartilage formation by fetal rat chondrocytes cultured in alginate beads: a proposed model for investigating tissue-biomaterial interactions. *J. Biomed. Mater. Res.* 42:213–222 (1998).
371. R. Glicklis, L. Shapiro, R. Agbaria, J. C. Merchuk, S. Cohen. Hepatocyte behaviour within three-dimensional porous alginate scaffolds. *Biotechnol. Bioeng.* 67:344–353 (2000).
372. J. J. Marler, A. Guha, J. Rowley, R. Koka, D. Mooney, J. Upton, J. P. Vacanti. Soft-tissue augmentation with injectable alginate and syngeneic fibroblasts. *Plast. Reconstr. Surg.* 105:2049–2058 (2000).
373. Y. Suzuki, M. Tanihara, K. Suzuki, A. Saitou, W. Sufan, Y. Nishimura. Alginate hydrogel linked with synthetic oligopeptide derived from BMP-2 allows ectopic osteoinduction in vivo. *J. Biomed. Mater. Res.* 50:405–409 (2000).
374. K. Ishikawa, Y. Ueyama, T. Mano, T. Koyama, K. Suzuki, T. Matsumura. Self-setting barrier membrane for guided tissue regeneration method: initial evaluation of alginate membrane made with sodium alginate and calcium chloride aqueous solutions. *J. Biomed. Mater. Res.* 47:111–115 (1999).
375. K. Suzuki, Y. Suzuki, M. Tanihara, K. Ohnishi, T. Hashimoto, K. Endo, Y. Nishimura. Reconstruction of rat peripheral nerve gap without sutures using freeze-dried alginate gel. *J. Biomed. Mater. Res.* 49:528–533 (2000).
376. A. Becker, F. Katzen, A. Puhler, L. Ielpi. Xanthan gum biosynthesis and application: a biochemical/genetic perspective. *Appl. Microbiol. Biotechnol.* 50:145–152 (1998).
377. B. T. Stokke, B. E. Christensen, O. Smidsrod. Macromolecule properties of xanthan. In: *Polysaccharides*, S. Dumitriu (Ed.). Marcel Dekker, New York, 1998, pp. 433–472.
378. E. Atkins. Structure of microbial polysaccharides using X-ray diffraction. In: *Novel Biodegradable Microbial Polymers*, E. A. Dawes (Ed.). NATO ASI Ser. SER E, Kluwer Academic, Dordrecht, The Netherlands, 1990, p. 371.
379. B. T. Stokke, A. Elgsaeter. Conformation, order-disorder conformational transitions and gelation of non-crystalline polysaccharide studies using electron microscopy. *Micron* 25:469 (1994).
380. V. J. Morris. Biotechnically produced carbohydrates with functional properties for use in food systems. *Food Biotechnol* 4:45 (1990).
381. S. Levy, S. C. Schuyler, R. K. Maglothin, L. A. Staehelin. Dynamic simulations of the molecular conformations of wild type and mutant xanthan polymers suggest that conformational differences may contribute to observed differences in viscosity. *Biopolymers* 38:251–272 (1996).
382. N. P. Yevlampieva, G. M. Pavlov, E. I. Rjumtsev. Flow birefringence of xanthan and other polysaccharide solutions. *Int. J. Biol. Macromol.* 26:295–301 (1999).
383. M. Milas, W. F. Reed, S. Printz. Conformations and flexibility of native and re-natured xanthan in aqueous solutions. *Int. J. Biol. Macromol.* 18:211–221 (1996).
384. A. Gamini, M. Mandel. Physicochemical properties of aqueous xanthan solutions: static light scattering. *Biopolymers* 34:783–797 (1994).
385. S. Takigami, M. Shimada, P. A. Williams, G. O. Phillips. E.S.R. study of the conformational transition of spin-labelled xanthan gum in aqueous solution. *Int. J. Biol. Macromol.* 15:367–371 (1993).
386. K. P. Shatwell, I. W. Sutherland, I. C. M. Dea, S. B. Ross-Murphy. The influence of acetyl and pyruvate substituents on the helix-coil transition behaviour of xanthan. *Carbohydr. Res.* 206:87–103 (1990).
387. I. C. M. Dea. In: *Industrial Polysaccharides: Genetic Engineering, Structure/Property Relations and Applications*, M. Yalpani (Ed.) Elsevier, Amsterdam, 1987, p. 209.
388. M. Tako, A. Asato, S. Nakamura. Rheological aspects of the intermolecular interaction between xanthan and locust bean gum in aqueous media. *Agric. Biol. Chem.* 48:2995 (1984).
389. M. Tako, S. Nakamura. Synergistic interaction between xanthan and guar gum. *Carbohydr. Res.* 138:207 (1985).
390. I. C. M. Dea. Specificity of interactions between polysaccharide helices and β-1,4-linked polysaccharides. *ACS Symp. Ser.* 150:439 (1981).
391. R. H. Marchessault, A. Buleon, Y. Deslandes, T. Goto. Comparison of X-ray diffraction data of galactomannans. *J. Colloid Interface Sci.* 71:375 (1979).
392. P. Cairns, M. J. Miles, V. J. Morris. Intermolecular binding of xanthan gum and carob gum. *Nature* 322:89 (1986).
393. P. Cairns, M. J. Miles, V. J. Morris, G. J. Brownsey. X-Ray fiber-diffraction studies of synergistic, binary polysaccharide gels. *Carbohydr. Res.* 160:410 (1987).
394. I. W. Sutherland. The role of acetylation in exopolysaccharides including those for food use. *Food Biochem.* 6: 75 (1992).
395. S. A. Frangou, E. R. Moris, D. A. Rees, R. K. Richardson, S. B. Ross-Murphy. Molecular origin of xanthan solution rheology: effect of urea on chain conformation and interactions. *J. Polym. Sci.* (Polym. Lett. Ed.) 20:531 (1982).
396. L. Lopes, C. T. Andrade, M. Milas, M. Rinaudo. Role of conformation and acetylation of xanthan on xanthan–guar interaction. *Carbohydr. Polym.* 17:121 (1992).
397. T. M. Bresolin, M. Milas, M. Rinaudo, F. Reicher, J. L. Ganter. Role of galactomannan composition on the binary gel formation with xanthan. *Int. J. Biol. Macromol.* 26: 225–231 (1999).
398. T. M. Bresolin, M. Milas, M. Rinaudo, J. L. M. Ganter. Xanthan–galactomannan interactions as related to xanthan conformations. *Int. J. Biol. Macromol.* 23:263–271 (1998).

399. K. P. Shatwell, I. W. Sutherland, S. B. Ross-Murphy, I. C. M. Dea. Influence of the acetyl substituent on the interaction of xanthan with plant polysaccharides-I. Xanthan-Locust Bean gum systems. *Carbohydr. Polym*. 14:29 (1991).

400. L. Lundin, A. M. Hermansson, Supramolecular aspects of xanthan-locust bean gum gels based on rheology and electronmicroscopy. *Carbohydr. Polym*. 26:129 (1995).

401. A. L. Ruissen, W. A. van der Reijden, W. van't Hof, E. C. Veerman, A. V. Nieuw Amerongen. Evaluation of the use of xanthan as vehicle for cationic antifungal peptides. *J. Controlled Release* 60:49–56 (1999).

402. M. M. Talukdar, I. Vinckier, P. Moldenaers, R. Kinget. Rheological characterization of xanthan gum and hydroxypropylmethyl cellulose with respect to controlled-release drug delivery. *J. Pharm. Sci*. 85:537–540 (1996).

403. M. C. Bonferoni, S. Rossi, M. Tamayo, J. L. Pedraz, A. Dominguez-Gil, C. Caramella. On the employment of λ-carrageenan in a matrix system. I. Sensitivity to dissolution medium and comparison with Na carboxymethylcellulose and xanthan gum. *J. Controlled Release* 26:119 (1993).

404. J. W. Anderson. Fiber and health: an overview. *Am. J. Gastroenterol*. 81:892 (1986).

405. J. W. Anderson, C. A. Bryant. Dietary fiber: diabetes and obesity. *Am. J. Gastroenterol*. 81:898 (1986).

406. J. W. Anderson, D. A. Deakins, T. L. Floore, B. M. Smith, S. E. Whitis. Dietary fiber and coronary heart disease. *Food Sci. Nutr*. 29:95 (1990).

407. R. A. Baker. Potential dietary benefits of citrus pectin and fiber. *Food Technol*. 48:133 (1994).

408. C. Rolin, B. U. Nielson, P. E. Glahn. Pectin. In: *Polysaccharides*, S. Dumitriu (Ed.). Marcel Dekker, New York, 1998, pp. 377–433.

409. A. Barcelo, J. Claustre, F. Moro, J. A. Chayvialle, J. C. Cuber, P. Plaisancie. Mucin secretion is modulated by luminal factors in the isolated vascularly perfused rat colon. *Gut* 46:218–224 (2000).

410. S. E. Schwartz, R. A. levine, A. Singh, J. R. Scheidecker, N. S. Track. Sustained pectin ingestion delays gastric emptying. *Gastroenterology* 83:812 (1982).

411. R. Kohen, V. Shadmi, A. Kakunda, A. Rubinstein. Prevention of oxidative damage in the rat jejunal mucosa by pectin. *Br. J. Nutr*. 69:789 (1993).

412. P. Bachmann. A composition for treating and preventing diarrhoea in humans and animals and a method of preparing same. WO 87/02243, 23.04.87 (1985).

413. T. Havelund, C. Aalykke, L. Rasmussen. Efficacy of a pectin-based anti-reflux agent on acid reflux and recurrence of symptoms and oesophagitis in gastro-oesophageal reflux disease. *Eur. J. Gastroenterol. Hepatol*. 9:509–514 (1997).

414. R. McPherson, Kay. Dietary fiber. *J. Lipid Res*. 23:221 (1982).

415. K. L. Roehrig. The physiological effects of dietary fiber—a review. *Food Hydrocoll*. 2:1 (1988).

416. K. Behall, S. Reiser. Effects of pectin on human metabolism. Chemistry and Function of pectins. *ACS Sympos. Ser*. 310:248 (1986).

417. S. Reiser. Metabolic effects of dietary pectins related to human health. *Food Technol*. 41:91 (1987).

418. K. Yamada, Y. Tokunaga, A. Ikeda, K. Ohkura, S. Mamiya, S. Kaku, M. Sugano, H. Tachibana. Dietary effect of guar gum and its partially hydrolyzed product on the lipid metabolism and immune function of Sprague-Dawley rats. *Biosc. Biotechn. Biochem*. 63:2163–2167 (1999).

419. M. F. Ismail, M. Z. Gad, M. A. Hamdy. Study of the hypolipidemic properties of pectin, garlic and ginseng in hypercholesterolemic rabbits. *Pharmac. Res*. 39:157–166 (1999).

420. K. Tazawa, H. Okami, I. Yamashita, Y. Ohnishi, K. Kobashi, M. Fujimaki. Anticarcinogenic action of apple pectin on fecal enzyme activities and mucosal or portal prostaglandin E2 levels in experimental rat colon carcinogenesis. *J. Exp. Clin. Cancer Res*. 16:33–38 (1997).

421. A. Hensel, K. Meier. Pectins and xyloglucans exhibit antimutagenic activities against nitroaromatic compounds. *Planta Medica* 65:395–399 (1999).

422. H. S. Taper, N. M. Delzenne, M. B. Roberfroid. Growth inhibition of transplantable mouse tumors by non-digestible carbohydrates. *Int. J. Cancer* 71:1109–1112 (1997).

423. H. S. Taper, C. Lemort, M. B. Roberfroid. Inhibition effect of dietary inulin and oligofructose on the growth of transplantable mouse tumor. *Anticancer Res*. 18:4123–4126 (1998).

424. H. S. Taper, M. Roberfroid. Influence of inulin and oligofructose on breast cancer and tumor growth. *J. Nutr*. 129(7 Suppl):1488S–1491S (1999).

425. C. Rolin, B. U. Nielson, P. E. Glahn. Pectin. In: *Polysaccharides*, S. Dumitriu (Ed.). Marcel Dekker, New York, 1998, pp. 377–433.

426. J. N. McMullen, D. W. Newton, C. H. Becker. Pectin gelatin complex coacervates. I: Determinants of microglobule size, morphology, and recovery as water-dispersible powders. *J. Pharmaceut. Sci*. 71:628 (1982).

427. N. McMullen, D. W. Newton, C. H. Becker. Pectin–gelatin complex coacervates. II: Effect of microencapsulated sulfamerazine on size, morphology, recovery, and extraction of water-dispersible microglobules. *J. Pharmaceut. Sci*. 73:1799 (1984).

428. I. Krogel, R. Bodmeier. Evaluation of an enzyme-containing capsular shaped pulsatile drug delivery system. *Pharm. Res*. 16:1424–1429 (1999).

429. M. Gonzalez, C. Rivas, B. Caride, M. A. Lamas, M. C. Taboada. Effects of orange and apple pectin on cholesterol concentration in serum, liver and faeces. *J. Physiol. Biochem*. 54:99–104 (1998).

430. M. M. Meshali, K. E. Gabr. Effect of interpolymer complex formation of chitosan with pectin or acacia on the release of behaviour of chlorpromazine HCl. *Int. J. Pharmaceut*. 89:177 (1993).

431. M. Ashford, J. Fell, D. Attwood, H. Sharma, P. Woodhead. An evaluation of pectin as a carrier for drug

targeting to the colon. *J. Controlled Release* 26:213 (1993).
432. M. Ashford, J. Fell, D. Attwood, H. Sharma, P. Woodhead. Studies on pectin formulations for colonic drug delivery. *J. Controlled Release* 30:225 (1994).
433. T. Takahashi, N. Nambu, T. Nagai. Dissolution of thioridacine hydrochloride from the coprecipitate with pectin. *Chem. Pharm. Bull*. 34:327 (1986).
434. H. W. T. Matthew, S. Basu, W. D. Peterson, S. O. Salley, M. D. Klein. Performance of plasma-perfused, microencapsulated hepatocytes: prospects for extracorporeal liver support. *J. Pediatr. Surg*. 28:1423 (1993).
435. L. A. Fransson, Mammalian glycosaminoglicans. In: *The Polysaccharides*, Vol. 3, G. O. Aspinall (Ed.). Academic Press, New York, 1985, pp. 337–415.
436. B. Casu. Heparin and Heparin-like polysaccharides. In: *Polymeric Biomaterials*, S. Dumitriu (Ed.). Marcel Dekker, New York, 1994, pp. 159–174.
437. E. Young, T. Venner, J. Ribau, S. Shaughnessy, J. Hirsh, T. J. Podor. The binding of unfractionated heparin and low molecular weight heparin to thrombin-activated human endothelial cells. *Thromb. Res*. 96:373–381 (1999).
438. O. Lev-Ran, A. Kramer, J. Gurevitch, I. Shapira, R. Mohr. Low-molecular-weight heparin for prosthetic heart valves: treatment failure. *Ann. Thoracic Surg*. 69:264–266 (2000).
439. A. Greinachen. Heparin-induced thrombocytopenia—pathogenesis and treatment. *Thrombosis & Haemostasis* 82(Suppl 1):148–156 (1999).
440. T. C. Wright, J. J. Castellot, J. R. Diamond, M. J. Karnowski. Regulation of cellular proliferation by heparin and heparan sulfate. In: *Heparin: Chemical and Biological Properties; Clinical Applications*, D. A. Lane, U. Lindahl (Eds.). Edward Arnold, London, 1989, pp. 295–316.
441. J. Folkman, D. E. Ingber. Angiogenesis: regulatory role of heparin and related molecules. In: *Heparin: Chemical and Biological Properties; Clinical Applications*, D. A. Lane, U. Lindahl (Eds.). Edward Arnold, London, 1989, pp. 317–333.
442. M. Baba, R. Pauwels, J. Balzarini, J. Arnaut, J. Desmyter, E. De Clerq. Mechanism of inhibitory effects of dextrin sulfate and heparin on replication of human immunodeficiency virus *in vitro*. *Proc. Natl. Acad. Sci. USA* 85:6132–6136 (1988).
443. S. E. Sakiyama, J. C. Schense, J. A. Hubbell. Incorporation of heparin-binding peptides into fibrin gels enhances neurite extension: an example of designer matrices in tissue engineering. *FASEB J*. 13:2214–2224 (1999).
444. G. Kratz, M. Back, C. Arnander, O. Larm. Immobilised heparin accelerates the healing of human wounds in vivo. *Scand. J. Plast. Reconstr. Surg. Hand Surg*. 32:381–385 (1998).
445. B. Casu. Structure and biological activity of heparin. *Adv. Carbohydr. Chem. Biochem*. 43:51–134 (1985).
446. M. N. Sanchez, C. Barker, J. Brosnan. Heparin-induced skin necrosis: nurses beware. *Dermatology Nursing* 10: 419–423,429 (1998).
447. P. J. Drew, M. J. Smith, M. A. Milling. Heparin-induced skin necrosis and low molecular weight heparins. *Ann. Roy. Coll. Surg. Eng*. 81:266–269 (1999).
448. G. Zugmaier, R. Favoni, R. Jaeger, N. Rosen, C. Knabbe. Polysulfated heparinoids selectively inactivate heparin-binding angiogenesis factors. *Ann. New York Acad. Sci*. 886:243–248 (1999).
449. S. C. Vasudev, T. Chandy, C. P. Sharma, M. Mohanty, P. R. Umasankar. Synergistic effect of released aspirin/heparin for preventing bovine pericardial calcification. *Artif. Organs* 24:129–136 (2000).
450. S. M. Bates, J. I. Weitz. Prevention of activation of blood coagulation during acute coronary ischemic syndromes: beyond aspirin and heparin. *Cardiovasc. Res*. 41:418–432 (1999).
451. G. Montalescot, V. Polle, J. P. Collet, P. Leprince, A. Bellanger, I. Gandjbakhch, D. Thomas. Low molecular weight heparin after mechanical heart valve replacement. *Circulation* 101:1083–1086 (2000).
452. W. E. Hennink, J. Feijen, C. D. Ebert, S. W. Kim. Covalently bound conjugates of albumin and heparin: synthesis, fractionation and characterization. *Thromb. Res*. 29:1–13 (1983).
453. W. E. Hennink, L. Dost, J. Feijen, S. W. Kim. Interaction of albumin–heparin conjugates preadsorbed surfaces with blood. *Trans. Am. Soc. Artif. Intern. Organs* 29:200–205 (1983).
454. W. E. Hennink, C. D. Ebert, S. W. Kim, W. Breemhaar, A. Bantjes, J. Feijen. Interaction of antithrombin III with preadsorbed albumin–heparin conjugates. *Biomaterials* 5: 264–268 (1984).
455. W. E. Hennink, S. W. Kim, J. Feijen. Inhibition of surface induced coagulation by preadsorption of albumin–heparin conjugates. *J. Biomed. Mater. Res*. 18:911–926 (1984).
456. G. H. M. Engbers, L. Dost, W. E. Hennink, P. A. Aarts, J. J. Sixma, J. Feijen. An in vitro study of the adhesion of blood platelets onto vascular catheters. Part I. *J. Biomed. Mater. Res*. 21:613–627 (1987).
457. G. W. Bos, N. M. Scharenborg, A. A. Poot, G. H. Engbers, W. G. Beugeling, T. van Aken, J. Feijen. Endothelialization of crosslinked albumin–heparin gels. *Thrombosis & Haemostasis* 82(6):1757–1763 (1999).
458. J. Chanda, R. Kuribayashi, T. Abe. Heparin coupling in inhibition of calcification of vascular bioprostheses. *Biomaterials* 20:1753–1757 (1999).
459. M. Tamim, M. Demircin, M. Guvener, O. Peker, M. Yilmaz. Heparin-coated circuits reduce complement activation and inflammatory response to cardiopulmonary bypass. *Panminerva Medica* 41:193–198 (1999).
460. M. Ishihara, Y. Saito, H. Yura, K. Ono, K. Ishikawa, H. Hattori, T. Akaike, A. Kurita. Heparin-carrying polystyrene to mediate cellular attachment and growth via interaction with growth factors. *J. Biomed. Mater. Res*. 50:144–152 (2000).
461. R. S. Stephens, F. S. Fawaz, K. A. Kennedy, K. Koshiyama, C. Nichols B. van Ooij, J. N. Engel. Eukaryotic cell uptake of heparin-coated microspheres: a model of host

cell invasion by *Chlamydia trachomatis. Infect. Immun.* 68:1080–1085 (2000).

462. Y. J. Kim, I. K. Kang, M. W. Huh, S. C. Yoon. Surface characterization and in vitro blood compatibility of poly (ethylene terephthalate) immobilized with insulin and/or heparin using plasma glow discharge. *Biomaterials* 21: 121–130 (2000).
463. S. Mathison, E. Bakker. Renewable pH cross-sensitive potentiometric heparin sensors with incorporated electrically charged H^+ ionophores. *Anal. Chem.* 71:4614–4621 (1999).
464. S. Mathison, E. Bakker. Improving measurement stability and reproducibility of potentiometric sensors for polyions such as heparin. *J. Pharm. Biomed. Anal.* 19:163–173 (1999).
465. P. A. Sandford, G. P. Hutchings. Chitosan—a natural cationic biopolymer. In: *Industrial Polysaccharides: Genetic Engineering, Structure/Properties Relations and Applications*, M. Yalpani (Ed.). Elsevier, Amsterdam, 1987, pp. 363–376.
466. R. Shepherd, S. Reader, A. Falshaw. Chitosan functional properties. *Glycoconjugate J.* 14:535–542 (1997).
467. T. Mitani, C. Nakalima, I. E. Sungkano, H. Ishii. Effects of ionic strength on the adsorption of heavy metals by swollen chitosan beads. *J. Environ. Sci. Health Part. A Environ. Sci. Eng. Toxic* 30:669–674 (1995).
468. E. Piron, A. Domard. Interactions between chitosan and alpha emitters. 238Pu and 241 Am. *Int. J. Biol. Macromole.* 23:121–125 (1998).
469. A. G. Imeri, D. Knorr. Effects of chitosan on yield and compositional data of carrot and apple juice. *J. Food Sci.* 53:1707–1710 (1988).
470. P. Stossel, J. L. Leuba. Effect of chitosan, chitin and some amino sugars on growth of various soilborne phytopathogenic fungi. *Phytopathology and Zoology* 111:82–90 (1984).
471. P. Gross, E. Konard, M. Hager. Patent application DE PS 262714 (1976).
472. J. Dutkiewicz, L. Judkiewicz, A. Papiewski, M. Kuchaeska, R. Ciszewski. Some uses of krill chitosan as biomaterial. In: *Chitin and Chitosan: Chemistry, Biochemistry, Physical Properties and Application*, G. Skjak-Braek, T. Anthonsen, P. Sandford (Eds.). Elesevier Applied Science, London, 1989.
473. G. G. Allan, L. C. Altman, R. E. Rensinger, D. K. Ghosh, Y. Hirabayashi, A. N. Neogi, S. Neogi. Biomedical application of chitin and chitosan. In: *Chitin, Chitosan and Related Enzymes*, J. P. Zikakis (Ed.). Academic Press, New York, 1984.
474. R. A. Muzzarelli. Human enzymatic activities related to the therapeutic administration of chitin derivatives. *Cell. Mol. Life Sci.* 53:131–140 (1997).
475. K. Nishimura, S. Nishimura, H. Seo, N. Nishi, S. Tokura, I. Azuma. Effect of multiporous microspheres derived from chitin and partially deacetylated chitin on the activation of mouse peritoneal macrophages. *Vaccine* 5:136–140 (1987).
476. H. Seo, K. Mitsuhashi, H. Tanibe. Antibacterial and antifungal fiber blended by chitosan. In: *Advances in Chitin and Chitosan*, C. J. Brine, P. A. Sandford, J. P. Zikakis (Eds.). Elsevier Scientific, London, 1992, pp. 24–40.
477. R. Muzzarelli, R. Tarsi, O. Filippini, E. Giovanetti, G. Biagini, P. E. Varaldo. Antimicrobial properties of N-carboxylbutyl chitosan. *Antimicro. Ag. Chemotherap.* 34:2019–2023 (1990).
478. T. Tanigawa, Y. Tanaka, H. Sashiwa, H. Saimoto, Y. Shigemasa. Various biological effects of chitin derivatives. In: *Advances in Chitin and Chitosan*, C. J. Brine, P. A. Sanford, J. P. Zikahis (Eds.), Elsevier Scientific, London, 1992, pp. 206–215.
479. P. Darmadji, M. Izumimoto. Effect of chitosan in meat preservation. *Meat. Sci.* 38:243–254 (1994).
480. Y. Uchiba. Antibacterial property of chitin. *Food Chem.* 2:22–29 (1988).
481. G. J. Tsai, W. H. Su. Antibacterial activity of shrimp chitosan against *Escherichia coli. J. Food Protection* 62:239–243 (1999).
482. R. Muzzarelli, G. Biagini, A. Pugnaloni, O. Filippini, V. Baldassarre, C. Castaldini, C. Rizzoli. Reconstruction of parodontal tissue with chitosan. *Biomaterials* 10:598–603 (1989).
483. S. Roller, N. Covill. The antimicrobial properties of chitosan in mayonnaise and mayonnaise-based shrimp salads. *J. Food Protection* 63(2):202–209 (2000).
484. J. Rhoades, S. Roller. Antimicrobial actions of degraded and native chitosan against spoilage organisms in laboratory media and foods. *Appl. Environ. Microbiol.* 66:80–86 (2000).
485. P. Artursson, T. Lindmark, S. S. Davis, L. Illum. Effect of chitosan on the permeability of monolayers of intestinal epithelial cells (Caco-2). *Pharm. Res.* 11:1358–1361 (1994).
486. N. G. M. Schipper, K. M. Varum, P. Artursson. Chitosans as absorption enhancers for poorly absorbable drugs. I. Influence of molecular weight and degree of acetylation on drug transport across human intestinal epithelial (Caco-2) cells. *Pharm. Res.* 13:1686–1692 (1996).
487. S. S. Koide. Chitin–Chitosan: properties, benefits and risks. *Nutr. Res.* 18:1091–1101 (1998).
488. M. Sugano, T. Fujikawa, Y. Hiratsuji, K. Nakashima, N. Fukuda, Y. Hasegawa. A novel use of chitosan as a hypocholesterolemic agent in rats. *Am. J. Clin. Nutr.* 33:787–793 (1980).
489. I. Ikeda, M. Sugano, K. Yoshida, E. Sasaki, Y. Iwamoto, K. Hatano. Effects of chitosan hydrolysates on lipid adsorption and on derum liver lipid concentrations in rats. *J. Agri. Food Chem.* 41:431–435 (1993).
490. D. J. Ormrod, C. C. Holmes, T. E. Miller. Dietary chitosan inhibits hypercholesterolaemia and atherogenesis in the apolipoprotein E-deficient mouse model of atherosclerosis. *Atherosclerosis* 138:329–334 (1998).
491. A. Razdan, D. Pettersson, J. Pettersson. Broiler chicken body weights, feed intakes, plasma lipid and small-intestinal bile acid concentrations in response to feeding of chitosan and pectin. *Br. J. Nutr.* 78:283–291 (1997).

492. S. B. Jing, L. Li, D. Ji, Y. Takiguchi, T. Yamaguchi. Effect of chitosan on renal function in patients with chronic renal failure. *J. Pharm. Pharmacol*. 49(7):721–723 (1997).
493. H. Chen, X. Tian, H. Zou. Preparation and blood compatibility of new silica–chitosan hybrid biomaterials. *Art. Cells Blood Substit. Immobilization Biotechnol*. 26:431–436 (1998).
494. S. H. Pangburn, P. V. Trescony, J. Heller. Lysozyme degradation of partially deacetylated chitin, its films and hydrogels. *Biomaterials* 3:105–108 (1982).
495. C. Yomoto, T. Komuro, T. Kimura. *Yakugaku Zasshi* 110: 442 (1990).
496. I. A. Zupanets, S. M. Drogozov, L. V. Yakovleva, A. I. Pavly, O. V. Bykova. *Fiziol. Sh*. 36:115 (1990).
497. L. V. Yakovleva, I. A. Zyupanets, S. M. Drogozov, A. I. Pavly. *Farmakol. Toxicol*. 3:70 (1988).
498. I. R. Setnikar, M. A. Cereda, L. Revel. Antireactive properties of glucosamine sulfate. *Arzneim. Forsch*. 41:157–161 (1991).
499. W. Malette, H. Quigley, E. Adickes. Chitosan effect in vascular surgery, tissue culture and tissue regeneration. In: *Chitin in Nature and Technology*. R. Muzzarelli, C. Jeuniaux, G. Gooday (Eds.). Plenum Press, New York, 1986, pp. 435–442.
500. K. Arai, T. Kinumaki, T. Fujita. *Bull. Tokai Reg. Fish Res. Lab*. 56:89–94 (1968).
501. M. Amiji, K. Park. Surface modification of polymeric biomaterials with poly(ethylene oxide), albumin and heparin for reduced thrombogenicity. In: *Polymeric Biomaterials: In Solution, as Interfaces, and as Solids*, S. L. Cooper, C. H. Bamford, T. Tsuruta (Eds.). VSP, The Netherlands, 1995, pp. 535–552.
502. T. Chandy, C. P. Sharma. Chitosan as a biomaterial. *Biomater. Artif. Cells Artif. Organs* 18:1–24 (1990).
503. S. Hirano, Y. Noishiki, J. Kinugawa, H. Higashima, T. Hayashi. Chitin and chitosan for use as a novel biomedical material. In: *Advances in Biomedical Polymers*, G. C. Gebelien (Ed.). Plenum Press, New York, 1987, pp. 285–297.
504. A. S. Hoffman. Blood–biomaterial interactions: an overview. In: *Biomaterials: Interfacial Phenomena and Applications*, Vol. 199, S. L. Copper, N. A. Peppas (Eds.). ACS, Washington, D.C., 1982, pp. 3–8.
505. M. A. Amiji. Synthesis of anionic poly(ethylene glycol) derivative for chitosan surface modification in blood-contacting applications. *Carbohydr. Polym*. 32:193–199 (1997).
506. F. Ferrari, S. Rossi, M. C. Bonferoni, C. Caramella, J. Karlsen. Characterization of rheological and mucoadhesive properties of three grades of chitosan hydrochloride. *Farmaco*. 52:493–497 (1997).
507. P. Tengamnuay, A. Sahamethapat, A. Sailasuta, A. K. Mitra. Chitosan as nasal absorption enhancers of peptides: comparison between free amine chitosans and soluble salts. *Int. J. Pharm*. 197, 53–67 (2000).
508. A. Bernkop-Schnurch. Chitosan and its derivatives: potential excipients for peroral peptide delivery systems. *Int. J. Pharm*. 194:1–13 (2000).
509. D. N. S. Hon. Chitin and chitosan: medical applications. In: *Polysaccharides in Medicinal Applications*, S. Dumitriu (Ed.). Marcel Dekker, New York, 1996, pp. 631–651.
510. R. A. A. Muzzarelli. Amphoteric derivatives of chitosan and their biological significance. In: *Chitin and Chitosan*, G. Skjak-Braek, T. Anthonsen, P. Sandford (Eds.). Elsevier Applied Science, London, New York, 1989, pp. 87–99.
511. J. F. Prudden. *Arch. Surg*. 89:1046 (1966).
512. J. Allen, J. F. Prudden. Histologic response to a cartilage powder preparation in a controlled human study. *Am. J. Surg*. 112:888 (1966).
513. H. Ueno, H. Yamada, I. Tanaka, N. Kaba, M. Matsuura, M. Okumura, T. Kadosawa, T. Fujinaga. Accelerating effects of chitosan for healing at early phase of experimental open wound in dogs. *Biomaterials* 20:1407–1414 (1999).
514. G. G. Allan, L. C. Altman, R. E. Bensinger, D. K. Ghosh, Y. Hirabayashi, A. N. Neogi, S. Neogi. Biomedical application of chitin and chitosan. In: *Chitin, Chitosan and Related Enzymes*, J. P. Zikakis (Ed.). Academic Press, New York, 1984.
515. W. K. Loke, S. K. Lau, L. L. Yong, E. Khor, C. K. Sum. Wound dressing with sustained anti-microbial capability. *J. Biomed. Mater. Res*. 53:8–17 (2000).
516. S. S. Koide. Chitin–Chitosan: properties, benefits and risks. *Nutr. Res*. 18:1091–1101 (1998).
517. Y. Okamoto, K. Shibazaki, S. Minami, A. Matsuhashi, S. Tanioka, Y. Shigemasa. Evaluation of chitin and chitosan on open wound healing in dogs. *J. Vet. Med. Sci*. 57:851–854 (1995).
518. K. Ono, Y. Saito, H. Yura, K. Ishikawa, A. Kurita, T. Akaike, M. Ishihara. Photocrosslinkable chitosan as a biological adhesive. *J. Biomed. Mater. Res*. 49:289–295 (2000).
519. Y. W. Cho, Y. N. Cho, S. H. Chung, G. Yoo, S. W. Ko. Water-soluble chitin as a wound healing accelerator. *Biomaterials* 20:2139–2145 (1999).
520. O. Felt, P. Buri, R. Gurny. Chitosan: a unique polysaccharide for drug delivery. *Drug Dev. Ind. Pharm*. 24:979–993 (1998).
521. L. Illum. Chitosan and its use as a pharmaceutical excipient. *Pharm. Res*. 15:1326–1331 (1998).
522. H. S. Kas. Chitosan: properties, preparations and application to microparticulate systems. *J. Microencapsulation* 14:689–711 (1997).
523. N. G. M. Scipper, S. Olsson, J. A. Hoogstraate, A. G. deBoer, R. M. Varum, P. Artursson. Chitosan as adsorption enhancers for poorly absorbed drugs. 2 Mechanism of absorption enhancement. *Pharm. Res*. 14:923–929 (1997).
524. A. F. Kotze, H. L. Luessen, A. G. de Boer, J. C. Verhoef, H. E. Junginger. Chitosan for enhanced intestinal permeability: prospects for derivatives soluble in neutral and basic environments. *Eur. J. Pharm. Sci*. 7:145–151 (1999).
525. G. C. Ritthidej, P. Chomto, S. Pummangura, P. Menasveta. Chitin and chitosan as disintegrates in pharmaceutical tablets. *Drug Dev. Ind. Pharm*. 20:2109–2134 (1994).
526. S. M. Upadrashta, P. R. Katikaneni, N. O. Nuessle. Chito-

san as a tablet binder. *Drug Dev. Ind. Pharm.* 18:1701–1708 (1992).
527. S. Sabnis, P. Rege, L. H. Block. Use of chitosan in compressed tablets of diclofenac sodium: inhibition of drug release in an acidic environment. *Pharm. Dev. Tech.* 2: 243–255 (1997).
528. K. D. Vorlop, J. Klein. Entrapment of microbial cells in chitosan. In: *Methods in Enzymology*, Vol. 135, K. Mosbach (Ed.). Academic Press, New York, 1987, p. 259.
529. K. D. Vorlop, J. Klein. *Biotechnol. Lett.* 3:9 (1981).
530. S. Dumitriu, E. Chornet. Functional versatility of polyionic hydrogels. In: *Chitin Enzymology*, Vol. 2, R. A. A. Muzzarelli (Ed.). Atec Edizioni, Italy, 1996, pp. 543–564.
531. Y. Kawashima, T. Handa, A. Kasai, H. Takenaka, S. Y. Lin, Y. Ando. Novel method for the preparation of controlled release theophylline granules coated with a polyelectrolyte complex of sodium polyphosphate chitosan. *J. Pharm. Sci.* 74:264–268 (1985).
532. S. Miyazaki, A. Nakayama, M. Oda, M. Takada, D. Attwood. Chitosan and sodium alginate based bioadhesive tablets for intraoral drug delivery. *Biol. Pharm. Bull.* 17: 745–747 (1994).
533. I. Genta, P. Perugini, F. Pavanetto. Different molecular weight chitosan microspheres: influence on drug loading and drug release. *Drug Dev. Ind. Pharm.* 24:779–784 (1998).
534. C. Remunan-Lopez, M. L. Lorenzo-Lamosa, J. L. Vila-Jato, M. J. Alonso. Development of new chitosan–cellulose multicore microparticles for controlled drug delivery. *Eur. J. Pharm. Biopharm.* 45:49–56 (1998).
535. M. L. Lorenzo-Lamosa, C. Remunan-Lopez, J. L. Vila-Jato, M. J. Alonso. Design of microencapsulated chitosan microspheres for colonic drug delivery. *J. Controlled Release* 52:109–118 (1998).
536. P. Calvo, C. Remunan-Lopez, J. L. Vila-Jato, M. J. Alonso. Chitosan and chitosan/ethylene oxide-propylene oxide block copolymer nanoparticles as novel carriers for proteins and vaccines. *Pharm. Res.* 14:1431–1436 (1997).
537. F. Bugamelli, M. A. Raggi, I. Orienti, V. Zecchi. Controlled insulin release from chitosan microparticles. *Arch. Pharm. Pharm. Med. Chem.* 331:133–138 (1998).
538. H. Tozaki, J. Komoike, C. Tada, T. Maruyama, A. Terabe, T. Suzuki, A. Yamamoto, S. Muranishi, Chitosan capsules for colon-specific drug delivery: improvement of insulin absorption from the rat colon. *J. Pharm. Sci.* 86:1016–1021 (1997).
539. Y. Kawashima, H. Yamamoto, H. Takeuchi, Y. Kuno. Mucoadhesive DL-lactide/glycolide copolymer nanospheres coated with chitosan to improve oral delivery of elcatonin. *Pharm. Dev. Tech.* 5:77–85 (2000).
540. Y. Nishioka, S. Kyotai, M. Okamura, M. Miyazaki, K. Okazaki, S. Ohnishi, Y. Yamamoto, K. Ito. Release characteristics of cisplatin chitosan microspheres and effect of containing chitin. *Chem. Pharm. Bull.* 38:2871–2873 (1990).
541. S. R. Jameela, A. Jayakrishnan. Crosslinked chitosan microspheres as a long acting biodegradable drug delivery vehicle: studies on the in vitro release of mitoxantrone and *in vivo* degradation of microspheres in rat muscle. *Biomaterials* 16:769–775 (1995).
542. Y. M. Wang, H. Sato, I. Adachi, I. Horikoshi. Optimization of the formulation design of chitosan microspheres containing cisplatin. *J. Pharm. Sci.* 85:1204–1210 (1996).
543. X. X. Tian, M. J. Groves, Formulation and biological activity of antineoplastic proteoglycans derived from Mycobacterium vaccae in chitosan nanoparticles. *J. Pharm. Pharmacol.* 51:151–157 (1999).
544. Y. S. Suzuki, Y. Momose, N. Higashi, A. Shigematsu, K. B. Park, Y. M. Kim, J. R. Kim, J. M. Ryu. Biodistribution and kinetics of holmium-166-chitosan complex (DW-166HC) in rats and mice. *J. Nucl. Med.* 39:2161–2166 (1998).
545. H. Tokumitsu, J. Hiratsuka, Y. Sakurai, T. Kobayashi, H. Ichikawa, Y. Fukumori. Gadolinium neutron-capture therapy using novel gadopentetic acid-chitosan complex nanoparticles: in vivo growth suppression of experimental melanoma solid tumor. *Cancer Lett.* 150:177–182 (2000).
546. K. Hojo, M. Maeda, Y. Mu, H. Kamada, Y. Tsutsumi, Y. Nishiyama, T. Yoshikawa, K. Kurita, L. H. Block, T. Mayumi, K. Kawasaki. Facile synthesis of a chitosan hybrid of a laminin-related peptide and its antimetastatic effect in mice. *J. Pharm. Pharmacol.* 52:67–73 (2000).
547. D. Quong, R. J. Neufeld. DNA protection from extracapsular nucleases, within chitosan- or poly L lysine coated alginate beads. *Biotech. Bioeng.* 60:124–134 (1998).
548. D. Quong, J. N. Yeo, R. J. Neufeld. Stability of chitosan and poly-L-lysine membranes coating DNA-alginate beads when exposed to hydrolytic enzymes. *J. Microencapsulation* 16:73–82 (1999).
549. J. Filipovic-Grcic, D. Voinovich, M. Moneghini, M. Becirevic-Lacan, L. Magarotto, I. Jalsenjak. Chitosan microspheres with hydrocortisone and hydrocortisone–hydroxypropyl–beta-cyclodextrin inclusion complex. *Eur. J. Pharmac. Sci.* 9:373–379 (2000).
550. F. C. Mooren, A. Berthold, W. Domschke, J. Kreuter, Influence of chitosan microspheres on the transport of prednisolone sodium phosphate across HT-29 cell monolayers. *Pharm. Res.* 15:58–65 (1998).
551. I. Orienti, K. Aiedeh, E. Gianasi, V. Bertasi, V. Zecchi. Indomethacin loaded chitosan microspheres. Correlation between the erosion process and release kinetics. *J. Microencapsulation* 13:463–472 (1996).
552. H. Tozaki, T. Fujita, T. Odoriba, A. Terabe, S. Okabe, S. Muranishi, A. Yamamoto. Validation of a pharmacokinetic model of colon-specific drug delivery and the therapeutic effects of chitosan capsules containing 5-aminosalicylic acid on 2,4,6-trinitrobenzenesulphonic acid-induced colitis in rats. *J. Pharm. Pharmacol.* 51:1107–1112 (1999).
553. S. Patashnik, L. Rabinovich, G. Golomb. Preparation and evaluation of chitosan microspheres containing bisphosphonates. *J. Drug Targeting* 4:371–380, (1997).
554. S. Magdassi, U. Bach, K. Y. Mumcuoglu, Formation of positively charged microcapsules based on chitosan–

lecithin interactions. *J. Microencapsulation* 14:189–195 (1997).
555. S. Nakatsuka, A. L. Andrady. Permeability of vitamin B12 in chitosan membranes—effect of cross-linking and blending with poly(vinyl alcohol) on permeability. *J. Appl. Polym. Sci*. 44:17–28 (1992).
556. K. Deyao, T. Peng, H. B. Feng, Y. Y. He. Swelling kinetics and release characteristic of cross-linked chitosan-polyether polymer network (semi-IPN) hydrogels. *J. Polym. Sci. Part A Polym. Chem*. 32:1213–1223 (1994).
557. G. A. F. Roberts, K. E. Taylor. The formation of gels by reaction of chitosan with glutaraldehyde. *Makromol. Chem*. 190:951–960 (1989).
558. S. Nakatsuka, A. L. Andrady. Permeability of vitamin B12 in chitosan membranes—effect of cross-linking and blending with poly(vinyl alcohol) on permeability. *J. Appl. Polym. Sci*. 44:17–28 (1992).
559. A. A. DeAngelis, D. Capitani, V. Crescenzi. Synthesis and 13C CP-MAS NMR characterization of a new chitosan-based polymeric network. *Macromolecules* 31:1505–1601 (1998).
560. V. Crescenzi, D. Imbiaco, C. I. Velasquez, M. Dentini, A. Ciferri. Novel types of polysaccharidic assemblies. *Macromol. Chem. Phys*. 196:2873–2880 (1995).
561. Y. M. Lee, S. S. Kim, S. H. Kim. Synthesis and properties of poly(ethylene glycol) macromer/beta-chitosan hydrogels. *J. Mater. Sci.-Mater. Med*. 8:537–541 (1997).
562. V. R. Pastel, M. M. Amiji. Preparation and characterization of freeze-dried chitosan–poly(ethylene oxide) hydrogels for site-specific antibiotic delivery in the stomach. *Pharm. Res*. 13:588–593 (1996).
563. X. Chen, W. J. Li, W. Zhong, Y. H. Lu, T. Y. Yu. pH sensitivity and ion sensitivity of hydrogels based on complex-forming chitosan/silk fibroin interpenetrating polymer network. *J. Appl. Polym. Sci*. 65:2257–2262 (1997).
564. H. F. Wang, W. J. Li, Y. H. Lu, Z. L. Wang. Studies on chitosan and poly(acrylic acid) interpolymer complex. I Preparation, structure, pH-sensitivity, and salt sensitivity of complex-forming poly(acrylic acid):chitosan semi-interpenetrating polymer network. *J. Appl. Polym. Sci*. 65: 1445–1450 (1997).
565. J. Kristl, J. Smid-Korbar, E. Struc, M. Schara, H. Ruppercht, Hydrocolloids and gels of chitosan as drug carriers. *Int. J. Pharm*. 99:13–19 (1993).
566. K. R. Holme, L. D. Hall. Chitosan derivatives bearing C10-alkyl glycoside branches: a temperature-induced gelling polysaccharide. *Macromolecules* 24:3828–3833 (1991).
567. L. Noble, A. I. Gray, L. Sadiq, I. F. Uchegbu, A non-covalently cross-linked chitosan based hydrogel. *Int. J. Pharm*. 192:173–182 (1999).
568. J. Cruz, M. Kawasaki, W. Gorski. Electrode coatings based on chitosan scaffolds. *Anal. Chem*. 72:680–686 (2000).
569. S. Miyazaki, K. Ishii, T. Nadai. The use of chitin and chitosan as drug carriers. *Chem. Pharm. Bull*. 29:3067–3069 (1981).
570. C. C. Leffler, B. W. Muller, Influence of the acid type on the physical and drug liberation properties of chitosan-gelatin sponges. *Int. J. Pharm*. 194:229–237 (2000).
571. C. Aimin, H. Chunlin, B. Juliang, Z. Tinyin, D. Zhichao. Antibiotic loaded chitosan bar. An in vitro, in vivo study of a possible treatment for osteomyelitis. *Clin. Orthopaedics Relat. Res*. 366:239–247 (1999).
572. A. Bernkop-Schnurch, C. Paikl, C. Valenta. Novel bioadhesive chitosan–EDTA conjugate protects leucine enkephalin from degradation by aminopeptidase. *N. Pharm. Res*. 14:917–922 (1997).
573. C. Valenta, B. Christen, A. Bernkop-Schnurch. Chitosan–EDTA conjugate: a novel polymer for topical gels. *J. Pharm. Pharmacol*. 50:445–452 (1998).
574. A. Bernkop-Schnurch, M. E. Krajicek. Mucoadhesive polymers as platforms for peroral peptide delivery and absorption: synthesis and evaluation of different chitosan–EDTA conjugates. *J. Controlled Release* 50:215–223 (1998).
575. A. Bernkop-Schnurch, C. Valenta. Novel bioadhesive chitosan–EDTA conjugate protects leucine enkephalin from degradation by aminopeptidase. *N. Pharm. Res*. 14:917–922 (1997).
576. C. J. Tsai, L. R. Hsu, Y. J. Fang, H. H. Lin. Chitosan hydrogel as a base for transdermal delivery of berberine and its evaluation in rat skin. *Biological Pharm. Bull*. 22: 397–401 (1999).
577. G. Kratz, C. Arnander, J. Swedenborg, M. Back, C. Falk, I. Gouda, O. Larm. Heparin–chitosan complexes stimulate wound healing in human skin. *Scand. J. Plas. Reconstr. Surg. Hand Surg*. 31:119–123 (1997).
578. A. S. Hoffman et al. *Proc. Int. Symp. Control. Rel. Bioact. Mater*. 24:633–634 (1997).
579. E. Tsuchida, K. Abe. Interactions between macromolecules in solution and intermacromolecular complexes. *Adv. Polym. Sci*. 45:1 (1982).
580. E. Bekturov, L. A. Bimendina. Interpolymer complexes. *Adv. Polym. Sci*. 41:100 (1980).
581. E. A. Bekturov, I. A. Bimendina. Complexing of statistical copolymers. *Izv. Akad. Nauk Kaz. SSR Ser. Khim*. 28:68 (1978).
582. V. A. Kabanov, I. M. Papisov. The formation of complexes between complementary synthetic polymers and oligomers in dilute solutions. *Vysokomol. Soyed*. A21:243 (1979).
583. A. B. Scranton, J. Klier, C. L. Aronson. Complexation of polymeric acids with polymeric base. *ACS Symp. Ser*. 480: 171 (1992).
584. S. N. Vinogradov, R. H. Linnell. *Hydrogen Bonding*. Van Nostrand Reinhold, New York, 1971.
585. C. L. McCornick, C. B. Johnson. Water-soluble copolymers. 29 copolymers of sodium 2-acrylamido-2-methylpropanesulfonate with (2-acrylamido-2-methyl) dimethylammonium chloride: solution properties. *Macromolecules* 21:694 (1988).
586. V. M. Monroy Soto, J. C. Galin. Poly(sulphopropylbetaines): 2. Dilute solution properties. *Polymer* 25:254 (1984).

587. A. B. Scranton, J. Klier, C. L. Aronson. Complexion of polymeric acids with polymeric bases. In: *Polyelectrolyte Gels: Properties, Preparation, and Applications*, R. S. Harland, R. K. Prud'homme (Eds.). ACS Symposium Series 480, American Chemical Society, Washington, D.C., 1992, pp. 171.
588. E. G. Kolawule, S. M. Mathieson, *J. Polym. Sci. Polym. Chem.* 15:2291 (1977).
589. S. K. Chatterjee, S. Gupta, K. R. Sethi. Study of copper (II)-methacrylic acid—methacrylamide copolymer interactions and formation of metal–copolymer complexes. *Angew. Makromol. Chem.* 147:133 (1987).
590. N. Kubota, Y. Kikuchi. Macromolecule complexes of chitosan. In: *Polysaccharides*, S. Dumitriu (Ed.). Marcel Dekker, New York, 1998, pp. 595–627.
591. 209. E. L. Cussler, M. R. Strokar, J. E. Varberg. Gels as size selective extraction solvents. *AIChE J.* 30:578 (1984).
592. A. J. Ribeiro, R. J. Neufeld, P. Arnaud, J. C. Chaumeil. Microencapsulation of lipophilic drugs in chitosan-coated alginate microspheres. *Int. J. Pharm.* 187:115–123 (1999).
593. W. R. Good, K. F. Mueller. Hydrogels and controlled delivery. *AIChE Symp. Ser.* 77:42 (1981).
594. R. A. Siegel, B. Firestone, M. Falamarzin. Paper 122H, AIChE 1988 Annual Meeting, Washington D.C., 1988.
595. J. A. Menjivar. The use of water-soluble polymers in oil field applications: hydraulic fracturing. Biotechnol. Mar. Polysaccharides, Proc. 3rd Ann. MIT Sea Grant Coll. Program Lect. Semin., R. R. Colwell, E. R. Pariser, A. J. Sinskey (Eds.). Hemisphere, Washington, D.C., 1985, p. 249.
596. A. D. Moradi-Araghi, H. Beardmore, G. A. Stahl. The application of gels in enhanced oil recovery: theory, polymers, and crosslinker systems. In: *Water-Soluble Polymers for Petroleum Recovery*, G. A. Stahl, D. N. Schulz (Eds.). Plenum Press, New York, 1988, p. 299.
597. J. K. Borchardt. Chemicals used in oil-field operations. *Am. Chem. Soc. Symp. Ser.* 396:3 (1989).
598. N. A. Peppas, A. G. Mikos. Preparation methods and structure of hydrogels. In: *Hydrogels in Medicine and Pharmacy*, Vol. 1, N. A. Peppas (Ed.). CRC Press, Boca Raton FL, 1986, pp. 1–25.
599. S. Dumitriu, P. F. Vidal, E. Chornet. Hydrogel based on polysaccharides. In: *Polysaccharides in Medicinal Applications*, S. Dumitriu (Ed.). Marcel Dekker, New York, 1996, pp. 125–241.
600. O. Gaserod, O. Smidsrod, G. Skjak-Braek. Microcapsules of alginate-chitosan. I. A quantitative study of the interaction between alginate and chitosan. *Biomaterials* 19: 1815–1825 (1998).
601. H. Takeuchi, T. Yasuji, H. Yamamoto, Y. Kawashima. Spray-dried lactose composite particles containing an ion complex of alginate-chitosan for designing a dry-coated tablet having a time-controlled releasing function. *Pharm. Res.* 17:94–99 (2000).
602. O. Gaserod, A. Sannes, G. Skjak-Braek. Microcapsules of alginate-chitosan. II. A study of capsule stability and permeability. *Biomaterials* 20:773–783 (1999).
603. D. D. Long, D. VanLuyen. Chitosan–carboxymethylcellulose hydrogels as supports for cell immobilization. *J. Macromol. Sci.-Pure Appl. Chem.* A33:1875–1884 (1996).
604. C. H. Chu, T. Sakiyama, T. Yano. pH-sensitive swelling of a polyelectrolyte complex gel prepared from xanthan and chitosan. *Biosci. Biotech. Biochem.* 59:717–719 (1995).
605. S. Dumitriu, E. Chornet. Inclusion and release of proteins from polysaccharide-based polyion complexes. *Adv. Drug Delivery Rev.* 31:223–246 (1998).
606. D. Thacharodi, K. Rao. Collagen-chitosan composite membranes for controlled release of propranolol hydrochloride. *Int. J. Pharm.* 120:115–118 (1995).
607. K. Oungbho, B. W. Muller. Chitosan sponges as sustained release drug carriers. *Int. J. Pharm.* 156:229–237 (1997).
608. C. C. Leffer, B. W. Muller. Influence of the acid type on the physical and drug liberation properties of chitosan–gelatin sponges. *Int. J. Pharm.* 194:229–237 (2000).
609. Y. Tabata Y. Ikada. Synthesis of gelatin microspheres containing interferon. *Pharm. Res.* 6:422–427 (1989.).
610. S. R. Jameela, T. V. Kumary, A. V. Lal, A. Jayakrishnan. Progesterone-loaded chitosan microspheres: a long acting biodegradable controlled delivery system. *J. Controlled Release* 52:17–24 (1998).
611. M. Izume, T. Taira, T. Kimura, T. Miyata. A novel cell culture matrix composed of chitosan and collagen complex. In: *Chitin and Chitosan*, S. B. Gudmund, T. Anthonsen, P. Sandford (Eds.). Elsevier Applied Science, London, 1989, p. 653.
612. S. Dumitriu, Cr. Dumitriu. Hydrogel and general properties of biomaterials. In: *Polymeric Biomaterials*, S. Dumitriu (Ed.), Marcel Dekker, New York, 1994, pp. 3–85.
613. S. Dumitriu, D. Dumitriu. Biocompatibility of polymers. In: *Polymeric Biomaterials*, S. Dumitriu, (Ed.). Marcel Dekker, New York, 1994, pp. 99–142.
614. S. Dumitriu, M. Dumitriu. Polymeric drug carriers. In: *Polymeric Biomaterials*, S. Dumitriu (Ed.). Marcel Dekker, New York, 1994, pp. 435–689.
615. S. Dumitriu, E. Chornet. Polysaccharides as support for enzyme and cell immobilization. In: *Polysaccharides*, S. Dumitriu (Ed.). Marcel Dekker, New York, 1994, pp. 629–748.
616. T. Shioya, C. K. Rha, S. Brand. Transmembrane permeability of chitosan/carboxymethyl-cellulose capsule. In: *Chitin and Chitosan*, I. G. Skjak-Braek, T. Anthonsen, P. Sandford (Eds.). Elsevier Applied Science, London, 1989, p. 627.
617. C. D. Melia, Hydrophilic matrix sustained release systems based on polysaccharide carriers. *Crit. Rev. Therap. Drug Carrier Systems* 8:395 (1991).
618. T. L. Bonfield, E. D. Colton, J. M. Anderson. Plasma protein adsorbed biomedical polymers: activation of human monocytes and induction of interleukin-1. *J. Biomed. Mater. Res.* 23:535–548 (1989).
619. K. M. Miller, V. Rose-Caprara, J. M. Anderson. Generation of IL1-like activity in response to biomedical polymer

implants: a comparison of in vitro and in vivo models. *J. Biomed. Mater. Res*. 23:1007–1026 (1989).

620. K. M. Miller, J. M. Anderson. Human monocyte/macrophage activation and interleukin-1 generation by biomedical polymers. *J. Biomed. Mater. Res*. 22:713–731 (1988).

621. D. Barkley, M. Feldmann, R. N. Maini. The detection by immunofluorescence of distinct cell populations producing interleukin-1 in activated human peripheral blood. *J. Immunol. Methods* 120:277–283 (1989).

622. D. J. Stuehr, M. A. Marletta. Mammalian nitrate biosynthesis: mouse macrophages produce nitrite and nitrate in response to *Escherichia coli* lipopolysaccharide. *Proc. Natl. Acad. Sci. USA* 82:7738 (1985).

623. F. Chellat, M. Tabrizian, S. Dumitriu, E. Chornet, P. Magny, C.-H. Rivard, L.-H. Yahia. *In vitro* and *in vivo* biocompatibility of chitosan–xanthan polyionic complex. *J. Biomed. Mater. Res*. 51:107–116 (2000).

623. C. A. Kienzle-Sterzer, D. Rodriguez-Sanchez, C. K. Rha. Biopolymers flow behavior of a cationic biopolymer, chitosan. *Polymer Bull*. 13:1 (1985).

624. J. G. Domoszy, G. A. F. Roberts. Ionic interactions between chitosan and oxidized cellulose. In: *Chitin in Nature and Technology*, R. Muzzarelli, Ch. Jeuniaux, G. W. Gooday (Eds.). Plenum Press, New York, 1986, p. 331.

625. C. Remunan-Lopez, R. Bodmeier. *Int. J. Polym*. 135:63 (1996).

626. Y. Kikuchi, N. Kubota. Macromolecular complexes consisting of poly(aluminium chloride), [2-(diethylamino) ethyl]dextran hydrochloride, and potassium poly(vinyl sulfate). *J. Appl. Polym. Sci*. 36:599 (1988).

627. F. L. Mi, S. S. Shyu, C. Y. Kuan, S. T Lee, K. T Lu, S. F. Jang. Chitosan—polyelectrolyte complexation for the preparation of gel beads and controlled release of anticancer drug. I. Effect of phosphorous polyelectrolyte complex and enzymatic hydrolysis of polymer. *J. Appl. Polym. Sci*. 74:1868–1879 (1999).

628. R. Fernandez-Urrusuno, P. Calvo, C. Remunan-Lopez, J. L. Vila-Jato, M. J. Alonso. Enhancement of nasal absorption of insulin using chitosan nanoparticles. *Pharm. Res*. 16:1576–1581 (1999).

629. L. Illum, N. D. Farraj, S. S. Davis. Chitosan as a novel nasal delivery system for peptide drugs. *Pharm. Res*. 11: 1186–1189 (1994).

630. T. J. Aspden, L. Illum, O. Skaugrud. Chitosan as a nasal delivery system: evaluation of insulin absorption enhancement and effect on nasal membrane integrity using rat model. *Eur. J. Pharm. Sci*. 4:23–31 (1996).

631. R. Muzzarelli, V. Baldassare, F. Conti, P. Ferrara, G. Biagini, G. Gazzanelli, V. Vasi. Biological activity of chitosan: ultrastructural study. *Biomaterials* 9:247–252 (1988).

632. S. Hirano, H. Tsuchida, N. Nagao. N-acetylation in chitosan and the rate of its enzymatic hydrolysis. *Biomaterials* 10:574–576 (1989).

633. F. F. Bartone, E. D. Adickes. Chitosan: effects on wound healing in urogenital tissue: preliminary report. *J. Urol*. 140:1134–1137 (1988).

634. A. E. Elcin, Y. M. Elcin, G. D. Peppas. Neural tissue engineering: adrenal chromaffin cell attachment and viability on chitosan scaffolds. *Neurological Res*. 20:648–654 (1998).

635. M. Kawase, N. Michibayashi, Y. Nakashima, N. Kurikawa, K. Yagi, T. Mizoguchi. Application of glutaraldehyde-crosslinked chitosan as a scaffold for hepatocyte attachment. *Biol. Pharm. Bull*. 20:708–710 (1997).

636. V. F. Sechriest, Y. J. Miao, C. Niyibizi, A. Westerhausen-Larson, H. W. Matthew, C. H. Evans, F. H. Fu, J. K. Suh. GAG-augmented polysaccharide hydrogel: a novel biocompatible and biodegradable material to support chondrogenesis. *J. Biomed. Mater. Res*. 49:534–541 (2000).

637. Y. J. Park, Y. M. Lee, S. N. Park, S. Y. Sheen, C. P. Chung, S. J. Lee. Platelet derived growth factor releasing chitosan sponge for periodontal bone regeneration. *Biomaterials* 21:153–159 (2000).

638. M. R. Mariappan, E. A. Alas, J. G. Williams, M. D. Prager. Chitosan and chitosan sulfate have opposing effects on collagen–fibroblast interactions. *Wound Repair Regeneration* 7:400–406 (1999).

639. K. Roy, H. Q. Mao, K. W. Leong. DNA–chitosan microspheres: transfection efficiency and cellular uptake. *Proc. Int. Symp. Control. Rel. Biaoct. Mater*. 24:673–674 (1997).

640. J. L. Murata, Y. Ohya, T. Ouchi. Possibility of application of quaternary chitosan having pendant galactose residues as gene delivery tool. *Carbohydr. Polym*. 29:69–74 (1996).

641. P. Erbacher, S. Zou, T. Bettinger, A.-M. Steffan, J. S. Remy. Chitosan-based vector/DNA complexes for gene delivery: biophysical characteristics and transfection ability. *Pharm. Res*. 15:1332–1339 (1998).

642. H-Q. Mao, K. Roy, V. L. Troung-Le, K. A. Janes, K. Y. Lin, Y. Wang, J. T. August, K. W. Leong. Chitosan-DNA nanoparticles as gene carriers: synthesis, characterization and transfection efficiency. *J. Control. Rel*. 70:399–421 (2001).

643. J. I. Murata, Y. Ohya, T. Ouchi. Possibility of application of quaternary chitosan having pendant galactose residues as gene delivery tool. *Carbohydr. Polym*. 29:69–74 (1996).

644. F. C. MacLaughlin, R. J. Mumper, J. Wang, J. M. Tagliaferri, I. Gill, M. Hinchcliffe, A. P. Rolland. Chitosan and depolymerized chitosan oligomers as condensing carriers for in vivo plasmid delivery. *J. Controlled Release* 56: 259–272 (1998).

645. G. Wu, C. Wu. Receptor-modified in vitro bene transformation by soluble DNA carrier system. *J. Biol. Chem*. 262:4429–4432 (1987).

646. O. Boussif, F. Lezoualch, M. A. Zanta, M. D. Mergny, D. Scherman, B. Demeneix, J. P. Behr. A versatile vector for gene and oligonucleotide delivery into cells in culture and in vivo: polyethyleneimine. *Proc. Natl. Acad. Sci. USA* 92:7297–7310 (1995).

647. J. S. Remy, B. Abdallah, M. A. Santa, O. Boussif, J. P. Behr, B. Demeneix. Gene transfer with lipospermines and polyethylenimines. *Adv. Drug Delivery Rev*. 30:85–95 (1998).

648. R. J. Mumper, J. G. Duguid, K. Anwer, M. K. Barron, H.

Nitta, A. P. Rolland. Polyvinyl derivatives as novel interactive polymers for controlled gene delivery to muscle. *Pharm. Res.* 13:701–709 (1996).
649. M. X. Tang, C. T. Redemann, F. C. Szoka. In vitro gene delivery by degraded polyamidoamine dendrimers. *Bioconjugate Chem.* 7:703–714 (1996).
650. S. C. W. Richardson, H. V. J. Koble, R. Duncan. Potential of low molecular mass chitosan as a DNA delivery system: biocompatibility, body distribution and ability to complex and protect DNA. *Int. J. Pharm.* 178, 231–243 (1999).
651. K. Y. Lee, I. C. Kwon, Y. H. Kim, W. H. Jo, S. Y. Jeong. Preparation of chitosan self-aggregates as a gene delivery system. *J. Controlled Release* 51:213–220 (1998).
652. K. Mao, S. K. Huang, K. W. Leong. Oral gene delivery with chitosan—DNA nanoparticles generates immunologic protection in a murine model of peanut allergy. *Nature Medicine* 5:387–391 (1999).
653. A. Bernkop-Schnurch. Chitosan and its derivatives: potential excipients for peroral peptide delivery systems. *Int. J. Pharm.* 194:1–13 (2000).
654. W. Baschong, D. Huglin, T. Maier, E. Kulik. Influence of N-derivatization of chitosan to its cosmetic activities. *SOFW J.* 125:22–24 (1999).
655. K. Aiedeh, M. O. Taha. Synthesis of chitosan succinate and chitosan phthalate and their evaluation as suggested matrices in orally administered, colon-specific drug delivery systems. *Acta Pharm. Pharm. Med. Chem.* 332:103–107 (1999).
656. I. F. Uchegbu, A. G. Schatzlein, L. Tetley, A. I. Gray, J. Sludden, S. Siddique, E. Mosha. Polymeric chitosan based vesicles for drug delivery. *J. Pharm. Pharmacol.* 50:453–458 (1998).
657. A. G. Nigalaye, P. Adusumilli, S. Bolton. Investigation of prolonged drug release from matrix formulation of chitosan. *Drug Dev. Ind. Pharm.* 16:449–467 (1990).
658. P. S. Adusumilli, S. M. Bolton. Evaluation of chitosan citrate complexes as matrices for controlled release formulations using a 32 full factorial design. *Drug Dev. Ind. Pharm.* 17:1931–1945 (1991).
659. T. Phaechamud, T. Koizumi, G. C. Ritthidei. Chitosan citrate as film former: compatibility with water-soluble anionic dyes and drug dissolution from coated tablet. *Int. J. Pharm.* 198:97–111 (2000).
660. S. Y. Lin, P. C. Lin. Effect of acid type, acetic and sodium carboxymethylcellulose concentrations on the formulation, micrometric, dissolution and floating properties of theophylline chitosan microcapsules. *Chem. Pharm. Bull.* 40:2491–2497 (1992).
661. A. Miwa, A. Ishibe, M. Nakano, T. Yamahira, S. Itai, S. Jinno, H. Kawahara. Development of novel chitosan derivatives as micellar carriers of taxol. *Pharm. Res.* 15: 1844–1850 (1998).
662. T. Ouchi, T. Banba, M. Fujimoto, S. Hamamoto. Synthesis and antitumor activity of chitosan carrying 5-fluorouracils. *Makromol. Chem.* 190:1817–1825 (1989).
663. Y. Ohya, M. Shiratani, H. Kobayashi, T. Ouchi. Release behaviour of 5-fluorouracil from chitosan-gel nanospheres immobilizing 5-fluorouracil coated with polysaccharides and their cell specific cytotoxicity. *J. Macromol. Sci.-Pure Appl. Chem.* A31:629–642 (1994).
664. S. Tokura, Y. Kaned, Y. Miura, Y. Uraki. Two-step hydrolyses of a polymeric drug under a model system. *Carbohydr. Polym.* 195:191 (1992).
665. M. Sato, H. Onishi, J. Takahara, Y. Machida, T. Nagai. *In vivo* drug release and antitumor characteristics of water-soluble conjugates of mitomycin C with glycol-chitosan and N-succinyl-chitosan. *Biol. Pharm. Bull.* 19:1170–1177 (1996).
666. L. Y. Lim, E. Khor, O. Koo. Gamma irradiation of chitosan. *J. Biomed. Mater. Res.* 43:282–290 (1998).

2

Biomimetics

Weiyuan John Kao
University of Wisconsin, Madison, Wisconsin

I. INTRODUCTION

Current biomaterial research efforts mainly focus on elucidating the specific interaction among material physicochemical properties, proteins/enzymes and various biological molecules, observed cellular functions, and physiological parameters such as mechanical forces in vitro and in vivo (1–10). Based on these investigations, biomaterial developments have begun to utilize the diversity and uniqueness of biology for potential methods to control biological behavior of a myriad of tissue engineering and biomedical applications. As the result of this novel approach and new technology, the field of biomimetic materials emerges. However, the root of biomimetics can be traced to the early 1970s when functional artificial molecules were designed and synthesized by mimicking the functional structure and synthesis pathway of natural compounds. For example, biomimetic methodologies were employed to selectively synthesize antileukemic triptolides (11), cyclical polyenes such as racemic 11α-hydroxyprogesterone (12–14), camptothecin chromophores (15), catechol estrogens (16), polyketone-derived phenols (17), D-glucose-derived (+)-biotin (18), and much more. In the 1980s and early 1990s, the field of biomimetics expanded dramatically to such areas as antitumor pharmaceutics (19–21), the development of receptor agonists and antagonists (22,23), the study of functional structures and syntheses of proteins and nucleotides (24), and the elucidation of biological pathways such as enzyme catalysis (25). In the late 1990s and into the new millennium, biomimetic methodologies have been employed as an engineering tool to develop novel material processing methods such as calcium phosphate coatings on metal alloys (26) and poly(L-lactic acid)–co-apatite composites (27), to formulate biofunctional materials for biomedicine and bioengineering such as synthetic peptide amphiphiles (28) and polymers derived from molecular and microbiology (29–31), and to continue to be utilized in probing the molecular fundamentals and mechanisms of biology.

Although the biomimetic approach has been employed by various scientific disciplines to develop novel molecules and materials, there is not a consensus in the definition of ''biomimetics'' that is accepted by various science communities, namely, engineering, chemistry, medicine, and life sciences. Nonetheless, the semantic definition is the ability to camouflage a substance with suitable contrivances similar to the environmental background with the aim to elude a hostile observer. However, such a definition is insufficient and narrow. Hence, the author argues that in the biological and engineering arena, the application of mimetism should be broad and hence dynamic and encompassing. The author will attempt to address the wide spectrum of biomimetics in this chapter by highlighting selected research investigations in which the knowledge of biological sciences has been adopted as a template in the study and development of novel molecules and materials. The author will emphasize the complementary interrelationship between biomimetic research and the pursuit of understanding the complex biological system. Biological systems fundamentally function at the angstrom scale of molecular atoms, the nano scale of macromolecules, the micron scale of cells and protein matrices, and the macro scale of tissues, organs, and macroorganisms (32). The uniqueness of biological materials and systems is the metastable microstructure formed from the molecular to macroscale in order to attain a combination of required properties (i.e., physical, chemical, biological, structural, etc.) for optimal function. Hence, the intricate micro- and macroarchitecture and multifunctional properties of biological systems are utilized as foundations in the investigation of biomimetic materials. By studying the hierarchical structure of biological systems and biological constituents at all possible scales of spatial resolution, the fundamentals of the unique structural designs can be deduced and mimicked by currently available techniques. Alternatively, the molecular synthesis and processing mechanism of biological systems and biological components can be elucidated and applied to develop novel materials (32). Both strategies embody the principle and the dynamic nature of biomimetism. Investigations adopting these broad research principles will be presented in this chapter as illustrative examples of biomimetic macromolecules and novel biomaterials with potential biomedical values.

II. BIOFUNCTIONAL MOLECULAR BIOMIMETICS

From the beginning of biomimetic research in the filed of chemistry, pharmaceutical sciences, and biochemistry, bioactive molecules have been synthesized by studying the native functional structure and natural synthesis pathways of model molecules. This approach continues to this day. Of particular interest to biomedicine is the investigation on superoxide dismutase (SOD) enzymes to catalyze oxygen radicals for the management of chronic inflammation, excessive oxidative metabolism, and tumor (19–21). For example, hemodialysis polymer membranes may activate the phagocyte oxidative metabolism resulting in the highly oxidized low-density lipoprotein found in uremic patients. Superoxide dismutase eliminates the highly reactive and hence destructive superoxide radicals and participates in the oxygen radical detoxification process. However, the native SOD molecule is highly unstable and has low bioavailability and high immunogenicity. The biomimetic approach to develop molecules with SOD activity was first pioneered by Yamamoto et al. (21) and continued by Ohse and Nagaoka, who demonstrated that the reaction of various metalloporphyrins and poly(styrene-co-maleic anhydride) copolymers resulted in the formulation of novel chimeras with pH sensible sites and SOD-like active sites (Fig. 1) (33). Using in vitro assay systems based on cytochrome-*c*, the low molecular weight metalloporphyrins catalyzed the dismutation of O_2^- with a rate constant of 10^7–10^8/M/s. The k_{cat} value for the multifunctional chimera of metalloporphyrins and poly(styrene-co-maleic anhydride) was found to be $1.1 \pm 0.1 \times 10^6$/M/s at 8.1 pH in vitro. In vivo studies demonstrated that the chimera construct of metalloporphyrins and poly(styrene-co-maleic anhydride) bound to the warfarin site of albumin and showed an enhanced half-life circulation time; while the SOD activity was retained.

In the design of biofunctional molecules to study and modulate host cell behavior (34), Kao et al. have chosen a family of active proteins, i.e., interleukin-1 (IL1), as a model in the design of biofunctional molecules. IL1β is a potent proinflammatory cytokine that upregulates cellular function upon ligation with IL1 receptor type I (IL1RI) on the extracellular membrane (35). IL1β-activated macrophages (a primary class of phagocytic leukocytes) may release high levels of IL1β, tumor necrosis factor α, and granulocyte/macrophage–colony stimulating factor (GM-CSF), which all have important roles in the healing process (1,4,9,36). Circulating IL1 receptor antagonist (IL1ra) is the natural antagonist for IL1β. IL1β ligates with Domain 1, 2, and 3 of IL1RI and initiates signaling pathways to upregulate cellular functions (37); whereas, IL1ra only binds with Domain 1 and 2 of IL1RI, resulting in no postligation signal transduction (38). It has been proposed that IL1β complexes with Domain 3 of IL1RI by a strong electrostatic interaction. Kao et al. designed macromolecules based on the functional architecture of IL1β and IL1ra. Antagonists were designed from known IL1β and IL1ra amino acid residues showing strong avidity toward Domain 1 and 2 of IL1RI (35,37–39). These residues are $R^{(4)}$ $L^{(6)}$ $F^{(46)}$ $I^{(56)}$ $K^{(103)}$ $E^{(105)}$ of IL1β and $W^{(17)}$ $Q^{(21)}$ $Y^{(35)}$ $Q^{(37)}$ $Y^{(148)}$ of IL1ra (35,37–40). Strong evidence suggests that these amino acids are essential in ligand–receptor interaction. The tertiary structures of the native IL1β and IL1ra molecule in solution were utilized as a guide in determining the spatial interrelationship between each residue. The minimum distance between each residue was approximated using the structural coordinates archived in the SwissProt Database®. Based on the measurement, a polyglycine sequence of approximately the same length was utilized to link each amino acid at all possible orientations. Kao et al. further hypothesized that the resulting antagonist sequences, which target Domains 1 and 2 of IL1RI, can be converted to agonists by coupling a strong electrostatic moiety (i.e., poly-lysine) to the terminal amino acid of the antagonist sequence, thus allowing the complexation between the peptide macromolecule and Domains 1, 2, and 3 of IL1RI to occur. To formulate agonists, a trimeric glycine linker was utilized to join the antagonist domain with

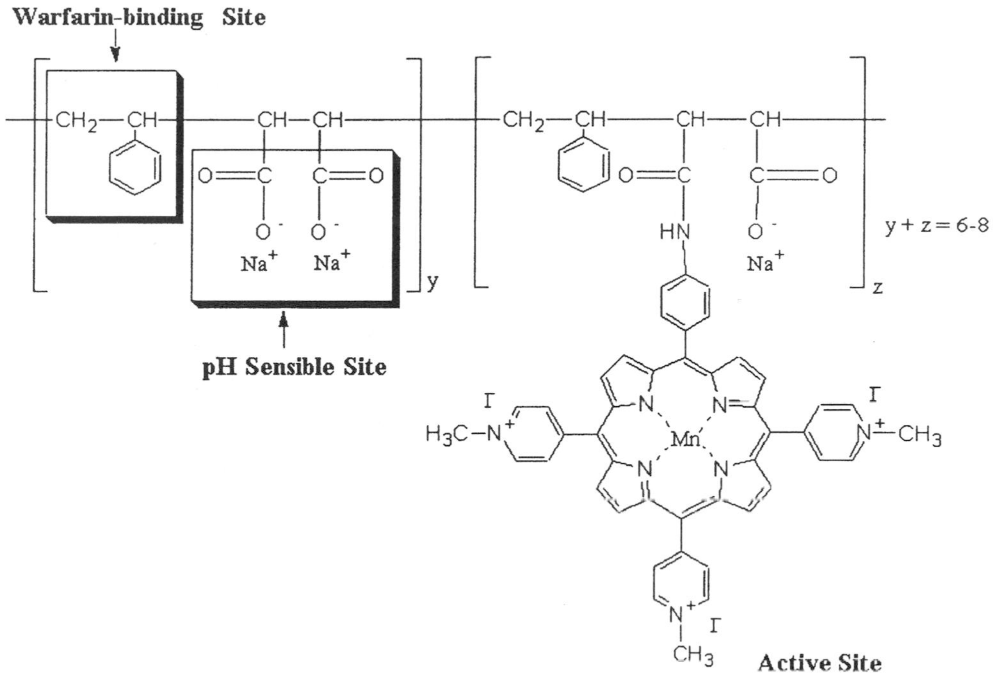

Figure 2.1 Structure of poly(styrene-co-maleic anhydride)–co-metalloporphyrin derivatives with superoxide dismutase activity. (Adopted from Ref. 33.)

the poly lysine domain. The trimeric glycine linker was designed to introduce spatial flexibility between the polylysine sequence and the antagonist oligopeptide domain since chain mobility may impact the dynamics of ligand–receptor association. Consequently, a library of linear oligopeptides were formulated and synthesized using solid-resin methods with standard 9-fluorenylmethyloxycarbonyl chemistry (41). Peptides thus formulated were designated as follows. Antagonists included those modeled after the IL1β molecule: (IL1iant) RGGLGGFGGIGKGGEG, (IL1iiant) FGRGGLGGGIGKGGEG, and (IL1iiiant) LGGRGFGGIGKGGEG; those modeled after the IL1ra molecule: (IL1ivant) WGGGQGGYGGQGGGYG and (IL1vant) WGGYGGQGGYGGQG; and that modeled after antagonists developed via combinatorial chemistry (40): (IL1viant) YWQPYALPL. Agonists included those modeled after the IL1β molecule: (IL1i) KKKGGGRGGL GGFGGIGKGGEG, (IL1ii) KKKGGGFGRGGLGGGIG KGGEG, and (IL1iii) KKKGGGLGGRGFGGIGKGGEG; those modeled after the IL1ra molecule: (IL1iv) KKKGGG WGGGQGGYGGQGGGYG and (IL1v) KKKGGGWGG YGGQGGYGGQG; and those modeled after antagonists developed via combinatorial chemistry (40): (IL1vi) KKK GGGYWQPYALPL and (IL1vii) YWQPYALPLGGGK KK. Human blood monocyte–derived macrophages were allowed to adhere on tissue culture polystyrene for 2 h in the presence of autologous serum in the culture medium. After 2 h, nonadherent cells were removed and adherent cells were challenged with free peptides at an optimal concentration of 50 pmol/mL that had been determined previously. Simultaneously, adherent cells were also challenged with or without recombinant human IL1β or IL1ra at 25 pmol/mL. Peptides with scrambled amino acid sequences were also employed as peptide controls. Cells were cultured thereafter in the presence of autologous serum and supernatants were collected at various time points for assay. At 4 h after the peptide challenge, no differences in GM-CSF release by adherent macrophages were observed among all test samples and controls. At 18 h after the free peptide challenge, IL1β-treated adherent macrophages on tissue-culture polystyrene (TCPS) showed a higher GM-CSF release than controls without treatment (Fig. 2a). Adherent macrophages treated with agonist IL1v showed a higher GM-CSF release than that treated with IL1β, IL1ra, none, and other agonists. It is known that IL1β upregulates GM-CSF release by human macrophages via the induction of AP-1 and NFκB gene expression factors. The result suggests that IL1v activates these gene expression factors resulting in an increased GM-CSF production. The GM-CSF level of cells treated with IL1v remained comparable when IL1ra was added simultaneously as the IL1v peptide indicating that the effect of IL1v in increasing GM-CSF release

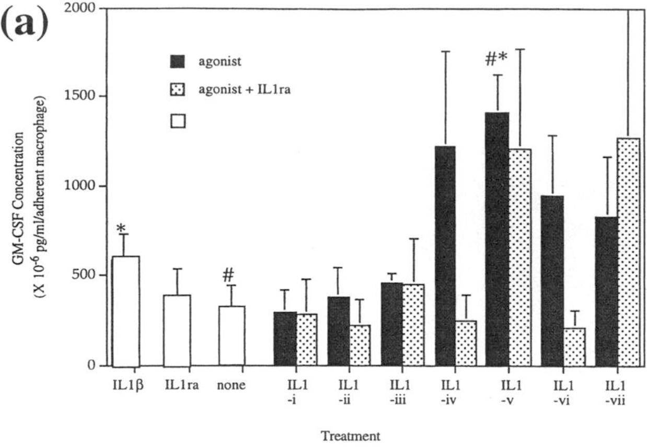

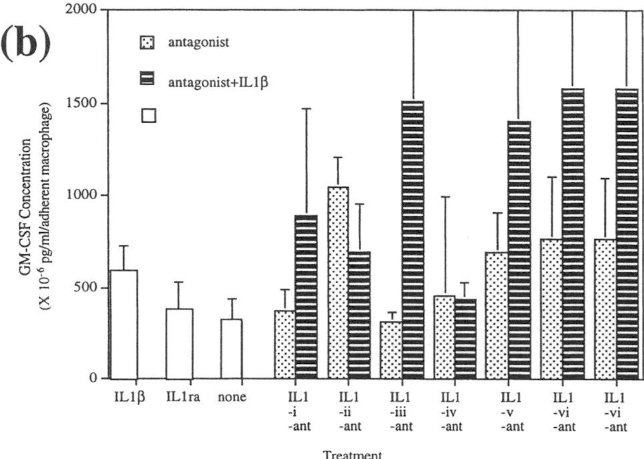

Figure 2.2 The release of GM-CSF by adherent human blood-derived macrophages (normalized to adherent macrophage density) on tissue-culture polystyrene after 18 h incubated with (a) biomimetic agonists derived from IL1-family proteins with or without recombinant human IL1ra (* represents $p < 0.05$ versus values of "none" control; # represents $p < 0.05$ versus values of IL1β control; IL1v values were larger than that of IL1i, IL1ii, IL1iii, IL1vi, and IL1vii at $p < 0.05$); or (b) biomimetic antagonists derived from IL1-family proteins with or without recombinant human IL1β (* represents $p < 0.05$ versus values of "none" control; # represents $p < 0.05$ versus values of IL1β control; no differences at $p > 0.95$ were found among each test group indicating that the GM-CSF release mediated by antagonists was not neutralized by the presence of IL1β). All values mean ± s.e.m., $n = 4$.

by adherent macrophages was not neutralized in the presence of the natural IL1β antagonist, namely, IL1ra. No significant differences in the GM-CSF release were observed between macrophages without treatment and those treated with both IL1β and one of the antagonists (Fig. 2b). These results indicate that the IL1-derived biomimetic antagonists neutralized the ability of IL1β in increasing the release of GM-CSF by adherent macrophages. Recombinant IL1β and IL1ra proteins have been investigated for their potential therapeutic values. For example, IL1β augments hematopoiesis and increases wound healing; whereas, IL1ra attenuates host inflammatory reaction. However, the clinical value of these regimens has yet to be demonstrated due to high cost, low bioavailability, deleterious side effects of the high dosage that is often required, and the presence of natural agonists, antagonists, and serum proteases. Hence, IL1-derived agonists and antagonists are currently under extensive investigation by pharmaceutical companies and basic science research laboratories to improve the therapeutic value of native IL1β and IL1ra moieties (42–44). IL1-derived agonists have demonstrated potential in increasing hematopoietic stem cell proliferation, increasing B cell proliferation, enhancing local wound healing, and modulating collagenase production. IL1-derived antagonists have shown promise in decreasing the extent of acute and chronic inflammation in diseases such as rheumatoid arthritis, treating septic shock, suppressing chronic leukemia, and modulating bone resorption by osteoclasts. These two examples illustrate the uniqueness of a biomimetic approach to develop synthetic molecules derived from naturally occurring precursors such as SOD and IL1 family proteins. These biomimetic molecules demonstrate enhanced biochemical and physical properties that can be employed to study biological systems and potentially improve clinical medicine.

III. NOVEL POLYMERIC MATERIALS CONTAINING PEPTIDE/PROTEIN BIOMIMETICS

Currently, several laboratories have taken the biomimetic approach to develop novel biomaterials containing biologically derived functional moieties to modulate specific cellular behavior in a variety of physiological systems such as cardiovascular, orthopedic, and nervous. The overall goal of these investigations is to obtain fundamental understanding of the complex biological–material interaction; the knowledge is necessary for the construction of tissue engineered devices and novel therapeutic biomedical products. Hubbell and West have focused on the study and development of enzymatically degradable hydrogels that are responsive to the local host environment (45,46). For example, West et al. utilized known protease substrate peptides (i.e., sequence LGPA from collagenase, sequence AAA from elastase, or sequence NRV from plasmin) copolymerized with polyethyleneglycols using dicyclohexylcarbodiimide-based chemistry to formulate BAB (peptide–polyethyleneglycol–peptide) block copolymers. The BAB copolymers were reacted with acryloyl chloride to provide photopolymerizable end groups. ABA (polyethyleneglycol–peptide–polyethyleneglycol) block copolymers were formed by reacting the peptide with two amine groups such as lysine residues at the C terminus with polyethyleneglycol–N-hydroxysuccinimide monoacrylate. Cell adhesive peptides such as RGDs were also reacted with polyethyleneglycol–N-hydroxysuccinimide monoacrylate and were subsequently grafted into the photopolymerized ABA hydrogel network. Hydrogel networks were formulated via photopolymerization with block copolymers, triethanolamine, and 2-2-dimethoxy-2-phenyl acetophenone photoinitiator in 1-vinyl-2-pyrrolidone (47). After the BAB (peptide–polyethyleneglycol–peptide)-containing hydrogels were hydrated to equilibrium, type IV collagenase, porcine pancreatic elastase, or plasmin at various concentrations was added to the hydrogel to commence degradation. Hydrogels were found to degrade in a protease-specific and dose-dependent fashion that increased with increasing treatment time. When the ABA (polyethyleneglycol–peptide–polyethyleneglycol)-containing hydrogel network grafted with the RGDs cell adhesion mediating motif was cultured with human dermal fibroblasts, cells were found to adhere and migrate throughout the hydrogel matrix. Since the polymer construct degrades enzymatically in the presence of a specific enzyme substrate, the degradation path within the hydrogel can be controlled spatially and temporally by varying the hydrogel chemical composition. These bioresponsive materials allow the study of cell migration mechanisms and function. Similar methodologies are currently been utilized by Hubbell et al. to investigate and to partly control the biomolecular mechanism of neurite extension and migration within a similar polyethyleneglycol-containing, bioresponsive hydrogel system immobilized with enzymatically degradable peptides (48,49).

In the development of biofunctional materials by mimicking ligand–receptor interactions, a clear understanding of the function–structure relationship between target proteins and cell membrane receptors is crucial. Furthermore, the ultimate challenge of eliciting specific cellular function by utilizing biofunctional materials lies within the normal host foreign body response, which might overcome the bioactive functionalities in the material intended to modulate cellular function. The major function of blood-derived phagocytic monocytes and macrophages is to mediate the normal host immune and inflammatory response against foreign objects such as microorganisms and biomaterials (1). Several characteristic macrophage functions are identified as critical events in the foreign body response. Macrophages recognize adsorbed proteins on the foreign surface via several ligand–receptor superfamilies (4). The process of adherent macrophage activation and fusion to form multinucleated foreign body giant cells (FBGC) is unique to the macrophage phenotype. The presence of FBGCs is utilized as a histopathology marker for chronic inflammation and the host foreign body reaction. Foreign body giant cells have been demonstrated on implanted biomaterials and the rate of material degradation underneath the giant cells has been shown to be markedly increased (50). However, the molecular mechanisms involved in FBGC formation remain unclear. Activated macrophages may release a variety of cytokines, growth factors, and other bioactive agents to modulate the function of other cell types in the inflammatory milieu (1). Several macrophage-derived cytokines and growth factors are utilized on a limited basis to alter the inflammatory response and promote the healing process. For example, IL1β and GM-CSF are given to injured patients and bone marrow transplant recipients to assist the recovery of hematopoiesis (42–44). Pharmaceutics based on the property, function, and structure of anti–tumor narcosis factor α are administered to patients suffering from chronic inflammation in order to enhance the wound healing process and attenuate inflammation (44). However, the long-term clinical efficacy of treatments involving the systemic administration of these bioactive agents has yet to be demonstrated. Hence, the localized delivery of selected cytokines and growth factors produced by endogenous inflammatory cells is desirable and may have potential therapeutic value in the fundamental processes of tissue healing, growth regulation, and biocompatibility.

Before the ability to control specific cellular function via biomaterials can be realized, a more detailed understanding of the interplay between material-bound ligands and receptors on the cell surface must be obtained. Furthermore, the redundancies that exist between receptors and target proteins must also be addressed. For instance, the tripeptide RGD sequence is found in several extracellular matrix proteins such as fibronectin, vitronectin, and fibrinogen. Several cell membrane receptors on different types of cells have been shown to complex with the RGD cell adhesion motif of these matrix proteins. For example, platelet surface glycoprotein IIb/IIIa recognizes the RGD sequence of fibrinogen, and integrin $\alpha_5\beta_1$ and $\alpha_{IIb}\beta_3$ receptors on macrophages also recognize the RGD motif of fibronectin. Other integrin receptors on a variety of cell types that recognize the RGD cell binding motif are $\alpha_{(2,3,4,5,7,8,v)}\beta_1$ and $\alpha_v\beta_{(1,3,5,6,8)}$. Some of the current biomaterial development and research utilize the RGD sequence to provide bioactive sites for cell adhesion. However, it is apparent that such a design strategy to elicit a specific cellular function is limited by the redundancies that exist between receptors and target proteins.

A clear elucidation of the mechanism involved in the interaction among cells, proteins/enzymes, materials, and various physiological parameters such as mechanical forces is critical for the development of biomimetic materials with specific cellular activity in the new field of cellular engineering (Fig. 3). The author acknowledges the critical role of this type of fundamental research at the molecular level between cells and material surfaces in the development of novel enabling technologies. The resulting technologies and know-how are vital for the future of tissue engineered products, the improvement of clinical biomaterials, and the construction of biofunctional materials.

Several methodologies were utilized to examine the interrelationship between material chemistry and cellular behavior at protein and cellular levels under in vitro and in vivo environments (1–10). Based on these findings, Kao et al. designed methodologies to probe the molecular mechanism of receptor–ligand recognition and postligation cellular function (51,52). In one of the studies, Kao et al. designed and employed biomimetic oligopeptides to probe the effect of primary and tertiary structures of a model protein (i.e., fibronectin) in ligand–receptor interaction (52). Synthetic peptides were formulated based on the primary and tertiary structures of human plasma fibronectin (53,54). Specifically, the amino acid sequences PHSRN (55) and RGD (56), which are located in the FIII-9 and the FIII-10 modules, respectively, of the central cell binding domain (Fig. 4) and the sequence PRRARV of the C terminal heparin binding domain of fibronectin (53) were chosen for exploration. RGD and PHSRN are present on adjacent loops of two connecting FIII modules and bind synergistically to the same integrin receptor (57). The heparin binding domain of fibronectin in which the sequence PRRARV resides also binds directly with integrin receptors (58,59); however, the precise mechanisms involved in this association remain unclear. The tertiary structure of fibronectin (60,61) was used as a guide in the design of peptides that included RGD and PHSRN. The distance between these sequences was approximated using the structural coordinates archived in the SwissProt Database (sequence FINC_HUMAN P02751). Based on the measurement, a hexamer of glycine (G_6) of approximately the same length was used to link the two bioactive sequences in both possible orientations. The combination of RGD and PRRARV was studied using a G_6 linker; although in this case the G_6 linker was not selected based on any structural considerations. A terminal trimeric glycine domain (G_3) was employed as a spacer in all peptides. Oligopeptides were synthesized using solid-resin methods with standard 9-fluorenylmethyloxycarbonyl chemistry (41). The following oligopeptides were synthesized: G_3RGDG, G_3PHSRNG, G_3PRRARVG, G_3RGDG$_6$PHSRNG, G_3PHSRNG$_6$RGDG, G_3RGDG$_6$PRRARVG, and G_3PRRARVG$_6$RGDG. A cyclic RGD peptide with an amino acid sequence of LNQEQVSPD(cRGDGRN) was utilized for comparison. LNQEQVSPD is a leader sequence and the cyclical RGDGRN sequence has been shown to bind with $\alpha_v\beta_3$ integrin with high specificity and affinity (62). Peptides were immobilized onto a polymer network terpolymerized with polyethyleneglycol, acrylic acid, and trimethylolpropanetriacrylate (51,52). The network had been shown to mediate low levels (<30 cells/mm^2) of adhesion by human macrophages, human neonatal foreskin dermal fibroblasts, and human umbilical vein endothelial cells in serum-containing culture medium up to 10 days. Macrophage ad-

Current Research:

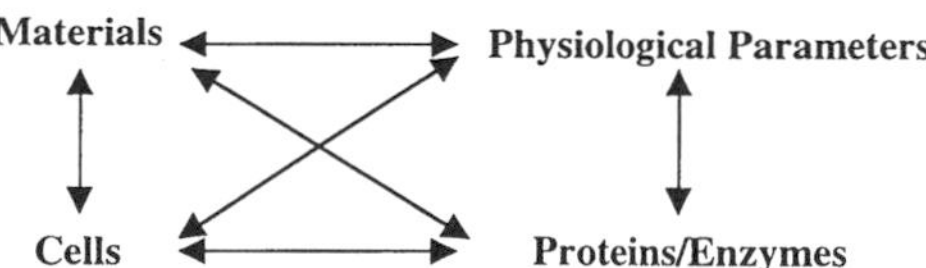

Future Research in "Cellular Engineering":

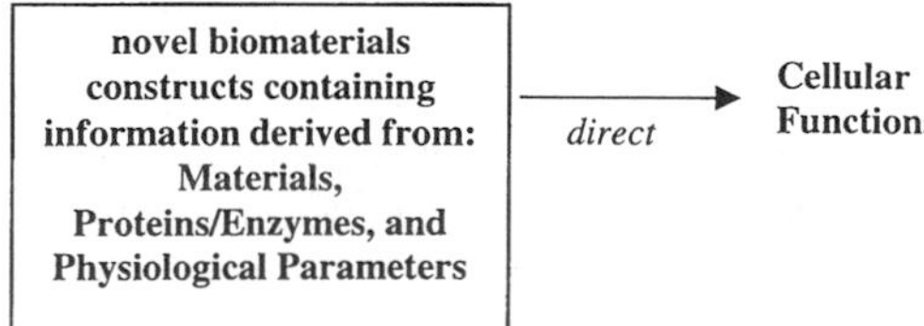

Figure 2.3 Current research focuses on elucidating the complex mechanisms in the interaction among cells, proteins/enzymes, materials, and various physiological parameters such as mechanical forces. Fundamental research at the molecular level results in the development of novel ''enabling technologies'' that are critical for the development of biomimetic materials with specific cellular activity in the new field of ''cellular engineering.'' Manipulation of cellular function is vital for the future of tissue-engineered products, the improvement of clinical biomaterials, and the pursuit of novel biofunctional materials.

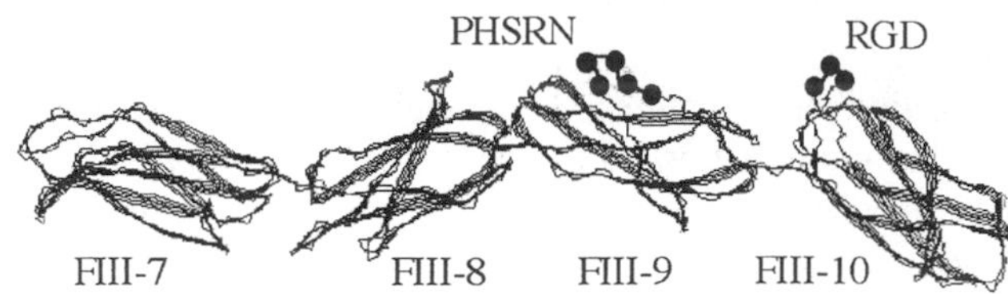

Figure 2.4 Amino acid sequences PHSRN and RGD (in ball-and-stick model) located in the FIII-9 and the FIII-10 modules, respectively, of the central cell-binding domain of human plasma fibronectin.

hesion and FBGC formation in vitro assays (63) were performed. Under the FBGC culture condition described, FBGCs containing up to 50 nuclei/cell formed consistently on tissue-culture polystyrene control surfaces. Briefly, freshly isolated human blood-derived monocytes were incubated with samples in RPMI containing 20% autologous serum. At 96 and 168 h, the medium was changed to RPMI with heat-inactivated autologous serum + 10 ng/mL of recombinant human interleukin-4 + 5 ng/mL of recombinant human GM-CSF. Each adherent cell with three or more nuclei per cell was defined as a FBGC. Competitive inhibition studies utilizing soluble free peptides and neutralizing antibodies were performed to ascertain ligand–receptor specificity and identification. Kao et al. (52) found that serum fibronectin modulated macrophage adhesion and the extent (i.e., size) of FBGC formation on control surfaces in the presence of serum proteins. Macrophages adhered to all peptide-grafted polymer networks with relatively subtle differences between adhesion mediated by RGD, PHSRN, PRRARV, or combinations thereof (Table 1) (52). β_1 integrin subunit was essential in macrophage adhesion to peptide-grafted networks in a receptor–peptide specific manner; whereas, β_3 integrin subunit was less important. Macrophage adhesion to PRRARV was mediated primarily by the direct interaction with integrins. RGD or PHSRN alone did not provide an adequate substrate for macrophage fusion to form FBGCs. However, the PHSRN synergistic site and the RGD site in a single oligopeptide provided a substrate for FBGC formation that was statistically comparable to that on the positive control material in the presence of serum proteins (Table 2) (52). This response was highly dependent upon the relative orientation between RGD and PHSRN. PRRARV alone or in tandem with RGD in a single peptide formulation did not support FBGC formation. Furthermore, neutralizing antibody was utilized in the FBGC culture assay to partly determine the role of integrins and fibronectin-derived oligopeptides in modulating the function of adherent macrophages to fuse and form FBGCs. Antibody isotype negative controls were also utilized for confirmation. Both integrin β_3 and β_1 subunits were found to play an importance role in mediating FBGC formation by macrophages adhered on peptide-grafted networks (Table 3) (52). Results showed that no antibody– and anti-β_3 neutralizing antibody–treated groups had a comparable FBGC density on TCPS (Table 3). On most of the peptide-grafted network, no FBGC formation was observed in the anti-β_3 antibody–treated group. One notable exception was the peptide that contains the PHSRN and the RGD domain in the optimal orientation, namely $G_3PHSRNG_6RGDG$, on which anti–integrin β_3 reduced FBGC formation by about 70%. When anti-β_1 neutralizing antibody was utilized, no FBGC formation was observed on any sample. Macrophages and FBGCs express α_2, α_4, β_1, and β_3 integrin subunits. It is also known that the first 160 amino acids of integrin β_1, specifically the α-helix formed by residues 141 to 160, are critical in integrin–ligand recognition. These findings suggest that the association between this region of the integrin β_1 receptor and $G_3PHSRNG_6RGDG$, but not G_3RGDG_6P HSRNG, results in the necessary binding characteristic that determine the subsequent cellular event leading to FBGC formation. Activated integrin β_1 or β_3 intracellular domains stimulate cell migration, modulate proliferation

Table 2.1 Adherent Macrophage Density on Polymer Networks Grafted with Fibronectin-Derived Biomimetic Peptides

Peptide	Culture time (h) 24	168
G_3RGDG	378 ± 59[a]	155 ± 83[a]
$G_3PHSRNG$	243 ± 30[a,b]	121 ± 70[a]
$G_3PRRARVG$	304 ± 46[a,b]	121 ± 48[a]
$G_3RGDG_6PHSRNG$	271 ± 28[a,b]	92 ± 25[a]
$G_3PHSRNG_6RGDG$	309 ± 34[a,b]	97 ± 29[a]
$G_3RGDG_6PRRARVG$	311 ± 39[a,b]	88 ± 30[a]
$G_3PRRARVG_6RGDG$	151 ± 24[a,b]	62 ± 24[a,b]
LNQEQVSPD(cRGDGRN)	290 ± 83[a,b]	220 ± 140[a]
No grafted peptides	69 ± 37[b]	24 ± 14[b]
Tissue-culture polystyrene	400 ± 49[a]	131 ± 22[a]

Note: All values ×10 macrophage/mm^2, mean ± s.e.m., n = 3 to 5.
[a] Value different at 95% confidence level ($p < 0.05$) versus no grafted peptide controls.
[b] Value different at 95% confidence level ($p < 0.05$) versus tissue-culture polystyrene controls.

Table 2.2 Macrophage-Derived FBGC Adherent Density and Size on Polymer Networks Grafted with Fibronectin-Derived Biomimetic Peptides at 240 h of the FBGC Assay

Peptide	FBGC/mm^2	Average FBGC Size (×100 μm^2)
G_3RGDG	18 ± 9	21 ± 8
$G_3PHSRNG$	14 ± 9	27 ± 9
$G_3PRRARVG$	15 ± 9	17 ± 4
$G_3RGDG_6PHSRNG$	16 ± 7	19 ± 5
$G_3PHSRNG_6RGDG$	88 ± 38[a]	32 ± 13
$G_3RGDG_6PRRARVG$	14 ± 7	20 ± 8
$G_3PRRARVG_6RGDG$	17 ± 9	19 ± 7
LNQEQVSPD(cRGDGRN)	20 ± 11	18 ± 5
Tissue-culture polystyrene	60 ± 20	162 ± 38

Note: All values expressed in mean ± s.e.m., n = 3 to 6.
[a] Values comparable ($p > 0.95$) versus that of tissue-culture polystyrene controls.

Table 2.3 Macrophage-Derived FBGC Adherent Density on Polymer Networks Grafted with Fibronectin-Derived Peptides at 240 h of FBGC Culture

Peptides	No antibody treatment[a]	Anti-integrin β_3 treatment	Anti-integrin β_1 treatment
G_3RGDG	18 ± 9	0 ± 0	0 ± 0
G_3PHSRNG	14 ± 9	0 ± 0	0 ± 0
G_3PRRARVG	15 ± 9	0 ± 0	0 ± 0
G_3RGDG$_6$PHSRNG	16 ± 7	0 ± 0	0 ± 0
G_3PHSRNG$_6$RGDG	88 ± 38	30 ± 30	0 ± 0
G_3RGDG$_6$PRRARVG	14 ± 7	0 ± 0	0 ± 0
G_3PRRARVG$_6$RGDG	17 ± 9	0 ± 0	0 ± 0
LNQEQVSPD(cRGDGRN)	20 ± 11	0 ± 0	0 ± 0
Tissue-culture polystyrene	60 ± 20	90 ± 30	0 ± 0

Notes: All values expressed in FBGC/mm^2, mean ± s.e.m., $n = 3$ to 6. Cells treated with anti-integrin antibody. To determine the role of integrins in mediating FBGC formation, neutralizing antibody [anti–human integrin β_1 (JB1a, purified IgG isotype ascite) or anti–human integrin β_3 (25E11, purified IgG2a isotype ascite) neutralizing antibody at 60 μg/mL] was added to the culture at 96 and 168 h of the FBGC assay.

[a] All values were significantly lower ($p < 0.05$) than that of tissue-culture polystyrene and networks grafted with G_6PHSRNG$_6$RGDG.

and gene expression, induce the assembly of F-actin cytoskeleton, and localize the activity of focal adhesion kinase pp125^{FAK}. These cellular events may contribute to the process of FBGC formation; however, the exact interrelationship between ligand–receptor architecture and association in activating intracellular signaling events resulting in the control of cellular behavior remains unclear.

To yield insights into the mechanisms coordinated by the interaction between integrins and fibronectin in mediating macrophage adhesion and FBGC formation, Kao et al. utilized the aforementioned biomimetic fibronectin-derived oligopeptides to probe the structure–function characteristic and signaling pathways of fibronectin–integrin association in modulating cellular function. Specifically, the key role played by RGD, PHSRN, and the spacing and orientation between the peptide sequences in modulating macrophage and FBGC behavior was investigated. Freshly isolated human blood-derived monocytes were preincubated with inhibitors of various signaling molecules at several concentrations to screen candidate signaling cascades in regulating macrophage behavior mediated by fibronectin and fibronectin-derived biomimetic oligopeptides. The signaling and transcriptional events and the corresponding inhibitor chosen for exploration included activated protein tyrosine kinases inhibitor AG82, lavendustin A, which inhibits Src-family kinases, activated protein serine/threonine kinases inhibitor H-7, protein kinase-A inhibitor 14-22 amide, PI-3K inhibitor wortmannin, PCK inhibitor EGF-R fragment, MAPK inhibitor PD98059, and NFκB inhibitor PSI. Activated integrin receptors have been shown to upregulate these selected signaling molecules under a variety of ligand–receptor associations. For example, Src is involved in integrin signaling upon ligation with extracellular matrix proteins, such as fibronectin, fibrinogen, or vitronectin, leading to macrophage adhesion and focal adhesion kinase formation. Treated cells were incubated with tissue-culture polystyrene or networks terpolymerized with polyethyleneglycols, acrylic acid, and trimethylolpropanetriacrylate. Tissue-culture polystyrene- and polyethyleneglycol-based networks were preadsorbed or immobilized with recombinant human fibronectin or fibronectin-derived biomimetic oligopeptides containing RGD and/or PHSRN domains. Surfaces without preadsorption nor immobilization were employed as negative surface controls. Inhibitor vehicle was also utilized as additional controls, and bovine serum albumin was employed as controls for fibronectin and peptides. From cell adhesion assays, the results showed the following. On networks immobilized with fibronectin-derived oligopeptides, macrophage adhesion was found to be independent of protein tyrosine kinases and protein serine/threonine kinases at 24, 48, and 120 h, except that on surfaces grafted with G_3PHSRNG and G_3PHSRNG$_6$RGDG where cell adhesion was dependent upon protein serine/threonine kinases at 48 h (Table 4). On networks immobilized with fibronectin or albumin, macrophage adhesion was found to be dependent of protein tyrosine kinases but not Src and dependent of protein serine/threonine kinases but not protein kinase-A at 24 and 48 h. Furthermore, the promotion of protein serine/threonine kinases, and specifically protein kinase-C, compen-

Table 2.4 Macrophage Adhesion on Polymer Networks and TCPS Immobilized with Proteins or Fibronectin-Derived Oligopeptides in Vitro

	24 h			48 h			120 h		
Networks/TCPS	Control	AG82	H7	Control	AG82	H7	Control	AG82	H7
Fibronectin	11 ± 5	4 ± 2[a]	3 ± 1[a]	5 ± 3	1 ± 1[a]	1 ± 1[a]	8 ± 6	2 ± 0	2 ± 0
Albumin	11 ± 3	5 ± 3[a]	4 ± 1[a]	5 ± 3	1 ± 1[a]	2 ± 1[a]	6 ± 4	3 ± 1	2 ± 1
G_3RGDG	6 ± 3	3 ± 1	2 ± 0	3 ± 2	2 ± 1	1 ± 0	7 ± 6	2 ± 2	1 ± 0
G_3PHSRNG	6 ± 4	3 ± 1	2 ± 1	4 ± 1	1 ± 1	1 ± 0[a]	7 ± 5	3 ± 0	1 ± 1
G_3RGDG$_6$PHSRNG	5 ± 3	2 ± 1	2 ± 1	6 ± 2	2 ± 1	2 ± 1	5 ± 2	1 ± 1	1 ± 1
G_3PHSRNG$_6$RGDG	6 ± 5	3 ± 1	2 ± 1	3 ± 1	2 ± 1	1 ± 1[a]	6 ± 4	2 ± 0	0 ± 0
TCPS	10 ± 6	6 ± 3	4 ± 2	7 ± 3	5 ± 3	3 ± 2[a]	4 ± 1	3 ± 1[a]	1 ± 1
TCPS + fibronectin	12 ± 4	7 ± 2[a]	6 ± 4[a]	10 ± 3	4 ± 2[a]	4 ± 3[a]	5 ± 2	3 ± 1	2 ± 0
TCPS + albumin	8 ± 4	7 ± 4	5 ± 2	11 ± 4	3 ± 2[a]	2 ± 1[a]	5 ± 1	3 ± 2	1 ± 0

Note: All values expressed in macrophage ×100/mm^2, mean ± sem, n = 3 to 6.
[a] $p < 0.1$ versus respective values of controls. AG82 inhibits PTK and H7 inhibits PSK. Tissue-culture polystyrene (TCPS).

sated protein tyrosine kinases inhibition in mediating macrophage adhesion at 24 h (Table 5). However, macrophage adhesion on networks immobilized with fibronectin or albumin was found to be independent of both protein tyrosine and serine/threonine kinases at 120 h. On tissue-culture polystyrene immobilized with fibronectin, macrophage adhesion was found to be dependent of protein tyrosine kinases but not Src and dependent of protein serine/threonine kinases but not protein kinase-A at 24 h. The promotion of protein serine/threonine kinases, and specifically protein kinase-C, did not compensate protein tyrosine kinases inhibition in mediating macrophage adhesion at 24 and 48 h. These results indicate that the crosstalk between protein tyrosine and serine/threonine kinases is different between adherent macrophages on fibronectin that was immobilized onto networks or tissue-culture polystyrene. Furthermore, macrophage adhesion on fibronectin-immobilized tissue-culture polystyrene was found to be dependent of protein tyrosine kinases but not Src and dependent of protein serine/threonine kinases, specifically protein kinase-C but not protein kinase-A, by 48 h. Cell adhesion on fibronectin-immobilized tissue-culture polystyrene was found to be independent of protein tyrosine and serine/threonine kinases by 120 h. Assays for FBGC demonstrated the following. On networks immobilized with fibronectin, FBGC formation was dependent of protein serine/threonine kinases, specifically protein kinase A, but was independent of Src (Table 6). It should be noted that macrophage adhesion was independent of protein kinase-A by 48 h and was independent of both protein tyrosine and serine/threonine kinases by 120 h. On tissue-culture polystyrene immobilized with fibronectin, FBGC formation was found to be independent of both protein tyrosine and serine/threonine kinases. Macrophage adhesion on tissue-culture polystyrene immobilized with fibronectin was independent of protein tyrosine and serine/threonine kinases by 120 h. Taken together, these results support the role of activated protein tyrosine kinases and serine/threonine kinases in integrin signalings leading to macrophage adhesion mediated by fibronectin–integrin association. Furthermore, RGD and PHSRN appears to be significant in mediating this receptor–ligand association resulting in the necessary signaling characteristic for macrophage adhesion and the subsequent development. Specifically, pro-

Table 2.5 Macrophage Adhesion on Networks and TCPS Immobilized with Proteins in Vitro

	2 h					24 h				
Networks/TCPS	Control	AG82	H7	AG82 + H7	AG82 + PMA	Control	AG82	H7	AG82 + H7	G82 + PMA
Fibronectin	16 ± 5	10 ± 7[a]	8 ± 5[a]	5 ± 0[a]	2 ± 1[a]	11 ± 5	4 ± 2[a]	3 ± 1[a]	3 ± 1[a]	5 ± 1
TCPS	12 ± 8	11 ± 1	13 ± 6	4 ± 1	6 ± 2	10 ± 6	6 ± 3	4 ± 2	5 ± 1[a]	8 ± 3
TCPS + FN	14 ± 6	9 ± 3	11 ± 2	3 ± 1[a]	4 ± 1[a]	12 ± 4	7 ± 2[a]	6 ± 4	4 ± 0[a]	5 ± 1[a]

Note: All values expressed in macrophage ×100/mm^2, mean ± sem, n = 3 to 6.
[a] $p < 0.1$ versus respective values of controls. AG82 inhibits protein tyrosine kinases, H7 inhibits protein serine/threonine kinases, and PMA promotes protein kinase-C of protein serine/threonine kinases.

Table 2.6 FBGC Formation on Networks and TCPS Immobilized with Fibronectin by 240 h of FBGC Assay in Vitro

Networks/TCPS	Control	Lavendustin A	PKI
Fibronectin	39 ± 16	21 ± 4	26 ± 17[a]
TCPS	25 ± 18	23 ± 15	23 ± 16
TCPS + fibronectin	9 ± 5	7 ± 2	17 ± 13

Note: All values expressed in FBGC/mm^2, mean ± sem, $n = 3$ to 6.
[a] $p < 0.1$ versus respective values of controls. Lavendustin A inhibits Src of protein tyrosine kinases and PKI inhibits protein kinase-A of protein serine/threonine kinases.

tein kinase-C of tyrosine kinases showed a significant role in modulating FBGC formation mediated by fibronectin, most importantly the RGD and PHSRN domains of fibronectin. Kao et al. showed the important role of RGD and PHSRN domains of fibronectin, and specifically the interpositional spacing between the motifs, in the complexation with integrin receptors to upregulate tyrosine kinases and serine/threonine kinases in mediating macrophage adhesion and FBGC formation. These findings represent a mechanistic correlation between the role of protein functional architectures in ligand–receptor recognition and the postligation signaling events that control cellular behavior. The fundamental understanding of these complex phenomena provides future researchers with necessary tools in the development of unique biomimetic enabling technologies that are vital for the advancement of cellular engineering and tissue engineered devices.

The incorporation of biospecific and biofunctional peptides into polymeric networks has been adopted in other systems in the management of various pathological conditions. Similar to the strategy outlined above by Kao et al., Healy et al. grafted RGD and FHRRIKA (putative heparin-binding) peptides onto interpenetrating polymer networks containing poly(acrylamide) and polyethyleneglycol (64–66). The resulting polymer network containing both RGD and FHRRIKA peptides mediated extensive adhesion, spreading, and mineralization of the extracellular matrix by human osteoblast–like cells when compared to that of homogenous peptide surfaces and controls. The cellular response was found to rely on the participation of integrin α_2, β_1, and α_v subunits in a temporally dependent manner. Specifically, $\alpha_2\beta_1$ and $\alpha_v\beta_3$ integrin receptors mediated the initial cell adhesion; whereas, only $\alpha_v\beta_3$ integrin receptors governed cell adhesion at a longer culture time. The cited investigations illustrate the diversity and capability of a biomimetic approach in the construction of biofunctional materials to study fundamental cell biology and to potentially manipulate local host environment for biomedical applications.

IV. PROTEIN MATRIX AND SUBCELLULAR BIOMIMETICS

In previous sections, examples were given to illustrate the utilization of macromolecules and proteins as templates to develop biomimetic molecules and novel materials. However, biological systems are composed hierarchically; hence, the impact of biological organization at the submicron and subcellular scales needs further elucidation. For example, most cells are intimately associated with basement membrane matrices that have a complex three-dimensional nanoarchitecture of 100–500 nm features. Hence, this submicron architecture may play an important role in mediating cell adhesion and function. A point of note, cellular focal adhesion sites are about 250 nm in size. Various three-dimensional fabrication methods such as inkjet, molding, impregnation, and laser perforation have been developed in an attempt to construct or reconstruct the unique topographical feature of protein layers, extracellular membranes, or extracellular matrices with high resolution and accuracy. Such topographical biomimetics can be utilized in tandem with current clinically utilized materials or novel biofunctional materials as a part of the construction of fully biomimetic material constructs (67–69). Goodman et al. utilized the inherent three-dimensional confinement of nonlinear photo-optical processes to develop three-dimensional freeform fabrication of proteins or synthetic polymers by conventional photochemically induced polymerization methods (70–73). A laser scanning confocal microscope was modified for near-infrared excitation to direct protein crosslinking by 2-photon photoactivation. Albumin, fibrinogen, and alkaline phosphatase were chosen as model proteins for fabrication in solution. Rose bengal was employed as the photoactivator. At the focal point of the lens where the photon density is sufficiently high for 2-photon excitation, crosslinking was induced by a mechanism where the photoactivator directly abstracts hydrogen atoms from protein molecules. Fabrication was then directed point by point and layer by layer within the solution and proceeded to achieve a desired, predetermined geometry and topography. Features of less than 200 nm, such as three-dimensional lattices, were synthesized by varying the type of lenses. The bioactivity of model proteins was maintained as determined by various biological functional assays. Goodman et al. demonstrated that human platelets extensively adhered to lines fabricated from fibrinogen molecules in vitro, and the lines were tenfold narrower than the diameter of a platelet cell body. A minimal level of platelet spreading was observed on substrates fabricated from the native albumin molecule in vitro. Furthermore, micron and submicron structures fabricated from alkaline phosphatase molecules demonstrated a high level of enzymatic activity when assayed with enzyme linked fluorescence imaging in vitro. These examples illustrate the feasibility of mimicking the complex hierarchical structure of biological systems that expands from the nano scale to subcellular levels in the study and the partial control of biological functions.

V. BIOMIMETIC-DERIVED MATERIAL PROCESSING METHODS

Currently, biomaterial development and synthesis have been focused on the utilization of a myriad of synthetic methodologies to obtain a wide range of polymers, ceramics, metal alloys, or biologically derived materials with a diverse physicochemical bulk and surface property. However, synthetic methodologies often result in heterogeneity and anisotropy (i.e., polydispersive molecular weight of polymers, large range of size distributions of polymer and ceramic particulates, phase separation of chemical and/or physical properties). Furthermore, current synthetic fabrication methods cannot incorporate or mimic the complex multilayer microstructure of biological systems to produce materials containing controlled microarchitecture and multifunctional properties present in biological materials, such as cellular membranes, extracellular matrices, or tissues. Several researchers are developing novel material synthesis and processing techniques to enhance the limitation of conventional synthesis and fabrication procedures.

By studying the complex interrelationship of structural organizations in biological systems, biomaterial researchers such as Baer and Hiltner pioneered the development of complex organic and inorganic composites that incorporated some of the ordered microstructure found in nature (32,74–76). For example, load-baring tissues such as tendon, comprise fibrous organic components embedded in a soft organic matrix that are analogous to fiber- or particle-reinforced polymeric composites. Baer et al. elucidated that tendon has six discrete levels of structure organized in a hierarchical manner from the molecular to the centimeter scale. Clearly, current composite materials lack the complexity and sophistication necessary to achieve highly ordered and multileveled hierarchical structure. Hence, composite material research has focused on elucidating the microstructure–property correlation that includes the coupling between the properties of an individual organic and inorganic component at various scales and in tandem at different levels of organization. The understanding of the structural organization and relationship between each discrete unit (i.e., molecules, macromolecules, cells, tissues, or extracellular matrix) that makes up a biological system is one important step in understanding basic structure–function relationships and the construction of complex biomimetic materials.

Due to the specificity of biological synthesis pathways such as biomineralization systems and protein and nucleic acid syntheses, microbiological and other biological techniques are currently being adopted and utilized in the construction and processing of novel biomimetic materials (77–83). For example, analysis of a variety of mineralizing biosystems leads to several guiding principles that have significant implications for biology and materials sciences. Biomineralization occurs within specific subunit compartments or microenvironments. The process produces specific minerals with defined crystal size, orientation, and ordered macroscopic growth. Many incremental units are packaged together to form unique composites (76). These complex systems are controlled by the structure and chemistry of the interfaces between the organic substrate, mineral, and medium. The presence of proteins such as osteopontin and catalysts is also critical for the control of biomineralization. Some examples of novel biomaterials synthesized by mimicking biomineralization systems include monolayer films, self-assembled monolayer films, self-assembled amphiphilic structures, and supramolecular templates (76). However, low stability, lack of control over local ordering, and inability to direct crystal growth in three dimensions are some of the existing technological hurdles that have yet to be overcome. One exciting area of research attempts to utilize microorganisms as a mean to synthesize calcium phosphate nanoparticles of a precise size distribution (84). Nanoparticles about 50 nm in diameter with a uniformed crystal size and structure were synthesized on the extracellular membrane of bacteria of the *Citrobacter* genus in the presence of calcium and phosphate ions. The *Citrobacter* genus contains membrane-bound phosphatase enzymes and has been utilized in the water purification process through the accumulation of heavy metal ions as cell-bound insoluble metal phosphates. The nature of the phosphate formed depends on the type of ions in the culture media. In the presence of calcium and phosphate ions in the culture medium, investigators showed that the crystals formed at the bacteria extracellular membrane were amorphous calcium phosphates produced by membrane-bound phosphatase enzymes via similar mechanisms as those involved in the extraction of heavy metals from water. The 50-nm particles within the bacteria were of the composition of $CaHPO_4$ and were produced through migration of calcium through the cell membrane. The exact mechanism of the $CaHPO_4$ transport across the extracellular membrane remains unclear. Nanoparticles were analyzed and characterized by Fourier transform infrared spectroscopy, x-ray diffraction, and scanning and transmission electron microscopy. $CaHPO_4$ can be further processed to hydroxyapatite and then the bacteria represents a self-constructed soft agglomerate of the particles. Hence, the utilization of *Citrobacter* provides a new approach to produce well-controlled nanoparticulates of optimal characteristics that are hydroxyapatite precursor powders. The use of the natural function of microorganisms in fabricating materials represents a unique ability of novel material synthesis and processing systems based on biomimetism.

As another example of utilizing the specificity and accuracy of biological systems to synthesize complex materials, genes that contain the instruction for the synthesis of desired macromolecules such as proteins and carbohydrates were incorporated into vectors that were utilized to transfect candidate cells such as bacteria. The new genetic code becomes a part of the genetic and biochemical machinery of the transfected cell. The marriage of microbiology and biomaterial development was pioneered by Tirrell et al. and continued in several other laboratories (29–31). The unique structural and functional properties of biological molecules such as proteins have been an area of intense investigation as potential building blocks for novel biomaterials that contain the multistructure and multifunctional characteristics of native biological molecules. For example, Tirrell et al. developed a series of methods for building proteinlike structural motifs that incorporate sequences of biological origin (28). In one of their studies, Tirrell et al. attached a lipophilic moiety onto the α-amine of a peptide chain resulting in a peptide–amphiphile. The alignment of amphiphilic compounds at the lipid–solvent interface was employed to facilitate peptide alignment and structure initiation and propagation while the lipophilic region adsorbed to hydrophobic surfaces. Peptide–amphiphiles containing triple-helical or α-helical structural motifs were synthesized and had been shown to be stable over physiological temperature ranges. These triple-helical peptide–amphiphiles were adopted to study surface modification and cell receptor binding.

VI. SUMMARY

As illustrated by the investigations discussed, the underlying basis of biomimetics is the need to better understand the structural and functional relationship of biological systems and organizations. Hence, the field of biomimetics represents a diverse and interdisciplinary approach to formulate functional compounds and materials with enhanced physicochemical and biological properties when compared to these synthesized via conventional formulation methodologies. Furthermore, the biomimetic approach offers unique tools that can, in time, be utilized to probe the fundamental questions of the complex biological systems and mechanisms.

ACKNOWLEDGMENTS

The author thanks the National Institutes of Health (HL 63686), the Whitaker Foundation (RG99-0285), the New Investigator Program Award of AACP, the Burroughs Wellcome Fund and the American Foundation for Pharmaceutical Education, and the American Cancer Society (IRG-58011423). The author also thanks Drs. Y. P. Liu, D. Lok, J. Li, and D. Lee of the University of Wisconsin—Madison for technical assistance. As the session organizer, the author also thanks the Society for Biomaterials (USA) and the presenters for establishing and participating in the Biomimetic Session of the 6th World Biomaterials Congress, May 15–20, 2000, in Kamuela, HI.

REFERENCES

1. J. M. Anderson. Mechanisms of inflammation and infection with implanted devices. *Cardiovasc. Pathol.* 2(3):33S–41S (1993).
2. J. M. Anderson. Inflammatory response to implants. *ASAIO* 11:101–107 (1988).
3. N. P. Ziats, K. M. Miller, J. M. Anderson. *In vivo* and *in vitro* interaction of cells with biomaterials. *Biomaterials* 9: 5–13 (1988).
4. R. Rudolph, D. Cheresh. Cell adhesion mechanisms and their potential impact on wound healing and tumor control. *Clin. Plast. Surg.* 17(3):457–462 (1990).
5. W. J. Kao, Q. H. Zhao, A. Hiltner, J. M. Anderson. Theoretical analysis of *in vivo* macrophage adhesion and foreign body giant cell formation on poly-dimethylsiloxane, low density polyethylene, and polyetherurethanes. *J. Biomed. Mater. Res.* 28:73–79 (1994).
6. S. S. Kaplan, R. E. Basford, E. Mora, M. H. Jeong, R. L. Simmons. Biomaterial-induced alterations of neutrophil su-

peroxide production. *J. Biomed. Mater. Res.* 26:1039–1051 (1992).

7. N. P. Ziats, D. A. Pankowsky, B. P. Tierney, Q. D. Ratnoff, J. M. Anderson. Adsorption of Hageman factor (factor XII) and other human plasma proteins to biomedical polymers. *J. Lab. Clin. Med.* 116(5):687–696 (1990).
8. S. Sapatnekar, W. J. Kao, J. M. Anderson. Leukocyte–biomaterial interactions in the presence of *Staphylococcus epidermidis*: flow cytometric evaluation of leukocyte activation. *J. Biomed. Mater. Res.* 15(35,4):409–420 (1997).
9. L. Tang, J. W. Eaton. Inflammatory responses to biomaterials. *Am. J. Clin. Pathol.* 103(4):466–471 (1995).
10. W. J. Kao, J. M. Anderson. The cage implant testing system, In: *Handbook of Biomaterials Evaluation: Scientific, Technical, and Clinical Testing of Implant Materials*, 2nd ed., A. F. von Recum (Ed.). Taylor & Francis, Philadelphia, 1999, pp. 659–671.
11. S. M. Kupchan, R. M. Schubert. Selective alkylation: a biomimetic reaction of the antileukemic triptolides. *Science* 185(153):791–793 (1974).
12. W. S. Johnson, G. E. DuBois. Biomimetic polyene cyclizations. Asymmetric induction by a chiral center remote from the initiating cationic center. 11alpha-methylprogesterone. *J. Am. Chem. Soc.* (letter) 98(4):1038–1039 (1976).
13. W. J. Johnson. Biomimetic polyene cyclizations. *Angew. Chem. Int. Ed. Engl.* 15(1):9–17 (1976).
14. W. S. Johnson, S. Escher, B. W. Metcalf. A stereospecific total synthesis of racemic 11alpha-hydroxyprogesterone via a biomimetic polyene cyclization. *J. Am. Chem. Soc.* (letter) 98(4):1039–1040 (1976).
15. C. R. Hutchinson, M. T. Hsia, A. H. Heckendorf, G. J. O'Loughlin. A biomimetic synthesis of the camptothecin chromophore. *J. Org. Chem.* (letter) 41(21):3493–3494 (1976).
16. P. W. Le Quesne, A. V. Durga, V. Subramanyam, A. H. Soloway, R. W. Hart, R. H. Purdy. Biomimetic synthesis of catechol estrogens: potentially mutagenic arene oxide intermediates in estrogen metabolism. *J. Med. Chem.* 23(3): 239–240 (1980).
17. G. E. Evans, M. J. Garson, D. A. Griffin, F. J. Leeper, J. Stauton. Biomimetic syntheses of phenols from polyketones. *Ciba Found. Symp.* 53:131–147 (1978).
18. T. Ogawa, T. Kawano, M. Matsui. A biomimetic synthesis of (+)-biotin from D-glucose. *Carbohydr. Res.* 57:C31–C35 (1977).
19. T. W. Kensler, D. M. Bush, W. J. Kozumbo. Inhibition of tumor promotion by a biomimetic superoxide dismutase. *Science* 221(4605):75–77 (1983).
20. P. A. Egner, T. W. Kensler. Effects of a biomimetic superoxide dismutase on complete and multistage carcinogenesis in mouse skin. *Carcinogenesis* 6(8):1167–1172 (1985).
21. S. Yamamoto, T. Nakadate, E. Aizu, R. Kato. Anti-tumor promoting action of phthalic acid mono-n-butyl ester cupric salt, a biomimetic superoxide dismutase. *Carcinogenesis* 11(5):749–754 (1990).
22. F. Guzman, M. Cain, P. Larscheid, T. Hagen, J. M. Cook, M. Schweri, P. Skolnick, S. M. Paul. Biomimetic approach to potential benzodiazepine receptor agonists and antagonists. *J. Med. Chem.* 27(5):564–570 (1984).
23. I. Berner, P. Yakirevitch, J. Libman, A. Shanzer, G. Winkelmann. Chiral linear hydroxamates as biomimetic analogues of ferrioxamine and coprogen and their use in probing siderophore-receptor specificity in bacteria and fungi. *Biol. Met.* 4(3):186–191 (1991).
24. S. Zbaida, R. Kariv. Biomimetic models for monooxygenases. *Biopharm. Drug Dispos.* 10(5):431–442 (1989).
25. S. Sasaki, K. Koga. Biomimetic studies using artificial systems. IV. Biomimetic peptide synthesis by using multifunctionalized crown ethers as a novel enzyme model: a new concept in mimicking of enzyme-catalyzed bond-forming reactions. *Chem. Pharm. Bull.* (Tokyo). 37(4):912–919 (1989).
26. F. Barrere, P. Layrolle, C. A. Van Blitterswijk, K. De Groot. Biomimetic calcium phosphate coatings on Ti6-Al4V: a crystal growth study of octacalcium phosphate and inhibition by Mg^{2+} and $HCO3^{-}$. *Bone* 25(2, suppl.):107S–111S (1999).
27. S. R. Winn, J. M. Schmitt, D. Buck, Y. Hu, D. Grainger, J. O. Hollinger. Tissue-engineered bone biomimetic to regenerate calvarial critical-sized defects in athymic rats. *J. Biomd. Mater. Res.* 45:414–421 (1999).
28. G. B. Fields, J. L. Lauer, Y. Dori, P. Forns. Y. C. Yu, M. Tirrell. Proteinlike molecular architecture: biomaterial applications for inducing cellular receptor binding and signal transduction. *Biopolymers (Peptide Science)* 47:143–151 (1998).
29. D. A. Tirrell. Polymer synthesis. For the living there is hope. *Nature* 390(6658):336–337,339 (1997).
30. J. C. van Hest, D. A. Tirrell. Efficient introduction of alkene functionality into proteins *in vivo*. *FEBS Lett.* 428(1-2):68–70 (1998).
31. W. A. Petka, J. L. Harden, K. P. McGrath, D. Wirtz, D. A. Tirrell. Reversible hydrogels from self-assembling artificial proteins. *Science* 281(5375):389–392 (1998).
32. M. Sarikaya. An introduction to biomimetics: a structural viewpoint. *Microscopy Res. Tech.* 27:360–375 (1994).
33. T. Ohse, H. Kawakami, S. Nagaoka. Superoxide scavenging by metalloporphyrin derivatives with SOD activity. *J. Biomater. Sci. Polym. Edn.* (in press, 2000).
34. W. J. Kao. Evaluation of protein-modulated macrophage behavior on biomaterials:designing biomimetic materials for cellular engineering. *Biomaterials* 20(23-24):2213–2221 (1999).
35. D. Boraschi, P. Bossu, G. Macchia, P. Ruggiero, A. Tagliabue. Structure–function relationship in the IL-1 family. *Frontiers in Bioscience* 1:d270–d308 (1996).
36. C. A. Dinarello. Biological basis for interleukin-1 in disease. *Blood* 87(6):2095–2147 (1996).
37. G. P. A. Vigers, L. J. Anderson, P. Caffes, B. J. Brandhuber. Crystal structure of the type-I interleukin-1 receptor complexed with interleukin-1beta. *Nature* 386:190–194 (1997).
38. H. Schreuder, C. Tardif, S. Trump-Kallmeyer, A. Soffientini, E. Sarubbi, A. Akeson, T. Bowlin, S. Yanofsky, R. W.

Barrett. A new cytokine-receptor binding mode revealed by the crystal structure of the IL-1 receptor with an antagonist. *Nature* 386:194–200 (1997).
39. R. J. Evans, J. Bray, J. D. Childs, G. P. Vigers, B. J. Brandhuber, J. J. Skalicky, R. C. Thompson, S. P. Eisenberg. Mapping receptor binding sites in interleukin (IL)-1 receptor antagonist and IL-1 beta by site-directed mutagenesis: identification of a single site in IL-1ra and two sites in IL-1 beta. *J. Biol. Chem.* 270(19):11477–11483 (1995).
40. S. D. Yanofsky, D. N. Baldwin, J. H. Butler, F. R. Holden, J. W. Jacobs, P. Balasubramanian, J. P. Chinn, S. E. Cirla, E. Peters-Bhat, E. A. Whitehorn, E. H. Tate, A. Akeson, T. L. Bowlin, W. J. Dower, R. W. Barrett. High affinity type I interleukin-1 receptor antagonists discovered by screening recombinant peptide libraries. *Proc. Natl. Acad. Sci. USA* 93:7381–7386 (1996).
41. G. Fields, R. Noble. Solid phase peptide synthesis utilizing 9-fluorenylmethoxycarbonyl amino acids. *Int. J. Peptide Protein Res.* 35:161–214 (1990).
42. S. Veltri, J. W. Smith. Interleukin-1 trails in cancer patients: a review of the toxicity, anti-tumor, and hematopoietic effects. *Stem Cells* 14(2):164–176 (1996).
43. R. Patarca, M. A. Fletcher. Interleukin-1: basic science and clinical applications. *Crit. Rev. Oncog.* 8(2-3):143–188 (1997).
44. M. D. Wewers. In: *Cytokines in Health and Disease*. Marcel Dekker, New York, 1992, pp. 235–256,327–352.
45. J. L. West, J. A. Hubbell. Polymeric biomaterials with degradation sites for proteases involved in cell migration. *Macromolecules* 32:241–244 (1999).
46. J. A. Hubbell. Bioactive biomaterials. *Cur. Opin. Biotechnol.* 10(2):123–129 (1999).
47. A. Gobin, A. Tsai, B. Mann, L. McIntire, J. West. Cell migration through biomimetic hydrogels. Biomimetic Session of the 6th World Biomaterials Congress, May 15–20, 2000, Kamuela, HI.
48. J. C. Schense, J. Bloch, P. Aebischer, J. A. Hubbell. Enzymatic incorporation of bioactive peptides into fibrin matrices enhances neurite extension. *Nat. Biotechnol.* 18(4): 415–419 (2000).
49. J. C. Schense, J. A. Hubbell. Three-dimensional migration of neurites is mediated by adhesion site density and affinity. *J. Biol. Chem.* 275(10):6813–6818 (2000).
50. Q. Zhao, N. Topham, J. M. Anderson, A. Hiltner, G. Lodoen, C. R. Payet. Foreign-body giant cells and polyurethane biostability: *in vivo* correlation of cell adhesion and surface cracking. *J. Biomed. Mater. Res.* 25:177–183 (1991).
51. W. J. Kao, J. A. Hubbell. Murine macrophage behavior on peptide-grafted polyethyleneglycol-containing networks. *Biotechnol. Bioeng.* 59(1):2–9 (1998).
52. W. J. Kao, D. Lee, J. C. Schense, J. A. Hubbell. Fibronectin modulates macrophage adhesion and FBGC formation: the role of RGD, PHSRN, PRRARV domains. *J. Biomed. Mater. Res.* (in press, 2000).
53. J. R. Potts, I. D. Campbell. Fibronectin structure and assembly. *Cur. Opin. Cell Biol.* 6(5):648–655 (1994).
54. J. R. Potts, I. D. Campbell. Structure and function of fibronectin modules. *Matrix Biol.* 15(5):313–320 (1996).
55. S. Aota, M. Nomizu, K. M. Yamada. The short amino acid sequence Pro-His-Ser-Arg-Asn in human fibronectin enhances cell-adhesive function. *J. Biol. Chem.* 269:24756–24761 (1994).
56. E. Ruoslahti. RGD and other recognition sequences for integrins. *Ann. Rev. Cell Dev. Biol.* 12:697–715 (1996).
57. R. P. Grant, C. Spitztaden, H. A. Altrott, I. D. Campbell, H. J. Mardon. Structural requirement for biological activity of the ninth and tenth FIII domains of human fibronectin. *J. Biol. Chem.* 272:6159–6166 (1997).
58. H. Mohri, J. Tanabe, K. Katoh, T. Okubo. Identification of a novel binding site to the integrin $\alpha_{IIb}\beta_3$ located in the C-terminal heparin-binding domain of human plasma fibronectin. *J. Biol. Chem.* 271:15724–15728 (1996).
59. H. Mohri, K. Katoh, A. Iwamatsu, T. Okubo. The novel recognition site in the C-terminal heparin-binding domain of fibronectin by integrin alpha 4 beta 1 receptor on HL-60 cells. *Exp. Cell Res.* 222(2):326–332 (1996).
60. D. J. Leahy, I. Aukhil, H. P. Erickson. 2.0 Å crystal structure of a four-domain segment of human fibronectin encompassing the RGD loop and synergy region. *Cell* 84:155–164 (1996).
61. C. Spitzfaden, R. P. Grant, H. J. Mardon, I. D. Campbell. Module–module interactions in the cell binding region of fibronectin: stability, flexibility, and specificity. *J. Mol. Biol.* 265:565–579 (1997).
62. E. Koivunen, B. Wang, E. Ruoslahti. Phage libraries displaying clclic peptides with different ring sizes: ligand specificities of the RGD-directed integrins. *Bio/Technology* 13:265–270 (1995).
63. A. K. McNally, K. M. DeFife, J. M. Anderson. Interleukin-4–induced macrophage fusion is prevented by inhibitors of mannose receptor activity. *Am. J. Pathol.* 149(3):975–985 (1996).
64. K. E. Healy, A. Rezania, R. A. Stile. Designing biomaterials to direct biological responses. *Ann. N.Y. Acad. Sci.* 875: 24–35 (1999).
65. A. Rezania, K. E. Healy. Integrin subunits responsible for adhesion of human osteoblast-like cells to biomimetic peptide surfaces. *J. Orthop. Res.* 17:615–623 (1999).
66. A. Rezania, K. E. Healy. Biomimetic peptide surfaces that regulate adhesion, spreading, cytoskeletal organization, and mineralization of the matrix deposited by osteoblast-like cells. *Biotechnol. Prog.* 15(1):19–32 (1999).
67. R. Langer. 1994 Whitaker Lecture: polymers for drug delivery and tissue engineering. *Anal. Biomed. Eng.* 23:101–111 (1995).
68. D. Williams. Biomimetic surfaces: how man-made becomes man-like. *Med. Device Technol.* 6(1):6–8,10 (1995).
69. A. S. Hoffman. Molecular bioengineering of biomaterials in the 1990s and beyond: a growing liaison of polymers with molecular biology. *Artif. Organs* 16(1):43–49 (1992).
70. J. D. Pitts, P. J. Campagnola, G. A. Epling, S. L. Goodman. Sub-micron multiphoton free form fabrication of proteins

and polymers: studies of reaction efficiencies and applications in sustained release. *Macromolecules* (in press, 2000).

71. P. J. Campagnola, D. M. DelGiudice, G. A. Epling, K. D. Hoffacker, A. R. Howell, J. D. Pitts, S. L. Goodman. Three-dimensional sub-micron polymerization of acrylamide by multi-photon excitation of xanthene dyes. *Macromolecules* (in press, 2000).
72. S. L. Goodman, P. A. Sims, R. M. Albrecht. Three-dimensional extracellular matrix textured biomaterials. *Biomaterials* 17(21):2087–2095 (1996).
73. R. G. Flemming, C. J. Murphy, G. A. Abrams, S. L. Goodman, P. F. Nealey. Effects of synthetic micro- and nano-structured surfaces on cell behavior. *Biomaterials* 20(6): 573–588 (1999).
74. E. Baer, J. J. Cassidy, A. Hiltner. Hierarchical structure of collagen and its relationship to the physical properties of tendon. In: *Collagen: Biochemistry and Biomechanics*, M. E. Mimi (Ed.). CRC Press, New York, pp. 177–199 (1988).
75. E. Baer, A. Hiltner, R. J. Morgan. Biological and synthetic hierarchical composites. *Physics Today*, October:60–67 (1992).
76. A. H. Heuer, D. J. Fink, V. J. Laraia, J. L. Arias, P. D. Calvert, K. Kendall, G. L. Messing, J. Blackwell, P. C. Rieke, D. H. Thompson, A. P. Wheeler, A. Veis, A. I. Caplan. Innovative materials processing strategies: a biomimetic approach. *Science* 255:1098–1105 (1992).
77. J. N. Cha, G. D. Stucky, D. E. Morse, T. J. Deming. Biomimetic synthesis of ordered silica structures mediated by block copolypeptides. *Nature* 403,6767: 289–292 (2000).
78. T. Pakalns, K. L. Haverstick, G. B. Fields, J. B. McCarthy, D. L. Mooradian, M. Tirrell. Cellular recognition of synthetic peptide amphiphiles in self-assembled monolayer films. *Biomaterials* 20(23-24):2265–2279 (1999).
79. C. Loty, J. M. Sautier, H. Boulekbache, T. Kokubo, H. M. Kim, N. Forest. *In vitro* bone formation on a bone-like apatite layer prepared by a biomimetic process on a bioactive glass-ceramic. *J. Biomed. Mater. Res.* 49(4):423–434 (2000).
80. R. Zhang, P. X. Ma. Porous poly(L-lactic acid)/apatite composites created by biomimetic process. *J. Biomed. Mater. Res.* 45(4):285–293 (1999).
81. U. B. Sleytr, D. Pum, M. Sara. Advances in S-layer nanotechnology and biomimetics. *Adv. Biophys.* 34:71–79 (1997).
82. I. A. Aksay, M. Trau, S. Manne, I. Honma, N. Yao, L. Zhou, P. Fenter, P. M. Eisenberger, S. M. Gruner. Biomimetic pathways for assembling inorganic thin films. *Science* 273:892–898 (1996).
83. K. M. Shakesheff, S. M. Cannizzaro, R. Langer. Creating biomimetic micro-environments with synthetic polymer–peptide hybrid molecule. *J. Biomater. Sci. Polymer Edn.* 9(5):507–518 (1998).
84. P. M. Marquis, H. Lugg, R. Sammons, L. Macaskie, P. Yong. Bacterial processing of nano particulate calcium phosphates. Biomimetic Session of the 6th World Biomaterials Congress, May 15–20, 2000, Kamuela, HI.

3

Silicones for Pharmaceutical and Biomedical Applications

Haissam S. El-Zaim and John P. Heggers
University of Texas Medical Branch, Galveston, Texas

I. INTRODUCTION

Until the early 1950s, the practice of medicine had not seen major changes for centuries. Before that time surgeons, for example, still used sutures processed from the intestinal walls of sheep or cows to sew wounds together. Around the mid-20th century, Eugene Rochow invented silicone, which proved to be one of the major inventions of the century. Silicone has some unique chemical and physical properties. It is inert, which implies low risk for interaction with the cells and chemicals in the body. It sustains high temperature, which suggests that silicone-based products could be easily sterilized. It was these properties that inspired scientists and physicians to consider the use of silicone for medical applications.

The first successful use of silicone in medical practice came about in 1956. At that time Doctors F. E. Nulsen and E. B. Spitz were cooperating with scientists from Dow Corning to develop a tube to be used in a hydrocephalic shunt procedure (68). In this procedure, an elastic tube is placed in the head of a child suffering from hydrocephalus in order to relieve the intracranial pressure caused by the excess of cerebrospinal fluid (CSF) in the cranial cavity. The shunt allows excess CSF to flow in one direction to the drainage site, which can be either intravenous, intrathoracic, or intra-abdominal. The tube had to be soft, flexible, elastomeric, and biocompatible with the human body. All of these requirements were met by the silicone elastomer that was successfully used in the procedure. Since then thousands of hydrocephalic shunt procedures have been performed using silicone elastomer. In other words, silicone is responsible for saving the lives of thousands of children borne with this birth defect (1 in 400–600 live birth), as well as improving the quality of life of survivors who otherwise would become mentally retarded. Hydrocephalic shunt is not the only procedure where the use of silicone has made a major contribution. Today, all intracranial indwelling tubing is made of silicone. Silicone is now used in many other surgical and nonsurgical procedures, and its use extends to a variety of medical instruments and indwelling medical devices.

Before we go into the details of the different applications for silicone in medical and pharmaceutical practice, let us briefly review the chemistry and physical properties of silicone.

II. CHEMICAL AND PHYSICAL PROPERTIES

Knowledge of the fundamental chemistry of the element silicon (Si) and understanding how Si can be integrated in biosystems are essential to understanding the role of Si in health and how it can be responsible for disease. Just like carbon, Si is capable of forming four bonds and is known for its ability to polymerize and form network covalent structures (9,134). However, unlike carbon, stable Si—Si bonds do not usually form. Silicon dioxide (SiO_2), also

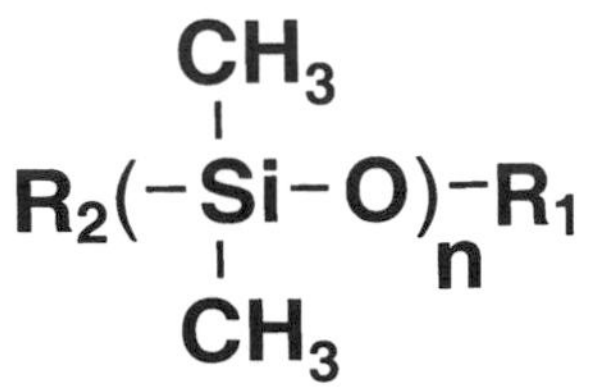

$$R_2(-\overset{CH_3}{\underset{CH_3}{Si}}-O)_n-R_1$$

Figure 3.1 The chemical structure of polydimethylsiloxane, the basic silicone compound used in medicine. (From Ref. 63b.)

$$2\ CH_3-\overset{CH_3}{\underset{CH_3}{Si}}-Cl + H_2O \longrightarrow CH_3-\overset{CH_3}{\underset{CH_3}{Si}}-O-\overset{CH_3}{\underset{CH_3}{Si}}-CH_3 + 2\ HCl$$

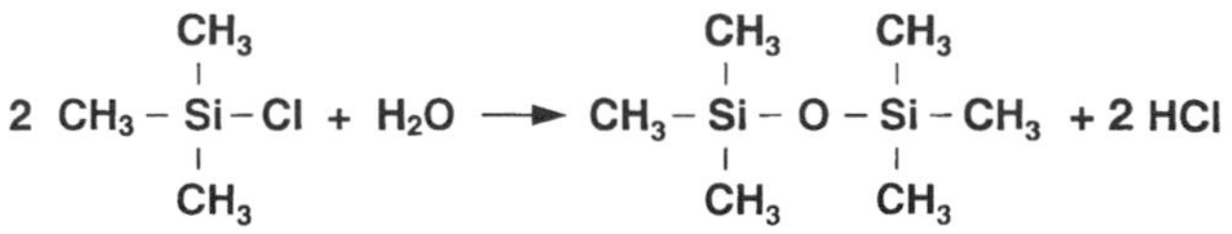

Figure 3.4 Condensation of two molecules of trimethylchlorosilane into hexamethyldisiloxane. (From Ref. 63b.)

called silica, is the most abundant mineral in the earth's crust and is the natural source of Si used in manufacturing silicone.

Silicone is simply a generic term that refers to siloxane polymers (polysiloxanes). In general, they are compounds of the structure $(R_1SiR_2O)_n$, where atoms of silicon are linked together by oxygen in a fashion analogous to the ether linkage of carbons by oxygen (36,89). The Si—O—Si bond is correctly termed a siloxane bond and thus the silicones are polysiloxanes (38,48,104,133). In polysiloxanes, the relatively stable Si—O bond is further stabilized by substitution of alkyl groups as the two side chain groups pensile from the Si molecule, which were CH_3 groups. Thus, the most common core structure of medical utility is polydimethylsiloxane (PDMS) (Fig. 1).

These highly hydrophobic compounds are chemically stable under physiologic conditions, can be synthesized with a variety of physical properties, and can be molded easily. Dialkylsiloxanes are prepared by the reaction of the dialkyldichlorosilane (generally dimethyldichlorosilane) with water (Figs. 2 and 3). Originally this structure was misinterpreted as oxygen linked to Si with a double bond, thus the term silicone. Further reaction with water yields silanolsdimethylsiloxanes with terminal hydroxyls. Polymerization usually yields a mixture of silanol oligomers with $n = 3$ to 9 and cyclic siloxanes.

The polysiloxanes then are ''capped'' by trimethylsiloxane groups, a process termed *equilibration* (50). This is first accomplished by condensing two molecules of trimethylchlorosilane to hexamethyldisiloxane via the addition of water (Fig. 4) and then reacting hexamethyldisiloxane with ailanol oligomers to produce silicones (Fig. 5). This process results in a fluid polysiloxane often termed an ''oil.'' Consequently, these compounds have widespread industrial and consumer uses as dielectric fluids, films, and coatings.

Polydimethylsiloxane, which has a specific gravity lower than that of water, is the most commonly used fluid in medical practice (133). The viscosity of the oil is a primary function of its molecular weight, but also of the ability of the molecules to slip past one another. The pensile methyl groups of silicones have little or no influence on this slippage. The polymerization reactions described can also be conducted using other alkyl or aromatic side chain–

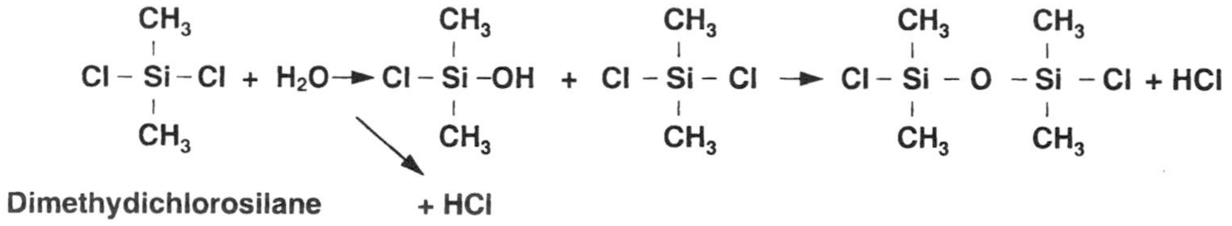

$$\underset{\text{Dimethydichlorosilane}}{Cl-\overset{CH_3}{\underset{CH_3}{Si}}-Cl} + H_2O \longrightarrow Cl-\overset{CH_3}{\underset{CH_3}{Si}}-OH\ (+\ HCl) + Cl-\overset{CH_3}{\underset{CH_3}{Si}}-Cl \longrightarrow Cl-\overset{CH_3}{\underset{CH_3}{Si}}-O-\overset{CH_3}{\underset{CH_3}{Si}}-Cl + HCl$$

Figure 3.2 Chemical reaction depicting the formation of dimethylsiloxane from dimethyldichlorosilane in the presence of water. (From Ref. 63b.)

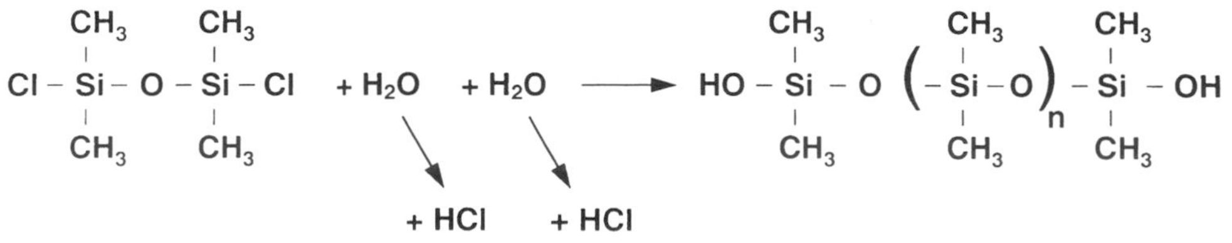

$$Cl-\overset{CH_3}{\underset{CH_3}{Si}}-O-\overset{CH_3}{\underset{CH_3}{Si}}-Cl + H_2O\ (+\ HCl) + H_2O\ (+\ HCl) \longrightarrow HO-\overset{CH_3}{\underset{CH_3}{Si}}-O\left(-\overset{CH_3}{\underset{CH_3}{Si}}-O\right)_n-\overset{CH_3}{\underset{CH_3}{Si}}-OH$$

Figure 3.3 Generation of silanols with terminal hydroxyl groups from dimethylsiloxane with water. (From Ref. 63b.)

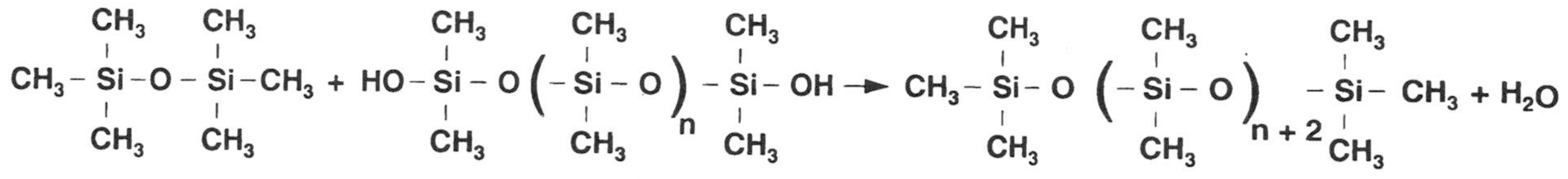

$$CH_3-\underset{CH_3}{\overset{CH_3}{Si}}-O-\underset{CH_3}{\overset{CH_3}{Si}}-CH_3 + HO-\underset{CH_3}{\overset{CH_3}{Si}}-O\left(-\underset{CH_3}{\overset{CH_3}{Si}}-O\right)_n-\underset{CH_3}{\overset{CH_3}{Si}}-OH \rightarrow CH_3-\underset{CH_3}{\overset{CH_3}{Si}}-O\left(-\underset{CH_3}{\overset{CH_3}{Si}}-O\right)_{n+2}-\underset{CH_3}{\overset{CH_3}{Si}}-CH_3 + H_2O$$

Figure 3.5 Production of silicone in a reaction between hexamthyldisiloxane and ailanol oligomers. (From Ref. 63b.)

modified chlorosilanes such as ethylmethyldichlorosilane or phenylmethyldichlorosilane as the feedstock and equilibrating with the homologous end groups. These will produce polysiloxanes, with increased viscosity because of the resistance of the more bulky pensile groups to slipping past each other. Silicone gels produced by this procedure consist of covalently crosslinked polysiloxane networks and interpenetrating polymer networks swollen in a silicone fluid. Commercial PDMS are not only produced as fluids or oils and gels, but also as ''gums,'' or rubber elastomers. The strength and elasticity of polysiloxane elastomers is a function of the length of the polymer chain and the degree of crosslinking. Elasticity occurs because polysiloxanes (like other polymers) exist in highly coiled conformers. As the material is stretched, the polymer unwinds. When tension is released the polymer recoils. Elasticity relies on the ability of adjacent polymer regions to slip past each other and therefore will be influenced by the presence of the bulky pendent groups. The physical rigidity of a siloxane polymer is increased by crosslinking the polymeric chains. Highly crosslinked polymers will lose elasticity because uncoiling will be inhibited as a result of the inability of the adjacent regions to slip past each other, but they will not deform easily and hence are more rigid. A lightly crosslinked polymer will deform easily but will have significant elasticity, with the ability to return to its original shape when stress is released.

The basic ingredient to make silicone elastomers, or silicone rubbers, is a clear and highly viscous PDMS with minor amounts of other substituents, particularly vinyl groups. Other radicals in the silicone rubber determine the physical properties of the resultant elastomer. The most common mechanism by which crosslinking occurs is by a radical attack on the alkyl groups pensile from the Si initiated by heat (heat vulcanization) or benzyl peroxide. The reactive intermediates containing carbon free radicals then combine to form the carbon–carbon covalent crosslinking bonds. In some cases, crosslinking is facilitated by substituting vinyl groups for occasional methyl groups because the double bond of this group is quite susceptible to free radical attack. In order to increase its tear and tensile strength, the prevulcanized silicone is usually mixed with a filler. A common filler used for this purpose is the submicroscopic fumed silica, which has a surface area of approximately 400 m^2/g. By changing the amount of filler added to a silicone elastomer, one can change its degree of hardness. This same property can be modified by changing the type of polymer used. Polydimethylsiloxane is used to prepare the soft grade silicone elastomers, whereas polysiloxane is used to prepare medium and hard grade silicone elastomers. Another type of silicone rubber is the room temperature–vulcanizing (RTV) silicone rubber. These silicones have much less tensile and tear strength than the high temperature–vulcanized materials. Unlike the other type of silicone elastomers, RTV silicones are provided as intermediate components which require mixing before their intended use (7,11).

III. BREAST AUGMENTATION IMPLANTS

Thomas Cronin and Frank Gerow, two plastic surgeons at the University of Texas, were looking for an alternative to the spongy breast implants commonly used for reconstruction of breasts among female patients undergoing mastectomy. The problem encountered with the spongy implants was that after a period of time they hardened and started to look and feel less natural. Around the year 1960, the two surgeons developed the first silicone breast implants in collaboration with Dow Corning (28,93). This prototype of silicone rubber sac filled with silicone gel was called a mammary prosthesis and was later marketed by Dow Corning after undergoing the first human trial. Soon after Dow Corning introduced this new product to the market, and to everybody's surprise, the demand for mammary prosthesis was much greater for breast augmentation procedures for cosmetic purposes (80%) rather than reconstruction procedures following mastectomy. It was estimated that by the year 1992, 1–2.5 million North American women and 100,000–150,000 British women received breast implants (62). During the 1960s and 1970s Dow Corning introduced many improvements to enhance the quality of the mammary prosthesis, which became a multimillion dollar product. Other companies joined Dow Corning to cash in on this rapidly growing market.

Silicone breast implants have been associated with complications such as capsular contracture, enlargement of lymph nodes draining the implant site, occasional rupture of the silicone gel sac, and bacterial contamination (26). Capsular contracture, which may be triggered by bleed of the silicone gel through the outer bag, is believed to be due to the natural attempt by the body to engulf the foreign material (18). The drying effect of silicone in the surrounding soft wet tissue enhances the scar formation and is another possible contributing factor to the formation of the capsule. This pseudocapsule is composed of fibrous tissue typically seen in the process of wound healing. Physical changes and alterations of the nature of the silicone implants, as well as use of wetting agents such as providone, have reduced the incidence of capsular contracture (45).

It was not these complications but rather the possible link of breast implants to a systemic connective tissue disease that fueled the debate over the safety of this procedure. Several reports appeared in the literature suggesting an association between breast implants and some ill-characterized immunological disorders. Most of the reports described cases of a poorly defined syndrome referred to as ''human adjuvant disease.'' Some reports claimed an association with systemic, probably autoimmune, connective tissue diseases (73,77). The first report of an association between silicone breast implant and connective tissue disease dates back to 1964, and the first documented cases were those of three patients in 1982. It is estimated now that the number of reported cases exceeds 290 patients (62). Reported symptoms and signs include fatigue, joint pain, bone pain, dry eyes, dry mouth, dry skin, cognitive dysfunction, myalgia, weakness, hair loss, nail changes, skin rashes, paresthesia, dysethesia, freckling, pigment change, headache, dizziness, nausea, foul taste, weight gain or weight loss, bruising, increased sensitivity to light, fever, chills, infections, diarrhea, constipation, periodontal disease, skin papules, muscle twitching, urinary symptoms, dysphagia, dysmenorrhea, blurred vision, tinnitus, drug reactions, emotional lability, insomnia, edema, hemangiomas, delayed wound healing, vascular abnormalities, partial hearing loss, reduced smell, tremor, mouth sores, tight skin, shortness of breath, wheezing, palpitations, seizures, parotid gland swelling, heat intolerance, and cancer. The range of symptoms was very large and many of the cases did not fulfill conventional clinical and laboratory criteria for a particular connective tissue disease. Scleroderma was the most common specific diagnosis (10,15, 16,30,76,99, 100).

In 1991, an American jury found that silicone breast implants were responsible for the plaintiffs' symptoms of a mixed connective tissue disease. The jury found that the company was responsible as they misrepresented the safety of their product. After this trial, public awareness of the issue rose steeply and the Food and Drug Administration, following the recommendations of two independent advisory panels, requested a halt on the use of silicone breast implants other than within clinical trials (27,94). In 1994, while the battle between the litigants and the manufacturers of the implants was raging inside the court and on the media front, a Department of Health advisory group in Great Britain reported that there was no evidence of an increased risk of connective tissue disease in patients with silicone breast implants (43). Analysis of published case series as well as several case–control and cohort studies suggested that the cumulative incidence of connective tissue disorders among recipients of breast implants is no different from that expected in the general population (101). Only one retrospective cohort study, by Hennekens et al., published in the *Journal of the American Medical Association* in 1996, pointed to a weak association (52). The authors of this study themselves concluded that an association between silicone breast implants and increased risk of major connective tissue disease is unlikely. Further, this study was criticized for the lack of certain diagnostic validity with potential bias due to differential over-reporting.

IV. OPHTHALMOLOGY PROCEDURES AND OPTIC DEVICES

Retinal detachment is a pathological process in which the neuronal layers of the retina, including the layer of photoreceptors (rods and cones), are separated from the pigment epithelium. Melanin in the pigment epithelium, located in the back of the neuronal retina, captures light that makes it through the photoreceptors layer and therefore prevents reflection of light off the sclera. In addition, the pigment epithelium is essential for the proper metabolism of the neuronal layer of the retina. It supplies the retina with nutrients and is involved in phagocytizing and clearing out the debris generated by the constant turnover of the outer segments of the photoreceptors containing the photopigments. Retinal detachment can be caused by numerous disease conditions. Cytomegalovirus-induced retinitis is a major cause of retinal detachment among patients with acquired immunodeficiency syndrome (AIDS) (56,57). Patients with retinal detachment are at serious risk of total blindness unless the problem is corrected. Traditionally, retinal reapposition was achieved by suturing a sponge or a band to the sclera to produce a ''buckle'' which closes the hole by external tamponade. With this procedure the retina resumes contact with the pigment epithelium. This procedure works well in uncomplicated cases; however, in complex retinal detachment, which requires removal of the preretinal membranes by vitrectomy, an external tamponade must be supplemented with an internal tamponade in order to successfully bring the retina and the pigment epithelium together. Investigators felt the need for a substance with stable surface properties at aqueous interfaces, that is biologically inert, and is amenable to surgical procedures to serve this purpose. Gases, such as air, sulfur, hexafluoride, and perfluoropenthrane were tested and proved to be effective in the short term. However, for longer-term maintenance of break closure and relief of retinal traction, a material that is insoluble in tissue fluid seemed to be more appropriate. Because of its long-term biotolerability and optical clarity, silicone seemed to be the perfect substance for this purpose. In 1962, Cibis was the first to report the use of liquid silicone (silicone oil) as a ''vitreous prosthesis'' in human eyes (23). Since then the use of silicone has been extended to include providing temporary or permanent internal tamponade, delaminating periretinal membranes, manipulating the retinal edge in giant tears, closing macular holes, and maintaining the surgeon's view of the back of the eye during surgery involving vascularized tissue. The use of intraocular silicone is considered a major breakthrough in the treatment of retinal detachment, which when left untreated can lead to blindness. Complex retinal detachments are now routinely repaired, and very few are deemed to be inoperable. Side effects due to intraocular silicone can be avoided or reduced by proper surgical technique. Foaming or emulsification is responsible for the main long-term complications such as glaucoma and impaired vision. This can be avoided with a more complete silicone fill (21,37,41,74,82,90,97).

In addition to the previous example, an experimental procedure to restore accommodation in primate eyes by refilling the lens capsule with injectable silicone compounds is currently being investigated (87,88). Last but not least, polysiloxane lenses coated with povidone are routinely used for patients after cataract surgery (58). Silicone is also used in manufacturing soft contact lenses.

V. TREATMENT OF URINARY AND FECAL INCONTINENCE AND GENITOURINARY DEVICES

Until the introduction of silicone-based prosthesis, attempts to replace any part of the urinary tract were plagued by the problem of phosphatic encrustation. Whereas silicone catheters may be left indwelling for 8 to 12 weeks with minimal phosphatic deposition, latex or plastic catheters have to be replaced within 3 to 4 weeks (130).

Urinary incontinence is a common problem affecting 10–25% of women between the ages of 15 and 60 years (64). Stress urinary incontinence, which is the involuntary loss of urine during physical activities that increase intra-abdominal pressure, is a common type. This condition could result from loss of pelvic support, and hence hypermobility of the vesicourethral junction (anatomic incontinence) or intrinsic sphincter deficiency (type III stress incontinence). The latter condition may be caused by myelodysplasia, sympathetic nerve injury, and surgical injury or trauma (75). Urinary incontinence may also occur as a result of spinal cord injury or following prostatectomy procedure in males (2,25,131).

As early as the year 1947, attempts to control urinary incontinence by applying a constricting cuff around the urethra via a cutaneous tunnel were reported. In 1973, Jo seph Kaufman described a procedure in which a silicone balloon is implanted around the crura of the penis to control incontinence (59). Upon distention the balloon would compress the urethra against the symphysis pubis. Following implantation the balloon can be further enlarged by injecting more fluid into it if necessary. That same year, Brantley Scott and his colleagues developed an artificial sphincter which consisted of a silicone rubber cuff, a small pump, a balloon, and a stainless steel control assembly (98). The cuff is surgically implanted around the neck of the bladder, or sometimes around the bulbous urethra of men. The balloon is positioned in the perivesical space and, for easy access, the pump is placed under the skin either in the scrotum in males or the labia minora in women. The inflated cuff cuts off the flow of urine in a manner similar to that of the natural sphincter muscle. By squeezing the pump several times, forcing the fluid out of the cuff and into the balloon, the patient could deflate the cuff and allow urine to flow. Within minutes the cuff refills automatically restoring continence. An overall success rate of 70% for an average of 3.5 years was reported with this device. Drawbacks of this procedure included cost and complications, such as periurethral fibrosis, tissue erosion, and infection.

Transurethral or periurethral injection of biomaterials into the intrinsic sphincter was developed as a less invasive technique for the treatment of incontinence (8). Various forms of bioinjectable materials have been developed for this purpose. Injection of polytetrafluoroethylene (Teflon) paste has had relatively good results in the short term (24). Long-term follow-up studies on patients treated with Teflon paste injections revealed a low success rate of only 38% and the occurrence of local side effects, such as fibrosis in the urethra and bladder granuloma balls (17). More recently, injection of autologous fat was tested and the results from this study were not impressive (6,20,95). Collagen has a good (46–77%) long-term (2 years) success rate and is currently the most widely used injectable (4). The major problem encountered with collagen use is the potential development of allergic reactions (3). Collagen, as well as autologous fat, are absorbed rather rapidly, which calls for repeated injections. Recently, reports on use of silicone particles for treatment of urinary incontinence appeared in the literature (61). According to these reports, injection of polydimethylsiloxane is a simple, safe, and effective procedure for the treatment of urinary incontinence. Another possible application for this procedure is in treating vesicoureteral reflux (61).

Soiling (fecal incontinence) is another distressing, demoralizing, and costly condition which may result from various causes. Although the pathophysiologic mechanisms are not fully understood, it is almost always associated with anorectal disorders that can deform the contour of the anus and anal canal. The condition is usually corrected by medical or surgical means; however, in certain situations reconstruction of the contour deformity of the anus is the only successful resort (86). Moderate success with sublevatorial implantation of a silicone ring was reported by Hansen in 1996 (47). Currently, perianal injection of PDMS paste is being investigated for treating soiling.

Silicone is also used in the making of penile implants used to treat male impotence. One model of these implants is a solid polysiloxane component that is inserted subcutaneously in the penis. Another model consists of an inflatable silicone bag connected to a reservoir which is used to pump fluid to the prosthesis when an erection is desired (92).

VI. ORTHOPEDIC AND RECONSTRUCTIVE SURGERY

Silicone prostheses play a major role in the treatment of degenerative arthritis and rheumatoid arthritis in the hand and wrist. Silicone devices are often used in orthopedic surgery, especially in operations to reconstruct the hand (45). Until the late 1950s, success rates were not impressive in attempts to achieve pain-free, stable implant arthroplasty with metacarpal cups or with soft tissue interpositions (8). The metal hinge implants for the metacarpophalangeal joint introduced in 1959 were not well tolerated due to bone resorption and metal corrosion (12,35). The search for better techniques and new biocompatible materials continued until silicone rubber prostheses were introduced in the late 1960s and early 1970s (13,84,85, 107,109–125). Niebauer and his colleagues introduced a hinged silicone joint with a dacron mesh covering for ingrowth fixation. A hinge spacer originally intended to improve the alignment and stability of resection and arthroplasty was introduced by Swanson. Swanson, who pioneered the field of silicone implant arthroplasty, introduced other designs used in the treatment of arthritis in the wrist, ulnar head, scaphoid, lunate, trapezium, and metacarpophalangeal and interphalangeal joints. Designs by Swanson and Niebauer became very popular soon after their introductions, but some of the problems encountered were implant fracture, persistent weakness of grip and pinch, and increasing deformity. This led to careful reconsideration of both prosthesis designs and materials. Today, prostheses based on those original designs continue to be widely used in arthroplasty. Another use for silicone in hand orthopedic surgery is in tendon transplantation (19). In this procedure, Hunter's Silastic rods are used to form a tunnel composed of a thin pseudocapsule in which tendon transplants are inserted later on. The purpose of this tunnel is to prevent the adhesion of the transplanted tendon to the adjacent structures in the hand, which allows for the gliding function of the tendon. Silicone is also used in reconstructing the wrist using silicone joint prostheses or bone prostheses made of silicone. Silicone can also be used to reconstruct larger joints such as the elbow (71,111,117–125). This may involve replacement of the head of the ulna with silicone components. However, reconstruction of large joints and long bones may require additional reinforcement.

The three major complications following silicone implant arthroplasty are fracture, subluxation, and synovitis. Fracture commonly occurs in the wrist and metacarpophalangeal implants, whereas subluxation occurs mainly with the carpal implants (63). Silicone synovitis is an inflammatory response to small particulate debris of 10–100 μm in diameter (5). This phenomenon is more commonly seen in association with high performance (HP) silicone elastomer than with the conventional silicone elastomer (125). Small particles shed from HP elastomer implants are ingested by phagocytes which fail to clear these foreign particles away. Frustrated macrophages release proinflammatory cytokines and attract more immune cells. This results in a chronic inflammatory response characterized by the presence of lymphocytes, eosinophils, and giant multinucleated epithelial cells (42). In very few cases, silicone lymphadenopathy has been reported following silicone implant arthroplasty (22).

In addition to hand orthopedic surgery, silicone is commonly used in maxillofacial reconstructive surgery. Silicone products are used to reconstruct the mandible, to augment the chin of patients with microgenia and to reconstruct the missing ear of patients with microtia (45,69).

VII. LARYNGEAL, TRACHEAL, AND ESOPHAGEAL PROSTHESES

A variety of devices used in the upper respiratory and digestive system are made of silicone. The silicone Safe-T-Tube is designed to maintain adequate tracheal airway and to provide support in the stenotic trachea. It can be kept in place for many years, if necessary. It is commonly used in cases with acute tracheal injuries, a need to support a reconstructed trachea or reconstituted trachea, as well as many other procedures (78). In cases such as obstructive sleep apnea, bilateral vocal cord paralysis, laryngeal carcinoma with glottic insufficiency, and neurological disorders with intermittent laryngeal insufficiency, a silicone device called the Tracheal Cannula System is usually used (79). By diverting the saliva into the distal esophagus, a silicone-made salivary bypass tube greatly reduces the healing time in cases with cervical esophageal and hypopharyngeal strictures and fistulas resulting from a variety of causes including malignancy, surgery, irradiation, trauma, and caustic indigestion (80). Esophageal tube is used to bridge the gap between the pharyngostome and esophagostome following laryngoesophagectomy and reconstructive surgery of the cervical esophagus (78,80,126,127). Silicone laryngeal stent can be used alone or in combination with the Safe-T-Tube to prevent and treat laryngeal stenosis, when the glottic stenosis involves the midglottis, posterior glottis, supraglottis, and subglottis (80). Laryngeal keel, an umbrella-shaped silicone prosthesis, is used to prevent and treat anterior laryngeal stenosis limited to the anterior commissure of the vocal cords (80). Silicone esophageal stents are used as a palliative therapy for tracheoesophageal fistula, a life-threatening late complication of carcinoma of the esophagus and lung (83,108). This procedure prevents food and gastric refluxate from entering the respiratory tract and restores the ability to swallow so proper nutrition intake can be maintained (54). Without treatment, most of these patients succumb to respiratory problems and die in a matter of few months (31,60).

Since the first laryngectomy was performed in 1866, several methods to restore speech have been attempted (1,44). This required an alternative vibratory source to replace the larynx in the reconstructed pharyngoesophageal region. This goal can be achieved by means of an artificial larynx that has an electric pharyngeal speech vibrator which can be externally applied to the neck. However, of all the speech restoration methods, the method of shunt esophageal speech by means of silicone tracheoesophageal valve prosthesis have the most consistently high success rate (102,132).

The major risks associated with the silicone devices are bacterial and fungal infections (55,72). The risk of infection can be greatly reduced by proper surgical techniques and by patient education on how to properly use, maintain, and clean their prostheses. Also, early intervention to disinfect or replace a contaminated device has been very efficient in preventing major complications.

VIII. CARDIOVASCULAR PROCEDURES AND DEVICES

With the advent of extracorporeal circulation in 1954 and the introduction of heart–lung machine by Gibbon (40), direct open treatment of valvular diseases became possible. Deformed, calcified valves could not be completely fixed by conservative procedures such as valvuloplasties and calcium debridement. Such procedures represented a major challenge to the cardiac surgeon (34). Early attempts to use artificial valves were not very successful in the long run due to the loss of flexibility of the material and to ingrowth of tissue (14). Charles Hufnagel was the first to implant a valvular prosthesis (46,53). His design consisted of a caged Lucite ball valve which he used to treat aortic insufficiency. Hufnagel's attempt set the stage for other investigators to develop more biocompatible and safer prostheses (33,49,106). Today, there are numerous heart valves available from different manufacturers and many of them consist of or contain silicone (34). Valvular endocarditis is the most dreaded complication of valve replacement (32). The source of bacteria in early nosocomial infection is believed to be patients themselves or their environment. Staphylococci and gram-negative bacilli are the leading cause of prosthetic valve infections (103). When the patient presents with symptoms of acute septicemia secondary to valvular endocarditis, death may ensue within few days. The best approach to treat early valvular endocarditis is the institution of appropriate antibiotics followed by surgery (70). The prognosis for late valvular endocardtitis is less severe and may be responsive to antibiotics alone.

The idea of externally stimulating the heart with an electric current was first conceived in 1952 by Paul Zoll (135). The first method of pacing was described in 1958 by Furman and Robinson (39). According to their method a catheter electrode with a metal guide wire was inserted into the right ventricle through a peripheral vein. The pacemaker itself was kept external to the body. Today, the transvenous approach is the most commonly applied procedure (91). In this method, a permanent pacemaker is implanted near the site where the electrode enters the cephalic vein just beneath the clavicle. In a pacemaker the only metal exposed to the body fluids is the tip of the electrode in contact with the heart muscle and the ground plate; all other parts are insulated with silicone rubber. Complications related to pacing leads include lead fracture, thrombosis and embolism, infection, lead failure, migration of epicardial leads, malposition, dislodgment, myocardial penetration and perforation, electrode corrosion, insulation failure, exit block, and silicone-induced endocarditis (128).

Other important silicone cardiovascular devices include silicone-coated tubing for extracorporeal circulation. Such tubing is important in open-heart surgery. The lower reactivity of silicone with blood reduces the thrombogenic effect of the polyethylene tubing during cardiopulmonary bypass, therefore making the use of silicones most advantageous in these circumstances. Catheters for intra-arterial and intravenous lines are encased in or coated with silicone to minimize thrombogenic activity. The silicone membrane in the artificial lining device allows the transport of oxygen and carbon dioxide (45). Silicone antifoams are essential in cardiopulmonary bypass procedures. They are used in bubble oxygenators to prevent gas bubble emboli (38).

IX. SILICONE FLUID LUBRICANTS

Silicone fluid lubricants facilitated the development of disposable hypodermic syringes and needles. In 1950, Darling and Spencer reported that silicone lubricant is chemically stable and withstands autoclave sterilization, and they recommended its use for lubrication of hypodermic syringes (29). Today, silicone remains an effective lubricant with both sterilization and time. Being chemically nonreactive and compatible with injectables, it is widely used in manufacturing disposable hypodermic needles and syringes.

X. MISCELLANEOUS DEVICES

Silicone is used in manufacturing dental impression materials, anti–gastric reflux devices, implantable drug-binding matrices for drug-release capsules such as implantable contraceptives, and peritoneal venous shunts. Silicone tubing is used routinely in kidney dialysis. Silicone net dressing is used on skin-grafted sites, and silicone cream occlusive dressing is used in the treatment of hypertrophic scars and keloid. Silicone is also used in manufacturing a variety of many other useful devices (65).

REFERENCES

1. P. W. Alberti. The evolution of laryngology and laryngectomy in the mid 19th century. *Laryngoscope* 85:288–298 (1975).
2. O. Alfthan. Treatment of postprostatectomy incontinence. *Annales of Chirugiae et Gynaecologiae* 71:244–249 (1982).
3. R. A. Appell. Collagen injection therapy for urinary incontinence. *Urol. Clin. North Am.* 21:177–182 (1994).
4. R. A. Appell, E. J. McGuire, P. A. DeRidder, A. H. Bennett, G. D. Webster, G. Badlani, et al. Summary of effectiveness and safety in the prospective, open, multicenter investigation of contigen implant for incontinence due to intrinsic sphincteric deficiency in females. *J. Urol.* 153: 418–420 (1994).
5. R. G. Aptekar, J. M. Davie, H. S. Cattell. Foreing body reaction to silicone rubber. *Clin. Orthop.* 98:231–232 (1974).
6. J. G. Balivas, D. Hertz, R. P. Santarosa, R. Dmochowski, K. Ganabathi, D. Roskamp, et al. Periurethral fat injection for sphincteric incontinence in women. *J. Urol.* 153:21–25 (1994).
7. S. Barley. The chemistry and properties of the medical-grade silicones. In: *Biomedical Polymers*, A. Rembaum, and M. Shen (Eds.). Marcel Dekker, New York, 1971, pp. 35–50.
8. R. D. Beckenbaugh, R. L. Linscheid. Arthroplasty in the hand and wrist. In *Operative Hand Surgery*, D. P. Green (Ed.). Churchill-Livingstone, New York, 1982, pp. 141–184.
9. G. Bendz, I. Lindqvist (Eds.). *Biochemistry of Silicon and Related Problems*. Plenum Press, New York, 1978.
10. D. Borenstein. Siliconosis: a spectrum of illness. *Semin. Arthritis Rheum.* 24(suppl. 1): 1–7 (1994).
11. J. W. Boretos. *Concise Guide to Biomedical Polymers*. Charles C. Thomas, Springfield, IL, 1973, pp. 72–73.
12. E. W. Brannon, G. Klein. Experiences with a finger-joint prosthesis. *J. Bone Joint Surg.* 41A:87–102 (1959).
13. R. M. Braun. Total joint replacement of arthritis at the base of the thumb. Presented at the Annual Meeting of the American Society for Surgery of the Hand. San Francisco, (1978).
14. N. S. Braunwald, A. G. Morrow. A late evaluation of flexible teflon prosthesis utilized for total aortic valve replacement. *J. Thorac. Cardiovas. Surg.* 49:485–496 (1965).
15. A. E. Brawer. Bones, groans, and silicones: beauty and the beast. *Arthritis Rheum.* 37(suppl.):R38 (1994).
16. A. J. Bridges, C. Connley, G. Wang, et al. A clinical and immunologic evaluation of women with silicone breast

implants and symptoms of rheumatic disease. *Ann. Int. Med.* 11:929–936 (1993).
17. J. F. Buckley, K. Lingam, R. N. Meddings, R. Scott. Injectable Teflon paste for female stress incontinence: long-term follow-up and results. *J. Urol.* 151:418–422 (1993).
18. B. R. Burkhardt. Capsular Contracture. In: *Silicone in Medical Devices*. Conference Proceedings. Baltimore, MD, 1991, pp. 187–193.
19. D. A. Caplin, V. L. Young. Silastic tendon passer. *Plast. Reconstr. Surg.* 13:529–536 (1981).
20. M. Cervigni, M. Panei. Periurethral autologous fat injection for type III stress urinary incontinence. *J. Urol.* 151: 403–406 (1993).
21. C. Chan, E. Okun. The question of ocular tolerance to intravitreal liquid silicone. A long-term analysis. *Ophtalmology* 93:651 (1986).
22. A. J. Christie, K. A. Weinberger, M. Dietrich. Silicone lymphadenopathy and synovitis. *JAMA* 237:1463–1464 (1977).
23. P. A. Cibis, B. Becker, E. Okun, S. Canaan. The use of liquid silicone in retinal detachment surgery. *Arch. Ophtalmol.* 68:590–599 (1962).
24. H. M. Cole. Diagnostic and therapeutic technology assessment (DATTA). *JAMA* 269:2975–2980 (1993).
25. D. J. Confer, M. E. Beall. Evolved improvements in placement of the silicone gel prosthesis for post-prostatectomy incontinence. *J. Urol.* 126:605–608 (1981).
26. R. R. Cook, M. C. Harrison, R. R. Levier. The breast implant controversy. *Arthritis Rheum.* 37:153–157 (1994).
27. C. Cooper, E. Dennison. Do silicone breast implants cause connective tissue disease? There is still no clear evidence that they do. *BMJ* 316:403–404 (1998).
28. T. D. Cronin, F. J. Gerow. Augmentation mammoplasty: a new ''natural feel'' prosthesis. Transactions of the Third International Congress of Plastic Surgery, Amsterdam. Excerpta Medica Foundation: 41–49 (1964).
29. G. H. Darling, J. G. C. Spencer. Silicone in syringe sterilization. *BMJ* Feb. 10:300 301 (1951).
30. J. Davis, J. Campagna, R. Perrillo, L. Criswell. Clinical characteristics of 343 patients with breast implants. *Arthritis Rheum.* 38(suppl): S263 (1995).
31. Y. S. Do, H. Song, B. H. Lee, et al. Esophagorespiratory fistula associated with esophageal cancer: treatment with a Gianturco stent tube. *Radiology* 187:673–677 (1993).
32. R. J. Duma (Ed.). Infection of prosthetic heart valves and vascular grafts. University Park Press, Baltimore, 1977.
33. W. S. Edwards. Prosthetic heart valves. In: *Surgery of the Chest*, W. B. Sunders Co., Philadelphia, 1962, pp. 872–879.
34. J. Fernandez. Prosthetic heart valves. In: *The Pacemaker and Valve Identification Guide*, D. Morse, R. M. Steiner (Eds.). MEBC, Garden City, NY, 1978, pp. 52–69.
35. A. E. Flatt. *The Care of the Rheumatoid Hand*, 3rd ed. C. V. Mosby, St. Louis, 1974.
36. S. Fordham. *Silicones*. George Newnes, London, 1960.
37. W. R. Freeman, D. E. Henderly, W. L. Wan, D. Causey, M. Trousdale, R. L. Green, N. A. Rao. Prevalence, pathophysiology, and treatment of rhegmatogenous retinal detachment in treated cytomegalovirus retinitis. *Am. J. Ophtalmol.* 103:527 (1987).
38. E. E. Frisch. Technology of silicone in biomedical applications. In: *Biomaterials in Reconstructive Surgery*, L. H. Rubin (Ed.). C. V. Mosby, St. Louis, 1983, pp. 73–90.
39. S. Furman, G. Robinson. The use of intracardial pacemaker in the correction of total heart block. *Surg. Forum* 9:245 (1958).
40. J. H. Gibbon, Jr. Application of a mechanical heart and lung apparatus to cardiac surgery. *Minnesota Med.* 37: 171–185 (1954).
41. M. Gonvers. Temporary silicone tamponade in the management of retinal detachment with proliferative vitreoretinopathy. *Am. J. Ophtalmol.* 100:239 (1985).
42. M. Gordon, P. G. Bullough. Synovial and osseous inflammation in failed silicone-rubber prosthesis. *J. Bone Joint Surg.* 64A:574–580 (1982).
43. D. M. Gott, J. J. B. Tinkler. Evaluation of evidence for an association between the implantation of silicone and connective tissue disease. Medical Devices Agency, London, 1994.
44. C. Gussenbauer. Ueber die erste durch th. Billroth am Menschen ausgefuhrte Kehlkopf-Extirpation und die Anwendung eines Kunstlichen kehkopfes. *Arch. Klin. Chir.* 17:343–356 (1874).
45. M. B. Habal. The biologic basis for the clinical application of the silicones. A correlate to the biocompatibility. *Arch. Surg.* 119:843–848 (1984).
46. E. Halkier, H. Fritz, I. Anderson. 1970. Aortic incompetence. The eventual outcome in a small series treated with Hufnagel's descending aorta ball-valve. *Scand. J. Thorac. Cardiovas. Surg.* 4:52–55 (1970).
47. H. Hansen. Surgical treatment of fecal incontinence. *Zentralblatt fur Chirugie*. 121:676–680 (1996).
48. B. B. Hardman, A. Torkelson. Silicones. In: *Kirk-Othmer Concise Encyclopedia of Chemical Technology*. John Wiley & Sons, New York, 1985, pp. 1062–1065.
49. D. E. Harken, H. S. Soroff, W. J. Taylor, et al. Partial and complete prostheses in aortic insufficiency. *J. Thorac. Cardiovas. Surg.* 40:744–762 (1960).
50. W. L. Hawkins. Polymer degradation and stabilization. In: *Polymers: Properties and Applications*. Springer-Verlag, New York, 1984.
51. J. P. Heggers, N. Kossovsky, R. W. Parsons, M. C. Robson, R. P. Pelley, T. J. Raine. Biocompatibility of silicone implants. *Ann. Plast. Surg.* 11(1):38–45 (1983).
52. C. H. Hennekens, I. Min Lee, N. Cook, P. Hebert, E. Karlson, F. LaMotte, et al. Self-reported breast implants and connective tissue diseases in female health professionals. *JAMA* 275:616–621 (1996).
53. C. A. Hufnagel, W. P. Harvey, P. J. Rabil, et. al. Surgical correction of aortic insufficiency. *Surgery* 35:673–683 (1954).
54. J. D. Irving, J. N. I. Simson. A new cuffed oesophageal prosthesis for the management of malignant oesophageo-

respiratory fistula. *Ann. Roy. Coll. Surg. Engl.* 70:13–15 (1988).
55. K. Izdebski, J. C. Ross, S. Lee. Fungal colonization of tracheoesophageal voice prosthesis. *Laryngoscope* 97: 594–597 (1987).
56. D. A. Jabs, C. Enger, J. G. Barlett. Cytomegalovirus retinitis and acquired immunodeficiency syndrome. *Arch. Ophtalmol.* 107:75 (1989).
57. D. A. Jabs, W. R. Green, R. Fox, B. F. Polk, J. G. Bartlett. Ocular manifestations of acquired immune deficiency syndrome. *Ophtalmology* 96:1092 (1989).
58. J. Katz, H. E. Kaufman, J. Valenti, et al. Corneal endothelium damage with intraocular lenses: contact adhesion between surgical materials and tissue. *Science* 198:525–527 (1977).
59. J. J. Kaufman. Treatment of post-prostatectomy urinary incontinence using a silicone gel prosthesis. *Br. J. Urol.* 45:646–653 (1973).
60. O. Kawashima, I. Yoshida, S. Ishikawa, K. Ohshima, Y. Morishita. Use of intratracheal silicone prosthesis (Dumaon type) for the treatment of tracheoesophageal fistula due to advanced lung cancer. *Surg. Today* 26:915–918 (1996).
61. R. T. Kershen, A. Atala. New advances in injectable therapies for the treatment of incontinence and vesicourethral reflux. *Urolog. Clinics North Am.* 26:81–94 (viii) (1999).
62. D. A. Kessler. The basis of the Food and Drug Administration's implants. *N. Engl. J. Med.* 3261:1713–1715 (1992).
63a. J. M. Kleinert, G. D. Lister. Silicone implants. *Hand Clinics* 2:271–290 (1986).
63b. B. Klitzman (Ed.). *Problems in General Surgery: Biomaterials*. J. B. Lippincott, Philadelphia, 1994.
64. H. Koelbel, V. Saz, D. Doerfler, G. Haeusler, C. Sam, E. Hanzal. Transurethral injection of silicone microimplants for intrinsic urethral sphincter deficiency. *Obstet. Gynecol.* 92:332–336 (1998).
65. N. Kossovsky. Pathophysiological basis of silicone bioreactivity. In: *Silicone in Medical Devices*. Conference proceedings, FDA, Baltimore, MD, 1991, pp. 45–58.
66. N. Kossovsky, J. P. Heggers, R. W. Parsons, M. C. Robson. Analysis of the surface morphology of recovered silicone mammary prostheses. *Plast. Reconstr. Surg.* 71:795–802 (1983).
67. N. Kossovsky, J. P. Heggers, R. W. Parsons, M. C. Robson. Acceleration of capsule formation around silicone implants by infection in a guinea pig model. *Plast. Reconstr. Surg.* 73:91–96 (1984).
68. H. La Fay. A father's last-chance invention saves his son. *The Reader's Digest* January 29–32 (1957).
69. D. Leake, J. E. Murray, M. B. Habal, et al. Custom fabrication for mandibular reconstruction. *Oral Surg.* 33:879–883 (1972).
70. E. A. Lefrak, A. Starr. Evaluation and management of postoperative problems. In: *Cardiac Valve Prostheses*. ACC, New York, 1979, pp. 371–391.
71. D. MacKay, B. Fitzgerald, J. H. Miller. Silastic replacement of the head of the radius. *Joint Surg. Br.* 61:494–497 (1979).
72. H. F. Mahieu, H. K. F. Van Saene, H. J. Rosingh, H. K. Schutte. Candida vegetation on silicone voice prostheses. *Arch. Otolaryngol. Head Neck Surg.* 112:321–325 (1986).
73. L. Martin. Silicone breast implants and connective tissue diseases: an ongoing controversy. *J. Rheumatol.* 22:198–200 (1995).
74. B. W. McGuen, E. Dejuan, M. B. Landers, R. Machemer. Silicone oil in vitreoretinal surgery. Part 2. Results and complications. *Retina* 5:198 (1985).
75. E. J. McGuire. Combined radiographic and manometric assessment of urethral sphincter function. *J. Urol.* 118: 632–635 (1977).
76. P. J. Mease, S. S. Overman, D. J. Green. Clinical symptoms/signs and laboratory features in symptomatic patients with silicone breast implants. *Arthritis Rheum.* 38(suppl.):S324.
77. K. Miyoshi, T. Miyaoka, Y. Kobayashi, K. Nishijo, M. Higashibara, et al. Hypergammaglobulinemia by prolonged adjuvanicity in man: disorders developed after augmentation mammoplasty. *Jpn. Med. J.* 2122:9–14 (1964).
78. W. W. Montgomery. Current modifications of the salivary bypass tube and tracheal T-tube. *Ann. Otol. Rhinol. Laryngol.* 95:121–125 (1986).
79. W. W. Montgomery. Silicone tracheal cannula. *Ann. Otol. Rhinol. Laryngol.* 89:521–528 (1980).
80. W. W. Montgomery, S. K. Montgomery. Manual for use of montgomery laryngeal, tracheal, and esophageal prostheses: update 1990. *Ann. Otol. Rhinol. Laryngol.* 99:22–28 (1990).
81. B. C. Murless. The injection treatment of stress incontinence. *J. Obstet. Gynaecol. Br. Emp.* 45:67–71 (1938).
82. J. G. Murray, A. D. Gean, R. M. Barr. Intraocular silicone oil for retinal detachment in AIDS: CT and MR appearances. *Clin. Radiol.* 51:415–417 (1996).
83. D. B. Nelson, A. M. Axelrad, D. E. Fleischer, R. A. Kozarek, S. E. Silvis, M. L. Freeman, S. B. Benjamin. Silicone-covered wallstent prototypes for palliation of malignant esophageal obstruction and digestive–respiratory fistulas. *Gastrointestinal Endoscopy* 45:31–37 (1997).
84. J. J. Niebauer, R. M. Landry. Dacron-silicone prosthesis for the metacarpophalangeal and interphalangeal joints. *Hand* 3:55–61 (1971).
85. J. J. Niebauer, J. L. Shaw, W. W. Shaw. Silicone-dacron hinge prosthesis. Design, evaluation, and application. *Ann. Rheum. Dis.* 28(suppl.):56–58 (1969).
86. P. H. Nijhuis, T. E. van den Bogaard, M. J. Daemen, C. G. Baeten. Perianal injection of polydimethylsiloxane (bioplastic implants) paste in the treatment of soiling: pilot study in rats to determine migratory tendency and locoregional reaction. *Dis. Colon Rectum* 41:624–629 (1998).
87. O. Nishi, K. Nishi. Accommodation amplitude after lens refilling with injectable silicone by sealing the capsule with a plug in primates. *Arch. Ophtalmol.* 116:1358–1361 (1998).

88. O. Nishi, K. Nishi, C. Mano, M. Ichihara, T. Honda. Controlling the capsular shape in lens refilling. *Arch. Ophthalmol.* 115:507–510 (1997).
89. W. Noll. *Chemistry and Technology of Silicones*. Academic Press, New York, 1968.
90. J. Orrelana, S. A. Teich, R. M. Lieberman, S. Restrepo, R. Peairs. Treatment of retinal detachments in patients with the acquired immune deficiency syndrome. *Ophthalmology* 98:939 (1991).
91. V. Parsonnet, et al. 1962. An intracardial bipolar electrode for interior treatment of complete heart block. *Ann. Cardiol.* 261 (1962).
92. J. P. Pryor. Surgery for impotence. *Br. J. Hosp. Med.* 139: 141–142 (1981).
93. J. Sanchez-Guerrero, P. H. Schur, J. S. Sergent, M. H. Liang. Silicone breast implants and rheumatic disease. *Arthritis Rheum.* 37:158–168 (1994).
94. J. Sanchez-Guerrero, M. H. Liang. Silicone breast implants and connective tissue diseases. No association has been convincingly established. *BMJ* 309:822–823 (1994).
95. R. P. Santarosa, J. G. Blaivas. Periurethral injection of autologous fat for the treatment of sphincteric incontinence. *J. Urol.* 151:607–611 (1994).
96. J. A. Sazy, D. J. Smith, J. D. Crissman, J. P. Heggers, M. C. Robson. Immunogenic potential of carpal implants. *Surgical Forum* 27:606–608 (1986).
97. J. D. Scott. Silicone oil as an instrument. In: *Retina*, Vol. 3, S. Ryan (Ed.). C. V. Mosby, St. Louis, 1989.
98. F. B. Scott, W. E. Bradley, G. W. Timm. Treatment of urinary incontinence by implantable prosthetic sphincter. *Urology* 1:252–259 (1973).
99. B. O. Shoaib, B. M. Patten, D. S. Calkins. Adjuvant breast disease: an evaluation of 100 symptomatic women with breast implants or silicone fluid injections. *Keio J. Med.* 43:79–87 (1994).
100. B. O. Shoaib, B. M. Patten. Human adjuvant disease: presentation as a multiple sclerosis–like syndrome. *South Med. J.* 89:179–188 (1996).
101. B. G. Silverman, S. L. Brown, R. A. Bright, R. G. Kaczmarek, J. B. Arrowsmith-Lowe, D. A. Kessler. Reported complications of silicone gel breast implants: an epidemiologic review. *Ann. Int. Med.* 124:744–756 (1996).
102. M. I. Singer, E. D. Blom. An endoscopic technique for restoration of voice after laryngectomy. *Ann. Otol. Rhinol. Laryngol.* 90:529–533 (1980).
103. L. Slaughter, J. E. Morris, A. Starr. Prosthetic valvular endocaditis: a 12-year review. *Circulation* 47:1319 (1973).
104. A. L. Smith (Ed.). *Analysis of Silicones*, Robert E. Krieger, Malabar, FL, 1983.
105. D. J. Smith, J. A. Sazy, J. D. Crissman, X. T. Niu, M. C. Robson, J. P. Heggers. Immunogenic potential of carpal implants. *J. Surgical Res.* 48:13–20 (1990).
106. A. Starr, M. L. Edwards. Mitral replacement: clinical experience with a ball-valve prosthesis. *Ann. Surg.* 154:726 (1961).
107. A. D. Steffe, R. D. Beckenbaugh, R. L. Linscheid, et al. The development, technique, and early clinical results of total joint replacement for the metacarpophalangeal joint of the fingers. *Orthopedics* 4:175–185 (1981).
108. D. H. Stemerman, D. F. Caroline, M. Dabezies, V. P. Mercader, B. Krevsky, R. A. Gatenby. Nonexpandable silicone esophageal stents for treatment of malignant tracheoesophageal fistulas: complications and radiographic appearances. *Abdom. Imaging* 22:14–19 (1997).
109. A. B. Swanson. Disabling arthritis at the base of the thumb. Treatment by resection of the trapezium and flexible (silicone) implant arthroplasty. *J. Bone Joint Surg.* 54A:456–471 (1972).
110. A. B. Swanson. Flexible implant arthroplasty for arthritic finger joints. Rationale, technique, and results of treatment. *J. Bone Joint Surg.* 54A:435–455 (1972).
111. A. B. Swanson. Flexible implant arthroplasty for arthritic disabilities of the radiocarpal joint. A silicone-rubber intramedullary stemmed flexible hinge implant for the wrist joint. *Orthop. Clin. North Am.* 4:383–394 (1973).
112. A. B. Swanson. Silicone rubber implants for replacement of arthritic or destroyed joints in the hand. *Surg. Clin. North Am.* 48:1113–1127 (1968).
113. A. B. Swanson, S. G. deGrott. Disabling osteoarthritis in the hand and its treatment. In: *AAOS Symposium on Osteoarthritis*. C. V. Mosby, St. Louis, 1974, pp. 196–232.
114. A. B. Swanson, S. G. deGrott. Flexible implant arthroplasty of the radiocarpal joint—surgical technique and long-term results. In: *AAOS Symposium on Total Joint Replacement of the Upper Extremity*, A. Inglis (Ed.). C. V. Mosby, St. Louis, 1982, pp. 301–316.
115. A. B. Swanson, S. G. deGrott. Joint replacement in the metacarpophalangeal joint. In: *AAOS Symposium on Total Joint Replacement of the Upper Extremity*, A. Inglis (Ed.). C. V. Mosby, St. Louis, 1982, pp. 217–237.
116. A. B. Swanson. Low-modulus force dampening materials for knee joint prostheses. *Acta Orthop. Belg.* 39:116–137 (1973).
117. A. B. Swanson. Flexible implant resection arthroplasty of the elbow. In: *Orthopaedic Surgery and Traumatology*. Proceedings of the 12th Congress of the International Society of Orthopaedic Surgery and Traumatology, Tel Aviv, October 1972, J. Delchef, R. deMarneffe, E. VanderElst (Eds.). Excerpta Medica, Amsterdam, 1973, pp. 894–895.
118. A. B. Swanson, J. H. Herndon. Arhtroplasty of the elbow. In: *The Elbow*, T. G. Wadsworth (Ed.). Churchill-Livingstone, London, 1982, pp. 303–345.
119. A. B. Swanson. Bipolar implant shoulder arthroplasty. In: *Surgery of the Shoulder*, J. E. Bateman, R. P. Welsch (Eds.). B. C. Decker, Toronto, 1984, pp. 211–223.
120. A. B. Swanson, G. deGroot Swanson, B. K. Maupin, J. N. Wei, M. A. Khalil. Bipolar implant shoulder arthroplasty. *Orthopedics* 9:343–351 (1986).
121. A. B. Swanson. Bipolar implant shoulder arthroplasty. *Rheum.* 12:103–123 (1989). In: *Rheumatoid Arthritis Surgery of the Shoulder*. A. W. F. Lettin, C. Peterson (Eds.). Karger, Basel.
122. A. B. Swanson, G. deGroot Swanson, A. B. Sattel, R. D.

Cendo, D. Hynes, J. N. Wei. Bipolar implant shoulder arthroplasty. *Clin. Orthop. Rel. Res.* 249:227–247 (1989).

123. A. B. Swanson, G. deGroot Swanson. Elbow joint reconstruction. In: *The CIBA Collection of Medical Illustrations*, Vol. 8, Part 2, *Musculoskeletal System*. CIBA-GEIGY, Summit, NJ, 1990, pp. 261–262.

124. A. B. Swanson, G. deGroot Swanson, K. Masada, M. Makino, P. R. Pires, D. M. Gannon, A. B. Sattel, V. A. Cestari. Constrained total elbow arthroplasty. *Rheum.* 15: 113–126 (1991). In: *Rheumatoid Arthritis Surgery of the Elbow*, M. Hamalainen, F. W. Hagena (Eds.). Krager, Basel.

125. A. B. Swanson. Silicone arthroplasty in the hand, upper extremity, and foot—25 years' experience. In: *Silicone in Medical Devices*. Conference Proceedings, FDA, Center for Devices and Radiological Health, Baltimore Maryland, February 1–2, 1991, pp. 233–259.

126. Y. Takimoto, N. Okumura, T. Nakamura, T. Natsume, Y. Shimizu. Long-term follow-up of the experimental replacement of the esophagus with a collagen-silicone composite tube. *ASAIO Journal* 39:M736–M739 (1993).

4

Biodegradable Polymers as Drug Carrier Systems

Abraham J. Domb, Neeraj Kumar, Tzviel Sheskin, Alfonso Bentolila, Joram Slager, and Doron Teomim
The Hebrew University of Jerusalem, Jerusalem, Israel

I. INTRODUCTION

Over the past decade the use of polymeric materials for the administration of pharmaceuticals and as biomedical devices has increased dramatically. The most important biomedical applications of biodegradable polymers are in the areas of controlled drug delivery systems (1–4) and in the form of implants and devices for fracture repairs (5,6), ligament reconstruction (7), surgical dressings (8), dental repairs, artificial heart valves, contact lenses, cardiac pacemakers, vascular grafts (9), tracheal replacements (10), and organ regeneration (11). The purpose of this chapter is to review the chemistry, properties, and formulation procedures of the different biodegradable polymers actually available and to describe some of their release kinetics, safety, and biocompatibility considerations.

II. BIODEGRADABLE POLYMERS

A. Polyesters

1. Lactide and Glycolide Copolymers

Linear polyesters of lactide and glycolide have been used for more than three decades for a variety of medical applications (5–8,12–14). Extensive research has been devoted to the use of these polymers as carriers for controlled drug delivery of a wide range of bioactive agents for human and animal use (15). They have been used for the delivery of steroids (16,17), anticancer agents (18), peptides and proteins (19), antibiotics (20), anesthetics (21,22), and vaccines (23). Injectable formulations containing microspheres of lactide/glicolide polymers have received the most attention in recent years.

a. Synthesis

Linear lactide- and glycolide-based polymers are most commonly synthesized by ring-opening melt polymerization of lactide and glycolide at 140–180°C for 2–10 h using a catalyst (24,25). When polymerization temperature is less than the melting point of polymer (~175°C), crystallization of the polymerizing polymer occurs, resulting in solid-state polymerization as with poly(glycolide) (PG). Solid-state polymerization has been found useful for very high molecular weight polymers, around 1000 kD (26). The polymerization reaction was studied and several mechanisms were proposed including cationic (27–30), anionic (31–34), and coordination–insertion (35,36). The common polymerization catalysts are tin derivatives such as tin octoate or tin hexanoate. A hypothetical mechanism of the ring-opening polymerization of lactide using a tin catalyst was suggested by Kissel et al. (Scheme 1) (37).

In this mechanism, Lewis acid character of the tin catalyst activates the ester carbonyl group in the dilactone (a, in Scheme 1). The activated species reacts with the alcohol initiator to form an unstable intermediate, which opens to the ester alcohol (b, in Scheme 1). The propagation reaction proceeds by tin catalyst activation of another dilactone carbonyl group and reaction with the hydroxyl end group.

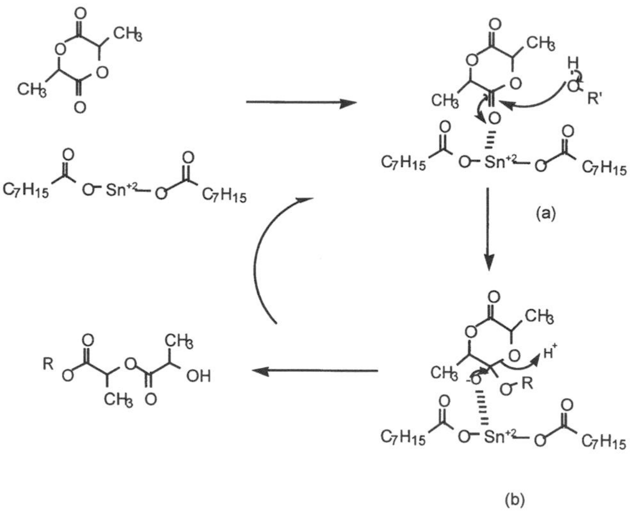

Scheme 1 Hypothetical mechanism of ring-opening polymerization using tin octoate as catalyst. (From Ref. 37).

Ring-opening rate for cyclic lactone can be increased by activation of a Zn- or Sn-based catalyst with carbonyl ester. Stannous octoate $S''[CO_2CH(''Bu)(Et)_2]_2$ is commonly used because it has FDA approval as a food stabilizer (38). Alternatively, resorbable Fe(II) salts have been utilized as initiators for lactide polymerization above 150°C (39). Zinc powder and CaH_2 have also been evaluated as potential nontoxic catalysts for copolymerization of poly(lactic acid) (PLA) with poly(ethylene oxide) (PEG) (40). Lauryl alcohol is generally added to control the molecular weight of the polymer (24,25). Polymers with a molecular weight as high as 500,000 can be obtained by the melt process when high purity monomers (>99.9%) are used (37). Since $Sn(OCT)_2^-$ promoted polymerizations are hardly controlled, a variety of organometallic derivatives, particularly metal alkoxides, are continuously tested as initiators in the polymerization of lactides (41,42). The aluminum alkoxide initiators are the most versatile and readily available. Both monoalkoxides (R_2-AlOR′) and trialkoxides ($Al(OR)_3$) were applied by several groups (43,44). By using only one form of the trialkoxide isopropyloxide ($Al(O\text{-}iPr)_3$), namely, its trimer, a perfect control of polymerization can be achieved.

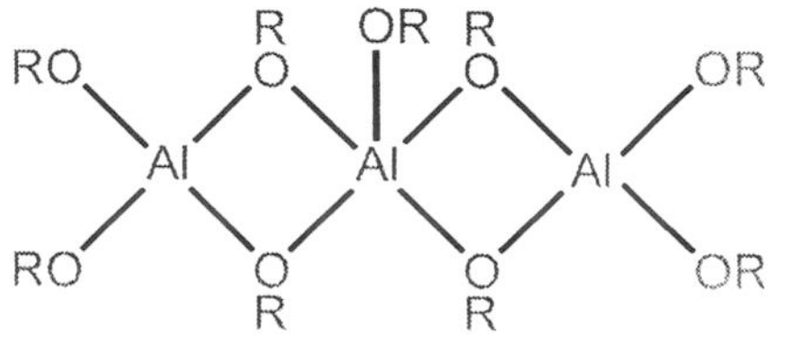

$R = CH(CH_3)_2$

Polymerization of L,L- and (L,L + D,D)-LA initiated with $Al(O\text{-}iPr)_3$ is claimed to be the first example of fully controlled synthesis of high molecular weight PLA (45). The aluminum trialkoxide–growing species belong to the most selective ones, in comparison with other metal (e.g., K, Sm, La, Fe, Sn, Ti) alkoxides.

Low molecular weight poly(lactic acid) (<3000 kD) can be synthesized by direct polycondensation of lactic or glycolic acid using phosphoric acid, *p*-toluene sulfonic acid, and antimony trifluoride as acid catalysts (46).

An interesting method of polymerization of α-hydroxy acids was offered by Tighe and coworkers (47–49), which did not receive much attention. In this method, the anhydrosulfite (I) and the anhydrocarboxylate (II) cyclic derivatives of α-hydroxy acids are polymerized in refluxing dioxane or nitrobenzene for 18–52 h with alcohol catalysis (Scheme 2).

The polymerization was preceded by thermal decomposition of (I) to give the sulfur dioxide and α-lactone (III). The α-lactone polymerizes by ring-opening polymerization using an alcohol catalysis to give exclusively poly(α-hydroxyalkanoic acid). The formation of (III) was rate controlling, and the subsequent polymerization was so rapid that sulfur dioxide and polymer were formed simultaneously. Pure polymers in >80% yield and molecular weights of MN = 20,000 were reported (49).

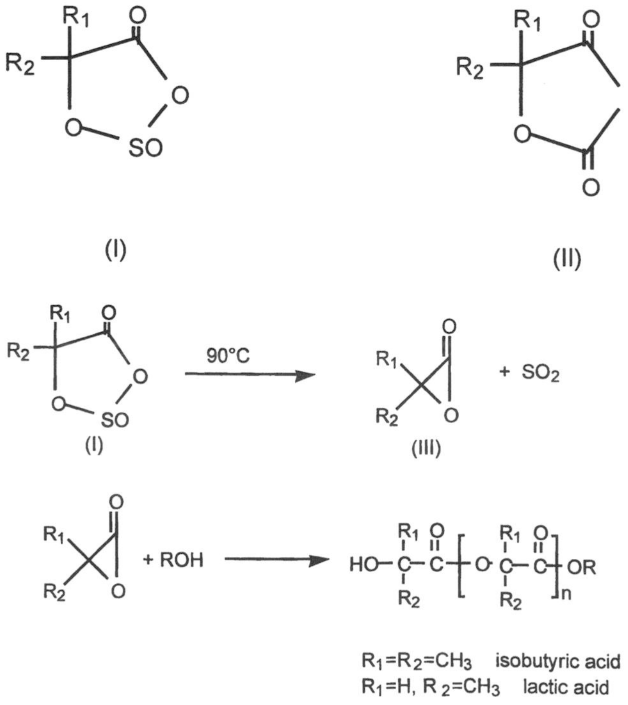

$R_1=R_2=CH_3$ isobutyric acid
$R_1=H$, $R_2=CH_3$ lactic acid

Scheme 2 Polymerization of α-hydroxy acids using anhydrosulfite cyclic derivatives of α-hydroxy acids.

Branched lactide polymers prepared from the polymerization of lactide and pentaerithritol as branching agent were reported. These branched polymers are degraded in a monophasic character, and their degradation is in general faster than for the linear polymers (50). Various polyol molecules were used to prepare a range of branched structures with different properties (37,51).

Copolymers of LA/GA 55:45 branched with 0.2–1 wt% of small polyols like manitol, cyclodextrine, pentaerithritol, and xylitol had a weight average molecular weight of 15,000 to 46,000. The copolymers branched with 1 wt% of poly(vinyl alcohol) of 3,000 to 72,000 molecular weight range resulted in a tenfold increase in molecular weight of 182,000 to 676,000, respectively. Star poly(ethylene oxide)-polylactide copolymers in a spherical form were obtained from the block copolymerization of lactic acid on the end groups of three or four arm poly(ethylene oxide) molecules (51).

Blends of low and high molecular weight poly(DL-lactide) were studied as carriers for drugs. As expected, the addition of low molecular weight polymer accelerated the release of drug from the blend formulation (52).

Different interactions are known to be involved in complex formation between two polymer molecules. It was found that PLA complex was formed in a solution of both enantiomers in chloroform. The reaction proceeds very slowly at room temperature (53). First, gel formation was noticed which turned into a precipitate only after about 30 days. For high molecular weight PLA the reaction takes even longer, up to 1 year. A quicker procedure was found by precipitating D-PLA and L-PLA together. This can be achieved by pouring a solution of D- and L-PLA in dichloromethane into a nonsolvent like methanol or ether. When a film is casted from a solution of low molecular weight isotactic PLA enantiomers in dichloromethane or chloroform, preferably stereocomplex crystals are formed. With PLA of higher molecular weight, less stereocomplex formation was noticed. An easy procedure to obtain the stereocomplex both for low and for height molecular weight PLA consists of heating a mixture of both enantiomers at 60°C in acetonitrile. Poly(lactic acid) was soluble in acetonitrile only at elevated temperatures (from 52°C upwards). Within several hours the solution became turbid and a precipitate was obtained after 3 days. It is suggested that complex formation takes place during precipitation from a solvent. Neither L-PLA or D-PLA gives a reaction on its own in various solvents under similar conditions (54,55). Differential scanning calorimetry (DSC) shows a shift of the melting temperature of about 50°C (from T_m = 180°C for D- or L-PLA, MW = 100,000, to T_m = 230°C for PLA stereocomplex of similar molecular weight) (54–56).

Scanning electron microscope (SEM) images revealed the formation of particles whose shape was influenced by the initial concentration of polymer. In concentrations up to 10 mg/mL discs were formed either with or without cavity in the middle (shape of a red blood cell). Polymers of higher molecular weight formed spheres consisting of a fiberlike structure. If polymer was used of 150,000 or higher molecular weight, a three-dimensional network could be noticed in which the previously mentioned spheres were absorbed (55).

The PLA stereocomplex appears to be soluble in dichloromethane or hexafluoroisopropanol (HFIP). By casting a low-concentrated solution on mica crystals PLA stereocomplexes were obtained. As opposed to isotactic PLA several reports indicate the formation of triangle-shaped crystals. There is a difference in opinion as to whether the PLA stereocomplexes consist of 3_1-helixes (β-form) or 10_3 helixes (α form). Brizollara et al. predicted 3_1 helix on the basis of computer calculations. The D- and L-helixes are intertwined forming ''double strand'' helices, which are packed in a hexagonal cell forming triangle-shaped crystals (57). On the other hand x-ray data suggested no change in the 10_3-helical conformation. ''Frustration'' of the packing of the L- and D-helixes was suggested to cause the triangular shape (58). According to the latter view, the helices are not intertwined, as was also suggested in earlier reports for stereocomplexes of other polymers (59,60). Recently it has been reported that both conformations exist in the PLA stereocomplex: 10_3 helix with a minimum of 11 units and 3_1-helices with a minimum of 7 monomers since only two helical turns are needed for complexation reaction (61). Also, stereocomplexes of PLA containing PEG have been synthesized and characterized by means of DSC and various microscopic techniques (62).

b. *Polymer Properties*

Polyesters based on poly(lactic acid), poly(glycolic acid), and poly(lactic-co-glycolic acid) are found as the best biomaterial with regard to design and performance. Among them, lactic acid contains an asymmetric α-carbon atom with three different isomers as D-, L-, and DL-lactic acid. The physiochemical properties of optically active homopolymers poly(D-lactic acid) (PDLA) and poly(L-lactic acid) (PLLA) are the same, whereas the racimic PLA has very different characteristics (63). Racimic PLA and PLLA have T_g of 57 and 56°C, respectively, but PLLA is highly crystalline with a T_m of 170°C, and racemic PLA is purely amorphous. The polymer characteristics are affected by the comonomer composition, the polymer architecture, and molecular weight. The crystallinity of the polymer is an important factor in polymer biodegradation and varies with the stereoregularity of the polymer. Sterilization using γ-irradiation decreases the polymer molecular weight by 30 to 40% (64). The irradiated polymers continue to decrease in molecular weight during storage at room temperature. This decline in molecular weight affects the mechanical properties and the release rate from the polymers. Poly(lactic acid) and its copolymers with less than 50% glycolic acid content are soluble in common solvents such as chlorinated hydrocarbons, tetrahydrofuran, and ethyl acetate. Poly(glycolic acid) is insoluble in common solvents, but soluble in hexafluoroisopropanol and hexafluoroacetone sesquihydrate (HFASH). In its highly crystalline form, PGA has a very high tensile strength (10,000–20,000 psi) and modulus of elasticity (~1,000,000 psi) (65).

The solubility parameters of several polymers were determined by Siemann (66). The solubility parameters were in the range of 16.2 and 16.8, which is comparable to those of polystyrene and polyisoprene (67).

A comparison study on the physicomechanical properties of several biodegradable polyesters was reported (68). The thermal properties, tensile properties, and the flexural storage modulus as a function of temperature were determined. The following polymers were compared: poly(L-lactic acid), poly(DL-lactic acid), poly(glycolic acid), poly(e-caprolactone), poly(hydroxybutirate) and copolymers with hydroxyvaleric acid, and poly(trimethylene carbonate). The thermal and mechanical properties of several of the polymers tested are summarized in Table 1 (68). A comprehensive review on the mechanical properties of several biodegradable materials used in orthopedic devices has been published (69). The tensile and flexural strength and modulus, as well as the biodegradation of various lactide/glycolide polymers, polyorthoester, and polycaprolactone have been summarized in a tabular or diagram format.

c. *Biodegradation*

Lewis (70), Gopferich (71), Holland et al. (72), and Tracy (73) reviewed the biodegradation behavior of lactide/glycolide polymers. The molecular weight and polydispersity as well as the crystallinity and morphology of the polymers are important factors in polymer biodegradation.

The factors that may affect the polylactide degradation include chemical and configurational structure, molecular weight and distribution, fabrication conditions, site of implantation, physical factors, and degradation conditions (74,75). The degradation of semicrystalline polymers proceeds in two phases: in the first phase the amorphous regions are hydrolyzed, and then the crystalline regions in the second. The polymers degraded by bulk hydrolysis of the ester bonds, which resulted in a decrease in molecular weight with no weight loss.

A comprehensive investigation on the hydrolysis of lactide polymers was described recently by Vert et al. (76). In these studies, a standardized set of experiments was designed. All specimens (2 × 10 × 15 mm) were prepared in a similar way and allowed to hydrolyze at 37°C in dis-

Table 4.1 Thermal and Mechanical Properties of Biodegradable Polyesters[a]

Polymer	MW	T_g (°C)	T_m (°C)	Tensile strength (MPa)	Tensile modulus (MPa)	Elongation	
						Yield (%)	Break (%)
Poly(lactic acid)							
L-PLA	50,000	54	170	28	1,200	3.7	6.0
L-PLA	100,000	58	159	50	2,700	2.6	3.3
L-PLA	300,000	59	178	48	3,000	1.8	2.0
DL-PLA	107,000	51	—	29	1,900	4.0	5.0
Poly(glycolic acid)							
PGA	50,000	35	210	NA	NA	NA	NA
Poly(β-hydroxybutyrate)							
PHB	370,000	1	171	36	2,500	2.2	2.5
P(HB-11%HV)	529,000	2	145	20	1,100	5.5	17
Poly(η-caprolactone)							
PCL	44,000	−62	57	16	400	7.0	80
Poly(trimethylene carbonate)							
PTC	48,000	−15	—	0.5	3	20	160
Poly(orthoesters)							
t-CDM:1,6-HD 35:65	99,000	55	—	20	820	4.1	220
t-CDM:1,6-HD 70:30	101,000	84	—	19	800	4.1	180

Source: Ref. 68.

tilled water or isotonic phosphate buffer. The changes in the polymer during hydrolysis were monitored by weighing for water uptake and weight loss, gel permeation chromatography (GPC) for molecular weight change, and DSC and x-ray scattering for thermal properties and crystallinity change, potentiometry and enzymatic assays for pH change and lactic acid release, and dynamic mechanical tests for changes in mechanical properties (76). The polymer, semicrystalline PLA lost about 50% of its mechanical strength after 18 weeks in buffer, with no weight loss until about 30 weeks of hydrolysis. The degradation of branched PLA was characterized as bulk erosion, like the linear polymers (37).

Vert et al. (77) demonstrated the complexicity of PLA, PGA, and PLGA degradation, and suggested that size dependency exists for hydrolytic degradation of PLA systems. Other research efforts suggest that PLA-derived microparticles will degrade faster than nanoparticles derived from PLA (78,79) and that this phenomenon is based on a diffusion mechanism.

The biodegradation of branched PLA with glucose or macromolecular polyol in rats is determined by weight loss (37). In vitro degradation was essentially the same, indicating minimal involvement of enzymatic degradation. The branched materials degraded much faster than the reference linear PLA. On the contrary, the linear PLA had a higher water uptake than the branched polymers. For example, after 36 weeks the linear PLA contained about 21 wt% water, while the corresponding branched PLA contained only 2%. No adequate explanation was given for this phenomenon.

The degradation of several aliphatic polyesters in the form of microspheres in phosphate buffer solution at 37 and 85°C was reported (80). Lower molecular weight polymers degraded faster than higher molecular weight polymers. Degradation at 85°C resulted in a similar degradation profile, but faster. The biodegradation of low molecular weight PLA used in tablets for oral delivery of drugs was also studied (81).

d. Other Copolymers of Lactic Acid

Low molecular weight homopolymers of D,L-mandelic acid (MA) and its copolymers with lactic acid were reported (82,83).

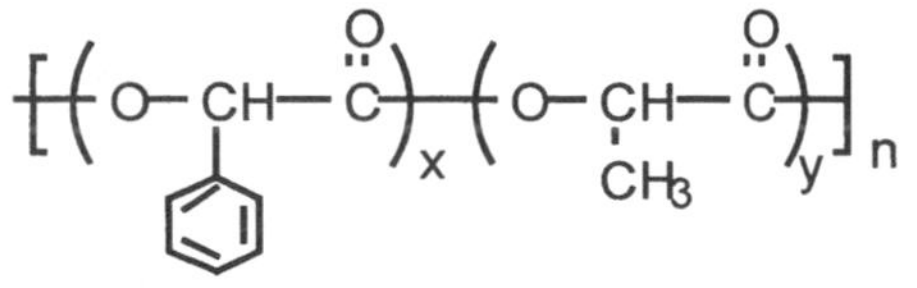

$$\left[\left(O-\underset{C_6H_5}{CH}-\overset{O}{\overset{\|}{C}}\right)_x\left(O-\underset{CH_3}{CH}-\overset{O}{\overset{\|}{C}}\right)_y\right]_n$$

copoly(lactide-mandelate)

The polymers were prepared by direct condensation of mandelic acid and lactic acid at 200°C under vacuum. The copolymers containing 15 to 100% MA were amorphous, and the degradation rate decreased with the increase in MA content with no degradation observed for the MA homopolymer after 15 weeks in buffer at 37°C.

Low molecular weight α-hydroxy acid copolymers composed of 70 mole% L-lactic acid and 30 mole% DL-hydroxy acids of the structure

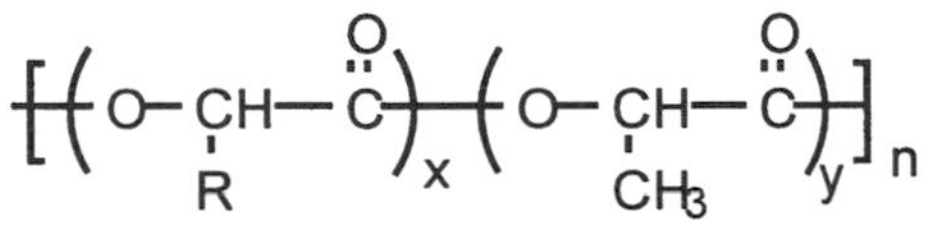

$$\left[\left(O-\underset{R}{CH}-\overset{O}{\overset{\|}{C}}\right)_x\left(O-\underset{CH_3}{CH}-\overset{O}{\overset{\|}{C}}\right)_y\right]_n$$

Monomer	R
DL-LA	$-CH_3$
DL-HBA	$-CH_2CH_3$
DL-HIVA	$-CH(CH_3)_2$
DL-HICA	$-CH_2CH(CH_3)_2$

were synthesized by direct polycondensation in the absence of catalysts (82,84). The polymers had a molecular weight of 5000 and T_g in the range of 20 to 37°C. These polymers were evaluated in vivo for their capabilities as biodegradable carriers for drug delivery systems. Small cylinders (2 × 10 mm) implanted subcutaneously in rats were 60 to 90% degraded in 15 weeks postimplantation. The copolymer L-LA/DL-HBA degraded the least with a lag time of no weight loss for 5 weeks. Polymer cylinders containing luteinizing hormone releasing hormone (LHRH) agonist were implanted in rats and released the drug for 15 weeks, with poly(L-LA/DL-HICA) being the most pharmacologically effective.

2. Polycaprolactones

The successful use of lactide and glycolide polymers in absorbable drug delivery systems and medical devices and absorbable sutures encouraged the evaluation of other polyesters for this purpose. The most studied polymers in this category are the polycaprolactones, polyhydroxybutyrates, and polymers of other α-hydroxy acids. These polymers were developed originally as synthetic plastics to be degraded by microorganisms in the environment (85,86).

a. Synthesis

Poly(ε-caprolactone) (PCL) has been synthesized from the anionic, cationic, and coordination polymerization of ε-caprolactone. The synthesis of polycaprolactones has been recently reviewed (87,88). A schematic description of caprolactone polymerization using these three types of initiators is shown in Scheme 3 (87).

Various initiators and polymerization conditions were reported for each type of catalyst. Effective anionic reaction systems are tertiary amines, alkali metal alkoxides, and carboxylates in tetrahydrofuran, toluene, and benzene (89). The anionic method of polymerization is most useful for the synthesis of low molecular weight hydroxy-terminated oligomers and polymers (90).

Known cationic catalysts in organic synthesis affect cationic polymerization and include protic acids, Lewis acids, acylating agents, and alkylating agents. The following agents, $FeCl_3$, BF_3, Et_2O, alkyl sulfonate, and trimethylsilyl triflate, have been used in 1,2-dichloroethane at 50°C to yield polymers with a molecular weight range of 15,000 to 50,000. High molecular weight homopolymers and random copolymers with lactides and other lactones were obtained using coordination catalysts such as di-n-butyl zinc, stannous octoate, and alkoxides and halids of Al, Sn, Mg, and Ti. Polymerization occurs at 120°C under argon to yield polymers with a narrow molecular weight distribution (MW/MN = 1.1) and molecular weights above 50,000 (91,92). Polycaprolactone was also obtained from the radical polymerization of 2-methylene-1,3-dioxepane (93).

In contrast to random copolymers, block and graft multicomponent systems are most often multiphase materials with properties that are different from the homopolymers or random copolymers (94). They are useful in improving the phase morphology, the interfacial adhesion and, accordingly, the ultimate mechanical properties of immiscible polymer blends. As an example, block copolymerization of ε-caprolactone (CL) and lactides (LA) allows the permeability of the PCL to be combined with the rapid biodegradation of PLA (95).

Block copolymers of CL and LA were synthesized by ring-opening polymerization of lactides (D,L and L,L) and ε-caprolactone using aluminum isopropoxide as initiator (96). Block copolymerization of CL and LA was also reported by Feng and Song using bimetallic (Al, Zn μ-oxo alkoxides) initiators (97,98). Because of the difference in reactivity, the sequential polymerization of these two monomers can only be achieved when CL is first polymerized followed by the lactide (99). Formation of large amounts of homo-PLA is observed and has been attributed to the increase in the mean degree in association of aluminum alkoxide in toluene from one to three in the presence of CL and LA, respectively. The homopolymer formation can be prevented by the addition of a small amount of an alcohol, like 2-propanol, or the use of Al derivative that bears only one alkoxide group (99).

Degradable block copolymers with polyethylene glycol, diglycolide, substituted caprolactones, and λ-valerolactone were also reviewed (87).

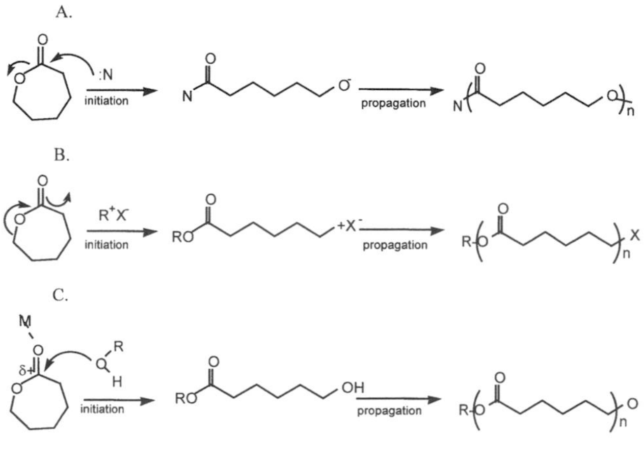

A.

B.

C.

Scheme 3 Polymerization of ε-caprolactone using (A) anionic, (B) cationic, and (C) coordination catalysts.

b. *Polymer Properties*

Polycaprolactone is soluble in chlorinated and aromatic hydrocarbons, cyclohexanone, and 2-nitropropane, and it is insoluble in aliphatic hydrocarbons, diethyl ether, and alcohols. The homopolymer melts at 59–64°C with a T_g of −60°C. Copolymerization with lactide increases the T_g with the increase in the lactide content in the polymer (100). The crystallinity of the polymer decreases with the increase in polymer molecular weight; polymer of 5000 is 80% crystalline, whereas the 60,000 polymer is 45% crystalline (101).

c. *Biodegradation*

The biodegradation of PCL has been extensively studied in the past 30 years and several reviews are available (87,102). Like the lactide polymers, PCL and its copolymers degrade both in vitro and in vivo by bulk hydrolysis (101), with the degradation rate affected by the size and shape of the device and additives.

The kinetic equivalency of the degradation of PCL in buffer and in rabbit was demonstrated by measuring the polymer intrinsic viscosity for 60 weeks (87). The polymers degrade in two phases. In the first phase a random hydrolytic chain scission occurs, which results in a reduction of the polymer molecular weight. In the second phase the low molecular fragments and the small polymer particles are carried away from the site of implantation by solubilization in the body fluids or by phagocytosis, which results in a weight loss. Complete degradation and elimination of PCL homopolymers may last for 2 to 4 years. The degradation rate is significantly increased by copolymerization or blending with lactide and glycolide. The rate of degradation can be increased also by the addition of oleic acid or tertiary amines to the polymer, which catalyzes the chain hydrolysis (103,104).

3. Poly(β-hydroxybutyrate)

a. Synthesis

Poly(β-hydroxybutyrate) (PHB) is made by a controlled bacterial fermentation. The producing organism occurs naturally. An optically active copolymer of 3-hydroxybutyrate (3HB) and 3-hydroxyvalerate (3HV) has been produced from propionic acid or pentanoic acid by *Alcaligenes eutrophus* (105):

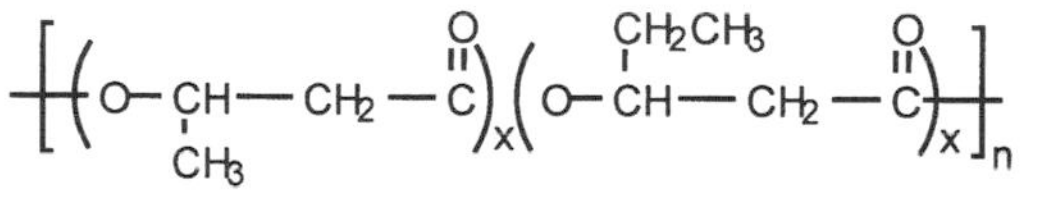

$$\left[\left(O-\underset{CH_3}{CH}-CH_2-\overset{O}{\overset{\|}{C}}\right)_x\left(O-\overset{CH_2CH_3}{CH}-CH_2-\overset{O}{\overset{\|}{C}}\right)_x\right]_n$$

3HB 3HV

The copolymer compositions (0 to 95 mole% 3HV content) can be controlled by the composition of the carbon sources. Random copolymers of 3HB and 4HV were produced from 4-hydroxybutyric acid and butyric acids by *Alcaligenes eutrophus* (106,107):

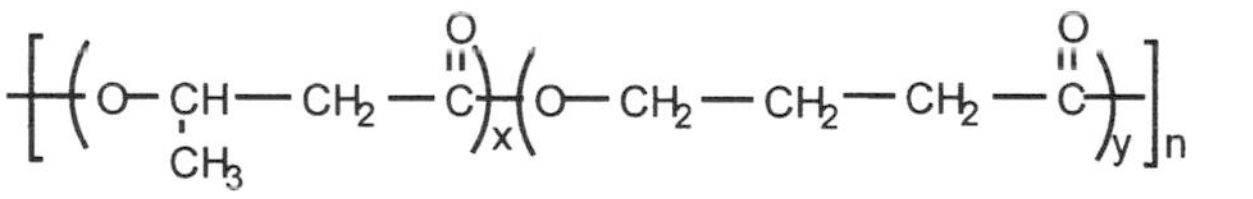

$$\left[\left(O-\underset{CH_3}{CH}-CH_2-\overset{O}{\overset{\|}{C}}\right)_x\left(O-CH_2-CH_2-CH_2-\overset{O}{\overset{\|}{C}}\right)_y\right]_n$$

3HB 4HB

b. Properties and Biodegradation.

The polymers are characterized as having a high molecular weight (>100,000, [n] > 3 dL/g) with a narrow polydispersity and a crystallinity of around 50%. The melting point depends on the polymer composition; P(3HB) homopolymer melts at 177°C with a T_g at 9°C, the 91 : 9 copolymer with 4HB melts at 159°C, and the 1 : 1 copolymer with 3HV melts at 91°C. The PHB properties in the living cells of *Alcaligenes eutrophus* were determined using x-ray and variable-temperature ^{13}C NMR relaxation studies (108). Polyhydroxybutyrate is an amorphous elastomer with a T_g around −40°C in its ''native'' state within the granules. The biodegradation of these polymers in soil and activated sludge show the rate of degradation to be in the following order: P(3HB-co-9% 4HB) > P(3HB) = P(3HB-co-50% 3HV).

The hydrolytic degradation of HB polymers was studied (109). Microspheres degraded slowly in phosphate buffer at 85°C, and after 5 months 20 to 40% of the polymer eroded under these conditions. Copolymers having a higher fraction of 3HV and low molecular weight polymers were more susceptible to hydrolysis.

4. Polycarbonates

Poly(ethylene carbonate) and poly(propylene carbonate) have been tested as biodegradable carriers for the delivery of 5-fluorouracil (5-FU) (110). They are linear thermoplastic polyesters of carbonic acid with aliphatic dihydroxy compounds.

a. Synthesis

The polymers are synthesized from the reaction of dihydroxy compounds with phosgene or with bischloroformates of aliphatic dihydroxy compounds by transesterification and by polymerization of cyclic carbonates:

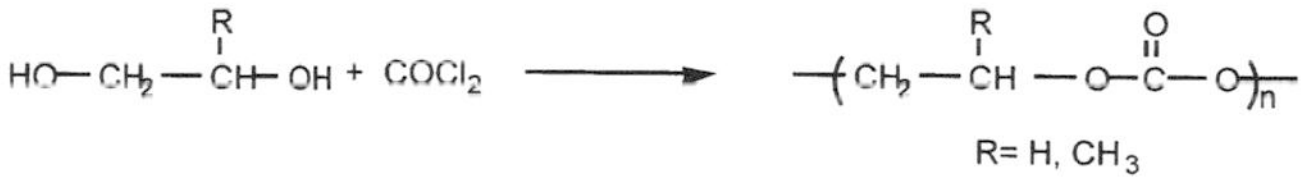

$$HO-CH_2-\overset{R}{CH}-OH + COCl_2 \longrightarrow -\left(CH_2-\overset{R}{CH}-O-\overset{O}{\overset{\|}{C}}-O\right)_n-$$

R= H, CH_3

These polymers have been synthesized from carbon dioxide and the corresponding epoxides in the presence of organometallic compounds as initiators (111).

b. Biodegradation

Since the carbonate linkage may be labile to hydrolysis, the biodegradability of polycarbonates has been studied (110,111). Pellets of poly(ethylene carbonate) and poly(propylene carbonate) were implanted into the peritoneal cavity of rats, and the toxicity and weight loss of polymer pellets were determined. Poly(ethylene carbonate) was completely eliminated 15 days postimplantation, while poly(propylene carbonate) remained intact after 60 days. When pellets of the polymers were incubated in phosphate buffer pH 7.4 and 37°C, both polymers did not degrade even after 40 days. These data indicate that poly(ethylene carbonate) was degraded by enzymes. No visible inflammatory reaction was noted at the implantation sites. In vitro release of 5-FU from poly(ethylene carbonate) pellets containing 20% 5-FU was poor, 20% of the drug was released in 2 h, and only an additional 15% of 5-FU were released during the following 60 days. Better release profiles were obtained when poly(propylene carbonate) was used.

Copolymers of aliphatic carbonates and lactide showed excellent biocompatibility and mechanical properties. Block copolymers of trimethylene carbonate (TMC) and lactide were synthesized from the reaction of the monomers with stannous octoate as catalyst at 160°C for 16 h:

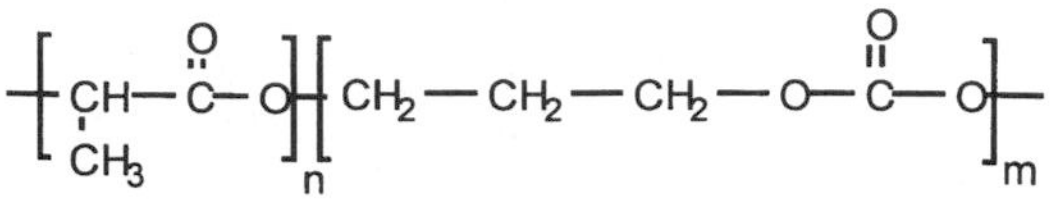

PLA TMC

The polymers were soluble in common organic solvents and had a weight average molecular weight of 90,000. They completely degraded in vivo in 1 year (112).

5. Other Polyesters

Poly(dihydropyrans) were developed for contraceptive delivery. The in vivo and in vitro release of contraceptive steroids and antimalarial agents from polymer matrices has been studied (113).

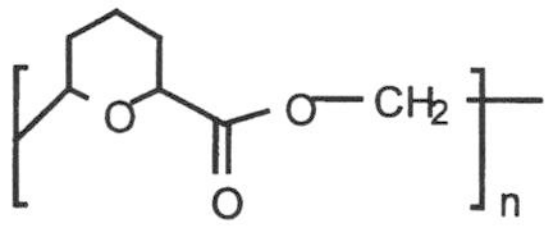

poly(dihydropyrans)

Poly(p-dioxanone) is clinically used as an alternative to poly(lactide) in absorbable sutures with similar properties to poly(lactide) with the advantage of better irradiation stability during sterilization (114):

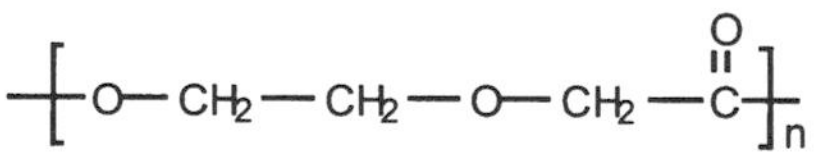

poly(p-dioxanone)

This polymer has not yet been developed as a carrier for controlled drug delivery.

Biodegradable polymers derived from naturally occurring, multifunctional hydroxy acids and amino acid have been investigated by Lenz and Guerin (115). The monomers, malic acid, and aspartic acid were polymerized into a polyester or polyamide using a ring-opening polymerization process as follows:

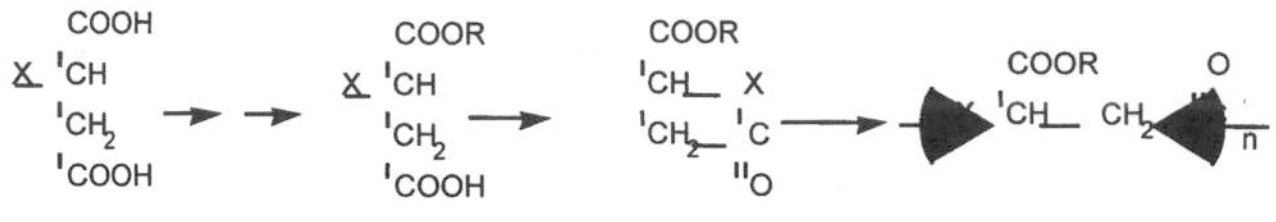

X= O (ester), X=NH (amide)

The molecular weights of the polymers were highly dependent on the purity of the cyclic monomers. Polymers of 50,000 were obtained in 93% polymerization yield when very pure monomers were used. Both polymers are water soluble; however, crystallinity and controlled number of acid groups in the polymer chains can be used to alter water swellability and solubility of the polymers. Biocompatibility evaluation in mice indicated that poly (β-malic acid) is nontoxic (115).

A series of polyesters, poly(propylene fumarate) (PPF), based on the reaction product of fumaric acid (116,117), and tartaric acid (118,119) with different aliphatic diols have also been evaluated as drug carriers. Poly(propylene fumarate) with acrylate and epoxide terminal groups was produced by Domb et al. (120). The polymers were crosslinked in a ratio of 30 wt% PPF to 70 wt% calcium carbonate–tricalcium phosphate mixture and were synthesized by a direct melt condensation of the acids with the diols with acid catalysis.

B. Poly(amides)

1. Synthesis and Biodegradation

The utilization of amide-based polymers, especially natural proteins, in the preparation of biodegradable matrices has been extensively investigated in recent years (121). Microcapsules and microspheres of crosslinked collagen, gelatin, and albumin have been used for drug delivery (122). The synthetic ability to manipulate amino acid sequences has seen its maturity over the last two decades, with new techniques and strategies continually being introduced. Poly(amides) such as poly(glutamic acid) and poly(lysine) and their copolymers with various amino acids have also been studied as drug carriers (123,124). Recently, Nathan and Kohn (125) have excellently reviewed the history of amino acid–derived polymers.

Pseudopoly(amino acids), a new approach for biomaterials based on amino acids, were first suggested by Kohn and Langer (126,127), who prepared a polyester from N-protected trans-4-hydroxy-L-proline, and poly(iminocarbonate) from tyrosine dipeptide as monomeric starting material. The structures of poly(N-acylhydroxyproline esters) and a homologous series of tyrosine-derived polymers is described in Scheme 4 (121).

The properties, biodegradability, drug release, and biocompatibility of this class of polymers have been reviewed (121,122). Biodegradable polyamides for use in controlled delivery of drugs were obtained from the reaction of 2,2′-bis[5(4HO-oxazolones] and alkane diamines (128). Drug release time from these polymers was short, less then 24 h.

Copolymers of glutamic acid and ethylglutamate were used for the delivery of naltrexone (129). The polymers were synthesized in three steps: (1) synthesis of N-carboxyanhydrides of γ-benzyl-L-glutamate and γ-ethyl-L-glutamate; (2) polyamidation of the benzyl blocked monomers; and, (3) debenzylation of the intermediate. Poly(amino acids) are generally hydrophilic, and their deg-

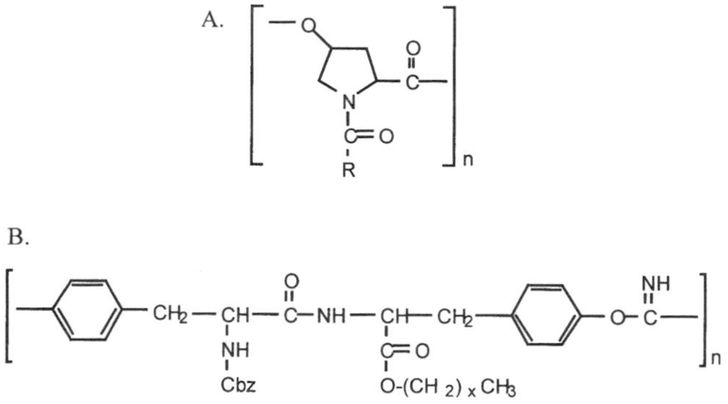

A. $\left[-O-\text{(pyrrolidine ring, N-C(=O)-R)}-C(=O)- \right]_n$

B. $\left[-C_6H_4-CH_2-CH(NH\text{-}Cbz)-C(=O)-NH-CH(C(=O)-O-(CH_2)_xCH_3)-CH_2-C_6H_4-O-C(=NH)- \right]_n$

Scheme 4 Molecular structure of pseudopoly(amino acids). (A) Poly(N-acylhydroxyproline esters); (B) Tyrosine-derived polyiminocarbonate, poly(CTTE), x = 1; poly(CTH), x = 4; poly(CTTP), x = 15.

radation rates are dependent upon hydrophilicity of the amino acids.

The degradation rate increases as the glutamic acid content increases. For example, copolymer containing 13 mole% glutamic acid remained intact for more than 79 days, while the 40% copolymer disintegrated in 7 days. Tubular capsules of 18:82 glutamic acid/ethylglutamate copolymer were used as a reservoir implant. The copolymer was biocompatible and completely degraded in 90 days. The degradation process involves hydrolysis of the ethyl esters followed by hydrolysis of the peptide bonds to produce glutamic acid, which enters the metabolic cycle. Amino acids are attractive due to their functionality by which they provide a polymer. Recently, poly(lactic acid-co-lysine) (PLAL) was synthesized using a stannous octoate catalyst from lactide and a lysine-containing monomer analogous to lactide (130). Inclusion of the amino acid lysine provides an amino group that allows for further modification of the PLAL system. In a recent report, Cook and coworkers (131) have successfully attached a peptide sequence which promotes cell adhesion to PLAL. The use of N-carboxyanhydride–activated amino acids was the first efficient method for production of amino acid homopolymers. In a further study, they have exploited the PLAL system by the reaction with lysine N-carboxyanhydride derivatives to increase the system's functionality with a poly(lysine) graft. Poly(lactic acid-co-lysine) has been formulated into microspheres that exhibit deep lung delivery from porous particles (132).

Copolymers of n-hydroxyalkyl-L-glutamine and γ-methyl-L-glutamate were synthesized from the reaction of poly(γ-methyl-L-glutamate) and 2-amino-1-ethanol or 5-amino-1-pentanol as follows (133):

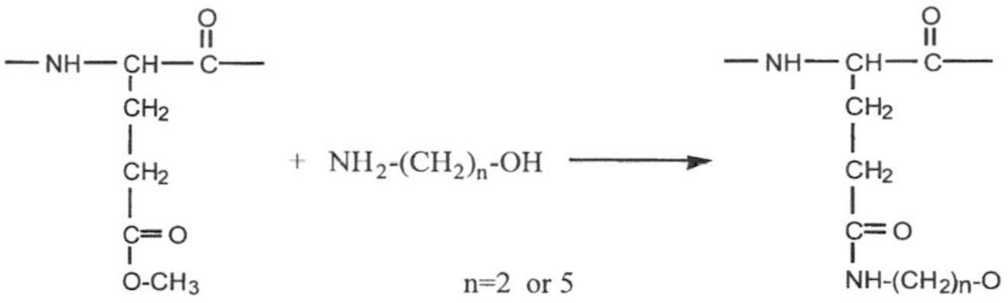

$$-NH-CH(CH_2-CH_2-C(=O)-O\text{-}CH_3)-C(=O)- \; + \; NH_2\text{-}(CH_2)_n\text{-}OH \longrightarrow -NH-CH(CH_2-CH_2-C(=O)-NH\text{-}(CH_2)_n\text{-}O)-C(=O)-$$

n=2 or 5

The in vitro enzymatic degradation using pronase E as protease showed that these polymers degrade from the surface, and the polymers were completely dissolved in from several hours to several days as a function of polymer structure and enzyme concentration. Negligible weight loss was detected without the enzyme.

Random copolymers of the α-amino acids N-(3-hydroxypropyl)-L-glutamine and L-leucine were synthesized and used as carriers for naltrexone (134). Naltrexone was covalently bound through the 3-phenolic or the 14-tertiary hydroxyls to the polymer hydroxyl side chains via a carbonate bond. Naltrexone was released from the polymer in a relatively constant way for 30 days, in both in vitro and in vivo experiments in rats (134). Tyrosine-derived poly(carbonates) are readily processible polymers that support the growth and attachment of the cells and have also shown a high degree of tissue biocompatibility (135,136). In vitro degradation of these compounds occurs due to hydrolysis of the pandent ester bonds and the imino–carbonate bonds of the backbone (136). Degradation goes over a period of months and rates are comparable to the degradation time of poly(L-lactic acid)

C. Poly(phosphate Esters)

Poly(phosphoester) has been studied as a potential biodegradable matrix for drug delivery (137,138). The polymers were synthesized from the reaction of ethyl or phenyl phosphorodichloridates and various dialcohols including bisphenol A and poly(ethylene glycol) of various molecular weights:

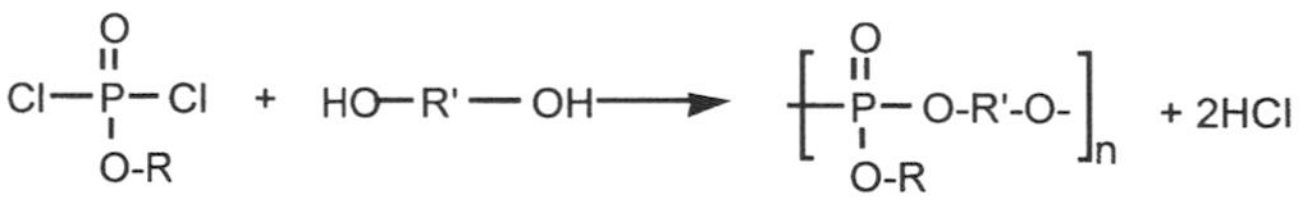

$$Cl-P(=O)(O\text{-}R)-Cl \; + \; HO-R'-OH \longrightarrow \left[-P(=O)(O\text{-}R)-O\text{-}R'\text{-}O- \right]_n \; + \; 2HCl$$

Interfacial condensation using a phase transfer catalyst and bisphenol A as comonomer yielded polymers with a weight average molecular weight around 36,000. Leong et al. have incorporated phosphoester groups into poly(urethanes) (139). Poly(urethanes) have been used as blood contacting biomaterials due to having a broad range of physical properties that can be obtained, from hard and brittle to soft and tacky. Leong et al. designed inert biomaterial for controlled release applications by introducing

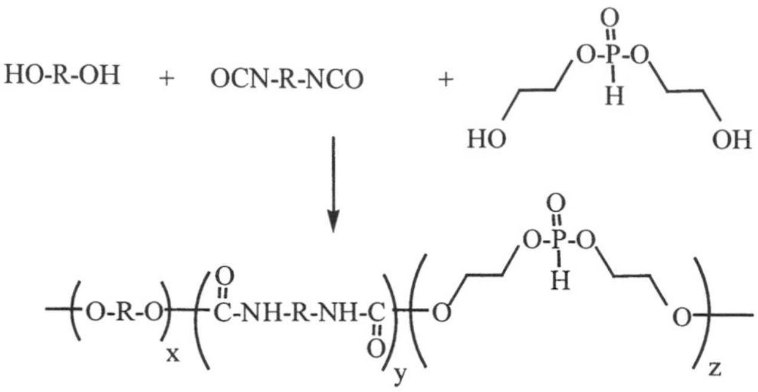

Scheme 5 Synthesis of poly(phosphoester-urethanes).

phosphoester linkage in poly(urethane). Introduction of phosphoester linkage does not change the mechanical properties inherent in the poly(urethanes) and provides an excellent biodegradable material.

Poly(phosphoester-urethanes) are obtained by the reaction of polydiols and di-isocyanates with phosphates as chain extenders as shown in Scheme 5.

Phosphoester bonds are readily cleaved under physiological conditions, and hydrolysis of poly(phosphoester-urethanes) leads to phosphates, alcohols, amines, and carbon dioxide, which makes it an excellent biomaterial for drug delivery applications. Leong et al. observed that the release kinetics of poly(phosphoester-urethanes) were influenced by the side chains attached via the phosphoester of the polymer backbone and thus pentavalency of the phosphorus contributes a site for future functionalization (140). The release mechanism was dependent combinedly on diffusion, swelling, and degradation of polymer.

The polymers based on bisphenol A release drug for a long period of time; 8 to 20% cortisone was released after 75 days in buffer solution. The degradation rate depends on the nature of the polymer side chain; polymers with phenyl side chains degrade much slower than those containing ethyl or ethoxyethyl side chains (138). The in vivo degradation of these polymers in rabbits was faster than in vitro.

D. Polyphosphazenes

The uniqueness of the polyphosphazenes stems from the inorganic backbone (N = P), which with certain organic side groups can be hydrolyzed to phosphate and ammonia. Several polymer structures have been used as matrix carriers for drugs (141,142) or as a hydrolyzable polymeric drug, where the drug is covalently bound to the polymer backbone and released from the polymer by hydrolysis (143). A comprehensive review on the synthesis, characterization, and medical applications of polyphosphazenes was published by Allcock (144).

1. Synthesis and Biodegradation

The polymers are most commonly synthesized by a substitution reaction of the reactive poly(dichlorophosphazene) with a wide range of reactive nucleophiles such as amines, alkoxides, and organometallic molecules (Scheme 6). The reaction is carried out in general at room temperature in tetrahydrofuran or aromatic hydrocarbone solutions. Polymers containing mixed substituent can be obtained from the sequential or simultaneous reaction with several nucleophiles (Scheme 6) (145).

The properties of the polymers depend on the nature of the side groups. Hydrolytically degradable polyphosphazene was obtained when amino acid and imidazole derivatives were used as substituents (142,146). The first bioerodible polymer was the ethylglycinato derivative ($R = NHCH_2COOEt$), which hydrolytically degrades to ethanol, glycine, phosphate, and ammonia (146). In vitro degradation studies on poly(phosphazenes) with different amino acid derivatives show that it takes several months to degrade the material depending on the amino acid present (147). Allcock et al. (148,149) synthesized three different poly[(amino acid ester)phosphazenes]: poly[di(ethylglycinato)phosphazene], poly[di(ethylalanato)phophazene], and poly[di(benzylalano)phosphazene]. According to the order of polymers, they show a tendency to decrease the molecular weight and mass loss with release of small molecules. The release of small molecules occurred through diffusion and decomposition of polymer.

Biodegradable poly(phosphazenes) that are insoluble in water prior to hydrolysis have been employed in the temporal controlled release of many drug classes including nonsteroidal anti-inflammatory agents and peptides (150–153). Amino acid ester–substituted polyphosphazenes have been used for controlled release of the covalently

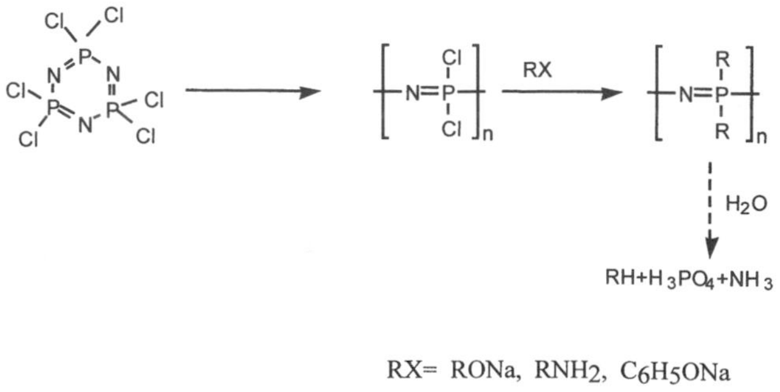

RX= RONa, RNH_2, C_6H_5ONa

Scheme 6 Synthesis and hydrolysis of polyphospazenes.

bonded anti-inflammatory agent naproxen. Steroids having a hydroxyl group were bound to the polymer chain through the hydroxyl group (154).

Poly(organophosphazenes) have been synthesized that possess amino acid side groups. The mechanical properties and rates of degradation have been controlled by appropriate selection of amino acid side chain groups (155). The versatility of these polymers has been demonstrated by the formation of 200-nm diameter poly(organophosphazene) nanoparticles that present covalently coupled poly(ethylene glycol) at their surface (156).

Imidazolyl-substituted polyphosphazenes are hydrolytically unstable and hydrolyze in room moisture (142). The rate of hydrolysis can be slowed by the incorporation of hydrophobic side groups such as phenoxy or methylphenoxy groups (141). The in vivo and in vitro release of progesterone and bovine serum albumin (BSA) from imidazole, 4-methylphenoxy–substituted polyphosphazene matrices were reported (141). Almost 90% of the loaded progesterone was released in 30 days with about 60% released in 8 days when placed in phosphate buffer pH 7.4 at 37°C. An in vivo release study in rats using radiolabelled progesterone demonstrated a zero-order release for 30 days. Various hydrophilic and hydrophobic polymers, hydrogels, water-soluble polymers, bioactive polymers, and various drug–polyphosphazene conjugates have been reviewed (144).

A number of approaches have been proposed to generate crosslinked polyphosphazene for temporal controlled release. Andianov and coworkers have synthesized poly [bis(carboxylatophenoxy)-phosphazene] crosslinked with Ca^{2+} ions for an ionically stabilized system (157). This polymer allowed drug molecules to be encapsulated into poly(phosphazene) microspheres under mild environmental conditions. Recently, pH-sensitive hydrogels have been synthesized by the formation of poly(phosphazenes) with oxybenzoate and methoxyethoxy side group. Swelling at different pH values was controlled by varying the ratios of the two side groups (158).

E. Poly(orthoesters)

Poly(orthoesters) were invented during the pursuit of developing a bioerodible polymer, subdermally implantable, that would release contraceptive steroids by close to zero-order kinetics for at least 6 months (159). An additional objective was that the polymer erosion and drug release should be concomitant so that no polymer remnants are present in the tissue after all the drug has been released. These objectives could only be met if the polymer were truly surface eroding. For a surface-eroding polymer, the erosion process at the surface of the polymer should be much faster than in the interior of the device. To exhibit such a phenomenon, the polymer has to be extremely hydrophobic with very labile linkages. Hence, it was envisioned that polymeric devices with an orthoester linkage in the backbone, which is an acid-sensitive linkage, could provide a surface-eroding polymer if the interior of the matrix is buffered with basic salts.

1. Synthesis

The first poly(orthoesters) were reported in a series of patents by Choi and Heller (160–163), assigned to Alza Corporation. These proprietary polymers were first designated as Chronomer and later as Alzamer. They were prepared by a transesterification reaction, as follows (164):

$$\text{(EtO)(OEt) cyclic orthoester} + \text{HO-R-OH} \longrightarrow \left[\text{O-(cyclic orthoester)-O-R} \right]_n + \text{EtOH}$$

The general synthesis involved the heating of the reaction mixture to 110 to 115°C for about 1.5 to 2 h and then further heated at 180°C and 0.01 torr for 24 h. The synthesis of such polymers, with a minimal amount of crosslinkage, requires an orthoester starting material in which one alkoxy group has a greatly reduced reactivity. This can be achieved by using a cyclic structure as shown in the reaction.

Hydrolysis of these polymers regenerates the diol and γ-butyrolactone (165). The γ-butyrolactone rapidly hydrolyzes to -hydroxybutyric acid. The production of -hydroxybutyric acid would further catalyze the breakdown of orthoester linkages leading to bulk erosion of the matrix. Thus it was decided to incorporate basic salts into the polymer matrix to neutralize the generated acid and keep the hydrolysis process under control. These Alzamer materials were investigated as bioerodible implants for the release of naltrexone and contraceptive steroids. Human clinical trials of the steroidal implant revealed local tissue irritation, and thus further work was discontinued (166). Recently the use of these polymers for the release of indomethacin in the prevention of reossification of experimental bone defects was reported (167).

Subsequently, another family of poly(orthoesters) was developed not related to Alzamer (168). These polymers are prepared by the addition of polyols to diketene acetals. The general reaction can be schematically represented as follows:

$$H_2C{=}C(OR)\text{-O-R'-O-}C(OR){=}CH_2 + \text{HO-R''-OH} \longrightarrow \left[\text{O-}C(OR)(CH_3)\text{-O-R'-O-}C(OR)(CH_3)\text{-O-R''} \right]$$

Initial work was conducted with the monomer, diketene acetal, derived form pentaerythritol, and 1,6-hexanediol. The reaction is exothermic and proceeds to completion virtually instantaneously. The reaction is as follows:

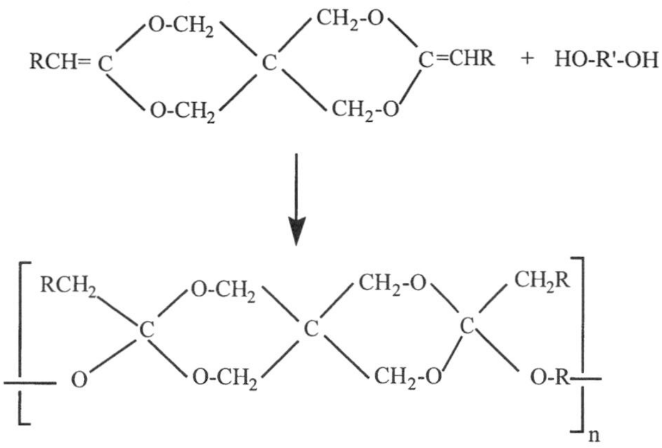

where R = H, for 3,9-bis(methylene 2,4,8,10-tetra oxaspiro [5,5] undecane). The diols investigated were 1,6-hexanediol trans-1,4-cyclohexane dimethanol, 1,6-cyclohexanediol, ethylene glycol, and biophenol A.

2. Polymer Properties

The molecular weight of poly(orthoesters) were significantly dependent on the type of diol and catalyst used for synthesis. A linear, flexible diol like 1,6-hexanediol gave molecular weights greater than 200 K, whereas bisphenol A in the presence of catalyst gave molecular weight only around 10,000 (169).

Mechanical properties of the linear poly(orthesters) can be varied over a large range by selecting various compositions of diols. It was shown that the glass transition temperature of the polymer prepared from DETOSU can be varied from 25 to 110°C by simply changing the amount of 1,6-hexanediol in trans-1,4-cyclohexane dimethanol from 100 to 0% (170). There seems to be a linearly decreasing relationship between the T_g and percentage of 1,6-hexanediol. One could take advantage of this relationship in selecting the polymer for in vivo applications because in vivo the T_g of the polymer would drop due to inbibition of water. This can result in the loss of stiffness and rigidity of the polymer.

3. Crosslinked Poly(orthoesters)

To prepare a crosslinked polymer, there should be at least one monomer which has a functionality greater than 2. In case of poly(orthoesters), it is possible for a ketene acetal or an alcohol to have a functionality greater than 2. Due to the difficulty in preparing trifunctional ketene acetals, triols were used to prepare the crosslinked polymer.

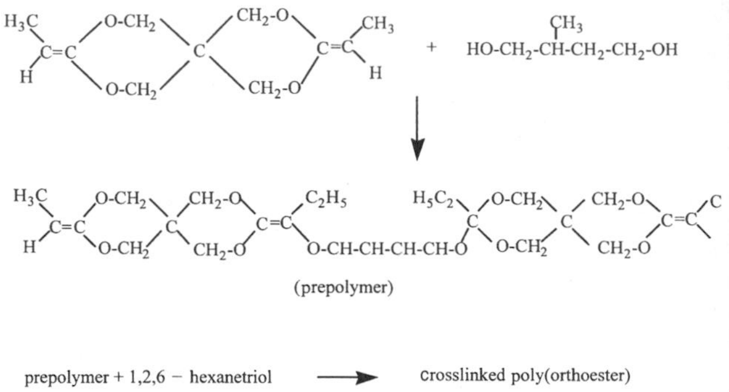

prepolymer + 1,2,6 – hexanetriol ⟶ crosslinked poly(orthoester)

Scheme 7 Synthesis of crosslinked polyorthoesters.

The general method used for the synthesis of crosslinked polymers was by reacting prepolymer with the triols or a mixture of diols and triols (Scheme 7) (171).

The prepolymer is an acetal with a diol and is a viscous liquid at room temperature. Thus, the compound of interest could be incorporated into the prepolymer along with the triol, and the mixture can be crosslinked at temperatures as low as 40°C. This can be a good method for incorporating thermolabile drugs. However, one should be cautious with using compounds with hydroxyl functionality.

In a series of experiments, Heller et al. (172) have reported the family of poly(orthoesters) which can be prepared by reacting a triol with two vicinal hydroxyl groups and one removed by at least three methylene groups with an alkyl orthoacetate as shown:

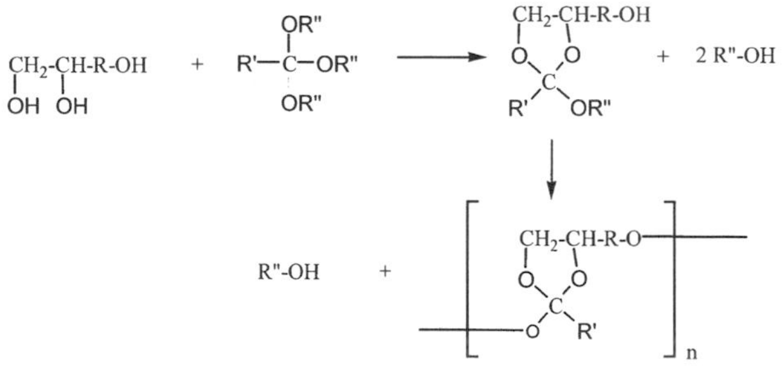

The use of flexible triols such as 1,2,6-hexanetriol produces highly flexible polymers that have ointmentlike properties even at relatively high molecular weights. However, properties such as viscosity, and hydrophobicity can be readily varied by controlling molecular weight and the size of the alkyl group R′. This polymer has an ointmentlike consistency at room temperature and is applicable where sensitive therapeutic agents such as proteins are incorporated into the polymer without use of solvent.

In another experiment, Heller et al. replaced the flexible triol to a rigid one such as 1,1,4-cyclohexanetrimethanol to obtain a solid poly(orthoester), as shown:

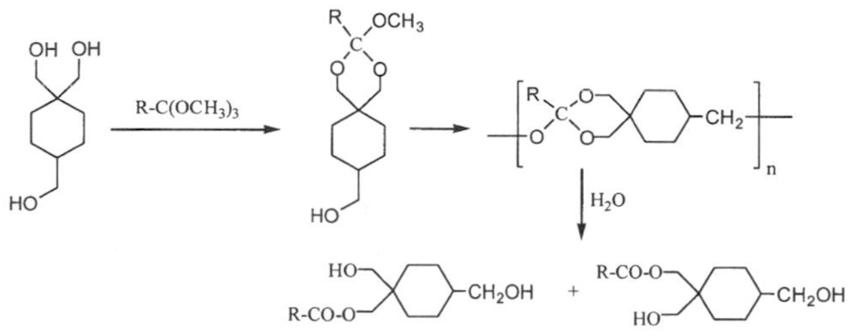

OH OH

$R\text{-}C(OCH_3)_3$ →

HO

R, OCH_3

HO

→ $-CH_2-$ n

H_2O ↓

HO– –CH_2OH + R-CO-O– –CH_2OH

R-CO-O– HO–

4. Polymer Hydrolysis

The primary mechanism for the degradation of poly (orthoesters) is via hydrolysis. Depending on the reactants used during the synthesis of the polymer, the hydrolysis products are diol, or a mixture of diols, and pentaerythritol dipropronate or diacetate if 3,9-bis(methylene-2,4,8,10-tetraoxaspiro [5,5]undecane) was used (144). The pentaerythritol esters hydrolyze at a slower rate to pentaerythritol and the corresponding acetic or propionic acid. The sequence of reaction is as follows:

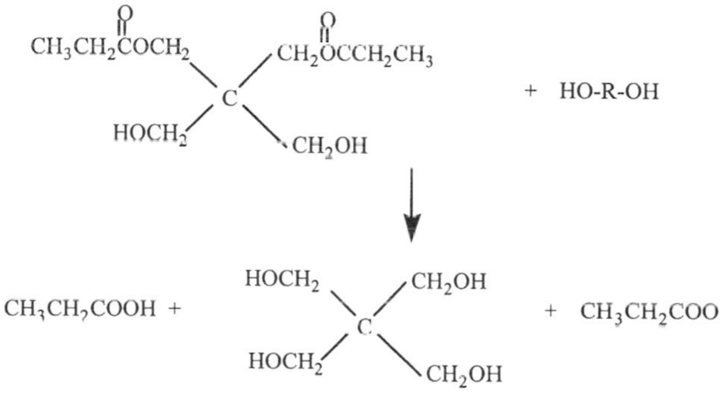

$CH_3CH_2\overset{O}{\overset{\|}{C}}OCH_2$, $CH_2O\overset{O}{\overset{\|}{C}}CH_2CH_3$, C, $HOCH_2$, CH_2OH + HO-R-OH

↓

CH_3CH_2COOH + $HOCH_2$, CH_2OH, C, $HOCH_2$, CH_2OH + CH_3CH_2COO

The difference in the sensitivity of the hydrolysis of orthoester linkages in an acid versus alkaline medium has been used to advantage in designing the orthoester-based delivery systems. Incorporating acid anhydrides into the matrix to accelerate the rate of hydrolysis uses this preferential sensitivity. While, on the other hand, a base is used to stabilize the interior of the matrix.

Acid-catalyzed hydrolysis of these polymers can be controlled by introducing acidic and basic excipients into the matrices. Rate of hydrolysis can be increased by the addition of acidic excipients, e.g., suberic acid, as demonstrated by the zero-order release of 5-fluorouracil over a 15-day period (173). Alternatively, basic excipients stabilize the bulk of the matrix but diffuse out of the surface region. This approach has been used in the temporal controlled release of tetracycline over a period of weeks in the treatment of periodontal disease (174). Ng et al. described the synthesis of self-catalyzed poly(orthoesters) that contain glycolide sequences and can be degraded hydrolytically without excipient catalysis (175).

Wuthrich et al. (176) proposed the hydrolysis of poly (orthoesters) derived from 1,2,6-hexanetriol as follows:

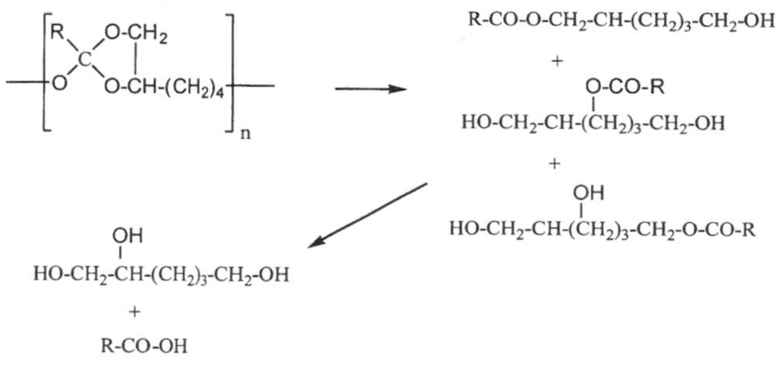

$[R, O\text{-}CH_2, C, O, O\text{-}CH\text{-}(CH_2)_4]_n$ →

$R\text{-}CO\text{-}O\text{-}CH_2\text{-}CH\text{-}(CH_2)_3\text{-}CH_2\text{-}OH$ + $HO\text{-}CH_2\text{-}CH(O\text{-}CO\text{-}R)\text{-}(CH_2)_3\text{-}CH_2\text{-}OH$ + $HO\text{-}CH_2\text{-}CH(OH)\text{-}(CH_2)_3\text{-}CH_2\text{-}O\text{-}CO\text{-}R$

→ $HO\text{-}CH_2\text{-}CH(OH)\text{-}(CH_2)_3\text{-}CH_2\text{-}OH$ + $R\text{-}CO\text{-}OH$

This initial hydrolysis is proceeded at a slow rate to produce a carboxylic acid and a triol, thus no autocatalysis is observed.

5. Polymer Processing

The orthoester linkage is inherently unstable in the presence of water. However, because of the polymer's highly hydrophobic nature, they can be stored without careful exclusion of moisture.

Even though the polymer is relatively stable in trace amounts of moisture, it is unstable to heat and undergoes disproportionation to an alcohol and ketene acetal. The combination of moisture and heat can be fatal for the processing of poly(orthoesters), which are designed to erode within days (177). Thus, if injection molding is necessary to fabricate the device, then moisture must be rigorously excluded during fabrication. One should also consider the interaction between the incorporated anhydride as catalysts, the polymer, and the drug during the thermal processing.

In one of the studies it was shown that phthalic anhydride reacted with the free hydroxyl end groups of the polymers. Using high-performance liquid chromatography and infrared spectroscopy the formation of half phthalate esters of 1,6-hexanediol and trans cyclohexane dimethanol was confirmed (178). The reaction led to the decrease in concentration of phthalic anhydride, leading to increased time for erosion of the matrix. The preferred method for fabricating the devices would be under low thermal and shear stresses (179). This would include solution mixing or powder blending followed by compression molding of the devices.

F. Polyanhydrides

A large number of biodegradable polymers have been investigated as carriers in the design of controlled drug delivery systems. It has been generally recognized that the matrix should undergo heterogeneous degradation to maximize the control over release process. In the early 1980s, Rosen et al. (180) envisioned the use of hydrophobic polyanhydrides in designing the surface-eroding matrix for applications in controlled drug delivery. However, the invention and development of polyanhydrides dates as far back as the early 1900s.

In 1909, Bucher and Slade (181) reported the development of aromatic polyanhydrides composed of isophthalic acid and terephthalic acid. Subsequently, Hill and Carothers (182,183) reported a series of aliphatic polyanhydrides. Systematic development of polyanhydrides as substitutes for polyesters in textile applications was undertaken initially by Conix (184,185) and then followed by Yoda and Miyake (186,187). They prepared and studied a number of aromatic and heterocyclic polyanhydrides. These polymers were not suitable for textile applications because of the extreme reactivity of anhydride linkage toward water. The fact that they are extremely hydrophobic and hydrolytically unstable renders them useful in drug delivery applications (187–189).

a. *Synthesis*

Melt polycondensation. The majority of the polyanhydrides are prepared by melt polycondensation. The sequence of reaction involves first the conversion of a dicarboxylic acid monomer into a prepolymer consisting of a mixed anhydride of the diacid with acetic anhydride. This is achieved by simply refluxing the diacid monomer with acetic anhydride for a specified length of time. The polymer is obtained subsequently by heating the prepolymer under vacuo to eliminate the acetic anhydride (190):

$$HOOC\text{-}(CH_2)_8\text{-}COOH \quad + \quad (CH_3CO)$$

$$\downarrow \text{reflux}$$

$$CH_3\text{-}\overset{O}{\overset{\|}{C}}\text{-}O\text{-}\overset{O}{\overset{\|}{C}}\text{-}(CH_2)_8\text{-}\overset{O}{\overset{\|}{C}}\text{-}O\text{-}\overset{O}{\overset{\|}{C}}\text{-}CH_3$$

$$10^{-4}\ \text{mmHg} \quad \downarrow \quad 180°C\text{ , }90\text{ min}$$

$$\left[\text{-}O\text{-}\overset{O}{\overset{\|}{C}}\text{-}(CH_2)_8\text{-}\overset{O}{\overset{\|}{C}}\text{-}O\text{-}\right]_n$$

This procedure was used by most of the early investigators. The polyanhydride thus obtained was of low molecular weight. For most practical applications high molecular weight polyanhydrides are desirable. Hence, a systematic study was undertaken to determine the factors that affected the polymer molecular weight (191). It was found that the critical factors were monomer purity, reaction time and temperature, and an efficient system to remove the byproduct, acetic anhydride. The highest molecular weight polymers were obtained using pure isolated prepolymers and heating them at 180°C for 90 min with a vacuum of 10^{-4} mmHg, using a dry ice/acetone trap. Molecular weights in excess of 125,000 were obtained.

Significantly higher molecular weights were obtained in shorter times by using coordination catalysts such as cadmium acetate, earth metal oxides, and $ZnEt_2 \cdot H_2O$ (159). The weight average molecular weight varied from 90,000 to 240,000 when the concentration of cadmium acetate was changed from 0.5 to 3 mole%, with the reaction time of less than 1 h. Other catalysts, such as titanium and iron, inhibited the polymerization reaction and the polymers were dark brown in color. Acidic catalysts, such as p-toluene sulfonic acid, did not show any effect on polymer molecular weight, while the basic catalyst 4-dimethyl amino pyridine caused a decrease in molecular weight.

Solution polymerization. Syntheses of polyanhydrides using melt polycondensation is useful to obtain high molecular weight polymers but is not useful if the monomers are thermolabile. Hence, methods were developed to synthesize polyanhydrides under ambient conditions for heat-sensitive monomers such as dipeptides and therapeutically active diacids.

The solution polymerization is carried out by the Scotten–Baumann technique. In this method the solution of diacid chloride is added dropwise into an ice-cooled solution of a dicarboxylic acid. The reaction is facilitated by using an acid acceptor such as triethylamine. Polymerization takes place instantly on contact of the monomers and is essentially complete within 1 h. The solvents employed can be a single solvent or a mixture of solvents like dichloromethane, chloroform, benzene, and ethyl ether. It was found that the order of addition is very important in obtaining relatively high molecular weight polyanhydrides. Addition of a diacid solution dropwise to the diacid chloride solution consistently produced high molecular weight polymers (192):

$$HOOC\text{-}R\text{-}COOH \ + ClOC\text{-}R'\text{-}COCl \xrightarrow{\text{base}} \text{-}(R\text{-}C(O)\text{-}O\text{-}C(O)\text{-}R'\text{-}C(O)\text{-}O\text{-}C(O))\text{-} \ + \text{base-}$$

The drawback of this homogeneous Schotten–Baumann condensation reaction in solution is that the diacid chloride monomer should be of very high purity. An alternative approach was the conversion of dicarboxylic acid monomer into the polyanhydride using a dehydrative coupling agent under ambient conditions. The dehydrative coupling agent, N′N-bis[2-oxo-3-oxazolidinyl]phosphonic chloride was the most effective in forming polyanhydrides with the degree of polymerization around 20 (193). It is essential that the catalyst be ground into fine particles before use and should be freshly prepared. A disadvantage of this method is that the final product contains polymerization byproducts which have to be removed by washing with protic solvents such as methanol or cold dilute hydrochloric acid. The washing by protic solvents may evoke some hydrolysis of the polymer.

Coupling agents such as phosgene and diphosgene could also be used for the polyanhydride formation. Polymerization of sebacic acid using either phosgene or diphosgene as coupling agents with the amine based heterogeneous acid acceptor poly(4-vinyl pyridine) produced higher molecular weights in comparison to nonamine heterogeneous base K_2CO_3 (194).

Ring-opening polymerization. Ring-opening polymerization (ROP) takes place in two steps; the first step is preparation of the cyclic monomer and the second is polymerization of the cyclic monomers. Albertsson and coworkers prepared adipic acid polyanhydride from cyclic adipic anhydride (Oxepane-2,7-dione) using cationic [e.g., $AlCl_3$ and $BF3{\cdot}(C_2H_5)_2O$] anionic (e.g., $CH_3COO^-K^+$ and NaH) and coordination-type inhibitors such as stannous-2-ethylhexanoate and dibutyltinoxide (195–197).

b. *Polymer Properties*

Almost all polyanhydrides show some degree of crystallinity as manifested by their crystalline melting points. An in depth x-ray diffraction analysis was conducted with the homopolymers of sebacic acid (SA); bis(carboxyphenoxy) propane (CPP); bis(carboxyphenoxy)hexane (CPH) and fumaric acid; and the copolymers of SA with CPP, CPH, and fumaric acid. The results indicated that the homopolymers were highly crystalline and the crystallinity of the copolymers was determined, in most cases, by the monomer of highest concentration. Copolymers with a composition close to 1:1 were essentially amorphous (198).

The melting point, as determined by differential scanning calorimetry, of aromatic polyanhydrides are much higher than the aliphatic polyanhydrides. The melting point of the aliphatic aromatic copolyanhydrides is proportional to the aromatic content. For this type of copolymer there is characteristically a minimum T_m between 5 to 20 mole% of the lower melting component (191).

The majority of polyanhydrides dissolves in solvents such as dichloromethane and chloroform. However, the aromatic polyanhydrides display much lower solubility than the aliphatic polyanhydrides. In an attempt to improve the solubility and decrease the T_m, copolymers of two different aromatic monomers were prepared. These copolymers displayed a substantial decrease in T_m and an increase in solubility compared to the corresponding homopolymers of aromatic diacids (199).

The data on mechanical properties of polyanhydrides are very limited. The fibers of poly[1,2-bis(p-carboxyphenoxy) ethane anhydride] showed a tensile strength of 40 kg/mm2 with an elongation of 17.2% and a Young's modulus of 505 kg/mm^2. A systematic study on the tensile strength of the copolymers of CPP and SA showed that increasing the CPP content in the copolymer or the molecular weight of the copolymer increased the tensile strength (177). Unsaturated polyanhydrides of the structure [—(OOC—CH=CH—CO)x—(OOC—R—CO)y]n were developed to improve the mechanical properties of the polymers. The advantage of the unsaturated polyanhydrides is that they can undergo secondary polymerization of the double bonds to create a crosslinked matrix (200).

c. *Polymer Hydrolysis*

Anhydride linkage is extremely susceptible to hydrolysis in presence of moisture to generate the dicarboxylic acids (201). Hydrolysis of monomeric anhydrides is catalyzed by both acid and base, the hydrolytic degradation rate of polyanhydrides increases with increase in pH. It is believed that the poor solubility of the oligomeric products, under low pH conditions, formed at the surface of the matrix impedes the degradation of the core. In general, the hydrophobic polymers such as P(CPP) and P(CPH) display constant erosion kinetics. The degradation rates of the polyanhydrides can be altered in a number of ways. The degradation rates can be enchanced by incorporating the aliphatic monomer, such as sebacic acid, into the polymer. The degradation can be slowed by increasing the methylene groups into the backbone of the polymer. For example, in the case of the poly[bis(p-carboxyphenoxy)alkane] series, increasing the methylene groups from one to six increased the hydrophobicity of the polymer, and the erosion rates underwent a decrease of three orders of magnitude.

To achieve a variety of degradation rates, aliphatic aromatic homopolyanhydrides of the structure $-(OOC-C_6H_4-O(CH_2)_x-CO-)_n$ were prepared with x varying from 1 to 10. Increasing the value of x decreases the erosion rates (202). Increased erosion rates were also observed when poly(sebacic acid) was branched with either 1,3,5-tricarboxylic acid or low molecular weight poly(acrylic acid) (203).

Apart from the reactivity of anhydride linkage toward water, aliphatic polyanhydrides and their copolymers are found to undergo self-depolymerization under anhydrous conditions in the solid state and in solution (204). The depolymerization reaction mainly affects the high molecular weight fraction of the polymer. Aromatic homopolymers show no sign of depolymerization when stored under anhydrous conditions. The depolymerization rate is found to follow a first-order kinetics, accelerate with temperature, and increase with polarity of the solvent (204). The depolymerized polymer can be repolymerized to yield the original polymer, suggesting that inter- or intramolecular anhydride interchange takes place during depolymerization.

d. Polymer Processing

Drug-incorporated matrices can be formulated either by compression or injection molding. The polymer and drug can be ground in a Micro Mill grinder, sieved into a particle size range of 90–120 μm and can be pressed into circular discs using a Carver press. Alternatively, the drug can be mixed into the molten polymer to form small chips of drug–polymer conjugate. These chips are fed into the injection molder to mold the drug–polymer matrix into the desired shaped device. One must consider the thermal stability of the polymer and potential chemical interaction between drug and polymer at the high temperatures of injection molding.

The preferred method of drug delivery, in many instances, is by injection. This requires the development of microcapsules or microspheres of the drug. Several different techniques have been developed for the preparation of microspheres from polyanhydrides, including hot-melt microencapsulation (205) and solvent removal technique (206).

e. Other Polyanhydrides

In addition to the previously discussed aliphatic aromatic and the copolyanhydrides of the respective diacids, several other modifications of the backbone of the polyanhydrides have been reported. These new polyanhydrides were developed to improve their physicochemical, mechanical, thermal, and hydrolytic properties.

One of these new polyanhydrides is polyanhydride-imides, also referred to as copolyimides (207). They showed good thermal resistance but were essentially insoluble in most organic solvents (208). In an attempt to improve the solubility in more polar solvents, polyanhydrides were synthesized using imide-diacids containing aliphatic aromatic characteristics. A systemic study was reported in which the starting monomers, imide-diacids, were prepared from aromatic acid anhydrides and x-amino acids. Varying the number of methylenic units in the α-amino acids provided the variability in the aliphatic character of the aromatic aliphatic monomer.

A typical example of such an aromatic aliphatic monomer is the one obtained by the reaction of trimellitic anhydride with glycine:

HOOC–(trimellitic anhydride) + $H_2N\text{-}CH_2\text{-}COOH$

↓

HOOC–(imide)–$N\text{-}CH_2\text{-}COOH$

For the synthesis of polyanhydride, the aliphatic aromatic diacid is first converted to the diacetyl derivative by refluxing the diacid in the presence of excess acetic anhydride. The diacetyl derivative is polymerized either by melt polycondensation or in solution. The polyanhydride-imides thus obtained are very soluble in polar organic solvents. However, they showed melt transitions at temperatures of 245°C and above (209). Along with that, insufficient data are available on the hydrolytic stability of these materials rendering them questionable materials as a carrier in drug delivery systems.

On similar lines, another research group has developed polyanhydrides containing amido groups. The polyanhydrides thus synthesized were of relatively low molecular weight, low melting points, and a glass transition higher than the room temperature (210). The same group also developed polyanhydrides containing ester groups (211). These polymers are in the early stages of development and have not been characterized with respect to their suitability in drug delivery systems.

Another class of polyanhydrides are based on natural fatty acids. The dimers of oleic acid and eurucic acid are liquid oils containing two carboxylic acids available for anhydride polymerization. The homopolymers are viscous liquids. Copolymerization with increasing amounts of sebacic acid forms solid polymers with increasing melting points as a function of SA content. The polymers are soluble in chlorinated hydrocarbones, tetrahydrofuran, 2-butanone, and acetone (212).

Polyanhydrides synthesized from nonlinear hydrophobic fatty acid esters based on ricinoleic, maleic, and sebacic acid possessed desired physicochemical properties such as low melting point, hydrophobicity, and flexibility to the polymer formed in addition to biocompatibility and biodegradability. The polymers were synthesized by melt condensation to yield film-forming polymers with molecular weights exceeding 100,000 (213).

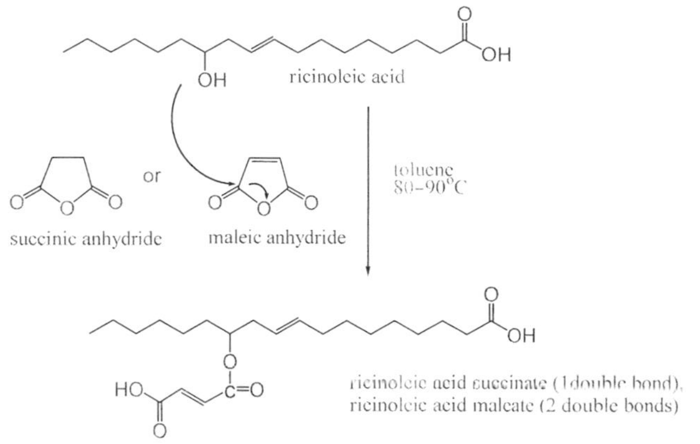

ricinoleic acid

succinic anhydride or maleic anhydride

toluene 80–90°C

ricinoleic acid succinate (1double bond), ricinoleic acid maleate (2 double bonds)

The properties of polyanhydrides were modified by the incorporation on long chain fatty acid terminals such as stearic acid in the polymer composition, which alters its hydrophobicity and decreases its degradation rate (214). Since natural fatty acids are monofunctional, they would act as polymerization chain terminators and control the molecular weight. A detailed analysis of the polymerization reaction shows that up to about 10 mole% content of stearic acid, the final product is essentially a stearic acid–terminated polymer. Whereas higher amounts of acetyl stearate in the reaction mixture resulted in the formation of increasing amounts of stearic anhydride byproduct with minimal effect on the polymer molecular weight, which remains in the range of 5000. Physical mixtures of polyanhydrides with triglycerides and fatty acids or alcohols did not form uniform blends.

III. BIOCOMPATIBILITY AND TOXICITY

In all the potential uses of polymeric material, a direct contact between the polymer and biological tissues is evident. Therefore, for the eventual human application of these biomedical implants and devices, an adequate testing for safety and biocompatibility of the specific polymer matrix used in each case is essential.

Whenever a synthetic polymer material is to be utilized in vivo, the possible tissue–implant interactions must be taken into consideration. In the case of biodegradable matrices, not only the possible toxicity of the polymer have to be evaluated, but also the potential toxicity of its degradation products.

The last section of this chapter reviews existing data about the biocompatibility and toxicity of the different polymers actually available for biomedical applications.

A. Polyesters

1. Lactide/Glycolide Copolymers

Biocompatibility of monomer is considered as the foundation for biocompatibility of degradable polymer systems, not the polymer itself. Thus, PLLA is found as an excellent biomaterial and safe for in vivo use because its degradation product L-lactic acid is a natural metabolite of the body. Even though PLGA is extensively used and represents the gold standard of degradable polymers, increased local acidity due to its degradation can lead to irritation at the site of polymer implant. Agrawal and Athanasiou have introduced a technique in which basic salts are used to control the pH in local environment of PLGA implant (215). The feasibility of lactide/glycolide polymers as excipients for the controlled release of bioactive agents is well proven, and they are the most widely investigated biodegradable polymers for drug delivery.

Most of the research work on the use of lactide/glycolide polymers as matrices for delivery systems has focused on the development of injectable microsphere formulations, although implantable rod and pellet devices are also being investigated.

The lactide/glycolide copolymers have been subjected to extensive animal and human trials without any significant harmful side effects (211). No evidence of inflammatory response, irritation, or other adverse effects have been reported upon implantation of lactide/glycolide polymer devices. However some limited incompatibility of certain macromolecules with lactide/glycolide was observed.

Lam and coworkers (216,217) studied the particles of PLA with particles of polytetrafluoroethylene (PTFE) as control. They injected intraperitoneally in mice as well as in rats. After up to 7 days, cells were harvested from the abdominal cavity. Microscopic examinations of cell morphology revealed the evidence of cell damage and death caused by phagocytosed PTFE particles. It was observed that there was more pronounced inflammatory response with PLA films than with PTFE films in subcutaneous tissues of rats.

Many conventional pharmaceutical agents formulated in lactide/glycolide polymer matrices were widely studied almost two decades ago, especially as injectable microsphere dosage forms (16–23). One of the most successful lactide/glycolide drug delivery formulations, in terms of clinical results obtained, is the steroid-loaded injectable microspheres for the controlled release of contraceptives

(218–221). Many animal and clinical trials with these systems were performed showing very good biocompatibility. For example a 90-day female contraceptive based on norethisterone as the active hormone and 85:15 DL-lactide/glycolide as the excipient has undergone successful Phase I and Phase II clinical trials in various geographic areas, demonstrating safety and efficacy of the formulation in about 300 subjects (222).

The success of the steroid microsphere system based on lactide/glycolide matrices is probably due to the combination of several factors: the reproducibility of the microencapsulation process, the in vivo drug-release performance, reliability in the treatment procedure, and the safety of the polymer (223).

Lactide/glycolide implants containing naltrexone and other narcotic antagonist agents have also been extensively studied (224–227). In one of these studies describing the clinical evaluation of a bead preparation containing 70% naltrexone and 30% of a 90:10 lactide/glycolide copolymer, a local inflammatory reaction at the site of implantation was reported in two of three subjects after subcutaneous implantation of the beads containing the drug (226). This finding prevented further clinical testing of that particular formulation. No similar problems were reported with other lactide/glycolide polymer preparations, and that incident was related to some unique aspect of that product (226).

Bergshma et al. (228) reported the use of a copolymer made of 96% L- and 4% D-lactic acid in subcutaneous tissues of rats for up to 52 weeks. Prior to implantation, discs made from the copolymer and homopolymer PLA were predegraded in vitro (229). The degradation of copolymer was faster than the homopolymer while inflammation was observed with both (230).

Good biocompatibility data were also reported with lactide/glycolide copolymer matrices containing antineoplastic drugs, antibiotics, and anti-inflammatory compounds (211).

The delivery of therapeutic molecules to the brain has been limited in part due to the presence of the blood–brain barrier. The biocompatibility of lactide/glycolide copolymer in the brain was examined regarding the gliotic response following implants of lactide/glycolide copolymer into the brains of rats. It was found that lactide/glycolide copolymer is well tolerated following implantation into the CNS and that the astrocytic response to lactide/glycolide copolymer is largely a consequence of the mechanical trauma that occurs during surgery (231).

Regarding the encapsulation of bioactive macromolecules as proteins, peptides, and antigens, the studies carried out so far show mixed biocompatibility results. Serious problems in achieving long-term release have been reported in several cases where the macromolecules lost bioactivity in vivo after a few days, as in the case of lactide/glycolide copolymer containing growth hormone. A complex interaction apparently occurs in vivo between the acidic polymer and the hormone. In vitro release studies have shown that growth hormone can become insoluble when incorporated in poly(lactide) films. Similar problems in maintaining biological activity for longer than 5–10 days were also observed with interferon–lactide/glycolide polymer formulation (211). On the other hand, promising results were obtained with luteinizing hormone–releasing hormone (LHRH) incorporated in lactide/glycolide polymer showing long-term delivery of the macromolecule (232,233). In general, hydrophilic polypeptides of low molecular weight (<5000) are considered quite stable in the presence of lactide/glycolide excipients and their acidic bioerosion byproducts.

Lactide/glycolide polymer implants, in the form of microbeads and pellets, containing insulin were reported to be effective in lowering blood glucose levels in diabetic rats for about 2 weeks, and no adverse effects, including inflammation at the implant site or deactivation of the macromolecule, were observed (234).

Toljan and Orthner (235) published details of the clinical use of an interference screw made from PGL copolymer. After 6 months, MRI data show the presence of local reactions at the implant site and a ''liquidization'' of the screw in 41% of the patients. Clinically, no febrile episodes or sterile effusions were observed.

Bezwada et al. (13) studied in vitro and in vivo biocompatibility and efficacy of block coplymer of poly(glycolide) and PCL in the form of MONOCRYL^{R} sutures. Using the size 2/0 monofilament sutures implanted subdermally in rats, MONOCRYL sutures retain 50% of their original strength after about 1 week, whereas homopolymeric PCL monofilament size 2/0 sutures retain 50% of their strength after about 52 weeks.

2. Poly(caprolactone)

The biocompatibility and toxicity of poly(caprolactone) have mostly been tested in conjuction with evaluations of Capronor™, which is an implantable 1-year contraceptive delivery system composed of a levonorgestrel-ethyl oleate slurry within a poly(caprolactone) capsule. In a preliminary 90-day toxicology study of Capronor in female rats and guinea pigs, except for a bland response at the implant site and a minimal tissue-encapsulating reaction, no toxic effects were observed (236).

The lack of an inflammatory response was confirmed

by implanting polyvinyl alcohol sponges impregnated with the powdered polymer in rats. In the Ames mutagenicity assay polycaprolactone was negatively tested (87).

The Capronor–polycaprolactone contraceptive delivery system was also tested implanted in rats and monkeys in a 2-year period (87). The results of this second, more extensive study, based on animal clinical and physical data such as blood and urine analysis, ophthalmoscopic tests, and histopathology after necropsy, showed no significant differences between the test and control groups.

Phase I and II clinical trials with Capronor were recently carried out in different medical centers (87).

B. Poly(amides)

1. Natural Polymers

The use of natural biodegradable polymers to deliver drugs continues to be an area of active research despite the advent of synthetic biodegradable polymers (237–239). Natural polymers remain attractive primarily because they are natural products of living organisms, readily available, relatively inexpensive, and capable of a multitude of chemical modifications (240).

Most of the investigations of natural polymers as matrices in drug delivery systems have focused on the use of proteins (polypeptides or polyamides) as gelatin, collagen, and albumin. Collagen is a major structural protein found in animal tissues where it is normally present in the form of aligned fibers. Because of its unique structural properties, collagen has been used in many biomedical applications such as absorbable sutures, sponge wound dressings, composite tissue–tendon allografts, injectables for facial reconstructive surgery, and as drug delivery systems especially in the form of microspheres (241).

Besides the collagen's biocompatibility and nontoxicity for most tissues (242), several factors such as the possible occurrence of antigenic responses, tissue irritation due to residual aldehyde crosslinking agents, and poor patient tolerance of ocular inserts have adversely influenced its use as a drug delivery vehicle (241). For example, 5-fluorouracil and bleomycin crosslinked sponges made from purified bovine skin collagen were implanted in rabbit eyes to test their possible use in preventing fibroblast proliferation following ophthalmic surgery, resulting in a chronic inflammatory reaction elicited by the sponges even in the absence of drug (243).

Noncollagenous proteins, particularly albumin and to a lesser extent gelatin, continue to be developed as drug delivery vehicles. The exploitable features of albumin include its reported biodegradation into natural products, its lack of toxicity, and its nonimmunogenicity (244,245).

Although many examples on the use of albumin microspheres were reported in the literature, there are only a few studies describing gelatin systems. Despite this, compared with albumin, gelatin offers the advantages of a good history in parenteral formulations and lower antigenicity. A biocompatibility study with gelatin microspheres reported no untoward effects when injected intravenously into mice over a 12-week period (246).

In a second study, when albumin microspheres were injected repeatedly into the knee joints of rabbits, pronounced rapid joint swelling occurred after the second and subsequent injections, compared to no swelling when gelatin microspheres were injected in the same way (247).

2. Pseudopoly(amino Acids)

The pseudopoly(amino acids), belonging to the poly(amides) group, represent one of the latest and therefore less advanced biodegradable polymers for medical use. Only a few of them have been synthesized and characterized, and no clinical tests have thus far been conducted. However, preliminary promising results were obtained showing no gross toxicity or tissue incompatibility upon subcutaneous implantation for the presently available pseudopoly(amino acids), and they are now being actively investigated in several laboratories for medical applications ranging from biodegradable bone nails to implantable adjuvants (121).

In an initial evaluation of hydroxyproline-derived polyesters, poly(N-palmitoyl hydroxyproline ester) was tested in a seriers of biocompatibility tests (248).

In the rabbit cornea bioassay, implantation of four small pieces of the polymer into four rabbit corneas elicited a pathological response in three of them and a very mild inflammatory response in one cornea. Histological examination of the corneas 4 weeks after implantation showed no invading blood vessels or migrating inflammatory cells in the area around the implants (249).

In a second biocompatibility study, 10-mg implants of poly(N-palmitoyl hydroxyproline ester) were inserted subcutaneously in mice in the dorsal area of the animals between the dermis and the adipose tissue layer. The mice responses were examined up to 1 year postimplantation (248).

No inflammatory reaction at the implantation site was observed in any of the mice during the first 7 weeks. A thin but distinct layer of fibrous connective tissue appeared around the implant by week 14 and, occasionally, a few multinucleated giant cells were associated with the fibrous connective tissue. Animal necropsies performed at 16 and 56 weeks postimplantation showed no histopathological abnormalities (248).

These preliminary biocompatibility results from two animal models, rabbit and mouse, indicate that the poly(N-palmitoyl hydroxyproline ester) elicits a very mild, local tissue reaction typical of a foreign body response, resulting in encapsulation of the foreign biomaterial without evidence for any significant pathological abnormalities. However, additional safety and toxicity tests need to be performed to evaluate possible systemic toxic effects, allergic reactions, or any mutagenic, teratogenic, and carcinogenic activities.

Pseudopoly(amino acids) containing aromatic side chains of tyrosine in their backbone structure were also developed recently to investigate whether biodegradable polymers that incorporate aromatic components would combine a high degree of biocompatibility with a high degree of mechanical strength (121).

In order to test the tissue compatibility of tyrosine-derived poly(iminocarbonates), solvent cast films of poly(CTTH) were subcutaneously implanted into the back of mice. In this study, conventional poly(L-tyrosine) served as a control (250). The data obtained from this biocompatibility assay were very similar to those observed with poly(N-palmitoyl hydroxyproline ester), showing no gross pathological changes from visual inspection of the implantation sites over a 1 year period.

Silver et al. (135) reported a comparative study on biocompatibility of solvent cast films of poly(desaminotyrosyl–tyrosine hexylester–iminocarbonate), or poly(DTH-iminocarbonate), in a subcutaneous rat model. In this study, high-density polyethylene (HDPE) and medical grade poly(D,L-lactic acid) served as controls. Considering the significantly faster degradation rate of poly(DTH-iminocarbonate), one would expect a different response from this material. After 7 days of postimplantation, a greater cell density and inflammatory response was noted for poly(DTH-iminocarbonate). However, at a later time point, the biological response observed for poly(DTH-iminocarbonate) was almost similar with the response observed for poly(DTH-iminocarbonate). The tissue response was characterized by a thin tissue capsule, absence of gaint cells, and a low inflammatory cell count and was not statistically different from the response observed for polyethylene and poly(lactic acid).

In vitro attachment and proliferation of fibroblasts on tyrosine-derived polycarbonates was a function of the pendent chain length. Ertal and Kohn fabricated poly(DTH-carbonate) pins and compared them to commercially available orthosorb^R pins made of polydioxanone (136). The pins were implanted transcortically in the distal femur and proximal tibia of New Zealand white rabbit for up to 26 weeks. In addition to routine histological evaluation of the implant sites, bone activity at the implant–tissue interface was visualized by UV illumination of sections labeled with fluorescent marker, and the degree of calcification around the implants was ascertained by backscattered electron microscopy. The bone tissue response was characterized by active bond remodeling at the surface of the degrading implant, the lack of fibrous capsule formation, and an unusually low number of inflammatory cells at the bone–implant interface. Poly(DTH-carbonate) exhibited very close bone apposition throughout the 26-week period of the initial study. A roughened interface was observed which was penetrated by new bone as early as 2 weeks postimplantation. Bone growth into the periphery of the implant material was visible at the 26-week time point.

C. Polyphosphazenes

Two different types of polyphosphazenes are of interest as bioinert materials: those with strongly hydrophobic surface characteristics and those with hydrophilic surfaces. Polyphosphazenes bearing fluoroalkoxy side groups are some of the most hydrophobic synthetic polymers known (251, 252). Such polymers are as hydrophobic as poly(tetrafluoroethylene) (Teflon), but unlike Teflon polyphosphazenes of this type are flexible or elastomeric, easy to prepare, and can be used as coatings for other materials.

Graft polymerization with dimethylaminoethylmethacrylate (DMAEM) onto the poly(phosphazene) surfaces highly enhances their biocompatibility. Surface modification of poly(trifluoroethoxy-phosphazenes) (PTFP) with polyethylene glycols was heading to materials with enhanced biocompatibility in comparision to nonmodified polymers (253). In the next study, the PTFP was grafted with hydrophilic monomers like dimethylacrylamide (DMAA) and acrylamide (AAm) (254). Large differences in the biocompatibility of these materials were found. Biocompatibility of AAm-grafted samples was greatly enhanced while an opposite behavior for DMAA-grafted samples was observed. All studies with these samples were done by intraperitoneal implantation of thin films in adult male wistar rats. Tissue around the implantation site was evaluated by histological examination after 25 days.

Biocompatibility and safety testings of these polymers by subcutaneous implantation in animals have shown minimal tissue response, similar in fact to the response reported with Teflon (255). The connection between hydrophobicity and tissue compatibility has been noted for classical organic polymers (256). Thus, these hydrophobic polyphosphazenes have been mentioned as good candidates for use in heart valves, heart pumps, blood vessel prostheses, and as coating materials for pacemakers or other implantable devices. However, more in vivo testing and clinical trials are needed (144).

In their bioerosion reactions polyphosphazenes display a uniqueness that stems from the presence of the inorganic backbone, which in the presence of appropriate side groups is capable of undergoing facile hydrolysis to phosphate and ammonia. The phosphate can be metabolized and the ammonia excreted. Theoretically, if side groups attached to the polymer are released by the same process being excretable or metabolizable, then the polymer can be eroded under hydrolitic conditions without the danger of a toxic response. Polyphosphazenes of this type are potential candidates as erodible biostructural materials for sutures or as matrices for controlled delivery of drugs (144).

Polyphosphazenes containing amino acid ester side groups were the first bioerodible polyphosphazenes synthesized (146). They are solid materials which erode hydrolitically to ethanol, glycine, phosphate, and ammonia. These polymers were tested in subcutaneous tissue response experiments showing no evidence of irritation, cell toxicity, giant cell formation, or tissue inflammation (143,255).

Imidazoyl groups linked to polyphosphazene chains are also hydrolized very easily, showing good biocompatibility tests (141,154). Langone et al. (257) studied poly[(ethyl-alanate)-co-(imidazole) phosphazene] derivatives for in vivo evaluation. Polymer films were subcutaneously implanted in rats. The animals were killed after 30 or 60 days. Biopsy samples of the implant zone were histologically examined. In both cases, animals were healthy and biological material surrounding to the polymers was found to correspond to fibroplast collagen with only a few monocytes in the internal site.

D. Polyorthoesters

As mentioned previously, the Chronomer polyorthoester material from Alza Corporation, or Alzamer, has been investigated as bioerodible inserts for the delivery of the narcotic antagonist naltrexone, and for the delivery of the contraceptive steroid norethisterone (258,259). The steroidal implant was tested in two separate human clinical trials causing local tissue irritation, and therefore further work with this formulation was discontinued (166,167). The reasons for the local irritation were never properly elucidated. New types of poly(orthoesters) were developed, but no data on the biocompatibility and safety of these materials were reported.

In vitro studies have shown that good control over release of tetracycline could be achieved and very good in vitro adhesion to bovine teeth was demonstrated (173). However, studies in beagle dogs with naturally occurring periodontitis were not successful because ointmentlike polymers with a relatively low viscosity are squeezed out of the pocket within about 1 day, despite good adhesiveness.

E. Polyanhydrides

Polyanhydrides are a novel class of biodegradable polymers under development as vehicles for the release of bioactive molecules including drug peptides and proteins (1). The polyanhydrides constitute so far the only class of surface-eroding polymers approved for clinical trials by the Food and Drug Administration.

A series of biocompatibility studies reported on several polyanhydrides have shown them to be nonmutagenic and nontoxic (189). In vitro tests measuring teratogenic potential were also negative. Growth of two types of mammalian cells in tissue culture was also not affected by the polyanhydride polymers (189); both the cellular doubling time and cellular morphology were unchanged when either bovine aorta endothelial cells or smooth muscle cells were grown directly on the polymeric substrate.

Subcutaneous implantation in rats of high doses of the 20:80 copolymer of bis-(p-carboxyphenoxy) propane and sebacic acid for up to 8 weeks indicated relatively minimal tissue irritation with no evidence of local or systemic toxicity (260). Since this polymer was designed to be used clinically to deliver an anticancer agent directly into the brain for the treatment of brain neoplasms, its biocompatibility in rat brain was also studied (261). The tissue reaction of the polymer was compared to the reaction observed with two control materials used in surgery, oxidized cellulose absorbable hemostat (Surgicel®, Johnson and Johnson) and absorbable gelatin sponge (Gelfoam®, Upjohn). The inflammatory reaction of the polymer was intermediate between the controls (261). A closely related polyanhydride copolymer poly(CPP)–SA 50:50 was also implanted in rabbit brains and was found to be essentially equivalent to Gelfoam in terms of biocompatibility evaluations (262). In a similar study conducted in monkey brains, no abnormalities were noted in the CT scans and magnetic resonance image, nor in the blood chemistry or hematology evaluations (263). No systemic effects of the implants were observed on histological examinations of any of the tissues tested (264).

New classes of polyanhydrides have been recently synthesized and are undergoing extensive preclinical testing, including a wide range of biocompatibility studies. Examples of these new materials are polymers of sebacic acid, poly(SA), and copolymers 1:1 of SA with fatty acid dimer (FAD) [poly(FAD:SA)], fumaric acid (FA) [poly(FA:SA)], and isophthalic acid [poly(ISO:SA)]. These materials were implanted in rabbits intramuscularly, subcutaneously, and in the cornea. Ocular and muscle irritation studies were performed compared to the material controls Gelfoam, Surgicel, and Vycryl®, a synthetic absorbable suture (Ethicon) (265). Detailed observations of toxicity, bleeding, swelling, or infection of the implantation site were conducted daily. At the end of the study the animals were sacrificed and a gross necropsy examination of the tissues surrounding the implant site was performed. No significant clinical signs or abnormalities of the incision sites were observed during the study period (4 weeks). No meaningful differences could be seen in reaction between the various polymer implants tested and the control materials (265).

In the rabbit cornea bioassay, no evidence of inflammatory response was observed with any of the implants at any time. On an average, the bulk of the polymers disappeared completely between 7 and 14 days after the implantation (265). The cornea is a very sensitive indicator of inflammatory reactions (266,267). The rabbit cornea possesses clear advantages over other implant sites for studying implant–host interactions due to the easy accessibility for frequent observations without having to gain surgical access to the implantation site. The transparency and avascularity of the cornea also enable the observer to distinguish among the different inflammatory characteristics such as edema, cellular infiltration, and ingrowth of blood vessels from the perifery of the cornea or neovascularization, which are strong indications that the biomaterial under testing is unsuitable for implantation (268).

In similar animal experiments in which polyanhydride matrices containing tumor angiogenic factor (TAF) were implanted in rabbit cornea, a significant vascularization response was observed without edema or white cells. Moreover, and most importantly from the biocompatibility standpoint, polymer matrices without incorporated TAF showed no adverse vascular response (268,269).

The biocompatibility of a new class of polyanhydrides based on ricinoleic acid as compared to Vycryl surgical suture and sham surgery was tested in rats (213). No evidence for tissue necrosis was detected in any of the treated animals upon evaluation of the tissues 21 days postimplantation. No indication for postimplantation test site contamination was noted.

Generally, in all tested groups there was a clear indication of a time-related healing process (i.e., there was a time-related reduction in the incidence and severity of necrosis and acute to subacute inflammatory reaction associated with increased fibroplasia).

The tissue reaction in the different treatment groups showed a clear trend of healing with time. In particular, comparison of the tissue reactivity along the three different time periods (3,7, and 21 days) indicated that the only noted remnants of tissue reaction at the 21-day time period were minimal subacute inflammation and mild fibrosis. In comparison, under the same conditions, Vycryl implant induced minimal fibrosis associated with the presence of minimal quantity of giant cells and encapsulated foreign material.

Based on the biocompatibility and safety preclinical studies carried out in rats (260,261), rabbits (262,266), and monkeys (263,264) reviewed here showing acceptability of the polyanhydrides for human use, a Phase I and II clinical protocol was instituted (270). In these clinical trials, a polyanhydride dosage form (Gliadel™), consisting of wafer polymer implants of poly(CPP-SA) 20:80 and containing the chemotherapeutic agent Carmustine (BCNU), was used for the treatment of glioblastoma multiforme, a universally fatal form of brain cancer. In these studies, up to eight of these wafer implants were placed to line the surgical cavity created during the surgical debulking of the brain tumor in patients undergoing a second operation for surgical debulking of either a grade III or IV anaplastic astrocytoma. In keeping with the results of the earlier preclinical studies suggesting a lack of toxicity, no central or systemic toxicity of the treatment was observed during the course of treating 21 patients under this protocol. Phase III human clinical trials have demonstrated that site-specific delivery of BCNU from a poly(CPP-SA) 20:80 wafer (Gliadel) in patients with recurring brain cancer (glioblastoma multiforme) significantly prolongs patient survival (271). Gliadel has finally won approval from the FDA as therapy for the treatment of brain tumors.

REFERENCES

1. A. J. Domb, O. Elmalak, V. R. Shastry, Z. Ta-Shma, D. M. Masters, I. Ringel, D. Teomim, R. Langer. Polyanhydrides. In: *Handbook of Biodegradable Polymers*, A. J. Domb, J. Kost, and D. M. Weiseman (Eds.). Hardwood Academic Publishers, Amsterdam, 1997, pp. 135–161.
2. R. Langer. Drug delivery and targeting. *Nature* 392(6679, suppl.):5–10 (1998).
3. E. Mathiowitz, J. S. Jacobs, Y. S. Jong, G. P. Carino, D. E. Chickering, P. Chaturvedi, C. A. Santos, K. Vijayaraghavan, S. Montgomery, M. Basset, C. Morrell. Biologically erodable microspheres as potential oral drug delivery systems. *Nature* 386(6623):410–414 (1997).
4. J. Van Brunt. Novel drug delivery systems. *Biotechnology* 7:127–130 (1989).
5. A. U. Daniels, M. K. O. Chang, K. P. Andriano, J. Heller. Mechanical properties of biodegradable polymers and composites proposed for internal fixation of bone. *J. Appl. Biomater*. 1:57 (1990).
6. P. Rokkanen, O. Bostman, S. Vainionpaa, K. Vihtonen, P. Tormala, J. Laiho, J. Kilpikari, M. Tamminmaki. Bio-

degradable implants in fracture fixation: early results of treatment of fractures of the ankle. *Lancet* 1:1442 (1985).
7. M. Bercovy, D. Goutallier, M. C. Voisin, D. Geiger, D. Blanquaert, A. Gaudichet, D. Patte. Carbon–PGLA prostheses for ligament reconstruction. Experimental basis and short-term results in man. *Clin. Orthop.* 159 (1985).
8. J. H. Brekke, M. Bresner, M. J. Reitman. Polylactic acid surgical dressing material. Postoperative therapy for dental extraction wounds. *Can. Dent. Assoc. J.* 52:599 (1986).
9. M. C. Wake, P. K. Gupta, A. G. Mikos. Fabrication of biodegradable polymer foams to engineer soft tissues. *Cell. Transplant* 5(4):465–463 (1996).
10. S. H. Mendak, Jr., R. J. Jensik, M. F. Haklin, D. L. Rose man. The evaluation of various bioabsorbable materials on the titanium fiber metal tracheal prosthesis. *Ann. Thorac. Surg.* 38:488 (1984).
11. J. P. Vacanti, R. Langer. Tissue engineering: the design and fabrication of living replacement devices for surgical reconstruction and transplantation. *Lancet* 354(suppl. 1): SI32–SI34 (1999).
12. F. A. Barber, B. F. Elrod, D. A. McGuire, L. E. Paulos. Preliminary results on absorbable interference screw. *Arthoscopy* 11:537–548 (1995).
13. R. S. Bezwada, D. D. Jamiolkowski, I. L. Lee, A. Vishvaroop, J. Persivale, S. Trenka-Benthin, E. Erneta, A. Y. Surydevara, S. Liu. Monocryl suture, a new ultrapliable absorbable monofilament suture. *Biomaterials* 16:1141–1148 (1995).
14. R. W. Bucholz, S. Henry, M. B. Henley. Fixation with bioabsorbable screws for the treatment of fractures of the ankle. *J. Bone Joint Surg.* 76-A:319–324 (1994).
15. R. Jain, N. H. Shah, A. W. Malick, C. T. Rhodes. Controlled drug delivery by biodegradable poly(ester) devices: different preparative approaches. *Drug Dev. Ind. Pharm.* 24(8):703–727 (1998).
16. D. R. Cowsar, T. R. Tice, R. M. Gilley, J. P. English. Poly(lactide-co-glycolide) for controlled release of steroids. *Methods Enzymol.* 112:101–116 (1985).
17. Y. Morita, A. Ohtori, M. Kimura, K. Tojo. Intravitreous delivery of dexamethasone sodium m-sulfobenzoate from poly(DL-lactic acid) implants. *Biol. Pharm. Bull.* 21(2): 188–190 (1998).
18. T. Kumanohoso, S. Natsugoe, M. Shimada, T. Aikou. Enhancement of therapeutic efficacy of bleomycin by incorporation into biodegradable poly-D,L-lactic acid. *Cancer Chemother. Pharmacol.* 40(2):112–116 (1997).
19. C. M. Agrawal, D. Best, J. D. Heckman, B. D. Boyan. Protein release kinetics of a biodegradable implant for fracture non-unions. *Biomaterials* 16:1255–1260 (1995).
20. J. D. Meyer, R. F. Falk, R. M. Kelly, S. J. Withrow, W. S. Dernell, D. J. Kroll, T. W. Randolph, M. C. Manning. Preparation and in vitro characterization of gentamycin-impregnated biodegradable beads suitable for treatment of osteomyelitis. *J. Pharm Sci.* 87(9):1149–1154 (1998).
21. M. G. Garry, D. L. Jackson, H. E. Geier, M. Southam, K. M. Hargreaves. Evaluation of the efficacy of a bioerodible bupivacaine polymer system on antinociception and inflammatory mediator release. *Pain* 82(1):49–55 (1999).
22. T. Gorner, R. Gref, D. Michenot, F. Sommer, M. N. Tran, E. Dellacherie. Lidocaine-loaded biodegradable nanospheres. I. Optimization Of the drug incorporation into the polymer matrix. *J. Controlled Release* 57(3):259–268 (1999).
23. M. Singh, X. M. Li, J. P. McGee., T. Zamb, W. Koff, C. Y. Wang, D. T. O'Hagan. Controlled release microparticles as a single dose hepatitis B vaccine: evaluation of immunogenicity in mice. *Vaccine* 15(5):475–481 (1997).
24. V. W. Dittrich, R. C. Schultz. Kinetics and mechanism of the ring opening polymerization of L-lactide. *Angew. Makromol. Chem.* 15:109 (1971).
25. A. J. Nijenhuis, D. W. Grijpma, A. J. Pennings. Lewis acid catalyzed polymerization of L lactide. Kinetics and mechanism of the bulk polymerization. *Macromolecules* 25:6419–6424 (1992).
26. J. E. Bergshma, F. R. Rozema, R. R. M. Bos, W. C. Bruijin. Foreign body reactions to resorbable poly(L-lactide) bone plates and screws used for the fixation of unstable zygometic fractures. *J. Oral Maxillofacial Surg.* 51:666–670 (1993).
27. E. Lilly, R. C. Schultz. H′- and ^{13}C-(H′)-NMR spectra of stereocopolymers of lactide. *Macromolec. Chem.* 175: 1901 (1975).
28. H. Cherdron, H. Ohse, F. Korte. Die polymerization von lactonen. Teil 1: homopolymerization 4-, 6- und 7-gliedrigerlactone meet kationischem Initiation. *Makromolek. Chem.* 56:179 (1962).
29. Y. Yamashita, T. Tsuda, O. Masahiko, S. Iwatsuky. Correlation of cationic copolymerization parametres of cyclic esters, formuls and esters. *J. Polym. Sci. A-1* 4:2121 (1966).
30. Y. Yamashita, K. Chiba, S. Kozawa. Cationic copolymerization of tetrahydrofuran with asctone. *Polym. J.* 3:389 (1972).
31. H. Cherdron, H. Ohse, F. Korte. Die polymerization von lactonen. Teil 2: homopolymerization 4-, 6- und 7-gliedrigerlactone meet kationischem Initiation. *Makromolek. Chem.* 56:1187 (1962).
32. Y. Yamashita, T. Tsuda, H. Ishida, A. Uchikawa, Y. Kuriyama. Anionic copolymerization of L-lactones in correlation with the mode of fission. *Makromolek. Chem.* 113: 139 (1968).
33. S. Somkowski, Panczek. Macroions and macroion pairs in the anionic polymerization of b-propiolactone (L-PL). *Macromolecules* 13:229 (1980).
34. P. Sigwalt. Ring opening polymerizations of heterocycles with organometallic catalysts. *Macromolek. Chem.* 94:161 (1981).
35. F. H. Kohn, J. G. Van Omnen, J. Feijen. The mechanisms of the ring-opening polymerization of lactide and glycolide. *Eur. Polym. J.* 12:1081 (1983).
36. W. Dittrich, R. C. Shultz. Kinetik und mechanismus der ringoffnenden polymeization von L-lactide. *Macromolek. Chem.* 15:109 (1971).

37. T. Kissel, Z. Brich, S. Bantle, I. Lancranjan, Nimmerfall, P. Vit. Parenteral depot-systems on the basis of biodegradable polyesters. *J. Controlled Release* 16:27 (1991).
38. W. H. Wong, D. J. Mooney. In: *Synthetic Biodegradable Polymer Scaffolds*, Atala, Mooney (Eds.). Birkhauser, Boston, 1997.
39. H. R. Kricheldorf, D.-O. Damrau. *Macromol. Chem. Phys.* 198:1767 (1997).
40. S. M. Li, I. Rashkov, J. L. Espartero, N. Manolova, M. Vert. Synthesis and chacterization and hydrolytic degradation of PLA/PEO/PLA triblock copolymers with long poly(L-lactic acid) blocks. *Macromolecules* 29:57–62 (1996).
41. P. Degee, P. Dubois, R. Jerome. *Macromol. Chem. Phys.* 198:1973 (1997).
42. W. M. Stevels, M. J. K. Ankone, P. J. Dijkastra, J. Feijen. *Macromolecules* 29:6132 (1996).
43. N. Spassky, M. Wisniewski, C. Pluta, A. LeBorgne. *Macromol. Chem. Phys.* 197:2627 (1996).
44. M. Bero, J. Kaspererczyk, G. Adamus. *Makromol. Chem.* 194:907 (1993).
45. A. Kowalski, Duda, S. Penczek. Polymerization of L,L-lactide initiate by aluminum isopropoxide trimer or tetramer. *Macromolecules* 31:2114–2122 (1998).
46. S. Hyon, K. Jamshidi, Y. Ikada. Synthesis of polylactides with different molecular weights. *Biomaterials* 18:1503–1508 (1997).
47. D. G. H. Ballard, B. J. Tighe. Studies of the reactions of the anhydrosulphites of α-hydroxy-carboxylic acids. Part I: polymerization of anhydrosulphite of α-hydroxyisobutyric acid. *J. Chem. Soc. B* 702:702, 976 (1967).
48. D. G. H. Ballard, B. J. Tighe. Studies of the reactions of the anhydrosulphites of α-hydroxy-carboxylic acids. Part II: Polymerization of glycolic and lactic acid anhydrosulphites. *J. Chem. Soc. B* 976 (1967).
49. J. Smith, B. J. Tighe. Ring opening polymerization. V. Hydroxyl initiated polymerization of phenyl-substituted 1,3-dioxolan-2,4-diones: a model study. *J. Polym. Sci. Polym. Chem.* 14:2293 (1976).
50. G. Caponetti, J. S. Hrkach, B. Kriwet, M. Poh, N. Lotan, P. Colombo, R. Langer. Microparticles of novel branched copolymers of lactic acid and amino acids: preparation and characterization. *J. Pharm. Sci.* 88(1):136–141 (1999).
51. K. J. Zhu, S. Bihai, Y. Shilin, "Super Microcapsules" (SMC). I. Preparation and characterization of star polyethylene oxide (PEO)-polylactide (PLA) copolymers. *J. Polym. Sci. Part A: Polym. Chem.* 27:2151 (1989).
52. R. Bodmeier, K. H. Oh, H. Chen. The effect of the addition of low molecular weight poly(DL-lactide) on drug release biodegradable poly(DL-lactide) drug delivery systems. *Int. J. Pharm.* 51:1 (1989).
53. Y. Ikada, K. Jamshidi, H. Tsuji, S. H. Hyon. Stereocomplex formation between enantiomeric poly(lactides). *Macromolecules* 20:904–906 (1987).
54. H. Tsuji, S.-H. Hyon, Y. Ikada. Stereocomplex formation between enantiomeric poly(lactic acids). 3. Calorimetric studies on blend films cast from dilute solution. *Macromolecules* 24:5651–5656 (1991).
55. H. Tsuji, S.-H. Hyon, Y. Idaka. Stereocomplex formation between enantiomeric poly(lactic acids). 5. Calorimetric and morphological studies on the stereocomplex formed in acetonitrile solution. *Macromolecules* 25:2940–2946 (1992).
56. M. Spinu, C. Jackson, M. Y. Keating, K. H. Gardner. Material design in poly(lactic acid) systems: block copolymers, star homo- and copolymers, and stereocomplexes. *J. Macromol. Sci. Pure Appl. Chem.* A 33:1497–1530 (1996).
57. D. Brizzolara, H.-J. Cantow, K. Diederichs, E. Keller, A. J. Domb. Mechanisms of the stereocomplex formation between enantiomeric poly(lactide)s. *Macromolecules* 29: 191–197 (1996).
58. L. Cartier, T. Okihara, B. Lotz. Triangular polymer single crystals: stereocomplexes, twins, and frustrated structures. *Macromolecules* 30:6313–6322 (1997).
59. H. Sakakihara, Y. Takahashi, H. Tadokoro, N. Oguni, H. Tani. Structural studies of isotactic poly(*tert*-butylethylene oxide). *Macromolecules* 6:205–212 (1973).
60. H. Matsubayashi, Y. Chatani, H. Tadokoro, P. Dumas, N. Spassky, P. Sigwalt. Crystal structures of optically active and inactive poly(*tert*-butylethylenesulfide). *Macromolecules* 10:996–1002 (1977).
61. S. J. de Jong, W. N. E. van Dijk-Wolthuis, J. J. Kettenes-van den Bosch, P. J. W. Schuyl, H. E. Hennink. Monodisperse enantiomeric lactic acid oligomers: preparation, characterization, and stereocomplex formation. *Macromolecules* 31:6397–6402 (1998).
62. D. Brizzolara, H.-J. Cantow, A. J. Domb. Crystallization and stereocomplexation governed self assembling of poly(lactide)-β-poly(ethyleneglycol) to mesoscopic structures.
63. K. A. Walter, R. Tamargo, A. Olivi, P. C. Burger, H. Brem. *Neurosurgery* 37:1129 (1995).
64. L. Montanari, M. Costantini, E. C. Sigoretti, L. Valvo, M. Santucci, M. Bartolomei, P. Fattibene, S. Onori, A. Faucitano, B. Conti, I. Genta. Gamma irradiation effects on poly(DL-lactide-co-glycolide) microspheres. *J. Controlled Release* 56(1–3):219–229 (1998).
65. BPI. *Technical Brochure: Lactel*[R] *Polymers*. Birmingham Polymers, Inc., Birmingham, AL, 1995.
66. U. Siemann. Densitometric determination of the solubility parameters of biodegradable of polyesters. *Proc. Int. Symp. Control. Rel. Bioact. Mater.* 12:53 (1985).
67. G. A. Bosell, R. M. Scribner. U.S. patent 3,773,919 (1973).
68. I. Engelberg, J. Kohn. Physicochemical properties of degradable polymers used in medical applications: a comparative study. *Biomaterials* 12:292 (1990).
69. U. Daniels, M. K. O. Chang, K. P. Andriano, J. Heller. Mechanical properties of biodegradable polymers and composites proposed for internal fixation of bone. *J. Appl. Biomater.* 1:57 (1990).
70. D. H. Lewis, Controlled release of bioactive agents from

lactide/glycolide polymers. In: *Biodegradable Polymers as Drug Deliver Systems*, M. Chasin, R. Langer (Eds.). Marcel Dekker, New York, 1990.

71. A. Gopferich. Mechanisms of polymer degradation and elimination. In: *Handbook of Biodegradable Polymers*, A. J. Domb, J. Kost, D. M. Weiseman (Eds.). Harwood Academic Publishers, Amsterdam, 1997, pp. 451–473.
72. S. J. Holland, B. J. Tighe, P. L. Gould. Polymers for biodegradable medical devices. 1. The potential of polyesters as controlled macromolecular release systems. *J. Controlled Release* 4:155–180 (1986).
73. M. A. Tracy. Development and scale-up of a microsphere protein delivery system. *Biotechnol. Prog.* 14:108–115 (1998).
74. S. M. Li, H. Garreau, M. Vert. Structure–property relationships in the case of the degradation of massive poly(α-hydroxy acids) in aqeous media. Part III: influence of the morphology of poly(L-lactic acid), PLA 100. *J. Mater. Sci. Mater. Medicine* 1:198–206 (1990).
75. T. S. Nakamura, S. Hitomi, S. Watanabe, Y. Shimizu, S. H. Jamashidi, Y. Ikada. Bioabsorption of polylactides with different molecular properties. *J. Biomed. Mater. Res.* 23:1115–1130 (1989).
76. M. Vert, S. M. Li, H. Garreau. More about the degradation of LA/GA-derived matrices in aqueous media. *J. Controlled Release* 16:15–26 (1991).
77. M. Vert, J. Mauduit, L. Suming. Biodegradation of PLA/GA polymers: increasing complexity. *Biomaterials* 15:1209–1213 (1994).
78. I. Grizzi, H. Garreau, S. Li, M. Vert. Hydrolytic degradation of devices based poly(DL-lactic acid) size dependence. *Biomaterials* 16:305–311 (1995).
79. T. G. Park. Degradation of poly(lactic-*co*-glycolic acid) microspheres: effect of copolymer composition. *Biomaterials* 16:1123–1130 (1995).
80. R. Lindhardt. Chapter 2. In: *Biodegradable Polymers for Controlled Release of Drugs*. Springer-Verlag, New York, 1988.
81. F. Moll, G. Koller. Biodegradable tablets having a matrix of low molecular weight poly-L-lactic acid and poly-DL-lactic acid. *Arch. Pharm.* (Weinheim) 323:887 (1990).
82. H. Fukuzaki, Y. Aiba. Synthesis of biodegradable poly (L-lactic acid -co-mandelic acid) with relatively low molecular weight. *Macromol. Chem.* 190:2407 (1989).
83. H. Fukuzaki, M. Yoshida M. Asano, M. Kumakura, K. Imasaka, T. Nagai, T. Mashimo, H. Yuasa, K. Imai, H. Yamanaka. Synthesis of biodegradable copoly(L-lactic acid/aromatic hydroxyacids) with relatively low molecular weight. *Eur. Polym. J.* 26:1273 (1990).
84. M. Yoshida, M. Asano, I. Kaetsu, et al. *Polym. J.* 18:287 (1986).
85. J. E. Potts, R. A. Cledining, W. B. Ackart, W. D. Niegisch. Biodegradability of synthetic polymers. *Polym. Sci. Technol.* 3:61 (1973).
86. S. Huang. Biodegradable polymers. In: *Encyclopedia of Polymer Science and Engineering*, Vol. 2, H. F. Mark, N. M. Bikales, C. G. Overberger, G. Menges, J. I. Kroschwitz (Eds.). John Wiley and Sons, New York, 1985, pp. 220–243.
87. C. G. Pitt. Poly(α-caprolactone) and its copolymers. In: *Biodegradable Polymers as Drug Delivery Systems*, M. Chasin, R. Langer (Eds.). Marcel Dekker, New York, 1990, pp. 71–120.
88. D. E. Perrin, J. P. English. Polycaprolactone. In: *Handbook of Biodegradable Polymers*, A. J. Domb, J. Kost, D. M. Weiseman (Eds.). Harwood Academic Publishers, Amsterdam, 1997, pp. 63–77.
89. K. Ito, Y. Yamashita. Propagation and depropagation rates in the anionic polymerization of actonecyclic oligomers. *Macromolecules* 11:68–72 (1978).
90. V. F. Jenkins. Caprolactone and its polymers. *Polym. Paint Colour J.* 167:622–627 (1977).
91. T. Ouhadi, C. Stevens, P. Teyssie. Study of poly(lactone) bulk degradation. *J. Appl. Polym. Sci.* 20:2963–2970 (1976).
92. A. Hamitou, T. Ouhadi, R. Jerome, P. Teyssie. Soluble bimetallic oxoalkoxides. VII. Characteristics and mechanism of ring-opening polymerization of lactones. *J. Polym. Sci. Part A Polym. Chem.* 15:865–873 (1977).
93. W. J. Bailey, Z. Ni, S. R. Wu. Synthesis of poly(actone via a free radical mechanism. Free radical ring-opening polymerization of 2-methylene-1,3-dioxepane. *J. Polym. Sci. Part A Polym. Chem.* 20:3021–3030 (1982).
94. C. Reiss, C. Hurtrez, P. Bahadur. Block copolymers. In: *Encyclopedia of Polymer Science and Engineering*, Vol. 2, H. F. Mark, N. M. Bikales, C. G. Overberger, G. Menges, J. I. Kroschwitz (Eds.). John Wiley and Sons, New York 1985, p. 398.
95. C. X. Song, H. F. Sun, X. D. Feng. *Polym. J.*, 19:485 (1987).
96. C. Jacobs, P. Dubois, R. Jerome, P. Teyessie. Macromolecular engineering of polylactones and polylactides. 5. Synthesis and characterization of diblock copolymers based on poly(ε-caprolactone) and poly(L,L or D,L)lactide by aluminium alkoxides. *Macromolecules* 24:3027 (1991).
97. C. X. Song, X. D. Feng. Synthesis of ABA triblock copolymers of ϵ-caprolactone and DL-lactide. *Macromolecules* 17:2764 (1984).
98. X. D. Feng, C. X. Song, W. Y. Chen. Synthesis and evaluation of biodegradable block copolymers of ε-caprolactone and DL-lactide. *J. Polym. Sci. Polym. Lett. Ed.* 21: 593 (1983).
99. M. Endo, T. Aida, S. Inoue. Immortal polymerization of initiated by aluminium porphyrin in the presence of alcohols. *Macromolecules* 20:2982–2988 (1987).
100. A. Schindler, R. Jeffcoat, G. L. Kimmel, C. G. Pitt, M. E. Wall, R. Zweidinger. Biodegradable polymers for sustained drug delivery. *Contemp. Top. Polym. Sci.* 2:251–289 (1977).
101. G. Pitt, F. I. Chasalow, Y. M. Hibionada, D. M. Klimas, A. Schindler. Aliphatic polyesters. I. The degradation of poly(lactone) in vivo. *J. Appl. Polym. Sci.* 26:3779–3787 (1981).

102. I. Engelberg, J. Kohn. Physico-mechanical prorerties of degradable polymers used in medical applications: a comparative study. *Biomaterials* 12:292–304 (1991).
103. G. Pitt, M. M. Gratzl, G. L. Kimmel, J. Surles, A. Schindler. Aliphatic polyesters. II. The degradation of poly(DL-lactide), poly(actone), and their copolymers in vivo. *Biomaterials* 2:215–220 (1981).
104. S. C. Woodward, P. S. Brewer, F. Moatmed, A. Schindler, C. G. Pitt. The intracellular degradation of poly(α-caprolactone). *J. Biomed. Mater. Res.* 19:437–444 (1985).
105. P. A. Holmes. Application of PHB—a microbially produced biodegradable thermoplastic. *Phys. Technol.* 16: 32–36 (1985).
106. Y. Doi, A. Tamaki, M. Kunioka, K. Soga. Production of copolyesters of 3-hydroxybutirate and 3-hydroxyvalerate by *Alcanigenes eutrophus* from butyric and pentanoic acids. *Appl. Microbiol. Biotech.* 28:330 (1988).
107. M. Kunioka, Y. Nakamura, Y. Doi. New bacterial copolyester produced in *Alcanigenese eutrophuse* from organic acids. *Polym. Commun.* 29:174–176 (1988).
108. S. R. Amor, T. Rayment, J. K. M. Sanders. Poly(hydroxybutirate) in vivo: NMR and X-ray characterization of the elastomeric state. *Macromolecules* 24:4583 (1991).
109. H. Brandl, R. A. Gross, R. W. Lenz, R. C. Fuller. *Appl. Environ. Microbiol.* 54:1977 (1988).
110. T. Kawaguchi, M. Nakano, K. Juni, S. Inoue, Y. Yoshida. Release profiles of 5-fluorouracil and its derivatives from polycarbonate in vitro. *Chem. Pharm. Bull.* 30:1517–1520 (1982).
111. T. Kawaguchi, M. Nakano, K. Juni, S. Inoue, Y. Yoshida. Examination of biodegradability of poly(ethylene carbonate) and poly(propylene carbonate) in the peritoneal cavity in rats. *Chem. Pharm. Bull.* 31:1400–1403 (1983).
112. A. Katz D. P. Mukherjee, A.L. Kaganov, S. Gordon. A new synthetic monofilament absorbable suture made from polytrimethylene carbonate. *Surg. Gynecol. Obstet* 166: 213–222 (1985).
113. N. B. Graham. *Proc. Int. Symp. Control. Rel. Bioact. Mater*. Florida, 1982, p. 16.
114. R. S. Bezwada, S. W. Shalaby, H. D. Newman, A. Kafrauy. Bioabsorbable copolymers of p-dioxane and lactide for surgical devices. *Trans. Soc Biomater* 13:194 (1900).
115. R. W. Lenz, P. Guerin. In: *Polymers in Medicine: Functional Polyesters and Polyamides for Medical applications of Biodegradable Polymers*, (E. Chiellini, P. Giusti (Eds.). Plenum Press, New York 1983.
116. A. J. Domb, C. T. Laurencin, O. Israeli, T. N. Gerhart, R. Langer. The formation of propylene fumarate oligomers for use in bioerodible bone cement composites. *J. Poly. Sci. Polym. Chem.* 28:973 (1990).
117. A. J. Domb, N. Manor, O. Elmalak. Biodegradable bone cement compositions based on acrylate and epoxide terminated poly (propylene fumerate) oligomers and calcium salt compositions. *Biomaterials* 17:411–417 (1996).
118. L. J. DiBenedetto, S. J. Huang. Biodegradable hydroxylated polymers as controlled release agents. *Polym. Preprints* 30:453 (1989).
119. S. J. Huang, O. Kitchen, L. J. DiBenedetto. Swellable biodegradable polymers as controlled release systems. *Polymeric Mater. Sci. Eng.* 62:804–807 (1990).
120. A. J. Domb, N. Manor, O. Elmalak. Biodegradable bone cement compositions based on acrylate and epoxide terminated poly(propylene fumarate) oligomers and calcium salt compositions. *Biomaterials* 17:411–417 (1996).
121. J. Kemnitzer, J. Kohn. In: *Handbook of Biodegradable Polymers: Degradable Polymers Derived from the Amino Acid L-Tyrosine*, A. J. Domb, J. Kost, D. M. Weiseman (Eds). Hardwood Academic Publishers, Amsterdam, 1997, pp. 251–272.
122. S. Bogdansky. In: *Biodegradable Polymers as Drug Delivery Systems: Natural Polymers as Drug Delivery Systems*, M. Chasin, R. Langer (Eds.). Marcel Dekker, New York, 1990, pp. 231–259.
123. L. J. Arnold, A. Dagan, N. O. Kaplan. In: *Targeted Drugs: Poly(L-Lysine) as an Antineoplastic Agent and a Tumor-Specific Drug Carrier*, E. P. Goldberg (Ed.). John Wiley and Sons, New York, 1983, pp. 89–112.
124. J. M. Anderson, K. L. Spilizewski, A. Hiltner. In: *Biocompatibility of Tissue Analogs*, Vol. 1: *Poly α-Amino Acids as Biomedical Polymers*, D. F. Williams (Ed.). CRC Press, Boca Raton, FL, 1985, pp. 67–88.
125. A. Nathan, J. Kohn. In: *Biomedical Polymers: Designed to Degrade Systems*, S. W. Salaby (Ed.). Hanser/Gardner, Cincinnati, OH, 1994.
126. J. Kohn, R. Langer. Polymerization reactions involving the side chains of α-L-amino acids. *J. Am. Chem. Soc.* 109:817–820 (1987).
127. J. Kohn. Design, synthesis, and possible applications of pseudo-poly(amino acids). *Trends. Polym. Sci.* 1(7):206–212 (1993).
128. F. Harris, R. Eury. Synthesis and evaluation of biodegradable non-peptide polyamides containing α-amino residues. *Polym. Preprints* 30:449 (1989).
129. K. R. Sidman, D. A. Schwope, W. D. Steber, S. E. Rudolph. Naltrexone. In: *Naltrexone: Research Monograph 28*, R. E. Willette, G. Barnett (Eds). National Institute of Drug Abuse, 1980.
130. D. Barrera, Zylstra, P. Lansbury, R. Langer. Synthesis and RGD peptide modification of a new biodegradable polymer: poly(lactic acid-*co*-lysine). *J. Am. Chem. Soc.* 115: 11010–11011 (1993).
131. A. D. Cook, J. S. Hrkash, N. N. Gao, I. M. Johnson, U. V. Pajvani, S. M. Cannizzaro, R. Langer. Characterization and development of RGD-peptide-modified poly(lactic acid-*co*-lysine) as an interactive, resorbable biomaterial. *J. Biomed. Mater. Res.* 35:513–523 (1997).
132. D. A. Edwards, J. Hanes, G. Caponetti, J. Hrkash, B.-J. Abdelaziz, M. L. Eskew, J. Mintzes, D. Deaver, N. Lotan, R. Langer. Large porous particles for pumonary drug delivery. *Science*. 276:1868–1872 (1997).
133. T. Hayashi, Y. Ikada. Enzymatic hydrolysis of copoly(N-hydroxyalkyl L-glutamine/γ-methyl L-glutamate) fibers. *Biomaterials* 11:409 (1990).
134. D. B. Bennett, X. Li, N. W. Adams, S. W. Kim, C. J. T.

Hoes, and J. Feijen. Biodegradable polymeric prodrugs of naltrexone. *J. Controlled Release* 16:43–52 (1991).
135. F. Silver, M. Marks, Y. Kato, C. Li, S. Purapura, J. Kohn. *Long-Term Effect Med. Implants* 1:329 (1992).
136. S. Ertal, J. Kohn. Evaluation of a series of tyrosine-derived polycarbonates for biomaterial applications. *J. Biomed. Mater. Res.* 28:919 (1994).
137. N. H. Li, M. Richards, K. Brandt, K. Leong. Poly(phosphate esters) as drug carriers. *Polym. Preprints* 30:454–455 (1989).
138. K. W. Leong, Alternative material for fracture fixation. *Conn. Tissue. Res.* 31(4):S69–S75 (1995).
139. K. W. Leong, Z. Zhao, B. I. Dahiyat. In: *Controlled Drug Delivery: Challenges and Strategies*, K. Park (Ed.). American Chemical Society, Washington, D.C. (1997).
140. B. Dahiyat, M. Richards, K. Leong. Controlled release from poly(phosphoester) matrices. *J. Controlled Release* 33:13–21 (1995).
141. T. Laurencin, M. E. Norman, H. M. Elgendy, S. F. El-Amin, H. R. Alcock, S. R. Pucher, A. A. Ambrosio. Use of polyphosphazenes for skeletal tissue regeneration. *J. Biomed. Mater. Res.* 27:963–973 (1993).
142. J. H. Goedemoed, K. de Groot, A. M. E. Klaessen, R. J. Scheper. Development of implantable antitumor devices based on polyphosphazene S-II. *J. Controlled Release* 17: 235–244 (1991).
143. H. R. Allcock, S. Kwon. Covalent linkage of proteins to surface-modified poly(organo-phosphazenes): imobilization of glucose-6-phosphatedehydrogenase and trypsin. *Macromolecules* 19:1502–1508 (1986).
144. H. R. Allcock. In: *Biodegradable Polymers as Drug Delivery Systems: Polyphosphazene as New Biomedical and Bioactive Materials*, M. Chasin, R. Langer (Eds.). Marcel Dekker, New York, 1990, pp. 163–193.
145. H. R. Allcock, R. L. Kugel. High molecular weight poly(diaminophosphazenes). *Inorg. Chem.* 5:1716 (1966).
146. H. R. Allcock, T. J. Fuller, D. P. Mack, K. Matsumura, K. M. Smeltz. Synthesis of poly[(amino acid alkyl ester) phosphazene]. *Macromolecules* 10:824 (1977).
147. H. R. Allcock, S. R. Pucher, A. G. Scopelianos. Synthesis of poly(organophosphazenes) with glycolic acid ester and lactic acid ester side groups: prototypes for new bioerodible polymers. *Macromolecules* 27:1–4 (1994).
148. H. R. Allcock, S. R. Pucher, A. G. Scopelianos. Poly [(amino acid ester) phosphazenes]: synthesis, crystallinity and hydrolytic sensitivity in solution and the solid state. *Macromolecules* 27:1071–1075 (1994).
149. H. R. Allcock, S. R. Pucher, A. G. Scopelianos. Poly [(amino acid ester) phosphazenes] as substrates for the controlled release of small molecules. *Biomaterials* 15: 563–569 (1994).
150. A. Conforti, S. Bertani, S. Lussignoli, L. Grigolini, M. Tarzi, S. Lora, P. Caliceti, F. Marsilio, F. M. Veronese. Anti-inflammatory activity of polyphosphozene-based naproxen slow-release systems. *J. Pharm. Pharmacol.* 48: 468–473 (1996).
151. E. Schacht, J. Vandorpe, S. Dejardin, Y. Lenmouchi, L. Seymour. Biomedical applications of degradable polyphosphazenes. *Biotechnol. Bioeng.* 52:102–108 (1996).
152. H. R. Allcock, S. R. Pucher, A. G. Scopelianos. Poly [(amino acid ester) phosphazene] as substrates for controlled release of small molecules. *Biomaterials* 15:563–569 (1994).
153. E. M. Ruiz, C. A. Ramirez, M. A. Aponte, G. V. Barbosa-Canovas. Degradation of poly[bis(glycine ethyl ester) phosphazene] in aqueous media. *Biomaterials* 14:491–496 (1993).
154. H. R. Allcock, T. J. Fyller. Phosphazene high polymers with steroidal side groups. *Macromolecules* 13:1388 (1990).
155. J. H. Crommen, E. Schacht. Biodegradable polymers. II. Degradation characteristics of hydrolysis-sensitive poly [(organo)phosphazenes]. *Biomaterials* 13:601–611 (1992).
156. J. Vandorpe, E. Schacht, S. Stolinik, M. C. Garnett, M. C. Davies, L. Illum, S. S. Davis. *Biotechnol. Bioeng.* 652:89 (1996).
157. A. K. Andianov, L. G. Payner, K. B. Visscher, H. R. Allcock, R. Langer. *J. Appl. Polym. Sci.* 112:7832 (1994).
158. H. R. Allcock, A. M. A. Ambrosio. Synthesis and characterization of pH-sensitive poly(organophosphazene) hydrogels. *Biomaterials* 17:2295–2302 (1996).
159. G. Benagiano, H. L. Gabelnick. Biodegradable systems for the sustained release of fertility-regulating agents. *J. Steroid Biochem.* 11:449 (1979).
160. N. S. Choi, J. P. Heller. Polycarbonates. U. S. patent 4,079,038 (1978).
161. N. S. Choi, J. Heller. Drug delivery devices manufactured from polyorthoesters and polyorthocarbonates. U. S. patent 4,093,709 (1978).
162. N. S. Choi, J. Heller. Structured orthoesters and orthocarbonate drug delivery devices. U. S. patent 4,131,648 (1978).
163. N. S. Choi, J. Heller. Erodible agent releasing device comprising poly(ortho esters) and poly(ortho carbonates). U. S. patent 4,138,344 (1979).
164. R. C. Capozza, E. E. Schmitt, L. R. Sendelbeck. Chapter 6, In: *National Institute of Drug Abuse Research Monograph: Development of Chronomer*™ for Narcotic Antagonists, R. Willete (Ed.). DHEW Publication (ADM) 76–296 (1975).
165. N. S. Choi. Drug delivery by bioerodible polymer systems. *Polymer* (Korea) 4:35 (1980).
166. H. L. Gabelnick. In: *Long Acting Steroids Contraception: Biodegradable Implants—Alternative Approaches*, (D. R. Mishel, Jr. (Ed.). Raven Press, New York, 1983, p. 149.
167. E. Solheim, E. M. Pinholt, R. Andersen, G. Bang, E. Sudmann. Local delivery of indomethacin by a polyorthoester inhibits reossification of experimental bone defects. *J. Biomed. Mater. Sci.* 29:1141 (1995).
168. J. Heller, D. W. H. Penhale, R. F. Helwing. Preparation of poly(ortho esters) by the reaction of ketene acetals and polyols. *J. Polym. Sci. Polym. Lett. Ed.* 18:82–83 (1980).
169. J. Heller, D. W. H. Penhale, B. K. Fritzinger, J. E. Rose,

R. F. Helwing. Controlled release of contraceptive steroids from biodegradable poly(ortho esters). *Contracept. Deliv. Syst.* 4:43–53 (1983).

170. J. Heller, B. K. Fritzinger, S. Y. Ng, D. W. H. Penhale. In vitro and in vivo release of levonorgestrel from poly(ortho esters). II. Crosslinked Polymers. *J. Controlled Release* 1: 233–238 (1985).
171. J. Heller, S. Y. Ng, D. W. H. Penhale, B. K. Fritzinger, L. M. Sanders, R. A. Burns, M. G. Gaynon, S. S. Bhosale. The use of poly(ortho esters) for the controlled release of 5-fluorouracyl and a LHRH analogue. *J. Controlled Release* 6:217–224 (1987).
172. J. Heller, S. Y. Ng, B. K. Fritzinger. Synthesis and characterization of a new family of poly(ortho esters). *Macromolecules* 25:3362 (1992).
173. L. W. Seymour, R. Duncan, J. Duffy, S. Y. Ng, J. Heller. Poly(ortho ester) matrices for controlled release of the antitumour agent 5-fluorouracil. *J. Controlled Release* 31: 201–206 (1994).
174. K. V. Roskos, B. K. Fritzinger, S. S. Rao, G. C. Armitage, J. Heller. Development of a drug delivery system for the treatment of a periodontal disease based on bioerodible poly(ortho ester). *Biomaterials* 16:313 (1995).
175. S. Y. Ng, T. Vandamme, M. S. Taylor, J. Heller. Synthesis and erosion studies of self catalyzed poly(ortho ester)s. *Macromolecules* 30:770 (1997).
176. P. Wuthrich, S. Y. Ng, B. K. Fritzinger, K. V. Raskos, J. Heller. Pulsatile and delayed release of lysozyme from ointment-like poly(ortho esters). *J. Controlled Release* 21: 191 (1992).
177. G. M. Zentner, S. A. Pogany, R. V. Sparer, C. Shih, F. Kaul. The design, fabrication and performance of acid catalyzed poly(ortho ester) erodible devices. Abstracts of the Third Annual Meeting, American Association of Pharmaceutical Scientists, Orlando, FL, 1988.
178. R. V. Sparer. Physicochemical characterization of phthalic anhydride catalyzed poly(ortho ester)/cyclobenzaprine devices. *J. Controlled Release* 1:23–32 (1984).
179. J. Heller, R. V. Sparer, G. M. Zentner. Poly(ortho esters). In: *Biodegradable Polymers as Drug Delivery Systems*, M. Chasin, R. Langer (Eds.). Marcel Dekker, New York, pp. 121–161 (1990).
180. H. B. Rosen, J. Chang, G. E. Wnek, R. J. Linhardt, R. Langer. Biodegradable poly(anhydrides) for controlled drug delivery. *Biomaterials* 4:131 (1983).
181. J. E. Bucher, W. C. Slade. The anhydrides of isophthalic and terephthelic acids. *J. Am. Chem. Soc.* 31:1319–1321 (1909).
182. J. Hill. Studies on polymerization and ring formation. VI. Adipic anhydride. *J. Am. Chem. Soc.* 52:4110–4114 (1930).
183. J. Hill, W. H. Carothers. Studies on polymerization and ring formation. XIV. Alinear superpolyanhydride and a cyclic dimeric anhydride from sebecic acid. *J. Am. Chem. Soc.* 54:1569 (1932).
184. A. Conix. *Macromol. Chem.* 24:96 (1957).
185. A. Conix. Aromatic poly(anhydrides): a new class of high melting fiber-forming polymers. *J. Polym. Sci.* 29:343–353 (1958).
186. N. Yoda, A. Miyake. *Bull. Chem. Soc. Jpn.* 32:1120 (1959).
187. N. Yoda. *Makromol. Chem.* 32:1 (1959).
188. K. W. Leong, B. C. Brott, R. Langer. *J. Biomed. Mater. Res.* 19:941 (1985).
189. K. W. Leong, P. D'Amore, M. Marletta, R. Langer. Bioerodible polyanhydrides as drug-carrier matrices: biocompatibility and chemical reactivity. *J. Biomed. Mater. Res.* 20:51–64 (1986).
190. N. Yoda. Synthesis of poly(anhydrides). Crystalline and high melting poly(amide), poly(anhydrides) of methylene bis(*p*-carboxy phenyl)amide. *J. Polym. Sci. A.* 1:1323 (1963).
191. A. Domb, R. Langer. Poly(anhydrides). I. Preparation of high molecular weight poly(anhydrides). *J. Polym. Sci.* 25:3373–3386 (1987).
192. R. Subramanyam, A. G. Pinkus. Synthesis of poly(terephthalic anhydride) by hydrolysis of terephthalolyl chloride triethyleneamine intermediate adduct: characterization of intermediate adduct. *J. Macromol. Sci. Chem. A* 22(1):23 (1985).
193. K. W. Leong, V. Simonte, R. Langer. Synthesis of poly (anhydrides): melt polycondensation, dehydrochlorination and dehydrative coupling. *J. Macromolecules* 20:705 (1987).
194. A. Domb, E. Ron, R. Langer. Poly(anhydride). II. One step polymerization using phosgene or diphosgene as coupling agents. *Macromolecules* 21:1925 (1988).
195. S. Lundmark, M. Sjoling, A.-C. Albertsson. Polymerization of oxipane-2,7-dione in solution and synthesis of block copolymer of oxipane-2,7-dione and 2-oxipanone. *J. Macromol. Sci. Chem. A* 28:15–29 (1991).
196. A.-C. Albertsson, M. Eklund. Influence of molecular structure on the degradation mechanism of degradable polymers: in vitro degradation of poly(trimethylene carbonate), poly(trimethylene carbonate-*co*-caprolactones) and poly(adipic anhydride). *J. Appl. Polym. Sci.* 57:87–103 (1995).
197. K. Stradsberg, A.-C. Albertsson. Ring opening polymerization of 1,5-dioxipan-2-one initiated by a cyclic tin alkoxide initiator in different solvents. *J. Polym. Sci. Part A Polym. Chem.* 37:3407–3417 (1999).
198. E. Mathiowitz, E. Ron, G. Mathiowitz, C. Amato, R. Langer. *Macromolecules* 23:3212 (1990).
199. A. J. Domb. Synthesis and characterization of bioerodible aromatic anhydrides copolymers. *Macromolecules* 25:12 (1992).
200. A. Domb, E. Mathiowitz, E. Ron, S. Giannos, R. Langer. Poly(anhydrides) IV. Unsaturated and crosslinked poly (anhydrides). *J. Polym. Sci. Polym. Chem.* 29:571–579 (1991).
201. A. J. Domb, R. Nudelman. In vivo and in vitro elimination of aliphatic polyanhydrides. *Biomaterials* 16:319–323 (1995).
202. A. Domb, C. F. Gallardo, R. Langer. Poly(anhydrides).

3. Poly(anhydrides) based on aliphatic-aromatic diacids. *Macromolecules* 22:3200–3204 (1989).
203. M. Maniar, X. Xie, A. Domb. Poly(anhydrides). V. Branched poly(anhydrides). *Biomaterials* 11:690 (1990).
204. A. Domb, R. Langer. Solid state and solution stability of poly(anhydrides) and poly(esters). *Macromolecules* 22: 2117–2122 (1989).
205. E. Mathiowitz, R. Langer. Poly(anhydrides) microspheres as drug carriers. I. Hot-melt microencapsulation. *J. Controlled Release* 5:13–22 (1987).
206. E. Mathiowitz, M. Saltzman, A. Domb, P. Dor, R. Langer. *J. Appl. Polym. Sci*. 35:755 (1988).
207. A. J. Domb. Biodegradable polymers derived from amino acids. *Biomaterials* 11:689 (1990).
208. A. Stabuli, E. Mathiowitz, M. Lucarell, R. Langer. Characterization of hydrolytically degradable amino acid containing poly(anhydride-co-imides). *Macromolecules* 24: 2283 (1991).
209. M. Hartmann, V. Schulz. Synthesis of poly(anhydrides) containing amino groups. *Makromol. Chem*. 190:2133 (1989).
210. P. Pinter, M. Hartmann. *Macromol. Chem. Rapid Commun*. 11:403 (1990).
211. D. H. Lewis. In: *Biodegradable Polymers as Drug Delivery Systems: Controlled Release of Bioactive Agents from Lactide/Glycolide Polymers*, M. Chasin, R. Langer (Eds.). Marcel Dekker, New York, pp 1–41 (1990).
212. A. J. Domb, M. Maniar. Absorbable biopolymers derived from dimer fatty acids. *J. Poly. Sci. Polym. Chem*. 31:1275 (1993).
213. D. Teomim, A. Nyska, A. J. Domb. Ricinoleic acid based biopolymers. *J. Biomed. Mater. Res*. 45:258–267 (1999).
214. D. Teomim, A. J. Domb. Fatty acid terminated polyanhydrides. *J. Polym. Sci. Polym. Chem*. 37:3337–3344 (1999).
215. C. M. Agrawal, K. A. Athanasiou. Technique to control pH in vicinity of biodegrading PLA-PGA implants. *J. Biol. Med. Mater. Res*. 38:105–114 (1997).
216. K. H. Lam, J. M. Schakenraad, H. Groen, H. Esselbrugge, P. J. Dijkstra, J. Feijen, P. Nieuwenhuis. The infuence of surface morphology and wettability on the inflammatory response against poly(L-lactic acid): a semiquantitative study with monoclonal antibodies. *J. Biomed. Mater. Res*. 29:929–942 (1995).
217. K. H. Lam, J. M. Schakenraad, H. Esselbrugge, J. Feijen, P. Nieuwenhuis. The effect of phagocytosis of poly(L-lactic acid) fragments on cellular morphology and viability. *J. Biomed. Mater. Res*. 27:1569–1577 (1993).
218. L. R. Beck, T. R. Tice. In: *Long-Acting Steroid Contraception*, R. Mishell, Jr. (Ed.). Raven Press, New York, 1983, pp. 175–199.
219. L. R. Beck, R. A. Ramos, C. E. Flowers, G. Z. Lopez, D. H. Lewis, D. R. Cowsar. Clinical evaluation of injectable biodegradable contraceptive system. *Am. J. Obstet. Gynecol*. 140:799 (1981).
220. L. R. Beck, V. Z. Pope, C. E. Flowers, Jr., D. R. Cowsar, T. R. Tice, D. H. Lewis, R. L. Dunn, A. R. Moore, R. M. Gilley. Poly(DL-lactide-co-glycolide)/norethisterone microcapsules: an injectable biodegradable contraceptive. *Biol. Reprod*. 28:186 (1983).
221. L. R. Beck, C. E. Flowers, Jr., V. Z. Pope, W. H. Wilborn, T. R. Tice. Clinical evaluation of an improved injectable microcapsule contraceptive system. *Am. J. Obstet. Gynecol*. 147:815 (1983).
222. Population Information Program. *Population Reports*. The Johns Hopkins University, Baltimore, No. 3, March 1987.
223. G. I. Zatuchni, A. Goldsmith, J. D. Shelton, J. J. Sciarra (Eds.). *Long-Acting Contraceptive Delivery Systems*. Harper and Row, Philadelphia, 1984.
224. D. L. Wise, A. D. Schwope, S. E. Harrigan, D. A. McCarty, J. F. Hower. In: *Polymeric Delivery Systems: Sustained Delivery of Narcotic Antagonists from Lactic/Glycolic Acid Copolymer Implants*, R. J. Kostelnik (Ed.). Gordon and Breach, New York, 1978.
225. A. D. Schwope, D. L. Wise, J. F. Hower. Lactic/glycolic acid polymers as narcotic antagonist delivery systems. *Life Sci*. 17:1877 (1975).
226. C. N. Chiang, L. E. Hollester, A. Kishimoto, G. Barnett. Kinetics of a naltrexone sustained-release preparation. *Clin. Pharmacol. Ther*. 36:51 (1984).
227. E. S. Nuwayser, M. H. Gay, D. J. DeRoo, P. D. Blaskovish. Sustained release injectable naltrexone microspheres. *Proc. Int. Symp. Control. Rel. Bioact. Mater*. 15 (1988).
228. J. E. Bergshma, F. R. Rozema, R. R. M. Bos, A. W. M. Van Rozendaal, W. H. De Jong, J. S. Teppema, C. A. P. Joziasse. Biocompatibility and degradation mechanisms of predegraded and non-predegraded poly(lactide) implants: an animal study. *J. Mater. Sci. Mater. Medicine* 6: 715–723 (1995).
229. J. E. Bergshma, F. R. Rozema, R. R. M. Bos, G. Boering, W. C. de Brruijin, A. J. Ppennings. In vivo degradation and biocompatibility study of in vitro pre-degraded as polymerized poly(lactides) particles. *Biomaterials* 16: 267–274 (1995).
230. J. E. Bergshma. Late complications using poly(lactide) osteosynthesis in vitro and in vivo tests. Dissertation, University of Groningen, The Netherlands (1995).
231. D. F. Emerich, M. A. Tracy, K. L. Ward, M. Figueiredo, R. Qian, C. Henschel, R. T. Bartus. Biocompatibility of poly(DL-lactide-co-glycolide) microspheres implanted into the brain. *Cell. Transplant* 8(1):47–58 (1999).
232. L. M. Sanders, J. S. Kent, G. I. McRae, B. H. Vickeryu, T. R. Tice, D. H. Lewis. Controlled release of an LHRH analogue from poly-DL-lactide co-glycolide microspheres. *J. Pharm. Sci*. 73:1294 (1984).
233. R. H. Asch, F. J. Rojas, A. Bartke, A. V. Schalley, T. R. Tice, H. G. Klemeke, T. M. Siler-Khodr, R. W. Bray, M. P. Hogan. Prolonged suppresion of plasma LH levels in male rats after a single injection of an LHRH agonist in DL-lactide/glycolide microcapsules. *J. Androl*. 6:83 (1985).
234. A. K. Kwong, S. Chou, A. M. Sun, M. V. Sefton, M. E. A. Goosen. In vitro and in vivo release of insulin from

poly(lactic acid) microbeads and pellets. *J. Controlled Release* 4:47 (1986).

235. M. A. Toljan, E. Orthner. Bioresorbable interference screws in ACL surgery. *Arthroscopy* 11:381 (1995).
236. C. G. Pitt, A. Schindler. In: *Long-Acting Contraceptive Delivery Systems: Capronor™: A Biodegradable Delivery System for Levonorgestrel*, G. I. Zatuchni, A. Goldsmith, J. D. Shelton, J. J. Sciarra (Eds.). Harper and Row, Philadelphia, 1984, pp 48–63.
237. H. B. Rosen, J. Kohn, K. Leong, R. Langer. In: *Controlled Release Systems: Fabrication Technology*, Vol. 2, D. Hsieh (Ed.). CRC Press, Boca Raton, FL, 1988.
238. L. Illum, S. S. Davis (Eds.). *Polymers in Controlled Drug Delivery*. Wright, Bristol, England, 1987.
239. K. Juni, M. Nakano. *Crit. Rev. Therapeut. Drug Carrier Syst*. 3:209 (1986).
240. S. Bogdansky. In: *Biodegradable Polymers as Drug Delivery Systems: Natural Polymers as Drug Delivery Systems*, M. Chasin, R. Langer (Eds.). Marcel Dekker, New York, 1990, pp 231–259.
241. A. Arem. *Clin. Plast. Surg*. 12:209 (1985).
242. F. DeLustro, R. A. Cendell, M. A. Nguyen, J. M. McPherson. *J. Biomed. Mater. Res*. 20:109 (1986).
243. J. S. Kay, B. S. Litin, M. A. Jones, A. W. Fryczkowsky, M. Chvapil, J. Herschler. *Ophthalmic Surg*. 17:796 (1986).
244. B. A. Rhodes, I. Zolle, J. W. Buchanan, H. N. Wagner, Jr. *Radiology* 92:1453 (1969).
245. S. S. Davis, L. Illum, J. G. McVie, E. Tomlinson (Eds.). *Microspheres and Drug Therapy: Pharmaceutical, Immunological and Medical Aspects*. Elsevier, Amsterdam, 1984.
246. J. J. Marty, R. C. Oppenheim. *Aust. J. Pharm. Sci*. 6:65 (1977).
247. K. T. Kennedy, M. Pharm. Thesis, Victorian College of Pharmacy, Australia (1983).
248. H. Yu. Pseudopoly(amino acids): a study of the synthesis and characterization of polyesters made from α-L-amino acids. Ph.D. Thesis, Massachusetts Institute of Technology, Cambridge, MA (1988).
249. R. Langer, H. Brem, D. Tapper. Biocompatibility of polymeric delivery systems for macromolecules. *J. Biomed. Mater. Res*. 15:267–277 (1981).
250. J. Kohn, S. M. Niemi, E. C. Albert, J. C. Murphy, R. Langer, J. G. Fox. Single step immunization using a controlled release biodegradable polymer with sustained adjuvant activity. *J. Immunol. Meth*. 95:31–38 (1986).
251. R. E. Singler, M. S. Sennett, R. A. Willingham. In: *Inorganic and Organometallic Polymers Phosphazene Polymers: Synthesis, Structure, and Properties*, M. Zeldin, K. J. Wynne, H. R. Allcock (Eds.). *ACS Symp. Ser*. 360: 360 (1988).
252. H. R. Penton. In: *Inorganic and Organometallic Polymers: Polyphosphazenes: Performance Polymers for Special Applications*, M. Zeldin, K. J. Wynne, H. R. Allcock (Eds.). ACS Symp. Ser. 360:278 (1988).
253. S. Lora, G. Palma, R. Bozio, P. Caliceti, G. Pezzin. Polyphosphazenes as biomaterials: surface modification of poly[bis(trifluoroethoxy)phosphazene] with polyethylene glycols. *Biomaterials* 14:430–436 (1993).
254. S. Lora, G. Palma, M. Carenza, P. Caliceti, G. Pezzin. Radiation grafting of hydrophilic monomers onto poly[bis (trifluoroethoxy) phosphazene] with polyethylene glycols. *Biomaterials* 14:430–436 (1993).
255. C. W. R. Wade, S. Gourlay, R. Rice, A. Hegyeli, R. Singler, J. White. In: *Organometallic Polymers: Biocompatibility of Eight Poly(organophosphazenes)*, C. E. Carraher, J. E. Sheats, C. U. Pittman (Eds.). Academic Press, New York, 1978, p. 289.
256. D. J. Lyman, K. Knutson. In: *Biomedical Polymers: Chemical, Physical, and Mechanical Aspects of Blood Compatibility*, E. P. Goldberg, A. Nakajima (Eds.). Academic Press, New York, 1980, p. 1.
257. F. Langone, S. Lora, F. M. Veronese, P. Caliceti, P. P. Parnigotto, F. Valenti, G. Palma. Peripheral nerve repair using a poly(organo)phosphazene tubular prosthesis. *Biomaterials* 16:347–353 (1995).
258. R. C. Capozza, L. Sendelbeck, W. J. Balkenhol. In: *Polymeric Delivery Systems: Preparation and Evaluation of a Bioerodible Naltrexone Delivery System*, R. J. Kostelnik (Ed.). Gordon and Breach, New York, 1978, pp. 59–73.
259. B. B. Pharriss, V. A. Place, L. Sendelbeck, E. E. Schmitt. Steroid systems for contraception. *J. Reprod. Med*. 17:91–97 (1976).
260. C. Laurencin, A. Domb, C. Morris, V. Brown, M. Chasin, R. McConell, N. Lange, R. Langer. Poly(anhydride) administration in high doses in vivo: studies of biocompatibility and toxicology. *J. Biomed. Mater. Res*. 24:1463–1481 (1990).
261. R. J. Tamargo, J. I. Epstein, C. S. Reinhard, M. Chasin, H. Brem. Brain biocompatibility of a biodegradable controlled-release polymer in rats. *J. Biomed. Mater. Res*. 23: 253–266 (1989).
262. H. Brem, A. Kader, J. I. Epstein, R. J. Tamargo, A. Domb, R. Langer, K. W. Leong. Biocompatibility of a biodegradable, controlled-release polymer in the rabbit brain. *Selective Ther*. 5:55 (1989).
263. H. Brem, H. Ahn, R. J. Tamargo, M. Pinn, M. Chasin. A biodegradable polymer for intracraneal drug delivery: a radiological study in primates. *Am. Assoc. Neurol. Surg*. 349:1988.
264. H. Brem, R. J. Tamargo, M. Pinn, M. Chasin. Biocompatibility of BCNU-loaded biodegradable polymer: a toxicity study in primates. *Am. Assoc. Neurol. Surg*. 381:1988.
265. M. Rock, M. Green, C. Fait, R. Geil, J. Myer, M. Maniar, A. Domb. Evaluation and comparison of biocompatibility of various classes of polyanhydrides. *Polym. Preprints* 32: 221–222 (1991).
266. S. B. Aaronson, R. S. Horton. Mechanisms of the host response in the eye. *Arch. Ophthalmol*. 85:306–308 (1971).
267. P. Henkind. Ocular neovascularization. *Am. J. Ophthalmol*. 85:287–301 (1978).
268. R. Langer, H. Brem, D. Tapper. Biocompatibility of polymeric delivery systems for macromolecules. *J. Biomed. Mater. Res*. 15:267–277 (1981).

269. R. Langer, D. Lund, K. Leong, et al. Controlled release of macromolecules: biological studies. *J. Controlled Release* 2:331–341 (1985).
270. A. J. Domb, M. Maniar, S. Bogdansky, M. Chasin. Drug delivery to the brain using polymers. *Crit. Rev. Therap. Drug Carrier Syst*. 8:1–17 (1991).
271. H. Brem, S. Piantadosi, P. C. Burger, M. Walker, R. Selker, N. A. Vick, K. Black, M. Sisti, S. Brem, G. Mohr, et al. Placebo-controlled trial of safety and efficacy of intraoperative controlled delivery by biodegradable polymers of chemotherapy for recurrent gliomas. *Lancet* 345(8956):1008–1012 (1995).

5

Biodegradable Biomaterials Having Nitric Oxide Biological Activity

C. C. Chu
Cornell University, Ithaca, New York

I. INTRODUCTION

The chemical incorporation of biologically active compounds into synthetic biodegradable biomaterials was successfully synthesized by using nitric oxide derivatives (e.g., 4-amino-2,2,6,6-tetramethylpiperidine-1-oxy) as the biochemical agents and synthetic aliphatic polyesters like polyglycolide as the biodegradable biomaterials. The resulting new biomaterial was characterized and its in vitro hydrolytic degradation property was studied to examine the release profiles of the chemically incorporated nitric oxide derivative from polyglycolide. The biological activity of this new class of biodegradable biomaterial was tested by examining its ability to retard the proliferation of human smooth muscle cell in vitro. It was found that this new class of biologically active biodegradable biomaterial has indeed one of the well-known biological functions of nitric oxide: retardation of the proliferation of smooth muscle cells. The potential biomedical applications of this new biomaterial include the treatment of hyperplasia in cardiovascular disorders, promoting wound healing in healing-impaired patients, and nitric oxide–related diseases.

The interests in biodegradable polymeric biomaterials for biomedical engineering use have increased dramatically during the past decade. This is because this class of biomaterials has two major advantages that nonbiodegradable biomaterials do not have. First, they do not elicit permanent chronic foreign body reaction due to the fact that they would be gradually absorbed by human body and do not permanently retain trace of residual in the implantation sites. Second, some of them have recently been found to be able to regenerate tissues, via so-called tissue engineering, through the interaction of their biodegradation with immunologic cells like macrophages. Hence, surgical implants made from biodegradable biomaterials could be used as temporary scaffold for tissue regeneration. This approach toward the reconstruction of injured, diseased, or aged tissues is one of the most promising fields in the future medicine.

Although the earliest and most commercially significant biodegradable polymeric biomaterials are originated from linear aliphatic polyesters like polyglycolide and polylactide from poly(-hydroxyacetic acids), recent introduction of several new synthetic and natural biodegradable polymeric biomaterials extends the domain beyond this family of simple polyesters. These new commercially significant biodegradable polymeric biomaterials include poly (orthoesters), polyanhydrides, polysaccharides, poly(esteramides), tyrosine-based polyarylates or polyiminocarbonates or polycarbonates, poly(D,L-lactide-urethane), poly(-hydroxybutyrate), poly(-caprolactone), poly[bis(carboxylatophenoxy) phosphazene], poly(amino acids), pseudopoly(amino acids), and copolymers derived from amino acids and non–amino acids.

The earliest and most successful and frequent biomedical application of biodegradable polymeric biomaterials has been in wound closure (1). All biodegradable wound closure biomaterials are based upon glycolide and lactide family. For example, polyglycolide (Dexon from American Cyanamid), poly(glycolide-L-lactide) random copolymer with 90-to-10 molar ratio (Vicryl from Ethicon), poly(ester-ether) (PDS from Ethicon), poly(glycolide-trimethylene carbonate) random block copolymer (Maxon from American Cyanamid), and poly(glycolide-caprolactone) copolymer (Monocryl from Ethicon). Some of these materials like Vicryl have been commercially used as surgical meshes for hernia and body wall repair. Besides wound closure application, biodegradable polymeric biomaterials that are commercially satisfactory include those for drug control/release devices. Some well-known examples in this application are polyanhydrides and poly (orthoester). Biodegradable polymeric biomaterials, particularly totally resorbable composites, have also been experimentally used in the field of orthopedics, mainly as components for internal bone fracture fixation like PDS pins. However, their wide acceptance in other parts of orthopedic implants may be limited due to their inherent mechanical properties and their biodegradation rate. Biodegradable polymeric biomaterials have been experimented as vascular grafts, vascular stents, vascular coupler for vessel anastomosis, nerve growth conduits, augmentation of defected bone, ligament/tendon prostheses, intramedullary plug during total hip replacement, anastomosis ring for intestinal surgery, and stents in ureteroureterostomies for accurate suture placement. The details of the biomedical applications of biodegradable polymeric biomaterials and their chemical, physical, mechanical biological, and biodegradation properties can be found in other recent reviews (1–7).

There is one common characteristic among all these biodegradable biomaterials: they do not ''actively'' participate in the process of wound healing, tissue regeneration and engineering. In other words, these biomaterials are not ''alive'' and cannot remodel and/or release cytokines upon stimulation like normal tissues. These biomaterials, however, elicit inflammatory and foreign body reactions and play a ''passive'' role in wound healing. It would be ideal if these synthetic biomaterials could be engineered so that they could become alive after implantation and hence actively participate in the biological functions with the surrounding tissues, such as the ability to modulate inflammatory reactions, to facilitate wound healing, or to mediate host defense system to combat diseases. In this chapter, we would describe the new biodegradable biomaterials having nitric oxide function.

II. BIODEGRADABLE BIOMATERIALS HAVING NITRIC OXIDE FUNCTION

Nitric oxide (NO·) is a very small but highly reactive and unstable free radical biomolecule with expanding known biological functions. This small biomolecule and its biological functions have recently become one of the most studied and intriguing subjects as recently reviewed by several investigators (8–18). Nitric oxide is extremely labile and short lived (about 6 to 10 s).

Nitric oxide and its radical derivatives have been known to play a very important role in a host of expanding biological functions, such as inflammation, neurotransmission, blood clotting, blood pressure, cardiovascular disorders, rheumatic and autoimmune diseases, antitumor activity with a high therapeutic index, antimicrobial property, sensitization or protection of cells and tissues against irradiation, oxidative stress, respiratory distress syndrome, and cytoprotective property in reperfusion injury, to name a few (8–30). Nitric oxide acts both as an essential regulatory agent to normal physiological activities and as cytotoxic species in diseases and their treatments. Nathan et al. reported that nitric oxide is a potent antiviral compound against two disfiguring poxvirus and herpes simplex virus type 1, which causes cold sores in humans (12). Levi et al. also found that nitric oxide could protect human heart against low oxygen supply, a condition known as myocardial ischemia, by widening blood vessels so that more oxygen-rich blood reaches the heart (14). Elliott et al. reported that a new NO·-releasing nonsteroidal anti-inflammatory drug has the benefits of accelerating gastric ulcer healing (31,32). It is important to know, however, that excessive introduction of NO· into body may have adverse effects like microvascular leakage, tissue damage in cystic fibrosis, septic shock, B cell destruction, and possible mutagenic risk, to name a few (18,19,27,33–35).

Nitric oxide and NO·-derived radicals are not normal biological messengers whose trafficking depend on specific transporters or channels. Instead, nitric oxide radicals released by cells like macrophage and endothelial cells would diffuse randomly in all directions from the site of release. Because of this unusual property, the only way to control the biological functions of nitric oxide is to control its site of synthesis. This suggests that the only way to deliver the desirable biological functions of nitric oxide is through nature. Existing science and technology are not able to modulate the release of nitric oxide according to our wish for a variety of therapeutic purposes.

A. Synthesis and Characterization

We recently used a patented chemical method to incorporate nitric oxide derivatives into a series of synthetic biodegradable biomaterials (36,37). Upon the hydrolytic degradation of the host biomaterials, nitric oxide derivatives could be released to the surrounding and the rate of release could be controlled by the nature of the biodegradable biomaterials. The amounts of the nitric oxide derivatives that can be incorporated into biodegradable biomaterials would depend on the molecular weight of the biomaterials. Figure 1 illustrates the chemical scheme of this patented method of incorporating 4-amino-2,2,6,6-tetramethylpiperidine-1-oxy (TAM) as the source of nitroxyl radicals nitric oxide into synthetic biodegradable polyesters like polyglycolide or polylactide.

Due to the free radical characteristic of Tempamine nitroxyl radicals, the radical incorporated polyglycolide (PGA) must exhibit an electron paramagnetic resonance (EPR) spectrum that has the characteristic of nitroxide. Figure 2 is such an EPR spectrum of TAM-PGA. This EPR spectrum shows a considerable broadening of linewidth when compared with an EPR spectrum of free TAM nitric oxide. An EPR characterization of nitroxyl radicals is based on the measurement of the signal intensity of an EPR spectrum. This measurement can provide fundamental information of free radicals, such as linewidth, which bears a relationship to the tumbling motion of free radicals; g value, which largely depends on the immediate environments of the free radicals; and hyperfine splitting constants, which describe the classical multiplicity of EPR spectram due to the interaction of the unpaired electron spins with nuclear spins.

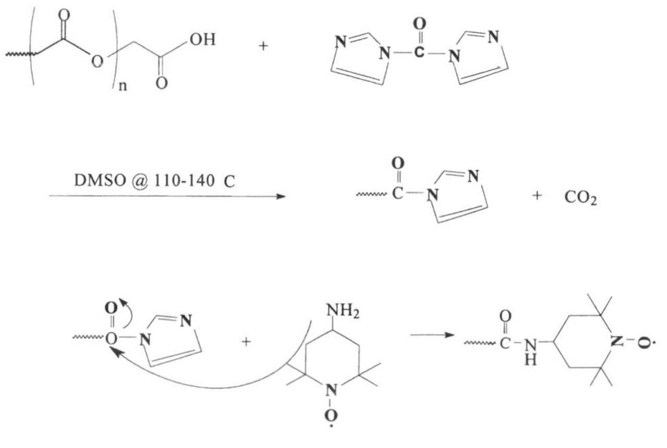

Figure 5.1 Chemical scheme for incorporating 4-amino-2,2,6,6-tetramethylpiperidine-1-oxy into the carboxyl chain ends of linear aliphatic polyesters like polyglycolide.

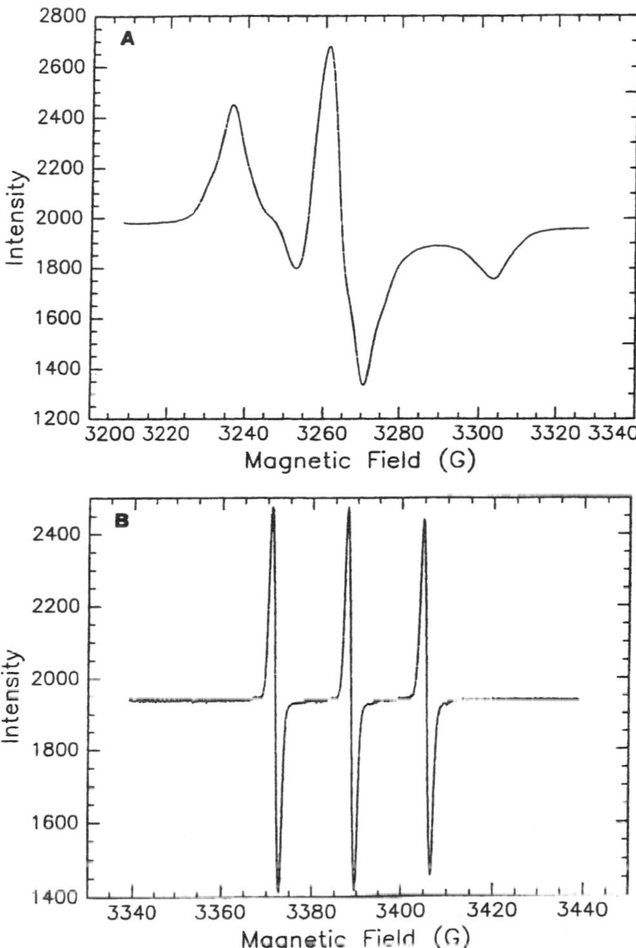

Figure 5.2 Electron paramagnetic resonance spectra of TAM radical (A) in conjugation with polyglycolide and (B) in free form.

The considerable broadening of the EPR spectrum of the TAM-PGA biomaterial (Fig. 2A) when compared with free TAM nitroxyl radical (Fig. 2B) is attributed to the viscous macromolecular environment surrounding TAM nitroxyl radicals that were chemically bound to PGA chain ends. These nitroxyl radicals cannot move as freely as free nitric oxide due to the restricted PGA chain segmental motion. This relationship between free radical motion and the characteristic of its immediate environment would provide a useful means to study the release pattern of nitroxyl radicals that were chemically bound to biodegradable substrates.

Because TAM nitroxyl radicals were incorporated only into the carboxylic chain ends of PGA macromolecules, their physical, thermal, and mechanical properties are insignificantly different from the parent PGA macromole-

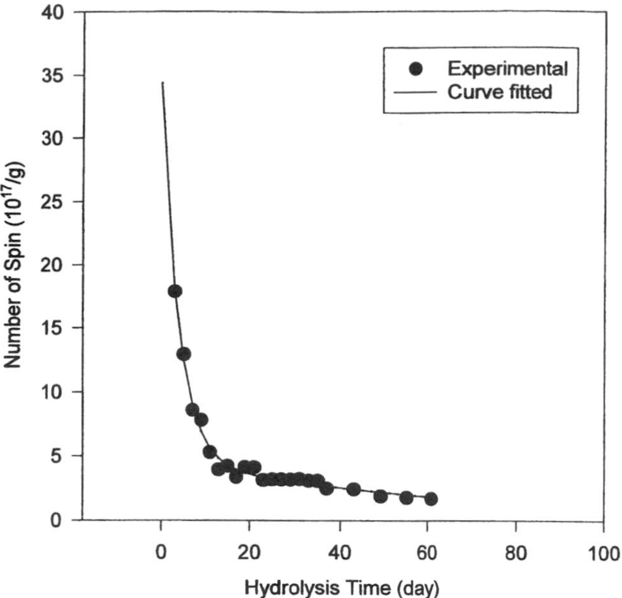

Figure 5.3 The kinetics of in vitro release of TAM nitroxyl radicals from TAM-PGA biomaterials in buffer media of original pH 7.44 at 37°C.

cules, such as similar melting temperature and heat of fusion. This lack of change in fundamental properties between TAM-PGA and parent PGA biomaterials should be beneficial because the similar processing conditions that have been used to fabricate PGA for a variety of clinical applications could also be used to fabricate the new TAM-PGA. In addition, the knowledge of the well-known biodegradation properties of PGA could be applied to estimate the release pattern of TAM nitroxyl radicals from PGA upon its biodegradation. The main difference between the parent and TAM-PGA, however, is their degradation products and their subsequent biological properties.

B. In Vitro Hydrolytic Degradation

Since the expected biological functions of the TAM-PGA must come from the nitroxyl radicals that would be released into the surrounding environment upon hydrolytic degradation of PGA, the amount and the rate of release of TAM nitroxyl radicals should have a direct impact on the applicability of the newly synthesized TAM-PGA biomaterials to medicine. Figure 3 illustrates such an in vitro release pattern of TAM radicals from PGA. The release of Tempamine nitroxyl radicals upon PGA hydrolysis fol-

(RNOH) (RNO·) (RNO+)

Figure 5.4 The reversible one-electron reduction and oxidation reactions of nitroxyl free radicals.

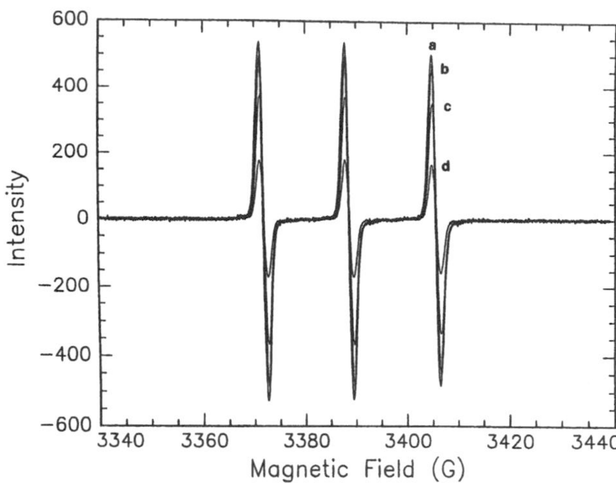

Figure 5.5 The effect of pH of the media on the EPR spectra intensity of TAM nitroxyl radical at 10 g/mL. The media was glycolic acid: (a) pH 7.44, (b) pH 4.0, (c) pH 3.5, (d) pH 3.0.

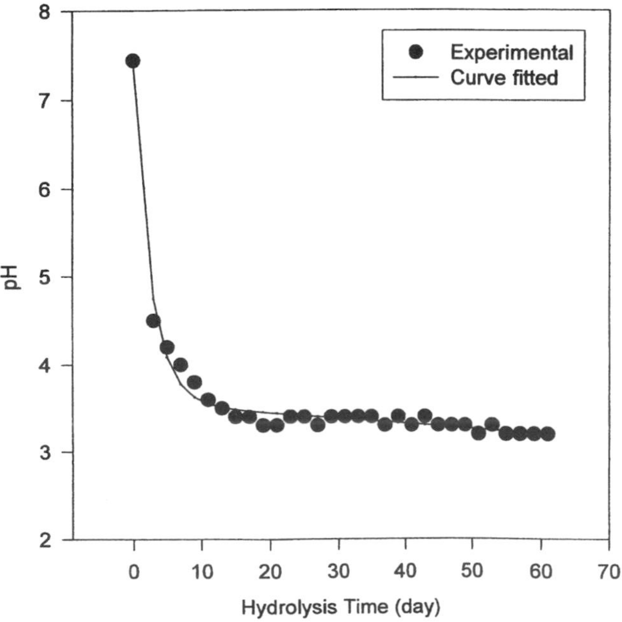

Figure 5.6 The change in pH of the buffer medium used for the in vitro release study of TAM-PGA at 37°C.

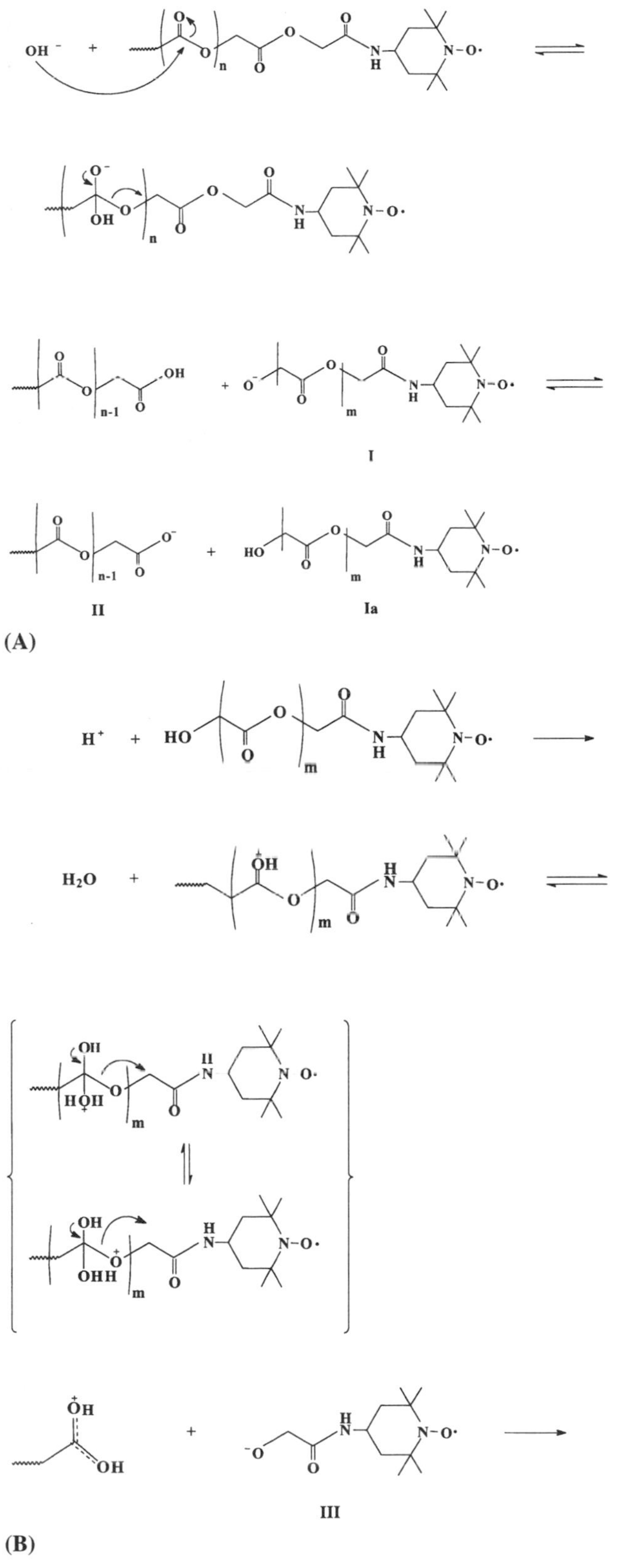

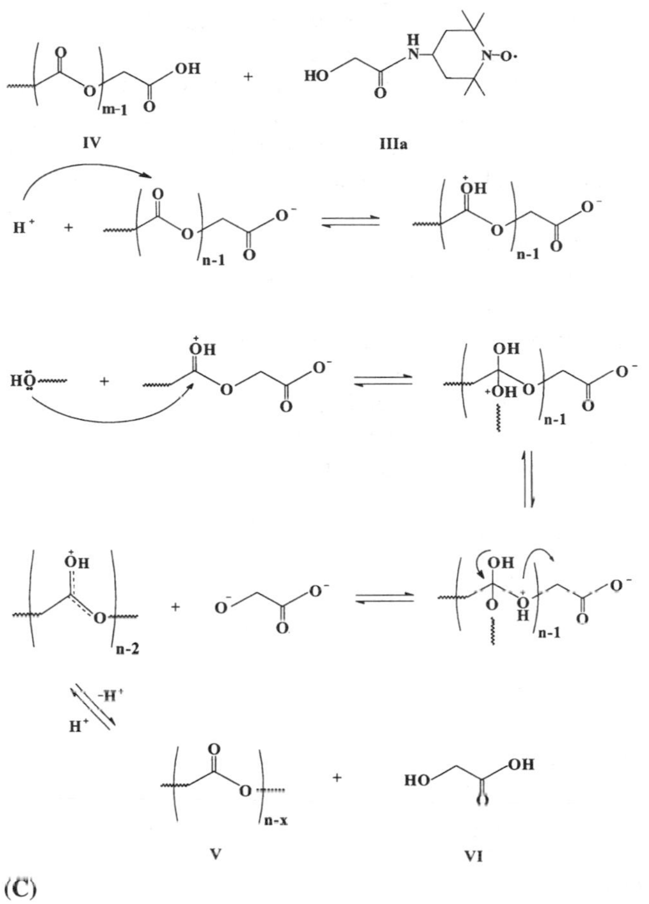

(A)

(B)

(C)

Figure 5.7 The hydrolytic release mechanism of TAM radicals from TAM-PGA biomaterials via alkaline and acid hydrolytic degradation. (A) Initial alkaline hydrolytic degradation of the ester linkages in polyglycolide backbone during the early stage of immersion. (B) Acidic hydrolytic degradation of the ester linkages in polyglycolide backbone during the middle stage of immersion. (C) Acidic hydrolytic degradation of the amide linkage to free TAM radical from the polyglycolide substrate during the late stage of immersion.

lowed a double exponential behavior in which significant amounts of nitroxyl radicals were released before 20 days followed by a gradual release thereafter.

It was well known that the nitroxyl free radicals could participate in one-electron reduction reaction in an acidic medium to yield relatively stable diamagnetic products, hydroxypiperidines structure (NOH), and *oxo*-piperidinium cations (NO^+). Figure 4 shows such one-electron reduction and oxidation reaction of free nitric oxide derivatives.

Because the conversion of TAM nitroxyl radical (i.e., chemical reduction of nitroxyl radical) to hydroxypiperine structure in an acidic environment would remove the free radical characteristic of TAM, EPR could not detect the presence of hydroxypiperine, and the spin numbers of TAM would continue decrease with time. The extent of this conversion and hence the number of spins of TAM would depend on the strength of acidity of the medium. Figure 5 illustrates such an effect of pH on the peak intensity of EPR spectra of TAM nitroxyl radicals in glycolic acid media. As the pH of the media decreased from 7.44 to 4.0 (EPR spectrum b in Fig. 5) and 3.0 (EPR spectrum d in Fig. 5), the EPR spectra peak intensities were reduced accordingly. This would suggest that more TAM nitroxyl radicals would be reduced to hydroxypiperine at a lower pH, and fewer TAM nitroxyl radicals would be remained in an acidic medium and exhibit weaker EPR signal.

In addition, the nitroxyl group could also be easily polarizable by acids toward its ionic resonance form resulting in an increased electron–nuclear coupling constant (hyperfine splitting constant) since the spin density of the nitroxyl bond has been influenced by the proton activity in the acidic media.

The acidifying of the degradation media can be demonstrated by the decrease in pH as shown in Fig. 6. The result indicated a similar double exponential reduction [$y(x) = 3.83\,e^{-0.39x} + 3.57\,e^{-0.0018x}$] in pH as for the number of spins versus time in Fig. 3. The pH of the medium was reduced from the initial 7.4 at day 0 to 3.5 during the first 14 days of hydrolysis, and there was a very little reduction in pH thereafter. A comparison between Figs. 3 and 6 showed a remarkable similarity between the release pattern of TAM nitroxyl radicals and the profile of the reduction in pH of the degradation media.

The observed in vitro release pattern of TAM nitroxyl radicals from the TAM-PGA biomaterial and the close similarity between the TAM release pattern and the media pH reduction profiles suggested that pH must play a major role. Based on the well-known hydrolytic degradation mechanism of PGA biomaterial, we postulated the hydrolytic degradation mechanism of TAM-PGA biomaterial as illustrated in Fig. 7. TAM-PGA biomaterials experienced both alkaline and acid catalyzed hydrolytic degradation as evident in the change of pH of the media (Fig. 3).

Alkaline hydrolysis would occur during the very early stage (within a few days) because of the initial pH of the buffer (7.4). The possible products of this early stage alkaline hydrolysis of TAM-PGA were expected to be carboxylic fragments of PGA segments (II in Fig. 7A) and the TAM-PGA fragments (Ia in Fig. 7A). As hydrolysis time proceeded further, the PGA fragments would eventually be degraded into glycolic acid or/and its cyclic dimers (glycolide) as the pH of the media became more acidic. The length of the PGA segment of the TAM-PGA frag-

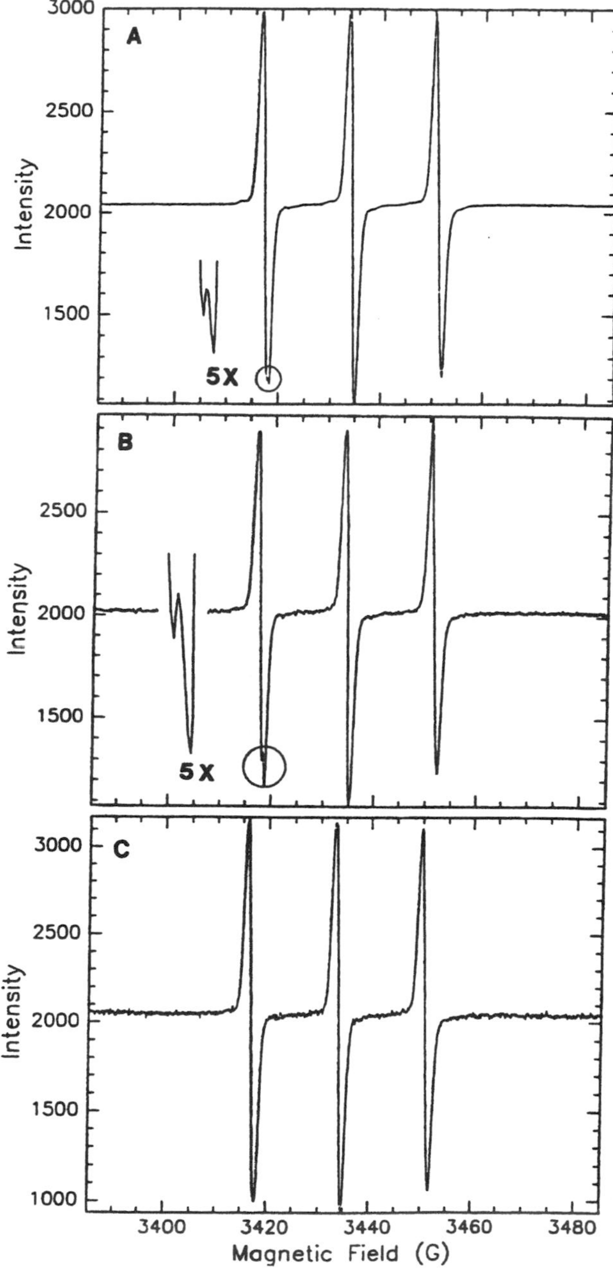

Figure 5.8 Electron paramagnetic resonance spectra of the degradation products from the in vitro hydrolytic degradation of TAM-PGA in buffer media over a period of 55 days at 37°C. (A) 3 days, (B) 23 days, (C) 55 days. Note the presence of split peaks in A and B that are not there in C.

ments (Ia) would also be reduced via ester hydrolysis in both alkaline and acidic media till all the ester linkages were scissioned (III or/and IIIa in Fig. 7B). Because the amide linkages where the TAM nitroxyl radicals were attached to PGA were far more hydrolytic resistant than ester linkages, we did not expect the formation of free TAM nitroxyl radicals via the scission of the amide linkage (VII in Fig. 7C) until at the late stage of hydrolysis.

Based on the proposed hydrolytic degradation mechanism of TAM-PGA biomaterials shown in Fig. 7, one should expect only one type of TAM nitroxyl radical (I or Ia in Fig. 7A) in the very early stage of hydrolysis, as hydrolysis proceeded further, a mixture of TAM nitroxyl radicals with different PGA chain length (i.e., different m in I or Ia, would be the predominant products in the degradation media). At the late stage of hydrolysis, all PGA segments of the TAM-PGA fragments were hydrolyzed and eventually the amide linkages where the TAM radicals were attached would be hydrolyzed to free TAM radicals (VII in Fig. 7C).

These different types of TAM radicals due to different PGA chain lengths attached might be evident in the EPR spectra shown in Fig. 8, which was a series of EPR spectra from three different stages of hydrolysis time.

At 3 days of hydrolysis, its EPR spectrum had a unique character on its first peak (i.e., a spectral feature with splitted peaks) that was characteristically different from the typical 3 EPR peaks of homogeneous nitroxyl radicals in solution. This spectral feature could imply that there was a mixture of TAM-PGA nitroxyl radicals with various PGA chain lengths (heterogeneity). In other words, there might be at least two different segmental lengths with different tumbling rate of TAM nitroxyl radicals (i.e., two different A tensor and g factors). The largest EPR spectral feature of the first peak was observed at 23 days hydrolysis. This EPR spectral feature persisted for a long period up to 40 days of hydrolysis. At the late stage of hydrolysis (55 days), however, this spectral feature of TAM radicals disappeared and the typical 3 EPR spectrum of nitroxyl radicals reappeared. This suggested that a more homogeneous type of nitroxyl radical, like VII or IIIa shown in Fig. 7, were present in the media. Due to the lack of restriction imposed by the PGA long chain segments, the tumbling motion of free TAM nitroxyl radicals in aqueous solution was easier than their polymeric counterparts and was reflected in the typical nitroxyl radical hyperfine spectrum.

C. In Vitro Biological Activity on Human Smooth Muscle Cells

The biological activity of TAM-PGA can best be illustrated by the level of retardation of smooth muscle cell (SMC) in cell culture. This is because NO has been theorized to be able to retard SMC proliferation in humans. As shown in Fig. 9, TAM-PGA showed profound retardation of the proliferation of SMC in vitro. There was virtually no change in the number of live SMCs in the culture medium having TAM-PGA over the entire period of cell culture, while the number of live SMCs in the culture medium control had been more than double (134% increase) during the same culture period. This level of SMC retardation of TAM-PGA biomaterial was found to be similar with free

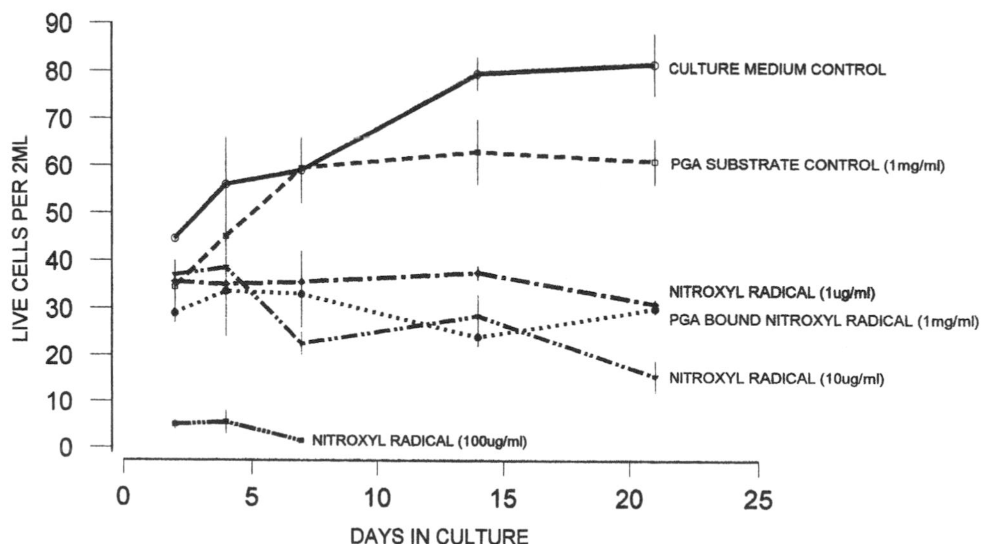

Figure 5.9 The effect of TAM-PGA on the proliferation of human smooth muscle. The cell density at day 0 was $5.0 \times 10^4/2$ mL.

TAM nitroxyl radicals at 1 μg/mL. Figure 9 also shows that any higher concentrations of TAM nitroxyl radicals in the culture media (>1 μg/mL) would appear toxic to SMC as evident in the reduction in SMC population from the initial number (0 day), particularly at 100 g/mL.

The preliminary in vitro SMC culture data suggest that the newly synthesized TAM-PGA biomaterials have the same biological function as free nitric oxide in terms of the retardation of SMC proliferation. Although the intermediate and final degradation products from the hydrolysis of TAM-PGA biomaterials (I/Ia, III/IIIa, and VII) are chemically different from pure nitric oxide they exhibited the same retardation of SMC proliferation as free TAM nitroxyl radicals. Thus, it appears that both the free TAM and the TAM-PGA nitroxyl radicals would have the same biological function as pure nitric oxide. The long PGA chain segments where the TAM nitroxyl radicals were attached appeared not to interfere with the biological functions of the nitric oxide portion of the TAM-PGA molecules. Since the level of retardation of SMC proliferation by TAM-PGA biomaterials at a concentration of 1 mg/mL was found to be similar to the pure TAM nitroxyl radicals at 1 μ/mL over the entire culture period, it appears that the amounts of TAM that were incorporated into PGA chain ends based on the stipulated chemical reaction conditions were adequate for this particular purpose.

III. POTENTIAL BIOMEDICAL APPLICATIONS

Some examples of the potential use of this new generation of biologically active biodegradable polymers are the treatment of intimal hyperplasia after balloon angioplastic procedures, anticancer drugs, wound closure materials with improved healing and antimicrobial capability, and synthetic vascular grafts that would not clot. In the anticancer drug area, the nitroxyl radical–incorporated biodegradable biomaterials could be used as the vehicles to precisely deliver the antitumor property of nitroxyl radicals to tumor sites via the biodegradation release of the incorporated nitroxyl radicals. The nitroxyl radical–incorporated biomaterials could also be used to improve the efficacy of radiation therapy in cancer because nitroxyl radicals are known to be able to considerably sensitize tumor cells toward radiation. The benefit will be lessened side effects of radiation therapy because a lower dosage of radiation could be used without compromising the therapeutic effect. In addition to the therapeutic effect and reconstruction of injured or diseased tissues, the nitroxyl radical–incorporated biomaterials may also be used as a useful tool for a fundamental study of a host of biochemical reactions involving free radicals and superoxide anions because these nitroxyl radical–labelled biomaterials could react with any reactive free radicals and neutralize them. These modified biomaterials could also be used to mimic the functions of superoxide dismutase, a naturally occurring enzyme to neutralize superoxide anions and other reactive radicals. Thus, the modified biomaterials may be used to control local inflammatory reaction induced by wounds or/and surgical implants.

ACKNOWLEDGMENTS

The author wishes to thank his former graduate student, Dr. Keun-Ho Lee, for his thesis upon which the bulk of this chapter is based. The in vitro smooth cell proliferation was conducted by Professor Fred Quimby and his research staff, Suzanne Klaessig, Department of Pathology, College of Veterinary Medicine, Cornell University.

REFERENCES

1. C. C. Chu, J. A. von Fraunhofer, H. P. Greisler. *Wound Closure Biomaterials and Devices*, CRC Press, Boca Raton, FL, 1996.
2. T. H. Barrows. *Clin. Mater.* 1:233–257 (1986).
3. Y. Kimura. *Biomedical Applications of Polymeric Materials*, T. Tsuruta, T. Hayashi, K. Kataoka, K. Ishihara, Y. Kimura (Eds.). CRC Press, Boca Raton, FL, 1993, pp. 164–190.
4. S. W. Shalaby. *Biomedical Polymers: Designed-to-Degrade Systems*. Hanser Publishers, New York, 1994.
5. J. O. Hollinger. *Biomedical Applications of Synthetic Biodegradable Polymers*, CRC Press, Boca Raton, FL, 1995.
6. M. Vert, J. Feijen, A. Albertsson, G. Scott, E. Chiellini. *Biodegradable Polymers and Plastics*. Royal Society of Chemists, Cambridge, England, 1992.
7. K. Park, W. S. W. Shalaby, H. Park. *Biodegradable Hydrogels for Drug Delivery*. Technomic Publishing, Lancaster, PA, 1993.
8. R. I. Zhdanov. *Bioactive Spin Labels*. Springer-Verlag, Berlin, Germany, 1992.
9. S. H. Snyder, D. S. Bredt. Biological roles of nitric oxide. *Sci. Amer.* 5:68 (1992).
10. S. Moncada, R. M. J. Palmer, E. A. Higgs. Nitric oxide: physiology, pathophysiology, and pharmacology. *Pharmacol. Rev.* 43:109–142 (1991).
11. D. S. Bredt, P. M. Hwang, C. E. Glatt. Cloned and expressed nitric oxide synthase structurally resembles cytochrome P-450 reductase. *Nature* 351:714–718 (1991).
12. G. Karupiah, Q. W. Xie, R. M. L. Buller. Inhibition of viral replication by interferon-gamma-induced nitric oxide synthase. *Science* 261:1445–1448 (1993).
13. H. Esumi, S. R. Tannenbaum, U.S.–Japan Cooperative Cancer Research Program, seminar on nitric oxide synthase and carcinogenesis, *Cancer Res.* 54:297–301 (1994).

14. K. H. Park. Nitric oxide is a mediator of hypoxic coronary vasodilatation. Relation to adenosine and cyclooxygenase-derived metabolites. *Circulation Res.* 71:992–1001 (1992).
15. V. Darley-Usmar, H. Wiseman, B. Halliwell. Nitric oxide and oxygen radicals: a question of balance. *FEBS Letters* 369:131 (1995).
16. S. M. Sagar, G. Singh, D. L. Hodson, A. C. Whitton. Nitric oxide and anti-cancer therapy. *Cancer Treatment Review* 21:159 (1995).
17. C. J. Morris, et al. *Handbook of Immunopharmacology: Immunopharmacology of Free Radical Species*. Academic Press, San Diego, CA, 1995, pp. 113–125.
18. E. Moilamen, et al. *Ann. Medicine* 27:359 (1995).
19. R. H. Liu, J. H. Hotchkiss. Potential genotoxicity of chronically elevated nitric oxide: a review. *Mutation Res.* 339: 73–89 (1995).
20. L. H. Pheng, C. Francoeur, M. Denis. The involvement of nitric oxide in a mouse model of adult respiratory distress syndrome. *Inflammation* 19:599–610 (1995).
21. A. R. Amin, P. Vyas, M. Attur, J. Leszcxynskapiziak, I. R. Patel, G. Weissmann, S. B. Abramson. The mode of action of aspirin-like drugs: effect on inducible nitric oxide synthase. *Proc. Natl. Acad. Sci.* 92:7926–7930 (1995).
22. R. A. Star, N. Rajora, J. J. Huang, R. C. Srock, A. Catania, J. M. Lipton. Evidence of autocrine modulation of macrophage nitric oxide synthase by alpha-melanocyte–stimulating hormone. *Proc. Natl. Acad. Sci.* 92:8016–8020 (1995).
23. M. Ikeda, M. Suzuki, K. Watarai, M. Sagai, T. Tomita. Impairment of endothelium-dependent relaxation by diesel exhaust particles in rat thoracic aorta. *Jpn. J. Pharmocol.* 68: 183–189 (1995).
24. P. J. Barnes. Nitric oxide and airway disease. *Ann. Medicine* 27:389–393 (1995).
25. M. B. Ganz, N. E. Kasner, R. J. Unwin. Nitric oxide alters cytosolic potassium in cultured glomerular mesangial cells. *Am. J. Physiol.* 268:F1081–F1086 (1995).
26. J. A. Lovchik, C. R. Lyons, M. F. Lipscomb. A role for gamma-interferon–induced nitric oxide in pulmonary clearance of *Cryptococcus neoformans*. *Am. J. Respir. Cell Mol. Biol.* 13:116–124 (1995).
27. G. J. Dusting. Nitric oxide in cardiovascular disorders. *J. Vasc. Res.* 32:143–161 (1995).
28. B. Halliwell. Oxygen radicals, nitric oxide and human inflammatory joint disease. *Ann. Rheumatic Dis.* 54:505–510 (1995).
29. D. A. Wink, J. A. Cook, M. C. Krishna, I. Hanbauer, W. Degraff, J. Gamson, J. B. Mitchell. Nitric oxide protects against alkyl peroxide-mediated cytotoxicity: further insights into the role nitric oxide plays in oxidative stress. *Arch. Biochem. Biophys.* 319:402–407 (1995).
30. E. M. Sotomayor, M. R. Dinapoli, C. Calderon, A. Colsky, Y. X. Fu, D. M. Lopez. Decreased macrophage-mediated cytotoxicity in mammary-tumor-bearing mice is related to alteration of nitric-oxide production and/or release. *Int. J. Cancer* 60:660–667 (1995).
31. S. N. Elliott, W. McKnight, G. Cirino, J. L. Wallace. A nitric oxide–releasing nonsteroidal anti-inflammatory drug accelerates gastric ulcer healing in rats. *Gastroenterology* 109:524–530 (1995).
32. J. L. Wallace, G. Cirino, G. W. McKnight, S. N. Elliott. Reduction of gastrointestinal injury in acute endotoxic shock by flurbiprofen nitroxybutylester. *Eur. J. Pharmacol.* 280:63–68 (1995).
33. H. H. Schmidt, U. Walter. NO at work. *Cell* 78:919–925 (1994).
34. A. K. Nussler, et al. *Modulation of the Inflammatory Response in Severe Sepsis*, Vol. 20, *Progress in Surgery*, J. M. Telladoo et al. (Eds.). S. Karger, Basel, 1995, pp. 33–55.
35. C. Francoeur, M. Denis. Nitric oxide and interleukin-8 as inflammatory components of cystic fibrosis. *Inflammation* 19:587–598 (1995).
36. K. H. Lee, C. C. Chu, J. Freed. U.S. patent 5,516,881 (1996).
37. K. H. Lee. Ph.D. thesis. Cornell University, Ithaca, New York, 1997.

6

Hydrogels for Biomedical and Pharmaceutical Applications

Akio Kishida
National Cardiovascular Center Research Institute, Osaka, Japan

Yoshito Ikada
Suzuka University of Medical Science, Suzuka-shi, Mie, Japan

I. INTRODUCTION

Hydrogels are a three-dimensional network of hydrophilic polymers held together by association bonds such as covalent bonds and weaker cohesive forces such as hydrogen and ionic bonds and intermolecular hydrophobic association. These networks are able to retain a large quantity of water within their structure without dissolving. Due to their superior chemical and physical properties, hydrogels have received much attention for preparing biomedical materials. The elastic nature of the hydrated hydrogels when used as implants has been found to minimize irritation to surrounding tissue. The low interfacial tension between the hydrogel surface and the aqueous solution has been found to minimize protein adsorption and cell adhesion. The research for an ideal biocompatible material was pioneered by Wichterle and Lim (1). They developed the first synthetic hydrogel and the subsequent development of therapeutic soft contact lenses. Since then the research for hydrogels increased year by year. There are some reviews and textbooks for historical and detailed information of hydrogels (2–4). In this chapter, we deal with a brief overview of hydrogels and focus on the ongoing researches.

II. FUNDAMENTALS

A. Classification

Hydrogels are classified in several ways (Table 1). The most simple classification is hydrogels obtained from natural products and hydrogels made of synthetic materials. Peppas classified hydrogels in three ways: method of preparation, ionic charge, and physical structure features (5). Based on the method of preparation, hydrogels are classified further into four groups: homopolymer hydrogels, copolymer hydrogels, multipolymer hydrogels, and interpenetrating polymeric hydrogels. Based on ionic charges, hydrogels are classified as neutral, cationic, anionic, or ampholytic hydrogels. Based on physical structural features, they are classified as amorphous hydrogels, semicrystalline hydrogels, or hydrogen-bonded structures.

Another classification is based on the crosslinking method. It basically involves two types of gel preparation. One is the chemical crosslinking, and the other is the physical crosslinking. In the former case, functional groups on polymer chains are bound with crosslinking agent or radiation to form insoluble gels. On the other hand, the physical method introduces physical crosslinks between polymer chains through intermolecular force such as van der Waals

Table 6.1 Classification of Hydrogels

Classification	Contents
Source	Natural
	Synthetic
Component	Homopolymer
	Copolymer
	Multipolymer
Preparation method	Simultaneous polymerization
	Crosslink of polymer
Electric charge	Nonion
	Anion
	Cation
	Zwitter ion
Physical structure	Amorphous
	Semicrystaline
	Hydrogen bonded
Crosslink	Covalent bond
	Intermolecular force
Functions	Biodegradable
	Stimuli responsive
	Superabsorbent
	Etc.

force and hydrophobic association, for example. Classification by function of hydrogel is also useful. A wide variety of functional hydrogels have been proposed such as biodegradable, stimuli-responsible, and superporous hydrogels.

B. Preparation

Hydrogels are prepared by various methods. In many situations, water may be present during the initial formation of the crosslinked structure. Typical methods of preparing hydrogels are irradiation, chemical reactions, and physical association. The radiation method utilizes electron beams, gamma rays, x-rays, or ultraviolet light to excite polymer chains to produce a crosslinking point (6). Chemical crosslinking requires a di- or trifunctional crosslinking agent and polymers with reactive functional groups in the side chain or the chain ends. Another chemical crosslinking method is a simultaneous copolymerization–crosslinking reaction between one or more monomers using polymerizable crosslinking agent. Physical crosslink association introduces physical crosslinks between polymer chains through intermolecular force.

A wide variety of synthetic hydrogels can be manufactured by selecting an appropriate polymer or monomer source. There are three basic types of monomers and polymers used to create hydrogels: neutral, acidic or anionic, and basic or cationic. Among the neutral monomers and polymers, hydrophilic or hydrophobic groups may be available. Most of the monomers and the polymer are hydrophilic because the water-soluble groups contribute to swelling. In order to improve the mechanical strength of hydrogels, hydrophobic portion is incorporated. For instance, when 2-hydroxyethyl methacrylate (HEMA) is copolymerized with hydrophobic monomers, hydrogels with superior tensile strength and machinability are obtained (7). To incorporate water into hydrogels, various hydrophilic groups can be applied, such as hydroxyl, amide, and ether group. Monomers with acidic groups are thought to minimize calcification when the hydrogels is implanted (8,9). Cationic hydrogels acquire positive charge, which has a favorable effect on the permeability of anions such as phosphate, a property important in hemodialysis (10).

C. Properties

1. Water Content and Swelling Ratio

The polymer chains of hydrogels interact with the solvent molecule and tend to expand to the fully solvated state. On the other hand, the crosslink structure works as the retractive force to pull back the polymer chain inside. This retractive force is described by the Flory rubber elasticity theory (11). The counterbalance of the expanding and retracting force attains to equilibrium in a particular solvent at particular temperature.

To describe the swelling behavior of hydrogels, their swelling ratio or water content is currently used in most cases. The water content of a hydrogel is expressed in terms of percentage of water by weight:

$$\text{water content} = \frac{\text{weight of water}}{\text{weight of water} + \text{weight of dry gel}} \times 100$$

For instance, most of hydrogel contact lenses have water content between 38 and 75%. When the water content of the hydrogel is over 90%, the hydrogel is called superadsorbent hydrogel. Another index which characterizes hydrogels is the swelling ratio, expressed by the ratio of the weight of swollen sample over that of the dry sample:

$$\text{swelling ratio} = \frac{\text{weight of swollen gel}}{\text{weight of dry gel}}$$

Peppas suggested that the swelling characteristic of hydrogels is a key for the use of hydrogels in biomedical and pharmaceutical applications, since the equilibrium swelling ratio influences the solute diffusion coefficient, surface wettability and mobility, and optical and mechanical properties of hydrogels.

2. Permeability

The permeability of target molecules is of utmost importance for medical application of hydrogels. For instance, oxygen permeation for contact lens, nutrient and immunological biosubstance transport for immunoisolation, and releasing drugs and proteins for drug delivery systems are core characteristics for each application.

III. MEDICAL APPLICATIONS OF HYDROGELS

Because of their extraordinary biocompatibility, hydrogels have been successfully used in a wide range of biomedical applications that include lubrication for surgical gloves, urinary catheters and surgical drainage systems, contact lenses, wound dressings, and drug delivery systems.

A. Lubricant

Hydrogels have been used extensively for lubricating the surfaces of biomaterials. As the dry surface of latex gloves and catheters exhibits a high faction coefficient, hydrogels were applied to provide a low friction surface (12). Similarly, drainage tubes used to evacuate collections of fluid within body cavities require surface lubricity to facilitate insertion and removal.

The mechanical friction between a catheter and the mucosa tissue may injure the urethra and may cause microhematuria (13). Because of this side effect, hydrogel coatings have been applied to catheters to protect the urethra and form a hydrophilic lubricious surface. Hydrogels swell on contact with water and retain a significant proportion of water, so that the hydrogel surface becomes slippery, providing a less frictional interface between the mucosa. Then the hydrogel-coated catheter provides protection to the contacting urethral epithelium during the catheter use.

Jackson and Fleming proposed five prerequisites for the optimal performance of a surgical drain: a slippery surface, antithrombogenicity, softness, radiopacity, and sterilizability (14). One example of this hydrogel formation is to react poly(vinylpyrrolidone) with an isocyanate prepolymer on a silicone drain (15).

B. Contact Lens

Contact lenses are used to correct the optical function of the eye with intimate contact to the eye (16,17). Contact lenses are classified into two groups: soft (flexible) and hard contact lenses. Hydrogel contact lenses are a member of soft contact lens group. The hydrogel lenses are made of slightly crosslinked hydrophilic polymers and copolymers. The original material for hydrogel contact lens was poly (2-hydroxyethyl methacrylate) (polyHEMA) (1); at the equilibrium swelling in physiological saline solution, it contains about 40% water of hydration.

Hydrogel contact lenses are more comfortable than other types and are easier to fit. On the other hand, they are more easily damaged and require more hygienic care than rubber or plastic lenses. The hydrogel contact lens is only comfortable when it has a large diameter, thin edges, and when they move on a limited region. Another factor to be considered in the comfort of a hydrogel contact lens is the surface quality. There has been little study about this problem; however, less interaction of eyelid with hydrogel contact lens is one of the possible reasons for the comfortability of this type of contact lens.

Wide varieties of hydrogel contact lenses have been developed, while the ''traditional'' material, HEMA, is still the most frequently used material for contact lenses. As the oxygen permeability of the original, rather thick polyHEMA hydrogel contact lenses was found to be insufficient for the corneal metabolism, improved hydrogel contact lenses were developed to enhance oxygen transport. One method was to increase the water content by copolymerizing with a more hydrophilic monomer, and the other method was to fabricate ultrathin lenses. The oxygen permeability coefficient of hydrogel materials increases exponentially with the water content. Ultrathin fabrication is applicable to any lens type. Generally, the oxygen flux through the lens will double when the thickness is halved.

Other hydrogel contact lens materials are methacrylic acid, 1-vinyl-2-pyrrolidone (NVP), dimethylacrylamide, glyceryl methacrylate, and so on. A variety of other monomers as well as a variety of crosslinking agents are used as minor ingredients in hydrogel contact lenses (Table 2) (18). Hydrogel lenses have been classified by the U.S. Food and Drug Administration (FDA) into four general groups (Table 2): low water (<50% H_2O) nonionic; high water (>50% H_2O) nonionic; low water ionic; and high water ionic. The ionic character is mainly due to the presence of methacrylic acid. This ionic character contributes to high water content, while it is responsible for higher surface protein binding to the contact lenses. High hydration is a desirable property for good oxygen permeability, but there are some disadvantages, such as low tear tolerance and protein penetration into the hydrogel. For daily-wear lenses, ultrathin low-water-content contact lenses are currently recommended.

Of all the studies for permeability, oxygen permeation through hydrogel has been investigated in detail for a long time. Most of the available contact lenses were developed with the important property of oxygen permeability in mind. The oxygen permeability coefficient, *Dk*, is a prop-

Table 6.2 Classification of Contact Lens by U.S. FDA

U.S. adopted name	Polymer	% H_2O	*Dk*
Group 1: low water (<50% H_2O), nonionic			
Polymacon	HEMA	38	9
Tetrafilcon	HEMA/MMA/VP	43	9
Crofilcon	DHPMMA/MMA	38	12
Group 2: high water (>50% H_2O), nonionic			
Lidofilcon A	VP/MMA	70	31
Lidofilcon B	VP/MMA	79	38
Group 3: low water, ionic			
Ocufilcon	HEMA/MAA	44	16
Bufilcon A	HEMA/NDOAAm	45	12
Group 4: high water, ionic			
Etafilcon	HEMA/MAA	58	28
Bufilcon B	HEMA/NDOAAm	55	16
Vifilcon	MAA/HEMA/VP	55	16

Note: HEMA: 2-hydroxyethyl methacrylate; MAA: methacrylic acid; VP: 1-vinyl-2-pyrrolidone; NDOAAm: N-(1,1-dimethyl-3-oxobutyl) acrylamide; DHPMMA: 2,3-dihydroxypropyl methacrylate.

erty characteristic of a material, where D is the diffusion coefficient and k is Henry's law solubility coefficient. The dimension of Dk is generally written in $\times 10^{-11}$ (cm_2/s)(mL O_2/mL mmHg). For a given contact lens, oxygen transmissibility (Dk/L) is defined as the oxygen permeability coefficient of the material divided by the average thickness of the lens (L, in cm) (19).

C. Dressing

Historically, gauze or nonwovens of cotton or wool have been used in medicine for a long time. Nowadays, there exist a plenty of polymeric wound covering materials. They are subdivided into polymer films, polymer foams, hydrogels, hydrocolloids, and alginates. The prerequisites for a successful wound covering material are flexibility, strength, nonantigenicity, and permeability of water and metabolites (20). As a barrier effect, a secure wound covering is also necessary to prevent infection (21). Hydrogels possess all of these characteristics except high mechanical strength. The swelling capacity of hydrogels in water, which gives them advantageous permeability and flexibility, reduces the tensile strength (21). In order to solve this contradiction, composite blends of hydrogels and other polymers have been created.

There are many successful examples of hydrogel composite wound dressings (22). The Hydron (23,24) (National Patent Hydro Med Science Division) was the first hydrogel developed for use as a wound covering material. Vigilon (25) (Bard Home Health), Gelperm (26) (Geistlich), Intrasite (Smith & Nephew), and Biolex (Catalina Biomedical) are currently available. All these hydrogel wound dressings are composed of hydrophilic monomers, such as HEMA, and other ingredients, such as poly(ethylene glycol), acrylamide, or agar. Some of these dressings have a multiple layer structure. The success of various hydrogel composites for wound covering depends on their ability to protect against infection and to promote wound healing by keeping a moist atmosphere.

IV. PHARMACEUTICAL APPLICATIONS OF HYDROGELS

A. Drug Delivery System

Hydrogels are currently being studied as controlled release carriers of drugs and proteins because of their good tissue compatibility, easy manipulation under swelling condition, and solute permeability (27,28). There are two general methods for loading drugs into hydrogels as drug carriers. In one method, a hydrogel monomer is mixed with drug, initiator, and crosslinker and is polymerized to entrap the drug within the matrix. In another method, a preformed hydrogel (in most cases lyophilized) is allowed to swell to equilibrium in a suitable drug solution. The release of these drugs from the hydrogel delivery system involves absorption of water into the polymer matrix and subsequent diffusion of the drugs as determined by Fick's Law. Kim et al. (27) reviewed the detailed consideration of swelling, drug loading, and drug release. They reported that factors that affect the drug release from hydrogels include the drug loading method, the local partition of drugs, the overall hydrophilic/hydrophobic balance, the osmotic effect of dissolved drugs, and the polymer chain elasticity. The preparation of hydrogel matrix for a specific drug carrier should be tailored considering the aforementioned effects as well as the physical properties of drugs, loading level, and release kinetics.

Due to an enormous number of drugs and proteins, the number of tailored hydrogels for specific drugs is also countless. Some typical examples are reviewed herein.

Pluronic polyols or polyoxamers are block copolymers of poly(ethylene oxide) and poly(propylene oxide). Some grades of pluronics have a unique character of reversible thermal gelation and good nondenaturing effects on proteins. Pluronic gels have been used as delivery systems for several proteins including IL-2 (29,30), urease (30), rat intestinal natriuretic factor (31). Poly(vinyl alcohol) (PVA) is used for releasing bovine serum albumin in vitro (32). The initial release of the drug was attributed to diffusion of drug through water-filled pores near the surface of the polymer matrix. Physically crosslinked PVA gels have been prepared by a freeze–thawing process, which causes structural densification of the hydrogel, and have been studied as protein-releasing matrices (33,34). One of the pioneer studies on the use of a chemically crosslinked hydrogel as a protein-delivery system were done with incorporation of chymotrypsin in poly(NVP) (PVP) hydrogels (35). Because of their high porosity, the protein underwent a rapid diffusional release within a time of 2–3 days, and the kinetics were reported to be difficult to control. Beck et al. used 3% methylcellulose gels to deliver transforming growth factor β1 (TGF-β1) both to topical skin wounds (36,37) and to bone defects (38). In both cases, the protein in the gel showed a significant enhancement in the healing of skin wounds or bone defects when compared to protein that was applied to the site in a saline buffer solution. It means that the incorporated TGF-β1 was not denatured and worked appropriately.

Hydrogels from natural products are also used as drug carriers, similar to synthetic polymer hydrogels. Alginate is one of most popular natural hydrogel matrices for drug release. Ionically crosslinked alginate hydrogels have been used to incorporate several different proteins for controlled release applications including TGF-β1 (39), basic fibroblast growth factor (bFGF) (39), tumor necrosis factor receptor (40), epidermal growth factor (EGF), and urogastrone (41). Most of these proteins are incorporated into alginate hydrogels via ionic bonding to the polymer, but sometimes the bioactivity of incorporated proteins was reduced by such ionic interaction. The addition of poly (acrylic acid) to the alginate hydrogel was shown to prevent the inactivation of proteins by alginate. Another method is preincorporation with other polymers. Basic FGF adsorbed to heparin-sepharose beads can keep its activity after incorporation into the alginate matrix (41).

Hyaluronic acid derivatives are a good example of naturally occurring polymers that have been modified to control the degradation and release rates. Hyaluronic acid has a high molecular weight (MW $= 5$ to 6×10^6) and exhibits excellent biocompatibility. In most cases, hyaluronic acid was applied after modification by chemical crosslinking and derivatization for enhancement of the rheological properties or for control of the degradation time (42–44). The controlled releases of insulin (45) and nerve growth factor (NGF) (46) from hyaluronic acid esters were reported.

Several groups have demonstrated the release of proteins from collagen matrices. Fujiwara et al. reported incorporation of IL-2 into a collagen pellet (47,48). A similar collagen system was used to release NGF (49). Collagen gels have been also reported to effectively deliver EGF (50) and TGF-β1 (51) to experimentally induced wounds in a mouse model. In both cases, the growth factors were shown to accelerate wound healing. The protein release from collagen gel is thought to be a diffusion-controlled system because the in vitro release rates were linear with the square root of time in many cases. Gelatin is a denatured substance of collagen and is soluble in water under the physiological condition. An appropriate crosslinking is necessary to obtain gelatin–hydrogel. Tabata et al. reported a gelatin microsphere delivery system containing IFN-α. The gelatin was often crosslinked with glutaraldehyde (52). Gelatin-based matrices are studied in many aspects, such as the release of insulin (53), and granulocyte macrophage colony stimulating factor (GM-CSF) (54,55), bFGF (56), and TGF-β1 (57). The gelatin system was shown to be a useful delivering matrix, but glutaraldehyde crosslinking still remains as a potential problem in clinical application if gelatin is crosslinked in the presence of protein pharmaceuticals.

V. FUTURE PERSPECTIVES

A. Immunoisolation

The basic concept of immunoisolation is very simple: to capsulate living cells with a semipermeable barrier which permits bidirectional passage of small molecules (nutrients, oxygen, and bioactive cell secretions) while restricting transport of larger molecules and host immunocytes. Several different approaches to immunoisolation have been proposed and evaluated (58). The current types of the system are cylindrical or planar diffusion chambers (macrocapsules) and dispersions of spherical beads (microcapsules). The objective cells may be allowed to float freely within a capsule but are mostly supported on a three-dimensional hydrogel matrix. In many studies, the immunoisolative barrier was made of hydrogels prepared from naturally occurring polysaccharides (alginate, agar, or chitosan). Modified hydrogel membrane barriers are fabricated either from weak polyelectrolytes, typically poly (lysine)-alginate, or from engineering thermoplastics, such as polysulfone, poly(acrylonitrile-vinyl chloride), or polyolefins (59–64).

Among the many requirements to these matrices, their permeability may be the most important issue. Oxygen and nutrients should be supplied at a sufficiently high rate through the membrane, though passage of components of the host immune system should be prevented by the semipermeability of the membrane. The properties required for the semipermeable membrane used in cell transplantation highly depend on the source of cells, whether allo- or xenogeneic.

An allograft is a graft between different individuals from the same species. It is believed that the predominant cause of allograft rejection is activation of cellular immunity by interactions of host T cells with a graft, while humoral immunity including antibodies and complement proteins is thought to play a major role in the rejection of xenografts. For the allograft applied to a recipient without preformed antibodies, a membrane which can physically inhibit the contact of host immune cells with the graft is expected to effectively protect the graft from rejection. For xenografts, the semipermeable membrane must be designed to be highly permeable to low molecular weight molecules, but be able to prevent permeation of high molecular weight biomolecules, such as antibodies and complement proteins for a long time. The development of immunoisolative membranes applicable to xenografts is challenging because of shortage of allogeneic donors. Although various membranes which allow xenotransplantation have been reported, there is still ambiguity in their long-term effectiveness.

A hydrogel membrane does not have a distinct molecular weight cutoff and its selectivity is characterized by the diffusion coefficient of solutes. Iwata et al. (65) employed a 5% agarose hydrogel to immunoisolate islets of Langerhans. Most xenogeneic recipients with microencapsulated hamster islets in 5% agarose hydrogel could not demonstrate normoglycemia for more than 10 days. The agarose gel cannot effectively protect xenogeneic cells from the humoral immunity. To solve this problem, they tried to control the complement cytolytic activities by using poly (styrene sulfonic acid) (65).

B. Tissue Engineering

Tissue engineering has been extensively studied by many researchers in a wide area. Tissue engineering, as well as cell transplantation, has been the most widely investigated strategy in the recent biomaterials field. Tissue engineering aims at restoring a tissue defect by inducing endogenous tissue regeneration and manipulating the cascades of cellular events during the healing process. Tissue engineering uses polymer devices with controlled macro- and microstructures and chemical properties to achieve organ regeneration. Combination of cells and an immunoisolative membrane forms biohybrid organs, which become a permanent part of the host organ by acting as the functional analog of the original organ by continuously supporting the organ.

Polymers play an important role in tissue engineering, providing a scaffold for cell adhesion and a space for cell proliferation. The interest in hydrogel as a biomaterial for soft tissue replacement existed in their ability to retain water within the polymer network and also in the possibility to retain the bioactive proteins such as growth factors without inactivating. The swelling characteristics of hydrogels provide the maintenance of a chemical balance with the surrounding tissue and allow the exchange between water in the hydrogel and ions and metabolites of tissue fluids. In addition, the viscoelastic behavior, low interfacial tension with biological fluids, and structural stability make hydrogels suitable for scaffold for tissue engineering (66–68).

In particular, the use of porous hydrogels to assist tissue repair and axonal regeneration in the brain (69–71) and the spinal cord (72–75) has been investigated. Hydrogels with a three-dimensional network of hydrophilic copolymers are well tolerated by living tissues and may serve as a substrate for tissue formation.

It has been widely recognized that growth factors greatly contribute to tissue regeneration at different stages of cell proliferation and differentiation. However, successful tissue regeneration by the use of growth factors has not always been achieved. One of the reasons for this is the very short half-life periods of growth factors in the body to sustain biological activities. Thus, it is highly necessary to contrive the dosage of growth factors for enhancing the in vivo efficacy.

Tabata et al. reported a series of growth factor release by using gelatin as a basic material (56,57,76–81). The skull bone regeneration induced by TGF-β1–containing gelatin hydrogels (TGF-β1–hydrogels) is significant (57). Gelatin hydrogels with a water content of 95 wt% that incorporated at least 0.1 μg of TGF-β1 induced significant bone regeneration at the rabbit skull defect site 6 weeks after treatment, whereas TGF-β1 in solution form was ineffective, regardless of the dose. The in vivo degradability of the hydrogels, which varied according to the water content, played an important role in skull bone regeneration induced by TGF-β1–hydrogels. In their hydrogel system, TGF-β1 is released from the hydrogels as a result of hydrogel degradation. When the hydrogel degrades too quickly, it does not retain TGF-β1 or prevent ingrowth of soft tissues at the skull defect site and does not induce bone regeneration at the skull defect. It is likely that hydrogel that degrades too slowly physically impedes formation of new bone at the skull defect. Following the treatment with 0.1 μg TGF-β1–hydrogel (95 wt%), newly formed bone remained at the defect site without being resorbed 6 and 12 months later. The histological structure of the newly formed bone was similar to that of the normal skull bone. Overgrowth of regenerated bone and tissue reaction were not observed after treatment with TGF-β1–hydrogels.

Tabata et al. also reported the sustained release of TGF-β1 by using a biodegradable hydrogel based on polyion complexation for the enhancement of bone regeneration activity (77). The TGF-β1 was adsorbed onto the biodegradable hydrogel of acidic gelatin with an isoelectric point of 5.0 by electrostatic interaction. The TGF-β1 could not be adsorbed onto the basic gelatin. When acidic gelatin hydrogels incorporating ^{125}I-labelled TGF-β1 were implanted into the back of mice, the radioactivity decreased with time and the in vivo retention of TGF-β1 was prolonged with a decrease in the water content of hydrogels. The higher the water content of hydrogels, the faster their biodegradation. The in vivo retention of TGF-β1 was correlated well with that of gelatin hydrogels, indicating that TGF-β1 was released from the gelatin hydrogel as a result of hydrogel biodegradation. The ability of TGF-β1 incorporated into acidic gelatin hydrogels to induce bone regeneration was evaluated in a rabbit calvarial defect model. Eight weeks after treatment, the gelatin hydrogels with water contents of 90 and 95 wt% induced significantly high bone regeneration compared with those with lower and higher water contents and free TGF-β1. This indicates that the sustained release of TGF-β1 from the hydrogel with suitable in vivo degradability is necessary to effectively enhance its osteoinductive function. Rapid hydrogel degradation will result in too short retention of TGF-β1 to induce bone regeneration. It is possible that the slow degradation of the hydrogel physically blocked TGF-β1–induced bone regeneration at the skull defect.

The polyion complexation is applicable for release of other growth factors. Acidic gelatin hydrogel was used as a biodegradable vehicle for bFGF release (78). This growth factor was incorporated by polyion complexation into a biodegradable hydrogel prepared by crosslinking acidic gelatin with the isoelectric point of 4.9. The dried hydrogel was hydrated with bFGF aqueous solution including different doses of bFGF (20, 50, 125, 250, and 500 ng) and implanted into a rabbit corneal pocket. Corneal angiogenesis was evaluated by biomicroscopy, corneal fluorescein angiography, and histology for 21 days. The hydrogel degraded with time after its implantation into the corneal pocket. Experimental eyes receiving the hydrogel containing more than 50 ng of bFGF demonstrated significant corneal angiogenesis. Control eyes and eyes receiving the hydrogel containing 20 ng of bFGF showed no corneal angiogenesis. Corneal angiogenesis, which occurred on the third or fourth day after implantation, reached maximal growth on about day 7 and regressed from day 10 after implantation. The area of angiogenesis showed a dose dependency on bFGF. The gelatin hydrogel itself induced neither angiogenesis nor inflammation. These results suggest that acidic gelatin hydrogel releases bioactive bFGF with its biodegradation, resulting in corneal neovascularization.

In vivo release of bFGF from a biodegradable gelatin hydrogel carrier was compared with the in vivo degradation of hydrogel (79). When gelatin hydrogels incorporating ^{125}I-labelled bFGF were implanted into the back of mice, the bFGF radioactivity remaining decreased with time, and the retention period was prolonged with a decrease in the water content of the hydrogels. The lower the water content of ^{125}I-labelled gelatin hydrogels, the faster both the weight of the hydrogels and the gelatin radioactivity remaining decreased with time. The decrement profile of bFGF remaining in hydrogels was correlated with that of the hydrogel weight and gelatin radioactivity, irrespective of the water content. Subcutaneous implantation of bFGF-incorporating gelatin hydrogels into the mice induced significant neovascularization. The retention period of neovascularization became longer as the water content of the hydrogels decreased. To study the decrease of activity of bFGF when implanted, bFGF-incorporating hydrogels were placed in a diffusion chamber and implanted in the mouse subcutis for certain periods of time. When hydrogels explanted from the mice were again implanted significant neovascularization was still observed, indicating that most of the biological activity of bFGF was retained in the hydrogels. It was concluded that biologically active bFGF was released as a result of in vivo degradation of the hydrogel. The release profile was controllable by changing the water content of hydrogels.

Other approaches for releasing growth factors from hydrogels were also reported. Among cytokines, bFGF is a well-known heparin-binding growth factor which has mitogenic activity for many cell types (82). It has been reported that heparin protects bFGF from proteolytic degradation and thermal inactivation (83), strongly potentiates its activity (84), and stabilizes its molecular conformation. Heparan sulfate and other sulfated glycosaminoglycans, which are functional analogs of heparin, are also present in the extracellular matrix and work as stabilizers for bFGF. In order to prepare the bFGF-reserving and -activating matrix on the inner surface of porous materials, the use of sulfated polysaccharide seems to be promising. Glucosyloxyethyl methacrylate (GEMA) is a novel methacryloyl monomer, and radical polymerization of GEMA gives highly water-soluble polymers [poly(GEMA)] and hydrogels with a crosslinking agent (85–87). The sulfation of poly(GEMA) provided poly(GEMA) with a heparinlike activity (88). Hence, poly(GEMA) was considered as one of polysaccharide model compounds. The reserving and activating ability of poly(GEMA)–sulfate, and its hydrogel for bFGF was investigated. It was revealed that poly (GEMA)–sulfate hydrogel has activity for activating bFGF and enhances the cell proliferative activity of bFGF.

C. Hydrogels of Stimuli-Responsive Polymers

1. Overview

Some polymers undergo strong conformational changes when only small changes occur in the environment, such as temperature, pH, and ionic strength. These polymers are called stimulus-responsive polymers. Nonlinear responses by the polymer systems have mainly been observed in water. An appropriate balance of hydrophobicity and hydrophilicity in the molecular structure of the polymer is required for the phase transition to occur. There have been a lot of reviews of this field (89–93).

2. Immobilized Biocatalysts

When an enzyme is incorporated in a hydrogel, phase transition of stimuli-responsive polymer can significantly affect the enzyme activity and substrate access to the enzyme molecule. A wide range of stimuli-responsive polymers have been used for the development of reversibly soluble biocatalysts whose solubility is controlled by pH (94) or temperature (95–99). A biocatalyst preparation sensitive to magnetic fields has been produced by immobilizing invertase and γ-Fe_2O_3 in a poly(N-isopropylacrylamide (NIPAAm)-co-acrylamide) hydrogel. The heat generated by the exposure of γ-Fe_2O_3 to a magnetic field causes the hydrogel to collapse, followed by a sharp decrease in the rate of sucrose hydrolysis.

3. Drug Delivery

The swelling or shrinking of stimuli-responsive hydrogel in response to small changes in pH or temperature can be used to control drug release (92). The development of a glucose-sensitive insulin releasing system for diabetes therapy was developed by using pH-responsive polymer, poly[(N,N-dimethylamino) ethyl methacrylate-co-ethylacrylamide]. The polymer was mixed and compressed with glucose oxidase, bovine serum albumin, and insulin. When the matrix was exposed to glucose, it was oxidized to form gluconic acid. As the hydration state was changed by a decrease in the pH, then insulin was released. A variety of designs for an insulin-delivery system that responds to glucose were also developed using glucose oxidase.

4. Biomimetic Actuators

Biomimetic actuators, which mimic the conversion of chemical energy into mechanical energy in living organisms, have become of major interest lately (100). Ionic hydrogels are known to show a discontinuous volumetric change above a certain threshold of an external stimulus such as pH, temperature, ionic strength, or concentration of organic solvent. The volume change of several hundred times over and under the threshold is utilized to exert a significant force. As a stimulus that induces a volumetric change in a hydrogel, an electric field is utilized. For example, a crosslinked hydrogel of poly(vinyl alcohol) chains entangled with poly(acrylic acid) chains has good mechanical properties and shows rapid electric field–associated bending deformation. Hydrogels capable of mechanical response to electric field have also been developed using the cooperative binding of positively charged surfactant molecules to the polyanionic polymer poly(2-acrylamido-2-methyl-1-propanesulfonic acid).

Copolymer gels consisting of NIPAAm and acrylic acid were studied for constructing biomimetic actuators. A pH-induced change in the -COOH ionization of acrylic acid alters the repulsive force. NIPAAm produced the attractive force by hydrophobic interactions over lower critical solution temperature (32°C). Glucose dehydrogenase was used here again for converting neutral glucose into gluconic acid. When this reaction occurs inside the gel, the repulsive force is eliminated owing to protonation of COO- groups. As a result, the attractive force dominates, followed by collapse of the gel. The problem of this system is extremely slow response to the stimuli.

D. Novel Hydrogels

Hubbell et al. have been studying applications of a novel hydrogel prepared by poly(ethylene glycol) (PEG) diacrylate derivatives (101–105). The hydrogel obtained by photopolymerization of poly(ethylene glycol) diacrylate was derivatized with Arg-Gly-Asp (RGD)–containing peptide sequences (101). Incorporation was achieved by functionalizing the amine terminus of the peptide with an acrylate moiety, thereby enabling the adhesion peptide to copolymerize rapidly with the PEG diacrylate upon photoinitiation. Hydrogels incorporated the peptide with a PEG spacer arm and incorporation of RGD-promoted fibroblast spreading. Water-soluble macromers based on block copolymers of PEG and poly(lactic acid) or poly(glycolic acid) with terminal acrylate groups were used for a nonadhesive barrier at the free surface on the treated wound site. These materials, photopolymerized in vivo in direct contact with tissues, appear to form an adherent hydrogel barrier that is highly effective in reducing postoperative adhesions in the models used.

Yui and coworkers have proposed new drug release systems: dual stimuli–responsive release (106,107) and pulsatile release systems (108). For the dual stimuli–responsive system, interpenetrating polymer network (IPN) –structured hydrogels of gelatin and dextran were prepared with lipid microspheres as a drug microreservoir. The IPN-structured hydrogel prepared below the transition temperature of gelatin exhibited a specific degradation-controlled lipid microsphere release behavior: lipid microsphere release from the hydrogel in the presence of either α-chymotrypsin or dextranase alone was completely hindered, whereas lipid microsphere release was observed in the presence of both enzymes. For the pulsatile release system, multilayered hydrogels consisting of PEG-grafted dextran (PEG-g-Dex) and ungrafted dextran were prepared. In these formulations, it is expected that the grafted PEG domains act as a drug reservoir dispersed in the dextran matrix based on aqueous polymer two-phase systems. The formulations exhibited surface-controlled degradation by dextranase, and insulin release was observed in a pulsatile manner because of the multilayered structure in which PEG-g-Dex hydrogel layers contained insulin, and dextran hydrogel layers did not.

E. Other Hydrogels

Some hydrogels are currently used for biomedical applications, but the use of hydrogels under physiological conditions is sometimes limited by calcification. The calcification of biomaterials is an important pathologic process, and has been observed not only in hydrogels but also in many kinds of biomaterials, for example, in devices (109), artificial blood pumps (artificial hearts) (110), heart valves (111,112), and soft contact lenses (113), as well as in polyurethane (109) and silicone rubber (114). As calcification lowered the performance of the materials by changing the bulk properties or causing thrombosis, a great deal of research has been done to make clear the calcification mechanism (115–118). On the other hand, calcification is useful as a biomedical material which requires bone bonding. Hydroxyapatite [HAp: $Ca_{10}(PO_4)_6(OH)_2$], which is one of the main components of calcified materials, is a biofunctional inorganic material. In some orthopedic and tissue adhesive applications, many researchers have attempted to create polymer–HAp composites by mixing (119), plasma spray (120,121), and coating (122,123); they reported that formation of composite was good. However, there have been few studies about hydrogel–HAp composites. If hydrogel–HAp composites can be formed, the use of hydrogels as implantable materials will increase. However, the calcification mechanism (HAp formation) has not yet been clarified.

Many possible factors, such as animal species, age, hormone level, type of materials, shear stress, surface defect, protein adsorption, hydrophilicity, and thrombus, may contribute to calcification of biomedical materials in vivo (124,125). In regard to studying calcification, the main obstacle was the lack of an adequate in vitro experimental system. Most studies of calcification have been done in vivo (118) because it is assumed that devitalized cells and cellular debris are keys to calcification. Recent research has revealed that the physicochemical aspect of materials is another essential factor for calcium phosphate deposition in the initial stage of calcification. Animal models can provide useful data, but it takes a long period of time to study calcification. In 1990, Kokubo et al. developed a biomimetic process and noted that a HAp layer with the desired thickness was formed on ceramic, metal, and polymer films at a normal temperature and pressure in vitro (126). Using this experimental procedure, some research (127) in regard to creating polymer–HAp composites has been done (122,128). Assuming the biomimetic process as an in vitro calcification model, HAp formation (which is a type of calcification) on/in hydrogels was analyzed. There are many kinds of hydrogels and many factors to study, including chemical formula, functional group, swelling ratio, and so on. Imai et al., who used PHEMA copolymer hydrogels, found that the amount of calcified deposits depends on the HEMA content in the hydrogels and that their chemical structure was a more important factor in calcification than hydrophilicity. Using various hydrogels that carried hydroxyl groups, the relationship between hydrogel characteristics and HAp formation was studied (115). It was revealed that there were two factors (the bound water content and the swelling ratio of a hydrogel) that affect HAp formation. The higher bound water content will provide a large number of nucleation sites for HAp on/in a hydrogel. On the other hand, the higher the swelling ratio becomes, the larger amount of ions is supplied into a hydrogel matrix from a solution in order to grow HAp nuclei on/in a hydrogel.

VI. SUMMARY

Hydrogels have been studied as biomaterials in the medical field, but the clinical applications are very limited. The continuous advances in biology and medicine will produce further demand for biomaterials, and hydrogels could be a practical answer, especially in the field of tissue engineering and drug delivery systems.

REFERENCES

1. O. Wichterle, D. Lim. Hydrophilic gels for biological use. *Nature* 185:117 (1960).
2. J. D. Andrade. *Hydrogels for Medical and Related Applications*. ACS Symposium Series, Vol. 31, American Chemical Society, Washington D.C., 1976.
3. N. A. Peppas. *Hydrogels in Medicine and Pharmacy*. CRC Press, Boca Raton, FL, 1987.
4. J. C. Wheeler, J. A. Woods, M. J. Cox, R. W. Cantrell, F. H. Watkins, R. F. Eldlich. Evolution of hydrogel polymers as contact lenses, surface coatings, dressings, and drug delivery systems. *J. Long-Term Effects Med. Implants* 6:207 (1996).
5. N. A. Peppas. Hydrogel. In: *Biomaterials Science*, B. D. Ratner, A. S. Hoffman, F. D. Schoen, J. E. Lemons (Eds.). Academic Press, San Diego, 1996, p. 60.
6. A. Chapiro. *Radiation Chemistry of Polymeric Systems*. Interscience, New York, 1962.
7. V. Kudela. Hydrogels. *Encyclopedia of Polymer Science and Engineering*, H. F. Mark, M. M. Bikals, C. G. Overburger, G. Menges, J. I. Kroschwitz (Eds.). John Wiley and Sons, New York, 1987, p. 783.

8. L. Pinchuk, E. C. Eckstein, M. R. Van De Mark. The interaction of urea with the generic class of poly(2-hydroxyethyl methacrylate) hydrogels. *J. Biomed. Mater. Res.* 18:671 (1984).
9. G. F. Klomp, H. Hashiguchi, P. C. Ursell, Y. Takeda, T. Taguchi, W. H. Dobelle. Macroporous hydrogel membranes for a hybrid artificial pancreas. II. Biocompatibility. *J. Biomed. Mater. Res.* 17:865 (1983).
10. O. Moise, S. Sideman, E. Hoffer, I. Rousseau, O. S. Better. Membrane permeability for inorganic phosphate ion. *J. Biomed. Mater. Res.* 11:903 (1977).
11. P. J. Flory. *Principles of Polymer Chemistry*. Cornell University Press, New York, 1953.
12. D. L. Podell, H. I. Podell. U.S. patent 3,813,695 (1974).
13. M. Ruutu, O. Alfthan, L. Heikkinen, A. Jarbinden, T. Lehtonen, E. Merikallia, C. G. Spamdertskjold-Mordenspam. Epodermic of acute urethal stricture after open-heart surgery. *Lancet* 1:218 (1982).
14. F. E. Jackson, P. M. Fleming. Jackson-Pratt brain drain. Use in general surgical conditions requiring drainage. *Int. Surg.* 57:658 (1972).
15. R. S. C. Pearce, L. R. West, G. T. Rodeheaver, R. F. Eldlich. Evaluation of a new hydrogel coating for drainage tubes. *Am. J. Surg.* 148:687 (1984).
16. R. B. Mandell. Contact Lens Practice, 4th ed., Charles C. Thomas, Springfield, IL, 1988.
17. ''Contact Lenses and Solutions Summary, Contact Lens Spectrum.'' Johnson & Johnson Vision Products, Inc., 1999, p. 3.
18. M. F. Refojo. Contact Lenses. In: *Kirk-Othmer Encyclopedia of Chemical Technology*, 3rd ed., Vol. 6, Wiley, New York, p. 720.
19. B. A. Holden, J. Newton-Homes, L. Winterton, I. Fatt, H. Hamano, D. La Hood, N. A. Brennan, N. Efron. The Dk project: an interlaboratory comparison of Dk/L measurements. *Optom. Vis. Sci.* 67:476 (1990).
20. Y. Kuroyanagi. Advances in wound dressing and cultured skin substitutes. *J. Artif. Organs* 2:97 (1999).
21. P. H. Corkhill, C. J. Hamilton, B. J. Toghe. Synthetic hydrogels VI. Hydrogel composites as wound dressings and implant materials. *Biomaterials* 10:3 (1989).
22. T. Helfman, L. Ovington, V. Falanga. Occlusive dressings and wound healing. *Clin. Dermatol.* 12:121 (1994).
23. A. S. Brown. Hydron for burns. *Plast. Reconstruct. Surg.* 67:810 (1981).
24. M. T. Husain, M. Akhtar, N. Akhtar. Report on evaluation Hydron as burn wound dressing. *Burns* 9:330 (1983).
25. D. W. Yates, J. M. Hadfield. Clinical experience with a new hydrogel wound dressing. *Injury* 16:230 (1984).
26. J. A. Myers. Delperm: a nontextile wound dressing. *Pharm. J.* 230:263 (1983).
27. S. W. Kim, Y. H. Bae, T. Okano. Hydrogels: swelling, drug loading, and release. *Pharmaceut. Res.* 9:283 (1992).
28. S.-J. Hwang, H. Park, K. Park. Gastric retentive drug-delivery systems. *Crit. Rev. Therap. Drug Carrier Syst.* 15:243 (1998).
29. K. Morikawa, O. Okada, M. Hosokawa, H. Kobayashi. Enhancement of therapeutic effects of recombinant interleukin-2 on a transplantable rat fibrosarcoma by the use of a sustained release vehicle, pluronic gel. *Cancer* 47:37 (1987).
30. K. A. Fults, T. P. Johnston. Sustained-release of urease from a polyoxamer gel matrix. *J. Parenter. Sci. Technol.* 44:58 (1990).
31. I. Juhasz, V. Lenaerts, P. Raymond, H. Ong. Diffusion of rat atrial natriuretic factor in thermoreversible polyoxamer gels. *Biomaterials* 10:265 (1989).
32. R. W. Korsmeyer, R. Gurny, E. Doelker, P. Biru, N. A. Peppas. Mechanisms of solute release from porous hydrophilic polymers. *Int. J. Pharm.* 15:25 (1983).
33. N. A. Peppas, J. E. Scott. Controlled release from poly(vinyl alcohol) gels prepared by freezing–thawing processes. *J. Controlled Release* 18:95 (1992).
34. B. J. Ficek, N. A. Peppas. Novel preparation of poly(vinyl alcohol) microparticles without crosslinking agent for controlled drug delivery of proteins. *J. Controlled Release* 27:259 (1993).
35. V. P. Torchillin, E. G. Tischenko, V. V. Smirnov, E. I. Chazoc. Immobilization of enzymes on slowly soluble carriers. *J. Biomed. Mater. Res.* 11:223 (1977).
36. S. L. Beck, T. L. Chen, P. Mikalauski, A. J. Amman. Recombinant human transforming growth factor beta 1 (rhTGF-β1) enhances healing and strength of granulation skin wounds. *Growth Factors* 3:267 (1990).
37. S. L. Beck, L. Deguzman, W. P. Lee, Y. Xu, L. A. McFatridge, E. P. Amento. TGF-β1 accelerates wound healing: reversal of steroid-impaired healing ion tars and rabbits. *Growth Factors* 5:295 (1991).
38. S. L. Beck, L. Deguzman, W. P. Lee, Y. Xu, L. A. McFatridge, N. A. Gillet, E. P. Amento. TGF-β1 induces bone closure in skull defects. *J. Bone Miner. Res.* 6:1257 (1991).
39. E. R. Edelman, E. Mathiowitz, R. Langer, M. Klagsbrun. Controlled and modulated release of basic fibroblast growth factor. *Biomaterials* 12:619 (1991).
40. S. Wee, W. R. Gombotz. Controlled release of recombinant human tumor necrosis factor receptor from alginate beads. *Proc. Int. Symp. Controlled Release Bioact. Mater.* 1:730 (1994).
41. E. C. Downs, N. E. N. E. Robertson, T. L. Riss, M. L. Plunkett. Calcium alginate beads as a slow-release system for delivering angiogenic molecules in vivo and in vitro. *J. Cell. Physiol.* 152:422 (1992).
42. R. Cortivo, P. Brun, A. Rastrelli, G. Abatangelo. In vitro studies on biocompatibility of hyaluronic acid esters. *Biomaterials* 2:727 (1991).
43. L. M. Benedetti, E. M. Topp, V. J. Stella. Microspheres of hyaluronic acid esters—fabrication methods and in vitro hydrocortisone release. *J. Controlled Release* 13:33 (1990).
44. J. A. Hunt, H. N. Joshi, V. J. Stella, E. M. Topp. Diffusion and drug release in polymer films prepared from ester derivatives of hyaluronic acid. *J. Controlled Release* 12:159 (1990).

45. L. Illum, N. F. Farraj, A. N. Fisher, I. Gill, M. Miglietta, L. M. Bendetti. Hyaluronic acid microspheres as a nasal delivery system for insulin. *J. Controlled Release* 29:133 (1994).
46. E. Ghezzo, L. M. Benedetti, M. Rochira, F. Biviano, L. Callegaro. Hyaluronic acid derivative microspheres as NGF delivery devices: preparation methods and in vitro release characterization. *Int. J. Pharm.* 87:21 (1992).
47. T. Fujiwara, K. Sakagami, J. Matsuoka, S. Shiozaki, K. Fujioka, Y. Takada, S. Uchida, T. Onoda, K. Orita. Augmentation of antitumor effect on syngeneic murine solid tumors by an interleukine 2 slow delivery system, the IL-2 mini-pellet. *Biotherapy* 3:203 (1991).
48. T. Fujiwara, K. Sakagami, J. Matsuoka, Y. Shiozaki, S. Uchida, K. Fujioka, S. Takada, T. Onoda, K. Orita. Application of an interleukine 2 slow delivery system to the immunotherapy of established murine colon 26 adeno carcinoma liver metastases. *Cancer Res.* 50:7003 (1990).
49. S. Yamamoto, T. Yoshimine, T. Fujita, R. Kuroda, T. Irie, K. Fujioka, T. Hayakawa. Protective effect of NGF atelocollagen mini-pellet on the hippocampal delayed neuronal death in gerbils. *Neurosci. Lett.* 141:161 (1992).
50. G. L. Brown, L. J. Curtsinger, M. White, R. O. Mitchell, J. Pietsch, R. Nordquist, A. von Fraunhofeer, G. S. Schultz. Acceleration of tensile strength incisions treated with EGF and TGF-β. *Ann. Surg.* 208:788 (1988).
51. T. A. Mustoe, G. F. Pierce, A. Thomason, P. Gramates, M. B. Sporn, T. F. Deuel. Accelerated healing of incisional wounds in rats induced by transforming growth factor-β. *Science* 237:1333 (1987).
52. Y. Tabata, Y. Ikada. Synthesis of gelatin microspheres containing interferon. *Pharm. Res.* 6:422 (1989).
53. B. G. Shinde, S. Erhan. Flexibilized gelatin film–based artificial skin model. II. Release kinetics of incorporated bioactive molecules. *Bio.-Med. Mater. Eng.* 2:127 (1992).
54. P. T. Golumbek, R. Azhari, E. Jafee, H. Levitsky, A. Lazenby, K. Leong, D. Pardoll. Controlled release, biodegradable sytokine depots: a new approach in cancer vaccine design. *Cancer Res.* 53:5841 (1993).
55. W. Shao, K. W. Leong. Microcapsules obtained from complex coacervation of collagen and chondroitin sulfate. *J. Biomater. Sci. Polym. Ed.* 7:389 (1995).
56. K. Yamada, Y. Tabata, K. Yamamoto, S. Miyamoto, I. Nagai, H. Kikuchi, Y. Ikada. Potential efficiency of basic fibroblast growth factor incorporated in biodegradable hydrogels for skull bone regeneration. *J. Neurosurg.* 86:871 (1997).
57. M. Yamamoto, Y. Tabata, L. Hong, S. Miyamoto, N. Hashimoto, Y. Ikada. Bone regeneration by transforming growth factor beta 1 released from biodegradable hydrogel. *J. Controlled Release* 64:133 (2000).
58. M. J. Lysaght, B. Frydel, F. Gentile, D. Emerich, S. Winn. Recent progeress in immunoisolated cell therapy. *J. Cellular Biochem.* 56:196 (1994).
59. T. M. S. Chang. Semipermeable microcapsules. *Science* 146:524 (1964).
60. F. Lim, A. M. Sun. Microencapsulated islets as bioartificial endocrine pancreas. *Science* 210:980 (1980).
61. M. F. A. Goosen, G. M. Oshea, H. M. Gharapetian, S. Chou, A. M. Sun. Optimization of microencapsulation parameters: semipermeable microcapsules as a bioartificial pancreas. *Biotecnol. Bioeng.* 27:146 (1985).
62. M. E. Sugamori, M. V. Sefton. Microencapsulation of pancreatic islets in a water insoluble polyacrylate. *Trans. Am. Soc. Artif. Organs* 35:791 (1989).
63. A. S. Michaels. U.S. patent 3,615,024 (1971).
64. W. Pusch, A. Walch. Synthetic membranes—preparation, structure, and applications. *Angew. Chem. Int. Ed. Engl.* 21:660 (1982).
65. H. Iwata, Y. Murakami, Y. Ikada. Control of complement activities for immunoisolation. *Ann. N. Y. Acad. Sci.* 875: 7 (1999).
66. B. D. Ratner, A. S. Hoffman. Synthetic hydrogels for biomedical applications. *Hydrogels for Medical and Related Applications*, J. D. Andrade (Ed.). ACS Symposium Series 31, American Chemical Society, Washington D.C., 1976, p. 1.
67. H. Park, K. Park. Hydrogels in Bioapplications. In: *Hydrogels and Biodegradable Polymers for Bioapplications*, R. M. Ottenbrite, S. J. Huang, K. Park (Eds.). ACS Symposium Series 627, American Chemical Society, Washington D.C., 1996, p. 2.
68. S. Dumitriu, C. Dumitriu-Medvichi. Hydrogel and general properties of biomaterials. In: *Polymeric Biomaterials*, S. Dumitriu (Ed.). Marcel Dekker, New York, 1994, p. 3.
69. S. Woerly, R. Marchand, C. Lavallee. Intracerebral implantation of synthetic polymer/biopolymer matrix: a new perspective for brain repair. *Biomaterials* 11:97 (1990).
70. S. Woerly. Porous hydrogels for neural tissue engineering. In: *Porous Materials for Tissue Engineering*, D.-M. Liu, V. Dixit (Eds.). Trans. Tech. Publications, Switzerland, 1997, p. 53.
71. S. Woerly, G. Maghami, R. Duncan, V. Subr, K. S. Ulbrich. Synthetic polymer derivatives as substrata for neuronal adhesion and growth. *Brain Res. Bull.* 30:423 (1993).
72. G. W. Plant, A. R. Harvey, T. V. Chirila. Axonal growth with poly(2-hydroxyethyl methacrylate) sponges infiltrated with Schwann cells and implanted into the lesioned rat optic tract. *Brain Res.* 671:287 (1997).
73. G. W. Plant, S. Woerly, A. R. Harvey. Hydrogels containing peptide or aminosugar sequences implanted into the rat brain: influence on cellular migration and axonal growth. *Exp. Neurol.* 143:287 (1997).
74. G. W. Plant, T. V. Chirila, A. R. Harvey. Implantation of collagen IV/poly(2-hydroxyethyl methacrylate) hydrogels containing Schwann cells into lesioned rat optic tract. *Cell Transpl.* 7:381 (1998).
75. S. Woerly, E. Pinet, L. De Robertis, M. Bousmina, G. Laroche, T. Roitback, L. Vargova, E. Sykova. Heterogeneous PHPMA hydrogels for tissue repair and axonal regeneration in the injured spinal cord. *J. Biomater. Sci. Polym. Ed.* 9:681 (1998).

76. K. Kawai, S. Suzuki, Y. Tabata, Y. Ikada, Y. Nishimura. Accelerated tissue regeneration through incorporation of basic fibroblast growth factor–impregnated gelatin microspheres into artificial dermis. *Biomaterials* 21:489 (2000).
77. L. Hong, Y. Tabata, S. Miyamoto, M. Yamamoto, K. Yamada, N. Hashimoto, Y. Ikada. Bone regeneration at rabbit skull defects treated with transforming growth factor-β 1 incorporated into hydrogels with different levels of biodegradability. *J. Neurosurg.* 92:315 (2000).
78. C. Yang, T. Yasukawa, H. Kimura, H. Miyamoto, Y. Honda, Y. Tabata, Y. Ikada, Y. Ogura. Experimental corneal neovascularization by basic fibroblast growth factor incorporated into gelatin hydrogel. *Ophthalmic Res.* 32: 19 (2000).
79. Y. Tabata, Y. Ikada. Vascularization effect of basic fibroblast growth factor released from gelatin hydrogels with different biodegradabilities. *Biomaterials* 20:2169 (1999).
80. Y. Tabata, A. Nagano, Y. Ikada. Biodegradation of hydrogel carrier incorporating fibroblast growth factor. *Tissue Eng.* 5:127 (1999).
81. Y. Tabata, A. Nagano, M. Muniruzzaman, Y. Ikada. In vitro sorption and desorption of basic fibroblast growth factor from biodegradable hydrogels. *Biomaterials* 19: 1781 (1998).
82. D. Gospodarowicz, G. Neufeld, L. Schweigerer. Molecular and biological characterization of fibroblast growth factor, an angiogenic factor which also controls the proliferation and differentiation of mesoderm and neuroectoderm derived cells. *Cell. Differ.* 19:1 (1986).
83. D. Gospodarowicz, J. Cheng. Heparin protects basic and acidic FGF from inactivation. *J. Cell. Physiol.* 128:475 (1986).
84. A. B. Schreiber, J. Kenney, J. Kowalski, R. Freisel, T. Mehlman. Interaction of endothelial cell growth factor with heparin: characterization by receptor and antibody recognition. *Proc. Natl. Acad. Sci. USA* 82:6138 (1985).
85. S. Kitazawa, M. Okumura, K. Kinomura, T. Sakakibara. Synthesis and properties of novel vinyl monomers bearing glycoside residues. *Chem. Lett.* 1733 (1990).
86. S. Kitazawa, M. Okumura, K. Kinomura, T. Sakakibara, K. Nakamae, Miyata, M. Akashi, K. Suzuki. Preparation and some properties of glycoside bearing polymer. *Carbohydrate as Organic Raw Materials* 2:115 (1994).
87. N. Fukudome, K. Suzuki, E. Yashima, M. Akashi. Synthesis of nonionic hydrogels bearing a monosaccharide residue and their properties. *J. Appl. Polym. Sci.* 52:1759 (1994).
88. M. Akashi, N. Sakamoto, K. Suzuki, A. Kishida. Synthesis and anticoagulant activity of sulfated glucoside-bearing polymer. *Bioconj. Chem.* 7:393 (1996).
89. R. S. Harland, R. K. Prud'homme (Eds.). *Polyelectrolyte Gels: Properties, Preparation and Applications*. American Chemical Society, Washington D.C., 1992.
90. K. Dusek (Ed.). *Responsive Gels: Volume Transitions I.* Advances in Polymer Science, Vol. 109. Springer-Verlag, Berlin, 1993.
91. K. Dusek (Ed.). *Responsive Gels: Volume Transitions II.* Advances in Polymer Science, Vol. 110. Springer-Verlag, Berlin, 1993.
92. J. Kost (Ed.). *Pulsed and Self-Regulated Drug Delivery.* CRC Press, Boca Raton, FL, 1990.
93. I. Y. Galaev, B. Mattiasson. 'Smart' polymers and what they could do in biotechnology and medicine. *Tibtech.* 17: 335 (1999).
94. J. P. Chen, K. C. Chang. Immobilization of chitinase on a reversibly soluble-insoluble polymer for chitin hydrolysis. *J. Chem. Technol. Biotechnol.* 60:133 (1994).
95. K. Hoshino, M. Taniguchi, T. Kitao, S. Morohashi, T. Sasakura. Preparation of a new thermo-responsive adsorbent with maltose as a ligand and its application to affinity precipitation. *Biotechnol. Bioeng.* 60:568 (1998).
96. Y. Sun, X. H. Jin, S. Y. Dong, K. Yu, X. Z. Zhou. Immobilized chymotrypsin on reversively precipitable polymerized liposome. *Appl. Biochem. Biotechnol.* 56:331 (1996).
97. A. Chilkoti, G. Chen, P. S. Stayton, A. S. Hoffman. Site-specific conjugation of a temperature-sensitive polymer to a genetically-engineered protein. *Bioconj. Chem.* 5:504 (1994).
98. G. Chen, A. S. Hoffman. Synthesis of carboxylated poly (NIPAAm) oligomers and their application to form thermo-reversible polymer-enzyme conjugates. *J. Biomater. Sci. Polym. Ed.* 5:371 (1994).
99. H. Kawaguchi, K. Fujimoto. Smart latexes for bioseparation. *Bioseparation* 7:253 (1998–1999).
100. E. Kokufuta, Y. Q. Zhang, T. Tanaka. Biochemo-mechanical function of urease-loaded gels. *J. Biomater. Sci. Polm. Ed.* 6:35 (1994).
101. D. L. Hern, J. A. Hubbell. Incorporation of adhesion peptides into nonadhesive hydrogels useful for tissue resurfacing. *J. Biomed. Mater. Res.* 39:266 (1998).
102. Y. An, J. A. Hubbell. Intraarterial protein delivery via intimally-adherent bilayer hydrogels. *J. Controlled Release* 64:205 (2000).
103. G. M. Cruise, O. D. Hegre, F. V. Lamberti, S. R. Hager, R. Hill, S. S. Scharp, J. A. Hubbell. In vitro and in vivo performance of porcine islets encapsulated in interfacially photopolymerized poly(ethylene glycol) diacrylate membranes. *Cell Transplant.* 8:293 (1999).
104. G. M. Cruise, O. D. Herge, D. S. Scharp, J. A. Hubbell. A sensitivity study of the key parameters in the interfacial photopolymerization of poly(ethylene glycol) diacrylate upon porcine islets. *Biotechnol. Bioeng.* 57:655 (1998).
105. G. M. Cruise, O. D. Herge, J. A. Hubbell. Characterization of permeability and network structure of interfacially photopolymerized poly(ethylene glycol) diacrylate hydrogels. *Biomaterials* 19:1287 (1998).
106. M. Kurisawa, M. Terano, N. Yui. Double-stimuli-responsive degradation of hydrogels consisting of oligopeptide-terminated poly(ethylene glycol) and dextran with an interpenetrating polymer network. *J. Biomater. Sci. Polym. Ed.* 8:691 (1997).
107. M. Kurisawa, N. Yui. Dual-stimuli-responsive drug release from interpenetrating polymer network-structures

hydrogels of gelatin and dextran. *J. Controlled Release* 54:191 (1998).
108. K. Moriyama, T. Ooya, N. Yui. Pulsatile peptide release from multi-layered hydrogel formulations consisting of poly(ethylene glycol)-grafted and ungrafted dextrans. *J. Biomater. Sci. Polym. Ed.* 10:1251 (1999).
109. H. Harasaki, R. Gerrity, R. Kiraly, G. Jacobs, Y. Nose. Calcification in blood pumps. *Trans. A.S.A.I.O.* 25:305 (1979).
110. R. J. Thoma, R. E. Phillips. The role of material surface chemistry in implant device calcification: a hypothesis. *J. Heart Valve. Dis.* 4:214 (1995).
111. W. N. Neethling, J. J. van den Heever, J. M. Meyer, H. C. Barnard. Processing factors as determinants of tissue valve calcification. *J. Cardiovasc. Surg.* (Torino) 33:285 (1992).
112. A. C. Fisher, G. M. Bernacca, T. G. Mackay, W. R. Dimitri, R. Wilkinson, D. Wheatley. Calcification modelling in artificial heart valves. *Int. J. Artif. Organs* 15:248 (1992).
113. P. J. Bucher, E. R. Buchi, B. C. Daicker. Dystrophic calcification of an implanted hydroxyethylmethacrylate intraocular lens. *Arch. Ophthalmol.* 113:1431 (1995).
114. R. J. Brockhurst, R. C. Ward, P. Lou, D. Ormerod, D. Albert. Dystrophic calcification of silicone scleral buckling implant materials. *Am. J. Ophthalmol.* 115:524 (1993).
115. J. M. Girardot, M. N. Girardot. Amide cross-linking: an alternative to glutaraldehyde fixation. *J. Heart Valve. Dis.* 5:518 (1996).
116. R. R. Joshi, T. Underwood, J. R. Frautschi, R. E. Phillips, F. J. Schoen, R. J. Levy. Calcification of polyurethanes implanted subdermally in rats is enhanced by calciphylaxis. *J. Biomed. Mater. Res.* 31:201 (1996).
117. G. Golomb, I. Lewinstein, V. Ezra, F. J. Schoen. Mechanical properties and histology of charge modified bioprosthetic tissue resistant to calcification. *Biomaterials* 13:353 (1992).
118. D. K. Han, K. D. Park, S. Y. Jeong, Y. H. Kim, U. Y. Kim, B. G. Min. In vivo biostability and calcification-resistance of surface-modified PU-PEO-SO3. *J. Biomed. Mater. Res.* 27:1063 (1993).
119. C. C. P. M. Verheyen, J. R. de Wijn, C. A. van Bliterswijk, K. de Groot, P. M. Rozing. Hydroxylapatite/poly(L-lactide) composites: an animal study on push-out strengths and interface histology. *J. Biomed. Mater. Res.* 27:433 (1993).
120. J. Weng, X. Liu, X. Zhang, K. de Groot. Integrity and thermal decomposition of apatite in coatings influenced by underlying titanium during plasma spraying and post-heat-treatment. *J. Biomed. Mater. Res.* 30:5 (1996).
121. T. Kitsugi, T. Nakamura, M. Oka, Y. Senaha, T. Goto, T. Shibuya. Bone-bonding behavior of plasma-sprayed coatings of BioglassR, AW-glass ceramic, and tricalcium phosphate on titanium alloy. *J. Biomed. Mater. Res.* 30: 261 (1996).
122. M. Tanahashi, T. Yao, T. Kokubo, M. Minoda, T. Miyamoto, T. Nakamura, T. Yamamuro. Apatite coated on organic polymers by biomimetic process: improvement in its adhesion to substrate by glow-discharge treatment. *J. Biomed. Mater. Res.* 29:349 (1995).
123. K. Kato, Y. Eika, Y. Ikada. Deposition of a hydroxyapatite thin layer onto a polymer surface carrying grafted phosphate polymer chains. *J. Biomed. Mater. Res.* 32:687 (1996).
124. D. L. Coleman. Mineralization of blood pump bladders. *Trans. A.S.A.I.O.* 27:708 (1981).
125. Y. Nose, H. Harasaki, J. Murry. Mineralization of artificial surfaces that contact blood. *Trans. A.S.A.I.O.* 27:714 (1981).
126. T. Kokubo. Biomimetic mineralization. *J. Noncryst. Sol.* 120:138 (1990).
127. M. Tanahashi, T. Matsuda. Surface functional group dependence on apatite formation on self-assembled monolayers in a simulated body fluid. *J. Biomed. Mater. Res.* 34:305 (1997).
128. M. Tanahashi, T. Kokubo, T. Nakamura, Y. Katsura, M. Nagano. Ultrastructural study of an apatite layer formed by a biomimetic process and its bonding to bone. *Biomaterials* 17:47 (1996).

7

Mucoadhesive Polymers

Andreas Bernkop-Schnürch
University of Vienna, Vienna, Austria

I. INTRODUCTION

In the early 1980s, academic research groups pioneered the concept of mucoadhesion as a new strategy to improve the therapeutic effect of various drugs. Mucoadhesive polymers are able to adhere on the mucus gel layer that covers various tissues of the body. These mucoadhesive properties are in many cases advantageous rendering such polymers interesting tools for various pharmaceutical reasons:

1. Mediated by mucoadhesive polymers, the residence time of dosage forms on the mucosa can be prolonged, which allows a sustained drug release at a given target site in order to maximize the therapeutic effect. Robinson and Bologna, for instance, have reported that a polycarbophil gel is capable of remaining on the vaginal tissue for 3 to 4 days and thus provides an excellent vehicle for the delivery of drugs such as progesterone and nonoxynol-9 (1).
2. Furthermore, drug delivery systems can be localized on a certain surface area for purpose of local therapy or for drug liberation at the ''absorption window,'' e.g., the absorption of riboflavin, which has its absorption window in the stomach as well as the small intestine, could be strongly improved in human volunteers by oral administration of mucoadhesive microspheres versus nonadhesive microspheres (2).
3. In addition, mucoadhesive polymers can guarantee an intimate contact with the absorption membrane providing the basis for a high concentration gradient as driving force of drug absorption; for the exclusion of a presystemic metabolism such as the degradation of orally given (poly)peptide drugs by luminally secreted intestinal enzymes (3,4); and for interactions of the polymer with the epithelium, such as a permeation enhancing effect (5,6) or the inhibition of brush border membrane-bound enzymes (7,8).

Because of all these benefits, research work in the field of mucoadhesive polymers has strongly increased within the last two decades resulting in numerous promising ideas, strategies, systems, and techniques based on a more and more profound basic knowledge. An overview reflecting the status quo as well as future trends concerning mucoadhesive polymers should provide a good platform for ongoing research and development in this field.

II. MUCUS GEL COMPOSITION

As mucoadhesive macromolecules adhere to the mucus gel layer, it is important to characterize first this polymeric network representing the target structure for mucoadhesive polymers. The most important component building up the mucus structure are glycoproteins with a relative molecular mass range of $1–40 \times 10^6$ Da. These so-called mucins possess a linear protein core, typically of high serine and threonine content that is glycosilated by oligosaccharide side chains that contain blood group structures. The protein core of many mucins exhibits furthermore N and/or C ter-

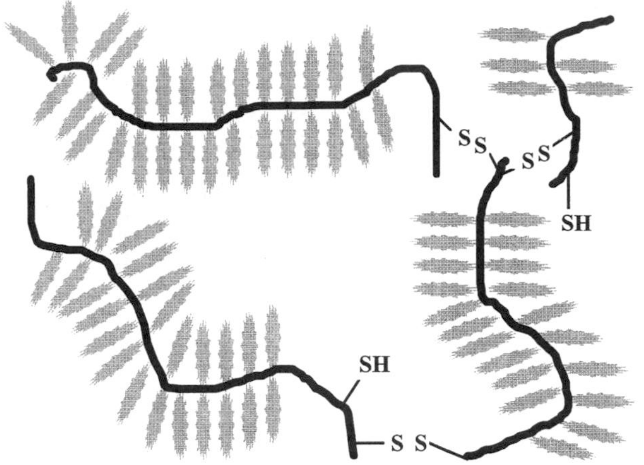

Figure 7.1 Schematic presentation of the three-dimensional network of the mucus gel layer; protein core; ▒; glycosidic side chains, ■.

minally located cysteine-rich subdomains, which are connected with each other via intra- and/or intermolecular disulfide bonds. This presumptive structure of the mucus is illustrated in Fig. 1. The water content of the mucus gel has been determined to be 83% (9). Generally, mucins may be classified into two classes, secretory and membrane-bound forms. Secretory mucins are secreted from mucosal absorptive epithelial cells as well as specialized goblet cells. They constitute the major component of mucus gels of the gastrointestinal, respiratory, ocular, and urogenital surface. The mucus layer based on secretory mucins represents not only a physical barrier but also a protective diffusion barrier (9,10). Membrane-bound mucins possess a hydrophilic membrane-spanning domain and are attached to cell surfaces. Up to now, eight different types of human mucins have been discovered and characterized, which are listed in Table 1.

Secreted mucins are continuously released from cells as well as glands undergoing immediately thereafter a polymerization process mainly based on an intermolecular disulfide bond formation. This so-formed mucus layer, on the other hand, is continuously eroded by enzymatic and mechanical challenges on the luminal surface. Although the turnover time of the mucus gel layer seems to be crucial in order to estimate how long a mucoadhesive delivery system can remain on the mucosa in maximum, there is no accurate information on this time scale available. A clue was given by Lehr et al., who determined the turnover time of the mucus gel layer of chronically isolated intestinal loops in rats to be in the order of approximately 1–4 h (21). Both mucus secretion and mucus erosion, however, are influenced by so many factors, such as mucus secretagogues, mechanical stimuli, stress, calcium concentration, and the enzymatic activity of luminal proteases leading to a highly variable turnover time a time scale is therefore quite complex and difficult to evaluate.

Table 7.1 Synopsis of Human Mucins

Mucin	Characteristics	High-level expression	Cysteine-rich subdomains	Reference
MUC1	Membrane-bound epithelial mucin	Breast, pancreas		12
MUC2	Secreted intestinal mucin	Intestine, tracheobronchus	Yes	13
MUC3	Secreted intestinal mucin	Intestine, gallbladder, pancreas	Yes	14
MUC4	Tracheobronchial, secretory mucin	Tracheobronchus, colon, uterine endocervix		15
MUC5A/C	Secretory mucin	Tracheobronchus, stomach, ocular, Uterine endocervix	Yes	16
MUC5B	Secretory mucin	Tracheobronchus, salivary	Yes	17
MUC6	Major secretory mucin of the stomach	Stomach, gallbladder	Yes	18
MUC7		Salivary		19
MUC8		Tracheobronchus, reproductive tract		20

Source: Ref. 11.

III. PRINCIPLES OF MUCOADHESION

Since the concept of mucoadhesion was introduced into the scientific literature, considerably many attempts have been undertaken in order to explain this phenomenon. So far, however, no generally accepted theory has been found. A reason for this situation can be seen in the fact that many parameters seem to have an impact on mucoadhesion (Fig. 2), which makes a unique explanation impossible. Although there are many controversies over which theories should be favored, at least two basic steps are generally accepted. In step I, the *contact stage*, an intimate contact between the mucoadhesive and the mucus gel layer is formed. In step II, the *consolidation stage*, the adhesive joint is strengthened and consolidated, providing a prolonged adhesion.

A. Chemical Principles

1. Formation of Noncovalent Bonds

Noncovalent chemical bonds include *hydrogen bonding*, which is based on hydrophilic functional groups such as hydroxylic groups, carboxylic groups, amino groups, and sulfate groups; *ionic interactions*, such as the interaction of the cationic polymer chitosan with anionic sialic acid moieties of the mucus; and *van der Waals forces*, based on various dipole–dipole interactions.

2. Formation of Covalent Bonds

In contrast to secondary bonds, covalent bonds are much stronger and are not any more influenced by parameters such as ionic strength and pH value. Functional groups that are able to form covalent bonds to the mucus layer are over all thiol groups. The way such functional groups can form covalent bonds with mucus glycoproteins is illustrated in Fig. 3. Thiolated polymers are able to mimic the naturally occurring mechanism whereby mucus glycoproteins are immobilized in the mucus.

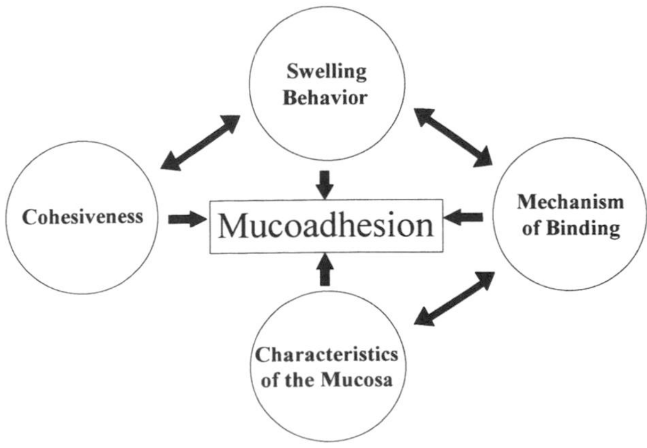

Figure 7.2 Schematic presentation of effects influencing mucoadhesion. (From Ref. 22.)

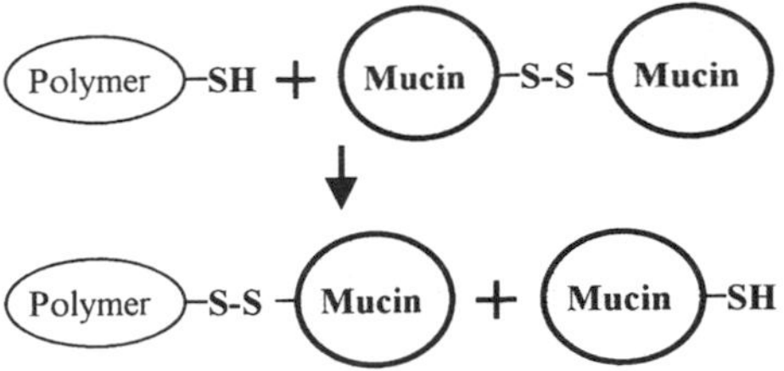

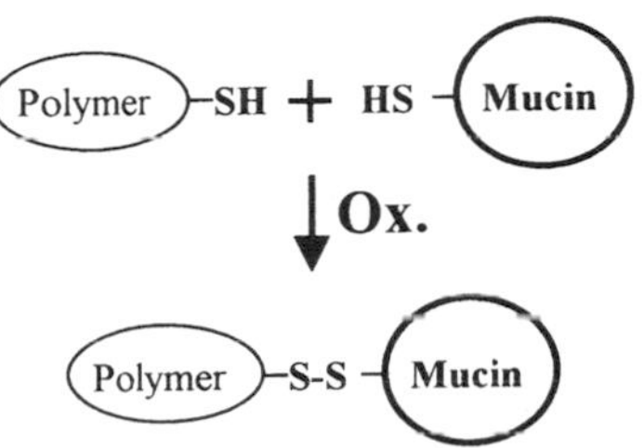

Figure 7.3 Thiolated polymers forming covalent bonds with the mucus layer.

B. Physical Principles

1. Interpenetration

One theory explaining the phenomenon of mucoadhesion is based on a macromolecular interpenetration effect. The mucoadhesive macromolecules interpenetrate with mucus glycoproteins as illustrated in Fig. 4. The resulting consolidation provides the formation of a strong stable mucoadhesive joint. The theory can be substantiated by the observation that chain flexibility favoring a polymeric

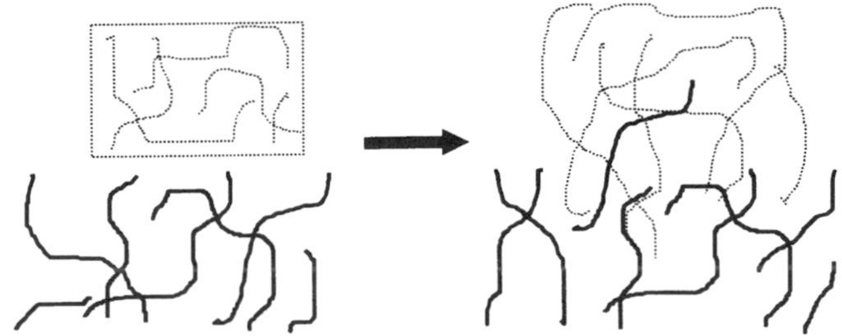

Figure 7.4 Interpenetration of a mucoadhesive matrix tablet (······) and the mucus gel layer (—).

interpenetration is a crucial parameter for mucoadhesion. The crosslinking of various polymers or the covalent attachment of large sized ligands (23) leading to a reduction in chain flexibility results in a strong decrease in mucoadhesive properties.

Rheological approaches (discussed in detail later on) demonstrate a synergistic increase in the resistance to elastic deformation by mixing mucus with mucoadhesive polymer. Attenuated total reflection Fourier transform infrared (ATR-FTIR) studies showing changes in the spectrum of a poly(acrylic acid) film because of interpenetrating mucin molecules within 6 min (24) provide further evidence for this theory.

Objections to the interpenetration theory, however, are provided mainly by electron microscopy studies giving no evidence of interpenetration in the micrometer range (25) and the strong adhesion of mucoadhesive polymers to plane solid surfaces where the opportunity for macromolecular interpenetration is obviously marginal.

2. Mucus Dehydration

Dehydration of a mucus gel layer increases its cohesive nature, which was shown by Mortazavi and Smart (26). Dehydration essentially alters the physicochemical properties of a mucus gel layer, making it locally more cohesive and promoting the retention of a delivery system. The theory can be substantiated by studies with dialysis tubings. Bringing dry mucoadhesive polymers wrapped in dialysis tubings into contact with a mucus layer leads to its dehydration rapidly (26). Dehydrating a mucus gel increases its cohesive nature and subsequently its adhesive behavior, which could be shown by tensile studies (26). An objection to the theory that no interpenetration but exclusively mucus dehydration occurs is given by various rheological studies and tensile studies carried out by Caramella et al. (27), who observed a significant increase in the total work of adhesion (TWA) by mixing the polymer directly with mucin before tensile measurements. It is therefore likely that glycoproteins of the mucus are carried with the flow of water into the mucoadhesive polymer which leads to the already described interpenetration, an explanation which allows the combination of both the interpenetration and mucus dehydration theory.

3. Entanglements of Polymer Chains

The mucoadhesive as well as cohesive properties of polymers can also be explained by physical entanglements of polymer chains. The difference in cohesive as well as mucoadhesive properties of lyophilized and precipitated mucoadhesive polymers gives strong evidence for this theory. Whereas precipitated polymers display high cohesive as well as mucoadhesive properties with a likely high number of polymer chain entanglements, these properties are comparatively lower for the corresponding lyophilized polymers, which do not exhibit such a high extent of entanglements (28).

IV. METHODS TO EVALUATE MUCOADHESIVE PROPERTIES

A. In Vitro Methods

The selection of the mucoadhesive material is the first step in developing a mucoadhesive drug delivery system. A screening of the adhesive properties of polymeric materials can be done by various in vitro methods such as visual tests, tensile studies, and rheological methods, which are often accompanied by additional spectroscopic techniques. Apart from these well-established methods, which are described in detail in this section, some novel methods such as magnetic (29) and direct force measurement techniques (30) have been introduced into the literature as well.

1. Visual Tests

a. Rotating Cylinder

In order to evaluate the duration of binding to the mucosa as well as the cohesiveness of mucoadhesive polymers, rotating cylinder seems to be an appropriate method. In particular, tablets consisting of the test polymer can be brought into contact with freshly excised intestinal mucosa (e.g., porcine) that has been spanned on a stainless steel cylinder (diameter, 4.4 cm; height, 5.1 cm; apparatus, 4-cylinder, USP XXII). Thereafter, the cylinder is placed in the dissolution apparatus according to the USP containing an artificial gastric or intestinal fluid at 37°C. The experimental set-up is illustrated in Fig. 5. The fully immersed cylinder is agitated at 250 rpm. The time needed for de-

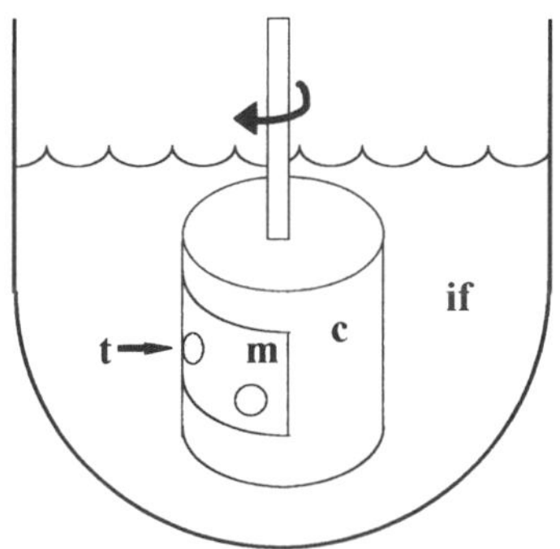

Figure 7.5 Schematic presentation of the test system used to evaluate the mucoadhesive properties of tablets based on various polymers. **c** = cylinder; **if** = intestinal fluid; **m** = porcine mucosa; **t** = tablet.

tachment, disintegration, and/or erosion of test tablets can be determined visually (22).

b. Rinsed Channel

At this method, freshly excised mucosa is spread out on a lopsided channel with the mucus gel layer facing upwards and placed in a thermostatic chamber, which is kept at 37°C. After applying the test material on the mucosa, the rinse is flushed with an artificial gastric or intestinal fluid at a constant flow rate, and the residence time of the mucoadhesive polymer is determined visually. The experimental set-up is illustrated in Fig. 6 (31,32).

2. Tensile Tests

a. Tensile Studies with Dry Polymer Compacts

Test polymers are thereby compressed to flat-faced discs. Tensiometer studies with these test discs are carried out on freshly excised mucosa. Test discs are therefore attached to the mucosa. After a certain contact time between test disc and mucosa, the mucosa is pulled at a certain rate (mm/s) from the disc. The total work of adhesion representing the area under the force–distance curve and the maximum detachment force (MDF) are determined. The experimental set-up is illustrated in Fig. 7. It represents one of the best established in vitro test systems used by numerous research groups (23,33,34).

b. Tensile Studies with Hydrated Polymers

In order to minimize the influence of an ''adhesion by hydration,'' tensile studies can also be carried out with hydrated polymers, as described by Ch'ng and coworkers (35). Hydrated test polymers are thereby spread in a uniform monolayer over excised mucosa which has been fixed on a flat surface. In an artificial gastric or intestinal fluid at 37°C, the hydrated polymer is brought in contact with the mucus layer of a second mucosa, which is fixed on a flat surface of a weight hanging on a balance. The TWA and MDF are then determined as already described.

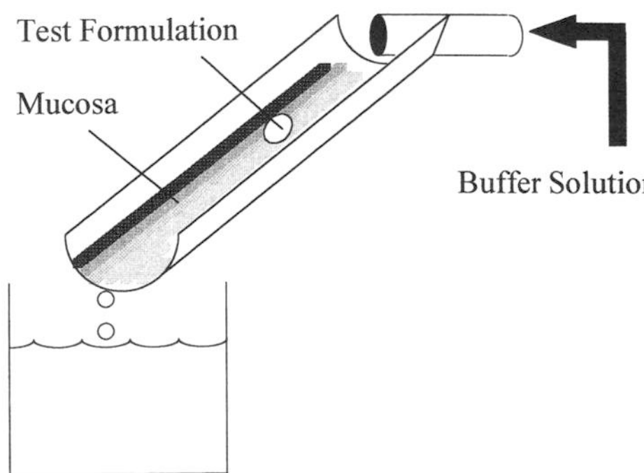

Figure 7.6 Experimental set-up in order to evaluate the mucoadhesive properties of test formulations on a freshly excised mucosa spread out on a lopsided channel.

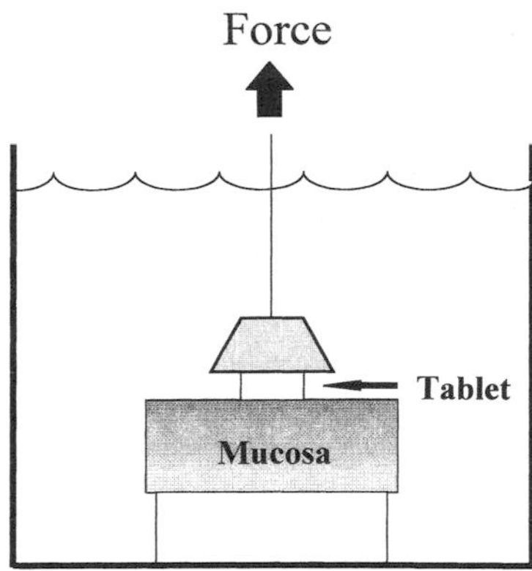

Figure 7.7 Experimental set-up for tensile studies with dry polymer compacts.

c. Tensile Studies with Microspheres

Tensile studies as described are not designed for measuring microscopic interactions such as those that may occur between microparticles and the mucus gel layer. Hence, a method was developed for measuring mucoadhesive properties of microspheres. In vivo interactions are thereby mimicked utilizing a miniature tissue chamber, which is heated by a water jacket. Thermoplastic microspheres are mounted to the tips of fine iron wires using a melting technique. The nonloaded ends of the wires are then attached to a sample clip and suspended in the microbalance enclosure. The freshly excised section of mucosa is clamped in the buffer-filled chamber at 37°C, and the microsphere is brought into contact with the tissue. To fracture the adhesive interactions, the tissue is pulled off the microsphere and certain mucoadhesive parameters are calculated and graphs of force versus position and time are plotted using appropriate software for the microbalance. The method provides valuable information concerning the adhesive properties of microspheres. So far, however, it is limited to microspheres not smaller than 300 μm (36).

3. Rheological Techniques

During the chain interpenetration of mucoadhesive polymers with mucin macromolecules, physical entanglements, conformational changes, and chemical interactions occur. Thereby, changes in the rheological behavior of the two macromolecular species are produced. An evaluation of the resulting synergistic increase in viscosity, which is supposed to be in many cases directly proportional to results obtained with tensile studies (37), can be achieved by mixing the mucoadhesive polymer with mucus and measuring viscosity. The rheological behavior can be determined either by a classical rotational viscometry test at a certain shear rate or by dynamic oscillatory measurements, which give useful information about the structure of the polymer–mucin network.

4. Spectroscopic Techniques

Since mucoadhesive properties of polymers have been investigated, these tests were accompanied by additional spectroscopic analyses (38). Kerr et al. (39), for instance, using ^{13}C nuclear magnetic resonance spectroscopy have provided evidence of hydrogen bonding between mucus glycoproteins and the carboxylic acid groups of polyacrylic acid. Moreover, Tobyn et al., using Fourier transform infrared spectroscopy, reported also interactions between the pig gastric mucus glycoproteins and the test mucoadhesive (40). Jabbari et al. could confirm the chain interpenetration theory by investigating a mucin and polyacrylic acid interface via attenuated total reflection Fourier transform infrared spectroscopy (24). Mortazavi et al., using infrared and ^{13}C NMR spectroscopy, suggested the formation of hydrogen bonds between the mucoadhesive polyacrylic acid and the terminal sugar residues on the mucus glycoprotein (41). In another study, the nature of interactions between the mucus gel and polyacrylic acid was investigated by tensile studies, dynamic oscillatory rheology, and ^{13}C NMR spectroscopy as well. The addition of hydrogen bond breaking agents resulted thereby in a decrease in mucoadhesive strength, a reduction in viscoelastic properties of polymer–mucin mixtures, and a positional change in the chemical shift of the polyacrylic acid signal (42).

B. In Vivo Methods

To date, the mucoadhesion of dosage forms in the body has been evaluated by direct observation and by gamma scintigraphy. For both methods, either the time period of mucoadhesion is determined or in case of the GI tract to what extent the transit time of the mucoadhesive dosage form can be prolonged.

The direct observation offers the advantage that radionuclides are not needed and that a pretreatment of the test formulation is in most cases not necessary. The technique can be used to evaluate mucoadhesion on various tissues in animal studies. Akiyama and Nagahara for instance, administered mucoadhesive microspheres orally to rats. After 2.5 h, the extent of the adhesion of these microspheres to the gastric mucosa was evaluated visually, demonstrating a high amount of mucoadhesive microspheres being present in the stomach compared to nonadhesive microspheres (2). For human studies, however, the direct observation is limited only to a few mucosal tissues such as the oral cavity. Bouckaert et al., for instance, determined the adhesion of tablets in the region of the upper canine. Test tablets were thereby fixed for 1 min with a slight manual pressure on the lip followed by moistening with the tongue to prevent sticking of the tablet to the lip. The adhesion time was defined as the time after which the mucoadhesive tablet was no longer visible under the lip upon control at 30-min intervals (43). In another study, which was carried out in the same way, volunteers participating in this study were asked to record the time and circumstances of the end of adhesion (erosion or detachment of the tablet) (44).

In contrast, no tissue limitations seem to exist for gamma scintigraphic methods. Using these techniques makes it even possible to evaluate the increase in the GI transit time of mucoadhesive formulations. Radionuclides used for imaging studies include $^{99}Tc^{m}$, $^{111}In^{m}$, $^{113}In^{m}$, and $^{81}Kr^{m}$. Among them technetium-99m represents the most commonly used radionuclide, as it displays no beta or alpha radiation and a comparatively short half-life of 6.03 h. Indium-113m, which has an even shorter half-life of 1.7 h has a different energy to technetium-99m and can therefore be used in double-labelling studies. For many applications the longer lived isotope indium-111m (half-life 2.8 days) seems to be more appropriate. Whereas a strongly prolonged GI transit time of mucoadhesive polymers was demonstrated in various animal studies (35), the same effect could not be shown in human volunteers (45,46).

V. MUCOADHESIVE POLYMERS

A system for mucoadhesive polymers can be based on their origin (e.g., natural/synthetic), the type of mucosa on which they are mainly applied (e.g., ocular/buccal), or their chemical structure (e.g., cellulose derivatives/polyacrylates). Apart from these approaches, mucoadhesive polymers can also be classified according to their mechanism of binding as shown within this chapter.

A. Noncovalent Binding Polymers

According to their surface charge, which is important for the mechanism of adhesion, they can be divided into anionic, cationic, nonionic, and ambiphilic polymers.

1. Anionic Polymers

For this group of polymers mainly —COOH groups are responsible for their adhesion to the mucus gel layer. Carbonic acid moieties are supposed to form hydrogen bonds with hydroxyl groups of the oligosaccharide side chains on mucus proteins. Further anionic groups are sulfate as well as sulfonate moieties, which seem to be more of theoretical than of practical relevance. Important representatives of this group of mucoadhesive polymers are listed in Table 2. Among them one can find the most adhesive noncovalent binding polymers such as polyacrylates and NaCMC (33,58). Because of their high charge density, these polymers display a high buffer capacity, which might be beneficial for various reasons as discussed in Section VI.C. In contrast to nonionic polymers, their swelling behavior, which is also crucial for mucoadhesion (Fig. 2), strongly depends on the pH value. The lower the pH value, the lower is the swelling behavior, leading to a quite insufficient adhesion in many cases. The correlation between the pH value of an anionic polymer and its swelling behavior is illustrated in Fig. 8. On the contrary, a too rapid swelling of such polymers at higher pH values can lead to an overswelling, which causes a strong decrease in the cohesive properties of such polymers. Even if the polymer sticks to the mucus layer, in this case, the drug delivery system won't be any more mucoadhesive, as the adhesive bond fails within the mucoadhesive polymer itself. This effect is illustrated in Fig. 9. A further drawback of anionic mucoadhesive polymers, however, is their incomatibility with multivalent cations like Ca^{2+}, Mg^{2+}, and Fe^{3+}. In the presence of such cations, these polymers precipitate and/or coagulate (61), leading to a strong reduction in their adhesive properties.

2. Cationic Polymers

The strong mucoadhesion of cationic polymers can be explained by ionic interactions between these polymers and

Table 7.2 Anionic Mucoadhesive Polymers

Polymer	Chemical structure	Additional information	References
Alginate	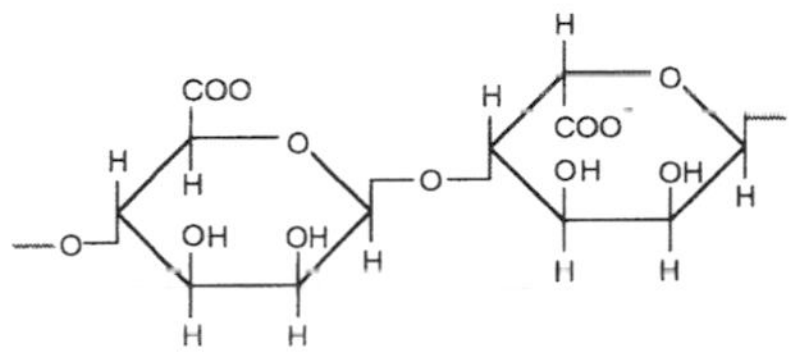		47–49
Carbomer		Crosslinked with sucrose	33, 49, 50
Chitosan-EDTA	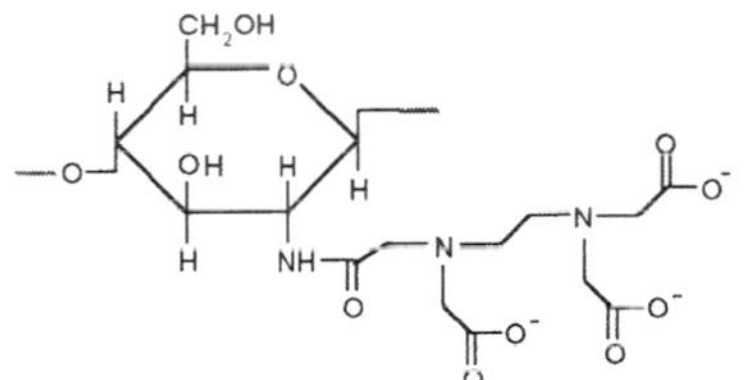	Optionally crosslinked with EDTA	51, 52
Hyaluronic acid	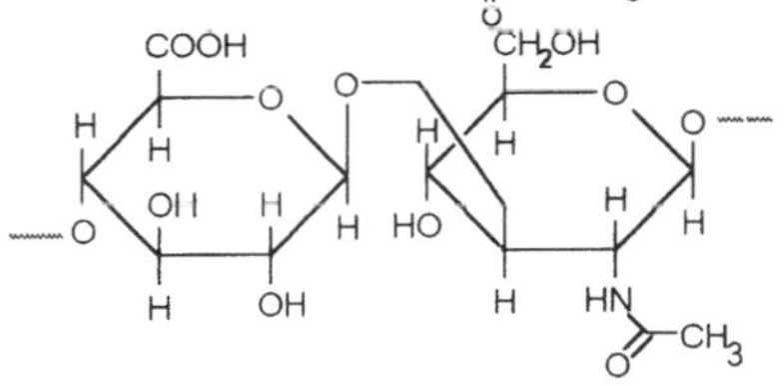		53–55
NaCMC	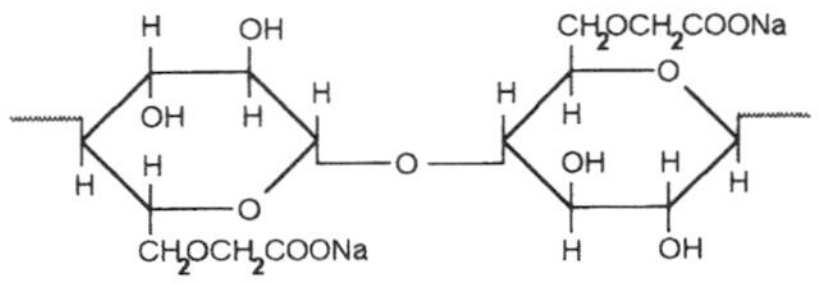	0.3–1.0 carboxymethyl groups per glucose unit	33, 49, 56
Pectins	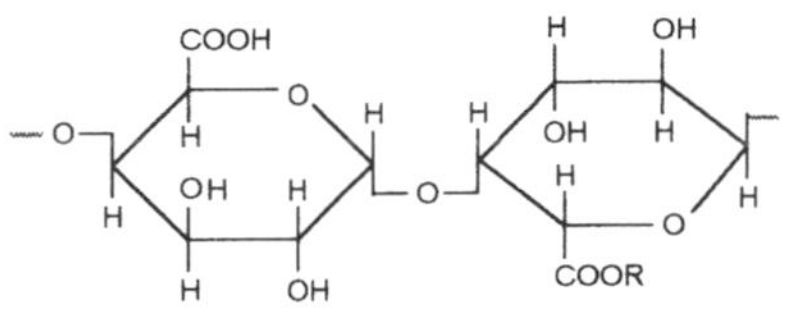	R=OH or methyl	57, 58
Polycarbophil	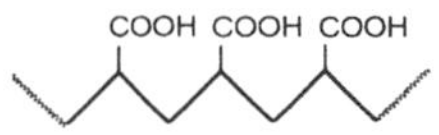	Crosslinked with divinylglycol	33, 56, 59, 60

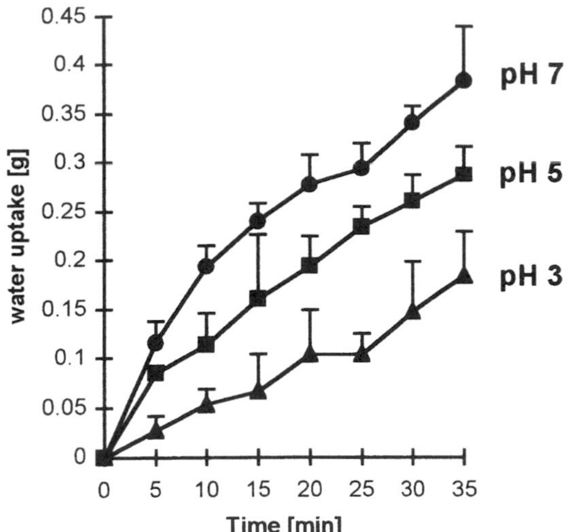

Figure 7.8 Correlation between the pH value of the anionic polymer polycarbophil and its swelling behavior. (From Ref. 22.)

anionic substructures such as sialic acid moieties of the mucus gel layer. In particular, chitosan, which can be produced in high amounts for a reasonable price, seems to be a promising mucoadhesive excipient. Apart from its mucoadhesive properties, chitosan is also reported to display permeation-enhancing properties (62,63). The most important cationic mucoadhesive polymers are listed in Table 3. Their swelling behavior is strongly pH dependent as well. In contrast to anionic polymers, however, their swelling behavior is improved at higher proton concentrations. Chitosan, for instance, is hydrated rapidly in the gastric fluid, leading to a strong reduction of its cohesive properties, whereas it does not swell at all at pH values above 6.5, causing a complete loss in its mucoadhesive properties.

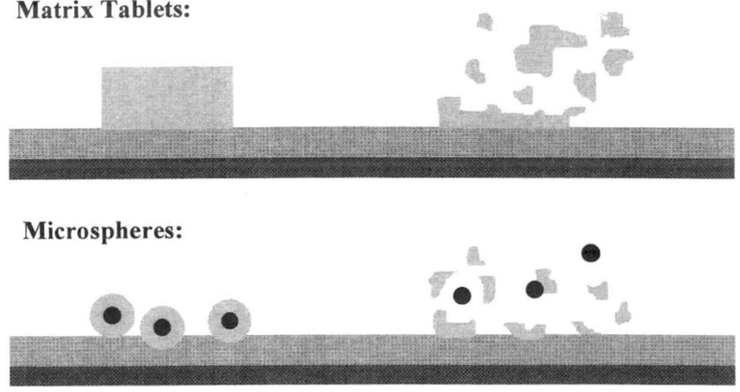

Figure 7.9 Adhesive bond failure in case of insufficient cohesive properties of the mucoadhesive polymer.

3. Nonionic Polymers

The adhesion of anionic as well as cationic polymers strongly depends on the pH value of the surrounding fluid. On the contrary, non ionic mucoadhesive polymers are mostly independent from this parameter. Whereas the formation of secondary chemical bonds due to ionic interactions can be completely excluded for this group of polymers, some of them such as poly(ethylene oxide) are capable of forming hydrogen bonds. Apart from these interactions, their adhesion to the mucosa seems to be rather based on an interpenetration followed by polymer chain entanglements. These theoretical considerations are in good accordance with mucoadhesion studies, demonstrating almost no adhesion of nonionic polymers if they are applied to the mucosa already in the completely hydrated form, whereas they are adhesive if applied in dry form (64). Hence, nonionic polymers are in most cases less adhesive than anionic as well as cationic mucoadhesive polymers. Well-known representatives of this group of mucoadhesive polymers are listed in Table 4. In contrast to ionic polymers, nonionic polymers are not influenced by electrolytes of the surrounding milieu. The addition of 0.9% NaCl, for instance, to carbomer leads to a tremendous decrease in its cohesiveness and subsequently to a strong reduction of its mucoadhesive properties, whereas these electrolytes have no influence on nonionic mucoadhesive polymers.

4. Ambiphilic Polymers

Ambiphilic mucoadhesive polymers display cationic as well as anionic substructures on their polymer chains. On one hand, mucoadhesion of the cationic polymers is said to be caused by electrostatic interactions with negatively charged mucosal surfaces. For anionic polymers, on the other hand, mucoadhesion can be explained by the formation of hydrogen bonds of carboxylic acid groups with the mucus gel layer. The combination of positive as well as negative charges on the same polymer, however, seems to compensate both effects, leading to strongly reduced adhesive properties of ambiphilic polymers. Mucoadhesion studies of chitosan–EDTA conjugates with increasing amounts of covalently attached EDTA can clearly show this effect. Whereas the exclusively anionic chitosan–EDTA conjugate (Table 2) exhibiting no remaining cationic moieties and the exclusively cationic polymer chitosan displayed the highest mucoadhesive properties, the mucoadhesion of chitosan–EDTA conjugates showing both cationic moieties of remaining primary amino groups and anionic moieties of already covalently attached EDTA was much lower. In addition, Luessen et al. could show a strongly increased intestinal buserelin bioavailability in

Table 7.3 Cationic Mucoadhesive Polymers

Polymer	Chemical structure	Additional information	References
Chitosan	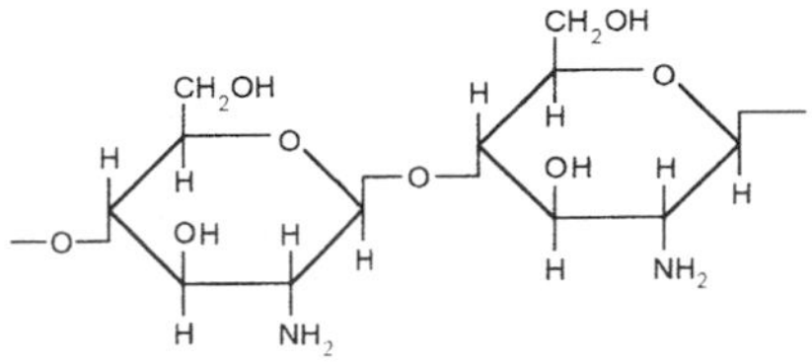	Primary amino groups can be acetylated to some extend	3, 60
Polylysine	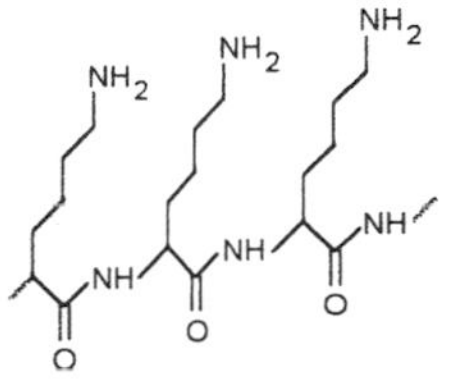		57

rats using chitosan HCl as mucoadhesive excipient. A mixture of this cationic polymer with the anionic polymer carbomer, however, led to a significantly reduced bioavailability of the therapeutic peptide (66). Representatives of this type of mucoadhesive polymer are mainly proteins such as gelatin, which is reported as mucoadhesive in various studies (58,67).

Due to the combination of cationic as well as anionic mucoadhesive polymers leading to ionic interactions, however, the cohesiveness of delivery systems can be strongly improved (53,57). If the adhesive bond of a delivery system fails within the mucoadhesive polymer itself rather than between the polymer and the mucus layer as illustrated in Fig. 9, this effect is more important than the mucoadhesive properties of the polymer in order to improve the adhesiveness of the whole dosage form.

B. Covalent Binding Polymers

Recently, a presumptive new generation of mucoadhesive polymers has been introduced into the pharmaceutical literature (68). Whereas the attachment of mucoadhesive polymers to the mucus layer has to date been achieved by noncovalent bonds, these novel polymers are capable of forming covalent bonds. The bridging structure most commonly encountered in biological systems—the disulfide bond—has thereby been discovered for the covalent adhesion of polymers to the mucus layer of the mucosa. Thio-

Table 7.4 Nonionic Mucoadhesive Polymers

Polymer	Chemical structure	Additional information	References
Hydroxypropyl–cellulose	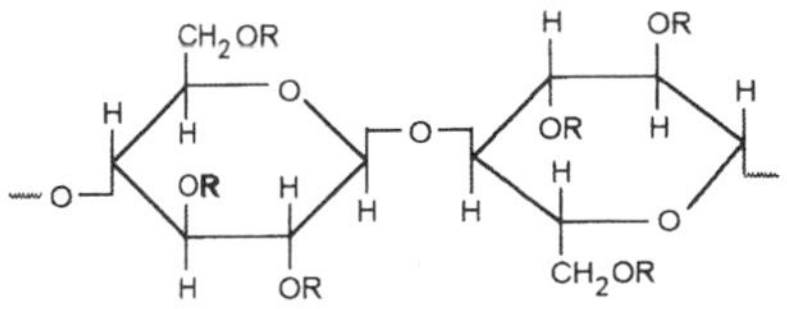	R=H or hydroxypropyl	34, 65
Hydroxypropyl–methylcellulose		R=H or methoxy or hydroxypropyl	49, 50
Poly(ethylene oxide)	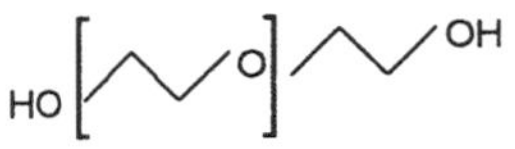		34, 49
Poly(vinyl alcohol)	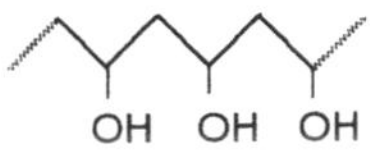		50, 58
Poly(vinyl pyrrolidone)	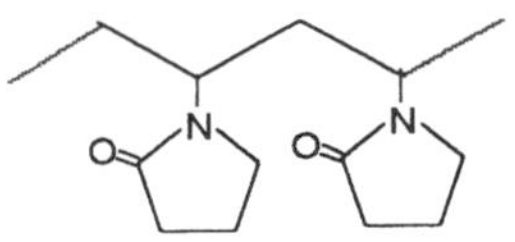		58

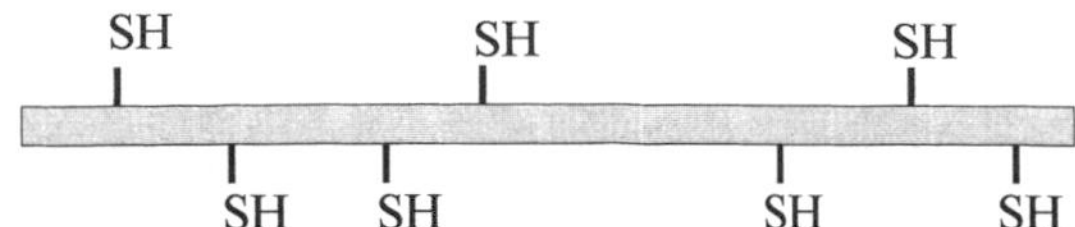

Figure 7.10 Thiomer (thiolated polymer): mucoadhesive polymers which display thiol moieties bearing side chains.

lated polymers, or so-called thiomers, are mucoadhesive basis polymers that display thiol-bearing side chains (Fig. 10). Based on thiol/disulfide exchange reactions, as illustrated in Fig. 11, and/or a simple oxidation process, disulfide bonds are formed between such polymers and cysteine-rich subdomains of mucus glycoproteins (see Table 1). Hence, thiomers mimic the natural mechanism of secreted mucus glycoproteins, which are also covalently anchored in the mucus layer by the formation of disulfide bonds. Due to the covalent attachment of cysteine to polycarbophil, for instance, the adhesive properties of this polymer could be strongly increased. Whereas the unmodified basis polymer displayed a total work of adhesion of 104 ± 21 μJ, it was 280 ± 67 μJ for a corresponding polymer–cysteine conjugate. The results of this study are shown in Fig. 12. Apart from these improved mucoadhesive properties, which could meanwhile also be shown for various other thiolated polymers, thiomers exhibit strongly improved cohesive properties as well. Whereas, for example, tablets consisting of polycarbophil disintegrate within 2 h, tablets based on the corresponding thiolated polymer remain stable even for days in the disintegration apparatus according to the Pharmacopoea Europea (70). The result, as shown in Fig. 13, can be explained by the continuous oxidation of thiol moieties on thiomers which takes place in aqueous solutions at pH values above 5. The decrease in sulfhydryl groups, i.e., the formation of inter- and/or intramolecular disulfide bonds within a sodium carboxymethylcellulose–cysteine conjugate, is illustrated in Fig. 14. Due to the high density of negative charges within anionic mucoadhesive polymers, they can also function as ion exchange resins displaying a high buffer capacity (see Section VI.C). According to this, the formation of disulfide bonds within polymer–cysteine conjugates can be controlled by adjusting the pH value of the system a priori. Adhesion of many quick swelling polymers, as already mentioned, is limited by an insufficient cohesion of the polymer, resulting in a break within the polymer network rather than

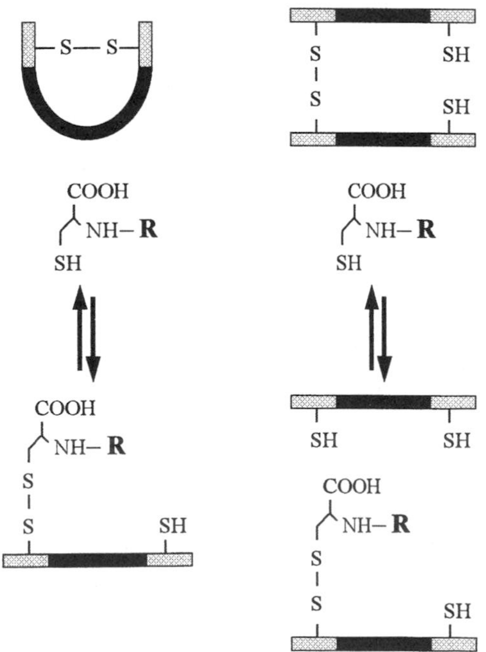

Figure 7.11 Schematic presentation of disulfide exchange reactions between a (poly)peptide and a cysteine derivative according to Snyder (69). The (poly)peptide stands here for a mucin glycoprotein of the mucus, and the cysteine derivative is a polymer–cysteine conjugate. (**R** = mucoadhesive basis polymer) (Adapted from Ref. 68.)

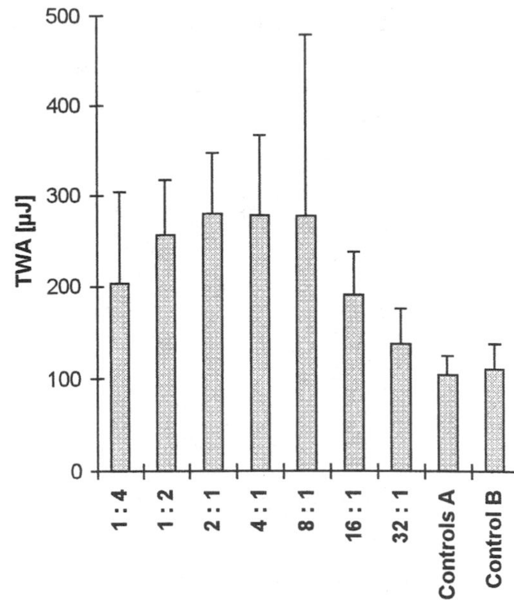

Figure 7.12 Comparison of the adhesive properties of polycarbophil–cysteine conjugates displaying increasing amounts of covalently attached cysteine (32:1 up to 1:4) and controls of unmodified polycarbophil. Represented values are means ± S.D. (n = 3–8) of the TWA determined in tensile studies at pH 6.8 with dry compacts of indicated test material. (From Ref. 68.)

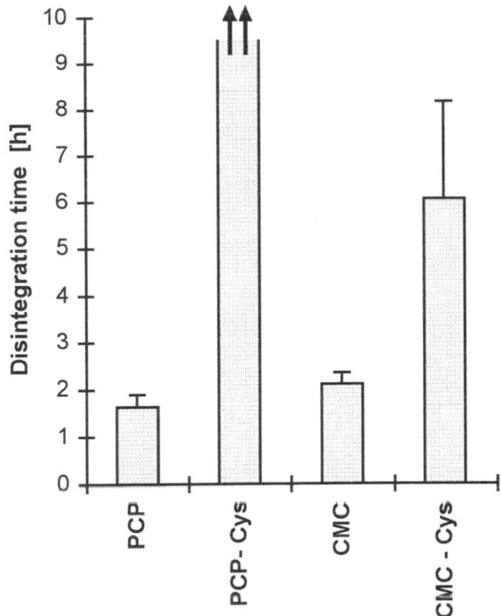

Figure 7.13 Comparison of the disintegration behavior of matrix tablets (30 mg; 5 mm i.d.) containing indicated lyophilized polymers. (PCP = polycarbophil; CMC = carboxymethylcellulose; PCP-Cys = polycarbophil–cysteine conjugate; CMC-Cys = carboxymethylcellulose–cysteine conjugate.) Studies were carried out with a disintegration test apparatus (Pharm. Eur.) in 50 mM TBS, pH 6.8, at 37°C. Indicated values are means ± S.D. of at least three experiments. Polycarbophil–cysteine tablets did not disintegrate even after 48 h of incubation. (From Ref. 70.)

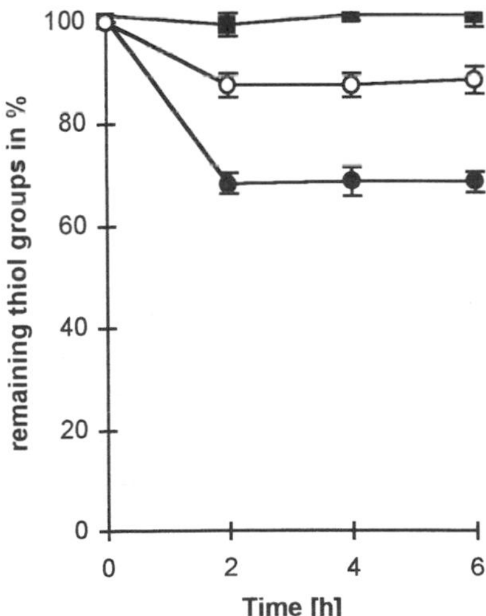

Figure 7.14 Decrease in thiol groups of 0.5% (m/v) (○) and 1.0% (m/v) NaCMC–cysteine conjugate (●, ■), in 50 mM phosphate buffer pH 6.8 (○, ●) and 50 mM acetate buffer pH 5.0 (■), at 37°C. Indicated values are means ± S.D. of at least three experiments. (From Ref. 70.)

between the polymer and mucus layer (Fig. 9). Although thiolated polymers are rapidly hydrated, they are able to form highly cohesive and viscoelastic gels due to the formation of additional disulfide bonds. The formation of an overhydrated slippery mucilage can thereby be excluded completely. Using the rotating cylinder (Section IV.A.1) in order to evaluate the mucoadhesive properties of various formulations, for instance, revealed a comparatively much longer adhesion of tablets consisting of thiolated polymers (22). Meanwhile, various anionic as well as cationic thiolated polymers have been synthesized, as listed in Table 5. They all display strongly improved mucoadhesive properties compared to the corresponding unmodified basis polymers.

VI. MULTIFUNCTIONAL MUCOADHESIVE POLYMERS

Whereas in the 1980s the adhesive properties of polymers played a central role in the field of mucoadhesion, numerous scientists began in the 1990s to focus their interest also on additional features of mucoadhesive polymers. These properties include enzyme inhibition, permeation enhancement, high buffer capacity, and controlled drug release. For some of these properties, mucoadhesion is substantial; for others, various synergistic effects can be expected due to adhesion.

A. Mucoadhesion and Enzyme Inhibition

Due to the great progress in biotechnology as well as gentechnology, the industry is capable of producing a large number of potential therapeutic peptides and proteins in commercial quantities. The majority of such drugs are most commonly administered by the parenteral routes that are often complex, difficult, and occasionally dangerous. Besides so-called alternative routes of application such as the nasal or transdermal route, there is no doubt that the peroral route is one of the most favored, as it offers the greatest ease of application. A presystemic metabolism of therapeutic peptides and proteins caused by proteolytic enzymes of the GI tract, however, leads to a very poor bioavailability after oral dosing. Attempts to reduce this barrier include the use of analogs; prodrugs; formulations such as nanoparticles, microparticles, and liposomes that shield therapeutic peptides and proteins from luminal enzymatic attack; and the design of delivery systems targeting the colon where the proteolytic activity is relatively low. Moreover, considerable interest is shown in the use of mucoadhesive polymers, since due to such excipients various in vivo studies could demonstrate a significantly

Table 7.5 Thiolated Mucoadhesive Polymers

Polymer	Chemical structure	Additional information	References
Chitosan–cysteine	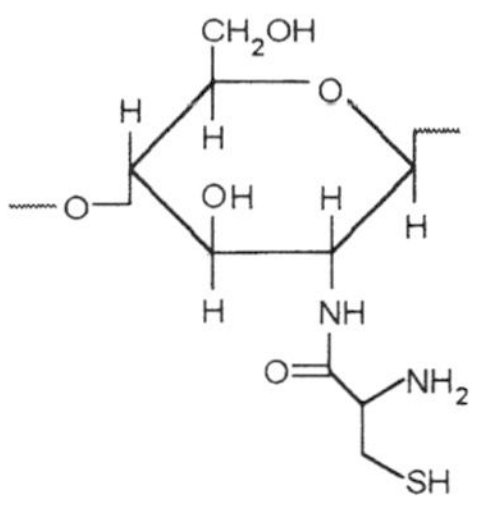	21 up to 100 μMol thiol groups per gram polymer	71
Chitosan–thioglycolic acid	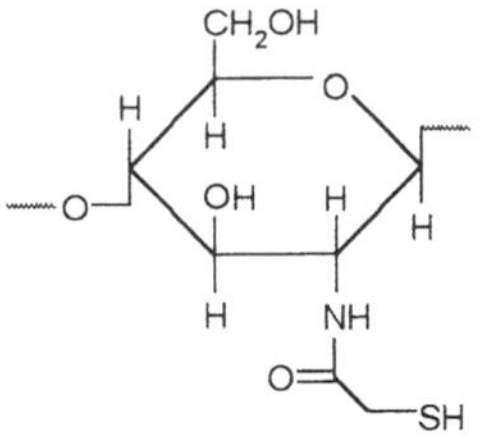	11 up to 25 μMol thiol groups per gram polymer	72
NaCMC–cysteine	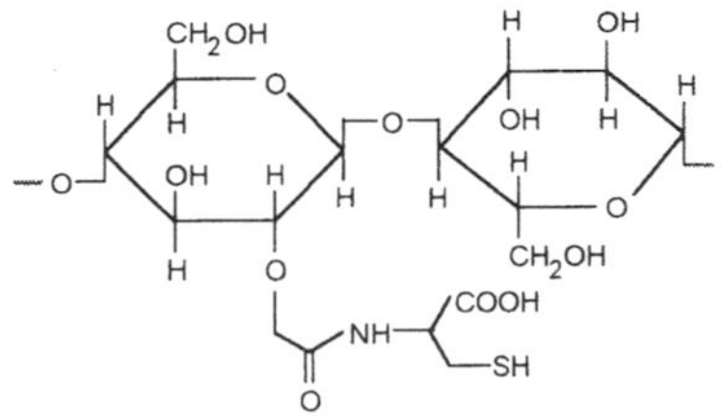	22 up to 1280 μMol thiol groups per gram polymer	22, 70
Polycarbophil–cystamine	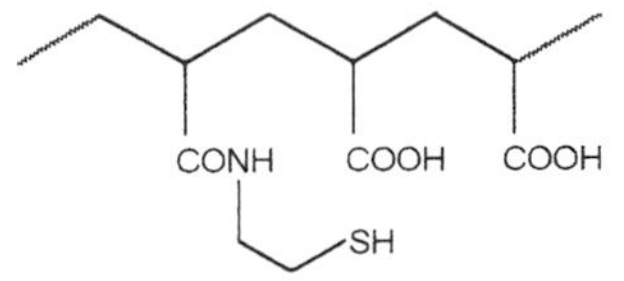	1 up to 20 μMol thiol groups per gram polymer	
Polycarbophil–cysteine	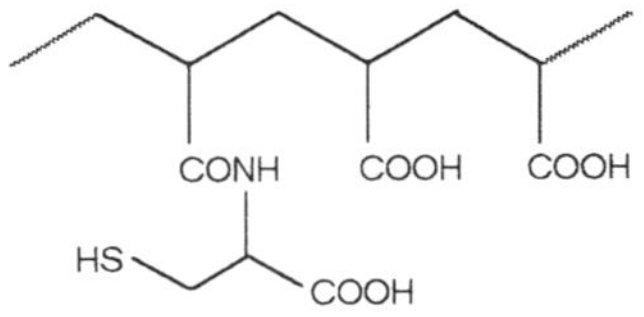	1 up to 142 μMol thiol groups per gram polymer	22, 68, 70, 73

improved bioavailability of peptide and protein drugs after oral dosing.

As formulations containing mucoadhesive polymers provide an intimate contact with the mucosa, a presystemic degradation of these drugs on the way between the delivery system and the absorbing membrane can be excluded. Takeuchi et al., for instance, demonstrated a significantly stronger reduction in the plasma calcium level after oral administration of calcitonin-loaded liposomes which were coated with a mucoadhesive polymer in comparison to the same formulation without the mucoadhesive coating (4).

In recent years it could be demonstrated by various studies that certain mucoadhesive polymers display also an enzyme inhibitory effect. In particular, poly(acrylic acid) was shown to exhibit a pronounced inhibitory effect toward trypsin (74,75). Additionally, the immobilization of enzyme inhibitors to mucoadhesive polymers acting only in a very restricted area of the intestine seems to be a promising approach in order to improve their enzyme inhibitory properties. Due to the immobilization of the inhibitor, it remains concentrated on the polymer, which should certainly make a reduced share of this auxiliary agent in the dosage form sufficient. Side effects of the inhibitor such as systemic toxicity, a disturbed digestion of nutritive proteins, and pancreatic hypersecretion caused by a luminal feedback regulation can be avoided. Hence, the covalent attachment of enzyme inhibitors to mucoadhesive polymers, such as shown in Fig. 15, represents the combination of two favorable strategies for the oral administration of (poly)peptide drugs, offering additional advantages

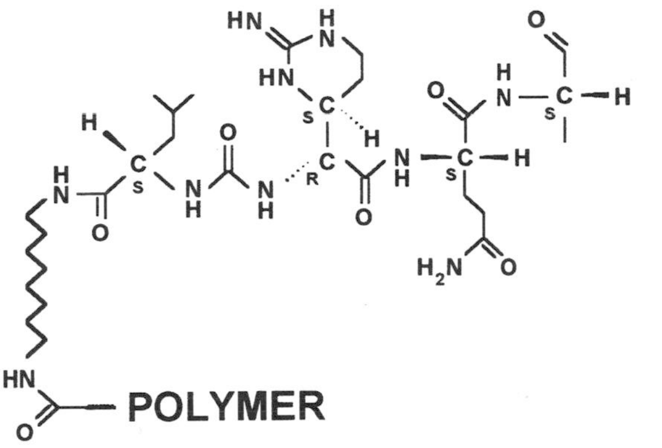

Figure 7.15 Example for a mucoadhesive polymer inhibitor conjugate. The elastase inhibitor elastatinal is thereby covalently attached via a C8 spacer to polymers like polycarbophil or NaCMC. (From Ref. 76.)

compared to a simple combination of both excipients without the covalent linkage. So far, various polymer–inhibitor conjugates have been generated as listed in Table 6. Their efficacy could be verified by in vivo studies in diabetic mice showing a significantly reduced glucose level after the oral administration of insulin tablets containing a CMC–Bowman Birk inhibitor conjugate and a CMC–elastatinal conjugate. The results of this study are shown in Fig. 16 (83).

B. Mucoadhesion and Permeation Enhancement

A number of mucoadhesive polymers also have promising effects on the modulation of the absorption barrier by opening of the intestinal intercellular junctions (63,84). In contrast to permeation enhancers of low molecular size such as sodium salicylate and medium-chain glycerides (85,86), these polymers might not be absorbed from the intestine, which should exclude systemic side effects of these auxiliary agents. Permeation studies, for instance, carried out in Ussing chambers on Caco-2 monolayers demonstrated a strong permeation-enhancing effect of chitosan and carbomer (84,87). This permeation-enhancing effect of these mucoadhesive polymers can even be significantly improved due to the immobilization of cysteine on these polymers (5,71). Results demonstrating this permeation-enhancing effect of a thiolated polymer are shown in Fig. 17. This improved permeation across the mucosa was accompanied by a decrease in the TEER, indicating a loosening of the tightness of intercellular junctions, i.e., the opening of the paracellular route across the epithelium for otherwise nonabsorbable hydrophilic compounds such as peptides. Mechanisms which are responsible for the permeation-enhancing effect of mucoadhesive polymers, however, are still unclear. As anionic polymers such as polyacrylates display a high binding capacity for Ca^{2+} ions, their permeation-enhancing effect might be based on the depletion of this cation. The theory relies on the knowledge that on the one hand many complexing agents like EDTA display a permeation-enhancing effect (88), and Ca^{2+}, on the other hand, plays an essential role concerning the gate fence function of the tight junctions being responsible for their closing (89).

In contrast, the permeation-enhancing effect of cationic polymers like chitosan might be based on the positive charges of these polymers which interact with the cell membrane, resulting in a structural reorganization of tight junction-associated proteins (90).

In case of thiolated polymers, protein tyrosine phosphatase might be involved in the underlying mechanism. This thiol-dependent enzyme mediates the closing process of tight junctions by dephosphorylation of tyrosine groups from the extracellular region (91). The inhibition of PTP by specific inhibitors such as vanadate or pervanadate causes an enhanced opening of the tight junctions. As it is also inhibited by sulfhydryl compounds such as glutathione forming a mixed disulfide with Cys 215 (92), it is likely that thiolated polymers might also lead to an inhibition of this enzyme.

C. Mucoadhesion and Buffer System

A further advantage of mucoadhesive polymers can be seen in the high buffer capacity of ionic polymers. As these polymers can act as ion exchange resins, they are able to maintain their pH value inside the polymeric network over a considerable period of time. Matrix tablets based on neutralized carbomer, for instance, can buffer the pH value inside the swollen carrier system even for hours in an artificial gastric fluid of pH 2 (93). The results of this study are shown in Fig. 18. This high buffer capacity seems to be highly beneficial for various reasons. For example, the epidermal growth factor (EGF) is recognized as an important agent for acceleration of ulcer healing and has a peculiar biological property to repair tissue damage by an enhanced proliferation and differentiation of epithelial tissues (94). Itoh and Matsuo demonstrated in a double-blind controlled clinical study the enhanced healing of rat gastric ulcers after oral administration of EGF. This effect could even be drastically increased by using the mucoadhesive polymer hydroxypropyl cellulose as drug carrier matrix (95). As EGF is strongly degraded by pepsin (96), the use of mucoadhesive polymers providing an additional protective effect toward pepsinic degradation might therefore be helpful. It can be achieved by a comparatively

Table 7.6 Comparison of Various Mucoadhesive Polymer–Inhibitor Conjugates

Polymer–inhibitor conjugate	Inhibited enzymes		Mucoadhesive properties	References
	Based on complexing properties	Based on competitive inhibition		
Carboxymethylcellulose–elastatinal conjugate		Elastase	+	76
Carboxymethylcellulose–pepstatin conjugate		Pepsin	n.d.	77
Chitosan–antipain conjugate		Trypsin	++	78, 79
Chitosan–chymostatin conjugate		Chymotrypsin	n.d.	79
Chitosan–elastatinal conjugate		Elastase	n.d.	79
Chitosan–ACE[a] conjugate		Trypsin, chymotrypsin, elastase	n.d.	79
Chitosan–EDTA	Aminopeptidase N, carboxypeptidase A/B		+++	8, 51
Chitosan–EDTA–antipain conjugate	Aminopeptidase N, carboxypeptidase A/B	Trypsin	n.d.	79
Chitosan–EDTA–chymostatin conjugate	Aminopeptidase N, carboxypeptidase A/B	Chymotrypsin	n.d.	79
Chitosan–EDTA–elastatinal conjugate	Aminopeptidase N, carboxypeptidase A/B	Elastase	n.d.	79
Chitosan–EDTA–ACE conjugate	Aminopeptidase N, carboxypeptidase A/B	Trypsin, chymotrypsin, elastase	+	79
Chitosan–EDTA–BBI[b] conjugate	Aminopeptidase N, carboxypeptidase A	Trypsin, chymotrypsin, elastase	+	80
Poly(acrylic acid)–Bowman–BBI conjugate		Chymotrypsin	++	81
Poly(acrylic acid)–chymostatin conjugate		Chymotrypsin	+++	23
Poly(acrylic acid)–elastatinal conjugate		Elastase	+++	76
Polycarbophil–elastatinal conjugate		Elastase	+++	76
Poly(acrylic acid)–bacitracin conjugate	Aminopeptidase N		n.d.	82
Polycarbophil-cysteine	Carboxypeptidase A/B		++++	73

Note: Mucoadhesive properties are classified as poor (+), good (++), very good (+++), and excellent (++++).
[a] Antipain/chymostatin/elastatinal.
[b] Bowman Birk inhibitor.

higher pH value inside the drug carrier matrix based on its high buffer capacity at which penetrating pepsin is already inactive.

Apart from this likely advantage for the (poly)peptide administration, the high buffer capacity of neutralized anionic polymer might also be highly beneficial in treatment of *Helicobacter pylori* infection in peptic ulcer disease, as common antibiotics such as amoxicillin or metronidazole display poor stability in the acidic pH of the stomach. The incorporation of these therapeutic agents in mucoadhesive polymers displaying a high buffer capacity might improve their stability in the acidic milieu.

D. Mucoadhesion and Controlled Release

If the therapeutic agent is incorporated in the mucoadhesive polymer, the excipient can act both as mucoadhesive and as a matrix system providing a controlled drug release. The release behavior of drugs embedded in mucoadhesive polymers depends thereby mainly on their molecular size and charge. According to the equation determining the diffusion coefficient, in which the radius of a molecule indirectly correlates with the diffusion coefficient, small-sized drugs will be released faster than larger ones. Apart from their size, the release of therapeutic agents from mucoadhesive polymers can be simply controlled by raising or lowering the share of the polymer in the delivery system. Whereas a low amount of the mucoadhesive polymer in the carrier matrix can guarantee a rapid drug release, a sustained drug release can be provided by raising the share of the polymer in the delivery system (77). In addition, the drug release from mucoadhesive polymers can also be controlled by the extent of crosslinking. The higher the polymer is crosslinked, the lower is the release rate of the

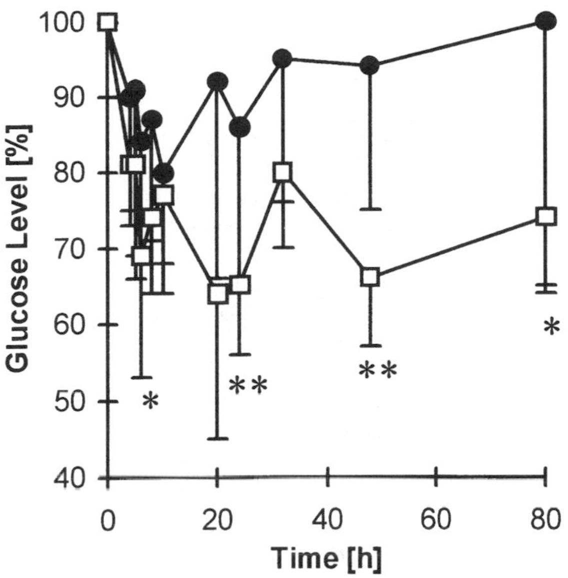

Figure 7.16 Blood glucose level of diabetic mice after oral administration of insulin microtablets containing mucoadhesive polymer inhibitor conjugates (□) versus control (●). Values are means ± S.D.; n = 10; * differs from control $p < 0.05$; ** differs from control $p < 0.001$. (From Ref. 83.)

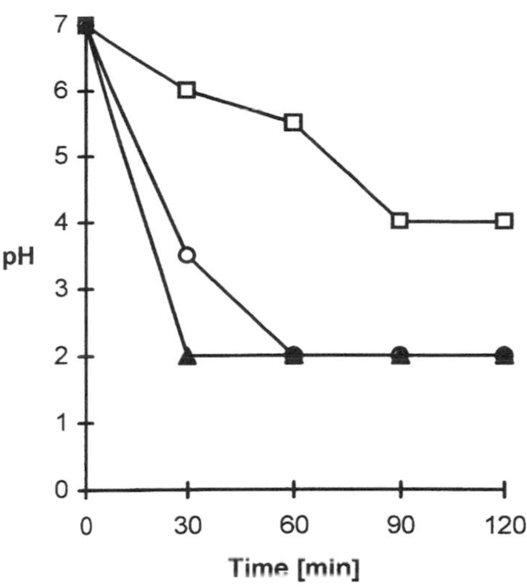

Figure 7.18 Buffer capacity of mucoadhesive polymers; matrix tablets consisting of carbomer (□); matrix tablets consisting of NaCMC (○); control (▲). (Adopted from Ref. 93.)

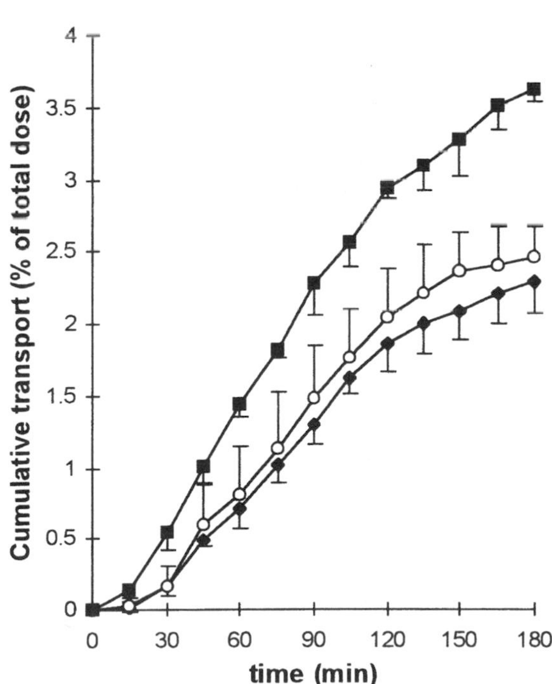

Figure 7.17 Permeation-enhancing effect of polycarbophil and thiolated polycarbophil; polycarbophil–cysteine conjugate (■); polycarbophil (○); control without polymer (◆). (From Ref. 5.)

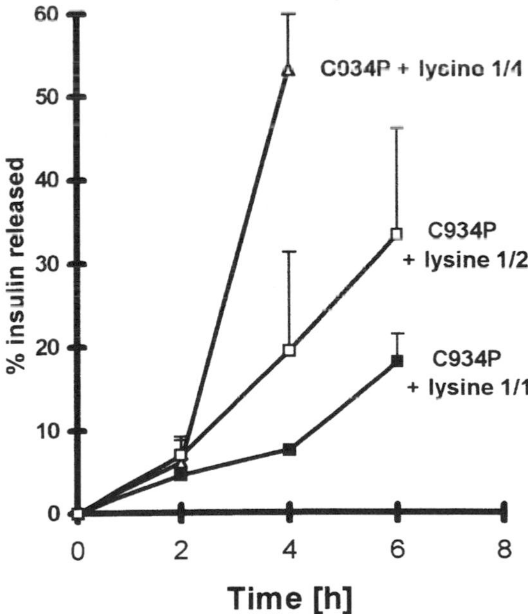

Figure 7.19 Release profile of insulin from tablets consisting of 3.3% insulin, 30% mannitol, and 66.7% carbomer (C934P), which had been crosslinked with lysine. Tablets were incubated in 10 mL release medium (20 mM Tris-HCl, pH 7.8) on a waterbath-shaker (100 rpm) at 37 ± 0.5°C. Indicated values are means ± S.D. of at least three experiments. (From Ref. 28.)

drug. Such a crosslinking can be achieved by the formation of covalent bonds, e.g., the crosslinking of gelatin with glutaraldehyde (67), or on the basis of ionic interactions. The release rate of insulin from matrix tablets consisting of the anionic mucoadhesive polymer carbomer (C934P), for instance, can be strongly reduced by the addition of the divalent cationic amino acid lysine (28). The results of this study are shown in Fig. 19. In the case of ionic drugs, a sustained release can also be guaranteed by the use of an ionic mucoadhesive polymer displaying the opposite charge of the therapeutic agent. On the basis of an ion exchange resin, for instance, a sustained release of the therapeutic agent α-lipoic acid over a period of more than 12 h can be provided by the incorporation of this anionic drug in the cationic mucoadhesive polymer chitosan (97).

VII. CONCLUDING REMARKS

Motivated by the great benefits that can be provided by mucoadhesive polymers, such as a prolonged residence time and an intimate contact of the dosage form on the mucosa, considerably intensive research and development has been performed in this field since the concept of mucoadhesion had been pioneered in the early 1980s. Merits of these efforts are the establishment of various useful techniques to evaluate the mucoadhesive properties of different polymers as well as the design of novel polymers displaying improved mucoadhesive properties. In addition, the development of multifunctional mucoadhesive polymers also exhibiting features such as enzyme inhibitory properties, permeation enhancing effects, high buffer capacity, and the possibility to control the drug release make them even more promising auxiliary agents. Although there are already numerous formulations based on mucoadhesive polymers on the market, the number of delivery systems making use of these advantages will certainly increase in the near future.

REFERENCES

1. J. R. Robinson, W. J. Bologna. Vaginal and reproductive system treatments using a bioadhesive polymer. *J. Controlled Release* 28:87 (1994).
2. Y. Akiyama, N. Nagahara. Novel formulation approaches to oral mucoadhesive drug delivery systems. In: *Bioadhesive Drug Delivery Systems*, E. Mathiowitz, D. E. Chickering, III, C.-M. Lehr (Eds.). Marcel Dekker, New York, 1999, p. 177.
3. H. Takeuchi, H. Yamamoto, N. Toshiyuki, H. Tomoaki, Y. Kawashima. Enteral absorption of insulin in rats from mucoadhesive chitosan-coated liposomes. *Pharm. Res.* 13: 896 (1996).
4. H. Takeuchi, Y. Matsui, H. Yamamoto, Y. Kawashima. Mucoadhesive liposomes coated with chitosan or carbopol for oral administration of peptide drugs. *Proceed. Int. Symp. Control. Rel. Bioact. Mater.* 26:6372 (1999).
5. A. Clausen, A. Bernkop-Schnürch. *In vitro* evaluation of the permeation enhancing effect of thiolated polycarbophil. *J. Pharm. Sci.* 89:1253 (2000).
6. N. G. M. Schipper, S. Olsson, J. A. Hoogstraate, A. G. de Boer, K. M. Varum, P. Artursson. Chitosans as absorption enhancers for poorly absorbable drugs. 2. Mechanism of absorption enhancement. *Pharm. Res.* 14:923 (1997).
7. H. L. Luessen, V. Bohner, D. Perard, P. Langguth, J. C. Verhoef, A. G. de Boer, H. P. Merkle, H. E. Junginger. Mucoadhesive polymers in peroral peptide drug delivery. V. Effect of poly(acrylates) on the enzymatic degradation of peptide drugs by intestinal brush border membrane vesicles. *Int. J. Pharm.* 141:39 (1996).
8. A. Bernkop-Schnürch, Ch. Paikl, C. Valenta. Novel bioadhesive chitosan–EDTA conjugate protects leucine enkephalin from degradation by aminopeptidase N. *Pharm. Res.* 14:917 (1997).
9. I. Matthes, F. Nimmerfall, H. Sucker. Mucusmodelle zur Untersuchung von intestinalen Absorptionsmechanismen. *Pharmazie* 47:505 (1992).
10. A. Bernkop-Schnürch, R. Fragner. Investigations into the diffusion behaviour of polypeptides in native intestinal mucus with regard to their peroral administration. *Pharm. Sciences* 2:361 (1996).
11. B. J. Campbell. Biochemical and functional aspects of mucus and mucin-type glycoproteins. In: *Bioadhesive Drug Delivery Systems*, E. Mathiowitz, D. E. Chickering, III, C.-M. Lehr (Eds.). Marcel Dekker, New York, 1999, p. 85.
12. S. J. Gendler, A. P. Spicer. Epithelial mucin genes. *Ann. Rev. Physiol.* 57:607 (1995).
13. J. R. Gum, J. W. Hicks, N. W. Toribara, E. M. Rothe, R. E. Lagace, Y. S. Kim. The human MUC2 intestinal mucin has cysteine-rich subdomains located both upstream and downstream of its central repetitive region. *J. Biol. Chem.* 267:21375 (1992).
14. J. R. Gum, J. W. Hicks, D. M. Swallow, R. L. Lagace, J. C. Byrd, D. T. A. Lamport, B. Siddiki, Y. S. Kim. Molecular cloning of cDNAs derived from a novel human intestinal mucin gene. *Biochem. Biophys. Res. Commun.* 171:407 (1990).
15. N. Porchet, N. V. Cong, J. Dufosse, J. Audie, V. Guyonnet-Duperat, M. S. Gross, C. Denis, P. Degand, A. Bernheim, J. P. Aubert. Molecular cloning and chromosomal localisation of a novel human tracheo-bronchial mucin cDNA containing tandemly repeated sequences of 48 base pairs. *Biochem. Biophys. Res. Commun.* 175:514 (1991).
16. V. Guyonnet-Duperat, J. P. Audie, V. Debailleul, A. Laine, M. P. Buisine, S. Galiegue-Zouitina, P. Pigny, P. Degand, J. P. Aubert, N. Porchet. Characterisation of the human mucin gene MUC5AC:A consensus cysteine rich domain for 11p15 mucin genes? *Biochem. J.* 305:211 (1995).

17. D. J. Thornton, M. Howard, N. Khan, J. K. Sheehan. Identification of two glycoforms of the MUC5B mucin in human respiratory mucus. Evidence for a cysteine-rich sequence repeated within the molecule. *J. Biol. Chem.* 272:9561 (1997).
18. N. W. Toribara, A. M. Roberton, S. B. Ho, W. L. Kuo, E. T. Gum, J. R. Gum, J. C. Byrd, B. Siddiki, Y. S. Kim. Human gastric mucin: identification of unique species by expression cloning. *J. Biol. Chem.* 268:5879 (1993).
19. L. A. Bobek, H. Tsai, A. R. Biesbrock, M. J. Levine. Molecular cloning, sequence, and specifity of expression of the gene encoding the low molecular weight human salivary mucin (MUC7). *J. Biol. Chem.* 268:20563 (1993).
20. V. Shankar, P. Pichan, R. L. Eddy, V. Tonk, N. Nowak, S. N. Sait, T. B. Shows, R. E. Shultz, G. Gotway, R. C. Elkins, M. S. Gilmore, G. P. Sachdev. Chromosomal localisation of a human mucin gene (MUC8) and cloning of the cDNA corresponding to the carboxy terminus. *Am. J. Respir. Cell Mol. Biol.* 16:232 (1997).
21. C.-M. Lehr, F. G. J. Poelma, H. E. Junginger. An estimate of turnover time of intestinal mucus gel layer in the rat in situ loop. *Int. J. Pharm.* 70:235 (1991).
22. A. Bernkop-Schnürch, S. Steininger. Synthesis and characterisation of mucoadhesive thiolated polymers. *Int. J. Pharm.* 194;239 (2000).
23. A. Bernkop-Schnürch, I. Apprich. Synthesis and evaluation of a modified mucoadhesive polymer protecting from α-chymotrypsinic degradation. *Int. J. Pharm.* 146:247 (1997).
24. E. Jabbari, N. Wisniewski, N. A. Peppas. Evidence of mucoadhesion by chain interpenetration at a poly(acrylic acid)/mucin interface using ATR-FTIR spectroscopy. *J. Controlled Release* 26:99 (1993).
25. C.-M. Lehr, J. A. Bouwstra, F. Spies, J. Onderwater, H. E. Junginger. Visualization studies of the mucoadhesive interface. *J. Controlled Release* 18:249 (1992).
26. S. A. Mortazavi, J. D. Smart. An investigation into the role of water movement and mucus gel dehydratation in mucoadhesion. *J. Controlled Release* 25:197 (1993).
27. C. M. Caramella, S. Rossi, M. C. Bonferoni. A rheological approach to explain the mucoadhesive behavior of polymer hydrogels. In: *Bioadhesive Drug Delivery Systems*, E. Mathiowitz, D. E. Chickering, III, C.-M. Lehr (Eds.) Marcel Dekker, New York, 1999, p. 25.
28. A. Bernkop-Schnürch, C. Humenberger, C. Valenta. Basic studies on bioadhesive delivery systems for peptide and protein drugs. *Int. J. Pharm.* 165:217 (1998).
29. B. A. Hertzog, E. Mathiowitz. Novel magnetic technique to measure bioadhesion. In: *Bioadhesive Drug Delivery Systems*, E. Mathiowitz, D. E. Chickering, III, C.-M. Lehr (Eds.). Marcel Dekker, New York, 1999, p. 147.
30. J. Schneider, M. Tirrell. Direct measurement of molecular-level forces and adhesion in biological systems. In: *Bioadhesive Drug Delivery Systems*, E. Mathiowitz, D. E. Chickering, III, C.-M. Lehr (Eds.). Marcel Dekker, New York, 1999, p. 223.
31. L. S. Nielsen, L. Schubert, J. Hansen. Bioadhesive drug delivery systems. I. Characterisation of mucoadhesive properties of systems based on glyceryl mono-oleate and glyceryl monolinoleate. *Eur. J. Pharm. Sci.* 6:231 (1998).
32. K. V. Rango Rao, P. Buri. A novel in situ method to test polymers and coated microparticles for bioadhesion. *Int. J. Pharm.* 52:265 (1989).
33. M. J. Tobyn, J. R. Johnson, P. W. Dettmar. Factors affecting in vitro gastric mucoadhesion II. Physical properties of polymers. *Eur. J. Pharm. Biopharm.* 42:56 (1996).
34. S. A. Mortazavi, J. D. Smart. An in-vitro evaluation of mucosa-adhesion using tensile and shear stresses. *J. Pharm. Pharmacol.* 45 (suppl.):1108 (1993).
35. H. S. Ch'ng, H. Park, P. Kelly, J. R. Robinson. Bioadhesive polymers as platforms for oral controlled drug delivery. II. Synthesis and evaluation of some swelling, water-insoluble bioadhesive polymers. *J. Pharm. Sci.* 74:399 (1985).
36. D. E. Chickering, III, C. A. Santos, E. Mathiowitz. Adaptation of a microbalance to measure bioadhesive properties of microspheres. In: *Bioadhesive Drug Delivery Systems*, E. Mathiowitz, D. E. Chickering, III, C.-M. Lehr (Eds.). Marcel Dekker, New York, 1999, p. 131.
37. E. E. Hassan, J. M. Gallo. A simple rheological method for the in vitro assessment of mucin-polymer bioadhesive bond strength. *Pharm. Res.* 7:491 (1990).
38. I. W. Kellaway. In vitro test methods for the measurement of mucoadhesion. In: *Bioadhesion Possibilities and Future Trends*, R. Gurny, H. E. Junginger (Eds.). Wissenschaftliche Verlagsgesellschaft, Stuttgart, 1990, p. 86.
39. L. J. Kerr, I. W. Kellaway, C. Rowlands, G. D. Parr. The influence of poly(acrylic) acids on the rheology of glycoprotein gels. *Proc. Int. Symp. Contr. Rel. Bioact. Mat.* pp. 122–123 (1990).
40. M. J. Tobyn, J. R. Johnson, S. A. Gibson. Investigations into the role of hydrogen bonding in the interaction between mucoadhesives and mucin at gastric pH. *J. Pharm. Pharmacol.* 44 (suppl.):1048 (1992).
41. S. A. Mortazavi, B. G. Carpenter, J. D. Smart. An investigation into the nature of mucoadhesive interactions. *J. Pharm. Pharmacol.* 45 (suppl.):1141 (1993).
42. S. A. Mortazavi. In vitro assessment of mucus/mucoadhesive interactions. *Int. J. Pharm.* 124:173 (1995).
43. S. Bouckaert, R. A. Lefebvre, J.-P. Remon. *In vitro/in vivo* Correlation of the bioadhesive properties of a buccal bioadhesive miconazole slow-release tablet. *Pharm. Res.* 10:853 (1993).
44. P. Bottenberg, R. Cleymaet, C. de Muynck, J.-P. Remon, D. Coomans, Y. Michotte, D. Slop. Development and testing of bioadhesive, fluoride-containing slow-release tablets for oral use. *J. Pharm. Pharmacol.* 43:457 (1991).
45. L. Khosla, S. S. Davis. The effect of polycarbophil on the gastric emptying of pellets. *J. Pharm. Pharmacol.* 39:47 (1987).
46. D. Harris, J. T. Fell, H. L. Sharma, D. C. Taylor. GI transit of potential bioadhesive formulations in man: a scintigraphic study. *J. Controlled Release* 12:45 (1990).
47. C. Witschi, R. J. Mrsny. In vitro evaluation of microparticles and polymer gels for use as nasal platforms for protein delivery. *Pharm. Res.* 16:382 (1999).

48. I. V. Evans. Mucilaginous substances from macroalgae: an overview. *Symp. Soc. Exp. Biol.* 43:455 (1989).
49. S. A. Mortazavi, J. D. Smart. An in-vitro method for assessing the duration of mucoadhesion. *J. Controlled Release* 31:207 (1994).
50. M. D. El Hameed, I. W. Kellaway. Preparation and in vitro characterization of mucoadhesive polymeric micropheres as intra-nasal delivery systems. *Eur. J. Pharm. Biopharm.* 44:53 (1997).
51. A. Bernkop-Schnürch, M. E. Krajicek. Mucoadhesive polymers as platforms for peroral peptide delivery and absorption: synthesis and evaluation of different chitosan–EDTA conjugates. *J. Controlled Release* 50:215 (1998).
52. A. Bernkop-Schnürch, J. Freudl. Comparative *in vitro* study of different chitosan-complexing agent conjugates. *Pharmazie* 54:369 (1999).
53. K. Takayama. M. Hirata, Y. Machida, T. Masada, T. Sannan, T. Nagai. Effect of interpolymer complex formation on bioadhesive property and drug release phenomenon of compressed tablets consisting of chitosan and sodium hyaluronate. *Chem. Pharm. Bull* 38:1993 (1990).
54. N. M. Hadler, R. R. Dourmashikin, M. V. Nermut, L. D. Williams. Ultrastructure of hyaluronic acid matrix. *Biochemistry* 79:307 (1982).
55. Y. D. Sanzgiri, E. M. Topp, L. Benedetti, V. J. Stella. Evaluation of mucoadhesive properties of hyaluronic acid benzyl esters. *Int. J. Pharm.* 107:91 (1994).
56. F. Madsen, K. Eberth, J. D. Smart. A rheological assessment of the nature of interactions between mucoadhesive polymers and a homogenised mucus gel. *Biomaterials* 19: 1083 (1998).
57. P. Liu, T. R. Krishnan. Alginate–pectin–poly-L-lysine particulate as a potential controlled release formulation. *J. Pharm. Pharmacol.* 51:141 (1999).
58. J. D. Smart, I. W. Kellaway, H. E. C. Worthington. An in vitro investigation of mucosa-adhesive materials for use in controlled drug delivery. *J. Pharm. Pharmacol* 36:295 (1984).
59. M. A. Longer, H. S. Ch'ng, J. R. Robinson. Bioadhesive polymers as platforms for oral controlled drug delivery. III. Oral delivery of chlorothiazide using a bioadhesive polymer. *J. Pharm. Sci.* 74:406 (1985).
60. C.-M. Lehr, J. A. Bouwstra, E. H. Schacht, H. E. Junginger. In vitro evaluation of mucoadhesive properties of chitosan and some other natural polymers. *Int. J. Pharm.* 78:43 (1992).
61. C. Valenta, B. Christen, A. Bernkop-Schnürch. Chitosan-EDTA conjugate: a novel polymer for topical use gels. *J. Pharm. Pharmacol.* 50:445 (1998).
62. P. Artursson, T. Lindmark, S. S. Davis, L. Illum. Effect of chitosan on the permeability of monolayers of intestinal epithelial cells (Caco-2). *Pharm. Res.* 11:1358 (1994).
63. H. L. Luessen, C.-O. Rentel, A. F. Kotzé, C.-M. Lehr, A. G. de Boer, J. C. Verhoef, H. E. Junginger. Mucoadhesive polymers in peroral peptide drug delivery. IV. Polycarbophil and chitosan are potent enhancers of peptide transport across intestinal mucosae *in vitro*. *J. Controlled Release* 45:15 (1997).
64. C.-M. Lehr. From sticky stuff to sweet receptors—achievements, limits and novel approaches to bioadhesion. *Eur. J. Drug Met. Pharmacok.* 21:139 (1996).
65. M. Rillosi, G. Buckton. Modelling mucoadhesion by use of surface energy terms obtained from the Lewis acid–Lewis base approach. II. Studies on anionic, cationic, and unionisable polymers. *Pharm. Res.* 12:669 (1995).
66. H. L. Luessen, B. J. de Leeuw, M. W. Langemeyer, A. G. de Boer, J. C. Verhoef, H. E. Junginger. Mucoadhesive polymers in peroral peptide drug delivery. VI. Carbomer and chitosan improve the intestinal absorption of the peptide drug buserelin *in vivo*. *Pharm. Res.* 13:1668 (1996).
67. S. Matsuda, H. Iwata, N. Se, Y. Ikada. Bioadhesion of gelatin films crosslinked with glutaraldehyde. *J. Biomed. Mater. Res.* 45:20 (1999).
68. A. Bernkop-Schnürch, V. Schwarz, S. Steininger. Polymers with thiol groups: a new generation of mucoadhesive polymers? *Pharm. Res.* 16:876 (1999).
69. G. H. Snyder. Intramolecular disulfide loop formation in a peptide containing two cysteines. *Biochemistry* 26:688 (1987).
70. A. Bernkop-Schnürch, S. Scholler, R. G. Biebel. Development of controlled drug release systems based on polymer–cysteine conjugates. *J. Controlled Release* 66:39 (2000).
71. A. Bernkop-Schnürch, U.-M. Brandt, A. Clausen. Synthese und *in vitro* Evaluierung von Chitosan–Cystein Konjugaten. *Sci. Pharm.* 67:197 (1999).
72. C. E. Kast, J. Freudl, A. Bernkop-Schnürch. Mucoadhesive thiolated polymers: synthesis and in vitro evaluation of chitosan-thioglycolic acid conjugates. *Proc. Int. Symp. Contr. Rel. Bioact. Mat.* 27:1222 (2000).
73. A. Bernkop-Schnürch, S. Thaler. Polycarbophil–cysteine conjugates as platforms for peroral (poly)peptide delivery systems. *J. Pharm. Sci.* 89:901 (2000).
74. H. L. Luessen, J. C. Verhoef, G. Borchard, C.-M. Lehr, A. G. de Boer, H. E. Junginger. Mucoadhesive polymers in peroral peptide drug delivery. II. Carbomer and polycarbophil are potent inhibitors of the intestinal proteolytic enzyme trypsin. *Pharm. Res.* 12:1293 (1995).
75. G. F. Walker, R. Ledger, I. G. Tucker. Carbomer inhibits tryptic proteolysis of luteinizing hormone-releasing hormone and N-α-benzoyl-L-arginine ethyl ester by binding the enzyme. *Pharm. Res.* 16:1074 (1999).
76. A. Bernkop-Schnürch, G. Schwarz, M. Kratzel. Modified mucoadhesive polymers for the peroral administration of mainly elastase degradable therapeutic (poly)peptides. *J. Controlled Release* 47:113 (1997).
77. A. Bernkop-Schnürch, K. Dundalek. Novel bioadhesive drug delivery system protecting (poly)peptides from gastric enzymatic degradation. *Int. J. Pharm.* 138:75 (1996).
78. A. Bernkop-Schnürch, I. Bratengeyer, C. Valenta. Development and in vitro evaluation of a drug delivery system protecting from trypsinic degradation. *Int. J. Pharm.* 157: 17 (1997).

79. A. Bernkop-Schnürch, A. Scerbe-Saiko. Synthesis and *in vitro* evaluation of chitosan–EDTA–protease-inhibitor conjugates which might be useful in oral delivery of peptides and proteins. *Pharm. Res.* 15:263 (1998).
80. A. Bernkop-Schnürch, M. Pasta. Intestinal peptide and protein delivery: novel bioadhesive drug carrier matrix shielding from enzymatic attack. *J. Pharm. Sci.* 87:430 (1998).
81. A. Bernkop-Schnürch, N. C. Göckel. Development and analysis of a polymer protecting from luminal enzymatic degradation caused by α-chymotrypsin. *Drug Dev. Ind. Pharm.* 23:733 (1997).
82. A. Bernkop-Schnürch, M. K. Marschütz. Development and *in vitro* evaluation of systems to protect peptide drugs from aminopeptidase N. *Pharm. Res.* 14:181 (1997).
83. M. K. Marschütz, P. Caliceti, A. E. Clausen, A. Bernkop-Schnürch. Design and in vivo evaluation of an oral delivery system for insulin. *Proc. Int. Symp. Contr. Rel. Bioact. Mat.* 27:1228 (2000).
84. G. Borchard, H. L. Luessen, J. C. Verhoef, C.-M. Lehr, A. G. de Boer, H. E. Junginger. The potential of mucoadhesive polymers in enhancing intestinal peptide drug absorption. III. Effects of chitosan–glutamate and carbomer on epithelial tight junctions *in vitro*. *J. Controlled Release* 39:131 (1996).
85. V. H. L. Lee. Protease inhibitors and permeation enhancers as approaches to modify peptide absorption. *J. Controlled Release* 13:213 (1990).
86. B. J. Aungst, H. Saitoh, D. L. Burcham, S. M. Huang, S. A. Mousa, M. A. Hussain. Enhancement of the intestinal absorption of peptides and non-peptides. *J. Controlled Release* 41:19 (1996).
87. L. Illum, N. F. Farraj, S. S. Davis. Chitosan as a novel nasal delivery system for peptide drugs. *Pharm. Res.* 11:1186 (1994).
88. M. Tomita, M. Hayashi, S. Awazu. Comparison of absorption-enhancing effect between sodium caprate and disodium ethylenediaminetetraacetate in Caco-2 cells. *Biol. Pharm. Bull.* 17:753 (1994).
89. M. S. Balda, L. Gonzalez-Mariscal, R. G. Contreras, M. Macias-Silva, M. E. Torres-Marquez, J. A. Gracia-Sainz, M. Cereijido. Assembly and sealing of tight junctions: possible participation of G-proteins, phospholipase C, protein kinase C and calmodulin. *J. Membr. Biol.* 122:193 (1991).
90. N. G. M. Schipper, S. Olsson, J. A. Hoogstraate, A. G. de Boer, K. M. Varum, P. Artursson. Chitosans as absorption enhancers for poorly absorbable drugs. 2. Mechanism of absorption enhancement. *Pharm. Res.* 14:923 (1997).
91. R. K. Rao, R. D. Baker, S. S. Baker, A. Gupta, M. Holycross. Oxidant-induced disruption of intestinal epithelial barrier function: role of tyrosine phosphorylation. *Am. J. Physiol.* 273:G812–G823 (1997).
92. W. C. Barret, J. P. DeGnore, S. Konig, H. M. Fales, Y. F. Keng, Y. Zhang, M. B. Yim, P. B. Chock. Regulation of PTP1B via glutathionylation of the active site cysteine 215. *Biochemistry* 38:6699 (1999).
93. A. Bernkop-Schnürch, B. Gilge. Anionic mucoadhesive polymers as auxiliary agents for the peroral administration of (poly)peptide drugs: influence of the gastric fluid. *Drug Dev. Ind. Pharm.* 26:107 (2000).
94. O. P. Skov, S. S. Poulsen, P. Kirkegaard, E. Nexo. Role of submandibular saliva and epidermal growth factor in gastric cytoprotection. *Gastroenterology* 87:103 (1984).
95. M. Itoh, Y. Matsuo. Gastric ulcer treatment with intravenous human epidermal growth factor: a double-blind controlled clinical study. *J. Gastroen. Hepatol.* 9:78 (1994).
96. B. L. Slomiany, H. Nishikawa, J. Bilski, A. Slomiany. Colloidal bismuth subcitrate inhibits peptic degradation of gastric mucus and epidermal growth factor *in vitro*. *Am. J. Gastroenterol.* 85:390 (1990).
97. A. Bernkop-Schnürch, H. Schuhbauer, I. Pischel. (1999). α-Liponsäure(-Derivate) enthaltende Retardform. Deutsche Patentschrift 1999-09-30.

8

Polymers for Tissue Engineering Scaffolds

Howard W. T. Matthew
Wayne State University, Detroit, Michigan

I. INTRODUCTION

A. Tissue Engineering and the Concept of the Scaffold

In general terms, the goal of tissue engineering is to develop materials and approaches which can be used to facilitate repair, regeneration, or replacement of damaged or diseased tissues. In some applications, reconstitution of tissue function outside the body is the desired goal. In the repair/regeneration scenario, the goal is to enhance or improve upon natural repair processes which often produce nonfunctional or poorly functional scar in place of normal tissue. In the replacement scenario, the tissue ''foundation'' may be assembled in vitro and subsequently implanted. This approach may employ a cellular component together with an appropriately shaped structural template or scaffold. Alternatively, a material scaffold may be implanted directly into the desired tissue site and colonized by the target cells. Most tissues are not merely collections of randomly arranged cells, but possess highly detailed organizational features which are closely tied to tissue function. The tissue engineer typically seeks to restore this tissue-specific architecture, and the use of a scaffold provides the means to this end. In order to achieve restoration of tissue architecture, the tissue scaffold may be required to perform a variety of tasks. A porous microstructure which allows cellular ingrowth and scaffold colonization is almost a universal requirement. Enhancements to the microstructure may include spatial variations in pore morphology to help orient cells or variations in material surface properties to facilitate selective cell adhesion and/or migration. In contrast, inhibition of cell adhesion may be an important performance characteristic for certain locations. Similarly, the scaffold material must be either overtly biodegradable or at least amenable to long-term integration with the host tissue. Biodegradability allows the gradual and orderly replacement of the scaffold with functional tissue, and also prevents the development of adverse chronic responses to the artificial structure. The scaffold material may be required to deliver biologically active agents (for example, growth factors or genes) to the target tissue. The material itself may be required to possess intrinsic biological activity to elicit specific migration or proliferation responses in adjacent cell populations. To this list of cell interaction characteristics may be added certain physical properties such as precise matching of the following: tissue mechanical properties, macromolecular permeability, protein binding or repulsion, tissue adhesion or lubricity, and ease of processing. Since restoration of normal tissue architecture and function is the ultimate goal, the scaffold is usually considered to be a temporary structure. Thus the ability to tune degradation rates to achieve an optimal degradation time profile may greatly broaden the applicability of a particular material.

Given the variety of target tissues, possible scaffold microstructures, and technical approaches, it is unlikely that a single polymeric material can meet the requirements of all systems. In keeping with this idea, a variety of polymers have been and continue to be evaluated for tissue engineering applications, and new or modified materials are constantly under development. In this chapter, the chemistry and properties of some of the more widely used or promising biodegradable materials are surveyed.

B. Polymers for Tissue Engineering

At the most fundamental level, biodegradable polymers used in engineered implants can be broadly classified as either natural or synthetic. In general terms there are advantages and disadvantages associated with both types of materials. For example, natural polymers provide the clear advantage of possessing specific molecular recognition features which generally translate into specific biological activities. They are usually enzymatically degradable to nontoxic subunits which can be utilized by adjacent cells. Furthermore, these materials are amenable to both cell-mediated remodeling and direct integration into neotissue. The disadvantages of such natural materials include source-associated variability and contamination; limited control over parameters such as molecular weight; the potential for adverse immunological responses; variations in degradation rates due to differences in host enzyme levels; and, in some cases, inferior mechanical properties. In contrast, synthetic polymers can be synthesized in pure form with complete control over molecular and physical properties. Degradation usually occurs via passive hydrolysis, thus ensuring uniform in vivo degradation rates from one host to another. These polymers also usually permit a wider range of processing options. However, they generally lack intrinsic biological activity. Introduction of bioactivity, by side chain grafting, for example, may also be limited by molecular properties or incompatibility. The degradation products may be toxic or may drastically alter the local microenvironment (e.g., pH). Surface hydrophobicity may also mediate protein denaturation in the vicinity of the implant, and may thus play a role in triggering fibrous encapsulation of the implant.

II. COLLAGENS

A. General Overview

The collagens are a family of fibrous proteins that occur in almost all mammalian tissues. They are particularly abundant in load-bearing tissues such as tendon and bone. Skin is also rich in this protein. The structure, sequence, and biology of the many types of collagen, as well as its use in tissue repair, have been extensively reviewed elsewhere (1–3). Because of the relative ease of extraction and its abundance, collagens in both their native and denatured forms have had a long history of use as implantable materials. Its use has been facilitated by the fact that its sequences are highly conserved across species boundaries. Thus, collagens from xenogeneic sources (e.g., bovine collagen) generally produce only mild immune response. To date, 14 types of collagen have been identified. The form that has been most investigated for tissue scaffold purposes is Type I collagen, which is also the most abundant form.

In its native form, collagen exists as three similar polypeptide chains arranged into a triple helical structure. Each polypeptide chain has the general sequence (glycine–X–Y)$_n$, where X and Y can be any amino acid but are often proline and hydroxyproline, respectively. Hydrogen bonding by these two amino acids serves to stabilize the helix. As synthesized, the ends of the polypeptide chains (i.e., the N and C termini) deviate from the three–amino acid repeat structure and therefore are not helical. These regions are up to 26 amino acid residues in length and are termed telopeptides. The triple helical collagen molecule is the main functional subunit, and has the unique ability to self-assemble into fibrils of extraordinary strength. The fibrils are further strengthened by the formation of intermolecular crosslinks between adjacent telopeptide regions. This crosslinking generally takes place as a postsecretion, enzymatic process in the extracellular space. Nonenzymatic crosslinking processes such as glycation (4–6) also contribute to postsecretion fibril strengthening. Fibrils may be further assembled into bundles to form collagen fibers, which are the main tensile load-bearing component in non-ligamentous tissues.

Collagen (Type I) is typically isolated by either enzymatic digestion or salt or acid extraction from collagenous animal tissues. Enzymatic digestion (e.g., via pepsin) cleaves the crosslinked telopeptide regions, thus producing soluble triple helices lacking the telopeptides; the so-called atelocollagen. Salt or acid extraction disrupts intra- and intermolecular hydrogen bonds, leading to swelling and subsequent dissolution of the structure. Reversion to a physiological environment restores conditions suitable for collagen fibril self-assembly. Hence an acidic solution of collagen can be induced to form fibrils in vitro by imposing conditions of physiological pH, temperature, and ionic strength. The formation of a fibril network is usually manifested by gelation of the solution.

B. Biological Properties

Since collagen serves as a structural component in almost all tissues and is present to some degree in all basement membranes, it stands to reason that most anchorage-dependent cells express receptors for one or more epitopes of the molecule. As a result, collagen gels and fibers are excellent substrates for cell attachment and migration. Numerous experimental studies have used collagen as a culture surface for in vitro maintenance of a wide variety of primary cell types (3,7–11). Cell attachment is usually mediated by integrins which recognize specific amino acid motifs of the molecule, such as the Arg-Gly-Asp (RGD) sequence. Attachment to this natural substrate and the associated transmembrane signalling events have been associated with improved cell metabolism, proliferation, and longevity, as well as maintenance of superior differentiated function. In particular, it has been demonstrated that embedding some highly differentiated cell types within a collagen gel allows the cells to maintain their differentiated phenotype, including high levels of metabolic function, for long durations in culture (12,13).

Collagen has long been recognized as a hemostatic agent, i.e., a material which stimulates blood coagulation. As such, it has enjoyed a long history of use during surgical procedures as a tool to stop hemorrhaging from injured tissue. Blood clotting is stimulated by platelet binding to collagen via a number of receptors (14–16). This is quickly followed by platelet activation and degranulation events leading to the classical coagulation pathway. The provisional matrix laid down during the first stages of wound healing consists mainly of crosslinked fibrin clot. As healing progresses this initial matrix is replaced with a more complex system using collagen fibers as the main structural element. Given this natural progression it is possible that tissue repair processes could be accelerated by intervening to install an advanced collagen-based tissue scaffold. This fundamental idea led to the use of collagen as tissue repair material in the form of sutures and sponges and as a biocompatible coating on nondegradable implants such as vascular prostheses.

C. Collagen Scaffolds

For tissue scaffold applications, collagen has been utilized in several forms, most notably, sponges, woven and nonwoven meshes, gels, decellularized tissue matrices, and porous composites. Porous collagen matrices have been prepared as both sponges and as woven and nonwoven (feltlike) fabric structures. These constructs have been applied both as classical tissue scaffolds for delivering cultured cells and generating new tissue in vivo, and as materials for physically repairing traumatized tissue (1,8,17–20). The porous structure provides a three-dimensional framework for developing a multilayered tissue from either a seeded single cell suspension or migrating cells from adjacent tissue, while simultaneously providing initial organization and mechanical integrity. Collagen gels have been utilized mainly as cell delivery and localization agents in cases where initial mechanical functionality is not a major concern. Because of its thermally induced gelation characteristics, this collagen form has the added advantage of conforming to fill available tissue spaces. As a result, gels can be used to engineer space-filling connective tissue structures without the need for prefabricated scaffold shapes. The high water content and low density of a typical collagen gel also facilitates cellular migration and matrix remodeling and reorganization. This type of remodeling is mainly accomplished by invading (or seeded) fibroblasts and may be associated with extensive contraction of the gel, resulting in significant shape and volume alterations postimplantation.

While collagenous tissues tend to have high strength, the lack of long-range fibril organization in synthetic collagen constructs usually results in these materials (in particular, gel-based systems) having fairly low mechanical strength. However tensile strength can be augmented through chemical crosslinking. For example, glutaraldehyde has long been used to strengthen collagen-based materials used in surgery (21–28). In particular, the nonenzymatic crosslinking mechanism of glycation has been successfully used to strengthen cell seeded, collagen based scaffolds without compromising cell viability or function (29). Glycation is a complex crosslinking reaction which occurs mainly in collagen-rich tissues and is mediated by reducing sugars (4–6,30,31).

Decellularized tissue matrices are prepared by a combination of enzymatic and detergent treatments of collagen-rich tissues such as porcine heart valves or small intestinal submucosa (SIS). The treatment removes cellular material as well as most of the potentially antigenic noncollagenous proteins. The remaining material is composed mainly of collagen and elastin with some associated extracellular matrix polysaccharides. The mechanical properties and degradation kinetics of this material can be subsequently modified by chemical crosslinking with an agent such as glutaraldehyde. Constructs of SIS have been studied for the repair of heart valves, blood vessels, and intestinal tracts (23,32–37).

Collagen in vivo is often found in association with extracellular matrix polysaccharides, termed glycosaminoglycans (GAGs). This association is mediated by both electrostatic and covalent linkages and is most notable in basement membranes and cartilage. In fact soluble collagens are known to form ionic complexes with soluble GAGs under appropriate conditions. This interaction has been used to generate collagen–GAG composite materials as tissue scaffolds. Porous microstructures were generated by controlled freezing of the hydrated complexes followed by lyophilization (38–41). The constructs were then stabilized by chemical crosslinking. The resultant scaffolds have been successfully applied as skin repair/regeneration aids and are being studied for cartilage repair and as conduits for nerve regeneration (18,42,43).

III. POLYSACCHARIDES

A. General Overview

While polysaccharides have enjoyed popularity for certain specialty applications, they have generally been underutilized in the biomaterials field. Recognition of the potential utility of this class of materials is growing however, and the field of polysaccharide biomaterials is poised to experience rapid growth. Four factors have specifically contributed to this growing recognition of polysaccharide potential. First is the longstanding recognition that the oligosaccharide components of most glycoproteins are absolutely essential for normal function of these molecules. This has been coupled with a large and growing body of information pointing to the critical role of saccharide moieties in numerous cell-signaling schemes and in the area of immune modulation in particular. Second has been the realization that the so-called extracellular matrix polysaccharides play critical roles in modulating the activities of signaling molecules as well as mediating certain intercellular signaling directly. Third is the recent development of powerful new synthetic techniques with the potential for automated synthesis of biologically active oligosaccharides. These techniques may allow researchers to finally decode and exploit the ''language'' of oligosaccharide signaling. The fourth factor is the explosion in tissue engineering research and the associated need for new materials with specific, controllable biological activity and biodegradability.

Polysaccharides in general share a number of features which make them as a class particularly desirable starting materials for a number of biomaterials applications. The numerous hydroxyl groups present on the typical molecule, in addition to providing the means for the stereospecific biological activity, also provide numerous sites for the attachment of side groups. Such groups can provide specific functionality and biological recognition features, or may serve simply to modulate mechanical or biological properties of the parent molecule. Polysaccharides are also hydrophilic, and many have the potential to be processed as hydrogels of various densities as well as denser structures. This feature has made some members of the family highly sought after as drug delivery matrices and components of drug formulations. The charged members of the polysaccharide family possess additional features which may facilitate the development of useful materials. Such materials may be held together by ionic interactions. The ionic functional groups also play a significant role in some of the more useful known biological activities involving charge-based binding of proteins ranging from growth factors to matrix proteins such as collagens.

With few exceptions, natural polysaccharides and their constituent monomers and degradation products can generally be considered to be either nontoxic or of low toxicity. Furthermore, while all polysaccharides can be considered biodegradable, mammals do not possess the enzymatic machinery to rapidly degrade many of the structural plant and marine varieties. However, most are susceptible to poorly understood, slow enzyme-mediated cleavage processes as well as free radical–based chain scission.

Polysaccharides can be classified into several groups in a number of different ways. This review will focus on polysaccharides of major significance in the biomaterials arena—in particular, marine polysaccharides of algal and animal origin and extracellular matrix polysaccharides of mammalian origin. It should be noted here that certain polysaccharides are synthesized by more than one type of organism, thus alginates are made by both algae and bacteria, and chitin/chitosan can be obtained from both arthropods and fungi.

Polysaccharides are polymers of five-carbon (pentose) or six-carbon (hexose) sugar molecules. As such, the term glycopolymers is often used to describe these substances. The polymers may be linear or branched depending on the polysaccharide, and linkage between subunits occurs via a condensation reaction between hydroxyl groups, resulting in the formation of an etherlike glycosidic bond.

Most nonsubstituted monosaccharides possess four or five hydroxyl groups capable of forming glycosidic linkages. Furthermore, each hydroxyl group can exhibit two optical isomeric configurations, leading to a large number of theoretically possible ways in which two different sugars can be linked to form a disaccharide. This feature, together with interactions mediated by charged substituent groups, leads to a range of chain configurations found in various polysaccharides. These configurations range from rigid rod conformation to flexible coil to random coil to globular. As expected, the chain configuration plays a major role in defining the physical properties of polysaccharide solutions, gels, and solids. The abundant hydroxyl groups in these molecules are capable of forming intermolecular hydrogen bonds and are the main contributor to both the solubility of many of these substances and their ability to form gels. It is the gel-forming capabilities that drew initial attention to this class of compounds for biotechnology applications. More recently, the cellular recognition features and the attendant biological activity have sparked renewed interest in their therapeutic and biomaterials potential.

Polysaccharide gel formation is generally of two types.

Hydrogen bonded gels are typical of molecules such as agarose and neutral chitosan, whereas ionically bonded gels are characteristic of alginates and carrageenans. The distinction is, however, limited since many charged polysaccharides will exhibit hydrogen bonded gel formation under neutralizing conditions.

The polar nature of the hydroxyl groups together with the presence of ionizable functional groups in some species means that polysaccharides are intrinsically hydrophilic and most are soluble in aqueous media. The exceptions to this rule tend to be species which form ordered structures with a high degree of crystallinity in the solid state. These exceptions include the crystalline polysaccharides chitin and cellulose. The conversion of these insoluble native forms to soluble derivatives generally involves the addition of charged or bulky side groups which disrupt the crystal structures and thus facilitate solvent penetration. Because of their high molecular weights, extended chain configurations, and the high degree of interchain hydrogen bonding potential, most polysaccharides form highly viscous solutions. This property ultimately limits the solution concentrations attainable in practice and also makes these molecules ideal for viscosity enhancement and gel formation.

The growing interest in polysaccharide-based drugs and biomaterials is partly fueled by the recognition that many biological recognition events at both the cellular and molecular levels are mediated or at least modulated by polysaccharide or oligosaccharide moieties. As a result, these molecular features and interactions form attractive targets for a variety of therapeutic interventions and also raise the possibility of developing polysaccharide biomaterials with a variety of useful biological activities. Within the tissue engineering paradigm there is a constant search for new materials which are biodegradable and which also possess targeted biological activity. In the sections which follow we describe the features and potential of some of the more promising polysaccharide biomaterials.

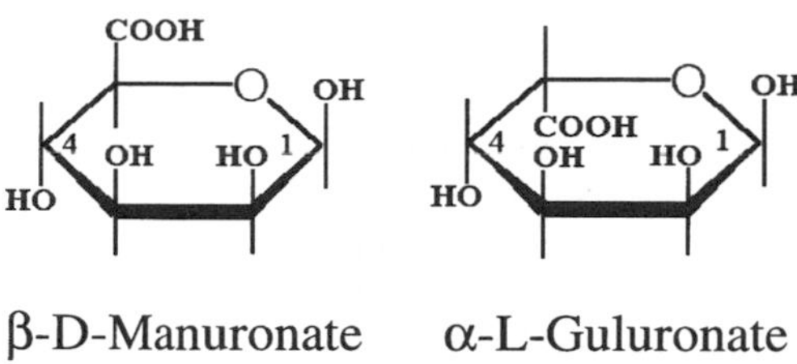

Figure 8.1 Alginate monosaccharide components.

B. Alginates

1. Chemical and Physical Properties

Alginic acids, or alginates, are isolated from several species of brown algae (e.g., *Macrocystis pyrifera*). Alginates are composed of two monomers, α-L-guluronic acid (G) and β-D-mannuronic acid (M) (Fig. 1). The polymer is structured as a block copolymer with blocks of G and M alternating. Within the blocks, residues are linked by α(1 → 4) and β(1 → 4) glycosidic bonds, respectively. In some regions single M and G units may alternate, giving rise to MG blocks. Within these MG blocks, the two bond types also alternate. Sizes of the three blocks can vary over a wide range, giving rise to alginates of different properties. Most notably, gels of alginate richer in the G blocks have a higher elastic modulus and also show higher solute diffusivities. These observations suggests a more open and more ordered polymer network for G-rich alginate gels. Average molecular weights range from 200 to 500 kDa.

Chain flexibility is a function of the block sizes and the relative quantities of the three block types. G blocks have the highest stiffness and MG blocks the lowest. Industrially, different alginates are often referred to by their M:G ratio. Thus, alginates with low M:G values (i.e., high G content) have chains which tend more toward the rigid rod conformation, while high M:G material tends to the flexible coil type conformation.

Within the biomaterials sphere, alginates have been extensively investigated for their gelation capabilities. The molecules have an extraordinarily high exclusion volume, which is evidenced by their high intrinsic viscosity. Alginates gel in the presence of divalent cations. The most widely used agents are calcium ions. Upon addition of Ca^{2+} to an alginate solution, the chains undergo rearrangement to form the so-called egg-box structures (Fig. 2), where groups of chain segments align with several intercalating Ca^{2+} ions. These structures are formed by the G blocks and serve to ionically crosslink the molecules in solution. As can be expected, the greater the G block con-

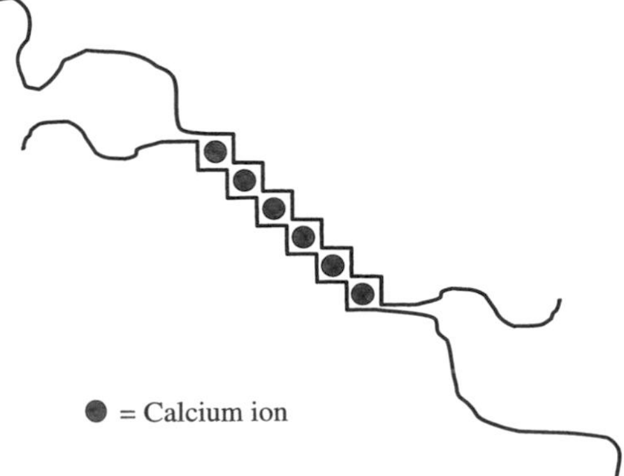

Figure 8.2 Alginate ‘‘egg box’’ structure.

tent of an alginate, the more crosslinks are formed and the stronger and stiffer the gel.

The excellent gellation ability has led to the use of alginates for entrapment of cells within gel beads. Cells so entrapped maintain viability and metabolic function and are capable of secreting a variety of protein products into the external medium. This scenario can be and has been used for the transplantation of cells for restoration of metabolic function. The strength of the techniques arises from the fact that xenogeneic entrapped cells can be protected from immune attack by controlling the density and permeability of the gel. Permeability parameters can be set to exclude immune cells and antibodies while allowing both nutrient access and the outward diffusion of smaller, desirable cell products. In an enhancement of the technique, the outermost layers of alginate in the beads are ionically crosslinked by reacting with a polycationic species such as poly-L-lysine (PLL) (44,45). The ionic complexes formed are stable over a wide range of pH and ionic strength conditions and are also calcium independent. This then allows the use of a calcium chelating agent such as EDTA or citrate to dissolve the inner gel regions of the bead, thus producing a hollow capsule enclosed in an ionic complex membrane. The complex formation provides an additional degree of freedom with regard to setting and controlling both the permeability and the strength of the capsule membrane. These capsules have been studied for use in a wide variety of cell transplantation and bioreactor design scenarios (44–46).

Alginates have been extensively researched for the encapsulation of pancreatic islet cells for controlled release of insulin to treat Type 1 diabetes. The research in this area has focused on modifying the properties of the semipermeable membrane coating the alginate capsule in an effort to reduce pericapsular fibrosis and to control the diffusion rate of nutrients to the islets and insulin from the islets. This application of alginates has been extensively reviewed (46–48). Encapsulation is conducted by suspending the islets in a solution of sodium alginate which is sprayed into a container of calcium chloride solution using a droplet-forming device. The droplets are then surface reacted with a poly-L-lysine solution to form a semipermeable complex membrane, after which the internal gel is dissolved by treatment with a calcium chelating solution. The alginate–PLL membrane allows small molecules such as insulin and glucose to pass into and out of the capsule, but presents a barrier to immune cells and large molecules such as immunoglobulins.

The mechanical properties of calcium alginate gels vary greatly with the composition of the alginate. The highest mechanical strengths are achieved using alginate with a G content higher than 70% and an average G block length of approximately 15 residues. For alginate molecular weights above some critical value, the mechanical strength of gels is determined mainly by chemical compositions and block structure and is independent of molecular weight. For low molecular weight alginates, however, molecular weight is still a major determinant of gel strength. The critical molecular weight depends on the concentration of alginate (48).

2. Drug Delivery Potential and Bioactivity

The ability to control alginate gel permeability means that these materials have significant potential for controlled release applications. Ionically crosslinked alginate gels have been used to incorporate a variety of proteins for controlled release. These have included TGF-β_1, the TNF receptor, and angiogenic molecules such as EGF and bFGF. Because alginate is an anionic polymer at physiological conditions (pH 7.4), proteins with a net positive charge can ionically bond to the polymer and thus exhibit reduced bioactivity upon incorporation into an alginate delivery system. TGF-β_1, with a pI of 9.8, is one such example (49). The porosity of alginate gels can be controlled by varying alginate concentration, M:G ratio, and calcium concentration. In this way the release kinetics of entrapped solutes can be controlled (50–53).

In general, alginates exhibit minimal intrinsic biological activity and therefore are employed mainly as inert matrices for delivery of cells or therapeutic agents. However, as with most polysaccharides, the hydroxyl and carboxyl groups provide attractive derivatization sites through which specific bioactive groups can be immobilized. Thus, bioactive peptides such as the fibronectin-derived adhesion sequence RGD have been covalently attached to alginate gels to provide cell adhesion properties (54).

3. Immunology and the Tissue Response

The biocompatibility of alginate materials is also dependent on the composition of the alginates. In vivo studies with empty alginate–PLL capsules have shown that the cellular overgrowth on capsules implanted intraperitoneally was composed of fibroblasts in cases of alginate with intermediate G content. But the overgrowth was mainly composed of macrophages when high-G alginate was used. There was no evidence of other immune cell elements such as B or T lymphocytes (55,56). In other studies, purified alginates with a high M content (68%) showed no fibrotic reaction after 3 weeks of implantation (57).

One of the issues facing the use of alginates and other polysaccharide biomaterials in vivo is the presence of impurities in commercial alginate preparations. These impurities may include contaminating algal or bacterial proteins, endotoxin, and other trace polysaccharides. It is quite possible that the biocompatibility difference between high and low guluronate alginates may in fact be partly due to different affinities for the contaminants exhibited by M and G blocks. Such contaminants have been implicated in the severity of the tissue response to alginate implants, fibrotic overgrowth, and general failure of the transplanted cells within the implant. It has been suggested that some of the contaminants are mitogenic in nature and may be directly responsible for fibrotic tissue overgrowth (58). Alternatively, an inflammatory response to contaminating elements would result in the local release of endogenous mitogens, again leading to cellular overgrowth of the implants. These issues have been addressed by the introduction of purification strategies involving the separation of alginates from more soluble contaminants using multistep precipitation and extraction procedures (56,59). The purified material was shown to elicit a minimal fibrotic response in vivo.

C. Chitosan

1. Chemical and Physical Properties

Chitosans are partially or fully deacetylated derivatives of chitin, the primary structural polymer in arthropod exoskeletons. From the biomaterials standpoint, chitosans are the most promising polysaccharides, with great potential for development of resorbable, biologically active, implant materials (60,61). The primary source for chitin/chitosan is shells from crab, shrimp, and lobster. Shells are ground, demineralized with HCl, deproteinized with a protease or dilute NaOH, and then deacetylated with concentrated NaOH. Structurally, chitosans are very similar to cellulose. The polymer is linear, consisting of $\beta(1 \rightarrow 4)$ linked D-glucosamine residues with a variable number of randomly located N-acetyl-glucosamine groups (Fig. 3). In essence, chitosan is cellulose with the 2-hydroxyl group replaced by an amino or acetylated amino group. Depending on the source and preparation procedure, molecular weight may range from 50 to 1000 kDa. Commercially available preparations have degrees of deacetylation ranging from 70 to 90%. As an additional method of property control, chitosan

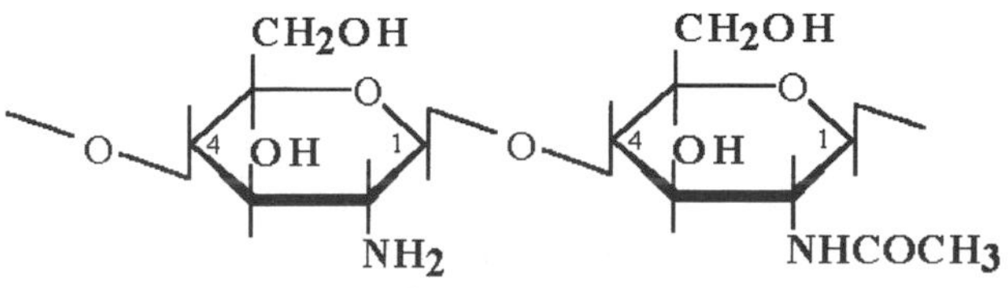

Figure 8.3 Structure of chitosan. One acetylated residue is shown linked to a deacetylated residue.

can be stoichiometrically depolymerized by reacting in solution with nitrous acid (62).

Chitosan is a crystalline polysaccharide and the degree of crystallinity is a function of the degree of deacetylation. Crystallinity is maximum for both chitin (i.e., 0% deacetylated) and 100% deacetylated chitosan. Minimum crystallinity is achieved at intermediate degrees of deacetylation. Because of the stable, crystalline structure, chitosan is normally insoluble in aqueous solutions above pH 7. However, in dilute acids, the free amino groups are protonated and the molecule becomes fully soluble below ~pH 5. The high charge density causes the linear chitosan chain to extend into a semirigid rod conformation. Upon raising the pH, amino groups are increasingly deprotonated and become available for hydrogen bonding. At some critical pH, the molecules in solution develop enough hydrogen bonds to establish a gel network. The pH at which this transition occurs is dependent upon both the degree of deacetylation and the average molecular weight. For a typical 86% deacetylated commercial chitosan with an average molecular weight of 300 kDa, gellation occurs at about pH 6. As the pH is raised further, deprotonation continues and the molecules establish additional hydrogen bonds, ultimately rearranging to establish miniature crystalline domains. This effect results in an increase in gel stiffness and can be associated with minor gel contraction. Chitosan's pH-dependent solubility provides a convenient mechanism for processing under mild conditions. Viscous solutions can be extruded and gelled in high-pH solutions or baths of nonsolvents such as methanol. Such gel fibers can be subsequently drawn and dried to form high strength fibers. The polymer has been extensively studied for industrial applications based on film and fiber formation, and the preparation and mechanical properties of these forms have been reviewed previously (63,64). Woven and nonwoven fiber-based structures and materials have potential for applications as implantable biomaterials and have been investigated to a limited extent (65).

2. Interactions with Biomolecules and Cells

Much of chitosan's biomaterial potential stems from its cationic nature and high charge density in solution. The charge density allows chitosan to form insoluble ionic complexes with a wide variety of water-soluble anionic polymers. Complex formation has been documented with anionic polysaccharides such as heparin and alginates, as well as synthetic polyanions like poly(acrylic acid) (66–69). Because the chitosan charge density is pH dependent, transfer of these complexes to physiological pH results in dissociation of a portion of the immobilized polyanion. This property can be used as a technique for local delivery of biologically active polyanions such as heparin and other glycosaminoglycans or even DNA. Heparin release from ionic complexes may enhance the effectiveness of growth factors released by inflammatory cells in the vicinity of an implant (70). In the case of DNA, complexation with chitosan has been shown to protect plasmids from degradation by nucleases and also facilitates cellular transfection by poorly understood interactions with cell membranes (71–73). Chitosan's positive charge also mediates nonspecific binding interactions with a variety of proteins. Since the majority of soluble proteins carry a negative charge at physiological pH, many of these molecules can be expected to bind with varying affinities to chitosan-based materials. The N-acetyl-glucosamine moiety in chitosan is also a structural feature found in the extracellular matrix polysaccharides termed glycosaminoglycans. These molecules are discussed later, but their properties include many specific interactions with growth factor receptors and adhesion proteins. This suggests that the analogous structure in chitosan may also have related bioactivity. In fact, chitosan oligosaccharides have been shown to have a stimulatory effect on macrophages, and the effect has been linked to the acetylated residues (74). Furthermore, both chitosan and its parent molecule chitin have been shown to exert chemoattractive effects on neutrophils in vitro and in vivo (75–78).

Chitosan is hydrolyzed enzymatically in vivo. The primary agent is lysozyme, which appears to target acetylated residues (79). The degradation products are chitosan oligosaccharides of variable length. The degradation kinetics appear to be inversely related to the degree of crystallinity. Crystallinity in the chitin–chitosan family of materials exhibits a minimum at intermediate levels of deacetylation, with chitin and fully deacetylated chitosan showing the highest levels of crystallinity. In fact, the low degradation rate of highly deacetylated chitosan implants (several months in vivo) is believed to be due to the inability of hydrolytic enzymes to penetrate the crystalline microstructure. This issue has been addressed by derivatizing the molecule with side chains of various types (80–84). Such treatments alter molecular chain packing and increase the amorphous fraction, thus allowing more rapid degradation. They also inherently affect both the mechanical and solubility properties.

Chitosan has been widely studied for applications in systems for the controlled release of various drugs and therapeutic agents. These applications have been extensively reviewed by others (49,64,85).

3. Immunology and the Tissue Response

A number of researchers have examined the tissue response to a variety of chitosan-based implants (60,61,69, 83,84,86–95). In general, these materials have been found to evoke a minimal foreign body reaction. In most cases, no major fibrous encapsulation developed. Formation of normal granulation tissue, often with accelerated angiogenesis, appears to be the typical course of healing. In the short term ($<$10 days), significant accumulation of neutrophils in the vicinity of the implants is often seen, but this dissipates rapidly and a chronic inflammatory response does not develop. The stimulatory effects of chitosan and chitosan fragments on immune cells mentioned, may play a role in inducing local cell proliferation and ultimately integration of the implanted material with the host tissue.

In the area of blood interactions, chitosan has been shown to have a strong hemostatic capacity. Blood coagulation occurs rapidly even under conditions of heparin anticoagulation. The process involves fibrin-independent aggregation of erythrocytes and appears to involve direct chitosan interactions with erythrocyte membranes (96–98). This type of reaction can be abrogated by acylating the free amino groups (81,99). In contrast, water-soluble, sulfated chitosan derivatives have been shown to have anticoagulant activity even higher than that of heparin (100–103).

As was discussed for alginates, contaminants within a chitosan preparation can dictate the nature of the tissue response, and an otherwise biocompatible material may exhibit adverse bioactivity in terms of a severe and possibly chronic inflammatory response accompanied by excessive fibrosis. The typical problem impurities are residual tissue proteins from the source organism and bacterial cell wall components. Thus care must be taken to ensure that chitosan for implant use is of a highly purified medical grade. Alternatively, the general purification methods used for alginates may be modified for application to chitosan. Chitosans also have the added feature of precipitating as gels with a relatively small change in pH. This characteristic can be used as an additional method of separating the polymer from more soluble contaminants. It should be noted that the highly deacetylated ($>$86%) chitosan forms are less likely to contain peptide contaminants since few peptides could withstand the harsh processing conditions needed to attain high degrees of deacetylation.

4. Formation of Tissue Scaffolds

One of chitosan's most promising features is its excellent ability to be processed into porous structures for use in cell transplantation and tissue regeneration. The tissue engineering approach to the repair and regeneration of damaged or diseased tissues relies on the use of polymer scaffolds which serve to support and organize the regenerating tissue (104–106). There is a continuous ongoing search for materials which either possess particularly desirable tissue-specific properties or which may have broad applicability and can be tailored to several tissue systems. Chitosan may be one such broadly applicable material. Porous chitosan structures can be formed by freezing and lyophilizing chitosan solutions in suitable molds (107). During the freezing process, ice crystals nucleate from solution and grow along the lines of thermal gradients. Exclusion of the chitosan salt from the ice crystal phase and subsequent ice removal by lyophilization generates a porous material whose mean pore size can be controlled by varying the freezing rate. Figure 4 illustrates tissue scaffold pore morphologies which can be produced by this simple technique. Pore orientation can be directed by controlling the geometry of the temperature gradients during freezing. The structures resulting from lyophilization of chitosan solutions are actually composed of the chitosan salt, for example, chitosan acetate if an acetic acid solution was used. The acid component must be neutralized or otherwise removed prior to rehydrating the scaffold. This can be accomplished by direct neutralization and rehydration in a basic solution such as dilute ammonia or NaOH.

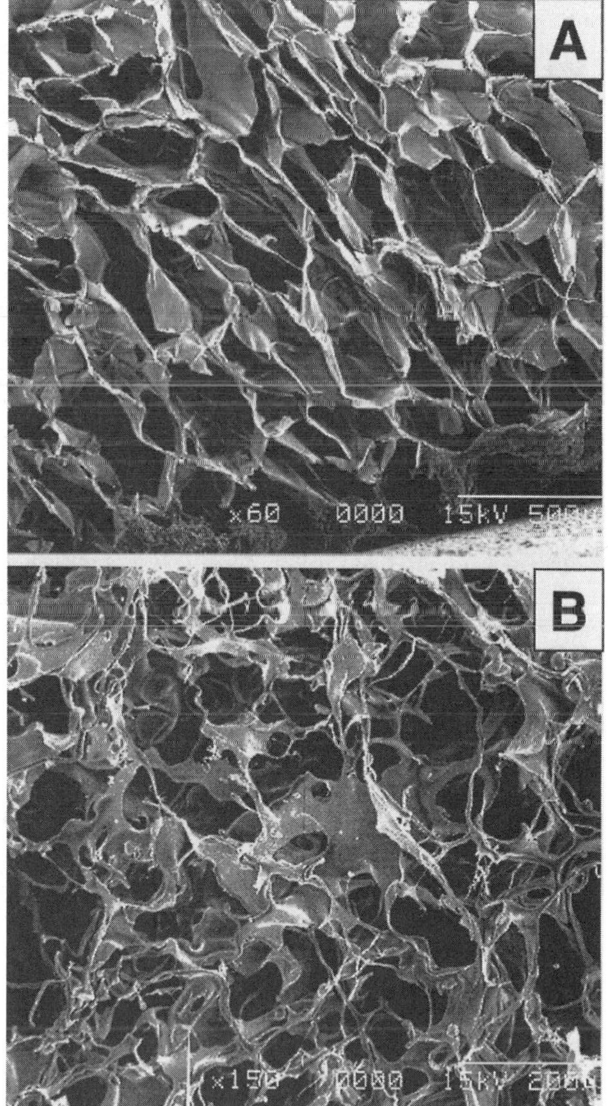

Figure 8.4 Porous chitosan tissue scaffolds produced by controlled rate freezing and lyophilization of chitosan solutions.

The mechanical properties of chitosan scaffolds formed in this way are mainly dependent on the pore sizes and pore orientations. Tensile testing of hydrated samples shows that porous membranes have greatly reduced elastic moduli (0.1–0.5 MPa) compared to nonporous chitosan membranes (5–7 MPa). The extensibility (maximum strain) of porous membranes varied from values similar to nonporous chitosan (~30%) to greater than 100% as a function of both pore size and orientation. The highest extensibility was obtained with a random pore orientation structure frozen rapidly at −78°C to give pores 120 μm in diameter. Porous membranes exhibited a stress–strain curve typical of composite materials with two distinct linear regions: a low modulus region at low strains and a transition to an approximately twofold higher modulus at high strains (Fig. 5). For all pore orientations, membranes formed at lower freezing rates exhibited higher moduli in both curve regions. The differences were most notable for samples with a random pore orientation. This suggests that in addition to larger pores, the lower freezing rate produces a more crystalline (and hence stiffer) chitosan phase. The tensile strengths of these porous structures were in the range 30–60 kPa.

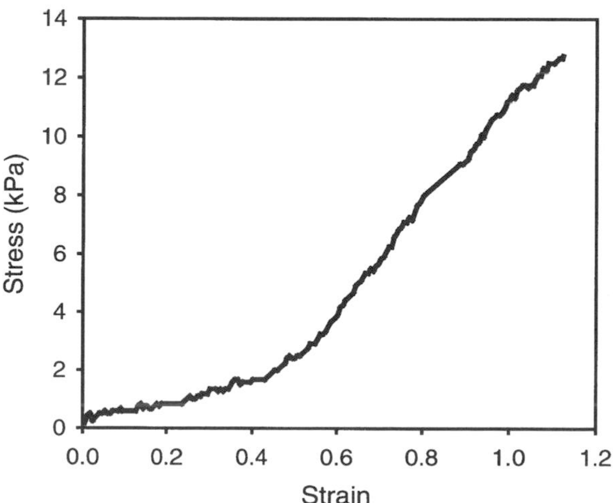

Figure 8.5 Tensile stress–strain plot for a porous chitosan membrane.

4. Modified Chitosans

Several chitosan derivatives have already been mentioned. Generally speaking, derivatization provides a powerful technique for generating new biological activities and for controlling chitosan's fundamental materials properties. The primary amino groups on the molecule are reactive and provide a means for side group attachment using very mild conditions. The general effect of side chain addition is to disrupt the material's crystal structure and hence increase the amorphous fraction. This modification generates a material with lower stiffness and often greater solubility. The precise nature of chemical and biological property changes will of course depend on the form of the side group. Thus, N-alkyl derivatives can exhibit reduced solubility and micellar aggregation in solution for alkyl lengths greater than five carbons. The character of chitosan as cationic, hemostatic, and insoluble at high pH can be completely reversed by sulfation, which, as described, can render the molecule anionic, anticoagulant, and water soluble. The variety of groups which can be attached to chitosan is almost unlimited, and side groups can be chosen to provide specific chemical functionality, alter biological properties, or modify physical properties. As an example, simply grafting on randomly distributed polyethylene glycol (PEG) side chains to the chitosan molecule reduces the elastic modulus of cast chitosan membranes in direct proportion to the PEG content (Fig. 6). Since PEG is a neutral, water-soluble, inert polymer with a large excluded volume, the grafted side chains serve to mask the chitosan surface in addition to reducing membrane stiffness. This masking effect has been shown to reduce the spreading and proliferation of endothelial and smooth muscle cells seeded onto PEG chitosan membranes as compared to unmodified chitosan membranes.

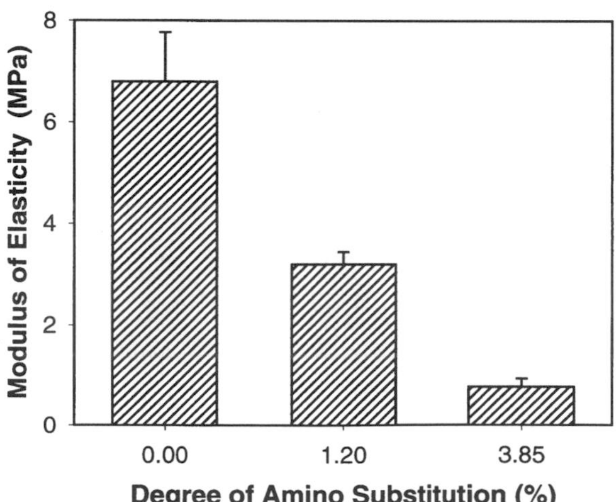

Figure 8.6 Effect of grafted polyethylene glycol side chains on the elastic modulus of dense chitosan membranes. Increasing PEG levels decreases membrane crystallinity, resulting in a reduction in elastic modulus. Membranes were tested in a hydrated state.

E. Glycosaminoglycans

1. General

Glycosaminoglycans are polysaccharides which occur ubiquitously within the extracellular matrix (ECM) of most animals. They are unbranched heteropolysaccharides, consisting of repeated disaccharide units, with the general structure [uronic acid–amino sugar]$_n$. In their native form, several GAG chains are covalently linked to a central protein core, and the protein–polysaccharide complex is termed a proteoglycan. Proteoglycans play a major role in organizing and determining the properties and functionality of the ECM, and the GAGs are the major factor in determining proteoglycan properties.

There are six different types of glycosaminoglycans: chondroitin sulfates, dermatan sulfate, keratan sulfate, heparan sulfate, heparin, and hyaluronic acid. Table 1 lists the component monosaccharides, primary source tissues, and molecular features of the commonly occurring GAGs. The monosaccharides in GAGs are sulfated to varying degrees, with the exception of hyaluronic acid, which is not sulfated. Unlike the other GAGs, hyaluronate exists as a free, high molecular weight molecule with no covalently attached protein. Figure 7 illustrates the repeating structures of hyaluronate and chondroitin-6-sulfate and a disaccharide segment of the variable heparin structure.

The GAGs are usually obtained as salts of sodium, potassium, or ammonia, and in this form they are all water soluble. The presence of strongly ionizing sulfate groups means that the charge density of these molecules is much less pH dependent than in the case of chitosan, and as a result these molecules are soluble over a wide pH range. The disaccharide repeat structure and the resultant alternating glycosidic bond types (analogous to the MG blocks of alginate) prevent this family of materials from forming high strength, crystalline structures in the solid state, and the dried or precipitated GAGs are essentially amorphous. The three-dimensional configurations which do form in solution tend to be flexible coil structures, extended as a result of the charge effects. Heparin for example is known to form single chain helical structures in solution. The only GAG with significant gel-forming ability is hyaluronate. This high molecular weight molecule is the primary gelling agent in the vitreous humor of the eye and can form very high viscosity solutions at concentrations as low as 0.3 wt%.

The biomaterials potential of the GAGs stems not from their intrinsic physical properties but from their biological

Table 8.1 Glycosaminoglycan Sources and Molecular Properties

Glycosaminoglycan	Source tissues	Molecular weight (kDa)	Charges per disaccharide	Component monosaccharides
Hyaluronate	Vitreous humor, cartilage	250–8000	1.0	N-acetyl-D-glucosamine D-glucuronic acid
Chondroitin 4-sulfate	Cartilage, cornea, skin, artery	5–50	1.1–2.0	N-acetyl-D-galactosamine D-glucuronic acid
Chondroitin 6-sulfate	Cartilage, cornea, skin, artery	5–50	1.2–2.3	N-acetyl-D-galactosamine D-glucuronic acid
Dermatan sulfate	Skin, heart valve, tendon, artery	15–40	2.0–2.2	N-acetyl-D-galactosamine L-iduronic acid D-glucuronic acid
Keratan sulfate	Cartilage, cornea, vertebral disks	2.5–20	0.9–1.8	N-acetyl-D-glucosamine D-galactose
Heparan sulfate	Lung, artery, cell surfaces	5–50	1.1–2.8	D-glucosamine N-acetylglucosamine L-iduronic acid D-glucuronic acid
Heparin	Mast cells in: lung, liver, skin, intestinal mucosa	5–25	3–5	D-glucosamine L-iduronic acid D-glucuronic acid

Source: Adapted from Ref. 156.

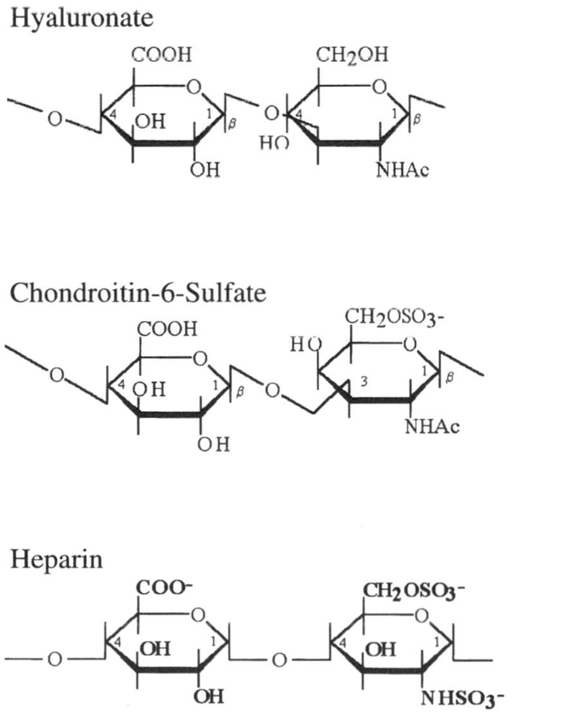

Hyaluronate

Chondroitin-6-Sulfate

Heparin

Figure 8.7 Repeating structures of representative glycosaminoglycans.

activity, which is a direct result of their numerous interactions with many important biomolecules. The biological activities described subsequently can be exploited by binding, complexing, or covalently linking GAG moieties to other polymers with superior structural or mechanical characteristics. Suitable polymers may include ECM proteins such as collagen, polysaccharides like chitosan, or a variety of synthetic polymers. Formation of composites of this type can endow the structural polymer with desirable tissue interaction properties and drastically improve its overall biocompatibility. Some of the better known GAG biological activities are described herein.

2. Interactions with Enzymes

These are the most extensively characterized of all GAG–protein interactions. In particular, the anticoagulant activity of heparin, via the activation of the protease inhibitor antithrombin III (ATIII), has been thoroughly characterized (108,109). Antithrombin III inactivates several proteases in the coagulation cascade, the most important of which is thrombin. Binding of heparin to ATIII increases the inactivation kinetics by as much as 2000-fold. Binding and activation of ATIII involves highly specific recognition of a pentasaccharide sequence found in both heparin and heparan sulfate. Heparin and heparan sulfate are also known to bind and activate the enzyme lipoprotein lipase (110–112). This property facilitates hydrolysis of triglycerides in lipoprotein particles in peripheral tissues. The fatty acids so formed are then available for uptake by adjacent cells. Lipoprotein lipase is normally found bound to heparan sulfate on endothelial cell surfaces where it is active. The soluble enzyme is inactive in the absence of GAG. This property has been used to reduce the levels of circulating lipoprotein by activating inactive lipase via injections of soluble GAG.

3. Interactions with Polypeptide Growth Factors

A wide variety of polypeptide growth factors bind avidly to heparin and heparan sulfate. Lower affinity binding also occurs with some other GAGs. Members of the fibroblast growth factor (FGF) family have been most extensively studied, but heparin/GAG affinity has been demonstrated with EGF, VEGF, a variety of interleukins (e.g., IL-3), as well as hematopoietic factors such as GM-CSF (113–122). Glycosaminoglycan binding can have a variety of effects on growth factors depending on the ratios of the two molecular species and the binding environment. Growth factors can be bound and sequestered in the extracellular matrix by GAGs. In the bound state, the polypeptides are often protected from proteolytic attack, and can be subsequently released from the ECM under the action of GAG lyases, matrix metaloproteinases, or changes in ionic strength. Growth factor binding to GAGs can also induce conformational changes which enhance receptor-growth factor binding. The GAGs may also enhance growth factor activity indirectly by mediating the clustering of growth factors or receptor–ligand complexes on the cell surface.

4. Interactions with Extracellular Matrix Proteins

Most ECM proteins have at least one GAG binding domain and many have several. These polypeptides bind heparin and heparan sulfate with the highest affinities, but binding of other GAGs is also known. The GAG–protein interaction may serve to facilitate the aggregation and organization of matrix proteins into ordered structures such as fibrils. Alternatively, the interaction may inhibit matrix protein organization. As an example, soluble GAG interactions can inhibit the assembly of Type I collagen fibrils from a collagen solution at physiological pH, while ionic GAG–collagen complexes are precipitated under acidic conditions. Binding of heparan sulfate to fibronectin may affect the conformation of the protein and thereby regulate its function. Furthermore, the binding of fibronectin to collagen in vitro is stabilized by the presence of heparin. Similarly, the apparent affinity of fibronectin for its cell surface receptors on rat hepatocytes and peritoneal macrophages is increased in the presence of heparan sulfate (123).

5. The Tissue Response to GAG-Containing Biomaterials

Because of their tissue origins, the GAGs are intrinsically biocompatible. Furthermore, because of their biological activity, interest in GAGs as components of implantable materials has been growing, particularly in recent years. Attention has focused mainly on heparin and hyaluronate. Heparin is being examined as the GAG of choice for tissue regeneration and wound healing applications primarily for two reasons. First, compared to other GAGs, it generally exhibits the highest level of biological activity with the broadest set of biological systems (124). Second, it is readily available at fairly low cost, in contrast to its less sulfated counterpart heparan sulfate. For wound healing and regeneration applications, heparin is generally complexed with or covalently linked to another biopolymer which serves as the main structural component. Both chitosan (70,94) and collagen (18,42) have been used for this purpose. Under these circumstances, the heparin can exert a number of influences dictated by the biological activities described previously. Thus growth factors released in the vicinity of the implant by neutrophils and macrophages may be sequestered and stabilized within the implant structure by immobilized heparin and subsequently released and utilized by ingrowing tissue cells. The bound heparin may also facilitate the binding and organization of deposited ECM components to the implant and consequently enhance the integration of the implant with neotissue and existing tissue. Alternatively, heparin released from an implant may activate and protect growth factors from degradation in the vicinity of the implant, a phenomenon which may lead to accelerated healing and/or enhanced angiogenesis. The typical outcome of these implantation studies is formation of new tissue in and around the implant which is histologically and functionally intermediate between normal tissue and scar. Although heparin has garnered the most interest for tissue repair, the other sulfated GAGs have also been applied with similar goals (69). In particular, crosslinked collagen–chondroitin sulfate complexes have demonstrated superior healing and regeneration in skin and nerve applications (18,42).

In contrast to heparin, hyaluronic acid is usually employed as a structural component in its own right because of its high molecular weight and gel-forming ability. The properties of the molecule may be broadly altered by chemical modification. For example, partial esterification of the carboxyl groups reduces the water solubility of the polymer and increases its viscosity. Extensive esterification generates materials that form water-insoluble films or swellable hydrogels (125). One of hyaluronate's main roles in tissue is as an organizer of the ECM. It appears to be particularly important during development of embryonic tissues. In addition, its importance in wound healing is emphasized by its transient elevation during the granulation phase (126). Ethyl- and benzyl-esterified hyaluronate membranes have demonstrated excellent healing responses and biodegradability in vivo. The fully esterified membranes had in vivo lifetimes of several months, whereas the partially esterified forms degraded within a few weeks. This type of degradation behavior has also been seen in vitro (127). The esterified hyaluronate derivatives undergo spontaneous de-esterification in aqueous environments. However, the hyaluronate backbone is very stable under the same conditions. As with all GAGs, degradation of the backbone occurs mainly by enzymatic depolymerization.

Hyaluronate-based materials may range from high viscosity solutions to low density hydrogels to high density hydrogels to dense but swellable hydrophilic polymers. The precise properties depend on the degree of esterification. Esters have been extruded into fibers and woven into fabrics for wound dressing applications. Dehydrated membranes and fibers may resemble similar structures made from synthetic polymers. They exhibit good mechanical and physical properties compared to other biological materials such as collagen. However, because the polysaccharide backbone retains its hydrophilic nature even at 100% esterification, these materials do exhibit some water absorption, swelling, and associated reductions in the mechanical properties.

IV. POLY(α-HYDROXY ACIDS)

Currently, the poly(α-hydroxy acids), specifically poly (lactic acid) (PLA), poly(glycolic acid) PGA (Fig. 8), and copolymers of the two are the most widely used synthetic polymers in the tissue engineering arena. These polymers have gained popularity because they offer the typical synthetic polymer advantages of high purity, convenient processing, and good mechanical properties, in addition to their biodegradability. Furthermore, since these materials have been used for over 25 years as resorbable sutures and fixation devices, they are already approved for in vivo human use in several forms and formulations. Polymerization can be induced either by direct condensation reactions between the acidic monomers or by catalytic ring-opening polymerization reactions of cyclic precursors.

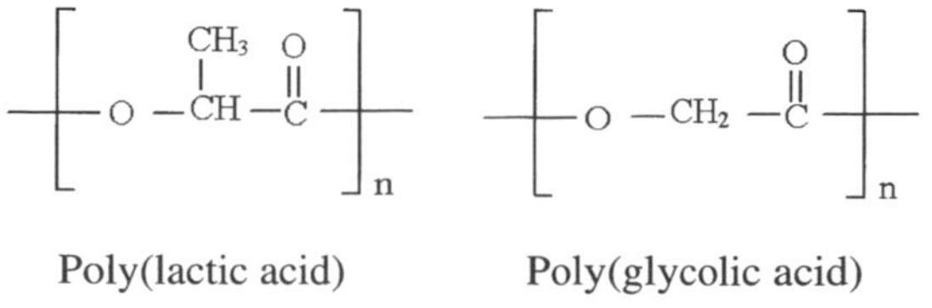

$$\left[-O-\underset{}{\overset{CH_3}{\overset{|}{CH}}}-\overset{O}{\overset{\|}{C}}- \right]_n \qquad \left[-O-CH_2-\overset{O}{\overset{\|}{C}}- \right]_n$$

Poly(lactic acid) Poly(glycolic acid)

Figure 8.8 Repeating structures of the two most commonly used poly(α-hydroxy acids).

Degradation of these polymers occurs by passive hydrolysis, with the monomeric acids as the final metabolizable products. The degradation rates are functions of polymer molecular weight, degree of crystallinity, and surface-to-volume ratio. Poly(glycolic acid) is generally available in a highly crystalline form. As such it possesses a high melting point (~300°C) and overall low solubility in organic solvents. However, its hydrophilic nature allows rapid absorption of water leading to rapid degradation within a few weeks. In contrast, poly(lactic acid) is available as fully crystalline or fully amorphous forms depending on the relative levels of the D and L optical isomers of lactic acid present in the molecule. Thus poly(D-lactide) and poly(L-lactide) are both highly crystalline, whereas the racemic copolymer poly(D,L-lactide) is completely amorphous. The presence of a methyl group makes PLA more hydrophobic than PGA, resulting in lower water absorption rates and correspondingly slower degradation, on the order of months. In both PLA and PGA, hydrolysis is accelerated by a reduction in pH. Thus, accumulation of acidic degradation products in the vicinity of the implant can have an autocatalytic effect, increasing the degradation rate. The effect is more pronounced in thicker samples since the outward diffusion of low molecular weight products is slower, leading to a drop in pH within the material and enhanced hydrolysis.

Neither PGA nor PLA possess intrinsic bioactivity. Cellular responses to implants of these polymers are mediated mostly by surface adsorbed proteins, surface topography, and the pH effects of degradation products. It is noteworthy that there is an effect of crystallinity on cell adhesion to these materials (128). The effects are believed to be a result of differences in protein conformation stemming from variations in binding to amorphous versus crystalline polymer surfaces. The chirality of crystalline regions may also play a role in controlling cellular response, but the detailed mechanisms of these effects are not fully understood.

Given the stability of these synthetic polymers, they are amenable to a wide variety of processing procedures for fabricating tissue scaffolds. They can be melt-spun into fibers which can be subsequently assembled into woven or nonwoven (i.e., felted) fabrics (129). Thin films can be directly cast from polymer melts or solutions in volatile organic solvents. Bulk porous scaffolds can be prepared using the porogen leaching procedure (130–132).

In this method, water-soluble porogen particles (e.g., sodium chloride crystals) are blended into a viscous polymer solution in a volatile organic solvent. The solution is then cast as a slab, and after evaporation of the solvent the slab is soaked in an aqueous solution to dissolve and leach out the porogen. Removal of the porogen leaves behind a porous matrix suitable for cellular ingrowth. Control of pore size and connectivity can be achieved by varying the porogen particle size and volume fraction, respectively.

An alternative fabrication procedure uses supercritical carbon dioxide as a foaming agent (133). In this approach, a nonporous slab of polymer is equilibrated with supercritical carbon dioxide under conditions of elevated temperature and pressure. Subsequent release of the pressure induces nucleation and expansion of gas bubbles within the polymer matrix, resulting in a closed cell polymer foam being formed. An open cellular architecture can be generated if the salt leaching method is combined with the gas foaming method.

More recently, freezing of polymer solutions followed by solvent removal under vacuum has been employed to generate macroporous scaffolds with a defined nanoscale fibrous architecture (134). This advance may represent a potentially useful technique for influencing cell behavior on these materals without altering the material chemistry or incorporating extrinsic biologically active molecules.

V. PSEUDOPOLY(AMINO ACIDS)

In this recently developed family of polymers, amino acid derivatives are assembled into polymers using either nonamide linkages for homopolymers or a combination of amide and nonamide bonds for heteropolymers. Unlike the previously developed poly(amino acids), which are in essence synthetic proteins, these materials can be generated as dimensionally stable thermoplastics. The most promising family of these materials examined to date are the tyrosine-based polymers. Desaminotyrosine is esterified with a tyrosine ester to form a diphenolic compound with an internal amide bond. This diphenol can be subsequently polymerized by linking the phenolic groups through a variety of chemistries (Fig. 9). Linkage of phenols via a carbonate group gives a tyrosine-derived polycarbonate such as poly(DTH carbonate) (135–140). Alternatively, a dicarboxylic acid may be used as the linking agent to give copolymers consisting of an alternating sequence of a diphenol and a diacid. By varying the length of the dicarboxylic acid and the type of tyrosine ester used to form the diphenol, a combinatorial library of related polymers was produced with physical and associated biological properties spanning a broad range (141). The chemical properties of the tyrosine-derived monomers impart thermoplastic behavior and water insolubility to this family of materials. In contrast to the poly(amino acids), the pseudopoly(amino acids) are thermally stable and can be processed using standard synthetic polymer techniques. They exhibit particularly high strength and stiffness and have been considered for use as bone substitutes and fixation devices (142). In animal studies, poly(DTH carbonate) scaffolds have exhibited good integration with regenerating bone without triggering a fibrous encapsulation response (142,143).

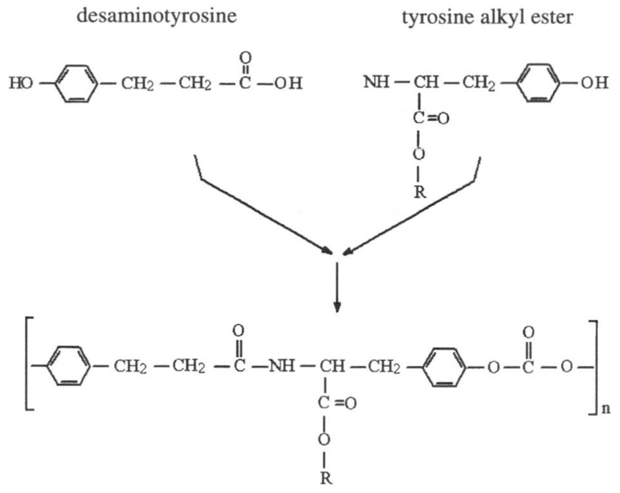

desaminotyrosine

tyrosine alkyl ester

Figure 8.9 Tyrosine derivatives and the repeating structure of the polycarbonate-type pseudopoly(amino acid) resulting from polymerization of their diphenolic dimer.

VI. POLYETHYLENE GLYCOL

Polyethylene glycol, also known as polyethylene oxide, is a petroleum-derived synthetic polymer that has garnered much attention in recent years. It is a linear polymer with the structure $HO–\{CH_2CH_2O\}_n–OH$. The presence of the backbone oxygen makes the molecule water soluble, and its high hydration volume and molecular mobility in solution cause it to exhibit negligible binding interactions with proteins. Thus, PEG-modified surfaces have been shown to resist protein adsorption and cellular attachment (144–147). The molecule is nontoxic and nondegradable under physiological conditions. Polyethylene glycol possesses no biological activity and in soluble form it is rapidly excreted by the kidneys without undergoing any transformation or chemical modifications. The terminal hydroxyl groups have low reactivity but can be activated under rigorous conditions. Thus they can be used as attachment sites for a variety of other more reactive or polymerizable groups. It is the end-modified PEG derivatives which have shown promise as tissue scaffolds. Derivatives such as PEG–diacrylate can be crosslinked under mild photoinitiated polymerization conditions to form crosslinked hydrogels which are nondegradable and completely inert. The mildness of the crosslinking procedures allows cells to be entrapped within these gels without toxic side effects. The power of this method stems from the fact that these inert hydrogels can be designed to incorporate cell adhesion factors such as the RGD peptide as well as other bioactive agents. This then allows the researcher complete and independent control over the physical gel properties as well as the biological properties. For example, biodegradable PLA oligomers or peptide sequences have been incorporated into PEG hydrogels to allow them to degrade or to allow cell migration through the gel structure (148–155). Apart from its direct use as a scaffolding material, PEG has also been used to modify a variety of biosurfaces. Its ability to resist protein binding led to its use as a modification for prolonging the lifetime of various peptide drugs in blood. Polyethylene glycol has also been examined as a nonfouling, covalently grafted surface layer on various medical materials (145). Such layers can improve the blood compatibility of procoagulant surfaces by preventing the adsorption and denaturation of plasma proteins. This type of modification also reduces or inhibits cell adhesion and migration on the modified surfaces and in principle can be used to modulate the level of interaction between tissue and implant materials.

REFERENCES

1. B. Chevallay, D. Herbage. Collagen-based biomaterials as 3D scaffold for cell cultures: applications for tissue engineering and gene therapy. *Med. Biol. Eng. Comput.* 38: 211–218 (2000).
2. D. R. Eyre. Collagen: molecular diversity in the body's protein scaffold. *Science* 207:1315–1322 (1980).
3. F. H. Silver, G. Pins. Cell growth on collagen: a review of tissue engineering using scaffolds containing extracellular matrix. *J. Long Term Eff. Med. Implants* 2:67–80 (1992).
4. H. K. Pokharna, L. A. Pottenger. Glycation induced cross-linking of link proteins, in vivo and in vitro. *J. Surg. Res.* 94:35–42 (2000).
5. R. Nagai, K. Matsumoto, X. Ling, H. Suzuki, T. Araki, S. Horiuchi. Glycolaldehyde, a reactive intermediate for advanced glycation end products, plays an important role in the generation of an active ligand for the macrophage scavenger receptor. *Diabetes* 49:1714–1723 (2000).
6. O. Devuyst, C. Van Ypersele De Strihou. Nitric oxide, advanced glycation end products, and uremia. *Kidney Int.* 58:1814–1815 (2000).
7. E. K. Yang, Y. K. Seo, H. H. Youn, D. H. Lee, S. N. Park, J. K. Park. Tissue engineered artificial skin composed of dermis and epidermis. *Artif. Organs* 24:7–17 (2000).
8. B. Hafemann, S. Ensslen, C. Erdmann, R. Niedballa, A. Zuhlke, K. Ghofrani, C. J. Kirkpatrick. Use of a collagen/elastin-membrane for the tissue engineering of dermis. *Burns* 25:373–384 (1999).
9. L. Germain, F. A. Auger, E. Grandbois, R. Guignard, M. Giasson, H. Boisjoly, S. L. Guerin. Reconstructed human cornea produced in vitro by tissue engineering. *Pathobiology* 67:140–147 (1999).
10. J. A. Chromiak, J. Shansky, C. Perrone, H. H. Vandenburgh. Bioreactor perfusion system for the long-term maintenance of tissue-engineered skeletal muscle organoids. *In Vitro Cell Dev. Biol. Anim.* 34:694–703 (1998).
11. S. Surapaneni, T. Pryor, M. D. Klein, H. W. Matthew. Rapid hepatocyte spheroid formation: optimization and long-term function in perfused microcapsules. *Asaino J.* 43:M848–M853 (1997).
12. J. C. Dunn, M. L. Yarmush, H. G. Koebe, R. G. Tompkins. Hepatocyte function and extracellular matrix geometry: long-term culture in a sandwich configuration. *FASEB J.* 3:174–177 (1989).
13. J. C. Y. Dunn, R. G. Tompkins, M. L. Yarmush. Long-term in vitro function of adult hepatocytes in a collagen sandwich configuration. *Biotechnol. Prog.* 7:237–245 (1991).
14. J. M. Pasquet, L. Quek, S. Pasquet, A. Poole, J. R. Matthews, C. Lowell, S. P. Watson. Evidence of a role for SHP-1 in platelet activation by the collagen receptor glycoprotein VI. *J. Biol. Chem.* 275:28526–28531 (2000).
15. M. Roest, J. J. Sixma, Y. P. Wu, M. J. Ijsseldijk, M. Tempelman, P. J. Slootweg, P. G. de Groot, G. H. van Zanten. Platelet adhesion to collagen in healthy volunteers is influenced by variation of both alpha(2)beta(1) density and von Willebrand factor. *Blood* 96:1433–1437 (2000).
16. M. Jandrot-Perrus, et al. Cloning, characterization, and functional studies of human and mouse glycoprotein VI: a platelet-specific collagen receptor from the immunoglobulin superfamily. *Blood* 96:1798–1807 (2000).
17. B. K. Tay, A. X. Le, M. Heilman, J. Lotz, D. S. Bradford. Use of a collagen–hydroxyapatite matrix in spinal fusion. A rabbit model. *Spine* 23:2276–2281 (1998).
18. D. L. Ellis, I. V. Yannas. Recent advances in tissue synthesis in vivo by use of collagen–glycosaminoglycan copolymers. *Biomaterials* 17:291–299 (1996).
19. S. Hakim, P. A. Merguerian, D. R. Chavez. Use of biodegradable mesh as a transport for a cultured uroepithelial graft: an improved method using collagen gel. *Urology* 44:139–142 (1994).
20. M. Chvapil, D. P. Speer, H. Holubec, T. A. Chvapil, D. H. King. Collagen fibers as a temporary scaffold for replacement of ACL in goats. *J. Biomed. Mater. Res.* 27: 313–325 (1993).
21. H. W. Kang, Y. Tabata, Y. Ikada. Fabrication of porous gelatin scaffolds for tissue engineering. *Biomaterials* 20: 1339–1344 (1999).
22. A. Bader, T. Schilling, O. E. Teebken, G. Brandes, T. Herden, G. Steinhoff, A. Haverich, Tissue engineering of heart valves—human endothelial cell seeding of detergent acellularized porcine valves. *Eur. J. Cardiothorac. Surg.* 14:279–284 (1998).

23. D. C. Gloeckner, M. S. Sacks, K. L. Billiar, N. Bachrach. Mechanical evaluation and design of a multilayered collagenous repair biomaterial. *J. Biomed. Mater. Res.* 52: 365–373 (2000).
24. A. P. Sclafani, T. Romo, III, A. A. Jacono, S. McCormick, R. Cocker, A. Parker. Evaluation of acellular dermal graft in sheet (AlloDerm) and injectable (micronized AlloDerm) forms for soft tissue augmentation. Clinical observations and histological analysis. *Arch. Facial Plast. Surg.* 2:130–136 (2000).
25. K. Inoue, T. Nakamoto, A. Usui, T. Usui. Endoscopic subureteral glutaraldehyde cross-linked collagen injection for the treatment of secondary vesicoureteral reflux: comparison with primary vesicoureteral reflux in adults. *J. Urol.* 164:336–339 (2000).
26. L. J. Southern, H. Hughes, P. V. Lawford, M. R. Clench, N. J. Manning. Glutaraldehyde-induced cross-links: a study of model compounds and commercial bioprosthetic valves. *J. Heart Valve Dis.* 9:241–248; discussion 8–9 (2000).
27. A. J. Bailey. Perspective article: the fate of collagen implants in tissue defects. *Wound Repair Regen.* 8:5–12 (2000).
28. M. Grabenwoger, F. Fitzal, C. Gross, D. Hutschala, P. Bock, P. Brucke, E. Wolner. Different modes of degeneration in autologous and heterologous heart valve prostheses. *J. Heart Valve Dis.* 9:104–109; discussion 10–11 (2000).
29. T. S. Girton, T. R. Oegema, R. T. Tranquillo. Exploiting glycation to stiffen and strengthen tissue equivalents for tissue engineering. *J. Biomed. Mater. Res.* 46:87–92 (1999).
30. Y. Iwashima, M. Eto, A. Hata, K. Kaku, S. Horiuchi, F. Ushikubi, H. Sano. Advanced glycation end products–induced gene expression of scavenger receptors in cultured human monocyte-derived macrophages. *Biochem. Biophys. Res. Commun.* 277:368–380 (2000).
31. A. W. Stitt, T. Bhaduri, C. B. McMullen, T. A. Gardiner, D. B. Archer. Advanced glycation end products induce blood–retinal barrier dysfunction in normoglycemic rats. *Mol. Cell Biol. Res. Commun.* 3:380–388 (2000).
32. S. F. Badylak, R. Tullius, K. Kokini, K. D. Shelbourne, T. Klootwyk, S. L. Voytik, M. R. Kraine, C. Simmons. The use of xenogeneic small intestinal submucosa as a biomaterial for Achilles tendon repair in a dog model. *J. Biomed. Mater. Res.* 29:977–985 (1995).
33. S. F. Badylak, B. Kropp, T. McPherson, H. Liang, P. W. Snyder. Small intestonal submucosa: a rapidly resorbed bioscaffold for augmentation cystoplasty in a dog model. *Tissue Eng.* 4:379–387 (1998).
34. S. F. Badylak, R. Record, K. Lindberg, J. Hodde, K. Park. Small intestinal submucosa: a substrate for in vitro cell growth. *J. Biomater. Sci. Polym. Ed.* 9:863–878 (1998).
35. S. Badylak, A. Liang, R. Record, R. Tullius, J. Hodde. Endothelial cell adherence to small intestinal submucosa: an acellular bioscaffold. *Biomaterials* 20:2257–2263 (1999).
36. S. Badylak, S. Arnoczky, P. Plouhar, R. Haut, V. Mendenhall, R. Clarke, C. Horvath. Naturally occurring extracellular matrix as a scaffold for musculoskeletal repair. *Clin. Orthop.* S333–S343 (1999).
37. B. P. Kropp, E. Y. Cheng. Bioengineering organs using small intestinal submucosa scaffolds: in vivo tissue-engineering technology. *J. Endourol.* 14:59–62 (2000).
38. N. Dagalakis, J. Flink, P. Stasikelis, J. F. Burke, I. V. Yannas. Design of an artificial skin. Part III. Control of pore structure. *J. Biomed. Mater. Res.* 14:511–528 (1980).
39. I. V. Yannas, J. F. Burke, Design of an artificial skin. I. Basic design principles. *J. Biomed. Mater. Res.* 14:65–81 (1980).
40. I. V. Yannas, J. F. Burke, P. L. Gordon, C. Huang, R. H. Rubenstein. Design of an artificial skin. II. Control of chemical composition. *J. Biomed. Mater. Res.* 14:107–132 (1980).
41. C. S. Chen, I. V. Yannas, M. Spector. Pore strain behaviour of collagen–glycosaminoglycan analogues of extracellular matrix. *Biomaterials* 16:777–783 (1995).
42. I. V. Yannas. Applications of ECM analogs in surgery. *J. Cell Biochem.* 56:188–191 (1994).
43. S. Nehrer, H. A. Breinan, A. Ramappa, S. Shortkroff, G. Young, T. Minas, C. B. Sledge, I. V. Yannas, M. Spector. Canine chondrocytes seeded in type I and type II collagen implants investigated in vitro. *J. Biomed. Mater. Res.* 38: 95–104 (1997).
44. J. P. Halle, S. Bourassa, F. A. Leblond, S. Chevalier, M. Beaudry, A. Chapdelaine, S. Cousineau, J. Saintonge, J. F. Yale. Protection of islets of Langerhans from antibodies by microencapsulation with alginate–poly-L-lysine membranes. *Transplantation* 55:350–354 (1993).
45. H. Wong, T. M. Chang. The microencapsulation of cells within alginate poly-L-lysine microcapsules prepared with the standard single step drop technique: histologically identified membrane imperfections and the associated graft rejection. *Biomater. Artif. Cells Immobilization Biotechnol.* 19:675–686 (1991).
46. H. A. Clayton, R. F. James, N. J. London. Islet microencapsulation: a review. *Acta Diabetol.* 30:181–189 (1993).
47. K. I. Draget, G. Skjak-Braek, O. Smidsrod. Alginate based new materials. *Int. J. Biol. Macromol.* 21:47–55 (1997).
48. O. Smidsrod, G. Skjak-Braek. Alginate as immobilization matrix for cells. *Trends Biotechnol.* 8:71–78 (1990).
49. W. R. Gombotz, D. K. Pettit. Biodegradable polymers for protein and peptide drug delivery. *Bioconjug. Chem.* 6: 332–351 (1995).
50. C. Y. Ko, V. Dixit, W. W. Shaw, G. Gitnick. Extensive in vivo angiogenesis from the controlled release of endothelial cell growth factor: implications for cell transplantation and wound healing. *J. Controlled Release* 44:209–214 (1997).
51. E. R. Edelman, E. Mathiowitz, R. Langer, M. Klagsbrun. Controlled and modulated release of basic fibroblast growth factor. *Biomaterials* 12:619–626 (1991).
52. N. H. Cho, S. Y. Seong, K. H. Chun, Y. H. Kim, I. C. Kwon, B. Y. Ahn, S. Y. Jeong. Novel mucosal immuniza-

tion with polysaccharide–protein conjugates entrapped in alginate microspheres. *J. Controlled Release* 53:215–224 (1998).
53. A. D. Sezer, J. Akbuga. Release characteristics of chitosan treated alginate beads. I. Sustained release of a macromolecular drug from chitosan treated alginate beads. *J. Microencapsul.* 16:195–203 (1999).
54. J. A. Rowley, G. Madlambayan, D. J. Mooney. Alginate hydrogels as synthetic extracellular matrix materials. *Biomaterials* 20:45–53 (1999).
55. P. De Vos, B. De Haan, R. Van Schilfgaarde. Effect of the alginate composition on the biocompatibility of alginate–polylysine microcapsules. *Biomaterials* 18:273–278 (1997).
56. P. De Vos, B. J. De Haan, G. H. Wolters, J. H. Strubbe, R. Van Schilfgaarde. Improved biocompatibility but limited graft survival after purification of alginate for microencapsulation of pancreatic islets. *Diabetologia* 40:262–270 (1997).
57. G. Klock, A. Pfeffermann, C. Ryser, P. Grohn, B. Kuttler, H. J. Hahn, U. Zimmermann. Biocompatibility of mannuronic acid–rich alginates. *Biomaterials* 18:707–713 (1997).
58. P. De Vos, G. H. Wolters, W. M. Fritschy, R. Van Schilfgaarde. Obstacles in the application of microencapsulation in islet transplantation. *Int. J. Artif. Organs* 16:205–212 (1993).
59. G. Klock, H. Frank, R. Houben, T. Zekorn, A. Horcher, U. Siebers, M. Wohrle, K. Federlin, U. Zimmermann. Production of purified alginates suitable for use in immuno isolated transplantation. *Appl. Microbiol. Biotechnol.* 40: 638 643 (1994).
60. T. Chandy, C. P. Sharma, Chitosan—as a biomaterial. *Biomater. Artif. Cells Artif. Organs* 18:1–24 (1990).
61. R. Muzzarelli, V. Baldassarre, F. Conti, P. Ferrara, G. Biagini, G. Gazzanelli, V. Vasi. Biological activity of chitosan: ultrastructural study. *Biomaterials* 9:247–252 (1988).
62. G. G. Allan, M. Peyron. Molecular weight manipulation of chitosan. I: Kinetics of depolymerization by nitrous acid. *Carbohydr. Res.* 277:257–272 (1995).
63. T. D. Rathke, S. M. Hudson. Review of chitin and chitosan as fiber and film formers. *Rev. Macromol. Chem. Phys.* C34:375–437 (1994).
64. Y. Qin, O. C. Agboh. Chitin and chitosan fibres. *Med. Device Technol.* 9:24–28 (1998).
65. S. Hirano, T. Midorikawa. Novel method for the preparation of N-acylchitosan fiber and N-acylchitosan-cellulose fiber. *Biomaterials* 19:293–297 (1998).
66. H. Fukuda, Y. Kikuchi. In vitro clot formation on the polyelectrolyte complexes of sodium dextran surface with chitosan. *J. Biomed. Mater. Res.* 12:531–539 (1978).
67. T. Takahashi, K. Takayama, Y. Machida, T. Nagai. Characteristics of polyion complexes of chitosan with sodium alginate and sodium polyacrylate. *Int. J. Pharmaceut.* 61: 35–41 (1990).
68. O. Gaserod, O. Smidsrod, G. Skjak-Braek. Microcapsules of alginate-chitosan. I. A quantitative study of the interaction between alginate and chitosan. *Biomaterials* 19: 1815–1825 (1998).
69. A. Denuziere, D. Ferrier, O. Damour, A. Domard. Chitosan–chondroitin sulfate and chitosan–hyaluronate polyelectrolyte complexes: biological properties. *Biomaterials* 19:1275–1285 (1998).
70. G. Kratz, C. Arnander, J. Swedenborg, M. Back, C. G. I. Falk, O. Larm. Heparin–chitosan complexes stimulate wound healing in human skin. *Scand. J. Plast. Reconstr. Surg. Hand Surg.* 31:119–123 (1997).
71. K. W. Leong, H. Q. Mao, V. L. Truong-Le, K. Roy, S. M. Walsh, J. T. August. DNA–polycation nanospheres as non-viral gene delivery vehicles. *J. Controlled Release* 53: 183–193 (1998).
72. F. C. MacLaughlin, R. J. Mumper, J. Wang, J. M. Tagliaferri, I. Gill, M. Hinchcliffe, A. P. Rolland. Chitosan and depolymerized chitosan oligomers as condensing carriers for in vivo plasmid delivery. *J. Controlled Release* 56: 259–272 (1998).
73. K. Roy, H. Q. Mao, S. K. Huang, K. W. Leong. Oral gene delivery with chitosan–DNA nanoparticles generates immunologic protection in a murine model of peanut allergy. *Nat. Med.* 5:387–391 (see comments) (1999).
74. G. Peluso, O. Petillo, M. Ranieri, M. Santin, L. Ambrosio, D. Calabro, B. Avallone, G. Balsamo. Chitosan-mediated stimulation of macrophage function. *Biomaterials* 15: 1215–1220 (1994)
75. Y. Usami, Y. Okamoto, T. Takayama, Y. Shigemasa, S. Minami. Chitin and chitosan stimulate canine polymorphonuclear cells to release leukotriene D4 and prostaglandin E2. *J. Biomed. Mater. Res.* 42:517–522 (1998).
76. Y. Shibata, L. A. Foster, W. J. Metzger, Q. N. Myrvik. Alveolar macrophage priming by intravenous administration of chitin particles, polymers of N-acetyl-D-glucosamine, in mice. *Infect. Immun.* 65:1734–1741 (1997).
77. Y. Usami, Y. Okamoto, S. Minami, A. Matsuhashi, N. H. Kumazawa, S. Tanioka, Y. Shigemasa. Migration of canine neutrophils to chitin and chitosan. *J. Vet. Med. Sci.* 56:1215–1216 (1994).
78. Y. Usami, Y. Okamoto, S. Minami, A. Matsuhashi, N. H. Kumazawa, S. Tanioka, Y. Shigemasa. Chitin and chitosan induce migration of bovine polymorphonuclear cells. *J. Vet. Med. Sci.* 56:761–762 (1994).
79. S. Hirano, H. Tsuchida, N. Nagao. N-acetylation in chitosan and the rate of its enzymic hydrolysis. *Biomaterials* 10:574–576 (1989).
80. G. Paradossi, E. Chiessi, M. Venanzi, B. Pispisa, A. Palleschi. Branched-chain analogues of linear polysaccharides: a spectroscopic and conformational investigation of chitosan derivatives. *Int. J. Biol. Macromol.* 14:73–80 (1992).
81. K. Y. Lee, W. S. Ha, W. H. Park. Blood compatibility and biodegradability of partially N-acylated chitosan derivatives. *Biomaterials* 16:1211–1216 (1995).
82. K. Kamiyama, H. Onishi, Y. Machida. Biodisposition characteristics of N-succinyl-chitosan and glycol- chitosan

in normal and tumor-bearing mice. *Biol. Pharm. Bull.* 22: 179–186 (1999).

83. R. A. Muzzarelli, G. Biagini, M. Bellardini, L. Simonelli, C. Castaldini, G. Fratto. Osteoconduction exerted by methylpyrrolidinone chitosan used in dental surgery. *Biomaterials* 14:39–43 (1993).
84. R. A. Muzzarelli, C. Zucchini, P. Ilari, A. Pugnaloni, M. Mattioli Belmonte, G. Biagini, C. Castaldini. Osteoconductive properties of methylpyrrolidinone chitosan in an animal model. *Biomaterials* 14:925–929 (1993).
85. O. Felt, P. Buri, R. Gurny. Chitosan: a unique polysaccharide for drug delivery. *Drug Dev. Ind. Pharm.* 24:979–993 (1998).
86. H. Onishi, Y. Machida. Biodegradation and distribution of water-soluble chitosan in mice. *Biomaterials* 20:175–182 (1999).
87. R. A. Muzzarelli, et al. Stimulatory effect on bone formation exerted by a modified chitosan. *Biomaterials* 15: 1075–1081 (1994).
88. R. A. Muzzarelli. Human enzymatic activities related to the therapeutic administration of chitin derivatives. *Cell Mol. Life Sci.* 53:131–140 (1997).
89. R. Muzzarelli, G. Biagini, A. Pugnaloni, O. Filippini, V. Baldassarre, C. Castaldini, C. Rizzoli. Reconstruction of parodontal tissue with chitosan. *Biomaterials* 10:598–603 (1989).
90. A. Eser Elcin, Y. M. Elcin, G. D. Pappas. Neural tissue engineering: adrenal chromaffin cell attachment and viability on chitosan scaffolds. *Neurol. Res.* 20:648–654 (1998).
91. Y. M. Elcin, V. Dixit, K. Lewin, G. Gitnick. Xenotransplantation of fetal porcine hepatocytes in rats using a tissue engineering approach. *Artif. Organs* 23:146–152 (1999).
92. O. Damour, P. Y. Gueugniaud, M. Berthin-Maghit, P. Rousselle, F. Berthod, F. Sahuc, C. Collombel. A dermal substrate made of collagen–GAG–chitosan for deep burn coverage: first clinical uses. *Clin. Mater.* 15:273–276 (1994).
93. K. Nishimura, S. Nishimura, N. Nishi, I. Saiki, S. Tokura, I. Azuma. Immunological activity of chitin and its derivatives. *Vaccine* 2:93–99 (1984).
94. G. Kratz, M. Back, C. Arnander, O. Larm. Immobilised heparin accelerates the healing of human wounds in vivo. *Scand. J. Plast. Reconstr. Surg. Hand Surg.* 32:381–385 (1998).
95. T. Hamano, A. Teramoto, E. Iizuka, K. Abe. Effects of polyelectrolyte complex (PEC) on human periodontal ligament fibroblast (HPLF) function. II. Enhancement of HPLF differentiation and aggregation on PEC by L-ascorbic acid and dexamethasone. *J. Biomed. Mater. Res.* 41:270–277 (1998).
96. S. B. Rao, C. P. Sharma. Use of chitosan as a biomaterial: studies on its safety and hemostatic potential. *J. Biomed. Mater. Res.* 34:21–28 (1997).
97. P. R. Klokkevold, D. S. Lew, D. G. Ellis, C. N. Bertolami. Effect of chitosan on lingual hemostasis in rabbits. *J. Oral Maxillofac. Surg.* 49:858–863 (1991).
98. P. R. Klokkevold, H. Fukayama, E. C. Sung, C. N. Bertolami. The effect of chitosan (poly-N-acetyl glucosamine) on lingual hemostasis in heparinized rabbits. *J. Oral Maxillofac. Surg.* 57:49–52 (1999).
99. S. Hirano, Y. Noishiki. The blood compatibility of chitosan and N-acylchitosans. *J. Biomed. Mater. Res.* 19:413–417 (1985).
100. R. L. Whistler, M. Kosik. Anticoagulant activity of oxidized and N- and O-sulfated chitosan. *Arch. Biochem. Biophys.* 142:106–110 (1971).
101. R. A. Muzzarelli, F. Tanfani, M. Emanuelli, D. P. Pace, E. Chiurazzi, M. Piani, Sulfated N-(carboxymethyl)chitosans: novel blood anticoagulants. *Carbohydr. Res.* 126: 225–231 (1984).
102. S. Hirano, Y. Tanaka, M. Hasegawa, K. Tobetto, A. Nishioka. Effect of sulfated derivatives of chitosan on some blood coagulant factors. *Carbohydr. Res.* 137:205–215 (1985).
103. C. Ishihara, S. Shimakawa, M. Tsuji, J. Arikawa, S. Tokura. A sulfated chitin, SCM-chitin III, inhibits the clearance of human erythrocytes from the blood circulation in erythrocyte-transfused SCID mice. *Immunopharmacology* 29:65–71 (1995).
104. R. Langer, J. P. Vacanti. Tissue engineering. *Science* 260: 920–926 (1993).
105. R. Langer. Tissue engineering: a new field and its challenges. *Pharm. Res.* 14:840–841 (1997).
106. J. A. Hubbell. Biomaterials in tissue engineering. *Biotechnology (N Y)* 13:565–576 (1995).
107. S. V. Madihally, H. W. Matthew. Porous chitosan scaffolds for tissue engineering. *Biomaterials* 20:1133–1142 (1999).
108. I. Bjork, U. Lindahl. Mechanism of the anticoagulant action of heparin. *Mol. Cell Biochem.* 48:161–182 (1982).
109. B. Bray, D. A. Lane, J. M. Freyssinet, G. Pejler, U. Lindahl. Anti-thrombin activities of heparin. Effect of saccharide chain length on thrombin inhibition by heparin cofactor II and by antithrombin. *Biochem. J.* 262:225–232 (1989).
110. T. Olivecrona, T. Egelrud, P. H. Iverius, U. Lindahl. Evidence for an ionic binding of lipoprotein lipase to heparin. *Biochem. Biophys. Res. Commun.* 43:524–529 (1971).
111. T. Olivecrona, G. Bengtsson, S. E. Marklund, U. Lindahl, M. Hook. Heparin-lipoprotein lipase interactions. *Fed. Proc.* 36:60–65 (1977).
112. G. Bengtsson, T. Olivecrona, M. Hook, J. Riesenfeld, U. Lindahl. Interaction of lipoprotein lipase with native and modified heparin-like polysaccharides. *Biochem. J.* 189: 625–633 (1980).
113. M. Belting, B. Havsmark, M. Jonsson, S. Persson, L. A. Fransson. Heparan sulphate/heparin glycosaminoglycans with strong affinity for the growth-promoter spermine have high antiproliferative activity. *Glycobiology* 6:121–129 (1996).
114. R. Flaumenhaft, D. Moscatelli, D. B. Rifkin. Heparin and heparan sulfate increase the radius of diffusion and action

of basic fibroblast growth factor. *J. Cell. Biol.* 111:1651–1659 (1990).

115. G. Klein, S. Conzelmann, S. Beck, R. Timpl, C. A. Muller. Perlecan in human bone marrow: a growth-factor-presenting, but anti-adhesive, extracellular matrix component for hematopoietic cells. *Matrix Biol.* 14:457–465 (1995).
116. W. Koopmann, M. S. Krangel. Identification of a glycosaminoglycan-binding site in chemokine macrophage inflammatory protein-1 alpha. *J. Biol. Chem.* 272:10103–10109 (1997).
117. G. S. Kuschert, A. J. Hoogewerf, A. E. Proudfoot, C. W. Chung, R. M. Cooke, R. E. Hubbard, T. N. Wells, P. N. Sanderson. Identification of a glycosaminoglycan binding surface on human interleukin-8. *Biochemistry* 37:11193–11201 (1998).
118. G. S. Kuschert, F. Coulin, C. A. Power, A. E. Proudfoot, R. E. Hubbard, A. J. Hoogewerf, T. N. Wells. Glycosaminoglycans interact selectively with chemokines and modulate receptor binding and cellular responses. *Biochemistry* 38:12959–12968 (1999).
119. H. Lortat-Jacob, P. Garrone, J. Banchereau, J. A. Grimaud. Human interleukin 4 is a glycosaminoglycan-binding protein. *Cytokine* 9:101–105 (1997).
120. S. Najjam, B. Mulloy, J. Theze, M. Gordon, R. Gibbs, C. C. Rider. Further characterization of the binding of human recombinant interleukin 2 to heparin and identification of putative binding sites. *Glycobiology* 8:509–516 (1998).
121. A. Wettreich, A. Sebollela, M. A. Carvalho, S. P. Azevedo, R. Borojevic, S. T. Ferreira, T. Coelho-Sampaio. Acidic pH modulates the interaction between human granulocyte-macrophage colony-stimulating factor and glycosaminoglycans. *J. Biol. Chem.* 274:31468–31475 (1999).
122. L. E. Wrenshall, J. L. Platt. Regulation of T cell homeostasis by heparan sulfate-bound IL-2. *J. Immunol.* 163:3793–3800 (1999).
123. E. Ruoslahti, E. Engvall. Complexing of fibronectin glycosaminoglycans and collagen. *Biochim. Biophys. Acta* 631:350–358 (1980).
124. D. J. Tyrell, S. Kilfeather, C. P. Page. Therapeutic uses of heparin beyond its traditional role as an anticoagulant. *Trends Pharmacol. Sci.* 16:198–204 (1995).
125. D. Campoccia, P. Doherty, M. Radice, P. Brun, G. Abatangelo, D. F. Williams. Semisynthetic resorbable materials from hyaluronan esterification. *Biomaterials* 19:2101–2127 (1998).
126. W. Y. Chen, G. Abatangelo. Functions of hyaluronan in wound repair. *Wound Repair Regen.* 7:79–89 (1999).
127. D. Campoccia, J. A. Hunt, P. J. Doherty, S. P. O. R. M. Zhong, L. Benedetti, D. F. Williams. Quantitative assessment of the tissue response to films of hyaluronan derivatives. *Biomaterials* 17:963–975 (1996).
128. A. Park, L. G. Cima. In vitro cell response to differences in poly-L-lactide crystallinity. *J. Biomed. Mater. Res.* 31:117–130 (1996).
129. A. G. Mikos, Y. Bao, L. G. Cima, D. E. Ingber, J. P. Vacanti, R. Langer. Preparation of poly(glycolic acid) bonded fiber structures for cell attachment and transplantation. *J. Biomed. Mater. Res.* 27:183–189 (1993).
130. M. C. Wake, P. K. Gupta, A. G. Mikos. Fabrication of pliable biodegradable polymer foams to engineer soft tissues. *Cell Transplant* 5:465–473 (1996).
131. L. Lu, S. J. Peter, M. D. Lyman, H. L. Lai, S. M. Leite, J. A. Tamada, J. P. Vacanti, R. Langer, A. G. Mikos. In vitro degradation of porous poly(L-lactic acid) foams. *Biomaterials* 21:1595–1605 (2000).
132. W. L. Murphy, D. H. Kohn, D. J. Mooney. Growth of continuous bonelike mineral within porous poly(lactide-co-glycolide) scaffolds in vitro. *J. Biomed. Mater. Res.* 50:50–58 (2000).
133. M. H. Sheridan, L. D. Shea, M. C. Peters, D. J. Mooney. Bioabsorbable polymer scaffolds for tissue engineering capable of sustained growth factor delivery. *J. Controlled Release* 64:91–102 (2000).
134. P. X. Ma, R. Zhang. Synthetic nano-scale fibrous extracellular matrix. *J. Biomed. Mater. Res.* 46:60–72 (1999).
135. S. I. Ertel, J. Kohn, M. C. Zimmerman, J. R. Parsons. Evaluation of poly(DTH carbonate), a tyrosine-derived degradable polymer, for orthopaedic applications. *J. Biomed. Mater. Res.* 29:1337–1348 (1995).
136. J. Fiordeliso, S. Bron, J. Kohn. Design, synthesis, and preliminary characterization of tyrosine-containing polyarylates: new biomaterials for medical applications. *J. Biomater. Sci. Polym. Ed.* 5:497–510 (1994).
137. S. I. Ertel, J. Kohn. Evaluation of a series of tyrosine-derived polycarbonates as degradable biomaterials. *J. Biomed. Mater. Res.* 28:919–930 (1994).
138. F. H. Silver, M. Marks, Y. P. Kato, C. Li, S. Pulapura, J. Kohn. Tissue compatibility of tyrosine-derived polycarbonates and polyiminocarbonates: an initial evaluation. *J. Long Term Eff. Med. Implants* 1:329–346 (1992).
139. S. Pulapura, J. Kohn. Tyrosine-derived polycarbonates: backbone-modified "pseudo" -poly (amino acids) designed for biomedical applications. *Biopolymers* 32:411–417 (1992).
140. V. Tangpasuthadol, A. Shefer, K. A. Hooper, J. Kohn. Thermal properties and physical ageing behaviour of tyrosine-derived polycarbonates. *Biomaterials* 17:463–468 (1996).
141. S. Brocchini, K. James, V. Tangpasuthadol, J. Kohn. Structure-property correlations in a combinatorial library of degradable biomaterials. *J. Biomed. Mater. Res.* 42:66–75 (1998).
142. J. Choueka, J. L. Charvet, K. J. Koval, H. Alexander, K. S. James, K. A. Hooper, J. Kohn. Canine bone response to tyrosine-derived polycarbonates and poly(L-lactic acid). *J. Biomed. Mater. Res.* 31:35–41 (1996).
143. K. James, H. Levene, J. R. Parsons, J. Kohn. Small changes in polymer chemistry have a large effect on the bone-implant interface: evaluation of a series of degradable tyrosine-derived polycarbonates in bone defects. *Biomaterials* 20:2203–2212 (1999).
144. M. J. Slepian. Polymeric endoluminal gel paving: thera-

peutic hydrogel barriers and sustained drug delivery depots for local arterial wall biomanipulation. *Semin. Interv. Cardiol.* 1:103–116 (1996).

145. E. Tziampazis, J. Kohn, P. V. Moghe. PEG-variant biomaterials as selectively adhesive protein templates: model surfaces for controlled cell adhesion and migration. *Biomaterials* 21:511–520 (2000).
146. J. L. West, J. A. Hubbell. Comparison of covalently and physically cross-linked polyethylene glycol-based hydrogels for the prevention of postoperative adhesions in a rat model. *Biomaterials* 16:1153–1156 (1995).
147. J. L. West, J. A. Hubbel. Separation of the arterial wall from blood contact using hydrogel barriers reduces intimal thickening after balloon injury in the rat: the roles of medial and luminal factors in arterial healing. *Proc. Natl. Acad. Sci. USA* 93:13188–13193 (1996).
148. D. K. Han, K. D. Park, J. A. Hubbell, Y. H. Kim. Surface characteristics and biocompatibility of lactide-based poly(ethylene glycol) scaffolds for tissue engineering. *J. Biomater. Sci. Polym. Ed.* 9:667–680 (1998).
149. C. Yu, J. Kohn. Tyrosine-PEG-derived poly(ether carbonate)s as new biomaterials. Part I: synthesis and evaluation. *Biomaterials* 20:253–264 (1999).
150. C. Yu, S. S. Mielewczyk, K. J. Breslauer, J. Kohn. Tyrosine-PEG-derived poly(ether carbonate)s as new biomaterials. Part II: study of inverse temperature transitions. *Biomaterials* 20:265–272 (1999).
151. A. Nathan, S. Zalipsky, S. I. Ertel, S. N. Agathos, M. L. Yarmush, J. Kohn. Copolymers of lysine and polyethylene glycol: a new family of functionalized drug carriers. *Bioconjug. Chem* 4:54–62 (1993) [published erratum appears in *Bioconjug. Chem.* 4(5):410 (1993)].
152. L. J. Suggs, R. S. Krishnan, C. A. Garcia, S. J. Peter, J. M. Anderson, A. G. Mikos. In vitro and in vivo degradation of poly(propylene fumarate-co-ethylene glycol) hydrogels. *J. Biomed. Mater. Res.* 42:312–320 (1998).
153. A. Gopferich, S. J. Peter, A. Lucke, L. Lu, A. G. Mikos. Modulation of marrow stromal cell function using poly(D,L-lactic acid)–block-poly(ethylene glycol)–monomethyl ether surfaces. *J. Biomed. Mater. Res.* 46:390–398 (1999).
154. L. J. Suggs, J. L. West, A. G. Mikos. Platelet adhesion on a bioresorbable poly(propylene fumarate-co-ethylene glycol) copolymer. *Biomaterials* 20:683–690 (1999).
155. K. S. Bohl, J. L. West. Nitric oxide–generating polymers reduce platelet adhesion and smooth muscle cell proliferation. *Biomaterials* 21:2273–2278 (2000).
156. U. Lindahl, M. Hook. Glycosaminoglycans and their binding to biological macromolecules. *Annu. Rev. Biochem.* 47:385–417 (1978).

9

Chitosan: Structure–Properties Relationship and Biomedical Applications

Alain Domard and Monique Domard
University Claude Bernard Lyon I, Villeurbanne, France

I. INTRODUCTION

A. Definition of Chitosan

Chitosan is a naturally occurring polysaccharide whose commercial forms are essentially produced from N-deacetylation of chitin. Chitin and chitosan can be represented by a unique structure, shown in Fig. 1. This structure corresponds to the series of the copolymers of β $(1 \rightarrow 4)$ linked glucosamine, and N-acetyl-glucosamine. They can be considered as belonging to the family of glycosaminoglycans (GAGs), a family to which also belong chondroitin sulfates, hyaluronic acid, and heparine. Glycosaminoglycans are particularly interesting since they seem to be alone among the polysaccharides that express the property of bioactivity.

The difference between chitin and chitosan is essentially related to the possibility to solubilize the polymer in dilute acidic media. Thus, when the structure can be dissolved in this kind of solvent, it corresponds to chitosan; in the reverse case, to chitin. Therefore the degree of acetylation (DA), which is related to the balance between the two kinds of residues, is essential to define these two terms. When chitin is deacetylated in heterogeneous conditions, the solubility in water is only achieved for DA generally below 30% (1). Nevertheless, on reacetylating chitosan in homogeneous conditions, it is possible to observe a solubilization up to DA close to 60% (2,3). As a consequence, the frontier between chitin/chitosan can be located at a DA of 60%. As we shall see, DA is also an interesting adaptable parameter whose simple modification allows us to provide this series of copolymers with a very great variety of properties. In numerous circumstances, this opportunity gives the advantage of avoiding some difficult and expensive chemical modifications.

B. Origin

The production of chitin in biomass has been evaluated to be as abundant as that of cellulose, with a yearly production within 10^{10}–10^{12} T (4,5). This polymer is present in most of the families of living species (5). Thus, it constitutes the structure polymer of the cuticles of all the arthropods and the endoskeletons of all the cephalopods. It is also very often present at the cell wall and in the extracellular matrix of most fungi. It is encountered in numerous microorganisms, in some algae, etc. Chitosan is much less present in living media and to date it has only been observed in some microorganisms, particularly of fungal nature (6). This lack of occurrence explains why the latter polymer, whose direct applications are much more important than that of the former, is essentially produced from chitin. It is noteworthy that both chitin and chitosan in their polymeric form are completely absent in numerous animals including mammals. Nevertheless, the N-acetyl-glucosamine unit is largely widespread in the chemical structure of the other GAGs and the glycoconjugates such as glycoproteins and glycolipids (7). On the other hand,

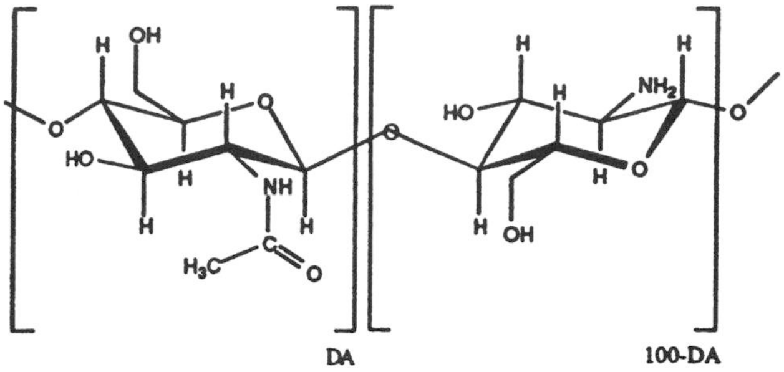

Figure 9.1 Chemical structure of chitin and chitosan.

the nearly total absence of chitosan in living media must be regarded as an interesting opportunity for numerous possible applications, especially if we consider that it corresponds to a more or less charged polycation, a chemical structure totally absent in nature.

A brief summary of the origin of chitin and chitosan from a historical point of view is also interesting. Chitin was discovered in 1811 by H. Braconnot during his studies on mushrooms and was termed *fungine* (8). This discovery occurred approximately 30 years before that of cellulose. The term *chitin* was first proposed by C. Odier in 1823 (9), who ignored the works of Braconnot and found chitin in the elytrum of the cock-chafer beetle. In 1859, C. Rouget discovered chitosan after he treated chitin in hot and concentrated KOH. He logically proposed to name this new product ''modified chitin'' (10). Unfortunately, in 1894, F. Hoppe-Seyler, ignoring the works of Rouget proposed to term this derivative *chitosan* (11). This name, which is now strongly anchored in the scientific language, is at the origin of an important problem of nomenclature since we have one chemical structure but, depending on DA, two names.

Surprisingly, the modern development of the research on chitin and chitosan was not initially related to an important discovery but to governments concerned with the production of food processing of crabs and shrimps, specifically to solve the problem of the significant amount of waste related to this process. Grants were offered to try to valorize this waste. Thus, the beginning of modern research on chitin/chitosan is situated at the beginning of the 1970s. The continuously increasing amount of research on these polymers, especially chitosan, signifies their importance. Before 1970 there were less than 10 papers per year abstracted in *Chemical Abstracts*. For 1999, they number more than 2000.

C. Production

Chitin is produced essentially from the cuticles of crustaceans, especially from crab and shrimp shells. Another interesting source is represented by the endoskeletons of cephalopods. The method is relatively simple and consists of two successive extractions. The first is a demineralization related to the presence of a more or less significant amount of calcium carbonate in the shells (5). This step is performed by means of mineral acids. It must be performed with care, particularly for medical applications for which the mineral part must be eliminated but also because polysaccharides are well known to be easily hydrolyzed in acidic media. The second step is a deproteinization, which consists of placing the material in alkaline media of sufficient concentration associated with a moderate heating (5). There also, the complete deproteinization is very important in the case of medical uses. Endoskeletons of calmars such as squid pens are particularly interesting for their very low mineralization with a content of calcium carbonate generally below 1% (w/w) (12). In this case, the demineralization step can be avoided. This source is probably the most interesting for the preparation of the polymers with the highest molecular weights. Some recent papers propose an enzymatic method for this step (13).

Chitosan is obtained from the N-deacetylation of chitin. All the methods are derived from the descriptions given in two patents (14,15) and consist in the use of highly concentrated solutions of sodium hydroxyde (30–50%) at temperatures over 90°C for times over 1 h. For mild conditions, these methods allow us to reach in one step DA close to 10–15%. If repeated one time, the deacetylation can be within 96–95%. Nevertheless, each step operated in relatively drastic conditions contributes to an important decrease of the molecular weight. Thus, if a complete deacetylation can be achieved with a three-step process, the chain length becomes relatively low at the end of the treatment. In order to avoid this problem, the use of sodium thiophenolate has been proposed, allowing both the catalysis of the reaction and the protection of the polymer chains from degradation. This method allows us also to considerably decrease the amount of sodium hydroxyde used (16). There also, endoskeletons of calmars are very interesting. Indeed, their crystalline parts are different from those of the cuticles of crustaceans. They are more accessible to reactives and thus the reaction of deacetylation is achieved more easily (12). For this additional reason, the molecular weight of chitosans produced from these sources are much higher.

II. THE THREE PHYSICAL STATES OF CHITOSAN

A. Solutions

Solutions of chitosan correspond to the most important physical state for this polymer, either from a fundamental point of view or for the very great number of applications obtained from this physical form. Indeed, in the dry solid state, although unknown, the temperatures of glass transition and melting of chitosan are certainly, as for cellulose, higher than its temperature of thermal decomposition, whose beginning is generally found to be close to 180°C (17). As a consequence, except for the native forms, the solid and gel states of chitosan are necessarily processed from solutions.

From a fundamental point of view, solution behavior of chitosan is informative of the properties of this polymer. Indeed, it reflects the role of the structural parameters and the molecular weight distribution, based on the interactions with the solvent and on the molecular and supramolecular organization. It also allows us to understand the various interactions in which these polymers can be involved. It is also very important to remember that the origin of the starting material and the quality of the industrial processes necessary for the production of these polymers play very important roles in determining their properties.

Considering the chemical structure schematized in Fig. 1, it is highly functionalized with a great variety of functions and atoms. When the glucosamine residues are in the free amino form, these structures are strongly involved in the formation of three-dimensional networks thanks to hydrogen bonding and hydrophobic and van der Waals interactions. The consequence is a full insolubility of this series of copolymers whatever the solvent or the degree of acetylation. Nevertheless, the solubilization can be achieved thanks to an influx of energy which depends on parameters such as the degree of acetylation, the molecular weight, and various environmental parameters.

1. Role of DA

When DA is below 60%, the polymer is soluble in dilute acidic solutions. This behavior is related to the fact that the protonation of the amino groups of glucosamine residues contributes to the disruption of hydrogen bonding, the solvation of the cationic sites, and then to the solubilization when the balance between solvent/polymer and polymer/polymer interactions becomes favorable. This limit is tightly related to the value of the intrinsic pKa (pK_0) of chitosan, which is found to be close to 6.5 for DA below 25% (18). For a given molecular weight (out of the range of the oligomers), the pH range where chitosan is soluble increases with the degree of acetylation when DA increases from 0 to 50% (19). This variation can be attributed to two contributions. One is related to the increase of the pK_0 value when DA becomes over 25%. This value should tend toward that of the monomer, close to 7.7 (20) and then contribute to an increase in solubility. The second is related to the increase of the stiffness of the polymer chain on increasing DA, in relation with an increase of the steric hindrance of acetylated residues and then to an increase of the excluded volume (3). The consequence is that when DA equals 0, which corresponds to a fully deacetylated structure, the polymer is only soluble in water for a range of pH within 2–6 (18), although when DA is close to 50%, the polymer becomes soluble whatever the pH.

When DA becomes over 60%, we enter the range of chitin and the chains become completely insoluble in water. This insolubility has to be related to the numerous hydrogen bonds occurring between the alcohol, amide, and ether functions distributed on the repeating units all along the polymer chains. They also correspond to hydrophobic interactions due to the presence of the methyl groups of the acetamide functions and to the —CH and —CH_2 of the glucosidic rings. Nevertheless, there also the solubilization can be achieved thanks to the use of hydrogen-bonding ruptors such as complexing agents of alcohol functions. Lithium ions are well known for this kind of interaction and contribute to destabilize the crystalline domains. The adjunction of some solvents such as N,N′-dimethylacetamide (DMAc) or N-methylpyrrolydone (NMP) allows the complete solubilization (21). Nevertheless, these solutions are not true solutions and contain an important amount of aggregates and microgels.

2. Role of Molecular Weight

For thermodynamic reasons, the solubility of neutral polymers is known to decrease with an increase of their molecular weight (22). In the case of chitosan, two factors play an additional role, the possibility of interchain association by hydrogen bonding and the polyelectrolyte character. The former behavior has been identified in the case of high molecular weight chitosan, whose aggregation capacity increases with molecular weight (23). The latter is classical and explains why chitosan is insoluble at low pH. Indeed, in these conditions, the ionic strength becomes sufficiently high to favor the condensation of the counterions. For these three reasons it is difficult to prepare high concentrations of chitosan chains. Another limitation is the increase of viscosity with the molecular weight, especially due to the electroviscous effects. Therefore, it is often difficult to prepare solutions with a concentration over 2–3% (w/w). On the contrary and for the same reasons, it is possible to increase the concentration on decreasing the molecular weight. In the case of very low molecular weight oligomers, for degree of polymerization (DP) below 8, the oligomers are highly soluble whatever the pH and DA (24). In this case, the increase of the pK_0 value (20) contributes to the enlargement of the pH range of solubility. In the case of intermediate molecular weights, for DP within 15–50, the behavior is similar to that of the high molecular weights as concerns the pH range of solubility, but the viscosity remains much lower and thus we can considerably increase the concentrations up to 20–30%. In the same range of DP, we also observe a typical behavior for fully deacetylated molecules. In this case, solutions preserve the memory of the crystalline organization in the solid state (25), especially when the molecules have been isolated in the free amino form. Thus, a monodisperse sample of a given molecular weight, after isolation in the free amino form and a storage of a few days, leads to a chromatographic analysis exhibiting a distribution showing several peaks corresponding to molecular weights multiple of the initial molecule (monomer, dimer, trimer, etc.) This behavior depends on the DP, the concentration, and the degree of crystallinity before dissolution. It does not change with time contrary to a classical process of agregation. It is interesting to remember that this kind of molecular weight crystallizes very easily (26) and has the dimensions that avoid the folding of the chains to build the crystallites. Therefore, in this case, the chains which have the same dimensions are packed in the crystallites in parallel. When these systems are placed in a solvent, the amorphous parts lead to isolated chains, although molecules constituting the crystalline parts are only partly dissociated and constitute arrangements of two, three, four chains, etc., which are either packed in parallel or attached together by their chain ends (25).

3. Role of Other Parameters

The solubility of chitosan is essentially governed by the possibility of formation of inter- and intramolecular hydrogen bonds. Then, all factors contributing to the contrary of this kind of interaction can contribute to solubilization. This is the case of a decrease of pH which allows the protonation of the amino sites. In the free amino form, we could consider the case of the complexation/solvation mentioned for chitin. Nevertheless, this situation does not operate with chitosan, certainly in relation with the too high polarity of water compared to solvents such as DMAc or NMP. We could also consider the formation of soluble complexes with some metals. Unfortunately, all the known complexes of chitosan in the free amino form are neutral, involving hydroxylated forms of metals which can also contribute to reinforce the network of hydrogen bonding (27).

Ionic strength is also a disfavorable parameter since it contributes to reduce the Debye atmosphere of the ionized sites of the polycationic chains and then to favor the ionic condensation, thus reinforcing the depletion of the polymer chains. This behavior is quite well illustrated when we study the hydrolysis of chitosan in acidic media. Indeed, initialy, the polymer chains are not soluble in relation with the polyelectrolyte behavior just described. On decreasing the molecular weight, the total ionic strength does not change, but the electrostatic potential of the chains decreases considerably and contributes to a critical molecular weight below which the chitosan molecules become soluble.

The dielectric constant of the media can also play an important role. Indeed, polar solvents favor dipolar and ionic interactions and should contribute to solubilize chitosan. The experimental results show that this parameter is not sufficient alone to induce the solubilization. The case of solvent blends is particuliarly interesting. In some conditions it can give rise to the gelation of chitosan instead of precipitation. This is particuliarly the case when solvents such as DMSO (16) or others (28) are added to a solution of chitosan in water. The decrease of the dielectric constant disfavors the ionization and ionic interactions, especially the protonation of the amino sites. This behavior is emphasized when a weak acid has been chosen to prepare the solution. The consequence is that, for a given concentration of the less polar solvent, the number of free amino groups becomes sufficient to allow the formation of a network of chitosan chains in which hydrogen bonds constitute the physical links. This situation is often metastable and turns to precipitation on increasing the concentration of the solvent of weaker polarity.

The solubility of chitosan can also be constrained by the use of acids which can interact with chitosan in a particular way. The most typical behavior is that of sulfate ions, which interact with chitosan according to two modes. The first is related to the very high salting out (dehydrating) power of these ions, as has been evaluated by the scale of Hofmeister (29). The second is that chitosan would form a typical complex with sulfate ions through intermolecular crosslinking (5). Although the latter point remains unclearly demonstrated, we can observe that chitosan is not soluble in the presence of sulfate ions.

The origin and the quality of the materials and processes can also play an important role on the solubility of chitosan. There are several sources of chitin and several recipes for the preparation of chitosan (5). Contrary to the wrong ideas published in the past, it is now definitely decided that the chemical structure is independent of the source. The only differences we can observe concern the values of DA and molecular weight. They can be related to the origin, but the most important parameter is the quality of the treatment. Thus, if the raw material is selected in order to eliminate the too highly pigmented part of the shells, the corresponding chitosan is less colored. The coloration can also be related to a too high temperature of deacetylation. As a consequence, the less colored chitosans are those obtained taking into account these parameters, and the best materials are logically those produced from endoskeletons of cephalopods which are not pigmented. The second problem concerns the presence of insoluble parts corresponding to particles or colloidal forms. The first case is due to impurities or a bad deacetylation; the second is essentially related to the possible presence of silica or lipids. There also, the quality of the process is important. A sufficient washing of the raw material can eliminate completely solid impurities. The elimination of organic colloids possibly due to the presence of residual lipids is more difficult to solve. Some methods used by producers consist of a washing with organic solvents or detergents before the last washing and drying. Endoskeletons of cephalopods are there also the most promising materials.

4. Molecular and Supramolecular Organizations in Solution

The conformation of chitosan in solution is not completely known. This polymer, depending on the parameters already discussed, is generally considered as adopting a disordered conformation with a certain stiffness (3). This stiffness is related to the relatively constrained β (1 → 4) backbone, less flexible than α (1 → 4) chains. Various models have been proposed in the literature to evaluate the stiffness. The Kuhn model consists in the definition of an equivalent statistical segment A_m (30), but is limited to conformations of relatively low stiffness. Beyond this limit, the ''wormlike'' model (31) defining a persistence length P_L is preferred. For moderate stiffnesses, the two models are equivalent. The values reported in the literature are generally deduced from viscometric measurements at various concentrations and ionic strengths. We can mention a value of the Kuhn segment found to be close to 23 nm (32) at a low ionic strength, and a value of the persistence length extrapolated near to 4.2 nm at infinite ionic strength (33). The stiffness of the chitosan chains is confirmed by the values of the Mark–Houwink–Sakurada exponent, which is located within 0.9–1.1 (23), values in agreement with semirigid conformations. As a conclusion, we can consider that the stiffness of chitosan in solution decreases with the deacetylation, ionic strength, and temperature (32), although no conformational transition order/disorder can be observed (33).

No ordered conformation has ever been observed in chitosan solutions at low concentration. Nevertheless, a kinetic process of aggregation has been identified by light scattering studies and steric exclusion chromatography (16). These aggregates are characterized by very high apparent molecular weights. The structure of these aggregates has not yet been described. Nevertheless, the presence of a supramolecular organization in these molecular associations has been reported (34).

5. Stability of Chitosan Solutions

We already mentioned the two limitations related to the polyelectrolyte character of chitosan in aqueous solution, in particular the possible precipitation at more or less high ionic strength and the viscosity which depends on the molecular weight and charge density.

As for all polysaccharides, the glycosidic linkage is sensitive to acidic hydrolysis, a phenomenon which depends on concentration, temperature, and pH (24). The consequence is a continuous decrease of the molecular weight of chitosan chains in solution. As for all natural polymers, there is also a possible contamination of the chitosan solutions by microorganisms specific to their chemical structure. As a consequence, in order to avoid these problems it seems better to limit the storage of these solutions and to work with the readily soluble forms of this polymer represented by chitosan salts.

6. Applications of Chitosan Solutions

Solutions of polysaccharides are necessary for the processing of articles for various uses either in solid form or as gels. Nevertheless, these solutions find many direct uses, and the literature is filled with several thousands of papers and patents reporting on these applications (5,35,36).

B. Hydrogels

Most of the polysaccharides are known to undergo gel formation. These gels are only obtained in the presence of an important proportion of water and belong to the family of hydrogels. They can be of either chemical or physical nature. In the first case, the three-dimensional network of the polymer chains responsible for the gel formation is due to covalent crosslinks and, as a consequence, these gels are not reversible. In the second case, the junctions between the chains are due to two kinds of interactions. In a first case, they can correspond to low energy interactions such as hydrogen bonding or van der Waals or hydrophobic interactions. In another case, they are due to the presence of some conformations stabilized by ionic interactions such as in the case of pectins and alginates (38). For most of the applications in which we are interested in this book, only physical gels can be considered as interesting. These gels are generally solvo- and thermoreversible.

1. Physical Gels

Physical gels have the advantage of being reversible in some conditions. They also correspond to molecular systems which have fully preserved their chemical structure of origin. Except in the case of gels stabilized by some ionic species, the biological properties of the polymer in the gel form are very close to that of the polymer in the solid or solution state. These gels are superabsorbants since they are generally prepared in conditions where the concentration of the polymer network is below 5% (w/w). They have a soft consistency and in numerous cases are highly transparent to the visible spectra.

2. Physical Gels Formed from Chitosan Solutions

a. Pure Chitosan Gels

If we consider chitosan solutions, the gel formation needs to be at a concentration sufficiently high to favor the chain entanglement useful for the formation of a continuous gel instead of microgels or aggregates. This critical concentration is well known as C* and is easily deduced from the inverse of the intrinsic viscosity [η].

The second parameter corresponds to the balance between attractive and repulsive interactions along the polymer chains and the ability to favor the formation of junction points by means of ordered domains. This balance can be influenced in different ways. We can play on the dielectric constant of the media. Then, a weak decrease of this parameter, associated with a control of the ionization state of the amine groups, favors the formation of hydrophobic interactions and hydrogen bondings. The latter constitute the junction points on chains where numerous hydrophilic interactions still exist and contribute to the existence of the chain segments between the junction points (28). The presence of ordered conformations as well as order–disorder transitions has never been identified in chitosan solutions (5). Moreover, the absence of any information on the presence of ordered domains in chitosan gels led us to suppose that gel formation from a chitosan solution could be due to a statistical reticulation.

The crosslinking of chitosan chains can also be achieved thanks to interchain ionic bridging. This behavior has been described in the case of various multivalent ions such as oxalates, sulfonates, or molybdates (38–40).

Whatever the case, in some circumstances, the gels are thermoreversible. They can be also fully regenerated in the form of pure chitosan gels after a complete exchange of the components of the initial media of their formation (38). Unfortunately, after this exchange, they become thermostable. This behavior must necessarily be related to the generation of other junction points than those initially responsible for gelation. These pure chitosan gels have then the same properties as chitosan both in solution and in the solid state.

b. Chitin Gels

The last possibility to play on the balance between segments/segments and segments/solvent interactions is to chemically modify the polymer chains by introduction of chemical groups as generators of hydrophobic interactions and hydrogen bonding. The most well-known physical gels formed by chemical modifications of chitosan in aqueous solutions are chitin gels. They were described for the first time in the literature by Hirano et al. (41). They observed their formation in the course of the reaction of acetylation of chitosan in aqueous media in the presence of acetic anhydride. These gels were described as rigid, thermostable, and soluble in formic acid. Unfortunately, in these conditions, both N- and O-acylations occurred. They were discarded to the benefit of true chitin gels discovered by the same authors. The presence of a hydroalcoholic media allowed them to avoid the O-acylation (42).

The formation of this kind of gel can only be achieved when chitosan is initially in the form of a salt of a weak acid such as acetic acid. Then, if chitosan is in the form of hydrochloride, the gel formation is not observed. This behavior must be related to the fact that the reaction of acylation needs the presence of free amino groups and that gelation only occurs at DA over 80%. In addition, as already mentioned, the value of pK_0 increases with DA. This is the reason why a weak acid favors gelation contrary to strong acids whose salts are always completely formed. As a consequence, the reacetylation in the case of a chitosan hydrochloride does not overcome 60% and then remains below the critical DA of gelation (43). The gel formation is obviously dependent on the molecular mobility of the polymer chains in solution and on the extent of acylation. This is to be related to the high DA necessary for gelation. It is interesting to notice that this value is independent of the nature of the alcohol (43). In fact, the role of the alcohol is only to reduce the dielectric constant of the media since it has no chemical function to play. According to its viscosity, it can operate in a role on the kinetics of gelation (43). The relatively high level of the acetylation degree necessary for gelation confirms a statistical process of gelation related both to the fact that no ordered conformation is generated and that no long-range homogeneously acetylated sequences are observed. It also shows that when the hydrophobic sites are short enough, the repulsive forces (electrostatic, steric hindrance) and the solvation by water molecules play very important roles. This kind of mechanism was confirmed by the fact that the initial value of the DA has no influence on the DA at which gelation occurs (43). Another confirmation was given by the fact that when the chain length of the acyl group increases, the value of the critical DA necessary for gelation decreases. It has also been shown that the molecular weight has no effect on the critical DA of gelation (43). Nevertheless, the kinetics of gelation increase with the molecular weight certainly in relation with a decrease of both the molecular mobility and the solubility parameter on increasing the molecular weight. It was also shown that below DP 280, only microgels were formed instead of a macrogel, and that no gel at all was observed for an oligomer of DP 7. These results confirm the mechanism of a statistical gelation with a probability to form a stable reticulation which is relatively weak when the concentration of polymer is below C*.

c. *Specific Properties of Chitin Gels*

Hydrogels are known to undergo the classic process of syneresis. This phenomenon is typically of thermodynamic origin and depends on the molecular mobility of chain segments between two junction points. As acetylation proceeds, the solubility of these segments decreases and the polymer network excludes progressively the solvent to favor segment–segment interactions with, as a consequence, a more or less important depletion of the gel. All parameters contributing to decrease this solubility contribute to this phenomenon (44). Syneresis is sensitive to the density of reticulation, which for reasons of van der Waals and electrostatic interactions disfavors this phenomenon. The consequence is that for high concentrations of polymer or low DA (near the limit of gelation), no syneresis is observed. External parameters such as pH, ionic strength, the dielectric constant, or temperature can also influence the solvent exclusion. In this case, the laws of electrostatic and osmotic mechanisms are generally followed. Then, according to the DA, the content of the gel, and the nature of the external media in which the gel is placed, we can observe either a swelling or a depletion of the gel (44). The relaxation of chain segments must also be considered for the understanding of the syneresis. Indeed, when the gel is formed, the chain segments between two reticulation points are necessarily stressed and then, for entropic reasons, they obey a relaxation process up to an equilibrium state with the thermodynamic parameters.

The mechanical properties of chitin gels are very interesting for their numerous possible applications. As for all gels, they obviously depend on the density of reticulation, the polymer concentration in the gel, and the molecular weight. The limitation comes from the fact that the best gels are necessarily obtained from the highest molecular weights, the highest DA, and the highest polymer concentrations. The consequence is that it is difficult to prepare chitin gels containing more than 2% polymer in relation with too high viscosities and kinetics of gelation.

These gels have the interesting property of being completely exchangeable in pure water with no major alteration of their properties (43,44). Thus, the initial hydroalcoholic media containing (after gelation) acetic acid is easily replaced by pure water after a sufficient contact with this solvent. This property reveals that the physical reticulation is not sensitive to this important change. It also signifies that these gels can be considered as having a very large potential of applications in the fields of cosmetics, biotechnology, biomaterials (45), pharmacology and so on.

These gels are certainly formed mainly due to the conjugation of hydrogen bonding and hydrophobic interactions. According to the value of DA, they also contain ionizable amino sites. From a physical point of view, in addition to their good mechanical properties, in relation to a relatively high polymer concentration, they necessarily have a high porosity (44). The consequence is that these gels are highly interactive and can be used to interact with a great variety of molecules.

C. Solid State

Chitosan exists in the solid state in various forms such as films, fibers, sponges, flakes, powders, and microparticles.

Its properties in this state depend on the origin of the material and on the method of processing. As for the other physical states, they also depend on DA and the molecular weight distribution. Moreover, they depend on whether chitosan is used as received from the manufacturer or it has been regenerated after solubilization. In the former case, the material has the memory of the crystalline organization in the cuticle since all along the industrial process of production the material has never been solubilized. In the latter case, the solubilization destroys this organization.

1. Crystallinity of Chitosan

The solid state of chitosan can vary first from a morphological point of view. This is obvious when we compare the micrographs of a sample of chitosan as received from the manufacturer to a lyophilizate of the same chitosan after dissolution with hydrochloric acid (Fig. 2) (46). In the case of the sample as received from the manufacturer, we observe a succession of dark and clear striations which correspond to the differences of opacity to electrons characteristic of the cholesteric geometry. This morphology is exactly that we observe in the cuticles of crab or shrimp. In this case, the material has never been dissolved during the industrial process and this result agrees with the fact that the material has maintained the supramolecular organization all along the process of extraction of chitin, then of deacetylation. Therefore, the material has physical behaviors re-

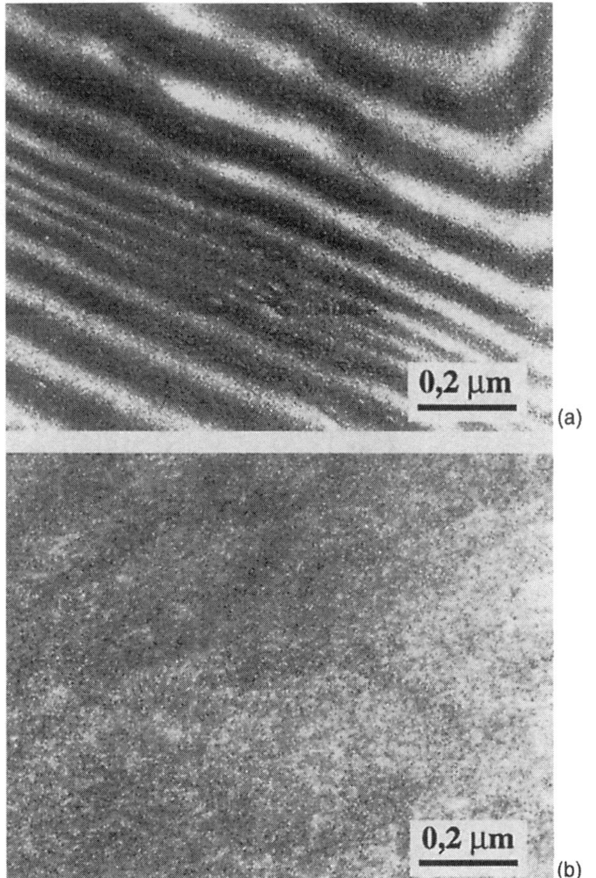

Figure 9.2 (a) Ultrafine cut of a sample of chitosan initially in the free amine form, as received from the manufacturer. TEM observation after contrasting with uranyl acetate (×30,000). (b) Ultrafine cut of a sample of chitosan lyophilized in the hydrochloride form, observed in the same conditions as for (a).

lated to this morphology. Thus, it remains highly crystalline with degrees of crystallinity close to 40% with relatively large crystallites (46). The consequence is a relatively low porosity of the material, which is essentially due to fractures of the supramolecular organization initially present in the cuticles or formed in the course of the industrial process. This kind of material has a relatively low accessibility and its hydration is particularly slow to be fully achieved (46).

In the case of the sample in the hydrochloride form, obtained after dissolution then lyophilization, the morphology represented in Figure 2.a shows the absence of a long-range supramolecular organization. The material appears as a sponge in which the periodic structure has completely disappeared. The volumic mass is only 0.016–0.02 g/cm^3 although it is measured as close to 0.12 g/cm^3 in the initial material (46). In the lyophilizate, the porosity is very high and the crystallinity is only 10%. This important decrease is also followed by a decrease of the length of the crystallites. The consequence of these important changes is an easy dispersion of this material in water with a relatively fast hydration certainly followed with a complete destruction of the crystallites (27).

The crystallographic structure of chitosan is necessarily related to the crystalline organization of chitin. Chitin exists under three allomorphs termed α, β, and γ. Only the first two are now completely accepted. α chitin, the most abundant structure, is present in the cuticles of all arthropods and fungi. It is characterized by an orthorhombic unit cell in which chains are arranged in an antiparallel manner following the *b* axis (47). This structure is stabilized thanks to numerous hydrogen bondings according to the three axes (*a, b*, and *c*) of the unit cell. β chitin is much less widespread in the biomass. The most important source is represented by the endoskeletons of cephalopods such as squid pens. The main difference with α chitin is the parallel arrangement of the polymeric chains whatever the axis, *a* or *b*. This organization contributes to reduce the length of the *b* axis and to slightly increase that of the *a* axis (48). We also notice an important difference due to the absence of hydrogen bondings between the chains, along the *b* axis. The consequence is that this chitin is chemically more accessible than α chitin (12).

According to the literature, the polymorphism of chitosan is more important than in the case of chitin. Its structure can be more or less modified by the insertion in the unit cell of small molecules such as water, some metals, or salts. Degree of acetylation and the salt or the free amine form are also important parameters (49). In the free amino form, two allomorphs are described. When chitosan is obtained from α chitin or when it has been regenerated after precipitation of a solution, the solid state corresponds to the allomorph ''tendon,'' which is a hydrated form (49). When this form, especially when it corresponds to a fully deacetylated chitosan, is heated up to 240°C, we observe another allomorph named '' annealed,'' whose main difference to the previous one is having no water in the unit cell. Nevertheless, this transition is never total and we always observe a mixture of the two allomorphs.

In the case of salt forms, two other kinds of allomorphs have been proposed. These salts can be obtained in sufficiently concentrated solutions of various acids thus avoiding the dissolution of chitosan. Salts of various acids have been tested, especially those prepared from HNO_3, HBr, HI, HF, HCl, and H_2SO_4 (50). For the first three acids, the structure is that of a helix 2_1 and contains no water. Nevertheless, these structures show some differences especially along the *a* or *b* axis, although the length of the *c* axis is preserved. This behavior must certainly be related to the fact that the protonation of the amine groups, its solvation, and the inclusion of anions in the structure only are contrary to the hydrogen bondings according to the *a* and *b* axes of the unit cell and have no great influence on the length of the chain axis essentially stabilized by a hydrogen bond between the —OH of C_2 and the heterocyclic oxygen atom. Another difference between the crystalline structures of these salts is that in the case of the two first salts, the unit cell is orthorhombic although it is monoclinic for the last one. In the case of the three last acids, the structures of the salts correspond to a monoclinic unit cell with polymer chains adopting a more extended conformation identified as a 8_3 helix. The cell unit, according to the salt, is more or less hydrated, but the number of water molecules included in the structure has not been really determined. For all the salts, the regeneration of the free amino form after dipping in a solution of sodium hydroxyde leads to a mixture of the allomorphs tendon and annealed in a proportion which depends on the nature of the acid. Thus, it seems that HF and H_2SO_4 favor the annealed structure more than the others. On the other hand, the proportion of the annealed structure seems to be dependent on the length of the chains. Thus, for DP within 30–50, it is possible to obtain monocrystals of the annealed structure (26).

The modification of the crystalline structure of α chitin during the process of deacetylation in NaOH at 95°C has been studyed by x-ray diffraction (51). The authors observed the progressive disappearance of the pics characteristics of the allomorph α to be progressively replaced by those specific of the allomorph tendon of chitosan. Nevertheless, in the course of this heterogeneous process of deacetylation one peak characteristic of the chitin structure persists up to degrees of deacetylation of 70% (51–53). Some authors have also observed a decrease of the degree of order during the conversion of chitin into chitosan (54) with, moreover, the quasicomplete disappearance of the 020 crystallographic plan associated to an anisotropic resistance of the network toward the stretching stresses. It has finally been proposed that the chains could be arranged differently whether the amino form is regenerated by precipitation of a solution or is obtained from a film dried in the salt form then neutralized in alkaline media (55,56).

This information on the morphology and the crystallography of chitosan is necessary for the understanding of the properties of this polymer in the solid state.

2. Solid Forms of Chitosan

a. *Flakes and Powders*

For a long time chitosan has been commercialized in the physical state achieved at the end of the process of deacetylation. This state corresponds to flakes of various forms and dimensions whose average length is generally close to 4–5 mm. These particles are the result of the treatment in heterogeneous media of the cuticles of shrimp or crab without any dissolution. These particles have the advantage of being easily manipulated, in particular in the case of their washing, filtration, collection, and drying. Their morphology and crystallinity have been described previously. This kind of particle presents several disadvantages. First, their volumic mass is relatively low and this fact contributes to increase the cost of their transportation. This form cannot be used for the filling of small volumes such as capsules. It is also disfavorable for direct uses of chitosan, especially those for which the specific area plays an important role. This is the case when chitosan is used for the sorption of chemical species such as metal ions (46). For these various reasons, all the manufacturers recently decided to commercialize chitosan in the solid state in the form of powders. In all cases, these powders are obtained by grinding the flakes described. In most processes, this grinding is not performed under a rigorous control of the local increase of temperature, which can be at the origin of a superficial thermal decomposition of the material associated with a yellowing. The best way consists of grinding in liquid nitrogen which completely avoids this problem. The particles can then be passed on sifters of different mesh allowing the preparation of particles of selected granulometry according to the application considered. Chitosan which has never been dissolved can be considered as a true porous material. Then, when particles of such material are used as sorbents of various types of molecules such as metal ions, the total specific area, which includes the sum of the porous area plus the superficial area, plays an important role in the kinetics of sorption (57). It can be easily increased by decreasing the dimensions of the particles. Nevertheless, there is a limit to the grinding related to the difficulty to collect small particles in a dispersed state. The specific area is certainly an important parameter for the circumstances in which chitosan has to be modified in heterogeneous media.

b. *Films*

Chitosan, as all members of the familly of linked β (1 → 4) polysaccharides, is known for its very good film-forming ability. The films are usually obtained from a solution of the polymer in the salt form. The solution is then cast on a surface such as a sheet of polystyrene. After evaporation of the solvent, we obain a film which is easily detached from the support. The film can be regenerated in the free amino form after it has been subjected to an alkaline solution. The thickness of these films can be controled either by the polymer concentration or by the height of the liquid to be evaporated.

The mechanic characteristics of chitosan films depend on various parameters. The most important is the hydration. Thus, for a chitosan film in anhydrous conditions, thc stress at break can be relatively high and close to 83 MPa (58) with a stretching close to 33% and a Young modulus near 3620 MPa. After an immersion one night in deionized water, we observe an important loss of the mechanical properties: 99% for the Young modulus and 88% for the stress at break. The behavior in aqueous media depends on the conditions of preparation of the membrane (59). Then, the elastic modulus increases with the concentration of the solution used for the preparation of the film. It also depends on the chemical structure of the acid used for the preparation of the solution. For the authors, the nature of the interactions, the concentration of chitosan, and the nature of the acid used could influence the conformation of the polymer chains during the formation of the film and, as a consequence, the density of entanglement of the chains in the solid state. Indeed, a densification of the material induces necessarily a decrease of the molecular mobility and then of the stress relaxation responsible of an increase of the elastic modulus. Nevertheless, the differences of behavior according to these parameters, in most cases, remain relatively low. It has also been demonstrated that a storage of the films at 23°C, 1 atmosphere, and 50% relative humidity had no particular influence on the mechanical properties of the films (60).

Membranes of chitosan show other physical properties, especially their very high impermeability to various gasses in their dehydrated form. The permeability depends on various parameters such as DA, the crystallinity, and the degree of hydration (28). These membranes of chitosan show a selectivity for the transport of some gasses, for example, toward CO_2 compared to O_2 (61). They are effective in applications of pervaporation of water–alcohol mixtures (62). The permeability of these membranes in aqueous media has also been studied, especially in the case of the diffusion of vitamin B12 (63).

All the properties of chitosan membranes associated with their physicochemical and biological properties are particularly interesting for various biomedical applications, such as for pharmaceuticals, wound coverings and tissue engineering.

c. *Chitosan Fibers*

The filmogenic properties of chitosan must necessarily be associated to its easy spinning. Due to the properties already mentioned, the spinning of chitosan is only possible by the method of wet spinning. Mild conditions of solutions of chitosan at concentrations close to 5–6% (w/w) prepared with acetic acid. This highly viscous collodion is pushed through a spinneret and the filaments are coagulated in an alkaline bath. After washing and drying we can obtain very nice chitosan fibers (64) whose tenacity in the dry state can be close to 2 g/Denier with an elongation at break near 14% and a Young modulus of 73.27 g/Denier (65). Chitosan fibers can then be processed in the form of yarns with direct applications or used to prepare woven or nonwoven fabrics with particularly interesting biological properties.

d. *Chitosan Sponges*

The filmogenic properties of chitosan can also be associated with its capability to form spongelike materials. These materials are easily formed from a solution or a dispersion of chitosan with a concentration located within 0.5–2%, followed by a lyophilization. These materials are highly expansed systems with open pores. Their volumic mass can be very low and close to only 0.016 g/cm^3 (46). Their prosity can be calculated from SEM studies, and parameters such as the concentration and the temperature of freezing considerably influence this parameter (66). It is then possible to prepare materials with average pore sizes varying within 20–250 μm. In fact, there are several levels of pore size depending on the scale of observation. Thus, the walls of the pores seen by SEM when observed by TEM reveal a secondary porosity due to the fibrous structure of these walls which is below 50 nm (46). When they are hydrated, these lyophilizates become very brittle and can be used as a source of particles when they are dispersed in water under heavy stirring. At higher densities, they can also constitute materials operating as a sponge with interesting biological properties.

e. *Microparticles of Chitosan*

Microparticles, especially microspheres of chitosan, can also be obtained from the method of emulsification, followed by an extraction/evaporation. Typically, the inner phase is constituted by a solution of chitosan in a mixed solvent such as acetic acid/acetone. This solution is emulsified in an outer phase constituted with a vegetal oil to which an emulsifier is added. Afterward, the solvent of the inner phase is evaporated and finally extracted up to completion leading to microspheres of diameter generally below 10 μm (67).

III. PHYSICOCHEMICAL PROPERTIES

If we consider the chemical structure of chitosan shown in Fig. 1, it bears atoms and functions allowing this structure to give rise to all the known interactions of chemistry.

The nature of these interactions depends particuliarly on two parameters, the degree of acetylation and the degree of ionization of the ionizable functions, i.e., the amine and alcohol functions.

A. Hydrogen Bonding

Alcohol, amine, amide, and ether functions present in the structure of chitosan can be involved in the formation of hydrogen bonds either with various substrates or by means of inter- and/or intramolecular bonding.

The most important substrate in the formation of this kind of interaction is water. It participates in the formation of hydrogen bonds responsible for the solubility of chitosan in water. Nevertheless, the competition between inter- and intramolecular bonding is always present. This competition is regulated by the balance between attractive and repulsive interactions. Here, the degree of ionization and DA can play an inportant role. When DA and the ionization degree are close to 0, the possibility of hydrogen bonding should be minimum. Nevertheless, we always observe the formation of aggregates. If DA increases, the stiffness increases but the second virial coefficient decreases in relation with the increase in the ability to form aggregates by intermolecular hydrogen bonds between the polymer chains. These aggregates have been described as corresponding to relatively ordered supramolecular organizations (35). Aggregation is a kinetic process which can be compared to crystallization (16). When the degree of ionization decreases, the repulsive interactions decrease, thus favoring the formation of hydrogen bonding leading to precipitation.

Other substrates can give rise to hydrogen bonding with chitosan. It is the case of some dyes (5) or polymers. In the latter case, we can mention the interaction of chitosan with collagen (68). Thus, when a very great excess of chitosan (over 200%, w/w) is added to a solution of collagen, we observe the formation of a precipitate in which the interactions between the two kinds of polymer chains are essentially due to hydrogen bondings. The complex can contain up to 85% (w/w) chitosan, which is largely beyond the maximum theoritical value (28.5%) calculated for a pure polyelectrolyte complex (PEC). The formation of such a complex is confirmed by IR spectroscopy analyses. It also allows us to show that this interaction is very strong and induces the denaturation of collagen characterized by the transformation of the threefold helixes into single helixes. This behavior is confirmed by the fact that this complex is not hydrolyzed by collagenases which specifically recognize the threefold helix (69).

Hydrogen bondings also plays an important role in the gelation process of chitosan (43).

In the solid state, the role of hydrogen bonding is very important. It is responsible for the stability of the ordered structures in the crystalline domains and of the interchain interactions which reinforce the stability of these domains. In some circumstances, hydrogen bonding with water can be observed in the unit cell (49).

B. Hydrophobic and van der Waals Interactions

This kind of interaction is often used to interpret some results especially in the case of amphiphilic polymers such as biopolymers. Thus, if we consider the structure of chitosan, the $-CH$, $-NH_2$, $-CH_2$, and $-CH_3$ groups are hydrophobic. The conformation of chitosan in solution is relatively stressed and the stiffness increases with DA. As a consequence, The deprotonation of amine groups and the increase of DA, temperature, and dielectric constant contribute to influence not only the hydrogen bonding, but also this kind of interaction. Thus, when we increase the temperature of chitosan, we contribute to decrease its solubility by the disruption of hydrogen bonding with water and increase the interactions mentioned previously. This behavior is emphasized when the dielectric constant of the media is decreased by addition of a solvent of lower dielectric constant such as an alcohol (43).

C. Electrostatic Interactions

1. General Aspects

These interactions are due to the ammonium form of the glucosamine residues and then to the charge density of the polymer chains. This property is quite well illustrated by the Katchalsky equation (70):

$$pK_a = pH + \frac{\log(1 - \alpha)}{\alpha} = pK_0 - \frac{\varepsilon\Delta\Psi(\alpha)}{KT} \tag{1}$$

The polyelectrolyte character of chitosan is illustrated by the second term on the right part of the equation which depends especially on the charge density of the polymer chains. From a theoretical point of view, if we assume the crystallographic parameters, the projection of the repeating unit on the chain axis is at least 0.512 nm (26). As a consequence, the charge density is relatively low and precludes the phenomena of ionic selectivity and ionic condensation.

The low value of pK_0 already mentioned (6.5) provides chitosan with interesting properties when it is in the salt form. Thus, in the case of a strong acid, the salt is quite stable and cannot be easily displaced. In the case of acetic acid, or weaker acids, the salt is not completely formed and then can be displaced especially in the pesence of a stronger acid or simply by elimination of the free acid. This behavior is illustrated especially when this salt is isolated in the solid form. In the case of acetate, the acid is weak, weakly soluble in water, and has a relatively low boiling point compared to other acids. The consequence is that if we prepare a film from a solution of chitosan in acetic acid, the solid obtained after evaporation is only partially in the acetate form, and if we let this film in ambient conditions, after a few days the free amine form is completely regenerated (17). This behavior is lowered in the case of higher molecular weight carboxylic acids in relation with the increase of their boiling points. A particular case is observed with α-hydroxyacids which can form stable complexes with chitosan, although their pK_a is higher than that of acetic acid (17).

2. Interaction with Lipids

The interaction of chitosan with lipids is particuliarly interesting if we consider the biological properties of chitosan or the role of this polymer as a food additive, allowing a decrease of the metabolization of lipids, especially fatty acids, triglycerides, and cholesterol.

First, salts of chitosan and lipids can be formed (71). When chitosan is added to solutions of lipids such as undecylenic acid, it induces an important decrease of the critical concentration of micellization, which is thus divided by three. In addition, the presence of chitosan is responsible for a phenomenon of flocculation dispersion of lipidic aggregates. In the case of undecylenic acid, the flocculation is maximum for a ratio glucosamine mole/fatty acid mole ranged near 0.25. On increasing the concentration of chitosan, we observe a strong redispersion which is much more stable than initially. The maximum of flocculation is then close to the maximum redispersion. This phenomenon is dependent on the ionic strength, pH, and concentration of lipid. Therefore, the maximum flocculation occurs sooner on increasing the ionic strength and decreasing the concentration. It is optimal at pH 5.8, which corresponds to the maximum electrostatic charge of the whole system. The particle size is minimum at the maximum of redipersion and depends also on these parameters (71).

It was also shown that for the optimal pH, the interaction between chitosan and the lipidic particles is essentially of electrostatic nature and is directly related to the DA of chitosan (72). The molecular weight plays another important role in the mechanism of flocculation dispersion. When the molecular weight increases, less chitosan is necessary to flocculate/redisperse the system, and the size of the redispersed particles increases. Adsorption studies based on the Langmuir model show that within a wide range of molecular weights, one major mechanism of interaction is involved in the process of flocculation, namely, electrostatic interactions of the mosaic type. Above a critical mass, another mechanism arises which corresponds to the polymer bridging with several particles (72).

Analyses by TEM and SEM (73) reveal that flocs obtained by means of high molecular weight chitosans correspond to catenary fractal structures in agreement with a theoretical model established for the flocculation of spherical particles by polymers. In addition, when chitosan chains are adsorbed on a polymeric surface such as a modified cellulose, the presence of lipids in solution allows the formation of an asymmetric membrane with one side constituted of pure chitosan and the other side of chitosan/lipid aggregates (73).

3. Polyelectrolyte Complexes

Another case of electrostatic interactions in which chitosan-can be involved corresponds to the formation of polyelectrolyte complexes (PECs) with polyanions. Polycationic polymers are particularly rare in nature and thus chitosan is a very interesting tool for this kind of interaction. This interaction plays an important role in living media and the problem of its possible appearance on introducing chitosan in such media is of particular interest. The formation of a PEC depends essentially on the charge density of each kind of polymer and on the pK_0 of the various ionic sites. If we represent the ammonium sites of the glucosamine residues by $-NH_3^+$, the optimal conditions for the formation of such complexes are summarized in the following equation:

$$-P^-, A^+ + -NH_3^+, B^- \Leftrightarrow -P\,NH_3 + A^+, B^- \quad (2)$$

where A^+ and B^- are the counterions of the polyanion and polycation, respectively. Two other reactions must be considered:

$$-P\,NH_3 + H^+ \Leftrightarrow -PH + -NH_3^+ \quad (3)$$

$$-P\,NH_3 + OH^- \Leftrightarrow -P^- + -NH_2 + H_2O \quad (4)$$

From these relations, we deduce that the maximum complexation occurs when the charge densities are maximum for each polymer and when their ionization state is maximum. In these conditions the complete formation of the PEC is achieved when the total amount of negative charges is exactly neutralized by the same amount of ammoium functions. At this point, the medium contains only the complex, which is electrically neutral and insoluble in water, and the stoechiometric amount of the AB salt liberated during the interaction (74). As shown by Eq. (3) and (4), the stabilty of the complex depends on the pK_0 of the anionic and cationic sites. Thus, when the ionic site corresponds to a weak carboxylic acid, the complex is easily destroyed on decreasing pH. This behavior explains why the PEC is only slightly formed when chitosan in the salt form is added to a solution containing the free acidic form of a weak polyanion. This situation is observed, for example, when chitosan is mixed with collagen at low pH (75). On the contrary, when the anionic sites correspond to strong acids such as sulfates, whatever the pH, these sites are always protonated and the complex is stable even at low pH as in the case of the sulfate sites of chondroitin sulfates (74). This behavior is also observed in the case of carboxylic sites of low pK as in the case of the uronic residues of hyaluronic acid whose pK_0 is close to 2.9 (76). In this case the complex can be formed even in acidic media (74). Furthermore, we observe a continuous decrease of the pH of the media during its formation related to a progressive deprotonation of the carboxylic sites. This behavior is possible provided that the pH of the media remains over the pK_0 of hyaluronic acid. When the pK_0 of the anionic sites increases, depending on both the concentration of the polymers and pK_0, we can observe an intermediate situation where only a part of the carboxylic sites initially in the free acidic form are deprotonated. This is the case of the carboxylic sites of chondroitin sulfates whose pK_0 is close to 3.83 (74).

IV. BIOLOGICAL PROPERTIES

A. Biodegradability

It is important to remember that chitosan is completely absent in mammals. The first interesting biological property of chitosan is its ability to be biodegradable and bioresorbable. Indeed its structure constitutes a source of carbon and nitrogen for various kinds of bacteria. Chitin deacetylases and enzymes hydrolyzing chitosan, such as chitinases, chitobiases, chitosanases, as well as glucosaminidases and N-acetyl-glucosaminidases, are now well known (36,37). If for some of them, the structure is clearly identified, the mechanisms of their activity remain obscure. In mammals, these enzymes seem to be completely absent. Nevertheless, when chitosan is implanted it disappears definitely after more or less time depending on the DA (77). Lysozyme, which is a nonspecific proteolytic enzyme widespread in this category of animals can hydrolyze chitosan, but this activity disappears rapidly when chitosan has a DA below 30% (78). Fully dacetylated chitosans are then completely insensitive to this enzyme. In addition, it seems that at least three consecutive N-acetylated residues are necessary to be recognized by this enzyme (79). The degradation of these polymers is nevertheless observed but the mechanism is not really elucidated. It is attributed to the presence of some oxidative agents produced by the immune system in the exudates (80).

It has been shown that whatever the circumstance, the biodegradation of chitosan is a phenomenon depending on several factors, especially the degree of acetylation, the molecular weight, the degree of crystallinity, the water content, and also the shape and the state of surface of the material (8,82).

B. Biocompatibility

1. Oral Administration

The consequence of oral administration of chitosan in humans has been studied in the case of dietetic applications (see Section IV.E). No adverse reaction has been observed for normal conditions of use. In animals, more information was obtained. Thus, in the case of the oral delivery of chitosan in rabbits, no particular adverse response of the host was noticed in normal conditions of administration (83). These conditions corresponded to the use of chitosan as a food additive with a ratio of 2% in the basal ration, for periods of 85 to 189 days. No significative difference of weight, height, color, fat deposition, or any pathology (lung, heart, stomach, intestin, or behavior) were observed. The LD_{50} was found as over 16 g/day/kg body weight, a dose which is higher than that of sucrose (12 g/day/kg body weight). The same experiments were performed on mice where the LD_{50} was lower, close to 10 g/day/kg body weight (84). The observation of the digestive organs revealed that chitosan flocculate with feed residues and remains in the duodenum and in the upper part of the jejunum. This observation confirms the effective dissolution of chitosan in the acidic conditions of the stomach (83). The perfect biocompatibility of chitosan after oral administration can also be related to the fact that there is no report mentioning the metabolization of this polymer through the intestinal cells.

2. Hemocompatibility

The hemocompatibility of chitosan is generally accepted. Added to blood, it induces the formation of a coagulum (85). In the case of solid forms of chitosan, the response is related to the surface properties (77). The first important results were obtained from experiments on dogs (86). It was shown that a graft of aorta covered with chitosan allowed a good integration of the synthetic material owing to a local hemostasis on the surface of the material created by the chitosan coating. Autopsies revealed that the graft was covered by normal epithelial cells. In vitro, the hemostatic properties of chitosan have been observed even in severe anticoagulating conditions (87) and in the case of an abnormal activity of platelets (88).

The consequences of direct injections of chitosan in blood have also been studied. In mice, the LD_{50} was found to be over 10 mg/day/kg body weight. In an experiment on rabbits, intravenous injections of chitosan oligomers (DP 2–8) were made at a dosage of 4.5 mg/kg body weight. No abnormal physiological symptom was noticed after treatment for 11 days (89). The serum hexosamine value increased on the second day after the last injection and decreased immediately on the next day. This kind of experiment revealed also the increase of the lysozyme activity in blood. Thus, for a dosage of 7.1–8.6 mg/day/kg body weight, the lysozyme activity was more than twice the normal value and was coming back to the normal 5 days after the last day of treatment. In the case of relatively high doses, severe diseases were observed. Thus, 200 mg/kg subcutaneous administration in dogs induces severe hemorrhagic pneumonia (90) and 10 mg/kg induces a systemic activation of chemiluminescence response (CL) in canine circulating polymorphonuclear cells (PMN). The increase of CL response was observed in the PMN recovered from the chitosan treated dogs after incubation with normal plasma and in the PMN of chitosan nontreated dogs after incubation with the plasma recovered from the chitosan treated dogs. When chitosan is administered to subcutaneous tissues of dogs and mice, in vitro studies (91) reveal that subcutaneous complement wastes rapidly by contact with chitosan material, due to the plasma C3 increase in these treated animals. On the contrary, when chitosan is administered to human plasma, C3 and C5 are decreased whereas C4 does not change. Therefore, there is no doubt that chitosan activates the complement alternative pathway. Nevertheless, in the alternative pathway, C3b is originated by a C3i enzymatic degradation of C3, but a formation of C3i due to chitosan remains unknown.

A last response concerns the production of interleukins by fibroblasts subjected to chitosan. The activation of the production of IL-8, in relation with the stimulation of both IL-1 and tumor necrosis factor-α (TNF-α), has been reported (91).

3. Cytocompatibility, Cell Adhesion and Cell Proliferation on Chitosan Films

The cytocompatibility of chitosan films, at physiological pH, toward keratinocytes, fibroblasts, or chondrocytes has been recently studied in vitro. In the case of the first two kinds of cells, the role of the degree of acetylation was also investigated (29). The results revealed that whatever the DA, all chitosan films are cytocompatible for these two kinds of cells. In the case of chondrocytes, there also, no cytoxocicity at all was detected in the case of a film with a DA of 2.5% (92). The adhesion tested on the two first types of cells showed that cell adhesion increases considerably on decreasing DA. This increase is important as long as the DA is low. In addition, for a given DA, fibroblasts appeared to adhere twice as much as keratinocytes on these films. Although alive, fibroblasts do not proliferate on chitosan films. This behavior is related to the too high adhesion of these cells on this surface, which inhibits their development. On the contrary, the proliferation is quite good in the case of keratinocytes. This proliferation increases when the DA of chitosan decreases. All these experiments contribute to show that electrostatic interactions between the negative charges of the cell membranes and the cationic charge of chitosan play the major roles. Thus, the adhesion seems to reach an optimum depending on these parameters.

Some studies were performed in vitro on the role of the addition of chitosan solutions to cell suspensions. In the case of fibroblasts, it was observed that the presence of chitosan at low concentration inhibited the proliferation of the cells but did not act as a cell killer (93). Although not claimed by the authors, it seemed that the microscopic views of the cell cultivations agreed with a mechanisms of cell agglutination in the presence of chitosan, which in turn could be the reason for the inhibition of cell proliferation without cell death induction (see the discussion in Section IV.B.4).

4. Bacterio- and Fungistatic Properties

Chitosan under various physical forms including solutions, powders, and films has been shown as inducing fungi- and bacteriostatic properties. These properties are relatively interesting since chitosan is not considered as toxic. As a consequence, they allow us to consider numerous applications for this polymer in various fields such as biomaterials, textiles, etc. Although extensively studied, these properties remain relatively unexplained with regard to the biological mechanisms involved.

Among the various types of bacteria and fungi, positive and significant results were obtained with the following families.

Bacteria: Escherichia (93), Pseudomonas (94), Lactobacillus (94), Staphylococcus (94,95), Micrococcus (94), Enterococcus (94), and Clostridium (94). In this field some antimitogenic activities were reported when chitosan was contacted to Salmonella associated to the mitogene Trp-P-1 (98)

Fungi and yeasts: Candida (97), Fusarium (94,97), Scombomorus (99), Pseudosciaena (99), Botritis, Saccharomyces (97), Pyricularia (97), and others

This kind of biological property must be first related to the physicochemical properties, in particular to the possibility for chitosan to flocculate dispersions of particles including cells and to be involved in hydrogen bond formation and polyelectrolyte interactions. The flocculating properties of cationic polyelectrolytes toward cell suspensions were demonstrated at least 30 years ago (100). An interesting study of the flocculation properties of chitosan in cell suspensions concerns the cases of *Escheriscia coli*, *Bacillus subtilis*, and *Zymomonas mobilis* (101). In this study, it was shown that the flocculation properties of chitosan were comparable to those of synthetic molecules known for their high potenty to induce this phenomenon. The role of pH showed that the maximum flocculation was located within 5.5–6 for the three kinds of microorganisms. It is interesting to notice that this optimal pH is quite similar to that observed in the flocculation of lipid aggregates (71). Unfortunately the study was limited at high pH by the precipitation of chitosan (over pH 6) where the authors observed a sudden decrease of the flocculation properties. The dose necessary to achieve the maximum property certainly depends on the nature of the cell in relation with the charge density of the cell surface but was comprised between 2 and 3.5 mg chitosan per gram of cells. Whatever the case, we can consider that the cell entrapment in chitosan flocs limits the cell mobility necessary for their division and then to their proliferation. This behavior is quite similar to that mentioned when epithelial cells such as fibroblasts adhere too much on the surface of a chitosan biomaterial. Nevertheless, in the conditions of maximum flocculation, the cells are maintained alive and then chitosan has ''cell-static'' properties. We can also consider as demonstrated in the case of plant cell suspensions that when chitosan concentration becomes sufficiently high to produce a redispersion of the flocs, the cell surface becomes highly positively charged and leads to the death of the cells (102). Fortunately, in relation with the physiological pH which is generally over 7, this situation cannot occur and chitosan can never be considered directly as a cell killer in vivo.

C. Bioactivity

Although considered as nontoxic, chitosan is often shown as being a strong elicitor of biological activity whether in plants or in animals. In both cases, the consequence of the contact between chitosan and living media has been extensively studied during the last 20 years. Some results in the case of plants could be used to understand some behaviors observed in animals.

1. ''Immune'' Response in Plants

Except one case mentioning the possible cell internalization of chitosan oligomers in plant cells (103), the crossing of the cell wall and the plasma membrane of these cells by chitosan has never really been demonstrated. As a consequence, the elicitation of various cell activities must be related to signals coming from the interactions between chitosan and the cell envelope. Among the elicitations reported in the literature, the induction of the production of callose has been the most studied. Thus, when chitosans of various degrees of acetylation and molecular weights were subjected to supension-cultured cells of *Catharanthus roseus*, the following results were observed (102): The elicitation of the formation of the β $(1 \rightarrow 3)$ glucan callose was not observed with low molecular weight oligomers. Callose synthesis increased with the degree of polymerization up to several thousands corresponding to a weight average molecular mass of 10^6 g/mol^1. At a comparable degree of polymerization, the increase of the acetylation degree contributed to decrease the elicitation. In the case of protoplasts, the biological response was greatly emphasized. These results showing the role of the charge density and molecular weight allowed the authors to consider that the primary interaction responsible of the biological signal may be attributed to an electrostatic interaction between the cationic charges of chitosan and the anionic charges of the phospholipidic structure of the cell membrane. This interaction is repressed by the electrostatic barrier constituted by the cell wall. In the case of another kind of cell such as those of *Rubus*, the elicitation of laminarinase was studied. Similar results were obtained with an additional interesting observation showing that the elicitation reached a maximum for a DP close to 3000 then decreased (104). This behavior has to be related to a change in the mechanism of interaction between chitosan and cells. In the first case, several chains can bind at the surface of one cell with a progressive lowering of the number of chains and cationic sites involved on each chain on increasing the molecular weight. In this case we observe a process of cell agglutination of mosaic type where the cell surface is certainly highly stressed. This stress increases with the molecular weight and then favors an increase of the elicitation process. Over a critical chain length, the mechanism becomes of interparticular type. In this situation, one chain can bind several cells. The consequence is a progressive decrease of the cell stress on increasing the molecular weight. In the case of *rubus* cells, a lectin specific for glucosamine oligomers especially of DP 4 was identified (105).

2. Immunological Properties in Animals

The antigenicity of chitosan has not already been demonstrated and antichitosan antibodies are not known. Nevertheless, this state of our knowledge does not signify that they do not exist. The immunologic activity of chitosan is particularly interesting and contributes to potentially very important applications of this polymer in the treatment of various tumoral afflictions and in the treatment of several pathologies of viral origin. Since chitosan can be degraded in living media, the question remains whether the biological activity is due to the monomer and oligomers or directly to the polymer. It seems that both kinds of involvement must be considered.

The activation of macrophages has been studied by several authors. Thus, the peritoneal exudate macrophage (PEM) chemotaxis has been demonstrated in vitro in the presence of chitosan. The experiments show that machrophages migrate in response to a positive gradient of chitosan, thus indicating a true chemotactic effect. Another consequence of this activation concerns the nitrite production by the PEM after preincubation with chitosan. This activity is related to the N-acetylglucosamine residues since the monomer glucosamine, contrary to NAG, has no influence on this response (106). This result is of great importance. It could signify that fully deacetylated chitosan is not biodegradable in mammals. Indeed, lysozyme has no action on it and nitrite production should not be elicited in its presence. On the contrary, the production of nitrite if it were confirmed in vivo should be associated with biocompatibility problems. Indeed, NO production contributes to amplify the inflammatory reaction leading to the destruction of both the polymer material and the tissue surrounding the implant. In vivo experiments performed by subcutaneous implantation of chitosan in mice reveal adverse inflammatory reactions at the level of the implants. After 2 weeks implantation, the foreign materials are loosely attached to the host tissues showing no integration at a macroscopic level. Histologic studies by TEM demonstrate the penetration and invasion of the cells into the chitosan material. Thus, many migrating macrophages and eosinophils are present inside the material or at the interface between this material and the host tissue in contact. Remnant particles of chitosan can be observed inside and outside the cells (106). Neverthless, vascularized neotissue is also observed. It can be considered that the complete degradation of the implant produces the progressive normalization of the histological picture with a final wound healing. This behavior explains the numerous results (see Section V.D.2) on the wound healing induced by chitosan following implantations of at least 2 months.

a. *Antitumoral and Antiviral Activity*

The antitumoral activity of glucosamine tested in vitro on human tumors shows a specific killer activity without any toxicity toward the normal tissues (107).

The antitumoral activity of chitosan oligomers especially the hexamer has been demonstrated in mice The use of this kind of oligomer is particularly interesting as an immunotherapeutic agent compared to other polysaccharidic systems. Indeed, their antigenicity has not been demonstrated and their solubility in water associated to their facile biodegradation allow them to be easily eliminated from the human body (108).

The antitumoral properties of the polymeric forms of chitosan are also demonstrated. They are related to the production of tumor necrosis factors, especially cytokines (109). They are also due to activation of macrophages leading to the stimulation of killer cells and T cells by the secretion of lymphokines. It is interesting to notice that it has no particular influence on the production of lymphokines by spleen cells (110).

The antitumoral activity of chitosan inducing the activation of killer cells, especially lymphocytelike cells, led us to consider that it can also act as a killer of virus-infected cells.

V. APPLICATIONS

A. Cosmetics

If we consider all the properties of chitosan mentioned, they necessarily predict numerous possible opportunities in the field of cosmetics. Thus, chitosan is a good moisturizer with no allergenic properties and possesses the two fundamental properties for this field of application (considering the nonmedical aspect of this domain). There are several hundred patents reporting on the possible applications of this polysaccharide in the various aspects of this branch. Thus, as mentioned, chitosan can be used as a lipidic dispersant and participates to the formulation of various cosmetic creams. Chitosan has filmogenic properties; it weakly interacts with proteins such as keratine and possesses cationic sites. Therefore, it can be used, especially in the case of low molecular weights, as aftershampoo giving a silky aspect associated with repulsive electrostatic properties. If the molecular weight is sufficiently low, the polymer film will be easily eliminated after a brushing or a new washing. The power to improve the glossiness associated to fixating and comb-out properties was claimed for formulations containing both chitosan and linear block copolymers of polysiloxane and polyoxyalkilene (111). The capability to act as a moisturizer can be illustrated by formulations containing chitosan and organic acids such as gluconic or succinic acids (112). The moisturizing properties of chitosan associated with an emulsifying character were claimed in the case of formulations corresponding to chitosan modified by acylation with fatty acid anhydrides or chlorides (113). The cationic character of chitosan was claimed in other cosmetic formulations for the cleaning and makeup remover of the skin and mucous membrane

(114). Chitosan has been proposed in toothpaste formulations, in the solid state for its abrasive properties associated to its possible anti–dental caries activity (115). Chitosan gels have good skin adhesive properties and could be used in the form of patches for the delivery of various substrates such as caffeine, vitamins, or nicotine. A complex between chitosan and iodine was proposed as a good disinfectant and wound healing system (116). Another interesting patent concerns the use of chitosan powder in the formulation of soaps. A soap was prepared from palm oil, fatty acids, tetrasodium edetate, and chitosan powder (1%, w/w). This preparation was then applied to patients with atopic dermatitis and revealed a decrease of their symptoms and no skin trouble formation (117).

Although the number of patents is very important in this field, surprisingly the number of papers describing the scientific background of cosmetic applications of chitosan is very poor. We can mention the use of chitosan associated with collagen and other glycosaminoglycans in a porous form as a dermal equivalent (DE) and a skin equivalent (SE) (118,119). The DE model was made from this three-dimensional matrix populated with human fibroblasts. The SE corresponded to the DE after seeding with normal human keratinocytes. These systems were then subjected to UVA in order to test in vitro the deleterious effects of this radiation. These systems were also tested in the presence of chemicals usually used for skin protection. A good accordance was obtained between the results on the cell viability and the IL-1 alpha release obtained both in vivo and in vitro with these models. As a consequence, they could be used to test the efficacy and toxicity of sunscreens. We can also note that chitosan could be possibly used itself as a sunscreen since UV spectra of chitosan solutions and films reveal a strong absorption below 200 nm, although it is quite transparent in the visible range (18).

Chewing gum supplemented with low molecular weight chitosans (3%) were used to evaluate the pH response of the human dental plaque (120). Thus, the role of these molecules was tested after an exposure to fermentable carbohydrates by means of a microelectrode. The fermentable substances corresponding to 5% glucose solution or sugared caramel were directly applied. Then, the subjects started chewing gums with or without chitosan. In the former case, the recovery of the normal pH (close to 5.5) was much faster than in the latter. This behavior can be simply related to the fact that chitosan as a polyamine has the potency to actively neutralize acidic media. A comparison was made between the capability of various buffering substances of different pK_a to inhibit the pH lowering of the dental plaque. Aspartame ($pK_a = 7.8$), phosphate buffer (7.1), low molecular weight chitosan (6.4), and maleate buffer (6.2) were tested after addition of lactic acid. It was shown that the pK_a value plays an important role since the most effective inhibition of pH lowering was observed both with chitosan and maleate buffer (121). The findings indicate that low molecular weight chitosans may be used as a food additive or, better, in the formulation of toothpastes to decrease the cariogenicity of foods.

B. Dietetics

Due to is polycationic character and its amphiphilic structure chitosan allows, as mentioned, the formation of all the known interactions in chemistry. Therefore, it can interact with numerous components of the diet, especially those of lipidic nature. It also exhibits some particular properties which in some cases could be at the origin of adverse responses.

1. Effect on Calcium and Vitamin Metabolism

The effect of chitosan intake was investigated in rats fed with diets containing 5% chitosan. The role of chitosan on the decrease of the whole-body retention of ^{47}Ca was demonstrated, although there was no significant increase of the calcium content in fecal excretions. In fact the ^{47}Ca elimination was identified by the increase of calcium in urinary excretions. These results suggest that dietary chitosan would affect the calcium metabolism in animals (122). This result was confirmed in the case of mice fed with a continuous and massive dose of chitosan. The decrease of calcium adsorption was characterized by an important decrease of the bone calcium content (123). In the case of the latter experiments, a marked decrease of the serum vitamin E level was also observed.

2. Effect on Intestinal Media

The effect of dietary chitosan on the induction and development of colonic precursor lesions was determined in CF1 mice fed with diets containing 2% of low or high molecular weight chitosans for a period of 6 weeks (124). Over a 2-week pretreatment period, one group of mice was treated with azoxymethane (four injections of 5 mg/kg body weight). Whatever the molecular weight of chitosans, an important reduction in the number of aberrant crypt foci cell proliferation, crypt height, and crypt circumference in the colonic epithelium was observed ($p < 0.05$). As a consequence, it seems that dietary chitosan reduces the incidence of early preneoplastic markers of colon carcinogenesis as evidenced by aberrant crypts and mitotic figures and altered crypt morphometrics in the murine colon. This property could be influenced by the molecular weight of chitosan.

3. Effect on Intestinal Immune System

Experiments made on rats fed with diets containing 5% chitosan revealed some changes in immune functions. Thus, the serum IgE was significantly lower while the IgM did not differ. IgA, IgG, and IgM in mesenteric lymph node (MLN) lymphocytes were unchanged, although the IgE concentration was also lower. The CD4+/CD8+ T cells in MLN lymphocytes were not affected nor the γ-interferon concentration or the tumor necrosis factor-α (125).

4. Hypocholesterolemic Properties

The most important studies on the use of chitosan as a food additive were performed on the effectiveness of this polymer to promote a hypocholesterolemic effect when it is associated with diet. The first results were published in 1978–1979 (126,127). The authors noticed that when rats were fed diets containing additional cholesterol, addition of chitosan induced a marked decrease of liver and plasma cholesterol. The magnitude of the reduction was comparable to that observed with cholestyramine but without any adverse secondary effects. The same authors enlarged their investigations to several series of experiments. They confirmed the first results and added that chitosan seemed to have no effect on bile acids and that the hypocholesterolemic activity had to be related to an increase of cholesterol in fecal excretions. They also mentioned the growth lowering of the rats fed with diets containing chitosan. The role of chitosan was effective from 2% (w/w diet) and the reduction of cholesterol was by 25–30% (128). In another study, these authors mentioned that on a large scale, the chain length does not play a major role. Nevertheless, they observed that the first terms of the series of the oligomers of glucosamine were ineffective (129). Similar experiments performed on rats (130) confirmed the previous results. In addition, it was mentioned that feeding with diets containing chitosan had no effect on hemoglobin and serum iron. The same authors observed that contrary to cholestyramine, there was no mucosal change in the gastrointestinal tract, except an increase of the distal small bowel. The hypocholesterolemic activity of chitosan was also shown in the presence of a hypocholesterolemic substance such as cystine (131). In the same manner, it was observed that the type of dietary fat influences the cholesterol absorption. Thus, the cholesterol uptake was more effective in the presence of sunflower oil than with palm oil (132). The results do not seem to vary considerably with the kind of animal. Thus, similar results were obtained with hens or rabbits where the decrease of the cholesterol was also observed in the blood plasma and liver (133,134). A particular case was mentioned in the second paper concerning rabbits. When these animals were fed with cholesterol enriched diet (up to 9 g/kg), the serum cholesterol and triacyglycerol levels of these high–serum cholesterol rabbits were affected neither by feeding with a 20 g/kg chitosan–supplemented diet nor by intravenous injection with chitosan of low molecular weight or chitosan oligomers daily. These results signify that chitosan has a hypocholesterolemic action only in the intestine.

The first human experimentation was reported in 1993 (135). Three to six grams per day of chitosan introduced in biscuits was given to eight healthy males. Their LDL and total cholesterol level decreased significantly. Moreover, serum HDL cholesterol was increased. In parallel, the excreted amounts of primery bile acids were significantly increased during ingestion of chitosan.

More recent studies concerned the treatment of 100, 90, 80, and 30 adults in four different places in Italy. Contrary to previous experiments, chitosan was directly ingested by means of tablets (four per day) containing 250 mg chitosan each. The experiments were performed in a randomized, double-blind, placebo-controled trial for 4 weeks. The results confirmed quite well those previously reported (135), but with more details and quantification. Lowering of LDL cholesterol up to 35.1% and increase of HDL cholesterol up to 11.12% were observed. Numerous other benefits related to chitosan ingestion were also mentioned, such as the decrease of arterial pressure and the increase of respiratory rate (136–139).

5. Fat Binding and Antilipidic Metabolism

In parallel with the ability to decrease the cholesterol level, numerous papers reported the capability of chitosan to bind fats and thus to decrease their metabolism by animals as well as humans. Thus, in a first report (128) a decrease of the triglyceride level in plasma and liver was observed in rats fed with diets containing chitosan. The inhibition of the absorption of fatty acids, bile acids, and triglycerides was claimed in a first patent in 1980 (140). A study on rats whose gut contents were examined 2 h after feeding with chitosan-containing diets gave interesting results. Thus, the ratio between the content of bile acids and phospholipids in the liquid phase and in the solid phase (after centrifugation) was significantly lowered compared to the control (without chitosan in diets). This result allowed the authors to show the uptake of bile acids and phospholipids by chitosan (141). When an emulsion containing various kinds of oils was introduced intragastrically in rats, the absorption of triglycerides was relatively low, showing that the gastric medium plays a major role on the uptake of triglycerides except for the very first terms of the series of the oligomers of glucosamine, which were inefficient (128). A synergistic effect related to the association of chitosan and ascorbic acid in diets was investigated in a study made on rats and by the observation of ileal contents (142). It was claimed that this association should allow the uptake of 3.4 g fat per 1 g chitosan. In the same kind of study, the same team showed that both the increase of the viscosity (equivalent to the molecular weight) and the deacetylation of chitosan favor an increase of the fat binding. They also noticed that their results were dependent on the nature of the fat. Thus, the excretion was more significant with corn oil than with lard (143). The hypolipidic capability of chitosan was also demonstrated on broiler chickens fed with diets containing chitosan. This behavior was also associated with a lower weight intake (144).

Human experiments on the lipid uptake by chitosan were more recently reported (136–139). The triglyceride reduction was 26.6% associated with a weight decrease of at least 7.9 kg after 4 weeks of chitosan treatment (four tablets containing 250 mg chitosan per day).

6. Mechanisms of Interaction Between Chitosan and Lipids

The mechanism of inhibition of lipid metabolization by chitosan can be attributed to either physicochemical interactions between chitosan and the substrate or to the inhibition of enzymatic mechanisms. Another question concerns the possible metabolization of chitosan and/or its degradation products. The intravenous injection of low molecular weight chitosan, as mentioned previously, has no influence on the cholesterol level in blood. In addition, although chitosan can be degraded by lysozyme produced by intestinal flora, up to this time there is no evidence of the metabolization of chitosan through the intestinal cells, whatever the molecular weight. Therefore, all the results reported in the literature agree with an effect only to the conditions of the gastrointestinal tract.

An important question concerns the possible involvement of the inhibition of pancreatic lipase by chitosan in the mechanism of antilipidic metabolism. A simple experiment (145) showed that chitosan inhibited the hydrolysis by lipase of triolein emulsified with phosphatidylcholine but not that of triolein emulsified with gum arabic and triton X-100. The conclusion is that the inhibition of lipid absorption by chitosan is essentially due to a physicochemical mechanism.

7. Conclusion

All the papers mentioned agree that chitosan has the capability to reduce significantly the intestinal metabolization of all kinds of lipids. When the absorption of chitosan does not exceed a critical amount, this activity seems to be achieved without any damage whether to the intestinal flora or to any tissue or mucous membrane in contact with the polymer. It is also interesting to note that chitosan seems to have no effect on the inhibition of lipid-specific enzymes.

The inhibition of lipid metabolism by chitosan seems essentially of physicochemical origin. In absence of chitosan, the metabolism is attributed to the micellization of lipids in the gastrointestinal tract thanks to the presence of micellizing substances such as biliary acids. These aggregates are stabilized by cholesterol and the whole can be then recognized by intestinal cells and transfered in the blood plasma.

When a solution of chitosan is added to a dispersion of lipids at 37°C (for example, oleic acid or various edible oils) in acidic conditions corresponding to the gastric media, no particular phenomena are observed and the dispersion is preserved after incubation during 30 min. If we increase the pH to intestinal conditions, we observe the formation of an impressive fatty coagulum contrary to the control (without chitosan). This fat clot corresponds to a coprecipitation of lipid aggregates formed in the presence of chitosan (see Section III.C.2) in a three-dimensional network. This clot obviously inhibits any possible recognition by lipases and intestinal cells. This mechanism agrees with all the results mentioned, especially the rejection of fatty acids, cholesterol, and triglycerides as well as biliary acids in feces. It also agrees with the noninhibition of lipases and the fact that the antilipidic activity of chitosan is not observed when chitosan is directly subjected to blood. It finally agrees with the fact that the molecular weight of chitosan is not very important except for molecular dimensions below a critical limit where the three-dimensional network cannot be formed. This mechanism, to be really observed, needs the prior solubilization of chitosan. In these conditions, we can observe in vitro a lipidic coagulum containing up to 30 g lipid per 1 g chitosan. It is important to remember that chitosan cannot be solubilized at pH below 2, in that case if chitosan is absorbed in a free amine form corresponding to a powder in capsules or tablets, the efficacy is very low and limited to the high intestine at the beginning of the increase of pH. This observation certainly aggrees with the low efficacy of most commercial formulations. It is also important to notice that all formulations including polyanions such as ascorbic acid, etc., necessarily contribute to limit this efficacy.

C. E-3-Pharmacology

Chitosan has very interesting physical and physicochemical poperties which provide it with numerous possible applications in the field of phamacolosy. Since another complete chapter in this book is devoted to this application, we will concentrate especially on the physicochemical and physical aspects. Thanks to its properties, chitosan has been extensively studied in the field of vectorization and sustained release of drugs. The bioactivity of this biopolymer has also been shown as favoring the targeting or the induction of some synergistic effects.

1. Vectorization

Chitosan is a biodegradable, bioresorbable, biocompatible, and bioactive polymer. Moreover, it is highly functionalized and as a consequence it can be considered as an interesting tool for the vectorization of drugs.

We first mention the simple case of the vectorization of anionic forms of acids (17). Indeed, chitosan is a weak alkali with a pK_0 close to 6.5. Thus, the salts prepared from weak acids and chitosan are easily displaced and anionic forms of acids are then liberated in the media. We have the particular case of α-hydroxyacids, which are weak acids but are strongly complexed with chitosan by means of hydrogen bonds between OH groups. This simple kind of vectorization can also be observed at physiologic pH where chitosan is essentially in the free amino form and is then a strong complexing agent of metals. These complexes are pending forms (27) and chitosan can be used for the vectorization of some transition or alkaline earth metals and actinides.

An important research activity has been developed during the last 10 years in Japan on the vectorization of 5-fluorouracyl. This drug is well known for its antitumoral activity but unfortunately also for its high toxicity. It was first bound covalently to a hexamethylene spacer by a carbamoyl bond (146), then to chitosan through an amide bond on the amine groups of the glucosamine residues. This system was successfully used to treat various tumors, especially fibrosarcoma cells. After inoculation of these cells to rats, the 5-fluorouracyl–chitosan conjugate was administered to animals at different doses and at various times. On day 30 after inoculation of the tumoral cells, the mice were sacrified. The inhibition ratio was then measured for the different cases. It is interesting to note that chitosan alone had an antitumoral activity. Also, a synergy was observed with the drug vector compared to treatment with the drug alone. In addition to the biological activity of chitosan, this behavior was attributed to the protecting effect brought about by the polymer avoiding the dilution effect of the drug in the organism and to the targeting property of chitosan toward the tumoral cells whose phospholipidic composition of their membrane is relatively different from that of healthy cells.

Magnetic microspheres of chitosan can be used to entrap some drugs. This complex system has the advantage of being easily transported under a magnetic field toward the site to be treated. We have a sort of a double vectorization. A combined emulsion/polymer crosslinking/solvent evaporation technique was used to obtain chitosan microspheres containing the anticancer drug oxantrazole (147). The particulates were crosslinked to minimize the biodegradation and reduce the porosity. It was possible to entrap up to 3% (w/w) oxantrazole. These particulates were administered to rats intra-arterially with solutions of the same drug as reference. After subjection of the animals to a magnetic field of 6000 G for 30 and 120 s, the observation of different tissues revealed a concentration of the drug in the brain 100-fold greater than that of reference with an ipsilateral repartition. Therefore, there was a targeting effect toward the brain tissues with a maximum efficacy achieved only after 30 s exposure to the magnetic field (148). Chitosan is a good tool for this kind of application involving magnetic particles. Indeed, as mentioned, this polymer has excellent complexing properties especially with metal oxides and hydroxides.

2. Chitosan as an Enhancer of the Permeability of Epithelial Tissues

Several experiments reveal the property of chitosan to enhance the permeability of various mucosal surfaces. This property has been tested on intestinal epithelial cells using Caco-2 cell monolayers. Measurements of the transepithelial electrical resistance (TEER) were performed at pH 6.2 after subjection to chitosan hydrochloride at a concentration within 0.25–1.5% (w/v). The presence of chitosan caused an immediate and pronounced lowering in TEER values of 84% after 2 h incubation. The permabilty was then tested by means of hydrophilic radioactive markers and revealed a large increase in the permeation of these markers. The increase of the transport of ^{14}C mannitol was increased 34-fold in the presence of chitosan. In addition, no deleterious effect to the cells could be demonstrated with tripan blue exclusion studies and confocal laser scanning microscopy. The conclusion is that chitosan, due to its chemical structure, especially in the presence of cationic charges even at physiologic pH, contributes to the opening of the tight junctions and allows the paracellular transport of drugs (149).

This kind of property contributed to the recent proposal of chitosan for the nasal drug delivery of insulin (150). Solutions or particulates of chitosan glutamate containing insulin were prepared and administered to sheep. The results showed that in the presence of chitosan, the delivery of insulin in the blood plasma was multiplied by a factor of 3 compared to the same treatment in absence of this polymer. Moreover, it was observed that the level of delivery was also multiplied by 4 when chitosan was administered in the solid form. These behaviors were attributed to the fact that chitosan can bind to the negative sites of the sialic residues in the nasal lining. Therefore, chitosan favors the paracellular transport of polar molecules thanks to a weakening of the tight junctions between epithelial cells. These interpretations are supported by an effect which increases when DA and pH decrease in relation with an increase of the charge density of the polymer chains.

3. Forms of Chitosan Containing Drug Delivery

The conjugation of all the properties of chitosan already mentioned point to this polysaccharide as a polymer of choice for drug delivery, via the two main routes, oral and injectable, or other pathways.

a. Oral Administration

Oral administration is certainly the most interesting for its easy processing by patients. Nevertheless, it exhibits numerous disadvantages, which in numerous circumstances limits its use. It is for this kind of drug administration that we will find the most important number of available physical forms of chitosan.

Chitosan can be proposed in the formulation of tablets in relation to several of its properties. Chitosan, especially in the salt form, is a strong water absorbent and provides tablets with the property of disintegrant in the gastrointestinal tract (151). Due to its capability to give rise to all the known chemical interactions, it constitutes an interesting binder for the processing of tablets (152). Its antiulcer properties related to its antiacid behavior in the free amino form allows its association to the formulations of tablets which can be used for the administration of ulcerogenic drugs such as aspirin (153).

Granules of chitosan-containing drugs have been proposed for their capability to slowly dissolve in the gastric media. The main difference between granules and beads is that in the former case, the disintegration is the major mechanism of drug delivery whereas diffusion is in the latter. These physical forms were tested by several authors in vitro and in vivo for various drugs such as acetaminophen, ampicillin, indomethacin, nifedipine, prednisolone, sulfadiazine, and theophiline (154–160). In vivo experiments all agree and reveal a delayed and prolonged effect and a high blood plasma level of drug quite different to that obtained with other means of administration.

The production of microspheres and microcapsules of chitosan-containing drugs has been extensively described in the literature (161–168). In these papers, many parameters were analyzed concerning their formation, their stability, and their release of drugs. There also, various drugs were tested including aspirin, diclofenac, furosemide, indomethacin, methotrexate, nifedipine, prednisolone, theophyline, and others.

Numerous other kinds of chitosan drug delivery systems were also proposed. Thus, as already mentioned, chitosan favors and stabilizes the micellization of lipids including phospholipids (71–73,169,170). These properties were applied to improve the oral absorption of insulin by means of chitosan-coated liposomes (171). Capsules of chitosan containing R68070 were recently used as antiinflammatory and for the treatment of ulcerative colitis, especially in the case of colon-specific delivery. A synergistic effect due to chitosan was observed in comparison with carboxymethylcellulose suspensions (172). We can also mention the studies concerning the immobilization of prednisolone (173) and indomethacin (174).

b. Administration by Other Routes

The oral route is limited by various problems, especially those related to strong irritations, the nonmetabolization through the intestinal mucosis, and the inactivation during transit in the gastrointestinal tract. The injectable, nasal, transdermal, and ophthalmic routes, can also be considered for chitosan-containing drugs. These routes are described elsewhere in this book.

D. Biomaterials

Chitosan can be processed according to a great variety of physical forms and as a general rule most of the forms used in the field of biomaterials can be processed from either solutions or gels. If we add to this possibility the fact that chitosan has very useful biological properties, we must necessarily also consider that it represents a very interesting tool for many applications in the field of biomaterials. Numerous results in the literature reveal that chitosan is a bioactive material which stimulates cell proliferation and contributes to tissue regeneration for either soft or hard tissues, whatever the nature of the cell constitution.

1. Cell Activation

The first important fact is that chitosan has an immunogenic activity, which has been already discussed. Cell activation is another important factor. Cytocompatibility and the cell adhesion observed in the presence of chitosan have also been addressed already. Administration of chitosan contributes to activate inflammatory cells leading to an enhancement of the antiphlogistic activity (175). In the presence of chitosan, mesenchymal cells generate a well-defined three-dimensional network. For most of the compartments of this scaffold an undifferentiated aspect is observed showing sometimes an endothelial appearance (176). In vivo studies on animals revealed that the organization of chitosan chains when submitted to living media where the polymer is insoluble contributes to stimulate the migration of polymorphonuclear and munonuclear cells (175). In the case of wounds, the presence of chitosan limits the exudates, thus demonstrating a positive response of the biological media (175). In this paper, it was also reported that chitosan is particuliarly effective in stimulating the formation of granulated tissues.

Chitosan has a chemical structure which is not so far from that of mammal glycosaminoglycans. Thus, it can constitute a material which mimics the extracellular matrix of numerous cells constituted partly of these molecules, alone or included in the structure of glycoproteins

We now examine various examples of applications of chitosan in the field of tissue regenaration.

2. Wound Healing of Soft Tissue

Artificial wounds created on animals, for example of 2 cm^2, were treated by a chitosan material. On day 28 after treatment, a lower level of inflammatory cells than in the control was observed. Moreover, the esthetic quality was much better (177). In another experiment performed on beagles, a cotton fiber type chitosan was applied on open skin wounds for 15 days. On day 3, a severe infiltration of PMN cells was observed with an increase of effusion. Granulation was more important in the presence of chitosan than in the control on days 9 and 15. An increase of the production of collagen III was also noticed. A mitotic effect was apparent on day 6 and was delayed compared to that of the control (178).

An interesting experiment was performed on a model of abcess created experimentally in dogs after a subcutaneous inoculation of *Staphylococcus aureus* T-6, with a 4-cm silk suture (179). After draining the pus, the abscess was treated with a suspention of finely granulated chitosan and other systems. After euthanasia on day 8, the contraction of the wound cavity was as high in the case of dogs treated with a sufficient dose of chitosan (>0.1 mg) than for those treated by ampicillin. Nevertheless the best results of healing on day 8 (90%) were obtained in the group of animals treated with chitosan (only 40% with ampicillin and no healing for the control groups or for those treated with a dose of chitosan below 0.01 mg). Moreover, histological observation showed the formation of highly vascularized granulation tissue in the group subjected to chitosan from 0.1 mg.

Chitosan was tested in vivo in other difficult conditions of wound healing. We can mention the case of the application of chitosan on wounds of the genitourinary system in dogs. Wounds were created in the kidney, ureter, and penile foreskin. Wound healing was normally acheived and a decrease of fibrosis was seen in the presence of chitosan compared to the control (180). Chitosan in the form of hydrogel obtained in the presence of ascorbic acid was also used for the regeneration of gingival tissue after an important surgical operation consecutive to the formation of dental pockets. After 2 months, the result was quite interesting and showed a very good tissue regeneration with the recovery of the gingival tissue over the level before treatment (181).

All the very good results mentioned concern epithelial tissue regeneration. In the case of endothelial tissues, the results are quite absent in the literature. This fact could be related to the absence of satisfying results in this field. This comment is corroborated by the experiments performed on the eyes of rabbits. Chitosan was applied topically after they had sustained central corneal wounds. Even after 21 days treatment, there were no apparent differences in corneal wound tensile strength between the chitosan-treated and the control eyes (182).

In the presence of cartilaginous material, chitosan can induce typical reactions. Thus, the influence of the injection of a chitosan solution in the rat knee articular cavity was studied up to 6 weeks after injection (0.2 mL of 0.1% chitosan solution). It was noticed that chitosan slows significantly the decrease in epiphyseal cartilage thickness and increases significantly articular cartilage chondrocyte densities. Nevertheless, also observed was an induction of the proliferation of fibrous tissue with abundant fibroblasts, fibrocytes, and monocytes (183). In fact, the latter problem can be related to a possible coprecipitation of chitosan and hyaluronane, proteins, etc., present in the joint cavity by PEC formation. This problem possibly does not exist when chitosan is introduced in the free amine form.

3. Osteogenesis by Chitosan

In vitro experiments can show the effect of chitosan on osteoblast differentiation and bone formation. Wells pretreated with chitosan solutions were seeded with stem cells. After 14 days of growth under optimal conditions, histological studies confirmed the bone formation. Image analysis led the authors to consider that chitosan potentiates the differentiation of osteoprogenitor cells and may facilitate the formation of bone (184).

In vivo experiments reveal the osteoinductive property of chitosan. Experiments were performed on rabbits by direct application of a paste made from chitosan and hydroxyapatite after removal of the osteomembrane. After 1 week treatment, immature slender trabecular bone was formed on the surface of the bone. At 2 weeks, the new bone tissue matrix showed a mosaic pattern and osteoblasts were arranged at the surrounding area. At 6 weeks the new bone tissue began to mature and continued its maturation up to 20 weeks. The new bone formation was observed to appear only at the interface between the biomaterial and the bone (185). Our experiment in this field (28) shows that the osteoinduction is due to chitosan and that hydroxyapatite has only a kinetic role on the tissue regeneration.

VI. CONCLUSION

In this chapter we gave an overview of the properties of chitosan, deliberately neglecting its derivatives. The reason for that is simple: there is no example in the literature showing better results obtained with chitosan derivatives in the field of biomedical applications.

It is very important to understand all the properties of this polymer in order to avoid misapplication or misinterpretation. Also, it is very important to consider that this polymer may interact with numerous substances when is introduced in the living media.

The results of the literature confirm quite well the interest in the use of such a polysaccharide for numerous applications in the life sciences. In the near future, we should see the application of this polymer in various physical forms on the corresponding markets. Some new directions solving the present limitations of its use should appear. To mention a few: the use of chitin and chitosan gels and true nanoparticles (no more than a few tenths of a nanometer).

Another field which is not sufficiently explored concerns the research on the immunogenicity of chitosan, which could give very interesting results from a molecule completely absent in mammals.

REFERENCES

1. T. Sannan, K. Kurita, Y. Iwakura. *Makromol. Chem.* 177: 3589 (1976).
2. K. Kurita, M. Kamiya, S.-I. Nishimura. *Carbohydrate Polymers* 16:83 (1991).
3. M. W. Anthosen, K. M. Vårum, Ø Smidsrød. *Carbohydrate Polymers* 22:193 (1993).
4. C. Jeuniaux, M. F. Voss-Foucart, M. Poulicek, J. C. Bussert. *Chitin and Chitosan*, G. Skjåk-Bræk, T. Anthonsen, P. Sandford (Eds.). Elsevier Applied Science, London, 1989, p. 3.
5. G. A. F. Roberts. *Chitin Chemistry*. Macmillan, London, 1992.
6. N. Hanson, G. Hanson. *Appl. Microbiol. Biotechnol.* 36: 618 (1992).
7. J. Montreuil. *Polysaccharides in Medicinal Applications*, S. Dimitriu (Ed.). Marcel Dekker., New York, 1996, p. 265.
8. H. Braconnot. *Ann. Chim. Paris*. 79:265 (1811).
9. A. Odier. *Mem. Soc. Histoire Natl. Paris* 1:29 (1923).
10. C. Rouget. *Comptes Rendus Acad. Sci.* 48:792 (1859).
11. F. Hoppe-Seyler. *Ber.* 27:3329 (1894).
12. K. Kurita, K. Tomita, T. Tada, S. Ishii, S.-I. Nishimura. *J. Polymer Sci. Part A: Polymer Chem.* 31:485 (1993).
13. K. Shirai, I. Guerrero-lagarreta, G. Rodriguez-Serrano, S. Huerta-Ochoa, G. Saucedo-Castaneda, G. M. Hall. *Advances in Chitin Science*, Vol. II, A. Domard, G. A. F. Roberts, K. M. Vårum (Eds.). Jacques André, Lyon, 1998, p. 56.
14. G. W. Rigby. U.S. patent US 204087 (1934).
15. E. I. du Pont de Nemours & Company. UK patent UK 458839 (1936).
16. A. Domard, M. Rinaudo. *Int. J. Biol. Macromol.* 5:49 (1983).
17. S. Demarger-André, A. Domard. *Carbohydrate Polymers* 23:211 (1994).
18. A. Domard. *Int. J. Biol. Macromol.* 9:98 (1987).
19. M. H. Ottøy, K. M. Vårum, Ø. Smidsrød. *Carbohydrate Polymers* 29:17 (1996).
20. S. Tsukuda, Y. Inoue. *Carbohydr. Res.* 88:19 (1981).
21. P. R. Austin. German patent Ger. 2707164 (1977).
22. P. J. Flory. *Principles of Polymer Chemistry*. Cornel University Press, Ithaca, New York, 1967.
23. M. W. Anthonsen. *PhD thesis*. Trondheim, Norway (1993).
24. A. Domard, N. Cartier. *Int. J. Biol. Macromol.* 11:297 (1989).
25. A. Domard, N. Cartier. *Int. J. Biol. Macromol.* 14:100 (1992).
26. N. Cartier, A. Domard, and H. Chanzy. *Int. J. Biol. Macromol.* 12:289 (1990).
27. E. Piron, A. Domard. *Int. J. Biol. Macromol.* 22:33 (1998).
28. A. Domard et al. (to be published).
29. W. Brümmer, G. Gunzer. *Biotechnologies*, J. F. Kennedy, H. J. Rehm, and G. Reed (Eds.), Weinheim, New York, 1987.
30. W. Kuhn. *Kolloid-Zeit.* 68:2 (1934).
31. O. Kratky, G. Porod. *Recueil Trav. Chim. Pays-Bas.* 68: 1106 (1940).
32. N. V. Pogodina, G. M. Pavlov, S. V. Bushin, A. B. Mel'nikov, Y. B. Lysenko, L. A. Nud'ga, V. N. Marsheva, G. H. Marchenko, V. N. Tsvetkov. *Vysokomol Soyed.* A28: 232 (1986).
33. A. Domard, M. Rinaudo. *Chitin and Chitosan*, G. Skjåk-Bræk, T. Anthonsen, P. Sandford (Eds.) Elsevier Applied Science, London, 1989, p. 71.
34. M. W. Anthonsen, K. M. Varum, A. M. Hermanson, Ø. smidsrød, D. Brant. *Carbohydr. Polym.* 25:13 (1994).

35. A. Domard, Ch. Jeuniaux, R. Muzzarelli, G. Roberts (Eds.) *Advances in Chitin Science*, Vol. I. Jacques André Press, Lyon, 1996.
36. A. Domard, G. A. F. Roberts, K. M. Vårum (Eds.) *Advances in Chitin Science*, Vol. II. Jacques André Press, Lyon, 1998.
37. J. M. Guenet (Ed.) *Thermoreversible Gelation of Polymers and Biopolymers*. Academic Press, New York, 1992.
38. S. Hirano, R. Yamaguschi, N. Fukui, M. Iwata. *Carbohydr. Res.* 201:145 (1990).
39. G. A. F. Roberts. *Chitin and Chitosan*, G. Skjåk-Bræk, T. Anthonsen, P. Sandford (Eds.). Elsevier Applied Science, London, 1989, p. 479.
40. K. L. Draget, K. M. Varum, E. Moen, H. Gynnild, Ø. Smidsrød. *J. Appl. Polym. Sci.*, 50:2021 (1993).
41. S. Hirano, R. Yamaguchi. *Biopolymers* 15:1685 (1976).
42. S. Hirano, Y. Ohe, H. Ono. *Carbohydr. Res.* 47:315 (1976).
43. L. Vachoud, N. Zydowicz, A. Domard. *Carbohydr. Res.* 302:169 (1997).
44. L. Vachoud, N. Zydowicz, A. Domard. (submitted for publication).
45. A. Domard, B. Grandmontagne, T. Karibian, G. Sparacca, H. Tournebise. French patent FR 95 52016 (1995).
46. E. Piron, M. Accominotti, A. Domard. *Langmuir* 13:1653 (1997).
47. R. Minke, J. Blackwell. *J. Mol. Biol.* 120:167 (1978).
48. J. Blackwell. *Biopolymers* 7:281 (1969).
49. K. Ogawa. *Agric. Biol. Chem.* 55:2375 (1991).
50. K. Ogawa, S. Inukai. *Carbohydr. Res.* 160:425 (1987).
51. G. L. Clark, A. F. Smith. *J. Phys. Chem.* 40:863 (1936).
52. K. Kurita, T. Sannan, Y. Iwakura. *Makromol. Chemie.* 178:3197 (1977).
53. S. I. Aiba. *Int. J. Biol. Macromol.* 13:40 (1991).
54. B. Focher, A. Naggi, G. Torri, A. Cosani, M. Terbojevitch. *Carbohydr. Polym.* 18:43 (1992).
55. K. Sakurai, M. Takagi, T. Takahashi. *Sen-i Gakkaishi* 40: T-246 (1984).
56. K. Sakurai, T. Shibano, K. Kimura, T. Takahashi. *Sen-i Gakkaishi* 41:T-361 (1985).
57. I. Saucedo, E. Guibal, C. Roulph, P. Le Cloirec. *Water SA* 19:119 (1993).
58. P. Wanichpongpan, S. Chandrkrachang. In: *Advances in Chitin Science*, Vol. II, A. Domard, G. A. F. Roberts, K. M. Vårum (Eds.). Jacques André Press, Lyon, 1998, p. 449.
59. C. A. Kienzle-Sterzer, D. Rodriguez-Sanchez, C. Rha. *Makromol. Chemie* 183:1353 (1982).
60. B. L. Butler, P. J. Vergano, R. F. Testin, J. M. Bunn, J. L. Wiles. *J. Food Sci.* 61:953 (1996).
61. R.-K. Bai, M.-Y. Huang, Y.-Y. Jiang. *Polym. Bull.* 20:83 (1988).
62. T. Iwatsubo, T. Masuoka, K. Mizoguchi. *J. Membrane Sci.* 65:51 (1992).
63. S. Nakatsuka, A. L. Andrady. *J. Appl. Polym. Sci.* 44:17 (1992).
64. J. Knaul, M. Hooper, C. Chanyil, K. A. M. Creber. *J. Appl. Polym. Sci.* 69:1435 (1998).
65. J. Knaul, S. M. Hudson, K. A. M. Creber. *J. Appl. Polym. Sci.* 72:1435 (1999).
66. S. V. Madihally, H. W. T. Matthew. *Biomaterials* 20:1133 (1999).
67. C. Remunian-Lopez, M. L. Lorenzo, A. Portero, J. L. Vilajato, M. J. Alonso. In: *Advances in Chitin Science*, Vol. II, A. Domard, G. A. F. Roberts, K. M. Vårum (Eds.) Jacques André Press, Lyon 1998, p. 600.
68. M. N. Taravel, A. Domard. *Biomaterials* 16:865 (1994).
69. M. N. Taravel, A. Domard. *Biomaterials* 17:451 (1995).
70. A. Katchalsky. *Pure Appl. Chem.* 26:159 (1971).
71. S. Demarger-André, A. Domard. *Carbohydr. Polym.* 22: 117 (1993).
72. S. Demarger-André, A. Domard. *Carbohydr. Polym.* 24: 177 (1994).
73. S. Demarger-André, A. Domard. *Carbohydr. Polym.* 27: 101 (1995).
74. A. Denuzière, D. Ferrier, and A. Domard. *Carbohydr. Polym.* 29:317 (1996).
75. M. N. Taravel, A. Domard. *Biomaterials* 14:1993 (1993).
76. R. L. Cleland, J. L. Wang, Detweiler. D. M. *Carbohydr. Res.* 50:169 (1982).
77. K. Y. Lee, W. S. Ha, W. H. Park. *Biomaterials* 16:1211 (1995).
78. S. Hirano, H. Tnui, N. Hutadilok, H. Kosaki, Y. Uno, T. Toda. *Polym. Mater. Sci. Eng.* 66:348 (1992).
79. S. Aiba. *Int. J. Biol. Macromol.* 14:225 (1992).
80. S. Minami, Y. okamoto, A. Matsuhashi, S. Tanioka, Y. Shigemasa. *Zeitai Zairyo* 13:112 (1996).
81. H. Struszcyk, D. Wawro, A. Niekraszewicz. In: *Advances in Chitin and Chitosan*, C. J. Brine, P. A. Sandford, J. P. Zikakis (Eds.). Elsevier Applied Science, New York, 1991, p. 580.
82. S. Tokura, S. I. Nishimura, N. Sakairi, N. Nishi. *Macromol. Symp.* 101:389 (1996).
83. S. Hirano, H. Seino, Y. Akiyama, I. Nonaka. *Polym. Mater. Sci. Eng.* 59:897 (1988).
84. K. Arai, T. Kinumaki, T. Fiyita. *Bull. Tokai. Reg. Fish. Res. Lab.* 56:89 (1968).
85. S. Hirano, Y. Noishiki, J. Kinugawa, H. Hihigashijima, T. Hayashi. *Polym. Sci. Technol. (Adv. Biomed. Polym.)* 35: 285 (1987).
86. W. Malette, H. Quigley, R. Gaines N. Johnson, W. Rainer. *Ann. Thor. Surg.* 36 (1983).
87. G. Fradet, S. Brister D. S. Mulder, J. Lough, B. L. Averbach. In: *Chitin in Nature and Technology*, R. Muzzarelli, C. Jeuniaux, G. W. Gooday (Eds.). Plenum Press, New York, 1986, p. 443.
88. P. Subar, P. Klokkevold. *Dentistry* (US) 12:18 (1992).
89. S. Hirano, H. Inui, M. Iwata, K. Yamanaka, H. Tanaka, T. Toda. In: *Progress in Clinical Biochemistry*, K. Miyai, T. Kanno, E. Ischikawa (Eds.). Elsevier Applied Science, London, 1992, p. 1009.
90. S. Minami, M. Oh-oka, Y. Okamoto, K. Miyatake, A.

Matsuhashi, Y. Shigemasa, Y Fukumoto. *Carbohydr. Polym.* 29:241 (1996).
91. S. Minami, Y. Okamoto, T. Mori, T. Fujinaga, Y. Shigemasa. In: *Advances in Chitin Science*, Vol. II, A. Domard, G. A. F. Roberts, K. M. Vårum (Eds.). Jacques André Press, Lyon, 1998, p. 633.
92. A. Denuzière, D. Ferrier, O. Damour, A. Domard. *Biomaterials* 22:1275 (1998).
93. W. G. Malette, H. J. Quigley, E. D. Adickes. In: *Chitin in Nature and Technology*, R. Muzzarelli, C. Jeuniaux, G. W. Gooday (Eds.). Plenum Press, New York, 1986, p. 435.
94. A. Taiho. Japan patent JP 62 83877 (1987).
95. W. Guanghua, Z. Yanwan. *Yu Fajiao Gongye* 2:1 (1992).
96. N. Qinxiang, W. Tuanghua, Z. Yanwan. *Proc. Int. Congr. Meat Sci. Technol. 37th* 2:60 (1991).
97. C. Van Minh, P. Hu'u Dien, D. Lan Hu'ong, T. Du'uc Hu'ng, H. Than Hu'ong. *Tap Chi Hoa Hoc* 34:29 (1996).
98. U. C. Guyuen, S. Takashi. Japan patent JP 08 81379 (1996).
99. Y. Guangli, L. Xiaohong, W. Shi, T. Xualin. *Zhongghuo Haiyang Yaowu* 13:45 (1994).
100. W. C. MacGregor, R. K. Finn. *Biotech. Bioeng.* 11:127 (1969).
101. J. Hugues, D. K. Ramsden, K. C. Symes. *Biotech. Tech.* 4:55 (1990).
102. H. Kauss, W. Jeblick, A. Domard. *Planta* 178:385 (1989).
103. D. F. Kendra, L. A. Hadwiger. *Exp. Mycol.* 8:276 (1984).
104. Y. Lienart, C. Gauthier, A. Domard. *Phytochemistry* 34: 621 (1993).
105. Y. Lienart, C. Gauthier, A. Domard. *Planta* 184:8 (1991).
106. G. Peluso, O. Petillo, M. Ranieri, M. Santin, L. Ambrosio, D. Calabro, B. Avallone, G. Balsamo. *Biomaterials* 15: 1215 (1994).
107. D. S. Matheson, B. J. Green, S. J. Friedman. *J. Biol. Resp. Modif.* 3:445 (1984).
108. K. Suzuki, T. Mikami, Y. Okawa, A. Tokoro, S. Suzuki. *Carbohydr. Res.* 151:403 (1986).
109. K. Nishimura. In: *Chitin Derivatives in Life Sciences*, S. Tokura (Eds.). Plenum Press, New York, 1986, p. 477.
110. K. Nishimura, S. Nishimura, H. Seo, N. Nishi, S. Tokura, I. Azuma. *J. Biomed. Mater. Res.* 20:1359 (1986).
111. C. Dupuis. French patent FR 9310967 (1967).
112. N. Hugues, French patent FR 8911170 (1991).
113. A. Huc, D. Antoni, E. Perrier. French patent FR 9100758 (1991).
114. A. Vanstraceele, C. Marion. French patent FR 9606287 (1996).
115. A. Yoichi, I. Takateru, Y. Giichi. Japanese patent JP 9249541 (1997).
116. A. De Rosa, A. Rossi, P. Affaitati. International patent WO 94/26788 (1994).
117. A. Toshiya. Japanese patent JP 11124324 (1997).
118. C. Augustin, C. Collombel, O. Damour. *Photochem. Photobiol.* 66:853 (1997).
119. C. Augustin, C. Collombel, O. Damour. *Photodermatol. Photoimmunol. Photomed.* 13:27 (1997).
120. K. Shibasaki, H. Sano, T. Matsukubo, Y. Takaesu. *Bull. Tokyo Dent. Coll.* 35:61 (1994).
121. K. Shibasaki, H. Sano, T. Matsukubo, Y. Takaesu. *Bull. Tokyo Dent. Coll.* 35:27 (1994).
122. M. Wada, Y. Nishimura, Y. Watanabe, T. Takita, S. Innami. *Biosci. Biotechnol. Biochem.* 61:1206 (1997).
123. K. Deuchi, O. Kanauchi, M. Shizukuishi, E. Kobayashi. *Biosci. Biotechnol. Biochem.* 59:1211 (1995).
124. T. L. Torzsas, C. W. Kendall, M. Sugano, Y. Iwamoto, A. V. Rao. *Food Chem. Toxicol.* 34:73 (1996).
125. B. O. Lim, K. Yamada, M. Nonaka, Y. Kuramoto, P. Hung, M. Sugano. *J. Nutr.* 127:663 (1997).
126. M. Sugano, T. Fujikawa, Y. Hiratsuji, Y. Hasegawa. *Nutr. Rep. Int.* 1:53 (1978).
127. T. Kobayashi, T. Fujikawa, M. Sugano. Japanese patent JP 79148090 (1979).
128. M. Sugano, T. Fujikawa, Y. Hiratsuji, K. Nakashima, N. Fukuda, Y. Hasegawa. *Am. J. Clin. Nutr.* 33:787 (1980).
129. Y. Sugano, S. Watanable, A. Kishi, M. Izume, A. Ohtakara. *Lipids* 23:91 (1998).
130. I. Ikeda, Y. Tomari, M. Sugano. *J. Nutr.* 119:1383 (1989).
131. Y. Aoyama, H. Matsumoto, T. Tsuda, E. Ohmara, A. Yoshida. *Agric. Biol. Chem.* 52:2811 (1988).
132. I. Ikeda, Y. Tomari, M. Sugano. *J. Nutr.* 119:1383 (1989).
133. M. Waki, M. Kataoka, R. Ino. *Kenkyu Hokoku-China-ken Chikusan Senta* 13:67 (1990).
134. S. Hirano, Y. Akiyama. *J. Sci. Food Agric.* 69:91 (1995).
135. Y. Maezaki, K. Tsuji, Y. Nakagawa, Y. Kawai, M. Akimoto, T. Tsugita, W. Takekawa, A. Terada, H. Hara, T. Mitsuoka. *Biosci. Biotechnol. Biochem.* 57:1439 (1993).
136. A. Giustina, P. Ventura. *Acta Toxicol. Ther.* 16:199 (1995).
137. A. M. Sciutto, P. Colombo. *Acta Toxicol. Ther.* 16:215 (1995).
138. G. Veneroni, F. Veneroni, S. Contos, S. Tripoli, M. De Bernardi, M. Guarino, M. Marletta. *Acta Toxicol. Ther.* 17:53 (1996).
139. G. Macchi. *Acta Toxicol. Ther.* 17:303 (1996).
140. I. Furda. U.S. patent US 4223023 (1980).
141. K. Ebihara, B. Schneeman. *J. Nutr.* 119:1100 (1989).
142. O. Kanauchi, K. Deuchi, Y. Imasato, M. Shizukuishi, E. Kobayashi. *Biosci. Biotech. Biochem.* 59:786 (1995).
143. K. Deuchi, O. Kanauchi, Y. Imasato, E. Kobayashi. *Biosci. Biotech. Biochem.* 59:781 (1995).
144. A. Razdan, D. Peterson. In: *Advances in Chitin Science*, Vol. I, A. Domard, C. Jeuniaux, R. Muzzarelli, G. Roberts (Eds.). Jacques André Press, Lyon, 1996, p. 422.
145. L. K. Han, Y. Kimura, H. Okuda. *Int. J. Obes. Metab. Disord.* 23:174 (1999).
146. T. Ouchi, T. Banba, T. Matsumoto, S. Suzuki, M. Suzuki. *J. Bioactive Comp. Polym.* 4:362 (1989).
147. E. E. Hassan, R. C. Parish, J. M. Gallo. *Pharm. Res.* 9: 390 (1992).
148. E. E. Hassan, J. M. Gallo. *J. Drug Target* 1:7 (1993).
149. A. F. Kotze, H. L. Luessen, B. J. De Leeuw, A. G. De

Boer, J. C. Verhoel, H. E. Junginger, *J. Controlled Release* 51:35 (1998).
150. M. Dornish, O. Skaugrund, L. Illum, S. S. Davis. In: *Advances in Chitin Science*, Vol. II, A. Domard, G. A. F. Roberts, K. M. Vårum, (Eds.). Jacques André Press, Lyon, 1998, p. 694.
151. A. G. Nigalaye, P. Andusumilli, S. Bolton. *Drug Dev. Ind. Pharm.* 16:449 (1990).
152. S. M. Upadrashta, P. R. Katikaneni, N. O. Nuessle. *Drug Dev. Ind. Pharm.* 18:1701 (1992).
153. I. W. Hillyard, J. Doczi, P. B. Kieman. *Proc. Soc. Exp. Biol. Med.* 115:108 (1964).
154. Y. Machida, T. Nagai, K. Inoue, T. Sannan. In: *Chitin and Chitosan*, G. Skjåk-Bræk, T. Anthonsen, P. Sandford (Eds.). Elsevier Applied Science, London, 1984, p. 693.
155. W. Hou, S. Miyazaki, M. Takada, T. Komai. *Chem. Pharm. Bull.* 33:3986 (1985).
156. S. Miyazaki, H. Yamagouchi, C. Yokouchi, M. Takada, W. Hou. *Chem. Pharm. Bull.* 36:4033 (1988).
157. T. Chandy, C. P. Sharma. *Biomaterials* 13:949 (1992).
158. T. Chandy, C. P. Sharma. *Biomaterials* 14:939 (1993).
159. R. Bodmeier, K. H. Oh, Y. Pramar. *Drug Dev. Ind. Pharm.* 15:1475 (1989).
160. R. Goskonda, S. M. Upadrashta. *Drug Dev. Ind. Pharm.* 19:915 (1993).
161. A. Berthold, K. Kremer, J. Kreuter. *S.T.P. pharm. Sci.* 6: 358 (1996).
162. B. C. Thanou, M. C. Sunny, A. Jayakrishnan. *J. Pharm. Pharmacol.* 44:283 (1992).
163. J. Akbuga, G. Durmaz. *Int. J. Pharm.* 111:217 (1994).
164. R. Narayani, K. Panduranga Rao. *J. Appl. Polym. Sci.* 58: 1761 (1995).
165. I. Orienti, K. Aiedeh, E. Gianasi, V. Bertasi, V. Zecchi. *J. Microencapsulation* 13:463 (1996).
166. J. Filipovic-Griec, M. Becirevic-Lacan, N. Skalko, I. Jalsenjak. *Int. J. Pharm.* 135:183 (1996).
167. A. Berthold, K. Cremer, J. Kreuter. *J. Controlled Release* 39:17 (1996).
168. M. Acigöz, H. S. Kas, C. P. Sharma. *J. Microencapsulation* 13:141 (1996).
169. I. Henriksen, G. Smistad, J. Karlsen. *Int. J. Pharm.* 101: 227 (1994).
170. E. Onsoyen. In: *Advances in Chitin and Chitosan*, C. J. Brine, P. A. Sandford, J. P. Zikakis (Eds.). Elsevier Applied Science, London, 1991, p. 479.
171. H. Takeuchi, H. Yamamoto, T. Niwa, T. Hino, Y. Kawashima. *Pharm. Res.* 13:893 (1996).
172. H. Tozaki, T. Fujita, T. Odoriba, A. Terabe, T. Suzuki, C. Tanaka, S. Okabe, S. Muranishi, A. Yamamoto. *Life Sci.* 64:1155 (1999).
173. M. Kanke, H. Katayama, S. Tsuzuki, H. Kuramoto. *Chem. Pharm. Bull.* 37:523 (1989).
174. S.-H. Lin and R.-I. Perng. *Chem. Pharm. Bull.* 40:1058 (1992).
175. S. Minami, Y. Okamoto, A. Matsuhashi, H. Sashiwa, H. Saimoto, Y. Shigemasa, T. Tanigawa, Y. Tanaka, S. Tokura. In: *Advances in Chitin and Chitosan*. C. J. Brine, P. A. Sandford, J. P. Zikakis (Eds.). Elsevier Applied Science, London, 1991, p. 61.
176. R. A. A. Muzzarelli. *Carbohydr. Polym.* 8:1 (1988).
177. Y. Okamoto, K. Shibazaki, S. Minami, A. Matsuhashi, S. Tanioka, Y. Shigemasa. *J. Vet. Med. Sci.* 57:851 (1995).
178. H. Ueno, H. Yamada, I. Tanaka, N. Kaba, M. Matsuura, M. Okamura, T. Kadosawa, T. Fuginaga. *Biomaterials* 20: 1407 (1999).
179. Y. Okamoto, T. Tomita, S. Minami, A. Matsuhashi, N. H. Kumazawa, S. Tanioka, Y. Shigemasa. *J. Vet. Med. Sci.* 57:765 (1995).
180. F. F. Bartone, E. D. Adickes. *J. Urol.* 140:1134 (1988).
181. R. A. A. Muzzarelli, G. Biagini, A. Pugnaloni, O. Filipini, V. Baldassare, C. Castaldini, C. Rizzoli. *Biomaterials* 10: 598 (1989).
182. K. N. Sall, J. K. Kreiter, R. H. Keates. *Ann. Ophthalmol.* 19:31 (1987).
183. J. X. Lu, F. Prudhomme, A. Meunier, L. Sedel, G. Guillemin. *Biomaterials* 20:1937 (1999).
184. P. R. Klokkevold, L. Vandemark, E. B. Kenney, G. W. Bernard. *J. Periodontol.* 67:1170 (1996).
185. T. Kawakami, M. Antoh, H. Hasegawa, T. Yamagishi, M. Ito, S. Eda. *Biomaterials* 13:759 (1992).

10

Chitosan-Based Delivery Systems: Physicochemical Properties and Pharmaceutical Applications

Radi Hejazi and Mansoor Amiji
Northeastern University, Boston, Massachusetts

I. INTRODUCTION

A. Chemical Structure of Chitosan

Chitosan [a (1 → 4) 2-amino-2-deoxy-β-D-glucan] is obtained by the alkaline deacetylation of chitin. Thus, the chitosan molecule is a copolymer of N-acetylglucosamine and glucosamine (Fig. 1) (1–3). The sugar backbone consists of β-1,4-linked glucosamine with a high degree of N-acetylation, a structure very similar to that of cellulose, except that the acetyl amino group replaces the hydroxyl group on the C-2 position. Thus, chitosan is poly(N-acetyl-2-amino-2-deoxy-D-glucopyranose), where the N-acetyl-2-amino-2-deoxy-D-glucopyranose (or Glu-NH_2) units are linked by (1 → 4)-β-glycosidic bonds (4,5).

B. Availability of Chitosan

Chitin is the second most abundant polysaccharide in nature, cellulose being the most abundant. Chitin is found in the exoskeleton of crustacea, insects, and some fungi. Chitosan has a rigid crystalline structure through inter- and intramolecular hydrogen bonding. The main commercial sources of chitin are the shell wastes of shrimp, lobster, krill, and crab. In the world, several million tons of chitin are harvested annually (4,6,18).

C. Chemical Methodology for the Preparation of Chitosan

Shrimp or crab shell proteins are removed by treatment with 3–5% (w/v) NaOH aqueous solution at 80–90°C for a few hours or at room temperature overnight. Afterward, the product's inorganic constituents are removed by treatment with 3–5% (w/v) HCl aqueous solution at room temperature giving a white to beige colored sample of chitin. The treatment of the sample with an aqueous 40–45% (w/v) NaOH solution at 90–120°C for 4–5 h results in N-deacetylation of chitin. The insoluble precipitate is washed with water to give a crude sample of chitosan. The conditions used for deacetylation will determine the polymer molecular weight and the degree of deacetylation. The crude sample is dissolved in aqueous 2% (w/v) acetic acid, then the insoluble material is removed giving a clear supernatant solution which is neutralized with NaOH solution, resulting in a purified sample of chitosan as a white precipitate (7). Further purification may be necessary to prepare medical and pharmaceutical grade chitosan.

D. Physicochemical and Biological Properties of Chitosan

The word chitosan refers to a large number of polymers which differ in their degree of N-deacetylation (40–98%)

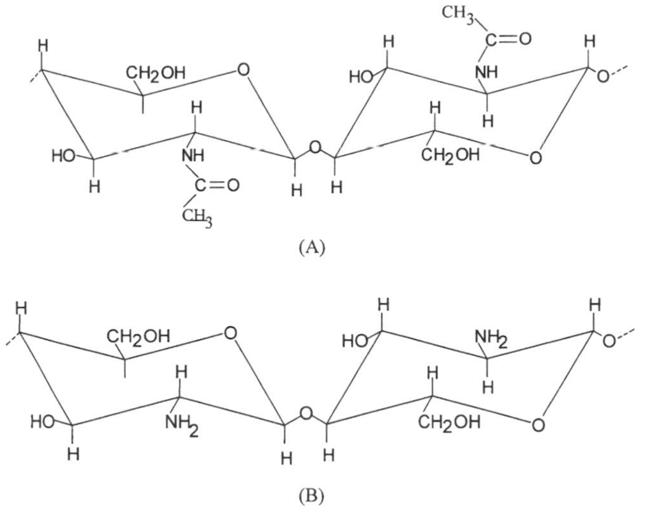

Figure 10.1 Structures of chitin (A) and chitosan (B).

and molecular weight (50,000–2,000,000 daltons). These two characteristics are very important to the physicochemical properties of the chitosans and, hence, they have a major effect on the biological properties (8).

Chitosan is a weak base with a pK_a value of D-glucosamine of about 6.2–7 and, therefore, insoluble at neutral and alkaline pH values. It does, however, make salts with inorganic and organic acids such as hydrochloric acid, acetic acid, glutamic acid, and lactic acid. In acidic medium, the amine groups of the polymer are protonated resulting in a soluble, positively charged polysaccharide that has a high charge density (one charge for each D-glucosamine unit). Chitosan can form gels by interacting with many multivalent anions (9–14).

Chitosan salts are soluble in water, the solubility depending on the degree of deacetylation (and thereby the pKa value of the D-glucosamine) and the pH. Chitosans with low degree of deacetylation (≤40%) are soluble up to a pH of 9, whereas highly deacetylated chitosans (≥85%) are soluble only up to a pH of 6.5. Addition of salts to the solution interferes with the solubility of chitosans. The higher the ionic strength, the lower the solubility as charge neutralization occurs by increasing concentrations of counterion. Increasing degree of deacetylation increases the viscosity. This can be explained by the fact that high and low deacetylated chitosan have different conformations. Chitosan has an extended conformation with a more flexible chain when it is highly deacetylated because of the charge repulsion in the molecule. However, the chitosan molecule has a rodlike shape or a coiled shape at low degree of deacetylation due to the low charge density in polymer chain (12,15). The viscosity of chitosan solution is also affected by factors such as concentration and temperature. As the chitosan concentration increases and the temperature decreases, the viscosity increases. The nearly linear negative dependence of viscosity on temperature could be explained by the enhanced chain flexibility and reduced root-mean-square unperturbed end-to-end distance of the polymer chains with increasing temperature (16).

Chitosan in solution exists in the form of quasiglobular conformation stabilized by extensive intra- and intermolecular hydrogen bonding. The hydrogen bonding that might attribute to the high viscosity of chitosan solutions exists mainly between the amine and hydroxyl groups of the polymer chains. Chitosans with lower degree of deacetylation have lower hydrogen bonding density because of the lower number of amino groups in the polymer chains. Since urea is capable of breaking the hydrogen bonding, the chain conformation of chitosan changes from a spherical structure in acetic acid–NaCl to a random coil conformation in acetic acid–NaCl–urea system (2). Intermolecular hydrogen bonding in chitosan is responsible for the film and fiber-forming properties of the polymer. According to Chen et al. (17), the degree of deacetylation of chitosan, which will determine the number of intermolecular hydrogen bonds, was found to affect the rigidity of the polymer film. For chitosan hydrogels, the extent of dissociation of the hydrogen bonding may affect the swelling kinetics of the gels. At low pH, the hydrogen bonding dissociates due to the protonation of the amine groups leading to faster swelling.

Because chitosan has favorable biological properties such as biodegradability (18) and biocompatibility (19,20), it has attracted a lot of attention in the pharmaceutical and medical fields. Chitosan has low oral toxicity with an LD_{50} in rats of 16 g/kg (20,21). Toxicity of chitosan might depend on different factors such as degree of deacetylation, molecular weight, purity, and route of administration. Further studies of toxicity need to be carried out for particular application of the polymer.

The pharmaceutical requirements of chitosan are as follows: particle size <30 μm, density between 1.35 and 1.40 g/cc, pH 6.5–7.5, insoluble in water, and partially soluble

Table 10.1 Chemical Properties of Chitosan

Cationic polyamine
High charge density at pH <6.5
Adheres to negatively charged surfaces
Forms gels with polyanions
Low to high molecular weight range
High to low viscosity
Chelates certain transition metals
Reactive amino/hydroxyl groups

Source: Adapted from Ref. 22.

Table 10.2 Biological Properties of Chitosan

Natural polymer
Biodegradable to normal body constituents
Safe and nontoxic
Hemostatic
Microbicidal
Spermicidal
Anticarcinogenic
Anticholesterolemic

Source: Adapted from Ref. 22.

in acids (11). The chemical and biological properties of chitosan are summarized in Tables 1 and 2.

E. Chitosan as a Biomaterial

A biomaterial is defined as a material that is capable of establishing an interaction with the surrounding biological environment without stimulating an adverse response by the host (23,24). Chitosan has a broad range of biological activities and thus has many applications. It could be used in wound healing by giving comfortable and painless wound surface protection. For burns, chitosan can be used for the development of an artificial skin that is devoid of postoperative scar formation. Chitosan could also be applied as a topical treatment for certain infections like athlete's foot infections. In purulent diseases, chitosan can accelerate the granulated tissue formation. Acting as a drug delivery system, chitosan is also used for tumor-selective drug delivery. Since chitosan acts as a chemotactic for polymorphonuclear leukocytes it is promising for treating mastitis. N-Acyl chitosan membranes and fibers are commonly used for artificial kidney. Since chitosan is highly permeable to oxygen, contact lenses made from chitosan could be worn for a long period of time. N-Octanoyl and N-hexanoyl derivatives of chitosan have blood coagulant and antithrombogenic effect. Chitosan acts as a hypocholesterolemic agent since it binds bile acids. The biopolymer has also found utility in agricultural, food, and cosmetic industries (7,25).

II. CHITOSAN AS A BIOADHESIVE

A. Definition of Bioadhesion

Bioadhesion is defined as the ability of a synthetic or biological material to adhere to a biological tissue for an extended period of time (26).

B. Importance of Bioadhesion

Bioadhesive polymers have attracted a lot of attention as they can increase the residence time and also can improve the localization of controlled release formulations at the mucosal sites of administration. Bioadhesive formulations have been used for administration of drugs to the gastrointestinal tract and buccal, nasal, ocular, and vaginal routes (27–29). Examples of bioadhesive polymers include cellulose derivatives, alginates, natural gums (30–32), polyacrylates (33), hyaluronic acid (34), scleroglucan (35), and pregelatinized and modified starches (36,37).

C. Chitosan as a Bioadhesive

In the cationic form, the D-glucosamine residue of chitosan could interact with the N-acetylneuraminic acid (sialic acid) residues of mucin by electrostatic interactions (32,38). The bioadhesive properties of chitosan may allow a prolonged interaction of the delivered drug with the membrane epithelia facilitating more efficient absorption. Increased absorption of drugs at mucosal sites by chitosan could be due to prolonged interaction with the membrane epithelia, inhibition of degrading enzymes, or opening of the tight junctions between cells to facilitate transport.

Table 3 illustrates the role of chitosan in increasing the absorption of drugs at mucosal sites.

1. Bioadhesive and Rheological Properties

Ferrari et al. (39) studied the rheological and bioadhesive properties of three viscosity grades of chitosan: high, medium, and low. They found that the temperature and the pH of the hydration medium could influence the rheological properties of the three grades of chitosan. The polymer solutions and polymer–mucin mixtures prepared at the same concentrations were subjected to mucoadhesive testing with a tensile stress tester. All of the grades of chitosan were found to strongly interact with mucin and can be excellent candidates for use in bioadhesive formulations. Table 4 illustrates the maximum force of detachment and the work of adhesion of polymer solutions and the polymer–mucin mixtures. For the high grade chitosan, the force of detachment was 1038.8 mN for polymer alone and 1420 mN for the polymer–mucin mixture. The low grade chitosan, on the other hand, had a force of detachment of 667.6 mN for polymer alone and 1345.9 mN for the polymer–mucin mixture.

2. Bioadhesive Bilayered Devices as Buccal Drug Delivery Systems

Bioadhesive bilayered devices were developed by Lopez et al. (40). Chitosan was used as the adhesive material,

Table 10.3 Chitosan as a Bioadhesive

Model compound	Function of chitosan	Reference
Nifedipine and propranolol	Mucoadhesive	40
Ampicillin	Mucoadhesive	41
Acyclovir	Mucoadhesive	42
Bovine serum albumin	Mucoadhesive	43
Insulin	Enzyme inhibition	44
Leucine enkephalin	Enzyme inhibition	45
ND[a]	Enzyme inhibition	46
ND[a]	Enzyme inhibition	47
Buserelin	Absorption enhancer	48
^{14}C mannitol	Absorption enhancer	49
9-Desglycinamide,8-L-arginine vasopressin	Absorption enhancer	50
Insulin	Absorption enhancer	52
Morphine-6-glucoronide	Absorption enhancer	54
^{14}C mannitol	Absorption enhancer	56

[a] Not described.

while ethylcellulose was used as an impermeable backing material. Adhesive capacity was estimated by measuring the maximum detachment force using a tensile tester. Bovine sublingual mucosa was used as the biological substrate. The detachment force of buccal bilayered devices based on chitosan was 0.5 N /cm^2, while for the buccal bilayered devices based on chitosan crosslinked with polycarbophil, the detachment force was 0.85 N/cm^2. Crosslinking of chitosan with polycarbophil allowed further swelling and hence enhanced adhesion. To evaluate these new devices as buccal drug delivery systems, a lipophilic drug, nifedipine, and a hydrophilic drug, propranolol hydrochloride, were entrapped. Almost 100% of propranolol was released within 4 h, while only 30% of nifedipine was released within 6 h. Propranolol release was faster than nifedipine because of the higher solubility of the drug in the release medium.

3. Poly(Vinyl Alcohol) Gel Spheres Containing Chitosan

Sugimoto et al. (41) formulated poly(vinyl alcohol) gel spheres that contained chitosan. When administered orally, the bioavailability of ampicillin, a poorly absorbable drug, was significantly improved. The bioavailability was further increased by the addition of chitosan. Seventy percent of the ampicillin oral solution was unabsorbed and entered the large intestine. The absorption of ampicillin from chitosan-containing gel spheres occurred for a longer period of time than from gel spheres without chitosan. The residence time of the gel spheres in the small intestine was increased by chitosan addition and the period of absorption of ampicillin was increased. Table 5 illustrates the pharmacokinetic parameters of ampicillin after oral administration of poly(vinyl alcohol) gel spheres and

Table 10.4 Force of Detachment and Work of Adhesion of the Three Viscosity Grades of Chitosan Hydrated in 0.1 M HCl

Chitosan grade	Force of detachment (mN)		Work of adhesion (μ joule)	
	Polymer alone	Polymer + mucin	Polymer alone	Polymer + mucin
Low	667.6	1345.9	71.4	182.5
Medium	784.8	1154.9	85.8	170.2
High	1038.8	1420.0	101.3	166.6

Source: Adapted from Ref. 39.

Table 10.5 Pharmacokinetic Parameters of Ampicillin After Oral Administration in Poly(Vinyl Alcohol) Gel Spheres and Chitosan–Poly(Vinyl Alcohol) Gel Spheres

Dosage form	AUC[a](μg · h/mL)	MRT[b](h)	MAT[c](h)	Bioavailability
Solution medium	6.91	1.77	1.26	0.316
Poly(vinyl alcohol) gel spheres	8.6	4.08	3.57	0.393
Chitosan–poly(vinyl alcohol) gel spheres	10.63	4.79	4.28	0.485

Note: The dose of ampicillin was 50 mg/kg.
[a] AUC is the area under the plasma concentration–time curve of the drug for the different formulations.
[b] MRT is the mean release time of the drug for the different formulations.
[c] MAT is the mean absorption time of the drug for the different formulations.
Source: Adapted from Ref. 41.

chitosan–poly(vinyl alcohol) gel spheres in rats. The bioavailability of ampicillin in solution medium is only 31.6%; however, the bioavailability of the drug was increased up to 48.5% using chitosan–poly(vinyl alcohol) gel spheres.

4. Acyclovir-Loaded Chitosan Microspheres

Since acyclovir is a relatively hydrophilic antiviral drug, its permeability through the ophthalmic epithelium is very low. A drug delivery system that can prolong the release of acyclovir and increases the residence time at the site of absorption will enhance the bioavailability of the drug in the eye. Acyclovir-loaded chitosan microspheres with a diameter of ≤20 μm were prepared (42). Following administration into the rabbit eye, the area under the aqueous humor concentration–time curve from 0 to 240 min of the drug-loaded chitosan microspheres was 172 μg · min/mL. However, for the control suspension formulation, the area under the aqueous humor concentration–time curve from 0 to 240 min was only 39.3 μg · min/mL. The total dose of acyclovir was 0.26 mg for both the microsphere formulation and the suspension. The maximum drug concentration in the aqueous humor was reached after 240 min from the drug-loaded chitosan microspheres. However, for suspension formulation, the drug was not detected after 120 min. Chitosan microspheres were able to prolong the residence time and increase bioavailability of acyclovir.

5. Bioadhesion of Hydrated Chitosans

Freshly excised bovine corneas were treated with tritiated [^{3}H] chitosan in solution. A two-cubic factorial design was used to find out the effect of parameters like pH, ionic strength, residence time, and chitosan molecular weight on the adsorption of chitosan on the corneal surface. All of the parameters were significant for the adsorption. Also, the interaction between these parameters was observed. In the in vivo study, liposomes and chitosan-coated liposomes, containing ^{125}I-labeled bovine serum albumin (BSA), were instilled into the eyes of anesthetized Wistar rats. Liposomes were produced using egg L-α-phosphatidylcholine and egg L-α-phosphatidyl-DL-glycerol in a ratio of 10:1. Free ^{125}I-BSA (7 μL approximating 300–600 nCi) or BSA loaded in liposomes or chitosan liposomes was applied to each eye. The eyelids and eyes were excised, and then the eyes were washed with 1 mL 1% Triton X-100 to remove excess ^{125}I-BSA on the surface. The retention of radioactivity in the eyes at 10, 30, and 90 min was measured. Table 6 shows the total recovery of free ^{125}I-BSA, or liposome and chitosan-coated liposome entrapped ^{125}I-BSA instilled into the eyes of anesthetized rats. Both coated and noncoated liposomes had significantly higher retention of ^{125}I-BSA than from solution. The retention of chitosan-coated liposomes, however, was not significantly

Table 10.6 Total Percentage Recovery of Free Albumin, Albumin Loaded in Liposome and Chitosan-Coated Liposome Instilled into the Eyes of Anesthetized Rats

Time (min)	Formulation	Average percentage in the eye
0	Free BSA	42.2
10	Free BSA	6.8
	Liposomes	17.3
	Chitosan-coated liposomes	23.9
30	Free BSA	8.1
	Liposomes	23.7
	Chitosan-coated liposomes	20.2
90	Free BSA	5.4
	Liposomes	10.3
	Chitosan-coated liposomes	11.0

Note: Bovine serum albumin was labeled with iodine-125 and administered at a dose of 7 μL, approximating 300–600 nCi.
Source: Adapted from Ref. 43.

different from noncoated liposomes (43). After 90 min, the percent retention in the eye of BSA in solution was 5.4%, while for liposomes and chitosan-coated liposomes the adhesive percentage was around 11%.

6. Chitosan-Coated Liposomes

Takeuchi et al. (44) developed polymer-coated liposomes as a novel drug carrier system for the oral administration of insulin. The polymers were chitosan, poly(vinyl alcohol) having a long alkyl chain, and poly(acrylic acid) that had cholesteryl groups. Isolated rat intestine was used for the in vitro evaluation of the bioadhesiveness of the polymer-coated liposomes. The number of liposome particles was measured with a Coulter counter after administering the liposome suspension in the isolated rat intestine. The adhesive percentage was calculated by dividing the difference between the number of liposomes before and after contacting with the intestine by the number of liposomes in suspension multiplied by 100. The particle counting method was adopted after confirming a linear relationship between the number of chitosan and the lipid concentration. Table 7 illustrates the adhesive percentage of liposomes coated with various polymers to the rat intestine. The polymer-coated liposomes showed higher adhesive percentage than the noncoated liposomes. The bioadhesive polymer layer on the surface of the liposomes resulted in the attachment of these liposomes to the intestinal mucosa. Chitosan-coated liposomes had the highest adhesive percentage among all of the polymer-coated liposomes tested. For in vivo study, rats were used for the drug absorption. For all the formulations administered orally, 24 IU was used as a dose of insulin. Insulin solution, at an equivalent dose, was used as a reference, and the blood glucose level without administering any formulation was kept as control. After the oral administration of chitosan-coated liposomes loaded with insulin, the reduction in the basal blood glucose level was more than that caused by noncoated liposomes or insulin solution for at least 12 h. For instance, a 10% reduction in the basal blood glucose level occurred following the administration of noncoated liposomes or insulin solution after 8 h. However, the reduction in the basal glucose level increased to 33% after the administration of chitosan-coated liposomes in the same time period.

Table 10.7 Adhesive Percentage of Liposomes Coated with Various Polymers to the Rat Intestine

Liposomes formulation	Adhesive percentage
Noncoated liposomes	3
Chitosan-coated liposomes	45
Poly(acrylic acid) with a cholesteryl group	12
Poly(vinyl alcohol) having a long alkyl chain	22

Notes: The polymers were dissolved in phosphate buffer solution (pH 7.4) at a concentration of 0.75% (w/v). Adhesive percentage was determined as discussed in the text.
Source: Adapted from Ref. 44.

7. Chitosan–Ethylenediaminetetraacetic Acid Conjugate

Schnurch et al. (45) have developed a chitosan-based bioadhesive system in which leucine enkephalin was protected from luminal degradation, particularly by aminopeptidase N, one of the most abundant membrane-bound peptidases. Ethylenediaminetetraacetic acid (EDTA), a chelating agent that can complex with divalent metal ions, was conjugated with chitosan. The conjugate was able to bind with zinc, a cofactor for aminopeptidase N activity. The hydrolysis of L-leucine-p-nitroanilide by aminopeptidase N was completely inhibited in the presence of the chitosan–EDTA conjugate. Schnurch et al. (46) showed that the detachment force of chitosan–EDTA conjugates with a D-glucosamine subunit of chitosan to EDTA at a molar ratio of 1:20 was 43.7 mN with mucin, while for unmodified chitosan the detachment force was only 33.1 mN. Also, it was noticed that chitosan–EDTA conjugates do not precipitate once they form complexes with low concentrations of bivalent cations.

Chitosan–EDTA conjugates were found to inhibit the enzymes carboxypeptidase A and aminopeptidase N. After 13 min, the substrate hippuryl-L-phenylalanine was completely hydrolyzed by carboxypeptidase A in the presence of 0.125% chitosan. However, in the presence of the chitosan–EDTA conjugate with a molar ratio of 1:20, 60% inhibition of the enzyme was observed. para-Nitroaniline is produced from the hydrolysis of L-leucine-p-nitroanilide by aminopeptidase N. In the presence of 1% chitosan, the concentration of p-nitroaniline after 120 min was almost 170 μM. This concentration was reduced to 60 μM in the presence of chitosan–EDTA conjugate. Chitosan–EDTA conjugate, therefore, was able to partially inhibit the proteolytic enzyme without compromising the bioadhesive property.

8. Chitosan–EDTA Linked with Bowman–Birk Inhibitor

In a separate study, Schnurch et al. (47) examined the bioadhesive polymer that can protect peptide drugs from degradation by the most abundant enzymes in the small intestine. The Bowman–Birk inhibitor (BBI), a protein of a molecular weight of 8000 daltons derived from soybean, inhibits serine proteases such as trypsin and chymotrypsin. The bioadhesive polymer chitosan–EDTA that was found to inhibit carboxypeptidase A and aminopeptidase N (45) was covalently linked to the BBI. Using native porcine mucosa, tensiometer studies showed that the bioadhesion of polymer-inhibitor conjugates was less than that of unmodified chitosan–EDTA. The loss in bioadhesion was proportional to the inhibitor concentration. The average adhesive strength of unmodified chitosan–EDTA was 54.4 mN, while for the comparably most adhesive polymer-inhibitor conjugate was 21.0 mN. Hydrolysis of the substrate N-α-benzoylarginine ethyl ester to N-α-benzoylarginine by trypsin was determined at 1-min intervals for 9 min at pH 6.8 and 20°C. After 9 min, using 0.055% of chitosan–EDTA conjugate, 10% of the enzyme activity was inhibited. Using the same concentration of chitosan–EDTA–BBI conjugate (5:1 ratio), 70% of the enzyme activity was inhibited. Hydrolysis of the substrate hippuryl-L-phenylalanine to L-phenylalanine and hippurate by chymotrypsin was also inhibited by chitosan–EDTA–BBI conjugate. Although this novel drug carrier matrix can protect peptides from the enzymatic attack in the intestine after oral administration, there was a compromise in the bioadhesion property of chitosan.

9. Chitosan as an Absorption Enhancer of Buserelin

Luessen et al. (48) showed that the intestinal absorption of buserelin, a synthetic nonapeptide with a molecular weight of 1240 daltons, was enhanced by chitosan in rats. Since chitosan did not inhibit the proteolytic enzyme activity, it was assumed that opening of the intercellular junctions by chitosan could lead to the enhancement of peptide absorption across the mucosa. Table 8 lists the pharmacokinetic parameters after intraduodenal administration of buserelin at a total dose of 500 μg. The bioavailability of buserelin was higher when delivered with chitosan polymer than that of the crosslinked polyacrylate. While the absolute bioavailability of buserelin was 0.6% using freeze-dried sodium salt of crosslinked poly(acrylic acid), the bioavailability increased to 5.1% for chitosan formulation when the same dose of buserelin was administered. The authors concluded that the enhancement of buserelin bioavailability was probably by opening of the intercellular junctions. The enzyme inhibitory activity was thought not to be as important as the opening of tight junctions, thus enhancing the permeability.

10. Chitosan Glutamate as an Absorption Enhancer of 9-Desglycinamide, 8-L-Arginine Vasopressin

Borchard et al. (49) used well-characterized colon carcinoma (Caco-2) cell culture system as an in vitro model for the intestinal epithelium. Chitosan glutamate influenced the structure of the epithelial tight junctions, thus enhancing the peptide uptake through the paracellular route. Transepithelial electrical resistance (TEER) is an indicator of the permeability of ionic molecules across the cell monolayer. Chitosan glutamate decreased TEER indicating that the polymer was able to affect the permeability of epithelial cell monolayers in vitro.

Using chitosan glutamate having a concentration of 1% (w/v), almost 1.2% of the total dose of 9-desglycinamide, 8-L-arginine vasopressin, a peptide containing eight amino acids, was transported in vitro across the Caco-2 cell monolayers within 4 h. While for the control only 0.2% of the total dose of 9-desglycinamide, 8-L-arginine vasopressin dissolved in the culture medium was transported in the same time period (50).

Table 10.8 Pharmacokinetic Parameters After Intraduodenal Administration of Buserelin

Polymer	T_{max} (min)	C_{max} (ng/mL)	Absolute bioavailability (%)
Buserelin in buffer medium[a]	60–90	6.7	0.1
C934P gel (0.5%)[b]	40–90	112.1	1.9
FNaC934P gel (0.5%)[c]	40–60	45.8	0.6
Chitosan–hydrochloride gel (1.5%)	40–90	364.0	5.1

Note: The dose of buserelin administered was 500 μg per rat.
[a] Buserelin in a buffer solution (pH 6.7) was used as control.
[b] C934P is the crosslinked form of poly(acrylic acid) available as Carbopol® 934P.
[c] FNaC934P is the freeze-dried sodium salt of crosslinked poly(acrylic acid).
Source: Adapted from Ref. 48.

11. Chitosan as an Absorption Enhancer of Insulin

It is believed that the structural reorganization of tight junction–associated proteins could happen as a result of the interaction of chitosans with the cell membrane. This leads to the enhanced transport through the paracellular pathway (51). Illum et al. (52) showed that chitosan can enhance the absorption of insulin, a polypeptide composed of 51 amino acids and having a molecular weight of 6000 daltons, from the nasal cavity in sheep. The area under the plasma insulin concentration–time curve was 1346 mIU/L following the nasal administration of 2 IU/kg sodium human insulin solution in sheep. The area under the plasma insulin concentration–time curve increased to 9809 mIU/L after the nasal administration of 2 IU/kg sodium human insulin solution with 0.5% (w/v) chitosan solution. The histological studies were conducted using light microscopy. Using the epithelium of the nostril of rat that was not treated with chitosan formulation as a control, the nasal epithelium of the other nostril was more disordered than the control after exposure to chitosan formulation. Normal permeability properties of the nasal membrane were regained after 60 min. Thus, chitosan acted as an absorption enhancer on the nasal membrane in a reversible fashion. Eventually, this chitosan delivery system enhanced the nasal uptake of insulin without using penetration-enhancing agents that could be damaging to the nasal mucosa. Further studies are necessary to investigate how chitosan affects the integrity of the nasal epithelial cells. In addition, opening the tight junction for 60 min might lead to certain complications such as the passage of foreign particles through the nasal mucosa that could be dangerous, especially if these particles reach other areas of the body.

12. Chitosan as an Absorption Enhancer of Morphine-6-Glucuronide

Although chitosan could affect the integrity of the tight junction between cells, it also can delay the mucociliary clearance. The mucociliary transport rates were depressed longer by increasing the molecular weight of chitosan (53). Chitosan, an absorption enhancer, was administered nasally in combination with morphine-6-glucuronide, a morphine metabolite that has a molecular weight of 461.5. Table 9 shows the pharmacokinetic parameters for morphine-6-glucuronide after intravenous and nasal administration to sheep. Following nasal administration, the maximum plasma concentration was obtained after 13 min. Relative to intravenous injection, the bioavailability of morphine-6-glucuronide after nasal administration in sheep was 31.4%. Nasal administration of morphine-6-glucuronide combined with chitosan could be a good route of administration for this analgesic compound for pain relief that necessitates a rapid onset of action. Unfortunately, in this study the authors have failed to show a comparative bioavailability associated with the nasal administration of morphine-6-glucuronide in saline solution without chitosan (54).

Table 10.9 Pharmacokinetic Parameters of Morphine-6-Glucuronide After Intravenous and Nasal Administration to Sheep

Pharmacokinetic parameter	Intravenous	Nasal
C_{max} (ng/mL)	50	132
T_{max} (min)	3	13.2
AUC (ng · min/mL^1 · kg^1)	2,928	9,314
Bioavailability (%)	100	31.4

Note: The doses of morphine-6-glucuronide administered intravenously and nasally were 0.015 mg/kg and 0.15 mg/kg, respectively, in 0.9% (w/v) sodium chloride solution.
Source: Adapted from Ref. 54.

13. N-Trimethyl Chitosan as an Absorption Enhancer

Chitosan salts, such as chitosan hydrochloride or glutamate, can enhance the absorption of large hydrophilic compounds through the mucosal surfaces. Chitosan salts, however, will only be soluble at pH <6.5. To enhance the solubility at neutral pH, N-trimethyl chitosan chloride was synthesized (55). Reduction in the TEER of Caco-2 cells by 25% was observed when chitosan derivative was applied at 0.25% (w/v) concentration. Using the same polymer concentration, the transport rate of [^{14}C]-mannitol was increased by 11-fold. Confocal laser scanning microscopy study showed that N-trimethyl chitosan chloride was able to open the tight junctions of the intestinal epithelial cells increasing the transport of hydrophilic compounds through the paracellular transport pathway. The authors, however, did not examine the bioadhesive properties of N-trimethyl chitosan for comparison with unmodified chitosan (56).

III. pH-SENSITIVE CHITOSAN HYDROGELS

A. Definition of Hydrogels

Hydrogels are defined as three-dimensional crosslinked polymeric material that can absorb significant amounts of water (57). Implants (58), contact lenses (59), and controlled-release drug delivery systems (60) are examples of different biomedical applications where hydrogels have been used. Changes in the environmental factors such as temperature, pH, light, and electrical field might cause a phase transition of certain hydrogels. These hydrogels are important as physiologically responsive drug delivery devices (61,62).

B. Interpenetrating Polymer Networks

Interpenetrating polymer networks (IPNs) are blends of polymers in a network configuration. They are intimate mixtures of two or more different crosslinked polymer networks without covalent bonding between the polymers. Thus, each polymer is crosslinked only with other molecules of the same polymer to provide the physical and chemical combination of two or more structurally dissimilar polymers. In this way, the properties of the polymers such as processing, flexibility, tensile and impact strength, and chemical and flammability resistance could be modified to meet specific needs (63). Semi-IPN is defined as a graft copolymer in which one of the polymers is crosslinked, while the other is linear, so different morphologies and properties are developed (64).

C. pH-Sensitive Chitosan Hydrogels

In acidic environment, the primary amine group of the D-glucosamine residue is protonated making chitosan positively charged. Like charges on the polymer chains lead to electrostatic repulsion and swelling in the network, probably due to the osmotic effect of bound counterions. In the neutral or alkaline environment, the decrease in cationic charge density leads to the collapse of the hydrogel network. The pH-sensitive swelling system would be useful for localized drug delivery in a specific region of the gastrointestinal tract (65).

Table 10 illustrates the different types of semi-IPNs of chitosan that have been evaluated.

1. pH-sensitive Semi-IPN by Crosslinking Chitosan with Glutaraldehyde

Chitosan–poly(propylene glycol) blends were crosslinked with glutaraldehyde by Schiff's base reaction mechanism to form novel pH-sensitive semi-IPNs (66). In the presence of the polyether, infrared analysis showed that the semi-IPN was formed by crosslinking the amine groups of chitosan with the aldehyde groups of glutaraldehyde. Also, there was hydrogen bonding between amino hydrogen and oxygen from polyether. The semi-IPN response to pH changes was reversible. At pH 1, the ionization of the amino groups occurred leading to dissociation of the hydrogen bonding in the network. At pH 13, however, the association of the hydrogen bonding occurred because of the increase in the unionized fraction of the amino groups.

2. Chitosan–Poly(Propylene Glycol) Semi-IPN

The pH-sensitive swelling behavior of chitosan–poly(propylene glycol) semi-IPN was examined by Yao et al. (67). The swelling kinetics experiments were conducted by placing the preweighted initially dry samples of semi-IPN gel in the buffer solution and the weight of the swollen samples was measured as a function of time. The degree of swelling at predetermined intervals was measured by dividing the difference between the weight of the swollen gel and the dry gel by the weight of the swollen gel multiplied by 100. At pH 3.19 the swelling was maximal was maximal having a swelling ratio of 10% after 50 min, while at pH 13.0 the swelling was minimal having a swelling ratio of 0.2%. Infrared analysis confirmed that when the hydrogels were transferred from an acidic medium to an alkaline medium, the $NH_3^+ \rightarrow NH_2$ transition was reversible.

3. Chitosan–Poly(Ethylene Glycol Diacrylate) Semi-IPN

Semi-IPN hydrogels composed of β-chitosan and poly(ethylene glycol diacrylate) macromer (PEGM) were synthesized by adding 10 wt% PEGM aqueous solution into 2 wt% β-chitosan solution that was prepared by dissolving β-chitosan in acetic acid aqueous solution. Hydrogel films were prepared by casting the PEGM–β-chitosan solution. The hydrogels were formed by 2,2-dimethoxy-2-phenyla-

Table 10.10 pH-Sensitive Swelling Hydrogels

Type of semi-IPN	Drug	Reference
Crosslinked glutaraldehyde chitosan semi-IPN	ND[a]	66
Chitosan–poly(propylene glycol) semi-IPN	Chlorhexidine acetate	67
Chitosan–poly(ethylene glycol diacrylate) semi-IPN	ND[a]	68
Chitosan–poly(ethylene oxide) semi-IPN	Riboflavin	69
Freeze-dried chitosan–poly(ethylene oxide) semi-IPN	Amoxicillin and metronidazole	71

[a] Not described.

cetophenone, a nontoxic photoinitiator, using ultraviolet radiation. High equilibrium water content in the range of 77–83% was observed for the photocrosslinked hydrogels and was mainly attributed to the free water content rather than the bound water, hydrogen bonded with components of the semi-IPN hydrogels. The crystallinity, thermal properties, and mechanical properties of semi-IPN hydrogels were examined as well. The crystallinity of chitosan–PEGM photocrosslinked hydrogels was reduced. Crosslinked PEGM segment in the semi-IPNs showed higher glass transition temperature than that of poly(ethylene glycol) itself. The number of crosslinkable acrylate end groups that contribute to the lower glass transition temperature decreased by increasing the β-chitosan content in the semi-IPNs. Since hydrogels in the wet state are more rubbery, the tensile strength values of semi-IPNs in the wet state were much lower than those in the dry state (68).

4. Chitosan-Poly(Ethylene Oxide) Semi-IPN

By crosslinking blends of chitosan and poly(ethylene oxide) (PEO) with an average molecular weight of 1,000,000 daltons with glyoxal, a pH-sensitive semi-IPN hydrogel system was prepared (69). The swelling of chitosan hydrogels was not significant in simulated gastric fluid (SGF) of pH 1.2 or simulated intestinal fluid (SIF) of pH 7.0. However, after 6 h in SGF, the swelling of chitosan–PEO semi-IPNs was about six times higher than that of chitosan hydrogels. The increase in the internal osmotic pressure of the network with the addition of high molecular weight PEO could explain the higher swelling ratio. In other words, at higher osmotic pressure, hydronium ions were able to diffuse into the network at a faster rate leading to a higher charge density. Also, the crystallinity of chitosan matrix might be decreased as a result of the intermolecular association between chitosan and PEO in the semi-IPN. In addition, the presence of PEO in the network may also facilitate the absorption of aqueous medium.

The swelling ratio of the hydrogel increased by increasing the molecular weight and the amount of incorporated PEO. High molecular weight PEO was important for the initial swelling of the hydrogels. Thus, PEO having molecular weight of 1,000,000 daltons was chosen to prepare chitosan–PEO semi-IPN. The swelling ratio of chitosan–PEO semi-IPN in SGF was ten times higher than that in SIF. After 6 h, almost 50% of riboflavin was released from chitosan–PEO semi-IPN in SGF. However, only 30% of the drug was released from chitosan–PEO semi-IPN in the intestinal fluid after 6 h. The drug release from chitosan hydrogels was significantly lower than from chitosan–PEO semi-IPN. Only 30.2 and 6.6% of riboflavin was released from chitosan hydrogels after 6 h in SGF and SIF, respectively (70).

5. Freeze-Dried Chitosan–Poly(Ethylene Oxide) Semi-IPN

For stomach-specific antibiotic delivery, freeze-dried chitosan and chitosan–PEO semi-IPN hydrogels containing PEO of molecular weight 1,000,000 daltons were prepared for faster swelling and drug release (71). Because of the rapid influx of simulated gastric fluid (pH 1.2) into the porous matrix by capillary action, the swelling of freeze-dried hydrogels was faster than for air-dried hydrogels. The swelling and drug release properties of the freeze-dried chitosan–PEO semi-IPN were affected by the porosity of the matrix, the ionization of glucosamine residues in acidic medium, and the presence of high molecular weight PEO. After 2 h in SGF, the freeze-dried chitosan–PEO semi-IPN released about 65 and 59% of the entrapped amoxicillin and metronidazole, respectively.

IV. CHITOSAN FOR GENE DELIVERY

The course of many hereditary diseases could be reversed by gene delivery. In addition, many acquired diseases such as multigenetic disorders and those diseases caused by viral genes could be treated by genetic therapy (72–75). Gene delivery systems include viral vectors, cationic liposomes, polycation complexes, and microencapsulated systems (76–82).

A. Advantages and Disadvantages of Viral Vectors

Viral vectors are advantageous for gene delivery because they have high efficiency and a wide range of cell targets. However, when used in vivo they might exert potential immune responses and oncogenic effects. In addition, it is sometimes difficult to reproducibly prepare viral vectors in large batches (83).

B. Nonviral Delivery Systems

To overcome the limitations of viral vectors, nonviral delivery systems are considered for gene therapy. Nonviral delivery systems have certain advantages like ease of preparation, cell/tissue targeting, low immune response, unrestricted plasmid size, and large-scale reproducible production (83).

C. Chitosan as a Gene Delivery Vehicle

Chitosan could be used as a carrier of DNA in gene delivery systems. It is biocompatible, biodegradable, and has an average amino group density of 0.837 per monosaccharide D-glucosamine unit. This amino group density depends on the degree of deacetylation. The cationically charged chitosan will form polyelectrolyte complexes with the negatively charged plasmid DNA. Chitosan could be modified by coupling with specific ligands, such as lactose, so the chitosan–DNA complex could be targeted to cells that express a galactose-binding membrane lectin. Chitosan–DNA complexes could be protected from DNase to improve the bioavailability of the plasmid DNA delivered into the body for gene therapy.

Table 11 summarizes the different types of chitosan or modified chitosan that have been used for gene delivery.

1. Chitosan as a Delivery System for Plasmids

Negatively charged plasmid DNA will form polyelectrolyte complexes with chitosan, a polycation. Chitosan was first described as a delivery vehicle for plasmids by Mumper et al. (84). They showed that the molecular weight of chitosan ranging from 2000 to 540,000 daltons had significant influence on the particle size of chitosan-condensed DNA particles. As the molecular weight decreased, a more monodisperse particle size distribution was obtained.

2. Self-Aggregates–Plasmid Complexes

Lee et al. (82) showed that self-aggregates, which were stable in aqueous media, were formed when chitosan was modified with deoxycholic acid. The hydrophobically modified chitosan self-aggregates were relatively small with a mean diameter of 160 nm and a unimodal size distribution. After mixing with plasmid of cytomegalo virus encoding chloramphenicol acetyltransferase (pCMV-CAT), the self-aggregates formed charge complexes and were used for in vitro transfection of genes into mammalian cells. The self-aggregates–plasmid complex formation was confirmed by gel electrophoresis. The transfection efficiency of self-aggregates–plasmid complexes in kidney-derived cells (COS-1) was measured by assaying chloramphenicol acetyltransferase activity. Chloramphenicol and the products of its enzymatic hydrolysis were analyzed by thin layer chromatography and autoradiography. The transfection efficiency of chitosan self-aggregates–plasmid complexes was slightly lower than that of the commercial cationic liposome transfecting agent Lipofectamine® but higher than the transfection efficiency of the plasmid in phosphate buffer solution (pH 7.2). Using Lipofectamine, 1-acetate, 3-acetate, and 1,3-acetate metabolites were formed, while with chitosan self-aggregates–plasmid complexes only 1-acetate and 3-acetate metabolites were formed.

3. Complex Coacervation of DNA with Chitosan

A nanosphere delivery system based on the complex coacervation of DNA with chitosan was developed by Leong et al. (83). The advantages of this gene delivery system are (1) the ability of conjugating these nanosphere with ligands that might help in targeting purposes or stimulating receptor-mediated endocytosis, (2) that DNA degradation could be reduced in the endosomal and lysosomal compartments by incorporating lysosomolytic agents, (3) the coencapsulation of bioactive agents or multiple plasmids, (4) the protection of the DNA from serum nuclease degradation, and (5) the stability of the nanosphere formulation after lyophilization. In addition, since polymers can reduce the degradation of plasmid by DNase, the DNA bioavailability could be improved. Plasmid–chitosan nanoparticles with a diameter of 200–300 nm were prepared using a complex coacervation method. Chitosan solution (0.02% w/v) in acetate buffer was vortexed with equal volume of DNA solution (100 μg/mL in 50 mM of sodium sulfate solution) at 55°C. Human embryonic kidney cells (HEK

Table 10.11 Chitosan as a Gene Delivery System

Carrier type	Reference
Chitosan	84,87,88
Deoxycholic acid–modified chitosan	82
Complex coacervation of DNA with chitosan	83,85
Chitosan and depolymerized chitosan oligomers	86
Modified chitosan microspheres with transferrin and PEG	89
Lactosylated chitosan	90
N,N,N-trimethyl chitosan–galactose conjugate	91

293) were transfected in vitro using these nanoparticles. The transfection efficiency was evaluated with luciferase activity and green fluorescence protein. The transfection efficiency of the chitosan nanospheres was compared to that of Lipofectamine reagent. The percentage of cells transfected with the nanospheres having 5 μg of DNA was 80% of that transfected with 1 μg of standard Lipofectamine (85).

4. Chitosan and Depolymerized Chitosan Oligomers

MacLaughlin et al. (86) showed that plasmid DNA containing cytomegalo virus promoter sequence and a luciferase reporter gene could be delivered in vivo by chitosan and depolymerized chitosan oligomers to express a luciferase gene in the intestinal tract. The morphology and size of plasmid–chitosan complexes were characterized by transmission electron microscopy. The transfection of kidney-derived cells (COS-1) with complexes was assayed by measuring the relative luciferase unit per milligram of protein. The expression level of chitosan complexes (chitosan molecular weight 102,000 daltons) was 250-fold less than that of the plasmid–Lipofectamine, a positive control. By incorporating an endosomolytic peptide GM227.3 (Genemedicine, Inc., The Woodlands, TX), the transfection was increased four times as compared with the complex without the peptide. A plasmid consisting of cytomegalo virus promoter sequence and a chloramphenicol acetyltransferase reporter gene was complexed with chitosan and administered in the upper small intestine or in the colon of female New Zealand rabbits. At 72 h postinstillation, the rabbits were sacrificed and tissues were collected for chloramphenicol acetyltransferase expression. The average chloramphenicol acetyltransferase expression after plasmid–chitosan–GM227.3 complex administration was 7.06 pg/mg in the mesenteric lymph nodes. However, for the control (plasmid in 10% lactose), there was no detectable chloramphenicol acetyltransferase activity in the mesenteric lymph nodes.

5. Biodistribution of Chitosan-Complexed DNA

The bioadhesion of intravenously injected chitosan and the ability to complex and prevent DNA degradation was investigated by Richardson et al. (87,88). The molecular weight of highly purified chitosans varied from <5000 to >10,000 daltons. For the low molecular weight ^{125}I-labelled chitosan (<5000), 26.5% of the dose was in the liver, 32.2% in the blood pool, and 19.1% in the urine after 1 h of administration. In contrast, 82.7% of the high molecular weight chitosan (>10,000) was in the liver and only 2.6% in the blood pool after the same time period. The hemolytic potential of chitosans in terms of percent hemoglobin released was evaluated spectrophotometrically at 550 nm. For all of the chitosans, less than 10% of hemolysis was observed even after 5 h of incubation. Agarose gel electrophoresis showed that all chitosans complexed with DNA. DNA degradation was assayed spectrophotometrically at 260 nm and the degradation by DNase II was reduced in vitro when condensed with chitosan.

6. Modified Chitosan Nanospheres

Chitosan nanospheres containing 2 μg of plasmid DNA encoding luciferase gene and containing respiratory synctial virus promoter sequence were modified with transferrin and poly(ethylene glycol) (PEG) (89). The size range of these microspheres was between 200 and 700 nm as determined by differential interference microscopy and photon correlation spectroscopy. The transfection of HEK 293 cells was measured by assaying the luciferase activity in permeabilized cell extracts. The transfection efficiency of chitosan–DNA nanospheres was not significantly higher than that of the Lipofectamine control. After lyophilization, the PEG-treated nanospheres were also capable of transfecting cells and the luciferase expression was 7×10^7 light units/min · mg protein. The transfection efficacy of the PEG-treated nanospheres was maintained for 1 month. The higher transfection of lyophilized PEG-treated nanospheres could be due to the fact that PEG might stabilize the nanospheres against aggregation.

7. Lactosylated Chitosan for Cell Targeting

One of the promising uses of chitosan for targeted gene delivery was described by Erbacher et al. (90). The authors developed a lactose conjugate of chitosan and condensed DNA to form stable complexes with a size of about 100 nm. The complex formation was confirmed by transmission electron microscopy. Interestingly, after the chemical modification by lactose, chitosan could be targeted to certain cells like hepatocytes since they have galactose-binding membrane lectins. The authors conducted several transfection experiments mediated by chitosan and lactosylated chitosans. Utrocervical carcinoma cells (HeLa) were transfected with luciferase gene with cytomegalo virus promoter sequence plasmid complexed with chitosan. The gene expression was measured by luciferase assay and expressed as relative luciferase unit (RLU) per milligram of protein in cell lysates. Chitosan–DNA complexes with the number of chitosan nitrogen per DNA phosphate ratio (N/P) of 3 gave the highest transfection efficiency of 3×10^7 RLU/mg protein after 72 h post-transfection. To selectively deliver a gene, lactosylated chitosans were used as vectors for targeted delivery to cells expressing a galactose-specific membrane lectin [normal mouse liver cells (BNL CL2) or human hepatoma cells (Hep G2)]. These cells were poorly transfected by lactosylated chitosan with the luciferase activities below 10^5 RLU/mg of protein. The low transfection efficacy could be explained by the fact that the lactosylated chitosan complexes do not have as many cationic amino groups. Lack of cationic charge on chitosan may decrease DNA complexation or charge-mediated interactions with the cell surface.

8. Quaternary Chitosan with Galactose Residues

Murata et al. (91) synthesized N,N,N-trimethyl chitosan–galactose conjugate, a quaternary chitosan with recognizable galactose residues. This conjugate was targeted to human hepatoma cells (HepG2) which express galactose binding membrane lectin receptors on the surface. After interacting with the plasmid, a formation of a neutral polyelectrolyte complex was confirmed by gel electrophoresis. After incubating the human hepatoma cells in a culture medium, the expression of β-galactosidase activity by the polycation–luciferase with simian virus promoter sequence plasmid complex was measured using a microplate reader at 420 nm. An absorbance value of 0.1, reflecting low β-galactosidase activity, was obtained by the conjugate. A similar β-galactosidase activity was observed for diethylaminoethyl dextran–complexed plasmid, used as a control.

V. CHITOSAN AS A VACCINE CARRIER AND MUCOSAL ADJUVANT

A. Vaccine Carriers

Many approaches to enhance the absorption of proteins and peptides across the gastrointestinal, urinogenital, and nasal mucosae have been developed. Among these approaches are the use of absorption enhancers and coadministration of enzyme inhibitors. However, the implication of this approach is to alter the membrane permeability, thus allowing foreign particles to go through. For this reason, a better approach is to deliver antigens entrapped in a carrier system. These carriers could be liposomes, poly(methyl methacrylate), poly(ethylene vinyl acetate), poly(lactide-co-glycolide), albumin, and starch (92).

Chitosan has interesting features that render it as a good vaccine carrier. This compound is mucoadhesive, biocompatible, and can enhance the penetration of proteins and peptides across the nasal, gastrointestinal, and urinogenital mucosae. Chitosan could be associated with proteins and vaccines by electrostatic interactions between the positive amine group of chitosan and the negative acidic groups of proteins. Vaccine carrier systems that are prepared with chitosan could maintain the stability of proteins, allowing them to be released in their active form (93).

The different types of chitosan that have been used as vaccine carriers are shown in Table 12.

1. Chitosan Nanoparticles as a Novel Carrier for Vaccines

Chitosan nanoparticles, prepared by a simple precipitation method with tripolyphosphate, were evaluated as carriers of proteins and vaccines by Calvo et al. (93). Transmission electron microscopy showed that chitosan nanoparticles were uniformly solid and had consistent structure. Proteins and vaccines were associated with chitosan nanoparticles probably by electrostatic interactions between the positive ammonium group of chitosan and the negative acid groups of proteins. Table 13 compares the uptake of bovine serum albumin (BSA), tetanus toxoid, and diphtheria toxoid to chitosan nanoparticles prepared at pH 5. The proteins were incorporated into the chitosan solution after they were dissolved in tripolyphosphate solution. Although the uptake of toxoids to the nanoparticles was efficient, BSA uptake was higher than that of toxoids. The isoelectric point (pI) of BSA is 4.8 and the pI of toxoids varies from 4 to 6 as the toxoids are a mixture of proteins. Thus, different ionization of the toxoids, caused by different pI values, will lead to different uptake by nanoparticles. However, the pH value of 5 is very close to the pI of BSA. Also the positively charged amino group of chitosan was cross-linked with the negative groups of tripolyphosphate, so positively charged amino groups of chitosan are not avail-

Table 10.12 Chitosan as a Vaccine Carrier

Carrier type	Antigens	Reference
Chitosan–ethylene oxide–propylene oxide block copolymer (Pluronic® F68)	Bovine serum albumin and tetanus and diphtheria toxoids	93
Glutaraldehyde crosslinked chitosan microspheres	Bovine serum albumin, diphtheria toxoid	94

Table 10.13 Entrapment Efficiency of Bovine Serum Albumin and Tetanus and Diphtheria Toxoids into Chitosan Nanoparticles

Protein	Molecular weight (daltons)	Protein/chitosan (%)	Isoelectric point	Entrapment efficiency (%)
BSA	69,000	12	4.5–4.8	100.0
Tetanus toxoid	150,000	6	4.4–5.9	53.30
		12		56.73
Diphtheria toxoid	62,000	6	4.1–6.0	55.10

Source: Adapted from Ref. 93.

able to provide electrostatic interactions. As a result, it could be concluded that other forces such as hydrogen bonding and hydrophobic interactions may have led to the high uptake of BSA into the nanoparticles. A prolonged release of the entrapped and active protein was obtained. Nanoparticles with BSA loading of 25% released the entrapped protein without burst effect and at a fairly constant rate. The amount released of BSA was almost 14% after 4 h and about 30% after 8 h. For tetanus toxoid, the release profile was biphasic. After incubating the particles for almost 9 days in an aqueous 5% (w/v) trehalose solution, a slower second phase release profile was achieved. The amount of tetanus toxoid released after 9 days was 16%, and an additional 2% was released for up to 18 days.

2. Crosslinked Chitosan Microspheres as Carriers of Macromolecular Drugs

Some biological macromolecules are sensitive to pH, organic solvents, surfactants, and temperature. Loading them passively into hydrophilic matrices would preserve the biological activity of the protein molecules. For this reason, bovine serum albumin and diphtheria toxoid were loaded by passive absorption from aqueous solutions into performed glutaraldehyde crosslinked chitosan microspheres (300–600 μm) (94). The drug release into phosphate buffer (pH 7.4) was modulated after coating the bovine serum albumin–loaded particles with paraffin oil or with a poly(L-lactic acid). Although there was an initial burst of drug seen at day 1 from poly(L-lactic acid)–coated particles, the release rate increased as a function of time for nearly 2 months. Diphtheria toxoid–loaded chitosan microspheres were injected intramuscularly in Wistar rats and showed a constant antibody titer for up to a 5-month period. Tissue compatibility was confirmed by histological studies after intramuscular injection of placebo chitosan microspheres (150–300 μm diameter) into rats. The crosslinked chitosan microspheres were retained in the rat muscle for up to 6 months without any degradation.

B. Chitosan as a Mucosal Adjuvant

Parenteral route is the most common method for the administration of vaccines. Since this route is invasive, administration of vaccines into mucosal surfaces such as the nasal, urinogenital, pulmonary, and peroral mucosae has been investigated (50,95–97). However, these routes have limitations because of the chemical and physical instability of vaccines, the high metabolic activity, and the low permeability of the mucosal barrier (98). These limitations can be overcome by using enzyme inhibitors, absorption enhancers, and colloidal carriers (99,100).

Adjuvants are substances that can enhance the immunogen effectiveness by enhancing the immune response. Sometimes antigens on their own may produce a weak specific immune response and hence are ineffective as vaccines. By adding adjuvants, the immune response can be elevated to an effective level. The production of viral vaccines in large quantities is very difficult and expensive. Growing and expanding the viral particles and the extracting and purifying of the effective antigens require sophisticated equipment, highly trained personnel, and prepared biochemicals. Thus, using adjuvants to elevate the immune response to an effective level would enhance the action of small doses of vaccine. Different adjuvants had been investigated for the immunogenicity enhancement of vaccines. The adjuvants include cholera toxin (101,102), poly(L-lactide-co-glycolide) microspheres (103–105), liposomes (106), pluronics surfactants (107), lipopolysaccharides, and aluminum salts (108).

Chitosan can affect the permeability of the epithelial membrane by opening the tight junctions between the cells leading to enhancement of absorption. Chitosan can act as an enzyme inhibitor and a good mucoadhesive. All the

Table 10.14 Chitosan as a Mucosal Adjuvant

Immune response enhancement	Reference
Mucosal and systemic immune response	108
Mucosal immune response	109

previous properties make chitosan a good candidate as a mucosal adjuvant to enhance the immune response by increasing the residence time of vaccine available at the mucosal site for specific immunity.

Table 14 summarizes the various examples of chitosan used in enhancing immune response.

1. Enhancement of the Immune Response

When antigens were administered nasally in rats, the immune response was enhanced by chitosan. Both serum and mucosal immune responses after intranasal immunization of mice with filamentous haemagglutinin (FHA) and pertussis toxin (rPT) doses of 2 μg combined with chitosan were higher than those of the solution of FHA and rPT administered intranasally. This could be explained by higher retention time of the antigen at the absorption site in the chitosan formulation. Two doses (2 μg/mL) of FHA combined with chitosan were also administered in mice nasally. While the mucosal response was 100 times higher than that produced after the nasal administration of FHA by itself, the systemic response was 60 times higher. The results show that chitosan can serve to enhance both mucosal and systemic immunity after mucosal administration of vaccine (108).

2. Mucosal Vaccination with Chitosan

The immune response to an antigen that is delivered through the mucosa could be enhanced by chitosan, as shown by Makin et al. (109). After nasal administration of purified surface antigen (PSA) of influenza B virus combined with chitosan to mice, the serum antibody response was similar to that produced after the subcutaneous immunization of PSA combined with aluminum hydroxide adjuvant. While the subcutaneous administration of the antigen did not cause any mucosal response, the nasal administration of PSA–chitosan formulation did. The results clearly showed that chitosan could enhance mucosal immunogenic response probably by increasing the residence time of the immunogen at the site of absorption or by increasing the fraction of dose absorbed across the membrane. The combined effect was to enhance the exposure of the antigen to the immune cells of the mucosal epithelia (108).

VI. MISCELLANEOUS APPLICATIONS OF CHITOSAN

A. Chitosan Microspheres for Controlled Release of Drugs

Microspheres in drug formulation offer many advantages such as controlled drug release, large surface area, protection against unpleasant tastes, and separation of incompatible drugs. Since microspheres have uniform distribution over absorption sites in the gastrointestinal tract, higher absorption and less irritation could be obtained (110,111).

Table 15 summarizes the various examples of chitosan microspheres that have been used for drug delivery. Different methods had been used to prepare chitosan microspheres, such as crosslinking with glutaraldehyde (112), interfacial acylation (113), precipitation (114), spray-drying technique (115), ionotropic gelation (116), solvent evaporation (117), and capillary extrusion (118).

1. Crosslinked Chitosan Microspheres

An aqueous acetic acid dispersion of chitosan containing progesterone was crosslinked—using glutaraldehyde—in a dispersion of liquid paraffin and petroleum ether (112). The resulting crosslinked chitosan microspheres were characterized by being smooth, spherical, and in the size range of 45–300 μm. The crosslinking density of the microspheres was an important factor affecting the extent of drug release. Microspheres prepared by using 4 mL of glutaraldehyde-saturated toluene were considered weakly crosslinked, while those prepared by adding 1 mL of aqueous glutaraldehyde were highly crosslinked. Seventy percent of the incorporated steroid was released in 40 days from the weakly crosslinked microspheres, whereas only 35% of the drug was released from the highly crosslinked spheres. After intramuscular injection of the microsphere formulation in rabbits, the plasma level of progesterone of 1 to 2 ng/mL was maintained for up to 5 months. The authors have failed to examine the biodegradation of microspheres after intramuscular administration.

2. Interfacial Acylation

To minimize stomach irritation, chitosan microspheres containing oxytetracycline were prepared by interfacial acylation method by Mi et al. (113). These microspheres had particle diameters between 420 and 590 μm and were used for intramuscular injection for controlled drug release property. In the simulated stomach medium, the oxytetracycline release was significantly reduced in the microsphere formulation. Using chitosan having a molecular

Table 10.15 Chitosan Microspheres for Controlled Release of Drugs

Microsphere preparation method	Drug	Reference
Crosslinked microspheres by glutaraldehyde	Progesterone	112
Spray-hardening and interfacial acylation	Oxytetracycline	113
Precipitation process	Prednisolone sodium phosphate	114
Spray-drying technique	Diclofenac sodium	115
Ionic gelation	Bovine serum albumin	116
Solvent evaporation	Diclofenac sodium, bovine serum albumin	117
Capillary extrusion	Diclofenac sodium	118

weight of 2,000,000 daltons, it took almost 9 h to release 50% of the drug. The release of the drug was prolonged in the stomach as acetylation of the amino group of D-glucosamine will reduce the charge density of the polymer leading to lower swelling of the microspheres.

3. Precipitation Technique

Using sodium sulfate as a precipitant, chitosan microspheres were prepared by Berthold et al. (114). Photon correlation spectroscopy and centrifugal sedimentation were used to measure the particle size, found to be between 1.5 and 2.5 μm. Side-by-side diffusion cells having a dialysis membrane in between made of cellulose acetate with a molecular weight cut-off of 12,000–14,000 daltons was used to measure the release of prednisolone sodium phosphate from the microsphere formulation. The drug/polymer ratio was a significant factor that affected the release of the drug from the microspheres. As the chitosan/drug ratio increased, the release rate decreased. For instance, 23% of prednisolone sodium phosphate was released after 6 h from microspheres having a chitosan/drug weight ratio of 1:1. However, when the chitosan/drug weight ratio increased to 7:1, the percentage of the drug released after the same period was only 13%.

4. Spray-Drying Technique

Using spray-drying technique, diclofenac sodium was loaded into chitosan microspheres. The drug-loaded microspheres were further microencapsulated with Eudragit L-100 and Eudragit S-100 using an oil-in-oil solvent evaporation method to form a multireservoir system (115). Eudragit L-100 and Eudragit S-100 are anionic copolymers of methacrylic acid and methyl methacrylate, used for enteric coating, with ratio of the free carboxyl groups to the ester groups of approximately 1:1 in Eudragit L-100 and about 1:2 in Eudragit S-100. By coating the chitosan microspheres with Eudragit, perfect pH-dependent release profiles were obtained. Although no release was detected at acidic pH, a continuous release for about 8–12 h was observed at the pH of Eudragit solubility. Eudragit L-100 is soluble at pH above 6, while Eudragit S-100 is soluble at pH above 7. The dissolution of chitosan gel cores, the dissolution of the Eudragit coating, and the swelling of the chitosan microspheres were factors that affected the release of diclofenac sodium. Controlled drug release was also achieved by the ionic interaction between the carboxyl groups of Eudragit and the amine groups of chitosan. Coating of chitosan microspheres with anionic polymers, such as Eudragit, could be useful for delivering drugs to specific regions of the gastrointestinal tract.

5. Ionic Gelation

Ionic gelation is a simple technique that involves the mixing of an aqueous phase of chitosan and a diblock copolymer of ethylene oxide and propylene oxide with an aqueous phase of sodium tripolyphosphate (116). The size of these microspheres was between 200 and 1000 nm. The particle size increased by increasing the concentration and molecular weight of the diblock copolymer of ethylene oxide and propylene oxide. Up to 80% of bovine serum albumin was entrapped within these microspheres. About 25% of BSA was released after 2 days, and after 6 days the percentage release of BSA increased up to 82%. Almost all of the entrapped drug was released in one week.

6. Solvent Evaporation

Chitosan microspheres, prepared by methylene chloride solvent evaporation technique, were entrapped in a hydrophobic polymer made of cellulose acetate butyrate and ethyl cellulose (117). The loading of diclofenac sodium was almost 100% in these microspheres. The loading of the drug was not affected by the amount of chitosan or the molecular weight of the polymer. One hundred percent of the drug was released after 8 h in a controlled release form from the chitosan microspheres coated with ethyl cellulose, and about 85% was released from chitosan microspheres coated with cellulose acetate butyrate in the same period of time. Since this method of microsphere preparation does not require a crosslinking agent, it could be an alternative method of chitosan microsphere preparation.

7. Capillary Extrusion

Diclofenac sodium was loaded into chitosan microspheres that were prepared by capillary extrusion (118). The microspheres were characterized by having a narrow size distribution with a particle of diameter 760 μm. Four different formulations were evaluated for oral administration. The formulations were (1) drug powder, (2) empty chitosan microspheres, (3) sustained-release pellets, and (4) optimum diclofenac sodium–containing chitosan microspheres. In vitro release study was carried out in simulated intestinal medium at a pH of 6.8 and 37°C. A slow release of diclofenac sodium from the optimum microsphere formulation was observed. Nearly 45% of the drug was released from the microspheres after 6 h. To evaluate the efficacy of the product in treatment of ulceration, rabbits were given a dose of 5 mg/kg body weight of the free drug or in the microspheres form once a day by gastric tubing for 21 days. After sacrificing the rabbits, the mucosa of the stomach was examined under a low power (20×) dissecting microscope. The degree of ulceration was determined qualitatively by using a scale of 0 to 4 (4 being highly ulcerative). The ulcerogenic index (UI) was calculated from the degree of ulceration and the percentage of the rabbits that had ulcer in the study. No difference in gastric lesions and ulcerogenic index was observed between the optimum formulation and the commercial product of diclofenac sodium. The microsphere formulation had a lower level of ulcerogenic index (UI 0.25) than the plain drug (UI 1.5), which suggests that chitosan could protect the gastric mucosa from ulceration by nonsteroidal antiinflammatory drugs.

B. Implantable Sustained Release Forms of Chitosan

Surgical implants devices in the body tissues could provide sustained release of a drug at the site of action (119). One of the most promising materials that could be used for implantable devices is a polymer that can degrade at the site of delivery. Biodegradable polymeric devices could be useful as a second surgery to remove the implant after drug release is avoided. The biodegradable polymer should also have good mechanical properties, not form harmful degradation products, have capability to be fabricated into devices with different geometry, and be sterilizable (120). Chitosan and chitosan derivatives have been used to develop implantable sustained release dosage forms.

Table 16 shows the different examples of chitosan being used for implantable sustained release systems and the drugs that have been incorporated.

1. Implantable Chitosan Using Uracil as a Model Drug

Chitosan with a degree of deacetylation of ≥85% and hydroxypropyl chitosan were used to prepare an implantable sustained release form of uracil (121). For in vitro studies, lysozyme was used to ensure the enzymatic degradation of chitosan and hydroxypropyl chitosan. The enzymatic degradation was confirmed by gel filtration chromatography. For in vivo studies, 100 mg spherical polymer with an approximate diameter of 5 mm was implanted subcutaneously in the backs of Wistar rats. After 1 month, the weight of chitosan and hydroxypropyl chitosan samples were reduced to 69 and 50% of the original weight, respectively. Extrusion-in-air method was used to design film- and stick-type implantable dosage forms of chitosan containing uracil. After the implantation of 50 mg stick type dosage form containing 5 mg of uracil into the back of a Wistar rat, controlled release of the drug was achieved for almost 3 days. Following subcutaneous injection of uracil, the peak drug concentration was 3.282 μg/mL after 1 h. No uracil was found after 24 h, and the area under the plasma concentration–time curve was only 9 μg · h/mL. After subcutaneous implantation of the chitosan-based device, however, the peak concentration of uracil was 0.422 μg/mL after 3 h, and the concentration was sustained at 0.024 μg/mL for up to 3 days. The area under the plasma concentration–time curve was also 9 μg · h/mL. Both methods of administration had the same area under the plasma concentration–time curve, meaning that the release of the drug was prolonged without changing the overall bioavailability. The results of this study showed that a chitosan-based implantable delivery system could be used for sustained release of the anticancer drug. The devices were able to release the drug for up to 3 days in vivo.

2. Local Implantation of Chitosan Microspheres Containing Bisphosphonates

Suberoylbisphosphonate, an agent that is used clinically to inhibit bone resorption in the treatment of hypercalcemia, which may lead to calcification of bioprosthetic heart valve tissue, was loaded into chitosan microspheres (122). The in vitro release rate of the drug in TRIS buffer (pH 7.4)

Table 10.16 Implantable Sustained Release Forms of Chitosan

Carrier type	Drug	Reference
Film- and stick-type implantable dosage form	Uracil	121
Chitosan microspheres prepared from water-in-oil emulsion	Pamidronate, suberoylbisphosphonate	122
N,O-Carboxymethyl chitosan	Human coagulation factor IX	123
Glutaraldehyde crosslinked microspheres	Mitoxantrone	125
N-Succinyl–chitosan–mitomycin C conjugate	Mitomycin C	126

was retarded in the chitosan microspheres. Almost 100% of the drug was released in 24 h from the microsphere formulation. Chitosan films were used to grow endothelial cells to prove biocompatibility. The results showed that the cells were viable on the polymer surface. After implanting suberoylbisphosphonate microspheres (500 μm diameter) containing 10% (w/w) of the drug subdermally in rats, the bioprosthetic heart valve tissue calcification was inhibited. Sixty-eight percent reduction in calcium concentration of the valve was achieved in 35 days.

3. Chitosan as a Delivery System for Coagulation Proteins

An aqueous formulation of human coagulation factor IX was developed for hemophilia treatment. As compared to intravenous dose, the bioavailability of 16% of human coagulation factor IX was achieved after bolus subcutaneous injection in rabbits. When osmotic pumps were implanted subcutaneously in rabbits, human coagulation factor IX reached a plasma plateau that is equivalent to 2.5% of the normal human levels. A chitosan derivative, N,O-carboxymethyl chitosan, was used to prepare hydrogels containing human coagulation factor IX. Using the same dose (8000 units/kg body weight) of intravenous or subcutaneous bolus injections as controls, higher levels of human coagulation factor IX were maintained in mice plasma following subcutaneous injection of the hydrogels. After 72 h, the concentration of human coagulation factor IX in mouse plasma was 0.1 units/mL from the hydrogel, while from the intravenous or subcutaneous bolus injection, the concentration was only 0.05 units/mL (123).

4. Crosslinked Chitosan Microspheres

Jameela et al. (124) prepared glutaraldehyde crosslinked chitosan microspheres. Chitosan microspheres, without the drug, were implanted in the skeletal muscle of rats. The tolerability of the microspheres in muscle tissue was confirmed by histological analysis. There was a moderate cellular response with a few macrophages encircling the beads. All of the loaded mitoxantrone was released into phosphate buffer (pH 7.4) in 1 day from the weakly crosslinked spheres that were made with 2 mL of of glutaraldehyde-saturated toluene followed by 0.2 g of additional glutaraldehyde. However, only 25% of the drug was released after 36 days from the highly crosslinked spheres prepared with 10 mL of glutaraldehyde-saturated toluene followed by 0.2 g of additional glutaraldehyde. For 3 months, the microspheres remained intact in the skeletal muscle of rats without any degradation.

Mitoxantrone-loaded chitosan microspheres were injected intraperitonealy in mice with Ehrlich ascites carcinoma. The mean survival time of mice treated with mitoxantrone-loaded chitosan microspheres containing 2 mg of the drug was 50 days. Mice receiving 2 mg of the drug in solution, however, had a mean survival time of only 2.1 days. Thus, mitoxantrone-loaded chitosan microspheres could enhance the therapeutic efficacy of anticancer drugs like mitoxantrone (125).

5. N-Succinyl–Chitosan–Mitomycin C as an Implant

Water-soluble carbodiimide was used as a coupling agent between N-succinyl–chitosan and mitomycin C (126). A total weight of 20 mg of the drug was incorporated in the conjugate N-succinyl–chitosan–mitomycin C as a tablet form. After implantation subcutaneously into the back of a male Wistar rat, mitomycin C was released gradually for almost 1 week. For tumor inhibition study, 0.1 mL of the ascitic fluid containing 1×10^7 Sarcoma cells was injected subcutaneously into the axillary tissue of 6-week-old mice. To investigate the antitumor effect, the tumor size was measured with calipers at appropriate times. The percentage growth ratio was calculated by dividing the volume of the tumor after 19 days of drug administration over the volume of the tumor after 7 days following inoculation. After intratumoral administration of 5 mg/kg of mitomycin C in the mice, the percentage of tumor growth ratio was 30.5%. However, the percentage of tumor growth ratio was reduced to 17.5% after the administration of the conjugate N-succinyl–chitosan–mitomycin C containing 15 mg equivalent mitomycin C/kg by the same route. The conjugate was more effective than free mitomycin C in reducing the tumor growth, but at a much higher dose.

C. Chitosan for Colon-Specific Drug Delivery

Following oral administration, peptide and protein drugs suffer from poor intestinal absorption due to susceptibility of these drugs to the proteolytic enzymes in the gastrointestinal tract and poor membrane permeability. The absorption of peptide and protein drugs could be enhanced in the colon because of the low activity of proteolytic enzymes there and the long residence time. In addition, it is more effective to treat colonic diseases such as ulcerative colitis, colorectal cancer, and Crohn's disease with direct delivery of drugs to the affected area (127).

Many pH-sensitive or bacterial degradable polymers have been used for colon-specific drug delivery. These polymers include chitosan (127), methacrylic acid or hydroxyethyl methacrylate (128), azoaromatic polymers (129), methacrylated inulin (130), and cellulose derivatives such as ethyl cellulose (131), hydroxypropyl methylcellulose acetate succinates, cellulose acetate phthalate (132), and cellulose acetate butyrate (133).

Chitosan was used in oral drug formulations to provide sustained release of drugs. Recently it was found that chitosan is degraded by the microflora that are available in the colon. As a result, this compound could be promising for colon-specific drug delivery (127).

1. Improvement of Insulin Absorption Using Chitosan

Hydroxypropyl methylcellulose phthalate, an enteric-coating material, was used to coat chitosan capsules loaded with insulin (127). After 2 h following oral administration, the capsules were eliminated from the stomach. After 2 to 6 h, the capsules moved into the small intestine and were in the large intestine after 6 to 12 h. For in vitro release medium, cecal content from rats was suspended in two volumes of bicarbonate buffer and the pH was adjusted to 7.0. The release of 5(6)-carboxyfluorescein loaded in chitosan capsules, a model water-soluble compound, was increased in the rat cecal content suspension as compared to simulated gastric juice (pH 1.2) and simulated intestinal juice (pH 6.8). While all of the drug was released after 12 h in the cecal suspension, only 20% was released in the simulated intestinal juice after 6 h, and none was released in the simulated gastric juice after 2 h. Degradation of chitosan capsules in cecal contents facilitated the release of 5(6)-carboxyfluorescein. Using male Wistar rats, insulin-containing chitosan capsules were administered orally with a total dose of 20 IU into the stomach with a polyethylene tubing. The hypoglycemic effect started 6 h after the administration, when the capsules were in the colon and lasted for 24 h. The peak plasma insulin concentration of 326 μunit/mL was noticed after 7 h of administration. To find out the insulin bioavailability, insulin solution (0.1 IU)

Table 10.17 Chitosan as a Pharmaceutical Excipient in Tablet

Drug	Chitosan DD[a] %	Tests conducted	Reference
No drug	92.7	Disintegration test	134
No drug	92.7	Disintegration test	135
Propranolol · HCl	92.7	In vitro drug release	138
No drug	92.7	Disintegration test	139
Chlorpheniramine maleate	ND[b]	In vitro drug release	140
Theophylline	ND	In vitro drug release	141
Ketoprofen	80	In vivo drug absorption, In	32
		vitro and in vivo adhesion	32
Diltiazem	80	In vitro drug release, In vitro	142
		adhesion	142

[a] DD refers to the degree of deacetylation of chitosan in the tablet formulation.
[b] ND = not determined.
Source: Adapted from Ref. 143.

was injected intravenously by bolus injection and the area under the plasma concentration–time curve was normalized according to the dose. The bioavailability of insulin from the chitosan formulation was 5.73% as compared to the intravenous one.

D. Chitosan as a Pharmaceutical Excipient in Tablet

Table 17 provides the various examples of chitosan when used in the preparation of tablet dosage forms.

1. Angle of Repose Reduction of a Powder Mixture

Angle of repose is defined as the base angle of the cone formed when a powdered or granular material falls freely on a flat surface from an orifice. By adding chitosan to other conventional tablet excipients such as mannitol, lactose, or starch, the angle of repose was decreased. Thus, the fluidity of the powder from the hopper into the die cavity in the tablet machine would be increased. The angle of repose decreased from 65° to 39° in the presence of 80% (w/w) chitosan. In the presence of microcrystalline cellulose, on the other hand, the angle of repose was reduced to 45° at the same concentration (134,135).

2. Chitosan as a Binder in Wet Granulation

When chitosan was used as a binder in wet granulation, the binding efficiency was less than that of hydroxypropyl methylcellulose, but more than that of methylcellulose or sodium carboxymethyl cellulose (136). The binding efficiency was evaluated based on the properties of tablet hardness, friability, and disintegration time. Thus, by dividing the hardness by the friability value, the hardness/friability ratio was obtained. The binder efficiency parameter was obtained by dividing the hardness/friability ratio by the disintegration time. Table 18 compares the binder efficiency parameter of chlorpheniramine maleate using 2% (w/v) granulating fluid of different polymers. The binding efficiency of chitosan was found to be 0.116 kg/s, and that of sodium carboxymethyl cellulose was 0.037 kg/s.

Table 10.18 Binding Efficiency Parameter of Chlorpheniramine Maleate

Polymer	Binding efficiency parameter (kg/s)
Chitosan	0.116
Hydroxypropyl methylcellulose	0.215
Methylcellulose	0.101
Sodium carboxymethyl cellulose	0.037

Note: The binding efficiency was measured using a 2% (w/v) solution of the various polymers.
Source: Adapted from Ref. 136.

Table 10.19 Disintegration Times of Paracetamol Tablets

Disintegrants	Disintegration time (min)
Chitosan	0.6
Corn starch	2.65
Microcrystalline cellulose	7.05

Note: Tablets were compressed at 600 pounds force using the various disintegrating agents.
Source: Adapted from Ref. 137.

3. Chitosan as a Disintegrant

When chitosan was used in tablets at a concentration above 5% (w/w), its disintegration efficiency was higher than starch and microcrystalline cellulose. Table 19 lists the disintegration times of paracetamol tablets compressed at two forces using different disintegrants (137). The disintegration time of chitosan tablet compressed at 600 pounds force was 0.6 min. The disintegration time of 7.05 min, however, was obtained from tablets made with microcrystalline cellulose.

VII. CONCLUSION

Chitosan is an abundant natural polymer, obtained by alkaline N-deacetylation of chitin. The physical and chemical properties of chitosan, such as inter- and intramolecular hydrogen bonding and the cationic charge in acidic medium, makes this polymer attractive for the development of conventional and novel pharmaceutical products. As a bioadhesive polymer, chitosan has been found to increase residence time of dosage forms at mucosal sites, inhibit enzymes, and increase the permeability of protein and peptide drugs across mucosal membranes. Chitosan networks can serve as pH-sensitive swelling system for region-specific drug delivery in the gastrointestinal tract. Chitosan–DNA complexes have been shown to facilitate transfection and inhibit degradation of DNA by DNase. A number of investigators have also shown that chitosan could be used as a mucosal vaccine carrier and may have adjuvant effect. The future of chitosan as a pharmaceutical aid looks very bright as a number of derivatives have been synthesized that will lead to more useful applications.

REFERENCES

1. F. Hoppe-Seiler. Chitin and chitosan. *Ber. Deutsch Chem. Ges.* 27:3329–3331 (1994).
2. G. A. F. Roberts. Solubility and solution behaviour of chitin and chitosan. In: *Chitin Chemistry*, G. A. F. Roberts (Ed.). Macmillan Press, Houndmills, England, 1992, pp. 274–329.
3. A. Domard, N. Cartier. Glucosamine oligomers. 4. Solid state-crystallization and sustained dissolution. *Int. J. Biol. Macromol.* 14:100–106 (1992).
4. G. A. F. Roberts. Structure of chitin and chitosan. In: *Chitin Chemistry*, G. A. F. Roberts (Ed.). Macmillan Press, Houndmills, England, 1992, pp. 1–53.
5. K. Kurita. Chemical modifications of chitin and chitosan. In: *Chitin in Nature and Technology*, R. A. A. Muzzarelli, C. Jeuniaux, G. W. Gooday (Eds.). Plenum Press, New York, 1986, pp. 287–293.
6. C. K. Rha, D. Rodriguez-Sanchez, C. Kienzle-Sterzer. Novel applications of chitosan. In: *Biotechnology of Marine Polysaccharides*, R. R. Colwell, E. R. Pariser, A. J. Sinskey (Eds.). Hemisphere Publishing, Washington, D.C., 1984, pp. 284–311.
7. S. Hirano. Chitin biotechnology applications. *Biotech. Ann. Rev.* 2:237–258 (1996).
8. T. Sannan, K. Kurita, Y. Iwakura. Studies on chitin. 2. Effect of deacetylation on solubility. *Macromol. Chem.* 177:3589–3600 (1976).
9. M. M. Amiji, V. R. Patel. *Proceed. Int. Symp. Control. Rel. Bioact. Mater.* 22.330–331 (1995).
10. J. Knapczyk, L. Krowczynski, J. Krzek, M. Brzeski, E. Nurnberg, D. Schenk, H. Struszczyk. Requirements of chitosan for pharmaceutical and biomedical applications. In: *Chitin and Chitosan Sources: Chemistry, Biochemistry, Physical Properties and Applications*, G. Skjak Braek, T. Anthonsen, P. Sandford (Eds.). Elsevier, London, 1989, pp. 657–664.
11. P. A. Sandford. Chitosan: commercial uses and potential applications. In: *Chitin and Chitosan Sources: Chemistry, Biochemistry, Physical Properties and Applications*, G. Skjak-Braek, T. Anthonsen, P. Sandford (Eds.). Elsevier, London, 1989, pp. 51–69.
12. N. Errington, S. E. Harding, K. M. Varum, L. Illum. Hydrodynamic characterization of chitosans varying in degree of acetylation. *Int. J. Biol. Macromol.* 15:113–117 (1993).
13. H. Fukuda. Polyelectrolyte complexes of chitosan with sodium carboxymethylcellulose. *Bull. Chem. Soc. Japan* 53: 837–840 (1980).
14. H. Fukuda, Y. Kikuchi. Polyelectrolyte complexes of chitosan with sodium carboxymethyldextran. *Bull. Chem. Soc. Japan* 51:1142–1144 (1978).
15. L. Illum. Chitosan and its use as a pharmaceutical excipient. *Pharm. Res.* 15:1326–1331 (1998).
16. R. A. A. Muzzarelli. Stereochemistry and physical characterization. In: *Chitin*, R. A. A. Muzzarelli (Ed.). Pergamon Press, Oxford, 1977, pp. 45–86.
17. R. H. Chen, J. H. Lin, M. H. Yang. Relationships between the chain flexibilities of chitosan molecules and the physical properties of their casted films. *Carbo. Polymers* 24: 41–46 (1994).
18. H. Struszczyk, D. Wawro, A. Niekraszewicz. Biodegradability of chitosan fibres. In: *Advances in Chitin and Chitosan*, C. J. Brine, P. A. Sandford, J. P. Zikakis. (Eds.). Elsevier Applied Science, London, 1991, pp. 580–585.
19. T. Chandy, C. P. Sharma. Chitosan—as a biomaterial. *Biomat. Art. Cells. Art. Org.* 18:1–24 (1990).
20. S. Hirano, H. Seino, Y. Akiyama, I. Nonaka. Chitosan: a biocompatible material for oral and intravenous administrations. In: *Progress in Biomedical Polymers*, C. G. Gebelein, R. L. Dunn (Eds.). Plenum Press, New York, 1990, pp. 283–290.
21. J. Knapczk, L. Krowczynski, B. Pawlik, Z. Liber. Pharmaceutical dosage forms with chitosan. In: *Chitin and Chitosan: Sources, Chemistry, Biochemistry, Physical Properties and Applications*, G. Skjak-Braek, T. Anthonsen, P. Sandford (Eds.). Elsevier Applied Science, London, 1984, pp. 665–669.
22. P. A. Sandford. Chitosan and alginate: new forms of commercial interest. *Am. Chem. Soc. Div. Polym. Chem.* 31: 628–628 (1990).
23. S. Palapura, J. Kohn. Trends in the development of bioresorbable polymers for medical application. *J. Biomater. Appl.* 6:216–250 (1992).
24. Q. Zhao, M. P. Agger, M. Fitzpatrick, J. M. Anderson, A. Hiltner, K. Stokes, P. Urbanski. Cellular interactions with biomaterials: in vivo cracking of prestressed pellethane 2363-80 A. *J. Biomed. Mater. Res.* 24:621–637 (1990).
25. Y. Shigemasa, S. Minami. Applications of chitin and chitosan for biomaterials. *Biotechnol. Gen. Eng. Rev.* 13: 383–420 (1995).
26. N. A. Peppas, P. Buri. Surface, interfacial and molecular aspects of polymer bioadhesion on soft tissues. *J. Controlled Release* 2:257–275 (1985).
27. M. A. Longer, J. R. Robinson. Fundamental aspects of bioadhesion. *Pharm. Int.* 9:114–117 (1986).
28. D. Duchene, F. Touchard, N. A. Peppas. Pharmaceutical and medical aspects of bioadhesive systems for drug administration. *Drug. Dev. Ind. Pharm.* 14:283–318 (1988).
29. H. E. Junginger. Bioadhesive polymer systems for peptide delivery. *Acta Pharm. Technol.* 36:110–126 (1990).
30. J. D. Smart, I. W. Kellaway, H. E. C. Worthington. An in vitro investigation of mucosa-adhesive materials for use in controlled/drug delivery. *J. Pharm. Pharmacol.* 36: 295–299 (1984).
31. R. Gurny, J. M. Meyer, N. A. Peppas. Bioadhesive intra oral release systems: design, testing and analysis. *Biomaterials* 5:336–340 (1984).
32. S. Miyazaki, A. Nakayama, M. Oda, M. Takada, D. Att-

wood. Chitosan and sodium alginate based bioadhesive tablets for intraoral drug delivery. *Biol. Pharm. Bull.* 17: 745–747 (1994).

33. G. Ponchel, F. Thouchard, D. Duchene, N. A. Peppas. Bioadhesive analysis of controlled release systems. I: Fracture and interpenetration analysis in poly(acrylic acid)-containing systems. *J. Controlled Release* 5:129–141 (1987).
34. M. F. Saettone, P. Chetoni, M. T. Torracca, S. Burgalassi, P. Giannaccini. Evaluation of muco-adhesive properties and in vivo activity of ophthalmic vehicles based on hyaluronic acid. *Int. J. Pharm.* 51:203–212 (1989).
35. P. Esposito, I. Colombo, M. Lovrecich. Investigation of surface properties of some polymers by a thermodynamic and mechanical approach: possibility of predicting mucoadhesion and biocompatibility. *Biomaterials* 15(3): 177–182 (1994).
36. P. Bottenberg, R. Cleymae, C. De Muynck, J. P. Remon, D. Coomans, Y. Michpotte, D. Slop. Development and testing of bioadhesive, fluoride containing slow release tablets for oral use. *J. Pharm. Pharmacol.* 43:457–464 (1991).
37. F. Ferrari, S. Rossi, A. Martini, L. Muggetti, R. De Ponti, C. Caramella. Technological induction of mucoadhesive properties on waxy starches by grinding. *Eur. J. Pharm. Sci.* 5:277–285 (1997).
38. C. M. Lehr, J. A. Bouwstra, E. Schacht, H. E. Junginger. In vitro evaluation of mucoadhesive properties of chitosan and some other natural polymers. *Int. J. Pharm.* 78:43–48 (1992).
39. F. Ferrari, S. Rossi, M. Cristina, B. Caramella, C. Caramella. Characterization of rheological and mucoadhesive properties of three grades of chitosan hydrochloride. *Il Farmaco* 52:493–497 (1997).
40. C. R. Lopez, A. Portero, J. L. Vila-Jato, M. J. Alonso. Design and evaluation of chitosan ethylcellulose mucoadhesive bilayered devices for buccal drug delivery. *J. Controlled Release* 55:143–152 (1998).
41. K. Sugimoto, M. Yoshida, T. Yata, K. Higaki, T. Kimura. Evaluation of poly(vinyl alcohol)-gel spheres containing chitosan as dosage form to control gastrointestinal transit time of drugs. *Biol. Pharm. Bull.* 21:1202–1206 (1998).
42. I. Genta, B. Conti, P. Perugini, F. Pavanetto, A. Spadaro, G. Puglisi. Bioadhesive microspheres for ophthalmic administration of acyclovir. *J. Pharm. Pharmacol.* 49:737–742 (1997).
43. I. Henriksen, K. L. Green, J. D. Smart, G. Smistad, J. Karlsen. Bioadhesion of hydrated chitosans: an in vitro and in vivo study. *Int. J. Pharm.* 145:231–240 (1996).
44. H. Takeuchi, H. Yamamoto, T. Niwa, T. Hino, Y. Kawashima. Enteral absorption of insulin in rats from mucoadhesive chitosan-coated liposomes. *Pharm. Res.* 13:896–901 (1996).
45. A. B. Schnurch, C. Paikl, C. Valenta. Novel bioadhesive chitosan–EDTA conjugate protects leucine enkephalin from degradation by aminopeptidase N. *Pharm. Res.* 14: 917–922 (1997).
46. A. B. Schnurch, M. E. Krajicek. Mucoadhesive polymers as platforms for peroral peptide delivery and absorption: synthesis and evaluation of different chitosan–EDTA conjugates. *J. Controlled Release* 50:215–223 (1998).
47. A. B. Schnurch, M. Pasta. Intestinal peptide and protein delivery: novel bioadhesive drug-carrier matrix shielding from enzymatic attack. *J. Pharm. Sci.* 87:430–434 (1998).
48. H. L. Luessen, B. J. de Leeuw, M. W. E. Langemeyer, A. G. de Boer, J. C. Verhoef, H. E. Junginger. Mucoadhesive polymers in peroral peptide drug delivery. VI. Carbomer and chitosan improve the intestinal absorption of the peptide drug buserelin in vivo. *Pharm. Res.* 13:1668–1672 (1996).
49. G. Borchard, H. L. Luessen, A. G. de Boer, J. C. Verhoef, C. M. Lehr, H. E. Junginger. The potential of mucoadhesive polymers in enhancing intestinal peptide drug absorption. III: Effects of chitosan-glutamate and carbomer on epithelial tight junctions in vitro. *J. Controlled Release* 39: 131–138 (1996).
50. H. L. Luessen, C. O. Rental, A. F. Kotze, C. M. Lehr, A. G. deBoer, J. C. Verhoef, H. E. Junginger. Mucoadhesive polymers in peroral peptide drug delivery. IV. Polycarbophil and chitosan are potent enhancers of peptide transport across intestinal mucosae in vitro. *J. Controlled Release* 45:15–23 (1997).
51. N. G. Schipper, S. Olsson, J. A. Hoogstraate, A. G. deBoer, K. M. Varum, P. Artursson. Chitosans as absorption enhancers for poorly absorbable drugs 2:mechanism of absorption enhancement. *Pharm. Res.* 14:923–929 (1997).
52. L. Illum, N. F. Farraj, S. S. Davis. Chitosan as a novel nasal delivery system for peptide drugs. *Pharm. Res.* 11: 1186–1189 (1994).
53. T. J. Aspden, J. T. Mason, N. S. Jones, J. Lowe, O. Skaugrud, L. Illum. Chitosan as a nasal delivery system: the effect of chitosan solutions on in vitro and in vivo mucociliary transport rates in human turbinates and volunteers. *J. Pharm. Sci.* 86:509–513 (1997).
54. L. Illum, S. S. Davis, M. Pawula, A. N. Fisher, D. A. Barrett, N. F. Farraj, P. N. Shaw. Nasal administration of morphine-6-glucuronide in sheep-a pharmacokinetic study. *Biopharm. Drug Dis.* 17:717–724 (1996).
55. A. F. Kotze, H. L. Luessen, Bas J. de Leeuw, B. G. de Boer, J. C. Verhoef, H. E. Junginger. N-Trimethyl chitosan chloride as a potential absorption enhancer across mucosal surfaces: in vitro evaluation in intestinal epithelial cells (Caco-2). *Pharm. Res.* 14:1197–1202 (1997).
56. A. F. Kotze, H. L. Luessen, B. J. de Leeuw, B. G. de Boer, J. C. Verhoef, H. E. Junginger. Comparison of the effect of different chitosan salts and N-trimethyl chitosan chloride on the permeability of intestinal epithelial cells (Caco-2). *J. Controlled Release* 51:35–46 (1998).
57. N. A. Peppas, A. G. Mikos. Preparation methods and structure of hydrogels. In: *Hydrogels in Medicine and Pharmacy*, Vol. I, *Fundamentals*, N. A. Peppas (Ed.). CRC Press, Boca Raton, FL, 1986, pp. 1–25.

58. B. D. Ratner, A. S. Hoffman. Synthetic hydrogels for biomedical applications. In: *Hydrogels for Medical and Related Applications*, J. D. Andrade (Ed.). *ACS Symposium Series* 31. American Chemical Society, Washington, D.C., 1976, pp. 1–36.
59. D. G. Pedley, P. J. Skelly, B. J. Tighe. Hydrogels in biomedical applications. *Br. Polym. J.* 12:99–110 (1980).
60. N. B. Graham, M. E. McNeill. Hydrogels for controlled drug delivery. *Biomaterials* 5:27–36 (1984).
61. T. Tanaka. Phase transitions of gels. In: *Polyelectrolyte Gels*, R. S. Harland, R. K. Prud'homme (Eds.). *ACS Symposium Series* 480. American Chemical Society, Washington, D.C., 1992, pp. 1–21.
62. J. Heller. Chemically self-regulated drug delivery systems. *J. Controlled Release* 8:111–125 (1988).
63. R. A. Siegel, B. A. Firestone. pH-dependent equilibrium swelling properties of hydrophobic polyelectrolyte copolymer gels. *Macromolecules* 21:3254–3259 (1988).
64. A. A. Donatelli, D. A. Thomas, L. H. Sperling. Poly(butadiene-co-styrene)/polystyrene IPN's, semi-IPN's and graft copolymers: staining behavior and morphology. In: *Recent Advances in Polymer Blends, Grafts, and Blocks*, L. H. Sperling (Ed.). Plenum Press, New York, 1974, pp. 375–393.
65. H. L. Frisch, J. Cifaratti, R. Palma, R. Schwartz, R. Foreman. Barrier and surface properties of polyurethane–epoxy interpenetrating polymer networks. In: *Polymer Alloys: Blends, Blocks, Grafts, and Interpenetrating Networks*, D. Klempner, K. C. Frisch (Eds.). Plenum Press, New York, 1977, pp. 97–112.
66. K. D. Yao, T. Peng, M. F. A. Goosen, J. M. Min, Y. Y. He. pH-sensitivity of hydrogels based on complex forming chitosan: polymer interpenetrating polymer network. *J. Appl. Polym. Sci.* 48:343–354 (1993).
67. K. D. Yao, T. Peng, H. B. Feng, Y. Y. He. Swelling kinetics and release characteristic of crosslinked chitosan: polyether polymer network (semi-IPN) hydrogels. *J. Polym. Sci. Part A: Polym. Chem.* 32:1213–1223 (1994).
68. Y. M. Lee, S. S. Kim, S. H. Kim. Synthesis and properties of poly(ethylene glycol) macromer/β-chitosan hydrogels. *J. Material. Sci. Materials Med.* 8:537–541 (1997).
69. M. M. Amiji, V. R. Patel. Chitosan-poly(ethylene oxide) semi-IPNs as a pH-sensitive drug delivery system. *Polymer Preprints* 35:403–404 (1994).
70. V. R. Patel, M. M. Amiji. pH-sensitive swelling and drug release properties of chitosan–poly(ethylene oxide) semi-interpenetrating polymer network. In: *Hydrogels and Biodegradable Polymers for Bioapplications*, R. Ottenbrite, S. Huang, K. Park (Eds.). *ACS Symposium Series* 627. American Chemical Society, Washington, D.C., 1996, pp. 209–220.
71. V. R. Patel, M. M. Amiji. Preparation and characterization of freeze-dried chitosan–poly(ethylene oxide) hydrogels for site-specific antibiotic delivery in the stomach. *Pharm. Res.* 13:588–593 (1996).
72. J. P. Behr. Synthetic gene transfer vectors. *Acc. Chem. Res.* 26:274–278 (1993).
73. R. C. Mulligan. The basic science of gene therapy. *Science* 260:926–932 (1993).
74. T. Friedmann. Human gene therapy—an immune genie, but certainly out of the bottle. *Nature Med.* 2:144–147 (1996).
75. R. G. Crystal. The gene as the drug. *Nature Med.* 1:15–17 (1995).
76. A. D. Miller. Human gene therapy comes of age. *Nature* 357:455–460 (1992).
77. Z. Q. Xiang, Y. Yang, J. M. Wilson, H. C. Ertl. A replication defective human adenovirus recombinant serves as a highly efficacious vaccine carrier. *Virology* 219:220–227 (1996).
78. M. R. Knowles, K. W. Hohnecker, Z. Zhou, J. C. Olsen, T. L. Noah, P. C. Hu, M. W. Leigh, J. F. Engelhardt, L. J. Edwards, K. R. Jones. A controlled study of adenoviral vector-mediated gene transfer in the nasal epithelial of patients with cystic fibrosis. *N. Engl. J. Med.* 333:823–831 (1995).
79. E. W. Alton, D. M. Geddes. Gene therapy for cystic fibrosis: a clinical perspective. *Gene Ther.* 2:88–95 (1995).
80. J. S. Kim, A. Maruyama, T. Akaike, S. W. Kim. In vitro gene expression on smooth muscle cells using a terplex delivery system. *J. Controlled Release* 47:51–59 (1997).
81. P. L. Chang, N. Shen, A. J. Westcott. Delivery of recombinant gene products with microencapsulated cells in vivo. *Hum. Gene Ther.* 4:433–440 (1993).
82. K. Y. Lee, I. C. Kwon, Y. H. Kim, W. H. Jo, S. Y. Jeong. Preparation of chitosan self aggregates as a gene delivery system. *J. Controlled Release* 51:213–220 (1998).
83. K. W. Leong, H. Q. Mao, V. L. Truong-Le, K. Roy, S. M. Walsh, J. T. August. DNA-polycation nanospheres as non-viral gene delivery vehicles. *J. Controlled Release* 53: 183–193 (1998).
84. R. J. Mumper, J. Wang, J. M. Claspell, A. P. Rolland. Novel polymeric condensing carriers for gene delivery. *Proceed. Int. Symp. Control. Rel. Bioact. Mater.* 22:178–179 (1995).
85. K. Roy, H. Q. Mao, K. W. Leong, DNA–chitosan nanospheres: transfection efficiency and cellular uptake. *Proceed. Int. Symp. Control. Rel. Bioact. Mater.* 24:673–674 (1997).
86. F. C. MacLaughlin, R. J. Mumper, J. Wang, J. M. Tagliaferri, I. Gill, M. Hinchcliffe, A. P. Rollad. Chitosan and depolymerized chitosan oligomers as condensing carriers for in vivo plasmid delivery. *J. Controlled Release* 56: 259–272 (1998).
87. S. Richardson, H. V. J. Kolbe, R. Duncan. Potential of low molecular mass chitosan as a DNA delivery system: biocompatibility, body distribution and ability to complex and protect DNA. *Int. J. Pharm.* 178:231–243 (1999).
88. S. Richardson, H. V. J. Kolbe, R. Duncan. Evaluation of highly purified chitosan as a potential gene delivery vector. *Proceed. Int. Symp. Control. Rel. Bioact. Mater.* 24: 649–650 (1997).
89. H. Q. Mao, K. Roy, V. Truong-Le, J. T. August, K. W. Leong. DNA–chitosan nanospheres: derivatization and

storage stability. *Proceed. Int. Symp. Control. Rel. Bioact. Mater.* 24:671–672 (1997).

90. P. Erbacher, S. Zou, T. Bettinger, A. M. Steffan, J. S. Remy. Chitosan-based vector/DNA complexes for gene delivery: biophysical characteristics and transfection ability. *Pharm. Res.* 15:1332–1330 (1998).
91. J. I. Murata, Y. Ohya, T. Ouchi. Possibility of application of quaternary chitosan having pendant galactose residues as gene delivery tool. *Carb. Polymers* 29:69–74 (1996).
92. D. T. O'Hagan. Microparticles as oral vaccines. In: *Novel Delivery Systems for Oral Vaccines*, D. T. O'Hagan (Ed.). CRC Press, Boca Raton, Florida, 1994, pp. 175–205.
93. P. Calvo, C. R. Lopez, J. L. Vila-Jato, M. J. Alonso. Chitosan and chitosan/ethylene oxide–propylene oxide block copolymer nanoparticles as novel carriers for proteins and vaccines. *Pharm. Res.* 14:1431–1436 (1997).
94. S. R. Jameela, A. Misra, A. Jayakrishnan. Cross-linked chitosan microspheres as carriers for prolonged delivery of macromolecular drugs. *J. Biomater. Sci. Polymer Edn.* 6:621–632 (1994).
95. S. Kobayashi, S. Kondo, K. Juni. Pulmonary delivery of salmon calcitonin dry powders containing absorption enhancers in rats. *Pharm. Res.* 13:80–83 (1996).
96. H. P. Merkle, G. Wolany. Buccal delivery for peptide drugs. *J. Controlled Release* 21:155–164 (1992).
97. J. C. Verhoef, N. G. M. Schipper, S. G. Romeijn, F. W. H. McMerkus. The potential of cyclodextrins as absorption enhancers in nasal delivery of peptide drugs. *J. Controlled Release* 29:351–360 (1994).
98. B. Lipka, J. Crison, G. L. Amidon. Transmembrane transport of peptide type compounds: prospects for oral delivery. *J. Controlled Release* 39:121–129 (1996).
99. D. A. Eppstein, J. P. Longenecker. Alternative delivery systems for peptides and proteins as drugs. *Crit. Rev. Ther. Drug Carrier Syst.* 5:99–139 (1988).
100. V. H. L. Lee, A. Yamamoto, U. B. Kompelia. Mucosal penetration enhancers for facilitation of peptide and protein drug absorption. *Crit. Rev. Ther. Drug Carrier Syst.* 8:91–192 (1991).
101. P. D. Reuman, S. P. Keely, G. M. Schiff. Similar subclass antibody responses after intranasal immunization with UV-inactivated RSV mixed with cholera toxin or live RSV. *J. Med. Virology* 35:192–197 (1991).
102. S. Tamura, Y. Samegai, H. Kurata, T. Nagamine, C. Aizawa, T. Kurata. Protection against influenza virus infection by vaccine inoculated intranasally with cholera toxin B subunit. *Vaccine* 6:409–413 (1988).
103. D. H. Jones, B. W. McBride, C. Thornton, D. T. O'Hagan, A. Robinson, G. H. Farrar. Orally administered microencapsulated Bordetella pertussis fimbriae protect mice from B. Pertussis respiratory infection. *Infect. Imm.* 64: 489–494 (1996).
104. R. Shahin, M. Leef, J. Eldridge, M. F. Hudson, R. Gilley. Adjuvanticity and protective immunity elicited by Bordetella pertussis antigens encapsulated in poly(DL-lactide-co-glycolide) microspheres. *Infect. Imm.* 63:1195–1200 (1995).
105. E. S. Cahill, D. T. O'Hagan, L. Illum, A. Bernard, K. H. J. Mills, K. Redhead. Immune responses and protection against Bordetella pertussis infection after intranasal immunization of mice with filamentous haemagglutinin in solution or incorporated in biodegradable microparticles. *Vaccine* 13:455–462 (1995).
106. A. de Haan, J. F. C. Tomee, J. P. Huchshorn, J. Wilschut. Liposomes as an adjuvant system for stimulation of mucosal and systemic antibody responses against inactivated measles virus administered intranasally to mice. *Vaccine* 13:1320–1324 (1995).
107. N. Spitzer, A. Jardim, D. Lippert, R. W. Olafson. Long-term protection of mice against Leishmania major with a synthetic peptide vaccine. *Vaccine* 17:1298–1300 (1999).
108. I. Jabbal-Gill, A. N. Fisher, R. Rappuoli, S. S. Davis, L. Illum. Stimulation of mucosal and systemic antibody responses against Bordetella pertussis filamentous haemagglutinin and recombinant pertussis toxin after nasal administration with chitosan in mice. *Vaccine* 16:2039–2046 (1998).
109. J. Makin, A. Bacon, M. Roberts, P. J. Sizer, I. Jabbal-Gill, M. Hinchcliffe, L. Illum, S. Chattfield. Carbohydrate biopolymers enhance antibody responses to mucosally delivered vaccine antigens. *Infect. Imm.* (to be published, 1998).
110. R. Arshady. Microspheres and microcapsules: a survey of manufacturing techniques. Part 1: Suspension cross-linking. *Polym. Eng. Sci.* 29:1746–1758 (1989).
111. P. B. Deasy. General introduction. In: *Microencapsulation and Related Processes*, P. B. Deasy (Ed.). Marcel Dekker, New York, 1984, pp. 1–19.
112. S. R. Jameela, T. V. Kumary, A. V. Lal, A. Jayakrishnan. Progesterone-loaded chitosan microspheres: a long acting biodegradable controlled delivery system. *J. Controlled Release* 52:17–24 (1998).
113. F. L. Mi, T. B. Wong, S. S. Shyu. Sustained-release of oxytetracycline from chitosan microspheres prepared by interfacial acylation and spray hardening methods. *J. Microencapsulation* 14:577–591 (1997).
114. A. Berthold, K. Cremer, J. Kreuter. Preparation and characterization of chitosan microspheres as drug carrier for prednisolone sodium phosphate as model for anti-inflammatory drugs. *J. Controlled Release* 39:17–25 (1996).
115. M. L. Lorenzo-Lamosa, C. Remunan-Lopez, J. L. Vila-Jato, M. J. Alonso. Design of microencapsulated chitosan microspheres for colonic drug delivery. *J. Controlled Release* 52:109–118 (1998).
116. P. Calvo, C. Remunan-Lopez, J. L. Vila-Jato, M. J. Alonso. Novel hydrophilic chitosan-polyethylene oxide nanoparticles as protein carriers. *J. Appl. Polym. Sci.* 63: 125–132 (1997).
117. C. Remunan-Lopez, M. L. Lorenzo-Lamosa, J. L. Vila-Jato, M. J. Alonso. Development of new chitosan–cellulose multicore microparticles for controlled drug delivery. *Eur. J. Pharm. Biopharm.* 45:49–56 (1998).
118. M. Acikgoz, H. S. Kas, Z. Hascelik, O. Milli, A. A. Hin-

cal. Chitosan microspheres of diclofenac sodium. II: In vitro and in vivo evaluation. *Pharmazie* 50:275–277 (1995).

119. G. M. Jantzen, J. R. Robinson. Sustained- and controlled-release drug delivery systems. In: *Modern Pharmaceutics*, G. S. Banker, C. T. Rhodes (Eds.). Marcel Dekker, New York, 1996, pp. 575–609.
120. R. L. Dunn, J. P. English, J. D. Strobel, D. R. Cowsar, T. R. Tice. Preparation and evaluation of lactide/glycolide copolymers for drug delivery. In: *Polymers in Medicine III*, C. Migliaresi (Ed.). Elsevier Science Publishers, Amsterdam, 1988, pp. 149–160.
121. Y. Machida, T. Nagai, M. Abe, T. Sannan. Use of chitosan and hydroxypropylchitosan in drug formulations to effect sustained release. *Drug Des. Del.* 1:119–130 (1986).
122. S. Patashnik, L. Rabinovich, G. Golomb. Preparation and evaluation of chitosan microspheres containing bisphosphonate. *J. Drug Targ.* 4:371–380 (1997).
123. Miekka *et al.* Novel delivery systems for coagulation proteins. *Haemophilia* 4:436–442 (1998).
124. S. R. Jameela, A. Jayakrishnan. Glutaraldehyde cross-linked chitosan microspheres as a long acting biodegradable drug delivery vehicle: studies on the in vitro release of mitoxantrone and in vivo degradation of microspheres in rat muscle. *Biomaterials* 16:769–775 (1995).
125. S. R. Jameela, P. G. Latha, A. Subramoniam, A. Jayakrishnan. Antitumor activity of mitoxantrone-loaded chitosan microspheres against ehrlich ascites carcinoma. *J. Pharm. Pharmacol.* 48:685–688 (1996).
126. Y. Song, H. Onishi, Y. Machida, T. Nagai. Drug release and antitumor characteristics of N-succinyl-chitosan-mitomycin C as an implant. *J. Controlled Release* 42:93–100 (1996).
127. Tozaki *et al.* Chitosan capsules for colon-specific drug delivery: improvement of insulin absorption from the rat colon. *J. Pharm. Sci.* 86:1016–1021 (1997).
128. S. Davaran, J. Hanaee, A. Khosravi. Release of 5-amino salicylic acid from acrylic type polymeric prodrugs designed for colon-specific drug delivery. *J. Controlled Release* 58:279–287 (1999).
129. C. L. Cheng, S. H. Gehrke, W. A. Ritschel. Development of an azopolymer based colonic release capsule for delivering proteins/macromolecules. *Methods Find. Exp. Clin. Pharmacol.* 16:271–278 (1994).
130. L. Vervoort, I. Vinckier, P. Moldenaers, G. Van den Mooter, P. Augustijns, R. Kinget. Inulin hydrogels as carriers for colonic drug targeting. Rheological characterization of the hydrogel formation and the hydrogel network. *J. Pharm. Sci.* 88:209–214 (1999).
131. S. Y. Lin, J. W. Ayres. Calcium alginate beads as core carriers of 5-aminosalicylic acid. *Pharm. Res.* 9:1128–1131 (1992).
132. M. Marvola, P. Nykanen, S. Rautio, N. Isonen, A. Autere. Enteric polymers as binders and coating materials in multiple-unit site-specific drug delivery systems. *Eur. J. Pharm. Sci.* 7:259–267 (1999).
133. M. Rodriguez, J. L. Vila Jato, D. Torres. Design of a new multiparticulate system for potential site-specific and controlled drug delivery to the colonic region. *J. Controlled Release* 55:67–77 (1998).
134. Y. Sawayanagi, N. Nambu, T. Nagai. Directly compressed tablets containing chitin or chitosan in addition to lactose or potato starch. *Chem. Pharm. Bull.* 30:2935–2940 (1982).
135. Y. Sawayanagi, N. Nambu, T. Nagai. Directly compressed tablets containing chitin or chitosan in addition to mannitol. *Chem. Pharm. Bull.* 30:4216–4218 (1982).
136. S. M. Upadrashta, P. R. Katikaneni, N. O. Nuessle. Chitosan as a tablet binder. *Drug Dev. Ind. Pharm.* 18:1701–1708 (1992).
137. G. C. Ritthidej, P. Chomto, S. Pummangura, P. Menasveta. Chitin and chitosan as disintegrants in paracetamol tables. *Drug Dev. Ind. Pharm.* 20:2109–2134 (1994).
138. Y. Sawayanagi, N. Nambu, T. Nagai. Use of chitosan for sustained-release preparations of water-soluble drugs. *Chem. Pharm. Bull.* 30:4213–4215 (1982).
139. T. Nagai, Y. Sawayanagi, N. Nambu. Application of chitin and chitosan to pharmaceutical preparations. In: *Chitin, Chitosan, and Related Enzymes*, J. P. Zikakis (Ed.). Academic Press, Orlando, FL, 1984 pp. 21–39.
140. C. J. Brine. Controlled release pharmaceutical applications of chitosan. In: *Chitin and Chitosan: Sources, Chemistry, Biochemistry, Physical Properties and Applications*, G. Skjak-Braek, T. Anthonsen, P. Sandford (Eds.). Elsevier Applied Science, London, 1984, pp. 679–691.
141. A. G. Nigalaye, P. Adusumilli, S. Bolton. Investigation of prolonged drug release from matrix formulations of chitosan. *Drug Dev. Ind. Pharm.* 16:449–467 (1990).
142. S. Miyazaki, A. Nakayama, M. Oda, M. Takada, D. Attwood. Drug release from oral mucosal adhesive tablets of chitosan and sodium alginate. *Int. J. Pharm.* 118:257–263 (1995).
143. O. Felt, P. Buri, R. Gurny. Chitosan: a unique polysaccharide for drug delivery. *Drug Dev. Ind. Pharm.* 24:979–993 (1998).

11

Immobilization of Active Biopolymers from Cold Plasma–Functionalized Surfaces for the Creation of Molecular Recognition and Molecular Manufacturing Systems

Ferencz Denes and Sorin Manolache
University of Wisconsin, Madison, Wisconsin

I. INTRODUCTION

Biomaterials can be broadly defined as specially designed inorganic materials (e.g., ceramics, metals) and organic polymers (e.g., natural or synthetic polymers or their surface-modified versions) whose surfaces are in direct contact with biological components such as blood, tissue fluid, proteins, cells, etc. The interaction of these substrates with biological components, including covalent coupling, and adsorption of these materials on polymer surfaces depend greatly on their surface properties, including surface energy, surface functionality, composition, and surface morphology.

Depending on the intensity of the interaction (chemical bonding or adsorption), biomaterials are involved in different applications. Biomaterials for in vivo applications should not adsorb any proteins and should not induce any activation of complementary systems. Similarly material surfaces involved in food processing industry should have antifouling characteristics in order to avoid biofilm formation. On the contrary, biosensor applications require functionalized surfaces (e.g., polymer surfaces), which should selectively interact (e.g., covalently link) with specific molecules, cells, etc. Significant biological effects can also be induced by strongly adsorbed or covalently attached specific biomolecules (e.g., albumin) on polymer surfaces.

Biosensors combine the selective interaction between the immobilized biological molecules and entities, such as DNA, enzymes, and antibodies as well as certain molecules (biological or nonbiological), cells, bacteria, etc., with detection systems based on electrical or optical signals. The transducer of the detection system recognizes the molecular interaction between the target molecules and the detector biomolecules and converts the electrical, electrochemical, optical, thermal, or mass changes of the nascent surface layers generated as a result of the interaction into electrical or optical signals.

The first step in the preparation of molecular recognition systems is the surface functionalization of specific substrates. The resulting active layers should anchor, in a next step, covalently or by physical adsorption, biomolecules with highly selective activity.

Molecular recognition between biological macromolecules, small molecules, and macromolecules and specific surfaces plays a crucial role both in the understanding of various biological systems and in the design of artificial, intelligent surfaces (molecular manufacturing systems) with tremendous practical potential. As soon as a primitive molecular assembler, which is able to self-replicate by atomic precision positional control is constructed, migration pathways will allow the generation of increasingly sophisticated assemblers.

In order to understand the molecular basis of the interaction, controlled immobilization of biomolecules must be performed by taking into account the chemical and physical (crystallinity, morphology) structure of the substrates, the nature of the bonding between the biologically active molecules and substrates, the surface density of immobilized molecules, and the distance between the anchored biomolecules and the substrate. By understanding and controlling of surface functionalization and the consequent anchoring reaction mechanisms, tailored and very specific reaction pathways (molecular recognition) can be developed.

The nature of surface topographies of ''host substrates'' plays a significant role both in the efficiency and selectivity of the immobilization processes. Biologically active molecules like enzymes and nucleotides usually favor a single, initial conformation to ''catalyze'' reactions. However, recently it has been found that over time an enzyme can adopt more than one functional conformation with different catalytic efficiencies (1). The bonds responsible for the existence of secondary, tertiary, and quaternary structures are based on noncovalent-type ionic, hydrogen, and van der Waals interactions. The high number of these relatively labile individual bonds involved in the interactions assure the stability of the bimolecular structures. However, the labile nature of individual linkages restricts significantly their handling parameters. Mild pH and temperature conditions are usually required.

Conformational changes of biomolecules as a result of replica interactions (presence of amorphous or crystalline surfaces) and multiple bonding between the surface of the substrate and the biomolecules can alter significantly the efficiencies of catalytic activities. As a consequence, the activities of immobilized biomolecules usually are modest (e.g., immobilized enzymes) in comparison to the free molecules. However, controlled substrate–molecule interactions might open up a novel route for generating optimal substrate surfaces for the immobilization of biomolecules. Molecularly imprinted polymer surfaces and the corresponding technologies are already a rapidly emerging area in which synthetic polymer surfaces are generated in the presence of the ''print molecules.''

The preparation of oligonucleotides for clinical trials will require kilogram to ton amounts of specific oligonucleotide derivatives, while the preparation of oligonucleotide libraries requires nanoscale beads that have high loading characteristics and optimal accessibility of their surfaces to biomacromolecules. Conventional rigid-lattice supports have their limitations due to the nonspecific inside/outside loading. Controlled pore glass (CPG) or macroporous polystyrene exhibit, for instance, nucleoside loading of 0.8–0.9 μmol/m^2 regardless of porosity or size dimensions. Swellable resins show at the same time diffusion-retarded kinetics in the diverse reactions of the elongation cycle. Hence, the yields from oligonucleotide chain extension mechanisms do not exceed 97%, which would be required for application purposes. A solution to this problem would be the preparation of surface-functionalized, nonporous, inert, inorganic and organic polymeric beads. These support systems would allow oligonucleotide extension yield of 99%, and solutionlike kinetics could be achieved, while having oligonucleotide capacities of 100–200 μmol/g in comparison to 50–70 μmol/g of the currently used systems. Thus, polymer supports of the core shell type may lead to a 3- to 4-fold increase in the output of therapeutic oligonucleotides per batch (2,3).

Polymer-bound oligonucleotides will find their applications in hybridization-based diagnostics and in the discovery of new therapeutics based on molecular recognition. Prenatal diagnostics of genetic aberrations, identification of virus-borne diseases, detection of mutations of regulatory proteins controlling carcinogenesis, and novel hybridization-based identification techniques oriented to forensic or archeology fields are only some of potential applications.

Molecular recognition will also play an essential role in areas other than medicine, pharmaceutics, and biotechnology. Development of ultraselective chemical sensors and absorbent surfaces is crucial for creating environmentally safe processes. Monitoring the quality of water is one of the major demands in this area. Bimolecular-based chemical sensors and filters for toxic chemicals and microorganisms (e.g., *Escherichia coli*) will play a significant role in future technologies.

Current accurate tests for DNA sequences are based on fluorescent signals. Analytical set-ups for quantitative evaluations usually are time consuming and expensive. Recently it has been demonstrated that hybridization (when a single DNA strand ''bumps'' into a complementary partner, it will hybridize) can be detected by changes in an electrochemical signal through a redox or conducting molecule, which behaves differently in the presence of a DNA hybrid than in the presence of a single strand (4). The advantage of using electrical signals (current or voltage) instead of fluorescence for detecting hybridization would be the rapidity of the assay and the possibility of using large area arrayed DNA-immobilized surfaces for the development of automated hybridization monitoring procedures. The electrochemical approach *ab initio* involves surfaces, e.g., DNA immobilized on solid electrodes. In many systems under investigation electroactive groups that mediate electron transfer (e.g., ferrocene complexes) are attached to molecules that preferentially bind double-stranded

DNA. Recently DNA-functionalized conducting polymers like polypyrolle, polytiophene, etc., have been employed as well.

Polyanilines have received increased attention due to their ease of synthesis and doping and stability in the doped state; their processability; and their great versatility. For example, polyanilines can be tuned to conductivities in the range from insulators to conductors, and there is a large variation in the chemical nature of the dopants that can be employed. Controlled thickness, ultrathin polyaniline films and self-assembled polyion-containing (e.g., protein-containing) polyaniline layers can conveniently be prepared. These polymeric structures are very promising candidates for mediating electric signal–based detection of hybridization mechanisms.

Most of the natural and synthetic polymeric substrates can easily be functionalized through polymer–analog reactions. Main chain and side group homogeneous reactions are the most common approaches in this field. The use of polymers as ''carriers'' or ''supports'' for chemical reagents, catalysts, or substrates represents a relatively new significant rapidly developing area. In this case the polymer is in the form of an insoluble inert substrate that may be a solvent-swollen crosslinked gel or a surface-active solid. This approach eases the separation of reagents or catalysts (e.g., enzymes) from the reaction products, permitting consequently the automation of the complex chemistry. However, the specific structure of the repeating units of the macromolecules often limit considerably the variety of polymer–analog reactions. These reactions are even more difficult to develop under heterogeneous environments. Natural and even some synthetic polymeric substrates can also undergo undesired chemical modifications and sometimes biodegradation during the polymer-supported organic reactions. Inert polymeric substrates, e.g., polyethylene (PE), polypropylene (PP), poly(ethylene terephthalate) (PET), polytetrafluoroethylene (PTFE) and inorganic supports (e.g., glass, silica), however, cannot be functionalized efficiently by using conventional wet chemistry approaches.

II. CONVENTIONAL FUNCTIONALIZATION AND COATING FOR BIOTECH APPLICATIONS

The immobilized affinity ligand technology (5,6) has grown rapidly in recent years. Starting as a purification method, it has now become a technique also for scavenging reagents to remove unwanted contaminants (7–11), for modification and catalysis to effect specific transformations (12–21), and for separation tools to develop sensitive and highly accurate analytical techniques (5,8,22).

The affinity ligand technique is based on the combination of a ligand with a matrix (a solid insoluble substance). The ligand usually is a well-known structure (e.g., oligonucleotide (7), biotin (9), enzyme (5,12–21), bacteria (10), fungi (11), etc.) and is firmly attached to the surface of the support by physical forces (23–25), covalent bonds (most used method), or autocoupling (26). The matrix can be any material to which a biospecific ligand may covalently be attached. Each component of the molecular assembly influences the partner molecules, and the entire combination matrix must function as a unique system in an environment. A ''matrix effect'' or an ''activation effect'' occurs when a change in the matrix material or activation chemistry will result in a radically different behavior of the same ligand from that seen on a reference support.

A. Natural and Synthetic Supports

The preferable functional group for conventional functionalization reactions is the primary hydroxyl group (5). It is amenable to various activation procedures, but does not contribute to nonspecific binding of nontarget molecules. Agarose, which has a natural abundance of primary and secondary OH groups, represents an ideal surface for functionalization: the secondary hydroxyls preserve the hydrophilic nature of the matrix even after the primary hydroxyls are consumed in activation or crosslinking process. The abundance of primary amine or carboxylic acid functionalities on supports like polyethyleneimine or carboxymethyl cellulose (15) improve the functionalization possibilities. However, the resulting support may exhibit nonspecific ion exchange effects.

1. Natural Supports

Natural polysaccharide matrices (7,12,13,24,25) such as agarose and cellulose or naturally occurring materials such as silica, alumina, and glass (6,9,14,23) must be processed before they can be exploited as affinity supports. Table 1 summarizes the chemical structure and the required engineering of natural materials in order to convert them into the desired affinity ligand supports.

2. Synthetic Supports

The synthetic supports (5,10,16–21) typically have superior physical and chemical durability in comparison to the natural matrices. Most synthetic polymer supports can withstand higher-pressure environments without structural collapse and can work under high-pressure gradient and

Table 11.1 Natural Supports and the Required Engineering for Their Modification

Support	Primary structure	Engineering	Observations
Agarose		Alternating residues of β-D-galactose and 3-anhydrogalactose Crosslinking (30–50% loss of hydroxyl groups) Crosslinked agarose stability: 110–120°C and pH 3–14 Working pH 4–9 Tolerates a wide range of water-miscible solvents such as alcohols, dioxane, tetrahydrofuran, dimethylformamide, dimethylsulfoxide, pyridine, etc.	Contains primary and secondary alcohol groups Secondary and tertiary structure can be described as single fibers spun into a yarn of multiple fibers Remarkable fabric with large accessible pores Beads can be dissolved by heating or by strong denaturants such as guanidine hydrochloride
Cellulose		Linear polymer of 1,4-β-D-glucose Easy derivatization (substitution) Stability: 115°C at pH 7 Working pH 3–10 Can withstand a wide range of solvents, denaturating, and mild oxidation conditions	Contains primary and secondary alcohol groups Simple secondary and tertiary structure Tangled wad of string Low level of effective porosity or extended surface area
Silica and controlled pore glass	Si—ONa Macromolecular anorganic structures	Glass with covalent coating such as glycerol Dissolved in deionized water Working buffers should be neutral or acids (pH lower than 8) and with at least 0.05 M salt Can withstand enormous pressure, any organic solvent, or temperature in excess of 300°C	Contain silanol groups Porous glass, silica, alumina, and zeolites relatively rigid

flow rate conditions. Various-geometry (spheres, tubes, sticks, plates, etc.) porous or nonporous polymeric materials have already been successfully involved in the immobilization and affinity ligand techniques. Commercially available synthetic supports for affinity chromatography applications are presented in Table 2.

B. Functionalization

Functionalization is a process of chemical modification of the matrix (support) in order to form a covalent bond with the ligand. The chemistry of the process must be compatible with both the support and the ligand. It is required that the activation of the matrix and the coupling of the ligand should not add any nonspecific characteristics to the system. The major objectives of the functionalization processes are to create a stable binding of ligand to the support without changing the porosity or the surface topography of the substrate, and without generating nonspecific effects, and to be rapid, efficient, and user friendly.

The functionalization process should be specific for the active groups present both in matrix and ligand. Table 3 summarizes the most used classical chemical derivatization techniques for commercial available affinity supports.

Table 11.2 Synthetic Support and the Required Engineering for Their Modification

Support	Primary structure	Engineering	Observations
Acrylamide derivatives			
Polyacrylamide	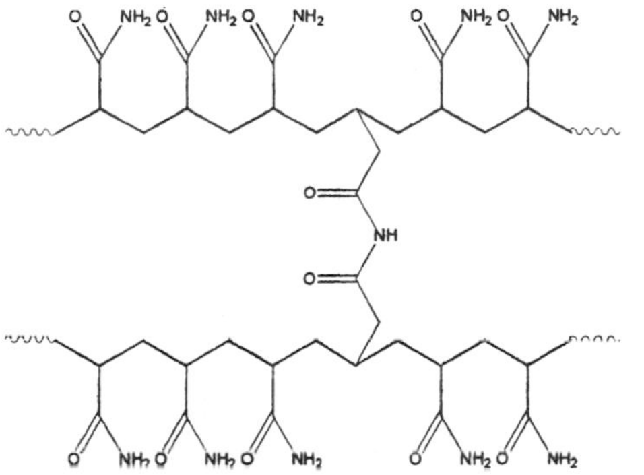	Copolymer of acrylamide and N,N′-methylene-bis (acrylamide) (crosslinking agent) Poor mechanical stability Low working flow rates Dimensional changes when working with various solvents (shrink or swell)	Resistance to microbial attack pH stability: 2–10 Excellent chemical stability Low nonspecific binding characteristics
Trisacryl	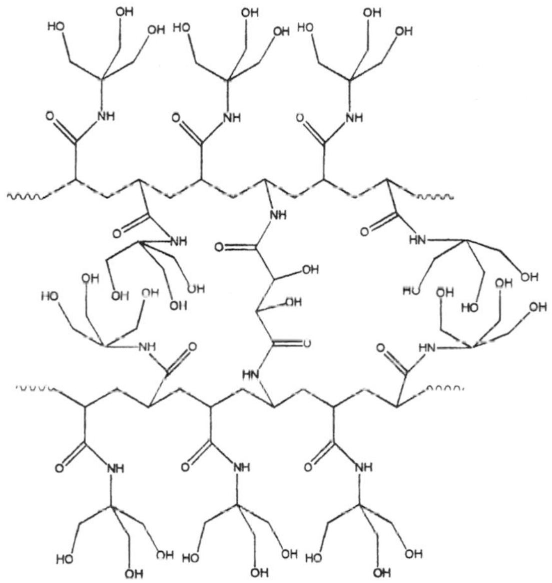	Copolymerimer of trihydroxylic derivative of acrylamide and N,N′-diallyltartra-diamide (crosslinking agent) Good mechanical properties Can withstand a wide range of solvents	Resistance to microbial attack pH stability: 2–11; more basic pH can slowly hydrolize it Excellent chemical stability Stability: −20 to 120°C
Sephacryl	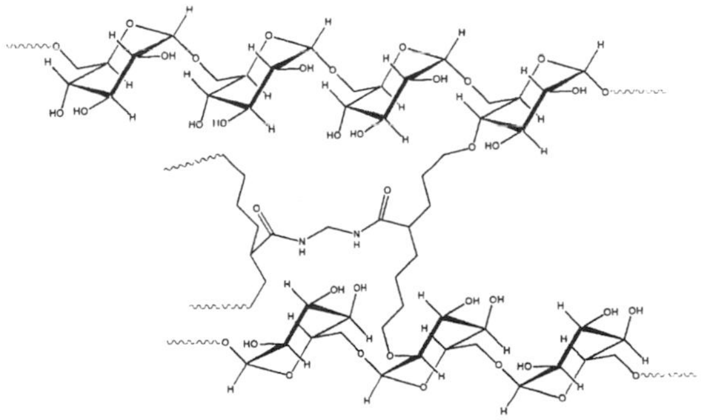	Copolymer of allyl dextran and N,N′-methylene-bis (acrylamide) (crosslinking agent) Increased mechanical properties Can withstand a wide range of organic solvents without shrinking or swelling	Resistance to microbial attack Stable to 0.5 N NaOH solution Excellent chemical stability Stable in autoclave conditions
Ultrogel	Complex mixtures of polyacrylamide and agarose polymers knitted together	Matrices resulting from a blend of agarose and polyacrylamide Denaturants or detergents can affect and destroy the integrity of macromolecular structure Some organic solvents may damage the gel	Working pH range: 3–10 Working temperature range: 2 to 36°C

Table 11.2 Continued

Support	Primary structure	Engineering	Observations
Azlactone	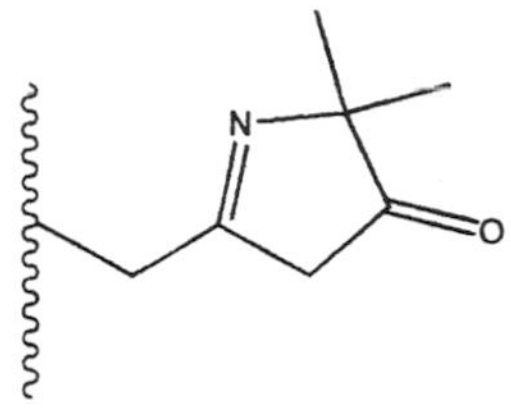	Copolymer of vinyldimethyl azlactone and N,N′-methylene-bis(acrylamide) (crosslinking agent) Slightly hydrophobic; for some proteins a lyotropic agent is required Allows high pressure and high flow rate operation	Stability, pH range: 1–14 Working pH rangc: 4–9 Proteins and small ligands are coupled in high yields and short times Maintains high coupling capacity even at elevated flows
Methacrylate derivatives			
TSK-gel toyopearl HW	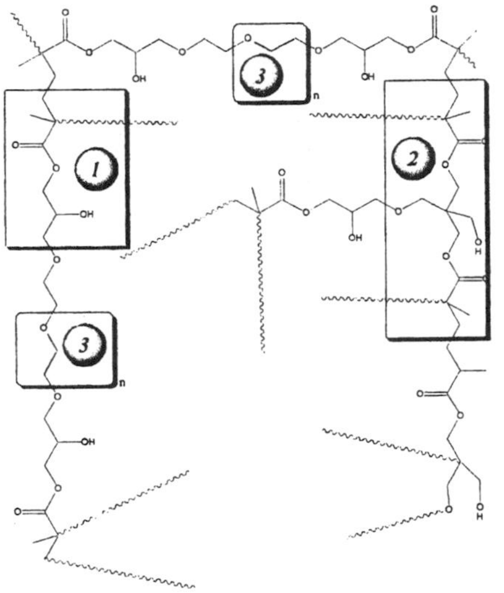	Copolymer of glycidyl methacrylate (*1*), pentaerythritol dimethacrylate (*2*), and polyethylene glycol (*3*) Surface mildly hydrophilic Hydrophobic properties under certain conditions Good mechanical stability	Stability, pH range: 2–12 Do not support microbial growth Stable in detergents, organic solvents, or concentrated denaturants solutions Originally designed for GPC
HEMA	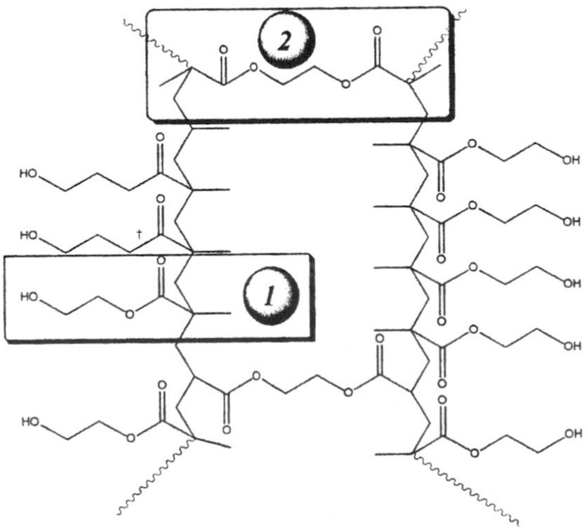	Copolymer of 2-hydroxyethyl methacrylate (*1*) and ethylene dimethacrylate (*2*) (crosslinking agent) Micropore and macropores structures Can withstand pressures exceeding 1,000 psi (4,400 psi for some products) and very high flow rates	Stability, pH range: 2–13 Can be washed with 2 N NaOH or 2 N HCl for short period of time Termal stability: 170°C Very resistant to detergents, organic solvents, or concentrated buffers solutions Contains numerous tertiary α-carbonyl esters of pivalic acid (most stable and least hydrolysable ester known)

Table 11.2 Continued

Support	Primary structure	Engineering	Observations
Eupergit	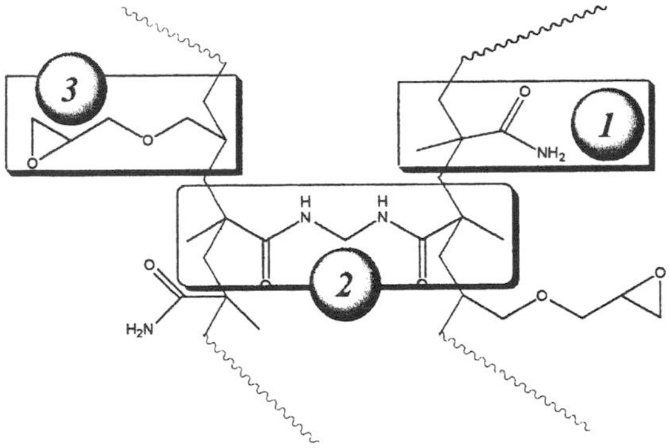	Copolymer of methacrylamide (*1*), N,N′-methylene-bis(methacrylamide) (*2*), and allyl glycidyl ether (*3*) or glycidyl methacrylate Can withstand pressures up to 4,000 psi	Very good chemical and physical stability Porous or nonporous varieties
Polystyrene derivatives			
Poros	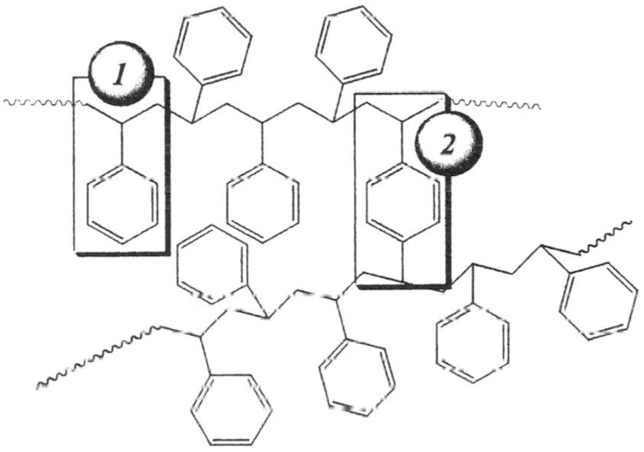	Copolymer of styrene (*1*) and divinylbenzene (*2*) (crosslinking agent) Maximum operational pressure: 3,000 psi Large pores passing through particles that allow high flow rates	Resistant to extreme pH: 1–14 Robust chemical and physical structures
Polystyrene balls, plates, and devices	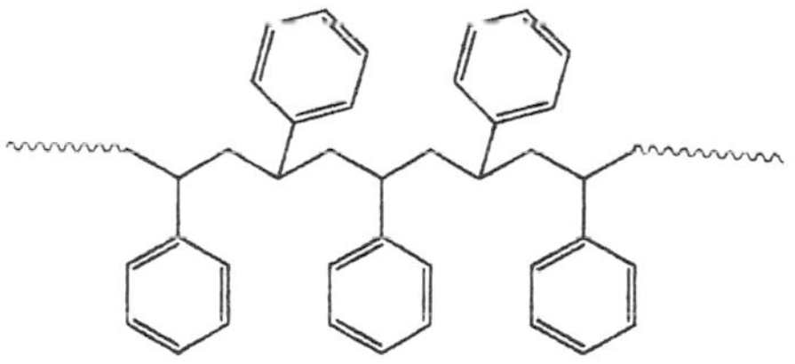	Nonporous particles with different geometries Surface etched to increase area for affinity interactions Polymer surface blocked by adsorption with a nonrelevant substance to prevent nonspecific hydrophobic interactions (e.g., bovine albumin)	Not suitable for chromatographic applications Can easily be mixed or separate from solvents Good optical properties in visible range for light absorption diagnostic using enzyme-linked assay

III. DEPOSITION OF ANTIFOULING LAYERS

The growth of living organisms and the attachment of biologically active molecules on solid surfaces exposed to aqueous media present one of the most important problems to be considered for food industry and medicine. Bacterial adherence to various solid surfaces is a naturally occurring process. Many bacterial species live and proliferate under ''attached'' conditions or embedded in polymeric matrices, called biofilms, generated by their own metabolism. Adherence of bacteria on various substrates and the subsequent biofilm formation can lead to the enzymatic degradation of the host materials or to the development of biofilm-related infections. It is recognized that adhesion of microorganisms to surfaces represents the initiating step in the development of biofilms and, as a consequence, of many infectious diseases. Biofilm environ-

(*text continues on p. 252*)

Table 11.3 Functionalization Processes in Respect to Matrix and Ligand Active Groups

Matrix	Ligand	Functionalization	Binding
~OH	$\sim NH_2$	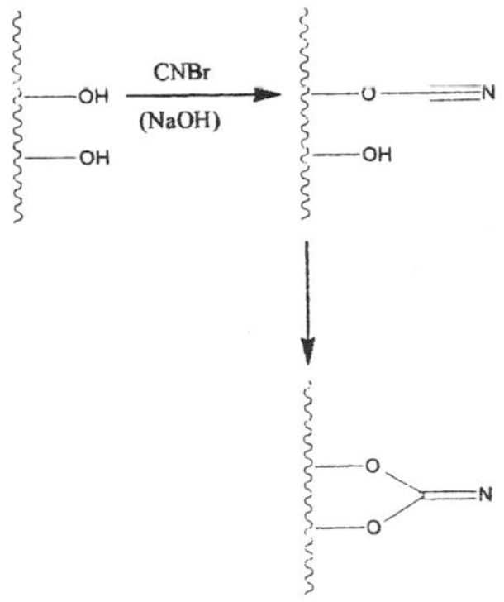 CNBr (NaOH) — Cyanogen bromide	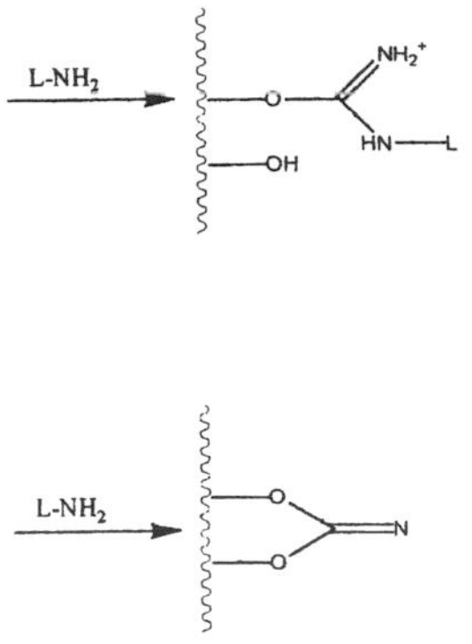 L-NH_2
		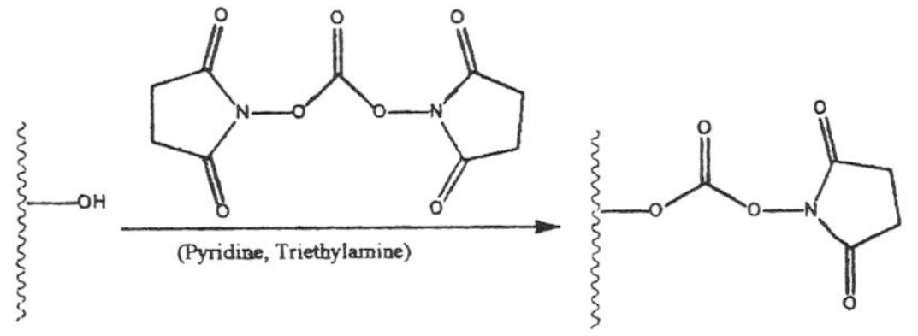 (Pyridine, Triethylamine) — N-Hydroxy succinimide esters (N,N′-disuccinimidylcarbonate)	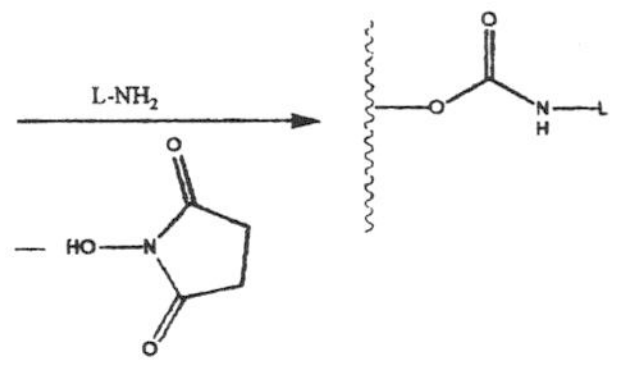 L-NH_2
		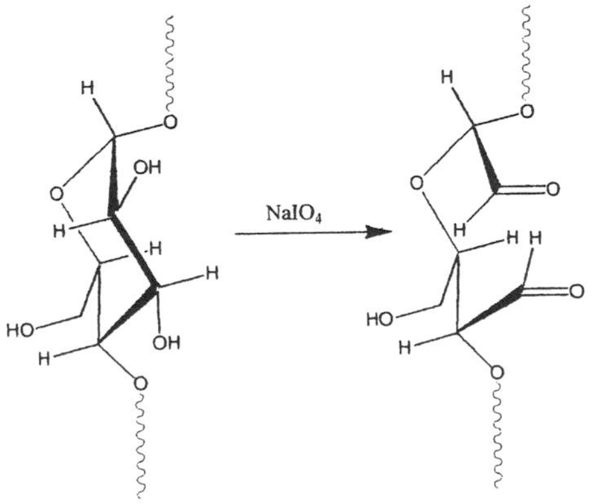 $NaIO_4$ — Sodium periodate	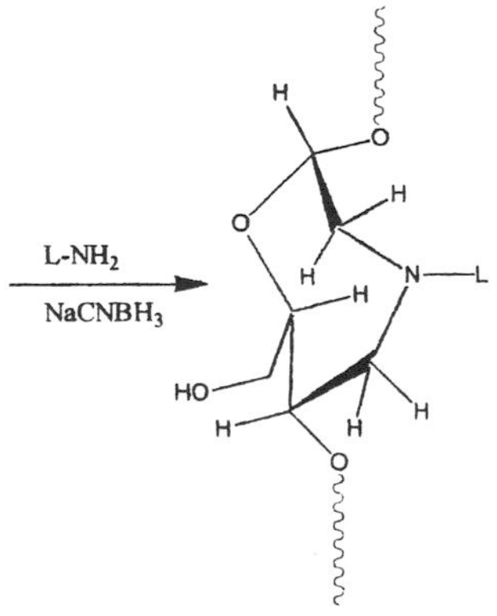 L-NH_2, $NaCNBH_3$ — Sodium borohydride
		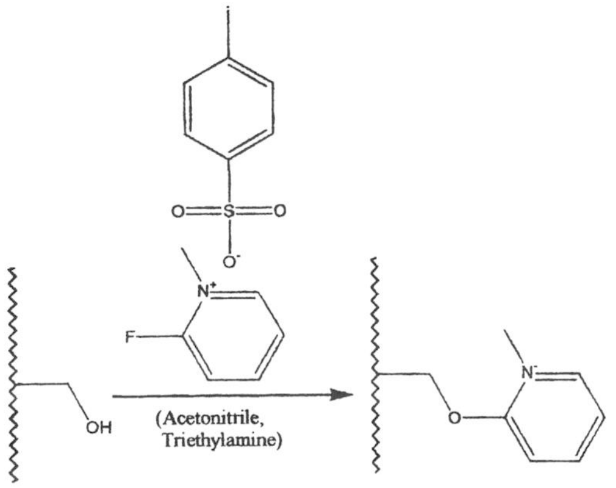 (Acetonitrile, Triethylamine) — 2-Fluoro-1-methylpyridinium toluene-4-sulfonate (FMP)	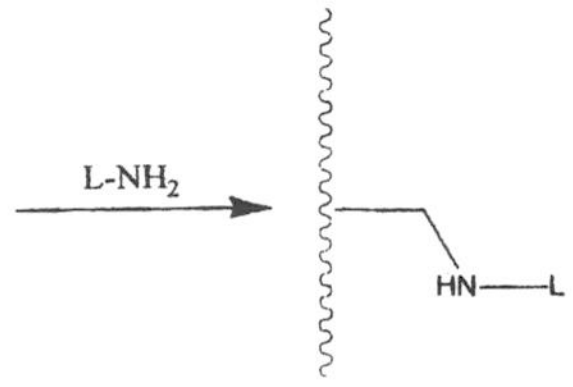 L-NH_2

Table 11.3 Continued

Matrix	Ligand	Functionalization	Binding
		p-Toluenesulfonyl chloride (tosyl chloride)	
		2,2,2-Trifluoroethanesulfonyl chloride (tresyl chloride)	
		Trichloro-s-triazine (TsT) (cyanuric chloride)	

Table 11.3 Continued

Matrix	Ligand	Functionalization	Binding
		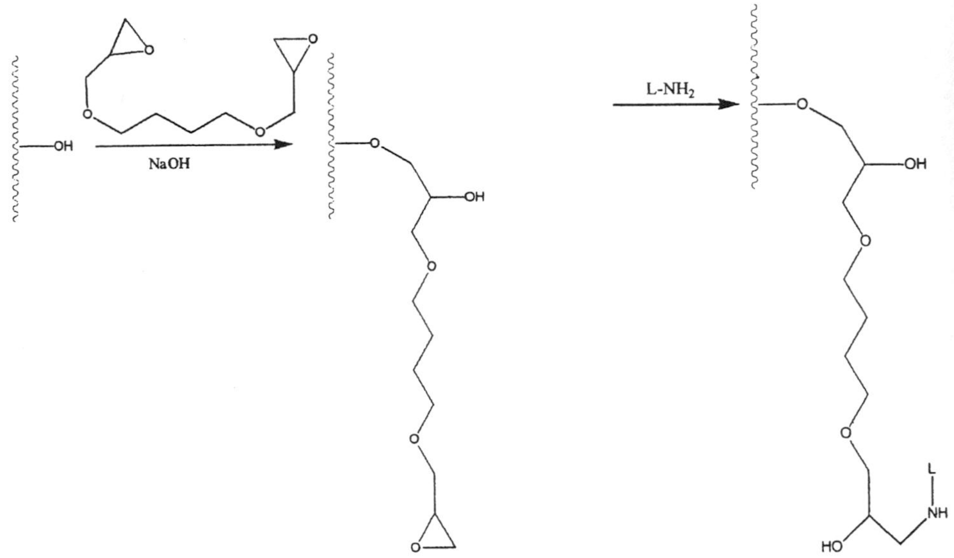 NaOH; 1,4-Butanedioldiglycidylether	L-NH_2
~OH	~OH	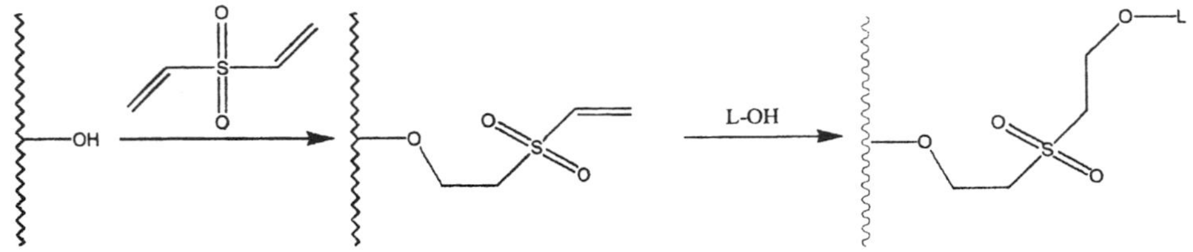 Divinyl sulfone	L-OH
~COOH	~NH_2	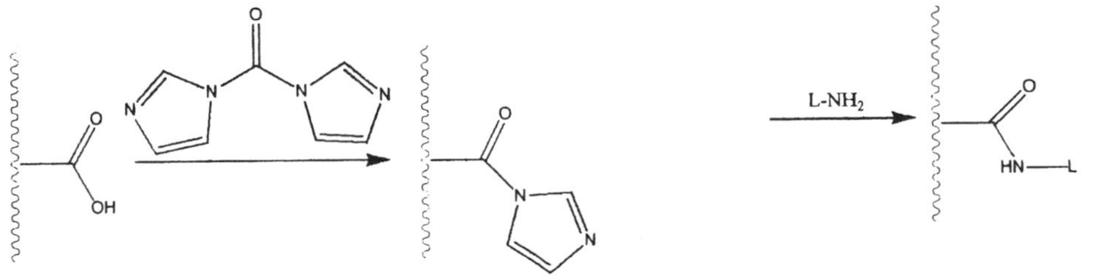 Carbonyldiimidazole	L-NH_2
		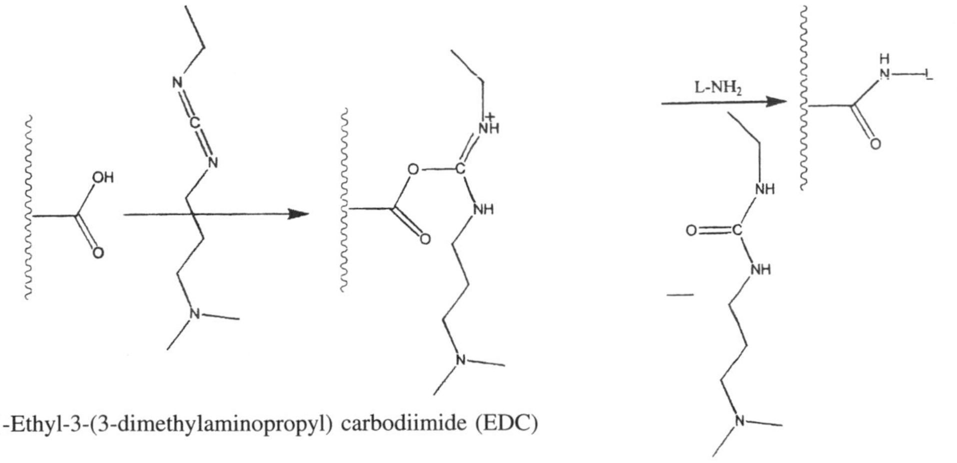 1-Ethyl-3-(3-dimethylaminopropyl) carbodiimide (EDC)	L-NH_2

Table 11.3 Continued

Matrix	Ligand	Functionalization	Binding
		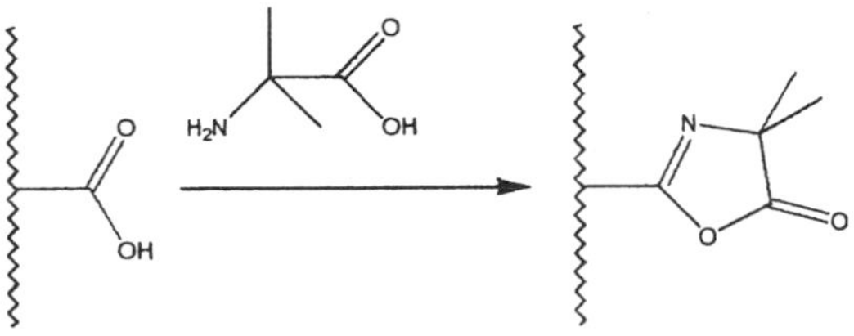 H_2N, OH, O; OH, O; N, O, O; Azlactone ring	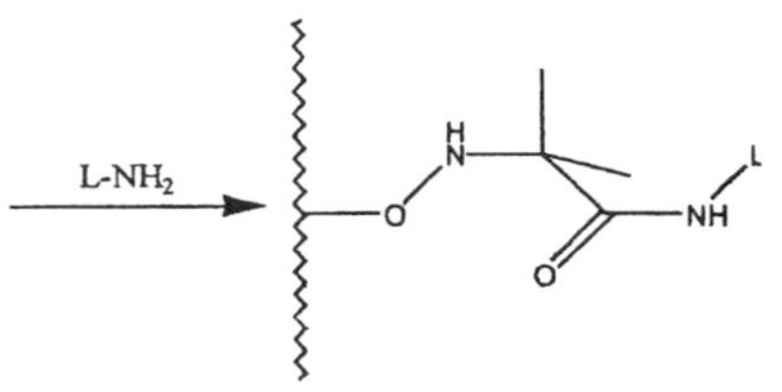 L-NH_2; O, H N, O, NH, L
~COOH	~SH	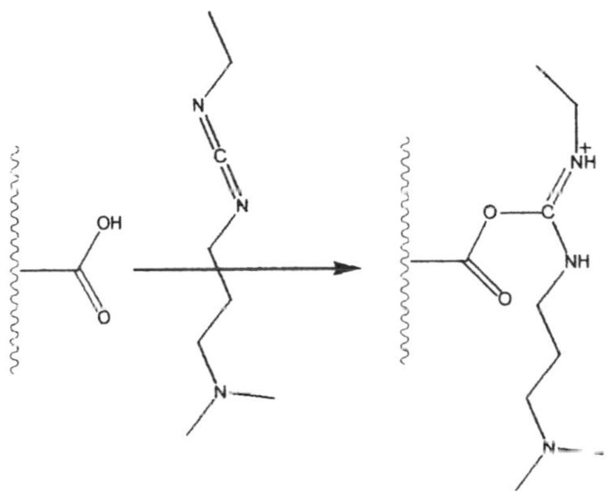 N, C, N, N; OH, O; NH, O, C, NH, O, N; 1-Ethyl-3-(3-dimethylaminopropyl) carbodiimide	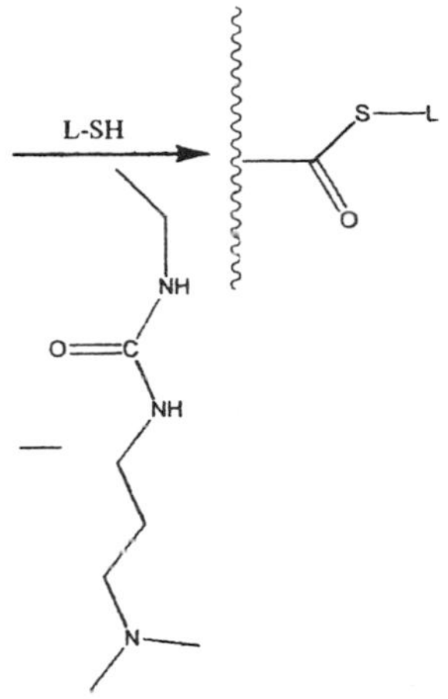 L-SH; S—L, O; NH, O=C, NH, N
~NH_2	~SH	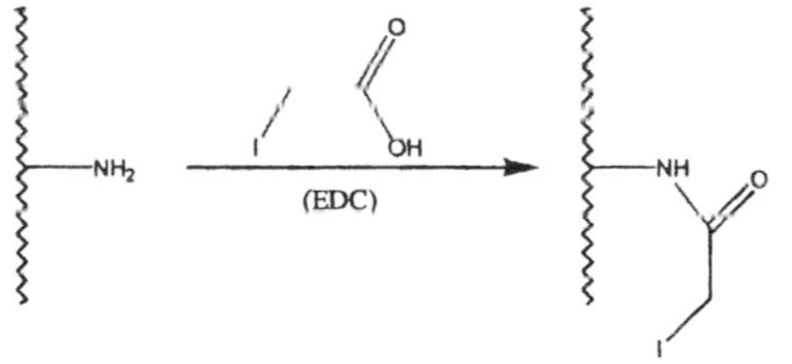 NH_2; I, O, OH (EDC); NH, O, I; Iodoacetic acid or bromoacetic acid	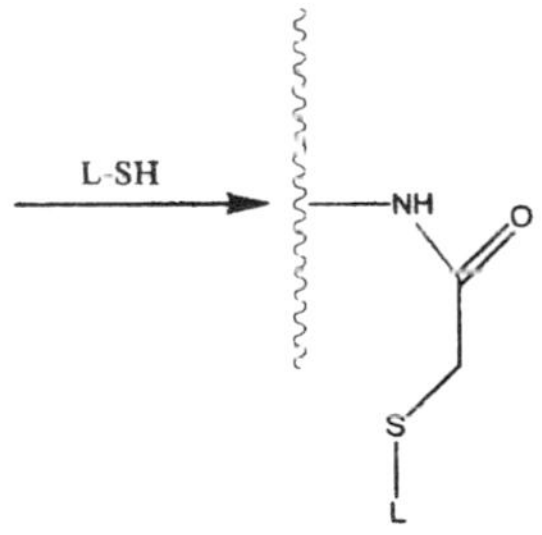 L-SH; NH, O, S, L
		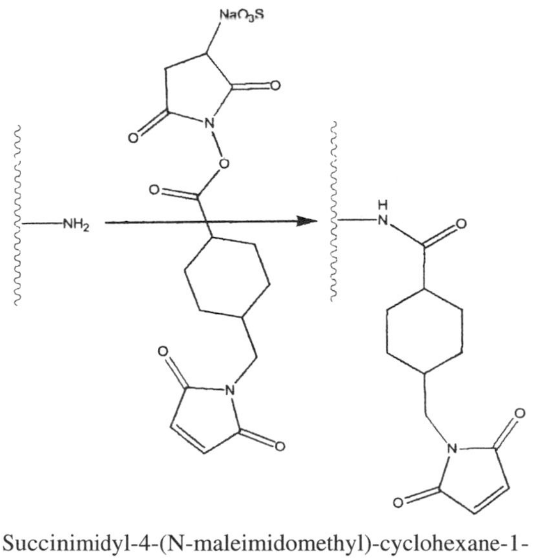 NaO_3S, O, N, O, O, O, NH_2, O, N, O; H N, O, O, N, O; Succinimidyl-4-(N-maleimidomethyl)-cyclohexane-1-carboxylate (SMCC)	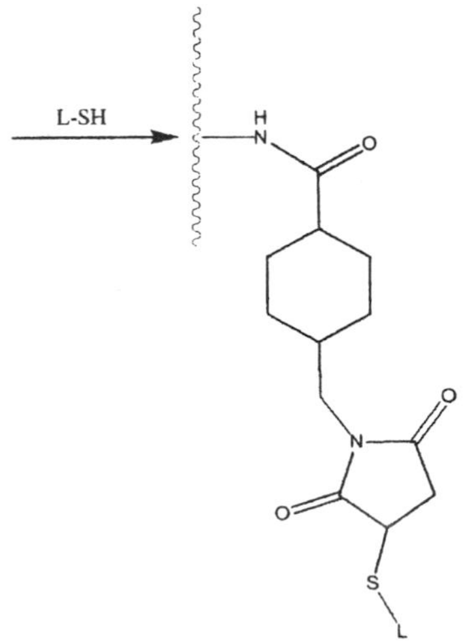 L-SH; H N, O, O, N, O, S, L

Table 11.3 Continued

Matrix	Ligand	Functionalization	Binding
		Traut's Reagent; 5,5-Dithio-bis-(2-nitrobenzoic acid) (Ellman's reagent)	L-SH
~NH_2	Phenol group	$Na_2S_2O_4$; $NaNO_2$, HCl; Diazonium	L–C₆H₄–OH

Table 11.3 Continued

Matrix	Ligand	Functionalization	Binding
			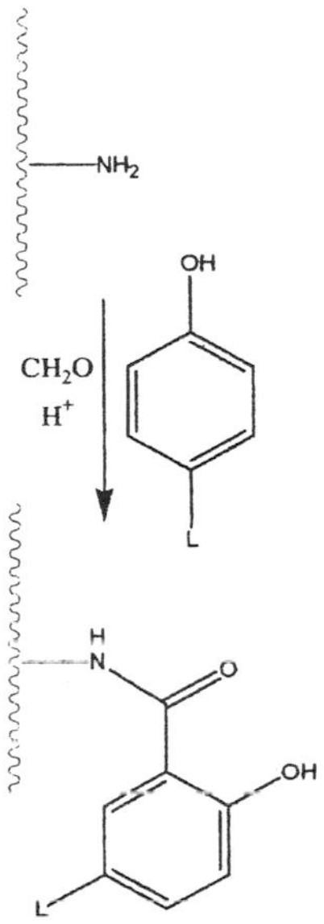 NH_2; CH_2O, H^+; OH, L; H N, O, OH, L Mannish condensation
~SH	~SH	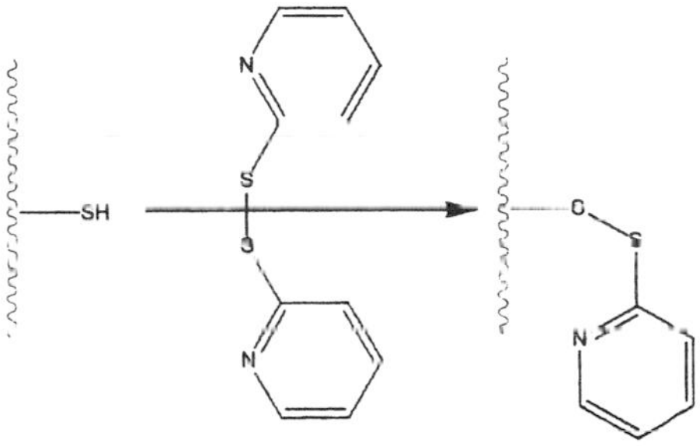 SH; N, S, S, N; S, S, N 2,2′-Dipyridyl disulfide	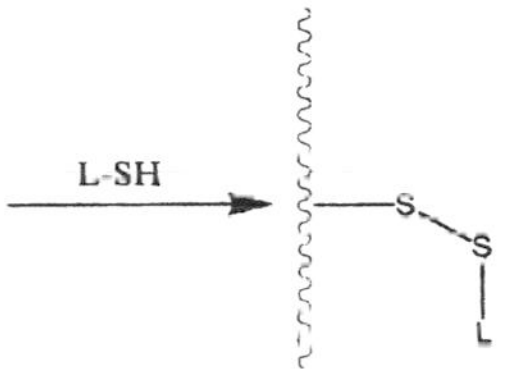 L-SH; S, S, L
		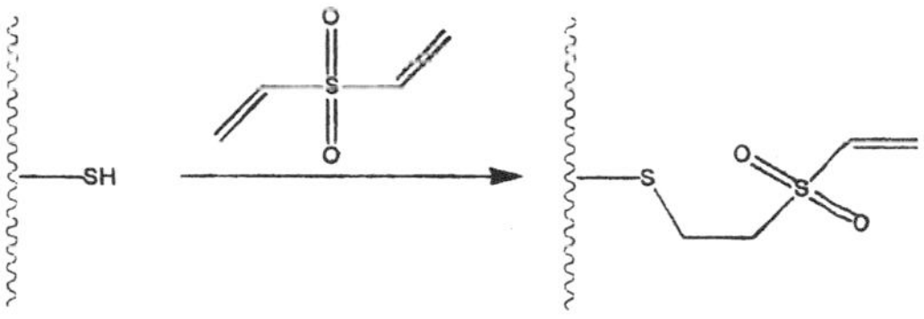 SH; O, S, O; S, O, S, O Divinyl sulfone	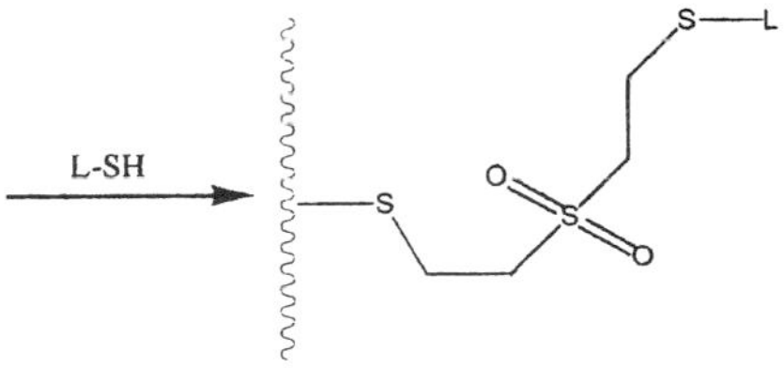 L-SH; S, O, S, O, S—L
		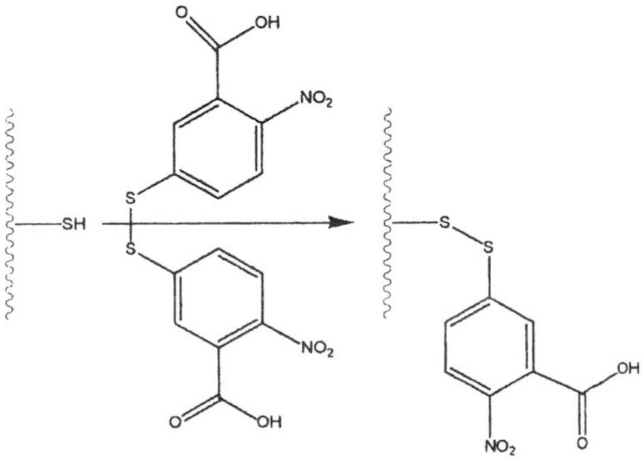 SH; O, OH, NO_2, S, S, NO_2, O, OH; S, S, OH, NO_2, O 5,5-Dithio-bis-(2-nitrobenzoic acid) (Ellman's reagent)	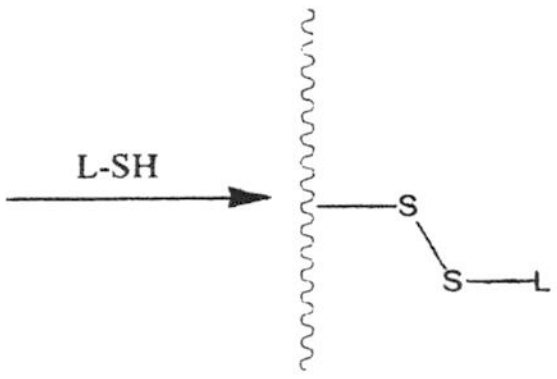 L-SH; S, S—L

Table 11.3 Continued

Matrix	Ligand	Functionalization	Binding
~CHO	~CHO	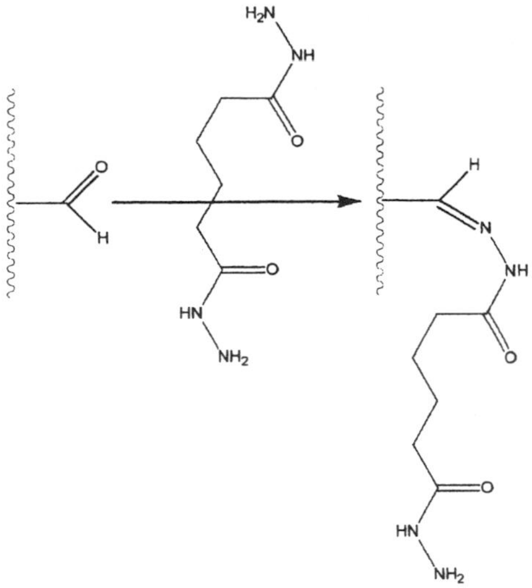 Adipic dihydrazide	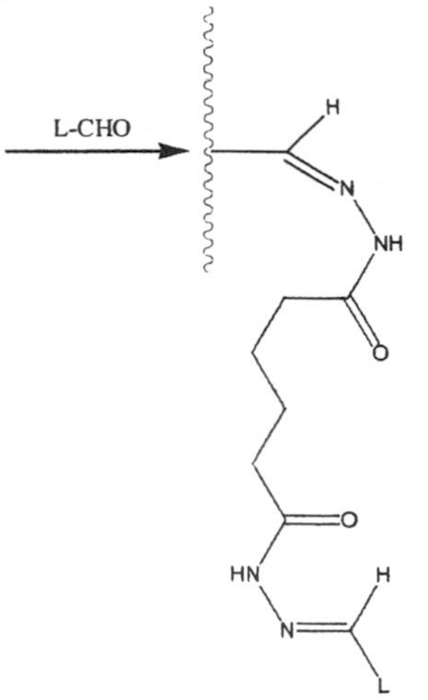 L-CHO
~CHO	$\sim NH_2$		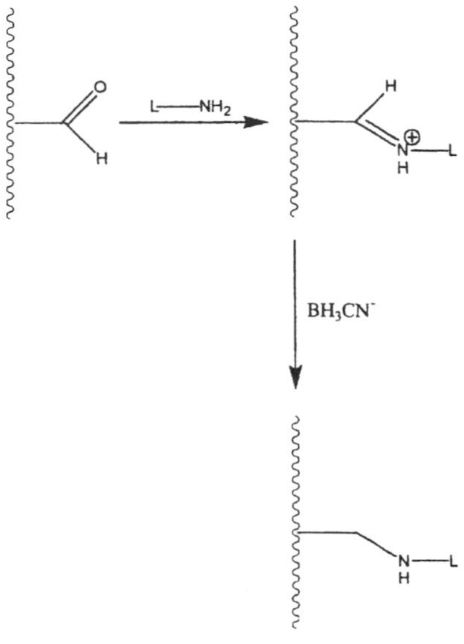 L—NH_2; BH_3CN^-; Cyanoborohydride

ments protect microorganisms, through the embedding phenomenon, against substrate-origin defense mechanisms and antibiotics, enhancing in this way the survival of microorganisms and making extremely difficult treatments against biofilm-origin infections.

When a typical material is placed in a biological fluid or in a microorganism-containing liquid environment, proteins are adsorbed rapidly and usually nonspecifically on its surface. It is suggested that this physicochemical protein adsorption process is mediated by entropic phenomena associated with the presence of a highly ordered water monolayer on protein molecule surfaces and electrostatic and van der Waals long- and short-range interactions. The adsorbed protein molecules act as ligands and the cell surface proteins and proteoglucans produced by the cells as the receptor counterparts in the molecular recognition mechanism.

Adherence of microorganisms to polymer or metal surfaces and subsequent colonization of the exposed surfaces are the major initial steps in the pathogenesis of foreign body infections. Two approaches have been developed for avoiding polymer surface–mediated infections (9): development of polymers with antiadhesive characteristics and development of polymers with antimicrobial properties. A large number of polymers were tested for antiadhesive properties. It has been found that bacterial strains usually exhibited a high adhesion to polymers with a high surface energy and having rough surface topographies, and that ionic groups are also involved in the adhesion mechanism. Most bacteria exhibit in an aqueous environment a negative surface charge; accordingly, it has been shown that negatively charged polymer surfaces can reduce bacterial adhesion. However, none of the attempts focused on the generation of antiadhesive polymeric surfaces resulted in a total reduction of bacterial attachment. It was also emphasized that the adhesion of bacteria to biomaterials in vivo is also mediated by bacterial adhesion. It was demonstrated, for instance, that preadsorption of albumin decreases bacterial adhesion, while fibronectin promotes adherence. By considering polymeric materials which selectively adsorb proteins, potential routes could be developed for the reduction of bacterial adhesion, and thus prevention of foreign body–mediated infections.

It has been shown that coating surfaces with noncharged hydrophilic polymers resulted in reducing protein and cell adsorption on a variety of surfaces due to the elimination of electrostatic attractive forces and the hydrophobic interactions between the solid surfaces (usually polymeric materials) and proteins in solutions. Most of the research in this area involved poly(ethylene glycol) or its derivatives; however, polysaccharides and nonionic cellulose ethers were also considered.

A. Conventional Approaches for Avoiding Protein Adsorption, Bacterial Attachment, and Biofilm Formation

Poly(ethylene oxide) (PEO), -[—CH_2—CH_2—O—$]_n^-$, [or polyethylene glycol (PEG) (when the molecular weight is less than 10,000 Daltons)] is highly water soluble and has a good structural fit with the water molecules, which explains the presence of a strong hydrogen bonding between the ether oxygen atom of PEG and hydrogen atoms of the water molecules. Its decreased solubility in water at elevated temperatures is also related to the hydrogen bonding; higher temperatures result in decreased hydrogen bonding and in an increase of hydrophobic interactions between the macromolecular chains.

Poly(ethylene oxide)–containing surfaces have been prepared by a number of approaches including physical adsorption and entrapment, covalent grafting, photo-induced grafting, high-energy radiation grafting, and glow discharge–mediated grafting (27–42). It has been shown that only relatively high molecular weight PEO (>100,000) can be effectively adsorbed on hydrophobic surfaces, and that the adsorption of PEO-containing amphiphatic block copolymers and multiblock copolymers (pluronics) would result in stable adsorbed layers. Homopolymers or copolymers of PEO containing shorter hydrophobic segments can be displaced by other macromolecules that have higher affinity for the specific base polymer surfaces. Entrapment of PEO chains or its copolymers bearing hydrophobic segments is achieved by surface swelling of polymeric supports, which will allow the entanglement of loosened macromolecular chains of the substrate with the hydrophobic segments of the PEO-based copolymer. The process is usually completed by immersing the system into a nonsolvent of the base polymer. The limitations of this approach are related to the toxicity of the solvents and their removal from the surface layers. Covalent grafting of PEO or its copolymers can only be achieved in the cases of functional polymer surfaces (e.g., PET, polyurethane, etc). Poly(ethylene oxide) can only be covalently linked to chemically inert substrates such as PE, PP, and Teflon when their surfaces are prefunctionalized. Photo-induced and high-energy radiation–mediated grafting and grafting/crosslinking processes represent an additional approach to this problem. However, the complex wet chemistry required for the generation of photoactive groups bearing PEG and the often undesired high-energy radiation-induced structural modification of the bulk of polymeric systems are limitations for application purposes.

Literature data indicate that most of approaches for the generation of PEO-rich polymeric surfaces involving physical adsorption or entrapment (including the generation of interpenetrating PEO-containing polymeric networks) and covalent grafting of PEO on a variety of substrate surfaces (PET, polyurethane, PVC, cellulose, functionalized glass, gold, etc.) have resulted in significantly reduced protein, platelet, and bacterial adsorption and in enhanced biocompatibility.

B. Silver as an Antimicrobial Molecular Recognition Agent

Another approach to prevent biofilm formation and consequently foreign body infections is to create antimicrobial polymeric materials by loading the substrates with antibiotics, detergents, disinfectants, iodine, or metals (e.g., silver) (27). These composite materials gradually release the antibacterial agents and can in this way prevent colonization. However, most of the organic bacteriostatic compounds incorporated in biomaterials have limited action due to the limited loading, and they only can be used for short-term biofilm prevention.

Silver was considered in many investigations due to its good antimicrobial activity and low toxicity. Silver is one of the most powerful natural disinfectants known. Research indicates that resistant strains do not appear in connection with the use of colloidal silver. Other metals also have germicidal effects, such as copper and mercury. However, most of these metals are toxic to mammalian cells. Bacteria have a strong ability to adapt to various substances. Antibiotics, for instance, destroy bacteria which are susceptible to them; however, they can become resistant by mutation.

Colloidal silver is a dispersion of tiny silver particles in water. The ultimate colloidal silver suspension would be at the molecular level. It has been suggested that the smaller the silver particles, the more effective the colloid is.

Silver occurs under several oxidation states. The most common are elemental silver (Ag^0) and the monovalent silver ion (Ag^+). However, higher oxidation states of silver are also known (Ag^{2+}, Ag^{3+}). It has been proposed that silver could be prepared in a molecular crystal form (an oxide lattice where one pair of silver ions in the molecule is trivalent and another pair of silver ions is monovalent) (43,44). This tetrasilver tetroxide is suggested to have biocidal properties through an electron release mechanism. The exact antimicrobial action of silver is not completely understood. Several possible mechanisms have been proposed which involve the interaction of silver with biological macromolecules, such as enzymes, DNA, etc. (45–49).

The deposition of silver or silver oxide on polymer (plastic) surfaces is rather difficult. Conventional silver coatings involve the preparation of silver-containing polymer solution (e.g., silicon, latex) that can be applied in the next step to polymer surfaces. However, most of the silver is buried into the bulk of the coating, and accordingly the resulting low silver surface area reduces significantly the efficiency of the molecular interaction. Sputtering and ion beam–assisted deposition of silver on plastic surfaces have also been evaluated. These processes are energy intensive and most of the silver is lost through undesired reactor wall deposition processes, and it can be applied only on substrates of high thermal stability.

These conventional approaches focused on the generation of antifouling layers on various substrate surfaces, including deposition of PEO and PEO-based copolymers. Incorporation of antimicrobial substances into common or modified polymeric structures and deposition of thin silver layers have the following limitations for their application:

Physical adsorption of PEO or PEO-based copolymers on polymeric surfaces usually results in the formation of labile films and even in less stable layers in the cases of low surface energy polymeric materials (e.g., PP, PE, Teflon).

PEO structures cannot be deposited on metal surfaces and exploited in water environments due to their high water solubility, while less-soluble PEO-based structures require complex procedures for their synthesis and deposition.

Conventional grafting processes developed for the covalent anchorage of PEO type structures on hydrophobic polymeric substrate surfaces require complex wet chemistry processes and often the presence of toxic solvents. The removal the trace solvent amounts or other chemicals (toxic materials incorporated into the surface layers of the substrates as a result of swelling processes) from the structure of surface layers of the base polymers is sometimes very difficult.

Photochemical grafting mechanisms have low efficiencies and the high-energy radiation processes usually alter the bulk characteristics of the materials.

Polymeric materials are not conductive electrically; consequently, thin silver layers cannot be deposited using conventional electrochemical techniques. Silver does not have stable, volatile organometallic compounds except a low–vapor pressure complex derivative: perfluoro-1-methyl propenyl silver (50), and as a result plasma-enhanced chemical vapor deposition (PECVD) is also less attractive. The sputtering approach is also excluded for polymeric materials; their surface structures would be damaged by the energetic (high temperature) metal particles. Sputtering is also usually limited to low surface area coatings.

IV. SURFACE FUNCTIONALIZATION OF BIOMATERIALS USING A COLD PLASMA APPROACH

Literature data accumulated in the last two decades indicate that active species of cold plasma environments can induce significant chemical reactions in the surface layers of various substrates including organic or inorganic materials. It has been demonstrated that surface characteristics like surface energy, surface roughness, and optical and electrical properties can conveniently be tailored under plasma conditions. Due to the fairly high energy levels of the plasma particles, even surface characteristics of materials as inert as quartz or Teflon can be altered by selecting the proper discharge environments (51–57).

A. Cold Plasma State

The plasma state, the fourth state of matter, is associated with the high energy content of the matter, and it can be broadly defined as a collection of an equal number of oppositely charged gas phase carriers with a net zero electrical charge. It is estimated that more than 95% of the known universe exists in the plasma state. All active celestial bodies are plasmas, and the interstellar space is also considered as rarefied plasma. Lightning, the Northern lights, and the ionosphere surrounding our planet are plasma states as well.

Manmade plasmas can be created by increasing the energy content of matter regardless of the nature of the energy source. Mechanical, chemical, thermal, radiant, and electrical energies can be employed for the creation of plasma state. Plasmas are losing energy toward the walls which confine them, and in order to sustain the discharge, energy must be supplied into the system continuously. The easiest way to supply energy into the plasma on a continuous manner is by using electrical energy. This explains why electrical discharges are the most common manmade plasma states.

Depending on the nature of the electric or electromagnetic fields that generate the plasma, DC, low- and high-frequency AC, and microwave glow discharges can be defined.

Close to totally ionized gases are termed as hot plasmas (near-equilibrium discharges). In these cases, the temperature of both the charged species of atomic or molecular origin and the electrons is extremely high.

Cold plasmas (glow discharges, silent discharges, and nonequilibrium discharges) are produced in matter with much lower energy contents. The degree of ionization is small (10^{-1} to 5×10^{-3}%), and the atomic and molecular charged and neutral species have low energies, while the electrons have relatively high energies. These electrical discharges are nonequilibrium plasmas and, owing to the low energy levels of the species composing the plasma, they alone are suitable for modifying organic matter. Electrical energy–initiated cold plasma states arise when the accidentally free electrons (e.g., created by cosmic rays) of a low pressure gaseous environment are accelerated by electric or electromagnetic fields to kinetic energy levels at which ionization, excitation, and molecular fragmentation processes are generated. These energies are usually expressed in energy or temperature units (1 eV = 11,600K) based on the kinetic theory. The mass of electrons being much smaller than those of molecular or atomic species, their velocities and consequently their kinetic energies will be much higher in comparison to ions or free radicals. Characteristic 3- to 4-eV electron energies of glow discharges would be equivalent to 34,800–46,400K. However, owing to the low heat capacity of electrons these energies will not raise significantly the temperatures of the surfaces which confine the plasma. These systems are characterized by the simultaneous presence of photons, electrons, ions of either polarity, neutral atoms and gas molecules and their corresponding excited species, and molecular fragments (mono and multiple free radicals). They can be considered as quasineutral gases exhibiting a collective behavior (local concentration of charges can influence, for instance, the motion of particles through long-range Coulomb effects) (58–61).

Depending on the nature of the electric or electromagnetic fields that generate the plasma, DC, low- and high-frequency AC (e.g., radiofrequency plasmas), and micro wave glow discharges can be defined.

B. Plasma Chemistry

The electron energy distribution of nonequilibrium (low pressure) plasmas can be often described by a Druyvesteyn approximation (Fig. 1) when the temperature of the electrons is considered much higher than that of the ions and when it is assumed that the only energy losses are by elastic collisions (the electric field strength in the plasma is sufficiently low to neglect inelastic collisions). However, at higher degrees of ionization the influence of electron density on the energy distribution can be significant (60,61). It can be observed that a small number of electrons have relatively high energies (5–15 eV), while the bulk of the electrons belong to the low energy electron range (0.5–5 eV). Since the ionization potentials of the atoms of com-

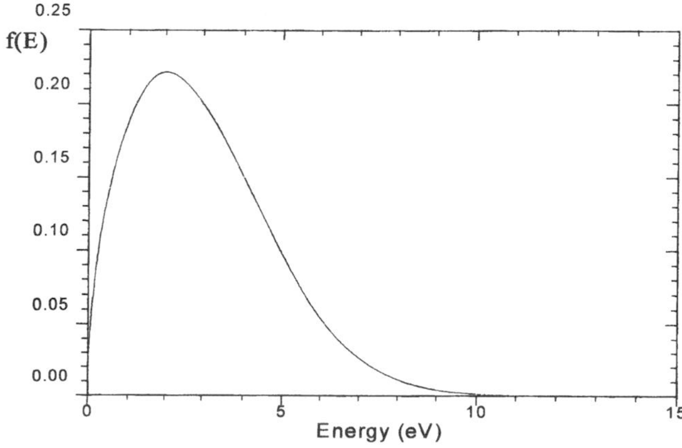

Figure 11.1 Druvesteyn electron energy distribution of a cold plasma (average electron energy: 3 eV).

mon organic structures belong to the tail region of the electron energy distribution (e.g., $C^+ = 11.26$ eV; $H^+ = 13.6$ eV; $O^+ = 13.6$ eV; $N^+ = 14.53$ eV, etc.), the low degrees of ionization of cold plasmas appear obvious.

However, it is important to note that the energy range of most of the electrons (2–8 eV) is intense enough to dissociate almost all chemical bonds involved in organic structures (62) and to create free radical species capable of reorganizing into macromolecular structures. As a consequence, all volatile organic compounds and organic derivatives containing main group elements can be fragmented and/or converted into high molecular weight compounds, even if they do not have the functionalities peculiar for common monomers (63,64). The plasma-generated macromolecular networks do not retain the structural characteristics of the starting materials, they are not based on repeating units. Consequently, the term plasma polymers which is often used is inappropriate for describing these structures. Higher energies are required only for the dissociation of unsaturated linkages and the formation of free radicals from a small number of atoms (CH:, SiCl:, etc.). Thus, it can be understood why plasma-generated macromolecular structures are usually characterized by unsaturation and branched and crosslinked architecture.

Besides the recombination mechanisms developed on the surfaces which limit the plasma, the active species of the discharge interact and continuously tailor the artificially exposed (reactor walls, various substrates, etc.) and self-generated (e.g., plasma-synthesized macromolecular structures) surface layers. The competition between the recombination–deposition processes and interaction of plasma species with the nascent macromolecular structures (etching) will control the intensities and the predominance of ablation, surface functionalization, or macromolecular film formation reactions.

Macromolecular plasma chemistry has developed in the last two decades in the following directions: plasma-enhanced synthesis (deposition and/or grafting) of thin layer macromolecular structures; conventional graft-polymerization reactions from plasma-activated substrate surfaces; surface functionalization of polymeric materials; and etching of inorganic or polymeric substrate surfaces.

C. Plasma Reactors

The most commonly used plasma reactors for the synthesis of macromolecular structures and surface modification of various substrates are radiofrequency (RF) installations (62,65). These discharges can be excited and sustained even by using insulated electrodes located inside or outside the reaction chamber. This is an important feature relative to DC plasmas because in the case of RF discharges dielectric material depositions on the electrodes during the plasma process will not significantly influence the regime of the discharge. Radiofrequency plasmas can also be sustained under lower pressure conditions and have higher ionization efficiencies than DC discharges. The energies of sample-bombarding ions of RF plasmas can conveniently be controlled by adjustable negative bias, while in the case of DC plasmas this possibility is limited by the breakdown voltage.

The main advantages of plasma chemistry versus conventional wet chemistry approaches in the areas of coating and modification (e.g., functionalization, ablation) of surfaces are

The energies of the active species of plasmas are comparable with the bond energies involved in all organic compounds and organic compounds containing main group elements, and as a result all volatile derivatives can be converted from their plasma-generated molecular fragments into thin layer macromolecular film depositions. This practically assures unlimited possibilities for the development of structure- and functionality-controlled coatings.

The controlled interaction of plasma state with surfaces allows the tailoring of surface characteristics (chemical functionalities) and morphologies of even the most inert substrates (e.g., Teflon, silica, etc.).

Cold plasma treatments are dry chemistry technologies. The reaction mechanisms are mediated by the interaction of gas phase active species of the discharge with solid surfaces (metal, inorganic, and organic macromolecular compounds).

Plasma chemistry technologies usually require moderate vacuum environments (100–10,000 mTorr) and the active species of the discharge alter only the very top layers of the substrates (deposition/etching). As a consequence, these processes require very low quantities of starting materials.

Plasma chemistry technologies are cost effective and environmentally friendly.

V. PLASMA TECHNOLOGIES FOR THE PREVENTION OF PROTEIN ADSORPTION, BACTERIAL ATTACHMENT, AND BIOFILM FORMATION

The cold plasma state can be used to create new surface functionalities, morphologies, and chemistries that will re-

Table 11.4 Cold Plasma–Enhanced Modification of Surfaces for the Control of Protein and Cell Adhesion

Objective	Experimental conditions	Results	Ref.
Covalent immobilization of PEO on allylamine (AA) and allyl alcohol (AAl) plasma–treated PET	13.56 MHz RF reactor with external moving (controlled speed of capacitor plates in same direction with vapor flow) copper plate electrodes	Successful cyanuric chloride–mediated immobilization of PEO from AA and AA plasma–coated PET surfaces. AA-treated PET surfaces exhibit higher PEO attachment than AAl-modified PET substrates.	70
Evaluation of *S. epidermis* adhesion to hydrophilic films deposited on PS from hydroxyethyl–methacrylate (HEMA) plasma and the effects of surface energetic parameters on the bacteria–synthetic surface interaction	Stainless steel, capacitively coupled, parallel plate reactor (RF power: 40–100 W; base pressure: 15 Pa; gap between electrodes; 10 cm), with samples located on the water-cooled ground electrode	The *S. epidermis* adhesion on PS discs coated with HEMA plasma films was significantly higher than on PS. The number of bacteria adhered to plasma-coated and oxygen plasma–treated PS was comparable. It is suggested that a significantly higher Lewis base characteristic associated with the plasma-modified substrates is responsible for this phenomenon.	71
Study of cell adhesion on ammonia, dimethyl acetamid, methanol, and methyl methacrylate RF plasma–coated fluoroethylenepropylene copolymer substrates evaluating the putative role of serum fibronectin (Fn) and vitronectin (Vn) in the initial attachment of human vein endothelia (HUVE) and human dermal fibroblast (HDF) to surfaces coated under oxygen- and nitrogen-containing vapor conditions	RF (125–375 kHz and 13.56 MHz) reactor; specially designed vapor flow system for the controlled introduction of vapors directly into discharge zone (4–15 sscm); tape type substrate; reversible, speed-controlled transport mechanism for the tape (0.05–0.4 m/min)	Surfaces resulting from oxygen-containing vapors do not bind sufficient Fn from fetal bovine serum (FBS) to support cell attachment, and the initial adhesion of HUEV cells has a high dependence upon FBS-origin Vn. Surfaces based on nitrogen-containing gases/vapors more effectively bind FBS-origin Fn, resulting in a higher dependence of cell attachment upon serum Fn. It was concluded that cell adhesion to the nitrogen-based substrates is dependent upon either Vn or Fn, but not exclusively to one of them.	72
Evaluation of the platelet-adhesion activity of fibrinogen adsorbed on tetrachloroethylene (TCE) and tetrafluoroehtylene (TFE) plasma–treated PET surfaces	Tubular Pyrex glass reactor; degassed monomers; substrates initially treated with Ar plasma (2.5 W; 3 cm^3/min; 0.1 mmHg; 3.3 mm/s)	Fibrinogen adsorbs strongly on both TFE- and TCE-treated PET. It is suggested that this tenacious binding behavior is related to the conformational rearrangement of fibrinogen, which results in a molecular state, that prevents its recognition and binding by platelet receptors.	73
Investigation of the adsorption and subsequent detergent elutability of fibrinogen and albumin on virgin and tetrafluoroethylene and perfluoropropane (PFP) plasma–treated polymer (PET) surfaces, evaluating their relative adsorption-related thromboresistance	13.56-MHz, 300-W RF capacitively coupled tubular glass reactor provided with external movable (1–4 mm/s electrodes (Ar etched substrates; TFE: 15 sscm; 200 μmHg; 30 W; 3.3 mm/s; and 3 sscm, 200 μmHg; 2.5 W; 3.3 mm/s; PFP: 3 sccm; 200 μmHg; 2.5 W; 1 mm/s)	The adsorption of fibrinogen and albumin to plasma-treated PET are comparable to their adsorption to PTFE and PET; however, their elutability with SDS is much lower in comparison to PTFE and PET.	74
Study of deposition of ultrathin poly(2-hydroxyethylmethacrylate (PHEMA) by simultaneous condensation and RF plasma exposure of HEMA vapors, comparing interaction of plasma and conventional PHEMA surfaces with ^{125}I-radiolabelled fibrinogen	Capacitively coupled, external electrode, tubular glass, RF plasma reactor provided with a liquid nitrogen cooling system of the substrate holder (0.2 Torr; 20 W; Ar plasma–activated substrate: PE; deposition time: 5 min for cooled and 15 min for noncooled substrates.	Plasma HEMA films deposited at low temperatures were similar in nature and interacted similarly with fibrinogen in comparison to conventional PHEMA structures.	75

Table 11.4 Continued

Objective	Experimental conditions	Results	Ref.
Study of nitrogen and oxygen plasma–mediated modification of PTFE for enhancement of adhesion of endothelial cells and serum fibronectin	Plasmafab 505 (Electrotech, Bristol, UK) plasma reactor (20 Pa for N_2; 9 Pa for O_2 plasmas; 250 W)	Oxygen and nitrogen plasma–treated PTFE have improved wettability and human endothelial cell adhesion (from serum-containing culture medium) comparable to tissue culture PS. The amounts of serum proteins including fibronectin adsorbed to plasma-modified substrates are larger than those adsorbed on virgin PTFE	76
Study of the platelet activation and adhesion to ethylene oxide (EO) and N-vinyl-2-pyrrolidone (VP) plasma–coated dimethyldichlorosilane (DDS)-treated glass substrates	13.56-MHz inductively coupled, Pyrex glass RF plasma reactor [Ar plasma pretreatment; EO plasma: 15–100 mTorr; 20–80 W; EO flow: 0.8–8.6 mL/min (STP); treatment time: 5–30 min; VP plasma: 30 mTorr; 0.1 mL/min (STP); 30 W].	Hydrophilic EO- and VP-based plasma-created macromolecular layers do not prevent platelet adhesion and activation. The film thickness does not influence significantly these parameters. Based on adhesion data resulting from Pluronic structures, it is suggested that steric repulsion might be involved in the protein and platelet repelling mechanisms	77
Evaluation of protein adsorption, cell attachment, and biocompatibility of 2-methacroyiloxyethyl phosphorylcholine–grafted silicon rubber surfaces (pMPC-SR), prepared by plasma-induced reaction mechanisms	Ar plasma–treated and oxygen-exposed SR membranes sealed in glass ampoules containing degassed MPC/ethanol solution and heated for a predermined time at 75°C	Protein adsorption of high MPC concentration pMPC-SR surfaces was reduced in comparison to SR. Epithelial cell attachment and growth onto the grafted samples were suppressed. Platelet adhesion data indicate good biocompatibility.	78
Evaluation of fibrinogen adsorption and host tissue responses to hydrophilic and hydrophobic plasma-functionalized PET substrates	RF pulsed allyl alcohol, allylamine, perfluorohexenes (PH), and hexamethyldisiloxane (HMDSO) plasma environments (0.1 ms on, 3 ms off; 200 W)	Surface functionalities do significantly influence both the adsorption and denaturation of adsorbed fibrinogen, and these surfaces provoke different degrees of acute inflammatory responses. It is suggested that surface-induced conformational changes of adsorbed fibrinogen may be critical to the recognition of biomaterial implants by phagocytes.	79
Analysis of the influence of free energy of polymers (PET, PE, PTFE, PDMS, and fluorocarbon-copolymer-coated PET) on resistance of adsorbed albumin to sodium dodecyl sulfate (SDS) elution for untreated and TFE, ethylene, and HMDSO RF plasma–exposed surfaces	Tubular Pyrex glass, capacitively coupled reactor with external, stationary or movable electrodes (2.5 W; 3 cm^3/min; 0.1 Torr; 3.3 mm/s) and presence or absence of Ar plasma precleaning	TFE-treated polymers retain a larger fraction of the adsorbed albumin than ethylene- and HMDSO-modified surfaces. It is suggested that the albumin retention is closely related to their surface free energy, and that the tight binding of albumin to the fluorocarbon polymers may be related to the strong hydrophobic interactions.	80

Table 11.4 Continued

Objective	Experimental conditions	Results	Ref.
Preparation of fouling-resistant PE, PTFE, TCPS, and glass surfaces by plasma deposition of macromolecular layers from tetraethylene glycol dimethyl ether (TEGDME)	Capacitively coupled, external electrode, RF tubular, heated glass reactor (Ar plasma; 40W; 0.175 mmHg; 5 min)	Adsorption of ^{125}I-labelled fibrinogen, albumin, and IgG from buffer and plasma on plasma-treated polymer surfaces was very low in comparison to unmodified substrata. Coated glass substrates adsorbed more protein than plasma-treated PE and PTFE. Short-term nonadhesiveness of platelet and endothelial cells to plasma-treated substrates was also observed.	81
Evaluation of five different gram-positive and gram-negative bacteria and the different strains'adhesion to plasma-treated Vicryl sutures	13.56-MHz RF tubular, capacitively coupled, external electrode plasma reactor (plasma vapors: dimethylaminoethylmethacrylate (DMAEMA) and acrylic acid (AAc); 10 W; 5 min; 0.5 Torr; monomer flow: 30 mL/min)	Hydrophobic bacteria attached more to hydrophobic (unmodified) Vicryl sutures. Both plasma treatments generated significant decrease in bacterial attachment. The effect of AAc plasma was more pronounced.	82
Study of adhesion of S. epidermis to plasma-deposited HEMA macromolecular layers deposited on PS substrates	Capacitively coupled, parallel plate reactor (gap between the electrodes: 10 cm); water cooled electrodes (HEMA plasma: 40–100 W; 15 Pa; monomer temperature: 75°C; monomer flow: 10 cc(STP)/min; oxygen plasma: 20 cc(STP)/min; 80 W; 20 Pa)	Bacteria adhered more to the plasma-deposited or plasma-treated surfaces than to PS. Experimental conditions did not have a significant influence on the adhesion efficiencies. It is suggested that electron donor–electron acceptor interactions control the adhesion mechanism.	83
XPS and SSIMS characterization of glass surfaces modified by plasma-modified oligo(glyme) films	13.56-MHz RF tubular, capacitively coupled (external copper electrodes), glass reactor (20–80 W, base pressure: 14–18 mT; oligo-glyme flow: 50 sccm; deposition time: 20 min)	The surfaces of oligo(glyme)-origin deposited layers appear to be populated with methyl-terminated oligo(ethylene oxide) chain ends. At higher RF power values, the oxygen-based functionalities are lost; the deposited layers have an unsaturated hydrocarbon type character.	84
Evaluation of controlled release of antibiotic (Ciprofloxacin) of triethylene glycol dimethyl ether (triglyme) and poly(butyl methacrylate) (pBMA) plasma–modified polyurethane membranes (PEU) to prevent bacterial (*Pseudomonas aeruginosa*) adhesion and biofilm formation	Glow discharge deposition of triglyme- and pBMA-origin layers (''monomer'' temperature: 60°C)	The rate of initial bacterial cell adhesion to triglyme-coated PEU was 0.77% and to the pBMA-coated PEU releasing ciprofloxacin was 6% of the observed adhesion rates for the control PEU.	85

sult in a different interaction of discharge-exposed surfaces with biomacromolecules and cells in comparison to the nonmodified materials. Literature data indicate that a large variety of plasma reactors, precursor plasma gases/vapors (selected for the creation of specific surface functionalities and surface layers), and experimental parameters has been employed to control protein and cell adsorption and consequently biofilm formation and foreign body–mediated infections (66–69).

A. Plasma-Enhanced Modification of Polymeric Substrates

Owing to this ''multidirectional approach'' the comparison and evaluation of experimental data and theoretical considerations is very difficult and often the results are apparently inconsistent. Accordingly, a tabular comparison of experimental parameter space, objectives, and findings appears more adequate (Table 4).

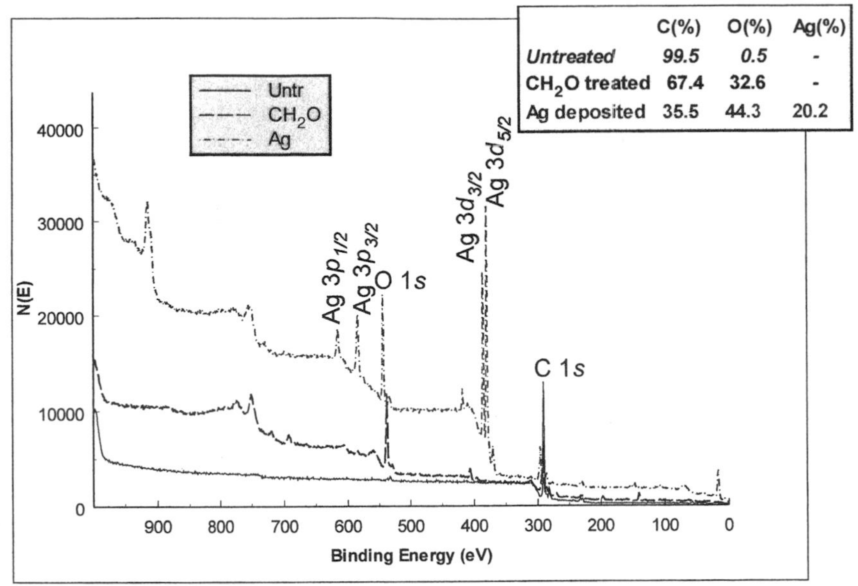

Figure 11.2 Survey ESCA diagram of polyethylene, formaldehyde plasma–treated polyethylene, and silver-coated polyethylene.

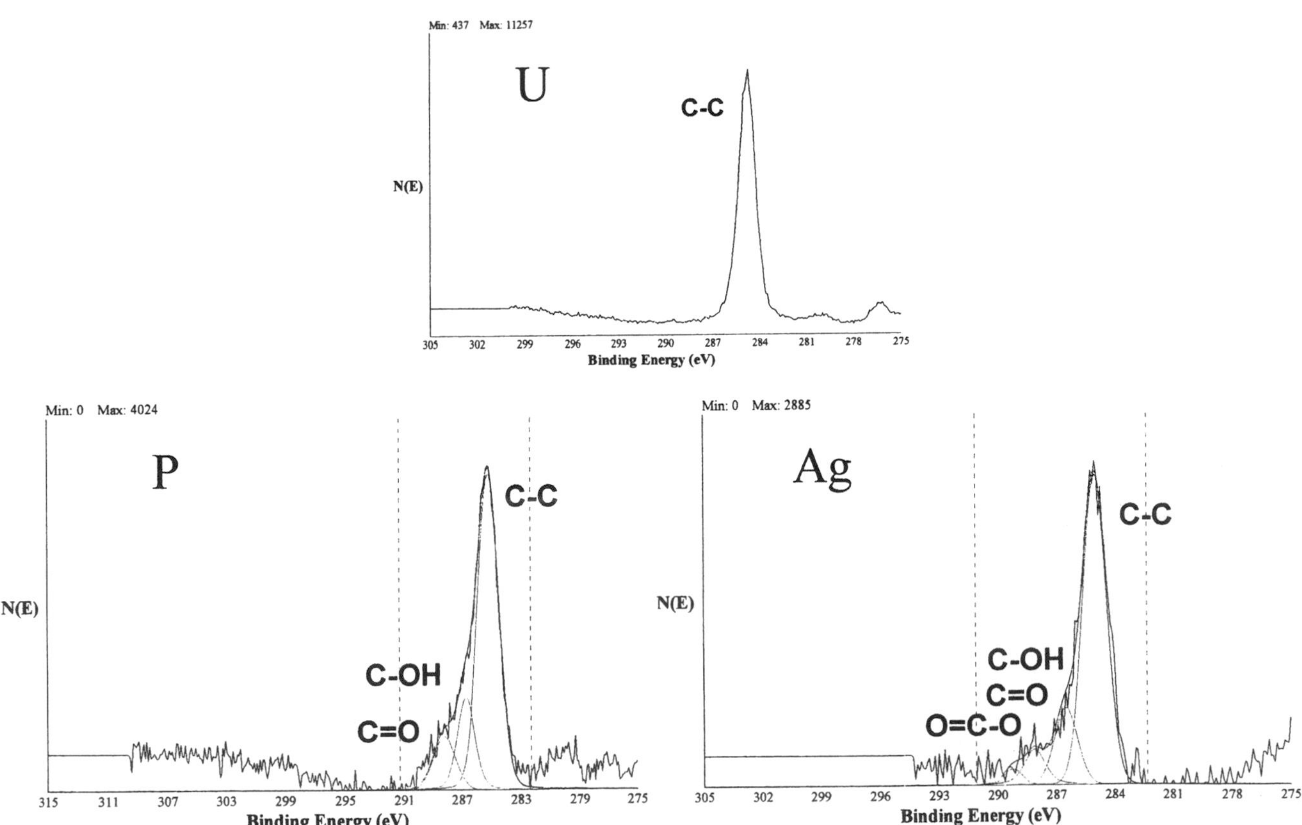

Figure 11.3 High resolution C*1s* ESCA diagrams of untreated polyethylene (U), formaldehyde plasma–treated polyethylene (P), and silver-coated polyethylene (Ag).

B. Silver Deposition on Polymers

Plasma-enhanced deposition of thin silver layers on polymer (dielectric) substrate surfaces and the production of large quantities of high-concentration (200–400 ppm) colloidal silver solutions under dense medium plasma (DMP) conditions have also been accomplished (86–88). Thin silver layers were deposited on formaldehyde plasma-functionalized PE surfaces by involving ex situ, second-stage Tollen's reaction in the absence of plasma, where diamminosilver (I) ions are reduced by the discharge-generated aldehyde functionalities to metallic silver according to the following reaction:

$$R\text{—}C(=O)H \xrightarrow[H_2O]{Ag(NH_3)_2^+} R\text{—}C(=O)O^- + Ag^0$$

The implantation of aldehyde functionalities was achieved using paraformaldehyde (PF)–origin formaldehyde–RF plasma environments. Formaldehyde is only available commercially in water solution, and other aldehydes (aldehyde group precursors) are usually less volatile and would also generate under plasma conditions undesired hydrocarbon-type fragments. Paraformaldehyde is a solid phase material that decomposes at relatively elevated temperatures into formaldehyde. Accordingly, a temperature-controlled ''formaldehyde feeder'' was designed and employed. Aldehyde functionalities were implanted onto PE surfaces under the following experimental conditions: stainless steel, cylindrical, parallel-plate, capacitively coupled reactor; frequency of the driving field: 40 kHz; base pressure: 50 mTorr; formaldehyde pressure in the absence of plasma: 200 mTorr; formaldehyde pressure in the presence of discharge: 400 mTorr; RF power dissipated to the

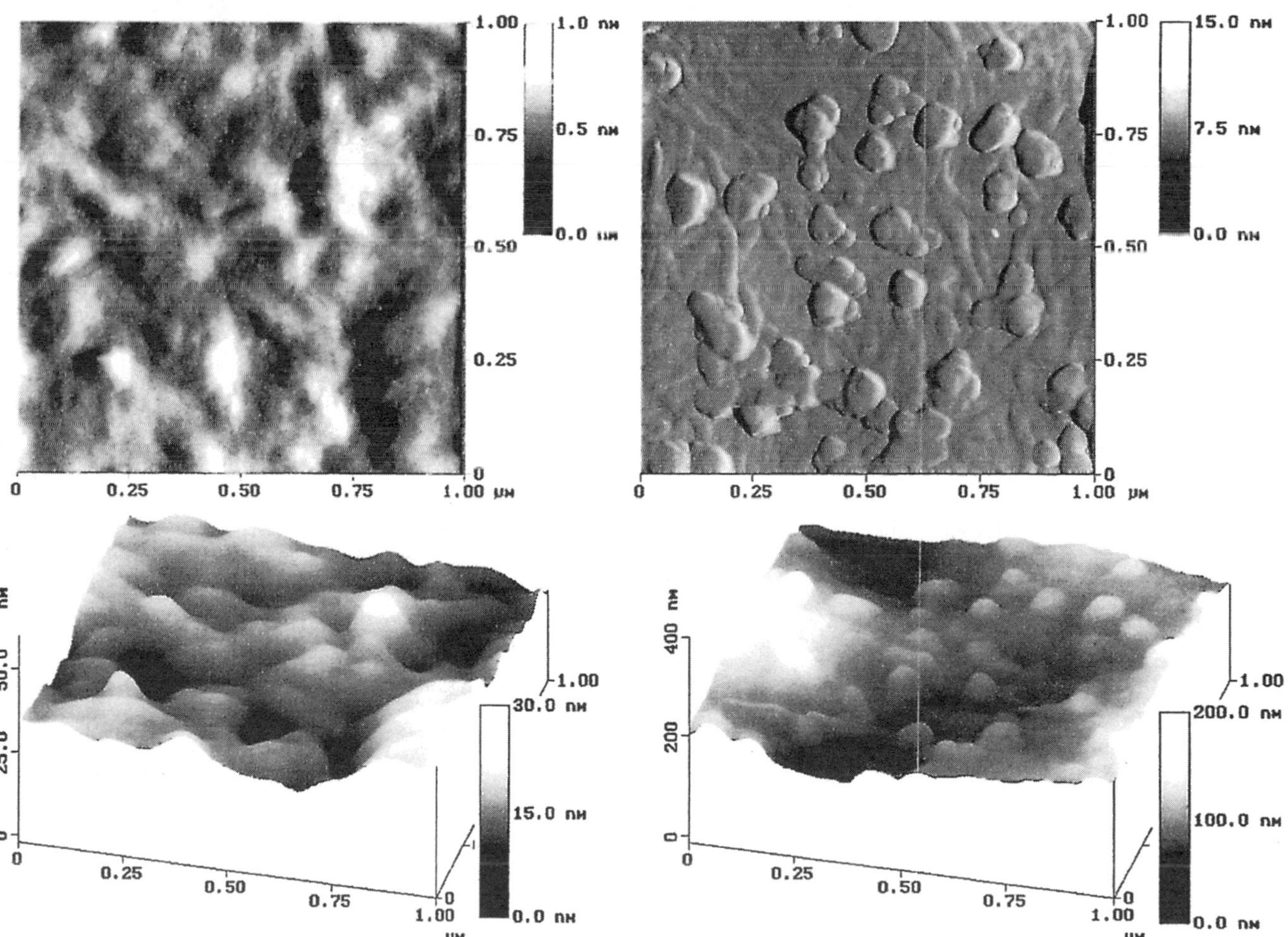

Figure 11.4 AFM images of untreated (left) and silver deposited (right) polyethylene.

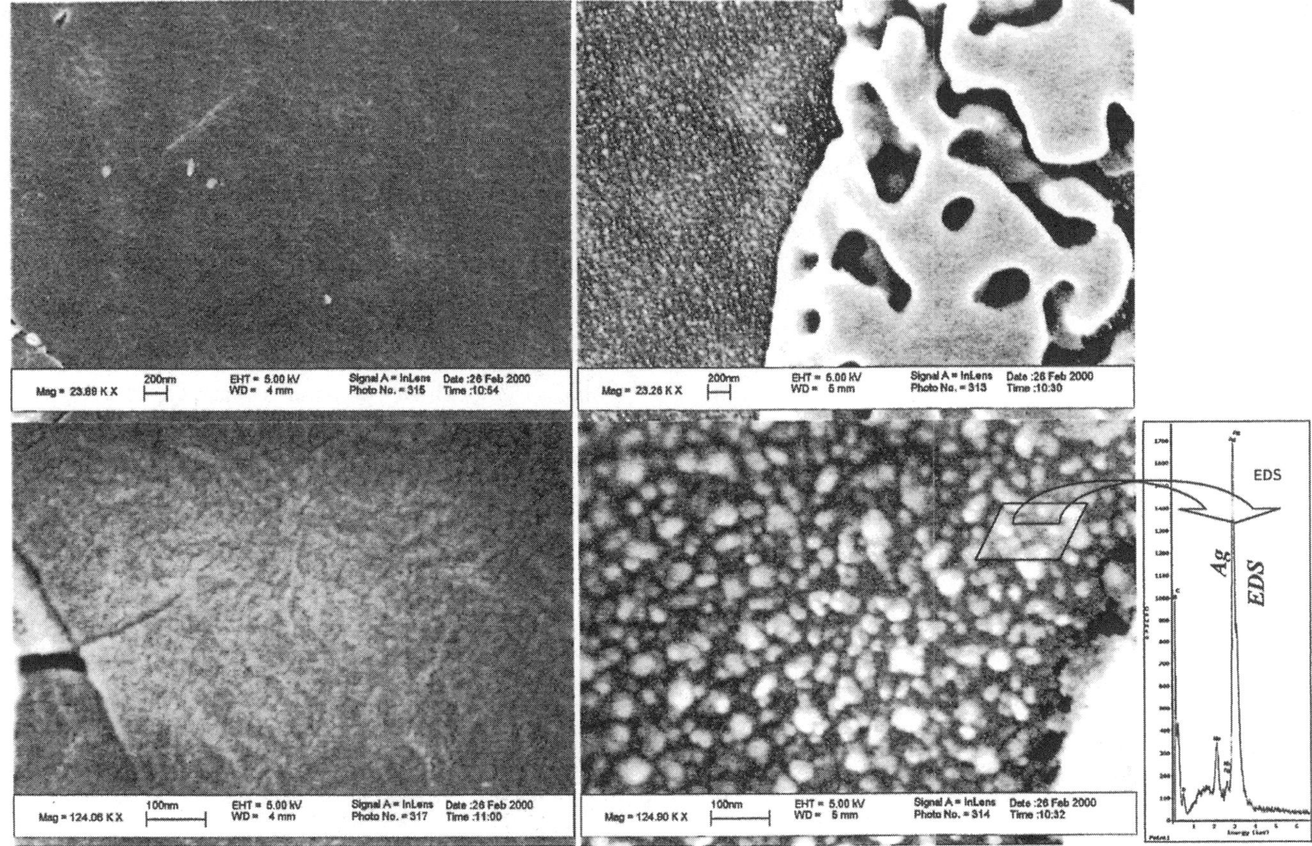

Figure 11.5 SEM images of untreated (left) and silver-coated (right) polyethylene. Magnification: ×24,000 (top); × 125,000 (bottom).

electrodes: 250 W; reaction time: 10 min; temperature of the PF container: 170°C. Survey (Fig. 2) and high resolution C*1s* (Fig. 3), O*1s*, and Ag*3d* ESCA data indicate the presence of silver on the surface of PE films and the formation of —COOH functionalities as a result of the Tollen's reaction. Visual observation (metallized surface appearance), AFM (Fig. 4), SEM, and energy dispersive x-ray spectroscopy (Fig. 5) substantiate the presence of silver particles of the plasma/Tollen's modification of PE.

C. DMP Plasma Generation of Colloidal Silver

Water-based colloidal silver nanoparticles were prepared in a redesigned version of DMP reactor. The DMP, a new plasma tool, developed at the Center for Plasma Aided Manufacturing (C-PAM), University of Wisconsin—Madison, allows the initiation and sustaining of discharges in coexisting liquid/vapor medium at atmospheric pressure and may offer a significantly higher efficiency for the processing of liquid phase materials in comparison to common plasma technologies.

The reactor (Fig. 6) is composed of a cylindrical glass chamber (7), provided with two stainless steel upper and bottom caps (9,17), and a cooling jacket (4). The rotating, cylindrical, stainless steel, upper electrode (19) is equipped with a quartz jacket for avoiding the penetration of the reaction media to the electrode-sustaining central shaft and bearings. The upper electrode has a cylindrically shaped, cross-section disc end, which is terminated in an interchangeable ceramic pin array (8) and holder (23). The lower electrode is hollow, has also an interchangeable conical cross-section end piece, and in addition is provided with channels (25) for the recirculation of the reaction media. Both the spirally located pin array and the interchangeable metallic part of the lower electrode can be made of different metals required by the specific plasma treatment. In this case, for the synthesis of colloidal silver solutions, both the spirally located pin array and the interchangeable metallic part of the lower electrode are made of pure silver. The distance between the pin array and the lower electrode can be varied by a micrometric (thimble) screw system. The reactor is vacuum tight (copper gasket sealings to

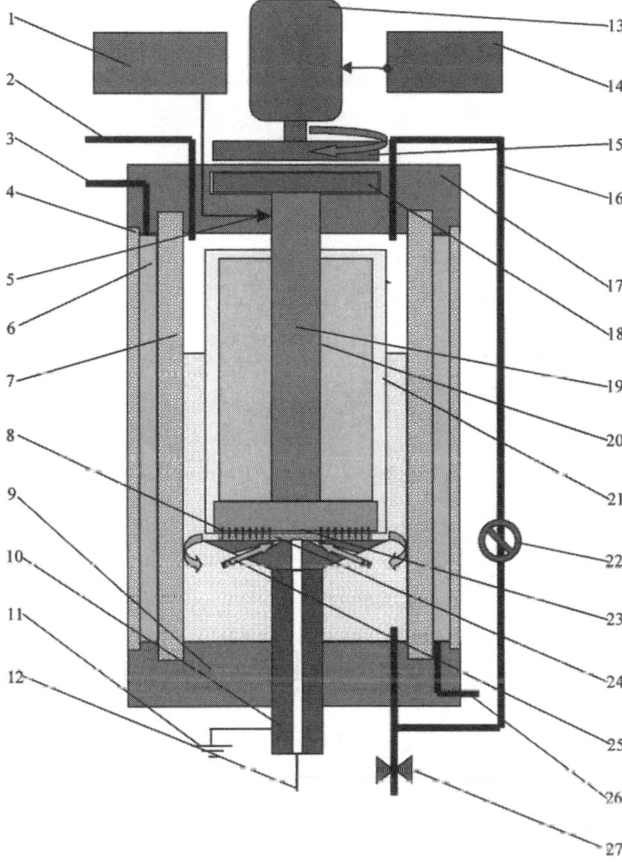

Figure 11.6 Redesigned version of dense medium plasma reactor. 1—DC power supply; 2—gases evacuation; 3, 26—coolant exit and inlet; 4, 7—glass cylinders; 5—electrical contact; 6—coolant; 8—ceramic pin array; 9, 17—caps; 10—nonrotating electrode; 11—ground; 12—gas inlet; 13—motor; 14—digital controler; 15, 18—magnetic coupling system; 16—liquid inlet; 19—rotating electrode; 20—sealed volume; 21—quartz isolator; 22—recirculating pump; 23—pins; 24—electrical discharges; 25—recirculated flux; 27—valve.

avoid the escape of the components of the reaction medium into the environment), and the rotation of the upper electrode is assured through magnetic coupling system (15,18). The reactor can be operated in a batch-type or continuous flow mode, depending on the specific application. The rotation of the upper electrode is digitally controlled in the range of 0–5000 rpm. The plasma (multitude of spark discharges) can be initiated and sustained using adjustable and commercially available DC or AC power supplies. Although the actual mechanism for electron emission and energy transport through the liquid is not well characterized at this time, the rotation of the electrode, and the spirally arranged pin array system (which acts as a high current density field emission arc source) will generate under DC or AC voltage conditions many microdischarges covering the whole area of the electrode surfaces.

Rotating the electrode serves several important purposes. The action spatially homogenizes the multiple microarcs, activating a larger effective volume of fluid. Spinning the electrode also simultaneously pumps fresh liquid and vapors into the discharge zone, and it thins the boundary layer between the emission tips and the bulk liquid.

Reactive or inert gases can also be introduced into the reaction media during the plasma process through the hollow, lower electrode. The simultaneous presence of a gas environment contributes also to the homogenization of the reaction system and enhances the microarc formation process owing to the lowered density of the media.

The DMP reactor shows an attractive potential for the creation of an efficient way for the synthesis of nanoparticle systems: the rapid spinning of the pin array electrode, and consequently the microarc system, essentially converts a surface process to a volume process. The individual arclets, under spin, sweep out a circular volume of plasma while simultaneously pumping fresh fluid into the reaction zone, thus thinning the diffusive boundary layer and enhancing mobility of free radicals into the bulk. Accordingly, all plasma-initiated and -sustained processes will be more efficient and will have a volume character.

Typical colloidal silver solutions were prepared under the following experimental conditions: initial liquid media: 200 mL pure water; angular speed of the upper electrode: 1000 rpm; DC voltage: 200 V; DC current during the process: 1.0–1.5 A; flow rate of Ar (bubbled through the liquid medium): 20 sccm; reaction time: 1 min; temperature of the reaction medium: 8°C; modus operandi: batch-type. SEM images of nanoparticle silver clusters formed by the evaporation of 100 μL colloidal silver solution (200 mL initial water; 20-s reaction time) show the nano-scale nature of the DMP-plasma generated silver particles (Fig. 7). It was demonstrated that the colloidal silver solutions have very strong bactericid characteristics (Table 5). Extremely short plasma treatment times are enough to kill high concentrations of bacterial populations. The following can be concluded:

Cold plasma reaction mechanisms allow the deposition of macromolecular layers, which, depending on the starting components and the selected plasma parameters, can enhance or diminish the adsorption of proteins and cells on various substrate surfaces.

For each application the specific plasma gases and reaction conditions should be carefully selected.

Systematic investigations targeting the deposition of antifouling layers should be developed for the specific PEG precursor material (influence of experimental

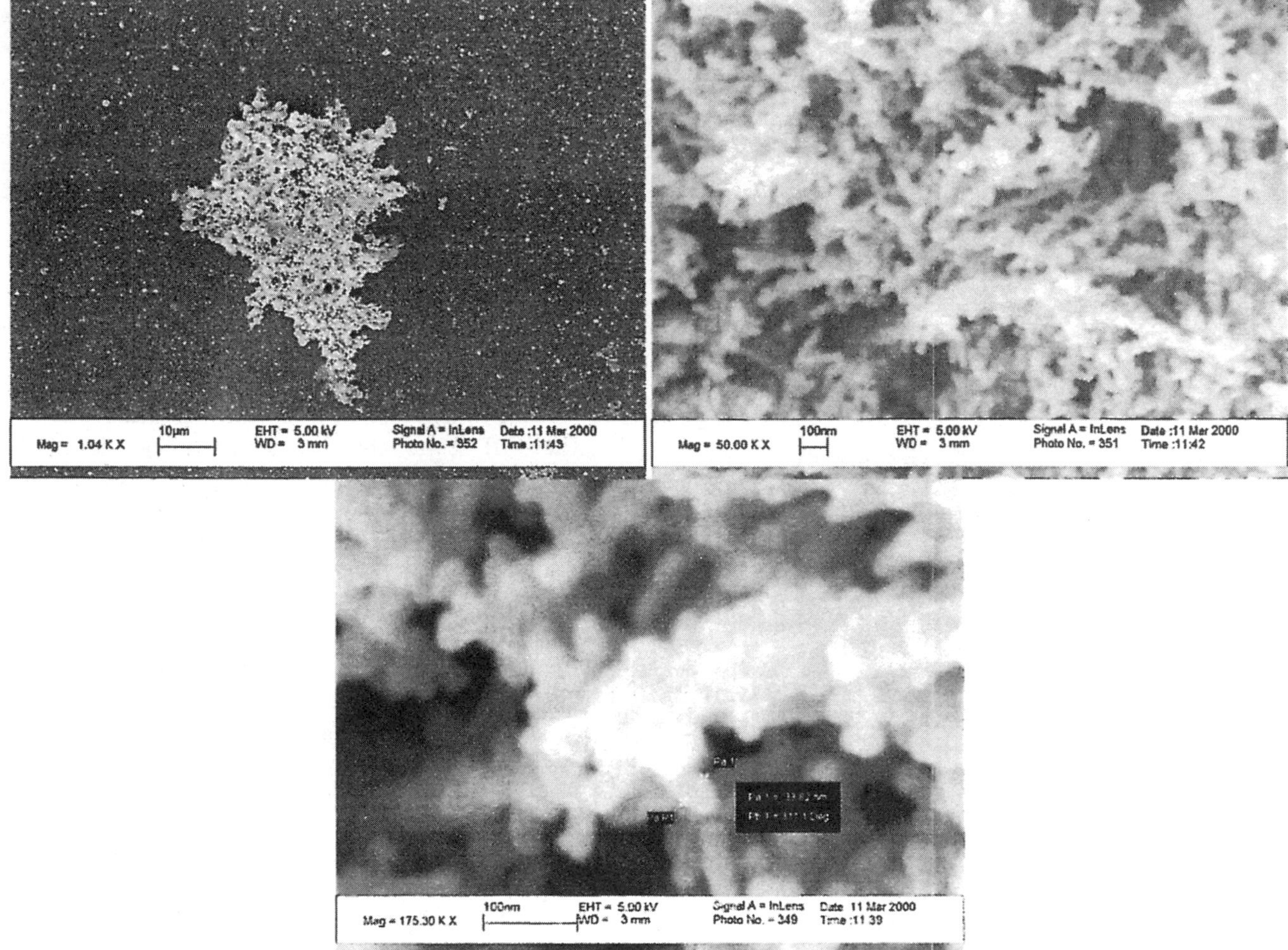

Figure 11.7 SEM image of nanoparticle silver clusters formed by the evaporation of 100µL colloidal silver solution. Magnification: × 1,000 (upper left), × 50,000 (upper right), × 175,000 (bottom). Measured diameter: 33 nm.

parameters on the molecular fragmentation and surface mediated recombination reaction mechanisms).

VI. PLASMA-MEDIATED IMMOBILIZATION OF ACTIVE BIOMOLECULES

The advantages of using enzymes in chemical synthesis are related to their very high specificity (regio- and stereospecificity) and versatility, to their mild reaction conditions (close to room temperature and to pH neutral media), and to their high reaction rates (89–91). However, due to the poor recovery yields and reusability of free enzymes, much attention has been paid to the development of efficient enzyme immobilization processes.

Immobilized enzymes are defined as enzymes physically confined in a certain region of a substrate with retention of their catalytic activities. The immobilization process includes the localization of enzymes chemically or physically on various insoluble (usually water-insoluble) organic or inorganic polymer matrices. This research area is currently generating increased interest among biochemists, biophysicists, chemists, and physicists due to its interdisciplinary nature. Two main reasons attracted the attention of scientists in this field: first, it was obvious that the results of these investigations can lead to potential for using immobilized enzymes as stable and renewable industrial catalysts, and second these investigations might help to understand the influence of heterogeneous environments on enzyme-catalyzed reactions. Most biologically active in

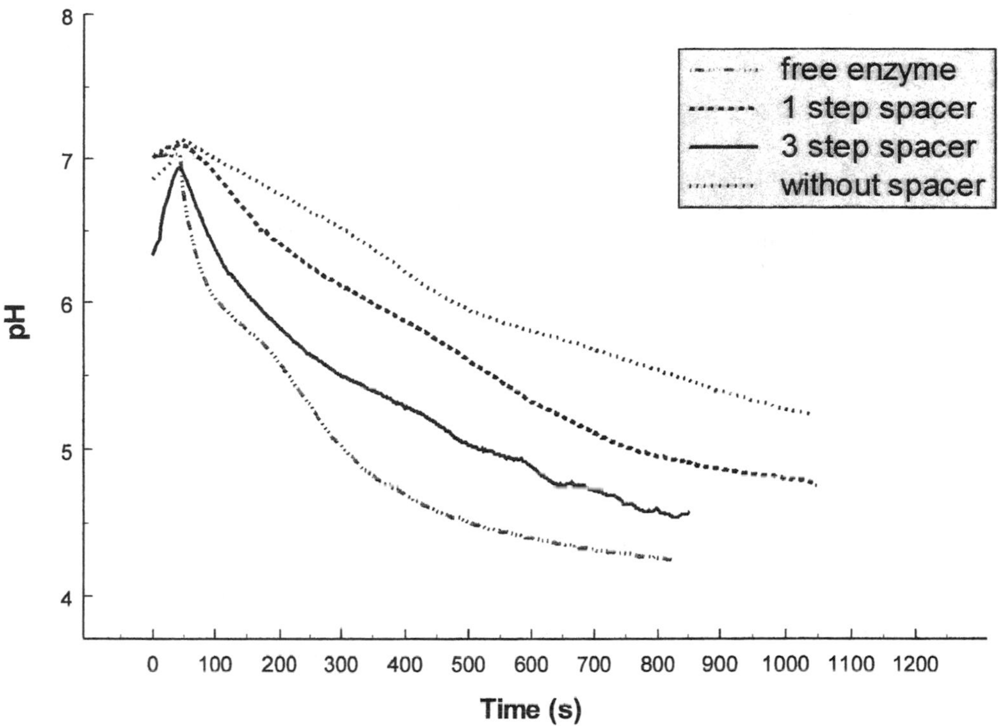

Figure 11.8 pH diagrams of enzyme assay involving free enzyme, one-step spacer-attached AC, three-step spacer-attached AC, and directly attached AC to the polyethylene substrate. Experimental conditions: *DS functionalization*: base pressure: 40 mTorr; pressure in the absence of plasma: 200 mTorr; pressure in the presence of plasma: 220 mTorr; RF power: 100 W; treatment time: 30 s; flow rate DS: 6 sccm. *Oxygen functionalization*: base pressure: 40 mTorr; pressure in the absence of plasma: 200 mTorr; pressure in the presence of plasma: 220 mTorr; RF power: 200 W; treatment time: 60 s; flow rate O_2: 6 sccm.

Table 11.5 Bacterial Inoculum Treated in DMP Reactor

		Plate counts of surviving bacteria	
No.	Sample	(log CFU/mL)	(CFU/mL)
1	Initial inoculum of water	5.73	537,032
2	Water held until treated samples were plated	5.41	257,040
3	Bacteria sample treated for 5 s	<1.0	0
4	Bacteria sample treated for 10 s	<1.0	0
5	Bacteria sample treated for 1 min	<1.0	0
6	Water ACS treated for 1 min and added 1:1 to bacteria sample	<1.0	0
7	1 mL of bacteria sample treated for 10 s added to 200 mL untreated bacteria sample	3.69	4,898

Note: Reactor: 200 V; 1 A.

vivo species, such as enzymes and antibodies, function in heterogeneous media. These environments are difficult to reproduce in vitro for applicative purposes. Immobilized enzyme systems are useful for experimental and theoretical research purposes for understanding the mechanisms of in vivo biocatalyzed reactions, and can offer solutions for use in batch-type reactions where there is a poor adaptability to various technological designs and recovery of the enzymes is difficult. Most of the intracellular enzymes are working under such environments and do not meet the Michaelis–Menten dilute solution conditions.

The molecular recognition ability and activity of enzymes (polypeptide molecules) are based on their complex three-dimensional structures containing sterically exposed, specific functionalities. The polypeptide chains are folded into one or several discrete units (domains) which represent the basic functional and three-dimensional structural entities. The cores of domains are composed of a combination of motifs which are combinations of secondary structure elements with a specific geometric arrangement. The molecular structure–driven chain-folding mechanisms generate three-dimensional enzyme structures with protein molecules orienting their hydrophobic side chains toward the interior exposing a hydrophillic surface. The —C(R)—CO—NH— based main chain is also organized into a secondary structure to neutralize its polar components through hydrogen bonds. These structural characteristics are extremely important, and they make the enzyme molecules very sensitive to the morphological and functional characteristics of the potential immobilizing substrates. High surface concentrations of enzyme-anchoring functionalities can result, for instance, in excessive enzyme densities or multipoint connections which can ''neutralize'' the active sites or can alter the three-dimensional morphologies of the enzyme molecules through their mutual interaction and their interaction with the substrate surfaces. These are just a few of the factors that may be responsible for the significantly lower activities of immobilized enzymes in comparison to the activities of free enzyme molecules. Rough substrate surface topographies or stereoregular surfaces (e.g., isotactic or syndiotactic polymers) might also influence, in a positive or negative way, the specific activities. Morphologically ordered surfaces might induce changes of the stereoregular shapes of protein molecules. Recently it has been found that enzymes can adopt more than one functional conformation other than its lowest potential energy state (92). Consequently, one of the potential solutions for achieving enhanced immobilized enzyme activities, comparable to that of the free enzymes, would be to intercalate selected length and densities of spacer molecules between the substrate and the enzyme molecules during the immobilization reaction processes (93–97).

Techniques for immobilizing enzymes can broadly be divided into two categories: chemical and physical methods. Chemical methods of immobilization involve the formation of at least one covalent bond between one or more enzyme molecules and the polymer matrix. These processes are usually irreversible. Physical methods include adsorption (e.g., electrostatic interaction) and entrapment within microcompartments (e.g., entrapment of enzyme within gel matrices or semipermeable microcapsules, etc.) and are usually based on reversible bond formations.

The classification of enzyme immobilization techniques can also be made according to the binding procedure: carrier binding, crosslinking, and entrapping (98). Carrier binding can be accomplished by physical adsorption, ionic bonding, and covalent bonding, while entrapping can be achieved by microencapsulation and lattice entrapping. According to the carrier binding technique, the amount of the enzyme bound to the carrier and the activity after immobilization depend on the nature of the inorganic or organic carrier. An increase of the densities of the hydrophilic groups on the carrier surfaces and the concentration of the bounded enzymes results in enhanced activity of the immobilized enzyme.

Covalent attachment of enzymes to a solid phase matrix must involve only functional groups of the enzyme that are not essential for catalytic action. This immobilization approach is much more complex and less mild than those based on physical adsorption and ionic binding procedures. As a consequence, covalent binding may alter the conformational structure and the active centers of the enzyme, resulting in diminished activity. However, due to the covalent bonding, the immobilized enzyme will not be released into the surrounding liquid environment even in the case of reactive substrates and under high ionic strength conditions.

Physical adsorption immobilization of enzymes causes little or no conformational change of the enzyme or annihilation of its active centers. However, due to the much weaker forces (van der Waals, hydrogen bonding, etc.) which retain the protein molecules on the carrier surfaces, the enzyme may be released from the carrier surface and diffuse into the surrounding solution.

The ion binding method relies on the attachment of the enzyme molecules by charged functionalities (e.g., ion exchange residues) present on the surfaces of water-insoluble carriers. This approach causes little change in the conformation and the active site of the enzyme, and as a consequence the activity of the immobilized enzyme is usually high. The enzyme-to-carrier linkages are significantly stronger in this case, in comparison to the physically adsorbed enzyme systems, but are less strong than in covalent binding. Dissociation of the enzyme from the carrier surface under high ionic strength solution conditions (specific pH values) is one of the weak points of this approach.

A. Plasma Functionalization of Natural and Synthetic Supports

Immobilization of active enzymes from 40 kHz and 13.56 continuous wave (CW) and pulsed MHz, capacitively coupled, RF plasma–functionalized natural and synthetic polymer surfaces has been accomplished in the absence and presence of spacer molecular chains intercalated between the polymeric substrates and the immobilized enzyme (99–102). It has been demonstrated that oxygen, hydrazine, and dichlorosilane RF plasma environments are proper for the implantation of carbonyl, primary amine, and halosilane functionalities onto various polymeric substrate surfaces, including poly(ethylene terephthalate), polypropylene, polyethylene, atactic and isotactic polystyrene (APS, IPS), and cellophane, and that α-chymotrypsin can conveniently be immobilized under ex situ plasma conditions directly to the substrate surfaces or by involving spacer chain molecules generated under in situ conditions in the absence of the discharge, according to the reaction mechanisms shown in Scheme 1 (99–101).

The activity of the immobilized enzymes was evaluated by monitoring the pH change as a result of the enzyme-induced hydrolysis of acetyl tyrosine ethyl ester. The pH values were recorded using a Virtual instrument (LabView) serial connected to a pH meter. The enzyme activity was also tested by reusing thoroughly washed enzyme-immobilized substrates for as many as five cycles.

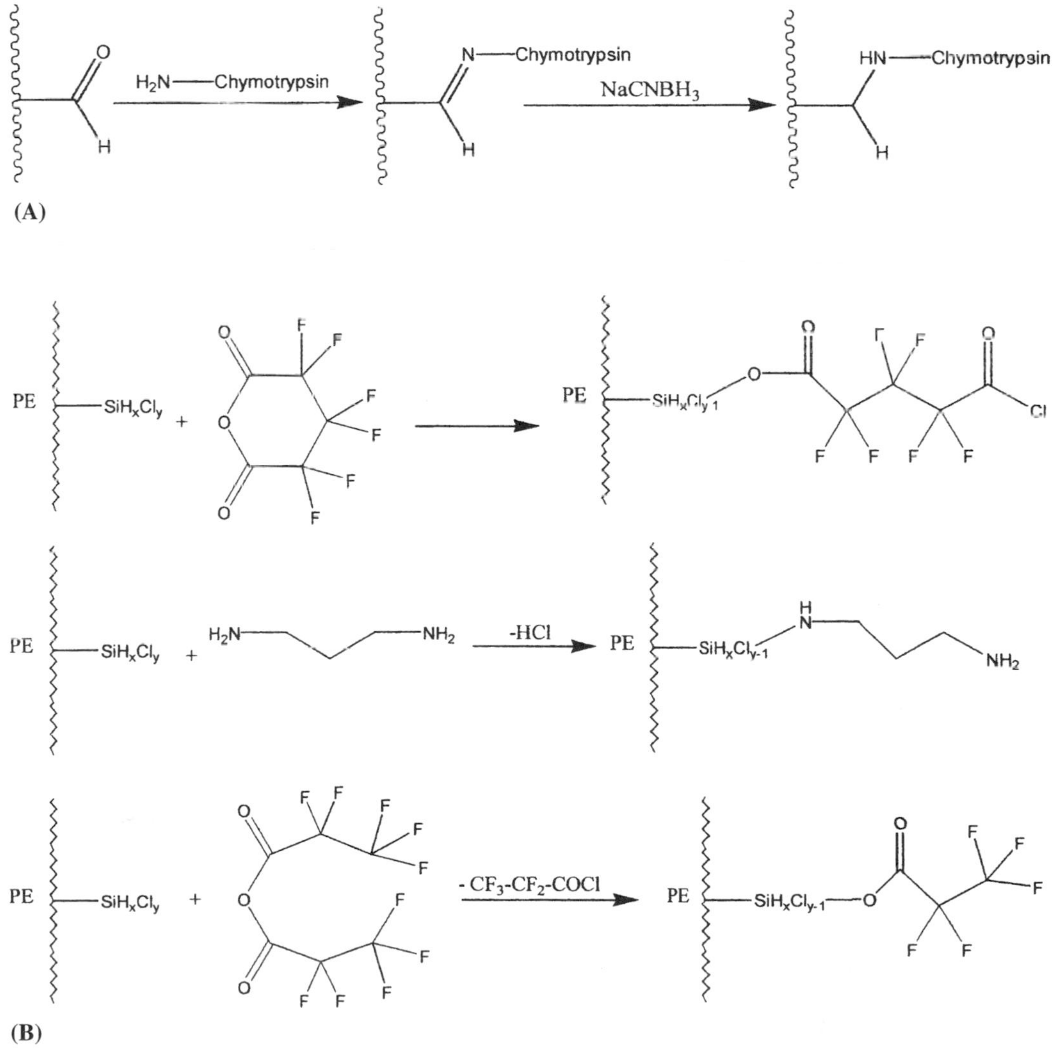

(A)

(B)

Scheme 11.1 (A) Direct immobilization of enzyme from oxygen plasma–functionalized PE (aldehyde and ketone groups). (B) Immobilization of enzyme involving SiH_xCl_y functionalities, from SiH_2Cl_2 RF plasma–functionalized PE. (C) Immobilization of enzyme involving spacer molecules from SiH_2Cl_2 RF plasma–functionalized PE.

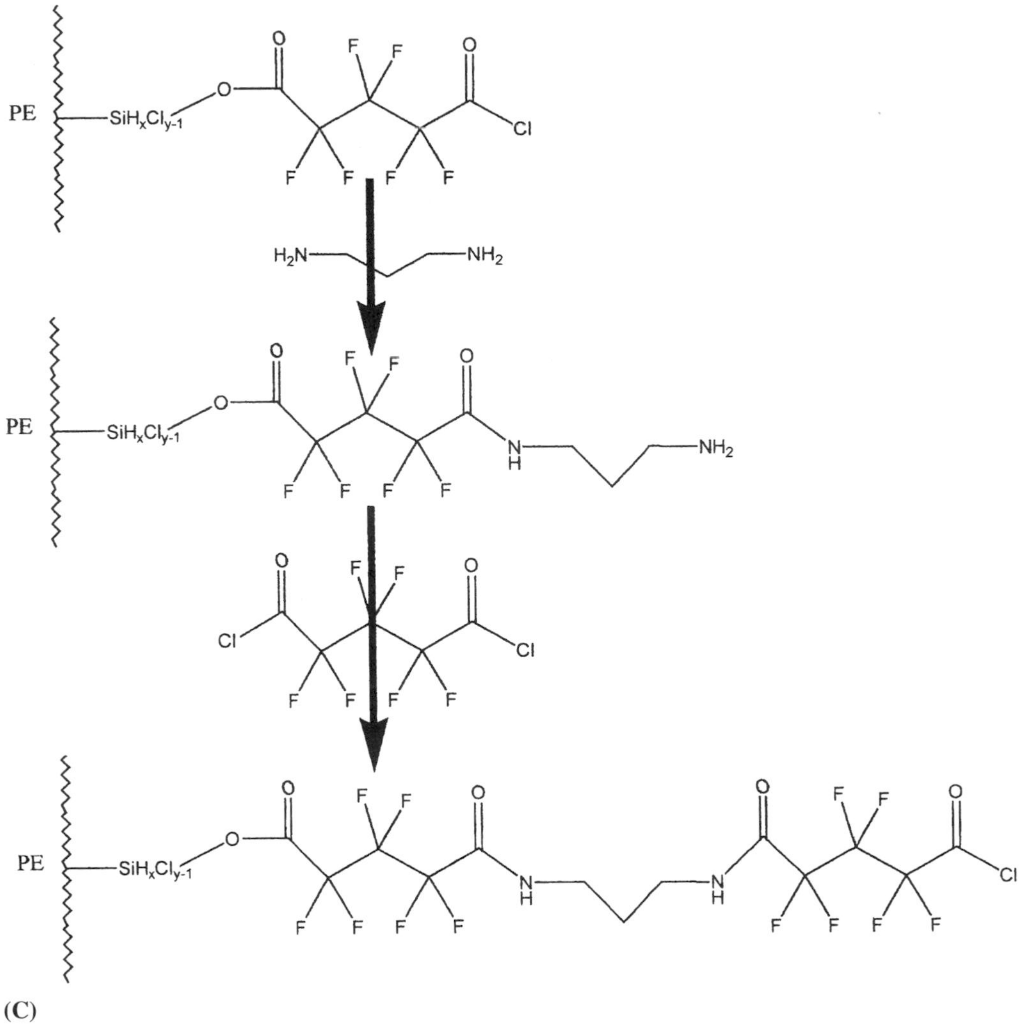

(C)

Scheme 11.1 Continued

The presence of the enzyme on the polymeric substrate surfaces and the covalent nature of the anchorage was demonstrated by x-ray photoelectron spectroscopy, attenuated total reflectance, Fourier transform IR spectroscopy, atomic force microscopy, and laser desorption Fourier transform ion cyclotron resonance mass spectroscopy techniques. It was shown that the tacticity of the polystyrene substrates do not influence the activity of the immobilized α-chymotrypsin; however, spacer molecules intercalated between the synthetic polymeric substrates and enzyme significantly increase the enzyme activity (Fig. 8) (99,100). It was suggested based on computer-aided conformational modeling of the substrate–spacer system (Fig. 9) that the enhanced enzyme activity might be related to the increased freedom of mobility of α-chymotrypsin as a result of the intercalation of spacer-chain molecules. The stability of

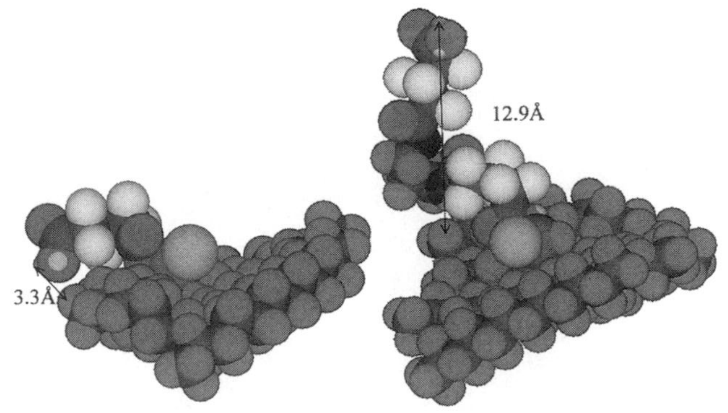

Figure 11.9 Space-filling models of the conformations of HFGA and HFGA-PD-HFGCl spacers attached onto polyethylene surface.

the immobilized enzyme was found to be fairly good in all cases; it retained most of its activity after several washing/assay cycles.

Cellophane (CE), a natural polymeric material which swells in water-based environments, has also been tested as a substrate for the plasma-enhanced immobilization of α-chymotrypsin (101). Results from 40 kHz and 13.56 MHz CW and pulsed hydrazine plasma treatments of CE revealed that the 40-KHz plasma environments generate significantly rougher cellophane surfaces in comparison to the pulsed 13.56-MHz plasma conditions (Figs. 10 and 11). It also has been shown that the longer the plasma treatment time (e.g., the longer the plasma-on interval of the duty cycle), the higher the surface roughness of the plasma-exposed CE, regardless of the low or high driving frequency conditions. Accordingly, most of the enzyme immobilization reactions were carried out using cellophane substrates exhibiting the lowest surface roughness in order to avoid the influence of the plasma-created cavity structures on the physical retention of the enzyme on the CE surface (Tables 6 and 7). Enzyme assay data indicate that the activities of the immobilized enzyme from CE—CO—$(CF_2)_3$— CO—OH and CE—[CO—CO—NH—$(CH_2)_2$—NH]$_n$ —CO—CO—Cl (where n = 1, 2, and 4) spacer-chain segments and that of the enzyme directly immobilized to the CE substrate from =C=O groups are comparable with the activity of the free enzyme (Fig. 12). It was also found that the presence of CE in the free enzyme solution reduces significantly the activity of the enzyme, and that the spacer molecular chains intercalated between the substrate and the enzyme do not enhance the activity of the immobilized enzyme in comparison to

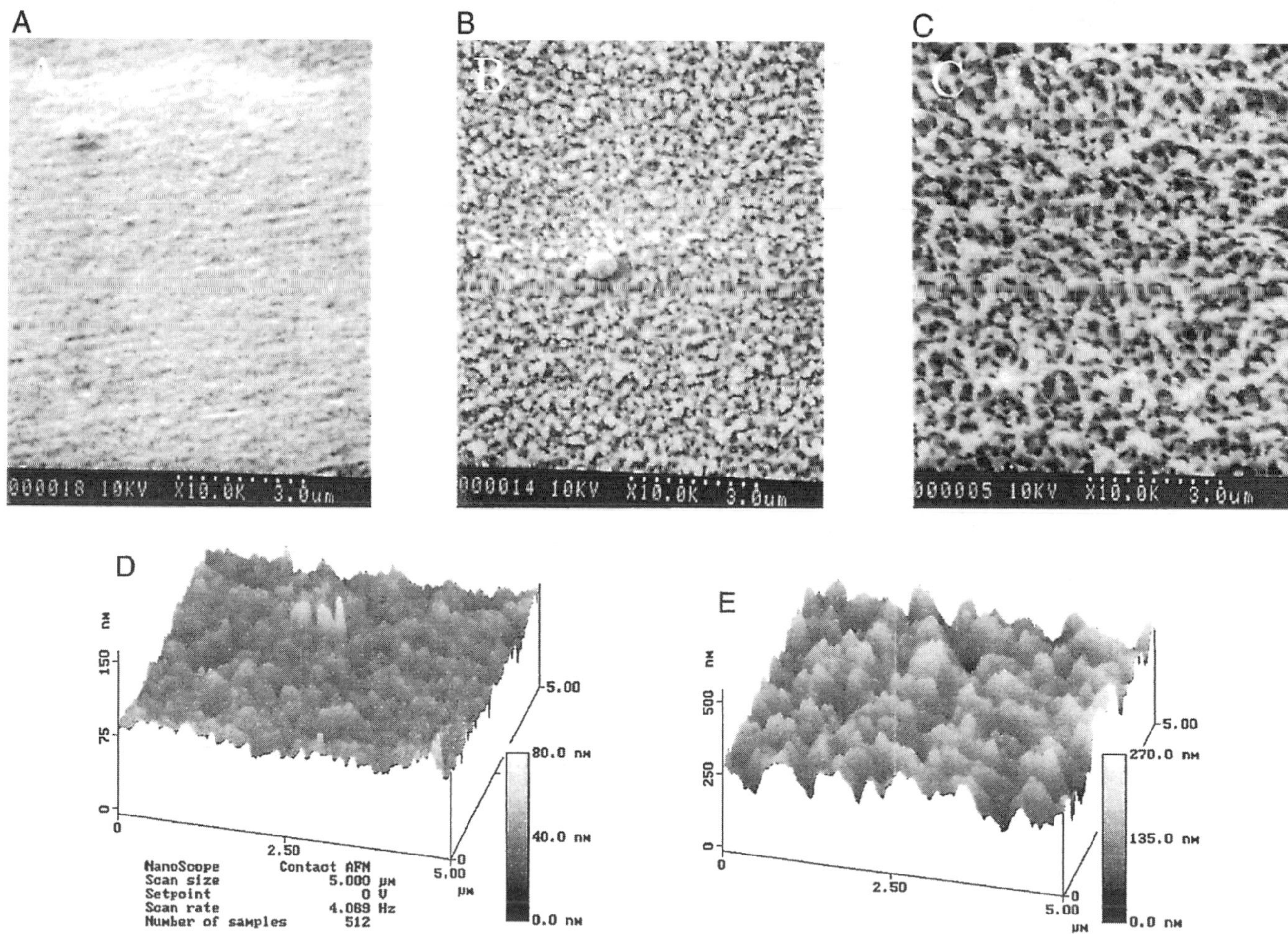

Figure 11.10 Surface topography of the untreated (A—SEM; D—AFM) and 40-kHz HY plasma–treated cellophane for the lowest (B—SEM; E—AFM) experimental conditions (power: 100 W; treatment time: 15 min) and the highest (C—SEM) experimental conditions (power: 150 W; treatment time: 25 min) crystallinity index samples.

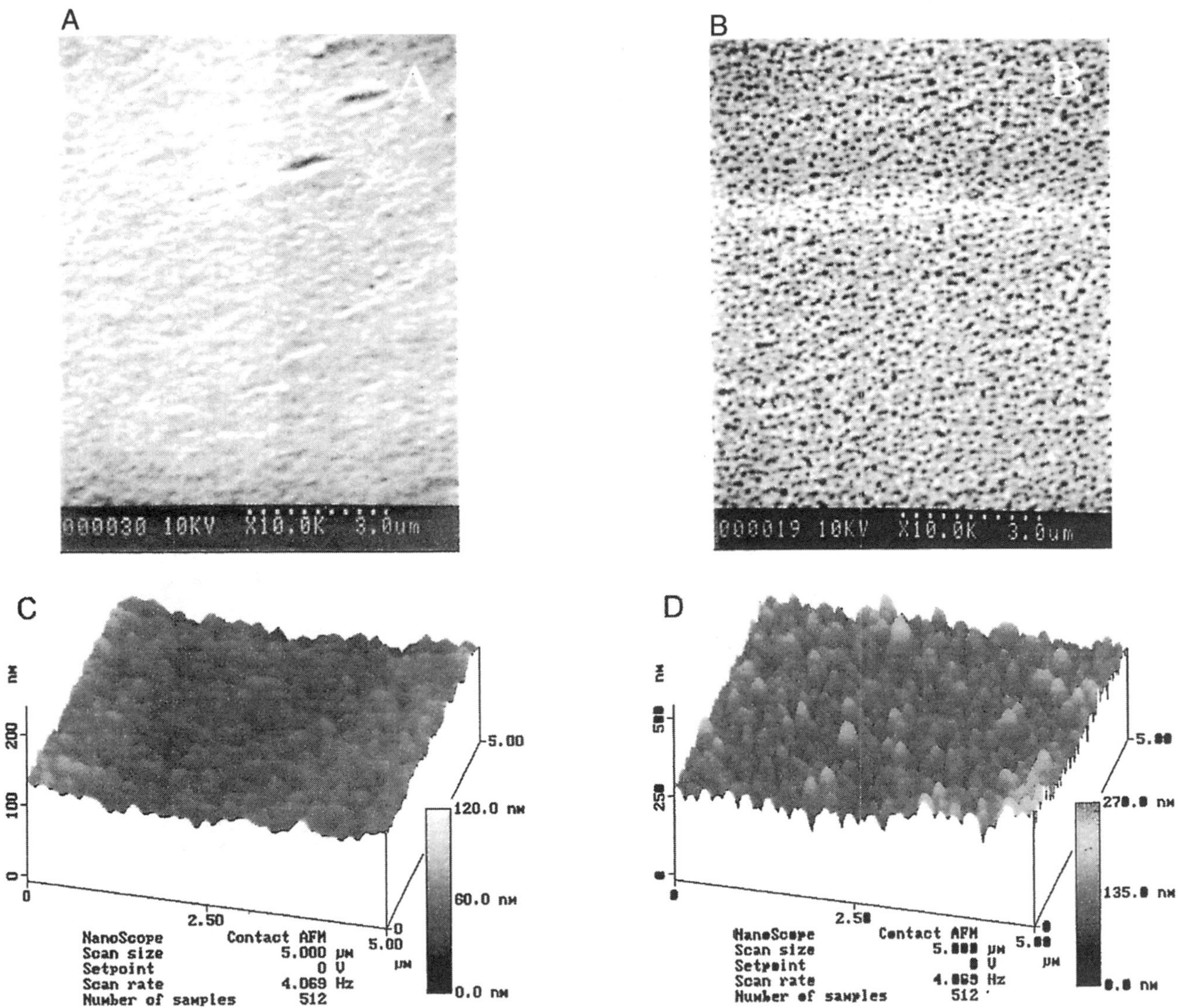

Figure 11.11 Surface topography of the 13.56-MHz HY plasma–treated cellophane surfaces for the lowest (A—SEM; C—AFM) experimental conditions (power: 140 W; treatment time: 22 min; pulse period: 700 µs; duty cycle: 35%) and the highest (B—SEM; D—AFM) experimental conditions (power: 140 W; treatment time: 22 min; pulse period: 300 µs; duty cycle: 65%) crystallinity index samples.

Table 11.6 Activity of Cellophane-Absorbed AC after Successive Washing

No.	Substrate	pH after 5 min	pH after 20 min	Ultrasonic washed	pH after 5 min	pH after 20 min	Ultrasonic washed	pH after 5 min	pH after 20 min
1.	Cellophane	6.3	5.3	None	5.6	5.6	None	5.7	5.6
2.	Cellophane	6.2	5.3	Yes	5.9	5.8	Yes	5.2	5.1

Table 11.7 Activity of AC Covalently Linked to Cellophane

No.	Sample	Plasma conditions: Gas	Plasma conditions: Power supply	Spacer	Enzyme	Activity pH after: 5 min	Activity pH after: 20 min
1.	Cellophane	—	—	—	—	7.1	7.0
2.	Free enzyme	—	—	—	Yes	4.5	4.1
3.	Free enzyme/ cellophane	—	—	—	Yes	4.9	4.8
4.	Cellophane	HY	140 W, 22 min, pulsing 700 μs, 35% duty	—	Yes	6.6	6.1
5.	Cellophane	HY	140 W, 22 min, pulsing 700 μs, 35% duty	HFGA	Yes	6.4	6.0
6.	Cellophane	HY	140 W, 22 min, pulsing 300 μs, 65% duty	OC	Yes	5.5	5.1
7.	Cellophane	HY	140 W, 22 min, pulsing 700 μs, 35% duty	OC	Yes	5.6	5.2
8.	Cellophane	HY	140 W, 22 min, pulsing 700 μs, 35% duty	OC—$(EDA\text{-}OC)_2$	Yes	5.1	4.7
9.	Cellophane	HY	140 W, 22 min, pulsing 700 μs, 35% duty	OC—$(EDA\text{-}OC)_4$	Yes	5.2	4.7
10.	Cellophane	O_2	140 W, 11 min, pulsing 700 μs, 35% duty	—	Yes	5.0	4.5
11.	Cellophane	O_2	140 W, 11 min, pulsing 300 μs, 65% duty	—	Yes	5.0	5.0

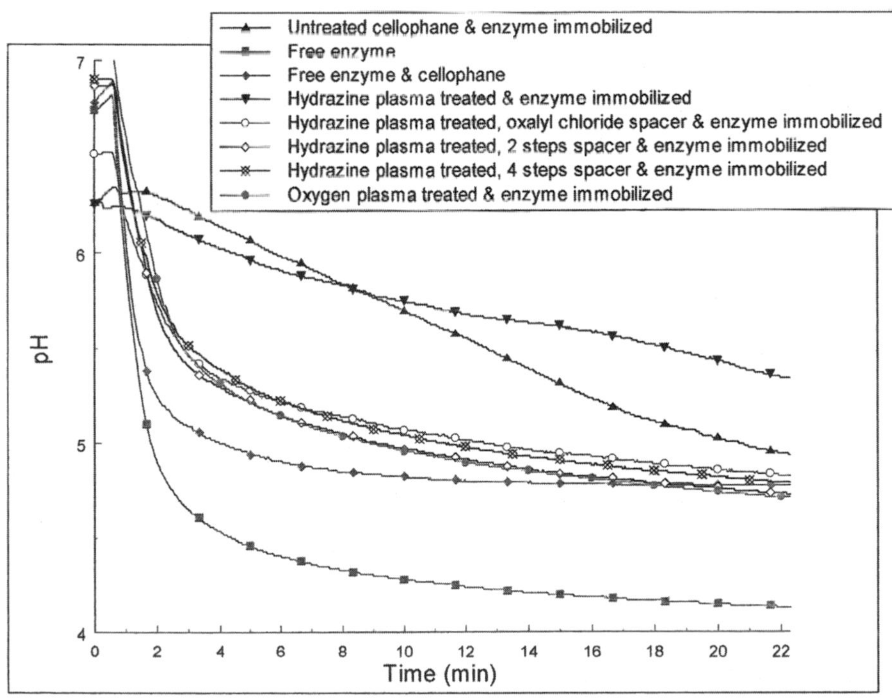

Figure 11.12 Activities of the free enzyme and the enzyme immobilized on cellophane samples expressed as time-dependent pH changes.

the enzyme directly, covalently attached in the absence of spacer molecules. It is suggested that the swollen state of the CE plays a significant role in the deactivation both of the free and immobilized enzyme molecules as a result of the entrapment of the enzyme molecules into the three-dimensional complex CE matrix. The incorporation of enzyme molecules into the swollen environment might result in reduced freedom of mobility and might also alter the supramolecular morphology of the enzyme through the development of multipoint connections (Scheme 2).

Papain has also been immobilized on RF plasma–functionalized polyethylene and glass substrate surfaces in the presence and absence of spacer molecules (Scheme 3) (102).

It has been shown that the activities of the papain, immobilized involving dichlorosilane (DS), propylenediamine (PD), oxallyl chloride (OC), and DS–PD–hexafluoroglutaric anhydride (HFGA)–based one- and multistep spacer chains, are comparable with the activity of the free enzyme, and that the activity of papain anchored directly to the oxygen plasma–functionalized PE substrate surfaces is significantly lower (Fig. 13). However, the papain covalently anchored to the PE surface by involving a longer spacer chain (PE—DS—[PD—OC]$_3$—PD—OC) has considerably lower activity (pH after 25 min: 5.7) in comparison to that of two-step spacer chain–linked papain (PE—DS—PD—OC—PD—OC) (pH after 25 min: 4.6) (Fig. 14). Computer-aided conformational modeling of the two- and four-step spacer chains revealed that the most probable conformations for two- and four-step systems render distances between the terminal active sites (—COCl) and the PE substrate surface of 5.3 and 3.1 Å, respectively (Fig. 15). Accordingly, it is suggested that the distance of the immobilized enzyme relative to the surface of the substrate might influence significantly the activity of the enzyme.

The influence of the concentration of N-α-benzoyl L-arginine ethyl ester (BAEE) substrate on the activities of free and immobilized enzyme were also evaluated. It was found that at higher BAEE concentrations (12 mM) the activities of immobilized enzymes are comparable with that of the free enzyme, and that at lower concentration values of BAEE the activity of the directly connected papain to the PE is significantly diminished relative to the free and spacer chain–anchored enzyme molecules. It is suggested that the modification of the conformational shape of papain as a result of the close vicinity to the PE substrate might be responsible for this phenomenon.

The Lineweaver–Burk diagrams plotted for the free papain and for the enzyme anchored directly and by involving spacer chains (Fig. 16; Table 8) indicate that the Michaelis constant, K_m, of spacer-immobilized papain are

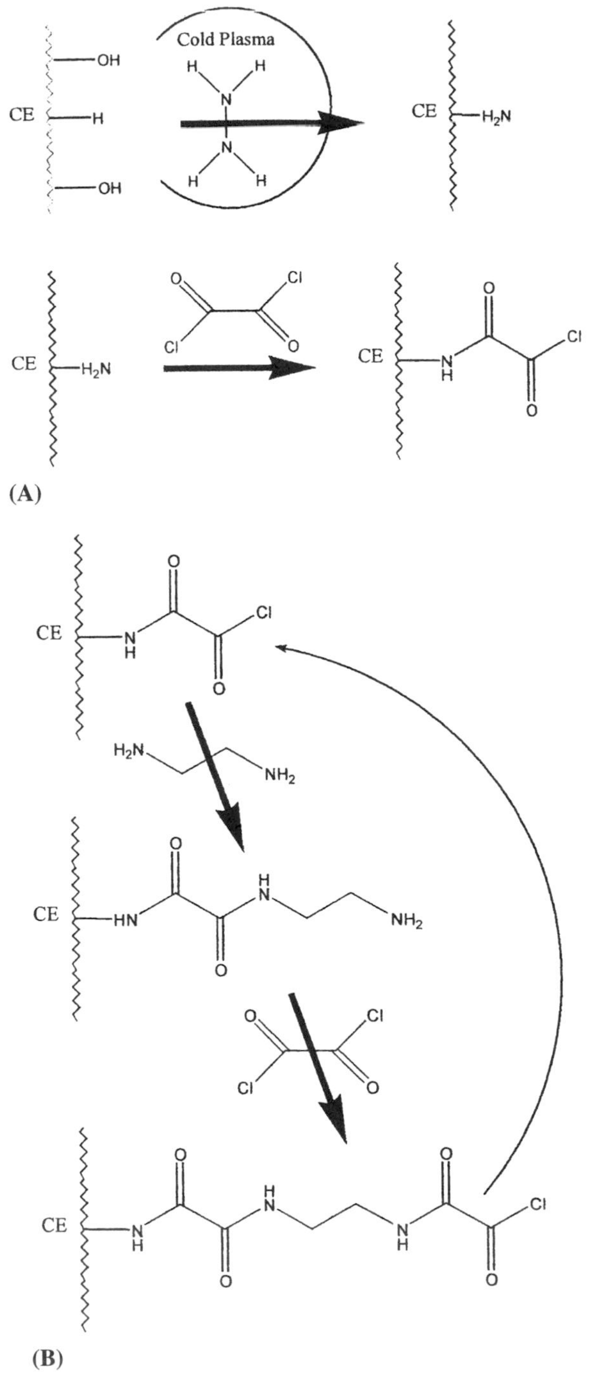

CE —OH —H —OH Cold Plasma H H N N H H CE —H_2N

CE —H_2N O Cl Cl O CE N H O Cl O

(A)

CE N H O Cl O H_2N NH_2 CE HN O O H N NH_2 O Cl Cl O CE N H O O H N N H O Cl O

(B)

Scheme 11.2 (A) Immobilization of enzyme involving NH—NH_2 and NH_2 functionalities from N_2H_4 RF plasma–functionalized PE. (B) Immobilization of enzyme involving spacer molecules from N_2H_4 RF plasma–functionalized PE.

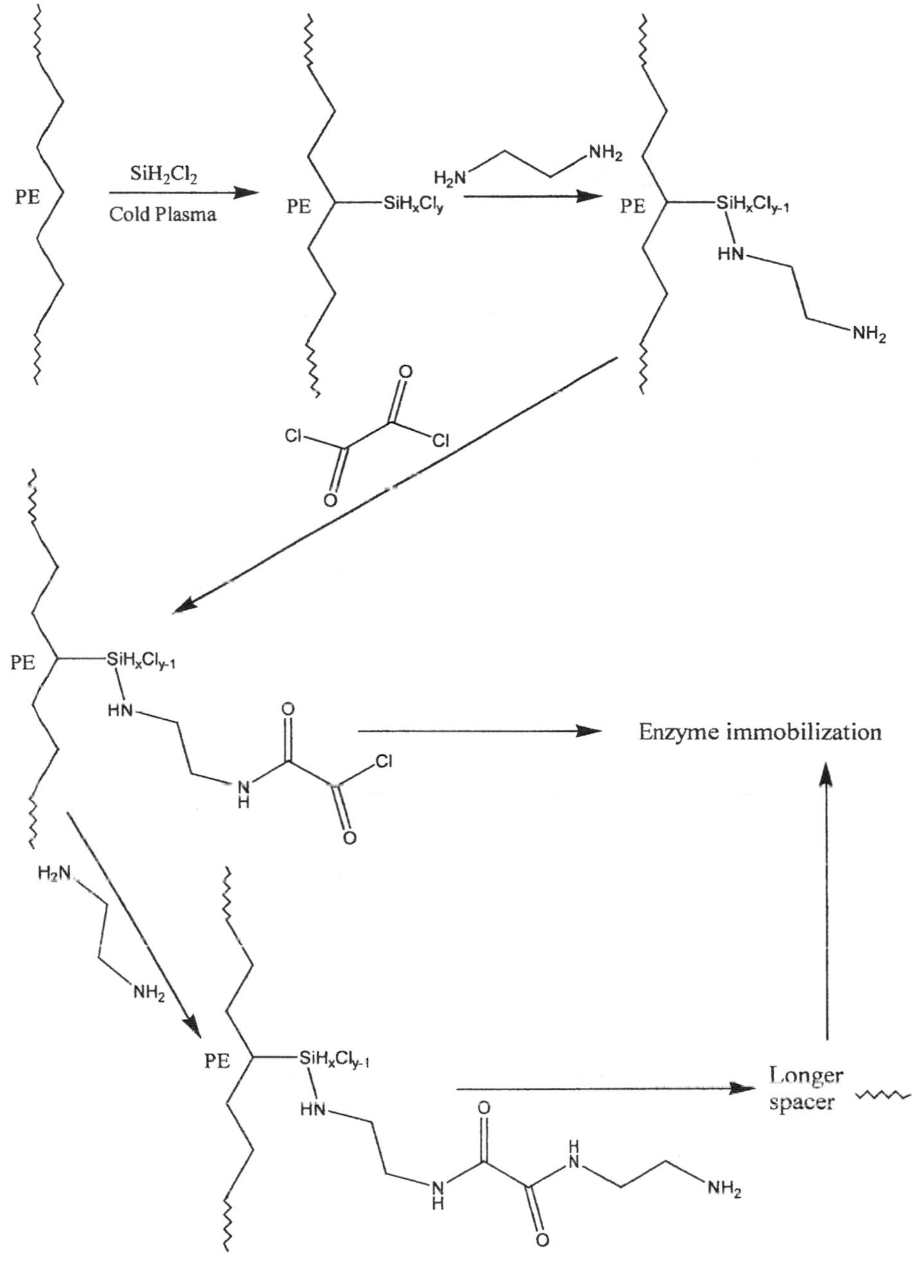

Scheme 11.3 Immobilization of papain involving RF plasma–functionalized polyethylene (PE).

higher in comparison to the K_m values of free and directly anchored enzyme. On the other hand, the maximum reaction velocity (V_m) value of spacer chain (one-step)–immobilized papain is comparable to that of the free enzyme, while the V_m value of the directly connected papain is substantially lower. It is suggested that the activity of the directly anchored papain to the PE was diminished significantly during the close-vicinity coupling.

Trypsyn (from bovine pancreas type III) was also covalently immobilized onto plasma-functionalized polytetrafluoroethylene substrate surfaces (103). The surface modification of PTFE was carried out in a two-step reaction. Soxhlet, methanol extracted, thin PTFE films (0.01 cm) were first pretreated under argon plasma conditions (bell jar type reactor; gap between the parallel plate electrodes: 8.0 cm; rotating substrate holder located between the elec-

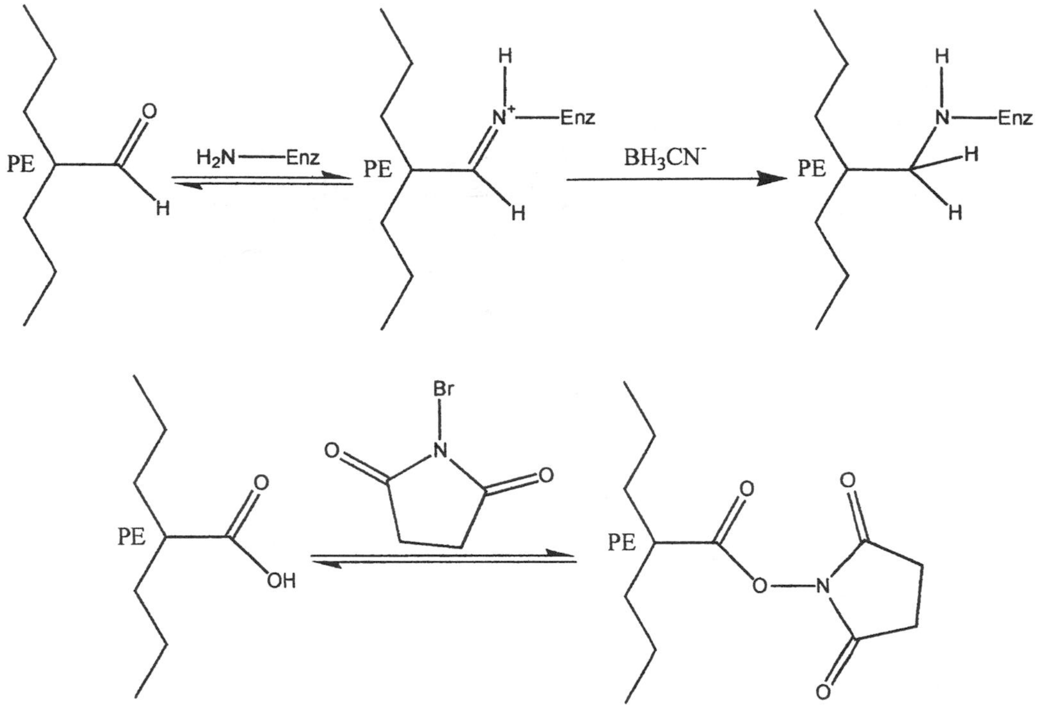

Scheme 11.3 Continued

trodes; 5 kHz; 28 W; 0.04 Torr) followed by the near-UV radiation–mediated graft polymerization reactions of acrylic acid (AAc), sodium salt of styrensulfonic acid (NaSS), and N,N-dimethacrylamide (DMAA). The AAc graft polymer films were further functionalized via covalent immobilization of trypsin. It has been shown that a stratified thin-layer microstructure is generated on the PTFE substrate surfaces and that the graft yields increased with the argon plasma pretreatment time and monomer concentration (Fig. 17) during the graft polymerization reactions. Acrylic acid and NaSS polymers grafted with PTFE substrates coated with thin polyaniline (emeraldine) layers resulted in semiconducting surfaces as a result of charge transfer interactions. The immobilized trypsin retained around 30% of its original activity, and the effective enzyme activity was dependent on the surface concentration of the grafted AAc polymer. The enzyme activities increased initially with increasing AAc graft polymer concentrations, then became saturated at higher AAc polymer concentration values.

B. Plasma-Assisted Development of Biosensors

The development of reliable and sensitive glucose biosensors will open up novel ways for the continuous in vivo monitoring of this clinically relevant compound. It is not surprising, then, that the immobilization of glucose oxidase (GOD) has become an increasingly active research area in recent years. Approaches for the achievements of efficient immobilization of GOD based on cold plasma techniques offer additional possibilities in this area.

Glucose oxidase was immobilized onto PE substrates functionalized according to a multistep reaction mechanism, involving argon plasma treatment followed by exposure to oxygen (for the generation of hydroperoxid groups) as a first step, and the activities of the immobilized enzymes were compared to that of the free GOD (Scheme 4) (104).

Glucose oxidase was anchored covalently to the modified PE surface in the absence and presence of poly(ethylene oxide) spacer molecules. It was demonstrated that the amount of immobilized GOD depends significantly on the initial GOD concentration in the low concentration range, and it remains constant at higher initial GOD concentration values (Fig. 18). Glucose oxidase immobilized onto polymeric films in the absence and presence of PEO spacer chains obeys the Michaelis kinetics. The Michaelis constant, K_m, was found to be larger for the immobilized GOD than for the free enzyme, while the V_m value was smaller for the immobilized enzyme (Table 9). The bioactivity of PEO-modified PAAc-grafted PE (PAAc-PEO-GOD) was also higher than that of PAAc-grafted PE (PAAc-GOD).

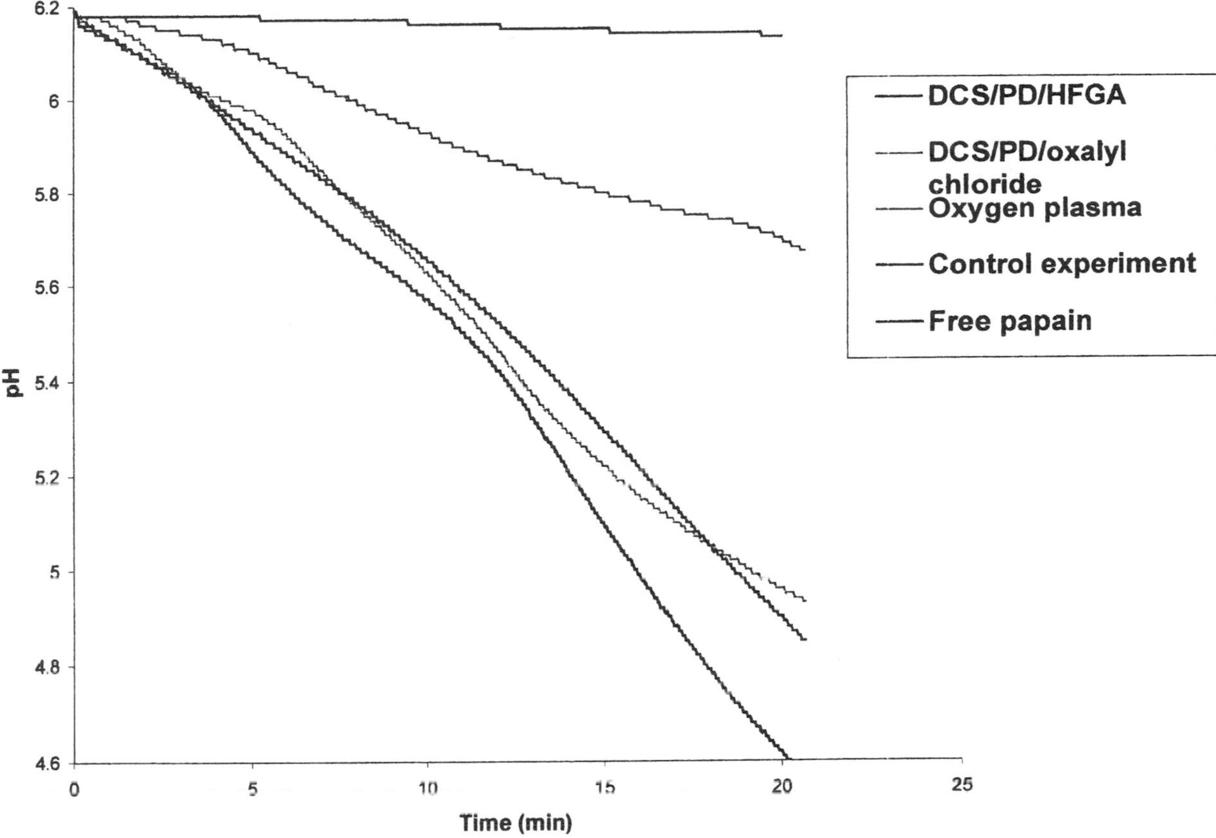

Figure 11.13 pH diagrams of enzyme assay involving free papain, DCS-PD-HFGA spacer-attached, DCS-PD-OC spacer-attached, and directly attached papain to the polyethylene substrate.

Both the thermal and pH stabilities of the immobilized enzymes prepared in the absence of spacer PEO were higher in comparison to the same characteristics of the immobilized enzyme involving spacer molecules.

Glucose oxidase was also immobilized onto RF plasma–modified poly(etherurethane urea) (PEUU), and the activities of the free and immobilized enzymes were evaluated by measuring the amount of hydrogen peroxide produced in the GOD-catalyzed redox reaction developed between the glucose and oxygen using cyclic voltametry and a specially designed sandwich-type thin layer electrochemical cell (105). Approximately 100-nm thick films of plasma-polymerized N-vinyl-2-pyrrolidone (PPNVP) were deposited onto PEUU films by RF glow discharge to PPNVP/PEUU layers. Surface hydroxyl groups were generated by the reduction of carbonyl functionalities present in the PPNVP layers with aqueous sodium borohydride, then the films were activated by cyanotransfer coupling using 1-cyano-4-dimethyl-aminopyridinium tetrafluoroborate. The first and second cycles of a typical cyclic voltammogram of a GOD-PPNVP/PEUU thin layer cell under air is presented in Fig. 19. It was concluded that the incorporation of GOD, immobilized onto a thin film of RF plasma–modified PEUU into the thin layer chamber allowed for the detection of immobilized enzyme activity at room temperature.

A glucose sensor was fabricated using semiconductor dry technology and plasma-deposited macromolecular layers, originating from ethylendiamine RF discharge (106). The sensor was prepared by the successive deposition of an intermediate adhesive dexadimethyldisiloxane (HMDSO) and a functional (amino group–bearing) ethylenediamine plasma layer on glass substrates, followed by patterned platinum sputtering. The plasma-enhanced depositions were carried out in a Pyrex glass, tubular reactor provided with an external inductive copper coil under the following experimental conditions: frequency of the driving field: 10 MHz; base pressure: 7×10^{-3} Pa; RF power: 40 W; pressure of the plasma vapors: 4.6 Pa; deposition time: 1 min. Immobilization of GOD was achieved by applying 2.5%

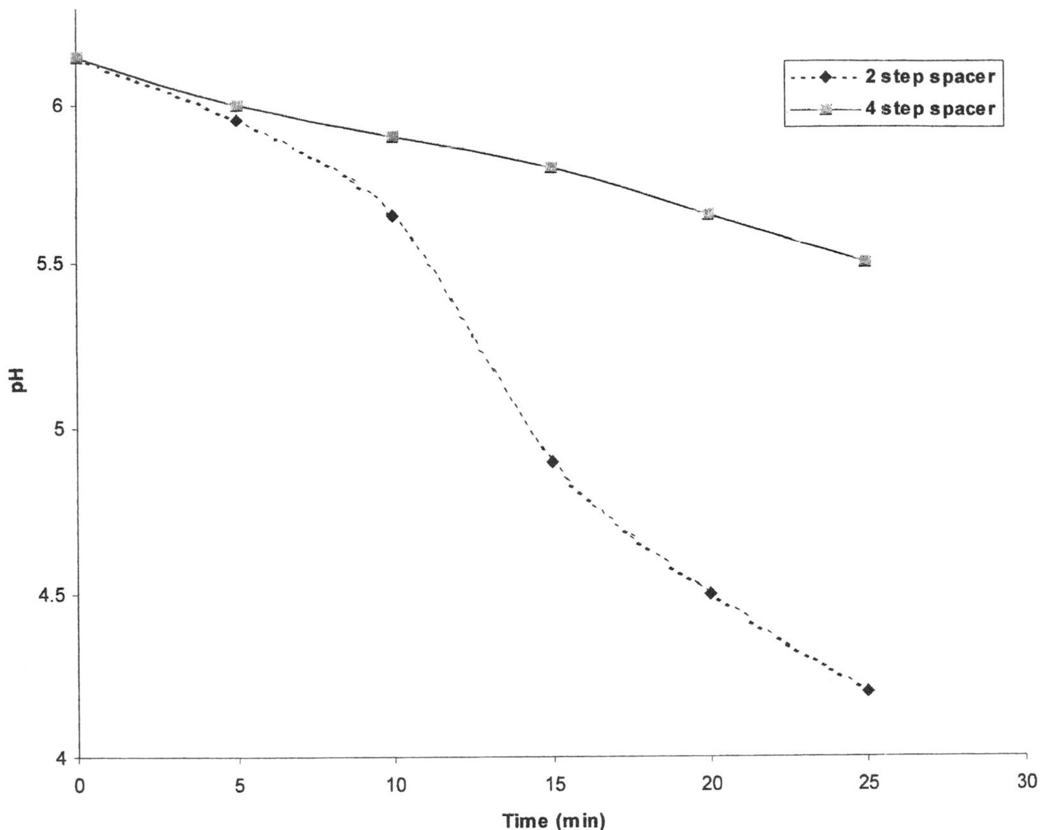

Figure 11.14 pH diagrams of enzyme assay involving two- and four-step spacer-attached papain onto the polyethylene substrate surfaces.

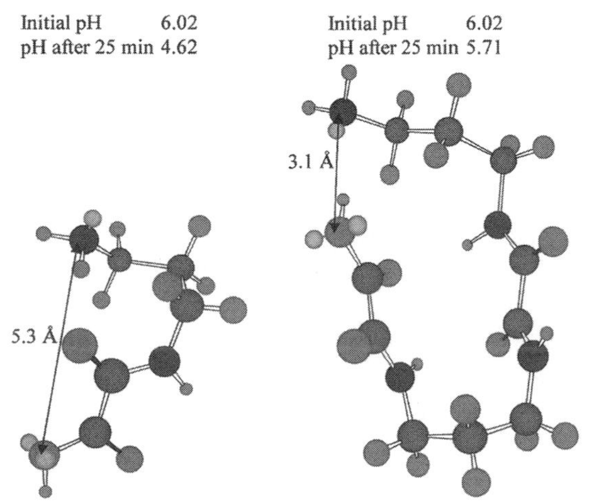

Figure 11.15 Influence of spacer chain length on the activity of the immobilized enzyme (lowest pH after 25 min, highest enzymatic activity). Molecular design for 2 and 4 units of alternating 1,3-propylene diamine and oxalyl chloride spacer chains are computed using Conformer and Chem3D.

aqueous solution of glutaraldehyde to the surface of plasma-deposited ethylenediamine-based film, followed by washing and by the exposure of the functionalized plasma film to a 0.1% GOD solution (phosphate buffer: pH 5.6). The evaluation of the activity of the immobilized enzyme was performed using amperometric technique. It was demonstrated that the biosensor has a short response time (12 s) and is effective for glucose concentration estimation in the range of 0 to 30 mM with a detection limit of 1 mM.

High-performance glucose biosensor was also developed using acetonitrile–RF plasma–deposited, cross-linked, thin macromolecular layers (107). The step-by-step schematic diagram of the biosensor fabrication is presented in Scheme 5.

Hexamethyldisiloxane (HMDSO) and acetonitrile (AN) RF plasma environments (*HMDSO plasma*—Pyrex glass bell jar type reactor; 13.56 MHz; inductive coupling; base pressure: 10^{-3} Pa; RF power: 100 W; flow rate HMDSO: 15 mL/min; HMDSO pressure: 4.6 Pa; deposition time: 1 min; resulting film thickness: 100 nm; *AN plasma*—base pressure: 10^{-3} Pa; RF power: 80 W; flow rate AN: 15 mL/

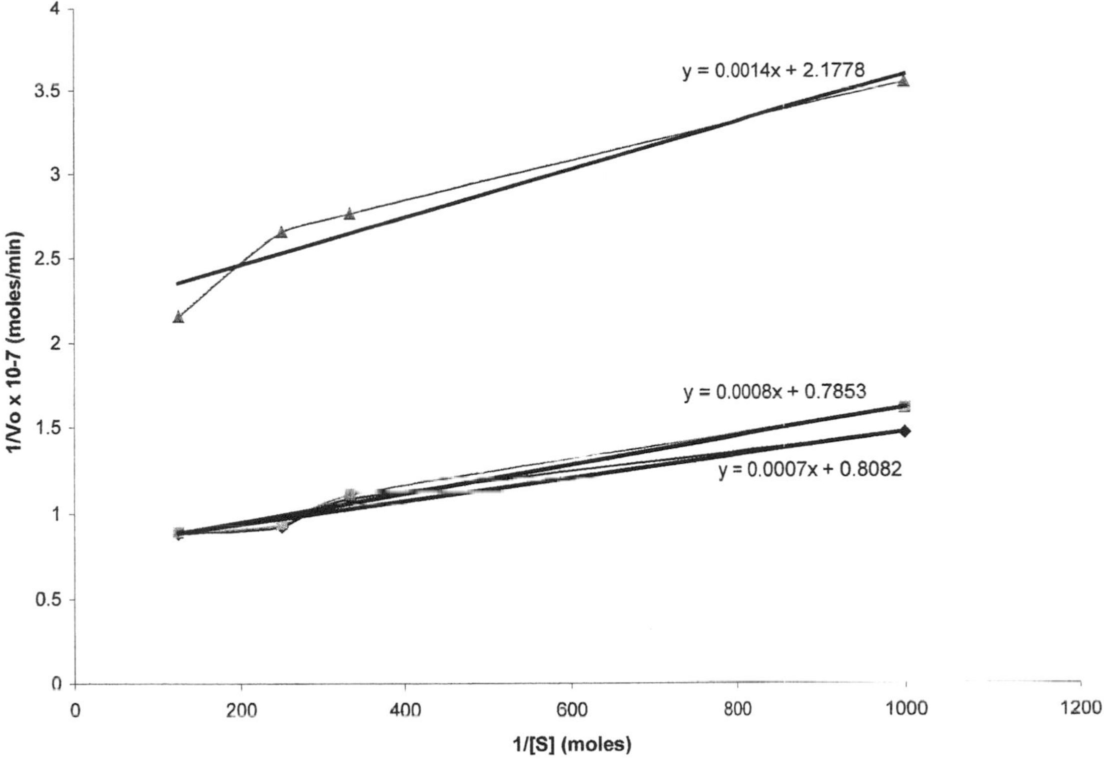

Figure 11.16 Lineweaver–Burk plot for (◆) free papain, (▲) direct, and (■) spacer-attached papain onto plasma-treated polyethylene.

min; AN pressure: 2.1 Pa; deposition time: 1 min; resulting film thickness: 118 nm) were employed both for the deposition of an intermediate layer between the glass substrate and the sputtered platinum electrode and for the generation of AN-based reactive surface. A second biosensor was also fabricated by substituting the AN plasma step (AN-based plasma deposition) for a conventional APTES step [functionalization via 3-(aminopropyl)triethoxysilane) (APTES) in the absence of plasma], and the performances of the two devices were compared. It is concluded based on steady-state amperometric responses (Fig. 20) that the sensors fabricated using AN plasma films are more reproducible, exhibit lower noise, and have a reduced effect of interference than the sensors made using conventional immobilization method (e.g., APTES). It is suggested that due to the highly crosslinked network structure of the plasma-deposited AN-based layers and their thin layer character, the amperometric response was significantly reduced.

Table 11.8 Michaelis Parameters K_m and V_m at pH 6.1 and 25°C

Sample	K_m(μM/g)	V_m(μM/min × g)
Free papin	12.3	0.00861
Immobilized papain on DCS plasma–treated PE and 1,3-DAP-OC–attached spacer	12.7	0.0101
Immobilized papain on oxygen plasma–treated PE	4.5	0.00642

Amperometric and colorimetric enzyme immunoassay biosensor was developed for the evaluation of concentration of insulin in serum and that of urinary human serum albumin (HAS) using as a molecular recognition element antibodies and enzymes immobilized on water vapor plasma–functionalized microporous, hydrophobic polypropylene films (108–110). The biaxially streched PP films were partially functionalized in a bell jar type reactor equipped with aluminum, parallel plate electodes (180 × 180 × 1 mm) with a gap between them of 180 mm. The PP film was sandwiched between two aluminum masks (180 × 180 × 4 mm), provided with 64 holes of 6 mm in diameter, and positioned symmetrically between the electrodes and exposed to the water vapor discharge (base pressure: 10^{-3} Torr; driving frequency: 5 kHz; power:42 W; treatment time: 5 min; pressure in the presence of discharge: 0.5 Torr) in order to generate the water-permeable 6-mm hydrophilic spots on the membrane. The hydrophilic spots were aldehyde groups functionalized by a conven-

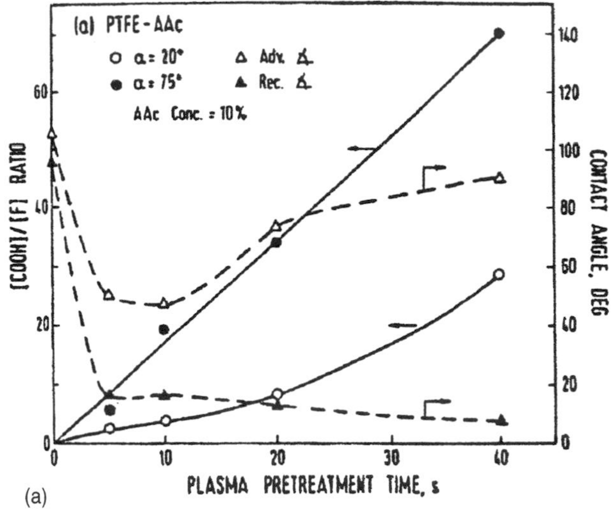

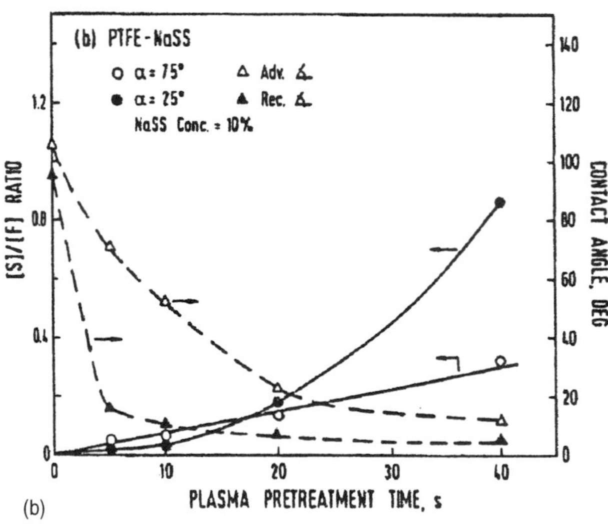

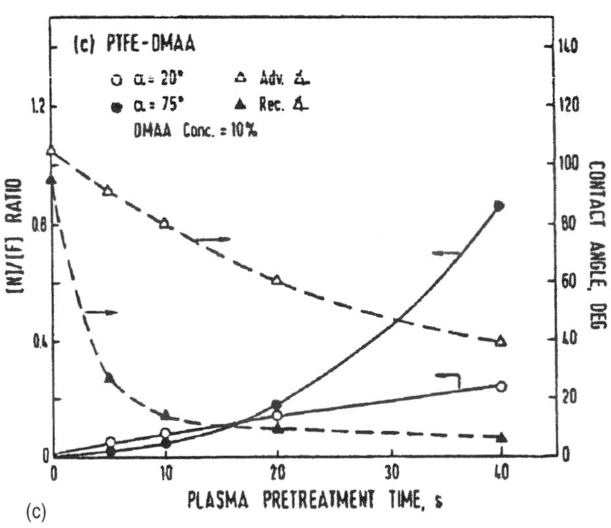

Figure 11.17 Effect of Ar plasma pretreatment time on the amount of surface-grafted (a) AAc, (b) styrenesulfonic acid, and (c) DMAA polymer on PTFE films (From Ref. 103.)

tional consecutive octamethylenediamine and glutaraldehyde treatment. The proteins were immobilized by dropping on each functionalized spot of the membrane 20 μL of antiserum, diluted 150-fold in 0.15 M phosphate buffered saline (PBS) (pH 7.2) and incubated at 4°C overnight. Then, 20 μL of 0.3% (w/v) BSA solution was applied at 25°C for 2 h. After each treatment with protein the film was washed with 0.15 M PBS (pH 7.2). Results from the immunoreactions performed on the antibody-immobilized membrane spots (a competitive reaction between GOD-HAS and HAS), amperometric determination of HAS, and colorimetric evaluation of HAS indicate that the measurable range of this system is between 0.5 to 100 mg/L HAS in urine, which is suitable for the prognosis of diabetic nephritis.

Urea biosensor was also assembled from ammonia plasma–treated PP membrane (111,112). One side of a porous Celgard PP (pore size: 0.04 μm; thickness: 25 μm) membrane was ammonia plasma treated in a bell jar type reactor (base pressure: 0.1 Torr; ammonia pressure: 0.2–0.9 Torr; driving field frequency: 13.56 MHz; RF power: 20–60 W; ammonia plasma exposure time: 1–6 min), and subsequently ex situ functionalized in the presence of a 3% glutaraldehyde aqueous solution (pH: 7.0; 20 h; 25°C). It is suggested that the ammonia plasma treatment resulted in the generation of nitrogen atom–based functionalities on the PP surface.

A urea sensor was assembled by substituting the gas-permeable membrane of an ammonia electrode for plasma-modified membrane containing the immobilized urease. It was demonstrated that there is a strong interdependence between the external plasma parameters (pressure, RF power, and reaction time) and the activity of the immobilized urease (Figs. 21–23). This might be explained by the retention of different enzyme densities on the PP membrane surfaces. It was shown that the urea sensor has response sensitivities ranging from 19 mV/decade to 30 mV/decade, depending on the plasma parameters employed, and it has a shorter response time in comparison to the corresponding conventional urea electrodes. It was emphasized that deamination of the plasma-modified PP membrane did not occur in aqueous solution, even after 12 days of operation.

Application of extremozymes, enzymes that are capable of surviving under harsh conditions, including low and high pH environments, extreme temperatures and pressure, organic solvent environments, etc., immobilized on ce-

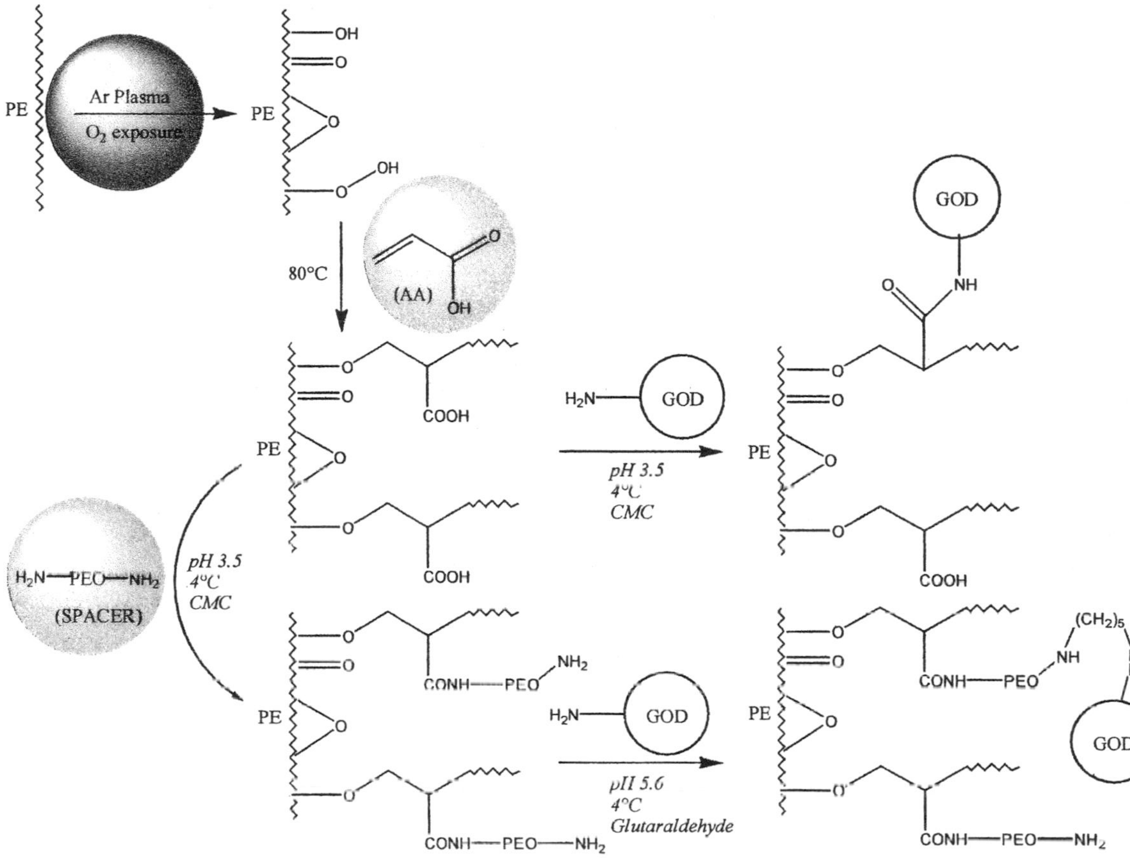

Scheme 11.4 Immobilization of glucose oxidase onto PE substrates functionalized according to a multistep reaction mechanism. (CMC—1-cyclohexyl-3-(2-morpholinoethyl) carbodiimide metho-p-toluene sulfonate.)

ramic substrate surfaces presents a special interest for bioreactor applications. Recently glucoamylase was immobilized on γ-aminopropyltriethoxysilane/water vapor corona discharge–modified ceramic membrane surfaces, and the activity of the covalently bound glucoamylase was evaluated (113). The schematic diagram of the corona discharge apparatus is presented in Fig. 24. The surface modification and enzyme immobilization reaction mechanisms are shown in Scheme 6.

The tubular alumina eramic membranes are placed between two rod type (exciting electrode) and coil copper (discharge electrode) electrodes, and the discharge is initiated and sustained in the vapor phase mixture, at atmospheric pressure at 9.5 kHz and 20 kV. It is emphasized that the corona discharge is a less aggressive technique for the fragmentation of gas phase molecules, and consequently a larger part of the structures of the starting components can be retained in the nascent plasma-deposited layers. It was demonstrated that the plasma-enhanced immobilization of glucoamylase is more efficient relative to the conventional technique, and that the corona-immobilized enzyme is active in a broader pH and temperature range in comparison to the free enzyme (Figs. 25–27). It also has been found that the operational stability of the immobilized enzyme increased with the number of plasma-enhanced surface modification processes.

The toxicity of organophosphorus derivatives (neurotoxins) and their use in agricultural and military applications stimulated the development of accurate detection methods for monitoring the concentrations of these neurotoxins. One of the promising approaches to this subject is to use organophosphorus hydrolase (OPH) for the development of efficient biosensors (114,115). Organophosphorus hydrolase is often used as an alternative to acethylcholinesterase, which requires inhibition-mode sensor operation, long sample incubation time, and a constant source of acethylcholine substrate. Organophosphorus hydrolase hydrolyzes a range of organophosphate esters (e.g., pesticides, parathion, soman, sarin, etc), and as a result protons are generated that can be measured and correlated to the

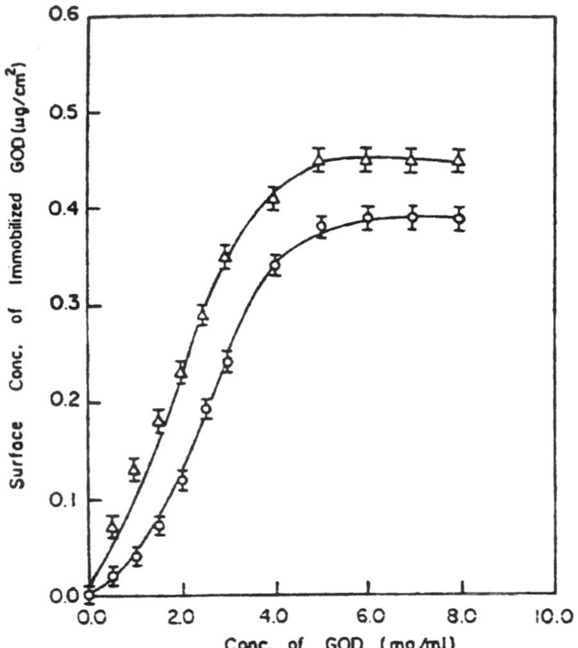

Figure 11.18 Effect of the GOD concentration on the amount of GOD immobilized onto (△) PAAc-grafted PE and (○) PEO spacer-modified PAAc-grafted PE at 4°C and 16 h. (From Ref. 104.)

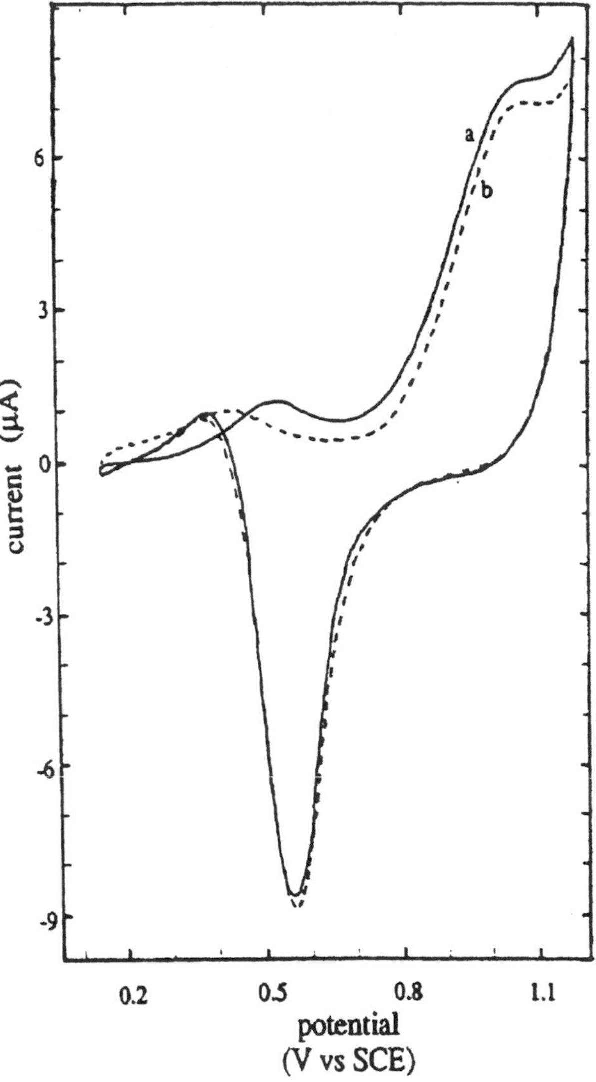

Figure 11.19 Cyclic voltammogram of a GO_x-PPNVP/PEUU thin-layer cell in 0.2 M sodium phosphate/5 mM β-D(+)-glucose (pH 5.2) under air. Sweep rate: 1.0 mV/s. (a) First cycle; (b) second cycle. (From Ref. 105.)

concentrations of the organophosphorus derivatives (OPDs). The hydrolysis of some of the OPDs produce specific chromophoric sites, which allow spectroscopic concentration evaluation of OPDs. Potential efficient plasma-enhanced immobilization routes of OPH can be envisaged for the fabrication of biosensors for the detection and monitoring of the concentration of OPDs.

Immobilization of tailored-structure synthetic, active biopolymers (e.g., synthetic peptides, polysaccharides) might play a significant role in the future for the study and the development of high quality biosensors. Recently it was suggested that ''side effects'' associated with the immobilization and use of natural biopolymers during preliminary studies could be avoided. For example, immobilized proteins may partly or entirely lose their activities through denaturation mechanisms, extracellular enzymes may rapidly break down a surface-anchored protein layer, and biological coating may generate safety hazards during in vivo testing. Current measures for diminishing these phenomena can be costly and time consuming; however, tailored-structure active synthetic biopolymers could be used instead (116). Accordingly, peptoid-containing [H—(Gly—Nleu—Pro)$_9$—NH_2; Ac—(Gly—Nleu—Pro)$_9$—NH_2; H—(Gly—Pro—Nleu)$_9$—NH_2; H—(Gly—Pro—Nleu)$_{10}$—Gly—Pro—NH_2, and H— (Gly—Nleu—Pro)$_{10}$—NH_2] collagenelike molecules (CLMs) were synthesized and immobilized onto cold plasma–functionalized (coated sur-

Table 11.9 Michaelis Parameters K_m and V_m at pH 5.6 and 30°C

Sample	K_m(mM)	V_m(mM/min)
GOD	23.58	5.62
PAAc-GOD	31.44	4.17
PAAc-PEO-GOD	27.33	4.75

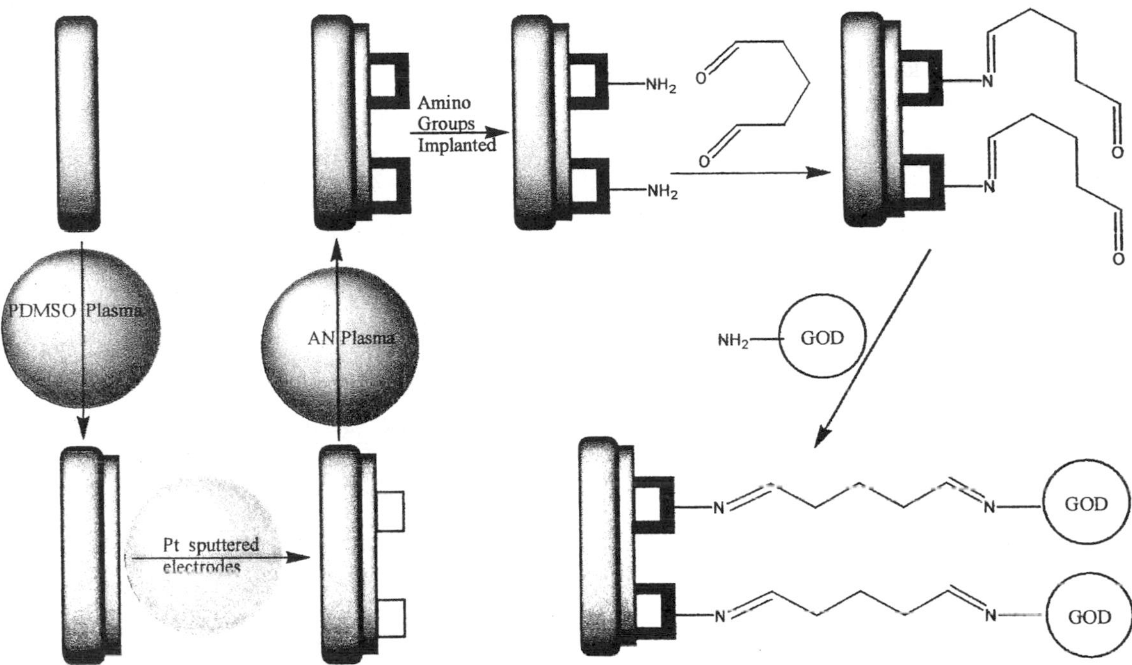

Scheme 11.5 Fabrication of high-performance glucose biosensor.

faces) perfluorinated ethylene–propylene (FEP) surfaces, and the biological performances of the modified surfaces were evaluated. For the generation of aldehyde and amine groups containing macromolecular layers on the FEP substrates, acetaldehyde (A), acrolein (Al), ethylbutyraldehyde (EBA), and n-heptylamine (HA) plasma deposition reactions were performed, respectively. All plasma exposures were carried out in a custom-built, cylindrical, glass plasma reactor provided with a substrate holder, circular grounded electrode, and a U-shaped powered electrode under the following plasma parameter conditions: driving-field frequency: 125 kHz for aldehyde discharges and 200 kHz for HA plasma; base pressure: 0.02 Torr; AA, Al, and EBA pressure: 0.3 Torr; HA pressure: 0.13 Torr; RF power for AA, Al, and EBA: 5 W for 1 min; RF power for HA: 20 W for 20 s. The immobilization of synthetic enzymes from aldehyde group–functionalized surfaces were carried out using a conventional sodium cyanoboronhydride chemistry, while the covalent anchorage of the enzymes from amine group–functionalized surfaces was achieved by a two-step process involving polycarboxylic acid derivatives (32-carboxy-terminated PAMAM starburst dendrimer and custom-synthesized carboxymethyldextran)/carbodiimide mechanism.

Based on XPS, MALDI-MS, and autoclaving data resulting from the substrates bearing the immobilized enzymes, it was shown that not all the surface-bound CLM molecules are attached through covalent bonding; some of them are only retained by the physical forces within the triple helical peptide assemblies. Initial cell attachment and growth assays indicate that the biological performance of the CLMs is related to specific amino acid sequences.

It was recently found that the surface protein binding interactions strongly influence the efficiency of protein ionization by MALDI technique (117). Poly(vinyliden fluoride) and poly(ethylene terephthalate) substrate surfaces were modified under pulsed allylamine–RF plasma environments for the generation of primary amine rich functionality surfaces and consecutively exposed to radiolabelled peptide (^{125}I-radiolabelled peptides including angiotensin I and porcine insulin) adsorption procedure. The peptide retention processes were carried out in a flow-through cell to avoid denaturation of the peptides at the air–water interface and to avoid formation of Langmuir–Blodgett film formation. The experimental results demonstrated that for the sample preparation method employed (α-cyano-4-hydroxycinnamic acid in methanol and 10% trifluoroacetic acid in water-based MALDI matrix), increases in the surface peptide binding affinity leads to decreases in the peptide MALDI ion signals. This phenomenon should be considered during MALDI evaluations.

Good antithrombocity of blood-contacting devices is a crucial characteristic for the development of advanced biomedical applications. To achieve these requirements, hepa-

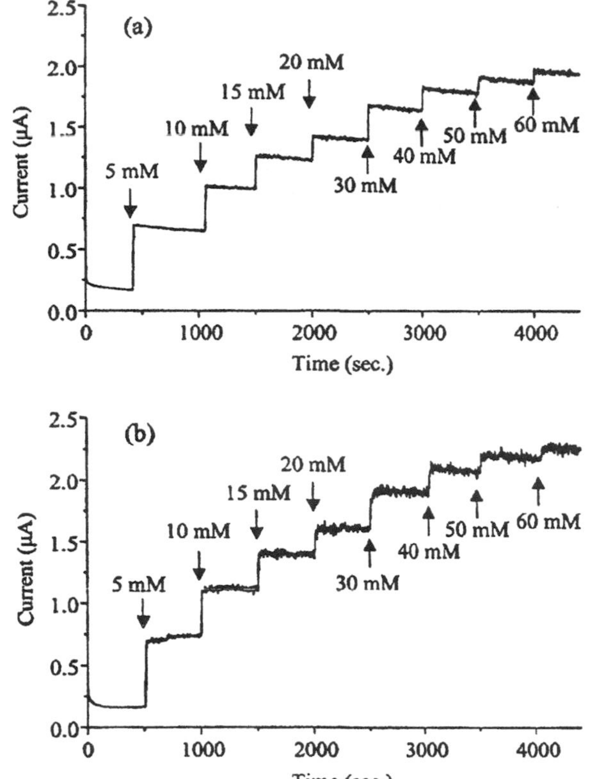

Figure 11.20 Time base measurement of glucose in 20 mM phosphate buffer (pH 7.4). Applied potential: +700 mV versus Ag/AgCl. Device based on (a) plasma-polymerized film (device 1) and (b) 3-(aminopropyl)-triethoxysilane (device 2). Sampling time: 0.5 s. (From Ref. 107.)

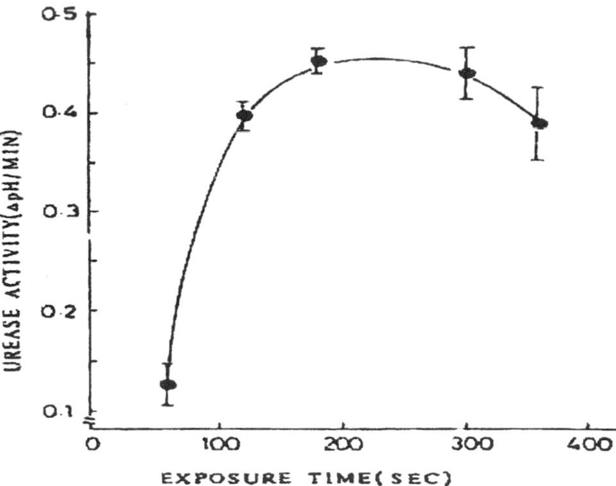

Figure 11.21 Effect of plasma exposure time on the extent of polypropylene membrane modification. The modified membrane was immersed in 2 mL of 0.01 M phosphate buffer (initial pH 7.0) containing 100 mM urea, and the extent of modification was analyzed by the rate of pH change. Each bar represents standard deviation from three experimental data points. Plasma treatment conditions were 60 W discharge power, 0.9 Torr ammonia gas pressure. (From Ref. 111.)

rin, a well-known anticoagulant, is often considered for surface coating of these devices. Polyurethanes are frequently used for the fabrication of catheters, artificial hearts, etc., due their remarkable mechanical properties.

Polyetherurethane urea (PU) was functionalized in a three-step process composed of an initial oxygen plasma treatment, followed by graft polymerization of 1-acryloylbenzotriazole (AB) and a subsequent substitution reaction of AB with sodium hydroxide and ethylene diamine (Scheme 7) (118).

The primary amine and carboxylic groups were further functionalized by coupling with heparin in the presence of carbodiimide. The plasma-modification step of PU was performed in a cylindrical stainless steel reactor equipped with two disc-shaped upper (stressed) and lower (grounded) 12-cm diameter parallel plate electrodes. The peroxide group concentrations on the plasma-exposed PU surfaces were evaluated by using 1,1-diphenyl-2-picrylhydrazyl. It was shown that the amounts of heparin covalently immobilized on the PU-NH_2 and PU-COOH surfaces were 2.0 and 1.4 μg/cm^2, respectively, and that the immobilized heparin exhibited high stability in physiological solution.

Various enzymes have been immobilized by the N_2, NH_3, or O_2 RF plasma treatment (parallel plate reactor; capacitive coupling) of dry mixtures of enzymes or mix-

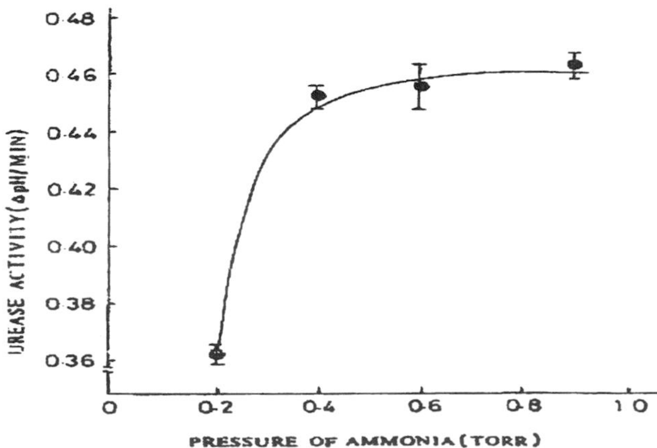

Figure 11.22 Effect of ammonia gas pressure on the extent of polypropylene membrane modification. Each bar represents standard deviation from three experimental data points. Plasma treatment conditions were 60 W discharge power, 2 min exposure. (From Ref. 111.)

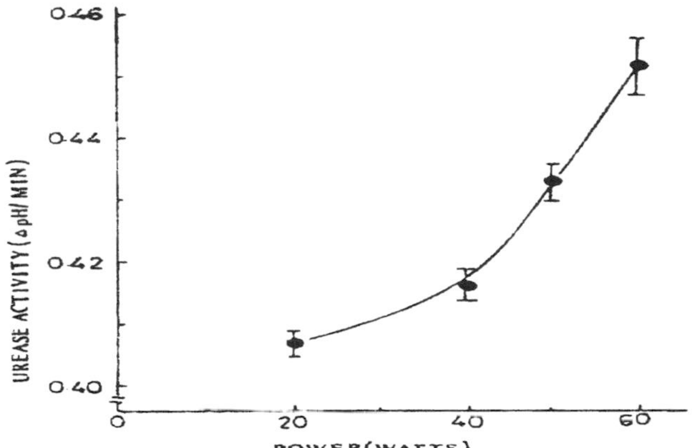

Figure 11.23 Effect of plasma discharge power on the extent of PP membrane modification. Each bar represents standard deviation from three experimental data points. Plasma treatment conditions were 0.9 Torr ammonia gas pressure, 2 min exposure. (From Ref. 111.)

tures predeposited from buffer (phosphate, tris, citric buffers) solution onto hydrophilic PE membrane surfaces. The authors claim (119) that various proteins, including glucose oxidase, lactate oxidase, 1-glutamate decarboxylase, 1-lysine decarboxylase and invertase, mutarotase, and glucose oxidase mixtures, can be covalently immobilized on various hydrophobic and hydrophilic polymeric materials, including polyolefins and cellulose acetate (e.g., cuprophane), and that the membranes (enzymes immobilized on polymeric supports) are stable and retain 50% of their activities after storing them at 4°C for 45 days in a glycerol buffer. The plasma exposures of the predeposited proteins/buffer layers were performed under the following plasma parameter range: substrate temperature: 0–25°C; pressure: 0.1–0.2 Torr, RF power: 50–300 W; driving frequency: 13.56 MHz. The authors do not present any evidence for the covalent coupling of the proteins to the polymeric substrates. This process might only resulted in the surface crosslinking of the solid-phase protein particles with the formation of an insoluble surface layer. This crosslinked protein surface layer (the plasma species interact only with

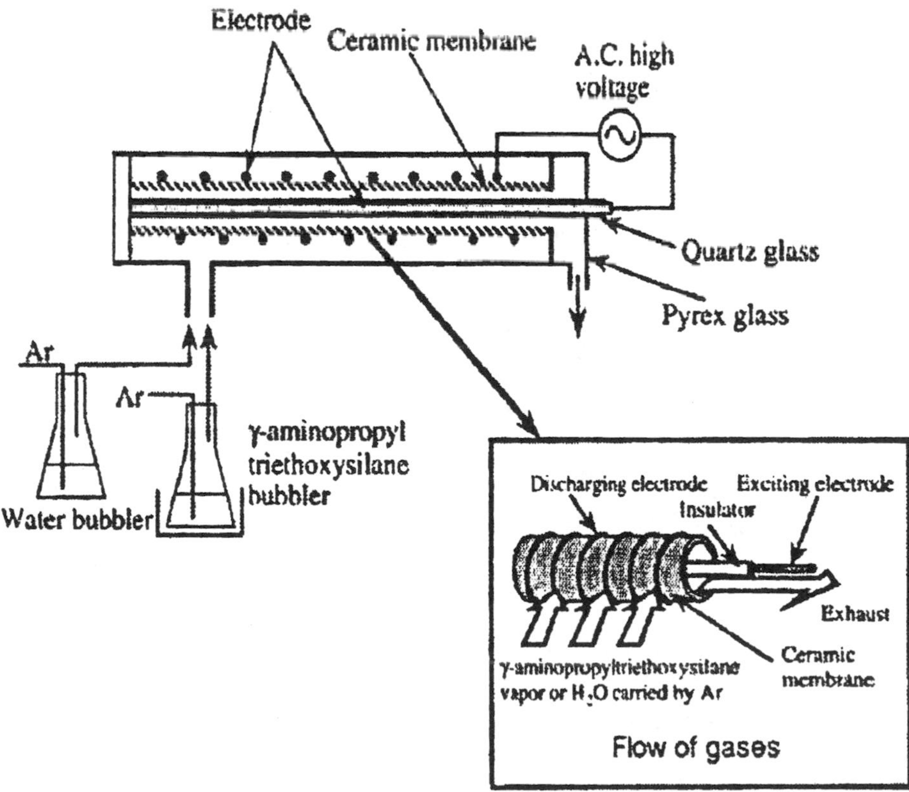

Figure 11.24 Experimental set-up for the surface modification of a ceramic membrane. (From Ref. 113.)

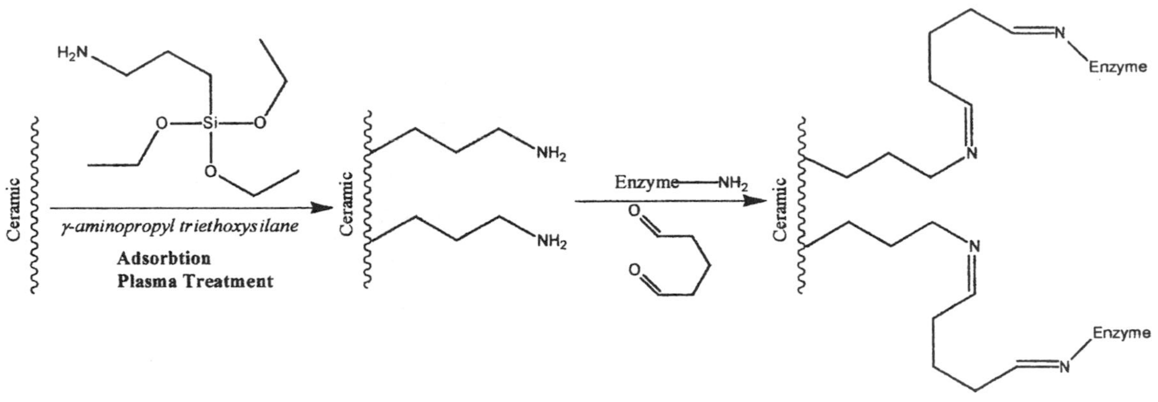

H_2N O—Si—O O Ceramic γ-aminopropyl triethoxysilane Adsorbtion Plasma Treatment Ceramic NH_2 NH_2 Enzyme—NH_2 O O Ceramic N N Enzyme N N Enzyme

Scheme 11.6 Surface modification and enzyme immobilization of glucoamylase on ceramic membrane surface.

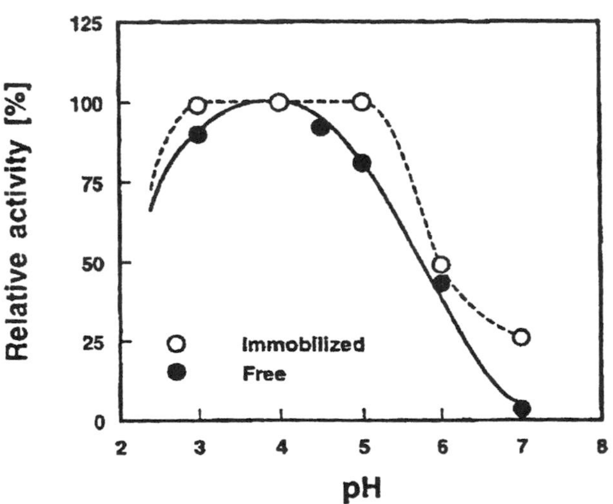

Figure 11.25 pH profiles of free and immobilized glucoamylase. (From Ref. 113.)

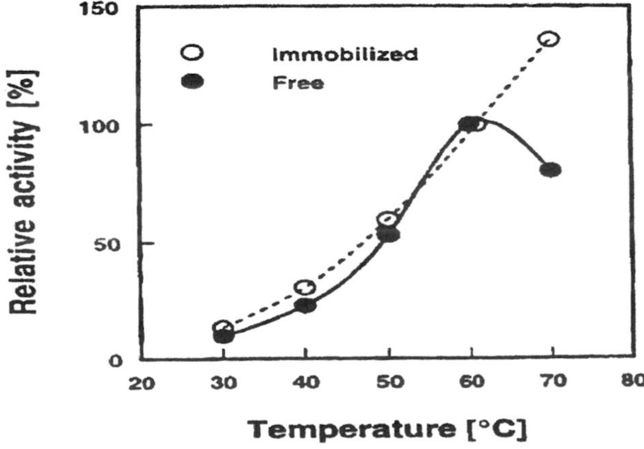

Figure 11.26 Temperature–activity profiles for free and immobilized glucoamylase. (From Ref. 113.)

the 100–150 Å depth of the exposed solid surfaces) might be responsible for the encapsulation of unmodified protein molecules and consequently for the retained enzyme activities.

The use of cold plasma technique for patterning of immobilized biopolymer layers for biosensor applications has recently attracted attention (120–122). It has been shown that plasma-mediated reaction mechanisms can be involved in different stages of the fabrication of biosensors both for accomplishing surface-functionalization reactions and for generating conveniently patterned surfaces for mo-

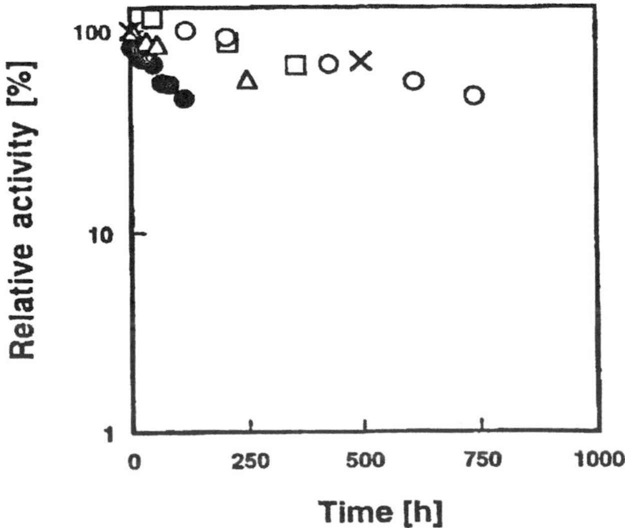

Figure 11.27 Operational stability of free and immobilized glucoamylase: (△) immobilized glucoamylase prepared by SPCP method (1-time treatment); (○) immobilized glucoamylase prepared by SPCP method (3-time treatment); (□) immobilized glucoamylase prepared by SPCP method (5-time treatment); (×) immobilized glucoamylase prepared by conventional method; (●) free glucoamylase. (From Ref. 113.)

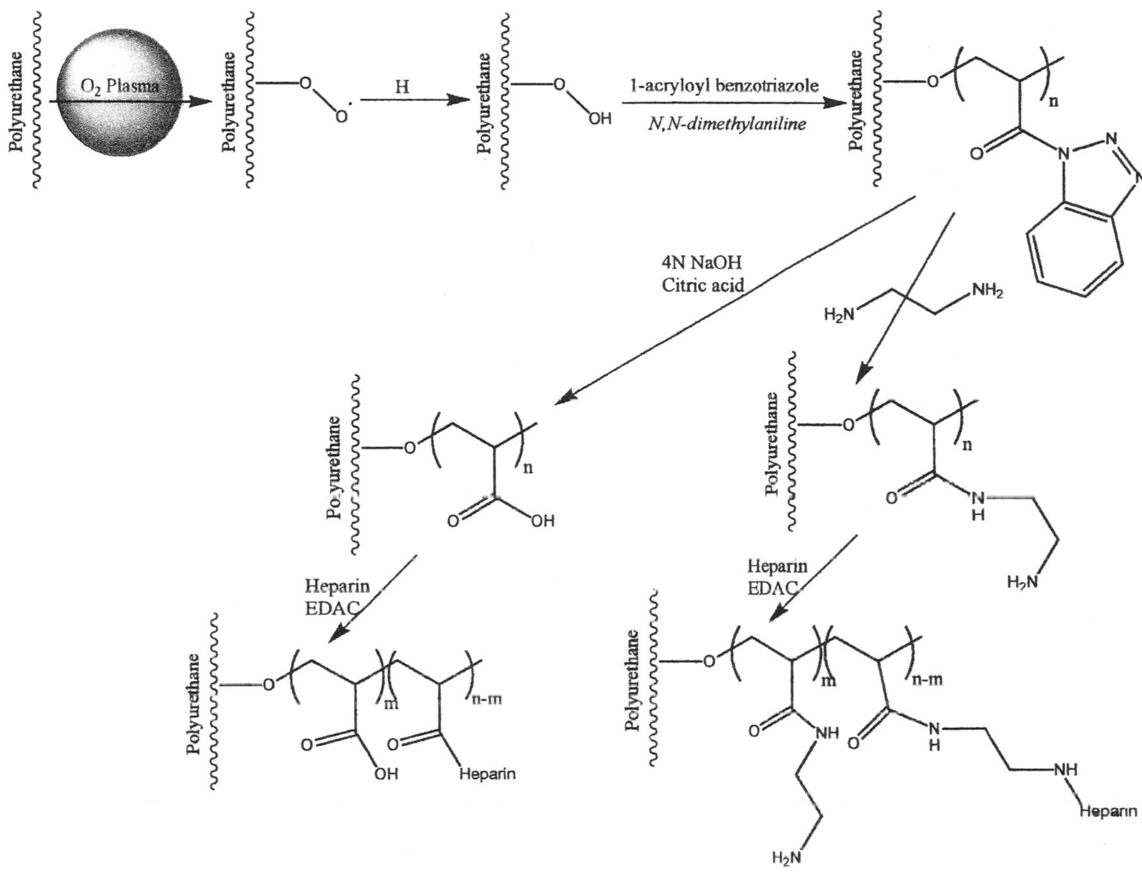

Scheme 11.7 Oxygen plasma treatment, graft polymerization, and substitution reaction of polyetherurethane urea.

lecular recognition processes. Protein, antibody, and conducting polymeric layers were successfully patterned by involving various cold plasma techniques and plasma gases including oxygen, water vapor, and others.

Hot plasma technique (plasma spraying) has also been considered for biotech applications. Owning to the high energy levels of the plasma species this approach can only be employed for the processing of inorganic and thermally stable starting materials, such as metals and inorganic oxides. Hydroxyapatite, a calcium phosphate ceramic, titanium and titanium in association with of alkaline phosphate has been extensively used as a bone replacement (123–125). The importance of plasma spray technique and the plasma-sprayed implants for the long-term proliferation of human osteoblastic cells have been emphasized.

C. Synthesis of Oligonucleotides from Plasma-Functionalized Surfaces

Dichlorosilane (DS) RF plasmas have successfully been used for the functionalization of both inorganic and organic polymeric substrates. Eight hundred–nm silica beads were exposed to DS 13.56 MHz discharges under RF rotating plasma reactor conditions for the implantation onto the particle surfaces $SiCl_xH_y$ functionalities and, consecutively, further in situ, derivatized in the presence of various volatile diamines (e.g., hexafluoropropylenediamine). The presence of Si—O— and C—NH_2 linkages were revealed with survey and high resolution ESCA and differential ATR-FTIR techniques and fluorescence labelling techniques. The surface-functionalized beads were used in the next step for synthesizing oligonucleotides according to the mechanism shown in Scheme 8.

Twenty-six-mer oligonucleotides were synthesized from —$SiCl_x$—NH—$(CF_2)_x$—NH_2 plasma-implanted functional spacer chains in the total absence of side reactions. From primary amine functionalized and cytidine coupled [loading 0.16 μmol/g, which is only 1% of the loading of a porous-controlled pore glass (CPG) support] silica bead surfaces, long chain (26 mer) oligonucleotides were synthesized using a Pharmacia Gene Assembler 4 Primers system according to the modified small scale pro-

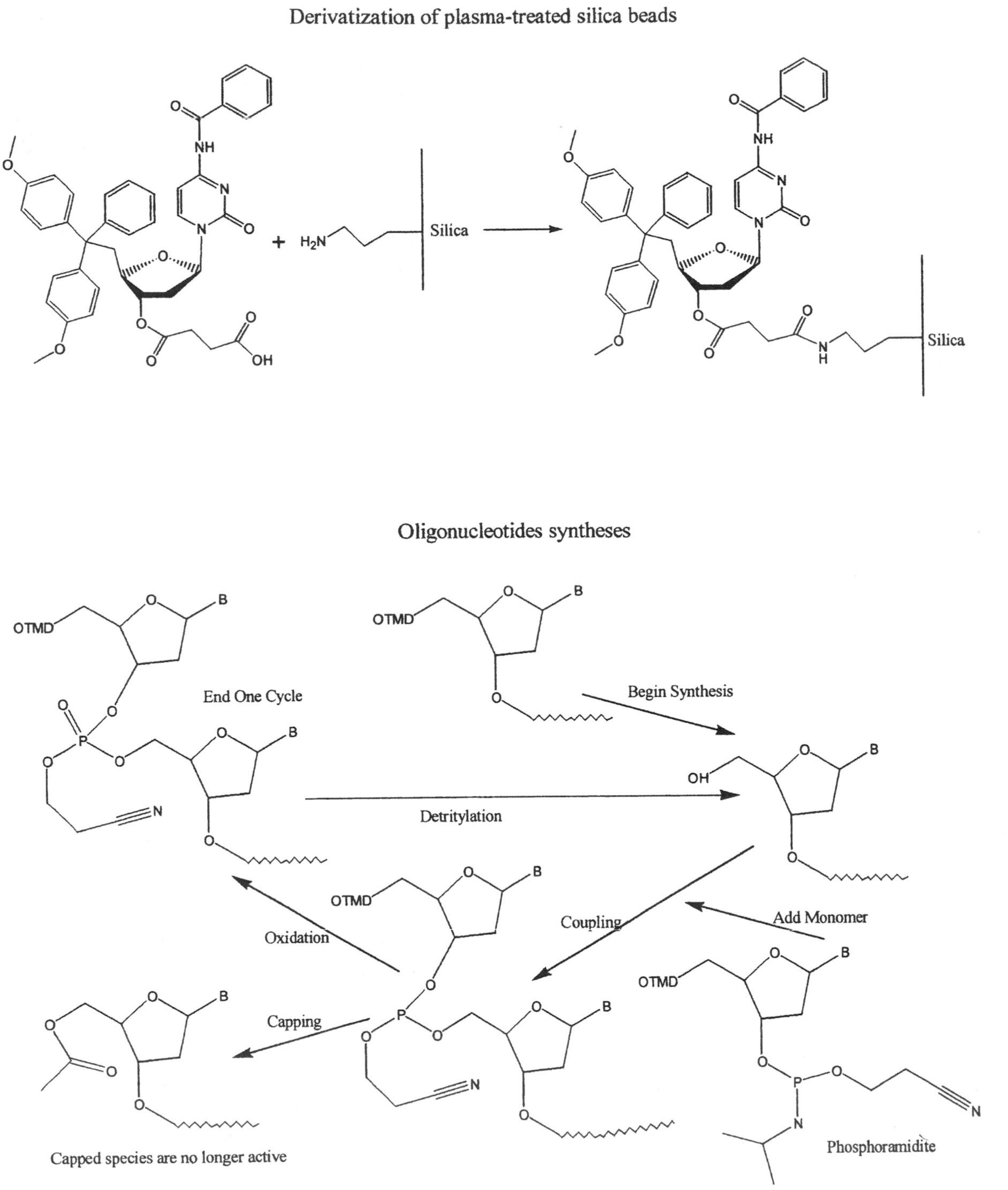

Derivatization of plasma-treated silica beads

NH
N
N
O
O
+ H_2N
Silica
OH

Oligonucleotides syntheses

OTMD
B
End One Cycle
Begin Synthesis
OH
N
Detritylation
Oxidation
Coupling
Add Monomer
Capping
Capped species are no longer active
Phosphoramidite

Scheme 11.8 Synthesis of oligonucleotides with surface-functionalized silica beads.

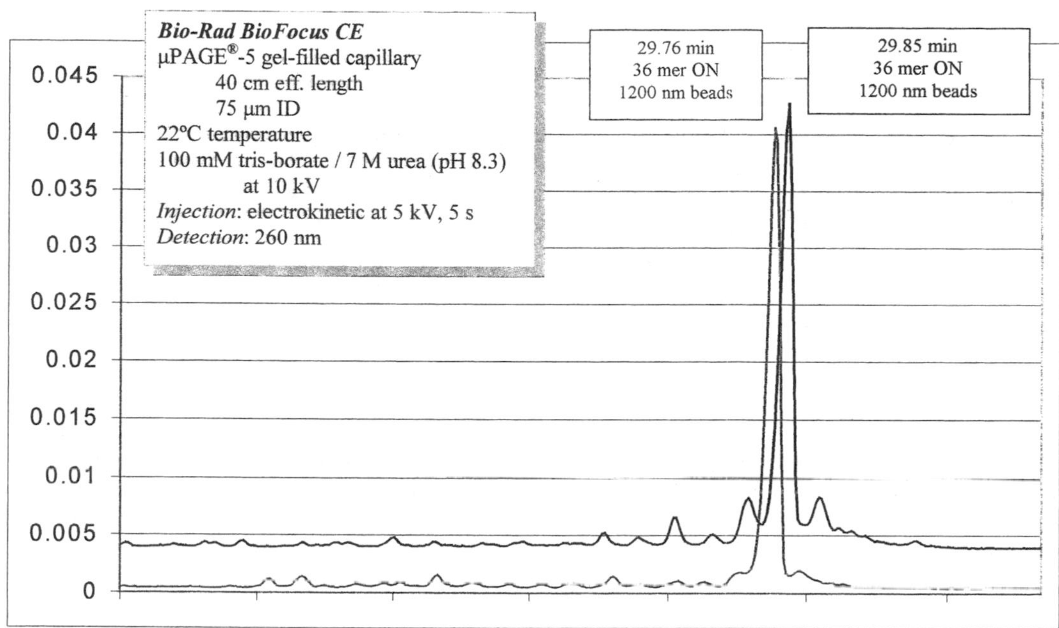

Figure 11.28 Capillary gel electrophoresis chromatograms of 36-mer oligonucleotides synthesized on plasma-functionalized 1200-nm beads.

tocols. It has been shown that the average coupling yield was as good as that of GPS with a higher than 99% loading (126, 128) (Fig. 28). These findings indicate that the surface has a good accessibility and that the support presents significant interest for small scale synthesis of long nucleotides and that the plasma-enhanced technique might open up novel highways for building up oligonucleotide libraries and for creating molecular recognition– and molecular machining–based advanced technologies.

VII. CONCLUSIONS

Nonequilibrium plasma technologies allow the deposition of tailored-structure, thin macromolecular layers on various substrate surfaces by involving monomers or common organic derivatives as plasma gases. The controlled interaction of plasma state with surfaces permits the predesigned modification of surface characteristics (implantation of the required chemical functionalities) and morphologies of even the most inert organic and inorganic materials (e.g., Teflon, silica, etc.).

Cold plasma approaches are environmentally friendly dry-chemistry technologies; the reaction pathways are mediated by the interaction of plasma-generated nascent charged and neutral active species with the surfaces, which confine the discharge. The importance of using plasma technologies, relies on the fact that they can take advantage of most of the technological developments associated with the semiconductor industry (e.g., patterning, etching, deposition, etc.).

Plasma-modified surfaces open up novel highways for the efficient anchoring covalently or by physical forces of active biomolecules on inorganic or organic substrate surfaces, under in situ or ex situ conditions, and in the presence or absence of spacer molecular chains.

These technologies will find their applications in the near future in the preparation of oligonucleotide libraries; the identification of motifs of single-strand nucleic acids to selectively interact with cellular (pathogenic) proteins, glycoproteins, etc.; the basic understanding of structural chemistry and/or biomedical diagnostics; the identification of hybridization processes based on electrical signal by plasma deposition of electrically conducting organic polymeric layers; the development of biopolymer-based molecular recognition sensors (biosensors, biochips) for the instant identification of the presence of specific biomolecules and toxins, for example, and in many other areas.

REFERENCES

1. E. S. Yeung, W. Tan. *Anal. Chem.* 69:4242 (1997).
2. H. Seliger, R. Bader, E. Birch-Hirschfeld, Z. Földes-Papp, K. H. Gührs, M. Hinz, R. Rösch, C. Scharpf. *Reactive & Functional Polymers* 26:119–126 (1995).

3. R. Bader, M. Hinz, B. Schu, H. Seliger. *Nucleosides & Nucleotides* 16(5&6):829–833 (1997).
4. *Chemical & Engineering News*, May 25, p. 47 (1998).
5. G. T. Hermanson, A. K. Mallia, P. K. Smith. *Immobilized Affinity Ligand Techniques*. Academic Press, San Diego, CA, 1992.
6. L. A. Potempa, M. Motie, B. Anderson, E. Klein, U. Baurmeister. *Clin. Mater.* 11(1–4):105–117 (1992).
7. T. Schluep, C. L. Cooney. *Bioseparation* 7(6):317–326 (1999).
8. S. N. Eremenko, D. M. Koulich, V. E. Petukhov, K. V. Maltsev, A. N. Wulfson, A. I. Miroshnikov. *Biotechnol. Tech.* 8(11):805–810 (1994).
9. M. K. Walsh, H. E. Swaisgood. *Biotechnol. Bioeng.* 44(11):1348–1354 (1994).
10. M. Meraz, O. Monroy, A. Noyola, K. Ilangovan. *Water Sci. Technol.* 32(8):243–250 (1995).
11. C. J. Banks, M. E. Parkinson. *J. Chem. Technol. Biotechnol.* 54(2):192–196 (1992).
12. A. K. M. Kabzinski, T. Takagi, K. Tanaka. *Chem. Anal.* (Warsaw) 44(5):805–823 (1999).
13. P. E. Gustavsson, K. Mosbach, K. Nilsson, P. O. Larsson. *J. Chromatogr. A* 776(2):197–203 (1997).
14. A. Serres, E. Legendre, J. Jozefonvicz, D. Muller. *J. Chromatogr. B—Biomed. Appl.* 681(2):219–226 (1996).
15. S. Dumitriu, M. Popa, V. Artenie, F. Dan. *Biotechnol. Bioeng.* 34:283–290 (1989).
16. M. D. Jonzo, A. Hiol, I. Zagol, D. Druet, L.-C. Comeau. *Enzyme and Microbial Technology* 27:443–450 (2000).
17. J.-T. Oh, J.-H. Kim. *Enzyme and Microbial Technology* 27:356–361 (2000).
18. M. Y. Arica, N. G. Alaeddinoglu, V. Nasirci. *Enzyme Microb. Technol.* 22(3):152–157 (1998).
19. M. Y. Arica, S. Senel, N. G. Alaeddinoglu, S. Patir, A. Denizli. *J. Appl. Polym. Sci.* 75:1685–1692 (2000).
20. A. Condo, T. Teshima. *Biotechnol. Bioeng.* 46(5):421–428 (1995).
21. T. Hayashi, Y. Ikada. *Biotechnol. Bioeng.* 36:593–600 (1990).
22. D. H. Kim, A. A. Garcia. *Biotechnol. Prog.* 11(4):465–467 (1995).
23. R. Gaffar, S. Kermasha, B. Bisakowski. *J. Biotechnol.* 75(1):45–55 (1999).
24. F. Azari, M. Nemat-Gorgani. *Biotechnol. Bioeng.* 62(2):193–199 (1999).
25. R. Madoery, C. G. Gattone, G. Fidelio. *J. Biotechnol.* 40(3):145–153 (1995).
26. O. Befani, M. T. Graziani, E. Agostinelli, E. Gripa, B. Mondovi, M. A. Mateescu. *Biotechnol. Appl. Biochem.* 28(2):99–104 (1998).
27. W. Kohnen, B. Jansen. *ZBL. Fact.* 283:175–186 (1995).
28. M. Amiji, K. Park. *J. Biomater. Sci. Polymer Edn.* 4(3):217–219 (1993).
29. T. Akizawa, K. Kino, S. Koshikawa, Y. Ikada, A. Kishida, M. Yamashita, K. Imamura. *Trans. Am. Soc. Artif. Organs* 35:333–335 (1989).
30. K. D. Park, Y. S. Kim, D. K. Han, Y. H. Kim, E. H. B. Lee, H. Suh, K. S. Choi. *Biomaterials* 19:851–859 (1998).
31. A. Z. Okkema, T. G. Grasel, R. J. Zdrahala, D. D. Solomon, S. L. Cooper. *J. Biomater. Sci. Polymer Edn.* 1(1):43–62 (1989).
32. L. K. Itsa, H. Fan, O. Baca, G. P. Lopez. *FEMS Microbiology Letters* 142:59–63 (1996).
33. S. J. Sofia, E. W. Merrill. Protein adsorption on poly (ethylene glycol)–grafted silicon surfaces. In: *Poly(Ethylene Glycol): Chemistry and Biological Applications*, J. M. Harris, S. Zalipsky (Eds.). American Chemical Society, Washington, D.C., 1997. Chapter 22.
34. M. Mrksich, G. M. Whitesides. Using self-assembled monolayers that present oligo (ethylene glycol) groups to control the interactions of proteins with surfaces. In: *Poly-(Ethylene Glycol): Chemistry and Biological Applications*, J. M. Harris, S. Zalipsky (Eds.). American Chemical Society, Washington, D.C., 1997. Chapter 23.
35. B. Wesslen, M. Kober, C. Freij-Larsson, A. Ljungh. *Biomaterials* 15:278–287 (1997).
36. M. Zhang, T. Desai, M. Ferrari. *Biomaterials* 19:953–960 (1998).
37. W. R. Gombotz, W. Guanghui, T. A. Horbett, A. S. Hoffman. Protein adsorption to and elution from polyether surfaces. In: *Poly(Ethylene Glycol) Chemistry Biotechnical and Biomedical Applications*, J. M. Harris (Ed.). Plenum Press, New York, 1992.
38. E. Osterberg, K. Bergstrom, K. Holmberg, T. P. Schuman, J. A. Riggs, N. L. Burns, J. M. Van Alstine, J. M. Harris. *J. Biomed. Mater. Res.* 29:741–747 (1995).
39. Y. C. Shin, D. K. Han, Y. H. Kim, S. C. Kim. *J. Biomater. Sci. Polymer Edn.* 6(3):281–295 (1994).
40. H. W. Roh, M. J. Song, D. K. Han, D. S. Lee, J. H. Ahn, S. C. Kim. *J. Biomater. Sci. Polymer Edn.* 10(1):1–21 (1998).
41. Y. C. Shin, D. K. Han, Y. H. Kim, S. C. Kim. *J. Biomater. Sci. Polymer Edn.* 6(2): 195–210 (1994).
42. K. Ishihara. *TRIPS* 5(12):401–407 (1997).
43. M. S. Antelman. *Int. Prec. Met. Inst.* 16:141 (1992).
44. D. G. Ahearn, L. L. May, M. M. Gabriel. *J. Industrial Microbiology* 15:372 (1995).
45. K. S. Rogers. *Biochim. Biophys. Acta* 263:309 (1972).
46. H. S. Rosenkranz, H. S. Carr. *Antimicrob. Agents Chemother.* 2:367 (1978).
47. R. Saruno, M. Tanaka, F. Kato. *Agric. Biol. Chem.* 43:2227 (1979).
48. R. M. Slawson, E. M. Lohmeier-Vogel, H. Lee, J. T. Trevors. *Biometals* 7:30 (1990).
49. Y. Yakabe, T. Sano, H. Ushio, T. Yasumaga. *Chem. Lett.* 4:373 (1980).
50. C. Oehr, H. Suhr. *Appl. Phys. A* 49:691–696 (1989).
51. M. Shen. *Plasma Chemistry of Polymers*. Marcel Dekker, New York, 1976.
52. H. V. Boenig. *Plasma Science and Technology*. Cornell University Press, Ithaca, NY, 1982.
53. H. Yasuda. *Plasma Polymerization*. Academic Press, New York, 1985.

54. R. d'Agostino. *Plasma Deposition, Treatment and Etching of Polymers*. Academic Press, New York, 1990.
55. D. T. Clark, A. Dilks, D. Shuttleworth. In: *Polymer Surfaces*, D. T. Clark and W. J. Feast (Eds.). John Wiley & Sons, New York, 1978.
56. H. V. Boenig. *Advances in Low Temperature Plasma Chemistry: Technology Applications, Vol. 3*. Technomic, Lancaster, PA, 1991.
57. H. Biederman, Y. Osada. *Plasma Polymerization Process: Plasma Technology, Vol. 3*, Elsevier, Amsterdam, 1992.
58. E. Nasser. *Fundamentals of Gaseous Ionization and Plasma Electronics*. Wiley Interscience, New York, 1971.
59. M. Venugopalan. *Reactions Under Plasma Conditions*, Vol.1, Wiley Interscience, New York, 1971, pp. 1–53.
60. M. A. Lieberman, A. J. Lichtenberg. *Principles of Plasma Discharges and Materials Processing*. Wiley Interscience, New York, 1994, pp. 552–553.
61. A. Grill. *Cold Plasma in Material Fabrication*. IEEE Press, 1994 p. 9.
62. F. Denes, *TRIP* 5(1).23–31 (1997).
63. F. D. Egitto, V. Vucanovic, G. N. Taylor. Plasma etching of polymers. In: *Plasma Deposition, Treatment, and Etching of Polymers*, R. d'Agostino (Ed.). Academic Press, New York, 1990, pp. 321–422.
64. K. Inagaki. *Plasma Surface Modification and Plasma Polymerization*. Technomic, Lancaster, PA. 1996, pp. 22–28.
65. F. Denes, R. A. Young. Surface modification of polysaccharides under cold plasma conditions. In: *Polysaccharides: Structural Diversity and Functional Versatility*, S. Dumitriu (Ed.). Marcel Dekker, New York, 1998.
66. A. S. Hoffman, D. Kiael, A. Safranj, T. A. Horbett, S. R. Hanson. Binding of proteins and platelets to gas discharge–deposited polymers. In: *Polymers in Medicine: Clinical Materials*, C. Migliaressi, E. Chiellini, O. P. Giusti, L. Nicolais (Eds.). (in press).
67. B. D. Ratner. *Biosensors & Bioelectronics* 10:797–804 (1995).
68. F. Poncin-Epaillard, G. Legea, J.-C. Brosse. *J. Appl. Polym. Sci.* 44:1513–1522 (1992).
69. B. D. Ratner. *J. Biomater. Sci. Polymer Edn.* 4(1):3–11 (1992).
70. W. R. Gombotz, W. Guanghui, A. S. Hoffman. *J. Appl. Polym. Sci.* 37:91–107 (1989).
71. M. Morra, C. Cassinelli. *J. Biomed. Mater. Res.* 31:149–155 (1996).
72. J. G. Steele, G. Johnson, C. McFarland, B. A. Dalton, T. R. Gengenbach, R. C. Chatelier, P. A. Underwood, H. J. Griesser. *J. Biomater. Sci. Polym. Edn.* 6:511–532 (1994).
73. D. Kiaei, A. S. Hoffman, T. A. Horbett. Platelet adhesion to fibrinogen adsorbed on glow discharge–deposited polymers. In: *Proteins at Interfaces II: Fundamentals and Applications*, T. A. Horbett and J. L. Bash (Eds.). ACS Symposium Series 602. American Chemical Society, Washington, D.C., 1994, 1995, pp. 450–462.
74. J. L. Bohnert, B. C. Fowler, T. A. Horbett, A. S. Hoffman. *J. Biomater. Sci. Polymer Edn.* 4:279–297 (1990).
75. G. P. Lopez, B. D. Ratner, R. J. Rapoza, T. A. Horbett. *Macromolecules* 26(13):3247–3253 (1993).
76. K. Dekker, T. Reitsma, A. Beugeling, J. Bantjes, J. Feijen, W. G. Aken. *Biomaterials* 12:130–138 (1991).
77. K. R. Kamath, M. J. Danilich, R. E. Marchant, K. Park. *J. Biomater. Sci. Polymer Edn.* 7(11):977–988 (1996).
78. G.-H. Hsiue, S.-D. Lee, P. C-T. Chang, C.-Y. Kao. *J. Biomaterials Res.* 42(1):143–147 (1998).
79. L. Tang, Y. Wu, R. B. Timmons. *J. Biomaterials Res.* 42(1):156-163 (1998).
80. D. Kiaei, A. S. Hoffman, T. A. Horbett. *J. Biomater. Sci. Polymer Edn.* 4(1):35–44 (1992).
81. G. P. Lopez, B. D. Ratner, C. D. Tidwell, C. L. Haycox, R. J. Rapoza, T. A. Horbett. *J. Biomaterials Res.* 26:415–439 (1992).
82. A. Y. Rad, H. Ayhan, E. Piskin. *J. Biomaterials Res.* 41(3):349–358 (1998).
83. M. Mora, C. Cassinelli. *J. Biomaterials Res.* 31:149–155 (1996).
84. E. E. Johnston, B. D. Ratner. XPS and SSIMS characterization of surfaces modified by plasma deposited oligo (glyme) films. In: *Surface Modification of Polymeric Biomaterials*, B. D. Ratner, D. G. Castner (Eds.). Plenum Press, New York, 1996, pp. 35–44.
85. S. K. Hendricks, C. Kwok, M. Shen, T. A. Horbett, B. D. Ratner, J. D. Bryers. *J. Biomed. Mater. Res.* 50:160–170 (2000).
86. F. Denes, S. Manolache. Plasma-enhanced technique for the implantation of aldehyde functionalities onto polymeric surfaces, and for the deposition of thin silver layers on polymer–substrate surfaces. Patent disclosure application, WARF (2000).
87. F. Denes, S. Manolache, N. Hershkowitz. Generation of large quantities of colloidal silver/water solutions and decomposition of organic contaminants (including microorganisms) in a modified dense-medium-plasma installation. Patent disclosure application, WARF (2000).
88. S. Manolache, F. Denes. *J. Photopolymer Sci. Technol.* 13(1):51–62 (2000).
89. T. Hayashi, Y. Ikada. *Biotechnol. Bioeng.* 35:518 (1990).
90. E. K. Katchalsky. Enzyme engineering. In: *Immobilized Enzyme Technology: Research and Applications*, H. H. Weetall, S. Suzuki (Eds.). Plenum Press, New York, 1975.
91. R. F. Taylor. *Protein Immobilization: Fundamentals and Applications*. Marcel Dekker, New York, 1991.
92. E. S. Yeung, W. Tan. *Anal. Chem.* 69:4242 (1997).
93. P. Lozano, A. Manjon, F. Romojaro, J. L. Iborra. *Eur. Congr. Biotechnol.* 2:52–55 (1987).
94. J. P. Telo, L. P. Candeias, J. M. A. Empis, J. M. S. Cabral, J. F. Kennedy. *Chim. Oggi.* 8(10):15–18 (1990).
95. T. Hayashi, Y. Ikada. *Abstr. Pap. Am. Chem. Soc.* 1990 Meeting, Pt. 2, PMSE:113 (1990).
96. T. Hayashi, Y. Ikada. *Biotechnol. Bioeng.* 35(5):518–524 (1990).
97. T. Hayashi, Y. Ikada. *Biotechnol. Polymers* 3:21–32 (1991).

98. H. Pham, L-M. Usher. http://www.esb.ucp.pt/bungah/immob/immob.htm.
99. R. Ganapathy, M. Sarmadi, F. Denes. *J. Biomater. Sci. Polymer Edn.* 9(4):389–404 (1989).
100. R. Ganapathy, S. Manolache, M. Sarmadi, W. J. Simonsick Jr., F. Denes. J. *Appl. Polym. Sci.* 78(10): 1783–1796 (2000).
101. A. J. Martinez, S. Manolache, V. Gonzales, R. A. Young, F. Denes. *J. Biomater. Sci. Polymer Edn.* 11(4):415–438 (2000).
102. R. Ganapathy, S. Manolache, M. Sarmadi, F. Denes. *J. Biomater. Sci. Polymer Edn.* (submitted for publication).
103. E. T. Kang, K. L. Tan, K. Kato, Y. Uyama, Y. Ikada. *Macromolecules* 29:6872–6879 (1996).
104. C.-C. Wang, G.-H. Hsiue. *J. Appl. Polym. Sci.* 50:11141–11149 (1993).
105. M. J. Danilich, D. Gervasio, R. E. Marchant. *Ann. Biomed. Eng.* 21:655–668 (1993).
106. A. Hiratsuka, H. Muguruma, S. Sasaki, K. Ikebukuro, I. Karube. *Electroanalysis* 11(15):1098–1100 (1999).
107. H. Muguruma, A. Hiratsuka, I. Karube. *Anal. Chem.* 72: 2671–2675 (2000).
108. S. Kaku, S. Nakanishi, K. Horiguchi. *Anal. Chim. Acta* 225:283 (1989).
109. S. Kaku, S. Nakanishi, K. Horiguchi, M. Sato. *Anal. Chim. Acta* 272:213 (1993).
110. S. Kaku, S. Nakanishi, K. Horiguchi. *Anal. Chim. Acta* 281:35–43 (1993).
111. Y. J. Wang, C. H. Chen, M. L. Yen, G. H. Hsiue, B. C. Yu. *J. Membrane Sci.* 53:275–286 (1990).
112. Y. J. Wang, C. H. Chen, G. H. Hsiue, B. C. Yu. *Biotechnol. Bioeng.* 40:446–449 (1992).
113. J.-I. Ida, T. Matsuyama, H. Yamamoto. *Biochem. Eng. J.* 5:179–184 (2000).
114. A. Mulchandani, S. Pan, W. Chen. *Biotechnol. Prog.* 15: 130–134 (1999).
115. A. W. Flounders, A. K. Singh, J. V. Volponi, S. C. Carichner, K. Wally, A. S. Simonian, J. R. Wild, J. S. Schoeniger. *Biosensors & Bioelectronics* 14:715–722 (1999).
116. H. J. Giesser, K. M. McLean, G. J. Beumer, X. Gong, P. Kingshott, G. Johnson, J. G. Steele. In: *Mater. Res. Soc. Symp. Proc*, Vol. 544, W. W. Lee, R. d'Agostino, M. R. Wertheimer, B. D. Ratner (Eds.). Materials Science Society, 1999, pp. 9–20.
117. A. K. Walker, Y. Wu, R. B. Timmons, G. R. Kinsel. *Anal. Chem.* 71:286–272 (1999).
118. I.-K. Kang, O. H. Known, Y. M. Moo, Y. Kiel. *Biomaterials* 17(8):841–847 (1996).
119. T.-T. Hsu, M.-T. Wang, K.-P. Hsiung, G. H. Hsiue, M.-S. Sheu. Plasma-induced protein immobilization on Polymeric surfaces. U.S. patent 5,306,768, Apr. 24, 1994.
120. L. Dai, H. J. Griesser, A. W. H. Mau. *J. Phys. Chem. B* 101:9548–9554 (1997).
121. I. Moser, T. Schalkhammer, E. Mann-Buxbaum, G. Hawa. *Sensors and Actuators*. B7:356–362 (1992).
122. A. W. Flounders, D. L. Brandon, A. H. Bates. *Biosensors and Bioelectronics* 12(6):447–456 (1997).
123. M. P. Ferraz, M. H. Fernandes, A. Trigo Cabral, J. D. Santos, F. J. Monteiro. *J. Mater. Sci. Mater. Medicine* 10: 567–576 (1999).
124. D. De Santis, C. Guerriero, P. F. Nocini, A. Ungersbock, G. Richards, P. Gotteand, U. Armato. *J. Mater. Sci. Mater. Medicine* 7:21–28 (1996).
125. A. Piattelli, A. Scarano, M. Corigliano, M. Piattelli. *Biomaterials* 17(14):1443–1449 (1996).
126. F. Denes, S. Manolache, R. A. Young. *J. Photopolymer Sci. Technol.* 12(1):27–38 (1999).
127. H. S, Seliger, M. Hinz, P. Jaisankar, F. Eisenbeib, S. Manolache, F. Denes, S. Gura, B. Nitzan, S. Margel. Proceedings of the Second International Symposium on Natural Polymers and Composites, ISNaPol 98, Atibaia-SP, Brazil, May 10–13, 1998, pp. 71–74.
128. F. Denes, S. Manolache, M. Hinz, H. Seliger. World Polymer Congress, Macro-98, 37th International Symposium on Macromolecules (Abstract), Gold Coast, Australia, July 13–17, 1998.

12

Advances in Designed Multivalent Bioconjugates with Dendritic Structure

Bogdan Comanita
Institute for Chemical Process and Environmental Technology, Ottawa, Ontario, Canada

I. INTRODUCTION TO DENDRIMERS

Dendrimer chemistry is a burgeoning field of macromolecular science that evolved over the last decade from the status of "aesthetically appealing molecules" to mainstream interdisciplinary research. The results of a key word search on the term "dendrimer*" carried out in scientific and technical databases serve to illustrate the exponential surge of publications and patents over the last couple of years (Fig. 1).

What sets dendrimers apart in the rich landscape of polymer chemistry and why have research groups all over the world become increasingly interested in these materials?

The progress of the macromolecular chemistry was marked by the discovery of three types of major macromolecular architectures that triggered the development of significant chemical industries (Fig. 2). New polymeric structures were therefore related to novel materials with important economical implications. The dendrimers have indeed a unique structure consisting of constantly branching segments from a central core, and they are expected to prompt the discovery of breakthrough, high value added products. Unlike older milestone discoveries in polymer science that had a crucial impact on the commodities market, the increased cost of the dendrimer synthesis (vide infra) will most likely confine these new molecular entities to the high technology sector such as pharmaceuticals and medical devices (Fig. 2) (1).

Worldwide producers of dendrimers are Dendritech (United States), DSM (Netherlands), and Perstorp (Sweden). The commercial availability of the polyamidoamine (PAMAM) and polypropylene amine (DAB) dendrimers, as well as the in depth study of the benzyl ether dendritic structures (2) have provided an important initial impetus to this field (Fig. 3).

The dynamics of the dendrimer chemistry have certainly lived up to the expectations and their applications are now spanning the entire spectrum of macromolecular science.

This chapter focuses on the use of dendrimers in the biomedical field and is intended as a cross-pollination point for professionals who are active at the interface of chemistry and biology. Other excellent reviews focusing on more specialized aspects of dendrimer chemistry are highlighted as an entry point to the literature for the interested reader (3–5).

A. The Dendritic Molecular Architecture

The etymology of the word *dendrimer* stems from the Greek words *dendron* and *meros*, meaning tree and part. Other common terminology such as Starburst® (6), Cascade, or Arborol (7) polymers are suggestive for the same

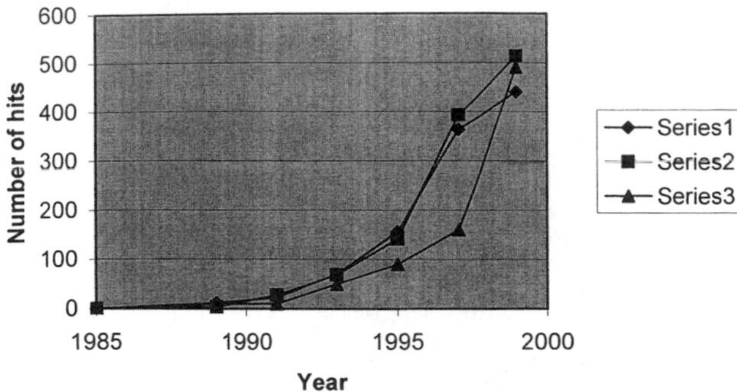

Figure 12.1 Dynamics of dendrimer chemistry: Series 1: Chemical Abstracts; Series 2: Science Citation Index; Series 3: U.S. patent and trademarks database (number of patents multiplied by a factor of 10).

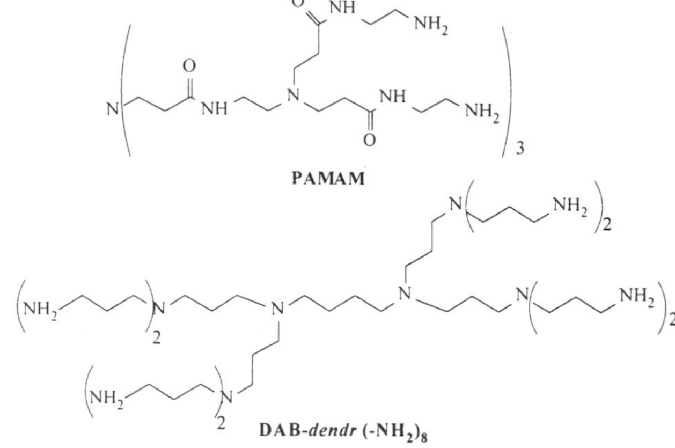

Figure 12.3 Commercially available dendrimers.

type of molecular structure consisting of (1) one common core (2) successive layers of concentric branched monomers, and (3) end groups (Fig. 4).

The core can be a single atom or a group of atoms and is defined by a branching functionality, which is given by the number of covalent bonds irradiating outward (e.g., the branching functionality of the core is 3 in Fig. 4). A first ''layer'' of branched monomeric units surrounds the core to form the first-generation dendrimer. Larger homologs can be built through a sequence of similar iterations leading to an onionlike structure of the molecule, each monomer layer corresponding to a new generation. Due to the branching of the monomer the number of segments in each layer grows exponentially from the core to the periphery

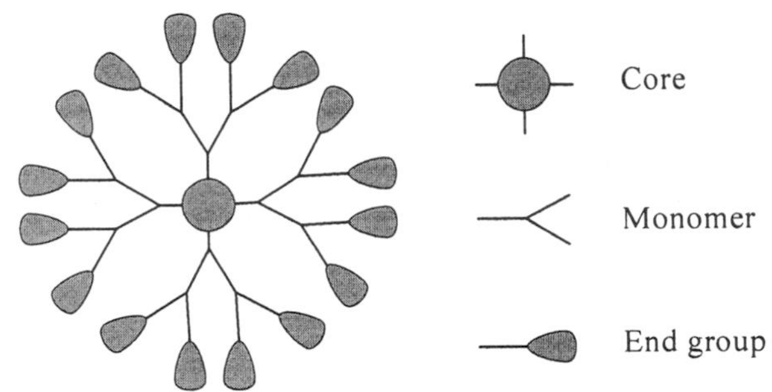

Figure 12.4 The main components of the dendritic architecture.

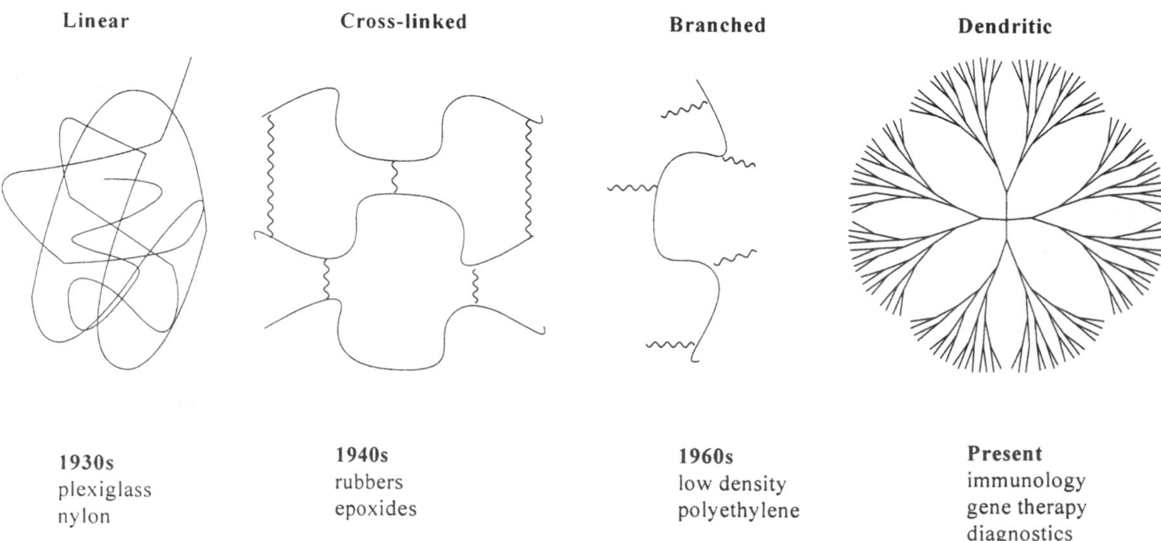

Figure 12.2 New macromolecular architecture is conducive to the discovery of novel materials.

of the dendrimers. Thus, the number of end groups (also known as peripheral, superficial, or terminal groups) will be given by the formula

$$N = f(b - 1)^G \quad (1)$$

where

N	=	number of end groups
f	=	branching functionality of the core
b	=	functionality of the monomer
G	=	generation number

The exponential nature of Eq. (1) translates into an accelerated increase of the molecular mass with the generation of the dendrimer. Consequently, it is possible to obtain macromolecules with molecular weight of 10^4–10^5 D and extremely low polydispersity by using only several monomer layers.

B. Theoretical Models and Experimental Characterization of Dendrimers

Many of the potential applications for dendritic materials depend directly on the organization and distribution of the end groups and internal segments throughout the volume of the molecule. In particular, the possibility of reduced density of the core (e.g., in drug delivery systems) and the location and accessibility of terminal end groups (e.g., in neoglycoconjugate synthesis) are crucial parameters for the successful utilization of these molecules in biomedical applications. The microstructure of the dendrimers has been therefore extensively investigated both theoretically and experimentally.

Maciejewski's ''dendritic box'' (8) hypothesis predicted that increased peripheral steric compression with the generation number would ultimately lead to a high-density outer shell with hollow intramolecular cavities. Theoretical calculations by Hervet and De Gennes (9) used a modified version of Edwards's self-consistent field (10) to confirm this molecular architecture. According to these authors, the inner branches become elongated at higher generations while the end groups lie on the dense surface of the globular construct. This is the dense shell model and is graphically described by the structure shown in Fig. 4.

At the other end of the spectrum Muthukumar and Lescanec generated dendritic structures using an off-lattice kinetic growth algorithm of self-avoiding walks to arrive at a density profile that decreases monotonically from the center to the periphery (11). The end groups are not necessarily positioned on the outer surface. This is the dense core model.

Theoretical Monte Carlo (12) and molecular dynamics (13) simulations also predict extensive back folding of the dendritic branches. According to these models, the segment density profile within the dendritic space is relatively constant with minimal values at the core and on the surface. The end groups are positioned preferentially throughout the outer shells, while the segments pertaining to the lower generations are located near the core. Similar results are obtained for ''flexible dendrimers'' by using a self-consistent mean-field model (14).

None of these generic molecular models address the chemical nature of the dendrimers. Consequently, it is not surprising that experimental measurements and computational simulations on chemically defined dendrimers have validated them all. The false dichotomy between these findings can be explained by the peculiarities of each case in point. It is important to realize that either the dense shell or the dense core architecture will be dictated not only by the branching functionality and the connectivity between the atoms, but also by the chemical nature of the layers within the dendrimer and the surrounding environment (pH, ionic strength, etc.).

The dense shell model will be favored when the back-folding of the dendritic arms is prevented. One way to achieve this end is by the use of ''stiff'' branches generated from acetylene- and phenylene-based monomers (15).

Back-folding can also be prevented by the incompatibility between the nature of the end groups and the core of the dendrimer. For example, the use of hydrophilic oligosaccharides as end groups on a hydrophobic carbosilane dendrimer will most likely lead to a more favorable presentation of the carbohydrate antigen on the surface, while the use of a hydrophilic dendritic core could favor back-folding (Fig. 5).

Recent data using this architectural principle offer strong proof of this concept. Theoretical studies on poly (propylene imine) dendrimers peripherally functionalized with N-*t*-BOC-l-phenylalanine confirm increasing interaction between the end groups of higher generation dendrimers (16) and a reduced degree of back-folding in these systems. Similar results were obtained for hydroxyl-terminated carbosilane dendrimers, although in this case hydrogen bond networking on the surface of the dendrimers could not be observed up to the fourth generation (17). These theoretical studies are confirmed by experimental evidence for perfluorinated end groups on polyaryl ether (18,19) and carbosilane dendrimers (20). The dendritic box hypothesis is also supported by experimental proof reporting on the successful inclusion of guest molecules inside the dendritic cavities (21), as well as by another body of evidence (22,23).

On the other hand ^{2}H and ^{13}C NMR[22] provided early experimental support for the back-folding of PAMAM dendrimers. Electron paramagnetic resonance (EPR) (24)

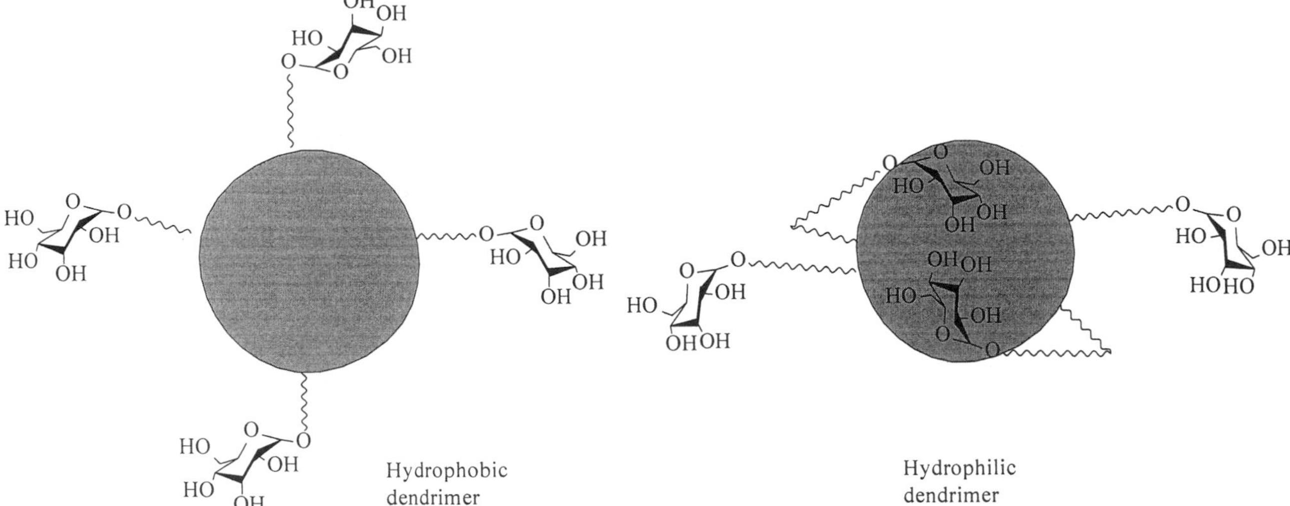

Figure 12.5 The chemical nature of the dendritic layers and the surrounding environment can favor the dense shell or the dense core models.

and fluorescence spectroscopy (25,26) have also been used to characterize with limited success the molecular dynamics of various dendrimers. More recently, small angle x-ray (SAXS) (27,28) and neutron (SANS) (29) scattering experiments provided crucial evidence that PAMAM, DAB, and Fréchet type (benzyl ether) dendrimers display a rather uniform internal segment density with partial backfolding of the dendritic branches. These experiments clearly establish that neither the dense shell nor the dense core model apply for these dendrimers under the reported experimental conditions.

C. Synthetic Strategies in Dendrimer Chemistry

''Retrosynthetic'' analysis (30) of the generic dendritic structure **1** (refer to Fig. 6) suggest two possible solutions for the synthesis of dendrimers. This is essentially a logical way to deconstruct a target molecule to simpler molecular entities, which in turn undergo the same analysis until simple raw materials are identified. The process follows a direction that is opposite to the actual synthesis and is therefore called retrosynthesis. Disconnection A leads to the generation n − 1 dendrimer **2** and the branched monomer synthon **3**. This rationale can be successively applied until the problem is reduced to the reaction of a core synthetic equivalent **4** with the branched monomer **3**. The process represents the divergent method (31) and is currently applied for the synthesis of the industrially available dendrimers DAB and PAMAM. From a synthetic perspective, the dendrimers are assembled from the core to the periphery.

Alternatively, along path B, the core synthon **4** can react with the branched ''dendron'' **5** (A dendron is a dendritic structure with a chemically reactive core—also called focal point—that will react to form a higher generation dendrimer.) to provide the target dendrimer **1**. After further iterative disconnection of dendron **5**, the problem is reduced to the reaction of two monomer synthons **6** and **3**. This is the convergent method (32), whereby the dendrimers are assembled from the periphery to the core.

The recent synthesis of the carbosilane-based dendrimers shown in Fig.7 (33) can illustrate the advantages and drawbacks of the divergent approach. The core tetravinylsilane is hydrosilylated with methyldichlorosilane followed by the reaction of the octachloro intermediate **8** (refer to Fig. 7) with vinylmagnesium bromide to yield the first-generation (8 arm) dendrimer **9**. Reiteration of the same sequence of reactions provides the second-generation (16 arms) compound **10**. Finally, the third-generation (32 arms) dendrimer **11** is obtained by the reaction with a hydrosilane containing the end group in a masked form. Assuming that all 16 hydrosilylation reactions in the final step occur with a 99% yield, one expects an 85% overall yield. The remaining 15% will be side products that are very similar in structure and molecular mass with the target dendrimer and therefore impossible to remove in the work-up. The divergent approach is thus suitable only for the very high yielding reaction sequences, and considerable re-

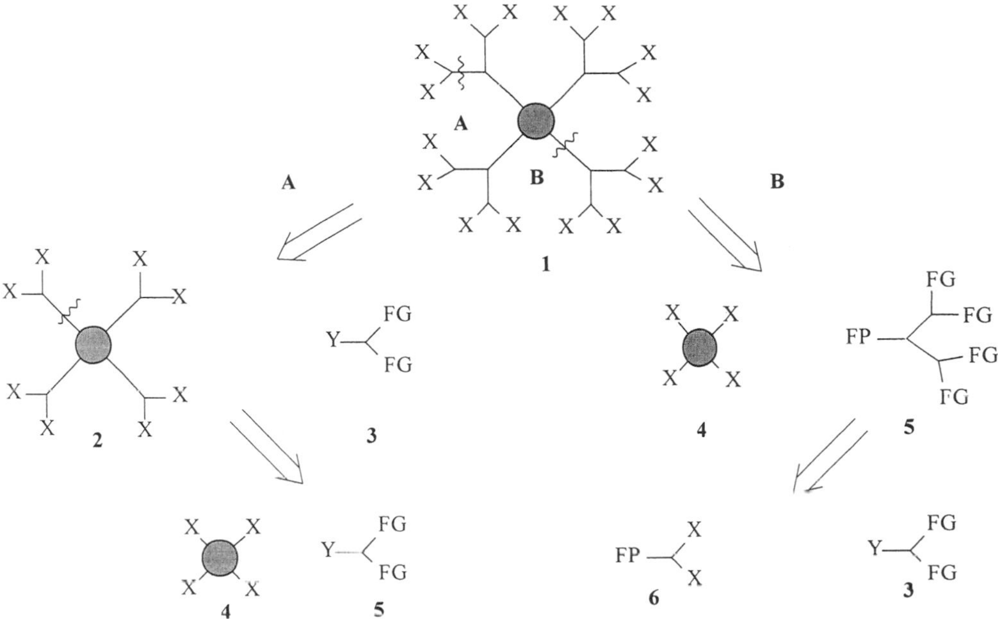

Figure 12.6 Retrosynthetic analysis of dendritic structures: FG, FP, and X,Y are respectively interconvertible functional groups.

search effort is required to optimize the system for industrial scale production. Nevertheless, this approach is applied today for the synthesis of the commercially available dendrimers produced by DSM (ASTRAMOL®) and Dendritech (PAMAM) (34–38).

In the convergent approach (39), the dendritic construct is synthesized from the periphery to the core, as exemplified in Fig. 8. Protection of the phenolic groups (that will eventually become the end groups of the dendrimer) is followed by activation of the benzyl alcohol **13** (refer to Fig. 8) to the benzyl bromide **14**. This highly reactive intermediate is further developed to generate dendron **16**. The focal point of this dendron is transformed back to a benzyl bromide **17** and the reaction sequence is reiterated to the branched structure **18**. Finally, this reacts with the trifunctional core **19** to obtain the desired dendrimer **20**. Any misfire of the reagents in the last step will result in a side product **21** that has considerably lower molecular weight than **20**. Consequently, purification of the product at the final and intermediate stages becomes possible, and this is the method of choice to produce defect-free dendrimers on a laboratory scale. Unfortunately, steric hindrance often limits the accessibility of higher generation dendrimers due to reduced reactivity of the focal point. Another important drawback of the convergent method is the exposure of the end groups to repeated chemical cycles. This can be a substantial restriction for dendrimers with sensitive end groups such as proteins or polyglycosides.

When radial block dendrimers are desired, the direct reaction of the end groups on a dendrimer with the focal group of a chemically different dendron can be achieved in a process called the double growth synthesis (40–46). This hybrid method also has the advantage of an accelerated growth due to increased overall convergence.

II. DENDRIMERS AS NANOSCOPIC SCAFFOLDS IN BIOLOGICAL APPLICATIONS

A. Three-Dimensional Control at the Nanometer Level

1. The β Factor, Practical Alternative to the Cooperativity α Factor

The 20th century has witnessed extraordinary scientific and technical achievements that will prefigure our advancement for forthcoming generations. Headline-making news such as the exploration of space and the creation of cyberspace was accompanied by the quieter but equally relevant biological revolution that provided us with spectacular insights in the microcosmos of life.

One of the most intriguing working hypotheses of modern molecular biology is the concept of polyvalent interactions (47). Multiple simultaneous interactions between various ligands and receptors have important functional consequences that could not be accomplished through a

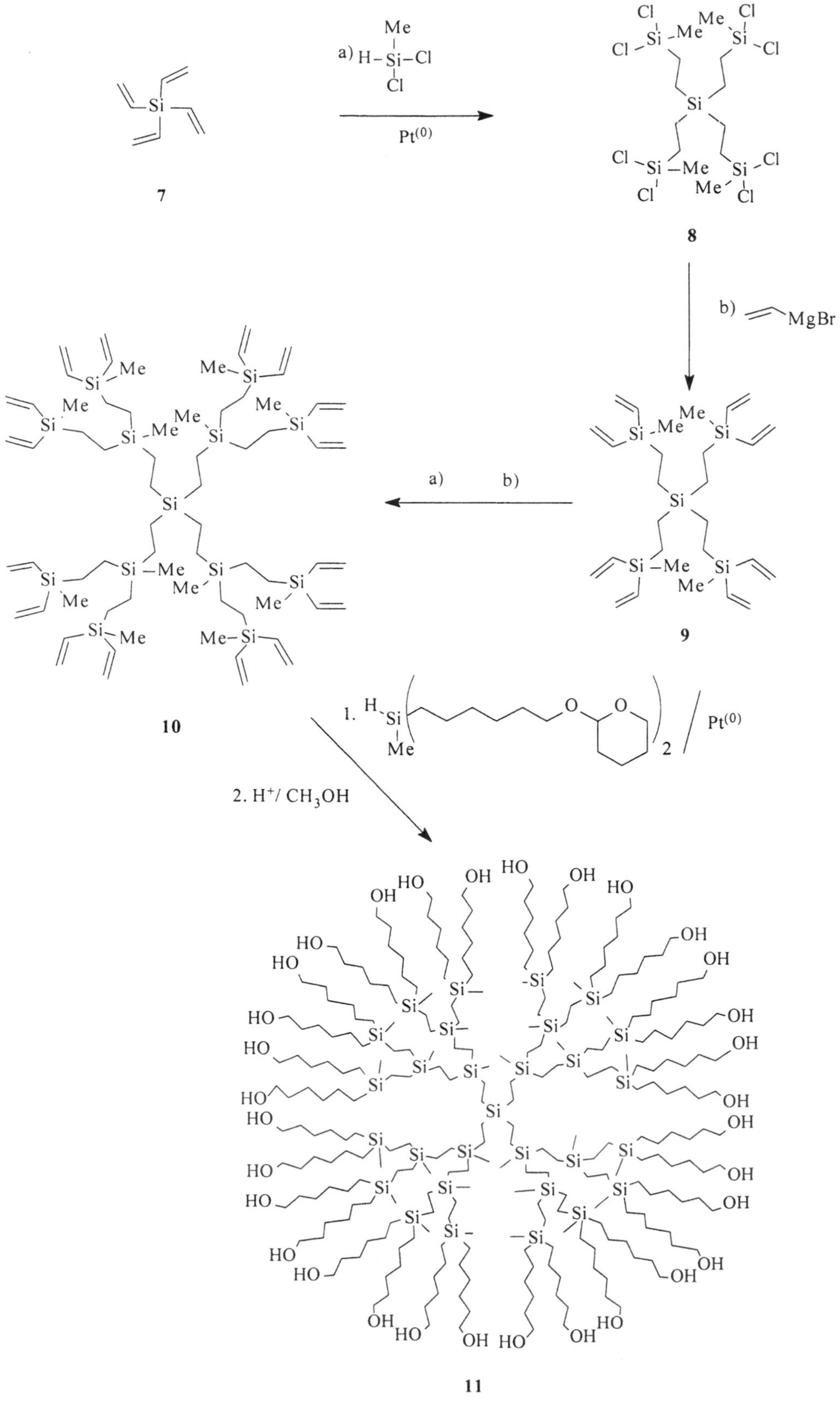

Figure 12.7 The divergent synthesis.

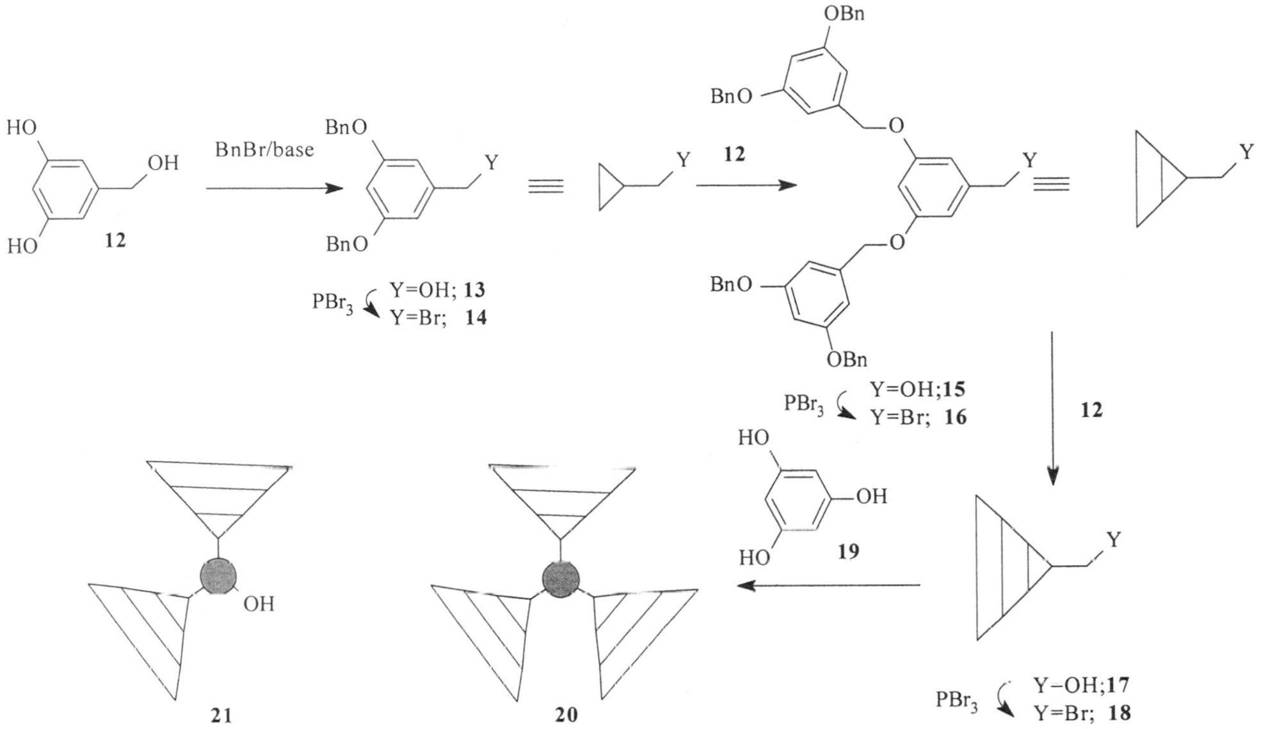

Figure 12.8 The convergent synthesis.

single or an equivalent number of independent binding events.

In the case of monovalent ligands (Fig. 9) each binding event has approximately the same constant of equilibrium, i.e., $K_1, K_2, ..., K_N$ in the figure are the same, provided that the receptors behave independently. (Often, the binding of the first ligand can trigger important conformational modifications that will accelerate the binding of the subsequent monovalent ligands. The classic example is binding of oxygen to hemoglobin. These are cooperative effects and do not involve polyvalency.) For polyvalent ligands however the overall affinity is given by the product of the individual binding constants $K_1, K_2, ..., K_N$. Thus, a considerably stronger binding can occur between the two multivalent entities.

The cooperative effect has been largely underappreciated in the rational design of new polyvalent drugs and research reagents for biochemistry and biology. However, a recent milestone article by Whitesides et al. (47) summarizes the latest developments in the field and sets the theoretical framework for using polyvalency as a key concept in the design of new biopharmaceutical products.

Starting from basic thermodynamic considerations, the article defines the free energy and equilibrium constant for two entities (cells, molecules, organelles, etc.) interacting through an N-order polyvalent interaction, that is, through N ligands and N receptors. See Table 1 and Eq. (2) to (5). The thermodynamic equations relating these metrics are the following:

$$\Delta G_{avg}^{poly} = \frac{\Delta G_N^{poly}}{N} \quad (2)$$

$$\Delta G_N^{poly} = -RT \ln K_N^{poly} \quad (3)$$

$$K_N^{poly} = (K_{avg}^{poly})^N \quad (4)$$

$$\Delta G^{mono} = -RT \ln K^{mono} \quad (5)$$

Positive cooperativity in the traditional sense (48) is defined by the sequence of binding events that display a higher average binding constant K_{avg}^{poly} than the corresponding monovalent binding constant K^{mono}. The degree of cooperativity, α, is a parameter that relates the monovalent free energy ΔG^{mono} to the observed polyvalent free energy ΔG_{avg}^{poly} through Eq. (6).

$$\Delta G_{avg}^{poly} = \alpha \cdot \Delta G^{mono} \quad (6)$$

A polyvalent interaction is noncooperative (additive) when $\alpha = 1$, positive (synergistic) for $\alpha > 1$, and negative (interfering) if $\alpha < 1$. It is important to realize that polyvalent interaction, even with interfering cooperativity, will

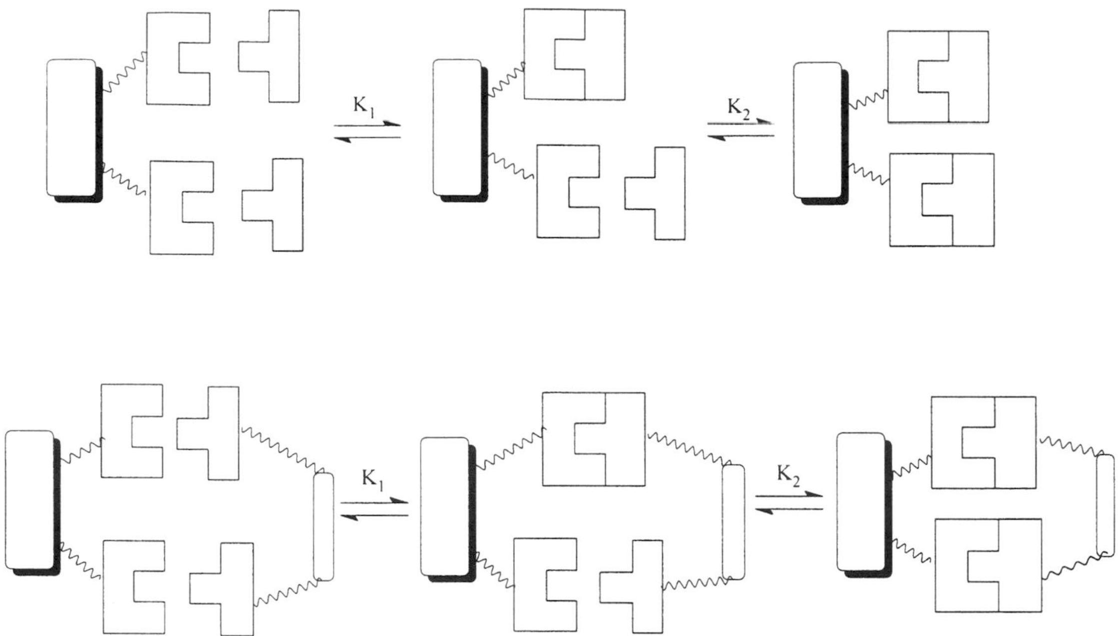

Figure 12.9 Polyvalent interactions versus monovalent interactions.

ultimately lead to an enhanced overall binding. For example, Lee et al. (49) reported in a classic reference the affinity of the divalent and trivalent galactose ligands which bind to C type lectines on the surface of hepatocites (Table 2).

Since $K_{avg}^{bivalent} = \sqrt{K_2^{poly}} < K^{mono}$ and $K_{avg}^{trivalent} = \sqrt[3]{K_3^{poly}} < K^{mono}$, the divalent and trivalent interactions both have negative cooperativity, and therefore $\alpha < 1$. Nevertheless, the binding of the trivalent ligand is still 2857 times stronger than the monovalent galactose ligand. Consequently, the cooperativity parameter α is not as useful a parameter in the case of the multivalent interactions as it is in the case of multiple monovalent, independent ligands. Moreover, from a practical point of view, the number of polyvalent interactions N is usually unknown and no statement about cooperativity α is even possible. Whitesides et al. propose therefore a new parameter, β, correlating the binding constant of the monovalent ligand, K^{mono}, and K^{Elisa} determined from a common Elisa type essay [Eq. (7)]:

$$\beta = \frac{K^{Elisa}}{K^{mono}} = \frac{K_N^{poly}}{K^{mono}} \tag{7}$$

This "enhancement factor" β can better serve the purpose of comparing various polyvalent systems and will emphasize the beneficial effect of polyvalency even for systems with negative cooperativity.

Table 12.1 Proposed Nomenclature for the Polyvalent Interactions

Symbol	Comments
ΔG_N^{poly}	Free energy of the polyvalent interaction
K_N^{poly}	Equilibrium constant of the polyvalent interaction
ΔG_{avg}^{poly}	Average free energy per each ligand–receptor interaction
K_{avg}^{poly}	Average equilibrium constant per each ligand–receptor interaction
ΔG^{mono}	Free energy for the monovalent ligand–receptor interaction
K^{mono}	Equilibrium constant of the monovalent ligand–receptor interaction

2. Enthalpy and Entropy in Multivalent Interaction

Another important aspect of multivalent interactions is the correlation between the enthalpic and the entropic components. Thus, in order to maximize the overall binding effect, it is necessary to use multivalent ligands that fit perfectly in the receptor (the well-known lock and key concept). The ideal solution requires a rigid spacer that places the multivalent ligands in a matching, precisely defined topology. Any deviation from this three-dimensional distribution will lead to a strained conformation at the binding site and diminish the favorable enthalpic component of the multivalent interaction. Since the configuration

Table 12.2 Binding Constant for Multivalent Galactose Ligands

Oligosaccharide	Binding constant (M^{-1})
Gal(β1)OMe	$K^{mono} = 7 \times 10^4$
Gal(β1,4)GlcNAc(β1,2) Gal(β1,4)GlcNAc(β1,4) → Man	$K_2^{poly} = 3 \times 10^7$
Gal(β1,4)GlcNAc(β1,2)Man(α1,6) Gal(β1,4)GlcNAc(β1,2)Man(α1,3) → Man Gal(β1,4)GlcNAc(β1,4)	$K_3^{poly} = 2 \times 10^8$

of the receptor is usually not known, one way to overcome strain is by using flexible spacers. This allows the ligands to probe the surface of the receptor and achieve interaction at the expense of an entropic price, which explains why flexible linkers often fail to provide a spectacular solution to the problem (50–51).

Although the entropic and the enthalpic effects of the multivalent interactions are difficult to delineate quantitatively, it is obvious that the end groups on the dendritic surface will have quite a different behavior than the functionality attached to the graft or star polymers. It can be speculated that the entropic loss of the dendritic multivalent ligands will be considerably diminished due to their restricted motion on a spherical surface, as opposed to the usual situation whereby the ligands move freely in the three dimensional space. At the same time, the fine tuning of the distance between ligands on the dendritic surface can be achieved from a combination of architectural parameters, such as monomer and core functionality, length of the monomer arm, and chemical nature of the constituent atoms. Consequently, it can be concluded that dendrimers can provide a unique compromise between the enthalpic and entropic components of the multivalent interactions warranting systematic future studies.

3. Shape and Size Control at the Nanoscopic Level

Shape and size control at the nanoscopic level represents an important desiderate from a technological point of view since it defines a theoretical threshold for miniaturization. The subnanoscopic level circumscribes the interaction within the molecules and is therefore irrelevant for practical applications in the field of life sciences. Typical dimensions for biological compounds range between a few nanometers (e.g., insulin and hemoglobin) to hundreds of nanometers (e.g., histones). It is therefore obvious that having precise control at this scale is of paramount importance for the rational design of the new drugs.

The dendritic structure offers peerless possibilities in the control of the size and shape of new molecular entities. For example, the PAMAM dendrimers increase in diameter by approximately 1 nm per generation (52). Even finer tuning can be achieved with the carbosilane dendrimers shown in Fig. 6, whose diameter can be incrementally increased by one-half of a nanometer from one generation to another (53).

By using dendrimers as building blocks in a molecular Lego® chemists have already synthesized macromolecules in various shapes such as spheres, ellipsoids, and rods, as shown schematically in Fig. 9. (See Ref. 54 for a recent review on the dimensions and shapes of dendrimers and dendrimer–polymer hybrids.) The possibility to design clefts and protrusions on these surfaces having a desired chemical functionality in a precise reciprocal position has therefore become accessible to a degree that had been previously encountered only in biological molecules (Fig. 10).

B. Dendrimer–Carbohydrate Conjugates

Carbohydrate–protein interaction constitutes a crucial biochemical event in a wide range of key cell–cell interactions and infectious diseases. The multivalent character of the carbohydrate ligands compensates the intrinsic weak nature of this binding and is lately generating growing interest in the design of novel therapeutics.

Dendritic carbohydrates, also known as glycodendrimers (55), fill the dimensional gap between high molecular weight neoglycoconjugates and small clusters. Typically, but not necessarily, they are confined to the range of nanometers up to tens of nanometers.

One important class of dendritic carbohydrate agent targets the competitive inhibition of the protein–carbohydrate binding phenomena involved, for example, in various pathogenic infections.

The immune response against bacterial infections can be mediated by the mannose binding proteins (MBPs) which tag the mannose-rich surface of the bacteria and

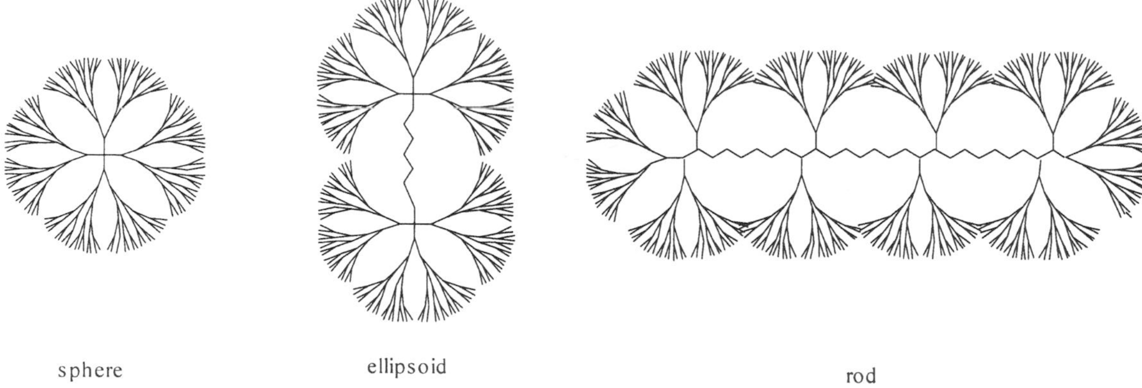

Figure 12.10 Shape and size control at the nanometer level with dendritic structures.

prompt the phagocytosis by macrophages. On the other hand, mannose-terminated glycoproteins serve as high affinity ligands for bacterial attachment, facilitating the invasion of the host cells. Since these interactions involve multivalent binding, the dendritic mannosides have potential in both therapeutic and drug targeting applications.

Multivalent mannosides 22 (refer to Fig. 11, only octavalent dendrimers shown) inhibited the binding of the yeast mannan (56) to concavalin A and pea lectins (57). Thus, it was shown that the binding potency of these inhibitors is influenced both by the dendritic chemical structure and its multivalency. The hexadecamannose polylysine (only 8-arm analog 22 shown) displayed a 2000-fold increase in overall binding (approximately 100-fold increase per sugar if compared to the monovalent mannose). A similar study was performed on the commercially available PAMAM dendrimers conjugated with mannose. The tetravalent and octavalent carbohydrate achieved the optimum binding with a 30-fold increase binding per sugar (58). As these interactions are at the origin of the host infections by fimbriated bacteria, mannoside dendrimers can form the basis of novel antiadhesin molecules.

Multivalent mannose glycomimetics have also been synthesized in order to inhibit the adhesion of *Escherichia coli* HB 101 (pPK14) to erythrocytes and yeast mannan (59). *E. Coli* strains are important pathogens in urogenital and gastrointestinal infections that use α-mannoside specific fimbriae to adhere to the cell surface carbohydrates of the host cell. In order to allow for a broad variation in the structure of the glycoclusters the authors have considered both different attachment sites at the sugar ring and different linkage types between the sugar and the dendritic substrate. It is known from various other lectins that the hydroxyl groups in the 2, 3, and 4 positions of the pyranosides are essential to binding, therefore these sites were not used for attachment. Finally, the role of the aglycone structure has also been investigated in this study. Based on the known affinity of the monomeric p-nitrophenyl mannoside aglycone to bind 80-fold stronger than the methyl glycoside it was expected that these clusters should display improved inhibitory properties.

The synthesis of the most potent inhibitor 31 is highlighted in Fig. 12. The six–amino methyl mannoside 30 was coupled with the tricarboxylic acid 29 and then debenzoylated under Zémplen conditions to afford the unprotected six-peptide bridged trimannoside 31 in quantitative yield. A structurally related inhibitor having a thiourea bridge instead of the peptide linker showed a considerably reduced inhibitory capacity, probably due to conformational effects.

The N-acetyllactosamine (LacNAc) and its simpler constituent N-acetylglucose (GlcNAc) characterize the lactosaminoglycans as tumor-associated antigenic carbohydrates. Various polylysine dendrimers bearing β-lactoside 25, N-acetyllactosaminide 26 (60), and β-*D*-N-acetylthioglucosamine 23 (61) (Fig. 11) were synthesized by chemical or enzymatic routes. All of these glycodendrimers showed enhanced affinities to their respective binding proteins.

One of the most widespread mammalian cell carbohydrate ligands is the N-acetylneuraminic acid (NeuAc), or sialic acid (62). Dendritic sialoside inhibitors of the Influenza virus hemagglutination to human erythrocytes have therefore been synthesized (24 in Fig. 11). The solid phase synthesis of sialidase-resistant N-linked α-sialodendrimers has also been reported (63).

The more complex tetrasaccharide NeuAcα2 → 3Galβ1 → 4[Fucα1 → 3]GlcNAc, or sLe^x, is a ligand for E-selectin endothelial receptors initiating the leukocyte adhesion, attachment, and extravasation in the inflammatory response. Traditional structure activity relationship (SAR) studies have led to the development of numerous sialyl Lewis^x (sLe^x) analogs without taking into account cluster

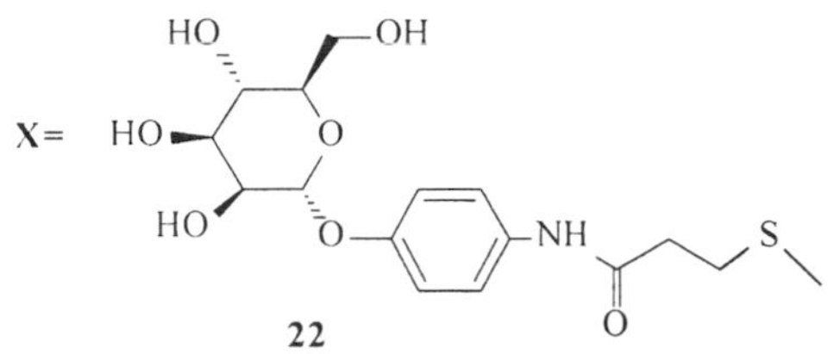

Figure 12.11 Polylysine-based dendrimer–carbohydrate conjugates (glycodendrimers).

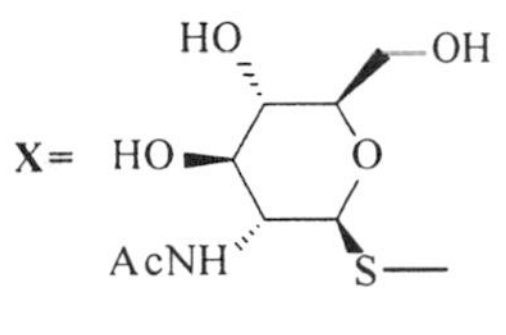

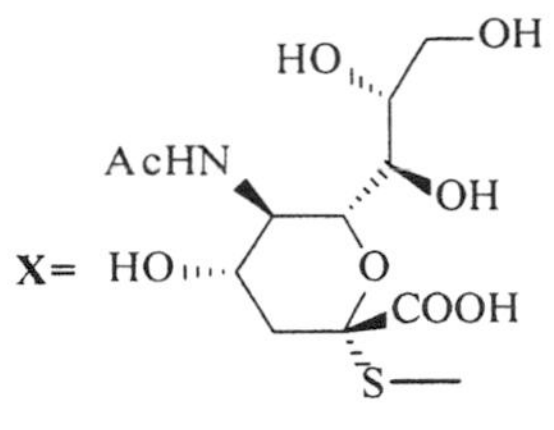

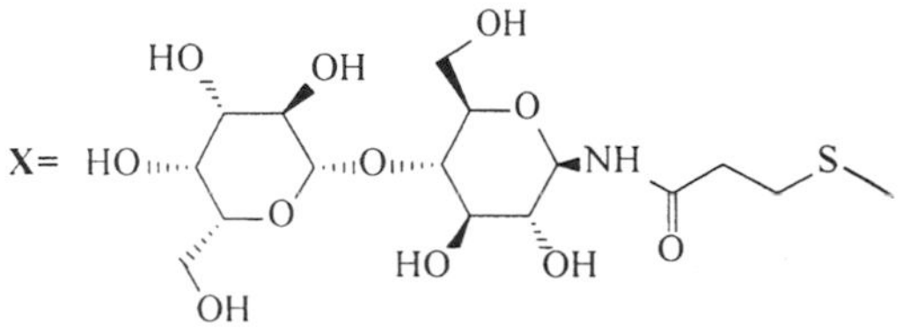

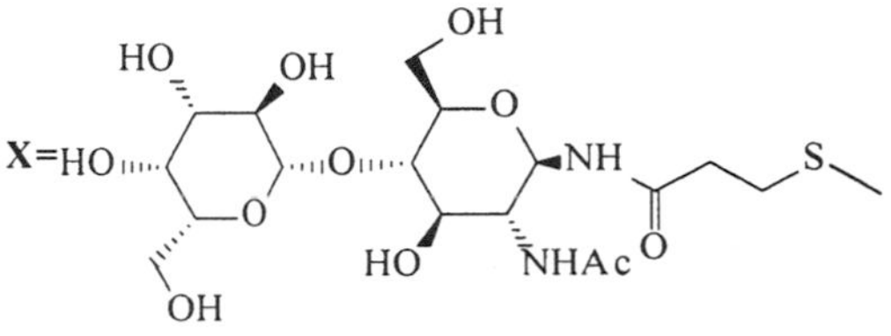

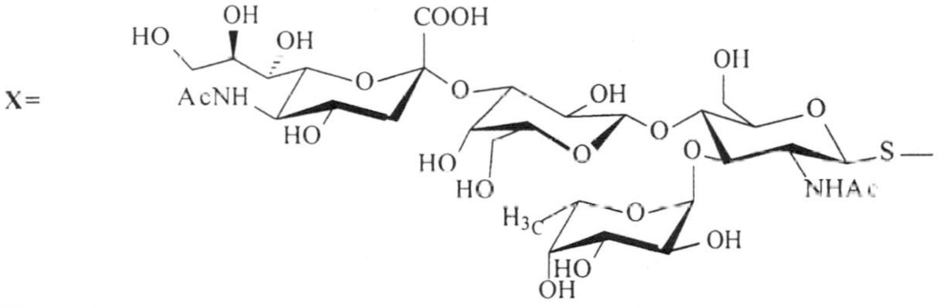

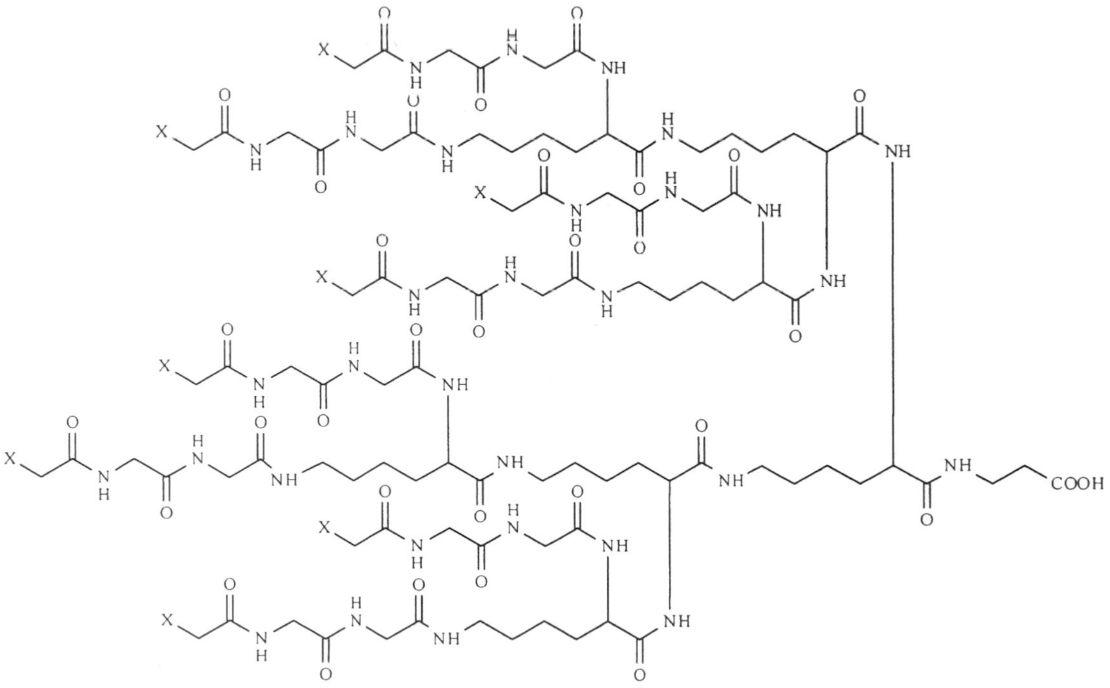

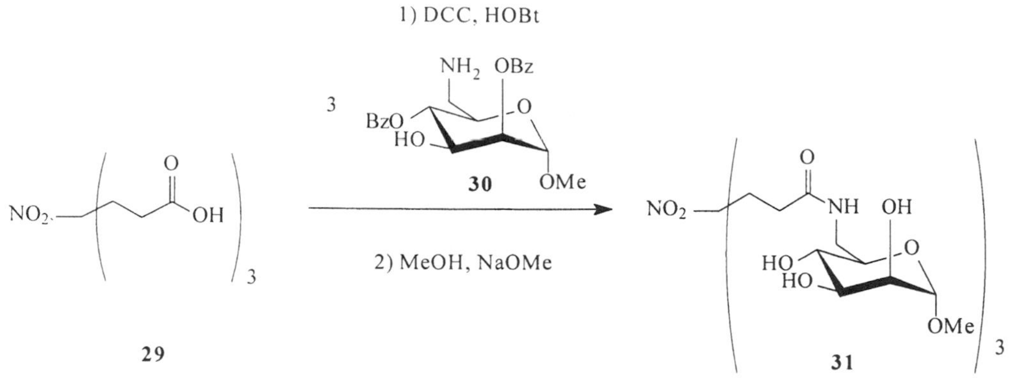

Figure 12.12 The synthesis of a multivalent mannose inhibitor for the adhesion of *Escherichia coli* HB 101 (pPK14).

or multivalency effects. The first example of enzymatic synthesis of sLe^x-functionalized polylysine dendrimers 27 has been reported (64). These dendrimers are currently investigated as selectin antagonists and could provide important information regarding the rational design of anti-inflammatory drugs.

In the quest for a better understanding of the role played by the interligand distances on binding of neoglycoconjugates, a series of polydentate dendritic disaccharides were constructed on a PAMAM scaffold (65). Thus, 4, 8, 16, and 32 ligand-bearing dendrimers were assessed as inhibitors for the anti-B_{di} immunoglobulin which causes rejection in pig to human xenotransplantation (32 in Fig 13). The results of molecular dynamic simulations suggested a match between the interligand distances of the 32-arm neoglycoconjugates and the antigen binding sites of the IgG immunoglobulin. Experimentally, it was found that the di-, tetra-, and octameric dendrimers bound as tight as the monomeric B_{di} ligand. The 16 and 32 glycodendrimers however had an IC_{50} about tenfold lower than the monomeric ligand, which may be indicative of a better fit in the topology of the ligand–receptor interactions.

While carbohydrates have mainly been used as end groups on dendrimers of a different chemical nature, fully synthetic carbohydrate dendrimers could also mimic a broad spectrum of polysaccharide structures and properties. The three-dimensional closely packed structure of highly branched dendritic oligosaccharides could be, for example, responsible for inclusion properties similar to those of cyclodextrines (66). Thus, the attachment of the heptasaccharide 33 to a central trifunctional core should afford a C_3-symmetrical glucodendrimer made entirely of carbohydrate moieties.

Carbohydrates can be also used as multivalent cores for dendritic molecules. Allyl α-D-glucopyranoside was O-perallylated to the corresponding tera-allyl derivative 34 (Fig. 14). The rich chemistry of the allyl ether group was subsequently exploited to provide uniformly functionalized spacer glycosides. The new ''octopus glycosides'' 35–38 are useful core molecules for the synthesis of glycoclusters and carbohydrate-centered dendrimers.

C. Dendrimer–Peptide Conjugates

Synthetic peptides conjugated to carrier proteins can stimulate the production of antibodies leading to an immune response to a certain amino acid sequence. The carrier protein has generally a low and random surface density of the peptide antigen, and the control of the protein functionalization is often difficult. Complications from cross-reactivity of the antibodies with the carrier protein can also often arise.

The concept of multivalent interaction has been successfully applied to elicit a stronger immune response to a certain peptide sequence. Multiple peptide–antigen systems (MAPS) (67) have been mounted on dendritic polylysine structures similar to 28 and yielded results normally obtained at much higher concentrations of the monomeric vaccine preparations (68). The peptides are presented with high surface density, and the homogenous nature of the dendrimers should ensure a highly reproducible effect. Remarkably, the branching polylysine dendrimers did not cause any immunological response (69), thus avoiding cross-reactivity complications. Unlike the peptide-carrier protein vaccines, the MAP system accounted for approximately 90% of the total molecular mass of the dendrimer–peptide conjugate.

Functional groups on the surface of the dendritic structures show increased chemical reactivity in comparison to other polymer homolog reactions (70). Biologically active

PAMAM

32

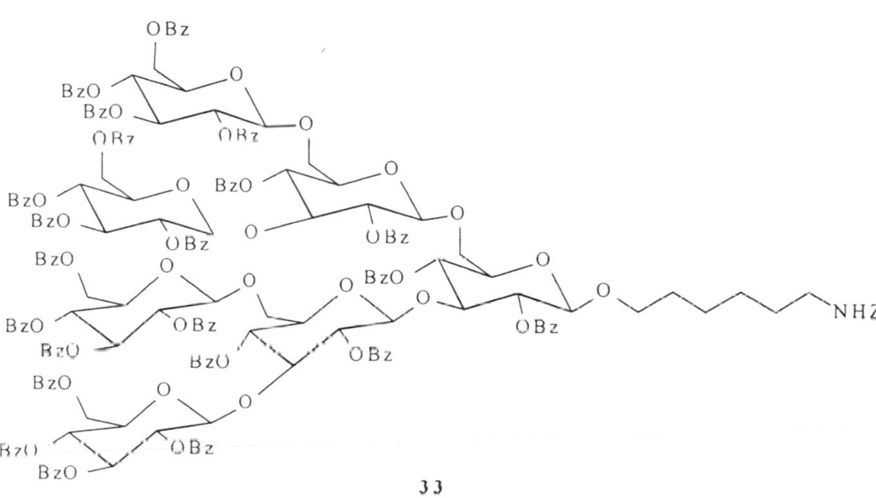

Figure 12.13 Examples of sugar-coated and sugar-made dendrimers.

molecules should therefore react under milder reaction conditions to yield various dendritic conjugates without diminishing their activity due to degradation.

This hypothesis has been confirmed by the synthesis of dendrimers bearing sensitive proteins (71). Thus, the activation of the surface groups of Starburst® dendrimers to various electrophilic and nucleophilic groups (as in 44 and 45, Fig. 15) was reported. Both the amino- and carboxyl-terminated dendrimers could be reacted with alkaline phosphatase to form a protein–dendrimer conjugate of type 41, which in turn was further conjugated with the Fab' fragment of an anticreatine kinase MB isoenzyme antibody to yield 43.

Antibodies are ideal vectors for targeting specific cells due to their selective interactions with antigens expressed on the surface of the cellular membrane. Consequently, antibody conjugates with radioisotopes are under active investigation for applications in radioimmunodetection and radioimmunotherapy of cancer.

PAMAM dendrimers were reacted with 2-(p-isothiocyanatobenzyl)-6-methyl-diethylene triamine pentaacetic acid (1B4M) to generate a dendrimer–chelating agent conjugate. This was further chemically bound to an anti-Tac (72) IgG(HuTac) monoclonal antibody and finally complexed with ^{111}In and ^{88}Y. Unfortunately, the immunoreactivity and the biodistribution of the radiolabelled dendrimer chelate and the antibody–dendrimer chelate did not compare favorably with the radiolabelled antibody alone due to significant accumulation in the liver, kidney, and spleen.

More encouraging results were obtained in the use of dendrimers for the ''pretargeting'' approach to cancer therapy (73). In this approach the tumor-selective targeting of monoclonal antibodies (mAbs) is retained, but the delivery of the radionuclide is relegated to another molecule, which is often a radiolabelled biotin, avidin, or streptavidin molecule (Fig. 16). Thus, it is hoped that the immunoreactivity and biodistribution of the mAbs would be better preserved, while the selective tumor localization of the radiolabelled biotin would be secured by the strong biotin–streptavidin binding. In addition to improving the pharmacokinetic distribution of the therapeutic radionuclide, the biotin/streptavidin-based reagents may provide a method for the step-

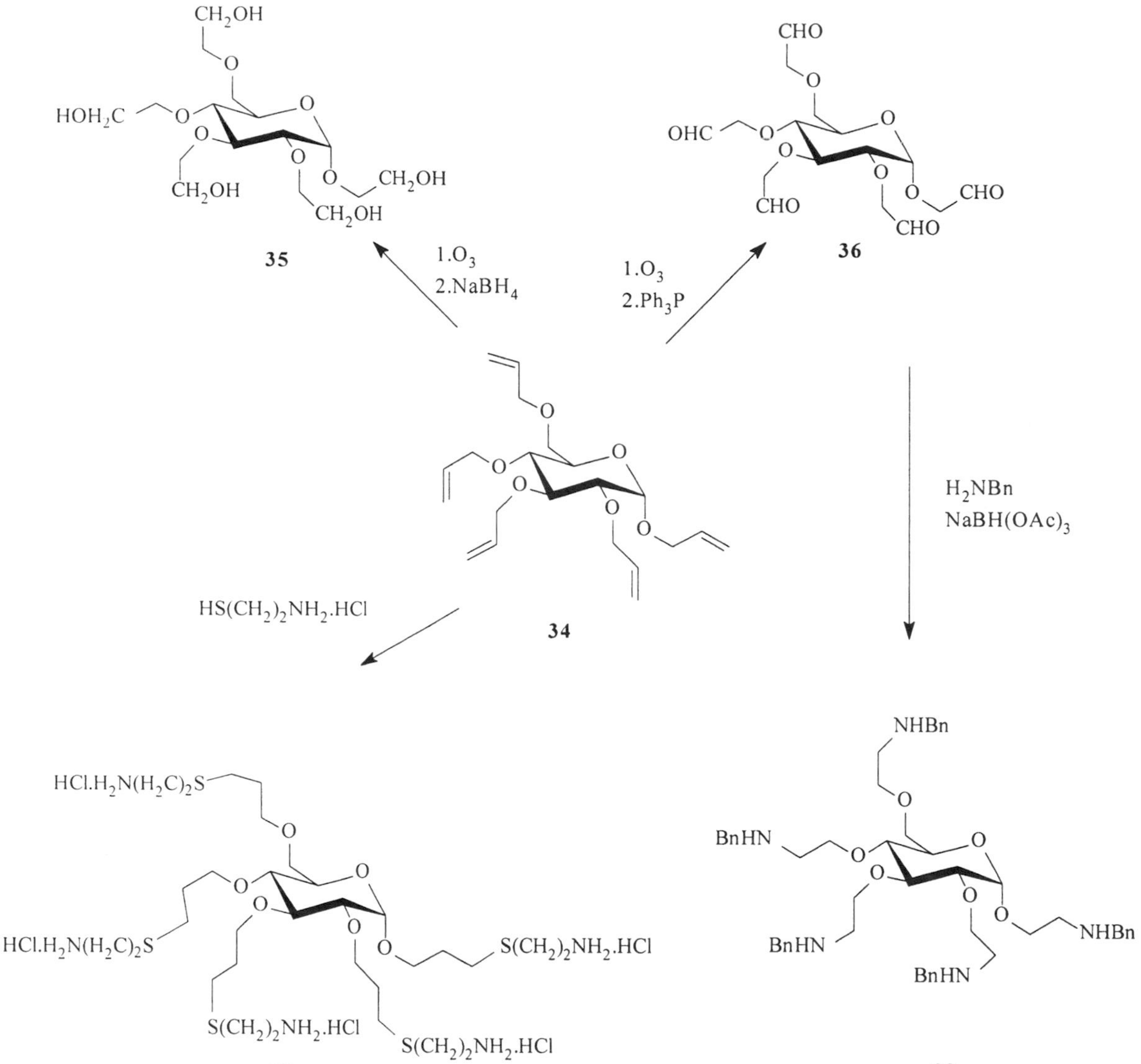

Figure 12.14 Octopus glycosides are useful core molecules for the synthesis of glycoclusters and carbohydrate centered dendrimers.

wise increase of the amount of radioactive metal bound to cancer cells in vivo.

In order to achieve this goal, PAMAM dendrimers conjugated with biotin are particularly attractive due to their compactness, multifunctionality, and size control (Fig. 17). These parameters are essential for reaching the optimal balance between tumor penetration and rapid increase in the bound radionuclide. Thus, various generations of Starburst® dendrimers were reacted with N-succinimidyl p-tributylstannyl benzoate **47** (in Fig. 17) followed by perbiotinylation with the new reagent **48**. Iododestannylation allowed the radiolabelling of the Starburst® dendrimer **49** (Fig. 17).

D. Dendrimer–Oligonucleotide and –Nucleic Acid Conjugates

PAMAM dendrimers have been investigated regarding their ability to transfer biomolecules into several mammalian cell lines (74). Their use as gene transfer vectors in achieving a highly efficient transfection exploits the electrostatic interactions that occur at physiological pH. In this environment, the protonated amino terminal groups interact with biologically relevant polyanions, such as nucleic acids. Thus, synthetic polycationic dendrimers can serve as mimics of naturally occurring macromolecules with a vital biological function, such as histones and spermidines.

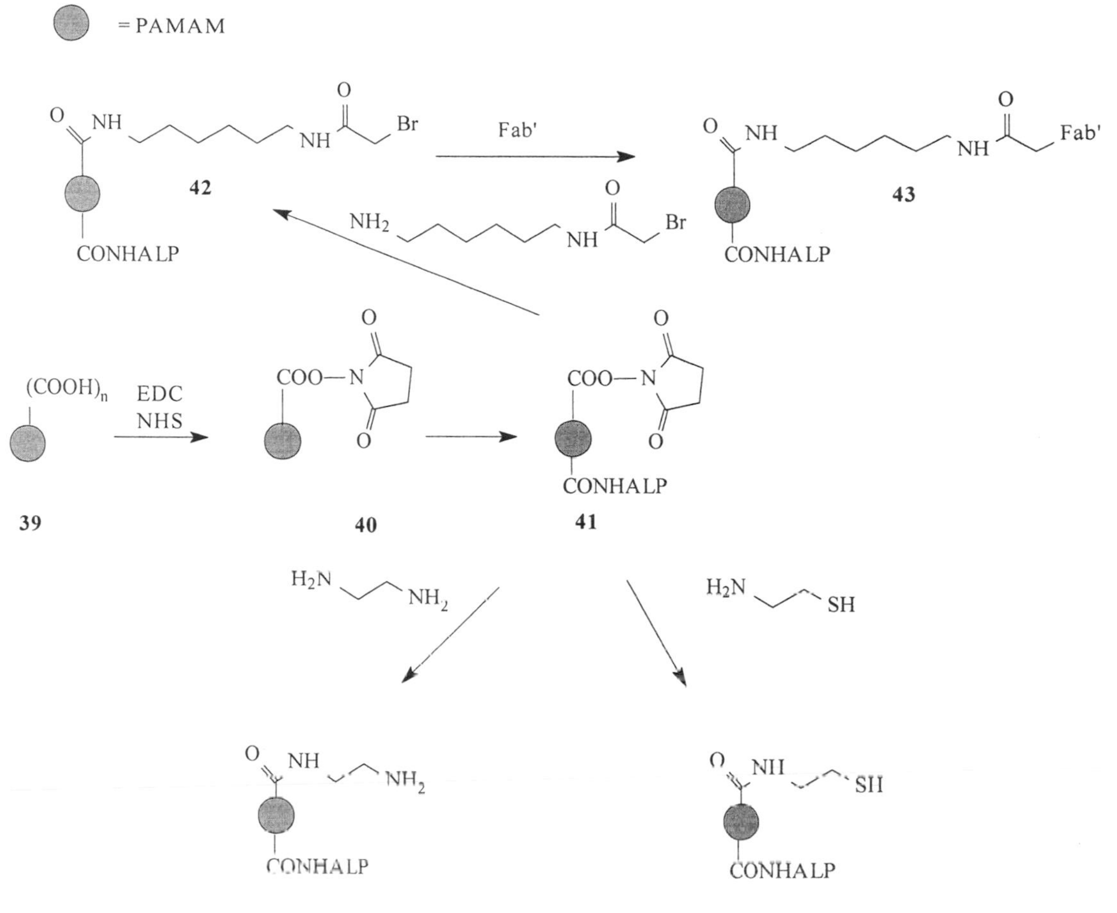

Figure 12.15 Activation and reactions of dendrimers containing terminal carboxyl groups.

New studies (75) analyzed the interactions between the Starburst® dendrimers and polynucleotides such as calf thymus DNA, poly(AT), poly(GC), and double-stranded oligonucleotides of 12 base pairs (DNA-12mer). These interactions were studied by EPR spectroscopy employing nitroxide radicals as spin labels (covalently bound to dendrimers). Computer-aided analysis of the EPR spectra provided information on the mobility of the labels and their partitions in different environments. This type of interaction is extremely important in understanding the mechanism of PAMAM-mediated gene transfer in mammalian cells. Spectral analysis of the investigated polynucleotides indicated that the DNA-12mer presented the strongest interaction, presumably due to its short length. Given the fact that dendrimer–DNA interactions depend on the degree of protonation of the external amino groups, their dependence on the pH of the environment was investigated. Small dendrimers at low pH showed significant interaction with the polynucleotides. At higher concentration of dendritic molecules the interaction decreased due to self-aggregation. By contrast, the large dendrimers suffer a partial swelling as a result of the electrostatic repulsion of the charged surface groups. This destabilization can be lowered by the interaction with the polyanionic DNA leading to much stronger interactions.

Studies have also shown plasmid DNA (76) conjugation with block copolymers containing methoxy-poly(ethylene glycol) (mPEG) and poly(L-lysine) dendrimer. The shape and particle size distributions of the copolymer–plasmid complex at various charge ratios were examined by atomic force microscopy. It was shown that this novel linear dendrimer–polymer block copolymer can self-assemble with plasmid DNA at physiological conditions, forming a compact and water-soluble polyionic complex. The nuclease resistance of the complex proved that the copolymer increased the stability of the plasmid DNA suggesting that these materials could be valuable for the delivery of genetic material such as antisense or plasmid DNA.

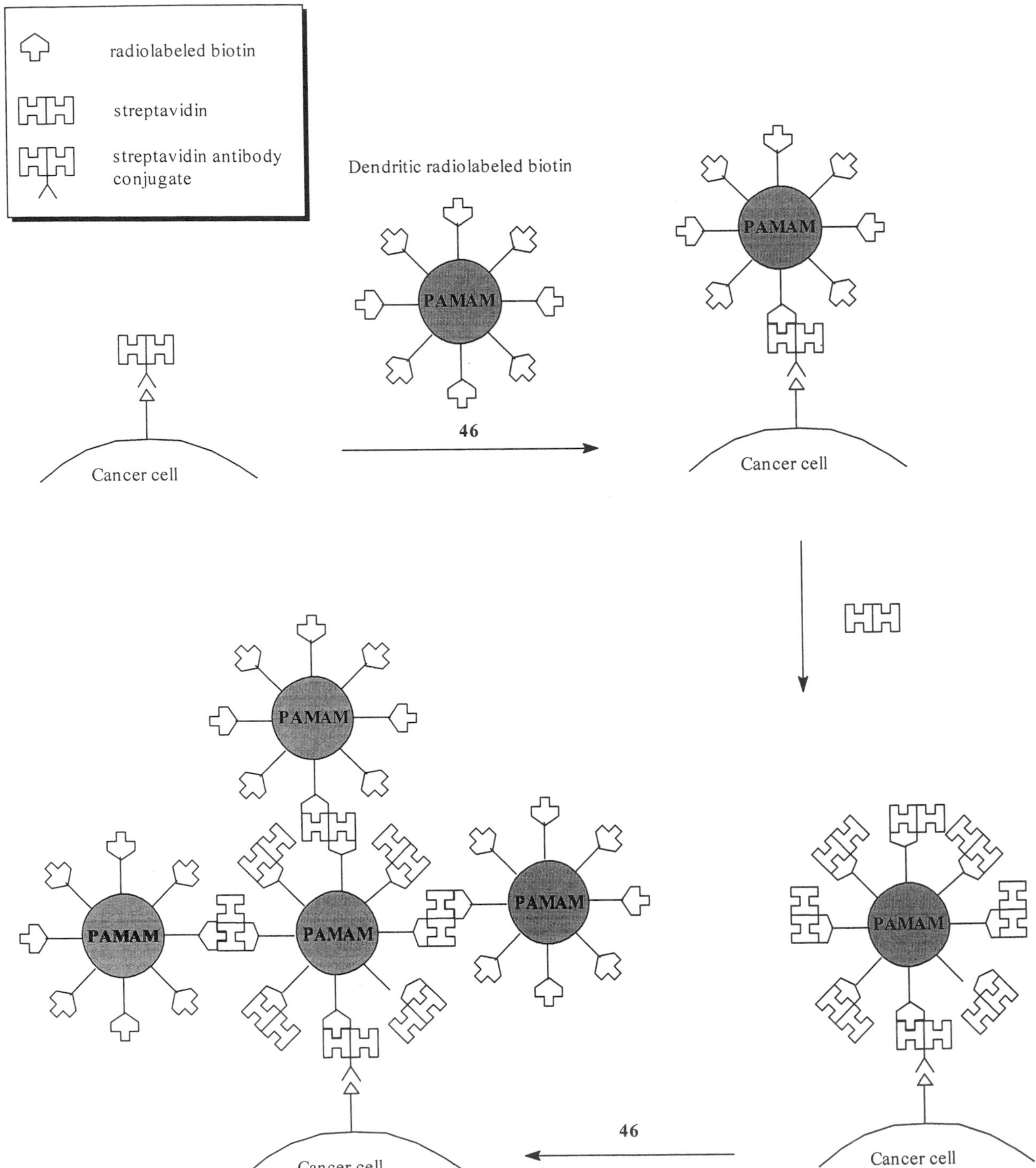

Figure 12.16 The pretargeting approach to cancer therapy. A streptavidin–antibody conjugate initially identifies the cancer cell and then a radiolabelled dendritic biotin binds strongly to this substrate. Thus, in just a few steps one can bind a lethal amount of radionuclide to the cancer cell.

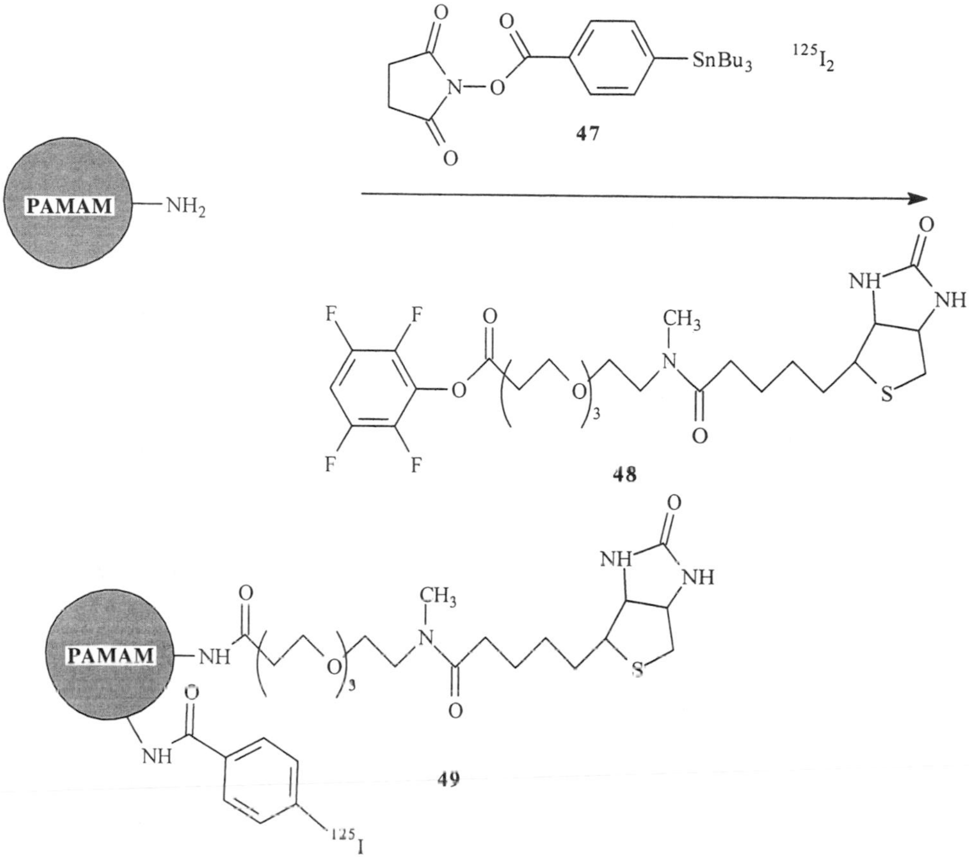

Figure 12.17 Synthesis of the radiolabelled dendritic biotin reagent for cancer therapy.

III. CONCLUSIONS

As it is increasingly evident that multivalent interactions play a key role in various biological phenomena, this field will continue to flourish in the years to come. Any biological surface interacting with another biological entity through multivalency can in principle become the target for the design of a new polyvalent drug.

Dendrimers can provide unique opportunities in the rational design of nanomolecular scaffolds for the attachment of active biological ligands. Thus, by controlling the chemical nature of the dendritic layers, the presentation of the end groups on the surface may change. The core functionality, the length, and the functionality of the monomer are important parameters controlling the surface density and the topology of the end groups.

The results obtained so far predict that breakthrough applications can occur in the design of inhibitors for the attachment of both viruses and bacteria. Other applications in the fields of gene therapy and new cancer treatment protocols based on multivalent dendritic conjugates may follow.

The field of dendrimer chemistry is currently in the exponential growth phase, as demonstrated by Fig. 1 of this brief review. Important applications will certainly mushroom across the spectrum of the medical sciences and have the potential to generate significant intellectual property rights.

REFERENCES

1. D. A. Tomalia, R. Esfand. Dendrons, dendrimers and dendrigrafts. *Chemistry & Industry*. 416 (1997).
2. C. J. Hawker, J. M. J. Fréchet. Preparation of polymers with controlled molecular architecture. A new convergent approach to dendritic macromolecules. *J. Am. Chem. Soc.* 112:7638 (1990).
3. A. W. Bosman, H. M. Jansen, E. W. Meijer. About dendrimers: structure, physical properties, and applications. *Chem. Rev.* 99:1665 (1999).
4. J. M. Gardiner. The therapeutic potential of synthetic multivalent carbohydrates. *Exp. Opin. Invest. Drugs* 7:405 (1998).
5. F. Zeng, S. C. Zimmerman. Dendrimers in supramolecular

chemistry: from molecular recognition to self-assembly. *Chem. Rev*. 97: 1681 (1997).

6. D. A. Tomalia, H. Baker, J. R. Dewald, M. Hall, G. Kallos, S. Martin, J. Roeck, J. Ryder, P. Smith. A new class of polymers—Starburst dendritic macromolecules. *Polym. J*. 17:117 (1985).
7. G. R. Newcome, Z-Q. Yao, G. R. Baker, V. K. Gupta, P. S. Russo, M. J. Saunders. Cascade molecules. 2. Synthesis and characterization of a benzene (9)3-Arborol. *J. Am. Chem. Soc*. 108:849 (1986).
8. M. J. Maciejewski. Concepts of trapping topologically by shell molecules. *J. Macromol. Sci. Chem*. A17:689 (1982).
9. P. J. De Gennes, H. J. Hervet. Statistics of Starburst polymers. *Phys. Let*. 44:351 (1983).
10. S. F. Edwards. The statistical mechanics of polymerized material. *Proc. Phys. Soc. London* 92:9 (1967).
11. R. L. Lescanec, M. Muthukuma. Configurational characteristics and scaling behavior of Starburst molecules: a computational study. *Macromolecules* 23:2280 (1990).
12. M. L. Mansfield, L. I. Klushin. Monte Carlo studies of dendrimer macromolecules. *Macromolecules* 26:4262 (1993).
13. M. Murat, G. S. Grest. Molecular dynamics study of dendrimer molecules in solvents of varying quality. *Macromolecules* 29:1278 (1996).
14. D. Boris, M. Rubinstein. A self-consistent mean field model of a Starburst dendrimer: dense core vs dense shell. *Macromolecules* 29:7251 (1996).
15. J. S. Moore. Shape-persistent molecular architectures of nanoscale dimensions. *Acc. Chem. Res*. 30:402 (1997).
16. P. Miklis, T. Cagin, W. A. Goddard. Dynamics of Bengal Rose encapsulated in the Meijer dendrimer box. *J. Am. Chem. Soc*. 119:7458 (1997).
17. S. Brownstein, B. Comanita, J. Roovers. Unpublished results.
18. V. Percec, G. Johansson, G. Ungar, J. Zhou. Fluorophobic effect induces the shelf-assembly of semifluorinated tapered monodendrons containing crown ethers into supramolecular columnar dendrimers which exhibit a homeotropic hexagonal columnar liquid crystalline phase. *J. Am. Chem. Soc*. 118:9855 (1996).
19. S. D. Hudson, H.-T. Jung, V. Percec, W.-D. Cho, G. Johansson, G. Ungar, V. S. K. Balagurusamy. Direct visualisation of individual cylindrical and spherical supramolecular dendrimers. *Science* 278:449 (1997).
20. B. Stark, B. Stühn, H. Frey, C. Lach, K. Lorentz, B. Frick. Segmental dynamics in dendrimers with perfluorinated end groups. A study using quasielastic neutron scattering. *Macromolecules* 3:5415 (1998).
21. J. F. G. A. Jansen, E. M. M. de Brabander-van den Berg, E. W. Meijer. Encapsulation of guest molecules into a dendritic box. *Science* 266:1226 (1994).
22. D. A. Tomalia, A. M. Naylor, W. A. Goddard. Starburst dendrimers. Molecular level control of size, shape, surface chemistry and flexibility from atoms to macroscopic matter. *Angew. Chem. Int. Ed. Engl*. 29:138 (1990).
23. M. C. Moreno-Bondi, G. Orellana, N. J. Turro, D. A. Tomalia. Photoinduced electron-transfer reactions to probe the structure of Starburst dendrimers. *Macromolecules* 23:912 (1990).
24. M. F. Ottaviani, F. Montalti, N. J. Turro, D. A. Tomalia. Characterization of Starburst dendrimers by the technique copper(II) ions binding full-generation. *J. Phys. Chem*. B101:158 (1997).
25. G. Caminati, N. J. Turro, D. A. Tomalia. Photophysical investigation of Starburst dendrimers and their interactions with anionic and cationic surfactants. *J. Am. Chem. Soc*. 112:8515 (1990).
26. R. Wilken, J. Adams. End-group dynamics of fluorescently labeled dendrimers. *Macromol. Rapid Commun*. 18:659 (1997).
27. T. J. Prosa, B. J. Bauer, E. J. Amis, D. A. Tomalia, R. Scherremberg. A SAXS Study of the internal structure of dendritic polymer systems. *Polymer Sci. Part B: Polym. Physics* 35:2913 (1997).
28. R. Kleppinger, H. Reiners, K. Desmedt, B. Forier, W. Dehaen, M. Koch, P. Verhaert. A small-angle x-ray scattering study of sizes and shapes of poly(benzyl ether) dendrimer molecules. *Rapid Commun*. 19:111 (1998).
29. R. Scherremberg, B. Coussens, P. van Vliet, G. Edward, J. Brackman, E. De Brabander. The molecular characterization of poly(propyleneimine) dendrimers as studied with small-angle neutron scattering. Viscosimetry and molecular dynamics. *Macromolecules* 31:456 (1998).
30. E. J. Corey, X. M. Cheng. *The Logic of Chemical Synthesis*. Wiley, New York, 1989.
31. D. A. Tomalia, H. D. Durst. Genealogically directed synthesis: Starburst cascade dendrimers and hyperbranched structures. *Topics in current chemistry* 165:193 (1993).
32. C. J. Hawker, J. M. J. Frèchet. Preparation of polymers with controlled molecular architecture. A new convergent approach to dendritic macromolecules. *J. Am. Chem. Soc*. 112:7638 (1990).
33. B. Comanita, J. Roovers. Synthesis of new carbosilane dendrimers with hydrophilic end-groups. Polyols. *Designed Monomers and Polymers* 2:112 (1999).
34. E. Buhleier, W. Wehner, F. Vögtle. ''Cascade'' and ''nonskid-chain-like'' syntheses of molecular cavity topologies. *Synthesis* 155 (1978).
35. D. A. Tomalia, H. Baker, J. Dewald, M. Hall, G. Kallos, S. Martin, J. Roeck, J. Ryder, P. Smith. A new class of polymers. Starburst dendritic macromolecules. *Polymer J*. 17:117 (1985).
36. C. Wömer, R. Mülhaupt. Polynitrile-functional and polyamine-functional poly(trimethylene imine) dendrimers. *Angew. Chem. Int. Ed. Engl*. 32:1306 (1993).
37. E. M. M. de Brabander-van den Berg, E. W. Meijer. Poly (propylene imine) dendrimers. Large-scale synthesis by heterogeneously catalized hydrogenations. *Angew. Chem. Int. Ed. Engl*. 32:1308 (1993).
38. E. M. M. de Brabander, J. Brackman, M. Mure-Mak, H. De Man, M. Hogeweg, J. Keulen, R. Scherrenberg, B. Coussens, Y. Mengerink, S. J. van der Wal, E. W. Meijer. Polypropyleimine dendrimers: improved synthesis and characterization. *Macromol. Symp*. 102:9 (1996).

39. T. M. Miller, T. X. Neenan. Convergent synthesis of monodisperse dendrimers based upon 1,3,5-trisubstituted benzenes. *Chem Matter* 2:346 (1990).
40. K. L. Wooley, C. J. Hawker, J. M. J. Fréchet. Hyperbranched macromolecules via a novel double-stage convergent growth approach. *J. Am. Chem. Soc.* 113:4252 (1991).
41. R. Spindler, J. M. J. Fréchet. 2-step approach towards the accelerated synthesis of dendritic macromolecules. *J. Chem. Soc. Perkin Trans* I:913 (1993).
42. J. W. Leon, M. Kawa, J. M. J. Fréchet. Isophthalate ester-terminated dendrimers. Versatile nanoscopic building-blocks with readily modifiable surface functionalities. *J. Am. Chem. Soc.* 118:8847 (1996).
43. K. L. Wooley, C. J. Hawker, J. M. J. Fréchet. A branched monomer approach for the rapid synthesis of dendrimers. *Angew. Chem. Int. Ed. Engl.* 33:82 (1994).
44. G. L'abbé, B. Forier, W. Dehaen. A fast double-stage convergent synthesis of dendritic polyethers. *Chem. Commun.* 2143 (1996)
45. H.-T. Chang, C.-T. Chen, T. Kondo, G. Siuzdal, K. B. Sharpless. Asymmetric dihydroxylation enables rapid construction of chiral dendrimers based on 1,2-diols. *Angew. Chem. Int. Ed. Engl.* 35:182 (1996).
46. T. Kawaguchi, K. L. Walker, C. L. Wilkins, J. S. Moore. Double exponential dendrimer growth. *J. Am. Chem. Soc.* 117:2159 (1995).
47. M. Mammen, S.-K. Choi, G. M. Whitesides. Polyvalent interactions in biological systems: implications for design and use of multivalent ligands and inhibitors. *Angew. Chem. Int. Ed.* 2755 (1998).
48. B. Perlmutter-Hayman. Cooperative binding to macromolecules. A formal approach. *Acc. Chem. Res.* 19:90 (1986).
49. Y. Lee, R. Lee. Carbohydrate–protein interactions—basis of glycobiology. *Acc. Chem. Res.* 28:321 (1995).
50. G. D. Glick, J. R. Knowles. Molecular recognition of bivalent sialosides by influenza-virus. *J. Am. Chem. Soc.* 113: 4701 (1991).
51. G. D. Glick, P. L. Toogwood, D. C. Wiley, J. J. Skehel, J. R. Knowles. Ligand recognition by Influenza-virus. The binding of bivalent sialosides. *J. Biol. Chem.* 266:3660 (1991).
52. D. A. Tomalia, R. Esfand. Dendrons, dendrimers and dendrigrafts. *Chemistry & Industry* 416 (1997).
53. S. Brownstein, B. Comanita, J. Roovers. (in press).
54. J. Roovers, B. Comanita. Dendrimers and dendrimer-polymer hybrids. In: *Advances in Polymer Science 142*, J. Roovers (Ed.). Springer-Verlag, Berlin, 1999, p. 179.
55. R. Roy, D. Zanini, S. J. Meunier, A. Romanowska. Solid-phase synthesis of dendritic sialoside inhibitors of Influenza-A virus hemagglutinin. *J. Chem. Soc. Chem. Commun.* 1869 (1993).
56. D. Pagé, S. Aravind, R. Roy. Synthesis and lectin binding properties of dendritic mannopyranoside. *J. Chem. Soc. Chem. Commun.* 1913 (1996).
57. D. Pagé, D. Zanini, R. Roy. Macromolecular recognition effect of multivalency in the inhibition of binding of yeast mannan to concavalin A and pea lectins by mannosylated dendrimers. *Bioorg. Med. Chem.* 4:1949 (1996).
58. D. Pagé, R. Roy. Synthesis and biological properties of mannosylated Starburst poly(amidoamine) Dendrimers. *Bioconjugate Chem.* 8:714 (1997).
59. S. Kötter, U. Krallmann-Wenzel, S. Ehlers, T. K. Lindhorst. Multivalent ligands for the mannose specific lectin type 1 fimbriae of *Escherichia coli*: synthesis and testing of the trivalent α-D mannoside clusters. *J. Chem. Soc., Perkin Trans.* I:2193 (1998).
60. D. Zanini, W. K. C. Park, R. Roy. Synthesis of novel dendritic glycosides. *Tetrahedron Lett.* 36:7383 (1995).
61. D. Zanini, R. Roy. Chemoenzymatic synthesis and selectin binding properties of dendritic N-acetyllactosamine. *Bioconjugate Chem.* 8:187 (1997).
62. J. C. Paulson. *The Receptors*, (Michael Conn (Ed.). Academic Press, Orlando, FL, 1985, p. 131.
63. M. Linares, R. Roy. Multivalent neoglycoconjugates solid-phase synthesis of N-linked α-sialodendrimers. *Chem. Commun.* 2119 (1997).
64. M. M. Palcic, H. Li, D. Zanini, R. S. Bhella, R. Roy. Chemoenzymatic synthesis of dendritic sialyl Lewis$_x$. *Carbohydr Res.* 305:433 (1998).
65. D. E. Tsevtkov, P. E. Cheshev, A. B. Tuzicov, G. V. Pazinina, N. V. Bovin, R. Rieben, N. E. Nifant'ev. Synthesis of neoglycoconjugate dendrimers. *Mendeleev Commun.* 47 (1999).
66. N. Jayaraman, S. A. Nepogodiev, J. F. Stoddart. Synthetic carbohydrates containing dendrimers. *Chem. Eur. J.* 3:1193 (1997).
67. J. P. Tam. Synthetic peptide vaccine design: synthesis and properties of a high-density multiple antigenic peptide system. *Proc. Natl. Acad. Sci. USA* 85:5409 (1988).
68. M. J. Francis, G. Z. Hastings, F. Brown, J. McDermed, Y. Lu, J. P. Tam. Immunological evaluation of the multiple antigen peptide (MAP) System using the major immunogenic site of foot-and-mouth disease virus. *Immunology* 73: 249 (1991).
69. J. P. Tam. High-density multiple antigen–peptide system for preparation of antipeptide antibodies. *Methods Enzymol.* 168:7 (1998).
70. J. M. J. Fréchet. Functional polymers and dendrimers: reactivity, molecular architecture and interfacial energy. *Science* 263:1710 (1994).
71. P. Singh. Terminal groups in Starburst dendrimers: activation and reaction with proteins. *Bioconjugate Chem.* 9:54 (1998).
72. H. Kobayaski, C. Wu, M.-K. Kim, H. C. Paik, J. A. Carrasquillo, M. W. Brechbiel. Evaluation of the in vivo biodistribution of indium-111 and yttrium-88 labeled dendrimer-1B4M-DTPA and its conjugation with anti-tac monoclonal antibody. *Bioconjugate Chem.* 10:103 (1999).
73. D. S. Wilbur, P. M. Pathare, D. K. Hamlin, K. R. Buhler, R. L. Vessella. Biotin reagents for antibody pretargetting. 3. Synthesis, radioiodination and evaluation of biotynylated Starburst dendrimers. *Bioconjugate Chem.* 9:813 (1998).
74. J. Haensler, F. C. Szoka, Jr. Polyamidoamine cascade poly-

mers mediate efficient transfection of cells in culture. *Bioconjugate Chem.* 4:372 (1993).

75. M. F. Ottaviani, B. Sacchi, N. J. Turro, W. Chen, S. Jockusch, D. A. Tomalia. An EPR study of the interactions between Starburst dendrimers and polynucleotides. *Macromolecules* 32:2275 (1999).

76. J. S. Choi, E. J. Lee, Y. H. Choi, Y. J. Jeang, J. S. Park. Poly(ethyleneglycol)-block-poly(L-lysine) dendrimer: novel linear polymer/dendrimer block copolymer forming a spherical water-soluble polyionic complex with DNA. *Bioconjugate Chem.* 10:62 (1999).

13

Biocompatibility of Elastomers

D. J. Chauvel-Lebret, P. Auroy, M. Bonnaure-Mallet
Université de Rennes, Rennes, France

I. ELASTOMERS

A. Introduction

Elastomers, like plastics, are polymers. The term *elastomer* is currently used to designate all rubbers, i.e., natural or synthetic macromolecular materials with rubberlike elasticity. Medical applications for elastomers appeared soon after they began being produced on an industrial scale (1) following the discovery of vulcanization (crosslinking) in 1839 by the American Charles Goodyear. Over the past 15 years, the literature on biomedical applications for elastomers has grown considerably and reports have appeared regularly (2,3). Yoda (1) recently published a detailed review.

1. Definition

While they are members of the broad family of polymers, elastomers behave differently than plastic materials or plastomers. Generally speaking, for materials to be considered elastomers they must be

- Flexible, i.e., have low rigidity (several megapascals)
- Highly deformable, i.e., are able to withstand strong deforming forces without rupturing and have an elongation at rupture over 200%., while possessing relatively high tensile strength at ultimate elongation
- Elastic or resilient, i.e., are able to return to their original shape and size after the deforming force is removed and quantitatively release the energy used to deform them (Table 1) (4).

Many so-called biomedical elastomers do not meet the second criterion. While it is difficult to set an elastic limit beyond which a polymer stops being a plastomer and becomes an elastomer, we will use the broadest medical definition and refer to all polymers with sufficient elasticity to return to their original shape after a substantial deformation as elastomers (5). However, it will occasionally be more appropriate to talk about elastic polymers than true elastomers.

2. Classification

a. *The Various Families of Elastomers*

There are some 15 families of elastomers, some of which include 10 to 20 different grades. Modern polymerization processes increasingly allow the manufacture of customized products that address specific application problems and provide specific elastomeric properties.

The current classification system separates elastomers into four categories (the acronyms are based on the ISO 1629:1995 standard):

- *General-use elastomers.* Includes natural rubber (NR), synthetic polyisoprene (IR), styrene butadiene copolymers (SBR), and polybutadienes (BR)
- *Special elastomers.* Includes ethylene propylene and diene copolymers (EPM, EPDM), isobutylene and chloro- and bromoisobutylene isoprene copolymers (IIR, BIIR, CIIR), nitrile butadiene copolymers (NBR), and polychloroprenes (CR)
- *Very special elastomers.* High thermal and/or chemical

Table 13.1 Some Commercially Available Elastomers Arranged by Tensile Strength

Elastomer	Trade name	Supplier	Tensile strength (PSI)	Tensile strength (MPa)	Ultimate elongation (%)
Butyl rubber (polyisobutylene)	Butyl	Polysar Inc., Akron, OH	700–1500	4.8–10.3	200–250
Polyisoprene	Santoprene	Monsanto Co., Akron, OH	1640	11.3	520
Silicone rubber	Silastic	Dow Corning, Midland, MI	500–1500	3.4–10.3	200–1000
	Med	McGhan NuSil Corp., Carpenteria, CA	750–1750	5.2–12.1	850
	SM	Sil-Med Corp., Taunton, MA	500–1500	3.4–10.3	250–900
Silicone rubber copolymers	C-Flex	Concept polymer, Inc., Clearwater, FL	1000–2000	6.9–13.8	200–600
Polyvinylchloride (PVC)	Goodtouch	B.F. Goodrich, Cleveland, OH	1700–1850	11.7–12.7	430–460
	Tygon	Norton Perform Plastic, Wayne, NJ	1200–2300	8.3–15.9	375–410
Ethylene vinyl acetate (EVA)	Ultrathene	U.S. Industrial Chemical Co., New York, NY	2000	13.8	700
Acrylonitrile butadiene	Krynac	Polysar Inc., Akron, OH	500–2250	3.4–15.5	150–650
Polyhexene	Hexsyn	Goodyear Tire Co., Akron, OH	1500–2400	10.3–16.5	250–350
Polychloroprene	Neoprene	Du Pont, Wilmington, DE	1000–2500	6.9–17.2	200–600
Fluoroelastomers	Fluorel	3M, St. Paul, MN	1500–2500	10.3–17.2	100–400
Styrene butadiene rubber (SBR)	Krayton	Shell Chemical Co., Houston, TX	750–2650	5.2–18.3	550–850
	Plioflex	Goodyear Tire Co., Akron, OH	200–2825	1.4–19.5	290–900
Ethylene vinyl acetate (EVA)	EVA	Du Pont, Wilmington, DE	3200	22.1	600–750
Natural rubber (cis-polyisoprene)	Natsyn	Goodyear Tire Co., Akron, OH	2200–3450	15.2–23.8	500–600
Polyester terephthalate glycolated	Hytrel	Du Pont, Wilmington, DE	4000–6800	27.6–46.9	350–700
Nylon (amorphous)	Pebax	Atochem Polymers, Philadelphia, PA	4200–7400	29.0–51.0	350–680
Thermoplastic polyether urethanes	Tecoflex	Thermedics, Inc., Woburn, MA	3000–6000	20.7–41.4	250–600
Thermoplastic polyester urethanes	Estane	B.F. Goodrich, Cleveland, OH	4000–8000	27.6–55.2	400–600
Thermoplastic polycarbonate urethane	Corethane	Corvita Corp., Miami, FL	4500–9000	31.0–62.1	250–500
Chlorotrifluoroethylene	Aclar	Allied-Signal, Morristown, NJ	7500–11000	51.7–75.8	200–300
	Saran	Dow Chemical, Midland, MI	8000–16800	55.1–115.8	70–90

Source: Ref. 4.

resistance elastomers, including silicone elastomers (VMQ), fluoroelastomers (FPM), chloropolyethylenes and chlorosulfonyl polyethylenes (CM, CSM), polyacrylates (ACM), ethylene vinyl acetate elastomers (EVA), ethylene methyl acrylate (EAM), hydrogenated nitrile elastomers (HNBR), and epichlorhydrin elastomers (CO, ECO, GECO)

Thermoplastic elastomers. A separate category indicating that they do not need to be vulcanized since they can be used like any thermoplastic material.

The first two families of elastomers account for 97% of total world consumption. The third family includes high thermal and/or chemical performance elastomers, which are experiencing strong market growth. The vast majority of elastomers are polymerized from petroleum monomer derivatives.

b. *Very Special Elastomers*

Polysiloxanes, or silicone elastomers, are the most abundantly produced very special elastomers. The backbone of these elastomers is not a chain of carbon atoms but an arrangement of silicon and oxygen atoms. Their most remarkable quality is their outstanding thermal performance. They retain their service properties from −85 to 250°C. However, their mechanical properties are relatively weak. They are used for a broad range of applications in the electrical, electronic, and automotive industries. Their physiological inertness makes them well suited for use in the biomedical field.

Fluorocarbons are another family of high-performance elastomers, containing 70% fluoride that gives them outstanding resistance to heat and chemicals. They can be used at temperatures up to 250°C. Their cold resistance, however, is less than that of silicones (−30°C). Because of their pasivity and their insolubility they are useful in the biomedical field.

Polyacrylates were specially designed to resist oils containing sulfur-based additives. They have high heat (180°C) and oil swell resistance. They are used as components of grafted or linear biomedical elastomers [polyurethane (PU), styrene–ethylene/butylene–styrene (SEBS)].

Chloropolyethylenes and chlorosulfonyl polythylenes compete with polychloroprenes because of very similar chemical compositions. The high chemical saturation of the polymer chain gives them better heat resistance (120 to 130°C), and they are often used for wires and cables.

Hydrogenated nitrile, one of the most recent elastomers to arrive on the market, has hydrogenated double bonds, which provide it with better heat resistance (150°C) than conventional nitrile rubber as well as with excellent resistance to oils and ozone.

c. *Thermoplastic Elastomers*

Polymers are commonly defined by their thermal properties, i.e., as thermoplastics and thermosetting materials. *Thermoplastics* are composed of independent, linear molecules (which may have side branches). They can be melted (e.g., polyethylene) and are processed by heating. *Thermosetting materials* are composed of networks of covalently linked chains. They are initially polymerized at room temperature or by heating and cannot be remelted once they have solidified (e.g., silicone elastomers). It is actually a little more complex since certain elastomers have intermediate structures and properties (Fig. 1).

Many types of thermoplastic elastomers (TPE) are being developed. They have certain advantages with respect to classic elastomers and are replacing them in some applications. Thermoplastic elastomers can be defined as materials that combine the flexibility or elasticity of rubber compounds with the thermoreversible versatility of thermoplastic polymers. Most require little or no formulation. If formulation is required, it can be performed by the producer or by an intermediary between the producer and the processor—the compounder. Thermoplastic elastomers are relatively simple to use and require fewer processing steps than thermosetting elastomers (no mixing or crosslinking), thus reducing costs. Processing time is also much shorter, reducing energy consumption. Processing methods are nonspecific and more numerous, being the same as those used for plastomers. Unlike classic elastomers, thermoplastics can be recycled. Manufacturing scraps and clippings can be reused with little or no change in their properties. Despite these numerous advantages, most of these

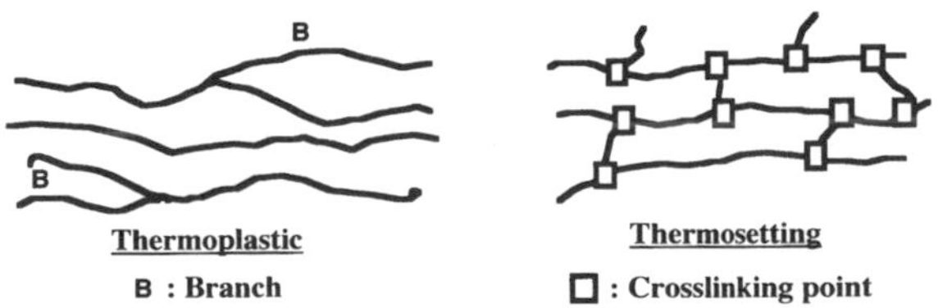

Figure 13.1 Schematic drawing of thermoplastic and thermosetting elastomer structure.

elastomers do have drawbacks, including their temperature sensitivity. In 1995, the international thermoplastic market totaled 900,000 tons (according to Rapra Technology Ltd.), including 329,000 in Europe. Only 2 to 3% were used for medical applications.

3. Chemistry and Synthesis

a. *Structure*

Elastomers are composed of macromolecular chains that occasionally exceed 100,000 atoms in length, fold back on themselves when at rest, and have a very high degree of self-cohesion (bond energies of several hundred kilojoules per mole). On the other hand, cohesion between different chains is weak because of the small number of bonds and their relatively low energy (covalent, van der Waals, polar bonds). Because of their ability to rotate around their carbon–carbon or silicon–silicon catenary backbones, the chains can twist into various shapes and are thus very flexible at their service temperature. When subjected to a deforming stress, these entangled macromolecular chains begin to unfold, sliding over each other, and separating along the direction of the stress, giving the material its elastomeric property. The greater the number of bonds between chains and the more chemically stable the bonds (crosslinking), the less likely the material is to deform under stress. What limits rotation of the catenary backbone and sliding is the formation of chemical bonds (generally covalent) between the chains, thereby leading to the formation of a stable, three-dimensional network. The greater the crosslinking, the more rigid the polymer. The formation of chemical bonds produces various active sites on the macromolecular chains (double bonds, hydrogen atoms, labile halogen atoms, etc.). From a thermodynamic perspective, when an elastomer is subjected to a tensile stress, it may stretch to several times its free length when the macromolecular chains unfold. The energy transferred to the elastomer makes it thermodynamically unstable. When the stress is removed, the macromolecules return to their initial, energetically stable, intertwined state, releasing most of the input energy.

b. *Polymerization*

True polymerization only involves the formation of macromolecules by the creation of multiple carbon–carbon bonds. However, in a larger sense, it simply means the formation of macromolecules. Monomers with accessible double bonds, i.e., bonds that are not obstructed by large substituents may from giant molecules. The double bonds can be rearranged by catalysts such as radiation, oxygen, and free radicals (like those produced by the dissociation of peroxides), giving a new radical that attacks another

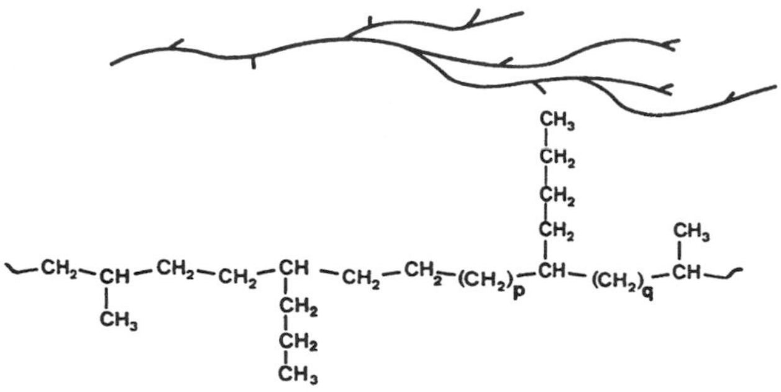

Figure 13.2 Example of branched homopolymer: polyethylene.

monomer, in a chain reaction. This is called *free radical polymerization*:

The catalyst may attack the main chain, producing side chains (Fig. 2). However, in principle, bridges never form between two chains (see, crosslinking subsection). Certain molecules with different functional groups can combine into a single molecule, e.g., acid and alcohol or aldehyde and amine. Molecules with different functional groups can combine into macromolecules. Molecules with identical functional groups can also react with other molecules with antagonistic functional groups to produce macromolecules. Siloxanes are an example of this. They condense to form polysiloxanes:

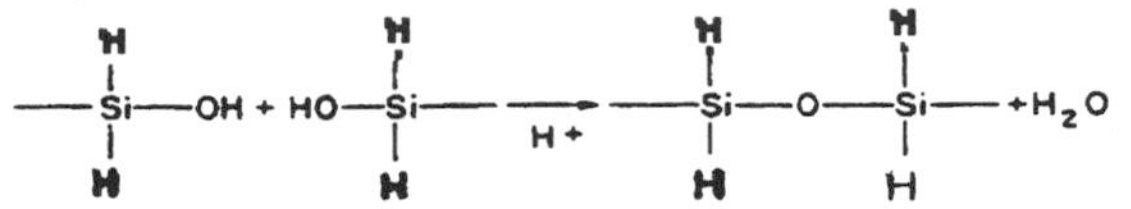

Silicon elastomer

Another example is carbonyl chloride (phosgene) and bisphenol A, which combine to produce polycarbonate:

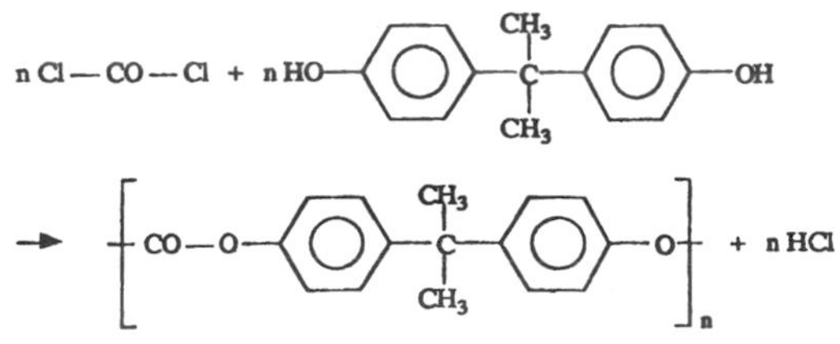

In the two examples, for each bond created a small molecule (H_2O, HCl) is lost in a reaction called *polycondensation*. In other cases, as with the formation of urethanes and other forms of polysiloxanes, no molecules are lost:

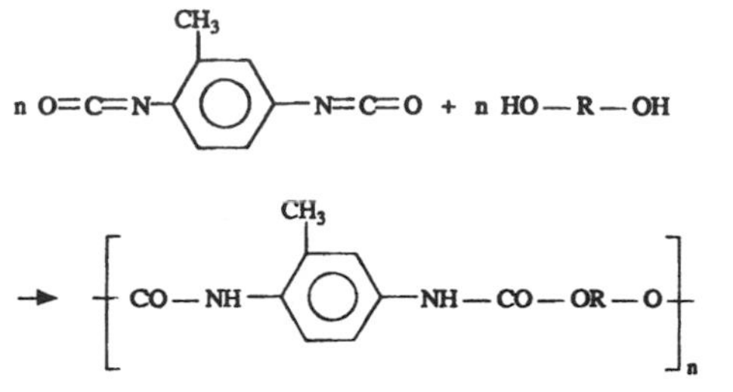

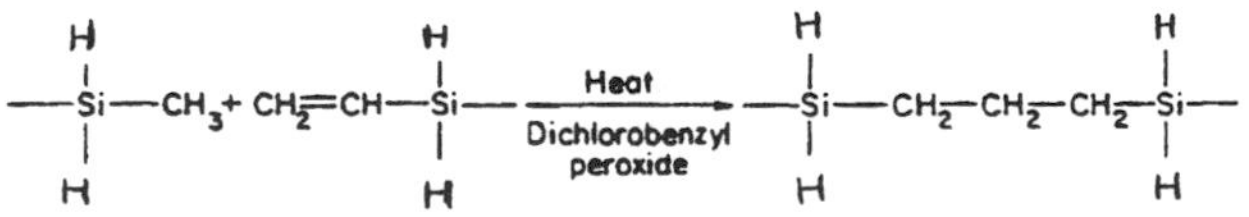

This is referred to as *polyaddition*.

A wide variety of catalysts (specific to each reaction) are involved in free radical polymerization reactions (polyaddition or polycondensation), all of which lead to long polymers that concatenate either the same monomer (homopolymers) or different monomers (copolymers).

Homopolymerization. The linear bonding of structurally identical small molecules (monomers) leads to the formation of homogeneous polymers. Most commonly used biomedical elastomers are homopolymers (Fig. 2). However, the possibility of mixing different monomers and/or small polymers is leading to the synthesis of new polymers with infinitely varied properties, which, because of their improved performance, are increasingly replacing homopolymers for specific applications.

Copolymerization. Different types of monomers can be copolymerized or mixed with small reactive polymers to obtain alternating (-A-B-A-B-A-B-), sequential (-A-A-A-B-B-B-A-A-A-B-B-B-), or random (A-B-B-A-B-A-A-B-) copolymers with different block structures containing two, or occasionally more, identical monomers (Fig. 3).

Other types of monomers or small polymers can also be attached to existing chains to create branched or grafted copolymers (Fig. 4). Alterations to the catenary or terminal structures of copolymers change their properties, creating new types of elastomers, some of which have potentially interesting biomedical applications. It would be more accurate to talk about bi- and terpolymers when a reaction involves two or three types of monomers, but the term co-

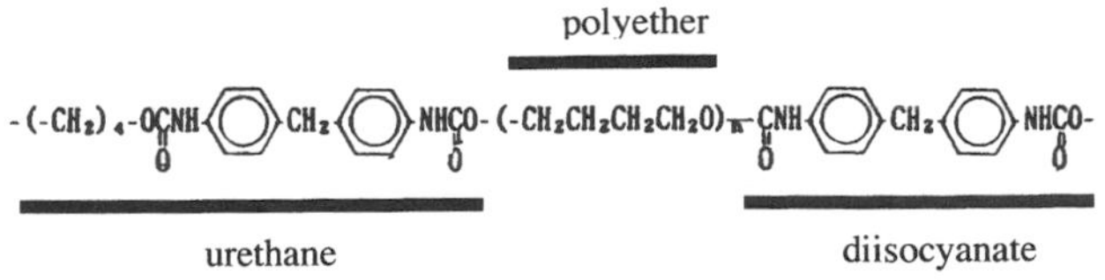

polyether

urethane

diisocyanate

Figure 13.3 Different blocks of a terpolymer segmented polyetherurethane: pellethane® (A-(Bn)-C).

```
CH2          COOH    COOH
   \CH—CH2—CH—CH2—CH~
CH2
   \CH2
CH2
   \CH2      COOH    COOH
CH2
   \CH—CH2—CH—CH2—CH~
CH2
   \CH2
CH2
```

acrylic polyacid grafted polyethylen

```
CH2
   \CH—CH2—CH2—CH2—CH3
CH2
   \CH2
CH2
   \CH2
CH2
   \CH2
CH2
   \CH—CH2—CH2—CH2—CH3
CH2
```

ethylene-hexene grafted copolymer

Figure 13.4 Monomers attached to existing chains to create branched or grafted copolymers.

polymer is used to describe all polymers containing two or more different types of monomers.

Crosslinking. When polymers have reactive sites within the chain (e.g., the double bonds of unsaturated polyesters) or at the terminal (e.g., the hydroxyl group of polyols), they can be reacted with a polyfunctional agent to create a network in a process referred to as crosslinking or vulcanization. Such agents are generally called catalysts, hardeners, or coupling agents, and the initial polymer is called a prepolymer (Fig. 5).

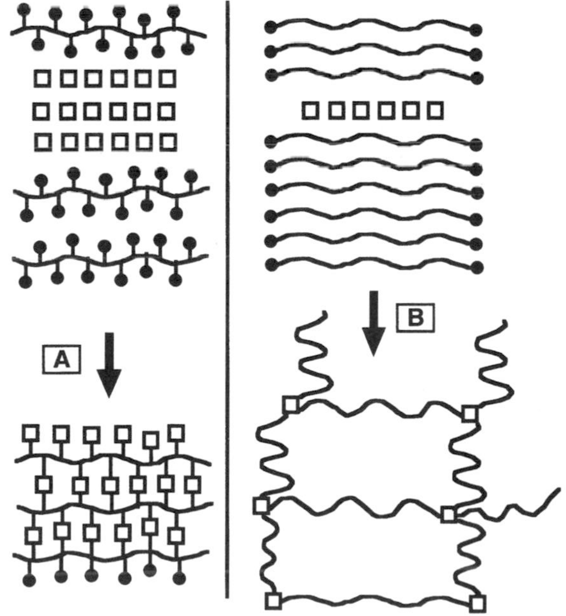

Figure 13.5 Crosslinking of a polymer. (A) Crosslinking of a polymer with reactive side groups; (B) crosslinking of a polymer with reactive terminal groups.

Crosslinking can be performed by free radical polymerization using organic peroxides, which greatly increases the thermal resistance of the network by creating very strong carbon–carbon bridges. Other types of crosslinking agents can be used, especially for elastomers with specific reactive groups (carboxyls, halogens, etc.).

d. Formulation

Apart from a few rare exceptions, the properties of ''green'' elastomers do not allow them to be used as is. Various additives, all with well-defined functions, must be incorporated. This is what is called formulation. Despite a certain empirical approach, formulation requires a great deal of know-how to compensate for the occasionally antagonistic properties required by certain applications. A standard formulation contains

- Elastomer(s). Specific properties, especially chemical properties, must address application specifications. A mixture of elastomers may be used.
- Filler(s). Fillers are used to provide superior mechanical properties, and also to improve handling and to lower production costs. Precipitated silica is the most common additive in biomedical elastomers. Chalks and kaolins are also occasionally used.
- Plasticizers. Plasticizers are used to make processing easier and improve low temperature flexibility. Heavy solvents, oils, and esters are generally used.
- Protective agents. These are used to protect the macromolecular chains from oxygen, ozone, and ultraviolet radiation. Amine and phenol derivatives are generally used.
- A crosslinker. These allow the formation of a stable, three-dimensional network without which the elastomer would be useless.
- Various other ingredients. Numerous other ingredients, such as colorants, conditioning agents, swelling agents for manufacturing cellular elastomers, with very specific functions may be used.

It is clear that the release of additives like plasticizers, antioxidants, initiators, catalysts, fillers, etc., by biomedical elastomers may have a significant impact on host tissues.

e. Elastomer Reinforcement

Reinforcement fillers are quite often added because elastomers have fairly low breaking strengths at ambient temperature. Adding more precipitated silica enhances breaking strength, but decreases flexibility. The amount of filler can be changed to generate variable stress–strain curves and produce elastomers with specific properties. Fillers like semireinforcing kaolins can be used to generate economical, clear mixtures. The same is true of chalks, which are considered as diluents. Elastomers are also occasionally reinforced by the addition of bulk fibers or tissues (glass, carbon, aramide) for very specialized aeronautics and medical applications for example.

B. Silicone Elastomers

1. Introduction

Silicones (a contraction of silicon ketone) make up a vast family of polymers with remarkable properties due to the presence of both silicon–oxygen and silicon–carbon bonds. Depending on the nature of the functional groups attached to the silicon atoms, silicone polymers may be resins, oils, or elastomers, making them well-suited for diverse applications in a wide variety of areas. Silicon elastomers are composed of very long polymers containing several thousand silicon atoms per molecule (6). The first biomedical applications for silicone elastomers can be traced back to the 1950s, while industrial-scale manufacturing dates to 1965 (7,8). Today, silicone elastomers are the most widely used polymers in medical applications (1) because of the strong, very mobile bonds of their Si—O—Si (siloxane) catenary backbone, which provide elevated chemical inertness and exceptional flexibility. They are also very stable over time and at body temperature, show little tissue reactivity, and are highly resistant to chemical attack and heat, which allows them to be autoclaved. They also have exceptional mechanical properties such as high tear strength, outstanding elasticity, and high gas permeability, which makes them suitable for many medical applications such as contact lenses, special dressings, and air/blood filter membranes.(Table 2) (9,10).

2. Manufacture

Chlorosilanes, especially dichlorosilanes and trichlorosilanes (where R is an organic radical), are the basic building blocks of silicone elastomers:

$$Cl-\underset{R}{\overset{R}{|}}{Si}-Cl \qquad Cl-\underset{Cl}{\overset{R}{|}}{Si}-Cl$$

The main organic radicals used are methyl (—CH_3) and phenyl (—C_6H_5) groups. Others (ethyl, vinyl, etc.) are used for special application silicones. Chlorosilanes are complicated and costly to produce. They react with water to give silanols, which, being unstable, lose water molecules and condense into polymers—polysiloxanes:

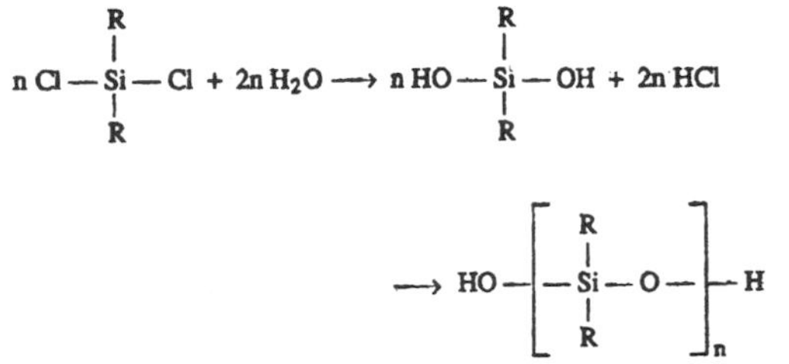

$$n\,Cl-\underset{R}{\overset{R}{|}}{Si}-Cl + 2n\,H_2O \longrightarrow n\,HO-\underset{R}{\overset{R}{|}}{Si}-OH + 2n\,HCl$$

$$\longrightarrow HO-\left[-\underset{R}{\overset{R}{|}}{Si}-O-\right]_n-H$$

Depending on the dichlorosilane/trichlorosilane ratio, different polymers are obtained. The greater the percentage of dichlorosilane, the more linear the polymer; and the greater the percentage of trichlorosilane, the more branched the polymer. Depending on the reaction conditions, polymers with varying molecular masses can be produced. The terminal hydroxyl groups can react with other components or with each other in the presence of catalysts. It is the termination reaction that produces the final silicone product:

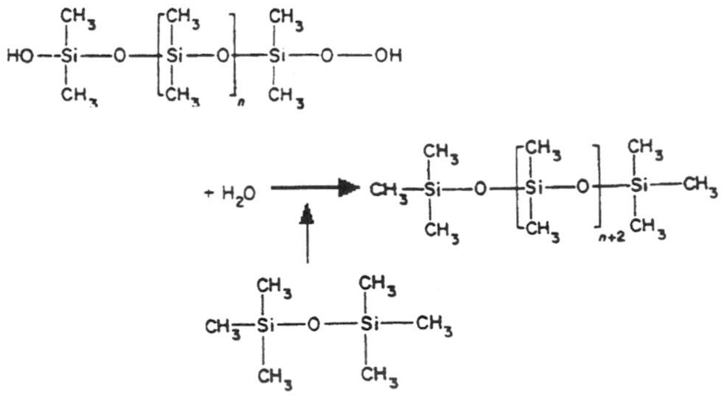

$$HO-\underset{CH_3}{\overset{CH_3}{|}}{Si}-O-\left[\underset{CH_3}{\overset{CH_3}{|}}{Si}-O\right]_n-\underset{CH_3}{\overset{CH_3}{|}}{Si}-O-OH$$

$$+\,H_2O \longrightarrow CH_3-\underset{CH_3}{\overset{CH_3}{|}}{Si}-O-\left[\underset{CH_3}{\overset{CH_3}{|}}{Si}-O\right]_{n+2}-\underset{CH_3}{\overset{CH_3}{|}}{Si}-CH_3$$

$$\uparrow$$

$$CH_3-\underset{CH_3}{\overset{CH_3}{|}}{Si}-O-\underset{CH_3}{\overset{CH_3}{|}}{Si}-CH_3$$

Table 13.2 Properties of Silicones for Use in Biomedical Applications

Good elastomeric and relatively uniform properties over a wide temperature range
Physiological indifference
Good low temperature resistance and stability at high temperatures
Excellent resistance to biodegradation
Excellent resistance to oxidation and ultraviolet light
Moderate biocompatibility
Outstanding resistance to aging
Excellent dielectric behavior over a wide range of temperatures

The properties of the finished product depend on the nature of the organic radicals grafted onto the catenary backbone, meaning that an infinite number of chemical structures—and properties—are possible.

The end products of this industrial process are polycondensates of basically linear or weakly branched polysiloxane molecules of varying molecular mass and consistency with terminal reactive groups such as —Cl or —OH. These reactive polycondensates or prepolymers serve as the raw materials to produce the final silicone products by room temperature vulcanization (RTV) or high temperature vulcanization (HTV).

3. Homopolymerization of Silicone Elastomers

Most commonly used biomedical silicone elastomers are homopolymers. Increasingly, however, various polysiloxane prepolymers as well as other types of prepolymers with varying properties are being combined to synthesize copolymers for the biomedical market. Different polymerization processes can be used to produce a variety of silicone elastomers.

a. One Component RTV Silicones

There result from a reaction between atmospheric moisture and a mixture of polycondensates and catalyst. Crosslinking thus occurs from the outside in, limiting the useful thickness:

$$-\mathrm{SiH_2}-\mathrm{OH} + \mathrm{HO}-\mathrm{SiH_2}- \xrightarrow{H^+} -\mathrm{SiH_2}-\mathrm{O}-\mathrm{SiH_2}- + \mathrm{H_2O}$$

Such silicones are only used for adhesives, coatings, and sealants. Their biomedical utility arises from the absence of fillers and additives, which gives them an excellent biocompatibility (11).

b. Two-Component RTV Silicones

These result from a reaction between a polycondensate and a crosslinker added to initiate the reaction. They are used in the medical field for limited production runs of molded products.

Crosslinking can occur by polycondensation of a relatively low molecular weight polydimethyl siloxane with a reactive hydroxyl group in the presence of tin octoate (i.e., condensation cure):

$$-\mathrm{SiH_2}-\mathrm{H} + \mathrm{HO}-\mathrm{SiH_2}- \xrightarrow{\text{Metal salt}} -\mathrm{SiH_2}-\mathrm{O}-\mathrm{SiH_2}- + \mathrm{H_2}$$

Or crosslinking can occur by polyaddition of a relatively low molecular weight vinylsiloxane, a polysiloxane with a terminal hydrogen, and an organometallic catalyst like chloroplatinum acid [e.g., RTV 71556 Silbione from Aventis formerly Rhône—Poulenc]:

$$-\mathrm{SiH_2}-\mathrm{H} + \mathrm{CH_2{=}CH}-\mathrm{SiH_2}- \xrightarrow{\text{Organo-metallic catalyst}} -\mathrm{SiH_2}-\mathrm{CH_2}-\mathrm{CH_2}-\mathrm{SiH_2}-$$

c. High Temperature Vulcanization Elastomers

These are produced by heat vulcanizing similar but higher molecular weight polycondensates premixed with crosslinkers (peroxides):

$$-\mathrm{SiH_2}-\mathrm{CH_3} + \mathrm{CH_2{=}CH}-\mathrm{SiH_2}- \xrightarrow[\text{Dichlorobenzyl peroxide}]{\text{Heat}} -\mathrm{SiH_2}-\mathrm{CH_2}-\mathrm{CH_2}-\mathrm{CH_2}-\mathrm{SiH_2}-$$

They are used in small- and large-scale industrial production runs by injection molding, extrusion, calendaring, etc. (e.g.,MDX 44210 Silastic from Dow Corning).

4. Copolymerization of Silicone Elastomers

Silicone elastomer copolymers can be produced by addition or condensation. Condensation generally involves low to medium molecular weight prepolymers. For example, vinyl polymethylsiloxane is condensed with hydroxy polymethylsiloxane to create a new prepolymer—siloxyethanol ether—which can in turn be condensed with di-isocyanate to produce silicone urethane copolymers whose blood compatibility and stiffness have been improved to the point where they can be used as blood catheters or blood pump coatings. Examples of different reactive Silicone prepolymers include the following (12):

$$^{*}\mathrm{HO}-\mathrm{Si(CH_3)_2}-\mathrm{O}[-\mathrm{Si(CH_3)_2}-\mathrm{O}]_n-\mathrm{Si(CH_3)_2}-\mathrm{OH}^{*}$$

Silanol terminated

$$^{*}\mathrm{CH}{=}\mathrm{Si(CH_3)_2}-\mathrm{O}[-\mathrm{Si(CH_3)_2}-\mathrm{O}]_n-\mathrm{Si(CH_3)_2}{=}\mathrm{CH}^{*}$$

Vinyl terminated

$$\mathrm{CH_3}-\mathrm{Si(CH_3)_2}-\mathrm{O}[-\mathrm{Si(CH_3)(CH{=}CH_2^{*})}-\mathrm{O}]_m-[-\mathrm{Si(CH_3)_2}-\mathrm{O}]_n-\mathrm{Si(CH_3)_2}-\mathrm{CH_3}$$

Vinyl backbone

$$\mathrm{CH_3}-\mathrm{Si(CH_3)_2}-\mathrm{O}[-\mathrm{Si(CH_3)(H^{*})}-\mathrm{O}]_n-\mathrm{Si(CH_3)_2}-\mathrm{CH_3}$$

Polymethyl hydrosiloxane

The copolymerization reaction is (12):

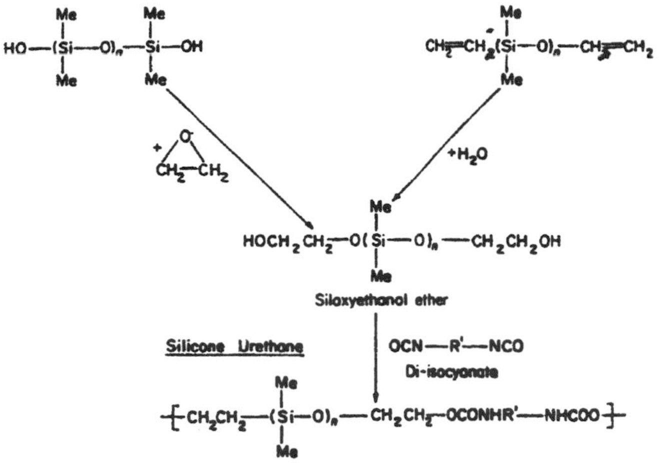

$HO-(SiMe_2-O)_n-SiMe_2-OH$ + ethylene oxide (CH_2-CH_2, O)

$CH_2=CH_2(SiMe_2-O)_n-CH_2=CH_2$ + H_2O

$HOCH_2CH_2-O(SiMe_2-O)_n-CH_2CH_2OH$

Siloxyethanol ether

Silicone Urethane

OCN—R′—NCO

Di-isocyanate

$-[CH_2CH_2-(SiMe_2-O)_n-CH_2CH_2-OCONHR'-NHCOO]-$

Polysiloxane copolymers—condensation

Condensation has also been used to produce silicone polycarbonate copolymers designed to transport ions (12):

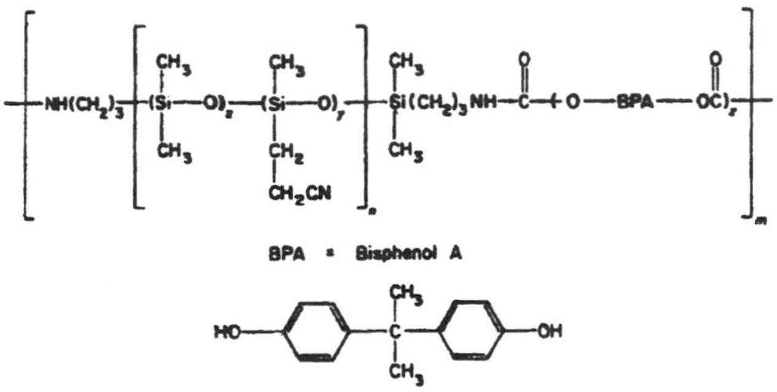

$-[NH(CH_2)_3-[(Si(CH_3)_2-O)_x-(Si(CH_3)(CH_2CH_2CN)-O)_y]_n-Si(CH_3)_2(CH_2)_3NH-C(=O)-(O-BPA-OC(=O))_z]_m-$

BPA = Bisphenol A

$HO-C_6H_4-C(CH_3)_2-C_6H_4-OH$

Their good mechanical resistance and high stability in aqueous environments make them useful for microelectrodes.

Certain linear silicone elastomer copolymers are also prepared by addition curing. These are generally higher molecular weight polymers that can be processed or molded using industrial processes. Siloxane and methacrylate copolymers were developed to improve the properties of elastomeric contact lenses by increasing their hardness, stiffness, transparency, wettability, and gas permeability by combining the properties of the various monomers used to produce the copolymer (12):

$CH_2=C(CH_3)CO(CH_2)_3Si(OSiMe_3)_3$

Methacryloxypropyltris(trimethylsiloxy)silane

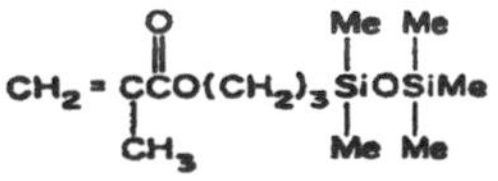

$CH_2=C(CH_3)CO(CH_2)_3SiMe_2OSiMe_3$

Methacryloxypropylpentamethyldisiloxane

$CH_2=C(CH_3)CO(CH_2)_3Si(OSiMe_2OSiMe_3)_3$

Methacryloxypropyltris(pentamethyldisiloxanyl) silane

Graft silicone elastomer copolymers can have two types of grafted side chains, one type on all the molecules in the elastomer and a second added by graft polymerization at the surface (see section I.B.) Silicone elastomers can be mixed relatively easily with numerous copolymers [styrene–butadiene–styrene (SBS), SEBS, Polypropylene (PP), SEBS, PU]. Silicone elastomer copolymers are all very biocompatible because the addition of silicone gives them the elasticity that only the incorporation of plasticizers (phthalates) would otherwise provide. They are suitable for large-scale production runs by molding or injection because of the thermoplasticity provided by the SBS, SEBS, and PP copolymers. They also combine the surface properties of silicones with the intrinsic mechanical performance of the added copolymers. Polyurethanes are good examples of copolymers with greatly enhanced blood compatibility and mechanical strength. Polyurethane products are produced by mixing a vinylsiloxane polycondensate with a curing agent (crosslinker) followed by injection into an appropriate mold. The components crosslink to form a thermoplastic silicone–polyurethane network with remarkable physical and biomedical properties. These biomedical silicone elastomers can be further modified by the addition of other components before the final cure. These components may or may not be involved in the crosslinking, and they can have an impact on the physical and chemical properties, and thus the biocompatibility of the final elastomeric products (13,14).

5. Fillers

Biomedical silicone elastomers may be considered composite materials in the sense that they are composed of organomineral polymer matrixes containing fillers that are more or less linked to the network. Fillers are used to improve the mechanical performance of elastomers, notably to increase tear and tensile strength. They are generally mineral in nature, but may also be organic, like high molecular weight polyacrylic acid (15). Amorphous silica is the most widely used filler and may be modified using a coupling agent that improves adherence to the polymer matrix (sizing). This is generally done by silanization because the most common coupling agents are silanes like $X_3Si(CH_2)_nY$, where n = 0 to 3, X = hydrolysable group, and Y = organic group selected on the basis of the polymer matrix. Silica with reactive Si—H covalent bonds can be obtained by a reaction with methyldichlorosilane for example (1). Other more specific treatments have been reported like the attachment of antithrombogenic heparinoid substances to amorphous silica (16,17). Fibers (glass, aramide, carbon, etc.) can also be used to reinforce the structure of biomedical silicone elastomers. Sizing the fibers ensures good adherence between the reinforcement and the matrix (18).

The fillers must never reduce material biocompatibility, either because they are not biocompatible themselves or they are released in situ due to insufficient bonding to the polymer.

6. Surface Treatments

Changes to the basic silicone elastomer may be insufficient for special applications. The desire to improve the materials has led to a number of surface treatments to optimize the biocompatibility of surfaces in contact with living tissue, to seal in undesirable residues or additives using a coating, and to regulate excretion and/or absorption by the sealed elastomer using a selectively permeable surface (19). Silicone elastomers are hydrophobic and have high coefficients of friction, which in combination with their low surface energy (10 to 18 dynes/cm) ensures low wettability (5,20). While these biophysical characteristics can be advantageous (no tissue adherence to mammary implants, no bacterial adherence, no sliding of urethral stents or blood catheters due to surface friction caused by contact with living tissue), they can cause numerous biological and technical problems in need of solving in other applications. Various surface treatments can be performed using a variety of techniques.

a. Plasma Deposition of Simple Substances or Composites

Heating a gas to a very high temperature using an electric arc (electric glow discharge) causes extreme atomic excitation resulting in partially or almost totally ionized gases, or plasmas. When the ions come into contact with a "cold" surface, they are deposited and become chemically bonded to the surface. Exposure of elastomers to such plasmas (He, Ar, etc.) allows the deposition and chemical bonding of monomers or polymers to the exposed surface in a process called plasma polymerization (or glow discharge polymerization to avoid confusion with blood plasma). The atomic, molecular, or macromolecular deposits obtained using this technique create a thin, smooth film several angstroms to several hundred angstroms thick. Most polymers with suitable biological or technical properties (polyethylene, polyvinylchloride, polytetrafluoroethylene, polycarbonate, polymethylmethacrylates, polysiloxanes, etc.) can be applied to silicone elastomers.

The plasma deposition of H_2^+, N_2^+, O_2^+, and Ne^+ ions breaks existing chemical bonds on the surface and creates new radicals: $=C=O$ using O^+ plasma, for example. The affinity of certain proteins (albumin, fibrin) for the surface of silicone elastomers modified in this way creates a layer of adsorbed protein that prevents the adherence of other blood constituents by steric repulsion, a well-known effect in colloidal chemistry (21). The result is a powerful antithrombogenic effect.

The deposition of ionic oxygen, argon, carbon dioxide, or ammonia makes the silicone elastomer surface more hydrophilic. However, the migration of free polymer to the elastomer surface affects the durability of such treatments (22,23). The deposition of these ions also allows for polymer grafting on the elastomer surface.

b. Surface Polymer Grafting

This is a more long-lasting solution, but more technically complicated. After activation by an argon plasma or a solvent, the silicone elastomer is placed in a reactor containing a monomer solution (acrylic acid, hydroxyethyl methacrylate, bis amino polyethylene oxide, etc.). Heating promotes chemical bonding of the monomers to the elastomer and crosslinking among the grafted monomers, producing a smooth, continuous coating of grafted copolymer (polyacrylic or polyethylene, for example) covering the elastomer surface. By alternating the chemical treatment of the various surfaces of an elastomer, a bifunctional, homobifunctional, or heterobifunctional material can be created (Fig. 6). Such treatments have been used to improve the adherence and growth of corneal epithelial cells by grafting 2-hydroxyethylmethacrylate to the outside surfaces of silicone elastomer corneas and to prevent adherence by grafting bis amino polyethylene oxide to the inside surfaces (24).

c. Denucleation of Silicone Elastomers

This treatment helps remove the micro air bubbles in the surface irregularities of the elastomer. The goal is to remove air/blood interfaces to prevent complement activation and platelet aggregation. The procedure is relatively long but technically simple. It involves rinsing the elastomer in double distilled water for 12 h, soaking in 99% ethanol for 24 h, soaking in a buffer solution with vacuum degassing for 6 h, then gradually replacing the buffer solution with ethanol over a 6-h period. The elastomer is then stored in the buffer solution until use (25). This treatment helps increase hemocompatibility without modifying the structure or composition of the silicone elastomer surface.

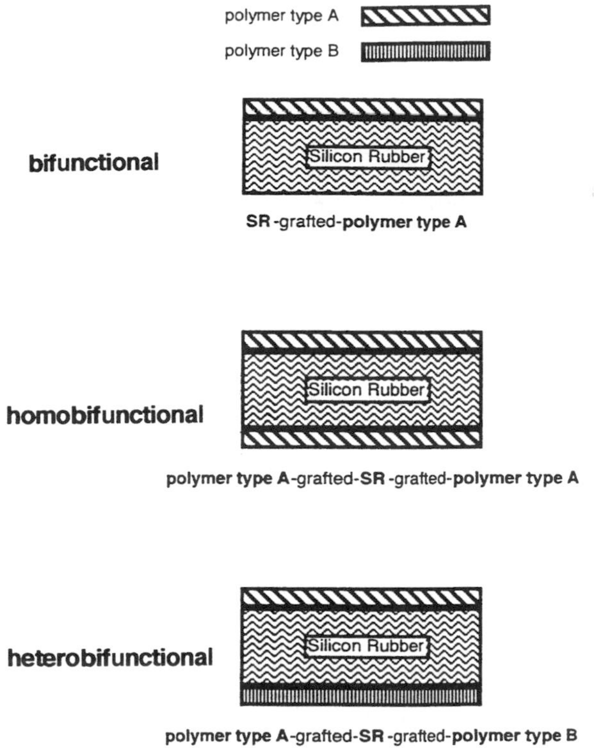

Figure 13.6 Schematic diagrams for grafted silicon rubber surfaces.

d. Acylation or Hydroxylation of the Silicon Elastomer Surface

This process increases the affinity for serum proteins (26). Hydroxylation is performed by an oxymercuration/demercuration reaction, or hydroboration, which transforms the polydimethylsiloxanes on the surface into hydroxymethylvinylsiloxane. Acylation of —OH groups is achieved through simple esterification of exposed carbons on the siloxane side chains (26).

C. Polyurethane Elastomers

1. Introduction

The first polyurethanes were developed by Otto Bayer in the late 1930s. They now make up the largest family of thermoplastic and thermosetting plastics. They range from very rigid to very flexible and from compact to cellular, and are used to produce varnishes, fibers, textiles, plastics, and elastomers. Polyurethane elastomers were studied in great detail in the 1950s and 1960s (27–29). Their use for biomedical applications was proposed by Pangman (30) and Boretos (31), not to mention Yoda (1). They have been widely used in the medical field for over 25 years now (31–36). They are the products of polyaddition reactions between polyisocyanates and polyalcohols (polyols). Broadly speaking, all elastomers are isocyanate derivatives.

2. Manufacture

Polyurethane elastomers are formed by reactions between isocyanate and substances with mobile hydrogen ions, generally polyols, amines, and water:

$$\mathrm{R{-}N{=}C{=}O + H{-}A \rightarrow R{-}NH{-}\underset{\displaystyle A}{\underset{|}{C}}{=}O}$$

Alcohols give urethanes:

$$\mathrm{R{-}N{=}CO + H{-}OR' \rightarrow R{-}NH{-}CO{-}OR'}$$

Nontertiary amines give ureins or substituted ureas:

$$\mathrm{R{-}N{=}CO + H{-}N\begin{matrix}R'\\R''\end{matrix} \rightarrow R{-}NH{-}CO{-}N\begin{matrix}R'\\R''\end{matrix}}$$

Water also give ureins together with carbon dioxide:

$$\mathrm{\begin{matrix}R{-}N{=}CO\\R{-}N{=}CO\end{matrix} + \begin{matrix}H\\H\end{matrix}\!\!>O \rightarrow \begin{matrix}R{-}NH\\R{-}NH\end{matrix}\!\!>CO + CO_2\uparrow}$$

In most cases, the components (i.e., the raw materials, the catalysts, and the adjuvants) are mixed as liquids and then injected into molds or cavities. Aromatic polyisocyanates are the most frequently used raw materials. Toluenediisocyanate (TDI) is used as is or is dimerized in order to synthesize elastomers with high tear resistance; diphenylmethanediisocyante (MDI) and naphthylenediisocyante (NDAI) are used to produce polyester elastomers:

2-4 2-6

Toluene diisocyanate

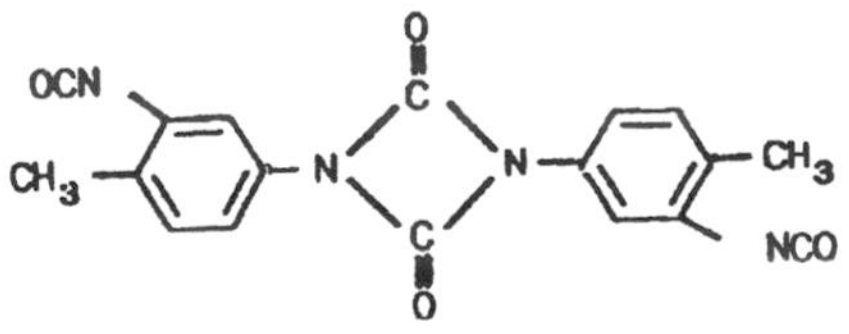

Dimerized toluene diisocyanate

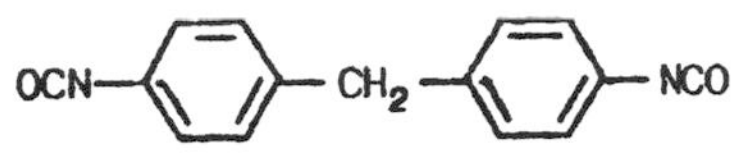

OCN–C₆H₄–CH₂–C₆H₄–NCO

Diphenylmethane 4-4′-diisocyanate

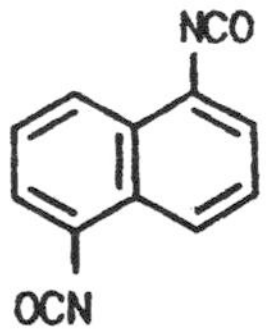

(naphthalene ring with NCO and OCN)

Naphthylene 1-5-diisocyanate

Polyethers are the most frequently used polyols and result in polyetherurethanes; polyesters are used less often but the resulting polyesterurethanes are much more chemically stable:

Formation of polyether polyols

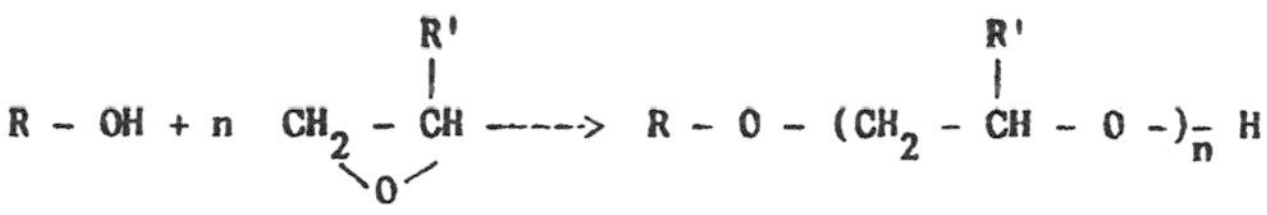

$$R-OH + n\ CH_2-CH(R')-O\ (\text{epoxide}) \dashrightarrow R-O-(CH_2-CH(R')-O-)_n H$$

Formation of polyester polyols

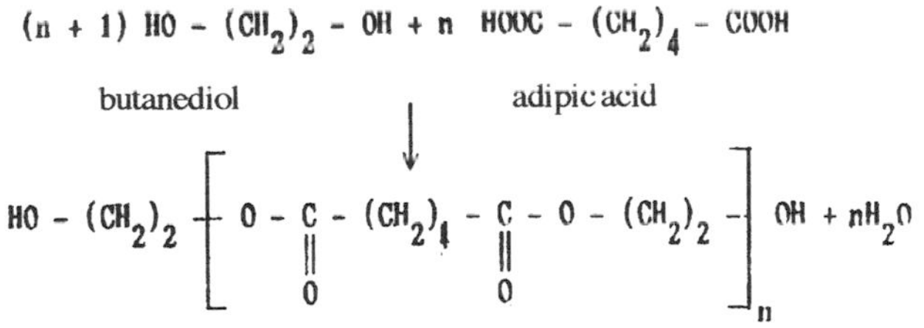

$$(n+1)\ HO-(CH_2)_2-OH + n\ HOOC-(CH_2)_4-COOH$$

butanediol adipic acid

$$\downarrow$$

$$HO-(CH_2)_2-\left[O-\underset{\|}{\overset{}{C}}(=O)-(CH_2)_4-C(=O)-O-(CH_2)_2-\right]_n OH + nH_2O$$

Pellethane® is a poly(ether urethane) that has been used for many years in the medical field, and is the result of a reaction between MDI and polytetramethyleneglycol (PTMG), which produces a soft segment ending in an isocyanate that reacts with the diol moiety of butanediol to extend the chain (Fig. 7) (1).

The greater the number of reactive groups on the polyols used to synthesize the polyurethane, the more the end product is crosslinked and thus rigid. The polyols with the highest molecular mass and the least crosslinking give the most flexible elastomers. To improve the mechanical resistance of certain polyurethane elastomers, all or part of the polyol may be replaced by a reactive diamine [e.g., methylene dianailine (MDA), methylene *bis*-orthochloroaniline (MOCA), ethylene dianiline (EDA)], giving rise to segmented polyetherurethaneurea elastomer such as Biomer® or Mitrathane®, which are used in cardiovascular applications. Prepolymer processing (foaming) using expansion agents produces cellular elastomers: either water is reacted with isocyanate to produce a urein and carbon dioxide (chemical expansion) or low boiling point inert liquids are added and subsequently expelled by degassing during high temperature polymerization (physical expansion). Adjuvants are also required for crosslinking: catalysts (tertiary amines, organic tin salts), surfactants (silicones), fillers, reinforcements, pigments, etc. During manufacture (whatever the process) and storage, polyurethanes must be kept in a dry environment because their affinity to and reactivity with water can result in changes to their composition and/or create defects in the end product.

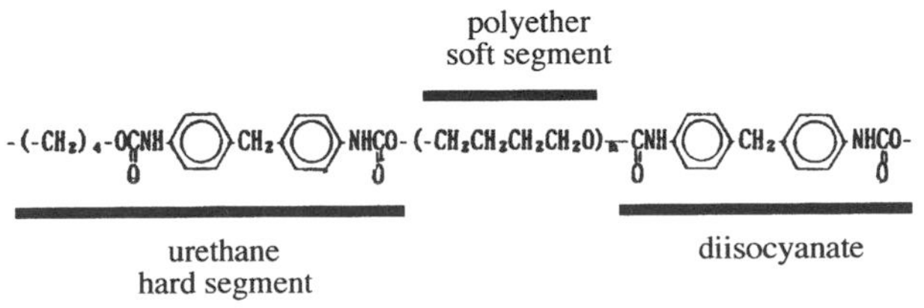

$$-(-CH_2)_4-O\overset{O}{C}NH-C_6H_4-CH_2-C_6H_4-NH\overset{O}{C}O-(-CH_2CH_2CH_2CH_2O)_n-\overset{O}{C}NH-C_6H_4-CH_2-C_6H_4-NH\overset{O}{C}O-$$

polyether soft segment

urethane hard segment

diisocyanate

Figure 13.7 Polyetherurethane-segmented Pellethane.

3. Mechanical Behavior of Polyurethane Elastomers

The hard segments crosslink among themselves within the mass of the elastomer to form agglomerations that act as fillers, improving the mechanical resistance of the material. The soft segments remain free and randomly arranged (Figs. 8 and 9). Under a tensile stress of 150% (29), the polyether soft segments line up along the elongation axis, displacing the urethane hard segments so that they are more or less perpendicular to the vertical axis. Soft segments brought into close proximity with each other by a stress crystallize, a process that becomes complete when elongation reaches approximately 250% (Fig. 10). When the elongation increases even more (~500%), the chemical bonds crosslinking the hard segments break down and the now separate hard segments align along the vertical axis. At this point, the soft segments relax because of the stretching due to the release of the hard segments (Fig. 11).

Segmented polyurethane elastomers thus have elastic behavior under low stress (deformation) conditions, which becomes plastic when the hard segment network breaks down. Deformed samples do not immediately return to their original shape following removal of the stress. However, rearrangement of the hard segments eventually results in a total or partial return to the initial shape, depending on the elastomer structure. The greater the concentration of hard segments, the more plastic the defor-

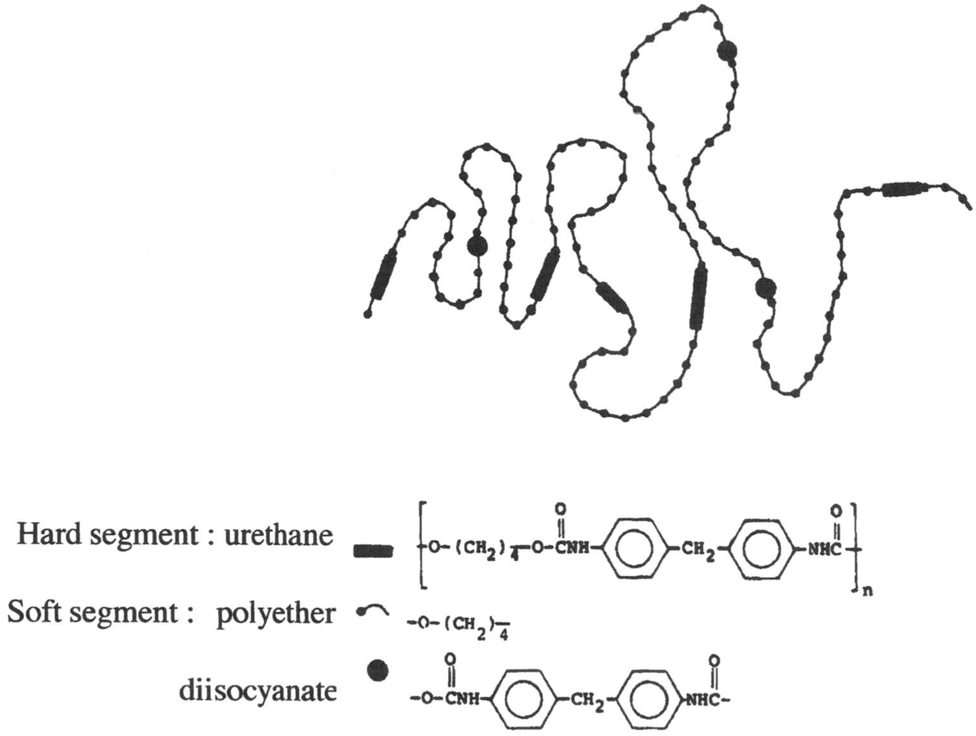

Hard segment : urethane $\left[-O-(CH_2)_4-O-\overset{O}{\overset{\|}{C}}NH-C_6H_4-CH_2-C_6H_4-NH\overset{O}{\overset{\|}{C}}- \right]_n$

Soft segment : polyether $-O-(CH_2)_4-$

diisocyanate $-O-\overset{O}{\overset{\|}{C}}NH-C_6H_4-CH_2-C_6H_4-NH\overset{O}{\overset{\|}{C}}-$

Figure 13.8 Linear polyetherurethane chain. (From Ref. 63.)

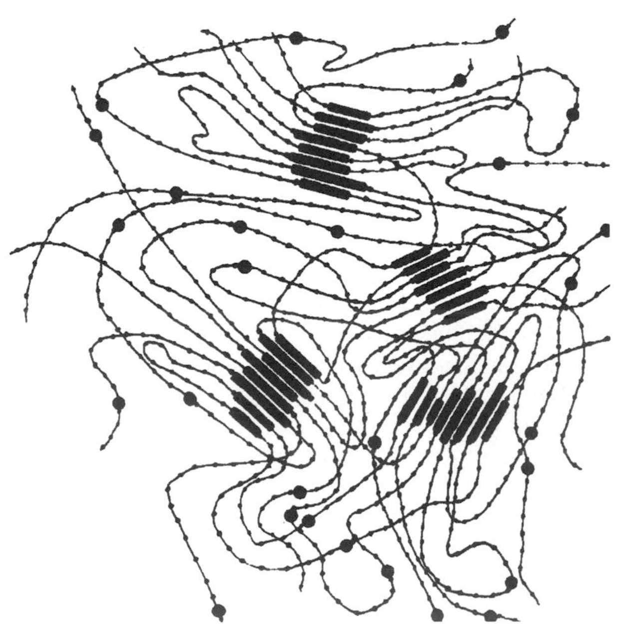

Figure 13.9 Distribution of hard and soft segments in a polyurethane elastomer. (From Ref. 63.)

mation. On the other hand, the lower the concentration of hard segments, the more elastic the behavior.

Chemical properties also depend on the soft/hard segment ratio. A large proportion of hard segments makes the material harder and provides better stress resistance, but decreases elasticity and resistance to abrasion (37). The presence of long polyether or polyester soft segments provides superior break elongation but makes the material more sensitive to oxidation and degradation by biological fluids and thus to failure under repeated stress (Table 3).

4. Biodegradation of Polyurethane Elastomers

In vivo degradation of polyurethane elastomers occurs via four main mechanisms: calcification, macromolecular chain breakdown (1), hydrolysis, and environmental stress cracking. Calcification involves the deposition of calcium hydroxyapatite on the surfaces of implanted elastomers (38). Macromolecular chain breakdown of polyurethane elastomers occurs via oxidation reactions with the soft segments and at the junctions between hard and soft segments (39). Corrodable metallic residues favor oxidation by acting as catalysts in vivo (4,40). Hydrolysis depends on the susceptibility of the urea and urethane bonds and the concentration of hydrolytic enzymes at the cell–polymer interface. The extreme sensitivity of polyesterurethanes to hy-

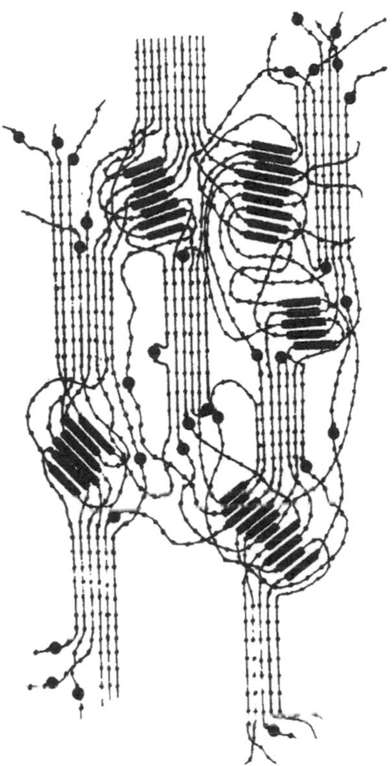

Figure 13.10 Crystallization of soft segments under a weak vertical stress. (From Ref. 63.)

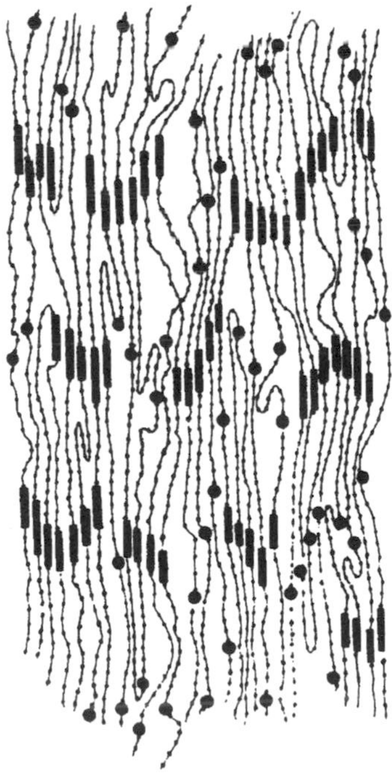

Figure 13.11 Reorientation of hard segments and relaxation of soft segments under a strong vertical stress. (From Ref. 63.)

drolysis limits their use in biomedical applications as opposed to polyetherurethanes, which are more resistant (1). However, polyurethane elastomers in general cannot be autoclaved because of the risk of hydrolysis.

Environmental stress cracking (ESC) occurs when mechanical stresses are combined with interactions with living tissue (41) and can lead to the complete destruction of the implanted material. Three phenomena must coincide for ESC to occur: the presence of enzymes as a result of an inflammatory response at the interface (42–45), a susceptible poly(ether urethane) elastomer, and a mechanical stress on the elastomer (41,46). Environmental stress cracking, which has never been observed with polyurethanes in vitro, does not occur if a single factor is missing (37), and seems to be due to the oxidation of the methyl group on the soft segment polyether chains (Figs. 12–14) (37).

Chemical properties are not the only factors at play. The hardness, mechanical resistance, and morphology or microstructure, as well as the affinity for proteins, also influence sensitivity to biodegradation (47). The microporous surfaces of polyetherurethane vascular devices promote tissue colonization, neointima formation, and fibrotic capsule formation that in turn favor biodegradation by increasing the size of the biological interface with respect to the volume of the device (48,49).

5. Biostable Polyurethane Elastomers

The impact of environmental stress cracking on polyetherurethanes has stimulated research into polyurethane elastomers that do not contain ether groups and that remain stable when subjected to environmental stresses in vivo, i.e., biostable polyurethanes (37,50,51). Coury et al. (52) have filed a patent for a polyurethane that contains no ether groups. An aliphatic complex replaces the soft segment polyethers with a very rigid biostable skeleton that makes the polymer much too hard for many applications. Numerous attempts have been made to modify or replace the soft segments in order to improve biostability. Soft segment polyethers have been replaced by polybutadienes and polymethylsiloxanes (39,53,54), polycarbonates (55,56), and aliphatic hydrocarbons (55). Polysiloxanes are attractive substitutes for polyethers because of their low toxicity, good thermal and oxidative stability, low coefficient of friction, good hemocompatibility, and that fact that they do not affect the mechanical properties of the polyurethanes because their modulus of elasticity is similar to that of the soft segment they are replacing. Polycarbonate soft segements are also less sensitive to oxidation, although the hydrolysis of carbonate linkages in vivo is always possible (Fig. 15) (4,55–60).

Table 13.3 Commercially Available Polyurethanes Used in Biomedical Applications

Type and molecular architecture	Processing	Trade name	Supplier
Segmented polyetherurethane			
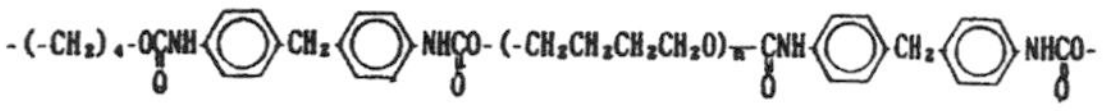 $-(-CH_2)_4-OCNH-C_6H_4-CH_2-C_6H_4-NHCO-(-CH_2CH_2CH_2CH_2O)_n-CNH-C_6H_4-CH_2-C_6H_4-NHCO-$ (C=O groups)	Extrusion Injection molding Solution casting	Pellethane™	Dow Chemical Co., United States
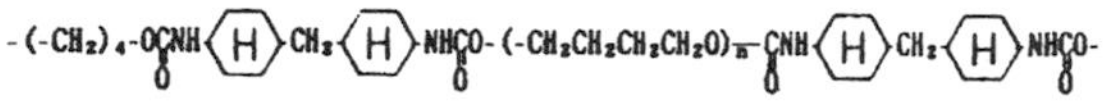 $-(-CH_2)_4-OCNH-\langle H\rangle-CH_2-\langle H\rangle-NHCO-(-CH_2CH_2CH_2CH_2O)_n-CNH-\langle H\rangle-CH_2-\langle H\rangle-NHCO-$ (C=O groups)	Extrusion Injection molding Solution casting	Tecoflex™	Thermedics Inc., United States
Segmented polyetherurethane urea			
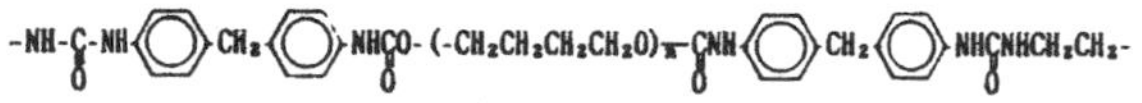 $-NH-C(O)-NH-C_6H_4-CH_2-C_6H_4-NHCO-(-CH_2CH_2CH_2CH_2O)_n-CNH-C_6H_4-CH_2-C_6H_4-NHCNHCH_2CH_2-$	Solution casting	Biomer™	Ethicon Co., United States
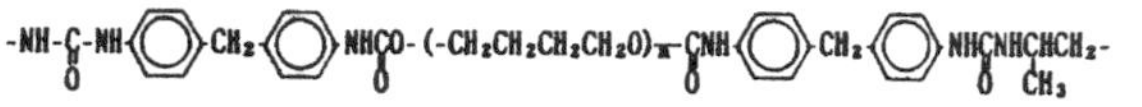 $-NH-C(O)-NH-C_6H_4-CH_2-C_6H_4-NHCO-(-CH_2CH_2CH_2CH_2O)_n-CNH-C_6H_4-CH_2-C_6H_4-NHCNHCH(CH_3)CH_2-$	Solution casting	TM3™, TM5™	Toyobo Co., Japan
Copolymer of segmented polyetherurethane and polydimethylsiloxane			
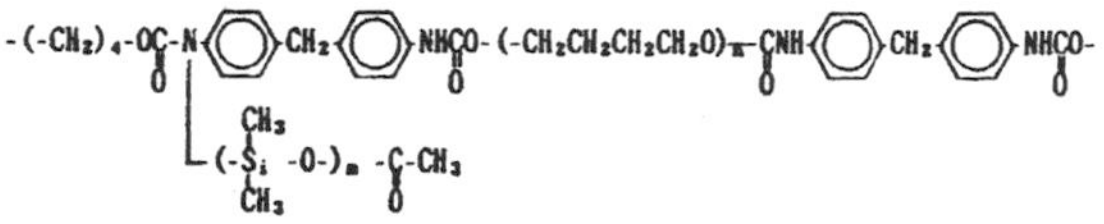 $-(-CH_2)_4-OC(O)-N(-)-C_6H_4-CH_2-C_6H_4-NHCO-(-CH_2CH_2CH_2CH_2O)_n-CNH-C_6H_4-CH_2-C_6H_4-NHCO-$, N-substituted by $(-Si(CH_3)_2-O-)_m-C(O)-CH_3$	Solution casting	Cardiothene51™	Kontron Inc., United States
			Nippon Zeon Co., Japan

Source: Ref. 1.

The complementary qualities of polyurethane and polyvinylsiloxane elastomers led to the idea of producing "alloys" in the hope of creating new polymers with enhanced resistance to biodegradation. The copolymerization of aromatic poly(ether urethane urea) and polydimethylsiloxane (acetoxy-terminated siloxane blocks linked to unsubstituted urethane nitrogen) produces a heterogenous elastomer composed of 10% water-vulcanizable polymethyldisiloxane and a 90% polyurethane matrix in which phase separation (soft/hard segments) of the polyurethane mass is prevented by the interpositioning of the silicone–urethanes formed during the polymerization process (1,61,62). This complex mixture of polyurethane and silicone (Cardiotane 51®) is difficult to process because it is neither soluble nor thermoplastic (63).

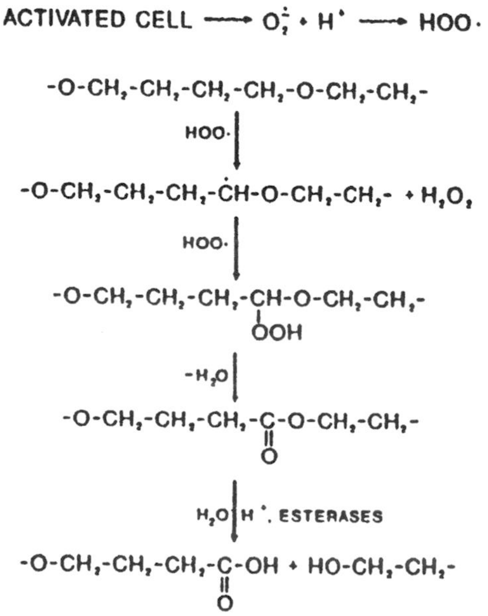

ACTIVATED CELL ⟶ $O_2^{\cdot -}$ + $H^{\cdot}$ ⟶ $HOO\cdot$

$-O-CH_2-CH_2-CH_2-CH_2-O-CH_2-CH_2-$

↓ $HOO\cdot$

$-O-CH_2-CH_2-CH_2-\dot{C}H-O-CH_2-CH_2-$ + H_2O_2

↓ $HOO\cdot$

$-O-CH_2-CH_2-CH_2-CH(OOH)-O-CH_2-CH_2-$

↓ $-H_2O$

$-O-CH_2-CH_2-CH_2-C(=O)-O-CH_2-CH_2-$

↓ H_2O | H^+, ESTERASES

$-O-CH_2-CH_2-CH_2-C(=O)-OH$ + $HO-CH_2-CH_2-$

Figure 13.12 The degradation mechanism of a polyurethane at the carbon alpha to the ether group as proposed by Anderson et al. (43).

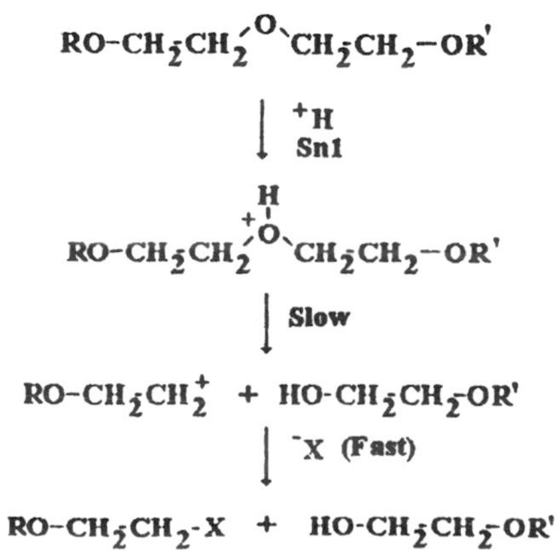

$RO-CH_2CH_2-O-CH_2CH_2-OR'$

↓ ^+H, Sn1

$RO-CH_2CH_2-O^+(H)-CH_2CH_2-OR'$

↓ Slow

$RO-CH_2CH_2^+$ + $HO-CH_2CH_2-OR'$

↓ ^-X (Fast)

$RO-CH_2CH_2-X$ + $HO-CH_2CH_2-OR'$

Figure 13.13 Cleavage of a polyether urethane by Sn1 type acid hydrolysis. (From Ref. 4.)

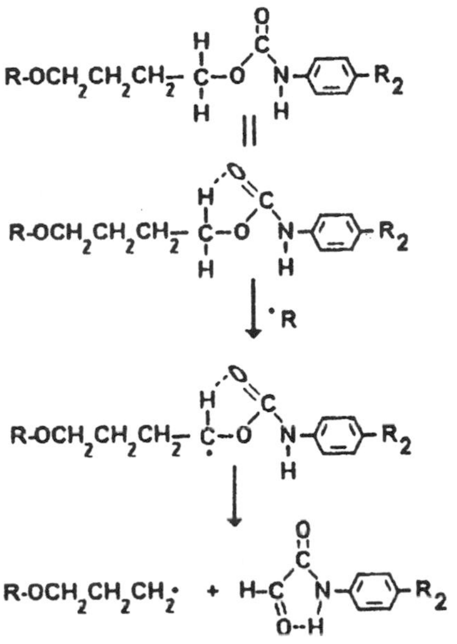

Figure 13.14 The degradation mechanism of a polyurethane urea at the carbon alpha to the urethane group as proposed by Tyler and Ratner. (From Ref. 4.)

6. Grafted Polyurethane Elastomers

The biomedical properties of polyurethane elastomers can be improved by encapsulation to isolate them from living tissue. Grafting polysiloxane molecules on the surface makes the material hydrophobic and protects the soft segments from oxidation and hydrolysis in vivo (64–66). Grafting alkyl or polyethylene [oligoethylene monoalkyl (aryl)alcohol ether] groups to the sodium atoms of urethane hard segments or incorporating sulfate groups by nucleophilic molecular displacement of N-H urethane groups creates sulfonate polyurethane anionomers with chemical properties that limit biodegradation and thrombogenicity (67–74). Researchers have also grafted proteins and other molecules on polyurethane elastomers in an attempt to improve hemocompatibility—albumin (75), heparin (76), adenosine diphosphate (77), fibronectin, and gelatin (78). The grafting processes, or impregnation in the case of gelatin, are complex and require numerous steps. Collagen grafted to carboxyl group–enriched polyurethanes via 1,2-bis(2,3-epoxypropoxy)ethane links allows better epithelial cell growth than raw polyurethanes (79).

Seeding the lumen of microvessel vascular prostheses prepared by glow discharge allows the in vivo proliferation of a multicell layer close to but distinct from the elastomer wall. The cell layer is histologically similar to a native artery with antithrombogenic endothelial cells on the surface (80).

Plasma processing can be used to modify surfaces by depositing metallic silver (81–85), fluorinated films (86), nitrogen, and hexamethyldisiloxan polymers (83) to, among other things, improve the antithrombogenic and antibacterial properties of cardiovascular implants.

7. Modified Polyurethane Elastomers

Solvent extraction (toluene, acetone, methanol, etc.) of small molecular weight surface polymers (which are different in composition from the bulk polyurethane) increases the molecular weight of the surface fraction and enriches it in hard segments, making it more polar and improving hemocompatibility (36,87,88).

Since biogradation mainly affects the ether bonds of soft segments, which are very sensitive to oxidation, certain biomedical polyetherurethanes contain high concentrations of antioxidants (89). Unfortunately, these elastomers (or their degradation products) may be toxic to host tissues (90). Researchers have recently begun examining

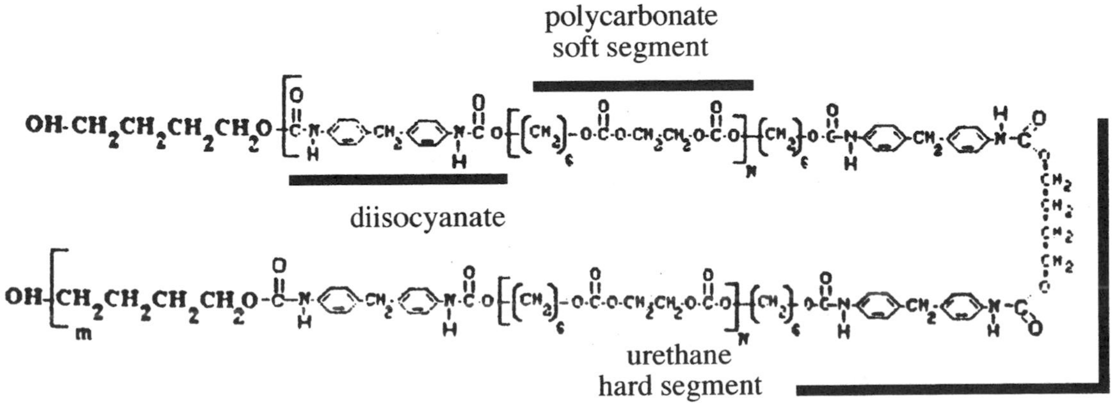

polycarbonate soft segment

diisocyanate

urethane hard segment

Figure 13.15 An example of a biostable polyurethane: Corethane® (polycarbonate urethane), produced by reacting poly(1,6-hexyl 1,2-ethyl carbonate)diol with MDI. (From Ref. 4.)

natural antioxidants like vitamin E as possible alternatives to industrial additives.

Attempts to inhibit tissue responses have also been made. Adding hydroepiandrosterone (DHEA) to poly (etherurethane urea) reduces biodegradation by limiting macrophage activation. However, the effect is dose-dependent, and since DHEA is very hydrosoluble, it rapidly diffuses away, making it useless for long-term protection (91).

Toyo Cloth Co. recently announced that it had prepared a high molecular weight poly(ether/urethane/amide) from polytetramethylene oxide, MDI, ethylene glycol, and Nylon 66. The addition of calcium chloride to this polyurethane–nylon polymer modifies its physical and biomedical properties (92) by increasing water permeability, hydrophilicity, biocompatibility, thrombogenicity, and elasticity, and decreasing breaking strength (tensile modulus). An affinity for calcium ions points to potential thrombogenic or antithrombogenic applications for heparinized versions of the elastomer (76).

While polyurethanes may themselves be encapsulated, they can also be used to coat medical devices like pacemakers and endovascular stents (93–95). Such coatings have various degrees of hemocompatibility (96) and occasionally shift because of a lack of adherence to vessel walls (95).

The development of biodegradable polymers is a major advance in the research on biomaterials. Designed to gradually degrade in vivo over the implantation period, these polymers are used in numerous medical and surgical applications. Biodegradable polyesterurethane have been developed in recent years (97–99). DegraPol/btc® is a polyesterurethane copolymer with polyhydroxybutyric acid hard segments and polycaprolactone soft segments (Fig. 16).

The hard and soft segments can be replaced by a variety of substitutes that make the end product more or less biodegradable in vivo. By selecting the right substitute, it is possible to design cellular polyesterurethanes with varying degrees of rigidity and with degradation rates suitable for osteoblast colonization of bone implants (98). They can lose half their mass within several months of implantation. They are not strictly speaking elastomers, but lengthening their soft segments may provide new mechanical properties and open the door to soft tissue applications.

D. Saturated Polyesters and Copolyesters

These thermoplastic polymers have long been used in textiles (Dacron®, Tergal®) and plastic films; polyethylene terephthalate (PET) holds the lion's share of the market:

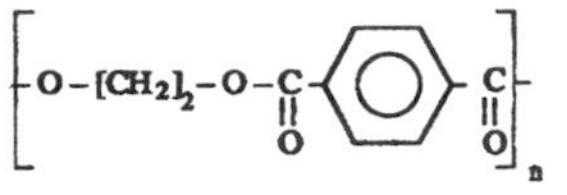

$$\left[-O-(CH_2)_2-O-\overset{O}{\overset{\|}{C}}-C_6H_4-\overset{O}{\overset{\|}{C}}-\right]_n$$

Polyethylene terephthalate

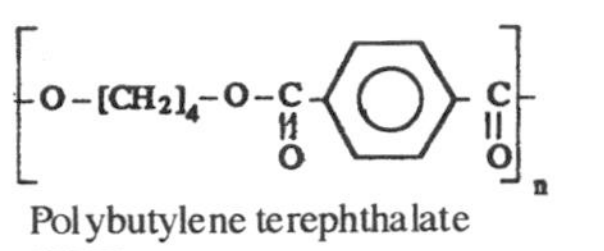

$$\left[-O-(CH_2)_4-O-\overset{O}{\overset{\|}{C}}-C_6H_4-\overset{O}{\overset{\|}{C}}-\right]_n$$

Polybutylene terephthalate (PBT)

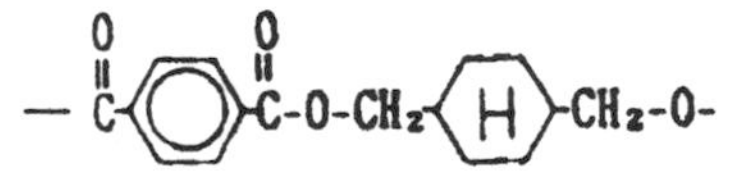

$$-\overset{O}{\overset{\|}{C}}-C_6H_4-\overset{O}{\overset{\|}{C}}-O-CH_2-C_6H_{10}-CH_2-O-$$

Poly(cyclohexylenedimethylene terephthalate) (PCT)

Polyethylene terephthalate was first used as a medical elastomer when Du Pont introduced Hytrel® copolymer in 1972. Together with PET, polybutylene terephthalate, which is used in injection molding, polycyclohexylenedimethylene terephthalate, and polyester–glycol copolymers (ethylenes and cyclohexane dimethanols) are excellent elastomers (polyesters and copolyesters) for biomedical applications. They are used as surgical field, bag, flask, bottle, and bandage-backing materials. Polyethylene terephthalate and PBT are especially suitable for making copolymers with polycarbonate, and are also available with fiberglass reinforcements (~30%). Both these polymers, which theoretically can be esterified on their diacid or dialcohol groups, are in practice obtained by transesterification of dimethylterephthalate by ethylene glycol and butylene glycol with the release of methanol:

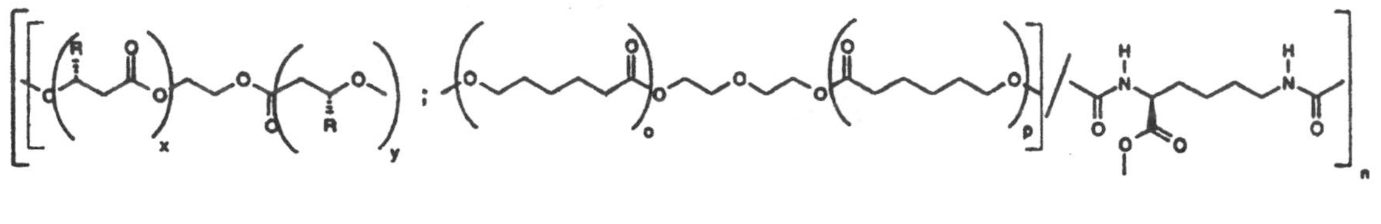

BLC: Poly{[α,ω-dihydroxy-(poly((R)-3-hydroxybutyrate-*co*-(R)-3-hydroxyvalerate)-block-ethylene glycol)-*block*-((S)-2,6-diisocyanato methyl caproate)]-*co*-[α,ω-dihydroxy-(poly(ε-caprolacton)-diethylene glycol-(ε-caprolacton)-*block*-((S)-2,6-diisocyanato methyl caproate)]}

Figure 13.16 Example of a biodegradable polyurethane. (From Ref. 99.)

Dimethylterephthalate Ethylene glycol

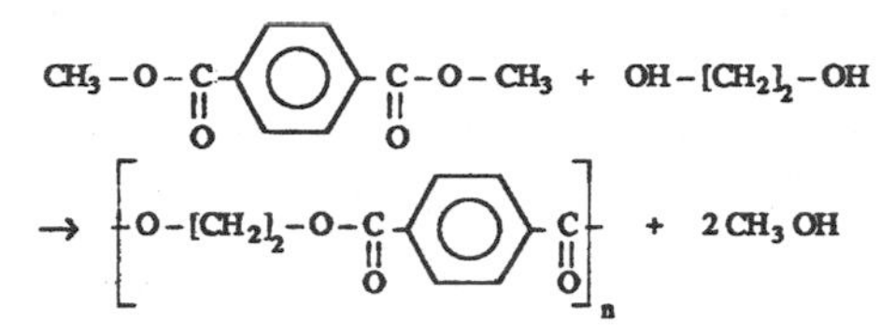

$$CH_3-O-\underset{\underset{O}{\|}}{C}-C_6H_4-\underset{\underset{O}{\|}}{C}-O-CH_3 + OH-[CH_2]_2-OH$$

$$\rightarrow \left[O-[CH_2]_2-O-\underset{\underset{O}{\|}}{C}-C_6H_4-\underset{\underset{O}{\|}}{C}\right]_n + 2\,CH_3OH$$

PET Methanol

The reaction is controlled by gradually increasing the temperature until the desired molecular mass is obtained. They are easy and cheap to produce on an industrial scale. They are sensitive to hydrolysis but are chemically resistant to a wide range of chemicals and solvents, are relatively nontoxic, contain no plasticizers, can be easily washed and sterilized (autoclave, gamma radiation), are very rigid, have the lowest coefficient of friction of any thermoplastic, and are very esthetic (smooth surface, very translucid). Thin films are permeable to water vapor and are very flexible and tear resistant (100). When copolymerized with polycarbonate, they are very easily colonized by endothelial cells and may see use as vascular prostheses in the future.

E. Polyvinyl Chlorides

Polyvinyl chloride (PVC) is a widely used polymer, and PVC products can be produced with a broad range of properties (flexible, rigid, cellular). It has been widely used in the medical field for some 40 years (gloves, tubes, tubing, catheters, bags, etc.) because of its low cost and ease of processing. The structure of PVC:

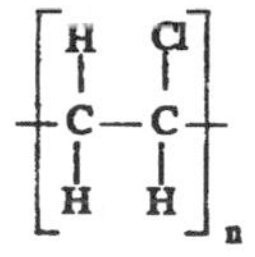

$$\left[-\underset{H}{\overset{H}{C}}-\underset{H}{\overset{Cl}{C}}-\right]_n$$

The high pressure polymerization of vinyl chloride monomer, which is a gas at room temperature and pressure, is activated by the addition of an initiator (peroxide) and controlled using temperature, mixing, and surfactants. The result is a white pulverulent polymer whose structure varies depending on the process (bulk, suspension, or emulsion polymerization). Vinyl chloride monomer lends itself to copolymerization, especially with vinyl chloroacetate (VCA). Copolymers PVC/VCA generally contain 5 to 15% vinyl chloroacetate. They can be processed at lower temperatures and pressures than PVC alone; more filler can be added; and they can be used to produce very transparent products, Polyvinyl chloride is almost never used on its own. Small amounts of specific adjuvants such as heat stabilizers (to protect the resin during processing) and lubricants (to facilitate or make the processing possible) are often added. Fillers, pigments, reinforcing agents, light stabilizers, antioxidants, etc., may also be added. In addition to the adjuvants mentioned above, PVC elastomers also contain plasticizers (5 to 70 parts per 100 parts of resin). Since conventional plasticizers (phthalates) are incompatible with biomedical applications, PVC elastomers intended for biomedical applications are specially designed not to release toxic products (101,102). Phthlate-free PVC elastomers either contain placticizers that are not released or are nontoxic (azelate, phosphate ester and polyester, citrate) or they are made without placticizer by combining PVC with other elastomers to produce copolymers (polyurethanes, vinyl chloroacetate) or simple phase mixtures. The elastomer phase plasticizes the rigid PVC (polyadipates, polyesters, polyacrylonitrile–butadiene) (103). The surface properties of PVC elastomers can be altered by plasma treatments and hydrophilic monomer grafts (carboxylation, acrylic and metacrylic acid, vinyl pyrrolidone) (104). Single-use antithrombogenic endovascular materials can be produced by grafting PU/SI copolymers on the surface of PVC tubing.

F. Polyolefins

Polyolefins are polymers produced by the polymerization of ethylene molecules where the hydrogens are replaced by various hydrocarbon groups. Ethylene (CH_2—CH_2) can be used to produce polyethylene, polypropylene, etc., by simple substitution of the side group (H, CH_3, C_2H_5, C_4H_9, etc.):

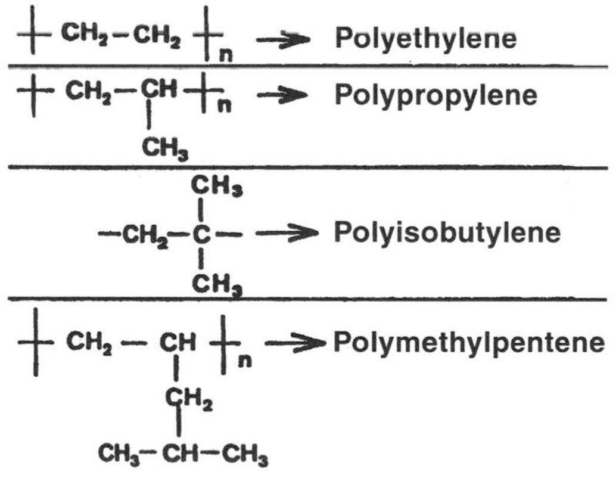

$$\left[CH_2-CH_2\right]_n \rightarrow \text{Polyethylene}$$

$$\left[CH_2-\underset{CH_3}{\underset{|}{CH}}\right]_n \rightarrow \text{Polypropylene}$$

$$-CH_2-\underset{CH_3}{\overset{CH_3}{C}}- \rightarrow \text{Polyisobutylene}$$

$$\left[CH_2-\underset{\underset{CH_3-CH-CH_3}{\underset{|}{CH_2}}}{\underset{|}{CH}}\right]_n \rightarrow \text{Polymethylpentene}$$

Polyethylenes are produced by polymerizing gaseous ethylene under different conditions. Polyisobutylenes (PIBs) for example are polymerized in the presence of catalysts. Propylene/ethylene copolymers (often called polypropylene copolymers) are elastomeric in nature and have improved shock resistance properties (block copolymers).

Random copolymers are especially well suited for manufacturing plastic films, hollow objects, plates, etc. Polymethylpentenes (PMPs) are obtained by high pressure polymerization in the presence of a catalyst. They have good water and gas permeability and superior physicochemical properties, making them resistant to autoclaving and inert to most pharmaceuticals. Numerous compounds may be added to facilitate processing. Polymethylpentenes must, however, be washed to eliminate potentially toxic residues. They are used for all sorts of packaging materials as well as for flexible tubing, stoppers, implants, catheters, etc. The excellent resistance to flexural fatigue of polypropylene copolymers makes them more suitable than other elastomers for certain applications (extracorporeal pump diaphragms and synthetic finger joints made of Hexsyn®, for example). Grafting various monomers (vinyl pyrrolidone, ethyl methacrylate, acrylamide) on the surfaces of polypropylenes can provide excellent wettability while conserving their superior mechanical performance (105,106). The recent production of polypropylene homopolymers containing alternating isotactic and atactic segments (relative configurations of successive asymmetric elements in the polymer chain) by a new metalocen polymerization process opens the way to new medical applications for polyolefins (1).

G. Styrene Copolymers

The styrene family contains a wide variety of polymers. Liquid styrene can be polymerized in two ways:

Bulk polymerization—delicate method
Suspension polymerization—fine droplets (0.1 to 10 μm) of styrene monomer dispersed in an aqueous phase

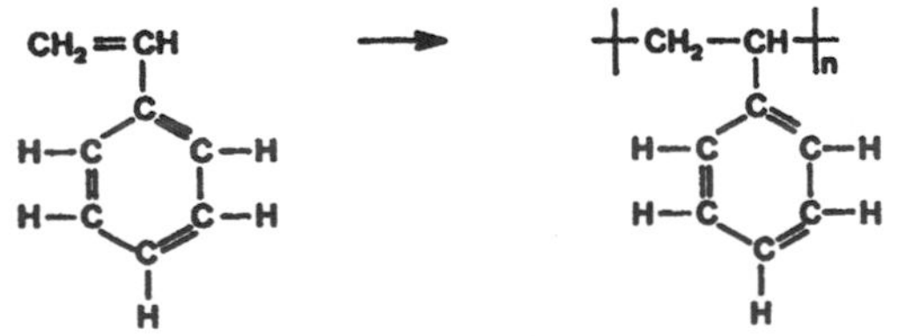

$$CH_2{=}CH{-}C_6H_5 \longrightarrow {-}\!\left[CH_2{-}CH(C_6H_5)\right]_n\!{-}$$

Styrene can be copolymerized with polybutadiene or ethylene/butylene to produce graft styrene–butadiene–styrene (SBS) and styrene–ethylene/butylene–styrene (SEBS), which are widely used in medical applications. These elastomers not only have improved shock resistance but also do not require additives. They are very flexible and pure and do not release any residues in situ (107). However, they are not very histocompatible and numerous attempts have been made to improve them. Polyhydroxyethylmethacrylate grafts improve blood compatibility; polysiloxane grafts impart the properties of silicone elastomers (108); while N^+, F^+, and Ar^+ ions improve cell adhesion (109). The SBS and SEBS elastomers can be sterilized by heat, vapor, gamma radiation, or ethylene oxide without losing their physical properties. They are used to manufacture catheters, stoppers, nonrigid containers, surgical fields, condoms, gloves, etc.

H. Natural Rubbers

Natural rubbers are the most elastic and resistant of all biomedical elastomers, but are also the least hemocompatible due to the release of accelerator (dithiocarbamate) residues (110). Many attempts have been made to improve their blood compatibility. Methylmethacrylate grafts have shown great promise because they make natural rubber more hemocompatible than many silicone elastomers (111). Another approach has been to vulcanize natural rubber by gamma radiation without additives. Natural rubbers polymerized in this way are very pure and demonstrate remarkably good histocompatibility. Untreated natural rubbers are mainly used to manufacture latex gloves, while treated natural rubbers are used to produce catheters and tubing. Latex gloves have been shown to prevent cross-infections (112). The well-balanced physical properties of natural rubbers (elasticity, tear resistance) make them ideal for this use. Hydrosoluble proteins are responsible for allergies associated with latex products. Simple washing is sufficient to completely remove the proteins and eliminate the risk of allergic reactions (113).

I. Hydrogels

Biomedical uses for hydrogels first appeared over 30 years ago (114,115). A number of polymers can be designed to act as hydrogels and retain water. Polyvinyl alcohol is the most widely used such polymer in ophthalmological applications (1). Polyvinylpyrrolidone is a relatively nontoxic, hydrosoluble polymer that can be impregnated with water or organic solvents once it has been crosslinked (116). It can be grafted on the surfaces of silicone polymers (115) and polyurethanes to improve thromboresistance (117). It is also suitable for use in soft contact lenses (1). Polyurethane hydrogels contain asymmetric hard segments that leave enough space for the absorbed liquids, if hard segment coalescence is avoided. The soft segments are generally longer than those of classic polyurethanes. Translucidity can be improved by using polypropylene glycol instead of polyethylene glycol soft segments. Hydroxyethylmethacrylates are hydrophilic methylmethacrylate copolymers (e.g., Miragel®) (115,118–120). Polysiloxane–hydroxyethylmethacrylate copolymers are used in drug delivery systems. In addition to contact lenses and drug delivery systems, hydrogels are very occasionally used in prosthetic devices. Their low strength means that they are frequently used to give more rigid elastomers (silicone, polyurethane, natural rubber) a thin hydrophilic coating (121).

J. Polypeptide and Collagen Elastomers

Polypeptide elastomers of varying composition have been synthesized to mimic natural biological elastic polymers with different physical properties and can be designed to be either biostable or biodegradable (1). They contain polypeptides (synthetic and/or natural) copolymerized with artifical polymers. Collagen is the most widely used substance for producing bulk or graft polypeptide elastomers. Elastin is also used to synthesize elastomeric biological tissues for numerous promising biomedical applications (Table 4).

K. Polyphosphazenes

Polyphosphazenes are polymers that have long flexible chains with alternating phosphorus and nitrogen atoms (1).

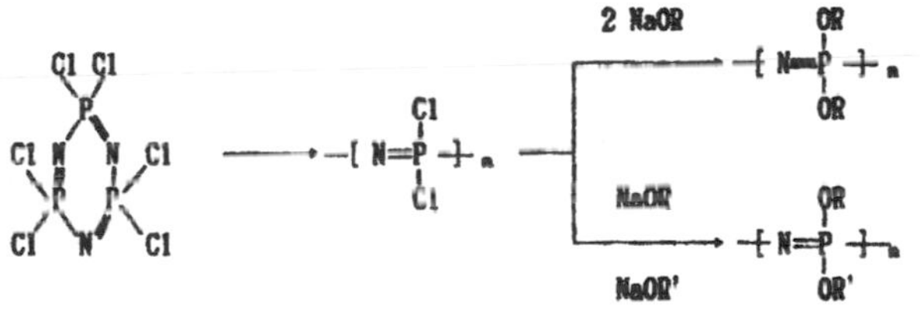

Synthesis of poly(dichlorophosphazene) and poly(organophosphazenes)

Several hundred different polyphosphazenes have been produced simply by changing the radical on the phosphorus atom. The flexibility of the chain provides good elasticity and superior flexibility. Fluorinated polyphosphazenes are very hydrophobic, which means they are theoretically weakly thrombogenic. Given their mechanical properties, they are used in external devices that come into contact with blood. Their good biocompatibility means that microporous polyphosphazene polymer matrices can be seeded with osteoblasts to serve as bone fillers or to induce bone regeneration. While they are also used to coat metallic vascular stents to improve biocompatibility, they can occasionally provoke strong tissue reactions (122). Certain polyphosphazenes with aminoester acid radicals are biodegradable (123).

L. Polyamides

Polyamides have an amide (—CO—NH—) bond that is the result of the condensation of an organic acid group on an amine. They make up a large family of textile fibers and widely used technical polymers. The first polyamide (Nylon 66), which was invented by Carothers, is a member of the group of polymers obtained by reactions between diacids [HO—CO—(CH_2)b—CO—OH] and diamines [NH_2—(CH_2)a—NH_2]. Two other types, derived from polycondensation reactions between neighboring caprolactam and amino acid molecules (—CH_2—COOH + H_2N—CH and —CH_2—CO—NH—CH_2— + H_2O), appeared a little later. Gradual heating to 250°C results in the elimination of the water and an increase in molecular mass. Diacid–diamine polyamides (e.g., nylon 66) are pro-

Table 13.4 Polypeptide Elastomers for Medical Applications

Elastomer	Composition
Polypentapeptide	$(\text{Val-Pro-Gly-Val-Gly})_n$
Polypeptide	Poly(ethylene-graft-γ-benzyl L-glutamate)
A-B-A type block copolymers	A: poly(γ-benzyl L-glutamate) B: polybutadiene
A-B-A type block copolymers	A: poly(γ-ethyl L-glutamate) B: polybutadiene
A-B-A type block copolymers	A: poly(γ-methyl L- or D,L-glutamate) B: polybutadiene
A-B-A type block copolymers	A: poly(ε-N-benzyloxycarbonyl L-lysine) B: polybutadiene
A-B-A type block copolymers	A: poly(γ-benzyl L-glutamate) B: polyisoprene
Collagen	Collagen–polyacrylates graft copolymer
Gelatin/chitin	Gelatin–carboxymethylchitin complex
Pseudopoly(amino acids)	Poly(desaminotyrosyltyrosine hexylester cabonate)

Source: Ref. 1.

duced by isolating and purifying a crystalline intermediate (nylon salt), which then undergoes polycondensation. A wide variety of copolymers can be produced in this way. Polyether block amides (PEBA's) are very large family of copolymers that have alternating sequences of crystalline polyamides and amorphous polyethers. These thermoplastic copolymers are intermediates between plastics and elastomers. Ferruti et al. (124) have produced numerous functional polyethylene glycol and polyamidoamine polymers that are used as grafts on a number of elastomeric biomaterials. Polyamide elastomers are used in a wide variety of applications: sterile packing material, cardiovascular surgery, and perfusion and transfusion materials.

M. Polyacrylics

Polyacrylics are the result of polymerization of methylmethacrylate:

Emulsion polymerization is catalyzed by adding peroxide to an aqueous monomer emulsion. The result is a low molecular weight white power ready for molding.

Molding polymerization involves heating a monomer and catalyst in an oven between two glass plates, which act as the mold. Thermoplastic objects are rarely obtained directly by polymerization. The higher molecular weight (compared to emulsion polymerization) facilitate thermomolding by increasing the temperature range of the rubber platform.

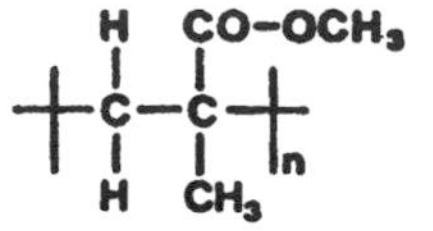

$$\left[-\overset{\displaystyle H}{\underset{\displaystyle H}{C}} - \overset{\displaystyle CO\text{-}OCH_3}{\underset{\displaystyle CH_3}{C}} - \right]_n$$

Polymethylmethacrylate

Polymethylmethacrylate (PMMA) is mainly known for its exceptional optical properties. This amorphous polymer is remarkably transparent (92% light transmission). A 3-m thick block of PMMA only absorbs 50% of incident light. This makes it ideal for use in ophthalmological devices (contact lenses, crystalline lenses). It is also very biocompatible (125,126). The addition of plasticizers (phthalates) provides PMMAs with elastomeric properties, enabling them to be used for odontological applications (maxillofacial prostheses, mouthguards) (127). They are also used in combination with other biomedical elastomers (polyurethanes, SEBS, etc.) to produce linear and graft copolymers. They can only be sterilized using ethylene oxide or gamma radiation (128) since they melt at low temperatures (80 to 100°C).

N. Fluorinated Elastomers

Fluorinated elastomers are a group of technical polymers with exceptional properties. Replacing the hydrogen with fluorine greatly improves heat and chemical resistance. Industrial fluorinated monomers are gases that liquefy at very low temperatures. They are polymerized at very high pressures using radical catalysts, generally by suspension or

Table 13.5 Grouping of Medical Devices

Type	Environment	Duration	Application
I	Internal devices	Less than 30 days	Intravenous catheters Drainage tubes
		More than 30 days	Hip implants Pacemakers Artificial heart valves
II	External devices	Less than 30 days	Devices that contact the skin such as gloves, tapes, dressings, and orthopedic casts
		More than 30 days	Devices that contact the mucous membranes such as urinary catheters and intravaginal devices
III	Indirect devices (do not contact the body)		Hypodermic syringes Transfusion assemblies Dialysis components
IV	Nonpatient contact devices (do not touch the body)		Dressing trays Packaging materials

Source: Ref. 1.

Table 13.6 Typical Research Concerning Small Diameter (<6 mm i.d.) Vascular Grafts

Material	Important design feature
1a. Polyetherurethane	Tube walls are porous and microfibrous three-dimensional mesh composed of polyurethane fibers of 1 μm diameter. Penetration depth of living tissue cell is limited to about 20–30 μm.
1b. Polyetherurethane	Graft with pore size 30–100 μm.
1c. Polyetherurethane	Grafted with albumin and adenosine-cyclic monophosphate. Surface has better blood compatibility. Modified the surface further with urokinase.
1d. Polyetherurethane	Microporous graft prepared by an excimer laser ablation technique. The pore size (100 μm) and the pore-to-pore distance (200 μm) are constant.
1e. Polyetherurethane	Made by a filament-winding technique comprising Lycra™ elastomeric fibers embedded in an elastomeric matrix of Pellethane™.
2a. Polyetherurethane urea	It is of sufficient porosity to permit tissue ingrowth. Its stress–strain curve is similar to that of the thoracic aorta.
2b. Polyetherurethane urea/silicones	Microporous replamine form graft (20–30 μm pore size) composed of silicones was coated with commercial Biomer™.
2c. Polyetherurethane urea	Fibrillar microporous hydrophobic graft made from commercial Mitrathane™.
3a. Degradable, aliphatic polyesterurethane	It has well-defined chemical and physical properties and controlled rates of degradation. It induces the growth of functional neoartery, which has a cellular structure similar to that of the native artery.
3b. Polyesterurethane (Vascugraft™)	Made from commercial polyesterurethane free from migrating additives such as catalysts and stabilizers. Microfibrous, open pore structure.
4. Expanded Teflon-coated with silicones	Platelet accumulation was reduced by coating the graft surfaces with a smooth layer of silicones.
5a. Polyurethane/poly(L-lactide)	Microporous, compliant, and biodegradable. Function as a temporary scaffold for the generation of small-caliber arteries.
5b. Polyurethane/poly(L-lactide)	Microporous and biodegradable graft consisted of two layers: the inner layer made from aliphatic polyetherurethane, the outer ply constructed by precipitating a physical mixture of polyesterurethane and poly(L-lactide).
5c. Polyurethane/poly(HEMA-block-styrene)	Polyurethane modified by coating with poly (HEMA-block-styrene) from solution.
5d. Polyurethane/endothelial cell	The hybrid artificial graft made of an endothelialized microporous polyurethane.
5e. Poly(carbonate urethane)/pentapeptide	Glycine or fibronectin are covalently bound to polyurethane surface by succinyl dichloride coupling.
5f. Biodurable poly(carbonate urethane) (ChronoFlex™)	Microporous grafts with excellent physical and mechanical characteristics. Self-sealing and maintenance of compliance.
6. Copolymers of MPC	Copolymers of 2-methacryloyloxyethyl phosphorylcholine (MPC) with hydrophobic alkyl methacrylates.
7. Polyester/human vein	Vascular graft made of human vein and a highly porous polyester fabric vascular prosthesis.

Source: Ref. 1.

Table 13.7 Elastomers Used for Making Artificial Hearts

Supplier or investigator	Type of blood pump	Elastomer
Novacor's LVAD	Pusher-plate	Biomer™
Symbion	Diaphragm	Biomer™
Aviomed	Tuberous	Angioflex™
Nimbus	Pusher-plate	Hexsyn™/gelatine
Thermedics LVAD	Tuberous	Tecoflex™
Nippon Zeon VAD	Sac	Cardiothane™
Totobo VAD	Diaphragm	TM-3™, TM-5™
Berlin University	Diaphragm	Pellethane™

Source: Ref. 1.

aqueous emulsion polymerization in order to control the strong exothermic reaction:

$$\underset{\textbf{Tetrafluoroethylene}}{F_2C=CF_2} \longrightarrow \underset{\textbf{Polytetrafluoroethylene (PTFE)}}{\left[-\overset{F}{\underset{F}{\overset{|}{\underset{|}{C}}}}-\overset{F}{\underset{F}{\overset{|}{\underset{|}{C}}}}- \right]_n}$$

The very strong electronegativity of the PTFE fluoride ions protects the carbon chain and gives these elastomers their extreme chemical stability. They are inert and insoluble (6,129). They are naphthalene-plasticized resins and are extruded and made microporous for vascular graft prostheses (Teflon®, Impra®, Goretex®) (130,131). Their vascular biocompatibility is adequate but less than that of polyurethanes (132). They are used in orthopedic surgery to reduce friction between metallic and/or composite joint prostheses (133) both alone and in combination with polyurethanes.

Table 13.9 Elastomers Used in Transdermal Therapeutic Systems

Classification	Elastomer
Hydrophobic polymers	Poly(ethylene-co-vinyl acetate)
	Silicone rubber
	Polyurethane
	Poly(vinyl chloride)
	Collagen-based materials
	Ethylcellulose
	Cellulose acetate
Hydrogels	Poly(hydroxyethyl methacrylate)
	Crosslinked poly(vinyl alcohol)
	Crosslinked poly(vinylpyrrolidone)
	Polyacrylamide

Source: Ref. 1.

O. Elastomeric Medical Devices

See Tables 5–10 and Fig. 17 for information on the use of elastomers in medical device applications.

II. BIOCOMPATIBILITY EVALUATION OF ELASTOMERS

A. Introduction

Most medical devices use synthetic biomaterials as their principal component. A biomaterial is a nondrug substance for inclusion in a physiological system that augments or replaces the functions of a bodily tissue or organ. Elastomers, like others biomaterials, must be compatible and inert, must interact with the assorted tissues and organs in a nontoxic manner, and must not destroy the cellular constit-

Table 13.8 Implant Applications for Medical Grade Silicone Elastomers

Treatment	Application
Plastic and reconstructive surgery	Reconstruction of nose, chin, ear armature, etc.
	Breast reconstruction
Ophthalmology	Correction of detached retina
	Prosthetric eye
	Repairing fracture of the floor of the orbit
Orthopedic surgery	Reconstruction of fingers, thumbs, wrists, elbows, feet, tendons, temporomandibular joint, etc.
	Maxillofacial prosthesis
	Penile prosthesis
Cardiovascular surgery	Ball in the ball-and-cage heart valve
	Coatings on pacemakers and leads/wires
	Construction in artificial hearts and heart assist devices

Source: Ref. 1.

Table 13.10 Typical Materials for Contact Lenses

Classification	Polymer and characteristics
Hydrogel soft lenses	Poly(hydroxyethyl methacrylate) (PHEMA). Generally crosslinked with ethylene glycol dimethacrylate, at a level of 26–42%, sufficient to give water content in equilibrium. It may be copolymerized with other hydrophilic monomers, such as N-vinyl pyrrolidone, acrylamide, and methacrylic acid.
Elastic soft lenses	Polydimethylsiloxane and its copolymers. They have good flexibility and very high oxygen permeability, but strong hydrophobicity results in considerable discomfort for the patient. Surface treatment may improve wettability.
	Perfluoropolyethers and its copolymers have attractive combinations of flexibility and gas permeability.
Rigid lenses	Poly(methyl methacrylate) (PMMA) has good properties: very clear optically, good biocompatibility, and good handling characteristics. The main problem relates to poor oxygen permeability. There have been attempts to introduce arrays of very small holes (15 μm) to facilitate oxygen transport.
Gas-permeable rigid lenses	Cellulose acetate butyrate and others. The butyrate is a highly oxygen permeable and relatively rigid material.
	The copolymerization of siloxany alkyl methacrylates of fluorosilicone methacrylates with methyl methacrylate provides a rigid methacrylate backbone with rubbery, highly permeable alkylsiloxane side groups.

Source: Ref. 1.

uents of the body fluid with which they interface (134). A battle takes place between the organism and the implanted elastomer. Possible outcomes include acceptance, slow digestion of the implant, or rapid ejection accompanied by undesired secondary effects (135). Medical devices must thus be designed to prevent rejection. This has led to the development of a new class of materials called biomaterials, which includes elastomers used in the medical field. The ability of biomaterials to fulfill their role in medical devices depends on their degree of biofunctionality and biocompatibility.

It is hard to talk about the biocompatibility of elastomers since they are only in contact with physiological systems in the form of medical devices. In addition, the bio-

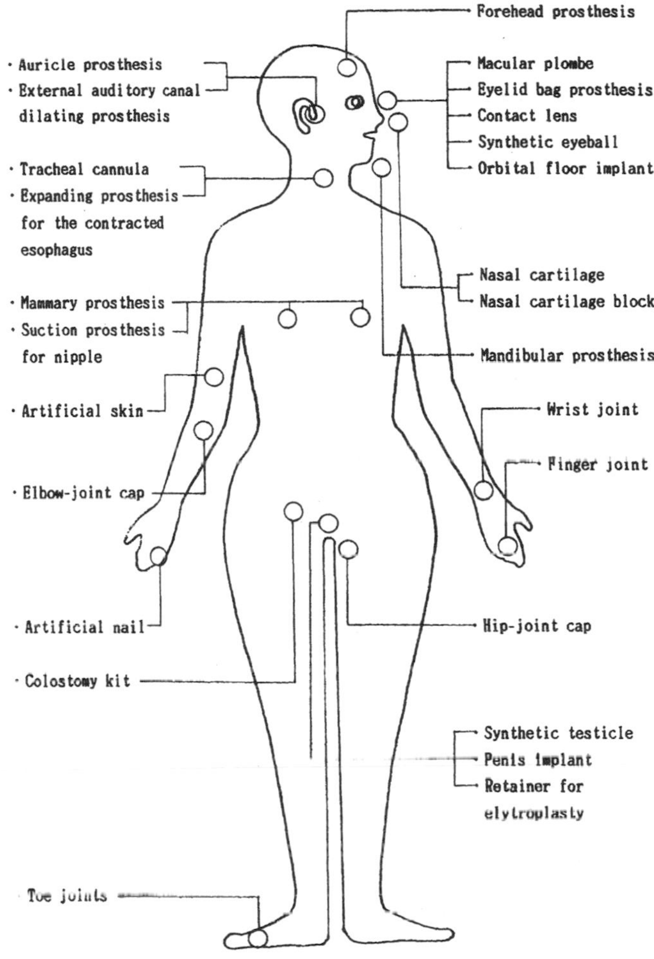

Figure 13.17 Medical device in the human body. (From Ref. 1.)

compatibility of elastomers depends on the type of device and its function. Theoretically, an evaluation of the biocompatibility of an elastomer would include

The behavior of the host tissue and the host itself in the presence of the biomaterial

The behavior over time of the biomaterial in the host environment

However, in practice, in vivo studies of all these parameters are difficult if not impossible to conduct, which is why the evaluation of biocompatibility begins with in vitro testing. Briefly, biomaterials have evolved through three generations:

1. Technical performance (noncytotoxic = tolerance)
2. Biological performance (cytocompatibility)
3. Biological performance (biofunctionality)

Four trends currently dominate the evolution of biomaterials:

- Providing materials with a surface that mimics the biological properties necessary for the intended application. This requires fundamental knowledge of cellular reactions at the material–tissue interface.
- Taking implantation in young subjects into account, which implies long-term use and the need for immunological, mutagenic, and carcinogenic evaluations.
- Decreasing animal testing, which requires the development of alternative methods.
- Setting up marketing regulations, which includes the development of standard tests.

Elastomers are very important in the manufacture of medical devices that may come into contact with host tissues for varying periods of time. Exposure times have been arbitrarily divided into three categories: limited exposure (<24 h), prolonged exposure (24 h to 30 days), and permanent exposure (>30 days). The tests used to evaluate biocompatibility depend on exposure time. It is thus easy to understand why elastomers used to make implants have to meet stricter standards than those used for dental impressions. This classification system is only a guide; since elastomers may be used for short, but repeated periods, biocompatibility testing must be based on the maximum exposure time. It must also take into account how the material will be used and the type of contact. Differing standards apply to medical devices that are

- In contact with surfaces like the skin or mucous membranes other than the skin
- In direct or indirect contact with other tissues, including blood
- Implantable devices

B. Testing

1. Cytotoxicity

In vitro cytotoxicity testing evaluates lysis, growth inhibition, and other impacts on cells by medical devices, component materials, and leachates using morphological, biochemical, and metabolic criteria. These techniques have well-defined standards for material preparation, sterilization, cell lines, and the type of testing required. Whatever methods are used, controls must be included.

a. Material Preparation

The shape of the material used for biocompatibility testing varies greatly from author to author. Generally speaking, however, elastomers are tested as discs of the size of multiwell plates used for cell cultures. Menconi et al. (136) used 10-cm^2 discs. Saad et al. (99) used 14-mm diameter 100-μm thick discs to evaluate polyesterurethanes. Other techniques involve producing 10-cm^3 blocks, which are then cut into 300-μm thick slices using a microtome (98). Special attention must be paid to the cut surface, which must not only be identical in test products and controls, but must also resemble the surface of the finished product. It is often preferable in the industrial research phase to test materials that correspond as closely as possible to the finished product. On the other hand, in the development and marketing phases, evaluations can be performed on randomly selected elastomeric ''parts'' to avoid testing bias.

Many studies have been conducted on the cytotoxicity of finished elastomeric products. Extracts are prepared on the basis of recommended normative standards. A range of extraction protocols are available: dynamic extraction, static extraction, and heat extraction.

The chemical substances extracted from elastomeric articles are usually residues from accelerators (e.g., mercaptobenzthiazole, tetramethylthiurim disulfite), activators (e.g., zinc oxide), inert filler materials (e.g., wax), plasticizers (e.g., oil), lubricants (e.g., stearic acid), and vulcanization products (e.g., zinc stearates and dithiocarbonate zinc salts). In order to minimize the amount of residue extracted from elastomeric items, the items must be either cleaned or coated, or both. Polyurethanes (PUR) and silicones are the elastomers from which the smallest amount of material is extracted (137).

b. Sterilization

The sterilization protocols for the various items used for cell cultures vary depending on the elastomer being evaluated. Most authors report using ethylene oxide (500–540 mg/L EO in 80% CO_2 at a relative humidity of over 35% for 3 h at 35–41 °C). Others have used gamma radiation (2.5 Mrad for 5 h) (86,98), dry heat or autoclaving (138), antiseptic solutions (136,139), antibiotics occasionally combined with UV radiation for 48 h (136), and immersion in 70% ethanol for 30 min followed by immersion in phosphate-buffered saline containing antibiotics (140). Albanese et al. (76) tested three different sterilization methods (gamma radiation, ethylene oxide, and formaldehyde) to evaluate the effect on the cytotoxic potential of PUPA both in native and heparinized form. PUPA and heparinized PUPA sterilized by gamma radiation are no more cytotoxic than materials sterilized by ethylene oxide or formaldehyde. Ethylene oxide gas is widely used for sterilizing elastomers since it can be employed at low temperatures with short exposure times. The exposure is followed by a degassing period, usually 10 h in a 55°C air chamber. Most elastomers are able to survive these conditions with minimal property loss. In recent years, sterilization by exposure to gamma irradiation has increased in popularity with device manufacturers because it does not produce toxic residues. All polymers are affected to some degree by irradiation, with crosslinking and molecular scission reported (141). Bouet et al. (142) compared the effect of sterilization on the cytotoxicity of PUR and various latexes and showed that while some differences may show up in vivo, no differences could be detected in vitro. While there have been reports that sterilization of polymers influences their performance in cell cultures, certain authors do not mention the sterilization method used. There is an international standard for the management of residues following sterilization by ethylene oxide, glutaraldehyde, and formaldehyde (ISO 10993—Parts 7 and 17 (143,144), but there are no standards regulating sterilization procedures per se.

c. *Cell Lines*

Fibroblast cell lines (3T3, L929) are frequently used in cytotoxicity tests (76,86,99,145,146). Tests based on the medical end uses of materials are conducted with cell lines that are as similar as possible to the cells or tissues that will be in contact with the elastomers in vivo. For example, MC.3T3.E1 osteoblast cell lines (98) and J774 macrophages (98) are used for testing materials that will be in contact with bone, and SV-HUC1 cells for materials that will be in contact with the urinary tract (147).

Primary cultures are more suitable for cytotoxicity testing. Bone cells from rat tibias (98), endothelial cells from cow thoracic aortas (136), and endothelial cells from microvessels in the abdominal fat of various animals (148), as well as human primary cultures of scaphoid bone cells (149), urothelial, ureter, and bladder cells (147), and vascular smooth muscle cells strained by inborn aortic coarctation surgery (109) have been successfully used to evaluate the toxicity of elastomers. While primary cultures, especially those derived from human tissues, provide superior biocompatibility testing results (137) because they are only one step removed from in vivo conditions, they do not allow interlaboratory comparisons to be made. Ideally, researchers should include both an established reference line and a primary culture in their studies, as was the case with Pariente et al. (147).

Certain authors have reported using organotypic cornea and skin models (23,150) to evaluate polyoxyethylene films. Others have used the culture supernatants of monocytes activated by biomedical polymers to evaluate cytotoxicity using fibroblast cultures.

d. *Tests*

The crystal violet, neutral red, or trypan blue cell viability screening tool developed over 30 years ago for evaluating the toxicity of plastics is the most commonly used test (151). Whatever the cell type, two methods are used: direct tests, where the cells are placed in direct contact with the material, and indirect tests, where samples are extracted in the medium used to culture the cells and the extract is then placed in contact with prepared cell culture either by diffusion across a membrane or an agar overlay (141,152). While the incubation period can vary from 24 h to 6 days, depending on the protocol, it also depends on the type of cells used. Generally speaking, the incubation period can be longer for established cell lines than for primary cultures. A tetrazolium-based colorimetric assay (MTT) is often used to evaluate polymer cytotoxicity, or more precisely the metabolic activity of cells in contact with the material. This assay has largely replaced those using ^{3}H-thymidine (153) and other isotopes. The MTT assay is simple, reproducible, and does not use radioisotopes. Tests can be carried out using a fully automated microtiter plate system, making it possible to economically test polymers of limited availability (154). It has also been used to evaluate polyurethanes (98), polyesterurethanes (99), elastomers (127), and urinary stents (147).

In addition to cell viability and metabolic activity assays, cell morphology is also verified to complete the first-phase screening of elastomers. The impact on cell morphology is generally determined by scanning electron microscopy, which is a relatively limited technique because it relies on obtaining images of cell adherence to the material and intracellular organization (79,136,155). In most studies, samples are dehydrated by critical point drying and then metallicized with colloidal gold. Controls involve elastomers that have not been in contact with the cells. While cell viability and morphology tests provide a first screening for cytotoxicity, many researchers have been working to examine the biofunctionality of cells in contact with elastomers and elastomer extracts. Various tests have been used, including protein synthesis and mRNA expression. Such tests obviously depend on the end use of the elastomer.

The very concept of biocompatibility is based on the interaction between a material and a biological environment, most often expressed as an inflammatory response. When an elastomer is implanted, the tissues provoke an inflammatory response, which initiates the process of tissue repair and regeneration. Several cell types are involved in tissue repair, including fibroblasts, macrophages, PMNs, and endothelial cells. Certain authors (99) have reported a more complex response to elastomers involving lymphoid and myeloid cells. Activated macrophages produce cytokines such as interleukin 1 (IL-1), interleukin 6 (IL-6), interleukin 10 (IL-10), and tumor necrosis factor α (TNFα), which are involved in regulating the inflammatory and repair processes. In addition to cytokines, activated macrophages also produce nitric oxides (NO). Macrophages are also involved in the biodegradation of foreign materials (156).

The size, surface morphology, wettability, physicochemical properties, and degradation of elastomers are key factors in the intensity and duration of the inflammatory response and repair process. Various in vitro tests have been developed to evaluate the inflammatory response and repair phases. The most frequently reported tests in the literature on the biocompatibility of polyurethanes are measurements of amounts of NO and TNFα, substances produced when macrophages and osteoblasts interact with polyesterurethanes (98,99), various types of fibronectin and collagen (24,98,99), and protein adsorption. Most of these tests are conducted by ELISA.

Osteoblast cultures are used to evaluate bone tissue biocompatibility. Alkaline phosphatase activity and vitamin D3 stimulation responses can be measured. Osteocalcine levels can be evaluated by RIA or immunofluorescence (98) and mineralization by SEM or Von Kossa staining for calcium (98,149). These assays can also be combined with more advanced, specific fluorescence techniques using devices with very low detection thresholds. Patel et al. (145) have proposed using fluorescence probe response (FPR) to evaluate biocompatibility. This technique can be used to discriminate between functional and dead cells, and is essentially based on differences in cytoplasmic membrane permeability (145,157). While these tests are commercially available (Molecular Probe, among others), they are yet not cited in the literature as reference techniques. Kirkpatrick and Mittermayer (141) published an article on the scope and limitations of in vitro systems and proposed some guiding principles for the in vitro evaluation of biocompatibility, including the use of relevant cells types and biological parameters, and stressed the need to use dynamic conditions.

Techniques based on molecular approaches are being developed to characterize the interactions between cells, biological fluids, and biomaterials. These mRNA-based techniques analyze cell functions, especially the ability to synthesize an extracellular matrix. Very detailed protocols have been published, especially that of Menconi et al. (136). Most of these techniques have become more accessible with the arrival of commercially available molecular biology kits.

e. *Controls*

The tests cannot be valid or objective unless internal controls are included. Tissue culture plastic is a commonly used negative control, although silicone rubber has also been used (158). Asbestos, phenol, and latex, which cause major disruptions to metabolic activity, are often used as positive or cytotoxic controls (147,155). It should be noted that latex has been used as both a negative and positive control. Its cytotoxicity arises from the presence of accelerators and oxidants used during the vulcanization process (137). As indicated by Park and Park (21), in the absence of clear criteria for evaluating biocompatibility, many materials were mistakenly thought to be biocompatible and were used as controls. Silicone rubber is a case in point. For a long time, silicone rubber was believed to be totally biocompatible. It is clearly necessary to reevaluate the biocompatibility of existing biomaterials and *a fortiori* the negative controls. Positive controls are not reported in many studies, which undermines the validity of the results and proposed evaluation technique.

2. Sensitization Assays of Irritation or Intercutaneous Reactivity from Implantation

Even today, there are no totally satisfactory in vitro elastomer evaluation systems. While such assays provide a first screening, in vivo testing on animals is required before implanting, injecting, or using elastomers in humans. A nonexhaustive search of the literature between 1990 and 1999 reveals that too many animals have been sacrificed just to study the implantation response of elastomers. It is vitally important that researchers meticulously plan their studies to extract the maximum amount of information from each animal.

a. *Material Preparation*

For solid implants, the physical properties (shape, hardness, surface finish) must be identical for each implant, except if one of the parameters is being singled out for study. The chemical and morphological aspects of elastomeric surfaces must be controlled (159). All contaminants must be removed from the implants, which must then be sterilized using the method that will be applied to the finished device. Tests on nonsolid materials such as powders, liquids, and particles (including mixtures) can be conducted in PTFE tubes. However, Williams (160) disputes the suitability of PTFE as a negative control, especially as regards surface properties. Control tubes must have the same physical properties as test tubes.

b. *Animals*

The choice of animal depends on the size of the implant, the test period, and known differences in biological responses of soft and hard tissues. Rats are the animal of choice for many studies. Implants are placed bilaterally in the back muscles. Two animals per material per implant period are used with a minimum of two implants per animal (101,102,130,139,161–163). When rabbits are used, 8 to 10 implants are placed in the back muscles (81,164). Dogs are used to study new materials for vasectomies and arteriovenous fistulas (95,121). Dogs are also used to test materials for temporary replacement organs. The number of animals used varies greatly depending on the type of research and the complexity of the procedure. Because of the differences, few studies can be compared.

c. *Implantation Period*

The implantation period varies greatly depending on the animals used and the material being evaluated. In the few chronic inflammation studies that have been reported, positive controls that could lead to a major inflammatory response within a 12-week period are missing. Generally speaking, observations are made over a 30-day (102) or 12-week (121,130,139) period. The implantation period must be chosen to obtain a stable biological reaction, which depends on the nature and properties of the material and the damage caused by the surgery.

d. *Evaluation of Biological Reactivity*

Evaluation of biological reactivity begins with an analysis of the state of health of the animal (vigilance, nutrition, coat, weight, temperature, etc.). The macroscopic and histopathological reactions are evaluated over time. The histopathological examination includes the degree of inflammation, the number and type of inflammatory cells (leukocytes, PMNs, lymphocytes, plasmocytes, macrophages, polynuclear cells, etc.), necrosis, and the presence of material debris, granuloma, fat, and calcified tissue.

e. *Irritation and Sensitization Assays*

Irritation is a nonimmunological inflammatory reaction. Sensitization assays (contact allergies) and delayed hypersensitivity reactions involve the immune system. These assays are vital to understanding acute toxicity. Rabbits are often used for skin irritation assays with observations at 1, 24, 48, and 72 h. The test period may be prolonged in the event of persistent lesions to determine whether the lesions are reversible or permanent, but must in no case exceed 14 days. Less than 0.2 mL of extract is injected intradermally. Irritation is scored based on the level of edema and erythema. Generally speaking, ten injections are made on the anterior and posterior portions of the trunk (including controls). Ocular and oral mucosal assays may also be conducted. To evaluate immune responses, tests developed for other purposes such as research on autoimmune disease animal models and conjunctive tissue pathologies are used. These tests are based on delayed hypersensitivity reactions induced by a given autoantibody (thyroglobulin), anticollagen (165), etc. In the research reported by Naim et al. (165), silicone gels were tested for toxicology, antigenic, and adjuvancy properties. According to Brautbar et al. (166), more rigorous in vivo animal testing is required before concluding there is a relationship between silicone breast implants and autoimmune disorders.

f. *Implantation*

All elastomers act as foreign bodies when they are implanted, leading to an acute inflammatory response and the accumulation of phagocytes. Elastomers are relatively inert, unreactive, and nontoxic. It is thus difficult to understand how they are detected by the immune system and how they provoke an inflammatory response. However, considering the mechanisms involved in such responses, a good starting point would be the initial interaction between the biomaterial and proteins in the surrounding host tissue. Immediately after implantation, hydrophobic polymers like polyurethanes, polyethylenes, polydimethylsiloxanes, and dacron become coated with host protein. Plasma and interstitial fluid proteins rapidly colonize the implant, binding together to form a disorganized matrix on the implant surface (167). The behavior is probably the result of a progressive denaturing of the proteins. After several hours, the proteins adsorbed to most biomaterials cannot be removed, even with powerful detergents. Implants are thus spontaneously coated with a random layer of partially or totally denatured proteins. Since proteins rapidly adsorb to the surface of the biomaterial, host inflammatory cells and fibroblasts cannot come into direct contact with the implant. This protein coat determines future cell reactions to the implant and it is undoubtedly this initial phase that is key in determining biocompatibility. Albumin, immunoglobulin G, and fibrinogen are the most important proteins coating the surfaces of implants. Conventional histological techniques involving hematoxylin, eosin, Von Gieson, and Schiff periodic acid stains are often used to study the inflammatory response. The inflammatory cells, which include macrophages, PMNs, and lymphocytes, are identified based on morphological criteria. The distribution of these cells around the implant provides a qualitative description of the response. A more quantitative evaluation is provided by making cell counts based on morphological criteria and scoring the response, generally on a scale of 1 to 5. This evaluation method depends on subjective evaluations by the operator, which can lead to cell typing and counting errors. A number of assays using computer assisted image analysis have been developed (163,168,169). Discriminating between different cell types has also greatly evolved with the arrival of specific monoclonal antibodies. Specific enzymatic assays and signal amplification kits (avidin, biotin, etc.) have considerably enhanced the ability to gauge implant inflammatory responses (163). These techniques have been used to evaluate numerous elastomers (81,95,121,130,139,161,162,168,169) using more or less complete cell screening panels. In addition to these histochemical staining, immunocytochemical, and counting techniques, certain authors have published in situ hybridization techniques to evaluate inflammatory protein production (TNFα) and detect the expression of TNFα mRNA (139). These techniques give a clear picture of the intensity of the inflammatory response. The reaction of TNFα and other growth factors or enzymes must reflect a balance between the tendency to promote biocompatibility and the tendency to cause inflammatory damage and implant rejection (139,170).

A number of researchers have analyzed the quality and quantity of the inflammatory exudate using the in vivo cage implant system originally developed by Marchant et al. (171). The exudate is removed before the animal is sacrificed and is analyzed using a hemocytometer. Different stains are used to discriminate between the various cells, including Wright staining and nonspecific esterase staining (66,101,161,162). A number of biochemical tests (protein analyses, albumin concentration, extracellular enzymes) are also possible (164). The exudate may also be analyzed by cytofluorometry for quantitative results (102,130). Flow cytometry appears to be particularly well suited for studying exudate cell composition. The literature on exudate cell responses is relatively meager compared to that on tissue reactions. To our knowledge, there have been no reports correlating tissue inflammation (as measured by histomorphometric image analysis) and cellular response (as measured by flow cytometry). The analysis of the inflammatory responses that occur during biocompatibility testing are now commonplace. All materials provoke a more or less intense response, and understanding the mechanisms at play in the initial colonization phase of the implanted material helps reduce the inflammatory response. McNally and Anderson (172) made a first step toward this understanding by demonstrating the participation of the C3 complement fraction in the adherence of monocytes to fluorinated, nitrogenated, and oxygenated surfaces.

3. Hemocompatibility and Blood Interaction Assays

a. General

Only materials in prolonged or permanent contact with blood pathways must be subjected to this type of testing, which can be divided into five categories depending on the primary process (thrombus, coagulation, platelet aggregation) or biological system (blood, immune) involved. Various standards determine which tests will be used to evaluate hemocompatibility depending on the type of device:

Devices whose surfaces are not in direct contact with the blood
Devices whose surfaces are in direct contact with the blood
Implantable devices

In certain cases, it is difficult to determine whether biocompatibility and/or hemocompatibility tests should be performed. In general, experience has shown that materials that are hemocompatible will also be biocompatible. However, not all biocompatible materials are also hemocompatible. The situation is even more complex because there has been a tendency to designate materials as being biocompatible on the basis of toxicological tests, while others have been designated as being hemocompatible on the basis of a single whole blood clotting test (173).

b. Tests

A given material may be adequate for one type of application but not for others. Differences between arterial and venous blood, varying blood flow patterns, and the design and mechanical operation of a medical device are among the parameters that influence the performance of materials when they are in contact with blood. Most biomaterial researchers would like to be able to predict biological performance using simplified methodologies. However, no single procedure is adequate for this purpose. Tests that merely measure the amount of thrombus give only the final result of the blood coagulation sequence without any insight into the initial events. A recently developed rheology-based test (71) now allows the sequence of events to be measured over time. ISO 10993 guidelines for the biological evaluation for implantable medical devices (174) list a number of recommended in vitro and in vivo tests that can be used to study hemocompatibility. The ISO guidelines recommend the hemolysis test as the standard screening method. However, hemolysis is mainly the result of mechanical blood cell trauma from high flow conditions that cause local turbulence and rupture of erythrocytes (175). Axisymmetric drop shape analysis—profile (ADSA-P), as described by Rakhorst et al. (175), is of special interest, especially for studying plasma protein–material interactions and is destined to become a standard test.

Many researchers use the acute ex vivo canine femoral anteriovenous series shunt technique, which allows the testing of a number of materials under similar physiological and hematological conditions for in vivo hemocompatibility studies (71,88,146,176,177). High levels of platelet and fibrinogen deposition are generally associated with a more thrombogenic polymeric surface. The morphology of the adherent platelets examined by SEM also provides interesting information. A measure of thrombogenicity is the tendency of a surface to cause adherent platelets to change shape and activate.

Hemocompatibility tests include evaluations of platelet adhesion, aggregation, activation, and release reactions under dynamic blood flow conditions together with measurements of protein adsorption and coagulation factor activation. The adhesion of human blood platelets to vascular catheters has been studied using a perfusion chamber. Polyurethane catheters were exposed to extracted human blood for different periods (up to 20 min) and at different wall shear rates (190–330 s^{-1}), and the rate of adhesion was determined using In-labelled platelets by electron spectroscopy for chemical analysis (ESCA) (178). The most common test for evaluating platelet adhesion uses fresh blood from healthy donors. The platelets are isolated and a standard microscopic analysis is performed (96,146). Platelet counts can also be determined using the petri disc model with flat sheet membranes. Complement activation and release reactions have also been explored. To evaluate the extent of in vitro complement activation by elastomers, the concentration of C3a des Arg was measured according to the method of Wagner and Hugli (179) using radioimmunoassay kits available from a number of manufacturers (103).

Tests evaluating contact activation, the intrinsic coagulation system, the Hageman factor (f XII) and dependent kinin formation and fibrinolysis pathways have been developed. A quantitative measurement of the activation of the contact system is required in studies examining the effect of artificial surfaces on blood. Van der Kamp and Van Oeveren (180) proposed analyzing these cascades by using the kallikrein inhibitor aprotinine to calculate the activity of factor f XII and kallikrein.

One approach to improving the hemocompatibility of biomaterials has been to immobilize anticoagulants like heparin at the interface (76). Most such attempts have used one of the following strategies set out by Dumitriu (135):

- Ionic binding of heparin to the material surface
- Covalent attachment of heparin to or in the material
- Chemical modification of the material surface to confer heparinoid properties

Tests including clotting assays (TT, reptilase time, and anti-XA activity), coagulation time, platelet counts, and resonance thrombography (RTG) have been used (76).

4. Biodegradation Assays

These assays are used to evaluate both biodegradation and biostability. Degradation products are generated by the decomposition or chemical degradation of a material. Biodegradation is the degradation of a biological material involving a loss of integrity and/or performance during exposure to a physiological or simulated environment. Elastomer degradation is studied using toxicokinetic modeling. Many types of surgically implantable devices and drug delivery systems that only function for a relatively short time in vivo can be made from polymers that are eliminated from the body by hydrolytic degradation and subsequent metabolism after serving their intended purpose. Biomaterials made of biodegradable polymers are designed to degrade in vivo in a controlled manner over a predetermined period. The main in vivo degradation mechanism of polymers is hydrolytic degradation in which enzymes may play a role. In vivo cell culture and animal models may also be used (181,182).

So-called nonbiodegradable biomaterials may be slowly degraded over time by the organism. One example is the failure of PU medical devices like pacemaker leads, which are manufactured from poly(ether) polymers. Failure of PU consistently occurs 5 to 10 years after implantation. The mechanism of this process is still not well understood. Although in vitro studies have been performed with single enzyme systems, which showed release of products, the in vivo situation involves complex biosystems that act synergistically (43). Other assays using sophisticated technology such as attenuated total reflection Fourier transform infrared spectroscopy (ATR-FTIR) (66), transmission Fourier transform infrared analysis (T-FTIR), and SEM (91) have also been developed. Using these techniques to evaluate inflammatory responses can throw light on the entire degradation process of a biomaterial.

Some research has been published reporting the use of autoradiographic techniques following the injection of animals with tritiated thymidine. This allows cell proliferation around the biomaterial (macrophages, foreign body giant cells, fibroblasts, bone cells) to be measured over time (172). Labow et al. (183) have studied on the effect of phospholipids on the biodegradation of polyurethanes by lysosomal enzymes. Labow et al. (183) have also studied the effect of enzymes like CE (cholesterol esterase) phospholipase A2 on biomaterial surfaces in vitro.

5. Reproduction and Embryo Development Assays

It is not known whether implant materials can have an impact on reproduction and embryo development. The few tests that have been conducted in this area involve the functional testing of biomaterials used for preparing sperm prior to artificial insemination, in vitro fertilization procedures, and embryo transfers (184). These assays can be conducted in vitro (e.g., spermatozoid cultures) or in vivo (e.g., sperm penetration tests using hamster ova and mouse embryo compatibility studies) (185). To our knowledge, no standards have been established for these assays.

6. Genotoxicity

ISO 10993 is a set of criteria based on the intended use of a product or device and establishes testing guidelines depending on the implantation period and the nature of the contact with the biomaterial. The criteria stipulate that any material or implantable device placed in contact with mucosal, bone, or oral tissue, when the contact exceeds 30 days, must be subjected to genotoxicity testing prior to commercialization. The tests are intended to detect genetic anomalies (mutations, chromosomal alterations) in cells (prokaryote, yeast, mammalian) placed in contact with the material. A review of the literature has revealed that four main techniques are used to evaluate genotoxicity: the sister chromatid exchange method, the micronucleus test, the Ames test, and the chromosomal aberration test (186,187). A study conducted in 1995 by Purves et al. (188) showed that the pharmaceutical industry uses this wide range of genotoxicity tests and protocols. New genotoxicity tests have been reported (189). However, few studies on elastomers have reported using eukaryotic cells to detect DNA damage. Further work in this area is thus required. Our team is currently investigating an in vitro test.

7. Biocompatibility Testing in Humans

Clinical tests are better indicators of biocompatibility and also help improve biomaterials. While implantable devices are supposed to be biocompatible based on in vitro testing, in vivo clinical testing is the only way to be sure. Human tests are conducted using ethical research practices. These tests can be included in randomized clinical research protocols (126,190) and in medium- to long-term epidemiological studies (133,191,192).

8. Biocompatibility Testing—Microbial Colonization

As for assays in the area of reproduction and embryo development, there are no bacteriological standards or compulsory tests. Elastomers used in implantable medical devices are by definition sterile. Nevertheless, certain authors have included microbial (including yeast) adherence and colonization in their battery of tests (13,14,193). Such testing is recommended for certain elastomers for which implantation failure may be caused by bacterial infections (194–196). The plasma proteins deposited on implant surfaces may mediate bacterial adherence, especially that of *Staphylococcus aureus*, a pathogen associated with recurrent infections. Bacterial adherence testing is conducted in radial flow chambers mounted on the motorized stage of a video microscopy system. Image processing software is used to perform automated data collection and image analysis (73). A simpler technique for quantifying adherence involves submerging the material in a bacterial suspension of known concentration. After 24 h, the material is rinsed until the rinsing liquid is sterile. The material is then placed in a culture medium (86). The agar disk diffusion test may also be used, notably for biomaterial extracts (82). Bacteria-induced infections in the presence of polymers have been studied in animals (168,169). It would be of interest to determine how bacteria adhere to polymer substrates and, in certain cases, to evaluate the efficacy of antibiotic treatments. A better understanding of bacterial genomes should also lead to a better understanding of virulence factors produced by bacteria. These virulence factors may vary depending on the biomaterial as suggested by changes in bacterial phenotypes (unpublished data).

III. BIOCOMPATIBILITY OF DIFFERENT CLASSES OF ELASTOMERS

A. Elastomers in General

Biomaterials are materials designed for use with living tissues and/or biological fluids in order to evaluate, treat, modify, or replace a tissue, organ, or bodily function (25). For materials to be biocompatible, they must not provoke allergic, inflammatory, or immunological reactions; they must be nonthrombogenic, nontoxic, and noncarcinogenic; and they must not damage surrounding tissues, plasma proteins, or enzymes (25,181). However, in most cases, it is not a lack of response that is important but rather an appropriate host response to the specific application for which the biomaterial was designed. Biocompatibility is thus defined by the response of the biological system to the biomaterial, which is seen as a foreign body and which provokes a cascade of interrelated reactions both systematically as well as locally at the interface with the biomaterial (181). Host responses to biomaterials can be divided into two categories: (1) acute, compulsory events or local reactions and (2) delayed, secondary responses or general reactions (Fig. 18). All implanted materials adsorb proteins and other macromolecules, which in turn mediate biological activities from cell attachment to inflammation to other physiopathological processes (13).

These interactions are influenced by the nature and use of the material as well as implantation site and duration (91,99). Biomaterials may thus lead to numerous undesirable side effects (inflammation, thrombosis, tissue necrosis, carcinogenesis, etc.). These undesirable side effects are

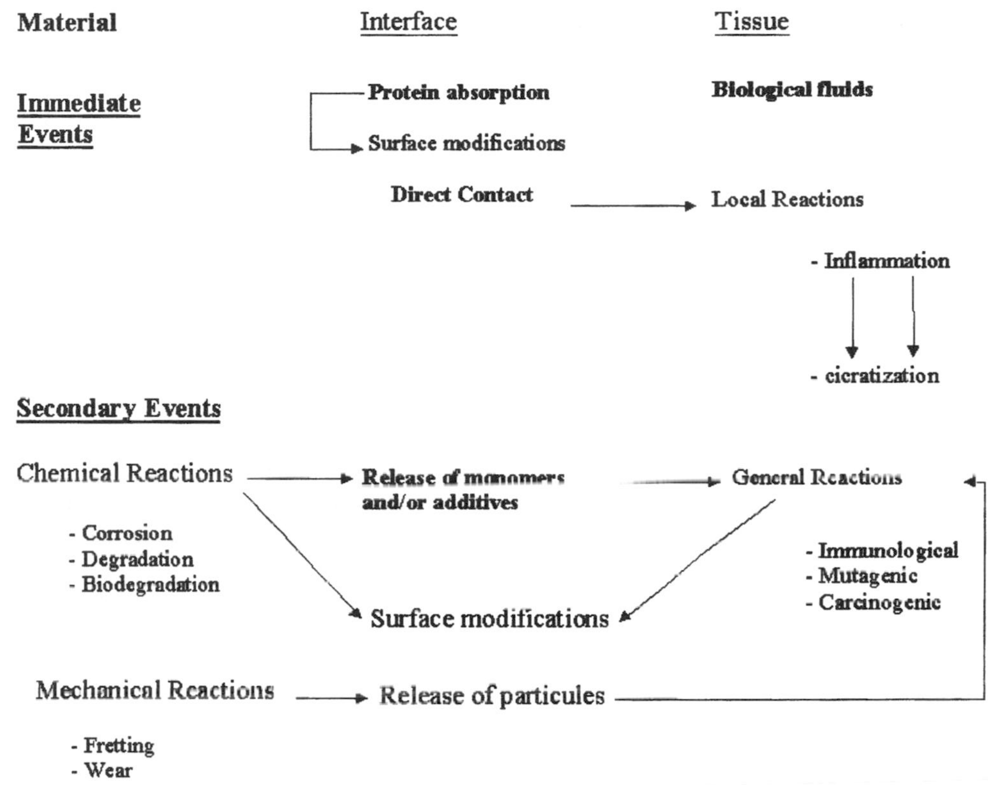

Figure 13.18 Interactions of biomaterial and tissue.

especially exacerbated by elastomers that, notwithstanding their surface biocompatibility, readily release additives (initiators, emulsifiers, plasticizers, etc.), unreacted monomers and/or oligomers (181), and degradation products in situ.

B. Local Reactions

When a biomaterial comes in contact with an intra- or extravascular system, the first interaction with host tissue is protein adsorption via noncovalent, nonspecific van der Waals forces and electrostatic bonds. Proteins compete to adsorb. The most strongly adsorbed then change their three-dimensional structures depending on the physiochemical surface properties of the biomaterial. The nature of the protein film determines the course of subsequent events such as the coagulation cascade, fibrin formation, complement activation, and the adherence and activation of inflammatory cells (72,181). Activated cells migrate and establish specific ligand-receptor bonds with the biomaterial and/or adsorbed proteins, the second step in the adhesion process. The bonding can occur in three ways: focal contacts with 20-nm separations, closed contacts with 50-nm separations, or extracellular matrix contacts with 100-nm separations (16). The host response to adherence is inflammation, an intense, local response that is the beginning of the cicatrization phase (21,162,181).

1. Protein Adsorption

Polymers, especially elastomers, are available in a wide range of shapes and formats. Depending on the end use, their surface properties such as wettability, hydrophobicity, surface charge and energy, polarity, distribution of reactive chemical groups, and surface state (smooth, rough, porous) can be chemically, physically, and/or biochemically altered. All these features, which are selected on the basis of the intended application, are subject to change under the pressure of the environmental conditions to which the biomaterial is exposed. They must, however, be preserved for the duration of the implant time because they play an essential role in the biocompatibility of the materials by directly influencing protein adsorption and host inflammatory responses (13,14,16,17,126,130, 162,181).

Albumin, IgG, fibrinogen, and fibronectin are the first proteins to adsorb (167). Albumin does not attract inflammatory cells because it does not have the peptide sequence required to bind to the cell membrane receptors. It also does not activate the coagulation or complement cascades because it cannot bind to the appropriate enzymatic sites (167). IgG antibodies produced by mature B lymphocytes or plasmocytes activate the complement system, the humoral mediator of inflammation. The complement cascade can follow either the classical pathway or the alternative pathway. Activation via the classical pathway requires an antigen–antibody complex where the antibody is either IgG or IgM. The alternative pathway is activated by microorganisms (viruses, bacteria, yeasts, parasites), tumors, nonimmune substances (bacterial polysaccharides and endotoxins, hemoglobin, etc.), or immune IgA or IgG complexes (197). When IgG adsorbs to an elastomer surface and activates the complement, an inflammatory response occurs, which can lead to increased vascular permeability, chemoattraction of polynuclear neutrophils and monocytes, leukocytosis, and/or phagocytosis (Fig. 19) (167,198).

Fibrinogen is indirectly activated by the Hageman factor (factor XII) when it comes in contact with the biomaterial (Fig. 20). The contact system is the first step in the coagulation cascade via the intrinsic pathway, which is activated when blood comes into contact with a surface other than that of an endothelial cell (subendothelial, foreign). The Hageman factor (XII), the Rosenthal factor (XI), and the prekallicrein/kallicrein and kininogen/kinin systems are involved in this activation pathway (199). Factor XII and prekallicrein (Fletcher factor), which are minor clotting factors in vivo, become very active when they interact with the surface of a biomaterial (Figs. 21 and 22) (25,180).

The Hageman factor, a serum enzyme, is activated in the presence of bacterial endotoxins, certain enzymes (trypsin, plasmin), immune complexes, collagen released by proteolysis, or by binding to subendothelial surfaces exposed by injury. Activated Hageman factor in turn activates prekallicrein, complement, coagulation, and fibrinolysis (Fig. 23) (200,201). Fibrinogen adsorbed to the surface of the biomaterial is converted into fibrin and fibrinopeptides by thrombin. The fibrinopeptides cause vasodilation and have a chemoattractant effect on leukocytes (167,200).

Fibronectin is intimately involved in cell–substrate interactions (16). Fibronectin is an adhesive protein that adsorbs to the surfaces of biomaterials. It is a high molecular weight glycoprotein that is soluble in plasma and insoluble in connective tissue. This adhesion protein enables cells to bind to each other and to adhere to artificial surfaces (13,14).

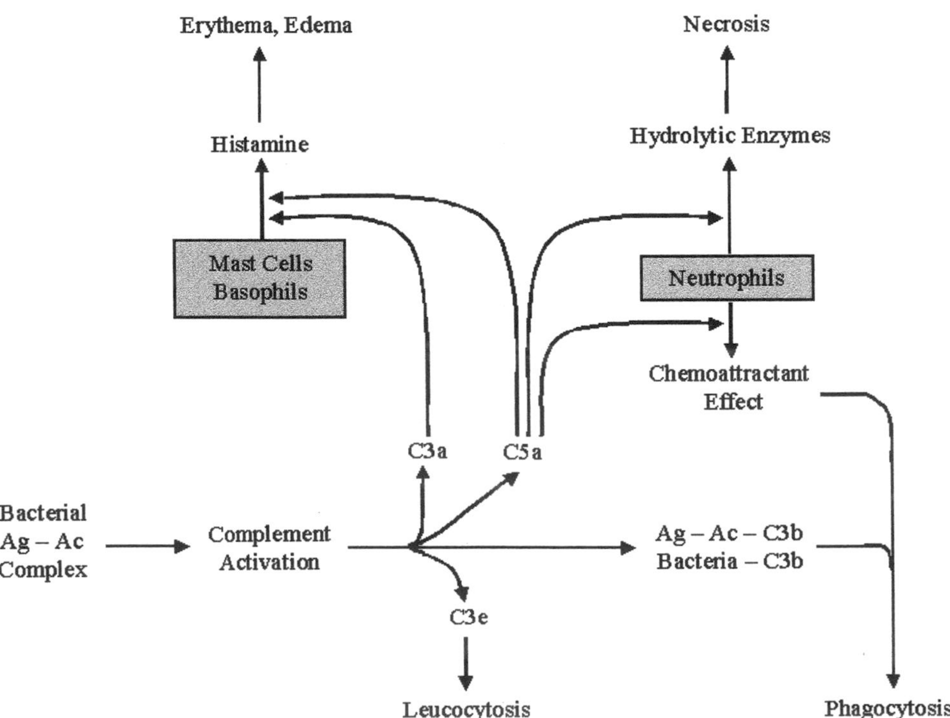

Figure 13.19 Role of complement in inflammation. (From Ref. 198.)

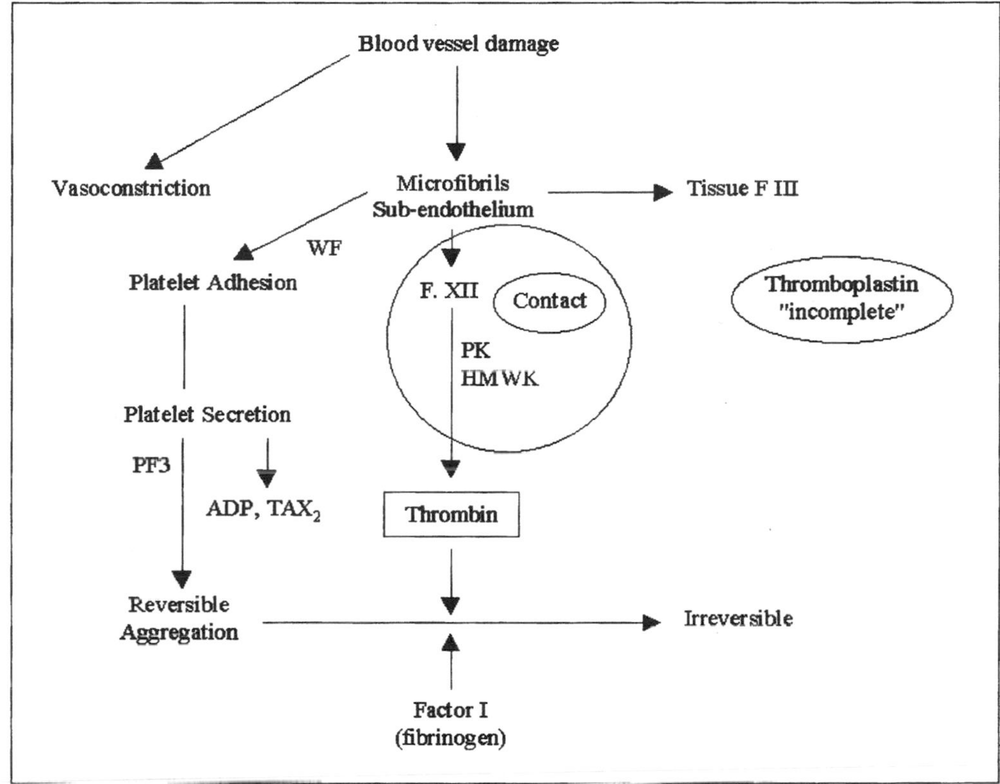

PF3 : Platelet factor 3
ADP : Adenosine diphosphate
TXA_2 : Thromboxane A2

WF : vonWillebrand Factor VII
PK : Prekallicrein
HMWK : High Molecular Weight Kininogen

Figure 13.20 Classic coagulation mechanism.

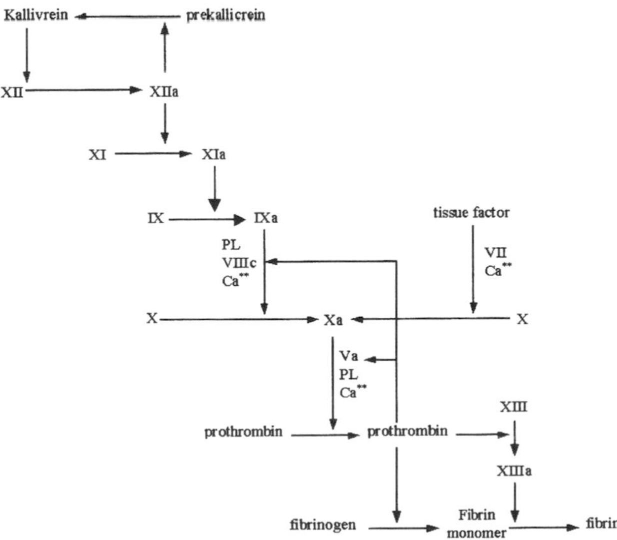

Figure 13.21 In vivo hemostasis. (From Ref. 199.)

2. Adherence and Activation of Inflammatory Cells

Leukocyte diapedesis at the implantation site is the last step in the vascular reactions involved in inflammation (91,200). Their in situ fate determines the intensity and duration of the inflammatory response because they release

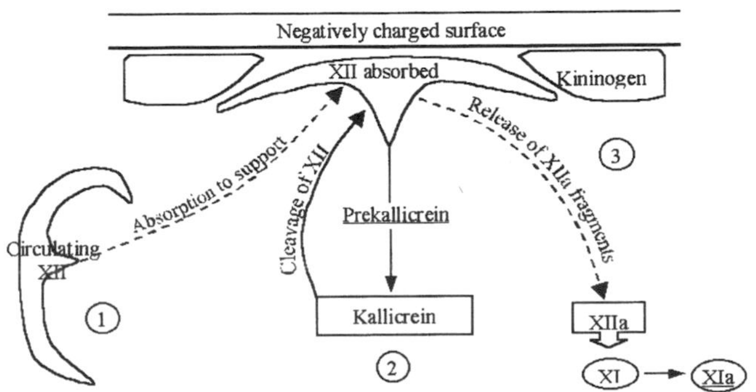

Figure 13.22 Coagulation contact phase.

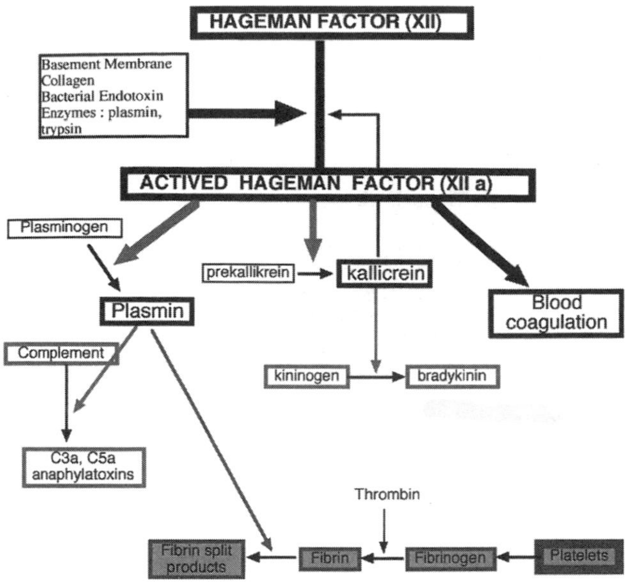

Figure 13.23 Role of activated Hageman factor in several systems.

and activate numerous cellular and humoral mediators (162). Polynuclear neutrophils and eosinophils appear in the early stages of inflammation and are followed by monocytes, which will become macrophages and lymphocytes. The macrophages activate the reticuloendothelial or reticulohistiocytic system, which helps remove foreign substances and triggers cell and humoral immune responses by activating lymphocytes (99,138,200). All phagocytes contain lysosomes, intracellular organites that contain a great variety of proteolytic enzymes. When liberated in situ, these enzymes can destroy the extracellular matrix supporting connective and epithelial tissues. Lysosomes also release vasodilating substances and activate the complement and kallicrein/kinin systems, coagulation, and fibrinolysis (25,175,180,200).

Platelets, which are very important in homeostasis, can act as inflammation mediators via substances in their intracytoplasmic granules (201). When a biomaterial is exposed to the intravascular system, protein adsorption to the surface of the substrate is followed by platelet adhesion and spreading (21). The substances released by the platelets promote interplatelet adhesion, the first, reversible step in aggregation. This process becomes irreversible when thrombin causes the formation of a fibrin network of varying thickness called a clot (21,181,199).

Connective tissue histiocytes, mast cells, and fibroblasts also participate in the host response to biomaterials (200). During cell–elastomer interactions, fibroblasts synthesize collagen and other matrix molecules (138). Under normal conditions, a fibrotic capsule quickly surrounds the implant (167). Host inflammatory response to contact with a biomaterial are normal reactions. In certain cases, local complications may occur, the most frequent being the formation of a thrombosis on contact with blood, the formation, of a fibrotic hyperplasia around the implant, and bacterial infections (21,167).

3. Local Complications

While platelet aggregation may be a normal reaction when a biomaterial comes in contact with blood, it may become iatrogenic with the formation of a thrombus and the risk of emboli (Fig. 24) (21,181).

Factor XII activates the complement system. Activated fractions C3a and C5a cause platelet aggregation. C3b promotes cell adherence, increasing the risk of thrombosis (25,72).

Certain authors have reported that hemocompatibility is influenced by the type of proteins adsorbed to the surface of the implant. Albumin seems to prevent the adhesion of platelets, thus providing acceptable hemocompatibility, while fibrinogen and gamma-globulin increase platelet adhesion, leading to bioincompatibility (135). It has been shown that albumin specifically adsorbs to hydrophilic domains while gamma-globulin adsorbs to hydrophobic domains (202). Hydrophilic biomaterials are thus more hemocompatible than hydrophobic polymers (203).

Contact with the vascular system may lead to large variations in the surface wettability of biomaterials. This change from a hydrophobic to a hydrophilic state has more of an impact on hemocompatibility than initial wettability (175). More hydrophilic surfaces tend to activate factor XII, the coagulation cascade promoter, while hydrophobic surfaces tend to show a preference for activating the kallicrein/kinin system (180). When vascular grafts are implanted, the response of endothelial cells is crucial because they play a role in maintaining the antithromobogenicity of the prosthesis. To achieve optimal integration of a vascular implant, the endothelialization of the prosthesis lumen must be quick and complete. A rapid migration of endothelial cells prevents the formation of blot clots and vascular occlusion. On the other hand, the proliferation of these cells must not be excessive in order to avoid hyperplasia and occlusion (132).

The formation of a fibrotic capsule is a normal reaction to the implantation of a foreign body. The process may become pathological if the fibroblasts proliferate too extensively and cause fibrotic hyperplasia. An intense host inflammatory response on exposure to a biomaterial stimulates phagocytes, especially macrophages, which are

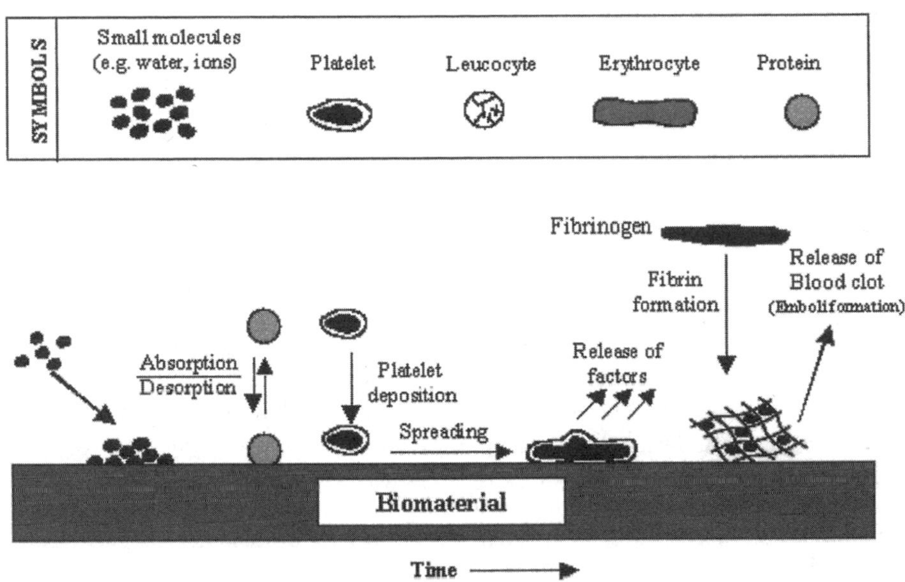

Figure 13.24 Blood response. (From Ref. 4.)

attracted in large number and produce a cytokine, interleukin 1. Interleukin 1 stimulates fibroblast proliferation and thus regulates the duration and extent of fibroblast activity and can also lead to fibrotic hyperplasia, and thus has a direct impact on graft success or failure (21,138,167).

A lack of implant–tissue adherence and the space between the two surfaces cause a certain amount of friction that results in the accumulation of serous fluid and inflammatory cells. The intense inflammatory response at the interface would explain the fibrotic hyperplasia (16). The contraction of the fibrotic capsule around the implant, which is more frequent with silicone elastomers than polyurethanes, is a common complication with implanted biomaterials (167). Inflammation and the various humoral and cell mediators play a clear role in the contraction of the fibrotic capsule. Good acceptance of an implant means a more moderate inflammatory response and less contraction. The structure of the implant also plays a role in this phenomenon. Contraction of capsules surrounding porous polyurethane implants occurs at a very low rate (1 to 5%) compared to smooth implants (70%). The phagocytic activity of macrophages and foreign body giant cells prevents the formation of linear capsules around implants. Fibroblasts migrate through the pores in implants, dissipating vectoral forces. Numerous microcapsules replace the linear, continuous fibrotic capsule, thus preventing contraction (167,191).

Bacterial infections are frequently associated with implanted biomaterials. Fibrinogen is known to mediate the adhesion of *Staphylococcus aureus*, the main pathogen involved in implant-associated infections. Fibrinogen can act as an opsonin and promote the aggregation of bacterial cells, especially Staphylococci and Streptococci (201). The access route for catheter-related infections remains a point of contention. Certain authors (204–206) have suggested that infections involving percutaneously inserted catheters are due to contamination of the external surfaces. Others have suggested that the contamination of the catheter hub can lead to intralumenal colonization and catheter-related septicemia (84). Bacteria and fungi colonize on the polymer surface, forming an adherent biofilm. The microorganisms embedded in these highly adherent films may in the long term secrete products that degrade certain substrates (169). In addition, biofilms make antibiotic treatments and host defense mechanisms less effective (13,14,73,82).

Silicone elastomers, however, do not seem to be affected by infections. Analyses of fibrotic capsules around silicone implants, whether or not there was a deliberately induced *Staphylococcus aureus* infection, did not reveal any differences in fibrotic capsule composition or thickness. Analogous results with porous polyurethane also showed that biocompatibility was not significantly altered by infections (169).

C. General Reactions

1. Degradation

Degradation is a complex mechanism that affects all elastomeric biomaterials and depends on the chemical, physical, and structural properties of the implant and the inflammatory response of the host (4). Degradation operates at the molecular, macromolecular, microscopic, and macroscopic levels (161). All implanted materials provoke an inflammatory response from the host when it attempts to attack, destroy, and reject the foreign body. Leukocytes, especially macrophages and FBGCs, are involved in this mechanism which leads to the degradation of the elastomer (37,167).

Macrophages appear to play a predominant role in the degradation of elastomers. They are involved in the pinocytosis and phagocytosis of cell and tissue debris as well as polymer contaminants on the surfaces of biomaterials. Phagolysosomes digest the particles they have ingested and the degradation products are then released into the extracellular environment together with lysosomal enzymes, peroxides, and hydrogen ions (37,91,162,167). Lysosomal enzymes have often been implicated in the degradation process despite the difficulties in analyzing this phenomenon in vivo (161). Two lysosomal enzymes have been shown to participate in the chronic inflammation phase following the implantation of polyurethane devices. They are cholestherolesterase (CE) and phospholipase A2 (PLA). Cholestherolesterase catalyzes the hydrolysis of amphipathic molecules (183). It was shown in a study on polyurethane elastomers that the products generated by the lysosomal enzymes (CE and PLA) as well as the physicochemical properties of the elastomer affect the biodegradation of polyurethanes (183). The release of lysosomal enzymes, peroxides, and hydrogen ions, which are very concentrated at the cell–polymer surface interface, leads to significant degradation at this site. The degradation is faster for porous elastomers than dense elastomers. Stress and tension combined with the chemical effects of the elastomer are also involved in the degradation process (91). Degradation leads to surface cracking and a loss of molecular weight, which in turn results in a loss of tensile strength (4).

2. Carcinogenic Reactions

The carcinogenic potential of materials implanted in rodents has been recognized since 1950. This process is called the Oppenheimer effect. Many substances have been shown to induce foreign body carcinogenesis at the implantation site (4). In 1975, Brand et al. (207) described the various steps in foreign body tumorigenesis:

Proliferation and tissue infiltration
Progressive formation of a fibrotic tissue capsule
Quiescence of the tissue reaction
Direct contact of clonal preneoplastic cells on the foreign body surface
Maturation of preneoplastic cells
Sarcomatous proliferation

Tumorigenesis is controlled by a number of very controversial factors. Foreign body carcinogenesis is influenced by physical features like shape, size, hardness, and porosity (4,34,158). Implant size is critical. Tumorigenesis does not occur unless the implant is of sufficient size. The bigger the implant, the greater the probability that preneoplastic clones will attach to the foreign body surface (207). Other researchers have reported that chemical properties play an important role in tumorigenesis (158). Nakamura (158) concluded that the chemical release of decomposition products and the hardness and physicochemical surface properties of biomaterials play a role in the long-term tissue response. Numerous studies have examined the incidence of tumor formation as a function of substrate and, in particular, polyurethanes and silicones. Polyurethanes appear more carcinogenic than silicones. Differences between the two elastomers in terms of cell adhesion and thus cell proliferation are important factors in the carcinogenic process (158,208,209). The higher thermal sensitivity and water absorption of polyurethanes with respect to silicones also play an important role in this process (34). In addition, a degradation product of polyurethane (MDA or 4,4′-methylene dianiline) has been implicated in the development of cancer in rats (4,158,167,210). Heat (autoclave) and gamma radiation sterilization have also been implicated in the formation of this toxic, carcinogenic compound (1). The mechanism remains very controversial. On the other hand, a 1995 study produced no evidence that sterilization leads to the formation of MDA (211). The tumorigenesis mechanism is not clear despite numerous studies on the carcinogenic potential of biomaterials. Nakamura and his team (158) have tried to provide an explanation by suggesting a two-step mechanism: initiation followed by promotion. The initiation step depends on active oxygen species produced by cells like macrophages and foreign body giant cells or on initiators in the biomaterial (208,209). This step may be triggered by the hard segment of the chemical moiety such as MDU, which is associated with leachable and biodegradable oligomers (158). The promotion step then occurs, with polyether soft segment moieties such as PTMO inhibiting gap junction intercellular communications. The soft segment moiety is derived from leachable oligomers and degradation by direct cell–material interactions (208,212). The role of surface properties in the promotion of tumors has been examined by gap junction intercellular communication inhibition studies using collagen immobilized on polyethylene film. The studies have shown that the inhibition of gap junction intercellular communication reduces the tumorigenic potential of polyethylene by decreasing tumor promotion activity (213).

D. Silicones

While the belief that silicone elastomers are inert is no longer a point of discussion, the biocompatibility of these elastomers remains an open question. Undesirable effects from silicone implants have been reported. Silicone particles have been shown to migrate to synovial fluid, bone tissue, and peripheral lymphatic organs via the vascular and lymphatic systems. Chronic inflammatory reactions, with and without fibrotic hyperplasia, have also been reported (214). Silicone elastomers have elevated coefficients of friction and low surface energies, which, combined with their hydrophobic nature, give them a low wettability index (5,20). As early as 1979, in vitro cytotoxicity studies provided contradictory results. Cytotoxicity is the result of the release of cytotoxic compounds and/or the surface state of the silicone elastomer. The hydrophobic/lipophilic behavior of silicone is a major problem for in vitro studies because it makes cell adhesion difficult. One study has shown that cell adhesion and growth increased when talc was applied to the elastomer surface to make it more hydrophilic. This was not due to a decrease in the release of toxic substance but improved adhesion resulting from a change to the surface of the biomaterial (215–218).

An analysis of epithelial cell adhesion to various elastomers has revealed similar results for all samples (silicones, polyurethanes). However, cell growth was poorer on silicone elastomers than on polyurethanes. These results tend to minimize the role of cell adhesion to the substrate. Cytotoxicity is more likely related to the release of toxic substances (219). Scanning electron microscopic analyses of the behavior of cells grown in contact with silicone elastomers have shown that the cells grow in aggregates with little spreading and produce numerous adherence structures linking them to one another, but few binding them to the substrate (127). The properties of silicon elastomers described affect cell adherence to the substrate and thus cell growth, but have no effect on viability.

In vivo studies on the biocompatibility of elastomer implants have shown that a moderate inflammatory response occurs with the appearance of low numbers of FGBC (129,220,221).

Hemocompatibility studies on silicones have shown that albumin and fibrinogen strongly adsorb to these elastomers and that platelets also adhere in large quantities. Wettability can be calculated using contact angle measurements. There is a very large decrease in the contact angle of silicone elastomers over time, changing them from hydrophobic to hydrophilic biomaterials. According to the authors of one study (175), variations in wettability have a greater impact on hemocompatibility than initial wettability. When the biomaterial comes in contact with blood, the intrinsic homeostasis pathway is triggered and the Hageman (XII) factor and prekallicrein/kallicrein and kininogen/kinin systems are activated. The Hageman factor stimulates coagulation while the kallicrein/kinin system causes vasodilatation (180,199). A study on these systems during biomaterial–blood contact has shown that silicones preferentially activate the kallicrein/kinin system. The ratio of Factor XII to the kallicrein/kinin system was 1:2. The wettability of biomaterials plays a role in the activation of these systems. The kallicrein/kinin system is more active on hydrophobic surfaces, which leads to vasodilatation, platelet aggregation, and an increase in permeability (180). Various treatments have been attempted to improve the hemocompatibility of silicones. The elimination of micro air bubbles on the surfaces of the elastomer by denucleation reduces complement activation and platelet aggregation (25). The affinity of albumin, which acts as an activation inhibitor of various biological reactions, can be increased by a number of different processes. This reduces inflammatory cell migration, greatly improves hemocompatibility, decreases complement activation, and prolongs contact time in vivo (26,222). The presence of fibers in polymethylvinylsiloxane does not modify hemocompatibility (223).

Free radicals released by polymers were implicated as early as 1960 in immune disorders caused when polymer particles are phagocytosed by macrophages. The macrophages can die or release antigenic substances, which lead to the production of antibodies in the reticuloendothelial system (6). Animal experiments have confirmed this and have reached the conclusion that injected silicone can precipitate an immune response. The silicone is transported by the lymphatic system to the liver, spleen, kidney, and pancreatic lymph glands of the reticuloendothelial system (166). Foreign macromolecules that are phagocytosed and presented as antigens may provoke a humoral or cellular immune response (13). This immune response is mediated by the nature of the protein film adsorbed to the surface of the substrate as well as the nature of the substrate itself. Bonfield and Anderson have shown that the adsorbed protein film has a major impact on the behavior of cultivated human mononuclear cells (224). Monocytes and macrophages secrete more functional IL-1β when grown on a protein film of fibrinogen and fibronectin than on a surface without a protein film. It has also been shown that the substrate itself plays a role in the production of IL-1β. In the presence of an identical protein film on different substrates, monocytes/macrophages produce more functional IL-1β when grown on silicone than on Teflon, Dacron, polyethylene, or Biomer (224). The nature of the substrate and the adsorbed protein film also plays a role in molecular conformation (13). Silicone elastomers are immunogenic, but the involvement of silicone mammary implants in immunological and conjunctive tissue diseases remains controversial (166,210,225–227). Chronic diseases, especially cancers, may be the result of immune system disruptions caused by silicone elastomers (210).

While silicone elastomers are hydrophobic, very thrombogenic, and occasionally result in catheters that are difficult to insert without breaking (20), they are very stable over time, are highly resistant to attacks by the host and microorganisms, provoke minimum tissue reactions, and can be sterilized by autoclaving (1).

E. Polyurethanes

Polyurethanes have found a number of medical uses because of their acceptable level of biocompatibility, their high mechanical resistance, and their elastomeric properties (1). Polyurethanes, together with silicones, are among the few elastomers than can be implanted for long periods in humans (4). Other authors prefer polyurethanes to silicones because of their often superior mechanical properties and comparable biocompatibility. They are materials of choice for soft tissue and cardiovascular applications, especially for catheters (20,37,132). Polyurethanes have also been cited as the best biomaterial for applications requiring a combination of hardness, durability, biocompatibility, and biostability (228). They are divided into two main families: polyetherurethanes and polyesterurethanes. Since the ester bonds of the polyesterurethanes are subject to hydrolysis (vascugraft), these elastomers are used less frequently in medical applications. Another class of polyurethanes includes poly(ether urethane urea), in which the diol group has been replaced by a diamine.

Numerous studies have compared the biocompatibilities of the various polyurethanes. In vitro biocompatibility analyses comparing polyurethane elastomers (polyether and polyester) and copolymers (polyether–polyester) have shown that epithelial cells have the same growth patterns and explant morphologies as cells grown on control substrates. Comparable results were obtained with the ''artificial aging'' technique (219). There is strong endothelial cell proliferation within 1 week despite slower spontaneous endothelialization compared to other polymers like PTFE (148). The biocompatibilities of a polyetherurethane and a porous copolymer (polyether–polyester) have been studied following implantation in rats. The degradation of these polyurethanes, which is less rapid than other elastomers, does not result in the release of toxic products. The proliferation of fibroblasts and the growth of fibrous and bone tissue are signs of acceptable implant fixation despite the presence of macrophages and foreign body giant cells (168).

A number of biocompatibility studies have been conducted to evaluate the role of wettability and implant structure (porosity) on cell adhesion, endothelial cell proliferation, inflammation, and cicatrization. The studies evaluated a polyesterurethane (vascugraft) with interconnected pores and two polyetherurethaneureas (Mitrathane), one hydrophilic with closed pores and the other hydrophobic with open pores. Following implantation in rats, the histological analysis revealed a slight cell reaction to the polyether and a medium reaction to the polyester. The fibrotic capsule surrounding the polyester implant pointed to good cicatrization (130). Cell adhesion to the polyester was weak, slightly stronger to the hydrophobic polyether, and equivalent to the control substrate for the hydrophilic polyether. Cell proliferation followed the same pattern, although a uniform monolayer of endothelial cells formed on the polyester, while multiple layers formed on the hydrophilic polyether (130,155). Similar results were reported in a study on hydrophilic and hydrophobic polyetherurethaneureas. The formation of cell monolayers on a hydrophobic substrate is facilitated more by the presence of interconnecting pores than wettability (131). A porous structure promotes fibroblast proliferation and the production of new collagen on polyester. Tissue growth is a function of material porosity (130,229). If there are no interconnecting pores, a thin but linear fibrotic capsule is deposited around the implant, but fibrotic tissue does not form within the implant (230,231). Open pores should allow the development of fibrotic tissue within the implant. A porous structure thus prevents deformation and reduces the risk of infection and contraction of the capsule around the implant (130,167,191). These reports show the importance of the physical properties of substrates such as porosity and surface wettability in elastomer biocompatibility.

Polyetherurethanes can be aromatic or aliphatic. When aliphatic they are produced using hydrogenated methylene diisocyanate (HMDI) rather than MDI, they no longer release 4-4′-methylene dianiline as a decomposition product, which has been shown to be carcinogenic in the rat. In addition, these aliphatic polyetherurethanes are not affected by ultraviolet radiation (1). As for hemocompatibility, platelet adherence is similar for both aromatic and aliphatic polyetherurethanes although more fibrinogen is deposited on the aliphatic (203). Aliphatic polyetherurethanes are more difficult to synthesize and have slightly weaker mechanical properties than the aromatic (1,203).

There are no differences in endothelialization or neointimal formation between aliphatic polyether-coated stents (Tecoflex) and uncoated stents. On the other hand, the inflammatory tissue response associated with aliphatic polyetherurethane–coated stents may be an indication of low biocompatibility (96). A new type of aliphatic polyetherurethane (UTA) has been shown to increase the speed of cell adhesion, equaling that of hydrophilic polyetherurethane urea. After 1 week, both the UTA and hydrophilic polyetherurethane surfaces were covered by multiple cell layers, while the hydrophobic polyetherurethane was only covered by a monolayer (131). Extraction cytotoxicity analyses revealed that this family of elastomers does not release toxic contaminants and has no clear superiority in terms of biocompatibility over the other elastomers tested (polyetherurethane urea and PTFE, Goretex, Impra) (6). The author of this study also pointed to the role of biomaterial hydrophobicity in cell adhesion.

The degradation of polyurethanes has been studied in great detail and involves three main mechanisms: mineralization, oxidation, and environmental stress cracking. Mineralization or calcification (161) involves the incorporation of crystalline inorganic substances on the surface of the implanted material. This mechanism often results in hardening, distortion, and mechanical degradation of the elastomer (228). This mechanism most often comes into play when elastomers are in contact with the bloodstream or are in zones subject to dynamic flexing. Microdefects in the material can catch small particles of cellular debris, which in turn can act as nuclei for the formation and growth of calcium phosphate–rich crystals (1,37,228).

Oxidation is induced by free radicals that break bonds, opening the elastomer up to attack by oxygen molecules. This can be influenced by factors such as the flexibility of the medical device, the type of soft segment, and the ability to interact with ions (4).

Environmental stress cracking is mediated by various stress and environmental factors acting on the polymers such as oxidation, enzymatic activity, and mineralization. Three conditions are required for ESC to occur: a polyetherurethane device, enzyme activity, and mechanical stress. While this biological mechanism has never been described for polyetherurethanes in vitro, Szycher et al. (37) have mentioned the notion of BI-ESC. Cell- and humoral-mediated inflammatory responses may also be involved in the surface degradation of polyurethanes. Environmental stress cracking first occurs in the direction of the stress and then takes on a three-dimensional aspect that may extend to the very heart of the material, leading to total failure (37). The problem with degradation has been minimized with the introduction of ''biostable'' polyurethanes. Coury (52) has synthesized polyurethanes with no ether functions. In vitro the absence of an ether bond removes a point of attack for enzymes and chemical oxidants, thus lowering the risk of degradation (4,37). Szycher et al. (37) detected no degradation, cracking, or failure in biostable PUs after 12 weeks of implantation.

Modified polyurethane elastomers (polycarbonate urethanes) have also been produced using polycarbonate glycol as a soft segment. While polyetherurethanes undergo significant degradation after 6 months of implantation, the surface of a polycarbonate urethane remains smooth and crack-free (4).

New elastomers such as polycarbonate polyurethanes and polyurethanes endcapped with polydimethylsiloxane have made their appearance. Studies have recently been published comparing the biocompatibility and biostability of these new PUs with classic polyetherurethanes and PTFEs. They show that all polyurethanes share the same biological reactions (acute and chronic inflammation). The hydrophobicity of polydimethylsiloxane leads to a smaller accumulation of FBGC on the material surface. Polydimethylsiloxane-endcapped PUs are less subject to biodegradation because the endcap provides protection against the oxygenated free radicals released by macrophages and FBGCs. The even greater biostability of polycarbonate polyurethanes has been attributed to their carbonate bonds (66). When used as vascular prostheses, polycarbonate polyurethanes become endothelialized much more quickly than PTFEs. In addition, chronic proliferation is lower than with PTFEs, decreasing the risk of hyperplasia and occlusion (132).

The biocompatibility of elastomers can be improved by modifying the surface or internal properties of existing polymers. However, these physical, chemical, and biological modifications must not affect their bulk properties. To quote Han et al. (74), ''practically every physical and chemical material property has been suggested as being important in biocompatibility for areas such as thrombus, calcification and infection'' (74).

The chemical and surface properties of biomaterials play a major role in cell and tissue responses on implantation. The production of factor TNFα, which regulates inflammation and cicatrization, has been measured using polyurethanes containing increasing quantities of sulfonate (10, 20, and 30%) and thus varying surface charges. Fewer neutrophils are observed from 2 to 12 weeks with the PU containing 20% sulfonate. However, after 12 weeks the situation is similar for all polyurethanes, with a few neutrophils and the same number of macrophages associated with the elastomer. This indicates that surface charge may play a role in the acute inflammatory response during implantation (139).

Other techniques have been developed to improve the biocompatibility of polyurethane elastomers. The incorporation of glycerophosphocholine as a chain extender in poly(tetramethylene oxide)–based polyurethane significantly decreases bacterial adhesion and protein adsorption. The incorporation of dehydroepiandrosterone (DHEA) in polyetherurethane urea decreases macrophage adhesion and FBGC formation for up to 7 days. After 7 days, the biocompatibility of all PUs (with and without DHEA) is similar (91). The incorporation of vitamin E (tocopherol) generates a very stable PU, something that is not observed with DHEA (91). Calcium chloride added to poly(ether urethane amide) improves biocompatibility, hydrophilicity, and plasticity. However, it also increases thrombogenicity (92). Grafting collagen onto PU increases endothelial cell growth, which is dependent on the morphology and diameter of the collagen fibers. However, the collagen may also activate and aggregate platelets, making the material thrombogenic (79). Coating the prosthesis (PU + collagen) with a confluent monolayer of endothelial cells should resolve this problem. All these studies clearly demonstrate the complexity of the biocompatibility of elastomers in general.

While polyurethane elastomers have been used in a broad range of applications, the presence of microemboli and microscopic thromboses requires them to be modified to make them less thrombogenic (1). Rakhorst et al. (175) have shown that less albumin and fewer platelets attach to polyurethane elastomers than to polydimethyl siloxane or PTFE.

Manufacturing processes and the composition of polyurethane as well as surface properties play a role in polyurethane–blood interactions (203). The roles of the various components of polyurethanes have been studied. The nature of the soft segment makes polyurethane more or less thrombogenic. It has been shown that polyethylene oxide is more thrombogenic than polypropylene oxide or polytetramethylene oxide, with greater platelet retention (203). The nature of the hard segment is also important in the blood response. Aliphatic PUs do not have the same properties as aromatic PUs. They both attract platelets to an equal degree, but there is an increased fibrinogen response with aliphatic PUs (148). The quantity of soft and hard segments influences hemocompatibility. Polyurethanes with large proportions of hard segments are subject to rapid cell adhesion and sustained cell proliferation (203). The lower the proportion of hard segments, the weaker the blood interactions (146). The molecular mass of the hard segment is important when such devices must be in contact with blood. It has been shown that PTMO 1000 is more hemocompatible than PTMO 2000 (146). This shows that both the concentration and conformation of the hard segment play a role in hemocompatibility (203). Two main types of chain extender can be used: diols (polyether urethanes) and diamines (polyetherurethane urea). Diamines provide enhanced thromboresistance and hydrophilicity (203).

Solvent casting affects the blood response by changing the orientation and/or conformation of the diisocyanate and the components of the chain extender (148). An analysis of the biocompatibility and the cell response of aliphatic PUs containing varying amounts of plasticizer has shown the important role played by plasticizers. The biocompatibility, durability, and adhesive properties of aliphatic PUs increase in line with the decrease in the amount of plasticizer (101).

To get around the problem of thromboresistance, researchers have tried adding and/or replacing components. Sulfonated polyurethanes (PEO SO_3) produced by various techniques exhibit a lower level of platelet adhesion and platelet factor IV release from the time they first come in contact with blood until up to 12 to 24 h later (74,232). Hemocompatibility parallels bacterial repellence; the more hemocompatible a polymer, the more resistant it is to infections (74).

The hemocompatibility of sulfonated polyurethanes modified by a surface derivation technique using 2-acrylamido-2-methyl-propane sulfonic acid (AMPS) varies depending on the species, with in vivo and in vitro studies providing very contradictory results (72). While very hemocompatible in vitro, sulfonated polyurethanes attract large numbers of neutrophils when implanted in mice, increase macrophage recruitment to unexpected levels, and lead to the formation of thromboses when implanted in dogs (72).

The heparinization of polymers to prevent the deposition of fibrin has been studied in great detail. Applying a layer of heparin to PUPA [PU + poly(amido amine) + HMDI] provides excellent antithrombogenic properties, thus increasing hemocompatibility (76).

To reduce the risk of infections associated with medical devices, a number of metallic ions known for their bactericidal effect, especially silver, have been investigated. Numerous studies have examined the biocompatibility, infection rates, and thrombogenicity of silver-impregnated or silver-coated polyurethanes and silicones (81,82,84,85, 233). Coating polyurethane, silicon, and Dacron catheters with silver ions prevents microbial colonization both in vitro and in vivo (82,84,234). An animal implantation study (233) has shown that infection rates dropped when silver-impregnated polyurethane catheters were used compared to untreated catheters. In vitro tests have shown that silver ions have no cytotoxic or thrombogenic effect and do not provoke an exaggerated inflammatory or tissue response. Biocompatibility also depends on the material to which the silver is bound (81,84,85).

F. Biodegradable Polyurethanes

The chemical properties of elastomers have also been modified to produce biodegradable products. Studies on biodegradable polyesterurethanes have shown that these elastomers have good cell compatibility and that cell–substrate interactions do not lead to the release of toxic substances or the activation of macrophages. Relatively strong adhesion and acceptable growth of macrophages and osteoblasts occur. Fibroblasts keep their phenotype for 12 days. The main problem seems to the actual degradation of the elastomer, which results in the production of crystalline poly(r-3-hydroxybutyric acid), or PHB-P, particles. Macrophages, and to a lesser degree osteoblasts, may internalize the particles by phagocytosis. At high concentrations, these particles may be toxic for macrophages and osteoblasts at concentrations of 200 pg PHB P/cell and 400 pg PHB-P/cell, respectively (98,99).

G. Polyesters

Dacron, or polyethylene terephthalate, which has been used since 1940 as a bone and cartilage substitute, is considered to be biocompatible (1,95). It is used in numerous medical applications, especially in vascular surgery and for stent coatings (95). But a chronic, systemic inflammation was reported following implantation of a Dacron device. Lymphocytes appear after 30 days and the size of giant cells increases over time (15,235–241). Their tendency to provoke intimal hyperplasia and platelet stimulation, and thus their thrombogenicity, limits their clinical use to blood vessels greater than 6 mm in diameter (1). Polytetraphthalate-coated stents are preferred to polyurethane stents, which are much more subject to dislocation (95).

H. Polyvinylchloride

Polyvinyl chloride is increasingly being used for medical applications, with an annual increase of 5.4%. It has been used in the healthcare field since 1940 because of its broad biocompatibility, ease of manufacture, and low cost (1). A study in 1991 reported chronic inflammation associated with PVC (164). Two days after implantation, the inflammatory exudate around the PVC contains mainly neutrophils, while that around PU or silicone contains a mixture of neutrophils and monocytes (102).

Additives have often been implicated in harmful host responses. Additives include plasticizers, antioxidants, pigments, and UV stabilizers. Plasticizers in particular can release substances that are incompatible with biological applications (1). A study in 1994 (101) on the biocompatibility of and cell response to PVC has shown that there is a clear correlation between the amount of plasticizer and the host inflammatory response. Various processes have been used in an attempt to make PVCs more biocompatible. The most commonly used plasticizer is phthalic ester, but it can be replaced by trioctyl trimellitate, azelate, or phosphate ester,which are nontoxic and are not released (1).

The effect of plasticizers on hemocompatibility has been studied using PVC blood tubing manufactured with various plasticizers (phthalate alone, trimellitate or TD 360, and phthalate coextruded with polyurethane). The concentration of C3a anaphylatoxin and protein morphology were examined by electron microscopy. Phthlate-free PVC is more biocompatible, and polyurethane-coated PVC releases less C3a anaphylatoxin, although hemocompatibility does not change (103).

I. Polytetrafluoroethylene

Polytetrafluoroethylene is used for many medical applications, especially for peripheral vascular surgery (148). In vitro extraction and direct contact analyses reveal no cytotoxic contaminants (131). In vivo, the inflammatory response resorbs within 9 weeks, indicating acceptable biocompatibility (130,131). Fibroblasts respond to vascular PFTE prostheses by forming a thick external fibrotic capsule and a thin layer coating the pores (131,155). Despite an initial quick, spontaneous endothelialization process, complete endothelialization of the lumen of a PTFE prosthesis can take up to 6 months because of the rough surface, which, compared to polyurethanes, makes endothelial cell proliferation more difficult. Chronic proliferation is, however, significant. After 6 months, the cells are still dividing, increasing neointimal formation and thus the risk of hyperplasia and thrombosis (132,148). Surface texture also influences cell migration. Cell migration, which can prevent thrombosis and occlusion, is slow on PTFE (132).

Following a 500-min contact between human plasma and a PTFE prosthesis, variations in contact angle are minimal and there is little fibrinogen adsorption, while platelet adhesion is elevated (175). Blood contact with hydrophobic materials preferentially activates the kallicrein/kinin system, leading to vasodilatation, platelet aggregation, and an increase in vascular permeability (180).

J. Hydrogels

Hydrogels have been used in a wide variety of medical applications for over 30 years, for contact lenses, prosthetic devices, and drug delivery systems (1). General properties include swelling in water, high mobility of surface chains, low interfacial tension, low surface friction, and appreciable elasticity, making them reasonably biocompatible. Their hydrophilicity and low surface tension decreases protein adsorption and cell adhesion, which makes them suitable for use in contact lenses. The low surface friction reduces fibroblast stimulation and prevents the formation of a fibrotic capsule (21). The high elasticity and low mechanical resistance means that they have to be combined with other more solid elastomers, providing hydrogels with excellent mechanical and biological properties.

K. Polyethylenes

A 1983 study evaluated the cell response to polyethylene following implantation in the peritoneal cavities of mice (129). The rough surface of the elastomer becomes covered with numerous macrophages and foreign body giant cells. The tissue reaction to polypropylene oxide includes an acute inflammatory response involving numerous macrophages and lymphocytes. Polyethylenes are rapidly degraded within 5 months of implantation, and the resulting reactions point to the toxicity of the degradation products. The study by Bakker et al. (169) shows that polypropylene oxide cannot be used for alloplastic tympanic membranes. Hemocompatibility studies have shown strong albumin adsorption and weak platelet adhesion to high density polyethylene. In addition, analyses of the kallicrein/kinin system and the Hageman factor have shown that both systems are activated to the same extent, indicating acceptable hemocompatibility (175,180).

L. Natural Rubbers

While natural rubbers are used in surgical gloves, urinary catheters, and tubing, their low hemocompatibility means they cannot be used in many other applications. Methylmethacrylate grafts can improve hemocompatibility (111,242). Natural rubber also cause cytotoxic and allergic reactions of various etiologies (113). The proteins in natural rubbers cause Type I, IgE-induced allergic reactions. It is possible to remove these proteins by simple washing. Gamma radiation vulcanization can be used to produce nitrosamine-free, low toxicity elastomers (1). It has also been shown that the cytotoxicity of natural rubbers is due to dithiocarbamates (110). Various improvements mean that natural rubbers could soon see more frequent use in medical applications.

ACKNOWLEDGEMENTS

The authors would like to thank Gene Bourgeau, Céline Allaire, and Christiane Péju for editorial assistance.

REFERENCES

1. R. Yoda. Elastomers for biomedical applications. *J. Biomater. Sci. Polym. Ed.* 9:561–626 (1998).
2. H. M. Leeper, R. M. Wright. Elastomers in medicine. *Rubber Chem. Technol.* 56:523–556 (1983).
3. C. R. McMillin. Elastomers for biomedical application. *Rubber Chem. Technol.* 67:417 (1994).
4. L. Pinchuk. A review of the biostability and carcinogenicity of polyurethanes in medicine and the new generation of biostable polyurethanes. *J. Biomater. Sci. Polym. Ed.* 6:225–267 (1994).
5. H. K. Mardis, R. M. Kroeger, J. J. Morton, J. M. Donovan. Comparative evaluation of materials used for internal ureteral stents. *J. Endourology* 7:105–115 (1993).
6. F. Bischoff. Organic polymer biocompatibility and toxicology. *Clin. Chem.* 18:869–894 (1972).
7. M. R. Toub. Technical innovations enhance commercial value of silicone rubber. *Elastomerics* 119:20–22 (1987).
8. D. P. Jones. High quality silicones still dominate biomedical market after three decades. *Elastomerics* 120:12–16 (1988).
9. J. L. Boone, S. A. Braley. Resistance of silicone rubbers to body fluids. *Rubber Chem. Technol.* 39:1293–1297 (1966).
10. P. Vondracek. Some aspects of the medical application of silicone rubber. *Int. Polym. Sci. Technol.* 8:16 (1981).
11. B. Ashar, R. S. Ward, L. R. Turcotte. Development of a silica-free silicone system for medical applications. *J. Biomed. Mater. Res.* 15:663–672 (1981).
12. M. J. Whitford. The chemistry of silicone materials for biomedical devices and contact lenses. *Biomaterials* 5: 298–300 (1984).
13. N. Kossovsky, C. J. Freilan. Review of physicochemical and immunological basis of silicone pathophysiology. *J. Biomater. Sci. Polym. Ed.* 7:101–113 (1995).
14. N. Kossovsky. Can the silicone controversy be resolved with rational certainty. *J. Biomater. Sci. Polym. Ed.* 7:97–100 (1995).
15. L. Maturri, A. Azzolini, G. L. Campiglio, E. Tardito. Are synthetic prostheses really inert? Preliminary results of a study on the biocompatibility of Dacron vascular prostheses and silicone skin expanders. *Int. Surg.* 76:115–118 (1991).
16. A. F. Von Recum, T. G. Van Kooten. The influence of micro-topography on cellular response and the implica-

tions for silicone implants. *J. Biomater. Sci. Polym. Ed.* 7:181–198 (1995).
17. A. F. Von Recum, M. LaBerge. Educational goals for biomaterials science and engineering: prospective view. *J. Biomater. Appl.* 6:137–144 (1995).
18. P. Auroy, P. Duchatelard, N. E. Zmantar, M. Hennequin. Hardness and shock absorption of silicone rubber for mouth guards. *J. Prosthetic Dentistry* 75:463–471 (1996).
19. K. D. Colter, M. Shen, A. T. Bell. Reduction of progesterone release rate through silicone membranes by plasma polymerization. *Biomater. Med. Dev. Artif. Organs* 5:13–24 (1977).
20. J. M. Brown. Polyurethane and silicone: myths and misconceptions. *J. Intravenous Nursing* 18:120–122 (1995).
21. H. Park, K. Park. Biocompatibility issues of implantable drug delivery systems. *Pharmaceut. Res.* 13:1770–1776 (1996).
22. E. P. Everaert, H. C. Van Der Mei, J. De Vries, H. J. Busscher. Hydrophobic recovery of repeatedly plasma-treated silicone rubber. I. Storage in air. *J. Adhesion Sci. Technol.* 9:1263 (1995).
23. G. H. Hsiue, S. D. Lee, P. C. Chang. Surface modification of silicone rubber membrane by plasma induced graft copolymerization as artificial cornea. *Artif. Organs* 20: 1196–1207 (1996).
24. S. D. Lee, G. H. Hsiue, C. Y. Kao, P. C. Chang. Artificial cornea: surface modification of silicone rubber membrane by graft polymerization of pHEMA via glow discharge. *Biomaterials* 17:587–595 (1996b).
25. P. G. Kalman, C. A. Ward, N. B. McKeown, D. McCullough, A. D. Romaschin. Improved biocompatibility of silicone rubber by removal of surface entrapped air nuclei. *J. Biomed. Mater. Res.* 25:199–211 (1991).
26. C. C. Tsai, M. L. Dollar, A. Constantinescu, P. V. Kulkarni, R. C. Eberhart. Performance evaluation of hydroxylated and acylated silicone rubber coatings. Presented at ASAIO Transactions, 1991.
27. K. Frisch, J. Saunders. *Polyurethanes: Chemistry and Technology*. Interscience, New York, 1962.
28. S. Cooper, A. Toblosky. Properties of linear elastomeric polyurethanes. *J. Appl. Polym. Sci.* 10:1837–1844 (1966).
29. R. Bonart. X-ray investigations concerning the physical structure of crosslinking in segmented urethane elastomers. *J. Macromol. Sci. Phys.* B7:115–138 (1968).
30. W. J. Pangman. U.S. patent 2,842,775 (1958).
31. J. W. Boretos, W. S. Pierce. Segmented polyurethane: a new elastomer for biomedical applications. *Science* 158: 1481–1482 (1967).
32. G. L. Wilkes, S. L. Samuels. Porous segmented polyurethanes: possible candidates as biomaterials. *J. Biomed. Mater. Res.* 7:541–544 (1973).
33. J. Autian. *Biological Models for the Testing of the Toxicity of Biomaterials*. Plenum Press, New York, 1974, pp. 181–203.
34. K. Stokes, K. Cobian. Polyether polyurethanes for implantable pacemaker leads. *Biomaterials* 3:225–231 (1982).
35. A. Coury, K. Cobian, P. Cahalan, A. Jevne. In: *Biomedical Uses of Polyurethanes*. Technomic Publishing, Lancaster, PA, 1984, pp. 133–139.
36. M. D. Lelah, S. L. Cooper. *Polyurethanes in Medicine*. CRC Press, Boca Raton, FL, 1986.
37. M. Szycher, A. M. Reed. Biostable polyurethane elastomers. *Medical Device Technol.* 3:42–51 (1992).
38. F. J. Schoen, H. Harasaki, K. M. Kim, H. C. Anderson, R. J. Levy. Biomaterial-associated calcification: pathology, mechanisms, and strategies for prevention. *J. Biomed. Mater. Res.* 22:11–36 (1988).
39. A. Takahara, A. J. Coury, R. W. Hergenrother, S. L. Cooper. Effect of soft segment chemistry on the biostability of segmented polyurethanes. I. *In vitro* oxidation. *J. Biomed. Mater. Res.* 25:341–356 (1991).
40. K. B. Stokes. Polyether polyurethanes: biostable or not? *J. Biomater. Appl.* 3:228–259 (1988).
41. R. E. Phillips, M. C. Smith, R. J. Thoma. Biomedical applications of polyurethanes: implications of failure mechanisms. *J. Biomater. Appl.* 3:207–227 (1988).
42. J. M. Anderson, K. M. Miller. Biomaterial biocompatibility and the macrophage. *Biomaterials* 5:5–10 (1984).
43. J. M. Anderson, A. Hiltner, Q. H. Zhao, Y. Wu, M. Reiner, M. Schubert M. Brunstedt, G. A. Lodoen, C. R. Payet. In: *Cell/Polymer Interactions in the Biodegradation of Polyurethanes*. Royal Society of Chemistry, Cambridge, 1992, pp. 122–136.
44. O. H. Zhao, J. M. Anderson, A. Hiltner, G. A. Lodoen, C. R. Payet. Theoretical analysis on cell size distribution and kinetics of foreign body giant cell formation *in vivo* on polyurethane elastomers. *J. Biomed. Mater. Res.* 26: 1019–1038 (1992).
45. W. J. Kao, A. Hiltner, J. M. Anderson, G. A. Lodoen. Theoretical analysis of *in vivo* macrophage adhesion and foreign body giant cell formation on strained poly(etherurethane urea) elastomers. *J. Biomed. Mater. Res.* 28:819–829 (1994).
46. R. P. Rambour. A review of crazing and fracture in thermoplastics. *J. Polym. Sci. Macromol. Rev.* 7:1–154 (1973).
47. Z. Zhang, M. W. King, R. Guidoin, M. Therrien, M. Pezolet, A. Adnot, P. Ukpabi, M. H. Vantal. Morphological, physical and chemical evaluation of the Vascugraft arterial prosthesis: comparison of a novel polyurethane device with other microporous structures. *Biomaterials* 15:483–501 (1994).
48. Y. Marois, R. Guidoin, D. Boyer, F. Assayed, C. J. Doillon, R. Paynter, M. Marois. *In vivo* evaluation of hydrophobic and fibrillar microporous polyetherurethane urea graft. *Biomaterials* 10:521–531 (1989).
49. R. A. White. The effect of porosity on the variability of the neointima. An histological investigation on implanted synthetic vascular prostheses. Presented at American Society for Artificial Internal Organs, 1988.
50. K. B. Stokes. *The Biostability of Various Polyether Polyurethanes Under Stress*. Medtronic, Inc., Minneapolis, MN, 1983.

51. K. B. Stokes, P. Urbanski, R. Cobian. In: *New Test Methods for the Evaluation of Stress Cracking and Metal Catalysed Oxidation in Implanted Polymers*. Elsevier, Amsterdam, 1987, pp. 109–127.
52. A. J. Coury, C. N. Hobot. Method for producing polyurethanes form poly(hydroxyalkyl urethane) (1991). U.S. Patent No. 5 001 210.
53. R. W. Hergenrother, X. H. Yu, S. L. Cooper. Blood-contacting properties of polydimethylsiloxane polyurea-urethanes. *Biomaterials* 15:635–640 (1994).
54. F. Lim, C. Z. Yang, S. L. Cooper. Synthesis, characterization and *ex vivo* evaluation of polydimethylsiloxane polyurea-urethanes. *Biomaterials* 15:408–416 (1994).
55. C. D. Capone. Biostability of a non-ether polyurethane. *J. Biomater. Applications* 7:108–129 (1992).
56. K. Stokes, R. McVenes, J. M. Anderson. Polyurethane elastomer biostability. *J. Biomater. Appl.* 9:321–354 (1995).
57. A. J. Coury, C. M. Hobot, P. C. Slaikeu, K. B. Stokes, P. T. Cahalan. A new family of implantable biostable polyurethanes. Presented at the 16th annual meeting for the Society for Biomaterials, Charleston, 1990.
58. L. Pinchuk, Y. P. Kato, M. L. Eckstein, G. J. Wilson, D. C. MacGregor. Polycarbonate urethanes as elastomeric materials for long-term implant applications. Presented at the 16th annual meeting for the Society for Biomaterials, Charleston, 1993.
59. M. Szycher, A. Reed, J. Potter. A solution grade biostable polyurethane elastomer: ChronoFlex AR. *J. Biomater. Appl.* 8:210–236 (1994).
60. R. S. Ward, K. A. White, R. S. Gill, F. Lim. The effect of phase separation and endgroup chemistry on the *in vivo* biostability of polyurethanes. Presented at ASAIO meeting, Washington, D.C., 1996.
61. E. Nyilas, R. S. Ward, Jr. Development of blood-compatible elastomers. V. Surface structure and blood compatibility of avcothane elastomers. *J. Biomed. Mater. Res.* 11: 69–84 (1977).
62. R. Iwamoto, K. Ohta, T. Matsuda, K. Imachi. Quantitative surface analysis of Cardiothane 51 by FT-IR-ATR spectroscopy. *J. Biomed. Mater. Res.* 20:507–520 (1986).
63. A. J. Coury, P. C. Slaikeu, P. T. Cahalan, K. B. Stokes, C. M. Hobot. Factors and interactions affecting the performance of polyurethane elastomers in medical devices. *J. Biomater. Appl.* 3:130–179 (1988).
64. R. S. Ward. Surface modification prior to surface formation: control of polymer surface properties via bulk composition. *Med. Plastics Biomater.* 34–41 Spring (1995).
65. R. S. Ward, K. A. White, R. S. Gill, C. A. Wolcott. Development of biostable thermoplastic polyurethanes with oligomeric polydimethylsiloxane end groups. Presented at the 21st meeting of the Society for Biomaterials, San Francisco, 1995.
66. A. B. Mathur, T. O. Collier, W. J. Kao, M. Wiggins, M. A. Schubert, A. Hiltner, J. M. Anderson. *In vivo* biocompatibility and biostability of modified polyurethanes. *J. Biomed. Mater. Res.* 36:246–257 (1997).
67. M. D. Lelah, J. A. Pierce, L. K. Lambrecht, S. L. Cooper. Polyetherurethane ionomers: surface property/ex vivoblood compatibility relationships. *J. Colloid Interface Sci.* 104:422–439 (1985).
68. T. G. Grasel, J. A. Pierce, S. L. Cooper. Effects of alkyl grafting on surface properties and blood compatibility of polyurethane block copolymers. *J. Biomed. Mater. Res.* 21:815–842 (1987).
69. T. G. Grasel, S. L. Cooper. Properties and biological interactions of polyurethane anionomers: effect of sulfonate incorporation. *J. Biomed. Mater. Res.* 23:311–338 (1989).
70. A. Z. Okkema, T. A. Giroux, T. G. Grasel, S. L. Cooper. Ionic polyurethanes: surface and blood contacting properties. Presented at MRS symposium on biomedical materials and devices, Boston, 1987.
71. F. Lim, X. H. Yu, S. L. Cooper. Effects of oligoethylene oxide monoalkyl(aryl) alcohol ether grafting on the surface properties and blood compatibility of a polyurethane. *Biomaterials* 14:537–545 (1993).
72. J. R. Keogh, M. F. Wolf, M. E. Overend, L. Tang, J. W. Eaton. Biocompatibility of sulphonated polyurethane surfaces. *Biomaterials* 17:1987–1994 (1996).
73. J. N. Baumgartner, C. Z. Yang, S. L. Cooper. Physical property analysis and bacterial adhesion on a series of phosphonated polyurethanes. *Biomaterials* 18:831–837 (1997).
74. D. K. Han, K. D. Park, Y. H. Kim. Sulfonated poly(ethylene oxide)–grafted polyurethane copolymer for biomedical applications. *J. Biomater. Sci. Polym. Ed.* 9:163–174 (1998).
75. J. R. Keogh, J. W. Eaton. Albumin binding surfaces for biomaterials. *J. Lab. Clin. Med.* 124:537–545 (1994).
76. A. Albanese, R. Barbucci, J. Belleville, S. Bowry, R. Eloy, H. D. Lemke, L. Sabatini. *In vitro* biocompatibility evaluation of a heparinizable material (PUPA), based on polyurethane and poly(amido-amine) components. *Biomaterials* 15:129–136 (1994).
77. W. W. Bakker, B. van der Lei, P. Nieuwenhuis, P. Robinson, H. L. Bartels. Reduced thrombogenicity of artificial materials by coating with ADPase. *Biomaterials* 12: 603–606 (1991).
78. F. Hess, R. Jerusalem, O. Reijnders, C. Jerusalem, S. Steeghs, B. Braun, P. Grande. Seeding of enzymatically derived and subcultivated canine endothelial cells on fibrous polyurethane vascular prostheses. *Biomaterials* 13: 657–663 (1992).
79. P. C. Lee, L. L. Huang, L. W. Chen, K. H. Hsieh, C. L. Tsai. Effect of forms of collagen linked to polyurethane on endothelial cell growth. *J. Biomed. Mater. Res.* 32: 645–653 (1996).
80. S. K. Williams, T. Carter, P. K. Park, D. G. Rose, T. Schneider, B. E. Jarrell. Formation of a multilayer cellular lining on a polyurethane vascular graft following endothelial cell sodding. *J. Biomed. Mater. Res.* 26:103–117 (1992).
81. A. Oloffs, C. Grosse-Siestrup, S. Bisson, M. Rinck, R. Rudolph, U. Gross. Biocompatibility of silver-coated poly-

urethane catheters and silver-coated Dacron material. *Biomaterials* 15:753–758 (1994).
82. B. Jansen, M. Rinck, P. Wolbring, A. Strohmeier, T. Jahns. *In vitro* evaluation of the antimicrobial efficacy and biocompatibility of a silver-coated central venous catheter. *J. Biomater. Appl.* 9:55–70 (1994).
83. D. Li, J. Zhao. Surface biomedical effects of plasma on polyetherurethane. *J. Adhesion Sci. Technol.* 9:1249–1261 (1995).
84. M. Boswald, M. Girisch, J. Greil, T. Spies, K. Stehr, T. Krall, J. P. Guggenbichler. Antimicrobial activity and biocompatibility of polyurethane and silicone catheters containing low concentrations of silver: a new perspective in prevention of polymer-associated foreign-body infections. *Zentralblatt fur Bakteriologie* 283:187–200 (1995).
85. J. Greil, T. Spies, M. Boswald, T. Bechert, S. Lugauer, A. Regenfus, J. P. Guggenbichler. Analysis of the acute cytotoxicity of the Erlanger silver catheter. *Infection* 27: S34–37 (1999).
86. A. Pizzoferrato, C. R. Arciola, E. Cenni, G. Ciapetti, S. Sassi. in vitro biocompatibility of a polyurethane catheter after deposition of fluorinated film. *Biomaterials* 16:361–367 (1995).
87. B. D. Ratner. Surface characterization of biomaterials by electron spectroscopy for chemical analysis. *Ann. Biomed. Eng.* 11:313–336 (1983).
88. T. G. Grasel, D. C. Lee, A. Z. Okkema, T. J. Slowinski, S. L. Cooper. Extraction of polyurethane block copolymers: effects on bulk and surface properties and biocompatibility, *Biomaterials* 9:383–392 (1988).
89. M. A. Schubert, M. J. Wiggins, K. M. DeFife, A. Hiltner, J. M. Anderson. Vitamin E as an antioxidant for poly (etherurethane urea): in vivo studies. *J. Biomed. Mater. Res.* 32:493–504 (1996).
90. M. Szycher. Biostability of polyurethane elastomers: a critical review. *J. Biomater. Applications* 3:297–402 (1988).
91. T. Collier, J. Tan, M. Shive, S. Hasan, A. Hiltner, J. Anderson. Biocompatibility of poly(etherurethane urea) containing dehydroepiandrosterone. *J. Biomed. Mater. Res.* 41:192–201 (1998).
92. K. Kawashima, H. Sato. Calcium effect on the membrane preparation of segmented poly(ether/urethane/amide) (PEUN) as a biomedical material. *J. Biomater. Sci. Polym. Ed.* 8:467–480 (1997).
93. D. R. Holmes, A. R. Camrud, M. A. Jorgenson, W. D. Edwards, R. S. Schwartz. Polymeric stenting in the porcine coronary artery model: differential outcome of exogenous fibrin sleeves versus polyurethane-coated stents. *J. Am. Coll. Cardiol.* 24:525–531 (1994).
94. I. K. De Scheerder, K. L. Wilczek, E. V. Verbeken, J. Vandorpe, P. N. Lan, E. Schacht, H. De Geest, J. Piessens. Biocompatibility of polymer-coated oversized metallic stents implanted in normal porcine coronary arteries. *Atherosclerosis* 114:105–114 (1995b).
95. F. Schellhammer, M. Walter, A. Berlis, H. G. Bloss, E. Wellens, M. Schumacher. Polyethylene terephthalate and polyurethane coatings for endovascular stents: preliminary results in canine experimental arteriovenous fistulas. *Radiology* 211:169–175 (1999).
96. E. Rechavia, F. Litvack, M. C. Fishbien, M. Nakamura, N. Eigler. Biocompatibility of polyurethane-coated stents: tissue and vascular aspects. *Catheterization Cardiovasc. Diagn.* 45:202–207 (1998).
97. T. D. Hirt, P. Neuenschwander, U. W. Suter. Telechelic diols from poly(r)-3-hydroxybutyric acid and poly(r)-3-hydroxybutyric acid-co-(r)-3-hydroxyvaleric acid. *Macromol. Chem. Phys.* 197:1609–1614 (1996).
98. B. Saad, S. Matter, G. Ciardelli, G. K. Uhlschmid, M. Welti, P. Neuenschwander, U. W. Suter. Interactions of osteoblasts and macrophages with biodegradable and highly porous polyesterurethane foam and its degradation products. *J. Biomed. Mater. Res.* 32:355–366 (1996).
99. B. Saad, T. D. Hirt, M. Welti, G. K. Uhlschmid, P. Neuenschwander, U. W. Suter. Development of degradable polyesterurethanes for medical applications: in vitro and in vivo evaluations. *J. Biomed. Mater. Res.* 36:65–74 (1997).
100. L. White. Clean TPEs find medical uses. *Eur. Rubber J.* 173:26–29 (1991).
101. E. Lindner, V. V. Cosofret, S. Ufer, R. P. Buck, W. J. Kao, M. R. Neuman, J. M. Anderson. Ion-selective membranes with low plasticizer content: electroanalytical characterization and biocompatibility studies. *J. Biomed. Mater. Res.* 28:591–601 (1994).
102. T. Fabre, J. Bertrand-Barat, G. Freyburger, J. Rivel, B. Dupuy, A. Durandeau, C. Baquey. Quantification of the inflammatory response in exudates to three polymers implanted *in vivo*. *J. Biomed. Mater. Res.* 39:637–641 (1998).
103. B. Branger, M. Garreau, G. Baudin, J. C. Gris. Biocompatibility of blood tubings. *Int. J. Artif. Organs* 13:697–703 (1990).
104. J. Singh, K. K. Agrawal. Modification of poly(vinyl chloride) for biocompatibility improvement and biomedical application. *Polym.-Plastics Technol. Eng.* 31:203–212 (1992).
105. A. A. Katbab, R. P. Burford, J. L. Garnett. Radiation graft modification of EPDM rubber. *Int. J. Radiat. Appl. Instr. Part C. Radiat. Phys. Chem.* 39:293–302 (1992).
106. H. Mirzadeh, A. A. Katbab, R. P. Burford. CO-pulsed laser induced surface grafting of acrylamide onto ethylene-propylene rubber. *Int. J. Radiat. Appl. Instr. Part C. Radiat. Phys. Chem.* 42:53–56 (1993).
107. C. Freij-Larsson, M. Kober, B. Wesslen, E. Willquist, P. Tengvall. Effects of a polymeric additive in a biomedical poly(ether urethaneurea). *J. Appl. Polym. Sci.* 49:815–821 (1993).
108. R. Deisler. New silicone modified TPEs for medical applications. *Rubber World* 24 (1987).
109. L. Bacakova, V. Svorcik, V. Rybka, I. Micek, V. Hnatowicz, V. Lisa, F. Kocourek. Adhesion and proliferation of cultured human aortic smooth muscle cells on polystyrene implanted with N^+, F^+ and Ar^+ ions: correlation with poly-

mer surface polarity and carbonization. *Biomaterials* 17: 1121–1126 (1996).
110. A. Nakamura, Y. Ikarashi, T. Tsuchiya, M. A. Kaniwa, M. Sato, K. Toyoda, M. Takahashi, N. Ohsawa, T. Uchima. Correlations among chemical constituents, cytotoxicities and tissue responses: in the case of natural rubber latex materials. *Biomaterials* 11:92–94 (1990).
111. M. T. Razzak, K. Otsuhata, Y. Tabata, F. Ohashi, A. Takeuchi. Modification of natural rubber tubes for biomaterials. I. Radiation induced grafting of N,N-dimetyl acrylamide onto natural rubber tubes. *J. Appl. Polym. Sci.* 36: 645 (1988).
112. M. G. Tucci, M. Mattioli Belmonte, E. Toschi, G. A. Pelliccioni, L. Checchi, C. Castaldini, G. Biagini, G. Piana. Structural features of latex gloves in dental practice. *Biomaterials* 17:517–522 (1996).
113. L. Cormio, K. Turjanmaa, M. Talja, L. C. Andersson, M. Ruutu. Toxicity and immediate allergenicity of latex gloves. *Clin. Exper. Allergy* 23:618–623 (1993).
114. Annual report of the medical devices applications program of the National Heart and Lung Institute. Bethesda, MD, 1971.
115. B. D. Ratner, T. Horbett, A. S. Hoffman, S. D. Hauschka. Cell adhesion to polymeric materials: implications with respect to biocompatibility. *J. Biomed. Mater. Res.* 9:407–422 (1975).
116. T. Vijayasekaran, T. V. Chirila, Y. Hong, G. Tahija, I. J. Constable, I. L. McAllister. Poly(1-vinyl-2-pyrrolidinone) hydrogels as vitreous substitutes: histopathological evaluation in the animal eye. *J. Biomater. Sci. Polym. Ed.* 7: 685 (1996).
117. A. Chapiro. Radiation grafting of hydrogels to improve the thromboresistance of polymers. *Eur. Polym. J.* 19:859 (1983).
118. A. Warren, F. Gould, R. Capulong, B. Glotfelty, S. Boley, W. Calem, S. Levowitz. Prosthetic applications of a new hydrophilic plastic. *Surg. Forum* 18:183–185 (1967).
119. L. Sprincl, J. Kopecek, D. Lim. Effect of porosity of heterogeneous poly(glycol monomethacrylate) gels on the healing-in of test implants. *J. Biomed. Mater. Res.* 5:447–458 (1971).
120. F. D'Hermies, J. F. Korobelnik, M. Savoldelli, D. Chauvaud, Y. Pouliquen. Miragel versus silastic used as episcleral implants in rabbits. An experimental and histopathologic comparative study. *Retina* 15:62–67 (1995).
121. J. C. Carroll, S. D. Schwaitzberg, A. A. Ucci, Jr., R. M. Schlesinger, D. Lauritzen, G. R. Sant. New matrix material for potential use in ''reversible'' vasectomy. Preliminary animal biocompatibility studies. *Urology* 41:34–37 (1993).
122. I. K. De Scheerder, K. L. Wilczek, E. V. Verbeken, J. Vandorpe, P. N. Lan, E. Schacht, J. Piessens, H. De Geest. Biocompatibility of biodegradable and nonbiodegradable polymer-coated stents implanted in porcine peripheral arteries. *Cardiovas. Interventional Radiol.* 18:227–232 (1995).
123. J. Crommen, J. Vandorpe, E. Schacht. Degradable polyphazenes for biomedical applications. *J. Controlled Release* 24:167–180 (1993).
124. P. Ferruti, E. Ranucci. New functional polymers for medical applications. *Polym. J.* 23:541–550 (1991).
125. W. L. Jongebloed, G. van der Veen, D. Kalicharan, M. V. van Andel, G. Bartman, J. G. Worst. New material for low-cost intraocular lenses. *Biomaterials* 15:766–773 (1994).
126. E. J. Hollick, D. J. Spalton, P. G. Ursell, M. V. Pande. Biocompatibility of poly(methyl methacrylate), silicone, and AcrySof intraocular lenses: randomized comparison of the cellular reaction on the anterior lens surface. *J. Cataract Refractive Surg.* 24:361–366 (1998).
127. D. Chauvel-Lebret, P. Pellen-Mussi, P. Auroy, M. Bonnaure-Mallet. Evaluation of the in vitro biocompatibility of various elastomers. *Biomaterials* 20:291–299 (1999).
128. T. J. Henry. *Guidelines for the Preclinical Safety Evaluation of Materials Used in Medical Devices*. Health Industry Manufactures Association, Washington, D.C., 1985.
129. R. S. Wortman, K. Merritt, S. A. Brown. The use of the mouse peritoneal cavity for screening for biocompatibility of polymers. *Biomater. Med. Dev. Artif. Organs* 11:103–114 (1983).
130. B. Huang, Y. Marois, R. Roy, M. Julien, R. Guidoin. Cellular reaction to the Vascugraft polyesterurethane vascular prosthesis: in vivo studies in rats. *Biomaterials* 13:209–216 (1992).
131. M. F. Sigot-Luizard, M. Sigot, R. Guidoin, M. King, W. W. von Maltzahn, R. Kowligi, R. C. Eberhart. A novel microporous polyurethane blood conduit: biocompatibility assessment of the UTA arterial prosthesis by an organo-typic culture technique. *J. Invest. Surg.* 6:251–271 (1993).
132. M. G. Jeschke, V. Hermanutz, S. E. Wolf, G. B. Koveker. Polyurethane vascular prostheses decreases neointimal formation compared with expanded polytetrafluoroethylene. *J. Vasc. Surg.* 29:168–176 (1999).
133. J. Defrere, A. Franckart. Teflon/polyurethane arthroplasty of the knee: the first 2 years preliminary clinical experience in a new concept of artificial resurfacing of full thickness cartilage lesions of the knee. *Acta Chirurgica Belgica* Vol. 5, 92:217–227 (1992).
134. M. Szycher, A. A. Siciliano, A. M. Reed. In: *Polyurethane Elastomers in Medicine*. Marcel Dekker, New York, 1990, pp. 234–244.
135. S. Dumitriu, D. Dumitriu. In: *Biocompatibility of Polymers*, Marcel Dekker, New York, 1990, pp. 100–158.
136. M. J. Menconi, T. Owen, K. A. Dasse, G. Stein, J. B. Lian. Molecular approaches to the characterization of cell and blood/biomaterial interactions. *J. Cardiac Surg.* 7:177–187 (1992).
137. H. Oshima, M. Nakamura. A study on reference standard for cytotoxicity assay of biomaterials. *Biomed. Mater. Eng.* 4:327–332 (1994).
138. K. M. Miller, J. M. Anderson. in vitro stimulation of fibroblast activity by factors generated from human mono-

cytes activated by biomedical polymers. *J. Biomed. Mater. Res.* 23:911–930 (1989).
139. J. A. Hunt, B. F. Flanagan, P. J. McLaughlin, I. Strickland, D. F. Williams. Effect of biomaterial surface charge on the inflammatory response: evaluation of cellular infiltration and TNF alpha production. *J. Biomed. Mater. Res.* 31:139–144 (1996).
140. C. Morrison, C. Macnair, C. Macdonald, A. Wykam, I. Goldie, M. H. Grant. *In vitro* biocompatibility testing of polymers for orthopaedic implants using cultured fibroblasts and osteoblasts. *Biomaterials* 16:987–992 (1995).
141. C. J. Kirkpatrick, C. Mittermayer. Theoretical and practical aspects of testing potential biomaterials *in vitro*. *J. Mater. Sci. Mater. Med.* 1:9–13 (1990).
142. T. Bouet, K. Toyoda, Y. Ikarashi, T. Uchima, A. Nakamura, T. Tsuchiya, M. Takahashi, R. Eloy. Evaluation of biocompatibility based on quantitative determination of the vascular response induced by material implantation. *J. Biomed. Mater. Res.* 25:1507–1521 (1991).
143. AFNOR. Norm. Biological evaluation of medical devices. Part 7: Ethylene oxide sterilization residuals, AFNOR, Paris, 1996.
144. AFNOR, Norm. Biological evaluation of medical devices. Part 16: Toxicokinetic study design for degradation product and leachables, AFNOR, Paris, 1997.
145. B. C. Patel, J. M. Courtney, J. H. Evans, J. P. Paul. Biocompatibility assessment: application of fluorescent probe response (FPR) technique. *Biomaterials* 12:722–726 (1991).
146. J. H. Chen, J. Wei, C. Y. Chang, R. F. Laiw, Y. D. Lee. Studies on segmented polyetherurethane for biomedical application: effects of composition and hard-segment content on biocompatibility. *J. Biomed. Mater. Res.* 41:633–648 (1998).
147. J. L. Pariente, L. Bordenave, R. Bareille, F. Rouais, C. Courtes, G. Daude, M. Le Guillou, C. Baquey. First use of cultured human urothelial cells for biocompatibility assessment: application to urinary catheters. *J. Biomed. Mater. Res.* 40:31–39 (1998).
148. R. Bschorer, G. B. Koveker, G. Gehrke, M. Jeschke, V. Hermanutz. Experimental improvement of microvascular allografts with the new material polyurethane and microvessel endothelial cell seeding. *Int. J. Oral Maxillofacial Surg.* 23:389–392 (1994).
149. A. Oliva, A. Salerno, B. Locardi, V. Riccio, F. Della Ragione, P. Iardino, V. Zappia. Behaviour of human osteoblasts cultured on bioactive glass coatings. *Biomaterials* 19:1019–1025 (1998).
150. H. Beele, H. Thierens, R. Deveux, E. Goethals, L. Ridder. Skin organ culture model to test the toxicity of polyoxyethylene networks. *Biomaterials* 13:1031–1037 (1992).
151. S. A. Rosenbluth, G. R. Weddington, W. L. Guess, J. Autian. Tissue culture method for screening toxicity of plastic materials to be used in medical practice. *J. Pharmaceut. Sci.* 54:156–159 (1965).
152. H. J. Johnson, S. J. Northup, P. A. Seagraves, P. J. Garvin, R. F. Wallin. Biocompatibility test procedures for materials evaluation *in vitro*. I. Comparative test system sensitivity. *J. Biomed. Mater. Res.* 17:571–586 (1983).
153. J. B. Ulreich, M. Chvapil. A quantitative microassay for in vitro toxicity testing of biomaterials. *J. Biomed. Mater. Res.* 15:913–922 (1981).
154. D. Sgouras, R. Duncan. Methods for the evaluation of biocompatibility of soluble synthetic polymers which have potential for biomedical use. I. Use of the tetrazolium-based colorimetric assay (MTT) as a preliminary screen for evaluation of in vitro cytotoxicity. *J. Mater. Sci. Mater. Medicine* 1:61–68 (1990).
155. R. Guidoin, M. Sigot, M. King, M. F. Sigot-Luizard. Biocompatibility of the Vascugraft: evaluation of a novel polyester urethane vascular substitute by an organotypic culture technique. *Biomaterials* 13:281–288 (1992).
156. Y. Tabata, Y. Ikata. Phagocytosis of polymer by macrophages. *Adv. Polym. Sci.* 94:108–141 (1990).
157. M. D. Smith, J. C. Barbenel, J. M. Courtney, M. H. Grant. Novel quantitative methods for the determination of biomaterial cytotoxicity. *Int. J. Artif. Organs* 15:191–194 (1992).
158. A. Nakamura, Y. Kawasaki, K. Takada, Y. Aida, Y. Kurokama, S. Kojima, H. Shintani, M. Matsui, T. Nohmi, A. Matsuoka. Difference in tumor incidence and other tissue responses to polyetherurethanes and polydimethylsiloxane in long-term subcutaneous implantation into rats. *J. Biomed. Mater. Res.* 26:631–650 (1992).
159. K. D. Chesmel, J. Black. Cellular responses to chemical and morphologic aspects of biomaterial surfaces. I. A novel in vitro model system. *J. Biomed. Mater. Res.* 29: 1089–1099 (1995).
160. D. F. Williams. *Techniques of Biocompatibility Testing*. CRC Press, Boca Raton, FL, 1986.
161. R. E. Marchant, J. M. Anderson, K. Phua, A. Hiltner. in vivo biocompatibility studies. II. Biomer: preliminary cell adhesion and surface characterization studies. *J. Biomed. Mater. Res.* 18:309–315 (1984).
162. R. E. Marchant, K. M. Miller, J. M. Anderson. in vivo biocompatibility studies. V. *In vivo* leukocyte interactions with biomer. *J. Biomed. Mater. Res.* 18:1169–1190 (1984).
163. D. G. Vince, J. A. Hunt, D. F. Williams. Quantitative assessment of the tissue response to implanted biomaterials. *Biomaterials* 12:731–736 (1991).
164. M. Jayabalan, N. S. Kumar, K. Rathinam, T. V. Kumari. *In vivo* biocompatibility of an aliphatic crosslinked polyurethane in rabbit. *J. Biomed. Mater. Res.* 25:1431–1432 (1991).
165. J. O. Naim, R. J. Lanzafame, C. J. V. Oss. The effect of silicone-gel on the immune response. *J. Biomater. Sci. Polym. Ed.* 7:123–132 (1995).
166. N. Brautbar, A. Campbell, A. Vojdani. Silicone breast implants and autoimmunity: causation, association, or myth? *J. Biomater. Sci. Polym. Ed.* 7:133–145 (1995).
167. L. Tang, J. W. Eaton. Inflammatory responses to biomaterials. *Am. J. Clin. Pathol.* 103:466–471 (1995).
168. D. Bakker, C. A. van Blitterswijk, S. C. Hesseling, H. K.

Koerten, W. Kuijpers, J. J. Grote. Biocompatibility of a polyether urethane, polypropylene oxide, and a polyether polyester copolymer. A qualitative and quantitative study of three alloplastic tympanic membrane materials in the rat middle ear. *J. Biomed. Mater. Res.* 24:489–515 (1990).

169. D. Bakker, C. A. van Blitterswijk, S. C. Hesseling, W. T. Daems, W. Kuijpers, J. J. Grote. The behavior of alloplastic tympanic membranes in *Staphylococcus aureus*–induced middle ear infection. I. Quantitative biocompatibility evaluation. *J. Biomed. Mater. Res.* 24:669–688 (1990).
170. M. Löbler, M. Sass, P. Michel, U. T. Hopt, C. Kunze, K. P. Schmitz. Differential gene expression after implantation of biomaterials into rat gastrointestine. *J. Mater. Sci. Mater. Medicine* 10:797–799 (1999).
171. R. Marchant, A. Hiltner, C. Hamlin, A. Rabinovitch, R. Slobodkin, J. M. Anderson. *In vivo* biocompatibility studies. I. The cage implant system and a biodegradable hydrogel. *J. Biomed. Mater. Res.* 17:301–325 (1983).
172. A. K. McNally, J. M. Anderson. Complement C3 participation in monocyte adhesion to different surfaces. *Proc. Nat. Acad. Sci. U.S.A.* 91:10119–10123 (1994).
173. S. D. Bruck. Problems and artefacts in the evaluation of polymeric materials for medical uses. *Biomaterials* 1: 103–107 (1980).
174. AFNOR. Norm. Biological evaluation of medical devices. Part 4: Selection of tests for interactions with blood, AFNOR, Paris, 1994.
175. G. Rakhorst, H. C. Van der Mei, W. Van Oeveren, H. T. Spijker, H. J. Busscher. Time-related contact angle measurements with human plasma on biomaterial surfaces. *Int. J. Artif. Organs* 22:35–39 (1999).
176. M. D. Lelah, C. A. Jordon, M. E. Pariso, L. K. Lambrecht, S. L. Cooper, R. M. Albrecht. Morphological changes occurring during thrombogenesis and embolization on biomaterials in a canine ex vivo series shunt. *Scanning Electron Microscopy* 4:1983–1984 (1983).
177. M. D. Lelah, L. K. Lambrecht, S. L. Cooper. A canine ex vivo series shunt for evaluating thrombus deposition on polymer surfaces. *J. Biomed. Mater. Res.* 18:475–496 (1984).
178. I. A. Feuerstein, B. D. Ratner. Adhesion and aggregation of thrombin prestimulated human platelets: evaluation of a series of biomaterials characterized by ESCA. *Biomaterials* 11:127–132 (1990).
179. J. L. Wagner, T. E. Hugli. Radioimmunoassay for anaphylatoxins: a sensitive method for determining complement activation products in biological fluids. *Anal. Biochem.* 136:75–88 (1984).
180. K. W. Van der Kamp, W. Van Oeveren. Factor XII fragment and kallikrein generation in plasma during incubation with biomaterials. *J. Biomed. Mater. Res.* 28:349–352 (1994).
181. E. Piskin. Review of biodegradable polymers as biomaterials. *J. Biomater. Sci. Polym. Ed.* 6:775–795 (1994).
182. M. Singh, A. R. Ray, P. Vasudevan, K. Verma, S. K. Guha. Potential biosoluble carriers: biocompatibility and biodegradability of oxidized cellulose. *Biomater. Med. Dev. Artif. Organs* 7:495–512 (1979).
183. R. S. Labow, J. P. Santerre, G. Waghray. The effect of phospholipids on the biodegradation of polyurethanes by lysosomal enzymes. *J. Biomater. Sci. Polym. Ed.* 8:779–795 (1997).
184. G. Grimaldi, R. L. Zobell, J. R. Scott, R. L. Urry. Pregnancies following implantation of an artificial fallopian tube in rabbits. *Artif. Organs* 16:213–216 (1992).
185. S. K. Hunter, J. R. Scott, D. Hull, R. L. Urry. The gamete and embryo compatibility of various synthetic polymers. *Fertility and Sterility* 50:110–116 (1988).
186. K. U. Schendel, L. Erdinger, G. Komposch, H. G. Sonntag, Neon-colored plastics for orthodontic appliances. Biocompatibility studies. *Fortschritte der Kieferorthopadie* Vol. 1, 56:41–48 (1995).
187. V. G. Nadeenko, I. R. Goldina, Z. D'Iachenko O, L. V. Pestova. Comparative informative value of chromosome aberrations and sister chromatid exchanges in the evaluation of metals in the environment. *Gigiena i. Sanitariia* Vol. 3, 10–13 (1997).
188. D. Purves, C. Harvey, D. Tweats, C. E. Lumley. Genotoxicity testing: current practices and strategies used by the pharmaceutical industry. *Mutagenesis* 10:297–312 (1995).
189. X. C. Le, J. Z. Xing, J. Lee, S. A. Leadon, M. Weinfeld. Inducible repair of thymine glycol detected by an ultrasensitive assay for DNA damage. *Science* 280:1066–1069 (1998).
190. E. J. Hollick, D. J. Spalton, P. G. Ursell. Surface cytologic features on intraocular lenses: can increased biocompatibility have disadvantages? *Arch. Ophthalmol.* 117:872–878 (1999).
191. M. Szycher, A. A. Siciliano. Polyurethane-covered mammary prosthesis: a nine year follow-up assessment. *J. Biomater. Appl.* 5:282–322 (1991).
192. J. P. Rubin, M. J. Yaremchuk. Complications and toxicities of implantable biomaterials used in facial reconstructive and aesthetic surgery: a comprehensive review of the literature. *Plastic Reconstr. Surg.* 100:1336–1353 (1997).
193. B. Gottenbos, H. C. Van Der Mei, H. J. Busscher, D. W. Grijpma, J. Feijen. Initial adhesion and surface growth of *Pseudomonas aeruginosa* on negatively and positively charged poly(methacrylates). *J. Mater. Sci. Mater. Med.* 10:853–855 (1999).
194. D. G. Maki. Epidemiology and prevention of intravascular device–related infections. Presented at Infection Control Symposium: Influence of Medical Device Design, Bethesda, MD, 1992.
195. C. R. Aricola, L. Radin, P. Alvergna, E. Cenni, A. Pizzoferrato. Heparin surface treatment of poly(methylmethacrylate) alters adhesion of *Staphylococcus aureus* strain: utility of bacterial fatty acid analysis. *Biomaterials* 14: 1161–1164 (1993).
196. J. M. Anderson. Mechanisms of inflammation and infec-

tion with implanted devices. *Cardiovasc. Pathol.* 2:33–41 (1993).
197. J. C. Homberg. *Immunologie Fondamentale*, Estem, Paris, 1999.
198. B. Genetet. *Immunologie*. Lavoisier, Paris, 1989.
199. D. Vignon. *Physiologie de l'Hémostase*. 22-009-D-20. Editions Techniques, Paris, 1995, pp. 1–8.
200. J. Diebold, J. P. Camilleri, M. Reynes, P. Callard. *Anatomie Pathologique Générale*. Baillière, Paris, 1977.
201. H. O. Trowbridge, R. C. Emling. *Inflammation: A Review of the Process*, 5th ed. Quintessence Books, Carol Stream, IL, 1997.
202. A. Baszkin, M. M. Boissonnade. Characterization and albumin absorption on surface oxidized polyethylene films in polymers in medicine. *Biomed. Pharmacolog. Appl.* 271–285 (1983).
203. M. D. Lelah, T. G. Grasel, J. A. Pierce, S. L. Cooper. *Ex vivo* interactions and surface property relationships of polyetherurethanes. *J. Biomed. Mater. Res.* 20:433–468 (1986).
204. D. G. Maki, C. E. Weise, H. W. Sarafin. A semiquantitative culture method for identifying intravenous catheter–related infection. *N. Engl. J. Med.* 296:1305–1309 (1977).
205. G. L. Cooper, C. C. Hopkins. Rapid diagnosis of intravascular catheter–associated infection by direct Gram staining of catheter segments. *N. Engl. J. Med.* 312:1142–1147 (1985).
206. D. G. Maki, S. J. Wheeler, S. M. Stolz. Study of a novel antiseptic-coated central venous catheter. *Crit. Care Med.* 19.99 (1991).
207. K. G. Brand, L. C. Buoen, K. H. Johnson, I. Brand. Etiological factors, stages, and the role of the foreign body in foreign body tumorigenesis. *Cancer Res.* 35:279–286 (1975).
208. T. Tsuchiya, H. Hata, A. Nakamura. Studies on the tumor-promoting activity of biomaterials: inhibition of metabolic cooperation by polyetherurethane and silicone. *J. Biomed. Mater. Res.* 29:113–119 (1995).
209. T. Tsuchiya, K. Fukuhara, H. Hata, Y. Ikarashi, N. Miyata, F. Katoh, H. Yamasaki, A. Nakamura. Studies on the tumor-promoting activy of additives in biomaterials: inhibition of metabolic cooperation by phenolic antioxidants involved in rubber materials. *J. Biomed. Mater. Res.* 29: 121–126 (1995).
210. L. A. Brinton, S. L. Brown. Breast implants and cancer. *J. Nat. Cancer Inst.* 89:1341–1349 (1997).
211. H. Shintani. Formation and elution of toxic compounds from sterilized medical products: methylenedianiline formation in polyurethane. *J. Biomater. Applications* 10:23–58 (1995).
212. T. Tsuchiya, R. Nakaoka, H. Degawa, A. Nakamura. Studies on the mechanisms of tumorigenesis induced by polyetherurethanes in rats: leachable and biodegradable oligomers involving the diphenyl carbamate structure acted as an initiator on the transformation of balb 3T3 cells. *J. Biomed. Mater. Res.* 31:299–303 (1996).
213. R. Nakaoka, T. Tsuchiya, K. Kato, Y. Ikada, A. Nakamura. Studies on tumor-promoting activity of polyethylene: inhibitory activity of metabolic cooperation on polyethylene surfaces is markedly decreased by surface modification with collagen but not with RGDS peptide. *J. Biomed. Mater. Res.* 35:391–397 (1997).
214. L. Needelman. The microscopic interaction between silicone and the surrounding tissues. *Clinics in Podiatric Medicine and Surgery* 12:415–423 (1995).
215. M. S. Lucas, D. J. Moore. Tissue culture and histologic study of a new elastomer. *J. Prosthetic Dentistry* 42:447–451 (1979).
216. A. Hensten-Pettersen, A. Hulterström. Assessment of *in vitro* cytotoxicity of four RTV-silicone elastomers used for maxillo-facial prostheses. *Acta Odontologica Scandinavica* 38:163–167 (1980).
217. G. L. Polyzois, A. Hensten-Pettersen, A. Kullman. Effects of RTC-silicone maxillofacial prosthetic elastomers on cell cultures. *J. Prosthetic Dentistry* 71:505–510 (1994).
218. G. L. Polyzois, A. Hensten-Pettersen, A. Kullmann. An assessment of the physical properties and biocompatibility of three silicone elastomers. *J. Prosthetic Dentistry* 71: 500–504 (1994).
219. D. Bakker, C. A. Van Blitterswijk, W. T. Daems, J. J. Grote. Biocompatibility of six elastomers *in vitro*. *J. Biomed. Mater. Res.* 22:423–439 (1988).
220. D. Kalicharan, W. L. Jongebloed, G. van der Veen, I. I Los, J. G. Worst. Cell-ingrowth in a silicone plombe. Interactions between biomaterial and scleral tissue after 8 years in situ: a SEM and TEM investigation. *Documenta Ophthalmologic.* 78:307–315 (1991).
221. J. F. Wolfaardt, P. Cleaton-Jones, J. Lownie, G. Ackermann. Biocompatibility testing of a silicone maxillofacial prosthetic elastomer: soft tissue study in primates. *J. Prosthetic Dentistry* 68:331–338 (1992).
222. C. C. Tsai, H. H. Huo, P. Kulkarni, R. C. Eberhart. Biocompatible coating with high albumin affinity. Presented at ASAIO, 1990.
223. S. W. Fountain, J. Duffin, C. A. Ward, H. Osada, B. A. Martin, J. D. Cooper. Biocompatibility of standard and silica-free silicone rubber membrane oxygenators. *Am. J. Physiol.* 236:H371–H375 (1979).
224. T. L. Bonfield, J. M. Anderson. Functional versus quantitative comparison of IL-1 beta from monocytes/macrophages on biomedical polymers. *J. Biomed. Mater. Res.* 27:1195–1199 (1993).
225. S. E. Gabriel, W. M. O'Fallon, L. T. Kurland, C. M. Beard, J. E. Woods, L. J. Melton. Risk of connective-tissue diseases and other disorders after breast implantation. *N. Engl. J. Med.* 330:1697–1702 (1994).
226. J. Sanchez-Guerrero, G. A. Colditz, E. W. Karlson, D. J. Hunter, F. E. Speizer, M. H. Liang. Silicone breast implants and the risk of connective-tissue diseases and symptoms. *N. Engl. J. Med.* 332:1666–1670 (1995).
227. C. H. Hennekens, I. M. Lee, N. R. Cook, P. R. Hebert, E. W. Karlson, F. Lamotte. Self-reported breast implants and connective-tissue diseases in female health professionals. A retrospective cohort study. *JAMA* 275:616–621 (1996).

228. A. J. Coury, K. B. Stokes, P. T. Cahalan, P. C. Slaikeu. Biostability considerations for implantable polyurethanes. *Life Support Systems* 5:25–39 (1987).
229. E. Pollock, E. J. Andrews, D. Lentz, K. Sheikh. Tissue ingrowth and porosity of biomer. Presented at American Society for Artificial Internal Organs, 1981.
230. C. L. Ives, J. L. Zamora, S. G. Eskin, D. G. Weilbaecher, Z. R. Gao, G. P. Noon, M. E. DeBakey. *In vivo* investigation of a new elastomeric vascular graft (Mitrathane). Presented at American Society for Artificial Internal Organs, 1984.
231. H. Martz, R. Paynter, J. C. Forest, A. Downs, R. Guidoin. Microporous hydrophilic polyurethane vascular grafts as substitutes in the abdominal aorta of dogs. *Biomaterials* 8:3–11 (1987).
232. T. J. McCoy, H. D. Wabers, S. L. Cooper. Series shunt evaluation of polyurethane vascular graft materials in chronically AV-shunted canines. *J. Biomed. Mater. Res.* 24:107–129 (1990).
233. M. Boswald, K. Mende, W. Bernschneider, S. Bonakdar, H. Ruder, H. Kissler, E. Sieber, J. P. Guggenbichler. Biocompatibility testing of a new silver-impregnated catheter *in vivo. Infection* 27:S38–S42 (1999).
234. G. Zhao, S. E. Stevens, Jr. Multiple parameters for the comprehensive evaluation of the susceptibility of *Escherichia coli* to the silver ion. *Biomaterials* 11:27–32 (1998).
235. D. S. Feldman, S. M. Hultman, R. S. Colaizzo, A. F. von Recum. Electron microscope investigation of soft tissue ingrowth into Dacron velour with dogs. *Biomaterials* 4: 105–111 (1983).
236. L. Granstroem, L. Backam, S. E. Dahlgren. Tissue reaction to polypropylene and polyester in obese patients. *Biomaterials* 7:455–458 (1986).
237. O. Sato, Y. Tada, A. Takagi. The biologic fate of Dacron double velour vascular prostheses: a clinicopathological study. *Jpn. J. Surg.* 19:301–311 (1986).
238. G. Cavallini, M. Lanfredi, M. Lodi, M. Govoni, M. Pampolini. Detection and measurement of a cellular immune-reactivity towards polyester and polytetrafluoroethylene grafts. Leukocyte adherence inhibition test. *Acta Chirurgica Scandinavica* 153:179–184 (1987).
239. U. Blum, M. Langer, G. Spillner, C. Mialhe, F. Beyersdorf, C. Buitrago-Tellez, G. Voshage, C. Duber, V. Schlosser, A. H. Cragg. Abdominal aortic aneurysms: preliminary technical and clinical results with transfemoral placement of endovascular self-expanding stent-grafts. *Radiology* 198:25–31 (1996).
240. J. Link, B. Feyerabend, M. Grabener, U. Linstedt, J. Brossmann, H. Thomsen, M. Heller. Dacron-covered stent-grafts for the percutaneous treatment of carotid aneurysms: effectiveness and biocompatibility-experimental study in swine. *Radiology* 200:397–401 (1996).
241. M. Lodi, G. Cavallini, A. Susa, M. Lanfredi. Biomaterials and immune system: cellular reactivity towards PTFE and Dacron vascular substitutes pointed out by the leukocyte adhrence inhibiyion (LAI) test. *Int. Angiology* 7:344–348 (1988).
242. M. T. Razzak, K. Otsuhata, Y. Tabata, F. Ohashi, A. Takeuchi. Modification of natural rubber tubes for biomaterials. II. Radiation induced grafting of N,N-dimethylaminoethylacrylate (DMAEA) onto natural rubber (NR) tubes. *J. Appl. Polym. Sci.* 38:829 (1989).

14

Control of Cell–Biomaterial Interactions

Danielle C. Giliberti, Kyle K. White, and Kay C Dee
Tulane University, New Orleans, Louisiana

I. INTRODUCTION

An ideal implanted prosthesis or device would be quickly integrated and stabilized within the surrounding tissue. If tissue integration occurs too slowly, or is compromised over time, common clinical problems (such as loosening of the implant, pain, and loss of tissue near the implant site) may lead to subsequent, difficult reimplantation. Many recent strategies for avoiding these adverse outcomes are derived from the fundamental premise that *macroscopic tissue-level events are ultimately derived from, and thus could be controlled by, cellular- and molecular-level events at the tissue–implant interface.* An ideal implant could be designed to encourage quick healing and maintenance of tissue by controlling cellular- and molecular-level events (such as cell adhesion, differentiation, matrix deposition, etc.) at the tissue–biomaterial interface. Understanding how cells interact with biomaterials is a necessary prerequisite for the development of novel methods to control cell–biomaterial and, eventually, tissue–biomaterial interactions.

Controlling cell–biomaterial interactions is an important and relevant goal for the development of polymeric biomaterials because techniques are available for the fabrication of polymeric biomaterials with a wide range of bulk and surface properties. However, influencing even the most fundamental cellular functions (such as adhesion and migration) requires an understanding of a variety of topics, including cell biology, surface chemistry, and physical properties. The purpose of this chapter is, therefore, to provide an overview of key concepts and current research on controlling cell–biomaterial interactions, with the hope of stimulating further interest in this rapidly growing field of research.

II. MEDIATING CELLULAR FUNCTIONS: BIOLOGICAL CONSIDERATIONS

Cellular anchorage (adhesion to a substrate) and migration are essential to the normal progression of many biological processes, including proper functioning of the immune system, embryogenesis, hemostasis, and maintenance of tissue integrity (1). The ability of cells to interact with each other and their surroundings in a coordinated manner depends on multiple cell–cell and cell–matrix adhesive interactions (2). For anchorage-dependent cells, adhesion to a substrate or matrix is a necessary prerequisite for viability and proliferation (3). Due to its fundamental biological importance, cell–substrate adhesion has been the focus of many studies, including much of the research on controlling cell–biomaterial interactions. Therefore, this chapter will focus primarily on cell adhesion, and will discuss biological, surface chemistry, and surface architectural/mechanical considerations involved in mediating cellular adhesion and related functions.

A. Cell Adhesion

Adhesion molecules are a versatile class of cell membrane receptors which function as links between cells and the extracellular environment. Cell adhesion molecules can transmit both mechanical and biochemical signals bidirectionally across the cell membrane (2), from adjacent cells and/or from the extracellular matrix, thereby regulating cellular functions. Signaling via cell adhesion receptors mediates a variety of processes such as gene expression, differentiation, apoptosis (2), and cellular migration (4). The interdependence between cell adhesion molecules, growth factor receptors, and the extracellular matrix (demonstrated through initiation of biochemical signaling cascades and formation of intracellular multiprotein complexes) helps explain the need for multiple cell membrane molecules to be activated for normal cellular function (2). The following discussion focuses on four major superfamilies of cell adhesion molecules that function in cellular adhesion and intercellular signaling and are the focus of much current research: cadherins, selectins, immunoglobulins, and integrins.

1. Cadherins

Cadherins are a family of transmembrane glycoproteins whose purpose is to organize solid tissues and ensure tissue integrity (5,6). Cadherins are specialized for the maintenance of stable adhesive interactions through long-term, almost irreversible binding (7) and constitute the major component of adherens junctions (2,4,8). Mutations in cadherins can cause tissue disorder and cellular dedifferentiation and may even lead to malignancy (9). Cadherins generally exhibit homotypic adhesion (2,7,8); however, E-cadherin has been shown to bind integrin ligands (7,8). The integrin-binding site on the cadherin molecule is spatially separate from the sites that mediate homotypic binding, suggesting that heterotypic and homotypic binding can occur simultaneously (7). Cadherin adhesion is calcium dependent (2,7). Calcium ions serve to rigidify cadherin structure, induce conformational changes needed for adhesion, and confer resistance to proteolysis (7). Cadherins have a rather large extracellular domain, a transmembrane segment, and a short cytoplasmic tail (7). The cytoplasmic domain indirectly interacts with the actin cytoskeleton via linker molecules called catenins (2,7) that bind to α-actinin and vinculin (4), which in turn interact with actin filaments. The cytoplasmic domains of cadherins are required for desmosome assembly and cytoskeletal filament anchorage (4).

2. Selectins

Unlike the stable adhesion demonstrated by cadherins, selectins provide transient adhesive interactions. For example, selectins are important for leukocyte–endothelium weak adhesion (8,10) and leukocyte signaling (11), and are therefore a focus for the development of anti-inflammatory agents (7). Selectins also play a role in platelet adhesion (7,12). Selectin-mediated adhesion can occur under flow (7) and is regulated by divalent calcium ions (2). Selectins have five domains: three extracellular, one transmembrane, and one intracellular, or cytoplasmic (8). Selectins bind to carbohydrate-containing ligands (7) as well as to integrin ligands (8).

3. Immunoglobulins

Many members of the immunoglobulin superfamily are involved in cell adhesion, and play roles in platelet–endothelial cell adhesion, neural cell–matrix adhesion (13,14), vascular cell–matrix adhesion, and general intracellular adhesion in a variety of cell types (8). Immunoglobulins exhibit homotypic binding as well as heterotypic binding to integrin and extracellular matrix ligands (2). Dysfunction in immunoglobulin-mediated adhesion has been implicated in a variety of pathological disorders (8,15,16).

4. Integrins

Integrins are a significant family of cell adhesion receptors as well as a major focus of current biological and biomaterials research (and, therefore, this discussion). Integrin-mediated cell adhesion and signaling have been cited as crucial to a wide variety of physiologic and pathologic processes, including but not limited to embryonic development, fertilization, maintenance of tissue integrity, wound healing and angiogenesis, leukocyte recirculation and recruitment, antigen presentation, cytotoxicity, and phagocytosis (2). Cell–matrix interactions that take place through integrin ligation directly participate in the control of cellular differentiation, morphogenesis, proliferation, migration, and apoptosis, influencing such processes as wound healing, inflammation, and metastasis (3,17,18). Integrins have therefore been targeted for the development of novel therapies for inflammatory, neoplastic, and infectious diseases (8). In other words, in order to fully understand and potentially control select clinically relevant, macroscopic-level events (metastasis, wound healing, tissue integrity) it is important to understand the mechanisms of integrin-mediated cellular- and molecular-level events (receptor activation, intracellular signaling, cell adhesion, etc.).

Integrins are perhaps most widely recognized as the primary cell adhesion receptors for many proteins of the extracellular matrix, such as fibronectin, laminin, vitronectin, and collagen (19). The integrin family contains sixteen α and eight β subunits which can combine to form at least 22 heterodimeric receptors. Both the α and β subunits are type-1 transmembrane glycoproteins (2,4,8), containing a large extracellular domain, a transmembrane segment, and a short cytoplasmic tail (Fig. 1) (8). Cell surface expression of the mature αβ heterodimer can be regulated by modulat-

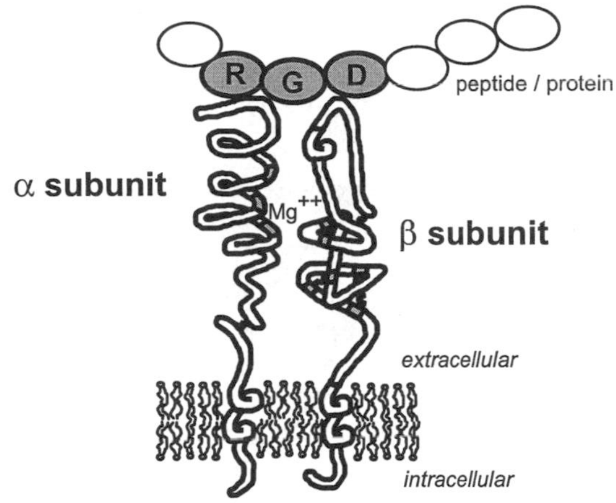

Figure 14.1 Schematic of an integrin receptor. Each receptor has an α and a β subunit, both of which span the cell membrane. The ligand binding site of the receptor is a ''pocket'' between the subunits; divalent ions also bind between the subunits. The ligand shown is a sequence of three amino acids, arginine–glycine–aspartic acid (RGD). The three-dimensional structure of the integrin subunits shown here is not definitive. (Adapted from Ref. 120.)

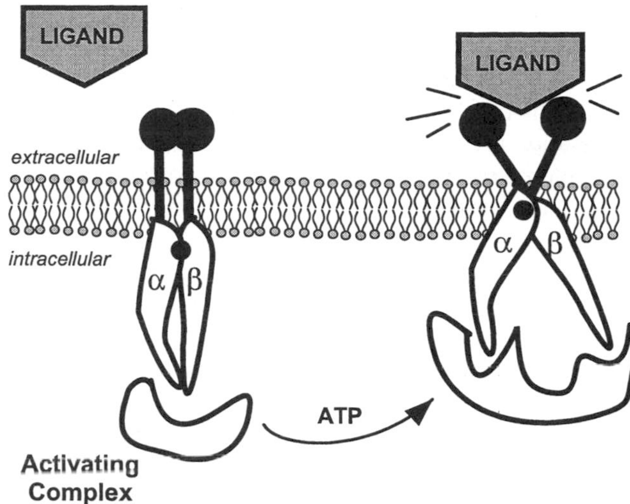

Figure 14.2 Inside-out signaling. An intracellular complex interacts with the cytoplasmic domain(s) of the receptor, changing the receptor conformation to one which allows ligand binding. While a scissorlike conformational change is depicted here for the sake of illustration, the actual spatial rotation/translation may be different. (Adapted from Ref. 25.)

ing the level of α chain expression (through, for example, chemical means such as inflammatory cytokines) (2). Cells can express more than one type of integrin receptor (4), and distinct integrin pairs can act to differentially regulate signals across the cell membrane (7,20). Due to interactions with the cytoskeleton, activated integrins can function as mechanotransducers between a cell and the surrounding microenvironment (4,7,21,22). Integrin activation leads to tyrosine and serine phosphorylation, increases in intracellular calcium, and changes in phosphoinositide turnover and gene expression (23,24).

a. *Activation and Conformation*

Integrins move between an inactive state, in which they cannot bind ligands, and an active state, in which ligation occurs. Altering integrin activation states through receptor conformation changes or spatial clustering, rather than by changing the number of receptors on a cell surface (2), can control adhesive properties of integrins and, therefore, the adhesive properties of a cell (7,25,26). Integrin *avidity* depends on the spatial arrangement of the receptors on the cell membrane (1,23); the *affinity* of integrin receptors for ligands is dependent upon receptor conformation (1,23). Conformational changes to the ligand-binding domain [which is a ''pocket'' between the α and β subunits (7)] of an integrin receptor modulates the affinity of the receptor (1,25) and occurs through the interaction of the integrin cytoplasmic domain with intracellular factors (21,26). This mechanism of affinity modulation is often termed ''inside-out'' signaling (Fig. 2) (2,3,8) and is thought to be primarily brought about by phosphorylation/dephosphorylation events, which subsequently allow association of other regulatory proteins (3). Calreticulin, for example, is a calcium-binding intracellular protein that is believed to function by binding to integrins, forcing the maintenance of a high affinity state (1,26).

There is a three-way interdependence between integrin conformation, divalent cation occupancy, and ligand binding (27). Integrin structure (and therefore function) is cation dependent, with cation specificity differing between integrins (2). Cation occupancy may regulate an opening of the integrin dimer to expose the ligand-binding domain between the α and β subunits (27). Mn^{2+} stimulates ligand binding; Mg^{2+} also stimulates ligand binding, but not as strongly; and Ca^{2+} inhibits ligand binding in most integrins (21,23,25–28). Mn^{2+} and Ca^{2+} noncompetitively inhibit ligation (two distinct binding sites exist for these cations), whereas Mg^{2+} and Ca^{2+} can competitively inhibit ligation by binding at the same site (26).

b. *Ligation*

Integrin receptors recognize ligands that are components of extracellular environments they would often encounter (e.g., connective tissue matrix, basement membranes, mineralized matrix) (22). Ligands may be cell surface members of the immunoglobulin superfamily, microorganisms, coagulation factors, or portions of extracellular matrix molecules (7,8,25). A classic example of an integrin ligand is the peptide arginine–glycine–aspartic acid, or RGD, which was the first ligand sequence identified for integrin recognition (7,17,21,22,28). The RGD peptide is present in a number of extracellular matrix components including fibronectin, fibrinogen, vitronectin, and von Willebrand's Factor (23).

Each integrin heterodimer type has a specific binding profile (2,4,7,8). Several regions from both the α and the β subunits constitute the ligand-binding domain; therefore a particular ligand may bind at two or more distinct sites within the dimer (26). Integrin ligation elicits a variety of intracellular events such as the stimulation of protein kinase C, the Na^+/H^+ antiporter, phosphoinositide hydrolysis, tyrosine phosphorylation of cellular proteins, changes in intracellular pH, and mitogen-activated protein kinase activation (18–21,29). The ligation of integrins also leads to an increase in intracellular calcium concentration (28). There is evidence that calcium signaling may be a "fine-tuning" mechanism in cell adhesion and migratory activity (28).

Cell adhesion to an extracellular matrix via integrin–ligand interactions is accompanied by integrin aggregation. Integrins are able to move laterally along the surface of the cell membrane to form localized clusters (1,23), often due to ligation of the β subunit (23). Signaling initiated through integrin clustering and ligation is often termed "outside-in" signaling (1) since the clustering can be viewed as a response to signals—the presence of ligands—from the extracellular matrix. "Inside-out" intracellular signaling, which remodels cytoskeletal linkages or alters receptor diffusion rates, can also regulate integrin clustering (1). Clustering appears to trigger increased tyrosine phosphorylation of a number of intracellular proteins (30) and may regulate many cellular processes (3), including gene expression, proliferation, and apoptosis (2,8). Integrins can cocluster with other membrane receptors and with intracellular proteins to form "focal adhesions" (18,24,30,31), which are essentially links between the cytoskeleton and the extracellular environment (Fig. 3) (26).

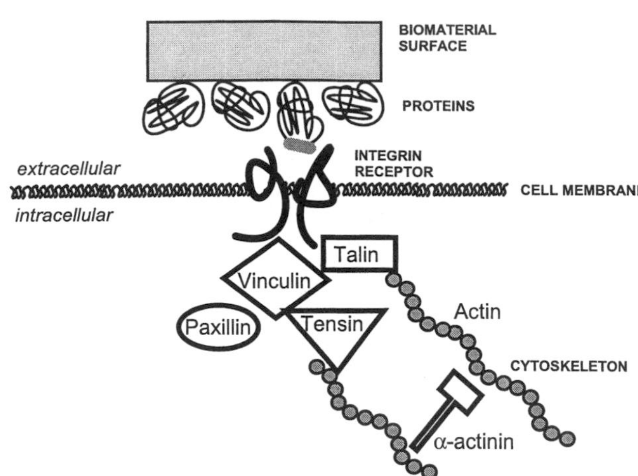

Figure 14.3 Focal adhesions and the cytoskeleton. When integrins bind to an extracellular ligand, such as a specific region of a protein adsorbed to the surface of a biomaterial, a number of intracellular proteins (only a few of which are shown) localize at the integrin cytoplasmic domains. This intracellular complex, or focal adhesion, can anchor the actin cytoskeleton to the location of integrin–ligand interaction. Focal adhesions therefore provide a direct link between the extracellular environment and the cytoskeleton.

c. *Focal Adhesions*

Integrins colocalize with the ends of stress fibers (actin filament bundles that span the interior of the cell) at focal adhesions (19,21,23,24). The formation of focal adhesions causes the formation of intracellular signaling complexes and the reorganization of the cytoskeleton to facilitate signal transfer (19).

At least 20 different intracellular proteins can be recruited to the ligand–integrin binding site to make up a focal adhesion (30). The intracellular proteins F-actin, filamin, paxillin, tensin, and vinculin have been shown to associate with integrin cytoplasmic domains (26,32), and α-actinin and talin have been shown to directly bind to these domains (23). Some focal adhesion proteins, such as paxillin, are phosphorylated in an integrin-dependent manner (24). Integrins lack intrinsic enzymatic activity; therefore, some integrin-mediated signaling is dependent upon nonreceptor kinases (17).

Focal adhesion kinase (FAK) is a nonreceptor tyrosine kinase that, as the name suggests, localizes in focal adhesions (17,18,29,33) and is tyrosine phosphorylated upon integrin binding (20,24,28,30,33). Focal adhesion kinase phosphorylation and activity appears to be triggered by the clustering of integrins (17,18,29,33), and FAK appears to play a large role in ligand binding (1,23,26,33). Talin, α-actinin, and FAK can interact directly with the β subunit to form a focal adhesion complex with the cytoskeleton (1,19,23). Focal adhesion kinase activity also leads to the activation of mitogen-activated protein kinase (MAPK) and Ras, a family of small GTPases (17,23,29,33). This (FAK, MAPK, Ras) signaling cascade is initiated as cytoskeletal elements assemble to anchor actin filaments during the formation of focal adhesions (Fig. 4) (29). In this way, integrin ligation can lead to direct activation of MAPK and Ras, both of which have been associated with

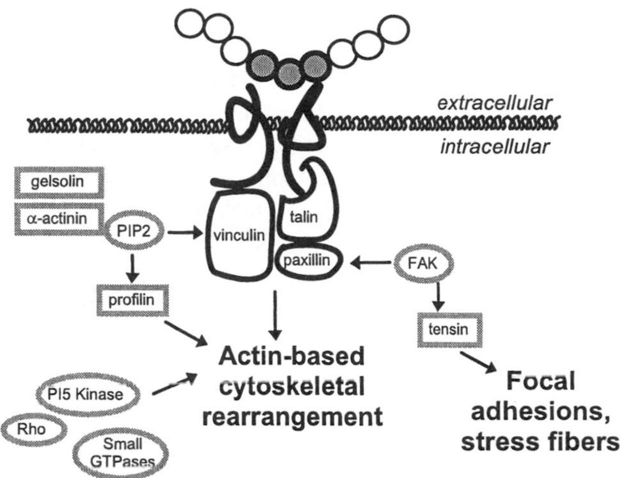

Figure 14.4 Integrins and cytoskeletal rearrangement. When integrins bind to an extracellular ligand, a number of signaling cascades are locally initiated, some of which participate in rearrangement and anchoring of the cytoskeleton.

the generation of mitogenic signals (Fig. 5). In other words, specific intracellular signaling pathways (such as FAK, MAPK, Ras) essentially constitute a line of communication between integrins and the nucleus (18,20,23,29). Activation of the MAPK signal transduction pathway is a common route leading to transcriptional regulation of genes (30). Anoikis (apoptosis that occurs when cells are detached from their matrix, and therefore integrins are detached from their ligands) is regulated through FAK and MAPK (3,25).

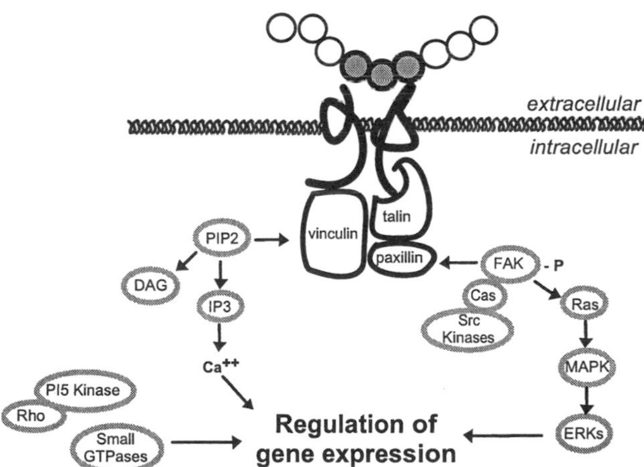

Figure 14.5 Integrins and gene expression. Some of the intracellular signaling cascades that are initiated upon integrin ligation are known to mediate gene expression and cell proliferation. Cytokines and growth factors mediate cell proliferation via similar pathways, providing plausible locations for signaling convergence and cooperative amplification.

The activation of MAPK leads to suppression of high affinity ligand binding by a number of integrins (30). The Ras family of small GTPases also modulates integrin affinity (1). Rho is a member of the Ras family that regulates actin stress fiber organization and assembly of the focal adhesion (17). The actin structures controlled by each Rho GTPases are associated with integrin clustering and other focal adhesion proteins (30). Activated Rho induces focal adhesion formation. However, integrin clustering can alter Rho localization (18,25). Rho GTPases are currently thought to be key parts of focal adhesion–dependent signaling pathways that control cell proliferation and gene transcription (30).

Formation of focal adhesions involves tyrosine phosphorylation events, clustering of integrins, integrin–ligand binding, and actin cytoskeletal integrity (20). Clustering of integrins without ligand binding localizes FAK and tensin to the integrin aggregation site. If tyrosine phosphorylation occurs at this point, signaling molecules such as Ras, MAPK, and PI3-kinase will also be found at the aggregation site. Clustering of integrins with ligand binding colocalizes talin, α-actinin, and vinculin in addition to FAK and tensin. If tyrosine phosphorylation occurs at this point, paxillin and actin filaments are also recruited to the site, in addition to a number of signaling molecules (18,24). Integrins can bind ligands in the absence of FAK. In this case, vinculin, α actinin, and talin may colocalize to form a link to the cytoskeleton (23). Therefore, clustering of integrins is sufficient for some signal transduction, but ligand binding is required for cytoskeletal assembly. These results suggest that integrin–ligand binding changes the conformation of the integrin cytoplasmic domain (20,24,25,29, 33), allowing the initiation of focal adhesions. The intracellular events associated with integrin–ligand binding and focal adhesion formation may explain why integrins seem to synergistically enhance the effects of growth factors on cells (2).

d. *Growth Factors*

Integrins may enhance the actions of growth factors by increasing receptor clustering (20,25). The actin cytoskeleton and focal adhesion proteins may act as a scaffold that coordinates interactions between integrin– and growth factor receptor–tyrosine kinases (30). Focal adhesion kinase can be activated by growth factors or by integrins, so FAK is a plausible location for integrin and growth factor signaling pathways to converge (30,33). Integrins cooperate with growth factors in the activation of the MAPK pathway by increased clustering and phosphorylation of FAK (20,21).

Extracellular growth factors activate different Rho family small GTPases that are essential for the formation of F-actin cytoskeletal structures (18). Signal transduction mediated by Rho and Rac, in response to growth factors and extracellular matrix proteins, is required for actin cytoskeletal reorganization and assembly of focal adhesions (19). Convergence of the adhesion and growth factor pathways is important for cell proliferation and cell cycle progression (3,19).

B. Cell Migration

Cell migration is dependent on rapid controlled changes in integrin-dependent cellular adhesion (1) and plays key roles in both normal physiology and disease. The ability of cells to migrate may be defined by four major factors: the amount of extracellular ligand present, the amount of integrins present, and the binding affinity and avidity of the integrins (34–36). The ligand concentration needed for a cell to migrate at its maximum speed has an inversely proportional relationship with the amount of integrin expression on the cell surface and the integrin–ligand binding affinity (36). Cellular migration may also be affected by the spatial representation or avidity of integrins on a cell surface. When integrins are clustered, the ligand density necessary for cell migration is greatly decreased (35). The attachment of integrins to the actin cytoskeleton couples integrins to the actomyosin contractile system, which is necessary for cell motility (37).

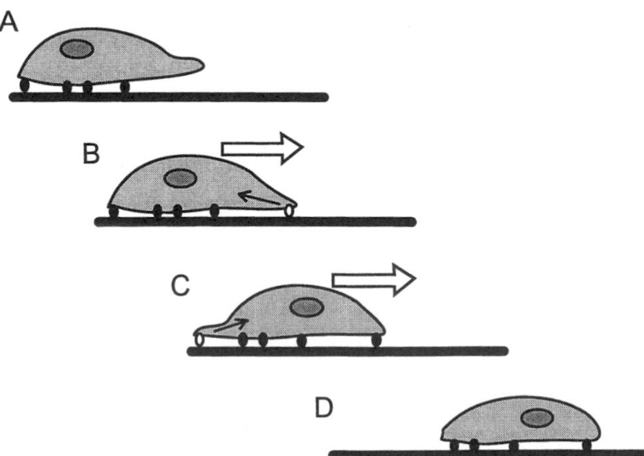

Figure 14.6 Cell migration depends on adhesion. (A) The cell forms a protrusion to the front. (B) The protrusion adheres to the substrate; the cell pulls against that adhesion site to move forward. (C) As the cell continues to move forward, it begins to pull against an adhesion in the rear. (D) The adhesion site in the rear of the cell has been pulled away from the substrate; the cell is in a new position and the process begins again.

In cellular locomotion, one end of the cell spreads while the other retracts; therefore this process may be considered to be a kind of polarized cell spreading (25). The basic mechanism behind cell locomotion is polymerization of the actin cytoskeleton (38). One main type of actin-based cell motility is known is amoeboid movement. Once a cell has adhered to a substrate, a lamellar protrusion is formed (Fig. 6) (38). Protrusion is tightly coupled with actin polymerization (38). This protrusion adheres to the substrate. The cell body and nucleus then move forward into this protrusion in a process known as traction (38), which involves the cooperation of actin and myosin cytoskeletal filaments. The cytoplasm left behind the cell body after movement is known as the tail. After traction occurs and the main cell body has moved forward, this tail de-adheres and retracts and the locomotion process begins again (38).

III. MODIFICATION OF BIOMATERIAL SURFACES: CHEMICAL CONSIDERATIONS

Following the rationale that cellular- and molecular-level processes (such as cell adhesion, proliferation, and intracellular signaling) ultimately direct the healing and maintenance of healthy tissue around an implant, a great deal of current biomaterials research is devoted to understanding and controlling the cellular-level processes described in Section II of this chapter. Many investigators are utilizing knowledge of receptor–ligand interactions, particularly the present understanding of integrin–ligand interactions, and modifying surface parameters of biomaterials in attempts to understand and control the behavior of cells. If the ''correct'' ligands are present on a biomaterial, either as distinct biomolecules chemically immobilized on the surface or as conformationally available domains of adsorbed proteins, then cell membrane receptors should interact with the ligands. If the ''appropriate'' types and numbers of receptor–ligand complexes are formed, signals will be sent to the cell nucleus via intracellular signaling cascades, and consequently the functions of the cells on the biomaterial could be regulated. Currently, determining the correct ligands and the appropriate receptor–ligand complexes for particular biomaterial applications is a challenge. This section reviewing recent work on the design of polymer surfaces in order to control cellular behavior is divided into three categories of chemical surface modification strategies: physicochemical methods, biochemical methods, and morphological methods (39).

A. Physicochemical Modifications

Cell adhesion, spreading, and migration on a substrate may be influenced by modifying the physicochemical surface properties of the material or by changing the surface charge, energy, and/or chemical composition of the material (39). One focus of much materials research is the hydrophilicity/hydrophobicity of surfaces, often characterized as wettability. A variety of cell lines including corneal epithelial cells, endothelial cells, fibroblasts, and bone-derived cells have demonstrated increased adhesion and spreading on hydrophilic surfaces (34,40–43). Webb et al. demonstrated that when materials were exposed to dilute serum, fibroblast attachment, spreading, and cytoskeletal organization were increased on hydrophilic (oxidized thiol–, quaternary amine–, and amine-modified glass) versus hydrophobic (thiol- and methyl-modified glass) surfaces (40). Differences in charge and wettability influenced only cell attachment on hydrophilic surfaces (40). In the absence of serum, positive charge had a large influence on the number of adherent cells, as demonstrated by the large number of cells adhered to the quaternary amine–modified glass and the small number of cells attached to the oxidized thiol–modified glass (40). Griesser et al. studied the effects of both surface hydrophilicity and nitrogen content (in the presence of serum) on human umbilical vein endothelial cell and human dermal fibroblast adhesion (43). Amine and amide compounds were polymerized onto fluorinated ethylene propylene copolymer using a low pressure gas plasma. The number of adherent cells increased with hydrophilicity, and in general the amide plasma polymers were the most adherent surfaces (although the individual monomer and the plasma parameters also affected cell attachment) (43). Dalton et al. studied the effects of surface hydrophilicity, in the presence and absence of serum proteins, on colonization and migration of human bone–derived cells (42). Vitronectin and fibronectin were used as model serum proteins, and the substrates tested were untreated polystyrene and sulfuric acid–treated polystyrene surfaces (possessing surface oxidative and sulfonic functional groups) (42). Results indicated that cells preferentially adhered to and migrated on hydrophilic surfaces; in fact, the cells would not migrate across a border between hydrophilic and hydrophobic surfaces (42). Vitronectin was found to stimulate cellular migration, while the same concentration of fibronectin was found to be insufficient for this activity (42). Evans and Steele studied the effect of hydrophilicity on corneal epithelial cell adhesion to tissue culture and untreated polystyrene, in the absence of fibronectin and vitronectin (41). A large amount of cell attachment and spreading was observed on the hydrophilic surfaces; however, a small number of cells also adhered to the moderately hydrophobic untreated polystyrene surface (41).

Many polymers designed for the control of cellular functions use copolymer grafting and polymer coatings to elicit specific cell behaviors. Mayes et al. studied two different surfaces for the control of cellular activity and of protein adsorption. First, using silicone wafers with either end-grafted star copolymers or linear polymers of poly (ethylene oxide), it was found that star polymers provided good resistance to protein absorption; however, a combination of star and linear copolymers gave complete resistance to protein absorption at relatively low densities of poly (ethylene oxide) (44). The second substrate examined was a comb copolymer with a poly(lactide) backbone and poly-(ethylene glycol) (PEG) teeth, which assumed a brush structure in water and showed excellent resistance to protein absorption (44). Due to its highly mobile chain conformation in water and large excluded volume, PEG is well known to repel protein adhesion and subsequent cell–surface interactions. Addition of bioactive ligands to PEG surfaces could be an opportunity to create surfaces which suppress nonspecific protein adsorption, yet still make the ligands of interest available for interactions with cells (44). Kao et al. reported that otherwise protein adhesion– and cell adhesion–resistant PEG networks could be grafted with interleukin 1-β or a component of complement C3a to elicit macrophage adhesion, or with C3a and fibronectin to elicit fibroblast adhesion (45). Drumheller and Hubbell have created a polymer system incorporating PEG throughout a hydrophobic, densely crosslinked matrix; this retained the bulk properties of the substrate while still providing a substrate that resisted protein and cell adsorption (46). Du et al. prepared a PEG-grafted phospholipid bilayer surface; the PEG grafting greatly decreased the amount of surface-bound fibronectin, albumin, and laminin and also greatly decreased the number of surface-adherent erythrocytes, lymphocytes, and macrophages (47). Iwasaki et al. investigated the effects of phospholipid grafts on platelet and protein adsorption to polymer substrates in the presence of serum (48). Three ω-methacryloxyalkyl phosphorylcholine (MAPC) polymers were grafted onto polyethylene backbones using a corona discharge treatment (48). The amount of fibrinogen adsorbed to the polymer surface decreased as the phospholipid density increased; the amount of absorbed fibrinogen decreased with an increase in the length of the methylene chain (increasing hydrophobicity) on the MAPC unit (48). The number of platelets in serum attached to the polymer decreased as the density of the MAPC grafts increased; however, there was no clear relationship between the platelet adhesion and MAPC chain length (48).

The coating of substrates such as silicone and glass with

polymer films can completely hide the nature of the underlying substrates, allowing the creation of a biologically active surface (49). By using self-assembled monolayers and appropriate monomers and initiators, many surface compositions can be created in a very controlled manner (49) using thermal or photopolymerization procedures. Ruhe et al. demonstrated that surfaces modified with poly(methacryloyl oxypropyl (trimethyl ammonium bromide)) and poly(methyl methacrylate) promoted neuron adhesion and outgrowth through interactions of the cells with the surface positive charges (49). Recently, Zhang et al. synthesized a novel nontoxic biodegradable lysine-di-isocyanate–based urethane polymer (50). In vitro, this polymer has been found to support bone marrow stromal cell attachment and viability, opening the door for future synthesis of biodegradable peptide-based urethane polymers and new applications in cellular engineering (50).

Modification of polymer surfaces can involve the deposition of nonpolymeric materials. Khang et al. have studied platelet adhesion on gold sputter-coated polymer surfaces (51). Platelet adhesion to the substrate was decreased and cell adhesion and growth was increased due to the alterations in surface wettability as well as to the chemistry of the gold (51). Another group, Kaibara et al., has investigated the effect of carbon deposition as a means of controlling cells on polymer surfaces (52). The adhesion and proliferation of endothelial cells on a carbon-deposited polymer surface was greatly improved, most likely due to serum proteins preferentially adsorbing to the surface (52).

B. Biochemical Modifications

The attachment of biomolecules to polymer surfaces could be an important tool for controlling cell–biomaterial interactions. Covalent immobilization of various integrin-binding peptides on a variety of materials has been shown to mediate the adhesion of many cell types [for example, osteoblasts (53–56) fibroblasts (57–63), endothelial cells (64–69), and neuronal cells (64,70–74)]. Increased cell adhesion, proliferation, and/or spreading has been obtained by grafting the specific peptide arginine–glycine–aspartic acid (RGD) to polymer surfaces (56,58,61,75–78). RGD is probably the most well-known integrin ligand and promotes adhesion of a number of cell types, including bone-derived cells (53,54,56), fibroblasts (58,61,75,77), macrophages (79), and aortic endothelial cells (76). Surfaces modified with peptides that mimic the adhesive activity of glycosaminoglycan-binding domains in cell adhesion proteins (and which are not integrin ligands) can also mediate cell adhesion via interactions with cell surface heparin or chondroitin sulfate proteoglycans (54,80,81). Biochemical surface modifications are not limited to the use of peptides; entire proteins may be immobilized on surfaces as well. For example, polyurethane substrates with surface-immobilized heparin demonstrate excellent blood compatibility in vitro and in vivo (82), and enhanced human umbilical vein endothelial cell adhesion to a variety of fibronectin-grafted polymers has been reported (83,84).

To date, biochemical surface modification strategies show a good deal of promise for controlling cellular functions at the tissue–implant interface. However, a number of questions must still be answered to fully evaluate the potential applications of this emerging area of research. Can integrin-binding peptides reproducibly control cell functions other than adhesion? Can surface modifications be developed to allow selective control of particular cell types, even when multiple cell types are present at the surface? How will biochemical surface modifications fare during the complex wound healing processes which occur at the tissue–implant interface in vivo? Continued development of surface modification approaches, as well as of analytical methodologies to quantitatively characterize these novel surfaces and the resulting cell/biological responses, will help answer these questions.

C. Morphological Modifications

Chemical micropatterning of polymer surfaces is a recent development in the pursuit of controlling cell–surface interactions. This approach allows very specific control of cell function, since micropatterned surfaces may contain areas supporting and areas inhibiting cellular adhesion (85). Cells will adhere preferentially to areas ''patterned'' to support cellular attachment and can migrate on these surfaces within the patterns supporting adhesion (85). Currently, most surfaces are morphologically modified using photolithographic patterning techniques. However, soft lithographic techniques such as microcontact printing, patterning using microfluidic channels, and laminar flow patterning (86) may provide advantages over photolithography since these techniques are relatively simple, inexpensive, and do not need extremely stringent environmental conditions (86). Most importantly, these techniques allow the patterning of surfaces with delicate ligands while providing control over surface chemistry and the cellular environment (86). Micropatterning techniques have been used in a number of studies of cell–biomaterial interactions. A few examples include investigations of human bone–derived cell preferential attachment and spreading on N-(2-aminoethyl)-3-aminopropyl trimethoxysilane–treated regions as opposed to dimethyldichlorosilane-treated areas (87) and the role of vitronectin on the attachment and distribution of these cells (88); the control of retinal pigment epithelial cell attachment, spreading, proliferation, and morphology on microcontact-printed surfaces (89); and the use of microtextured membranes for cardiac myocyte adhesion and spatial orientation (90).

IV. MODIFICATION OF BIOMATERIAL SURFACES: ARCHITECTURAL AND MECHANICAL CONSIDERATIONS

The majority of current efforts to understand and influence cell–biomaterial interactions have focused on controlling the chemistry of the biomaterial surface. This approach is certainly logical, since it builds on the knowledge that cells respond to chemical stimuli; however, this approach neglects an important and broad class of cellular stimuli. In vivo, cells are typically exposed to, and thus must be able to respond to, a mechanically dynamic environment. Depending on cell type and maturation state, cells are able to detect and respond to mechanical strain, fluid shear, electromagnetic forces, etc. Researchers have, in some cases, chosen to exploit the natural cellular responses to mechanical stimuli by carefully choosing the mechanical and structural parameters of biomaterials presented to cells.

A. Surface Architexture

The term ''architexture'' is intended to refer to planned, three-dimensional surface structure and topography at the micron scale. Cellular responses to microtopographical features of substrates in vitro have been extensively studied over the past decade. It is widely accepted that the microtexture–cell interaction can be a powerful stimulus, especially in a controlled in vitro cell culture environment. Surface microtexture has been shown to influence cellular shape, proliferation, motility, and adhesion (91,92). Many attempts to control architextural parameters have involved adding microgrooves or micropores to the substrata used in cell culture (93,94), while other studies have attempted to correlate a general surface roughness to cellular responses (95). For an extensive discussion of cellular responses to topographical phenomena, see the excellent review by Curtis and Wilkinson (92).

1. Microgrooved and Microporous Surfaces

In the presence of microgrooves, many cells undergo a phenomenon known as contact guidance. That is, the cells align or orient themselves in the direction of the groove. Various studies have attempted to determine optimal groove width (96), groove depth (97), and groove ridge (98) dimensions to influence cellular functions. Because the contact guidance response varies according to cell type (99), recommended values for these surface structure parameters will probably vary depending on the intended application of the biomaterial.

The incorporation of micropores in tissue engineered scaffolds for corneal epithelial cells has been shown to allow essential nutritional fluxes (94). Researchers are currently investigating cellular and tissue responses to various physical parameters of microporous surfaces. For example, Nakayama et al. conducted an in vivo canine study which correlated pore density with neointimal tissue thickness and thrombus formation in polyurethane cardiovascular stents (100). Pore diameter was held constant in this study at 30 μm; increased micropore density (percentage of total pore area per unit area: 0.3, 1.1, and 4.5%) aided tissue ingrowth, reduced the thickness of thrombus layer, and resulted in a thinner neointimal layer (100). Cell responses to materials incorporating micropores of various sizes (ranging from 0.1 to 30 μm) have also been investigated (94,100–102). For fibroblasts in vitro, proliferation and adhesion increased as the micropore size decreased from 8 to 0.2 μm (102). This result is consistent with other reports of cell interactions with microscale surface topographical phenomena (91), suggesting that the scale, or size, of architextural features has a greater influence than the actual geometrical structure or shape in governing cell–biomaterial interactions.

2. Surface Roughness

Surface roughness, though difficult to define and quantify, has been used to influence cellular functions on biomaterials. Roughness can be characterized by using a scanning mechanical microscope or profilometer to quantify surface topological aspects including the average surface height, distance between highest and lowest points, and increase in total surface area made possible by the roughness (as compared to a smooth surface of equal cross-sectional area) (103). Using corneal and vascular tissue from 14-day chick embryos, Lampin et al. showed a positive correlation between surface roughness of poly(methyl methacrylate) and cell adhesion. This increase was attributed to increased surface hydrophobicity and subsequent serum protein adhesion rather than directly to specific topological parameters of the surfaces (103). Compared to satin-polished titanium surfaces, roughened titanium surfaces enhanced osteoblast adhesion as well as production of extracellular matrix and subsequent mineralization (95). Surface roughness has been shown to be inversely correlated with MG63 osteoblast-like cell proliferation (104,105). However, MG63 osteoblast-like cell alkaline phosphatase activity and matrix production (indicators of cell differentiation) were positively correlated with increased surface roughness (104,105). These results suggest that in vitro topographical parameters of surface roughness (or secondary effects of surface roughness) may be used to direct cell–material interactions.

3. In Vivo Evaluations of Microtextured Biomaterials

In vivo studies concerning microtextured material–cell interactions have had mixed results. Studies of microgrooved silicone substrates implanted subcutaneously in the guinea pig (97) and of microgrooved polystyrene implants in the goat (106) both concluded that the microgrooved surfaces produced the same cellular and tissue response as smooth surfaces: fibrous capsules separated from the implant by a layer of inflammatory cells. The microtextures did not aid the subcutaneous wound healing response, perhaps as a result of differences between the material properties of the implant and the subcutaneous soft tissue surrounding the implant (106). It has also been suggested that these results were due to the influence of inflammatory cells present at the implant surface, which formed an amorphous layer between the surface architexture and the connective tissue. Because inflammatory cells do not respond to the contact guidance phenomenon (107), the cells in the connective tissue may have been unable to sense and respond to the microtopography (97).

However, in other areas or applications, microtextures may prove to be beneficial. Roughened surfaces have been reported to yield better integration of breast and subcutaneous implants as well as enhanced osteointegration and less fibrous encapsulation when compared to the performance of smooth surfaces (108). As previously described, the addition of micropores to a cardiovascular stent has been reported to aid tissue ingrowth, reduce thrombus thickness, and yield a thinner neointimal layer (100). Moreover, it is possible that the architexture of a surface could be planned to maximize the addition of bioactive proteins or ligands (97), creating a biomaterial which uses dual (both chemical and structural/topological) strategies for controlling cell–biomaterial interactions.

B. Special Application: Scaffolds for Tissue Engineering

Biomaterials used for tissue engineering present special challenges to materials scientists and engineers. On the macroscopic scale, a material scaffold for tissue engineering must have an appropriate shape and sufficient bulk mechanical properties for the intended application. On the microscopic scale, the scaffold must have microstructural and surface characteristics which are conducive to cell growth. Many tissue engineering efforts currently utilize biodegradable, porous matrices as scaffolds; these materials provide a temporary three-dimensional structure to which cells can adhere and, ideally, maintain their normal differentiated function. Over time, the scaffold can be broken down in vivo, either by hydrolysis or enzymatic activity, and replaced by the extracellular matrix (or tissue) deposited by the cells. A variety of scaffold microstructures have been developed for tissue engineering applications, including fiber-based scaffolds and sponge scaffolds with interconnected pores. Benefits of fiber-based scaffolds include extremely high porosity, up to 97% (109), and a high surface-to-volume ratio. One drawback to fiber-based scaffolds are a potential inability to withstand compressive loads (109), which could cause the scaffold structure to collapse in vivo. Fiber bonding techniques can strengthen scaffolds with respect to compressive loading (110), but may make it difficult to independently control porosity and pore size within the scaffolds (111).

To overcome some of the drawbacks associated with fiber-based scaffolds, a solvent-casting particulate leaching technique has been developed which produces sponge scaffolds, also referred to as foams (112). This technique allows control of porosity, pore size, surface-to-volume ratio, and crystallinity. The highly controllable nature of these foams makes them desirable scaffolds for tissue engineering applications.

1. Composition

Poly(L-lactic acid) (PLA), poly(glycolic acid) (PGA), and poly(D, L-lactic-co-glycolic acid) (PLGA) belong to the family of poly(α-hydroxy esters) and are commonly used polymers for the production of synthetic foam scaffolds for a number of reasons. First, these polymers seem to provide a good substrate for cell adhesion and proliferation. Second, because poly(α-hydroxy ester) polymers are degraded primarily via hydrolysis of the polyester bond, the degradation rate in vivo should not vary significantly from one individual to another. Moreover, the degradation rate can be modified via the selection of molecular weight during the fabrication process. In the case of PLGA, the degradation rate can also be controlled by varying the relative composition of the two polymers. Third, the degradation products of these polymers (lactic acid or glycolic acid) can be removed through naturally occurring metabolic pathways, primarily respiration. Finally, PLA, PGA, and PLGA have demonstrated biocompatibility, and the FDA has approved their use for a variety of different applications (113).

One drawback to the use of the poly(α-hydroxy ester) polymers is that the lactic and glycolic acid degradation products may lower the pH in the immediate vicinity of the polymer (111). In the worst case, the microenvironment may become nonviable for the cells, creating necrosis. This is especially likely in the event that the implant undergoes autocatalysis (111). Since degradation increases with increasing acidity, the release of lactic and glycolic acid could positively feed back and cause the breakdown and release of even more acid products. In addition to cellular and tissue damage, a buildup of degradation products also has the potential to shorten the expected functional life of the scaffold matrix (114,115).

2. Porosity

Porosity and pore size are key factors for tissue engineering scaffolds since these parameters help determine a number of performance aspects of the scaffold. Depending on the fabrication technique used, the porosity can range from 0 to 95% (116–118), while pore sizes typically range from ~10 to ~700 μm in diameter (114,117). Porosity can affect the overall initial mechanical strength of the polymer foam. Poly(L-lactic acid) and PLGA have been shown capable of supporting compressional forces of up to 10 kPa when fabricated with a porosity ranging from 80 to 93% (116). Porosity may also affect the rate of foam degradation. An increase in the polymer surface area exposed to the biological environment increases the rate of hydrolysis of the polymer units. Additionally, pore number and size delineate the surface area available for cell attachment, usually maximized as much as possible in tissue engineering constructs (but limited by the mechanical strength required for the specific application and the fabrication techniques). Finally, highly porous constructs can possess thinner pore walls and highly interconnected pore spaces. Interconnection of the pore spaces is important as this allows cells to penetrate deeper into the foam.

3. Design for Mechanical Properties

Historically, materials have been selected for implantation applications because they possessed adequate mechanical properties. The mechanical properties of polymer scaffolds vary widely depending on the foam fabrication technique used (112,117,119), even while the biomaterial is kept constant. Additionally, the same degradative properties which make these materials desirable for tissue engineering applications cause changes in the mechanical properties over time (118). The mechanical properties and structure of a tissue engineering scaffold mediate how mechanical stimuli from the biological environment will be transferred to cells within the scaffold. It is well-known that mechanical stimuli have an important role in governing the phenotype and functions of many cell types. Therefore, it is becoming increasingly important to be able to characterize scaffold mechanical properties to determine if they are appropriate for use in engineered tissues. It is also important to look closely at points of cellular adhesion on the scaffold surfaces to see how and to what degree the macroscopically induced mechanical stimuli are being transferred to the microscopic environment of the cell. An increased fundamental understanding of how cells respond to mechanical influences may provide impetus to design polymeric biomaterials such that a desired micromechanical environment is provided to cells as another means of controlling cell–biomaterial interactions.

V. CONCLUSIONS

Even the most basic cellular functions, such as adhesion, are the result of complex interactions between a cell and the extracellular environment. Controlling cellular functions on a biomaterial surface is a nontrivial undertaking. First, the correct type and amount of extracellular chemical ligand (or, potentially, mechanical stimuli) must be identified and reproducibly supplied at the biomaterial surface for communication with the cell. This could be accomplished by chemically modifying the biomaterial surface in a number of ways, by creating surface architexture, or by designing the biomaterial to transduce macroscopically applied mechanical loads into an appropriate mechanical microenvironment. Second, the necessary cell membrane receptors need to be identified and strategically targeted to suitably interact with the signal(s) at the biomaterial surface. The receptors must perform adequately, meaning that issues including but not limited to receptor conformation, activation, and ligation need to be addressed. Once a ''link'' is established between the intracellular and the extracellular environment via receptor ligation, the subsequent intracellular processes—often complex and overlapping chemical signaling cascades—must predictably cause desirable cellular functions. Enough beneficial cellular interactions with the biomaterial must occur to sum to a beneficial tissue interaction with the biomaterial.

Many of the scientific and engineering issues outlined herein can and should be modeled and tested under the controlled conditions possible with in vitro cell culture experiments. Computational models and analytical methodologies can help quantitatively characterize and confirm the interactions of cells and proteins with biomaterial surfaces, as well as the influences of mechanical stimuli on cell functions. Carefully designed in vivo studies can then follow to perform tests in the presence of conditions that cannot be replicated in vitro (e.g., stages of the wound healing process), but which are crucial to establishing the potential of a biomaterial for clinical use. Fundamental studies in a number of research areas can contribute to the development of strategies for controlling cell and, ultimately, tissue interactions with biomaterials in a clinically desirable manner.

REFERENCES

1. P. E. Hughes, M. Pfaff. Integrin affinity modulation. *Trends Cell Biol.* 8:359–364 (1998).
2. C. D. Buckley, G. E. Rainger, P. F. Bradfield, G. B. Nash, D. L. Simmons. Cell adhesion: more than just glue. (Review). *Mol. Membr. Biol.* 15(4):167–176 (1998).
3. J. L. Jones, R. A. Walker. Integrins: a role as cell signalling molecules. *J. Clin. Pathol.: Mol. Pathol.* 52(4):208–213 (1999).
4. F. M. Pavalko, C. A. Otey. Role of adhesion molecule cytoplasmic domains in mediating interactions with the cytoskeleton. *Proc. Soc. Exp. Biol. Med.* 205(4):282–293 (1994).
5. K. Vleminckx, R. Kemler. Cadherins and tissue formation: integrating adhesion and signaling. *Bioessays* 21(3):211–220 (1999).
6. T. Rowlands et al. Cadherins: crucial regulators of structure and function in reproductive tissues. *Rev. Reprod.* 5(1):53–61 (2000).
7. M. J. Humphries, P. Newham. The structure of cell-adhesion molecules. *Trends Cell Biol.* 8(2):78–83 (1998).
8. A. I. Rojas, A. R. Ahmed. Adhesion receptors in health and disease. *Crit. Rev. Oral Biol. Med.* 10(3):337–358 (1999).
9. J. Behrens. Cadherins and catenins: role in signal transduction and tumor progression. *Cancer Metastasis Rev.* 18(1):15–30 (1999).
10. D. Vestweber, J. Blanks. Mechanisms that regulate the function of the selectins and their ligands. *Physiol. Rev.* 79(1):181–213 (1999).
11. E. Crockett-Torabi, J. Fantone. The selectins: insights into selectin-induced intracellular signaling in leukocytes. *Immunol. Res.* 14(4):237–251 (1995).
12. R. Andrews, J. Lopez, M. Berndt. Molecular mechanisms of platelet adhesion and activation. *Int. J. Biochem. Cell. Biol.* 29(1):91–105 (1997).
13. K. Crossin, L. Krushel. Cellular signaling by neural cell adhesion molecules of the immunoglobulin superfamily. *Dev. Dyn.* 218(2):260–279 (2000).
14. Y. Yoshihara, S. Oka, J. Ikeda, K. Mori. Immunoglobulin superfamily molecules in the nervous system. *Neurosci. Res.* 10(2):83–105 (1991).
15. R. McMurray. Adhesion molecules in autoimmune disease. *Semin. Arthritis Rheum.* 25(4):215–233 (1996).
16. H. Jaeschke. Cellular adhesion molecules: regulation and functional significance in the pathogenesis of liver diseases. *Am. J. Physiol.* 273(3, Pt. 1):G602–G611 (1997).
17. M. G. Coppolino, S. Dedhar. Bi-directional signal transduction by integrin receptors. *Int. J. Biochem. Cell Biol.* 32(2):171–188 (2000).
18. B. Katz, K. Yamada. Integrins in morphogenesis and signaling. *Biochimie* 79(8):467–476 (1997).
19. G. E. Hannigan, S. Dedhar. Protein kinase mediators of integrin signal transduction. *J. Mol. Med.* 75(1):35–44 (1997).
20. E. H. J. Danen, R. M. Lafrenie, S. Miyamoto, K. M. Yamada. Integrin signaling: cytoskeletal complexes, MAP kinase activation, and regulation of gene expression. *Cell Adhesion Commun.* 6(2,3):217–224 (1998).
21. A. N. Garratt, M. J. Humphries. Recent insights into ligand binding, activation and signalling by integrin adhesion receptors. *Acta Anatomica* 154(1):34–45 (1995).
22. M. J. Humphries. Integrin cell adhesion receptors and the concept of agonism. *Trends Pharmacol. Sci.* 21(1):29–32 (2000).
23. M. J. Humphries. Towards a structural model of an integrin. *Biochem. Soc. Symp.* 65:63–78 (1999).
24. S. K. Akiyama. Integrins in cell adhesion and signaling. *Human Cell* 9(3):181–186 (1996).
25. M. A. Schwartz, M. D. Schaller, M. H. Ginsberg. Integrins: Emerging paradigms of signal transduction. *Ann. Rev. Cell Dev. Biol.* 11:549–599 (1995).
26. C. Fernandez, K. Clark, L. Burrows, N. R. Schofield, M. J. Humphries. Regulation of the extracellular ligand binding activity of integrins. *Frontiers Biosci.* 3:d684–d700 (1998).
27. M. J. Humphries. Integrin activation: the link between ligand binding and signal transduction. *Curr. Opin. Cell Biol.* 8(5):632–640 (1996).
28. C. M. Longhurst, L. K. Jennings. Integrin-mediated signal transduction. *Cell. Mol. Life Sci.* 54(6):514–526 (1998).
29. E. Brown, N. Hogg. Where the outside meets the inside: integrins as activators and targets of signal transduction cascades. *Immunol. Lett.* 54(2,3):189–193 (1996).
30. N. J. Boudreau, P. L. Jones. Extracellular matrix and integrin signalling: the shape of things to come. *Biochem. J.* 339(3):481–488 (1999).
31. J. C. Porter, N. Hogg. Integrins take partners: cross-talk between integrins and other membrane receptors. *Trends Cell Biol.* 8(10):390–396 (1998).
32. G. M. O'Neill, S. J. Fashena, E. A. Golemis. Integrin signalling: a new Cas(t) of characters enters the stage. *Trends Cell Biol.* 10(3):111–119 (2000).
33. D. D. Schlaepfer, T. Hunter. Integrin signalling and tyrosine phosphorylation: just the FAKS? *Trends Cell Biol.* 8: 151–157 (1998).
34. B. D. Ratner, A. S. Hoffman, F. J. Schoen, J. E. Lemons (Eds.). *Biomaterials Science: An Introduction to Materials in Medicine.* Academic Press, San Diego, 1996.
35. G. Maheshwari, G. Brown, D. A. Lauffenburger, A. Wells, L. G. Griffith. Cell adhesion and motility depend on nanoscale RGD clustering. *J. Cell Sci.* 113(10):1677–1686 (2000).
36. S. P. Palecek, J. C. Loftus, M. H. Ginsberg, D. A. Lauf-

fenburger, A. F. Horwitz. Integrin–ligand binding properties govern cell migration speed through cell–substratum adhesiveness. *Nature* 385(6616):537–540 (1997).
37. D. R. Critchley. Focal adhesions—the cytoskeletal connection. *Curr Opin. Cell Biol.* 12(1):133–139 (2000).
38. T. J. Mitchison, L. P. Cramer. Actin-based cell motility and cell locomotion. *Cell* 84:371–379 (1996).
39. D. A. Puleo, A. Nanci. Understanding and controlling the bone–implant interface. *Biomaterials* 20(24):2311–2321 (1999).
40. K. Webb, V. Hlady, P. A. Tresco. Relative importance of surface wettability and charged functional groups on NIH 3T3 fibroblast attachment, spreading, and cytoskeletal organization. *J. Biomed. Mater. Res.* 41(3):422–430 (1998).
41. M. D. M. Evans, J. G. Steele. Polymer surface chemistry and a novel attachment mechanism in corneal epithelial cells. *J. Biomed. Mater. Res.* 40(4):621–630 (1998).
42. B. A. Dalton, C. D. McFarland. T. R. Gengenbach, H. J. Griesser, J. G. Steele. Polymer surface chemistry and bone cell migration. *J. Biomater. Sci. Polym. Ed.* 9(8):781–799 (1998).
43. H. J. Griesser, R. C. Chatelier, T. R. Gengenbach, G. Johnson, J. G. Steele. Growth of human cells on plasma polymers: putative role of amine and amide groups. *J. Biomater. Sci. Polym. Ed.* 5(6):531–554 (1994).
44. A. M. Mayes, D. J. Irvine, L. G. Griffith. Tailoring polymer surfaces for controlled cell behavior. *Mater. Res. Soc. Symp. Proc.* 530:73–84 (1998).
45. W. J. Kao, J. A. Hubbell. Murine macrophage behavior on peptide-grafted polyethyleneglycol containing networks. *Biotech. Bioeng.* 59(1):2–9 (1998).
46. P. D. Drumheller, J. A. Hubbell. Densely crosslinked polymer networks of poly(ethylene glycol) in trimethylolpropane triacrylate for cell-adhesion-resistant surfaces. *J. Biomed. Mater. Res.* 29(2):207–215 (1995).
47. H. Du, P. Chandaroy, S. W. Hui. Grafted poly-(ethylene glycol) on lipid surfaces inhibits protein adsorption and cell adhesion. *Biochim. Biophys. Acta* 1326(2):236–248 (1997).
48. Y. Iwasaki, K. Ishihara, N. Nakabyashi, G. Khang, J. H. Jeon, J. W. Lee, H. B. Lee. Platelet adhesion on the gradient surfaces grafted with phospholipid polymer. *J. Biomater. Sci., Polym. Ed.* 9(8):801–816 (1998).
49. J. Ruhe, R. Yano, J.-S. Lee, P. Koberle, W. Knoll, A. Offenhausser. Tailoring of surfaces with ultrathin polymer films for survival and growth of neurons in culture. *J. Biomater. Sci., Polym. Ed.* 10(8):859–874 (1999).
50. J. Y. Zhang, E. J. Beckman, N. P. Piesco, S. Agarwal. A new peptide-based urethane polymer: synthesis, biodegradation, and potential to support cell growth in vitro. *Biomaterials* 21(12):1247–1258 (2000).
51. G. Khang, J. H. Jeon, J. W. Lee, H. B. Lee. Platelet and cell interactions on gold sputter-deposited polymeric surfaces. *Bio-Med. Mater. Eng.* 8:299–309 (1998).
52. M. Kaibara, H. Iwata, H. Wada, Y. Kawamoto, M. Iwaki, Y. Suzuki. Promotion and control of selective adhesion and proliferation of endothelial cells on polymer surface by carbon deposition. *J. Biomed. Mater. Res.* 31(3):429–435 (1996).
53. K. C. Dee, D. C. Rueger, T. T. Andersen, R. Bizios. Conditions which promote mineralization at the bone–implant interface: a model *in vitro* study. *Biomaterials* 17(2):209–215 (1996).
54. K. C. Dee, T. T. Andersen, R. Bizios. Design and function of novel osteoblast-adhesive peptides for chemical modification of biomaterials. *J. Biomed. Mater. Res.* 40(3): 371–377 (1998).
55. K. C. Dee, T. T. Andersen, R. Bizios. Osteoblast population migration characteristics on substrates modified with immobilized adhesive peptides. *Biomaterials* 20:221–227 (1999).
56. A. Rezania, C. H. Thomas, A. B. Branger, C. M. Waters, K. E. Healy. The detachment strength and morphology of bone cells contacting materials modified with a peptide sequence found within bone sialoprotein. *J. Biomed. Mater. Res.* 37(1):9–19 (1997).
57. A. Rezania, K. E. Healy. Biomimetic peptide surfaces that regulate adhesion, spreading, cytoskeletal organization, and mineralization of the matrix deposited by osteoblast-like cells. *Biotechnol. Prog.* 15(1):19–32 (1999).
58. P. D. Drumheller, J. A. Hubbell. Polymer networkds with grafted cell adhesion peptides for highly biospecific cell adhesive substrates. *Anal. Biochem.* 222(2):380–388 (1994).
59. S. P. Massia, J. A. Hubbell. Covalent surface immobilization of Arg-Gly-Asp–and Tyr-Ile-Gly-Ser-Arg containing peptides to obtain well-defined cell-adhesive substrates. *Anal. Biochem.* 187(2):292–301 (1990).
60. S. P. Massia, J. A. Hubbell. An RGD spacing of 440 nm is sufficient for integrin $\alpha v\beta 3$-mediated fibroblast spreading and 140nm for focal contact and stress fiber formation. *J. Cell Biol.* 114(5):1089–1100 (1991).
61. J. A. Neff, P. A. Tresco, K. D. Caldwell. Surface modification for controlled studies of cell–ligand interactions. *Biomaterials* 20:2377–2393 (1999).
62. K. Kugo, M. Okuno, K. Masuda, J. Nishino, H. Masuda, and M. Iwatsuki. Fibroblast attachment to Arg-Gly-Asp peptide–immobilized poly(gamma-methyl L-glutamate). *J. Biomater. Sci. Polym. Ed.* 5(4):325–337 (1994).
63. M. D. Pierschbacher, J. W. Polarek, W. S. Craig, J. F. Tschopp, N. J. Sipes, J. R. Harper. Manipulation of cellular interactions with biomaterials toward a therapeutic outcome: a perspective. *J. Cell Biochem.* 56(2):150–154 (1994).
64. N. Patel, R. Padera, G. H. W. Sanders, S. M. Cannizzaro, M. C. Davies, R. Langer, C. J. Roberts, S. J. B. Tendler, P. M. Williams, K. M. Shakesheff. Spatially controlled cell engineering on biodegradable polymer surfaces. *FASEB* 12(4):1447–1454 (1998).
65. K. C. Dee, T. T. Andersen, R. Bizios. Enhanced endothelialization of substrates modified with immobilized bioactive peptides. *Tissue Eng.* 1:135–145 (1995).
66. S. P. Massia, J. A. Hubbell. Human endothelial cell inter-

actions with surface-coupled adhesion peptides on a nonadhesive glass substrate and two polymeric biomaterials. *J. Biomed. Mater. Res.* 25(2):223–242 (1991).

67. H. B. Lin, C. Garcia-Echeverria, S. Asakura, W. Sun, D. F. Mosher, S. L. Cooper. Endothelial cell adhesion on polyurethanes containing covalently attached RGD-peptides. *Biomaterials* 13(13):905–914 (1992).
68. H. B. Lin, W. Sun, D. F. Mosher, C. Garcia-Echeverria, K. Schaufelberger, P. I. Lelkes, S. L. Cooper. Synthesis, surface, and cell-adhesion properties of polyurethanes containing covalently grafted RGD-peptides. *J. Biomed. Mater. Res.* 28(3):329–342 (1994).
69. J. J. Grzesiak, M. D. Pierschbacher, M. F. Amodeo, T. I. Malaney, J. R. Glass. Enhancement of cell interactions with collagen/glycosaminoglycan matrices by RGD derivatization. *Biomaterials* 18(24):1625–1632 (1997).
70. R. B. Patterson, A. Messier, R. F. Valentini. Effects of radiofrequency glow discharge and oligopeptides on the attachment of human endothelial cells to polyurethane. *Asaio J.* 41(3):M625–M629 (1995).
71. S. Saneinejad, M. S. Shoichet. Patterned glass surfaces direct cell adhesion and process outgrowth of primary neurons of the central nervous system. *J. Biomed. Mater. Res.* 42(1):13–19 (1998).
72. T. G. Vargo, E. J. Bekos, Y. S. Kim, J. P. Ranieri, R. Bellamkonda, P. Aebischer, D. E. Margevich, P. M. Thompson, F. V. Bright, J. A. Gardella, Jr. Synthesis and characterization of fluoropolymeric substrata with immobilized minimal peptide sequences for cell adhesion studies. I. *J. Biomed. Mater. Res.* 29:767–778 (1995).
73. J. P. Ranieri, R. Bellamkonda, E. J. Bekos, T. G. Vargo, J. A. Gardella, Jr., P. Aebischer. Neuronal cell attachment to fluorinated ethylene propylene films with covalently immobilized laminin oligopeptides YIGSR and IKVAV. II. *J. Biomed. Mater. Res.* 29(6):779–785 (1995).
74. J. P. Ranieri, R. Bellamkonda, E. J. Bekos, J. A. Gardella, Jr., H. J. Mathieu, L. Ruiz, P. Aebischer. Spatial control of neuronal cell attachment and differentiation on covalently patterned laminin oligopeptide substrates. *Int. J. Dev. Neurosci.* 12(8):725–735 (1994).
75. P. Banerjee, D. J. Irvine, A. M. Mayes, L. G. Griffith. Polymer latexes for cell- resistant and cell-interactive surfaces. *J. Biomed. Mater. Res.* 50(3):331–339 (2000).
76. A. D. Cook, J. S. Hrkach, N. N. Gao, I. M. Johnson, U. B. Pajvani, S. M. Cannizzaro, R. Langer. Characterization and development of RGD-peptide–modified poly-(lactic acid-co-lysine) as an interactive, resorbable biomaterial. *J. Biomed. Mater. Res.* 35(4):513–523 (1997).
77. S. P. Massia, J. A. Hubbell. Covalently attached GRGD on polymer surfaces promotes biospecific adhesion of mammalian cells. *Ann. N. Y. Acad. Sci.* 589:261–270 (1990).
78. K. M. Shakesheff, S. M. Cannizzaro, R. Langer. Creating biomimetic microenvironments with synthetic polymer–peptide hybrid molecules. *J. Biomater. Sci. Polym. Ed.* 9(5):507–518 (1998).
79. W. J. Kao. Evaluation of protein-modulated macrophage behavior on biomaterials: designing biomimetic materials for cellular engineering. *Biomaterials* 20:2213–2221 (1999).
80. S. P. Massia, J. A. Hubbell. Immobilized amines and basic amino acids as mimetic heparin-binding domains for cell surface proteoglycan-mediated adhesion. *J. Biol. Chem.* 267(14):10133–10141 (1992).
81. K. E. Healy, B. Lom, P. E. Hockberger. Spatial distribution of mammalian cells dictated by material surface chemistry. *Biotechnol. Bioeng.* 43:792–800 (1994).
82. S. W. Kim, H. Jacobs. Design of nonthrombogenic polymer surfaces for blood-contacting medical devices. *Blood Purification* 14:357–372 (1996).
83. V. D. Bhat, B. Klitzman, K. Koger, G. A. Truskey, W. M. Reichert. Improving endothelial cell adhesion to vascular graft surfaces: clinical need and strategies. *J. Biomater. Sci. Polym. Ed.* 9(11):1117–1135 (1998).
84. E. Imbert, A. A. Poot, C. G. Figdor, J. Feijen. Different growth behavior of human umbilical vein endothelial cells and an endothelial cell line seeded on various polymer surfaces. *Biomaterials* 19(24):2285–2290 (1998).
85. S. Zhang, L. Yan, M. Altman, M. Lassle, H. Nugent, F. Frankel, D. A. Lauffenburger, G. M. Whitesides, A. Rich. Biological surface engineering: a simple system for cell pattern formation. *Biomaterials* 20(13):1213–1220 (1999).
86. R. S. Kane, S. Takayama, E. Ostuni, D. E. Ingber, G. M. Whitesides. Patterning proteins and cells using soft lithography. *Biomaterials* 20:2363–2376 (1999).
87. C. D. McFarland, C. H. Thomas, C. DeFilippis, J. G. Steele, K. E. Healy. Protein adsorption and cell attachment to patterned surfaces. *J. Biomed. Mater. Res.* 49(2):200–210 (2000).
88. C. H. Thomas, C. D. McFarland, M. L. Jenkins, A. Rezania, J. G. Steele, K. E. Healy. The role of vitronectin in the attachment and spatial distribution of bone-derived cells on materials with patterned surface chemistry. *J. Biomed. Mater. Res.* 37(1):81–93 (1997).
89. L. Lu, L. Kam, M. Hasenbein, K. Nyalakonda, R. Bizios, A. Gopferich, J. F. Young, A. G. Mikos. Retinal pigment epithelial cell function on substrates with chemically micropatterned surfaces. *Biomaterials* 20:2351–2361 (1999).
90. J. Deutsch, D. Motlagh, B. Russell, T. A. Desai. Fabrication of microtextured membranes for cardiac myocyte attachment and orientation. *J. Biomed. Mater. Res. (Appl. Biomater.)* 53(3):267–275 (2000).
91. A. M. Green, J. A. Jansen, J. P. van der Waerden, A. F. von Recum. Fibroblast response to microtextured silicone surfaces: texture orientation into or out of the surface. *J. Biomed. Mater. Res.* 28(5):647–653 (1994).
92. A. Curtis, C. Wilkinson. Topographical control of cells. *Biomaterials* 18(24):1573–1583 (1997).
93. E. T. den Braber, J. E. de Ruijter, H. T. Smits, L. A. Ginsel, A. F. von Recum, J. A. Jansen, Effect of parallel surface microgrooves and surface energy on cell growth. *J. Biomed. Mater. Res.* 29(4): 511–518 (1995).

94. M. D. Evans, B. A. Dalton, J. G. Steele. Persistent adhesion of epithelial tissue is sensitive to polymer topography. *J. Biomed. Mater. Res.* 46(4):485–493 (1999).
95. B. Groessner-Schreiber, R. S. Tuan. Enhanced extracellular matrix production and mineralization by osteoblasts cultured on titanium surfaces in vitro. *J. Cell. Sci.* 101(Pt.1):209–217 (1992).
96. B. D. Boyan, T. W. Hummert, D. D. Dean, Z. Schwartz. Role of material surfaces in regulating bone and cartilage cell response. *Biomaterials* 17(2):137–146 (1996).
97. X. F. Walboomers, J. A. Jansen. Microgrooved silicone subcutaneous implants in guinea pigs. *Biomaterials* 21(6):629–636 (2000).
98. E. T. den Braber, J. E. de Ruijter, L. A. Ginsel, A. F. von Recum, J. A. Jansen. Quantitative analysis of fibroblast morphology on microgrooved surfaces with various groove and ridge dimensions. *Biomaterials* 17(21):2037–2044 (1996).
99. J. L. Ricci, J. Charvet, R. Chang, C. Howard, W. S. Green, L. Weiser, H. Alexander. In vitro effects of surface microgeometry on colony formation by fibroblasts and bone cells. Proceedings of the 20th annual meeting of the Society of Biomaterials, Boston, 1994, p. 401.
100. Y. Nakayama, S. Nishi, H. Ishibashi-Ueda, T. Matsuda. Surface microarchitectural design in biomedical applications: in vivo analysis of tissue ingrowth in excimer laser-directed micropored scaffold for cardiovascular tissue engineering. *J. Biomed. Mater. Res.* 51(3):520–528 (2000).
101. J. G. Steele, G. Johnson, K. M. McLean, G. J. Beumer, H. J. Griesser. Effect of porosity and surface hydrophilicity on migration of epithelial tissue over synthetic polymer. *J. Biomed. Mater. Res.* 50(4):475–482 (2000).
102. J. H. Lee, S. J. Lee, G. Khang, H. B. Lee. Interaction of fibroblasts on polycarbonate membrane surfaces with different micropore sizes and hydrophilicity. *J. Biomater. Sci. Polym. Ed.* 10(3):283–294 (1999).
103. M. Lampin, R. Warocquier-Clerout, C. Legris, M. Degrange, M. F. Sigot-Luizard. Correlation between substratum roughness and wettability, cell adhesion, and cell migration. *J. Biomed. Mater. Res.* 36(1):99–108 (1997).
104. J. Y. Martin, Z. Schwartz, T. W. Hummert, D. M. Schraub, J. Simpson, J. Lankford, Jr., D. D. Dean, D. L. Cochran, B. D. Boyan. Effect of titanium surface roughness on proliferation, differentiation, and protein synthesis of human osteoblast-like cells. (MG63). *J. Biomed. Mater. Res.* 29(3):389–401 (1995).
105. J. Lincks, B. D. Boyan, C. R. Blanchard, C. H. Lohmann, Y. Liu, D. L. Cochran, D. D. Dean, Z. Schwartz. Response of MG63 osteoblast-like cells to titanium and titanium alloy is dependent on surface roughness and composition. *Biomaterials* 19(23):2219–2232 (1998).
106. X. F. Walboomers, H. J. Croes, L. A. Ginsel, J. A. Jansen. Microgrooved subcutaneous implants in the goat. *J. Biomed. Mater. Res.* 42(4):634–641 (1998).
107. J. Meyle, K. Gultig, W. Nisch. Variation in contact guidance by human cells on a microstructured surface. *J. Biomed. Mater. Res.* 29(1):81–88 (1995). [erratum appears in *J. Biomed. Mater. Res.* 29(7):905 (1995).]
108. Y. Ito. Surface micropatterning to regulate cell functions. *Biomaterials* 20:2333–2342 (1999).
109. B. S. Kim, D. J. Mooney. Development of biocompatible synthetic extracellular matrices for tissue engineering. *Trends Biotechnol.* 16(5):224–230 (1998).
110. A. G. Mikos, Y. Bao, L. G. Cima, D. E. Ingber, J. P. Vacanti, R. Langer. Preparation of poly(glycolic acid) bonded fiber structures for cell attachment and transplantation. *J. Biomed. Mater. Res.* 27(2):183–189 (1993).
111. R. C. Thomson, M. C. Wake, M. J. Yaszemski, A. G. Mikos. Biodegradable polymer scaffolds to regenerate organs. *Adv. Polym. Sci.* 122:245–274 (1995).
112. A. G. Mikos et al. Preparation and characterization of poly(L-lactic acid) foams. *Polymer* 35(5):1068–1077 (1994).
113. W. H. Wong, D. J. Mooney. Synthesis and properties of biodegradable polymers used as synthetic matrices for tissue engineering. In:*Synthetic Biodegradable Polymer Scaffolds*, A. Atala, D. Mooney (Eds.). Birkhauser, Boston, 1997, pp. 51–82.
114. S. L. Ishaug-Riley, G. M. Crane-Kruger, M. J. Yaszemski, A. G. Mikos. Three-dimensional culture of rat calvarial osteoblasts in porous biodegradable polymers. *Biomaterials* 19(15):1405–1412 (1998).
115. D. E. Thompson, C. M. Agrawal, K. A. Athanasaiou. The effects of dynamic compressive loading on biodegradable implants of 50–50% polylactic acid–polyglycolic acid. *Tissue Eng.* 2(1):61–74 (1996).
116. M. C. Peters, D. J. Mooney. Synthetic extracellular matrices for cell transplantation. *Mater. Sci. Forum* 250:43–52 (1997).
117. C. M. Agrawal, K. A. Athanasiou, J. D. Heckman. Biodegradable PLA-PGA polymers for tissue engineering in orthopaedics. *Mater. Sci. Forum* 250:115–128 (1997).
118. K. A. Athanasiou, J. P. Schmitz, C. M. Agrawal. The effects of porosity on in vitro degradation of polylactic acid–polyglycolic acid implants used in repair of articular cartilage. *Tissue Eng.* 4(1):53–63 (1998).
119. K. Whang, C. H. Thomas, K. E. Healy, G. Nuber. Novel method to fabricate bioabsorbable scaffolds. *Polymer* 36(4):837–842 (1995).
120. R. H. Kramer, J. Enenstein, N. S. Waleh. Integrin structure and ligand specificity in cell–matrix interactions. In: *Molecular and Cellular Aspects of Basement Membranes*, D. H. Rohrbach, R. Timpl (Eds.). Academic Press, New York, 1993, pp. 239–258.

15

Polymeric Systems for Ophthalmic Drug Delivery

O. Felt, S. Einmahl, and R. Gurny
University of Geneva, Geneva, Switzerland

P. Furrer
University of Lausanne, Lausanne, Switzerland

V. Baeyens
Centre Interuniversitaire de Recherche et d'Enseignement, Archamps, France

I. INTRODUCTION

The main routes of administration for ocular therapeutics include topical, periocular, and intraocular routes (Fig. 1) (1). In addition, some ocular diseases may be treated by systemic dosage forms, either by oral ingestion or parenteral injection (2). However, the systemic route presents the great disadvantage of exposing all the organs of the body to the action of the drug, thus leading to unwanted side effects. Since such oral and parenteral systems are largely beyond the scope of this review, the systemic route will not be treated in this chapter.

For evident convenience reasons, the administration of medications to the eye by topical instillation of eyedrops into the lower cul-de-sac is undoubtedly the most popular method. The local instillation of drugs may be used to treat both surface and intraocular pathological conditions (1). However, efficient protective mechanisms such as solution drainage, lachrymation, and diversion of exogenous substances into the systemic circulation via the conjunctiva reduce considerably the availability of the applied drugs. Hence, the success of the therapy is based on frequent instillations for most of the pathological states encountered, leading to poor patient compliance. One of the most common methods to improve this parameter is to increase the precorneal residence time of topical formulations by using polymers for the preparation of delivery systems such as hydrogels, micro- and nanoparticles, and inserts. The use of such systems for ocular applications has been recently reviewed by Baeyens et al. (3) and by Felt et al. (4). Though some of these systems provided interesting results for local diseases when compared with classical solutions, they remain unsatisfactory in most cases for the treatment of pathologies affecting the posterior segment of the eye. This is mainly due to the structure of the cornea, which is relatively impermeable to most drugs, and to the tight junctional complexes (*zonulae occludentes*) of the retinal pigment epithelium and retinal capillaries, a barrier which inhibits the penetration of substances from the blood flow into the posterior segment of the eye. To circumvent these barriers, different routes of intraocular drug delivery have been investigated.

The first part of this chapter is devoted to the description of some ocular diseases, describing briefly disorders that may be treated by local applications and those which necessitate intraocular interventions. The different polymeric systems (hydrogels, microparticulate systems, and solid dosage forms) allowing ocular drug delivery after local or intraocular administration are discussed in the second and third part of this chapter, respectively.

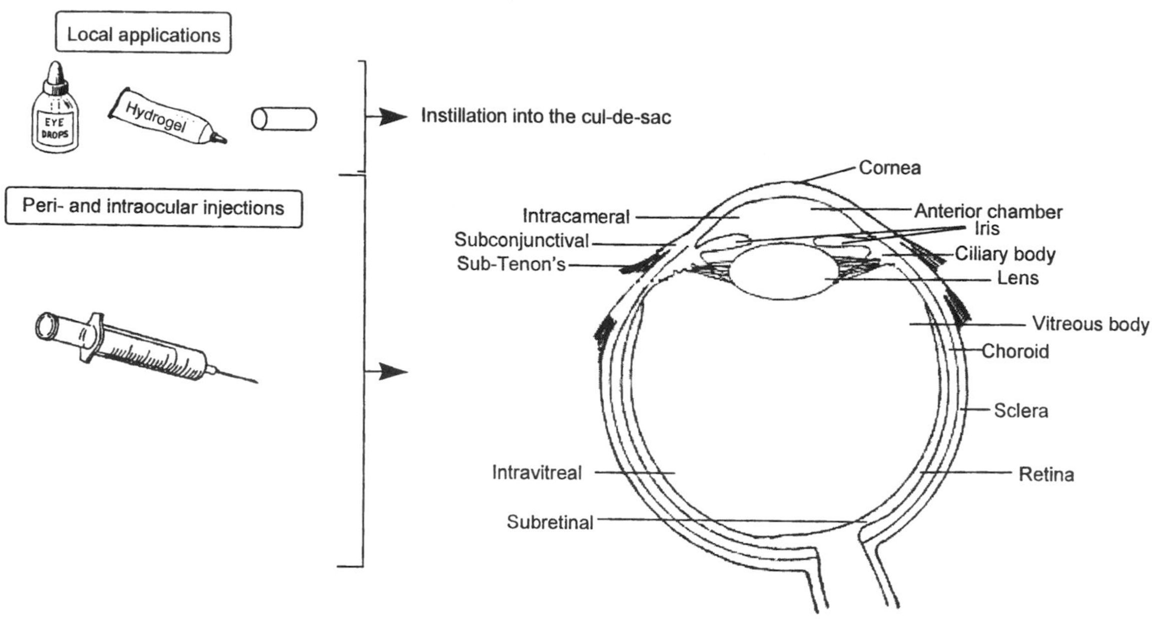

Figure 15.1 Routes for the administration of ophthalmic therapeutics.

II. PATHOLOGIES

A. Topically Treated Pathologies

1. Inflammation

Inflammation is one of the most common ocular disorders and can have several etiologic causes including trauma, viral or bacterial infections, immunorelated phenomena, or corneal ulceration (3). Blepharitis and conjunctivitis represent common cases of inflammation. Blepharitis describes a variety of inflammatory conditions of the eyelid margins, where usual symptoms are redness, thickening with scales and crusts, and shallow ulcers (5). It may be caused by bacteria, local allergic reactions, or sebaceous secretions from the skin (5). The inflammation of the conjunctiva can be the consequence of bacterial, fungal, or viral infections or of an allergic reaction. In the case of bacterial conjunctivitis, the main causative agents include *Staphylococcus aureus*, *Pneumococcus* species, and *Haemophilus aegyptius* (5).

Treatments of inflammation vary with the etiologic agent and may be based on the use of a topical steroidal active compound associated with nonsteroidal anti-inflammatory drugs such as diclofenac and flurbiprofen, immunosuppressive agent (azathioprine), antimicrobial agents (tobramycin, ofloxacin), and mydriatic-cycloplegic drugs (pilocarpine) (3).

2. Keratoconjunctivitis Sicca

The terms dry eye and keratoconjunctivitis sicca (KCS) refer to a group of disorders of the ocular surface connected with reduction or instability of the precorneal tear film (6). The symptoms usually identified in such diseases include itchy and burning sensations, photophobia, reflex tearing, redness, blurred vision, mucous strands, and intolerance to contact lenses (7). Diagnosis of dry eye may be based on physical examination as well as on laboratory tests (Table 1). The treatment of KCS relies mainly on the

Table 15.1 Diagnosis Tools for the Investigation of KCS

Physical examination	Laboratory tests
Rose bengal staining	Lysozyme measurement
Fluorescein staining	Goblet cell count
Tear film breakup time	Tear osmolarity
Schirmer test	Tear mucin measurement

Source: Ref. 6.

use of artificial tears to fulfill the physiological role of normal tears, e.g., lowering the surface tension of the tear film, forming a hydrophilic layer, and enhancing the tear volume. Others possible treating agents include Vitamin A derivatives, mucolytic (N-acetylcysteine) and tear stimulating products (bromhexine) (8).

3. Glaucoma

Glaucoma is a consequence of the impairment of the aqueous humor outflow. An increased intraocular pressure, progressive optic neuropathy, visual impairment, and in severe cases blindness are common characteristics of this disease (3). The therapy of glaucoma aims either to reduce the rate of aqueous humor secretion or to increase the aqueous humor outflow. The first approach is achieved by using beta-adrenergic agonists (epinephrine) and antagonists (timolol) or carbonic anhydrase inhibitors (acetazolamide), whereas the second approach is based on the use of parasympathomimetic agents (pilocarpine) (9). However, when such medical therapy is unsuccessful argon laser trabeculoplasty or surgical actions are required (see Section II. B) (9).

B. Intraocular Pathologies

1. Glaucoma Filtering Surgery Failure

Glaucoma filtration surgery entails the construction of a fistula between the anterior chamber and the subconjunctival space as an alternative or enhanced passage for the drainage of the aqueous humor from the eye in order to reduce the intraocular pressure and preserve vision in severe cases of glaucoma. The outcomes of filtration surgery are affected by wound healing, a normal event. The healing response is initiated at the moment of injury, when the tissue architecture is disrupted. A complex cascade of events characterized by coagulation, inflammation, angiogenesis, extracellular matrix deposition, and epithelialization occur and fibroblasts proliferate, leading to the obstruction of the filtration site (10). Pharmacological agents have been used to control certain stages of the wound healing response; these include anti-inflammatory agents, antimetabolites or antiproliferative agents, and collagen crosslinking inhibitory agents (10).

2. Proliferative Vitreoretinopathy

Proliferative vitreoretinopathy (PVR) is a pathologic condition occurring in part as a complication of retinal detachment, in which cells originating from the retina and transdifferentiating into fibroblasts proliferate, inducing the formation of retractile membranes on both surfaces of the detached retina (11). The aim of the pharmacological treatment of PVR is to intervene at different stages of disease progression (12), that is, inflammation and cellular proliferation. In the inflammatory phase, long-effect steroids are more suitable, while antifibroblastic drugs can be used in the proliferative phase. Proliferative vitreoretinopathy is a recurrent disease; even if surgical removal of the contractile membranes enables flattening of the detached retina, the proliferation process can reappear (13). The controlled release of anti-inflammatory and antiproliferative drugs can both prevent the development of PVR in high risk eyes and hinder PVR recurrence while minimizing toxic side effects and improving patient comfort.

3. Cytomegalovirus Retinitis

Cytomegalovirus (CMV) retinitis is an opportunistic infection frequently seen in patients with acquired immune deficiency syndrome (AIDS), occurring in 15–40% of patients (14), and those with other forms of immune suppression. Ganciclovir was the first virustatic agent that showed an activity against CMV in patients with AIDS (15). Another drug used in the treatment of CMV retinitis is foscarnet, an alternative to ganciclovir if resistant strains of cytomegalovirus develop (16). Both drugs are administered systemically or, if an intolerance develops, intravitreally.

A promising new treatment of CMV retinitis consists of the intravitreal administration of a phosphorothioate oligonucleotide, fomivirsen (17). Its activity is very selective against the virus; its half-life of elimination from the vitreous is long; and it triggers no relevant toxicity.

4. Endophthalmitis

Endophthalmitis is an inflammatory response to the invasion of the internal structures of the eye by replicating microorganisms, i.e., bacteria, fungi, or parasites (18). This disease is most often a complication of ocular surgical procedures such as cataract extraction, corneal transplantation, glaucoma filtering surgery, or retinal detachment surgery. Another cause is a penetrating ocular trauma, which can occur in up to 20–30% of the cases. The vitreous body and the retina are generally involved, and depending on the virulence of the infectious agent this disease can lead to retinal necrosis and blindness within a very short period of time.

The current therapeutic strategy for presumed endophthalmitis includes the intravitreal injection of an antibiotic or more frequently a combination of broad-spectrum antibiotics (19). Besides antibiotics, intravitreal corticosteroids such as dexamethasone play an important role in the management of bacterial endophthalmitis because of their beneficial effect in reducing the inflammation that accompanies the infection. Therapy for fungal endophthalmitis is similar to that for bacterial endophthalmitis except that antifungals, rather than antibacterials, are administered.

5. Posterior Uveitis

Posterior uveitis, whether it is endogenous and chronic or associated with systemic diseases such as Behçet's disease, is a pathology characterized by a vitreal inflammation and a retinal vasculitis. In the majority of patients, the etiologic agent is not known. It has been postulated that in many cases uveitis may be autoimmune in origin. Uveitis usually has a low response to treatments with steroids or cytotoxic drugs, and it may require long-term chronic treatment. If untreated, the outcome is generally severe, eventually leading to progressive blindness. The immunomodulating drug cyclosporine was investigated and found effective in treating experimental autoimmune uveitis and acute inflammation unresponsive to the usual treatment (20).

6. Posterior Capsule Opacification

Cataract extraction with the implantation of a posterior chamber intraocular lens has become the most popular technique for cataract surgery. However, one of its most frequent complications is posterior capsule opacification (PCO), which appears in 50% of the cases between the third and fifth year postoperatively (21). It is caused by migration, proliferation, and metaplasia to fibroblasts of the epithelial cells of the anterior capsule on the posterior capsule surface. The prevention of PCO, other than by surgical procedures such as cryocoagulation, consists of the administration of antiproliferative drugs in the intraocular infusion fluids during surgery (21).

7. Retinal Degenerative Diseases

Age-related macular degeneration (AMD) is the leading cause of irreversible visual loss in the United States in the elderly people (22). Blindness associated with AMD is caused by degeneration of visual cells, a result of degenerative changes in the retinal pigment epithelium (RPE). Whether the disease takes the form of localized degeneration without the complications of vascular invasion or whether cells are destroyed by the disruptive effects of neovascularization, degeneration of the RPE precedes or accompanies death of the associated rods and cones. Laser photocoagulation of choroidal neovascular membranes (CNVM) is currently the only well-studied and widely accepted treatment modality. However, it is beneficial for only a small minority of patients and is associated with an unacceptably high CNVM persistence and recurrence rate. Consequently, investigators have attempted to develop new modalities for treatment of CNVM: subretinal surgery, radiation therapy, photodynamic therapy, and pharmacological inhibition of CNVM formation (23). Pharmacological therapy has focused on interfering with inflammation, oxidative stress, and angiogenesis. Inflammation is often associated with neovascular membranes, and steroids have been shown to have a direct antiangiogenic activity, in addition to their anti-inflammatory effects. Free radical production is partly the result of the high oxygen consumption by the retina. Antioxidants have been proposed as a preventative measure to reduce lipid peroxidation. Antiangiogenic drugs are the latest addition to the list of suggested pharmacological treatments for AMD. These drugs may be useful in treating and preventing the neovascular phase of this disorder. Local administration would be possible with the advent of slow-release technologies for chronic drug delivery.

III. TOPICAL DELIVERY SYSTEMS

A. Polymeric Gels

Hydrogels, also called aqueous gels, are defined as colloidal dispersions of high molecular weight polymers which have the ability to swell, but not dissolve, in water or aqueous solvent systems (24–26). Macromolecules in the hydrogels form a swollen, crosslinked network of twisted matted strands bound together by various electrochemical interactions like covalent bonds or somewhat weaker intermolecular forces (van der Waals forces and hydrogen bonds, electrostatic forces, ion bridging interactions, or hydrophobic interactions). The network contains crystalline and amorphous regions throughout the entire system (27,28).

The desired properties of polymers used in ophthalmic hydrogels as drug delivery systems are the following (29):

- Chemical compatibility with the other excipients and the drug
- Stability after prolonged storage and, if possible, heat stability permitting sterilization by autoclaving
- Absence of ocular irritation or toxicity
- Little or no problem for vision
- Promotion of precorneal retention due to viscosity or bioadhesive properties
- Physicochemical properties (pH, osmolality) compatible with the ocular use

Polymeric hydrogels used in ophthalmology are generally classified in two distinct groups (Fig. 2). The first category includes systems which are administered as viscous

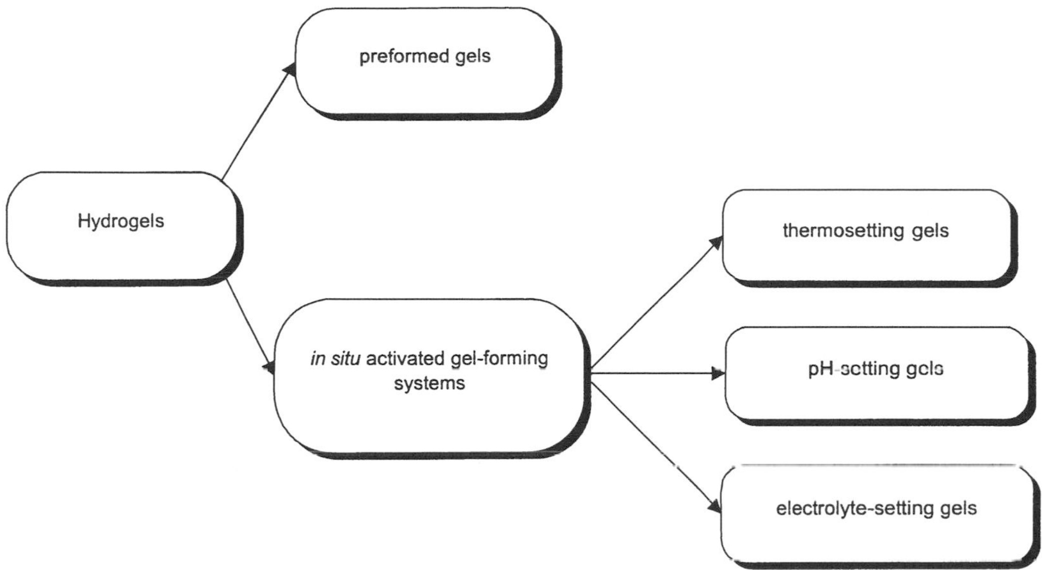

Figure 15.2 Classification of hydrogels as sustained ocular drug delivery systems.

preparations which do not undergo any other transformation on the eye such as an eventual dilution by lacrymation or gradual elimination by lachrymal drainage. These hydrogels are called preformed gels, i.e., they are structured before ocular application. The second group refers to polymeric systems that are topically applied on the ocular surface as solutions or suspensions and which shift from a sol to a gel phase as they are exposed to ocular physiological conditions like pH, temperature, or the presence of ions. This second group of hydrogels is called in situ activated gel-forming systems. Both polymeric gels—preformed as well as in situ forming gels—are used for the ocular route to improve drug bioavailability by extending drug residence on the eye (30). The main advantage of in situ activated gels over preformed hydrogels is that they facilitate the administration of the ophthalmic preparation, since the preparation is conveniently delivered in a drop form and undergoes a transition from a solution to a gel on the ocular surface, thus permitting an administration of accurate and reproducible quantities in contrast to already gelled preparations (31).

The mechanism of drug release from hydrogels involves a combination of diffusion and erosion of the gel surface (30). Due to the hydrophilic nature of the gels, tears readily diffuse into the gel interior thereby rapidly leaching out water-soluble drugs such as pilocarpine salts.

There are two general methods for the loading of hydrogels as drug carriers (25). In the first method, the hydrogel monomer is polymerized in the presence of the drug by addition of chemicals—initiator and crosslinker—so that the drug remains trapped in the polymeric matrix. This method has the disadvantage that some polymerization conditions may have deleterious effects on the drug and that some chemicals are toxic and difficult to extract from the polymeric device. In the second method, a polymer is allowed to swell in a suitable drug solution or the drug is added to a preformed hydrogel and diffuses through the polymeric network.

It should be noted that in the ophthalmic field, hydrogels are not all loaded with drug, for example, in the case of artificial tears, where the polymeric hydrogels are themselves used as a treatment of tear film disorders (32). This point is further developed subsequently.

From a medical point of view, ophthalmic hydrogels aim to optimize ocular drug delivery of an active ingredient by prolonging the corneal contact time of the drug and decreasing its drainage rate, thus increasing its ocular bioavailability. Furthermore, the hydrogels may also lower the frictional resistance between the cornea and the eyelids which occurs with each blink, thereby exerting a lubricating effect beneficial in tear film disorders (33).

1. Preformed Hydrogels

a. Cellulose Derivatives

Semisynthetic water-soluble cellulosic polymers have been widely used in ophthalmic preparations such as eye drops, artificial tears, and contact lens solutions as viscosity-increasing agents. The first cellulosic derivative used in ophthalmology was methylcellulose (MC), a topically inert material able to provide gels of high viscosity (34,35). Other cellulosic compounds used in ocular preparations include hydroxyethylcellulose (HEC), hydroxypropylcellulose (HPC), hydroxypropylmethylcellulose (HPMC), and sodium carboxymethylcellulose (CMC Na). These viscolysers differ with respect to their degree of substitution and the nature of the chemical substituents, but present common properties include (36)

Wide range of viscosity (400 to 15,000 mPas)
Compatibility with many topically applied drugs
Stability to autoclaving (120°C)
Good ocular tolerance

The viscosity of cellulose derivative–based hydrogels depends on the molecular weight grade and on the substitution (nature, degree of substitution) of the derivative (37).

Methylcellulose and HPMC hydrogels are prepared by addition of hot (80–90°C) water to the polymer powder under strong agitation, followed by cooling for 24 h in the refrigerator and swelling of the polymer. Afterward, the final volume is adjusted. The preparation of HEC can be made with cold water. Carboxymethylcellulose and HPC hydrogels are obtained by sprinkling the polymer on the water surface and mixing and boiling the solution.

Cellulosic polymers have been extensively used for ocular drug delivery as viscous solutions in low concentrations rather than as hydrogels in higher concentrations. Some representative examples of clinical and pharmacological studies with these polymers are listed in Table 2. Several investigations have shown the efficacy of cellulosic derivatives in increasing ocular drug availability, by extending the corneal residence time, with respect to simple saline solutions (38–40). Ludwig and Van Ooteghem (41) used slit lamp fluorophotometry to compare the effect of different cellulosic viscolysers, namely, HEC, HPC, and HPMC, on the ocular retention of a fluorescent tracer in human subjects. They concluded that HEC is more comfortable and thus is superior to the two other polymers in reducing the ocular elimination rate of the viscous eye drops. Other authors have compared the performance of cellulosic derivatives with more recently developed polymers in the field of ocular drug delivery. Generally these polymers, such as polyacrylic acid, sodium hyaluronate, and poly(vinyl alcohol), give better results than cellulose (40,42–45).

Currently, a large number of hydrogels on the market are formulated with cellulosic viscolysers, among them CMC Na (Celluvise® from Allergan, Irvine, CA), HEC (Adsorbotear® from Alcon, Fort Worth, TX), HPC (Lacrisert® from Merck Sharp & Dohme, West Point, PA), HPMC (Lubritears® Solution from Bausch & Lomb, Tampa, FL), MC (Methocel® from Ciba Vision, Niederwangen, Switzerland).

b. *Polyacrylic Acid*

Polyacrylic acids, also called carbomers or carboxyvinyl polymers, are acrylic acid–based polymers which are available in different molecular weights and different structures (linear, branched, or crosslinked). Carbopol® resins (BFGoodrich Specialty Chemical, Cleveland, OH) are high molecular weight, crosslinked carbomers. Carbopol was introduced in the mid- to late-1950s as a group of synthetic, hydrophilic colloidal gums with improved and constant thickening properties at relatively low concentrations compared with natural gums (51).

Rheograms of carbomers exhibit non-Newtonian flow behavior. A pseudoplastic (decrease of the viscosity with increasing shear rates) or plastic (rheogram with a yield

Table 15.2 Examples of Ocular Hydrogels Based on Cellulosic Derivatives

Cellulose derivative	Concentration (%)	Species	Drug	Ref.
Methylcellulose	NS	R	Tritiated pilocarpine	38
	1.0	R + H	Nonmedicated hydrogel	45
Hydroxyethylcellulose	1.1, 1.5, 3.4	H	Sodium fluorescein	46
	0.325, 0.5	R + H	Pilocarpine nitrate	39
Hydroxypropylcellulose	0.35–3.4	H	Sodium fluorescein	41
	1.2, 5.0	R + H	Pilocarpine nitrate	47
Hydroxypropylmethylcellulose	0.5	H	Nonmedicated hydrogel	48
	NS	R	Cromolyn sodium	49
Sodium carboxymethylcellulose	1.0	R	Timolol	50
	1.63	R + H	Tropicamide	40

Notes: NS, not specified; H, human; R, rabbit.

value) nonthixotropic (none or negligible hysteresis loop in the rheological curves) flow has been described (44,51–55). Carbopol gels have shown elastic properties (56).

Several mechanisms have been suggested to account for the formation of Carbopol hydrogels. Carbomer molecules in their dry state are tightly coiled, thus limiting their thickening capability (Fig. 3). When dispersed in water, the molecules begin to hydrate and partially uncoil. A cloudy acidic dispersion of low consistency is formed (51). Neutralizing the carbomer resin with a suitable base generates negative charges along the polymer backbone. Repulsion of these negative charges causes the polymer to completely uncoil in an extended rigid structure of high viscosity. Neutralization increases consistency and reduces turbidity. Another method for the thickening of Carbopol gels without neutralization involves hydrogen bonding of solvent molecules along the polymer chain to produce a rigid chain (57). The pH of such systems tends to be acidic. A third mechanism explaining the hydrogel formation involves the absorption of water by the crosslinked polymer to form a gel consisting of macroscopic, swollen particles (58,59). There is no evidence for any one mechanism, and it seems likely that gelation occurs by a combination of these processes (51).

The simplest way to prepare Carbopol hydrogels is to disperse the resin in water at room temperature and to neutralize the polymer with bases such as triethanolamine, sodium hydroxide, or even an active alkaline compound. The other method to form carbomer hydrogels is to add hydroxyl donors such as polyols (e.g., glycerine, propylene glycol), sugar alcohols (e.g., mannitol), or polyethylene oxide to carbomer dispersions.

Carbomers present many advantages for the ocular drug therapy, including (54,55,60):

Good flow and good thickening properties at low concentration
Relatively good patient compliance
Excellent appearance
Compatibility with many active ingredients
Possibility to neutralize the polymer with a drug
Bioadhesive properties

The bioadhesive properties are very useful for a controlled drug delivery to the eye because they enable a prolonged contact time between the eye and the drug and because they delay drug elimination (61). The mucoadhesion of carbomer involves several mechanisms such as chain interpenetration of the polymer in the mucus (since both carbomer and mucin are macromolecular expanded networks and both exhibit a significant hydration in aqueous media), hydrogen bonds (both have carboxyl groups), electrostatic interactions (both are ionic polymers), and hydrophobic interaction (both polymer have various hydrophobic functional groups) (62,63). Besides carbomer, other polymers having good mucoadhesive properties include hydroxypropylcellulose, carboxymethylcellulose, carbophil, gelatin, and hyaluronic acid (54,64). Davies and coworkers (65) have emphasized the importance of mucoadhesion for improving the ocular bioavailability of pilocarpine in rabbits and shown the superiority of bioadhesive Carbopol 934P over the nonbioadhesive poly(vinyl alcohol) (PVAL). Dittgen et al. (66) also found Carbopol to be superior to PVAL.

Several authors have examined the stability of carboxyvinyl polymers at various storage conditions and demonstrated the good stability of Carbopol 940 and 941, whose viscosity decreased the least (67,68). Comparing various Carbopol types (910,934,940,941), Ünlü et al. (54) concluded that Carbopol 934 had the worst thickening properties and that Carbopol 940 had an excellent appearance and clarity when compared with the other types.

Many studies have explored the potential of carbomer in extending the corneal residence time and increasing the bioavailability of active compounds such as pilocarpine (43,44,69); some representative examples are listed in Table 3.

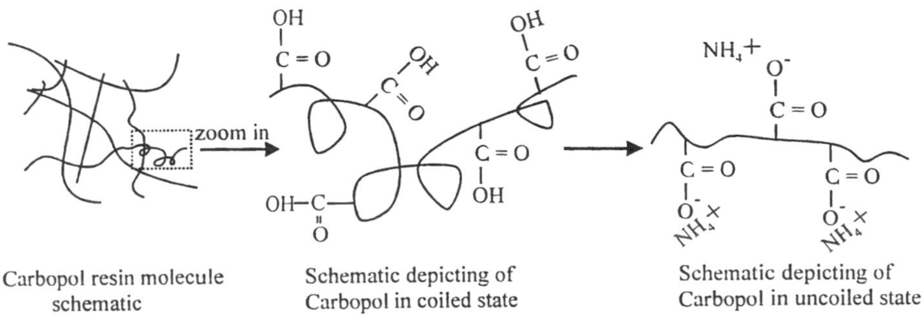

Carbopol resin molecule schematic

Schematic depicting of Carbopol in coiled state

Schematic depicting of Carbopol in uncoiled state

Figure 15.3 Schematic representation of the mechanism of the sol–gel transition of carbomers (From Ref. 51.)

Table 15.3 Examples of Topical Hydrogels Based on Carbopol

Carbopol type	Concentration (%)	Drug	Species	Effects	Ref.
940	0.05, 0.1, 0.2	Tropicamide	H	Increased bioavailability	70
940	0.3	Tropicamide	R	Increased bioavailability compared to an ointment and a HPC hydrogel	52
940 and 910	0.6 1.5	Pilocarpine	R + H	2–3 × increase of the drug bioavailability; effects more pronounced in humans than in rabbits	69
940	NS[c]	Prednisolone (acetate)	R	4.5× increase of the aqueous humor levels; 4× increase of the concentration in the cornea	71
934	NS[c]	Prednisolone (sodium phosphate)	R	5.5× increase of the aqueous humor levels; 10.6× increase of the concentration in the cornea	71
940	1; 2; 4; 6	Pilocarpine	R	Extended duration of miosis	43
941	6	Pilocarpine	R	Extended duration of miosis	43
940	1.0–4.0	Flurbiprofen	R	More increased bioavailability than with pluronic F 127 hydrogels	72
934	0.49	Pilocarpine	R	Increased bioavailability as compared to a 6% PVAL hydrogel	65
940	0.1, 0.2	Fluorescein	H	Increased corneal retention	53
940	0.9–5.0	Pilocarpine	R	Extended duration of miosis	44

Notes: H, human; R, rabbit; HPC, hydroxypropylcellulose, 0.7% w/v; NS, not specified.

A large number of commercial products are based on carbomer, including tears substitutes like Lacrinorm® with carbomer 940 (Chauvin, Montpellier, France) and eye drops with active compounds such as Pilopine® (Alcon).

c. *Sodium Hyaluronate*

Hyaluronic acid (HA) is a high molecular weight linear unbranched polysaccharide consisting of repeating disaccharide units of glucuronic acid and N-acetylglucosamine (73). It is a glycosaminoglycan and has a flexible open coil double helix conformation (74). The carboxyl groups of HA are dissociated at physiological pH, thus conferring a polyanionic character to the polymer. In the human body, HA plays an important role in retaining water in the intercellular matrix of connective tissues. It is present in diverse tissues including skin, tendons, cartilage, synovial fluid, and umbilical cords, but also in ocular fluids, particularly in the vitreous and the aqueous humor. Thus, it has been suggested that HA might be used in eye surgery as a substitute for aqueous humor or vitreous either to maintain the shape of the anterior chamber during cataract surgery or to protect corneal endothelium during lens implantation (75).

Until recently, HA had been produced by tissue extraction from umbilical cord of rooster comb, which made its ophthalmic solutions very expensive. However, HA can now be produced by microbial fermentation (76). A highly purified sterile form can be produced in bulk with a variety of molecular weight ranges and at a cost considerably less than the tissue-extracted material. Healon® (Kabi Pharmacia, Sweden) is a patented ultrapurified fraction of the sodium salt of HA (77).

The use of hyaluronic acid in ocular therapy has been reviewed by Bernatchez et al. (78). The concentration of HA commonly used in eye drops varies between 0.1 to 0.3%.

Sodium hyaluronate solutions exhibit a typical pseudoplastic rheological behavior with a low viscosity at high shear rate (46). Its viscosity is shear dependent, and the hydrogel exhibits a drop in viscosity at high shear rates such as during blinking (shear thinning). This rheological property is particularly advantageous in ophthalmic applications because the high viscosity hydrogel can be administered with the advantage of longer ocular surface retention times but without the associated sensation of a foreign body, blurred vision, and patient discomfort as encountered with high viscosity Newtonian solutions (79). Furthermore, the low viscosity of HA hydrogels favors a homogeneous distribution on the whole corneal surface after each blink.

The performance of HA in extending the precorneal residence time of several ophthalmic drugs such as pilocarpine, gentamicin, and tropicamide is summarized in Table 4. The HA 0.25% vehicle led to a twofold increase of the bioavailability of topically administered gentamicin in

Table 15.4 Ophthalmic Formulations Based on Sodium Hyaluronate

Concentration (% w/v)	Drug	Species	Effects	Ref.
0.25	Gentamicin sulfate	H	Increased bioavailability for at least 10 min	80
0.125, 0.2, 0.75	Pilocarpine HCl	R	Extended duration of miosis; increased corneal residence time and intraocular absorption	81, 82
5.0, 15.0	Pilocarpine nitrate	R	Prolonged miosis	83
	Tropicamide		Prolonged mydriasis	
	Sodium fluorescein		Extended corneal residence	
	Nonmedicated gel		Good mucoadhesion	
0.125, 0.25	Nonmedicated gel	H	No effect of the 0.125% gel on the ocular clearance; extended corneal residence with 0.25%	84
0.125	Nonmedicated gel	R	Extended corneal residence	85
0.125, 0.25	Nonmedicated gel	H	Extended corneal residence	84
0.1, 0.19, 0.25	Sodium fluorescein	H	Extended corneal residence	46

Notes: H, human; R, rabbit; HA; hyaluronic acid (sodium salt).

the human eye for at least 10 min after instillation compared to a saline isotonic buffered solution (80). Gamma-scintigraphic studies of a nonmedicated HA hydrogel also showed an extended corneal residence time at the surface of the eye (84,85).

d. Poly(Vinyl Alcohol)

Poly(vinyl alcohol) is a synthetic long-chain polymer obtained by condensation of vinylacetate and partial or full hydrolysis of the resulting poly(vinyl acetate) (86). It is available in various grades that differ in the degree of polymerization and residual acetate content, thus providing different solubilities (87). The molecular weight average has a significant influence on the viscosity: very high molecular weight grades show a very marked change in viscosity for a small change in concentration, whereas low molecular weight grades have little influence on the viscosity of the solution (76). Aqueous solutions of PVAL approximate a Newtonian rheological behavior (47,88).

Since PVAL was first introduced for ophthalmic use in the early 1960s (89) a lot of time and effort have been expended on exploring the influence on ocular bioavailability of various PVAL hydrogels compared mostly to other hydrogels such as cellulose preparations. Some representative examples of ophthalmic formulations based on PVAL and their subsequent effects on ocular bioavailability are presented in Table 5. There have been many contradictory results about the superiority of PVAL in prolonging the corneal contact time. On one hand, PVAL (1.4%) was reported to be superior to MC (0.5%) based on rabbit studies (89). On the other hand, PVAL (1.4%) has exhibited a significantly shorter elimination time than HPMC (0.25–2.5%) (90). Other authors concluded that cellulose derivatives were superior to 1.4% PVAL (48,91,92). The rationale for these conflicting findings is explained by differences in flow properties and in the viscosity of the tested hydrogel. Indeed, when comparing hydrogels of different polymers, it is important to take into account not the concentration used of the polymers, but rather their viscosity. The importance of the viscosity was pointed out by Patton and Robinson (88), who demonstrated that two nonmucoadhesive hydrogels based on PVAL and MC with Newtonian rheological behavior and with the same viscosity range show comparable effects on the ocular residence time.

Attempts have been made to find the optimal concentration of PVAL in terms of decreased drainage of the hydrogel from the precorneal area. Patton and Robinson (88), testing PVAL concentration range between 0 and 5.0%, concluded that a 3% PVAL solution was the optimum because beyond this concentration large increase in viscosity resulted in only minor decreases in drainage rate and also resulted in a blockage of the drainage duct. However, the concentration of PVAL commonly used commercially is 1.4%, the rationale for this choice of concentration is the visual comfort, concentrated solutions leading to the sensation of foreign body and blurred vision (88). Kassem et al. (96) have reported that PVAL was the most effective vehicle, compared to other cellulosic derivatives, for increasing the ocular bioavailability of phenylephrine, a drug lowering the intraocular pressure (IOP) and added to a betamethasone preparation in order to balance the side effect of the steroid (elevation of the IOP).

Table 15.5 Effects of Polyvinyl Alcohol on the Precorneal Residence Time of Hydrogel Loaded or Not with Drugs Compared to Saline Solution

Concentration (%)	Type of PVAL	Drug	Species	Effects	Ref.
0–10.0	Various grades (20–90)	Pilocarpine nitrate	R	Delayed corneal drainage rate; prolonged aqueous humor levels	88
1.0–5.0	Polyviol W40/140	Nonmedicated (tracer ^{99m}Tc)	R	Prolonged residence time	93
1.2	High molecular weight (90 000)	Pilocarpine nitrate	R + H	Extended miosis time	47
1.4	NS	Nonmedicated (tracer ^{99m}Tc)	R + H	Diminution of the drainage rate	45
1.4, 2.8, 4.2		Sodium fluorescein	H	Prolonged residence time; 4.2%: discomfort, blurred vision	94
3.0–6.5	Polyviol W40/140	Nonmedicated (tracer ^{99m}Tc)	R + H	Prolonged residence time (more pronounced in rabbit than in man)	95
6.0	Fully hydrolyzed (MW 86 000)	Pilocarpine nitrate	R	Delayed corneal drainage rate; increased bioavailability	65
4	NS	Sodium fluorescein	H	Delayed corneal drainage rate	42

Notes: PVAL, polyvinylalcohol; R, rabbit; H, human; NS, not specified.

Poly(vinyl alcohol) is a common viscolyser used in many marketed eye drops such as Tears Plus® (Allergan), and Hypotears® (Ciba Vision).

e. Other Polymers

Various natural and synthetic polymers have been investigated as potential vehicles for ocular drug delivery and are still in development. These polymers include poly(vinyl pyrrolidone) (PVP), chondroitin sulfate, carrageenans, dextrans, xanthan gum, and chitosan. The synthetic polymer PVP has found interesting applications in ophthalmic solutions, eye drops, and contact lens solutions (40,97,98). They will be briefly described. Chondroitin sulfate is a natural mucopolysaccharide (glycosaminoglycan) acting as a flexible connecting matrix in cartilage and connective tissues (99). It has been examined as a potential polymeric vehicle, but the results were not conclusive in that chondroitin hydrogels failed to enhance the bioavailability of pilocarpine and cyclopentolate (100). Carrageenans are a group of water-soluble sulfated galactans extracted from red seaweed. The main gel-forming carrageenans fractions are the kappa and iota, while the lambda is known as non-gelling, because of its higher level of ester sulfate. Carrageenan hydrogels exhibit pseudoplastic rheological behavior and viscoelastic properties (101). Verschueren et al. (101) have reported the good acceptability of carrageenan hydrogels in human volunteers. Xanthan gum is an anionic branched polysaccharide composed of glucose, mannose, and glucuronic acid. The polymer is obtained by fermentation of the bacterium *Xanthomonas campestris*. It has been examined for use in eye products and has shown interesting properties: Meseguer et al. (102–104) showed by gamma-scintigraphy that xanthan hydrogels are able to delay corneal clearance by tear flow in the human eye. Xanthan gum has been proposed as a polymer in artificial tears (105). Chitosan is a polycationic natural polymer which exhibits favorable properties for ocular administration such as biodegradability, nontoxicity, biocompatibility, and bioadhesion (106–108). Gamma-scintigraphic evaluation has shown the superior efficiency of chitosan for prolonging precorneal residence time of tobramycin with respect to a commercial collyrium containing the same antibiotic at a concentration of 0.3% (Tobrex®) (Fig. 4) (109).

2. In Situ Forming Gels

Gelling systems, also called phase transition systems, consist of a mixture of the drug with a polymeric material that has low viscosity in the dispensing bottle but changes to a very viscous solution or a gel once in contact with the eye (110). The sol–gel transition of in situ activated gel-forming systems occurs as a result of a chemical/physical change induced by the physiological ocular environment. The transition can by triggered by a shift in the pH, by changing the temperature, or by the presence of cations in

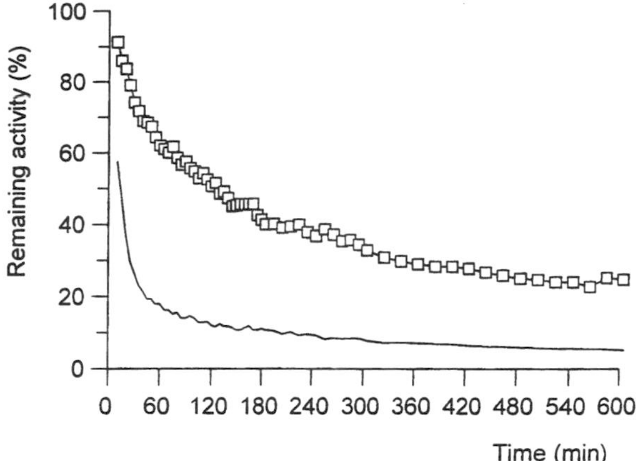

Figure 15.4 Precorneal drainage of ^{99m}Tc-DTPA in formulations containing 0.3% tobramycin and 0.5% of high molecular weight chitosan (□) as well as Tobrex® (−) (From Ref. 109.)

tears (Fig. 5). It is assumed that the formation of the gel in the conjunctival cul-de-sac could lead to a prolonged release system for the drug (111).

a. Gelling Activated by pH Change

The concept of using hydrogels based on pH-sensitive polymers for the ocular route began in the 1980s (112). These hydrogels are based on latex. The term latex defines a low viscosity, stable polymeric dispersion of particles, generally in an aqueous solvent (113). According to the manufacturing method, latexes are globally classified in three categories: natural latexes like rubber, synthetic latexes obtained by polymerization of insoluble monomers emulsified in water, and artificial latexes, also called pseudolatexes, which are prepared by dispersion of a pre-existing polymer in an aqueous medium (114). In situ gelling pseudolatexes are pH-sensitive aqueous colloidal dispersions which undergo spontaneous coagulation and gelation

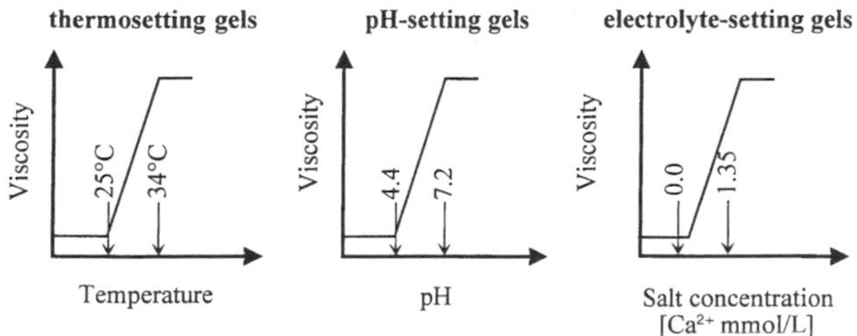

Figure 15.5 Schematic viscosity behavior of in situ activated gel forming systems: thermosetting gels, pH-setting gels and electrolyte-setting gels (From Ref. 112.)

after instillation in the conjunctival cul-de-sac because of the pH increase (85,115).

The preparation of ophthalmic pseudolatexes involves an emulsification of an organic solvent containing the polymer with an aqueous solution leading to an oil-in-water (O/W) emulsion (112). There are two general methods of preparation (Fig. 6) (116,117). In the first method, the pseudolatexes result from an evaporation of solvents (removal of all organic solvent and a fraction of water). The internal phase of the emulsion contains a mixture of organic solvents not miscible with water and the polymer. The external phase is composed of an aqueous solution with an emulsifier. The bioactive material, e.g., pilocarpine, is added to the dispersion, where it is partially adsorbed onto the polymer. The drug can also be introduced at the beginning of the preparation, in one or other phase before emulsification. In the second method, the dispersion is prepared by a salting out process: the gradual addition of water breaks the equilibrium between the internal phase of the emulsion based on polymer in organic solvent miscible with water and the external phase containing salts in high concentration and hydrocolloids, so that polymeric particles are formed. In this second case, the bioactive material, e.g., pilocarpine, is blended either to the organic phase before emulsification or to the dry powder obtained after lyophilization of the particles (118).

The gel-forming polymers have to be carefully selected with respect to their physicochemical properties and biocompatibility. Cellulose acetate phthalate (CAP) has been shown to be suitable for pseudolatexes containing pilocarpine for several reasons: the compatibility of the polymer with the drug, the stabilizing effect of the polymer at relatively low pH avoiding hydrolysis of pilocarpine, and the relatively low buffer capacity of the polymeric system in the conjunctival cul-de-sac (112). The latex formulation described has a total polymer content (CAP) of 30% w/v and an average particle size of 250 nm. The CAP pseudolatexes form a free-running dispersion of acidic pH (pH = 4.4) and coagulate within a few seconds when in contact with the lachrymal fluid (pH = 7.2). Gamma-scintigraphy was used to monitor formulations based on pilocarpine–CAP pseudolatexes and showed a tenfold increase in precorneal residence time (residence half-life = 400 s) compared to a solution of pilocarpine (residence half-life = 40 s) (110,119). The pharmacological response of pilocarpine pseudolatexes has also been evaluated in terms of miosis (diminution of the pupil diameter), and it has been concluded that the bioavailability of the drug was substantially increased, the pseudolatexes maintaining a constant miosis in the rabbit for up to 10 h compared to 4 h with pilocarpine eye drops (120).

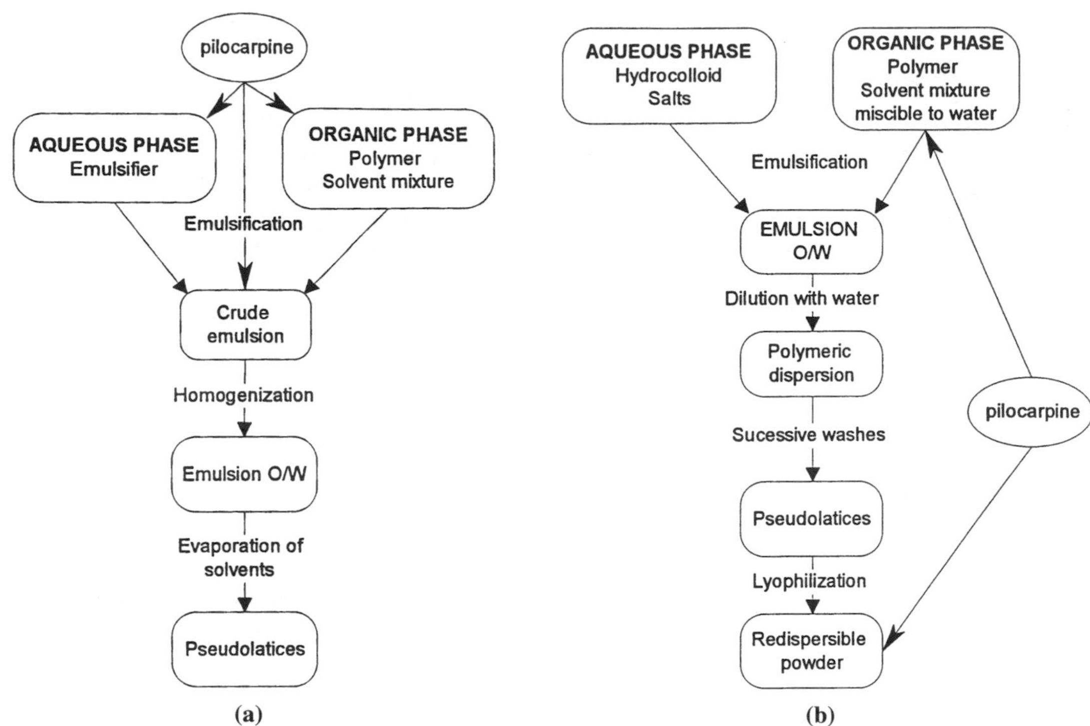

Figure 15.6 Methods for the preparation of pseudolatexes for the ocular route: (a) evaporation of organic solvent in the internal phase of an O/W emulsion; (b) dilution of the external phase of an O/W emulsion. (From Ref. 112.)

It should be underlined that the long-acting latex, once coagulated, has no impact on vision (112), although a sensation of discomfort seems to be unavoidable immediately after administration as in the case of any preparation of suspension type.

b. Gelling Activated by Temperature Change

Another method for causing phase transitions of the ophthalmic polymeric preparations on the ocular surface is based on thermogelation. Temperature-sensitive polymeric gels, also called thermoreversible hydrogels, are liquid at room temperature (20–25°C) and undergo gelation when at body temperature. This thermosensitive property makes these hydrogels useful in controlled drug delivery either in topical administration on the skin or mucosa (nasal, ocular, or rectal route) or in parenteral administration (121). Miller and Donovan (122) first mentioned this in situ gel formation for the ocular route. The most common thermosensitive polymers suitable for ophthalmic preparations are poloxamers.

Poloxamers are synthetic nonionic surfactants and are water-soluble triblock copolymers made of poly(oxy ethylene) (PEO) and poly(oxy propylene) (POP) (123,124). These polymers consist of a central hydrophobic nucleus of POP surrounded by hydrophilic sequences of PEO, conferring surface active properties to the copolymer. Chemically, poloxamers can be classified as polyethers of alcohol ethers, but generically they are referred to copolymer polyols. Different types of poloxamers exhibit different gelation properties and vary over a wide range of molecular weights and the relative proportions of the two constituents (125). Poloxamers are defined by a number, the first two digits of which, when multiplied by 100, correspond to the approximate average molecular weight of the POP portion of the molecule. The last digit, times 10, gives the percentage by weight of the PEO portion.

The series of poloxamers is commercially available as Pluronic® (BASF-Wyandotte) (124). The most commonly used poloxamer, not only in ophthalmology but also in dermatology and cosmetics, is poloxamer 407, also known as Lutrol F127 (BASF, Ludwigshafen, Germany). This is explained by several reasons: first, poloxamer 407 gives colorless and transparent gels, an important parameter for the ocular route; second, it has a very low toxicity and it is also inert toward mucosa; third, the polymer has a high solubilizing capacity, and, finally, poloxamer 407 requires the lowest concentration for gel formation (123,126,127).

The concentration at which poloxamer solutions exhibits reversible gelation depends on the type of polymer: poloxamers 184 to 188 form gels from concentrations of 50–60% w/v; poloxamers 234 to 288 form gels from concentrations of 40%; poloxamers 333 to 338 give hydrogels by a concentration of 25%; whereas poloxamer 407 forms a gel at only 20% concentration in water at 25°C. For a given type of polymer, as the concentration is increased, the gel becomes harder and harder, and the sol–gel transition temperature decreases (126,128). An increasing concentration of the polymer also results in an increased corneal contact time (128).

Several mechanisms have been proposed to explain the thermogelification, including micellar aggregation forming an intermicellar network and gradual desolvation of the polymer breaking the hydrogen bonds between water molecules and polymer oxygen atoms. But the most recognized mechanism is that thermogelification results from the interaction between the different molecules of poloxamer (124,129–131). The increase in temperature modifies the hydration spheres around the hydrophobic units of POP, which in turn induces greater interactions between these different units, leading to an increased entanglement of the polymeric network (130,132). A schematic representation of the gelation process proposed by Waring and Harris (133) is shown in Fig. 7.

The phenomenon of thermogelling is perfectly reversible: the viscosity of poloxamer solutions increases when the temperature is raised to the eye temperature (33–34°C) from a critical temperature (23–26°C) (130,134). The hydrogel reverts to a solution by placing the preparation at refrigerator temperature and gels again at ambient temperature. The gel can be cooled down and/or warmed up innumerable times with no change in properties (126). The rheological behavior of a poloxamer thermoreversible hydrogel at 20% was shown to be Newtonian below the transition temperature and non-Newtonian above this temperature. Furthermore, the viscosity of the poloxamer hydrogel is affected by the presence of additives: for example, preservatives such as benzoic acid and parabens produce a decrease in the poloxamer gelation temperature (126,135).

Hydrogels of poloxamer can be prepared by two methods. In the cold water process, the polymer is stirred into water at 5–10°C; gelling sets in on heating to room temperature. In the hot water process, the required amount of polymer is dissolved in water under gentle shaking at about 90°C. By this hot method, it takes about 4 h for the polymer to go completely into solution (126). Both methods yield gels with the same properties.

Saettone et al. (136) demonstrated that poloxamers are useful not only for modified ocular drug delivery, but are also outstanding ophthalmic solubilizing vehicles for poorly water-soluble drugs such as tropicamide. Miller and Donovan (122) have evaluated the pharmacological activity of pilocarpine nitrate–poloxamer 407 gels in rabbits and reported an enhanced activity of the hydrogel compared to a simple solution. This increased therapeutic effect was explained by the longer permanence and contact of the drug with the ocular surface. Conflicting results were obtained by Dumortier et al. (137), who showed that a thermoreversible poloxamer 407 hydrogel did not prolong the release kinetics of morphine over a simple morphine salt solution in rabbits. Comparing the anti-inflammatory effect of flurbiprofen hydrogel formulated in Carbopol 940 2% and Pluronic F127 20% of the same viscosity, Mengi and Deshpande (72) reported that Carbopol provided a more sustained action than hydrogel with poloxamer. They suggested that bioadhesion of carbopol could play an important role in prolonging the contact of the drug with the

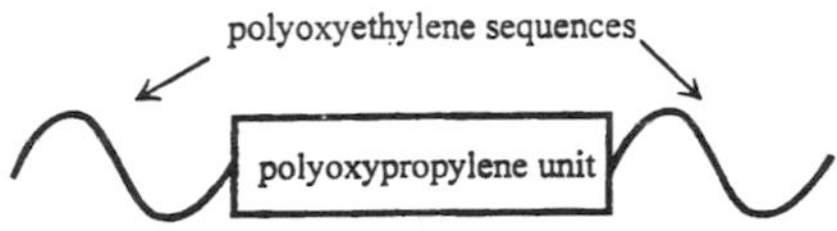

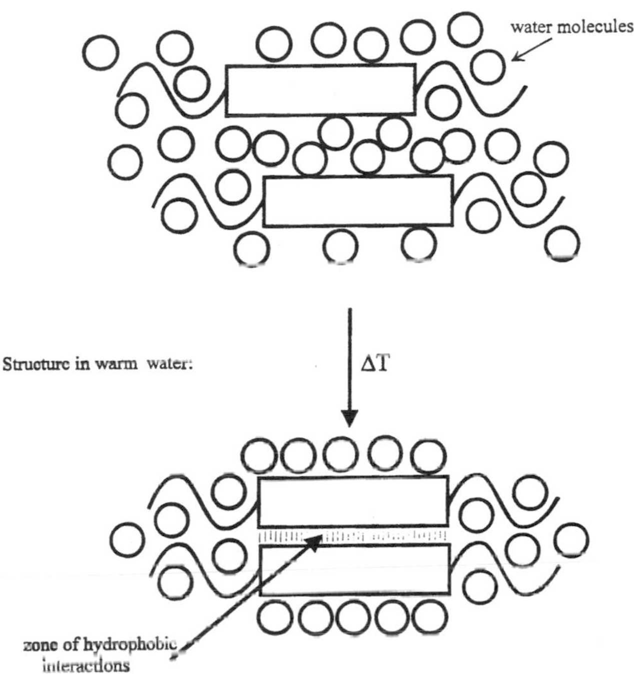

Figure 15.7 Scheme of sol–gel transition of poloxamer hydrogels (From Ref. 133.)

ocular surface. Some commercial products, particularly for contact lenses, are based on poloxamers including Clerz® and Miraflow® (Ciba Vision).

Some other ophthalmic applications of thermoreversible hydrogels have recourse to other polymers such as the poloxamine marketed as Tetronic® (138,139). These are copolymers derived from an ethylene diamine derivative (125).

Despite their undeniable qualities, thermoreversible gels are associated with an important disadvantage, namely, the risk of gelling before use by an increased ambient temperature due to improper packing or storage conditions.

c. *Gelling Activated by Electrolyte Composition Change*

A possible way to overcome the drawback of critical storage conditions of thermosensitive hydrogels is to use a more specific physiological parameter to trigger gelation, for example, the chemical composition of tears. Gellan gum shows interesting gelling properties in that it forms a gel in the presence of certain mono- or divalent cations present in tears, such as sodium, calcium, and potassium ions (140,141).

Gellan gum is a natural anionic extracellular polysaccharide obtained by fermentation of the bacteria *Pseudomonas elodea*. The anaerobic fermentation conditions are controlled to produce a high yield, the isolation of the polymer from the fermentation liquor is followed by a deacetylation step and a subsequent purification (142). Chemically, this heteropolysaccharide is a linear polymer with a tetrasaccharide repeating unit of L-rhamnose, D-glucose, and D-glucuronate partially O-acetylated in the molar ratios 1:2:1 (143–145).

Deacetylated gellan gum is available commercially under the tradename of Gelrite® (Kelco Division of Merck and Co), also called low-acetyl gellan gum. Gelrite is obtained by saponification and clarified by filtration (146).

The gelling mechanism of the deacetylated gellan gum is based on the formation of double-helical segments, which leads to a three-dimensional network and which is induced to form by cations (147–149). Divalent cations, in particular divalent metallic ions such as calcium and magnesium, can coordinate to carboxylate groups to stabilize the gel formed.

Gelrite is capable of forming an optically clear dispersion in water at low concentrations (<2% w/v), whose apparent viscosity can be easily increased by increasing the pH or the polysaccharide concentration (150,151). Another parameter that influences the phase transition is the nature and the concentration of cations. Mono- as well as divalent cations promote the gelation of the gellan gum. Sodium, a monocation whose concentration in tears approximates 113 mmol/L, causes the gelation of the material when topically instilled into the conjunctival sac (152). However, divalent ions such as magnesium (0.45 mmol/L in tears) or calcium (1.35 mmol/L) are superior in promoting the gelation of the polymer (153,154).

Osmolality is another critical parameter that governs the salt uptake of gellan gum and hence the rate of gelling (155). Owing to an osmotic gradient between the hydrogel and the surrounding environment, hypotonic hydrogel undergoes a rapid gelling and prolongs ocular contact time longer than isotonic or hypertonic gels.

Finally, the presence of ionic additives in gellan gum hydrogel may affect the rheological behavior. For instance, benzalkonium chloride, a common cationic preservative for ophthalmic preparations, increases the elasticity, probably by forming micellar aggregates at the oppositely charged carboxylic groups of the polysaccharide which participate in the intermolecular network of the gel (155).

Heat treatment including autoclaving for sterilization enables the full dispersion of the polymer without loss of apparent viscosity: the hydrogel then has a good thermal stability (150). Thus, the usual method to prepare gellan gum hydrogel (containing generally 0.6% of polymer) consists of dispersing the polymer in deionized water and autoclaving the mixture at 120°C for 10 min to obtain a sterile optically clear hydrogel (152). Additives such as the buffer system (e.g., maleate buffer), tonicity adjusting agent (e.g., mannitol) or preservatives, (e.g., benzalkonium chloride) can be dissolved in water before autoclaving.

Gellan gum hydrogels show pseudoplastic flow properties devoid of a static yield value and little thixotropy (150).

Comparing the methylprednisolone (MP) ester of gellan gum with MP suspension in gellan gum hydrogel, Sanzgiri et al. (156) made the assumption that the release mechanism of the drug is influenced by three factors, namely, the dissolution of the drug, its diffusion through the polymeric network, and the possible erosion of the matrix.

The suitability of gellan gum hydrogel as an ophthalmic vehicle has been investigated in different randomized clinical trials. Levy and Alsbury (157) demonstrated that gellan gum prolonged the pharmacological action of timolol so that a single daily instillation of the hydrogel was comparable to a solution used twice a day to lower the intraocular pressure of glaucomatous patients. This finding is consistent with two other studies. Gunning et al. (158) reported that a Gelrite hydrogel slightly prolonged the duration of action of two new antiglaucoma drugs, and Vogel et al.

(159) observed a twofold decrease of the intraocular pressure in patients who were given a Gelrite-based hydrogel of timolol compared to simple eye drops.

The performance of formulations based on gellan gum, in terms of improved residence time or in terms of enhanced bioavailability, has been assessed either by noninvasive techniques such as fluorometry (155,160) and gamma-scintingraphy (102,161) or by invasive methods involving the dosage of drugs in different parts of the eye (154,162). All techniques showed the prolonged contact time and improved bioavailability of the drug formulated in gellan gum hydrogel when compared with other solutions containing the same drug. Meseguer (102) has demonstrated the performance of Gelrite in delaying the ocular clearance of pilocarpine hydrogel. Despite the extended ocular residence and enhanced ocular bioavailability, gellan gum preparations do not have pronounced side effects caused by systemic absorption.

Dickstein and Aarsaland (163) demonstrated that timolol formulated in gellan gum was significantly less absorbed than a timolol solution and caused less heart rate response.

Ophthalmic preparations containing Gelrite are marketed under the tradename of TimopticXE® (Merck Sharp & Dohme, France).

3. Artificial Tears

Besides their use in ocular drug delivery systems, polymers are also commonly employed as wetting agents for contact lens solutions to avoid friction between the eye and the contact lens and as tear substitutes, also called artificial tears, to replace natural lachrymal secretions in cases of dry eye syndrome or KCS. One of the important criteria for selecting a polymer for artificial tears is an extended residence time on the cornea that enables reduction in the instillation frequency of artificial tears. Sodium hyaluronate (SH) exhibits specific properties that makes it an interesting vehicle for the preparation of artificial tears. Indeed, SH has chemical, structural, and rheological similarities with mucin (164). Furthermore, SH has been shown to protect against corneal damage by benzalkonium chloride, a preservative often added in tear substitutes and well known for its potential cytotoxicity (165–169). Thus, SH has been proposed as a vehicle of choice in the preparation of artificial tears (99,170–172). Snibson et al. (173) demonstrated that the corneal residence times of 0.2 and 0.3% SH hydrogels were significantly prolonged for patients with dry eye syndrome compared to healthy volunteers, probably because the tear mucin in dry eyes is altered and binds more strongly to the SH.

Poloxamer 407 hydrogel, with its protective and mucomimetic properties, as well as its optical transparency, can also be used for the treatment of dry eye (133,174). Flow Base®, a preparation containing 18% of poloxamer 407 and sodium chloride together with potassium chloride, has been shown to provide clinical advantages as a tear substitute (133). Poly(vinyl alcohol) can lower the surface tension of water, reduce interfacial tension at an oil–water interface, and enhance tear film stability. These interfacial properties have led to the widespread use of PVAL in artificial tear preparation (42,89). It is, for example, contained in the artificial tears Tears Plus and Liquifilm Forte® (Allergan). Cellulosic derivatives are also used in medication for dry eye: Ultra Tears® (Alcon) contains HPMC; Murocel® (Bausli & Lomb, Rochester, NY) contains MC; and Celluvisc® (Allergan) is based on CMC Na.

4. Ocular Tolerance of Hydrogels

Ocular tolerance is an essential parameter in the development of ophthalmic preparations, which have to be safe, i.e., they neither cause any irritation nor lead to any toxicity. The assessment of ocular tolerance involves either in vitro methods or in vivo tests (175–177). The acceptability of ocular hydrogels in human volunteers can be evaluated on the basis of a subjective evaluation: the volunteers are asked to answer a standard questionnaire with a grading scale (41,46,178). The pain sensation, lacrymation, blurring of vision, irritation, blinking rate, and crusting around the eye are the main evaluated parameters. Visual acuity and reading speed are also indicative of discomfort (178).

Tolerance tests have shown that generally hydrogels are better tolerated than ointments by patients (155,179). Indeed, blurring of vision is common with ointments, whereas hydrogels cause a blurring of short duration even with hydrogels designed for prolonged drug delivery (158). For instance, Gelrite was reported to form an opaque material immediately after administration into the eye (156). However, it does not seem to cause undue irritation. Nelson et al. (180) investigated the ocular tolerance of gellan gum hydrogel in a double-masked, randomized, crossover study where the patient rated the medication using a visual analog scale at different time intervals. The gellan gum hydrogel was found to impair visual function more than the saline solution, but these effects were brief and transitory.

Ludwig and coworkers (53) reported that some human volunteers complained of blurred vision after administration of Carbopol hydrogels. These findings are consistent with a study on human volunteers that reported irritation with artificial tears based on a carbomer 940 gel (105). Despite these negative results based on a subjective appreciation, carbomer hydrogels do not induce evidence of irritation in rabbits (68).

The ocular tolerance of hydrogels can be related to their rheological behavior. Polymer solutions generally exhibit either a Newtonian behavior or a non-Newtonian behavior (36). Hydrogels with Newtonian flow properties, where the viscosity is independent of the shear rate, are poorly tolerated in the eye because of the transmitted shearing forces associated with blinking and rapid eye movements. Hydrogels displaying non-Newtonian properties, in which the viscosity decreases with increasing shear rate, show much better acceptance because they offer less resistance to the movement of the lid over the ocular globe than viscous Newtonian formulations (76,181).

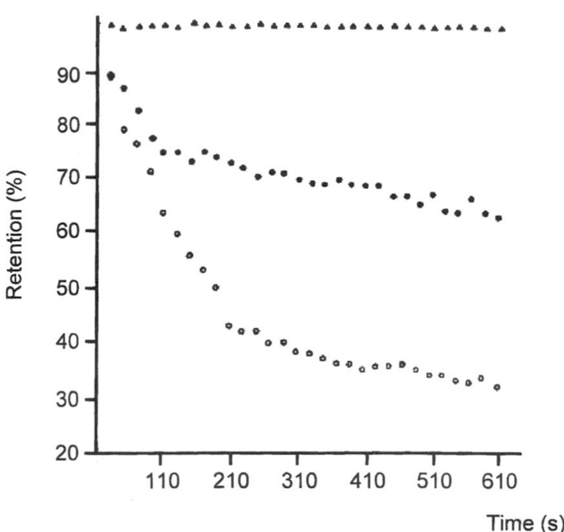

Figure 15.8 Precorneal retention of topically instilled particles: (●) 25 μL dose of a 3 μm suspension; (○) 50 μL dose of a 3 μm suspension; (▲) 25 and 50 μL doses of a 25 μm suspension (From Ref. 186.)

B. Dispersed Systems

Nano- and microparticles are polymeric drug carriers in the submicron and micron size range. The only difference between these particles is based on their size. Particles in the micrometer size range (>1 μm) are called microparticles or microspheres, whereas those in the nanometer range (<1 μm) are called nanoparticles (182). These systems were first developed to resolve solubility problems of poorly soluble drugs, as well as for long-acting injectable formulations and for improved drug targeting (182). An important potential use for these systems was to improve classical aqueous eye drop formulations, which have several disadvantages such as a rapid elimination of the drug from the precorneal site (183). Consequently, colloidal suspensions were designed to combine ophthalmic sustained action with the ease of application of eye drop solutions. Upon topical instillation, these particles are expected to be retained in the cul-de-sac and the drug to be slowly released from the particles through diffusion, chemical reaction, or polymer degradation or by an ion exchange mechanism (184). When using particulate systems for topical ocular purposes, an upper size limit of 10 μm is generally respected in order to prevent patient discomfort (185). The instillation of larger particles could increase lacrymation rate and hence accelerate their elimination from the precorneal area with the consequence of dramatically reducing drug bioavailability. However, Sieg and Triplett (186) demonstrated in the early 1980s that a smaller size limit should also be considered. In fact, comparing the performances of polystyrene microspheres of different sizes by scintigraphy they found that the precorneal elimination of small particles (3 μm) (1) was dependent on the volume instilled and (2) had an accelerated washout phase when compared to larger particles (25 μm) (Fig. 8) (186).

Among the several manufacturing techniques generally used for preparing particulate systems (e.g., emulsion polymerization, desolvation of proteins, heat denaturation, electrocapillary emulsification, and interfacial polymerization), emulsion polymerization is the most common for producing nanoparticles for ophthalmic drug delivery (183). The active principles are dissolved, entrapped, encapsulated, or linked to these colloidal carriers (187).

To date synthetic polymers such as poly(alkyl cyanoacrylates) (PACA), polyacrylamides, and polymethylmethacrylates (PMMA) have been most frequently used for the preparation of colloidal formulations for ocular applications (188). The polymer PACA has been the most extensively studied since it has certain advantages over the other two: it is well tolerated, biodegradable, and mucoadhesive (185,188), and it is not histotoxic (189). Poly(alkyl cyanoacrylate) nanoparticles are rapidly biodegrade with a degradation rate depending on the chain length of their ester side chain (182). Two metabolization pathways may take place, depending on the environmental conditions of the nanoparticles. The first possible mechanism corresponds to an erosion of the polymer leading to the formation of formaldehyde (190). The second way of metabolization is the hydrolysis of the ester bonds resulting in a water-soluble polymer backbone and a corresponding alcohol (191) (Fig. 9). The latter mechanism is predominant in vivo due to the presence of esterases, which is quite convenient considering the toxicity of formaldehyde produced during the first degradation process.

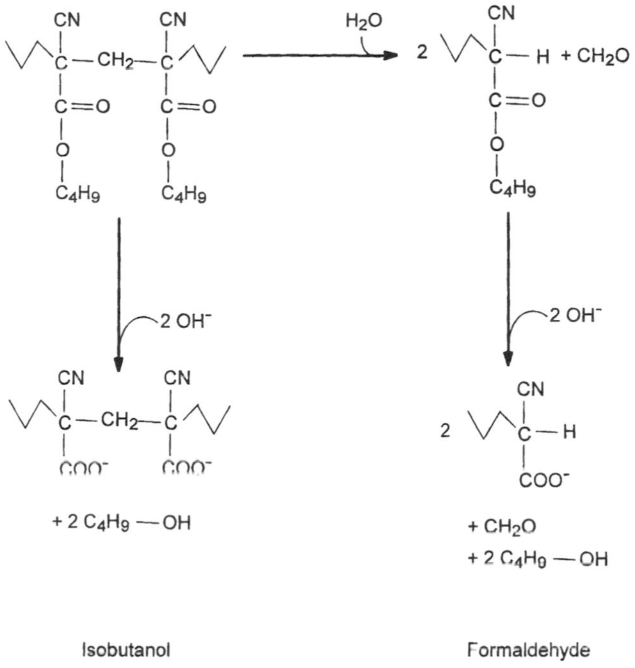

Isobutanol

Formaldehyde

Figure 15.9 Poly(isobutyl cyanoacrylate) degradation mechanisms (From Ref. 191.)

Besides acute and subacute tests of toxicity, acrylic polymers have also been evaluated regarding their tolerability after topical application to the eye. It was concluded that empty poly(butylcyanoacrylate) nanoparticles in saline did not cause any adverse reactions to human volunteers (185). Some other acrylic derivatives, i.e., poly(isobutylcyanoacrylate) and poly(hexylcyanoacrylate), have also been shown to be nontoxic after topical iterative application into rabbit eyes (185).

As shown in Table 6, the main field of application of colloidal systems is in glaucoma therapy, studying in vitro or in vivo drugs such as betaxolol, metipranolol, and especially pilocarpine. Evaluating the release of pilocarpine in vivo, Harmia et al. (192) pointed out the importance of the manufacturing method for the pharmacological efficacy of the drug. In fact, they compared the behavior of two particulate systems differing by the localization of pilocarpine, which was either adsorbed or incorporated into poly(butylcyanoacrylate) nanoparticles. Their results showed that the miotic activity was prolonged from about 200 to 270 min when compared with a simple solution for the nanoparticles with adsorbed pilocarpine unlikely for nanoparticles with the incorporated drug. Such an improvement of the miotic response using nanoparticles instead of conventional collyrium was confirmed some years later (Fig. 10) (193). However, regarding the performance of the same formulations in terms of bioavailability of the drug, it appeared that this parameter was only slightly increased, by a factor 1.2 (193).

So far, two nanoparticulate systems have been developed for ophthalmic drug delivery under the trade names of Piloplex® and Glaupex®, both containing pilocarpine (185). In the Piloplex delivery system, pilocarpine is bound to charged poly(methylmethacrylate–acrylic acid) copolymer particles, whereas Glaupex is based on PACA nanoparticles.

In conclusion, most of the results subsequent to the use of colloidal systems for topical ocular application have been promising. However, the commercial development of these products remains limited for various reasons, including (1) stability problems after sterilization, (2) only slight improvement of the pharmacological and pharmacokinetical performances respective of the long manufacturing processes involved when compared to classical solutions.

C. Inserts

Ophthalmic inserts are defined as preparations with a solid or semisolid consistency, whose size and shape are especially designed for ophthalmic application (i.e., rods or shields) (221). Historically, the first solid medication, precursors of the present insoluble inserts, consisted of squares of dry filter paper, previously impregnated with drug solutions (e.g., atropine sulfate, pilocarpine hydrochloride) (222). Small sections were cut and applied under the eyelid. Later, lamellae, the precursors of the present soluble inserts, were developed. They consisted of glycerinated gelatin containing different ophthalmic drugs (222). However, the use of lamellae ended when more stringent requirements for sterility of ophthalmic preparations were enforced. Nowadays, growing interest is observed for ophthalmic inserts as demonstrated by the increasing number of publications in this field in recent years. The uses of ocular inserts have been extensively reviewed by Felt et al. (4), Bawa (223), Saettone et al. (222,224,225), Khan et al. (226), and Shell (227,228). The patent literature on ophthalmic inserts has been reviewed by Gurtler and Gurny (221), while Baeyens et al. (3) have described the use of inserts as ocular drug delivery devices in veterinary medicine.

Inserts are placed in the lower fornix and, less frequently, in the upper fornix or on the cornea. They are usually composed of a polymeric vehicle containing the drug and are mainly used for topical therapy. Despite several advantages (3) such as accurate dosing, absence of preservatives, and increased shelf-life, they have one significant disadvantage, their solid consistency, which means that they are perceived by patients as a foreign body in the

Table 15.6 Some Polymers Used for the Preparation of Micro- and Nanoparticles for Topical Ocular Administration

Polymer	Drug	Ref.
Albumin	Hydrocortisone	194
	Pilocarpine	195–198
Chitosan	Acyclovir	107
Gelatin	Pilocarpine	195
Polyamide	Pilocarpine	199
Poly(ε-caprolactone)	Betaxolol	200
	Carteolol	201
	Indomethacin	108, 202, 203
	Metipranolol	204
Poly(lactic acid)	Chloramphenicol	205
	Pilocarpine	206
Poly(lactic-co-glycolide)	Betaxolol	200
Polyphthalamide	Pilocarpine	207
Poly(alkyl cyanoacrylate)	Timolol	208
Poly(butyl cyanoacrylate)	Amikacin	209
	Pilocarpine	192, 193, 210–213
	Progesterone	214
Poly(hexyl cyanoacrylate)	Pilocarpine	193, 212
	Progesterone	215
Poly(isobutyl cyanoacrylate)	Betaxolol	200, 216
Poly(methyl methacrylate)	Pilocarpine	212
Poly(methyl methacrylate-*c*-acrylic acid)	Pilocarpine	217–220

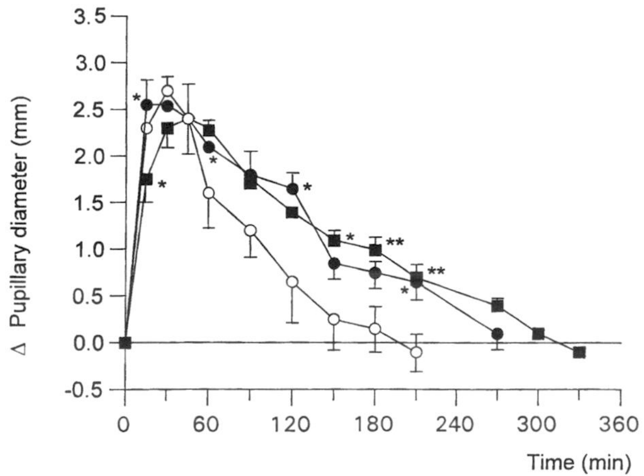

Figure 15.10 Miotic effect of 2% pilocarpine formulations in untreated rabbits with normotone IOP; (○) aqueous reference solution; (●) nanoparticle suspensions (Poloxamer 188/BCA ratio 1.2:1); (■) nanoparticle suspensions (Poloxamer 188/BCA ratio 1.2:8) (From Ref. 193.)

eye (224). Besides the initial discomfort, other potential disadvantages arising from their solid state are possible movement around the eye, occasional inadvertent loss during sleep or while rubbing the eyes, and interference with vision and difficult placement (and removal in the case of insoluble inserts) (224). Most of the ongoing research is therefore dedicated to improving ocular retention and to ensure an easy placement by the patient, while reducing the foreign body sensation in the eye.

Ophthalmic inserts are generally classified according to their solubility behavior and their possible bioerodibility, depending on the polymer(s) incorporated in the device.

1. Soluble Inserts

Inserts described in the literature are most frequently based on soluble polymers (Table 7). Their main advantage relies on their complete solubility compared with their insoluble counterparts, so that they do not need to be removed from the eye after therapy. They are usually divided into two categories according to the origin of the polymer. The first type is based on natural polymers, and the second is derived from synthetic or semisynthetic polymers. Drug release from soluble inserts is generally characterized by a

Table 15.7 Currently Investigated Polymers for Soluble Ocular Inserts

Carrier	Drugs	Main conclusions	Species	Ref.
Alginate, MC	Pilocarpine	Increased duration of the miotic activity as compared to solutions in the presence of MC.	R	230
Collagen	Gentamicin	3 h after administration, collagen insert gives the highest tear film and tissue concentration compared to topical ointment, aqueous solution, as well as subconjunctival application.	R	231
Collagen	Gentamicin	Concentrations in the precorneal tear film approximate the MIC during the 24 h after insertion.	B	232
Collagen	Gentamicin + dexamethasone	Collagen shields are comparable to the subconjunctival delivery of the same drugs over a 10-h period.	R	233
Collagen	Tobramycin	No significant difference between collagen and aqueous solution in the treatment of induced keratitis.	R	234
Copolymers of N-vinyl-pyrrolidone	Erythromycin	Complete suppression of a *Chlamydia trachomatis* infection. Inserts remain in the eye for 7 days.	M	235, 236
HA, HAE	Pilocarpine	Increased bioavailability of pilocarpine compared to a standard aqueous vehicle.	R	83
HPC	Pilocarpine	Delayed and decreased peak concentration in general circulation compared to aqueous solution.	R	237
HPC/D-lactose/glyceryl palmitostearate/Eudragit® RS	Pilocarpine	Increased lipophily as well as coating result in an increased pilocarpine bioavailability.	R	238
HPC/EC/Carbopol® 934P (BODI®)	Gentamicin	Concentrations in the precorneal tear film approximate the MIC during the 72 h after insertion.	R, D	239
HPC/EC/Carbopol 934P (Chrono-Logic®)	Gentamicin + dexamethasone	Release of dexamethasone and gentamicin over 24 and 48 h, respectively. Increased lacrymal availability for both drugs when compared to commercial solution.	R, D	240–242
HPC/Eudragit RS	Timolol	Coated inserts antagonized isoproterenol-induced ocular hypotension significantly more than Timolol eyedrops and uncoated inserts; sustained release of Timolol in tear fluid and decreased systemic peak concentration with coated and uncoated inserts compared to the aqueous solution.	R	243
HPC, PVAL, PVAL/Carbopol 940	Timolol	Drug release from the insert decreases in the order PVAL > HPC > PVAL/C940. Reduction in peak plasma Timolol concentration 2–5 times compared to the aqueous solution.	R	244
HPM	Tilisolol	Release by a non-Fickian mechanism.	R	245
PVAL (NODS®)	Chloramphenicol	Increased bioavailability with respect to standard eyedrop formulations.	R	246
PVAL (NODS)	Pilocarpine	Increased bioavailability with respect to standard eyedrop formulations.	R	246
PVAL/XG/HPMC/HA/GB/Eudragit RS 30 D	Pilocarpine	Coated inserts show a sustained release over 9 to 10 h; shift of the peak time to 120–240 min; over 3-fold increased AUC over the standard aqueous solution.	R	247
PVAL, HPC, EC, CAP, Eudragit	Dexamethasone	Increased concentration in eye tissues compared to suspension.	R	248
PVAL, HPC, PVAL/Carbopol 940	Timolol and prodrugs of Timolol	Release from the insert decreased in the order PVAL > HPC > PVAL/C940. Release rate much slower for Timolol prodrugs compared to the Timolol-containing inserts.	R	249
PVP	Pilocarpine	2.5-fold increased AUC over the aqueous solution.	R	250

Notes: CAP, cellulose acetophtalate; EC, ethylcellulose; GB, glyceryl behenate; HA, hyaluronic acid; HAE, ethyl ester hyaluronic acid; HPC, hydroxypropylcellulose; HPM, poly(2-hydroxypropyl methacrylate); HPMC, hydroxypropylmethylcellulose; MC, methylcellulose; MIC, minimum inhibitiory concentration; PVAL, poly(vinyl alcohol), PVP, poly(vinyl pyrrolidone). poly(vinylpyrrolidone); XG, xanthan gum. B, bovine; D, dog; M, monkey; R, rabbit; AUC, area under the curve.

diffusion process occurring in two steps (221,229). The first corresponds to the penetration of tear fluid into the insert, which induces a rapid diffusion of the drug and forms a gel layer around the core of the insert. This external gelification induces the second phase corresponding to a decreased release rate, again controlled by diffusion. The major problems of these soluble inserts are the rapid penetration of the lachrymal fluid into the device, the blurred vision caused by the solubilization of insert components, and the glassy constitution of the insert increasing the risk of expulsion.

a. *Natural Polymers*

Natural polymers include collagen, which was the first ophthalmic insert excipient described in the literature. Collagen-based inserts were first developed by Fyodorov (223,251) as a corneal bandage used after surgical operations and various eye diseases. Collagen shields were later suggested also as potential drug carriers (231). The therapeutic agent is generally absorbed by soaking the collagen shield in a solution containing the drug. Once placed in the eye, the drug is gradually released from the interstices between the collagen molecules, as the collagen dissolves. Accordingly, the residence time of drugs (252) such as antibacterials (253,254), anti-inflammatory agents (255,256), antivirals (257,258), or combination drugs (233) was increased when compared to traditional eye drops. For example, Bloomfield et al. (231) compared the levels of gentamicin in tears, cornea, and sclera of the rabbit eye after application of a collagen insert, drops, an ointment or by subconjunctival administration. After 3 h, they found that the collagen insert gave the highest concentration of gentamicin in the tear film and in the tissues when compared with the other types of formulations. The corneal shields, currently available for clinical use, do not contain drugs and are designed specifically as disposable therapeutic corneal bandages (224). For example, Bio-Cor® (Bausch and Lomb, Clearwater, FL) is made of porcine scleral collagen, while Medilens® (Chiron Ophthalmics, Irvine, CA) and ProShield® (Alcon) are prepared from bovine corium tissue (258–260). The main problem arising from the use of collagen is that it may cause an inflammatory reaction in the ocular tissues. Also, if shields are not used in association with antibacterials, a secondary infection may occur (257). Nowadays, the use of such devices is controversial due to possible prion-based infection.

b. *Synthetic and Semisynthetic Polymers*

Ophthalmic inserts containing synthetic [i.e., poly(vinyl) alcohol (247,248)] and semisynthetic [i.e., cellulosic derivatives (230,237–239,243,247,248)] polymers are frequently described in the literature. This stems in part from their advantage of being based on products well adapted for ophthalmic use and their ease of manufacture by conventional methods, including extrusion (239), compression (238), and compression molding (261). Ethylcellulose, a hydrophobic polymer, can be incorpored in the formulation to decrease insert deformation and therefore prevent blurred vision (239,241,248). Regarding the risk of expulsion, several authors (239,241,242,244,249) have incorporated a carbomer (Carbopol) which is, at low concentrations, a strong and rather well-tolerated bioadhesive polymer. Recently, Baeyens et al. (240–242) and Gurtler et al. (239,262) have used cellulose acetate phthalate in combination with gentamicin sulfate to decrease drug solubility. Prolonged release of the drug above the minimum inhibitory concentration (MIC) was obtained for more than 50 h, while gentamicin incorporated without CAP was released over less than 24 h. Subsequently, a new insert (ChronoLogic®), providing release of gentamicin and dexamethasone at different rates was developed. The prolonged release of gentamicin—an antibacterial agent—is combined with the immediate release of dexamethasone—an anti-inflammatory agent—against structural damage that can be caused by the inflammatory reaction (Fig. 11).

The release rate can also be decreased by using a copolymer of methacrylic acid (Eudragit®) as coating agent (238,243,247). Saettone et al. (238) have observed in rabbits that Eudragit®RS-coated inserts containing pilocarpine induced a miotic effect of longer duration compared to the corresponding uncoated products (Fig. 12).

Lacrisert® is a soluble insert that was commercialized by Merck Sharp and Dohme in 1981 (223). The device

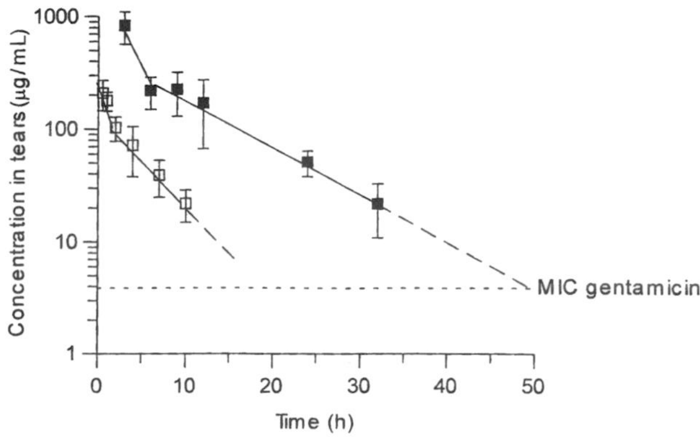

Figure 15.11 Combined release of gentamicin sulfate (■) and dexamethasone phosphate (□) at different release rates from a single insert containing 5.0 mg of gentamicin sulfate and 1.0 mg of dexamethasone phosphate (tested in rabbit, mean ± SEM, n = 6). (From Ref. 241.)

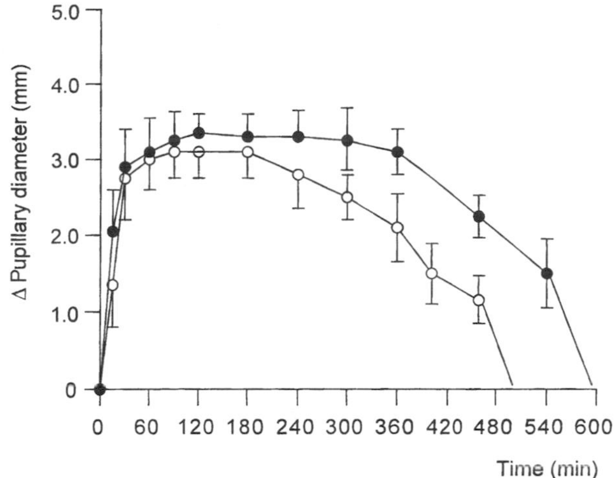

Figure 15.12 Miotic activity profiles of uncoated (○) and coated (●) ophthalmic inserts. (Adapted from Ref. 238.)

weighs 5 mg, measures 1.27 mm in diameter with a length of 3.5 mm, and is composed of hydroxypropylcellulose and is useful in the treatment of dry eye syndrome. The device is placed in the lower fornix where it slowly dissolves over 6–8 h to stabilize and thicken the tear film (263).

New Ophthalmic Delivery System (NODS®), originally patented by Smith & Nephew Pharmaceuticals Ltd. in 1985, consists of a medicated flag (4 mm × 6 mm; thickness, 20 μm; weight, 0.5 g) which is attached to a paper-covered handle by means of a short (0.7 mm) and thin (3–4 μm) membrane (246). All components (flag, membrane, and handle) are made of the same grade water-soluble poly(vinyl) alcohol. For use, the flag is touched onto the surface of the lower conjunctival sac. The membrane dissolves, rapidly releasing the flag which swells and dissolves in the lachrymal fluid to deliver the drug. When evaluated in vivo, NODS produced an increase in bioavailability for pilocarpine and chloramphenicol with respect to standard eyedrop formulations (246).

2. Insoluble Inserts

Inserts made up of insoluble polymer can be classified into two categories (242): reservoir and matrix systems. Each class of insert shows different drug release profiles. The reservoir systems can release drug either by diffusion or by an osmotic process. It contains, respectively, a liquid, a gel, a colloid, a semisolid, a solid matrix, or a carrier containing drug. Carriers are made of hydrophobic, hydrophilic, organic, natural, or synthetic polymers. The second category, matrix systems, is a particular group of insoluble ophthalmic devices mainly represented by contact lenses. The main disadvantage of the insoluble devices is their insolubility, since they need to be removed from the eye after treatment.

a. *Reservoir Inserts*

Reservoir inserts based on a diffusional release mechanism of the drug consist of a central reservoir of drug enclosed in specially designed semipermeable or microporous membranes allowing the drug to diffuse from the reservoir at a precisely determined rate. Drug release from such a system is controlled by the lacrimal fluid permeating through the membrane until a sufficient internal pressure is reached to drive the drug out of the reservoir. These diffusional systems prevent a continuous decrease in release rate by the use of a barrier membrane of fixed thickness, resulting in a zero order release pattern.

Ocusert® (developed by Alza Corporation, Palo Alto, CA) is undoubtedly the most commonly described insoluble insert in the literature (Fig. 13) (64,224,228,264–269). This flat, flexible elliptical device consists of a pilocarpine reservoir with alginic acid, a mixture surrounded on both sides by a membrane of ethylene vinyl acetate copolymer. The device is encircled by a retaining ring impregnated with titanium dioxide. The dimensions of the elliptical device are as follows: major axis, 13.4 mm; minor axis, 5.7 mm; thickness, 0.3 mm. Two types of Ocusert are available for humans: Pilo-20® and Pilo-40®, providing two different release rates for pilocarpine (20 and 40 μg/h, respectively) over a period of 7 days (264). In rabbits, Sendelbeck et al. (264) have compared the distribution of pilocarpine in ocular tissues after administration by eyedrop or by the Ocusert system. After the administration of eyedrops, pilocarpine levels in ocular tissues rose and fell within each 6-h intervals between eyedrops. On the other hand, pilocar-

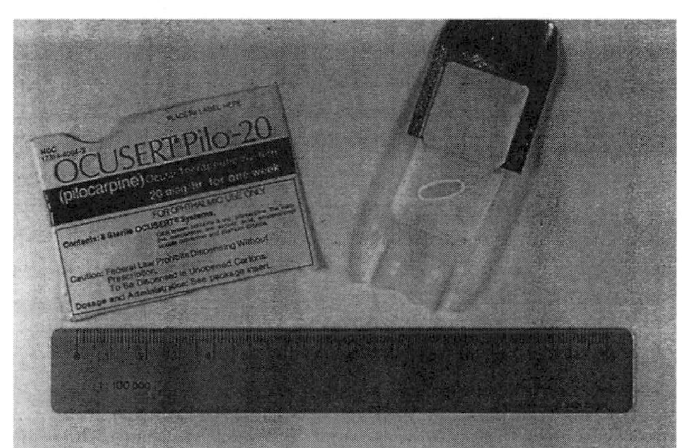

Figure 15.13 Ocusert® system.

pine levels remained constant over a 2- to 8-day period with the delivery system.

Reservoir inserts based on an osmotic release mechanism of the drug are mostly described in the patent literature, and in vivo tests are rarely reported (221). They are generally made up of a central core surrounded by a peripheral component. The central part can consist of a single reservoir or two distinct compartments (Fig. 14). The single reservoir contains the drug, with or without an additional osmotic solute dispersed through a water-permeable polymeric matrix (e.g., ethylene vinyl esters copolymers) so that the drug is surrounded by the polymer as discrete small deposits. If the central part is made up of two compartments, the drug and the osmotic solute are placed in separate chambers, the drug reservoir being surrounded by an elastic impermeable membrane and the osmotic solute reservoir by a semipermeable membrane (e.g., cellulose acetate derivatives). The peripheral part of these osmotic inserts comprises in all cases a covering film made of an insoluble semipermeable polymer.

The majority of patents deals with systems having a unique central reservoir (270,271). The release of drug from these inserts can occur via two release mechanisms, an osmotic and a diffusional release (221). When the insert is placed in the ocular environment, this starts the osmotic release, meaning tear fluid diffuses to the peripheral deposits through the semipermeable polymeric membrane, wets them, and induces their dissolution. The solubilized deposits generate a hydrostatic pressure against the polymer matrix triggering its rupture in the form of pores releasing the drug. This osmotic part of the release is characterized by a zero order release profile. Ideally, the drug is continuously

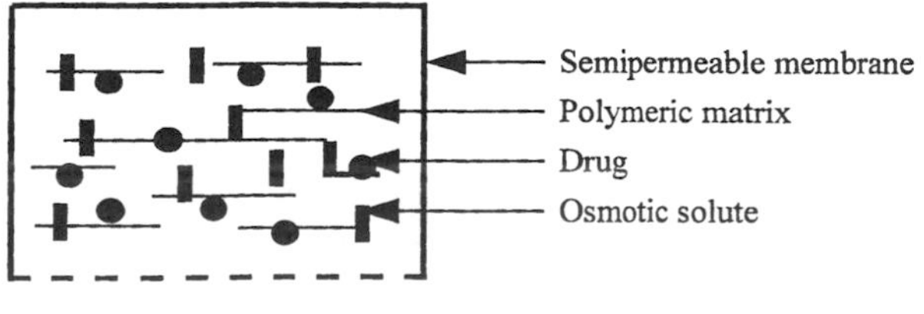

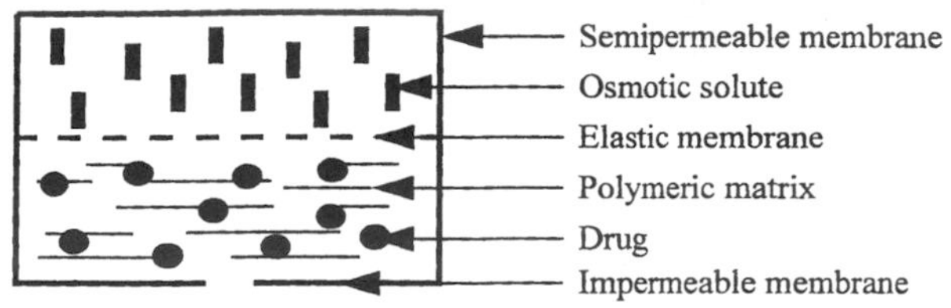

Figure 15.14 Schematic representation of two types of osmotic inserts.

released from the osmotic insert by the increasing formation of apertures in the device forming a lattice of composition dispensing paths in the polymer on all sides of the inserts. In fact, these paths are interconnected, forming tortuous microchannels of irregular shapes and size, causing a second drug release corresponding to a diffusional nonconstant release.

The release of drug from systems having a central reservoir subdivided into two compartments (272,273) starts as soon as the device is placed in the eye. Tears diffuse into the osmotic compartment, inducing osmotic pressure that stretches the elastic membrane and contracts the compartment containing the drug, so that the active compound is forced through the single drug release aperture. Thus, these systems are characterized by two distinct compartments and a single aperture having a very small diameter.

A constant zero order release can also be reached by using reservoir devices based on silicone polymers. The prolonged release of drugs can be controlled either by a diffusion process (274,275) or by an osmotic mechanism (276,277). Recently, Chetoni and coworkers (277) manufactured bioadhesive polydimethylsiloxane rod-shaped inserts maintaining lacrimal concentrations of oxytetracycline above the MIC for at least 3 days.

b. *Matrix Inserts*

As mentioned previously, the matrix insoluble inserts are mainly represented by contact lenses. The initial use of contact lenses was for vision correction. Their use has been extended as potential drug delivery devices by presoaking them in drug solutions. The main advantage of this system is the possibility of correcting vision and releasing drug simultaneously. Contact lenses are composed of a hydrophilic or hydrophobic polymer which swells by absorbing water. The swelling, caused by the osmotic pressure of the polymer segments, is opposed by the elastic retroactive forces arising along the chains as crosslinks are stretched until a final swelling (equilibrium) is reached.

Refojo (278) has proposed a subdivision in five groups of contact lenses, namely, rigid, semirigid, elastomeric, soft hydrophilic, and biopolymeric. Rigid contact lenses have the disadvantage of being composed of polymers [e.g., poly(methylmethacrylic) acid] hardly permeable to moisture and oxygen, a problem which has been overcome by using gas permeable polymers such as cellulose acetate butyrate. However, these systems are not suitable for prolonged delivery of drug to the eye and their rigidity makes them very uncomfortable to wear. For this reason, soft hydrophilic contact lenses were developed for prolonged release of drugs such as pilocarpine (279), chloramphenicol and tetracycline (280), and prednisolone sodium phosphate (281). The most commonly used polymer in the composition of these types of lenses is hydroxyethylmethyl metacrylic acid copolymerized with poly(vinylpyrrolidone) or ethylene glycol dimethacrylic acid (EGDM). Poly(vinylpyrrolidone) is used for increasing water of hydration, while EGDM is used to decrease the water of hydration.

Shell et al. (229) and Khan et al. (226) have shown that drug release from presoaked contact lenses was extremely rapid, with an in vivo residence time in general not longer than 24 h. In addition, preservatives, such as benzalkonium chloride, cannot be avoided and they have greater affinity with the hydrophilic contact lens material than for the aqueous drug solution (223). Bawa (282,283) has described other approaches to decrease drug release rate from contact lenses. These include the introduction of the drug into the monomer mixture followed by polymerization of the monomers in the presence of the drug. This procedure removes the need for preservatives and consequent eye sensitization, since the drug is added in the matrix as a solid (282,283). However, the main problem associated with all contact lenses is their high cost of manufacture and the difficulty of incorporating a precise amount of drug into the matrix. Disposable contact lenses have been commercially available for many years, and the continued progress made in polymer chemistry should facilitate the development of this type of insert in the future.

3. Bioerodible Inserts

These inserts are formed by bioerodible polymers (e.g., crosslinked gelatin derivatives, polyesters) which undergo hydrolysis of chemical bonds and hence dissolution (Table 8) (237,248,264,284,285). The great advantage of these inserts is the possibility of modulating their erosion rate by modification of their chemical structure during synthesis and by addition of anionic or cationic surfactants. A crosslinked gelatin insert was proposed by Attia et al. (248) in order to increase bioavailability of dexamethasone in the rabbit eye. The dexamethasone levels in the aqueous humor were found to be fourfold greater compared to a dexamethasone suspension. More recently, absorbable gelatin sponges (Gelfoam®) described in *USP XXIII* were used to deliver drugs to the eye (286–289). The device consists of a crosslinked gelatin sponge. The drug is incorporated in this polymeric system by soaking it in a solution containing the active compound. In vivo results showed that the Gelfoam device is more effective than conventional

Table 15.8 Currently Investigated Polymers for Bioerodible Ocular Inserts

Carrier	Drugs	Effects	Species	Ref.
Gelatin	Dexamethasone	Increased dexamethasone concentration in eye tissues compared to suspension.	R	248
Gelatin	Gentamycin	Ocular irritation.	B	285
Gelatin	Pilocarpine	2-fold increase in duration of effect as compared to pilocarpine administered as a solution of 1.0 and 60 cps viscosity.	R	290
Gelatin (Gelfoam®)	Phenylephrine + tropicamide	Mydriatic response is larger and longer lasting than that produced by eyedrops with an equivalent amount of drugs.	R	289
Gelatin (Gelfoam)	Pilocarpine	Gelfoam device is more effective than conventional pilocarpine dosage forms in prolonging the duration of the pilocarpine.	R	286
Polypeptide	Idoxuridine	Better results than drops and ointment in the treatment of herpes simplex keratitis.	R	284
PVMMA	Pilocarpine	Delayed and decreased peak concentration of pilocarpine in general circulation compared to aqueous solution.	R	237
PVMMA	Timolol	2- to 5-fold reduction of peak plasma Timolol concentration compared to the aqueous solution.	R	244
PVMMA	Timolol and prodrugs of Timolol	Release rate much slower for Timolol prodrugs compared to the Timolol-containing inserts.	R	249

Notes: PVMMA, n-butyl half ester of poly(methyl vinyl ether/maleic anhydride); B, bovine; R, rabbit.

pilocarpine eyedrops and various other gel formulations in prolonging the action of pilocarpine (286), phenylephrine (289), and tropicamide (289).

However, erodible systems can have significantly variable erosion rates based on individual patient physiology and lacrimation patterns, while degradation products and residual solvents used during the polymer preparation can cause inflammatory reaction.

In conclusion, a wide range of polymers can be used to manufacture ocular inserts. Inserts are a promising alternative administration route because of their various advantages compared to classical dosage forms. However, only a few of these formulations have been commercialized. This can be attributed to the reluctance of ophthalmologists and patients to replace traditional ophthalmic solutions as well as the cost and the need to train both the prescribers and the patients to place the inserts correctly in the eyes. In the future, the use of ophthalmic inserts will certainly increase because of the development of new polymers, the emergence of new drugs having short biological half-lives or systemic side effects, and the need to improve the efficacy of ophthalmic treatments by ensuring an effective drug concentration in the eye for several days.

IV. INTRAOCULAR DELIVERY SYSTEMS

Different routes of intraocular drug delivery have been investigated. Periocular injections may either be subconjunctival or sub-Tenon's. Subconjunctival injections are made underneath the conjunctiva, whereas sub-Tenon's injections are made beneath the Tenon's capsule. Following periocular injection, the administered drug passes through the sclera and into the eye by diffusion. Penetration of drugs into the vitreous following periocular injection is generally negligible. This route of administration is often employed to treat severe infections of the anterior portion of the eye or to administer drugs as an adjunctive treatment to glaucoma filtering surgery (2). Intraocular injections may be intracameral, intravitreal, intracapsular, or subretinal (see Fig. 1). The intracameral injection, i.e., in the anterior chamber, may be sought for intraocular antibiotic therapy as a last resort in case of severe infections of the eye. Injections within the posterior chamber might be beneficial after cataract surgery to prevent complications. Intravitreal injection of drugs represents a direct way of attaining effective drug concentrations in the vitreous cavity (291). The vitreous humor is a hydrophilic gel, containing 99% water and 1% of organic substances such as collagen and hyaluronic acid, through which drugs diffuse freely as they would in water (292). After injection into the vitreous body, drugs may be eliminated through the retina or by diffusion into the aqueous humor of the posterior chamber with subsequent removal by the normal egress of fluid from the anterior chamber. The intravitreal dosage route was initially reserved primarily for the treatment of endophthalmitis; nowadays this route is also used for the treatment of other sight-threatening intraocular diseases.

Intracapsular administration is achieved by dissolving drugs into the intraocular infusion fluids used during surgery. This route allows the treatment of afflictions of the capsular bag such as posterior capsular opacification, a frequent complication of cataract surgery (21).

Subretinal injections allow drug delivery directly to the retina, or more precisely to the retinal pigment epithelium cells. Conventional delivery routes cannot achieve adequate intracellular concentrations of drugs or other substances such as oligonucleotides or genes to modulate functions of RPE cells effectively. Subretinal injections are performed by a transvitreal approach (293) or by creating a tunnel in the sclera.

Successful treatment of most intraocular diseases requires multiple injections to maintain therapeutically effective drug concentrations for a desired period of time. Additionally, repeated intraocular injections can cause several complications, such as increasing risk of ocular infection, vitreal hemorrhage, retinal detachment, and cataract. Moreover, the initial peak level of drug achieved immediately after a bolus injection may result in direct toxicity to ocular tissues. Because of the rapid clearance of drugs from the eye and the variable toxicity encountered in various studies, researchers have been encouraged to develop new systems of intraocular drug delivery that would sustain a therapeutic level of the drug over an extended period of time. Thus, the ''peak and valley'' effects could be minimized and drug concentrations could be maintained at an effective therapeutic level for a prolonged period of time.

A. Intraocular Implants

1. Bioerodible Polymers

Bioerodible polymers are particularly suitable in the case of intraocular implants since their use eliminates the step of removing the implant after the drug has been released, which in some applications can represent a significant advantage over other systems.

a. Poly(Lactic Acid), Poly(Glycolic Acid) and Their Copolymers

Polymers and copolymers of lactic and glycolic acids are poly(α-hydroxy acids) (PLA and PLGA) that are generally synthesized by a condensation reaction at high temperature via ring-opening polymerization of the corresponding lactide and glycolide (294). The mechanism of degradation of these polymers is a simple chemical hydrolysis predominantly occurring in the core of the matrix (295). The polymeric chains are first cleaved by hydrolysis, thereby releasing acidic monomers which are rapidly eliminated by a physiological pathway. The degradation rate can be varied by many factors: the composition of the polymer used, its conformation, and also the physicochemical characteristics of the incorporated drugs (296).

Intraocular biodegradable poly(α-hydroxy acid)–based implants have been developed as new drug delivery systems for the treatment of glaucoma filtering surgery failure or vitreoretinal pathologies. These implants are generally in the form of small cylinders which can be implanted in the eye by means of a syringe with a conventional needle (19- or 20-gauge). Table 9 summarizes the different experiments using these polymers.

Various experiments were carried out concerning implantation of PLGA-based drug delivery systems containing 5-fluorouracil (5-FU) for the management of glaucoma filtering surgery outcome. In the first studies in 1994 (298,299), efficacy and biocompatibility of such implants were assessed. After subconjunctival implantation, filtration fistulae survival time was significantly lengthened. A

Table 15.9 Poly(Lactic Acid) and Poly(Lactic and Glycolic Acid) Implants for Intraocular Administration

Polymer	Drug	Implant type	Site	Animal	Ref.
PLA	Fluorescein	Implant	Intravitreal	Rabbit	297
Glaucoma filtering surgery failure					
PLGA	5-Fluorouracil	Disc	Subconjunctival	Rabbit	298
PLGA	5-Fluorouracil	Disc	Subconjunctival	Rabbit Guinea pig	299
PLGA	5-Fluorouracil	Implant	Subconjunctival	Rabbit	300
PLGA	5-Fluorouracil	Implant	Subconjunctival	Rabbit	301
Proliferative vitreoretinopathy					
PLGA	5-Fluorouracil	Cylinder	Intravitreal	Rabbit	302
PLGA	Doxorubicin	Scleral plug	Intravitreal	Rabbit	303
PLA	Doxorubicin	Scleral plug	Intravitreal	Rabbit	304
PLGA	5-FUrd, t-PA, triamcinolone	Cylinder	—	In vitro	305
CMV retinitis					
PLA/PLGA	Ganciclovir	Scleral plug	Intravitreal	Rabbit	306
PLGA	Ganciclovir	Scleral plug	Intravitreal	Rabbit	307
PLGA	ODN	Pillar	—	In vitro	308
Endophthalmitis					
PLA	Ciprofloxacin	Disc	Intravitreal	Rabbit	309
PLGA	Fluconazole	Scleral plug	Intravitreal	Rabbit	310
PLA	Dexamethasone Sulfobenzoate	Rod	Intravitreal	Rabbit	311
Posterior capsule opacification					
PLGA	Indomethacin	Disc	Intracapsular	Rabbit	312
PLA	Daunorubicin, Indomethacin	Ring	Intracapsular	Rabbit	313
PLGA	EDTA	Disc	Intracapsular	Rabbit	312
PLA/PVP	Fluorometholone	Rod	Intracameral	Rabbit	314
PLGA	Dexamethasone	Tablet	Intracameral (posterior chamber)	Human	315
PLGA	Dexamethasone	Tablet	Intracameral (posterior chamber)	Human	316

Notes: PLA, poly(lactic acid); PLGA, poly(lactic and glycolic acid); 5-FUrd, 5-fluorouridine; t-PA, tissue plasminogen activator; ODN, oligodeoxynucleotide; EDTA, ethylenediaminetetraacetic acid; PVP, poly(vinyl pyrrolidone).

mild host response inflammatory reaction was triggered in both guinea pig and rabbit eyes 1 week after implantation, and slight inflammation persisted for at least 6 weeks. Steady-state concentrations of 5-FU were achieved in the conjunctiva and the sclera 24 h after implantation and continued for about 200 h (300,301). The advantage of such a system is its reversibility during the first 14 days in case a contraindication to the use of 5-FU appears postoperatively.

Rubsamen et al. (302) investigated PLGA-based cylinders containing 5-FU to treat experimental PVR in rabbit eyes. The devices could deliver therapeutic levels of 5-FU for almost 3 weeks and reduce the incidence of tractional retinal detachment. No adverse mechanical or toxic effects were apparent during the time of the study. However, because these devices were not fixed in the vitreous cavity, they might move and come into contact with the retina. Consequently, a stable localization of the device in the vitreous base through a suture fixation or an alternative mode of fixation at the site of implantation might be desirable. In 1994, Hashizoe et al. designed a new device that is implanted at the pars plana and releases drugs directly into the vitreous (303). These scleral plugs, made of PLGA, gradually released doxorubicin in a concentration maintained within the therapeutic levels for 1 month, without any notable toxicity to the retina. In another study with PLA-based scleral plugs with doxorubicin, drug release was still prolonged and showed less initial burst (304).

Sustained release of ganciclovir for the treatment of cytomegalovirus retinitis was also investigated (Fig. 15) (306,307). The concentration of ganciclovir was maintained within the therapeutic range for over 3 months in the vitreous and for over 5 months in the retina and the choroid. Concerning retinal toxicity, the authors concluded that the biocompatibility of the device might be clinically acceptable. A promising study was carried out by Yamakawa concerning the in vitro release of phosphorothioate oligonucleotide from PLGA matrices (308). Pseudo–zero order release of oligonucleotide lasted more than 20 days.

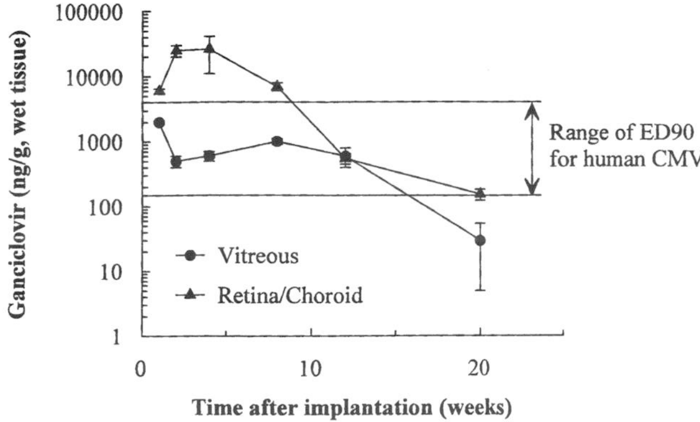

Figure 15.15 Concentrations of ganciclovir in the vitreous and the retina/choroid after implantation of ganciclovir-loaded PLGA scleral implant (From Ref. 306.)

Treatment of endophthalmitis was investigated with antibiotics (309), antifungals (310), and anti-inflammatory agents (311). In all cases, the intravitreal sustained delivery device showed great promise as a means of delivering therapeutic intravitreal levels of drugs over an extended period of time.

In order to prevent posterior capsule opacification (PCO) after cataract surgery, several investigations have been made to inhibit proliferation of lens epithelial cells. Poly(lactic and glycolic acid) with indomethacin, implanted in the capsular bag, significantly decreased intraocular inflammation, but did not reduce PCO (317). On the other hand, the implantation of PLA implants containing indomethacin and daunorubicin reduced PCO formation by approximately 50%. The administration of EDTA, which complexes with calcium, disrupted the interaction between the posterior capsule and migrating lens epithelial cells by inactivating the adhesion molecule integrin, significantly reducing cell migration onto the posterior capsule (312). Implants with fluorometholone (314) as well as a new dexamethasone delivery system (Surodex®) (315,316) were found to be a safe and effective treatment method to reduce intraocular inflammation after cataract surgery.

b. Polycaprolactones

Other polymers from the polyesters family have been evaluated as drug delivery systems. Poly-ε-caprolactone (PCL) is synthesized by a ring-opening polymerization of the monomer ε-caprolactone at high temperature. It is semicrystalline and rather hydrophobic compared to poly (α-hydroxy acids) (318).

As with PLA and PLGA, degradation of PCL produces ε-hydroxycaproic acid by cleavage of the polymeric chains at the ester linkage, leading to an initial decrease of the molecular weight without any significant weight loss. When the molecular weight reaches about 5000, small polymeric fragments begin to diffuse from the matrix and weight loss occurs; subsequently, fragments undergo phagocytosis (318).

Intraocular application of polycaprolactone-based delivery systems has been investigated by Borhani et al. (319). Devices of PCL loaded with 5-FU and implanted intravitreally exhibited a slow release and thus proved to be efficient against PVR by providing a constant concentration of drug during the active period of the disease. An inflammatory cell reaction was found in the sclera near the implant site; however, no implant-related inflammation was observed (319).

Peyman et al. (320) developed a PCL-based biodegradable porous implant. Nontoxic water-soluble inorganic salts such as sodium or potassium chloride were used as pore-forming agents and incorporated into PCL by a melting process. Biodegradable porous reservoir devices were prepared and loaded with different substances, i.e., 5-FU, ganciclovir, foscarnet, and the dye carboxyfluorescein. When the implants are immersed in water, the inorganic salts dissolve, leaving pores in the PCL devices. A concentration gradient across the wall of the polymer is created that forces the flow of the drug from the inner reservoir system to the surrounding aqueous medium. The release pattern followed zero order kinetics without any initial burst and lasted more than 6 months (320).

c. *Polyanhydrides*

Polyanhydrides can be synthesized by melt-polycondensation, dehydrochlorination, or dehydrative coupling (321). The first method leads to the formation of high molecular weight polymers, and the two last methods to polymers of lower molecular weight. The most frequently investigated polyanhydride is a copolymer of 1,3-bis(carboxyphenoxypropane) (PCPP) with sebacic acid (SA), a more hydrophilic monomer. Pure PCPP has an extremely long lifetime (over 3 years), whereas after copolymerization with 80% SA, this is reduced to a few days (322). The ratio of the two monomers plays an important role in the rate of hydrolysis of the resulting polymer: when the proportion of sebacic acid increases, the hydrophobicity of the polymer decreases, resulting in a faster rate of hydrolysis of the polymeric chains at the surface. It should also be noted that polyanhydrides are pH sensitive: they undergo faster breakdown at high pH and are more stable in acidic media (322).

Due to their surface erosion and excellent biocompatibility (323), polyanhydrides have been particularly useful for the controlled delivery of drugs for the prevention of glaucoma filtering surgery failure (Table 10).

Implants made of bis(p-carboxyphenoxy)hexane and sebacic acid, loaded with 5-FU, were implanted intraoperatively at the site of filtration surgery in rabbits (324). Intraocular pressures were lower in the experimental groups during the second postoperative week, but eventually both experimental and control eyes returned to preoperative levels. Filtration blebs lasted longer in experimental eyes; implant disappearance occurred after intraocular pressure and bleb failure. Experimental eyes showed more postoperative complications, such as hyphema, implant extrusion, corneal pigmentation, and haze. Eventually, the filtration surgery failed in both the experimental and control rabbit eyes (324). When using bis(p-carboxyphenoxy)propane instead of PCPH (325), the filtration surgery also failed in both experimental and control rabbit eyes. These negative results are probably due to the fact that the polymer becomes depleted of drug before its erosion is complete. To prolong the duration of drug release, a more hydrophobic polymeric carrier may be required to slow down the diffusion of the hydrophilic 5-FU. Jampel et al. implanted PCPP–SA devices containing 5-fluorouridine in monkeys (326). The duration of success of the filtration surgery was significantly longer in the eyes that received the implant than in the controls. Since the anatomy and physiology of the monkey eye resemble closely the human eye, the assessment of the implants efficacy has much clinical relevance.

Table 15.10 Subconjunctival Applications of Polyanhydrides in Glaucoma Filtering Surgery Failure

Polymer	Drug	Implant type	Animal	Ref.
PCPH-SA	5-Fluorouracil	Cylinder	Rabbit	324
PCPP-SA	5-Fluorouracil	Disc	Rabbit	325
PCPP-SA	5-Fluorouracil	Disc	Monkey	326
PCPP-SA	Mitomycin C	Disc	Rabbit	327
PCPP-SA	Taxol, etoposide	Disc	In vitro	328
PCPP-SA	Taxol, etoposide	Disc	Monkey	329
PCPP-SA	Etoposide	Disc	Rabbit	330
PCPP-SA	Daunorubicin	Disc	Rabbit	331

Notes: PCPH, bis (p-carboxyphenoxy)hexane.

The intraoperative use of mitomycin C (MMC) is common among ophthalmic surgeons; however, it is associated with multiple complications such as loss of the tissue vascularization, which leads to leaky filtering blebs that have thinner epithelium, bleb-associated endophthalmitis, and scleritis (332). A localized and sustained delivery of this agent would allow reduction of side effects. Incorporation of MMC into polyanhydrides enhanced efficacy and reduced toxicity (327).

Polyanhydrides containing taxol and etoposide, which are hydrophobic drugs, showed favorable release kinetics (328,330) and were thus implanted in monkey eyes (329). Use of disks containing taxol, but not etoposide, had a marked beneficial effect on intraocular pressure and bleb appearance after filtration surgery. The difference between the two drugs may result from the greater antiproliferative potency of taxol and its greater duration of release from the polymer (329).

2. Nonbiodegradable Polymers

a. Ethylene Vinyl Acetate

The ethylene vinyl acetate (EVA) copolymers are a family of materials with a number of promising advantages. This thermoplastic lipophilic polymer is mainly used as a membrane in reservoir systems. Ethylene vinyl acetate is permeable to certain lipophilic substances such as pilocarpine but sufficiently hydrophobic to be relatively impermeable to hydrophilic drugs as well as water. In 1992, Ashton and coworkers developed an implantable sustained release device to treat chronic disorders of the eye (333). This device consists of a central core of drug entirely coated in polyvinyl alcohol, a permeable polymer. The surface area available for release, and consequently the release rate, is then controlled by additional layers of ethylene EVA, which is impermeable to the drugs used, and PVAL. Water diffuses into the device and dissolves part of the pellet, forming a saturated drug solution. The drug diffuses then out of the device, and as long as the solution inside the device is saturated, the release rate is constant (334).

This device has been implanted subconjunctivally and intravitreally in animal and human eyes for the treatment of various vitreoretinal disorders (Table 11). As it is nondegradable, this system is almost devoid of any intraocular inflammatory response. However, it has to be removed or replaced to prevent any risk of fibrous encapsulation.

Table 15.11 Intraocular Administration of the EVA/PVAL Devices

Drug	Model	Site of implantation	Ref.
Glaucoma filtering surgery failure			
5-Fluorouracil	In vitro	—	335
5-Fluorouracil	Rabbit, monkey	Subconjunctival	336
5-Fluorouracil	Human	Subconjunctival	337
CMV retinitis			
Ganciclovir	Rabbit	Intravitreal	334
Ganciclovir	Human	Intravitreal	338
Ganciclovir	Human	Intravitreal	339
Ganciclovir	Human	Intravitreal	340
Ganciclovir	Human	Intravitreal	341
Uveitis			
Dexamethasone	Rabbit	Intravitreal	342
Dexamethasone	Rabbit	Intravitreal	343
Cyclosporine	Rabbit	Intravitreal	344
Cyclosporine, dexamethasone	Rabbit	Intravitreal	345
Cyclosporine	Rabbit	Intravitreal	346

The first devices were developed to prevent glaucoma filtering surgery failure. Blandford et al. (336) implanted subconjunctival implants containing 5-FU. Devices prolonged the reduction of intraocular pressure. Although 5-FU was released from the devices for 2 weeks, intraocular pressure remained low for 3 months. These implants have also been implanted in patients undergoing filtering surgery (337); the mean follow-up was 2.5 years. In three of the four patients, intraocular pressure was controlled with stabilization of the visual field. The other patient had early failure. No untoward events were linked to the placement of the implant.

Treatment of intraocular inflammation and posterior uveitis with EVA/PVAL devices was also investigated. Implants were prepared, containing dexamethasone (342,343). Dexamethasone was released continuously for 3.5 months at an effective concentration. Furthermore, the devices were well tolerated with no indication of toxicity or inflammation. Cyclosporine devices were implanted by Pearson et al. (344). Data suggest that due to the high potency of cyclosporine and the very slow release rate from the device, these implants would provide sustained intravitreal levels of the drug for up to 10 years. However, cyclosporine levels produced by these implants appeared to be toxic in the rabbit, resulting in focal lens opacification. This toxicity was reversible and resolved after the device was removed. In a rabbit model of uveitis (346), cyclosporine implants appeared to effectively suppress ocular inflammation. Intravitreal cyclosporine remained at therapeutic levels for at least 6 months, and histopathologic examination showed that cyclosporine-treated eyes had preserved architecture and greatly reduced inflammatory cells. Devices containing both cyclosporine and dexamethasone were also prepared (345). Based on the fact that cyclosporine and corticosteroids reduce inflammation by different mechanisms, a device which contains both drugs in combination would be more effective than a device containing either agent alone. These implants maintained therapeutic levels of both drugs in the vitreous over 10 weeks, and the presence of dexamethasone did not seem to change the release characteristics for cyclosporine (345).

However, the main application of EVA/PVAL devices is the treatment of CMV retinitis with ganciclovir. First developed by Smith et al. in 1992 (334), this system has been widely experimented in both animal and human eyes (338–340,347–352). The ganciclovir device was approved by the Food and Drug Administration in 1996 and is commercially available under the name Vitrasert™ (Bausch and Lomb Surgical), each implant containing a minimum

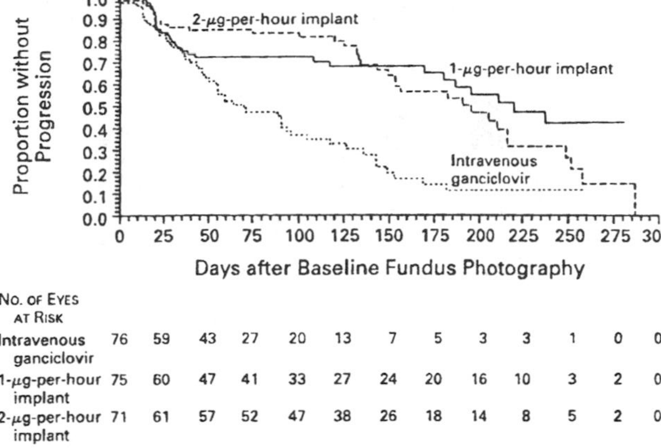

Figure 15.16 Progression of CMV retinitis in eyes affected at baseline (From Ref. 353.)

of 4.5 mg of ganciclovir and being designed to release the drug over a 5- to 8-month period of time. The international AIDS Society has promulgated recommendations concerning Vitrasert (341), which offers a number of advantages over systemic ganciclovir therapy, including the longest median time of control of retinitis reported to date (Fig. 16) (340,353). Better control of retinitis from higher intraocular drug levels may reduce the risk of emergence of viral resistance. The disadvantages include pain and discomfort from surgery, a transient decrease in visual acuity, and the risk of complications inherent in the surgical procedure, which may increase with multiple operations. Because of the lack of systemic anti-CMV effect of the ganciclovir implant, patients remain at risk for developing CMV disease elsewhere (340). Thus, concomitant oral ganciclovir therapy is usually recommended.

b. *Polysulfone*

Polysulfone is a water-impermeable polymer. A polysulfone capillary fiber (PCF) has been investigated as a slow-release delivery system. It is an anisotropic ultrafiltration fiber mainly developed for cell culture purposes. Polysulfone capillary fibers have the following important features: they have deep macrovoids in its outer membrane, which increase the surface area for drug diffusion and release; they are permeable to lipophilic as well as hydrophilic compounds; the PCF–drug device can be readily sterilized; the device prevents dispersion of drug formulations, such as liposomes or microspheres, in the vitreous; and most important the bioactivity of the drug is unchanged by utilizing the PCF device since the fabrication process does not require chemical reaction, heat, or solvent exposure. The obvious disadvantage of this device is the need for surgical removal of the drug-depleted device since it is not a biodegradable system.

Rahimy et al. (354) found that the PCF system is capable of maintaining a constant intravitreal level of the hydrophilic dye carboxyfluorescein (CF) for up to 45 days. Investigation of the in vitro release suggests that the PCF device is a reservoir diffusion-controlled system. Carboxyfluorescein diffusion through the PCF membrane is mainly driven by the concentration gradient between the PCF core and the incubation medium. The transport of CF molecules across the PCF membrane remains constant if the core is maintained in a saturated state, such as a solid or suspension form. However, such a state is difficult to maintain in this case since CF is highly water soluble. Intravitreal implantation of this device into rabbit eyes showed sustained-release kinetics for up to 45 days in the vitreous cavity. Furthermore, both clinical and histopathological data demonstrated that the PCF device is well tolerated by the rabbit eye (354).

Promising results have been obtained with daunomycin-charged PCF for the treatment of PVR (355). Devices were implanted intravitreally in rabbit eyes and results showed that the release of daunomycin was maintained at a therapeutic concentration in the vitreous for at least 21 days. The devices prevented tractional retinal detachment in an experimental model of PVR and were well tolerated without causing toxicity to ocular tissues (355).

B. Viscous Nonaqueous Injectable Preparations

1. Third Generation Poly(Ortho Esters)

Poly(ortho esters) (POE) are hydrophobic and bioerodible polymers whose main characteristics are a release of the drug by surface erosion, following zero order kinetics. Since the late 1970s, four generations of POE have been synthesized to produce bioerodible carriers for drug delivery. Among them, the third generation has been recently described (356) and characterized by Merkli et al. (357).

This viscous, ointmentlike, material is prepared by a transesterification between trimethyl orthoacetate and 1,2,6-hexanetriol (357). Drugs can be incorporated by simple mixing at room temperature, without the use of a solvent. This characteristic is of considerable interest with respect to peptide and protein delivery, as well as for other thermolabile drugs. Poly(ortho esters) can be injected using a syringe with an appropriate hypodermic needle without any other surgical intervention, which is a significant advantage when compared with solid devices that must be placed either with a trocar or through a surgical procedure.

The release of drugs from semisolid POE occurs via a combined erosion and diffusion mechanism. Because POE contains pH-sensitive linkages in the polymeric backbone, the erosion rate can be controlled by incorporating acidic substances into the polymer matrix to accelerate the erosion rates or basic ones to stabilize the polymer backbone (358,359).

Degradation occurs in two steps: the first reaction is a rapid hydrolysis of the labile ortho ester bonds. This initial reaction is followed by a slower hydrolysis which produces a carboxylic acid and a triol (357). Degradation rate is also influenced by the physicochemical properties of the incorporated substances (359).

POE has been investigated as an injectable bioerodible polymeric implant for controlled subconjunctival release of antimetabolites such as 5-fluorouracil and mitomycin C as an adjunctive treatment to glaucoma filtering surgery (360). The biotolerance of the semisolid biodegradable POE was investigated subcutaneously and with the cage-implant system (361): the biocompatibility appeared to be acceptable. The subconjunctival biocompatibility in rabbits was studied by Bernatchez et al. (362). The acute inflammation resolved rapidly and no chronic inflammation developed. After further purification of the polymer and using an aseptic preparation method instead of γ-sterilization, Zignani et al. (363,364) further improved the biocompatibility of the polymer. Extended biocompatibility investigations have also been carried out both in the anterior and posterior parts of the rabbit eye (365). After intracameral injection, polymer biocompatibility appeared to the dependent on the amount injected in the anterior chamber. When a small quantity (typically 50 μL) was administered, the polymer degraded within 2 weeks with a good biocompatibility, without harming any ocular tissues nor increasing the intraocular pressure. After intravitreal administration, POE was well tolerated and no inflammatory reaction developed during the study. The polymer degraded slowly, appearing as a round and whitish bubble in the vitreous cavity (Fig. 17).

Since POE appears to be a biocompatible material, the efficacy of a system containing 5-FU has been tested in a rabbit model of glaucoma filtering surgery. Results appear encouraging because POE systems lowered intraocular pressure and maintained filtration blebs for more than 4 weeks (unpublished results).

2. Fourth Generation Poly(Ortho Esters)

A fourth generation of POE has been very recently developed (366,367). These POE are said to be self-catalyzed since they possess short dimers of glycolic or lactic acid within the polymeric chain, allowing an excellent control of the polymer degradation rate.

The exact mechanism of hydrolysis of self-catalyzed POE has recently been elucidated (369). It proceeds in three consecutive steps: in the first step, the lactic acid dimer segment hydrolyzes to generate a polymer fragment containing a carboxylic acid end group which will catalyze the ortho ester hydrolysis. A second cleavage produces free lactic acid which also catalyzes hydrolysis of the ortho ester links (369).

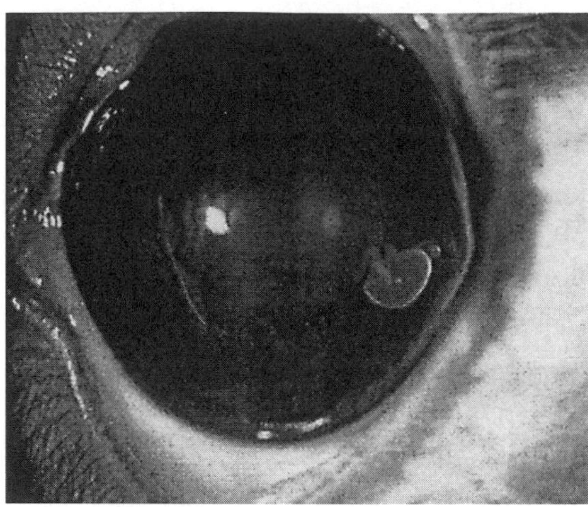

Figure 15.17 External photograph of rabbit eye 5 days after intracameral injection of 50 μL POE (From Ref. 365.)

The biocompatibility of this last generation of POE has been assessed subconjunctivally (unpublished results). These polymers are far more hydrophobic than the third generation, which means that polymer lifetime under the conjunctiva was much longer. Almost no hyperemia and chemosis was triggered, which suggests that POE are promising candidates for ophthalmic applications (unpublished data).

C. Dispersed Systems

1. Nanoparticles and Microspheres

a. Nanoparticles

Concerning the ophthalmic potential applications of nanoparticles, they are mainly used as a topical treatment for glaucoma, inflammation, or external infections of the eye since they are administered as drops and provide a sustained action (see Section III.B). A few studies have been conducted concerning the intraocular application of nanoparticles. El-Samaligy and coworkers (370) described the intravitreal injection of acyclovir and ganciclovir nanoparticles made of biodegradable polymers, i.e., bovine serum albumin (BSA), polyethylcyanoacrylate, and chitosan. All of these polymers exhibited interesting characteristics during in vitro evaluation (drug loading capacity, particle size, homogeneity of nanosphere formation, drug release). Polyethylcyanoacrylate was chosen for in vivo studies in rabbits because it demonstrated the best drug loading and the smallest particle size. Therapeutic levels of ganciclovir were detected in the vitreous for up to 10 days following injection of the PECA nanospheres. Lens opacification and vitreous turbidity appeared after the injection of this colloidal opaque dispersion, certainly due to the drug formulation.

b. Microspheres

The polymers used for the preparation of microspheres are mainly based on lactic and glycolic acids. These bioerodible systems present the advantage that they do not have to be removed after degradation of the microspheres and delivery of the drug. Following administration of the suspension in the eye, the particles remain at the delivery site and the drug is released from the polymeric matrix through diffusion, polymer degradation, erosion, or a combination of these processes.

Besides topical ophthalmic application, which have been widely studied (184) (see Section III.B), some authors have investigated microspheres for intraocular injections to provide sustained release of drugs (Table 12).

Subconjunctival injection of microspheres for the prevention of glaucoma filtering surgery failure was investigated by incorporating adriamycin (372). This delivery system using PLA microspheres was found to reduce significantly intraocular pressure as well as prolong filtering bleb persistence. When injecting fluorescein-loaded microspheres, transient conjunctival inflammation was found 6 to 7 days after injection. Clinically, a granulomatouslike reaction appeared, and one eye showed necrosis at the site of injection. The conjunctival reaction disappeared in a few days with no further complication (371).

In 1991, Khoobehi et al. (373) investigated the intravitreal injection of microspheres containing fluorescein as a drug model. High concentrations of the dye were detected within 24 h and gradually decreased after 5 days; lower concentrations were still measurable for 16 days. Moritera et al. (374) entrapped 5-FU in PLA and PLGA microspheres and injected them intravitreally in rabbits. The drug was released over a period up to 7 days. The microspheres were observed as a white mass in the vitreous cavity and were completely cleared out after 48 days. Administration of microspheres had no adverse effects on the ocular tissues, even though the degradation products of PLA and PLGA are acidic. Peyman et al. (375) demonstrated the efficacy of microspheres in increasing the half-life of encapsulated antimetabolites when injected intravitreally in primate eyes. Both cytarabine and 5-FU were entrapped in PLGA microspheres; both were still detect-

Table 15.12 Microspheres for Intraocular Applications

Polymer	Drug	Animal model	Route of administration	Ref.
Glaucoma filtering surgery failure				
PLA/PLGA	Fluorescein	Rabbit	Subconjunctival	371
PLA	Adriamycin	Rabbit	Subconjunctival	372
PVR				
PLA/PLGA	Fluorescein	Rabbit	Intravitreal	373
PLA/PLGA	5-Fluorouracil	Rabbit	Intravitreal	374
PLGA	5-Fluorouracil, cytarabine	Monkey	Intravitreal	375
PLA	Adriamycin	Rabbit	Intravitreal	376
PLGA	Retinoic acid	Rabbit	Intravitreal	377
PLGA	None	Rabbit	Intravitreal	378
CMV retinitis				
PLA/PLGA	Acyclovir	Rabbit	Intravitreal	379
PLGA	Ganciclovir	Rabbit	Intravitreal	380
Retinal degenerative diseases				
PLA	POPOP	In vitro	—	381
PLA/PLGA	Rhodamine	In vitro	—	382
PLA/PLGA	Rhodamine	Rabbit	Subretinal	293
Gelatin	bFGF	Rabbit	Subretinal	383
Gelatin	bFGF	Rabbit	Subretinal	384

Notes: POPOP, 1,4-bis[2-(5-phenyloxazolyl)]-benzene; bFGF, basic fibroblast growth factor.

able in the eye 11 days after injection and the clearance kinetics were similar for both drugs. Moritera injected PLA microspheres with adriamycin in rabbits (376). This drug, 50 times more potent than 5-FU, was released at a continuous rate for over 14 days and was effective in preventing the rate of retinal detachment after 4 weeks in an experimental model of PVR and showed no toxicity to the retina even at a relatively high dose. Giordano et al. (377) entrapped retinoic acid, a highly hydrophobic drug, in PLGA microspheres and obtained a constant release of the drug. This system could provide therapeutic levels of retinoic acid for 40 days. Giordano et al. assessed biodegradation and tissue reaction to intravitreal PLGA microspheres (378). A mild localized, nonprogressive foreign body reaction was observed. The cell reaction was composed mostly of glial cells and fibroblasts. The choroid and retina were normal. No clinical inflammatory signs were observed 4 days postoperatively and thereafter (378).

Antiviral drugs such as acyclovir and ganciclovir were also encapsulated into microspheres and administered intravitreally (379,380). In both cases, drugs were released at a sustained rate and appeared to be promising for the treatment of viral retinitis.

As mentioned, there is no effective therapy available for achieving sufficient intracellular drug concentrations to modulate functions of retinal pigment epithelium cells. Since RPE cells have phagocytic activity, drug delivery could be targeted with the use of microspheres. A first study by Moritera et al. (381) showed that gelatin-coated PLA microspheres containing a fluorescent dye, 1,4-bis[2-(5-phenyloxazolyl)]-benzene (POPOP), were phagocytosed by RPE cells in vitro. In further work by Kimura et al. (382), PLA microspheres were loaded with a fluorescent dye, rhodamine. Ogura et al. (293) injected these rhodamine-loaded microspheres subretinally via a transvitreal approach with a glass micropipette in the rabbit. Microspheres were degraded in the cytoplasm of the RPE cells, but fragments were observed for up to 4 weeks. The retinal architecture overlying the delivery site was well preserved, which suggests that direct delivery of substances to the RPE cells is feasible, without damaging the neural retinal structure (Fig. 18) (293).

Another application of subretinal administration of microspheres is the creation of a model of choroidal neovascularization. By administrating gelatin microspheres im-Rev 16.01egnated with basic fibroblast growth factor (bFGF), a reproducible subretinal neovascularization was triggered, which represents a useful model for testing drugs against retinal diseases such as age-related macular degeneration (383,384).

In conclusion, most intraocular diseases, including sight-threatening endophthalmitis, devastating CMV retinitis, proliferative vitreoretinopathy and uveitis, as well as glaucoma filtering surgery failure and degenerative diseases of the retina, were treated by systemic and topical administration of drugs in the past. These routes of treatment led, in the majority of cases, to the loss of the eye due to the poor penetration of the various drugs into the vitreous cavity. The means of bypassing physiological and anatomical barriers and achieving high concentrations of drugs within the eye is through direct intraocular administration. The use of intraocular drugs has formerly been restricted to the treatment of endophthalmitis. Nowadays, it is also considered for the disorders described in this chapter. As all these clinical conditions require frequent and repeated intraocular injections of drugs, this may cause severe complications such as retinal detachment, vitreous hemorrhage, or intraocular infections. That is why sustained delivery systems have been developed, in order to prolong the therapeutic intraocular levels of drugs and thus diminish the frequency of the injections.

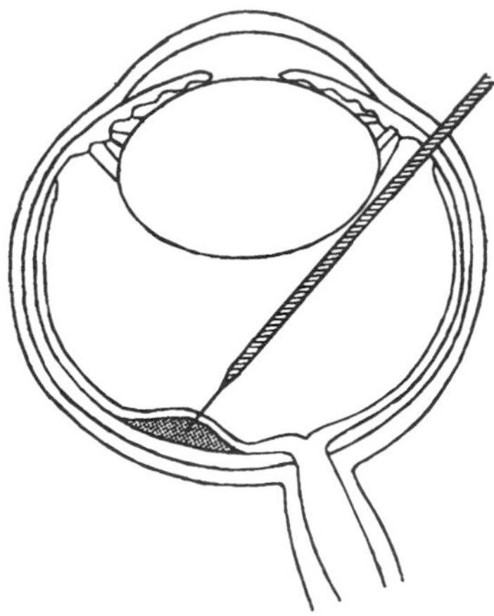

Figure 15.18 A glass micropipette is inserted transvitreally to administer the microspheres subretinally. (From Ref. 293.)

Biodegradable polymers have an advantage over other controlled release systems in obviating the need to surgically remove the drug depleted device. Potentially, biodegradable matrix systems also enjoy a number of other advantages in terms of simplicity of design and predictability of release if this is controlled solely by the erosion of the matrix. In many cases, however, the release occurs also by diffusion through the matrix, making the process more difficult to control, particularly if the matrix is hydrophilic and thereby absorbs water, promoting degradation in the interior of the matrix. To maximize control over the release process, it is desirable to have a polymeric system which degrades only from the surface and delays the permeation of drug molecules. Achieving such a heterogeneous degradation requires the rate of hydrolytic degradation on the surface to be much faster than the rate of water penetration into the bulk. The ideal polymer would have a hydrophobic backbone, but with a water-labile linkage. Many classes of polymers have been designed with these considerations in mind. Among them, polyanhydrides and poly(ortho esters) erode from the surface. Taking advantage of the pH dependence of the rate of ortho ester cleavage, preferential hydrolysis at the surface is obtained by either addition of basic substances to suppress degradation in the bulk or incorporation of acidic catalysts to promote degradation on the surface.

Devices that are not biodegradable such as the EVA-PVAL implants are of clinical utility for chronic, sight-threatening diseases such as CMV retinitis and uveitis. These conditions are likely to require long-term drug therapy; therefore any device may need to be replaced once depleted of drug. On the other hand, in diseases such as glaucoma filtering surgery failure or PVR, which require relatively short-term therapy (weeks or months), a biodegradable system is clearly more desirable.

Microspheres have been shown to increase the efficacy, reduce the toxicity, prolong the activity, and provide site-specific delivery for many drugs when administered to the eye. Microspheres are very stable; their production is reproducible; and the release rate is controllable. Yet, the injection of microspheres in the vitreous is always followed by an impairment of the visual acuity.

V. CONCLUSION

The ideal ophthalmic drug-release system should fulfill criteria which are identical for local as well as intraocular application: the systems should provide a sustained therapeutic concentration of drug in a reliable and predictable way over an adequate period to obtain the desired clinical response; the concentration of drug administered should be uniform over the administration period with no early or late excessive release of the drug; the formulation should be easily handled and administered either by the patient or by medical personnel; the device should have a long shelf-life and be easily sterilized; and there should be no toxic effect from sustained exposure to the drug or degradation products of the device.

None of the sustained delivery systems introduced to date and discussed in this chapter fulfill all these criteria, though they represent great strides when compared with classical solutions. Viscous formulations of the hydrogel type provide an increased bioavailability, but do not allow a significant sustained and controlled release of the incorporated drugs. Colloidal systems are often associated with stability problems and are difficult to sterilize. Solid systems are not always well tolerated by the patients.

Some new approaches, such as the use of lectins or cyclodextrins for topical formulations and iontophoresis to allow enhanced penetration of drugs to the posterior segment of the eye, may be promising alternatives to the various polymeric systems discussed in this chapter.

ACKNOWLEDGMENTS

The authors wish to express their thanks to Annette Meier-Nabholz (Galenica Distribution AG, Schöubühl, Switzerland) for providing information on marketed hydrogels.

REFERENCES

1. I. K. Reddy, M. G. Ganesan. Ocular therapeutics and drug delivery: an overview. In: *Ocular Therapeutics and Drug Delivery: A Multi-Disciplinary Approach*, I. K. Reddy (Ed.). Technomic, Lancaster, PA, 1996, p. 3.
2. D. M. Maurice, S. Mishima. Ocular pharmacokinetics. *Pharmacology of the Eye*, M. L. Sears (Ed.). Springer-Verlag, Berlin, 1984, p. 19.
3. V. Baeyens, C. Percicot, M. Zignani, A. A. Deshpande, V. Kaltsatos, R. Gurny. Ocular drug delivery in veterinary medicine. *Adv. Drug Del. Rev.* 28:335 (1997).
4. O. Felt, V. Baeyens, M. Zignani, P. Buri, R. Gurny. Mucosal drug delivery, ocular. In: *The Encyclopedia of Controlled Drug Delivery*, E. Mathiowitz (Ed.). Wiley and Sons, New York, 1999, p. 605.
5. M. A. Bartletta, I. K. Reddy. Clinical pharmacology of the anterior segment of the eye. In: *Ocular Therapeutics and Drug Delivery*, I. K. Reddy (Ed.). Technomic, Lancaster, PA, 1996, p. 213.
6. S. P. Kulkarni, B. Bhanumathi, S. O. Thoppil, C. P. Bhogte, D. Vartak, P. D. Amin. Dry eye syndrome. *Drug Dev. Ind. Pharm.* 23:465 (1997).
7. E. Woolard, M. Frank. Management of dry eye. *On Cont. Pract.* 18:41 (1991).
8. F. J. Holly, M. A. Lemp. Tear physiology and dry eyes. *Surv. Ophthalmol.* 22:69 (1977).
9. M. Leino, A. Urtti. Recent developments in anti-glaucoma drug research. In: *Ocular Therapeutics and Drug Delivery*, I. K. Reddy (Ed.). Technomic, Lancaster, PA, 1996, p. 245.
10. M. R. Chang, Q. Cheng, D. A. Lee. Basic science and clinical aspects of wound healing in glaucoma filtering surgery. *J. Ocul. Pharmacol.* 14:75 (1998).
11. P. Wiedemann, K. Heimann. Proliferative vitreoretinopathy. *Curr. Opin. Ophth.* 3:357 (1992).
12. M. S. Blumenkranz, M. K. Hartzer. The mechanism of action of drugs for the treatment of vitreoretinal scarring. In: *Retina*, S. J. Ryan (Ed.). Mosby, St Louis, MO, 1989, p. 401.
13. H. Lewis, T. M. Aaberg. Causes of failure after repeat vitreoretinal surgery for recurrent proliferative vitreoretinopathy. *Am. J. Ophthalmol.* 111:15 (1991).

14. D. E. Henderley, W. R. Freeman, D. M. Causey, N. A. Rao. Cytomegalovirus retinitis and response to therapy with ganciclovir. *Ophthalmol.* 94:425 (1987).
15. J. A. Schulman, G. A. Peyman, M. B. Horton, J. Liu, R. G. Fiscella, J. S. Pulido, J. C. Barber, P. de Miranda. Intraocular 9-([2-hydroxy-1-(hydroxymethyl) ethoxy] methyl) guanine levels after intravitreal and subconjunctival administration. *Ophthalmic Surg.* 17:429 (1986).
16. W. L. Drew, R. C. Miner, D. F. Busch, S. E. Follansbee, J. Gullett, S. G. Mehalko, S. M. Gordon, W. F. Owen, T. R. Matthews, W. C. Buhles, B. DeArmond. Prevalence of resistance in patients receiving ganciclovir for serious cytomegalovirus infection. *J. Inf. Dis.* 163:716 (1991).
17. C. M. Perry, J. A. Balfour. Fomivirsen. *Drugs* 57:375 (1999).
18. T. A. Meredith. Clinical microbiology of infectious endophthalmitis. In: *Retina*, S. J. Ryan (Ed.). Mosby, St Louis, MO, 1989, p. 183.
19. G. A. Peyman, S. S. Bassili. A practical guideline for management of endophthalmitis. *Ophthalmic Surg.* 26:294 (1995).
20. R. B. Nussenblatt, A. G. Palestine, C. C. Chan. Cyclosporin A therapy in the treatment of intraocular inflammatory disease resistant to systemic corticosteroids and cytotoxic agents. *Am. J. Ophthalmol.* 96:275 (1983).
21. J. M. Ruiz, M. Medrano, J. L. Alio. Inhibition of posterior capsule opacification by 5-fluorouracil in rabbits. *Ophthalmic Res.* 22:201 (1990).
22. N. M. Bressler, S. B. Bressler. Preventative ophthalmology. Age-related macular degeneration. *Ophthalmol.* 102: 1206 (1995).
23. T. A. Ciulla, R. P. Danis, A. Harris. Age-related macular degeneration: a review of experimental treatments. *Surv. Ophthalmol.* 43:134 (1998).
24. S. Burgalassi, P. Chetoni, M. F. Saettone. Hydrogels for ocular delivery of pilocarpine: preliminary evaluation in rabbits of the influence of viscosity and of drug solubility. *Eur. J. Pharm. Biopharm.* 42:385 (1996).
25. S. W. Kim, Y. H. Bae, T. Okano. Hydrogels: swelling, drug loading, and release, *Pharm. Res.* 9:283 (1992).
26. S. Dumitriu, P. F. Vidal, E. Chornet. Hydrogels based on polysaccharides, *Polysaccharides in Medicinal Applications*, S. Dumitriu (Ed.). Marcel Dekker, New York, 1999, p. 125.
27. S. B. Ross-Murphy, H. McEvoy. Fundamentals of hydrogels and gelation. *Br. Polym. J.* 18:2 (1986).
28. A. Martin, J. Swarbrick, A. Cammarata, A. H. C. Chun. Coarse dispersions. In: *Physical Pharmacy, Physical Chemical Principles in the Pharmaceutical Sciences*. Lea & Febiger, Philadelphia, 1983, p. 544.
29. R. Bawa, M. Nandu. Physico-chemical considerations in the development of an ocular polymeric drug delivery system. *Biomaterials* 11:724 (1990).
30. V. H. L. Lee, J. R. Robinson. Review: topical ocular drug delivery: recent developments and future challenges. *J. Ocul. Pharmacol.* 2:67 (1986).
31. C. Le Bourlais, L. Acar, H. Zia, P. A. Sado, T. Needham, R. Leverage. Ophtalmic drug delivery systems—recent advances. *Progress in Retinal and Eye Research* 17:33 (1997).
32. L. M. Kralian. What's artificial in artificial tears. *Am. J. Optom. Physiol. Optics* 63:304 (1986).
33. S. D. Desai, J. Blanchard. Ocular drug formulation and delivery. In: *Encyclopedia of Pharmaceutical Technology/Nuclear Medicine and Pharmacy to Permeation Enhancement Through Skin*, J. Swarbrick et al. (Eds.), Marcel Dekker, New York, 1995, p. 43.
34. K. C. Swan. Use of methylcellulose in ophthalmology. *Arch. Ophthalmol.* 33:378 (1945).
35. J. L. Mims. Methylcellulose solutions for ophthalmic use. *Arch. Ophthalmol.* 46:664 (1951).
36. C. A. Le Bourlais, L. Treupel-Acar, C. T. Rhodes, P. A. Sado, R. Leverge. New ophthalmic drug delivery systems. *Drug Dev. Ind. Pharm.* 21:19 (1995).
37. E. Doelker. Water-swollen cellulose derivatives in pharmacy. In: *Hydrogels in Medicine and Pharmacy*, Vol. II, N. Peppas (Ed.). CRC Press, Boca Raton, FL, 1987, p. 116.
38. S. S. Chrai, J. R. Robinson. Ocular evaluation of methylcellulose vehicle in albino rabbits. *J. Pharm. Sci.* 63:1218 (1974).
39. G. Meseguer, R. Gurny, P. Buri, A. Rozier, B. Plazonnet. Gamma scintigraphic study of precorneal drainage and assessment of miotic response in rabbits of various ophthalmic formulations containing pilocarpine. *Int. J. Pharm.* 95:229 (1993).
40. M. F. Saettone, B. Giannaccini, S. Ravecca, F. La Marca, G. Tota. Polymer effects on ocular bioavailability—the influence of different liquid vehicles on the mydriatic response of tropicamide in humans and in rabbits. *Int. J. Pharm.* 20:187 (1984).
41. A. Ludwig, M. M. Van Ooteghem. Influence of viscolysers on the residence of ophthalmic solutions evaluated by slit lamp fluorophotometry. *S. T. P. Pharma Sci.* 2:81 (1992).
42. D. A. Benedetto, D. O. Shah, H. E. Kaufman. The instilled fluid dynamics and surface chemistry of polymers in the preocular tear film. *Invest. Ophthalmol. Vis. Sci.* 14:887 (1975).
43. S. G. Deshpande, S. Shirokolar. Sustained release ophthalmic formulations of pilocarpine. *J. Pharm. Pharmacol.* 41:197 (1989).
44. R. D. Schoenwald, R. L. Ward, L. M. DeSantis, R. E. Roehrs. Influence of high-viscosity vehicles on miotic effect of pilocarpine. *J. Pharm. Sci.* 67:1280 (1978).
45. R. Harderger, C. Hanna, C. M. Boyd. Effects of drug vehicles on ocular contact. *Arch. Ophthalmol.* 93:42 (1975).
46. A. Ludwig, M. Van Ooteghem. Evaluation of sodium hyaluronate as viscous vehicle for eye drops. *J. Pharm. Belg.* 44:391 (1989).
47. M. F. Saettone, B. Giannaccini, P. Savagni, N. Tellini. Vehicle effects on ophthalmic bioavailability: the influence of different polymers on the activity of pilocarpine in rabbit and human. *J. Pharm. Pharmacol.* 34:464 (1982).

48. J. H. Trueblood, R. M. Rossomondo, W. H. Carlton, L. A. Wilson. Corneal contact times of ophthalmic vehicles. Evaluation by microscintigraphy. *Arch. Ophthalmol.* 93: 127 (1975).
49. V. Hon-Kin Li, J. R. Robinson. Solution viscosity effects on the ocular disposition of cromolyn sodium in the albino rabbit. *Int. J. Pharm.* 53:219 (1989).
50. K. Järvinen, E. Vartiainen, A. Urtti. Optimizing the systemic and ocular absorption of timolol from eye-drops. *S. T. P. Pharm. Sci.* 2:105 (1992).
51. B. W. Barry, M. C. Meyer. The rheological properties of Carbopol gels. I. Continuous shear and creep properties of Carbopol gels. *Int. J. Pharm.* 2:1 (1979).
52. M. F. Saettone, B. Giannaccini, F. Barattini, N. Tellini. The validity of rabbits for investigations on ophthalmic vehicles: a comparison for four different vehicles containing tropicamide in humans and rabbits. *Pharm. Acta Helv.* 57:3 (1982).
53. A. Ludwig, N. Unlu, M. Van Ooteghem. Evaluation of viscous ophtalmic vehicles containing carbomer by slit-lamp fluorophotometry in humans. *Int. J. Pharm.* 61:15 (1990).
54. N. Unlü, M. M. Van Ooteghem, A. A. Hincal. A comparative rheological study on carbopol viscous solutions and the evaluation of their suitability as the ophthalmic vehicles and artificial tears. *Pharm. Acta Helv.* 67:5 (1992).
55. N. Unlü, A. Ludwig, M. M. Van Ooteghem, A. A. Hincal. Formulation of carbopol 940 ophthalmic vehicles, and *in vitro* evaluation of the influence of stimulated lacrimal fluid on their physico-chemical properties. *Pharmazie* 46: 784 (1991).
56. B. W. Barry, M. C. Meyer. The rheological properties of Carbopol gels. II. Oscillatory properties of Carbopol gels. *Int. J. Pharm.* 2:27 (1979).
57. W. Lang. Thickening cosmetics without neutralisation. *Drug Cosmet. Ind.* 110:52 (1972).
58. N. W. Taylor, E. B. Bagley. Dispersions of solutions. A mechanism for certain thickening agents. *J. Appl. Polymer. Sci.* 18:2747 (1974).
59. N. W. Taylor, E. B. Bagley. Rheology of dispersions of swollen gel particles. *J. Polymer. Sci.* 13:1133 (1975).
60. K. Park, J. R. Robinson. Bioadhesive polymers as platforms for oral-controlled drug delivery: method to study bioadhesion. *Int. J. Pharm.* 19:107 (1984).
61. E. M. Slovin, J. R. Robinson. Bioadhesives in ocular drug delivery. In: *Biopharmaceutics of Ocular Drug Delivery*, P. Edman. (Ed.). CRC Press, Boca Raton, FL, 1993, p. 145.
62. J. R. Robinson, G. M. Mlynek. Bioadhesive and phase-change polymers for ocular drug delivery. *Adv. Drug Del. Rev.* 16:45 (1995).
63. H. Park, J. R. Robinson. Mechanisms of mucoadhesion of poly(acrylic acid) hydrogels. *Pharm. Res.* 4:457 (1987).
64. J. R. Robinson. Ocular drug delivery. Mechanism(s) of corneal drug transport and mucoadhesive delivery systems. *S. T. P. Pharma. Sci.* 5:839 (1989).
65. N. M. Davies, S. J. Farr, J. Hadgraft, I. W. Kellaway. Evaluation of mucoadhesive polymers in ocular drug delivery. I. Viscous solutions. *Pharm. Res.* 8:1039 (1991).
66. M. Dittgen, S. Oestereich, D. Eckhardt. Influence of bioadhesion on the elimination of drugs from the eye and on their penetration ability across the pig cornea. *S. T. P. Pharm. Sci.* 2:93 (1992).
67. M. Schrenzel. Hydrogels a base de polymeres de l'acide acrylique. Essais galeniques et pharmacodynamiques. Iere communication. *Pharm. Acta Helv.* 39:546 (1963).
68. J. Giroux, M. Schrenzel. Hydrogels a base de polymeres de l'acide acrylique. Essais galeniques et pharmacodynamiques. 2e communication. *Pharm. Acta Helv.* 39:615 (1964).
69. M. F. Saettone, B. Giannaccini, A. Guiducci, P. Savigni. Semisolid ophthalmic vehicles. III. An evaluation of four organic hydrogels containing pilocarpine. *Int. J. Pharm.* 31:261 (1986).
70. M. F. Saettone, B. Giannaccini, P. Savigni, A. Wirth. The effect of different ophthalmic vehicles on the activity of tropicamide in man. *J. Pharm. Pharmacol.* 32:519 (1980).
71. R. D. Schoenwald, J. J. Boltralik. A bioavailability comparison in rabbits of two steroids formulated as high-viscosity gels and reference aqueous preparations. *Invest. Ophthalmol. Vis. Sci.* 61 (1979).
72. S. Mengi, S. G. Deshpande. Development and evaluation of flurbiprofen hydrogels on the breakdown of the blood/aqueous humor barrier. *S. T. P. Pharm. Sci.* 2:118 (1992).
73. E. A. Balazs, P. Band. Hyaluronic acid: its structure and use. *Cosmetics and Toiletries* 99:65 (1984).
74 I. C. Dea, R. Moorhouse, D. A. Rees, S. Arnott, J. M. Guss, E. A. Balazs. Hyaluronic acid: a novel, double helical molecule. *Science* 179:560 (1973).
75. F. M. Polack. Healon (Na hyaluronate). *Cornea* 5:81 (1986).
76. J. L. Greaves, O. Olejnik, C. G. Wilson. Polymers and the precorneal tear film. *S. T. P. Pharm. Sci.* 2:13 (1992).
77. E. A. Balazs. Ultrapure hyaluronic acid and the use thereof. U.S. patent 4,141,973 (1979).
78. S. F. Bernatchez, O. Camber, C. Tabatabay, R. Gurny. Use of hyaluronic acid in ocular therapy. In: *Biopharmaceutics of Ocular Drug Delivery*, P. Edman (Ed.). CRC Press, Boca Raton, FL, 1993, p. 105.
79. A. J. Bron. Prospects for the dry eye. *Trans. Ophthalmol. Soc. UK* 104:801 (1985).
80. S. F. Bernatchez, C. Tabatabay, R. Gurny. Sodium hyaluronate 0.25% used as a vehicle increases the bioavailability of topically administered gentamicin. *Graefe's Arch. Clin. Exp. Ophthalmol.* 231:157 (1993).
81. O. Camber, P. Edman, R. Gurny. Influence of sodium hyaluronate on the miotic effect of pilocarpine in rabbits. *Curr. Eye Res.* 6:779 (1987).
82. O. Camber, P. Edman. Sodium hyaluronate as an ophthalmic vehicle: some factors governing its effect on the ocular absorption of pilocarpine. *Curr. Eye Res.* 8:563 (1989).
83. M. F. Saettone, P. Chetoni, M. T. Torracca, S. Burgalassi,

B. Giannaccini. Evaluation of muco-adhesive properties and *in vivo* activity of ophthalmic vehicles based on hyaluronic acid. *Int. J. Pharm.* 51:203 (1989).

84. R. Gurny, J. E. Ryser, C. Tabatabay, M. Martenet, P. Edman, O. Camber. Precorneal residence time in humans of sodium hyaluronate as measured by gamma scintigraphy. *Graefe's Arch. Clin. Exp. Ophthalmol.* 228:510 (1991).
85. R. Gurny, H. Ibrahim, A. Aebi, P. Buri, C. G. Wilson, N. Washington, P. Edman, O. Camber. Design and evaluation of controlled release systems for the eye. *J. Control. Rel.* 6:367 (1987).
86. N. A. Peppas, Hydrogels of poly(vinyl alcohol) and its copolymers. In: *Hydrogels in Medicine and Pharmacy*, N. A. Peppas (Ed.). CRC Press, Boca Raton, FL, 1987, p. 1.
87. A. A. Deshpande, J. Heller, R. Gurny. Bioerodible polymers for ocular drug delivery. *Crit. Rev. Ther. Drug Carrier Syst.* 15:381 (1998).
88. T. F. Patton, J. R. Robinson. Ocular evaluation of polyvinyl alcohol vehicle in rabbits. *J. Pharm. Sci.* 64:1312 (1975).
89. N. Krishna, F. Brow. Polyvinyl alcohol as an ophthalmic vehicle. *Am. J. Ophthalmol.* 57:99 (1964).
90. M. L. Linn, L. T. Jones. Rate of lacrimal excretion of ophthalmic vehicles. *Am. J. Ophthalmol.* 65:76 (1973).
91. S. R. Waltman, T. C. Patrowicz. Effects of hydroxypropyl methylcellulose and polyvinyl alcohol on intraocular penetration. *Invest. Ophthalmol. Vis. Sci.* 9:966 (1970).
92. F. C. Bach, G. Riddel, C. Miller, J. A. Martin, J. D. Mullins. *Am. J. Ophthalmol.* 68:659 (1970).
93. C. G. Wilson, O. Olejnik, J. G. Hardy. Precorneal drainage of polyvinyl alcohol solutions in the rabbit assessed by gamma scintigraphy. *J. Pharm. Pharmacol.* 35:451 (1983).
94. A. Ludwig, M. M. Van Ooteghem. Influence of the viscosity and the surface tension of ophthalmic vehicles on the retention of a tracer in the precorneal area of human eyes. *Drug Dev. Ind. Pharm.* 14:2267 (1988).
95. I. Zaki, P. Fitzgerald, J. G. Hardy, C. G. Wilson. A comparison of the effect of viscosity on the precorneal residence of solutions in rabbit and man. *J. Pharm. Pharmacol.* 38:463 (1986).
96. M. A. Kassem, M. A. Attia, F. S. Habib, A. M. Mohamed. Effect of phenylephrine hydrochloride on betamethasone side effect in relation to viscosity ophthalmic preparations. I. Viscous solutions. *Drug Dev. Ind. Pharm.* 15:253 (1989).
97. M. Oechsner, S. Keipert. Polyacrylic acid/polyvinylpyrrolidone biopolymeric systems. I. Rheological and mucoadhesive properties of formulations potentially useful for the treatment of dry-eye syndrome. *Eur. J. Pharm. Biopharm.* 47:113 (1999).
98. A. H. Abd-El-Gawad, E. M. Ramadan, U. A. Seleman. The effect of ointments, bases and gels on release and ocular disposition of co-trimoxazole in rabbits' eyes. *Pharm. Ind.* 54:977 (1992).
99. M. B. Limberg, C. McCaa, G. E. Kissling, H. E. Kaufman. Topical application of hyaluronic acid and chondroitin sulfate in the treatment of dry eyes. *Am. J. Ophthalmol.* 103: 194 (1987).
100. M. F. Saettone, D. Monti, M. T. Torracca, P. Chetoni. Mucoadhesive ophthalmic vehicles: evaluation of polymeric low-viscosity formulations. *J. Ocular. Pharmacol.* 10:83 (1994).
101. E. Verschueren, L. Van Santvliet, A. Ludwig. Evaluation of various carrageenans as ophthalmic viscolysers. *S. T. P. Pharm. Sci.* 6:203 (1996).
102. Meseguer G, P. Buri, B. Plazonnet, A. Rozier, Gurny R. Gamma Scintigraphy comparison of eyedrops containing pilocarpine in healthy volunteers. *J. Ocular Pharmacol.* 12:481 (1996).
103. G. Meseguer, R. Gurny, P. Buri. Gamma scintigraphic evaluation of precorneal clearance in human volunteers and in rabbits. *Eur. J. Drug Metabol. Pharmacokin.* 18: 190 (1993).
104. G. Meseguer, R. Gurny, P. Buri. *In vivo* evaluation of dosage forms: application of gamma scintigraphy to nonenteral routes of administration. *J. Drug Targ.* 2:269 (1994).
105. P. D. Amin, C. P. Bhogte, M. A. Deshpande. Studies on gel tears. *Drug Dev. Ind. Pharm.* 22:735 (1996).
106. O. Felt, P. Buri, R. Gurny. Chitosan: a unique polysaccharide for drug delivery. *Drug Dev. Ind. Pharm.* 24:979 (1998).
107. I. Genta, B. Conti, P. Perugini, F. Pavanetto, A. Spadaro, G. Puglisi. Bioadhesive microspheres for ophthalmic administration of acyclovir. *J. Pharm. Pharmacol.* 49:737 (1997).
108. P. Calvo, J. L. Vila-Jato, M. J. Alonso. Evaluation of cationic polymer-coated nanocapsules as ocular drug carriers. *Int. J. Pharm.* 153:41 (1997).
109. O. Felt, P. Furrer, J. M. Mayer, B. Plazonnet, P. Buri, R. Gurny. Topical use of chitosan in ophtalmology: tolerance assessment and evaluation of precorneal retention *Int. J. Pharm.* 180:185 (1999).
110. R. Gurny, H. Ibrahim, P. Buri. The development and use of in situ formed gels triggered by pH. In: *Biopharmaceutics of Ocular Drug Delivery*, P. Edman (Ed.). CRC Press, Boca Raton, FL, 1993, p. 82.
111. D. M. Maurice. Prolonged-action drops. *Int. Ophthalmol. Clin.* 33:81 (1993).
112. R. Gurny, H. Ibrahim, T. Boye, P. Buri. Latices and thermosensitive gels as sustained delivery systems to the eye. In: *Ophthalmic Drug Delivery. Biopharmaceutical, Technological and Clinical Aspects*, M. S. Sattone et al. (Eds.). Fidia Research Series. Liviana Press, Padua, 1987, p. 27.
113. M. S. El-Aasser. Formation of polymer latexes by direct emulsification. Advances in emulsion polymerization and latex technology, 10th annual short course, Lehigh University, Bethlehem, PA, 1979, p. 1.
114. R. Gurny. Systèmes thérapeutiques à base de latex. In: *Formes Pharmaceutiques Nouvelles*, P. Buri et al. (Eds.). Techniques et Documentation, Lavoisier, Paris, 1985, p. 657.

115. R. Gurny, T. Boye, H. Ibrahim. Ocular therapy with nanoparticulate systems for controlled drug delivery. *J. Control. Rel.* 2:353 (1985).
116. H. Ibrahim, C. Bindschaedler, E. Doelker, P. Buri, R. Gurny. Concept and development of ophthalmic pseudolatexes triggered by pH. *Int. J. Pharm.* 77:211 (1991).
117. H. Ibrahim, C. Bindschaedler, E. Doelker, P. Buri, R. Gurny. Aqueous nanodispersions prepared by a salting-out process. *Int. J. Pharm.* 87:239 (1992).
118. H. Ibrahim. Concept et évaluation de systèmes polymériques dipersés (pseudo-latex) à usage ophtalmique. Ph.D Thesis, University of Geneva (1989).
119. T. Boye. Développement de nouvelles formes opthalmiques à libération prolongée et évaluation de la durée de leur activité pharmacologique. Ph.D Thesis, University of Geneva (1986).
120. R. Gurny, T. Boye, H. Ibrahim, P. Buri. Recent developments in controlled drug delivery to the eye. *Proc. Int. Symp. Control. Rel. Bioact. Mater.* 12:300 (1985).
121. A. S. Hoffman. Applications of thermally reversible polymers and hydrogels in therapeutics and diagnostics. *J. Control. Rel.* 6:297 (1987).
122. S. C. Miller, M. D. Donovan. Effect of poloxamer 407 gel on the miotic activity of pilocarpine nitrate in rabbits. *Int. J. Pharm.* 12:147 (1982).
123. G. Dumortier, J. L. Grossiord, M. Zuber, G. Couarraze, J. C. Chaumeil. Rheological study of thermoreversible morphine gel. *Drug Dev. Ind. Pharm.* 17:1255 (1991).
124. P. Alexandridis, J. F. Holzwarth, T. A. Hatton. Micellization of poly(ethylene oxide)–poly(propylene oxide)–poly(ethylene oxide) triblock copolymers in aqueous solutions: thermodynamics of copolymer association. *Macromolecules* 27:2414 (1994).
125. R. L. Henry, I. R. Schmolka. Burn wound coverings and the use of poloxamer preparations. *Crit. Rev. Biocomp.* 5: 207 (1989).
126. I. R. Schmolka. Artificial skin. I. Preparation and properties of Pluronic F-127 gels. *J. Biomed. Mater. Res.* 6:571 (1972).
127. BASF. Pluronic polyols toxicity and irritation data. BASF. Wyandotte, MI (1967).
128. K. Edsman, J. Carlfors, R. Petersson. Rheological evaluation of poloxamer as an in situ gel for ophthalmic use. *Eur. J. Pharm. Sci.* 6:105 (1998).
129. J. C. Gilbert, C. Washington, M. C. Davies, J. Hadgraft. The behaviour of Pluronic F127 in aqueous solution studied using fluorescent probes. *Int. J. Pharm.* 40:93 (1987).
130. M. Vadnere, G. Amidon, S. Lindebaum, J. L. Haslam. Thermodynamic studies on the gel–sol transition of some pluronic polyols. *Int. J. Pharm.* 22:207 (1984).
131. V. Lenaerts, C. Trinqueneaux, M. Quarton, F. Rieg-Falson, P. Couvreur. Temperature-dependent rheological behaviore of Pluronic F-127 aqueous solutions. *Int. J. Pharm.* 39:121 (1987).
132. S. C. Miller, B. R. Drabik. Rheological properties of poloxamer vehicles. *Int. J. Pharm.* 18:269 (1984).
133. G. O. Waring, R. R. Harris. Double-masked evaluation of a poloxamer artificial tear in keratoconjunctivitis sicca. In: *Symposium on Ocular Therapy*, I. H. Leopold et al. (Eds.). John Wiley, New York, 1979, p. 127.
134. A. S. Hoffman, A. Afrassiabi, L. C. Dong. Thermally reversible hydrogels. II. Delivery and selective removal of substances from aqueous solutions. *J. Control. Rel.* 4:213 (1986).
135. J. C. Gilbert, J. L. Richardson, M. C. Davies, K. J. Palin, J. Hadgraft. The effect of solutes and polymers on the gelation properties of Pluronic F-127 solutions for controlled drug delivery. *J. Control. Rel.* 5:113 (1987).
136. M. F. Saettone, G. Giannaccini, V. Delmonte, G. Campighi, G. Tota, F. La Marca. Solubilization of tropicamide by poloxamer: physicochemical data and activity data in rabbits and humans. *Int. J. Pharm.* 43:(1988).
137. G. Dumortier, M. Zuber, F. Chast, P. Sandouk, J. C. Chaumeil. Comparison between a thermoreversible gel and an insert in order to prolong the systemic absorption of morphine after ocular administration. *S. T. P. Pharm. Sci.* 2(1):111 (1992).
138. A. Al-Saden, A. T. Florence, T. L. Whateley. Novel poloxamer (pluronic) and poloxamine (tetronic) hydrogels: swelling and drug release. *J. Mater. Sci.* 9:1815 (1974).
139. C. Koller, P. Buri. Propriétés et intérêt pharmaceutique des gels thermoréversibles à base de poloxamers et poloxamines. *S. T. P. Pharm. Sci.* 3:115 (1987).
140. C. Mazuel, M.-C. Friteyre. Ophthalmological composition of the type which undergoes liquid-gel phase transition. U.S. patent 4,861,760 (1989).
141. C. Mazuel, M.-C. Friteyre. Composition pharmaceutique de type à transition de phase liquide-gel. F. patent 2,588,189 (1987).
142. R. Moorhouse, G. T. Colegrove, P. A. Sandford, J. Baird, K. S. Kang. A new gel forming polysaccharide. *A.C.S. Symp. Ser.* 150:111 (1981).
143. P.-E. Jansson, B. Lindberg. Structural studies of gellan gum, an extracellular polysaccharide elaborated by *Pseudomonas elodea. Carbohydr. Res.* 124:135 (1983).
144. M. A. O'Neill, R. R. Selvendran, V. J. Morris. Structure of the acidic extracellular gelling polysaccharide produced by *Pseudomonas elodea. Carbohydr. Res.* 124:123 (1983).
145. M. Milas, X. Shi, M. Rinaudo. On the physicochemical properties of gellan gum. *Biopolymers* 30:451 (1990).
146. K. S. Kang, G. T. Veeder, P. J. Mirrasoul, T. Kaneko, I. W. Cottrell. Agar-like polysaccharide produced by a *Pseudomonas* species: production and basic properties. *Appl. Environm. Microbiol.* 43:1086 (1982).
147. H. Grasdalen, O. Smidsroed. Gelation of gellan gum. *Carbohydr. Polym.* 7:371 (1987).
148. R. Chandrasekaran, R. P. Millane, S. Arnott, E. D. T. Atkins. The crystal structure of gellan gum. *Carbohydr. Res.* 175:1 (1988).
149. R. P. L. C. Chandrasekaran, K. L. Joyce, S. Arnott. Cation interaction in gellan: an X-ray study of the potassium salt. *Carbohydr. Res.* 181:23 (1988).
150. P. B. Deasy, J. Quigley. Rheological evaluation of deacetylated gellen gum (Gelrite) for pharmaceutical use. *Int. J. Pharm.* 73:117 (1991).

151. M. Paulsson, H. Hägerström, K. Edsman. Rheological studies of the gelation of deacetylated gellan gum (Gelrite®) in physiological conditions. *Eur. J. Pharm. Sci.* 9: 99 (1999).
152. A. Rozier, C. Mazuel, J. Grove, B. Plazonnet. Gelrite: a novel, ion-activated, in-situ gelling polymer for ophthalmic vehicles. Effect on bioavailability of timolol. *Int. J. Pharm.* 57:163 (1989).
153. H. Ibrahim, P. Buri, R. Gurny. Composition, structure et parametres physiologiques du systeme lacrymal impliques dans la conception des formes ophtalmiques. *Pharm. Acta Helv.* 63:146 (1988).
154. A. Rozier, J. Grove, C. Mazuel, B. Plazonnet. Gelrite solutions: novel ophthalmic vehicles that enhance ocular drug penetration. *Proc. Int. Symp. Control. Rel. Bioact. Mater.* 16:109 (1989).
155. J. Carlfors, K. Edsman, R. Petersson, K. Jörnving. Rheological evaluation of Gelrite® in situ gels for ophthalmic use. *Eur. J. Pharm. Sci.* 6:113 (1998).
156. Y. D. Sanzgiri, S. Mashi, V. Crescenzi, L. Callegaro, E. M. Topp, V. J. Stella. Gellan-based systems for ophthalmic sustained delivery of methylprednisolone. *J. Control. Rel.* 26:195 (1993).
157. N. S. Levy, C. Alsbury. Evaluation of timolol in gellan gum—a new vehicle to extend its duration of action. *Ann. Ophthalmol.* 26:166 (1994).
158. F. P. Gunning, E. L. Greve, A. M. Bron, J. M. Bosc, J. G. Royer, J. L. George, P. Lesure, D. Sirbat. Two topical carbonic anhydrase inhibitors sezolamide and dorzolamide in Gelrite vehicle: a multiple-dose efficacy study. *Graefe's Arch. Clin. Exp. Ophthalmol.* 231:384 (1993).
159. R. Vogel, S. F. Kulaga, J. K. Laurence, R. L. Gross, B. G. Haik, D. Karp, M. Koby, T. J. Zimmerman. The effect of a Gelrite vehicle on the efficacy of low concentrations of timolol. *Invest. Ophthalmol. Vis. Sci.* 31:404 (1990).
160. D. M. Maurice, S. P. Srinivas. Use of fluorometry in assessing the efficacy of a cation-sensitive gel as an ophtalmic vehicle: comparison with scintigraphy. *J. Pharm. Sci.* 81:615 (1992).
161. J. L. Greaves, C. G. Wilson, A. Rozier, J. Grove, B. Plazonnet. Scintigraphic assessment of an ophtalmic gelling vehicle in man and rabbit. *Curr. Eye Res.* 9:415 (1990).
162. A. Rozier, J. Grove, C. Mazuel, B. Plazonnet. Gelrite solutions: novel ophthalmic vehicles that enhance ocular drug penetration. *Proc. Int. Symp. Control. Rel. Bioact. Mater.* 16:109 (1989).
163. K. Dickstein, T. Aarsland. Comparison of the effects of aqueous and gellan ophthalmic timolol on peak exercise performance in middle-aged men. *Am. J. Ophthalmol.* 121:367 (1996).
164. H. Bothner, Waalwe T, Wik O. Rheological characterization of tear substitutes. *Drug Dev. Ind. Pharm.* 16:755 (1990).
165. Y. S. Wysenbeek, N. Loya, I. Ben Sira. The effect of sodium hyaluronate on the corneal epithelium. An ultrastructural study. *Invest. Ophthalmol. Vis. Sci.* 29:194 (1988).
166. A. M. Tonjum. Effects of benzalkonium chloride upon the corneal epithelium studied with scanning electron microscopy. *Acta Ophthalmol.* 53:358 (1975).
167. A. J. Duncan, W. S. Wilson. Some preservatives in eyedrop preparations hasten the formulation of dry spots in the rabbit cornea. *Brit. J. Pharmacol.* 56:359 (1976).
168. A. R. Gasset. Benzalkonium chloride toxicity to the human cornea. *Am. J. Ophthalmol.* 84:169 (1977).
169. W. S. Wilson, A. J. Duncan, J. L. Jay. Effect of benzalkonium chloride on the stability of the precorneal tear film in rabbit and man. *Br. J. Ophthalmol.* 59:667 (1975).
170. G. R. Snibson, J. L. Greaves, N. D. W. Soper, J. M. Tiffany, C. G. Wilson, A. J. Bron. Ocular surface residence times of artificial tear solutions. *Cornea* 11:288 (1992).
171. C. Tabatabay. Instillation d'acide hyaluronique à 0,1% lors de kératite sèche sévère. *J. Fr. Ophtalmol.* 8:513 (1985).
172. J. G. Orsoni, M. Chiari, A. Guazzi, M. De Carli, D. Guidolin. Efficacité de l'acide hyaluronique en collyre dans le traitement de l'oeil sec. *Ophtalmologie* 2:355 (1988).
173. G. R. Snibson, J. L. Greaves, N. D. W. Soper, J. I. Prydal, C. G. Wilson, A. J. Bron. Precorneal residence times of sodium hyaluronate solutions studied by quantitative gamma scintigraphy. *Eye* 4:594 (1990).
174. M. A. Lemp. Artificial tear solutions. *Int. Ophthalmol. Clin.* 13:221 (1973).
175. L. H. Bruner. Ocular irritation. In: In Vitro *Toxicity Testing: Applications to Safety Evaluation*, J. M. Frazier (Ed.). Marcel Dekker, New York, 1992, p. 149.
176. B. Ballantyne. Ophthalmic toxicology. In: *General & Applied Toxicology*, B. Ballantyne et al. (Eds.). Macmillan, London, 1995, p. 503.
177. G. P. Daston, F. E. Freeberg. Ocular irritation testing. In: *Dermal and Ocular Toxicology: Fundamentals and Methods*, D. W. Hobson (Ed.). CRC Press, Boca Raton, FL, 1991, p. 509.
178. O. Dudinski, B. C. Finnin, B. L. Reed. Acceptability of thickened eye drops to human subjects. *Curr. Ther. Res.* 33:322 (1983).
179. F. Bottari, B. Giannaccini, B. Cristofori, M. F. Saettone, N. Tellini. Semisolid ophthalmic vehicles. Part I. A study of eye irritation in albino rabbits of a series of gel type aqueous bases. *Farmaco* 33:434 (1978).
180. M. D. Nelson, J. D. Bartlett, T. Karkkainen, M. Voce. Ocular tolerability of timolol in Gelrite in young glaucoma patients. *J. Am. Opt. Assoc.* 67:659 (1996).
181. M. Zignani, C. Tabatabay, R. Gurny. Topical semi-solid drug delivery: kinetics and tolerance of ophthalmic hydrogels. *Adv. Drug Deliv. Rev.* 16:51 (1995).
182. J. Kreuter. Particulates (nanoparticles and microparticles), In: *Ophthalmic Drug Delivery Systems*, A. K. Mitra. (Ed.). Marcel Dekker, New York, 1993, p. 275.
183. A. Zimmer, J. Kreuter. Microspheres and nanoparticles used in ocular delivery systems. *Adv. Drug Del. Rev.* 16: 61 (1995).
184. A. Joshi. Microparticulates for ophthalmic drug delivery. *J. Ocular. Pharmacol.* 10:29 (1994).
185. J. Kreuter. Nanoparticles as bioadhesive ocular drug deliv-

ery systems. In: *Bioadhesive Drug Delivery Systems*, V. Lenaerts et al. (Eds.). CRC Press, Boca Raton, FL, 1990, p. 203.
186. J. W. Sieg, J. W. Triplett. Precorneal retention of topically instilled micronized particles. *J. Pharm. Sci.* 69:863 (1980).
187. J. Kreuter. Evaluation of nanoparticles as drug delivery systems. I. Preparation methods. *Pharm. Acta Helv.* 58: 196 (1983).
188. A. Joshi. Microparticulates as an ocular delivery system. In: *Ocular Therapeutics and Drug Delivery*, I. K. Reddy (Ed.). Technomic, Lancaster, PA, 1996, p. 441.
189. P. Couvreur, L. Grislain, V. Lenaerts, F. Brasseur, P. Guiot, A. Bienacki. Biodegradable polymeric nanoparticles as drug carriers for antitumor drugs. In: *Polymeric Nanoparticles and Microspheres*, P. Guiot et al. (Eds.). CRC Press, Boca Raton, FL, 1986, p. 27.
190. W. R. Vezin, A. T. Florence. *In vitro* degradation rates of biodegradable poly-n-alkyl cyanoacrylates. *J. Pharm. Pharmacol.* 30:5P (1978).
191. V. Lenaerts, P. Couvreur, D. Christiaens-Leyh, E. Joiris, M. Roland, B. Rollmann, P. Speiser. Degradation of poly (isobutylcyanoacrylate) nanoparticles. *Biomaterials* 5.65 (1984).
192. T. Harmia, J. Kreuter, P. Speiser, T. Boye, R. Gurny, A. Kubis. Enhancement of the myotic response of rabbits with pilocarpine-loaded polybutylcyanoacrylate nanoparticles. *Int. J. Pharm.* 33:187 (1986).
193. A. Zimmer, E. Mutschler, G. Lambrecht, D. Mayer, J. Kreuter. Pharmacokinetic and pharmacodynamic aspects of an ophthalmic pilocarpine nanoparticle–delivery system. *Pharm. Res.* 11:1435 (1994).
194. A. K. Zimmer, P. Maincent, P. Thouvenot, J. Kreuter. Hydrocortisone delivery to healthy and inflamed eyes using a micellar polysorbate 80 solution or albumin nanoparticles. *Int. J. Pharm.* 110:211 (1994).
195. S. E. Leucuta. The kinetics of *in vitro* release and the pharmacokinetics of miotic response in rabbits of gelatin and albumin microspheres with pilocarpine. *Int. J. Pharm.* 54: 71 (1989).
196. A. K. Zimmer, J. Kreuter, M. F. Saettone, H. Zerbe. Size dependency of albumin carrier systems on *in vivo* effects of pilocarpine in the rabbit eye. *Proc. Int. Symp. Control. Rel. Bioact. Mater.* 18:493 (1991).
197. A. K. Zimmer, H. Zerbe, J. Kreuter. Evaluation of pilocarpine-loaded albumin particles as drug delivery systems for controlled delivery in the eye. I. *In vitro* and *in vivo* characterisation. *J. Control. Rel.* 32:57 (1994).
198. A. K. Zimmer, P. Chetoni, M. F. Saettone, H. Zerbe, J. Kreuter. Evaluation of pilocarpine-loaded albumin particles as controlled drug delivery systems for the eye. II. Co-administration with bioadhesive and viscous polymers. *J. Control. Rel.* Vol. 33:19, 53:31 (1995).
199. M. E. Evans, N. E. Richardson, D. A. Norton. The preparation and properties of polyamide microcapsules containing pilocarpine in solutions. *J. Pharm. Pharmacol.* (1981).
200. L. Marchal-Heussler, H. Fessi, J. P. Devissaguet, M. Hoffman, P. Maincent. Colloidal drug delivery systems for the eye. A comparison of the efficacy of three different polymers: polyisobutylcyanoacrylate, polylactic-co-glycolic acid, poly-epsilon-caprolacton. *S.T.P. Pharm. Sci.* 2:98 (1992).
201. P. Maincent, L. Marchal-Heussler, D. Sirbat, P. Thouvenot, M. Hoffman, J. A. Vallet. Polycaprolacton colloidal carriers for controlled ophthalmic delivery. *Proc. Int. Symp. Control. Rel. Bioact. Mater.* 19:226 (1992).
202. V. Masson, P. Billardon, H. Fessi, J. P. Devissaguet, F. Puisieux. Tolerance studies and pharmacokinetic evaluation of indomethacin-loaded nanocapsules in rabbit eyes. *Proc. Int. Symp. Control. Rel. Bioact. Mater.* 19:423 (1992).
203. P. Calvo, J. L. Vila-Jato, M. J. Alonso. Cationic polymer-coated nanocapsules as ocular drug carriers. *Proc. Int. Symp. Control. Rel. Bioact. Mater.* 24:97 (1997).
204. C. Losa, M. J. Alonso, J. L. Vila, F. Orallo, J. Martinez, J. A. Saavedra, J. C. Pastor. Reduction of cardiovascular side effects associated with ocular administration of metipranolol by inclusion in polymeric nanocapsules. *J. Ocular. Pharmacol.* 8:191 (1992).
205. J. W. Shell. Ophthalmic bioerodible drug dispensing formulation. *U.S. patent* 4,001,388 (1977).
206. V. Vidmar, S. Pepeljnjak, I. Jalsenjak. The *in vivo* evaluation of poly(lactic acid) microcapsules of pilocarpine hydrochloride. *J. Microencapsulation* 2:289 (1985).
207. M. Beal, N. E. Richardson, B. J. Meakin, D. J. G. Davies. The use of polyphthalamide microcapsules for obtaining extended periods of therapy in the eye. In: *Microspheres and Drug Therapy: Pharmaceutical Immunological and Medical Aspects*, Davis et al. (Eds.). 1984, p. 347.
208. T. Harmia-Pulkinnen, A. Ihantola, A. Tuomi, E. Kristoffersson. Nanoencapsulation of timolol by suspension and micelle polymerization. *Acta Pharm. Fennica* 95:89 (1997).
209. M. J. Alonso, C. Losa, B. Seijo, D. Torres, J. L. Vila-Jato. New ophthalmic drug release systems, formulation and ocular disposition of amikacin-loaded nanoparticles. *5ème Congr. Int. Technol. Pharm.* 1:77 (1989).
210. R. Diepold, J. Kreuter, J. Himber, R. Gurny, V. H. L. Lee, J. R. Robinson, M. F. Saettone, O. E. Schnaudigel. Comparison of different models for the testing of pilocarpine eyedrops using conventional eyedrops and a novel depot formulation (nanoparticules). *Graefe's Arch. Clin. Exp. Ophthalmol.* 227:188 (1989).
211. T. Harmia, P. Speiser, J. Kreuter. A solid colloidal drug delivery system for the eye: encapsulation of pilocarpin in nanoparticles. *J. Microencapsulation* 3:3 (1986).
212. T. Harmia, P. Speiser, J. Kreuter. Optimization of pilocarpine loading onto nanoparticles by sorption procedures. *Int. J. Pharm.* 33:45 (1986).
213. J. Kreuter. Nanoparticles and liposomes in ophthalmic drug delivery; biopharmaceutical, technological and clinical aspects. In: *Ophthalmic Drug Delivery*, M. F. Saettone et al. (Eds.). Liviana Press, Padua, 1987, p. 101.

214. V. H. K. Li, R. W. Wood, J. Kreuter, T. Harmia, J. R. Robinson. Ocular drug delivery of progesterone using nanoparticles. *J. Microencapsulation* 3:213 (1986).
215. J. Kreuter, V. H. K. Li, R. W. Wood, J. R. Robinson, T. Harmia, P. Speiser. Ocular disposition of polyhexylcyanoacrylate nanoparticles and of nanoparticles-bound progesterone. *Proc. Int. Symp. Control. Rel. Bioact. Mater.* 12:304 (1985).
216. L. Marchal-Heussler, P. Maincent, M. Hoffman, J. Spittler, P. Couvreur. Antiglaucomatous activity of betaxolol chlorhydrate sorbed onto different isobutylcyanoacrylate nanoparticle preparations. *Int. J. Pharm.* 58:115 (1990).
217. Z. Mazor, U. Ticho, U. Rehany, L. Rose. Piloplex, a new long-acting pilocarpine polymer salt. B: Comparative study of the visual effects of pilocarpine and Piloplex eye drops. *Br. J. Ophthalmol.* 63:48 (1979).
218. C. Andermann, G. de Burlet, C. Cannet. Etude comparative de l'activité antiglaucomateuse de glaupex 2 et du nitrate de pilocarpine sur le glaucome expérimental à l'alphachymotrypsine. *J. Fr. Ophtalmol.* 5:499 (1982).
219. H. Z. Klein, M. Lugo, M. Shields, M. B. Leon, E. Duzman. A dose-response study of piloplex for duration of action in humans. *Am. J. Ophthalmol.* 99:23 (1985).
220. U. Ticho, M. Blumenthal, S. Zonis, A. Gal, I. Blank, Z. W. Mazor. Piloplex, a new long-acting pilocarpine polymer salt. A: Long-term study. *Br. J. Ophthalmol.* 63:45 (1979).
221. F. Gurtler, R. Gurny. Patent litterature review of ophthalmic inserts. *Drug Dev. Ind. Pharm.* 21:1 (1995).
222. M. F. Saettone. Solid polymeric inserts/disks as drug delivery devices. In: *Biopharmaceutics of Ocular Drug Delivery*, P. Edman (Ed.). CRC Press, London, 1993, p. 61.
223. R. Bawa. Ocular inserts. In: *Ophthalmic Drug Delivery Systems*, A. K. Mitra (Ed.). Marcel Dekker, New York, 1993, p. 223.
224. M. F. Saettone, L. Salminen. Ocular inserts for topical delivery. *Adv. Drug Deliv. Rev.* 16:95 (1995).
225. M. F. Saettone, S. Burgalassi, P. Chetoni. Ocular bioadhesive drug delivery systems. In: *Bioadhesive Drug Delivery Systems: Fundamentals, Novel Approaches, and Development*, E. Mathiowitz et al. (Eds.). Marcel Dekker, New York, 1999, p. 601.
226. M. A. Khan, M. J. Durrani. Polymers in ophthalmic drug delivery systems. In: *Ocular Therapeutics and Drug Delivery. A Multi-Disciplinary Approach*, I. K. Reddy (Ed.). Technomic, Lancaster, PA, 1996, p. 405.
227. J. W. Shell. Ophthalmic drug delivery systems. *Drug Dev. Res.* 6:245 (1985).
228. J. W. Shell. Ophthalmic drug delivery systems. *Surv. Ophthalmol.* 29:117 (1984).
229. J. W. Shell, R. W. Baker. Diffusional systems for controlled release of drugs to the eye. *Ann. Ophthalmol.* 6: 1037 (1974).
230. S. P. Loucas, M. Heskel, M. Haddad. Solid state ophthalmic dosage system. II. Use of polyuronic acid in effecting prolonged delivery of pilocarpine in the eye. *Metab. Ophthalmol.* 1:27 (1976).
231. S. E. Bloomfield, T. Miyata, M. W. Dunn, N. Bueser, K. H. Stenzel, A. L. Rubin. Soluble gentamycin ophthalmic inserts as a drug delivery system. *Arch. Ophthalmol.* 96: 885 (1978).
232. D. H. Slatter, N. D. Costa, M. E. Edwards. Ocular inserts for application of drugs to bovine eyes—*in vivo* and *in vitro* studies on the release of gentamycin from collagen inserts. *Austr. Vet. J.* 59:4 (1982).
233. J. K. Milani, I. Verbukh, U. Pleyer, H. Sumner, S. A. Adamu, H. P. Halabi, H. J. Chou, D. A. Lee, B. J. Mondino. Collagen shields impregnated with gentamicin–dexamethasone as a potential drug delivery device. *Am. J. Ophthalmol.* 116:622 (1993).
234. K. K. Assil, S. R. Zarnegar, B. D. Fouraker, D. J. Schanzlin. Efficacy of tobramycin-soaked collagen shields vs. tobramycin eyedrop loading dose for sustained treatment of experimental pseudomonas aeruginosa induced keratitis in rabbits. *Am. J. Ophthalmol.* 113:418 (1992).
235. S. Hosaka, H. Ozawa, H. Tanzawa. *In vivo* evaluation of ocular inserts of hydrogel impregnated with antibiotics for trachoma therapy. *Biomaterials* 4:243 (1983).
236. H. Ozawa, S. Hosaka, T. Kunitomo, H. Tanzawa. Ocular inserts for controlled release of antibiotics. *Biomaterials* 4:170 (1983).
237. A. Urtti, L. Salminen, O. Miinalainen. Systemic absorption of ocular pilocarpine is modified by polymer matrices. *Int. J. Pharm.* 23:147 (1985).
238. M. F. Saettone, P. Chetoni, M. T. Torraca, B. Giannaccini, L. Naber, U. Conte, M. E. Sangalli, A. Gazzaniga. Application of the compression technique to the manufacture of pilocarpine inserts. *Acta Pharm. Technol.* 36(1):15 (1990).
239. F. Gurtler, V. Kaltsatos, B. Boisramé, R. Gurny. Long-acting soluble bioadhesive ophthalmic drug insert (BODI) containing gentamicin for veterinary use: optimization and clinical investigation. *J. Control. Rel.* 33:231 (1995).
240. V. Baeyens, V. Kaltsatos, B. Boisramé, M. Fathi, R. Gurny. Evaluation of soluble ophthalmic inserts for prolonged release of gentamicin: lachrymal pharmacokinetics and ocular tolerance. *J. Ocular Pharmacol. Ther.* 14:263 (1998).
241. V. Baeyens, V. Kaltsatos, B. Boisramé, E. Varesio, J.-L. Veuthey, M. Fathi, L. P. Balant, M. Gex-Fabry, R. Gurny. Optimized release of dexamethasone and gentamicin from a soluble insert for the treatment of external ophthalmic infections. *J. Control. Rel.* 52:215 (1998).
242. V. Baeyens. Développement et évaluation d'inserts ophtalmiques solubles destinés au traitement des infections bactériennes du segment antérieur de l'oeil. PhD Thesis, Université de Genève, No. 2992 (1998).
243. Y. Yamamoto, Y. Kaga, T. Yoshikawa, A. Moribe. Ultraviolet-hardenable adhesive. U.S. patent 5,145,884 (1992).
244. V. H. L. Lee, S. Yong Li, H. Sasaki, M. F. Saettone, P. Chetoni. Influence of drug release rate on systemic timolol absorption from polymeric ocular inserts in the pigmented rabbit. *J. Ocular. Pharmacol.* 10:421 (1994).

245. H. Sasaki, C. Tei, K. Nishida, J. Nakamura. Drug release of an ophthalmic insert of a beta-blocker as an ocular drug delivery system. *J. Control. Rel.* 27:127 (1993).
246. M. C. Richardson, P. H. Bentley. A new ophthalmic delivery system. In: *Ophthalmic Drug Delivery Systems*, A. K. Mitra (Ed.). Marcel Dekker, New York, 1993, p. 355.
247. M. F. Saettone, M. T. Torraca, A. Pagano, B. Giannaccini, L. Rodriguez, M. Cini. Controlled release of pilocarpine from coated polymeric ophthalmic inserts prepared by extrusion. *Int. J. Pharm.* 86:159 (1992).
248. M. A. Attia, M. A. Kassem, S. M. Safwat. *In vivo* performance of 3-H-dexamethasone ophthalmic film delivery system in the rabbit eye. *Int. J. Pharm.* 47:21 (1988).
249. V. H. L. Lee, S. Li, M. F. Saettone, P. Chetoni, H. Bundgaard. Systemic and ocular absorption of timolol prodrugs from erodible inserts. *Proc. Intern. Symp. Control. Rel. Bioact. Mater.* 18:291 (1991).
250. L. Salminen, A. Urtti, H. Kujari, M. Juslin. Prolonged pulse-entry of pilocarpine with a soluble drug insert. *Graefe's Arch. Clin. Exp. Ophthalmol.* 221:96 (1983).
251. S. N. Fyodorov, S. N. Bagrov, T. S. Amstislavskaya, I. A. Maklakova, S. V. Maslenkov. Ophthalmological collagen covering. U.S. patent 4,913,904 (1990).
252. M. L. Friedberg, U. Pleyer, B. J. Mondino. Device drug delivery to the eye. Collagen shields, iontophoresis, and pumps. *Ophthalmology* 98:725 (1991).
253. R. B. Phinney, D. Schwartz, D. A. Lee, B. J. Mondino. Collagen-shield delivery of gentamicin and vancomycin. *Arch. Ophthalmol.* 106:1599 (1988).
254. M. R. Sawusch, T. P. O'Brien, J. D. Dick, J. D. Gottsch. Use of collagen corneal shields in the treatment of bacterial keratitis. *Am. J. Ophthalmol.* 106:279 (1988).
255. D. G. Hwang, W. H. Stern, P. H. Hwang, L. A. Macgowan-Smith. Collagen shield enhancement of topical dexamethasone penetration. *Arch. Ophthalmol.* 107:1375 (1989).
256. M. R. Sawusch, T. P. O'Brien, S. A. Updegraff. Collagen corneal shields enhance penetration of topical prednisolone acetate. *J. Cataract Refract. Surg.* 15:625 (1989).
257. J. R. Gussler, P. Ashton, W. S. Van Meter, T. J. Smith. Collagen shield delivery of trifluorothymidine. *J. Cataract. Refract. Surg.* 16:719 (1990).
258. J. M. Hill, R. J. O'Callaghan, J. A. Hobden, H. E. Kaufman. Corneal collagen shields for ocular delivery. In: *Ophthalmic Drug Delivery Systems*, A. K. Mitra (Ed.). Marcel Dekker, New York, 1993, p. 261.
259. R. S. Shofner, H. E. Kaufman, J. M. Hill. New horizons in ocular drug delivery. *Ophtal. Clin. North Am.* 2:15 (1989).
260. B. J. Mondino. Collagen shields. *Am. J. Ophthalmol.* 112: 587 (1991).
261. R. J. Harwood, J. B. Schwartz. Drug release from compression molded films preliminary studies with pilocarpine. *Drug Dev. Ind. Pharm.* 8(5):663 (1982).
262. F. Gurtler, V. Kaltsatos, B. Boisramé, J. Deleforge, M. Gex-Fabry, L. P. Balant, R. Gurny. Ocular availability of gentamicin in small animals after topical administration of a conventional eye drop solution and a novel long acting bioadhesive ophthalmic drug insert. *Pharm. Res.* 12:1791 (1995).
263. D. W. Lamberts, D. P. Langston, W. Chu. A clinical study of slow-releasing artificial tears. *Ophthalmology* 85:794 (1978).
264. L. Sendelbeck, D. Moore, J. Urqhart. Comparative distribution of pilocarpine in ocular tissues of the rabbit during administration by eyedrop or by membrane-controlled delivery systems. *Am. J. Ophthalmol.* 80:274 (1975).
265. B. G. Clerc, C. Robberechts. Le traitement de glaucome par injection intr-vitréenne de gentamicine (compte rendu de recherche clinique). *Rec. Med. Vet.* 168:97 (1992).
266. S. M. Drance, D. W. A. Mitchell, M. Schulzer. The duration of action of pilocarpine ocusert on intraocular pressure in man. *Canad. J. Ophthalmol.* 10:450 (1975).
267. I. P. Pollack, H. A. Quigley, T. S. Harbin. The Ocusert pilocarpine system: advantages and disadvantages. *South Med. J.* 69(10):1296 (1976).
268. R. L. Friederich. The pilocarpine Ocusert: a new drug delivery system. *Ann. Ophthalmol.* 6:1279 (1974).
269. R. K. Arnold. Ocular drug delivery device. U.S. patent 4,014,335 (1977).
270. R. M. Gale, M. Ben-Dor, N. Keller. Ocular therapeutic system for dispensing a medication formulation. U.S. patent 4,190,642 (1980).
271. R. M. Gale. Vorrichtung zur verzögerten Freisetzung von Arzneimittel und Verfahren zu deren Herstellung. Ger. Offen. P 26 33 987.7 (1977).
272. S. Darougar. Ocular insert. *Eur. Pat. Appl.* 262 893 A2 (1988).
273. S. Darougar. Ocular insert for the fornix. U.S. patent 5,147,647 (1992).
274. A. Urtti, J. D. Pipkin, G. Rork, A. J. Repta. Controlled drug delivery devices for experimental ocular studies with timolol. 1. Invitro release studies. *Int. J. Pharm.* 61:235 (1990).
275. A. Urtti, J. D. Pipkin, T. Sendo, U. Finne, A. J. Repta. Controlled drug delivery devices for experimental ocular studies with timolol. 2. Ocular and systemic absorption in rabbits. *Int. J. Pharm.* 61:241 (1990).
276. R. Sutinen, A. Urtti, R. Miettunen, P. Paronen. Water-activated and pH-controlled release of weak bases from silicone reservoir devices. *Int. J. Pharm.* 62:113 (1990).
277. Chetoni, P., G. Di Colo, M. Grandi, M. Morelli, M. F. Saettone, S. Darougar. Silicone rubber/hydrogel composite ophthalmic inserts: preparation and preliminary *in vitro*/*in vivo* evaluation. *Eur. J. Pharm. Biopharm.* 46:125 (1998).
278. M. F. Refojo. Polymers in contact lenses: an overview. *Curr. Eye Res.* 4(6):719 (1985).
279. Y. T. Maddox, H. N. Bernstein. An evaluation of the bionite hydrophilic contact lens for use in a drug delivery system. *Ann. Ophthalmol.* 4:789 (1972).
280. R. Praus, I. Brettschneider, L. Krejci, D. Kalvodova. Hydrophilic contact lenses as a new therapeutic approach for the topical use of chloramphenicol and tetracycline. *Ophthalmologica* 165:62 (1972).

281. D. S. Hull, H. F. Edelhauser, R. A. Hyndiuk. Ocular penetration of prednisolone and the hydrophilic contact lense. *Arch. Ophthalmol.* 92:413 (1974).

282. R. Bawa. Sustained-release formulation comprising a hydrophobic polymer system. *Eur. Pat. Appl.* 219-207-A2 (1987).

283. R. Bawa. Sustained-release formulation comprising a hydrophobic polymer. *Eur. Pat. Appl.* 219-207-B1 (1987).

284. D. Pavan-Langston, R. H. S. Langston, P. A. Geary. Idoxuridine ocular insert therapy: use in treatment of experimental herpes simplex keratitis. *Arch. Ophthalmol.* 93: 1349 (1975).

285. P. I. Punch, D. H. Slatter, N. D. Costa, M. E. Edwards. Investigation of gelatin as a possible biodegradable matrix for sustained delivery of gentamicin to the bovine eye. *J. Vet. Pharmacol. Ther.* 8:335 (1985).

286. P. Simamora, S. R. Nadkarni, Y.-C. Lee, H. S. Yalkowsky. Controlled delivery of pilocarpine. 2. *In vivo* evaluation of Gelfoam device. *Int. J. Pharm.* 170:209 (1998).

287. Y. Lee, S. H. Yalkowsky. Effect of formulation on the systemic absorption of insulin from enhancer-free ocular devices. *Int. J. Pharm.* 185:199 (1999).

288. Y. C. Lee, S. H. Yalkowsky. Ocular devices for the controlled systemic delivery of insulin: *in vitro* and *in vivo* dissolution. *Int. J. Pharm.* 181:71 (1999).

289. Y. C. Lee, J. Millard, G. J. Negvesky, S. I. Butrus, S. H. Yalkowsky. Formulation and *in vivo* evaluation of ocular insert containing phenylephrine and tropicamide. *Int. J. Pharm.* 182:121 (1999).

290. G. M. Grass, J. Cobby, M. C. Makoid. Ocular delivery of pilocarpine from erodable matrices. *J. Pharm. Sci.* 73:618 (1984).

291. V. H. L. Lee, K. J. Pince, D. A. Frambach, B. Martini. Drug delivery to the posterior segment. In: *Retina*, S. J. Ryan (Ed.). Mosby, St Louis, MO, 1989, p. 483.

292. D. M. Maurice. The exchange of sodium between the vitreous body and the blood and aqueous humour. *J. Physiol.* 137:110 (1957).

293. Y. Ogura, H. Kimura. Biodegradable polymer microspheres for targeted drug delivery to the retinal pigment epithelium. *Surv. Ophthalmol.* 39:S17–S24 (1995).

294. D. H. Lewis. Controlled release of bioactive agents from lactide/glycolide polymers. In: *Biodegradable Polymers as Drug Delivery Systems*, M. Chasin et al. (Eds.). Marcel Dekker, New York, 1990, p. 1.

295. M. Vert, S. M. Li, H. Garreau. More about the degradation of LA/GA-derived matrices in aqueous media. *J. Control. Rel.* 16:15 (1991).

296. S. Li, S. Girod-Holland, M. Vert. Hydrolytic degradation of poly(DL-lactic acid) in the presence of caffeine base. *J. Control. Rel.* 40:41 (1996).

297. H. Kimura, Y. Ogura, M. Hashizoe, H. Nishiwaki, Y. Honda, Y. Ikada. A new vitreal drug delivery system using an implantable biodegradable polymeric device. *Invest. Ophthalmol. Vis. Sci.* 35:2815 (1994).

298. G. E. Trope, Y. L. Cheng, H. Sheardown, G. S. Liu, I. A. Menon, J. G. Heathcote, D. S. Rootman, W. J. Chiu, L. Gould. Depot drug delivery system for 5-fluorouracil after filtration surgery in the rabbit. *Can. J. Ophthalmol.* 29:263 (1994).

299. L. Gould, G. Trope, Y. L. Cheng, J. G. Heathcote, H. Sheardown, D. Rootman, G. S. Liu, I. A. Menon. Fifty: fifty poly(DL glycolic acid-lactic acid) copolymer as a drug delivery system for 5-fluorouracil: a histopathological evaluation. *Can. J. Ophthalmol.* 29:168 (1994).

300. P. Hostyn, F. Villain, N. Malek-Chehire, F. Kühne, Y. Takesue, R. K. Parrish, J. M. Parel. Implant biodégradable à libération contrôlée de 5-FU dans la chirurgie du glaucome. Etude expérimentale. *J. Fr. Ophtalmol.* 19:133 (1996).

301. G. Wang, I. G. Tucker, M. S. Roberts, L. W. Hirst. *In vitro* and *in vivo* evaluation in rabbits of a controlled release 5-fluorouracil subconjunctival implant based on poly(D,L-lactide-co-glycolide). *Pharm. Res.* 13:1059 (1996).

302. P. E. Rubsamen, P. A. Davis, E. Hernandez, G. E. O'Grady, S. W. Cousins. Prevention of experimental proliferative vitreoretinopathy with a biodegradable intravitreal implant for the sustained release of fluorouracil. *Arch. Ophthalmol.* 112:407 (1994).

303. M. Hashizoe, Y. Ogura, H. Kimura, T. Moritera, Y. Honda, M. Kyo, S. H. Hyon, Y. Ikada. Scleral plug of biodegradable polymers for controlled drug release in the vitreous. *Arch. Ophthalmol.* 112:1380 (1994).

304. M. Hashizoe, Y. Ogura, T. Takanashi, N. Kunou, Y. Honda, Y. Ikada. Implantable biodegradable polymeric device in the treatment of experimental proliferative vitreoretinopathy. *Curr. Eye Res.* 14:473 (1995).

305. T. Zhou, H. Lewis, R. E. Foster, S. P. Schwendeman. Development of a multiple-drug delivery implant for intraocular management of proliferative vitreoretinopathy. *J. Control. Rel.* 55:281 (1998).

306. N. Kunou, Y. Ogura, M. Hashizoe, Y. Honda, S. H. Hyon, Y. Ikada. Controlled intraocular delivery of ganciclovir with use of biodegradable scleral implant in rabbits. *J. Control. Rel.* 37:143 (1995).

307. M. Hashizoe, Y. Ogura, T. Takanashi, N. Kunou, Y. Honda, Y. Ikada. Biodegradable polymeric device for sustained intravitreal release of ganciclovir in rabbits. *Curr. Eye Res.* 16:633 (1997).

308. I. Yamakawa, M. Ishida, T. Kato, H. Ando, N. Asakawa. Release behavior of poly(lactic acid-co-glycolic acid) implants containing phosphorothioate oligodeoxynucleotide. *Biol. Pharm. Bull.* 20:455 (1997).

309. D. P. Hainsworth, J. D. Conklin, J. R. Bierly, D. Ax, P. Ashton. Intravitreal delivery of ciprofloxacin. *J. Ocul. Pharmacol.* 12:183 (1996).

310. H. Miyamoto, Y. Ogura, M. Hashizoe, N. Kunou, Y. Honda, Y. Ikada. Biodegradable scleral implant for intravitreal controlled release of fluconazole. *Curr. Eye Res.* 16: 930 (1997).

311. Y. Morita, A. Ohtori, M. Kimura, K. Tojo. Intravitreous delivery of dexamethasone sodium m-sulfobenzoate from poly(DL-lactic acid) implants. *Biol. Pharm. Bull.* 21:188 (1998).

312. O. Nishi, K. Nishi, I. Saitoh, K. Sakanishi. Inhibition of migrating lens epithelial cells by sustained release of ethylenediaminetetraacetic acid. *J. Cataract. Refract. Surg.* 22(Suppl. 1):863 (1996).
313. M. R. Tetz, M. W. Ries, C. Lucas, H. Stricker, H. E. Volcker. Inhibition of posterior capsule opacification by an intraocular-lens-bound sustained drug delivery system: an experimental animal study and literature review. *J. Cataract. Refract. Surg.* 22:1070 (1996).
314. Y. Morita, H. Saino, K. Tojo. Polymer blend implant for ocular delivery of fluorometholone. *Biol. Pharm. Bull.* 21:72 (1998).
315. D. T. Tan, S. P. Chee, L. Lim, A. S. Lim. Randomized clinical trial of a new dexamethasone delivery system (Surodex) for treatment of post-cataract surgery inflammation. *Ophthalmology* 106:223 (1999).
316. D. F. Chang, I. H. Garcia, J. D. Hunkeler, T. Minas. Phase II results of an intraocular steroid delivery system for cataract surgery. *Ophthalmology* 106:1172 (1999).
317. O. Nishi, K. Nishi, T. Morita, Y. Tada, E. Shirasawa, K. Sakanishi. Effect of intraocular sustained release of indomethacin on postoperative inflammation and posterior capsule opacification. *J. Cataract. Refract. Surg.* 22(Suppl. 1):806 (1996).
318. C. G. Pitt. Poly-ε-caprolactone and its copolymers. In: *Biodegradable Polymers as Drug Delivery Systems*, M. Chasin et al. (Eds.). Marcel Dekker, New York, 1990, p. 71.
319. H. Borhani, G. A. Peyman, M. H. Rahimy, H. Thompson. Suppression of experimental proliferative vitreoretinopathy by sustained intraocular delivery of 5-FU. *Int. Ophthalmol.* 19:43 (1995).
320. G. A. Peyman, D. Yang, B. Khoobehi, M. H. Rahimy, S. Y. Chin. In vitro evaluation of polymeric matrix and porous biodegradable reservoir devices for slow-release drug delivery. *Ophthalmic Surg. Lasers* 27:384 (1996).
321. K. W. Leong, V. Simonte, R. Langer. Synthesis of polyanhydrides: melt-polycondensation, dehydrochlorination, and dehydrative coupling. *Macromolecules* 20:705 (1987).
322. K. W. Leong, B. C. Brott, R. Langer. Bioerodible polyanhydrides as drug-carrier matrices. I. Characterization, degradation, and release characteristics. *J. Biomed. Mater. Res.* 19:941 (1985).
323. K. W. Leong, P. A. D'Amore, M. Marletta, R. Langer. Bioerodible polyanhydrides as drug-carrier matrices. II. Biocompatibility and chemical reactivity. *J. Biomed. Mater. Res.* 20:51 (1986).
324. D. A. Lee, R. A. Flores, P. J. Anderson, K. W. Leong, C. Teekhasaenee, A. W. De Kater, E. Hertzmark. Glaucoma filtration surgery in rabbits using bioerodible polymers and 5-fluorouracil. *Ophthalmology* 94:1523 (1987).
325. D. A. Lee, K. W. Leong, W. C. Panek, C. T. Eng, B. J. Glasgow. The use of bioerodible polymers and 5-fluorouracil in glaucoma filtering surgery. *Invest. Ophthalmol. Vis. Sci.* 29:1692 (1988).
326. H. D. Jampel, K. W. Leong, G. R. Dunkelburger, H. A. Quigley. Glaucoma filtration surgery in monkeys using 5-fluorouridine in polyanhydride disks. *Arch. Ophthalmol.* 108:430 (1990).
327. J. B. Charles, R. Ganthier, M. R. Wilson, D. A. Lee, R. S. Baker, K. W. Leong, B. J. Glasgow. Use of bioerodible polymers impregnated with mitomycin in glaucoma filtration surgery in rabbits. *Ophthalmology* 98:503 (1991).
328. H. D. Jampel, P. Koya, K. W. Leong, H. A. Quigley. In vitro release of hydrophobic drugs from polyanhydride disks. *Ophthalmic Surg.* 22:676 (1991).
329. H. D. Jampel, D. Thibault, K. W. Leong, P. Uppal, H. A. Quigley. Glaucoma filtration surgery in nonhuman primates using taxol and etoposide in polyanhydride carriers. *Invest. Ophthalmol. Vis. Sci.* 34:3076 (1993).
330. P. Uppal, H. D. Jampel, H. A. Quigley, K. W. Leong. Pharmacokinetics of etoposide delivery by a bioerodible drug carrier implanted at glaucoma surgery. *J. Ocul. Pharmacol.* 10:471 (1994).
331. J. H. Rabowsky, A. J. Dukes, D. A. Lee, K. W. Leong. The use of bioerodible polymers and daunorubicin in glaucoma filtration surgery. *Ophthalmology* 103:800 (1996).
332. S. Fourman. Scleritis after glaucoma filtering surgery with mitomycin C. *Ophthalmology* 102:1569 (1995).
333. P. Ashton, D. L. Blandford, P. A. Pearson, G. J. Jaffe, D. F. Martin, R. B. Nussenblatt. Review: implants. *J. Ocul. Pharmacol.* 10:691 (1994).
334. T. J. Smith, P. A. Pearson, D. L. Blandford, J. D. Brown, K. A. Goins, J. L. Hollins, E. T. Schmeisser, P. Glavinos, L. B. Baldwin, P. Ashton. Intravitreal sustained-release ganciclovir. *Arch. Ophthalmol.* 110:255 (1992).
335. R. E. Wyszynski, J. B. Vahey, L. Manning, W. E. Bruner, K. M. Morgan, E. N. Burney. Sustained release of 5-fluorouracil from ethylene acetate copolymer. *J. Ocul. Pharmacol.* 5:141 (1989).
336. D. L. Blandford, T. J. Smith, J. D. Brown, P. A. Pearson, P. Ashton. Subconjunctival sustained release 5-fluorouracil. *Invest. Ophthalmol. Vis. Sci.* 33:3430 (1992).
337. T. J. Smith, P. Ashton. Sustained-release subconjunctival 5-fluorouracil. *Ophthalmic Surg. Lasers* 27:763 (1996).
338. G. E. Sanborn, R. Anand, R. E. Torti, S. D. Nightingale, S. X. Cal, B. Yates, P. Ashton, T. Smith. Sustained-release ganciclovir therapy for treatment of cytomegalovirus retinitis. Use of an intravitreal device. *Arch. Ophthalmol.* 110:188 (1992).
339. R. Anand, S. D. Nightingale, R. H. Fish, T. J. Smith, P. Ashton. Control of cytomegalovirus retinitis using sustained release of intraocular ganciclovir. *Arch. Ophthalmol.* 111:223 (1993).
340. D. F. Martin, D. J. Parks, S. D. Mellow, F. L. Ferris, R. C. Walton, N. A. Remaley, E. Y. Chew, P. Ashton, M. D. Davis, R. B. Nussenblatt. Treatment of cytomegalovirus retinitis with an intraocular sustained-release ganciclovir implant. A randomized controlled clinical trial. *Arch. Ophthalmol.* 112:1531 (1994).
341. D. F. Martin, J. P. Dunn, J. L. Davis, J. S. Duker, R. E. J. Engstrom, D. N. Friedberg, G. J. Jaffe, B. D. Kuppermann, M. A. Polis, R. J. Whitley, R. A. Wolitz, C. A.

Benson. Use of the ganciclovir implant for the treatment of cytomegalovirus retinitis in the era of potent antiretroviral therapy: recommendations of the International AIDS Society—USA panel. *Am. J. Ophthalmol.* 127:329 (1999).

342. C. K. Cheng, A. S. Berger, P. A. Pearson, P. Ashton, G. J. Jaffe. Intravitreal sustained-release dexamethasone device in the treatment of experimental uveitis. *Invest. Ophthalmol. Vis. Sci.* 36:442 (1995).
343. D. P. Hainsworth, P. A. Pearson, J. D. Conklin, P. Ashton. Sustained release intravitreal dexamethasone. *J. Ocul. Pharmacol.* 12:57 (1996).
344. P. A. Pearson, G. J. Jaffe, D. F. Martin, G. J. Cordahi, H. Grossniklaus, E. T. Schmeisser, P. Ashton. Evaluation of a delivery system providing long-term release of cyclosporine. *Arch. Ophthalmol.* 114:311 (1996).
345. L. B. Enyedi, P. A. Pearson, P. Ashton, G. J. Jaffe. An intravitreal device providing sustained release of cyclosporine and dexamethasone. *Curr. Eye Res.* 15:549 (1996).
346. G. J. Jaffe, C. S. Yang, X. C. Wang, S. W. Cousins, R. P. Gallemore, P. Ashton. Intravitreal sustained-release cyclosporine in the treatment of experimental uveitis. *Ophthalmology* 105:46 (1998).
347. P. Ashton, J. D. Brown. P. A. Pearson, D. L. Blandford, T. J. Smith, R. Anand, S. D. Nightingale, G. E. Sanborn. Intravitreal ganciclovir pharmacokinetics in rabbits and man. *J. Ocul. Pharmacol.* 8:343 (1992).
348. R. Anand, R. L. Font, R. H. Fish, S. D. Nightingale. Pathology of cytomegalovirus retinitis treated with sustained release intravitreal ganciclovir. *Ophthalmology* 100:1032 (1993).
349. M. G. Morley, J. S. Duker, P. Ashton, M. R. Robinson. Replacing ganciclovir implants. *Ophthalmology* 102:388 (1995).
350. R. E. Engstrom, G. N. Holland. Local therapy for cytomegalovirus retinopathy. *Am. J. Ophthalmol.* 120:376 (1995).
351. N. C. Charles, G. C. Steiner. Ganciclovir intraocular implant. A clinicopathologic study. *Ophthalmology* 103:416 (1996).
352. J. L. Marx, M. A. Kapusta, S. S. Patel, L. D. LaBree, F. Walonker, N. A. Rao, L. P. Chong. Use of the ganciclovir implant in the treatment of recurrent cytomegalovirus retinitis. *Arch. Ophthalmol.* 114:815 (1996).
353. D. C. Musch, D. F. Martin, J. F. Gordon, M. D. Davis, B. D. Kuppermann. Treatment of cytomegalovirus retinitis with a sustained-release ganciclovir implant. The Ganciclovir Implant Study Group. *N. Engl. J. Med.* 337:83 (1997).
354. M. H. Rahimy, G. A. Peyman, S. Y. Chin, R. Golshani, C. Aras, H. Borhani, H. Thompson. Polysulfone capillary fiber for intraocular drug delivery: *in vitro* and *in vivo* evaluations. *J. Drug Targ.* 2:289 (1994).
355. M. H. Rahimy, G. A. Peyman, M. L. Fernandes, S. H. El-Sayed, Q. Luo, H. Borhani. Effects of an intravitreal daunomycin implant on experimental proliferative vitreoretinopathy: simultaneous pharmacokinetic and pharmacodynamic evaluations. *J. Ocul. Pharmacol.* 10:561 (1994).
356. J. Heller, S. Y. Ng, B. K. Fritzinger, K. V. Roskos. Controlled drug release from bioerodible hydrophobic ointments. *Biomaterials* 11:235 (1990).
357. A. Merkli, J. Heller, C. Tabatabay, R. Gurny. Synthesis and characterization of a new biodegradable semi-solid poly(ortho ester) for drug delivery systems. *J. Biomater. Sci. Polym. Edn.* 4:505 (1993).
358. S. Y. Ng, T. Vandamme, M. S. Taylor, J. Heller. Synthesis and erosion studies of self-catalyzed poly(ortho esters). *Adv. ACS Abstracts* (1997).
359. S. Einmahl, M. Zignani, E. Varesio, J. Heller, J. L. Veuthey, C. Tabatabay, R. Gurny. Concomitant and controlled release of dexamethasone and 5-fluorouracil from poly(ortho ester). *Int. J. Pharm.* 185:189 (1999).
360. A. Merkli, J. Heller, C. Tabatabay, R. Gurny, Semi-solid hydrophobic bioerodible poly(ortho ester) for potential application in glaucoma filtering surgery. *J. Control. Rel.* 29: 105–112 (1994).
361. S. B. Bernatchez, A. Merkli, C. Tabatabay, R. Gurny, Q. H. Zhao, J. M. Anderson, J. Heller. Biotolerance of a semisolid hydrophobic biodegradable poly(ortho ester) for controlled drug delivery. *J. Biomed. Mater. Res.* 27:677 (1993).
362. S. B. Bernatchez, A. Merkli, T. Le Minh, C. Tabatabay, J. M. Anderson, R. Gurny. Biocompatibility of a new semi-solid bioerodible poly(ortho ester) intended for the ocular delivery of 5-fluorouracil. *J. Biomed. Mater. Res.* 28:1037 (1994).
363. M. Zignani, A. Merkli, M. B. Sintzel, S. B. Bernatchez, W. Kloeti, J. Heller, C. Tabatabay, R. Gurny. New generation of poly(ortho esters): synthesis, characterization, kinetics, sterilization and biocompatibility. *J. Control. Rel.* 48:115 (1997).
364. M. Zignani, S. B. Bernatchez, T. Le Minh, C. Tabatabay, J. M. Anderson, R. Gurny. Subconjonctival biocompatibility of a viscous bioerodible poly(ortho ester). *J. Biomed. Mater. Res.* 39:277 (1998).
365. S. Einmahl, F. F. Behar-Cohen, C. Tabatabay, M. Savoldelli, F. D'Hermies, D. Chauvaud, J. Heller, R. Gurny. A viscous bioerodible poly(ortho ester) as a new biomaterial for intraocular application. *J. Biomed. Mater. Res.* 50:566 (2000).
366. M. B. Sintzel, J. Heller, S. Y. Ng, M. S. Taylor, C. Tabatabay, R. Gurny. Synthesis and characterization of self-catalyzed poly(ortho ester). *Biomaterials* 19:791 (1998).
367. K. Schwach-Abdellaoui, J. Heller, R. Gurny. Synthesis and characterization of self-catalyzed poly(ortho-ester) based on decanediol and decanediol-lactate. *J. Biomater. Sci. Polym. Edn.* 10:375 (1999).
368. K. A. Hooper, N. D. Macon, J. Kohn. Comparative histological evaluation of new tyrosine-derived polymers and poly(L-lactic acid) as a function of polymer degradation. *J. Biomed. Mater. Res.* 41:443 (1998).
369. K. Schwach-Abdellaoui, J. Heller, R. Gurny. Hydrolysis and erosion studies of autocatalyzed poly(ortho esters)

containing lactoyl-lactyl acid dimers. *Macromolecules* 32: 301 (2000).

370. M. S. El-Samaligy, Y. Rojanasakul, J. F. Charlton, G. W. Weinstein, J. K. Lim. Ocular disposition of nanoencapsulated acyclovir and ganciclovir via intravitreal injection in rabbit's eye. *Drug Deliv.* 3:93 (1996).
371. B. Khoobehi, M. O. Stradtmann, G. A. Peyman, O. M. Aly. Clearance of fluorescein incorporated into microspheres from the cornea and aqueous after subconjunctival injection. *Ophthalmic Surg.* 21:840 (1990).
372. H. Kimura, Y. Ogura, T. Moritera, Y. Honda, R. Wada, S. H. Hyon, Y. Ikada. Injectable microspheres with controlled drug release for glaucoma filtering surgery. *Invest. Ophthalmol. Vis. Sci.* 33:3436 (1992).
373. B. Khoobehi, M. O. Stradtmann, G. A. Peyman, O. M. Aly. Clearance of sodium fluorescein incorporated into microspheres from the vitreous after intravitreal injection. *Ophthalmic Surg.* 22:175 (1991).
374. T. Moritera, Y. Ogura, Y. Honda, R. Wada, S. H. Hyon, Y. Ikada. Microspheres of biodegradable polymers as a drug-delivery system in the vitreous. *Invest. Ophthalmol. Vis. Sci.* 32:1785 (1991).
375. G. A. Peyman, M. Conway, B. Khoobehi, K. Soike. Clearance of microsphere-entrapped 5 fluorouracil and cytosine arabinoside from the vitreous of the primates. *Int. Ophthalmol.* 16:109 (1992).
376. T. Moritera, Y. Ogura, N. Yoshimura, Y. Honda, R. Wada, S. H. Hyon, Y. Ikada. Biodegradable microspheres containing adriamycin in the treatment of proliferative vitreoretinopathy. *Invest. Ophthalmol. Vis. Sci.* 33:3125 (1992).
377. G. G. Giordano, M. F. Refojo, M. H. Arroyo. Sustained delivery of retinoic acid from microspheres of biodegradable polymer in PVR. *Invest. Ophthalmol. Vis. Sci.* 34: 2743 (1993).
378. G. G. Giordano, P. Chevez-Barrios, M. F. Refojo, C. A. Garcia. Biodegradation and tissue reaction to intravitreous biodegradable poly(D,L-lactic-co-glycolic) acid microspheres. *Curr. Eye Res.* 14:761 (1995).
379. B. Conti, C. Bucolo, C. Giannavola, G. Puglisi, P. Giunchedi, U. Conte. Biodegradable microspheres for the intravitreal administration of acyclovir: *in vitro*/*in vivo* evaluation. *Eur. J. Pharm. Sci.* 5:287 (1997).
380. A. A. S. Veloso, Q. Zhu, R. Herrero-Vanrell, M. F. Refojo. Ganciclovir-loaded polymer microspheres in rabbit eyes inoculated with human cytomegalovirus. *Invest. Ophthalmol. Vis. Sci.* 38:665 (1997).
381. T. Moritera, Y. Ogura, N. Yoshimura, S. Kuriyama, Y. Honda, Y. Tabata, Y. Ikada. Feasibility of drug targeting to the retinal pigment epithelium with biodegradable microspheres. *Curr. Eye Res.* 13:171 (1994).
382. H. Kimura, Y. Ogura, T. Moritera, Y. Honda, Y. Tabata, Y. Ikada. In vitro phagocytosis of polylactide microspheres by retinal pigment epithelial cells and intracellular drug release. *Curr. Eye Res.* 13:353 (1994).
383. H. Kimura, T. Sakamoto, D. R. Hinton, C. Spee, Y. Ogura, Y. Tabata, Y. Ikada, S. J. Ryan. A new model of subretinal neovascularization in the rabbit. *Invest. Ophthalmol. Vis Sci.* 36:2110 (1995).
384. H. Kimura, C. Spee, T. Sakamoto, D. R. Hinton, Y. Ogura, Y. Tabata, Y. Ikada, S. J. Ryan. Cellular response in subretinal neovascularization induced by bFGF-impregnated microspheres. *Invest. Ophthalmol. Vis. Sci.* 40:524 (1999).

16

Dental and Maxillofacial Surgery Applications of Polymers

A. Bascones, J. M. Vega, N. Olmo, J. Turnay, J. G. Gavilanes, and M. A. Lizarbe
Universidad Complutense, Madrid, Spain

I. INTRODUCTION

The oral cavity is an external environment where a wide range of conditions are in constant change (i.e., pH, temperature, abrasion forces, presence of bacteria, etc.). However, under physiological conditions, the oral cavity is very tolerant and resistant to external challenges. In order to achieve this goal, the teeth and their associated structures, such as the jaws and gums, must display harmonious behavior and interrelationships. The anatomical design of teeth and associated tissues provides resistance to this hostile environment. However, several factors (diseases, physiologically unacceptable conditions, trauma, etc.) may alter the function and structure of the oral cavity elements. Clinical treatments directed to restore and maintain the normal functions of the oral structures (teeth and supporting tissues) are grouped in what is known as dentistry.

Dentistry involves the repair and/or removal of decayed teeth, malocclusion correction, straightening of the teeth, and the design and manufacture of false teeth or other prosthetic devices. Studies within the dental biomaterial field are focused on the clinical functionality and longevity of dental implant devices in order to improve human dentition health, esthetics, and quality of life.

There are four main groups of widely used materials in dentistry: metals and alloys, ceramics, synthetic organic polymers and biopolymers (derived from natural tissues), and composites (an organic matrix of different kinds of polymers filled with inorganic fine particles). Although several metallic biomaterials are widely used due to their good properties, they show a slow and progressive decrease in specific applications which is concomitant to an increase in the use of polymers, composites, and inorganic materials. This might have different explanations: (1) nonprecious metal alloys can be chemically unstable in the oral environment (etching, oxidation, pigmentation); (2) a general negative feeling against visible alloys in dentistry has appeared in recent years based on esthetic criteria; (3) many of the nonprecious metal alloys frequently show biological complications (allergenicity, cytotoxicity, carcinogenicity); (4) complex equipment is required for metal processing, which usually makes its dental application too expensive; (5) the price of many metals is not balanced and depends on economic or political crisis. For these and other reasons, the materials used for dentistry have been modified in recent years. These changes in use tendency are characterized by (1) increased use of polymers, composites, and ceramics; (2) development of new techniques for covering metal surfaces with polymer or ceramic materials; (3) development of composites for specific dental applications; and (4) improvement of adhesiveness of all these materials to dental structures. However, the use of polymers in dentistry has to be combined with the use of other types of biomaterials due to the special characteristics of the oral biological environment where different soft and hard tissues are found. Thus, for example, endosteal or endosseous devices placed into the bone tissue are mainly metals and alloys (1).

In a short historical overview, it must be mentioned that oral disease treatment and tooth replacement can be found in ancient civilizations. Dead animal parts or minerals were used to replace damaged structures. In the 17th century, the implants used for oral treatment became more similar to the missing or damaged part: teeth extracted from cadavers or from living persons were used. In the 18th and 19th centuries, increasing attention was given to the basic properties of the implanted materials. A systematic investigation of the physical, mechanical, and biological properties of dental implant devices was carried out. The technological and scientific advances in the 20th century provided the basis for the design of new biomaterials with a better behavior and an adequate durability for dental applications.

The use of polymers for dentistry was initiated in the 19th century. Thus, gutta-percha (leathery material derived from the latex of certain trees) was used as a material for dental impressions in 1848 (it is still an important material for root canal filling); vulcanized caoutchouc was employed in 1854 for denture bases; and celluloid (developed in 1869 from a homogeneous colloidal dispersion of cellulose nitrate and camphor) was used arround 1870 for dental prosthetics. Figure 1 shows an old dental prosthetic device (a caoutchouc base with porcelain teeth) that has been in continuous use for 45 years (from 1940) and finally substituted by another polymethyl methacrylate (PMMA) device. However, the modern age of polymers began in 1909 with bakelite (a phenol–formaldehyde resin), a material scarcely used in dentistry. The first rigid polymer employed in odontology was PMMA, which has been used from 1930 for different purposes (denture bases, artificial teeth, removable orthodontics, surgical splinting, and even for esthetic filling in anterior teeth). In the 1930s many other polymers appeared for dental and surgical uses (polyamide, polyester, polyethylene, etc.), prepared in different forms (rigids, softs, fibers, adhesives, etc.), and exhibiting several applications (dentures, veneering of crowns and bridges, fillings, mouth protectors, sutures, implants, etc.). Bis-GMA, a combination of bis-phenol and glycidyl methacrylate, appeared in 1956 as an important innovation in the field of rigid polymers for dental uses. From that time on, there has been continuous and important research concerning the use of polymers and composite materials for dentistry.

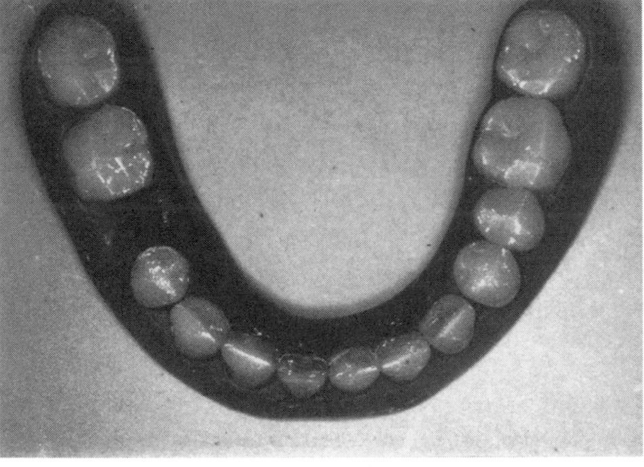

Figure 16.1 Old caoutchout (prepared around 1940) complete dental lower prosthesis with porcelain teeth.

II. ORAL BIOLOGY AND BIOCHEMISTRY

When a material is used for any surgical purpose (implant, suture, etc.), the organism reacts in order to restore the integrity of the affected tissues. There are many general processes following any kind of injury, such as inflammation and development of inflammatory tissue, cell proliferation, and necrosis of damaged cells. But the specific response depends on the biology of the particular affected regions. Thus, the use of restorative materials for biomedical applications requires a good knowledge of the biology and biochemistry of the specific location considered. Oral surgery repair treatments move into a highly specialized biological environment. A significant part of the functions performed at the maxillofacial level shows a mechanical component that exhibits high stress. The specific properties of the biological components at this location must accomplish such requirements. Most treatments involve the use of nonnatural materials that must be able to perform specifically designed functions in that particular biological theater. Therefore, the characteristics of the biology and physical chemistry of this neighborhood must be analyzed when considering the use of biomaterials at the maxillofacial level. Human dentition consists of 32 teeth (candidates to implant sites); there are four incisors, two canines, four premolars, and six molars in each jaw (mandible and maxilla). The anatomical shape of these teeth is heterogeneous reflecting their specific functions; the incisors and canines are used for cutting and tearing, while premolars are used for crushing, and the molars for grinding. Three different zones can be distinguished in each tooth: the root, attached to the tooth-bearing bone of the jaws by a fibrous ligament (periodontal membrane), the neck, embraced by the fleshy gum tissue, and the crown, that projects into the mouth. Figure 2 shows a longitudinal section of a tooth with its major structures.

After an oral surgical operation, in most cases the picture is characterized by the presence of a hard material surrounded by soft tissues. All of these components must be adequately interfaced in order to function properly, and

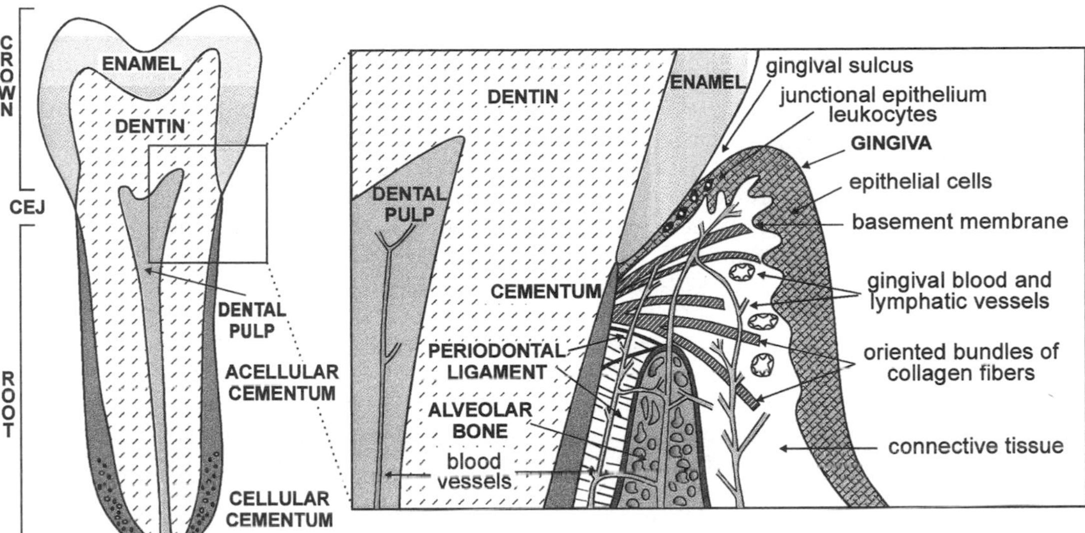

Figure 16.2 Schematic diagram of the major tooth compartments, connective tissues, and cells in the periodontal unit. CEJ: cementun–enamel junction.

this requires the cooperation of many different tissue cells and biopolymers. The main role corresponds to connective tissues forming the periodontium, which is responsible for the support and attachment of the tooth in the bones of the jaw. Many different connective tissues (ranging from mineralized to fibrous) can be distinguished in the oral cavity (Fig. 2; Table 1): alveolar bone, enamel, dentin, cementum, periodontal ligament, and gingiva (2–18).

The basic constituents of human enamel and dentin, their composition, and several other parameters (mechanical properties, elastic modulus, and viscoelastic properties) have been reported (18–20). The enamel is the very hard, dense, and brittle cover of the tooth's crown that is located above the gumline. Enamel is highly mineralized and contains prismlike structures comprised of hydroxyapatite crystallites and very little organic material. Its properties make dental enamel a good material for cutting and masticating food, processes that involve friction and wear. Mature enamel is an acellular nonliving structure that cannot be renewed after damage.

The bulk of the tooth is made up by dentin, which defines its shape. Dentin, a continuous nonuniform mineralized tissue, is similar in composition to bone but is tougher and less hard than enamel (9,10). It is formed by a highly mineralized matrix (peritubular dentin) that surrounds the dentinal tubules. These tubules are distributed throughout the dentin matrix in a more or less uniform manner. In contrast to enamel, dentin is a living tissue. A layer of dentin producing cells, the odontoblasts, surrounds the pulp cavity of the tooth sending projections into the dentinal tubules. Secondary dentin, a less organized form of tubular dentin, is produced throughout life. Tertiary, or reparative, dentin is formed when the enamel of the crown is damaged. The dentin is nourished by the pulp in which cells, tiny blood vessels, and a nerve are enclosed in the pulp chamber. Below the gumline, from the cervix to the apex of the root, dentin is covered by the cementum to which the periodontal ligaments bind attaching the tooth to the alveolar bone (17). Cementum is less hard than dentin, showing properties very similar to those of bone.

The connective tissues of the oral cavity are involved in several specific and diverse functions, including speech, mastication, and the maintainance of face esthetics. Moreover, the periodontium represents one of the biological barriers against the external environment (21). Thus, it can be affected by a variety of diseases producing alterations of its connective tissues and requiring frequent clinical treatments (22–28).

From a general point of view, the connective tissues can be considered as composed of cells and extracellular matrix (ECM). In such tissues there are many types of cells performing different functions, e.g., biosynthesis of the ECM components, lipid storage, biological protection, etc. Table 1 shows some of the different tissues and cell types from the periodontal unit. Fibroblasts, one of the principal cells of the periodontal connective tissues, are involved in maintaining the connective tissue matrix (29). After an alteration of the periodontal connective tissues, following injury, fibroblasts migrate into the wound site. Also epithelial cells appear to be related to the formation of epithelial

Table 16.1 Tooth Compartments and Associated Tissues

Tissue and cells	Characteristics and function	Ref.
Mineralized		
Enamel		
Ameloblasts	Enamel-producing cells; differentiated and polarized epithelial cells. Secretion of amelogenins (proline-rich proteins; 5–28 kDa) and enamelins (more acidic and lower proline content than amelogenins; 42, 56, and 72 kDa). Noncollagenous fibrils. Mineral content: 95–98%. Mature enamel: nonliving tissue.	2
Dentin		
Odontoblasts	Dentin-producing cells; differentiation of mesenchymal cells. Single layer of fully differentiated and polarized cells; located at the boundary between the dentin and pulp chamber. Mineral content: 65%. Secretion of extracellular matrix components, uncalcified matrix: predentin and control of its mineralization. Synthesis of noncollagenous proteins.	2–10
Cementum		
Cementoblasts	Secretion of collagen and noncollagenous proteins. Type I (>90%) and type III (<5%) collagen.	2–4,8–17
Alveolar bone		
Osteoblasts	Matrix-producing cells. Formation of mineral bone tissue. Synthesis of type I collagen (100%) and noncollagenous proteins. High levels of alkaline phosphatase. Synthesis of collagenolytic enzymes and their inhibitors.	2,3
Osteoclasts	Specialized resorptive cells. Role in remodeling processes. Resorption of the degraded organic matrix by osteoblasts.	2,3,18
Osteocytes	Embedded in the mineralized matrix in the lacunae space. Derived from osteoblasts. Role in the calcium exchange between mineralized tissue and blood. Matrix-producing cells; resorptive properties but to a lower extent than osteoclasts and osteoblasts.	2,3
Nonmineralized		
Dental pulp		
Dental pulp cells	Loose connective tissue enclosed in the dentin. Nerves and vascular elements permeate the tissue and provide nutrients entry, supporting metabolic activity via the blood vessels and capillaries.	2,3,5,11,13
Periodontal ligament		
Fibroblasts	Collagen-producing cells. Dense collagenous structure. Synthesis of type I (>90%), type III (<18%), and type V (<1%) collagens. Presence of other cells from blood and lymphatic vessels.	2–4,11
Gingiva		
Fibroblasts, epithelial cells	Collagen-producing cells. Type I and type III collagens represent 60% of the total tissue protein. Synthesis of type IV and type V collagens (<1%).	2–4,10,13
Mast cells, leukocytes, plasma cells	Primary defense against invading microorganisms. They are present in gingiva and periodontal ligament.	2,4

attachments during wound healing after periodontal therapy. On the other hand, the ECMs can be divided into two groups: those forming interstitial connective tissue and those from the basement membrane (30–36). However, both types of ECM comprise many different macromolecular structures, which can be grouped into collagens, glycoproteins, and proteoglycans, attending to their major components (37–45).

Collagens and glycosaminoglycans (one of the components of proteoglycans) are discussed elsewhere in this book since these molecules are used as natural biopolymers in the biomaterials field. The periodontal tissues are highly collagenous. The family of collagens accounts for more than 50% of the total tissular protein. Collagens are well-known proteins whose chemical and tridimensional structures are clearly elucidated (37–42). The fibrillar type I collagen is the one present in the largest amount. However, this type of collagen is only the representative molecule of a broad family of proteins which mainly have a structural function in connective tissues. But in spite of such a structural role, the relative proportions of the different collagen types are important factors in determining biological functions at the ECM. In particular, the periodontal structures contain high proportions of type I collagen, about 90% of the collagen present in the periodontium, whereas type III collagen only accounts for about 10%, and other minor collagen types are only present in about 1% (4).

Based on this information, it is perfectly comprehensible that a significant part of the diseases and misfunctions at a maxillofacial location will be related to collagens. For instance, a destruction of collagen bundles occurs during gingivitis (22), and extensive degradation of connective tissues with a concomitant reduction in the amount of collagen appears in chronic periodontitis (4), considering only two of the most common diseases at the periodontal level. Also physical trauma induced lesions, which are frequent at the oral cavity, may affect collagen deposition and fibroblasts migration, although cellular and biochemical studies concerning this topic are scarce.

The ECM is a structurally stable association of biopolymers surrounding the connective tissue cells. But this matrix is not a simple inert supporting material (46–48). It exhibits close functional relationships with the tissue cells; specific cell receptors interact with ECM components modulating cell behavior (43,49). These functional components have been grouped into a heterogenous family of adhesion glycoproteins that include some types of collagens and several noncollagenous ECM proteins (43,49,50). Particular matrix components can associate in a diversity of macromolecular structures defining specific ECMs with whom different types of cells can interact. Noncollagenous proteins present in oral cavity tissues are involved in several processes, as the initiation and formation of apatite crystals (mineralized tissues) and the regulation of cell proliferation and tissue growth and development (16,17,51). In the dentin matrix, a mineralized tissue, there are several noncollagenous ECM proteins (Table 2) (52,53). The main characteristic of these proteins is their acidic nature, some of them presenting the capacity of being phosphorylated. Several of these proteins are also found in the bone matrix (osteonectin, osteopontin, osteocalcin, matrix Gla protein, proteoglycans, etc.) (53–65). However, some of them appear to be dentin specific such as dentin phosphophoryn, dentin sialoprotein, and dentin matrix protein-1. Some characteristics of these proteins from the mineralized tissues are summarized in Table 2. Bone morphogenetic proteins (BMPs), which are present in dentin, deserve special attention since they can induce bone formation (66–68).

As already mentioned, some natural polymers such as collagen are used in different applications in biomaterials science (69). Other noncollagenous glycoproteins from the ECM that also play important roles in the control of cell behavior should be taken into consideration in dentistry. A brief description of some of these proteins is shown in Table 3 (70–85). In this way, fibronectin has been used in the treatment of periodontal disease (27,28), and laminin-5 is used to promote cell attachment of gingival epithelial cells to titanium alloy implants (86).

When studying biomaterials, one should never forget that they are going to be in close contact with the surrounding tissue cells. Cells are in constant communication with their external environment. Several cell surface receptors are involved in cell-to-cell association and cell–ECM interactions (87–89). These receptors are grouped into different families based on similarity of structure and function, such as integrins, selectins, cadherins, etc. Integrins are heterodimers comprised by two different transmembrane subunits (α and β) that play a critical role in cell-to-cell and cell–ECM association. They are capable of transducing information from the external environment to the inside of the cell (90–92). The maintainance of a normal cell phenotype is dependent on the signals received from the ECM. Thus, the posible alteration of the external cell environment is an important aspect to be considered in order to understand host response to biomaterials. The binding site or receptor recognition sequences, among them the arginine–glycine–aspartic acid sequence (RGD), have been described in several adhesive ECM proteins (Tables 2 and 3) (93,94).

However, many other natural molecules are also involved in the complex mechanisms that control cell behavior. Thus, polypeptide growth factors have a determinant role in stimulating migration and growth of cells in wound healing and repair. Interesting studies are still in progress concerning this topic (26).

These brief comments are intended to present the complexity of the biology of the maxillofacial region where cells (fibroblasts, endothelial cells, bone cells and nerve cells, among other cell types), many biopolymers (collagens, proteoglycans, and glycoproteins), and growth factors play specific roles resulting in a highly specialized structure. All of them must be considered when studying the response of a biomaterial after any use at this location.

Table 16.2 Noncollagenous Proteins in Mineralized Tissues

	Bone	Dentin
Glycoproteins/sialoproteins	Osteonectin (SPARC) Protein of 35 kDa. Acidic secreted protein, cystein rich. Affinity for collagen, gelatin and hydroxyapatite. Calcium binding protein. Role in bone matrix assembly. 13% carbohydrate. (2,10,54) Bone sialoprotein (BSP) 70–80 kDa; 20% sialic acid. (2,55,56) Osteopontin (sialoprotein I) 32 kDa; 5% sialic acid, high content of phosphate. Contains the -GRGDS-sequence (2,18,57,56,58) Bone sialoprotein II 13% sialic acid. (2,9,10) Bone acidic glycoprotein (BPG-75) (2,10,59)	Osteonectin Similar to those from bone. Osteopontin Similar to those from bone. Dentin matrix protein-1 (Dmp-1) Specific of dentin. Aspartic acid and serine rich protein. Expression in odontoblasts and osteoblasts. Binding to collagen. Contains one RGD sequence. Role in matrix calcification. (10,62) Dentin sialoprotein (DSP) Specific of dentin. 53 kDa. Synthetized and secreted by odontoblast cells of the dental papila. (10,63)
Phosphoproteins	Phosphoserine and phosphothreonine proteins 5–70 kDa. Some are glycoproteins and/or sialoproteins. (2,9,10)	Phosphophoryns 155 kDa. Specific of dentin. More than 50% of noncollagenous dentin proteins. 75% of serine and aspartic acid and 85% of serine residues are phosphorylated. They are able to bind calcium ion with high affinity. Higher phosphorylation and molecular mass than bone phosphophoryns. Heterogeneity between dentin of different species. (7,10,64,65)
γ-Carboxy glutamic acid–containing proteins	Osteocalcin (BGP) Major endogenous protein constituent of the bone (5.2–5.9 kDa). It contains three Gla residues; present in mineralized tissues and also found in plasma. Its plasma content is related to metabolic bone disease. Specific role in matrix mineralization. (2,9,10,58,60) Matrix Gla protein (MGP) 10 kDa. Five Gla residues. (2,9,60)	Osteocalcin (BGP) Similar to those from bone. Matrix Gla protein (MGP) Similar to those from bone.
Proteoglycans	Biglycan Core of 38 kDa and two GAG chains of 40 kDa. Containing primarily chondroitin 4-sulfate. (10,61) Decorin Core of 38 kDa and only one GAG chain of 40 kDa. Containing primarily chondroitin 4-sulfate. (10,61)	Biglycan and decorin (major proteoglycans) Proteoglycans represent about 2.5% of the organic matrix. Small core proteins and GAG chains of 35–40 kDa. All the GAG have been detected but they are composed predominantly of chondroitin 4-sulfate. The content in predentin is higher than in dentin and is enriched in chondroitin-6-sulfate. (5,7,44)
Serum-derived proteins	α_2HS-Glycoprotein, serum albumin, fibronectin	α_2HS-Glycoprotein, immunoglobulin, serum albumin, transferrin

Notes: Some serum derived proteins trapped or binding to the mineralized tissues are also indicated. Gla: γ-carboxy glutamic acid; GAG: glycosaminoglycans. References in brackets.

Table 16.3 Main Noncollagenous Proteins (Elastin, Glycoproteins, and Proteogycans) of the Extracellular Matrix

Component	Characteristics	Ref.
Elastin	Single polypeptide chain of 72 kDa. Glycine rich protein (33%). High alanine, valine, and proline contents; presence of hydroxyproline. Elastin fibers form a network with two different components: amorphous elastin and microfibrils (10–12 nm diameter). Covalent crosslinks. Responsible for tissue elastic properties. Synthesis by fibroblasts, chondrocytes, and endothelial and smooth muscle cells.	70,71
Glycoproteins		
Fibronectin	Variants of a dimer (440–550 kDa) of two polypeptide chains (230 kDa) disulfide bonded near the C terminus. Each chain is composed of a different number of repeats (type I, type II, and type III repeats). Three regions with alternative splicing; 4–10% carbohydrate. High degree of sequence homology between fibronectin from various species. Interactions with collagen, glycosaminoglycans, fibrin, heparin, cell surface receptors, etc. Cell attachment site of -Arg-Gly-Asp-(Ser)- recognized by cell surface receptors of the integrin family. Functions: control of cell adhesion, cell spreading, cellular motility, differentiation, opsonization, and wound healing. Serum, stroma, and particular basement membrane. Synthesis by fibroblasts and other cell types.	72–74
Laminin	Family of trimeric basement membrane glycoproteins with multiple domains, structures, and functions. Three disulfide-linked polypeptide chains (850–900 kDa) are assembled in a cruciform structure with three short arms containing globular domains and a long arm with a distal globular domain. Different isoforms according with tissues and origins. Major noncollagenous protein of basement membrane. They show different cell biological activities controlling cell adhesion, growth, morphology, differentiation, and migration; have role in matrix assembly. Repeats with similar sequences to the EGF-like domains. Cell laminin receptors are mainly of the integrin family.	75–78
Tenascin	Six-armed structure (1900 kDa). Three arms are assembled forming a trimer (disulfide bonded) and two trimers are bound forming the hexamer molecule through disulfide bonds. Isoforms. The molecule contains three structural domains: a terminal globule at each arm (C terminus), thick distal, and thin proximal segments. Repeats with similar sequences to the EGF-like domains and domains similar to the type III repeats of fibronectin. Tendons, ligaments, bone, cartilage, and smooth muscle. Synthesis by fibroblasts and glial cells.	79,80
Thrombospondin	Three polypeptide chains (180 kDa) crosslinked by interchain disulfide bonds near the N terminal domains (420 kDa). The molecule is organized into globular domains connected by thin regions of polypeptide. Large globular structure at the N terminal (three chains assembled) and three globular domains each corresponding to the C terminal polypeptide chain. Binds to heparin, fibronectin, fibrinogen, type V collagen, and calcium. Thrombin-sensitive protein. Associated to extracellular matrix. Synthesis by platelets.	81,82
Proteoglycans	Family of molecules composed of different core polypeptide chains where one or more types of glycosaminoglycans are attached by O-glycosidic linkages. Widespread distribution on cell surfaces, in basement membranes, and in the extracellular matrix of most tissues. They are involved in cell–cell interactions, cell migration, differentiation, tissue morphogenesis, cell–substrate attachment, and in the ionic control of filtration through basement membranes.	83–85

III. TESTING AND EVALUATION OF MATERIALS FOR DENTISTRY

Many factors must be considered in the evaluation of the potential use of a material in a biological environment. Every process resulting from the presence of such a material is a consequence of a reciprocal set of reactions, i.e., host response against the material and vice versa. These responses cannot be independently analyzed since both arise from the presence of the material considered and are closely related. The main events occurring related to the material–host reactions are summarized in Fig. 3.

The behavior of a material in a biological environment would first depend on its intrinsic properties related to its physicochemical structure. However, these properties are modulated by the surrounding tissues, with the cellular and biochemical components of that particular site playing a determinant role. Depending on the characteristics of this local double response, the material might be cataloged as biocompatible. However, in spite of these short-range responses, the potential long-range consequences arising from the presence of the material should be also considered. For instance, products liberated from the material may produce significant effects in distant tissues or organs (95). Such possibilities should be taken under consideration when designing assays for testing and evaluating a material for biomedical applications.

A biomaterial behaves as a foreign body in terms of the tissue response it evokes (96). Therefore, there is a sequence of events, ranging from the molecular scale to the cellular and tissue level, that can be expected following any biomaterial application. But the particular characteristics of the whole response are also dependent on the surgeon's skill, the patient's condition, and, more importantly from the present point of view, the material properties. Thus, the chemical, physicochemical, and mechanical properties of the material must be considered. On the other hand, the biodegradability of the biomaterial represents another point of interest. Nonbiodegradable materials constitute a permanent trauma to the adjacent tissues. But absorbable products result in scar tissues whose presence would depend on the material removal rate. In summary, there are many properties of biomaterials to be considered in terms of tissue reactions against their presence. Among these properties, some of the most relevant are surface roughness and porosity, interfacial forces, surface charge, hydrophilicity, chemical composition, leaching of chemicals into the surrounding tissues, shape, size, and permeability (97).

The surface roughness determines the contact area of

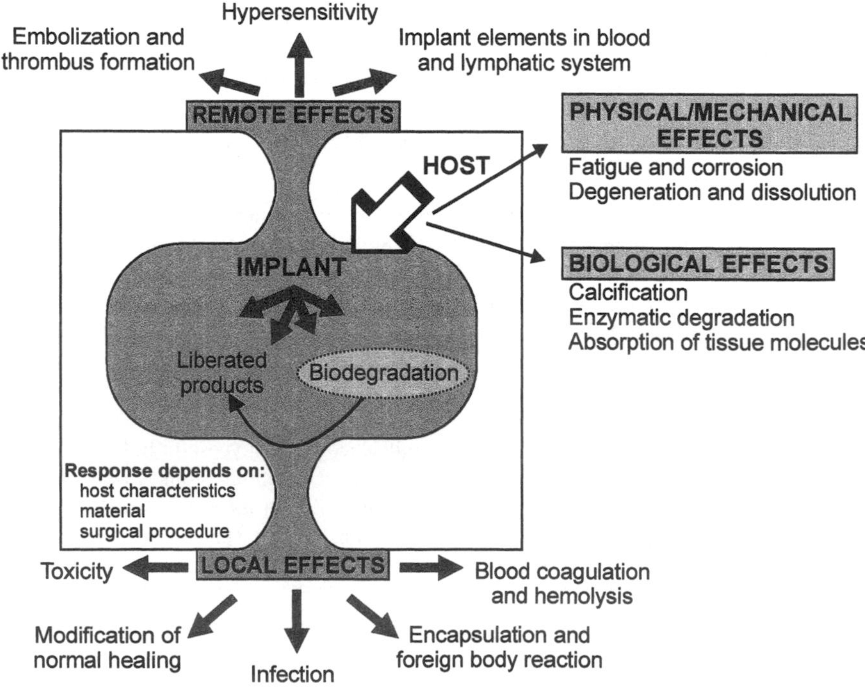

Figure 16.3 Summary of the general host–implant reactions.

the material with the surrounding tissue and the extent of the tissue–material interactions at the physicochemical level. The porosity of the material, pore size, and distribution affect the ingrowth, attachment of the adjacent tissues, and protein deposition (98). The interfacial forces arising from the surface properties of the product can also determine to a large extent interactions between the biomaterials and the tissue or physiological fluids. Surface charge must be also considered. This charge may result from the processing of the biomaterial or from either surface defects or adsorbed charged molecules. The hydrophilicity of the material and the distribution of polar and nonpolar groups in its surface are also important since the material will be located in a polar environment, and these factors determine the thermodynamical parameters of the biomaterial–tissue interactions. All the surface properties of the biomaterial may be involved in processes of macromolecules adsorption (98,99).

Most of the nonnatural products to be used as biomaterials can be considered as inert in terms of liberation of component products. However, the biological environment is highly aggressive for any foreign material, and consequently leaching of chemicals into the surrounding tissue may occur during degradation of the used product. This is especially important when the liberated products are monomers from the polymerization process due to their potentially high toxic effect (100). Thus, the chemical composition of the biomaterial is also something to be considered. Size and shape are also important physical factors when considering the material–tissue relationships. In fact, most of the mentioned effects are related to the relative surface of the biomaterial, which correlates with size and shape. But these two factors are also intrinsically important. The size and shape of the device must be compatible with the anatomical and histological characteristics of the implantation site. Moreover, in general, large and sharp devices can induce more reactions in the surrounding tissue than small and flat structures.

On the other hand, any biomaterial is potentially subjected to etching processes. Thus, corrosion would affect metal devices, or dissolution would affect ceramics. Polymers are subjected to degradation. They can be partially or totally destroyed by different causes, e.g., heat, oxidation, mechanical forces, radiation, etc. But biological degradation of polymers is also possible, e.g., enzymic action or phagocytic activity of macrophages (101). This is very important when considering the tissue reaction to an implanted polymer. Sometimes biological degradation is a required characteristic of resorbable implants. However, it can also result in secondary toxic effects produced by the liberated components. Therefore, the synthesis of the polymeric biomaterial should tentatively control all of these specific requirements and precautions.

In spite of all the previous considerations concerning the properties that polymeric material should exhibit to be used for biomedical applications, its response against biological environments must be tested prior to human clinical uses. Thus, after manufacture of a polymer according to the structural characteristics required for its desired function, the resulting product must tested both in vitro and in vivo.

Most of the in vitro tests involve cell culture in the presence of either the potential biomaterial or products resulting from its degradation (97,102,103). The validity of these assays can be extended by using tissue or organ culture, a more reliable test but involving highly specific methodologies. There are many different in vitro tests (104–110). They usually deal with the cell morphology, cell adhesion and spreading, cell migration patterns and rates, cell growth kinetics, and DNA and protein synthesis in the presence of the material to be tested (105,106,111–113). These cellular characteristics are representative of the cell response against nonbiological materials. Thus, their analyses give information concerning the potential biological performance of the material.

Results from in vitro tests are obtained in relatively short periods of time and normally with a low cost. However, they do not completely mimic the complexity of the potential responses after implantation.

The in vivo tests intend to overcome such limitations. But they require a defined standardization in allowing a clear explanation of the obtained data. Sometimes the results appearing in the literature are difficult to interpret because the experiments have been performed under very different conditions. In fact, the in vivo tests are mainly related to implantation in animals, and the selected animal species is of crucial importance, as well as the animal age and the dietary and laboratory living conditions, for generalization purposes.

The effects of implanted materials on biological fluids must also be considered. After implantation, the blood and cellular exudates contact the material and many physiological reactions can occur. These are of great importance when related to blood. Among these potential effects, immune response and thrombogenicity are of special significance. Immune responses should be considered under in vivo testing conditions of an implant. This assay should be extended to products potentially arising from the degradation of the implanted material. Remote site immune responses, e.g., urticaria or asthma, may also appear after implantation. Concerning thrombogenicity, even chemically inert materials can produce thrombus on contact with blood (114,115). The surface of the implanted material may adsorb plasma proteins resulting in altered local protein concentrations and modified blood coagulation.

In order to evaluate materials for dentistry, all of the mentioned steps should be considered. First, if the material to be used is a new one, its toxicity must be tested by in vitro assays (e.g., Fig. 4; Table 4) (105). Further, implant devices should be tested in vivo. Subcutaneous and/or intramuscular implantation in experimentation animals can be the first step since the host response, including cellular and immunological data, can thus be measured for relatively short periods of time (e.g., Fig. 5) (116–120). The bacteriological effects are also very important when materials for dentistry are being considered. First, alterations on the local immune response, modifying the resistance of the oral environment to bacteriological contamination, may result from the presence of the biomaterial. On the other hand, the product would be exposed to bacterial degradation at the maxillofacial unit, which should also be taken into account (Fig. 5). The next step is to consider the device implantation at the maxillofacial unit of a suitable animal model. Probably primates are the most preferred experimental animals because their physiological and immunological responses are very similar to those in humans. However, some materials used in maxillofacial surgery are not adequately tested in monkeys. The behavior of this type of animal as well as its anatomy sometimes do not allow these clinical tests. For instance, these animals can remove the devices to be tested with their fingers. Thus, small and calm beagle dogs are sometimes used for these purposes. The in vivo analyses require histopathological and radiographic examinations and should be performed for long periods of time, even at the order of years, to evaluate the functional abilities of the implant. When studying the use of a biomaterial as dental implant, its osteoconductive and osteoinductive abilities should also be taken under consideration when bone regeneration is required.

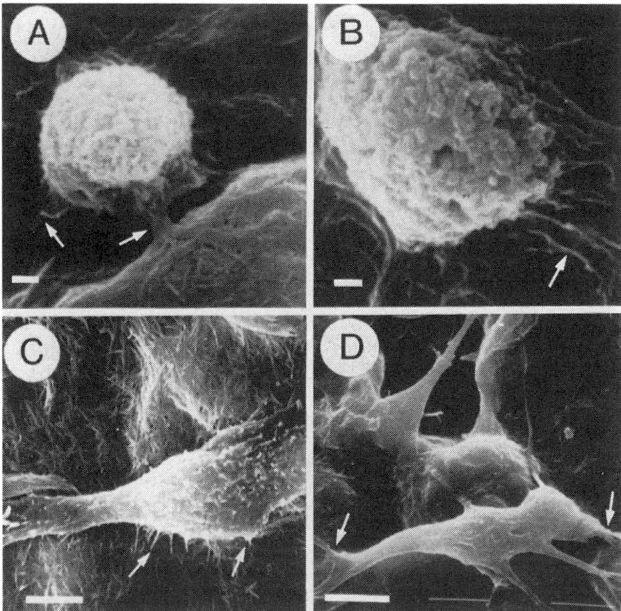

Figure 16.4 Sepiolite (clay)–collagen complex has been tested by scanning electron microscopy for biocompatibility. The adhesion and spreading of human skin fibroblasts have been studied as a function of time: (A) 20 min, ×5000; (B) 30 min, ×5000; (C) 60 min, ×2500; (D) 120 min, ×2500. Cell-to-sepiolite–collagen complex attachments are indicated by arrows. The clay–type I collagen complex is a suitable substrate for fibroblast attachment (see Ref. 105). According to this analysis fibroblasts exhibit a normal morphogenetic process when sepiolite–collagen complexes are used as substrate. The cell spreading seems to be essential for fundamental biological properties such as mobility, macromolecule biosynthesis, or proliferation. Bars = 1 μm.

IV. OPERATIVE DENTISTRY AND ENDODONTIC TREATMENT

Specific dental surgical treatments using polymer materials are shown in Fig. 6. Some dental defects are shown as well as the division of dentistry dealing with them and the currently used methods of treatment.

A. Operative Dentistry

Operative dentistry deals mainly with the removal of tooth caries, a localized disease that begins at the surface of the tooth and may progress through the dentin into the pulp cavity. Caries is one of the most widespread health problems in humans, with a concomitant economic relevance. Nowadays three main topics can be distinguished in operative dentistry: (1) floor and wall protection of the carved cavity after caries removal; (2) filling and restoration of the cavity by using composite materials; and (3) bonding between composites and dental structures.

1. Bases, Liners, and Varnishes for Cavities

There is a large diversity of organic and inorganic materials for these purposes. They can be used as a barrier against other materials with agressive pH, for thermal or electrical insulation, or to provide hardness and mechanical resistance (121,122). These materials are nowadays under revision since in the last years special attention has been paid to adhesion mechanisms between biomaterials and dental tissues (123). The materials most frecuently used are zinc polycarboxylate cement, glass-ionomer cement, and different varnishes.

Table 16.4 Summary of the Results Obtained from Proliferation Curves of Human Skin Fibroblasts Grown on Sepiolite–Collagen Complex

Substrate	DNA[a] (pg/cell)	Protein[a] (pg/cell)	Saturation cell density[a] ($cells/cm^2$)	Doubling time[b] (days)
Plastic	8.81 ± 0.19	454 ± 49	60,000 ± 4,500	1.40 ± 0.08
Sepiolite–collagen	8.88 ± 0.29	451 ± 37	66,000 ± 5,435	1.36 ± 0.05

[a] Values determined for cells at the stationary phase.
[b] Time required to double the cell population determined in the logarythmic phase.
Notes: Sepiolite–collagen complex does not modify the extent of DNA and protein contents nor the kinetic parameters of the cell growth when compared to a standard substrate, like culture plastic dishes, under the conditions used in this in vitro assay. Normal adhesion and growth of fibroblasts and other so-called anchorage-dependent cell types in culture require suitable substrates. Thus, sepiolite–collagen complex exhibits biocompatibility with regard to cell proliferation. Data are expressed as averages ± SD.
Source: Ref. 106.

Zinc polycarboxylate (or polyacrylate) cement is prepared by mixing zinc oxide and the polymer solution, a highly viscous water solution of polyacrylic acid (25,000–50,000 molecular weight). Carboxyl groups of the polymer structure are crosslinked by the metal ion or form a complex with the calcium ions in the surface of the calcified tissue. The resulting product is not only highly resistant, with similar properties to dental zinc phosphate cement, and has an excellent biocompatibility, but also shows adhesion to the enamel and dentin layers of the tooth (124).

Glass-ionomer cement also results from a two-component mixture. Instead of zinc oxide, fluorocalcium aluminosilicate glass powder is used and, as the second component, a solution of a variety of acids, mainly homopolymers or copolymers of acrylic acid with different proportions of

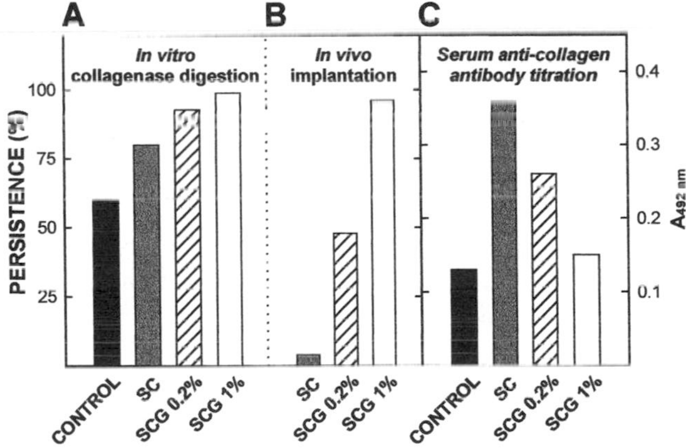

Figure 16.5 (A) Degradation of collagen in type I collagen–sepiolite complex by *Clostridium histolyticum* collagenase (see Ref. 116). The degradation of collagen by the bacterial enzyme decreases when the protein is bound to the clay; the degradation of collagen fibers by collagenase is lower than that produced in collagen solutions (control) (see Ref. 117). This would be in agreement with the longitudinally ordered structure of collagen in the protein-clay complex (see Ref. 118). (B) Evaluation of the subcutaneous implantation of sepiolite–collagen complex. Persistence of the biomaterial determined as percentage of (^{14}C)-acetylated collagen remaining at the implantation site after two months. (C) Titration of anti-type I collagen antibodies by ELISA tests (measurements of absorbance at 492 nm, A_{492}) in rat sera 2 months after implantation of sepiolite–collagen complex (SC) and sepiolite–collagen complex treated with 0.2% (v/v) and 1% (v/v) glutaraldehyde (SCG).

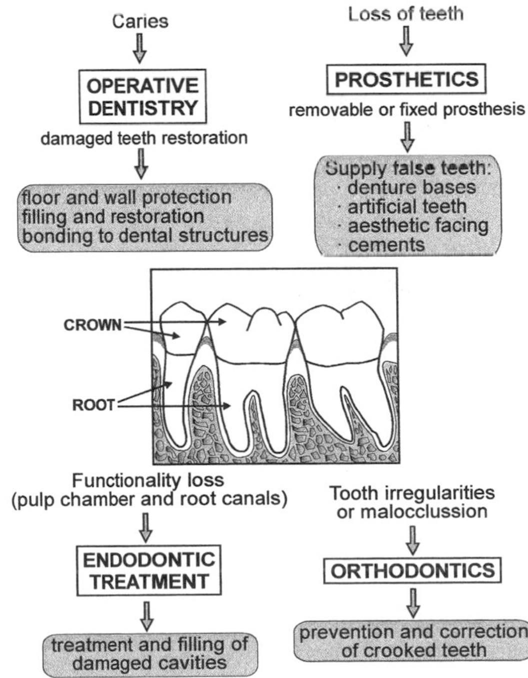

Figure 16.6 Scheme of different areas of dentristry.

tartaric, itaconic, or maleic acid. Polyacid groups interact with calcium and aluminum ions from the glass. The resulting product is a more rigid cement, exhibiting a high mechanical resistance and adherence to enamel and dentin as the zinc polycarboxylate cement. It can also release fluoride ions, which induce the formation of fluorapatite in the teeth. Glass-ionomer cement can be also used as restorative material and luting cement (125–130). These materials present some advantages over acrylic bone cements: absence of monomer, lack of exotherm during setting, and improved release of incorporated therapeutic agents. They integrate with the bone influencing its growth and development. They have been used successfully in several biomedical applications; however, adverse effects have been described on neural tissues (131).

The varnishes are constituted by natural (copal, colophony, etc.) or artificial (polystyrene) resins dissolved in conventional organic solvents like chloroform, ether, or some alcohols. A thin (few micrometers) film of the polymer appears after application of these solutions on the dental surface. These products are used as barriers against several irritants. When protection against thermal changes or rigid support are required, a suitable cement is also used (122,124).

2. Filling and Restorative Materials

a. *Composite Resins*

Composite resins are made up of an organic matrix (polymeric structure) and an inorganic particulate or fibrous (small fibers) filling. They are held together by using a coupling or binding agent. Additionally, some colorants, stabilizers, or primers for the polymerization reaction are added.

Dental amalgam has traditionally been employed as material for cavity filling; but the use of this material is controversial (esthetic problems, eventual toxicity, environmental pollution by mercury, etc.) (132–134). Glass-ionomer is a product also now employed for restorative purposes (129,130). Polymethyl methacrylate resins have also been used as filling materials, but they have several disadvantages such as low molecular weight of the monomer, high polymerization shrinkage, marginal percolation (caries relapse), nonadhesion to dental structures, pulp toxicity, and low color stability.

Bis-GMA represented an improvement over PMMA mainly due to its higher monomer molecular weight (135,136). In order to reduce the viscosity of the original monomer solutions, many others monomers have been considered (bis-phenol dimethacrylate, aromatic or aliphatic urethane dimethacrylates, diethylene glycol dimethacrylate, triethylene glycol dimethacrylate, decamethylene glycol dimethacrylate) (137–139). Therefore, there are many different products available that result from mixtures of these monomers and exhibit a diversity of properties (rheological, mechanical, polymerization contraction capacity, etc.). The potential risk of the use of bisphenols (used in some industrial sectors) is nowadays being considered since estrogenic effects can be induced as a result of the degradation of the organic matrix (140,141). Recently, some composite resins have appeared for their use as dental filling, bis-GMA being replaced by monomers of polysiloxane containing photopolymerizable methacrylate groups. The addition of small particles of inorganic fillers (silica, aluminum oxide, different kinds of glasses, etc.) has produced the modern dental composite resins which are characterized by high hardness, resistance to wearing, and low polymerization shrinkage. The size and amount of the inorganic particles present in these filling materials are dependent on the required properties. Products composed of large particles (several micrometers) are more resistant, whereas the presence of small particles (average size lower than 1 μm) results in a lower resistance but an easier polish. The first type is indicated when high load and resistance to wearing are required (molars) and the second one when the esthetic criteria are essential (anterior teeth) (137,138). There are many different commercially available composite resins containing particles of variable size (142): macroparticles from 8 to 25 μm, miniparticles from 1 to 8 μm, microparticles from 0.04 to 0.20 μm, and blends of different sizes, thus covering many different clinical requirements (Fig. 7).

The properties of these materials are improved when the surface of the inorganic particles is covered by an agent that increases the adhesion between the inorganic compo-

Figure 16.7 Scanning electron microscopy of a contemporary dental composite resin.

nent and the polymeric (organic) matrix of the composite. Silanes (i.e., γ-methacryloxypropyltrimethoxysilane) have been used for such a purpose (143).

The polymerization of these filling materials can be produced by either chemical (autopolymerization or autocuring) or photochemical (photopolymerization or light-curing) mechanisms. Auto-cured filling composites are formed by two components—initiator (i.e., benzoyl peroxide) and activator (usually a tertiary amine, i.e., dihydroxyethyl-p-toluidine)—that are mixed prior to their application (139). Light-cured composites are now widely used. They are available in a single-component form (usually a paste-containing syringe). Some of them contain benzoin ether (which absorbs at 365 nm) and are activated by long-wavelength ultraviolet light (from 340 to 380 nm). Others can be cured by visible light (around 470 nm wavelength) and contain diketones and amines (e.g., dimethylaminoethyl methacrylate). Visible light–cured composites allow larger polymerization thickness than the UV light–cured ones (139).

Clinical applications of composite resins have increased significatively in recent years. Their most extended use is in direct plastic restorations (fillers) using them with different strength and densities. Besides, they are also used as rigid restoration (inlays, onlays, laminate veneers); cementing agents (fluids with little inorganic component); fissure and tooth enamel damage sealers (prevention of caries in cases of higher susceptibility); artificial teeth; covering of metal surfaces; etc.

Research is going on as this group of materials is the most dynamic in the dental biomaterials field. Thus, new nomenclature is arising in order to categorize the new composite resins. The new names incorporate either new manufacturing technologies or new properties, such as: *ormocer* (*or*ganic *mo*dified *cer*amic), *ceromer* (*cer*amic *o*ptimized poly*mer*), and *Polyglass* (glass polymers); others incorporate buffering capacities (i.e., liberation of fluoride, hydroxyl, or calcium ions if the pH conditions vary during restoration).

b. *Compomers*

The term compomer refers to some restorative versatile materials composed by a mixture of composite resins and glass-ionomer cements (144). However, this term is not yet universally accepted, nor defines all the possibilities that this type of material offers. Compomers contain a bifunctional monomer, which is able to react with the methacrylate groups of other monomers as well as with the cations liberated by the glass particles (145). In some cases, they are formed by a composite resin modified by a certain amount of ionomer, while in others glass-ionomer is the main component. Compomers combine the most positive attributes of both components with a concomitant decrease in their individual possible incoveniences. Composite resins contribute mainly mechanical and wearing resistance (low in ionomers), while ionomers increase the adhesive capacity to the dental structures (composite resins do not adhere due to their hydrophobic nature) and release fluoride ions.

3. Bonding to Dental Structures

Composite resins lack adherence to the dental surfaces. Thus, specific attachment treatments are required for the establishment, by physicochemical processes, of a continuity between the tooth mineralized tissues and the implant (123,146). The two main approaches currently in use are bonding to either enamel or dentin, since bonding to cementum is difficult and is under study.

Enamel is mainly an inorganic structure (contains about 95% inorganic material, 1% collagen, and water). To obtain this bonding, the enamel surface is etched by a diluted acid solution (acid-etch technique). The most commonly used is a treatment with 37% orthophosphoric acid for 15–25 s. It causes a partial demineralization at the end of the enamel prisms, either at the central (type I) or interprismatic (type II) zones. Longer periods of treatment increase the depth of the demineralized zone and the resulting adherence is lower (type III). The acid treatment generates a wide irregular surface with plenty of microretention tags that are filled by a fluid resin. Finally, the process is concluded when the composite is applied. After polymerization the resulting product is firmly attached via micromechanical bonds (138,147).

Enamel and dentin present significant differences in composition and structure that do not allow similar bonding treatments for both locations. Dentin contains a large amount of collagen making it hydrophilic, whereas most composites are hydrophobic. Moreover, dentin has odontoblastic extensions that project into the dentinal tubules. The smear layer produced during the cavity carving protects the dentinary canals and is sometimes used to form a network with polymers and adhesives which bonds to dentin. In other cases, such a smear layer is partially or completely removed prior to any other treatment. Therefore, there are many different bonding products containing "conditioners," "cleansers," or "primers" to be used as required (148).

Normally, complex bifunctional reagents, or adhesives, are required to promote composite–dentin bonding. One molecule end (usually a functional methacrylate group) reacts with the composite, and the other end is able to interact with the inorganic (calcium, hydroxyapatite) or the organic

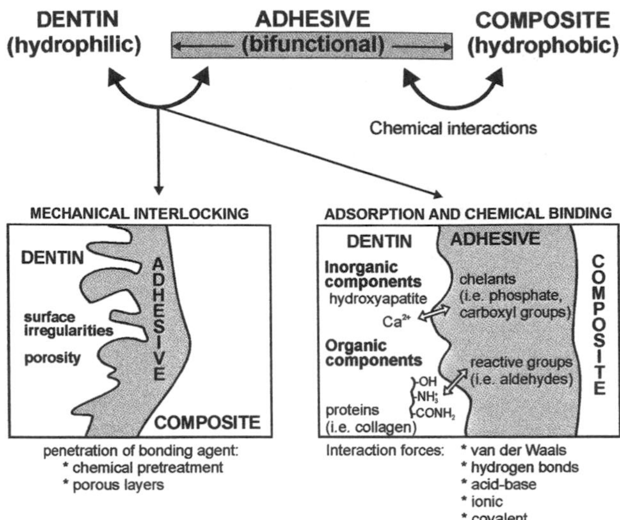

Figure 16.8 Scheme of the interactions between dentin and adhesives.

(reactive protein groups) or both components of the dentin (Fig. 8). Those chemicals for the inorganic component would also react with the inorganic part of the enamel. The resulting structures are maintained by different types of interactions, including covalent bonds, ionic interactions, hydrogen bonds, and van der Waals forces. Some of the most widely used agents added to the resins, such as dichlorophosphate, phenylphosphate, glycerolphosphate, etc., interact with dentin calcium through their phosphate groups. A similar interaction in which carboxyl groups are involved is found in polyacrylic acid–containing glass-ionomers (125). These chemicals can also react with the amine and hydroxy collagen groups (137,147). The interaction of the adhesive with the hydrophilic matrix of dentin is also established by the addition of hydrophilic polymers such as hydroxyethyl metacrylate (HEMA) (148). Binding can be also obtained by the presence of isocyanate groups which react with the hydroxy, carboxy, amino, or amide groups of collagen as well as with hydroxyapatite (137,149); in other cases the reagents used are aldehydes and oxalates, e.g., 4-methacryloxyethyl trimellitate anhydride (4-META) (137,150).

The adhesive can penetrate into the dentin surface irregularities or pores, thus achieving a micromechanical interlocking between the adhesive and the dentin surface or dentinary canals, as illustrated in Fig. 8 (147,151). In order to increase the roughness of the dentin surface, the use of primers (chemical pretreatments) is normally used. When an acid treatment is used on the dentin surface, a partial demineralization is obtained, making collagen molecules more accesible to the reactive groups of the adhesive. Thus, a hybrid layer is formed where, in addition to a mechanical interlocking, a more favorable chemical binding is achieved.

In summary, the study of bonding materials to dentin is an active field characterized by widespread research. The apparent clinical success that is sometimes obtained still requires conclusive evaluation (123). The constant progress in this field has introduced the appearance of different generations of adhesive procedures. Briefly, the first adhesives used were hydrophobic liquid resin applied to the etched enamel surface; over this layer, the restorative resin was attached without manipulation of the dentin. In a second generation, dentin begins to be involved and the adhesive is composed by a mixture of hydrophobic and hydrophilic resins that interact with the smear layer. These treatments evolved into the use of hydrophilic resins applied on cleared dentin surfaces. The next step was the use of hybrid layers (partially demineralized dentin in which exposed collagen fibers interact chemically with the adhesive) (152). Nowadays, the trend is to use simpler techniques, reducing the number of involved steps, and trying to apply the different components simultaneously.

B. Endodontic Treatment

Necrosis of the tissues at the pulp chamber and the root canals of the teeth occur by deep caries or other aggressions (Fig. 6). Treatment of the infection requires the removal of the damaged tissues that cannot be regenerated. Therefore, the resulting pulp cavities, previously enlarged, cleaned, disinfected, and dried, must be filled by using different materials and techniques. These materials should provide an increased tooth stability by improving the mechanical force transfer from the tooth to the underlaying bone. The long-term stability of these fillers depends on the proper sealing of the tooth root apex area.

The nature of the materials employed is very important since they contact internal tissues through the root apex. They can be plastics (cements, pastes, etc.) or solid pieces (thin cones). Among those of the first group there is a wide diversity: antiseptic pastes, alkaline pastes (usually containing calcium hydroxyde), medical cements, etc. Many of these cements contain synthetic resins, i.e., polyvinyl, polyethylene, epoxy, polyacrylate, polycarbonate, etc., which contribute to the hardness of the final product and also seal the internal part of the canals (153–155). However, solid or semisolid materials with a small enough diameter to be passed through the tooth root canal and with different shapes (sharp cones, long pins or screws) can be used for filling purposes (Fig. 9). Silver was first used; but gutta-percha (resulting from the polymerization of the trans isomer of the isoprene) mixed with a cement is now

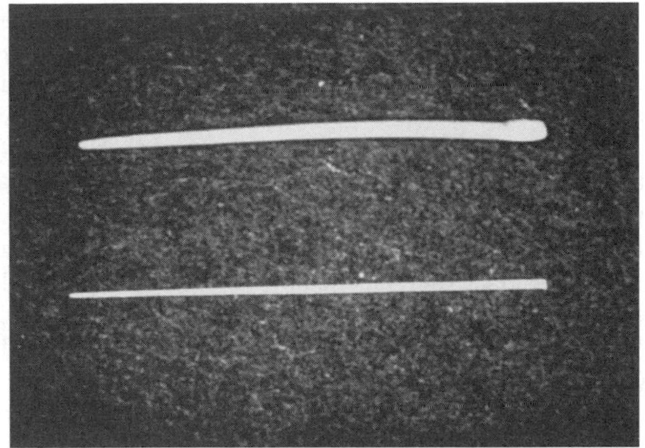

Figure 16.9 Gutta-percha cones of two different sizes.

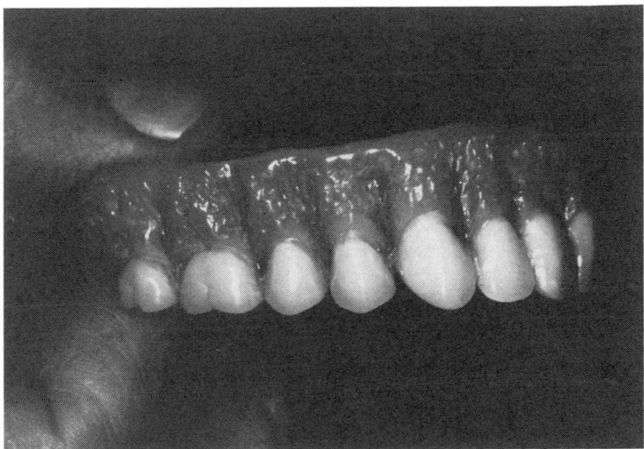

Figure 16.10 Complete dental upper prosthesis prepared with acrylic resin.

the most employed sealing material. In order to improve the properties of this polymer for dental purposes, several different additives have been included—zinc oxide, fillers, plasticizers, radiopaque agents, etc. Nowadays it is also used under the form of soft gutta-percha, prepared by heating the thermoplastic product immediately prior to its use (154,155).

Finally, other polymers used for the endodontic treatment are silicones, hydrophilic polymers (e.g., HEMA), epoxy, and so forth (155).

V. PROSTHETICS

Dental extraction of teeth considered hopeless because of the extent of their pathology is one of the most frequent surgical procedures performed upon humans. In order to maintain a proper oral functionality, prostheses are required (Fig. 6). Patients can use two types of dental prostheses: removable or fixed. The first group of devices are classified into complete/total (contacting to the oral mucosa) and partial (supported by the teeth and the oral mucosa). Complete prostheses are constituted by a base (a rigid polymer) and the artificial teeth (porcelain or polymers) (Fig. 10). Removable partial prostheses used to be formed by a metal alloy base (cobalt–chromium) and a rigid polymer that simulates the gum, and the artificial teeth (Fig. 11).

Fixed prosthesis requires carving of the natural teeth, either the one to be treated (preparation of a dental crown) or those adjacent to the toothless zone (preparation of a dental bridge). Afterward a metal structure is prepared that is covered by a polymer or ceramic for esthetic purposes. Finally, the fixed prosthesis will be cemented to the natural carved teeth (anchor). Fiber-reinforced composites (i.e., based on a light-polymerized bis-GMA matrix) are now under study in order to make metal-free prostheses with better esthetic qualities (156–158).

A. Removable Dental Prosthetic Bases

The most frecuently used removable denture bases are rigid; however, for specific cases resilient ones are recommended.

Rigid bases for complete or partially removable prostheses are normally prepared with PMMA. This polymer is known to be an ideal base material because of its stability in the oral environment although it absorbs some amount of water and is fragile. It can be used in either thermopolymerized systems (by using benzoyl peroxide as

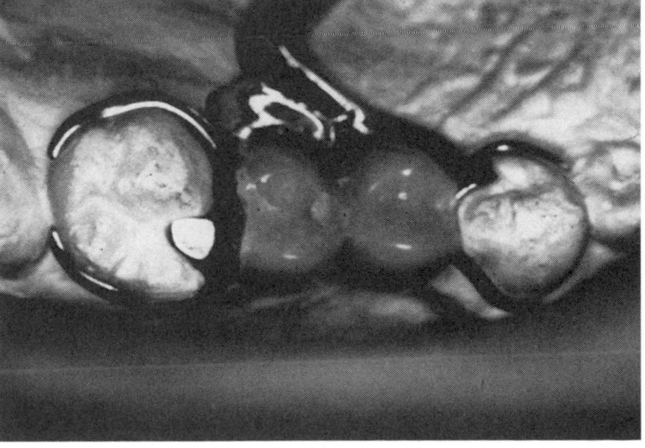

Figure 16.11 Detail (metal structure and acrylic teeth) of a removable partial prosthesis.

heat activated initiator) or light-cured systems, as those described in the operative dentistry section. The mechanical properties of PMMA, mainly its fragility, are improved by including small amounts of ethyl methacrylate, butyl methacrylate, etc.; or small particles of an artificial rubber (butadiene styrene); or aluminum, magnesium, and zirconium oxide powders; or by forming copolymers with vinyl acetate and vinyl chloride (158–160).

Other polymers used for this sort of base are copolymers of HEMA and methyl methacrylate, which exhibit a good adhesion between the prosthesis and the oral mucosa due to the adsorption and elasticity properties of HEMA. Polystyrene and polycarbonate can also be used, but their processing is difficult and expensive (159).

There are clinical cases related to highly sensible oral tissues where the use of soft and resilient bases is recommended (e.g., after surgery). Some acrylate derivatives mixed with plasticizers (e.g., dibutylphthalate), which are soft materials exhibiting a slow polymerization rate, are used in these cases (161).

B. Artificial Teeth, Esthetic Facing, and Cements

Porcelain has been used for the construction of artificial teeth since the end of the 18th century. Polymethyl methacrylate has also been used for this purpose. Now PMMA copolymers and certain vinyl or styrene resins are used. Nevertheless, the selection of the material employed for the artificial teeth is influenced by clinical considerations as well as the wearing of the material. The development of new composites, as described in the operative dentistry section, has resulted in new materials for the elaboration of artificial teeth (e.g., dimethacrylate urethane resins with microfilling particles) (159).

Veneering in crown and bridge prosthetics (esthetic facing) present a metal structure (precious or nonprecious alloys attached to the anchor teeth) covered by a PMMA resin or porcelain. Initially, thermocured acrylic derivatives were also used. But vinyl acrylic and acrylate derivatives, thermo- or light-cured composites (such as those described for operative dentistry), are now in wide use due to their improved resistance against wearing (159). However, these compounds have no adhesion to the underlying metal structure. Therefore, some adhesive must be also included (162); 4-META has been used to perform this function. Pyrolytic recovering of the metal surface with a thin glass layer is also employed since the resin can be attached to this layer by a silane treatment (150,163,164).

More recently, the appearance of ceromers has made it possible to introduce metal-free veneers, inlays/onlays, and short bridges in regions without heavy mechanical stress (165,166). In this way, metallic materials are substituted by fiber-reinforced ceromers forming an inner layer that is then recovered by additional layers of conventional or modified composite resins.

The prosthetic elements are usually fixed to the carved teeth by either an inorganic (e.g., zinc phosphate) or a polymer cement (e.g., zinc polycarboxylate or glass ionomer). In recent years, prosthetics directly attached to the teeth have appeared. They are mainly attached to the dental enamel through composite resins (e.g., Rochette bridge, Maryland bridge, etc.). The device is subjected to physical (perforation, blasting, abrassive treatment, etc.), chemical (acid treatments), or electrolytic procedures which allow a further surface retention of the composite (122,167,168).

VI. ORAL AND MAXILLOFACIAL SURGERY

There are many oral and maxillofacial surgical treatments where polymers of natural or artificial origin are being tested or used. Some of these applications are shown in Table 5.

A. Bone Loss Replacement, Fracture Immobilization, and Facial Outward Prostheses

At the end of the 19th century, vulcanized caoutchouc was introduced among jaw fragments to avoid maxillomandible deviations or deformities after surgical resections. Further, PMMA implants were used since they faithfully reproduced the removed mandibulary fragment. Silicon or metal alloys have been also employed for bone loss replacement. But now autologous or conservated bone are preferred for these purposes (169).

Table 16.5 Applications of Polymers in Oral and Maxillofacial Surgery

Bone loss replacement
Dental implants
Facial outward prostheses
Fracture immobilization
Hemostasis
Mouth protectors and habit suppresion devices
Obturating prostheses for abnormal communications
Plastic and reconstructive operations
Preprosthetic surgery
Surgical dressing
Sutures
Temporomandibular joint surgery

Regarding fracture immobilization, two different groups of lesions are distinguished: small dentary and/or alveolodentary fractures with luxation of a short dental group and complex maxillomandible fractures with deviation of fragments. Concerning the first group of lesions, composites such as those used in operative dentistry can be employed either to reconstruct a tooth fragment lost or to immobilize the affected dental group by using splints attached to the enamel after acid etching. Auto-cured PMMA can be also used for these splints. In maxillomandible fractures, osteosynthesis via the use of metal miniplates or steel wires is required (169,170).

Palliative treatments of the lesions resulting from complex fractures, with vast loss of facial soft tissues, were studied and developed after the world wars. Many different materials have been used for these treatments but there is not an ideal product. There are rigid or elastic materials, some of which can admit colorants. Rubbers, polyvinyl chloride, PMMA, vinyl resins with plasticizers, RTV (room temperature–vulcanized) silicone elastomers, polyurethanes, methacrylamide, butylacrylate or methylmethacrylate are some of the products used (159,171).

B. Plastic, Reconstructive, Preprosthetic, and Temporomandibular Joint Surgery

Some facial congenital or acquired deformities (sinking, asymmetries, etc.) as well as rhinoplasty, mentoplasty, ocular displacement due to orbit floor fracture, and cyclide surgery in facial paralysis treatment require corrective procedures involving implanted polymers. Paraffin was used at the beginning of the 20th century for remodeling of small facial defects. But tumor formation and migration of the implant were two main problems. Liquid silicone injections were used, manufactured with differing viscosity degree. However, migration of the implanted material has been observed (microspherical particle deposits in liver, lymph nodes, and kidney have been reported) (169). Elastic solid silicones manufactured under the form of blocks or sheets are preferred. Polyvinyl alcohol, polytetrafluoroethene (PTFE), polyester, and polyurethane have been also tested (169,172).

The aim of preprosthetic surgery is the increase of the maxillar or mandibular alveolar ridge by using an implant. This treatment is mainly recommended for patients exhibiting atrophy where conventional denture base prostheses cannot be used. The most tested materials have been polyvinyl alcohol, acrylate-amide sponge, silicones, silicone polyester, and Proplast (173–176). Further, hydroxylapatite and other ceramic materials have also been tested successfully (177,178).

Relapsing luxation treatment, insertion of materials in ankylosis treatment, condyle or meniscus replacement, etc., are included among temporomandibular joint (TMJ) surgical operations. Polyethylene, acrylic cups, PTFE, silicone discs, etc., were formerly used for this surgery. Also, Proplast was employed. It is a mixture of PTFE and glass carbon fibers, which can be invaded and covered by connective tissue, thus acting as a remodeling material (173–175). Autologous grafts or lyophilized bone are also used for these treatments.

C. Sutures

Suture threads can be classified taking into account different criteria: according to the type of material from which they are made (natural or synthetic) or considering the permanence of the material (resorbables and nonresorbables) (179). Although metals can be employed, polymers are the most commonly used materials. Depending on the geometry of the thread fibers, they can be considered as monofilamentary or polyfilamentary plaited. This is an interesting difference since the space between fibers of plaited threads can be filled by exudation or oral secretions. Thus, plaited threads must be processed in order to increase their hydrophobic character (e.g., silicone treatment).

Resorbable sutures have many uses in oral and maxillofacial surgery to suture deep tissues. Catgut and collagen are used to elaborate natural resorbable sutures. Collagen is obtained from animal tendons; it is rarely used, although it seems to be less of an irritant than catgut. Submucous and serous layer of bovine intestine are used to elaborate catgut. This traditional material is a monofilamentary thread but exhibits a disadvantage: its diameter increases in only a few hours due to its absorption properties. Catgut can be simple or chromic. The latter is absorbed at a slower rate and is less of an irritant than simple catgut (167,180–182).

Polyglycolic acid is a resorbable synthetic homopolymer. There are also copolymers of glycolic and lactic acids. Polygluconate and polydioxanone are also employed as resorbable suture threads. The tissue reaction produced against these synthetic materials is lower than that appearing for those of a natural origin. This is due to the different degradation processes involved in the resorption mechanism: physicochemical reactions for the polymers and enzymatic degradation for collagen and catgut (101,180,181). New materials for intra- and extraoral applications are appearing combining reduced tissue reaction with fast-absorbing properties (183).

Among the nonresorbable threads, silk is the most used suture in oral and maxillofacial surgery. Capillarity phenomena are its main disadvantage. Polyamides, polyester, polypropylene, polyacrylonitrile, etc., are also used as nonresorbable threads. All these threads tend to slacken slightly; thus more knots are required in order to tighten and reinforce the sutures (182).

D. Other Applications

Obturating prostheses are used for abnormal, temporary or permanent, communication treatments between the oral cavity and other cavities (nasal, sinus, orbit, etc.) due to congenital (cleft palate) or acquired (post-tumoral resections, complex injuries, etc.) lesions. Polymethyl methacrylate is usually employed and is attached to the teeth by metal wires (Fig. 12) (184).

So far polymers have no application as dental implants. Nevertheless, there are some old reports about the use of acrylic intraosseous implants for the immobilization of teeth in periodontal disease.

In several surgical dental procedures, the control of hemorrhage is critical. This is the case of apical sealing in endodontic surgery, which requires a dry root end cavity to insert the filling material. Bovine collagen and bone wax (with oils, wax mixtures, antiseptics, etc.) are used as hemostatic agents for some bone or periosteum hemorrhages. A mixture of surgical wax and fibers of calcium alginate permits an easy placement of a root end filling under sterile and nontoxic conditions (185).

Surgical cement is sometimes required during wound healing following some surgery in gingiva. Most of these cements used for surgical dressing are inorganic substances containing fatty acids and many additives, such as antiseptics, sedatives, etc. Some light-cured materials, containing urethane and acrylate resins are currently under use.

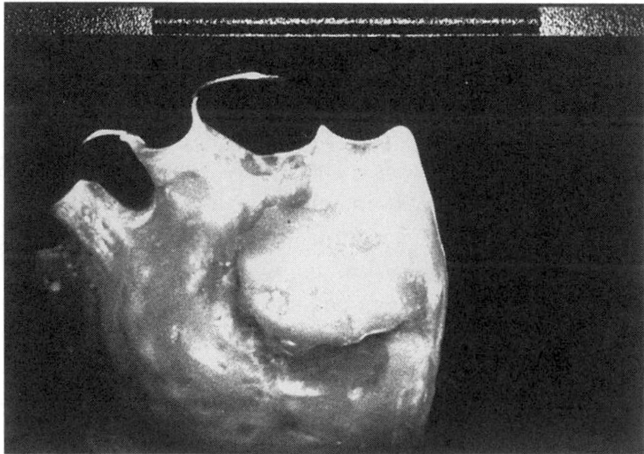

Figure 16.12 Obturating acrylic device to avoid the communication between oral and nasal cavities after surgical removing of a palatal tumour.

Different kinds of intra- or extraoral mouth protectors and habit suppression devices are frequently used for the prevention of dental or bone fractures and complex facial injuries, TMJ lesions, dental abrasions, etc., which can result from violent activities related to sports, bruxism, or certain habitual behaviors. Polymers are the most used materials: silicones, PMMA, caoutchouc, soft acrylic resins, polyvinyl chloride, vinyl polyacetate, polyethylene copolymers, etc. Many of these are thermoplastics whereas others are prepared by auto- or thermo-cure mechanisms (159,171,184).

VII. PERIODONTOLOGY

Periodontal disease is caused by specific hardened bacteria which adhere to the teeth inducing inflammation of the soft tissues surrounding the teeth. A major consequence of periodontal disease is the loss of periodontal support with destruction of the fibrous connective tissue attachment. This is the support of the periodontal ligament which connects tooth root cementum and surrounding alveolar bone. If this process goes on, destruction of cementum and bone may occur, leading eventually to the loss of the teeth. The ultimate goal of periodontology is the development and improvement of therapies for the prevention, diagnosis, and treatment of periodontal disease (Fig. 13). There are many different groups of periodontology treatments using biomaterials, as summarized in the following sections. However, basic research is now focused in periodontal regeneration, trying to induce the formation of new bone, new cementum, and supportive periodontal ligament (68,186).

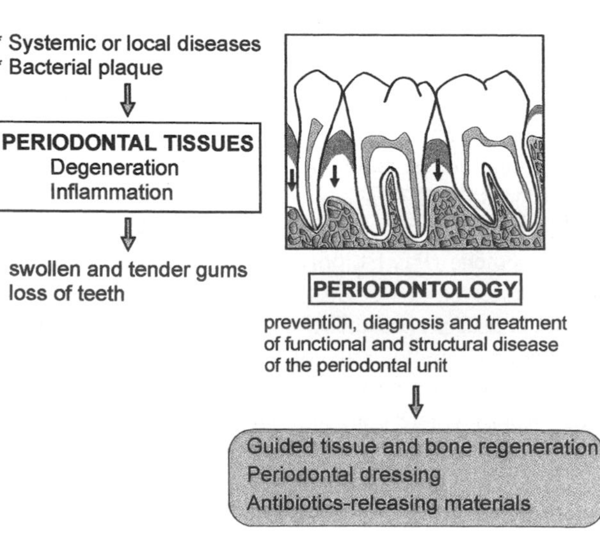

Figure 16.13 Origin of periodontal disease and different treatments used in periodontology.

This approach is based on the regenerative wound healing potential of progenitor stem cell populations from the periodontal tissues.

A. Guided Tissue Regeneration

Guided tissue regeneration (GTR) procedures have been used for regeneration of the attachment apparatus on teeth (187–193). This technique has also been considered as a surgical procedure for either ridge augmentation or implantation of osteointegrated materials to support a prosthesis.

The possibility of periodontal tissue regeneration is of particular importance in the treatment of furcation-involved teeth. It has been demonstrated that GTR therapy accounts for a high rate of success and predictability in the treatment of second-degree furcation involvement of mandibular molars (194). The GTR approach has also been used in the treatment of class II furcation defects in combination with other modalities designed to achieve regeneration, such as bone grafting, root demineralization, attachment proteins, and/or coronally positioned flaps (195).

Prevention of the epithelial migration along the cemental wall of the pocket has been treated by placing barriers of different types covering the bone and the periodontal ligament (Fig. 14), these membranes being removed 5 weeks after the operation. Millipore, Teflon, and Gore-Tex (expanded polytetrafluorethylene) membranes have been used for these purposes (196). However, there are no ideal universal characteristics for these materials, such as pore size and polymer type, and different criteria must be applied in each particular clinical case (197,198).

Bioresorbable type I collagen barrier membranes have also been used for GTR (69,113,199). The results obtained have demonstrated that sites treated with a collagen barrier exhibit significantly better healing when compared to control sites over a 1-year period of study. Subgingivally placed membranes of the type used for treatments of defects in monkey may predictably promote the formation of a new connective tissue attachment in recession-type defects (190,191). A combination of the bone inductive protein (osteogenin or bone morphogenetic protein-3) plus type I collagen as well as a combination of osteogenin with coralline hydroxyapatite have been used for treatment of critical size calvaria defects in rats (200). The results from histometric measurements of healing indicate that osteogenin plus collagen has the greatest potential for regenerating calvarial bone defects (190).

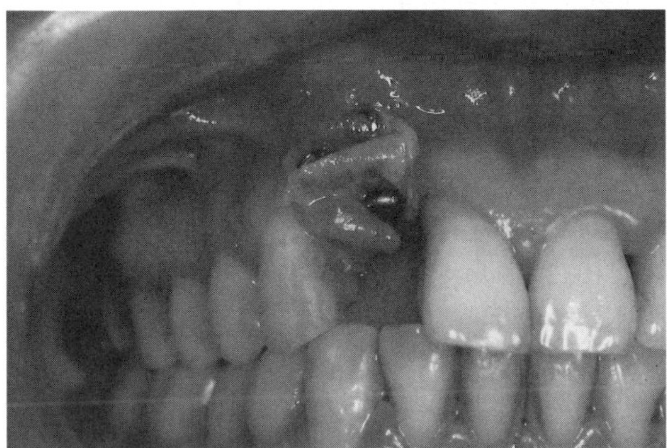

Figure 16.14 Implantation of expanded polytetrafluoroethylene membrane (Gore-Tex).

One of the new trends in the research involves the development of bioresorbable grafting materials chemically based on synthetic polymers (201). Specific membrane porosities do not seem to be required for bone formation, but optimal pore sizes are advantageous regarding nutrient flow, wound stabilization, and peripheral sealing to prevent ingrowth of soft tissue forming cells. The best known groups of polymers used for medical and dental purposes are collagen, mentioned previously, and aliphatic polyesters. The most frequently used synthetic membranes are homopolymers or copolymers of polyglycolide and/or polylactide (202). Occlusive bioresorbable membranes made of polylactic or polyglycolic acid are equally succesful as expanded polytetrafluoroethylene membranes in regenerating bone in transosseous defects in the rabbit mandibular ramus (203). Clinical results using these bioresorbable barrier membranes in the treatment of interproximal intrabony periodontal defects show comparable results to bioresorbable type I collagen membranes (204).

One apparent advantage of poly(α-hydroxy acids) is their degradation by hydrolysis, resulting in decomposition products that are mostly metabolized to carbon dioxide and water. The degradation rate is dependent on the presence of enzymes, bacteria (infection), mechanical strain, and pH. The addition of D-lactide or glycolide results in a faster degradation. However, the hydrolysis of these polymers releases lactic and/or glycolic acids; if the concentration of these monomers goes over a certain threshold they may be toxic to cells and present osteolytic effects.

Guided tissue regeneration has also been considered in relation to osteointegrated implants. Following extraction of the natural tooth, the edentulous ridge begins to resorb palatally, lingually, and toward the apex. Therefore, the implants must be placed immediately into the extraction sites and covered with a membrane. This technique has been reported as valid for preservation of the bone at the site of implantation.

Bone grafts have been used in periodontal and implant

surgery with the aim of increasing the bone volume in the defect area. Bone substitute can be classified into two groups: xenogenic and autogenic materials. The second is the most frequently used. Demineralized freeze-dried bone represents an allograft material (from the same species, but not the same individual) (205). Bone substitutes should exhibit biocompatible material properties, being tolerated and integrated by the host tissues and gradually replaced by newly formed bone (206,207). Recently, several synthetic bone grafts have also been used for clinical applications and the obtained results are comparable to those obtained with natural bone (208). Guided bone regeneration (GBR) seems to be more efficient when autogenic bone grafts are used in combination with adequate protection by barrier membranes (209).

The future of GTR and GBR promises to be exciting; new materials or modified ones, especially in combination with biological factors, will become available.

B. Periodontal Dressings

Periodontal dressings are mainly used to achieve and maintain a close adaptation of the mucosal flaps to the underlying bone, for the comfort of the patient and to protect the wound prepared during surgery. The most commonly used periodontal dressings may be divided into two groups: eugenol-containing and non–eugenol-containing dressings. They are formed by a solid component—zinc oxide, resins, tannic acid, cellulose fibers, and zinc acetate—and a liquid one that may contain eugenol, vegetable oils, thymol, and color additives. Cyanoacrylates have also been used as periodontal dressings with varying success. They are directly applied as a liquid onto the wound or sprayed over the wound surface. Antibacterial agents should be incorporated in periodontal dressings to prevent bacterial growth in the wound area.

C. Antibiotic-Releasing Materials

Antibiotic-containing implants can be used to suppress or prevent postsurgical infections after biomaterial implantation. The design of these systems must include consideration of the controlled diffusion of the associated drug (antibacterial agent, anti-inflammatory drugs, etc.) (206,210). These drugs could also be mixed with the cement (e.g., polymethyl methacrylate) at the insertion time.

Different materials, such as acrylic strips, monolithic fibers, and dialysis tubings, have been used for sustained local release of antimicrobial agents (192). Bioabsorbable materials can also be used for in situ applications. This eliminates the risk of disturbing a healing site after therapy. The most frequently used bioabsorbable materials are Surgical (hemostatic gauze made of oxidized regenerated cellulose), Colla Cote (a collagen wound dressing), and Tisell (a fibrin sealant).

Hydroxypropylcellulose has been used as matrix for degradable films containing chlorhexidine or tetracycline (211). Polycaprolactone, another degradable polymer, and ethylene vinyl acetate copolymers have served as a matrices for monolithic fibers loaded with tetracycline for local treatment of periodontal pockets or subgingival root surfaces (212). Most of the intrapocket controlled release devices are composed of nondegradable matrices.

Treatment of marginal periodontitis may include the use of local antibiotics. Surgical bioabsorbable materials (e.g., Tisell and Colla Cote) and acrylic strips have been reported to be capable of an in vivo prolonged release of doxycycline (213). This mode of delivery increases the concentration of antibiotic at the site of action by incorporating the drug into different devices for insertion into periodontal pockets. In contrast to irrigation, these devices may assure an antimicrobial effect for a prolonged period.

VIII. ORTHODONTICS AND OTHER APPLICATIONS

Removable intraoral devices, composed of PMMA plates and wires (in the form of springs, arches, gibs, screws, elastics, etc.), are used in orthodontics to obtain tooth movement or habit supression (Fig. 6). These devices tend to give tone to the perioral muscles. Fixed devices are also employed. Orthodontic tooth movement is finally achieved by a remodeling of alveolar bone and the response of the periodontal ligament to a mechanical force (214). The dental movement is obtained by different arches and wires forming bands and brackets. Bands are cemented to the teeth and brackets can be welded to the bands or directly bonded to the enamel. For esthetic reasons, polycarbonate or ceramic bases for the brackets have also been employed; but they exhibit a significant fragility that is sometimes reduced by using plastic-covered metal bases. Attachment of bands is obtained by using zinc phosphate, polycarboxylate, or glass-ionomer cement, whereas slightly fluid composites and enamel acid etching have been used for the bracket attachment. Rubber elastic rings are also elements employed for the orthodontic treatments (215,216).

Other applications of polymers in dentistry should also be mentioned; among them are those in which polymers are used as impression materials or for the construction of work models. Elastic dental impression materials, hydrocolloids and elastomers, are very useful in dentistry. Agar, a linear polymer of D-galactose, was the first hydrocolloid used for these purposes. Later on, alginates (alginic acid salts; polymer of D-mannuronic and L-guluronic acids) were introduced in this field and are nowadays widely used. Polysulfide rubber materials, polysiloxanes, and polyethers are among the elastomeric impression materials (217). Polyvinyl siloxane is used in different indirect procedures in prosthodontics and restorative dentistry due to its favorable handling properties, good patient acceptance, and physical properties (218).

Calcium sulfate and dental stone gypsum were the first materials used to construct work models, but epoxy resins are also now employed due to their hardness and dimensional stability (219).

Finally, waxes and thermoplastics are also used in dentistry. They are natural and synthetic waxes used in occlusion registration, pattern elaboration, and many other auxiliary applications in dentistry. Complex mixtures, composed of waxes, resins, fillers, and plasticizers are also sometimes employed (220,221).

IX. NEW PERSPECTIVES

Much research remains to be done concerning oral biology and biochemistry in connection with biomaterial applications in dentistry. Polymers and composites are widely used in this field and their applications are under continuous evolution. Due to the understanding of their structure and properties, these compounds have a bright future in the field of dentistry. Connective tissue adherence to these materials, replacement of hard structures, achievement of proper cell–biomaterial interaction controlling cell proliferation and differentiation after implantation are some of the main biological problems to be treated over the next few years.

Polymers and polymer-containing biomaterials require specific considerations. The chemical composition of a macromolecule determines its conformation, which is further responsible for most of the properties determining its behavior as an implant. An ideal biomaterial must provide not only a scaffolding or three-dimensional architecture for tissue regeneration, but also must modulate specific cell functions. Thus, the goal in the development of a polymeric biomaterial is the synthesis of an adequate chemical structure, designed to be nonresorbable, partially resorbable, or fully resorbable, with specific surface properties in order to promote cell adhesion and to enhance tissue adaptation. In this way, a more stable implant-to-tissue interface, according with the demands of each particular application, could be obtained. New approaches in the synthesis of new biomaterials have been reviewed (222,223).

Besides the traditional chemical methods of polymer synthesis, a combination of biological and chemical synthesis is now being considered in order to improve polymer synthesis (e.g., polyphenylenes) (223). Moreover, an alternative synthesis of polymeric materials has been described using bacteria and plants in large-scale production of biodegradable plastics. Polyhydroxyalkanoates (polyesters), a family of bacterial biodegradable plastics and elastomers, are polymers that have properties ranging from stiff and brittle plastics to rubberlike materials, and possess inherent biodegradability (224). Another example is the production of poly-3-hydroxybutyrate in *Saccharomyces cerevisiae* transformed with the bacterial polyhydroxybutyrate synthase gene (225).

The great advances in genetic engineering have also contributed to the development of new biopolymers or novel proteins encoded by artificial genes. This promising approach allows the synthesis of new biopolymers (natural or quimeric proteins) with specific sequences, structures, and properties by using recombinant cDNA technology. DNA sequences coding for specific molecular arrangements of proteins have been selected and transfected into bacterial systems, and the protein was produced in large amounts. In this way, different quimeric proteins, containing sequences of natural fibrous proteins (silk, elastin, and collagen) have been prepared using either bacterial or yeast systems. Artificial fibrous proteins containing repetitive sequences have been successfully synthesized including silklike polymers, elastin analogs, polymers containing both silklike and elastinlike blocks, and collagenlike proteins (226,227). Another example is the Pronectin F polymer—a commercially available molecule that is the product of an artificial gene containing sequences of silk and fibronectin. Silk contributes a high tensile strength, even stronger than high tensile steel, and fibronectin apports the RGD receptor recognition amino acid sequences that enable the attachment of several types of mammalian cells (93,94,228).

It is well known that growth factors and other mitogenic and/or differentiating agents can be used in different systems to modify the response to a material (12,67,186,229). Cellular and molecular biology basic research is studying the specific proteins in the oral cavity presenting mitogenic or differentiation activity that may be useful for periodontal and alveolar bone regeneration. It is well known that bone morphogenetic proteins (BMPs) are excellent molecules for the stimulation of oral bone formation (67,68,230,231). Potential applications of these proteins in the restoration of oral cavity defects have been studied, allowing periodontal ligament regeneration. Other therapeutic candidates are enamel matrix proteins that have been reported to stimulate acellular cementum formation, also allowing functional periodontal ligament formation (232). The therapeutic use of the functionally active proteins requires their large-scale production. This has been achieved by the use of recombinant cDNA technology. As an example, recombinant BMP-2 has been extensively studied in animal models and is currently being tested in human clinical trials (231). Porous synthetic polymers (e.g., poly-D,L-lactide-co-glycolide particles) are beginning to be used as carriers for BMPs, promoting the induction of new bone formation (233).

REFERENCES

1. J. E. Lemons. Dental implants. In: *Biomaterials Science: An Introduction to Materials in Medicine*, B. D. Ratner, A. S. Hoffman (Eds.). Academic Press, New York, 1996, pp. 308–319.
2. A. Veis. Bones and teeth. In: *Extracellular Matrix Biochemistry*, K. A. Piez, A. H. Reddi (Eds.). Elsevier, New York, 1984, pp. 329–374.
3. C. A. Shuttleworth, J. W. Smalley. Periodontal ligament. *Int. Rev. Connect. Tissue Res.* 10:211–247 (1983).
4. A. S. Narayanan, R. C. Page. Connective tissue of the periodontium: a summary of current work. *Collagen Rel. Res.* 3:33–64 (1983).
5. M. Bouvier, A. Joffre, H. Magloire. In vitro mineralization of a three-dimensional collagen matrix by human dental pulp cells in the presence of chondroitin sulphate. *Arch. Oral Biol.* 35:301–109 (1990).
6. H. J. Höhling, R. H. Barckhaus, E. R. Krefting, J. Althoff, P. Quint. Collagen mineralization: aspects of the structural relationship between collagen and the apatitic crystallites. In: *Ultrastructure of Skeletal Tissues*, E. Bonucci, P. M. Motta (Eds.). Kluwer Academic, Dordrecht, The Netherlands, 1990, pp. 41–62.
7. W. T. Buttler, M. Bhown, M. T. Dimuzio, A. Linde. Noncollagenous proteins of dentin. Isolation and partial characterization of rat dentin proteins and proteoglycans using a three-step preparative method. *Collagen Rel. Res.* 1: 187–199 (1981).
8. D. J. Carmichael, A. Veis, E. T. Wang. Dentin matrix collagen: evidence for a covalently linked phosphoprotein attachment. *Calcif. Tiss. Res.* 7:331–334 (1971).
9. W. T. Butler. Dentin-specific proteins. *Methods Enzymol.* 145:291–303 (1987).
10. A. Veis. Dentin. In: *Extracellular Matrix*, W. D. Comper (Ed.). Harwood Academic, Amsterdam, 1996, pp. 41–76.
11. M. H. Veron, M. L. Couble, G. Caillot, D. J. Hartmann, H. Magloire. Expression of fibronectin and type I collagen by human dental pulp cells and gingiva fibroblasts grown on fibronectin substrate. *Arch. Oral Biol.* 35:565–569 (1990).
12. B. McAllister, A. S. Narayanan, Y. Miki, R. C. Page. Isolation of a fibroblast attachment factor from cementum. *J. Periodontal. Res.* 25:99–105 (1990).
13. M. J. Somerman, B. Shroff, W. S. Argraves, G. Morrison, A. M. Craig, D. T. Denhardt, R. A. Foster, J. J. Sauk. Expression of attachment proteins during cementogenesis. *J. Biol. Buccale* 18:207–214 (1990).
14. S. Sengupta, F. E. Shovlin, A. Slomiany, B. L. Slomiany. Identification of laminin receptor in gingival tissue and its interaction with tooth cementum laminin. *Int. J. Biochem.* 23:115–121 (1991).
15. M. J. Somerman, J. J. Sauk, R. A. Foster, K. Dickerson, K. Norris, W. S. Argraves. Cell attachment activity of cementum: bone sialoprotein II identified in cementum. *J. Periodontal Res.* 26:10–16 (1991).
16. S. Pitaru, N. Savion, H. Hekmati, S. Olson, S. A. Narayanan. Molecular and cellular interactions of a cementum attachment protein with periodontal cells and cementum matrix components. *J. Periodontal Res.* 28:560–572 (1993).
17. R. L. MacNeil, J. Berry, J. D'Errico, C. Strayhorn, M. J. Somerman. Localization and expression of osteopontin in mineralized and nonmineralized tissues of the periodontum. In: *Osteopontin: Role in Cell Signalling and Adhesion*, D. T. Denhardt, W. T. Butler, A. F. Chambers, D. R. Senger (Eds.). Annals of the New York Academy of Sciences. Vol. 760, 1995, pp. 166–176.
18. F. P. Reinholt, K. Hultenby, A. Oldberg, D. Heinegard. Osteopontin—a possible anchor of osteoclasts to bone. *Proc. Natl. Acad. Sci. U.S.A.* 87:4473–4475 (1990).
19. K. E. Healy. Dentin and enamel. In: *Handbook of Biomaterial Properties*, J. Black, G. Hastings (Eds.). Chapman & Hall, London, 1998, pp. 24–39.
20. N. E. Waters. Some mechanical and physical properties of teeth. *Symp. Soc. Exp. Biol.* 34:99–135 (1980).
21. A. Mariotte. The extracellular matrix of the periodontium: dynamic and interactive tissues. *J. Periodontol. 2000* 3: 39–63 (1993).
22. R. C. Page, H. E. Schroeder. Biochemical aspects of connective tissue alterations in inflammatory gingival and periodontal disease. *Int. Dent. J.* 23:455–469 (1973).
23. Y. Takagi, A. Veis, J. J. Sauk. Relation of mineralization defects in collagen matrices to noncollagenous protein components. Identification of a molecular defect in dentinogenesis imperfecta. *Clin. Orthop.* 176:282–289 (1983).
24. P. M. Bartold, R. C. Page. Hyaluronic acid synthesized by fibroblasts from normal and chronically inflamed gingiva. *Collagen Rel. Res.* 6:365–377 (1986).
25. P. L. Lukinmaa, H. Ranta, A. Vaheri. Osteogenesis imperfecta: fibronectin in dentin matrix. *J. Cran. Genet. Dev. Biol.* 8:75–82 (1988).
26. V. P. Terranova, M. Jendresen, F. Young. Healing, regeneration, and repair: prospectus for new dental treatment. *Adv. Dent. Res.* 3:69–79 (1989).
27. R. G. Caffesse, C. E. Nasjleti, G. B. Anderson, D. E. Lopatin, B. A. Smith, E. C. Morrison. Periodontal healing following guided tissue regeneration with citric acid and fibronectin application. *J. Periodontol.* 62:21–29 (1991).
28. Y. L. Kapila, H. Lancero, P. W. Johnson. The response of periodontal ligament cells to fibronectin. *J. Periodontol.* 69:1008–1019 (1998).
29. A. E. Postlethwaite, A. H. Kang. Fibroblasts and matrix proteins. In: *Inflammation: Basic Principles and Clinical*

Correlates, J. I. Gallin, I. M. Goldstein, R. Snyderman (Eds.). Raven Press, New York, 1992, pp. 747–773.

30. M. Aumailley, T. Krieg. Structure and function of the cutaneous extracellular matrix. *Eur. J. Dermatol.* 4:271–280 (1994).
31. T. Scott-Burden. Extracellular matrix: the cellular environment. *NIPS* 9:110–115 (1994).
32. J. Ph. Kleman, C. Giry-Loringuez, M. van der Rest. Diversity and modularity of extracellular matrix molecules at the gene and protein levels, *J. Hepatol.* 22:3–9 (1995).
33. M. Aumailley, B. Gayraud. Structure and biological activity of the extracellular matrix. *J. Mol. Med.* 76:253–265 (1998).
34. P. D. Yurchenco, J. J. O'Rear. Basal lamina assembly. *Curr. Opin. Cell. Biol.* 6:674–681 (1994).
35. A. Lindblom, M. Paulsson. Basement membranes. In: *Extracellular Matrix*, W. D. Comper (Ed.). Harwood Academic, Amsterdam, 1996, pp. 132–174.
36. R. Timpl. Macromolecular organization of basement membranes. *Curr. Opin. Cell. Biol.* 8:618–624 (1996).
37. M. van der Rest, R. Garrone. Collagen as multidomain proteins. *Biochemie* 72:473–484 (1990).
38. E. Vuorio, B. DeCrombrugghe. The family of collagen genes. *Annu. Rev. Biochem.* 59:837–872 (1990).
39. C. M. Kielty, I. Hopkinson, M. E. Grant. The collagen family: structure, assembly and organization in the extracellular matrix. In: *Connective Tissue and Its Heritable Disorders*, P. M. Royce, B. Steinmann (Eds.). Wiley-Liss, New York, 1993, pp. 103–147.
40. R. Mayne, R. G. Brewton. New members of the collagen superfamily. *Curr. Opin. Cell. Biol.* 5:883–890 (1993).
41. J. C. Brown, R. Timpl. The collagen superfamily. *Int. Arch. Allergy Immunol.* 107:484–490 (1995).
42. B. R. Olsen. New insights into the functions of collagens from genetic analysis. *Curr. Opin. Cell. Biol.* 7:720–727 (1995).
43. K. von der Mark, S. Goodman. Adhesive glycoproteins. In: *Connective Tissue and Its Heritable Disorders*, P. M. Royce, B. Steinmann (Eds.). Wiley-Liss, New York, 1993, pp. 211–236.
44. C. W. Prince. Proteoglycans of mineralized matrices. *Methods Enzymol.* 145:261–269 (1987).
45. A. J. Fosang, T. E. Hardingham. Matrix proteoglycans. In: *Extracellular Matrix*, W. D. Comper (Ed.). Harwood Academic, Amsterdam, 1996, pp. 200–229.
46. J. C. Adams, F. M. Watt. Regulation of development and differentiation by the extracellular matrix. *Development* 64:1183–1198 (1993).
47. H. K. Kleinman, M. C. Kibbey, H. W. Schnaper, M. A. Hadley, M Dym, D. S. Grant. Role of basement membrane in differentiation. In: *Molecular and Cellular Aspects of Basement Membranes*, D. H. Rohrbach, R. Timpl (Eds.). Academic Press, San Diego, 1993, pp. 309–326.
48. T. N. Wight, M. G. Kinsella, E. E. Qwarnström. The role of proteoglycans in cell adhesion, migration and proliferation. *Curr. Opin. Cell. Biol.* 4:793–801 (1992).
49. R. A. Brown, C. D. McFarland. Cell–matrix adhesive proteins: further therapeutic applications. *Exp. Opin. Ther. Patents* 6:11–19 (1996).
50. A. Stockmann, S. Hes, P. Declerck, R. Timpl, K. T. Preissner. Multimeric vitronectin. Identification and characterization of conformation-dependent self-association of the adhesive proteins. *J. Biol. Chem.* 268:22874–22882 (1993).
51. S. Johansson. Non-collagenous matrix proteins. In: *Extracellular Matrix*, W. D. Comper (Ed.). Harwood Academic, Amsterdam, 1996, pp. 68–94.
52. M. Silbermann, K. von der Mark, D. Heinegard. An immunohistochemical study of the distribution of matricial proteins in the mandibular condyle of neonatal mice. 2. Non-collagenous proteins. *J. Anatomy* 170:23–31 (1990).
53. A. Lindle, M. Bhown, W. T. Butler. Noncollagenous protein of dentin. A re-examination of proteins from rat incisor dentin utilizing techniques to avoid artifacts. *J. Biol. Chem.* 255:5931–5942 (1980).
54. R. W. Romberg, P. G. Werness, P. Lollar, B. L. Riggs, K. G. Mann. Isolation and characterization of native adult osteonectin. *J. Biol. Chem.* 260:2728–2736 (1985).
55. R. L. MacNeil, N. Sheng, C. Strayhorn, L. W. Fisher, M. J. Somerman. Bone sialoprotein is localized to the root surface during cementogenesis. *J. Bone Miner. Res.* 9: 1597–1606 (1994).
56. J. Sodek, J. Chen, S. Kasugai, T. Nagata, Q. Zhang, M. D. McKee, A. Nanci. Elucidating the functions of bone sialoprotein and osteopontin in bone formation. In: *Chemistry and Biology of Mineralized Tissues*, H. Slavkin, P. Price (Eds.). Excerpta Medica, New York, 1992, pp. 297–307.
57. M. N. Helder, A. L. J. J. Bronckers, J. H. M. Woltgens. Dissimilar expression patterns for the extracellular matrix proteins osteopontin (OPN) and collagen type 1 in dental tissues and alveolar bone of the neonatal rat. *Matrix* 13: 415–425 (1993).
58. M. P. Mark, W. T. Butler, C. W. Prince, R. D. Findelman, J.-V. Ruch. Developmental expression of 44-kDa bone phosphoprotein (osteopontin) and bone γ carboxyglutamic acid (Gla)–containing protein (osteocalcin) in calcifying tissues of rat. *Differentiation* 37:123–136 (1988).
59. M. Sato, W. Grasser, S. Harm, C. Fullenkamp, J. P. Gorski. Bone acidic glycoprotein 75 inhibits resorption activity of isolated rat and chicken osteoclasts. *FASEB J.* 6: 2966–2976 (1992).
60. P. A. Price, J. W. Pser, N. Raman. Primary structure of the γ-carboxyglutamic acid–containing protein from bovine bone. *Proc. Natl. Acad. Sci. U.S.A.* 73:3374–3375 (1976).
61. L. W. Fisher, J. D. Termine, M. F. Young. Deduced protein sequence of bone small proteoglycan I (byglycan) shows homology with proteglycan II (decorin) and several non-connective tissue proteins in a variety of species. *J. Biol. Chem.* 264:4571–4576 (1989).
62. A. George, R. Silberstein, A. Veis. In situ hybridization shows Dmp1 to be a developmentally regulated dentin

specific protein produced by mature odontoblasts. *Connective Tissue Res.* 33:67–72 (1995).
63. H. H. Ritchie, H. Hou, A. Veis, W. T. Butler. Cloning and sequencing determination of rat dentin sialoprotein, a novel dentin protein. *J. Biol. Chem.* 269:3698–3702 (1994).
64. W. G. Stetler-Stevenson, A. Veis. Bovine dentin phosphophoryn: composition and molecular weight. *Biochemistry* 22:4326–4335 (1983).
65. R. Fujisawa, T. Takagi, Y. Kuboki, S. Sasaki. The characterization of collagen–phosphophoryn complex in bovine dentin. In: *Chemistry and Biology of Mineralized Connective Tissues*, A. Veis (Ed.). Elsevier, New York, 1981, pp. 483–487.
66. C. L. Strayhorn, J. S. Garret, R. L. Dunn, J. J. Benedict, M. J. Somerman. Growth factors regulate expression of osteoblast-associated genes. *J. Periodontol.* 70:1345–1354 (1999).
67. R. D. Finkelman, S. Mohan, J. C. Jennings, A. K. Taylor, S. Jepson, D. J. Baylink. Quantitation of growth factors IGF-I, SGF/IGF-II, and TGF-β in human dentin. *J. Bone Mineral Res.* 5:717–723 (1990).
68. U. Ripamonti, A. H. Reddi. Periodontal regeneration: potential role of bone morphogenetic proteins, *J. Periodontal Res.* 29:225–235 (1994).
69. J. S. Mattson, L. L. McLey, M. H. Jabro. Treatment of intrabony defects with collagen membrane barriers. Case reports. *J. Periodontol.* 66:635–645 (1995).
70. J. Rosenbloom, W. R. Abrams, R. Mecham. Extracellular matrix: the elastic fiber. *FASEB J.* 7:1208–1218 (1993).
71. J. Rosenbloom. Elastin. In: *Connective Tissue and Its Heritable Disorders*, P. M. Royce, B. Steinmann (Eds.). Wiley-Liss, New York, 1993, pp. 167–188.
72. C. Mendieta, C. Caravana, D. H. Fine. Sorption of fibronectin to human root surfaces in vitro. *J. Periodontol.* 61: 254–260 (1990).
73. J. R. Potts, I. D. Campbell. Fibronectin structure and assembly. *Curr. Opin. Cell. Biol.* 6:648–655 (1994).
74. K. Beck, I. Hunter, J. Engel. Structure and function of laminin: anatomy of a multidomain glycoprotein. *FASEB J.* 4:148–160 (1990).
75. E. Engvall. Laminin variants: why, where and when? *Kidney Int.* 43:2–6 (1993).
76. R. Timpl, J. C. Brown. The laminins. *Matrix Biol.* 14:275–281 (1994).
77. E. Engvall, U. M. Wewer. Domains of laminin. *J. Cell Biochem.* 61:493–501 (1996).
78. K. M. Malinda, H. K. Kleinman. The laminins. *Int. J. Biochem. Cell Biol.* 28:957–959 (1996).
79. D. E. Nies, T. J. Hemesath, J. H. Kim, J. R. Gulcher, K. Stefansson. The complete cDNA sequence of human hexabrachion (tenascin). A multidomain protein containing unique epidermal growth factor repeats. *J. Biol. Chem.* 266:2818–2823 (1991).
80. H. P. Erickson. Tenascin-C, tenascin-R and tenascin-X: a family of talented proteins in search of functions. *Curr. Opin. Cell. Biol.* 5:869–876 (1993).
81. P. G. Robey, M. F. Young, L. W. Fisher, T. D. McClain. Thrombospondin is an osteoblast-derived component of mineralized extracellular matrix. *J. Cell. Biol.* 108:719–727 (1989).
82. P. Bornstein. Thrombospondins: structure and regulation of expression. *FASEB J.* 6:3290–3299 (1992).
83. P. M. Bartold, R. R. Boyd, R. C. Page. Proteoglycans synthesized by gingival fibroblasts derived from human donors of different ages. *J. Cell Physiol.* 126:37–46 (1986).
84. L. Kjellén, U. Lindahl. Proteoglycans: structures and interactions. *Annu. Rev. Biochem.* 60:443–475 (1991).
85. J. E. Scott. Supramolecular organization of extracellular matrix glycosaminoglycans, in vitro and in the tissues. *FASEB J.* 6:2639–2645 (1992).
86. R. N. Tamura, D. Oda, V. Quaranta, G. Plopper, R. Lambert, S. Glaser, J. C. Jones. Coating of titanium alloy with soluble laminin-5 promotes cell attachment and hemidesmosome assembly in gingival epithelial cells: potential application to dental implants. *J. Periodontal. Res.* 32:287–294 (1997).
87. M. Hormia, J. Ylanne, I. Virtanen. Expression of integrins in human gingiva. *J. Dent. Res.* 69:1817–1823 (1990).
88. E. Ruoslahti. Integrins. *J. Clin. Invest.* 87:1–5 (1991).
89. S. Sengupta, F. E. Shovlin, A. Slomiany, B. L. Slomiany. Identification of laminin receptor in gingival tissue and its interaction with tooth cementum laminin. *Int. J. Biochem.* 23:115–121 (1991).
90. A. Howe, A. E. Aplin, S. K. Alahari, R. I. Juliano. Integrin signaling and cell growth control. *Curr. Opin. Cell Biol.* 10:220–231 (1998).
91. N. Olmo, M. A. Lizarbe. Signal transduction through laminin receptors. Effects of extracellular matrix on BCS-TC2 adenocarcinoma cells. In: *Cell Signal Transduction, Second Messengers and Protein Phosphorylation in Health and Disease*, A. Martín-Municio, M. T. Miras-Portugal (Eds.). Plenum, New York, 1994, pp. 227–249.
92. S. Dedhar, G. E. Hanningan. Integrin cytoplasmic interactions and bidirectional transmembrane signalling. *Curr. Opin. Cell Biol.* 8:657–669 (1996).
93. E. Ruoslahti. RGD and other recognition sequences for integrins. *Ann. Rev. Cell Devel. Biol.* 12:697–715 (1996).
94. K. Rubin, D. Gullberg, B. Tomasini-Johansson, R. K. Reed, C. Rydèn, T. K. Borg. Molecular recognition of the extracellular matrix by cell surface receptors. In: *Extracellular Matrix*, W. D. Comper (Ed.). Harwood Academic, Amsterdam, 1996, pp. 262–309.
95. L. L. H. Huang-Lee, D. T. Cheung, M. E. Nimni. Biochemical changes and cytotoxicity associated with the degradation of polymeric glutaraldehyde derived crosslinks. *J. Biomed. Mater. Res.* 24:1885–1201 (1990).
96. T. J. Krizek. The normal body defenses against foreign implants. In: *Biomaterials in Reconstructive Surgery*, L. R. Rubin (Ed.). C. V. Mosby, 1983, pp. 9–16.
97. L. L. Hench, E. C. Ethridge. In: *Biomaterials: An Interfacial Approach.* Academic Press, New York, 1982.
98. E. T. den Braber, J. E. de Ruijter, L. A. Ginsel, A. F. von

Recumm, J. A. Jansen. Orientation of ECM protein deposition, fibroblast cytoskeleton, and attachment complex components on silicone microgrooved surfaces. *J. Biomed. Mater. Res.* 40:291–300 (1998).
99. C. H. Thomas, C. D. McFarland, M. L. Jenkins, A. Rezania, J. G. Steele, K. E. Healy. The role of vitronectin in the attachment and spatial distribution of bone-derived cells on materials with patterned surface chemistry. *J. Biomed. Mater. Res.* 37:81–93 (1997).
100. R. Langer, L. G. Cima, J. A. Tamada, I. E. Wintermantel. Future directions in biomaterials. *Biomaterials* 11:738–745 (1990).
101. A. J. Coury. Degradation of materials in the biological environment. In: *Biomaterials Science: An Introduction to Materials in Medicine*, B. D. Ratner, A. S. Hoffman (Eds.). Academic Press, New York, 1996, pp. 243–259.
102. H. Kawahara, A. Yamagami, M. Nakamura, Jr. Biological testing of dental materials by means of tissue culture. *Int. Dent. J.* 18:443–467 (1968).
103. J. Black. Biological performance of materials. In: *Fundamentals of Biocompatibility: Third Edition, Revised and expanded*, J. Black (Ed.). Marcel Dekker, New York, 1999.
104. W. M. Murphy. An *in vitro* model for testing biocompatibility of endodontic materials to bone. *Biomaterials* 6: 427–430 (1985).
105. M. A. Lizarbe, N. Olmo, J. G. Gavilanes. Adhesion and spreading of fibroblasts on sepiolite–collagen complex. *J. Biomed. Mater. Res.* 21:137–144 (1987).
106. N. Olmo, M. A. Lizarbe, F. J. Turnay, K. P. Müller, J. G. Gavilanes. Cell morphology, proliferation and collagen synthesis of human fibroblasts cultured on sepiolite–collagen complexes. *J. Biomed. Mater. Res.* 22:257–270 (1988).
107. I. B. Rosanova, B. P. Mischenko, U. V. Zaitsev, S. L. Vasin, V. I. Sevastianov. The effect of cells on biomaterial calcification—experiments with in vivo diffusion chambers. *J. Biomed. Mater. Res.* 25:277–280 (1991).
108. M. Maeda, M. Kimura, S. Inoue, K. Kataoka, T. Okano, Y. Sakurai. Adhesion behaviour of rat lymphocyte subpopulations (B cell and T cell) on the surface of polystyrene/polypeptide graft copolymer. *J. Biomed. Mater. Res.* 20:25–35 (1986).
109. C. N. Sukenik, N. Balachander, L. A. Culp, K. Lewandowska, K. Merritt. Modulation of cell adhesion by modification of titanium surfaces with covalently attached self-assembled monolayers. *J. Biomed. Mater. Res.* 24:1307–1323 (1990).
110. D. E. Steflik, R. S. Corpe, T. R. Young, K. Buttle. In vivo evaluation of the biocompatibility of implanted biomaterials: morphology of the implant–tissue interactions. *Implant. Dent.* 7:338–350 (1998).
111. A. Schedle, A. Franz, X. Rausch-Fan, A. Spittler, T. Lucas, P. Samorapoompichit, W. Sperr, G. Boltz-Nitulescu. Cytotoxic effects of dental composites, adhesive substances, compomers and cements. *Dent. Mater.* 14:429–440 (1998).
112. E. H. Rompen, G. H. Goffinet, B. Nusgens. Human periodontal ligament fibroblast behavior on chemically conditioned dentine: an in vitro study. *J. Periodontol.* 70:1144–1152 (1999).
113. B. Unsal, G. Ozcan, G. Tuter, B. Kurtis, M. Yalim. Evaluation of initial attachment of human gingival fibroblast cells to biodegradable membranes in vitro by light and scanning electron microscopy. *J. Oral Sci.* 41:57–60 (1999).
114. J. A. Hayward, D. Chapman. Biomembrane surfaces as models for polymer design: the potential for haemocompatibility. *Biomaterials* 5:135–142 (1984).
115. P. W. Heyman, C. S. Cho, J. C. McRea, D. B. Olsen, S. W. Kim. Heparinized polyurethanes: in vitro and in vivo studies. *J. Biomed. Mater. Res.* 19:419–436 (1985).
116. N. Olmo, M. A. Lizarbe, J. G. Gavilanes. Biocompatibility and degradability of sepiolite–collagen complexes. *Biomaterials* 8:67–69 (1987).
117. D. T. Cheung, M. E. Nimni. Mechanism of cross-linking of proteins by glutaraldehyde. II. Reaction with monomeric and polymeric collagen. *Connect. Tissue Res.* 10: 201–216 (1982).
118. N. Olmo, A. Martinez del Pozo, M. A. Lizarbe, J. G. Gavilanes. Interaction of type I collagen with sepiolite (magnesium silicate). *Collagen Rel. Res.* 5:9–16 (1985).
119. N. Olmo, J. Turnay, J. G. Gavilanes, M. A. Lizarbe. Subcutaneous and intramuscular implantation of sepiolite-collagen complexes. *J. Mater. Sci.: Mater. Med.* 3:239–244 (1992).
120. J. I. Herrera, N. Olmo, J. Turnay, A. Sicilia, A. Bascones, J. G. Gavilanes, M. A. Lizarbe. Implantation of sepiolite–collagen complexes in surgically created rat calvaria defects. *Biomaterials* 16:625–631 (1995).
121. J. C. Mitchem. In:*Dental Materials in Clinical Dentistry*, M. H. Reisbick (Ed.). John Wright, Guilford, 1982.
122. B. G. N. Smith, P. S. Wright, D. Brown. *The Clinical Handling of Dental Materials*. Wright, Bristol, 1988, pp. 143–164.
123. M. A. Latta, W. W. Barkmeier. Dental adhesives in contemporary restorative dentistry. *Dent. Clin. North Am.* 42: 567–577 (1998).
124. G. Bourdairon. *Abrégé des Biomateriaux Dentaires*. Masson, Paris, 1990, pp. 267–277.
125. G. J. Mount. *An Atlas of Glass-Ionomer Cements. A Clinician's Guide*. Martin Dunitz, London, 1990, pp. 1–24.
126. G. J. Mount. *An Atlas of Glass-Ionomer Cements. A Clinican's Guide*, Martin Dunitz, London, 1990, pp. 34–65.
127. T. Wang, T. Nikaido, N. Nakabayashi. Photocure bonding agent containing phosphoric methacrylate. *Dent. Mater.* 7: 59–62 (1991).
128. R. R. Welbury, A. W. G. Walls, J. J. Murray, J. F. McCabe. The 5-year results of a clinical trial comparing a glass polyalkenoate (ionomer) cement restoration with amalgam restoration. *Brit. Dent. J.* 5:177–181 (1991).
129. J. W. Nicholson, T. P. Croll. Glass-ionomer cements in restorative dentistry. *Quintessence Int.* 28:705–714 (1997).

130. J. F. McCabe. Resin-modified glass-ionomers. *Biomaterials* 19:521–527 (1998).
131. I. M. Brook, P. V. Hatton. Glass-ionomers: bioactive implant materials, *Biomaterials* 19:565–571 (1998).
132. R. J. Simonsen. The amalgam controversy. *Quintessence Int.* 4:241–242 (1991).
133. R. V. Katz. The safety of dental amalgam: the classic problem of early questions and premature conclusions. *Quintessence Int.* 4:243–246 (1991).
134. S. C. Bayne. The amalgam controversy. *Quintessence Int.* 4:248 (1991).
135. R. L. Bowen. Use of epoxy resins in restorative materials. *J. Dent. Rest.* 35:360–369 (1965).
136. R. L. Bowen, J. A. Barton, A. L. Mullineaux. Composite restorative materials. Dental Materials Research. National Bureau of Standards special publication 93 Gaithersburg, MD, 1972.
137. H. F. Albers. In: *Tooth Colored Restoratives. A text for Selection, Placement and Finishing*, H. Albers (Ed.). Santa Rosa, CA, 1985.
138. R. E. Jordan, A. J. Gwinnett. In: *Esthetic Composite Bonding: Techniques and Materials*, R. E. Jordan (Ed.). B. C. Decker Inc., Burlington, Ontario, 1986.
139. D. C. Smith. Posterior composite dental restorative materials: materials development. In: *Posterior Composite Resin Dental Restorative Materials*, G. Vanherle, D. C. Smith (Eds.). International Symposium in Minnesota Peter Szule Publishing, The Netherlands, 1985, pp. 47–60.
140. J. L. Baños, J. M. Vega. Resinas compuestas en odontología. In: *Materiales en Odontología, fundamentos biológicos, clínicos, biofísicos y fisico-químicos*, J. M. Vega (Ed.). Ediciones Avances, Madrid, 1996, pp. 291–312.
141. N. Olea et al. Estrogenicity of Resin-based composites and sealants used in dentistry. *Environmental Health Perspectives* 104:298–305 (1996).
142. R. G. Craig. Overview of posterior composite resins for use in clinical practice. In: *Posterior Composite Resin Dental Restorative Materials*, G. Vanherle, D. C. Smith, (Eds.). International Synposium in Minnesota. Peter Szule Publishing, The Netherlands, 1985, pp. 199–211.
143. K. Soderholm. Filler systems and resin interface. In: *Posterior Composite Resin Dental Restorative Materials*, G. Vanherle, D. C. Smith (Eds.). International Symposium in Minnesota. Peter Szule Publishing, The Netherlands, 1985, pp. 139–159.
144. D. L. Lambert. Clinical versatility of new compomer restorative technology. *Signature* Winter:14–17 (1996).
145. J. M. Meyer, M. A. Cattani-Lorente, V. Dupuis. Compomers: between glass-ionomer cements and composites. *Biomaterials* 19:529–539 (1998).
146. N. G. Nuñez, E. G. Priotto. Restauraciones con resinas compuestas en el sector anterior. In: *Operatoria Dental: Ciencia y Práctica*, J. Uribe (Ed.). Avances S. L., Madrid, 1990, pp. 231–306.
147. D. C. Smith. Adhesives and sealants. In: *Biomaterials Science: An Introduction to Materials in Medicine*, B. D. Ratner, A. S. Hoffman (Eds.). Academic Press, New York, 1996, pp. 319–328.
148. K. Nakanuma, T. Hayakawa, T. Tomita, M. Yamazaki. Effect of the application of dentin primers and a dentin bonding agent on the adhesion between the resin-modified glass-ionomer cement and dentin. *Dent. Mater.* 14:281–286 (1998).
149. C. Q. Bowles CQ, R. G. Miller, C. C. Chappelow, C. S. Pinzino, J. D. Eick. Design, formulation, and evaluation of isocyanatoacrylate copolymer dental adhesives. *J. Biomed. Mater. Res.* 48:496–503 (1999).
150. R. L. Cooley, J. W. McCourt, T. E. Train. Bond strength of resin to amalgam as affected by surface finish. *Quintessence Int.* 20:237–239 (1989).
151. F. A. Rueggeberg. Substrate for adhesion testing to tooth structure—review of the literature. *Dent. Mater.* 1:2–10 (1991).
152. S. E. Bishara, V. V. Gordan, L. VonWald, J. R. Jakobsen. Shear bond strength of composite, glass ionomer, and acidic primer adhesive systems. *Am. J. Orthod. Dentofacial. Orthop.* 115:24–28 (1999).
153. J. I. Ingle, J. F. Taintor. *Endodontics*. Lea & Febiger, Philadelphia, 1985.
154. L. Spanberg, K. Langeland. Biologic effects of dental materials: 1. Toxicity of root canal filling materials on HeLa cells in vitro. *Oral Surg. Oral Med. Oral Path.* 3:402–414 (1973).
155. B. G. N. Smith, P. S. Wright, D. Brown. *The Clinical Handling of Dental Materials*. Wright, Bristol, 1986, pp. 177–181.
156. M. A. Freilich, A. C. Karmaker, C. J. Burstone, A. J. Goldberg. Development and clinical applications of a light-polymerized fiber-reinforced composite. *J. Prosthet. Dent.* 80:311–318 (1998).
157. J. C. Meiers, J. P. Duncan, M. A. Freilich, A. J. Goldberg. Preimpregnated, fiber-reinforced prostheses. Part II. Direct applications: splints and fixed partial dentures. *Quintessence Int.* 29:761–768 (1998).
158. A. G. Zuccari, Y. Oshida, B. K. Moore. Reinforcement of acrylic resins for provisional fixed restorations. Part I: Mechanical properties. *Biomed. Mater. Eng.* 7:327–343 (1997).
159. F. Simionato. Resine sintetiche per protesi. In: *Tecnologie dei Materiali Dentali*. Piccin, Padua, 1985, pp. 825–887.
160. B. Taner, A. Dogan, T. Tincer, A. E. Akinay. A study on impact and tensile strength of acrylic resin filled with short ultra-high molecular weight polyethylene fibers. *J. Oral. Sci.* 41:15–18 (1999).
161. A. III. Koran, B. R. Lang. In: *An Outline of Dental Materials and Their Selection*, J. O'Brien, G. Ryge (Eds.). W. B. Saunders, Philadelphia, 1980.
162. P. Magne, W. H. Douglas. Porcelain veneers: dentin bonding optimization and biomimetic recovery of the crown. *Int. J. Prosthodont.* 12:111–121 (1999).
163. I. Barzilay, M. L. Myiers, L. B. Cooper, G. Graser. Mechanical and chemical retention of laboratory cured composite to metal surfaces. *J. Prosthet. Dent.* 59:131–137 (1988).
164. T. Tanaka, K. Nagata, M. Takeyama, N. Nakabayashi, E.

Masuhara. 4-META opaque resin—a new resin strongly adhesive to nickel-chromium alloy. *J. Dent. Rest.* 60: 1697–1705 (1981).

165. T. Trinkner. Achieving functional restorations utilizing a new Ceromer system. *Signature* 4:12–17 (1997).
166. M. J. Koczarski. Utilization of ceromer inlays/onlays for replacement of amalgam restorations. *Pract. Periodontics. Aesthet. Dent.* 10:405–412 (1998).
167. B. G. N. Smith, P. S. Wright, D. Brown. *The Clinical Handling of Dental Materials*. Wright, Bristol, 1988, pp. 236–248.
168. D. R. Gratton. In: *Esthetic Composite Bonding: Techniques and Materials*, R. E. Jordan (ed.). B. C. Decker Inc., Burlington, Ontario, 1986.
169. D. O. Maisels. Implants in plastic and reconstructive surgery. In: *Implants in Surgery*, D. F. Williams and R. Roaf (Eds.). W. B. Saunders, London, 1973, pp. 537–568.
170. P. Horsted-Bindslev, I. A. Mjor. Periodontal and oclusal considerations in operative dentistry. In: *Modern Concepts in Operative Dentistry*, P. Horsted-Bindslev, I. A. Mjor (Eds.). Munksgaard, Copenhagen, 1988, pp. 302–318.
171. A. C. Roberts. Silicones for facial prosthesis. *The Dental Practitioner* 21:276–284 (1971).
172. J. B. Brown. Investigations of and use of dimethyl-siloxanes, halogenated carbons and polyvynil-alcohol as subcutaneous prosthesis. *Ann. Surg.* 152:534–547 (1968).
173. C. A. Homsy. Bio-compatibility in selection of materials for implantation. *J. Biomed. Mater. Res.* 4:341–356 (1970).
174 C. A. Homsy. Implant stabilization chemical and biochemical considerations. Symposium on interposition and implant arthroplasty. *Orthopedic Clinics of North America* 4:295–311 (1973).
175. E. C. Hinds. Use of a biocompatible interface for binding tissues and prosthesis in temporomandibular joint surgery. *Oral Surg. Oral Med. Oral Path.* 38:512–519 (1974).
176. E. P. Hennefer, T. A. McFall, D. C. Hauschild. Acrylate-amide sponge for repair of alveolar bone defects. *J. Oral Surg.* 26:577–581 (1968).
177. L. L. Hench. Ceramic implants for humans. *Adv. Ceram. Mater.* 1:306–324 (1986).
178. J. W. Boretos, Advances in Bioceramics, *Adv. Ceram. Mater.* 2:15–30 (1986).
179. D. Goupil. Sutures. In: *Biomaterials Science. An Introduction to Materials in Medicine*, B. D. Ratner, A. S. Hoffman (Eds.). Academic Press, New York, 1996, pp. 356–360.
180. W. R. Wallace. Comparison of polyglycolic acid suture to black silk, chromic and plain catgut in human oral tissues. *J. Oral Surg.* 10:739–742 (1970).
181. J. O. Dawson. Surgical ligatures and sutures. In: *Surgical Dressings, Ligatures and Sutures*, F. Fish, J. O. Dawson (Eds.). William Heinemann, Medical Books, London, 1964, p. 115–117.
182. D. F. Williams. The response of the body environment to implants. In: *Implants in Surgery*, D. F. Williams and R. Roaf (Eds.). W. B. Saunders, London, 1973, pp. 203–297.
183. D. Aderriotis, G. K. Sandor. Outcomes of irradiated polyglactin 910 Vicryl Rapide fast-absorbing suture in oral and scalp wound. *J. Can. Dent. Assoc.* 65:345–347 (1999).
184. E. T. Silverman. Speech rehabilitation habits and myofunctional therapy in restorative procedures. In: *A Dynamic Approach to Restorative Dentistry*, J. Seidel (Ed.). W. B. Saunders, Philadelphia, 1980, p. 615–677.
185. G. Sauveur, F. Roth, M. Sobel, Y. Boucher. The control of haemorrhage at the operative site during periradicular surgery. *Int. Endod. J.* 32:225–228 (1999).
186. D. L. Cochran, J. M. Wozney. Biological mediators for periodontal regeneration. *Periodontol. 2000* 19:40–58 (1999).
187. T. Karring, S. Nyman, J. Lindhe. Healing following implantation of periodontitis affected roots into bone tissue. *J. Clin. Periodontol.* 45:725–730 (1980).
188. S. Nyman, J. Lindhe, T. Karring, H. Rylander. New attachment following surgical treatment of human periodontal disease. *J. Clin. Periodontol.* 9:290–296 (1982).
189. S. Nyman, J. Gottlow, T. Karring, J. Lindhe. The regenerative potential of the periodontal ligament. *J. Clin. Periodontol.* 9:257–265 (1982).
190. J. Gottlow, S. Nyman, J. Lindhe, T. Karring, J. Wennstrom. New attachment formation in the human periodontium by guided tissue regeneration. *J. Clin. Periodontol.* 13:604–616 (1986)
191. J. Gottlow, T. Karring, S. Nyman. Guided tissue regeneration following treatment of recession type defects in the monkey. *J. Periodontol.* 61:680–685 (1990).
192. R. O'Neal, H.-L. Wang, R. L. MacNeil, M. J. Somerman. Cells and materials involved in guided tissue regeneration. *Cur. Opin. Periodontol.* 1994:141–156 (1994)
193. H. L. Wang, R. L. MacNeil. Guided tissue regeneration. Absorbable barriers. *Dent. Clin. North Am.* 42:505–522 (1998).
194. R. Pontoriero, S. Nyman, J. Lindhe, E. Rosenberg, F. Sanavi. Guided tissue regeneration in the treatment of furcation defects in man. *J. Clin. Periodontol.* 14:618–620 (1987).
195. W. Becker, B. E. Becker, J. F. Prichard, R. Caffesse, E. Rosenberg, J. A. Gian-Grasso. A surgical and suturing method: three case reports. *J. Periodontol.* 12:819–826 (1987).
196. B. G. Kersten, A. D. Chamberlainm, S. Khorsandi, U. M. Wikesjo, K. A. Selvig, R. E. Nilveus. Healing of the intrabony periodontal lesion following root conditioning with citric acid and wound closure including an expanded PTFE membrane. *J. Periodontol.* 63:876–882 (1992).
197. R. Caffesse, B. A. Smith, B. Duff, E. C. Morrison, D. Merrill, W. Becker. Class II furcation treated by guided tissue regeneration in humans: case reports, *J. Periodontol.* 61: 510–514 (1990).
198. B. K. Bartee. Evalution of a new polytetrafluoroethylene guided tissue regeneration membrane in healing extration sites. *Compend. Contin. Educ. Dent.* 19:1256–1258 (1998).
199. K. M. Chung, L. M. Salkin, M. D. Stein, A. L. Freedman. Clinical evaluation of a biodegradable collagen membrane in guided tissue regeneration. *J. Periodontol.* 61:732–736 (1990).
200. B. A. Doll, H. J. Towle, J. O. Hollinger, A. H. Reddi, J. I. Mellonig. The osteogenic potential oif two composite

graft systems using osteogenin. *J. Periodontol.* 61:745–750 (1990).

201. B. WallKamm, J. Schmid, C. H. F. Hammerle, S. Gogolewski, N. P. Lang. Effect of a bioresorbable Goam (Polygoam on experimental bone neoformation. *J. Dent. Res.* (in press; IADR abstract).
202. D. Hutmacher, M. B. Hurzeler, M. Schliephake. A review of material properties of biodegradable polymers and devices for GTR and GBR applications. *Int. J. Oral Maxillofac. Implants* 11:667–678 (1976).
203. E. Sandberg, C. Dahlin, A. Linde. Bone regeneration by the osteopromotion technique using bioabsorbable membranes. An experimental study in rats. *J. Oral Maxillofac. Surg.* 51:1106–1114 (1993).
204. J. S. Mattson, S. J. Gallagher, M. H. Jabro. The use of 2 bioabsorbable barrier membranes in the treatment of interproximal intrabony periodontal defects. *J. Periodontol.* 70:510–517 (1999).
205. C. H. Hammerle, T. Karring. Guided bone regeneration at oral implant sites. *Periodontology 2000* 17:151–175 (1998).
206. D. D. Lee, A. Tofighi, M. Aiolova, P. Chakravarthy, A. Catalano, A. Majahad, D. Knaack. Alpha-BSM: a biomimetic bone substitute and drug delivery vehicle. *Clin. Orthoped.* 367:S396–S405 (1999).
207. J. S. Gross. Bone grafting materials for dental applications: a practical guide. *Compend. Contin. Educ. Dent.* 18: 1013–1018 (1997).
208. A. G. Haris, G. Szabo, A. Ashman, T. Divinyi, Z. Suba, K. Martonffy. Five-year 224-patient prospective histological study of clinical applications using a synthetic bone alloplast. *Implant. Dent.* 7:287–299 (1998).
209. O. T. Jensen, R. O. Geer, L. Johnson, D. Kassebaum. Vertical guided bone-graft augmentation in a new canine mandibular model. *Int. J. Oral Maxillofac. Implants* 10:335–344 (1995).
210. D. Corry, J. Moran. Assessment of acrylic bone cement as a local delivery vehicle for the application of non-steroidal antiinflammatory drugs. *Biomaterials* 19:1295–1301 (1998).
211. T. Noguchi, K. Izumizawa, M. Fukuda, S. Kitamura, Y. Suzuki, H. Ikura. New method for local drug delivery using resorbable base material in periodontal therapy. *Bull. Tokyo Med. Dent. Univ.* 31:145 (1984).
212. S. L. Morrison, C. M. Cobb, G. M. Kazakos, W. J. Killoy. Root surface characteristics associated with subgingival placement of monolithic tetracycline-impregnated fibers. *J. Periodontol.* 63:137–143 (1992).
213. T. Larsen. In vitro release of doxycicline from bioabsorbable materials and acrylic strips. *J. Periodontol.* 61:30–34 (1990).
214. M. Redlich, S. Shoshan, A. Palmon. Gingival response to orthodontic force. *Am. J. Orthod. Dentofacial Orthoped.* 116:152–158 (1999).
215. D. R. Stirrups. A comparative clinical trial of a glass ionomer and zinc phosphate cement for securing orthodontic bands. *Brit. J. Orthod.* 18. 1:15–20 (1991).
216. N. R. Smith, I. R. Reynolds. A comparison of three bracket bases: an in vivo study. *Brit. J. Orthod.* 18. 1:29–35 (1991).
217. G. Burdairon. *Abrégé des Biomateriaux Dentaires.* Masson, Paris, 1990, pp. 153–180.
218. M. N. Mandikos. Polyvinyl siloxane impression materials: an update on clinical use. *Aust. Dent. J.* 43:428–434 (1998).
219. F. Simionato. Materiali per modelli e monconi. In: *Tecnologie dei Materiali Dentali.* Piccin, Padua, 1985, pp. 538–579.
220. F. Simionato. Materiali da impronta. In: *Tecnologie dei Materiali Dentali.* Piccin, Padua, 1985, pp. 443–534.
221. F. Simionato. Cere dentali. In: *Tecnologie dei materiali dentali.* Piccin, Padua, 1985, pp. 581–603.
222. N. A. Peppas, R. Langer. New Challenges in Biomaterials. *Science* 263:1715–1720 (1994).
223. J. G. Tirrell, M. J. Fournier, T. L. Mason, D. A. Tirrell. Biomolecular Materials. *C&EN* 19:40–51 (1994).
224. Y. Poirier, C. Nawrath, C. Somerville. Production of polyhydroxyalkanoates, a family of biodegradable plastics and elastomers, in bacteria and plants. *Biotechnology (N.Y.)* 13:142–150 (1995).
225. T. A. Leaf, M. S. Peterson, S. K. Stoup, D. Somers, F. Srienc. *Saccharomyces cerevisiae* expressing bacterial polyhydroxybutyrate synthase produces poly-3-hydroxybutyrate. *Microbiology* 142:1169–1180 (1996).
226. J. P. Anderson, M. Stephen-Hassard, C. Martin. *Silk Polymers: Materials Science and Biotechnology*, D. Kaplan, W. W. Adams, B. Farmer, C. Viney (Eds.). ACS Symposium Series 544. American Chemical Society, Washington D.C., 1994, p. 37.
227. H. Heslot. Artificial fibrous proteins: a review. *Biochimie* 80:19–31 (1998).
228. M. Kantlehner, D. Finsinger, J. Meyer, P. Schaffner, A. Jonczyk, B. Diefenbach, B. Nies, H. Kessler. Selective RGD-mediated adhesion of osteoblasts at surfaces of implants. *Angew. Chem. Int. Ed.* 38:417–572 (1999).
229. R. G. Caffesse, C. R. Quiñones. Polypeptide growth factors and attachment proteins in periodontal wound healing and regeneration. *J. Periodontol. 2000* 1:69–79 (1993).
230. E. M. Wikesjo, P. Guglielmoni, A. Promsudthi, K. S. Cho, L. Trombelli, K. A. Selvig, L. Jin, J. M. Wozney. Periodontal repair in dogs: effect of rhBMP-2 concentration on regeneration of alveolar bone and periodontal attachment. *J. Clin. Periodontol.* 26:392–400 (1999).
231. D. L. Cochran, R. Schenk, D. Buser, J. M. Wozney, A. A. Jones. Recombinant human bone morphogenetic protein-2 stimulation of bone formation around endosseous dental implants. *J. Periodontol.* 70:139–150 (1999).
232. L. Hammarstrom, L. Heijl, S. Gestrelius. Periodontal regeneration in a buccal dehiscence model in monkeys after application of enamel matrix proteins. *J. Clin. Periodontol.* 24:669–677 (1997).
233. B. D. Boyan, C. H. Lohmann, A. Somers, G. G. Niederauer, J. M. Wozney, D. D. Dean, D. L. Carnes, Z. Schwartz. Potential of porous poly-D, L-lactide-co-glycolide particles as a carrier for recombinant human bone morphogenetic protein-2 during osteoinduction in vivo. *J. Biomed. Mater. Res.* 46:51–59 (1999).

17

Biomaterials in Burn and Wound Dressings

Robert L. Sheridan, Jeffrey R. Morgan, and Rashid Mohammad
Shriners Burns Hospital, Massachusetts General Hospital, Boston, Massachusetts, and Harvard Medical School, Cambridge, Massachusetts

I. INTRODUCTION

In few areas have new biomaterials been incorporated into daily practice as in the management of burns and wounds. Burn management has changed dramatically over the past two or three decades, thanks to a host of critical care and surgical innovations (1), and new biomaterials are playing an increasingly important role as patients with larger and more complex wounds are surviving. The initial improvements in survival were due largely to an understanding of fluid resuscitation that developed in the 1950s and 1960s (2,3). The next major hurdle to overcome was the routine occurrence of wound sepsis. This was eliminated as a common problem by the development and practice of early and accurate identification of deep wounds and prompt and effective excision and closure (4–7). Various biomaterials have materially contributed to the success of this important progress in burn care.

Refinement of the surgical operations for excision of large wounds combined with improvements in critical care techniques have extended our ability to support patients who have suffered serious burns (8). Burn physical and occupational therapy and burn reconstruction have developed in parallel, facilitating our ability to deliver increasingly satisfying long-term outcomes (9). Further progress is hampered severely by the lack of a durable skin substitute. The successful development of a permanent skin substitute will have an enormous impact on the care of patients with serious burns. Multiple clinical research groups and biomaterials scientists continue to work on this difficult problem.

Conceptually, skin substitutes are temporary or permanent; epidermal, dermal, or composite; and biologic or synthetic. Biologic components are xenogenic, allogenic, or autogenic. From the perspective of the practicing clinician, skin substitutes are either temporary or permanent. The objective of this chapter is to review the current state of skin substitutes and to speculate on future directions in this important area of research. Whenever possible, proprietary product names will not be used; any products not mentioned are not purposefully excluded.

II. STRUCTURE AND FUNCTION OF THE SKIN

Skin, the body's largest organ, is incredibly complex. Functionally there are two layers with a highly specialized and effective bonding mechanism. The epidermis, consisting of the strata basale, spinosum, granulosum, and corneum, provides a vapor and bacterial barrier. The dermis provides strength and elasticity. The thin epidermal layer is constantly refreshing itself from its basal layer, with new keratinocytes undergoing terminal differentiation over approximately 4 weeks to anuclear keratin-filled cells that make up the stratum corneum, which provides much of the barrier function of the epidermis. The basal layer of the epidermis is firmly attached to the dermis by a complex bonding mechanism containing collagen types IV and VII. When this bond fails, serious morbidity results, as demonstrated by the disease processes of toxic epidermal necrolysis (10) and dystrophic epidermolysis bullosa (11).

III. TEMPORARY SKIN SUBSTITUTES

Temporary skin substitutes are used to provide a number of potential benefits to healing: transient physiologic wound closure, pain control, absorption of wound exudate, and prevention of wound dessication. Physiologic wound closure implies a degree of protection from mechanical trauma, vapor transmission characteristics similar to skin, and a physical barrier to bacteria. These membranes attempt to foster a moist wound environment with a low bacterial density. There are four common uses for temporary skin substitutes in burn care: (1) as a dressing on donor sites to facilitate pain control and epithelialization from skin appendages, (2) as a dressing on clean superficial wounds to a similar end, (3) to provide temporary physiologic closure of deep dermal and full thickness wounds after excision while awaiting autografting or healing of underlying widely meshed autografts, and (4) as a ''test'' graft in questionable wound beds. There are a large number of such membranes in common use.

A. Human Allograft

Human allograft, applied as a split thickness graft after procurement from organ donors remains the gold standard of temporary dressings in burns (Fig. 1) (12–14). This material is applied in a viable state after storage in the refrigerated or frozen state. It can be refrigerated for 7 days or less, but can be stored for extended periods when cryopreserved. Viable split thickness allograft will vascularize and will reliably provide durable biologic cover until it is rejected by the host, usually within 3 or 4 weeks. Prolongation of allograft survival, through the use of antirejection drugs, has been done clinically (15), but is not generally practiced for fear that this will result in excessive infection (16). When modern screening techniques are followed, the viral disease transmission risk appears to be vanishingly small.

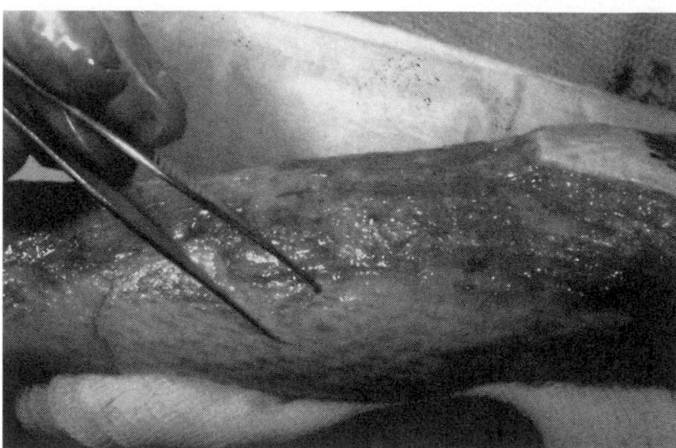

Figure 17.1 Human allograft, applied as a split thickness graft after procurement from organ donors, remains the gold standard of temporary dressings in burns.

B. Human Amnion

Human amniotic membrane is used in many parts of the world to cover clean superficial wounds, including partial thickness burns, donor sites, and freshly excised burns awaiting donor site availability (17,18). This inexpensive membrane material is generally obtained fresh and used immediately or after brief refrigerated storage (19,20). Like porcine xenograft, it has been combined with silver to facilitate control of bacterial overgrowth (21). Although amnion does not vascularize (22), it can provide coverage that, in optimal circumstances, is nearly as good as that of allograft in some wounds (22). The principal concern with amnion in the developed world has been the difficulty in screening donors for viral diseases. In North America, human skin allografts are placed into frozen storage awaiting the return of numerous laboratory tests allowing one to safely exclude the possibility of viral disease transmission. When using amnion, the risks of disease transmission must be balanced against the clinical need and the known characteristics of the donor in individual circumstances.

C. Xenograft

Skin from various animals has been used for many years to provide temporary coverage of wounds (23). The only example of this practice still widely seen is the use of porcine xenograft (Fig. 2) (24). Porcine xenograft is often used as a reconstituted product consisting of homogenized porcine dermis which is fashioned into sheets and meshed (25). It is generally used for temporary coverage of clean wounds such as superficial second degree burns and donor sites (26). Its has been used in patients with toxic epidermal necrolysis (10,27), another superficial but extensive

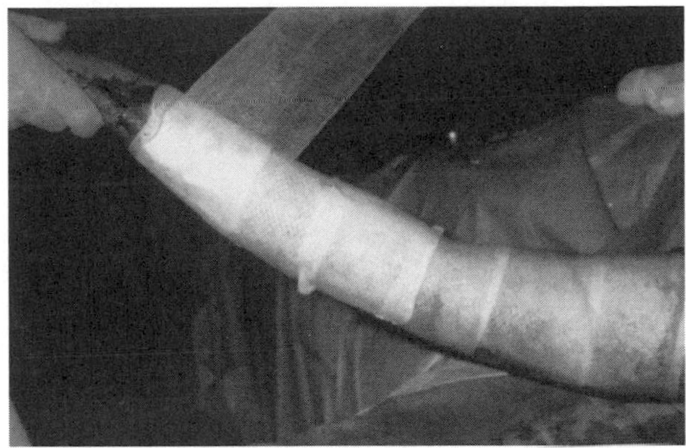

Figure 17.2 Skin from various animals has been used for many years to provide temporary coverage of wounds. The only example of this practice still widely seen is the use of porcine xenograft, which remains a very effective and safe temporary dressing.

wound. Porcine xenograft has been combined with silver to suppress wound colonization (28,29). Porcine xenograft does not vascularize, but it will adhere to a clean superficial wound and can provide excellent pain control while the underlying wound heals.

D. Synthetic Membranes

An increasing number of proprietary semipermeable membrane dressings provide a vapor and bacterial barrier and control pain while the underlying superficial wound or donor site re-epithelializes. Most consist of a single semipermeable layer that provides a mechanical barrier to bacteria and has physiologic vapor transmission characteristics (30,31). These can be used with good effect on clean superficial wounds and split thickness donor sites. Biobrane® (Dow-Hickham, Sugarland, TX) is a two-layer membrane consisting of an inner layer of nylon mesh that allows fibrovascular ingrowth and an outer layer of silastic that serves as a vapor and bacterial barrier (32). It has been used to good effect in clean superficial burns and donor sites.

Hydrocolloid dressings generally consist of three layers and attempt to create a moist wound environment while absorbing exudate. A moist wound environment has been found to favor wound healing in experimental and clinical trails (33). There is commonly a porous gently adherent inner layer, a methylcellulose absorbent middle layer, and a semipermeable outer layer. Also available are a number of pastes and powders made from hydocolloid materials that are well applied to the control of wound exudate and maintenance of a moist wound environment.

All synthetic membranes are more or less occlusive. As such they must be used with caution if wounds are not clearly clean and superficial. If placed over devitalized tissue, submembrane purulence can occur with potentially disastrous results (34).

E. Combined Allogenic and Synthetic Membranes

There are an increasing number of growth factors thought to play important roles in wound healing: epidermal growth factor, transforming growth factor-beta, insulinlike growth factor, platelet-derived growth factors, and fibroblast growth factors (35,36). In an effort to apply some of these topically to wounds, investigators have placed both viable and nonviable allogenic cell types into temporary dressings which are then placed on superficial wounds and donor sites (37). These cells persist for no more than 14 days, but it is hoped that factors secreted by the allogenic cells, or released upon their dissolution, will provide signals to the host that enhance wound healing. Scientific support for this presumption remains elusive.

Perhaps the first group to attempt the manufacture of a biologic composite skin substitute to this end was that led by Bell, who developed a completely allogenic dermal/epidermal product that used a collagen lattice as scaffold for culturing both cell types (38). After it was demonstrated in an athymic mouse model that this device would successfully engraft (39,40), it went into clinical trials. Although it has not been demonstrated to have a clinical role in burn patients, the device is being explored for utility in chronic ulcers of the lower extremity (41–45). Allogenic fibroblasts grown into the nylon inner layer of Dow-Hickam's Biobrane is undergoing evaluation in wound healing (46–50). Although stimulation of wound healing by topical application of mixed growth factors in this fashion is an intriguing concept, convincing evidence of the concept's general validity is awaited. Viral transfection has been used to modify keratinocytes so that they overexpress platelet-derived growth factor, human growth hormone, insulinlike growth factor-1, and other growth factors (51). It is likely that such cell lines will be applied to wounds as components of wound membranes over the next few years (52).

IV. PERMANENT SKIN SUBSTITUTES

A durable permanent skin substitute will make an enormous difference to patients with burns and other difficult wounds. The ''perfect'' substitute is described in Table 1. Currently, no such substitute exists. However, there are a

Table 17.1 The Perfect Skin Substitute

Prevents water loss
Barrier to bacteria
Inexpensive
Long shelf-life
Flexible
Conforms to irregular wound surfaces
Can be used ''off the shelf''
Does not require refrigeration
Cannot transmit viral diseases
Does not incite inflammatory response
Durable
Easy to secure
Grows with a child
Can be applied in one operation
Does not become hypertrophic

number of partial substitutes in clinical use that are valuable and may be the forerunners of this hypothetical ideal. In this section, dermal, epidermal, and composite substitutes will be described.

A. Dermal Substitutes

A functioning dermis is essential for normal skin durability and function. Perhaps the first dermal substitute used clinically was Integra® ''artificial skin'' (Integra LifeSciences Corp., Plainsboro, NJ). This material was developed in the 1980s by a biomaterials research team from the Massachusetts General Hospital and Massachusetts Institute of Technology and was recently released for general use (53). The research team, lead by Burke and Yannas, conceived a membrane that would both provide a temporary vapor and bacterial barrier and serve as a scaffold for dermal regeneration. The material was designed for use in freshly excised burn wounds and is now approved for clinical use in patients with life-threatening burns. The inner layer of this material is a 2-mm thick combination of fibers of collagen isolated from bovine tissue and the glycosaminoglycan chondroitin-6-sulfate, which has a pore size of 70 to 200 μm and a structure that allows fibrovascular ingrowth, after which it is designed to biodegrade (54,55). To fabricate this device requires precipitation of glycosaminoglycan and collagen fibers, which are then freeze-dried and cross-linked by gluteraldehyde. The outer layer is 0.009 in. polysiloxane polymer with vapor transmission characteristics similar to epithelium. This material is placed on excised full thickness burns, and the outer silicone membrane is replaced with an ultrathin epithelial autograft 2 to 3 weeks later, after fibrovascular ingrowth has occurred into the inner layer (Fig. 3) (56). Clinical reports in patients with large burns have been favorable (57–59), and post marketing trials of Integra artificial skin are in progress. As in any occlusive wound dressing, submembrane purulence must be watched for and promptly treated.

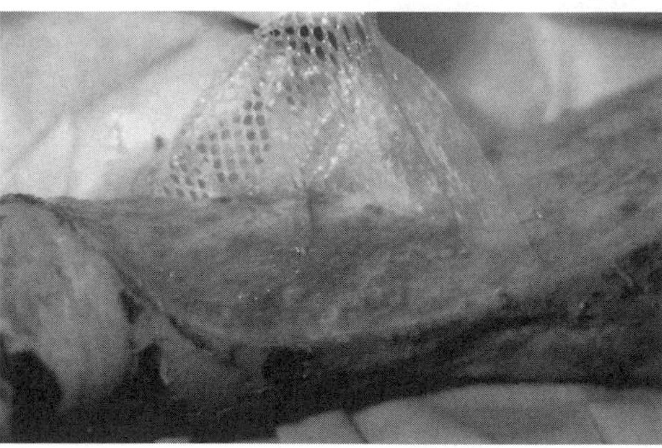

Figure 17.3 Perhaps the first dermal substitute used clinically was Integra® ''artificial skin.'' The material was designed for use in freshly excised burn wounds and is now approved for clinical use in patients with life-threatening burns. The device is placed on excised full thickness burns and the outer silicone membrane replaced with an ultrathin epithelial autograft 2 to 3 weeks later, after fibrovascular ingrowth has occurred into the inner layer.

Biodegradeable polyglactin mesh, seeded with allogenic fibroblasts from neonatal foreskin, has also been explored as a dermal analog in burn wounds. The carrier material biodegrades by hydrolysis, while the allogenic fibroblasts are hoped to facilitate the formation of a neodermis. This material, like Integra, is designed to be combined with an ultrathin epidermal autograft (60–62). Its clinical utility remains to be demonstrated.

Cryopreserved allogenic dermis is another strategy designed to contribute to permanent coverage of wounds by providing for a replacement of the lost dermis. This material, designed to be combined with a thin epithelial autograft at the time of application, is marketed as AlloDerm® (LifeCell Corporation, The Woodlands, TX) (63,64). Split thickness skin allograft skin is procured from cadaver donors, properly screened for disease. Using hypertonic saline, the epithelial elements of the grafts are removed and the remaining dermis is treated in a detergent to inactivate any viruses. The material is then freeze-dried. This process results in a nonantigenic complete dermal scaffold with basement membrane proteins, including laminin and type IV and VII collagen. The material is rehydrated and applied to wounds with an overlying ultrathin epithelial autograft (Fig. 4). Clinical experience with this material in

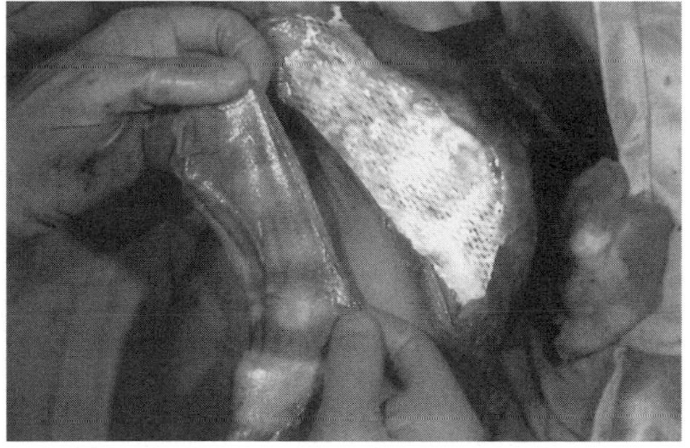

Figure 17.4 Cryopreserved allogenic dermis is another strategy designed to contribute to permanent coverage of wounds by providing for a replacement of the lost dermis. This material, designed to be combined with a thin epithelial autograft at the time of application, is marketed as AlloDerm®.

acute and reconstructive burn wounds has been favorably reported (65,66).

B. Epidermal Substitutes

In the 1970s, Rheinwald and Green developed a method of culturing epithelial cells from a small skin biopsy (67,68). This technique has become the basis for the widespread clinical use of cultured epithelial grafts. From a full thickness skin biopsy, epithelial cells are separated with trypsin. The resulting epithelial cell suspension is plated in culture dishes containing culture medium with fetal calf serum, insulin, transferrin, hydrocortisone, epidermal growth factor, and cholera toxin, overlying a layer of murine fibroblasts that have been treated with a nonlethal dose of radiation that prevents them from multiplying. Isolated colonies of epithelial cells then expand into broad sheets of undifferentiated epithelial cells. These cultures are treated with trypsin, and the cells are taken to secondary culture using the same techniques. The resulting sheets are removed from the dishes after treatment with dipase, which digests the proteins attaching the epithelial cells to the dish. The sheets of epithelial cells are attached to a petrolatum gauze carrier to make it easy to handle in the operating room. Shortly after they were developed, epithelial cultures were used in patients with large burns (69–71). Epithelial grafts are now commercially available, most notably by Genzyme Tissue Repair of Cambridge, MA. With more frequent use of epithelial grafts, certain liabilities have been seen (72,73). These include suboptimal engraftment rates and long-term durability. However, when faced with a very large wound and minimal donor sites, epithelial cell wound closure is a valuable adjunct to overall patient management. It is generally felt that at least the latter liability is the result of the absence of a dermal element.

C. Composite Substitutes

Combining epithelial grafts with a dermal analog is a logical objective. This can be done either on the wound, at the time of application of epithelial grafts, or in the laboratory prior to surgery. The first clinical report of the former concept was that by Cuono and coworkers. They described a technique in which epithelial cells are grafted on wounds closed initially with vascularized allograft. Subsequently the allogenic epithelial cells were removed by dermabrasion or tangential excision, leaving behind a vascularized but theoretically nonantigenic allogenic dermal layer (74–76). The method has been favorably reported (77,78), but the technique has not been universally effective and has not been widely adopted. Perhaps the epithelial excision either leaves behind nests of antigenic epithelial cells if too superficial, or removes the epidermal–dermal attachment structures if too deep. The application of epithelial cells onto Integra or AlloDerm has been attempted without known reliability, and no nonanecdotal data exists.

Combining epithelial cells with a dermal analog in the laboratory prior to use on human wounds also seems a logical approach. This effort began perhaps with the work of Boyce and Hansbrough, who have produced a completely biologic composite skin substitute, culturing human fibroblasts in a collagen–glycosaminoglycan membrane upon which are grown keratinocytes (79,80). Although this com-

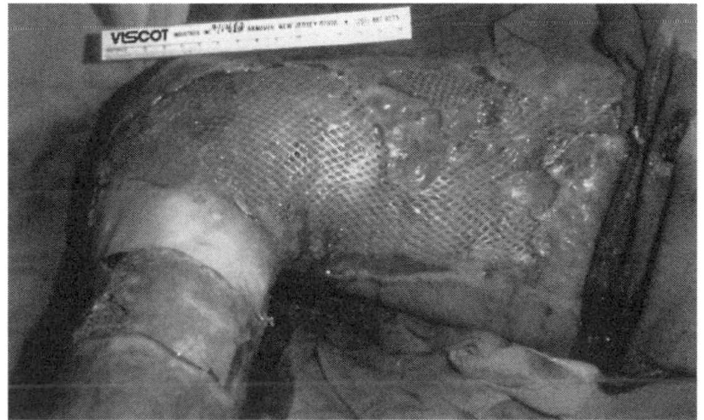

Figure 17.5 An example of efforts to develop a composite skin substitute includes culturing autogenic epithelial cells are onto allogenic dermis.

posite membrane would successfully engraft in a nude mouse model (81), engraftment rates were found to be suboptimal in a small clinical series (82). Further investigations of this potentially exciting technology continue (83). Another example of this approach is work going on at the Shriners Hospital in Boston, where autogenic epithelial cells are being cultured onto allogenic dermis (84). This material was also successful in an animal model and is in early clinical pilot trials (Fig. 5).

V. CONCLUSIONS

Although it has saved lives, the increasing success of burn resuscitation and supportive care has created a growing clinical problem: the need for temporary and permanent closure of wounds. As the burn and wound care field has evolved, the biomaterials sector has kept pace, with a series of products that have enormously facilitated the care of these patients. The future looks increasingly bright, with new temporary and permanent wound membranes on the horizon. These new products may include populations of allogeneic epithelial cells genetically modified to express essential growth factors to facilitate underlying wound healing. However the problem of definitive wound closure is solved, it is certain that biomaterials scientists will play an essential role in the solution.

REFERENCES

1. M. J. Friedrich. Polymer scientists engineer better remedies. *JAMA* 2000 Apr 19. 1947; 283:1943, 1947.
2. R. L. Sheridan, J. P. Remensnyder, J. J. Schnitzer, J. T. Schulz, C. M. Ryan, R. G. Tompkins. Current expectations for survival in pediatric burns. *Arch. Pediatr. Adolesc. Med.* 154:245–9 (2000).
3. R. L. Sheridan, M. I. Hinson, M. H. Liang, et al. Long-term outcome of children surviving massive burns. *JAMA* 283:69–73 (2000).
4. F. D. Moore. The body-weight burn budget. Basic fluid therapy for the early burn. *Surgical Clinics of North America* 50:1249–1265 (1970).
5. C. P. Artz, J. A. Moncrief. The burn problem. In: *The Treatment of Burns*, C. P. Artz, J. A. Moncrief (Eds.). W. B. Saunders, Philadelphia, 1969, pp. 1–22.
6. Z. Janzekovic. A new concept in the early excision and immediate grafting of burns. *J. Trauma* 10:1103–1108 (1970).
7. J. F. Burke, W. C. Quinby, Jr., C. C. Bondoc. Primary excision and prompt grafting as routine therapy for the treatment of thermal burns in children. 1976 [classical article]. *Hand Clinics*. 6:305–317 (1990).
8. J. F. Burke, W. C. Quinby, Jr., C. C. Bondoc. Primary excision and prompt grafting as routine therapy for the treatment of thermal burns in children. *Surgical Clinics of North America*. 56:477–494 (1976).
9. D. N. Herndon, D. Gore, M. Cole, et al. Determinants of mortality in pediatric patients with greater than 70% full-thickness total body surface area thermal injury treated by early total excision and grafting. *J. Trauma* 27:208–212 (1987).
10. D. M. Heimbach, L. H. Engrav, J. A. Marvin, T. J. Harnar, B. J. Grube. Toxic epidermal necrolysis. A step forward in treatment. [Published erratum appears in *JAMA* 258(14): 1894 (1987).] *JAMA* 257:2171–2175 (1987).
11. J. D. Fine, L. B. Johnson, D. Cronce, et al. Intracytoplasmic retention of type VII collagen and dominant dystrophic epidermolysis bullosa: reversal of defect following cessation of or marked improvement in disease activity. *J. Invest. Dermatol.* 101:232–236 (1993).
12. C. C. Bondoc, J. F. Burke. Clinical experience with viable frozen human skin and a frozen skin bank. *Ann. Surg.* 174: 371–382 (1971).
13. D. N. Herndon. Perspectives in the use of allograft. J. Burn Care *Rehabil.* 18:S6 (1997).
14. S. R. May, J. M. Still, Jr., W. B. Atkinson. Recent developments in skin banking and the clinical uses of cryopreserved skin. [Review]. *J. Med. Assoc. Georgia* 73:233–236 (1957).
15. J. F. Burke, J. W. May, Jr., N. Albright, W. C. Quinby, P. S. Russell. Temporary skin transplantation and immunosuppression for extensive burns. *N. Engl. J. Med.* 290:269–271 (1974).
16. J. F. Bale, Jr., G. P. Kealey, C. L. Ebelhack, C. E. Platz, J. A. Goeken. Cytomegalovirus infection in a cyclosporine-treated burn patient: case report. *J. Trauma* 32:263–267 (1992).
17. K. M. Ramakrishnan, V. Jayaraman. Management of partial-thickness burn wounds by amniotic membrane: a cost-effective treatment in developing countries. *Burns* 23 (Suppl. 1):S33–S36 (1997).
18. M. Subrahmanyam. Amniotic membrane as a cover for microskin grafts. *Br. J. Plastic Surg.* 48:477–478 (1995).
19. M. A. Ganatra, K. M. Durrani. Method of obtaining and preparation of fresh human amniotic membrane for clinical use. *J. Pakistan Med. Assoc.* 46:126–128 (1996).
20. P. D. Thomson, D. H. Parks. Monitoring, banking, and clinical use of amnion as a burn wound dressing. *Ann. Plastic Surg.* 7:354–356 (1981).
21. M. Haberal, Z. Oner, U. Bayraktar, N. Bilgin. The use of silver nitrate–incorporated amniotic membrane as a temporary dressing. *Burns, Incl. Thermal Injury* 13:159–163 (1987).
22. W. C. Quinby, Jr., H. C. Hoover, M. Scheflan, P. T. Walters, S. A. Slavin, C. C. Bondoc. Clinical trials of amniotic membranes in burn wound care. *Plast. Reconstr. Surg.* 70: 711–717 (1982).
23. I. C. Song, B. E. Bromberg, M. P. Mohn, E. Koehnlein.

Heterografts as biological dressings for large skin wounds. *Surgery* 59:576–583 (1966).
24. R. A. Elliott, Jr., J. G. Hoehn. Use of commercial porcine skin for wound dressings. *Plastic Reconstr. Surg*. 52:401–405 (1973).
25. R. A. Ersek, H. J. Hachen. Porcine xenografts in the treatment of pressure ulcers. Ann. *Plast. Surg*. 5:464–470 (1980).
26. D. S. Chatterjee. A controlled comparative study of the use of porcine xenograft in the treatment of partial thickness skin loss in an occupational health centre. *Curr. Med. Res. Opin*. 5:726–733 (1978).
27. J. A. Marvin, D. M. Heimbach, L. H. Engrav, T. J. Harnar. Improved treatment of the Stevens-Johnson syndrome. *Arch. Surg*. 119:601–605 (1984).
28. R. A. Ersek, J. A. Navarro. Maximizing wound healing with silver-impregnated porcine xenograft. *Today's OR-Nurse*. 12:4–9 (1990).
29. R. A. Ersek, D. R. Denton. Silver-impregnated porcine xenografts for treatment of meshed autografts. *Ann. Plast. Surg*. 13:482–487 (1984).
30. R. E. Salisbury, D. W. Wilmore, P. Silverstein, B. A. Pruitt, Jr. Biological dressings for skin graft donor sites. *Arch. Surg*. 106:705–706 (1973).
31. R. E. Salisbury, R. W. Carnes, D. Enterline. Biological dressings and evaporative water loss from burn wounds. *Ann. Plas. Surg*. 5:270–272 (1980).
32. R. H. Demling. Burns. *N. Engl. J. Med*. 313:1389–1398 (1985).
33. P. Vanstraelen. Comparison of calcium sodium alginate (KALTOSTAT) and porcine xenograft (E-Z DERM) in the healing of split-thickness skin graft donor sites. *Burns* 18: 145–148 (1992).
34. E. A. Bacha, R. L. Sheridan, G. A. Donohue, R. G. Tompkins. Staphylococcal toxic shock syndrome in a paediatric burn unit. *Burns* 20:499–502 (1994).
35. D. G. Greenhalgh. The role of growth factors in wound healing. [Review]. *J. Trauma* 41:159–167 (1996).
36. N. T. Bennett, G. S. Schultz. Growth factors and wound healing: biochemical properties of growth factors and their receptors. [Review]. *Am. J. Surg*. 165:728–737 (1993).
37. R. G. Teepe, R. Koch, B. Haeseker. Randomized trial comparing cryopreserved cultured epidermal allografts with tulle-gras in the treatment of split-thickness skin graft donor sites. *J. Trauma* 35:850–854 (1993).
38. E. Bell, H. P. Ehrlich, D. J. Buttle, T. Nakatsuji. Living tissue formed in vitro and accepted as skin-equivalent tissue of full thickness. *Science* 211:1052–1054 (1981).
39. C. J. Nolte, M. A. Oleson, J. F. Hansbrough, J. Morgan, G. Greenleaf, L. Wilkins. Ultrastructural features of composite skin cultures grafted onto athymic mice. *J. Anat*. 185:325–333 (1994).
40. J. F. Hansbrough, J. Morgan, G. Greenleaf, M. Parikh, C. Nolte, L. Wilkins. Evaluation of Graftskin composite grafts on full-thickness wounds on athymic mice. *J. Burn Care Rehabil*. 15:346–353 (1994).
41. G. D. Gentzkow, S. D. Iwasaki, K. S. Hershon, et al. Use of dermagraft, a cultured human dermis, to treat diabetic foot ulcers [see comments]. *Diabetes Care* 19:350–354 (1996).
42. M. S. Sacks, C. J. Chuong, W. M. Petroll, M. Kwan, C. Halberstadt. Collagen fiber architecture of a cultured dermal tissue. *J. Biomechan. Eng*. 119:124–127 (1997).
43. T. P. Economou, M. D. Rosenquist, R. W. Lewis, II, G. P. Kealey. An experimental study to determine the effects of Dermagraft on skin graft viability in the presence of bacterial wound contamination. *J. Burn Care Rehabil*. 16:27–30 (1995).
44. V. Falanga, M. Sabolinski. A bilayered living skin construct (APLIGRAF) accelerates complete closure of hard-to-heal venous ulcers. *Wound Repair Regen*. 7:201–207 (2000).
45. P. Wickware. Progress from a fragile start. *Nature* 403:466 (2000).
46. J. F. Hansbrough, J. Morgan, G. Greenleaf, J. Underwood. Development of a temporary living skin replacement composed of human neonatal fibroblasts cultured in Biobrane, a synthetic dressing material. *Surgery* 115:633–644 (1994).
47. J. Hansbrough. Dermagraft-TC for partial-thickness burns: a clinical evaluation. *J. Burn Care Rehabil*. 18:S25–S28 (1997).
48. S. T. Parente. Estimating the economic cost offsets of using Dermagraft-TC as an alternative to cadaver allograft in the treatment of graftable burns. *J. Burn Care Rehabil*. 18: S18–S24 (1997).
49. R. L. Spielvogel. A histological study of Dermagraft-TC in patients' burn wounds. *J. Burn Care Rehabil*. 18:S16–S18 (1997).
50. G. F. Purdue. Dermagraft-TC pivotal efficacy and safety study. *J. Burn Care Rehabil*. 18:S13–S14 (1997).
51. J. R. Morgan, Y. Barrandon, H. Green, R. C. Mulligan. Expression of an exogenous growth hormone gene by transplantable human epidermal cells. *Science* 237:1476–1479 (1987).
52. J. R. Morgan, M. L. Yarmush. Bioengineered skin substitutes. *Sci. Med*. July/August:6–15 (1997).
53. R. G. Tompkins, J. F. Burke. Progress in burn treatment and the use of artificial skin. [Review]. *World J. Surg*. 14: 819–824 (1990).
54. I. V. Yannas, J. F. Burke, M. Warpehoski, et al. Prompt, long-term functional replacement of skin. *Trans. Am. Soc. Artif. Internal Organs* 27:19–23 (1981).
55. I. V. Yannas, J. F. Burke, D. P. Orgill, E. M. Skrabut. Wound tissue can utilize a polymeric template to synthesize a functional extension of skin. *Science* 215:174–176 (1982).
56. R. G. Tompkins, J. F. Hilton, J. F. Burke, et al. Increased survival after massive thermal injuries in adults: preliminary report using artificial skin. *Crit. Care Med*. 17:734–740 (1989).
57. T. M. Scalea, H. M. Simon, A. O. Duncan, et al. Geriatric blunt multiple trauma: improved survival with early invasive monitoring. *J. Trauma* 30:129–136 (1990).
58. R. L. Sheridan, M. Heggerty, R. G. Tompkins, J. F. Burke.

Artificial skin in massive burns—results at ten years. *Eur. J. Plast. Surg.* 17:91–93 (1994).

59. D. Heimbach, A. Luterman, J. Burke, et al. Artificial dermis for major burns. A multi-center randomized clinical trial. *Ann. Surg.* 208:313–320 (1988).
60. M. L. Cooper, J. F. Hansbrough, R. L. Spielvogel, R. Cohen, R. L. Bartel, G. Naughton. In vivo optimization of a living dermal substitute employing cultured human fibroblasts on a biodegradable polyglycolic acid or polyglactin mesh. *Biomaterials* 12:243–248 (1991).
61. J. F. Hansbrough, M. L. Cooper, R. Cohen, et al. Evaluation of a biodegradable matrix containing cultured human fibroblasts as a dermal replacement beneath meshed skin grafts on athymic mice. *Surgery* 111:438–446 (1992).
62. J. F. Hansbrough, C. Dore, W. B. Hansbrough. Clinical trials of a living dermal tissue replacement placed beneath meshed, split-thickness skin grafts on excised burn wounds. *J. Burn Care Rehabil.* 13:519–529 (1992).
63. D. J. Wainwright. Use of an acellular allograft dermal matrix (Alloderm) in the management of full-thickness burns. *Burns* 21:243–248 (1995).
64. D. Wainwright, M. Madden, A. Luterman, et al. Clinical evaluation of an acellular allograft dermal matrix in full-thickness burns. *J. Burn Care Rehabil.* 17:124–136 (1996).
65. R. L. Sheridan, R. J. Choucair. Acellular allograft dermis does not hinder initial engraftment in burn resurfacing and reconstruction. *J. Burn Care Rehabil.* 18:496–499 (1997).
66. R. L. Sheridan, R. J. Choucair. Acellular allodermis in burn surgery: 1-year results of a pilot trial. [Abstract]. *J. Burn Care Rehabil.* 19:528–530 (1998).
67. J. G. Rheinwald, H. Green. Serial cultivation of strains of human epidermal keratinocytes: the formation of keratinizing colonies from single cells. *Cell* 6:331–343 (1975).
68. H. Green, O. Kehinde, J. Thomas. Growth of cultured human epidermal cells into multiple epithelia suitable for grafting. *Proc. Nat. Acad. Sci. U.S.A.* 76:5665–5668 (1979).
69. H. Green. Cultured cells for the treatment of disease. [Review]. *Sci. Am.* 265:96–102 (1991).
70. G. G. Gallico, III, N. E. O'Connor, C. C. Compton, O. Kehinde, H. Green. Permanent coverage of large burn wounds with autologous cultured human epithelium. *N. Engl. J. Med.* 311:448–451 (1984).
71. G. G. Gallico, III, N. E. O'Connor, C. C. Compton, J. P. Remensnyder, O. Kehinde, H. Green. Cultured epithelial autografts for giant congenital nevi [see comments]. *Plast. Reconstr. Surg.* 84:1–9 (1989).
72. R. L. Sheridan, R. G. Tompkins. Cultured autologous epithelium in patients with burns of ninety percent or more of the body surface. *J. Trauma* 38:48–50 (1995).
73. L. W. Rue, III, W. G. Cioffi, W. F. McManus, B. A. Pruitt, Jr. Wound closure and outcome in extensively burned patients treated with cultured autologous keratinocytes. *J. Trauma* 34:662–667 (1993).
74. R. C. Langdon, C. B. Cuono, N. Birchall, et al. Reconstitution of structure and cell function in human skin grafts derived from cryopreserved allogeneic dermis and autologous cultured keratinocytes. *J. Invest. Dermatol.* 91:478–485 (1988).
75. C. B. Cuono, R. Langdon, N. Birchall, S. Barttelbort, J. McGuire. Composite autologous-allogeneic skin replacement: development and clinical application. *Plast. Reconstr. Surg.* 80:626–637 (1987).
76. C. Cuono, R. Langdon, J. McGuire. Use of cultured epidermal autografts and dermal allografts as skin replacement after burn injury. *Lancet* 1:1123–1124 (1986).
77. W. L. Hickerson, C. Compton, S. Fletchall, L. R. Smith. Cultured epidermal autografts and allodermis combination for permanent burn wound coverage. *Burns* 20(Suppl. 1): S52–S56 (1994).
78. C. C. Compton, W. Hickerson, K. Nadire, W. Press. Acceleration of skin regeneration from cultured epithelial autografts by transplantation to homograft dermis. *J. Burn Care Rehabil.* 14:653–662 (1993).
79. S. T. Boyce, D. J. Christianson, J. F. Hansbrough. Structure of a collagen-GAG dermal skin substitute optimized for cultured human epidermal keratinocytes. *J. Biomed. Mater. Res.* 22:939–957 (1988).
80. S. T. Boyce, J. F. Hansbrough. Biologic attachment, growth, and differentiation of cultured human epidermal keratinocytes on a graftable collagen and chondroitin-6-sulfate substrate. *Surgery* 103:421–431 (1988).
81. M. L. Cooper, J. F. Hansbrough. Use of a composite skin graft composed of cultured human keratinocytes and fibroblasts and a collagen-GAG matrix to cover full-thickness wounds on athymic mice. *Surgery* 109:198–207 (1991).
82. J. F. Hansbrough, S. T. Boyce, M. L. Cooper, T. J. Foreman. Burn wound closure with cultured autologous keratinocytes and fibroblasts attached to a collagen-glycosaminoglycan substrate. *JAMA* 262:2125–2130 (1989).
83. J. F. Hansbrough, J. L. Morgan, G. E. Greenleaf, R. Bartel. Composite grafts of human keratinocytes grown on a polyglactin mesh-cultured fibroblast dermal substitute function as a bilayer skin replacement in full-thickness wounds on athymic mice. *J. Burn Care Rehabil.* 14:485–494 (1993).
84. R. L. Sheridan, J. R. Morgan. Initial clinical experience with an autologoug composite skin substitute. *J. Burn Care Rehabil.* 21:S214 (2000).

18

Dermocosmetic Applications of Polymeric Biomaterials

P. Corvi Mora
EUPHAR Group, Piacenza, Italy

P. G. Baraldi
Ferrara University, Ferrara, Italy

I. INTRODUCTION

The Federal Food, Drug, and Cosmetic Act (FD&C Act) defines cosmetics as "articles other than soap which are applied to the human body for cleansing, beautifying, promoting attractiveness, or altering the appearance" (1). The Food and Drug Administration (FDA) has classified cosmetics into 13 categories: skin care (creams, lotions, powders, and sprays), fragrances, eye makeup, manicure products, makeup other than eye (e.g., lipstick, foundation, and blusher), shampoos, permanent waves and other hair products, deodorants, shaving products, baby products (e.g., shampoos, lotions, and powders), bath oils and bubble baths, mouthwashes, and tanning products.

It is against the law to distribute cosmetics that contain poisonous or harmful substances that might injure users under normal conditions. Manufacturing or holding cosmetics under insanitary conditions, using nonpermitted colors, or including any filthy, putrid, or decomposed substance is also illegal.

Except for color additives and a few prohibited ingredients, a cosmetic manufacturer may use any ingredient or raw material and market the final product without government approval. The prohibited ingredients are: biothionol, exachlorophene, mercury compounds (except under certain conditions as preservatives in eye cosmetics), vinyl chloride and zirconium salts in aerosol products, halogenated salicylanilides, chloroform, and methylene chloride. Manufacturers must test color additives for safety and gain FDA approval for their intended use.

It is important for both consumers and manufacturers to understand the difference between cosmetics and drugs. Different regulations are applied to each type of product. The FDA's Office of Cosmetics and Colors, which is part of the Center for Food Safety and Applied Nutrition, handles issues related to cosmetics and color additives. The agency's Center for Drug Evaluation and Research handles issues related to drugs.

The FD&C Act defines drugs as "articles intended for use in the diagnosis, cure, mitigation, treatment, or prevention of disease . . . and articles (other than food) intended to affect the structure or any function of the body of man or other animals." Over-the-counter (OTC) drugs are drugs that can be purchased without a doctor's prescription. The agency is conducting a review of all OTC drugs to establish monographs (rules) under which the drugs are generally recognized as safe and effective, and not misbranded. These rules are established on a class-by-class basis (e.g., fluoride dentifrices, cough suppressants, antihistamines). OTC drugs must meet the requirements of the appropriate class once that rule is published as a final regulation.

OTC drugs are often marketed side by side with cosmetics, and some products qualify both as cosmetics and as OTC drugs. This may happen when a product has two intended uses, with ingredients intended to do two different things. For instance, a shampoo is a cosmetic, since its intended use is to cleanse the hair. An antidandruff treatment is a drug, since its intended use is to treat dandruff. Consequently, an antidandruff shampoo is both a cosmetic and a drug. Among other cosmetic/drug combinations are toothpastes that contain fluoride, deodorants that are also antiperspirants, and moisturizers and makeup marketed with sun-protection claims. Certain claims may cause a product to qualify as a drug, even if the product is marketed as if it were a cosmetic. Such claims establish the product as a drug because the intended use is to treat or prevent disease or otherwise affect the structure or functions of the human body. Some examples are claims that products will restore hair growth, reduce cellulite, treat varicose veins, or revitalize cells. The same is true of essential oils in fragrance products. A fragrance marketed for promoting attractiveness is a cosmetic. But a fragrance marketed with ''aromatherapy'' claims such as assertions that the scent will help the consumer sleep or quit smoking meets the definition of a drug because of its intended use.

In light of these considerations, and in respect of the actual regulation, cosmetics have purely the main function to maintain the skin in *good condition*. For dermocosmetics we need knowledge of skin biology and physiology. To allow the cosmetic to support the skin's natural physiological state, the cosmetic function has to be involved in the biological and physiological processes of the skin structure.

The scientific developments derived from biological research applied to dermatology have profoundly changed numerous concepts in skin physiology and physiopathology (2). Ideas of skin homeostasis and skin purely as a protective barrier have been corrected with the idea of the skin as a dynamic *equilibrium* of a complex system: the skin function as a barrier is a result of the balance between proliferation, differentiation, and loss of keratinocyte cells. Moreover, experimental data has shown that the keratinocytes, melanocytes, and Langerhans cells are functionally coordinated by autocrine and/or paracrine mechanisms; their actions confer a functional autonomy on the epidermis, so removing it, at least in part, from the conditioning of the dermis.

II. BIOCHEMICAL SECTION

A. Introduction

Studying and searching for new functional cosmetics require deep knowledge about the normal function of cells and structural elements in skin. Moreover, the terms that describe the skin are used repeatedly in discussing the goals of skin-directed products.

The skin divides conveniently into four regions or zones, layered one on top of the other:

Epidermis
Basement membrane zone
Dermis
Subcutaneous tissue

In certain parts of the body these regions and zones are highly modified, reflecting very special needs. For example, the scalp is ordinarily covered with thick hair (a product of the epidermis) and the palms are covered with a highly thickened epidermis. On the other hand, the face contains large numbers of sebaceous oil glands that are susceptible to inflammation (acne), and the face exhibits a special susceptibility to the development of basal cell carcinomas (3).

1. Epidermis

The upper layer, the epidermis, is an unusual structure, being composed almost entirely of cells (keratinocytes, melanocytes, Langerhans cells). These cells are locked together into a tight membrane that covers the entire skin surface. The most common epidermal cell is the keratinocyte, making up 95% of the skin surface. Concepts of the activities of keratinocytes have expanded greatly in the last ten years to go far beyond their capacity to protect skin through barrier formation. In fact, keratinocytes are now known to produce a wide spectrum of cytokines and to express important adhesion molecules.

Two other resident populations of epidermal cells include melanocytes and Langerhans cells, each with a specialized function. Melanocytes manufacture the pigment melanin, which is then released and taken up into the adjacent keratinocytes to provide a protective barrier against ultraviolet light. Langerhans cells, on the other hand, are components of the immune system and serve as the most remote aspect of foreign antigen recognition in the skin. Importantly, keratinocytes are able to influence the functions of these two cells through cytokines and adhesion molecules.

2. Basement Membrane Zone

The basement membrane zone is a noncellular attachment zone that links the epidermis to the dermis. The proteins within the basement membrane zone are manufactured by cells on both sides, epidermis and dermis.

3. Dermis

The underlying dermis provides nutrition through blood vessels and strength through large bundles of the protein collagen. Cells that reside normally in the dermis include fibroblasts, endothelial cells (in blood vessels), and mast cells.

4. Subcutis

Deep beneath the dermis one finds the subcutis, which contains large blood vessels, large nerves, and fat cells.

5. Specialized Structures

Hair, sweat glands, and sebaceous glands are all specialized structures that are produced through the coordinated efforts of both epidermis and dermis.

B. Epidermis

The epidermis is a multilayered structure (stratified epithelium) that renews itself continuously by keratinocyte division in its deepest layer, the basal layer. The cells produced by cell division in the basal layer constitute the prickle cell layer, and as they ascend toward the surface they undergo a process known as keratinization, which involves the synthesis of the fibrous protein keratin. The total epidermal renewal time is 52–75 days. The cells on the surface of the skin, forming the horny layer (*stratum corneum*), are fully keratinized dead cells, which are gradually abraded by day-to-day wear and tear from the environment (4–6).

The basal layer is composed of columnar cells that are anchored to a basement membrane (this lies between the epidermis and the dermis). The basement membrane is a multilayered structure from which anchoring fibrils extend into the superficial dermis. Interspersed amongst the basal cells are melanocytes, large dendritic cells responsible for melanin pigment production.

The prickle cell layer acquires its name from the spiky appearance produced by intercellular bridges (desmosomes) that connect adjacent cells. Scattered throughout the prickle cell layer are numbers of dendritic cells called Langerhans cells. Like macrophages, Langerhans cells originate in the bone marrow and have an antigen-presenting capacity.

Above the prickle cell layer is the granular layer, which is composed of rather flattened cells containing numerous darkly-staining particles known as keratohyaline granules. Also present in the cytoplasm of cells in the granular layer are organelles known as lamellar granules (Odland bodies). Lamellar granules contain lipids and enzymes, and they discharge their contents into the intercellular spaces between the cells of the granular layer and the *stratum corneum*, providing something akin to ''mortar'' between the cellular ''bricks.'' In the granular layer the cell membranes become thickened as a result of deposition of dense material on their inner surfaces.

The cells of the stratum corneum are flattened keratinized cells that are devoid of nuclei and cytoplasmic organelles. These cellular components degenerate in the upper granular layer. Adjacent cells overlap at their margins, and this locking together of cells, together with intercellular lipid, forms a very effective barrier. The stratum corneum varies in thickness depending on the region of the body, being thickest over the palms of the hands and soles of the feet.

The rate of cell production in the germinative compartment of the epidermis must be balanced by the rate of cell loss at the surface of the stratum corneum. The control mechanism of epidermopoiesis consists of a balance of stimulatory and inhibitory signals. Wound healing provides a model to examine the changes in growth control that occur in establishing a new epidermis. Wounding of the skin is followed by a wave of epidermal mitotic activity, which represents the effects of diffusible factors spreading from the wound into the surrounding tissue. These factors include cytokines and growth factors. Their production is not limited to immune cells, as they are produced by keratinocytes in vitro and can be found in physiological amounts in normal human skin.

a. *Regulation of Epidermopoiesis: Stimulatory Factors*

The growth factors that stimulate the epidermal cells include epidermal growth factor (EGF), transforming growth factor alpha (TGFalpha), interleukins (IL) and other immunological cytokines, and basic fibroblast growth factor (bFGF).

EGF binds to specific cell-surface receptors (EGFR, a transmembrane glycoprotein receptor) present in the basal layer of the human epidermis. Following binding of EGF to EGFR, the receptor is internalized and carries EGF into an intracellular cycle within the cytoplasm and the nucleus to mediate all its effects. EGF has been shown to increase the growth and persistence of epidermal keratinocytes and to promote wound healing in vitro. EGF transcripts are not found in the epidermis, but in salivary glands and the intestinal tract.

TGFalpha was the first growth factor known to be produced by keratinocytes. Its mRNA predominates in the basal compartment of the epidermis. TGFalpha is related to EGF; it binds to and activates the EGF receptor. It stimulates keratinocyte growth.

The normal epidermis also contains large amounts of Interleukin-1. There are two forms, alpha and beta, and unlike macrophages, the epidermis largely produces IL-1 alpha. IL-1 has been shown to be mitogenic for keratinocytes (other effects include fibroblast proliferation and synthesis of collagenase, stimulation of IL-2 production, stimulation of B-cell function, and fever induction). IL-1 releases IL-6 from keratinocytes. IL-6 appears to stimulate growth of keratinocytes and can be detected in epidermal cells. Keratinocytes also synthesize IL-3, IL-4, IL-8 (neutrophil activating protein), and granulocyte-macrophage colony stimulating factor.

Thus the epidermal keratinocytes can, under activation conditions, secrete a large number of cytokines, which can modulate lymphocyte activation and granulocyte function. These factors do not work in isolation but have complex interactions and may be synergistic or antagonistic. The factors controlling synthesis and secretion of these factors may be important in the pathogenesis of skin disease as well as epidermal growth control.

The regulation of the effects of growth factors includes the control of expression of the specific growth-factor receptors. The epidermal cell cycle is also controlled by the intracellular concentrations of the cyclic nucleotides: cAMP and cGMP. These are small molecules that are formed and broken down intracellularly as a response to external signals acting on the cell membrane. Cyclic AMP is believed to be the intracellular agent or ''second messenger'' of those hormones, i.e., catecholamines and polypeptides, which do not themselves penetrate the surface of cells. Cyclic AMP inhibits epidermal cell division, while cGMP stimulates it. Epidermal mitosis exhibits a circadian rhythm that is inversely related to blood adrenaline levels.

Steroid hormones like testosterone enter the target cells. Epidermal keratinocytes contain a 5 alpha-reductase enzyme, and they can convert testosterone to 5 alpha-dihydrotestosterone (DHT). DHT binds to specific cytoplasmic receptors, which then translocate to the nucleus, thereafter, altering protein synthesis via messenger RNA. Androgens and vitamin A stimulate epidermal mitosis, while glucocorticoid hormones inhibit it.

Prostaglandins, which are metabolic products of arachidonic acid, can affect nucleotide metabolism. Prostaglandins of the D and E series can increase cAMP, although not all such components are present in the epidermis. The main prostaglandin formed in the epidermis is PGE2. On the other hand, lipoxygenase products of arachidonic acid metabolism, namely HETE (12-hydroxy-eicosa-tetraenoic acid) and the leucotrienes, are capable of inducing epidermal cell proliferation in vitro. Polyamines, including spermidine, putrescein, and spermine, stimulate mitosis. Ornithine decarboxylase is a particularly important enzyme for the generation of this group of substances.

b. Regulation of Epidermopoiesis: Inhibitory Factors

Growth inhibitors for keratinocytes include chalones, transforming growth factor beta (TGF beta), alpha and gamma interferons (IFN-gamma), and tumor necrosis factor (TNF).

Chalones are polypeptides produced by suprabasal cells which slow basal mitosis.

TGF beta stimulates fibroblast growth and increases fibrosis but inhibits the growth of keratinocytes. Thus although it may have an inhibitory effect on epidermal growth, the effect on wound healing is complex, because of mesenchymal effects (on fibroblasts), and it has been reported to stimulate wound healing.

Alpha and gamma interferons have cytostatic effects on keratinocytes both in vivo following systemic administration and in vitro. Following stimulation with IFN-gamma, keratinocytes express class II antigens, predominantly HLA-DR. High doses of interferon-gamma are cytotoxic.

Thirty percent of administered TNF localizes in the epidermis, suggesting the presence of many TNF binding sites. Keratinocytes also secrete TNF. TNF can cause release of IL-1. It stimulates fibroblast proliferation and cytokine production. TNF has also been shown to be reversibly cytostatic to keratinocytes.

1. Sensory System

The skin is innervated with around one million afferent nerve fibers. Most terminate in the face and extremities; relatively few supply the back. The cutaneous nerves contain axons with cell bodies in the dorsal root ganglia. Their diameters range from 0.2 to 20 μm. The main nerve trunks entering the subdermal fatty tissue each divide into smaller bundles. Groups of myelinated fibers fan out in a horizontal plane to form a branching network from which fibers ascend, usually accompanying blood vessels, to form a mesh of interlacing nerves in the superficial dermis. Throughout their course, the axons are enveloped in Schwann cells, and as they run peripherally, an increasing number lack myelin sheaths. Most end in the dermis; some penetrate the basement membrane but do not travel far into the epidermis.

Sensory endings are of two main kinds: corpuscular, which embrace nonnervous elements, and ''free,'' which do not. Corpuscular endings can, in turn, be subdivided into encapsulated receptors, of which a range occurs in the dermis, and nonencapsulated, exemplified by Merkel's ''touch spot,'' which is epidermal.

Each Merkel's touch spot is composed of a battery of Merkel cells borne on branches of a myelinated axon. A Merkel cell has a lobulated nucleus and characteristic granules; it is embedded in the basal layer of epidermal cells, with which it has desmosomal connections; it contains intermediate filaments composed of low molecular weight keratin rather than neurofilament protein. The Pacinian corpuscle is one of the encapsulated receptors. It is an ovoid structure about 1 mm in length, which is lamellated in cross section like an onion and is innervated by a myelinated sensory axon that loses its sheath as it traverses the core. The Golgi–Mazzoni corpuscle found in the subcutaneous tissue of the human finger is similarly laminate but of much simpler organization. These last two lamellated end organs are movement and vibration detectors.

The Krause end bulb is an encapsulated swelling on myelinated fibers situated in the superficial layers of the dermis. Meissner corpuscles are characteristic of the papillary ridges of glabrous (hairless skin) skin; they are touch receptors; they have a thick lamellated capsule 20–40 μm in diameter and up to 150 μm long. Ruffini endings in the human digits have several expanded endings branching from a single myelinated afferent fiber; the endings are directly related to collagen fibrils; they are stretch receptors.

''Free nerve endings,'' which appear to be derived from nonmyelinated fibers, occur in the superficial dermis and in the overlying epidermis; they are receptors for pain, touch, pressure, and temperature. Hair follicles have fine nerve filaments running parallel to and encircling the follicles; each group of axons is surrounded by Schwann cells; they mediate touch sensation.

a. *Physiology of Sensory Receptors*

Adaptation. When a maintained stimulus of constant strength is applied to a receptor, the frequency of the action potentials in its sensory nerve declines over time. There are two types of receptors: (1) tonic slowly adapting receptors: as the nociceptors (pain receptors) that continue to transmit impulses to the brain as long as the stimulus is applied, thus keeping the CNS continuously informed about the state of the body, and (2) phasic rapidly adapting receptors: as Pacinian corpuscles, these receptors adapt rapidly and cannot be used to transmit a continuous signal to the CNS; they are stimulated only when the stimulus strength is changed.

Touch sensation is provoked by a harmless stimulus to the skin allowing us to distinguish between hard and soft objects; touch receptors belong to the class of mechanoreceptors, and many of them can be found around hair follicles, so that removal of hair decreases touch sensitivity; the tips of the fingers and lips are rich in touch receptors.

Tickle and itch. The heparin-containing tissue cells called mast cells have a high histamine content in their granules. They also contain serotonin. Mast cells are particularly numerous in the skin (about 7,000 mast cells per cubic millimeter in normal skin in the subpapillary dermis) and play an important role in type I immediate hypersensitivity reation (IgE-mediated anaphylactic reaction).

Temperature sensation. Receptors for warmth and cold are specialized free nerve endings; a rise in skin temperature above body temperature causes a sensation of warmth, while a fall in skin temperature below body temperature is experienced as a cold sensation; pain is felt if skin temperature increases above 45°C or decreases below 10°C; the mucous membrane of the mouth is less sensitive than the skin, so tea can be drunk at a temperature that is painful to fingers. Paradoxical cold: cold receptors are stimulated by intrinsic heat (e.g., shivering that occurs with fever).

Pain is evoked by nonspecific stimuli (chemical, mechanical, thermal, or electrical) of an intensity that can produce tissue damage. Pain is a high-threshold sensation. The nociceptors (pain receptors) are free nerve endings. Cutaneous pain may be sharp and localized or dull and diffuse. A painful stimulus causes at first sharp pain followed by dull aching pain. Reflex withdrawal movements also occur, with an increase in heart rate and blood pressure. Fast sharp (pricking) pain is mediated by nociceptors innervated by group A delta thick myelinated nerve fibers, which transmit pain impulses at a velocity of 20 m/s. Slow, chronic (dull aching or burning) pain is mediated by nociceptors innervated by group C thin unmyelinated nerve fibers that conduct pain at a low velocity of 1 m/s.

2. Vascular System

Circulation through the skin serves two functions, nutrition of the skin tissue and regulation of body temperature by conducting heat from the internal structures of the body to the skin, where it is lost by exchange with the external environment (by convection, conduction, and radiation). (See also ''Sweat Glands'' below.)

The cutaneous circulatory apparatus is well-suited to its functions. It comprises two types of vessels, (1) the usual nutritive vessels (arteries, capillaries, and veins) and (2) vascular structures concerned with heat regulation. The latter include an extensive subcutaneous venous plexus that can hold large quantities of blood (to heat the surface of the skin) and arteriovenous anastomoses, which are large, direct vascular communications between arterial and venous plexuses. Arteriovenous anastomoses are only present in some skin areas that are often exposed to maximal cooling, as the volar surfaces of hands and feet, the lips, the nose, and the ear.

The specialized vascular structures just mentioned bear strong muscular coats innervated by sympathetic adrenergic vasoconstrictor nerve fibers. When constricted, blood flow into the subcutaneous venous plexus is reduced to almost nothing (minimal heat loss); while, when dilated, an extremely rapid flow of warm blood into the venous plexus is allowed (maximal heat loss).

The cutaneous circulation also serves as a blood reservoir. Under conditions of circulatory stress, e.g., exercise and hemorrhage, sympathetic stimulation of subcutaneous venous plexus forces a large volume of blood (5–10% of the blood volume) into the general circulation.

Reactive hyperemia occurs if one, for example, sits on one portion of his skin for 30 minutes or more and then removes the pressure. In such conditions, the individual will notice intense redness of the skin at the site of previous pressure, which resulted from the accumulation of vasodilator metabolites at that site (due to decreased availability of nutrients to the tissues during compression).

3. Pigmentary System

The melanin pigmentary system is composed of functional units called epidermal melanin units. Each unit consists of a melanocyte that supplies melanin pigment to a group of keratinocytes (about 36). Pigmentation is determined primarily by the amount of melanin transferred to the keratinocytes.

The melanocyte is a dendritic cell present in the basal layer of the epidermis with no desmosomes (intercellular bridges) or tonofilaments. Melanocytes arise from the neural crest as melanoblasts and migrate to the dermis, hair follicles, leptomeninx, uveal tract, and retina. By the eighth week of intrauterine life, they start to migrate from the dermis to the epidermis. Although full melanocyte migration is normally completed prior to birth, residual dermal melanocytes are sometimes left (clinically appearing as mongoloid spots in the sacral area of oriental and black infants).

Melanosomes are membrane-bound organelles located in the cytoplasm of melanocytes and bearing tyrosinase enzyme. They are responsible for melanin synthesis and pigment transfer from the melanocyte to the surrounding keratinocytes. During their passage from the perinuclear area of the melanocyte to the dendrites, the melanosomes show four stages of development: I and II (with no melanin deposition), III (with high levels of tyrosinase activity and is partially obscured by melanin deposition), and IV (with low levels of tyrosinase activity and is completely obscured by melanin deposition). Pigment transfer occurs by keratinocyte phagocytosis of melanosome-containing dendritic tips. As squamous cells differentiate, the melanosomes within them are degraded by lysosomal enzymes.

The differences in racial pigmentation are not due to differences in the number of melanocytes but rather to differences in melanocyte activity. In black skin, there is greater production of melanosomes, a higher degree of melanization of melanosomes, and larger unaggregated melanosomes showing a slow rate of degradation.

a. Melanin: Types, Synthesis, and Hormonal Regulation

Melanin is a brown-black light-absorbing pigment that protects the skin against ultraviolet rays. Two major forms of melanin exist in humans: eumelanin, a brown-to-black pigment synthesized from indole 5,6-quinone and found within the ellipsoid melanosomes, and phaeomelanin, a yellow-red sulfur-containing pigment found within the spherical melanosomes (Fig. 1).

Tyrosinase is a copper-containing enzyme that can catalyse two distinct reactions, the orthohydroxylation of a monophenol (hydroxylase activity) and the conversion of an *o*-diphenol to the corresponding *o*-quinone (oxidase activity). The two enzymatic activities are intimately coupled and are also referred to as cresolase or phenolase and catecholase of diphenolase activities, respectively (Fig. 2). Typical substrates of tyrosinase are mono- and *o*-diphenols (catechols) devoid of bulky groups or crowded substituent patterns adjacent to the hydroxyl group(s) on the aromatic ring. Traditionally, studies on the biosynthesis of melanins have showed that all steps following the tyrosinase-catalyzed formation of dopaquinone are envisaged as proceeding more or less spontaneously, without any further enzymatic assistance. There are, however, a number of biological observations suggesting that tyrosinase, though essential for melanogenesis to occur, is not the sole requirement (16–18). Once dopaquinone is generated, its metabolic fate is ultimately dependent upon the sulfhydril content within melanocytes. Thus, by implication, enzymes that influence the redox state of the glutathione system can indirectly affect the level of melanization, even if tyrosinase is normally expressed (19–20). In addition, evidence has been accumulating indicating the existence of a number of other enzymes and cofactors that can play a critical role in the later stages of melanogenesis. There are receptors on the surface of melanocytes for melanocyte-stimulating hormone (MSH). MSH, ACTH (adrenocorticotropin, similar to MSH in the arrangement of first 13 amino acids), estrogen, androgens, and progesterone all

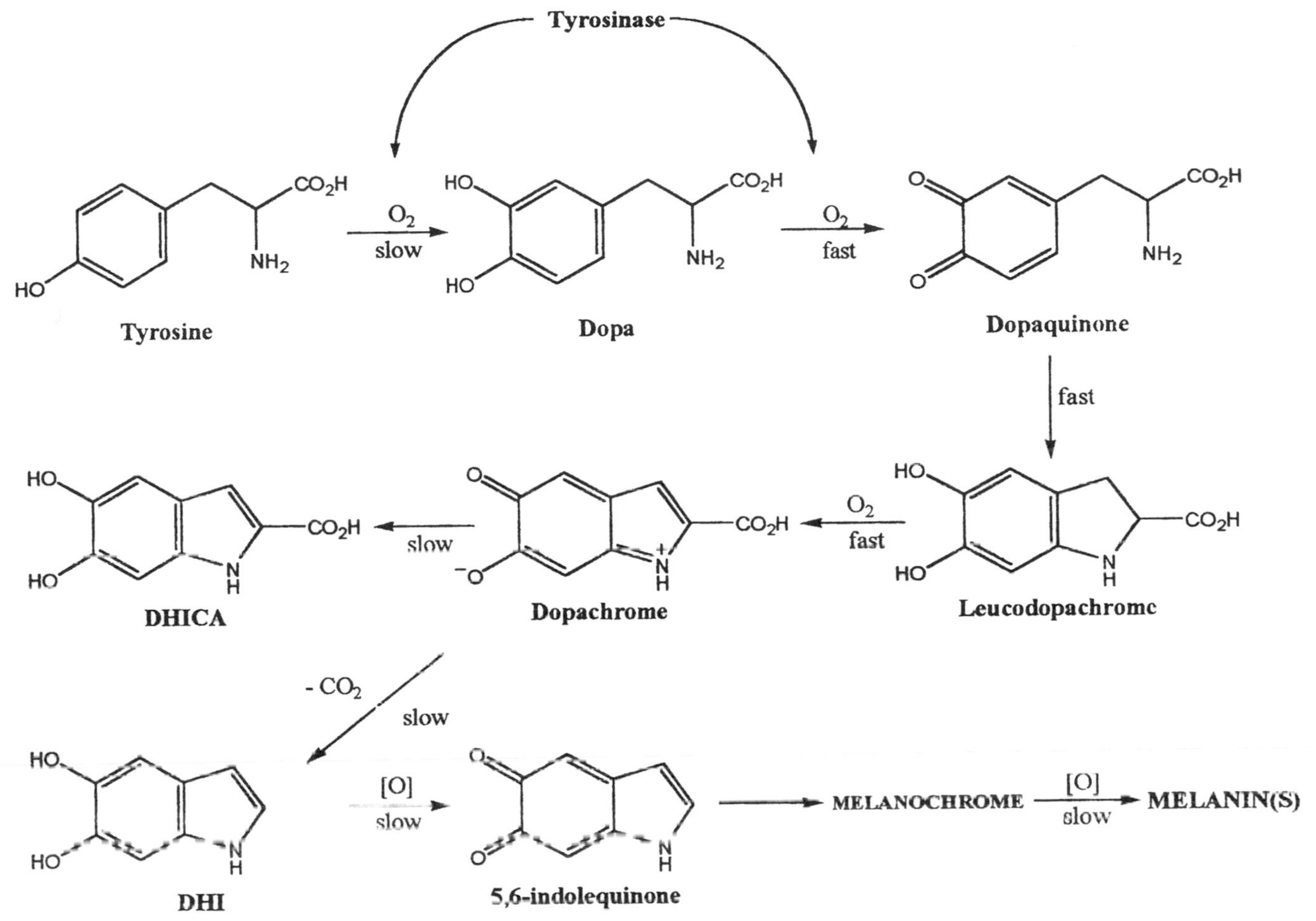

Figure 1 Scheme of melanogenesis.

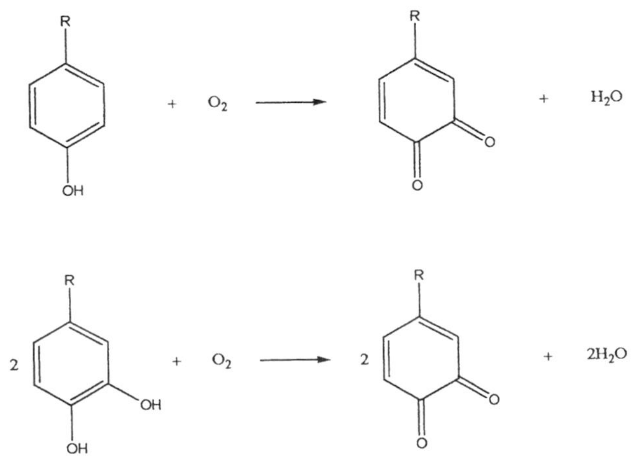

Figure 2 Enzymatic activity of tyrosinase.

stimulate pigmentation through increasing cAMP and tyrosinase activity, resulting in increased melanin formation and transfer.

b. *Photobiology and Photochemistry of Melanogenesis*

Exposure of human skin to solar radiation or UV light from artificial sources results in a profound alteration of metabolism, structure, and function of epidermal melanocytes. The visible outcome is a marked increase in skin pigmentation, commonly known as tanning (21). When viewed from the standpoint of photobiology, skin tanning involves two distinct phenomena: (1) immediate tanning (IT), sometimes referred to as immediate pigment darkening (IPD); and (2) delayed tanning (DT). This latter is but one of a complex cascade of events beginning with cell injury and leading to inflammation (erythema) and enhanced keratinocyte proliferation (22). The end result is an increased protection of skin against the harmful effects of UV radiation. Multiple exposures further increase this protection, but excessive doses of UV radiation result in the long term in a number of pathological changes in skin, such as atrophy, aging, actinic keratoses, and development of many cutaneous cancers, for example, squamous cell carcinoma and basal cell carcinoma. A number of authors have also expressed concern that UV exposure might be responsible for the increased incidence of malignant melanomas (23–24). However, this view has not reached general consensus (25–26).

Our incomplete knowledge of skin photobiology makes it difficult to attempt to explain individual susceptibility to the harmful effects of UV radiation. What seems clear is that skin with high melanogenic activity, be it constitutive or UV induced (facultative), is by far less susceptible to actinic damage than white unpigmented skin. Thus, for example, albinotic and vitiliginous skins, which lack melanin, sunburn easily and exhibit little or no ability to tan. Redheads also show a pronounced sensitivity to sunlight, but whether this results from the presence of pheomelanins in the skin is controversial.

Kaidbey et al. (27), using spectroscopic techniques, found that the transmission and protection factors of the stratum corneum of black and white skin toward UVB rays (280–320 nm) are quite similar, but the dark pigmented epidermis transmits much less and protects much more than the white epidermis. Thus, since the stratum corneum and epidermis are of essentially equal thickness in all races, it follows that melanin is the most important skin photoprotective agent. However, the picture is not so simple, since other products of melanocyte activity may be implicated to account for the apparent relationship between skin color and photoprotection.

c. *Skin Phototypes*

The concept of sun-reactive skin typing was introduced by Fitzpatrick (28) for the specific purpose of selecting the correct initial doses of UV light for photo- and chemotherapy. Based on the individual's tendency to sunburn and tan, six main phototypes were recognized. In the proposed classification, persons of skin types I and II have less constitutive pigmentation and less capacity to tan than individuals with skin types III, IV, or V, in that order, while type VI is represented by African and American blacks. Another classification, proposed by Cesarini (29), takes also into consideration some genetic markers, such as freckles and hair color, which allow a more precise definition of skin phototypes.

Though these classifications serve the purpose for Caucasian subjects, they do not work well when applied to darker populations. For example, in Japan, it is necessary to define a local skin typing system that allows for a burning reaction to UV in the minimal range and a tanning reaction in the moderate to intense range (30). Currently, when a more objective determination of sensitivity of skin to UV radiation is desired, the usual method is measurement of the minimal erythemal dose (MED), as evaluated either visually or more accurately by chromameters (31). In terms of MED, it has been estimated that the protection offered by melanin pigmentation reaches a maximum value of 10–15 for very black individuals, whereas for Hispanics or Kuwaitis or dark mediterraneans it reaches a value of 2.5 (32). Yet within each skin there may be a wide range of variability in the photobiological response to UVB, suggesting the involvement of another variable that has not been identified so far.

d. *Melanin, Melanogenesis, and Skin Photoprotection*

The role of melanin in photoprotection is either obvious or wholly obscure. Certainly, pigmented skin is by far less susceptible than white skin to sunburn and phototoxic reactions. However, whether this is solely related to the epidermal content of melanin or other products of melanocyte activity is not clear. Confusion also exists as to the actual mechanism by which melanin would provide photoprotection.

A widely accepted view, which was first proposed in 1820 by Sir Everald Home, is that melanin absorbs UVR and therefore acts as a passive screening filter to protect the underlying tissue against actinic damage. Advocates of this view often refer, as supporting evidence, to the umbrellalike distribution of the melanin granules over the keratinocytes and the increased skin photoprotection following suntanning. However, if we analyze the epidermal response to UVR without prejudices, we find that it has all the features of a nonspecific hyperplasia in response to damage and death of cells, as hyperpigmentation of skin occurs in response to a variety of other traumas such as ionizing radiation, heat, abrasion, and inflammation. In any case, suntanning comes about too long after the UV radiation has done its worst damage. In passing, it may be worth noting that black people tend to avoid sunning because they are not exempt from sunburn, although their skin is heavily pigmented (33). Kaidbey and Kligman (34) exposed eight volunteers to UVA twice a week for 8 weeks and compared the photo protection 1 week after the last irradiation against UVA and sunlight. Although much pigment was generated, the minimal protection against a MED was only 2 to 3 times higher with respect to control unpigmented skin.

These observations and others in both human and animal models (35) suggest that melanin alone may not be such a good photoprotective agent. In fact, when compared with conventional sunscreens, natural or synthetically produced eumelanins appear to have little effectiveness in protecting skin from UVR exposure (L. J. Wolfram, unpublished). It could be argued, however, that a melanin-containing cream or lotion applied over the skin is far from a reproduction of the potential photoprotection afforded by the strategic distribution and aggregation of melanosomes in the epidermis. In any case, a role of melanin as a neutral sunscreen is unlikely and would not be in keeping with the insoluble character of the pigment granules, giving rise to scattering rather than absorption of light.

An alternative interpretation of the photoprotective role of melanins comes from recent studies, discussed earlier, showing that melanin is by no means an inert passive material, as previously believed, but a rather unstable chemical entity exhibiting an impressive range of physicochemical properties. Among these, perhaps the most relevant from the physiological viewpoint is the ability to scavenge oxygen-derived radicals, such as superoxide anion and hydrogen peroxide, which are normally formed during biochemical or photochemical processes. Both species are known to produce several biological effects, most of which are deleterious to cell homeostasis. While host defenses are adequate to remove these toxic species under physiological conditions, on exposure to UVR or other inflammatory stimuli their concentration increases, and thus the presence of an in situ residing quencher such as melanin becomes important.

In this context, the possible contribution of other products of melanocyte activity should not be overlooked. In a preliminary study at Naples (G. Prota et al, unpublished), it was found that 5,6-dihydroxyindole (DHI) is capable of inhibiting lipooxygenase-induced oxidation of arachidonic acid, the major unsaturated fatty acid present in mammalian phospholipids. DHI and 5,6-dihydroxindole-2-carboxylic acid (DHICA) are also endowed with excellent antioxidant properties and are capable of capturing oxygen radicals, themselves being eventually converted to melanin. Viewed in this way, melanin pigmentation could be regarded as a visible indicator of the ability of DHI and related colorless metabolites to counteract noxious effects induced by internal or external stimuli, including UV light. A further contribution to this picture follows from several studies showing that 5,6-dihydroxyindoles and other melanogenic precursors for example cysdopas, undergo photodestruction when irradiated with biologically relevant UV light (i.e., with wavelengths greater than 300 nm). As a result of their photochemical reactivity, a number of free-radical species can be generated. Whether such photodegradation products of melanin precursors are formed in vivo is not yet documented; however, in vitro studies suggest that they may play a role in the acute and chronic response of human skin to sunlight.

4. Functions of the Skin

The most obvious function of skin is to create a barrier between humans and their environment. The barrier to infectious organisms and toxic chemicals is created by the surface cells of skin, keratinocytes, which divide slowly but continuously throughout life. After cell division, keratinocytes move to the surface where they flatten into a very tough and durable membrane. As these cells slowly come off of the surface, they are replaced by the underlying dividing cells.

Establishing a barrier to ultraviolet light is a special requirement of skin. In fact, ultraviolet light from the sun has the capacity to cause skin cancer and to produce accelerated skin aging. One of the defense mechanisms against ultraviolet light in skin is the elaboration of protective melanin by melanocytes.

a. *The Role of Lipids in the Structure and Junctional Organization of the Horny Layer*

The skin is structured to prevent loss of essential body fluids, and to protect the body against the entry of toxic environmental chemicals. In the absence of a stratum corneum we would lose significant amounts of water to the environment and rapidly become dehydrated. The stratum corneum, with its overlapping cells and intercellular lipid, makes diffusion of water into the environment very difficult (7–8).

In the horny layer the intercellular spaces contain formations of overlayed lamellae derived from the lamellar granules localized close to the cell membrane of the living cells below. When the granule membrane fuses with the cell membrane by the process of exocytosis, the contents of the granules are expelled into the intercellular spaces. Initially they are arranged in discoidal form deposited between the cells of the horny layer, thus being transformed into wide lamellae in the terminal phase of the keratinocyte differentiation process (9).

The following play an important role in the assembly of the granules:

Acylglucosyl ceramides, during the dispersion of the lamellar discs, mediated by the removal of the saccharide residue

A Ca-dependent phospholipase, which hydrolyses the phospholipids present in the granular layer as substrate

The concomitant increase of fatty acids in the zone, which separates the granular layer from the horny layer

Also the cholesterol esters have a functional as well as a structural role. In fact, the desquamation of horny layer cells, which constitutes the final act of epidermal differentiation, is accompanied by hydrolysis of cholesteryl sulphate.

The epidermal lamellar bodies, rich in glycosphingolipids, sterols, and fatty acids, are the source of all the intercellular lipids of the horny layer. These ellipsoid organelles are assembled in the spinous and deep granular layers. In the outer areas of this latter layer, the lamellar bodies arrange themselves along the lateral and apical surfaces so that they can be quickly exocytosed. Their secretion brings about not only the release of the lipids they contain but also the dispatch of numerous hydrolytic enzymes into the intercellular spaces of the horny layer (acid lipases, phospholipases, sphingomyelinases, glycosphingohydrolases, acid phosphatases, and carboxypeptidases).

Consequently, in the granular and horny layers an intense metabolic activity takes place on the complex lipids present in the lamellar bodies and initially ''deposited'' in the intercellular spaces. This accounts for the structural changes and qualitative and quantitative variations in the composition of the epidermal lipids (Table 1).

The ceramides constitute almost 50% of the total lipids present in the horny layer. Several molecular species have been identified, different as regards their hydrophobic sections: sphingosine, phytosphingosine, and fatty acids, these last often being hydroxylated. The hydroxyl can be esterified with linoleic acid residues (which are sometimes also present in the glucose residues of glucosyl ceramides present in the cellular layers). Particular to the horny layer is a ceramide linked to a very long chain hydroxyacid (30–32 carbon atoms), in its turn esterified with a fatty acid whose major component is linoleic acid.

In conclusion, the characteristics of the horny layer and the associated lipid ''mantel'' organized in a double layer can be summarized as follows:

An almost complete absence of phospholipids

The structural and functional substitution of phospholipids (as the essential fatty acid reserve) by neutral glycolipids

Large quantities of nonesterified cholesterol, which by esterification reactions allow the regulation of lipid layer fluidity

The presence of free fatty acids, many of which have peculiar structures not detectable in other organs and tissues

The need for both saturated and unsaturated fatty acids to be present in a precise ratio in order to ensure the optimal barrier effect

The presence of water, therefore, making it more appropriate to talk of a hydrolipid barrier and of a lamellar structure composed of various layers

Table 1 Lipidic Composition of the Epidermis (% Distribution)

Class of lipid	Layers with nucleate cell	Horny layer
Phospholipids	40	traces
Sphingolipids	10	35
Cholesterol sulphate	—	2–5
Cholesterol	15	20
Triglycerides	25	traces
Fatty acids	5	25
Sterol esters, squalene	5	10

The skin is also part of the innate immunity (natural resistance) of the body against invasion by microorganisms. The dryness and constant desquamation of the skin, the normal flora of the skin, the fatty acids of *sebum* and lactic acid of sweat, all represent natural defense mechanisms against invasion by microorganisms. Langerhans cells present in the epidermis have an antigen-presenting capacity and might play an important role in delayed hypersensitivity reactions. They also play a role in immunosurveillance against viral infections. Langerhans cells interact with neighboring keratinocytes, which secrete a number of immunoregulating cytokines.

Melanin pigment of the skin protects the nuclear structures against damage from ultraviolet radiation.

The skin is also a huge sensory receptor for heat, cold, pain, touch, and tickle. Parts of the skin are considered as erogenous zones. The skin has great psychological importance at all ages. It is an organ of emotional expression and a site for the discharge of anxiety. Caressing favors emotional development, learning, and growth of newborn infants.

The skin is a vital part of the body's temperature regulation system, protecting us against hypothermia and hyperthermia, both of which can be fatal (specialized vascular structures of the dermis, insulation by fat in subcutaneous tissue, evaporation of sweat).

The skin plays an important role in calcium homeostasis by contributing to the body's supply of vitamin D. Vitamin D3 (cholecalciferol) is produced in the skin by the action of ultraviolet light on dehydrocholesterol. It is then hydroxylated in the liver and kidneys (needs parathyroid hormone to activate alpha-hydroxylase) to 1,25 dihydroxycholecalciferol, the active form of vitamin D. This antirachitic vitamin acts on the intestine, increasing calcium absorption (through stimulation of the synthesis of calcium-binding proteins in the mucosal cells of the intestine), as well as on the kidneys, promoting calcium reabsorption.

b. *Keratin*

Electron-microscopic examination of cells from all tissues reveals that they contain a complex, heterogenous, intracytoplasmic system of filaments. The components of this system include actin, myosin, and tubulin, whose diameters average approximately 60Å, 150Å, and 250Å, respectively. In addition, other intracytoplasmic filaments were noted, and since the diameter of these latter structures was found to be between 70 and 100Å, they were called intermediate filaments (10–12).

Intermediate filaments form a major part of the cytoskeleton of most cells and fulfill a variety of roles related to cell shape, spatial organization, and perhaps informational transfer. The nucleus contains structures related to these intermediate filaments, and many intracellular components including polyribosomes, mitochondria, nucleic acids, enzymes, and cyclic nucleotides are attached to the cytoskeleton.

Based on their biochemical, biophysical, and antigenic properties, a number of classes of intermediate filaments can be recognized in different cell types: desmin (skeletin) in muscle cells, glial fibrillary acidic filaments in glial cells, neurofilaments in neurons, vimentin in mesenchymal cells, and keratin in epithelial cells. In cultured epidermal cells, keratins account for up to 30% of the cellular protein, while in the stratum corneum, keratin accounts for up to 85% of the cellular protein.

At least 19 keratin proteins can be identified ranging in molecular weight from approximately 40,000 to 68,000 micrograms. The polypeptide structures of all the intermediate filaments have a similar skeletal part composed of structural blocks of polypeptide subunits. The number of subunits varies between one and 30. The proteins are products of separate genes, which by cross-hybridization fall into two gene subfamilies: type 1 (basic) and type 2 (acidic). The molecular weights of the members of one (the basic subfamily) are larger than those of the members of the other (the acidic subfamily).

The fundamental polypeptide chain of keratin is a classical alpha helix with repeated groups of seven aminoacids. It is composed of four separate helical zones linked by interhelical sequences and is preceded and followed by nonhelical carboxyterminal and aminoterminal sequences. During keratinization two polypeptide chains (one basic and one acidic) are assembled as a heterodimer called a protofilament. Two protofilaments then assemble into larger protofilaments so forming a protofibril, the basic element of the keratin fiber heterodimer. In epithelial cells, keratin filaments radiate from the perinuclear region to the internal face of the plasma membrane. Bundles of these filaments are associated with specialized structures of the plasma membrane called desmosomes. Looking at their chemical structure it is evident that keratin filaments are heterogeneous. This heterogeneity explains how the filaments perform diverse functions in different tissues, i.e, there are different types of keratin for keratinized epidermis, hyperproliferative epidermis of palms and soles, corneal epithelium, stratified epithelium of the esophagus and cervix, and simple epithelium of the epidermal glands. As mentioned before, keratin is the main structural protein of the epidermis (13). The keratinocytes in the basal layer and prickle cell layer synthesize keratin filaments (tonofilaments), which aggregate into bundles (tonofibrils). In the cells of the stratum corneum, these bundles of keratin filaments form a complex intracellular network embedded in an amorphous protein matrix. The matrix is derived from the keratohyaline granules of the granular layer; filagrin is the substance produced by keratohyalin granules, which links the tonofilaments and allows their alignment into microfibers. Epidermal keratinization results in the production of a barrier that is relatively impermeable to substances passing into or out of the body (14,15).

C. Ultrastructure of the Dermoepidermal Junction

The most superficial component of the junction is the basal plasma membrane of keratinocytes, melanocytes, and Merkel cells.

Hemidesmosomes have a complicated ultrastructure that resembles half a desmosome. Hemidesmosomes consist of an intracellular component, the attachment plaque, which is associated with tonofilaments, and an extracellular component, known as the subbasal dense plate. This latter structure is located in the lamina lucida (see below) and resembles a fine, dense line parallel to and just beneath the plasma membrane. Hemidesmosomes are important in maintaining adhesion between dermis and epidermis.

Immediately beneath the basal plasma membrane is the basement membrane, which consists of three layers: the lamina lucida, the lamina densa, and the lamina fibroreticularis (sublamina densa).

Distributed throughout the lamina lucida are anchoring filaments. Anchoring filaments are very fine structures that are oriented vertically between the lamina densa and the basal plasma membrane.

The lamina densa is an electron-dense amorphous layer that lies parallel to and below the lamina lucida.

Anchoring fibrils are the major constituent of the fibroreticular layer of the basement membrane. These are short, often curved structures, with an irregular cross-banding, that insert into the lamina densa and extend into the upper part of the dermis. They may also insert into amorphous bodies in the superficial dermis known as anchoring plaques, or curve back to have a second insertion in the lamina densa.

Another component of the lamina fibroreticularis is the elastic microfibril bundles, each consisting of many microfibrils that extend into the dermis and may enmesh with the microfibrillar system of dermal elastic fibers.

Laminin, a high molecular weight glycoprotein required for cell adhesion, and fibronectin have been immunolocalized to the lamina lucida.

Type IV collagen antigen has been immunolocalized to the lamina densa. Type VII collagen antigen has been immunolocalized to anchoring fibrils and plaques. Type VII collagen has a role in normal dermoepidermal adherence.

D. Dermis

The dermis accounts for approximately 90% of the cutis. It can be considered an aqueous gel ''held'' in a structure composed of collagen and elastin fibers. Compared to the relatively rapid turnover of the epidermis, the connective tissue of the dermis is regenerated more slowly even though it is more highly vascularized, being pervaded by a woven network of arterioles, venules, and capillaries. Like all differentiated connective tissues, it is composed of an abundant intercellular matrix and specialized cells or fibroblasts. Other cell types found in the dermis are the macrophages and mast cells, and in addition lymphocytes and plasma cells frequently populate the dermis in response to injury or other stimuli.

1. Principal Function of the Cells Present in the Dermis and the Characteristic of the Molecules They Synthesize

Fibroblasts are bound to the matrix by fibronectin, and their main function is the synthesis and remodeling of connective tissue proteins. The remodeling process is mediated by specific enzymes, collagenase and elastase, whose presence is particularly high in the dermal papillae found immediately below the basal lamina and which therefore delimit the epidermal region. The main product of fibroblast synthesis in the papillary layer is type III collagen, characterized by high hydroxyproline content and a small percentage of hydroxylysine, which has few glycosylation sites. Type I collagen is synthesized by fibroblasts of the underlying reticular dermis, organized in large fibrillary bundles interspersed by bundles of mature elastin fibers. Macrophages represent the final stage of differentiation of progenitor cells, which originate in the bone marrow and mature into monocytes in the blood circulation. Their principal function is the processing and presentation of antigens to the immunocompetent lymphoid cells. They synthesize and secrete interleukin 1, growth factors, prostaglandin, interferon, as well as components of the complement system and diverse hydrolytic enzymes. Mast cells are present in all layers of the dermis but are found in highest concentrations around the blood vessels. They synthesize granules, which contain vasoactive or chemotactic factors for the neutrophils and eosinophils that are released in response to stimuli of various kinds. These cells are able to respond to stimuli such as light, cold, acute trauma, vibrations, variations in pressure, and other stimuli of chemical and immunological nature.

2. Biosynthetic Processes of the Dermis

The main classes of molecules biosynthesized by the dermis are collagen and elastin, which constitute the fibrous proteins; and proteoglycans and structural glycoproteins, which are the main components of the ground substance.

The timing of biosynthesis of these molecules is different but occurs in a manner coordinated with the morphogenesis of the dermis fibrous network and has the following characteristics:

1. The first characteristic is determined by the affinity ''coded'' within the structure of the classes of macromolecules mentioned previously and in particular by the structural glycoproteins, which govern the interactions between fibroblasts and collagen fibers, and by proteoglycans, which determine the three-dimensional arrangement of the collagen fibers themselves.
2. Considerable changes are seen with age following modifications in the biosynthetic program, alterations in the relationship between synthesis/degradation and under the influence of hormonal, environmental, and nutritional factors.

3. Biochemical-Molecular Aspects of Biosynthesis of Collagen and Elastin

Collagen is an extracellular protein, but it is synthesized intracellularly in the form of a precursor, which undergoes extensive posttranslation modifications before becoming a mature collagen fibril. The molecule initially synthesized at the ribosome is preprocollagen (37). The main metabolic steps that lead to the synthesis of mature collagen and the intra- and extracellular sites where it occurs are schematically reported in Table 2. Elastin is present in the skin in lower concentrations than collagen and is the principal component of elastic fibers. It is synthesized and ''refined'' by a process essentially similar to that described above. The amino acid composition of elastin is highly analogous to that of collagen; it is rich in glycine and proline residues, although it contains little hydroxyproline, no hydroxylysine, and a high quantity of alanine and other nonpolar aliphatic amino acids.

Mature elastin contains many cross-bonds due to numerous lysine-containing domains separated by two-three alanine residues that can give rise to two types of cross-links:

1. Condensation of an oxidated residue, with an unmodified residue (norleucine–lysine)
2. Condensation of four different residues—three oxidated and one nonoxidated residue to give a compound called desmosine (Fig. 3).

These cross links are typical of elastin and allow the elastic fibers to regain their proper form and dimension after any stretching that they may be subjected to.

Elastin biosynthesis must be coordinated with that of the structural glycoproteins, as the two classes of molecules associate to give the ''young'' elastin fibers rich in microfibrils. As the process of maturation proceeds, the concentration of polymerized elastin increases following the cross-linking of tropoelastin chains and desmosine residues.

From a biomolecular point of view, the genes that code for these molecules constitute a many-gene family. In the case of collagen, for example, about 20 genes have so far been identified that code for at least 11 different types of

Table 2 Sequence and Localization of the Collagen Synthesis Process

Nucleus and ribosomes	
Site of process	Functional significance
Gene selection	Determines the type of protein that will be synthesized
Transcription	Formation of primary mRNA transcript
Processing	Formation of final mRNA
Translation	Formation of polypeptides and the initiation of their assembly
Control of expression	Determines the quantity of polypeptides synthesized
Endoplasmic reticulum and Golgi apparatus [cotranslational and posttranslational modifications of repetitive sequences (Gly-X-Y)n]	
Site of process	Functional significance
Removal of signal peptide	Necessary for secretion into the extracelluar space
Hydroxylation of 4-proline residues	Still not clear but probably the signal for subsequent hydroxylations
Hydroxylation of 3-proline residues	
Hydroxylation of lysine residues	Necessary to mark lysine residues for glycosylation
Glycosylation	
Formation of intra- and interchain S–S bonds	Start of tertiary structure assembly
Formation of procollagen triple helices	
Extracellular environment (further posttranslational modifications)	
Site of process	Functional significance
Proteolysis of N and C terminal fragments	Elimination of alignment peptides and start of final assembly
Oxidative deamination of lysine, unglycosylated and glycosylated hydroxylysine	Prerequisite for the formation of cross bonds
Formation of interchain cross bonds mediated by aldonic condensation and by the formation of Schiff's bases	
Degradation by specific collagenases	Removal of old damaged extracellular fibers

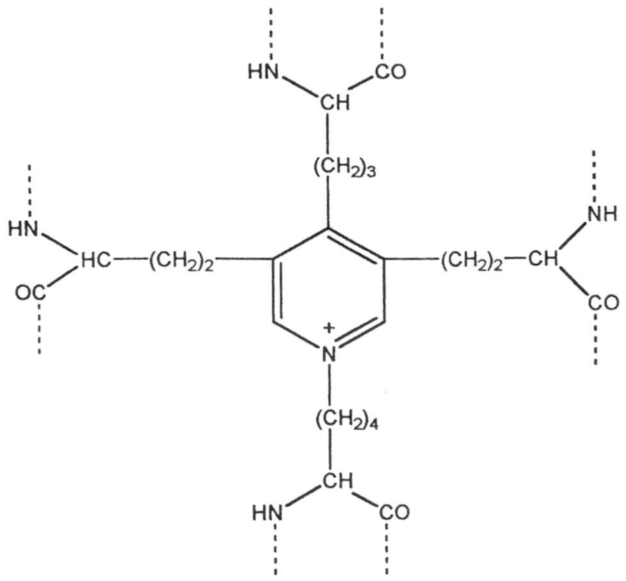

Figure 3 Structure of desmosine.

collagen protein. Observed structural differences reflect the different functions that the collagens perform in the various tissues and cell sections in which they are found: tendons and ligaments, the skin, renal glomeruli, bones, and teeth.

At least two pertinent molecular biological problems are still to be clarified:

1. How the genes necessary for the synthesis of a certain type of collagen in a particular cell are selected (in other words, how cells of a given tissue produce a structurally defined type of collagen)
2. How the expression of these genes is regulated (in other words, has a given cell the capacity to produce more than one collagen protein, and it only expresses one of them as a function of time, or does the synthesis of a certain type of collagen contemporaneously inhibit the production of other chains?)

Concerning elastin, the domains in which the cross-links are concentrated are rich in glycine, proline, and valine and are highly conserved structures easily forming beta spirals. Elucidation of the three-dimensional structure of these regions in relation to the elastic characteristics of the protein and the control of expression of the genes coding for the sequences mentioned above are still objects of in-depth research.

4. Fibrous Proteins and Their Degradation: Collagenases and Elastases

From a metabolic point of view collagen is not an ''inert'' protein. In an individual's lifetime collagen undergoes processes of maturation, remodeling, cicatrization and aging, processes in which the mature and/or aged fibers are progressively substituted by young fibers. The degradation begins with the intervention of specific proteases, which transform polymeric collagen into monomers, which are then hydrolyzed by collagenases. Collagenases, whose activity is regulated by specific protein factors, are derived from an inactive precursor by the proteolytic removal of the amino terminal portion.

Vertebrate collagenases are metal-enzymes containing Ca^{2+} and Zn^{2+} and are synthesized and secreted by fibroblasts, leucocytes, polymorphonucleates, macrophages, cutaneous epithelial cells, and condrocytes. Their synthesis is controlled by hormones (parathyroid hormone is a positive effector in contrast to progesterone and hydrocortisone). Hormonal action is probably mediated by a specific prostaglandin and other factors, heparin for example.

The situation with elastin is different: the thin elastic fibers present in the subpapillary region disappear at about 40 years of age, when the first lysis of elastin appears. With advancing age, the dermis fibroblasts secrete large quantities of elastases, which degrade the fibers of both the dermis and the epidermis, fibers that cannot be replaced because the activity of the lysyl oxidases diminishes drastically.

5. Ground Substance Biosynthesis and the Role of Structural Proteins, Glucosaminoglycans, and Proteoglycans

As previously stated, the structural glycoproteins (GPS) are synthesized in situ by mesenchymal cells of the connective tissue and are found in the basal membrane and intercellular spaces, either free or associated with other matrix proteins. Some of these can be imbricated with collagen or elastin, others with proteoglycans, thus playing an important role in the three-dimensional organization of the fibrous framework and in the positioning of cells, especially the fibroblasts, in the body of the matrix itself. Therefore they are directly involved in fibrillogenesis, a process that is accelerated in vitro by the addition of GPS to a tropocollagen solution. Given their high affinity for lipids, it is likely that the GPSs have an important role in the aging process, in a characteristic way in the case of diabetes and atherosclerosis, conditions in which the dermis accumulates lipids, especially cholesterol esters.

From a chemical point of view, the GPSs contain many

dicarboxylic amino acid residues and also aromatic (triptophan), sulphurated (cysteine), and aliphatic (glycine, proline, and valine) amino acids. The saccharide content (glucose, galactose, mannose; *N*-acetyl esosamine, fucose, and sialic acid) varies from one molecular species to another. Generally they are not very soluble and are often composed of subunits linked by disulphide bridges. Their biosynthesis is not dissimilar to that of other glycoproteins.

Fibronectin is one of the best-studied GPSs synthesized by fibroblasts. It is found either associated with the cell membrane or to the collagen of the intercellular matrix. It is most probably responsible for the "anchorage" of fibroblasts to the fibrous framework. The binding both to fibroblasts and to collagen and proteoglycans is mediated by specific amino acid sequences. Fibronectin is composed of two polypeptide chains linked by disulphide bridges close to the carboxyl terminal end. The gene for this GPS contains more than 50 exons, which code for "variations" in three principal sequences. In the protein the domain, that binds to collagen possesses three modules: one that performs a different specific function (that of binding fibrin) and two with another type of amino acid sequence. The module, which contains the sequence *Arg-Gly-Asp-Ser*, seems to be responsible for binding to the intercellular matrix and/or to certain proteoglycans.

Experimental data indicate that cancer cells or cells transformed by oncogenic viruses lose the ability to fix fibronectin to their membrane surfaces, and this could be responsible for their inability to form organized tissues and their progression to malignancy.

Finally it should be mentioned that a single mRNA primary transcript, which undergoes different splicing, codes these "different" fibronectins.

Laminin is another adhesion glycoprotein of the extracellular matrix and allows the epithelial cells to adhere to the connective tissue that surrounds them. It is able to bind to type IV collagen, the characteristic component of the basal membrane. Like fibronectin it possesses different domains with specific binding functions.

The connective tissue is also rich in *proteoglycans*, which are essentially composed (approximately 90%) by polysaccharide residues. These are important in the determination of the skin's viscoelastic properties in that being negatively charged they are able to bind water and different cations. The proteoglycan heteropolysaccharide chains are formed by glucosaminoglycans (GAG) composed of repetitive disaccharide units containing an amino sugar (glucosamine or galactosamine) and a sugar containing a carboxylic or sulphate group (38).

Principal GAGs are hyaluronic acid, the chondroitin sulphates, dermatan sulphate, and heparan sulphate. In proteoglycans the GAGs are covalently linked to a peptide matrix ("core protein"). Approximately 140 protein residues are then linked at regular intervals of approximately 300 Å to a long filament of hyaluronic acid, thus forming high molecular weight complexes. The interaction with hyaluronic acid is mediated by two small peptides: one links the protein, the other links hyaluronic acid. The schematic structure of a generic proteoglycan in its monomeric form is shown in Fig. 4

The proteoglycans represent the final product of a complex "assembly" process that involves both the protein synthesis apparatus of the ribosomes and the Golgi system, where posttranslational modifications of splicing and glycosylation occur. The schematic view of the synthesis of a generic proteoglycan in monomer form, not yet conjugated with a hyaluronic acid, is shown in Fig. 7. Functionally, GAG and proteoglycans are strictly associated with tropocollagen molecules, which "control" the level of polymerization. They also contribute to the correct "alignment" of the collagen and elastin fibrils and change the chemical and physical state of the dermis from a fibrous structure to a gel containing a high percentage of water.

Hyaluronic acid itself contributes not only to this hydrating effect but also to the elastic properties of the dermis. The "viscosity" level of the gel can be reduced following depolymerization of hyaluronic acid and the subsequent "disorganization" of the aggregates. This may occur under the action of hyaluronidases, the influence of chemical agents (cysteine, ascorbic acid, and hydroqui-

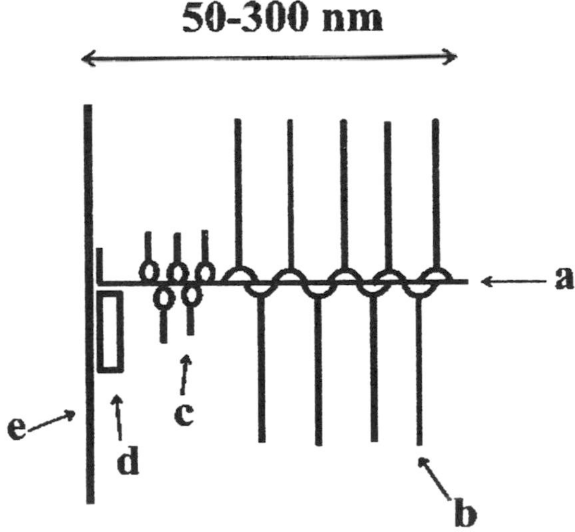

Figure 4 Schematic structure of a generic proteoglycan in its monomeric form. a, core protein; b, chondroitin sulphates; c, keratan sulphates; d, peptide linker (junction); e, hyaluronic acid.

nones), and finally due to the presence of free radicals produced by ultraviolet radiation on the skin.

E. Skin Annexes

1. Hair

Hairs grow out of tubular invaginations of the epidermis known as follicles, and a hair follicle and its associated sebaceous glands are referred to as a pilosebaceous unit. Hair follicles extend into the dermis at an angle. A small bundle of smooth muscle fibers, the arrector pili muscle, extends from just beneath the epidermis and is attached to the side of the follicle at an angle. Arrector pili muscles are supplied by adrenergic nerves, and are responsible for the erection of hair during cold or emotional stress (gooseflesh). The sebaceous gland is attached to the follicle just above the point of attachment of the arrector pili (36).

At the lower end of the follicle is the hair bulb, part of which, the hair matrix, is a zone of rapidly dividing cells that is responsible for the formation of the hair shaft. Hair pigment is produced by melanocytes in the hair bulb. Cells produced in the hair bulb become densely packed, elongated, and arranged parallel to the long axis of the hair shaft. They gradually become keratinized as they ascend in the hair follicle.

The average rate of growth of human scalp hair is 0.37 mm per day. In women, scalp hair grows faster and body hair grows more slowly than in men. The rate of growth of body hair is undoubtedly increased by androgens, since it can be reduced by treatment with antiandrogenic steroids.

2. Hair: Types and Growth Cycle

The type of hair produced by any particular follicle can change. The most striking example is the replacement of vellus by terminal hair at puberty, which starts in the pubic regions. This leads us to the definition of androgen-dependent hair. It is obvious from the events of puberty that pubic, axillary, facial, and body hair are hormone dependent. So, paradoxically, is pattern baldness (male), in which terminal hair is replaced by fine, short hair resembling vellus. The growth of the male beard depends on testicular hormones. The action of testosterone in general involves its reduction to 5 alpha-dihydrotestosterone and binding to an intracellular receptor.

The most important feature of hair follicles is that their activity is intermittent (cyclical). As the hair reaches a definitive length, it is shed and replaced by a new hair. Thus a hair follicle will pass into three stages: an active (anagen) stage, a resting (catagen) stage, and a telogen stage wherein the hair stops growing and is finally shed. In human scalp hair, the anagen stage takes about three years, the catagen stage takes three weeks, and the telogen phase takes three months. The hair cycle occurs in different hair follicles asynchronously, i.e., at a given time, each individual hair follicle is at a different stage of the hair cycle.

3. Nails

The nail acts as a protective covering to the end of the digit and assists in grasping small objects. The nail has also a cosmetic function. The major part of this appendage is the hard nail plate, which arises from the matrix (see below). The nail plate is roughly rectangular and flat in shape but shows considerable variation in different persons. The pink color of the nail bed results from its extensive vascular network and can be seen because of the transparency of the plate.

Usually in the thumbs, uncommonly in other fingers and in the large toenails, a whitish crescent-shaped lunula is seen projecting from under the proximal nail folds. The lunula is the most distal portion of the matrix and determines the shape of the free edge of the nail plate. Its color is due in part to the effect of light scattered by the nucleated cells of the matrix and in part to the thick layer of epithelial cells making up the matrix.

As the nail plate emerges from the matrix, its lateral and proximal borders are enveloped by folds of the skin termed the lateral and proximal nail folds. The skin underlying the free end of the nail is referred to as the hyponychium and is contiguous with the skin on the tip of the finger.

The nail plate is formed by a process that involves flattening of the basal cells of the matrix, fragmentation of the nuclei, and condensation of cytoplasm to form horny flat cells that are strongly adherent to one another.

4. Sebaceous Glands

Sebaceous glands are found on all areas of the skin with the exception of the palms, soles, and dorsa of the feet. They are holocrine glands, i.e., their secretion is formed by complete destruction of the cells.

Most sebaceous glands have their ducts opening into hair follicles (pilosebaceous apparatus). Free sebaceous glands (not associated with hair follicles) open directly to the surface of the skin, e.g., Meibomian glands of the eyelids, Tyson's glands of the prepuce, and free glands in the female genitalia and in the areola of the nipple.

The production of sebum is under hormonal control, and sebaceous secretion is a continuous process. Sebaceous gland development is an early event in puberty, and the prime hormonal stimulus for this glandular development is androgen. Although the sebaceous glands are very small throughout the prepubertal period, they are large at the time of birth, probably as a result of androgen stimulation in utero, and acne may be seen in the neonatal period. It should be noted that (1) sebum production is low in children; (2) in adults, sebum production is higher in men than in women; (3) in men, sebum production falls only slightly with advancing age, whereas in women it decreases significantly after the age of 50. Orchidectomy causes a marked decrease in sebum production. Therefore it can be assumed that testicular androgen maintains sebum production at high levels in men. The role of adrenal androgens is also important, especially in women where they play a contributory role in sebum production together with the ovaries.

Estrogens have a profound effect on sebaceous gland function that is opposite of that of androgens. In both sexes, estrogen administration decreases the size of the sebaceous glands and the production of sebum.

The sebum is composed of triglycerides and free fatty acids, wax esters, squalene, and cholesterol. The sebum controls moisture loss from the epidermis. The water-holding power of cornified epithelium depends on the presence of lipids. The sebum also protects against fungal and bacterial infections of the skin due to its contents of free fatty acids. Ringworm of the scalp becomes rare after puberty.

5. Sweat Glands

Generalized sweating is the normal response to exercise or thermal stress by which human beings control their body temperature through evaporative heat loss. Failure of this mechanism can cause hyperthermia and death. (See also ''Vascular System'' above).

Humans have several million eccrine sweat glands distributed over nearly the entire body surface (except the labia minora and the glans penis). The total mass of eccrine sweat glands roughly equals that of one kidney, i.e., 100 g. A person can perspire as much as several liters per hour and 10 liters per day, which is far greater than the secretory rates of other exocrine glands such as the salivary and lacrimal glands and the pancreas.

Each eccrine sweat gland consists of a secretory coil deep in the dermis and a duct that conveys the secreted sweat to the surface. The secretory activity of the human eccrine sweat glands consists of two major functions: (1) the secretion of an ultrafiltrate of a plasmalike precursor fluid by the secretory coil in response to acetylcholine released from the sympathetic nerve endings, and (2) reabsorption of sodium in excess of water by the duct, thereby producing a hypotonic skin surface sweat. Under extreme conditions, where the amount of perspiration reaches several liters a day, the ductal reabsorptive function assumes a vital role in maintaining homeostasis of the entire body.

In addition to the secretion of water and electrolytes, the sweat glands serve as excretory organs for heavy metals, organic compounds, and macromolecules. The sweat is composed of 99% water, electrolytes, lactate (provides an acidic pH to resist infection), urea, ammonia, proteolytic enzymes, and other substances.

The term apocrine glands was given to sweat glands present in the axillae and anogenital area that are under the control of sex hormones, mainly androgens. But nowadays by electron microscopy, these apocrine glands (apocrine = apical part of the cell is destroyed during the process of secretion) proved to be merocrine in nature (merocrine = no destruction of the cell during the process of secretion). The ''apocrine'' sweat of humans has been described as milky (because it is mixed with sebum due to shared ducts) and viscid, without odor when it is first secreted. Subsequent bacterial action is necessary for odor production. Unlike eccrine glands, which have a duct that opens onto the skin surface independently of a hair follicle, apocrine glands have a duct that opens into a hair follicle.

III. DERMOCOSMETIC SECTION

A. Introduction

The skin can be defined as healthy when its physical, biochemical, and structural components are all in equilibrium. The use of a cosmetic sometimes does not affect this equilibrium and has purely aesthetic functions, while at other times it can and must provide correction and prevention of changes in the skin brought about by both internal and external factors.

In a cosmetic, both the active principle and the excipient can perform this function. The active principles are all those substances or mixtures of substances used in cosmetics to obtain a defined nutritional, hydrating, stimulative, or protective action. Naturally, the use of active principles in cosmetics must be directed exclusively at the treatment of healthy skin, and when substances with pharmacological activity are used (e.g., vitamins or antimicrobials), the percentage content must be lower than the dosage utilized in the treatment of pathological changes.

B. Skin and Hydration

As water is life for the entire body, so it is for the skin. Hydration represents the most important parameter in the health of our skin. Numerous factors determine the water content of the skin, although overall it is directly related to the ambient humidity, and the skin can only retain adequate concentrations of water at a relative humidity of 60%.

The skin's factors that govern the maintenance of its hydration are epidermal lipids, surface hydrolipidic film, natural moisturizing factors (NMF), the horny layer, and organic substances (salts, amino acids, hyaluronic acid).

Epidermal lipids have a fundamental role both in binding water and in occluding the intercellular spaces. Their action is expressed in the correct structure of the lamellae in the horny layer, which otherwise would not be able to retain water.

The hydrolipidic film covers the epidermis with a thin protective layer that softens it, slows down its desquamation, and defends it against both chemical and bacteriomicotic agents. Removal of the film by excessive cleansing, for example, leads to damage of the horny layer, which desquamates and loses intercellular lipids, so opening the door to the entry of germs and harmful substances.

The NMF, are a series of substances produced by the epidermal cells and by sweating, which function to bind water both intracellular and extracellular together with the intercellular lipids. Of the NMF components the most important are pyrrolidon carboxylic acid and urea, but many other organic substances and mineral salts are part of the group.

Below the epidermis, in the dermis, there is a larger quantity of water. It is this water of the dermis that hydrates the epidermis by two principal mechanisms, sweating and transpiration (perspiratio insensibilis). The latter is essential for life; the body must continuously exchange the heat produced by the energy reactions that occur uninterruptedly in all its cells.

In the dermis, the majority of water is bound to glycosaminoglycans and proteoglycans. Sugar complexes with organic acids and proteins form them, respectively. Of particular interest is hyaluronic acid, which is able to bind many molecules of water. All substances that bind water are defined as hygroscopic, while the property of being soluble in water is defined as hydrosolubility. Hyaluronic acid is also present in the epidermis where, in addition to hydration, it is also involved in other activities connected with the life of keratinocytes.

Many specific and nonspecific hydrating substances are used in the cosmetic sector.

Hyaluronic acid (HA) is the main component of a polysaccharide family defined as mucopolysaccharides or glycosaminoglycans, which all have similar structure and behavior. The HA (39) molecule is composed of a repeating disaccharide (beta-*N*-acetyl-D-glucosamine-beta-glucuronic acid)n organized as a random coil. Its molecular weight can approach 10 million daltons. It takes on a huge volume of water of hydration that makes it a very expansive molecule. It can expand its solvent domain up to 10,000 times its actual polymer volume. Hyaluronic acid is the simplest of the GAGs, the only one not covalently linked to a core protein, and it is the only nonsulfated GAG.

Hyaluronic acid maintains skin moisture, organizes the extracellular matrix, and has an immunosuppressive function. HA also lubricates joints and maintains the necessary turgor of the eye. Understanding the metabolism of HA, its reactions within skin, and the interactions of HA with other skin components, will facilitate the ability to modulate skin moisture in a rational manner.

Recent progress in the details of the metabolism of HA has also clarified the long-appreciated observations that premature aging of the skin has two causes, chronic inflammation and sun damage caused by ultraviolet light (UV). These processes, as well as normal aging, all use similar mechanisms that cause loss of moisture and changes in HA distribution (40). UV light causes dramatic changes in the arrangement and distribution of HA in skin (41). In addition, HA has some unusual features that, at present, cannot be explained. The turnover rate of HA in the body as well as in the skin is very rapid, much of it being degraded and renewed every hour (42).

The HA molecule is critical from the onset of human development, and it promotes cell movement and cell proliferation. When HA is removed by hyaluronidases, cells are able to make contact with each other, undergo compaction or condensation, and begin to interact with each other. Cell movement slows, permitting these units to differentiate into tissues and organs (43).

Thus it seems that HA must be removed by hyaluronidases before commitment and differentiation can begin. This also documents how important HA is as a signalling and informational molecule. It is not just a sugar polymer that feels and hydrates space.

The HA molecule belongs to the class of compounds termed glycosaminoglycans (GAGs). Other members of this class include heparan, dermatan, keratan, and chondroitin sulfates. In skin, however, HA is the major GAG.

It is present in the dermis and in the epidermis (44). In the epidermis, the HA provides a means of cell motility. It creates an environment for cell movement by swelling the extracellular space. HA molecules simultaneously endow cells with motility by binding with receptors that interact with the cytoskeleton. That action initiates cascades of signal transduction events (45,46). The basal keratinocytes, in cultures, in the presence of low calcium, remain undifferentiated and proliferate rapidly. They also synthesize large volumes of HA. When calcium is added to this medium, these cells cease growing, begin differentiation, and become stratified, so that they appear like the layers of true skin epidermis. They also start making great amounts of hyaluronidase. The HA becomes degraded, and the cells remain relatively depleted of HA. The HA of the dermis is in contiguity with the body's bloodstream and lymphatic system. It is the dermal fibroblasts that make dermal HA. However, when such cells are grown in tissue culture, they do not demonstrate the calcium sensitivity exhibited by epidermal keratinocytes. Here is another example to justify the claim that dermis and epidermis are different organs, even though they lie immediately adjacent to each other.

Why 50% of HA in the body is located in the dermis is not known, other than that it serves as a control of interstitial fluid volume.

The HA of the dermis is loosely bound to tissue, in equilibrium with the rest of the body, and is easily removed by the lymphatic. On the other hand, the HA of the epidermis is tightly bound, not easily removed, and appears to be tightly regulated in conjunction with differentiation of epidermal skin cells. HA synthesis, degradation, and the attendant control mechanisms are distinct and separate in epidermis and dermis. It is necessary to begin to think about attempts to modulate HA deposition in skin, therefore, in relation to two separate compartments. Whether the desired moist, soft, and supple appearance of skin is a reflection of HA in one or the other, or possibly in both compartments, is yet to be established.

1. Implications for the Cosmetic Industry

The HA field is a rapidly growing area of research, and continued research is assured. A number of discoveries should be noted that might have impact on the cosmetics industry and among skin-care researchers. Attempts are frequently made to enhance the moisture and appearance of skin by using agents topically that stimulate HA content, increasing the level and the length of time HA is present in skin, preserving optimal chain length of this sugar polymer, and inducing expression of the best profile of HA binding proteins to decorate the molecule. Agents that stimulate HA production should be separated, distinguishing those that act upon epidermal HA and those whose target is the HA of the dermis.

The size of the HA polymer is critical in vivo. In fact, opposite biological effects can be achieved from two different chain lengths of HA. High molecular weight (HMW) HA, in the range of millions of daltons, is present in cartilage, in the vitreous of the eye, and in the synovial fluid of normal joints. This HMW material inhibits the growth of blood vessels (47).

On the other hand, HA fragments are highly angiogenic and are potent stimulators of blood vessel growth (48). Small HA fragments can also stimulate the inflammatory response. In skin pathology, for example, the accumulation of fragmented HA molecules in dermal papillae supports the growth of psoriatic lesions, which stimulate the growth of capillaries and attract inflammatory cells (49).

In cosmetics, one might consider that the size of the HA molecule might be important. However, little, if any, of the HA molecules actually get through the horny layer when applied topically. Therefore HA size may be of little importance in cosmetic preparations.

Following application to the skin, HA forms a thin, light-permeable, viscoelastic surface film. This fixes the moisture on the surface of the skin and supports the skin's natural protective mechanism. The HA regulates water balance and osmotic pressure, functions as an ion-exchange resin, and regulates ion flow. It functions as sieve, to exclude certain molecules, to enhance the extracellular domain of cell surfaces (particularly the lumen surface of endothelial cells), to stabilize structures by electrostatic interactions, and also to act as a lubricant. HA also acts as an organizer of the extra cellular matrix. It is the central molecule around which other components of the extracellular matrix distribute and orient themselves (50).

Since it is an excellent water reservoir and an ideal lubricant, HA, when incorporated in cosmetics, leads to a perceptible and visible improvement in skin condition.

The anomalous ability of HA to be both hydrophobic and hydrophilic, to associate with itself, with cell surface membranes, with proteins, or with other GAGs speaks to the versatility of this remarkable molecule.

HA as well as other glycosaminoglycans can be extracted from animals, e.g., fish cartilage, or alternatively from plants. For example, the seed extract of the St. John's bread tree (*Ceratonia siliqua, Leguminosae*) is locust bean gum, a branched galactomannane. This gum has moisturizing and skin-smoothing properties just like HA and is now in use for cosmetic preparations (Fig. 5). A sulfated heteropolysaccharide, rich in glucuronic acid, is extracted from a green seaweed (*Codium tomentosum*) and has the natural function of moisturizing agent.

Another class of molecules of cosmetic interest is that of molecules with a high affinity to keratin, that form a moisturizing, semiocclusive, protective, antiwrinkle film on the skin surface. For example, from the mulberry silkworm (*Bombix mori*) is obtained a water-soluble biopolymer, a high molecular weight glycoprotein that shows high affinity with keratin of the skin and of the hair, to form a protective film, imparting an immediate, long-lasting, smooth, silky feeling.

C. Skin and Defense System

The skin is the first line of the body's nonspecific defense system. Unlike the immune system, which defends by

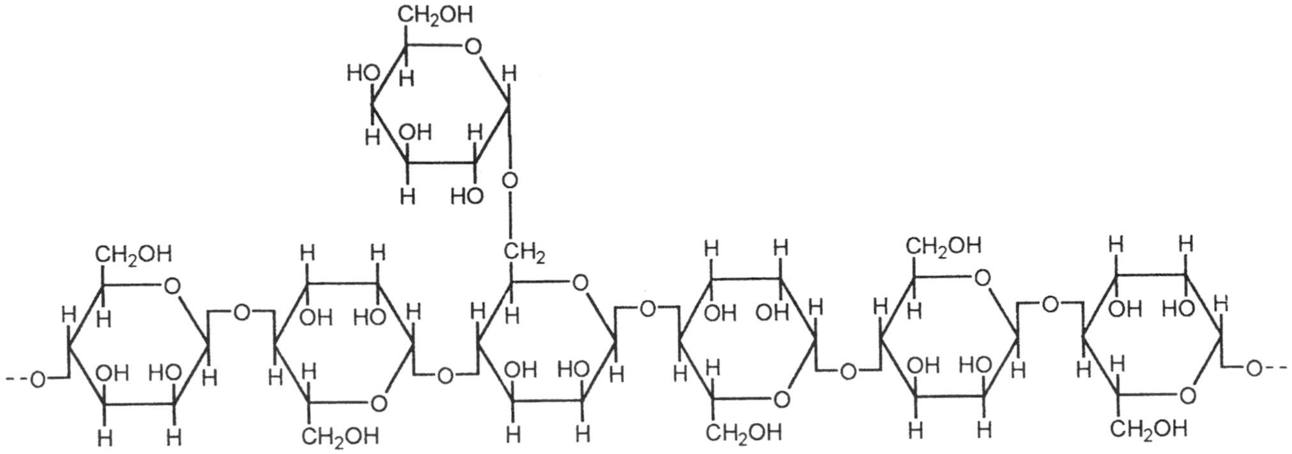

Figure 5 Chemical structure of locust bean gum.

forming antibodies against specific invaders, the skin provides a nonspecific barrier of cells and an environment that most bacteria, viruses, and other pollutants cannot penetrate (51). Skin is no different from any other organ in the body with a whole host of immune cells. Perhaps the most important of these defenses is the activity of the Langerhans cells, the skin resident macrophages. Activating the macrophages sets up a cascade reaction in the skin, resulting in fibroblast activation and the production of cytokines and epidermal growth factor (EGF), which aid the healing of wounds. In aged or wrinkled skin, increased production of EGF increases the production of collagen and elastin, thus improving the skin's appearance and causing fine lines and wrinkles to disappear (52). In the last two decades the role of *beta-glucan* in stimulating the immune system of the human body has been confirmed (53,54).

Glucans are natural polysaccharides found in oats, barley, wheat, yeast, and fungi. The term glucan refers generically to a variety of naturally occurring homopolysaccharides or polyglucoses, including polymers such as cellulose, amylose, glycogen, and starch. Glucan encompasses branched and unbranched chains of glucose units linked by 1–3, 1–4, and 1–6 glucosidic bonds that may be of either the alpha or beta type.

Polybranched beta-1,3-D-glucan is a naturally occurring polysaccharide that can be found in a variety of fungal cells including cell walls of yeast, i.e., *Saccharomyces cerevisiae*. Out of different glucans, the beta-1,3-D-glucan configuration has been shown to act as a nonspecific immune activator (55).

A macrophage cell surface receptor specific for a small oligosaccharide of the beta-(1,3)-D-glucan series has been identified (56). This receptor is a protein complex that appears to be present throughout the whole differentiation cycle of macrophages, starting in the bone marrow. Mature macrophages are found in virtually all the tissues including the central nervous system. When a macrophage encounters beta-1,3-D-glucan, it becomes activated. All the functions, including phagocytosis, release of certain cytokines, and the processing of antigens are improved and brought up to date.

1. Implications for the Cosmetic Industry

Glucan is an excellent wound healer. In experiments, glucan-treated wounds showed a higher number of macrophages in the early, inflammatory stage of repair, with fewer polymorphonuclear neutrophilic leukocytes than did control wounds. Both reepithelization and the onset of fibroplasia commenced at an earlier stage in glucan-treated wounds than in control wounds (57).

In humans, topical glucan treatment resulted in a 73% improvement in chronic decubitus ulcers with complete closure and epitalization in 27% of treated ulcers. All wounds remained clean, and no infections occurred during this treatment (58).

Considering the data above, a topical combination of an antibiotic and beta-1,3-D-glucan as an adjuvant for wound healing applications seems to be appropriate.

An interesting effect of the topical application of glucan was observed in regard to nonwounded aged skin. Revitalizing, such as reducing the number, depth, and length of wrinkles, thickening, and reducing roughness and dryness of the skin was shown in a group of female volunteers.

Applied topically, glucan activates epidermal macrophages (Langerhans cells). This mechanism plus its free-radical scavenging effect makes it a photoprotective agent. Glucan application resulted in the reduction of after-UV erythema and preservation of the amount of Langerhans cells in the epidermis (59), so that a combination of a sunscreen and glucan is suggested.

The anti-irritant effect of beta-(1,3)-D-glucan was also shown in combination with otherwise severe irritation caused by high level of lactic acid. Glucan also has a synergistic effect with another antiaging topical ingredient: retinoic acid. Similar to corticosteroids, retinoic acid significantly increases the number of beta-glucan receptor sites on phagocytic cells.

Beta-glucan of the *Schizophyllum commune* mushroom has a beta-(1,6) branch on every third glucose residue of the beta-(1,3) main chain (Fig. 6). Moo-Sung et al. (60) show that beta-(1,6)-branched beta-(1,3)-glucan from *Schizophyllum commune* is an active ingredient that can help to increase skin cell proliferation, collagen biosynthesis, and sunburn recovery. It can effectively reduce skin irritation. In addition, this beta-glucan is water soluble and stable under conditions of changing pH and temperature. The authors suggest that beta-(1,6)-branched beta-(1,3)-glucan from *S. commune* is an effective ingredient for use in antiaging and anti-irritant dermocosmetic formulations.

Carboxymethylated beta-(1,3)-Glucan (CM-Glucan) (61) enhances the renewal rate of the stratum corneum significantly compared to untreated skin (62). CM-Glucan also offers a concentration-dependent protection against UV-A irradiation. Skin pretreatment with a formulation containing 0.2% CM-Glucan showed an almost complete inhibition of lipid peroxidation compared to placebo. Protection against lipid peroxidation induced by UV-A irradiation is usually only observed by the application of antioxidants (63). Colin et al. (64) reported an inhibition of about 90% upon the application of 0.2% D-alpha-tocopherol but only an inhibition of less than 25% by the application of the cosmetically stable vitamin E acetate at the same concentration. Since CM-Glucan is neither an antioxidant nor an iron chelator, it must use mechanisms other than extracellular radical scavenging or activities related to it. CM-Glucan appears to stimulate cells to produce endogenous factors that protect the skin against oxidative stress and other environmental insults.

The pretreatment of skin with cosmetic formulations containing CM-Glucan at different concentrations showed substantial protecting effects against skin damage caused by detergent challenge. In a concentration-dependent manner, CM-Glucan protected the skin against the decrease of skin humidity and the increase of transepidermal water loss (65).

The treatment of aged skin (age of volunteers >60 years) with an oil-in-water emulsion containing 0.04% CM-Glucan clearly improved skin firmness and eye wrinkle depth even compared to the placebo control, and not

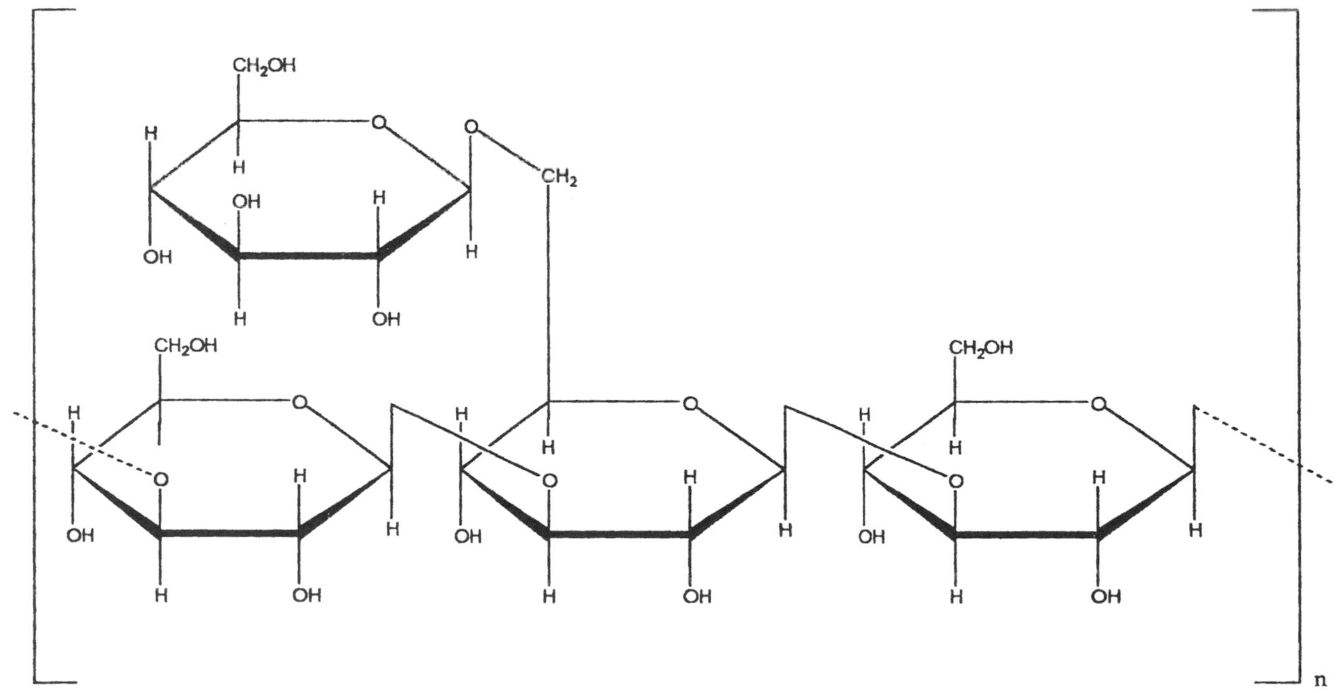

Figure 6 Structure of a monomer of beta-glucan in *Schizophyllum commune*.

only to untreated skin. However, the exact mode of CM-Glucan activity has not been fully elucidated and warrants further investigations. The increase of skin firmness is related to the reduction of wrinkles. It seems that CM-Glucan stimulates the production of skin factors, which improve the elasticity and smoothness of the skin. In addition, the film-forming properties and the UV-A protecting effects of CM-Glucan could also contribute to the observed results.

Considerable attention has been recently given to *Aloe Vera* extract, which was found to carry one of the macrophage activating polysaccharides, mannan or polymannose. While mannan has some macrophage activating potential, it is very slight compared to that of glucan (66).

D. Biopolymers as Hydrocolloids

Gums, hydrophilic colloids, mucilages, and water-soluble polymers are but a few designations for materials that have the ability to thicken or gel aqueous systems (67,68). These materials were first found in exudates from trees or bushes, or were extracts from plants or seaweeds, flours from seeds or grains, gummy slimes from fermentation processes, and many other natural products. In more recent times, new and modified gums have been made by chemical modification and derivation of many of the natural gums. In addition, some very new gums were developed by complete chemical synthesis to yield new polymers having completely new hydrophilic properties. The origin of these hydrocolloids is shown in Table 3, and a classification of natural and synthetic polymers is shown in Table 4.

Natural gums have a long history of usage. Many have been used in food applications, so their safety is well documented. There are several drawbacks to this product category: a standardization problem due to their natural origin (i.e., the difficulty in predicting viscosity buildup), possible contamination by microorganisms and environmental pollutants, their natural color and odor (which may bring formulation problems), and the great number of cellulose derivatives with different structures and qualities (that make the proper choice difficult), and possible incompatibilities with other ingredients of the formulation (69).

a. *Functional Properties*

The utility and importance of hydrocolloids are based upon their functional properties. Hydrocolloids are long-chain polymers that dissolve or disperse in water to give a thickening or viscosity-producing effect. The water-thickening property is common to all hydrocolloids; the degree of thickening varies between gums. The rheological characteristics of a hydrocolloid solution are useful aids to the selection of the proper gum for a specific formulation.

Table 3 Origins of Biopolymers

Exudates from specific trees or bushes in teardrop or flake shapes, then ground to a powder.
Extracts from specific seaweed species are made by heating with water and separating from the insoluble plant material. Then the aqueous liquid is purified, dried, and ground to a fine powder. In similar fashion, pectin is isolated as an aqueous extract of citrus fruit peel, and gelatin is extracted not only from cowhides, pigskins, or animal bones, but also from plants.
Flours separated by purely mechanical means from the cereal or plant seed. Purified or clarified material can be made by dissolving the gums in water, filtering the solution, then drying.
Fermentation or *biosynthesis* of pure polymers produced by controlled bacterial strains and substrates.
Chemical modification and derivatization of natural polymers to remove inherent limitations and deficiencies that restrict their overall utilization. While normal guar gum is quite soluble in cold water, its solubility can be greatly increased by forming the hydroxypropyl guar derivative, while simultaneously giving a greatly increased viscosity. Pure cellulose is completely insoluble in water as well as poorly absorptive in its native form. By chemical treatment to form cellulose ether compounds—such as methyl cellulose and hydroxypropyl cellulose—water solubility can be imparted, thus making a useful series of water-soluble functional polymers.
Chemical synthesis of new and better synthetic polymers.

While all hydrocolloids thicken and impart stickiness to aqueous solutions, a few biopolymers also have another major property of being able to form a gel. Gellification is the phenomenon involving the association of cross-linking of the polymer chains to form a three-dimensional continuous network that traps or immobilizes the water within it to form a rigid structure that is resistant to flow under pressure.

In addition to thickening and gelling, hydrocolloids have many other functional properties in cosmetics: they are conditioners, emulsifiers, film formers, foam stabilizers, stabilizers, suspending agents, and swelling agents.

The properties of these biopolymers are largely due to physical effects, primarily those dealing with their interaction with water.

Most hydrocolloids are polysaccharides. The physical effects of these materials derive from the interaction of the polysaccharide molecules both with themselves and with the molecules in their environments. Environmental molecules, with which polysaccharides may interact, range from protein molecules to provide suspension and solution stability, to lipid molecules to provide viscous and emulsification effects. In almost all uses, however, polysaccharides

Table 4 Natural and Synthetic Polymers in Cosmetics

Natural Polymers
1. *Vegetal Origin*
Starch
Cellulose and cellulose derivatives
Seaweed gum
Extracts from leaves and flowers
Chondroitin sulfate and hyaluronic acid of marine origin
Branched galactomannan derived from the seeds of the St. John's bread tree (*Ceratonia siliqua, Leguminosae*)
Sulftated heteropolysaccharide, rich in glucuronic acid synthetized by the green seaweed *Codium.*
Seaweed proteins of the alga *Ulva lactuca.*
Vegetal substitute of elastin
Glycoprotein obtained from mulberry silkworm (*Bombix mori*)
Collagen of vegetal derivation
2. *Animal Origin*
Hyaluronic acid
Glycosamminoglycans (from fish cartilage)
Glycogen
Sodium carboxymethyl betaglucan (extracted from e.g. *Saccharomyces cerevisiae*)
Collagen from bovine tendons, bones, skin, and gelatine
Elastin from bovine tendons
Casein
Synthetic Polymers
Acrylic polymers, specifically the carboxyvinyl polymers (Carbomers) (INCI)
Acrylates
Glycerylacrylates
Polyacrylamides
Polyvinylpyrrolidone (PVP). Copolimeri PVP/PVA e PVA/acido crotonico
Polyvinyl methylethers. Copolymers methylvinylether/maleic anhydride (MVE/MA). Polyvinyl quaternary
PVM/MA decadiene crosspolymer (INCI) or sodium PVM/VA decadiene cross-polymer (INCI of the neutralized molecule)

exist in an environment rich in water molecules. Thus polysaccharides must primarily and continuously react with water molecules. By such interactions, polysaccharides perform their function to provide viscosity, solution stability, suspending ability, emulsifying action, and gelation.

b. *Structure*

Polysaccharides are either branched or linear (70). The structure of the polymer is of great importance in determining its application, because the structure controls the functional properties. Linear polysaccharides generally have high viscosity, because they will come into contact with each other more easily, and increase the friction or viscosity characteristics of the solution at much lower dilutions than will highly branched molecules. Linear uncharged polymers usually tend to produce unstable solutions because of their behavior in solution. As linear molecules, such as starch amylose, move about in solution, they collide, shear off absorbed water molecules, and stick together. Other molecules come into contact in the same manner, adhere, and consequently build up, one molecule on another, so that a colloidal particle and eventually a particle large enough to precipitate from solution develops. The solubility of linear molecules can be improved by changes in the molecules that would reduce or prevent the fit or planar association of one molecule to another, or by introducing charges that, by Coulombic repulsion, would facilitate solution of the molecules and while in solution prevent their extensive attraction so as to produce a precipitate. Highly branched structures tend to give solutions with lower viscosities than those of linear molecules of equal molecular weights; however, these solutions tend to be quite stable and will not retrograde or precipitate out of solution. Highly branched molecules, because of their high affinity for water, also have excellent adhesive properties (gum arabic or amylopectin).

Native polymers, such as guar gum, have extending side chains, which do prevent close polymeric association and thus give these polymers greatly improved solubility properties in water. Cellulose and starch polymers, deficient in side chains, have been reacted with various reagents to introduce specific side chain groups and thus produce solubilization.

A classification of polysaccharides is

A. *Homopolysaccharides*

Linear—Polysaccharides formed from a single sugar unit and having a straight chain (unbranched) structure. These substances are most often found in structural or cell wall materials of plants and lower animals (Table 5).

Branched—Single sugar unit polysaccharides having a branched structure. For the most part, these polysaccharides serve as energy reserves and not as structural elements, although mannans are known to have structural functions. Branched homopolysaccharides usually have a

Table 5 Representative Linear Homopolysaccharides

Polysaccharides	Sugar component and linkage
Cellulose	Glucose, β 1 $\rightarrow$ 4
Amylose	Glucose, α 1 $\rightarrow$ 4
Chitin	*N*-Acetylglucosamine, β 1 $\rightarrow$ 4
Galactan (pectin)	Galactose, β 1 $\rightarrow$ 4

Table 6 Representative Branched Homopolysaccharides

Polysaccharides	Sugar components and linkages
Amylopectin[a]	Glucose, α 1 $\rightarrow$ 4, 6 $\leftarrow$ 1 α
Glycogen	Glucose, α 1 $\rightarrow$ 4, 6 $\leftarrow$ 1 α
Dextrans	Glucose, α 1 $\rightarrow$ 6, 4 $\leftarrow$ 1 α, α 1 $\rightarrow$ 6, 3 $\leftarrow$ 1 α
Galactan	Galactose, β 1 $\rightarrow$ 6, 3 $\leftarrow$ 1 β
Mannan	Mannose, α 1 $\rightarrow$ 2 and α 1 $\rightarrow$ 3, 6 $\leftarrow$ 1 α

[a] There are numerous representatives of each of these types throughout animal and plant systems. Differences will lie mainly in branching frequency, but occasionally unusual linkages or other minor sugar components are present.

linear main chain with uniform linkages with respect to position and configuration. At the branch points, new chains of the same general type are initiated with all branch points alike (Table 6).

B. Heteropolysaccharides

Linear—Polysaccharides of this group contain two or more sugar units. Mostly, they are connective tissue polysaccharides, such as chondroitin-4-sulfate and hyaluronic acid. Many carbohydrates in this group are linked covalently to proteins or lipids.

Branched—These complex polysaccharides may contain up to six different sugar units and, like the linear variety, may be covalently bound to protein or fat. The polysaccharides of this class may be ''type specific'' for microorganisms or animals (blood group substances) or may serve a recognition function for cell surfaces (Table 7).

Table 7 Representative Heteropolysaccharides

Polysaccharides	Sugar components and linkages
Pectin	Galacturonic acid, galacturonic acid methyl ester, α 1 $\rightarrow$ 4
Alginic acid	D-mannuronic acid, L-glucuronic acid[a] β 1 $\rightarrow$ 4
Hyaluronic acid	Glucuronic acid, *N*-acetyl glucosamine α 1 $\rightarrow$ 3, β 1 $\rightarrow$ 4
Chondroitin sulfate C	Glucuronic acid, *N*-acetyl galactosamine 6-*O*-sulfate, β 1 $\rightarrow$ 3, β 1 $\rightarrow$ 4
Chondroitin sulfate B (Dermatan-SO_4)	L-iduronic acid, *N*-acetyl galactosamine 4-*O*-sulfate, α 1 $\rightarrow$ 3, β 1 $\rightarrow$ 4
Keratosulfate	D-galactose, *N*-acetyl galactosamine 6-*O*-sulfate, β 1 $\rightarrow$ 3
Heparitin sulfate	D-glucuronic acid, *N*-acetyl galactosamine 6-*O*-sulfate, α 1 $\rightarrow$ 4
Heparin	D-glucuronic acid, *N*-sulfoglucosamine ester sulfate, α 1 $\rightarrow$ 4

[a] C-5 epimer of D-mannuronic. Contrast with L-iduronic in dermatan sulfate.

1. Vegetal Biopolymers

a. Gums

Gum arabic. This is the dried exudate of the acacia tree (*Acacia senegal* or related species of *Acacia* Fam. Leguminosae). Gum arabic is a complex mixture of Ca, Mg, and K salts of arabic acid (a complex, highly branched polysaccharide colloid) that contains galactose, rhamnose, glucuronic acid, 4-O-methylglucuronic acid, and arabinose residues; it shows a broad molecular weight distribution, e.g., 260,000 to 1,160,000 (Fig. 7). The gum is highly soluble in water, and solutions of up to 50% gum concentration can be prepared; the solution is slightly acidic. Physically, gum arabic is considered to be a complex, highly branched, globular molecule, which is closely packed rather than linear, thus accounting for its low viscosity. High viscosities are not obtained with gum arabic until concentrations of about 40 to 50% are obtained. Rheologically, gum arabic solutions exhibit typical Newtonian behavior at concentrations up to 40%. Above 40%, solutions become pseudoplastic, as is shown by a decrease in viscosity with increasing shearing stress (71).

In cosmetic products, gum arabic acts as binder and stabilizer of O/W emulsions (72), and it facilitates the smoothness and fluency of the products and softens the epidermis forming a protective film.

In high doses it constitutes a good fixative for hair, not greasy. In smaller doses it is useful in detergents for foam stabilizing and in pastes and toothpastes as a gingival emollient.

Gum tragacanth. Gum tragacanth is the exudation of small shrub-like plants of the *Astragalus* species (Leguminosae). Structurally, gum tragacanth is a heterogeneous, highly branched polymer composed of a major acidic polysaccharide and a minor neutral polysaccharide plus insignificant amounts of associated cellulose, protein, and starch. The acidic, water-swellable component, tragacanthic acid (formerly called bassorin) accounts for 60 to 70% of the polymer and upon acid hydrolysis yields D-xylose, L-fucose, D-galacturonic acid, D-galactose, and a very small amount of L-rhamnose. This acidic portion of the molecule is associated with calcium, magnesium, and potassium cations. The neutral, water-soluble polysaccharide component of tragacanth (also called tragacanthin) is an arabinogalactan in which L-arabinose is the preponderant sugar. The highly branched arabinogalactan is believed to consist of a core of D-galactose residues to which highly ramified chains of L-arabinofuranose residues are attached.

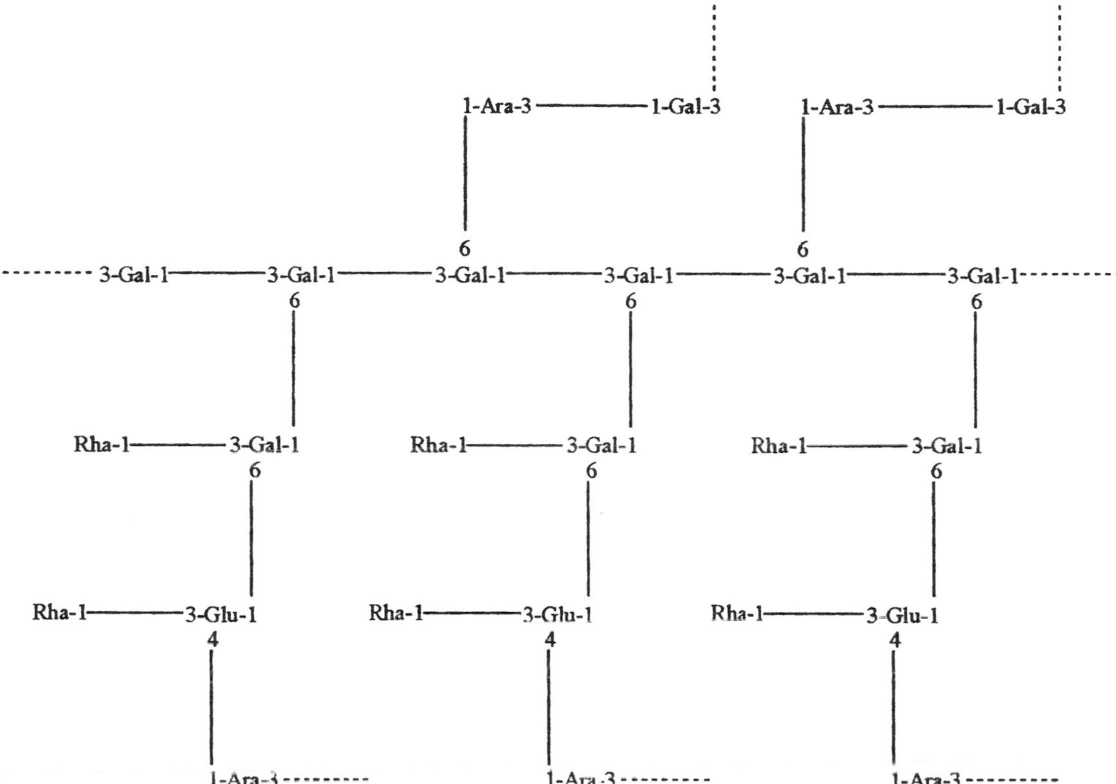

Figure 7 Schematic structure of gum arabic. Ara = arabinose, Rha = rhamnose, Gal = galactose, Glu = glucuronic acid.

Gum tragacanth swells in water to give thick viscous dispersions or pastes similar in texture to soft gels. Gum tragacanth solutions are acidic, usually in the pH range of 5 to 6. Its maximum initial viscosity is at pH 8, but its maximum stable viscosity was found to be near pH 5; the stability is definitely decreased at a pH below 4 or above 6 (73). Gum tragacanth has well-defined surface activity properties and produces a rapid lowering of the surface tension of water at low concentrations (0.25–0.50%). This property when used in O/W mixtures lowers the interfacial tension between the oil and water surfaces and accounts for its effective functioning as an emulsifying agent.

In cosmetic applications, this vegetal biopolymer is used to form gels and protective films. In toothpastes, it acts as a softener and gingival emollient.

Gum karaja. Gum karaja is the dried gummy exudates from *Sterculia urens* and other species of *Sterculia* (Fam. Sterculiaceae).

Gum karaja occurs naturally as a complex, partially acetylated branched polysaccharide having a very high molecular weight of about 9,500,000 (74). It contains about 37% uronic acid residues and approximately 8% acetyl groups. Gum karaja contains a central chain of D-galactose, L-rhamnose, and D-galacturonic acid units (in the molecular ratio of 6:4:5) with some side chains containing D-glucuronic acid. The exudates occur in a salt form containing calcium and magnesium ions.

Gum karaja is water-swellable rather than water-soluble and adsorbs water very rapidly to form viscous colloidal dispersions at low concentrations. In dilute solutions of gum karaja, the viscosity increases linearly with concentration up to about 0.5%; thereafter karaja dispersions behave as non-Newtonian solutions.

In cosmetic applications, gum karaja, at concentrations above 2 to 3% (viscosity 5,000 to 10,000 cps), forms thick, nonflowing pastes resembling spreadable gels. This natural biopolymer has film-forming properties when plasticized with compounds such as glycols to reduce its brittleness (it has found some use in hair-setting preparations). At higher concentrations of 20 to 50%, gum karaja exhibits strong wet-adhesive properties, which find major applications in denture adhesion products where strong bonding properties are required.

Gum ghatti. Gum ghatti, also known as Indian gum, is an amorphous, translucent exudate of the *Anogeissus latifolia* tree of the Combretaceae family. Structurally, gum ghatti is a mixed calcium and magnesium salt of a complex polysaccharidic acid. Complete acid hydrolysis has shown the polymer to consist of L-arabinose, D-galactose, D-mannose, D-xylose, and D-glucuronic acid in a molar ratio of 10:6:2:1:2 plus traces (less than 1%) of 6-deoxyhexose (75). The gum contains alternating 4-*O*-substituted and 2-*O*-substituted alpha-D-mannopyranose units (76) and chains of 1,6 linked beta-D-galactopyranose units, which are mostly single L-arabinofuranose residues but may also be short chains of L-arabinose sugar units (77). Gum ghatti does not form true aqueous solutions but forms viscous dispersions in water at a concentration of about 5% or higher and exhibits typical non-Newtonian behavior common to most hydrocolloids. It is a moderately viscous gum, lying intermediate between arabic and karaja. Gum ghatti normally forms viscous dispersions having a pH of about 4.8. Viscosity increases sharply with pH up to a maximum at about pH 5 to 7 and then drops off gradually down to pH 12. At all pHs, upon aging, the viscosity decreases noticeably over time (78).

The main cosmetic application of gum ghatti is to stabilize O/W emulsions, and thus has the same uses as gum arabic.

b. *Extracts*

Hydrocolloids extracted from seaweeds are very important because of the large quantity of extracts that can be obtained and their wide field of application in cosmetics. The seaweed extracts are divided into two groups: extracts from red algae and from brown algae.

Agar. Agar is extracted from several genera of marine algae of the class red algae (*Rhodophyceae*). Agarose, the gelling portion of agar, has a double helical structure, forming a three dimensional framework that holds the water molecules within the interstices. Chemically, agarose consists of chains of alternate beta-1,3-linked-D-galactose and alfa-1,4-linked 3,6-anhydro-L-galactose. 6-*O*-Methyl-D-galactose may also be present in variable amounts from about 1 to 20% depending upon the algal species (79) (Fig. 8). Agaropectin is probably a mixture of polysaccharides. It contains sulphated residues (3 to 10% sulfate), glucuronic acid, and in some species a small proportion of pyruvic acid linked in acetal linkage.

Agar is insoluble in cold water but soluble in boiling water. The viscosity at temperatures above its gelation point is relatively constant (80). Agar finds some cosmetic uses as a consistency agent in shampoos and toothpastes (81), in hand gels, and in eye makeup.

Carrageenans. Carrageenan is a structural polysaccharide found in red seaweed (*Rhodophyceae*), mainly *Chondrus crispus* L., *Eucheuma*, and *Gigartina stellata*.

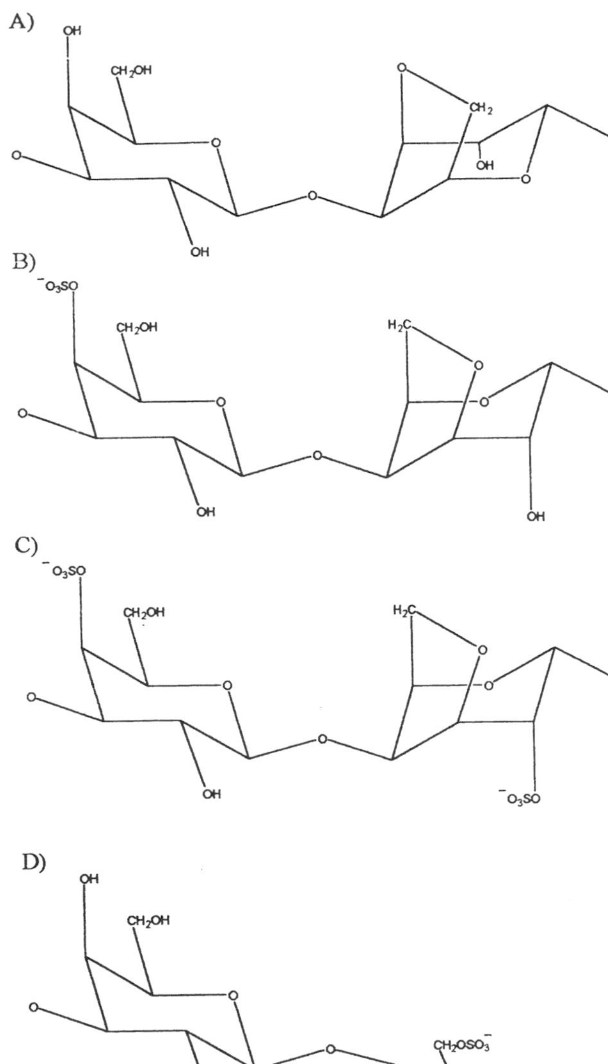

Figure 8 Basic structures of agarose and carrageenans. (A) Agarose = D-galactose + 3,6-anhydro-L-galactose (left and right, respectively); (B) kappa carrageenan = D-galactose-4-sulphate + 3,6-anhydro-D-galactose (left and right, respectively); (C) iota carrageenan = D-galactose-4-sulphate + 3,6-anhydro-D-galactose-2-sulphate (left and right, respectively); (D) lambda carrageenan = D-galactose-2-sulphate + D-galactose-2,6-disulphate (left and right, respectively).

Carrageenans are sulphated polymers made up of D-galactose residues linked alternately alpha-(1,3) and beta-(1,4) as a main chain (Fig. 8). The molecular weight is 100,000 to 1,000,000. According to the half-ester sulfate group linkages to one or more of the galactose OH-groups, iota-, kappa-, lambda- micro-, and nu-carrageenans can be distinguished, of which only the first two form gels. Carrageenan is soluble in hot water, forming viscous solutions, which gel on addition of cations such as K^+, NH_4^+, Ca^{2+} (72); the presence of alcohol can lead to precipitation.

In cosmetic products, carrageenans lead in O/W emulsions to high viscosity, suspensive ability, and emulsion stabilizing. They are used as foam stabilizers of beard creams and have immense applications, under gel shapes, as excipients in lotions and creams for the face and hands, beauty masks, and above all in toothpastes. The gels of carrageenans introduce moreover one interesting property: they can be incorporated without breaking off the gel in numerous products like alcohols, detergents, solutions, and also weak acids.

Alginates. Algin occurs in all members of the class *Phaeophycea*, i.e., brown seaweed, as a structural component of the cell walls in the form of the insoluble mixed Ca, Mg, Na, and K salts of alginic acid. Algin (INCI) is the sodium salt of alginic acid. The molecular weight of alginic acids is approximately 240,000. Alginates can be attached from microorganisms, and so they have to be preserved.

In cosmetics they are used as coemulsifiers and thickeners for gels [the viscosity of these gels can be regulated with Ca salts (81)], and in lotions, emulsions, shaving foams, toothpastes, and shampoos; they create also excellent protective films.

Pectins. Pectin is a generic term for a group of mucilagineous polysaccharides. For extraction, we use materials from some plants particularly rich in pectins like the rinds of some citrus fruits (the lemon, for example, family of the Rutaceae) or apple pulp (*Pyrus*, family of the Rosaceae). Pectins are primarily composed of (1,4)-linked alpha-D-galactopyranosyluronic acid units, partially esterified with methanol. They contain some L-rhamnose residues and some galactose, arabinose, and xylose units. Pectins are readily soluble in water to give viscous, stable solutions; they possess some surfactant properties. Pectins possess strong water-retaining capacities and form on the skin a refreshing and easily absorbable styptic film (81).

In cosmetics, pectins, especially the high methoxylated ones, are used mainly for gels (72,81) (Fig. 9); they give high viscosity, stability, and suspend ability to emulsions and toothpastes. Because of their emulsifier property, these natural biopolymers have been proposed as natural substitutes for synthetic detergents to use in cosmetic preparations.

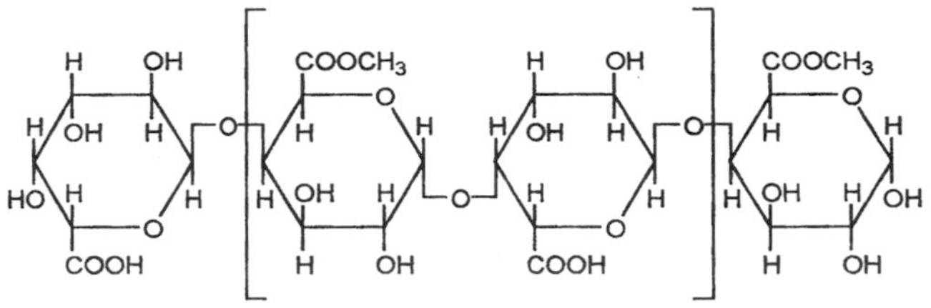

Figure 9 Structure of pectin.

c. *Mucilages from Plants*

Althea officinalis, Calendula officinalis, Linum usitatissimum, Malva sylvestris, Plantago psillium. The hydrocolloids extracted from these plants are a mix of polysaccharides and polyuronic acids; after hydrolysis, they provide galacturonic acid, rhamnose, galactose, xylose, arabinose, glucose, and glucuronic acid.

Mucilage extracted from *Althea* contains galacturonic acid, glucuronic acid, rhamnose, and galactose; it associates with starch and pectin. Mucilage from *Calendula* contains glucuronic acid, rhamnose, arabinose, and galactose. Mucilage from *Linum* contains galacturonic acid, rhamnose, xylose, and galactose; it associates with a high percentage of oil. Mucilage from *Malva* contains galacturonic acid, rhamnose, arabinose, and galactose. Mucilage extracted from *Plantago* contains galacturonic acid, xylose, and arabinose (85).

These extracts can be used in cosmetics to soothe, soften, refresh, hydrate, and create a protective film.

Aloe barbadensis. The biopolymer obtained from the plant contains mainly uronic acid, glucose, mannose, and in smaller quantity xilose, arabinose, rhamnose, and galactose. These last carbohydrates seem to give to Aloe mucilage great soothing properties. In Aloe extract are also present amino acids, citric acid, succinic acid, malic acid, sterols, and partially acetylated glucomannans. These last compounds give to this mucilage its strong gel-forming capacity.

Aloe gel is used in moisturizing, after-sun, and soothing cosmetic preparations.

d. *Fermentation Gums*

Xanthan gum. Xanthan gum is a secondary metabolite polysaccharide, produced by industrial fermentation of a carbohydrate under aerobic conditions with the microorganism *Xanthomonas campestris*. The xanthan gum polymer backbone consists of 1,4 linked beta-D-glucose units as in cellulose. At the 3-position of the alternate glucose monomer units trisaccharide side chains branch out consisting of one beta-D-glucuronic acid and two 2-alpha-mannose residues. To about half of the terminal alpha-D-mannose residues a pyruvate group is linked by an acetal linkage to the 4- and 6-carbon positions.

The D-mannose units adjacent to the main chain contain an acetyl group at position 6 (86). The presence of a glucuronic acid unit and a pyruvate group imparts an anionic character to xanthan gum. Xanthan gum dissolves readily in water at ambient temperature and over broad pH ranges and salt concentrations. It is soluble in glycerol at 65°C but is insoluble in other organic solvents or cosmetic oils. Nevertheless, once dissolved in water, up to 40% acetone, ethanol, or isopropanol can be added without precipitating the gum. Concentrations used typically range between 0.3 and 1.0%. The characteristics of aqueous solutions of xanthan gum are

1. High viscosity at low concentration
2. Pseudoplasticity with yield value
3. Insensitivity of the viscosity to heat, electrolytes, prolonged shearing or pH
4. Pleasing skin-feel when used in cosmetics
5. Imparting freeze/thaw stability

e. Cellulose Derivatives

Cellulose is the most ubiquitous structure-polysaccharide. It constitutes half of the cell-membrane material of wood and other vegetable products. Cotton consists of practically pure cellulose. Cellulose consists of chains of beta-anhydroglucose units, joined together by acetal linkages. Cellobiose (Fig. 10) and cellotetraose are obtained by the controlled hydrolysis of cellulose. The molecular weight of cellulose is 300,000 to 500,000, which corresponds to 3,000 to 5,000 glucose units. Cellulose consists of amorphous and crystalline regions. It is the prototype unbranched linear colloid which is resistant even to hot water and has to be derivatized, e.g., carboxymethylated or hydroxyalkylated to become water soluble (82,83). For this purpose, cellulose is dissolved in soda lye and reacted with monochloroacetic acid, ethylene oxide, propylene oxide, methylchloride, or methylchloride/propylene oxide mixture, etc. Thus water or organosoluble cellulose derivatives are obtained. The solution characteristics of these cellulose derivatives depend on the type of substituent R, the average chain length or degree of polymerization, and the degree of substitution. They can be used in the pH range 3 to 12.

Ethyl cellulose. This is the cellulose ethyl ether (C_2H_5 Cl); the percentage of C_2H_5 groups is 40–50%. The substitution degree is 2–3. Ethyl cellulose in cosmetics is used as a film-creating agent in lipsticks, enamels for nails, and hair fixing.

Ethylmethyl cellulose. This is the cellulose ethyl methyl ether (C_2H_5 Cl + CH + Cl); it is water soluble and confers high viscosity and stability to cosmetic emulsions and gels.

Hydroxyethyl cellulose. This is the cellulose hydroxy ethyl ether ($(H_2COCH_2)_n$), a nonionic polymer that is water soluble. It forms solutions with gel and has film-forming capacity. The viscosity of hydroxyethyl cellulose solutions depends upon the degree of polymerization.

In cosmetic preparations, hydroxyethyl cellulose solution with high viscosity is utilized at low concentration in toothpastes; solutions with average viscosity are used at higher dosages to suspend and to make viscous emulsions; low viscous solutions of hydroxyethyl cellulose are used at maximal concentration to create a protective film over the skin.

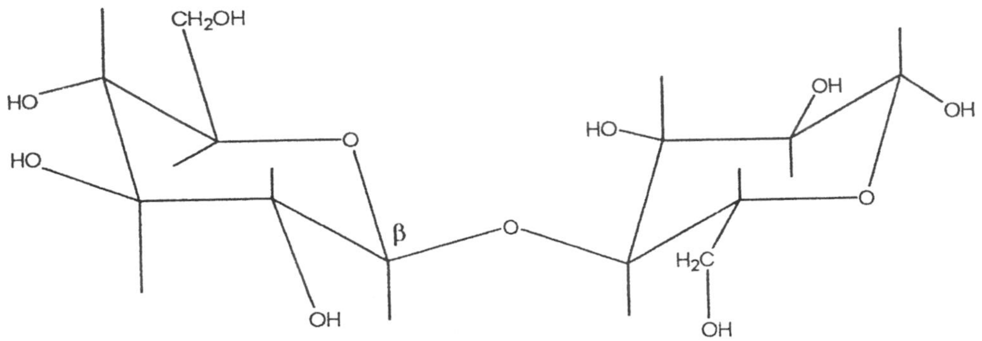

Figure 10 Chemical structure of cellobiose, the disaccharide units of cellulose.

Methyl cellulose, methyl-hydroxyethyl cellulose, methyl-hydroxypropyl cellulose, and hydroxypropyl cellulose. Methyl cellulose is the cellulose methyl ether; it is a protective colloid with the ability to stabilize emulsions, suspend, and disperse.

Methyl-hydroxyethyl cellulose has a property similar to hydroxyethyl cellulose and methyl cellulose. Methyl-hydroxypropyl cellulose is a nonionic polymer with the ability to protect, stabilize emulsions, suspend, and disperse.

Hydroxypropyl celluloses have the capacity to increase remarkably the viscosity of anionic mixtures and reduce them in nonionic solutions.

f. *Starch and Starch Derivatives*

Natural starch consists of 20% of a water-insoluble linear amylose and 80% water-soluble branched amylopectin. By stepwise acid or enzymatic hydrolysis, starch is hydrolyzed to dextrins (polysaccharides of lower molecular weights), (+)-maltose and finally to D-(+)-glucose. Because of the property of amylose to swell in water, it is used in the manufacture of tablets. A starch-ether called amylum non mucilaginosum (ANM) is a modified starch used in skin powders for the treatment of acne, because of its strong liquid-absorbing power, without the inconvenience of swelling (84).

2. Animal Biopolymers

a. *Polysaccharides*

Hyaluronic acid, glycosaminoglycans, sodium carboxymethyl beta-glucan: these polysaccarides have been treated in Sections III.B and III.C.

b. *Polypeptides*

Protein hydrolysates are added to hair-care products (to ''repair'' broken hair thanks to their strong affinity to keratin), soaps, bath gels, creams, etc. These peptides derive from the hydrolysis mainly from collagen, keratin, elastin, milk, wheat, almond, and silk. They are particularly important for the water-binding capacity which regulates the skin's moisture content, and also for the high affinity for skin proteins resulting in a tightening, antiwrinkle effect. Protein hydrolysates form a uniform substantive semiocclusive film.

c. *Gelatin*

Gelatin is a natural protein colloid, possessing gelling and stabilizing properties in aqueous phases. It is obtained by the partial hydrolysis (molecules with an average molecular weight of 500–1,500) of the collagen contained in skin, conjunctive tissues, and bones of animals. In common with all other proteins, gelatin comprises L-amino acids linked by peptide bonds. Gelatin contains all the amino acids essential to man with the exception of tryptophane (87,81). Technical application of gelatin is in facial masks and protective and conditioning creams for hair and hand care. Vegetal substitutes of gelatin have been proposed.

d. *Elastin*

Elastin from bovine tendons is hydrolyzed to obtain peptides with a molecular weight that ranges from 5,000 to 85,000. It is used in cosmetic formulations as a moisturizing and conditioning agent. Recently, vegetal substitutes of elastin have been studied. An extract from the seaweed *Ulva lactuca* seems to have some properties similar to those of animal elastin, or *perhaps* inhibiting elastase activity.

e. *Keratin*

The controlled lysis of hair, nails, and other keratinic materials (of bovine origin usually) supplies polypeptides characterized by the presence of sulphured amino acids (above all cystein), typical horny structures. Thanks to the presence of this reservoir, these keratinlike polypeptides show more analogy to hair and epidermis than those of collagen.

Keratinic hydrolysates are added to hair-care products, soaps, bath gels, creams, etc.

f. *Melanin*

Throughout nature, melanin is used in such diverse areas as protection from ultraviolet radiation. It is insoluble and difficult to work with, making it impractical for inclusion in creams and lotions. A great result is the discovery of methods for creating melanin derivatives that dissolve readily in water and, when incorporated into cosmetic products, can be spread evenly on the skin to instantly produce a tan.

Sunscreens are most efficient at absorbing and thus protect from the burning UV-B rays. There are many conventional sunscreens available. However, the ideal is to have a sunscreen containing the best natural protectant nature has to offer: melanin. Melanin naturally protects the skin from cancer induced by ultraviolet light. Perhaps synthetic melanin would do the same. Skin cancers are the most prevalent type of cancer in the world, so even a small reduction in their incidence would make an impact on a significant number of people.

Melanin-containing skin-care and hair-care products make melanin available for topical application to the skin. Soluble melanin formulated into cosmetics protects the skin against harmful UVR and also has the photoprotective ability to scavenge free radicals.

These specially formulated products will also moisturize the skin to help protect against dryness and dehydration. Many manufacturers are even putting sunscreening ingredients in their cosmetic products (foundation, eye makeup, lipstick, and face creams) to provide the skin with even further protection.

Pre-tan accelerators, applied at least 24 hours before exposure, usually contain tyrosine, an amino acid that helps induce the melanocytes into producing additional melanin prior to actual sun exposure.

Sunscreens are also being added to hair styling products including shampoos, sprays, gels, mousses, and conditioning packs. Melanin-based hair-care cosmetics nourish the scalp with a special combination of pH-balanced ingredients. For strong, healthy hair, they also minimize dryness, brittleness, and split ends.

Shampoo containing a combination of amino acids and melanin cleanses without stripping the hair shaft of its natural moisture.

Conditioning hair and scalp treatment, enriched with proteins, vitamins, minerals, and melanin, penetrates the hair follicles to moisturize and protect the hair, leaving it shiny, healthy and more manageable. On the scalp, it soothes and nourishes it, while reducing itchy and flaking scalp.

Moisturizing leave-in-conditioner containing a combination of vitamins and melanin protects the hair from the damaging effects of the sun's UVR.

Hair balm containing a combination of oils plus melanin leaves hair with luster, body, and sheen as nature intended.

E. Conclusion

The state of the art in cosmetology has evolved in the last few years, enriched by knowledge acquired from wider and wider fields of research, from chemistry to biochemistry, from medicine to biophysics and biotechnology. This has allowed the pursuit of research into new formulations with intrinsic cosmetic functionality that is identified and tested in its mechanisms. This functionality expresses itself on the biological structure of the cutaneous tissue often modifying it in a significant way. These are in synthesis the needs of dermocosmetics:

Research into new active ingredients
Research into new sources of natural active ingredients (biotechnologies)
Knowledge of skin biochemical mechanisms
Knowledge of the way cosmetically functional substances work

Often, new sanitary emergencies force a rapid evolution in the research and utilization of alternative cosmetic ingredients. The potential contamination of bovine collagen with bovine spongiforme encephalopathy (BSE, the fatal mad cow disease) and its human variant, Creuzfeldt–Jakob disease, has led to numerous efforts (1) to define effective methods of eliminating any risk of BSE in material containing collagen and (2) to research new sources of natural collagen (for example, biotechnology has developed a recombinant human collagen).

REFERENCES

1. Anonymous. On the approximation of the laws of the Member States relating to cosmetic products, Council Directive 76/768/EEC of July 1976, *O. J. E. C.* n. L 262/169 (and following amendments) (27.9.1976).
2. R. H. Champion, J. L. Burton, F. J. G. Ebling, eds. *Textbook of Dermatology* (Rook). 5th ed. Blackwell, Oxford, 1992, Vol. 4.
3. T. B. Fitzpatrick, A. Z. Eisen, K. Wolff, I. M. Freedberg, K. F. Austen, eds. *Dermatology in General Medicine*. 3d ed. McGraw-Hill, New York, 1987, Vol. 2.
4. W. F. Ganong. *Review of Medical Physiology*. 14th ed. Appleton & Lange, Connecticut, 1989.
5. R. Graham-Brown, T. Burns. *Lecture Notes on Dermatology*. 6th ed. Blackwell, Oxford, 1990.
6. B. Berra, S. Rapelli. Carbohydrate, protein and lipid metabolism in the skin; biochemical and molecular aspects. *Skin Pharmacology and Toxicology* (C. L. Galli et al., eds.). Plenum Press, 1990, p. 37.
7. B. Berra. Detergenza e biochimismo della cute: problemi e prospettive. *Cosmesi dermatologica*, 21:8 (1988).
8. P. M. Elias. Epidermal lipid membranes and keratinization. *Int. J. Derm.*, 20:1 (1981).
9. L. Landmann. Epidermal permeability barrier: transformation of lamellar granule-disk into intracellular sheets by a membrane-fusion process; a freeze-fraction study. *J. Invest. Dermatol.*, 87:202 (1986).
10. E. Lazarides. Intermediate filaments: a chemically heterogeneous, developmentally regulated class of proteins. *Ann. Rev. Biochem.*, 51:219 (1982).
11. P. Steinert, D. Parry, Intermediate Filaments, *Ann. Rev. Cell. Biol.*, I:4166 (1985).
12. P. Steinert, D. Roop. Molecular and cellular biology of intermediate filaments. *Ann. Biochem.*, 57:593 (1988).
13. E. Fuchs, H. Green. Changes in keratin gene expression during terminal differentiation of the keratinocyte. *Cell*, 19: 1033 (1980).

14. L. A. Goldschmidt. *Biochemistry and Physiology of the Skin*. Oxford University Press, Oxford, 1983.
15. L. A. Goldschmidt. *Physiology, Biochemistry and Molecular Biology of the Skin*. Oxford University Press, Oxford, 1992.
16. T. P. Dryja, M. O'Neil-Dryja, J. M. Pawelek, D. M. Albert. Demonstration of tyrosinase in the adult bovine uveal tract and retinal pigment epithelium. *Invest. Ophthalmol. Visual Sci*., 17:511 (1978).
17. G. Imokawa, Y. Mishima. Isolation and biochemical characterization of tyrosinase rich GERL and coated vesicle in the melanin synthesizing cells. *Br. J. Dermatol*., 104:160 (1981).
18. Y. Mishima, G. Imokawa. Role of glycosilation in initial melanogenesis: post inhibition dynamics. In: *Pigment Cell 1985: Biological and Molecular Aspects of Pigmentation* (J. Bagnara, S. N. Klaus, E. Paul, M. Scharl, eds.) University of Tokyo Press, Tokyo, 1985, p. 17.
19. G. Prota. Recent advances in the chemistry of melanogenesis in mammals. *J. Invest. Derm*., 75:122 (1980).
20. G. Prota. Melanin and pigmentation. In: *Coenzymes and Cofactors* (D. Dolphin, R. Paulson, and O. Abramovic, eds.) John Wiley, New York, 1989, Vol. 3, p. 441.
21. M. A. Pathak, K. Jimbow, T. B. Fitzpatrick. Photobiology of pigment cell. In: *Pigment Cell 1981: Phenotypic Expression in Pigment Cell* (M. Seiji, ed.). University of Tokyo Press, Tokyo, 1981, p. 655.
22. J. Parrish. Photobiology of melanin pigmentation. In: *Psoralens in Cosmetics and Dermatology* (J. Cahn, P. Forlot, C. Grupper, A. Meybeck, F. Urbach, eds.). Pergamon Press, Paris, 1981, p. 17.
23. R. M. Mackie, J. M. Elwood, J. L. M. Hawk. Links between exposure to ultraviolet radiation and skin cancer. *J. Royal Coll. Phys*., London, 21:91 (1987).
24. D. E. Elder. Human melanocytic neoplasms and their etiologic relationship with sunlight. *J. Invest. Dermatol*., 928: 297 (1989).
25. N. Cascinelli, R. Marchesini. Increasing incidence of cutaneous melanoma, ultraviolet radiation and the clinician. *Photochem. Photobiol*., 50:497 (1989).
26. J. F. Dore, F. Anders. Contributions of experimental models to understanding of human melanoma. In: *Cutaneous Melanoma: Biology and Management* (N. Cascinelli, M. Santinami, U. Veronesi, eds.). Masson, Milan, 1990, p. 63.
27. K. H. Kaidbey, P. P. Agin, R. M. Sayre, A. M. Kligman. Photoprotection by melanin—a comparison of Black and Caucasian skin. *J. Am. Acad. Dermatol*., 1:249 (1979).
28. T. B. Fitzpatrick. The validity and praticality of sun-reactive skin types I through VI. *Arch. Dermatol*., 124:869 (1988).
29. J. P. Cesarini. Photo-induced events in the human melanocytic system: photoaggression and photoprotection. *Pigment Cell Res*., 1:223 (1988).
30. A. Kawava. UVB-induced erythema, induced tanning, and UVA-induced immediate tanning in Japanese skin. *Photodermatol*., 3:372 (1986).
31. S. Shono, M. Imura, M. Ota, O. Satoshi, K. Toda. The relationship of skin color, UVB-induced erythema, and melanogenesis. *J. Invest. Dermatol*., 84:265 (1985).
32. N. Kollias, R. M. Sayre, L. Zeise, M. R. Chedekel. Photoprotection by melanin. *J. Photochem. Photobiol. B. Biol*., 9:135 (1991).
33. I. Illis. Photosensitivity reactions in black skin. *Dermatologis Clinics*, 6:369 (1988).
34. K. H. Kaidbey, A. M. Kligman. Sunburn protection by long wave ultraviolet light upon human skin. *Arch. Dermatol*., 114:46 (1978).
35. H. Z. Hill. The relationship of the photobiology of skin cancer and melanins to the radiation biology of melanosomas: a selective review. *Comments Mol. Cell. Biophys*., 6:141 (1989).
36. L. A. Goldshmidt. *Biochemistry and Physiology of the skin*. Oxford University Press, Oxford, 1983.
37. E. Solomen. The collagen gene family. *Nature*, 286:656 (1980).
38. McK. J. Snowden, D. A. Swann. Effects of GAG and proteoglycans on the in vitro assembly and thermal stability of collagen fibrils. *Biopolymers*, 19:767 (1980).
39. M. M. Rapport, B. Weissman, A. Linkerand K. Meyer. Isolation of a crystalline disaccharide, hyalobiuronic acid, from hyaluronic acid. *Nature* 168996 (1951).
40. L. J. Meyer, R. Stern. Age-dependent changes of hyaluronan in human skin. *J. Invest. Dermatol*., 102:385 (1994).
41. E. F. Bernstein, C. B. Underhill, P. J. Hahn, D. B. Brown, J. Uitto. Chronic sun exposure alters both the content and distribution of dermal glycosaminoglycans. *Br. J. Dermatol*., 135:255 (1996).
42. R. K. Reed, U. B. Laurent, J. R. Fraser, T. C. Laurent. Removal rate of [3H]hyaluronan injected subcutaneously in rabbits. *Am. J. Physiol*., 259:H532 (1990).
43. R. Stern, G. I. Frost, S. Shuster, V. Shuster, J. Hall, T. Wong, P. Gakunga. Hyaluronic acid and skin. *Cosmet. Toil*., 113:43 (1998).
44. R. K. Reed, K. Lilja, T. C. Laurent. Hyaluronan in the rat with special reference to the skin. *Acta Physiol. Scand*., 134:405 (1998).
45. L. Y. Bourguignon, N. Lida, C. F. Welsh, D. Zhu, A. Krongrad, D. Pasquale. Involvement of CD44 and its variant isoforms in membrane-cytoskeleton interaction, cell adhesion and tumor metastasis. *J. Neurooncol*., 26:201 (1995).
46. J. Entwistle, C. L. Hall, E. A. Turley. HA receptors: regulators of signalling to the cytoskeleton. *J. Cell Biochem*., 61: 569 (1996).
47. R. N. Feinberg, D. C. Beebe. Hyaluronate in vasculogenesis. *Science*, 220:1177 (1983).
48. D. C. West, I. N. Hampson, F. Arnold, S. Kumar. Angiogenesis induced by degradation products of hyaluronic acid. *Science*, 228:1324 (1985).
49. R. Tammi, K. Paukkonen, C. Wang, M. Horsmanheimo, M. Tammi. Hyaluronan and CD44 in psoriatic skin, intense staining for hyaluronan on dermal capillary loops and reduced expression of CD44 and hyaluronan in keratinocyte-leukocyte interfaces. *Arch. Dermatol. Res*., 286:21 (1994).
50. T. N. White, D. K. Heinegard, V. C. Hascall. Proteogly-

cans, structure and function. In: *Cell Biology of the Extracellular Matrix* (E. D. Hay, ed.). New York, Plenum Press, 1991, p. 45.
51. A. V. Benedeto. The environment and skin aging. *Clinics in Dermatology*, 16:129 (1998).
52. P. W. A. Mansell. Polysaccharides in skin care. *Cosmet. Toil.*, 109(9):67 (1994).
53. R. C. Goldman. Biological response modification by beta-glucans. *Annual Reports in Medicinal Chemistry*, 30:129 (1995).
54. J. A. Bohn, J. N. BeMiller. (1-3)-Beta-glucans as biological response modifier: a review of structure-functional activity relationships. *Carbohydrate Polymers*, 28:3 (1995).
55. S. Yoshio, H. Katsuhiko, M. Kazumasa. Augmenting effect of sizofiran on the immunofunction of regional lymph nodes in cervical cancer. *Cancer*, 69(5):1188 (1992).
56. J. K. Czop, J. Kay. Isolation and characterization of beta-glucan receptors on human mononuclear phagocytes. *J. Exp. Med.*, 173:1511 (1991).
57. S. J. Leibovich, D. Danon. Promotion of wound repair in mice by application of glucan. *J. Reticuloendothel.* Soc., 27:1 (1980).
58. Di Luzio. Enhanced healing of decubitus ulcers by topical application of particulate glucan. Tulane University School of Medicine, *Research Summary* (1984).
59. C. A. Elmets. Photoprotective effects of sunscreens in cosmetics on sunburn and Langerhans cell photodamage. *Photodermatol. Photoimmunol. Photomed.*, 9:113 (1992).
60. K. Moo-Sung, P. Kyung-Mok, C. Ih-Seop, K. Hak-Hee, S. Young-Chul. Beta-(1,6)-branched beta-(1,3)-glucan in skin care, *Cosmet. Toil.*, 115(7):79 (2000).
61. F. Zülli, F. Suter. Patent pending, Mibelle AG.
62. F. Zülli, F. Suter, H. Biltz, H. P. Nissen, M. Birman. Carboxymethylated β-(1-3)-Glucan. *Cosmet. Toil.*, 111:91–98 (1996).
63. F. Zülli, F. Suter, H. Biltz, H. P. Nissen, M. Birman. Carboxymethylated beta-(1-3)-Glucan. *Cosmet. Toil.*, 111:91 (1996).
64. C. Colin, B. Boussouira, D. Bernard, D. Moyal, and Q. L. Nguyen. Non invasive methods of evaluation of oxidative stress induced by low doses of ultra violet in humans. IFSCC Congress Venezia A 105, 50–72 (1994).
65. F. Zülli, F. Suter, H. Blitz, H. P. Nissen. CM-glucan: a biological response modifier from baker's yeast for skin care. *SÖFW Journal* 123:535–541 (1997).
66. R. Goldman. Characteristic of the β-glucan receptor of murine macrophages. *Exp. Cel. Res.*, 174:481 (1988).
67. H. D. Graham. *Food Colloids*. A VI Publishing Co., Westport, CT (1977).
68. Y. L. Meltzer. *Water Soluble Resins and Polymers*. Technology and Applications. Noyes Data Corporation, Park Ridge, NJ (1976).
69. K. Klein. Improving emulsion stability. *Cosmet. Toil.*,99: 121 (1984).
70. I. Danishefsky, R. L. Whistler, F. A. Bettelheim. Introduction to polysaccharide chemistry. In: *The Carbohydrates: Chemisty and Biochemistry* (W. Pigman, D. Horton, eds.). Vol. 2A. Academic Press, New York, 1970, p. 375.
71. B. Warburton. The rheology and physical chemistry of some Acacia systems. In: *The Chemistry and Rheology of Water Soluble Gums and Colloids*. Soc. Chem. Ind. (London) Monograph, 24:118 (1966).
72. Kirk-Othmer, Encyclopedia of Technical Chemistry, Vol. 12. 3d ed. Wiley-Interscience, New York, 1980, p. 45.
73. T. W. Schwarz, G. Levy, H. H. Kawagoe. Tragacanth solutions. III. The effect of pH on stability. *J. Am. Pharm. Assoc. Sci. Ed.*, 47:695 (1958).
74. J. V. Kubal, N. Gralen. Physiochemical properties of karaja gum and locust bean mucilage. *J. Colloid Sci.*, 3:457 (1948).
75. G. O. Phillips, G. Pass, M. Jeffries, R. Morley. *Gelling and Thickening Agents in Foods* (H. Neukom, W. Pilnik, (Eds.). Forster, Zurich, 1980, p. 135.
76. G. O. Aspinall, J. M. McNab. Anogeissus leiocarpus gum. Part III. Interior chains of leiocarpan A. *J. Chem. Soc.*, C: 845 (1969).
77. G. O. Aspinall, V. P. Bhavanadan, J. B. Christensen. Gum ghatti (Indian gum). Degradation of the periodate-oxidized gum. *J. Chem. Soc.*, Part V: 2677 (1965).
78. G. Meer, W. A. Meer, T. Gerard. Gum ghatti. In: *Industrial Gums*, 2[d] ed. (R. L. Whisler, Ed.). Academic Press, New York, 1973.
79. G. G. Allan, P. G. Johnson, Y. Z. Lai, K. V. Sarkanen. Marine polymers. Part I. A new procedure for the fractionation of agar. *Carbohydr. Res.*, 17:234 (1971).
80. H. H. Selby, T. A. Selby. Agar. In: *Industrial Gum* (R. L. Whisler, ed.). Academic Press, New York, 1959, p. 15.
81. L. Trüger. *Chemie in der Kosmetik*. 2. Aufl. Hüthig Verlag, Heidelberg, 1989.
82. L. Fieser, M. Fieser. *Organische chemie*. Verlag Chemie, Weinheim/Bergstraße, 1965.
83. J. D. Roberts, M. C. Caserio. *Basic Principles of Organic Chemistry*. Benjamin, New York, 1964.
84. K. F. De Polo. *A Short Textbook of Cosmetology*. Verlag für Chemische Industrie, H. Zlolkowsky, Augsburg, 1998, p. 173.
85. G. Proserpio. *Cosmesi Funzionale*. Sinerga ed., Milan, 1988, p. 231.
86. Mero Rousselot Satia.*Food-Grade Xanthan Gum*. 15 Av. D'Eylau, F-75116 Paris Satiaxan, 1986.
87. Sanofi Bio-Industries. Hydrocolloids. Technical Brochure. 66, av. Marceau, F-75 008 Paris.

19

Textile-Based Biomaterials for Surgical Applications

C. C. Chu
Cornell University, Ithaca, New York

I. INTRODUCTION

Textile-based biomaterials have the longest history in bio materials. In ancient China and Egypt as far back as 2000 B.C., natural fibers like linen, silk, bark, horsehair, and dried guts were used as suture materials for wound closure. The introduction of steel wire and synthetic fibers like polypropylene, nylon, and polyester during and after World War II brought the first revolution of medical textiles and greatly expanded the chemical composition of textile-based biomaterials beyond natural sources. Owing to their precisely controlled manufacturing processes and uniform and reproducible properties, these synthetic fibrous biomaterials have received a great deal of attention from both surgeons and researchers. However, the basic textile structure has not kept up the pace of the development of materials.

The advancement of textile-based biomaterials made another major breakthrough in the early 1970s after the successful introduction of two synthetic absorbable wound closure biomaterials, i.e., Dexon® and Vicryl®. The availability of this class of absorbable aliphatic polyester-based biomaterials opened a new chapter for medical textiles, particularly in wound closure, cardiovascular implants, and body wall repair. Today, surgeons can choose among a large number of textile-based biomaterials with various chemical, physical, mechanical, textile structural, and biological properties for their specific clinical applications.

Textile-based biomaterials are different from other biomaterials largely in their physical form and anisotropic mechanical behavior. They are the only class of polymeric biomaterials that could have a wide range of three-dimensional porous structure. The four most widely used textile structures as biomaterials are woven, knitted, nonwoven, and braided. Within each of these four classes of textile structure, there are many other variations, such as weft knit vs. warp knit. These structural features and the wide range of structural possibilities and mechanical strength orientation provide the basis for their numerous applications in biomedical fields, such as (1) the requirement of tissue ingrowth for fastening devices, as with the suture ring of an artificial heart valve, (2) the requirement for integration with surrounding tissues for achieving better tissue biocompatibility, as with vascular grafts, (3) the requirement of large surface and pore area in a 3D structure, as with tissue engineering scaffolds, (4) the requirement to fill void space and provide mechanical support, as with surgical meshes for body wall repair, (5) the requirement to carry physiological force anisotropically, as with wound closure, and tendon/ligament prostheses, and (6) the requirement of mechanical strength, as with fiber and fabric reinforced composites in a variety of surgical implants such as total artificial hearts, penile implants, bone fixation devices, to name a few.

In addition to the textile-based implants, medical textiles have been used outside the human body in hospital linens, surgical gowns and drapes, antiembolism stockings, the reinforced components of casts, etc. In this chapter, we focus on the use of medical textiles that human lives are dependent on, particularly in the areas of wound closure, cardiovascular implants, and surgical meshes for body wall repair. Among these three areas of application, the textile biomaterials for wound closure are the most important not only because they are used in virtually every surgery for the fastening of surgical implants but also because other textile-based biomaterials like those for vascular grafts and surgical meshes have largely been based upon the same fibrous polymers as suture materials.

II. WOUND CLOSURE BIOMATERIALS

Wound closure biomaterials are generally divided into three major categories: suture materials, tissue adhesives, and staplers. Only the suture materials are in textile form, and they have received the most research and development attention. They are also the most widely used in wound closure.

A suture, by definition, is a strand of textile material, natural or synthetic, used to ligate blood vessels and to draw tissues together. It consists of a fiber with a metallic needle attached at one of the fiber ends. An ideal suture should

- Handle comfortably and naturally.
- Show minimum tissue reaction.
- Have adequate tensile strength and knot security.
- Be unfavorable for bacterial growth and easily sterilized.
- Be nonelectrolytic, noncapillary, nonallergenic, and noncarcinogenic.
- If it is absorbable, its tensile strength loss must match the healing rate of the tissues to be closed, and the degradation products must be biocompatible and be metabolized by normal body metabolic mechanisms.

Suture materials are commonly classified by absorbability in biological tissues, size, physical configuration (monofilament vs. multifilament), and type of coating materials used to facilitate handling properties. Suture materials have vast differences in chemistry, physical/mechanical properties, and biological characteristics.

The four most important properties of suture materials are physical and mechanical properties, handling properties, biological properties, and biodegradation properties. Table 1 summarizes individual properties under each of the four categories. It is important to recognize that these properties are interrelated. For example, the capillarity of a suture material under physical/mechanical characteristics is closely related to the ability of the suture to transport bacteria, which is a biological characteristic. The modulus of elasticity under physical/mechanical characteristics is frequently used to relate to pliability of sutures under handling characteristics. A brief description of each of those essential properties will be given below and they are listed in Tables 2 through 4. The wide range of data in these tables indicates the complexity of the issues and the difficulty of drawing general conclusions. Nevertheless, the information in these tables should provide readers with an overall view of the various essential properties of suture materials. Readers should be aware that the data in the tables vary depending on specific clinical and/or physical environments that suture materials are subject to and constant refining of the manufacturing processes by suture manufacturers. Readers are referred to a recently published book for its comprehensive review of these four major properties (1).

A. Chemical Structure and Manufacturing Processes

Suture materials are generally classified into two broad categories: absorbable and nonabsorbable. Absorbable suture materials lose their entire tensile strength within two to

Table 19.1 Four Major Categories of the Characteristics of Suture Materials

Physical/mechanical	Handling	Biocompatibility	Biodegradation
USP vs. EP	Pliability	Inflammatory reaction	Tensile breaking strength and mass loss profiles
Size (diameter)	Memory	Propensity toward wound infection, calculi formation, thrombi formation, carcinogenicity, allergy	Biocompatibility of degradation products
Mono vs. multifilament	Packing		
Tensile breaking strength and elongation	Knot tie-down		
Modulus of elasticity	Knot slippage		
Bending stiffness	Tissue drag		
Stress relaxation and creep			
Capillarity			
Swelling			
Coefficient of friction			

Table 19.2 Relative Order of Mechanical Properties of Absorbable Sutures[a]

Class (chemical name)	Commercial name	Break strength straight pull (MPa)	Break strength knot pull (MPa)	Elongation to break (%)	Young's modulus (GPa) (psi)[b]
Catgut		310–380	110–210	15–35	2.4(358,000)[b]
Regenerated collagen					
Poly(*p*-dioxanone)	PDS®, PDSII®	450–560	240–340	30–38	1.2–1.7 (211,000)[b]
Poly(glycolide-co-trimethylene car-	Maxon™	540–610	280–480	26–38	3.0–3.4 (380,000)[b]
bonate)		570–910	300–400	18–25	7–14
Poly(glycolide-co-lactide) or poly- glactin-910	Vicryl®				
Polyglycolide-co-ε-caprolactone or	Monocryl®	654–882		67–96	(113,000)[a]
polyglecaprone 25	Dexon S®	760–920	310–590	18–25	7–14
Poly(glycolic acid) or polyglyco- lide	Dexon Plus®				

[a] Mechanical properties presented are typical for sizes 0 through 3–0 but may differ for finer or larger sizes.
[b] Data in parentheses are in psi unit of 2/0 size from R. S. Bezwada. et al. Monocryl, a new ultrapliable absorbable monofilament suture derived from caprolactone and glycolide. *Biomaterials 16*:1141 (1995).
Partial Source: D. J. Casey, and O. G. Lewis. Absorbable and nonabsorbable sutures. In: *Handbook of Biomaterials Evaluation: Scientific, Technical, and Clinical Testing of Implant Materials* A. F. Von Recuin, ed.). Macmillan, New York, 1986, Chap. 7.

three months; those which retain their strength longer than two to three months are nonabsorbable. This definition has recently become questionable because a new absorbable suture, Panacryl®, from Ethicon can retain some strength longer than six months. The absorbable suture materials are catgut (collagen sutures derived from sheep intestinal submucosa), reconstituted collagen, polyglycolide (PGA), poly(glycolide-lactide) random copolymers (Vicryl® and Panacryl®), poly-*p*-dioxanone (PDS®, PDSII®), poly (glycolide-trimethylene carbonate) block copolymer (Maxon®), poly(glycolide-ε-caprolactone) (Monocryl®), and Gycolide-trimethylene carbonate block copolymer (Biosyn®). The nonabsorbable sutures are divided into the natural fibers (i.e., silk, cotton, linen), and manmade fibers

Table 19.3 Relative Order of Mechanical Properties of Nonabsorbable Sutures

Class (chemical name)	Commercial name	Break strength straight pull (MPa)	Break strength knot pull (MPa)	Elongation to break (%)	Young's modulus (GPa)
Silk	Silk, Surgical Silk®, Dermal®, Virgin Silk®	370–570	240–290	9–31	8.4–12.9
Polypropylene	Surgilene®, Prolene®	410–460	280–320	24–62	2.2–6.9
Nylon 66 & nylon 6	Surgilon®, Dermalon®, Nurolon®, Ethilon®, Supramid®	460–710	300–330	17–65	1.8–4.5
Poly[(tetramethylene ether)terephthalate-co-tetramethylene terephthalate]	Novafil®	480–550	290–370	29–38	1.9–2.1
Poly(Butylene terephthal- ate)	Miralene®	490–550	280–400	19–22	3.6–3.7
Poly(ethylene) terephtha- late	Dacron®, Ethiflex®, Ti.Cron®, Polydek®, Ethibond®, Tevdek®, Mersilene®, Mirafil®	510–1,060	300–390	8–42	1.2–6.5
Stainless steel	Flexon®, stainless steel, surgical s.s.	540–780	420–710	29–65	200

* Mechanical properties presented are typical for sizes 0 through 3–0 but may differ for finer or larger sizes.
Source: D. J. Casey, and O. G. Lewis. Absorbable and nonabsorbable sutures. In: *Handbook of Biomaterials Evaluation: Scientific, Technical, and Clinical Testing of Implant Materials* A. F. von Recuin, (ed.). Macmillan, New York, 1986, Chap. 7.

Table 19.4 Relative Tissue Reactivity to Sutures

Relative tissue reactivity	Nonabsorbable	Absorbable
Most	Silk, cotton	Catgut
	Polyester coated	Dexon and Vicryl
	Polyester uncoated	Maxon, PDS, Monocryl, Biosyn
	Nylon	
Least	Polypropylene, Gore-Tex	

Source: R. G. Bennett. Selection of wound closure materials. *J. Am. Acad. Dermatology 18*(4):619–657 (1988).

[i.e., polyethylene, polypropylene, polyamide, polyester, polytetrafluoroethylene (Gore-Tex®), poly (hexafluoropropylene-VDF) (Pronova®), and stainless steel]. Table 5 summarizes all commercial suture materials that are available mainly in the United States, Europe and the Pacific, their generic and trade names, physical configurations, and manufacturers (2). Figure 1 shows scanning electron micrographs of these commercial suture materials.

1. Catgut and Reconstituted Collagen Sutures

Multifilament twisted plain catgut was the first natural absorbable suture and was described as early as A.D. 175. The basic constituent is collagen. Collagen is a protein consisting of three polypeptides interweaved in a left-hand triple-helical structure. Each polypeptide chain has the general amino acid sequence of $(\text{-Gly-X-Y-})_n$, where X is frequently proline (Pro) and Y is frequently hydroxyproline (Hyp) (3). One important aspect of collagen is its essential electrical neutrality under physiological conditions due to the approximately equal number of acidic (glutamic and aspartic acids) and basic (lysine and arginine) side groups (4). Because the pKs for amino and carboxyl groups are about 10 and 4, respectively, the electrostatic interactions among the acidic and basic groups are expected to be significantly disturbed at pH either <4 or >10 (5). Thus any pH change would lead to a weakening of the inter- or intramolecular electrostatic interactions in collagen fibers. Such a weakening in fiber structure due to pH change is evident in the observed swelling of the fibers, which would be ultimately reflected in the observed accelerated loss of strength and mass of catgut sutures at highly acidic or alkaline conditions. The use of intermolecular cross-linking agents like formaldehyde or glutaldehyde could stabilize fiber structure against the pH-induced change.

Catgut suture is derived from the submucosa of sheep intestines or serosa of bovine intestine. The jejunum and ileum portions of the intestine of sheep or cattle are split into two or more longitudinal ribbons, and then the mucosa muscularis and other unwanted layers are removed by chemical and mechanical treatments. The remaining portion is treated in diluted formaldehyde to block the —OH and —NH_2 groups on collagen in order to increase the strength and the resistance to enzyme attack. Several ribbons are then twisted into strands, dried, machine ground, and polished by a centerless grinder to a correct and smooth size. This grinding and polishing process can produce unpredictable amounts of weak points and local tearing of fibrils. Thus fibrils could fray during use. Reproducible strength is also difficult to achieve.

The resulting untreated catgut suture is called plain catgut. If the plain catgut is further tanned in a bath of chromium trioxide, it is called chromic catgut, a variant first developed by Lister in 1840. There are two types of chromicizing processes: tru and surface chromicizings. The former is done to each ribbon before it is spun into strands, while the latter is conducted on the finished strand. This treatment changes the color of plain catgut from a yellowish tan shade to a darker shade of brown. Depending on the concentration of the chromic bath and the duration of chromicizing, mild and extra chromic catgut are available. Of course, the degrees of absorption and tissue response are also affected by the severity of the tanning process. Chromic catgut suture is generally more resistant to absorption and causes less tissue reaction than plain catgut suture. Catgut sutures are packaged in alcohol solution like ethanol or isopropanol to retain their flexibility, and the packages are sterilized by either Co^{60} γ-irradiation or ethylene oxide.

Davis and Geck introduced a glycerin-coated chromic catgut (Softgut®) to eliminate the need for alcohol in packaging catgut and to improve handling qualities (6). The glycerin-treated chromic catgut sutures have a smoother and more uniform surface appearance than untreated catgut, and, as a result of glycerin treatment, the sutures are thicker. Davis and Geck, Ethicon, and Deknatel all have plain gut and chromic gut sutures.

Collagen can be reconstituted either from enzymatic digestion of native collagen-rich tissues or via the extraction of these tissues with salt solutions to form reconstituted collagen sutures. Reconstituted collagen, however, exhibits various polymorphic aggregated forms that are different from the native collagen. Piez reported that the formation of polymorphic aggregates of collagen depends on the environment of reconstitution (7). Reconstituted collagen sutures are prepared from the long flexor tendons of cattle. The tendon is cleaned, frozen, sliced, treated with ficin, and then swollen in dilute cyanoacetic acid. The resulting viscous gel is extruded through a spinneret into an acetone

Table 19.5 A List of Commercially Available Suture Materials

Generic name	Trade name	Physical configuration	Surface treatment	Manufacturer
Natural Absorbable Sutures				
Catgut	Catgut or Surgical gut Surgical gut	twisted multifilament	plain & chromic	A, E, D/G, SSC
Catgut	Surgigut®	twisted multifilament	plain & chromic	USS
Catgut	Softgut®	twisted multifilament	glycerin-coated	D/G
Reconstituted collagen	Collagen	twisted multifilament	plain & chromic	E
Synthetic Absorbable Sutures				
Polyglycolic acid	Dexon "S"®	braided multifilament	None	D/G
Polyglycolic acid	Dexon Plus®	braided multifilament	poly(oxyethylene-oxypropylene)	D/G
Polyglycolic acid	Dexon II®	braided multifilament	polycaprolate	D/G
Polyglycolic acid	Medifit®	braided multifilament	None	JPS
Poly(glycolide-lactide) (Polyglactin 910)	Vicryl®	braided multifilament	Polyglactin 370 and calcium stearate	E
Poly(glycolide-L-Lactide)	Panacryl®	braided multifilament	None	E
Poly(glycolide-L-Lactide)	Polysorb®	braided multifilament	coated	USS
Poly-*p*-dioxanone	PDS II®	monofilament	None	E
Poly(glycolide-co-tri methylene carbonate)	Maxon®	monofilament	None	D/G
Poly(glycolide-co-ε-Caprolactone) (poliglecaprone 25)	Monocryl®	monofilament	None	E
Glycomer 631	Biosyn®	monofilament	None	USS
Nonabsorbable Sutures				
Silk	Surgical Silk®	braided multifilament	tru-permanizing	E
Silk	Dermal®	twisted multifilament	tanned gelatin (or other proteins)	E
Silk	Virgin Silk®	twisted multifilament	None	E
Silk	Silk	braided multifilament	silicone	E & D/G
Silk	Sofsilk®	braided multifilament	coated	USS
Silk	Silk	braided multifilament	paraffin wax	SSC
Cotton	Surgical cotton	twisted multifilament	None	E
Cotton	Cotton	twisted multifilament	None	D/G
Linen	Linen	twisted multifilament	None	SSC & E
Polyester	Ethibond®	braided multifilament	Polybutilate	E
Polyester	Mersilene®	braided multifilament	None	E
Polyester	Ethiflex®	braided multifilament	teflon	E
Polyester	Dacron®	braided multifilament	None	D/G
Polyester	Ti-cron®	braided multifilament	silicone	D/G
Polyester	Surgidac®	braided & monofilament	coated with braid	USS
Polyester	Silky Polydek®	braided multifilament	teflonized	SSC
Polyester	Sterilene®	braided multifilament	teflonized	SSC
Polyester	Tevedek®	braided multifilament	teflonized	SSC
Polyester	Astralen®	braided multifilament	teflonized	A
Polyester	Polyviolene®	braided multifilament	None	L
Polyester	Mirafil®	monofilament	None	BM
Polyester	Novafil®	monofilament	None	D/G

Table 19.5 Continued

Generic name	Trade name	Physical configuration	Surface treatment	Manufacturer
Nonabsorbable Sutures				
Polyamide (Nylon 6 & 66)	Ethilon®	monofilament	None	E
Polyamide (Nylon 6 & 66)	Nurolon®	braided multiflament	coated	E
Polyamide (Nylon 66)	Surgilon®	braided multifilament	silicone	D/G
Polyamide (Nylon 66)	Dermalon® Bralon®	monofilament	None	D/G
Polyamide (Nylon 66)	Monosof®	braided monofilament	coated	USS
Polyamide	Sutron®	monofilament	None	SSC
Polyamide (Nylon 6)	Supramid®	core-sheath	None	A
Polyamide (Nylon 6)	Perlon®	core-sheath	None	PTA
Polypropylene	Prolene®	monofilament	None	E
Polypropylene	Surgilene®	monofilament	None	D/G
Polypropylene	Surgipro®	monofilament	None	USS
Polyethylene		monofilament	None	D/G
Poly(tetrafluoro-ethylene)	Gore-Tex®	monofilament	None	Gore
Poly(Hexafluoro-propylene-VDF)	Pronova®	monofilament	None	E
Stainless steel	Surgical stainless steel	mono & twisted multi-filament	None	E
Stainless steel	Flexon®	twisted multifilament	None	D/G
Stainless steel	Stainless steel	mono & twisted multi-filament	None	D/G
Stainless steel	Steel	monofilament	None	USS
Stainless steel	Stainless steel	mono & twisted multi-filament	None	A, SSC

A: Astra; BM: Braun Melsungen; E: Ethicon; PS: Japan Medical Supplies; L: Look; USS: US Surgical; D/G: Davis and Geck; Gore: W. L. Gore and Associates; SSC: Société Steril Catgut.

bath for coagulation. The coagulated fibril is stretched, twisted, and dried or treated with chromic salts before twisting and drying. These sutures are similar in appearance to catgut. They are made only in fine sizes and thus are almost exclusively used in microsurgery.

2. Polyglycolic Acid Suture (Dexon®) and Poly(glycolide-lactide) Copolymer Suture (Vicryl®)

Polyglycolic acid (PGA) was the first synthetic absorbable suture introduced in the early 1970s (8–10). It was developed by Davis and Geck under the trade name Dexon®. PGA can be polymerized either directly or indirectly from glycolic acid. The direct polycondensation produces a polymer of M_n less than 10,000 because of the requirement of a very high degree of dehydration (99.28% and up) and the absence of monofunctional impurities (11,12). For PGA of molecular weight higher than 10,000 it is necessary to proceed through the ring-opening polymerization of the cyclic dimers of glycolic acid. Numerous catalysts are available for this ring-opening polymerization. For biomedical applications, stannous chloride dihydrate, stannous octoate, or trialkyl aluminum are preferred. The resulting PGA polymer having M_w from 20,000 to 140,000 is suitable for fiber extrusion and suture manufacturing. Dexon suture fibers are made through the melting spinning of PGA chips. The fibers are stretched to several hundred percent of their original length at a temperature above the glass transition temperature (about 36°C), heat-set for improving dimensional stability and inhibiting shrinkage, and subsequently braided into final multifilament braid suture forms of various sizes. Before packaging, all Dexon sutures are subject to heat under vacuum to remove residual unreacted monomers or very low molecular weight volatile oligomers according to a patented procedure (13,14). De-

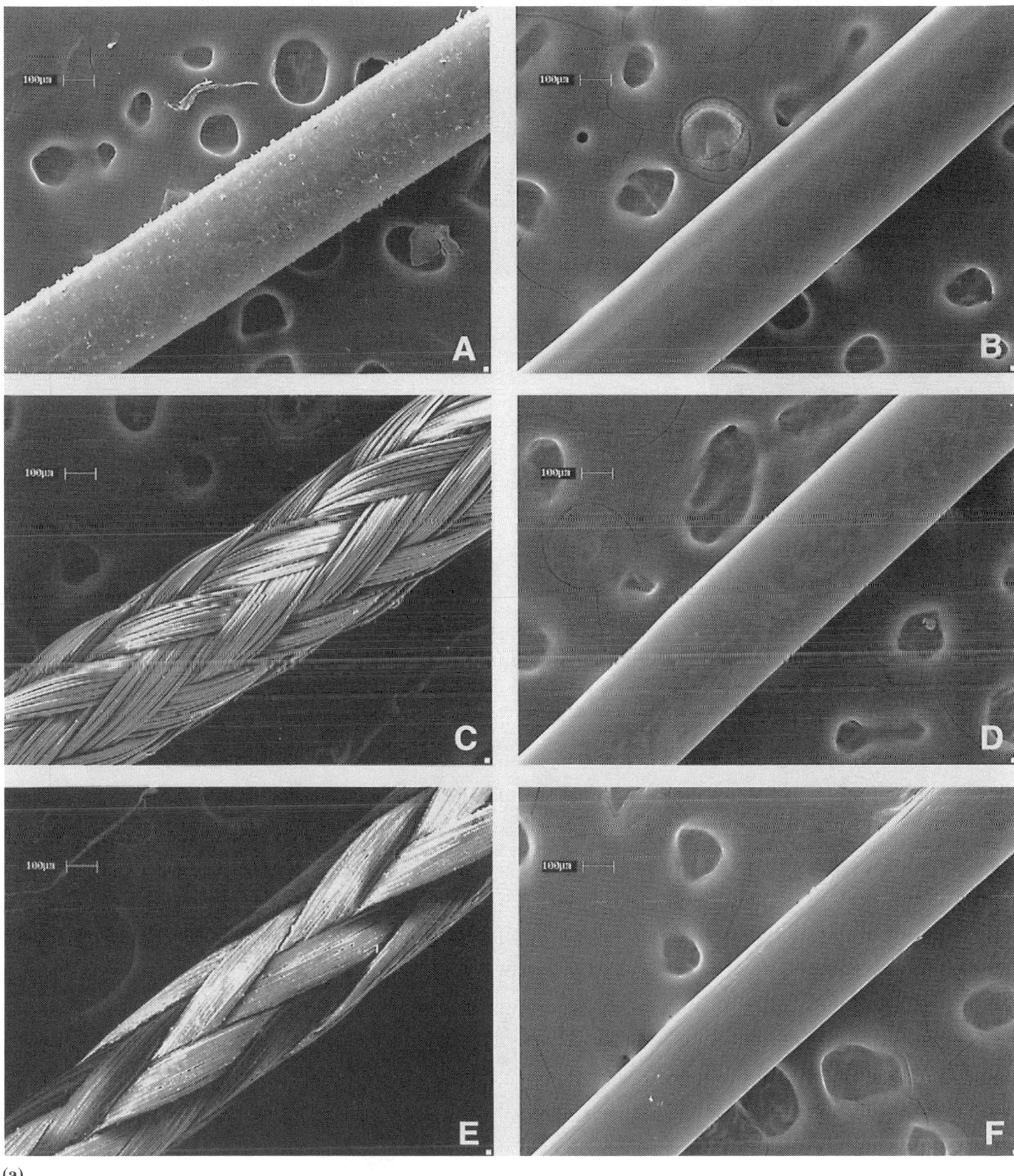

(a)

Figure 19.1 Scanning electron micrographs of commercial suture materials. (a) 2/0 size absorbable sutures: (A) chromic catgut, (B) Maxon, (C) Dexon, (D) Monocryl, (E) Vicryl, (F) PDSII. (b) nonabsorbable sutures: (A) 1/0 Silk, (B) polypropylene (2/0 Prolene), (C) polyester (2/0 Mersilene), (D) monofilament nylon (2/0 Ethilon), (E) multifilament nylon (2/0 Nurolon), (F) expanded polytetrafluoroethylene (Gore-Tex), (G) stainless steel (3/0 surgical steel).

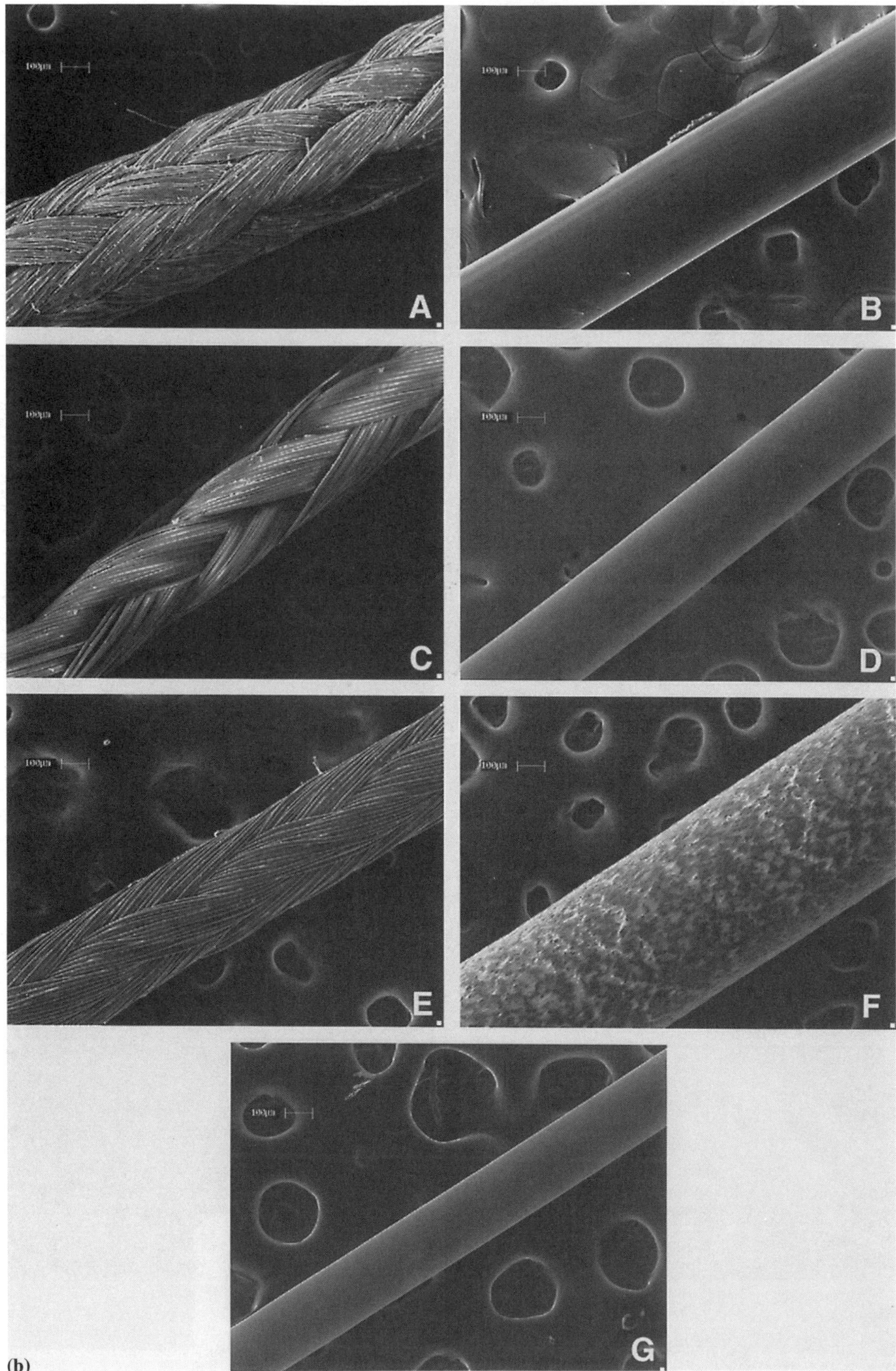

Figure 19.1 Continued

xon sutures are sterilized by ethylene oxide because of the well-known adverse effect of γ-irradiation, i.e., accelerated loss of tensile strength.

The glycolide-L-lactide random copolymer suture material (Vicryl), sometimes called polyglactin-910, is also copolymerized in the same manner as PGA. For suture use, the glycolide-L-lactide copolymers must have a high concentration of glycolide monomer (90/10 molar ratio of glycolic to L-lactic acids) for achieving proper mechanical and degradation properties. This is because, for wound closure use, synthetic absorbable sutures must have some level of crystallinity to achieve a proper tensile strength and its retention during biodegradation in vivo. Absorbable polymers of very low crystalline or totally amorphous structure would biodegrade too fast to be useful as wound closure biomaterials. Figure 2 illustrates such a relationship between the rate of degradation in vivo and the composition of glycolide to L-lactide (15). Glycolide-lactide copolymers with L-lactide composition between 25 and 75% are totally amorphous and hence would degrade the fastest. For this reason, a 90:10 molar ratio of glycolide to L-lactide has been used as the optimal comonomer ratio for Vicryl suture use. The very recent introduction of Panacryl absorbable suture, however, breaks away from this high glycolide-to-L-lactide composition ratio approach. Panacryl has a glycolide-to-L-lactide ratio almost opposite to that of the Vicryl suture. Please refer to Sec. II.A.6 Panacryl for details.

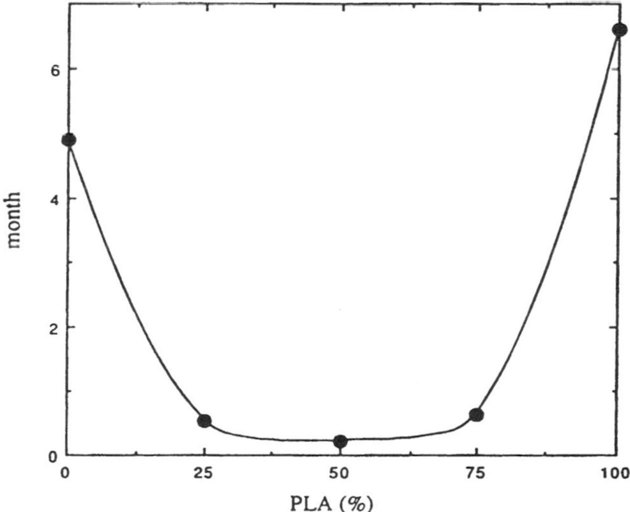

Figure 19.2 The effect of glycolide-to-lactide composition on the in vivo degradation rate of polyglactin implanted under the dorsal skin of rat. The degradation rate was expressed in terms of the time for 50% weight loss of the implant. (From P. A. Miller, J. M. Brady, and D. E. Cutright. *J. Biomed. Mater. Res. 11*:719 (1977). With permission.)

If DL-instead of L-lactide is used as the comonomer, however, the U-shape relationship between the level of crystallinity and the glycolide composition disappears. This is because polylactide from 100% DL-lactide composition is totally amorphous.

Multifilament braided Vicryl sutures are coated with 2–10% of a 50:50 mixture of an amorphous polyglactin 370 (a 65/35 mole ratio of lactide–glycolide copolymer) and calcium stearate. Vicryl sutures are sterilized by ethylene oxide like other synthetic absorbable sutures.

3. Poly-*p*-dioxanone Suture (PDS® & PDS II®)

Unlike the multifilament absorbable sutures (Dexon and Vicryl), PDS and PDSII sutures are the first commercially available monofilament absorbable sutures that derive from the glycolide family. They are poly(ester-ethers). These synthetic degradable monofilament sutures are polymerized from *p*-dioxanone monomers. The monomer 1,4-dioxanone-2,5 dione or *p*-dioxanone is obtained from the reaction of chloroacetic acid with metallic sodium dissolved in a large excess of ethylene glycol (16). The resulting *p*-dioxanone is purified by multiple redistillation and crystallization. Fiber grade high-molecular-weight PDS polymer is made from the ring-opening polymerization of highly purified *p*-dioxanone (>99%) in the presence of organometallic catalysts such as Et_2Zn or zirconium acetylacetonate (17). The resulting polymer has an inherent viscosity of 0.70 (0.1% polymer solution in tetrachloroethane at 25°C) and a degree of crystallinity of 37%. Poly-*p*-dioxanone polymer has $T_g = -16$ to -10°C and $T_m = 110$–115°C. Monofilament PDS sutures are made from the melt-spinning of the dried polymer chips through a spinneret into monofilaments of any desired suture diameter (16). The extruded fibers are then drawn about five times at $T > 43$°C and heat set to orient the molecules for better physical and mechanical properties. The drug/cosmetic violet dye #2 is added to make violet-colored PDS sutures.

PDS and PDSII are chemically identical, but the PDSII suture differs from the PDS in fiber morphology resulting from different fiber spinning conditions. PDSII sutures are made by subjecting the melt-spun PDS fibers to a short period of annealing at a temperature above T_m of PDS (about 125°C) (18). This additional heated drawing treatment, not used in the PDS suture, partially melts the surface layer of PDS fibers and subsequently modifies the near-surface crystalline structure of the PDS monofilament suture. Thus a distinctive skin-core morphology that PDS sutures do not have is observed in PDSII sutures (18). The core of the PDSII suture has a more highly ordered and larger spherulitic crystal structure than the surrounding annular area characterized by smaller crystals, while the untreated PDS suture shows a relatively even crystalline structure throughout its cross section. In general, PDSII sutures have a lower modulus of elasticity with improved flexibility and handling characteristics. Recently, there are a few varieties of PDS-related copolymers reported in the literature that may have use as wound closure biomaterials (19–22). A copolymer of PDS and PGA (20%) has an absorption profile similar to Dexon and Vicryl sutures, but it has compliance similar to PDS. A copolymer of PDS and PLLA (15%) results in a more compliant suture than homopolymer PDS but with similar absorption profiles as PDS.

PDS sutures are sterilized by ethylene oxide in the same way as other synthetic absorbable sutures, and no coating is used. Based on the PDS chemical structure, i.e., ether and ester linkages separated by a methylene group (CH_2), PDS is expected to be and has been found to be sensitive to oxidative, photo-oxidative, and γ-irradiation degradation.

4. Poly(glycolide-trimethylene carbonate) Copolymer Suture (Maxon®)

Another commercially available monofilament absorbable suture, Maxon, is made from block copolymer of glycolide and 1,3-dioxan-2-one (trimethylene carbonate or GTMC). It consists of 32.5% by weight (or 36 mol%) of trimethylene carbonate (23,24). Maxon is a poly(ester-carbonate). The polymerization process of Maxon suture is divided into two stages. The first stage is the formation of the middle block, which is a random copolymer of glycolide and 1,3-dioxan-2-one (trimethylene carbonate). Diethylene glycol is used as an initiator, and stannous chloride dihydrate ($SnCl_2 \cdot 2H_2O$) serves as the catalyst. The polymerization is conducted at about 180°C. The weight ratio of glycolide to trimethylene carbonate in the middle block is 15:85. After the synthesis of the middle block, the temperature of the reactive bath is raised to about 220°C to prevent the crystallization of the copolymer, and additional glycolide monomers as the end blocks are added into the reaction bath to form the final triblock copolymer. The undyed Maxon has a natural clear appearance, while green Maxon is dyed by green DG#6 with $<0.3\%$ by weight. The maxon suture is sterilized by ethylene oxide, and no coating is used.

5. Glycolide-Caprolactone Copolymer Suture (Monocryl®)

Monocryl is a monofilament absorbable copolymer suture of glycolide and ε-caprolactone. The Monocryl suture is a segmented block copolymer consisting of both soft and hard segments. The purpose of having soft segments in the copolymer is to provide good handling properties like pliability, while the hard segments are used to provide adequate strength. Monocryl is made in a two-stage polymerization process (25). In the first stage, soft segments of prepolymer of glycolide and ε-caprolactone are made. This soft-segmented prepolymer is further polymerized with glycolides to provide hard segments of polyglycolide. Monocryl has a composition of 75% glycolide and 25% ε-caprolactone.

An aspect of Monocryl monofilament suture is its pliability as claimed by Ethicon (25). The force required to bend a 2/0 suture is only about 2.8×10^4 lb-in^2 (193 MPa) for Monocryl, while the same size PDSII and Maxon monofilament sutures require about 3.9 and 11.6×10^4 lb-in^2 (269 and 800 MPa) force, respectively. This inherent pliability of Monocryl is due to its low glass transition temperature resulting from the ε-caprolactone comonomer unit. Its T_g is expected to be between 15 and -36°C. Unlike Maxon and PDSII, the Monocryl suture appears to have less out-of-package memory, which improves its handling characteristics.

6. Poly (glycolide-L-lactide) Copolymer Suture (Panacryl®)

The latest addition to the arsenal of synthetic absorbable sutures is Panacryl, a glycolide-L-lactide copolymer absorbable suture from Ethicon, Inc. It is the first and only braided synthetic absorbable suture that provides long-term mechanical strength support over six months. Panacryl is made from the same copolymer components as the Vicryl suture, i.e., glycolide and L-lactide; but the composition ratio of these two components was almost exactly opposed to that of Vicryl, i.e., a 10/90 molar ratio of glycolide to L-lactide. Since poly-L-lactide biodegrades at a significantly longer time than polyglycolide, Panacryl is expected to retain its strength and mass much longer than Vicryl. Ethicon claimed that Panacryl can retain about 80 and 60% of its original tensile strength at the end of 3 and 6 months in vivo, respectively. Based on the tensile strength loss profile, the Panacryl suture appears to bridge the gap between traditional absorbable and nonabsorbable sutures. The coating of Panacryl is ε-caprolactone/glycolide copolymer.

Panacryl has a unique braided construction that allows for excellent handling and knot tying and provides knot security at four to five throws. Because it is coated, it passes smoothly through tissue and allows for 100% successful knot slides. The undyed Panacryl has a white color that may provide good visibility in the presence of blood and tissue.

The Panacryl absorbable suture is suggested for use in general soft tissue approximation and/or ligation, and tissues that heal slowly like tendons and ligaments, and for reattachment to bone, or patents with compromised wound healing capability due to diabetes, malignancy, deficiency in the immune system, obesity, and malnutrition. Due to its pending patents, there is virtually no published technical information about the Panacryl suture at this time.

7. Glycolide-Trimethylene Carbonate Triblock Copolymer Suture (Biosyn®)

Biosyn is a triblock copolymer, Glycomer 631 from US Surgical (26). This block copolymer consists of glycolide (60%), dioxanone (14%), and trimethylene carbonate (26%). The center block is a random copolymer of 1,3 dioxanone-2-one (65% by weight) and 1,4 dioxane-2-one (35% by weight). The two ends of this center block are capped by a block copolymer consisting mainly of glycolide (>50%) that could be copolymerized with lactide, trimethylene carbonate, *p*-dioxanone, or ε-caprolactone.

The purpose of using triblock polymer chemistry to design Biosyn is to make synthetic absorbable monofilament sutures having good handling properties like flexibility without compromising other mechanical properties, like knot strength and security and biodegradation properties. As we shall describe later (B.S, Handling Properties), sutures must be bent for packaging. Due to the viscoelastic nature of synthetic polymers, the bent sutures during storage would introduce "packaging memory" to the sutures. Therefore, before use, the packaged suture material must be straightened out from its bent position upon its removal from the package. Such an ability to remove the kinks of a suture after a long period of shelf life is a very important handling characteristic that surgeons like to have during wound closure.

Biosyn has the next lowest modulus of elasticity (145×10^3 psi) among all existing synthetic absorbable suture materials, and its strain energy at both 5 and 10% strain (0.84 and 2.76 kg-mm, respectively) is about half of the corresponding size Maxon sutures (26). The rate of tensile strength loss of Biosyn sutures is similar to that of Vicryl suture in vitro with about 8% of its original tensile breaking strength remaining at the end of 28 days.

8. Silk Suture

Silk is one of the three major fibrous proteins (the others are wool and collagen). Silk fiber is semicrystalline and consists of two major constituents, a fiber protein called fibroin and the gummy substance called sericin that holds the fibers together. As it is spun by the silkworm, silk fiber consists of two triangular filaments stuck together with sericin. The most significant feature of the amino acid composition of silk fibroins is the high concentration of glycine, alanine, and serine (27). Together, these three small amino acids take up about 80 to 85% of the total amino acids in *Bombyx mori* and *Anaphe pernyi*. Such a high concentration of these three amino acids having small and simple side groups in silk fibroins permits the arrangement of polypeptide molecules into an orderly and crystalline manner and is responsible for the desirable mechanical, physical, and chemical properties of silk fibers. The amino acids with bulky side groups like tyrosine consist of a smaller portion of silk fibroins and cannot be accommodated within the three-dimensional, ordered crystalline structure. Thus disordered amorphous regions coexist with the crystalline ones. Silk fibers have about a 60% level of crystallinity (28,29). This arrangement of crystalline to amorphous domains in fibroin bears an important relationship to properties, such as moisture absorption and biodegradation, that occur mainly in the amorphous regions of silk. Cross-sectional views of silk fiber shows somewhat irregular triangular shape.

Sericin is the gummy protein that holds the fibroin filaments together in cocoon silk and consists of about 25% of the cocoon by weight. The hydrophilic nature of sericin derives from the high concentration of amino acids with side polar groups like —OH and —COOH. In most silk sutures, except Virgin Silk for microsurgery, sericin is removed.

The most common form of silk suture is produced by the silkworm *Bombyx mori*. The process of making silk sutures is similar to that for traditional silks. It involves the collection of the raw silk fibers from the cocoon and the removal of the natural waxes and sericin (called degumming). The cleaned silk fibers are then either twisted or braided to form the suture strands. Silk has a relatively high standard moisture regain of 11%. Various types of surface treatments are used to render it noncapillary, serum proof, incapable of the ingrowth of tissues, or all of these. Wax or silicone has been used as the coating material. Tissue ingrowth is prevented by encasing the twisted silk fibers in a nonabsorbable coating of tanned gelatin or other protein substances. The trade name of this specially treated silk suture is Dermal® from Ethicon. Other trademarks of silk sutures are braided Surgical silk and twisted Virgin Silk from Ethicon, and braided Protein Silk (Davis and Geck). With the very fine size silk sutures for microsurgery and ophthalmology, the sericin component of the Virgin Silk is not removed during manufacturing processing. Silk sutures are sterilized by either Co^{60} γ-irradiation or ethylene oxide.

The mechanical properties of silk closely correlate with the fraction of bulky side groups present and thus with the crystalline-amorphous ratio as well as the conformational arrangement of chain segments in the amorphous domains. Earland and Robins (30) and Zuber (31) reported that peptide chain segments located in the amorphous domains are interconnected among themselves and to the crystallites via disulfide and ester linkages and hydrogen and ionic bonds where they are appropriate. Silk sutures have a relatively very high modulus of elasticity. This high elastic modulus is attributed to the strong inter- and intramolecular interactions among peptide chain segments in the amorphous domains as well as interactions between these randomly oriented amorphous and highly ordered crystalline peptide chain segments through various types of H-bonds, ionic bonds, and ester and disulfide linkages. If silk sutures are stretched to beyond their yield point, these inter- or intramolecular bonds are broken first, and randomly oriented amorphous peptide chain segments are thus allowed to be extended. At this phase of the stress–strain curve, a characteristic low modulus yield plateau is evident. The degree of yield plateau depends on the strength and amount of inter- and intramolecular force among the amorphous peptide chain segments and between the amorphous and crystalline peptide chain segments. A weaker inter- and intramolecular force would result in a pronounced yield plateau.

The macromolecular structure of silk fibers is also responsible for their well-known loss of tensile strength in vivo. Any loss of fiber tensile strength could be attributed to either the scission of primary bonds in the backbone macromolecules or/and the breakage of secondary bonds such as H-bonds due to a reactive species like water. The moisture regain of a typical silk fiber is about 9.9% due to the high concentration of polar side groups in amino acid residues. Water molecules absorbed by silk fibers would reside in the amorphous domains and compete with the amino acids in the amorphous chain segments for inter- and intramolecular interaction (27). This competition by water molecules would lead to a more open amorphous structure by the replacement of relatively strong inter- and intramolecular interaction in the amorphous domains with water. As a result, a lower tensile strength is observed. The observed in vivo strength loss of silk sutures might be a combination of the scissions of both primary and secondary bonds; however, the breakage of secondary bonds within silk fibers in water probably plays a more important role than the scission of primary bonds. This is because the high crystallinity level found in most silk fibers should retard the diffusion of relatively large proteolytic enzymes (32) and hence render silk fibers resistant to proteolytic enzymatic hydrolysis.

Because of the undesirable in vivo loss of tensile strength and the somewhat higher tissue reactions and ingrowth to silk sutures, Shalaby et al. reported the impregnation of silk with hydrophobic thermoplastic elastomers to improve the performance of silk sutures (33). The rationale behind the approach is to provide an inert barrier between silk sutures and the surrounding tissue so that tissue reactions and cellular invasion could be minimized. This modified silk suture was prepared by treating a multifilament silk suture with a solution of a highly flexible hydrophobic and deformable polymer in a solvent and heating the moving suture to obtain a continuous impregnation of the silk with the elastomer. This elastomer with a molecular weight >10,000 daltons is a segmented polyetherester and consists of soft segments, poly(polyxytetramethylene) terephthalate, and crystallizable hard segments, polybutylene terephthalate. The elastomer has 5 to 50% by weight of the composite silk suture. In vivo evaluation of this composite silk suture in Sprague Dawley rats indicated that the impregnation of silk sutures indeed not only provided better tensile breaking strength retention over a period of 56 days but also significantly reduced tissue reactions and cellular invasion.

9. Polyester-Based Sutures

There are three types of nonabsorbable polyester-based sutures: poly(ethylene terephthalate) (PET), poly(butylene terephthalate) (PBT), and the copolymer of poly(tetramethylene ether) terephthalate and poly(tetramethylene terephthalate) (polybutester). PET based polyester sutures are braided Dacron®, TI-CRON®, from Davis and Geck, Ethibond®, Mersilene®, Ethiflex®, from Ethicon, Surgidac®, from US Surgical, Polydek® and Tevdek® from Deknatel, and monofilament Mirafil®, from B. Braun. PBT-based polyester sutures are monofilament ® from B. Braun. The monofilament copolymeric polybutester suture, Novafil® from Davis and Geck, is the most recent polyester-based suture.

PET sutures are made of polyethylene terephthalate, which in turn is polymerized from ethylene glycol and terephthalic acid (or dimethyl terephthalate in the case of Terylene®). Polymerization is conducted in a vacuum at a high temperature. The first stage involves the formation of low-molecular-weight oligomers through ester interchange. The oligomers are then polycondensed further to build up high-molecular-weight polyester. The second stage is carried out at about 270°C and reduced pressure (0.5 torr). The resulting polyester chips are melt-spun into round cross-sectional shaped filaments. The filaments are then hot-stretched (above T_g 69°C) to about five times their original length. Further crystallization occurs during the hot drawing. Depending on the degree of drawing, polyester yarns of either normal strength or high tenacity can be obtained. The molecular weight of PET that is capable of making fibers is of the order of 20,000.

Due to the benzene ring, PET forms a single rigid structure. Consequently, the PET chain is rigid and less flexible than nylon and polyethylene. This accounts for the slow rate of crystallization and high melting point (265°C). The melt-spun PET filaments are thus largely amorphous before drawing. The maximum rate of crystallization of amorphous PET occurs at 190°C. The molecular structure of PET is also winding-speed-dependent during the melt-spinning process. PET yarns wound at a speed <3,000 meters/minute has a low level of crystallinity, while at a higher speed, an increasingly higher level of crystalline structure develops but reaches a plateau at 7,000 meters/min (34). A range of crystal density from 1.455 to 1.515 g/cm^3 has been reported, but the amorphous density has a narrower range from 1.335 to 1.336 g/cm^3

Novafil® is a relatively new monofilament copolyester suture (Polybutester) and is made from a block copolymer of poly(butylene terephthalate) as the hard segment and poly(tetramethylene ether) glycol terephthalate as the soft segment (35). Novafil suture fibers made from polybutester copolymer are drawn in two stages from 6× to 8× at temperature from 120 to 165°C to provide desirable tensile strength, knot strength, security, flexibility, fatigue life, and low tissue drag. The number average molecular weight ranges from 25,000 to 30,000. The ratio of the hard and soft segments can be adjusted to achieve the desirable handling properties. Novafil suture has the ratio of hard to soft segments of 84/16% (36). Titanium dioxide, carbon black, or iron oxide could be used to color the suture. Novafil sutures can be sterilized by ethylene oxide or γ-irradiation, but they should not be sterilized by heat because the elastomeric property of the sutures could be adversely affected.

Because of its elastomeric character, Novafil has a distinctly different stress–strain behavior than PET based polyester sutures and other synthetic monofilament sutures like nylon and polypropylene. The stress–strain curve of Novafil suture exhibits a biphasic pattern, while both PET-based polyester sutures and polypropylene and nylon monofilament sutures show a monophasic pattern (36,37). Novafil suture elongates to about 10% easily under tension; a steep rise in force is required for further elongation. This mechanical behavior may be advantageous in management of wound edema because the suture loop easily stretches instead of cutting through the edematous tissue. After the resolution of the edema, a Novafil suture loop is expected to return to its original diameter more easily than those sutures exhibiting monophasic stress–strain behavior. As a result of this elastomeric character, Novafil is expected to be less stiff than other monofilament sutures, as is evident in the reported stiffness coefficient of Novafil (14.81 ± 1.14), which is about two folds less than equivalent size nylon (Dermalon®) and polypropylene (Surgilene®) based monofilament sutures. The elastomeric characteristic of polybutester block copolymer also renders Novafil less susceptible to package memory than polypropylene and nylon based monofilament sutures.

Although aromatic polyester based sutures are not considered to be degradable and there is no reported clinical case of failed polyester sutures due to the hydrolytic scission of their ester linkages, the presence of these ester linkages inherently provides the opportunity for their eventual hydrolytic scissions. The relatively hydrophobic nature of the polyester sutures and their high glass transition temperatures (relative to body temperature) are responsible for the observed in vivo stability of this class of suture materials. Under alkaline conditions, the hydrolytic degradation of polyester sutures is a surface phenomenon with very little change in molecular weight, and the rate of hydrolysis is inversely proportional to the diameter of the fibers. This is because highly ionized reagents like NaOH cannot readily diffuse into the relatively nonpolar polyesters. Because fiber diameter would decrease with alkaline hydrolysis, tensile strength (force/cross-sectional area) of the fiber would not change. However, the rate of alkaline hydrolysis would increase significantly if the alkaline agent could readily diffuse into the fiber by the addition of chemicals like quaternary ammonium compounds to serve as carriers for —OH anions. Acid catalyzed hydrolysis of polyester fibers is usually significantly slower that alkaline catalyzed hydrolysis. Recent studies have also suggested that Novafil suture may degrade if it is used in areas exposed to sunlight, such as in cataract surgery. Hence the long-term stability of Novafil sutures in ophthalmologic surgery has been reported to be a concern.

10. Polyamide Based Sutures

The most successful aliphatic polyamide that polymer scientists have synthesized is nylon. Nylon is polymerized either from polycondensation of a dicarboxylic acid and a diamine, or through a ring-opening polymerization of appropriate lactams. Although there are numerous types of nylon, such as nylon 3, 4, 5, 6, 7, 8, 9, 11, 12, 66, and 610, only 66 and 6 are used to make suture materials. The former is more popular in the United States, while the latter is used in Europe. The tradenames of nylon sutures are braided Surgilon® and monofilament Dermalon® from Davis and Geck, braided Nurolon® and monofilament Ethilon® from Ethicon, monofilament Monosof® and braided Bralon® from US Surgical, and sheath-core structure Supramid® from S. Jackson. Surgilon nylon suture is coated with silicone, while Nurolon nylon suture is coated with wax. Bralon is coated with a proprietary material. All nylon sutures are sterilized with Co^{60} γ-irradiation.

Nylon 66 is made from adipic acid and hexamethylene diamine. These two chemicals, dissolved separately in methanol, are mixed to form nylon salt as precipitate. Equal moles of adipic acid and diamine are important for achieving high molecular-weight nylon 66. Nylon salt is then melted under an inert gas atmosphere to prevent discoloration. Acetic acid is added as a stabilizer during polycondensation. No catalyst is needed for polymerization; suitable conditions are four hours at a temperature of 280°C. The molten polymer is extruded in ribbon form, which is quenched with cold water in order to reduce the crystal size, and then cut into chips. The extent of polymerization is determined by the residual moisture content. A vacuum should be applied if high-molecular-weight nylon is desirable (38). Nylon 66 can also be made by interfacial polycondensation (38). The reaction takes place at the interface between a diamine water solution and a dicarboxylic acid chloride in a water-insoluble solvent. The reaction is rapid, and a precise amount of the reactants is unnecessary. This method results in high-molecular-weight nylon 66. Nylon 66 of molecular weight 12,000 to 20,000 is suitable for melt-spinning into fibers.

Nylon 6 is made from caprolactam. Two alternative methods are used (39). The first method involves the liquefying and heating of caprolactam under high pressure. The resulting nylon 6 polymer chain consists of an average of 200 repeating units. The second method needs about 10% water and is carried out at a high temperature with a controlled release of steam. There are three reactions in the second method of polymerization: addition, condensation, and hydrolysis (38).

The resulting nylon 66 or 6 chips are then melt-spun into a cooling chamber in which nylon filaments form. In the case of nylon 66, the filaments are run through a steam chamber to wet them before they are wound. This treatment eliminates the undesirable extension of the yarns after they reach equilibrium by absorbing moisture from the air. The wound yarn is further cold-drawn by stretching about 400% in order to acquire better strength. During this cold-drawing process, vegetable oil is applied to the yarn as a lubricant and is washed off afterwards.

In the case of nylon 6, the same process of melt-spinning as for nylon 66 is used except that the steam chamber is not suitable. The relatively high residual concentration of monomers in nylon 6 would make the filaments sticky and adhere to one another if steamed. The filaments are normally drawn to 350–400%. Higher drawing can be achieved if a stronger fiber is needed. It is well recognized in fiber technology that by increasing the draw ratio (the ratio of the speed of output and input feed rollers), the tensile strength and elastic modulus increase, while the elongation at break decreases. If the molecular weight of nylon 6 is between 20,000 and 25,000, hot-drawing is used.

Unlike PET, both nylon fibers are fairly crystalline when they are spun. The crystal structures of nylon 66 and 6 fall into α, β, and γ phases. Almost all the commercially important nylons exist in either the α or the β phase.

Because of the inherent susceptibility of the amide linkage to hydrolytic degradation, nylon sutures have been reported to lose strength after implantation. Thus nylon sutures should not be used for fastening implants.

11. Polypropylene Sutures

Polypropylene (PP) suture materials are made from isotactic polypropylene, which is polymerized from propylene with a Ziegler–Natta catalyst (40). The Z-N catalyst consists of a transition metal halide (e.g., $TiCl_3$) with a reducing agent (e.g., AIR_3). Ziegler–Natta catalyst systems are quite complex, and the exact structure of the catalysts cannot be determined precisely (41,42). Many factors could influence the activity of the catalyst, such as the crystal structure, the molar ratio of the components, the aging of the complex, impurities, and the temperature of preparation (42a). In addition to isotactic polypropylene suture, a syndiotactic PP suture has recently been reported by Liu et al. (42b).

PP usually has a wide range of molecular weight distribution (MWD) ranging from 2 to 12 polydispersity (M_w/M_n). Recently, narrower MWD PP has become available and is called controlled rheology (CR) PP (43). The melting and glass transition temperatures of PP are about 165 and −15°C, respectively.

PP sutures made from isotactic polypropylene have a molecular weight of about 80,000 or a melt flow rate between 3 and 35. The crystallinity is 50% before melt-spinning and decreases to 33% after spinning and before drawing, and increases to 47% after drawing. Annealing increases the crystallinity further to 68% (43). The currently available monofilament polypropylene sutures are Surgilene® from Davis and Geck, Prolene® from Ethicon, and Surgipro® from US Surgical. PP sutures, in general, do not have any coating material applied, and they are sterilized by ethylene oxide because of their sensitivity toward Co^{60} γ-irradiation sterilization.

PP fibers are inherently unstable to both heat and light, but they are totally hydrolytically resistant due to the lack of ester and amide linkages. The heat and light instability arise from the high temperature during fiber spinning, which frequently leads to the formation of oxygen-containing functional groups like >C=O in the molecule. This PP fiber instability may be the cause of the observed clinical failure of PP sutures in ophthalmology.

12. Polytetrafluoroethylene Sutures (Gore-Tex®)

A nonabsorbable monofilament suture, Gore-Tex®, is made from a highly crystalline linear polytetrafluoroethylene (PTFE). This fully fluorinated thermoplastic polyolefin is an addition polymer and is polymerized through a free radical polymerization route in aqueous dispersion under pressure with persulfates and hydrogen peroxide as initiators. PTFE has the highest enthalpy and entropy of polymerization (−156 KJ/mol and −112 J/mol-deg, respectively) in vinyl polymerization (44). Its molecular weight can be as high as 5×10^6.

Due to the extremely stable C—F bond, PTFE has a very high melting temperature, 327°C, which makes the fabrication of PTFE very difficult. This difficulty is further compounded by a high viscosity above T_m due to restricted bond rotation and high molecular weight. Thus PTFE is usually fabricated by a combination of heat and pressure, namely sintering, which is usually applied to metal and ceramics. PTFE powders are preformed at high pressure (2,000–10,000 psi, 14–70 MPa) at room temperature and then sintered above their melting point (>365°C) for a brief period. The resulting products can subsequently be machined to the desirable shape and size. Microfibrous PTFE is made from wet-spinning of a mixture of an aqueous PTFE dispersion and cellulose xanthate to provide fibers that are subsequently sintered at 385°C by contact with a metal roll to develop fibers of low but useful strength.

The monofilament suture made from PTFE (Gore-Tex®), however, is different from molded PTFE in morphological structure. The Gore-Tex suture is an expanded PTFE characterized by two distinctive components: nodes held together by fine fibrils 5–10 μm in diameter and >17 μm long. The most unique property of Gore-Tex is its microporous structure with up to about 9 billion pores/in^2. Hence the Gore-Tex suture has >50% air by volume. The pore size varies but is big enough for ingrowth of fibroblasts and leukocytes similar to those observed with Gore-Tex vascular grafts. The size of the Gore-Tex suture does not follow USP classification and is designated by CV (i.e., cardiovascular). Its suture diameter is measured in its pre-expanded dimension, and the actual diameter of a Gore-Tex suture is obviously much larger because it contains >50% air by volume. For example, the CV-4 Gore-Tex suture has a similar diameter (0.35 mm) to a 2-0 Prolene (0.303 mm).

Because of its large pore volume, the Gore-Tex suture has a unique property that other sutures do not have, namely a needle-to-suture diameter ratio of one (45). Other commercial sutures have a needle-to-thread ratio >1.0, and they frequently range from 2.0 to 3.0. As a result of this high needle-to-suture diameter ratio, the thread portion of a suture cannot fill up the hole generated by a needle in wound closure. Bleeding through this unfilled space at needle holes has been a common problem associated with sutures other than Gore-Tex. Miller et al. reported that a 5-0 Gore-Tex suture had about 1/3 of the needle hole blood leakage of the same size Prolene in the abdominal aorta of mongrel dogs (45). Thus Miller et al. recommended that the Gore-Tex suture should be a better choice for wound closure in multiple arterial anastomoses like complicated extra-anatomic bypasses, fully heparinized patients, and aortic surgery, where larger needle holes are made. The unique porous structure of the Gore-Tex suture also results in very low bending stiffness (46,47). Dang et al. reported that the CV-4 Gore-Tex suture has a bending stiffness coefficient of 1.21, while a 2-0 size Prolene is more than 100 times stiffer with bending stiffness coefficient of 180.07 (46). Similar findings were reported by Chu and Kizil (47)

13. Poly (Hexafluoropropylene-VDF) Suture (Pronova®)

Pronova® is the latest new synthetic nonabsorbable monofilament suture from Ethicon and is a copolymer from polyvinylidine fluoride homopolymer (PVDF) and polyvinylidine fluoride hexafluoropropylene copolymer over a wide range of composition ratio. It appears that Pronova is similar to the PVDF monofilament suture developed by Peters Laboratoire Pharmaceutique (Bobigny, France). PVDF was designed to provide a wound closure biomaterial in vascular surgery that would have very good antithrombogenicity, with the same satisfactory handling characteristics as polypropylene sutures, and yet be as durable as polyester sutures. It was the developer's hope to replace polypropylene with PVDF in vascular surgery for some of its unique properties not available with isotactic polypropylene sutures. Combining PVDF with polypropylene into one single entity like Pronova should combine the merits of PVDF and polypropylene and achieve a better nonabsorbable monofilament suture. Ethicon claims that Pronova has excellent handling properties like pliability and low package memory with very good tensile breaking strength and resistance to fraying and damage done by surgical instrument. One of the properties of Pronova is that its chemical composition of PVDF to polyvinylidine fluoride hexafluoropropylene copolymer depends on the suture size.

Pronova sutures of smaller size (4/0 to 10/0) have the polymer composition ratio of 80/20 for the purpose of retaining adequate tensile strength without the expense of their handling characteristics, since finer sutures are inherently weaker than larger ones. Larger Pronova sutures (2/0 to 2 size), however, have a 50/50 composition ratio of the two constituent polymers to achieve better handling properties without the expense of their tensile strength, because larger monofilament sutures are well known to be more rigid and less pliable than finer ones. Pronova sutures are designed mainly for cardiovascular, ophthalmic, and neurological surgeries.

Although the detailed technical information of Pronova is very limited at this time, an examination of PVDF sutures should shine some light on this relatively new nonabsorbable monofilament Pronova suture.

The chemical, physical, mechanical, morphological, and biocompatible properties of 5/0 and 6/0 PVDF sutures (Teflene®) were recently reported by Urban et al. (48). PVDF and polypropylene sutures were found to be similar in tensile breaking force and biocompatibility in blood vessels. The three properties that most differentiate PVDF from polypropylene sutures are creep behavior, the extent of iatrogenic trauma by a needle holder, and sterilization by γ-irradiation. Over a period of 10^3 minutes there was about a 10% dimensional increase in PVDF sutures, while polypropylene (Prolene) sutures had more than a 50% dimensional increase. Thus it appears that PVDF sutures have a better resistance to creep than polypropylene sutures, and hence they are expected to be more dimensionally stable. However, PVDF exhibited more dimensional change than polypropylene during the initial 30 min of creep testing.

PVDF sutures appeared to sustain the damage from needle holders better than polypropylene sutures, at least from the surface morphological point of view. PVDF sutures showed some flattening with a roughened surface, but they did not have the longitudinal cracks and fibrillar formation found with polypropylene sutures. This morphological difference between PVDF and polypropylene sutures, however, was not reflected to a marked degree in their tensile breaking force values. In other words, the iatrogenic trauma done by needle holders on polypropylene and PVDF sutures did not reduce their tensile breaking force significantly.

PVDF and polypropylene sutures have very similar melting temperatures (165–175°C), but distinctively different levels of crystallinity. PVDF has a level of crystallinity 59 ± 7%, while polypropylene has 43 ± 3%. Because of the lack of an a alkyl group, PVDF can be sterilized by the conventional γ-irradiation method, while polypropylene requires the use ethylene oxide gas. Thus PVDF can take advantage of the efficiency and convenience of γ-irradiation sterilization. Like the polypropylene suture, the PVDF suture should not have any O_2 in its chemical structure. However the surfaces of PVDF and polypropylene sutures showed oxidation products as confirmed by electron spectroscopy for chemical analysis. The amounts of O_2 on the surface of PVDF and polypropylene sutures were 7.4 and 7.9%, respectively. However, bulk FTIR data failed to reveal such oxidation products. This suggests that oxidation of PVDF and polypropylene is introduced during melt-spinning of fibers and is mainly restricted to the surface of suture fibers.

PVDF sutures showed a similar histological response as a polypropylene suture. At the end of 6 months' implantation in adult mongrel dogs, PVDF sutures were encapsulated by a thin layer of newly formed connective tissue with the absence of inflammatory cells. The explanted and cleaned PVDF sutures revealed no visible surface damage or degradation.

14. Stainless Steel Sutures

The use of iron base metallic sutures started as early as 1666. The most commonly used metallic suture now is stainless steel based. Stainless steel is an alloy of mainly iron, chromium, and nickel, but due to the wide range of possible alloy compositions, the incorporation of trace elements, and fabrications, there are many different grades of stainless steel. Table 6 lists some examples of stainless steels and their structure (49). The three-digit classification in the table is based upon the main composition in the alloy stipulated by the American Iron and Steel Institute (AISI). Series 200 has mainly chromium, nickel, and manganese, Series 300 has mainly chromium and nickel, Series 400 has mainly chromium, and Series 500 has low chromium.

Only 304, 316, and 316L wrought stainless steels are used as sutures. Series 300 stainless steel has high chromium and nickel contents with an austenitic structure and is characterized by high corrosion resistance and ductility. The L in 316L stainless steel indicates extra low carbon content, since a high level of C can result, under certain circumstances, in the precipitation of chromium as chromium carbides ($Cr_{23}C_6$), which depletes the chromium content of the matrix, making the Cr depleted areas more susceptible to corrosion. Thus the use of extra low C would reduce this undesirable precipitation effect. Presently, extra low C stainless steel is made by the electroslag-remelted (ESR) or the vacuum-remelted (VM) processes. Different metallurgical processing conditions and finishing have a large effect on the mechanical properties and degree of corrosion resistance of stainless steel.

Stainless steel sutures are fabricated from cold-worked

Table 19.6 Representative Stainless Steel Compositions[a] (%) and Structure

Alloy designation (wrought)	Carbon	Manganese	Phosphorus	Sulfur	Silicon	Chromium	Nickel	Molybdenum	Others	Structure
302	0.15	2.00	0.045	0.03	1.00	17.0–19.0	8.0–10.0			Austenitic
303	0.15	2.00	0.020	≤0.15	1.00	17.0–19.0	8.0–10.0	0.6[b]		Austenitic
304	0.08	2.00	0.045	0.03	1.00	18.0–20.0	8.0–10.5			Austenitic
305	0.12	2.00	0.45	0.30	1.00	17.0–19.0	10.5–13.0			Austenitic
316	0.08	2.00	0.045	0.03	1.00	16.0–18.0	10.0–14.0	2.0–3.0		Austenitic
316L	0.03	2.00	0.045	0.03	1.00	16.0–18.0	10.0–14.0	2.0–3.0		Austenitic
317	0.08	2.00	0.045	0.03	1.00	18.0–20.0	11.0–15.0	3.0–4.0		Austenitic
431	0.20	1.00	0.04	0.03	1.00	15.0–17.0	1.25–2.50			Martensitic
Precipitation hard-enable 17-7 PH[c]	0.09	1.00	0.04	0.03	1.00	16.0–18.0	6.5–7.75		0.75–1.5 Al	Austenitic/ martensitic
Cast CF-8M	0.08	1.50	0.04	0.04	2.00	18.0–21.0	9.0–12.0	2.0–3.0		Austenitic/ ferritic

[a] Single values are maximum values unless otherwise noted.
[b] Optional.
[c] Trademark of the Armco Steel Corporation.
Source: E. J. Sutow. Iron-based alloys. In: *Concise Encyclopedia of Medical and Dental Materials* (D.F. Williams, ed.). Pergamon Press, New York, 1990, pp. 232–240.

mill products to enhance their mechanical properties so that the resulting wire would not fail during bending and twisting when tying knots. It is important, however, to ensure the generation of as few surface defects as possible during cold-working because an irregular surface could promote crevice and fretting corrosion and/or stress concentration effects. Any debris from fabrication is removed by a process called ''passivation treatment'' in which the surface is treated with, typically, a strong acid like nitric acid. The acid treatment results in an oxide film on the surface, which is considered to be more stable than the natural air-formed film. Whether this passivated oxidative film should not be disturbed during subsequent handling is still debatable. Stainless steel is sterilized by heat or steam with heat. It was reported that steam sterilization also increases its resistance to corrosion (49).

Currently the available stainless steel sutures are twisted Flexon® and monofilament Stainless Steel® from American Cyanamid, twisted or monofilament Surgical Stainless Steel from Ethicon, and monofilament steel from US Surgical.

B. Size Classification of Sutures

Suture materials are also classified according to their size. Currently, two standards are used to describe the size of suture materials: the USP (United States Pharmacopoeia) and the EP (European Pharmacopoeia) (2). Table 7 summarizes EP and USP standards. The USP standard is more commonly used. In the USP standard, the size is represented by a series combination of two arabic numbers: a zero and any number other than zero, like 2-0 (or 2/0). The higher the first number, the smaller the suture material is. Sizes greater than O are denoted by 1, 2, 3, etc. This standard size also varies with the type of suture material.

In the EP standard, the code number ranges from 0.1 to 10. The corresponding minimum diameter (mm) can be easily calculated by taking the code number and dividing by 10. The EP standard does not separate natural from synthetic absorbable sutures as the USP does.

Because a range of diameters is permitted for each USP suture size, the tensile strength (force/cross-sectional area) of sutures of the same USP size may be different from each other. For example, two polypropylene sutures of the same USP size from two different manufacturers having the same tensile breaking load may have different tensile strengths because of a possible difference in suture cross-sectional area due to slightly different diameters. The polypropylene suture with a smaller diameter may have a higher tensile breaking strength than the one with a larger diameter. A recent study by von Fraunhofer et al. reported that some types of sutures have consistently less variability in diameter than others (50). They found that Prolene has the greatest variability in diameter, particularly in the 2/0 and 3/0 sizes among six tested sutures (chromic catgut, Softgut, Dexon Plus, Vicryl, PDS, and Prolene) of USP sizes ranging from 3/0, 2/0, 1/0, to 1. However, Prolene has overall the lowest diameter at all four USP sizes (50). Softgut, a glycerin-coated catgut from Davis/Geck, has a significantly greater diameter than all other five sutures tested at all four USP sizes. The diameters of chromic gut, Dexon Plus, Vicryl, and PDS are comparable at most USP

Table 19.7 Suture Size Classification

USP size codes		EP size codes	
Nonsynthetic absorbable	Nonabsorbable and synthetic absorbable	Absorbable and nonabsorbable materials	Diameter (mm) Min. Max.
	11/0	0.1	0.01–0.019
	10/0	0.2	0.02–0.029
	9/0	0.3	0.03–0.039
	8/0	0.4	0.04–0.049
8/0	7/0	0.5	0.05–0.069
7/0	6/0	0.7	0.07–0.099
6/0	5/0	1	0.10–0.14
5/0	4/0	1.5	0.15–0.19
4/0	3/0	2	0.20–0.24
3/0	2/0	2.5	0.25–0.29
2/0	0	3	0.30–0.39
0	1	4	0.40–0.49
1	2	5	0.50–0.59
2	3	6	0.60–0.69
3	4	7	0.70–0.79
4	5	8	0.80–0.89
5	6	9	0.90–0.99
6	7	10	1.00–1.09

sizes, as shown in Fig. 3, although PDS is a monofilament and the rest are multifilaments.

C. Physical Configuration and Surface Treatments

In terms of the physical configuration of suture materials, they can be classified into monofilament, multifilament, twisted, and braided. Suture materials made of nylon, polyester and stainless steel are available in both multifilament and monofilament forms. Catgut, reconstituted collagen, and cotton are available in twisted multifilament form, while PGA, Vicryl, Panacryl, silk, polyester based, and polyamide based suture materials are available in the braided multifilament configuration. PDS, Maxon, Monocryl, Biosyn polypropylene based, Gore-Tex, and Pronova suture materials exist in monofilament form only. Stainless steel metallic suture materials can be obtained in either monofilament or twisted multifilament configurations. Another unique physical configuration of suture material is polyamide (nylon 6), which has the trade name Supramid®. It has a twisted core covered by a jacket of the same material.

Suture materials are frequently coated to facilitate their handling properties, particularly for a reduction in tissue drag when passing through the needle tract and ease of sliding knots down the suture during knotting (i.e., knot tie-down). Although nonabsorbable beeswax, paraffin, silicone, and polytetrafluoroethylene are the traditional coating materials, new coating materials have been reported, particularly those that are absorbable. This is because the coating materials used for absorbable sutures must be absorbable, and traditional nonabsorbable coating materials like wax are not appropriate for absorbable sutures (51–

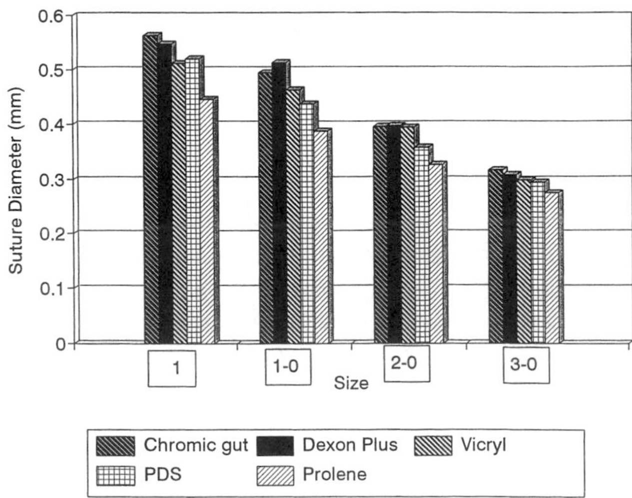

Figure 19.3 Variation in suture diameter of five sutures in four different USP sizes. (From J. A. von Fraunhofer, R. S. Storey, I. K. Stone, and B. J. Materson. *J. Biomed. Mater. Res.* *19*:595 (1985). Data recompiled with permission.)

53). Furthermore, absorbable coating materials should have better tissue biocompatibility because of the lack of chronic tissue reaction.

The trend of suture coating is to develop coating materials that have a chemical property similar to the suture to be coated. There are basically two types of absorbable coating materials: water soluble and water insoluble. Water-insoluble coating materials have similar chemical constituents to the suture, and they are broken down by hydrolysis. They remain on the suture surface longer than water-soluble coatings. A typical example is polyglactin 370 used for the Vicryl suture. Dexon II sutures have a polycaprolate coating, which is water insoluble. Water-soluble coating materials dissolve promptly to reveal the underlying uncoated suture after wound closure. A typical example is poloxamer 188 found on Dexon Plus. Multifilament sutures are more commonly provided with coating materials than monofilament sutures. For example, multifilament Vicryl and Dexon Plus or II have coating materials applied, while monofilament PDS and Maxon sutures have no coatings.

Although coating of suture materials facilitates easy passage through tissue and handling properties, it frequently results in poor knot security. For example, Dexon Plus and coated Vicryl require four or five square throws to form secure square knots, while the uncoated Dexon and Vicryl sutures form secure knots with only two throws (54,55). Water-soluble coating materials do not have the same adverse effects as water-insoluble ones on suture knot security. This is because the dissolution of water-soluble coating material reveals the underlying substrate, i.e., uncoated sutures, which have better knot security (55).

There is, however, one technical concern about using water-soluble coating material in actual wound closure. Suture materials are frequently soaked in saline after their removal from packages before wound closure. Some or the bulk of the water-soluble coating materials might be removed by this routine soaking practice. Thus it is important to minimize the time of soaking when dealing with water-soluble coated suture materials.

Besides using water soluble coating materials to improve knot security and handling properties, there are several other patented procedures and materials reported recently to improve either knot tie-down performance (the ease of sliding a knot down the suture into place during knotting) or/and knot security (the ability of a knot to hold after knotting) (56–59). In general, a coating designed to improve knot tie-down would reduce knot security. It is difficult to achieve both ease of knot tie-down and enhanced knot security of sutures. There are very few reported treatments that would achieve these two contradictory and mutually exclusive properties. One of them is the use of a combination of both coating and textured yarns reported by Kawai et al. (56) who used sucrose fatty acid ester, beeswax, paraffin, poly(oxyethylene-co-oxypropylene), polyglactin-370, gelatin, silicone, and polytetrafluoroethylene as coating materials. Textured yarns could be achieved by standard texturing processes employed in the textile industry, and they include false-twist, knit-deknit, stuffing box, and crimping gear methods.

Other recently reported absorbable but not water-soluble coating materials that could improve knot tie-down and knot security are high-molecular-weight poly-ε-caprolactone, copolymer of at least 90% by weight of caprolactone and 10% at most of other biodegradable monomers like glycolide, lactide, and their derivatives (57,58) or a random copolymer of 25–75% by weight of glycolide and the remaining trimethylene carbonate (59). The improved knot tie-down and knot security were attributed to deep penetration and even distribution of the coating materials into the interstices of the suture filaments. The patented random copolymer of glycolide and trimethylene carbonate coating material was suggested to have the advantages of not flaking off from the substrate sutures because of its high molecular weight and low glass transition temperature and of retaining its lubricant property even when the coated suture is wet.

There is relatively little research and development of new coating materials for natural absorbable sutures like catgut and reconstituted collagen sutures. Gaillard recently reported that the widely reported relatively poor in vivo biocompatibility and performance of collagen based sutures could be improved by using biodegradable polymeric coating materials instead of the conventional chromium salts (60). The rationale behind his approach is that a biodegradable protective coating could temporarily shield the substrate catgut sutures from enzymatic biodegradation during the early stage of wound healing, so that the coated catgut sutures should be able not only to retain better strength in this critical period of wound healing but also to delay and reduce the well-known adverse tissue reactions by delaying the onset of enzymatic biodegradation beyond the initial stage of wound healing. The use of synthetic biodegradable polymers as the coating for catgut sutures would also eliminate the known toxicity of chromium salts and improve knot tie-down and security.

D. Physical and Mechanical Properties

The physical and mechanical characteristics of suture materials are probably the most important criteria for suture functions, i.e., to close wounds and to carry the physiologic load during healing. They include those related to strength, stiffness, viscoelasticity, coefficient of friction, compliance, size, form (monofilament or multifilament), fluid absorption and transport, etc. Mechanical properties include knotted and unknotted (straight pull) tensile strength, modulus of elasticity (relating to stiffness), elongation at break, and toughness (the area under the stress–strain curve, which relates to the ability of a material to absorb impact energy).

There is a wide variety of mechanical properties among existing commercial suture materials. The tensile strength property is the most frequently reported and studied mechanical characteristic of suture materials. Because strength is expressed in terms of the cross-sectional area of a material, it is normalized based on the dimension of the material and hence could be used to compare sutures having different chemical structure or/and sizes. Tensile breaking force, however, does not take into account suture size (i.e., diameter). Thus a larger suture would have a higher tensile breaking force than the same suture of a smaller size, even though the two sutures may have the same tensile strength. Therefore a meaningful comparison of tensile breaking forces of sutures should be done under the same suture size (diameter) and physical form. In addition, knotted tensile strength or breaking force is frequently lower than for the unknotted suture. Strength values are obtained in either dry or wet conditions. Among these physical and mechanical properties, viscoelasticity, bending stiffness, and compliance are the least studied and understood and are worthy of brief comments below.

Viscoelasticity relates to the ability of materials to change either their dimension under a constant force (i.e., creep) or their strength under a constant dimension (i.e., stress relaxation). This ability is believed to be related to the quality of a suture to hold a wound. For example, a suture with a high creep phenomenon will become ''relaxed'' after the recession of wound edema, which could lead to a suture knot loop becoming too loose to appose the wound edges properly. Viscoelasticity of suture materials is also the least studied and understood (61,62).

Bending stiffness is a complex mechanical phenomenon that closely relates to the handling characteristics of suture materials, particularly knot security. There is very little reported data describing the bending stiffness of sutures. Most reported stiffness data were derived from the modulus of elasticity obtained from the tensile strength test. Because a knot involves the bending of suture strands, stiffness data based on the modulus of elasticity do not adequately represent the performance of knot strength, security, or tie-down. There are two reported studies of the bending stiffness of sutures (47,63).

The study of bending stiffness by Chu et al. (47) was based on the force required to bend a suture to a predetermined angle. The measured bending force was converted to flexural stiffness in pounds/in^2 according to an ASTM formula. Sutures with a braided structure were generally more flexible than those of a monofilament structure, irrespective of their chemical constituents. Coated sutures had a significantly higher bending stiffness than the corresponding uncoated ones. This is particularly true when polymers rather than wax or silicone were used as the coating material. This coating-related increase in bending stiffness is attributed to the loss of mobility of constituent fibers under bending force. An increase in suture size significantly increased their stiffness, and the magnitude of increase depended on the chemical constituents of the suture. The large porous volume inherent in the Gore-Tex monofilament suture was the reason for its lowest flexural stiffness. These quantitative bending stiffness data were consistent with the reported semiquantitative stiffness data based on torsional mode (63).

The torsional bending stiffness study of sutures [Tomita et al. (63)] used the technique reported by Scott and Robbin (64). In principle, a constant weight was attached to each end of a suture of fixed length (25.5 cm), and the distance between these two ends was measured after one minute loading. The stiffness coefficient (G) was calculated by the formula $G = TD^2/8$, where T is the applied force in dynes and D was the stiffness index, which was the average distance between the two parallel ends of the suture. The bending stiffness data from Tomita et al. generally agree with the data of Chu et al. that braided sutures are generally more flexible than monofilaments, and Gore-Tex suture has the lowest bending stiffness.

Suture compliance is a mechanical property that closely relates to the ease of a suture to elongate under a tensile force. It is believed that the level of suture compliance should contribute to the compliance of tissues at the anastomotic site. Suture compliance is particularly important in surgery where there is a tubular anastomosis, such as vascular anastomoses. There is only one reported study that examined the effect of suture compliance on the compliance of arterial anastomotic tissues closed with sutures (65). Compliance mismatch between a vascular graft and host tissue has long been suggested as one of the several factors contributing to graft failure (66). Compliance mismatch at the anastomotic site constitutes a major component of overall compliance mismatch associated with vascular grafts. Since sutures are the only foreign materials at the anastomotic site, it is expected that a wide range of suture compliance might result in different levels of anastomotic compliance. Megerman et al. (65) very recently tested this hypothesis by using two 6/0 sutures with a vast difference in suture compliance, Novafil and Prolene. Novafil is an elastomeric suture made from polybutester and is characterized by a high elongation at low tensile force, a low modulus of elasticity, and high hysteresis, while the Prolene suture has a relatively higher modulus of elasticity, low elongation at low tensile force, and low hysteresis. They reported that arterial anastomoses closed with a more compliant suture like Novafil produced arterial anastomotic compliance on average over 75% more than those closed with a less compliant Prolene suture.

Capillarity of a suture describes the ease of a suture to transport liquid along the suture strand and is an inherent physical property of multifilament sutures due to the available interstitial space. Capillarity is thus related to the ability of a suture to transport or spread microorganisms and hence is an important property that relates to wound infection. Bucknall (67) reported that a braided nylon suture could take up three times as many microorganisms as monofilament nylon. There are two methods to evaluate the capillarity of sutures, qualitative according to USP XVII (68) and quantitative as developed by Blomstedt and Osterberg (69). The former method determines whether a 0.1% methylene blue dye solution can transport through a fixed length of a suture weighing under 2 grams to reach the end marked by a white cotton within 8 h. If the cotton becomes blue within 8 h, the suture is classified as capillary; otherwise it is noncapillary.

Table 8 summarizes capillarity data of some size 0 sutures according to the USP XVII qualitative method (69). None of the monofilament sutures exhibited capillarity, while braided polyester (Mersilene®), twisted polyamide with cover (Supramid®), and twisted linen showed capillarity. It was surprising that braided silk with wax coating and catgut (plain and chromic) did not have any capillarity. It appears that the wax treatment on silk was able to reduce capillarity. Although the qualitative USP XVII method is a fast and easy way to evaluate suture capillarity, it does not distinguish the rate of capillarity.

Blomstedt et al. developed a quantitative method to determine the rate of capillarity of sutures (69). The method is based on the conductivity of suture strands when they are wet via capillarity. There was a more than 10-fold difference in the rate of fluid transport (the slope of the curve) among the three sutures having capillarity based on the USP XVII method. For example, Supramid, Mersilene, and twisted linen had the fluid transport rate constants 0.24, 0.04, and 0.01 cm^2/s, respectively.

Fluid absorption of sutures, a property relating to capillarity, may also be responsible for the spread of microorganisms in tissues. Table 9 summarizes the level of absorption of saline and blood plasma of several sutures (69). In general, both the chemical nature and the physical structure of the sutures determine the level of fluid absorption; however, it appears that the chemical nature is more important than the physical structure for the wide range of fluid absorption. Synthetic sutures have much lower fluid absorption capability than natural sutures, because the former are more hydrophobic than the latter. Within each type of suture, multifilament sutures have a higher fluid absorption than their monofilament counterparts due to the additional capillarity effect associated with multifilament sutures. Among the sutures tested, plain and chromic catgut sutures showed the highest level of fluid absorption in both saline and blood plasma media. Obviously, bulk fluid absorption in addition to surface fluid absorption is required to reach such a high level of fluid absorption. The protein nature of catgut sutures provides many bonding sites for fluid molecules to attach to, and such attachments are not lim-

Table 19.8 Capillarity of Different Suture Materials According to the USP XVII Method

Material	Capillarity
Polypropylene, monofilament	−
Polyamide, monofilament	−
Polyamide, braided waxed	−
Polyamide, twisted with cover	+
Polyester, braided	+
Linen, twisted	+
Silk, braided waxed	−
Catgut, plain	−
Catgut, chromic	−

+ = capillary, − = noncapillary.
Source: B. Blomstedt, and B. Osterberg. Fluid absorption and capillarity of suture materials. *Acta Chir Scand. 143*:67–70 (1977).

Table 19.9 Absorption of Saline and Blood Plasma in Different Suture Materials

	Fluid absorption in percent of strand dry weight ± S.E.	
Material	Saline, $n = 20$	Blood plasma, $n = 20$
Polypropylene, monofilament	0.4 ± 0.1	0.2 ± 0.1
Polyamide, monofilament	7.8 ± 0.1	10.1 ± 0.2
Polyamide, braided waxed	11.9 ± 0.5	16.1 ± 0.4
Polyamide, twisted with cover	23.6 ± 0.3	27.8 ± 0.8
Polyester, braided	13.9 ± 0.4	16.0 ± 0.3
Linen, twisted	84.5 ± 1.2	92.2 ± 1.3
Silk, braided waxed	59.5 ± 1.0	65.2 ± 0.7
Catgut, plain	96.8 ± 0.5	100.1 ± 0.7
Catgut, chromic	85.2 ± 0.7	100.1 ± 0.8

Source: B. Blomstedt, and B. Osterberg. Fluid absorption and capillarity of suture materials. *Acta Chir Scand. 143*:67–70 (1977).

ited to the surface of catgut sutures. In contrast, almost all absorbed fluid in synthetic sutures was retained on the surface of the sutures.

E. Handling Properties

Handling characteristics describe those properties that relate to the feel of suture materials by surgeons during wound closure. These suture characteristics are the most difficult to evaluate objectively. The handling properties of sutures are pliability (or stiffness), knot tie-down, knot security, packaging memory, surface friction, viscoelasticity, tissue drag, etc., and hence are directly and indirectly related to the physical/mechanical characteristics of sutures. For example, the term pliability is a subjective description of how easily a person could bend the suture and hence relates to the surgeon's feel of a suture during knot tying. It is directly related to the bending modulus and indirectly to the coefficient of friction. Packaging memory, another handling property that indirectly relates to pliability, is the ability to retain the kink form after unpacking. The ability to retain such kink form makes surgeons' handling of the sutures more difficult during wound closure, particularly when tying a knot. This is because sutures with high memory, like nylon, polypropylene, PDS and Maxon, tend to untie their knots as they try to return to their kink form. Thus packaging memory should be as low as possible. In general, monofilament sutures have more packaging memory than braided ones. The four exceptions are the newly available Monocryl, Biosyn, Gore-Tex, and Pronova sutures, which are reported to have exceptionally low packaging memory. The easiest means of evaluating the packaging memory is to hang the sutures in air and measure the time required to straighten out the kink.

Knot tie-down and security describe how easily a surgeon can slide a knot down to the wound edge and how well the knot will stay in position without untying or slippage. These handling characteristics relate to surface and mechanical properties of sutures. A relatively smooth surface like that of monofilament or coated braided sutures will have a better knot tie-down than a suture with a rough surface such as an uncoated braided suture, if everything else is equal. The coefficient of friction of sutures also relates to knot tie-down and security. A linear relationship between knot security and the coefficient of friction was reported by Herman (70). A high coefficient of friction will make knot tie-down difficult but will lead to a more secure knot. This is because a high-friction suture can provide additional frictional force to hold the knot together. This high-friction suture surface also makes the passage of suture strands difficult during knot tie-down. It thus appears that knot tie-down and knot security are two contradictory requirements. There is no reported standard test for evaluating knot tie-down capacity.

Tomita et al. (63) recently reported a method to quantify objectively the knot tie-down capacity of 2/0 silk, polyester sutures, Gore-Tex, and an experimental ultrahigh-molecular-weight polyethylene suture (Nesplon®). Their method was based on the technique developed by Kobayashi (71). The pullout friction test measures the frictional resistances produced by both surface friction and cross-sectional deformation of sutures when tying and holding a knot securely. In principle, a suture thread was wound around a sponge tube, tied with a square knot, and placed in a tensile testing machine. Table 10 summarizes the knot tie-down resistance and roughness as well as the knot security of the four tested sutures. The two monofilament sutures (Gore-Tex and UHMW-PE) as a group had a much lower knot tie-down resistance than the braided sutures (Surgical Silk and Ethibond). A higher contact surface area among braided suture strands may be responsible for the high resistance to knot sliding. Silk suture was found to have the largest difference between static (for knot security) and dynamic (for knot tie-down) resistances. These silk data are consistent with their well-known excellent handling characteristics.

The high knot tie-down resistance of braided sutures also led to their higher knot security as evident in the lower number of throws that was required to achieve a secure knot (defined as one that broke without slipping >10 mm) in Table 10. Tomita et al. attributed the high knot security of these braided sutures not only to the surface friction force but also to the resistance resulting from cross-sectional deformity.

Tissue drag describes how easy it is to pull a suture through tissue during wound closure and suture removal. It hence gives an indication of the extent of tissue tear in needle holes. Obviously, tissue drag should be maintained at a minimum for easy passage of sutures through tissue, which is reflected in a surgeon's feel about a suture. Tissue

Table 19.10 Knot Properties of USP Size 2 Sutures

Suture material	Knot security[a]	Tie-down resistance ($\times 10^3$ N) (n = 3)	Tie-down roughness ($\times 10^3$ N) (n = 3)
Silk	3	11.7 ± 0.4	0.58 ± 0.14
PET	3–4	12.7 ± 3.6	2.2 ± 0.40
E-PTFE(CV2)	—	3.6 ± 0.4	0.15 ± 0.01
UHMW-PE	6–	7.3 ± 1.5	1.1 ± 0.60

Source: N. Tomita, S. Tamai, T. Morihara et al. *J. Appl. Biomater. 4*: 61–65 (1993)

[a] Number of throws required to achieve a secured knot defined as the one broken without slipping >10 mm.

Table 19.11A Maximum Withdrawal Stress (g/cm[a])

Suture size	A: Silk plain		B: Silk/silicone		C: PE/teflon		D: Nylon		E: Polyethylene		F: Polypropylene	
	Mode I	Mode II	Mode I	Mode II	Mode I	Mode II	Mode I	Mode II	Mode I	Mode II	Mode I	Mode II
2–0	36.2	146.9[b]	25.5	118.1[b]	14.2	85.3[b]	10.3	34.4	13.8	49.3	13.2	45.3
4–0	20.8	146.9[b]	12.5	118.1[b]	9.3	85.3[b]	8.6	73.7	5.7	35.5	5.5	32.4
6–0	17.4	146.9[b]	6.3	118.1[b]		85.3[b]	6.6					

drag of a suture is closely related to both the surface physical roughness and the coefficient of friction. In general, coated or/and monofilament suture materials have less tissue drag than their uncoated or/and braided counterparts. This is because a coating material can smooth a rough surface and also change the coefficient of friction.

There are very few reported studies that examine the relationship between the structure/physical configuration of sutures and tissue drag (72,73). These published studies focus on the evaluation of suture withdrawal stress and work after predetermined periods of implantation. It is known that healing around a suture in a wound leads to ingrowth of fibrous connective tissues into the interstitial space of the suture. This tissue ingrowth can exert a significant resistance during suture withdrawal. Table 11 summarizes the maximum suture withdrawal stress (g/cm) and work (g-cm/cm) of six commercial sutures subdermally implanted in dogs for up to 14 days (72). Model II, which used the test sutures to close three 1 inch long cutaneous incisions via a short curved needle and a continuous stitch, was a better model for simulating clinical trauma proximal to the suture location. In general, monofilament sutures (e.g., nylon, polypropylene, and polyethylene) showed significantly lower maximum withdrawal stress values than multifilament braid sutures like silk (plain or silicone-coated) and Tevdek® (a PTFE-coated polyester). Plain silk had the highest maximum withdrawal stress (146.9 g/cm) among the six types of sutures. The duration of implantation appeared not to exert a strong effect. The data of withdrawal work were consistent with the maximum withdrawal stress. A subsequent detailed analysis of the withdrawal stress profile vs. the length of suture withdrawn in cutaneous tissues of pigs indicated a unique pattern (73). A sharp maximum withdrawal stress along the length of the suture withdrawn was the most characteristic pattern for silicone-coated silk sutures over all three periods of implantation (1, 2, and 3 weeks), while the synthetic monofilament polyethylene, polypropylene, and nylon as well as Tevdek® sutures showed relatively flat withdrawal stress curves (less profound maximum withdrawal stress) with the length of suture withdrawn.

F. Biological Properties

Biocompatibility of suture materials describes how sutures, which are foreign materials to the body, affect surrounding tissues and how the surrounding tissues affect the properties of sutures. Thus biocompatibility is a two way relationship. The extent of tissue reactions to sutures depends largely on the chemical nature of the sutures and their degradation products if they are absorbable. Sutures from natural sources like catgut and silk usually provoke more tissue reactions than synthetic ones. This is due to the availability of enzymes to react with natural biopolymers. Besides the most important chemical factors, physical form, the amount and stiffness of suture materials have been reported to elicit different levels of tissue reactions. For example, a stiff suture would result in stiff projecting

Table 19.11B Withdrawal Work (g-cm/cm)

Suture size	A: Silk plain		B: Silk/silicone		C: PE/teflon		D: Nylon		E: Polyethylene		F: Polypropylene	
	Mode I	Mode II	Mode I	Mode II	Mode I	Mode II	Mode I	Mode II	Mode I	Mode II	Mode I	Mode II
2–0	3.9	47.1[b]	3.7	40.6[b]	2.7	43.8[b]	3.6	13.2	4.1	33.5	4.1	22.3
4–0	3.8	47.1[b]	2.6	40.6[b]	2.7	43.8[b]	3.4	32.7	2.4	15.1	2.4	10.6
6.0	1.4	47.1[b]	1.4	40.6[b]		43.8[b]	3.1					

[a] Averages for all implant duration periods.
[b] No trend with suture size
Source: C. A. Homsy, K. E. McDonald, and W. W. Akers. *J. Biomed. Mater. Res.* 2:215 (1968). With permission.

ends in a knot where cut. These stiff ends could irritate surrounding tissues through mechanical means, a problem associated with some monofilament sutures but generally not found in braided multifilament sutures.

Because the quantity of a buried suture relates to the extent of tissue reaction, it is a well-known practice in surgery that one should use as little suture material as possible, such as a smaller knot or a smaller size, to close wounds. The use of a smaller size for wound closure without the expenses of adequate support to wounds and cutting through wound tissue is due to the square relationship between diameter (D) and volume (V) ($V = \pi D^2 \times$ length), which suggests that a slight increase in suture size or diameter would increase its volume considerably.

There are two basic means to study the biocompatibility of suture materials, cellular response and enzyme histochemistry. The former is the most frequently used and provides information about the type and density of inflammatory cells at a suture site. The latter provides information on what these inflammatory cells would do and is based on the fact that any cellular response is always associated with the presence of a variety of enzymes and is particularly useful in the study of the mechanism of absorption of absorbable sutures. In the cellular response approach, histological stains with a variety of dyes like the most frequently used H & E are the standard methods of evaluation of cellular activity at the suture sites. Figure 4 is a typical example of histological photomicrographs of PDS and Maxon sutures at 35 days postimplantation in a variety of tissues (74). In addition to a qualitative description of cellular activities, tissue response could be graded by the most frequently used and accepted Sewell et al. method or its modification (75).

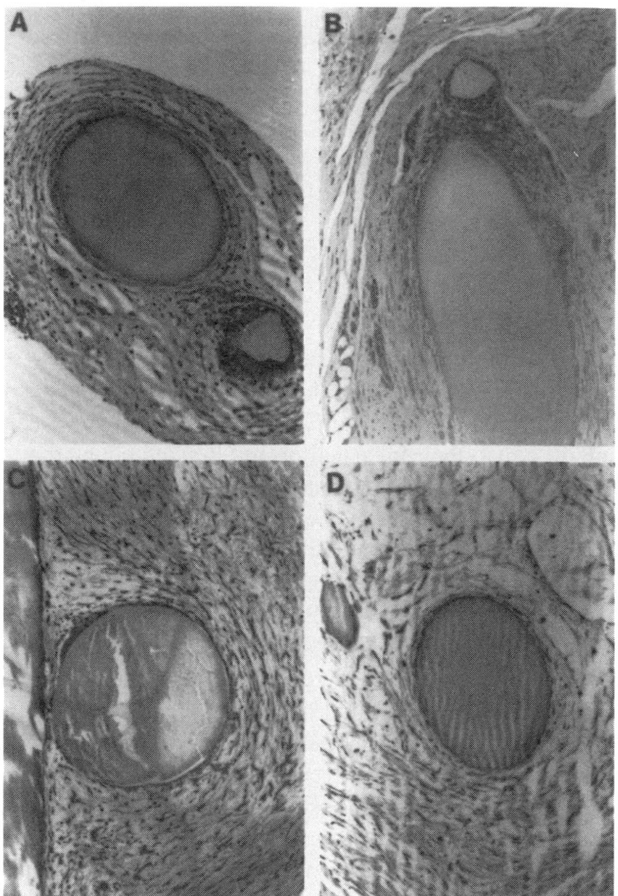

Figure 19.4 Light histological photomicrographs of tissue adjacent to PDS and Maxon sutures 3 and 5 days after implantation in a variety of tissues of New Zealand White rabbits. ×130. (A) Peritoneum/PDS, (B) Fascia/PDS, (C) Peritoneum/Maxon, and (D) Fascia/Maxon. (From S. A. Metz, N. Chegini, and B. J. Masterson. *J. Gynecol. Surg.* *5*(1):37–46 (1989). With permission.)

Most biocompatibility studies of suture materials have been performed in rat gluteal muscle. This implantation site has given a very consistent and reproducible cellular response for valid comparisons, even though it is not a common site for suture in surgery. However, Walton recently raised the question of using this common test procedure, particularly in orthopaedic surgery (76). His arguments were based on the observed inflamed nature of the postoperative synovial tissue and the mechanically stressed nature of the suture.

The second means for the study of suture biocompatibility is the use of enzyme histochemistry. It is a more objective, quantitative, consistent, and reproducible method than cellular response, which is based on a more subjective histological evaluation. Enzyme histochemistry is, however, more tedious and requires more sophisticated facilities and better experience. The data obtained provide additional insight into the functions of those cells that appear during various stages of wound healing. The enzymatic activity of a suture implant site is quantified by microscopic photometry of a cryostat section of the tissue. Figure 5 is a typical finding of an enzyme histochemical study of suture biocompatibility (77–79). The high level of cellular response to silk suture observed from a histological study is confirmed in this enzyme histochemical study. Enzyme histochemistry is also useful for studying the biodegradation mechanism of absorbable sutures because not only natural absorbable sutures are degraded through the enzymatic route but also the degradation products must be metabolized via enzyme activity.

The normal tissue reaction to sutures has three stages, according to the time for the appearance of a variety of inflammatory cells (77,79–81). They are the initial infiltration of polymorphonuclear leukocytes, lymphocytes,

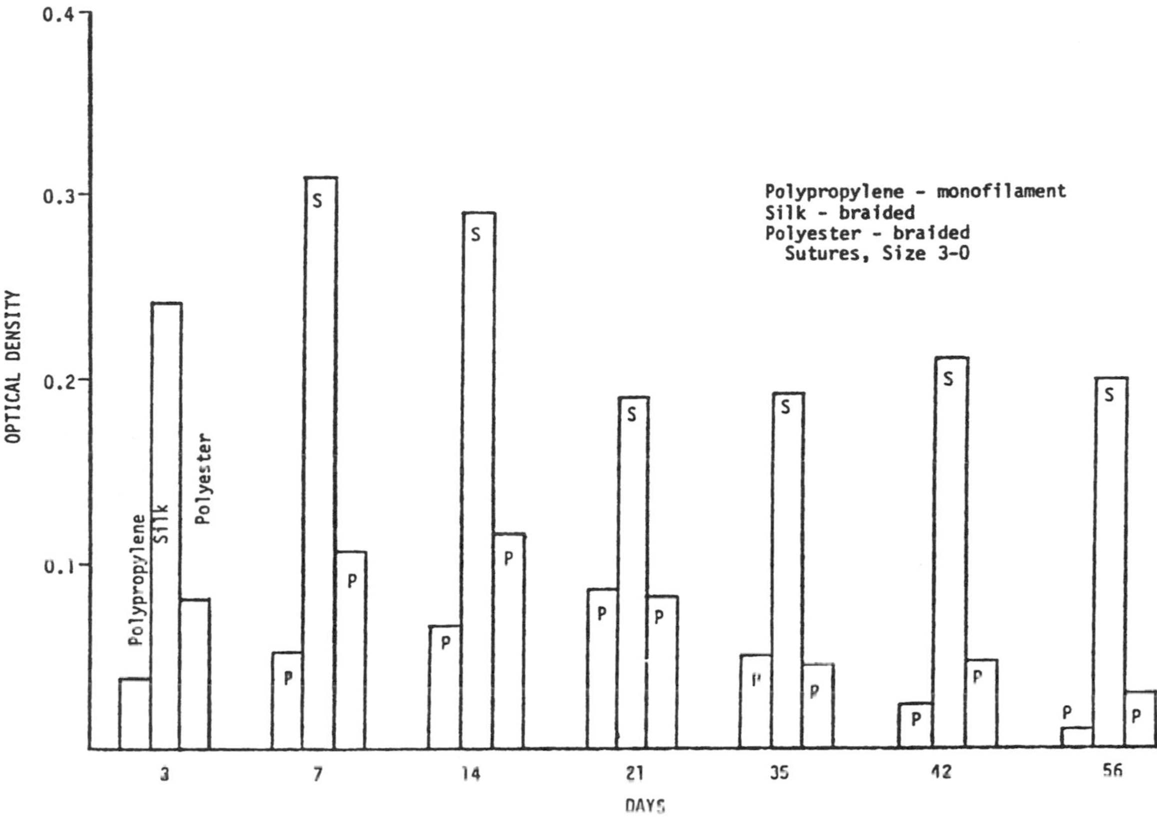

Figure 19.5 Enzyme histochemical evaluation of macrophage hydrolase activity of monofilament polypropylene and braided silk and polyester sutures. (From W. van Winkle, and T. N. Salthouse. Ethicon, Somerville, N.J., 1976. With permission.)

and monocytes during the first 3 or 4 days (i.e., acute response); the appearance of macrophages and fibroblasts from day 4 to day 7; and the beginning of maturation of fibrous connective tissue formation with chronic inflammation after the 7th to the 10th day. During the first 7 days after implantation, there is virtually no difference in normal tissue reaction between synthetic absorbable and nonabsorbable sutures. However, a slightly higher inflammatory reaction to synthetic absorbable sutures could persist for an extended period until they are completely absorbed and metabolized, while synthetic nonabsorbable sutures, in general, are characterized with a minimal chronic inflammatory reaction with a thin fibrous connective tissue capsule surrounding the sutures usually by 28 days after implantation.

Due to this normal tissue reaction, fibrous and/or epidermic tissue ingrowth into sutures may pose a problem during the removal of the sutures, particularly for those sutures placed through the cutaneous surface due to the ingrowth of epidermis in addition to fibrous connective tissue. This problem is particularly profound in multifilament sutures because of the available interstitial space within these sutures for tissue infiltration. The formation of a perisutural cuff due to a downward growth of epidermis along the suture path has been found to be responsible for 70 to 85% of the force required to remove the suture (72,73,80). Among the multifilament sutures, silk is the worst offender. A very recently reported U.S. patent disclosed the approach of using a composite suture to reduce tissue ingrowth into silk sutures through encapsulation of the silk suture with biocompatible polymeric resin (33).

Monofilament sutures are considered to be a better choice than multifilament ones in closing contaminated wounds. This is because not only do multifilament sutures elicit more tissue reactions, which may lessen tissue ability to deal with wound infections, but also multifilament sutures have a capillary effect, which could transport microorganisms from one region of the wound to another. The reason that multifilament sutures generally elicit more tissue reactions than their monofilament counterparts is that inflammatory cells are able to penetrate into the interstitial space within a multifilament suture and invade each filament. Such an invasion by inflammatory cells, well evident in histological pictures, does not occur in monofilament sutures. Thus the available surface area of a suture to tissue should bear a close relationship to the level of tissue reaction that a suture could elicit.

In addition to the normal tissue reactions to sutures, there are several adverse tissue reactions that are suture and site specific. Some examples include urinary stone or calci formation, granuloma formation, thrombogenicity, propensity toward wound infection and recurrence of tumor after radical surgery, and allergy.

G. Biodegradation and Absorption Properties

Absorbable sutures require consideration of their biodegradation and absorption, factors that are not relevant for most nonabsorbable sutures. This biodegradation characteristic is the reason that absorbable sutures do not elicit the permanent chronic inflammatory reactions found with nonabsorbable sutures. The most important characteristics in biodegradation and absorption of sutures are the strength and mass loss profiles and the biocompatibility of degradation products. Although there is a wide range of strength and mass loss profiles among the available absorbable sutures, they have one common characteristic: strength loss always occurs much earlier than mass loss, as shown in Fig. 6. This suggests that absorbable sutures retain a large portion of their mass in tissue, while they have already lost all their designated function to provide support for the wound tissue. Thus an ideal absorbable suture should have its mass loss and strength loss profiles synchronized. Such an ideal absorbable suture is not available due to the inherent relationship between strength and fiber structure.

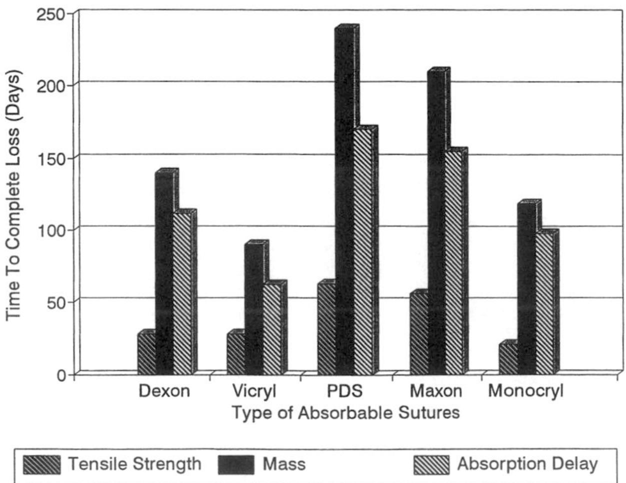

Figure 19.6 Time to complete loss in tensile breaking strength and mass profiles of five commercial absorbable sutures.

Biodegradation properties are usually examined in vitro or/and in vivo. In the in vitro environment, the most commonly used medium is physiological saline phosphate buffer of pH 7.44 at 37°C. However, other buffers, like Tris, or body fluids like urine, bile, or synovial fluids, have been used. Occasionally, microorganisms were deliberately incorporated into these media to examine the effect of microorganisms on biodegradation properties of absorbable sutures. In the in vivo environment, unstressed absorbable sutures are normally implanted in rat gluteal muscle for predetermined periods of implantation. As described above in Sec. B.6, Walton raised the point that the use of unstressed sutures in the gluteal muscle site may not represent the real clinical environment that absorbable sutures normally experience (76). The sutures retrieved at various periods of immersion or implantation are then subject to the evaluation of their mechanical and physical properties to assess their changes with time. The degree of absorption in vivo is evaluated by the change in suture cross-sectional area, while the level of tissue reaction is assessed by either the histological method or by the enzyme histochemistry method previously described.

The vast amounts of published information about the biodegradation phenomena of synthetic absorbable sutures since the 1970s show that these synthetic absorbable sutures can be degraded by a hydrolytic mechanism via the scission of ester linkages in the polymeric backbone (1,2). The observed wide range of strength and mass loss profiles of absorbable sutures is attributable not only to the chemical differences among the absorbable sutures but also to a variety of intrinsic and extrinsic factors, such as pH, electrolytes, stress applied, temperature, γ-irradiation, microorganisms, and tissue type, to name a few.

Very recently, an interesting study of the effect of superoxide ion on the degradation of absorbable sutures has been reported (82). Superoxide ion can act as an oxygen nucleophile agent to attack the ester linkage in absorbable suture polymers. In their study of five commercial 2/0 size synthetic absorbable sutures at 25°C (Dexon®S, Maxon®, Vicryl®, PDS II®, and Monocryl®), Lee and Chu (82) observed that there was a significant superoxide-ion-induced degradation effect on the mechanical properties, thermal properties, and surface morphology of these absorbable sutures. Among the five absorbable sutures and over the concentration range of that study, the monofilament Monocryl suture was the most sensitive toward superoxide-ion-induced degradation, followed by Maxon, Vicryl, Dexon, and PDSII sutures, which had relatively the least effect of superoxide-ion-induced hydrolytic degradation, as shown in Table 12. The amount of tensile breaking force loss ranged from as low as 3% to as high as 80%, depending

Table 19.12 The Tensile Breaking Force (Kg) and Its Percentage Retention of Five Absorbable Sutures as a Function of Superoxide Ion Concentration and Hydrolysis Reaction Time at 25°C

	Superoxide ion concentration (molar) 0	Superoxide ion concentration (molar) 0.0005		Superoxide ion concentration (molar) 0.005		Superoxide ion concentration (molar) 0.01	
	Hydrolysis time	Hydrolysis time		Hydrolysis time		Hydrolysis time	
	0 h	2 h	24 h	2 h	24 h	2 h	24 h
PDS II®	4.75 ± 0.01 (100.0%)[a]	4.59 ± 0.01 (97%)	4.59 ± 0.02 (97%)	4.57 ± 0.01 (96%)	3.09 ± 0.01 (65%)	0	0
Dexon®	6.23 ± 0.01 (100.0%)	5.73 ± 0.01 (92%)	5.59 ± 0.01 (90%)	5.32 ± 0.01 (85%)	2.89 ± 0.02 (46%)	0	0
Vicryl®	6.26 ± 0.01 (100.0%)	5.22 ± 0.01 (85%)	5.13 ± 0.01 (82%)	4.90 ± 0.01 (78%)	2.39 ± 0.01 (38%)	0	0
Maxon®	7.42 ± 0.01 (100.0%)	6.20 ± 0.01 (84%)	5.40 ± 0.02 (73%)	6.16 ± 0.01 (83%)	1.86 ± 0.01 (25%)	0	0
Monocryl®	5.66 ± 0.01 (100.0%)	4.40 ± 0.01 (78%)	3.70 ± 0.01 (65%)	3.01 ± 0.01 (53%)	1.14 ± 0.01 (20%)	0	0

[a] Retention of original (no superoxide and 0 hr) tensile strength.
Source: K. H. Lee and C. C. Chu. The effect of superoxide ions in the degradation of five synthetic absorbable suture materials. *J. Biomed. Mater. Res.*, *49*(1):25 (2000).

on the type of absorbable sutures, reaction time, and superoxide ion concentration. Regardless of degradation reaction time (up to 24 h), the measurement of tensile breaking force of these five absorbable sutures at the highest superoxide ion concentration (0.01 molar) was impossible because all absorbable sutures were so degraded that they completely lost their physical integrity as fibers and became almost invisible. For the purpose of comparison, none of these five absorbable sutures lost any measurable tensile strength in the presence of buffer medium within 24 h.

Along with the drastic reduction in mechanical properties, all these five absorbable sutures showed significant reductions in both peak melting temperature (T_m) and glass transition temperature (T_g). For examples, Dexon® suture showed the most significant change in T_g at both superoxide ion concentrations ($\Delta T_g = 9.0$°C, 19.5°C) followed by Vicryl® ($\Delta T_g = 12.6$°C). Among the monofilament absorbable sutures, Monocryl® had the highest T_g reduction ($\Delta T_g = 7.4$°C, 9.7°C), while PDS II® was the most resistant to superoxide-ion-induced T_g change with the smallest ΔT_g among all five synthetic absorbable suture materials ($\Delta T_g = 0.8$°C, 1.2°C).

Unlike the surface morphological change of the absorbable sutures in buffer solutions, the effects of superoxide-ion-induced degradation on the surface morphological change of the five absorbable sutures were unique and unusual (Fig. 7), particularly the moon-crater shaped impressions of various sizes and depth found in Monocryl and Maxon sutures that defied the anisotropic characteristic of the fibers.

Lee and Chu suggested that the cleavage of the ester bonds in the backbone of these absorbable suture materials was via a nucleophilic attack of superoxide ion. As shown in Scheme 1, this nucleophilic attack was predominantly an S_N2 type of reaction yielding mainly anionic end groups of different chain lengths. These active anionic species could subsequently attack the main backbone chains via transesterification with a rapid reduction in molecular weight until thermodynamic equilibrium was reached. The resulting degradation products would eventually lead to cylic and/or linear oligomers. The nucleophilic attack of the superoxide ion on polymers occurs primarily in Steps 1 and 2 of Scheme 1, which lead to the fragmentation of polyester chains and the formation of two different types of species: lower molecular weight macromolecules with peroxy radical chain end groups and anionic catalyzed chain end groups, respectively. Unlike the base catalyzed hydrolysis of simple aliphatic polyesters in an alkaline buffer solution in which the alcoholic anions were stabilized by hydrogen abstraction from the carboxylic end group to produce an alcohol group and a relatively stable carboxylic anion end group, the alcoholate end group from Step 2 may subsequently attack the main backbone chains

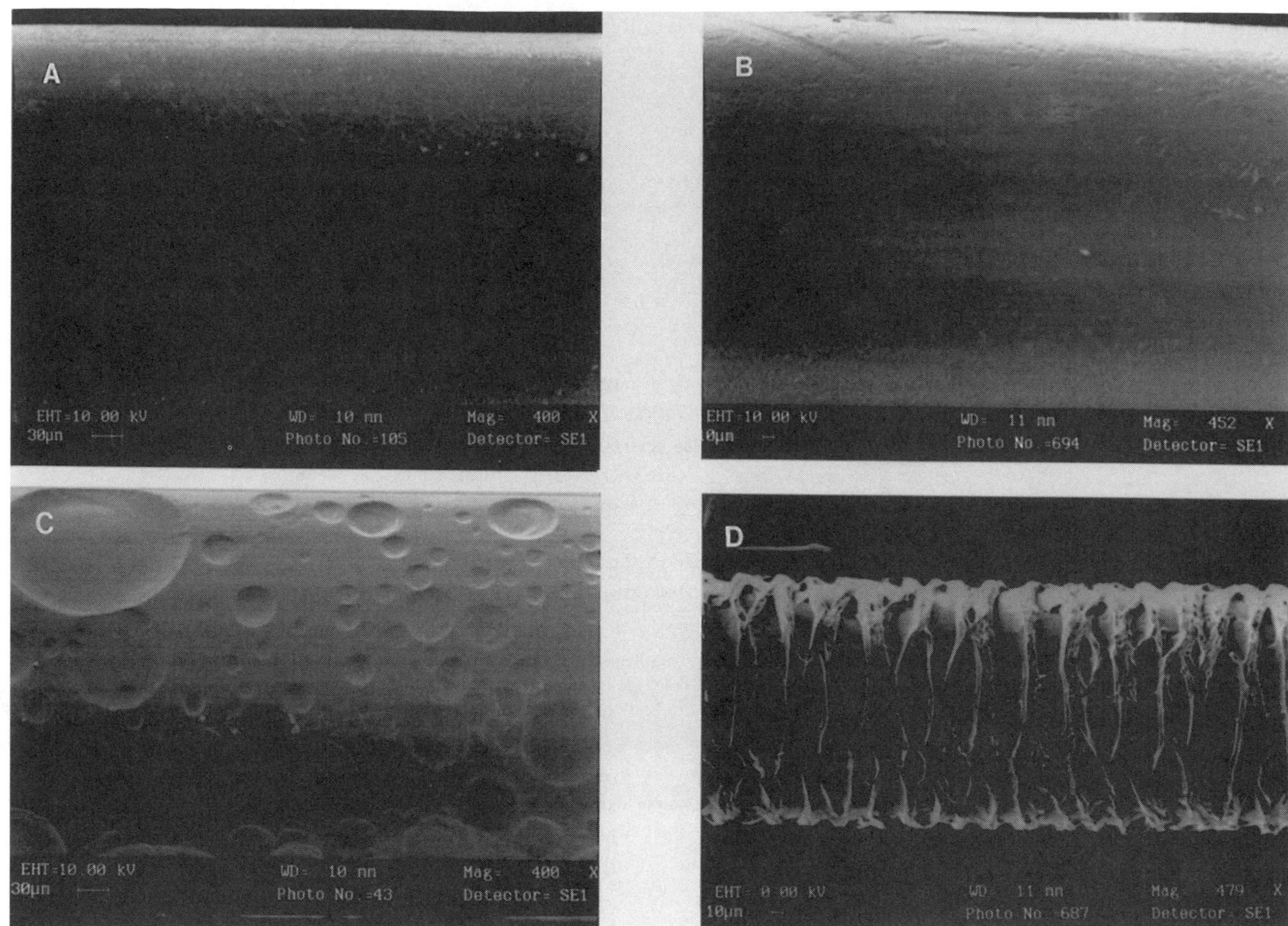

Figure 19.7 The surface morphology of 2-0 Monocryl suture upon superoxide-ion-induced hydrolytic degradation at 25°C for 24 h. (A) Original control, (B) 0.0005 molar, (C) 0.0025 molar, (D) 0.005 molar superoxide ion concentration.

via intra- and intermolecular transesterification reactions as illustrated in Step 3 of Scheme 1. This additional chain scission in Step 3 not only reduces the molecular weight of linear aliphatic polyesters like Dexon, Maxon, and Vicryl further but also produces another anionic species that repeats Step 3 again to accelerate the degradation further. Step 4 is a typical electron transfer reaction from the superoxide ion to the peroxy radical that leads to the formation of the carboxyl anionic intermediates, which are expected to be stable. Therefore their reactivity towards the polyester chain as Step 3 is very low or unlikely. Any excess superoxide ion would react with water as shown in Step 5 to form a hydroxyl anion and other species. Although it appeared to have competitive reactions between the hydroxyl anion attacking the ester linkage of the polyester for additional chain scission and the hydroxyl anion reacting with the carboxyl peroxide anion end group of the polymer chain (Step 6) to produce the carboxylic anion without chain scission, it was less feasible for the former (i.e., hydroxyl anion to attack ester linkages) because of the relatively low temperature (25°C). Step 6 was preferable during the final step of the termination reaction. Finally, it is possible to exchange potassium counter anions with hydrogen via diluted hydrochloric acid solution used for the termination reaction.

The biocompatibility of degradation products is usually not a problem because all existing absorbable sutures are made from the well-known biocompatible glycolide, lactide, and their derivatives. However, biocompatibility of degradation products also depends on the rate of their accumulation in the surrounding tissues. This implies that the ability of the surrounding tissues to actively metabolize degradation products is essential. Such a metabolism depends on the extent of blood circulation in the tissue. A

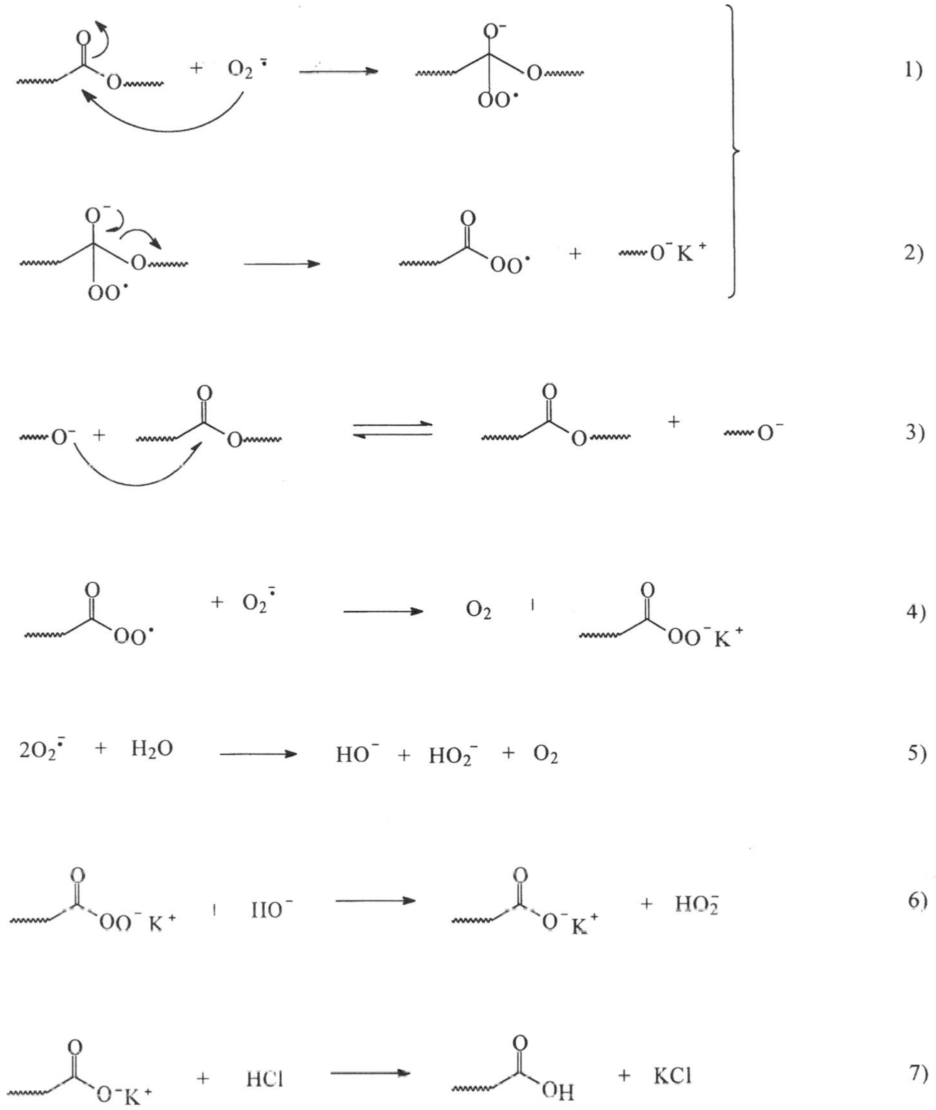

$$2O_2^{\cdot -} + H_2O \longrightarrow HO^- + HO_2^- + O_2 \qquad 5)$$

Scheme 1 Reaction mechanism of superoxide-ion-induced degradation of aliphatic polyesters. (From K. H. Lee and C. C. Chu. The effect of superoxide ions in the degradation of five synthetic absorbable suture materials. *J. Biomed. Mater. Res. 49*(1):25 (2000).)

well vascularized tissue could remove degradation products as fast as they are released from an absorbable suture and subsequently metabolized, which could minimize tissue reactions to degradation products.

Due to their ability to release degradation products, absorbable sutures have recently been studied as a vehicle to deliver a variety of biochemicals like growth factors to facilitate wound healing or antibiotics to combat wound infection. This new approach will increase the value of absorbable sutures and extend their function beyond the traditional role of wound closure.

III. SURGICAL MESHES

Another common medical textile product that has a long history of saving human lives is surgical mesh for body wall repair. Defects of the body's tissues are the direct result of the radical resection of a cancer tumor, congenital lesions, infection and subsequent necrosis and debridement, violent crimes, accidents, or large hernias. One in every 200 U.S. adults has experienced a hernia problem. Attempts to close these large defects are essential to save the patient's life, as well as to restore normal body functions. For those large defects lacking adequate tissue to permit primary repair, it is highly desirable to use materials of either a synthetic or a natural origin to reinforce tenuous aponeurotic closures and/or to bridge large defects in the abdominal or chest wall.

A wide variety of biological and synthetic textile materials have been used to repair the defective body wall. Biological materials include autologous, homologous, and heterologous transplants of fascia lataj skin or dermis, and dura mater (83–86). Synthetic textile materials for mesh use include the metallic based (e.g., tantalum, stainless steel) and the polymeric based. The polymeric based meshes include the nonabsorbable type, which includes polyamide (nylon), polyester (Dacron®), polytetrafluoroethylene (PTFE) or expanded polytetrafluoroethylene (Gore-Tex®), and polypropylene (Marlex®), and the absorbable type, which includes PGA (chemically identical to Dexon suture) and polyglactin-910 (chemically identical to Vicryl suture) (87–106). Very recently, there are a few new mesh products introduced for a variety of biomedical applications. For example, Genzyme introduced two unusual absorbable surgical meshes that are not based upon synthetic absorbable suture materials. These two absorbable meshes, Seprafilm® and Sepramesh®, are made from carboxymethycellulose and sodium hyaluronate (107,108). Klinge et al. (109) reported a new hybrid mesh (Soft Hernia Mesh) that consists of both nonabsorbable (i.e., polypropylene) and absorbable (i.e., polyglactin-910) components. They also introduced a polyglactin-910 coated polypropylene mesh for improving the handling property during surgery (110). Regardless of the nature and type of fibers used, all these synthetic surgical meshes are fabricated in either a woven or a knitted structure and exhibit various levels of porosity and strength.

In addition to the traditional use of surgical meshes for the repair of hernias and abdominal body wall defects, there are some innovative uses in many surgical areas, particularly for the protection of internal organs, that are worth of mentioning here briefly. They include pelvic floor reconstruction (108,111), isolation of the abdominal tissues from the denuded and possibly infected pelvis after radical pelvic surgery (112), the repair of the fractured orbital floor (113,114), the wrapping of severe traumatized hepatic tissues for controlling hemorrhage that otherwise would be a life-threatening problem (115,116), the management of the enlarged liver of an infant with stage 4S neuroblastoma that required decompressive laparotomy (117), transabdominal rectopexy in the treatment of rectal prolapse (118–120), wrapping of injured spleen during splenorrhaphy (121–126) or kidney during renorrhaphy (127) after blunt and penetrating trauma, mesh envelope of en bloc pediatric kidneys to facilitate the handling and positioning of the kidney in renal transplantation (during vascular anastomoses) with minimal vascular complications and lower risk of torsion of renal pedicles (127), gaining better abdominal access in patients with necrotizing pancreatitis by the use of absorbable mesh (129), retropubic sling procedures in women with urinary stress incontinence (130), repair of defects of the right atrium of pigs (131), use as a dural substitute (132), use in the absorbable mesh/fibrin glue technique to prevent cerebrospinal fluid leakage via the suture line along ePTFE mesh as dural substitute (133), the repair of tracheal defects (134), the reconstruction of chest wall defects following en bloc resection of lung carcinomas (135), the surgical management of esophageal perforation by absorbable mesh and fibrin glue (136), as the implant for chin augmentation (137), and finally as the treatment for periodontal defects (138).

A. Nonabsorbable Surgical Meshes

To reach a better understanding of the failure modes of surgical meshes, in terms of their properties, and to aid the development of new and better surgical meshes, Chu et al. (139) conducted a systematical and quantitative evaluation of the physical, mechanical, and morphologic properties of three most widely used nonabsorbable synthetic mesh materials, Marlex, Mersilene, and PTFE. Tables 13–16 and Figures 8–11 summarize the textile characteristics of the three most commonly used surgical meshes. It is clearly evident that there is a wide variation in structure and performance among these three.

Tables 13 and 14 show that the three meshes have distinctively different yarn and fabric structures. PTFE meshes are the heaviest and thickest and have the highest filament count and the lowest stitch density. Mersilene meshes are the lightest and thinnest and rank second in terms of filament count and stitch density. Marlex meshes, in terms of weight and thickness, rank second; this monofilament yarn has the highest stitch density. Overall fabric weights of the three surgical meshes are different because of variations in the linear density of the constituent fibers and fabric structure.

Table 15 shows that marked differences exist in tensile properties, not only among the three tested mesh specimens but also between the wale and course directions of the same mesh material. The difference in bursting strength and breaking strength and elongation at the wale direction among the three mesh fabrics was statistically significant according to Student's t-test ($p < 0.01$). However, the breaking strength and elongation at the course direction of Marlex and PTFE mesh fabrics were not statistically

Table 19.13 Yarn Characterization of Three Commercial Nonabsorbable Surgical Meshes

Trade name	Chemical component	Type	Texture	Linear density[a] (Tex[b])	Filaments per yarn	Filament linear density (Tex[b])	Filament cross section	Filament diameter (cm)
Mersilene	Poly(ethylene terephthalate)	Multifilament	Flat	2.5	12	0.212	Circular	0.0014
Marlex	Polypropylene	Monofilament	Flat	21	1	21	Circular	0.017
Teflon	Polytetrafluoroethylene	Multifilament	Flat	62	52	1.2	Circular	0.0026

[a] Calculated from filament diameter and filament count assuming fiber densities of 1.38, 0.91, and 2.30 g/cm^3 for Dacron, Polypropylene, and Teflon, respectively.
[b] Tex is defined as the weight in grams of a yarn 1000 meters long.
Source: C. C. Chu, B. Pourderyhm, and L. Welch. Characterization of morphologic and mechanical properties of surgical mesh fabrics. *J. Biomed. Mater. Res. 19*:903–916 (1985).

Table 19.14 Fabric Characterization of Three Commercial Nonabsorbable Surgical Meshes

Trade name	Type	Wales/ 10 cm	Courses/ 10 cm	Fabric count (sum of wales and courses in 10 cm^2)	Stitch density per cm^{2a}	Relative porosity (%)	Thickness (cm)	Weight (g/10 cm^2)	Density g/cm^3)	Packing[b] factor (%)
Mersilene	Warp knit	87	47	134	41	51	0.023	0.0432	0.19	13.8
Marlex	Warp knit	70	65	135	45	32	0.065	0.1522	0.23	25.3
Teflon	Warp knit	48	31	79	15	32	0.068	0.3165	0.49	21.3

[a] Calculated from wales/cm × courses/cm.
[b] Calculated from fabric density/fiber density × 100.
Source: C. C. Chu, B. Pourderyhm, and L. Welch. Characterization of morphologic and mechanical properties of surgical mesh fabrics. *J. Biomed. Mater. Res. 19*:903–916 (1985).

Table 19.15 Strength Evaluation of Three Commercial Nonabsorbable Surgical Meshes[a]

Mesh	Tensile properties: Breaking strength (kg) Wale	Breaking strength (kg) Course	Breaking elongation (%) Wale	Breaking elongation (%) Course	Bursting strength (kg)	Bursting strength (kg/cm^2)
Mersilene	12.2 ± 1.3	6.9 ± 0.9	45.2 ± 3.0	103.9 ± 3.1	19.9 ± 0.3	1.3 ± 0.02
Marlex	34.6 ± 1.1	16.5 ± 1.4	74.3 ± 2.3	203.2 ± 3.9	68.9 ± 1.9	4.5 ± 0.12
Teflon	22.7 ± 2.4	16.8 ± 1.5	159.4 ± 12.7	200.8 ± 2.6	35.1 ± 2.0	2.3 ± 0.13

[a] Calculated at 95% confidence limit.
Source: C. C. Chu, B. Pourderyhm, and L. Welch. Characterization of morphologic and mechanical properties of surgical mesh fabrics. *J. Biomed. Mater. Res. 19*:903–916 (1985).

significant. Marlex meshes have the highest tensile breaking strength among the three surgical meshes, particularly in the wale direction. In this direction the breaking strength of Mersilene mesh is found to be only about one-third of that of Marlex. Mersilene mesh also exhibits the least extensibility in both the wale and course directions, while PTFE mesh shows the highest breaking elongation. In terms of bursting strength a similar pattern was also observed. Marlex mesh exhibits the highest bursting strength, almost twice that of PTFE mesh and more than three times that of Mersilene mesh.

Table 16 shows large differences in the flexural rigidities of the three surgical meshes. While Mersilene and PTFE meshes are both extremely flexible, Marlex mesh is found to be extremely rigid, particularly in the course direction. Its overall rigidity is about 70 times greater than that of Mersilene mesh. In terms of wrinkle recovery, as shown in Table 17, Marlex mesh shows the lowest recov-

Table 19.16 Flexual Rigidity of Three Commercial Nonabsorbable Surgical Meshes (mg-cm)

Trade name	G_w, wale rigidity[a]	G_c, course rigidity[a]	G_0, fabric rigidity[a]
Mersilene	16.07	3.58	30.40
Marlex	78.07	825.22	2242.68
Teflon	16.20	6.83	42.34

[a] Calculated from the following formula: G_w (flexural rigidity in the wale direction) = $W \times (O/2)^3_w$; G_c (flexural rigidity in the course direction) = $W \times (O/2)^3_c$; $G_0 = (G_w \times G_c)^{1/2}$; W = weight per unit area in mg/cm^2. O = length of overhang in cm.

Source: C. C. Chu, B. Pourderyhm, and L. Welch. Characterization of morphologic and mechanical properties of surgical mesh fabrics. *J. Biomed. Mater. Res. 19*:903–916 (1985).

ery in the course direction. In the wale direction, however, it exhibits a slightly higher wrinkle recovery than both Mersilene and PTFE meshes. PTFE mesh has almost identical wrinkle recovery in both wale and course directions when measured immediately.

The wide range of observed characteristics and properties of these three surgical meshes is the combined result of both the constituent fibers and the yarn and fabric structure of each mesh. Based on fabric structure, all three surgical meshes examined are warp knitted; the yarns lie predominantly in the lengthwise direction. Knitted fabrics have a more open structure than woven. The size and shape of the pores are a function of the arrangement of the movement of needles or the fabric design and can therefore be easily controlled.

Figure 8 shows the way in which the yarns of the three mesh fabrics are interlooped around each other. Marlex mesh has a totally different structure from those of Mersilene and PTFE. The pores in Mersilene and PTFE meshes are mainly located between the two adjacent sets of yarns, while the pores in the Marlex are largely in the centers of the loops. The yarns in the Marlex mesh fabrics, however, are interlooped on almost every course. This would undoubtedly result in a higher strength but would also lead to a higher degree of stiffness as well as a reduced extensibility.

The shapes of the pores of the three surgical meshes are also different, as is evident in Figs. 9–11. The PTFE

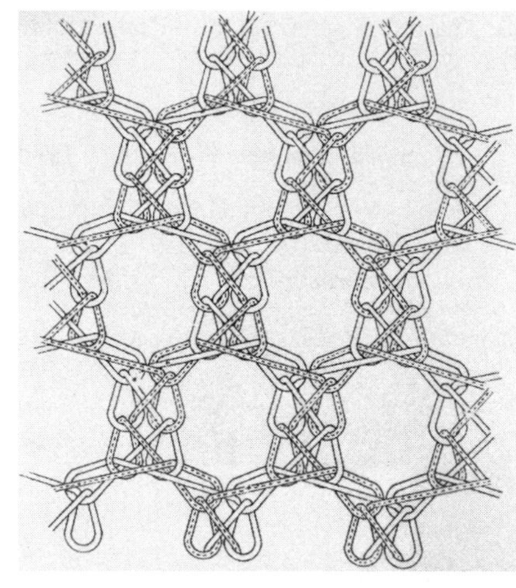

(a)

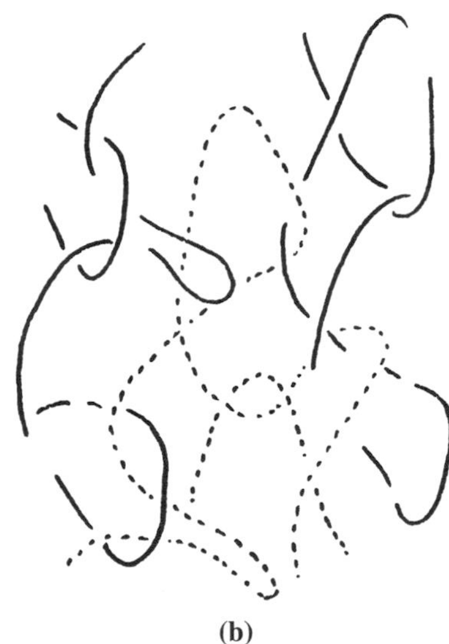

(b)

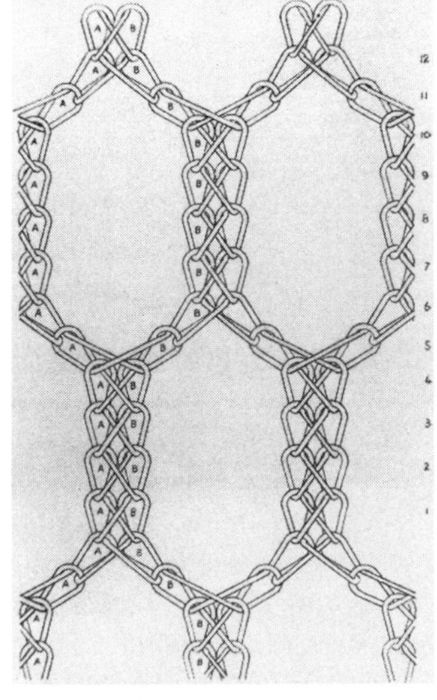

(c)

Figure 19.8 Schematic drawing of the actual configuration of the pore structure of the three nonabsorbable surgical mesh fabrics. (a) Mersilene, (b) Marlex, and (c) Teflon. (From C. C. Chu, B. Pourderyhm, and L. Welch. Characterization of morphologic and mechanical properties of surgical mesh fabrics. *J. Biomed. Mater. Res. 19*:903–916 (1985).)

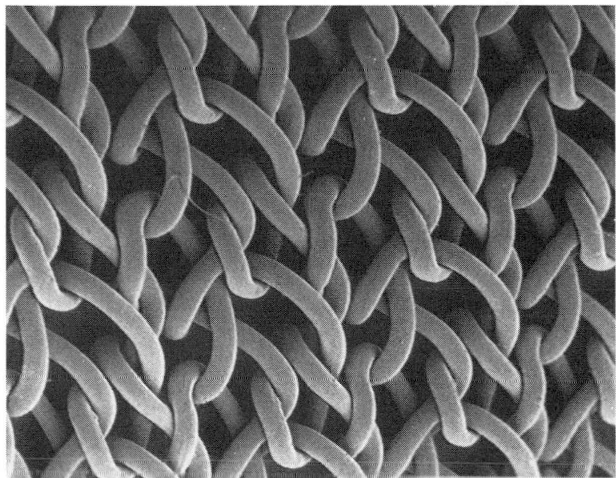

Figure 19.9 Scanning electron micrograph of Marlex mesh at 22×. (From C. C. Chu, B. Pourderyhm, and L. Welch. Characterization of morphologic and mechanical properties of surgical mesh fabrics. *J. Biomed. Mater. Res. 19*:903–916 (1985).)

mesh has nearly circular pores, the Mersilene mesh has hexagonal pores, and the Marlex mesh has irregular pores. The number of pores, as well as the pore size and shape, is important to the surgical mesh's ability to reinforce wounds. The porous area of the mesh serves as a scaffold for the subsequent ingrowth of a dense infiltrate of fibrous tissue, which, in turn, contributes to the strength of the wound (140). Arnaud et al. (141) used Wistar rats as models to show that porous mesh materials induced more than twice the fibroblastic cell infiltration than nonporous fabric materials, therefore exhibiting higher bursting strength of the abdominal wound. Among the three surgical meshes tested, Mersilene has the highest porosity (or the lowest packing factor) and thus may induce more ingrowth of fibrous tissue than the other two meshes. Ingrowth of fibrous tissue, however, also depends on other factors, such as the pore size and shape, the chemical nature of the constituent fibers, and the mechanical force on a porous material–fibrous tissue interface. The pore sizes of the three surgical meshes are different; Mersilene has a pore size of 120 μm × 85 μm measured by taking the two longest perpendicular axes of the pore. PTFE has a similar pore size to that of Mersilene mesh, 157 μm × 67 μm, while Marlex has the smallest pores, ranging from 68 μm × 32 μm to as small as 23 μm × 23 μm. Bobyn et al. (142) have dem-

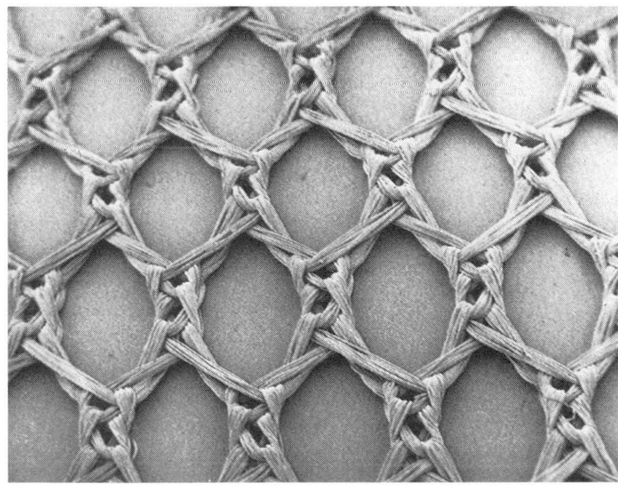

Figure 19.10 Scanning electron micrograph of Mersilene mesh at 22×. (From C. C. Chu, B. Pourderyhm, and L. Welch. Characterization of morphologic and mechanical properties of surgical mesh fabrics. *J. Biomed. Mater. Res. 19*:903–916 (1985).)

Figure 19.11 Scanning electron micrograph of PTFE mesh at 22× (From C. C. Chu, B. Pourderyhm, and L. Welch. Characterization of morphologic and mechanical properties of surgical mesh fabrics. *J. Biomed. Mater. Res. 19*:903–916 (1985).)

Table 19.17 Wrinkle Recovery of Three Commercial Nonabsorbable Surgical Meshes (Degrees)

	Wale		Course	
Trade name	Immediately	After 5 minutes	Immediately	After 5 minutes
Mersilene	143.7	149.5	159.5	163.0
Marlex	143.5	163.2	112.7	142.7
Teflon	138.5	158.2	137.0	167.2

Source: C. C. Chu, B. Pourderyhm, and L. Welch. Characterization of morphologic and mechanical properties of surgical mesh fabrics. *J. Biomed. Mater. Res. 19*:903–916 (1985).

onstrated that the optimal pore size for the strongest mechanical attachment of ingrowth for fibrous tissues should range from 50 to 200 μm, with an average pore size of 90 μm. Chvapil et al. (143) also reported that a pore size greater than 100 μm was needed for the rapid ingrowth of vascularized connective tissue. Smaller pore size is inadequate for the ingrowth of connective tissues, because of the lack of proper space for capillary penetration, as is discussed by Taylor et al. (144). These reported pore sizes for optimal fibrous tissue ingrowth are within the pore sizes of Mersilene and PTFE meshes but not Marlex mesh. Usher and Gannon (89) reported that PTFE meshes, however, stimulated less fibrosis than Marlex meshes in dogs and hence were not well bonded to the surrounding fascia. This lack of ingrowth of fibrous tissue in PTFE mesh, regardless of its porosity and pore size, is thought to be attributed to the chemical nature of the PTFE fibers. Because of a very low critical surface tension, 18.5 dyn/cm (145), it is difficult for fibrocells to attach and spread on PTFE surfaces.

The observed wide range of mechanical properties of these three surgical meshes are no doubt the result of the distinct difference in yarn and fabric structure. Marlex has monofilament yarns, whereas Mersilene and PTFE have multifilament yarns with 12 and 52 filament counts, respectively. The linear densities and the sizes of the yarns are also different. These differences in fabric and yarn structure, together with differences due to chemical properties of the fibers, influence the tensile and bursting strengths, the flexural rigidities, and the wrinkle recovery properties of these fabrics.

Regardless of the magnitude of strength measurements, all three surgical meshes have one common characteristic—greater strength in the wale direction and more extensibility in the course direction. While this is a typical characteristic of a knitted fabric structure, this unique anisotropic property is important in determining the orientation of the mesh during implantation. The wale direction of the mesh implant should be parallel to the maximum physiological stress. Among the three tested materials, PTFE has the highest elongation at both wale and course directions. This property is not considered to be an advantage in hernia repair because too much stretch in a mesh material can either separate the underlying tissue before the mesh resistance can be brought into operation (93) or pinch the ingrowth tissue by a narrowing of the pore size, which can result in tissue necrosis. A certain amount of extensibility, however, is desirable, because it can act as a stimulant to orient the connective tissue parallel to the direction of the physiological force. Another common observation of the three tested surgical meshes is that the mesh with higher tensile strength also exhibits higher bursting strength, because the force experienced by the mesh fabrics in bursting tests was of tensile mode. Marlex meshes exhibit the highest tensile and bursting strength, due mainly to the structure of the constituent yarn of this fabric.

A wrinkled mesh may cause more adhesion problems with viscera than a stretched mesh. Usher et al. (89) reported that there were adhesions of the bowel and omentum to the peritoneal side of the Marlex mesh in dogs, and that this might be due to the observed transverse wrinkling of the mesh. A wrinkled mesh can also result in the uneven distribution of stress on the mesh and cause an undesirable distortion and possibly, a premature weakening of the mesh. From the morphological point of view, a wrinkled mesh may result in bursas on the mesh material and hence predispose the wound to infection. The results of the wrinkle recovery test performed by Chu et al. (139) indicate that the recovery of a surgical mesh fabric depends on the direction of the mesh fabric (i.e., wale or course) and the time allowed for recovery. The exception was the PTFE mesh, which showed no difference of direction when measured immediately. Among the three surgical meshes, PTFE and Mersilene have better overall wrinkle recovery than Marlex meshes, and this can be attributed to different yarn and fabric structure.

Because of the difficulty of conforming to the topology of the surrounding tissues, a stiff and rigid mesh can cause the erosion or perforation of intraabdominal viscera, particularly with small children. Clinical cases of this complication have been previously reported (146). Moazam et al. (147) also reported that a pliable mesh will rarely develop problems with enteric fistulae, even when it is used over a prolonged period. Among the three types of meshes tested in this study, Marlex meshes have a significantly higher flexural rigidity than the other two. This is due to a high yarn stiffness and a bending modulus resulting from both the physical configuration of the monofilament yarn and the chemical structure of the high linear density polypropylene fiber. It is evident, therefore, that both the selection of the chemical nature of the constituents of the fibers and the selection of the yarn and fabric structures contribute to the properties of each mesh. These factors should be considered in the future design of surgical mesh fabrics to obtain the necessary balance between strength and properties such as porosity, stiffness, and wrinkle recovery. The availability of these characterization data can also serve as the foundation for a surgeon's selection of the appropriate commercial surgical mesh fabric for each particular case, and for its subsequent evaluation and comparison in vivo.

Among the synthetic nonabsorbable biomaterials, metallic meshes have been shown to become work-hardened, inflexible, friable, and fragmented with time (84,85,148,

149). They may protrude through the skin or erode into adjacent tissue or blood vessels.

The polymeric based meshes are far more successful than metallic ones, but they still exhibit various problems. Mersilene meshes produce good results in tissue repair but often produce seromas as the most common complication (150). Meshes made of PTFE are generally satisfactory, particularly in the presence of infection, but have ''too much stretch'' (93) and do not stimulate sufficient fibroplasia to be incorporated into the tissue (89). Meshes made of Marlex, the most commonly used, have the lowest frequency of related complications (151). However, they are relatively stiff and occasionally injure the underlying abdominal viscera (146). Long-term complications associated with Marlex, such as mesh extrusion and/or entric fistulae following coverage by split-thickness skin, have been reported (152). Recently, a five-year follow-up study of the use of Marlex mesh for parastomal hernia by Morris-Stiff and Hughes (153) indicated poor results due to high recurrent hernia (29%), serious complications (obstruction and dense adhesion to the intra-abdominal mesh (57%), and mesh-related abscess (15%) (153).

The most recent comprehensive evaluation of biological tissue response to nonabsorbable meshes in animals was reported by Ortiz-Oshiro et al. (154). They found that polypropylene mesh in rats over a period of 15 weeks provoked higher acute inflammatory reaction (e.g., higher cell density and polymorphonuclear response) and connective tissue formation (e.g., fibrohistocytic response) than Mersilene mesh, which induced higher foreign-body reaction. Similar findings were reported in the meshes retrieved from humans (106). In this study, Klinge et al. reported that polypropylene mesh had the highest partial volume of inflammatory cells (32%), followed by Gore-Tex (12%), Mersilene (8%), and ''reduced'' polypropylene mesh (7%). The extent of this inflammatory reaction correlated well with the level of connective tissue formation. They also concluded that these persistent foreign body reactions to nonabsorbable meshes could lead to chronic wound complications.

In order to reduce or eliminate the reported clinical disadvantages of nonabsorbable meshes, one approach was to reduce the amounts of nonabsorbable components in a surgical mesh to <30 g/m^2 via increasing mesh pore size. This reduced polypropylene mesh consists of multifilament yarn instead of the conventional monofilaments and has a larger pore size than nonreduced polypropylene mesh. This approach is based on the fact that tissue reaction to implants depends also on the amounts of the foreign materials presented in a biological environment. Thus the term ''reduced'' polypropylene mesh referred to a nonabsorbable mesh having smaller amounts of polypropylene than conventional Marlex for the purpose of reducing tissue reaction. As reported by Klinge et al. (106), reduced polypropylene indeed showed the lowest partial volume of inflammatory cells and numbers of macrophages at the interface between meshes and tissues among Marlex, Mersilene, and Gore-Tex meshes.

Since this reduced polypropylene mesh was too soft and lumpy for proper surgical handling, polyglactin-910 absorbable biomaterial was either coated onto or/and blended (in multifilament form) with the reduced polypropylene mesh to increase its initial stiffness to facilitate its manipulation during surgery (110). The polyglactin-910-coated reduced polypropylene mesh, however, showed a remarkably different tissue reaction from the uncoated ones in rats over 90 days by the preferential formation of fibroconnective tissue capsule, surrounding the whole mesh. Such capsule formation prevents a proper incorporation of the mesh with the surrounding tissue.

Nonabsorbable meshes, particularly the polypropylene based, are known to have a strong tendency to form intra-abdominal adhesion between the mesh material and the intraperitoneal viscera. Such adhesion is due to the high integration of the surrounding tissues into the porous structure of the mesh and may lead to the obstructions of intestine and enterocutaneous fistulae. One promising reported approach to solve this intra-abdominal adhesion problem is to use absorbable meshes with or without the presence of nonabsorbable meshes. The details of this approach will be discussed in the next section.

In addition, all the above polymeric based meshes share one common disadvantage—they must be removed, in most cases, if infection develops. Absorbable surgical meshes have shown better chronic tissue tolerance because they are biodegraded completely so that no traces of a foreign body remain.

B. Absorbable Surgical Meshes

Most absorbable surgical meshes are made from the same absorbable fibers as the absorbable suture materials. Among synthetic absorbable fibers, PGA (Dexon®) and polyglactin-910 (Vicryl®) are the two most popular fibrous biomaterials for making absorbable surgical meshes. The major advantage of absorbable surgical meshes over nonabsorbable ones is the absence of permanent foreign-body reaction, which could predispose to late complications like chronic infection, adhesion, mechanical irritation, drainage, erosion, and bleeding. The disadvantage of absorbable surgical meshes is the possible recurrence of hernia after the disappearance of the mesh.

Absorbable meshes are also particularly useful in closing contaminated abdominal wall defects. Dayton et al.

(101) reported that those patients who had prior infected polypropylene mesh or a chronically infected scar were treated by a single sheet of PGA mesh without the long-term complications of nonabsorbable meshes. However, six of the eight patients developed recurrent abdominal wall hernia at the site of the PGA mesh from 3–18 months follow-up. They suggested that this complication should balance against the more serious complications associated with nonabsorbable meshes, and the use of absorbable meshes as a temporary abdominal wall support in contaminated wound sites could lead to subsequent successful implantation of a nonabsorbable mesh after the infection was cleared. The recurrence of hernia due to absorbable meshes was also reported by Tyrell et al. (102) in their animal (rabbits) study and by Greene et al. (103) in their clinical study. Tyrell et al. found that all those rabbits repaired by either PGA or polyglactin-910 meshes exhibited ventral hernias by the 10th week after the operation, while those repaired by Marlex or PTFE showed no sign of hernias. In the Greene et al. study, PGA knitted meshes were used in 59 critically ill patients to bridge the abdominal wall defects after celiotomy; however, hernia defects were found in most patients 4 to 6 months after the operation. The time of hernia recurrence coincided with the complete disappearance of the PGA meshes (2 or 3 months). This suggests that the lack of proper mechanical support due to the complete biodegradation of the absorbable mesh was the cause for the hernia appearance. Since most trauma patients enrolled in emergency rooms have contaminated wounds, absorbable meshes were suggested as a better temporary alternative adjunct to fascial closure until the complete resolution of acute complications (105).

Absorbable meshes also have the advantage over nonabsorbable ones in terms of adhesion formation between prosthetic meshes and surrounding tissues. This advantage is supported by several reported studies (108,155–157). For example, Dasika and Widmann (156) implanted polyglactin-910 mesh alone or placed this absorbable mesh between Marlex mesh and the intraperitoneal viscera of rats for 1 to 3 months. Both polyglactin-910 mesh alone and the combination of polyglactin-910 and Marlex meshes showed a similar adhesion score (1.5), while the Marlex mesh had the worst adhesion score (2.75 with 3 the most severe). The control (midline laparotomy with suture closure only) had little adhesion and the lowest adhesion score (0.25 with 0 no adhesion). This difference in adhesion formation became the greatest at the longest implantation period (i.e., 3 months). The ability for absorbable meshes to reduce tissue adhesion also depends on the level of trauma. For example, Sondhi et al. (155) found that, in normal surgical trauma conditions, polyglactin-910 mesh could significantly reduce postoperative adhesion in the extraocular muscle surgery of rabbits, but no significant reduction in adhesion was found in extensive trauma.

In addition to the synthetic absorbable aliphatic polyester based meshes, Genzyme's new Seprafilm has also been used along with nonabsorbable polypropylene mesh to prevent tissue adhesion problem (108,157). In their study, Alponat et al. (108) placed two Seprafilms over the abdominal viscera of an incisional hernia in rats before the surgical repair of the defect with polypropylene mesh, and fewer adhesion formations were observed than with polypropylene mesh only at the end of 30 days. Significant reduction in adhesion formation in rabbits by placing Seprafilms between tissues and polypropylene mesh was also reported by Dinsmore and Calton (157).

Although most reports of using absorbable mesh showed promising results toward the reduction of intra-abdominal adhesion, the findings were not unanimous. Using mice, Baykal et al. (158) recently demonstrated that, based on adhesion grading and tissue hydroxyproline levels, PGA mesh actually increased intestinal adhesion formation at 90 days after operation than the control (no mesh) and polypropylene mesh groups. They also found that tissue adhesion correlated well with the level of tissue hydroxyproline formation. They attributed these unexpected findings to the higher inflammatory and foreign-body reactions associated with the absorption process of PGA mesh.

There is one innovative use of surgical meshes that has nothing to do with their traditional function to bridge body wall defects. Surgical meshes, particularly absorbable ones, have been used in postoperative radiation therapy for gynecological (e.g., pelvic) and rectal-related malignant tumors. The function of surgical meshes in this application is to lift up patients' small bowels away from the radiation target region (e.g., pelvic or rectal) during postoperative radiation therapy to prevent the devastating radiation-induced small bowel injury that could predispose to infection, adhesion, and obstruction. There is a high frequency of this type of small bowel injury due to postoperative radiation therapy following pelvic cancer surgery (up to 50%). Because of this concern, postoperative radiation dosage cannot be raised to a more effective tumoricidal dosage level and is frequently limited to $< 5{,}000$ cGy. In this nontraditional application of surgical meshes, absorbable ones are better than the nonabsorbable because there is no need to remove the mesh afterwards and there is no permanent foreign-body reaction.

Several investigators have reported the use of absorbable surgical meshes like PGA or polyglactin-910 for such a nontraditional purpose in humans and animals with promising results (159–166). For example, Devereux et al. (159) used PGA mesh as an intestinal sling to elevate the small bowel of 60 patents (18 with gynecologic malignancies and 42 with rectal carcinomas) away from the sites operated on and irradiated (a mean 5,500 rad dosage). After 28 months, there was no single case of radiation enteritis or complication associated with the implanted PGA meshes. Dasmahapatra et al. (163) reported the lack of incidence of clinical radiation-associated small bowel injury after a 12–53 months follow-up of the use of both PGA and polyglactin-910 meshes in 45 patients having rectal carcinoma, except for two patients with polyglactin-910 mesh that developed obstructions due to adhesions. Rodier et al. (165) reported that the PGA mesh was completely absorbed in such an application within 3 to 5 postoperative months. In a controlled animal study, Devereux et al. (161) found that cebus monkeys that had PGA mesh slings before irradiation showed normal small bowel function and histological data within 12 months, while the untreated 10 monkeys all died of small bowel necrosis upon 2,000 rads (single dose). PGA mesh disappeared completely by 6 months.

Therefore the use of absorbable surgical meshes to lift up small intestine bowels after surgical removal of tumors and before radiation therapy appears to be a promising technique to eliminate radiation-induced small bowel injury without significant mesh-related complications. Such application also permits the use of more effective radiation dosages for tumoricidal effect.

IV. VASCULAR GRAFTS

Vascular diseases have always been associated with human beings, as recorded history shows. They frequently result in either the loss of lives or limbs. The development of textile based vascular grafts of either synthetic or natural origin has been one of the most important biomedical applications. This development has made significant progress and allowed the reconstruction of obstructed or injured blood vessels with remarkable success. However, some serious problems still remain unsolved (167). While homografts are still considered to be the preferred arterial replacement for small-diameter vessels (below the knee), they are used in limited quantity and are insufficient to meet the increasing needs for vascular replacement, due mainly to an inadequate supply, nonuniform properties, and difficulty in preparation (168,169).

A. Introduction

In the long search for an ideal graft replacement, various synthetic materials have been experimented with. They include rigid nonporous tubes made of gold, silver, aluminum, glass, and polyethylene. However, these nonporous materials completely failed due to the lack of porosity and compliance. These two important requirements of an ideal vascular graft have been partially fulfilled by the use of fibrous based polymers and textile structure. The two most important polymeric biomaterials used for making vascular fabrics are poly(ethylene terephthalate) and expanded poly(tetrafluoroethylene). The former includes Dacron and has the largest market share. The latter is the same as Gore-Tex. In addition to these two common polymeric biomaterials as the major source for fabricating vascular grafts, the use of polyurethane based elastomeric fibers like Spandex has been tried for the purpose of improving the compliance of vascular grafts (170).

In contrast to a solid tube design, the inherent porous nature of textile materials has provided the required space for the fibrous connective tissue to grow into for achieving full-wall healing (defined by Sauvage as ''incorporation of the entire graft within a fibrous tissue matrix whose flow surface is covered with endothelium'') (167).

Currently, there are three basic textile structures for vascular grafts: woven, knitted, and nonwoven. These three are different from each other in porosity, mechanical properties, and other properties. All woven and knitted vascular fabrics are made from poly(ethylene terephthalate), while the nonwoven vascular fabrics are made from expanded poly(tetrafluoroethylene), the same polymer for the nonabsorbable Gore-Tex suture.

The woven structure provides strength with high dimensional stability and low permeability to blood, and it is less prone to kinking. It is used mainly in large-diameter vessels like the aorta and major arteries from which uncontrolled bleeding could lead to fatality. The main disadvantages of woven vascular fabrics are their very low healing porosity leading to poor healing, difficulty in suturing, fraying of cut edges, and poor compliance. In order to improve the healing porosity and compliance of woven vascular fabrics, velour woven vascular fabrics were introduced. These new fabrics were made by floating portions of the weft or/and warp yarns so that the numbers of intersections along both warp and weft directions were reduced considerably from the conventional 1×1 woven structure. Since the numbers of intersections of both weft and warp yarns are closely related to the mechanical and physical properties of the resulting fabrics, a significant reduction in these numbers would make the fabrics more flexible and porous. The floated portions of the yarns have another advantage: they produce three-dimensional loose surface structures as the framework for tissue attachment and ingrowth that conventional woven fabrics do not have.

The knitted vascular fabrics have a much higher porosity than woven grafts and hence show better full-wall healing. However, due to their high porosity, they are difficult to preclot during the time of implantation and therefore impose the problem of blood leakage. As a result, knitted vascular fabrics are not normally used in large arteries where bleeding is a major problem. Knitted vascular fabrics are also softer and more flexible than woven vascular grafts and show better compliance. There are far more varieties of knitted than of woven structure. The current commercially available knitted vascular fabrics are mainly warp knitted with a single or double velour surface structure.

All woven and knitted vascular grafts have one common appearance, a crimped structure. A crimping process introduces ''hills and valleys'' along the longitudinal direction of the grafts. The purpose of crimping these vascular grafts is to prevent kinking of the graft at its bending site. A kinked graft would block the flow of blood at the point of bending. Since the crimped process would destroy the smooth and even flow surface of the fabric, the wells in the valley portions could be prone to thrombus deposits due to the stagnation of the blood flow in these regions.

The most nonconventional textile structure in vascular fabrics is Gore-Tex. It consists of nodes that are connected by fine fibrils. Because of this unique structure, it is highly porous. The details of Gore-Tex have been previously described in Sec. II.A.12.

B. Two Important Criteria for Designing Vascular Fabrics

The two most important criteria in designing vascular fabrics are porosity and nonthrombogenic surface. These two criteria are closely related to each other because vascular fabrics require porous space for tissue ingrowth, which would ultimately lead to the formation of nonthrombogenic surfaces. The generation of a nonthrombogenic surface on synthetic vascular graft materials is one of the most important goals in the repair of damaged or diseased vascular systems. Incomplete healing imposes the continued risk of thrombotic occlusion, particularly in medium- and small-caliber blood vessels. There is only one reported case of a complete full-wall healing (defined by Sauvage as the incorporation of the entire graft within a fibrous tissue matrix whose flow surface is covered with endothelium) in humans at points distant from the suture line (171).

Because all of the existing commercial vascular grafts are constructed from a single type of nonabsorbable fiber (e.g., Dacron), their porosity is relatively constant and does not change with time. It is thus impossible to vary the porosity of these single-component fabrics so that they would be very tight during implantation (i.e., low bleeding porosity) to prevent the occurrence of blood leakage, and be very porous during healing (i.e., high healing porosity) to promote fibrous tissue ingrowth for a full-wall healing.

There are several approaches to try to design ideal vascular fabrics that would meet these two most important criteria. Among them, the use of absorbable fibers as the sole component or as one of two components of a fabric and the theoretical prediction of the porosity of a fabric based on some mathematical formula are quite intriguing and appear to be promising for the design of the next generation of vascular fabrics.

The rationale of using absorbable fibers as components of vascular fabrics is threefold. First, since the absorbable component would eventually be biodegraded, the resulting vascular fabrics would have high healing porosity due to the new porous space generated from the disappearance of the absorbable components. This high healing porosity would induce the growth of an endothelial cell lining within the graft luminal surface. The formation of an endothelial lining within a vascular graft could be achieved through pannus ingrowth, capillary ingrowth, attachment and growth of blood-borne pluripotential cells, and the artificial endothelial cell-seeding at the time of implantation. All the ingrowth mechanisms would require adequate porosity of the fabrics, and the use of absorbable fibers appears to be the most promising approach to achieve such a goal without compromising bleeding porosity during implantation. Secondly, absorbable vascular grafts would not elicit permanent foreign-body reactions as nonabsorbable fibers do and hence are more biocompatible and less prone to chronic inflammatory reactions that lead to infection. Finally, absorbable vascular grafts could regenerate pseudoarterial tissues with nonthrombogenic surfaces during the biodegradation process of the absorbable grafts. This approach of tissue regeneration via absorbable scaffolds is being pursued with some very exciting and promising results.

C. Absorbable Based Vascular Fabrics

Among absorbable based vascular fabrics, there are two kinds, totally and partially absorbable fabrics. The totally absorbable vascular fabrics do not have any nonabsorbable components, and the most representative researches are by Wesolowski et al., Bowald et al., and Greisler et al. (172–191). The partially absorbable vascular fabrics have both absorbable and nonabsorbable components, and the most representative researches are from Chu et al. (192–194).

1. Totally Absorbable Based Vascular Grafts

Greisler et al. have used the same absorbable fibers as the synthetic absorbable suture materials for the completely absorbable based vascular grafts, namely polyglycolic acid (PGA), polyglactin-910 (Vicryl), and poly-*p*-dioxanone (PDS). Their extensive in vivo and in vitro studies have documented active stimulatory or inhibitory effects of various absorbable materials on the regeneration of myofibroblast, vascular smooth muscle cells, and endothelial cells, and have shown a transinterstitial migration to be their source when lactide/glycolide copolymer prostheses are used (185). The rate of tissue ingrowth parallels the kinetics of macrophage-mediated prosthetic absorption in all polylactide/polyglycolide-based grafts (179,180,182, 183). Macrophage phagocytosis of these absorbable biomaterials is observed histologically as early as one week following implantation in the case of a rapidly absorbable material, such as PGA or polyglactin-910, and is followed by an extensive increase in the myofibroblast population and neovascularization of the inner capsules (179,180, 195). Autoradiographic analyses demonstrated a significantly increased mitotic index within these inner capsular cells, that mitotic index paralleling the course of the absorption of absorbable biomaterials (196). Polyglactin-910, for example, resulted in a mitotic index of 20.1 ± 16.6% 3 weeks following implantation, progressively decreasing to 1.2 ± 1.3% after 12 weeks. The more slowly absorbable PDS based vascular grafts demonstrated a persistently elevated mitotic index, 7.1 ± 3.8%, 12 weeks after implantation, a time in which the prosthetic material is still being biodegraded. The nonabsorbable Dacron vascular graft, however, never achieved greater than a 1.2 ± 1.3% mitotic index (196). These mitotic indices correlated closely with the slopes of the inner capsule thickening curves, suggesting that myofibroblast proliferation contributed heavily to this tissue deposition. The labeled cells are found in the deeper zones of the inner capsule in proximity to the macrophages and prosthetic graft material (196) in the area in which the myofibroblasts demonstrate ultrastructurally synthetic phenotypes, a phenotype more commonly found in cells actively cycling (179).

Inner capsular collagen deposition similarly follows the kinetics of both myofibroblast proliferation and macrophage-mediated biomaterial absorption. Explanted inner capsule tissues within polyglactin-910, PDS, and Dacron grafts showed extensive deposition of collagen within the inner capsules of both polyglactin-910 and PDS absorbable graft biomaterials, more rapidly within the more rapidly absorbed polyglactin-910. Collagen contents at one month were 35.5 ± 9.9 mg collagen/mg dry weight in polyglactin-910, 25.6 ± 5.0 in PDS, and 12.0 ± 0.9 in Dacron. By three months, PDS inner capsular collagen content had increased to 44.6 ± 6.3 mg/mg, while the polyglactin-910 and Dacron exhibited little increase. A normal rabbit aortic collagen content was 22.5 ± 1.2 mg/mg.

The research of Greisler et al. suggested that lactide/glycolide copolymer based absorbable vascular grafts could stimulate an extensive tissue ingrowth via their interactions with macrophages. Similarly, they showed that Dacron actively inhibited such tissue ingrowth (195). They reported that the presence of as little as 30% Dacron within a compound yarn could effectively abolish tissue ingrowth and inner capsule cellularity, resulting in inner capsules composed primarily of fibrin coagula. A compounded graft consisting of an absorbable yarn with the nonresorbable polypropylene, however, did not result in a similar blockade of the absorbable polymer induced tissue ingrowth (181,197).

Greisler et al. utilized this Dacron inhibitory activity to perform a study evaluating the source of regenerating tissues (185). Three 10-mm-long compound prostheses (30 mm × 4 mm I.D.) were constructed in such a fashion that Dacron segments were on both ends and polyglactin-910 between. These compounded grafts were implanted into adult NZW rabbit aortas. During the 4 month postimplantation period, all grafts showed 100% patency. Inner capsule thickness in the central polyglactin-910 segments increased from 2 weeks to 2 months, being significantly thicker than either Dacron segment at 1 and 2 months. By 1 month, inner capsules of polyglactin 910 segments were predominantly myofibroblasts and collagen, while the inner capsules in Dacron segments remained fibrin coagula beyond 2 mm from either anastomotic sites. This corresponded to the observations of capillary infiltration of inner capsules with capillary communication with regenerated aortic lumens (182) and suggested that a transinterstitial ingrowth rather than a transanastomotic pannus ingrowth would be the primary source of cells replacing absorbable vascular prostheses.

To demonstrate the humoral nature of the signal for cell proliferation, Greisler et al. reported that the freshly explanted polyglactin-910, PDS, and Dacron prosthesis/tissue complexes harvested 2–12 weeks following implantation were incubated in a physiologic salt solution (100 mg/mL) to elute growth factor activity from these tissues. The mitogenic activity of the eluates was then evaluated. The eluates prepared from the polyglactin-910 explants resulted in significantly more stimulation of DNA synthesis in the 3T3 cells than was seen in response to the Dacron or PDS prepared eluates (198). Radioimmunoassay for PDGF B chain in these eluates was negative, and likely other growth factors are the humoral mediators.

Greisler et al. have also demonstrated that the implantation of the same polylactide/polyglycolide vascular prostheses into rabbits with diet-induced arteriosclerosis results in a markedly attenuated regenerative response with significantly fewer endothelial cells and myofibroblasts in the inner capsule tissues when these absorbable prostheses are implanted into rabbits previously made atherogenic by dietary manipulation (199). Histologically, 3-week explants showed only small areas of neointima with myofibroblasts and endothelial cells; the outer capsules were infiltrated by lipid-laden macrophages. This is in contrast to the thicker inner capsule with abundant myofibroblasts and collagen in polyglactin-910 grafts implanted into normal NZW rabbits by 3 weeks.

Based on the data from absorbable based vascular grafts, Greisler et al. suggested that the macrophage–biomaterial interactions are the basis for the observed vascular tissue regeneration over absorbable scaffolds. These interactions are expected to yield a differential activation of the macrophage resulting in a differential release of biologically active mediators that are both stimulatory and inhibitory to myofibroblast, smooth muscle cell, and endothelial cell migration and proliferation. Chu advanced this hypothesis by suggesting that different rates of absorption of absorbable biomaterials induce different levels of macrophage–biomaterial interactions and hence different degrees of vascular tissue regeneration.

Recent cell culture studies by Greisler et al. appeared to support the hypothesis that the macrophage–biomaterial interactions control the level of vascular tissue regeneration (200–202). They reported the greater ability of lactide/glycolide copolymers to induce macrophages to release growth factors, including bFGF, capable of stimulating DNA synthesis in cultured murine capillary lung endothelial cells, human umbilical vein endothelial cells, BALB/c 3T3 mouse fibroblasts, and rabbit aortic smooth muscle cells. On the contrary, by exposure the same macrophage population to nonabsorbable Dacron resulted in an inhibition of endothelial cell proliferation and a diminution in the stimulation of fibroblast and smooth muscle cell proliferation.

Smooth muscle cells of normal rabbit aortas (SMC) in the presence of the media consisting of normal diet macrophages that had been exposed to polyglactin-910 were stimulated to synthesize DNA significantly more (4–9×) when compared to the same macrophage media exposed either to Dacron or to neither biomaterial. The Dacron-exposed macrophage media yielded minimal mitogen release. It thus appears that the absorbable biomaterial conditioned macrophage media stimulated SMC proliferation. It further suggested that polyglactin-910 induced more mitogen release from the biomaterial conditioned macrophage media than did Dacron. These results are consistent with the in vivo observations of greater tissue ingrowth and mitotic indices following implantation of polyglactin-910 prostheses into normal rabbits as compared to Dacron prostheses. These data are also consistent with the greater tissue ingrowth elicited by polyglactin-910 prostheses implanted into normal rabbit aortas as compared to artherosclerotic rabbit aortas.

Bioassays utilizing quiescent LE-II endothelial cells similarly showed an increase in LE-II proliferation induced by the macrophage polyglactin-910 conditioned media, an effect not seen in response to Dacron exposure. Conversely, exposure of the macrophage Dacron conditioned media yielded less LE-II proliferation even than the PBS negative control, suggesting the release of a growth inhibitor due to the presence of Dacron. Greisler et al. suggested that the extensive transinterstitial ingrowth of myofibroblasts and endothelial cells following implantation of absorbable grafts in normal animals and the attendant elevated mitotic index of the myofibroblasts may be mediated by macrophage activation and release of growth factors promoting the mitogenic response.

2. Partially Absorbable Based Vascular Grafts

Because of the complete disappearance of the totally absorbable based graft materials, there is the common theoretical concern that the degradation of the component yarns in the totally absorbable vascular fabrics (i.e., loss of strength), before the regeneration of tissues becomes strong enough to withstand the pulsatile force, may make the totally absorbable vascular grafts more prone to aneurysmal dilation and subsequent failure. Aneurysmal dilation has been found in biological grafts like bovine heterografts (203). The current commercial remedy for the problem of aneurysmal dilation associated with degradable biologic grafts is the use of either a bioprosthetic (a combination of synthetic and natural materials) (204) or a better cross-linking process (203).

One solution to the aneurysmal dilation problem associated with synthetic vascular grafts before the arrival of ideal synthetic absorbable polymers is to use bicomponent fabric design. In this approach, absorbable yarns are blended with nonabsorbable ones with various compositions. The purpose is to provide minimal strength of the grafts (from nonabsorbable components) during and after the biodegradation of the absorbable components of the grafts. Theoretically, this type of design could take the advantage of the property of absorbable polymers (i.e., stimulating the regeneration of vascular tissue and reducing chronic foreign-body reaction by reducing the amount of permanent graft mass in a biologic environment) without the expense of making vascular grafts too weak to be useful in the clinic situation of delayed wound healing.

Chu et al. reported a comprehensive study of these bicomponent vascular fabrics (192–194) in terms of (1) how these bicomponent vascular fabrics hydrolytically de-

Table 19.18 Fabric Characteristics of Bicomponent Woven Vascular Fabrics

Fabrics	Ends/inch	Picks/inch	Thickness (cm)	Weight (g/cm³)	Fabric density (g/cm³)	Bending stiffness (mg-cm)	Bursting strength (Kg/cm²)	Water permeability (mL/cm²/min/mm Hg)
W1	168	104	0.034 ± 0.001	0.0205 ± 0.0026	0.6029	G_{w1} = 2640 G_{w2} = 1430	15.0	300 ± 25
W2	78	128	0.050 ± 0.002	0.0300 ± 0.0022	0.6000	G_{w1} = 823 G_{w2} = 670	11.61	650 ± 70
W3	78	136	0.053 ± 0.002	0.0360 ± 0.0017	0.6792	G_{w1} = 420 G_{w2} = 810	13.51	880 ± 20
W4	156	128	0.052 ± 0.008	0.0300 ± 0.0022	0.5769	G_{w1} = 1045	22.3	400 ± 10
DeBakey woven	135	94	0.025	0.017	0.68	G_{w1} = 1895 G_{w2} = 1345	26.3	250 ± 3

[a] Bending stiffness measured by FRL cantilever Bending Tester, Model TMI-79-10, G_{w1}, in warp direction, G_{w2} in weft direction.
Source: T. J. Yu, and C. C. Chu. Bicomponent vascular grafts consisting of synthetic biodegradable fibers. Part I. In Vitro Study. *J. Biomed. Mater. Res. 27*:1329–1340 (1993).

graded in the in vitro environment and hence the generation of porosity with time, (2) whether the use of partially absorbable vascular grafts in vivo would improve the nonthrombogenic performance of these grafts, (3) whether these partially absorbable grafts would be more prone to aneurysmal dilation and subsequent failure in vivo, and finally (4) to find out the relationships, if any, between these in vivo and the previously reported in vitro data with an emphasis on how the in vitro changes in fabric structure and properties related to the in vivo data.

The bicomponent vascular fabrics of Chu et al. consisted of polyethylene terephthalate (Dacron as the nonabsorbable component) and polyglycolic acid (PGA as the absorbable component). Two types of vascular graft fabrics were fabricated: woven and weft-knitted fabrics with different water permeability. In the woven type (the W group), Dacron yarns were used to make the framework of the fabrics, while PGA yarns were incorporated in the warp, the weft, or both directions of the fabrics. Such a design would retain the structural integrity of the fabrics even after the complete absorption of PGA yarns. The composition of PGA yarns in the woven group ranged from 38% to 82% by weight. In the weft-knitted vascular fabrics (the K group), both Dacron and PGA yarns were fed through a circular knitting machine, and tubular bicomponent fabrics of various diameters could be made. The composition of PGA yarns ranged from 24 to 57% by weight. Tables 18 & 19 summarize the compositions of the eight types of bicomponent vascular fabrics. Depending on the type of vascular fabric construction, the location and composition of PGA yarns in the bicomponent fabrics, and the denier of Dacron yarns, bicomponent fabrics with a range of mechanical and physical properties could be made.

Table 19.19 Fabric Characteristics of Bicomponent Knitted Vascular Fabrics[a]

Fabrics	Course/inch	Wales/inch	Thickness (cm)	Weight (g/cm²)	Stitch density (#stitch/inch²)	Loop length (Cm)	Fabric density (g/cm³)	Bending stiffness[b] (Kg)	Bursting strength (kg/cm²)	Water permeability (mL/cm²/min/mm Hg)
K2	24	40	0.0576 ± 0.0059	0.0179 ± 0.0018	960	0.1071	0.3108	0.649 ± 0.055	19.95 ± 0.45	1700 ± 30
K3	24	32	0.0660 ± 0.00	0.2020 ± 0.0001	768	0.1667	0.3067	2.123 ± 0.608	22.75 ± 0.09	1800 ± 0
K6	28	36	0.0553 ± 0.0011	0.0189 ± 0.0005	1,008	0.1389	0.3433	1.016 ± 0.071	20.07 ± 0.37	1950 ± 100
K8	26	32	0.0772 ± 0.0003	0.0223 ± 0.0004	832	0.1607	0.2893	n/a	25.78 ± 1.28	1500 ± 20
Cooley knitted graft	50	70	0.0445 ± 0.0006	0.0274 ± 0.0006	3,500	n/a	0.6157	n/a	27.47 ± 0.39	2470 ± 30

[a] Dacron and PGA yarns blend together and count as one yarn.
[b] Bending stiffness obtained from the circular bend procedures in ASTM D4032-81. The amount of force required for a 5.3 mm diameter flat tip plunger to push the fabric through a 8.75 mm diameter hole.
Source: T. J. Yu and C. C. Chu. Bicomponent vascular grafts consisting of synthetic biodegradable fibers. Part I. In Vitro Study. *J. Biomed. Mater. Res. 27*:1329–1340 (1993).

Compared with commercial vascular fabrics, the porosities of these bicomponent fabrics are controlled not only by the conventional means (e.g., the number of warp and weft yarns per unit length and the denier of the yarns) but also by the amount and location of the absorbable yarns incorporated into the fabrics. Figure 12 exhibits the most important advantage of these bicomponent vascular grafts: a significant increase in water porosity with time that none of the current commercial vascular fabrics could achieve. Almost all bicomponent vascular fabrics had final water porosity 10 times higher than their initial values, and reached the lower limit of the suggested optimum porosity (i.e., 10,000 mL/cm^2/min of the graft suggested by Wesolowski et al. (173). On the contrary, the water porosities of the two commercial woven vascular fabrics did not change with time and were significantly lower than the bicomponent fabrics at the end of 90 days. This suggests that the bicomponent fabrics may meet the two contradictory and mutually exclusive requirements of an ideal vascular graft: high healing porosity without the expense of low bleeding porosity.

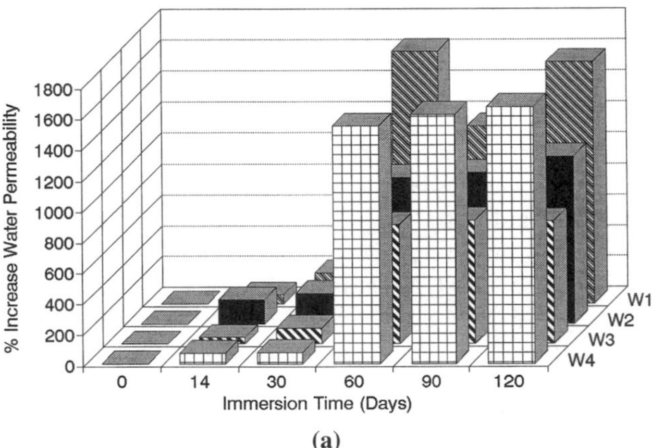

(a)

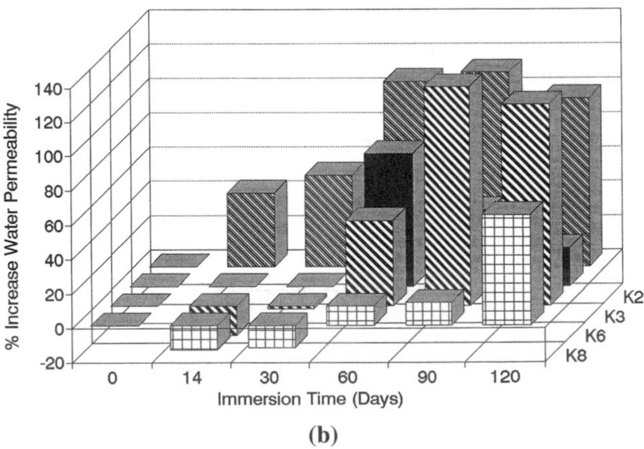

(b)

Figure 19.12 The percentage of increase in water permeability of bicomponent fabrics as a function of duration of in vitro hydrolysis in phosphate buffer solutions of pH 7.44 at 37°C. (a) Woven fabrics, (b) knitted fabrics. (From T. J. Yu, and C. C. Chu. Bicomponent vascular grafts consisting of synthetic biodegradable fibers. Part I. In vitro study. *J. Biomed. Mater. Res. 27*:1329–1340 (1993).)

The observed significant increase in water porosity with time in the bicomponent vascular fabrics is also confirmed by their morphological structure change upon immersion time as shown in Figure 13. Both W and K bicomponent fabrics became more porous and open as immersion (i.e., hydrolysis) time increased. This relaxation of the original tight fabric structure is due to the hydrolytic degradation of the absorbable component of the bicomponent fabrics and was the basic cause for the increasing water porosity and decreasing fabric weight and density with immersion time. The K group showed an isotropic and uniform structural change with time. The W group, however, exhibited unusual fabric surface morphology with time. Depending on the location of the absorbable component yarns, the tightness of the initial W fabrics would relax along the direction of a fabric without absorbable PGA yarns. The findings of Chu et al. suggest that the reduction in weft yarn size through the degradation and absorption of the PGA weft yarns is the primary requirement for generating a loose and velourlike surface morphology in bicomponent woven fabrics. The reduction in warp yarn size via the degradation of the PGA warp yarns, however, did not produce such a unique velourlike surface morphology. Therefore in addition to the advantage of increasing porosity with time, the generation of velourlike surfaces on woven vascular fabrics would provide a more open three-dimensional framework for tissue ingrowth.

Another important advantage of these bicomponent fabrics was their significantly lower initial and final stiffness than the corresponding commercial DeBakey woven vascular fabric. As shown in Fig. 14, the bending stiffness of all four bicomponent woven vascular fabrics decreased with time of immersion. Most of the final stiffness of these bicomponent vascular fabrics was only of the order of 1/10th of the values of the commercial woven grafts. A soft and compliant graft would certainly improve its performance and patency (205).

The knitted vascular fabrics made by Chu had the same textile structure as DeBakey Standard, DeBakey Ultralight Weight, and DeBakey Vascular D. They also consisted of both polyglycolic acid and Dacron fibers in various ratios (194). It was found that these vascular fabrics, in general, had lower initial water porosity than the commercial prod-

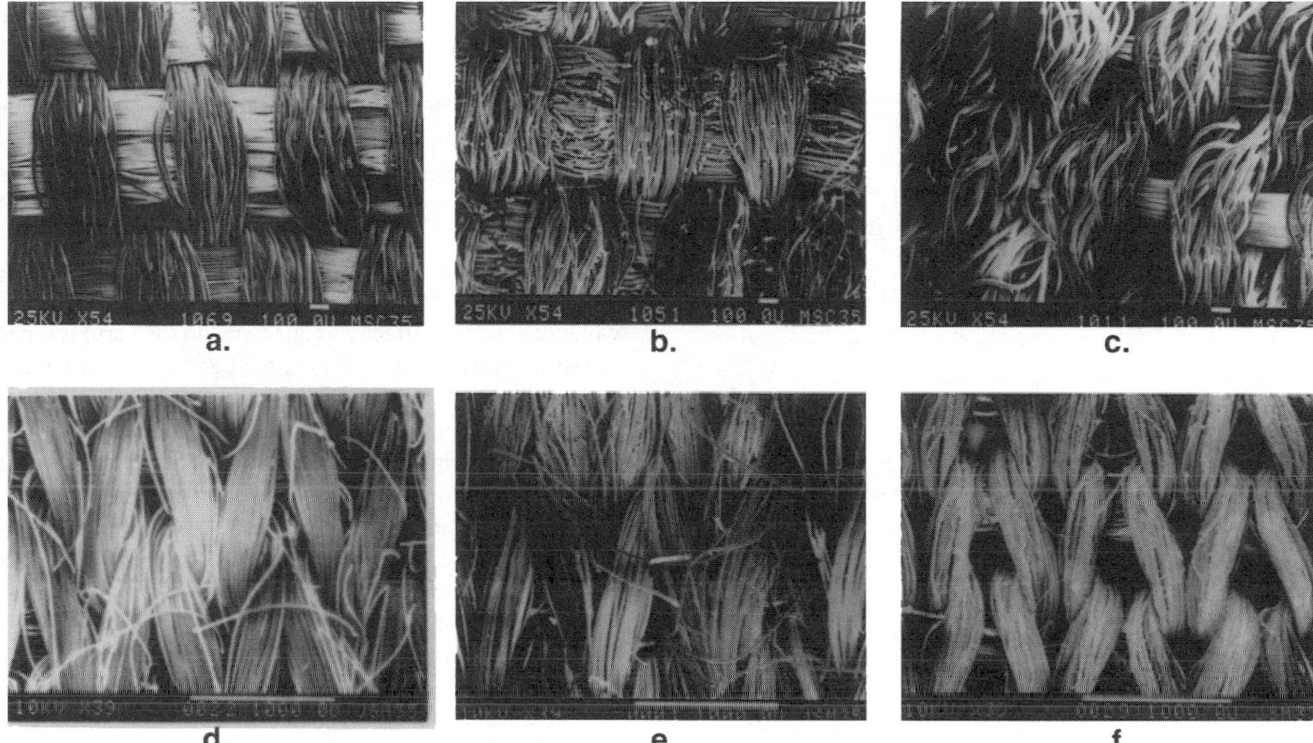

Figure 19.13 Scanning electron micrographs of bicomponent vascular fabrics as a function of in vitro hydrolysis time. (a) W3/0 d, (b) W3/30 d, (c) W3/90 d, (d) K3/0 d, (e) K3/30 d, (f) K3/90 d. (From T. J. Yu and C. C. Chu. Bicomponent vascular grafts consisting of synthetic biodegradable fibers. Part I. In vitro study. *J. Biomed. Mater. Res.* 27:1329–1340 (1993).)

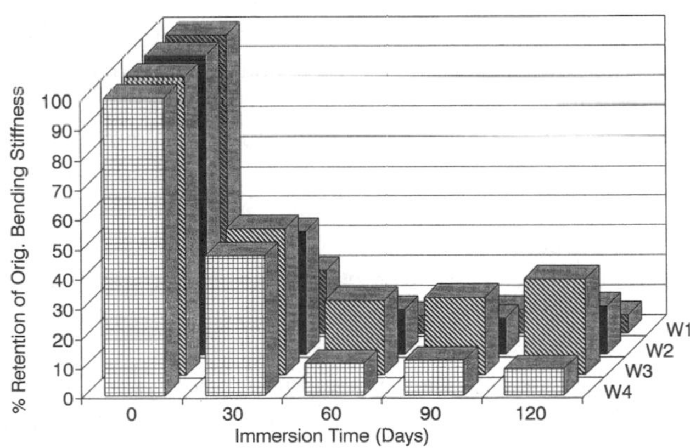

Figure 19.14 The change of fabric stiffness of bicomponent woven fabrics as a function of the duration of in vitro hydrolysis in phosphate buffer solutions of pH 7.44 at 37°C. (From T. J. Yu and C. C. Chu. Bicomponent vascular grafts consisting of synthetic biodegradable fibers. Part I. In vitro study. *J. Biomed. Mater. Res.* 27:1329–1340 (1993).)

ucts and exhibited an increase in water porosity ranging from 22 to 116%, while the corresponding commercial weft knitted vascular fabrics showed no change in water porosity during the same period.

Did these bicomponent vascular fabrics provide adequate mechanical strength support to relieve the concern of the lack of mechanical strength of absorbable based vascular fabrics at a later period of implantation? Figure 15 shows that both W and K bicomponent fabrics still exhibit a minimal more than 30% retention of their original bursting strength at the end of 30 days immersion. Most of the reduction in bursting strength occurred during the first 14 days of immersion time, and no appreciable reduction in bursting strength was observed beyond 14 days, an indication that the absorbable PGA component yarns had already lost the majority of their strength and the remaining nonabsorbable Dacron yarns became solely responsible for the observed fabric strength remaining. Therefore a minimum bursting strength level was retained during the whole period of study (120 days), depending on the relative concentration of absorbable to nonabsorbable components. Along

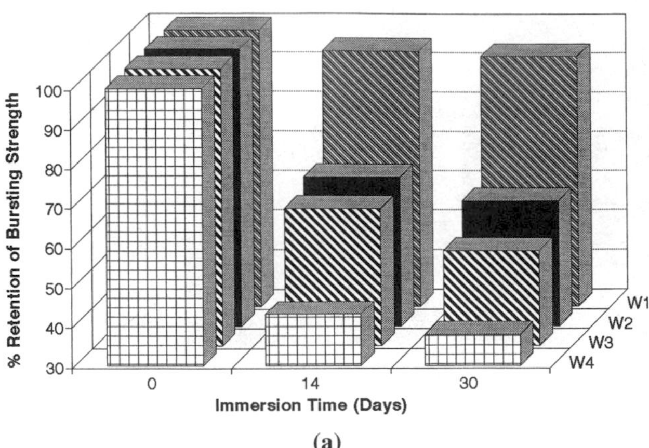

(a)

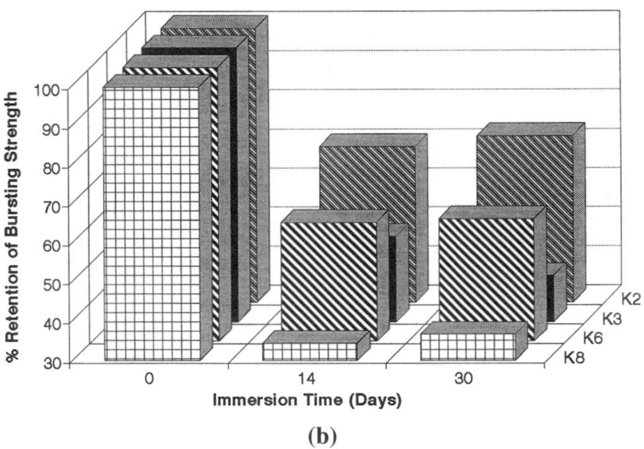

(b)

Figure 19.15 The change of fabric bursting strength as a function of in vitro hydrolysis in phosphate buffer solution of pH 7.44 at 37°C. (a) Woven, (b) knit. (From T. J. Yu and C. C. Chu. Bicomponent vascular grafts consisting of synthetic biodegradable fibers. Part I. In vitro study, *J. Biomed. Mater. Res. 27*:1329–1340 (1993).)

with this reduction in bursting strength with time, the weight of these bicomponent vascular fabrics also decreased. Regardless of the composition ratio of the bicomponent fabrics, their biggest weight loss occurred between 14 and 60 days, depending on the polyglycolic acid composition.

An in vivo trial of these bicomponent vascular fabrics in dogs for 4 months was reported by Yu et al. (194). They found that these bicomponent vascular fabrics resulted in a full-wall healing in the thoracic aorta of dogs. All bicomponent vascular grafts in surviving dogs exhibited a 100% patency rate, free of thrombus and aneurysmal formation, no hematoma or seroma around the grafts, and free of fibrin coagula in the inner capsules. The gross morphology of the regenerated tissues was visually very similar to the adjacent original arterial tissue. Histologically, the luminal surface was lined with a layer of endothelial cells with myofibroblasts, fibroblasts, and collagens underneath. The extent of the full-wall healing depended on the type of fabric structure, the concentration of absorbable yarns, the location of absorbable yarns (for the woven group only), and initial water permeability. It is believed that the concentration effect was related to the level of macrophage activation from the degradation products of the absorbable yarns, while the location effect was attributed to the various types of fabric structure change upon the degradation of the absorbable yarns. The knitted (K) group, in general, was better than the woven (W) group. The incorporation of absorbable yarns in the weft direction of the bicomponent fabrics resulted in a velourlike loose and porous fabric surface for facilitating tissue ingrowth. The placement of absorbable yarns in the warp direction, however, did not show this unique surface morphology. Calcification was, however, occasionally observed in the W samples with low initial water permeability. The observed in vivo performance correlated well with their in vitro study. This leads Chu et al. to conclude that the amount of an absorbable component in the bicomponent vascular graft appears to be the dominant factor for determining the healing characteristics of grafts consisting of absorbable components.

3. Mathematical Modeling of Porosity of Vascular Fabrics

In order to maintain a viable vessel after implantation, the vessel must be highly permeable to allow proper tissue ingrowth for full-wall healing and subsequent adequate fluid transport for minimizing the degeneration and calcification of the ingrown tissue in the pseudoneointima. At the time of surgery, however, the implant needs to be impermeable so that hemorrhage does not occur. The present commercial synthetic vascular grafts cannot achieve this dynamic nature needed for proper performance. Therefore an optimal balance needs to be determined for the porosity of the vascular fabric to allow healing of the vessel with tissue ingrowth, yet minimize blood leakage.

An industry standard to determine the ''porosity'' of vascular fabric is the water porosity test first developed by Wesolowski (206). However, Guidoin et al. recently pointed out that this term is incorrect in describing the relevant properties (207–209). They suggested that the permeability of a fabric describes how well a fluid passes through its fibers and gives an arbitrary parameter for surgeons to assess the need to preclot before implantation, while the porosity just describes the amount of void space within a fabric, and much of this space is not continuous throughout the structure and inadequately characterizes the fabric's ability to allow fluid transport. Porosity, instead of water permeability, is thus approximately related to the capability of a fabric toward tissue ingrowth. Contrary to nonwoven fabrics, Guidoin et al. found no close relationship between water permeability and porosity of woven and knitted vascular fabrics (207).

The water permeability test, as it should be known, uses water as the fluid passing through the vascular grafts. Maintaining the pressure of a flow through a sample at physiological conditions (120 mm Hg), the flow rate measured can lead to a coefficient of permittivity (mL/cm^2 min/mm Hg) (207). This coefficient describes the water permeability of a graft and will be used as such. However, the testing procedures are lengthy and time-consuming.

Although water permeability data of vascular fabrics are critical to their successful performance in vivo, the standard practice in the designing of new vascular fabrics with proper water permeability is still based on trial and error, that is, one makes the samples first and then evaluates their water permeability. If the water permeability does not fall into the desirable range, manufacturers must go back to change the design and make new samples for the water permeability test. This is a very tedious and time-consuming process. There is a strong need to determine reliably the water permeability of vascular fabrics before they are made and to simulate the water permeability test of the vascular fabrics to be manufactured. The availability of such a theoretical prediction of the water permeability of vascular fabrics would allow vascular graft manufacturers to design their products intelligently with a desirable water permeability without tedious and time-consuming trial and error.

Chu et al. (210), with the help of mathematical formulae and a computer program that they developed, created a new means to predict instantaneously the water permeability of a hypothetical vascular graft design, given the material and geometric properties of the fabric. Any user can input the specified fabric and fiber parameters of a hypothetical vascular graft and obtain its computed water permeability. If the computed water permeability is outside the desirable range, the user can adjust the input fiber and fabric parameters until the desirable water permeability is obtained. These desirable fiber and fabric parameters could then be used to construct real vascular fabrics, and their water permeability would be experimentally determined. This approach would save considerable resources and time by eliminating the need to run lengthy experiments to determine the correct fiber and fabric specifications for achieving a desirable water permeability of any given surgical fabric.

The mathematical formulae of Chu et al. for predicting water permeability of both woven and knitted vascular fabrics are

$$Q_w = \frac{A^{0.5}}{V_w}(-302.715 - 4.782\pi_2 + 28.062\pi_3 - 10489.354\pi_4 - 2.701\pi_6)$$

$$Q_k = \frac{A^{0.5}}{V_w}(851.694 - 70.644\pi_2 + 81.555\pi_3 - 23417.858\pi_4 + 292.568\pi_6)$$

where $\pi_2 = \eta A^{0.5}$, $\pi_3 = \delta A^{0.5}$, $\pi_4 = T_f/A^{0.5}$, $\pi_6 = D_t/D_f$, D_t = fiber density, D_f = fabric density, A = fabric cross-sectional area, V_w = water viscosity, Q_w = water permeability of woven fabrics, and Q_k = water permeability of knitted fabrics.

The above formulae, Q_w and Q_k, were then used to calculate the predicted water permeabilities, Q_{prd}, of all 25 vascular fabrics, and they were compared with their corresponding experimental water permeabilities, Q_{exp}. Table 20 summarizes Q_{exp}, Q_{prd}, and the percentage of error, $\Delta = \{[Q_{prd} - Q_{exp}]/Q_{exp}\} \times 100$, for determining the accuracy of the mathematical formulae of Chu et al. to predict the water permeability of various types of vascular fabrics. A close match between Q_{prd} and Q_{exp} was defined as those whose Δ were less than or equal to 10%.

In general, the accuracy of the predicted water permeability, Q_{prd}, of Chu et al. to the experimental water permeability, Q_{exp}, depended on the type of vascular fabrics (i.e., woven vs. knitted) and the source of Q_{exp}. The difference between Q_{prd} and Q_{exp} ranged from as small as 0.27% to as high as 74.2%. As a group, woven vascular fabrics, particularly those not commercially available (i.e., experimental), appeared to have a better match between Q_{prd} and Q_{exp} than knitted fabrics. If proper sources of Q_{exp} were chosen for comparison, we found that 86% of woven fabrics (6/7) and 77% of knitted fabrics (14/18) had their Q_{prd} within 10% of their Q_{exp}. This high percentage of close matches (within 10%) between Q_{prd} and Q_{exp} should be considered satisfactory, because the experimental error for obtaining Q_{exp} is generally higher than 10%. All experimental woven (4/4) and 63% (5/8) of knitted vascular fabrics exhibited less than 10% difference between Q_{prd} and Q_{exp}.

The data in Table 20 illustrate that mathematical modeling of water permeability of surgical fabrics for a reasonable prediction of their water permeability is feasible and that its level of reliability depends on many factors that are not fully understood at this time. The source of Q_{exp} had a big influence on the percentage of accuracy of our predicted water permeability formulae. For example, several knitted vascular fabrics like Weaveknit and Vasculour II had Q_{prd} within 1% of Q_{exp} if one set of Q_{exp} (211) was

Table 19.20 Comparison of Experimental and Predicted Water Permeabilities of Testing Vascular Fabrics

Vascular Grafts	Q_{exp} (mL/cm^2 min	Q_{prd} mm Hg)	$\Delta = (Q_{exp} - Q_{prd})/Q_{exp} \times 100$ (%)
Woven Fabrics			
Woven Cooley	120[a]	124	3.33
	122	124	1.61
Woven Cooley Verisoft	250[a]	252	0.80
	199	252	26.63
Woven DeBakey	232	184	−20.56
	350[a]	184	−47.42
W1	300	273	−9.18
W2	650	698	7.36
W3	880	834	−5.27
W4	400	419	4.71
Knitted Fabrics			
DeBakey Standard	2020	2771	37.19
	2530[a]	4771	9.52
DeBakey Ultralt Wt.	4080	3763	−7.78
	2650*	3763	42.00
Milliknit	3410	3401	−0.27
	5300[a]	3401	−35.83
Weaveknit	1670	2910	74.22
	2920[a]	2910	−0.34
Cooley Double Velour	1870	1946	4.04
	1660[a]	1946	17.23
DeBakey Vasculour D	3380	2327	−31.14
	2250[a]	2327	3.42
Lopor	2960	2734	−7.64
	3000[a]	2734	−8.86
Microvel	2580	2754	6.74
	2400[a]	2754	14.75
Rhodergon	5020	3344	−33.39
	5790[a]	3344	−42.25
Vasculour II	2150	2784	29.48
	2800[a]	2784	−0.57
K1	2800	3235	15.53
K2	1700	2565	50.86
K3	1800	1929	7.19
K4	2550	1534	−39.83
K5	4000	3753	−6.18
K6	1950	1961	0.57
K7	1800	1928	7.08
K8	1500	1604	6.90

[a] From M. King, P. Blais, R. Guidoin, E. Prowse, M. Marcois, C. Gosselin, and H. P. Noel. Polyethylene terephthalate (Dacron) vascular prostheses—material and fabric construction aspects. In: *Biocompatibility of Clinical Implant Materials* (D. F. Williams, Ed.). Vol. II. CRC Press, Boca Raton, FL, 1981, Chap. 8. The remaining Q_{exp} from R. Guidoin, M. King, D. Marceau, A. Cardou, D. de la Faye, J.-M. Legendre, and P. Blais. Textile arterial prostheses: is water permeability equivalent to porosity. *J. Biomed. Mat. Res. 21*:65–87 (1987).

Source: C. C. Chu, and J. Rawlinson. Mathematical modeling of water permeability of surgical fabrics for vascular use. *J. Biomed. Mater. Res. 28*(4):441–448 (1994).

used for comparison, while a difference as large as 74.2% (Weaveknit) was observed if a different set of Q_{exp} (208) was used. In the commercial knitted vascular fabrics, only two (Lopor and Rhodergon) were not affected by the source of Q_{exp} on the accuracy of our predicted water permeability formula. Both of them are velour, but Lopor is a weft knit while Rhodergon is a warp knit. These two different sets of Q_{exp} data were reported by the same authors (207,211). It is not known why the two sets of Q_{exp} data from the same lab are different from each other. The time dependence of Q_{exp} might be one of the factors contributing to this discrepancy. In their published study of water permeability of 34 commercial vascular fabrics, Guidoin et al. (207) reported that Q_{exp} decreased with time, irrespective of fabric construction type. The rate of reduction in Q_{exp}, however, was found to be less with woven structures than with knitted ones, presumably due to the rigid, compact, and dense packing of yarns and fibers in woven structures, which leaves very little room for them to swell with time under hydrostatic pressure. On the contrary, the loose knit structure provides plenty of interstitial space for the swelling of constituent fibers and yarns during water permeability tests. This higher degree of swelling would subsequently reduce the amount of water flowing through the fabric during the later stage of the water permeability test. This time dependence of Q_{exp} suggests that the Q_{exp} data from Guidoin et al. (207) (without * in Table 20) are the water permeabilities of virgin vascular fabrics estimated from the linear extrapolation of experimental water permeabilities obtained at time t to $t = 0$. If these estimated Q_{exp} of virgin vascular fabrics were used for determining the accuracy of the formulae of Chu et al., half of the tested commercial knitted vascular fabrics (DeBakey Ultralight Weight, Milliknit, Cooley Double Velour, Microvel, and Rhodergon) and one woven commercial fabric (Woven DeBakey) showed considerable improvement. For example, DeBakey Ultralight Weight showed a considerable reduction in the difference between Q_{exp} and Q_{prd} from as high as 42% to 7.78%. However, the time dependence of Q_{exp} is not the sole factor contributing to the dependence of the formulae on the source of Q_{exp}. This is because most of the remaining commercial woven and knitted vascular fabrics (i.e., those fabrics showed very poor agreement between Q_{exp} and Q_{prd} using estimated Q_{exp} at $t = 0$ or virgin Q_{exp}) exhibited a close match between Q_{exp} and Q_{prd}, if the data set of Q_{exp} from King et al. (211) was used. It is not clear whether this Q_{exp} data set (211) took into consideration the time dependence of experimental water permeability. Assuming this set of Q_{exp} represented the water permeability after t minutes of testing, then the observed close match between Q_{prd} and Q_{exp} found in Woven Cooley Verisoft, DeBakey Standard, Weaveknit, and DeBakey Vasculour D could not be attributed to the time dependence of Q_{exp} alone.

Although woven fabrics as a group showed relatively good agreement between Q_{prd} and Q_{exp}, Woven DeBakey was the only one of the commercial woven vascular fabrics that neither estimated Q_{exp} at $t = 0$ nor Q_{exp} at $t > 0$ min could narrow Δ to less than 10%. There must be some fabric and/or yarn characteristics in Woven DeBakey that the formula of Chu et al., Q_w, fails to take into consideration. An examination of the published fabric characteristics of the three tested commercial woven vascular fabrics (211) indicates they all have taffeta construction with 1/1 plain weave. However, Woven DeBakey has the lowest packing factor (47%) and highest thickness (0.25 mm). In addition, Woven DeBakey fabric is the only one in these commercial groups that uses textured yarns. This may suggest that the water permeability formula developed by Chu et al. for woven fabrics has its limitation and cannot accurately predict water permeability of those woven fabrics having a low packing factor and high thickness and is made from textured yarns. Whether it is the combination of all three of these differences in fabric and yarn characteristics, or whether it is one or two of them, that results in the larger discrepancy between Q_{prd} and Q_{exp} is not known at this time.

In the case of the commercial knitted fabrics, Lopor fabric was the only one that exhibited a close match between Q_{prd} and Q_{exp}, irrespective of which Q_{exp} was chosen. On the other hand, Rhodergon was the only knitted commercial vascular fabric that neither Q_{exp} would result in a $\Delta < 10\%$. An examination of their fabric and yarn characteristics did not reveal any apparent fabric and/or yarn characteristics that are responsible for the successful (i.e., Lopor) or failed (i.e., Rhodergon) prediction of their water permeability by our formulae.

Because of the lack of other sets of Q_{exp} for the experimental vascular fabrics, the above-mentioned Q_{exp} source problem does not apply. The Q_{exp} of these experimental fabrics were obtained after water flowed through the fabrics for one minute. Hence the reported accuracy of the predicted water permeability formula was not based on the estimated Q_{exp} of virgin fabrics like those that Guidoin et al. studied (207). Since the estimated Q_{exp} at $t = 0$ is always greater than the Q_{exp} obtained at t minutes, Chu et al. thus predict that the accuracy of their formulae would change if Q_{exp} at $t = 0$ were used, and the direction of change would depend on the relative magnitude of Q_{exp} to Q_{prd}. If $Q_{prd} > Q_{exp}$ at $t = 1$ min as with W2 and W4 (i.e., positive Δ), the use of estimated Q_{exp} at $t = 0$ would further narrow the difference between the predicted and the experimental water permeability. On the contrary, those experimental woven fabrics that showed initial negative difference ($Q_{prd} < Q_{exp}$), like W1 and W3, would exhibit a larger difference than the Δ based on Q_{exp} at $t = 1$ min. In the experimental knitted fabrics, it is interesting to know that most of the close match between Q_{prd} and Q_{exp} occurred in those experimental knitted fabrics ranging from K5 to K8. The main difference between these (K5–K8) and other (K1–K4) groups was the size of Dacron yarns (193). The K5 group had 200 denier Dacron yarn, while the K1 group had 250 denier Dacron yarn. It is known that heavier denier yarns are thicker and result in fabrics with more coverage for a given loop length. Thus the K1 group exhibited generally a higher packing factor than the K5 group. Whether this difference in yarn size really contributed to the observed larger error found in the K1 group is not known.

As with any mathematical formula for the purpose of prediction or forecasting, there is a limitation on what it can do. The less than completely satisfactory prediction of water permeability by the formula of Chu et al. in the knitted grafts, particularly in the commercial ones, would be expected with the lower R^2 value (0.492). This finding suggests that a nonlinear, and more complex, relationship among the Buckingham π groups must exist in the knitted fabrics. In addition, the far more complex fabric structure found in the commercial knitted vascular grafts than woven ones could result in additional parameters that have not been included in our modeling. Some examples of such new parameters will be the mathematical description of the surface morphology of velours and textured yarns. Obviously, future refinement of these formulae is required for improving their accuracy in predicting water permeability of various types of medical textiles.

It is important to know that the water permeability formulae of Chu et al. will not replace the experimental water permeability test. The formulae, however, will help investigators or manufacturers in their R/D of medical fabrics by providing some meaningful guidelines for them to work with. With the help of a personal computer, scientists could use the formulae to find out instantaneously how changes in fabric and fiber parameters could affect the water permeability of the resulting fabrics without actually manufacturing and testing them. The formulae also allow scientists to study the effects of a wide range of fabric and fiber parameters on water permeability before they choose a few of them for actual experimental trials. This would save considerable time and resources.

REFERENCES

1. C. C. Chu, A. von Fraunhofer, and H. G. Greisler. *Wound Closure Biomaterials and Devices*. CRC Press, Boca Raton, FL, 1997.
2. C. C. Chu. Survey of clinically important wound closure

biomaterials. In: *Biocompatible Polymers, Metals, and Composites* (M. Szycher, ed.). Technomic Publishing, Lancaster, PA, 1983.
3. S. T. Li. Biologic biomaterials: tissue-derived biomaterials (collagen); In: *Biomedical Engineering Handbook* (J. D. Bronzino, ed.). CRC Press, p. 627 (in press).
4. S. T. Li, and E. P. Katz. An electrostatic model for collagen fibrils. The interaction of reconstituted collagen with Ca++, Na+, and Cl−. *Biopolymers 15*:1439 (1976).
5. S. T. Li, E. Golub, and E. P. Katz. Electrostatic properties of reconstituted collagen fibrils, *J. Mol. Biol. 98*:835 (1975).
6. I. K. Stone, J. A. von Fraunhofer, and B. J. Masterson. A comparative study of suture materials: chromic gut and chromic gut treated with glycerin, *Amer. J. Obstet. Gynecol. 151*:1087 (1985).
7. K. A. Piez. Molecular and aggregate structures of the collagens. In: *Extracellular Matrix Biochemistry* (K. A. Piez, and A. H. Reddi, eds.). Elsevier, New York, 1984, p. 41.
8. E. J. Frazza, and E. E. Schmitt. A new absorbable suture. *J. Biomed. Mater. Res. Symp. 1*:43 (1971).
9. E. E. Schmitt, and R. A. Polistina. Surgical sutures. U. S. patent 3,297,033, American Cyanamid, 1967.
10. A. R. Katz, and R. J. Turner. Evaluation of tensile and absorption properties of polyglycolic acid sutures. *Surg. Gynecol. Obstet. 131*:701 (1970).
11. E. M. Filachione, and C. H. Fisher. Lactic acid condensation polymers—preparation by batch and continuous methods. *Ind. Eng. Chem. 36*:223 (1944).
12. D. L. Wise, T. D. Fellmann, J. E. Sanderson, and R. L. Wentworth. Lactic/glycolic acid polymers. In: *Drug Carriers in Biology and Medicine*. Academic Press, New York, 1979, p. 237.
13. D. Casey, and O. G. Lewis. Absorbable and nonabsorbable sutures. In: *Handbook of Biomaterials Evaluation: Scientific, Technical, and Clinical Testing of Implant Materials* (A. von Recum, ed.). Macmillan, New York, 1986, Chap. 7, p. 86.
14. A. Glick, and J. B. McPherson, Jr. Absorbable polyglycolic acid suture of enhanced in vivo strength retention. U.S. patent 3,626,948, American Cyanamid, 1971.
15. P. A. Miller, J. M. Brady, and D. E. Cutright. Degradation rates of oral resorbable implants (polylactates and polyglycolates): rate modification with changes in PLA/PGA copolymer ratios. *J. Biomed. Mater. Res. 11*:711 (1977).
16. N. Doddi, C. C. Versfelt, and D. Wasserman. Bioresorbable polydiozanone. In: *Fiber-Forming Polymers. Recent Advances* (J. S. Robinson, ed.). Noyes Data Corp., Park Ridge, NJ, 1990, p. 341.
17. N. Doddi, C. C. Versfelt, and D. Wasserman. Synthetic absorbable surgical devices of polydioxanone. U.S. Patent 4,052,988, 1977.
18. E. Broyer. Thermal treatment of theraplastic filaments for the preparation of surgical sutures. U.S. patent 5,294,395, Ethicon, 1994.
19. S. W. Shalaby. Synthetic absorbable polyesters. In: *Biomedical Polymers: Designed-to-Degrade Systems* (S. W. Shalaby, ed.). Hanser, New York, 1994, Chap. 1.
20. S. W. Shalaby, and D. F. Koelmel. U.S. patent 4,441,496, 1984.
21. R. S. Bezwada, S. W. Shalaby, and H. D. Newman. U.S. Patent 4,653,497, 1987.
22. R. S. Bezwada, S. W. Shalaby, H. D. Newman, and A. Kafrauy. U.S. patent 4,643,191, 1987.
23. A. Katz, D. P. Mukherjee, A. L. Kaganov, and S. Gordon. A new synthetic monofilament absorbable suture made from polytrimethylene carbonate. *Surg. Gynecol. Obstet. 161*:213 (1985).
24. D. J. Casey, and M. S. Roby. Synthetic copolymer surgical articles and method of manufacturing the same. U.S. patent 4,429,080, American Cyanamid, 1984.
25. R. S. Bezwada, D. D. Jamiolkowski, M. Erneta, J. Persivale, S. Trenka-Benthin, I. Y. Lee, J. Suryadevara, A. Yang, V. Agarwal, and S. Liu. Monocryl suture, a new ultra-pliable absorbable monofilament suture. *Biomaterials 16*:1141 (1995).
26. M. S. Roby, S. L. Bennett, and C. K. Liu. Absorbable block copolymers and surgical articles fabricated therefrom. U.S. patent, 5,403,347, April 4, 1995.
27. R. M. Robson. Silk: composition, structure, and properties. In: *Fiber Chemistry* (M. Lewin, and E. M. Pearce, eds.). Handbook of Fiber Science and Technology, Vol. IV. Marcel Dekker, New York, 1985, Chap. 7.
28. F. Lucas, and K. M. Rudall. *Comprehensive Biochemistry, Vol. 26B* (M. Florkin, and E. H. Stotz, eds.). Elsevier, Amsterdam, 1968.
29. R. D. B. Fraser, and T. P. MacRae. *Conformation in Fibrous Proteins and Related Synthetic Polymer*. Academic Press, New York, 1973, p. 293.
30. C. Earland, and S. P. Robins. Cystine residues in Bombyx mori and other silks. *Int. J. Protein Res. 5*:327 (1973).
31. H. Zuber. Silk fibroin III. *Kolloid-Z. 179*:100 (1961).
32. E. Mercer. The fine structure and biosynthesis of silk fibroin. *Aust. J. Sci. Res. B5*:366 (1952).
33. S. W. Shalaby, M. Stephenson, L. Schaap, and G. H. Hartley. Composite sutures of silk and hydrophobic thermoplastic elastomers. U.S. patent 4,461,298, 1984.
34. H. M. Heuvel, and R. Huisman. Effect of winding speed on the physical structure of as-spun poly(ethylene terephthalate) fibers, including orientation-induced crystallization. *J. Appl. Polym. Sci. 22*:2229 (1978).
35. D. S. Kaplan. Surgical suture derived from segmented polyether-ester block copolymers. U.S. patent 4,246,904, American Cyanamid, 1981.
36. G. T. Rodeheaver, D. C. Borzelleca, J. G. Thacker, and R. F. Edlich. Unique performance characteristics of Novafil. *Surg. Gynecol. Obstet. 164*:230 (1987).
37. J. A. von Fraunhofer, R. J. Storey, and B. J. Masterson. Tensile properties of suture materials. *Biomaterials 9*:324 (1988).
38. M. I. Kohan. Preparation and chemistry of nylon plastics. *Nylon Plastics* (M. I. Kohan, ed.). Wiley-Interscience, New York, 1973, p. 13.
39. R. W. Moncrieff. Nylon. In: *Man-Made Fibers*. John Wiley, New York, 1975, p. 333.

40. G. J. Listner. Polypropylene monofilament sutures. U.S. patent 3,630,205, Ethicon, 1971.
41. M. Goodman. *Topics in Stereochemistry, Vol. 2* (N. L. Allinger, and E. L. Eliel, eds.). Wiley-Interscience, New York, 1967, p. 73.
42. H. S. Kaufman, and J. J. Falcetta. *Introduction to Polymer Science and Technology: An SPE Textbook*. Wiley-Interscience, New York, 1977, p. 92.
43. M. Wishman, and G. E. Hagler. Polypropylene fibers. In: *Fiber Chemistry. Handbook of Fiber Science and Technology, Vol. IV* (M. Lewin, and E. M. Pearce, eds.). Marcel Dekker, New York, 1985, Chap. 4.
44. R. M. Joshi, and B. J. Zwolinski. Heats of polymerization and their structural and mechanistic implications. In: *Vinyl Polymerization, Vol. 1. Part I* (G. E. Ham, ed.). Marcel Dekker, New York, 1967.
45. C. M. Miller, P. Sangiolo, and J. J. Jacobson II. Reduced anastomotic bleeding using new sutures with a needle-suture diameter ratio of one. *Surgery 101*:156 (1987).
46. M. C. Dang, J. G. Thacker, J. C. S. Hwang, G. T. Rodeheaver, S. M. Melton, and R. F. Edlich. Some biomechanical considerations of polytetrafluoroethylene sutures. *Arch. Surg. 125*:647 (1990).
47. C. C. Chu, and Z. Kizil. Quantitative evaluation of stiffness of commercial suture materials. *Surg., Gynecol. Obstet. 168*:233 (1989).
48. E. Urban, M. W. King, R. Guidoin, G. Laroche, Y. Marois, L. Martin, A. Cardou, and Y. Douville. Why make monofilament sutures out of polyvinylidene fluoride. *Am. Soc. Artif. Intn. Org. 40*:145 (1994).
49. E. J. Sutow. Iron-based alloys. In: *Concise Encyclopedia of Medical and Dental Materials* (D. F. Williams, ed.) Pergamon Press, New York, 1990, p. 232.
50. J. A. von Fraunhofer, R. S. Storey, I. K. Stone, and B. J. Masterson. Tensile strength of suture materials. *J. Biomed. Mater. Res. 19*:595 (1985).
51. J. Conn, Jr., and J. M. Beal. Coated Vicryl synthetic absorbable sutures. *Surg. Gynecol. Obstet. 140*:377 (1975).
52. F. V. Mattei. Absorbable coating composition for sutures. U.S. patent 4,201,216, Ethicon, 1980.
53. D. J. Casey, and O. G. Lewis. Absorbable and nonabsorbable sutures. In: *Handbook of Biomaterials Evaluation: Scientific, Technical, and Clinical Testing of Implant Materials* (A. F. von Recum, ed.). Macmillan, New York, 1986, Chap. 7.
54. G. T. Rodeheaver, J. G. Thacker, J. Owen, et al. Knotting and handling characteristics of coated synthetic absorbable sutures. *J. Surg. Res. 35*:525 (1983).
55. G. T. Rodeheaver, J. G. Thacker, and R. F. Edlich. Mechanical performance of polyglycolic acid and polyglactin 910 synthetic absorbable sutures. *Surg. Gynecol. Obstet. 153*:835 (1981).
56. T. Kawai, T. Matsuda, and M. Yoshimoto. Coated sutures exhibiting improved knot security. U.S. patent 4,983,180, Japan Medical Supply, 1991.
57. K. A. Messier, and J. D. Rhum. Caprolactone polymers for suture coating. U.S. patent 4,624,256, Pfizer Hospital Products Group, 1986.
58. R. S. Bezwada, A. W. Hunter, and S. W. Shalaby. Copolymers of ε-caprolactone, glycolide and glycolic acid for suture coatings. U.S. patent 4,994,074, Ethicon, 1991.
59. D. W. Wang, D. J. Casey, and L. T. Lehmann. Surgical suture coating. U.S. patent 4,705,820, American Cyanamid, 1987.
60. B. D. Gaillard. Sheathed surgical suture filament and method for its preparation. U.S. patent 4,506,672, Assut S. A. Switzerland, 1985.
61. J. A. von Fraunhofer, and W. J. Sichina. Characterisation of surgical suture materials using dynamic mechanical analysis. *Biomaterials 13*:715 (1992).
62. S. A. Metz, J. A. von Fraunhofer, and B. J. Masterson. Stress relaxation of organic suture materials. *Biomaterials 11*:197 (1990).
63. N. Tomita, S. Tamai, T. Morihara, K. Ikeuchi, and Y. Ikada. Handling characteristics of braided suture materials for tight tying. *J. Appl. Biomater. 4*:61 (1993).
64. G. V. Scott, and C. R. Robbin. Stiffness of human hair fibers. *J. Soc. Cosmet. Chem. 29*:469 (1978).
65. J. Megerman, G. Hamilton, T. Schmitz-Rixen, and W. M. Abbott. Compliance of vascular anastomoses with polybutester and polypropylene sutures. *J. Vasc. Surg. 18*:827 (1993).
66. W. M. Abbott, J. Megerman, J. E. Hasson, G. L'Italien, D. Warnock. Effect of compliance mismatch upon vascular graft patency. *J. Vasc. Surg. 5*:376 (1987).
67. T. E. Bucknall. Factors influencing wound complications: a clinical and experimental study. *Ann. Royal Coll. Surg. 65*:71 (1983).
68. *United States Pharmacopeia*, 17th revision, 1965, p. 693.
69. B. Blomstedt, and B. Osterberg. Fluid absorption and capillarity of suture materials. *Acta Chir. Scand. 143*:67 (1977).
70. J. B. Herman. Tensile strength and knot security of surgical suture materials. *Am. Surgeon 37*:209 (1971).
71. H. Kobayashi. Suture materials (in Japanese). *GEKA Mook No.4*, Kinbara Syuppan, Toyko, Japan, 1976, pp. 1–14.
72. C. A. Homsy, K. E. McDonald, and W. W. Akers. Surgical suture—canine tissue interaction for six common suture types. *J. Biomed. Mater. Res. 2*:215 (1968).
73. B. S. Freeman, C. A. Homsy, J. Fissette, and S. B. Hardy. An analysis of suture withdrawal stress. *Surg. Gynecol. Obstet. 131*:441 (1970).
74. S. A. Metz, N. Chegini, and B. J. Masterson. In vivo tissue reactivity and degradation of suture materials: a comparison of Maxon and PDS. *J. Gynecol. Surg. 5*:37 (1989).
75. W. R. Sewell, J. Wiland, and B. N. Craver. A new method of comparing sutures of bovine catgut with sutures of bovine catgut in three species. *Surg. Gynecol. Obstet. 100*: 483 (1955).
76. M. Walton. Strength retention of chromic gut and monofilament synthetic absorbable suture materials in joint tissues. *Clin. Orthop. Related Res. 242*:303 (1989).

77. W. Van Winkle, and T. N. Salthouse. *Biological Response to Sutures and Principles of Suture Selection*. Ethicon, Somerville, NJ, 1976.
78. T. N. Salthouse, and B. F. Matlage. Significance of cellular enzyme activity at nonabsorbable suture implant sites: silk, polyester and polypropylene. *J. Surg. Res. 19*:127 (1975).
79. T. N. Salthouse. Biocompatibility of sutures. In: *Biocompatibility in Clinical Practice, Vol. 1* (D. F. Williams, ed.). CRC Press, Boca Raton, FL, 1981.
80. R. G. Bennett. Selection of wound closure materials. *J. Am. Acad. Dermatol. 18*:619 (1988).
81. E. T. Madsen. An experimental and clinical evaluation of surgical suture materials, I and II. *Surg. Gynecol. Obstet. 97*:73 (1953).
82. K. H. Lee, and C. C. Chu. The effect of superoxide ions in the degradation of five synthetic absorbable suture materials. *J. Biomed. Mater. Res. 49(1)*:25 (2000).
83. J. E. Hamilton. The repair of large or difficult hernia with mattressed onlay grafts of fascia lata: a 21-year experience. *Ann. Surg. 167*:85 (1968).
84. R. Smith. The use of prosthetic materials in the repair of hernias. *Surg. Clin. N. Am. 51*:1387 (1971).
85. L. M. Zimmerman. The use of prosthetic materials in the repair of hernias. *Surg. Clin. N. Am. 48*:143 (1968).
86. B. M. Rodgers, J. W. Maher, and J. L. Talbert. The use of preserved human dura for closure of abdominal wall and diaphragmatic defects. *Ann. Surg. 193*:606 (1981).
87. A. R. Koontz. Further experiences with the use of tantalum mesh in the repair of large ventral hernias. *Southern Med. J. 42*:455 (1949).
88. C. C. Guy, C. Y. Werelius, and L. B. Bell, Jr. Symposium on techniques and procedures in surgery: 5 years experience with tantalum mesh in hernia repair; results in 302 hernias. *Surg. Clin. N. Am. 35*:175 (1955).
89. F. C. Usher, and J. P. Gannon. Marlex mesh, a new plastic mesh for replacing tissue defects. I. Experimental studies. *Arch. Surg. 78*:131 (1959).
90. L. D. Gibson, and C. E. Stafford. Synthetic mesh repair of abdominal wall defects: follow up and reapprisal. *Am. Surg. 30*:481 (1964).
91. R. H. Adler, and C. N. Firme. The use of nylon prostheses for diaphragmatic defects. *Surg. Gynecol. Obst. 104*:669 (1957).
92. K. G. Callum, R. L. Doig, and J. B. Kimmonth. The results of nylon darn repair for inguinal hernia. *Arch. Surg. 108*: 25 (1974).
93. R. H. Adler, M. Mendz, and C. Darby. Effects of implanted mesh on the strength of healing wounds. *Surgen. J. 52*:898 (1962).
94. E. J. Cerise, R. W. Busuttil, C. C. Craighead, and W. W. Ogden. The use of Mersilene mesh in repair of abdominal wall hernias: a clinical and experimental study. *Ann. Surg. 181*:728 (1975).
95. R. H. Adler. An evaluation of surgical mesh in the repair of hernias and tissue defects. *Arch. Surg. 85*:836 (1962).
96. C. Johnson-Nurse, and D. H. R. Jenkins. The use of flexible carbon fiber in the repair of experimental large abdominal incisional hernias. *Br. J. Surg. 67*:135 (1980).
97. W. Sher, D. Pollack, C. A. Paulides, and T. Matsumoto. Repair of abdominal wall defects: Gore-Tex vs. Marlex graft. *Am. Surg. 46*:618 (1981).
98. F. C. Usher, et al. Use of Marlex mesh in the repair of incisional hernias. *Am. Surg. 24*:969 (1958).
99. S. D. Jenkins, T. W. Klamer, J. J. Parteka, and R. E. Condon. A comparison of prosthetic materials used to repair abdominal wall defects. *Surgery 94*:392–398 (1983).
100. L. M. Marmon, C. D. Vinocur, S. B. Standiford, C. W. Wagner, J. M. Dunn, and W. H. Weintraub. Evaluation of absorbable polyglycolic acid mesh as a wound support. *J. Pediatr. Surg. 20*:737–742 (1985).
101. M. T. Dayton, B. A. Buchele, S. S. Shirazi, and L. B. Hunt. Use of an absorbable mesh to repair contaminated abdominal-wall defects. *Arch. Surg. 121*:954–960 (1986).
102. J. Tyrell, H. Silberman, P. Chandrasoma, J. Niland, and J. Shull. Absorbable versus permanent mesh in abdominal operations. *Surg. Gynecol and Obstet. 168*:227–232 (1989).
103. M. A. Greene, R. J. Mullins, M. A. Malangoni, P. D. Feliciano, J. D. Richardson, and H. C. Polk, Jr. Laparotomy wound closure with absorbable polyglycolic acid mesh. *Surg. Gynecol. Obstet. 176*:213–218 (1993).
104. A. Pans and C. Desaive. Use of an absorbable polyglactin mesh for the prevention of incisional hernias. *Acta Chir. Belg. 95*:265–268 (1995).
105. J. R. Buck, J. J. Fath, S. K. Chung, V. J. Sorensen, H. M. Horst, and F. N. Obeid. Use of absorbable mesh as an aid in abdominal wall closure in the emergent setting. *Am. Surg. 61*:655–657 (1995).
106. U. Klinge, B. Klosterhalfen, M. Muller, and V. Schumepelick. Foreign body reaction to meshes used for the repair of abdominal wall hernias. *Eur. J. Surg. 165*:665–673 (1999).
107. M. Kusunoki, H. Yanagi, Y. Shoji, M. Noda, H. Ikeuchi, and T. Yamamura. Reconstruction of the pelvic floor using absorbable mesh with a bioresorbable membrane (Seprafilm) after abdominoperineal rectal excision. *J. Surg. Oncol. 70*:261–262 (1999).
108. A. Alponat, S. R. Lakshminarasappa, N. Yavua, and P. M. Goh. Prevention of adhesion by Seprafilm, an absorbable adhesion barrier: an incisional hernia model in rats. *Am. Surg. 63*:818–819 (1997).
109. U. Klinge, B. Klosterhalfen, J. Conze, W. Limberg, B. Obolenski, A. P. Ottinger, and V. Schumpelick. Modified mesh for hernia repair that is adapted to the physiology of the abdominal wall. *Eur. J. Surg. 164*:951–960 (1998).
110. U. Klinge, B. Klosterhalfen, M. Muller, M. Anurov, A. P. Ottinger, and V. Schumpelick. Influence of polyglactin-coating on functional and morphological parameters of polypropylene-mesh modifications for abdominal wall repair. *Biomaterials 20*:613–623 (1999).
111. H. J. Buchsbaum, W. Christopherson, S. Lifshitz, and S. Bernstein. Vicryl mesh in pelvic floor reconstruction. *Arch. Surg. 120*:1389–1391 (1985).

112. D. L. Clarke-Pearson, J. T. Soper, and W. T. Creasman. Absorbable synthetic mesh (polyglactin 910) for the formation of a pelvic ''lid4'' after radical pelvic resection. *Am. J. Obstet. Gynecol. 158*:158–61 (1988).
113. J. A. Maunello, B. Wasserman, and R. Kraut. Use of Vicryl (polyglactin-910) mesh implant for repair of orbital floor fracture causing diplopia: a study of 28 patients over 5 years. *Ophthal. Plast. Reconstr. Surg.* 9:191–195 (1993).
114. R. H. Haug, E. Nuveen, and T. Bredbenner. An evaluation of the support provided by common internal orbital reconstruction materials. *J. Oral Maxillofac. Surg. 57*:564–570 (1999).
115. S. B. Frame, B. L. Enderson, U. Schmidt, and K. I. Maull. Intrahepatic absorbable fine mesh packing of hepatic injuries: preliminary clinical report. *World J. Surg. 19*:575–579 (1995).
116. F. C. Bakker, F. Wille, P. Patka, H. J. Haarman. Surgical treatment of liver injury with an absorbable mesh: an experimental study. *J. Trauma 38*:891–894 (1995).
117. E. D. McGahren, B. M. Rodgers, and P. E. Waldron. Successful management of stage 4S neuroblastoma and severe hepatomegaly using absorbable mesh in an infant. *J. Pediatr. Surg. 33*:1554–1557 (1998).
118. M. Arndt, and W. Pircher. Absorbable mesh in the treatment of rectal prolapse. *Internl. J. Colorectal Disease 3*: 141–143 (1988).
119. G. Winde, B. Reers, H. Nottberg, T. Berns, J. Meyer, and H. Bunte. Clinical and functional results of abdominal rectopexy with absorbable mesh-graft for treatment of complete rectal prolapse. *Eur. J. Surg. 159*:301–305 (1993).
120. Y. Galili, and M. Rabau. Comparison of polyglycolic acid and polypropylene mesh for rectopexy in the treatment of rectal prolapse. *Eur. J. Surg. 163*:445–448 (1997).
121. H. M. Delany, A. Z. Rudavsky, and S. Lan. Preliminary clinical experience with the use of absorbable mesh splenorrhaphy. *J. Trauma 25*:903–913 (1985).
122. C. G. Tribble, A. W. Joob, G. W. Barone, and B. M. Rodgers. A new technique for wrapping the injured spleen with polyglactin mesh. *Am. Surg. 53*:661–613 (1987).
123. D. A. Lange, P. Zaret, G. J. Merlotti, A. P. Robin, C. Sheaff, and J. A. Barrett. The use of absorbable mesh in splenic trauma. *J. Trauma 28*:269–275 (1988).
124. F. B. Rogers, N. E. Baumgartner, A. P. Robin, and J. A. Barrett. Absorbable mesh splenorrhaphy for severe splenic injuries: functional studies in an animal model and an additional patient series. *J. Trauma 31*:200–204 (1991).
125. A. Fingerhul, P. Oberlin, J. L. Cotte, L. Aziz, J. C. Etienne, B. Vinson-Bonnett, J. D. Aubert, and S. Rea. Splenic salvage using an absorbable mesh: feasibility, reliability and safety. *Br. J. Surg. 79*:325–327 (1992).
126. P. Vanderschot, P. H. Cuypers, P. Rommens, and P. Broos. Splenic function after splenic rupture treated with an absorbable mesh. *Unfallchirugr. 96*:248–252 (1993).
127. R. A. White, S. M. Ramos, and H. M. Delany. Renorrhaphy using knitted polyglycolic acid mesh. *J. Trauma 27*: 689–690 (1987).
128. U. Kumar, and I. S. Gill. Absorbable mesh envelope facilitates handling of pediatric en bloc kidneys during transplantation. *Tech. Urol. 5*:195–197 (1999).
129. A. T. Gentile, P. D. Feliciano, R. J. Mullins, R. A. Crass, L. R. Eidemiller, and B. C. Sheppard. The utility of polyglycolic acid mesh for abdominal access in patients with necrotizing pancreatitis. *J. Am. Coll. Surg. 186*:313–318 (1998).
130. S. Fianu and G. Soderberg. Absorbable polyglactin mesh for retropublic sling operations in female urinary stress incontinence. *Gynecol. Obstet. Invest. 16*:45–50 (1983).
131. S. Bowald, C. Busch, I. Eriksson, and T. Aberg. Repair of cardiac defects with absorbable material. *Scand. J. Thorac. Cardiovasc. Surg. 15*:91–94 (1981).
132. P. K. Maurer, and J. V. McDonald. Vicryl (polyglactin 910) mesh as a dural substitute. *J. Neurosurg. 63*:448–452 (1985).
133. K. Nagata, S. Kawamoto, J. Sashida, T. Abe, A. Mukasa, and Y. Imaizumi. Mesh-and-glue technique to prevent leakage of cerebrospinal fluid after implantation of expanded polytetrafluoroethylene dura substitute—technical note. *Neuro. Med. Chir (Tokyo) 39*:316–318 (1999).
134. N. Okumura, T. Nakamura, Y. Takimoto, T. Kiyotani, Y. H. Lee, Y. Shimizu, K. Tomihata, Y. Ikada, and K. Shir aki. The repair of tracheal defects using absorbable mesh. *ASAIO J. 38*:M555–559 (1992).
135. A. Janni, M. Lucchi, F. Melfi, G. Menconi, and C. A. Angeletti. The utility of polyglactin 910 mesh in the plastic reconstruction of the chest wall after on-bloc resection. *Eur. J. Surg. Oncol. 22*:377–380 (1996).
136. E. Bardaxoglou, D. Manganas, B. Meunier, S. Landen, G. J. Maddern, J. P. Campion, and B. Launois. New approach to surgical management of early esophageal thoracic perforation: primary suture repair reinforced with absorbable mesh and fibrin glue. *World J. Surg. 21*:618–621 (1997).
137. E. G. McCollough, D. B. Hom, M. T. Weigel, and J. R. Anderson. Augmentation mentoplasty using Mersilene mesh. *Arch. Otolaryngol. Head Neck Surg. 116*:1154–1158 (1990).
138. A. H. Gager, and A. J. Schultz. Treatment of periodontal defects with an absorbable membrane (polyglactin 910) with and without osseous grafting: case reports. *J. Periodontol. 62*:276–683 (1991).
139. C. C. Chu, B. Pourderyhm, and L. Welch. Characterization of morphologic and mechanical properties of surgical mesh fabrics. *J. Biomed. Mater. Res. 19*:903–916 (1985).
140. M. Wagner. Evaluation of diverse plastic and cutis prostheses in a growing host. *Surg. Gynecol. Obst. 130*:1077 (1970).
141. J. P. Arnaud, R. Eloy, M. Adloff, and J. F. Grenier. Invivo exploration of the tensile strength of the abdominal wall after repair with different prosthetic materials. *Eur. Surg. Res. 11*:1–7 (1979).
142. J. D. Bobyn, G. J. Wilson, D. C. MacGregor, R. M. Pilliar, and G. C. Weatherly. Effect of pore size on the peel strength of attachment of fibrous tissue to porous-surfaced implants, *J. Biomed. Mater. Res. 16*:571–584 (1982).

143. M. Chvapil, R. Holusa, K. Kliment, et al. Some chemical and biological characteristics of a new collagen–polymer compound material. *J. Biomed. Mater. Res. 3*:315–322 (1969).
144. D. F. Taylor, and F. B. Smith. Porous methyl methacrylate as an implant material. *J. Biomed. Mater. Res. 6*:467–475 (1972).
145. W. A. Zisman. Relation of equilibrium contact angle to liquid and solid constitution. *Advances in Chemistry Series #43, Contact Angle, Wettability, and Adhesion*. The 144th Meeting of the American Chemical Society, Los Angeles, CA, April 2–3, 1963.
146. J. L. Talbert, M. B. Rodgers, et al. Surgical management of massive ventral hernias in children. *J. Pediatr. Surg. 12*:63 (1977).
147. F. Moazam, B. M. Rodgers, and J. L. Talbert. Use of Teflon mesh for repair of abdominal wall defects in neonates. *J. Pediatr. Surg. 14*:347 (1979).
148. J. P. Arnaud, R. Eloy, et al. Critical evaluation of prosthetic materials in repair of abdominal wall hernias: new criteria of tolerance and resistance. *Am. J. Surg. 1233*:338 (1977).
149. A. R. Koontz. On the need for prostheses in hernia repair. *Am. Surg. 28*:342 (1962).
150. J. G. Durden, and L. B. Pemberton. Dacron mesh in ventral and inguinal hernias. *Am. Surg. 40*:662 (1974).
151. J. H. Johnson. Use of polypropylene mesh as a prosthetic material for abdominal hernias in horses. *J. Am. Vet. Med. Assoc. 155*:1589 (1969).
152. C. R. Voyles, J. D. Richardson, K. I. Bland, G. R. Tobin, L. M. Flint, and H. C. Polk, Jr. Emergency abdominal wall reconstruction with polypropylene mesh. Short term benefits versus long term complications. *Ann. Surg. 194*:219 (1981).
153. G. Morris-Stiff, and L. E. Hughes. The continuing challenge of parastomal hernia: failure of a novel polypropylene mesh repair. *Ann. Royal Coll. Surg. Engl. 80*(3):184–187 (1998).
154. E. Ortiz-Oshiro, G. C. Villalta, V. Furio-Bacete, J. M. Martinez, D. Ortega-Lopez, and J. A. Fdez-Represa. Nonabsorbable prosthetic meshes: which is the best option in the repair of abdominal wall defects? *Internl. Surg., 84*(3): 246–250 (1999).
155. N. Sondhi, F. D. Ellis, L. M. Hamed, and E. M. Helveston. Evaluation of an absorbable muscle sleeve to limit postoperative adhesions in strabismus surgery. *Ophthalmic. Surg. 18*:441–443 (1987).
156. U. K. Dasika, and W. D. Widmann. Does lining polypropylene with polyglactin mesh reduce intraperitoneal adhesions? *Am. Surg. 64*:817–819 (1998).
157. R. C. Dinsmore, and W. C. Calton, Jr. Prevention of adhesions to polypropylene mesh in a rabbit model. *Am. Surg. 65*:383–387 (1999).
158. A. Baykal, D. Onat, K. Rasa, N. Renda, and I. Sayek. Effects of polyglycolic acid and polypropylene meshes on postoperative adhesion formation in mice. *World J. Surg. 21*:579–582 (1997).
159. D. F. Devereux, J. J. Chandler, T. Eisenstat, and L. Zinkin. Efficacy of an absorbable mesh in keeping the small bowel out of the human pelvis following surgery. *Diseases Colon Rectum 31*:17–21 (1988).
160. D. F. Devereux. Protection from radiation-associated small bowel injury with the aid of an absorbable mesh. *Seminars in Surg. Oncol. 2*:17–23 (1986).
161. D. F. Devereux, D. Thompson, L. Sandhaus, W. Sweeney, and A. Haas. Protection from radiation enteritis by an absorbable polyglycolic acid mesh sling. *Surgery 101*:123–129 (1987).
162. J. T. Soper, D. L. Clarke-Pearson, and W. T. Creasman. Absorbable synthetic mesh (910-polyglactin) intestinal sling to reduce radiation-induced small bowel injury in patents with pelvic malignancies. *Gynecol. Oncol. 29*: 283–289 (1988).
163. K. S. Dasmahapatra, and A. P. Swaminathan. The use of a biodegradable mesh to prevent radiation-associated small-bowel injury. *Arch. Surg. 126*:366–369 (1991).
164. J. B. Trimbos, T. Sanders-Keilholz, and A. A. Peters. Feasibility of the application of a resorbable polyglycolic-acid mesh (Dexon mesh) to prevent complications of radiotherapy following gynaecological surgery. *Eur. J. Surg. 157*: 281–284 (1991).
165. J. F. Rodier, J. C. Janser, D. Rodier, J. Dauplat, P. Kauffman, G. Le Bouedec, B. Giraud, and G. Lorimier. Prevention of radiation enteritis by an absorbable polyglycolic acid mesh sling. A 60 case multicentric study. *Cancer 68*: 2545–2549 (1991).
166. C. Ottosen, and E. Simonsen. The use of an absorbable mesh to avoid radiation-associated small-bowel injury in the treatment of gynaecological malignancy. *Acta Oncol. 33*:703–705 (1994).
167. S. A. Wesolowski, C. C. Fries, R. Domingo, W. Liebig, and P. Sawyer. The compound prosthetic vascular graft: a pathologic survey. *Surgery 53*:19–44 (1963).
168. W. A. Dale. Arterial Grafts: 1900–1978. In: *Graft Materials in Vascular Surgery* (Herbert Dardik ed.). Symposia Specialists Miami, FL, 1978.
169. R. S. Lord. The Search for an Ideal Arterial Substitute. *Australia-New Zealand J. Surg. 44*:362–369 (1974).
170. N. Rosenberg, A. Simpson, and R. Brown. A circumferentially elastic arterial prosthesis: three year studies of a Dacron-Spandex graft in the dog. *J. Surg. Res. 34*:7–16 (1983).
171. L. R. Sauvage, K. Berger, S. J. Wood, et al. Grafts for the 80's. Bob Hope International Heart Research Institute, Seattle, Washington, 1980.
172. L. R. Sauvage, K. Berger, L. B. Beilin, J. C. Smith, S. J. Wood, and P. B. Mansfield. Presence of endothelium in an axillary femoral graft of knitted Dacron with an external velour surface, *Ann. Surg. 182*:749–753 (1975).
173. S. A. Wesolowski, W. Liebig, W. Golaski, and C. C. Fries. Knitted arterial prosthesis with optimal characteristics. *Circulation 28*:825–826 (1936).
174. S. A. Wesolowski, C. C. Fries, A. Martinez, and J. D. McMahon. Arterial prosthetic materials. *Ann. New York Acad. Sci. 153*:325–344 (1968).

175. S. A. Wesolowski. Vascular prostheses: the need for standard—historical and surgical perspectives. In: *Vascular Graft Update: Safety and Performance* (H. E. Kambic, A. Kantrowitz, and P. Sung, eds.). ASTM Special Technical Publication 898, ASTM, Philadelphia, PA, 1986, pp. 253–277.
176. R. J. Ruderman, A. F. Hegyeli, B. G. Hattler, and F. Leonard. A partially biodegradable vascular prosthesis. *Trans. Amer. Soc. Artif. Int. Org. 28*:30–33 (1972).
177. S. Bowald, C. Busch, and I. Eriksson. Arterial regeneration following polyglactin 910 suture mesh grafting *Surgery 86*:722–729 (1979).
178. S. Bowald, C. Busch, and I. Eriksson. Absorbable material in vascular prosthesis *Acta Chir. Scand. 146*:391–395 (1980).
179. H. P. Greisler, D. U. Kim, J. B. Price, and A. B. Voorhees. Arterial regenerative activity after prosthetic implantation. *Arch. Surg. 120*:315–323 (1985).
180. H. P. Greisler. Arterial regeneration over absorbable prostheses. *Arch. Surg. 117*:1425–1431 (1982).
181. H. P. Greisler, D. U. Kim, J. W. Dennis, J. J. Klosak, K. A. Widerborg, E. D. Endean, R. M. Raymond, and J. Ellinger. Compound polyglactin 910/polypropylene small vessel prostheses. *J Vasc Surg. 5*:572–583 (1987).
182. H. P. Greisler, J. Ellinger, T. H. Schwarcz, J. Golan, R. M. Raymond and D. U. Kim. Arterial regeneration over polydioxanone prostheses in the rabbit. *Arch Surg. 122*: 715–721 (1987).
183. H. P. Greisler, E. D. Endean, J. J. Klosak, J. Ellinger, J. W. Dennis, K. Buttle, and D. U. Kim. Polyglactin 910/polydioxanone bicomponent totally resorbable vascular prostheses. *J. Vasc. Surg. 7*:697–705 (1988).
184. H. P. Greisler. Macrophage-biomaterial interactions with bioresorbable vascular prostheses. *Trans. ASAIO 34*: 1051–1059 (1988).
185. H. P. Greisler, J. W. Dennis, E. D. Endean, and D. U. Kim. Derivation of neointima of vascular grafts. *Circulation Supplement I. 78*:I6–I12 (1988).
186. H. P. Greisler, C. W. Tattersall, J. J. Kloask, et al. Partially bioresorbable vascular grafts in dogs. *Surgery 110*:645–655 (1991).
187. H. F. Sasken, L. A. Trudell, P. M. Galletti, et al. Histopathologic characterization of tissue—material interactions in bioresorbable vascular prostheses. *Trans. Soc. Biomaterials Ann. Meeting*, Vol. VI, Birmingham, AL, April 27–May 1 (1983), p. 82.
188. P. M. Galletti, L. A. Trudell, T. H. Chiu, et al. Coated bioresorbable mesh as vascular graft material. *Trans. Amer. Soc. Artif. Int. Org. 26*:257–263 (1985).
189. S. Gogolewski, and A. J. Pennings. Growth of a neo-artery induced by a biodegradable polymeric vascular prosthesis. *Makromol. Chem. Rapid Commun. 4*:213–219 (1983).
190. G. Uretzky, Y. Appelbaum, H. Younes, R. Udassin, P. Nataf, et al. Long-term evaluation of a new selectively biodegradable vascular graft coated with polyethylene oxide–polylactic acid for right ventricular conduit. *J. Thorac Cardiovas Surg. 100*:769–780 (1990).
191. Hans-Peter Zweep, S. Satoh, B. van der Lei, W. L. J. Hinrichs, F. Dijk, J. Feijen, and C. R. H. Wildevuur. Autologous vein supported with a biodegradable prosthesis for arterial grafting. *Ann. Thorac. Surg. 55*:427–433 (1993).
192. T. J. Yu, and C. C. Chu. Bicomponent vascular grafts consisting of synthetic biodegradable fibers. Part I. In vitro study. *J. Biomed. Mater. Res. 27*:1329–1340 (1993).
193. C. C. Chu, and L. E. Lecaroz. Design and in vitro testing of newly-made bicomponent knitted fabrics for vascular surgery. *Advances in Biomedical Polymers* (Charles G. Gebelein, ed.). Polymer Science and Technology, Vol. 35. Plenum Press, New York, 1987, pp. 185–214.
194. T. J. Yu, Donald M. Ho, and C. C. Chu. Bicomponent vascular grafts consisting of synthetic biodegradable fibers. Part II. In vivo healing response. *J. Investigative Surg. 7*:195–211 (1994).
195. H. P. Greisler, T. H. Schwarcz, J. Ellinger, and D. U. Kim. Dacron inhibition of arterial regenerative activity. *J. Vasc. Surg. 3*:747–756 (1986).
196. H. P. Greisler. Macrophage activation in bioresorbable vascular grafts. In: (NATO ASI Ser.: Life Sciences Vol. 208) *Vascular Endothelium: Physiological Basis of Clinical Problems.* (J. D. Catravas, A. D. Callow, C. N. Gillis, and U. Ryan, eds.) Plenum Publishing, New York, 1991, pp. 253–254.
197. E. D. Endean, D. U. Kim, J. Ellinger, S. Henderson, and H. P. Greisler. Effects of polypropylene's mechanical properties on histological and functional reactions to polyglactin 910/polypropylene vascular prostheses. *Surgical Forum 38*:323–325 (1987).
198. H. P. Greisler, T. M. Lam, C. W. Tattersall, J. Ellinger, and K. A. Joyce. Modulation of cell proliferation by im planted biomaterials. Transactions of the VIth International Symposium of the Biology of Vascular Cells 1990, p. 61.
199. H. P. Greisler, J. J. Klosak, E. D. Endean, J. F. McGurrin, and D. U. Kim. Effects of hypercholesterolemia on healing of vascular grafts. *J. Investigative Surgery 4*(3):299–312 (1991).
200. H. P. Greisler, J. Ellinger, S. C. Henderson, A. M. Shaheen, W. H. Burgess, and T. M. Lam. The effects of an atherogenic diet on macrophage/biomaterial interactions. *J. Vasc. Surg. 14*(1):10–23 (1991).
201. H. P. Greisler, J. W. Dennis, E. D. Endean, J. Ellinger, R. Friesel, and W. H. Burgess. Macrophage/biomaterial interactions—the stimulation of endothelialization. *J. Vasc. Surg. 9*:588:593 (1989).
202. H. P. Greisler. The role of the macrophages in intimal hyperplasia. *J. Vasc. Surg. 10*:566–568 (1989).
203. R. Baier. Properties and characteristics of bioprosthetic grafts. *Vascular Graft Update: Safety and Performance* (H. E. Kambic, A. Kantrowitz, and P. Sung, eds.). ASTM Special Technical Publication 898, ASTM, Philadelphia, PA, 1986, pp. 95–107.
204. H. Dardik. Lower extremity revascularization with the glutaraldehyde stabilized human umbilical cord vein graft. *Vascular Graft Update: Safety and Performance* (H. E.

Kambic, A. Kantrowitz, and P. Sung, eds.). ASTM Special Technical Publication 898, ASTM, Philadelphia, PA, 1986, pp. 50–59.

205. W. M. Abbott, and D. J. Bouchier-Hayes. The role of mechanical properties in graft desing. In: *Graft Materials in Vascular Surgery* (H. Dardik, editor). Symposia Specialists, Miami, FL, 1978, pp. 59–78.

206. S. A. Wesolowski, C. C. Fries, K. E. Karlson, M. De-Bakey, and P. N. Sawyer,. Porosity, primary determinant of the ultimate fate of synthetic vascular grafts. *Surgery* *50*:91–96 (1961).

207. R. Guidoin, M. King, D. Marceau, A. Cardou, D. de la Faye, J.-M. Legendre, and P. Blais. Textile arterial prostheses: is water permeability equivalent to porosity. *J. Biomed. Mat. Res. 21*:65–87 (1987).

208. R. Guidoin, C. Gosselin, L. Martin, M. Marois, et al. Polyester prostheses as substitutes in the thoracic aorta of dogs. I. Evaluation of commercial prostheses. *J. Biomed. Mater. Res. 17*:1049–1077 (1983).

209. D. de la Faye, J. M. Legendre, R. Guidoin, B. Bene, M. King, et al. Premier contact sang/prothèse artérielle. Étude de la rétention des éléments sanguins. *J. Chir. (Paris) 121*: 253–261 (1984).

210. C. C. Chu, and J. Rawlinson. Mathematical modeling of water permeability of surgical fabrics for vascular use. *J. Biomed. Mater. Res. 28*(4):441–448 (1994).

211. M. King, P. Blais, R. Guidoin, E. Prowse, M. Marcois, C. Gosselin, and H. P. Noel. Polyethylene terephthalate (Dacron) vascular prostheses—material and fabric construction aspects. In: *Biocompatibility of Clinical Implant Materials* (D. F. Williams, ed.), Vol. II. CRC Press, Boca Raton, FL, 1981, Chap. 8.

20

Bioabsorbable Polymers for Medical Applications with an Emphasis on Orthopedic Surgery

Pentti U. Rokkanen
Helsinki University Central Hospital and University of Helsinki, Helsinki, Finland

I. INTRODUCTION

Synthetic materials used for biological applications (biomaterials) can be classified as metals, ceramics, polymers,and composites. Polymeric materials and their composites have been used for years, typical applications being tissue replacement, augmentation and support of tissues, and the delivery of drugs. Based on their behavior in living tissue, polymeric biomaterials can be divided into biostable, bioabsorbable (biodegradable or bioresorbable), and partially bioabsorbable materials. Biostable polymers are inert, cause minimal response in the surrounding tissue, and retain their properties for years. Biostable polymers, e.g., polyethylene and polypropylene, are used as endoprostheses and sutures.

In many cases biomaterial needs to be in place only temporarily, and in such cases bioabsorbable or partially bioabsorbable polymeric materials are more appropriate than biostable ones. Bioabsorbable surgical materials are therefore most suited for temporary internal fixation when the tissue has been traumatized, as the bioabsorbable implant preserves the structure of the tissue at the early stage of the healing of, for example, bone, tendon, or skin. After that, the implant gradually decomposes, and stresses are gradually transferred to the healing tissue. The time taken depends on the material, its molecular weight and structural properties, and the place where it is inserted. Bioabsorption of a material means degradation induced by the metabolism of the organism. Therefore bioabsorbable surgical devices need no removal, which substantially benefits the individual and produces social savings.

This chapter deals mainly with the properties of bioabsorbable synthetic polymeric materials and their present and possible future surgical applications. It is based on the research work done by the research groups of Professor Pertti Törmälä and myself in traumatology and orthopedic surgery as well as on the published literature. Our investigations have, so far, led to 23 medical academic dissertations (1–23) and more than 1000 different papers (original scientific articles, reviews, patents, abstracts), books, or booklets.

This review deals with the bioabsorbable synthetic polyglycolide (PGA), polylactide (PLA), and polydioxanone (PDS) polymeric devices in the fixation of procedures in trauma or orthopedic disease mainly based on own investigations and experience in more than 3200 operations since 1984 (24).

II. BIOABSORBABLE SYNTHETIC POLYMERS AND THEIR PROPERTIES

Many polymers have been found to be bioabsorbable in living tissue. The most important surgical bioabsorbable polymers are aliphatic polyesters (polymers and copolymers) of α-hydroxy acid derivatives. Most of them are thermoplastic, partially crystalline, or totally amorphous polymers.

Polyglycolid acid (PGA), with a high molecular weight, is a hard, tough crystalline polymer melting at about 224–226°C with a glass transition temperature of 36°C. Unlike closely related polymers such as polylactic acid (PLA) and poly-beta-hydroxybutyric acid (PHBA), PGA is insoluble in most of the common polymer solvents. Polylactic acid is a pale polymer melting at about 174–184°C with a glass transition temperature of 57°C. Such polymers can be processed using different manufacturing processes into fibers, films, rods, screws, plates, clamps, etc. Polydioxanone (PDS) is a colorless crystalline polymer produced by polymerizing para-dioxanone by a ring-opening method in the presence of a suitable catalyst (25). The melting and glass transition temperatures of PDS are 110°C and −16°C, respectively, which means that at room temperature PDS is very flexible and elastic. One advantage of polymeric materials compared to metals and ceramics is that they are easy and cheap to make.

Bioabsorbable materials should fulfill certain criteria and also specific requirements related to any special application of the material or implant. The bioabsorbable materials must be nonmutagenic, nonantigenic, noncarcinogenic, nontoxic, nonteratogenic, antiseptic, and tissue-compatible. They should not cause morbidity and must provide adequate mechanical stiffness and strength. The degradation products should be water-soluble, comprise small molecules, and be naturally occurring metabolites. The degradation should preferably occur by hydrolysis in aqueous media, though it is faster in the presence of certain enzymes (26). The products should dissolve in the extracellular fluid and be excreated via the kidneys and lungs. The breakdown products of polyglycolic acid, CO_2 and water, are examples. Accordingly, this is one mechanism of biodegradation for this major polymer. Another more common mechanism is phagocytosis by macrophages, neutrophils, and giant cells.

No macroscopic inflammatory foreign-body reactions have been seen in own animal experiments. Nevertheless, bioabsorbable implants elicit a histopathologically recognizable nonspecific foreign-body type of tissue response. This seems to be a phenomenon inherent to the degradation and absorption processes of these polymers in the tissues. With regard to the phagocytic and clearing capacity of the tissues, the most demanding phase is the decomposition stage, when the gross geometry of the implant is rapidly lost. At that time, the production rate of polymeric debris particles may exceed the critical limits of the tissue tolerance and transport potential of the anatomical site concerned.

No signs of toxic or teratogenic effects have been recorded with the use of bioabsorbable implants. Sarcomatous changes have been reported to develop at PLA implants placed in the soft tissues of rats, but in other studies no difference in the tumorigenicity was found between the implants made of PLA and those of polyethylene (27). Probably, such tumors merely represent nonspecific response to any foreign material known in rodents. It is even observed that PLA inhibits carcinoma cell growth in vitro (28).

III. SURGICAL APPLICATIONS OF BIOABSORBABLE IMPLANTS

The bioabsorbable implants in current use are sutures and fiber constructions, porous composites, and drug delivery systems. Some are partially bioabsorbable devices, while the others are totally bioabsorbable. Although this review concentrates on bioabsorbable devices used in the fixation of fractures, osteotomies, and soft tissues, it will open with a short remark on other devices used in orthopedics.

A. Suture Materials

Suture materials provide support until the collagen is synthesized and woven into a scar. In 1954 Higgins (29) reported on the use of polyglycolide, and in 1967 Schmitt and Polistina (30) introduced PGA sutures, which three years later led to the first commercial synthetic biodegradable suture (Dexon®). Copolymerization of a small amount of lactide with glycolide (PGA/PLLA) led to the development of the second commercial synthetic bioabsorbable suture (Vicryl®) (31).

When bioabsorption occurs in vivo, PGA is hydrolyzed to glycolic acid (32). When exposed to glycolate oxidase, the glycolic acid molecules are transformed into glyoxylate, which reacts with glycine transaminase, thus producing glycine. Glycine may be used in protein synthesis or the synthesis of serine, which may be employed in the tricarboxylic acid cycle after transformation into pyruvate (33). PLA undergoes hydrolytic de-esterification into lactic acid, which is incorporated into the tricarboxylic acid cycle and subsequently excreted by the lungs as carbon dioxide (33,34).

Bioabsorbable sutures are widely used in the closure of soft tissue wounds. Several clinical studies have been published on the use of PGA, PGA/PLLA, PDS, and other sutures or bands that are manufactured for use in the repair of tendons and ligaments. Some studies have reported on the use of bioabsorbable sutures in the fixation of bone fractures and osteotomies. Cutright et al. (35) used polylevolactic acid (PLLA) sutures for internal fixation of the symphysis disruption in Macaca rhesus monkeys. Roed-Petersen (36) described two cases in which bioabsorbable polyglycolic acid sutures were used in the fixation of severely dislocated mandibular fractures in young patients. The stability of the fixation was, however, secured by intraoral arch bars with rigid intermaxillary fixation. Vihtonen (5) studied the fixation of distal femoral osteotomy in rabbits with PGA sutures, and the results were satisfactory.

B. Porous Composites

New possibilities for developing biomaterial for different purposes have been opened up by the technique of combining bioabsorbable polymers in porous or nonporous materials. Hydroxyapatite powders and blocks have applications in bone surgery, e.g., to fill in defects (37). Since porous ceramics are brittle, attempts have been made to increase their toughness by combining them with polymers. Tencer et al. (38) made coralline hydroxyapatite (CHAG)/polymer composites by coating porous CHAG externally or internally with PLA and polymethylmethacrylate (PMMA). Internal microcoatings of PLA and PMMA increased the strength of CHAG three- and tenfold, respectively.

C. Drug Delivery Systems

Numerous studies have been done on the use of biostable and bioabsorbable polymers for releasing drugs (39). Polymeric devices for the controlled release of drugs and antibiotics have several advantages over traditional repeated dosage methods. For instance, a slow-release mechanism can provide continuous administration without the variable physiological factors associated with gastrointestinal absorption presystemic metabolism. It can save patients from being exposed to greater amounts of the drug than is necessary at the desired site of action.

D. Partially Bioabsorbable Devices

The reinforcement of bioabsorbable polymeric matrices with biostable fibers yields strong, partially bioabsorbable materials. The first of such materials, plates, rods, ligaments, tendons, scaffolds, etc., were developed by Parsons et al. (40) and by Alexander et al. (41); they comprised PLA matrix reinforced with carbon fibers. Belykh et al. (42) developed partially bioabsorbable composites for bone surgery; they comprised a copolymer of methylmethacrylate and *N*-vinylpyrrolidone reinforced with polyamide fibres. These materials had relatively good mechanical properties (43). Clinical studies have also been in progress for many years in various Soviet medical centers (44). The long-term response of living tissues to the biostable (slowly eroding) polyamide fibers of these materials has, however, not been reported.

E. Totally Bioabsorbable Devices

The disadvantages of rigid plate fixation to bones, for example the reduction of mechanical strength owing to atrophy of bone and the risk of infection, prompted the experimental use of bioabsorbable devices in fixing fractures of the long bones, as suggested by Schmitt and Polistina (45). Despite some positive experimental results, the materials were not applied clinically. Early attempts to develop bioabsorbable devices were based on the use of implants manufactured with traditional melt-molding techniques, such as extrusion and injection molding. Secure fixation material for bone fractures must have a relatively high tensile strength, but these requirements are not fulfilled with nonreinforced bioabsorbable polymeric implants manufactured with melt-molding techniques. In the late 1970s another approach was made to the fundamental questions raised by bioabsorbable fixation. It was concluded that the requirements for high tensile strength of any fixation material could be best fulfilled by developing reinforced, bioabsorbable composites mostly of polyglycolic and polylactic acid (2). In fact, small PDS implants have been used also in the fixation of small fractures (46).

A review on the bioabsorbable implants was published eight years ago (47). Since then our knowledge has greatly increased, allowing new critical examinations. Many reviews have been published in journals or separately, e.g., on bioabsorbable implants in orthopedic surgery in 1992 (48), on applications in podiatry in 1995 (49), on PGA membranes in 1996 (50), on bioabsorbable polyglycolide devices in 1997 (51), on sports medicine in 1998 (52), and on the use of bioabsorbable implants in traumatology and orthopedics in 1996 (53), in 1997 (54–57), and in 1999 (58).

IV. DEVELOPMENT OF SELF-REINFORCED BIOABSORBABLE COMPOSITES AND THEIR PROPERTIES

Experimental background. Many macromolecular compounds are bioabsorbable, but only a few possess the properties necessary for an internal bone fixation device. Some of the most suitable materials have been PGA, PLA, and PDS. In addition to these homopolymers, various copolymers of PGA and PLA have been tested. PLA occurs in different stereoisomeric forms, including polylevolactic acid (PLLA) and various stereocopolymers of polylevo- and polydextrolactic acid (P(L/DL)LA). PGA : PLA copolymer in a ratio of 90 : 10 and polydioxanone are used as bioabsorbable sutures. Initially, the implants studied were prepared by casting the polymers into sheets or films, which permitted basic investigations on the biological behavior of the compounds in bone tissue but did not make the devices suitable for fracture fixation (47). The fabrication of implants was accomplished by melt-molding and extrusion into rods. The strength characteristics of implants (rods, screws, and tacks) have been improved by the fiber-reinforced composite texture in which the polymer matrix is reinforced with the same material (self-reinforced, SR) (59–62).

In our own early studies, different types of partially or totally bioabsorbable (biodegradable) fiber-reinforced composites were developed, mainly based on different poly-α-hydroxy acids. The most promising results were achieved using self-reinforced (SR) bioabsorbable polymeric composites. These are polymeric materials in which the reinforcing elements and matrix material contain the same chemical elements. The following paragraphs will deal with the technical and medical properties of self-reinforced bioabsorbable polymeric composites.

One method of manufacturing SR composites is based on sintering and/or partial melting of polymer fibers and on shaping the sample under pressure. When the temperature profile of the process is carefully controlled, an SR structure comprising strong fibers embedded in polymeric matrix is obtained. According to Törmälä, the most effective way to create the self-reinforced structure into the bioabsorbable polymer is the mechanical deformation of nonreinforced material. Different models of deformation like free drawing, die-drawing, shearing, rolling, etc. are possible and result in different kinds of microstructures. However, a common feature in all deformation processes is the induction of oriented, self-reinforced structures into the polymeric matrix (63). Our own data on the initial strength of SR-PGA and SR-PLLA rods 3.2 mm in diameter were, respectively, 430 and 300 MPa (bending strength) and 255 and 200 MPa (shear strength). The implants lose most of their strength after 30 to 120 days and are resorbed, respectively, during 6 to 12 months (SR-PGA) or 2 to 5 years (SR-PLLA) in vivo depending on the size and composition of the implant. Certain experiments suggest an osteostimulatory potential of bioabsorbable polyester implants, but transient local osteolytic lesions also occur (64,65). The degree of the final restoration of the original tissue architecture within the implant track varies for reasons not yet fully understood (66). The degradation of these polymers occurs mainly by hydrolysis and, to a lesser extent, through nonspecific enzymatic action (67). The degradation rates vary, PLLA having the longest degradation time. The molecular weight, the crystallinity, the thermal history of the sample, as well as the geometry of the implant influence the degradation. In experimental studies by light microscopy, PGA totally disappears from the tissues within 36 weeks (64). For copolymers the degradation rate is dependent on the ratio between the constituent homopolymers. It is stated that PLLA implants can be visualized in the bone with magnetic resonance imaging (68,69), though histologic examination is the most reliable method in this respect.

Prior to and parallel with the clinical studies of the bioabsorbable fracture fixation implants, more than 6200 animals were operated on in our experimental studies (e.g., 2–5,7–10,12,13,15–17,21,23). In these studies, macroscopic, radiographic, microradiographic, histologic, histomorphometric, oxytetracycline labeling, computed tomography, quantitative computed tomography, and magnetic resonance imaging studies were done, and the strength measurements of the implants and fixations were performed in different situations. These studies have made the basis for our clinical investigations and use of self-reinforced implants in the clinical trials and praxis. Osteotomies of the distal femur in 19 rabbits were operatively fixed with totally biodegradable implants, and the experimental operations were started on January 15, 1982 (2).

The use of PGA/PLA rods, SR-PGA, and SR-PLLA rods, screws, and other devices (Figs. 1 and 2) in fixing cancellous bone fractures has been documented in more than 3200 patients during the past 15 years at our department (e.g., 24,70,54) (Tables 1 and 2). The healing results are comparable or even better than those obtained with metallic fixation. SR-PGA rods and screws (Bionx devices) are routinely used nowadays or are being studied in clinical research in about 20 countries, Western European countries, the USA, Japan, etc., or have been used in more than 300,000 cases.

Since living cells help to absorb biomaterials, an important question is which factors affect the intensity and duration of the cellular response. Clinically it is desirable to keep any reaction as mild as possible so as not to delay healing. As with the clinical use of bioabsorbable PGA and PGA/PLA sutures (71), macroscopic symptoms of foreign-body reaction, such as local fluid accumulation (LFA) and transient sinus formation (TSF), sometimes occur but

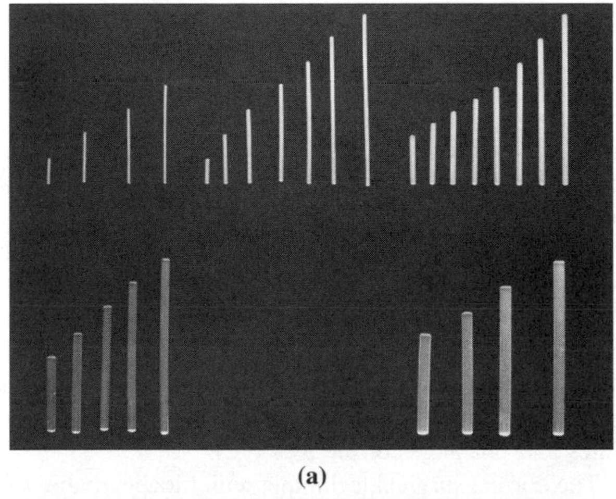

(a)

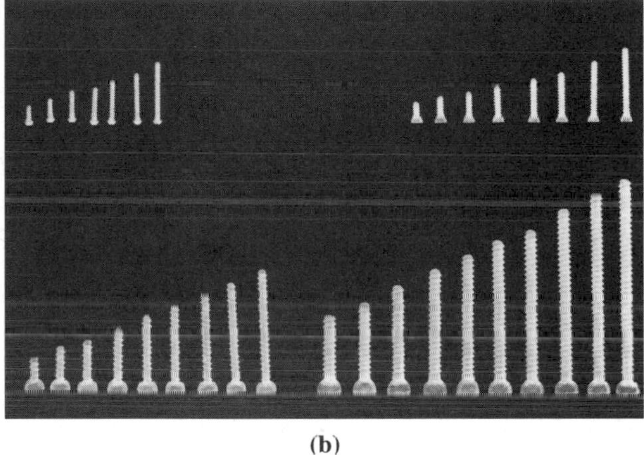

(b)

Figure 20.1 Self-reinforced poly-L-lactide (SR-PLLA) rods (a) and screws (b) routinely used in bone fixations.

without any adverse affect on the healing of the fracture (6,11).

SR-PGA is a widely used bioabsorbable fixation material about which qualitative conclusions can be drawn regarding factors contributing to LFA or TSF reactions and their treatment (Table 3). A general hypothesis, based on the above observations and on the experience of other research groups, is that this reaction is correlated to the relation between the local accumulation of polymer debris from the implant and the capacity of the surrounding tissues to eliminate the debris. If this capacity is low, e.g., due to poor circulation or low metabolic activity, the polymer debris may accumulate locally, leading to local transient disturbances, such as an increase of osmotic pressure, with subsequent manifestations of LFA or TSF. The correct surgical technique, including the use of the right implantation depth (21), is important to prevent sinuses.

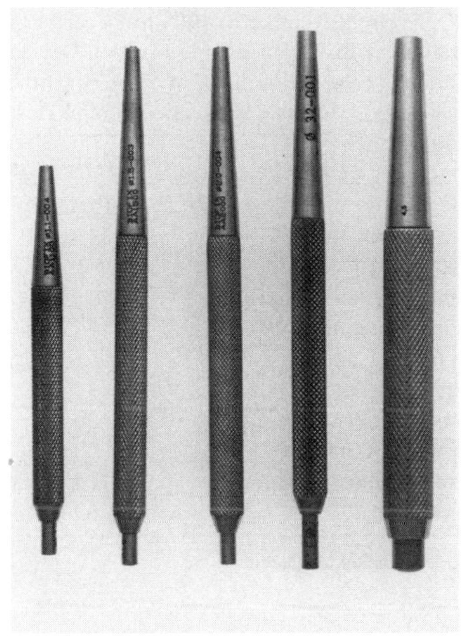

(a)

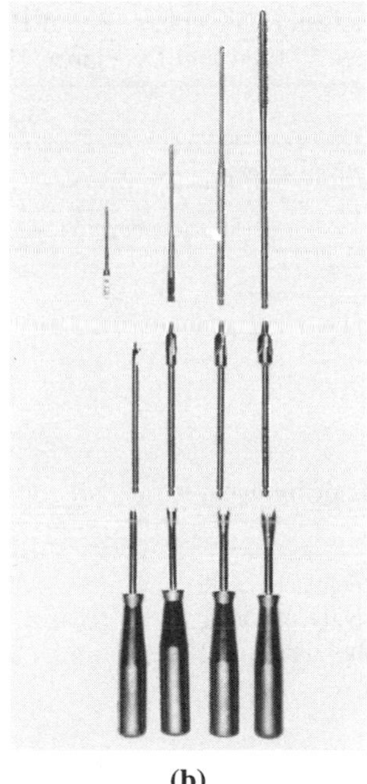

(b)

Figure 20.2 Applicators for insertion of rods (a). Instrumentation for insertion of screws (taps, countersinks, and screwdrivers) (b).

Table 20.1 From November 5, 1984 to December 31, 1998, 3241 Operations with Bioabsorbable Fixation Devices Have Been Performed at the Department of Surgery (Orthopaedics and Traumatology), Helsinki University Central Hospital

Issue	Number	%
Operation	3241	100
Age (mean; years)	38.3	—
Sex (female/male)	1738/1503	54/46
Operation time (mean; minutes)	52.1	—
Hospital stay (mean; days)	4.1	—
Sick leave (days)	68.6	—
Postoperative course		
Infections (mainly superficial)	136	4.2
Deep venous thrombosis	38	1.2
Failure of fixation[a]	129	4.0
Sinus formation	69	2.1

[a] A new operation is seldom needed.

Table 20.2 Bioabsorbable Implants Used at the Department of Surgery (Orthopaedics and Traumatology), Helsinki University Central Hospital Between November 5, 1984 and December 31, 1998

Implant	Operations
PGA/PLA[1] rod (copolymer)	53
SR-PGA[2] rod	1049
SR-PGA[3] screw	645
SR-PLLA rod	358
SR-PLLA screw	238
SR-PLLA tack	236
SR-PLLA plug	61
SR-PLLA arrow	32
SR-PLLA wire	3
At least two kinds of implants	566
Total	3241

[1] Polyglycolic acid/polylactic acid.
[2] Self-reinforced polyglycolic acid.
[3] Self-reinforced poly-L-lactic acid.

Table 20.3 Treatment of Transient Fluid Accumulation

Small and painless fluid accumulation can be observed.
Large or painful accumulations must be aspirated with a needle.

The first experimental operation with a PGA/PLA implant was done at our department on January 15, 1982, the first clinical operation on November 5, 1984 (Rokkanen-Böstman), and the first operation with the use of an SR-PLLA implant on July 27, 1986, and on August 19, 1988, respectively.

V. BIOABSORBABLE FIXATION IN FRACTURE TREATMENT

Bioabsorbable fixation (Table 4) has been used in several kinds of fractures in adults: glenoidal rim fractures, fractures of the proximal humerus, fractures of the lateral humeral condyle, fractures of the medial condyle of the humerus, fractures of the olecranon, fractures of the radial head, fractures of the distal radius, fractures of the hand, fractures of the femoral head and neck, fractures of the femoral condyles, fractures of the patella, fractures of the tibial condyles, malleolar fractures, fractures of the talus, fractures of the calcaneus, and fractures of the metatarsal bones and phalanges of the toes (72).

The correct surgical technique with bioabsorbable rods is essential. The fracture should be reduced exactly and fixed with clamps or fingers. Holes are drilled for implants, and the channel is flushed and measured. The rod is inserted into the drilled, measured, and flushed hole with an applicator. The rod is introduced into the channel by hammering at the piston of the applicator. Using this surgical technique, polyglycolide rods were clinically used in the fixation of several types of cancellous bone fractures and osteotomies of the upper and lower limbs. In these studies,

Table 20.4 The Most Common and Typical Indications for Application of Bioabsorbable Rods, Screws, Tacks, Wires, and Arrows in Traumatology and Orthopedics at the Department of Surgery (Orthopaedics and Traumatology), Helsinki University Central Hospital between November 5, 1984 and December 31, 1998[a]

Fresh fracture of	
Humeral condyles	112
Olecranon	84
Radial head and neck	180
Hand	84
Distal femur or proximal tibia	77
Patella	66
Ankle	1364
Ligament injury	
Rupture of the ulnar collateral ligament	214
Orthopedic diseases	
Osteochondritis dissecans	31
Hallux valgus	372
Recurrent shoulder dislocation	64
Osteoarthritis in the ankle, subtalar joint or in the small joints of the hand	51
Rheumatoid arthritis	89
Pediatric orthopaedics	6
Operations together	3241

[a] The clinical course was uneventful in 82% of the patients.

for example, 289 patients were operated on using mainly SR-PGA rods in the fixation of displaced ankle fractures (6). Other research groups have used PLA implants for the fixation of ankle fractures (73–75) and of several kinds of small-fragment fractures (76,77). In radial head fractures the first prospective study of 24 patients showed encouraging results (6). These results were confirmed (19). Eight patients with radial head fractures were also successfully treated with PDS pins (78). Similarly, the retention properties of bioabsorbable pins by transarticular fixation seemed to be successful in the treatment of fractures of the humeral capitellum (79). In fractures of the olecranon the use of both bioabsorbable pins and screws showed equal and favorable results (11). Open reduction and internal fixation with bioabsorbable SR-PGA or SR-PLLA devices were used in the treatment of 32 fractures of the hand at our department and also at another department in 46 cases (80). Promising results were obtained, with the exception of the Bennett's fractures, where the outcome was less favorable in our series. The same conclusion can be drawn from the fixation of scaphoid delayed union or nonunion at Matti-Russe operation (19). Promising clinical reports have been published on the use of PDS rods for fracture fixation in the hand (81).

An increasing number of clinical reports have been published on the growing use of polylactic acid implants. Thirty-five successful operations out of 40 were performed using PLA 98 rods or screws for osteosynthesis in cancellous bone fractures or orthopedic conditions (82). In one multicenter study, nonreinforced Poly-L/DL-lactide 70/30 pins were used in small fragment indications, but resulting in 4/57 redislocations (83). SR-PLLA pins were studied in 27 patients with small-fragment fractures or osteotomies treated by internal fixation without redisplacements (84).

The correct operation technique with bioabsorbable screws and with drill bit, tap, countersink, screwdriver, and screws is essential. The fracture should be reduced exactly and fixed with clamps. Holes are drilled and tapped with a special tap. The hole is countersunk for the screw head, and the channel is measured and flushed. The screw is placed in a screwdriver and inserted into the screw hole. The screw is tightened. The implants can be shortened during the operation with a small oscillating saw. The excess part of the screw head is removed with a small oscillating saw.

In a prospective study, 319 operations were performed between 1987 and 1991 (11). SR-PLLA screws alone were used in 38 fixations, SR-PLLA and SR-PGA screws in 14 fixations, and SR-PGA screws in 247 fixations (Fig. 3). SR-PGA rods were used as controls in 20 fixations. The trauma indications were 41 olecranon fractures and 259 ankle fractures. The method showed subjectively excellent or good results in more than 90% of the patients with a fracture. SR-PGA screws can also be used in osteoporotic bone (22), but they are not recommended for fixation of displaced ankle fractures in alcoholics because of the risk of poor cooperation and redislocation (6/16 reoperations in our series) (22). Six patients with displaced split-depres-

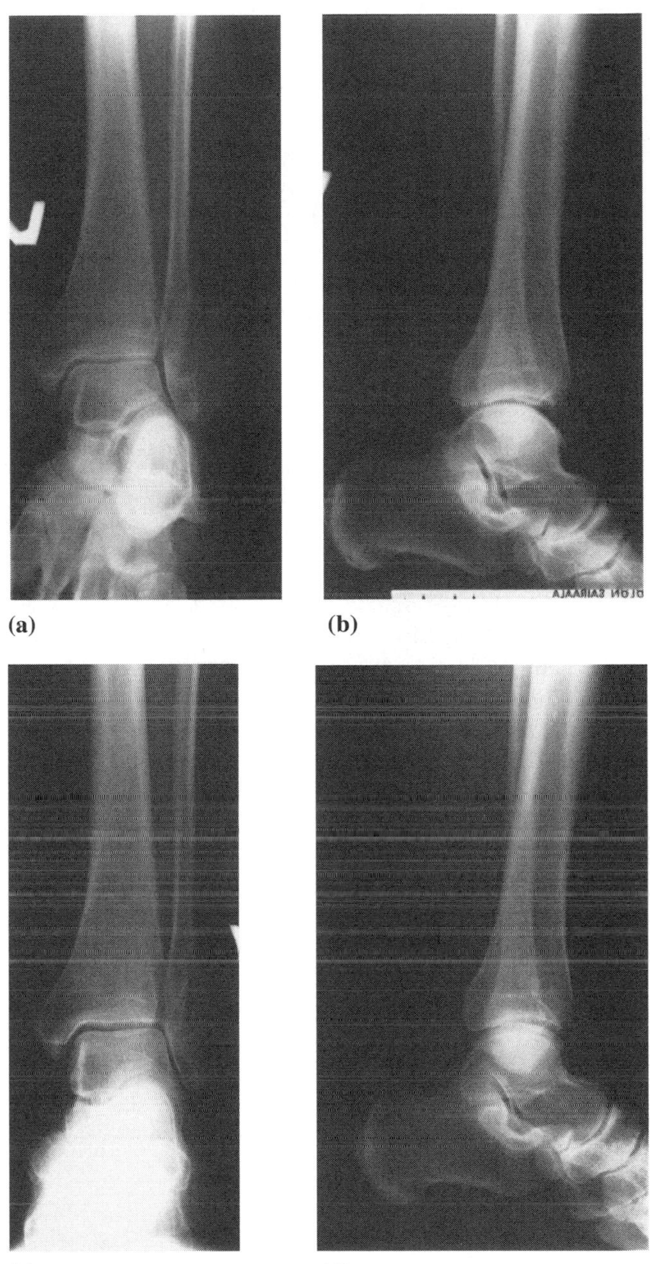

Figure 20.3 Displaced fracture of the ankle in a 59-year-old female. Radiographs taken on admission (a, b). Radiographs after open reduction and internal fixation with two SR-poly-L-lactide screws three and a half year later (c, d).

sion-type tibial condylar fractures were treated with SR-PGA screws. One patient underwent reoperation due to an unacceptable primary fracture reduction unrelated to the implant material. The others healed well (22). In 4/28 patients with proximal tibial osteotomies and fractures, redisplacement was noted (85). Good indications for bioabsorbable fixation are also fractures of the talus (22) and calcaneus (22).

In one clinical study, 19 ankle fractures were fixed by plates and screws made of bioabsorbable polylactic acid (86). Fracture union was achieved within 6 weeks, though several fractures demonstrated an aseptic soft tissue problem. In seven patients, volume-reduced plates and screws with flat heads were applied without any soft tissue reaction (86). Patients with a closed, displaced fracture of the ankle were managed with medial malleolar fixation using either 4.0-millimeter orientruded polylactide screws (83 patients) or 4.0-millimeter stainless steel screws (72 patients) (87). It was concluded that polylactide screws are safe and effective. Bioabsorbable screws, pins, and nails of orientated polylactide were developed for fixation of bone grafts, fractures, and osteotomies, and their use was evaluated in 143 patients with bony union in all except one (88).

The SR-PLLA screw 6.3 mm in diameter was found as good as the metal screw to fix a bone-patellar tendon-bone graft for the anterior cruciate ligament in a bovine experimental model (89). SR-PLLA lag screws 6.3 mm in diameter can also be used safely to fix subcapital femoral neck fractures in Garden Stage I and II fractures and in younger patients with Garden III fractures (90).

VI. BIOABSORBABLE FIXATION IN ORTHOPEDIC SURGERY

Bioabsorbable fixation of bone in osteotomies, arthrodeses, and in other reconstructive orthopedic surgery (Table 4) is valuable, as metallic fixation can never be an ideal fixation modality, due to corrosion, stiffness, and stress shielding. Mainly rods and screws are used in orthopedic procedures for bones.

The distal chevron osteotomy of the first metatarsal bone for hallux valgus is widely used. After osteotomy, lateralization of the distal part of the first metarsal bone is performed. The drill channel is made for a rod 2.0 mm in diameter. The osteotomy is fixed with one rod (Fig. 4). Bioabsorbable SR-PLLA pins (84), SR-PGA pins (91), and PDS-pins (92) can be used reliably to fix distal chevron osteotomy and PLLA staples in Akin osteotomy (93) as well as widely in foot surgery (94,95). In proximal osteotomy of the first metarsal bone for severe hallux valgus a wedge osteotomy is done one centimeter from the tarso-

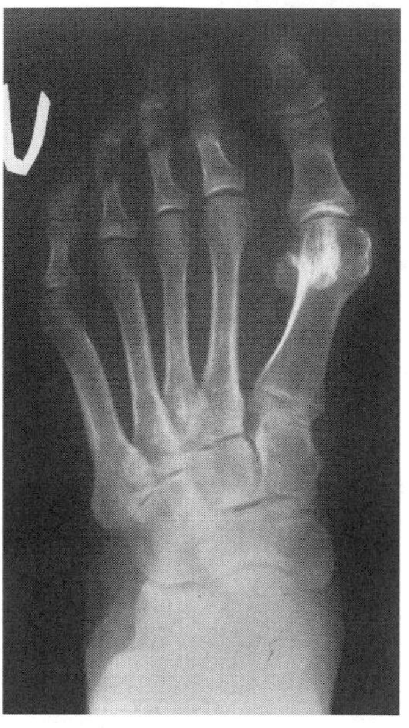

(a)

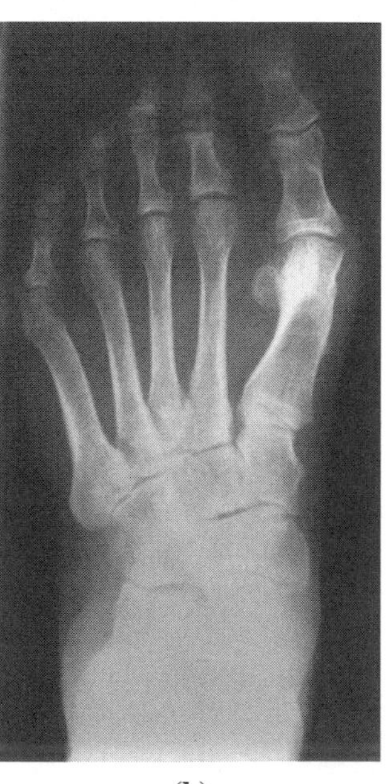

(b)

Figure 20.4 Hallux valgus of the left foot in a 45-year-old female. Radiography taken on admission (a). Radiography after chevron osteotomy fixed with one SR-polyglycolide rod one year later (b).

metatarsal joint distally. The osteotomy is fixed temporarily by a clamp. The tapped drill channel is made from the distal metatarsus to the proximal through the osteotomy. After countersinking the head one SR-PLLA screw (4.5 mm in diameter and 30–40 mm in length) is inserted to fix the osteotomy or the first tarsometatarsal arthrodesis. By using proximal osteotomy, statistically slightly better results were achieved in the treatment of metatarsus primus varus and hallux valgus (96).

In talocrural arthrodesis the articular surface from all parts of the joint is removed. Tibia, talus, and fibula are compressed together with clamps. One drill channel is drilled from the fibula through the tibia into the talus, one from the medial side of the tibia into the talus, and one anteriorly from the tibia to the talus. The arthrodesis is fixed with three SR-PLLA screws. The union rate achieved has been over 90% (11). Using this technique one redislocation out of 25 operations was observed. In a study of 11 patients, 12 arthrodeses of the ankle joint were performed by using bioabsorbable SR-PLLA (or SR-PGA) screws (Fig. 5). Solid fusion was achieved in 11 of the 12 cases in nine weeks (11). In subtalar arthrodesis, due to malposition after injury, osteoarthritis, or rheumatoid arthritis, the articular surfaces from the talocalcaneal, the talonavicular, and the calcaneocuboidal joint are removed. The surfaces are compressed tightly together and fixed with three SR-PLLA screws: one from the talus into the calcaneus, one from the navicular bone into the talus, and one from the calcaneus into the cuboid bone. The results, especially after subtalar arthrodesis, have been excellent (18). In severe destruction of the talocrural (TC) joint in osteoarthritis (or rheumatoid arthritis), operative fusion is required. In severe destruction of the subtalar joints in osteoarthritis (or rheumatoid arthritis), operative fusion is required for painless weight-bearing. The postoperative course was uneventful in all 21 cases operated at our department.

In the proximal tibial osteotomies for osteoarthritis of the knee joint the fixation has been done with three screws (54). Two screws are directed from the proximal tibia to

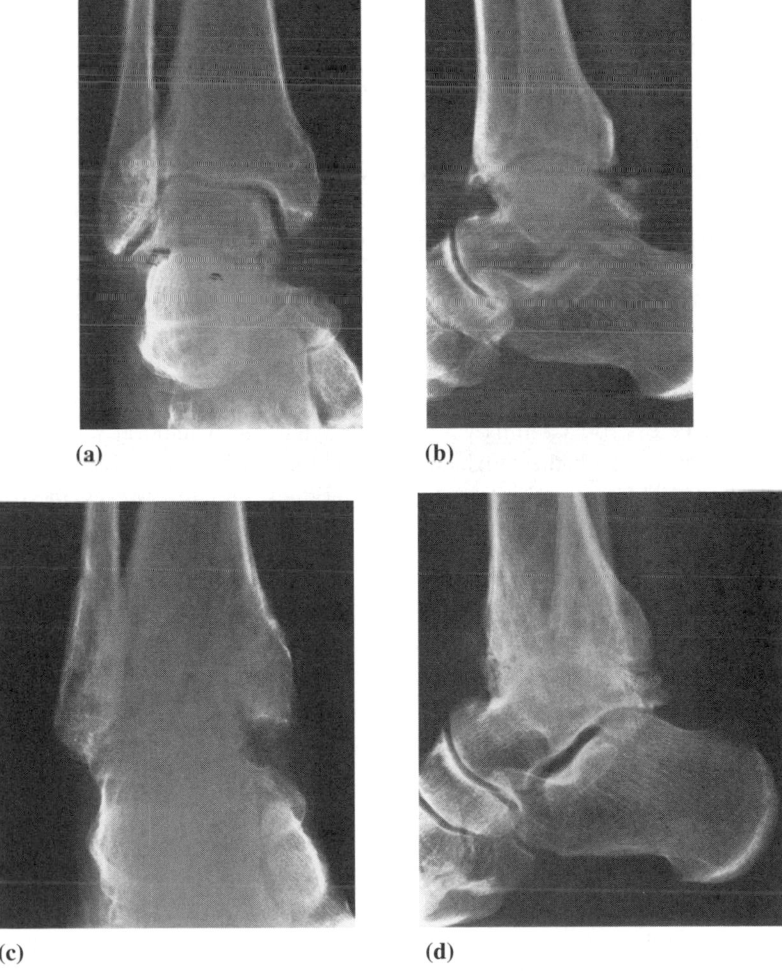

Figure 20.5 Posttraumatic osteoarthritis of the ankle in a 55-year-old female. Radiographs taken on admission (a, b). Radiographs after arthrodesis fixed with one SR-polyglycolide and two SR-poly-L-lactide screws one and one-half years later (c, d).

the distal from the lateral side and one from the diaphysis to the medial condyle. The fixation should be done using either SR-PLLA screws of 4.5 mm in diameter or SR-PLLA lag screws of 6.3 mm in diameter.

In the fixation of osteochondritis dissecans (OCD) of the knee, bioabsorbable rods of 1.1, 1.5, or 2.0 mm in diameter are used as a fixation device. In earlier operations the open technique was used, and the OCD fragment was reduced and usually fixed with one or two rods in the knee. In the arthroscopic technique a rod, usually 30 mm in length, can be inserted to fix the fragment. In our series the results were excellent or good in 19/24 patients (97). In one series it is concluded that the poly-L-lactide pin is safe and useful in the fixation of three osteochondrites and two osteochondral fragments in the knee joint (98).

It is possible to prevent or delay the course of the rheuma disease, improve the function, and eliminate or diminish the pain with operations. Bioabsorbable devices can be used in the following operative procedures for rheuma patients when fixing bone, joint, or ligament: osteotomy, arthrodesis (especially PIP, DIP, IP, and base joint of the thumb, trapezio-, metacarpo-, scaphotrapezio-, wrist-, subtalar, and TC joint), and refixation (especially the ulnar collateral ligament of the I metacarpophalangeal joint). Polylactide (not polyglycolide) devices must be used in rheuma patients.

A destroyed wrist mostly caused by rheumatoid arthritis is an indication for arthrodesis. In this surgical technique the synovial membrane, cartilage, and sclerotic bone from the radiocarpal and midcarpal joints are removed. With a drill bit of 3.2 mm a channel is drilled in the capitate bone and radius. Some of the bone between the scaphoid and lunate bones is removed. An SR-PLLA rod 3.2 mm by 50–70 mm is inserted first in the radius with the other end pushed back in the capitate bone. The rod will retain the position until fusion has occurred. Fifty-three arthrodeses (18 in the wrist, 18 in the hand, 6 in the talocrural joint, and 11 in the subtalar-calcaneocuboid-talonavicular joint) were performed on 47 patients using SR-PLLA screws and rods with two nonunions in the TC arthrodeses (18).

Arthrodesis is indicated in the osteoarthritic or rheumatic destruction of the joint in the hand if it is painful and causes continuous inconvenience for the patient. In the surgery, the synovial membrane and the rest of the cartilage are removed until the proper angle is obtained. Refreshed and molded surfaces are compressed and held together with hands or clamps. The channels are drilled for two crossed SR-PLLA rods with drillbits of the same size, and the rods are inserted. In a study, 24 metacarpophalangeal arthrodeses fixed together with polylactide and metallic rods were performed with union in 22/24 cases (99).

A destroyed I metatarsophalangeal joint owing to the osteoarthritis or rheumarthritis is an indication for operation. Resection arthroplasty and arthrodesis are the operative methods used. In a randomized study, the first metatarsophalangeal joint was fused in 39 patients with rheumatoid arthritis using either bioabsorbable SR-PLLA (3/20 nonunion) rods or Kirschner (0/19 nonunion) wires (100).

VII. BIOABSORBABLE FIXATION IN TRAUMATOLOGY AND ORTHOPEDIC SURGERY IN CHILDREN

Certain fractures involving the growth plate need specific consideration due to a growth disturbance often related to them. Accurate reduction and internal fixation are essential especially in types 3 and 4 according to the classification of Salter and Harris (101). Metallic pins have been used in the fixation of physeal fractures in children. In addition, osteotomies and arthrodeses have been fixed with metallic pins and with metallic screws. When bioabsorbable implants were experimentally used in young rabbits, it was found that the destruction of 7% of the cross-sectional area of the growth plate caused permament growth disturbance and shortening at the femur but that of 3% did not (4). In one study, two SR-PGA pins 1.1 mm in diameter were used in the fixation of experimental distal femoral physeal fractures in rabbits (102), giving sufficient stability for healing without any signs of impaired function of the growth plate. Accordingly, SR-PLLA pins 1.1 mm in diameter provided sufficient stability for healing of experimental distal femoral physeal fractures of five-week-old rabbits (103). In addition to small-fragment fractures, osteotomies, and arthrodeses, the self-reinforced implants can also be used in the intramedullary fixation of growing bone fractures even in the diaphyseal area (54).

Most physeal fractures and small-fragment fractures requiring open reduction and internal fixation are suitable for bioabsorbable rod fixation. Rod fixation with small-diameter (1.5–2.0 mm) SR-PGA rods appears to be safe even after transphyseal placement of the fixation rods. The operation technique is partly the same as that with metallic fixation. Our own series of 140 fractures in children treated successfully with bioabsorbable rod fixation included mostly elbow fractures (Fig. 6), ankle fractures, and fractures of the phalanges and metacarpal bones (104). The complications recorded were loss of reduction (2.8%), inaccurate position of fixation or minor redisplacement (1.4%), and superficial infection (0.7%). Transient reactions (fluid accumulations, not sinuses) seem to be quite

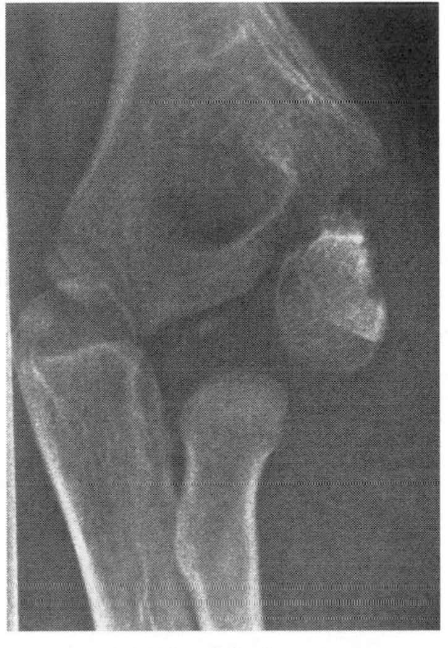

(a)

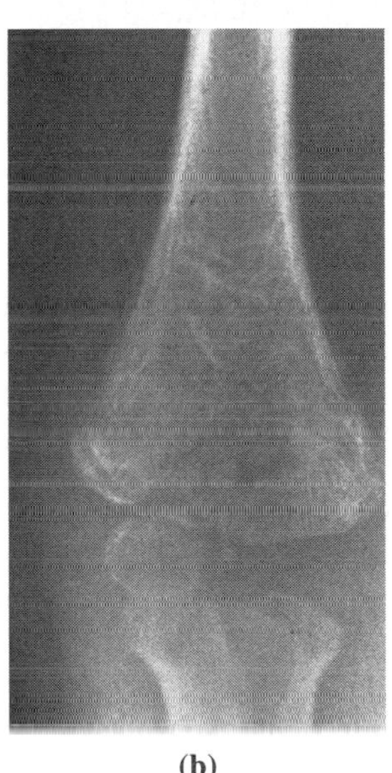

(b)

Figure 20.6 Displaced fracture of the lateral humeral condyle in a 7-year-old girl. Radiography taken on admission (a). Radiography (b) after open reduction and internal fixation with an SR-PGA rod one year later.

rare in children (2.1%) and always mild in character. There are many other investigations on pediatric fractures with successful results (105–108).

Bioabsorbable implants made of PGA and PLLA have been used in the fixation of metaphyseal osteotomies of small bones, such as the base and distal part of the first metatarsal bone in the treatment of hallux valgus. SR-PLLA screws are suitable for fixation of pelvic osteotomy (Chiari), whereas ultra-high molecular weight PLA screws are well suited in rotational acetabular osteotomies to transfix the acetabulum to the grafted bone and the pelvis in children and adults (28 hips) (109) and in another 13 cases (hips, total 41) (110) in the treatment of congenital hip dysplasia. "There were four postoperative complications. A small subcutaneous abscess occurred around the nonabsorbable sutures in 2 hips within 1 year after operation that healed after removal of the suture material. Another hip had developed mild thrombophlebitis in the lower leg of the operated side, which healed with aspirin. A 55-year-old female patient developed itching local dermatitis of unknown origin, 3 cm in diameter, on the skin incision, 8 months after surgery. The result of the patch test was negative for each of the following materials: lactide, low-molecular weight poly-L-lactide and extract of poly-lactide." In addition, SR-PLLA screw fixation has been successfully used to stabilize subtalar extra-articular arthrodesis in the treatment of severe flatfoot and hindfoot valgus deformity (111).

VIII. BIOABSORBABLE FIXATION IN SUPPORTING SOFT TISSUE INJURIES

The common injury of skiers is the total rupture of the ulnar collateral ligament of the first metacarpophalangeal joint of the thumb. The fixation of ruptures of the ulnar collateral ligament of the thumb is important (112). It offers a fast, simple, and safe possibility to treat this common sports injury. Seventy patients with clinically total rupture were treated fixing the ligament with SR-PLLA minitacks. Sixty-nine out of 70 patients healed without complications. The results of this method were confirmed in another study, which included 140 patients (113). Late ligament instability in the ankle can be treated safely also by using bioabsorbable tacks as a fixation device.

Another indication, the fixation of knee meniscus bucket-handle lesions with SR-PLLA arrows, has presently gained more importance (114). The arthroscopic technique has made it possible to fix a certain type of meniscal tear instead of using the former suturing technique. The method to fix meniscus bucket-handle ruptures with SR-PLLA arrows was developed (114,115) and proved to be safe and the fastest way to treat these injuries (116). In the operation, standard arthroscopic portals are used for meniscus fixation in the posterior 2/3 of the meniscus. In the anterior lesion (1/3), accessory portals may be necessary. The rupture is freshened with an arthroscopic rasp and reduced. The cannula with a blunt obturator inside is inserted. The obturator is removed. A channel is made with a special needle through the meniscus. Irrigation fluid is retracted. The needle is retracted. The arrow is pushed to the surface of the meniscus with the obturator via the cannula. The implant is inserted by hammering into the meniscus. The T-shaped head is left on as the surface of the meniscus. The entire procedure is repeated at another site (distance 5–10 mm). Two to four arrows are needed.

A third indication is the use of SR-PLLA plugs in connection with reconstruction operations on the habitual dislocation of the shoulder (117) and the Bankart lesion with special tacks using the arthroscopical technique. There are many different procedures to prevent the recurrent dislocation of the shoulder joint. The Bristow–Latarjet operation is one of the techniques in which the top of the coracoid process is transferred. The fixation of the coracoid bone block conjoined the tendon of coracobrachialis and the short head of the biceps in the drill hole in the scapula neck through the subscapularis tendon was tested with SR-PLLA expansion plugs (117). In 33 patients, no dislocation of the coracoid bone blocks was noticed. For a Bankart lesion, it is possible to do the fixation with special tacks using the arthroscopical technique (118–120).

IX. COMPLICATIONS OF BIOABSORBABLE IMPLANTS

The general complication rates with bioabsorbable fixation seem to be of the same magnitude as with metallic fixation. Several bioabsorbable materials form small debris particles during the last stages of degradation. Therefore it is important to know whether the resorption itself causes complications. Comprehensive studies of the complications associated with bioabsorbable orthopedic devices are limited, but extensive clinical experience with follow-up times of up to 15 years is available.

When wound infection associated with bioabsorbable or metallic devices used in the fixation of fractures, arthrodeses, and osteotomies was studied in 2114 operations, the infection rate was 0.7% with pure SR-PLLA implants but 4% with pure SR-PGA implants (121). When comparison of the infection rates was done between metallic (2073 patients) and bioabsorbable fracture fixation devices (1012 patients) in displaced ankle fractures, the infection rates were 4.1% and 3.2%, respectively.

The cytological analysis of the material aspirated from the effusion, which occasionally develops around a polyglycolic acid (PGA) osteosynthesis implant, showed a predominance of inflammatory monocytes and, in particular, lymphocytes (122). This suggested that PGA is an immunologically inert implant material that induces inflammatory mononuclear cell migration and adhesion, leading to a slight nonspecific lymphocyte activation.

Out of 216 patients with displaced malleolar fractures operated on using bioabsorbable polyglycolide screws, 24 developed a transient local nonbacterial inflammatory reaction on average 3 months after the operation. Upon histopathologic examination, these tissue responses were found to be nonspecific foreign-body reactions. No deleterious effect of these tissue responses on the union of the fractures could be detected (123). A series of 286 patients with unimalleolar or bimalleolar fractures were treated by open reduction and internal fixation using cylidrical rods made of polyglycolide (124). Among them 18 nonbacterial inflammatory tissue responses (6.3% of the total) occurred, requiring surgical drainage. When Kirschner wires were compared to PGA rods in the treatment of distal radial fractures (Frykman types I, II, V, and VI) in 15 human cases, PGA was associated with more foreign-body ractions at 3 to 6 months; in these cases the technique could be a causative factor, as the rods were placed partly outside the bone (125). Nine out of 40 Colles fractures fixed with SR-PGA rods developed sinus formation (126). FrØkjaer and MØller (1992) (127) could not recommend SR-PGA rod fixation for ankle fractures due to complications and osteolysis. However, osteolysis or decreased density is a transient phenomenon (70,128,129). Also, the cases of knee synovitis observed by Tegnander et al. (130) (1994) were too early to result from PLA degradation. Such complications can be explained with the rod insertion technique or with the lesion and operation itself. Tegnander et al. also found high levels of Csa des Arg in plasma incubated in the presence of polylactic acid with conclusions against the biocompatibility of polylactic acid. However, the test procedure has been found inadequatedly designed and the conclusion not justified (131). Two cases of severe aseptic synovitis of the knee were reported 8 and 13 weeks after fixation of a fracture of the intercondylar eminence with PGA rods (132). Both knees were treated by surgical revision and synovectomy. Histologic examination revealed a severe foreign-body type of reactive synovitis in the absence of infection. Local complication after fixation of displaced ankle fracture with PGA rods was reported in two cases (133).

Although complications with the use of bioabsorbable fixation are similar to those of any method of internal fixation, polyglycolide polymer may cause a clinical foreign-body reaction. This occurs 8–16 weeks after the operation but has nothing to do with bone healing. Aspiration with a needle of 1.1 mm is necessary in painful or reddish fluid accumulation and an incision and antibiotics in sinuses.

Complete healing is reached in a few weeks. The incidence of the fluid accumulation when using polyglycolide implants has been recently lower than 2% and has hardly ever occurred with the use of polylactide implants (58). To avoid this complication the accurate operative technique is essential, which is also emphasized in this article.

X. THE FUTURE OF BIOABSORBABLE SELF-REINFORCED COMPOSITES IN ORTHOPEDICS AND TRAUMATOLOGY

The advantages of bioabsorbable implants are significant. Osteoporosis and other harmful consequences associated with rigid metallic implants can be avoided. The avoidance of implant removal procedures results in financial benefits and psychological advantages. The bioabsorbable devices are especially suitable for arthroscopical and other minimum-invasive surgical techniques. The advantages of these implants are not only limited to the mechanical function. They themselves have osteogenic potential. With these polymeric implants it is also possible to combine physiologically active components which accelerate or facilitate tissue healing and prevent or heal infections.

The application of bioabsorbable SR composites such as rods, screws, and tacks is expanding rapidly. The annual number of operations with devices manufactured from these materials has increased steadily over the past years. This means that surgeons have accepted the use of these implants in certain well-defined circumstances. The main goals are to improve the properties of implants and develop surgical techniques suitable for use with bioabsorbable implants. Research is continuing in several units where it has been started already years ago (134–137).

The latest SR composites are strong enough for fractures of load-bearing cancellous bones to be fixed without a plaster cast and with an early mobilization of the patient. Cortical bone osteotomies have also been securely fixed on an experimental basis and clinically in hand and foot surgery.

Bioabsorbable fixation devices have been in clinical use in orthopedic surgery for 15 years. The cumulative number of operations with self-reinforced bioabsorbable implants in surgery has exceeded 300,000. In addition, other bioabsorbable implants than SR bioabsorbable implants increase the total number operations performed. The number of surgical applications will continue to increase in orthopedics and also in other surgical specialities, so that in the future these implants will have a more significant part in modern surgical technique.

ACKNOWLEDGMENTS

The studies of our research group have been financially supported by the Finnish Parliament, the Academy of Finland, the University of Helsinki, Helsinki University Central Hospital, Research Foundation of Orthopaedics and Traumatology, and Tampere University of Technology. The support has been of prime value to the investigations of our research group.

REFERENCES

1. J. Vainio. Experimental studies on the use of bis-GMA composite resin as an implant material in bone. Academic dissertation, University of Tampere, Tampere, Finland, 1980.
2. S. Vainionpää. Biodegradation and fixation properties of biodegradable implants in bone tissue. An experimental and clinical study. Academic dissertation, University of Helsinki, Helsinki, and Tampere University of Technology, Tampere, Finland, 1987.
3. P. B. Axelson. Fixation of cancellous bone and physeal canine and feline fractures with biodegradable self-reinforced polyglycolide devices. A clinical study. Academic dissertation, College of Veterinary Medicine, Helsinki, Finland, 1989.
4. E. A. Mäkelä. Fixation properties and biodegradation of absorbable implants in growing bone. An experimental and clinical study. Academic dissertation, University of Helsinki, Helsinki, and Tampere University of Technology, Tampere, Finland, 1989.
5. K. Vihtonen. Fixation of cancellous bone osteotomies with biodegradable polyglycolic acid thread. Comparison of combined cyanoacrylate, bone cement and biodegradable thread fixation to fixation with biodegradable thread alone. An experimental study. Academic dissertation, University of Helsinki, Helsinki, Finland, 1990.
6. E. Hirvensalo. Absorbable synthetic self-reinforced polymer rods in the fixation of fractures and osteotomies. A clinical study. Academic dissertation, University of Helsinki, Helsinki, Finland, 1990.
7. J. Vasenius. Biocompatibility, biodegradation, fixation properties and strength retention of absorbable implants. An experimental study. Academic dissertation, University of Helsinki, Helsinki, Finland, 1990.
8. R. Suuronen. Biodegradable self-reinforced polylactide plates and screws in the fixation of osteotomies in the mandible. An experimental study. Academic dissertation, Helsinki University, College of Veterinary Medicine, Helsinki, and Tampere University of Technology, Tampere, Finland, 1992.
9. A. Majola. Biodegradation, biocompatibility, strength re-

tention and fixation properties of polylactic acid rods and screws in bone tissue. An experimental study. Academic dissertation, Helsinki University, Helsinki, and Tampere University of Technology, Tampere, Finland, 1992.

10. H. Miettinen. Fracture of the growing femoral shaft. A clinical and experimental study with special reference to treatment using absorbable osteosynthesis. Academic dissertation, Kuopio University, Kuopio, Finland, 1992.
11. E. K. Partio. Absorbable screws in the fixation of cancellous bone fractures and arthrodeses. A clinical study of 318 patients. Academic dissertation, Helsinki University, Helsinki, Finland, 1992.
12. J. Räihä. Biodegradable self-reinforced polylactic acid (SR-PLA) implants for fracture fixation in small animals. An experimental and clinical study. Academic dissertation, College of Veterinary Medicine, Helsinki, Finland, 1993.
13. M. J. Manninen. Self-reinforced polyglycolide and poly-L-lactide devices in fixation of osteotomies of weight-bearing bones. An experimental mechanical study. Academic dissertation, Helsinki University, Helsinki College of Veterinary Medicine, Helsinki, and Tampere University of Technology, Tampere, Finland, 1993.
14. M. Lapinsuo. Bone fixation with plastics. An experimental study. Academic dissertation, Helsinki University, Helsinki, Tampere University, and Tampere University of Technology, Tampere, Finland, 1993.
15. H. Pihlajamäki. Absorbable self-reinforced poly-L-lactide pins and expansion plugs in the fixation of fractures and osteotomies in cancellous bone. An experimental and clinical study. Academic dissertation, Helsinki University, Helsinki, and Tampere University of Technology, Tampere, Finland, 1994.
16. O. Laitinen. Biodegradable polylactide implant as an augmentation device in anterior cruciate ligament repair or reconstruction. An experimental and clinical study. Academic dissertation, Helsinki University, Helsinki College of Veterinary Medicine, and Tampere University of Technology, Tampere, Finland, 1996.
17. N. Ashammakhi. Effect of absorbable polyglycolide membrane on the bone. An experimental study. Academic dissertation, Helsinki University, Helsinki, and Tampere University of Technology, Tampere, Finland, 1996.
18. T. Juutilainen. Absorbable self-reinforced poly-L-lactide wires, screws, and rods in the fixation of fractures and arthrodeses and mini tacks in the fixation of ligament ruptures. A clinical study. Academic dissertation, Helsinki University, Helsinki, Finland, 1997.
19. K. Pelto-Vasenius. Reliability of fixation of fractures and osteotomies with absorbable implants. A clinical study. Academic dissertation, Helsinki University, Helsinki, Finland, 1998.
20. T. Tarvainen. Porous-coated glassy carbon as a bone substitute. Experimental study on rats and rabbits, Academic dissertation, Tampere University, Tampere University of Technology Tampere, and Helsinki University, Helsinki, Finland, 1998.
21. K. Koskikare. Absorbable self-reinforced poly-L-lactide plates in the fixation of osteotomies of the distal femur with special references to the site of implantation. An experimental study. Academic dissertation, Helsinki University, Helsinki, and Biomaterials Laboratory, Institute of Plastics Technology, Tampere University of Technology, Tampere, Finland, 1998.
22. J. Kankare. Absorbable internal fixation of cancellous bone fractures in the lower extremity. A clinical study. Academic dissertation, Helsinki University, Helsinki, Finland, 1999.
23. J. Kettunen. Carbon fibre liquid crystalline polymer composite as an intramedullary bone implant. An experimental study on rabbits and dogs. Academic dissertation, Kuopio University, Kuopio, Finland, 1999.
24. P. Rokkanen, O. Böstman, S. Vainionpää, K. Vihtonen, P. Törmälä, J. Laiho, J. Kilpikari, M. Tamminmäki. Biodegradable implants in fracture fixation: early results of treatment of fractures of the ankle. *Lancet I*: 1422–1424 (1985).
25. N. Doddi, C. C. Versfelt, D. Wasserman. Synthetic absorbable surgical devices of poly-dioxanone. U.S. patent 4,052,988 (1977).
26. D. F. Williams. Enzymic hydrolysis of polylactic acid. *Eng. Med. 10*:5–7 (1981).
27. T. Nakamura, Y. Shimizu, N. Okumura, T. Matsui, S. H. Hyon, T. Shimamoto. Tumorigenicity of poly-L-lactide (PLLA) plates compared with medical-grade polyethylene. *J. Biom. Mater. Res. 28*:17–25 (1994).
28. J. H. Campbell, L. Edsberg, A. E. Meyer. Polylactide inhibition of carcinoma cell growth in vitro. *J. Oral. Maxillofac. Surg. 52*:49–51 (1994).
29. N. A. Higgins. Condensation polymers of hydroyacetic acid. U.S. patent 2,676,945 (1954).
30. E. E. Schmitt, R. A. Polistina. Surgical sutures. U.S. patent 3,297,033 (1967).
31. D. Wasserman, C. C. Versfelt. Use of stannous octoate catalyst in the manufacture of L(−) lactide-glycolide copolymer sutures. U.S. patent 3,839,297 (1974).
32. E. J. Frazza, E. E. Schmitt. A new absorbable suture. *J. Biomed. Mater. Res. 1*:43–58 (1971).
33. J. O. Hollinger. Preliminary report on the osteogenic potential of a biodegradable copolymer of polylactide (PLA) and polyglycolide (PGA). *J. Biomed. Mater. Res. 17*:71–82 (1983).
34. R. K. Kulkarni, K. C. Pani, C. Neuman, F. Leonard. Polylactic acid for surgical implants. *Arch. Surg. 93*:839–843 (1966).
35. D. E. Cutright, E. E. Hunsuck, J. D. Beasley III. Fracture reduction using a biodegradable material, polylactid acid. *J. Oral. Surg. 29*:393–397 (1971).
36. B. Roed-Petersen. Absorbable synthetic suture material for internal fixation of fractures of the mandible. *Int. J. Oral. Surg. 3*:133–136 (1974).
37. R. W. Bucholz, A. Carlton, R. E. Holmes. Hydroxyapatite and tricalcium phosphate bone graft substitutes. *Orthop. Clin. N. Am. 18*:323–334 (1987).

38. A. Tencer, P. Woodward, J. Swenson, K. Brown. Bone ingrowth into polymer coated porous synthetic coralline hydroxyapatite. *J. Orthop. Res.* *5*:275–282 (1987).
39. E. E. Schmitt, R. A. Polistina. Controlled release of medicaments using polymers from glycolic acid. U.S. patent 3,991,766 (1976).
40. J. R. Parsons, H. Alexander, S. F. Corcoran, A. B. Weiss. Development of a variable stiffness, absorbable bone plate. Transactions of the 25th Annual Congress of the Orthopaedic Research Society, San Fransisco, California, 1979, p. 168.
41. H. Alexander, J. R. Parsons, I. D. Strauchler, S. F. Corcoran, O. Gona, C. Mayott, A. B. Weiss. Canine patellar tendon replacement with a polylactic acid polymer-filamentous carbon degrading scaffold to form new tissue. *Orthop. Rev.* *11*:41–51 (1981).
42. S. I. Belykh, A. B. Davydov, G. L. Khoromov, A. D. Moschensky, G. I. Movskovich, G. I. Roitberg, G. L. Voskresensky, G. G. Pershin, V. A. Moskvitin. Biodestructive material for bone fixaton elements. U.S. patent 4,263,185 (1981).
43. V. Skondia, J. Merendino, M. Dales. The biocompatible orthopaedic polymer (BOP). Transaction of the 5th European Congress of Biomaterials, 1985, p. 185.
44. J. Merendino, G. Sertl, V. Skondia. Use of biocompatible orthopaedic polymer for fracture treatment and reconstructive orthopaedic procedures. *Int. Med. Res.* *12*:351–354 (1984).
45. E. E. Schmitt, R. A. Polistina. Polyglycolic acid prosthetic devices. U.S. patent 3,463,158 (1969).
46. L. Claes, C. Burri, H. Kiefer, W. Muntschler. Resorbierbare implantate zur refixierung von osteochondralen fragmenten in gelenkflachen. *Akt. Traumatol.* *16*:74–77 (1986).
47. O. M. Böstman. Current concepts review. Absorbable implants for the fixation of fractures. *J. Bone. Joint. Surg.* *73A*:148–153 (1991).
48. G. O. Hofmann. Biodegradable implants in orthopaedic surgery—a review on the state-of-the-art. *Clin. Mater.* *10*: 75–80 (1992).
49. K. A. Athanasiou, G. G. Niederauer, C. M. Agrawal, A. S. Landsman. Applications of biodegradable lactides and glycolides in podiatry. *Clin. Pod. Med. Surg.* *12*:475–495 (1995).
50. N. A. Ashammkhi. Neomembranes: a concept review with special reference to self-reinforced polyglycolide membranes. *J. Biomed. Mater. Res.* *33*:297–303 (1996).
51. N. Ashammakhi, P. Rokkanen. Review. Absorbable polyglycolide devices in trauma and bone surgery. *Biomaterials* *18*:3–9 (1997).
52. R. S. Maitra, J. C. Brand, D. N. M. Caborn. Biodegradable implants. *Sport. Med. Arthr. Rev.* *6*:103–117 (1998).
53. P. Rokkanen, P. Törmälä. Absorbable polylactide implants in the fixation of fractures. osteotomies, arthrodeses, and ligament injuries. A review focused on self-reinforced implants (P. Rokkanen, P. Törmälä, eds.). University of Helsinki, Department of Orthopaedics and Traumatology, and Tampere University of Technology, Biomaterials Laboratory, Tampere, Copy Shop Oy, Tampere, Finland, 1996.
54. P. Rokkanen, O. Böstman, E. Hirvensalo, E. A. Mäkelä, E. K. Partio, H. Pätiälä, K. Vihtonen, P. Törmälä. Bioabsorbable fixation in traumatology and orthopaedics (BFTO). Definitive fracture care at one operation (P. Rokkanen, ed.). University of Helsinki, Department of Surgery, and Kuopio University Hospital/Copy Center, Kuopio, Finland, 1997.
55. P. Rokkanen, P. Törmälä. Current concept review bioabsorbable implants in the fixation of fractures, (P. Rokkanen, P. Törmälä, eds.). University of Helsinki, Department of Orthopaedics and Traumatology, and Tampere University of Technology, Institute of Biomaterials, Domus-Offset Oy, Tampere, Finland, 1997.
56. J. A. Simon, J. L. Ricci, P. E. di Cesare. Bioresorbable fracture fixation in orthopaedics: a comprehensive review. Part I. Basic science and preclinical studies. *Am. J. Orhop* *26*:665–671 (1997).
57. J. A. Simon, J. L. Ricci, P. E. di Cesare. Bioresorbable fracture fixation in orthopaedics: a comprehensive review. Part II. Clinical studies. *Am. J. Orthop.* *26*:754–762 (1997).
58. P. Rokkanen, O. Böstman, E. Hirvensalo, E. K. Partio, E. A. Mäkelä, H. Pätiälä, K. Vihtonen. Bioabsorbable implants in orthopaedics. *Curr. Orthop.* *13*:223–228 (1999).
59. P. Törmälä, S. Vainionpää, J. Kilpikari, P. Rokkanen. The effects of fibre reinforcement and gold plating on the flexural and tensile strength of PGA/PLA copolymer materials in vitro. *Biomaterials* *8*:42–45 (1987).
60. P. Törmälä. Biodegradable self-reinforced composite materials: manufacturing structure and mechanical properties. *Clin. Mat.* *10*:29–34 (1992).
61. P. Törmälä. Ultra-high strength, self-reinforced absorbable polymeric composites for applications in different disciplines of surgery. *Clin. Mat.* *13*:35–40 (1993).
62. P. Törmälä, T. Pohjonen, P. Rokkanen. Bioabsorbable polymers: materials technology and surgical applications. *Proc. Instn. Mech. Engrs.* *212 Part II*:101–111 (1998).
63. P. Törmälä, P. Rokkanen, J. Laiho, M. Tamminmäki, S. Vainionpää. Material for osteosynthesis devices. U.S. patent 4,743,257 (1988).
64. O. Böstman, U. Päivärinta, E. Partio, J. Vasenius, M. Manninen, P. Rokkanen. Degradation and tissue replacement of an absorbable polyglycolide screw in the fixation of rabbit femoral osteotomies. *J. Bone. Joint. Surg.* *74A*: 1021–1031 (1992).
65. A. Weiler, H. J. Helling, U. Kirch, T. K. Zirbes, K. E. Rehm. Foreign-body reaction and the course of osteolysis after polyglycolide implants for fracture fixation. Experimental study in sheep. *J. Bone. Joint. Surg.* *78B*:369–376 (1996).
66. O. Böstman, U. Päivärinta. Restoration of tissue components after insertion of absorbable fracture fixation devices of polyglycolide through the articular surface: an experimental study in the distal rabbit femur. *J. Orthop. Res.* *12*: 403–411 (1994).

67. D. F. Williams. Some observations on the role of cellular enzymes in the in vivo degradation of polymers. *Spech. Tech. Publ. 684*:61–75 (1979).
68. H. Pihlajamäki, J. Kinnunen, O. Böstman. In vivo monitoring of the degradation process of bioresorbable polymeric implants using magnetic resonance imaging. *Biomaterials 18*:1311–1315 (1997).
69. H. K. Pihlajamäki, P. T. Karjalainen, H. J. Aronen, O. M. Böstman. MR imaging of biodegradable polylevolactide osteosynthesis devices in the ankle. *J. Orthop. Traum. 11*: 559–564 (1997).
70. O. Böstman, S. Vainionpää, E. Hirvensalo, E. A. Mäkelä, K. Vihtonen, P. Törmälä, P. Rokkanen. Biodegradable internal fixation for malleolar fractures. A prospective randomised trial. *J. Bone. Joint. Surg. 69B*:615–619 (1987).
71. N. Gammelgaard, J. Jensen. Wound complications after closure of abdominal incisions with Dexon® or Vicryl®. A randomized double-blind study. *Acta. Chir. Scand. 147*: 505–508 (1983).
72. P. U. Rokkanen. Bioabsorbable fixation devices in orthopaedics and traumatology. *Ann. Chir. Gyn. 87*:13–20 (1998).
73. J. T. Waston, W. D. Hovis, R. W. Bucholz. Bioabsorbable fixation of ankle fractures. *Tech. Orthop. 13*:180–186 (1998).
74. D. B. Thordarson. Fixation of the ankle syndesmosis with bioabsorbable screws. Tech. Orthop. 13:187–191 (1998).
75. E. A. Melamed, D. Seligson. Fixation of ankle fractures with biodegradable implants. *Tech. Orthop. 13*:192–200 (1998).
76. A. Chen, C. Hou, J. Bao, S. Guo. Comparison of biodegradable and metallic tension-band fixation for patella fractures. 38 patients followed for 2 years. *Acta. Orthop. Scand. 69*:39–42. (1998).
77. B. Bloss, T. B. Rogers, D. Seligson. Bioabsorbable pins and screws for the fixation of distal radial fractures. *Tech. Orthop. 13*:153–159 (1998).
78. R. Jahn, D. Diederichs, B. Friedrich. Resorbierbare Implantate und ihre Anwendung am Beispiel der Radiusköpfchenfraktur. *Akt. Traumatol. 19*:281–286 (1989).
79. E. Hirvensalo, O. Böstman, E. K. Partio, P. Törmälä, P. Rokkanen. Fracture of the humeral capitellum fixed with absorbable polyglycolide pins. 1-year follow-up of 8 adults. *Acta. Orthop. Scand. 64*:85–886 (1993).
80. S. M. Kumta, P. C. Leung. The technique and indications for the use of biodegradable implants in fractures of the hand. *Techn. Orthop. 13*:160–163 (1998).
81. H. G. Haas. PDS-Splinte zur Frakturbehandlung. *Handchirurgie 18*:295–297 (1986).
82. N. van Nieuwenhuyse, P. Merloz, S. Plaweski. Ostéosynthèse par vis bio-résorbables: 40 premiers cas. *Actualités en Biomatériaux, Vol. 2* (D. Mainard, M. Merle, J. P. Delagoutte, J. P. Louis, eds.). Romillat, Paris, France, 1993, pp. 44–46.
83. K. E. Rehm, H. J. Helling, L. Claes. Bericht der Arbeitsgruppe Biodegradable Implantate. *Akt. Traumatol. 24*:70–73 (1994).
84. H. Pihlajamäki, O. Böstman, E. Hirvensalo, P. Törmälä, P. Rokkanen. Absorbable pins of self-reinforced poly-L-lactic acid for fixation of fractures and osteotomies. *J. Bone. Joint. Surg. 74-B*:853–857 (1992).
85. P. Tuompo, E. K. Partio, P. Rokkanen. Bioabsorbable fixation in the treatment of proximal tibial osteotomies and fractures. A clinical study. *Ann. Chir. Gyn. 88*:66–72 (1999).
86. J. Eitenmüller, A. David, A. Pommer, G. Muhr. Operative Behandlung von Sprunggelenksfrakturen mit biodegradablen Schrauben und Platten aus Poly-L-Lactid. *Chirurg 67*:413–418 (1996).
87. R. W. Bucholz, S. Henry, M. B. Henley. Fixation with bioabsorbable screws for the treatment of fractures of the ankle. *J. Bone. Joint. Surg. 76-A*:319–324 (1994).
88. T. Yamamuro, Y. Matsusue, A. Uchida, K. Shimada, E. Shimozaki, K. Kitaoka. Bioabsorbable osteosynthetic implants of ultra high strength poly-L-lactide. A clinical study. *Int. Orthop. 18*:332–340 (1994).
89. P. Kousa, T. L. N. Järvinen, T. Pohjonen, P. Kannus, M. Kotikoski, M. Järvinen. Fixation strength of a biodegradable screw in anterior cruciate ligament reconstruction. *J. Bone. Joint. Surg. 77B*:901–905 (1995).
90. K. Jukkala-Partio, E. K. Partio, P. Helevirta, T. Pohjonen, P. Törmälä, P. Rokkanen. Fixation of subcapital femoral neck fractures with bioabsorbable or metallic screw fixation. A preliminary report. *Ann. Chir. Gyn. in press* (1999).
91. K. Pelto-Vasenius, E. Hirvensalo, J. Vasenius, P. Rokkanen. Osteolytic changes after polyglycolide pin fixation in chevron osteotomy. *Foot. Ankle. Int. 18*:21–25 (1997).
92. L. H. Gill, D. F. Martin, J. M. Coumas, G. M. Kiebzak. Fixation with bioabsorbable pins in chevron bunionectomy. *J. Bone. Joint. Surg. 79-A*:1510–1518 (1997).
93. F. Barca, R. Busa. Resorbable poly-L-lactic acid mini-staples for the fixation of Akin osteotomies. *J. Foot. Ankle. Surg. 36*:106–111 (1997).
94. A. E. Burns, J. Varin. Poly-L-lactic acid rod fixation results in foot surgery. *J. Foot. Ankle. Surg. 37*:37–41 (1998).
95. A. E. Burns. Absorbable fixation techniques in forefoot surgery. *Tech. Orthop. 13*:201–206 (1998).
96. E. K. Partio, E. Hirvensalo, A. Joukainen, P. Rokkanen. Proximal or distal osteotomy for hallux valgus fixed with absorbable implants. *Acta. Orthop. Scand. (suppl. 280)69*: 43 (1998).
97. P. Tuompo, V. Arvela, E. K. Partio, P. Rokkanen. Osteochondritis dissecans of the knee fixed with biodegradable self-reinforced polyglycolide and polylactide rods in 24 patients. *Int. Orthop. 21*:355–360 (1997).
98. Y. Matsusue, T. Nakamura, S. Suzuki, R. Iwasaki. Biodegradable pin fixation of osteochondral fragments of the knee. *Clin. Orthop. 322*:166–173 (1996).
99. P. Voche, M. Merle, H. Membre. Approche expérimentale et clinique de l'ostéosynthèse résorbable en chirurgie de la main. *Actualités en Biomatériaux, vol. 2* (D. Mainard, M. Merle, J.P. Delagoutte, J.P. Louis, eds.). Romillat, Paris, France, 1993, pp. 38–43.

100. R. O. Niskanen, M. Y. Lehtimäki, M. M. J. Hämäläinen, P. Törmälä, P. U. Rokkanen. Arthrodesis of the first metatarsophalangeal joint in rheumatoid arthritis. Biodegradable rods and Kirschner-wires in 39 cases. *Acta. Orthop. Scand. 64*:100–102 (1993).
101. R. B. Salter, W. R. Harris. Injuries involving the epiphyseal plate. *J. Bone. Joint. Surg. 45A*:587–622 (1963).
102. E. A. Mäkelä, S. Vainionpää, K. Vihtonen, M. Mero, J. Laiho, P. Törmälä, P. Rokkanen. Healing of physeal fracture after fixation with biodegradable self-reinforced polyglycolic acid pins. An experimental study on growing rabbits. *Clinical. Mater. 5*:1–12 (1990).
103. E. A. Mäkelä. Fixation properties and biodegradation of absorbable implants in growing bone. In: *Self-Reinforced Bioabsorbable Polymeric Composites in Surgery* (P. Rokkanen, P. Törmälä eds.). Department of Orthopaedics and Traumatology, Helsinki University, and Biomaterials Laboratory, Tampere University of Technology, Finland, 1995, pp. 175–198.
104. E. A. Mäkelä. Absorbable pin fixation of physeal fractures in children. In: *Seminar on Absorbable Fixation Devices in Bone and Ligament Surgery* (E. A. Mäkelä ed.). Tallinn, Estonia, August 28 and 29, 1997, pp. 29–33.
105. P. J. Svensson, P. M. Janary, G. Hirsch. Internal fixation with biodegradable rods in pediatric fractures: one-year follow-up of fifty patients. *J. Pediatr. Orthop. 14*:220–224 (1994).
106. D. Rovinsky, R. C. Durkin, N. Y. Otsuka. The use of bioabsorbables in the treatment of children's fractures. *Tech. Orthop. 13*:130–138 (1998).
107. G. Hirsch, A. Boman. Osteochondral fractures of the knee in children and adolescents—treatment with open reduction and osteosynthesis using biodegradable pins. *Tech. Orthop. 13*:139–142 (1998).
108. R. D. Blasier, R. R. White. The treatment of tibial tubercle avulsion in adolescence with resorbable implants. *Tech. Orthop. 13*:170–176 (1998).
109. S. Nakamura, S. Ninomiya, Y. Takatori, S. Morimoto, I. Kusaba, T. Kurokawa. Polylactide screws in acetabular osteotomy. 28 dysplastic hips followed for 1 year. *Acta. Orthop. Scand. 64*:301–302 (1993).
110. S. Nakamura, Y. Takatori, S. Morimoto, T. Umeyama, M. Yamamoto, T. Moro, S. Ninomiya. Rotantional acetabular osteotomy using biodegradable internal fixation. *Int. Orthop. 23*:148–149 (1999).
111. E. K. Partio, J. Merikanto, J. T. Heikkilä, P. Ylinen, E. A. Mäkelä, J. Vainio, P. Törmälä, P. Rokkanen. Totally absorbable screws in fixation of subtalar extra articular arthrodesis in children with spastic neuromuscular disease. Preliminary report of randomized prospective study of fourteen arthrodeses fixed with absorbable or metallic screws. *J. Pediatr. Orthop. 12*:646–650 (1992).
112. K. Vihtonen, T. Juutilainen, H. Pätiälä, P. Rokkanen, P. Törmälä. Reinsertion of the ruptured ulnar collateral ligament of the metacarpophalangeal joint with an absorbable self-reinforced polylactide tack. *J. Hand. Surg. 18B*:200–203 (1993).
113. T. Juutilainen, K. Vihtonen, H. Pätiälä, P. Rokkanen, P. Törmälä. Reinsertion of the ruptured ulnar collateral ligament of the metacarpophalangeal joint of thumb with an absorbable self-reinforced polylactide mini tack. *Ann. Chir. Gyn. 85*:364–368 (1996).
114. P. Albrecht-Olsen, G. Kristensen, P. Törmälä. Meniscus bucket-handle fixation with an absorbable Biofix tack: development of a new technique. *Knee. Surg. Sports. Traumatol. Arthroscopy. 1*:104–106 (1993).
115. G. Kristensen, P. M. Albrecht-Olsen. Meniscus repair as an all-inside method with an absorbable arrow. *Orthopaedic Product News Mar/Apr*:35–36 (1998).
116. B. Cohen, J. Tasto. Meniscal arrow. *Tech. Orthop. 13*: 164–169 (1998).
117. H. Pihlajamäki, O. Böstman, P. Rokkanen. A biodegradable expansion plug for fixation of the cortical bone block in the Bristow-Latarjet operation. *Int. Orthop. 18*:66–71 (1994).
118. K. P. Speer, R. F. Warren. Arthroscopic shoulder stabilization. A role for biodegradable materials. *Clin. Orthop. 291*:67–74 (1993).
119. K. P. Speer, R. F. Warren, M. Pagnani, J. J. P. Warner. An arthroscopic technique for anterior stabilization of the shoulder with a bioabsorbable tack. *J. Bone. Joint. Surg. 78-A*:1801–1807 (1996).
120. U. Väätäinen. Arthroscopic repair of bankart lesion using PLA-tacks. Operation technique. In: *Seminar on Absorbable Fixation Devices in Bone and Ligament Surgery* (E. A. Mäkelä ed.), Tallinn, Estonia, August 28 and 29, 1997, pp. 35–36.
121. I. Sinisaari, H. Pätiälä, O. Böstman, E. A. Mäkelä, E. Hirvensalo, E. K. Partio, P. Törmälä, P. Rokkanen. Wound infections associated with absorbable or metallic devices used in the fixation of fractures, arthrodeses and osteotomies. *Eur. J. Orthop. Surg. Traumatol. 5*:41–43 (1995).
122. S. Santavirta, Y. T. Konttinen, T. Saito, M. Grönblad, E. Partio, P. Kemppinen, P. Rokkanen. Immune response to polyglycolide acid implants. *J. Bone. Joint. Surg. 72-B*: 597–600 (1990).
123. O. Böstman, E. Partio, E. Hirvensalo, P. Rokkanen. Foreign-body reactions to polyglycolide screws. Observations in 24/216 malleolar fracture cases. *Acta. Orthop. Scand. 63*:173–176 (1992).
124. O. M. Böstman. Intense granulomatous inflammatory lesions associated with absorbable internal fixation devices made of polyglycolide in ankle fractures. *Clin. Orthop. 278*:193–199 (1992).
125. P. P. Casteleyn, E. Handelberg, P. Haentjens. Biodegradable rods versus Kirschner wire fixation of wrist fractures. A randomised trial. *J. Bone. Joint. Surg. 74-B*:858–861 (1992).
126. R. Hoffmann, C. Kretter, N. Haas, H. Tscherne. Die distale radiusfraktur. Frakturstabilisierung mit biodegradablen Osteosynthese-Stiften (Biofix®). Experimentelle untersuchungen und erfahrungen. *Unfallchirurg 92*:430–434 (1989).
127. J. Frøkjaer, B. N. Møller. Biodegradable fixation of ankle

fractures. Complications in a prospective study of 25 cases. *Acta. Orthop. Scand. 63*:434–436 (1992).

128. O. M. Böstman. Osteolytic changes accompanying degradation of absorbable fracture fixation implants. *J. Bone. Joint. Surg. 73-B*:679–682 (1991).
129. R. K. Fraser, W. G. Cole. Osteolysis after biodegradable pin fixation of fractures in children. *J. Bone. Joint. Surg. 74-B*:929–930 (1992).
130. A. Tegnander, L. Engebretsen, K. Bergh, E. Eide, K. J. Holen, O. J. Iversen. Activation of the complement system observed by use of biodegradable pins of polylactic acid (Biofix®) in osteochondritis dissecans. *Acta. Orthop. Scand. 65*:472–475 (1994).
131. P. Mainil-Varlet. Correspondence. Polylactic acid pins. *Acta. Orthop. Scand. 66*:573–574 (1995).
132. G. Barfod, R. N. Svendsen. Synovitis of the knee after intra-articular fracture fixation with Biofix®. Report of two cases. *Acta. Orthop. Scand. 63*:680–681 (1992).
133. J. Poigenfürst, M. Leixnering, M. B. Mokhtar. Lokalkomplikationen nach implantation von Biorod. *Akt. Traumatol. 20*:157–159 (1990).
134. M. Vert, F. Chabot, J. Leray, P. Christel. Stereoregular bioresorbable polyesters for orthopaedic surgery. *Makromol. Chem. Suppl. 5*:30–41 (1981).
135. L. E. Claes. Mechanical characterization of biodegradable implants. *Clin. Mater. 10*:41–46 (1992).
136. S. Gogolewski. Resorbable polymers for internal fixation. *Clin. Mater. 10*:13–20 (1992).
137. D. C. Tunc. In vivo degradation and biocompatability study of in vitro pre-degraded as-polymerized polylactide particles. Letters to the editor. *Biomaterials 17*:2109–2112 (1996).

21

Polymers for Artificial Joints*

Naohide Tomita and Hiroshi Fujita
Kyoto University, Kyoto, Japan

Kazuya Nagata
Industrial Technology Center of Okayama Prefecture, Okayama, Japan

I. ARTIFICIAL JOINTS

A. History of Artificial Joints

A typical view of a joint is shown in Fig. 1a; it is composed of bone, cartilage, a joint capsule, joint fluid, and ligaments. When the cartilage has large lesions as shown in Fig. 1b, some replacement or regeneration of the joint is required (1). Several regenerative treatments, including bone osteotomy, autograft, and cultured cell implantation, have been developed, but joint replacement with an artificial joint is the most common and effective treatment for the aged. The use of biological and inorganic materials for joint arthroplasty became popular in the early twentieth century. Deformed or ankylosed joint surfaces were contoured and an interpositional layer inserted to resurface the joint and allow motion, as shown in Fig. 2a. Fascia lata grafts, periarticular soft tissues, and gold foils were tried as the interpositional layer. Then cup authroplasty and joint head replacement, as shown in Fig. 2c, were also used for hip and finger joints as shown in Fig. 2b and c. They became the standard operation for hip reconstruction until the advent of modern total hip arthroplasty. The artificial joint in the restricted sense should have an artificial articulating joint mechanism as shown in Fig. 2d. The modern total joint arthroplasty was pioneered by McKee (1951)

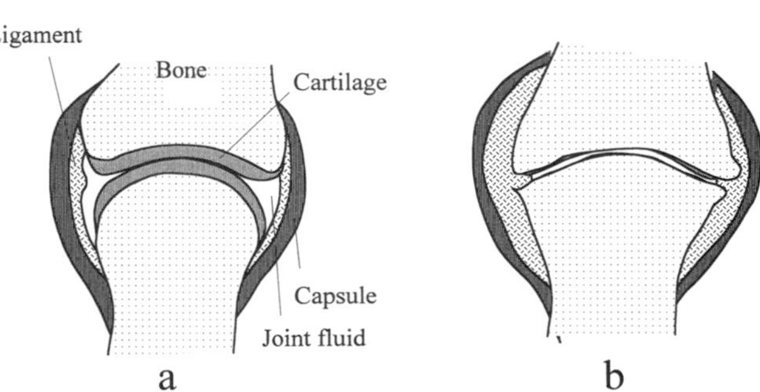

Figure 21.1 Typical view of joints. (a) Normal joint; (b) arthrotic deformity.

* Part I of this chapter is by Nagata and Tomita; Part II is by Fujita.

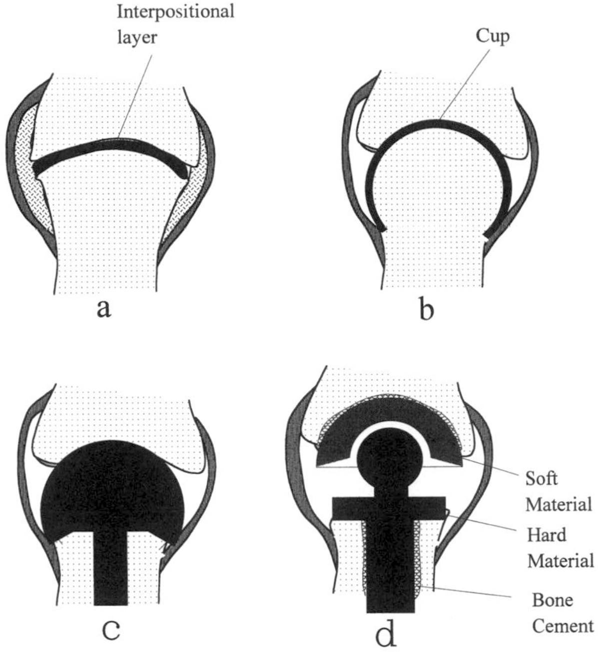

Figure 21.2 Typical view of arthroplasty for deformed joint.

and by Charnley (1961) (2,3,4) after several abortive attempts. Figure 2d shows schematics for modern total arthroplasty established by Sir John Charnley. The artificial joint has a sliding interface using a combination of a hard material against a soft material. He used a metallic femoral head as the hard material and a polytetrafluororthylene shell as the soft material. He then cemented the stem and cup with cold-curing acrylic cement (polymethylmethacrylate) to fix the components securely in the bone and to transfer stress more uniformly. The polytetrafluoroethylene was replaced by high-density polyethylene (HDP) and later by ultrahigh molecular weight polyethylene (UHMWPE) because of its excessive wear and tissue reaction. The metal-on-metal total hip arthroplasty did not prove satisfactory at that time because friction and metal wear could not be prevented.

Typical shapes for artificial hip and knee prostheses are shown in Fig. 3 and Fig. 4. Motion in hip prostheses is universal but there is only one center for bending and rotating, whereas knee prostheses should bear more complicated motion, including flexion, rotation, and anterior–posterior sliding. The contact area on the sliding surface of knee prostheses is narrower than that of hip prostheses because of the difference in movement.

Figure 5 shows a rough estimation for the prognosis of total hip replacement. Incidence of infection has been diminished because of improvements in surgical equipment and technique. Obviously, loosening is the most serious problem in modern total joint replacement. Loosening is caused by mechanical loading, but the UHMWPE wear particles generated at the articulating surfaces are thought to accelerate the loosening process.

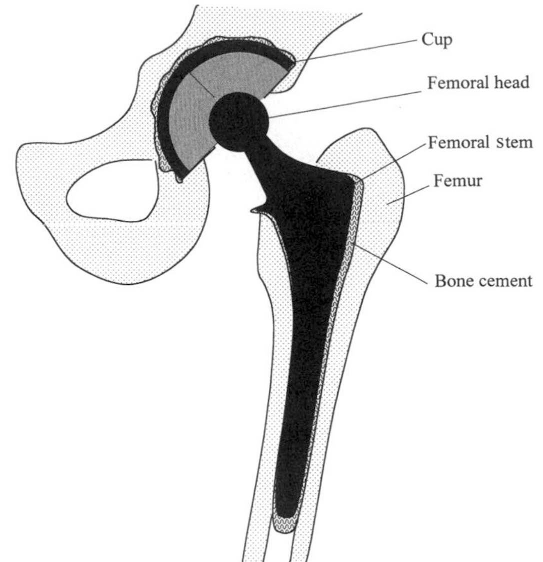

Figure 21.3 Schematic drawing of a typical artificial hip joint.

B. Ultra-High Molecular Weight Polyethylene for Articulating Surfaces

Ultra-high molecular weight polyethylene (UHMWPE) has been used as the bearing material in total joint prostheses (hip, knee, ankle, shoulder, elbow, and wrist) implanted since the 1960s. UHMWPE was chosen because of its unique material properties, such as a low friction coeffi-

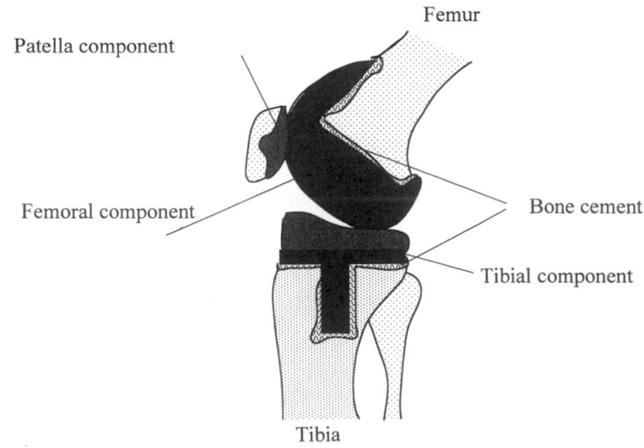

Figure 21.4 Schematic drawing of a typical artificial knee joint.

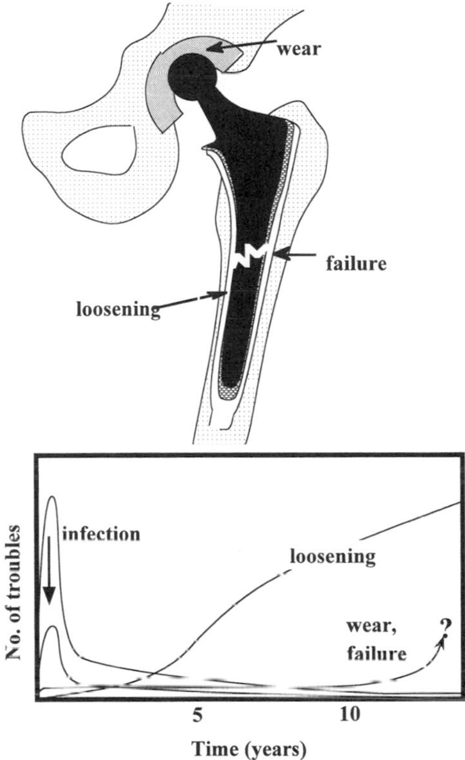

Figure 21.5 Rough estimation for prognoses of total hip replacement.

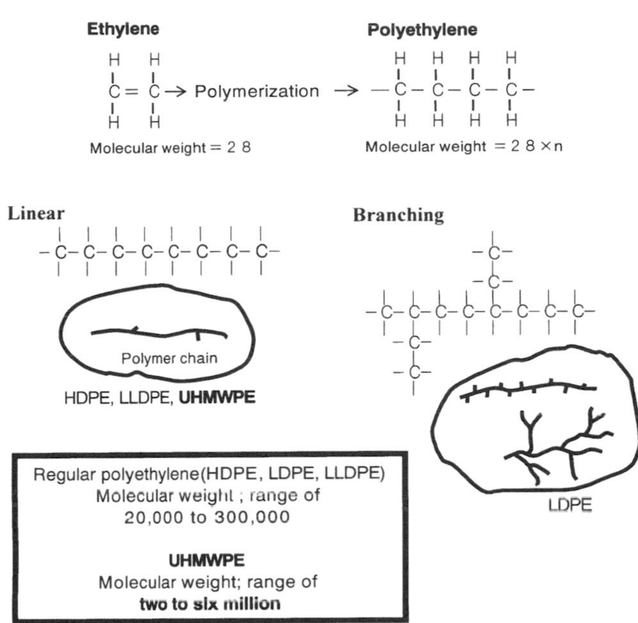

Figure 21.6 Chemical structure of UHMWPE.

cient, high resistance to wear, high impact resistance, high ductility, and stability in the body (5). However, producing the UHMWPE implant is complicated due to its high molecular weight. Research and development efforts in the manufacturing process of UHMWPE are the key points for improvements in quality.

1. Structure and Properties of UHMWPE

UHMWPE is a linear type of polyethylene that has a branching chain loosely contained in one chain (Fig. 6). The molecular weight of UHMWPE is extremely high, normally in the range of two to six million. These values are in contrast to 20,000 to 300,000 for regular polyethylenes such as high-density polyethylene (HDPE), low-density polyethylene (LDPE), and linear low-density polyethylene (LLDPE) (6). The UHMWPE molecular chain arrangement in the solid state is HDPE, which consists of highly ordered chain-folded crystallite regions and a relatively unoriented amorphous region. Long tie molecules in the amorphous region connect the crystalline with the surrounding crystalline. The major difference between the morphology of UHMWPE and that of HDPE is that the number and chain length of the tie molecules in UHMWPE is much higher than that in HDPE (7). This is the reason that UHMWPE exhibits superior physical properties and resistance to wear in regular polyethylene and various types of polymers (Table 1, Fig. 7) (7,8). UHMWPE also exhibits superior impact strength. Figure 8 shows a characteristic curve correlating impact strength with the molecular weight of polyethylene. The impact strength increases with increasing molecular weight initially, reaches a maximum, and decreases slightly upon a further increase in molecular weight in the ultra-high molecular weight region (9,10).

The density of polyethylene decreases with the increasing molecular weight of polyethylene as shown in Fig. 9.

Table 21.1 Physical Properties of Polyethylene

Property	High-density polyethylene	Ultra-high molecular polyethylene
Molecular weight/millions $g \cdot mol^{-1}$	0.05–0.2	2.6
Melting point/°C	130–137	125–135
Density/$g \cdot cm^{-3}$	0.952–0.965	0.930–0.945
Tensile yield/MPa	26.2–33.1	19.3–23
Elongation at break/%	10–1200	200–450
Tensile modulus/MPa	400–4000	600–1500
Izot impact/$J \cdot m^{-1}$	21–210	<1070–no break
Shore-D hardness	67–73	60–65

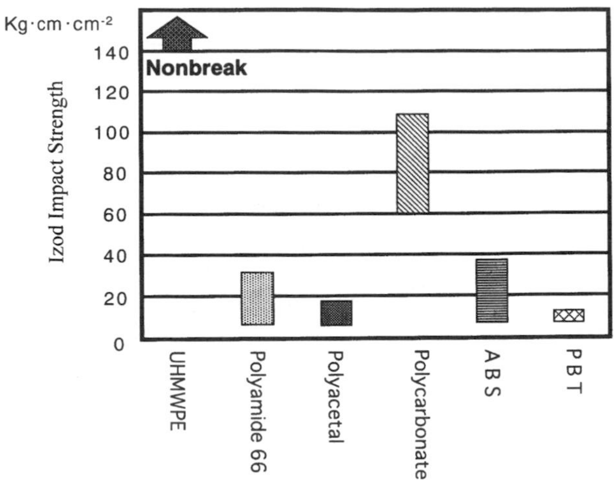

Figure 21.7 Izod impact strength of various polymers.

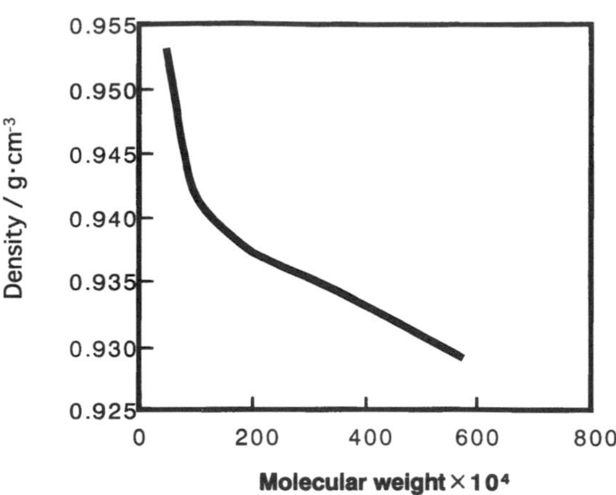

Figure 21.9 Relationship between molecular weight and density of polyethylene.

There always exists a correlation between crystallinity and density for crystalline polymers. The crystallinity and density of UHMWPE are lower than that of HDPE, due to the high molecular weight and chain structure. The crystallinity of UHMWPE, estimated with the use of differential scanning calorimetry, lies in the range of 45 to 65%, while that of HDPE is 70% or higher (10,13).

2. Synthesis of UHMWPE

Figure 10 shows the manufacturing process of UHMWPE for artificial joints. The synthesis of UHMWPE is conducted with use of a Ziegler and Natta coordination catalyst under low pressure. The catalyst is made from titanium chloride and an aluminum alkyl compound. The polymerization reaction is a heterogeneous system involving a hydrocarbon solvent (hexane), ethylene gas, and the catalyst. The polymerization is typically conducted at pressures between 0.4 and 0.6 MPa, at a temperature between 66 and 90°C (6,8). The polymerization results in a fine granular white powder. In some grades, a very fine calcium stearate powder is added after isolation and drying of the reactant powder. There are mainly two companies manufacturing UHMWPE for medical applications: Ticona (Hoechst, Hostalen GUR) and Montell (Himont, HiFax). The powder properties of typical grades of UHMWPE are shown in Table 2.

3. Processing of UHMWPE

When UHMWPE is melted, it does not flow easily, as low molecular weight polyethylenes such as LDPE and HDPE

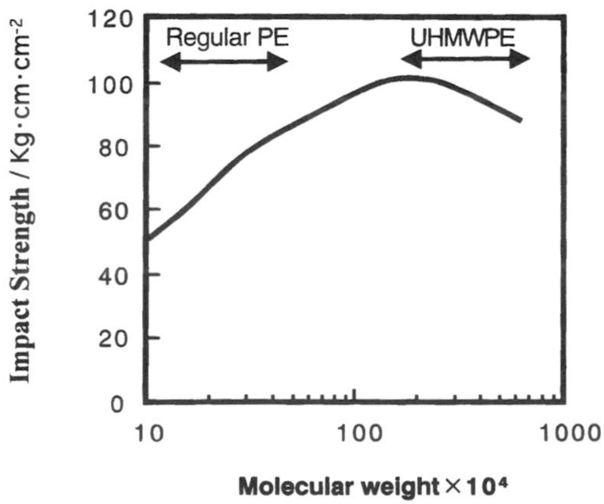

Figure 21.8 Relationship between molecular weight and impact strength of polyethylene.

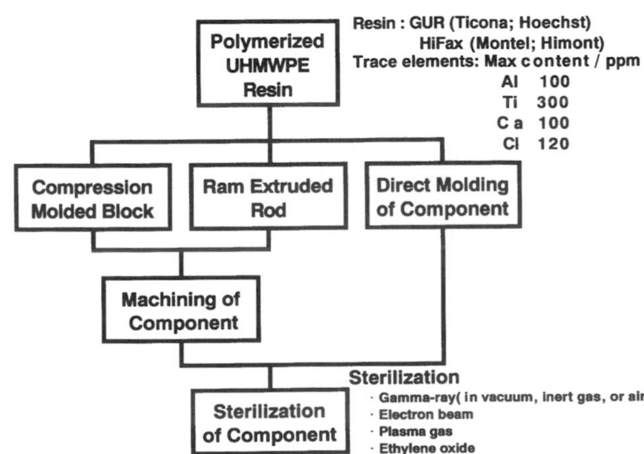

Figure 21.10 Manufacturing process of UHMWPE for artificial joints.

Table 21.2 Powder Properties of Typical UHMWPE Grades (Ticona; Hoechst, Hostalen GUR)

Grade	Mean particle size/μm	Intrinsic viscosity η/dl · g^{-1}	Molecular weight[a] × 10^4	Calcium stearate
GUR1120	120	1920	308	<50 ppm
GUR1150	120	2700	490	<50 ppm
GUR1020	120	1920	308	<5 ppm
GUR1050	120	2700	490	<5 ppm

[a] ASTM method: Molecular weight = $5.37 \times 10^4 \times [\eta]^{1.37}$

do, owing to its high molecular weight and excessive entanglement (Fig. 11) (9). When UHMWPE is heated above its crystalline melting temperature (125–135°C), the resin becomes rubbery but does not flow. Scanning electron microscopy (SEM) images of UHMWPE powder heated at 160, 180, and 200°C are shown in Fig. 12. A specimen heated to 200°C shows a highly granular structure in which the original powder particles are clearly evident. The method of fabricated forms for commonly used thermoplastics cannot be applied to process UHMWPE due to this extremely high melt viscosity. The processing of UHMWPE requires a proper combination of temperature, high pressure, and time to achieve complete plastification. Three methods are currently used to make artificial joints with UHMWPE. These methods are ram extrusion, compression molding, and direct compression molding (Figs. 13, 14, and 15). For each method the goal is to apply enough temperature and pressure to UHMWPE powder to insure that the particles are fully sintered.

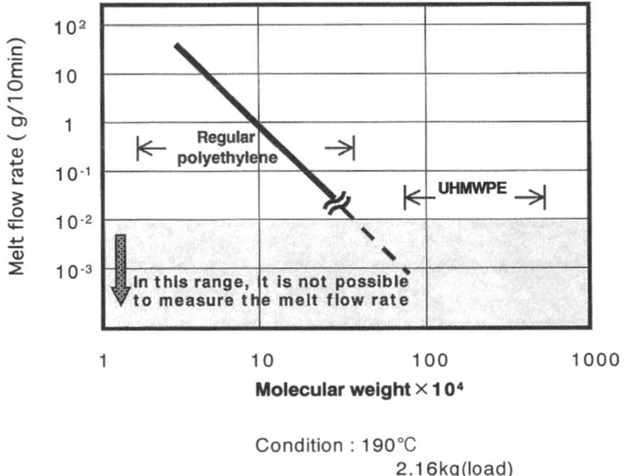

Figure 21.11 Relationship between molecular weight of polyethylene and melt flow rate.

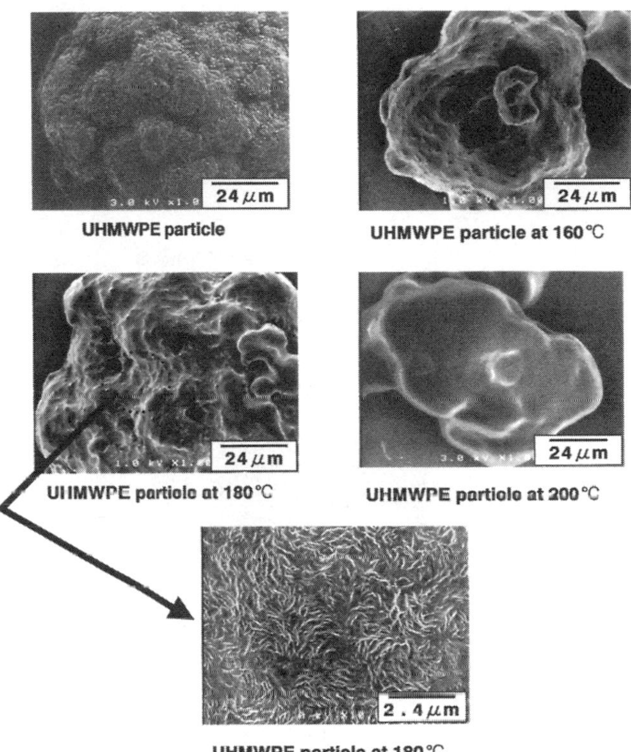

Figure 21.12 Scanning electron micrograph of UHMWPE particles at various temperatures.

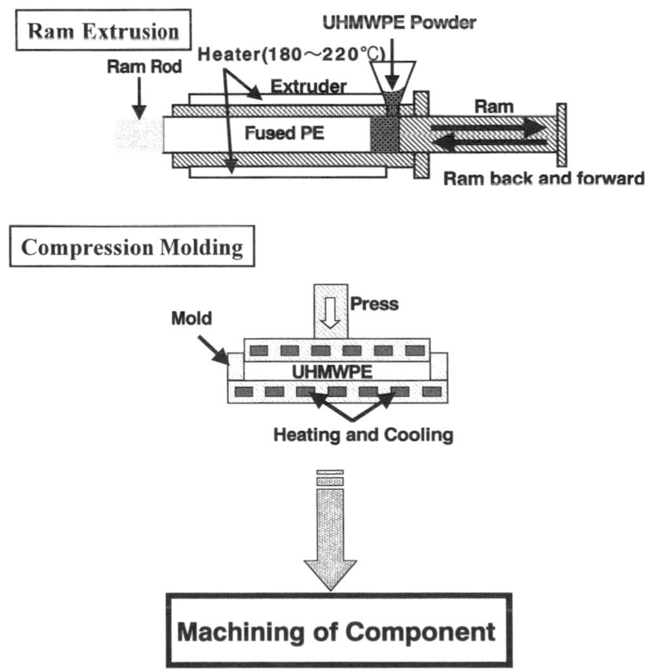

Figure 21.13 Manufacturing process of ram extrusion and compression molding for UHMWPE implants.

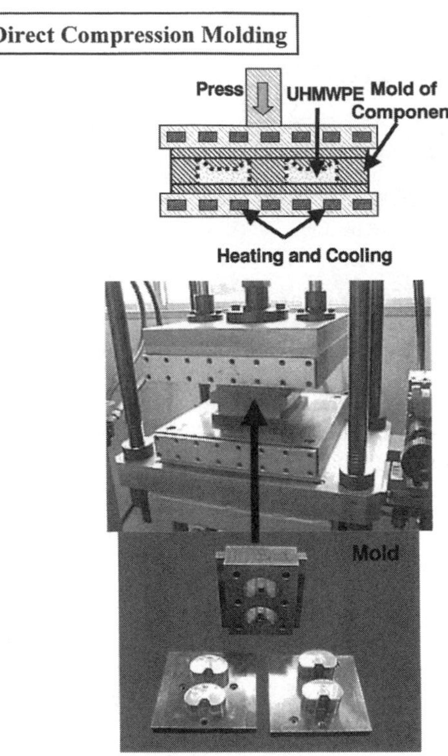

Figure 21.14 Manufacturing process of direct compression molding for UHMWPE implants.

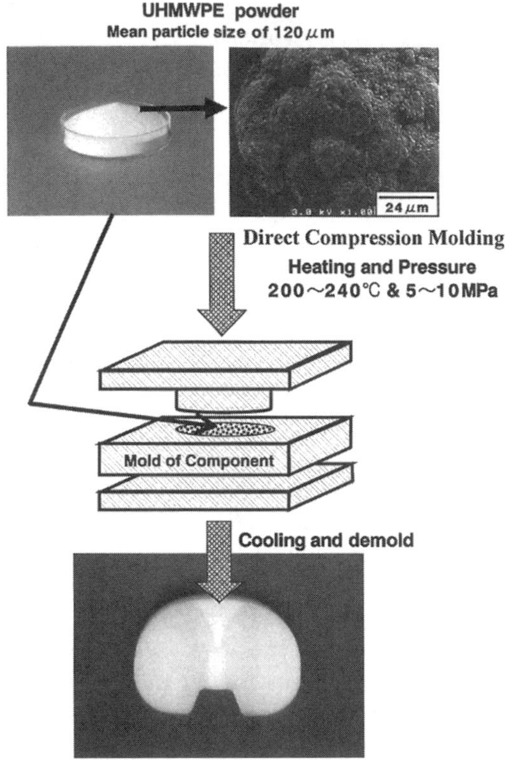

Figure 21.15 Manufacturing process of the flow of direct compression molding for UHMWPE knee implants.

a. *Ram Extrusion*

Ram extrusion of UHMWPE is essentially a continuous compression. A ram extruder consists of a hydraulic ram and a heated die. Rods are produced by feeding UHMWPE powder into an electrically heated die (from 5 to 15 cm in diameter, 180–220°C), in which a plunger driven by an oil hydraulic mechanisms is moved backward and forwards for compression (8,9). Following ram extrusion, the rods are annealed at a temperature slightly above the melting temperature of UHMWPE in order to improve the material dimensional stability. The implant is then machined from this bar stock. Over 90% of all the UHMWPE fabricated forms for surgical implants are prepared by this method.

b. *Compression Molding*

Compression molding is used to manufacture sheets or blocks. A metal mold, made of steel, stainless steel, or aluminum, is filled with UHMWPE powder. The mold is then inserted into a press that has heating and cooling capabilities. Heating at temperatures of 200 to 230°C is continued under pressures of 2 to 10 MPa until the powder has plasticized completely. After plastization is completed, the hot mold is cooled gradually at a controlled rate. The sintered UHMWPE sheet or block is demolded. The implant is then machined from the sheet or block.

c. *Direct Compression Molding*

UHMWPE can also be directly compression molded into final inserts. Direct compression molding results in a sintered insert that is fully consolidated and exhibits largely uniform properties. A metal mold designed for surgical implants is filled with UHMWPE powder. The powder is then heated under pressure until the powder has plasticized completely. After plastization is completed, the hot mold is cooled gradually at a controlled rate. The sintered implants are demolded.

In direct molding, the UHMWPE powder is usually heated at temperatures of 200 to 240°C under pressure from 5 to 10 MPa (Fig. 16). There are manufacturing defects such as fusion defects, macro voids, and grain boundary in sintered implants that were fabricated under insufficiently plastisized conditions such as inadequate temperature, pressure, or time (Fig. 17). The defects are formed by incomplete particle consolidation between some adjacent powder particles. The scanning electron micro-

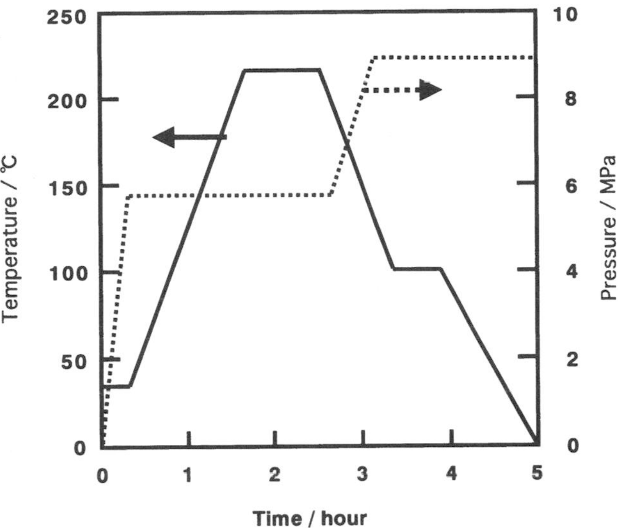

Figure 21.16 Typical processing conditions of direct compression molding.

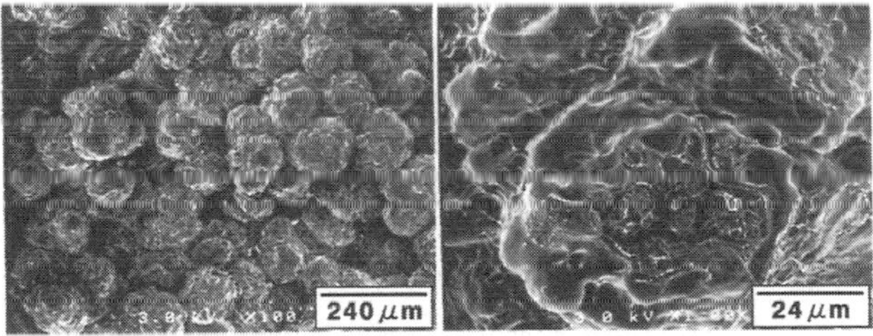

Direct compression molding
Condition
Heating and Pressure
220°C & <1kg·cm⁻²

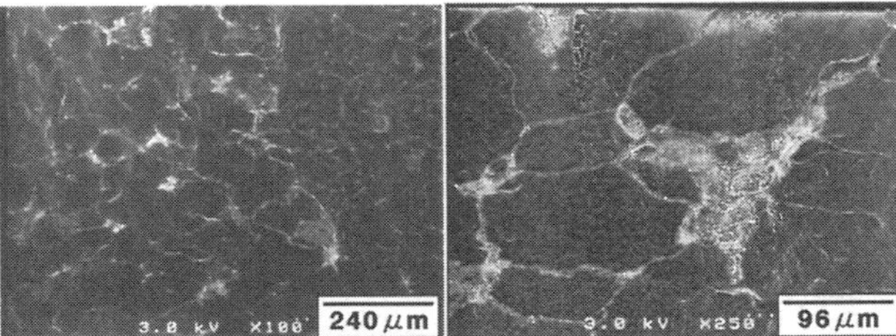

Direct compression molding
Condition
Heating and Pressure
220°C & <10kg·cm⁻²

Figure 21.17 Scanning electron micrograph of a fractured surface of UHMWPE plate fabricated by direct compression molding under insufficient condition.

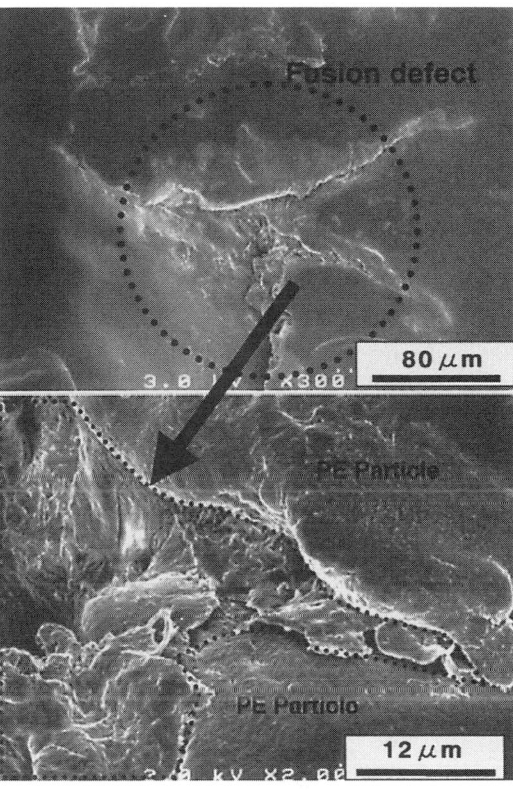

Figure 21.18 Scanning electron micrograph of fractured surface of UHMWPE knee implant fabricated by ram extrusion.

graph of the fracture surface of an unused UHMWPE knee implant produced by a ram extruder is shown in Fig. 18. Several fusion defects, other than rigid surfaces that are fully consolidated between adjacent powder particles, are observed in the implant. The defects may lower the mechanical properties of UHMWPE implants and make them more susceptible to fatigue damage (14,15). Figure 19 shows a photograph of a retrieved knee implant. Damage

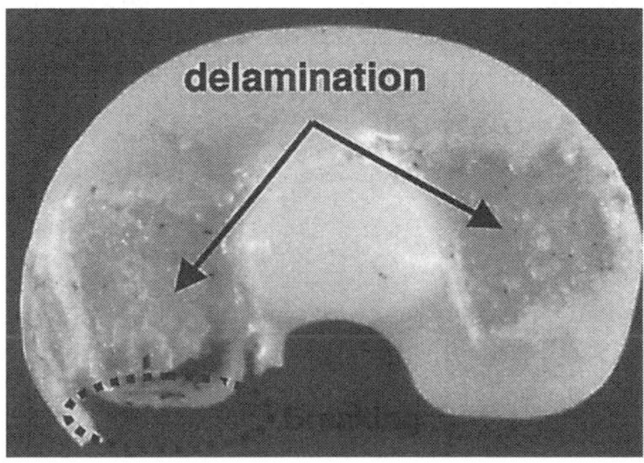

Figure 21.19 Photograph of a retrieved knee implant.

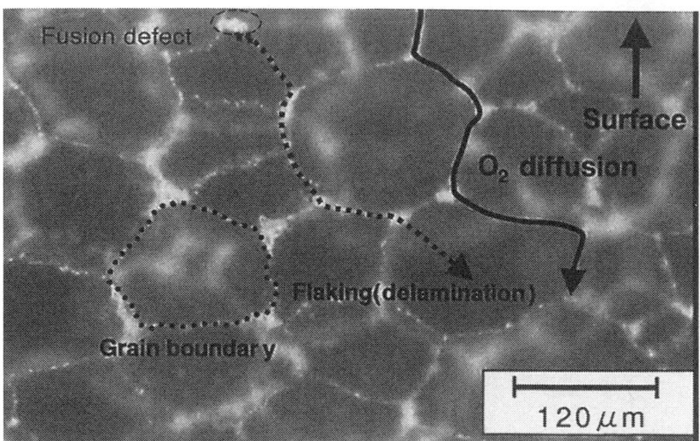

Figure 21.20 Photograph of a microtomized thin section of a UHMWPE implant.

Molecular weight; 100 millon
Calcium stearate; 0.12%

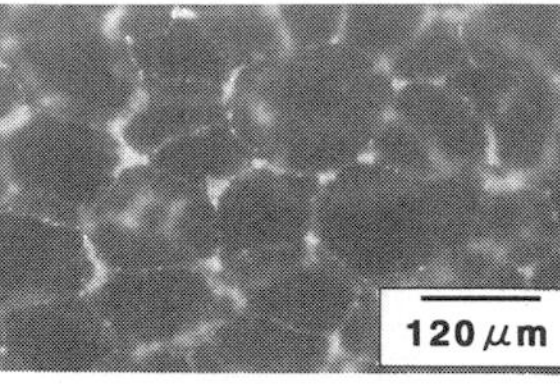

Molecular weight; 100 millon
Calcium stearate; 1.00%

Molecular weight; 600 millon
Calcium stearate; 0.12%

Molecular weight; 600 millon
Calcium stearate; $<0.05\%$

Figure 21.21 Photograph of a microtomized thin section of a plate fabricated by direct compression molding using various grades of UHMPWE powder.

such as a delamination and breaking is likely due to fatigue correlated by material processing and oxidation of the UHMWPE implants. The defects may become a diffusion path for oxidants such as oxygen in the air or hydrogen peroxide in body fluids, causing oxidation of UHMWPE (Fig. 20).

d. Calcium Stearate

Calcium stearate is added to UHMWPE for neutralization of the catalyst, improvement of melt flow, and reduction of wear and tear of the tool dies used to shape the final implants. The grade containing calcium stearate is standard for industrial use. However, it has been reported that fusion defects are associated with the presence of calcium stearate, and that use of UHMWPE source resin, which was free of calcium stearate results in improved material consolidation and wear resistance (16,17,18). Figure 21 shows the optical micrographs of microtoming thin sections of a plate prepared by direct compression molding using UHMWPE source resin of different calcium stearate contents. The visible grain boundaries in the plate become obvious with an increase in calcium stearate content. In spite of the high molecular weight, the plate prepared with calcium-stearate-free resin has no visible grain boundaries or fusion defects.

e. Sterilization

Sterilization is performed after packaging. There are several sterilization methods for surgical implants, such as dry heat, steam under pressure, steam–formaldehyde, ethylene oxide (EtO) or other chemicals, and irradiation (^{60}Co, ^{137}Cs, accelerated electron). Some of these procedures including steam autoclave are not recommended because the high temperature may affect UHMWPE properties and product dimensions. Currently, γ-ray irradiation from a cobalt source and EtO sterilization are commonly used as the sterilization methods for UHMWPE. The FDA requires that a minimum dose of 2.5 Mrad be used for γ-ray sterilization. The sterilization of EtO does not affect the mechanical properties or wear resistance of UHMWPE implants. γ-ray sterilization, in the presence of oxygen, can result in property changes in UHMWPE implants. γ-ray energy that destroys microbes on the implant surfaces also causes chemical reactions, such as oxidative chain scission, recombination, and cross-linking, to occur within the polyethylene (15). It is reported by many researchers that the oxidation of UHMWPE implants from γ-ray sterilization, in the presence of oxygen, may reduce the mechanical properties and wear resistance of UHMWPE implants (8,20,21).

f. Oxidation

It has been reported that UHMWPE oxidizes after γ-ray sterilization and that the process of oxidation continues after implantation (8,19,21–27,29–31). The γ-ray energy causes a chemical reaction from the breaking of carbon–hydrogen and carbon–carbon bonds, which results in the production of free radicals in the polyethylene molecule. These free radicals react with oxidants, such as oxygen in the air or hydrogen peroxide in body fluids. The free radicals also undergo reactions of recombination and cross-

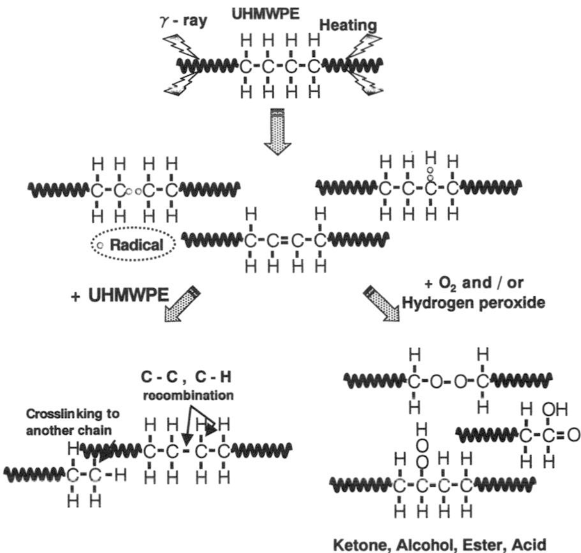

Figure 21.22 Predictive model of UHMWPE oxidation.

linking to other polyethylene molecules (19,27) (Fig. 22). During heating at temperatures exceeding the crystalline melting point of UHMWPE, particle fusion and crosslinking to other chains and chain scission occur simultaneously. These chemical cause changes in the physical properties of the UHMWPE implant.

Oxidation of polyethylene is a chemical attack that leads to a combination of chain scission and the introduction of oxygen containing groups, such as keton, alcohol, ester, and hydroxylic acid, into the molecules. Chain scissions, which take place during oxidation, lead to a relevant reduction of molecular weight. When the original UHMWPE has a molecular weight of 4 million, five chain scissions depress the molecular weight to 70,000, which is the typical molecular weight of regular polyethylene (HDPE, LDPE, and LLDPE). Figure 23 shows optical micrography of a UHMWPE knee component. The knee component was fabricated by ram extrusion, sterilized by γ-ray irradiation in air, and then stored for 7 years. A white area, a so called white band, can be seen along the edges of a section about 0.5 mm in depth and 2 mm in width. The chemical results of oxidation are the formation of carbon–carbon double bonds and the introduction of oxygen-containing groups such as ketone, ester, or hydroxylic acid into the polymer. Fourier transform infrared spectroscopy (FT-IR) has been shown to be a convenient and effective method for the direct detection and determination of relative amounts of these types of chemicals (30) (Fig. 24). The IR peaks of oxygen-containing groups appear in the range of 1600 cm^{-1} to 1800 cm^{-1} (19,23). The integrated peak area in this range is proportional to the amount of oxidized UHMWPE and can be used to quantify the extent of oxidation. The oxidation index—the oxidized peak area is divided by the peak area of 1464 cm^{-1} assigned to the CH_2 group—is often used (10) in order to have a valid comparison from sample to sample. Figure 25 shows a typical oxidation profile for a knee implant fabricated by ram extrusion after γ-ray sterilization in air and shelf aging for 7 years (30). The oxidation index shows a maximum at the subsurface area, which corresponds to the depth of the white band area seen in the section of oxidized UHMWPE implant. UHMWPE implants sterilized with γ-irradiation are known to undergo oxidative degradation, both while being stored on the shelf and after implantation. The oxidative degradation of implants results in both changes in chemical properties and changes in physical properties such as density, mechanical properties, and wear resistance. Oxidation would be ex-

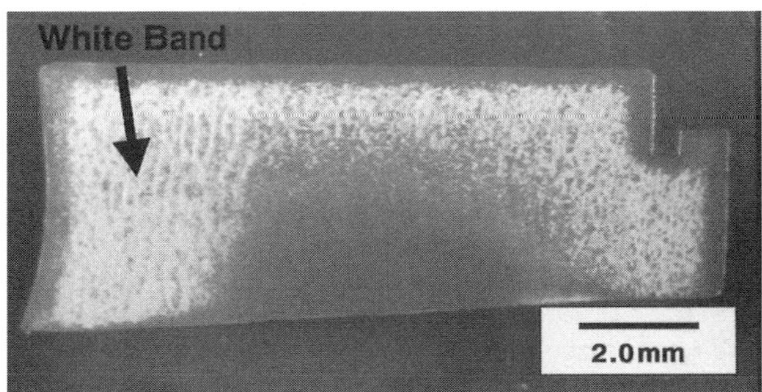

Figure 21.23 Photograph of a microtomized thin section of a UHMWPE knee implant fabricated by ram extrusion and sterilized in air.

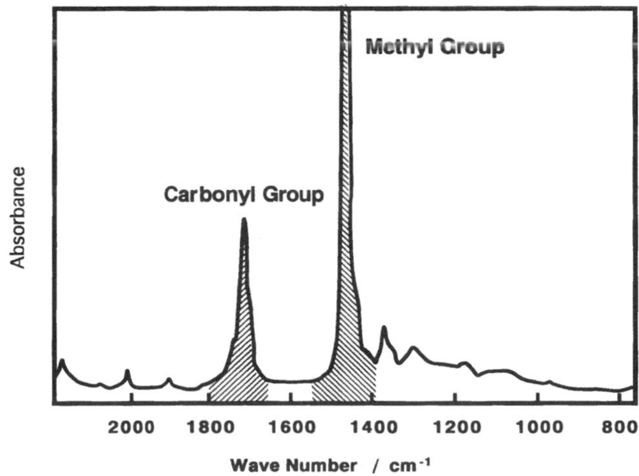

Figure 21.24 IR spectra of a microtomized thin section of a UHMWPE knee implant fabricated by ram extrusion and sterilized in air.

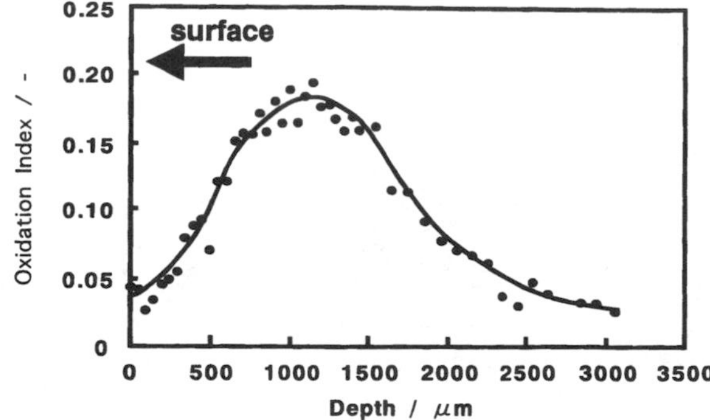

Figure 21.25 Depth profile of oxidation index for a UHMWPE knee implant fabricated by ram extrusion and sterilized in air.

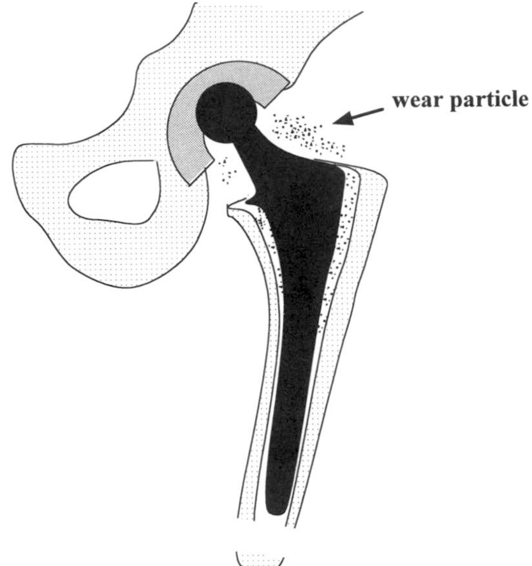

Figure 21.26 Schematic illustration of an artificial hip joint. Wear particle produced at sliding surface accelerates the loosening.

pected to increase the modulus and brittleness of polyethylene. Increases in density near surfaces (subsurfaces) of an implant cause major increases in the stress associated with wear damage (delamination) that occurs on and near the surfaces. The oxidative degradation of implants may lower the mechanical properties of UHMWPE and make them more susceptible to fatigue damage owing to chain scission, effectively turning them into regular polyethylene. In order to prevent the oxidative degradation, most manufacturers currently use γ-irradiation in a vacuum or inert gas to sterilize their products. The reduction of oxidative degradation leads to a minimum amount of chain scission and an increase in cross-linking, which preserves the mechanical properties associated with wear resistance.

4. Wear Properties of UHMWPE

Medical grade ultra-high molecular weight polyethylene (UHMWPE) is the current material of choice for use as a bearing surface in artificial joints because it offers high toughness, a low friction coefficient, and low wear in the body. However, UHMWPE also remains as the limiting factor for the longevity of joint replacement. Artificial hip joints do not generally wear out, but UHMWPE wear particles generated at the articulating surfaces can enter the tissues surrounding the prosthesis and cause adverse cellular reactions as shown in Fig. 26. Since loosening of the stem is thought to be accelerated by the cellular reactions, tribological performance is one of the most important characteristics of UHMWPE.

Figure 27 shows a typical classification of polyethylene wear for artificial joints proposed by the Leeds University group (32). Microscopic wear of UHMWPE is caused by abrasion against hard material, such as the metal surface of a joint, and is associated with very small asperities or smooth counterfaces (less than 0.2 micrometers). The mi-

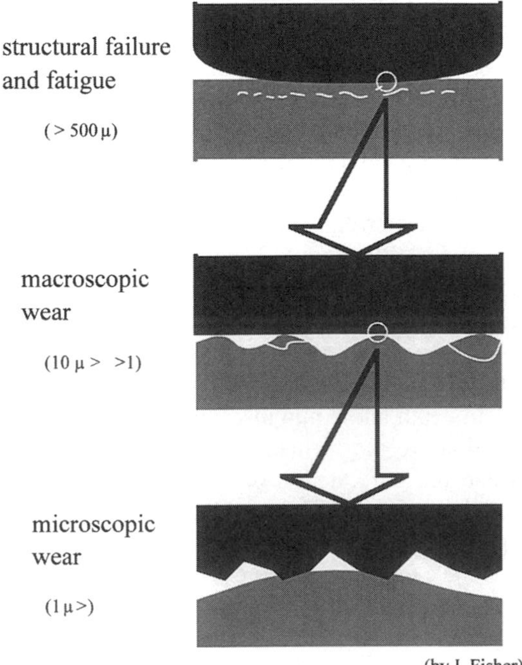

Figure 21.27 Classification of wear for artificial joints.

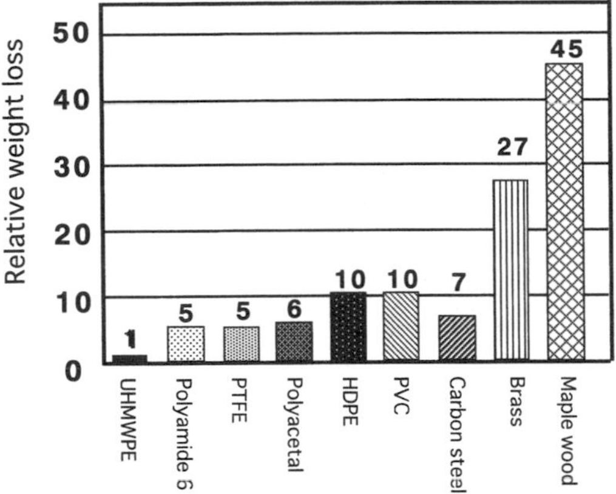

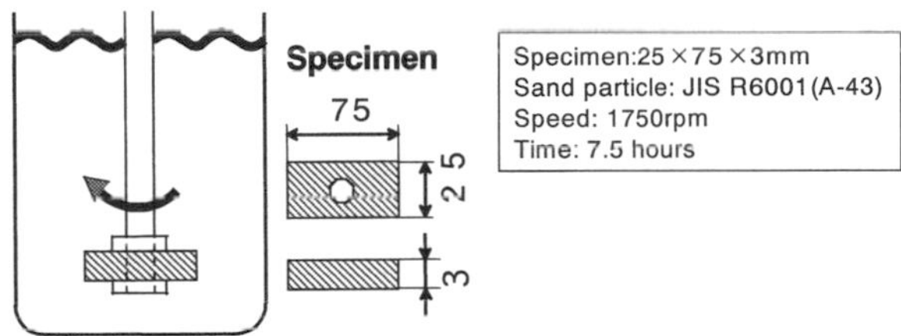

Specimen
75
25
3
Specimen:25×75×3mm
Sand particle: JIS R6001(A-43)
Speed: 1750rpm
Time: 7.5 hours
Relative weight loss = loss weight of materials / loss weight of UHMWPE

Figure 21.28 Comparison of abrasion properties between UHMWPE and other materials.

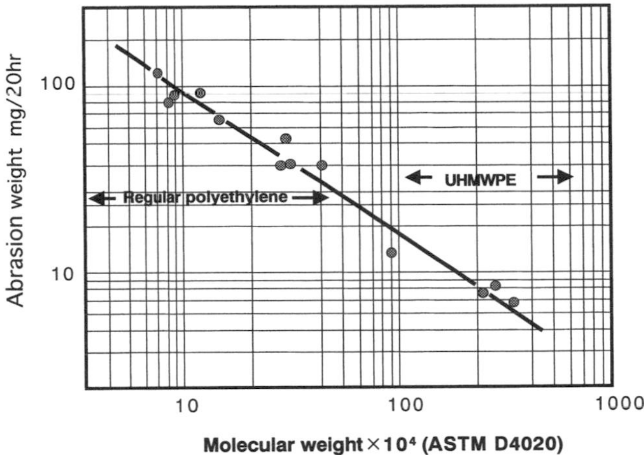

Figure 21.29 Relationship between molecular weight of polyethylene and abrasion weight loss.

croscopic wear is characterized by the formation of particles with diameters smaller than 1 micron. Macroscopic polymer asperity wear processes are associated with stress concentrations under the much larger peaks in the UHMWPE surface (amplitude less than 10 micrometers). Structural failure and fatigue is called delamination wear. This failure is initiated by high stress at the subsurface of UHMWPE, followed by sudden explosive destruction and the formation of particles with diameters greater than 0.5 mm.

Generally, microscopic wear corresponds to abrasive wear and macroscopic wear corresponds to adhesive wear. UHMWPE shows high resistance against abrasive wear as shown in Fig. 28, where the sand slurry abrasive wear test for various materials was performed. The abrasion resistance of UHMWPE is the highest of the various materials (2,5). The abrasive wear resistance is seen to increase with linearly increasing molecular weight as shown in Fig. 29. It is reported that the abrasive wear rate for a variety of polymers increased with the inverse of the product of the ultimate tensile strength and the tensile elongation at break and toughness.

The mechanism for the macroscopic wear seen in the body has not been clarified yet. UHMWPE and polytetrafluoroethylene demonstrated very low coefficients of friction as shown in Table 3 (5). The low coefficient of friction can be explained by transfer film lubrication: a very thin film of polymer is transferred to the opposing surface and this leads to the resultant coefficient of friction being very low. However, the formation of the transfer film cannot be observed in the body or in wear testing using bovine serum as lubricant.

There has been no simple test device that would yield wear rates and wear mechanisms similar to those found in retrieved prosthetic joint components. The wear factors obtained for UHMWPE against polished Co-Cr alloy in serum are of the order of 10–8 mm^3/Nm or almost negligible by pin-on-disk testing in which a polymer pin slides against a unidirectionally rotating or reciprocating plate. However, the clinical wear factors measured from retrieved acetabular cups of total hip prostheses are of the order of 10–6 mm^3/Nm or more. Some of the joint simulator testing yields wear rates and mechanisms similar to those in clinical studies but were expensive and had poor reproducability.

A simple testing method is required for wear and friction studies of prosthetic joint material. Recently, the effect

Table 21.3 Coefficient of Friction of Various Polymers (ASTM1894)

Polymer	Coefficient of kinetic friction
Polycarbonate	0.36
Polypropylene	0.38
Polyamide 6	0.36
Polyacetal	0.18
HDPE	0.12
Polytetrafluoroethylene	0.04–0.10
UHMW-PE	0.10–0.22

of multidirectional motion on the wear behavior of UHMWPE has been reported by several groups. Besong et al. reported that the number of wear particles produced in multidirectional wear testing was 26 times higher than that in unidirectional testing. Some of them advocated the importance of molecular orientation at the surface layer. Sambasivan et al. reported that unidirectionaly rubbed UHMWPE shows higher average molecular orientation than in cross-shearing rubbed UHWPE. Although the mechanisms for the effect of multidirectional sliding have not been clarified, multidirectional testing may open up a road to evaluating wear performance in the body.

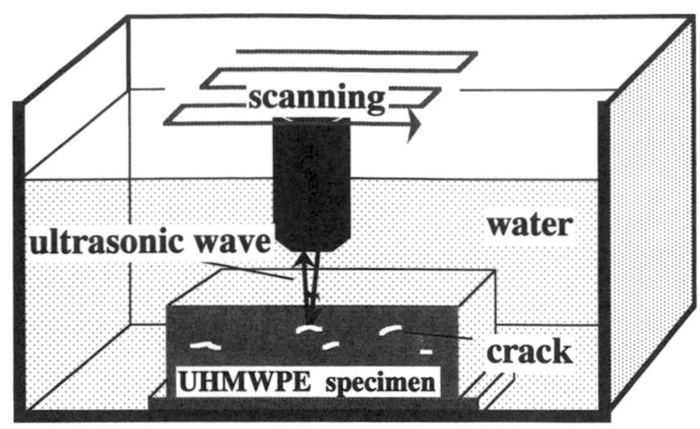

Figure 21.31 Schematic illustration of the scanning acoustic tomography (SAT) system 6.

5. Fatigue Properties of UHMWPE

Flaking type wear, so-called delamination, is often observed in polyethylene knee components as shown in Fig. 19. This wear can be classified as structural wear (32) caused by structural failure and fatigue. This failure will be initiated by high stress at the subsurface of the UHMWPE, followed by sudden explosive destruction accompanying the formation of particles with diameters greater than 0.5 mm as shown in Fig. 27. This destruction is thought to occur due to crack or defect formation and propagation and is accelerated by oxidation at the grain boundaries. However, this fatigue process has not yet been studied in detail, because the UHMWPE components have a complicated structure along with defects such as grain boundary and crystalline amorphous phase interface as shown in Fig. 30. Moreover, the methods available for fatigue study are quite limited, and detailed investigation of subsurface crack formation and propagation is obstructed by the difficulty of cleanly cutting UHMWPE products to produce a specimen suitable for outer surface and subsurface study. Scanning acoustic tomography (SAT, Fig. 31) is one possible method for estimating the crack or defect formation and propagation in UHMWPE (33–35). Figure 32 shows a photograph

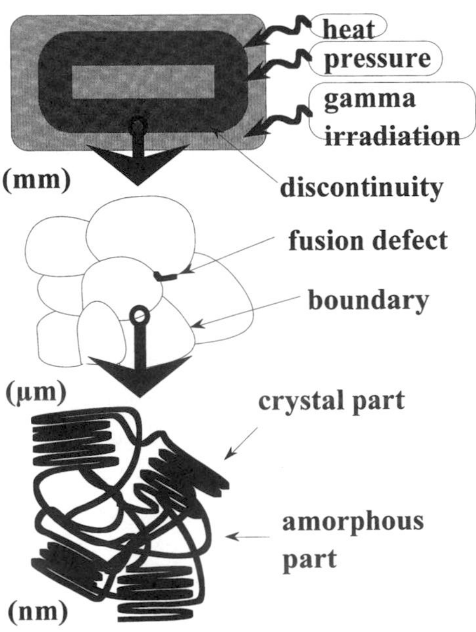

Figure 21.30 Structure of UHMWPE.

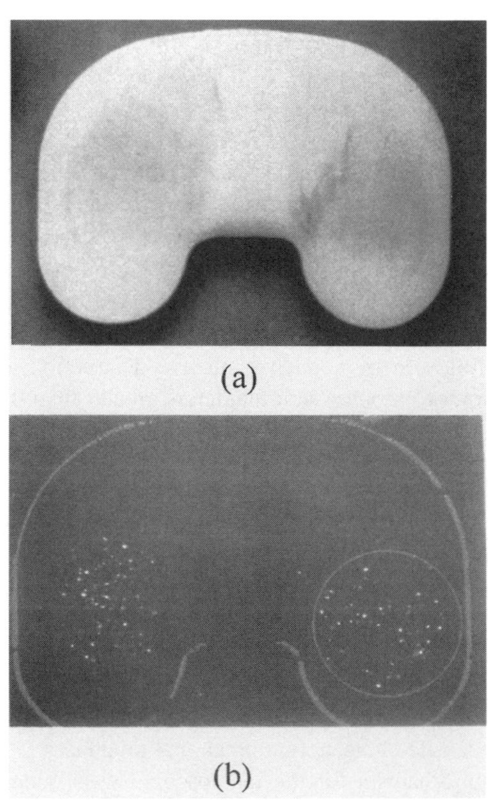

Figure 21.32 Photograph (a) and SAT image (b) of retrieved UHMWPE knee component.

and SAT image of a tibial component that has neither apparent severe wear nor flaking on the surface. However, it is obvious that subsurface cracks or defects are represented in the SAT image as dots at the loaded site. The accumulated area of subsurface cracks or defects evaluated by SAT is shown in Fig. 33, where the delamination destruction was reproduced by sliding fatigue testing. Fatigue seems to develop slowly during the early and middle stages after implantation, followed by sudden explosive destruction. A cross-sectional view of the UHMWPE sample showed some amount of looseness at the grain boundary of UHMWPE.

Numerous studies have reported a significant influence of gamma-irradiation on the UHMWPE delamination (36–43). Sun et al. (44) found that the maximal subsurface oxidation level of gamma-sterilized UHMWPE components peaked at a depth ranging from 0.2 to 0.8 mm below the surface. This depth is almost identical to where maximal residual strain was stored, found by analysis using a constitutive equation for cyclic plasticity (45,46). Probably, oxidation caused by gamma-irradiation has accelerated the fatigue process of polyethylene. Indeed, our study on explants revealed that flaking-like destruction or apparent propagation of subsurface cracks occurred in UHMWPE components sterilized by gamma-irradiation.

Figures 34–36 (47) shows results obtained by a two-dimensional sliding fatigue testing machine, where UHMWPE specimens were fixed on a computer-controlled X–Y table and loaded by a round rod. The end of the rod

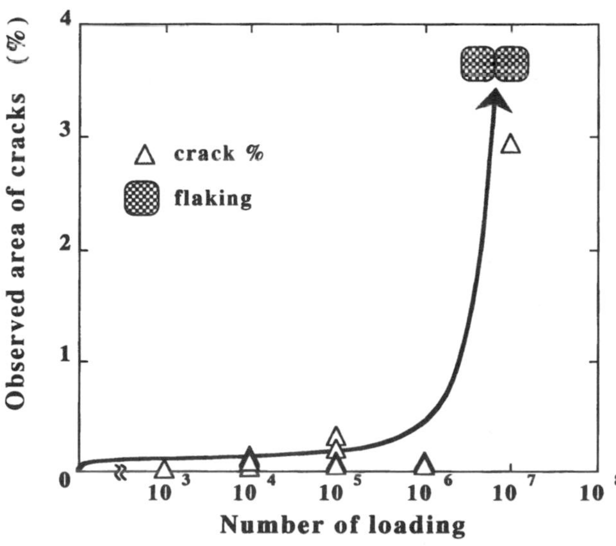

Figure 21.33 Cracks observed in UHMWPE loaded with Si3N4 balls (r = 5 mm) sliding on the specimen (load = 1000 N).

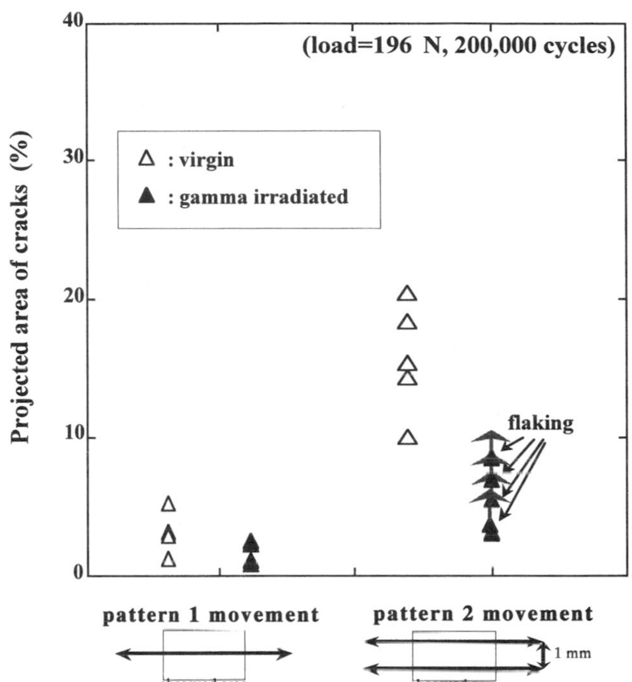

Figure 21.34 Cracks observed in UHMWPE loaded by a rounded rod (r = 3 mm) moving two different locis.

was made into a half ball with a curvature radius of 3 mm polished with cloth into a surface roughness value (Ra) of 0.3–0.4 μm. A pressure of 196 N was applied to the specimen through the rod. Twenty minutes was adapted as the stationary period before the start of the sliding movement. The end of the rod was sunk into the specimen during the 20-minute stationary period. Two simple patterns of sliding movement were adopted; simple reciprocating (pattern 1) or switched reciprocating (pattern 2), movement which consisted of two parallel lines at a distance of 1 mm and was switched at every other cycle of the reciprocating movement. Surrounding water in the chamber was maintained at 37°C, and the number of loadings in the observed area was 200,000 cycles. Subsurface cracks that formed in the middle portion of the loaded area were measured by SAT.

Figure 34 clearly demonstrates that complicated sliding motion (pattern 2) accelerates crack formation and/or propagation below the surface. This result suggests that the locus of the sliding movement is an important factor. It is likely that multidirectional fatigue loading causes tensile strain at grain boundaries to accelerate the fatigue process. Gamma-irradiated UHMWPE specimens demonstrated lower rates of crack formation than unirradiated specimens under pattern 1 movement as shown in Fig. 35, whereas flaking-like destruction was observed in the gamma-irradi-

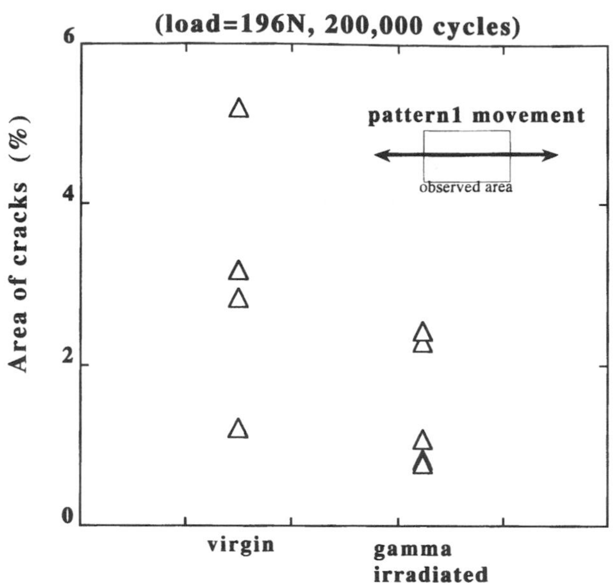

Figure 21.35 Cracks observed in UHMWPE loaded by a rounded rod (r = 3 mm) moving on pattern 1 locus.

ated UHMWPE under pattern 2 movement as shown in Fig. 36. One out of five gamma-irradiated UHMWPE specimens exhibited relatively low crack area, suggesting that the flaking-like destruction would occur suddenly from a low-cracked state. These results suggest that the gamma

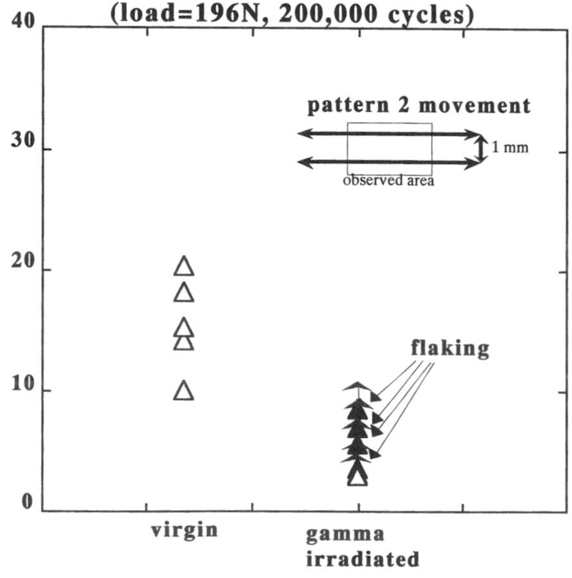

Figure 21.36 Cracks observed in UHMWPE loaded by a rounded rod (r = 3 mm) moving on pattern 2 locus.

irradiation reduces crack formation but increases crack propagation under multidirectional sliding.

6. New Processing of UHMWPE

The aim of many researchers in high performance implants has been to alter the properties of UHMWPE by changing its morphological characteristics by various manufacturing processes. It is reported that crystallinity in polyethylene can be increased without causing degradation by using temperatures greater than 250°C and pressure greater than 2,800 atmospheres (8,31,49). It is also reported that UHMWPE with crystallinity over 80% can be prepared by recrystallization of solid UHMWPE with crystallinity of 50% without degradation of the original material.

a. Cross-Linking

UHMWPE is cross-linked by low-dose (2–5 Mrads) γ-ray sterilization in a vacuum or inert gas and then stored in an oxygen-free environment or heat-treated at a temperature below the melting point in a vacuum or inert gas owing to the inactivation of free radicals (50,51). UHMWPE sterilized by this favorable cross-linking has a low density, low crystallinity, and a higher insoluble fraction than UHMWPE that is γ-sterilized in air. It is reported that the increase in the favorable cross-linking of UHMWPE results in a significant improvement in wear resistance as compared to both γ-ray sterilization in air and EtO sterilization; hip simulator tests indicated a 49% decrease in wear rate for the samples irradiated in a vacuum relative to samples irradiated in air (Fig. 37), and a 36% decrease in wear rate

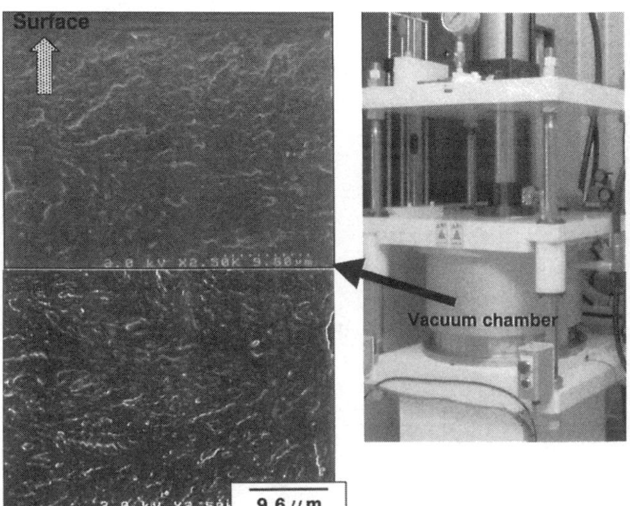

Figure 21.37 Direct compression molding in vacuum.

for samples irradiated in a vacuum relative to samples sterilized using EtO (52). The UHMWPE implants treated in a vacuum have the greatest wear resistance, presumably due to the increased cross-linking in the polyethylene.

UHMWPE is cross-linked by low-dose γ-ray irradiation (~2.5 Mrad) in a vacuum and then directly compression-molded into final inserts at a temperature of 200°C under pressure of 20MPa owing to orientation of the polyethylene chains and inactivation of free radicals. The wear factor of the samples significantly decreased with an increase in the degree of orientation of the UHMWPE (53). This suggests that the wear resistance of implants is improved by the cross-linked orientation of UHMWPE.

Many kinds of cross-linked UHMWPE have been developed and used in clinical applications (54–56).

b. *Vitamin E Addition*

The effect of vitamin E on crack formation and/or propagation in UHMWPE was evaluated using two-dimensional sliding fatigue testing and microindenter testing. Figure 38 (47) shows the effect of vitamin E on the delamination destruction. All the irradiated vitamin E–containing specimens did not demonstrate flaking-like destruction. Both crack formation and flaking-like destruction were significantly prevented by adding vitamin E. The amount of vitamin E at 0.1 and 0.3% demonstrated no apparent difference in this experiment. The effect of vitamin E on the delamination destruction is thought to be caused by many factors. However, it is possible that the presence of vitamin E prevents crack propagation in part due to reduced hardness at grain boundaries. Results of the microindentation testing are shown in Fig. 39 (47), where dynamic hardness at the grain boundary was higher than in the grain and was increased by gamma-irradiation. This hardening at the grain boundary was reduced by adding vitamin E in UHMWPE. The gamma-irradiated vitamin E–containing UHMWPE is a promising material to prevent flaking-like destruction of polyethylene joint components.

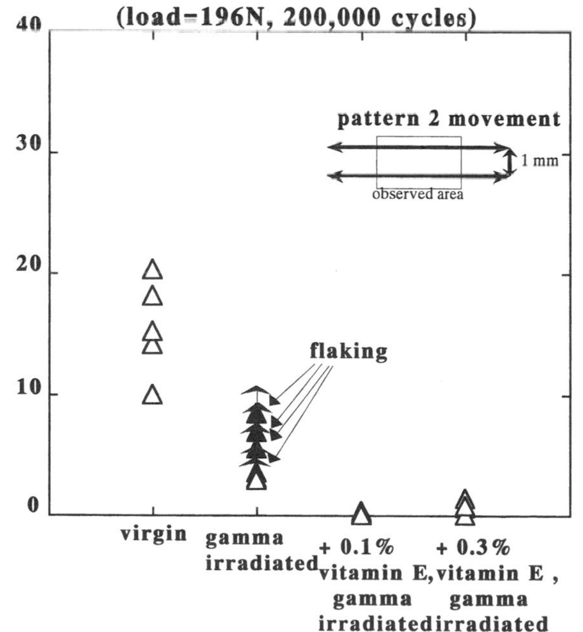

Figure 21.38 Cracks observed in vitamin E added UHMWPE loaded by a rounded rod (r = 3 mm) moving on pattern 2 locus.

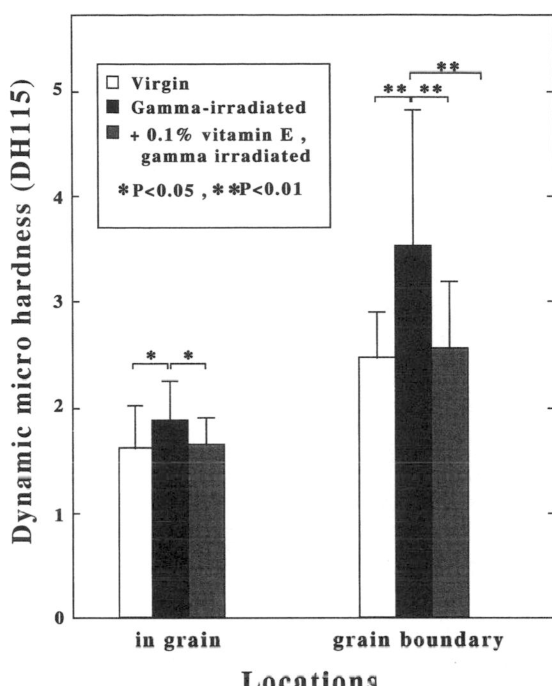

Figure 21.39 Dynamic hardness at the surface of a UHMWPE specimen.

II. BONE CEMENT

Surgeons must thoroughly understand the background of self-curing acrylic cement to appreciate the mechanism of fixation and the biological interaction between bone and cement and to deal with problems arising from defective fixation of a total hip prosthesis. Space limits a complete account of self-curing cement; the reader is strongly urged to review Charnley's original report and the appropriate references (57,58) at the end of this chapter. The goal of this chapter is to review the general aspect of the available acrylic cements as well as modern cementing techniques. Our new challenges of improving the bone cement and intramedullary plug are also discussed.

A. History of Bone Cement

The successful use of acrylic cement for fixation of total hip implants was one of Sir John Charnley's major contributions. At the suggestion of Dr. Dennis Smith of the materials laboratory of the Manchester University Dental School, he began to work with polymethyl methacrylate (PMMA), which was being used for the construction of temporary dentures.

The industrial form of polymethyl methacrylate is familiar as Plexiglas or Lucite. The material was developed in the years before and during World War II when it could be substituted for glass as a clear material. It was originally used as a denture base material in 1937 and it subsequently found use in a wide variety of medical applications, especially for cranioplasties. In 1946, the Judet hip was introduced with a heat-cured acrylic substitute for the femoral head (59). Although a number of applications were found in orthopedics, Charnley was the first to utilize the material in a self-curing form for fixation of an intramedullary stem in 1958 (60).

Our current understanding of acrylic bone cement stems from much of the original investigative work done by Charnley and other workers in the 1960s and 1970s when bone cement was first introduced. Charnley (61) first identified such issues as the porosity of bone cement and the effect of air and blood inclusions. Building on the extensive experience with PMMA in dental biomaterials, the essential properties and limitations of the material were identified as applied to orthopedics. After over 40 years of clinical use, acrylic cement is still the most frequently used method and material for the fixation of total joint prostheses.

B. Chemical Composition

Polymethyl methacrylate is one of a group of widely known polymers such as polyethylene, polyvinyl chloride (PVC), and polystyrene. These are all repeating carbon-based units of varying chain length. The methyl methacrylate molecule, which has a molecular weight of 100, is seen in Fig. 40. The polymer consists of repeating units of the basic molecule with the double carbon bond broken and replaced with single carbon bonds to the adjacent subunits. The ultimate chain length of the polymer is dependent on the manufacturing process as well as the initiators and activators that cause the reaction to occur when the mixture is prepared in surgery. A longer ultimate chain length will be reflected in a higher molecular weight and increased material properties and will cause a greater degree of viscosity during the mixing and curing process. While the PMMA used in orthopedic surgery is an example of a single-link polymer chain, much stronger materials are available in dentistry, such as bisphenol-A-glycidyl methacrylate (BIS-GMA), where significant cross-linking occurs between the subunits of different chains (62).

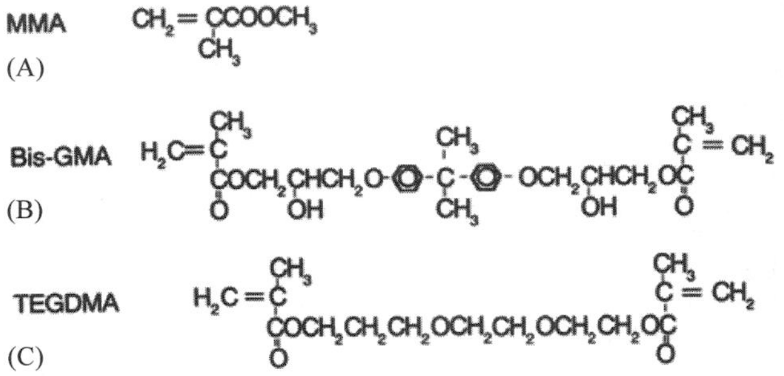

MMA $CH_2=C(CH_3)COOCH_3$
(A)

Bis-GMA $H_2C=C(CH_3)COCH_2CHCH_2O-\phi-C(CH_3)_2-\phi-OCH_2CHCH_2OC-C(CH_3)=CH_2$ (with OH on each CH, =O on each carbonyl C)
(B)

TEGDMA $H_2C=C(CH_3)COCH_2CH_2CH_2OCH_2CH_2OCH_2CH_2OC-C(CH_3)=CH_2$ (with =O on each carbonyl C)
(C)

Figure 21.40 Chemical formula of acrylic bone cement. (a) methyl methacrylate, (b) bisphenol-A-glycidyl dimethacrylate (bis-GMA), and (c) triethylene-glycol dimethacrylate (TEGDMA).

The cements used in orthopedic surgery rely on combining prepolymerized solid particles (powder) and the liquid monomer (liquid). A process of suspension polymerization that allows the manufacturer to control both particle size and the molecular weight or mean polymer chain length of the polymer produces the prepolymerized powder. The powder particles are generally spheres from 30 to 150 μm in diameter. There is a wide distribution of the molecular weight in a range from 20,000 to 2 million with a mean weight of 172,000 g/mol (63).

Whereas the methyl methacrylate monomer will polymerize on its own if exposed to light or heat, the self-curing process relies on chemical agents for the polymerization. The liquid monomer also generally contains the activator *N,N*-dimethyl-*p*-toluidine (DMPT). For the reaction to occur, the powder component of prepolymerized PMMA beads needs to contain an initiator, which is most commonly dibenzoyl peroxide (BP). To prevent the monomer from polymerizing, the liquid generally contains an inhibitor or retardant, hydroquinone, which functions by absorbing any free radicals that may occur and causing the breakup of the double carbon bonds and initiating the polymerization.

When the liquid monomer and prepolymerized powder are mixed together, a chemical reaction begins with the activator (DMPT) and the initiator (BP) combining and releasing a benzoyl peroxide free radical, which reacts with the methacrylate monomer molecule to begin the polymerization process. Chains are then formed of repeating subunits in the range of 20–20,000 (63). As the chains form with the double carbon bonds being converted to single carbon bonds, heat is generated as an exotherm of 130 calories/g of liquid monomer, which further accelerates the reaction as the cement cures into a hard mass.

In addition to these basic components, many of the cements also contain copolymers (polymers derived from more than one type of monomer) that can influence the mixing and handling characteristics of the cement. Simplex P cement, for example, includes a styrene copolymer in its composition. All of the cements also have an opacifier, usually barium sulfate ($BaSO_4$) or zirconium oxide (ZrO_2).

The exact composition and relative proportions of the components of the individual orthopedic bone cements influence both the final physical properties and the handling and flow characteristics of the cement during its preparation and application. Variations in the composition as well as the mixing method will affect total set time, amount of heat generated, expansion and shrinkage during the curing, the flow characteristics over time (liquid phase through dough phase), and the degree of polymerization or amount of residual monomer that may be present after the reaction has finished.

C. Problems of PMMA Bone Cement

Since the introduction of polymethylmethacrylate (PMMA) bone cement (60), PMMA bone cement is considered to have become one of the most effective means for the fixation of hip prostheses. Aseptic loosening, however, remains the most serious long-term drawback on a multifactorial basis. Biochemical (64–66) and mechanical (67, 68) factors are considered to be responsible for the loosening of hip prostheses. In the acetabular component, particulate polyethylene may invade the interfacial membrane between the bone and the cement, where phagocytosis by macrophages and foreign body giant cells leads to release of bone resorbing factors, resulting in osteolysis (66). In the femoral component, the mechanical (67,68) and biochemical factors mentioned above are thought to be responsible.

D. Bioactive Filler–Impregnated Bis-GMA-Based Resin

1. Introduction

To improve the mechanical properties of PMMA bone cement (69) and to avoid cement fracture, centrifugation (70) and pressurization of the cement (71) are now used. To improve the biochemical properties of PMMA bone cement, bioactive glass ceramic particles have been added (72). However, the mechanical properties of these composites have been shown to be inferior to the mechanical properties of PMMA bone cement, resulting in failure. Taken together, these problems warrant the development of a new bioactive bone cement (BABC) that has better mechanical and biochemical properties than those of PMMA bone cement. One important type of BABC is made by mixing bioactive powder with resin (72–76).

The apatite and wollastonite-containing glass ceramic (AW glass ceramic) developed at Kyoto University (77–80) is one of the most promising bioactive glass ceramic materials (81) and is now used clinically (82). The mother glass of AW glass ceramic (83) before crystallization is thought to have high bioactivity because of its glass phase. In 1993, Kawanabe et al. reported on a BABC consisting of $CaO\text{-}SiO_2\text{-}P_2O_5\text{-}CaF_2$ glass powder and a bisphenol-A-glycidyl dimethacrylate-based resin (Bis-GMA-based resin) (84). This new BABC showed direct contact with living bone through a calcium and phosphorous-rich layer and a higher mechanical strength than PMMA bone cement. To achieve better clinical results for fixation of orthopaedic implants, we have developed new BABCs consisting of $CaO\text{-}MgO\text{-}SiO_2\text{-}P_2O_5\text{-}CaF_2$ glass or a glass ceramic powder and Bis-GMA-based resin (85–88). These cements also showed direct contact with living bone through a calcium phosphorous rich layer and had improved mechanical strength.

However, these BABCs may demonstrate different characteristics when used for implant fixation and when subjected to greater mechanical stress. Senaha et al. (89) worked with dog femora and reported that a BABC consisting of bioactive glass powder and a Bis-GMA-based resin achieved good implant stability under weight-bearing conditions. Matsuda et al. (90) reported that a BABC consisting of bioactive glass powder and Bis-GMA-based resin showed a higher bonding strength than PMMA bone cement for up to 6 months after surgery when used for canine total hip arthroplasty (THA).

To achieve a higher mechanical strength and better handling properties, silica glass powder was added as a second filler to a BABC consisting of bioactive glass ceramic powder and Bis-GMA-based resin (91). We also reported that BABC with the same components showed a greater intrusion volume in 5 mm holes than PMMA bone cement when used for acetabular cup fixation in a simulated acetabular cavity (92). In this study, we performed canine THA using a BABC with the same components and followed up the outcome to 2 years (93). The outcomes were compared with the results of PMMA bone cement. Under load-bearing conditions, we subsequently compared the bonding strength and histological findings at the bone–cement interface produced by these two types of bone cement.

2. Materials and Methods

Total hip arthroplasties (THAs) were performed in beagle dogs using a bioactive bone cement (BABC) consisting of a silane-treated apatite and wollastonite-containing glass ceramic (AW glass ceramic) powder and a silica glass powder as the filling particles, and a bisphenol-A-glycidyl dimethacrylate-based resin (bis-GMA-based resin) as the organic matrix. Bis-GMA-based resin was prepared from equal weights of bis-GMA and triethylene-glycol dimethacrylate (TEGDMA) (Fig. 40).

BABC is composed of two pastes, type B paste and type T paste. Type B paste consists of fillers, bis-GMA-based resin, and benzoyl peroxide (BP) in a proportion of 1.2% per unit weight of the resin in type B paste. Type T paste consists of fillers, bis-GMA-based resin, and *N,N*-dimethyl-*p*-toluidine (DMPT) in the proportion of 1.1% per unit weight of the resin in type T paste, and phenothiazine, as an inhibitor of the polymerization reactions, at a proportion of 300 ppm of the resin in type T paste. We prepared two different compositions of BABC; the proportion by weight of the powder mixed into the cement was 85% for the acetabular side (dough type) and 79% for the femoral side (injection type), because the contemporary cemented THA uses the manual cementing technique for the acetabular side and the injection cementing technique for the femoral side. The viscosity of the dough type is high, which enables, manual handling of the cement, whereas that of the injection type is low, which enables injection into the femoral canal with a syringe. The two pastes (each weight is 40 g, 80 g in total) were packed and sterilized separately in ethylene oxide gas. In use, they were kneaded together for about one minute using a high-vacuum mixing system, Mixevac II (Stryker Co. Ltd., USA). Both cements polymerized within 8 minutes.

3. Results

The mechanical properties of the BABC were stronger than those of PMMA bone cement. The compressive strength of the two types of BABC was approximately 2.5 times stronger than the compressive strength of PMMA bone cement. The bending strength and the fracture toughness of both types of BABC were stronger than the bending strength of PMMA bone cement. The elastic modulus of the BABC was 3 to 5 times higher than the elastic modulus of PMMA bone cement.

All dogs were able to bear their body weight within 1 week and walked without a limp by 3 weeks. There was no postoperative hip dislocation. Two deep infections, one in one dog in the BABC group and one in one dog in the PMMA bone cement group, were shown on gross examination at the time of sacrifice at 24 months after implantation, so these two dogs were excluded from the present study. In the other 30 dogs, neither infection nor abnormal inflammatory reaction of the hip was shown on gross examination, and all the implants appeared to be fixed securely to the skeletons. Radiographs for both groups showed neither loosening of the hip prosthesis nor abnormal bone resorption around the cement mantle of the femur at 1 to 24 months (Fig. 41).

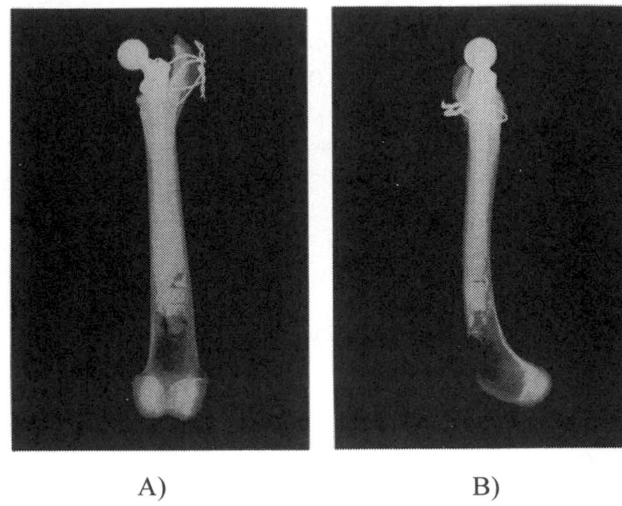

A) B)

Figure 21.41 Radiographs taken 24 months after implantation using BABC. (A) Anteroposterior view; (B) lateral view of the femur. The radiograph showed neither loosening of the hip prosthesis nor abnormal bone resorption around the cement mantle of the femur.

Histological examination showed direct bonding between the BABC and the femoral bone for up to 24 months after implantation (Fig. 42A). However, with PMMA bone cement, an intervening soft tissue layer was observed consistently at the bone–cement interface (Fig. 42B). Direct bonding at the interface between the BABC and the bone through a calcium phosphorous layer, 30 μm thick, was revealed by scanning electron microscopy (Fig. 43). Cement fractures of the BABC were observed on the acetabular side 24 months after implantation. The bonding strength of the BABC to bone in the dogs' femora increased with time and reached 3.7 MPa at 24 months after implantation, whereas that of PMMA bone cement was 2.0 MPa ($p < 0.05$).

4. Discussion

It is well known that oxygen is responsible for the formation of an uncured layer on the surface of a resin that has

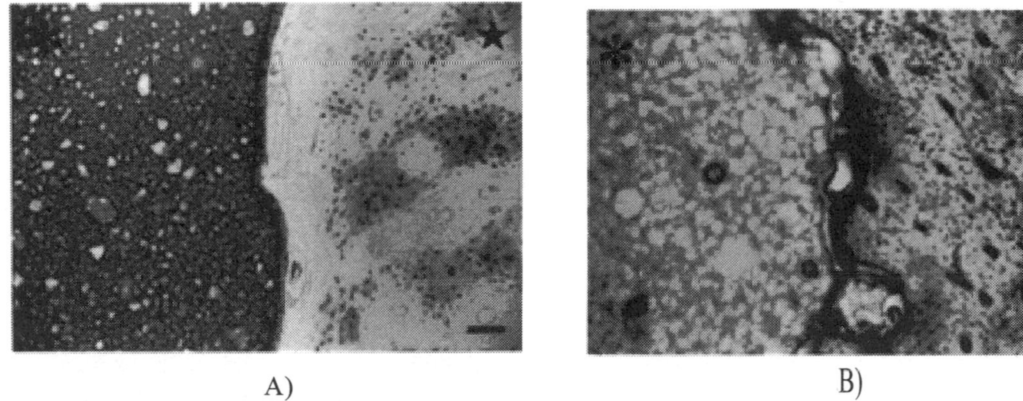

Figure 21.42 Giemsa surface staining of the femur at the bone cement interface of the (A) BABC and (B) PMMA bone cement at 24 months after implantation. Original magnification, ×100; scale = 100 μm. The star indicates the bone and the asterisk indicates the cement. (A) Direct bonding between the BABC and the bone was observed at the bone–cement interface. (B) An intervening soft tissue layer approximately 10 to 400 μm thick was consistently observed at the bone–cement interface.

been in contact with the atmosphere or with fluids, which include oxygen (94). Oxygen consumes radicals produced by benzoyl peroxide and *N,N*-dimethyl-*p*-toluidine and stops the polymerization chain, which leads to the formation of an incompletely polymerized thin layer (an uncured layer) on the surface of the resin.

Miyaji et al. reported that when a composite resin containing bis-GMA and TEGDMA, with a bioactive filler, was soaked in simulated body fluid (SBF) (95), incompletely polymerized resin, mainly TEGDMA, dissolved from the uncured layer into the solution within 1 hour, and glass grains exposed to SBF were converted into silica gel (96). The silica gels then induce apatite nucleation between the spaces of the grains, which induces the ability to bond directly to living bone. Okada et al. reported that the uncured layer was completely filled with newly formed bone-like tissue by their transmission electron microscopic (TEM) study when BABC was implanted in rat tibiae (97). They also revealed that this layer consisted of calcium, phosphorus, and some silicon by energy dispersive x ray microanalysis (EDX). In the scanning electron microscopy of the present study, direct bonding between the cement

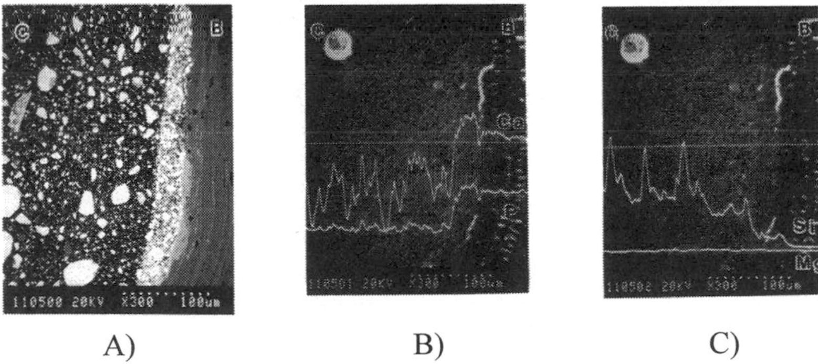

Figure 21.43 Back-scattered scanning electron micrograph of the femur at the bone–cement interface of the BABC group 24 months after implantation (A); electron probe microanalysis (B and C). Original magnification, ×300. The dotted line indicates 100 μm. B indicates the bone and C indicates the cement. (A) BABC showed direct bonding to bone without any intervening soft tissue layer. The surface of the cement was covered with a calcium phosphorus layer 30 μm thick, through which direct bonding between the cement mantle and host bone was accomplished. (B) The calcium level did not change, and the phosphorus level increased slightly across the bone–cement interface. (C) The silicon level decreased, and the magnesium level did not change across the bone cement–interface. A calcium phosphorus layer approximately 30 μm thick was present on the surface of the BABC, and the cement bonded directly to the host bone through this layer. No marked degradation of the fillers was observed in the cured region of the BABC.

and the host bone, both acetabular and femoral, was observed through the calcium phosphorous layer formed on the surface of the BABC.

Using a dog model, Pilliar et al. reported that bone ingrowth can occur in the presence of small movement (up to 28 μm) (98). In the present study, BABC bonded directly to bone under load-bearing conditions for up to 6 months and maintained this bond for up to 24 months and 12 months on the femoral and acetabular sides, respectively. This implies the high bioactivity of this material and indicates that adequate initial fixation was performed.

The failure on the acetabulum side at 24 months can be explained by several factors. Detachment between the BABC and the acetabular component occurred at 12 months, followed by cement fracture at the edge of the groove on the surface of the acetabular component. Although the loading mechanism in the acetabulum is not precisely known, it is believed to be a combination of compression, shear, and torsion. As the acetabulum of beagle dogs is shallow and the range of motion is wide, an excessive force may have loaded on the acetabular component. Moreover, the thickness of the acetabular component used in this study was 3 mm. This thinness of polyethylene also poses a concern in regard to creep of the cup, which can affect stress in the cement and the bone. An excessive load may have been applied to the bone–cement interface, which could have led to the breakage of the bone–cement interface. This breakage consistently occurred in the calcium phosphorus layer. In addition, the evaluation of the bone–cement interface after the push-out test demonstrated that breakage consistently occurred in the calcium phosphorus layer. Taken together, it has been revealed that the mechanically weakest layer of the bone–cement interface was the calcium phosphorus layer formed on the surface of the BABC.

In general, it is assumed that biomaterials with higher bioactivity tend to be more degradable and can show marked degradation after long-term implantation in the human body. In this study, cement fractures were observed on the acetabular side at 24 months after implantation, although no apparent degradation of the BABC was observed on the femoral side. The thin cement mantle of the acetabular side may have accelerated the degradation of the BABC.

BABCs are composite materials of glass ceramic and polymers, with their elastic moduli (dough type 14.4 GPa, and injection type 9.7 GPa) being lower than those of metals (stainless steel 193 GPa), higher than those of cancellous bone (0.2 to 0.5 GPa) and UHMWPE (0.012 GPa), and almost identical to those of cortical bone (10 to 20 GPa) (99). In the present study, the elastic moduli of the BABCs were higher than those of PMMA bone cement, which had an elastic modulus of 2.7 GPa due to the large amount of filler, mixed with bis-GMA-based resin in the BABCs.

On the acetabular side, bone cement lies between cancellous bone and UHMWPE. It is generally known that a discrepancy in an elastic modulus leads to stress concentration at the interface. Hori and Lewis reported that an intervening fibrous tissue layer is very compliant and deformable and can withstand large strains from a load applied to the bone–cement interface (100). As BABC bonds directly with bone, no shock absorber exists between BABC and cancellous bone, which leads to breakage inside the calcium and phosphorous layer. On the other hand, in the case of PMMA bone cement, an intervening soft tissue layer lies consistently between PMMA bone cement and the cancellous bone, which may act as a shock absorber.

On the femoral side, bone cement lies between the stainless steel and the cortical bone. As BABC and cortical bone have an almost identical elastic modulus, the bonding between BABC and the bone could be maintained for 24 months. Moreover, the bonding strength between metals and BABC or PMMA bone cement is stronger than the bonding strength between UHMWPE and BABC or PMMA bone cement (our unpublished data).

Wear debris is now considered to be one of the major factors responsible for osteolysis, resulting in aseptic loosening of the implant. Several attempts at resolving this problem are currently in progress, one of which involves improving the characteristics of the bone–cement interface (74,76). The improvement of the characteristics at the bone–cement interface can be achieved by using BABC, which has an ability to bond directly with bone. In the present study, the affinity indices of BABC for the femur was higher than that of PMMA bone cement at 12 and 24 months after implantation. This may result in closer adhesion and less micromotion at the bone–cement interface of the BABC group compared with the PMMA bone–cement group. The BABC used in the present study produced a calcium phosphorous layer at the bone–cement interface, and through this layer, direct bonding between the BABC and the host bone was observed. Although the affinity indices of the BABC for the femur were approximately 40% at 6, 12, and 24 months after implantation, the number of cells that can phagocytise wear debris of polyethylene and/or metal may be reduced at the bone–cement interface.

In the present study, the effectiveness of the BABC for fixation of the hip prosthesis was examined. This cement showed good bioactivity and bone-bonding ability on the femoral side for up to 24 months after implantation. Combination of weak bonding between BABC and UHMWPE, relatively high elastic characteristics of BABC, and weakness of the calcium phosphorous layer formed on the surface of this cement led to failure on the acetabular side at 24 months. We are now conducting a further fundamental study to produce a cement with a thinner uncured layer to strengthen further the bone–cement interface. Currently, we are also conducting studies to modify the elastic characteristics by changing the composition of the polymer. Further investigations will be necessary before a BABC is finally developed that is suitable for human THA.

E. Cementing Technique

Since its introduction by Charnley (60) in 1960, polymethyl methacrylate (PMMA) bone cement has generally been considered to be one of the most effective means for fixing hip prostheses. However, aseptic loosening remains a most serious long-term drawback. There have been a number of advances in the methods of cement fixation of the femoral component in total hip arthroplasty (THA), including pulsatile lavage (101), intramedullary plugging (102,103), vacuum mixing (104,105), retrograde insertion by cement gun, and final compaction. These techniques have been shown to enhance the fixation of cement to bone, and superior clinical results have been documented using them (106). An essential prerequisite to this procedure is adequate plugging of the distal femoral canal.

F. Intramedullary Plug

Various methods for plugging the canal have been reported, such as doughy acrylic cement (107), bone plugs (108,109), polyethylene plugs (110), biodegradable copolymer (111,112), and hydroxyapatite (113). There are advantages and disadvantages with each method. In general, plugging converts the femoral canal into a closed space, facilitates preparation of bone surfaces, prevents introduction of cement into the distal canal, and may reduce the potential hazard of embolization of canal contents into the venous circulation (114,115).

Cancellous bone chips obtained from resected femoral heads have generally been used in our institute for primary THAs. However, when it is difficult to obtain bone chips, as in revision surgery, porous apatite-wollastonite-containing glass ceramic (AW-GC) has been used for intramedullary plugging.

Autografts have been used widely for repairing bone defects. This technique has been successful, but the amount of available bone is limited, and harvesting it necessitates two operations. Implantation of allografts and xenografts does not require a second operation but does cause problems of antigenicity. Therefore artificial bone substitutes, such as alumina, hydroxyapatite, tricalcium phosphate, Bioglass^R, and AW-GC (78) have been investigated.

Dense AW-GC has been reported to have high mechanical strength as well as the capacity to form strong chemical bonds with bone tissue (80,81). Ijiri et al. (116) investigated porous AW-GC, which induced ectopic bone formation combined with bone morphogenic protein (BMP) and collagen.

The efficacy and biocompatibility of porous AW-GC as an intramedullary plug in canine total hip arthroplasty (THA) was evaluated for up to 2 years. Cylindrical porous AW-GC rods (70% porosity, 200 μm mean pore size) were prepared and provided by Nippon Electric Glass Co. Ltd. (Otsu, Japan) (Fig. 44). The chemical composition of the porous AW-GC was 4.6% MgO, 44.7% CaO, 34.0% SiO_2, 16.2% P_2O_5, and 0.5% CaF_2, and the crystallized glass-ceramic consisted of 28% residual glass, 38% apatite [$Ca_{10}(PO_4)_6(O, F_2)$], and 34% wollastonite ($SiO_2 \cdot CaO$), as described previously (79), being the same as that of dense AW-GC. Its compressive strength was 17.54 ± 3.82 MPa (mean ± S.D.) (16), which is identical with that of cancellous bone. Porous AW-GC has been used for bone defect filler in Japan. Twenty-two adult beagle dogs underwent unilateral THAs, and 4 dogs were sacrificed at 1, 3, 6, and 12 months each and 6 dogs at 24 months after implantation. Radiological evaluation confirmed the efficacy of po-

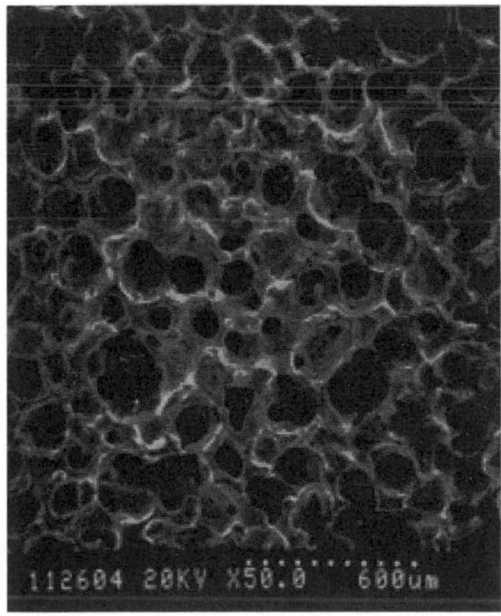

Figure 21.44 Scanning electron micrograph of the porous AW-GC used in the present study. Original magnification, ×50. The dotted line indicates 600 μm.

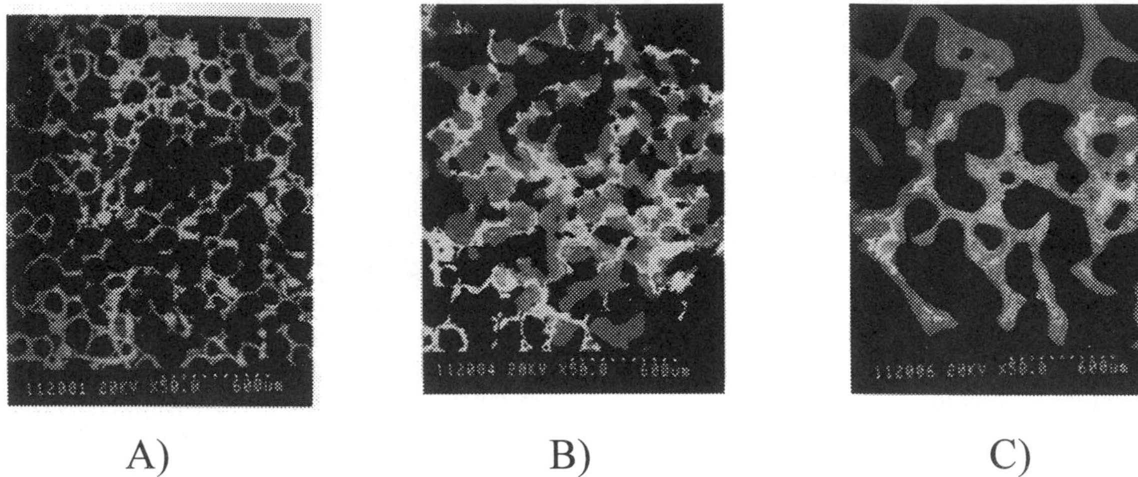

Figure 21.45 Back-scattered scanning electron micrograph of the femur at the bone–implant interface at (A) 1 month, (B) 6 months, and (C) 24 months after implantation. Original magnification, ×50; the dotted line indicates 600 μm. (A) Newly formed bone in the pores of the porous AW-GC was observed at one month. (B) The amount of bone increased until 6 months. (C) At 24 months, the walls of AW-GC pores were almost totally resorbed and replaced by newly formed bone.

rous AW-CG as an intramedullary plug. Histological evaluation showed osteoconduction at 1 month and resorption of porous AW-GC, which was replaced by newly formed bone, at 24 months (117) (Fig. 45).

There are various methods for plugging femoral canals, each with advantages and disadvantages. PMMA plugs have the disadvantage of a prolonged operation time plus additional mass cement handling and instruments (107). Also, the insertion of a cancellous bone plug taken from the removed femoral head requires experience and specialized instruments (108,109). Moreover, the femoral head is not always available, as in revision surgery. Polyethylene plugs are easier to use but are deficient in achieving adequate and consistent occlusion of the femoral canal because of its elliptical shape (103,111). Although cylindrical porous AW-GC may be insufficient to occlude the elliptical shape of the femoral canal, insertion of several spherical AW-GCs and compacting them is thought to achieve complete plugging of the femoral canal without any migration or leakage.

Raut et al. reported excellent clinical results from one-stage revision of an infected total hip arthroplasty with discharging sinuses using antibiotic-soaked PMMA bone cement (118). Kawanabe et al. reported on the efficacy of antibiotic-soaked AW-GC blocks as a new drug delivery system (DDS) for osteomyelitis in vitro and in vivo (119). They also showed in their pilot clinical study that antibiotic-soaked AW-GC blocks appeared to be useful for treating chronic osteomyelitis and infected arthroplasties with bone defects. On the basis of these two reports, we conclude that porous AW-GC is suitable as a DDS and that the combination of porous AW-GC plugs and antibiotics will be effective for preventing and treating infections at THA combined with antibiotic-impregnated PMMA bone cement.

Neo et al. reported that particles (100 to 220 μm in diameter) of AW-GC implanted in rat tibiae have not been resorbed completely, even 96 weeks after implantation (120). They reported that the resorbed or replaced width of the surface of AW-GC was less than 50 μm per year. In the present study, the thickness of the wall of porous AW-GC was 10 to 30 μm, and it was reasonable that the porous AW-GC was subtotally resorbed within 2 years (117). At revision surgery, we always remove all remaining cement, including the plug, because of possible infection. In many cases it is very difficult to remove the plug through the medullary canal. Giardino et al. reported that intramedullary plugs made of poly (D,L-lactic acid) (PDLLA) completely disappeared in the femoral medullary cavity of rabbits 26 weeks after implantation (112). Although the resorption rate of porous AW-GC was lower than bioresorbable materials including PDLLA, use of porous AW-GC plugs will also overcome this problem because of their resorption within 2 years. We are currently conducting clinical trials using porous AW-GC with a special design and have met with good preliminary results.

We conclude that porous AW-GC is effective as an intramedullary plug. AW-GC plugs appear to be promising for clinical application, having good biocompatibility and DDS capacity and showing resorption within 2 years.

ACKNOWLEDGMENTS

The authors would like to thank Dr. Eiichiro Watanabe (School of Medicine, Chiba University), Akiko Mori (Nakashima Propeller Co., Ltd.), and Sintaro Araki (Mitsui Chemicals, Inc.) for various data and information.

REFERENCES

1. Robert E. Tooms, James W. Harkess. Arthroplasty (Introduction and Overview). In: Campbell's Operative Orthopaedics (A. H. Crenshaw, ed.). Vol. 1, Chap. 14 (1992).
2. J. Charnley. Surgery of the hip-joint: present and future developments. Br. Med. J. 5176, 821 (1960).
3. J. Charnley. Arthoplasty of the hip: a new operation. Lancet 1, 1129 (1961).
4. J. Charnley. Total hip replacement by low-friction arthroplasty. Clin. Orthop. 72, 7 (1970).
5. D. F. Farrar, A. A. Brain. The microstructure of ultra-high molecular weight polyethylene used in total joint replacements. Biomaterials 18, 1677 (1997).
6. H. Ito. Ultra-high molecular polyethylene. Plastics (in Japanese) 35, No. 3, 28 (1984).
7. A Howmedica Research and Development Technical Monograph: Part Three of a Series on Ultra-High Molecular Weight Polyethylene. A Comparative Analysis of the Properties of Standard and "Enhanced" Ultra-High Molecular Weight Polyethylene.
8. S. Li, A. H. Burstein. Current concepts review: ultra high molecular polyethylene. J. Bone. Joint Surgery 76-A, No. 7, 1080 (1994).
9. T. Shiraki. Application and new processing of ultra-high molecular polyethylene. Plastics (in Japanese) 28, No. 5, 57 (1977).
10. A Howmedica Research and Development Technical Monograph: Part Two of a Series on Ultra-High Molecular Weight Polyethylene. Material Properties, Product Quality Control, and Their Relation to UHMWPE Performance.
11. R. M. Rose, E. L. Radin. A prognosis for ultra high weight polyethylene. Biomaterials 11, 63 (1990).
12. Hoechst Celanese Hostalen GUR UHMWPE Technical Data.
13. X. Y. Wang, R. Salovey. Melting of ultra high molecular polyethylene. J. Appl. Polym. Sci. 36, 593 (1987).
14. R. M. Rose, A. Crugnola, M. Ries, W. R. Cimino, I. Paul, E. L. Radin. On the origins of high in vivo wear rate in polyethylene components of total joint prostheses. Clin. Orthop. Rel. Res. 145, 277 (1992).
15. M. Wrona, M. B. Mayor, J. P. Collier, R. E. Jensen. The correlation between fusion defects and damage in tibial polyethylene bearings. Clin. Orthop. Rel. Res. 299, 92 (1994).
16. M. G. Tanner, L. A. Whiteside, S. White. Effects of polyethylene quality on wear in total knee arthroplasty. Clin. Orthop. 317, 83 (1995).
17. S. P. James, K. R. Lee, G. P. Beauregard, E. D. Rentfrow, J. R. Mclaughlin. Clinical wear of 63 ultrahigh molecular weight polyethylene acetabuular components: effect of starting resin and forming method. J. Biomed. Mater. Res. 48(3), 374 (1999).
18. H. A. Mckellop, F. W. Shen, P. Campbell, T. Ota. Effect of molecular weight, calcium stearate, and sterilization methods on the wear of ultra high molecular weight polyethylene acetabular cups in a hip joint simulator. J. Orthop. Res. 17(3), 329 (1999).
19. L. Costa, M. P. Luda, L. Trossarelli, E. M. Brach del Prever, M. Crova, P. Gallinaro. Oxidation in orthopaedic UHMWPE sterilized by gamma-radiation and ethylene oxide. Biomaterials 19, 659 (1998).
20. H. J. Nusbaum, R. M. Rose. The effect of radiation sterilization on the properties of ultrahigh molecular polyethylene. J. Biomed. Mater. Res. 13, 557 (1979).
21. P. Eyerer, Y. C. Ke. Property changes of UHMW polyethylene hip endoprostheses during implantation. J. Biomed. Mater. Res. 18, 1137 (1984).
22. C. M. Rimnac, R. W. Klein, F. Betts, T. M. Wright. Post-irradiation aging of ultra-high molecular polyethylene. J. Bone Joint Surgery 76-A, No. 7, 1052 (1994).
23. D. L. Tabb, J. J. Sevick, J. L. Koening. Fourier transform infrared study of the effects of irradiation on polyethylene. J. Polym. Sci. Polym. Phys. Ed. 13, 815 (1975).
24. R. J. Roe, E. S. Grood, R. Shastri, C. A. Gosselin, F. R. Noyes. Effect of radiation sterilization and aging on ultra-high molecular weight polyethylene. J. Biomed. Mater. Res., 15, 209 (1981).
25. E. S. Grood, R. Shastri, C. N. Hopson. Analysis of retrieved implants: crystallinity changes in ultrahigh molecular weight polyethylene. J. Biomed. Mater. Res. 16, 399 (1982).
26. A. Shinde, R. Salovey. Irradiation of UHMWPE. J. Polym. Sci. Polym. Phys. Ed. 23, 815 (1985).
27. V. Permnath, W. H. Harris, M. Jasty, E. W. Merrill. Gamma sterilization of UHMWPE articular implants: an analysis of the oxidation problem. Biomaterials 17(18), 1741 (1996).
28. E. Watanabe, K. Nagata, T. Tsuchida, H. Ohta, T. S. Kim, K. Kuramoto, H. Moriya. Clinical study of revision after total knee arthroplasty—wear of polyethylene insert. J. Jpn. Orthop. Assoc. 72(2), S243 (1998).
29. E. Watanabe, K. Nagata, T. Kaneeda, T. Tsuchida, H. Ohta, T. S. Kim, K. Kuramoto, H. Moriya. Analysis of the mechanism of breakage on UHMWPE tibial insert—comparing SEM findings on the fractured surface and oxidation. J. Jpn. Orthop. Assoc. 72(8), S1283 (1998).
30. E. Watanabe, T. Tsuchida, K. Nagata, T. S. Kim, T. Tamai, H. Moriya. New approach for assessing the characteristic of UHMWPE components. Transactions of the 45th Annual Meeting of the Orthopaedic Research Society, Anaheim California, 832 (1999).
31. D. C. Bassett, S. Block, G. J. Piermarini. A high pressure phase of polyethylene and chain extended growth. J. Appl. Phys 45, 4146 (1974).

32. J. R. Cooper, D. Dowson, J. Fisher. Macroscopic and microscopic wear mechanisms in ultra-high molecular weight polyethylene. Wear 162–164, 378–384 (1993).
33. M. Ohashi, N. Tomita, Y. Ikada, K. Ikeuchi. An observation of subsurface defects in ultra high molecular weight polyethylene due to rolling contact. Bio-Med. Mater. Eng. 9, 1–11 (1996).
34. M. Ohashi, N. Tomita, K. Ikeuchi. Contribution of grain boundary for subsurface crack to form in UHMWPE component of knee prostheses: morphology of cracks in UHMWPE under experimental and clinical condition. submitted to Wear.
35. N. Tomita, M. Oohashi, K. Ikeuchi, Y. Ikada, M. Endoh, T. Nagatoshi, E. Aoyama. The effect of compresive load on formation of fatigue crack in UHMWPE. J. Jpn. Soc. Clin. Biomech. Relat. Res. 18, 459–461 (1997).
36. S. K. Bhateja. Radiation-Induced crystalinity changes in linear polyethylene: influence of aging. J. Appl. Polymer Sci. 28, 861–872 (1983).
37. M. S. Jahan, C. Wang, G. Schwartz, J. A. Davidson. Combined chemical and mechanical effects on free radicals in UHMWPE joints during implantation. J. Biomed. Mater. Res. 25, 1005–1017 (1991).
38. S. Li, J. D. Chang, E. G. Barrena, B. D. Furman, T. M. Wright, E. Salvait. Nonconsolidated polyethylene particles and oxidation in Charnley acetabular cups. Clin. Orthop. Related Res. 319, 54–63 (1995).
39. S. E. White, R. D. Paxson, M. G. Tanner, L. A. Whiteside. Effects of sterilization on wear in total knee arthroplasty. Clin. Orthop. Related Res. 331, 164–171 (1996).
40. M. D. Ries, K. Weaver, R. M. Rose, J. Guther, W. Sauer, N. Beals. Fatigue strength of polyethylene after sterilization by gamma irradiation or ethylene oxide. Clin. Orthop. Related Res. 333, 87–95 (1996).
41. W. L. Sauer, K. D. Weaver, N. B. Beals. Fatigue performance of ultra-high molecular-weight polyethylene: effect of gamma radiation sterilization. Biomaterials 17, 1929–1935 (1996).
42. R. S. Pascaud, W. T. Evans, P. J. J. McCullagh, D. P. FitzPatrick. Influence of gamma-irradiation sterilization and temperature on the fracture toughness of ultra-high-molecular-weight polyethylene. Biomaterials 18, 727–735 (1997).
43. C. J. Bell, P. S. Walker, M. R. Abeysundera, J. M. Simmons, P. M. King, G. W. Blunn. Effect of oxidation on delamination of ultrahigh-molecular-weight polyethylene tibial components. J. Arthroplasty 13-3, 280–290 (1998).
44. M. Goldman, L. Pruitt. Comparison of the effects of gamma radiation and low temperature hydrogen peroxide gas plasma sterilization on the molecular structure, fatigue resistance, and wear behavior of UHMWPE. J. Biomed. Mater. Res. 40–43, 378–384 (1998).
44. D. C. Sun, G. Schmidig, C. Stark, J. H. Dumbleton. On the origins of a subsurface oxidation maximum and its relationship to the performance of UHMWPE implants. In: Abstract of the 21st Annual Meeting of the Society for Biomaterials, March 18–22, 1995, San Francisco.
45. H. Fujiki, H. Ishikawa, K. Yasuda. The effects of thickness and fixation of ultra high-molecular-weight polyethylene and of traction on wear behavior in artificial knee joint. Trans. Jpn. Soc. Mech. Eng. 60A–576, 159–65 (1993).
46. H. Fujiki, H. Ishikawa, K. Yasuda. Contact mechanics of ultrahigh-molecular-weight polyethylene for artificial knee joint. Trans. Jpn. Soc. Mech. Eng. 59A–567, 328–335 (1993).
47. N. Tomita, T. Kitakura, N. Onmori, Y. Ikada, E. Aoyama. Prevention of fatigue cracks in ultrahigh molecular weight polyethylene joint components by the addition of vitamin E. J. Biomed. Mater. Res. 48, 474–478 (1999).
49. D. C. Bassett, B. Turner. Pressure quenching and chain extended crystallization of polyethylene. J. Polym. Sci. Polym. Phys. Ed. 13, 1501 (1975).
50. F. W. Shen, H. A. Mckellop, R. Salovey. Irradiation of chemically crosslinked UHMWPE. J. Polym. Sci. Part B Polym. Phys. 11(34), 1063 (1996).
51. E. Jewett, F. Kurtz. Direct correlation of abrasive wear resistance of irradiation-crosslinked UHMWPE with large-deformation mechanical behavior determined at the articulating surface. Trans. Soc. Biomat. 21, 220 (1998).
52. Johnson & Johnson Professional Inc., Technical Data.
53. M. Ohta, S.-H. Hyon, Y.-B. Kang, S. Tsutsumi, S. Murakami, S. Kohjiya. Novel molding method of ultra-high molecular weight polyethylene for artificial joint. Polym. Preprints Jpn. 48, No. 11, 2705 (1999).
55. H. Oonishi, M. Saito, Y. Ikadoya. Wear of high-dose gamma irradiated polyethylene in total joint replacement: long term radiological evaluation In: Transactions of the 44th Annual Meeting, Orthopaedic Research Soc., 23 (1998).
56. D. C. Sun, C. Stark. Non-oxidizing polymeric medical implant. In: U.S. patent 5,414,049.
57. N. S. Eftekhar. Acrylic cement: properties and application. In: N. S. Eftekhar, ed. Total Hip Arthroplasty. St. Louis: Mosby Year Book, 1993, pp. 223–314.
58. R. L. Wixon, E. P. Lautenschlager. Methyl methacrylate. In: J. J. Callaghan, A. G. Rosenberg, H. E. Rubash, eds. The Adult Hip. Philadelphia: Lippincott-Raven, 1998, pp. 135–157.
59. J. Judet, R. Judet. The use of an artificial head for arthroplasty of the hip joint. J. Bone. Joint Surg. 1950; 32B:166.
60. J. Charnley. Anchorage of femoral head prosthesis to shaft of femur. J. Bone. Joint Surg. 1960; 42-B:28–30.
61. J. Charnley. Acrylic Cement in Orthopaedic Surgery. Baltimore: Williams Wilkins and, 1970.
62. J. Vainio, Rokkanen P. A. composite resin as an implant material in bone. Histologic, radiologic, microradiologic and oxytetracycline fluorescence examination of rats. Arch. Orthop. Traumat. Surg. 1978; 92:165–168.
63. E. P. Lautenschlager, S. I. Stupp, J. C. Keller. Structure and properties of acrylic bone cement. In: Duchaynep

Hasting GW, ed. Functional behavior of orthopaedic biomaterials, Vol. 2, Applications. CRC Series in Structure–Property Relationships of Biomaterials. Boca Raton, FL: CRC Press, 1984.

64. M. A. R. Freeman, G. W. Bradley, P. A. Revell. Observation upon the interface between bone and polymethylmethacrylate cement. J. Bone. Joint Surg. (Br) 1982; 64-B:489–493.
65. S. R. Goldring, A. L. Schiller, M. Roelke, C. M. Rourke, D. A. O'Neill, W. H. Harris. The synovial-like membrane at the bone–cement interface in loose total hip replacements and its proposed role in bone lysis. J. Bone. Joint Surg. (Am) 1983; 65-A:575–583.
66. T. P. Schmalzried, L. M. Kwong, M. Jasty, et al. The mechanism of loosening of cemented acetabular components in total hip arthroplasty. Analysis of specimens retrieved at autopsy. Clin. Orthop. 1992; 274:60–78.
67. M. Jasty, W. J. Maloney, C. R. Bragdon, T. Haire, W. H. Harris. Histomorphological studies of the long-term skeletal responses to well fixed cemented femoral components. J. Bone. Joint Surg. (Am) 1990; 72-A:1220–1229.
68. W. J. Maloney, M. Jasty, A. Rosenberg, W. H. Harris. Bone lysis in well-fixed cemented femoral component. J. Bone. Joint Surg. (Br) 1990; 72-B:966–970.
69. S. Saha, S. Pal. Mechanical properties of bone cement: a review. J. Biomed. Mater. Res. 1984; 18:435–462.
70. J. P. Davies, D. O. O'Conner, D. W. Burke, M. Jasty, W. H. Harris. The effect of centrifugation on the fatigue life of bone cement in the presence of surface irregularities. Clin. Orthop. 1988, 229, 156–161.
71. R. D. Mulroy Jr., W. H. Harris. The effect of improved cementing techniques on component loosening in total hip replacement. An 11-year radiographic review. J. Bone. Joint Surg. (Br) 1990; 72-B:757–760.
72. W. Henning, B. A. Blencke, K. K. Deutscher, A. Gross, W. Ege. Investigation with bioactivated polymethylmethacrylates. J. Biomed. Mater. Res. 1979; 13:89–97.
73. T. Kokubo, S. Yoshihara, N. Nishimura, T. Yamamuro, T. Nakamura. Bioactive bone cement based on $CaO\text{-}SiO_2\text{-}P_2O_5$ glass. J. Am. Ceram. Soc. 1991; 74:1739–1741.
74. R. Labella, M. Braden, S. Deb. Novel hydroxyapatite-based dental composites. Biomaterials 1994; 15:1197–1200.
75. J. Raveh, H. Stich, P. Schawalder, C. Ruchti, H. Cottier. Biocement—a new material. Acta. Otolaryngol. 1982; 94: 371–384.
76. M. Saito, A. Muraoka, T. Mori, N. Sugano, K. Hino. Experimental studies on a new bioactive bone cement: hydroxyapatite composite resin. Biomaterials 1994; 15:156–160.
77. T. Kitsugi, T. Yamamuro, T. Nakamura, et al. Bone bonding behavior of three kinds of apatite containing glass ceramic. J. Biomed. Mater. Res. 1986; 20:1295–1307.
78. T. Kokubo, S. Ito, M. Shigematsu, S. Sakka, T. Yamamuro. Mechanical properties of a new type of apatite-containing glass-ceramic for prosthetic application. J. Mater. Sci. 1985; 20:2001–2004.
79. T. Kokubo, S. Ito, S. Sakka, T. Yamamuro. Formation of a high-strength bioactive glass-ceramic in the system $MgO\text{-}CaO\text{-}SiO_2\text{-}P_2O_5$. J. Mater. Sci. 1986; 21:536–540.
80. T. Nakamura, T. Yamamuro, S. Higashi, T. Kokubo, S. Ito. A new glass-ceramic for bone replacement: evaluation of its bonding ability to bone tissue. J. Biomed. Mater. Res. 1985; 19:685–698.
81. T. Yamamuro, J. Shikata, H. Okumura, T. Kitsugi, Y. Kakutani, T. Matsui, T. Kokubo. Replacement of lumbar vertebrae of sheep with ceramic prosthesis. J. Bone. Joint Surg. (Br) 1990; 72-B:889–893.
82. T. Yamamuro. A-W glass-ceramic: clinical applications. In: Hench LL, Wilson J, eds. An Introduction to Bioceramics. Singapore:World Scientific, 1993, pp. 89–103.
83. T. Kitsugi, T. Yamamuro, T. Nakamura, T. Kokubo. Bone bonding behavior of $MgO\text{-}CaO\text{-}SiO_2\text{-}P_2O_5\text{-}CaF_2$ glass (mother glass of A-W glass ceramic). J. Biomed. Mater. Res. 1996; 23:631–648.
84. K. Kawanabe, J. Tamura, T. Yamamuro, T. Nakamura, T. Kokubo, and S. Yoshihara. A new bioactive bone cement consisting of BIS-GMA resin and bioactive glass powder. J. Appl. Biomat. 1993; 4:135–141.
85. H. Fujita, Nakamura, J. Tamura, M. Kobayashi, Y. Katsura, T. Kokubo, T. Kikutani. Bioactive bone cement: effect of the amount of glass-ceramic powder on bone-bonding strength. J. Biomed. Mater. Res. 1998; 40:145–152.
86. J. Tamura, K. Kawanabe, T. Yamamuro, T. Nakamura, T. Kokubo, S. Yoshihara, T. Shibuya. Bioactive bone cement: the effect of amounts of glass powder and histological changes with time. J. Biomed. Mater. Res. 1995, 29. 551–559.
87. J. Tamura, K. Kawanabe, M. Kobayashi, T. Nakamura, T. Kokubo, S. Yoshihara, T. Shibuya. Mechanical and biological properties of two types of bioactive bone cements containing $MgO\text{-}CaO\text{-}SiO_2\text{-}P_2O_5\text{-}CaF_2$ glass and glass-ceramic powder. J. Biomed. Mater. Res. 1996; 30:85–94.
88. J. Tamura, T. Kitsugi, H. Iida, H. Fujita, T. Nakamura, T. Kokubo, S. Yoshihara. Bone bonding ability of bioactive bone cements. Clin. Orthop. 1997; 343:183–191.
89. Y. Senaha, T. Nakamura, J. Tamura, K. Kawanabe, H. Iida, T. Yamamuro. Intercalary replacement of canine femora using a new bioactive bone cement. J. Bone. Joint Surg. (Br) 1996; 78-B:26–31.
90. Y. Matsuda, K. Ido, T. Nakamura, H. Fujita, T. Yamamuro, M. Oka, T. Shibuya. Prosthetic replacement of the hip in dogs using bioactive bone cement. Clin. Orthop. 1997; 336:263–277.
91. M. Kobayashi, T. Nakamura, J. Tamura, H. Fujita, T. Kokubo, T. Kikutani. Mechanical and biological properties of bioactive bone cement containing silica glass powder. J. Biomed. Mater. Res. 1997; 37:68–80.
92. H. Fujita, H. Iida, K. Kawanabe, Y. Okada, M. Oka, T. Masuda, Y. Kitamura, T. Nakamura. Pressurization of bioactive bone cement in vitro. J. Biomed. Mater. Res. (Appl Biomater) 1999; 48:43–51.
93. H. Fujita, K. Ido, Y. Matsuda, H. Iida, M. Oka, Y. Kitamura, T. Nakamura. Evaluation of bioactive bone cement

in canine total hip arthroplasty. J. Biomed. Mater. Res. 2000; 49:273–288.
94. F. A. Bovey, I. M. Kolthoff. The mechanism of emulsion polymerization. III. Oxygen as a comonomer in the emulsion polymerization of styrene. J. Am. Chem. Soc. 1947; 69:2143–2153.
95. T. Kokubo, H. Kushitani, S. Sakka, T. Kitsugi, T. Yamamuro. Solutions able to reproduce in vivo surface-structure changes in bioactive glass-ceramic A-W. J. Biomed. Mater. Res. 1990; 24:721–734.
96. F. Miyaji, Y. Morita, T. Kokubo, T. Nakamura. Surface structural change of bioactive inorganic filler-resin composite cement in simulated body fluid: effect of resin. J. Biomed. Mater. Res. 1998; 42:604–610.
97. Y. Okada, M. Kobayashi, H. Fujita, Y. Katsura, H. Matsuoka, H. Takadama, T. Kokubo, T. Nakamura. Transmission electron microscopic study of interface between bioactive bone cement and bone: comparison of apatite and wollastonite glass-ceramic filler with hydroxyapatite and β-tricalcium phosphate fillers. J. Biomed. Mater. Res. 1999; 45:277–284.
98. R. M. Pilliar, J. M. Lee, C. Maniatpoulos. Observations on the effect of movement on bone ingrowth into porous-surfaced implants. Clin. Orthop. 1986; 208:108–113.
99. J. Black. Properties of natural materials. In Black J, ed. Orthopaedic Biomaterials in Research and Practice. New York: Churchill Livingstone, 1988, pp. 105–131.
100. R. Y. Hori, J. L. Lewis. Mechanical properties of the fibrous tissue found at the bone–cement interface following total joint replacement. J. Biomed. Mater. Res. 1982; 16: 911–927.
101. W. R. Krause, W. Krug, B. Eng. Strength of the cement-bone interface. Clin. Orthop. 1982; 163:290–299.
102. I. Oh, C. E. Carlson, W. W. Tomford, W. H. Harris. Improved fixation of the femoral component after total hip replacement using a methacrylate intramedullary plug. J. Bone. Joint Surg. (Am) 1978; 60-A:608–613.
103. J. Johnson, J. Johnston, R. E. Hawary, et al. Occlusion and stability of synthetic femoral canal plugs used in cemented hip arthroplasty. J. Appl. Biomater. 1995; 6:213–218.
104. L. Lidgren, J. Moller Drar, Strength of polymethylmethacrylate increased by vacuum mixing. Acta. Orthop. Scand. 1988; 55:536–541.
105. E. W. Fritsch. Static and fatigue properties of two new low-viscosity PMMA bone cements improved by vacuum mixing. J. Biomed. Mater. Res. 1996; 31:451–456.
106. W. H. Harris, W. A. McGann. Loosening of the femoral component after use of the medullary-plug cementing technique. J. Bone. Joint Surg. (Am) 1986; 68-A:1064–1066.
107. I. Oh, C. E. Carlson, W. W. Tomford, W. H. Harris. Improved fixation of the femoral component after total hip replacement using a methacrylate intramedullary plug. J. Bone. Joint Surg. (Am) 1978; 60-A:608–613.
108. B. M. Wroblewski, A. Rijt. Intramedullary cancellous bone block to improve femoral stem fixation in Charnley low-friction arthroplasty. J. Bone. Joint Surg. (Br) 1984; 66-B:639–644.
109. W. E. Knight. Femoral plugging using cancellous bone. Clin. Orthop. 163:167–169, 1982.
110. T. H. Mallory. A plastic intermedullary plug for total hip arthroplasty. Clin. Orthop. 155:37–40, 1981.
111. S. K. Bulstra, R. G. T. Geesink, D. Bakker, et al. Femoral canal occlusion in total hip replacement using a resorbable and flexible cement restrictor. J. Bone. Joint Surg. (Br) 1996; 78-B:892–898.
112. R. Giardino, M. Fini, N. N. Aldini, et al. Experimental evaluation of a resorbable intramedullary plug for cemented total hip replacement. Biomaterials 18:907–913, 1997.
113. M. Maruyama. In vivo properties of an intramedullary hydroxyapatite plug to improve femoral stem fixation. Arch. Orthop. Trauma. Surg. 116:396–399, 1997.
114. J. Christie, C. M. Robinson, B. Singer, D. C. Ray. Medullary lavage reduces embolic phenomena and cardiopulmonary changes during cemented hemiarthroplasty. J. Bone. Joint Surg. (Br) 1995; 77-B:456–459.
115. A. W. McCaskie, M. R. Barnes, E. Lin, W. H. Harper, P. J. Gregg. Cement pressurization during hip replacement. J. Bone. Joint Surg. (Br) 1997; 79-B:379–384.
116. S. Ijiri, T. Nakamura, Y. Fujisawa, M. Hazama, S. Komatsudani. Ectopic bone induction in porous apatite-wollatonite-containing glass ceramic combined with bone morphogenic protein. J. Biomed. Mater. Res. 19:421–432, 1997.
117. H. Fujita, H. Iida, K. Ido, Y. Matsuda, M. Oka, T. Nakamura. Porous A-W glass-ceramic as an intramedullary plug. J. Bone. Joint. Surg. (Br), in press.
118. V. V. Raut, P. D. Siney, B. M. Wroblewski. One-stage revision of total hip replacements with discharging sinuses. J. Bone. Joint Surg. 76B:721–724, 1994.
119. K. Kawanabe, Y. Okada, Y. Matsusue, H. Iida, T. Nakamura. Treatment of osteomyelitis with antibiotics-soaked porous glass ceramic. J. Bone. Joint Surg. (Br) 1998; 80-B:527–530.
120. M. Neo, T. Nakamura, C. Ohtsuki, R. Kasai, T. Kokubo, T. Yamamuro. Ultrastructual study of the A-W GC-bone interface after long-term implantation in rat and human bone. J. Biomed. Mater. Res. 28:365–372, 1994.

22

Polymeric Occluders in Tilting Disc Heart Valve Prostheses

G. S. Bhuvaneshwar
Sree Chitra Tirunal Institute of Medical Sciences and Technology, Thiruvananthapuram, India

A. V. Ramani
T.T.K. Pharma, Ltd., Bangalore, India

K. B. Chandran
University of Iowa, Iowa City, Iowa

I. ARTIFICIAL HEART VALVE PROSTHESES

Replacement of diseased human heart valves with prostheses has become a common treatment modality today. The trileaflet aortic valve between the left ventricular outflow tract and the aorta, as well as the bicuspid mitral valve between the left atrium and the left ventricle, are subjected to high pressure generated due to the contraction of the heart. Valvular heart diseases and hence valve replacement predominantly occur for these two valves (1,2). With the advent of blood oxygenators and cold potassium cardioplegia to arrest the heart and perform open-heart surgery about four decades ago, many different artificial valves have been implanted. The valve prostheses commonly available today for replacement can be broadly classified into mechanical valves and biological tissue valves (1).

The biological tissue valves consist of either a porcine aortic valve or a trileaflet valve made of bovine pericardial tissue. The aortic valve of pigs, which has an anatomy closely resembling the human aortic valve, is treated with a preservative and mounted in flexible supporting struts for implantation. Similarly, treated bovine pericardial tissue is used to make a trileaflet valve and mounted in supporting struts for implantation. Tissue valves, being subjected to relatively large stresses as they open and close, fail due to fatigue stresses and need to be replaced about 10 to 12 years after implantation. Further, in patients below 40 years of age, tissue valves fail in 3 to 5 years due to calcification of the leaflets. On the other hand, mechanical valves made with high-strength biocompatible material are durable and have a long-term functional capability. However, mechanical valves are subjected to thrombus deposition and subsequent complications resulting from emboli, and patients with implanted mechanical valves need to be in long-term anticoagulant therapy.

The designs of the early mechanical valves were of centrally occluding caged ball or caged disc valves. In 1960, Harken et al. reported the successful implantation of a caged ball valve in the subcoronary position for severe aortic insufficiency (3), while later that year Starr performed the first long-term successful mitral valve replacement with a caged ball valve (4). During the following five years, thousands of Starr–Edwards mitral and aortic valves made of a Stellite® cage and a silicone rubber ball were implanted throughout the world with good overall results.

Following the initial successes, reports of problems with the caged ball design started. In some patients there was outflow tract obstruction, while in others there were late deaths due to ventricular fibrillation induced by septal irritation by the rubbing of the cage apex. Autopsy specimens demonstrated bulky and oversized ball valves projecting into a contracted left ventricular cavity, proving this point. This resulted in the search for a low-profile prosthesis in the mid-1960s. Hufnagel and Conrad described a polypropylene and silicone rubber caged disc valve. Many others designed similar disc valves with polymeric disc occluders. The notable ones are the Cross–Jones, Kay–Shiley, Harken–Cromie, Beall–Surgitool and Starr–Edwards models (4).

The early clinical trials established the vulnerability of the caged disc valves to catastrophic malfunction, high pressure drops, and thrombus formation behind the disc. The polymeric discs had a high propensity for degradation, wear, and getting stuck due to small thrombi on the struts. In this context, the Starr–Edwards group carried out extensive in vitro and in vivo testing to find a compatible combination of cage and disc material. Testing of 14 different polymers to determine the best abrasion-resistant material revealed ultra-high molecular weight polyethylene (UHMW PE, Hifax®) and Delrin® (acetal homopolymer, Dupont) to be superior to the others in wear resistance. Because of the possibility of water absorption by Delrin, Hifax® was chosen (4). However, due to poor hemodynamics and a higher propensity for the disc to get stuck, this model soon fell into disuse with the introduction of tilting disc valves at the end of the 1960s.

Wada was the first to introduce clinically a tilting disc design using a notched Teflon® (polytetra fluoroethylene, Dupont) disc, which engaged in an another pair of notches in the cage (5). The stepped occluder with notches was not free to rotate. This design permitted a large orifice diameter to be used when compared to equal-sized caged ball or disc valves, thus reducing the forward pressure drops considerably. The quest for a valve with optimal hemodynamics inspired Björk to analyze the cardiac catheterization data on early caged ball and disc valves (6,7). He switched over to the Wada–Cutter valve, impressed by its low pressure drops. Clinical data during the next couple of years indicated catastrophic thrombus formation around the notches and wear of the Teflon disc leading to severe valvular regurgitation or fatal disc embolization (8). From these studies, it became clear that any type of hinge created an area of low flow and stasis, which led to thrombus formation around it. Hence, the need was to do away with the hinge and make the disc free floating.

The year 1969 saw the introduction of two truly hingeless free floating disc valves, the Björk–Shiley (9) and the Lillehei–Kaster (10). The Björk-Shiley valve made use of a depression in the disc and two welded wire struts in the cage to retain the disc. This permitted the disc to tilt open and close, while it was free to rotate around its center. The Lillehei–Kaster valve used a pair of projecting arms on the outlet side and a pair of stubs on the inlet side to retain and allow the disc to tilt and rotate freely as well. The free rotation of the disc distributed the wear on the complete surface of the disc and allowed it to be washed well at the same time, thus preventing thrombus buildup on it.

The main features of these tilting disc valves compared to the caged ball and disc valves were

1. A large orifice diameter for a given tissue annulus diameter, resulting in very low pressure drops.
2. With an opening angle of 60° or more, the flow was more central.
3. The disc was free to rotate around its center, thus preventing any buildup of thrombus.

The Björk–Shiley valve used a Stellite-21 welded cage, a Delrin® disc, and a Teflon® sewing ring. The Lillehei–Kaster valve consisted of a pivoting disc suspended in a titanium housing encircled by a knitted Teflon fabric sewing ring. The disc was flat and consisted of a graphite substrate coated all around with a 250-micron-thick layer of carbon–silicon alloy (Pyrolite®) developed by Gulf Energy and Environmental Systems. A couple of years later, the disc material of the Björk–Shiley valve was also changed to Pyrolite®, as Delrin® had a propensity to swell during autoclaving (11,12).

Since its introduction, pyrolytic carbon has proven to be a very durable and blood-compatible material for use in prosthetic heart valves. Valve designs fabricated with this material and with a single tilting disc or two leaflets are the most popular mechanical valves in use today. The notable ones are the Hall–Kaster tilting disc valve (presently called the Medtronic-Hall valve), the St. Jude Medical bileaflet valve, the Carbomedics, and the Sorin tilting disc and bileaflet models. The bileaflet designs emerged as having superior forward flow dynamics in comparison to the tilting disc ones. However, clinical results indicate only a marginal reduction in the complication rates with the use of these models.

In spite of these many developments in prosthetic valve design and materials, problems of thrombosis, thromboembolism, and tissue overgrowth continue in all these valves. The thromboembolic complications associated with mechanical valves have been correlated with the forward flow dynamics (1,2). The flow characteristics suggested as factors causing thrombus deposition include turbulent stresses on the downstream side with flow across a fully open valve, and regions of relative stasis and flow separation.

Recently, a number of studies have associated thrombus initiation in the mechanical valves with the stresses induced during valve closure. Almost all currently available tilting disc and bileaflet mechanical valves have relatively rigid pyrolytic carbon occluders. At the instant of the impact of the occluder on the seat stop or the seating lip, the occluder tip comes to a sudden stop, resulting in a water hammer effect. Large positive pressure transients are induced near the leaflet tip close to the valve on the downstream side of the valve. Similarly, large negative pressure transients have been recorded on the upstream side of the valve close to the leaflet tip. This water hammer effect can be the source of several fluid dynamically induced stresses that can initiate hemolysis and platelet activation. Further, in many cases the negative pressure transients fall below the vapor pressure of blood near the occluder edge on the upstream side of the leaflet, leading to vapor phase cavitation. Numerous in vitro studies have demonstrated the formation of cavitation bubbles during valve closure (13–21). The cavitation bubbles visualized near the leaflets for a period of less than 1 ms in this region correspond to the regions where the negative pressure transients have also been recorded.

In recent in vivo studies, negative pressure transients have been recorded in the atrial chamber very close to the mitral valve in vivo and they were also remarkably similar to those measured with a single closing event of the same valve in vitro. Therefore it is clear that there is a *potential* for cavitation to occur in vivo with implanted mechanical valves. Even in the absence of cavitation bubbles, the negative pressure transients can result in the viscoelastic expansion of the formed elements in blood, resulting in thrombus initiation.

Similar valve closure studies were repeated with the Chitra tilting disc valve with a relatively flexible UHMW PE occluder, both in vitro and in vivo. Under the same closing conditions, the amplitude of the pressure transients with the Chitra valve was significantly smaller compared to that with a rigid leaflet valve. Since the negative pressures with the polyethylene leaflet valve never reached magnitudes close to the vapor pressure for the blood analog used in these studies, no cavitation bubbles were observed with these valves. These studies suggest that during valve closure with the flexible leaflets, the yielding of the flexible leaflets absorbs part of the energy during impact. Thus the fluid dynamically induced stresses in these regions during closing can be anticipated to be significantly smaller than those with the currently available mechanical valves. Whether this will lead to reduced incidence of thrombosis and embolism in the clinical situation needs to assessed by well-controlled clinical studies.

In the light of these data, the development and evaluation of the Chitra tilting disc valve prosthesis using UHMW PE occluder is described in this chapter. The current data available from a multicenter clinical study is presented. The closing mechanics of the valve is compared with that of the other clinical models with rigid occluders, which clearly brings out the advantage of a relatively flexible occluder.

II. DESIGN OF THE TILTING DISC VALVE WITH A ULTRA-HIGH MOLECULAR WEIGHT POLYETHYLENE DISC OCCLUDER

One major need for heart valve surgery in India and other developing countries continues to be rheumatic valvular disease. Most of these patients are from low socioeconomic levels and suffer lack of access to diagnosis and prophylaxis. In view of this, the need for valve replacement continues to be large and insistent. The high cost of imported valves was and continues to be one factor limiting the number of replacements being performed. Therefore a project was initiated in the late 1970s at the Sree Chitra Tirunal Institute for Medical Sciences and Technology (SCTIMST), India, for the development of an indigenous heart valve prosthesis, at a time when very little expertise and few facilities for biomedical research existed in the country. As in other valve development programs, the road was rough and had to go through a couple of failures. These details have been presented earlier (22), and the valve is now available under the name of TTK-Chitra Tilt ing Disc Heart valve prosthesis (manufactured by TTK Pharma Ltd., Chennai, India). For convenience however, the valve will continue to be referred here as the Chitra heart valve (CHV).

A. Choice of Design and Materials

1. Choice of Valve Type

As per the 1981 census, there were 200 million children in the age group of 5 to 15 years, and the Indian Council for Medical Research has estimated that 6 out of 1000 children in this age group were at risk of acquiring rheumatic disease. Hence the average age of valve replacement patients was below 30 years, and this could well be expected to continue for a few more decades. This ruled out the choice of a bioprosthetic valve, despite its advantage of not requiring long-term anticoagulant therapy. Further, the beef and pork industries were so poorly organized and emotionally charged that a manufacturing program based on such tissue collection was impractical. Thus the choice was limited to a mechanical model. In view of the lack of experience in valve development, and to ensure a reasonable chance of success, the choice was limited to the time-

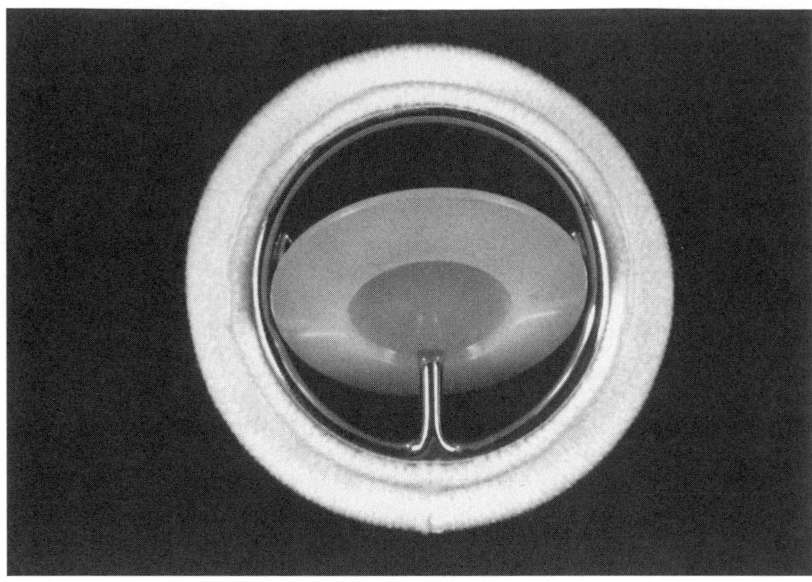

Figure 22.1 Photograph of a Chitra valve with a UHMW PE occluder.

tested ones, viz., the caged ball, the caged disc, and the tilting disc designs. With the majority being mitral valve replacements, the caged ball design with its high profile and relatively higher pressure drops was discarded. Between the other two, the tilting disc with its superior hemodynamics (lower pressure drops and more central flow) was considered the best available choice.

B. Design Features

The Chitra valve in current clinical use is shown in Fig. 1. It has an ultra-high molecular weight polyethylene (UHMW PE) disc, a Haynes-25 alloy (Haynes International, USA) cage, and a polyester suture ring. To start with, some of the basic design features of the Chitra tilting disc valve were drawn from the established designs, while others were modified to suit the current situation. It featured:

Free floating disc able to rotate on its center to avoid the problems of thrombosis around a hinge and to distribute the wear over its surface.

Tilt axis at one-fourth of the disc diameter (i.e., at the quarter chord point).

Cage entrance and exit curvature: bell mouthing of the inlet and outlet edges of the cage to permit smooth entrance and exit of blood.

Disc shape: A planoconvex disc with inlet side flat, increasing the inflow area into the minor orifice and making the fabrication of the cage and disc easier.

Opening angle: 70° opening angle for reduced pressure drops.

Cage structure: The struts machined integral with the ring from a solid block. The main feature of this design was that it enabled the cage to be machined with the indigenous capability at that time without the need for expensive CNC techniques.

C. Choice of Material

Four material combinations have been tried during this developmental program (Table 1). The first one, using a titanium cage and a polyacetal disc, was given up, as this plastic could swell and distort during steam sterilization (23). The second model used a titanium nitride (TiN) coated Haynes-25 alloy cage and a synthetic sapphire disc. This model successfully completed the accelerated durability test. In animal evaluation, the sapphire disc fractured in 5 out of 14 animal implants. In spite of this, sapphire proved to be an excellent blood-compatible material (24).

With ceramic materials being subject to unpredictable failures and LTI carbon being not available in the country, the search for a new disc material from a range of engineering plastics was initiated. Based on the need for low wear, high fatigue resistance, high toughness, and known biocompatibility, the choice of polymer for the disc was headed by UHMW PE and followed by Delrin ST® and Delrin AF® (Delrin super-tough and Delrin–PTFE mixture of Dupont, USA) and EKONOL® (composite of PTFE and a high-temperature polyester, Carborundum, USA). These

Table 1 Materials Used in the Different Models of the Chitra Valve

Valve model	Cage		Disc		Sewing ring
	Material	Fabrication	Material	Fabrication	
1	Titanium	Integral major strut; electron beam welded minor strut	Polyacetal (Delrin)	Injection molded	Polyester knitted fabric
2	Haynes-25 alloy with TiN coating	Integrally machined struts	Single crystal synthetic sapphire	Machined and polished	Polyester knitted fabric
3	Titanium	Integrally machined struts	UHMW PE	Solid state polishing and cryomachining	Polyester knitted fabric
4	Haynes-25 alloy	Integrally machined struts	UHMW PE	Solid state polishing and cryomachining	Polyester knitted fabric

four polymers were screened for their water absorption and adhesive and abrasive wear resistance (25).

Water absorption was measured by soaking five specimens 25 mm in diameter and 2 mm thick in distilled water for periods of up to 15 days. The samples were carefully dried and weighed at intervals of 24 hours, 96 hours, and 360 hours. Figure 2 shows that the water absorption of UHMW PE was extremely low and was less than one-tenth that of the Delrin-based polymers. The very low water absorption of UHMW PE ensured that the disc was unlikely to swell and change shape like the Delrin disc occluder of the first Bjork–Shiley model during steam sterilization (23).

Adhesive and *abrasive* wear are the two main types of wear occurring in conditions like those of artificial heart valves. Adhesive wear is predominant when highly polished surfaces in contact articulate against each other (26). Table 2 shows the data from these studies of the four polymers and clearly indicates that UHMW PE had the lowest wear rates.

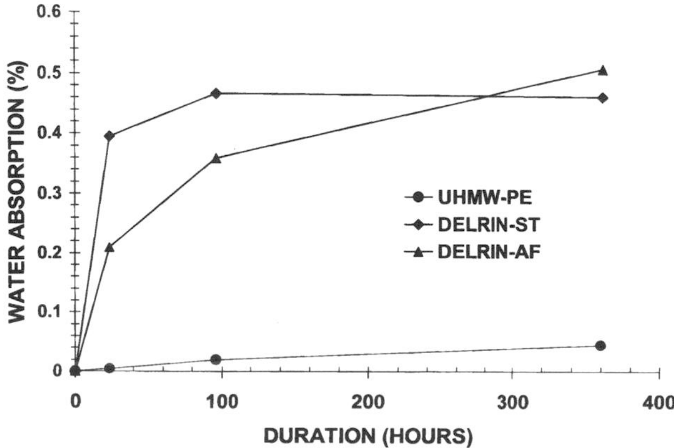

Figure 22.2 Comparison of water absorption for the various polymers evaluated as the occluder material.

The processing of UHMW PE for valve discs involves many thermal cycling stages such as compression molding, die polishing, annealing, and autoclaving of the valve for sterilization. UHMW PE specimens were thermally cycled as listed below and then tested for their wear resistance by both the methods.

Autoclaved once
Autoclaved thrice
Heated to 135°C
Heated to 160°C

These results, given in Table 3, establish that any thermal cycling of a material above 135°C can cause deterioration in its wear properties (the reported glass transition temperature of UHMW PE is 136–138°C). These tests showed that steam sterilization does not considerably increase the wear properties of UHMW PE.

Prototype valves were fabricated and evaluated as per the guidelines of the American National Standard for Cardiac Valve Prosthesis, ANSI/AAMI CVP3-1981, initially and later based on the guidelines of the International Stan-

Table 2 Results of Screening Wear Tests

Disc material	Pin-on-wheel test (volume loss in microliters)	Sand slurry test (volume loss in microliters)
UHMW PE	0.75	0.8
Delrin AF	5.68	4.1
Delrin ST	6.22	3.2
Delrin 500	23.18	6.3
Eknol PTFE	60.75	10.7
Mild steel	—	5.7

Table 3 Effect of Thermal Cycling of UHMW PE on Its Wear

Thermal cycle	Pin-on-wheel test (volume loss in microliters)	Sand slurry test (volume loss in microliters)
Virgin	0.75	0.80
Steam autoclaved once	0.82	—
Steam autoclaved thrice	0.84	—
Heated to 135°C	1.26	0.82
Heated to 160°C	9.65	4.30
Delrin 500 (control)	23.2	6.35

dard—Cardiovascular implants—Cardiac valve prostheses, ISO: 5840: 1989. The studies covered in vitro evaluation for hydraulic performance and accelerated durability followed by in vivo evaluation in animals and clinical trials.

III. IN VITRO FLOW DYNAMIC STUDIES

The hydraulic performance of the Chitra heart valve has been evaluated under different flow conditions in 1:1 model test flow channels and its performance compared with some of the standard models in current clinical use.

A. In Vitro Steady Flow Studies

The flow channels proximal and distal to the aortic and mitral valves (Fig. 3) were patterned on published work and provide cross-sectional areas analogous to the flow tracts in the human heart (27,28). The pressure drop measurements were made at tap P_1 located one valve diameter upstream and tap P_2 four diameters downstream of the test valve. This ensured that full pressure recovery had taken place at the point of measurement (29).

The steady flow system (30) consisted of the mitral test flow channel connected on its inlet side to a constant level tank via a venturi flow probe and a 1.5-meter-long straight entrance section. The outlet was connected to a flow control valve, which discharged into the main reservoir. The pressure drops across the test valves (at wall taps P_1 and P_2 as shown in Fig. 3) were measured using a signal conditioned semiconductor differential pressure transducer. The flow rate was accurately measured using a venturi tube, which was designed as recommended by the American Society for Mechanical Engineers (31).

The hydraulic performance of the 27 mm Chitra mitral valve (CHV) was compared with that of four popular valves of the same size, the Björk–Shiley standard (BSS), the Björk–Shiley monostrut (BSM), the Medtronic-Hall valve (MHV), and the St. Jude Medical bileaflet valve (SJM) under steady flow rates of 0 to 35 L/min.

Figure 4 shows the pressure drop of the 27 mm mitral valves under steady flow. The Björk–Shiley standard valve showed the highest pressure drop among the tilting disc valves due to its lower opening angle of 60°. A mild flutter was noticed at high flow rates, showing that for this disc shape, the center of pressure is just at the pivot point at full opening. The Medtronic-Hall valve also showed a similar flutter at its opening angle of 70°, while the Björk–Shiley monostrut with its convexoconcave disc did not show any. Woo and Yoganathan (32) have reported a similar observation, that the occluder of the MHV was oscillating during

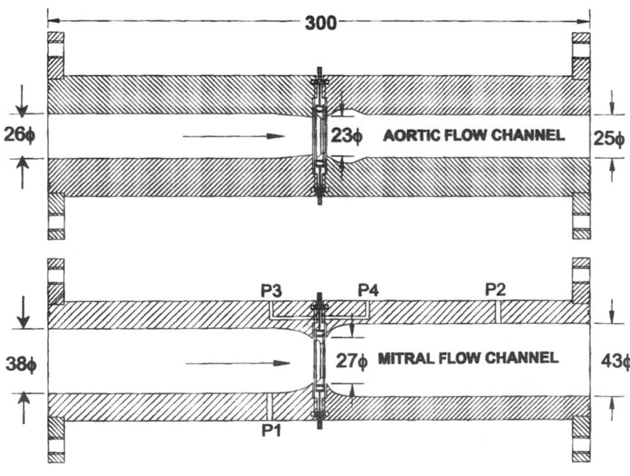

Figure 22.3 The aortic and mitral valve flow chambers used in the in vitro experiments.

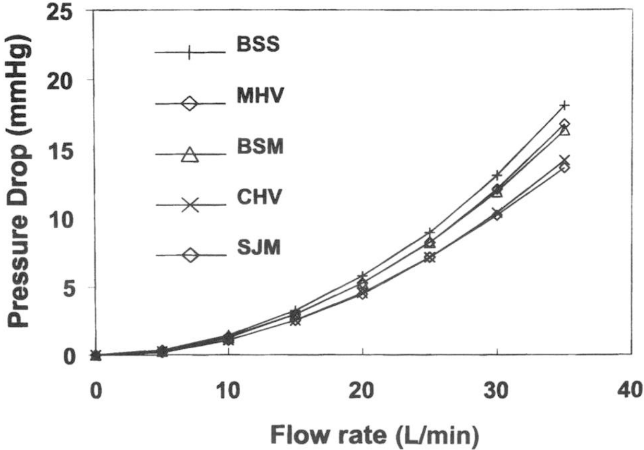

Figure 22.4 Comparison of steady flow pressure drop data for the various mechanical valves. BSS, Björk–Shiley standard; MHV, Medtronic-Hall; BSM, Björk–Shiley Monostrut; CHV, Chitra; SJM, St. Jude Medical.

diastole under pulsatile testing in the mitral position of a pulse duplicator, and they indicate that the flow could become nonstationary. The MHV is able to open to 70° in spite of its more central pivot location and a flat disc shape due to the presence of disc sliding. As shown by many workers, the St. Jude valve exhibited the lowest pressure drop, while the Chitra valve showed marginally higher magnitudes. The higher pressure drop of the BSM valve than the Chitra valve is probably due to the use of thicker struts.

B. Relative Flow Between Major and Minor Orifice Areas

The relative flow rates through the major and minor orifice areas of tilting disc valves were measured under steady flow rates of 15 to 25 L/min using the steady flow setup. Pressure drops across the fully open test valves were carefully measured at three flow rates of 15, 20, and 25 L/min; the flow rate being set using the venturi flow meter. Then with the disc in the fully open position, the minor orifice was blocked by sewing a piece of Teflon sheet over it and then filling the well thus formed in the valve with RTV silicone adhesive. In this state, the flow rate through the major orifice was measured using the venturi meter at the same pressure drop as measured first with the fully open valve. This procedure was then repeated with the major orifice similarly blocked. The percentage of the total flow rate for each orifice was then calculated and averaged over the three flow rates. The results are shown in Table 4.

The basic assumption in this technique is that the potential field across the valve is not significantly disturbed due to the blocking of the major or the minor orifice and remains nearly the same as when the valve is fully open. This of course is not strictly true and hence results in some error. Over this flow rate range, it can be seen that the technique yields fairly good estimates of the relative flow through the two orifices, and the effect of the differences in design can be appreciated.

Table 4 Relative Flow in Major and Minor Orifices of Tilting Disc Valves

	Percent of total flow through the valve	
Valve model	Major orifice	Minor orifice
Medtronic-Hall	69	29
Björk–Shiley monostrut	70	28
Björk–Shiley standard	80	19
Chitra heart valve	75	23

The present model of the Chitra valve exhibits a 4% improvement over the BSS due to the use of the planoconvex disc. The good improvement of flow through the minor orifices of the newer generation tilting disc valves, viz., Medtronic-Hall and Björk–Shiley Monostrut over the Björk–Shiley standard design can also be seen. These improvements in the minor orifice flow for the BSM and MHV have been documented by laser Doppler velocimetry by a number of investigators (28,29,32–34). However, there are no reports of such quantification of volumetric flow rates, which could be more meaningful to the cardiac surgeon.

C. In Vitro Pulsatile Flow Evaluation

Pulse duplicator setup. Many workers (35–39) have described the requirements and concepts for the design of a left heart pulse duplicator. The variety of systems that have been used reflects the problems and constraints involved in modeling the left ventricular dynamics and systemic circulation (27,28,36–38,40–42). The test system used (22) was an improvement of an earlier system developed here (25). The mitral and aortic flow channels were the same as those used in the steady flow test. The design of the systemic impedance was based on earlier work and was simulated by means of hydraulic elements consisting of two capacitances and resistances (38,43).

The blood analog fluid was a 35% solution of glycerol in water having a specific gravity of 1.06–1.08 at room temperature (28–30°C). The test parameters were

Cycling rate: 70 per minute
Cycle time: 856 ms
Systolic duration: 300 ms
Mean aortic pressure: 100 ± 2 mm Hg

The mean atrial pressure varied between 12 and 18 mm Hg depending on the mean flow. Adjusting the variable systemic resistance and controlling the pressure of the driving air set the mean aortic pressure at 100 mm Hg. Tests were conducted at mean flow rates of 2.5, 3.0, 4.0, and 5.25 LPM. Signals were acquired at 1000 samples/s for 32 consecutive cycles. Two such ensembles of data were acquired at each flow rate and averaged separately. The closing volume, the mean diastolic pressure drop across the valve, and the root mean square (RMS) diastolic flow rate were all calculated from the ensemble averaged data.

The effective orifice area (EOA) at each test flow rate for each valve was calculated using the well-known equation (28). The mean EOA for each valve and its standard deviation were obtained from these calculated values.

$$\mathrm{EOA} = \frac{Q}{51.6\sqrt{\Delta p}}$$

where

Q_{rms} = RMS systolic/diastolic flow rate or steady flow rate in cm^3/s and

Δp = mean pressure drop in mm Hg (systolic/diastolic in pulsatile).

The EOAs under both steady and pulsatile flow are given in Table 5. Among the mitral valves under steady flow, the St. Jude had the maximum EOA. The Chitra valve came a close second. Under pulsatile flow, all the valves showed comparable EOAs.

D. Closing Volume Measurement

Willshaw et al. (44) have used an artificial heart and a pneumotachograph to obtain more accurate measurements of the closing volumes and leaks of tilting disc valves. In this system, there is no dynamic pulsatile flow of the fluid at a given mean flow rate. The system essentially works with only the test valve, thus ensuring more controlled conditions of measurement. The pulse duplicator setup was modified. The ventricular pumping chamber and the mitral flow channel were used. The aortic side of the pumping chamber was sealed, and the valve under test was placed in the mitral position in the conventional manner.

The system was filled with the test fluid to a level of 10 cm above the valve, leaving a column of air space above. Inflating and deflating the rubber bag using compressed air generated the pumping action. A pneumotachograph was designed, fabricated, and calibrated (30). The device acts like a resistance and produces a differential pressure proportional to the rate of airflow through it. The principle of measurement in this system is based on the fact that any volume of water that passes through the valve under test must displace an equal volume of air through the pneumotachograph probe. As the differential pressure across the probe is kept small by design, the volume change due to the compressibility of air is negligible in comparison to the total flow.

This setup functions well with valves whose occluders are denser than the test fluid, like LTI carbon discs. In the case of the Chitra valve with its UHMW PE disc (specific gravity of UHMW PE = 0.94), the disc opens down fully at the start of forward flow. However, since the forward flow is small (equal to the total regurgitant volume), the flow rate falls to zero towards the end of the filling phase. At this point, the disc floats up and is in a semi closed position. At the start of the systole (bag inflation here), the valves close rapidly, exhibiting a small closing volume.

In order to test the UHMW PE disc valves more fully, the system was modified as shown in Fig. 5 with the disc opening upwards. The UHMW PE discs float up to a fully open position, while the denser discs tend to move down to a small opening angle. In this mode, the pneumotachograph probe was connected to a compliance chamber as shown in the figure. This compliance could be pressurized by pumping air into it. The valve mounting direction was reversed to open upwards in a manner similar to the aortic valve of the pulse duplicator. The mitral test chamber was used for all the tests. During inflation, the fluid flows up, displacing air into the compliance chamber. During deflation of the bag, the air pressure in the compliance chamber

Table 5 Effective Orifice Area

Valve	Steady flow (cm^2)	Pulsatile flow (cm^2)
St. Jude Medical	3.07 ± 0.05	2.51 ± 0.10
Medtronic-Hall	2.80 ± 0.02	2.41 ± 0.09
Björk–Shiley Monostrut	2.82 ± 0.02	2.57 ± 0.13
Björk–Shiley Standard	2.68 ± 0.01	2.51 ± 0.10
Chitra UHMW PE	2.98 ± 0.02	2.47 ± 0.14

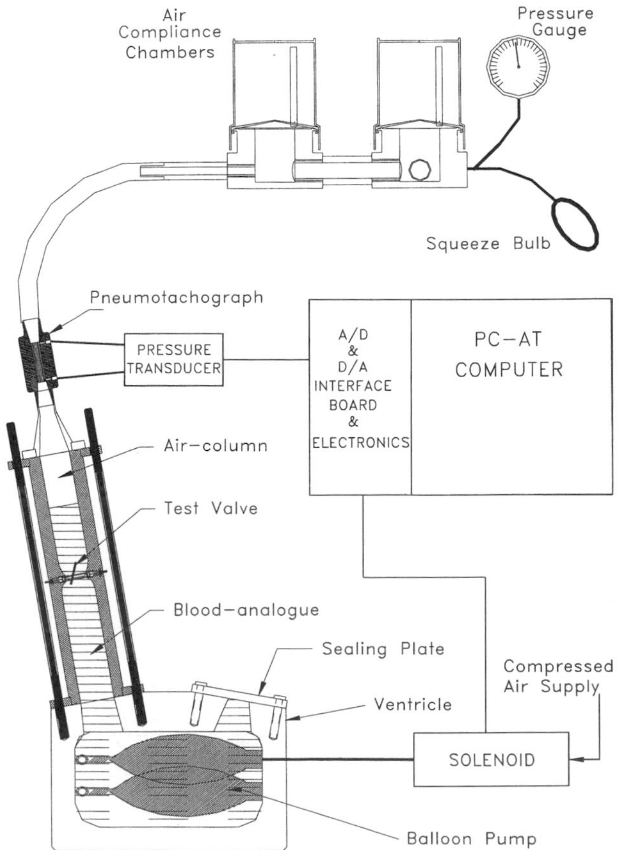

Figure 22.5 Schematic of the closing volume and impact force experimental setup: valve opening upwards.

forces the valve to close; the movement of air being equal to the closing volume as before. The cycling rate, timings, and data collection were all carried out as before

Figure 6 (top) shows typical ventricular pressure and air flow rate ensemble averaged waveforms for the valve opening downwards. In this figure, four phases can be seen. Phase 1 consists of a delay between the starting pulse from the computer (time zero) and the moment at which fluid begins to flow through the valve (point a). The delay, approximately 20 ms, was caused by the time required for the solenoid valve actuation and the buildup of air pressure in the ventricular bag. Phase 2 was the closing phase, during which water accelerated rapidly through the valve, reaching a peak flow rate that was abruptly terminated by valve closure. This abrupt termination produces resonance in the air space above the water column, which was phase 3. The start and finish of the closing phase was detected by the change in sign of the values. The valve closure was defined as the first zero crossing of the resonant phase (point b in the figure).

Figure 6 (bottom) shows typical ensemble averaged waveforms in the case of the valves opening upwards. As before, phase 1 was the delay for the inflation of the bag to start. Here phase 2 was the forward flow through the valve, when the air above the valve was displaced into the compliance chamber with consequent buildup of pressure there. At the start of deflation, the forward flow was terminated by the fall in ventricular pressure, resulting in valve closure due to the higher pressure in the compliance chamber—this was phase 3, the closing phase of the valve. The end of valve closure (point b) was followed by phase 4, when the air column resonates. The closing volume was the area under the curve during the closing phase, i.e., between points a and b. The closing volume, (CV in mL) was determined by integrating the flow signal from the start to the finish points. The results are given in Table 6.

Closing volumes under dynamic pulsatile flow conditions were obtained by integrating the area under the EM flowmeter signal for the duration of the closure period as defined in the international standard. The results are also given in Table 6. The St. Jude valve with its minimal leaflet travel shows the lowest closing volume, while that of the Chitra valve was marginally higher. Other investigators have shown that the closing volumes for the SJM are comparable to others like MHV and BSM (28,45). Knott et al. indicate that closure volumes are mainly dependent on the opening angles (42).

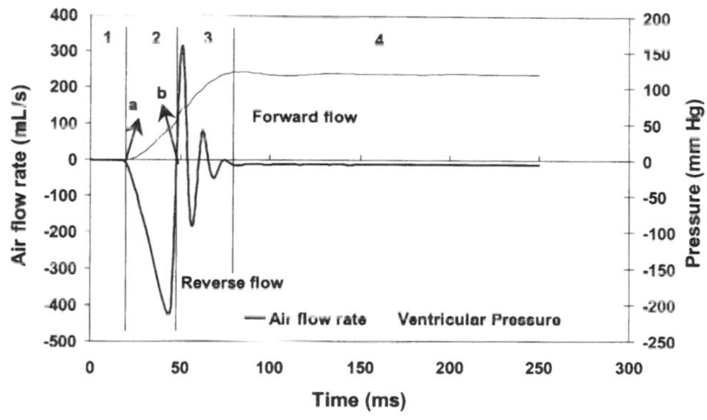

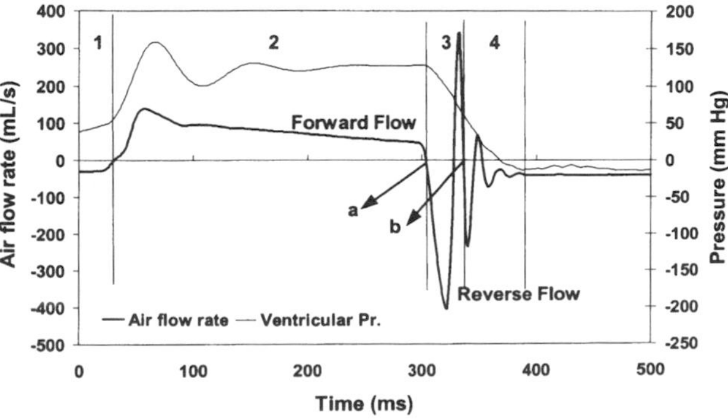

Figure 22.6 Ensemble averaged flow rate waveform: (top) valve opening downwards; (bottom) valve opening upwards.

E. Closing Impact Force Measurements

For the efficient working of any prosthetic heart valves it is necessary that the valve close quickly at the start of reverse flow. To achieve this quick closure, the occluder has to accelerate from its fully open position to the closed position and thereby impact with the cage mechanism before coming to rest. This impact at closure results in an impact force of short duration being transmitted through the cage, sewing ring, and sutures to the heart muscle. Large impact forces can naturally cause problems: suture dehiscence, fatigue fracture, wear of components, and probably tissue overgrowth due to chronic inflammatory response of the host tissue.

Table 6 Closing Volume and Impact Force Results

Valve model	Closing volume (mL)	Closing impact force (newtons)	Dynamic closing volume (mL)
Björk–Shiley monostrut	6.4	67.5	6.7
Björk–Shiley standard	5.4	51.1	5.7
Medtronic-Hall	6.2	66.0	6.5
St. Jude Medical	3.6	50.6	4.8
Chitra heart valve	4.9	34.3	5.2

The closing impact forces were measured in the same setup as for the closing volume measurements. The air flow probe was removed. Tests were conducted with the valves opening down for the denser discs and opening up for UHMW PE discs for the same reasons as before. The test and data acquisition conditions were also similar. The test valve was sutured to a force transducer plate. The transducer plate was a specially made stainless steel valve holder with strain gauges mounted for measuring the forces. The details of the design and calibration have been described earlier (30).

In the case of valves opening downwards, the initial force is zero with the valve open. At the onset of inflation, the valve closes rapidly with an impact followed by a short ringing. The force signal then settles at a high value due to the ventricular pressure during systole. In the case of valve opening upwards, the closure occurs at the start of deflation.

The closing impact force was taken as the peak value of the force waveform. The peak values were averaged over the 64 cycles. Figure 7 shows typical ventricular pressure and force waveforms for the two cases. The results of the closing impact force measurements are given in Table 6. The BSM showed the maximum impact force, followed by the MHV, BSS, and SJM. The Chitra valve showed the lowest impact force. This is due to the use of the soft UHMW PE occluder, which considerably absorbs the shock of impact at closure. This data correlates well with the closing mechanics for this model described in the following section.

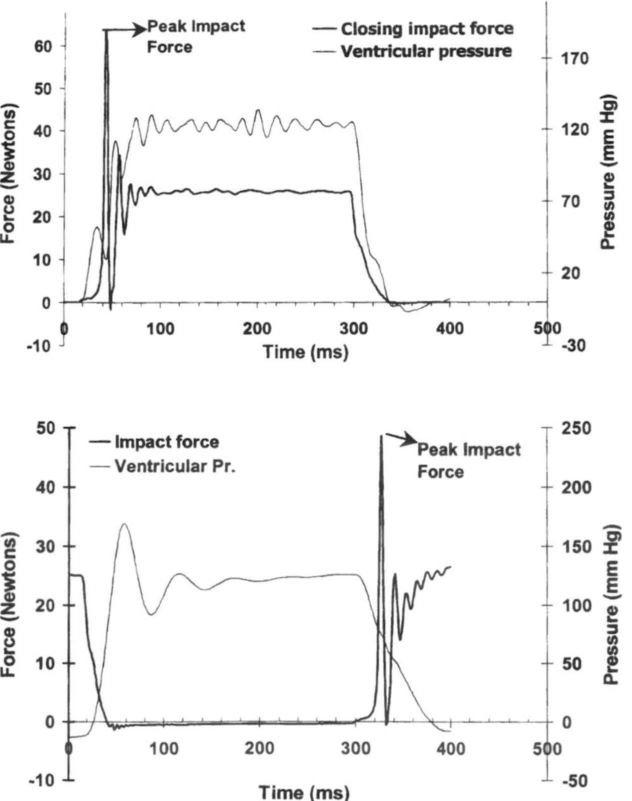

Figure 22.7 Closing impact force plots: (top) valve opening downwards; (bottom) valve opening upwards.

IV. FLEXIBLE (UHMW PE) LEAFLETS AND VALVE CLOSING DYNAMICS

A. Mechanical Valve Closing Dynamics and Induced Stresses

The thromboembolic complications associated with mechanical valve implants have been correlated with the forward flow dynamics since the introduction of mechanical valves (1,2). The flow characteristics suggested as factors causing thrombus deposition include turbulent stresses on the downstream side with flow across a fully open valve, and regions of relative stasis and flow separation. In spite of design improvements to reduce fluid induced stresses in the mechanical valve function, thromboembolic complications continue to be significant with these implants. In the last decade, problems associated with pitting and erosion in mechanical valves implanted in the artificial heart (46) as well as with a bileaflet valve implanted in animals and patients (47) have been reported. Structural failure with mechanical valves have included leaflet fracture (48) and fracture of pivot components and housing (49–52). Recently, a number of studies have associated thrombus initiation in mechanical valves with the stresses induced during valve closure. The occluder of a mechanical valve moves towards the closing position due to the adverse pressure gradient (larger pressure on the downstream side than on the upstream side) during the valve closing phase. All currently available tilting disc and bileaflet mechanical valves have relatively rigid pyrolytic carbon occluders. At the instant of impact of the occluder on the seat stop or the seating lip, the occluder tip comes to a sudden stop, resulting in a water hammer effect. Large positive pressure transients are induced near the leaflet tip close to the valve on the downstream side of the valve. Similarly, large negative pressure transients have been recorded on the upstream side of the valve close to the leaflet tip. This water hammer effect can be the source of several fluid dynamically induced stresses that can initiate hemolysis and platelet activation. The large pressure gradient induced across the valve leaflet, even though for a fraction of a second, can force blood through the clearance region between the leaflet and the valve housing. Model studies have demonstrated that relatively large wall shear stresses and local negative pressures can develop in this region at the instant of valve closure (53–56). In the simulation performed by Aluri (55,56), a simplified geometry of the leaflet and valve housing was employed to study the stresses induced by the flow through the clearance region during valve closure. In order to delineate the effect of leaflet motion during the valve closure, the first simulation involved the leaflet in the fully closed position with the ventricular pressure rise. The second simulation started with the leaflet in the open position and the leaflet moving towards the closed position with the ventricular pressure rise. A moving boundary for the rigid leaflet motion was included in the computational analysis for the latter simulation. At the instant of valve closure, the second simulation demonstrated the presence of a large negative pressure on the downstream side of the leaflet. The computed wall shear stress with the moving leaflet at the instant of valve closure was about 4000 Pa compared to about 580 Pa when the leaflet was held in the fully closed position. This study clearly demonstrated that the moving leaflet and the sudden closure of the rigid leaflet induced wall shear stresses of an order of magnitude larger than that when the leaflet was held stationary in the closed position. The clearance region between the leaflet and the valve housing in the periphery is also the region in which relative stasis and flow separation are usually observed during the forward flow phase of the cardiac cycle. Hence it is logical to assume that activated platelets subjected to relatively large shear stresses during valve closure can induce thrombus deposition at these sites in the subsequent forward flow phase.

The flow through the clearance region during valve closure has also been shown to induce regurgitant jets, resulting in relatively large turbulent stresses near the edge of the leaflet (57). The turbulent stresses measured in the regurgitant jets were significantly larger than the magnitudes measured downstream to the valve during the forward flow phase. Relative damage to blood during the forward and reverse flow phases of the cardiac cycle were studied in a steady flow system (58). In this study, damage due to steady forward flow through a fully open mitral valve was measured periodically at flow rates of 3 to 7 L/min. The blood damage was assessed by the measurement of plasma free hemoglobin in blood. Reversed flow across a closed mitral valve was also assessed at flow rates of 0.1 to 0.2 L/min in the same steady flow set up (by reversing the orientation of the valve). The levels of hemolysis observed were the same for both the forward and reverse flow phases, even though the flow rates in the reverse flow were an order of magnitude smaller. Significantly higher levels of hemolysis were also observed in pulsatile flow simulation. Even though pulsatile flow studies could not delineate the amount of blood damage between the forward and reverse flows through the valve, the significant increase in blood damage during pulsatile flow can be attributed to the valve closure induced stresses.

Cavitation is the rapid formation and collapse of vapor-filled bubbles caused by a transient reduction in local pressure to below the liquid vapor pressure. For glycerol solution, the vapor pressure is −743 mm Hg (59), and for whole blood, the magnitude is −713 mm Hg. Should the negative pressure transient (NPT) reach magnitudes below the vapor pressure for blood for a sufficient duration, cavitation bubble formation and collapse can result. The implosion of the bubbles can damage the blood cells and activate the platelets. The pitting and erosion observed on the occluder and housing surfaces with implanted mechanical valves suggest cavitation damage (46,47). It is not surprising that the suggestion of cavitation damage on valve surfaces has resulted in numerous in vitro studies on the visualization of cavitation bubbles during valve closure (13–21), the measurement of negative pressure transients at the instant of valve closure and cavitation initiation (13–16,18–21), and the velocity of the occluder tip at the instant of valve closure (14,60–64). These studies have demonstrated the presence of negative pressure transients with pressure below the vapor pressure of blood near the occluder edge on the upstream side of the leaflet. The cavitation bubbles visualized near the leaflets for a period of less than 1 ms in this region correspond to the regions where the negative pressure transients have also been recorded. Typical negative pressure transients recorded with tilting disc and bileaflet mechanical valves at the instant of valve closure, and the corresponding cavitation bubbles visualized, are shown in Fig. 8. Mechanisms that have been suggested to initiate cavitation bubbles during valve closure include the water hammer effect and the presence of negative pressure transients (13–15), vortices generated at the edge of the leaflet during valve closure (18), and the additional pressure drop induced by the velocity of fluid being squeezed between the leaflet and the seat stop at the instant of valve closure (16). In comparing the closure mechanics with valves of the same geometry (e.g., tilting disc valves), it was demonstrated that even with the presence of negative pressure transients, initiation of cavitation bubbles was dependent on the design of the valves. In valves where there was an interaction between the occluder and a seat stop at the instant of valve closure, bubbles were present, whereas no bubbles were observed in valves without such interaction (13,15). These studies suggest that the local flow dynamics in the vicinity of the leaflet edge and the squeeze flow effect are important in cavitation initiation. Numerical simulation of the squeeze flow effect has been reported (65–67), and these studies suggest that the fluid velocity

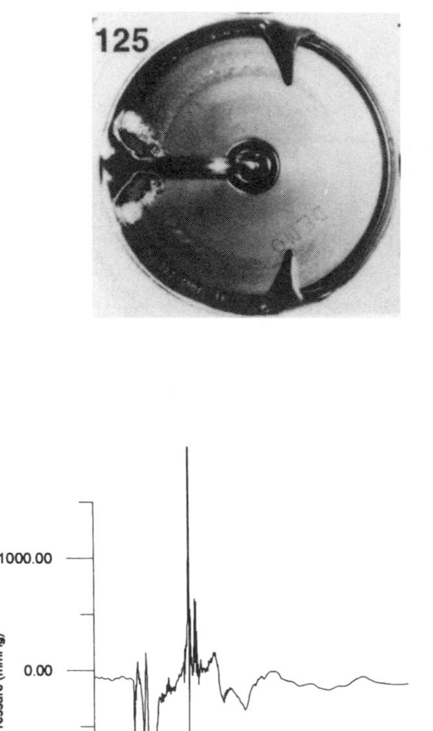

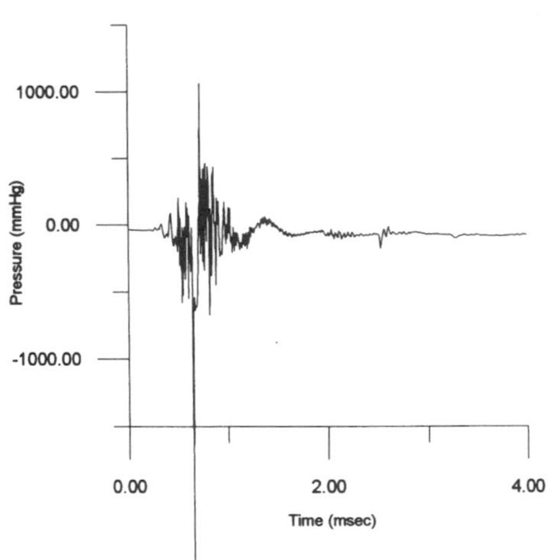

Figure 22.8 Negative pressure transients recorded in vitro with mechanical valve closure and the cavitation bubbles visualized in vitro. The inserts on the photographs indicate the time in microseconds after the instant of valve closure when the bubbles were visualized.

induced by the squeeze flow effect are of the order of several m/s and hence can induce additional local pressure drop that may initiate cavitation bubbles. Simulations have also suggested that the relatively rigid pyrolytic carbon leaflets in the mechanical valves rebound after the initial impact with the valve seat and hence induce additional negative pressure transients as well.

Studies performed on cavitation bubble visualization in pulse duplicators with the mechanical valve in the mitral position have used the peak ventricular pressure rise (dp/dt_{max}) as an index for the threshold for cavitation. However, the motion of the leaflet during valve closure will be governed by the ventricular pressure rise during the valve closure (about 30 ms after the beginning of the ventricular pressure rise), whereas the peak ventricular dp/dt occurs after the valve is fully closed in the mitral position. Hence the ventricular pressure rise rate during the leaflet closure, dp/dt_{cl}, has been suggested as the appropriate parameter to compare the closure mechanics of the various valve designs (13–17) and has been adopted by the FDA in mechanical valve closure studies. These studies were performed with controlled pressure rise rates during valve closure, and measurements of pressure transients, valve leaflet velocity, and bubble visualization were performed with a single closing event of the valve rather than simulating the periodic closure that the normal valve undergoes. However, a comparison of the cavitation bubbles visualized, as well as the characteristics of the pressure transients, in the studies with a single closing event with those obtained in the pulse duplicator have been observed to be similar.

The pressure transients in the above-referenced studies were from in vitro studies, generally in rigid flow chambers with the valve rigidly mounted in the mitral position. The effect of the flexibility of the mount (to simulate the valve sutured in the mitral orifice in vivo) has been suggested as important (64). However, studies with a single closing event with rigid and flexible valve holders have shown that this effect is not important. The water hammer effect and the local flow dynamics inducing cavitation bubbles will be due to the relative velocity between the valve housing and the leaflet and hence should not be affected by the relative rigidity of the valve holder (17).

Questions have also been raised about the presence of such negative pressure transients with mechanical valves implanted in vivo, since the in vitro studies do not simulate the distensibility characteristics of the atrial chamber. The only indirect evidence that cavitation is possible in vivo with implanted valves is the pitting and erosion observed in explanted valves that are characteristic of cavitation damage. At present, there are no techniques to visualize the presence of cavitation bubbles in vivo. However, if negative pressure transients similar to those demonstrated in the in vitro studies can be recorded with implanted mechanical valves in vivo, one can assert that at least a *potential* exists for cavitation to occur in vivo. Studies were performed in an animal model in which mechanical valves were implanted in the mitral position in sheep and a high-fidelity transducer was positioned very close to the leaflet in the closed position in the atrial chamber. The ventricular pressure rise was also recorded for a range of ventricular contractions with the aid of pharmacological interventions. These studies demonstrated that the ventricular pressure rise measured during mitral valve closure in vivo (68,69) were in the same range as those that were used in the in vitro studies earlier (13). Furthermore, negative pressure transients were recorded in the atrial chamber very close to the mitral valve in vivo, and they were also remarkably similar to those measured with a single closing event of the same valve in vitro. These studies further demonstrate that there is a potential for cavitation to occur in vivo with implanted mechanical valves. Typical negative pressure transients measured in vivo with the mechanical valve implanted in the mitral position are compared with those obtained in vitro with valves of the same design in Fig. 9. It should also be pointed out that these studies have demonstrated that the presence of negative pressure transients is not the only necessary condition for cavitation to be initiated; local fluid dynamics based on the valve designs play an important role. Since structural failure with implanted mechanical valves is not significant, one can assume that cavitation damage occurred in a very small fraction of mechanical valves and is due to the local flow dynamics related to the interaction of the leaflets and the valve housing in certain designs. Even in the absence of cavitation bubbles, the negative pressure transients can result in the viscoelastic expansion of the formed elements in blood, resulting in thrombus initiation. All these fluid dynamically induced stresses, including high shear stresses in the clearance region in the vicinity of the leaflet edge and the housing, significant regurgitant jet turbulent stresses, stresses induced by cavitation bubble collapse, and viscoelastic expansion of formed elements in the negative pressure region, can result in the activation of platelets and thrombus initiation. This region near the valve housing is also the region of flow reversal and relative stasis in the subsequent forward flow phase; hence induced thrombus will likely deposit and grow in this region. Thus mechanical valve closure and fluid dynamic stresses resulting during this phase of the cardiac cycle may be the dominant factors in thrombus deposition with implanted valves in the absence of anticoagulation.

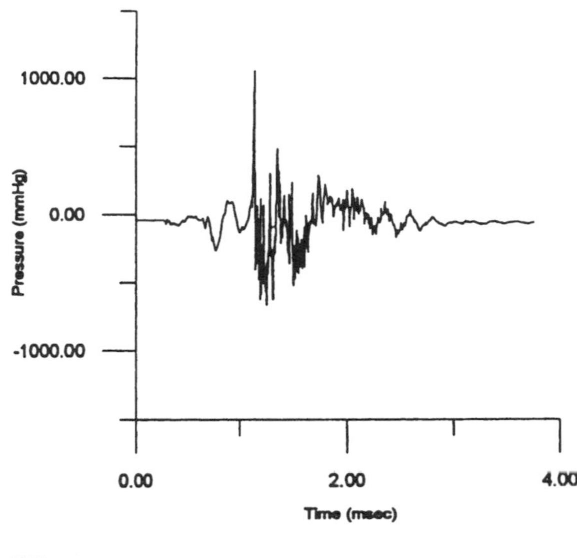

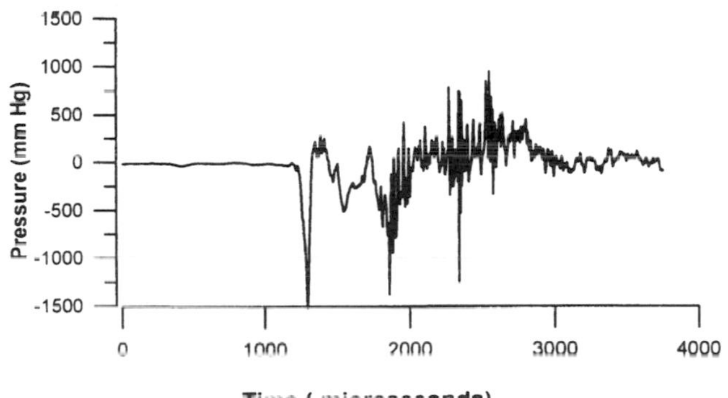

Figure 22.9 Comparison of negative pressure transients recorded during valve closure in vitro (top) with valve implanted in the mitral position of an animal model to that recorded for the same valve in vivo (bottom).

B. Occluder Flexibility and Valve Closing Dynamics

The studies described above have concentrated on the currently available mechanical valve prostheses with relatively rigid pyrolytic carbon leaflets. As such a leaflet impacts against the seat stop at the instant of valve closure, the water hammer effect results, since the leaflet does not yield significantly to the impact nor absorb part of the impact energy. Studies have shown that the leaflets rebound after impact and create additional negative pressures on the upstream side of the leaflets. Polymeric valves have flexible leaflets, which can absorb part of the energy during impact, and hence one can expect that the negative pressure transient amplitudes will be significantly smaller. Studies were done with the Chitra valve (with its relatively flexible UHMW PE leaflet) both in vitro and in vivo. Comparisons were made on the leaflet tip velocity at the instant of valve closure, negative pressure transients, and presence of cavitation bubbles in vitro under similar closing loads between mechanical valves with rigid leaflets and the Chitra valve (13–15). For a valve with the same design (e.g., tilting disc geometry) and the same size, the leaflet tip velocity at the instant of valve closure was the same irrespective of the rigidity of the leaflet. However, under the same closing conditions, the amplitude of the pressure transients with the Chitra valve was significantly smaller than those with a rigid leaflet valve. Since the negative pressures with the polyethylene leaflet valve never reached magnitudes close to the vapor pressure for the blood analog used in these studies, no cavitation bubbles were observed with these valves. Similar results were obtained from the in vivo studies as well. A comparison of the pressure transients of mechanical valves with rigid leaflets with those of the Chitra valve in vivo is shown in Fig. 10. These studies suggest that during valve closure with flexible leaflets, the yielding of the leaflets absorbs part of the energy during impact. Thus the fluid dynamically induced stresses in these regions during closing can be anticipated to be significantly smaller than those with the currently available mechanical valves. A controlled study on thrombus initiation with rigid and flexible leaflet mechanical valves is necessary to evaluate whether flexible leaflets will help minimize the problem.

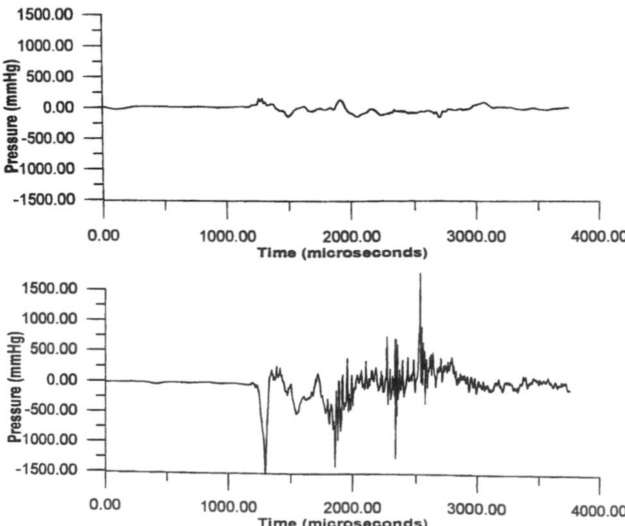

Figure 22.10 Comparison of negative pressure transients with a rigid leaflet mechanical valve with the flexible leaflet Chitra valve from in vitro and in vivo experiments.

V. ACCELERATED DURABILITY TESTING FOR WEAR AND FATIGUE RESISTANCE

The mechanical durability of the heart valves in terms of their wear resistance and fatigue properties was determined by accelerated life cycling testing. The test protocol is based on the guidelines of the ISO standard Cardiovascular implants—Cardiac valve prostheses: ISO: 5840:1989. Testing at normal heart rates in mechanical systems or animal models for over 10 years is unrealistic. Several methods of accelerated testing of prosthetic heart valves have been tried and used (70). These include pneumatic cycling and several variations and combinations of mechanical and hydraulic cycling.

The test system in use for the past 20 years has been described earlier (25,30). The hardware of the system presently in use is third generation, and many problems with the reliability of the equipment have been minimized so that the test results can be interpreted with reasonable accuracy. Five equally spaced valves are mounted on a stationary housing and covered with an acrylic chamber. The test fluid (distilled water treated with trace amounts of copper sulfate to control fungal growth) is supplied through the center inlet of the test housing. A DC motor with a variable speed control drives the rotor inside the housing. A rotor sequentially actuates each of the five test valves and distributes the fluid at a pressure of 200 mm Hg (26.7 kPa). The test valves are mounted with their sewing rings stitched to silicone rubber supports to isolate and damp their vibrations during cycling.

Two combinations of test valves (Haynes-25/UHMW PE: three each of size 27 mm and 23 mm; Titanium/UHMW PE: two of size 23 and one of size 27 mm) and two 27 mm Björk–Shiley standard valves were tested in this system for over 350 million cycles each. The valves were cycled at 800–840 times a minute, the rate being periodically checked using a strobe. The valves were removed after 1, 2, 5, 10, 20, and 40 million cycles and thereafter at intervals of 40 million cycles. In these instances, the assemblies were removed from their sewing ring holders, cleaned, degreased, and dried before weighing.

Wear of the valve components was measured by weighing them (or the cage/disc assembly during the test) to an accuracy of ± 0.1 mg in a single-pan electronic balance. The cage and disc were weighed separately before assembly. Control weights of titanium (coated with titanium nitride to prevent erosion) were used to monitor the reliability and consistency of the weighing balance over the prolonged period of the test.

The end of the cycling of any valve was determined when

1. There was a failure of any its components.
2. Excessive wear was noticed and the valve was expected to fail very soon.
3. The valve reached 400 million cycles without failure.

At the end of the test, the valves were dismantled and the components weighed again. They were also inspected for signs of wear and other degradation. The volume of wear was calculated from the weight loss and the density of the material. During cycling, since the cage and disc cannot be weighed separately, the weight loss was attributed to the component that wears most. The error due to this was found to be small as borne out by the final weighing.

Figure 11 shows the results of these tests. The volume of wear was calculated from the weight loss and the density of material (UHMW PE = 0.94 and LTI carbon disc = 2.0 g/mL). The bulk of the wear was in the UHMW PE disc. These rates are marginally higher than those of the Björk-Shiley standard. No mechanical failure or signs of excessive wear were encountered in any of the test valves.

The extrapolation of the wear data obtained from accelerated wear testing to predict actual implant durability depends on the relation of the wear occurring in vivo to that measured in the test system. To assess this factor, valves implanted in sheep have been carefully monitored for weight loss. The number of cycles that the valve has gone through is estimated by assuming that on an average, the heart of the animal beats at 70 per minute. Table 7 shows the weight loss of valves recovered after implantation and the corresponding weight loss interpolated from accelerated wear data. The clinical valve was recovered at reoperation for a thrombosed valve and cleaned well including enzymatic treatment to remove adherent proteins. It should be noted that the accuracy of the electronic balance over such long periods of time is at best ±0.2 mg, even with the careful calibration that was followed. The data clearly indicate that there are no signs of either excessive wear or absorption of body fluids; the polymer seems to stabilize inside the body as expected.

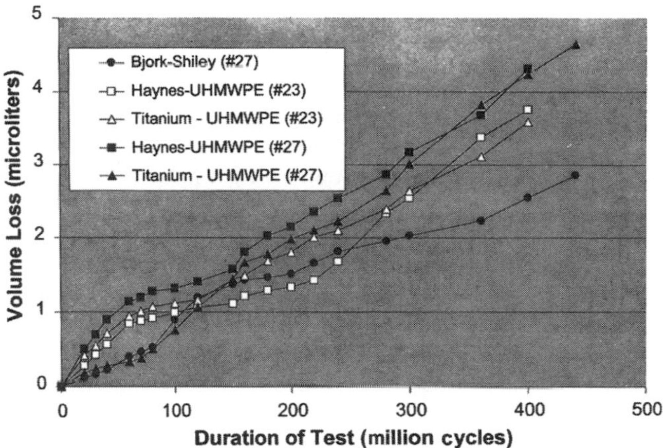

Figure 22.11 Volume loss plots from the accelerated wear tests.

Accelerated testing inherently imposes unrealistically severe conditions because of

1. The inability to achieve adequate system damping
2. Increased pressures that are required to achieve full valve excursion
3. Inferior lubricating properties of the test fluid
4. High frequencies imposing stresses that are not found at lower rates

Accurate prediction of device durability is of considerable importance in the development of life-saving devices like prosthetic heart valves. The reliability and durability of the test apparatus can be a significant problem, when one considers that it must outlive a relatively simplistic device designed to withstand an equivalent of 38 million cycles per year for over 50 years or more. Hence the design of the test system and the subsequent results obtained with a certain model of valve need to be validated with data from explanted devices, either animal or clinical.

The system used here has proved to be reliable in terms of long-term performance. Table 7 shows that this system gives a reasonably good 1:1 correspondence to the wear rates obtained from explanted valves.

The Björk–Shiley standard valves have been in clinical use since 1971 and have had an excellent record of durability. Their durability based on wear rates was estimated to be 400 years (70). The wear rate of UHMW PE disc valves

Table 7 Explant Data from Animal Trials

Implant duration (Days)	Estimated cycles (millions)	Wt. loss (mg)	Reasons for explantation	Accelerated test: Interpolated wt. loss (mg)
			Titanium cage—UHMW PE disc valve—animal explant	
210	21	0.2	Elective terminated—good healing	0.42
220	22	0.5	Elective terminated—good healing	0.44
418	42	0.4	Animal died—tissue overgrowth at the inflow side—valve disc got stuck	0.70
454	46	0.8	Elective sacrifice—excellent healing	0.75
498	50	0.3	Animal weak—terminated—evidence of myocardial infarction	0.81
			Haynes-25 cage—UHMW PE disc valve—animal explant	
96	9.7	0.2	Elective sacrifice	0.17
175	18	0.2	Elective sacrifice—good healing	0.28
1457	148	0.9	Animal died—bronchopneumonia	1.30
2765	280	2.1	Animal died—viral infection	2.48
3213	322	2.2	Animal died—valve thrombosis	2.60
			Haynes-25 cage—UHMW PE disc valve—clinical explant	
758	76	0.3	Reoperation for thrombosed valve	0.78

is marginally higher than that of the Björk–Shiley valve. The wear of the plastic disc gets distributed over its surface as it rotates during working. Hence, for a given volume of wear loss, the thickness of the component worn out is very much smaller than it would be if the cage struts were to wear. Clinical use of UHMW PE in artificial hip joints over the last 35 years has shown this material to be extremely stable in the body environment. Considering all this, valves with UHMW PE discs can be expected to last for over 50 years of implant life at a minimum.

VI. IN VIVO EVALUATION

A. Animal Studies

Adult sheep in the weight range of 30 to 40 kg are the animal model for this study. It was carried out with the authorization and supervision of the Institute's Animal Care Committee. *The International Guiding Principles for Biomedical Research Involving Animals* of the Council for International Organization of Medical Sciences (CIOMS) were strictly followed (71). Sheep were chosen as the model for this trial for the following reasons:

1. Adult sheep in the weight range of 30 to 40 kg had a heart size suitable for implantation of a 23 mm valve in the mitral position.
2. Well-established procedures for the conditioning, anesthesia, surgery, and postoperative care already existed following the development of a bubble oxygenator.
3. Earlier experiments showed pigs, dogs, and calves to be poor candidates for open-heart surgery in comparison.

The details of these studies for the clinical model of the Chitra valve have been published earlier (22). The mitral valve of the sheep was replaced with the test valve (23 mm sewing diameter) using standard open-heart surgical procedures.

Postoperatively, the animals were given antiplatelet agents for the first 14 days to reduce platelet aggregation and deposition. No anticoagulants were given. They were closely watched for signs of ill health such as loss of weight and appetite and general alertness. Valve sounds were regularly auscultated.

All animals that either died or were electively terminated were subjected to a detailed autopsy, covering the following areas:

1. Cause of death
2. Inspection of the valve for thrombus/platelet-fibrin deposit and tissue buildup
3. Evidence of thromboembolism by gross and histo-

pathological examination of the brain, kidneys, spleen, and liver
4. Explant analysis of the valve covering
5. Physical changes: Weight loss of the components as a measure of wear

Five mitral valve replacements of each of the Ti/UHMW PE and Haynes-25/UHMW PE models (all of 23 mm size) were carried out successfully. Three animals from the Ti group at 7 months and two from the Haynes-25 group at 3 and 6 months were electively terminated (Table 7). In the titanium group one animal died at 1 year and 2 months due to tissue overgrowth on the inflow side of the valve, leading to immobilization of the disc. The last animal became very weak and was electively terminated at 1 year and 4 months. An autopsy revealed evidence of myocardial infarction—the valve was clean, the sewing ring was well healed, and the disc was moving freely. Whether the infarction was due to a thromboembolism from the valve could not be determined at autopsy.

In the Haynes-25 group, three animals died at various periods as given in the Table 7. One died at 4 years due to bronchopneumonia and the second at 7 years and 7 months due to a viral infection. In both the cases the sewing ring was well healed all round and the disc was moving freely. In the third animal, which survived for the longest duration, 8 years and 9 months, a large thrombus was noticed on the valve at autopsy, which effectively immobilized the disc. It should be noted that except for the first 14 days following the valve implantation, the animals were not given any antiplatelet or anticoagulant medication.

The pressure drop across the valve was measured by transthoracic cannulation of the left atrium and the left ventricle in three anaesthetized animals (two from the titanium group and one from the Haynes group). The cannulae were connected to a semiconductor differential transducer using extension tubing and three-way stopcocks. ECG and the differential pressure signals were digitized and acquired using a data acquisition system. The end diastolic pressure drop was read for 10 consecutive cycles and averaged. The mean end diastolic pressure drop for the three animals was 3 mm Hg.

The hemolytic effect of the valve was determined by measuring free hemoglobin in plasma, the reticulocyte count, and the lactose dehydrogenase (LDH) before and after surgery and then after the third month. Hemolysis due to the valve could not be detected (22).

The explanted valves in general (except for the two noted earlier) showed excellent healing of the suture ring. No physical changes, including dimensional changes or degradation of the polymer, were observed. Weight loss of the cages was not measurable. The weight loss of the discs is given in Table 7. The good correlation between the weight loss in animals and the equivalent estimates interpolated from the accelerated wear data even at 7 and 8 years clearly indicate the stability of the polymer and its durability.

B. Clinical Studies

Based on engineering and animal data, it was clear that both combinations, titanium/UHMW PE and Haynes-25/UHMW PE, were equally good. Polishing of Haynes-25 with its high chromium content was found to be easier than with titanium. Obtaining a consistently high quality of surface finish with titanium alloys is more difficult than with chromium-containing alloys like stainless steels and the cobalt-chromium alloys. In view of this, it was decided to use a Haynes-25 cage and a UHMW PE disc for further clinical trials and commercial production.

On the basis of the engineering and in vivo animal data, the Ethics Committee of the Sree Chitra Tirunal Institute approved the evaluation of the Chitra valve in patients on October 27, 1990. The study was based on a common protocol developed at a joint meeting of the investigators with a senior statistician and approved by the institutional ethics committees. In the first phase, from December 1990 to November 1991, forty valves were implanted at the Institute. In the second phase, from January 1992 to January 1995, five additional centers across the country were included in the study. After a total of 306 valve replacements (101 aortic and 205 mitral), an analysis was carried out in February 1995 for the Monitoring Committee of the multicenter study, and the results have been reported previously (22).

A detailed follow-up was carried out from September 1997 to September 1998 for this same group of patients. This clinical data analyzed separately for aortic and mitral valves as per the American Association for Thoracic Surgery (AATS) guidelines of 1996 are being reported in detail (72). The overall analysis of all the patients, which brings out the clinical performance of this valve model, is highlighted in this section.

1. Patients

As mentioned, between December 1990 and January 1995, 306 patients underwent isolated mitral (=205) and aortic (=101) valve replacement. On discharge, written and verbal instructions on anticoagulation were given to the patients and families. Anticoagulation was achieved by Dicoumarol or Nicoumalone to maintain prothrombin time at 1½ to 2 times the control value.

The mean age was 28.9 years, the youngest being 6 and the oldest 58 with a male-to-female ratio of 1.2:1. Over

Table 8 Preoperative Clinical Profile

Population	306 patients
Age (years)	
Mean	28.87
Range	6–58
Male/female ratio	1.20 (167/139)
NYHA class	
II	99 (32.4%)
III	172 (56.2%)
IV	35 (11.4%)
Heart rhythm	
Atrial fibrillation	88 (28.7%)
Sinus rhythm	211 (68.9%)
Unknown	7 (2.4%)
Valve lesions	
Stenosis	71 (23.2)
Insufficiency	140 (45.8%)
Mixed	95 (31.0%)
History of CCF	85 (27.8%)
History of embolism	8 (2.6%)

two-thirds belonged to NYHA classes III and IV and shared a poor socioeconomic background (Table 8). Sixty-five percent of AVR patients had aortic insufficiency as the main lesion, while insufficiency and mixed lesions dominated (72%) the MVR group.

Early mortality (defined as death within 30 days of surgery or before discharge from the hospital, whichever is later) was 6.9 ± 1.5% (21 patients). While one patient died from anticoagulation related hemorrhage and another succumbed to infective endocarditis, four patients (two each of AVR and MVR) died suddenly at home within a month of the operation. Non-valve-related causes claimed 15 patients (4.9 ± 1.3%).

2. Postoperative Valve Function

In the first phase unicentric trial, postoperative valve function was assessed by Doppler echocardiography before discharge from the hospital. Eighteen aortic (all size 23 mm) and 20 mitral (all size 25 mm) valves were studied. Systolic or diastolic pressure drop and valve orifice areas were measured by the pressure half-time method. Regurgitation was assessed qualitatively.

This assessment showed that the systolic pressure drops were 8–26 mm Hg (mean 20) with the 23 mm aortic valves. With the 25 mm mitral valves, the diastolic pressure drops were 2–5 mm Hg (mean of 3) and the valve areas ranged from 2 to 3.4 cm^2 (mean 2.5). Doppler showed mild intravalvular regurgitation in most aortic and some mitral valves. This was acceptable and consistent with the leak allowed by the nonseating disc.

3. Follow-Up

Two hundred eighty five patients were reviewed at their respective institutions, initially at 1, 3, 6, and 12 months and thereafter at approximately yearly intervals. Between December 1997 and March 1998, 75 patients with no follow-up prior to September 1997 were vigorously traced. About 50% responded to direct letters and two-thirds of the remaining recalcitrant patients could be traced by direct house visits. At the end of the study period in September 1998, 13 patients remained lost to follow-up (defined as not having a checkup or any other form of contact (if not confirmed to have died) during the 12 months of the study period. Thus the net follow-up was 272/285 or 95.4% in terms of patients. The duration of follow-up ranged from 1 month to 7.5 years with a mean of 4.25 years; the minimum follow-up for survivors was 3 years. Total follow-up years actually observed was 1212 patient years.

4. Data Analysis

The linearized incidences and the actuarial probabilities of survival were calculated according to the method of Grunkemeier and Starr (73) and Lefrak and Starr (74). Percentages are expressed as percent ± 1 standard error.

5. Late Mortality

Late deaths occurred in 52 patients (4.3 ± 0.6% per patient year), 12 due to known valve-related causes (Table 9) and 17 due to non-valve-related causes. Twenty-three patients died at home between 2 months and 8 years after the operation, and the cause of death could not be determined. These unknown deaths have been treated as valve-related in all analyses.

6. Postoperative NYHA Status (Figure 12)

The majority of the operative survivors (258 of 285) moved up to NYHA Class I following valve implantation

Table 9 Valve Related Late Events and Deaths

Event	No. of patients	%/patient year	No. of deaths
Valve thrombosis	11 (13)[a]	1.1 ± 0.3	5
Systemic embolism	22 (22)	1.8 ± 0.4	1
AC bleeding	3 (6)	0.5 ± 0.2	3
Infective endocarditis	6 (6)	0.5 ± 0.2	3
Unknown deaths		2.4 ± 0.44	23
Total			35

[a] The figure in brackets gives the number of events in that category. AC = anticoagulant related.

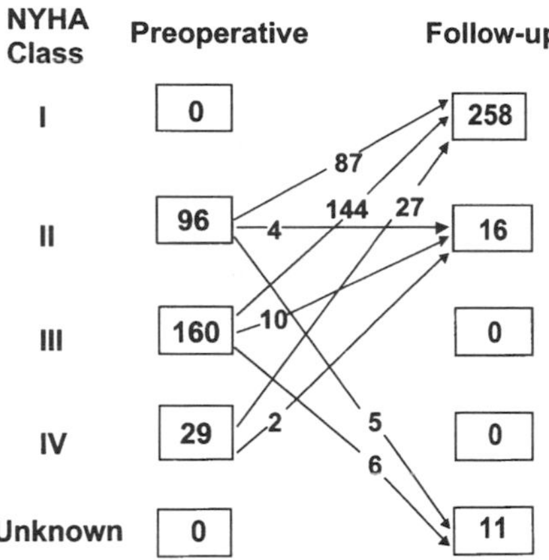

Figure 22.12 Postoperative status of the patients implanted with the Chitra valve.

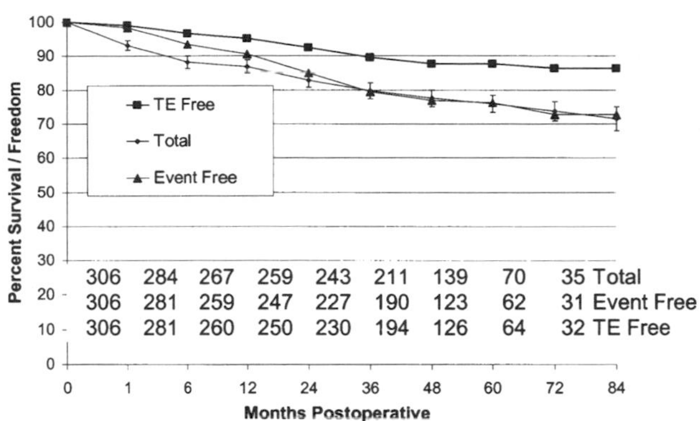

Figure 22.13 Long-term actuarial survival rates for the 285 survivors with Chitra valves.

and full recovery, while none remained in the preoperative levels of cardiac dysfunction (NYHA III and IV).

7. Valve-Related Mortality and Late Events

Twelve patients died due to confirmed valve-related complications (Table 9). To this figure was added the 23 unknown deaths, which resulted in a linearized incidence of 2.9 ± 0.5% per patient-year. Among late events during follow-up (Table 9), 13 episodes of valve thrombosis occurred in 11 patients (1.1 ± 0.3% per patient-year). Ten of these patients had discontinued anticoagulants for periods ranging from a few days to 3 months. Five patients in this group died before hospitalization became available to them. Four responded well to thrombolytic therapy, and the other two underwent successful valve re-replacement, one with a Starr–Edwards valve and the other with a CHV. Embolic episodes occurred in 22 patients (1.8 ± 0.4% per patient-year) of whom four had discontinued anticoagulants. Twenty-one patients in this group made a full recovery without residual neurological deficit, while one died. Six patients had anticoagulant-related bleeding (0.5 ± 0.2% per patient-year) of whom three succumbed to intracerebral bleeding. Of six patients who developed infective endocarditis (0.5 ± 0.2% per patient-year), three died.

The long term actuarial survival rates for the 285 survivors are shown in Fig. 13. At 5 years, the survival rate was 75.8 ± 2.6%, and at 7 years it was 71.5 ± 3.5%. Figure 13 also shows the actuarial data for thromboembolism free survival, which includes both valve thrombosis and systemic embolism. At 5 years, the freedom from thromboembolism was 87.7 ± 2.0%, and at 7 years it was 86.4 ± 2.4%. The freedom from all valve related mortality and morbidity was 76.3 ± 2.6% at 5 years and 72.6 ± 3.2% at 7 years.

There has so far been no report of mechanical failure, either in this study or elsewhere. Interestingly, there has been no incidence of paravalvular leak in this study. Also, there have been no complaints of discomfort due to the valve sounds. The closing sounds are soft and dull due to use of the UHMW PE disc. Even in the generally young and lean patients, the valve sounds are not easily heard nearby.

The important features of this patient profile are advanced disability (68% in NYHA classes III and IV) and poor socioeconomic background. The advanced preoperative functional disability would adversely influence operative mortality and long-term survival. The poor socioeconomic background is a major cause for late attendance at the hospital and, above all, lack of compliance in taking anticoagulants and other medications. It was not rare for poor patients to skip medications for financial reasons or because of nonavailability in a remote village. Nor was it rare for patients with valve thrombosis to delay reporting to the hospital in spite of progressive dyspnea, until they were in the terminal stage of pulmonary edema. These problems are encountered in varying degrees in third-world populations (75).

The low incidence of anticoagulant related bleeding in the present series probably reflects the less aggressive policy of physicians in the use of anticoagulants. The absence of any report of paravalvular leak is noteworthy and could be due to the low closing impact forces of this valve design. This needs further confirmation via larger and longer follow-up of patients.

Table 10 shows a comparison of the valve related complications, mortality rates, and actuarial estimates of the present series with those reported by Antunes *et al.* (75) with the use of the Medtronic-Hall valve in a third-world population. The striking similarities in the complications

Table 10 Comparison of Mortality, Complication, and Actuarial Values

Mortality/complication/ actuarial value		Incidence in %/patient-year	
		CHV	Antunes et al.
1	Late mortality	4.3 ± 0.6	6.3 ± 0.5
2	Valve related mortality	2.9 ± 0.5	2.9 ± 0.3[a]
3	Thrombosis	1.1 ± 0.3	1.2 ± 0.2
4	Embolism	1.8 ± 0.4	3.3 ± 0.3
5	Paravalvular leak	0.0	2.2 ± 0.2
6	Infective endocarditis	0.5 ± 0.2	0.7 ± 0.2
7	Anticoagulant related hemorrhage	0.5 ± 0.2	0.7 ± 0.2
8	Freedom from thrombo-embolism at 5 years	87.7 ± 2.0	85 ± 2.0
9	Freedom from all valve related mortality and morbidity at 5 years	76.3 ± 2.6	75 ± 3.0
10	Total survival at 5 years	75.8 ± 2.6	75 ± 2.0

[a] This excludes 41 of 64 cases of sudden or unknown death.

and survival rates highlight the comparable performance of the CHV and the Medtronic-Hall valve.

Bain and Nashef (76), in their review of the performance of different models of titling disc valves, have quoted the incidence of bleeding at 0.3 to 1.1% per patient-year, paravalvular leak at 0.7 to 2.0% per patient-year, and infective endocarditis of 0.8% per patient-year. They have also quoted the following figures for valve thrombosis and embolism for Bjõrk–Shiley valves (all models): Embolism for aortic valves at 0.4 to 1.6% per patient-year and for mitral valves at 1.5 to 3.0% per patient-year; thrombosis varies between 0.9 to 1.1% per patient-year. The data in this study are comparable to these. This study confirms the earlier observation (22) that the CHV is mechanically sound, structurally reliable, and hemodynamically and clinically comparable to other tilting disc valves of similar design.

SUMMARY AND CONCLUSIONS

In this chapter, the concept, design, development, and evaluation of a mechanical valve with a flexible polymeric occluder has been described. The development of a UHMW PE occluder enabled the technology to be indigenous to countries like India and proved to be cost-effective in treating a significant percentage of patients who cannot afford implantation of imported valve prostheses. The in vitro and accelerated wear evaluation assured that the prototype valve will have performance characteristics comparable to the currently available mechanical valve prostheses in the United States. Careful analysis of the valve closing dynamics and the stresses induced in this phase of the cardiac cycle suggests that valves with flexible occluders have significantly less fluid dynamic stresses, which may prove to be advantageous. The in vivo evaluation both in the animal model and in implants in patients proves to be promising. A carefully controlled study comparing the implant experience with the Chitra valve and the presently available mechanical valves with the pyrolytic carbon occluders is necessary for further investigation.

REFERENCES

1. K. B. Chandran. *Cardiovascular Biomechanics*. New York University Press, New York, 1992, Chap. 7, p. 294.
2. A. P. Yoganathan. Cardiac valve prostheses. In: *The Biomedical Engineering Handbook* (J. D. Bronzino, ed.). CRC Press, Boca Raton, Florida, 1995, Chap. 123, p. 184.
3. D. E. Harken, H. S. Soroff, W. J. Taylor, A. A. Lefemine, S. K. Gupta, S. Lunzer. Partial and complete prostheses in aortic insufficiency. *J. Thorac. Cardiovasc. Surg. 40*:744, 1960.
4. E. A. Lefrak, A. Starr. *Cardiac Valve Prostheses*. Appleton-Century-Crofts, New York, 1979.
5. J. Wada. Knotless suture method and Wada hingeless valve. *Jpn. J. Thorac. Surg. 15*:88, 1967.
6. L. Rodriguez. Haemodynamic and angiographic findings in patients with isolated aortic valvular disease before and after insertion of a Starr–Edwards aortic ball valve prosthesis. *Scand. J. Thorac. Cardiovasc. Surg*. suppl 5: 1, 1970.
7. V. O. Björk, C. Olin, H. S. Anstrom. Results of aortic valve replacement with the Kay–Shiley disc valve. *Scand. J. Thorac. Cardiovasc. Surg. 3*:93, 1969.
8. V. O. Björk. Experience with the Wada–Cutter valve prosthesis in the aortic area: one year follow-up. *J. Thorac. Cardiovasc. Surg. 60*:26, 1970.
9. V. O. Björk. A new tilting disc valve prosthesis. *Scand. J. Thorac. Cardiovasc. Surg 3*:1, 1969.
10. R. L. Kaster. Discussion. In: *Prosthetic Heart Valves* (L. A. Brewer III, ed.). Thomas, Springfield, 1969, p. 325.
11. V. O., Björk. Delrin as implant material for valve occluders. *Scand. J. Thorac. Cardiovasc. Surg. 6*:103, 1972.
12. V. O. Björk. The pyrolytic carbon occluder for the Björk–Shiley tilting disc valve prosthesis. *Scand. J. Thorac. Cardiovasc. Surg. 6*:109–113, 1972.
13. K. B. Chandran, C. S. Lee, L. D. Chen. Pressure field in the vicinity of mechanical valve occluders at the instant of valve closure: correlation with cavitation initiation. *J. Heart Valve Dis. 3*(suppl. 1): S65, 1994.
14. K. B. Chandran, S. Aluri. Mechanical valve closing dynamics: relationship between velocity of closing, pressure transients, and cavitation initiation. *Annals Biomed. Eng. 25*: 926, 1997.
15. C. S. Lee, K. B. Chandran, L. D. Chen. Cavitation dynamics of mechanical heart valve prostheses. *Artif. Org. 18*:758, 1994.

16. C. S. Lee, K. B. Chandran, L. D. Chen. Cavitation dynamics of Medtronic-Hall mechanical heart valve prosthesis: fluid squeezing effect. *ASME J. Biomech. Eng. 118*:97, 1996.
17. C. S. Lee, S. Aluri, K. B. Chandran. Effect of valve holder flexibility on cavitation initiation with mechanical heart valve prostheses: an in vitro study. *J. Heart Valve Dis. 5*: 104, 1996.
18. L. A. Garrison, T. C. Lamson, S. Deutsch, D. B. Geselowitz, R. P. Gaumond, J. M. Tarbell. An in vitro investigation of prosthetic heart valve cavitation in blood. *J. Heart Valve Dis. 3*(suppl. 1): S8, 1994.
19. C. M. Zapanta, E. G. Liszka Jr., T. C. Lamson, D. R. Stinebring, S. Deutsch, D. B. Geselowitz, J. M. Tarbell. A method for real-time in vitro observation of cavitation on prosthetic heart valves. *ASME J. Biomech. Eng. 116*:460, 1994.
20. T. Graf, H. Reul, W. Dietz, R. Wilmes, G. Rau. Cavitation at mechanical heart valves under simulated physiological conditions. *J. Heart Valve Dis. 1*:131, 1992.
21. T. Graf, H. Reul, C. Detlefs, R. Wilmes, G. Rau. Causes and formation of cavitation in mechanical heart valves. *J. Heart Valve Dis. 3*(suppl. 1): S49, 1994.
22. G. S. Bhuvaneshwar, C. V. Muraleedharan, G. Arthur Vijayan Lal, R. Sankar Kumar, M. S. Valiathan. Development of the Chitra tilting disc heart valve prosthesis. *J. Heart Valve Dis. 5*:448, 1996.
23. K. I. Larmi Teuvo, Pentti Karkola. Shrinkage and degradation of the Delrin occluder in the tilting disc valve prosthesis. *J. Thorac. Cardiovasc. Surg. 68*:66, 1974.
24. G. S. Bhuvaneshwar, C. V. Muraleedharan, G. Arthur Vijayan Lal, A. V. Ramani, M. S. Valiathan. Synthetic sapphire as an artificial heart valve occluder—promise and problems. *Trans. Indian Ceramic Soc. 50*:87, 1991.
25. G. S. Bhuvaneshwar, C. V. Muraleedharan, A. V. Ramani, M. S. Valiathan. Evaluation of materials for artificial heart valves. *Bull. Mater. Sci. 14*:1363, 1991.
26. E. Rabinowicz. *Friction and Wear of Materials*. John Wiley, New York, 1965, pp. 113.
27. D. W. Weiting, W. C. Hall, D. Liotta, M. E. DeBakey. Dynamic flow behavior of artificial heart valves. In: *Prosthetic Heart Valves* (L. A. Brewer III, ed.) Thomas, Springfield, 1969, pp. 34.
28. A. P. Yoganathan, W. Letzing. *Prosthetic Heart Valves: A Study of In-Vitro Performance, Phase II* (Report). Reproduced by National Technical Information Service, U.S. Department of Commerce, Springfield, 1983.
29. K.-H. Bruss, H. Reul, J. V. Gilse, E. Knott. Pressure drop and velocity fields at four mechanical heart valve prosthesis: Björk–Shiley standard, Björk–Shiley concave-convex, Hall–Kaster and St. Jude Medical. *Life Support Systems 1*: 3, 1983.
30. G. S. Bhuvaneshwar. *Design optimisation of the Chitra tilting disc heart valve prosthesis*. Ph.D. thesis, Sree Chitra Tirunal Institute for Medical Sciences and Technology, Trivandrum, India, 1993.
31. H. S. Bean, ed. *ASME Research Committee Report. Fluid Meters* (6th ed.). American Society of Mechanical Engineers, New York, 1971.
32. Y.-R. Woo, A. P. Yoganathan. In-vitro pulsatile flow velocity and shear stress measurements in the vicinity of mechanical mitral heart valve prostheses. *J. Biomechanics 19*: 47, 1986.
33. H. Reul, M. Giersiepen, E. Knott. Laboratory testing of prosthetic heart valves. *Proc. Heart Valve Engineering*. Institution of Mechanical Engineers, London, 1986, p. 3.
34. K. B. Chandran, T. V. Ferguson, C.-J. Chen, B. Khalighi. Experimental study of flow dynamics behind valve prosthesis. *Am. Soc. Artif. Int. Organs J. 6*:146, 1983.
35. D. W. Weiting. Dynamic flow characteristics of heart valves. Ph.D. thesis, Univ. of Texas, Austin, 1969.
36. J. T. M. Wright, L. J. Temple. An improved method for determining the flow characteristics of prosthetic mitral heart valves. *Thorax 26*:81, 1969.
37. W. M. Swanson, R. E. Clark. Cardiovascular system simulation requirements. *J. Bioeng. 1*:121, 1976.
38. T. R. P. Martin, M. M. Black. Problems of in-vitro testing of heart valve replacements. *Proc. European Soc. Artif. Organs 3*:131, 1976.
39. H. Reul. In-vitro evaluation of artificial heart valves. *Adv. Cardiovasc. Phys. 5* (part 4):16, 1983.
40. M. Klain, K. H. Letiz, W. J. Kolff. Comparative testing of artificial heart valves in a mock circulation. In: *Prosthetic Heart Valves* (L. A. Brewer III, ed.). Thomas, Springfield, 1969, p. 114.
41. W. M. Swanson, R. E. Clark. A simple cardiovascular system simulator and performance. *J. Bioeng. 1*:135–145, 1976.
42. E. Knott, H. Reul, M. Knoch, U. Steinsiefer, G. Rau. In-vitro comparison of aortic heart valve prostheses—Part 1: Mechanical valves. *J. Thorac. Cardiovasc. Surg. 96*:952, 1988.
43. H. Reul, H. Minammitani, J. Runge. A hydraulic analog of the systemic and pulmonary circulation for testing artificial hearts. *Proc. European Soc. Artif. Organs 120*, 1975.
44. P. Willshaw, M. Biagetti, R. H. Pichel. A comparative in-vitro study of the closing characteristics of Björk–Shiley and Bicer–Val tilting mitral valve prostheses. *J. Biomed. Eng. 8*:43, 1986.
45. K. C. Dellsperger, D. W. Weiting, D. A. Baehr, R. J. Bard, J.-P. Brugger, E. C. Harrison. Regurgitation of prosthetic heart valves: dependence on heart rate and cardiac output. *Am. J. Cardiol. 51*:321, 1983.
46. L. Leuer. Dynamics of the mechanical valves in the artificial heart. *Proce. 40th ACEMB*, 1987, p. 82.
47. R. Kafesjian, M. Howanec, G. D. Ward, L. Diep, L. S. Wagstaff, R. Rhee. Cavitation damage of pyrolytic carbon in mechanical heart valves. *J. Heart Valve Disease 3* (suppl. 1):S2–S7, 1993.
48. F. E. Deuvaert, J. Devriendt, J. Massaut. Leaflet escape of a mitral Duromedics prosthesis (case report). *Acta. Chir. 89*:15, 1989.
49. W. R. Dimitri, B. T. Williams. Fracture of the Duromedics

mitral valve housing with leaflet escape. *J. Cardiovasc. Surg. 31*:41, 1990.

50. E. Hjelms. Escape of a leaflet from a St. Jude medical prosthesis in the mitral position. *J. Thoracic. Cardiovasc. Surg. 31*:310, 1983.
51. W. Klepetko, A. Moritz, G. Mlzoch, H. Schurawitzki, E. Domanis, E. J. Wolner. Leaflet fracture in Edwards-Duromedics bileaflet valves. *J. Thorac. Cardiovasc. Surg. 97*: 90, 1989.
52. N. Kumar, S. Balasundaram, M. Rickard, N. al Halees, C. M. Duran. Leaflet embolization from Duromedics valves: a report of two cases. *Thoracic. Cardiovasc. Surgeon 39*: 382, 1991.
53. C. S. Lee, K. B. Chandran. Instantaneous backflow through peripheral clearance of Medtronic-Hall disc valve at the moment of closure. *Ann. Biomed. Eng. 22*:371, 1994.
54. C. S. Lee, K. B. Chandran. Numerical simulation of instantaneous backflow through central clearance of bileaflet mechanical heart valves at the moment of closure: shear stress and pressure fields within the clearance. *Med. Biol. Eng. Comput. 33*:257, 1995.
55. S. Aluri, K. B. Chandran. Numerical simulation of mechanical heart valve closure. *1999 Bioengineering Conference, ASME BED, 42*:565, 1999.
56. S. Aluri. Hemolysis induced by mechanical heart valve closure. Doctoral thesis, University of Iowa, 1999.
57. J. T. Baldwin, J. M. Tarbell, S. Deutsch, D. B. Geselowitz. Mean velocities and Reynolds stresses within regurgitant jets produced by tilting disc valves. *Trans. ASAIO 37*: M348, 1991.
58. T. C. Lamson, G. Rosenberg, D. B. Geselowitz, S. Deutsch, D. R. Stinebring, J. R. Frangos, J. M. Tarbell. Relative blood damage in the three phases of prosthetic heart valve flow cycle. *Trans. ASAIO* (suppl 1):M626, 1993.
59. J. W. Lawrie. *Glycerol and The Glycols*. Chemical Engineering Catalog Co., New York, 1928.
60. G. X. Guo, C. C. Xu, N. H. C. Hwang. The closing velocity of Baxter Duromedics heart valve prostheses. *Trans. ASAIO 36*:M529, 1990.
61. G. X. Guo, P. Adlparvar, M. Howanec, J. Roy, R. Kafesjian, C. Kingsbury. Effect of structural compliance on cavitation threshold measurements of mechanical heart valves. *J. Heart Valve Dis. 3* (suppl. 1):S77, 1994.
62. Z. J. Wu, B. Y. Wang, N. H. C. Hwang. Occluder closing behavior: a key factor in mechanical heart valve cavitation. *J. Heart Valve Dis. 3*:S25, 1994.
63. Z. J. Wu, M. C. S. Shu, D. R. Scott, N. H. C. Hwang. The closing behavior of Medtronic-Hall mechanical heart valves. *ASAIO J. 40*:M702, 1994.
64. Z. J. Wu, B. Z. Gao, N. H. C. Hwang. Transient pressure at closing of a monoleaflet valve prostheses: mounting compliance effect. *J. Heart Valve Dis. 4*:553, 1995.
65. D. Bluestein, S. Einav, N. H. C. Hwang. A squeeze flow phenomenon at the closing of a bileaflet mechanical heart valve prosthesis. *J. Biomechanics 27*:1369, 1994.
66. V. B. Makhijani, H. Q. Yang, A. K. Singhal, N. H. C. Hwang. An experimental-computational analysis of MHV cavitation: effects of leaflet squeezing and rebound. *J. Heart Valve Dis. 3* (suppl. 1):S35, 1994.
67. G. J. Cheon, K. B. Chandran. Dynamics of a mechanical monoleaflet heart valve prosthesis in the closing phase: effect of squeeze film. *Ann. Biomed. Eng. 23*:189, 1995.
68. K. B. Chandran, E. U. Dexter, S. Aluri, W. E. Richenbacher. Negative pressure transients with mechanical heart-valve closure: correlation between in vitro and in vivo results. *Ann. Biomed. Eng. 26*:546, 1998.
69. E. U. Dexter, S. Aluri, R. R. Radcliffe, H. Zhu, D. D. Carlson, T. E. Heilman, K. B. Chandran, W. E. Richenbacher. In vivo demonstration of cavitation potential of a mechanical heart valve. *ASAIO J. 45*:436, 1999.
70. B. E. Fettel, D. R. Johnston, P. E. Morris. Accelerated life testing of prosthetic heart valves. *Medical Instrumentation 14*(3):161, 1980.
71. N. Howard-Jones. ŸA CIOMS ethical code for animal experimentation. *WHO Chronicle 39*(2):52, 1985.
72. R. Sankar Kumar, G. S. Bhuvaneshwar, R. Magotra, S. Muralidharan, R. S. Rajan, D. Saha, K. V. S. K. Subba Rao, M. S. Valiathan, S. Radhakrishna, A. V. Ramani. Chitra heart valve: results of a multi-centre study. *J. Heart Valve Dis*. (in press).
73. G. L. Grunkemeier, A. Starr. Actuarial analysis of surgical results: rationale and methods. *Ann. Thor. Surg. 24*:404, 1977.
74. E. A. Lefrak, A. Starr (with the assistance of G. L. Grunkemeier). Data base and methodology. In: *Cardiac Valve Prosthesis*, Appleton-Century-Crofts, New York, 1979, Sec. 1, Chap. 2: pp. 38–63.
75. M. J. Antunes, A. Wessels, R. G. Sadowski, J. G. Schutz, K. M. Vanderdonck, J. M. Oliveira, L. E. Fernandes. Medtronic-Hall valve replacement in a third-world population group. *J. Thor. Cardiovasc. Surg. 95*:980, 1988.
76. W. H. Bain, S. A. M. Nashef. Tilting disk valves. In:*Replacement Cardiac Valves* (E. Bodnar, R. Frater, eds.). Pergamon Press, New York, 1991, p. 187.

23

Blood-Contacting Polymers

T. Avramoglou, J. Jozefonvicz, and M. Jozefowicz
Université de Paris, Villetaneuse, France

I. INTRODUCTION

When a foreign body is brought into contact with blood, it is confronted with two complex biological systems, coagulation and immunity. Although these protective mechanisms are essential in man, they create problems when a foreign body is intentionally brought into contact with blood for medical purposes. However, since the early 1950s, polymers have been widely used for a large range of medical applications, from long-term implants to short-term dressings, in all fields of medicine and surgery. Examples of practical devices that implicate blood contact polymers are (1):

- Extracorporal blood-circulating devices
- Catheters
- Blood bags and tubing used for blood transfusion
- Membranes, hollow fibers, and tubing used for dialysis devices, plasmapheresis and plasma detoxification, and oxygenators
- Cardiac valves and blood vessel replacement such as aortic bypasses
- Drug delivery systems
- Plasma expanders and blood substitutes
- Contrast agents
- Embolization agents

The medical use of such devices has increased during the second half of the 20th century due to improvements in the quality of the devices and that of the polymers used to produce them. These polymers are increasingly prescribed for use in cardiac surgery and in the treatment of renal and cardiovascular disease. Without the use of polymers, some of the major improvements made in the past decades, in medical and surgical practice, would have been impossible.

The global market for medical devices was approximately $120 billion in 1995. The market of blood-contacting polymers is relatively large and has an annual increase of approximately 5% (2). Synthetic materials are used in contact with blood in millions of medical devices and diagnostic systems. The number of devices used worldwide each year is impressive: 100,000 heart valves, 500,000 vascular grafts, >250,000 stents, 200,000 pacemakers, 200,000 blood oxygenators, 50,000,000 blood bags, and millions of catheters (3).

The polymers commonly used to produce blood-contacting devices are essentially either classical synthetic polymers or natural polymers (1) (Table 1) (Fig. 1). Poly (vinyl chloride) (PVC) is the most extensively used polymer for all short-term devices, such as extracorporal blood-circulating devices, catheters, and blood bags. Silicone rubber and polyethylene present alternative materials for these devices. Cellulose and cellulose derivatives, polyamides, polypropylene, polyacrylonitrile, polysulfone, and polyesters are the basic materials for membranes and hollow fibers for dialysis. Commercially available vascular grafts and cardiac valves are essentially made from polyesters, mainly poly(ethylene terephthalate) (Dacron®) and

Table 23.1 Main Blood-Contacting Polymers

Polymer	Device
Cellulose and derivatives	Membranes for dialysis
Cross-linked collagen	Heart valves
Dextran	Plasma expanders
Human albumin	Plasma expanders
Polyacetal	Heart valves
Polyacrylonitrile	Membranes for dialysis
Polyamides	Catheters, heart valves, membranes for dialysis
Polycarbonates	Syringes, catheters, heart valves
Poly(ethylene terephthalate)	Vascular prostheses
Polypropylene	Syringes, catheters
Polysulfones	Membranes for dialysis
Poly(tetrafluoroethylene)	Vascular prostheses, artificial hearts, catheters
Polyurethanes	Vascular prostheses, artificial hearts, catheters
Poly(vinyl chloride)	Blood bags, catheters
Silicones	Catheters, artificial hearts

poly(tetrafluoro ethylene) (Teflon®). Finally, polyurethanes and related polymers such as poly(ether-urethane) have also been developed as biomaterials (4,5). Biopolymers have been developed simultaneously, mainly for cardiovascular prosthetic devices. These materials are essentially collagen derivatives from animals, highly cross-linked by glutaraldehyde or other agents. Cross-linking improves the mechanical properties of the material and depresses its immunogenicity, which is important as collagen degradation products are known to be inflammatory (6–8).

The major improvement of the past decade, in this field, is the use of so-called medical grade polymers, i.e., polymeric compounds that release neither toxic nor carcinogenic products into the bloodstream. Moreover, the degradation products of such materials are nontoxic and noncarcinogenic and do not accumulate in the body (9–12).

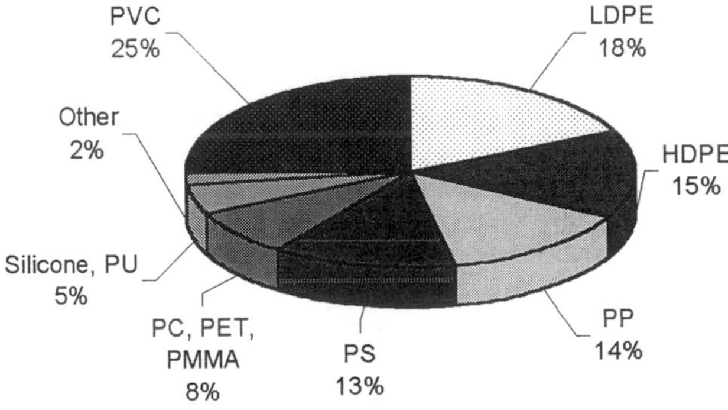

Figure 23.1 Weight distribution for blood contacting polymers. (From Ref. 357.)

A further improvement has been achieved through the modification of the processing technique employed to produce the polymers. The following desirable properties have been attained (13):

- Increased permeability and mechanical strength of the membranes and hollow fibers used in dialysis.
- Increased mechanical strength and relative porosity in vascular grafts. The porosity of the wall allows the healing process to occur within the walls of the vascular graft.

Despite their extensive use, the polymers available are essentially unsatisfactory, because of the undesirable events produced when blood comes into contact with the polymer. The first problem is the rapid adsorption of protein (14), which is the trigger for blood coagulation and leads to platelet adhesion. The second problem arises from the large difference between the mechanical compliance of the polymers and that of the natural blood vessel wall (15). Turbulence in the bloodstream results from the described variation, which in turn induces hemolysis, platelet activation, and aggregation (16–18). Moreover, the rugosity of the surface in contact with the blood may also produce similar undesirable effects (19). It should be noted that for long-term uses the loss of the mechanical strength of mobile devices (e.g., artificial heart membranes) might be a cause of failure of the device itself. It appears that these events are related to calcification processes occurring within the polymer matrix, after several months of exposure to blood (20). The third problem is created when the compounded polymer releases some of the adjuvant, stabilizers, and plasticizers into the bloodstream. This may induce blood damage (21). A literature survey carried out the early 1970s examined commercially available PVCs. These polymers, even when of medical grade, exhibit different blood compatibilities with respect to their ability to induce the blood coagulation process. This can be attributed to the release of plasticizers or stabilizers. The fourth undesirable event is produced by the degradation of the material. This may result from the long-term exposure of the polymers to blood. In this case nontoxic and noncarcinogenic products are released, which induce blood coagulation. This degradation effect is not necessarily related to the mechanical failures previously described. Finally, the exposure of polymers to blood may promote both a cellular and a humoral immune response, and may also carry a risk of bacterial infection. All these events can be catastrophic, and they sometimes require surgical intervention to replace the implanted device. This is the case for biovalves made of cross-linked collagen, which have a replacement rate of 25% after 10 years in the body (22).

In order to overcome the thrombotic problems encountered when using blood-contacting polymers, the medical professional can utilize an anticoagulant long-term adjuvant therapy, based on heparin and a vitamin K antagonist (coumarin, etc.). However, this type of therapy may induce some undesirable side effects.

The grafting of prostheses, made of synthetic materials, is adequate for the treatment of large arteries. However, the low hemocompatibility of these prostheses in small-diameter blood vessels does not allow their use in coronary or below-the-knee arterial replacement. The absence of an endothelium, in conjunction with the thrombogenic properties of the synthetic material, results in a high percentage of thrombosis when synthetic prostheses are grafted at these anatomic sites. To overcome this problem, it is possible to seed endothelial cells on expensed PTFE and then implant the endothelialized prosthesis. Although this technique is time-consuming, requiring approximately one month between the sampling of the autologous cell and the grafting, it leads to excellent results (23). Superior results are achieved with total vessel rebuilding (24). The latter technique is highly complex.

A considerable amount of research, during the past decades, has been devoted to the synthesis and tailoring of new biomaterials suitable for use as blood-contacting polymers. These new materials should not induce either the coagulation of blood or the immune response. Furthermore, they should not allow bacterial adhesion. With the availability of new materials, considerable progress in the field of blood-contacting artificial devices is expected in the near future.

In this chapter the following subjects will be reviewed:

- The physiological aspects of hemostasis.
- The events that occur when blood contacts a polymer, with particular attention to the coagulation of blood and the immune response. As a result, the blood compatibility of polymers will be defined.
- The factors that may affect the blood compatibility of polymers and their relevance to experimental data. In addition, the suitability of both commercial and noncommercial blood-compatible polymers for the manufacture of blood-contacting devices will be considered.
- The new concept of ''bioactive biomaterials.''

The response of blood to the artificial surfaces of polymers is inevitable, and it is possible to list the different interactions as follows: protein adsorption, platelet reaction, intrinsic coagulation, fibrinolytic activity, complement activation, and interaction with circulating cells.

II. BLOOD COAGULATION

Normal coagulation, or hemostasis, prevents vital internal fluids leaking, but pathological and inappropriate coagulation results in thrombosis. The hemostatic mechanism appeared at a very early stage of evolution (25,26), even before the formation of the vascular system (27,28).

Coagulation enables man to survive hemorrhaging following a trauma. It has to occur in a restricted area around the cut to avoid disturbing the general circulation of the blood. When the vascular breach is repaired, the blood clot will dissolve so that the circulation of the blood in the previously damaged vessel can resume: this is fibrinolysis (29). This mechanism prevents the permanent obstruction of damaged vessels. Thrombosis, which is the obstruction of undamaged vessels, is also avoided by fibrinolysis.

Coagulation represents the transformation of soluble fibrinogen into an insoluble fibrin network under the influence of thrombin, the key enzyme. It is a complex phenomenon and involves several enzymes, cofactors, and phospholipidic surfaces. All these elements are termed coagulation factors. The latter circulate as inactive precursors, of which some are zymogens and others are cofactors. Each zymogen is converted to an active form (a serine protease), which in turn activates the next coagulation factor in the sequence. However, this coagulation ''cascade'' is not a linear process; it implies intricate positive and negative feedback, in which thrombin affects its own formation and its own breakdown. Numerous steps in the hemostasis process are heterogeneous catalysis implying calcium ions and negatively charged phospholipids. Comprehension of the physiological nature of the coagulation is essential in order to elaborate a blood-compatible polymer.

Initiation of blood coagulation has been divided into two pathways (Fig. 2), the intrinsic and the extrinsic. Both pathways culminate in the generation of thrombin. The intrinsic cascade is initiated when blood comes into contact with an anionic surface and as only dependant on factors intrinsic to the flowing blood. The extrinsic pathway is initiated by tissue factor (TF or thromboplastin or factor III). This protein only becomes exposed to the bloodstream when a vascular wall is damaged (30).

A. Extrinsic Pathway and Primary Hemostasis

Primary hemostasis is a physiological process necessary to preserve homeostasis, as well as being the first step of the process of tissue repair. Primary hemostasis involves all of the complex interactions between the vascular walls, platelets, and coagulation factors described above, as well as hemorheological factors such as blood viscosity and flow.

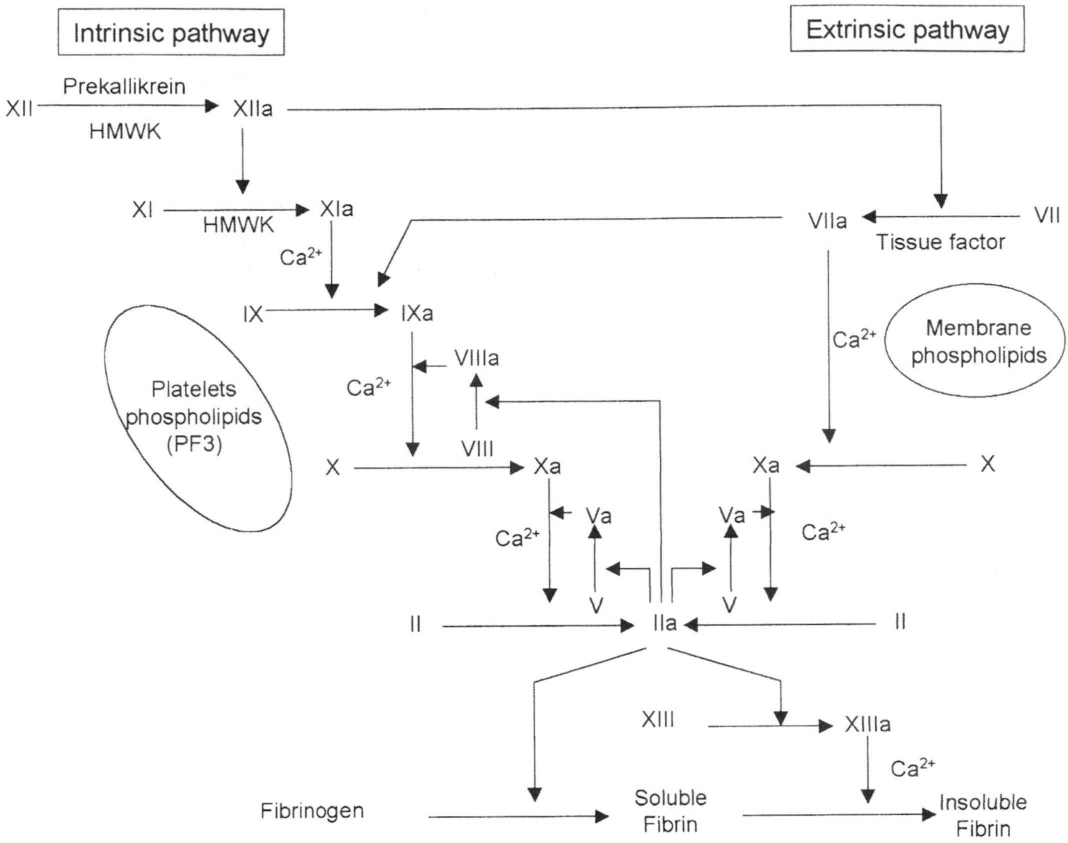

Figure 23.2 Overview of blood coagulation. The coagulation factors are designated with roman numerals, and the suffix a indicates a proteolytically activated factor.

Primary hemostasis in small blood vessels is determined by the standardized measurement of bleeding time. In man, the bleeding of sectioned small blood vessels stops after approximately four minutes. This process takes place in several stages. Cutting the vascular wall causes it to contract, which then modifies the local hemodynamic conditions and exposes the subendothelial conjunctive tissue. The surface of the endothelial cells is thromboresistant, whereas the collagen-rich subendothelium is thrombogenic. The platelets adhere to the subendothelium when submitted to hemodynamic forces and in the presence of the von Willebrand factor (factor VIII). This is the first stage of hemostasis. Both the coagulation pathways are activated, in conjunction with the activation of platelets. Upon vascular injury, the thromboplastin is exposed to blood and readily forms a one-to-one stoichiometric complex with the circulating factor VII (convertin). Complex formation with thromboplastin has two functions. It augments the activation of the factor VII to VIIa, and it enhances the proteolytic activity of factor VIIa with respect to its substrates. The factor VIIa/TF complex, then, initiates blood coagulation by proteolytically activating its substrates, factors IX and X, leading to the rapid formation of factor IIa (thrombin) (31). The subendothelium collagen activates the factor XII (Hageman factor) and leads to the slower, intrinsic formation of thrombin.

B. Intrinsic Pathway and the Contact System

The initiation of the intrinsic pathway involves four proteins forming a system termed a ''contact system'' (32). This pathway is triggered when factor XII (Hageman factor) comes into contact with collagen present in the exposed subendothelium. The activated factor XII (factor XIIa) then activates factor XI, which, in the presence of calcium ions, cleaves a peptide from factor IX. The latter reaction produces the activated form, factor Xa, in the presence of the phospholipids of the platelet membrane, factor VIII, and calcium ions. The activated factor X together with factor V bind to the platelet phospholipids in the presence of calcium. This complex, referred to as prothrombinase, catalyses the conversion of prothrombin (factor II) to thrombin (factor IIa). The effect of cofactors V and VIII is greatly enhanced by positive feedback created by the action of the thrombin.

C. Platelet Activation

In case of endothelium damage, the procoagulant subendothelium is exposed to the circulating platelets. Platelets, which are small nonnucleated cells, are very important for blood coagulation, because they provide procoagulant phospholipids and factor V.

The platelets are activated by the subendothelium, to which they adhere, and by the localized formation of trace thrombin. Platelets adhere to the subendothelium through their surface receptor (glycoprotein GP Ib-V-IX) via the von Willebrand factor (a polymer of factor VIII) and are activated to change shape and spread, undergo the release reaction, and stimulate the arachidonate pathway. Released adenosyl diphosphate (ADP) and thromboxane A2 (TxA2) cause the circulating platelets to change shape, express the fibrinogen receptor (glycoprotein GP IIb-IIIa), and stick to other platelets. Simultaneously, the platelets rearrange their phospholipidic membrane in such a way that the procoagulant lipids such as phosphatidyl serine, that normally remain inside the membrane, appear on its outer surface. Thus the growth of the thrombus is initiated. In addition, thrombin generated at the site of vascular injury causes further release and aggregation of platelets and transforms fibrinogen into fibrin, thereby contributing to the growth of the thrombus and its stabilization with a fibrin mesh (33). The surface of this thrombus exhibits a phospholipoprotein (designated platelet factor 3 or PF3). The latter promotes and accelerates the generation of the factor X (Stuart factor) and thrombin at the surface. It also enables these elements to be protected from the inhibiting action of antithrombin (AT). Furthermore, the platelet factor 4 (PF4), which has a great affinity for heparin, is released by activated platelets (see below). Therefore PF4 inactivates heparin, rendering it unavailable to catalyze the inhibition of thrombin by antithrombin (34).

D. Thrombin and Clot Formation

When materials are in contact with blood, the cascade of enzymatic reactions described above leads to the formation of thrombin, which plays numerous important roles in the coagulation of blood (35). Thrombin is able to induce both platelet aggregation and release. This enzyme has the possibility of catalyzing its own formation. Thus when small amounts of thrombin become available, increasing quantities of the enzyme will be readily formed. Thrombin is able to activate fibrinogen in a hydrolysis reaction that produces fibrinopeptides A and B and the soluble fibrin polymer (36). The cross-linking of the soluble fibrin is catalyzed by factor XIII (37), and this results in the formation of the insoluble fibrin clot. It should be noted that aggregated platelets, other blood cells, and plasma proteins are entrapped in the cross-linked fibrin network.

E. The Fibrinolytic System

Fibrinolysis is a physiological phenomenon that results in the enzymatic destruction of the thrombus (29,36). This process involves plasmin, which is a serine protease produced from an inactive precursor: the plasminogen. All the physiological activators of fibrinolysis are all serine proteases. The latter are formed from zymogens by limited pro teolysis after clotting or by the formation of a thrombus, which can occur in two different ways (Fig. 3). The activation of fibrinogen is dependent on the contact system, coagulation, circulating activators, and the endothelium of the vessel. The intrinsic pathway is initiated when the factor XII is activated to XIIa by interaction with a negatively charged foreign surface in the presence of high molecular weight kininogen and prekallikrein. The extrinsic pathway, however, seems to be more relevant to the physiological changes in fibrinolysis. There are two kinds of physiological activators, one of which is the tissue plasminogen activator (t-PA) synthesized within and released from the endothelial cell. The tissue plasminogen activator is predominately responsible for the lysis of clots in the vessels. The other type of activator is the urinary plasminogen activator (u-PA), which was initially identified in human urine and was later found in plasma (29). It should be noted that some foreign proteins, for example streptokinase and staphylokinase, and proteins from bacteria, have an enzymatic ability to form a complex with plasminogen and to promote plasmin formation (38).

The powerful inhibitor alpha-2-antiplasmin immediately inhibits any plasmin that may appear in the blood. It can be concluded that fibrinolysis is a localized phenomenon occurring at the surface of the thrombus or blood clot when it is confined in a closed cavity. In reality, plasmin is only observed in circulating blood during thrombolytic treatment using streptokinase.

Plasmin degrades the fibrous network of cross-linked fibrin as well as the fibrinogen. The products of this enzymatic digestion are called the fibrin degradation products (FDP). The quantity of the latter is an indirect measure of the action of the plasmin.

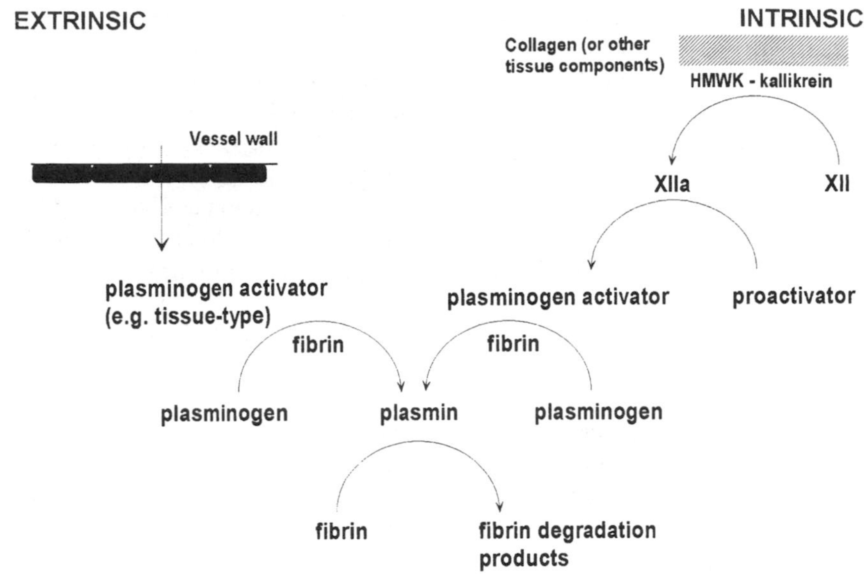

Figure 23.3 Schematic diagram showing the activation pathways of the fibrinolytic system.

F. Coagulation Control

From the discussion of the whole coagulation process, it would appear that the formation of the clot should occur when blood contacts any surface, including the normal vessel endothelium. However, this would result in the general thrombosis of the blood circulation system. As this is not the case in reality, it is obvious that other control systems are involved. Moreover, the normal healthy vessel endothelium is able to control the coagulation cascade. Thus blood coagulation is carefully controlled in vivo by several anticoagulant mechanisms, which ensure that clot propagation does not lead to occlusion of the vascular system (39,40).

1. Circulating Coagulation Inhibitors

Thrombin plays a central role in hemostasis, so there must be enough to ensure hemostasis but not so much as to cause thrombosis. As in all the important biological systems, thrombin generation is ensured by a delicate system of positive and negative feedback, in which thrombin influences both its own formation and its own breakdown. There are three types of coagulation inhibitors: the ''enzyme kidnappers'' (such as alpha-2-macroglobulin), the inhibitors that block active sites (such as antithrombin), and the serine proteases (such as alpha-1-antitrypsin) Table 2. These inhibitors are generally able to form complexes with several proteases, at variable rates. For instance, alpha-2-macroglobulin reacts readily with kallikrein but slowly with thrombin. In plasma, about 70% of the thrombin generated is inhibited by antithrombin, 20% by alpha-2-macroglubulin, and the rest by other inhibitors such as alpha-1-antitrypsin.

The most potent plasma inhibitor is antithrombin, which is able to form inactive stable complexes with serine proteases including factors IIa, IXa, Xa, XIa, and kallikrein. The reactions are irreversible and slow. They suggest the formation of a chemical bond between the serin active site of the protease and an arginyl residue of the inhibitor (41). These reactions are more or less subject to be catalyzed by heparin and heparin analogs that might be present in some subendothelial or endothelial tissue (42).

Another thrombin inhibitor, the second heparin cofactor (HC II), is also able irreversibly to form a stable complex with thrombin (43–45). However, unlike antithrombin,

Table 23.2 Coagulation Inhibitors in Human Plasma

Inhibitor	Concentration in plasma (g/L)
Alpha-1-antitrypsin	2–4
Alpha-2-antiplasmine	0.085
C1 inactivator	0.25
Alpha-1-antichymotrypsin	0.48
Alpha-2-macroglobulin	3
Antithrombin	0.15–0.25
Heparin cofactor II	0.079

this inhibitor is not able to react with other proteases involved in the coagulation cascade. In this case, the most effective catalysts appear to be dermatan sulfate and pentosan polysulfate rather than heparin (46–48).

2. Endothelial Cell Control

The endothelial cells associated with thrombin employ several different pathways to achieve the control of hemostasis.

Welsler et al. (49) showed that thrombin stimulates the endothelial cells to secrete prostacycline (PGI2), a prostaglandin that is a very potent inhibitor of platelet aggregation and hence restricts the growth of the thrombus.

Intact endothelial cell membranes contain a receptor called thrombomodulin. In the presence of the latter, thrombin loses its procoagulant properties and catalyzes the hydrolysis of protein C (50–52). The resulting activated protein C (aPC) is able to bond to platelet membranes. It has been suggested that activated protein C is able to catalyze the hydrolysis of factors Va (53) and VIIIa (54) when fixed onto the platelet membrane. As a result, the rate of activation of the factors X and II is decreased in the vicinity of the endothelial cells. During this process, the protein S also plays a role. Activated protein C, together with protein S, inactivates both tenase (a complex of factor Ixa, factor VIIIa, and calcium ions on phospholipid, activating the factor X) and prothrombinase, hence stopping the formation of thrombin. Activated protein C is also a profibrinolytic (55).

Thrombin also stimulates the production of the plasminogen activator by the endothelial cells, which initiates the fibrinolytic degradation of thrombi (56). Other mechanisms are also involved in the control of thrombosis by endothelial cells. For example, endothelial cells can synthesize and release nitric oxide (NO), which is a potent vasodilator and also an inhibitor of platelet aggregation (57).

All of these preceding pathways, and probably others, are involved in the control of clot formation (58,59). Their failure may contribute to atherothrombosis and therefore appears to be one possible manner by which general thromboses occur as a consequence of this disease. Other plasmatic control systems are involved in the prevention of general thrombosis.

3. Glycosaminoglycans

The ability of certain sulfated polysaccharides (such as glycosaminoglycans, the best known of which is heparin) to interfere with blood coagulation is well known (60,61). More than 80 years after its discovery, heparin remains an important antithrombotic agent in clinical use. The main effect of heparin (and heparan sulfate) is to accelerate the inactivation of coagulation enzymes by antithrombin, the most potent serine protease inhibitor (serpin) (42,62). Another serpin, the heparin cofactor II, which is activated by heparin and also by another glycosaminoglycan, dermantan sulfate, selectively inactivates thrombin (47,63). Control of blood coagulation is *via* the protein C pathway and involves a glycosaminoglycan-containing molecular species as the protein C activator thrombomodulin, which is a proteoglycan. The exact chemical nature of the catalytic site on heparin was a matter of controversy until Choay *et al.* (64,65) proved that a synthesized pentasaccharide was a replica of the catalytic site (Fig. 4). The conformational change of antithrombin bonded to the heparin is now well described (66).

The endothelial cells of blood vessels synthesize a minor subpopulation of proteoglycans, which exhibit heparan sulfate with the appropriate pentasaccharide to bind antithrombin (67–71). This interaction serves as the basis of the natural anticoagulant mechanism of the vascular system (72).

G. Procoagulant Regulation

It is important to ensure a fast initial thrombus formation and stabilization to prevent bleeding. Thus, in plasma, a specific inhibitor inhibits the anticoagulant properties of

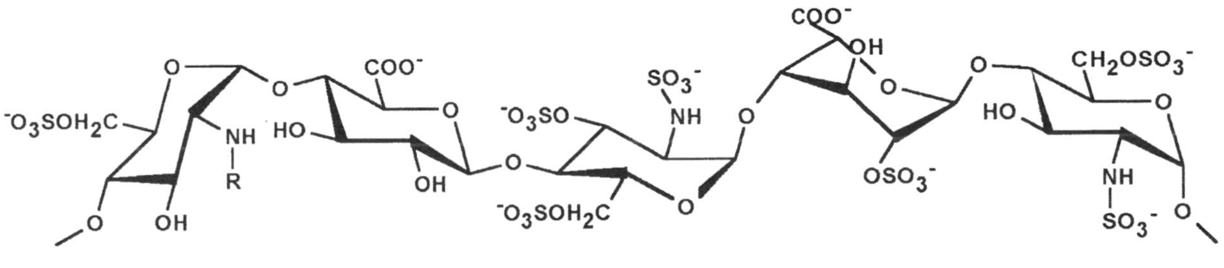

Figure 23.4 Pentasaccharide corresponding to the minimal sequence in heparin for binding to antithrombin.

the activated protein C: the protein C inhibitor (PCI). PCI is a member of the serine protease inhibitor family (serpin) (73). The reaction rate between aPC and its inhibitor is increased in the presence of heparin or dextran sulfate (74). The fibrinolytic system is controlled by two proteins: plasminogen activator inhibitors 1 and 2 (PAI-1 and PAI-2), which are also serine protease inhibitors (75,76).

The complement inhibitor S protein, which is identical to the adhesive protein vitronectin, functions as a heparin neutralizing factor by protecting thrombin and activated factor X against rapid inactivation by antithrombin. The vitronectin counteracts the anticoagulant activity of heparin and pentosan polysulfate but not that of dermatan sulfate (77). Vitronectin also plays another role in hemostasis, as it can strongly bond with a high affinity to PAI-1. The formation of the complex with vitronectin not only increases the half-life of PAI-1 in circulation but also enables the active inhibitor to be stabilized at sites of vascular injury and initial platelet plug formation (78).

H. Conclusion

Clot formation appears to be the normal consequence of the contact between blood and polymers, unless the latter have been specifically designed and tailored to prevent this catastrophic event. Clots initiated by contact may develop in different ways depending on the nature of the surface. The final result is always thrombin formation. The latter may well be the pivotal event in coagulation. Pro- and anticoagulant mechanisms generally favor anticoagulation under physiological conditions. However, the anticoagulant system is restrained, and procoagulant forces predominate at sites of vascular injury.

Blood and the natural endothelium of the vessel wall are able to control the coagulation process. One of the most potent control systems available is the heparin-catalyzed formation of the antithrombin–serine protease complex.

It must be emphasized that the coagulation of blood is a complicated process. Its kinetics cannot be completely described by the schematic models given above, as the formation of a clot is a complex combination of each of the reactions involved. Moreover, these kinetics are dependent on the flow rate of the circulating blood. This adds one further parameter to the multiparameter system controlling the coagulation of blood.

III. BLOOD COAGULATION ON FOREIGN SURFACES

It has been known for a long time that prothrombin time (a coagulation test) is shorter in glass than in plastic. This is a consequence of a relationship between the contact system and extrinsic coagulation: factor VII can be activated by factor XIIa. This is an indication of the importance of the surface composition in the coagulation process. The blood reaction to a foreign surface is driven by interfacial phenomena that are determined by the surface properties of the material (79) and blood flow characteristics (80).

A. Role of the Adsorption of Protein

As a result of different interactive forces (Tables 3 and 4), a competitive adsorption of proteins and glycoproteins occurs at the polymer surface, and they form a complex protein coating (81). Some of these adsorption processes are partially or completely reversible (82,83). Depending on the nature of the surface, some of the deposited proteins may initiate coagulation. Therefore in order to improve the blood compatibility of insoluble polymers, it is necessary to understand the phenomena that govern the adsorption of the proteins. The latter is the first essential step to triggering coagulation or to promoting complement activation by a foreign surface.

When a foreign surface (solid, liquid, or gas) is brought into contact with a protein solution like blood, a certain

Table 23.3 Fundamental Interaction Forces Between Materials and Blood Components

Interaction forces	Description
van-der-Waals forces including	Attractive dipole–dipole interaction considerably weaker than ionic or covalent bonds. These forces decrease rapidly with the distance (by the factor $1/d^7$ for London forces and by the factor $1/d^4$ for others).
Keesom forces (between two permanent dipoles)	
Debye forces (between permanent dipole and induced dipole)	
London forces (between instantaneous dipole and induced (dipole)	The typical energy is about 2 $kJ \cdot mol^{-1}$
Hydrogen bonds	Attractive, a special case of very strong dipole–dipole interaction. Typical energy is about 20 $kJ \cdot mol^{-1}$
Ionic interactions	Attractive or repulsive, these forces decrease by the factor $1/d^2$. Typical energy is about 250 $kJ \cdot mol^{-1}$

Table 23.4 Resultant Interaction Forces Between Materials and Blood Components

Interaction forces	Description
Electrostatic double layer forces	Attractive or repulsive long-range forces, depending on the electrolyte content of the medium
Hydration forces	Repulsive
Hydrophobic interactions	Attractive long-range forces
Polymer-induced forces	Sterical repulsion or bridging attraction
Specific forces	Attractive short-range (e.g., acid–base, Ca^{2+} bridges

Source: Adapted from Ref. 357.

amount of the dissolved protein will be adsorbed on the surface. The amount of protein adsorbed and the composition of this protein layer depend mainly on the nature of the foreign surface. To a great extent, the biological properties of the surface depend on the adsorbed protein. Unfortunately, the hemocompatibility is difficult to predict from the characteristics of the surface.

During the last 30 years, it has been established that the first event to occur when blood contacts a foreign body is the adsorption of plasma proteins to the surface. Vroman (84,85) has shown that a protein layer is formed on the surface within a few seconds of being in contact with blood. In the same period, Dutton *et al.* (86) showed that there is a protein layer between the substrate and the adhered platelets or thrombi. Hence the comprehension of the protein adsorption mechanism is essential in order to understand polymer/blood compatibility. The composition of the adsorbed layer is dependent on the chemical nature of the surface of the material (87,88). Furthermore, the conformation adopted by the adsorbed proteins plays an important role in the biological behavior of the surface (89,90).

Protein molecules in aqueous solutions are compact structures with hydrophilic amino acids at the surface, whereas hydrophobic ones are inside. In contact with a polymeric surface, hydrogen bonds, hydrophobic interactions, and ionic and polar forces can induce a conformational change of the adsorbed protein, sometimes leading to the unfolding of the macromolecular chain (91,92).

Prevention of protein adsorption is crucial for blood-contacting devices, including catheters, dialyzers, vascular grafts, blood containers, and oxygenators. When a material surface is brought into contact with blood, the initial stage of the adsorption of serum proteins successively triggers the thrombogenesis and then enhances the complement activation via the classical pathway (Sec. IV.A).

Experimental evidence shows that surface grafting can minimize both protein adsorption and thrombogenesis. The grafting effect lasts for approximately a month, which is sufficiently long for the specific purpose of the devices, such as dialysis and oxygenation, to be accomplished. However, the long-term durability of the grafting effect is still under examination (93). Permanent implants, such as vascular grafts and catheters for blood access, require a nonfouling effect over a long period of time.

Various water-soluble polymers have been used for surface grafting to create nonfouling surfaces. They include nonionic hydrophilic polymers such as polyacrylamide (PAAm), poly(*N,N*-dimethylacrylamide) (PDMAAm), poly(ethylene glycol) (PEG), ethylene–vinyl alcohol copolymer (EVA), and poly(2-hydroxyethyl methacrylate) (PHEMA). A polymer containing phosphorylcholine, a cell membrane containing zwitterions, has also been employed. Grafting of these polymers can be achieved by surface graft polymerization following a low-temperature plasma treatment (94,95), or with a coupling reaction (96–99).

Fujimoto et al. demonstrated, with an ex vivo adsorption experiment using radio-labeled immunoglobuline G (IgG), that IgG adsorption onto a PU film was considerably reduced by grafting of PAAm to the surface, employing a glow discharge treatment (100) and ozone oxidation (101). Interaction between the PU surface and the platelets was greatly reduced by this modification. The latter was assessed with an ex vivo arteriovenous shunt experiment in rabbits (100,101). Ruckert and Geuskens (102) reported that surface graft polymerization of PVP effectively reduced the adsorption of fibronectin onto a styrene-(ethylene-co-butene)-styrene triblock copolymer film. They also reported that the surface of PU catheters with tethered PDMAAm chains remained unfouled even 3 weeks after implantation in the inferior vena cava of a rabbit (103). Recently, Kishida *et al.* (104) used a reverse transcription-polymerase chain reaction (RT-PCR) to examine the expression of interleukine-1 beta mRNA secreted by macrophage-like cells (HL-60). The latter were cultured on a grafted PE surface as an index of inflammatory stimulation. They observed that the cells cultured on the PAAm-grafted surface had a low level of IL-1beta mRNA, indicating the nonfouling capacity of the grafted surface.

The prevention of protein adsorption to the outermost grafted surface is attributed to a steric hinderance effect created by the tethered chains. A grafted surface in contact with an aqueous medium, which is a good solvent of chains, exhibits a diffuse structure (105). The reversible deformation of the tethered chains, due to the invasion of mobile protein molecules into this layer, produces a repulsive force. This force is governed by the equilibrium between the entropic elasticity of the chains and the osmotic pressure resulting from a rise in the concentration of segment. The repulsive force prevents direct contact between the protein molecules and the substrate surface. It is interesting to note that the extent of protein repulsion is related to the polymer graft density (100).

1. Role of Platelets and Rheology

As previously described, a clot is generally produced when blood comes into contact with polymers. The clot is a consequence of the induction of the coagulation cascade. The mechanism is initiated by

1. The polymer/blood contact.
2. The products released in the bloodstream.
3. The platelet activation and hemolysis produced by turbulence in the bloodstream. This is the result of the poor mechanical compliance of the prosthesis.

The interaction of platelets with various stimuli both soluble and surface bound leads to activation with shape change from disc to sphere with pseudopodia. Activated platelets spread on the surface and aggregate with other platelets (33). Initially, the von Willebrand factor (a polymer of factor VIII) controls the bonding of platelets to the surface. Some of the internal constituents are then released from the aggregated platelets, into the plasma. These transformations, which are controlled by prostaglandins, calcium ions, cyclic adenyl monophosphate, and the concentration of adenyl diphosphate, are essential as they make both specific proteins and platelet membrane phospholipids available. The latter elements are of prime importance for the acceleration and control of the coagulation process. Thrombin adenyl diphosphate and other reactants also initiate platelet activation. It can also be produced by turbulence in the blood flow and is a function of the flow rate (106–110).

2. Conclusion

Coagulation on artificial surfaces is a complex phenomenon involving several systems: the intrinsic pathway of the coagulation cascade, triggered by the contact phase and controlled by antithrombin; the adhesion and activation of the platelets; and finally the fibrinolytic system. Therefore there are different methods for obtaining a hemocompatible polymer:

- To stimulate the catalytic inhibition of the thrombin by the antithrombin. This is the case of heparin-bearing or heparinlike polymers (108,111–114).
- To avoid the adhesion of proteins and platelets. This is the case for nonfouling materials, which can be, for instance, produced by grafting a hydrophilic polymer with a large hydrodynamic volume onto a polymeric substrate (20,98,99,115).
- To stimulate fibrinolysis. This is the case for plasminogen-bearing materials (116).

All of these possibilities are currently being explored.

B. Contact System and the Intrinsic Pathway

When artificial surfaces are placed in contact with blood, the activation of the coagulation process is induced by the activation of the contact phase. This process involves four proteins: high molecular weight kininogen (HMWK), prekallikrein/kallikrein (Fletcher factor), factor XII (Hageman factor), and factor XI [plasma thromboplastin antecedent (PTA)], and their complexes. The proteins are bound to the surface in a cascade of enzymatic reactions, which result in increasing amounts of factor XIa attached to the surface.

In the intrinsic pathway, thrombin (factor IIa) formation results from the enzyme activation cascade triggered by the contact between factor XII (Hageman factor) and an electronegative surface. This intrinsic pathway is self-accelerated as the formation of the enzymes subsequently activates the zymogens. In a purified medium, the rate of the latter reactions is slow and calcium dependent. In contrast, these reactions are accelerated when platelet factor 3 (PF3), from the platelet membranes, and cofactors VIIIa and Va are present.

Nowadays, coagulation is divided into two stages rather than two pathways: an ''initiation'' stage, which is controlled by the tissue factor–dependent pathway, and an ''augmentation'' stage, which is controlled by components of the intrinsic pathway (30). Both pathways are able to generate fibrin, but the tissue factor stage is impeded by the tissue factor pathway inhibitor-1 (TFPI-1) soon after its initiation.

IV. THE INFLAMMATORY RESPONSE

Polymers promote an inflammatory response when placed in contact with living systems. This response can be acute and/or chronic. These responses may be both beneficial and deleterious to the host (9,10,12). For example, inflammation is a major protection mechanism for the host, which uses it to rid itself of foreign materials. However, when a host is seriously impaired, the inflammatory response becomes a risk of infection. The acute inflammatory response appears to increase the rate of reaction, whereas the chronic inflammatory response may increase the infection rate (12).

The inflammatory response is a series of complex reactions involving various types of cells whose functions are controlled by various endogenous and exogenous mediators. The first phase of the acute inflammatory response is initiated when the formation of a blood clot occurs in the space around the biomaterial, after the latter has been exposed to the living system. This process is initiated by changes in the permeability in the adjacent vessel, which, in turn, are controlled by the intrinsic and extrinsic coagulation systems, the complement system, the fibrinolytic system, the kinin-generating system, and the platelets. It has been established that leukocytes have receptors for P-selectin, an alpha-granule membrane protein, expressed on the surface of the activated platelets (117,118).

Following these events there is preferential migration of the neutrophils towards the site of bonding. The differentiation of these cells into macrophages is therefore the first step of normal wound healing. This is then followed by the proliferation of the fibroblast, the deposition of collagen, and the capillary endothelial cell proliferation.

Various agents control monocyte and macrophage movement towards the site of injury within the tissue. This movement is defined as either chemotaxis or chemokinesis.

1. The Complement System

The complement system is one of the most important agents controlling chemotaxis. Like the coagulation system, the complement system consists of a series of plasma glycoproteins and inhibitors. Its activation can be initiated by antigen–antibody complexes, bacterial polysaccharides, viruses, endotoxins, and synthetic polymers. Once initiated, activation proceeds along one of two pathways, classical or alternative (Fig. 5). The cascades of enzymatic events generate components that produce an inflammatory or immune response.

To summarize, the contact between living systems and polymers induces a blood-controlled inflammatory response that may result in

1. Increased ability to resist infections
2. The isolation of the implant by fibrous tissue, resulting in the rejection of the implants
3. Acute inflammatory response with a hypersensitivity reaction

The activation of the complement system is one of the major agents controlling the inflammatory response. It consists of more than twenty components and plasma proteins, which are involved in the recognition of foreign substances and in both specific (immune) and natural (nonimmune) host defense (119). It appears necessary to control

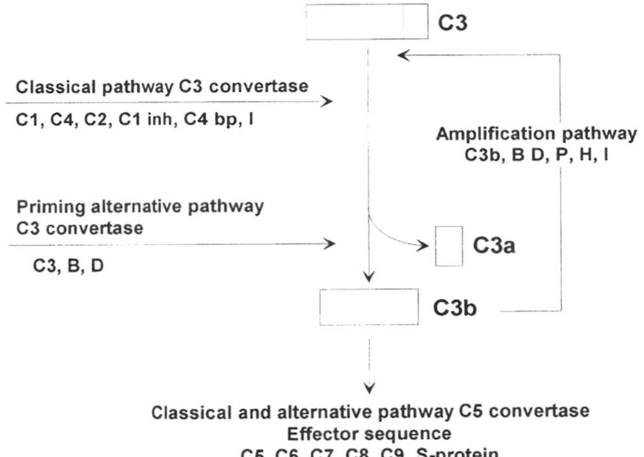

Figure 23.5 Pathways of complement activation.

this inflammatory response, and this can be achieved through the control of the complement system (120–122).

Both classical and alternative pathways of the activation of the complement system involve the formation of potent C3-convertases. These convertases catalyze the activation of zymogen C, a plasma protein, into a potent enzyme. This enzyme then mediates the formation of the C5–C9 complex, which is the final step of the activation of the complement system. Both alternative and classical C3-convertases can be inhibited by natural plasma inhibitors like H and I proteins (123).

This inhibition effect is promoted by heparin (124) either in solution or when bound to activating polymeric surfaces such as Zymosan or Cuprophan (125), a potent complement-activating polysaccharide. This control mechanism of the complement system, which includes the action of a potent plasma inhibitor called C1 inhibitor (C1inh), has already been employed to control both the inflammatory response and particularly the hypersensitivity reactions induced by dextrans, as described in this chapter.

The small fragments generated by the cleavage of the proteins C3 and C5 are the anaphylatoxins C3a and C5a. These anaphylatoxins induce smooth muscle cell contraction, the enhancement of the vascular permeability, and the release of histamine. The C5a is also a potent chemotactic factor for neutrophils. The activities of C3a and C5a characterize the common postcoagulative reactions associated with tissue injury. The larger portions of the activated complement factors C3 and C5 (C3b and C5b) participate in the amplification phase of the humoral immune system, serving as enzyme modulators, opsonins, and effectors of cellular recognition.

The first phase described above is followed by the production of small molecules such as cytokines. The latter modulate the inflammatory response by playing the role of a mediator between the cells of the immune system and those of the vascular system.

V. INFECTION

In spite of good hemocompatibility, a polymeric prosthesis may fail because of bacterial contamination. Most prosthesis-related infections are initiated during the surgical procedure, when bacteria-carrying particles from the air contaminate the host. Bacterial colonization is promoted by high hydrophobicity and the irregularity of the surface of the biomaterial surfaces (126). Glycocalyx, a slime layer produced by the bacteria, represents a factor of adhesion and plays an important role in the pathogenesis of the infection (127). Increased exopolysaccharide production produces a biofilm that protects the bacteria from leukocytes, antibodies, and the action of antibiotics (128,129).

As a result of the immune response, infection in the vicinity of the implanted material may occur, which results in the failure of the device. It has been demonstrated that 10^6 organisms cause infection in normal human skin, whereas only 10^2 organisms are necessary to produce the same effect when a suture is present (12).

One of the features of deep-tissue implants, in neurosurgical, cardiovascular, orthopaedic, and ophthalmic patients, is delayed infection. It develops 6 months or a year after surgery in patients who show successful rehabilitation. This delayed infection at the implant site is often caused by organisms of low virulence and is observed in total joint infections, sternal osteomyelitis, and heart valve infections as well as at the site of urinary catheters (12).

The mechanism by which the implanted materials alter the ability of the host to resist infection is not well understood. It has been suggested that the inflammatory response developed by the host as a protective mechanism increases the risk of infection. The acute inflammatory response appears to reduce the infection rate, whereas the chronic inflammatory response may increase the rate of infection (12).

Intravascular catheters are frequently used in hospitals. Several studies have demonstrated that catheter-associated infections are a significant source of nosocomial morbidity and mortality (130). The data reported in the literature indicates the complexity of biomaterial-associated infections. It appears increasingly probable that there is not one simple explanation for the latter. A series of events occur at the time of implantation. Bacterial adherence to the biomaterial occurs initially, followed by the colonization and the formation of an adherent biofilm. The biofilm then implants the bacteria onto the surface of the material, rendering any antibiotic treatment and host defense mechanism ineffective (131,132). Therefore this zone can be assumed to be a site of continuous infection.

There is no correlation between the physical properties (e.g., hydrophilicity, zeta potential, surface charge, rugosity) of the surface of the biomaterial and the bacterial attachment. Hence more specific interactions between the chemical groups, on both the surface of the bacteria and the biomaterials, need to be considered (133).

VI. FATE OF SYNTHETIC POLYMERS IN THE BODY

A. Insoluble Polymers

Living tissues present a very aggressive environment. Polymers, which appear to be very stable in vitro, can be dramatically degraded in vivo. The in vivo failure of polymeric cardiovascular devices has been attributed to calcification, hydrolysis, oxidation, and environmental stress cracking (ESC) (134).

1. Calcification

Deposition of calcium containing apatite mineral has been associated with the implantation of cardiovascular medical devices, such as bioprosthetic heart valves, aortic homografts, and trileaflet polymeric valve prostheses (8,135). Calcification causes the failure of bioprosthetic (136) and polymeric materials (137). Several methods to reduce calcification have been explored, namely, heparin coupling on glutaraldehyde-treated porcine pericardium (138,139) and the chemical modification (phosphonatation) of polyurethane (140).

2. Biodegradation

Biodegradation, including hydrolysis and oxidation, has been extensively studied for polyurethanes and their derivatives (141–144). The oxidation of polyurethane is catalyzed by the presence of metallic ions, especially cobalt, which is utilized in numerous biomedical devices (145). In vivo polymer degradation results mainly from the synergistic action of enzymatic components in the body fluids, oxidative agents, and stress (146). Alpha-2-macroglobulin (147), cholesterol esterase (148), phospholipase A2 (149), proteinase K (150), and cathepsin B (151) are all enzymes that promote the biodegradation of polyurethane.

To prevent oxidation, it is possible to employ a synthetic antioxidant such as Santowhite® or Irganox®. However, particularly for biomedical applications, it is more appropriate to use a natural antioxidant such as vitamin E (152,153). The oxidation of a polymeric biomaterial can also be prevented by avoiding the presence of unsaturated bonds and ether groups. Coury was one of the first researchers to develop an aliphatic polyurethane with a polymer backbone free of ether (154). This polymer, although biodurable, proved too stiff for medical application. Szycher used these fundamental studies to develop a novel aliphatic ether-free polyurethane using a polycarbonate-based macroglycol (155). This polyurethane was shown to be resistant to environmental stress cracking both in vitro (156) and in vivo (157).

B. Soluble Polymers

Any substance introduced into the body must be considered a foreign body. The interaction of the latter with the components of the biological environment determines its rate of elimination or storage. Sufficient understanding of its behavior and its fate in the organism is therefore necessary for a reliable evaluation of the benefits and the risks of employing the material.

In pharmacokinetics, it is useful to represent the body as a system of compartments (Fig. 6) (158). Polymers entering into the organism cannot move randomly, as anatomical and physiological barriers control their movement. Radioactive labeling is a very sensitive method for following the progression of the polymer and its possible metabolites in the bulk of the biological material. Most of the data on the fate of soluble polymers in the living body was obtained after their direct intravascular administration. An injected polymer is rapidly (within a few minutes) distributed to all vascularized parts of the body, and its transport to other compartments of the body starts immediately (158). A polymer introduced into the bloodstream circulates in a closed system of blood vessels. It can be cleared from this compartment either by the endocytic activity of specialized cells (cells of the reticuloendothelial system) or by passage through the endothelial wall, particularly at capillary level. The cells of the reticuloendothelial system (RES) are the cells of the kidney tubular epithelium and the liver hepatocytes. Generally, nonbiodegradable polymers of high molecular weight present a risk of accumulation in the RES (159). Mehvar et al. demonstrated that the disposition of dextran in serum and tissues is molecular weight dependent (160). The low molecular weight dextrans (4 kD) are rapidly excreted into urine with negligible accumulation in the liver, but the larger (150 kD) are mainly accumulated in the liver (161–163). The collection of nonbiodegradable polymers in the body suggests the uptake of the polymer by pinocytosis and its storage in lysosomes. The excessive storage of polymers may provoke an

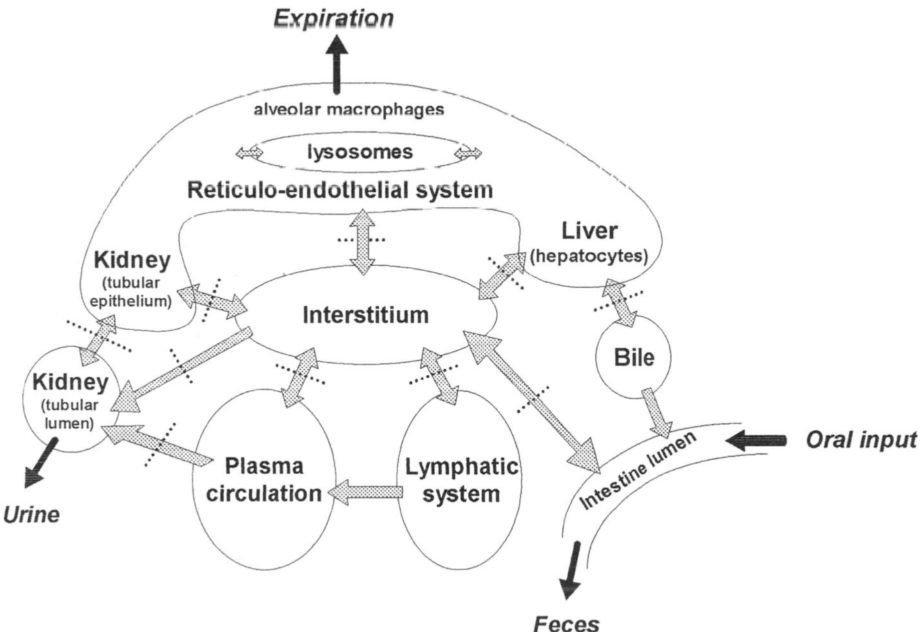

Figure 23.6 Multicompartment model of organism. Areas depict the body compartments. The dashed lines symbolize the restricted transfers between compartments. (Adapted from Ref. 158.)

adverse reaction. Therefore biodegradability becomes an imperative criterion for the polymer intended for use as plasma expanders or for other purposes, when large and repeated doses must be employed.

VII. USE OF BLOOD-CONTACTING POLYMERS

Polymers are not only used in insoluble forms such as tubes or containers, they also have a large array of applications in soluble forms, mainly for therapeutic aims. Polymers can also be employed in coating, for example to improve the hemocompatibility of intravascular grafts (164,165). The following section briefly presents a few applications in this field.

A. Drug Targeting and Drug Delivery

An important approach to increasing the therapeutic efficiency of bioactive agents while decreasing their toxicity has involved chemically bonding these agents to either synthetic or naturally occurring macromolecules. The rationale behind this approach was to design systems that would undergo hydrolysis or enzyme-catalyzed cleavage when placed in the body, in order to release the agent at a predetermined rate.

Synthetic polymers are widely used as drug delivery vehicles because of the diversity of their composition and their good biocompatibility. Water-soluble polymers are of particular interest as drug carriers, as the polymer dissolves when the drug is released. For example, Kopecek have synthesized and extensively investigated poly(hydroxypropyl methacrylamide) (PHPMA) as a drug carrier (166). Kim and his colleagues examined soluble pH- and temperature-sensitive linear polymers for protein and peptide delivery (167). Xu and Lee also investigated soluble polymers as drug delivery vehicles (168).

Mononuclear phagocytes play a central role in the defense of the human body against important pathologies, particularly infections. However, some of the agents of infection may survive phagocytosis, thus producing intracellular infection. The low permeability of cell membranes to several classes of antibiotics results in drug resistance. Temporary binding of a drug to a macromolecular carrier of natural or synthetic origin was suggested many years ago as a method to overcome the membrane barrier. However, a macromolecule–drug conjugate cannot reach its target unless there is target recognition. This is achieved through specific interactions with a membrane receptor. The conjugate must then be internalized by endocytosis. The use of mannosyl ligands covalently bound to a hydrosoluble bioresorbable drug carrier 'poly(L-lysine citramide imide) has been proposed (169). This conjugate targets drugs towards macrophages via the lectin-type specific receptor present on the macrophage membrane.

B. Polymers with Antitumor Activity

Most of the drugs currently used in conventional cancer chemotherapy do not intrinsically select towards tumor cells and are therefore toxic for normal tissues. Many natural products (e.g., anthracyclins, vinca alkaloids, taxanes, adriamycin) are utilized clinically as anticancer agents. However, their use is frequently restricted by dose-limiting toxicity, poor water solubility, or fast elimination due to metabolism or excretion. The conjugation of an active drug with a polymeric carrier through a biodegradable spacer yields a ''polymer therapeutic.'' This represents a new class of antitumor compounds (170,171). Two factors underlie the improved therapeutic index of such macromolecular prodrugs. First, many solid tumors possess vasculature that is hyperpermeable to macromolecules, and they usually lack effective lymphatic drainage, so they selectively accumulate circulating macromolecules (172). The second improvement is attained with the use of a polymer–drug linker that is stable in circulation but is degradable intracellularly by tumor-enhanced enzymes (173) or by acidic hydrolysis in a lysosomal compartment (67).

5-Fluorouracil (5-FU) is a phase-specific antimetabolite; it is commonly used in cancer chemotherapy to treat solid tumors, particularly those of a gastrointestinal origin. However, this treatment is often accompanied by adverse reactions that limit the dosage and hence the efficiency of the drug. Various macromolecular prodrugs of 5-FU have been proposed to reduce these side effects (174–176). In vitro and in vivo studies performed with these derivatives have shown that 5-FU is slowly released by nonspecific chemical hydrolysis. Schacht et al. have synthesized a macromolecular prodrug of 5-FU by covalent bonding to a polymeric carrier such as PEG or dextran. The in vitro hydrolysis study has shown that these derivatives are stable in the presence of serum peptidases but can be selectively activated by tumor-associated enzymes such as collagenase type IV or cathepsin B, to free 5-FU at the tumor site (175). The free radical polymerization of various types of 5-fluorouracil (5-FU) vinyl derivatives has also been studied by Akashi et al. (174).

Other polymers that are not prodrugs can be endowed with anticancer activity in vitro (177–179) and in vivo (180). This is the case for a dextran derivative that prevents tumor growth and tumor angiogenesis (181,182). The mode of action of these polysaccharides having carboxylic and benzylamide chemical functions is still under investigation. But it appears that these bioactive polymers exert an inhibitory effect on the growth of tumoral cell lines by interfering with specific autocrine and paracrine growth factors (181–183).

C. Immunoadjuvant

The interest in polyphosphazenes for biomedical uses, especially for exploitation in drug delivery and controlled release technology, has increased dramatically in recent years (184). Poly[di(carboxylatophenoxy)phosphazene] (PCPP) appears to be of particular interest as a biomaterial primarily because of its powerful immunoadjuvant properties but also due to its ability to form hydrogel microcapsules under physiological conditions. Polyphosphazene with its carboxylic and ester alkyl functions stimulates the immune response of mice infected by influenza virus (185).

D. Polymer Conjugates as Contrast Agents

Contrast agents may be categorized as nonspecific or specific. Nonspecific agents diffuse freely in the extracellular and extravascular compartments with the exception of the brain, where the contrast agent can only pass through brain barrier lesions. In the specific agent group, a new class of products has been developed. These are blood pool contrast agents, which are distributed throughout the whole intravascular volume and are slowly cleared from the blood. Diffusion through the healthy capillary wall is limited and depends on both the pathological state of the endothelial permeability tissue of the organ examined and on the characteristics of the contrast agent (size, charge, and molecular shape, etc.). The diagnostic efficacy of perfusion imaging, including cerebral perfusion, is modulated by the pharmacokinetic profile of the blood pool contrast agent. One way to improve the vascular residence time consists in bonding a vector, such as a synthetic polymer or a biological macromolecule, and a metal ion like Mn^{2+}, Gd^{3+}, or Dy^{3+} (186–189).

Organometallic compounds of gadolinium are used in medical imaging, but the macromolecular compounds (molecular weight 30,000 daltons) are more advantageous than those with a low molecular weight (550 daltons). Dynamic magnetic resonance imaging of breast neoplasms using a polymeric agent may help to differentiate better between benign and malignant lesions as it highlights the enlarged interstitial space and the increased capillary permeability in carcinomas (190).

E. Thrombogenic Blood-Contacting Polymers

Hemocompatibility of the polymers used in contact with blood is not always required. In one case, the coagulation of blood is desirable. Thus polymer beads are utilized to embolize the blood vessels above a solid tumor in order to asphyxiate the latter (191–193). Various polymeric embolic materials are currently in clinical use, such as ethylene–vinyl alcohol copolymer mixtures, poly(vinyl acetate), cellulose acetate polymer, and poly(vinyl alcohol) (194).

VIII. MODIFICATION OF THE POLYMER SURFACE

A. Chemical Modification of Polymers

The dextran molecule is a polymer formed by glycosidic bonds linking the number 1 and 6 carbon atoms of consecutive α-D-glucopyranose rings. This polysaccharide is relatively rigid. This property increases with the number of substituents with bonds axial to the glycosidic oxygen and in the presence of bulky equatorial groups adjacent to the linkage (195). The pyranose ring structure of the dextran chain is elastic. It is has been suggested that this elasticity plays an important role in accommodating mechanical stresses and modulating ligand bonding in biological systems (196).

Biospecific molecular recognition in living systems is known to be based on the lock and key principle as proposed by Emil Fischer. Based on this concept, Jozefonvicz and Jozefowicz (197) postulated that biospecificity can be achieved by random substitution of a preformed polymer with suitable chemical groups or random copolymerization of suitable functional monomers. Such polymers contain arrangements of the chemical functions, which mimic natural biospecific sites, and the probability of occurrence of such arrangements will depend on the average composition of the polymer. According to this principle, they have developed several functional random copolymer systems that possess a variety of biological properties depending on the type of chemical function. Examples are polymers possessing anticoagulant properties similar to those of heparin (198–202) and polymers that interact specifically with components of the immune system (203–207). These randomly functionalized polymers belong to the new concept of ''bioactive polymers'' possessing bioactivity through biological recognition (208,209).

B. Modification of the Surface of Polymers by Grafting

The modification of the surface of polymers can be achieved by grafting (210). Most of the untreated surfaces of polymers used in industry are hydrophobic. Polymer surfaces generally adsorb proteins when brought into direct contact with blood. The attachment of protein to the surface triggers a subsequent series of mostly adverse biological reactions with respect to the polymeric materials. Therefore the technologies for surface modification of polymers or regulation of the polymer surface interaction with other substances have been of prime importance in biomedical polymer applications (93,211). Theoretically, there is a large difference in the properties of the surface and the bulk of the material. Generally, only the outermost surface needs to be taken into consideration. However, this is not the case for polymer surfaces, as the physical structure of the outermost polymer surface changes continuously with time due to the microscopic Brownian motion of polymer segments. The polymer surface generally has high segmental mobility even at room temperature, in contrast to the rigid surfaces of metals and ceramics. There are two methods for producing grafted surfaces: direct coupling of existing polymer molecules to the surface and graft polymerization of monomers to the surface.

1. Polymer Coupling

There are two theories behind the grafting of a polymer to a surface: the first conjecture is to separate the blood proteins from the surface by grafting a hydrophilic polymer with a high hydrodynamic volume (for example poly(ethylene glycol)); the second hypothesis is to confer the biological properties of the grafted macromolecule (for example heparin) to the surface.

Kishida et al. directly immobilized poly(ethylene glycol) (PEG) chains onto a cellulose surface through esterification (96). Prior to the coupling reaction, they derivatized the terminal hydroxyl group of PEG molecules to carboxylic acid using succinic anhydride. This PEG molecule with terminal carboxylic acid was then chemically immobilized to the hydroxyl group on the surface using carbodiimide in nonaqueous media like toluene. The molecular weight of the grafted polymer has a significant influence on the blood compatibility. Desay and Hubbell (212) have covalently grafted poly(ethylene oxide) (PEO) of molecular weights between 5,000 and 100,000 g/mol on polyethylene terephthalate (PET) films. It was concluded that PEO of molecular weights of 18,500 g/mol and above was effective in reducing protein adsorption and cellular interactions on the surface. Saito et al. have grafted photoreactive α-propylsulfate-poly(ethylene oxide) (PEO-SO_3), one end of which was capped with an azidophenyl group, onto polyurethane surfaces employing a photochemical technique (115).

Han et al. (213) have studied the adsorption of fibrinogen, albumin, and gamma globulin from plasma onto surface-modified polyurethanes [polyurethane grafted with poly(ethylene oxide) (PU-PEO)], sulfonated polyurethane (PU-SO_3), and polyurethane grafted with sulfonated poly (ethylene oxide) (PU-PEO-SO_3). Adsorbed fibrinogen at steady state decreased in the order PU-SO_3 > PU > PU-PEO-SO_3 > PU-PEO, suggesting that sulfonate groups have a specific high affinity to fibrinogen. The intermediate fibrinogen adsorption on PU-PEO-SO_3 can be explained by the compensatory effect between the low protein binding affinity of the PEO chain and the high fibrinogen binding affinity of the sulfonate group. In addition, PU-PEO-SO_3 showed a very fast fibrinogen adsorption due to the high accessibility of the sulfonate group to fibrinogen by the poly(ethylene oxide) spacer. The kinetic profiles of their surfaces showed that as the adsorption time increases, fibrinogen initially adsorbed was decreased and a plateau reached, demonstrating that all the surfaces exhibited the Vroman effect. PU-PEO showed the least fibrinogen and albumin adsorption among polyurethanes, confirming the known nonadhesive property of PEO chains. It is very interesting that PU-PEO-SO_3 exhibited the highest adsorption of albumin and the lowest adsorption of IgG. The authors conclude that such adsorption behaviors of proteins to PU-PEO-SO_3 contribute to improved blood compatibility (213).

2. Graft Polymerization

This type of modification involves the radical polymerization of acrylic or vinylic monomers onto the polymeric surface, the latter having previously been oxidized to produce peroxides. A variety of methods for graft polymerization onto different substrate surfaces are available. They include direct chemical modification (214), ozone (215,216), gamma rays (217), electron beams (218), glow discharge (=low temperature plasma) (92,219,220), corona discharge (105,221), and UV irradiation (94).

Both glow and corona discharges are currently used for the modification of the surface of polymeric materials. The corona discharge treatment is far simpler than the glow discharge treatment, as the former technique does not require the evacuation of the discharge system. However, the corona treatment can result in greater damage of the substrate polymers, and the discharge conditions are more difficult to control than those of the plasma treatment. The latter treatment is generally carried out under reduced gas pressure. UV irradiation appears to be an excellent method for the surface grafting of polymers because it is both simple and clean.

IX. HEMOCOMPATIBILITY OF POLYMERS

When a polymer comes into contact with blood, it appears that

1. The polymeric compound may release components into the bloodstream that may be toxic, carcinogenic, blood coagulation–activating, and/or inflammatory response–promoting (222–226).
2. The polymeric compound may degrade into components that may induce blood coagulation, the inflammatory response, carcinogenesis, and the toxic response (149,150,227,228).
3. The polymers may not have the desired mechanical compliance, which may promote turbulence in the blood flow. As a consequence, hemolysis, platelet activation, and the inflammatory response may occur, all of which are undesirable effects (101,229–231). Long-term implantation provokes calcification, which leads to the mechanical failures of the devices (20,232,233).
4. The polymeric compound itself may induce the activation of the blood coagulation system. As a consequence, thrombosis on the site of implantation or thromboembolism may occur.
5. The presence of the polymeric compound may promote an inflammatory response (21,234–236), the effects of which can be either beneficial or prejudicial to the host as previously discussed.
6. The polymer may promote an infection in the host either alone or when mediated by the inflammatory response (237,238). Similarly, it may also induce tumorigenesis (10), although this phenomenon has never been accurately described.

The ideal blood-compatible polymer must not exhibit these undesirable properties, though in the case of heart valves and vascular grafts, the occurrence of failure is still high. Morbidity and mortality due to valve-related causes are around 25% of cases within 10 years. The thrombotic occlusion, especially in small diameter synthetic grafts, remains a serious problem, where 36–54% are occluded at 1 year and 50–62% are occluded by 5 years (22). Recently, significant improvement has been achieved with a Dacron® prosthesis coated with nonthrombogenic polymeric materials (239). There is a large array of laboratory tests used by hematologists to explore virtually each step of the coagulation system (240). Most of the tests are only applicable for the study of the hemocompatibility of soluble polymers, but some techniques are available to test the blood compatibility of insoluble materials (241).

It appears that blood compatibility is a multiparametric function of the characteristics of the polymeric compound. At the present time, only polymers that partially fill blood compatibility requirements are available. However, by employing medical grade polymers, nontoxicity of the released components and the degradation products has been achieved.

The noncarcinogenicity of the materials, their released adjuvants, and the degradation products remains controversial. Large-scale epidemiological studies are needed in order to determine whether the long-term systemic effects of the polymeric implants provoke carcinogenesis (9,10).

The mechanical compliance of the polymer is generally believed to be critical, and considerable efforts have been made to improve this characteristic. As a result, it appears that vascular grafts with improved mechanical properties are now commercially available. These grafts are made from polyesters (polyethyleneterephthalate, commercialized under the name Dacron®), expanded polytetrafluoro ethylene (commercialized by Gore Tex), and polyurethanes (242–244). These materials, for various structural reasons, match the mechanical compliance of the vessel wall.

The physical characteristics of the surface also appear to be critical. Various hypotheses have been proposed. One theory claims that the ideal surface is perfectly smooth and is best achieved by polyurethane grafting. Another widely used method employs either knitted or woven grafts of Dacron® or porous grafts of Teflon®. In the latter case, it has been proved that the ingrowth of tissue and the subsequent encapsulation of the implant by fibrous healing tissue provides a suitable smooth blood-compatible surface (245, 246).

The design of the device also affects the overall blood compatibility of the polymeric compound, as it is one of the parameters that controls the hemodynamic characteristics (1). The importance of this parameter is highlighted by the considerable research currently devoted to the improvement of the design of cardiac valves (247,248). Numerous types of cardiac valves are available that differ not only in their design and also in the chemical nature of the polymers used in their manufacture. The remarkable results obtained with biovalves may also result from their superior design. This is also illustrated by the fact that aortic vascular grafts of large diameter are currently successful, whereas small-diameter vascular grafts (of less than 4 mm diameter) fail. Aortic grafts are successful because the blood flow is high. Therefore the blood–polymer surface contact time is restricted, resulting in minimal activation of the coagulation of blood. Hence, only microthrombi of low adherence are formed and embolyzed. Therefore fibrinolysis is rapid, as the rate of fibrinolysis at a surface is inversely proportional to the diameter of the thrombus. This reasoning can also be applied to explain the failure of small-diameter vascular grafts, which have a lower blood flow rate. The other parameters that control the coagulation process at the blood/polymer interface must also be considered. These parameters include the relationship between the chemical nature of the polymer and that of the blood, and its influence on the formation of a clot at the interface. The inflammatory process can also be considered as a function of the chemical nature of the polymers.

A. Blood Compatibility and Chemical Physics of Polymeric Surfaces

The coagulation process, which is initiated at the blood/polymer interface, has been considered in relationship to the chemical structure of the polymer. Originally it was assumed that the formation of a clot was controlled by the physical properties of the surface, such as hydrophilicity or electrical potential. These properties are dependent on the chemical structure of the polymers.

Initially, it was postulated that hydrophilic polymers would be suitable blood-compatible materials, i.e., they should not induce the coagulation process. This assumption originated from the hypothesis that the blood compatibility of the natural vascular wall was a result of its ability either to bind water or to be hydrophobic.

Numerous studies have been devoted to the assessment of this assumption, which remains a matter of controversy. However, no clear correlation has been established experimentally between the affinity of water to the polymers and their blood compatibility. The various testing methods employed to determine blood compatibility are briefly discussed in this chapter.

Another hypothesis suggested that the electrical properties of the surface of polymers, as measured by polarizability, net surface electrical charge, or overall surface potential, control their blood compatibility (249,250). Again the numerous attempts made to check the validity of this hypothesis failed to establish any general correlation. Moreover, in 1977, Boffa et al. (251) studied the blood compatibility of polytetrafluoroethylene-polyvinylpyrrolidone copolymers of various chemical compositions and hydrophilicity. They showed that the ability to induce the formation of a clot of blood plasma is independent of the composition of the copolymers. In order to confirm this observation, the same authors checked the blood compatibility of various purified polymers including polystyrene, polytetrafluoroethylene, silicone, polyvinylchloride, polymethyl methacrylate, etc., either in powder or in film form. The authors concluded that the blood compatibility of the polymer, as assessed by the in vitro tests, did not vary greatly although their hydrophilicity and the polarizability of the monomeric units were markedly different.

In the early 1980s (252–255) it was postulated that the regular stretching of the hydrophilic and hydrophobic zones on the polymer surface would confer blood compatibility to the material if the distribution of the zones was appropriate. It was claimed that a distribution of hydrophilic zones of 5 μm diameter in a hydrophobic matrix would prevent the activation and aggregation of the platelets. Despite the considerable research, the theory that the blood compatibility of polymers is related to the physical properties of the surfaces has not been proved.

B. Blood Compatibility and Protein Adsorption

When a material is exposed to blood or blood plasma, adsorption of proteins occurs. This results in the formation of a complex protein layer at the interface which is strongly bonded to the surface.

Extensive studies of this process have been used to determine the blood compatibility of polymeric materials. It has been postulated, from the first results obtained, that the materials with an affinity for human serum albumin (HSA) should be blood-compatible, whereas the polymers with a high affinity for fibrinogen (Fg) should be thrombogenic (250,256,257).

However, studies of the adsorption of these proteins from purified protein plasma solution and whole blood, under either static or dynamic conditions (250,256,258–263), demonstrated that the process is not as simple as was initially suggested. The results of these studies can be summarized as follows:

1. Adsorption of proteins is a rapid process that may initially be reversible. This phase is followed by an irreversible adsorption process (258).
2. Competitive adsorption of the proteins occurs, which result from the reaction of numerous parameters such as the composition of the protein and the flow rate (256,264).
3. Blood cells, such as red blood cells, platelets, and white blood cells, influence the adsorption of HSA and Fg and their relative quantities in the blood layer (265).
4. Transient adsorption processes are observed with HSA, Fg, and immunoglobulins (266).

Therefore it can be concluded that it is not possible to assign a given value for the affinity of a polymeric material for HSA and Fg; hence these affinities cannot be employed as a control parameter of blood compatibility. Moreover, by considering these results, Vroman et al. (258) were able to show that the adsorption of traces of high molecular weight kininogen (HMWK), and other proteins that play a major role in the coagulation process, may be the determinant phenomenon in the blood compatibility of polymeric surfaces. Furthermore, Brash (256) observed that when a polymer is exposed to blood, the protein layer adsorbed onto its surface is complex. It includes the proteins HSA and Fg, which are always present, but also other proteins including low molecular weight peptides, which may control the coagulation process induced at the blood–material interface.

C. Antithrombogenic Materials

From the above data it would appear that there are two concepts on which the tailoring of blood-compatible polymers should be based. First, the designed biomaterial should not activate the coagulation process. Unfortunately, as previously described, the theories proposed to guide the tailoring of blood-compatible polymers have not resulted in completely successful biomaterials. Moreover, the vascular endothelium itself may activate the coagulation cascade. Second, the surface of the biomaterial is a blood coagulation activator, as is the case with all known surfaces. In order to prevent thrombus formation at the blood–material interface, the polymer should be able to inhibit the coagulation process.

Moreover, biomaterials should not act as inhibitors. If this were the case, the biomaterial would act as a reagent in an irreversible inhibition reaction, which would then induce the irreversible transformation of the surface of the biomaterial. This reaction would provoke the loss of the inhibiting properties of the biomaterials. Therefore a suitable anticoagulant biomaterial should act as a catalyst of the inhibition reactions of the coagulation process. The surface of such a material should catalyze some of the control mechanisms of the coagulation of the blood.

Several possibilities exist for the design of such biomaterials, and each one has been the basis of active research and development during the past two decades. Historically, the first attempt to produce antithrombogenic materials involved the ionic bonding of heparin onto a polymeric surface by Gott (267). The resulting material enabled the anticoagulant drug to be released slowly into the bloodstream. Since then, the same strategy has been applied in numerous studies but with different potent anticoagulant drugs.

D. Anticoagulant Drug-Releasing Materials

The Gott strategy has been applied to biomaterials that release prostacyclin (PGI2) entrapped in a macromolecular network (268–270). When placed in contact with blood, these biomaterials prevent the aggregation and release of platelets, thus controlling the coagulation process. Such materials are potent while they are able to release PGI2. Unfortunately, PGI2 is an expensive product and moreover is unstable (its lifetime, after hydrolysis under biological conditions, is less than 1 minute). Therefore this type of biomaterial has no practical uses at present. Other platelet antiaggregating surfaces, based on the incorporation of antiaggregating drugs, such as dipyridamole, in polymers, such as cellulose, cellulose triacetate, nylon, *N*-vinylpyrrolidone, and polyethyleneterephthalate, have been prepared (271–275). The resulting materials are claimed to be potent antithrombogenic materials even when implanted in dogs.

The strong anionic character of heparin enables it to bond ionically to cationic surfaces. This phenomenon has been widely researched. In each case, polycationic polymers have been prepared from polystyrene and its derivatives, cellulose, silicone rubber, epoxy resins, polyurethanes, and from copolymers, such as styrene or acrylonitrile and acrylate derivatives; these polycationic polymers were treated with heparin in various ways. Heparin has been also incorporated into cross-linked collagen. An alternative route was to incorporate heparin and prostaglandins simultaneously, in order to combine the respective thrombin and platelet aggregation-inhibiting effects of the released drugs.

Whatever the technique used, the resulting biomaterials have been proved potent if they release the ionically bonded heparin into the bloodstream. During this release process, the concentration of heparin near the surface is sufficiently high to prevent the formation of a thrombus for several days. Unfortunately, after this period of time, the release rate decreases. Therefore these biomaterials are only suitable for use in short-term devices, examples of which are extracorporal circulation and catheters (276–278). They were developed for temporary use by Japanese, American, Swedish, and French manufacturers for the production of catheters and peritoneovenous shunts.

An alternative approach to the ionic coating of heparin onto materials, to produce antithrombogenic surfaces, involves the synthesis of polymeric gels, which contain heparin. This can be achieved either by ionically bonding the heparin to the polymer, followed by the chemical or radiation-induced cross-linking of the resulting network, or by trapping heparin in a preformed hydrogel (251,279–282). The latter procedure has been employed for polymeric gels prepared using radiation processing treatments involving vinyl alcohol or vinyl acetate and *N*-vinyl-2-pyrrolidone monomers. Cobalt-60 γ rays were employed to graft chloromethylstyrene onto polysilicone samples followed by quaternization with pyridine. As in the preceding case, the uptake of heparin from the solutions occurs by exposure of the polymeric gel.

All these gel procedures have been proved to be effective, i.e., they have allowed the preparation of antithrombogenic biomaterials. When compared to previous procedures, the heparin release level was reduced but was adequate to prevent blood clotting in ex vivo assays. However, the potency of the devices is still insufficient to allow their use as permanent implants (283).

E. Covalent Binding of Anticoagulant Drugs to Polymers

Various physiologically active substances have been immobilized onto polymeric surfaces in order to confer blood compatibility (284–286). The use of fibrinolysis to control the thrombogenicity of biomaterials has been studied. Sugitachi et al. (287) attached urokinase or streptokinase onto polyvinylchloride and silicone rubber tubes. Kishida et al. have immobilized recombinant human thrombomodulin onto a polyurethane surface; the resulting polymer exhibits a very good anticoagulant activity (288,289). These surface treatments significantly improve the blood compatibility properties. Unfortunately, cleavage of the fixed enzyme may occur when the surface is exposed to blood, preventing the long-term use of the biomaterials. However, short-term use is possible.

Since platelet activation is a crucial step in the hemostasis, some authors have immobilized an inhibitor of platelet activation, the dipyridamole, onto a medical grade polyurethane. These experiments revealed that the clotting time was substantially prolonged (290). Other authors have covalently bonded a peptidic thrombin inhibitor extracted from leeches, the hirudin (291), on degradable polymer (292).

Although a heparinized surface, because of its dense negative charge, can initiate the intrinsic pathway of blood coagulation (293), an extensively explored way to obtain antithrombogenic biomaterials is the covalent binding of heparin onto polymeric materials. In principle, these biomaterials will not release their heparin content when exposed to the bloodstream. If the covalently bonded heparin remains active, the biomaterials will remain potent longer than those with ionically bonded heparin. The covalent bonding of heparin onto surfaces has been extensively studied during the past decade.

Covalent bonding of heparin onto polymers can be achieved in different ways. Heparin has been coupled to preformed hydrogels by employing various classical activation techniques. Hydrogels, derived from polyvinyl alcohol onto which heparin was fixed employing acetal links, were prepared (294). Alternatively, heparin has been bonded to agarose and Sepharose using cyanogen bromide or carbodiimide activation (295). The same coupling reagents has been utilized to bond heparin covalently either onto radiation-grafted hydrogels derived from hydroxyethyl methacrylate and methacrylic acid or onto the same hydrogels radiation-grafted onto the surface of a silicone rubber (296,297).

In a second general procedure, a polymer is chemically or radiochemically activated and then allowed to react chemically with heparin. This procedure was developed to fix isocyanate groups onto polystyrene and then produce a reaction between the resulting polymer and heparin. Mucopolysaccharide was also bonded to silicone in a two-step procedure. Analogous techniques have been developed to bond heparin onto modified polyvinyl alcohol hydrogels, elastomers, polyhydroxyethyl methacrylate-glycidyl methacrylate copolymers, or cellulosic membranes. Alternatively, heparin may be bonded to suitably prepared substrates to produce thromboresistant materials. This approach has been utilized by Goosen and Sefton (298) to develop a high-strength styrene-butadiene-styrene block copolymers. The same approach has been used by Larsson *et al.* (299), who adsorbed colloidal particles composed of heparin and acetylamine hydrochloride onto sulfated polyethylene and then reacted the mixture with glutaraldehyde. The authors claimed that in the resulting biomaterial heparin was covalently bonded, and the biomaterial was thromboresistant.

In a third general procedure, chemical or radiochemical treatment of heparin induces the formation of a macroradical. The latter then induces the polymerization of a monomer producing a copolymer with a heparin moiety covalently bonded to a synthetic polymeric sequence. This procedure was first carried out by Labarre *et al.* (300) using the cerium (IV) peroxidation of heparin, followed by a reaction with acrylic and methacrylic monomers (300,301). Baquey *et al.* (302,303) described a similar method, using the radicals resulting from the γ-ray irradiation of heparin instead of cerium (IV) oxidation. An end point immobilization of heparin can be achieved by partial depolymerization of heparin, followed by bonding on a cold plasma treated surface.

The heparin immobilization can also be achieved through hydrophilic spacer (304), and in a defined orientation by coupling through its reducing end. This is expected to mimic the attachment of heparin to the protein core in the naturally occurring proteoglycan and impart better ligand binding efficiency by exposing all the binding sites available in the naturally occurring heparin. The coupling chemistry was accomplished by modifying heparin at its reducing end to introduce reactive functionality that can react with appropriately functionalized supports (305). Partially nitrite-degraded heparin allows an end point attachment on the substrate (108,306). When the polymeric substrate is devoid of chemically reactive groups, it is useful to activate the former by radio-frequency plasma (307). Heparin has also been immobilized on polylactide for application to degradable biomaterials in contact with blood (308). Kim *et al.* (276,309) have compared heparin immobilized on polyurethane surfaces with heparin adsorbed on polyurethane. Immobilized heparin retained its ability to bind and inactivate thrombin and Factor Xa. Immobilized heparin–polyurethane catheters implanted in canine femoral and jugular veins for 1-hour periods exhibited significant reduction in thrombus formation compared with untreated polyurethane contralateral controls. The heparin releasing polyurethane provided even greater improvement in antithrombogenicity.

The properties and antithrombogenic activity of the covalently bonded heparin surfaces have been proved (300,310) to be similar to those of soluble heparin. An alternate method of achieving this consists of modifying the heparin so that it can polymerize or copolymerize. These biomaterials have been shown to be anticoagulant.

The above methods (311) have allowed the preparation of biomaterials with heparin covalently bonded to the surface. However, the anticoagulant heparinlike capacity of these materials depends on the method employed to fix the heparin to the polymer matrix. The activation of the platelets occurs when the surfaces to which the heparin is covalently bonded are exposed to blood or a platelet suspension. However, the preincubation of these surfaces with either aqueous solution of antithrombin or plasma passivates the surface with respect to platelet activation.

In conclusion it has been suggested that the thromboresistance of biomaterials to which heparin is covalently bonded might be due to their heparinlike anticoagulant activity and/or their ability to be passive to platelet activation by the bonding of antithrombin. Nevertheless, the use of such biomaterials is restricted for two reasons. The first of these is economic; the second results from the degradation of heparin by heparinases when exposed to blood. If this is the case, the use of such biomaterials as long-term implants may never be possible. This probably explains the current interest in the synthesis of anticoagulant and heparinlike materials.

F. Heparinlike Materials

As early as 1951, Lovelock and Porterfield (312) described the anticoagulant properties of polystyrenesulfonates, produced from a reaction of polystyrene tubing with sulfuric acid. The anticoagulant properties of sulfonated soluble polymers, and particularly polystyrene and polyethylene sulfonates, have since been described by Gregor (313). Furthermore, Machovich and Horvath (314) have shown that polymethacrylic acid exhibits an anticoagulant activity in aqueous solutions.

It has been shown that sulfonate and sulfamate chemical groups are essentials for the anticoagulant activity of the heparin (315–317). Thus it is not surprising to notice that many heparinlike polymers are bearing sulfate groups. Anticoagulant materials have been obtained by reacting *N*-chlorosulfonylisocyanate with an unsaturated polymer, followed by a suitable basic hydrolysis resulting in the formation of a water-soluble polyelectrolyte with *N*-sulfate and carboxylate groups. For example, a soluble polyelectrolyte derived from 1,4-cis-polyisoprene was shown to possess 10% of the anticoagulant activity of heparin (318,319). The anticoagulant activity of the latter polymer, just as with heparin, is related to molecular weight and *N*-sulfate content (320). Akashi *et al.* have synthesized a poly(glucosyloxyethyl methacrylate) sulfate [poly(GEMA) sulfate] that contains sulfated glucoside residues (321). The total human blood clotting time in the presence of poly(GEMA) sulfate is prolonged by increasing the dose or the sulfate content of the polymer (322). Other heparinlike polymers were obtained with bonding sulfate derived groups on various substrates such as polyurethane (323–326), polyethylene (327), hyaluronic acid (328,329), and vinylic polymers (330). Jozefowicz *et al.* (331–339) have obtained resins with significant antithrombogenic activity by bonding sulfonate, carboxylic, amino acid sulfamide, or amide groups onto cross-linked polystyrene, polysaccharides, or polystyrene–polyethylene graft copolymers. The dextran derivatives bearing carboxymethyl, benzylamide, and sulfonate chemical groups (CMDBS family) (Fig. 7) have been extensively studied (200,201,340). They have proved that this activity was antithrombic and anti-Xa, and that it involved plasmatic components (antithrombin and heparin cofactor II), by utilizing tests performed on platelet-poor plasma suspensions of the polymers.

This activity has been established to be on the surface of the polymers. Direct kinetic studies, performed on suspensions of the polymers in aqueous solutions of antithrombin and proteases, were able to show that the anticoagulant activity is heparinlike, i.e., the polymers act as a heterogeneous catalyst with respect to the antithrombin/protease-inhibiting reaction. Thermodynamic studies of the adsorption of proteins on the surface of the resins allowed the authors to propose a mechanism for these catalytic reactions (333–335).

The anticoagulant activities of the materials are strongly dependent on the content and the nature of the substituting groups, which can have either a cooperative or an additive effect. From these observations, a correlation between the anticoagulant activities and the compositions of the heparinlike resins has been established, as shown in Table 5.

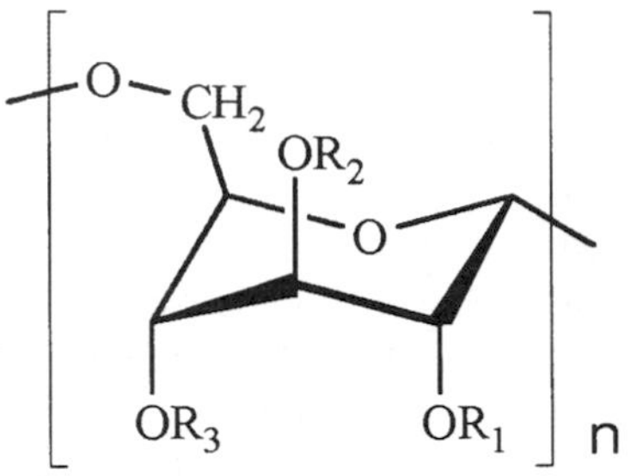

$R_1, R_2, R_3 = H$

$= {}^{-}CH_2COONa$

$= {}^{-}CH_2CONHCH_2C_6H_5$

$= {}^{-}CH_2CONHCH_2C_6H_4SO_3Na$

$= {}^{-}SO_3Na$

Figure 23.7 Schematic structure of the derivatized dextrans. Hydroxyl units of α-*D*-glucose are either unsubstituted or randomly substituted with carboxymethyl, carboxymethyl benzylamide, carboxymethyl benzylamide sulfonate, and sulfate groups.

Table 23.5 Activity Coefficient of Substituents Linked to Polystyrene Resins

Substituent	Activity coefficient (meq^{-1}) for 1 NIH thrombin unit
SO_3^-	50–70
SO_2NH butyl	0
SO_2 11-aminoundecanoic acid	40–80
SO_2 alanine	60–80
SO_2 glycine	100–120
SO_2 hydroxyproline	120–150
SO_2 proline	120–150
SO_2 methionine	120–150
SO_2 threonine	120–150
SO_2 *N*-benzyloxycarbonyllysine	120–150
SO_2 α-alanine	140–200
SO_2 β-aminocaproic acid	300–350
SO_2 glutamic acid	350–400
SO_2 δ-aminovaleric acid	400–450
SO_2 γ-aminobutyric acid	500–600
SO_2 aspartic acid	500–600

Based on these results, it has been suggested that it is the isolated functional groups of the polysaccharide chain of heparin that are responsible for its interaction with the antithrombin or the protease rather than the secondary or tertiary structure of this mucopolysaccharide. The comparison and discussion of the estimated activity parameters substantiate the hypothesis that carboxylic functions are essential for heparinlike activity.

Ex vivo animal experiments have been performed on surface-treated small-diameter tubing made of polystyrene–polyethylene copolymers. No significant platelet adhesion or aggregation was observed on the wall of the tubing, after it had been pretreated with plasma or antithrombin.

G. Polymers and the Immune Response

The formation of thrombin that is initiated when polymers contact blood is the most extensively studied phenomenon in the blood–polymer contact field. The inflammatory response induced by the contact of materials with blood has only recently been studied, and consequently the comprehension of this response is limited.

As with the blood coagulation process, the inflammatory response depends on the mechanical features and the shape of the implanted devices. Many polymers will wear when implanted, and there is evidence of giant cell reaction to the wear products. The inflammatory response also depends on the chemical nature of the polymers (12). Little is known about the relation between this characteristic and the immune response.

The complement system is an effective mechanism of the immune system and accomplishes many tasks. The tasks include recognition of infectious agents and foreign surfaces in the tissues, targeting of an object to enhance leukocyte migration and activation processes (e.g., phagocytosis and degranulation), and finally induction of cytolysis. The complement activation process includes an enzymatic cascade, which involves approximately 20 different proteins. The activation of the complement cascade is usually considered a characteristic of the hemocompatibility of the material in direct contact with blood.

Several enzyme immunoassay kits are available to investigate the various levels of complement activation. Thus it is possible to evaluate the production of C4d, Bd, iC3b, and SC5b-9 (341). C4b is a marker for the activation of the classical pathway, Bb for the alternative pathway, and iC3b and SC5b-9 for the final lytic pathway. A radioimmune assay for C3a explores the early steps of complement activation in both classical and alternative pathways. It allows us to monitor the newly formed C3a when C3 is cleaved (342). It should be noticed that measuring only

C3a remaining in the fluid phase leads to a wrong value of complement activation, since a part of the C3a may be absorbed on the polymer surface (343).

Zymosan, a polysaccharide, has been reported to be a complement activator. Cellulose and cellulose derivatives have been demonstrated to be responsible for the activation of both pathways of the complement system, in human patients treated by dialysis for renal diseases. In the latter case, cellulose and cellulose derivative membranes and hollow fibres were employed (344–346).

Moreover, soluble polysaccharides and polysaccharide derivatives are activators of the complement system. Dextran induced hypersensitivity reactions in over 40,000 cases when used as a plasma expander in humans. Glycosaminoglycans, such as heparin, may inhibit the complement activation either in solution or when bonded to zymosan. This activity is independent of the anticoagulant one (346).

Kazatchkine, Jozefowicz *et al.* (347) showed that Sephadex is able to activate both the classical and the alternative pathways of the complement system, upon exposure to human blood serum. On the other hand, the same research teams (348,349) showed that the substituted soluble dextrans have the ability to inhibit the formation of the alternative pathway C3 convertase. They demonstrated that a family of dextran derivatives, the carboxymethyl benzylamide sulfonate dextrans (CMDBS) (340), has a high anticomplement activity, which is dependent on the carboxylic and sulfonate group content of the polymer (343, 350,351). Thus these authors were able to synthesize dextran derivatives with high anticomplement and low anticoagulant activities, as shown in Fig. 8. Therefore, these soluble biomaterials are promising candidates for plasma expanders, as they are expected to prevent hypersensitivity reactions. The insoluble analogs of CMDBS are substituted Sephadex bearing carboxymethyl (CM), CM-benzylamide (CMB), and CMB-sulphonate (CMBS) groups. CM substitution decreases and can suppress the complement activation of Sephadex. Low CMB substitution also decreases the complement activation, whereas high CMB substitutions increase it again after a minimum (352).

In the case of blood substitutes, the infusion of large amounts of modified hemoglobin can potentially result in hypersensitivity and anaphylactic reactions (353,354). The same phenomenon is observed for the perfluorocarbon (PFC) emulsions usually called ''artificial blood'' (355).

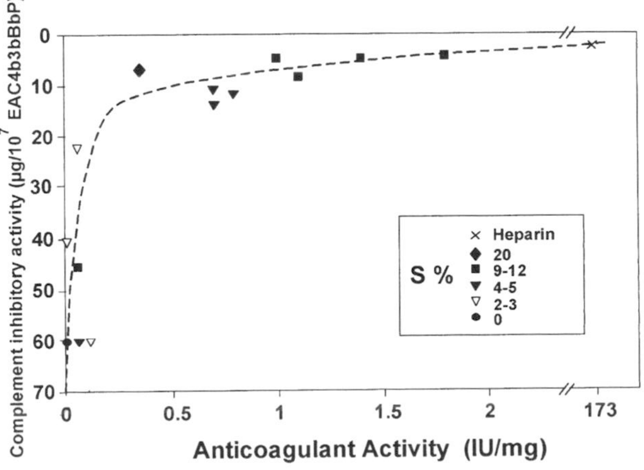

Figure 23.8 Relationship between anticomplementary activity and anticoagulant activity of the modified dextran.

X. CONCLUSION

Blood-contacting polymers have often been chosen by empirical observation. Nonetheless, cardiovascular devices are now in common use and represent an ever-increasing market. Still, there are limitations to the use of polymers. At the beginning of the 1970s, these limits appeared to be the blood coagulation process. More recently, the immune response of the host to the implants, and its consequence on long-term uses, i.e., delayed infection and calcification of the prosthetic devices, have been examined. Immediate hypersensitivity reactions to blood-contacting polymers were also observed in some patients. At present, nearly all the commercially available blood-contacting polymers are employed in conjunction with a permanent adjuvant anticoagulant therapy, which in turn leads to long-term medical complications.

Therefore considerable research has been devoted to understanding the scientific basis of these limitations. The data available for blood–polymer interactions show that the limitations lie in the fact that any surface, natural (including the healthy blood vessel endothelium) or artificial, probably activates the blood coagulation cascade by contact.

Improved knowledge of the parameters controlling this phenomenon has lead to improvements in the design of the artificial devices. It has permitted also the design and tailoring of antithrombogenic polymers, for example heparinlike polymers, which may be suitable for the production of cardiovascular devices in the near future.

Discussion of the available data suggests that fundamental studies of blood–material interfacial reactions are necessary in order to ascertain the structure of the materials, the coagulation factors, and the adsorption relationships responsible for the observed overall thrombogenicity. As a result, the tailoring of new polymers that are able to control the coagulation process initiated at the interface may be possible (356).

It can be concluded from observations made that the immune response may be a controlling parameter in the overall biocompatibility of blood-contacting polymers. At present it is not possible to describe completely the consequences of the host immune response on an implant, but from the available data it appears that two complications are of concern:

In the long term, it appears that the presence of an implant increases the risk of infection.

Blood-contacting polymers may induce sensitivity and short-term reactions.

Recent advances show that significant improvements in the tailoring of polymers enable the immune response to be controlled, and undoubtedly similar results will be obtained if sufficient research is devoted to the field of blood-contacting polymers. An example of significant advance in this field is the synthesis of substituted dextrans that control the activation of the complement system, which renders them promising candidates as plasma expanders. New methodologies for the modification and characterization of the biomaterial surfaces have been developed. In order to improve the biocompatibility of biomaterials, new bioactive biomaterials have been produced based on the principles of biological recognition in order to regulate biological pathways. Such biomaterials can be designed using approaches based on the lock and key concept proposed by Emil Fischer a hundred years ago. Oligopeptides such as the Arg-Gly-Asp (RGD) sequence, which are capable of being a binding domain with high specific affinity for cells, have been grafted to polymers. Other bioactive polymers have been developed based on the incorporation of growth factors. The delivery of the growth factors depends on the chemical and physical characteristics of the polymeric carrier. Some polymers also play the role of protecting agents when they are in contact with a living system. Based on Emil Fischer's concept, bioactive polymers have been synthesized by attaching biospecific ''keys'' along the polymer chain. It has been shown that the random attachment of functional groups, which are analogous to the active groups in the natural messenger molecule (such as heparin), results in materials endowed with biological specificity. Extensive work has been performed to prepare functionalized polymers capable of mimicking some of the actions of heparin. The latter are able to stimulate the healing of a wound in various in vivo models. These polymers may also constitute a family of tissue repair agents because of their protecting and potentiating effects with heparin binding growth factors. These polymers generally affect both cell proliferation and metabolism.

A large amount of work has been also devoted to the development of new methodologies for the characterization and modification of the biomaterial surface. For example, a chemical characterization of the surface is often performed with electron spectroscopy for chemical analysis (ESCA). Modification of the biomaterial surface can be achieved by the use of a gas plasma discharge. This method creates reactive chemical groups onto the surface capable of binding bioactive molecules.

Blood-contacting polymers play an important role in the current market for biomaterial engineering. However, the period of time between the concept and the clinical application is long. It is evident that the penetration of new products onto the market depends on the high cost of development, the extended length of time needed for the approval of the new product, and the collaboration between manufacturers and physicians and surgeons. The latter are often hesitant to try new products with which they are unfamiliar. However, there is no doubt that the market will expand rapidly in the next 10 years if the safety and functionality of new blood contacting polymers compensates for the high cost of their development. Blood-contacting polymers have been widely employed, and their use will continue to extend for the benefit of man.

ACKNOWLEDGMENTS

The experimental work performed by the authors was partially funded by grants from the Centre National de la Recherche Scientifique (CNRS) and the Ministère de l'Education Nationale, de la Recherche et de la Technologie. We are grateful to Stéphanie Cahuzac for assistance in writing the manuscript.

REFERENCES

1. M. N. Helmus, J. A. Hubbell. Materials selection. *Cardiovasc. Pathol. 2*:53S (1993).
2. M. Szycher. Review of cardiovascular devices. *J. Biomater. Appl. 12*:321 (1998).
3. B. D. Ratner. Issues in blood interaction at the end of the 20th century. ESB 99. Fifteenth European Conference on Biomaterials, Proceedings, September 8–12, 1999, Arcachon, France.
4. *Polyurethanes in Biomedical Applications*. CRC press, Boca Raton, 1997.
5. R. J. Zdrahala, I. J. Zdrahala. Biomedical applications of polyurethanes: a review of past promises, present realities, and a vibrant future. *J. Biomater. Appl. 14*:67 (1999).
6. D. L. Ellis, I. V. Yannas. Recent advances in tissue synthesis in vivo by use of collagen-glycosaminoglycan copolymers. *Biomaterials 17*:291 (1996).

7. E. Khor. Methods for the treatment of collagenous tissues for bioprostheses. *Biomaterials 18*:95 (1997).
8. F. J. Schoen, H. Harasaki, K. M. Kim, H. C. Anderson, R. J. Levy. Biomaterial-associated calcification: pathology, mechanisms, and strategies for prevention. *J. Biomed. Mater. Res. 22*:11 (1988).
9. J. M. Anderson, K. M. Miller. Biomaterial biocompatibility and the macrophage. *Biomaterials 5*:5 (1984).
10. J. Black. Systemic effects of biomaterials. *Biomaterials 5*: 11 (1984).
11. R. E. Marchant, K. M. Miller, J. M. Anderson. In vivo biocompatibility studies. V. In vivo leukocyte interactions with Biomer. *J. Biomed. Mater. Res. 18*:1169 (1984).
12. K. Merritt. Role of medical materials, both in implant and surface applications, in immune response and in resistance to infection. *Biomaterials 5*:47 (1984).
13. G. W. Hastings. Structural considerations of plastics materials. In: *Macromolecular Biomaterials* (G. W. Hastings, P. Ducheyne, eds.). CRC Press, Boca Raton, 1984, pp. 3–17.
14. R. M. Cornelius, J. L. Brash. Adsorption from plasma and buffer of single- and two-chain high molecular weight kininogen to glass and sulfonated polyurethane surfaces. *Biomaterials 20*:341 (1999).
15. D. Annis, A. C. Fisher, T. V. How, L. de Cossart. The effect of varying compliance on the long-term patency of a polyetherurethane arterial prosthesis. In: *Polymers in Medicine* (C. Migliaresi, ed.). Plenum Press, New York 1986.
16. H. L. Goldsmith, T. Karino. Microscopic considerations: the motions of individual particles. *Ann. NY Acad. Sci. 283*:241 (1977).
17. Z. M. Ruggeri. Mechanisms of shear-induced platelet adhesion and aggregation. *Thromb. Haemost. 70*:119 (1993).
18. S. M. Slack, Y. Cui, V. T. Turitto. The effects of flow on blood coagulation and thrombosis. *Thromb. Haemost. 70*: 129 (1993).
19. E. W. Merrill. Properties of materials affecting the behavior of blood at their interface. *Ann. NY Acad. Sci. 283*:6 (1977).
20. S. C. Vasudev, T. Chandy, C. P. Sharma. The antithrombotic versus calcium antagonistic effects of polyethylene glycol grafted bovine pericardium. *J. Biomater. Appl. 14*: 48 (1999).
21. T. Fabre, J. Bertrand-Barat, G. Freyburger, J. Rivel, B. Dupuy, A. Durandeau, C. Baquey. Quantification of the inflammatory response in exudates to three polymers implanted in vivo. *J. Biomed. Mater. Res. 39*:637 (1998).
22. S. S. Kaplan. Biomaterial–host interactions: consequences, determined by implant retrieval analysis. *Med. Prog. Technol. 20*:209 (1994).
23. P. Zilla, M. Deutsch, J. Meinhart. Endothelial cell transplantation. *Semin. Vasc. Surg. 12*:52 (1999).
24. N. L'Heureux, S. Paquet, R. Labbe, L. Germain, F. A. Auger. A completely biological tissue–engineered human blood vessel. *FASEB J 12*:47 (1998).
25. R. F. Doolittle. The evolution of vertebrate blood coagulation: a case of Yin and Yang. *Thromb. Haemost. 70*:24 (1993).
26. S. Iwanaga. Primitive coagulation systems and their message to modern biology. *Thromb. Haemost. 70*:48 (1993).
27. J. Leroy. Origines de l'hémostase. In: *Le sang et les vaisseaux* (J. Caen, ed.). Hermann, Paris, 1987, pp. 215–233.
28. S. Nakamura, T. Morita, S. Harada, S. Iwanaga, K. Takahashi, M. Niwa. A clotting enzyme associated with the hemolymph coagulation system of horseshoe crab (*Tachypleus tridentatus*): its purification and characterization. *J. Biochem. (Tokyo) 92*:781 (1982).
29. A. Takada, Y. Takada, T. Urano. The physiological aspects of fibrinolysis. *Thromb. Res. 76*:1 (1994).
30. E. Camerer, A. B. Kolsto, H. Prydz. Cell biology of tissue factor, the principal initiator of blood coagulation. *Thromb. Res. 81*:1 (1996).
31. L. C. Petersen, S. Valentin, U. Hedner. Regulation of the extrinsic pathway system in health and disease: the role of factor VIIa and tissue factor pathway inhibitor. *Thromb. Res. 79*:1 (1995).
32. Y. T. Wachtfogel, R. A. DeLa Cadena, R. W. Colman. Structural biology, cellular interactions and pathophysiology of the contact system. *Thromb. Res. 72*:1 (1993).
33. J.-P. Cazenave. Interaction of platelets with surfaces. In: *Blood-Surface Interactions: Biological Principles Underlying Haemocompatibility with Artificial Materials* (J.-P. Cazenave et al., eds.). Elsevier, Amsterdam, 1986, pp. 89–105.
34. M. L. Tiffany, J. A. Penner. Heparin and other sulfated polyanions: their interaction with the blood platelet. *Ann. NY Acad. Sci. 370*:662 (1981).
35. M. T. Stubbs, W. Bode. A player of many parts: the spotlight falls on thrombin's structure. *Thromb. Res. 69*:1 (1993).
36. B. Blomback. Fibrinogen and fibrin—proteins with complex roles in hemostasis and thrombosis. *Thromb. Res. 83*: 1 (1996).
37. L. Muszbek, V. C. Yee, Z. Hevessy. Blood coagulation factor XIII: structure and function [In Process Citation]. *Thromb. Res. 94*:271 (1999).
38. D. C. Berridge, J. J. Earnshaw, J. C. Westby, G. S. Makin, B. R. Hopkinson. Fibrinolytic profiles in local low-dose thrombolysis with streptokinase and recombinant tissue plasminogen activator. *Thromb. Haemost. 61*:275 (1989).
39. B. Furie, B. C. Furie. The molecular basis of blood coagulation. *Cell 53*:505 (1988).
40. E. W. Davie, K. Fujikawa, W. Kisiel. The coagulation cascade: initiation, maintenance, and regulation. *Biochemistry 30*:10363 (1991).
41. T. C. Laurent, A. Tengblad, L. Thunberg, M. Hook, U. Lindahl. The molecular-weight-dependence of the anticoagulant activity of heparin. *Biochem. J. 175*:691 (1978).
42. R. D. Rosenberg, G. M. Oosta, R. E. Jordan, W. T. Gardner. The interaction of heparin with thrombin and antithrombin. *Biochem. Biophys. Res. Commun. 96*:1200 (1980).
43. G. F. Briginshaw, J. N. Shanberge. Identification of two

distinct heparin cofactors in human plasma: separation and partial purification. *Arch. Biochem. Biophys. 161*:683 (1974).

44. D. M. Tollefsen, N. K. Blank. Detection of a new heparin-dependent inhibitor of thrombin in human plasma. *J. Clin. Invest. 68*:589 (1981).
45. R. M. Bertina, van der Linden, L. Engesser, H. P. Muller, E. J. Brommer. Hereditary heparin cofactor II deficiency and the risk of development of thrombosis. *Thromb. Haemost. 57*:196 (1987).
46. A. M. Fischer, M. D. Dantzenberg, M. H. Aurousseau, F. J. Behal. Comparison between the effect of pentosan polysulfate, heparin, and antithrombin III injections in antithrombin III deficient patients. *Thromb. Res. 37*:295 (1985).
47. D. M. Tollefsen, M. E. Peacock, W. J. Monafo. Molecular size of dermatan sulfate oligosaccharides required to bind and activate heparin cofactor II. *J. Biol. Chem. 261*:8854 (1986).
48. R. M. Maaroufi, M. Jozefowicz, J. Tapon-Bretaudière, J. Jozefonvicz, A. M. Fischer. Mechanism of thrombin inhibition by heparin cofactor II in the presence of dermatan sulphates, native or oversulphated, and a heparin-like dextran derivative. *Biomaterials 18*:359 (1997).
49. B. B. Welsler, A. J. Marcus, E. A. Jaffe. Synthesis of prostaglandin I_2 (prostacyclin) by cultured human and bovine endothelial cells. *Proc. Natl. Acad. Sci. USA 74*:3922 (1977).
50. M. C. Boffa, M. C. Bourin. Endothelial cell membrane and plasma protein interaction, the thrombomodulin protein C system. *Blood Transf. Immunohaematol. 6*:807 (1984).
51. C. T. Esmon. Molecular events that control the protein C anticoagulant pathway. *Thromb. Haemost. 70*:29 (1993).
52. B. Dahlback. Inherited thrombophilia: resistance to activated protein C as a pathogenic factor of venous thromboembolism. *Blood 85*:607 (1995).
53. W. Kisiel, N. Canfield, L. H. Ericsson, E. W. Davie. Anticoagulant properties of bovine plasma protein C following activation by thrombin. *Biochemistry 16*:5824 (1977).
54. G. Vehar, E. W. Davie. Preparation and properties of bovine factor VIII (antihemophilic factor). *Biochemistry 19*: 401 (1980).
55. Y. Sakata, S. Curriden, D. Lawrence, J. H. Griffin, D. J. Loskutoff. Activated protein C stimulates the fibrinolytic activity of cultured endothelial cells and decreases antiactivator activity. *Proc. Natl. Acad. Sci. USA 82*:1121 (1985).
56. E. G. Levin, D. M. Stern, P. P. Nawroth, R. A. Marlar, D. S. Fair, J. W. Fenton, L. A. Harker. Specificity of the thrombin-induced release of tissue plasminogen activator from cultured human endothelial cells. *Thromb. Haemost. 56*:115 (1986).
57. M. W. Radomski, S. Moncada. Regulation of vascular homeostasis by nitric oxide. *Thromb. Haemost. 70*:36 (1993).
58. P. C. Comp. Heparin–protein C interaction. *Nouv. Rev. Fr. Hematol. 26*:239 (1984).
59. S. Ecke, M. Geiger, B. R. Binder. Glycosaminoglycans regulate the enzyme specificity of protein C inhibitor. *Ann. NY Acad. Sci. 667*:84 (1992).
60. M. C. Bourin, U. Lindahl. Glycosaminoglycans and the regulation of blood coagulation. *Biochem. J. 289*:313 (1993).
61. U. Lindahl, K. Lidholt, D. Spillmann, L. Kjellen. More to "heparin" than anticoagulation. *Thromb. Res. 75*:1 (1994).
62. R. D. Rosenberg. Coagulation-fibrinolytic mechanism and the action of heparin. *Adv. Exp. Med. Biol. 52*:217 (1975).
63. D. M. Tollefsen, M. M. Maimone, E. A. McGuire, M. E. Peacock. Heparin cofactor II activation by dermatan sulfate. *Ann. NY Acad. Sci. 556*:116 (1989).
64. J. Choay, M. Petitou, J. C. Lormeau, P. Sinay, B. Casu, G. Gatti. Structure–activity relationship in heparin: a synthetic pentasaccharide with high affinity for antithrombin III and eliciting high anti-factor Xa activity. *Biochem. Biophys. Res. Commun. 116*:492 (1983).
65. M. Petitou, J. P. Herault, A. Bernat, P. A. Driguez, P. Duchaussoy, J. C. Lormeau, J. M. Herbert. Synthesis of thrombin-inhibiting heparin mimetics without side effects. *Nature 398*:417 (1999).
66. L. Jin, J. P. Abrahams, R. Skinner, M. Petitou, R. N. Pike, R. W. Carrell. The anticoagulant activation of antithrombin by heparin. *Proc. Natl. Acad. Sci. USA 94*:14683 (1997).
67. W. M. Choi, P. Kopeckova, T. Minko, J. Kopecek. Synthesis of HPMA copolymer containing adriamycin bound via an acid-labile spacer and its activity toward human ovarian carcinoma cells. *J. Bioact. Compat. Polym. 14*: 447 (1999).
68. S. Colliec-Jouault, N. W. Shworak, J. Liu, A. I. de Agostini, R. D. Rosenberg. Characterization of a cell mutant specifically defective in the synthesis of anticoagulantly active heparan sulfate. *J. Biol. Chem. 269*:24953 (1994).
69. N. W. Shworak, M. Shirakawa, S. Colliec-Jouault, J. Liu, R. C. Mulligan, L. K. Birinyi, R. D. Rosenberg. Pathway-specific regulation of the synthesis of anticoagulantly active heparan sulfate. *J. Biol. Chem. 269*:24941 (1994).
70. L. Mourey, J. P. Samama, M. Delarue, J. Choay, J. C. Lormeau, M. Petitou, D. Moras. Antithrombin III: structural and functional aspects. *Biochimie. 72*:599 (1990).
71. U. R. Desai, M. Petitou, I. Björk, S. T. Olson. Mechanism of heparin activation of antithrombin. Role of individual residues of the pentasaccharide activating sequence in the recognition of native and activated states of antithrombin. *J. Biol. Chem. 273*:7478 (1998).
72. J. A. Marcum, C. F. Reilly, R. D. Rosenberg. The role of specific forms of heparan sulfate in regulating blood vessel wall function. *Prog. Hemost. Thromb. 8*:185 (1986).
73. K. Suzuki, Y. Deyashiki, J. Nishioka, K. Kurachi, M. Akira, S. Yamamoto, S. Hashimoto. Characterization of a cDNA for human protein C inhibitor. A new member of the plasma serine protease inhibitor superfamily. *J. Biol. Chem. 262*:611 (1987).
74. K. Suzuki, J. Nishioka, H. Kusumoto, S. Hashimoto.

Mechanism of inhibition of activated protein C by protein C inhibitor. *J. Biochem. (Tokyo) 95*:187 (1984).
75. D. Belin. Biology and facultative secretion of plasminogen activator inhibitor-2. *Thromb. Haemost. 70*:144 (1993).
76. I. Juhan-Vague, M. C. Alessi. Plasminogen activator inhibitor 1 and atherothrombosis. *Thromb. Haemost. 70*:138 (1993).
77. K. T. Preissner, P. Sie. Modulation of heparin cofactor II function by S protein (vitronectin) and formation of a ternary S protein-thrombin-heparin cofactor II complex. *Thromb. Haemost. 60*:399 (1988).
78. K. T. Preissner, D. Seiffert. Role of vitronectin and its receptors in haemostasis and vascular remodeling. *Thromb. Res. 89*:1 (1998).
79. R. Banerjee, K. Nageswari, R. R. Puniyani. Hematological aspects of biocompatibility—review article. *J Biomater Appl. 12*:57 (1997).
80. N. P. Rhodes, T. V. Kumary, D. F. Williams. Influence of wall shear rate on parameters of blood compatibility of intravascular catheters. *Biomaterials 17*:1995 (1996).
81. J. L. Brash, T. A. Horbett. Proteins at interfaces: an overview. In: *Proteins at Interfaces II: Fundamentals and Applications* (T. A. Horbett, J. L. Brash, eds.). ASC Symposium Series, Washington DC, 1995, pp. 1–23
82. S. M. Slack, T. A. Horbett. The Vroman effect: a critical review. In: *Proteins at Interfaces II: Fundamentals and Applications* (T. A. Horbett, J. L. Brash, eds.). ASC Symposium Series, Washington DC, 1995, pp. 112–128.
83. W. Norde, C. A. Haynes. Reversibility and the mechanism of protein adsorption. In: *Proteins at Interfaces II: Fundamentals and Applications* (T. A. Horbett, J. L. Brash, eds.). ASC Symposium Series, Washington DC, 1995, pp. 26–40.
84. L. Vroman, A. L. Adams. Identification of rapid changes at plasma–solid interfaces. *J. Biomed. Mater. Res. 3*:43 (1969).
85. E. F. Leonard, L. Vroman. Is the Vroman effect of importance in the interaction of blood with artificial materials? *J. Biomater. Sci. Polym. Ed. 3*:95 (1991).
86. R. C. Dutton, A. J. Webber, S. A. Johnson, R. E. Baier. Microstructure of initial thrombus formation on foreign materials. *J. Biomed. Mater. Res. 3*:13 (1969).
87. T. A. Horbett, P. K. Weathersby. Adsorption of proteins from plasma to a series of hydrophilic–hydrophobic copolymers. I. Analysis with the in situ radioiodination technique. *J. Biomed. Mater. Res. 15*:403 (1981).
88. Y. Noishiki. Application of immunoperoxidase method to electron microscopic observation of plasma protein on polymer surface. *J. Biomed. Mater. Res. 16*:359 (1982).
89. L. Stanislawski, H. Serne, M. Stanislawski, M. Jozefowicz. Conformational changes of fibronectin induced by polystyrene derivatives with a heparin-like function. *J. Biomed. Mater. Res. 27*:619 (1993).
90. J. L. Brash. Behavior of proteins at interfaces. *Curr. Opin. Colloid Interface Sci. 1*:682 (1996).
91. S. I. Jeon, J. H. Lee, J. D. Andrade, P. G. de Gennes. Protein–surface interactions in the presence of polyethylene oxide I. Simplified theory. *J. Colloid. Interf. Sci. 142*:149 (1991).
92. H. Y. Chuang, J. D. Andrade. Immunochemical detection by specific antibody to thrombin of prothrombin conformational changes upon adsorption to artificial surfaces. *J. Biomed. Mater. Res. 19*:813 (1985).
93. Y. Ikada. Surface modification of polymers for medical applications. *Biomaterials 15*:725 (1994).
94. E. T. Kang, K. L. Tan, K. Kato, Y. Uyama, Y. Ikada. Surface modification and functionalization of polytetrafluoroethylene films. *Macromolecules 29*:6872 (1996).
95. M. Suzuki, A. Kishida, H. Iwata, Y. Ikada. Graft copolymerization of acrylamide onto a polyethylene surface pretreated with a glow discharge. *Macromolecules 19*:1804 (1986).
96. A. Kishida, K. Mishima, E. Corretge, H. Konishi, Y. Ikada. Interactions of poly(ethylene glycol)-grafted cellulose membranes with proteins and platelets. *Biomaterials 13*:113 (1992).
97. K. Bergstrom, E. Osterberg, K. Holmberg, A. S. Hoffman, T. P. Schuman, A. Kozlowski, J. H. Harris. Effects of branching and molecular weight of surface-bound poly (ethylene oxide) on protein rejection. *J. Biomater. Sci. Polym. Ed. 6*:123 (1994).
98. D. K. Han, S. Y. Jeong, K. D. Ahn, Y. H. Kim, B. G. Min. Preparation and surface properties of PEO-sulfonate grafted polyurethanes for enhanced blood compatibility. *J. Biomater. Sci. Polym. Ed 4*:579 (1993).
99. K. D. Park, W. K. Lee, J. Y. Yun, D. K. Han, S. H. Kim, Y. H. Kim, H. M. Kim, K. T. Kim. Novel anti-calcification treatment of biological tissues by grafting of sulphonated poly(ethylene oxide). *Biomaterials 18*:47 (1997).
100. K. Fujimoto, H. Tadokoro, Y. Ueda, Y. Ikada. Polyurethane surface modification by graft polymerization of acrylamide for reduced protein adsorption and platelet adhesion. *Biomaterials 14*:442 (1993).
101. K. Fujimoto, Y. Takebayashi, H. Inoue, Y. Ikada. Ozone-induced graft polymerization onto polymer surface. *J. Polym. Sci. A 31*:1035 (1993).
102. D. Ruckert, G. Geuskens, P. Fondu, S. van Erum. Surface modification of polymers—III. Photo-initiated grafting of water soluble vinyl monomers and influence on fibrinogen adsorption. *Eur. Polym. J. 31*:431 (1995).
103. H. Inoue, K. Fujimoto, Y. Uyama, Y. Ikada. Ex vivo and in vivo evaluation of the blood compatibility of surface-modified polyurethane catheters. *J. Biomed. Mater. Res. 35*:255 (1997).
104. A. Kishida, S. Kato, K. Ohmura, K. Sugimura, M. Akashi. Evaluation of biological responses to polymeric biomaterials by RT-PCR analysis. I. Study of IL-1 beta mRNA expression. *Biomaterials 17*:1301 (1996).
105. A. Kishida, H. Iwata, Y. Tamada, Y. Ikada. Cell behaviour on polymer surfaces grafted with non-ionic and ionic monomers. *Biomaterials 12*:786 (1991).
106. W. F. Ip, W. Zingg, M. V. Sefton. Parallel flow arteriovenous shunt for the ex vivo evaluation of heparinized materials. *J. Biomed. Mater. Res. 19*:161 (1985).

107. W. F. Ip, M. V. Sefton, Patency of heparin-PVA coated tubes at low flow rates. *Biomaterials 10*:313 (1989).
108. T. Lindhout, R. Blezer, P. Schoen, G. M. Willems, B. Fouache, M. Verhoeven, M. Hendriks, L. Cahalan, P. T. Cahalan. Antithrombin activity of surface-bound heparin studied under flow conditions. *J. Biomed. Mater. Res. 29*:1255 (1995).
109. A. Podias, T. Groth, Y. Missirlis. The effect of shear rate on the adhesion/activation of human platelets in flow through a closed-loop polymeric tubular system. *J. Biomater. Sci. Polym. Ed 6*:399 (1994).
110. N. P. Rhodes, M. Zuzel, D. F. Williams, M. R. Derrick. Granule secretion markers on fluid-phase platelets in whole blood perfused through capillary tubing. *J. Biomed. Mater. Res. 28*:435 (1994).
111. I. Noh, M. A. Lovich, E. R. Edelman. Mechanisms of heparin transport through expanded poly(tetrafluoroethylene) vascular grafts. *J. Biomed. Mater. Res. 49*:112 (2000).
112. V. Migonney, C. Fougnot, M. Jozefowicz. Heparin-like tubings. I. Preparation, characterization and biological in vitro activity assessment. *Biomaterials 9*:145 (1988).
113. V. Migonney, C. Fougnot, M. Jozefowicz. Heparin-like tubings. II. Mechanism of the thrombin–antithrombin III reaction at the surface. *Biomaterials 9*:230 (1988).
114. V. Migonney, C. Fougnot, M. Jozefowicz. Heparin-like tubings. III. Kinetics and mechanism of thrombin, antithrombin III and thrombin–antithrombin complex adsorption under controlled-flow conditions. *Biomaterials 9*:413 (1988).
115. N. Saito, C. Nojiri, S. Kuroda, K. Sakai. Photochemical grafting of alpha-propylsulphate-poly(ethylene oxide) on polyurethane surfaces and enhanced antithrombogenic potential. *Biomaterials 18*:1195 (1997).
116. C. A. Bense, K. A. Woodhouse. Plasmin degradation of fibrin coatings on synthetic polymer substrates. *J. Biomed. Mater. Res. 46*:305 (1999).
117. G. de Gaetano, C. Cerletti, V. Evangelista. Recent advances in platelet–polymorphonuclear leukocyte interaction, *Haemostasis 29*:41 (1999).
118. J. Yang, B. C. Furie, B. Furie. The biology of P-selectin glycoprotein ligand-1: its role as a selectin counterreceptor in leukocyte–endothelial and leukocyte–platelet interaction. *Thromb Haemost 81*:1 (1999).
119. M. D. Kazatchkine. Activation of the complement system. In: *Blood–Surface Interactions: Biological Principles Underlying Haemocompatibility with Artificial Materials* (J.-P. Cazenave et al., eds.). Elsevier, Amsterdam, 1986, pp. 75–87.
120. R. J. Johnson. Complement activation during extracorporeal therapy: biochemistry, cell biology and clinical relevance. *Nephrol Dial Transplant 9 suppl 2*:36 (1994).
121. J. Gong, R. Larsson, K. N. Ekdahl, T. E. Mollnes, U. Nilsson, B. Nilsson. Tubing loops as a model for cardiopulmonary bypass circuits: both the biomaterial and the blood–gas phase interfaces induce complement activation in an in vitro model. *J. Clin. Immunol. 16*:222 (1996).
122. L. Tang, L. Liu, H. B. Elwing. Complement activation and inflammation triggered by model biomaterial surfaces. *J. Biomed. Mater. Res. 41*:333 (1998).
123. E. Fischer, M. D. Kazatchkine. Surface-dependent modulation by H of C5 cleavage by the cell bound alternative pathway C5 convertase of human complement. *J. Immunol. 130*:2821 (1983).
124. F. Maillet, M. D. Kazatchkine, D. Glotz, E. Fischer, M. Rowe. Heparin prevents formation of the human C3 amplification convertase by inhibiting site for B on C3b. *Mol. Immunol. 20*:1401 (1983).
125. W. L. Hinrichs, H. W. ten Hoopen, G. H. Engbers, J. Feijen. In vitro evaluation of heparinized Cuprophan hemodialysis membrances. *J. Biomed. Mater. Res. 35*:443 (1997).
126. Y. H. An, R. J. Friedman. Concise review of mechanisms of bacterial adhesion to biomaterial surfaces. *J. Biomed. Mater. Res. 43*:338 (1998).
127. M. R. Brunstedt, S. Sapatnekar, K. R. Rubin, K. M. Kieswetter, N. P. Ziats, K. Merritt, J. M. Anderson. Bacteria/blood/material interactions. I. Injected and preseeded slime-forming Staphylococcus epidermidis in flowing blood with biomaterials. *J. Biomed. Mater. Res. 29*:455 (1995).
128. P. Fernandez. Clinical use of nuclear medicine in orthopedic and vascular biomaterials infection. Fifteenth European Conference on Biomaterials, Proceedings, September 8–12, 1999, Arcachon, France.
129. K. Merritt, A. Gaind, J. M. Anderson. Detection of bacterial adherence on biomedical polymers. *J. Biomed. Mater. Res. 39*:415 (1998).
130. G. Lopez-Lopez, A. Pascual, E. J. Perea. Effect of plastic catheter material on bacterial adherence and viability. *J. Med. Microbiol. 34*:349 (1991).
131. K. Merritt, V. M. Hitchins, A. R. Neale. Tissue colonization from implantable biomaterials with low numbers of bacteria. *J. Biomed. Mater. Res. 44*:261 (1999).
132. P. A. Suci, J. D. Vrany, M. W. Mittelman. Investigation of interactions between antimicrobial agents and bacterial biofilms using attenuated total reflection Fourier transform infrared spectroscopy. *Biomaterials 19*:327 (1998).
133. A. Yousefi, H. Ayhan, E. Piskin. Adhesion of different bacterial strains to low-temperature plasma treated biomedical silicon catheter surfaces. *J. Bioact. Compat. Polym. 13*:81 (1998).
134. E. Hennig, A. John, F. Zartnack, W. Lemm, E. S. Bucherl, G. Wick, K. Gerlach. Biostability of polyurethanes. *Int. J. Artif. Organs 11*:416 (1988).
135. R. J. Levy, F. J. Schoen, H. C. Anderson, H. Harasaki, T. H. Koch, W. Brown, J. B. Lian, R. Cumming, J. B. Gavin. Cardiovascular implant calcification: a survey and update. *Biomaterials 12*:707 (1991).
136. J. Chanda, K. Kondoh, K. Ijima, M. Matsukawa, R. Kuribayashi. In vitro and in vivo calcification of vascular bioprostheses. *Biomaterials 19*:1651 (1998).
137. G. M. Bernacca, T. G. Mackay, R. Wilkinson, D. J. Wheatley. Polyurethane heart valves: fatigue failure, cal-

cification, and polyurethane structure. *J. Biomed. Mater. Res. 34*:371 (1997).

138. J. Chanda, R. Kuribayashi, T. Abe. Heparin in calcification prevention of porcine pericardial bioprostheses. *Biomaterials 18*:1109 (1997).
139. J. Chanda, R. Kuribayashi, T. Abe. Heparin coupling in inhibition of calcification of vascular bioprostheses. *Biomaterials 20*:1753 (1999).
140. R. R. Joshi, J. R. Frautschi, R. E. J. Phillips, R. J. Levy. Phosphonated polyurethanes that resist calcification. *J. Appl. Biomater 5*:65 (1994).
141. A. Takahara, A. J. Coury, R. W. Hergenrother, S. L. Cooper. Effect of soft segment chemistry on the biostability of segmented polyurethanes. I. In vitro oxidation. *J. Biomed. Mater. Res. 25*:341 (1991).
142. L. Pinchuk. A review of the biostability and carcinogenicity of polyurethanes in medicine and the new generation of 'biostable' polyurethanes. *J. Biomater. Sci. Polym. Ed. 6*:225 (1994).
143. S. J. McCarthy, G. F. Meijs, N. Mitchell, P. A. Gunatillake, G. Heath, A. Brandwood, K. Schindhelm. In-vivo degradation of polyurethanes: transmission-FTIR microscopic characterization of polyurethanes sectioned by cryomicrotomy. *Biomaterials 18*:1387 (1997).
144. R. S. Labow, E. Meek, J. P. Santerre. Synthesis of cholesterol esterase by monocyte-derived macrophages: a potential role in the biodegradation of poly(urethane)s. *J. Biomater. Appl. 13*:187 (1999).
145. K. Stokes, P. Urbanski, J. Upton. The in vivo auto-oxidation of polyether polyurethane by metal ions. *J. Biomater. Sci. Polym. Ed. 1*:207 (1990).
146. S. Faré, P. Petrini, A. Motta, A. Cigada, M. C. Tanzi. Synergistic effects of oxidative environments and mechanical stress on in vitro stability of polyetherurethanes and polycarbonateurethanes. *J. Biomed. Mater. Res. 45*:62 (1999).
147. Q. H. Zhao, A. K. McNally, K. R. Rubin, M. Renier, Y. Wu, V. Rose-Caprara, J. M. Anderson, A. Hiltner, P. Urbanski, K. Stokes. Human plasma alpha 2-macroglobulin promotes in vitro oxidative stress cracking of Pellethane 2363-80A: in vivo and in vitro correlations. *J. Biomed. Mater. Res. 27*:379 (1993).
148. J. P. Santerre, R. S. Labow, D. G. Duguay, D. Erfle, G. A. Adams. Biodegradation evaluation of polyether and polyester-urethanes with oxidative and hydrolytic enzymes. *J. Biomed. Mater. Res. 28*:1187 (1994).
149. R. S. Labow, J. P. Santerre, G. Waghray. The effect of phospholipids on the biodegradation of polyurethanes by lysosomal enzymes. *J. Biomater. Sci. Polym. Ed. 8*:779 (1997).
150. R. S. Labow, E. Meek, J. P. Santerre. The biodegradation of poly(urethane)s by the esterolytic activity of serine proteases and oxidative enzyme systems. *J. Biomater. Sci. Polym. Ed. 10*:699 (1999).
151. S. K. Phua, E. Castillo, J. M. Anderson, A. Hiltner. Biodegradation of a polyurethane in vitro. *J. Biomed. Mater. Res. 21*:231 (1987).
152. M. A. Schubert, M. J. Wiggins, K. M. DeFife, A. Hiltner, J. M. Anderson. Vitamin E as an antioxidant for poly (etherurethane urea): in vivo studies. Student Research Award in the Doctoral Degree Candidate Category, Fifth World Biomaterials Congress (22nd Annual Meeting of the Society for Biomaterials), Toronto, Canada, May 29–June 2, 1996. *J. Biomed. Mater. Res. 32*:493 (1996).
153. M. A. Schubert, M. J. Wiggins, J. M. Anderson, A Hiltner. Comparison of two antioxidants for poly(etherurethane urea) in an accelerated in vitro biodegradation system. *J. Biomed. Mater. Res. 34*:493 (1997).
154. A. J. Coury, C. N. Hobot. Method for producing polyurethanes from poly(hydroxyalkyl urethane). U.S. patent 5,001,210,19 (1991).
155. M. Szycher, A. M. Reed. Biostable polyurethane products. U.S. patent 5,254,662 (19-10-1993).
156. R. J. Carson, A. Edwards, M. Szycher. Resistance to biodegradative stress cracking in microporous vascular access grafts. *J. Biomater. Appl. 11*:121 (1996).
157. M. Szycher, A. M. Reed, A. A. Siciliano. *In vivo* testing of biostable polyurethane. *J. Biomater. Appl. 6*:110 (1991).
158. J. Drobnik, F. Rypacek. Soluble synthetic polymers in biological systems. *Adv. Polym. Sci. 57*:1 (1984).
159. R. Mehvar, T. L. Shepard. Molecular-weight-dependent pharmacokinetics of fluorescein-labeled dextrans in rats. *J. Pharm. Sci. 81*:908 (1992).
160. R. Mehvar, M. A. Robinson, J. M. Reynolds. Molecular weight dependent tissue accumulation of dextrans: in vivo studies in rats. *J. Pharm. Sci. 83*:1495 (1994).
161. R. Mehvar, M. A. Robinson, J. M. Reynolds. Dose dependency of the kinetics of dextrans in rats: effects of molecular weight. *J. Pharm. Sci. 84*:815 (1995).
162. R. Mehvar. Effects of storage and homogenization methods on the hepatic recovery of dextrans determined by size-exclusion chromatography. *J. Pharm. Biomed. Anal. 14*: 801 (1996).
163. R. Mehvar. Kinetics of hepatic accumulation of dextrans in isolated perfused rat livers. *Drug. Metab. Dispos. 25*: 552 (1997).
164. J. Lahann, D. Klee, H. Thelen, H. Bienert, D. Vorwerk, H. Hocker. Improvement of haemocompatibility of metallic stents by polymer coating. *J. Mater. Sci. Mater Med 10*: 443 (1999).
165. T. Peng, P. Gibula, K. D. Yao, M. F. Goosen. Role of polymers in improving the results of stenting in coronary arteries. *Biomaterials 17*:685 (1996).
166. J. Kopecek. Targetable polymeric anticancer drugs. Temporal control of drug activity. *Ann. NY. Acad. Sci. 618*: 335 (1991).
167. A. Serres, M. Baudys, S. W. Kim. Temperature and pH-sensitive polymers for human calcitonin delivery. *Pharm. Res. 13*:196 (1996).
168. X. Xu, P. I. Lee. Programmable drug delivery from an erodible association polymer system. *Pharm. Res. 10*:1144 (1993).
169. O. Henin, M. Boustta, M. Domurado, J. Coudane, D. Domurado, M. Vert. Covalent binding of mannosyl ligand via 6-O position and glycolic arm to target a PLCA-type

degradable drug carrier toward macrophages. *J. Bioact. Compat. Polym. 13*:19 (1998).

170. T. Minko, P. Kopeckova, J. Kopecek. Efficacy of the chemotherapeutic action of HPMA copolymer-bound doxorubicin in a solid tumor model of oavian carcinoma. *Int. J. Cancer. 86*:108 (2000).

171. T. Minko, P. Kopeckova, J. Kopecek. Comparison of the anticancer effect of free and HPMA copolymer-bound adriamycin in human ovarian carcinoma cells. *Pharm. Res. 16*:986 (1999).

172. L. W. Seymour. Passive tumor targeting of soluble macromolecules and drug conjugates. *Crit. Rev. Ther. Drug. Carrier. Syst. 9*:135 (1992).

173. R. Duncan. Drug-polymer conjugates: potential for improved chemotherapy. *Anticancer Drugs 3*:175 (1992).

174. A. Kishida, H. Goto, K. Murakami, K. Kakinoki, M. Akashi, T. Endo. Polymer drugs and polymeric drugs IX. Synthesis and 5-fluorouracil release profiles of biodegradable polymeric prodrugs gamma-poly(alpha-hydroxymethyl-5-fluorouracil-glutamate). *J. Bioact. Compat. Polym. 13*:222 (1998).

175. M. Nichifor, E. H. Schacht. Cytotoxicity and anticancer activity of macromolecular prodrugs of 5-fluorouracil. *J. Bioact. Compat. Polym. 12*:265 (1997).

176. M. Akashi, K. Takemoto. New aspects of polymer drugs. *Adv. Polym. Sci. 97*:107 (1990).

177. J. F. Morere, D. Letourneur, P. Planchon, T. Avramoglou, J. Jozefonvicz, L. Israel, M. Crépin. Inhibitory effect of substituted dextrans on MCF7 human breast cancer cell growth in vitro. *Anticancer. Drugs. 3*:629 (1992).

178. R. Bagheri-Yarmand, J. F. Morere, D. Letourneur, J. Jozefonvicz, L. Israel, M. Crépin. Inhibitory effects of dextran derivatives in vitro on the growth characteristics of premalignant and malignant human mammary epithelial cell lines. *Anticancer. Res. 12*:1641 (1992).

179. R. Bagheri-Yarmand, P. Bittoun, J. Champion, D. Letourneur, J. Jozefonvicz, S. Fermandjian, M. Crépin. Carboxymethyl benzylamide dextrans inhibit breast cell growth [letter]. *In Vitro Cell Dev. Biol. Anim. 30A*:822 (1994).

180. R. Bagheri-Yarmand, Y. Kourbali, A. M. Rath, R. Vassy, A. Martin, J. Jozefonvicz, C. Soria, H. Lu, M. Crépin. Carboxymethyl benzylamide dextran blocks angiogenesis of MDA-MB435 breast carcinoma xenografted in fat pad and its lung metastases in nude mice. *Cancer Res. 59*:507 (1999).

181. R. Bagheri-Yarmand, Y. Kourbali, J. F. Morere, J. Jozefonvicz, M. Crépin. Inhibition of MCF-7ras tumor growth by carboxymethyl benzylamide dextran: blockage of the paracrine effect and receptor binding of transforming growth factor beta 1 and platelet-derived growth factor-BB. *Cell Growth Differ. 9*:497 (1998).

182. R. Bagheri-Yarmand, Y. Kourbali, C. Mabilat, J. F. Morere, A. Martin, H. Lu, C. Soria, J. Jozefonvicz, M. Crépin. The suppression of fibroblast growth factor 2/fibroblast growth factor 4-dependent tumour angiogenesis and growth by the anti-growth factor activity of dextran derivative (CMDB7). *Br. J. Cancer 78*:111 (1998).

183. R. Bagheri-Yarmand, J. F. Liu, D. Ledoux, J.-F. Morère, M. Crépin. Inhibition of human breast epithelial HBL100 cell proliferation by a dextran derivative (CMDB7): interference with the FGF2 autocrine loop. *Biochem. Biophys. Res. Commun. 239*:424 (1997).

184. Y. Lemmouchi, E. Schacht, S. Dejardin. Biodegradable poly[(amino acid ester)phosphazenes] for biomedical applications. *J. Bioact. Compat Polym. 13*:5 (1998).

185. A. K. Andrianov, J. R. Sargent, S. S. Sule, M. P. Le Golvan, A. L. Woods, S. A. Jenkins, L. G. Payne. Synthesis, physico-chemical properties and immunoadjuvant activity of water-soluble phosphazene polyacids. *J. Bioact Compat. Polym. 13*:243 (1998).

186. S. Benderbous, B. Bonnemain. Magnetic resonance contract agents and perfusion imaging. *J. Mal. Vasc. 21*:16 (1996).

187. S. Fallis, J. Beaty-Nosco, R. B. Dorshow, K. Adzamli. Polyethyleneglycol-stabilized manganese-substituted hydroxylapatite as a potential contrast agent for magnetic resonance imaging: particle stability in biologic fluids. *Invest. Radiol. 33*:847 (1998).

188. F. D. Knollmann, R. Sorge, A. Muhler, J. Maurer, P. Muschick, J. C. Bock, R. Felix. Hemodynamic tolerance of intravascular contrast agents for magnetic resonance imaging. *Invest. Radiol. 32*:755 (1997).

189. M. Sovak, J. G. Douglass, R. C. Terry, J. W. Brown, F. Bakir, T. S. Wasden. Blood pool radiopaque polymers. Design considerations. *Invest. Radiol. 29 suppl. 2*:S271 (1994).

190. G. Adam, A. Mühler, E. Spuntrup, J. M. Neuerburg, M. Kilbinger, H. Bauer, L. Fucezi, W. Kupper, R. W. Gunther. Differentiation of spontaneous canine breast tumors using dynamic magnetic resonance imaging with 24-Gadolinium-DTPA-cascade-polymer, a new blood-pool agent. Preliminary experience. *Invest. Radiol. 31*:267 (1996).

191. D. Horak, F. Jelinek, D. Krajickova. Artificial emboli based on poly(2-hydroxyethyl methacrylate) particles in animal experiments. *J. Biomater. Sci. Polym. Ed. 10*:455 (1999).

192. D. Horak, F. Svec, A. Adamyan, M. Titova, N. Skuba, O. Voronkova, N. Trostenyuk, V. Vishnevskii, K. Gumargalieva. Hydrogels in endovascular embolization. V. Antitumour agent methotrexate-containing p(HEMA). *Biomaterials 13*:361 (1992).

193. D. Horak, M. Cervinka, V. Puza. Hydrogels in endovascular embolization. VI. Toxicity tests of poly(2-hydroxyethyl methacrylate) particles on cell cultures. *Biomaterials 18*:1355 (1997).

194. Y. Matsumaru, A. Hyodo, T. Nose, T. Hirano, S. Ohashi. Embolic materials for endovascular treatment of cerebral lesions. *J. Biomater. Sci. Polym. Ed. 8*:555 (1997).

195. M. Yalpani. Polysaccarides. In: *Syntheses, Modifications and Structure/Property Relations* (M. Yalpani, ed.). Elsevier, Amsterdam, 1988, pp. 83–141.

196. P. E. Marszalek, A. F. Oberhauser, Y. P. Pang, J. M. Fernandez. Polysaccharide elasticity governed by chair–boat

transitions of the glucopyranose ring. *Nature 396*:661 (1998).

197. M. Jozefowicz, J. Jozefonvicz. Randomness and biospecificity: random copolymers are capable of biospecific molecular recognition in living systems. *Biomaterials 18*: 1633 (1997).
198. F. Chaubet, J. Champion, O. Maïga, S. Mauray, J. Jozefonvicz. Synthesis and structure-anticoagulant property relationships of functionalized dextrans: CMDBS. *Carbohydr. Polym. 28*:145 (1995).
199. E. de Raucourt, S. Mauray, F. Chaubet, O. Maiga-Revel, M. Jozefowicz, A. M. Fischer. Anticoagulant activity of dextran derivatives. *J. Biomed. Mater. Res. 41*:49 (1998).
200. L. Krentsel, F. Chaubet, A. Rebrov, J. Champion, I. Ermakov, P. Bittoun, S. Fermandjian, A. Litmanovich, N. Platé, J. Jozefonvicz. Anticoagulant activity of funtionalized dextrans. Structure analysis of carboxymethylated dextran and first Monte Carlo simulations. *Carbohydr. Polym. 33*: 63 (1997).
201. O. Maïga-Revel, F. Chaubet, J. Jozefonvicz. New investigations on heparin-like derivatized dextrans: CMDBS, synergistic role of benzylamide and sulfate substituents in anticoagulant activity. *Carbohydr. Polym. 32*:89 (1997).
202. S. Mauray, E. de Raucourt, F. Chaubet, O. Maiga-Revel, C. Sternberg, A. M. Fischer. Comparative anticoagulant activity and influence on thrombin generation of dextran derivatives and of a fucoidan fraction. *J. Biomater. Sci. Polym. Ed. 9*:373 (1998).
203. C. Boisson- Vidal, J. Jozefonvicz, J. L. Brash. Interactions of proteins in human plasma with modified polystyrene resins. *J. Biomed. Mater. Res. 25*:67 (1991).
204. M. P. Carreno, F. Maillet, D. Labarre, M. Jozefowicz, M. D. Kazatchkine. Specific antibodies enhance Sephadex-induced activation of the alternative complement pathway in human serum. *Biomaterials 9*:514 (1988).
205. B. Montdargent, D. Labarre, M. Jozefowicz. Interactions of functionalized polystyrene derivatives with the complement system in human serum. *J. Biomater. Sci. Polym. Ed.* 2:25 (1991).
206. B. Montdargent, F. Maillet, M. P. Carreno, M. Jozefowicz, M. Kazatchkine, D. Labarre. Regulation by sulphonate groups of complement activation induced by hydroxymethyl groups on polystyrene surfaces. *Biomaterials 14*: 203 (1993).
207. H. Thomas, F. Maillet, D. Letourneur, J. Jozefonvicz, M. D. Kazatchkine, E. Fischer. A synthetic dextran derivative inhibits complement activation and complement-mediated cytotoxicity in an in vitro model of hyperacute xenograft rejection. *Transplant. Proc. 28*:593 (1996).
208. J. A. Hubbell. Bioactive biomaterials. *Curr. Opin. Biotechnol. 10*:123 (1999).
209. B. D. Ratner. The engineering of biomaterials exhibiting recognition and specificity. *J. Mol. Recognit 9*:617 (1996).
210. Y. Uyama, K. Kato, Y. Ikada. Surface modification of polymers by grafting. *Adv. Polym. Sci. 137*:1 (1998).
211. H. Jacobs, D. Grainger, T. Okano, S. W. Kim. Surface modification for improved blood compatibility. *Artif. Organs 12*:506 (1988).
212. N. P. Desai, J. A. Hubbell. Biological responses to polyethylene oxide modified polyethylene terephthalate surfaces. *J. Biomed. Mater. Res. 25*:829 (1991).
213. D. K. Han, K. D. Park, G. H. Ryu, U. Y. Kim, B. G. Min, Y. H. Kim. Plasma protein adsorption to sulfonated poly (ethylene oxide)-grafted polyurethane surface. *J. Biomed. Mater. Res. 30*:23 (1996).
214. M. Ceuk, M. Sacak. Grafting of acrylamide–methacrylic acid mixture onto poly(ethylene terephthalate) fibers by azobisisobutyronitrile. *J. Appl. Polym. Sci. 59*:609 (1996).
215. F. C. Loh, K. L. Tan, E. T. Kang, K. G. Neoh, M. Y. Pun. Near-UV radiation induced surface graft copolymerization of some O_3-pretreated conventional polymer films. *Eur. Polym. J. 31*:481 (1995).
216. J. Buchenska. Modification of polyamide fibers (PA6) by grafting polyacrylamide (PAM). *J. Appl. Polym. Sci. 58*: 1901 (1995).
217. Y.-E. Fang, X. P. Lu, S. Z. Wang, X. Zhao, F. Fang. Study of radiation-induced graft copolymerization of vinyl acetate onto ethylene-co-propylene rubber. *J. Appl. Polym. Sci. 62*:2209 (1996).
218. S. Tsuneda, K. Saito, S. Furusaki, T. Sugo. High-throughput processing of proteins using a porous and tentacle anion-exchange membrane. *J. Chromatogr A. 689*: 211 (1995).
219. S. D. Lee, G. H. Hsiue, C. Y. Kao, P. C. Chang. Artificial cornea: surface modification of silicone rubber membrane by graft polymerization of pHEMA via glow discharge. *Biomaterials 17*:587 (1996).
220. A. S. Hoffman, Ionizing radiation and gas plasma (or glow) discharge treatments for preparation of novel polymeric biomaterials. *Adv. Polym. Sci. 57*:141 (1984).
221. J. H. Lee, H. W. Jung, I. K. Kang, H. B. Lee. Cell behaviour on polymer surfaces with different functional groups. *Biomaterials 15*:705 (1994).
222. A. Christensson, L. Ljunggren, C. Nilsson-Thorell, B. Arge, U. Diehl, K. E. Hagstam, M. Lundberg. In vivo comparative evaluation of hemodialysis tubing plasticized with DEHP and TEHTM. *Int. J. Artif. Organs. 14*:407 (1991).
223. N. A. Hoenich, J. Thompson, J. McCabe, D. R. Appleton. Particle release from haemodialysers. *Int. J. Artif. Organs. 13*:803 (1990).
224. N. M. Lamba, J. M. Courtney, J. D. Gaylor, G. D. Lowe. In vitro investigation of the blood response to medical grade PVC and the effect of heparin on the blood response. *Biomaterials 21*:89 (2000).
225. J. H. Lee, K. O. Kim, Y. M. Ju. Polyethylene oxide additive-entrapped polyvinyl chloride as a new blood bag material. *J. Biomed. Mater. Res. 48*:328 (1999).
226. V. Manojkumar, N. Padmakumaran, A. Santhosh, K. V. Deepadevi, P. Arun, L. R. Lakshmi, P. A. Kurup. Decrease in the concentration of vitamin E in blood and tissues caused by di(2-ethylhexyl) phthalate, a commonly used

plasticizer in blood storage bags and medical tubing. *Vox Sang 75*:139 (1998).

227. P. Amin, J. Wille, K. Shah, A. Kydonieus. Analysis of the extractive and hydrolytic behavior of microthane poly(ester-urethane) foam by high pressure liquid chromatography. *J. Biomed Mater Res. 27*:655 (1993).
228. A. Edwards, R. J. Carson, M. Szycher, S. Bowald. In vitro and in vivo biodurability of a compliant microporous vascular graft. *J. Biomater. Appl. 13*:23 (1998).
229. J. N. Mulvihill, A. A. Shah, J.-P. Cazenave. Interaction of biomaterials with blood platelets and plasma proteins. A method of evaluation under controlled hemodynamic in capillary tubes. In: *Blood Compatible Materials and Their Testing* (S. Dawids, A. Bantjes, eds.) Martinus Nijihoff, Boston, 1986, pp. 115–128.
230. N. Uchida, H. Kambic, H. Emoto, J. F. Chen, S. Hsu, S. Murabayshi, H. Harasaki, Y. Nose. Compliance effects on small diameter polyurethane graft patency. *J. Biomed Mater Res. 27*:1269 (1993).
231. D. J. Lyman, F. J. Fazzio, H. Voorhees, G. Robinson, D. J. Albo. Compliance as a factor effecting the patency of a copolyurethane vascular graft. *J. Biomed Mater Res. 12*: 337 (1978).
232. G. M. Bernacca, T. G. Mackay, R. Wilkinson, D. J. Wheatley. Calcification and fatigue failure in a polyurethane heart valve. *Biomaterials 16*:279 (1995).
233. M. E. Nimni, D. Myers, D. Ertl, B. Han. Factors which affect the calcification of tissue-derived bioprostheses. *J. Biomed Mater Res 35*:531 (1997).
234. J. A. Chinn, J. A. Sauter, R. E. J. Phillips, W. J. Kao, J. M. Anderson, S. R. Hanson, T. R. Ashton. Blood and tissue compatibility of modified polyester: thrombosis, inflammation, and healing. *J. Biomed Mater Res. 39*:130 (1998).
235. R. D. Hagerty, D. L. Salzmann, L. B. Kleinert, S. K. Williams. Cellular proliferation and macrophage populations associated with implanted expanded polytetrafluoroethylene and polyethyleneterephthalate. *J. Biomed. Mater. Res. 49*:489 (2000).
236. G. Janvier, C. Baquey, C. Roth, N. Benillan, S. Bélisle, J. F. Hardy. Extracorporeal circulation, hemocompatibility, and biomaterials. *Ann. Thorac. Surg. 62*:1926 (1996).
237. P. Vaudaux, R. Suzuki, F. A. Waldvogel, J. J. Morgenthaler, U. E. Nydegger. Foreign body infection: role of fibronectin as a ligand for the adherence of Staphylococcus aureus. *J. Infect. Dis. 150*:546 (1984).
238. E. Vinard, R. Eloy, J. Descotes, J. R. Brudon, H. Guidicelli, P. Patra, R. Streichenberger, M. David. Human vascular graft failure and frequency of infection. *J. Biomed. Mater. Res. 25*:499 (1991).
239. T. Yoneyama, K. Ishihara, N. Nakabayashi, M. Ito, Y. Mishima. Short-term in vivo evaluation of small-diameter vascular prosthesis composed of segmented poly(etherurethane)/2-methacryloyloxyethyl phosphorylcholine polymer blend. *J. Biomed. Mater. Res. 43*:15 (1998).
240. R. M. Schmidt. Section I: Hematology. CRC Handbood Series in Clinical Laboratory Science. CRC Press, Boca Raton, Florida, 1980.
241. S. Dawids, A. Bantjes. *Blood Compatible Materials and Their Testing*. Martinus Nijhoff, Boston, 1986.
242. K. Doi, Y. Nakayama, T. Matsuda. Novel compliant and tissue-permeable microporous polyurethane vascular prosthesis fabricated using an excimer laser ablation technique. *J. Biomed. Mater. Res. 31*:27 (1996).
243. K. Doi, T. Matsuda. Significance of porosity and compliance of microporous, polyurethane-based microarterial vessel on neoarterial wall regeneration. *J. Biomed Mater Res. 37*:573 (1997).
244. K. Fujimoto, M. Minato, S. Miyamoto, T. Kaneko, H. Kikuchi, K. Sakai, M. Okada, Y. Ikada. Porous polyurethane tubes as vascular graft. *J. Appl. Biomater 4*:347 (1993).
245. B. D. Ratner, T. Horbett, A. S. Hoffman, S. D. Hauschka. Cell adhesion to polymeric materials: implications with respect to biocompatibility. *J. Biomed Mater Res. 9*:407 (1975).
246. B. D. Ratner, A. B. Johnston, T. J. Lenk. Biomaterial surfaces. *J. Biomed Mater Res. 21*:59 (1987).
247. T. G. Mackay, D. J. Wheatley, G. M. Bernacca, A. C. Fisher, C. S. Hindle. New polyurethane heart valve prosthesis: design, manufacture and evaluation. *Biomaterials 17*:1857 (1996).
248. C. J. Underwood, W. F. Tait, D. Charlesworth. Design considerations for a small bore vascular prosthesis. *Int. J. Artif. Organs. 11*:272 (1988).
249. J. L. Brash, D. J. Lyman. Adsorption of plasma proteins in solution to uncharged, hydrophobic polymer surfaces. *J. Biomed. Mater. Res. 3*:175 (1969).
250. A. Schmitt, R. Varoqui, S. Uniyal, J. L. Brash, C. Pusineri. Interaction of fibrinogen with solid surfaces of varying charge and hydrophobic–hydrophilic balance. I: Adsorption isotherm. *J. Colloid Interf Sci. 92*:25 (1983).
251. G. A. Boffa, N. Lucien, A. Faure, M. C. Boffa, J. Jozefonvicz, A. Szubarga, M. J. Larrieu, P. Mandon. Polytetrafluoroethylene-*N*-vinylpyrrolidone graft copolymers: affinity with plasma proteins. *J. Biomed. Mater. Res. 11*: 317 (1977).
252. T. Okano, S. Nishiyama, I. Shinohara, T. Akaike, Y. Sakurai, K. Kataoka, T. Tsuruta. Effect of hydrophilic and hydrophobic microdomains on mode of interaction between block polymer and blood platelets. *J. Biomed. Mater. Res. 15*:393 (1981).
253. T. Okano, T. Aoyagi, K. Kataoka, K. Abe, Y. Sakurai, M. Shimada, I. Shinohara. Hydrophilic–hydrophobic microdomain surfaces having an ability to suppress platelet aggregation and their in vitro antithrombogenicity. *J. Biomed Mater Res. 20*:919 (1986).
254. M. Shimada, M. Unoki, N. Inaba, H. Tahara, I. Shinohara, T. Okano, Y. Sakurai, K. Kataoka. Effect of adsorbed protein on the adhesion behaviour of platelet to the microdomain surface of 2-hydroxyethyl methacrylate-stryrene block copolymer. *Eur. Polym J. 19*:929 (1983).
255. K. Kataoka, T. Okano, Y. Sakurai, T. Nishimura, S. Inoue, T. Watanabe, A. Maruyama, T. Tsuruta. Differential reten-

tion of lymphocyte subpopulations (B and T cells) on the microphase separated surface of polystyrene/polyamine graft copolymers. *Eur. Polym. J. 19*:979 (1983).
256. J. L. Brash. Mechanism of adsorption of proteins to solid surfaces and its relationship to blood compatibility. In: *Biocompatible Polymers, Metals, and Composites* (M. Szycher, ed.). Technomic, Lancaster PA, 1983, pp. 35–52.
257. A. L. Adams, G. C. Fischer, P. C. Munoz, L. Vroman. Convex-lens-on-slide: a simple system for the study of human plasma and blood in narrow spaces. *J. Biomed. Mater. Res. 18*:643 (1984).
258. L. Vroman, A. L. Adams, G. C. Fischer, P. C. Munoz. Interaction of high molecular weight kininogen, factor XII, and fibrinogen in plasma at interfaces. *Blood 55*:156 (1980).
259. S. Uniyal, J. L. Brash. Patterns of adsorption of proteins from human plasma onto foreign surfaces. *Thromb Haemost 47*:285 (1982).
260. J. L. Brash, P. ten Hove. Effect of plasma dilution on adsorption of fibrinogen to solid surfaces. *Thromb Haemost 51*:326 (1984).
261. A. H. Schmaier, L. Silver, A. L. Adams, G. C. Fischer, P. C. Munoz, L. Vroman, R. W. Colman. The effect of high molecular weight kininogen on surface-adsorbed fibrinogen. *Thromb. Res. 33*:51 (1984).
262. L. Vroman. Problems in the development of materials that are compatible with blood. *Biomater. Med. Devices Artif. Organs 12*:307 (1984).
263. D. L. Coleman, D. E. Gregonis, J. D. Andrade. Blood–materials interactions: the minimum interfacial free energy and the optimum polar/apolar ratio hypotheses. *J. Biomed. Mater. Res. 16*:381 (1982).
264. L. Vroman. Protein/surface interaction. In: *Biocompatible Polymers, Metals, and Composites* (M. Szycher, ed.). Technomic, Lancaster PA, 1983, pp. 81–88.
265. T. A. Horbett, J. L. Brash. Proteins at interfaces: current issues and future prospects. In: *Proteins at Interfaces. Physicochemical and Biochemical Studies* (J. L. Brash, T. A. Horbett, eds.). ASC Symposium Series, Washington DC, 1987, pp. 1–33.
266. B. W. Morrissey. The adsorption and conformation of plasma proteins: a physical approach. *Ann. NY Acad Sci. 283*:50 (1977).
267. V. L. Gott, J. D. Whiffen, R. C. Dutton. Heparin bonding on colloidal graphite surfaces. *Science 142*:1297 (1963).
268. C. D. Ebert, E. S. Lee, S. W. Kim. The antiplatelet activity of immobilized prostacyclin. *J. Biomed. Mater. Res. 16*: 629 (1982).
269. G. A. Grode, J. Pitman, J. P. Crowleg, R. I. Leininger, R. D. Falb. Surface immobilized prostaglandin as a platelet protective agent. *Trans Am. Soc. Artif. Intern Organs 20*: 38 (1974).
270. S. W. Kim, C. D. Ebert, J. Y. Lin, J. C. McRea. Non thromobogenic polymers: pharmaceutical approaches. *Am. Soc. Artif. Intern. Organs 6*:76 (1983).
271. C. H. Bamford, I. P. Middleton, K. G. Al-Lamee. Polymeric inhibitors of platelet aggregation. Synergistic effects and proposals for a new mechanism. *Biochim. Biophys. Acta. 886*:109 (1986).
272. C. H. Bamford, I. P. Middleton, K. G. Al-Lamee. Polymeric inhibitors of platelet aggregation. II. Copolymers of dipyridamole and related drugs with *N*-vinylpyrrolidone. *Biochim. Biophys. Acta. 924*:38 (1987).
273. C. H. Bamford, I. P. Middleton, K. G. Al-Lamee. Influence of molecular structure on the synergistic action of theophylline or dipyridamole derivatives in the prostaglandin-type inhibition of platelet aggregation. *J. Biomater. Sci. Polym. Ed. 2*:37 (1991).
274. C. H. Bamford, K. G. Al-Lamee. Chemical methods for improving the haemocompatibility of synthetic polymers. *Clin. Mater. 10*:243 (1992).
275. W. Marconi. New nonthrombogenic polymer compositions. *Makromol. Chem. Suppl. 5*:15 (1981).
276. P. W. Heyman, S. W. Kim. Blood compatibility of heparinized polyurethanes. *Makromol. Chem. Suppl. 9*:119 (1985).
277. Y. Mori, S. Nagaoka, Y. Masubuchi. The effect of released heparin from the heparinized hydrophilic polymer (HRSD) on the process of thrombus formation. *Trans. Am. Soc. Artif. Intern. Organs 24*:736 (1978).
278. C. Pusineri, R. Eloy, J. Baguet, J. Paul, J. Belleville, P. Leconte. In vivo evaluation of the heparinized cathethers in dogs, *Eur. Soc. Artif. Organs Proc. 1*:305 (1982).
279. W. E. Hennick, L. Dost, J. Feijen, S. W. Kim. Interaction of albumin–heparin conjugate preadsorbed surfaces with blood. *Trans. Am. Soc. Artif. Intern. Organs 29*:200 (1983).
280. R. Barbucci, M. Benvenuti, G. Casini, P. Ferruti, M. Nocentini. Heparinizable materials. V: Preparation and FTIR characterization of polyurethane surfaces grafted with heparin complexing poly(aminoamine) chains. *Makromol. Chem. 186*:2291 (1985).
281. R. Barbucci, M. Benvenuti, G. Casini, P. Ferruti, F. Tempesti. Surface grafting of heparin-complexing poly(aminoamide) on poly(ethylene terephtalate). (Dacron). *Makromol. Chem. Suppl. 9*:233 (1985).
282. O. Larm, R. Larsson, P. Olsson. A new non-thrombogenic surface prepared by selective covalent binding of heparin via a modified reducing terminal residue. *Biomater. Med. Devices. Artif. Organs. 11*:161 (1983).
283. M. F. Goosen, M. V. Sefton. Heparinized styrene-butadiene-styrene elastomers. *J Biomed. Mater. Res. 13*:347 (1979).
284. L. I. Valuev, N. Platé. Chemical modification of polymeric materials by physiologically active substances. *Adv. Mater. 2*:405 (1990).
285. L. D. Uzhinova, N. Platé, S. M. Krasovskaja. Interactions of biospecific sorbents with physiologically active substances. *J. Mater. Sci.: Mater Med 2*:189 (1991).
286. F. Sidouni, N. Nurdin, P. Chabrecek, D. Leonard, H. J. Mathieu, D. Lohmann, P. Descouts. Surface modification of a biomedical polyurethane to improve its long-term haemocompatibility, North Sea Biomaterials: 14th Euro-

pean Conference on Biomaterials, Proceedings, The Hague, The Netherlands, September 15–18, 1998, p. 115.

287. A. Sugitachi, M. Tanaka, Y. Kawahara, N. Kitamura, K. Takagi. A new type of drain tube. *Artif Organs 5*:69 (1981).

288. A. Kishida, Y. Ueno, I. Maruyama, M. Akashi. Immobilization of human thrombomodulin on biomaterials: evaluation of the activity of immobilized human thrombomodulin. *Biomaterials 15*:1170 (1994).

289. A. Kishida, Y. Ueno, N. Fukudome, E. Yashima, I. Maruyama, M. Akashi. Immobilization of human thrombomodulin onto poly(ether urethane urea) for developing antithrombogenic blood-contacting materials. *Biomaterials 15*:848 (1994).

290. Y. B. Aldenhoff, L. H. Koole. Studies on a new strategy for surface modification of polymeric biomaterials. *J. Biomed. Mater. Res. 29*:917 (1995).

291. F. Markwardt. The development of hirudin as an antithrombotic drug. *Thromb. Res. 74*:1 (1994).

292. B. Seifert, P. Romaniuk, T. Groth. Covalent immobilization of hirudin improves the haemocompatibility of polylactide-polyglycolide in vitro. *Biomaterials 18*:1495 (1997).

293. R. Blezer, B. Fouache, G. M. Willems, T. Lindhout. Activation of blood coagulation at heparin-coated surfaces. *J Biomed. Mater. Res. 37*:108 (1997).

294. E. W. Merrill, E. W. Salzman, P. S. L. Wong, T. P. Ashford, A. H. Brown, W. G. Austen. Polyvinyl alcohol–heparin hydrogel ''G''. *J. Appl. Physiol. 29*:723 (1970).

295. I. Danishefski, F. Tzeng. Preparation of heparin-linked agarose and its interaction with plasma. *Thromb. Res. 4*: 237 (1974).

296. A. S. Hoffman, G. Schmer, C. Harris, W. G. Kraft. Covalent binding of biomolecules to radiation-grafted hydrogels on inert polymer surfaces. *Trans. Am. Soc. Artif. Intern. Organs 18*:10 (1972).

297. A. S. Hoffman, D. Cohn, S. R. Hanson, L. A. Harker, T. Horbett, B. D. Ratner, L. O. Reynolds. Application of radiation-grafted hydrogels as blood-contacting biomaterials. *Radiat. Phys. Chem. 22*:267 (1983).

298. M. F. Goosen, M. V. Sefton. Properties of a heparin-poly (vinyl alcohol) hydrogel coating. *J. Biomed. Mater. Res. 17*:359 (1983).

299. R. Larsson, P. Olsson, U. Lindahl. Inhibition of thrombin on surfaces coated with immobilized heparin and heparin-like polysaccharides: a crucial non-thrombogenic principle. *Thromb. Res. 19*:43 (1980).

300. D. Labarre, M. Jozefowicz, M. C. Boffa. Properties of heparin–poly(methyl methacrylate) copolymers. II. *J. Biomed. Mater. Res. 11*:283 (1977).

301. D. Labarre, M. C. Boffa, M. Jozefowicz. Preparation and properties of heparin–poly(methacrylate) copolymers. *J. Pharm. Sci. Symp. 47*:131 (1974).

302. C. Baquey, C. Darnez, P. Blanquet. ESR study of gamma-irradiated sodium heparinate. *Radiat. Res. 70*:82 (1977).

303. C. Baquey, A. Beziade, D. Ducassou, P. Blanquet. Intérêt du greffage radiochimique de monomères vinyliques pour améliorer l'hémocompatibilité de matériaux artificiels. *Innov. Tech. Biol. Med. 2*:379 (1981).

304. Y. Byun, H. A. Jacobs, S. W. Kim. Heparin surface immobilization through hydrophilic spacers: thrombin and antithrombin III binding kinetics. *J. Biomater. Sci. Polym. Ed. 6*:1 (1994).

305. V. D. Nadkarni, A. Pervin, R. J. Linhardt. Directional immobilization of heparin onto beaded supports. *Anal. Biochem. 222*:59 (1994).

306. J. Sanchez, G. Elgue, J. Riesenfeld, P. Olsson. Inhibition of the plasma contact activation system of immobilized heparin: relation to surface density of functional antithrombin binding sites. *J. Biomed. Mater. Res. 37*:37 (1997).

307. P. V. Narayanan. Surface functionalization by RF plasma treatment of polymers for immobilization of bioactive-molecules. *J. Biomater. Sci. Polym. Ed. 6*:181 (1994).

308. B. Seifert, T. Groth, K. Herrmann, P. Romaniuk. Immobilization of heparin on polylactide for application to degradable biomaterials in contact with blood. *J. Biomater. Sci. Polym. Ed. 7*:277 (1995).

309. P. W. Heyman, C. S. Cho, J. C. McRea, D. B. Olsen, S. W. Kim. Heparinized polyurethanes: in vitro and in vivo studies. *J. Biomed. Mater. Res. 19*:419 (1985).

310. M. C. Boffa, D. Labarre, M. Jozefowicz, G. A. Boffa. Interactions between human plasma proteins and heparin-poly(methylmethacrylate) copolymer. *Thromb. Haemost. 41*:346 (1979).

311. C. Fougnot, D. Labarre, J. Jozefonvicz, M. Jozefowicz. Modifications of polymer surfaces to improve blood compatibility. In: *Macromolecular Biomaterials* (G. W. Hastings, P. Ducheyne, eds.). CRC Press, Boca Raton, 1984, p. 215.

312. J. E. Lovelock, J. S. Porterfield. Blood coagulation: its prolongation in vessels with negatively charged surfaces. *Nature 167*:39 (1951).

313. H. P. Gregor. Anticoagulant activity of sulfonate polymers and copolymers. *Polym. Sci. Technol. USA 7*:51 (1975).

314. R. Machovich, I. Horvath. Heparin-like effect of polymethacrylic acid on the reaction between thrombin and antithrombin III. *Thromb. Res. 11*:765 (1977).

315. J. M. Walenga, M. Petitou, M. Samama, J. Fareed, J. Choay. Importance of a 3-O-sulfate group in a heparin pentasaccharide for antithrombotic activity. *Thromb. Res. 52*:553 (1988).

316. U. Lindahl, G. Backstrom, L. Thunberg, I. G. Leder. Evidence for a 3-O-sulfated D-glucosamine residue in the antithrombin-binding sequence of heparin. *Proc. Natl. Acad. Sci. USA 77*:6551 (1980).

317. U. Lindahl, G. Backstrom, L. Thunberg. The antithrombin-binding sequence in heparin. Identification of an essential 6-O-sulfate group. *J. Biol. Chem. 258*:9826 (1983).

318. T. Beugeling, L. van der Does, A. Bantjes, W. L. Sederel. Antithrombin activity of a polyelectrolyte synthesized from cis-1,4-polyisoprene. *J. Biomed. Mater. Res. 8*:375 (1974).

319. L. van der Does, T. Beugeling, P. E. Froehling, A. Bantjes.

Synthetic polymers with anticoagulant activity. *J. Pharm. Sci. Symp. 66*:337 (1979).

320. L. C. Sederel, L. van der Does, J. F. van Duijl, T. Beugeling, A. Bantjes. Anticoagulant activity of a synthetic heparinoid in relation to molecular weight and *N*-sulfate content. *J Biomed Mater Res 15*:819 (1981).
321. M. Akashi, N. Sakamoto, K. Suzuki, A. Kishida. Synthesis and anticoagulant activity of sulfated glucoside-bearing polymer. *Bioconjug. Chem. 7*:393 (1996).
322. N. Sakamoto, A. Kishida, I. Maruyama, M. Akashi. The mechanism of anticoagulant activity of a novel heparinoid sulfated glucoside-bearing polymer. *J. Biomater. Sci. Polym. Ed. 8*:545 (1997). Erratum in J Biomater Sci Polym Ed 8(10):815 (1997).
323. J. P. Santerre, P. ten Hove, N. H. VanderKamp, J. L. Brash. Effect of sulfonation of segmented polyurethanes on the transient adsorption of fibrinogen from plasma: possible correlation with anticoagulant behavior. *J. Biomed. Mater. Res. 26*:39 (1992).
324. D. K. Han, K. B. Lee, K. D. Park, C. S. Kim, S. Y. Jeong, Y. H. Kim, H. M. Kim, B. G. Min. In vivo canine studies of a Sinkhole valve and vascular graft coated with biocompatible PU-PEO-SO3. *ASAIO J. 39*:M537 (1993).
325. T. G. Grasel, S. L. Cooper. Properties and biological interactions of polyurethane anionomers: effect of sulfonate incorporation. *J. Biomed. Mater. Res. 23*:311 (1989).
326. J. H. Silver, A. P. Hart, E. C. Williams, S. L. Cooper, S. Charef, D. Labarre, M. Jozefowicz. Anticoagulant effects of sulphonated polyurethanes. *Biomaterials 13*:339 (1992).
327. J. P. Lens, J. G. Terlingen, G. H. Engbers, J. Feijen. Preparation of heparin-like surfaces by introducing sulfate and carboxylate groups on poly(ethylene) using an argon plasma treatment. *J. Biomater. Sci. Polym. Ed. 9*:357 (1998).
328. A. Magnani, A. Albanese, S. Lamponi, R. Barbucci. Blood-interaction performance of differently sulfated hyaluronic acids. *Thromb. Res. 81*:383 (1996).
329. R. Barbucci, S. Lamponi, A. Magnani, D. Renier. The influence of molecular weight on the biological activity of heparin like sulphated hyaluronic acids. *Biomaterials 19*: 801 (1998).
330. N. Sakamoto, T. Yumura, M. Akashi. Synthesis and anticoagulant properties of a novel heparinoid *N*-sulfate-bearing vinylpolymer. *J. Bioact. Compat. Polym. 14*:150 (1999).
331. C. Fougnot, J. Jozefonvicz, M. Samama, L. Bara. New heparin-like insoluble materials. I. *Ann. Biomed. Eng. 7*: 429 (1979).
332. C. Fougnot, M. Jozefowicz, M. Samama, L. Bara. New heparin-like insoluble materials. II. *Ann. Biomed. Eng. 7*: 441 (1979).
333. C. Fougnot, M. Jozefowicz, R. D. Rosenberg. Affinity of purified thrombin or antithrombin III for two insoluble anticoagulant polystyrene derivatives: I. In vitro adsorption studies. *Biomaterials 4*:294 (1983).
334. C. Fougnot, M. Jozefowicz, R. D. Rosenberg. Adsorption of purified thrombin or antithrombin III for two insoluble anticoagulant polystyrene derivatives: II. Competition with the other plasma proteins. *Biomaterials 5*:89 (1984).
335. C. Fougnot, M. Jozefowicz, R. D. Rosenberg. Catalysis of the generation of thrombin–antithrombin complex by insoluble anticoagulant polystyrene derivatives. *Biomaterials 5*:94 (1984).
336. M. Mauzac, N. Aubert, J. Jozefonvicz. Antithrombic activity of some polysaccharide resins. *Biomaterials 3*:221 (1982).
337. C. Fougnot, M. Jozefowicz, L. Bara, M. Samama. Interactions of anticoagulant insoluble modified polystyrene resins with plasmatic proteins. *Thromb. Res. 28*:37 (1982).
338. M. Mauzac, J. Jozefonvicz. Anticoagulant activity of dextran derivatives. Part I: Synthesis and characterization. *Biomaterials 5*:301 (1984).
339. A. M. Fischer, M. Mauzac, J. Tapon-Bretaudiere, J. Jozefonvicz. Anticoagulant activity of dextran derivatives. Part II: Mechanism of thrombin inactivation. *Biomaterials 6*: 198 (1985).
340. D. Logeart-Avramoglou, J. Jozefonvicz. Carboxymethyl benzylamide sulfonate dextrans (CMDBS), a family of biospecific polymers endowed with numerous biological properties: a review. *J. Biomed. Mater. Res. 48*:578 (1999).
341. D. Campoccia, P. Doherty, M. Radice, P. Brun, G. Abatangelo, D. F. Williams. Semisynthetic resorbable materials from hyaluronan esterification. *Biomaterials 19*:2101 (1998).
342. T. E. Hugh. Interrelationships between coagulation and complement activation. In: *Perspectives in Hemostasis* (J. Fareed et al., eds.). Pergamon Press, New York, 1981, pp. 59–69.
343. B. Montdargent, J. Toufik, M. P. Carreno, D. Labarre, M. Jozefowicz. Complement activation and adsorption of protein fragments by functionalized polymer surfaces in human serum. *Biomaterials 13*:571 (1992).
344. P. R. Craddock, J. Ferh, A. P. Dalmasso, K. L. Bringham, H. S. Jacobs. Hemodialysis leukopenia: pulmonary vascular leukostasis resulting from complement activation by dialyser cellophane membrane. *J. Clin. Invest. 59*:879 (1977).
345. G. A. Herlinger, D. H. Bing, R. Stein, R. D. Cumming. Quantitative measurement of C3 activation at polymer surfaces. *Blood 57*:764 (1981).
346. M. D. Kazatchkine, M. P. Carreno. Activation of the complement system at the interface between blood and artificial surfaces. *Biomaterials 9*:30 (1988).
347. M. P. Carreno, D. Labarre, M. Jozefowicz, M. D. Kazatchkine. The ability of Sephadex to activate human complement is suppressed in specifically substituted functional Sephadex derivatives. *Mol. Immunol. 25*:165 (1988).
348. B. Crepon, F. Maillet, M. D. Kazatchkine, J. Jozefonvicz. Molecular weight dependency of the acquired anticomplementary and anticoagulant activities of specifically substituted dextrans. *Biomaterials 8*:248 (1987).
349. M. Mauzac, F. Maillet, J. Jozefonvicz, M. D. Kazatchkine.

Anticomplementary activity of dextran derivatives. *Biomaterials 6*:61 (1985).

350. H. Thomas, F. Maillet, D. Letourneur, J. Jozefonvicz, E. Fischer, M. D. Kazatchkine. Sulfonated dextran inhibits complement activation and complement-dependent cytotoxycity in an in vitro model of hyperacute xenograft rejection. *Mol. Immunol. 33*:643 (1996).
351. H. Thomas, F. Maillet, D. Letourneur, J. Jozefonvicz, M. D. Kazatchkine. Effect of substituted dextran derivative on complement activation in vivo. *Biomaterials 16*:1163 (1995).
352. J. Toufik, M. P. Carreno, M. Jozefowicz, D. Labarre. Activation of the complement system by polysaccharidic surfaces bearing carboxymethyl, carboxymethylbenzylamide and carboxymethylbenzylamide sulphonate groups. *Biomaterials 16*:993 (1995).
353. T. M. Chang, C. Lister. A preclinical screening test for modified hemoglobin to bridge the gap between animal safety studies and use in human. *Biomater. Artif. Cells. Immobilization. Biotechnol. 20*:565 (1992).
354. M. Feola, J. Simoni, P. C. Canizaro, R. Tran, G. Raschbaum, F. J. Behal. Toxicity of polymerized hemoglobin solutions. *Surg. Gynecol. Obstet. 166*:211 (1988).
355. L. A. Sedova, N. I. Kochetygov, M. V. Berkos, N. N. Pjatowskaja. Side reaction caused by the perfluorocarbon emulsions in intravenous infusion to experimental animals. *Artif. Cells. Blood. Substit. Immobil. Biotechnol. 26*: 149 (1998).
356. J. M. Anderson. Tissue engineering in cardiovascular desease: a report. *J. Biomed. Mater. Res. 29*:1473 (1995).
357. M. Ragaller, C. Werner, J. Bleyl, S. Adam, H. J. Jacobasch, D. M. Albrecht. Blood compatible polymers in intensive care units: state of the art and current aspects of biomaterials research. *Kidney. Int Suppl. 64*:S84 (1998).

24

Surface Modification of Dacron Vascular Grafts: Incorporation of Antithrombin and Mitogenic Properties

Matthew D. Phaneuf, William C. Quist, and Frank W. LoGerfo
Beth Israel Deaconess Medical Center, Boston, Massachusetts

Martin J. Bide
University of Rhode Island, Kingston, Rhode Island

I. INTRODUCTION

Medium (6–8 mm) and small (<5 mm) internal diameter (I.D.) prosthetic arterial grafts continue to have unacceptably high failure rates when used in the clinical setting. The two major complications associated with these grafts are acute thromboses and incomplete, unregulated cellular healing. Currently available biomaterials for vascular grafts do not emulate the multitude of biological and reparative processes that occur in a normal pulsatile arterial wall. There have been exhaustive studies aimed at creating a novel biomaterial surface by nonspecific binding of a biologically active agent, covalent linkage of an agent with a broad spectrum of activity, or altering the biomaterial surface. Thus far, none of these technologies has resulted in a clinically used prosthetic vascular graft. Based on our clinical and research observations, our hypothesis is that the ultimate prosthetic graft will have to possess multiple structural and biological properties that mimic those processes inherent to the native vessel. The structure of the prosthetic graft will have to possess long-term biodurability (strength over the lifetime of the implant), porosity (permits cellular ingrowth in order to promote complete healing), and excellent handling characteristics (ease of suturing). Biologically, the surface will have to possess both localized antithrombin activity at the blood/graft interface and specific endothelial cell growth promoting factors (to increase transwall and translumen cellular growth). This chapter will examine the use of polyester (Dacron) as a prosthetic vascular graft, complications associated with graft implantation, our clinical and research observations, previous efforts to develop a novel biocompatible material surface, and our preliminary studies developing technology for covalent linkage of antithrombin and mitogenic proteins onto a Dacron surface.

II. POLYESTER (DACRON): CHARACTERISTICS AND APPLICATIONS

Polyester (chemical nomenclature—polyethylene terephthalate) fibers were first characterized in 1941 and have become the most widely produced synthetic fibers in the world. They are most familiarly known by the DuPont commercial name Dacron®. The polymer is synthesized by a condensation reaction of derivatives of ethylene glycol and terephthalic acid, resulting in molecules that contain 80 to 100 repeat units (1). These molecules are then extruded through a plurality of holes (a spinneret) to produce multifilament yarns. Dacron yarns are further processed into various structures such as warp-knit, weft-knit, and woven fabrics (2) and have excellent resiliency as well as resistance to a wide range of chemical and biological challenges.

Dacron is utilized in items ranging from clothing to medical implants. Dacron yarn was first sewn into a tubular form and utilized as a large-diameter vascular graft in the mid-1950s (3). Since that point, Dacron has been incorporated into both large- and medium-bore vascular grafts in knitted and woven form. These grafts have shown excellent long-term biodurability, handling characteristics, and capsular tissue incorporation (4).

III. COMPLICATIONS ASSOCIATED WITH PROSTHETIC ARTERIAL GRAFTING

Failure of all prosthetic arterial grafts including Dacron can be grouped into three periods: acute, delayed, and late. Acute thrombosis (hours to days) is the result of low blood flow through the graft permitting activation of the coagulation cascade at the blood/graft interface with subsequent occlusive thrombus formation. Delayed failure (weeks to months) is the result of an incomplete cellular lining of the luminal graft surface (failure of complete neointimal formation) and associated unregulated smooth muscle cell proliferation at the anastomotic sites causing focal stenosis, decreased blood flow through the graft, and secondary occlusive thrombus formation. Late graft failure (years) is infrequent and most often due to progression of atherosclerosis in the inflow or outflow vessels.

In order to examine these phenomena, an understanding of how humans incorporate and/or heal prosthetic vascular grafts is essential. The capsular surface (non–blood flow) is immediately identified by the host immune system as foreign, thereby eliciting a mixed inflammatory response with ultimate foreign body giant cell reaction. This is subsequently reinforced by collagen-rich tissue, forming what is essentially a scar. Whether the result of inflammatory cytokines, growth inhibitors from the forming luminal pseudointima, or physical factors, cellular growth through the graft interstices from the host tissues does not occur in humans.

The luminal surface (blood flow surface) of all bare prosthetic conduits is inherently thrombogenic, resulting in the formation of a controlled mural thrombus when exposed to flowing blood. The thickness and composition of this mural thrombus is initially a function of biomaterial composition and design, with the final thickness/composition ultimately determined by blood flow velocity through the graft and the thrombogenic potential of the blood. This thrombotic acellular layer has been termed the pseudointima. The body under normal circumstances recognizes thrombus and organizes the coagulum by proliferation of the underlying blood vessel endothelium and smooth muscle cells. In the case of prosthetic grafts, this recanalization is not possible, since the underlying structure is synthetic and not biological. Recent studies have also determined that the pseudointimal lining itself contains growth-inhibiting compounds, further impeding the development of a cellular neointima (5). While several strategies to diminish or circumvent this incomplete healing response have been investigated (6,7), none has resulted in clinical success.

Thus all prosthetic conduits are lined by highly reactive pseudointima. This surface has the ability to activate platelets and provides for continuous low-level thrombotic events, generating biologically active molecules such as thrombin within the pseudointimal matrix, thereby promoting additional fibrin and platelet recruitment. This healing pattern is unique to humans and canines.

The final site to consider in graft healing is the anastomotic region. Both endothelial and smooth muscle cells from the host artery proliferate into the graft. The stimulus for this cellular migration has yet to be completely defined, but it is related both to the acute injury to the arterial wall at the time of surgery and subsequently to the deposition of pseudointima on the graft surface. Thrombin embedded within the pseudointima is a powerful smooth muscle and endothelial cell mitogen, providing one of the many stimuli for organizing in vivo arterial thrombi (8). The organization of perianastomotic pseudointima with subsequent cellular proliferation occurs approximately 1 cm into the graft and abruptly stops. Cellular ingrowth at an anastomosis, or pannus ingrowth, is no more than approximately 1 cm into the graft body regardless of how long a graft is implanted.

Following this initial proliferative cycle, a period of matrix production ensues creating a localized lesion (stenosis) at the anastomotic site. This lesion impacts blood flow through the graft, further enhancing the blood/pseudointima interaction. The observed healing responses of these three domains (capsule, luminal surface, anastomotic region) on graft incorporation is unique to human beings. The only animal model that mimics this healing response is the canine model (9).

IV. CLINICAL AND RESEARCH OBSERVATIONS: BASIS FOR RESEARCH

In previous studies from our clinical and laboratory experience, mechanisms of prosthetic arterial graft failure have been evaluated. Observations from these studies, which direct our research, are as follows:

Prosthetic arterial grafts, regardless of the material composition, fail to develop a functional cellular neointima within the mid-portion of the graft (incomplete healing). A relatively thrombogenic pseudointima composed of thrombin, platelets, fibrin, and serum proteins forms the majority of the flow surface to which the circulating blood is exposed for the duration of graft patency (Fig. 1) (10).

Pseudointima formation results in low-level chronic activation of platelets and proteins of the coagulation cascade and to some extent cellular and humoral aspects of the complement system (11). The only available cellular source that can respond to this activation is the distal anastomotic site and the adjacent arterial segment (12).

Unregulated smooth muscle cell proliferation and gene expression at the anastomosis results in focal stenosis. Anastomotic stenosis with subsequent lowering of blood flow through the graft amplifies the blood/graft/pseudointima contact time and potentiates the activation of platelets and the coagulation cascade (13). This results in increased pseudointima formation, which is essentially a controlled mural thrombosis or, in severe cases, occlusive graft thrombosis.

Providing a cellular source via increasing graft interstices (porosity) and direct luminal seeding with endothelial cells have been ineffective in forming a stable functional cellular neointima, which is necessary to modulate the blood/materials interaction.

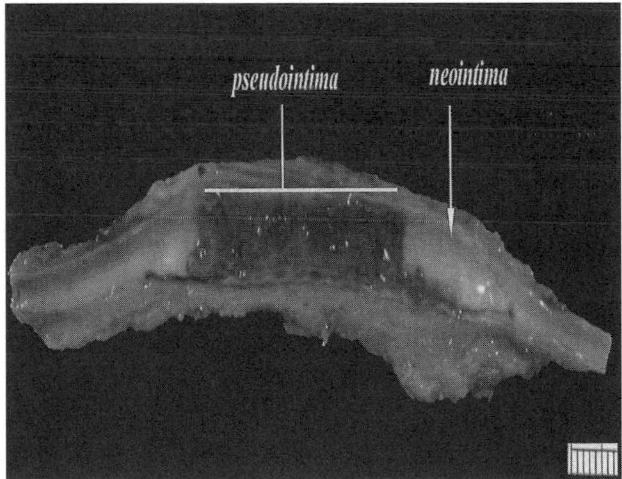

Figure 24.1 A longitudinal cross-section of a Dacron vascular graft after 30-day explantation from a canine carotid. A neointima composed of endothelial and smooth muscle cells proliferates into the graft and abruptly ends 1 cm from each anastomosis. A pseudointima composed of platelets, thrombin, fibrinogen, red blood cells, and serum proteins occupies the mid portion of the graft.

Systemic antiplatelet, anti-inflammatory, and antithrombin agents alone do not affect the onset of intimal hyperplasia or the long-term patency of the graft.

All biomaterials are excluded from direct blood interaction within minutes of establishing blood flow by an amorphous pseudointima composed of platelets, fibrin, fibrinogen, and serum proteins. Thus a biomaterial surface must be developed that will alter pseudointima composition and formation.

Studies evaluating the effects of circumferential compliance alone on graft healing/patency have not yielded conclusive findings.

V. HISTORICAL APPROACH TO SURFACE MODIFICATION

A. Prevention of Surface Thrombus Formation

A complication of all implantable biomaterials is incompatibility between blood and the biomaterial surface. The initial interaction of blood and the foreign surface results in a myriad of responses: platelet activation and adhesion (13,14), activation of the intrinsic pathway of the coagulation cascade, resulting in formation of active thrombin (15), leukocyte activation (14), and the release of complement and kallikrein (15,16). If unregulated, these responses lead to surface thrombus formation with subsequent failure of the implanted biomaterial.

Numerous attempts have been made to create a more biocompatible surface by establishing a new biologic lining on the luminal surface that would ''passivate'' this acute initial reaction. These efforts have ranged from nonspecifically binding albumin to the surface followed by heat denaturation (17) to nonspecifically cross-linking gelatin (18–20) and collagen (21–23). Covalent or ionic binding of the anticoagulant heparin albumin (14,24,25), alone (26–34), in conjunction with other biologic compounds (35–40), or with spacer moieties (41–43), as well as covalent linkage of thrombomodulin (44) have also been performed. Other studies have focused on modifying the composition of the biomaterial either by increasing hydrophilicity via incorporation of polyethylene oxide groups (45) or by creating an ionically charged surface (46).

Each of these methodologies has had limited success in creating a durable, biologically active surface. There are several limitations associated with these surface modifications: (1) thrombin is not directly inhibited, therefore fibrinogen amounts remain constant on the material surface, permitting platelet adhesion, (2) heparin-coated biomaterials may be subject to heparitinases, limiting the long-term use of these materials, (3) nonspecifically bound compounds are desorbed from the surface, which is under shear stresses, thereby re-exposing the thrombogenic biomaterial surface, (4) rapid release of nonspecifically bound compounds may create an undesired systemic effect, and (5) charge-based polymers may be covered by other blood proteins such that anticoagulant effects are masked.

One study that has shown promise involved left ventricular assist devices (LVAD) implanted in a sheep model that were either uncoated or contained covalently bound heparin (47). The heparin-coated LVAD had minimal thrombus formation (i.e., microthrombi, platelet accumulation) with a thin protein layer after implantation for 3 months as compared to uncoated control LVAD that had gross thrombus formation at 1 month. Heparin activity, which was monitored via chromogenic assay, was not detectable after 15 days, thus suggesting that altering the acute bioactivity of thrombin creates a long-term ''passivated'' flow surface. Despite these promising results, each of these studies has failed to change appreciably the pathobiology of biomaterial failure.

B. Promotion of an Endothelial Cell Lining

Another proposed mechanism to prevent thrombus formation on a graft surface was to develop a uniform layer of endothelial cells across the lumen. Endothelial cells play a pivotal role in mediating blood interaction with the arterial wall. These cells maintain hemostasis and also synthesizing growth mediators that block abnormal smooth muscle cell proliferation. Ideally, prosthetic grafts should promote endothelial cell adherence and growth on the luminal surface while permitting direct host tissue incorporation at the capsular surface. This type of cellular incorporation does not occur in actuality, thereby predisposing these grafts to infection (48,49), thrombosis (50,51), perigraft seromas (52), and delayed graft failure (53). Thus the failure of appropriate cell type growth and development to these biomaterials significantly limits their expanded use (48).

Endothelial cell adhesion to prosthetic grafts using endothelial cell seeding techniques has been extensively employed (54–56). Adhesive proteins such as fibronectin, fibrinogen, vitronectin, and collagen have served well in graft seeding protocols (57,58). The cell attachment properties of these matrices can also be duplicated by short peptide sequences such as RGD (Arg-Gly-Asp) (59). These adhesive proteins, however, have several drawbacks: (1) bacterial pathogens recognize and bind to these sequences (60,61), (2) nonendothelial cell lines also bind to these sequences (56,62), (3) patients requiring a seeded vascular graft have few donor endothelial cells, so that cells must be grown in culture (63), and (4) endothelial cell loss to shear forces remains a significant obstacle (64).

Modification of the surface has also been employed to modify host response to the foreign body, serving as an approach for improving endothelial cell adherence to Dacron. Endothelial cells after seeding have been shown to attach and grow on a variety of protein substrates coated onto vascular graft materials (58,65). Bioactive oligopeptides (66,67) and cell growth factors (68) have been immobilized onto various polymers and demonstrated to affect cell adherence and growth. Additional studies have described the incorporation of growth factors into a degradable protein mesh, resulting in the formation of capillaries into the graft wall (69). Utilizing these techniques to incorporate growth factors, however, does have limitations: (1) growth factor is rapidly released from the matrix, (2) matrix degradation reexposes the thrombogenic surface, such that endothelialization is not uniform, and (3) release of nonendothelial specific growth factor is not confined to the biomaterial matrix, thereby exposing the ''normal'' distal artery to the growth factor.

VI. MODIFICATION OF DACRON: CREATION OF ''ANCHOR'' SITES

The first step in developing a biocompatible surface was to create a surface to which biologically active agents could be covalently attached. Dacron was selected over other biomaterials due to its clinical acceptance, long-term biodurability, and potential for chemical manipulation. Our objective was to modify Dacron so that reactive sites could be created without significantly altering the physical characteristics of the biomaterial (70). Creation of carboxyl and hydroxyl groups on the Dacron surface was achieved via alkaline hydrolysis. Hydrolysis is typically utilized to break ester bonds on the outer periphery of the fiber surface (71). Applying this technique in the textile industry results in a finer, softer fabric with improved soil release due to the increased hydrophilicity of the material. If, however, concentrations of sodium hydroxide as well as temperature and time parameters are increased, extensive cleavage and loss of low molecular weight products occurs, thereby decreasing the strength and weight of the material. Reduction of tensile strength and fiber weight of a Dacron arterial graft could lead to aneurysmal formation due to the constant pressure exerted on a luminal surface that is under intense flow conditions once implanted. Thus it was imperative that creation of carboxyl groups on the Dacron surface via hydrolysis did not significantly modify the physical properties of the fiber.

Dacron segments were cut from a large fabric sheet and washed in a scouring solution containing Na_2CO_3 and Tween 20 detergent for 30 minutes at 60°C. Samples were then rinsed for 30 minutes at 60°C. Dacron segments were then oven dried for 15 minutes at 60°C to remove moisture, weighed (pretest weight) and combined into groups by closeness in weight. These groups were then exposed to the following solutions at 100°C for 30 minutes: (1) distilled H_2O (Control Dacron or CD), (2) 0.5% NaOH, (3) 1.0% NaOH, (4) 2.5% NaOH, and (5) 5.0% NaOH (referred to HD). These pieces were then rinsed with distilled H_2O (room temperature), oven dried for 15 minutes at 60°C, and weighed (posttest weight). Differences between pre and post weights were calculated and averaged (Fig. 2). Examination of the CD and HD for fiber weight loss resulted in the 1.0% (3.1%), 2.5% (11%), and 5.0% (27%) hydrolyzed materials demonstrating significant weight loss as compared to control. The 0.5% HD did not exhibit any significant weight loss.

CD and HD segments, used in determining weight loss, were then tested for tensile strength. An Instron T-1000 Apparatus (Instron Corporation) was calibrated according to manufacturer's specifications in a temperature-controlled environment (room temperature 31°C). Each segment was then placed into two air compressor clamps spaced 1″ apart, and the instrument was returned to baseline. Material stretching was then initiated and terminated upon observation of visible tearing. The pounds of force required to tear untreated and NaOH hydrolyzed materials were determined (Fig. 3). Evaluation of tensile strength of CD and HD materials revealed no significant difference between control, 0.5%, and 1.0% hydrolyzed materials. The 2.5% (20%) and 5.0% (40%) HD, however, had significant strength loss as compared to CD.

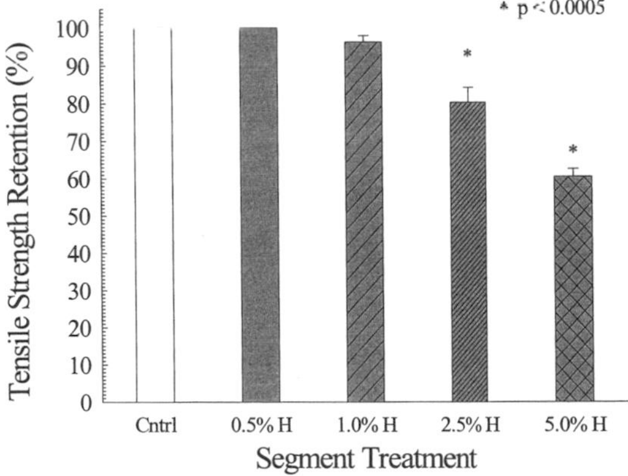

Figure 24.2 Fiber weight loss (%) of variously hydrolyzed Dacron segments versus untreated Dacron. The 1.0% (3.1, $p <$ 0.001), 2.5% (11%, $p < 0.0001$), and 5.0% (27%, $p < 0.0001$) hydrolyzed materials had significant weight loss in comparison to untreated Dacron. The fiber weight loss of the 0.5% hydrolyzed Dacron was not significant compared to untreated Dacron.

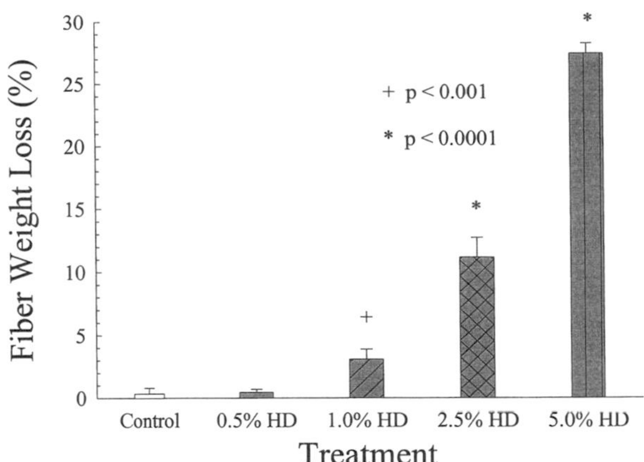

Figure 24.3 Strength retention (%) of variously hydrolyzed Dacron segments versus untreated Dacron. The 2.5% (20%, $p <$ 0.0003) and 5.0% (40%, $p < 0.0001$) hydrolyzed Dacron showed significant strength loss as compared to untreated Dacron, whereas the 0.5% and 1.0% hydrolyzed Dacron were not significantly different in strength.

Thus the 0.5% hydrolyzed Dacron was the only hydrolyzed material in which tensile strength and fiber weight loss were not significantly less than untreated Dacron. No extensive chain scission within the fiber occurred, as was displayed by the close correlation of fiber strength and weight loss (summation of strength retention and weight loss data for each assayed material is approximately equal to 100%). Any evidence of strength loss resulting from the various hydrolysis conditions was solely due to reduction in the surface fiber mass.

VII. SELECTION OF PROTEIN FOR "BASECOAT" LAYER ON DACRON SURFACE

Covalent linkage of a protein to a biomaterial surface in order to create a "basecoat" layer has numerous beneficial advantages. Nonspecific or covalent attachment of a protein coating has been shown to "passivate" a surface that is relatively thrombogenic, thereby decreasing adhesion of blood products such as platelets, red blood cells, and fibrinogen (17,26). Also, protein incorporated as a basecoat layer has been used as a "scaffolding" in order to promote a specific response, such as linkage of RGD peptides to promote cell adhesion (72). From our foundation studies, utilization of a basecoat layer permitted significant amplification of potential binding sites for secondary protein attachment via heterobifunctional cross-linkers (73–75). Thus utilization of a basecoat creates a biomaterial surface with distinct properties for a specific application.

A. Albumin Basecoat for Thromboresistant Surface

The effects of albumin coating on Dacron have been extensively examined both in vitro and in vivo. In vitro, albumin nonspecifically cross-linked around the fibers of Dacron via gluturaldehyde was shown to reduce platelet adhesion and coagulation cascade activation, release of platelet factor 4 and β-thromboglobulin, fibrinopeptide A formation, and leukocyte adhesion (14). In vivo analysis of aortocaval fistulae was performed between preclotted Dacron and Dacron that was immersed into a 25% albumin solution and autoclaved (17). Decreases of 5.2-, 6.0-, and 1.8-fold in red blood cell adhesion, platelet accumulation, and fibrinogen accumulation, respectively, were observed for the albumin-treated Dacron as compared to preclotted Dacron. Thus albumin was selected as the protein to bind covalently to HD due to the numerous beneficial results demonstrated in these early studies.

B. Type IV Collagen as Basecoat for Cellular Adhesion

In selecting a basecoat protein that would promote cellular adhesion, the composition of native blood vessels was examined. All mammalian epithelia are separated at their interface by specialized extracellular matrix proteins (basement membrane) composed predominantly of Type IV collagen networks with minor components of elastin, fibrillins, thrombospondins, laminin, fibronectin, and glycoaminoglycans. Endothelial cell growth and differentiation (morphogenesis) has an absolute requirement of this highly structured extracellular matrix environment for normal cell growth and differentiation (76). Additionally, matrix proteins have been shown to be crucial in binding signaling proteins such as fibroblast growth factor (FGF) and transforming growth factor-b (TGF-b) (77).

Type IV collagen has several advantages over albumin when considering basecoat proteins for a vascular graft surface. Type IV collagen is a basement membrane protein that is directly involved in cell adhesion as well as regulation of cell function (78). Conversely, albumin is a serum protein that has been shown to have beneficial results on protein adhesion when bound to a graft surface. However, albumin has no cell adhesion properties, an area that may be necessary for neointimal formation. Similar to albumin, Type IV collagen has been shown to provide numerous cross-linking sites suitable for covalent attachment of biologically active agents. Therefore Type IV collagen is a better candidate than albumin to serve as the basecoat protein for cellular adhesion studies.

VIII. RATIONALE FOR SELECTION OF ANTITHROMBIN AND MITOGENIC AGENTS

A. Recombinant Hirudin (rHir)

Thrombin, a pivotal enzyme in the blood coagulation cascade, is the primary agent responsible for thrombus formation. The principal function of thrombin is the cleavage of fibrinogen to fibrin (8). Additionally, thrombin also functions as a smooth muscle cell mitogen (79), is chemotactic for monocytes and neutrophils (80,81), and is an aggregator of lymphocytes (81). This enzyme has also been shown to bind to endothelial cells, inducing the release of platelet-derived growth factor (PDGF)-like growth factors (82) and has been shown to be a potent platelet aggregant, stimulating the release of platelet factors (83). Thus thrombin, beyond its role in clot formation, has tremendous secondary effects that induce inflammation at the site of synthesis and enhance cellular proliferation or hyperplasia by various proposed activation mechanisms, all of which are beneficial in wound healing but are extremely deleterious to prosthetic graft patency.

rHir, a 6,965 Da recombinant protein synthesized from the leech protein hirudin, is the most potent specific inhibitor of thrombin (84). rHir has demonstrated inhibition of thrombin's enzymatic, chemotactic, and mitogenic properties (8,85,86). rHir has also been shown to have potent antithrombin activity when covalently immobilized onto a Dacron surface (75) or to another biomolecule (87–89). This leech anticoagulant holds several advantages over heparin: (1) rHir inhibits thrombin directly, whereas heparin requires antithrombin III, (2) heparin enhances platelet aggregation (90), (3) rHir inhibits the uptake of thrombin into fibrin clots (91,92), and (4) heparin is regulated by platelet function (93). Thus rHir is an attractive agent for covalent attachment in order to reduce or eliminate the pseudointima formation on graft surface. Further, inhibition of thrombin via surface bound rHir may prevent smooth muscle cell proliferation and matrix production at the anastomoses.

B. Recombinant Vascular Endothelial Growth Factor (VEGF)

The rationale for selecting a growth factor in order to promote cellular migration and proliferation was based upon the specificity of the growth factor for endothelial cells. VEGF, a 38 kDa homodimeric glycoprotein, has been shown to be a potent endothelial cell mitogen and vasopermeability factor (94–96). VEGF is produced by many different cell types both in tissue culture and in vivo. VEGF binds to plasma membrane receptors on endothelial cells only with an extracellular transmembrane glycoprotein linked to an intracellular tyrosine kinase domain (94); thus VEGF is an endothelial-specific mitogen. These cell-specific, high-affinity receptors have structural similarities to both forms of the platelet-derived growth factor (PDGF) receptors (97). VEGF has been subsequently found in many tumors related to vascular angiogenesis (98), is responsible for vascular proliferation in the presence of hypoxia seen in diabetic retinopathy (99), and is shown to be involved in the normal development of the vasculature (100).

In vitro, VEGF expression can be increased by several mechanisms including stimulation by growth factors [basic FGF, TGF-b, and PDGF (101)] and the cytokines tumor necrosis factor-α and interleukin-1-β (102). VEGF mRNA expression is also increased by hypoxia in numerous cell lines, including endothelial cell and vascular smooth muscle cells (103). In addition, vascular growth factors PDGF and bFGF have synergistic effects on the induction of VEGF at moderate levels of hypoxia in vascular smooth muscle cells (104). Hypoxic induction of VEGF has been postulated to account for sustaining angiogenesis in tumor growth and in the induction of ischemic retinopathies such as diabetic retinopathy and central retinal vein occlusion (99).

In vivo, VEGF infused intra-arterially has been shown to improve collateral flow in a hindlimb model of ischemia (105) and mitigates the development of intimal hyperplasia observed after balloon catheter–induced arterial injury (106). The beneficial effect of VEGF following balloon denudation injury appears to result from an earlier, more extensive and more histologically normal reendothelialization of the vessel. Those portions of the vessel that were repopulated by endothelial cells had little or no neointimal growth (106). Of note, studies by Banai and Ibukiyama demonstrated the efficacy of human VEGF at stimulating proliferation of canine endothelial cells (107,108), an important factor for future implant studies. In autologous vein grafts, VEGF mRNA expression has been shown to increase 2.5-fold 48 hours after bypass grafting as compared to control veins (109). This may, in part, explain the superior patency rates achieved with vein grafts over prosthetic arterial grafts in small vessel arterial reconstruction.

Thus VEGF is specific for endothelial cell activation and proliferation, resulting in endothelial cell infiltration in several clinical conditions. These characteristics establish VEGF as an ideal candidate for covalent linkage to a vascular graft in an attempt to facilitate neointima formation throughout the luminal surface of the conduit.

IX. COVALENT BINDING OF rHIR TO THE DACRON SURFACE: IN VITRO AND IN VIVO RESULTS

A. Covalent Linkage of ^{125}I-BSA with Various Sulfo-SMCC Ratios to HD Surfaces Via EDC Cross-linker

The second phase of the surface modification studies was to determine if a basecoat protein with binding sites could be linked to the surface (75). HD segments were placed in an EDC solution and reacted for 30 minutes at room temperature. CD segments were placed into 100% ethanol without EDC cross-linker and reacted under the same conditions. Concurrently, seven ^{125}I-BSA solutions were prepared in PBS. One sample was utilized to establish the maximum binding control for this experiment as defined in our previous study (70), the incubation of HD segments with EDC followed by the addition of ^{125}I-BSA with no sulfo-SMCC cross-linker. The other six samples were reacted with sulfo-SMCC in molar ratios of 1:5, 1:10, 1:20, 1:30, 1:40, and 1:50 for 20 minutes at 37°C and purified via gel filtration. The final concentration of the ^{125}I-BSA and ^{125}I-BSA-SMCC samples was adjusted to 14.8 μM. CD and HD segments were then removed from their respective solutions after 30 minutes, shaken to remove excess ethanol/cross-linker, and incubated with the ^{125}I-BSA/^{125}I-BSA-SMCC solutions for 2 hours at room temperature on the inversion mixer. Dacron segments were then individually sonicated in PBS with 0.05% Tween 20. This procedure was repeated three times, changing the wash buffer between sonications to remove any weakly adherent ^{125}I-BSA on the surface of each segment. The segments with nonspecifically bound (CD + ^{125}I-BSA and CD + ^{125}I-BSA-SMCC) and covalently bound (HD-^{125}I-BSA and HD-^{125}I-BSA-SMCC) ^{125}I-BSA were then gamma counted, and the amount of ^{125}I-BSA bound (ng) per mg Dacron was determined.

The amount of ^{125}I-BSA-SMCC covalently bound to HD segments via EDC was not significantly altered by any of the sulfo-SMCC ratios examined as compared to HD that was incubated with only EDC and ^{125}I-BSA (Fig. 4). The amount of ^{125}I-BSA covalently bound to the HD segments via EDC was significantly greater (6.7-fold difference in the highest ^{125}I-BSA : sulfo-SMCC ratios to 11.4-fold difference for no cross-linker reacted with the ^{125}I-BSA) than the amount of ^{125}I-BSA or ^{125}I-BSA-SMCC complex nonspecifically bound to CD segments. The amount of nonspecifically bound ^{125}I-BSA to HD segments that were not incubated with EDC (data not shown) was slightly greater than the amount of ^{125}I-BSA nonspecifically bound to the CD segments.

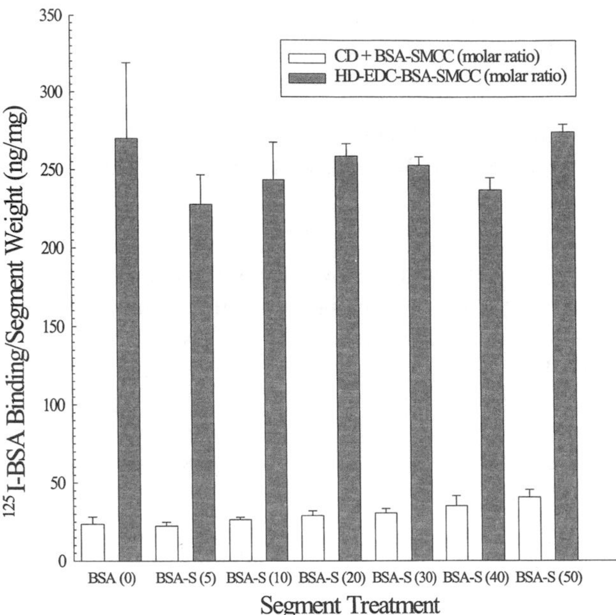

Figure 24.4 ^{125}I-BSA and ^{125}I-BSA-SMCC (1:5–1:50 ratios) binding to HD in the absence or presence of EDC. The amount of ^{125}I-BSA covalently bound to HD segments via EDC was not significantly altered by any of the sulfo-SMCC ratios examined.

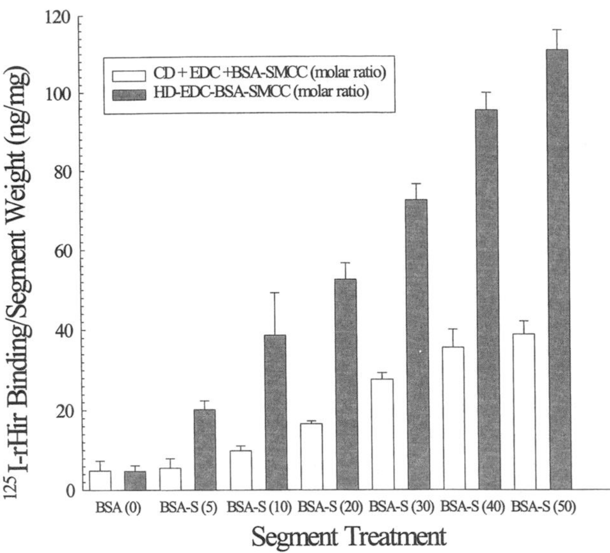

Figure 24.5 ^{125}I-rHir binding versus HD-BSA-SMCC with various SMCC ratios. The amount of ^{125}I-rHir-SH covalently bound to HD-BSA-SMCC segments increased with increasing sulfo-SMCC cross-linker.

B. Covalent Linkage ^{125}I-rHir to Dacron-BSA-SMCC Segments

In a separate series of experiments, the CD and HD segments were prepared without radiolabeling BSA in order to determine the amount of ^{125}I-rHir either nonspecifically or covalently bound to the surface. CD and HD segments, which were not exposed to BSA, also served as controls. ^{125}I-rHir was reacted in a 1:2 molar ratio with Traut's reagent for 20 minutes at 37°C in order to create sulfhydryl groups on rHir and purified via gel filtration. This ^{125}I-rHir-SH compound was then reacted with the modified Dacron segments overnight at 4°C. The segments were then washed and sonicated as described above. The nonspecifically bound ^{125}I-rHir segments (CD + ^{125}I-rHir-SH, HD + ^{125}I-rHir-SH, CD + BSA + ^{125}I-rHir-SH, and HD-BSA + ^{125}I-rHir-SH) and covalently linked ^{125}I-rHir segments (HD-BSA-SMCC-S-^{125}I-rHir with various sulfo-SMCC ratios) were then gamma counted in order to determine the amount of ^{125}I-rHir bound (ng) per mg Dacron.

The amount of ^{125}I-rHir-SH covalently bound to the HD-BSA-SMCC segments increased with increasing sulfo-SMCC cross-linker (20, 39, 53, 77, 95, and 111 ng rHir/mg Dacron) (Fig. 5). These ng amounts of ^{125}I-rHir corresponded to 0.23, 0.45, 0.61, 0.89, 1.1, and 1.3 antithrombin units (ATU)/mg) Dacron, with 1 ATU defined as the amount of rHir necessary to inhibit 1 NIHU of thrombin. Conversely, CD (not shown), CD + BSA, and HD-BSA segments incubated with the ^{125}I-rHir-SH bound minimal but comparable amounts of the inhibitor (5 ng or 0.06 ATU/mg Dacron). Covalently bound ^{125}I-rHir on the HD-BSA-SMCC-S-^{125}I-rHir segments resulted in 4-, 8-, 11-, 16-, 20-, and 22-fold greater ^{125}I-rHir binding as compared to control segments. The HD segments without BSA bound to the surface (not shown) had 1.7-fold greater (9 ng or 0.10 ATU/mg Dacron) nonspecific ^{125}I-rHir-SH binding than the other controls. Thus completion of these experiments resulted in a Dacron surface with a significant amount of ^{125}I-rHir bound to the surface. The next step was to determine which surface possessed optimum antithrombin activity.

C. Initial Determination of Antithrombin Activity by Modified Dacron Surfaces

Determination of thrombin inhibition by these modified Dacron segments was then assayed using the chromogenic substrate for thrombin S-2238, a synthetically derived peptide that possesses an affinity for thrombin comparable to that of the natural substrate fibrinogen. Upon cleavage of S-2238 by thrombin, the chromophore para-nitro aniline (4-nitroaniline) or pNA is released, giving rise to color. Thus color change is directly proportional to proteolytic activity of thrombin.

CD + BSA + ^{125}I-rHir-SH, HD-BSA + ^{125}I-rHir-SH or HD-BSA-SMCC-S-^{125}I-rHir (sulfo-SMCC ratios of 1:5 to 1:50) segments were placed flat into the bottom of 1 cm cuvettes. Thrombin (0.2 NIHU) was then added and equilibrated for 5 minutes at 37°C in a Beckman spectrophotometer. The thrombin assay was initiated by the addition of 1 mL of 50 μM S-2238 and the change in absorbance per minute monitored at 15-second intervals for 10 minutes at 410 nm, thus measuring residual proteolytic activity of thrombin. Using this data, the amount of thrombin inhibited (NIHU) by surface-bound rHir was determined. These segments were then rinsed with PBS with 0.05% Tween 20, placed into a new cuvette, and rechallenged with another 0.2 NIHU thrombin.

The CD + BSA + rHir-SH and HD-BSA + rHir-SH controls inhibited 0.022 and 0.046 NIHU thrombin, respectively, upon the initial 0.2 NIHU thrombin challenge. These control segments had minimal remaining antithrombin activity (0.010 and 0.014 NIHU, respectively) when rechallenged with 0.2 NIHU, Conversely, the HD-BSA-SMCC-S-rHir samples that contained various sulfo-SMCC ratios (1:5, 1:10, 1:20, 1:30, 1:40, and 1:50) had substantial antithrombin activity (0.134, 0.142, 0.170, 0.172, 0.176, and 0.186 NIHU) with the 1:50 HD-BSA-SMCC-S-rHir segments having the greatest antithrombin activity. Upon rechallenging these segments with 0.2 NIHU, thrombin inhibition occurred but decreased in all segments (0.106, 0.130, 0.150, 0.158, 0.166, and 0.172 NIHU). Overall, the 1:50 segments possessed the best antithrombin activity and were further evaluated.

D. Characterization of Antithrombin Activity of the 1:50 HD-BSA-SMCC-S-rHir Segments

CD + BSA + rHir-SH, HD-BSA + rHir-SH and the 1:50 HD-BSA-SMCC-S-rHir segments were prepared as outlined above and evaluated for ^{125}I-thrombin inhibition versus thrombin concentration and incubation time. ^{125}I-thrombin was prepared and graft segments were challenged against 0.66, 2.80, 6.12, 10.10, 22.40, and 43.50 NIHU of ^{125}I-thrombin. Test segments were reacted with ^{125}I-thrombin for 5, 15, 30, or 60 minutes using a 37°C water bath shaker and transferred to the spectrophotometer for a 5-minute equilibration period. An increased concentration of chromogenic substrate (1 mL of 100 μM S-2238) was utilized to initiate the assay in order to compensate for increased ^{125}I-thrombin concentrations.

The maximum ^{125}I-thrombin inhibited by either of the control segments (CD + BSA + rHir-SH and HD-BSA + rHir-SH) was 1.4 NIHU in the 2.80 NIHU challenge (Table 1). ^{125}I-thrombin inhibition in the lower concentrations increased with increasing incubation times. ^{125}I-thrombin inhibition by the control segments reacted with 6.12, 10.10, 22.40, or 43.50 NIHU samples was below the detectable limits of the assay. In the higher ^{125}I-thrombin concentrations (6.12–43.50 NIHU), increasing incubation time did not alter the results.

In contrast, the 1:50 HD-BSA-SMCC-S-rHir segments showed significant thrombin inhibition over the 0.66–22.40 NIHU ^{125}I-thrombin concentrations assayed with the inhibition greatest at the highest incubation times (30 and 60 minutes). The maximum ^{125}I-thrombin inhibited by these segments was 20.43 NIHU, 14.6-fold greater than control segments. In the 5- and 15-minute incubation times for 22.4 NIHU challenge as well as for all the incubation times at 43.50 NIHU, accurate assessment of ^{125}I thrombin inhibition could not be determined due to saturation of the chromogenic assay with uninhibited ^{125}I-thrombin.

After assaying the CD + BSA + rHir-SH, HD-BSA + rHir-SH, or HD-BSA-SMCC-S-rHir segments for ^{125}I-thrombin inhibition, segments were removed from the cuvette, dipped into PBS with 0.05% Tween 20 in order to remove any weakly adherent ^{125}I-thrombin, and gamma counted. Samples then underwent a more vigorous washing for 10 minutes in PBS with 0.05% Tween 20 on an inversion mixer and again gamma counted. The amount of ^{125}I-thrombin remaining after washing was ^{125}I-thrombin specifically bound to the material and the difference in ^{125}I-thrombin binding between pre- and postwash samples was ^{125}I-thrombin released.

^{125}I-thrombin binding to HD-BSA-SMCC-S-rHir segments was 5.4-, 5.2-, 4.8-, 6.5-, and 8.3-fold greater than the CD + BSA + rHir-SH and HD-BSA + rHir-SH control segments at their respective ^{125}I-thrombin concentration. ^{125}I-thrombin binding to control segments was independent of the ^{125}I-thrombin concentration applied, whereas binding to the HD-BSA-SMCC-S-rHir segments directly related to the ^{125}I-thrombin concentrations applied between 0.66 and 22.4 NIHU. The maximum amount of ^{125}I-thrombin specifically bound to this surface was 20.43 NIHU. Increasing the ^{125}I-thrombin concentration to 43.5 NIHU yielded no further increase in binding, exhibited by an overlap of the 22.4 and 43.5 NIHU curves, suggesting saturation of available rHir surface binding sites.

Nonspecific binding to and release from the control segments was significantly greater than the HD-BSA-SMCC-S-rHir segments. The percent ^{125}I-thrombin released from the controls after washing ranged from 26 to 62% over the various ^{125}I-thrombin concentrations assayed. In contrast, ^{125}I-thrombin release from the 1:50 segments was significantly lower, ranging from 1 to 12% across the ^{125}I-thrombin concentrations assayed, and was independent of incubation time. These data demonstrated the highly spe-

Table 24.1 ^{125}I-Thrombin Inhibition (NIHU) Versus Incubation Time (Minutes) and ^{125}I-Thrombin Applied (NIHU) by CD + BSA + rHir-SH, HD-BSA + rHir-SH and HD-BSA-SMCC-S-rHir Segments

CD + BSA + rHir-SH	Thrombin inhibition (NIHU)			
Thrombin applied (NIHU)	5 minutes	15 minutes	30 minutes	60 minutes
0.66	0.19 ± 0.13	0.16 ± 0.14	0.11 ± 0.10	0.35 ± 0.09
2.80	0.84 ± 0.44	0.60 ± 0.60	1.36 ± 0.92	1.36 ± 0.92
6.12	0.0 ± 0.0	0.0 ± 0.0	0.0 ± 0.0	0.0 ± 0.0
10.10	0.0 ± 0.0	0.0 ± 0.0	0.0 ± 0.0	0.0 ± 0.0
22.40	0.0 ± 0.0	0.0 ± 0.0	0.0 ± 0.0	0.0 ± 0.0
HD-BSA + rHir-SH	**Thrombin inhibition (NIHU)**			
Thrombin applied (NIHU)	5 minutes	15 minutes	30 minutes	60 minutes
0.66	0.22 ± 0.06	0.11 ± 0.12	0.16 ± 0.12	0.53 ± 0.50
2.80	0.28 ± 0.24	0.24 ± 0.44	0.40 ± 0.72	0.80 ± 0.92
6.12	0.0 ± 0.0	0.0 ± 0.0	0.0 ± 0.0	0.0 ± 0.0
10.10	0.0 ± 0.0	0.0 ± 0.0	0.0 ± 0.0	0.0 ± 0.0
22.40	0.0 ± 0.0	0.0 ± 0.0	0.0 ± 0.0	0.0 ± 0.0
HD-BSA-SMCC-S-rHir	**Thrombin inhibition (NIHU)**			
Thrombin applied (NIHU)	5 minutes	15 minutes	30 minutes	60 minutes
0.66	0.58 ± 0.03	0.64 ± 0.00	0.62 ± 0.005	0.66 ± 0.01
2.80	2.29 ± 0.12	2.35 ± 0.12	2.58 ± 0.12	2.78 ± 0.04
6.12	4.85 ± 0.24	5.49 ± 0.08	5.71 ± 0.40	5.97 ± 0.08
10.10	6.78 ± 0.48	9.51 ± 0.80	9.19 ± 0.16	9.49 ± 0.16
22.40	0.0 ± 0.0	0.0 ± 0.0	17.01 ± 0.96	20.43 ± 0.34

cific nature of ^{125}I-thrombin binding to surface immobilized rHir in comparison to the control segments in which ^{125}I-thrombin was easily removed.

E. In Vitro Flow Study

The in vitro pulsatile flow model was employed to determine structural stability of surface bound ^{125}I-rHir and ^{131}I-BSA on Dacron grafts (110). Both radiolabeled proteins were covalently linked to Cooley® Woven Dacron vascular grafts (3 cm × 0.6 cm). The grafts were perfused at a mean flow rate of 400 mL/min. Grafts were gamma counted at various times over seven days in order to determine pressure head to provide steady flow from the perfusate reservoir, through the heating coil that is used to maintain perfusate flow at 37°C, to the graft segment. Perfusate, which was 400 mL of a 5% albumin/PBS solution containing 200 units/mL penicillin, 0.2 mg/mL streptomycin, and ^{131}I-thrombin (0.5–0.8 NIHU/mL), ^{125}I-rHir and ^{131}I-BSA loss from the graft surface.

Using pulsatile flow to assess structural stability of surface bound ^{125}I-rHir and ^{131}I-BSA at high-flow and high-shear arterial flow conditions, without ^{131}I-thrombin present, resulted in 49.1% of the ^{125}I-rHir and 21.6% of the ^{131}I-BSA being removed from the surface after 7 days, with a majority of the protein release occurring after day 3. With a mean initial ^{125}I-rHir density of 51.8 ATU/cm^2, 26.4 ATU/cm^2 still remained covalently attached to the surface. Subjecting the ^{125}I-rHir and ^{125}I-BSA surfaces to constant flow did not significantly alter the total amount of protein released (data not shown).

The pulsatile flow system was then converted to a constant flow system, with only slight modifications, in order to evaluate ^{131}I-thrombin interaction with each of these surfaces. The constant flow perfusion system (flow rate = 400 mL/min) employs a gravity then flowed through a Cooley® Woven Dacron vascular graft (3 cm × 0.6 cm) with either covalently bound ^{125}I-rHir or ^{125}I-BSA. These grafts were in the same inert encasement chamber as described earlier. Recirculation of the perfusate in the system was maintained via peristaltic pump. A second fluid reservoir was employed to isolate the test graft from the significant pressure and flow oscillations generated by the peristaltic pump. Grafts were gamma counted at various time intervals for 27 hours in order to determine the stability of ^{125}I-rHir and ^{125}I-BSA on the surface and the interaction of ^{131}I-thrombin with each surface. Concurrently, 2 mL of the perfusate was drawn and assayed for ^{131}I-thrombin activity using the chromogenic substrate assay previously described.

The effects of ^{131}I-thrombin circulating under constant flow on surface-bound ^{125}I-rHir and ^{125}I-BSA were then evaluated (Table 2). Dacron grafts with covalently linked ^{125}I-rHir on the surface bound significantly greater amounts of ^{131}I-thrombin from the perfusate (23.8 NIHU) versus ^{125}I-BSA coated grafts (3.9 NIHU). The ^{125}I-rHir coated grafts were also more effective at inhibiting ^{131}I-thrombin during the perfusate period (125 NIHU) versus the ^{125}I-BSA coated grafts (3 NIHU). ^{125}I-BSA removal from the graft surface was relatively similar in both the presence and the absence of ^{131}I-thrombin. Conversely, ^{125}I-rHir removal from the surface within the 27 hour perfusion period (36.4%) when exposed to ^{131}I-thrombin was slightly less than the total ^{125}I-rHir loss over the 7 day perfusion (49.1%).

A potential mechanism for increased ^{125}I-rHir removal in the presence of ^{131}I thrombin may be a greater propensity for ^{125}I-rHir to dissociate following ^{131}I-thrombin binding due to the conformational changes that ^{175}I-rHir undergoes upon binding. These changes may, in turn, reduce the stability of the sulfur bond between the Traut's and the sulfo-SMCC cross-linkers, thereby enhancing the release of ^{125}I-rHir. The reduced stability may also be a result of the increased fluid drag forces associated with thrombin binding, which increases the mass of bound protein from 6.9 kDa to 43.5 kDa (3.4 fold). Overall, the ^{125}I-rHir coated grafts significantly inhibited and bound greater amounts of ^{131}I-thrombin from the perfusate. Even though ^{125}I-rHir was lost from the graft, the surface still possessed significant antithrombin activity after removal from the flow system.

F. In Vivo Assessment of a Dacron-BSA-SMCC-S-^{125}I-rHir Surface in a Thoracic Aorta Patch Model

In order to evaluate if a surface with covalently bound ^{125}I-rHir warranted long-term in vivo evaluation, an in vivo model was developed that would subject surface bound ^{125}I-rHir to increased thrombus and shear force challenges within a relatively short time period (111). The thoracic aorta patch model is a nonheparinized high-flow and high-shear stress model that poses such a severe challenge to the ^{125}I-rHir surface. ^{125}I-rHir (test) and BSA (control) were covalently bound onto the surfaces of 3 cm × 2 cm Dacron patches. Prior to implantation, Dacron-BSA-SMCC-S-^{125}I-rHir patches were gamma counted to determine the initial amount of ^{125}I-rHir on each patch. For each animal, an unimplanted control containing covalently bound ^{125}I-rHir was prepared in order to compare the postexplantation antithrombin activity of the patch.

Canines were tranquilized, intubated, and placed on halothane. A left thoracotomy was performed in order to expose the thoracic aorta. The aorta was then loosely dissected and major side branches ligated. A Satinski partial occluding vascular clamp was used to occlude a 2 cm × 4 cm segment of the thoracic aorta. At this point, a 3 cm × 2 cm Dacron patch containing either covalently bound BSA or ^{125}I-rHir was sutured to the artery using 3-0 Prolene. The clamp was then removed and the patch area packed until bleeding was controlled. The clamp was then placed distally and the remaining patch was implanted.

Table 24.2 ^{125}I-rHir or ^{125}I-BSA Loss From the Graft Surface, ^{131}I-Thrombin Binding to the Dacron Graft and ^{131}I-Thrombin Inactivation in Perfusate Under Constant Flow Conditions

	^{125}I-rHir grafts (n = 6)	^{125}I-BSA grafts (n = 3)	*p* value
Graft weight (mg)	200.0 ± 6.3	188.8 ± 7.2	
Initial ^{125}I-rHir concentration (ATU/cm^2)	38.2 ± 1.8	—	
Final ^{125}I-rHir concentration (ATU/cm^2)	28.7 ± 4.7	—	
Initial ^{125}I-rHir concentration (μg/cm^2)	—	7.4 ± 0.9	
Final ^{125}I-rHir concentration (μg/cm^2)	—	7.2 ± 0.7	
^{125}I-rHir or ^{125}I-BSA loss from surface (%)	36.4 ± 5.4	1.9 ± 2.4	$p < 0.01$
^{131}I-thrombin bound to graft (NIHU/cm^2)	3.08 ± 0.61	0.64 ± 0.04	$p < 0.01$
Total ^{131}I-thrombin bound to graft (NIHU)	23.8 ± 2.6	3.9 ± 0.3	$p < 0.001$
Total ^{131}I-thrombin inactivation in perfusate (NIHU)	125 ± 8	3 ± 14	$p < 0.005$

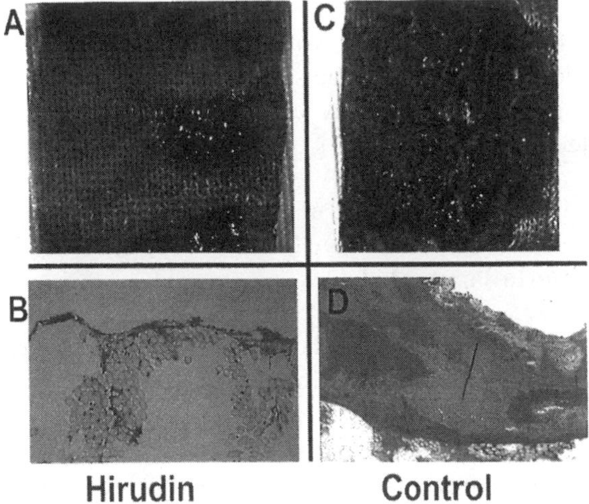

Figure 24.6 Macroscopic evaluation of the ^{125}I-rHir patches after explantation (A); the patches had virtually no gross thrombus formation on the luminal flow surface. Control BSA patches had a dense carpet of thrombus that lined the entire flow surface (C). Microscopic assessment of these patches confirmed these results. The ^{125}I-rHir surface had primarily a thin layer of platelets sealing the pores of the patches with minimal fibrin formation (B). Control BSA surfaces had dense, well-formed thrombus composed of fibrin and platelets (D).

After the second patch was implanted, blood was permitted to flow across the patch surfaces for 2 hours. After two hours, the aorta proximal and distal to both patches was clamped and the patches were retrieved. The ^{125}I-rHir patches were then gamma counted. Both patches were cut in half: one half was placed into formalin for routine histology and the other half was placed into saline. The saline patch was assayed for residual antithrombin activity via the chromogenic assay previously described. Both unimplanted and explanted patch halves were cut into 1 cm^2 segments and challenged with thrombin concentrations of 1, 5, 10, and 15 NIHU for 1 hour at 37°C.

Upon macroscopic evaluation of the ^{125}I-rHir patches after explantation, the patches had virtually no gross thrombus formation on the luminal flow surface (Fig. 6A). Some thrombus was observed near the suture site. In contrast, the control BSA patches had dense carpets of thrombus that lined the entire flow surface (Fig. 6C). These results were confirmed by routine histology that showed the ^{125}I-rHir surface had primarily a thin layer of platelets sealing the pores of the patches with minimal fibrin formation (Fig. 6B). The control BSA surfaces had dense, well-formed thrombus composed of fibrin and platelets (Fig. 6D). The magnification used to identify the presence of fibrin on the surface of the ^{125}I-rHir patches was significantly greater than the magnification used on the control BSA patches.

Covalently bound ^{125}I-rHir was released (20% ± 6.7, $n = 3$) from the patch surface subjected to 2 hours of this nonheparinized high-flow and high-shear thoracic aorta model. The ^{125}I-rHir loss is comparable to the results seen in the in vitro flow studies. Evaluation of the ^{125}I-rHir patches after explantation for antithrombin activity using the in vitro chromogenic assay resulted in 7 NIHU of thrombin inhibited in the 10 NIHU challenge that was incubated for 1 hour. The 15 NIHU challenge did not have any ''detectable'' thrombin inhibition. In contrast, the unimplanted ^{125}I-rHir patch segments had significant antithrombin activity throughout all thrombin concentrations assayed. Overall, the ^{125}I-rHir patches had platelet accumulation with minimal fibrin formation, whereas the BSA-coated patches had significant thrombus formation. ^{125}I-rHir was released from the patch, but significant antithrombin activity remained on the surface. Thus covalently bound ^{125}I-rHir had a significant impact on the interaction between the nonheparinized blood and the Dacron surface.

X. COVALENT BINDING OF VEGF TO DACRON: IN VITRO AND IN VIVO MITOGENIC PROPERTIES

A. Creation of Soluble VEGF Conjugate for Biological Activity Studies

Canine serum albumin (CSA) was utilized in the soluble conjugate studies in order to determine if covalent linkage of VEGF to another biomolecule would alter VEGF biologic activity (112). CSA, which has no cell adhesion properties, was selected for these studies over collagen in order to examine *only* the effects of modified VEGF. Creation of a soluble CSA-^{125}I-VEGF conjugate required the use of both sulfo-SMCC and Traut's reagent. CSA was reacted with sulfo-SMCC in a 1:50 molar ratio, and ^{125}I-VEGF was reacted with Traut's reagent in a 1:60 molar ratio (^{125}I-VEGF-SH). Both reactions were incubated for 20 minutes at 37°C with occasional mixing and purified via gel filtration. The two intermediate compounds were then mixed, incubated overnight at 4°C on an inversion mixer, and concentrated to 100 μL using a Filtron 10K concentrator. The CSA-SMCC-S-^{125}I-VEGF conjugate was then purified from the intermediate CSA-SMCC and ^{125}I-VEGF-SH compounds utilizing a Superdex 75 HR10/30 gel filtration column. Fractions were collected and gamma counted in order to determine which peak fractions contained ^{125}I-VEGF.

The CSA-SMCC-S-^{125}I-VEGF conjugate eluted from the column as a broad peak at 15.8 minutes, with a

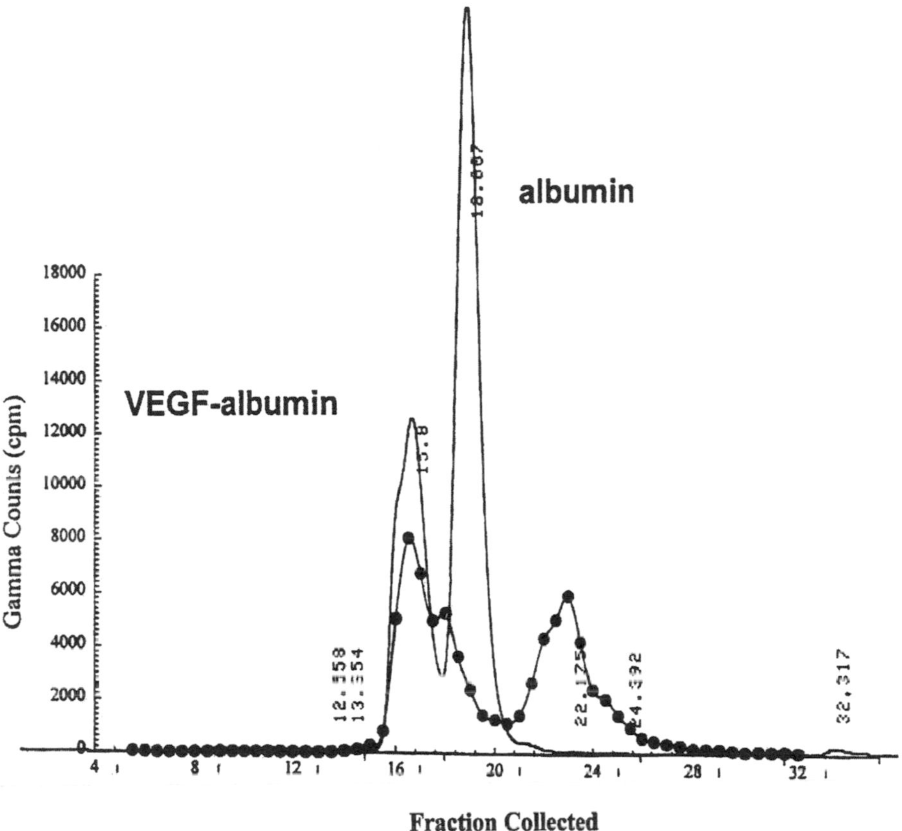

Figure 24.7 Profile of CSA-SMCC-S-^{125}I-VEGF conjugate purification using a Superdex 75 HR 10/30 gel filtration column. The CSA-SMCC-S-^{125}I-VEGF conjugate eluted from the column as a broad peak at 15.8 minutes, with a shoulder off the front portion of the peak. A shift in the ^{125}I-VEGF peak confirmed formation of the complex.

shoulder off the front portion of the peak (Fig. 7). The broadness of the peak suggests that multiple ratios of ^{125}I-VEGF to CSA (i.e., 1:1, 2:1) were created. The CSA-SMCC (67 kDa) peak, which eluted at 18 minutes, was comparable to the CSA standard (data not shown), which eluted at 18.2 minutes. The unbound ^{125}I-VEGF-SH peak (38 kDa), due to concentration, showed as a minor peak at 22.2 minutes. This retention time was in specific relation to the ovalbumin standard (43 kDa, data not shown), which eluted earlier at 20.4 minutes due to a 5 kDa difference in weight. Gamma counting the elution fractions confirmed that ^{125}I-VEGF linkage to CSA did occur, as shown by the shift in the ^{125}I-VEGF-SH peak from 38 kDa (fractions 22 to 24) to greater than 100 kDa (fractions 16 to 18). The total counts from this peak represented 54% of the applied ^{125}I-VEGF. The free ^{125}I-VEGF peak also correlated with the 22.2-minute absorbance peak. Thus VEGF could be covalently bound to another biomolecule using the described cross-linking techniques.

B. Comparison of Mitogenic Activity Between VEGF, VEGF-SH, and CSA-VEGF

The mitogenic activity of complex bound VEGF was then assessed. Bovine aortic endothelial cells (BAEC, passages = 1) were plated and grown to confluence in 10 cm dishes overnight in DMEM containing 10% calf serum. The media was changed to DMEM plus 2% calf serum for 24 hours prior to stimulation. The BAECs were stimulated for 5 minutes with 25 ng/mL of VEGF, VEGF-SH, or VEGF-CSA. Another well of BAECs was stimulated with 250 ng/mL VEGF. BAECs cultured without VEGF served as the negative control. The BAECs were then lysed in Laemmli buffer and boiled for 5 minutes. The lysates were centrifuged in an Eppendorf centrifuge at 14,000 rpm for 10 minutes. Aliquots of each cell extract, which had equivalent protein concentrations, were then loaded onto a 10% SDS polyacrylamide gel and separated. The separated proteins on the gel were then transferred to a nitrocellulose filter paper. Nonspecific binding of the antiphosphotyrosine anti-

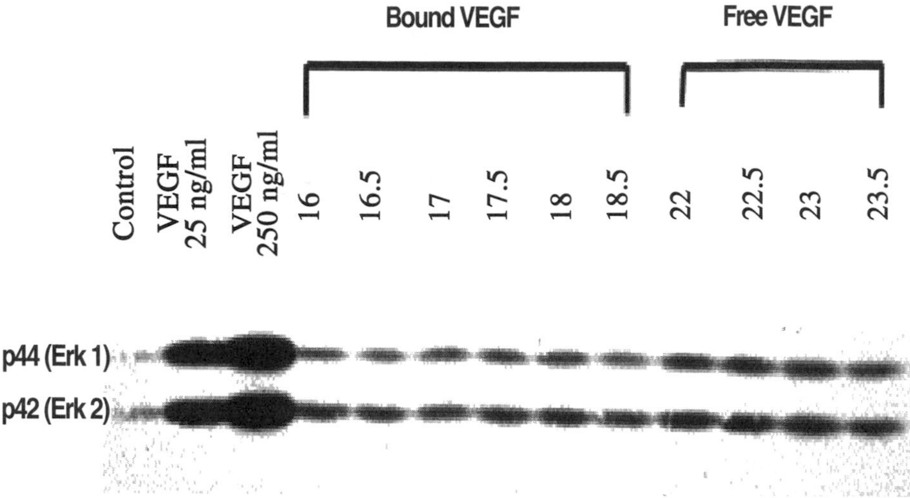

Figure 24.8 Determination of mitogenic activity using a MAP-kinase assay. BAECs exposed to PBS had minimal phosphorylation. In contrast, VEGF covalently linked to BSA (fractions 16–18.5) bound to and activated the transmembrane VEGF receptor as shown by intracellular phosphorylation by Erk 1 and 2 (MAP-kinases). VEGF, which was modified with Traut's reagent (fractions 22–23.5), also stimulated phosphorylation.

body was blocked via filter preincubation for 1 hour at room temperature using a blocking buffer that contained casein and Tween-20. The nitrocellulose blot was then incubated with the antiphosphotyrosine antibody (1 μg/mL), which was diluted using the blocking buffer, for 1 hour at room temperature. The bound antibodies were detected using an ECL kit in conjunction with a phosphoimager.

BAECs exposed to PBS had minimal phosphorylation (Fig. 8). In contrast, VEGF covalently linked to BSA (fractions 16–18.5) bound to and activated the transmembrane VEGF receptor as shown by intracellular phosphorylation by Erk 1 and 2 (MAP-kinases), a mechanism that is indicative of intracellular signaling activation. VEGF, which was modified with Traut's reagent (fractions 22–23.5), also stimulated phosphorylation comparable to the covalently bound VEGF. Native VEGF, the positive control, stimulated phosphorylation to a greater degree than both VEGF-SH and VEGF-CSA, possibly due to either lower concentrations of VEGF-SH/VEGF-CSA than calculated or a reduction in the mitogenic properties due to a modification of the VEGF.

C. Comparison of Migration Properties of Native Versus Complex Bound VEGF

The chemotactic properties of complex bound VEGF were then examined using a Boyden Chemotaxis Chamber (113). This 48-independent well chamber consists of two sections. The lower section is employed to incorporate the stimulant or inhibitor agent of choice. The cell type to be investigated is added to the upper section. A porous membrane is enclosed between these sections. Cells that migrate from the upper section onto the opposite side of the membrane were then counted. Advantages of this system are that sterility is not an issue since these are short-term studies and the chamber can be repeatedly used. Additionally, this type of apparatus permits simultaneous analysis of multiple agents at various concentrations.

For our study, BAECs were grown in DMEM with 10% FBS and then starved for 24 hours in DMEM containing 2% FBS. The BAECs were then trypsinized and diluted to a concentration of 100,000 cells/mL, and 50 μl was added to each well in the upper section. Complex bound and native VEGF concentrations ranging from 50 to 200 ng/mL were added to bottom section. Wells containing DMEM with 2% FBS were used as a negative control. The BAECs were incubated 4 hours at 37°C (5% CO_2). The apparatus was then disassembled, washed, fixed in ethanol, and stained with hematoxylin. BAECs bound to the porous membrane were then counted at 40×.

BAECs exposed to DMEM with 2% FBS had an average 17 adherent cells/high-power field, demonstrating the migration properties of FBS (Fig. 9). Native VEGF had an average 23.9, 35.3, and 49.1 cells/high-power field for increasing concentrations of the growth factor. Complex bound VEGF had 25.9, 39.1, and 69 cells/high-power field over similar concentrations, results comparable to native VEGF. These results demonstrate that VEGF maintains biological activity when covalently bound to another biomolecule.

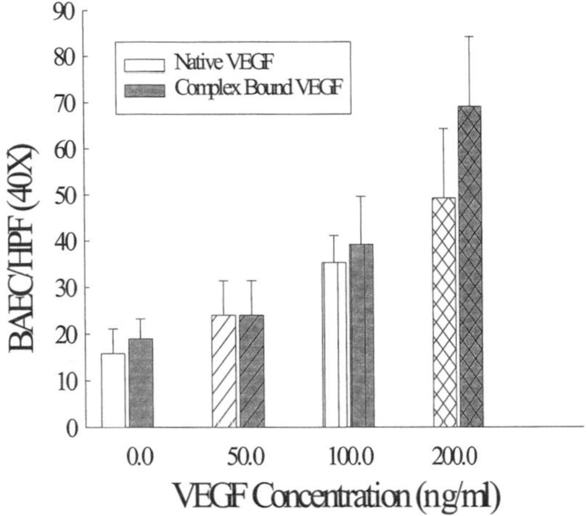

Figure 24.9 Endothelial cells/high power field for ^{125}I-VEGF and complex bound ^{125}I-VEGF using the Boyden chemotaxis chamber. BAECs exposed to DMEM with 2% FBS had an average 17 adherent cells/high power field. Native VEGF had an average 23.9, 35.3, and 49.1 cells/high power field for increasing concentrations of the growth factor, comparable to complex bound VEGF (25.9, 39.1, and 69 cells/high power field).

D. Covalent Linkage of ^{125}I-Collagen/ ^{125}I-Collagen-SMCC to HD Segments

To evaluate if Type IV collagen bound in a similar fashion to albumin, a group of collagen binding experiments was performed. CD and HD segments (1 cm^2) were cut and weighed. A stock ^{125}I-collagen solution was prepared and divided in half. One half of the ^{125}I-collagen solution was reacted with sulfo-SMCC in a 1:5 molar ratio. The other ^{125}I-collagen solution had an equal volume of PBS added in order to maintain consistency between the two ^{125}I-collagen solutions. Both ^{125}I-collagen solutions were then placed into a 37°C water bath, incubated for 20 minutes with occasional mixing, and purified by gel filtration. Both the ^{125}I-collagen/^{125}I-collagen-SMCC solutions, taking into account an approximate 50% loss in protein due to adhesiveness of the protein, were then brought up to a final concentration of 6.4 μM. Simultaneously, CD and HD controls were placed into 3 mL of 100% EtOH. The other set of HD segments was placed into 3 mL of a 10 mg/mL EDC solution in 100% EtOH. All segments were reacted for 30 minutes at room temperature on an inversion mixer. The CD and HD control segments as well as the HD test segment were then removed from their respective solutions after 30 minutes, shaken to remove excess solvent/cross-linker, and placed into 2.4 mL of a 6.4 μM ^{125}I-collagen. Another set of control and test segments was treated in a similar fashion, but these segments were placed into 2.4 mL of a 6.4 μM ^{125}I-collagen-SMCC. These segments were incubated for 2 hours at room temperature on an inversion mixer. The CD and HD segments were removed and washed/sonicated in PBS with 0.05% Tween 20 for 5 minutes. This procedure was repeated three times, changing the wash buffer between sonications, thereby removing any weakly adherent ^{125}I-collagen on the surface of each segment. Segments with nonspecifically bound (CD + ^{125}I-collagen, HD + ^{125}I-collagen, CD + ^{125}I-collagen-SMCC, HD + ^{125}I-collagen-SMCC) and covalently bound (HD-^{125}I-collagen and HD-^{125}I-collagen-SMCC) ^{125}I-collagen/ ^{125}I-collagen-SMCC were then gamma counted. Utilizing the specific activity, the amount of ^{125}I-collagen bound (ng) per mg Dacron was determined.

Nonspecific binding of ^{125}I-collagen to the CD segments (90 ng/mg) was significantly lower than the amount nonspecifically bound to the HD (429 ng/mg), thus creating a negative charge on the Dacron surface alone increased the binding by the positively charged ^{125}I-collagen (Fig. 10). Nonspecific ^{125}I-collagen-SMCC binding to CD (153 ng/mg) was 74% greater than with ^{125}I-collagen. In contrast, nonspecific ^{125}I-collagen-SMCC binding was 41% lower than with ^{125}I-collagen alone (304 ng/mg), suggesting that removal of the positive charge groups by sulfo-SMCC linkage reduced the net charge of the protein, thereby decreasing ^{125}I-collagen affinity for the HD surface. HD with EDC had significantly greater ^{125}I-collagen linkage (531 ng/mg) and ^{125}I-collagen-SMCC (571 ng/mg)

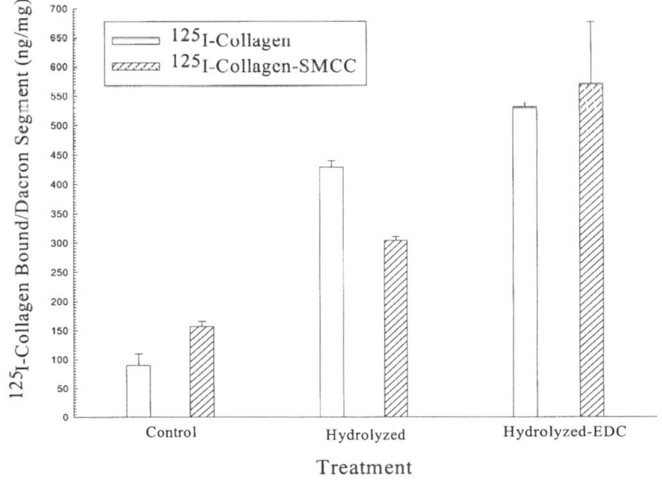

Figure 24.10 ^{125}I-collagen/^{125}I-collagen-SMCC binding (ng/mg Dacron) versus CD, HD, and HD reacted with EDC. Covalent binding of ^{125}I-collagen to the HD surface was greatest utilizing the EDC cross-linker, and linkage of the heterobifunctional cross-linker sulfo-SMCC did not alter ^{125}I-collagen immobilization to the HD surface.

than the HD controls and CD controls. No statistical difference was observed between the HD incubated with EDC in the presence of ^{125}I-collagen or ^{125}I-collagen-SMCC ($p = 0.48$). Therefore covalent binding of ^{125}I-collagen to the HD surface was greatest utilizing the EDC cross-linker, and linkage of the heterobifunctional cross-linker sulfo-SMCC did not alter ^{125}I-collagen immobilization to the HD surface.

E. Covalent Linkage of ^{125}I-VEGF to HD Segments

CD and HD segments were cut and weighed. CD and one group of HD segments were incubated with EDC followed by incubation with 6.4 μM collagen. The other set of HD segments was incubated with EDC followed by incubation in 6.4 μM collagen-SMCC. Removal of nonspecifically bound collagen/collagen-SMCC was performed via wash and sonication. ^{125}I-VEGF (4.5 μM) was then reacted with 2-iminothiolane hydrochloride (Traut's) reagent in a 1:60 (mole:mole) ratio for 20 minutes at 37°C in order to create sulfhydryl groups on ^{125}I-VEGF (^{125}I-VEGF-SH), purified by gel filtration and brought to a final concentration of 0.52 μM. This ^{125}I-VEGF-SH complex was then reacted with the Dacron segments overnight at 4°C followed by the wash/sonication procedure. Nonspecifically bound (CD + EDC + collagen + ^{125}I-VEGF-SH and HD-collagen + ^{125}I-VEGF-SH) and covalently linked (HD-collagen-SMCC-S-^{125}I-VEGF) ^{125}I-VEGF segments were then gamma counted. The amount of ^{125}I-VEGF bound (ng) per mg Dacron was determined.

^{125}I-VEGF binding to the CD (2.9 ng/mg) and HD (7.0 ng/mg) controls was 5.6- and 2.3-fold less than the HD test segments (16.4 ng/mg) (Fig. 11). The HD control segments had 2.4-fold greater ^{125}I-VEGF nonspecifically bound to the surface as compared to the CD segments, which could be the result of charge binding between the nonspecifically bound collagen and ^{125}I-VEGF.

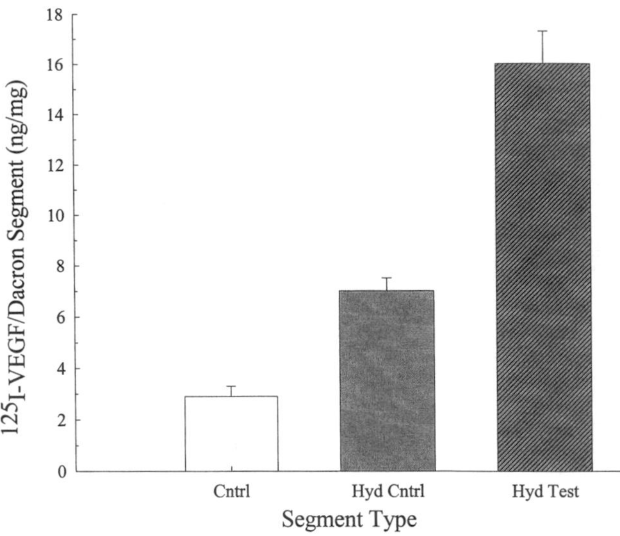

Figure 24.11 ^{125}I-VEGF binding (ng/mg Dacron) for CD, HD, and HD reacted with EDC. Linkage to the CD (2.9 ng/mg) and HD (7.0 ng/mg) control segments was 5.6- and 2.3-fold less than the HD test segments (16.4 ng/mg).

F. Assessment of In Vivo Mitogenic Properties of Surface-Bound VEGF

A rat dorsal subcutaneous implant model was employed in order to evaluate the in vivo mitogenic properties of covalently immobilized VEGF (Fig. 12) (114). Segments (1 cm^2) of HD with covalently bound ^{125}I-collagen (109 ± 26 ng/mg Dacron) and HD with covalently bound ^{125}I-VEGF (76 ± 18 ng/mg Dacron) were prepared as previously outlined. CD segments were also utilized. Rats were anesthetized by intraperitoneal injection of pentobarbital, dorsum shaved, and prepared with betadine. A longitudinal incision was made for the subcutaneous insertion of the

Figure 24.12 An illustration of the rat dorsal subcutaneous implantation model. This model permitted simultaneous evaluation of 8 segments, with each animal serving as its own control.

various Dacron segments. These segments were placed in random fashion by a single surgeon. Care was used to avoid subcutaneous hemorrhage, and skin was closed with staples. Rats were sacrificed at either 15 or 30 days after implantation via anesthetic overdose. Segments were explanted, gamma counted to determine the remaining amount of surface-bound protein, and fixed in buffered formalin for 24 hours prior to histological processing.

Formalin-fixed segments were processed for conventional histology, with sample sections performed every 0.5 mm in a serial manner. A total of six tissue sections were cut and analyzed for each Dacron segment, with 6 mm^2 of each Dacron segment represented. Sections were stained with hematoxylin phloxin saffron. Angiogenesis was quantified by counting the number of capillaries present in each segment under 40× magnification. The average number of capillaries per unit area (mm^2) was then determined.

Both time intervals showed a significant loss of protein from the Dacron surface. At 15 days, HD patches with ^{125}I-VEGF lost 91% of the protein from the surface. Similarly, loss from the HD patches with ^{125}I-collagen was significant at 88%. At 30 days, 99.7% of ^{125}I-VEGF and 97% of ^{125}I-collagen was released. Protein loss may have been caused by immune response to the immobilized human proteins (collagen and VEGF) as well as to the Dacron material. This hypothesis was further validated by morphological assessment of these segments, which showed granulation tissue and a mild inflammatory response that was evident in all segment types by 30 days. Gross microscopic observation of these segments revealed the greatest capillary formation in the HD-collagen-VEGF segments (Fig. 13).

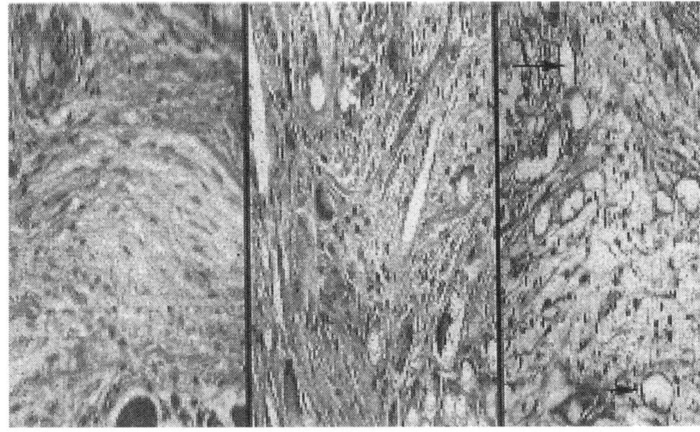

Figure 24.13 Histological examination of CD, HD-^{125}I-collagen, and HD-collagen-SMCC-^{125}I-VEGF segments after explant at 30 days. Segments with covalently bound ^{125}I-VEGF possessed the greatest amount of capillaries per area as compared to CD and HD-^{125}I-collagen segments. All segments showed granulation tissue and a mild inflammatory response.

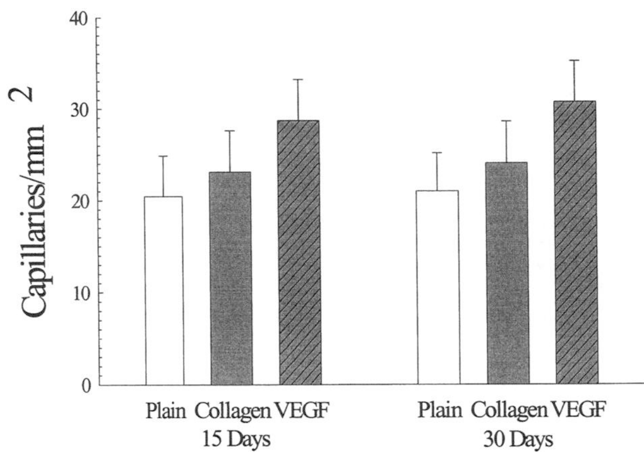

Figure 24.14 Quantification of the number of capillaries/mm^2 for CD, HD-^{125}I-collagen, and HD-collagen-SMCC-^{125}I-VEGF segments. There was significant capillary ingrowth for HD with covalently bound VEGF at both 15 and 30 days (29/31 capillaries/mm^2) as compared to HD-collagen (23/24 capillaries/mm^2) and CD (20/21 capillaries/mm^2).

Quantification of new microvessel formation confirmed the gross observations. There was significant capillary ingrowth for HD with covalently bound VEGF at both 15 and 30 days (29/31 capillaries/mm^2) as compared to HD-collagen (23/24 capillaries/mm^2) and CD (20/21 capillaries/mm^2) (Fig. 14). The number of capillaries increased slightly for all groups at 30 days, with the segments with covalently bound VEGF possessing the greatest amount of capillaries. Therefore covalent linkage of VEGF to a Dacron surface can stimulate capillary formation from the surrounding tissue.

XI. CONCLUSIONS FROM RESEARCH

Overall, this study demonstrates that a clinically accepted biomaterial (Dacron) can be modified using established textile techniques without altering the physical and chemical properties of the original biomaterial. This modification results in the creation of functional groups that can serve as ''anchor'' sites for covalent attachment of a specific basecoat protein, in an attempt to mask the adverse biomaterial properties. Using this basecoat layer in conjunction with heterobifunctional cross-linking agents, numerous target sites for a biologically active protein(s) can be created under controlled stoichiometric conditions. Antithrombin (rHir) and mitogenic (VEGF) agents have been independently attached to Dacron segments and have maintained biological activity in both in vitro and in vivo studies. Extensive in vivo studies still need to be performed in a prosthetic grafting model in order to determine the ultimate utility of this technology. Protein loss from these surfaces in our preliminary in vivo trials suggests altering the cross-linking methodology or increasing species specificity of the immobilized proteins for the animal model in order to reduce the immune response to the surface. Additionally, a surface with both biological moieties immobilized onto the Dacron surface will also be investigated.

XII. POTENTIAL FUTURE OF BIOMATERIAL RESEARCH

Thus far, most of research has focused on either altering an individual aspect of graft interaction within the vascular tree (i.e., thrombus formation, platelet activation, and cellular proliferation) or synthesizing a novel material in an attempt to promote host incorporation of the conduit. These studies have yet to yield a clinically suitable biocompatible prosthetic vascular graft. One possible reason that this problem has not been solved could be related to addressing this complex phenomenon with a single approach or so-called ''magic bullet'' theory. Also, research by a majority of these groups has been limited to their respective fields of expertise, so no collaborative exchange occurs in an area that requires a multidisciplinary approach. Our ultimate goal, using these foundation studies as well as a collaborative effort between academia and industry, is to design a prosthetic vascular graft that will possess multiple structural and biological properties that mimic those processes inherent to the native vessel.

ACKNOWLEDGMENTS

This work was funded, in part, by grants from the National Institutes of Health (R01 21796) and the Whitaker Foundation. The authors would like to express their gratitude to Boston Scientific and Farmitalia for the generous donations of Dacron vascular grafts and rHir, respectively. Additionally, the authors would especially like to acknowledge Dr. Mauricio Contreras, Dr. Scott Berceli, Dr. Stephen Kubaska III, Dr. Eva Rzucidlo, Dr. David Stone, and Dr. Lloyd Aiello for their considerable time and effort dedicated to this extensive research project.

REFERENCES

1. Joseph ML. Polyester fibers. In: Joseph M L (ed). *Introductory to Textile Science*. 5th ed. CBS College Publishing, New York, 1986, 114.
2. Cook JG. Polyester fibers. In: Cook JG (ed). *Handbook of Textile Fibers: Man-Made Fibers*. Merrow, Durham, England, 1984, 328.
3. Harrison JH. Synthetic materials as vascular prostheses: a comparison study in small vessels of nylon, Dacron, Orlon, Ivalon sponge and Teflon (I and II). Am J Surg 1958; 95:3.
4. Brewster DC. Prosthetic grafts. In: Rutherford RB (ed). *Vascular Surgery*. 4th ed. WB Saunders, Philadelphia, PA, 1995, 492.
5. Nagae T, Tsuchida H, Ishimaru S, Wilson SE. Enhanced neointima formation on the high-porosity inner surface of modified PTFE vascular grafts. J Invest Surg 1995; 8(4): 235.
6. Jankowski RJ, Severyn DA, Vorp DA, Wagner WR. Effect of retroviral transduction on human endothelial cell phenotype and adhesion to Dacron vascular grafts. J Vasc Surg 1997; 26(4):676.
7. Ziats NP, Anderson JM Human vascular endothelial cell attachment and growth inhibition by type V collagen J Vasc Surg 1993; 17(4):710.
8. Fenton JW Regulation of thrombin generation and functions. Semin Thromb Hemost 1988; 14:234.
9. Abbott W, Callow A, Moore W, Rutherford R, Veith F, Weinberg S. Evaluation and performance standards for arterial prostheses. J Vasc Surg 1993; 17(4):746.
10. LoGerfo FW, Quist WC, Nowak MD, Crawshaw H, Haudenschild CC. Downstream anastomotic intimal hyperplasia: a mechanism of failure in Dacron arterial grafts. Ann Surg 1993; 197:479.
11. Ozaki CK, Contreras M, Phaneuf M, Sheppeck RA, Rutter CM, Quist WC, LoGerfo FW. Platelet activation by healing ePTFE grafts. J Biomed Res 1995; 29(5):647.
12. Cantelmo NL, Quist WC, LoGerfo FW. Quantitative analysis of intimal hyperplasia in paired Dacron and PTFE grafts. Cordiovasc Surg 1989; 30:910.
13. Ito R, Brophy CM, Contreras MA, Tsoukas A, LoGerfo FW. Persistent platelet activation by passivated graft. J Vasc Surg 1991; 13:822.
14. Kottke-Marchant K, Anderson J, Umemura Y, Marchant R. Effect of albumin coating on the in vitro blood compatibility of Dacron arterial prostheses. Biomaterials 1989; 10: 147.
15. Coleman R, Hirsh J, Marder V, Salzman E. *Hemostasis and Thrombosis: Basic Principals and Clinical Practices*. (2d ed.) Philadelphia, PA, Lippincott, 1987.
16. Shepard A, Gelfand J, Callow A, O'Donnell T. Jr. Complement activation by synthetic vascular prostheses. J Vasc Surg 1984; 1:829.
17. Rumisek J, Wade C, Kaplan K, Okerberg C, Corley J, Barry M, Clarke J. The influence of early surface thromboreactivity on long-term arterial graft patency. Surgery 1989; 105:654.
18. Bascom J. Gelatin sealing to prevent blood loss from knitted arterial grafts. Surgery 1961; 50:947.
19. Drury J, Ashton T, Cunningham J, Maini R, Pollock J. Experimental and clinical experience with a gelatin im-

pregnated Dacron prosthesis. Ann Vasc Surg 1987; 1(5): 542.

20. Jonas R, Ziemer G, Schoen F, Britton L, Castaneda A. A new sealant for knitted Dacron prostheses: minimally cross-linked gelatin. J Vasc Surg 1988; 7:414.
21. Guidoin R, Marceau D, Couture J, Rao T, Merhi Y, Roy P-E, Faye DDL. Collagen coatings as biologic sealants for textile arterial prostheses. Biomaterials 1989; 10:156.
22. Quinones-Baldrich W, Moore W, Ziomet S, Chvapil M. Development of a ''leak proof'' knitted Dacron vascular prosthesis. J Vasc Surg 1986; 3:895.
23. Freischlag J, Moore W. Clinical experience with a collagen-impregnated knitted Dacron vascular graft. Ann Vasc Surg 1990; 4:449.
24. Slimane SB, Guidoin R, Merhi Y, King M, Domurado D, Sigot-Luizard M.-F. In vivo evaluation of polyester arterial grafts coated with albumin: the role and importance of cross-linking agents. Eur Surg Res 1988; 20:66.
25. Merhi Y, Roy R, Guidoin R, Hebert J, Mourad W, Slimane SB. Cellular reactions to polyester arterial prostheses impregnated with cross-linked albumin: in vivo studies in mice. Biomaterials 1989; 10:56.
26. Barbucci R, Magnani A. Conformation of human plasma proteins at polymer surfaces: the effectiveness of surface heparinization. Biomaterials 1994; 15(12):955.
27. Hardhammar PA, Van Beusekom HMM, Emanuelsson HU, et al. Reduction in thrombotic events with heparin-coated Palmaz–Schatz stents in normal porcine coronary arteries. Circulation 1996; 93:423.
28. Lindon JN, Salzman EW, Merrill EW, et al. Catalytic activity and platelet reactivity of heparin covalently bonded to surfaces. J Lab Clin Med 1985; 105:219.
29. Elgue G, Blomback M, Olsson PJ, Riessenfeld J. On the mechanism of coagulation inhibition on surfaces with end point immobilized heparin. Thromb Haemost 1993; 70(2): 289.
30. Tay SW, Merrill EW, Salzman EW, Lindon J. Activity toward thrombin–antithrombin of heparin immobilized on two hydrogels. Biomaterials 1989; 10:11.
31. Nemets EA, Sevastianov VI. The interaction of heparinized biomaterials with human serum, albumin, fibrinogen, antithrombin III, and platelets. Artif Organs 1991; 15: 381.
32. Ito Y, Imanishi Y, Sisido M. In vitro platelet adhesion and in vivo antithrombogenicity of heparinizd polyetherurethanes. Biomaterials 1988; 9:235.
33. Lyman DJ, Kim SW Interactions at the blood polymer interface. Fed Proc 1971; 30:1658.
34. Van Der Lei B, Bartels DF, Wildevuur Ch RH. The thrombogenic characteristics of small-caliber polyurethane vascular prostheses after heparin bonding. Trans Am Soc Artif Intern Organs 1985; 31:107.
35. Maksimenko AV, Torchitin VP Water soluble urokinase derivations of combined action. Thromb Res 1985; 38: 277.
36. Bakker WW, Van Der Lei B, Nieuwenhuis P, et al. Reduced thrombogenicity of artificial materials by coating with ADPase. Biomaterials 1991; 12:603.
37. Wilson JE. Hemocompatible polymers: preparation and properties. Polym Plast Technol Conf 1986; 25:233.
38. Hennick WE, Feyen J, Ebert CD, Kim SW. Covalently bound conjugates of albumin and heparin: synthesis fractionation and characterization. Thromb Res 1983; 29:1.
39. Jacobs H, Kim SW. In vitro bioactivity of a synthesized prostaglandin E, heparin conjugate. J Pharm Sci 1986; 75: 172.
40. Ma X, Mohammad SF, Kim SW. Heparin binding on poly(l-lysine) immobilized surface. J Coll Interface Science 1991; 147:251.
41. Nojiri C, Park KD, Grainger DW, et al. In vivo non-thrombogenicity of heparin immobilized polymer surfaces. ASAIO Trans 1990; 36:M168.
42. Vulic I, Okano T, Kim SW. Synthesis and characterization of polystyrene-poly(ethylene oxide)-heparin block copolymers. J Polym Science Part A: Polym Chem 1988; 26:381.
43. Lin SC, Jacobs HA, Kim SW. Heparin immobilization increased through chemical amplification. J Biomed Mater Res 1991; 25:791.
44. Kishida A, Ueno Y, Fukudome N, et al. Immobilization of human thrombomodulin onto poly(ether urethane urea) for developing antithrombogenic blood-contacting materials. Biomaterials 1994; 15(10):848.
45. Grainger DW, Okano T, Kim SW. Protein adsorption from buffer and plasma onto hydrophilic–hydrophobic poly (ethylene oxide)-polystyrene multiblock copolymers. J Coll Interface Science 1989; 132:161.
46. Silver JH, Hart AP, Williams EC, et al. Anticoagulant effects of sulphonated polyurethanes. Biomaterials 1992; 13(6):339.
47. Nojiri C, Kido T, Sugiyama T, et. al. Can heparin immobilized surfaces maintain nonthrombogenic activity during in vivo long-term implantation? ASAIO J 1996; 42:M468.
48. Gristina A. Biomaterial-centered infection: microbial adhesion versus tissue integration. Science 1987; 237:1588.
49. Sugarman B, Young EJ. Infections associated with prosthetic devices: magnitude of the problem. Inf Dis Clin N Am 1989; 3(2):187.
50. Craver JM, Ottinger LW, Darling RC, et al. Hemorrhage and thrombosis as early complications of femoropopliteal bypass grafts: causes, treatment and prognostic implications. Surgery 1973; 74(6):839.
51. Griesler HP, Kim DU. Vascular grafting in the management of thrombotic disorders. Sem Thromb Hemostas 1989; 15(2):206.
52. Blumenburg RM, Gelfand ML, Dale WA. Perigraft seromas complicating arterial grafts. Surgery 1985; 97(2):196.
53. Echave V, Koornick A, Haimov M, Jacobsen JH II. Intimal hyperplasia as a complication of the use of the polytetrafluoroethylene graft for femoral–popliteal bypass. Surgery 1979; 89(6):791.
54. Herring MB. Endothelial cell seeding. J Vasc Surg 1991; 13(5):731.
55. Griesler HP. Endothelial cell transplantation onto synthetic vascular grafts: panacea, poison or placebo. In:

Griesler HP. (ed). *New Biologic and Synthetic Vascular Prostheses*. RG Landes, Austin, TX, 1991.

56. Magometschnigg H, Kadletz M, Vodrazka M, et al. Prospective clinical study with in vitro endothelial cell lining of expanded polytetrafluoroethylene grafts in crural repeat reconstruction. J Vasc Surg 1992; 15(3):527.
57. Zilla P, Fasol R, Preiss P, et al. Use of fibrin glue as a substrate for in vitro endothelialization of PTFE vascular grafts. Surgery 1989; 105(4):515.
58. Schneider A, Melmed RN, Schwalb H, et al. An improved method for endothelial cell seeding of polytetrafluoroethylene small caliber vascular grafts. J Vasc Surg 1992; 15(4):649.
59. Pierschbacher MD, Ruoslahti E. Cell attachment activity of fibronectin can be duplicated by small synthetic fragments of the molecule. Nature 1984; 309:30.
60. Kuusela P. Fibronectin binds to Staph aureus. Nature 1978; 276:718.
61. Speziale P, Hook M, Wadstrom T, Timpl R. Binding of the basement membrane protein laminin to Escherichia coli. FEBS Let 1982; 146(1):55.
62. Visser MT, van Bockel H, van Muijen NP, van Hinsbergh VM. Cells derived from omental fat tissue and used for seeding vascular prostheses are not endothelial in origin. J Vasc Surg 1991; 13(3):373.
63. Radomski JS, Jarrell BE, Pratt KJ, Williams SK. Effects of in vitro aging on human endothelial cell adherence to Dacron vascular graft material. J Surg Res 1989; 47(2): 173.
64. Rosenman JE, Kempczinski RF, Pearce WH, Silberstein EB. Kinetics of endothelial cell seeding. J Vasc Surg 1985; 2(6):778.
65. Lindblad B, Burkel WE, Wakefield TW, et al. Endothelial cell seeding efficiency onto expanded polytetrafluoroethylene grafts with different coatings. Acta Chir Scan 1986; 152:653.
66. Massia SP, Hubbell JA. Human endothelial cell interactions with surface-coupled adhesion peptides on a nonadhesive glass substrate and two polymeric biomaterials. J Biomed Mater Res 1991; 25:223.
67. Hirano Y, Kando Y, Hayashi T, et al. Synthesis and cell attachment activity of bioactive oligopeptides: RGD, RGDS, RGDV and RGDT. J Biomater Res 1991; 25: 1523.
68. Ito Y, Liu SQ, Imanishi Y. Enhancement of cell growth on growth factor-immobilized polymer film. Biomaterials 1991; 12:449.
69. Gray JL, Kang SS, Zenni GC, et al. FGF-1 stimulates ePTFE endothelialization without intimal hyperplasia. J Surg Res 1994; 57:596
70. Phaneuf MD, Quist WC, Bide MJ, LoGerfo FW. Modification of polyethylene terepthalate (Dacron) via denier reduction: effects on material tensile strength, weight, and protein binding capabilities. J Applied Biomater 1995; 6: 289.
71. Kissa E. Soil release finishes-alkali treatment of polyester fibers (Sec. 4.8). In: Lewin M, Sello S (eds). *Handbook of Fiber Science and Technology*. Marcel Deckker, New York, 1984, 265.
72. Lin H-B, Sun W, Mosher DF, et al. Synthesis, surface, and cell-adhesion properties of polyurethanes containing covalently grafted RGD-peptides. J Biomed Mater Res 1994; 28:329.
73. Phaneuf MD, Szycher M, Berceli SA, Dempsey DJ, Quist WC, LoGerfo FW. Covalent linkage of recombinant hirudin to a novel poly(carbonate urea) urethane polymer with protein binding sites: determination of surface antithrombin activity. Artif Organs 1998; 22(8):657.
74. Phaneuf MD, Dempsey DJ, Bide MJ, Szycher M, Quist WC, LoGerfo FW. Bioengineering of a novel small-diameter polyurethane vascular graft with covalently bound recombinant hirudin. ASAIO J 1998; 44:M653.
75. Phaneuf MD, Berceli SA, Bide MJ, Quist WC, LoGerfo FW. Covalent linkage of recombinant hirudin to polyethylene terepthalate (Dacron): creation of a novel antithrombin surface. Biomaterials 1997; 18(10):755.
76. Mooney D, Hansen L, Vacanti J, et al. Switching from differentiation to growth in hepatocytes: control by extracellular matrix. J Cell Physiol 1992; 151:497.
77. Olsen BR. Matrix molecules and their ligands. In: Lanza R, Langer R, Chick W (eds). *Textbook of Tissue Engineering*. RG Landes Company, Academic Press, In press.
78. Hay ED. (ed.). *Cell Biology of Extracellular Matrix*. 2d ed. Plennum Press, New York, 1991.
79. Walz DA, Anderson GF, Fenton JW. Responses of aortic smooth muscle to thrombin and thrombin analogues. Ann NY Acad Sci 1986; 485:323.
80. Bar-Shavit R, Kahn A, Wilner G, Fenton J. Monocyte chemotaxis: stimulation by specific exosite region in thrombin. Science 1983; 220:728.
81. Bizios R, Lai L, Fenton J, Malik A. Thrombin-induced chemotaxis and aggregation of neutrophils. J Cell Physiol 1986; 128(3):485.
82. Kaplan JE, Moon DG, Weston LK, et al. Platelets adhere to thrombin-treated endothelial cells in vitro. Am J Physiol 1989; 257:H423.
83. Tollefsen DM, Feagler JR, Majerus PW. The binding of thrombin to the surface of human platelets. J Biol Chem 1974; 249:2646.
84. Markwardt F. Pharmacology of hirudin: one hundred years after the first report of the anticoagulant agent in medicinal leeches. Biochim Acta 1985; 44:1007.
85. Obberghen-Schilling EV, Perez-Rodriguez R, Pouyssegur J. Hidin, a probe to analyze the growth-promoting activity of thrombin in fibroblasts; reevaluation of the temporal action of competence factors. Biochem Biophys Res Comm 1982; 106:79.
86. Fenton JW, Bing DH. Thrombin active-site regions. Sem Thromb Hemost 1986; 12(3):200.
87. Phaneuf MD, Ito RK, LoGerfo FW. Synthesis and characterization of a recombinant hirudin-albumin complex. Blood Coagulation and Fibrinolysis 1994; 5:641.
88. Phaneuf MD, Ozaki CK, Johnstone MT, Loza JP, Quist

WC, LoGerof FW. Covalent linkage of streptokinase to recombinant hirudin: a novel thrombolytic agent with antithrombotic properties. Thromb Haemostas 1994; 71(4): 481.

89. Ito RK, Phaneuf MD, LoGerfo FW. Thrombin inhibition by covalently bound hirudin. Blood Coagulation and Fibrinolysis 1991; 2:77.

90. Salzman EW, Rosenberg RD, Smith MH, et al. Effect of heparin and heparin fractions on platelet aggregation. J Clin Invest 1980; 65:64.

91. Hogg PJ, Jackson CM. Fibrin monomer protects thrombin from inactivation by heparin-antithrombin III: implications for heparin efficacy. Proc Natl Acad Sci USA 1989; 86:3619.

92. Weitz J, Hudoba M, Massel D, et al. Clot-bound thrombin is protected from inhibition by heparin antithrombin III but is susceptible to inactivation by antithrombin III-independent inhibitors. J Clin Invest 1990; 86:385.

93. Oosta GM. Favreau LV, Beeler DL, Rosenberg RD. Purification and properties of human platelet heparitinase. J Biol Chem 1982; 257:11249.

94. Leung DW, Cachianes G, Kuang WJ, et al. Vascular endothelial growth factor is a secreted angiogenic mitogen. Science 1989; 246:1306.

95. Ferrara N, Houck KA, Jakeman LB, et al. Molecular and biologic properties of the vascular endothelial growth factor family of proteins. Endocr Rev 1992; 13:18.

96. Berse B, Brown LF, Van de Water L, et al. Vascular permeability factor (vascular endothelial growth factor) gene is expressed differentially in normal tissues, macrophages and tumors. Mol Biol Cell 1992; 3:211.

97. De Vries C, Escobedo H, Veno K, et al. The FMS-like tyrosine kinase, a receptor for vascular endothelial growth factor. Science 1992; 255:989.

98. Dvorak H, Sioussat T, Brown LF, et al. Distribution of vascular permeability factor (vascular endothelial growth factor) in tumors; concentration in tumor blood vessels. J Exp Med 1991; 174: 1275102.

99. Aiello LP, Northrup J, Keyt B, et al. Hypoxic regulation of vascular endothelial growth factor in retinal cells. Arch Ophthalmol 1995; 113:1538.

100. Millauer B, Wizigmann-Voos S, Schnurch H, et al. High affinity VEGF binding and developmental expression suggests FLK-1 as a major regulator of vasculogenesis and angiogenesis. Cell 1993; 72:835.

101. Brogi E, Wu T, Namiki A, Isner J. Indirect angiogenic cytokines upregulate VEGF and bFGF expression in vascular smooth muscle cells whereas hypoxia upregulates VEGF expression only. Circulation 1994; 90:649.

102. Frank S, Hubner G, Brier G, et al. Regulation of vascular endothelial growth factor expression in cultured keratinocytes. J Biol Chem 1995; 270:12607.

103. Shweiki D, Itin A, Soffer D, et al. Vascular endothelial growth factor induced by hypoxia may mediate hypoxia-initiated angiogenesis. Nature 1992; 359:843.

104. Stauri G, Zachary IC, Baskerville P, et al. Basic fibroblast growth factor upregulates the expression of vascular endothelial growth factor in vascular smooth muscle cells. FEBS Letters 1995; 358:311.

105. Takeshita S, Zheng L, Brogi E, et al. Therapeutic angiogenesis: a single intra-arterial bolus of vascular endothelial growth factor augments re-vascularization in a rabbit ischemic hind limb model. J Clin Invest 1994; 93:662.

106. Asahara T, Bauters C, Pastore C, et al. Local delivery of vascular endothelial growth factor accelerates re-endothelialization and attenuates intimal hyperplasia in balloon-injured rat carotid artery. Circulation 1995; 91:2793.

107. Ibukiyama C. Angiogenesis therapy using fibroblast growth factors and vascular endothelial growth factors for ischemic vascular lesions. Jpn Heart J 1996; 37:285.

108. Banai S, Jaklitsch MT, Shou M, et. al. Angiogenic-induced enhancement of collateral blood flow to ischemic myocardium by vascular endothelial growth factors in dogs. Circulation 1994; 89:2183.

109. Hamdan AD, Aiello LP, Misare BD, Contreras MA, King GL, LoGerfo FW, Quist WC. Vascular endothelial growth factor expression in canine peripheral vein bypass grafts. J Vasc Surg 1997; 26:79.

110. Berceli SA, Phaneuf MD, LoGerfo FW. Evaluation of a novel hirudin-coated polyester graft to physiologic flow conditions: hirudin bioavailability and thrombin uptake. J Vasc Surg 1998; 27:1117.

111. Wyers MC, Phaneuf MD, Rzucidlo EM, Contreras MA, Quist WC, LoGerfo FW. In vivo assessment of a Dacron surface with covalently bound recombinant hirudin. Cardiovasc Pathol 1999; 8(3):153.

112. Kubaska SM III, Phaneuf MD, Rook SL, Aiello LP, LoGerfo FW, Quist WC. Characterization of covalently bound vascular endothelial growth factor: creation of a novel Dacron prosthetic graft surface. Surgical Forum 1998; 49:322.

113. Stone DH, Phaneuf MD, Rohan DI, Sivamurthy N, LoGerfo FW, Quist WC. In vitro study of vascular endothelial growth factor in tissue culture for designing improved prosthetic vascular grafts. Presented at the 18th Annual Association of International Vascular Surgeons, March 2000.

114. Contreras MA, Phaneuf MD, Quist WC, LoGerfo FW. Promotion of angiogenesis in vivo utilizing VEGF covalently bound to Dacron patches. FASEB J 2000; 14(4): A710.

25

Antithrombin–Heparin Complexes

Leslie R. Berry, Maureen Andrew, and Anthony K. C. Chan
Hamilton Civic Hospitals Research Centre, Hamilton, Ontario, Canada

I. INTRODUCTION

Antithrombin is a serine protease inhibitor (serpin) that inhibits a number of plasma proteases. In particular, antithrombin functions as one of the major natural anticoagulants by irreversibly inhibiting a number of the enzymes formed during activation of the coagulation cascade in vivo (1). Antithrombin can form serpin–protease inhibitor complexes with activated factor (F) XII (FXIIa), FXIa, FIXa, FXa, and thrombin (1–3). Within this group of coagulant proteases, antithrombin has the fastest rate of reaction with thrombin, its preferred reactant (4).

Systemic generation of thrombin in vivo is dependent on a number of factors. These include plasma concentrations of pro- and anticoagulants, the presence of cell surface–associated stimulators (such as phospholipid, tissue factor) and inhibitors (thrombomodulin, tissue factor pathway inhibitor), and interactions with molecules in subendothelial and extravascular spaces (5). Ultimately, activation or inhibition of proenzymes or activated factors within the coagulation cascade affects the generation of thrombin from prothrombin, which, in turn, affects the conversion of fibrinogen to fibrin monomer that polymerizes to form a fibrin clot (see Fig. 1). Thrombin is the pivotal enzyme in the coagulation pathway (6). Once the initial amounts of thrombin are generated, thrombin causes feedback activation of its own formation by proteolytic cleavage of FV, FVIII, and FXI (Fig. 1) to produce FVa, FVIIIa, and FXIa, respectively. Furthermore, thrombin reacts with FXIII, leading to the covalent cross-linking of fibrin monomers by FXIIIa. The resultant cross-linked fibrin polymer has more structural stability and integrity within the site of injury (7). Additionally, thrombin has other procoagulant functions within the vasculature. Thrombin activates platelets, thus facilitating their cell–cell or cell–clot interactions (8), and causes inhibition of fibrinolysis by reaction with thrombin activatable fibrinolysis inhibitor (9–13). Alternatively, when thrombin is bound to endothelial cell associated thrombomodulin, it can activate protein C, the active form of which (in association with protein S) inactivates FVIIIa and FVa, thus limiting thrombin production (14). Summarily, inhibition of thrombin is a critical step in the regulation of coagulant activities in vivo.

The reaction of thrombin with antithrombin can be significantly accelerated by the action of heparin and heparan sulfate glycosaminoglycans (GAGs). In fact, the rate of inhibition of thrombin by antithrombin is increased 1,000-fold in the presence of native unfractionated heparin (UFH) (15). The increase in inhibition rate is due to two reasons. First, UFH molecules bind to antithrombin via a specific GAG sequence (16). The binding of UFH causes an allosteric change in antithrombin that results in a conformation that is more reactive with thrombin. Second, UFH can also bind to thrombin, which allows for a combination of the serpin and protease in a tertiary complex. In effect, UFH can bridge both antithrombin and thrombin. Once an irre-

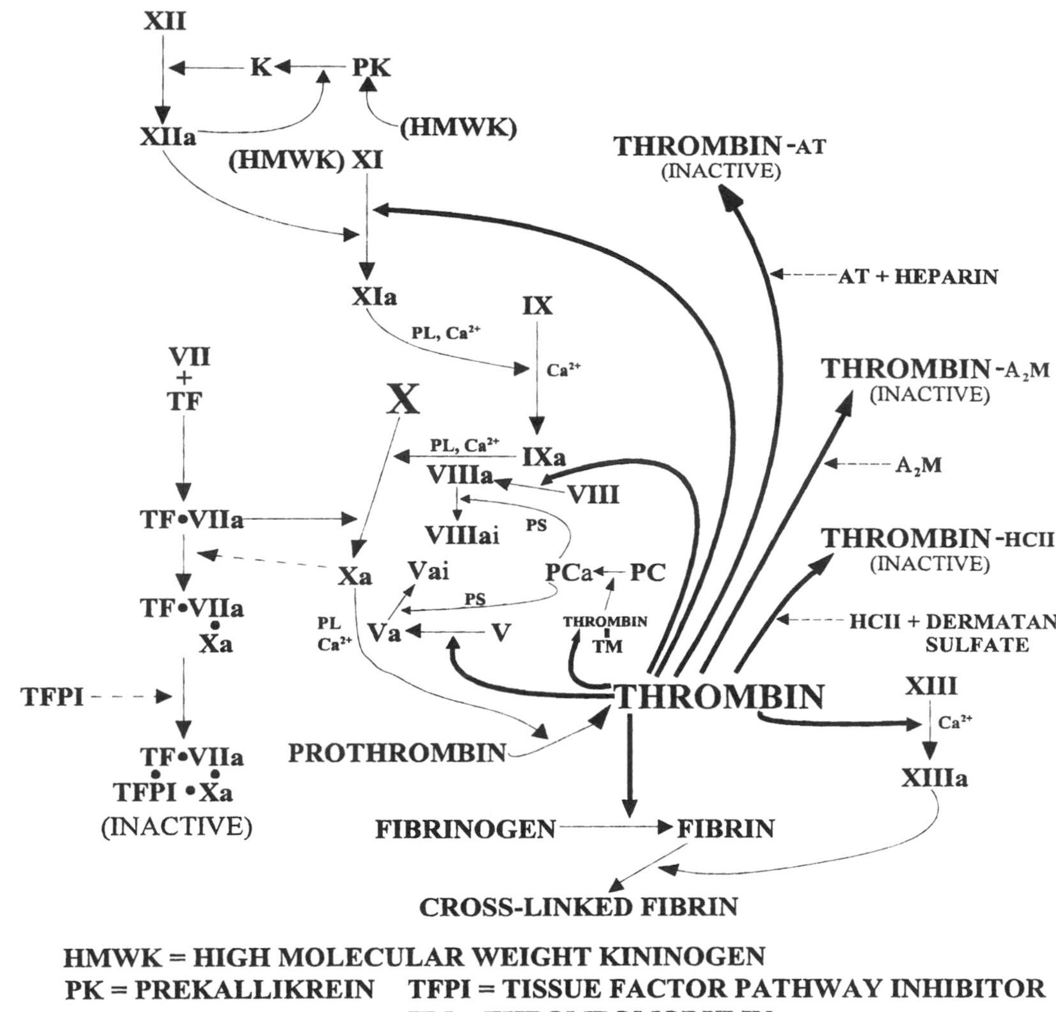

Figure 25.1 Plasma coagulation cascade for thrombin generation and inhibition.

versible covalent complex between thrombin and antithrombin (TAT) has been formed, the affinity of the antithrombin moiety for UFH decreases, which allows the GAG to dissociate and repeat the cycle of tertiary complex formation (17). By increasing the velocity of thrombin's reaction with antithrombin, UFH facilitates the inhibition of thrombin formed shortly after activation of coagulation. Thus the addition of UFH-like GAGs to plasma would significantly reduce the concentration of free, active thrombin, which in turn would decrease the generation of thrombin due to feedback activation of the cascade.

Unfortunately, UFH has a number of limitations that are related to its pharmacokinetic and biophysical properties. The pharmacokinetic limitation of UFH is that it has a short, dose-dependent, intravenous half-life (18). UFH's rapid clearance from the circulation is due, in part, to nonspecific protein binding (19). As a consequence, basic plasma and cell surface–associated proteins compete with antithrombin for noncovalent binding to UFH. The concentrations and distribution of the UFH-binding proteins vary widely between individuals, which results in UFH having an unpredictable anticoagulant effect in vivo (20). In addition, UFH can readily pass through tissue layers and is lost from the circulation due to its small size, which prevents its sequestration within various vascular spaces (21). One of the biophysical limitations of UFH relates to its inability to inhibit a number of coagulation factors bound to surfaces. The antithrombin–UFH complex is ineffective at inactivating thrombin bound to fibrin (22,23) and factor Xa bound to phospholipid (24,25). Clot propagation is due, in part, to the activity of this clot-bound thrombin (23–25). Clinical evidence supports the importance of clot-bound thrombin and the consequences of the inability to inactivate this focus for coagulation. The early recurrence of unstable coronary artery syndromes after discontinuation of heparin is likely due to this mechanism (26). Finally, the use of high in vivo concentrations of heparin results in bleeding complications. Previously, it has been shown that UFH anti–factor Xa activity is directly proportional to bleeding time (27). Attempts to anticoagulate vascular devices by coating surfaces with heparin derivatives have also lead to problems. Leaching of weakly attached heparin, resulting in undesired systemic anticoagulation and reduced activity of the modified heparin bound to the surface, is one example of the difficulties involved with heparin coating of biomaterials. Furthermore, only ~1/3 of starting commercial UFH preparations have anticoagulant activity because 2/3 of the molecules do not have high-affinity binding sites for antithrombin (28). Therefore biomaterials coated with heparin from mixtures derived from UFH would have the majority of their surface area covered by heparin with no antithrombin activity.

In order to address some of the problems associated with the clinical use of heparin, covalent complexes of antithrombin and heparin derivatives have been prepared. The rationale for construction of covalent antithrombin–heparin (ATH) is multifold. First, if heparin was irreversibly bonded to antithrombin, the serpin molecule should be fixed permanently in the active conformation. Thus the rate of thrombin inhibition will be increased compared to noncovalent antithrombin–heparin mixtures. Second, since the heparin component of ATH cannot dissociate from antithrombin, the intravascular pharmacokinetics would be expected to move towards that of antithrombin. Thus the half-life of the anticoagulant ATH should be significantly longer than heparin. Another aspect related to the transport and metabolism of ATH is that of endogenous protein binding in vivo. Because the antithrombin in ATH will interact noncovalently with a significant portion of the heparin moiety (particularly in the case of ATH molecules which contain relatively short heparin chains), binding of plasma or cell surface proteins (or other molecules) with the heparin in ATH should be reduced compared to free heparin. A decreased heparin binding by intra- and extravascular proteins might lead to a more consistent clearance pattern and, consequently, a more predictable anticoagulant response. Furthermore, if ATH binding to cell surface receptors is decreased compared to uncomplexed heparin, risk of hemorrhagic side effects may be reduced. Third, both the technology and the resultant polymeric surfaces coated with ATH may be significantly improved compared to that for biomaterials coated with heparin. Attachment of heparin onto a variety of polymers would be facilitated by linking antithrombin to the GAG. Immobilization of ATH could be more readily optimized due to the increased range of functional groups and chemistries found in the amino acid R groups of the antithrombin. Also, surface linkage of ATH through the antithrombin moiety may allow for a high number of attached molecules in which the heparin chain is oriented outwards from the surface into the fluid phase. Again, since all antithrombin species in ATH should be permanently activated, ATH bound to surfaces of biomedical devices may all be active, whereas only a maximum of 1/3 of GAGs on UFH coated surfaces would possess anticoagulant activity. Given the potential number of desirable properties of ATH products compared to other UFH derivatives, methods for covalent complexation of antithrombin and heparin and analysis of the covalent conjugates have been investigated.

In this chapter, the development of permanently linked antithrombin–heparin complexes will be assessed. The format for this review will be as follows. Previous work on the structure–function relationships and clinical use of antithrombin will be discussed. Heparin's structure, activities, in vivo occurrence, and biochemistry will be covered in light of the long history of medical application. Analysis of the research on antithrombin and heparin will be followed by a broad introduction to the production of covalent ATH complexes. The overview of ATH complexes will include assessments of clinical advantages for conjugation of antithrombin and heparin, novel applications of ATH products, and an introduction to the range of ATH conjugates that are available. A detailed analysis follows for the various types of ATH that have been reported. Each ATH complex will be reviewed according to its synthetic chemistry, physicochemical properties, effect of conjugation on anticoagulant activities, performance in animal models, and particular clinical advantages (including possible use for anticoagulating materials that come into contact with blood). Discussion of the various ATH compounds that have been produced will appear in the chronological order in which they have been reported. Finally, possible future directions for research on covalent serpin-GAG complexes will be presented.

II. ANTITHROMBIN

A. Antithrombin Chemical Structure

Antithrombin is a glycoprotein whose polypeptide moiety shares structural and functional homology with members of a large family of serpins (29–31). The degree of homology across the various proteins in this family is ~30%, indicating that the appearance of this molecular species may have been as early as 500 million years ago. In addition to primary sequence similarities, the serpins have a number of their tertiary structural features in common (32).

Antithrombin is produced within hepatocytes of the liver in mammalians. The gene for antithrombin has been reported to reside entirely within the long arm of chromosome 1 (33). In addition, it has been shown that within this region, the human antithrombin gene extends over an unbroken stretch of ~19 kb including 7 exons and 6 introns (34,35). The open reading frame for human antithrombin contains 1396 nucleotides. At the 5′ end of the reading frame there is a section of 96 nucleotides that code for a 32 amino acid segment called the signal peptide (35). This N-terminal peptide is removed prior to release from the cell. Within mammalian species 10% to 15% variation in primary sequence has been reported, but high conservation has been observed in the critical reactive site regions (36).

The antithrombin molecule is a single chain plasma glycoprotein with an approximate molecular mass of 60,000 daltons (37–39). The polypeptide chain in human antithrombin is composed of 432 amino acid residues (40). Upon folding into its native configuration, three disulfide bonds are formed between three pairs of cysteine residues in the polypeptide chain (40). Two of the disulfide bonds join a relatively unstructured stretch of 45 amino acid residues at the N-terminus to the third and fourth α-helices of the molecule. Overall, the protein has a neutral to basic pI due to a preponderance of arginyl and lysyl residues, as opposed to the acidic residues. These positively charged amino acids contribute to the UFH binding regions on the serpin molecule (32,41). Analysis of the tertiary structure, as determined by peptide modeling of the primary amino acid sequence, has shown that human antithrombin has 31% α-helix, 16% β-sheet, 9% β-turn and 44% random coil (42,43). Antithrombin has been crystallized recently, and the x-ray crystal structure determined to 3 Å resolution (44). Conformational studies of the topographic structure of antithrombin using x-ray diffraction show that the serpin can exist in two forms, one active and the other inactive (45). Analyses were made of dimers containing one active and one inactive molecule. Results showed that the active antithrombin consists of nine α-helices and three β-sheets. A composite model for the three-dimensional structure of antithrombin, devised from various x-ray diffraction and chemical analyses that have been reported, is shown in Fig.2.

N-linked glycosylation occurs at specific asparagine residues that are flanked by sequences in β-sheet regions that are recognized by a glycosyl transferase during post-

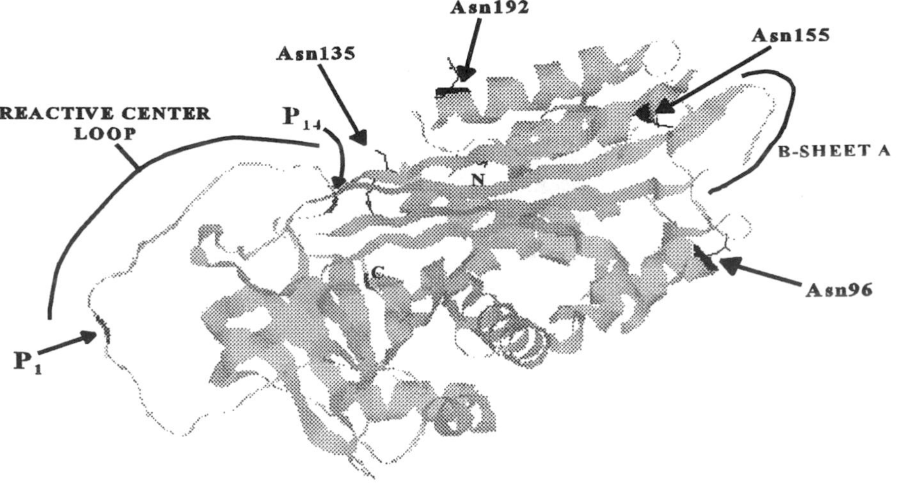

Figure 25.2 Antithrombin tertiary structure.

translational modification in vivo. The number of glycan chains that are present on the native antithrombin polypeptide varies from three (46) to four (47), although up to five glycan chains have been reported (48). In addition, conditions such as carbohydrate-deficient glycoprotein syndrome have been shown to result in reduced levels of functional antithrombin due to decreased posttranslational glycosylation of the protein (49). Isoforms of plasma antithrombin, occurring as the three and four glycan chain containing molecules, have been designated as α- and β-antithrombin, respectively. The fully glycosylated α-antithrombin has glycan chains linked through N-amido glycosidic bonds to the R groups of asparagine residue numbers 96, 135, 155, and 192 (50). In the case of β-antithrombin, the glycan at position 135 is missing.

Post translational modification of antithrombin occurs in the Golgi apparatus by transfer of tetra-antennary, branched high mannose glycan structures from dolicolphosphate onto the amino acid R group. These high mannose glycans are further processed by endoglycosidases to a core chitobiose-trimannose stub, which is the substrate for addition of further monosaccharide residues to produce the final ''complex type'' glycan structures. In addition to the major heterogeneity of glycan number described above, variations in structure within the glycans themselves (termed microheterogeneity) are observed. Complex N-linked glycans have been shown to occur in a number of forms that vary according to the number of branches extending out from the core region of the chain. Antithrombin molecules have been shown to exist as a group of populations in which the branching within glycans at the different glycosylation sites within the same polypeptide chain can range from mono- to bi-, tri-, and tetra-antennary forms (51,52). The glycans of antithrombin have been shown to vary in the level of N-acetylneuraminic acid residues present at the termini of the glycan chains (53). Additional groups may by added to the glycan structures in a heterogeneous fashion. For example, although fucosyl residues are present on the glycans at positions 96 and 192, a portion of glycans at position 155 was not fucosylated (51). Variations due to heterogeneity in antithrombin's carbohydrate moieties have also been identified as having significant effects on its functional activities, as will be outlined below.

B. Antithrombin Functional Biochemistry

The structural components that give rise to antithrombin's functional activities have been delineated. Early studies showed that a circulating anticoagulant existed because thrombin, when added to plasma, lost its activity over time (38,54). This plasma activity of the antithrombin molecule was given the name ''progressive antithrombin activity.'' With the subsequent isolation and clinical use of heparin, it was observed that heparin could inhibit thrombin rapidly when in the presence of an unidentified protein in plasma. The plasma protein responsible for this rapid thrombin inhibitory activity was designated ''heparin cofactor'' (55). Classification of the biological antithrombin activities was proposed in the 1950s by Seegers (56). Removal of plasma thrombin activity by binding to a fibrin clot was given the notation antithrombin I, while heparin cofactor and progressive antithrombin activities were called antithrombin II and antithrombin III, respectively. Isolation and further characterization of the native plasma thrombin inhibitor in the 1960s and 1970s showed that the protein responsible for the progressive antithrombin activity and that responsible for the heparin cofactor activity were identical (38). Afterwards, the term antithrombin III was used for the progressive antithrombin with heparin cofactor activity, which later became shortened to antithrombin.

Antithrombin, as well as a most of the other serpins, inhibits serine proteases (such as thrombin) by a stress-release mechanism. Within the antithrombin molecule, a reactive center loop exists that is towards the C-terminus of the polypeptide chain. Thrombin interacts noncovalently with antithrombin through the recognition of certain amino acids close to the reactive center. The thrombin binding region has been mapped out by the natural occurrence of amino acid substitution mutations in the protein, which give rise to expression of molecules with reduced antithrombin activity. The mutations $ala^{382} \rightarrow thr^{382}$ (antithrombin Hamilton) (57) and $pro^{407} \rightarrow leu^{407}$ (antithrombin Utah) (34) have indicated that the thrombin-binding region extends to at least include amino acid residues 382 and 407. Alternatively, investigations using thrombin mutants have illuminated structural components of the enzyme that are involved in the noncovalent interaction. Gly^{226} in the thrombin polypeptide has been shown to protrude into the antithrombin specificity pocket, since substitution of a valine at this position removed inhibitory activity (58). Molecular modeling indicated that the larger valine side chain would not allow enough space for the antithrombin P_1 arg^{393} R group. Once thrombin binding through weak interactions occurs, a more stable covalent bond is formed. Members of the serpin family have specific peptide bonds in the reactive center that are susceptible to enzymatic cleavage by the target protease. Two amino acids make up the reactive center peptide bond, and the peptide is designated as P_1-P_1'. The P_1 amino acid residue in antithrombin (arg^{393}) provides the inhibitor with selectivity towards thrombin and the other coagulation factors (31). Other residues to the N-terminal side of the P_1 arg^{393} give further enzyme selectivity and structural elements, such as in the case of the P_2 gly^{392}, which is preferred for reaction with FXa and is important for preventing release of FXa from the inhibitor complex (59). The role of the P_1' in antithrombin (ser^{394}) is not entirely clear.

Thrombin initially treats antithrombin as a substrate by attacking the inhibitor's arg^{393} C-terminal amide carbonyl via the protease active serine residue. Thus thrombin forms a covalent ester bond between the hydroxyl oxygen of the active serine and the arg^{393} carboxyl carbonyl. The active noncleaved antithrombin exists in an S-configuration where the two amino acids of the reactive center are in the middle of a stressed loop (60). Within antithrombin, the reactive center loop connects strand 4 of β-sheet A (central sheet in the serpin ordered numbering system) to strand 1 of β-sheet C (61). In this stressed configuration, the reactive center loop of antithrombin is under some tension and the resultant surface topology is such that the reactive center is readily available for reaction with thrombin. Once the arg^{393}–ser^{394} has been cleaved, leaving thrombin and antithrombin remaining linked by a covalent ester bond, a radical conformational change occurs leading to a relaxed or R-configuration (60). Several studies of thrombin–antithrombin inhibitor complexes and postcomplex cleaved antithrombin have been done that have helped elucidate the structural properties of the relaxed configuration of the antithrombin within the complex. Antibodies have been produced that do not bind to native antithrombin but recognize either the consumed inhibitor or the inhibitor when it is bound to a tetradecapeptide corresponding to the P_{14} to P_1 residues of the reactive center loop (61–68). These experiments, in conjunction with x-ray studies, strongly suggested that upon reaction of antithrombin with thrombin, the reactive center loop becomes embedded into the β-sheet A, which exposes new epitopes that are not present at the surface of the intact inhibitor. Model-building studies based on crystal structures and biochemical analyses of protease complexes with normal and mutant antithrombins have suggested that the reactive center loop is inserted into β-sheet A as far as P_{12}, strand s1C is absent from the β-sheet C, and the conformation of the C-terminus has changed so that it interacts with thrombin (69). In fact, an intact noninhibitory (latent) conformer of antithrombin exists with a structure where the P_{14} to P_3 residues of the reactive center are completely inserted into β-sheet A (70). However, experiments with an antibody, that did not bind to native, latent, or reactive center cleaved antithrombins, recognized antithrombin neoepitopes in either stable thrombin–antithrombin complexes or antithrombin complexed to a synthetic peptide corresponding to the P_{14} to P_9 sequence. Therefore only part of the reactive center may be imbedded in the antithrombin β-sheet A within the inhibitor complex (71). Regardless, insertion of the reactive center loop positions the thrombin reactive site close to or within the β-sheet A pocket. Thus the thrombin moiety becomes irreversibly trapped as a reaction intermediate covalently linked to antithrombin (72).

Interaction of antithrombin with heparin and heparan sulfate glycosaminoglycans accelerates the thrombin inhibition reaction. Activation of antithrombin by heparin binding has been verified by the concomitant increase in protein intrinsic fluorescence, as a result of a conformational change. Research involving antithrombin trp $\rightarrow$ phe mutants has shown that trp^{225} and trp^{307} each account for ~37% of the heparin-induced fluorescence enhancement (73). A spectral shift in trp^{49} towards the blue suggests partial burial due to contact with heparin, while a red shift for trp^{225} (along with fluorescence enhancement) indicates the increased access to the solvent due to heparin-induced movement of contact residue ser^{380}. The main heparin binding region on antithrombin has been shown (according to mutant studies) (74–77) to reside towards the N-terminus of the molecule (32,78). A second heparin-binding region on antithrombin was proposed to exist at residues from positions 107 to 156 (79). Ultimately, high resolution of the heparin-binding site was achieved by alanine scanning mutagenesis in a baculovirus expression system, which gives a normal product that is highly similar to plasma β-antithrombin (80). Mutant antithrombins were screened for heparin affinity by gradient elution from heparin columns. It was determined that only a subset of residues (lys^{11}, arg^{14}, arg^{24}, arg^{47}, lys^{125}, arg^{129}, and arg^{145}), which line in a 5 nm groove along antithrombin's surface, are critical for heparin binding (80). A few other residues have been proposed to be significant points for heparin binding (70,72). Molecular modeling has suggested that interaction of heparin with these amino acid R groups would induce the breakage of salt bridges between α-helix D and β-sheet B, thus facilitating movement of s123AhDEF to give a species that is conformationally primed for reactive center loop uptake by β-sheet A (80). X-ray studies have confirmed that binding of heparin pentasaccharide to certain of the key antithrombin residues gives an increased affinity between the GAG and antithrombin, which accompanies the change in conformation (70). In addition, peptides representing P_{14}-P_3 are able to inhibit formation of the conformation that would be induced by heparin binding to antithrombin (81). Finally, once antithrombin has been cleaved by thrombin, heparin affinity for the serpin in the inhibitor complex dramatically decreases. Model studies using immobilized heparin have shown that although native antithrombin readily bound to the heparin, upon covalent linkage to thrombin (cleavage) the serpin (as part of an inhibitor complex) dissociated from the GAG (17,82). Undoubtedly, conformational changes resulting from insertion of the reactive loop peptide into the β-sheet A of cleaved antithrombin disturb the heparin binding site channel.

Variation in glycosylation on the antithrombin molecule has substantial effects on the ability of heparin to catalyze antithrombin's inhibition of thrombin. It has been shown previously that human β-antithrombin, which lacks the carbohydrate side chain at asn^{135}, has higher affinity for heparin than the fully glycosylated (4 glycan) α-antithrombin (83). Previously, glycoforms of antithrombin produced in either BHK or CHO cells that had a 10-fold difference in heparin-binding affinity (52,84) were found to vary in the glycosylation at asn^{155} (85). Increased binding of heparin to β-antithrombin might be expected, given that the glycan position at asn^{135} is directly within the heparin-binding site residues lys^{125}, arg^{129}, and arg^{145} of the serpin. Rapid kinetic studies of heparin binding to α- and β-antithrombin demonstrated that, although the asn^{135}-glycan moderately interfered with the weak initial binding of heparin, the rate constant for the conformational change induced by heparin was significantly lower for α-antithrombin than for β-antithrombin (83). Thus there is a higher energy required for inducing the activated conformation in α-antithrombin, leading to a decrease in heparin-binding affinity. Since heparin-like GAGs exist in vivo, the variation in binding affinities by the antithrombin glycoforms may have physiological importance. In fact, evidence has been presented showing that β-antithrombin tends to be bound to vessel wall heparin/heparan sulfate, while α-antithrombin resides more commonly in the plasma phase (46). These data suggest specialized functions for the antithrombin glycoforms where β-antithrombin may be vital for controlling thrombogenic events arising from vessel wall injury, while α-antithrombin may be largely responsible for inhibition of fluid phase thrombin.

III. HEPARIN

A. Heparin Chemical Structure

Heparin is a member of the GAG family of molecules that occur not only in mammalians but in most multicellular, as well as some single-cell, organisms (86). This group of molecules includes heparin, heparan sulfate, dermatan sulfate, chrondroitin-4-sulfate, chondroitin-6-sulfate, keratan sulfate, hyluronic acid, and a number of minor subclasses (87). GAGs are straight chain polysaccharides that are composed of repeating uronic acid–hexosamine disaccharide units (88). Hyaluronic acid is of extremely high molecular weight and forms the structural matrix for many tissues, as well as molecules associated with the other GAGs. Heparin and heparan sulfate contain glucosamine derivative residues, whereas dermatan sulfate and the other chondroitins contain galactosamine (88). The uronic acid present in GAGs are either glucuronic or iduronic. Generally, both types of uronic acids appear in heparin and heparan sulfate, while dermatan sulfate and the other chondroitins contain solely iduronic acid or glucuronic acid, respectively (88,89). Saccharides in GAG chains are extensively modified during and after glycosidic polymerization. Galactosamines in dermatan sulfate and chondroitins are N-acetylated and variously O-sulfated. In the case of heparin and heparan sulfate, glucosamine residues can be N-acetylated or N-sulfated. However, while >80% of the glucosamines are N-sulfated in heparin, approximately equal amounts of N-sulfated and N-acetylated glucosamines have been detected in various sources of heparan sulfate (90). Furthermore, whereas there are ≥2 O-sulfates per disaccharide unit in heparin (91), O-sulfation per disaccharide in heparan sulfate has been found to range from 0.2 to 0.75 (90). Thus, from these and other structural observations, it has been concluded that heparin and heparan sulfate represent separate groups of N-sulfated GAGs.

GAGs are ubiquitous in mammalians. Heparin as well as other GAGs are produced in mast cells by biosynthetic attachment to a protein core. Each core protein may contain as many as 10 heparin chains, of molecular weights ranging from 60,000 to 100,000, linked via O-glycosidic bonds to serine residues in a glycine–serine sequence repeat (92). Heparin GAG is synthesized by glycosyl-transferase addition of monosaccharide residues. An initial xylose-galactose-galactose-glucuronic acid linkage region sequence is built up, with the terminal xylose linked through serine to the polypeptide backbone (93). Within the growing polysaccharide, uronosyl-β-1 → 4-glucosaminosyl units are joined by α-1 → 4 bonds (94). During polysaccharide chain formation, numerous functional group modifications are performed, in a somewhat concerted fashion, on the monosaccharide residues within the nascent oligosaccharide. Glycosyltransferases (95), N-acetylases (96), N-deacetylases (96), O-sulfotransferases (97), N-sulfotransferases (98), and glucuronosyl-5-epimerases (97) have been isolated and characterized which are involved in anabolism towards the final heparin product in vivo. Segments of glucuronosyl–glucosaminosyl oligosaccharide chain are initially produced which are partially N-unsubstituted, N-acetylated, or N-sulfated (99). Following (and during) (99) this stage, the growing chain is acted upon by a combination of a N-deacetylase and a N-sulfotransferase to yield a much greater N-sulfate containing stretch (96,99). Shortly after N-sulfation, a glucuronosyl-C5-epimerase causes conversion of the chain-end glucuronosyl residue to an iduronosyl residue (97). However, the epimerase reaction is one in which an equilibrium exists between glucuronosyl and iduronosyl residues, with the net equilibrium being towards the iduronosyl form. If O-sulfation occurs at either the C6 of the glucosaminosyl or the C2 of the iduronosyl residue in question, conversion of the iduronosyl back to a glucuronosyl does not occur (97). Chronologically, C2-O-sulfation of iduronosyl residues is carried out in the absence of C6-O-sulfation of neighboring glucosaminosyl residues, while C6-O-sulfation of glucosaminosyls occurs readily in the presence of 2-O-sulfates on adjacent iduronic acid units (100). However, if further glycosyl transfer occurs before C2-O-sulfation of the terminal uronic acid in the nascent chain, that uronic acid remains nonsulfated throughout subsequent modification reactions (100). Analyses of heparin samples has shown that ~78% of uronic acids are in the form of iduronic acid (94), of which ~75% are 2-O-sulfated (94,101). No significant amount of 2-O-sulfated glucuronic acid residues have been detected (94). Termination of the biosynthesis of each polysaccharide on the core protein varies, which yields (as mentioned above) side chains ranging in length (molecular weight). Chain size and monosaccharide modifications are not directed by transcriptional expression but are affected by factors within the milieu, such as substrate availability and cell status (102,103).

Studies of heparin metabolism in mastocytoma cells have shown that the newly synthesized heparin proteoglycan chains are partially depolymerized by an endoglucuronidase and stored in cytoplasmic granules (104,105). Due to this cellular processing, commercial heparin prepared from intestinal mucosa or lung mast cells are free GAG chains ranging in molecular weight from 5,000 to 30,000. Primary source commercial heparin, without isolation of any subpopulations, is called unfractionated heparin (UFH, as above). Within the last few decades, low molecular weight heparins (LMWHs) have been prepared via various methods (106). Partial depolymerization of UFH has been carried out by limited treatments with HNO_2, heparinases, heparitinases, and base elimination following partial esterification of uronic acid carboxyls (106,107). Molecular weights of LMWHs produced vary from 1,800 to 12,000 (106). Apart from variation in chain size, LMWHs have a number of pharmacokinetic and biological properties that separate them from their UFH parent compound.

B. Heparin Functional Biochemistry

Heparin (both UFH and LMWH) provides two major functions in vivo. First, heparin (and other GAGs), mainly in the proteoglycan form, acts as an extracellular matrix component for structural organization and as a chemoattractant in tissue (108,109). Second, via particular sequences, heparin can operate as an anticoagulant. Heparin's anticoagulant activities are based on its ability to bind to either plasma heparin cofactor II or antithrombin and catalyze their inhibition of thrombin (heparin cofactor II or antithrombin) or other coagulation factors [antithrombin (see above)] (110).

Reaction of antithrombin with various coagulation factors is catalyzed by heparin in vivo (111). Heparin's physiological anticoagulant activity resides particularly in facilitating antithrombin's inhibition of FXa and thrombin (112,113). However, heparin's augmentation of thrombin inhibition by antithrombin has been shown to be a major basis for the clinical use of heparin (particular UFH) (114). Regulation of in vivo FXa and thrombin inhibition by heparin relies on the presence of a particular pentasaccharide sequence in the GAG molecule that binds to antithrombin (115). This pentasaccharide antithrombin binding site has been shown to occur in the GAG chains of the proteoglycan form of heparin obtained from rat skin mast cells (which conserve intact proteoglycan) (116). Analyses indicated that while most proteoglycans contain no heparin chains with antithrombin binding sites, a small proportion of proteoglycans had chains with heparin pentasaccharide sequences numbering from 1 to 5 (average of 3) per polysaccharide unit (117). In the case of commercial UFH, on the average, only about 1/3 of the molecules have been found to have high-affinity antithrombin binding (118). Interestingly, though, some molecules of commercial UFH have been shown to contain two high-affinity antithrombin binding sites per molecule (119). On the other hand, pentasaccharide antithrombin binding sites in LMWH are reduced in number due to the degradative cleavages used to make LMWH molecules (112,120). It has been shown that the pentasaccharide antithrombin binding sequence occurs somewhat randomly along the chain in UFH molecules (121). However, reports have indicated that there may be a bias for the antithrombin binding sites to be located towards the nonaldose half of UFH chains (122,123).

The biosynthetic and structural aspects of the pentasaccharide antithrombin binding site in heparin have been elucidated. Previously, it was reported that an unusual 3-O-sulfate group was present on an internal glucosamine group within the pentasaccharide sequence (124). Further work has indicated that the 3-O-sulfated glucosamine residue was critical to the high-affinity antithrombin binding and anti-FXa activity (125). Final characterization and chemical synthesis of the complete pentasaccharide sequence has been accomplished (126) and is shown in Fig. 3. Important elements of the structure are the appearance of nonsulfated glucuronic acid and 3-O-sulfated glucos-

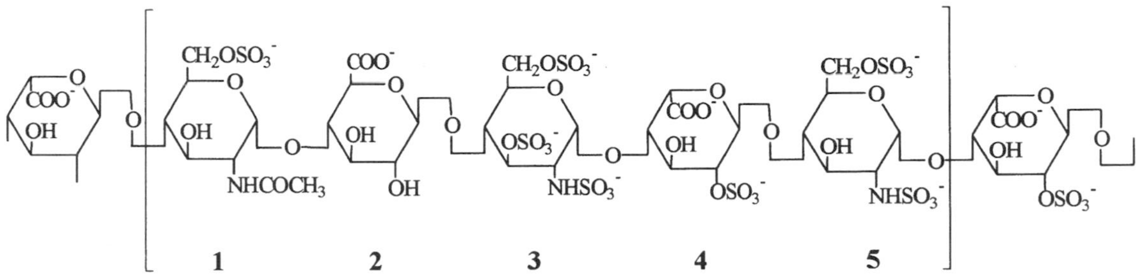

Figure 25.3 Heparin pentasaccharide high-affinity antithrombin binding sequence.

amine at positions 2 and 3 from the nonaldose terminus of the sequence. Binding of heparin to antithrombin, through the pentasaccharide, involves a two-stage mechanism that corresponds to the structural activation events that occur in the serpin. Evidence for the mechanism of activation of antithrombin by heparin pentasaccharide has come from experiments involving binding of various mono-, di-, tri-, and tetrasaccharide derivatives to antithrombin (127–129). Initially, residues 1, 2, and 3 (Fig. 3) of the heparin pentasaccharide bind to antithrombin via charge and hydrogen-bond interactions. Binding of the first three residues from the nonaldose end of the pentasaccharide is relatively weak but induces a conformational change in the antithrombin that is very similar to the heparin activated form (127). The 2-O-sulfated iduronic acid (residue 4, Fig. 3) has been shown to have a flexible capability to convert from a chair to a skew boat conformation (129). Upon binding of pentasaccharide residues 1–3 to antithrombin, movement of the polypeptide due to the trisaccharide-induced conformational change causes amino-acid R groups, particularly arg[47] (130), to come into contact with the remaining pentasaccharide residues 4 and 5. The skew boat conformation of the 2-O-sulfated iduronic acid also allows the 3-O-sulfate on the central glucosamine and a putative charge cluster on pentasaccharide residues 4 and 5 to interact with the polypeptide after antithrombin's change in conformation (129). This binding of charge groups from pentasaccharide residues 4 and 5 gives the high affinity binding with antithrombin that locks the GAG and serpin in place (128,129). No additional conformational activation of antithrombin occurs due to the locking interactions with residues 4 and 5 of the pentasaccharide (130).

It is now well understood that, although activation of antithrombin by heparin pentasaccharide binding is fairly sufficient to accelerate inhibition of FXa, a larger stretch of heparin chain is required for maximal inhibition of thrombin (115). For heparin catalysis of antithrombin's reaction with thrombin, both the serpin and the enzyme must bind to the GAG. It has been determined that effective ternary complexes of antithrombin, thrombin, and heparin require a heparin pentasaccharide sequence to bind the inhibitor and a total chain length of 18 to 22 monosaccharide residues in order also to accommodate interaction with thrombin (131).

Heparin binding to thrombin involves an anion binding exosite on the protease (132). However, although significant negative heparin charge density is important (133), a specific binding sequence in heparin for thrombin has never been found. Since binding to both antithrombin and thrombin by heparin has a minimum chain length requirement, certain LMWH molecules would be unable to catalyze thrombin inhibition. This has been borne out by the fact that LMWH preparations have a lower antithrombin-to-anti-FXa activity ratio. The importance of heparin's ability to catalyze thrombin activity has been implicated by reduced in vivo antithrombotic activity of various LMWHs compared to UFH (112). In fact, the lowest antithrombotic activity (as indicated by in vivo fibrin deposition) has been found with the pentasaccharide (134). These findings would appear to suggest a reduced clinical usefulness for LMWHs compared to UFH. However, the intravenous half-life of LMWHs is significantly longer than UFH (135). The difference in pharmacokinetics is due to two main reasons. First, nonspecific protein binding by LMWH molecules is reduced compared to UFH (136). Since LMWH has reduced affinity for other proteins in vivo, there is a reduction in pathways by which LMWH activity can be either pacified or removed from the circulation. Second, although UFH can be either metabolized by uptake into the liver or lost through the kidneys, plasma disappearance of LMWH occurs only via renal elimination (131). This single-phase elimination of LMWH is much slower and less variable than the concave–convex pattern seen with UFH. As a result of the reduction of pathways for LMWH's pharmacokinetics in vivo, LMWH produces a more predictable anticoagulant response than UFH, which gives a decreased risk of bleeding (137,138). Nevertheless, since different preparations of LMWH exhibit differing amounts of activity against thrombin (possibly due to contaminating higher molecular weight chains, varying charge density, etc.) (139), a narrow window of dosage and regimen exists for their application that must be evaluated for each type of LMWH (107). Furthermore, unlike UFH, LMWHs cannot be completely neutralized by protamine (administered to prevent bleeding if plasma levels become too high during treatment) (140). Thus development of a heparinoid with the thrombin inhibitory potency of UFH and the increased half-life of LMWH would be desirable.

IV. COVALENT ANTITHROMBIN–HEPARIN COMPLEXES: OVERVIEW

A. Limitations of Currently Available Anticoagulants

Limitations in the control of thrombin by heparin administration, as well as major adverse side effects induced by heparinization, have lead to the development of new anticoagulants for clinical use. Deficiencies in heparin's efficacy for thrombin regulation stem from either loss of heparin activity within the plasma compartment or the inability of heparin to accelerate inhibition of thrombin bound to different surfaces. Loss of plasma heparin activity occurs due to removal of heparin molecules from the circulation (disappearance) or neutralization. Pharmacokinetic loss of UFH is due to binding and uptake of heparin by cells, such as hepatocytes of the liver, or by glomerular filtration through the kidneys (131). In both elimination mechanisms, removal of UFH from the circulation requires the dissociation of UFH from antithrombin. That antithrombin UFH complexes break up prior to UFH elimination in vivo, is borne out by the fact that noncovalent mixtures of UFH and antithrombin have significantly different intravenous half-lives when injected as a bolus (118). It has been determined that the intravenous half-life of UFH in humans is dose dependent and ranges from 0.3 to 1 hour (131). UFH's short half-life makes it necessary to administer UFH either by intravenous infusion (which assures a constant delivery of UFH) or by subcutaneous injection (which provides a depot of heparin for slow release into the intravascular space) (141).

Subcutaneous UFH injection gives peak plasma concentrations at 4 hours (142). LMWHs have been shown to have longer intravenous half-lives than UFH (143). Interestingly, LMWH administered subcutaneously in humans at therapeutic doses gave peak plasma levels by 3 hours with plasma activity being undetectable after 12 hours (137). Apart from the disappearance of heparin from the plasma compartment, heparin's activity can be altered due to binding to other fluid phase or cell surface molecules within the lumen. It has been shown that plasma proteins (144,145), platelets (146), and endothelium (147) can bind UFH. Binding to plasma proteins, other than antithrombin and heparin cofactor II, is nonselective and based on charge (148). When bound to plasma proteins, UFH exhibits reduced activity (144), which can be regained following displacement from these basic heparin-binding proteins (149). However, dissociation of UFH from nonanticoagulant plasma proteins (part of the heparin rebound effect) results in variation of heparin activity levels and the regimen used to stop treatment by administration of protamine (149). Again, binding of UFH to basic plasma proteins involves dissociation of the heparin from antithrombin. As stated earlier, heparin has reduced efficacy at inhibiting fibrin-bound thrombin (22,23). In fact, at therapeutic levels, heparin promotes the binding of thrombin to fibrin polymer (150), which, in turn, protects thrombin from inactivation by antithrombin · UFH noncovalent complexes (151). The mechanism by which fibrin accretion of thrombin protects the protease from inactivation by antithrombin + heparin has been investigated. Fibrin, thrombin, and UFH interact to form a ternary complex (152). Once in this ternary complex, the action of thrombin against its substrates is altered (153), which causes decreased reaction of the protease with incoming antithrombin-heparin (151). Although LMWH has been found to have reduced nonselective plasma protein binding (154,155), both UFH and LMWH are unable to inactivate clot-bound thrombin (156). Numerous adverse side effects have been associated with heparin administration. Approximately 3% of adults receiving heparin develop heparin-induced thrombocytopenia (a condition in which antibodies are developed by the patient against platelet-bound heparin molecules) (157), and the use of UFH on a long-term basis results in osteoporosis in 17–36% of patients (158,159). Review of heparin use has indicated that the most significant problem is the risk of bleeding (160). Thus careful monitoring is required to ensure that UFH's anticoagulant activities remain within the therapeutic range to minimize the risk of recurrent disease while preventing hemorrhagic complications (161). Since heparin is rapidly cleared from the circulation, constant infusion or frequent subcutaneous injections are required to control the efficacy/bleeding ratio (162). From data in acute coronary syndrome patients, it has been suggested that LMWH may have a lower risk of heparin-induced thrombocytopenia than UFH (160). However, since LMWHs cross-react with 80% of antibodies generated during UFH exposure, LMWH is not recommended for the treatment of established heparin-induced thrombocytopenia (160). Furthermore, in clinical practice, there is no evidence that bleeding complications are reduced with LMWHs compared to UFH (163,164).

Non-heparin-related anticoagulants have been developed to improve on clinical therapy and to overcome problems with treatments involving heparin. Examples of new and effective anticoagulants include LMWHs (as discussed above), hirudin, hirulog, and PPACK (165–168). Hirudin is a small (7000 MW) direct, reversible thrombin inhibitor produced by the leech (169), and hirulog is the C-terminal portion (residues 53–65) of hirudin covalently linked to a short peptide active site inhibitor (170). Investigations have determined that hirudin can inhibit clot-bound thrombin, since clot growth did not occur even long after hirudin had been cleared from the circulation (171). However, the intravenous half-lives of hirudin (β-phase $\leq$1h, terminal half-life = 2.8 h) (172) and hirolog (36 min) (173) in humans are not significantly longer than that of UFH. Furthermore, it has been shown that persistent formation of thrombin occurs during declining plasma levels of hirudin after administering potent dosages, suggesting that thrombin generation is not blocked (174). PPACK (phe-pro-arg-chloromethyl ketone) is a modified tripeptide substrate that reacts selectively and rapidly with thrombin. Again, PPACK is very effective at irreversibly inhibiting clot-bound thrombin, as has been demonstrated in animal models (175). Given its small size, it is not surprising to note that PPACK's half-life is extremely short (176), which has lead to bleeding risk at therapeutic treatment doses (177). Thus the potency/bioavailability/bleeding issues that are concerns for heparin use have not been ameliorated by the development of alternative anticoagulant agents.

B. Potential Advantages of Covalent Antithrombin–Heparin Complexes

To address the aforementioned limitations of UFH, LMWH, and other anticoagulants, researchers have studied the possibility of stabilizing the interaction of antithrombin with heparin by covalent bonds. The rationale for producing a covalent antithrombin–heparin conjugate (ATH) was severalfold. First, if ATH molecules could be produced in which the antithrombin in the complex was able to interact with a pentasaccharide on the heparin component of the same complex, the ATH antithrombin would be activated for rapid reaction with thrombin. Reaction of ATH with thrombin should be faster than noncovalent mixtures of heparin + saturating amounts of antithrombin because the reaction step of antithrombin + heparin binding would be eliminated. In fact, previous workers have shown that antithrombin + heparin binding is the rate determining step in heparin-catalyzed inhibition of thrombin by antithrombin (178). Second, if antithrombin cannot dissociate from the heparin, the antithrombin would be permanently activated. Thus, wherever and whenever ATH appears at locations in the body, the antithrombin would always be in the conformation of a very potent anticoagulant towards thrombin, FXa, and other coagulant molecules of the cascade. Permanently activated antithrombin would not rely on any particular conditions to react rapidly with thrombin, which may be necessary for the prevention of noncovalent antithrombin · heparin complex dissociation. Third, depending on the methodology used, covalent ATH complexes may be prepared in which a selection takes place (prior to covalent linkage) for only the heparin molecules that have pentasaccharide sequences. That is, a mechanism for ATH synthesis may be possible in which antithrombin first binds heparin ionically, through the high-affinity sites on the GAG, followed by covalent bonding. In this way, the ATH formed would be a preparation in which all of the molecules would contain serpin species activated by heparin pentasaccharide sites. The heparin in ATH produced by preselection of the heparin component by the antithrombin component would have GAG that is a much more potent anticoagulant than the starting heparin, since only ~1/3 of commercial UFH (and even fewer for LMWH) have pentasaccharide containing molecules (118). Fourth, ATH may have the capability to inhibit fibrin-bound thrombin. Since the antithrombin and heparin in ATH do not dissociate, ternary complexes of fibrin/heparin/thrombin cannot form. Given that the antithrombin must remain attached to the heparin chain, it is expected that the heparin moiety in ATH will be less likely also to accommodate fibrin and thrombin compared to free heparin. Fifth, if antithrombin were permanently linked to heparin, the heparin moieties would not dissociate and be lost from the circulation through a renal mechanism. Given the increased size of an ATH complex, glomerular filtration of such a compound should be prohibited, or at least vastly reduced, compared to the much smaller low molecular weight species in polydisperse UFH (even more so for LMWH preparations). Sixth, since the antithrombin in ATH may cover (by noncovalent interaction or steric hinderance) a significant portion of the heparin chain in the complex, binding of the ATH GAG moiety by other plasma and cell surface proteins could be reduced. This reduction in nonselective protein binding would decrease both the removal of active compound from the circulation (via cell surface interactions, cell uptake, and cell matrix binding) and neutralization of anticoagulant activity (basic plasma protein, platelet factor 4 receptor, etc.). Seventh, close interaction of the heparin in ATH complexes with the conjugated antithrombin should sterically inhibit attack by proteases or glycosidases that may be released into the plasma environment. Therefore, by maintaining antithrombin and heparin as a complex, the serpin and GAG may, to a degree, inhibit degradation. Eighth, immune response against ATH may be much less compared to other anticoagulant modalities. For example, if human antithrombin is used to make ATH, no antibodies should be raised directly against the serpin during clinical application (unlike hirudin, hirulog, and PPACK). Also, since part of the heparin molecule permanently interacts with the antithrombin in ATH, the ATH heparin chain may interact with other proteins in a manner similar to that of LMWH (only a small part of the heparin in ATH is available for binding). Thus heparin-induced thrombocytopenia may be decreased for ATH (as it is for LMWH) (160), compared to UFH. Additionally, if binding between the antithrombin and heparin in an ATH preparation is very close, linking agents may be sterically hidden as epitopes for recognition for immune response. Ninth, ATH may not promote osteoporosis as readily as UFH. It has been shown previously that LMWH causes less osteopenia than UFH because it only decreases the rate of bone formation and does not increase bone resorption (179). ATH, again, would only have a portion of its heparin component available (free from the bound antithrombin) and therefore may interact poorly with osteoclasts to effect bone loss. Tenth, for the same reasons as those given for increased inhibition of fibrin-bound thrombin and reduced promotion of osteoporosis, ATH molecules may induce fewer hemorrhagic side effects than either UFH or LMWH. Platelet binding by ATH may be reduced compared to free heparin, since the covalently linked antithrombin may not allow heparin either to bridge receptors on the membrane or to interact with single protein molecules via the required geometry. Finally, ATH may combine many of the positive attributes exhibited by either UFH or LMWH, without a number of their deficiencies. For example, if ATH is prepared from UFH, the product should have high activity against thrombin (unlike LMWH) but have an increased intravenous half-life (decreased cell surface protein binding compared to UFH). In effect, ATH may be the optimum heparinoid.

C. Potential Uses of Covalent Antithrombin–Heparin Complexes

ATH has characteristics that make it attractive for use in a range of clinical indications. The rapid inhibition of thrombin, that should occur if antithrombin is permanently activated by covalently linked heparin, would be highly advantageous for anticoagulant prophylaxis. The low levels of thrombin, which need to be inactivated during ongoing prophylactic treatment, would be quickly complexed by ATH before feedback activation of FV, FVIII, or FXI can occur. Also, since ATH is likely to have a prolonged half-life compared to free heparin, it may be possible to give a single intravenous bolus injection of the conjugate to achieve safe protection against thrombosis. Monitoring of the ATH may be less critical if ATH has a reduced bleeding risk profile. Another possible application of ATH is administration of the complex as an antithrombotic treatment. As suggested above, although UFH and, to a lesser extent, LMWH have reduced activity against clot-bound thrombin, ATH may have good reactivity with thrombin associated with fibrin clots. Thus, in addition to an increased half-life for in vivo inhibition of thrombin generation, ATH may be able to pacify the procoagulant activity on thrombi. Rapid inhibition of clot-based procoagulant activity is critical to the prevention of myocardial infarction and stroke, and ATH may be a treatment that could reduce this risk in two ways. By inactivating thrombin on the surface of the polymerized fibrin, clot extension should not occur, since there will be no further activation of fluid phase coagulation by the clot (23–25). Thus fibrin accretion would be reduced if not eliminated. In addition, due to its increased potency and half-life compared to UFH and LMWH, ATH could be used at dosages that would not cause significant risk of hemorrhage. Another novel utilization of ATH involves the likelihood that its permeability from one compartment to another would be reduced compared to free heparin. The molecular size of ATH is much greater than that of heparin alone, which would prevent its percolation through small pores in cross-linked sections of the extracellular tissue matrix. Therefore ATH should be retained within the compartment or space in which it is placed. This property can be advantageous in situations where an anticoagulant must have potent direct activity in a confined space, without it being lost from that compartment. An example of a disease that would be best treated by sequestration of covalent serpin–GAG complex is respiratory distress syndrome (RDS). RDS, particularly in neonates, is characterized by intrapulmonary coagulation. Neonatal and adult RDSs are typified by leakage of plasma proteins of varying sizes into the airspace (180–182), which leads to interstitial and intra-alveolar thrombin generation with subsequent fibrin deposition. In fact, extravascular fibrin deposition is a hallmark of RDS (183,184), a common complication of premature birth. Furthermore, the success of surfactant therapy for RDS has not resulted in decreasing either the incidence or the severity of chronic bronchopulmonary dysplasia, which continues to be a problem for survivors of prematurity (185–191). Thus if an agent could be placed in the alveolar airspace that would eliminate thrombin activity, without being lost into the intravascular system, neonates could be treated prophylactically so that chronic effects of RDS would be prevented. ATH is a strong candidate for RDS treatment and prevention of chronic fibrotic lung disease. Coagulation in the lung space may occur where there is a paucity of antithrombin for thrombin inhibition. ATH would provide potent direct inhibition of pulmonary thrombin generation, even in the absence of antithrombin from the patient (a requirement for UFH and LMWH). Due to its size, ATH should slowly, if at all, disappear from the lung into the circulation. Therefore there would be little bleeding risk systemically from intratracheally instilled ATH. Finally, once ATH forms a complex with thrombin generated in the alveolar airspace, the serpin–GAG conjugate itself would be neutralized and unable to participate in any further, undesired reactions. In conclusion, prophylaxis, systemic antithrombotic treatment, and anticoagulation of selected compartments by sequestration are significant applications in which the particular characteristics of ATH may make it a superior agent compared to other available drug technologies.

D. Concepts for Producing Covalent Antithrombin–Heparin Complexes

There are only three general approaches to permanent linkage of antithrombin and heparin. Bonding can occur if heparin is activated to make it reactive, followed by interaction with antithrombin to effect covalent bond formation. Conversely, antithrombin can be preactivated and then incubated with heparin to obtain a stable complex. Finally, antithrombin and heparin can be conjugated by allowing the two macromolecules to interact noncovalently, followed by addition of a bifunctional reagent, one end of which bonds to the serpin and the other end of which reacts covalently with the GAG. The three synthetic schemes for ATH preparation are outlined in Fig. 4. In the first two

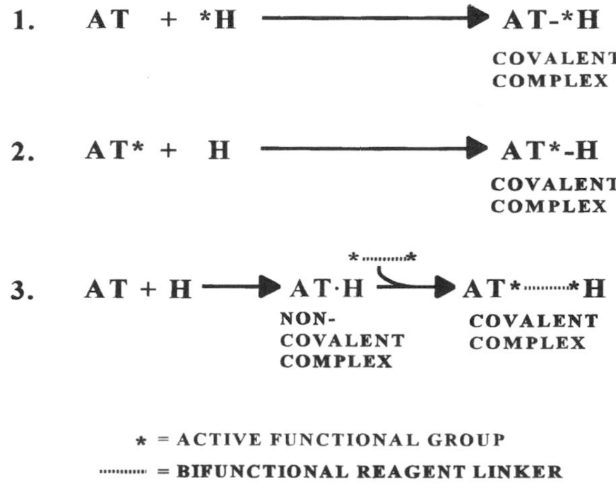

Figure 25.4 General approaches for covalent linkage of antithrombin and heparin.

methodologies, it is theoretically possible that the heparin or antithrombin may possess groups that, under the appropriate conditions, are already reactive enough to form a bond with the other macromolecule. Furthermore, in a number of procedures, care must be taken (by selective chemistry, appropriate reagent stoichiometries, or particular reaction conditions) to prevent linkage of either heparin to itself or antithrombin to itself.

Functional groups on heparin that may be activated (or have active or activatable groups attached to them) are carboxyl, sulfonyl, hydroxyl, amino, and acetal. Apart from the acetal and amino groups, attachment through heparin's other functional groups would be within the chain of the molecule. Only linkage through acetals, or aldehydes formed at the reducing terminus, would result in end-point attachment of heparin to antithrombin, which more closely resembles the structure of the glycosidically bonded chains in the proteoglycan form of heparin (93). An example of conjugation that occurs by reaction of antithrombin with a preactivated heparin follows. Heparin carboxyl groups are reacted with a carbodiimide followed by incubation with a diaminoalkane to form amide-linked groups with a free amino. Further reaction of the amino-substituted heparin with a diisothiocyanate then gives a heparin derivative with a free isothiocyanate group that can form thiourea bonds with the N-terminal or lys -amino groups of antithrombin. This type of mechanism for joining heparin to antithrombin has been carried out previously and will be discussed in further detail later (192).

Activation of antithrombin for covalent reaction with heparin is more problematic. Proteins not only have a greater number of different functional R group types than those on heparin, but the same type of amino acid R group can exist in a range of environments, either on the surface or within the tertiary structure. Thus more finely tuned chemistries may be required to obtain significant yields of purified ATH (containing linkages with heparin that are at the same point on the antithrombin molecule) resulting from the initial activation of the antithrombin. One possible example for this type of synthesis would involve activation of antithrombin with N-hydroxysuccinimidyl-4-azidobenzoate (formation of benzoyl amide with antithrombin lys ε-amino), followed by interaction with heparin in the presence of light to cause photochemical linkage between the benzoyl derivative and heparin. Additionally, the antithrombin carbohydrate residues could be modified to act as linkage points for heparin conjugation. Reaction of antithrombin with $NaIO_4$, under mild conditions, only produces aldehydes on the terminal sialic acids of the glycans. Aldehyde-containing antithrombin can be reacted with diaminoalkane + $NaBH_3CN$ to form alkyl-amino antithrombin. Reaction of the alkyl-amino groups on the antithrombin with N-hydroxysuccinimidyl-4-azidobenzoate (as above) would then be preferred, compared to other amino groups on the antithrombin, since the amino groups are very prominent (essentially on a spacer arm). Conjugation of such a modified antithrombin with heparin should then be possible. The likely advantages for linkage of antithrombin to heparin through the serpin glycan termini are that the number of linkage points on the antithrombin may be more controlled and there would be no perturbation due to modification of the polypeptide.

ATH synthesis involving simultaneous bonding of antithrombin and heparin with an intermediate agent is likely to be more sophisticated than procedures in which initial activation of either heparin or antithrombin is carried out. All of the functional groups that are present on heparin exist on antithrombin as well. Therefore it would be fortuitous to use heterobifunctional agents that have one functional end that would only react with a functional group found solely on antithrombin. An example would be use of a linking agent that contains one reactive group that would selectively link with the guanidinyl group of antithrombin arg residues (vicinal dione functional group) while the other end of the linker has a photoactivatable group ($—N_3$) with a neighboring plus-charged group ($—N(CH_3)_3^+$) to form a covalent bound with the negatively charged heparin chain. One disadvantage of this approach is that more than one link between antithrombin and heparin may occur, which may result in denaturation of the protein. On the other hand, allowing native antithrombin and heparin to interact prior to covalent modification would give the opportunity for the most active noncovalent antithrombin–heparin complexes to form prior to conjugation.

Investigations into ATH synthesis have been reported before. The vast majority involve the first method (discussed above and illustrated in Fig. 4) in which heparin is activated prior to interaction with antithrombin. The evolution of ATH compound investigation is reviewed below.

V. COVALENT ANTITHROMBIN–HEPARIN COMPLEXES: DEVELOPMENT

A. Chemistry, Physicochemical Properties, and In Vitro Activity

The chronological development of ATH products shows an evolution over time of more sophisticated, selective chemistries that are concomitant with higher anticoagulant activities. A number of studies have reported preparations in which noncovalent complexes of antithrombin and heparin have been produced (193–196). Such complexes are useful for short-term treatment but are not practical for long-term thrombin regulation or anticoagulation at surfaces where binding by the heparin moiety may cause dissociation of the antithrombin-heparin. As the various attempts to join antithrombin and heparin are described, the type of synthetic methods utilized will be related to the structure of the final product in order to explain the observed physical properties and anticoagulant activities.

Initial investigations in the construction of a permanently activated antithrombin relied on chemical linkage of GAGs, particularly heparin, to chromatographic matrices and other macroscopic surfaces. Conjugation of heparinoids to albumin has been accomplished in a manner similar to the methods employed for preparation of ATH compounds, but that literature will not be covered in detail. However, covalent attachment of heparin to other proteins (such as albumin) (197,198), natural polysaccharides (199), and synthetic polysaccharides (199) has added to the knowledge base for coupling of heparin to macromolecules. Modification of functional groups on heparin has been done over most of this century, and characterization of the broad range of reactions used eventually contributed to devising final methods for the conjugation of antithrombin and heparin. Some of the earliest attempts to form ATH were carried out by Ceustermans et al. (192). UFH was activated with CNBr at pH 10 according to the procedure of Cuatrecasas et al. (200), followed by reaction with antithrombin. These experiments failed to produce significant amounts of conjugate. Activation of heparin with CNBr may involve the formation of iminocarbonate or cyanate groups with the hydroxyl groups on heparin but, more likely, CNBr reaction requires the presence of amino groups, of which there are fewer than one per heparin molecule on the average (192). In addition, the amino groups on the heparin molecules may be at locations distal to the antithrombin that is interacting with the GAG pentasaccharide binding site. Further experiments were undertaken to form conjugate through the bifunctional reagent, 1,5-difluoro-2,4-dinitrobenzene (192), according to the methods of Zahn and Meienhofer (201,202). Again, no ATH was produced. Since 1,5-difluoro-2,4-dinitrobenzene would require the presence of heparin amino groups for nucleophilic substitution on the benzene ring to take place, lack of product was probably due to the same reasons as those for CNBr experiments. Modification of heparin with 4-fluoro-3-nitrophenyl azide or 4-azido-phenacyl bromide, followed by reaction with antithrombin via photoactivation of the $—N_3$ group, has been carried out according to the experimental conditions reported previously (203–206), but with negative results.

Successful linkage of antithrombin and heparin was finally obtained by Ceustermans et al. (192) using tolylene-2,4-diisocyanate according to the procedure of Clyne et al.

(207–209). In order to increase reactivity of aminos (or hydroxyls) with the isocyanate groups, reactions were carried out under basic conditions. However, the amount of conjugate obtained was ≤5%. The yield of ATH decreased rapidly due to the facile hydrolysis of isocyanate groups on the substituted heparin during incubation on ice in pH 9.5 borate buffer, just prior to interaction with the antithrombin. Thus the more stable tolylene-2,4-diisothiocyanate (TDTC) reagent was substituted in further attempts to couple antithrombin and heparin. Heparin was modified with TDTC using the procedure of Edman and Henschen (210), followed by reaction with antithrombin in pH 8.5 bicarbonate buffer (192). A yield of 30% was obtained for ATH using the heparin modified with TDTC.

In order further to improve yield and quality of the product with the TDTC reagent, ATH was synthesized following TDTC reaction with heparin that had amino groups introduced into it, either by deblocking amino groups on the glucosamine residues or by modifying the carboxyl groups with an amino-containing spacer arm (192). The resultant product was purified and structurally characterized, and its functional activities were assessed. A scheme for the reaction used to produce ATH with antithrombin linked to amino group modified heparin is shown as reaction 1 in Fig. 5. The high-affinity fraction of UFH passed over an immobilized antithrombin column was used as starting material for all of the ATH prepared from amino-substituted heparin. Partial N-desulfation was performed by incubation of the pyridinium salt of the acid form of heparin (prepared by passage of heparin through Dowex 50) in 95% dimethylsulfoxide for 0.5 h at 23°C (211,212), followed by adjustment of the pH to 9.5 and dialysis versus H_2O for 2 days. Introduction of aminos by carboxyl modification was achieved by incubation of high-affinity heparin with 1,6-diamino-hexane in the presence of a carbodiimide condensing agent. Either of the two amino-modified heparins was reacted with excess TDTC (to prevent cross-linkage of the GAG), after which the activated heparin was reacted with antithrombin via the remaining isothiocyanate group. Chemical analyses of the modified heparins and corresponding ATH products are found in Table 1. Partial N-desulfation of heparin resulted in an approximate two-fold increase in isothiocyanate groups incorporated per activated GAG molecule. Isothiocyanate substitution was improved by a factor of 2.9 in (hexyl-amino)-containing heparin compared to reactions with starting heparin. Yield of ATH (in terms of either starting heparin or antithrombin) was ~30% using either partially N-desulfated heparin or hexyl-amino heparin. Furthermore, up to 25% of the ATH obtained contained two antithrombin molecules per heparin molecule (heparin/antithrombin molar ratio Table 1) (192). Activity measurements of N-desulfated or hexyl-amino heparin were determined using an activated partial thromboplastin time (APTT) clotting time measurement or anti-FXa inhibition assay (Table 1). After partial N-desulfation, heparin APTT activity decreased by 35%, and anti-FXa activity was reduced by 24%. Hexyl-amino heparin had 44% and 39% lower APTT and anti-FXa activities, respectively, compared to starting high-affinity heparin (Table 1). The anti-FXa activities represent the ability of the heparin species to catalyze the inhibition of FXa by exogenous antithrombin (excess added antithrombin that is not covalently bound to heparin). Significant loss of activity occurred in both partially N-desulfated heparin and heparin with diamino-hexyl groups linked to the uronic acid residues, which is not surprising given that it has long been known that the N-sulfate and carboxyl groups are critical for heparin's anticoagulant activity (213). Anti-FXa measurements of the ATH complexes produced using partially N-desulfated heparin were not reported (192). Direct, noncatalytic reaction of the hexyl-amino containing ATH complexes with FXa was studied. Determinations of the reaction velocity revealed that the second-order rate constant for this ATH (Table 1) was $2.1 \times 10^6\ M^{-1}s^{-1}$ (~3 times lower than that for noncovalent mixtures of unmodified high-affinity heparin + saturating amounts of antithrombin) (192). Addition of heparin to the ATH doubled the rate of direct FXa inhibition, suggesting that not all ATH heparin was coupled so that the antithrombin was in the active conformation (192). Thus exogenous heparin increased the activity for these nonactivated molecules. ATH prepared from the amino-modified heparins had been purified by combinations of anion exchange, gel filtration, and sepharose-antithrombin affinity chromatographies (192). Further purification of hexyl-amino heparin-produced ATH on sepharose-concanavalin A allowed removal of residual free heparin and possibly some inactive complexes (214). The highly purified ATH was tested in reactions with thrombin, and a second-order rate constant of $6.7 \times 10^8\ M^{-1}s^{-1}$ was measured (Table 1). A bimolecular rate constant for reaction with thrombin of $2.2 \times 10^8\ M^{-1}s^{-1}$ was determined with high-affinity heparin in the presence of saturating amounts of antithrombin (compared to $2.5 \times 10^8\ M^{-1}\ s^{-1}$ for inhibition of thrombin by ATH) (214). Thus ATH that was an isolate of highly active covalent complexes of one amino-hexyl-thiourea-tolylene-isothiocyanate heparin molecule and one antithrombin molecule had direct inhibitory reactivity with thrombin that was essentially the same as that for inactivation of thrombin with noncovalent mixtures of heparin + saturating amounts of antithrombin. No specific antithrombin activity values have been reported for the hexyl-tolylene linkage-containing ATH. Hoylaerts et al. have also prepared ATH from high-affinity hexyl-amino containing heparin using

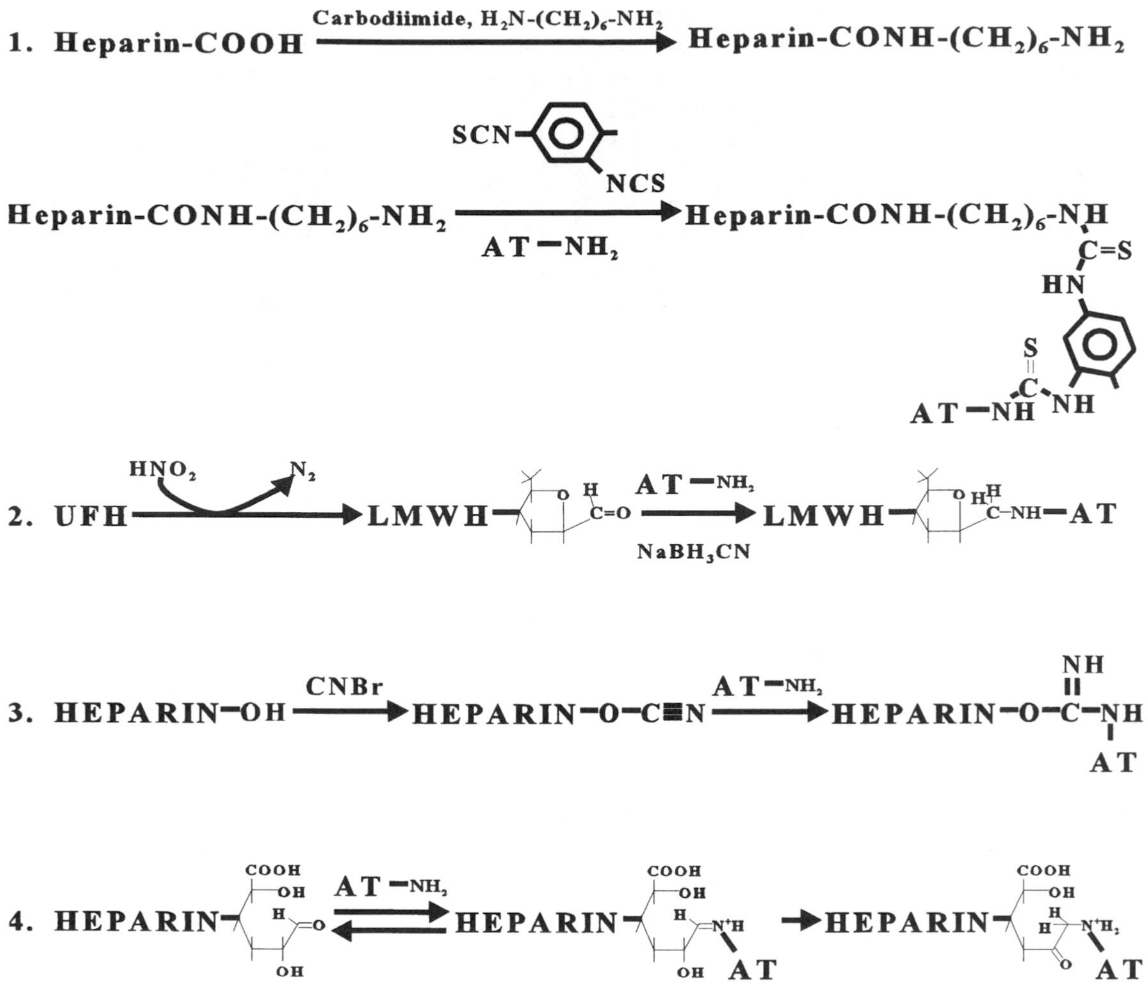

METHODS:

1. **Ceustermans** ***et al.***

2. **Björk** ***et al.***

3. **Mitra and Jordan**

4. **Chan** ***et al.***

Figure 25.5 Synthetic methods used to prepare various covalent antithrombin–heparin complexes.

Table 25.1 Physicochemical Properties and Activities of Heparin and Covalent Antithrombin–Heparin Complexes Prepared by Various Methods

Synthetic method	Start. heparin	Activating groups per heparin (mole/mole)	ATH H/AT (mole/mole)	Mod. heparin APTT (% start. heparin)	Activity: Anti-FXa catalytic activity (U/mg) Start. heparin	Mod. heparin	ATH	Anti-IIa catalytic activity (U/mg) Start. heparin	Mod. heparin	ATH	ATH + FXa k ($M^{-1}s^{-1}$)	ATH + IIa k ($M^{-1}s^{-1}$)
1	High affinity UFH	0.9 ± 0.3	0.9	65 ± 5	250	191 ± 28	ND	ND	ND	ND	ND	ND
2	High affinity UFH	1.9 ± 1.4	0.8	56 ± 7	250	153 ± 34	270	ND	ND	ND	2.1×10^6	6.7×10^8
3	UFH modified to LMWH	1	0.7	ND	~170 (UFH)	140 (LMWH)	5.2	~ 170 (UFH)	0 (LMWH)	0	ND	ND
4	UFH	≥1	≥1	ND	168	ND	low	168	ND	low	ND	6.7×10^7
5	UFH	0 – 0.18	1.1	100	209	209	861	198	198	754	3.8×10^6	3.1×10^9

Definitions: Start. heparin = Starting source heparin used for experiments; ATH = antithrombin–heparin covalent complex; H = heparin; AT = antithrombin; Mod. heparin = Starting heparin after chemical activation but before conjugation with antithrombin; APTT = activated partial thromboplastin time; Anti-FXa = antifactor Xa heparin activity (catalysis of the FXa + AT reaction); Anti-IIa = antithrombin heparin activity; k = rate constant (bimolecular or second order); High affinity = Fraction with highest binding strength to antithrombin; UFH = commercial unfractionated heparin; LMWH = low molecular weight heparin obtained by partial depolymerization of UFH with HNO_2; ND = Not determined.

Synthetic Methods:

1. Partially N-desulfated high-affinity heparin conjugated to antithrombin with tolylene-2,4-diisothiocyanate. (*Source*: Ref. 192.)
2. Hexyl-amino substituted high-affinity heparin conjugated to antithrombin with tolylene-2,4-diisothiocyanate. (*Source*: Refs. 192, 236, 214.)
3. Reductive alkylation of antithrombin with high-affinity anhydromannose-terminating heparin and $NaBH_3CN$. (*Source*: Refs. 215, 217.)
4. Conjugation of antithrombin to CNBr-activated heparin. (*Source*: Refs. 218, 219.)
5. Incubation of antithrombin with aldose-terminating unfractionated heparin to form a Schiff base between an antithrombin lysyl amino group and the heparin aldose aldehyde, which undergoes an Amadori rearrangement. (*Source*: Refs. 118, 230, 233.)

LMWH obtained by HNO_2 treatment of UFH [followed by size fractionation to give LMWH with chain lengths of 10–14 monosaccharide units (M_r = 3,200) or 14–18 monosaccharide units (M_r = 4,300)] (214). However, second-order rate constants for reaction with thrombin were significantly reduced for the LMWH containing ATH (3×10^5 $M^{-1}s^{-1}$ and 2×10^7 $M^{-1}s^{-1}$ for 3,200 and 4,300 M_r heparin containing ATH, respectively), compared to the corresponding ATH with standard molecular weight (M_r = 15,000) heparin.

Almost coincident to the ATH prepared by Ceustermans et al., Björk et al. produced ATH using end-group attachment to the heparin moiety (synthetic scheme 2, shown in Fig. 5) (215). UFH was treated with HNO_2 to cause partial deaminative cleavage. High-activity LMWH fragments were obtained by immobilized-antithrombin affinity chromatography, followed by covalent linkage to antithrombin by reduction of the Schiff base formed between the aldehyde group of the LMWH anhydromannose termini and lysyl amino groups on the protein (reductive alkylation) (215). Conjugate was isolated by gel filtration to separate ATH and antithrombin from unreacted LMWH, followed by heparin agarose affinity chromatography to separate ATH from free antithrombin. After completion of the purification procedures, determinations showed that ~40% conversion of the original antithrombin to ATH occurred. The molecular weight of heparin prepared by this partial deaminative cleavage method has been shown to range from 3,700 (215) to 10,000 (216). Unlike heparin in the hexyl-tolylene linked ATH, the LMWH generated by HNO_2 treatment of UFH has only one active group per molecule located at the terminal anhydromannose residue (Table 1). Therefore only one type of orientation of the heparin on the antithrombin can occur (end-point attachment). Analyses showed that ATH preparations, formed by reductive alkylation of antithrombin with high-affinity anhydromannose-terminating heparin, had an average of 0.7 moles of GAG chain conjugated per antithrombin (Table 1). LMWH obtained by HNO_2 for conjugation (prior to isolating the high-affinity fraction) was determined to have an anti-FXa activity of 140 units/mg, which is slightly lower than the UFH starting material (217). Titration with FXa showed that the Björk et al. ATH complex had an activity which was 98% of that for the high-affinity LMWH material used in the preparation (Table 1). Interestingly, although the anti-FXa activity of noncovalent mixtures of LMWH + antithrombin could be neutralized by high NaCl concentrations (1 M) or polybrene, reactivity of the Björk et al. ATH with FXa was only slightly decreased by high salt or polybrene (215). Therefore it may be concluded that covalent linkage of the HNO_2-produced LMWH to antithrombin prevents removal of the activating heparin species from the serpin by ionic media. Thus the Björk et al. ATH complex is at the same time potent in a variety of conditions and difficult to reverse if used for clinical treatment. Unfortunately, both the LMWH prepared using HNO_2 and the resultant ATH had no measurable antithrombin activity (217). The lack of activity against thrombin is not surprising, given that short-chain LMWH cannot provide the template required for binding both antithrombin and thrombin (115,131).

Further development of ATH by conjugation of heparin and antithrombin using CNBr was accomplished by Mitra and Jordan (reaction scheme 3, Fig. 5) (218). UFH was activated at pH 10.7 with ~ a 57-fold molar excess of CNBr for 40 min at 23°C, after which excess CNBr was removed by dialysis under basic conditions. A cyanate/iminocarbonate/N-nitrile active group–containing heparin was then incubated with 0.02 moles of antithrombin per mole of GAG at pH 9.4 for 18 h at 5°C. Separation of free heparin from ATH and free antithrombin was accomplished by chromatography on immobilized Concanavalin A, and isolation of ATH from antithrombin was done using a heparin-sepharose column (218). A yield for ATH of 40% (in terms of antithrombin) was estimated. Complexes with antithrombin were also produced using LMWH (obtained by fractionation of UFH on gel filtration columns), but no further characterization of the products was done. Data on the products of this CNBr-based synthetic procedure are shown in Table 1. No information on the number of activating groups formed on the heparin molecules by CNBr (cyanate/iminocarbonate/N-nitrile) was given (218). However, given the molar ratios involved and that Mitra and Jordan suggest the use of glycine to block any excess active groups after conjugation to antithrombin (218,219), it is likely that more than one reactive group per GAG chain was obtained and that ≥1 heparin was conjugated to each antithrombin in the ATH complexes. Titrations of Mitra and Jordan ATH with thrombin and FXa suggested that essentially all of the complex had rapid, direct inhibitory activity. Also, the increased intrinsic fluorescence of heparin-activated antithrombin was observed with the Mithra and Jordan. ATH, which was not further enhanced by addition of UFH. The ability of CNBr-conjugated ATH to catalyze inhibition of either FXa or thrombin by added antithrombin (exogenous to the antithrombin in the covalent complex) was not clearly investigated. However, description of thrombin and FXa inhibition assays of the complex in plasma suggests that the inhibitory capacity was only that of the covalent complex (no significant catalytic activity) (218,219). The rate of reaction of the Mitra and Jordan ATH with thrombin was investigated by generating inactivation time curves with equimolar enzyme and inhibitor concentrations (which allow for bimolecular rate constant calculation) (218). Results showed that the conjugate reacted with thrombin at a bimolecular rate of $6.7 \times 10^7\ M^{-1}s^{-1}$ (Table 1). Interestingly, noncovalent mixtures of antithrombin + high affinity heparin gave a bimolecular rate constant for reaction with thrombin of $2.0 \times 10^7\ M^{-1}s^{-1}$, which was 3.3 times lower than that for ATH. The authors suggested the possibility that the ATH may react with thrombin at a faster rate than noncovalent antithrombin + heparin because certain equilibrium constraints which apply to the free component system (such as association of the serpin with the GAG) are circumvented by the covalent complex. Unfortunately, since the rate measurements were not done under first-order conditions (reactants not at 5-to-10-fold higher concentrations with respect to the enzyme being inhibited), comparison of absolute values for reaction rates of the noncovalent and covalent complexes cannot be done. No measurement of the rate of inhibition of FXa by the Mitra and Jordan ATH was reported.

More recently, a new approach has been put forward by Chan et al. (118) that utilizes the chemistry of glucose and plasma proteins occurring in diabetics (see Fig. 5). It has long been known that hemoglobin and various plasma proteins undergo Schiff base formation between the aldehyde function on C_1 of plasma glucose and the ε-amino of lysyl residues of the protein (220). This intermediate product is metastable and in equilibrium with the starting materials. However, as shown in Fig. 5, over time, a tautomeric rearrangement (Amadori) between protons and groups on C_1 and C_2 occurs to form an ene-ol-amine and finally a stable keto-amine (220). The formation of Amadori rearrangement products is dependent on glucose concentration, pH, temperature, and availability of amino functions. However, once formed, nonenzymatically glycated proteins have been shown to survive in diabetic tissue from weeks to years after hyperglycemic exposure (220). Reports have indicated that nonenzymatic glycation can have effects on protein function (221). In the case of antithrombin, brief increases in blood glucose concentration have been shown to result in short alterations of antithrombin activity without loss of protein (likely due to nonenzymatic labile Schiff base formation) (222). Significant decreases in antithrombin activity due to longer term glycation has been noted (223), which may explain some of the pathogenesis for thrombosis in severe diabetics (224). Moreover, nonenzymatic glycation of antithrombin in vitro (225) and in vivo (225) has been found to inhibit heparin-catalyzed antithrombin activity. Even more fascinating was that glycation-induced effects on heparin-catalyzed antithrombin activity were completely reversed by incubation with excess Na heparin before assay (225). Therefore reaction of glucose with a particular antithrombin lysyl residue may be occurring to produce Schiff base/Amadori products that can be displaced by a large excess of heparin. Further evidence of selective glycation of antithrombin can be seen in that even in the presence of up to 5 mM glucose for 10 days, only 0.6 moles of glucose can be covalently bound per mole of protein (224). Although other literature has shown that a number of regions of the antithrombin molecule may form the keto-amine adducts, only one glucose per protein molecule could be yielded during incubations with up to 0.5 mM sugar (226)! Thus once one glycation site is affected, the probability of further glycation is drastically reduced. Given the effects of heparin on glycated antithrombin, the possibility arises that heparin molecules containing aldose residues of their own might be responsible for reversing (competing with) the Schiff base/Amadori structures being formed. Thus if UFH contained significant numbers of aldose-terminating molecules, it may be possible to form covalent ATH by simple incubation of unmodified UFH and antithrombin under the appropriate conditions. Previous work had shown that Schiff base/Amadori rearrangement of insulin by the disaccharide maltose can proceed (227). However, numerous workers had concluded that even Schiff base formation between polysaccharide aldoses and amino groups would be extremely limited because the aldose terminus represented a very small part of the sugar residues in the polymer, the equilibrium would be highly in favor of the hemiacetal (masked aldehyde) form, and significant steric hindrance for the approach of macromolecular amines seemed likely (228).

Chan et al. decided to test the hypothesis of ATH formation via Schiff base/Amadori rearrangement of UFH and antithrombin. It had been determined before that in commercial UFH preparations, approximately 10% of all molecules terminated in a free aldose (229). Since it would be desirable to have ATH complexes in which the antithrombin was activated by the covalently linked heparin, incubation conditions would need to be designed so that antithrombin molecules could have the opportunity to select for heparin with noncovalent pentasaccharide binding sequences that were close to the aldose end of the chain (118). Feasibility studies showed that nonenzymatic glycation of polypeptides by polysaccharide aldose was possible, but yields were greatly variable for a number of reasons (230). A format for ATH production via Schiff base/Amadori rearrangement was established. UFH and antithrombin from various sources were mixed in physiological buffer and heated for up to 16 days. Purification of the ATH formed was by a two-step procedure of hydrophobic chromatography (to remove unbound heparin), followed by separation of unbound antithrombin on DEAE sepharose (118). Since it had been shown that incubation of antithrombin with monosaccharide aldose could increase the temperature for thermal denaturation (Td) (231), interactions at different temperatures were studied. Yields of conjugate (in terms of antithrombin) were observed to be ~5% at 37°C, while ~50% conversion of antithrombin to complex was observed at 40°C (230). Properties of the Chan et al. ATH are found in Table 1. Analyses indicated that ~ one molecule of heparin was combined with each antithrombin in the complex (118). This would be in agreement with the binding of a heparin chain to the antithrombin binding site (through the high affinity pentasaccharide), followed by Schiff base/Amadori rearrangement of the aldose terminus with a lysyl residue in the vicinity. Since no prior modification of the heparin was used before conjugation, the number of activating groups, anti-FXa activity, and anti-IIa activity of the heparin in the coupling reaction were those of the starting UFH. Surprisingly, the Chan et al. ATH was found to possess catalytic activity for the inhibition of either FXa or thrombin by exogenous antithrombin (Table 1) (118). Furthermore, the catalytic anticoagulant activity of the covalent complex was greater than that for either the starting UFH or heparin with high affinity for antithrombin (118). This startling result was proposed to be due to the selection, during incubation with antithrombin, of not only heparin chains with a pentasaccharide but some GAG molecules with more than one pentasaccharide unit. The presence of more than one high-affinity antithrombin binding site in a subpopulation of UFH molecules (1–3%) has been described before (119,232). Selection of this minor subpopulation of UFH chains by antithrombin was made possible by using a vast molar excess (>200 fold) of UFH to antithrombin during the synthesis (118). It was posited that, since the interaction of incubating antithrombin and UFH would initially be via noncovalent binding to pentasaccharide sequences, heparin chains with more than one high-affinity site might be preferred because the intramolecular mean distance of diffusion would be less than the intermolecular mean free distance of diffusion, and there is a higher probability that at least one site may be close to the aldose terminus (where covalent linkage would occur). Further studies confirmed that, in fact, the Chan et al. ATH did contain a significant number of molecules with more than one pentasaccharide (233). Given that formation of Chan et al. ATH involved simple incubation of unmodified heparin with the antithrombin, it was not surprising to find that direct reaction of the inhibitor with thrombin proceeded with a second-order rate constant (3.1×10^9 $M^{-1}s^{-1}$) that was one of the fastest ever recorded (118).

B. In Vivo Performance

Further studies have been accomplished that investigated the in vivo properties of many of the ATH compounds described above. Investigations ranged from determination of pharmacokinetics to antithrombotic treatment of disease models in animals. Data for the different ATH products are given in Table 2.

ATH produced from the high-affinity fraction of UFH, according to the technique of Ceustermans et al. (192), has been tested for a number of characteristics in animals. Initial studies showed that, in rabbits, ATH joined to partially N-desulfated high-affinity heparin disappeared from plasma in a biphasic pattern with a short first phase (α) and a longer second phase ($\beta = 0.65 \pm 0.06$ h) (192). Further studies were carried out to characterize ATH made by joining hexyl-amino-substituted high-affinity heparin to antithrombin using diisothiocyanate. Using antithrombin linked to high-affinity hexyl-amino containing heparin (from UFH), disappearance from the circulation in rabbits could be described by a single phase with half-life of 0.89 ± 0.26 h (192). Neither of these ATH complexes had half-lives that were more than ~3-fold longer than UFH in rabbits (single phase = 0.23 ± 0.03 h), and both were well short of the half-life of antithrombin (11.0 ± 0.4 h) (192). As with UFH, covalent complexes of the hexyl-amino containing heparin with antithrombin may be eliminated by binding of the heparin moiety to basic plasma and cell surface proteins; since it was shown that these conjugates could also bind to histidine-rich glycoprotein (148). In order to extend longevity in the circulation, ATH from high-affinity hexyl-amino substituted LMWH (3,200 and 4,300 MW, produced by partial depolymerization with nitrous acid) was prepared and the turnover parameters in rabbits measured (234,235). Both LWMH ATH complexes were lost from plasma with β-phase half-lives of ~7–8 h (Table 2). This disappearance rate for the high-affinity hexyl-amino LMWH ATH complexes was 30 times longer than that for UFH in rabbits (0.25 ± 0.04 h). However, the half-lives of the unbound 3,200 and 4,300 starting high-affinity heparins were 2.35 ± 0.17 h and 2.50 ± 0.18 h, respectively, only ~ 3 times shorter than the corresponding conjugates with antithrombin (234). In addition, it was claimed by the authors that the covalent complexes (Ceustermans et al. ATH prepared from high-affinity UFH or LMWH) possessed the ability to catalyze reactions with the animal's own antithrombin. Thus, even with an increased half-life, any antithrombotic properties might be determined by their rate of consumption by activated coagulation factors, as opposed to how long the compound or its metabolites remained in the circulation. The antithrombotic properties of ATH prepared from either high-affinity hexyl-amino UFH or LMWHs were tested in a Wessler rabbit thrombus prevention model (235,236). Anesthetized rabbits, injected with test compounds, were injected with

Table 25.2 In Vivo Characteristics of Covalent Antithrombin–Heparin Compounds

Compound	Intravenous half-life (h)		Basic plasma protein binding	Bleeding characteristics	Antithrombotic activity (given equal mass concentration doses)
	α-Phase	β-Phase			
1	<0.5	0.65 ± 0.06	ND	ND	ND
2	0.89 ± 0.26	—	Yes	ND	Same as noncovalent mixtures
3	1	7.85 ± 0.18	ND	ND	Same as noncovalent mixtures
4	1	7.63 ± 0.75	ND	ND	Same as noncovalent mixtures
5	>UFH?	—	ND	ND	ND
6	<3	—	ND	ND	ND
7	2.6	13	Low	Similar to UFH	Efficacy superior to UFH

Definitions: UFH = Commercial unfractionated heparin; ND = Not determined.
Synthetic Methods:
1. Partially N-desulfated high-affinity heparin conjugated to antithrombin with tolylene-2,4-diisothiocyanate. (*Source*: Ref. 192.)
2. Hexyl-amino substituted high-affinity heparin conjugated to antithrombin with tolylene-2,4-diisothiocyanate. (*Source*: Refs. 192, 214, 236.)
3. Hexyl-amino substituted high-affinity low molecular weight heparin (MW = 3,200) conjugated to antithrombin with tolylene-2,4-diisothiocyanate. (*Source*: Refs. 234, 235.)
4. Hexyl-amino substituted high-affinity low molecular weight heparin (MW = 4,300) conjugated to antithrombin with tolylene-2,4-diisothiocyanate. (*Source*: Refs. 234, 235.)
5. Antithrombin reductively alkylated with anhydromannose-terminating heparin. (*Source*: Refs. 215, 217.)
6. Antithrombin conjugated to CNBr-activated heparin. (*Source*: Refs. 218, 219.)
7. Antithrombin coupled to heparin by Schiff base/Amadori rearrangement. (*Source*: Refs. 118, 230, 233.)

glass-activated plasma and stasis of a 2 cm length of isolated jugular vein induced for 10 min, followed by semi-quantitative analysis of the number and size of thrombi expressed within the vessel segment (236). Both effects of dosage and time delay from treatment to insult were examined. Results, up to 45–60 min after administration of anticoagulant, showed that all covalent complexes showed similar antithrombotic effects compared to that for injection of the same mass concentrations of the corresponding noncovalent mixtures of antithrombin + heparinoid. However, antithrombotic effects (reduction in size or number of thrombi) could be detected up to 60–120 min for the conjugates, due to the increased half-lives of ATH compounds compared to the corresponding free heparins (236). Hemorrhagic properties of Cuestermans et al. ATH produced from either UFH or LMWH have not been reported.

Information of in vivo use of Björk et al. ATH compounds is scarce. Dawes et al. have prepared ATH according to the method of Björk et al. and injected it, by intravenous bolus, into 12-week-old mice (217). Although no in vivo pharmacokinetic data were reported, administration of only three injections of small amounts of conjugate were necessary to obtain antibodies against the human antithrombin in the complex. Furthermore, interaction of radiolabelled ATH with hybridoma cells in vitro showed very low nonspecific binding to cell surface molecules (217). These data may suggest that the Björk et al. ATH complex has a low affinity for binding to proteins in vivo, which would enhance the half life. Unfortunately, assay of the anti-FXa activity of the conjugate gave a very low activity (5.2 U/mg) and no definite indication exists of its antithrombotic properties.

Mitra and Jordan have tested their ATH in vivo (218,219). Rabbits were given intravenous infusions of 84 or 200 anti-FXa U/kg using either ATH or UFH, respectively, at rates of 2 mL/kg. On completion of infusion, blood samples were taken over time and the supernatant plasma tested for anticoagulant activity. It was noted that, compared to the UFH group, animals receiving ATH had a significant prolongation of functional activity (as measured by activated partial thromboplastin time assays) for up to 3 h after discontinuation of treatment (218). This result would suggest that, at lower doses, a longer period of anticoagulant prophylaxis can be achieved with the Mitra and Jordan covalent ATH conjugate compared to use of free UFH alone. It is likely that this effect would be due to a longer intravenous half-life of ATH prepared from CNBr-activated heparin, since elimination of UFH from the rabbit circulation at this dosage would be less than 3 h (237), and lower activity of ATH than UFH was administered. Nevertheless, direct determination of the loss of ATH compound by mass analyses of the plasma time samples were not reported, and antithrombotic efficacy was not investigated.

A significant number of experiments have been performed to evaluate the in vivo character of ATH produced by Chan et al. Bolus injection of the (keto-amine)-linked ATH into rabbits gave an intravenous α-phase (two-compartment model) half-life of 2.5–2.6 h for both anticoagulant activity and mass (compared to 0.32 and 13 h for UFH and antithrombin, respectively) (118). Using a three-compartment two-exponential decay model, β-phase half-lives of 13 h and 69 h were calculated for loss of the ATH and antithrombin, respectively, from rabbit plasma. This terminal phase half-life was the longest ever reported for any ATH complex in vivo. Single subcutaneous injection of Chan et al. ATH resulted in systemic appearance of complex, which had peak concentrations at 24–30 h and could be detected up to 96 h after administration (118). However, only a small portion of the injected ATH was recovered in the plasma, indicating a very poor bioavailability of the subcutaneously administered Chan et al. compound. Given the long intravenous half-life of Chan et al. ATH, a likely explanation for the low total plasma levels of the subcutaneously administered adduct is that the large molecular size of ATH drastically reduced its permeability through extravascular tissue. This hypothesis led to the suggestion that Chan et al. ATH may be useful as an anticoagulant in sequestered spaces. One possibility is for anticoagulant treatment of respiratory distress syndrome in the lung (180,238–241), where administration of heparin is not useful due to ready loss of GAG to the circulation (21). Intratracheal instillation of keto-amine conjugated ATH into the lung airspace in rabbits led to significant anticoagulant activities and antithrombin antigen concentrations for up to 48 h after administration (118). Furthermore, no ATH protein could be measured in the plasma of the treated rabbits, indicating that ATH was retained in the lung without any leakage systemically. Functional activity of Chan et al. ATH as an anticoagulant on the alveolar surface of the lung is likely. Inhibition of thrombin generation in plasma on the surface of fetal distal lung epithelial cells in vitro was shown to be much more efficient with Chan et al. ATH than with similar concentrations of noncovalent mixtures of UFH and antithrombin (242). The superior efficacy of ATH on inhibition of epithelial-based plasma thrombin generation was due, in part, to the very rapid direct inactivation of initial thrombin feedback activation for the conversion of prothrombin to thrombin (242). Further tests to evaluate the antithrombotic potential of Schiff base/Amadori prepared ATH were carried out. ATH was compared to noncovalent mixtures of UFH + antithrombin in a rabbit treatment model of preformed jugular vein thrombi. Results showed that administration of a single intravenous bolus of Chan et al. ATH caused a reduction in clot mass and fibrin accretion, whereas similar (by mass) injections of noncovalent mixtures of UFH + antithrombin led to an increase in clot size (237). Parallel experiments using a rabbit bleeding ear model showed no significant difference in cumulative blood loss between covalent complex and noncovalent mixtures of UFH and antithrombin (237). Thus Chan et al. ATH was not only a highly potent antithrombotic but could be used effectively at safe doses. Data from the thrombosis treatment model were suggestive of the possibility that the conjugate tested may function via the direct noncatalytic inhibition of clot-bound thrombin. As stated earlier, noncovalent antithrombin · UFH has significantly reduced activity against fibrin-bound thrombin compared to that against fluid-phase thrombin. Further investigations with soluble fibrin have shown that the keto-amine adduct of antithrombin and heparin can directly inhibit fibrin-bound thrombin with a similar rate constant to that of thrombin in buffer only (243). This finding has major implications regarding possible indications for the clinical use of ATH. One rationale for the lack of effect of fibrin on the inhibition of thrombin by Chan et al. ATH is that binding of the ATH heparin chain to fibrin is reduced, due to steric hindrance by the covalently linked antithrombin. In that case, ATH heparin would not be able to form a ternary complex with fibrin and thrombin. Additionally, it has been noted, from comparison of the catalytic anti-FXa and antithrombin activities of ATH in either plasma or buffer, that the effects of nonselective plasma protein binding on the anticoagulant properties of Chan et al. ATH are reduced compared to those of UFH (private communication from the author). Given the Schiff base/Amadori mechanism by which antithrombin and commercial UFH combine to form Chan et al. ATH, the fascinating possibility presented itself that such a reaction may, in fact, occur spontaneously during the clinical use of UFH. Attempts were made to isolate ATH from plasmas of rabbits and a human injected subcutaneously with UFH (230). Using preparative procedures similar to that employed by Chan et al. for purification of their ATH, covalent adducts of heparin and antithrombin were detected in plasma samples of UFH-treated rabbits and humans (230). Thus given that endogenous, nonprotein linked heparin chains are known to circulate in humans (244–246), it is apparent that Chan et al. ATH may be a natural product.

VI. COVALENT ANTITHROMBIN–HEPARIN COMPLEXES: ATTACHMENT TO POLYMER SURFACES

A. Biochemical and Chemical Overview

Biomaterials that come into contact with blood have a number of limitations. First, cells such as platelets and polymorphonucleocytes can bind to the foreign surfaces and become activated (247,248). Second, binding of plasma proteins can cause modification of the surface properties towards other molecules or cells. The noncovalent plasma protein absorption encourages cells and other materials to bind in a layered effect that may even lead to constriction or change in the laminar flow characteristics of the blood. Third, interaction of some biomaterials with the blood cells or surrounding tissue can invoke inflammatory processes (249). In this case, inflammation can be the result of a plethora of factors such as providing a surface for infectious agents that induce inflammation, damaging blood cells or tissue that respond by arachadonic acid metabolism, and generating products at the biomaterial surface that are secondary to the inflammatory pathway. Fourth, the polymer composition or functional groups within the matrix of biomaterials may either contain epitopes or react with the blood to produce epitopes that cause an immunogenic response. Induction of complement activation by surface membranes, such as those used in dialysis, may actually exacerbate recovery of the patient (250). Finally, the surface may be procoagulant. Of all the problems listed above, surface-induced activation of the coagulation cascade is probably the most common difficulty encountered with blood-contacting materials. Materials are often coated on the polymeric surface in order to decrease the procoagulant nature of the foreign agent. However, such coatings frequently suffer from either incomplete anticoagulation of the entire surface or long-term instability, whereby the coating becomes modified to render it inactive, or some of the coated molecules are lost to the circulation (leaving part of the material unprotected and producing a systemic anticoagulant effect).

Covalently coupled antithrombin and heparin is an agent that may provide a surface covering that could overcome a number of the inherent blood incompatibilities of synthetic products. First, since antithrombin is often found noncovalently bound to the heparan sulfate (heparinlike) proteoglycans on the luminal surface of blood vessels (251), ATH could be a very good model for the native vein or artery. Second, since the heparin portion of ATH already has permanently attached antithrombin, it is less likely that absorption of other proteins in the blood will occur, for steric reasons. Third, both antithrombin and heparin have been shown to have anti-inflammatory properties. Heparin coated surfaces have long been known to possess reduced inflammatory effects due to decreased humoral, cellular, and complement activation by this very negative polymer (252). More recently, however, antithrombin has also been identified as having an anti-inflammatory effect in such conditions as septic shock (253). Fourth, since both antithrombin and heparin are natural products occurring in the human vascular system, very little immune response should be engendered by their combination. Low antigenicity should certainly be possible for the ATH produced by Chan et al., given that it is produced by spontaneous Schiff base/Amadori rearrangement of unmodified, native antithrombin and UFH. As would be expected, ATH should have potent anticoagulant activity for the direct inhibition of thrombin. However, once the covalently linked antithrombin in ATH is consumed by thrombin, FXa, etc., any further anticoagulant activity of a biomaterial coated with the conjugate would have to come from the heparin moiety. Thus not all ATH products may be as effective in providing a long-term, highly potent anticoagulant surface. Again, the high catalytic activity of the Chan et al. ATH towards inhibition of thrombin by plasma antithrombin in the patient would be a significant advantage for ongoing pacification during insults leading to thrombin generation near or at a biomaterial surface.

One ancillary advantage for the coating of biomaterials with ATH compared to heparin involves technical issues. Since antithrombin has a greater variety of functional groups than heparin, there is a wider range of chemistries available for covalent as well as noncovalent attachment of the serpin than that of the GAG. Carbamate, guanidinium, imidazole, indole, and phenolic groups on the protein allow for selective linkage of ATH through antithrombin, instead of the heparin chain. In addition, antithrombin's 37 lys residues (human) provide a very enriched amino content compared to heparin (192). Thus coupling reactions involving NH_2 functions would be strongly biased towards selection of ε-amino groups on antithrombin, as opposed to the serine amino or the small number of glucosamine residues on heparin. Since the NH_2 group has itself a large number of reactivities, connection of the surface or surface-spacer arm to ATH via primary amino functions on antithrombin would be an easy operation with the number of commercial reagents and conditions available. As alluded to above, noncovalent binding of biomaterial surfaces to the antithrombin of ATH is feasible. It has been shown that the Chan et al. ATH and antithrombin can be separated from free heparin by hydrophobic chromatography on butyl-agarose (118), which verifies that hydrophobic surface absorption of ATH would be through the serpin and not the GAG component. Again, this is advantageous in that it is preferable to bind ATH to the blood-contacting surfaces through the antithrombin, thereby allowing the heparin to be directed out into the fluid phase. Stronger attachment of active molecules to biomaterials, using antithrombin as the link point, would result if multiple covalent bonds from the surface to ATH are formed. The number and variety of amino acid R groups in the ATH protein lend a greater efficacy for joining the conjugate to the surface by several bonds, which would give a more stable coating that is less likely to be removed in vivo. Moreover, if heparin were joined to the biomaterial surface through many link-points, the likelihood that the pentasaccharide sequence will be adversely affected would increase significantly. Also, surface binding of heparin by bonds at several positions will almost certainly cause steric problems for the approach of either plasma antithrombin or thrombin during catalytic inhibition (the orientation of the heparin is improper). In fact, this is the major reason for the significantly superior commercial performance of blood-contacting products that have end-point-linked heparin, compared to those that do not. One other factor that makes coating with ATH advantageous is, simply, the dual functioning of the conjugate. Covering polymer surfaces can be done with either direct (hirudin) or indirect (heparin) thrombin inhibitors. In favor of the direct (noncatalytic) inhibitors is the rapid reaction with thrombin that is not limited by the requirement of a secondary molecule (is not affected by the metabolic environment of the patient to supply an inhibitor or cofactor). On the other hand, surface-bound indirect (catalytic) inhibitors would not be limited to a single knockout of thrombin but could, essentially, enable the inhibition of limitless amounts of ongoing thrombin generation, as long as they are supplied with plasma phase inhibitor molecules. Chan et al. ATH combines both very rapid noncatalytic and potent catalytic inhibition of thrombin in the same molecule. Therefore coatings with Chan et al. conjugate may, in the short term, quickly eliminate small amounts of thrombin that cause feedback activation when biomaterials are first in contact with blood, as well as catalyze the reaction of plasma antithrombin + thrombin for long-term patency.

B. Characteristics of Coated Surfaces

At present the only ATH product that has been attached to a biopolymer is the Schiff base/Amadori rearrangement or keto-amine linked conjugate prepared by Chan et al. Preliminary testing of the material, in vitro and in vivo, has shown promise for the conjugate when covalently linked to grafts composed of synthetic polymers that are found in commercial use.

ATH has been linked to two types of polyurethane. One is a polymer of polycarbonate with urethane extenders and the other material is an ester-grade polyurethane. The polycarbonate urethane tubing used was composed of a finely threaded material that in some cases was woven around a flexible stainless-steel wire mesh in order to provide further elasticity for application as an endoluminal vascular graft. This polycarbonate urethane came in several diameters and was produced by Corvita Corporation under the trade name Corethane™. The other material utilized in investigations was composed of a pure polyurethane that contained no plasticizers and low levels of extractables.

Pure polyurethane tubing was a clear, flexible, impermeable product that had resistance to fuels, oils, and a number of hydrophobic solvents. Developed by Nalgene, pure polyurethane conduit came in several wall thicknesses and was ideal for quick assessment of results from the different chemistries tested for graft polymers prepared to covalently link ATH to devices, etc.

Chan et al. ATH was attached to polymer surfaces and tested for various properties (254). Attachment was covalent and involved linkage to a graft poly/oligomer on the two types of polyurethane tubing described above. Chemical bonding was performed in three stages and is shown schematically in Fig. 6. Polycarbonate urethane was activated by reaction with NaOCl, after which the resultant N-chloro urethane groups were incubated with an initiator (i.e., $Na_2S_2O_4$) and a functional group containing monomer (allyl glycidyl ether). After reaction with the activated urethane surface, the tubing was washed and there was incubation with ATH for covalent linkage between the graft epoxide groups and lysyl amino groups of antithrombin in the ATH. Under the conditions employed, it has been shown previously that oligomers of the grafted species containing from 1 to 4.6 monomer units per urethane group can be obtained on the average (255,256). This polymer chemistry is based on the general mechanisms of radical-initiated polymerization on poly(N-chloroamide) donor

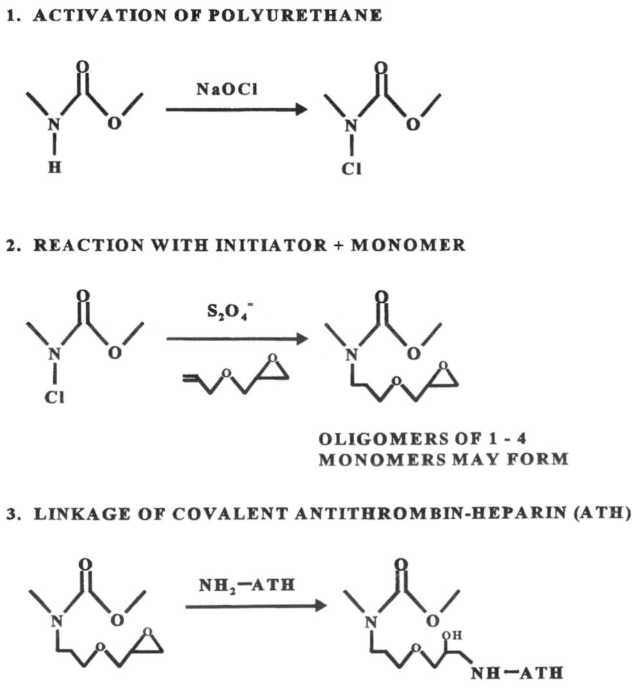

1. ACTIVATION OF POLYURETHANE

NaOCl

2. REACTION WITH INITIATOR + MONOMER

$S_2O_4^-$

OLIGOMERS OF 1 - 4 MONOMERS MAY FORM

3. LINKAGE OF COVALENT ANTITHROMBIN-HEPARIN (ATH)

NH_2—ATH

NH—ATH

Figure 25.6 Covalent coating of antithrombin–heparin conjugates onto polyurethane.

surfaces (257). Literature reviews have concluded that surface coating relies on a number of parameters such as surface area and mode of drug attachment (258). Therefore all steps in the process must be appraised. Preliminary tests on the pure polyurethane Nalgene surfaces verified that grafting of the epoxide materials had occurred. After reaction had taken place with the allyl glycidyl ether, the pure polyurethane tubing became cloudy, and a fine film could be observed that was strongly bound to the surface. Prereaction of NaOCl activated polyurethane followed by addition of the monomer (no initiator) did not result in coated surfaces nor did omission of any one of the reaction components. Epoxide group activity on the pure polyurethane grafted with propyl glycidyl ether was evident by the quenching of colored 2-nitro-5-thio-benzoic acid. Thus active surfaces could be prepared for covalent coupling of ATH. Coating of polycarbonate urethane with ATH was verified by staining of the protein or GAG present on the surface (259).

Immobilization of heparin and hirudin (one of the most potent direct thrombin inhibitors known) onto surfaces has been studied for some time. Hirudin has been bound to clinically used biomaterials by a number of methods (260,261). Significant coating densities with hirudin have been achieved with a surface character that gave efficient thrombin binding and inhibition. However, because hirudin–thrombin complexes are not covalent and only one thrombin can be inactivated per hirudin molecule, attempts have also been made to investigate the properties of heparinized biomaterials. Very ingenious methodologies have been developed to adapt heparin for use in a wide range of synthetic clinical devices. Only one such example is the bonding of heparin to bileaflet valves using a coating of graphite-carbon and benzalkonium chloride (262). However, the most viable structures for anticoagulating surfaces with heparin have been shown to require end-point attachment of heparin chains (263). This conclusion is consistent with the concept that active heparin molecules must be directed away from the surface in order to prevent surface activation of coagulants from the blood flow or surrounding tissue. In fact it has been deduced that the critical action of heparinized biomaterials is to mediate inhibition of the coagulation cascade leading to prothrombin activation (thrombin feedback reactions?) (263). Unfortunately, simple end-point binding of heparin chains to the polymer surface is not sufficient for good thrombin inhibition. A series of experiments have demonstrated that heparin that was directly covalently linked to polystyrene containing hydrophilic groups could not efficiently catalyze formation of thrombin–antithrombin complexes (17). In order to effect optimum anticoagulant activity of heparin on the surfaces, end-point attachment of the GAG to spacer arms

(>2,000 MW) is necessary (17). As a final caveat, in an attempt further to fortify heparinized surfaces, either free heparin has been coimmobilized on the same surface with antithrombin (264), or surfaces with covalently bonded heparin have been presaturated with free antithrombin (265). Although a mild improvement for thrombin inhibition has been noted, optimum arrangement of the antithrombin and heparin have not been achieved. Thus surface coating with a significant density of antithrombin and heparin that yields rapid, high-capacity inhibition of thrombin is desirable. Analysis of the ATH coated polycarbonate urethanes presented an excellent opportunity for testing the viability of attaching high-activity heparin to the surface through antithrombin itself.

Higher substitution of Chan et al. ATH on polycarbonate urethane was observed than that with either heparin alone (259) or hirudin (254). This verifies that covalent ATH has superior chemical features for attachment of the heparin containing anticoagulant. Storage properties of the ATH coated surface are favorable, since treated biomaterial could be utilized as an endoluminal graft in vivo after incubation in sterile saline at 4°C for minutes to months, with similar results (259). Even though ATH was likely covalently bonded through the antithrombin moiety to the propyl glycidyl linker on the surface, the immobilized ATH exhibited significant direct noncatalytic activity against thrombin (254). Thus polymer surfaces could be created with ATH coatings that have long-term rapid inhibitory action towards thrombin, which does not require a supply of antithrombin from the patient. However, binding experiments with labeled antithrombin indicated that significant surface concentrations of pentasaccharide sites were readily accessible on polycarbonate urethane coated with the Chan et al. ATH (259). High-affinity binding of antithrombin to the surface strongly suggests that ATH coated material can potentiate the reaction of plasma antithrombin with thrombin. Given the high catalytic anticoagulant activity reported for the keto-amine conjugated ATH, significant surface-associated catalysis of antithrombin reaction with thrombin would be expected from ATH coated biomaterials, provided that the coating process had oriented the heparin away from the polymer. The exogenous antithrombin binding results suggested that the ATH covered polycarbonate urethane product likely possesses the desired catalytic capacity (259).

In vivo experiments were performed to examine if the ATH coated endoluminal grafts had biological effectiveness that corresponded to the substitution density and antithrombin activity results observed in vitro. Endoluminal tubing (treated and nontreated) was deployed by a wire plunger through a narrow-gauge catheter into the jugular veins of anesthetized rabbits. No anticoagulant was administered to the animals and, after 3 h, the tubing segments were recovered and the weight of clot adherent to the device determined. Significant improvement with respect to tubing anticoagulant properties resulted from coating with ATH. A four- to fivefold reduction in weight of clot generated on the tubing was observed with ATH coated compared to nontreated surfaces (254). This improvement in antithrombogenicity of the biomaterial was not gained due to the presence of grafted propyl glycidyl groups, since tests with these control surfaces (no ATH attached) were negative (259). Furthermore, in vivo experiments with polycarbonate urethane coated with antithrombin gave results similar to those with surfaces that were reacted with allyl glycidyl ether alone. Thus, not surprisingly, the heparin moiety of the surface bound ATH was critical. Finally, hirudin was bonded to polycarbonate urethane in a fashion similar to that of ATH, followed by testing in the rabbit jugular vein. In vitro analyses of the tubing showed that most of the hirudin bound to the endoluminal graft could inhibit thrombin directly. However, significantly more clot was generated in the hirudin tubing than in the ATH tubing in vivo. Therefore ATH coated surfaces have greater potency than surfaces that are coated only with highly active direct thrombin inhibitors. The dual noncatalytic and catalytic anticoagulant activities of ATH may represent a new generation of pacified biomaterials.

VII. THE FUTURE

ATH is an agent that, either in fluid phase or surface immobilized form, pacifies the coagulation system in vivo. The fact that, with optimum methodologies, the highest activity heparin can be selected by antithrombin during conjugation, gives a significant advantage for prevention and treatment of coagulation, as well as the construction of highly anticoagulant surfaces for clinical use. A number of investigations are necessary in order to realize fully the potential applications of ATH. Long-term use of fluid-phase ATH in animals needs to be evaluated for its application as a prophylactic. Furthermore, a number of clinical procedures could be studied to determine the value of ATH treatment during surgery. ATH may also be efficacious for the treatment of deep vein thrombosis and RDS. Animal models of these pathological states exist that could be used to challenge the viability of ATH as a treatment agent. Linkage of the conjugate to a variety of polymers must be done to determine the optimum type of biomaterials for coating with ATH. In conjunction with the type of surface material is the investigation of other more sophisticated chemistries that could link to the antithrombin in ATH via amino acid R groups other than that of lysine. Although the lysine ε-amino is a standard for protein immobilization, the heparin in ATH may tie up some of the key residues. Once the chemical details on ATH attachment to biomaterial surfaces have been determined, in vivo experiments in a number of animal models will be required to evaluate the best indications for use. Finally, clinical trials will be necessary before commercial use of either fluid-phase ATH or ATH coated biomaterials is possible.

REFERENCES

1. I. M. Nilson. Coagulation and fibrinolysis. *Scand. J. Gastroenterol. 137*:11 (1987).
2. H. R. Buller, T. Ten Cate. Coagulation and platelet activation pathways. A review of the key components and the way in which these can be manipulated. *Eur. Heart. J. 16*: 8 (1995).
3. S. Butenas, C. van't Veer, K. G. Mann. ''Normal'' thrombin generation. *Blood 94*:2169 (1999).
4. R. E. Jordan, G. M. Oosta, R. T. Gardner, R. D. Rosenberg. The kinetics of hemostatic enzyme–antithrombin interactions in the presence of low molecular weight heparin. *J. Biol. Chem. 255*:10081 (1980).
5. B. Risberg, S. Andreasson, E. Eriksson. Disseminated intravascular coagulation. *Acta Anaesthesiol. Scand. 95*:60 (1991).
6. J. Choay, M. Petitou, J. Lormeau, P. Sinay, B. Casu, G. Gatti. Structure–activity relationship in heparin: a synthetic pentasaccharide with high affinity for antithrombin III and elicitin high anti factor Xa activity. *Biophy. Res. Comm. 116*:492 (1983).
7. L. Muszbek, R. Adany, H. Mikkola. Novel aspects of blood coagulation factor XIII. I. Structure, distribution, activation, and function. *Crit. Rev. Clin. Lab. Sci. 33*:357 (1996).
8. D. V. Devine, P. D. Bishop. Platelet-associated factor XIII in platelet activation, adhesion, and clot stabilization. *Thromb. Haemost. 22*:409 (1996).
9. W. Wang, M. B. Boffa, L. Bajzar, J. B. Walker, M. E. Nesheim. A study of the mechanism of inhibition of fibrinolysis by activated thrombin activable fibrinolysis inhibitor. *J. Biol. Chem. 273*:27176 (1998).
10. L. Bajzar, M. Nesheim, J. Morser, P. B. Tracey. Both cellular and soluble forms of thrombomodulin inhibit fibrinolysis by potentiating the activation of thrombin activable fibrinolysis inhibitor. *J. Biol. Chem. 273*:2792 (1998).
11. M. Nesheim, W. Wang, M. Boffa, M. Nagashima, J. Morser, L. Bajzar. Thrombin, thrombomodulin, and TAFI in the molecular link between coagulation and fibrinolysis. *Thromb. Haemostas. 78*:386 (1997).
12. M. B. Boffa, W. Wang, L. Bajzar, M. E. Nesheim. Plasma and recombinant thrombin activable fibrinolysis inhibitor (TAFI) and activated TAFI compared with respect to glycosylation, thrombin/thrombomodulin-dependent activation, thermal stability, and enzymatic properties. *J. Biol. Chem. 273*:2127 (1998).
13. K. Kokame, X. Zheng, J. E. Sadler. Activation of thrombin activable fibrinolysis inhibitor requires epidermal growth factor-like domain 3 of thrombomodulin and is inhibited competitively by protein C. *J. Biol. Chem. 273*:12135 (1998).
14. B. Dahlback. The protein C anticoagulant system: inherited defects as basis for venous thrombosis. *Thromb. Res. 77*:1 (1995).
15. A. R. Rezaie. Tryptophan 60-D in the B-insertion loop of thrombin modulates the thrombin–antithrombin reaction. *Biochem. 35*:1918 (1996).
16. B. Mille, J. Watton, T. W. Barrowcliffe, J. Mani, D. Lane. Role of N- and C-terminal amino acids in antithrombin binding to pentasaccharide. *J. Biol. Chem. 269*:29435 (1994).
17. Y. Byun, H. A. Jacobs, S. W. Kim. Mechanism of thrombin inactivation by immobilized heparin. *J. Biomed. Mater. Res. 30*:423 (1996).
18. J. W. Estes, E. W. Pelikan, E. Kruger-Thiemer. A retrospective study of the pharmacokinetics of heparin. *Clin. Pharmacol. Ther. 10*:329 (1969).
19. E. Young, M. Prins, M. N. Levine, J. Hirsh. Heparin binding to plasma proteins, an important mechanism for heparin clearance. *Thromb. Haemostas. 67*:639 (1992).
20. J. Hirsh, W. van Aken, A. Gallus, C. Dollery, J. Cade, W. Yung. Heparin kinetics in venous thrombosis and pulmonary embolism. *Circulation 53*:691 (1976).
21. L. B. Jaques, J. Mahadoo, L. W. Kavanagh. Intrapulmonary heparin: a new procedure for anticoagulant therapy. *Lancet 7996*:1157 (1976).
22. P. Hogg, C. Jackson. Fibrin monomer protects thrombin from inactivation by heparin–antithrombin III: implications for heparin efficacy. *Proc. Natl. Acad. Sci. USA 86*: 3619 (1989).
23. J. Weitz, M. Hudoba, D. Massel, J. Maraganore, J. Hirsh. Clot-bound thrombin is protected from inhibition by heparin–antithrombin III but is susceptible to inactivation by antithrombin III independent inhibitors. *J. Clin. Invest. 86*: 385 (1990).
24. P. Eisenberg, J. Siegel, D. Abendschein, J. Miletich. Importance of factor Xa in determining the procoagulant activity of whole-blood clots. *J. Clin. Invest. 91*:1877 (1993).
25. N. A. Prager, D. R. Abendschein, C. R. McKenzie, P. R. Eisenberg. Role of thrombin compared with factor Xa in the procoagulant activity of whole blood clots. *Circulation 92*:962 (1995).
26. P. Theroux, D. Waters, J. Lam, M. Juneau, J. McCans. Reactivation of unstable angina after the discontinuation of heparin. *N. Engl. J. Med. 327*:192 (1992).
27. M. Palm, C. Mattsson, C. M. Svahn, M. Weber. Bleeding times in rats treated with heparin, heparin fragments of high and low anticoagulant activity and chemically modified heparin fragments of low anticoagulant activity. *Throm. Haemostas. 64*:127 (1990).
28. M. W. C. Hatton, L. R. Berry, R. Machovich, E. Regoeczi. Tritiation of commercial heparins by reaction with NaB3H4: chemical analysis and biological properties of the product. *Anal. Biochem. 106*:417 (1980).

29. L. T. Hunt, M. O. Dayhoff. A surprising new protein superfamily containing ovalbumin, antithrombin-III, and alpha 1-protease inhibitor. *Biochem. Biophys. Res. Comm. 95*:864 (1980).
30. R. W. Carrell, D. R. Boswell, S. O. Brennan, M. C. Owen. Active site of al-antitrypsin: homologous site in antithrombin-III. *Biochem. Biophys. Res. Comm. 93*:399 (1980).
31. T. Chandra, R. Stackhouse, V. Kidd, K. J. Robson, S. L. Woo. Sequence homology between human alpha-1-antichymotrypsin, alpha-1-antitrypsin and antithrombin III. *Biochemistry 22*:5055 (1983).
32. R. W. Carrell, P. B. Christey, D. R. Boswell. Serpins: antithrombin and other inhibitors of coagulation and fibrinolysis: evidence from amino acid sequences. In: *Thrombosis and Haemostasis* (M. Vertsraete, ed.). Leuven University Press, Leuven, Belgium, 1987, p. 1.
33. T. Takano, Y. Yamanouchi, Y. Mori, S. Kudo, T. Nakayama, M. Sugiura, S. Hashira, T. Abe. Interstitial deletion of chromosome 1q [del(1)(q24q25.3)] identified by fluorescence in situ hybridization and gene dosage analysis of apolipoprotein A-II, coagulation factor V, and antithrombin III. *Am. J. Med. Genet. 68*:207 (1997).
34. S. C. Bock, J. A. Marrinan, E. Radziejewska. Antithrombin III Utah: proline-407 to leucine mutation in a highly conserved region near the inhibitor reactive site. *Biochemistry 27*:6171 (1988).
35. E. V. Prochownik, S. C. Bock, S. H. Orkin. Intron structure of the human antithrombin III gene differs from that of other members of the serine protease inhibitor superfamily. *J. Biol. Chem. 260*:9608 (1985).
36. R. W. Niessen, A. Sturk, P. L. Hordijk, F. Michiels, M. Peters. Sequence characterization of a sheep cDNA for antithrombin III. *Biochim. Biophys. Acta 1171*:207 (1992).
37. U. Abildegaard. Purification of two progressive antithrombins of human plasma. *Scand. J. Clin. Lab. Invest. 19*:190 (1967).
38. R. D. Rosenberg, P. S. Damus. The purification and mechanism of action of human antithrombin–heparin cofactor. *J. Biol. Chem. 248*:6490 (1973).
39. J. Miller-Andersson, H. Borg, L. O. Andersson. Purification of antithrombin III by affinity chromatography. *Thromb. Res. 5*:439 (1974).
40. A. E. Manson, R. C. Austin, F. Fernandez-Rachubinski, R. A. Rachubinski, M. A. Blajchman. The molecular pathology of inherited human antithrombin III deficiency. *Transf. Med. Rev. 3*:264 (1989).
41. M. N. Blackburn, R. L. Smith, J. Carson, C. C. Sibley. The heparin binding site of antithrombin III. Identification of a critical tryptophan in the amino acid sequence. *J. Biol. Chem. 259*:939 (1984).
42. G. B. Villaneuva. Predictions of the secondary structure of antithrombin III and the location of the heparin-binding site. *J. Biol. Chem. 259*:2531 (1984).
43. T. E. Petersen, G. Dudek-Wojciechowska, L. Sottrup-Jensen. Primary structure of antithrombin III (heparin cofactor). Partial homology between al-antitrypsin and antithrombin III. In: *The Physiological Inhibitors of Coagulation and Fibrinolysis* (D. Collen, ed.). Elsevier North-Holland Biomedical Press, Amsterdam, The Netherlands, 1979, p. 43.
44. R. W. Carrell, P. E. Stein, G. Fermi, M. R. Wardell. Biological implications of a 3 A structure of dimeric antithrombin. *Structure 2*:257 (1994).
45. M. R. Wardell, J. P. Abrahams, D. Bruce, R. Skinner, A. G. Leslie. Crystallization and preliminary X-ray diffraction analysis of two conformations of intact human antithrombin. *J. Mol. Biol. 234*:1253 (1993).
46. M. R. Witmer, M. W. Hatton. Anithrombin III-beta associates more readily than antithrombin III-alpha with uninjured and de-endothelialized aortic wall in vitro and in vivo. *Arterioscler. Thromb. 11*:530 (1991).
47. T. H. Carlson, M. R. Kolman, M. Piepkorn. Activation of antithrombin III isoforms by heparan sulphate glycosaminoglycans and other sulphated polysaccharides. *Blood Coagul. Fibrinol. 6*:474 (1995).
48. S. O. Brennan, J. Y. Borg, P. M. George, C. Soria, J. Soria, J. Caen, R. W. Carrell. New carbohydrate site in mutant antithrombin (7 Ile---Asn) with decreased heparin affinity. *FEBS Lett. 237*:118 (1988).
49. H. Stibler, U. Holzbach, B. Kristiansson. Isoforms and levels of transferrin, antithrombin, alpha(1)-antitrypsin and thyroxine-binding globulin in 48 patients with carbohydrate-deficient glycoprotein syndrome type I. *Scand. J. Clin. Lab. Invest. 58*:55 (1998).
50. V. Picard, E. Ersdal-Badju, S. C. Bock. Partial glycosylation of antithrombin III asparagine 135 is caused by the serine in the third position of its N-glycosylation consensus sequence and is responsible for production of the beta-antithrombin III isoform with enhanced heparin affinity. *Biochemistry 34*:8433 (1995).
51. L. Garone, T. Edmunds, E. Hanson, R. Bernasconi, J. A. Huntington, J. L. Meagher, B. Fan, P. G. Gettins. Antithrombin–heparin affinity reduced by fucosylation of carbohydrate at asparagine 155. *Biochemistry 35*:8881 (1996).
52. I. Bjork, K. Ylinenjarvi, S. T. Olson, P. Hermentin, H. S. Conradt, G. Zettlmeissl. Decreased affinity of recombinant antithrombin for heparin due to increased glycosylation. *Biochem. J. 286*:793 (1992).
53. A. Borsodi, T. R. Narasimhan. Microheterogeneity of human antithrombin III. *Br. J. Haematol. 39*:121 (1978).
54. W. H. Howell. The Coagulation of Blood. *The Harvey Lectures*. Lippincott, Philadelphia, PA, 1918, p. 272.
55. K. Brinkhous, H. P. Smith, E. D. Warner. The inhibition of blood clotting: an unidentified substance which acts in conjunction with heparin to prevent the conversion of prothrombin to thrombin. *Am. J. Physiol. 125*:683 (1939).
56. W. H. Seegers, J. F. Johnson, C. Fall. An antithrombin reaction related to prothrombin activation. *Am. J. Physiol. 176*:97 (1954).
57. R. Devraj-Kizuk, D. Chui, E. Prochownik, C. Carter, F. Ofosu, M. Blajchman. Antithrombin III-Hamilton: a gene with a point mutation (guanine to adenine) in codon 382 causing impaired serine protease reactivity. *Blood 72*:1518 (1988).

58. H. C. Whinna, F. C. Church. Interaction of thrombin with antithrombin, heparin cofactor II, and protein C inhibitor. *J. Protein Chem. 12*:677 (1993).
59. Y. J. Chuang, P. G. Gettins, S. T. Olson. Importance of the P2 glycine of antithrombin in target proteinase specificity, heparin activation, and the efficiency of proteinase trapping as revealed by a P2 gly → pro mutation. *J. Biol. Chem. 274*:28142 (1999).
60. R. W. Carrell, M. C. Owen. Plakalbumin, a2-antitrypsin, antithrombin and the mechanism of inflammatory thrombosis. *Nature 317*:730 (1985).
61. A. J. Schulze, R. Huber, W. Bode, R. A. Engh. Structural aspects of serpin inhibition. *FEBS Lett. 344*:117 (1994).
62. K. Skriver, W. R. Wikoff, P. A. Patston, F. Tausk, M. Schapira, A. P. Kaplan, S. C. Bock. Substrate properties of C1 inhibitor Ma (alanine 434---glutamic acid). Genetic and structural evidence suggesting that the P12-region contains critical determinants of serine protease inhibitor/substrate status. *J. Biol. Chem. 266*:9216 (1991).
63. A. J. Schulze, U. Baumann, S. Knof, E. Jaeger, R. Huber, C. B. Laurell. Structural transition of alpha-1-antitrypsin by a peptide sequentially similar to beta-strand s4A. *Eur. J. Biochem. 194*:51 (1990).
64. A. J. Schulze, P. W. Frohnert, R. A. Engh, R. Huber. Evidence for the extent of insertion of the active site loop of intact alpha 1 proteinase inhibitor in beta-sheet A. *Biochemistry 31*:7560 (1992).
65. I. Bjork, K. Nordling, I. Larsson, S. T. Olson. Kinetic characterization of the substrate reaction between a complex of antithrombin with a synthetic reactive-bond loop tetradecapeptide and four target proteinases of the inhibitor. *J. Biol. Chem. 267*:19047 (1992).
66. I. Bjork, K. Ylinenjarvi, S. T. Olson, P. E. Bock. Conversion of antithrombin from an inhibitor of thrombin to a substrate with reduced heparin affinity and enhanced conformational stability by binding of a tetradecapeptide corresponding to the P1 to P14 region of the putative reactive bond loop of the inhibitor. *J. Biol. Chem. 267*:1976 (1992).
67. S. Debrock, P. J. Decleck. Characterization of common neoantigenic epitopes generated in plasminogen activator inhibitor-1 after cleavage of the reactive center loop or after complex formation with various serine proteinases. *FEBS Lett. 376*:243 (1995).
68. K. Nordling, I. Bjork. Identification of an epitope in antithrombin appearing on insertion of the reactive-bond loop into the A beta-sheet. *Biochemistry 35*:10436 (1996).
69. J. Whisstock, A. M. Lesk, R. Carrell. Modeling of serpin–protease complexes: antithrombin–thrombin, alpha-1-antitrypsin (358Met → Arg)–thrombin, alpha-1-antitrypsin (358Met → Arg)–trypsin, and antitrypsin–elastase. *Proteins 26*:288 (1996).
70. R. Skinner, J.-P. Abrahams, J. C. Whisstock, A. M. Lesk, R. W. Carrell, M. R. Wardell. The 2.6 A structure of antithrombin indicates a conformational change at the heparin binding site. *J. Mol. Biol. 266*:601 (1997).
71. V. Picard, P.-E. Marque, F. Paolucci, M. Aiach, B. F. Le Bonniec. Topology of the stable serpin–protease complexes revealed by an autoantibody that fails to react with the monomeric conformers of antithrombin. *J. Biol. Chem. 274*:4586 (1999).
72. L. Jin, J. P. Abrahams, R. Skinner, M. Petitou, R. N. Pike, R. W. Carrell. The anticoagulant activation of antithrombin by heparin. *Proc. Natl. Acad. Sci. USA 94*:14683 (1997).
73. J. L. Meagher, J. M. Beechem, S. T. Olson, P. G. Gettins. Deconvolution of the fluorescence emission spectrum of human antithrombin and identification of the tryptophan residues that are responsive to heparin binding. *J. Biol. Chem. 273*: 23283 (1998).
74. T. Koide, S. Odani, K. Takahashi, T. Ono, N. Sakuragawa. Replacement of arginine-47 by cysteine in hereditary abnormal antithrombin III that lacks heparin-binding ability. *Proc. Natl. Acad. Sci. USA 81*:289 (1984).
75. M. C. Owen, J. Y. Borg, C. Soria, J. Soria, J. Caen, R. W. Carrell. Heparin binding defect in a new antithrombin III variant: Rouen, 47 Arg to His. *Blood 69*:1275 (1987).
76. J. Y. Borg, M. C. Owen, C. Soria, J. Soria, J. Caen, R. W. Carrell. Proposed heparin binding site in antithrombin based on arginine 47: a new variant Rouen-II, 47 Arg to Ser. *J. Clin. Invest. 81*:1292 (1988).
77. J. Y. Chang, T. H. Tran. Antithrombin III Basel. Identification of a Pro-Leu substitution in a hereditary abnormal antithrombin with impaired heparin cofactor activity. *J. Biol. Chem. 261*:1174 (1986).
78. H. L. Fitton, R. Skinner, T. R. Dafforn, L. Jin, R. N. Pike. The N-terminal segment of antithrombin acts as a steric gate for the binding of heparin. *Protein Sci. 7*:782 (1998).
79. J. W. Smith, D. J. Knauer. A heparin binding site in antithrombin III. *J. Biol. Chem. 262*:11964 (1987).
80. E. Ersdal-Badju, A. Lu, Y. Zuo, V. Picard, S. C. Bock. Identification of the antithrombin III heparin binding site. *J. Biol. Chem. 272*:19393 (1997).
81. R. Skinner, W. S. Chang, L. Jin, X. Pei, J. A. Huntington, J. P. Abrahams, R. W. Carrell, D. A. Lomas. Implications for function and therapy of a 2.9 A structure of binary-complexed antithrombin. *J. Mol. Biol. 283*:9 (1998).
82. M. W. Hatton, L. R. Berry, E. Regoeczi. Inhibition of thrombin by antithrombin III in the presence of certain glycosaminoglycans found in the mammalian aorta. *Thromb. Res. 13*:655 (1978).
83. T. B. Brieditis, S. C. Bock, S. T. Olson, I. Bjork. The oligosaccharide side chain on Asn-135 of alpha-antithrombin, absent in beta-antithrombin, decreases the heparin affinity of the inhibitor by affecting the heparin-induced conformational change. *Biochemstry 36*:6682 (1997).
84. B. Fan, B. C. Crews, I. V. Turko, J. Choay, G. Zettlmeissl, P. Gettins. Heterogeneity of recombinant human antithrombin III expressed in baby hamster kidney cells. Effect of glycosylation differences on heparin binding and structure. *J. Biol. Chem. 268*:17588 (1993).
85. S. T. Olson, A. M. Frances-Chmura, R. Swanson, I. Bjork, G. Zettlmeissl. Effect of individual carbohydrate chains of recombinant antithrombin on heparin affinity and on the generation of glycoforms differing in heparin affinity. *Arch. Biochem. Biophys. 341*:212 (1997).

86. K. Lidholt. Biosynthesis of glycosaminoglycans in mammalian cells and in bacteria. *Biochem. Soc. Trans. 25*:866 (1997).
87. D. M. Templeton. Proteoglycans in cell regulation. *Crit. Rev. Clin. Lab. Sci. 29*:141 (1992).
88. S. Ernst, R. Langer, C. L. Cooney, R. Sasisekharan. Enzymatic degradation of glycosaminoglycans. *Crit. Rev. Biochem. Mol. Biol. 30*:387 (1995).
89. M. Kobayashi, G. Sugumaran, J. Liu, N. W. Shworak, J. E. Silbert, R. D. Rosenberg. Molecular cloning and characterization of a human uronyl-2-sulfotransferase that sulfates iduronyl and glucuronyl residues in dermatan/chondroitin sulfate. *J. Biol. Chem. 274*:10474 (1999).
90. J. T. Gallagher, A. Walker. Molecular distinctions between heparin sulphate and heparin. Analysis of sulphation patterns indicates that heparin sulphate and heparin are separate families of N-sulphated polysaccharides. *Biochem. J. 230*:665 (1985).
91. J. E. Shively, H. E. Conrad. Nearest neighbor anaylsis of heparin: identification and quantitation of the products formed by selective depolymerization procedures. *Biochemistry 15*:3943 (1976).
92. H. C. Robinson, A. A. Horner, M. Hook, S. Ogren, U. Lindahl. A proteoglycan form of heparin and its degradation to single-chain molecules. *J. Biol. Chem. 253*:6687 (1978).
93. M. Iacomini, B. Casu, M. Guerrini, A. Naggi, A. Pirola, G. Torri. ''Linkage Region'' sequences of heparins and heparan sulfates: detection and quantification by nuclear magnetic resonance spectroscopy. *Anal. Biochem. 274*:50 (1999).
94. U. Lindahl, O. Axelsson. Identification of iduronic acid as the major sulfated uronic acid of heparin. *J. Biol. Chem. 246*:74 (1971).
95. T. Lind, U. Lindahl, K. Lidholt. Biosynthesis of heparin/heparan sulfate. Identification of a 0-kDa protein catalyzing both the D-glucuronosyl- and the N-acetyl-D-glucosaminyltransferase reactions. *J. Biol. Chem. 268*:20705 (1993).
96. J. Riesenfeld, M. Hook, U. Lindahl. Biosynthesis of heparin. Assay and properties of the microsomal N-acetyl-D-glucosaminyl-N-deacetylase. *J. Biol. Chem. 255*:922 (1980).
97. I. Jacobsson, U. Lindahl, J. W. Jensen, L. Roden, H. Prihar, D. S. Feingold. Biosynthesis of heparin. Substrate specificity of heparosan N-sulfate D-glucuronosyl 5-epimerase. *J. Biol. Chem. 259*:1056 (1984).
98. R. R. Miller, C. J. Waechter. Partial purification and characterization of detergent-solubilized N-sulfotransferase activity associated with calf brain microsomes. *J. Neurochem. 51*:87 (1988).
99. K. Lidholt, U. Lindahl. Biosynthesis of heparin. The D-glucuronosyl- and N-acetyl-D-glucosaminyltransferase reactions and their relation to polymer modification. *Biochem. J. 287*:21 (1992).
100. I. Jacobsson, U. Lindahl. Biosynthesis of heparin. Concerted action of late polymer-modification reactions. *J. Biol. Chem. 255*:5094 (1980).
101. A. B. Foster, R. Harrison, T. D. Inch, M. Stacey, J. M. Webber. Amino-sugars and related compounds. Part IX. Periodate oxidation of heparin and some related substances. *J. Am. Chem. Soc. 85*:2279 (1963).
102. K. Lidholt, L. Kjellen, U. Lindahl. Biosynthesis of heparin. Relationship between the polymerization and sulphation processes. *Biochem. J. 261*:999 (1989).
103. L. Toma, P. Berninsome, C. B. Hirschberg. The putative heparin specific N-acetylglucosaminyl N-deacetylase/N-sulfotransferase also occurs in non-heparin-producing cells. *J. Biol. Chem. 273*:22458 (1998).
104. S. Ogren, U. Lindahl. Cleavage of macromolecular heparin by an enzyme from mouse mastocytoma. *J. Biol. Chem. 250*:2690 (1975).
105. S. Ogren, U. Lindahl. Metabolism of macromolecular heparin in mouse neoplastic mast cells. *Biochem. J. 154*:605 (1976).
106. J. Fareed, K. Fu, L. H. Yang, D. A. Hoppensteadt. Pharmacokinetics of low molecular weight heparins in animal models. *Semin. Thromb. Hemostas. 25*:51 (1999).
107. J. Fareed, S. Haas, A. Sasahar. Past, present and future considerations on low molecular weight heparin differentiation: an epilogue. *Semin. Thromb. Hemostas. 25*:145 (1999).
108. N. S. Jaikaria, L. Rosenfeld, M. Y. Khan, I. Danishefsky, S. Newman. Interaction of fibronectin with heparin in model extracellular matrices: role of arginine residues and sulfate groups. *Biochemistry 30*:1538 (1991).
109. T. Zak-Nejmark, M. Krasnowska, R. Jankowska, M. Jutel. Heparin modulates migration of human peripheral blood mononuclear cells and neutrophils. *Arch. Immunol. Ther. Exp. (Warsz.) 47*:245 (1999).
110. W. Jeske, J. Fareed. Antithrombin III- and heparin cofactor II-mediated anticoagulant and antiprotease actions of heparin and its synthetic analogues. *Semin. Throm. Hemostas. 19*:241 (1993).
111. R. D. Rosenfeld. Role of heparin and heparinlike molecules in thrombosis and atherosclerosis. *Fed. Proc. 44*:404 (1985).
112. T. W. Barrowcliffe, B. Mulloy, E. A. Johnson, D. P. Thomas. The anticoagulant activity of heparin: measurement and relationship to chemical structure. *J. Pharm. Biomed. Anal. 7*:217 (1989).
113. S. Frebelius, U. Hedin, J. Swedenborg. Thrombogenecity of the injured vessel wall—role of antithrombin and heparin. *Thromb. Haemost. 71*:147 (1994).
114. L. Liu, L. Dewar, Y. Song, M. Kulczycky, M. Blajchman, J. W. Fenton II, M. M. Andrew, M. Delorme, J. Ginsberg, K. T. Preissner. Inhibition of thrombin by antithrombin III and heparin cofactor II in vivo. *Thromb. Haemost. 73*:405 (1995).
115. Y. I. Wu, W. P. Sheffield, M. Blajchman. Defining the heparin-binding domain of antithrombin. *Blood Coag. Fibrinol. 5*:83 (1994).
116. A. A. Horner, E. Young. Asymmetric distribution of sites with high affinity for antithrombin III in rat skin heparin proteoglycans. *J. Biol. Chem. 257*:8749 (1982).

117. K.-G. Jacobsson, U. Lindahl, A. A. Horner. Location of antithrombin-binding regions in rat skin heparin proteoglycans. *Biochem. J. 240*:625 (1986).
118. A. K. Chan, L. Berry, H. O'Brodovich, P. Klement, M. Mitchell, B. Baranowski, P. Monagle, M. Andrew. Covalent antithrombin–heparin complexes with high anticoagulant activity: intravenous, subcutaneous and intratracheal administration. *J. Biol. Chem. 272*:22111 (1997).
119. R. D. Rosenberg, R. E. Jordan, L. V. Favreau, L. H. Lam. Highly active heparin species with multiple binding sites for anithrombin. *Biochem. Biophys. Res. Comm. 86*:1319 (1979).
120. L. Thunberg, U. Lindahl, A. Tengblad, T. C. Laurent, C. M. Jackson. On the molecular-weight-dependence of the anticoagulant activity of heparin. *Biochem. J. 181*:241 (1979).
121. L. G. Oscarsson, G. Pejler, U. Lindhahl. Location of the antithrombin-binding sequence in the heparin chain. *J. Biol. Chem. 264*:296 (1989).
122. T. C. Laurent, A. Tengblad, L. Thunberg, M. Hook, U. Lindahl. The molecular-weight-dependence of the anti-coagulant activity of heparin. *Biochem. J. 175*:691 (1978).
123. L. R. Berry, M. W. Hatton. Controlled depolymerization of heparin: anticoagulant activity and molecular size of the products. *Biochem. Soc. Trans. 11*:101 (1983).
124. M. Kusche, G. Backstrom, J. Riesenfeld, M. Petitou, J. Choay, U. Lindahl. Biosynthesis of heparin. O-sulfation of the antithrombin-binding region. *J. Biol. Chem. 263*: 15474 (1988).
125. M. Petitou, P. Duchaussoy, I. Lederman, J. Choay, P. Sinay. Binding of heparin to antithrombin III: a chemical proof of the critical role played by a 3-sulfated 2-amino-2-deoxy-D-glucose residue. *Carb. Res. 179*:163 (1988).
126. M. Petitou, P. Duchaussoy, I. Lederman, J. Choay, P. Sinay, J. C. Jacquinet, G. Torri. Synthesis of heparin fragments. A chemical synthesis of the pentasaccharide O-(2-deoxy-2-sulfamido-6-O-sulfo-alpha-D-glucopyranosyl)-(1-4)-O-(beta-D-glucopyranosyluronic acid)-(1-4)-O-(2-deoxy-2-sulfamido-3,6-di-O-sulfo-alpha-D-glucopyrano syl)-(1-4)-O-(2-O-sulfo-alpha-L-idopyranosyluronic acid)-(1-4)-2-deoxy-2-sulfamido-6-O-sulfo-D-glucopyranose decasodium salt, a heparin fragment having high affinity for antithrombin III. *Carb. Res. 147*:221 (1986).
127. M. Petitou, T. Barzu, J. P. Herault, J. M. Herbert. A unique trisaccharide sequence in heparin mediates the early step of antithrombin III activation. *Glycobiology* 7:323 (1997).
128. U. R. Desai, M. Petitou, I. Bjork, S. T. Olson. Mechanism of heparin activation of antithrombin. Role of individual residues of the pentasaccharide activating sequence in the recognition of native and activated states of antithrombin. *J. Biol. Chem. 273*:7478 (1998).
129. U. R. Desai, M. Petitou, I. Bjork, S. T. Olson. Mechanism of heparin activation of antithrombin: evidence for an induced-fit model of allosteric activation involving two interaction subsites. *Biochemistry 37*:13033 (1998).
130. V. Arocas, S. C. Bock, S. T. Olson, I. Bjork. The role of Arg46 and Arg47 of antithrombin in heparin binding. *Biochemistry 38*:10196 (1999).
131. R. J. Kandrotas. Heparin pharmacokinetics and pharmacodynamics. *Clin. Pharmacokin. 22*:359 (1992).
132. J. Ye, A. R. Rezaie, C. T. Esmon. Glycosaminoglycan contributions to both protein C activation and thrombin inhibition involve a common arginine-rich site in thrombin that includes residues arginine 93, 97, and 101. *J. Biol. Chem. 269*:17965 (1994).
133. L. C. Petersen, M. Jorgensen. Electrostatic interactions in the heparin-enhanced reaction between human thrombin and antithrombin. *Biochem. J. 211*:91 (1983).
134. M. Lozano, A. Bos, P. G. de Groot, G. van Willigen, D. G. Meuleman, A. Ordinas, J. J. Sixma. Suitability of low molecular weight heparin(oid)s and a pentasaccharide for an in vitro human thrombosis model. *Arterioscler. Thromb. 14*:1215 (1994).
135. D. Bergqvist. Low molecular weight heparins. *J. Intern. Med. 240*:63 (1996).
136. E. Young, P. Wells, S. Holloway, J. Weitz, J. Hirsh. Ex vivo and in vitro evidence that low molecular weight heparins exhibit less binding to plasma protein than unfractionated heparin. *Thromb. Haemostas. 71*:300 (1994).
137. L. Bara, E. Billaud, G. Gramond, A. Kher. Comparative pharmacokinetics (PK 10169) and unfractionated heparin after intravenous and subcutaneous administration. *Thromb. Res. 39*:631 (1985).
138. J. Hirsh, M. Levine. Low molecular weight heparin. *Blood 79*:1 (1992).
139. J. Fareed, D. Hoppensteadt, W. Jeske, R. Clarizio, J. M. Walenga. Low molecular weight heparins: are they different? *Can. J. Cardiol. 14*:28E (1998).
140. M. Wolzt, A. Weltermann, M. Nieszpaur-Los, B. Schneider, A. Fassolt, K. Lechner, H. G. Eichler, P. A. Kyrle. Studies on the neutralizing effects of protamine on unfractionated and low molecular weight heparin (Fragmin) at the site of activation of the coagulation system in man. *Thromb. Haemost. 73*:439 (1995).
141. H. F. Schran, D. W. Bitz, F. J. DiSerio, J. Hirsh. The pharmacokinetics and bioavailability of subcutaneously administered dihydroergotamine, heparin and the dihydroergotamine–heparin combination. *Thromb. Res. 31*:51 (1983).
142. L. Briant, C. Caranobe, S. Saivin, P. Sie, B. Bayrou, G. Houin, B. Boneu. Unfractionated heparin and CY 216: pharmacokinetics and bioavailabilities of the antifactor Xa and IIa effects after intravenous and subcutaneous injection in the rabbit. *Thromb. Haemost. 61*:348 (1989).
143. M. D. Laforest, N. Colas-Linhart, S. Guiraud-Vitaux, B. Bok, L. Bara, M. Samama, J. Marin, F. Imbault, A. Uzan. Pharmacokinetics and biodistribution of technium 99m labelled standard heparin and a low molecular weight heparin (enoxaparin) after intravenous injection in normal volunteers. *Br. J. Haematol. 77*:201 (1991).
144. L. Manson, J. I. Weitz, T. J. Podor, J. Hirsh, E. Young. The variable anticoagulant response to unfractionated heparin in vivo reflects binding to plasma proteins rather than clearance. *J. Lab. Clin. Med. 130*:649 (1997).
145. E. Young, T. J. Podor, T. Venner, J. Hirsh. Induction of the acute-phase reaction increases heparin-binding pro-

teins in plasma. *Arterioscler. Thromb. Vasc. Biol. 17*:1568 (1997).

146. H. L. Messmore, B. Griffin, J. Fareed, E. Coyne, J. Seghatchian. In vitro studies of the interaction of heparin, low molecular weight heparin and heparinoids with platelets. *N.Y. Acad. Sci. 556*:217 (1989).
147. W. A. Patton 2nd, C. A. Granzow, L. A. Getts, S. C. Thomas, L. M. Zotter, K. A. Gunzel, L. J. Lowe-Krentz. Identification of a heparin-binding protein using monoclonal antibodies that block heparin binding to porcine aortic endothelial cells. *Biochem. J. 311*:461 (1995).
148. H. R. Lijnen, M. Hoylaerts, D. Collen. Heparin binding properties of human histidine rich glycoprotein: mechanism and role in the neutralization of heparin in plasma. *J. Biol. Chem. 258*:3803 (1983).
149. K. H. Teoh, E. Young, C. A. Bradley, J. Hirsh. Heparin binding proteins. Contribution to heparin rebound after cardiopulmonary bypass. *Circulation 88*:420 (1993).
150. P. J. Hogg, C. M. Jackson. Heparin promotes the binding of thrombin to fibrin polymer. Quantitative characterization of a thrombin–fibrin polymer–heparin ternary complex. *J. Biol. Chem. 265*:241 (1990).
151. P. J. Hogg, C. M. Jackson. Fibrin monomer protects thrombin from inactivation by heparin-antithrombin III: implications for heparin efficacy. *Proc. Natl. Acad. Sci. USA 86*:3619 (1989).
152. P. J. Hogg, C. M. Jackson, J. K. Labanowski, P. E. Bock. binding of fibrin monomer and heparin to thrombin in a ternary complex alters the environment of the thrombin catalytic site, reduces affinity for hirudin and inhibits cleavage of fibrinogen. *J. Biol. Chem. 271*:26088 (1996).
153. P. J. Hogg, C. M. Jackson. Formation of a ternary complex between thrombin, fibrin monomer, and heparin influences the action of thrombin on its substrates. *J. Biol. Chem. 265*:248 (1990).
154. E. Young, P. Wells, S. Holloway, J. Weitz, J. Hirsh. Ex-vivo and in-vitro evidence that low molecular weight heparins exhibit less binding to plasma proteins than unfractionated heparin. *Thromb. Haemost. 71*:300 (1999).
155. B. Cosmi, J. C. Fredenburgh, J. Rischke, J. Hirsh, E. Young, J. I. Weitz. Effect of nonspecific binding to plasma proteins on the antithrombin activities of unfractionated heparin, low-molecular-weight heparin, and dermatan sulfate. *Circulation 95*:118 (1997).
156. P. Bendayan, H. Boccalon, D. Dupouy, B. Boneu. Dermatan sulphate is a more potent inhibitor of clot-bound thrombin than unfractionated and low molecular weight heparins. *Thromb. Haemostas. 71*:576 (1994).
157. T. Warkentin, M. Levine, J. Hirsh, P. Horsewood, R. Roberts, M. Gent, J. Kelton. Heparin-induced thrombocytopenia in patients treated with low molecular weight heparins or unfractionated heparin. *N. Engl. J. Med. 332*:1330 (1995).
158. T. Dahlman, N. Lindvall, M. Hellgren. Osteopenia in pregnancy during longterm heparin treatment: a radiological study post partum. *Br. J. Obstet. Gynaecol. 97*:221 (1990).
159. L. A. Barbour, S. D. Kick, J. F. Steiner, M. E. LoVerde, L. N. Heddleson, J. L. Lear, A. E. Baron, P. L. Barton. A prospective study of heparin-induced osteoporosis in pregnancy using bone densitometry. *Am. J. Obstet. Gynecol. 170*:862 (1994).
160. M. Cohen. Heparin-induced thrombocytopenia and the clinical use of low molecular weight heparins in acute coronary syndromes. *Semin. Hematol. 36*:33 (1999).
161. R. Hull, G. Raskob, D. Rosenbloom, J. Lemaire, G. Pineo, B. Baylis, J. Ginsberg, A. Panju, P. Brill-Edwards, R. Brant. Optimal therapeutic level of heparin therapy in patients with venous thrombosis. *Arch. Intern. Med. 152*: 1589 (1992).
162. J. Hirsh, E. W. Salzman, V. J. Marder. Treatment of venous thromboembolism. In: *Hemostasis and Thrombosis: Basic Principles and Clinical Practice* (R. W. Colman, ed.). Lippincott, New York, 1994, p. 1346.
163. M. M. Koopman, P. Prandoni, F. Piovella, P. A. Ockelford, D. F. Brandjes, J. van der Meer, A. S. Gallus, C. Simoneau, C. H. Chesterman, M. H. Prins. Treatment of venous thrombosis with intravenous unfractionated heparin administered in the hospital as compared with subcutaneous low molecular weight heparin administered at home. The Tasman Study Group. *N. Engl. J. Med. 334*: 682 (1996).
164. M. Levine, M. Gent, J. Hirsh, J. Leclerc, D. Anderson, J. Weitz, J. Ginsberg, A. J. Turpie, C. Demers, M. Kovaks. A comparison of low molecular weight heparin administered primarily at home with unfractionated heparin administered in the hospital for proximal deep venous thrombosis. *N. Engl. J. Med. 334*:677 (1996).
165. J. Hirsh, S. Siragusa, B. Cosmi, J. Ginsberg. Low molecular weight heparins (LMWH) in the treatment of patients with acute venous thromboembolism. *Thromb. Haemostas. 74*:360 (1995).
166. GUSTO IIb investigators. A comparison of recombinant hirudin with heparin for the treatment of acute coronary syndromes. The Global Use of Strategies to Open Occluded Coronary Arteries (GUSTO) IIb investigators. *N. Engl. J. Med. 335*:775 (1996).
167. I. Sarembock, S. Gertz, L. Thome, K. McCoy, M. Ragosta, E. Powers, J. Maraganore, L. Gimple. Effectiveness of hirulog in reducing restenosis after balloon angioplasty of atherosclerotic femoral arteries in rabbits. *J. Vasc. Res. 33*: 308 (1996).
168. M. Verstraete. New developments in antiplatelet and antithrombotic therapy. *Eur. Heart. J. 16*:16 (1995).
169. W. E. Marki. The anticoagulant and antithrombotic properties of hirudins. *Thromb. Haemostas. 64*:344 (1991).
170. U. Egner, G. A. Hoyer, W. D. Schleuning. Rational design of hirulog-type inhibitors of thrombin. *J. Comput. Aided Mol. Des. 8*:479 (1994).
171. G. Agnelli, R. Cinzia, J. I. Weitz, G. G. Nenci, J. Hirsh. Sustained antithrombotic activity of hirudin after its plasma clearance: comparison with heparin. *Blood 80*:960 (1992).
172. J. M. Cardot, G. Y. Lefevre, J. A. Godbillon. Pharmacokinetics of rec-hirudin in healthy volunteers after intrave-

nous administration. *J. Pharmacokinet. Biopharm. 22*:147 (1994).
173. I. Fox, A. Dawson, P. Loynds, J. Eisner, K. Findlen, E. Levin, D. Hanson, T. Mant, J. Wagner, J. Maraganore. Anticoagulant activity of hirulog, a direct thrombin inhibitor, in humans. *Thromb. Haemost. 69*:157 (1993).
174. P. Zoldhelyi, J. Bichler, W. G. Owen, D. E. Grill, V. Fuster, J. S. Mruk, J. H. Chesebro. Persistent thrombin generation in humans during specific thrombin inhibition with hirudin. *Circulation 90*:2671 (1994).
175. W. A. Schumacher, T. E. Steinbacher, C. L. Heran, J. R. Megill, S. K. Durham. Effects of antithrombotic drugs in a rat model of aspirin-insensitive arterial thrombosis. *Thromb. Haemost. 69*:509 (1993).
176. N. A. Scott, G. L. Nunes, S. B. King 3rd, L. A. Harker, S. R. Hanson. Local delivery of an antithrombin inhibits platelet-dependent thrombosis. *Circulation 90*:1951 (1994).
177. S. Hollenbach, U. Sinha, P. H. Lin, K. Needham, L. Frey, T. Hancock, A. Wong, D. Wolf. A comparative study of prothrombinase and thrombin inhibitors in a novel rabbit model of non-occlusive deep vein thrombosis. *Thromb. Haemost. 71*:357 (1994).
178. C. H. Pletcher, G. L. Nelsestuen. The rate-determining step of the heparin-catalyzed antithrombin/thrombin reaction is independent of thrombin. *J. Biol. Chem. 257*:5342 (1982).
179. J. M. Muir, J. Hirsh, J. Weitz, M. Andrew, E. Young, S. G. Shaughnessy. A histomorphometric comparison of the effects of heparin and low molecular weight heparin on cancellous bone in rats. *Blood 89*:3236 (1997).
180. F. Brus, W. van Oeveren, A. Heikamp, A. Okken, S. B. Oetomo. Leakage of protein into lungs of preterm ventilated rabbits is correlated with activation of clotting, complement, and polymorphonuclear leukocytes in plasma. *Pediatr. Res. 39*:958 (1996).
181. J. Holter, J. Weiland, E. Packt, J. Gadek, W. Davis. Protein permeability in the adult respiratory distress syndrome. Loss of size selectivity of the alveolar epithelium. *J. Clin. Invest. 78*:1513 (1986).
182. C. L. Sprung, W. M. Long, E. H. Marcial, R. M. H. Schein, R. E. Parker, T. Schomer, K. L. Brigham. Distribution of proteins in pulmonary edema. *Am. Rev. Respir. Dis. 136*: 957 (1987).
183. M. Bachofen, E. Weibel. Structural alterations of lung parenchyma in the adult respiratory distress syndrome. *Clin. Chest. Med. 3*:35 (1982).
184. D. Gitlin, J. Craig. The nature of the hyaline membrane in asphyxia of the newborn. *Pediatrics 17*:64 (1956).
185. A. H. Jobe. Pulmonary surfactant therapy. *N. Engl. J. Med. 328*:861 (1993).
186. W. Long, T. Thompson, H. Sundell. Effects of two rescue doses of a synthetic surfactant on mortality rate and survival without bronchopulmonary dysplasia. *J. Pediatr. 118*:595 (1991).
187. The OSIRIS Collaborative Group. Early versus delayed neonatal administration of a synthetic surfactant—the judgement of OSIRIS. *Lancet 340*:1363 (1992).
188. E. A. Liechty, E. Donovan, D. Purohit. Reduction of neonatal mortality after multiple doses of bovine surfactant in low birth weight neonates with respiratory distress syndrome. *Pediatrics 88*:19 (1991).
189. J. Coalson, V. Winter, D. Gerstmann, S. Idell, R. King, R. Delemos. Pathophysiologic, morphometric and biochemical studies of the premature baboon with bronchopulmonary dysplasia. *Am. Rev. Respir. Dis. 145(4 Pt. 1)*: 872 (1992).
190. T. Saldeen. Fibrin-derived peptides and pulmonary injury. *Ann. N.Y. Acad. Sci. 384*:319 (1982).
191. Y. Fukuda, M. Ishizaki, Y. Masuda, G. Kimura, O. Kawanami, Y. Masugi. The role of intraalveolar fibrosis in the process of pulmonary structural remodelling in patients with diffuse alveolar damage. *Am. J. Pathol. 126*:171 (1987).
192. R. Ceustermans, M. Hoylaerts, M. DeMol, D. Collen. Preparation, characterization, and turnover properties of heparin-antithrombin III complexes stabilized by covalent bonds. *J. Biol. Chem. 257*:3401 (1982).
193. R. E. Jordan. Antithrombin–heparin complex and method for its production. U.S. patent 4,446,126, Cutter Laboratories, 1984.
194. J. Eibl, E. Hetzl, Y. Linnau. Method of producing an antithrombin III-heparin concentrate or antithrombin III-heparinoid concentrate. Immuno Aktiengesellschaft für chemisch-medizinische Produkte. U.S. patent 4,510,084, 1985.
195. M. Spannagl, R. Keller, W. Schramm. A new AT III–heparin-complex preparation: in vitro and in vivo characterisation. *Folia Haematol. 116*:879 (1989).
196. M. Spannagl, H. Hoffman, M. Siebeck, J. Weipert, H. P. Schwarz, W. Schramm. A purified antithrombin III–heparin complex as a potent inhibitor of thrombin in porcine endotoxin shock. *Thromb. Res. 61*:1 (1991).
197. W. E. Hennink, J. Feijen, C. D. Ebert, S. W. Kim. Covalently bound conjugates of albumin and heparin: synthesis, fractionation and characterization. *Thromb. Res. 29*:1 (1983).
198. G. Huhle, J. Harenberg, R. Malsch, D. L. Heene. Comparison of three heparin bovine serum albumin binding methods for production of antiheparin antibodies. *Sem. Thromb. Hemostas. 20*:193 (1994).
199. A. N. Teien, R. Odegard, T. B. Christensen. Heparin coupled to albumin, dextran and ficoll: influence on blood coagulation and platelets, and in vivo duration. *Throm. Res. 7*:273 (1975).
200. P. Cuatrecasas, S. Fuchs, C. B. Anfinsen. Cross-linking of aminotyrosyl residues in the active site of staphylococcal nuclease. *J. Biol. Chem. 244*:406 (1969).
201. H. Zahn, J. Meienhofer. Reaktionen von 1,5-difluor-2,4-dinitrobenzol mit insulin. 1. Mitt. synthese von modellverbindungen. *Makromol. Chem. 26*:126 (1958).
202. H. Zahn, J. Meienhofer. Reaktionen von 1,5-difluor-2,4-dinitrobenzol mit insulin. 2. Mitt. versuche mit insulin. *Makromol. Chem. 26*:153 (1958).
203. G. W. J. Fleet, R. R. Porter, J. R. Knowles. Affinity labelling of antibodies with aryl nitrene as reactive group. *Nature 244*:511 (1969).

204. J. V. Staros, F. M. Richards. Photochemical labeling of the surface proteins of human erythrocytes. *Biochemistry 13*:2720 (1974).
205. M. D. Bregman, D. Levy. Labeling of glucagon binding components in hepatocyte plasma membranes. *Biochem. Biophys. Res. Comm. 78*:584 (1977).
206. S. H. Hixson, S. S. Hixson. P-azidophenacyl bromide, a versatile photolabile bifunctional reagent. Reaction with glyceraldehyde-3-phosphate dehydrogenase. *Biochemistry 14*:4251 (1975).
207. D. H. Clyne, S. H. Norris, R. R. Modesto, A. J. Pesce, V. E. Pollak. The preparation of intermolecular conjugates of horseradish peroxidase and antibody and their use in immunohistology of renal cortex. *J. Histochem. Cytochem. 21*:233 (1973).
208. A. F. Schick, S. J. Singer. On the formation of covalent linkages between two protein molecules. *J. Biol. Chem. 236*:2477 (1961).
209. S. J. Singer, A. F. Schick. The properties of specific stains for electron microscopy prepared by the conjugation of antibody molecules with ferritin. *J. Biophys. Biochem. Cytol. 9*:519 (1961).
210. P. Edman, A. Henschen. *Protein Sequence Determination* (S. B. Needleman, ed.). Springer-Verlag, Berlin, Germany, 1975, p. 232.
211. Y. Inque, K. Nagasawa. Selective N-desulfation of heparin with dimethyl sulfoxide containing water or methanol. *Carb. Res. 46*:87 (1976).
212. K. Nagasawa, H. Yoshidome. Solvent catalytic degradation of sulfamic acid and its N-substituted derivatives. *Chem. Pharm. Bull. 17*:1316 (1969).
213. B. Casu. Structure and biological activity of heparin and other glycosaminoglycans. *Pharmacological Res. Comm. 11*:1 (1979).
214. M. Hoylaerts, W. G. Owen, D. Collen. Involvement of heparin chain length in the heparin catalyzed inhibition of thrombin antithrombin III. *J. Biol. Chem. 259*:5670 (1984).
215. I. Björk, O. Larm, U. Lindahl, K. Nordling, M. E. Riquelme. Permanent activation of antithrombin by covalent attachment of heparin oligosaccharides. *FEBS Lett. 143*:96 (1982).
216. A. P. Halluin. Protein heparin conjugates. U.S. patent 5,308,617, Halzyme, 1994.
217. J. Dawes, K. James, D. A. Lane. Conformational change in antithrombin induced by heparin probed with a monoclonal antibody against the 1C/4B region. *Biochemistry 33*:4375 (1994).
218. G. Mitra, R. E. Jordan. Covalently bound heparin–antithrombin III complex. U.S. patent 4,689,323, Miles Laboratories, 1987.
219. G. Mitra, R. E. Jordan. Covalently bound heparin–antithrombin III complex, method for its preparation and its use for treating thromboembolism. European patent 84111048.9 (0 137 356), Miles Laboratories, 1985.
220. H. Vlassara, M. Brownlee, A. Cerami. Nonenzymatic glycosylation: role in the pathogenesis of diabetic complications. *Clin. Chem. 32*:B37 (1986).
221. P. Hall, E. Tryon, T. F. Nikolai, R. C. Roberts. Functional activities and non-enzymatic glycosylation of plasma proteinase inhibitors in diabetes. *Clin. Chim. Acta 160*:55 (1986).
222. A. Ceriello, D. Giugliano, A. Quatraro, A. Stante, G. Consoli, P. Dello Russo, F. D'Onofrio. Daily rapid blood glucose variations may condition antithrombin III biologic activity but not its plasma concentration in insulin-dependent diabetes. A possible role for labile non-enzymatic glycation. *Diabete. Metab. 13*:16 (1987).
223. A. Ceriello, P. Dello Russo, C. Zuccotti, A. Florio, S. Nazzaro, C. Pietrantuono, G. B. Rosato. Decreased antithrombin III activity in diabetes may be due to non-enzymatic glycosylation — a preliminary report. *Thromb. Haemost. 50*:633 (1983).
224. G. B. Villanueva, N. Allen. Demonstration of altered antithrombin III activity due to non-enzymatic glycosylation at glucose concentration expected to be encountered in severely diabetic patients. *Diabetes 37*:1103 (1988).
225. M. Brownlee, H. Vlassara, A. Cerami. Inhibition of heparin-catalyzed human antithrombin III activity by non-enzymatic glycosylation. Possible role in fibrin deposition in diabetes. *Diabetes 33*:532 (1984).
226. T. Sakurai, J. P. Boissel, H. F. Bunn. Non-enzymatic glycation of antithrombin III in vitro. *Biochim. Biophys. Acta 964*:340 (1988).
227. M. Brownlee, A. Cerami. A glucose-controlled delivery system: semi-synthetic insulin bound to lectin. *Science 206*:1190 (1979).
228. J. Hoffman, O. Larm, E. Scholander. A new method for covalent coupling of heparin and other glycosaminoglycans to substances containing primary amino groups. *Carb. Res. 117*:328 (1983).
229. M. W. Hatton, L. R. Berry, R. Machovich, E. Regoeczi. Tritiation of commercial heparins by reaction of NaB3H4: chemical analysis and biological properties of the product. *Anal. Biochem. 106*:417 (1980).
230. L. Berry, A. K. C. Chan, M. Andrew. Polypeptide–polysaccharide conjugates produced by spontaneous non-enzymatic glycation. *J. Biochem. 124*:434 (1998).
231. T. F. Busby, K. C. Ingham. Thermal stabilization of antithrombin III by sugars and sugar derivatives and the effects of non-enzymatic glycosylation. *Biochim. Biophys. Acta 799*:80 (1984).
232. R. E. Jordan, L. V. Favreau, E. H. Braswell, R. D. Rosenberg. Heparin with two binding sites for antithrombin or platelet factor 4. *J. Biol. Chem. 257*:400 (1982).
233. L. Berry, A. Stafford, J. Fredenburgh, H. O'Brodovich, L. Mitchell, J. Weitz, M. Andrew, A. K. Chan. Investigation of the anticoagulant mechanisms of a covalent antithrombin–heparin complex. *J. Biol. Chem. 273*:34730 (1998).
234. M. Hoylaerts, E. Holmer, M. De Mol, D. Collen. Covalent complexes between low molecular weight heparin fragments and antithrombin III—inhibition kinetics and turnover parameters. *Thromb. Haemost. 49*:109 (1983).
235. D. J. Collen. Novel composition of matter of antithrombin III bound to a heparin fragment, KabiVitrum AB. U.S. patent 4,623,718, 1986.

236. C. Mattsson, M. Hoylaerts, E. Holmer, R. Uthne, D. Collen. Antithrombotic properties in rabbits of heparin and heparin fragments covalently coupled to human antithrombin III. *J. Clin. Invest. 75*:1169 (1985).
237. A. K. Chan, L. Berry, P. Klement, J. Julian, M. Mitchell, J. Weitz, J. Hirsh, M. Andrew. A novel antithrombin–heparin covalent complex: antithrombotic and bleeding studies in rabbits. *Blood Coag. Fibrinol. 9*:587 (1998).
238. K. K. Singhal, L. A. Parton. Plasminogen activator activity in preterm infants with respiratory distress syndrome: relationship to the development of bronchopulmonary dysplasia. *Pediatr. Res. 39*:229 (1996).
239. S. Idell, A. Kumar, K. B. Koenig, J. J. Coalson. Pathways of fibrin turnover in lavage of premature baboons with hypoxic lung injury. *Am. J. Respir. Crit. Care Med. 149*: 767 (1994).
240. S. Idell, K. K. James, E. G. Levin. Local abnormalities in coagulation and fibrinolytic pathways predispose to alveolar fibrin deposition in the adult respiratory distress syndrome. *J. Clin. Invest. 84*:695 (1989).
241. P. Bertozzi, B. Astedt, L. Zenzius, K. Lynch, F. Lemaire, W. Zapol, H. Chapman Jr. Depressed bronchoalveolar urokinase activity in patients with adult respiratory distress syndrome. *N. Engl. J. Med. 322*:890 (1990).
242. A. K. Chan, L. Berry, L. Mitchell, B. Baranowski, H. O'Brodovich, M. Andrew. Effect of a novel covalent antithrombin–heparin complex on thrombin generation on fetal distal lung epithelium. *Am. J. Physiol. 274*:L914 (1998).
243. D. L. Becker, J. C. Fredenburgh, A. R. Stafford, J. I. Weitz. Exosites 1 and 2 are essential for protection of fibrin-bound thrombin from heparin-catalyzed inhibition by antithrombin and heparin cofactor II. *J. Biol. Chem. 274*: 6226 (1999).
244. N. Volpi, M. Cusmano, T. Venturelli. Qualitative and quantitative studies of heparin and chondroitin sulfates in normal human plasma. *Biochim. Biophys. Acta 1243*:49 (1995).
245. L. Mitchell, R. Superina, M. Delorme, P. Vegh, L. Berry, H. Hoogendoorn, M. Andrew. Circulating dermatan sulphate/heparin sulfate proteoglycan(s) in children undergoing liver transplantation. *Thromb. Haemostas. 74*: 859 (1995).
246. M. Andrew, L. Mitchell, L. Berry, B. Paes, M. Delorme, F. Ofosu, R. Burrows, B. Khambalia. An anticoagulant dermatan sulfate proteoglycan circulates in the pregnant woman and her fetus. *J. Clin. Inst. 89*:321 (1992).
247. C. H. Bamford, K. G. Al-Lamee. Chemical methods for improving the haemocompatibility of synthetic polymers. *Clin. Mater. 10*:243 (1992).
248. C. Karlsson, H. Nygren, M. Braide. Exposure of blood to biomaterial surfaces liberates substances that activate polymorphonuclear granulocytes. *J. Lab. Clin. Med. 128*: 496 (1996).
249. L. Tang, J. W. Eaton. Inflammatory responses to biomaterials. *Am. J. Clin. Pathol. 103*:466 (1995).
250. R. Vanholder, N. Lameire. Does biocompatibility of dialysis membranes affect recovery of renal function and survival? *Lancet 354*:1316 (1999).
251. J. Sanchez, P. Olsson. On the control of the plasma contact activation system on human endothelium: comparisons with heparin surface. *Thromb. Res. 93*:27 (1999).
252. H. P. Wendel, G. Ziemer. Coating-techniques to improve the hemocompatibility of artificial devices used for extracorporeal circulation. *Eur. J. Cardiothorac. Surg. 16*:342 (1999).
253. G. Dickneite, B. Leithauser. Influence of antithrombin III on coagulation and inflammation in porcine septic shock. *Arterioscler. Thromb. Vasc. Biol. 19*:1566 (1999).
254. A. K. C. Chan, L. Berry, L. Mitchell, M. Andrew, P. Klement. Evaluation of antithrombin-heparin covalent coated endoluminal grafts in rabbits. Transactions of the Society for Biomaterials, Providence, Rhode Island, 1999, p. 102.
255. H. H. Hoerl, D. Nussbaumer, E. Wuenn. Surface grafting of microporous, nitrogen-containing polymer membranes and membranes obtained thereby. German patent DE 3,929,648, Sartorius, 1990.
256. H. H. Heinrich, D. Nussbaumer, E. Wuenn. Grafting of unsaturated monomers on polymers containing nitrogen. German patent DE 4,028,326, Sartorius A.-G., 1991.
257. K. van Phung, R. C. Schulz. Schöpfung von vinylverbindugen auf polyamide. *Makromol. Chem. 180*:1825 (1979).
258. V. K. Raman, E. R. Edelman. Coated stents: local pharmacology. *Semin. Interv. Cardiol. 3*:133 (1998).
259. A. K. C. Chan, L. Berry, M. Andrew, P. Klement. Antithrombin–heparin covalent complex: a novel approach for improving thromboresistance of cardiovascular devices. Proceedings of the International Biomaterials Congress, Honolulu, Hawaii, 2000.
260. M. D. Phaneuf, S. A. Berceli, M. J. Bide, W. C. Quist, F. W. LoGerfo. Covalent linkage of recombinant hirudin to poly(ethylene terephthalate) (Dacron): creation of a novel antithrombin surface. *Biomaterials 18*:755 (1997).
261. M. D. Phaneuf, M. Szycher, S. A. Berceli, D. J. Dempsey, W. C. Quist, F. W. LoGerfo. Covalent linkage of recombinant hirudin to a novel ionic poly (carbonate) urethane polymer with protein binding sites: determination of surface antithrombin activity. *Artific. Organs 22*:657 (1998).
262. V. L. Gott, R. L. Daggett. Serendipity and the development of heparin and carbon surfaces. *Ann. Thorac. Surg. 68*:S19 (1999).
263. G. Elgue, M. Blomback, P. Olsson, J. Riesenfeld. On the mechanism of coagulation inhibition on surfaces with end point immobilized heparin. *Thromb. Haemost. 70*:289 (1993).
264. Y. Miura, S. Aoyagi, F. Ikeda, K. Miyamoto. Anticoagulant activity of artificial biomedical materials with co-immobilized antithrombin III and heparin. *Biochimie 62*:595 (1980).
265. P. Cahalan, T. Lindhout, B. Fouache, M. Verhoeven, L. Cahalan, M. Hendriks, R. Blezer. Method for making improved heparinized biomaterials. U.S. patent 5,767,108, Medtronic, 1998.

26

Adhesives for Medical Applications

Iain Webster and Peter J. West
Smith & Nephew Group Research Centre, York, England

I. INTRODUCTION TO MEDICAL ADHESIVES

The area of medical adhesives is very broad and encompasses a variety of technologies that can be applied to a wide range of medical products and procedures. This technology already has a long and well-documented history, but, not surprisingly, as a result of the advances in health care, there are many new adhesives emerging to meet our demands for the future.

In this chapter we discuss medical adhesives for use in products that are applied directly to the body or used in surgery inside the body. This covers the areas of pressure-sensitive adhesives (PSAs) and tissue adhesives, which can be further divided into synthetic and biological adhesives. The scope of this chapter is shown in Fig. 1.

Investigations in pressure-sensitive adhesive technology have concentrated in recent years on materials that can do more than simply adhere medical dressings to a wound. Since the gradual replacement of rubber resin–based adhesives for medical applications with synthetic polymers, researchers have utilized the wide range of possible formulations to tailor adhesive properties to meet specific needs. Having a greater understanding of the skin as an adhesive substrate, formulators have been able to produce PSAs that form a vital part of modern wound dressings.

The use of surgical tissue adhesives in medicine has developed considerably over the last 40 years, and the list of applications is now increasing every year. Traditionally, the area of tissue reattachment or repair following surgery or trauma has been dominated by sutures, staples, and wiring, but recently the huge commercial potential for tissue adhesives has sparked a minirevolution in medical practices. Ultimately the general acceptance of new tissue adhesives will mark a step change in clinical practice.

The approach to developing new adhesives relies on understanding the different and often complicated requirements of various tissue types and a large number of surgical procedures. This understanding can only be gained by researchers working closely with medical professionals in hospitals and clinics. Equally, medical practitioners need to be prescriptive about their needs and be willing to contribute to, and ultimately accept and promote, new and exciting developments. With this challenge in mind, it is becoming possible to provide better products for us all as potential patients.

II. PRESSURE-SENSITIVE ADHESIVES

A. Introduction

Typically, pressure-sensitive adhesives have been used for adhering wound dressings to skin. PSAs have glass transition temperatures (Tg) or softening temperatures in the range $-20°C$ to $-60°C$, which means they are soft materials at room or skin temperature. These soft polymers are able to flow and ''wet out'' on to a surface and are formulated to be permanently tacky to enable adherence to that surface. The bond formed between PSA and substrate is not permanent and can be broken with a measurable force,

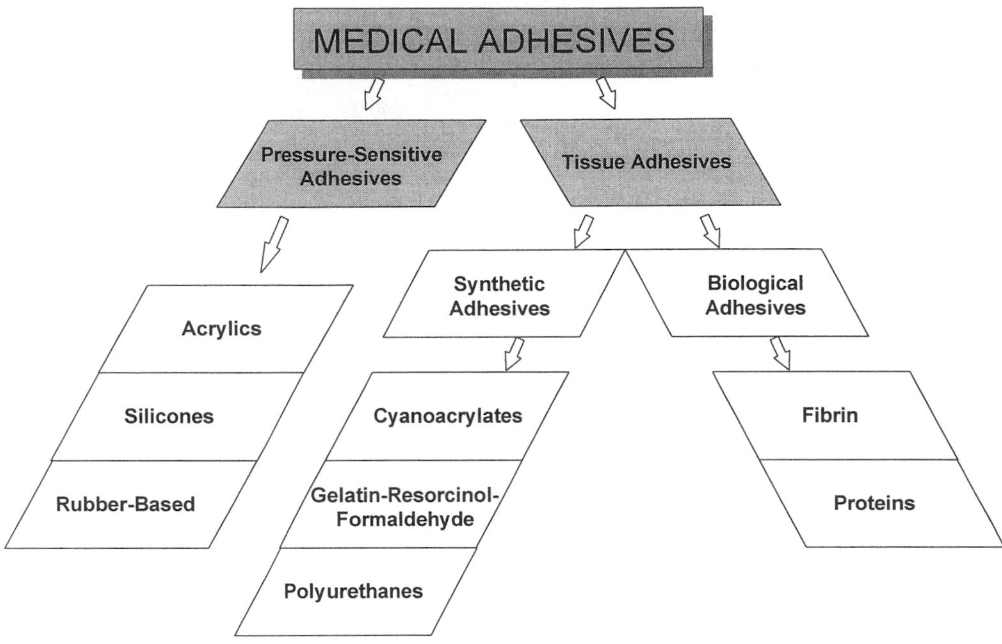

Figure 26.1 The scope of medical adhesives.

leaving the substrate relatively free from damage. Since the mid-19th century, when the first adhesive plasters were reported (1), the hospital and first aid applications of dressings have become more demanding, and the polymers used for PSAs have consequently undergone significant development (2) (Fig. 2). Early adhesive formulations were based on blends of natural rubber and resins derived from wood rosin, which act as tackifiers. However, a PSA is required to be more than just sticky. The adhesive should be permanently and aggressively tacky, adhere with only

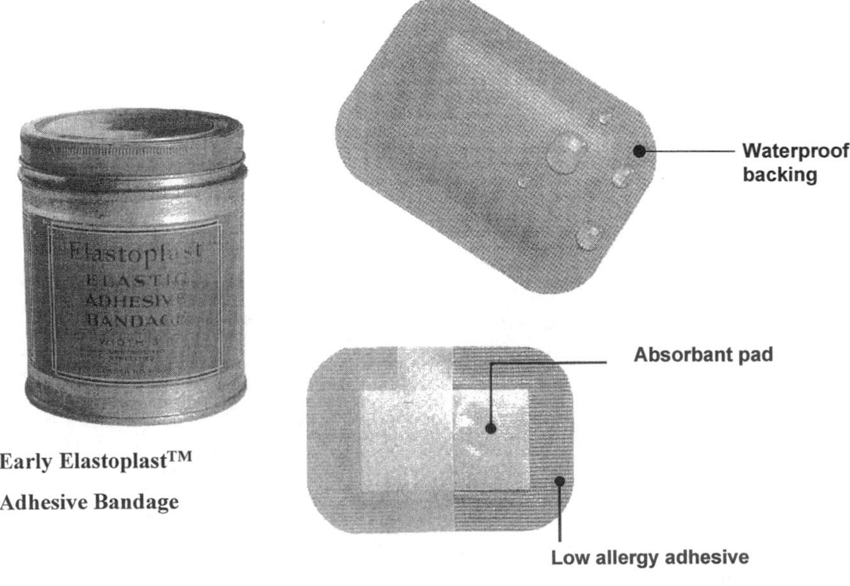

Figure 26.2 The development of first aid dressings; from early elastic adhesive bandage sold under the trade name *Elastoplast* to the modern day waterproof, breathable, absorbant, conformable dressings that also act as bacterial barriers.

slight finger pressure, form a strong bond with surfaces, and have sufficient cohesiveness that it can be removed without leaving a residue.

The properties for medical PSAs are made yet more demanding by the requirements for skin contact. Foremost is the need to be chemically and biologically acceptable to the skin, with the adhesive causing no irritation or sensitization. Skin is also very variable, both between individuals and even between different sites on the same individual. The roughness, elasticity, and surface energy of the skin must be taken into account, which means that adhesives must have sufficient flow to ensure intimate surface contact (3). Additionally, medical adhesives must be able to cope with moisture and sebum exuded by the skin without severely compromising performance. Finally, once securely adhered to the skin, medical PSAs should be easily removed with minimal trauma to the skin. Although this may appear to be a contradiction, recent developments in PSA technology have endeavored to solve this problem.

The last thirty years have seen the gradual replacement of rubber adhesives with acrylic copolymers, which now dominate the medical adhesive market. These acrylic copolymers are naturally tacky materials and have eliminated the need for the large number of components present in rubber PSAs, such as natural rosins. This reduction in potential for introducing irritant and sensitizing agents has led to a broad acceptance of acrylic PSAs within health care due to their natural hypoallergenicity (4).

Although polyacrylates are used widely and generally for medical PSAs, typically the polymer selected for a given PSA is directed by the application for which it is required (5) (Table 1). This section therefore describes the range of indications that require formulated medical grade adhesives and the properties required of the polymeric materials used. We will also detail the polymer types that are used to formulate these adhesives.

B. Indications and Adhesive Requirements

1. Adhesion To Skin

Skin is a very demanding and variable substrate for adhesive bonding, and consequently the properties of medical grade PSAs have to be adaptable. Skin is a rough surface, which requires the PSA to have good flow characteristics to enable it to form a good bond. Skin also flexes, which results in the adhesive being subject to various stresses as well as changing the angle of peel when tapes are removed. Furthermore, sebum and moisture exuded by skin may affect not only its surface energy but also its adhesive properties. Many studies have been made of the physical properties required of a PSA to produce good skin adhesives (3,6–8). The main properties that dictate adhesive performance are the surface energy of the PSA compared to that of skin and also the bulk rheological properties of the adhesive polymer.

a. Surface Properties

In order to provide a satisfactory bond, an adhesive must wet the skin surface. This requires that the surface energy

Table 26.1 Polymers Used in the Formulation of Medical Grade PSAs

Polymer	Application	Property
Silicone	Transdermal drug delivery	Chemically inert
		Biocompatible
		High drug permeability
Polyvinyl ether	Skin patches/surgical dressings	Moisture permeability
Polyvinyl pyrrolidone	Ostomy	Moisture absorption
Polyacrylate	Transdermal drug delivery	Chemically inert
		Good drug release properties
	First-aid dressings	Quick adhesion
		Adhere during normal daily activity
	Electromedical devices	Good, long-term adhesion
	Surgical dressings	Moisture permeability
	Incise drapes	Sterilizable
		Wet stick
	Surgical tape	Sterilizable
Hydrophilic gels	Electromedical applications	Electrical conductivity
		Moisture absorption
Rubber resins	First-aid dressings	Quick adhesion
		Adhere during normal daily activity

Table 26.2 Variation of Surface Energy of Human Skin with Temperature and Relative Humidity

Temperature (°C)	Relative humidity (%)	Total surface energy (J/m^2)
23	34	0.038
23	50	0.042
28	60	0.054
32	56	0.051
33	51	0.057
36	50	0.057

of an adhesive must be lower than that of the skin. The surface energy of skin increases with increasing temperature and relative humidity, as sweating produces a high-energy surface (8). This is illustrated in Table 2. The surface energy of a commercially available medical grade acrylic PSA is shown in Table 3 for comparison; it can be seen that humidity does not significantly affect the adhesive. The incorporation of polar monomers into acrylic PSAs has been shown to provide improved adhesion to skin during sweating, when the surface energy is higher (8). This is thought to be enhanced due to interfacial interactions between the adhesive and the skin surface.

b. Adhesive Peeling from Skin

The peeling of a PSA spread tape from a standard substrate is the most common assessment made of adhesives. This involves the measurement of the force required to peel an adhesive, spread onto a flexible backing, from a substrate following adhesion under controlled conditions of pressure and buildup time. The force required is dependent on the peel angle (typically 180° or 90°) and peel rate (estimated for removal of PSA tapes from skin as between 12 and 200 inches per minute). Although this type of test allows differences between adhesives to be identified, it is unrepresentative of how any given PSA will peel from skin.

Table 26.3 Variation of Surface Energy of Acrylic Medical-Grade PSA with Relative Humidity

Relative humidity (%)	Total surface energy (J/m^2)
10	0.027
20	0.027
42	0.028
52	0.026
81	0.027
100	0.025

Skin, as a peeling substrate, is rough, has a variable surface energy, and is flexible, whereas most data for peel adhesion has been determined using stainless steel as the substrate. The difficulty in reproducing skin as an in vitro substrate is demonstrated by the lack of suitable models developed over the years. Attempts have been made to model the surface energy and roughness using embossed polypropylene plaques (3) or a hydrated cross-linked gelatin/lipid substrate (9) with varying degrees of success.

c. Skin Stripping and Trauma

The requirements that a medical-grade PSA should adhere well to skin and yet be easily removed with minimal trauma are obviously contradictory. When an adhesive tape is removed, a layer of skin is also removed, because the separation is not at the skin–adhesive interface but deeper into the stratum corneum (4,10). The stripping of this surface barrier compromises the skin's ability to regulate fluid balance as well as its ability to protect from chemical, mechanical, and microbial effects (11). Research has been carried out for many years to devise a PSA that can be removed without causing damage or trauma to the skin while giving sufficient adhesion when required. However, a fundamental understanding of the interaction between the skin and adhesive is required to allow solutions to be explored (12).

Studies carried out using occlusive and nonocclusive tapes have shown that the buildup of moisture behind occlusive tapes results in hydration and therefore weakening of the stratum corneum. This means that the levels of adhesion between the outer layers and the adhesive are greater than the internal strength of the stratum corneum. Hence when tapes are peeled, layers of skin are also removed with the adhesive (4). Repeated application of a tape to the same site produces higher peel forces as the deeper epidermal cells have increased cohesion (Fig. 3). When more porous tapes and adhesives are used (i.e., changing from a hydrophobic rubber resin to a more hydrophilic acrylic adhesive), the moisture content of the skin is maintained and therefore so is its strength. This type of approach has enabled the development of adhesive dressings that can be removed with less trauma to the skin based on thin polyurethane films coated with moisture vapor permeable adhesives. Dressings based on this theory, typically surgical drapes and tapes, do not therefore cause maceration of the skin but do allow moist wound healing.

2. Wound Management Indications

The largest segment of the medical adhesive market is that for the protection of wounds. This diverse sector includes adhesive bandages, surgical drapes, ostomy bag mounts,

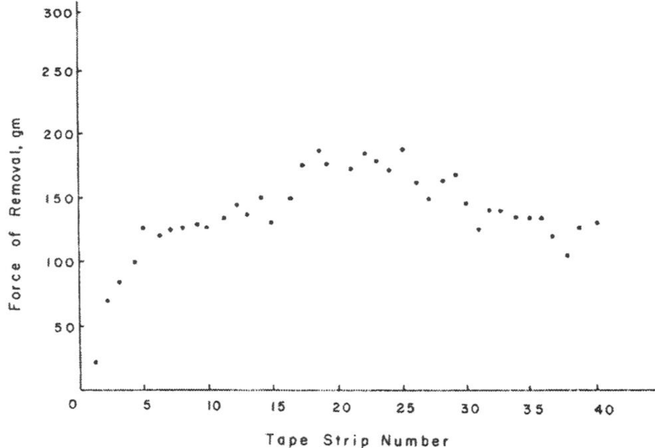

Figure 26.3 Force of removal of adhesive tapes peeled from a defined area of human back skin. (Graph reproduced from Ref. 4.)

intravenous needle attachments, electromedical procedures, microporous tapes, and transdermal drug delivery (5) (covered more extensively in Section II.B.3). The adhesives used in the majority of wound care applications are based on acrylic copolymers, although rubber-based PSAs (natural or synthetic) are still used where high levels of adhesion are required and skin irritation is not a major concern. The backing materials used with the adhesives range from foams to porous materials such as nonwovens, cotton and fabric, and polymeric films such as polyethylene and polyurethane.

Ostomy appliances require a PSA to perform two roles, firstly to provide a good seal around the device to prevent leakage of fluids around the stoma, and secondly to secure the device to the body. The adhesive must also be designed to allow prolonged and repeated application to the same site without causing significant skin deterioration by mechanical or chemical irritation. Acrylic adhesives were often found to be unsuitable for this application, as they did not provide a good seal around the stoma (5). However, several patents exist that describe soft acrylic PSA copolymers that provide good adhesion over prolonged wear times (13,14). The advent of acrylic hot melt PSAs has provided new materials suitable for this purpose. Other adhesive types include blends of acrylic/rubber components (15), polyisobutylenes (16), and vinyl ether copolymers (17).

PSAs for electromedical applications are used to provide an electrical connection between the skin and an electromedical device (5). Traditionally, a conductive gel in contact with a metal electrode was used and held in contact with the skin by a pressure-sensitive tape. However, more recent developments have seen the adhesive designed to both hold the electrode in place and provide a conductive medium (18,19). Typically, adhesives for this application are cross-linked hydrophilic polymers swollen with water (20). Examples include an ultraviolet curable PSA synthesized from acrylic acid and a multifunctional acrylate as cross-linking agent (21), an electron beam curable composition of polyvinyl pyrrolidone, polyethylene glycol, and water (22), and a PSA comprising cross-linked sodium carboxymethyl cellulose and polyisobutylene swollen with water (23). A hydrophilic PSA has also been described that comprises a zwitterionic polymer blended with a plasticiser (24). It is claimed that the zwitterionic functionality imparts high conductivity to the adhesive, making it suitable also for use as a matrix for iontophoretic delivery of drugs (Section II.B.3).

The acceptance of moist wound healing as most appropriate for rapid healing (25) has lead to the development of occlusive thin film dressings capable of maintaining the correct wound environment (26). The first dressing of this type was introduced in 1971 by Smith & Nephew under the tradename *OpSite*™, which comprised a polyurethane film spread with a vinyl ether adhesive (27). Other semiocclusive dressings were developed along similar lines based on polyurethane or copolyester backing films with an acrylic PSA. Moist wound healing was found to enhance the rate of experimentally induced wound resurfacing by up to 40% compared to air exposed wounds (28). It was therefore important that the dressing was capable of striking the balance between exudate handling and maintaining a moist environment. The moisture vapor transmission rate (MVTR) of the dressing and therefore the PSA played an important role in achieving this, as the thin film dressing could not absorb much moisture from the wound. The adhesive was required to have a MVTR (measured according to ASTM E96) greater than that of skin, which varies depending on the body site and skin condition (Table 4) (28). Changes in the polarity, porosity, and thickness of the ad-

Table 26.4 MVTR for Intact Skin and Varying Wound Types

	MVTR ($g/m^2/hr$)		
Body site/wound type	22°C	27°C	30°C
Legs	8.0	9.0	10
Abdomen	8.5	10	12
Arms	8.5	9.5	12
Head	18	22	35
Hands	50	78	90
Open granulating wound	214		
Full thickness burn	143		

Table 26.5 MVTR of Commercially Available Thin Film Dressings Determined Using Payne Cup Method at 37°C, 10% RH

Dressing	Manufacturer	MVTR ($g/m^2/24hr$)	
		Dry (upright)	Wet (inverted)
Bioclusive™	Johnson & Johnson	470	490
Bioclusive™ MVP	Johnson & Johnson	2200	4500
OpSite™ IV 3000	Smith & Nephew	1150	2900
Tegaderm™ HP	3M	450	1300

hesive provide control of the MVTR. Dressings based on the moist wound healing principle were required to cope with both lightly exuding wounds, requiring a low MVTR film and PSA, and high levels of moisture such as that experienced with invasive catheter fixation (Table 5) (28). *OpSite*™ IV 3000 (Fig. 4) used a specially developed emulsion-based acrylic PSA that was pattern spread to cover 75% of the film area to ensure optimum MVTR while maintaining good skin adhesion.

3. Transdermal Drug Delivery

PSAs have been used in transdermal delivery devices since the 1970s and have been found to offer many advantages over conventional methods for the controlled delivery of drugs (29). Transdermal drug delivery involves the delivery of an active ingredient through the skin and into the blood vessels before delivery to the target organ. This provides a constant rate of drug delivery that leads to stable levels within the blood that correlate more closely to those obtained for intravenous drug infusion (30). This technique allows the use of lower doses of drug than with oral medication and avoids the variable blood level profiles typically seen with other systems. Transdermal drug delivery therefore allows more effective and safe dosing to achieve the desired therapeutic result. Patient compliance is also likely to be higher, as the dosing regime may require only a single dressing change per week, and the delivery can be stopped by removing the adhesive patch.

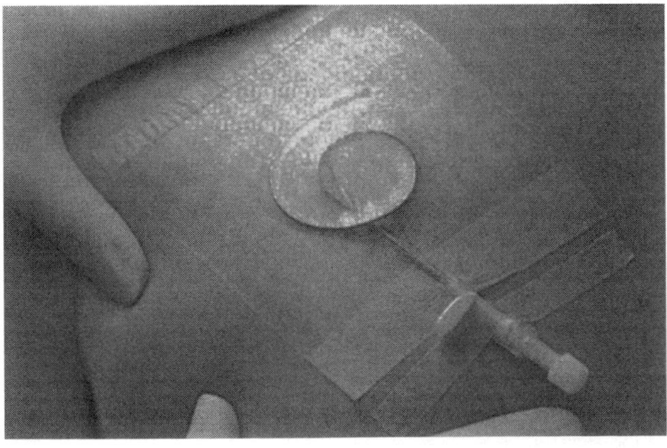

Figure 26.4 Catheter fixation using OpSite™ IV 3000* film dressing. (*Trademarks of Smith & Nephew.)

Conventional uses of transdermal delivery systems have included patches for the release of nitroglycerin for antianginal products (31), scopolamine (motion sickness), clonidine (hypertension), glycol salicylate (analgesia), and nicotine patches for smoking cessation. Such developments produced an industry generating annual sales of $1.7 billion in 1998, sales are estimated to continue growing as new drugs are proposed for transdermal delivery (29).

Transdermal drug delivery patches may be split into two basic types (29,30,32) with the drug being held either in a reservoir by a permeable membrane and adhered to the skin using a PSA, or in a matrix system where the drug is incorporated directly into the adhesive polymer (Fig. 5).

The adhesive chosen for transdermal applications must not only display the characteristics of a good skin adhesive but also be compatible with the drug so as not to alter its characteristics and, in the case of the matrix design (Fig. 5b), have the correct drug/adhesive solubility relationship. The polymers that meet these criteria tend to be silicones, acrylic copolymers (33), and polyisobutylene (29,32). Silicones, although the more expensive of these options, tend to offer the highest drug diffusion rates as well as being easily formulated to give the right adhesive properties. Acrylate copolymers can also be easily modified to optimize drug solubility, by controlling the hydrophilic nature of the polymer, diffusion, and adhesion. Polyisobutylenes provide more of a challenge to tailor their properties to transdermal drug delivery, but they are particularly suited for use with drugs of low solubility or polarity.

Many studies have compared the relative performance of these three polymer types for their suitability as drug delivery vehicles. Firstly, the ability of an adhesive to provide long-term adhesion is essential in order to complete the dosing regime required (34). Typical values of tack and

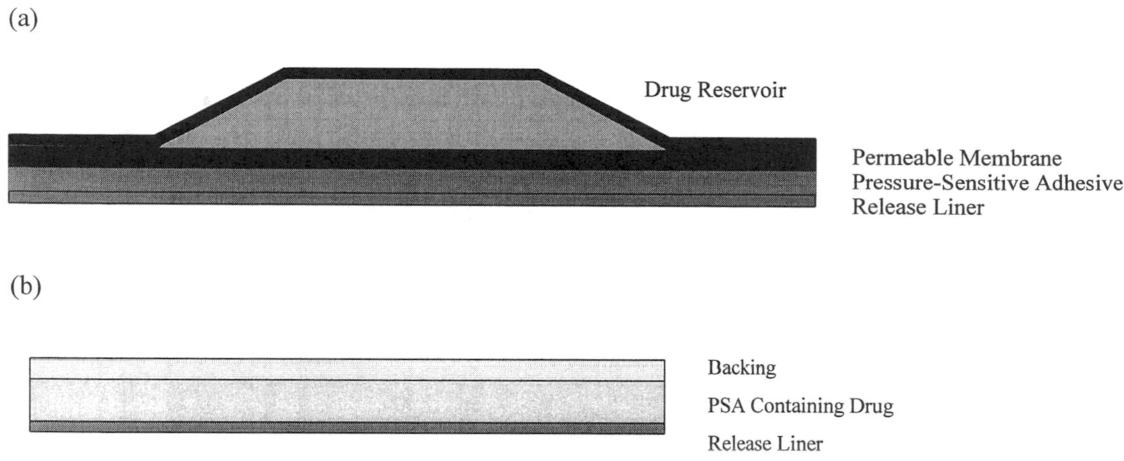

Figure 26.5 Transdermal drug delivery patch designs. (a) Reservoir system; (b) matrix system.

peel adhesion of PSAs loaded with drugs (progesterone and indomethacin) have been given as 60–70 g/cm^2 and 500–520 g/cm respectively (35). The efficiency of adhesives to release drugs has also been widely studied with respect to drug solubility in the adhesive, drug release kinetics, and skin permeation (36–38). Drug solubility may be related to the hydrophilicity of the adhesive; trends such as those shown in Table 6 are observed for examples of the three adhesive types.

Drug release is dependent on the Tg of the PSA and the functionalities on the adhesive polymer chain. The low Tg of silicone PSAs, for example, results in a very flexible and open polymer structure that gives high permeability values for a variety of molecules. Figure 6 illustrates the release profiles of four drugs from acrylic copolymers of 2-ethylhexyl acrylate with either acrylic acid or acrylamide, a silicone PSA, and a PSA blend of high and low molecular weight polyisobutylenes (38). These profiles demonstrate how the good flow characteristics of the silicone PSA results in a high rate of release for all of the drugs studied, whereas the potential for interaction between the acrylics and the drug produced varying release profiles depending on the drug type. The polyisobutylene gave fairly consistent release results for all drugs, as these polymers have no functional groups to cause differentiation.

Skin permeation of drugs does not however always follow the trends established for drug release in water. Skin permeation data, generated using human cadaver skin for the determination of skin flux of fentanyl, showed that the two properties do not necessarily give the same ranking (Table 7) (37). When released into phosphate buffer solution, the general ranking of fentanyl release by polymer type was acrylate > silicone > polyisobutylene. This order was changed, however, for skin permeation with a ranking of silicone 2920 (Dow Corning) > polyisobutylene (Exxon Chemical) > silicone 2675 (Dow Corning) > acrylate (Gelva-737, Monsanto) (silicone 2920 is moderately more hydrophobic than silicone 2675).

The future of transdermal drug delivery technology will be directed by the development of current delivery systems to allow extended wear times (39), biphasic drug delivery profiles, generic drug patches, and combination drug patches (29). A further challenge lies in the need to deliver higher molecular weight drugs through a skin barrier designed to prevent penetration of foreign molecules. Therefore external stimuli systems such as

Iontophoresis, where a conductive PSA and miniature battery set up an electrical potential to admit ionically charged drug into the skin (40,41)

Table 26.6 Drug Solubility (mg/mL) in Acrylic, Silicone, and Polyisobutylene PSAs

PSA	Fentanyl	Aminopyrine	Dipropylphthalate	Lidocaine	Ketoprofen
Acrylic copolymer	21.9	95.1	223.7	438.6	61.3
Silicone	20.0	39.6	12.9	81.0	4.2
Polyisobutylene	0.8	48.3	62.7	62.4	0.8

Source: Refs. 37 and 38.

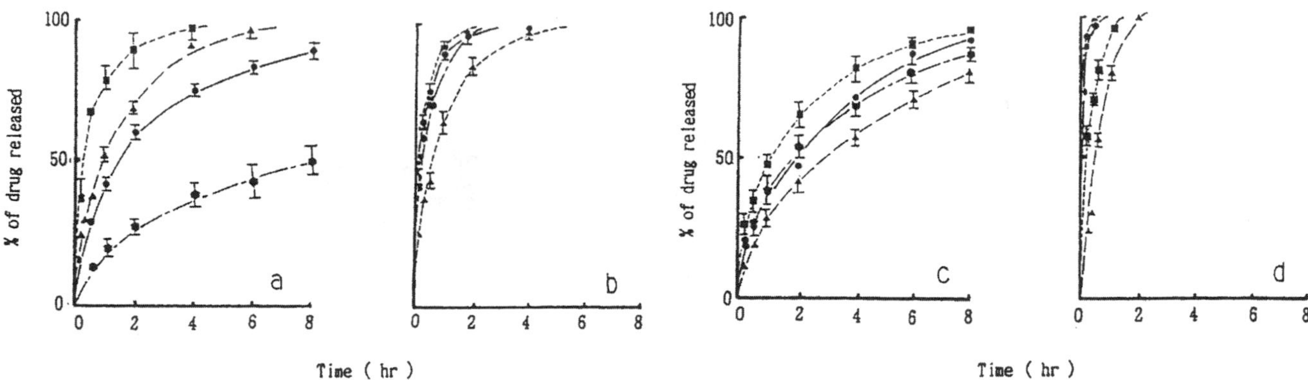

Figure 26.6 Release profiles of four drugs from different PSA formulations: (a) 2-ethylhexyl acrylate/acrylic acid copolymer; (b) 2-ethylhexyl acrylate/acrylamide copolymer; (c) polyisobutylene; (d) silicone. ● Dipropylphthalate, ■ Aminopyrine, ▲ Ketoprofen, ⬢ Lidocaine. (Graph reproduced from Ref. 33.).

Sonophoresis, where sound waves disrupt the stratum corneum to increase drug delivery rates

Electroporation, where pores within the skin are opened to create channels of low resistance

are being developed to allow the controlled delivery of high molecular weight compounds (29). The development of more hydrophilic PSAs such as hydrogels, polyurethanes, and modified acrylics has also led to enhanced drug delivery rates (42).

4. Trauma-Free Removal

A number of approaches have been taken to producing adhesives that initially adhere well but can be removed easily without causing trauma to the skin. Typically these have involved deactivation of the adhesive by some stimulus. One such development by Smith & Nephew has involved the use of light to deactivate, or switch off, an acrylic adhesive by deadening its tack and reducing the level of adhesion (43,44). The PSA was formulated using acrylic monomers (*n*-butyl acrylate, 2-ethylhexyl acrylate, and acrylic acid) copolymerized with either itaconic anhydride or vinyl azlactone. These adhesives were then further reacted with hydroxyethyl methacrylate to produce methacrylate functional polymers capable of free-radical cross-linking (Fig. 7). When formulated with a suitable visible light photoinitiator, the adhesive cross-links on exposure to light, resulting in reduced peel forces on removal from a substrate.

Analysis of their performance on skin, as part of a volunteer trial, demonstrated the benefits of switching (deactivating) the adhesive. Results showed that 60% reductions in peel strength were possible when the adhesive was exposed to light. This in turn gave significant reductions in perceived pain, skin erythema, and skin stripping (45) (Fig. 8).

Other methods for deactivating adhesives to give trauma-free removal of dressings include the use of water and temperature. Adhesives Research Inc. have developed a PSA for use on wound care products that can be deactivated when more than 25–30% water is absorbed (46). The adhesive is based on a hydrophilic polymer comprising a mixture of a vinyl ether/maleic acid ester copolymer (e.g., Fig. 9) and a polyvinyl alkyl ether, such as polyvinyl methyl ether. The base polymer is blended with a water-soluble tackifier to produce a PSA that is coated onto a nonwoven backing. When the adhesive absorbs excess moisture, the tackifier separates from the base polymer, causing a loss of adhesion (Fig. 10).

Adhesives are also known that perform as well as conventional PSAs but can be thermally deactivated by either heating or cooling during removal (47). This technology uses side chain crystallizable polymers combined with conventional acrylic medical grade PSAs to produce materials that have their adhesion controlled by the ambient temperature. The ''Cool Off'' adhesive was shown to lose 60% of its initial adhesion to skin when switched with ice or a cold compress (48). Similarly, the ''Warm Off'' adhesive gave reductions in peel strength of 70–90% with little stripping of the stratum corneum.

Table 26.7 Release Rates into Aqueous Medium and Skin Permeation of 2% Fentanyl from Various PSAs

PSA type	Release rate constant k ($min^{-1/2}$)	Skin flux ($\mu g/cm^2/hr$)
Acrylic copolymer	3.6	0.9 ± 0.2
Silicone 2675	1.3	1.1 ± 0.2
Silicone 2920	2.1	6.3 ± 0.7
Polyisobutylene	1.3	3.1 ± 0.3

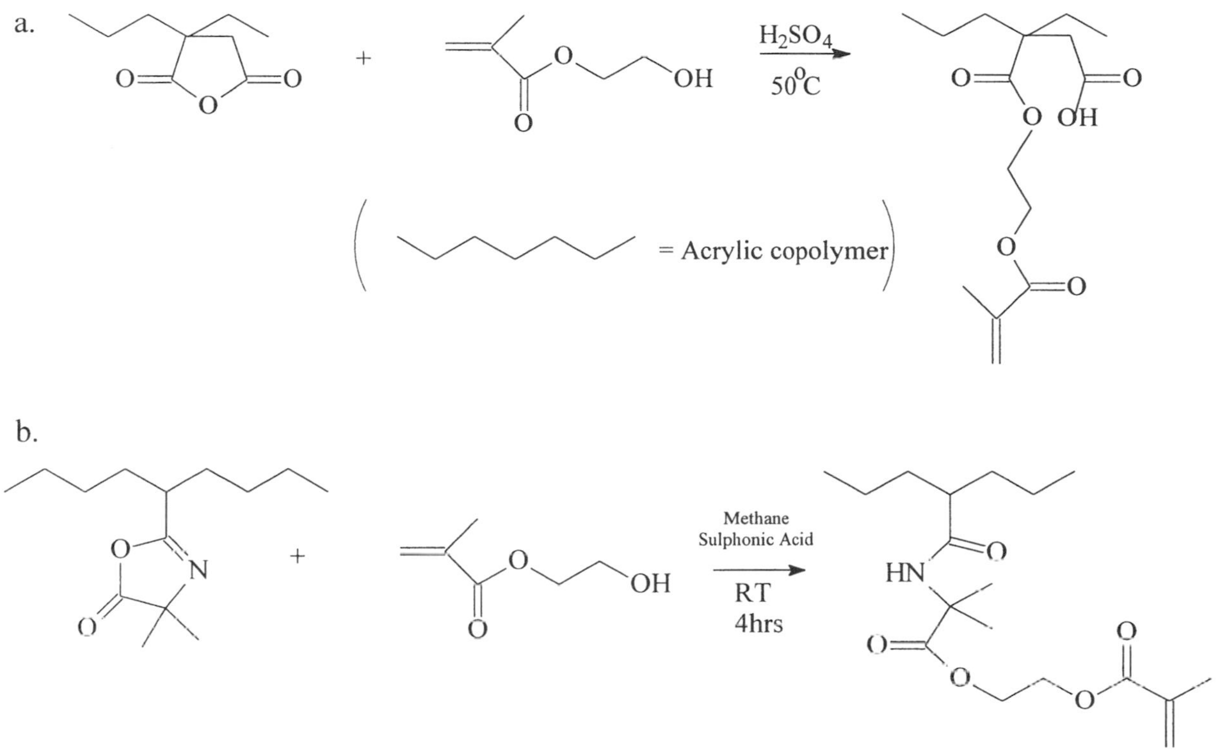

Figure 26.7 Synthesis of methacrylate functional PSAs using (a) itaconic anhydride and (b) vinyl azlactone containing copolymers reacted with hydroxyethyl methacrylate for the formation of light switchable PSAs

Many medical grade PSAs are designed to overcome the problem of trauma on removal by the use of cross-linking (covalent, hydrogen bonding or physical) or incorporation of hard monomers, such as acrylic acid (12,49–53). This is designed to control the buildup in adhesion once a sufficient bond has been formed between the adhesive and the skin. Therefore, by limiting the long term flow of the adhesive polymer, the ultimate bond strength is controlled and the cohesive strength of the adhesive is also increased. This results in reduced pain on removal with little or no adhesive residue left on the skin (54).

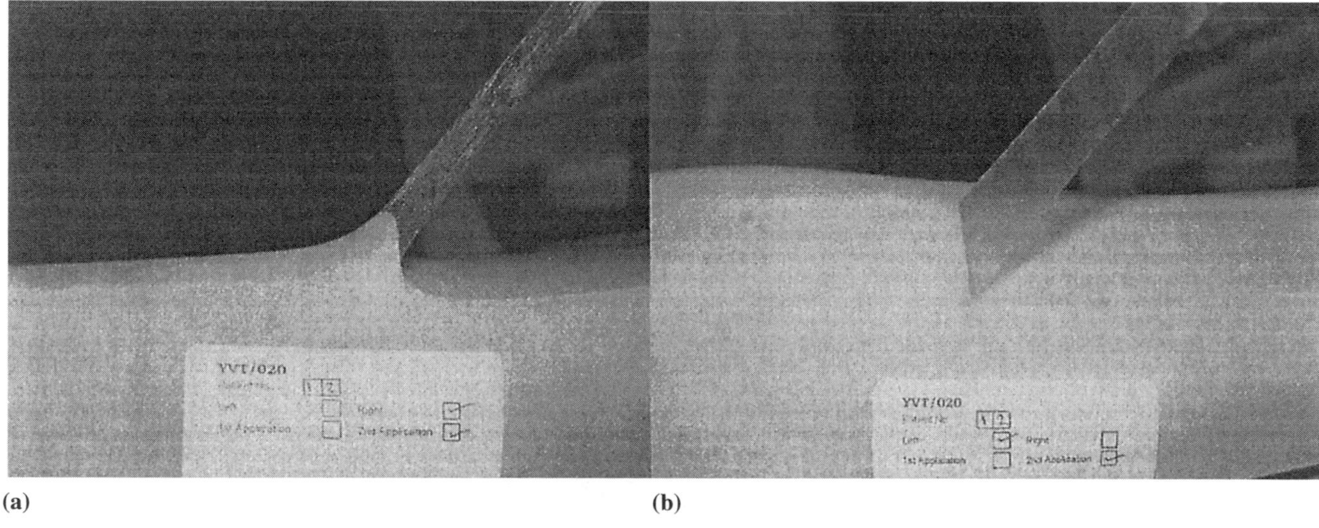

(a) (b)

Figure 26.8 Peeling of light switchable PSA tapes from volunteers showing (a) unswitched tape, (b) switched tape.

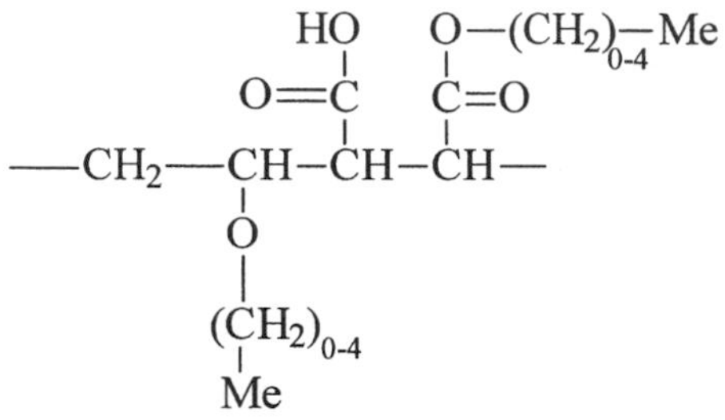

$$-CH_2-CH(O(CH_2)_{0-4}Me)-CH(COOH)-CH(C(=O)O(CH_2)_{0-4}Me)-$$

Figure 26.9 Vinyl ether/maleic acid ester copolymer used in the synthesis of moisture switchable PSAs.

Table 26.8 Glass Transition Temperatures of Acrylic Homopolymers Derived from Typical Monomers Used in Medical PSAs

Monomer	Homopolymer Tg	Remarks
n-Butyl acrylate	−54°C	Soft segment
2-Ethylhexyl acrylate	−70°C	Soft segment
Acrylic acid	+106°C	Hard segment
Vinyl acetate	+30°C	Hard segment
n-Butyl methacrylate	+20°C	Hard segment

C. Adhesive Types

1. Acrylic Polymers

Acrylic polymers are the most widely used materials for medical PSAs due to their hypoallergenicity, natural tackiness, and wide scope for formulation/property tailoring. They are typically copolymers composed of ''hard'' monomers that produce homopolymers with a high Tg and ''soft'' monomers that yield low Tg homopolymers (Table 8) (55,56). Acrylates are esters of acrylic acid and have the general structure shown in Fig. 11. The nature of the alkyl group, R, can be used to dictate the adhesive properties by varying the chain length and the hydrophilic/hydrophobic nature of the group (57). The length of the R group affects the Tg as shown in Fig. 12, where it can also be seen that the addition of a CH_3 group (R′) to produce a methacrylate leads to increased steric hinderance and therefore higher Tgs (56). The contribution made to adhesive properties by varying the polarity of the acrylic monomers is also highlighted in Table 9.

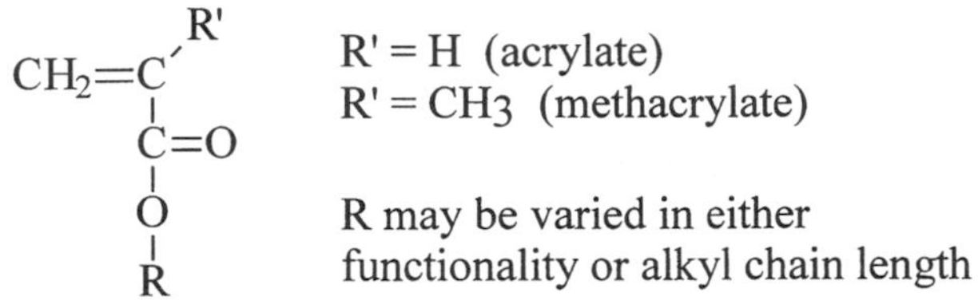

$$CH_2=C(R')-C(=O)-O-R$$

R' = H (acrylate)
R' = CH_3 (methacrylate)

R may be varied in either functionality or alkyl chain length

Figure 26.11 General structure of (meth)acrylic monomers.

Polymerization is normally carried out by a free-radical mechanism producing random copolymers of molecular weight typically in the range 200,000–1,000,000. The Tg of the resultant polymer can be controlled by the ratio of hard and soft monomers, which will in turn affect the adhesive properties. Varying the functionality of the monomers allows particular properties to be imparted on the adhesive:

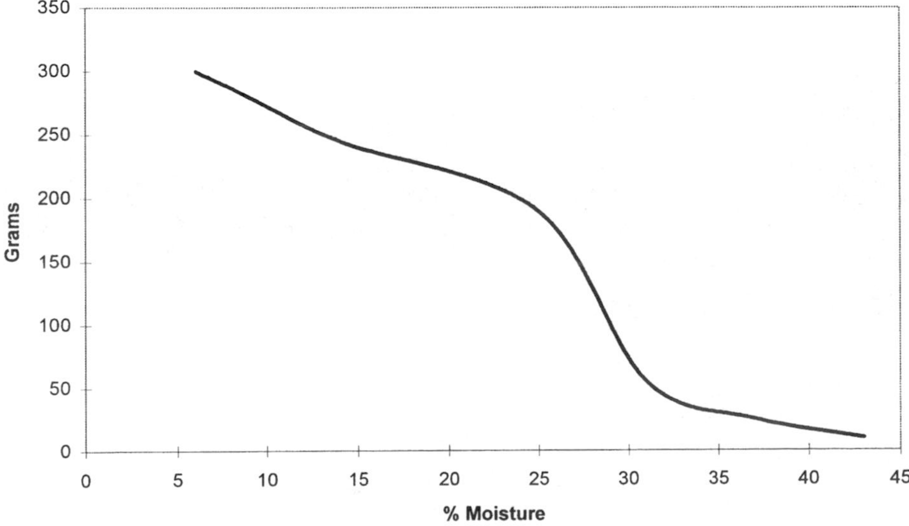

Figure 26.10 Relationship between tack of a moisture switchable PSA and moisture content. (Adapted from U.S. patent 5,032,637 assigned to Adhesives Research Inc.)

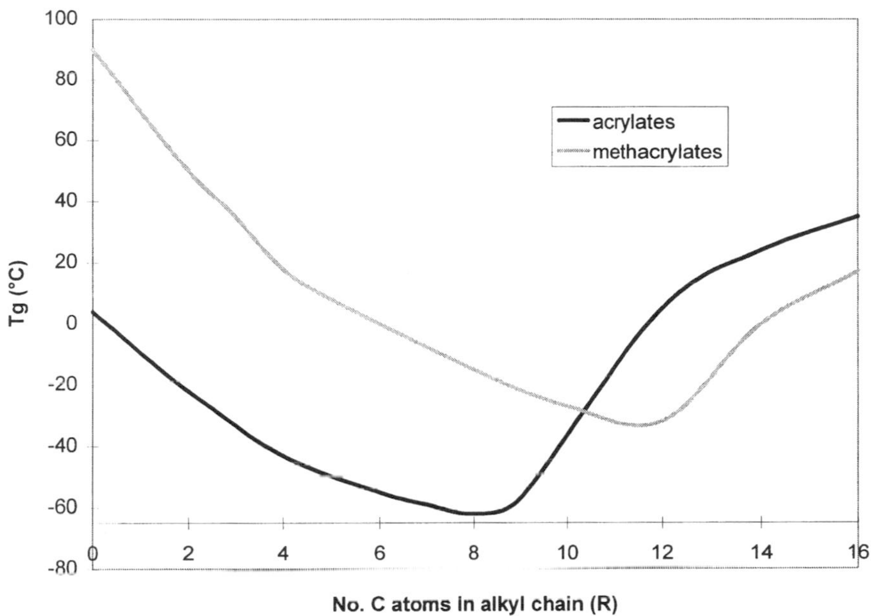

Figure 26.12 Effect of alkyl chain length on the Tg of poly(alkyl) acrylates and methacrylates. (Adapted from Ref. 56.)

High MVTR. A high proportion of ether oxygen atoms can be incorporated into the polymer to increase the MVTR. Monomers such as methoxybutyl and butoxyethyl acrylate therefore give increased MVTR to the PSA, as does decreasing the alkyl chain length (58–60).

Sebum tolerance. An alkyl chain length of at least C_6 is required to dissolve sebum, e.g., 2-ethylhexyl acrylate.

Tack. Low Tg acrylic monomers are required with chain lengths of C_4–C_{12} (61). These acrylates have lower Tgs than corresponding methacrylates (Fig. 12).

Wet stick. Changing the surface energy by varying the adhesive polarity (e.g., acrylic acid content) may affect the wet stick ability of the PSA (62–64). Increasing the alkyl chain length (e.g., *n*-butyl to *n*-lauryl acrylate) may also increase the wet stick ability.

Cohesive strength. Increasing the molecular weight by chain propagation or slight cross-linking will increase the cohesive strength of the adhesive.

Table 26.9 Adhesive Properties of Acrylic Monomers

Monomers	Adhesive property
Hydrophobic	*Present as a large proportion of the copolymer in order to*
2-Ethylhexyl acrylate	Impart low Tg
Isooctyl acrylate	Provide pressure sensitivity
n-Butyl acrylate	Impart sebum tolerance
Hydrophilic	*Present as a smaller proportion of the copolymer in order to*
(Meth)acrylic acid	Increase cohesive strength
Vinyl acetate	Provide surface polarity
Methyl acrylate	Enhance wear performance on skin
Vinyl caprolactam	
2-Hydroxyethyl methacrylate	
Methoxybutyl acrylate	
Butoxyethyl acrylate	

Although the majority of acrylic adhesives are solvent based, there is a growing move towards solvent-free systems such as water-borne or hot-melt acrylics. Developments in water-borne PSAs have generally concentrated on acrylic dispersions, which offer several advantages over solvent-based PSAs such as higher coating speeds, reduced impact on the environment (although water has a higher evaporation enthalpy than most solvents used), and greater resistance to aging.

The major development in this area concerns the surfactants used for emulsion polymerization. Typically, non-polymerizable surfactants were used in PSA formation, which left the adhesives sensitive to water due to extraction of the surfactant. However, in order to maintain good adhesion to skin it is essential that water does not significantly affect adhesive properties. To this end, polymerizable surfactants based on acrylic esters have been developed that overcome this problem (Fig. 13) (65). Therefore a PSA based on 2-ethylhexyl acrylate, *n*-butyl acrylate, and *n*-butyl methacrylate was produced that adhered well to human skin under dry and wet conditions as well as to sweaty skin.

Hot-melt adhesives are increasing in importance due to environmental pressure on solvent-based systems and the performance shortcomings of aqueous systems. The use of

$$CH_2{=}C(CH_2{-}C({=}O){-}OR){-}C({=}O){-}O{-}(CH_2)_3{-}S(O_2){-}OM$$

Figure 26.13 Polymerizable surfactant based on acrylic esters (M = metal).

hot-melt coating technology also offers superior economics to solvent- or water-based coatings.

The major health care applications of hot-melt PSAs (HMPSAs) are in adhesive tapes, bandages, and transdermal drug delivery systems. HMPSAs are particularly suited for use as combined adhesive/drug matrix media for drug delivery, as they can be formulated to contain little or no chemical functionality (66). This reduces the possibility of medication/adhesive interactions. Hot melts are also less prone to swelling when in contact with the alcohols used in certain medications, this being a particular problem with acrylics.

The major disadvantage of using hot-melt adhesives in skin contact applications is the high peel force required for removal due to their low cohesive strength and hence aggressive nature. Various examples do exist, however, of high cohesive strength HMPSAs formulated for medical use. Adhesives composed of soft acrylate (such as *n*-butyl acrylate) and hard methacrylate (such as isobutyl methacrylate) polymers containing methyl-3-oxa-4-oxo-5-azanonyl methacrylate as comonomer have been found to have high adhesive strength to human skin but only negligible adhesive transfer to the skin on removal (67,68). This is believed to be due to the ability of the comonomer to form hydrogen bonds and thus enhance cohesive strength.

Another approach to improving the cohesive strength of the adhesives is to introduce cross-links. The technique most suitable for cross-linking hot melts is radiation-induced cross-linking, as thermally activated cross-linking would occur prematurely during premelting. The polymerization of an ultraviolet (UV) photoinitiator into an acrylic PSA formulation has provided a novel route to a UV curable hot melt PSA (69). The copolymerized acrylic esters with substituted benzophenone terminal groups cross-link on UV irradiation to produce PSAs with a range of properties that remain constant with no post-curing (Fig. 14).

2. Rubber-Based PSAs

Rubber-based PSAs only find limited use in the modern medical field, as acrylic adhesive copolymers have been

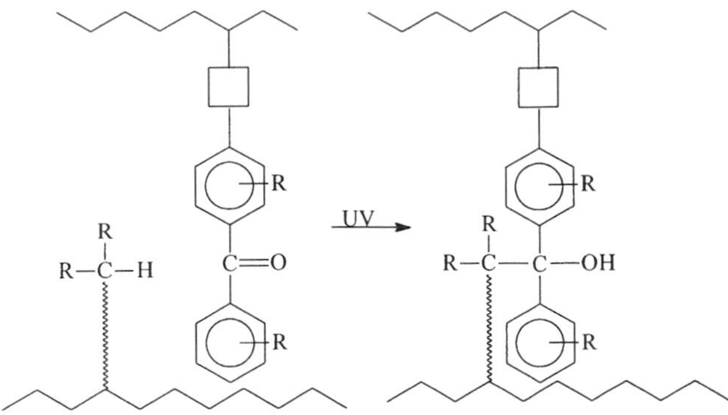

Figure 26.14 Cross-linking mechanism of UV curable hot melt PSA.

developed. Early medical adhesives, however, were most commonly based on natural rubber, an early composition is described as (1)

India rubber	5 lb
Gum of southern pine	1 lb
Balsam of Peru	6 oz
Ground litharge	1 lb
Extract of Capsicum annuum	4 oz
Solvent–turpentine	

A more general formulation for a surgical tape adhesive is given as (70)

	Parts by weight
Natural rubber	100
Colophony or ester gum	100
Lanolin	20
Zinc oxide	50

These natural rubber–based formulations have largely been superseded by synthetic elastomers such as polyisoprene and polyisobutylene (Fig. 15). Polyisobutylene tends to pack tightly on a molecular level, which results in low air and moisture permeability. The low Tgs of these polymers produce flexible materials that are naturally tacky, but because of their low polarity, adhesion of polyisobutylenes to many surfaces is weak and tackifiers are often used to impart a degree of polarity to the formulation (42). Low molecular weight polyisobutylenes may also be incorpo-

$$\left[C(CH_3)_2{-}CH_2 \right]_n$$

Figure 26.15 Repeat unit of polyisobutylene.

rated into the formulation in order to give suitable adhesive properties. These lower molecular weight polyisobutylenes (~50,000–100,000 molecular weight) are naturally tacky materials and provide the PSA with softness and flexibility, allowing the polymer to wet out on the skin surface. A typical polyisobutylene-based PSA formulation is described as (71)

Polyisobutylene (80,000)	50
Factice	29
Glyceryl ester of hydrogenated rosin	24
Liquid polyisobutylene	35
Lanolin	20
Beeswax	9
Titania	14
Hydrated alumina	62
Antioxidant	1.5

(Factice is used to limit the cold flow of the adhesive mass.)

3. Silicones

Silicone-based PSAs have been used in medical applications since the mid-1960s and have been utilized for tapes, dressings, and bandages. More recently, silicone adhesives have found use in transdermal drug delivery systems, which offer controlled entry of a pharmaceutical into the blood. Silicone PSAs are typically formulated from silicone resins and polydimethyl siloxane gum (72) (Fig. 16). In order to impart sufficient cohesive strength to these systems, the polymer and resin are cross-linked to form a network. The properties of the final adhesive can be controlled by varying the ratio of components and the cross-link density (73).

The properties attainable from silicones are summarized in Table 10 and illustrate why they are suitable for transdermal drug delivery. Firstly, silicones have excellent skin compatibility as they contain no tackifiers, plasticisers, antioxidants, or stabilizers and therefore have low levels of potential extractables (74). As well as satisfying the standard requirements for a medical PSA, it is especially important for drug delivery devices that the adhesive is biocompatible so as to be nonirritating and nonsensitizing as well as to produce no systemic toxicity. This is enhanced in silicone polymers by their surface activity and hydrophobic nature, which limits their interactions with body fluids (73). However, it is their high permeability to a wide variety of pharmaceuticals and their chemical resistance that allow controlled rates of delivery of an active ingredient to a patient (35).

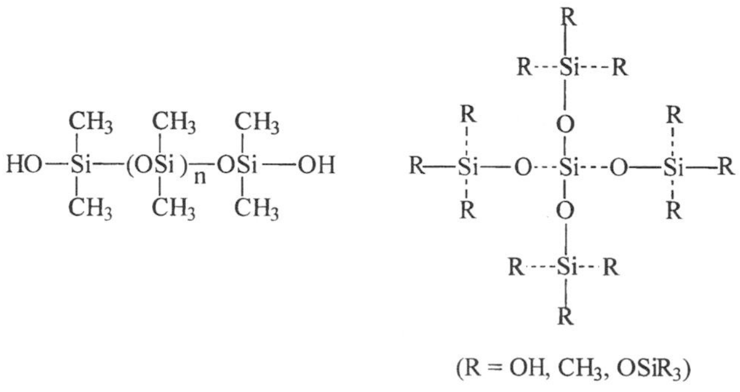

$HO{-}Si(CH_3)_2{-}(OSi(CH_3)_2)_n{-}OSi(CH_3)_2{-}OH$

(R = OH, CH_3, $OSiR_3$)

Figure 26.16 Silicone PSA components—polydimethyl siloxane polymer and silicate resin.

Table 26.10 Typical Properties of Silicone Polymers for Transdermal Drug Delivery

Biocompatibility	Biologically inert Low level of extractables Nonsensitizing and nonirritating
Adhesion	Low glass transition temperatures Good formulation control
Permeability	High gas permeability Permeable to a variety of pharmaceuticals
Chemical stability	Chemically inert to many substances Compatible with pharmaceuticals

III. TISSUE ADHESIVES

A. Introduction

1. Techniques for Tissue Reattachment

Before reviewing the various tissue adhesive technologies, it is worth considering the current practices with which any new tissue bonding adhesive would have to compete. Alternatives are dominated by the traditional approach offered by mechanical devices such as sutures and staples and more recently tissue welding with lasers. Sutures and staples are effective, well established, and widely used wound closure devices. Laser welding can also be used to join tissue defects. The worldwide market in which all of these technologies are to be found is worth in excess of US$2.5 billion per annum (75).

a. Sutures and Staples

For many hundreds of years suturing has been the primary means adopted by surgeons for most wound closure procedures, both postsurgery and posttrauma (Fig. 17). The materials used in the manufacture of sutures have become increasingly more high-tech, and it is now commonplace for these materials to have a predictable and consistent resorption profile to match the requirements of the wound. It

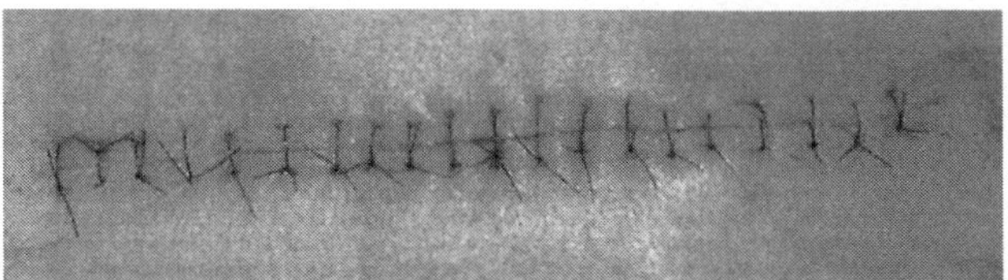

Figure 26.17 A typical suture line.

therefore comes as no surprise that sutures represent an annual worldwide market worth in excess of US$1.6 billion (75). Despite the recent search for alternative technologies to replace suturing for difficult procedures involving soft tissues such as liver and kidney, and harder tissues such as bone and cartilage, the market for this simple product continues to expand.

The suture market is dominated by Johnson & Johnson's Ethicon, which retains approximately three-quarters of the U.S. suture market worth US$550 million per annum (75). The remainder of the market is evenly divided between Sherwood Davis & Geck and U.S. Surgical. Stainless steel staples were first invented in 1969, and although they represent a sizeable market, staples have not replaced sutures.

b. *Laser Welding*

Tissue defects can be repaired with the use of lasers (76) in conjunction with minimally invasive surgery (MIS) via a fiber-optic arm. The physical changes in the tissue are dependent on the radiation parameters employed, such as the wavelength, spot size, power, and exposure. The overall efficiency of welding can be considerably enhanced with the use of protein solders such as fibrinogen or albumin-hyaluronic acid to reduce the damage to the irradiated tissue (77,78). Laser welded tissue has the potential to produce watertight closures with less trauma to surrounding tissues and much reduced operating times compared to suturing and stapling. This should also be the case for many of the emerging tissue adhesives.

2. Key Requirements of a Tissue Adhesive

Each individual key requirement for a successful tissue adhesive may appear obvious, but each one represents a significant challenge to any scientist working in this field. Only products that satisfy all these criteria can really expect to receive full recognition and widespread acceptance by the surgical community. Table 11 lists the main requirements of any tissue adhesive.

Clearly there is a need for a product that can deliver all these characteristics. Procedures that are inherently complicated, such as certain types of fracture repair involving small bone fragments, are difficult to repair using traditional pins, screws, plates, and wiring. Hence there is a need for a new type of adhesive product to be available to surgeons around the world.

Table 26.11 The Key Requirements of a Tissue Adhesive

In use, the adhesive must mimic the mechanical performance of the undamaged tissue.
The adhesive should provide sufficient tack for "primary" fixation with the opportunity for manipulation and realignment prior to setting strongly.
The adhesive must not elicit any toxic response by the surrounding healthy tissue and should facilitate the regrowth of new tissue where possible.
The adhesive should not liberate harmful degradation products.
Any exothermic process involved in the curing of the adhesive should not damage the surrounding tissue.
The adhesive should degrade, and as it does so, it should be replaced by new tissue with minimal scarring.
Any biodegradation products should not accumulate in the body but should be eliminated naturally either by excretion or incorporation into the natural biochemical cycle.

3. Factors Influencing the Choice of Tissue Adhesives

In selecting the most appropriate wound closure technique, the surgeon has a choice of technologies, which may be used alone or in combination. A surgeon will base the decision on a number of factors, the relative importance of which will vary from one patient or medical procedure to another. These are listed in Table 12. In addition it should be appreciated that the priority placed on delivering these would vary with the surgeon, the patient, and the manufacturer. The aim must therefore be to deliver a product that satisfies all of their needs.

Table 26.12 Factors Influencing the Use of Tissue Adhesives

Cost (in terms of materials and time spent in applying them to a patient)
Availability to the customer
Ease and reliability of the procedure
Properties and local effects of using the material
Applicability to the specific tissue site
Robustness of the treatment
Time to healing
Quality of the repair
Patient acceptability
Regulatory clearance

B. Synthetic Tissue Adhesives

There are two main types of synthetic polymer systems that readily adhere to moist tissue, synthetic and biological adhesives (79). The alkyl-2-cyanoacrylates have been evaluated as tissue adhesives quite widely and probably now lead the field, inasmuch as they attract the most interest from industries looking to enter this market. Other synthetic tissue adhesives, which include gelatin and resorcinol mixtures that are cross-linked by formaldehyde, have also been investigated, but the cross-linking reaction involved is very complex, and this has made investigation into the nature of the adhesion very difficult. Polyurethane prepolymers have received much attention, as this chemistry is well established, and the properties of the adhesive can be varied by choice of the polyol component and the chemical nature of the polyisocyanate. There are also a number of emerging technologies gaining interest from many companies as encouraging results are reported. Each will now be considered in turn.

1. Cyanoacrylates

a. Development and History of Cyanoacrylates

New developments in cyanoacrylate tissue adhesives are beginning to compete in the wound closure market, which has been traditionally dominated by sutures and staples. For example, Ethicon has a distribution agreement with Closure Medical, whose product, trademarked *Dermabond*, was unanimously recommended for approval by the FDA's General and Plastic Surgery Devices Panel at the beginning of 1998. *Dermabond* is the first synthetic cyanoacrylate adhesive product to join the U.S. market for closure of skin lacerations caused by trauma or small surgical incisions that would normally have been repaired by sutures or staples.

The earliest patents on the preparation of cyanoacrylates were issued in 1949 to Alan Ardis (80,81) and assigned to the B. F. Goodrich Company. These patents refer only to the applications of α-cyanoacrylates for the production of ''hard, clear, glass-like resins,'' which were simply obtained by heating the monomer. The adhesive properties of α-cyanoacrylates were discovered by accident during the investigation of a series of 1,1-disubstituted ethylenes in the Eastman Kodak laboratories. In attempting to measure the refractive index of a sample of ethyl-α-cyanoacrylate, a drop of the material was placed between the prisms of an Abbé refractometer with the result that these prisms became firmly bonded together.

Between the years 1955 and 1957, a number of patents were issued to Eastman Kodak including the use of cyanoacrylates as adhesives (82). The material only had speciality uses at first, partly due to its expense, although because of the tiny amounts of material necessary to form a strong bond, this adhesive was destined not to remain a curiosity for much longer.

b. The Role of Cyanoacrylates in the Medical Market

Most people have come across cyanoacrylates because of their widespread availability and common usage around the home as ''superglues,'' but cyanoacrylates also find numerous applications in, among others, the automotive and construction industries. The cyanoacrylate adhesive family is now enjoying increasing popularity as fast-setting ''instant'' adhesives. The makers of these household superglues have always warned the users that the glue can bond skin in seconds, but today this well-known hazard is being turned into a strong commercial selling point as cyanoacrylates find uses in medicine. The properties of cyanoacrylates that set them apart from nearly all other adhesives are that they are single-component, catalyst-free adhesives capable of bonding at room temperature within just a few seconds. Cyanoacrylates require no external initiation and for the vast majority of applications rely solely on the small amounts of water adsorbed on the surfaces of the substrates for their almost universal adhesion.

Cyanoacrylates were first used for a clinical application in the early 1960s (83,84), and among all the other synthetic polymers that have been considered for medical applications, the cyanoacrylates have received the most attention. Indeed, cyanoacrylates are now commonplace in many accident and emergency departments in hospitals as doctors and patients begin to realize their widespread appeal and clinical advantages. In the early literature, Eastman 910 monomer (methyl-2-cyanoacrylate) was the most noteworthy (the general structure of a cyanoacrylate monomer is shown in Fig. 18). Unlike sutures, which leave behind small openings in the wound (often called dehiscence), cyanoacrylates form a continuous sealing of the wound, along which the load is evenly distributed. This leads to formation of a bacterial barrier and also decreased scarring around the site.

Skin closure is potentially a major application area for cyanoacrylate tissue adhesives, and this has received ex-

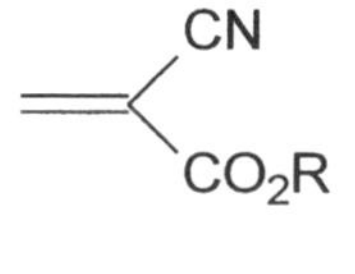

CN

CO_2R

(Where R = Me, Et, Bu etc)

Figure 26.18 General structure of cyanoacrylate monomer.

tensive evaluation. Indeed, many of the products marketed are aimed specifically at surface wound closure. There are also literature reports documenting the use of cyanoacrylates to bond bone and cartilage (85–96), nervous tissue (97–100), vascular tissue (97,98,101,102), and intestines (103).

Many other cyanoacrylates have been experimented with surgically. In early applications, methyl cyanoacrylate was used almost exclusively both as the monomer and as a commercial preparation with addition of a plasticizer and a thickener. Later, long chain alkyl cyanoacrylates were also used, especially isobutyl cyanoacrylate. The longer chain cyanoacrylates are generally considered to be less toxic due to their slower degradation when compared to the short side chain versions. The issues of toxicity will be addressed in a subsequent section.

Cyanoacrylates have been available commercially in the U.S. since the 1960s but were removed from the medical market due to concerns over carcinogenicity. However, products such as *Histoacryl* (trademark of B. Braun) have been available on the European market for more than two decades for external medical indications. *Histoacryl*, which is an *n*-butyl-2-cyanoacrylate monomer colored with a blue dye, has been widely used in hospital accident and emergency departments for the treatment of pediatric patients and traumatic skin incisions in adults. *Histoacryl* tissue adhesive has been compared with silk sutures as a means for closing incisions (104). The silk sutures were found to cause a severe inflammatory and giant cell reaction compared to *Histoacryl*, and the healing process was slower with the sutures. The cyanoacrylate caused less scarring and allowed shorter operation times by virtue of its hemostatic properties.

Closure Medical's *Dermabond* contains a different cyanoacrylate monomer, octyl-2-cyanoacrylate, that is less brittle than the butyl derivative and is used as a flexible seal over wound edges. This monomer is slower to polymerize than its butyl counterpart, taking up to 1 minute to polymerize sufficiently for a wound to hold itself together without support. Typically, cyanoacrylates used for these conditions would remain in place for about 7 days and then slowly slough off as the skin cells regenerate themselves and the natural healing process takes over. It is easy to appreciate that the use of cyanoacrylate tissue adhesives in the operating theater offers a much faster procedure when compared to the use of sutures, resulting in a far more acceptable cosmetic effect and, most appealing to younger patients, there is no need to make a second visit to the hospital to have the stitches removed.

Quinn et al. examined the use of *Dermabond* in the treatment of 136 lacerations in humans and recorded excellent results (105). The cyanoacrylate was reported to close the lacerations in a relatively rapid and painless manner. The authors stressed the importance of ensuring that the wound edges be well apposed to prevent the cyanoacrylate getting between the wound edges and penetrating into the underlying tissues, which may result in foreign body reactions and impaired healing. When used topically, octyl cyanoacrylate can result in fewer foreign body reactions than sutures and can decrease the infection rates in contaminated wounds.

Loctite (UK) has developed a product based on the same chemical monomer, butyl-2-cyanoacrylate, but in a clear formulation that does not tattoo the skin. This product has been granted a CE mark and is sold in Europe under the trademark *Indermil*. The product is sterilizable by gamma irradiation (106) and may be applied through a cannula attached to an electronic peristaltic pump (107) that helps deliver the adhesive in precise doses to the surgical site. The composition comprises a cyanoacrylate and a combination of an anionic stabilizer and a free-radical stabilizer in amounts effective to stabilize the composition during irradiation and to stabilize the sterilized composition during storage prior to use. The glue will form strong bonds almost immediately once the surgeon has applied it to the tissue. In their current formulations, cyanoacrylate monomers are not approved for use inside the body, although resorbable versions are no doubt the target of much research.

Many of the developments in cyanoacrylate technology focus on altering its stability and reactivity with the accurate use of inhibitors and stabilizers. Now researchers and numerous health care companies are turning their attention to developing cyanoacrylates for the medical market. This places additional demands on the cyanoacrylates, and new issues must be addressed that were not pertinent for their more conventional applications, such as toxicity, biodegradation profile, and sterilization issues.

c. *Mechanism of Bonding*

Adhesion is achieved through two independent mechanisms, molecular interaction (covalent bonding) and the mechanical interlocking of the poly(cyanoacrylate) with the substrate.

Primary chemical or covalent bonds are obtained when the adhesive undergoes a chemical reaction with the substrate. There are many functional groups present in the tissue substrate, particularly in proteins which are suitable for covalent bonding, most notably amine groups.

Physical or mechanical bonds are formed when the adhesive penetrates the substrate and forms interlocks when the material sets. This can occur very easily with tissue, and low-viscosity fluids such as cyanoacrylate monomers are particularly suited to this mode of action. The adhesive "enters" the substrate via cracks and channels and quickly becomes locked and held strongly once the polymerization process commences. This results in strong mechanical bonds between closely apposed surfaces. In this case, bond strengths would depend on the surface morphology of the substrates and the mechanical properties of the specific adhesive polymer.

In practice, many of the successful tissue bonds are achieved via a combination of mechanical and chemical bonds. Hence it is a combination of the properties of the adhesive and the substrate that determines the bond strength of the adhesive.

d. *The Chemistry of Cyanoacrylates*

Why are cyanoacrylates unique? The cyanoacrylate monomer is extremely reactive due to the electron withdrawing capacity of the nitrile and the ester groups, which results in the double bond being very polarized in the presence of even very weak nucleophiles. This effect can also be caused by water, which can be responsible for the initiation of the polymerization reaction. The properties of the adhesive can be controlled by varying the length of the alkyl group, by careful use of acidic substances to delay the polymerization, and also by the addition of other polymers to act as thickening agents. These minor components, such as stabilizers, plasticizers, and other additives that improve the viscosity, heat stability, bond strength, and shock resistance, generally account for less than 10% of the total formulation.

Like all acrylates, cyanoacrylates are capable of being polymerized via a free-radical mechanism. However they polymerize preferentially via an anionic mechanism that is catalyzed by bases (including weak nucleophiles such as water) so that the curing process of the single-component monomer is fast on contact with moist tissue. The anionic polymerization reaction is shown simply in Fig. 19.

The α-cyanoacrylates are prepared by depolymerization of the corresponding polycyanoacrylate formed by the Knoevenagel condensation reaction of an α-cyanoacetate with formaldehyde. This depolymerization occurs at high temperatures under high vacuum conditions, and pure monomer is distilled from the reaction. Water and bases must be removed or neutralized prior to the depolymerization reaction in order to prevent the polymer reforming during the distillation process. Polymerization inhibitors are usually added immediately to the purified monomer.

Medical-grade cyanoacrylates are far purer than "superglues" used elsewhere but have been formulated to contain specific additives, for example to enable sterilization. Cyanoacrylates would obviously not survive some of

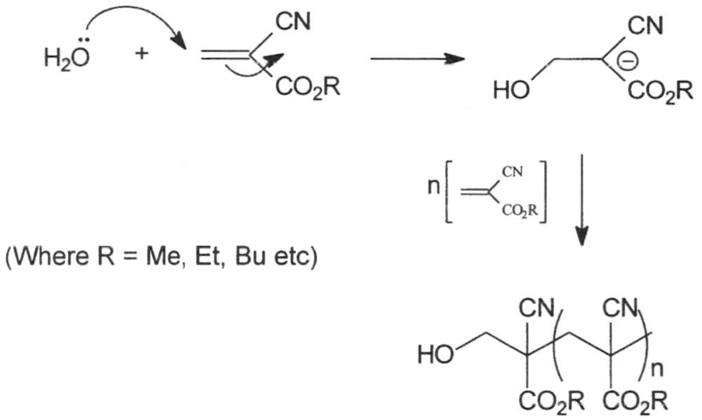

$H_2\ddot{O}$ + CN / CO_2R → HO CN ⊖ CO_2R

n[CN / CO_2R] ↓

(Where R = Me, Et, Bu etc)

HO CN CN CO_2R CO_2R n

Figure 26.19 Anionic polymerization mechanism of alkylcyanoacrylate.

the routine sterilization cycles used in the manufacture of many medical products, and care must be taken to choose a suitable procedure that excludes the possibility of unwanted free radical polymerization.

In the process of polymerization, a bond is created between adjacent tissue surfaces. The poly(cyanoacrylates) are considered to be relatively nonbiodegradable because they break down very slowly both in water and in tissue media. The main degradation products of poly(alkyl cyanoacrylates) are formaldehyde and the corresponding alkyl cyanoacetate. These degradation products, the monomer itself, or other impurities in the monomer may give rise to toxic species. The issues surrounding the toxicity of cyanoacrylates will be discussed in the next section.

The success of the adhesive joint is often judged by how well the tissue is replaced. It is therefore important that the polymeric material is degraded and eliminated from the body leading to the most widely debated area surrounding the use of cyanoacrylates as medical adhesives, namely toxicity.

e. *Issues of Toxicity in Cyanoacrylates*

All cyanoacrylate tissue adhesives are classed as medical devices and are required to be subjected to rigorous safety appraisal to prove that they are safe in use. It is therefore no longer sufficient for a medical adhesive product to rely on its mechanical performance.

Cyanoacrylates are used to bond two nonadherent surfaces and, in the case of external tissue, the adhesive either falls off or can be washed off after a period of time. The polymerization of the monomer at the tissue site can cause localized heat-induced tissue damage, but with only small volumes of adhesive being required, the temperatures reached as a result of the polymerization reaction should not be sufficient to cause necrosis. The greatest in vivo temperature rises for methyl cyanoacrylate have been reported to be 4°C (108); higher homologues, the hexyl and decyl esters, produce temperature rises of 2°C. It is, however, the chemical breakdown of the adhesive that may result in toxic species.

Cyanoacrylates belong to a class of monomers consisting of the alkyl esters of 2-cyanoacrylic acid. The choice of the alkyl side group greatly affects the properties of the adhesive formed when the monomer polymerizes. The methyl ester is volatile and lachrymatory and polymerizes extremely rapidly to give a rigid polymer. With increasing size of the alkyl group, volatility, polymerization rate, and rate of polymer degradation decrease, whereas the flexibility increases. As examples, methyl (102), ethyl (109), *n*-butyl and isobutyl (103,110–113) and more recently octyl (105) cyanoacrylates have been investigated for surgical use. The study involving the methyl derivative concluded that it caused tissue necrosis and did not spread well on tissues. As a result of this negative finding, much research has focused on the higher homologues, particularly butyl cyanoacrylate. It was believed that increasing the length of the alkyl ester side chain was associated with a decrease in histotoxicity. However, it has been found that the polymers disappear or degrade very slowly and still have too low a flexibility compared with wound tissues and hence are inadequate for use routinely as soft tissue adhesives.

Cyanoacrylates have been used as adhesives in soft tissues (110), arteriovenous malformations (97), opthalmology, and as drug carriers (114). In practice, few surgeons use cyanoacrylates except in unusual circumstances or in emergency cases. Inflammatory response (113), delayed healing and necrosis (113,115), thrombosis (97,101), and inhibition of bone formation (116) have been noted, whereas other researchers have found relative compatibility of the polymerized cyanoacrylates with the surrounding tissues, resulting in normal healing processes following injury (103,111,117).

It has been proposed that the toxic effects of cyanoacrylate polymers are due to the release of the products of their degradation (118). Cyanoacrylate degradation occurs by the breakdown of the polymer backbone, and it occurs because the methylene hydrogen in the polymer is highly activated inductively by the electron withdrawing neighboring groups. Water associated with the tissue can induce cyanoacrylate hydrolysis by a reverse Knoevenagel reaction liberating formaldehyde and alkyl cyanoacetate.

The degradability of cyanoacrylate polymers has been demonstrated to decrease with an increase in the size of the alkyl side chain, as a result of steric hindrance (119). Degradation also depends upon the molecular weight of the polymer; lower molecular weight cyanoacrylate polymers exhibit a more rapid release of formaldehyde. The reason for this is that for a fixed mass of polymer, lower molecular weight materials will contain more chain ends resulting in a higher probability of formaldehyde release (Fig. 20).

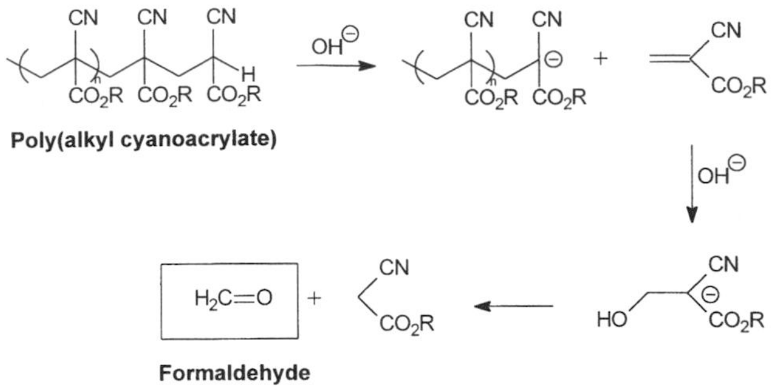

CN CN CN; H; CO_2R CO_2R CO_2R; $OH^{\ominus}$; CN CN; $\ominus$; +; CN; CO_2R CO_2R; CO_2R

Poly(alkyl cyanoacrylate)

$OH^{\ominus}$

$H_2C{=}O$ + CN; CO_2R ← HO; CN; $\ominus$; CO_2R

Formaldehyde

Figure 26.20 Degradation of cyanoacrylate polymer via a reverse Knoevenagel reaction.

An alternative degradation mechanism has been proposed whereby the cyanoacrylates degrade by hydrolysis of the ester group to produce cyanoacrylic acid and alcohol. Lenaerts et al. (114) studied degradation of isobutyl cyanoacrylate nanoparticles at pH 7 in phosphate buffer and found that the amount of formaldehyde produced after 1 day was 5% of the theoretical quantity that would have been produced if the polymer had been entirely degraded by this pathway. At pH 12 (in 0.01 N NaOH), this value reached 7%. Conversely, isobutanol production was 85% of the theoretical quantity in the alkaline medium.

This mechanism may occur to some extent in the physiological environment and may be catalyzed by enzymatic processes. Wade and Leonard (120) have shown the poor contribution of the formaldehyde producing pathway to the degradation of methyl cyanoacrylate in vivo and suggested that both hydrolytic chain scission and ester hydrolysis may be involved in the in vivo degradation of cyanoacrylate polymers. This is shown in Fig. 21.

The liberation of formaldehyde inside the body represents a significant issue regarding the acceptance of cyanoacrylates as internal tissue bonding adhesives. This is not a major issue in the use of cyanoacrylates to close external

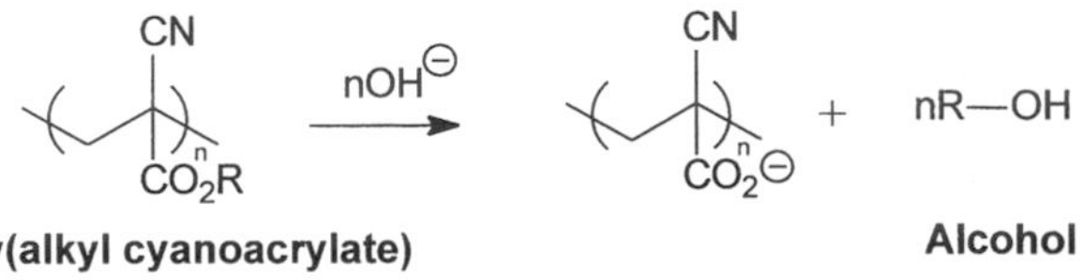

CN; n; CO_2R; $nOH^{\ominus}$; CN; n; $CO_2^{\ominus}$; + nR—OH

Poly(alkyl cyanoacrylate) Alcohol

Figure 26.21 Ester hydrolysis of cyanoacrylate polymers.

wounds, as care is taken not to use the adhesive inside the wound but only on the surface. Although formaldehyde is a product of the degradation of cyanoacrylate adhesives, it may not be the agent solely responsible for observed histotoxic effects. It is possible that formaldehyde levels released from cyanoacrylates in vivo are low enough to be processed by tissue metabolic systems and/or cleared by the normal flow of physiological fluids over the site, thus preventing buildup. If this is the case, the toxicity is attributable to other mechanisms.

Formaldehyde is a physiological metabolite and vital to the synthesis of essential biochemical substances in man, and it is not toxic at very low concentrations (121). It should be noted, however, that, due to formaldehyde's highly reactive carbonyl group, it undergoes typical aldehyde reactions: it readily reacts with free amino groups in proteins and forms methylol adducts with nucleic acids, histones, proteins, and amino acids. These reactions may be responsible for its toxic effects at higher concentrations. Epidemiological evidence has arisen to suggest that formaldehyde may be a human carcinogen. This has led to its regulation by U.S. federal agencies as a probable human carcinogen (122).

f. *Future Developments*

To date, the use of cyanoacrylates in medicine has been limited because of their physical properties and reports of histotoxicity. The currently available cyanoacrylate tissue bonding adhesives are only approved for external use in most territories, but there are many internal medical procedures that could be revolutionized by this type of product. Any internal procedure involving the use of a biomaterial inevitably relies on the polymer degrading safely inside the body (or not degrading at all and remaining inert), and it is well known that cyanoacrylate polymers are capable of undergoing hydrolytic degradation to produce formaldehyde and the corresponding alkyl cyanoacetate. Internal procedures therefore do not benefit from the adhesive being able to slough off the skin as in the case of an external application.

Clearly the development of a cyanoacrylate adhesive with a somewhat lower acute toxicity and faster degradability than butyl cyanoacrylate would go a long way towards replacing sutures and provide a faster and more efficient method of tissue approximation and wound closure. Further desirable design features would include flexibility of the adhesive polymer, appropriate viscosity (although this can easily be modified by the use of thickeners), and a set time that can be tailored to the needs of the surgeon.

In summary, although the use of cyanoacrylates is becoming more widespread, they are not the ideal tissue adhesive in their current form. Toxicity issues, improvement in their degradation rate, and to a lesser extent flexibility and viscosity will all need to be addressed before the use of these materials becomes commonplace and they gain full regulatory approval for the majority of surgical procedures.

2. Other Synthetic Medical Adhesives

a. *Polyurethane Adhesives*

A polyurethane is formed by the addition polymerization reaction of a diisocyanate and an oligomeric diol to form a series of urethane bonds. If both ends of the prepolymer that is formed are isocyanate functionalized, then the prepolymer will rapidly undergo gel formation in contact with water or moist tissue. Such a prepolymer may be suitable to act as a tissue adhesive, providing that the gel itself and the products of biodegradation are nontoxic.

Polyurethane prepolymer was first applied to tissue bonding in the late 1950s (123), and polyurethane adhesives were first used for repairing bone fractures (124,125). The adhesive, *Ostamer*, consisted of a prepolymer and an accelerator that were mixed prior to use. Curing occurred in an unacceptable time of 25 to 30 minutes, and maximum strength was not obtained for 1 to 2 days. The mechanisms of adhesion of polyurethane to living tissue have not been fully investigated.

It is possible to synthesize oligomeric diols that are readily biodegradable that can be built into a polyurethane adhesive. As an example, Kobayashi et al. (126) synthesized a copolymer of ϵ-caprolactone and D,L-lactide (P(CL-co-LA)) using ethylene glycol or poly(ethylene glycol) as an initiator to obtain hydroxyl terminated biodegradable polyesters. This reaction was carried out at 140°C under 1×10^{-3} mm Hg for six hours. These polyesters were allowed to react with an excess of diisocyanate such as hexamethylene diisocyanate (HMDI), tolylene diisocyanate (TDI), or diphenylmethane diisocyanate (MDI) in order to produce polyesters with terminal isocyanate functionalities. These materials were water curable, and their synthesis is shown in Fig. 22. Various formulations were prepared and evaluated in terms of set time, mechanical properties, and in vitro and in vivo performance (126). The set times were found to be greatly dependent on the composition of the polyesters and the nature of the isocyanate. More hydrophilic systems, such as those containing more polyethylene glycol, tended to cure more rapidly, as did those possessing the more reactive isocyanate groups of MDI. The in vivo degradation rate was almost the same as, or slightly more enhanced than, the in vitro degradation rate, with some adverse tissue responses being observed. In summary, although it was concluded that the incorpora-

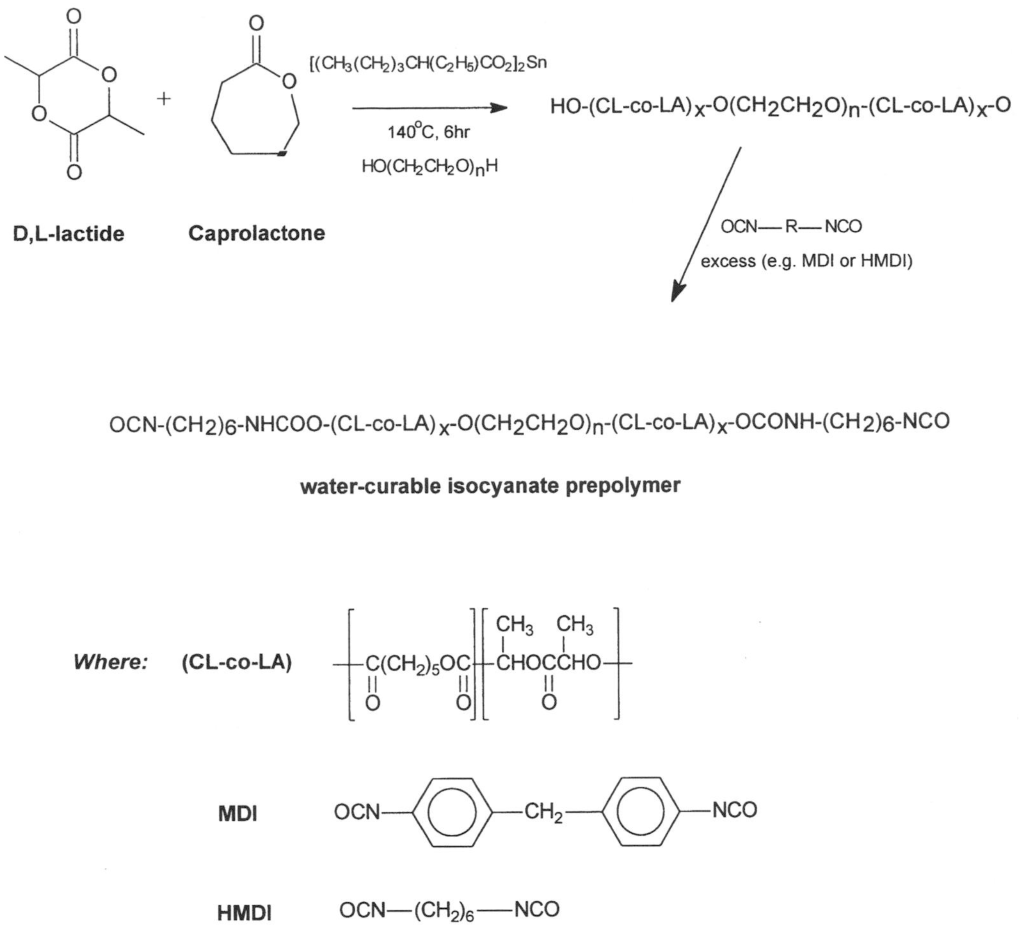

D,L-lactide + Caprolactone

$[(CH_3(CH_2)_3CH(C_2H_5)CO_2]_2Sn$, 140°C, 6hr, $HO(CH_2CH_2O)_nH$

$HO\text{-}(CL\text{-}co\text{-}LA)_x\text{-}O(CH_2CH_2O)_n\text{-}(CL\text{-}co\text{-}LA)_x\text{-}O$

OCN—R—NCO excess (e.g. MDI or HMDI)

$OCN\text{-}(CH_2)_6\text{-}NHCOO\text{-}(CL\text{-}co\text{-}LA)_x\text{-}O(CH_2CH_2O)_n\text{-}(CL\text{-}co\text{-}LA)_x\text{-}OCONH\text{-}(CH_2)_6\text{-}NCO$

water-curable isocyanate prepolymer

Where: **(CL-co-LA)**

MDI OCN—C_6H_4—CH_2—C_6H_4—NCO

HMDI OCN—$(CH_2)_6$—NCO

Figure 26.22 Synthesis of biodegradable polyurethanes.

tion of hydrophilic units was essential for high curing and degradation rates, because of the adverse tissue responses observed, further improvements to the prepolymers are necessary if these adhesives are to be used widely in surgery.

b. Gelatin–Resorcinol–Formaldehyde (GRF) Adhesives

Certain aqueous polymer solutions are capable of setting through gel formation in a few minutes when a cross-linking agent is added to the solution. Gelatin is an excellent example of such a polymer that is also bioresorbable. The use of cross-linked gelatin as a tissue adhesive was first reported in 1966 (127). A gelatin–resorcinol mixture cross-linked with a combination of formaldehyde (GRF) and also glutaraldehyde (GRFG) was also reported to give satisfactory bond strengths in hepatic and renal tissues (128).

In 1968, Cooper and Falb studied gelatin cross-linked with formaldehyde as a potential tissue adhesive (129). They also added resorcinol to the gelatin–formaldehyde mixtures in order to improve the bond strengths. Formaldehyde is able to condense with resorcinol to give a three-dimensional cross-linked resin. Possible mechanisms for the curing of gelatin–resorcinol–formaldehyde adhesive are shown in Fig. 23 (130).

This polymerized gelatin chemical glue contains no blood components and, outside the U.S., is a commercially available product. The glue consists of gelatin, resorcinol, and distilled water that is supplied in a sterile tube (usually sterilized by ethylene oxide) that is warmed to 45°C prior to use and applied to the operative site; a small volume of formaldehyde (glutaraldehyde can be substituted for formaldehyde) is then added to polymerize the adhesive.

Gelatin–resorcinol–formaldehyde glue was used to treat aortic dissections for the first time in the late 1970s (131). Although no toxicity problems were observed at the time, because of the presence of formaldehyde, questions have always remained about possible toxic and carcino-

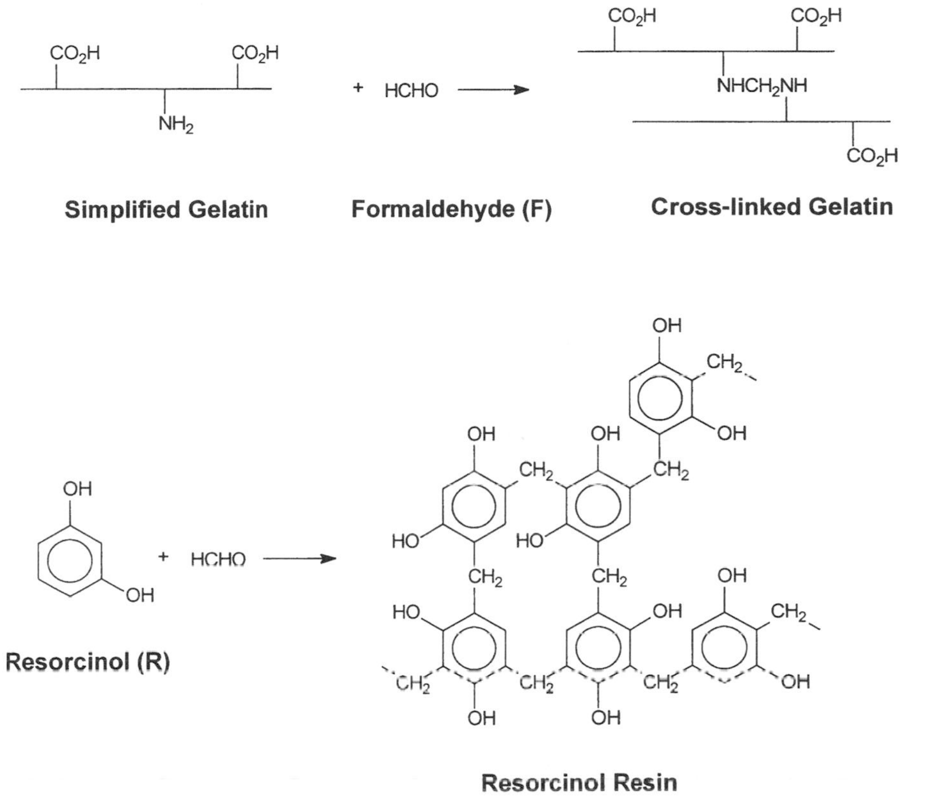

Figure 26.23 Possible mechanisms for the curing of gelatin resorcinol formaldehyde adhesives.

genic effects (132). Even if all the formaldehyde molecules are consumed in the condensation reactions with gelatin and resorcinol, formaldehyde will still be released from the cured adhesive when the gelatin undergoes enzymatic degradation. These concerns have limited its acceptance.

These adhesives bond well even to moist surfaces and have the advantage that they undergo a moderate degradation rate. The curing profile of these adhesives can be altered by careful adjustment of the ratios of the components used. However, the formaldehyde released can be responsible for the considerable inflammatory response, which, for a GRF adhesive, would interfere significantly with the healing process of the surrounding tissues. This echoes the primary concern with cyanoacrylate tissue bonding adhesives, which is their ability to liberate toxic formaldehyde on degradation.

Concerns about carcinogenicity of formaldehyde have prompted the development of gelatin–resorcinol–formaldehyde tissue glues that do not contain the formaldehyde component, this being replaced with both glutaraldehyde and glyoxal (133). These adaptations are not possible for cyanoacrylates.

A modification of GRF glue, so-called GR-DIAL glue has also been developed to reduce the toxicity of GRF by removing the formaldehyde component and replacing it with two less toxic aldehydes, pentane-1,5-dial and ethanedial (134). This type of adhesive is commercially available from Fehling Medical AG under the trademark *Gluetiss*.

3. Emerging Technologies

a. Bioabsorbable Synthetic Hydrogels as Surgical Sealants

To address the medical need for a biocompatible sealant for use in surgery a bioabsorbable hydrogel has been developed. Hydrogel-based adhesives have several potential advantages over cyanoacrylates such as a faster rate of degradation and the ability to act as space-filling adhesives. One such adhesive system is *FocalSeal* (trademark of Focal Inc.).

The hydrogel is formed by photopolymerization of water-soluble macromers. Poly(ethylene glycol) (PEG) is linked on both ends to hydrolyzable units, trimethylene carbonate (TMC) or lactate (LA) oligomeric segments, and end-capped with polymerizable acrylate groups (AA). In one embodiment, the hydrogel is polymerized in situ using two parts, a primer and a sealant, as a laminate forming a highly adherent coating of any desired thickness on tissue. The laminate provides both the bonding to the tissue (primer layer) and the desired mechanical properties (the sealant layer).

The materials are applied to the tissue in aqueous solution formulations. The primer solution is first brushed onto the tissue, and then the sealant is mixed with the primer using the brush to provide a transition layer between the primer and sealant layers. The final thicker sealant layer is then flowed over the application area and the laminate is photopolymerized. The primer and sealant macromers are shown in Fig. 24, and the synthesis of these macromers has been described (135) and is shown in Fig. 25.

Typically, photopolymerization is accomplished by formulating the macromers in buffered saline solutions containing triethanolamine and eosin Y as the photoinitiator. Visible light illumination from a xenon arc lamp (470–520 nm) at an intensity of 100 mW cm^{-2} is used for 40 seconds to initiate the polymerization (136). The general concern of heat related tissue necrosis is minimized in that the formulation contains a high proportion (>80%) of water.

Once implanted, the hydrogel breaks down by hydrolysis, releasing biocompatible components, as shown in the mechanism, that are metabolized or cleared by the kidneys. Tissue healing has been shown to occur unimpeded in the presence of the sealant.

Burst strengths have been measured for Focal Seal at 377 ± 98 mm Hg and compared to fibrin sealants at 23 ± 17 mm Hg. The degradation of the sealant has been measured at various temperatures. At 67°C there is 100% mass loss after 40 days.

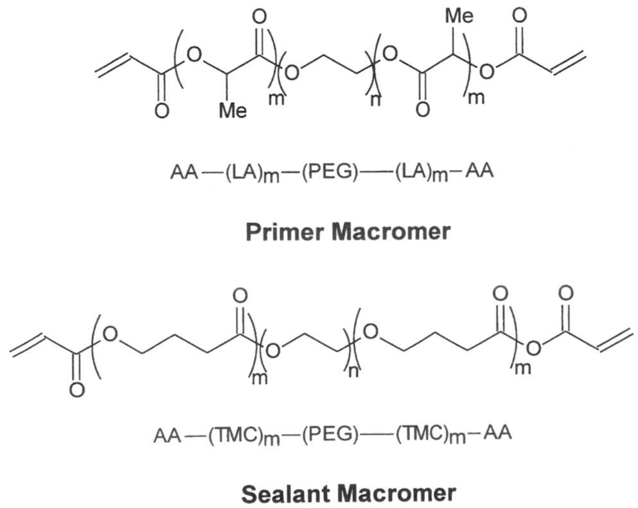

AA—(LA)m—(PEG)—(LA)m—AA

Primer Macromer

AA—(TMC)m—(PEG)—(TMC)m—AA

Sealant Macromer

Figure 26.24 Primer and sealant macromers used in photopolymerizable adhesive.

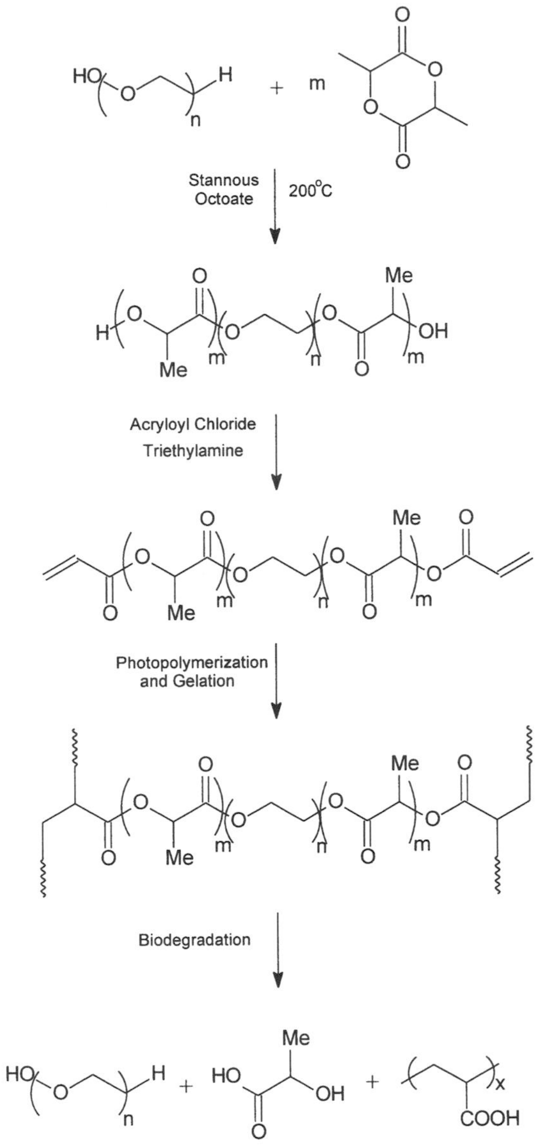

Stannous Octoate 200°C

Acryloyl Chloride Triethylamine

Photopolymerization and Gelation

Biodegradation

Figure 26.25 Synthesis of poly(ethylene glycol)-co-poly(D,L-lactide) diacrylate photopolymerizable monomer.

C. Biological Adhesives

1. Fibrin Tissue Adhesives

The majority of biological glues are essentially protein adhesives composed mainly of fibrin generated from fibrinogen activated by thrombin. These glues were first used in cerebral surgery by Grey (137) in 1915, though the hemostatic properties of these materials had already been reported in 1909 (138). Although the biochemical process of fibrin polymerization and cross-linking is well documented, studies in the mechanism of how these biological adhesives actually bond to various tissues are not as widely discussed.

a. *General Introduction to Fibrin as a Biomaterial*

Fibrin glues, or adhesives, have been used extensively in Europe since 1978 for joining tissue in many surgical procedures through products sold under trademarks such as *Tissucol, Biocol*, and *Beriplast*. Many of the concerns surrounding fibrin products center on the possibilities of viral transmission. All such products are therefore subjected to rigorous screening procedures, and companies have to go to great lengths to ensure that the risk of viral transmission is minimal. This potential problem is also being addressed by the use of fully recombinant materials that eliminate the use of, for example, pooled human blood products. These issues relating to their safety and efficacy had until recently prevented the U.S. Food and Drug Administration from approving fibrin adhesives (139). However, during 1998 approval was given to a human pooled plasma derived sealant manufactured by Immuno AG and sold under the trademark *Tisseel. Tisseel* has now been cleared for use in indications such as to control bleeding during heart bypass surgery, colon surgery, and traumatic spleen surgery, all situations that could involve hemorrhage originating from difficult-to-seal blood vessels. The product is marketed in the U.S. under the trademark *Hemaseel*. These biodegradable and biocompatible materials therefore realize three major clinical goals: reducing hemorrhage, increasing tissue adherance, and allowing the delivery of drugs and biologics, which has resulted in a current worldwide market for fibrin sealants worth in excess of US$300 million.

b. *Components of Fibrin Tissue Adhesives and Methods of Manufacture*

Fibrin adhesives are made from a number of components produced from pooled human plasma that enable the adhesive to mimic the final stages of blood clotting. The main constituents are fibrinogen, factor XIII, and thrombin, which are typically supplied as two components and then mixed in the presence of calcium ions. Component 1 typically contains concentrates of fibrinogen, factor XIII, and an antifibrinolytic agent, such as aprotinin or in a few cases tranexamic acid, with component 2 containing thrombin and calcium chloride solution. Table 13 outlines a typical formulation of a fibrin sealant product (140).

Fibrinogen. Fibrinogen is the polymeric precursor to fibrin monomer with a molecular weight of about 330,000 (141). The main source of fibrinogen is pooled plasma of human origin from screened volunteers or plasma from an autologous source. Blood is centrifuged to separate plasma from cellular components and frozen at −80°C for 24 hours, after which the plasma is thawed and centrifuged. The cryoprecipitate is concentrated in fibrinogen, which may be used immediately or stored under the appropriate conditions (142). Risks associated with possible viral transmission when using large-pool fibrinogen may be overcome using autologous sources (143–145), although this option is not always practical. The strength of fibrin glues is largely determined by the concentration of fibrinogen in component 1 of the adhesive (146).

Table 26.13 Typical Components of a Fibrin Sealant

Component	Concentration	Origin
Fibrinogen (clottable)	80–120 g/L	Human plasma
Factor XIII	10–30 IU/mL	Human plasma or human placenta
Fibronectin (antigen)	5–20 g/L	Human plasma
Thrombin (activity)	300–600 NIH-U/mL	Human plasma or bovine plasma
Aprotinin	3000 KIU/mL	Bovine lung
Calcium chloride	40–60 mM	Inorganic

Thrombin. Thrombin is an enzyme that acts to generate fibrin monomer in the initial stages of the clotting process. Until recently, commercial fibrin adhesives used thrombin of bovine origin, but human derived thrombin is now the preferred source, which, if virally inactivated, will prevent possible transmission of infectious agents (140). The concentration of thrombin in the formulation affects the speed of clotting, with increased concentrations giving reduced clotting times (147).

Factor XIII. Factor XIII is a cross-linking agent that, when activated by thrombin and calcium ions to factor XIIIa, catalyzes cross-linking of the fibrin polymer. This hardens the clot and thereby increases its mechanical strength and reduces susceptibility to proteolytic degradation by plasmin (148).

Fibrinolytic Inhibitors. Most fibrin adhesive compositions contain a fibrinolytic inhibitor that acts to stabilize the fibrin clot against degradation by naturally occurring proteolytic enzymes (140). Typically, an extract from bovine lung called aprotinin is used, although other agents such as tranexamic acid or aminocaproic acid have been used (149). Studies have shown that tranexamic acid decreases early fibrin digestion, while aprotinin is more ef-

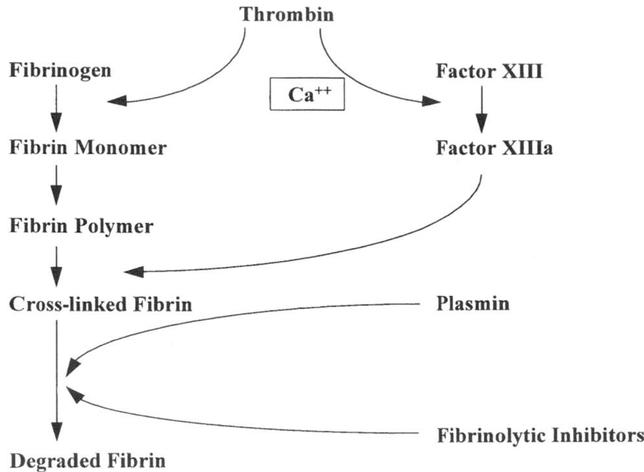

Figure 26.26 Mechanistic cascade of fibrin tissue adhesive clotting and degradation.

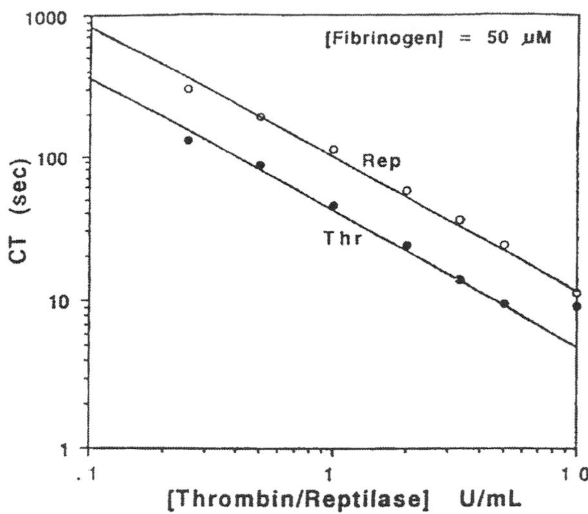

Figure 26.27 Clotting time (CT) of 2 mg/mL pure fibrinogen induced by various levels of thrombin or reptilase. (Reprinted from Ref. 153 with permission from Technomic Publishing Co. Inc., copyright 1996.)

fective over longer time periods by increasing fibrin recovery (150).

c. Mechanism of Action

Fibrin adhesives mimic the final step of the physiological coagulation cascade (Fig. 26). Thrombin acts proteolytically to cleave fibrinopeptide A and B from fibrinogen to produce fibrin monomer, which then polymerizes by hydrogen bonding and electrostatic reactions to form and unstable, soft clot. Thrombin also activates factor XIII to XIIIa in the presence of calcium ions, which catalyzes the cross-linking of the fibrin through the formation of amide links between glutamine and lysine residues in a variety of proteins (147,148,151). The cross-linking reaction stabilizes the clot as well as increasing clot resistance to proteolytic degradation by plasmin (152). Cross-linking between fibrin and adhesive glycoproteins (collagen, fibronectin, von Willebrand factor) as well as cellular glycoproteins anchors the clot to the site of the injury (140,148).

Clotting time is dependent on the concentration of both fibrinogen and thrombin (153). Figure 27 shows that the gelation rate is inversely proportional to the concentration of thrombin at a fixed level of fibrinogen. The dependency on fibrinogen concentration is more complex, showing a biphasic relationship at constant thrombin concentration (Fig. 28). Studies have also shown the strong influence of ionic strength and pH on clotting time (153).

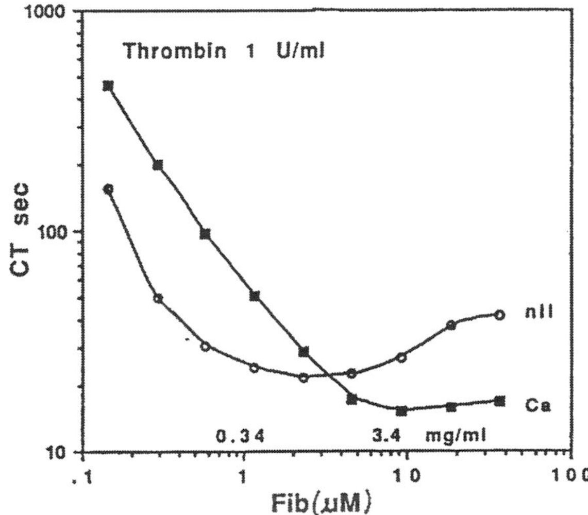

Figure 26.28 Clotting time of various concentrations of fibrinogen induced by 1 U/mL thrombin (with and without 2.5 mM Ca^{2+}). (Reprinted from Ref. 153 with permission from Technomic Publishing Co. Inc., copyright 1996.)

d. Mechanical and Adhesive Properties

Fibrin sealants have typically been used for hemostasis in surgical applications, but their use as tissue joining agents is a relatively new application. Within this area, other techniques, such as staples and soldering, provide alternative solutions. Therefore we must consider both the mechanical and the adhesive properties when characterizing fibrin adhesives.

The tensile strength of fibrin tissue adhesive is increased with increasing thrombin concentration and also

Table 26.14 Mean Tensile Strength of Wound Closures Using Male Sprague–Dawley Rats

Technique	Tensile strength (g/cm²) Day 0	Day 4
Staples	—	530
Fibrin sealant (thrombin activated cryoprecipitate)	99	733
Laser-activated cryoprecipitate	570	779
Laser-activated multicomponent solder	482	668

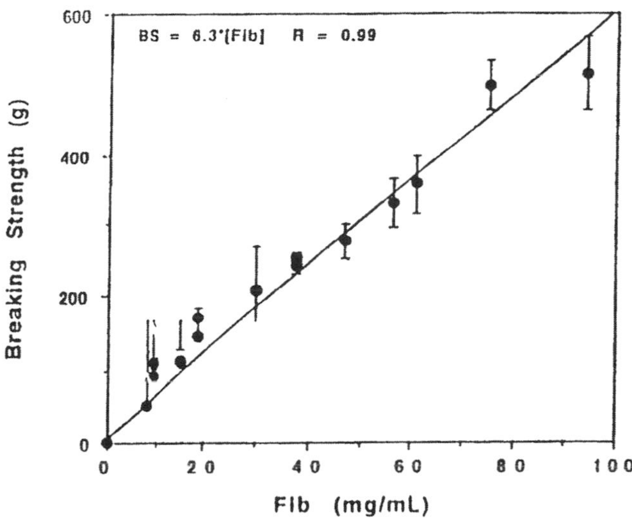

Figure 26.29 Breaking strength (g to break a 0.4 cm^2 cross section) of fibrin adhesive at varying fibrinogen concentrations. (Reprinted from Ref. 153 with permission from Technomic Publishing Co. Inc., copyright 1996.)

increases proportionally with fibrinogen concentration (149,154). When compared to other techniques, such as soldering, fibrin glues suffer from low initial mechanical strength, but after several days the strengths are comparable (155) (Table 14). The internal breaking strength of fibrin adhesives has also been shown to increase linearly with fibrinogen concentration at constant thrombin (5 U/ml) and Ca^{2+} (10 mM) levels (153) (Fig. 29). In the absence of Ca^{2+}, the breaking strengths were reported to be significantly lower, illustrating the importance of Ca ions in factor XIII–induced cross-linking. Trends have been identified between the breaking strength and the pH/ionic strength. The strength of the adhesive increases in relation to both parameters until a plateau is reached at 0.1 M and pH 7.4. This indicates that the fibrin adhesives have their highest strength at physiological pH.

Assessment of restored tissue has shown that cryoprecipitated fibrinogen concentrates compare favorably with suturing (156). Over a period of 90 days, wound closures on Wistar rats using sutures and a fibrin sealant were compared to a nonincised control. The results show that tensile break strength, tensile modulus, and tensile elongation for fibrin healed wounds provided almost 100% restoration of the biomechanical integrity of the control (Table 15).

a. *Fibrin Adhesives as Delivery Vehicles*

Fibrin tissue adhesives also find use as matrixes for the controlled delivery of drugs and biologics. Fibrin acts as an excellent carrier due to its hemostatic properties, natural biocompatibility, and ability to break down as part of the body's healing process.

Table 26.15 Tensile Data Comparing Fibrin Adhesive, Sutured, and Nonincised Sealing

Sample	Retrieval (days) 4	7	14	28	60	90
Tensile break force to failure (kg/cm²)						
Control	13.65	14.71	13.96	14.44	16.17	13.92
Suture	0.104	0.386	0.738	2.770	8.649	8.763
Fibrin	0.089	0.461	0.949	3.306	9.292	14.56
Tensile modulus to failure (kg/cm²)						
Control	—	73.40	65.12	68.01	80.89	81.73
Suture	1.267	1.471	3.759	12.46	39.49	44.49
Fibrin	1.422	1.760	4.421	15.07	45.76	77.44
Tensile elongation to failure (% (cm/cm × 100))						
Control	—	149.3	177.1	149.2	136.1	124.5
Suture	—	33.2	50.6	72.5	98.5	94.0
Fibrin	—	52.3	42.4	68.7	86.8	113.0

One of the first delivery matrix applications of fibrin sealants was for antibiotics (157–159). While the fibrin clot forms during bonding, a suitable medium for microbial growth is also formed. Therefore the controlled release of antibiotics from the glue provides a local antimicrobial effect during wound healing (160). Requirements for the delivery of antibiotics must include biodegradability under physiological conditions and ensure that the delivery matrix remains at the site of application. Several examples exist of the successful delivery of antibiotics (160), as either a solution or a solid dispersed within the matrix, including tetracycline (161), ampicillin, carbenicillin, gentamycin, and dibekacin. Some antibiotics, however, were found either to increase the clotting time or to reduce the strength of the resultant fibrin clot (149). Other issues surround the small loadings of antibiotic that may be incorporated into the matrix and also the relatively fast rate at which they are delivered. This rate may be controlled to some degree by the use of insoluble forms of antibiotic.

Fibrin has also been used for the delivery of chemotherapeutic agents (157,162), osteoinducers (163), and various growth factors (157).

f. Clinical Applications of Fibrin Adhesives

Fibrin adhesives have found use in many surgical applications due to their ability to join tissue, their hemostatic properties, and their ability to act as a barrier to fluid. Delivery of these adhesives to the wound site can be achieved by a dual syringe system, spray technique (164) or endoscopically (165). In reconstructive and aesthetic surgery, fibrin glues have been shown to give high success rates for skin graft take (166). For difficult wound sites, such as on the neck, shoulder, buttocks, and back, the use of fibrin glue to improve early adherence of skin grafts has reduced the amount of postoperative care (149,167–169).

Other indications for fibrin tissue adhesives include otolaryngology (140,170), nerve repair (171), cardiovascular surgery (140,172), thoracic surgery, opthalmology, dentistry (173), and general traumatology.

2. Protein Adhesives

The interest in synthetic adhesives, and the realization of their associated problems, has prompted scientists to look at how nature uses adhesion to perform various functions. For example, in order to develop even better adhesives, chemists have been looking at how common mussels adhere to objects underwater. None of the currently available commercial glues even come close to the incredible sticking power of the natural ''glue'' used by many molluscs. There are many other adhesives found in nature, many of which rely on the action of proteins on specific substrates.

a. Mussel Adhesive Proteins

Marine organisms including mussels and barnacles are capable of producing remarkable moisture-resistant adhesives. Since these function in moist environments, often underwater, they are potentially useful as medical adhesives.

The byssus threads of a mussel will hold it strongly to rocks, the hulls of ships, wood, or plant life. The adhesives are even more remarkable in that they function perfectly well over a wide range of temperatures and in fluctuating salinities despite the rigors of waves, tides, and currents in all marine environments.

In order for chemists to mimic the action of naturally occurring adhesives, they must first understand how marine organisms such as the mussel make their bonds. By placing its foot very close to a surface and expelling all the water and creating a partial vacuum it pumps liquid protein into the area through special glands in its foot. This liquid immediately forms a sticky foam and then numerous tiny threads of protein form within the foam. This is very similar to the silk threads formed by spiders and silkworms. It consists of highly insoluble polymeric fibers, much of which is keratin-like (the same protein which makes our fingernails and horses' hooves). Numerous detailed reviews have been published on marine bioadhesives, most notably those of Waite (174,175).

Scientists in laboratories all over the world have had limited success in making adhesives to match the performance and characteristics of the mussel's byssus threads. BioPolymers Inc. markets a glue based on mussel derived proteins under the trademark *Cell-Tak*, which was launched in 1986. This adhesive is used by scientists to attach submerged cells to the surface of tissue culture dishes. Another company, Gentex, has cloned the mussel gene that codes for the preliminary proteins that make byssus. They have been able to insert this gene into yeast cells, which are then able to produce small batches of the protein resin. They hope to make an adhesive that can be used instead of sutures, and that can even reunite severed nerves.

While this substance might make an excellent adhesive for use in wet environments, isolation of the uncured adhesive from mussels for commercial use is not practical since the extraction of 1 kg of the adhesive substance would require the use of 5 to 10 million mussels. A totally synthetic alternative would be a more realistic proposal.

Much of the initial work, however, in this area did concentrate on harvesting the polyphenolic proteins from common mussels, *Mytilus edulis*, and purifying and stabilizing these by isolating the high molecular weight fractions of acid-soluble extracts. These proteins are digested in trypsin to yield a specific decapeptide sequence. These decapeptides are then used as building blocks to make larger polyphenolic molecules possessing the adhesive capabilities of the native bioadhesive protein comprising about 1 to 1000 decapeptide repeating units using amino acid, oligopeptide, or other bifunctional spacers. Methods for purifying this protein and its uses have been published in U.S. patents (176–180).

The natural protein itself is rich in proline/hydroxyproline, serine, threonine, lysine, tyrosine/3,4-dihydroxyphenylalanine (DOPA), and alanine, a composition that suits the protein for interaction with biological surfaces, as shown in Table 16 (181).

The natural adhesive can be mimicked by focusing on the ten-acid sequence. This decapeptide is repeated many times in the long chain polypeptide and is one of the keys to the adhesive characteristics of these materials. In addition to the decapeptide sequence, other significant features identified include the presence of lysine and tyrosine or DOPA residues, which provide the essential cross-linking sites. Furthermore, there is an unusually frequent appearance of proline or hydroxyproline residues and an essentially complete absence of glycine residues, i.e., the decapeptide sequence is not typical of connective tissue proteins such as collagens.

Mytilus edulis, which has perhaps received the most attention, is not the only marine organism that has evolved an elaborate mechanism for adhering itself to surfaces. The Atlantic mussel, *Geukensia (Modiolus) demissa*, also possesses an ability to bond in a marine environment. Its proteinaceous adhesive is distinct from that of the common mussel. The Atlantic ribbed mussel adhesive has glycine as one of its most abundant amino acid residues, similar to collagen (though not as plentiful), and one of the novel sequences in this adhesive is based upon a tripeptide fragment that occurs with great frequency in the natural adhesive. Other novel peptide sequences include octapeptide and nonapeptides that contain the characteristic tripeptide "tail." The tripeptide is GLY-DOP-LYS, where GLY is glycine, DOP is dihydroxyphenylalanine, and LYS is lysine. These specific sequences are covered in a U.S. patent (182).

Table 26.16 Potential Adhesive Interactions of Frequently Occurring Amino Acids in *Mytilus edulis* Adhesive Protein

Amino acid	Chemical bond
Serine	Hydrogen
Threonine	Hydrogen
Lysine	Ionic
Tyrosine	Hydrogen
Hydroxyproline	Hydrogen
DOPA	Metal complex formation

b. *Synthetic Analogues of Protein Adhesives*

The direct chemical synthesis in the laboratory of a polypeptide identical to those found in mussels represents a huge challenge. Already, analogues of the marine mussel extract have been formed using recombinant technology as disclosed in U.S. patents (183,184).

Synthetic polymers (branch copolymers) have been prepared with a free-amine-rich backbone onto which were grafted variations of the natural decapeptide. These compounds are the subject of a patent by BioPolymers Inc. (185). A formulation based on polyphenolic proteins extracted from the marine mussel has already entered the market under the trademark *Cell-Tak*. *Cell-Tak* has been tested on cartilage surfaces and was found to be of very low adhesive strength (186).

In order to produce large quantities of mussel adhesive proteins for studying mechanisms and developing materials for medical applications, microbial systems have been developed (Genex Corporation) to produce the protein (187). The expression system was based on yeast but contains neither DOPA nor hydroxyproline residues. DOPA residues are unusual in proteins, and it has been suggested that these are crucial for moisture-resistant adhesion to surfaces and for cross-linking to build cohesive strength. Studies with genetically engineered protein support this and show that cross-linking and moisture-resistant adhesion require the presence of the reactive, oxidized form of DOPA, namely quinone. The DOPA adhesive proteins are stable in an acidic solution or as lyophilized powders. The DOPA protein is the preadhesive form of the protein. The adhesive properties are generated when the DOPA residues are oxidized to quinone. In order to achieve moisture-resistant adhesion, there is a need for tyrosine to be hydroxylated. The sequence in Fig. 30 shows the posttranslational modification of tyrosine residues in the *Mytilus edulis* protein (181). This reaction can be catalyzed by tyrosinases derived from mushrooms and bacteria.

Adherence of the engineered mussel protein in an aqueous environment to various substrates including polystyrene, glass, hydrogel, and collagen has been tested. In each case activation to the quinone form of the protein was required for good surface adhesion. The cohesive strength

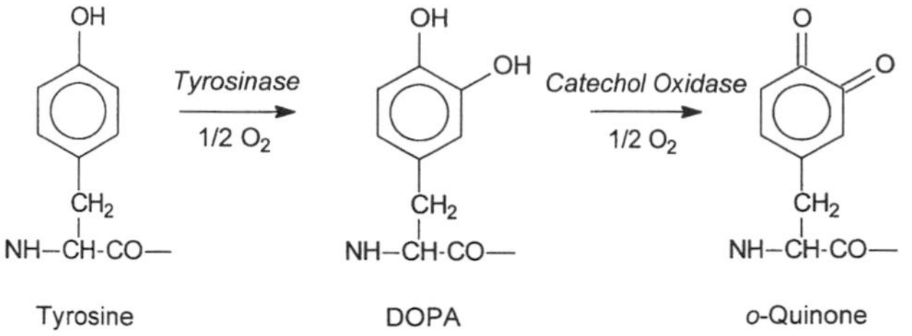

Figure 26.30 Posttranslational modifications of tyrosine residues in the *Mytilus edulis* adhesive protein.

of the engineered adhesive protein depends upon intermolecular cross-linking, which requires formation of quinone residues. The nature of the chemical bond involved in the cross-linking has yet to be determined.

It has also been found that the stereochemistry of the decapeptide oligomer is not essential to the adhesive behavior of the protein; the protein need only contain a certain amount of DOPA rather than the entire decapeptide sequence.

A synthetic approach has recently been adopted by Deming and Yu (188), who tested the premise that functionality, and not amino acid sequence, was the only feature necessary for moisture-resistant adhesion. Using the known composition of the natural adhesive proteins, sequentially random copolypeptides through copolymerization of a few select α-amino acid N-carboxyanhydrides (NCAs) were prepared. NCAs are readily prepared from amino acids by phosgenation and can be polymerized into high molecular weight polypeptides via successive ring-opening addition reactions that liberate carbon dioxide. The resulting copolymers were oxidized under a variety of conditions, and the effect of the oxidant on both cross-linking and adhesive capabilities was evaluated.

Simple copolypeptides of L-lysine and DOPA containing different compositions of the two monomers were prepared as shown in Fig. 31. L-Lysine was chosen as the other major component since (1) it is present in large quantities in marine adhesive proteins, (2) it is thought to be involved in protein cross-linking reactions, and (3) it should provide good water solubility to the copolymers when the side chain is ionized.

Many factors affect the bond strength in these synthetic systems including copolymer ratio, cure time, temperature, oxidant, and substrate. This system uses inexpensive oxidizing materials, and there is a route defined to produce a supply of adhesive polymers. Through adjustment of the copolymer composition, molecular weight, or curing conditions, the adhesive properties of these synthetic copolypeptides can, it is claimed, be tuned for specific applications (188).

Typical bond strengths for these materials have been measured (188) on steel, aluminum, glass, poly(methyl methacrylate), polystyrene, and poly(ethylene). The tensile strengths range from 2–4 MPa for the steel, aluminum, and glass substrates; the synthetic polymers were considerably less. These results, representing a totally synthetic ap-

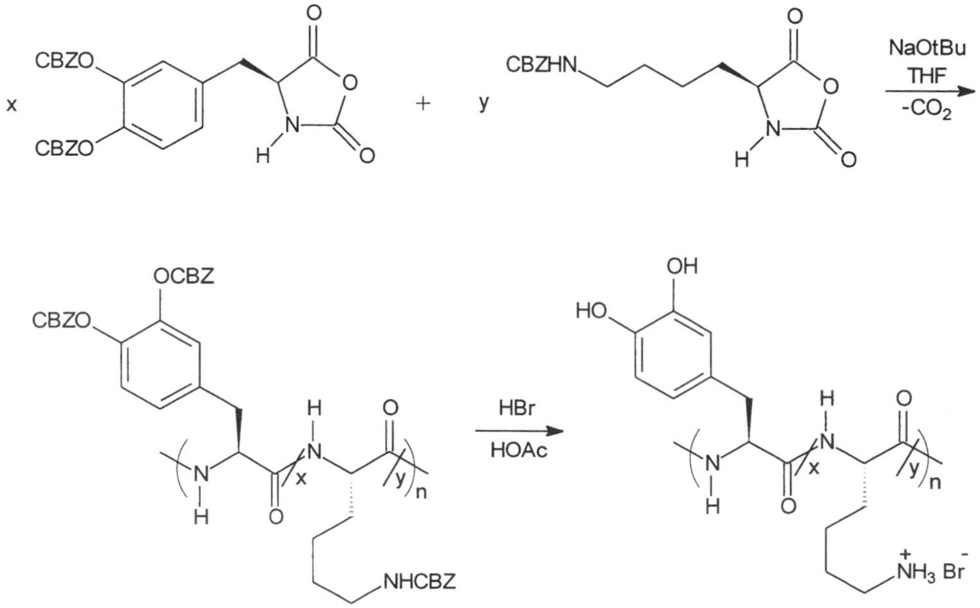

Figure 26.31 Preparation of lysine • HBr/DOPA copolymer.

Table 26.17 Characteristics of Currently Available Tissue Adhesives

	Fibrin glue	Cyanoacrylate	GRF
Handling	Excellent[a]	Poor[b]	Poor
Set time	Medium	Short	Medium
Tissue bonding	Poor	Good	Excellent
Pliability	Excellent	Poor	Poor
Toxicity	Low[c]	Medium[d]	High
Resorbability	Good	Poor	Poor
Cell infiltration	Excellent	Poor	Poor

[a] Spray type.
[b] Low viscous type.
[c] Not autologous.
[d] Long alkyl chains.
Source: Ref. 130.

proach, compare favorably with natural precursor protein. Natural adhesive precursor protein has been extracted from the blue mussel *Mytilus edulis*, which has the sequence Ala-Lys-Pro/Hyp-Ser-Tyr/DOPA-Hyp-Hyp-Thr-DOPA-Lys. A polydecapeptide having the sequence of the adhesive *Mytilus edulis* (Mw ca. 140,000) was reported to give a tensile strength of 2.75 MPa when cured in water on steel (189).

More recently, Deming has published experimental data showing that the adhesion and cross-linking chemistry of mussel adhesive model polymers is due primarily to the DOPA residues in the materials (190). The studies concluded that the catechol functionality is primarily responsible for moisture-resistant adhesion, and the oxidized *o*-quinone functionality is mainly responsible for cross-linking.

A new generation of biocompatible adhesive materials could soon be developed with the potential to become key components in surgical procedures. Opportunities lie in identifying novel peptide sequences that can be synthetically engineered to produce a strong adhesive. Initial results by key groups indicate that these characteristics are being investigated and optimized.

D. Tissue Adhesive Summary

The general properties of currently available tissue adhesives are summarized in Table 17 (130). Each adhesive type has characteristics making it suitable for specific applications. For example, cyanoacrylates give excellent bond strengths (186) (Table 18), and fibrin is rapidly resorbed. These relative benefits and side effects are associated with the use of all tissue adhesives, which are summarized in Table 19 (191).

Table 26.18 Mean Bond Strengths Between Different Tissue Types for Butt Joints

	Bone strength (MPa)		
	Cartilage	Bone	Skin
n-Butyl-2-cyanoacrylate	1.0	1.4	1.2
GRF	0.15	0.20	0.07
Fibrin	0.0049	0.011	0.019

Source: Ref. 186.

IV. CONCLUSIONS

The increasing understanding of wound management has largely been responsible for the development and formulation of polymers for PSAs that can facilitate the wound healing process. Highly moisture vapor permeable adhesives enable dressings to control wound exudate more effectively, the management of trauma has seen advances made in ''smart'' PSAs that can change adhesive properties by use of external stimuli, and the controlled delivery of active ingredients using transdermal delivery systems have all required novel PSAs to further this technology.

There are still shortcomings in the currently available tissue adhesives, and their widespread acceptance has been inhibited by the often demanding requirements of many clinical procedures. An ideal tissue adhesive should cure rapidly in situ, form strong bonds to a variety of tissue types, and ultimately be degraded to yield safe, nontoxic by-products. Cyanoacrylates have been used in surgery, but the cured polymers release formaldehyde on biodegradation, which is believed to be responsible for toxicity to-

Table 26.19 Relative Benefits/Side Effects of Current Tissue Adhesives

Property	Ranking
Adhesive power	Cyanoacrylate > GRF glue > Fibrin sealant
Tensile strength	Cyanoacrylate > GRF glue > Fibrin sealant
Flexibility	Fibrin sealant > GRF glue > Cyanoacrylate
Thrombogenicity	Cyanoacrylate > GRF glue > Fibrin sealant
Resorption rate	Fibrin sealant (10 days) > GRF glue (6 weeks) > Cyanoacrylate (12 months)
Chronic inflammation and stimulation of fibrosis	Cyanoacrylate > GRF glue > Fibrin sealant

Source: Ref. 190.

wards tissue. Butyl-2-cyanoacrylate is now being used as a safe cyanoacrylate for skin closure, although it is always stressed that the monomer should not be allowed to penetrate below the level of the skin. Fibrin has been more widely used in surgery than cyanoacrylates, not only as an adhesive but also as a hemostatic agent and surgical sealant, although fibrin only produces very weak adhesive bonds.

Significant developments have therefore been made in the area of medical adhesives over the last 20 years. Although PSA technology is well established, there have been many advances that have expanded their potential beyond a simple method of fixation. Research into tissue adhesives, however, has yet to identify the ideal surgical adhesive, although several candidate materials show promise and are currently under development.

ACKNOWLEDGMENTS

We are grateful to the following colleagues for useful discussions and technical assistance: Andrew Jackson, Chris Ansell, Robin Chivers, Sam Boyer, Kevin Yeomans, Judi Gledhill, and Gill Mawby.

REFERENCES

1. Day, H., Shecut, W. Improvements in adhesive plasters. U.S. patent 3,965 (1845).
2. Webster, I. Recent developments in pressure-sensitive adhesives for medical applications. *Int. J. Adhesion Adhesives*, 17(1), 69–73 (1997).
3. Chivers, R. A. Physical properties of pressure sensitive adhesives for medical applications. *Adhesion '93 Conference Proc.*, 305–309 (1993).
4. Bothwell, J. W. Adhesion to skin: effects of surgical tapes on skin. In: *Adhesion in Biological Systems* (Manly, R. S., ed.). Academic Press, Ch. 14, 215–221 (1970).
5. Pierson, D. G. An overview of skin contact applications for pressure-sensitive adhesives. *Tappi J.*, 101–107 (1990).
6. Walker, M. Instrumental techniques for the measurement of skin adhesion. *Analytical Proceedings*, 29(9), 391–393 (1992).
7. Satas, D. Pressure-sensitive adhesives—effect of rheological properties. In: *High Performance Biomaterials* (Szycher, M., ed.). Ch. 13, 185–189 (1991).
8. Kenney, J. F., Haddock, T. H., Sun, R. L., Parreira, H. C. Medical-grade acrylic adhesives for skin contact. *J. Appl. Polymer Sci.*, 45, 355–361 (1992).
9. Charkoudian, J. C. A model skin surface for testing adhesion to skin. *J. Soc. Cosmet. Chem.*, 39(4), 225–234 (1988).
10. O'Brien, J. M., Reilly, N. J. Comparison of tape products on skin integrity. *Advances in Wound Care*, 8(6), 26–30 (1995).
11. Mayrovitz, H. N., Carta, S. G. Laser-Doppler imaging assessment of skin hyperemia as an indicator of trauma after adhesive strip removal. *Advances in Wound Care*, 9(4), 38–42 (1996).
12. Schiraldi, M. T. Peel adhesion of tapes from skin. *Polymers, Laminations and Coatings Conference, TAPPI Proceedings*, 63–70 (1990).
13. Sun, R. L., Kenney, J. F. Vinyl caprolactam containing hot melt adhesives. European patent EP 175,562 (1988).
14. Cunningham, B. C., Shaw, C. J., Robinson, A. Pressure sensitive adhesives. British patent GB 2,198,441 (1988).
15. Rhodes, J., Douglas, W. Adhesive composition. U.S. patent 4,222,923 (1980).
16. Chen, J. L., Cilento, R. D., Hill, J. A., La Via, A. L. Ostomy adhesive. U.S. patent 4,192,785 (1980).
17. Russell, G. S., Pelesko, J. D. Pressure sensitive adhesive compositions and elements made therefrom. World patent WO 91/09883 (1991).
18. Dietz, T. P., Asmus, R. A., Uy, R. Biomedical electrode provided with two-phase composites conductive, pressure-sensitive adhesive. World patent WO 93/09713 (1993).
19. Dietz, T. M., Itoh, S. K., Uy, R. Adhesive comprising hydrogel and crosslinked polyvinyllactam. German patent DE 4,238,263 (1993).
20. Kantner, S. S., Kuester, W. Hydrophilic pressure sensitive adhesive composition. World patent WO 97/24378 (1997).
21. Zhang, Z., Cao, J., Su, Z. Development of UV-cured pressure-sensitive electrically conductive adhesives for medical electrodes. *Huadong Ligong Daxue Xuebao*, 20(2), 259–263 (1994).
22. Sieverding, D. L. Hydrophilic, elastomeric, pressure-sensitive adhesive. U.S. patent 4,699,146 (1987).
23. Thompson, J. A. Adhesive composites for biomedical electrodes. U.S. patent 4,830,776 (1989).
24. Everaerts, A. I., Kantner, S. S., Koski, N. L., Li, K., Nielsen, K. E., Koski, N. I. Hydrophilic pressure sensitive adhesive composition. World patent WO 97/34947 (1997).
25. Winter, G. D. Formation of the scab and the rate of epithelization of superficial wounds in the skin of the young domestic pig. *Nature*, 193, 293–294 (1962).
26. Thomas, S. Semipermeable film dressings. In: *Wound Management and Dressings*. Pharmaceutical Press, Ch. 4, 25–34 (1990).
27. Foreman-Peck, J. Research, pharmaceuticals and plastics. In: *Smith & Nephew in the Health Care Industry*. Edward Elgar, Ch. 9, 161–182 (1995).
28. Dabi, S., Haddock, T., Hill, A. S. Adaptive transparent film dressings. *Journal of Biomaterials Applications*, 9, 14–29 (1994).
29. Barnhart, S., Carrig, T. Critical role of PSAs in transdermal drug delivery. In: *Adhesives and Sealants Industry*, 26–30 (April 1998).
30. Wick, S. M. Developing a drug-in-adhesive design for

transdermal drug delivery. *Adhesive Age*, 18–24 (September 1995).

31. Cordes, G., Santoro, A. Setnikar, I. Transdermal drug delivery system for delivery of nitroglycerin to humans. World patent WO 98/25591 (1998).
32. Questel, J. PSAs for transdermal drug delivery. *Adhesives and Sealants Industry*, 28–36 (April/May 1995).
33. Kokubo, T., Sugibayashi, K., Morimoto, Y. Diffusion of drug in acrylic-type pressure-sensitive adhesive matrices. I. Influence of physical property of the matrices on the drug diffusion. *J. Controlled Release*, 17, 69–78 (1991).
34. Smith, S. E., Venkatraman, S. S. Contact adhesive extends wear time on skin. World patent WO 95/05138 (1995).
35. Sobieski, L. A., Tangrey, T. J. Silicone pressure sensitive adhesives. In: *Handbook of Pressure Sensitive Technology* (Satas, D., ed.). Ch. 18, 508–517 (1989).
36. Cheng, Y.-H., Hosoya, O., Sugibayashi, K., Morimoto, Y. Effect of skin surface lipid on the skin permeation of lidocaine from pressure sensitive adhesives. *Biol. Pharm. Bull.*, 17(12), 1640–1644 (1994).
37. Roy, S. D., Gutierrez, M., Flynn, G. L., Cleary, G. W. Controlled transdermal delivery of fentanyl: characterisation of pressure-sensitive adhesives for matrix patch design. *J. Pharmaceutical Sciences*, 85(5), 491–495 (1996).
38. Kokubo, T., Sugibayashi, K., Morimoto, Y. Interaction between drugs and pressure-sensitive adhesives in transdermal therapeutic systems. *Pharmaceutical Research*, 11(1), 104–107 (1994).
39. Smith, S. E., Venkatraman, S. S. Contact adhesive extends wear time on skin. World patent WO 95/05138 (1995).
40. Jevne, A. H., Holmblad, C., Phipps, J. B., Howland, W. W. Amphoteric hydrogel for medical devices. World patent WO 91/15250 (1991).
41. Holmblad, C., Howland, W. W., Jevne, A. H., Phipps, J. B. Iontophoretic drug delivery device. World patent WO 91/15260 (1991).
42. Tan, H. S., Pfister, W. R. Pressure-sensitive adhesives for transdermal drug delivery systems. *Pharmaceutical Science Technology Today*, 2(2), 60–69 (1999).
43. Webster, I. The development of a pressure-sensitive adhesive for trauma-free removal. *International J. Adhesion Adhesives*, 19, 29–34 (1999).
44. Webster, I. Adhesives. World patent WO 97/06836 (1997).
45. Webster, I. The development of a light switchable pressure-sensitive adhesive for trauma-free wound dressings. *Spring Medical Device and Technology Conference Proceedings*, 241–246 (1999).
46. Therriault, D. J., Workinger, J. E. A liner, dried to give a pressure-sensitive adhesive, and then laminated to a backing material. U.S. patent 5,032,637 (1991).
47. Schmitt, E. E., Tsugita, R. Temperature zone specific pressure-sensitive adhesive compositions and adhesive assemblies and methods of use associated therewith. World patent WO 92/13901 (1992).
48. Clarke, R., Larson, A., Schmitt, E. E., Bitler, S. P. Temperature switchable pressure sensitive adhesives. *Adhesives Age*, 9, 39–41 (1993).
49. Lucast, D. H., Taylor, C. W. Crosslinked acrylate adhesives for use on skin. In: *Polymers, Laminations and Coatings Conference Proceedings*, 721–725 (1989).
50. Beier, H., Petereit, H.-U., Bergmann, G. Water-soluble pressure sensitive skin adhesive. European patent EP 394,956 (1990).
51. Hashimoto, M., Soga, W., Fukuda, M. Medical adhesive for adhesive plaster. Japanese patent JP 04,178,322 (1992).
52. Appelt, M. R., Grosh, S. K. Crosslinked pressure-sensitive adhesives tolerant of alcohol-based excipients used in transdermal delivery devices and method of preparing same. European patent EP 455,458 (1991).
53. Delgado, J., Goetz, R. J., Lucast, D. H., Silver, S. F. Low trauma wound dressing having improved moisture vapour permeability. World patent WO 96/14094 (1996).
54. Kenney, J. F., Sun, R. L. Surgical pressure-sensitive sheet products of high cohesive strength. U.S. patent 4,879,178 (1989).
55. Venkatraman, S., Gale, R. Skin adhesives and skin adhesion 1. Transdermal drug delivery systems. *Biomaterials*, 19, 1119–1136 (1998).
56. Gehman, D. R. Acrylic adhesives. In: *Handbook of Adhesives*. 3d ed. (Skeist, I., ed.). Van Nostrand Reinhold, New York, Ch. 25, 437 (1990).
57. Kenney, J. F., Haddock, T. H., Sun, R. L., Parreira, H. C. Medical-grade acrylic adhesives for skin contact. *Proceedings of 137*th Meeting of the Rubber Division, American Chemical Society, Paper No. 33 (1990).
58. Brunsveld, G. H., Minnigh, J. T. Water vapor permeable, pressure sensitive adhesive composition. European patent EP 501,124 (1992).
59. Shah, K. R. Water vapor permeable pressure sensitive adhesives incorporating modified acrylate copolymers. U.S. patent 4,510,197 (1985).
60. Sugii, T., Wada, S., Konno, M. Dermal pressure-sensitive adhesive sheet material. European patent EP 369,092 (1990).
61. Krampe, S. E., Moore, C. L., Taylor, C. W. Macromer reinforced pressure sensitive skin adhesive. European patent EP 202,831 (1986).
62. Kishi, T., Kamiyama, F. Resin for medical pressure sensitive adhesive. Japanese patent JP 1,290,956 (1986).
63. Haddock, T. H. Pressure sensitive adhesive. European patent EP 130,080 (1985).
64. Cliento, R. D. Hydrophilic acrylic adhesive. European patent EP 297,769 (1989).
65. Howes, J. G. B. Emulsion polymers. European patent EP 194,881 (1986).
66. Jaros, S. E. Hot melts in pressure-sensitive and heat-seal applications. *Tappi Journal*, 72(12), 79–83 (1989).
67. Sun, R. L., Kenney, J. F. Hot melt pressure sensitive adhesives. U.S. patent 4,762,888 (1988).
68. Sun, R. L., Kenney, J. F. Surgical pressure sensitive adhesive sheet products. U.S. patent 4,879,178 (1989).

69. Auchter, G., Barwick, J., Rehmer, G., Jager, H. Developing UV-crosslinkable acrylic hot melt PSAs. *Adhesives Age*, 7, 20–25 (1994).
70. Gazely, K. F., Wake, W. C. Natural rubber adhesives. In: *Handbook of Adhesives*, 3d ed. (Skeist, I., ed.). Van Nostrand Reinhold, New York, Ch. 9, 167–184 (1990).
71. Satas, D. Hospital and first aid products. In: *Handbook of Pressure-Sensitive Adhesive Technology*. (Satas, D., ed.). Van Nostrand Reinhold, New York, Ch. 20, 419–425 (1982).
72. Pfister, W. R., Wilson, J. M. Silicone pressure sensitive adhesive compositions for transdermal drug delivery devices and related medical devices. European patent EP 524,776 (1993).
73. Ulman, K. L., Thomas, X. Silicone pressure sensitive adhesives for healthcare applications. In: *Advances in Pressure Sensitive Adhesive Technology* (Satas, D., ed.). Satas & Associates, Ch. 6, 133–157 (1995).
74. Pfister, W. R. Silicone adhesives for transdermal drug delivery systems. *Drug Cosmet. Ind*, 143(4), 44–52 (1988).
75. Medical Data International Inc. Sealing the market for surgical adhesives. *MedPro Month*, 84–88, March (1998).
76. Dew, D. K., Supik, L., Darrow, C. R., Price, G. F. Tissue repair using lasers: a review. *Orthopaedics*, 16(5), 581–587 (1993).
77. Bass, L. S., Libutti, S. K., Oz, M. C., Rosen, J., Williams, M. R., Nowygrod, R., Treat, M. R. Canine choledochotomy with diode-laser activated fibrinogen solder. *Surgery*, 115(3), 398–401 (1994).
78. Oz, M. C., Johnson, J. P., Parangi, S., Chuck, R. S., Marboe, C. C., Bass, L. S., Nowygrod, R., Treat, M. R. Tissue soldering by use of indocyanine green dye-enhanced fibrinogen with the near infrared diode laser. *J. Vasc. Surg.*, 11(5), 718–725 (1990).
79. Wang, P. Y. Surgical Adhesives and Coatings. In: *Medical Engineering* (Ray, C., ed.). Yearbook Medical, Chicago, 1123 (1975).
80. Ardis, A. E. Preparation of monomeric alkyl alpha-cyanoacrylates. U.S. patent 2,467,926 (1949).
81. Ardis, A. E. Preparation of monomeric alkyl alpha-cyanoacrylates. U.S. patent 2,467,927 (1949).
82. Joyner, F. B., Hawkins, G. F. Method of making α-cyanoacrylates. U.S. patent 2,721,858 (1955).
83. Carton, C. A., Kessler, L. A., Seidenberg, B., Hurwitt, E. S. *World Neurol.*, 1, 356, (1960).
84. Healy, J. E., Brooks, B. J., Gallager, H. S. *J. Surg.*, 1, 267 (1963).
85. Harper, M. C., Ralston, M. Isobutyl 2-cyanoacrylate as an osseous adhesive in the repair of osteochondral fractures. *J. Biomed. Mater. Res.*, 17(1), 167–177 (1983).
86. Harper, M. C. Viscous isoamyl 2-cyanoacrylate as an osseous adhesive in the repair of osteochondral osteotomies in rabbits. *J. Orthop. Res.*, 6, 287–292 (1988).
87. Duck-Kyoon-Ahn, Sims, C. D., Randolph, M. A., O'Conner, D., Butler, P. E., Amarante, M. T. J., Yaremchuk, M. J. Craniofacial skeletal fixation using biodegradable plates and cyanoacrylate glue. *Plast. Reconstr. Surg.*, 99, 1508–1517 (1997).
88. Brown, P. N., McGuff, S., Noorily, A. D. Comparison of N-octyl cyanoacrylate vs sutures in the stabilisation of cartilage grafts. *Arch. Otolaryngol. Head Neck Surg.*, 122, 873–877 (1996).
89. Toriumi, D. M., Raslan, W. F., Friedman, M. Histotoxicity of cyanoacrylate tissue adhesives: a comparative study. *Arch. Otolaryngol. Head Neck Surg.*, 116, 546–550 (1990).
90. Toriumi, D. M., Raslan, W. F., Friedman, M., Tardy, M. E. Variable histotoxicity of histoacryl when used in a subcutaneous site: an experimental study. *Laryngoscope*, 101, 339–343 (1991).
91. Kerr, A. G., Smyth, G. D. L. Experimental evaluation of tympanoplasty methods. *Arch. Otolaryngol.*, 91, 327–333 (1970).
92. Guilford, F. R., Shaver, E. F., Halpert, B. Incus repositioning in dogs. *Arch. Otolaryngol.*, 84(3), 316–319 (1966).
93. Kerr, A. G., Smyth, G. D. L. Experimental evaluation of tympanoplasty methods. *Arch. Otolaryngol.* 94, 129–131 (1971).
94. Seidentop, K. H. Tissue adhesive histoacryl (2-cyano-butyl-acrylate) in experimental middle ear surgery. *Am. J. Otolaryngol.*, 2, 77–87 (1980).
95. Sachs, M. E. Eubucrilate as cartilage adhesive in augmentation rhinoplasty. *Arch. Otolaryngol.*, 111, 389–393 (1985).
96. Fung, R. O., Ronis, M. L., Mohr, R. M. Use of butyl-2-cyanoacrylate in rabbit auricular cartilage. *Arch. Otolaryngol.*, 111, 459–464 (1985).
97. Coe, J. E., Bondurant, C. P. Late thrombosis following the use of autogenous fascia and a cyanoacrylate (Eastman 910 monomer) for the wrapping of an intracranial aneurysm. *J. Neurosurg.*, 21, 884–886 (1964).
98. Hood, T. W., Mastri, A. R., Chou, S. N. Neural and vascular tissue reaction to cyanoacrylate adhesives: a further report. *Neurosurgery*, 11, 363–366 (1982).
99. Cain, J. E., Dryer, R. F., Barton, B. R. Evaluation of dural closure techniques. *Spine*, 13, 720–725 (1988).
100. Zumpano, B. J., Jacobs, L. R., Hall, J. B., Margolis, G., Sachs, E. Bioadhesive and histotoxic properties of ethyl-2-cyanoacrylate. *Surg. Neurol.*, 18, 452–457 (1982).
101. Hoppenstein, R., Weissberg, D., Goetz, R. H. Fusiform dilation and thrombosis of arteries following the application of methyl 2-cyanoacrylate (Eastman 910 monomer). *J. Neurosurg.*, 23, 556–564 (1965).
102. Robicsek, F., Reilly, J. P., Marroum, M. C. The use of cyanoacrylate adhesive (Krazy Glue) in cardiac surgery. *J. Card. Surg.*, 9, 353–356 (1994).
103. Tebala, G. D., Ceriati, F., Ceriati, E., Vecchioli, A., Nori, S. The use of cyanoacrylate tissue adhesive on high risk intestinal anastomoses. *Jpn. J. Surg.*, 25, 1069–1072 (1995).
104. Ayton, J. M. Polar hands: spontaneous skin fissures closed with cyanoacrylate (histoacryl blue) tissue adhesive in antarctica. *Antarctica Arctic Med. Res.*, 52, 127–130 (1993).
105. Quinn, J., Wells, G., Sutcliffe, T., Jarmuske, M., Maw, J., Stiell, I., Johns, P. A. Randomized trial comparing octyl-

cyanoacrylate tissue adhesive and sutures in the management of lacerations. *JAMA*, 277, 1527–1530 (1997).
106. McDonnell, P. F. Sterilized cyanoacrylate adhesive composition and a method of making such a composition. European patent EP 659,441 (1993).
107. Rauh, E. Hose pump for the exact dosing of small quantities of liquids. U.S. patent 5,693,020 (1997).
108. Al-Khawam, E. M., Brewis, D. M., Glasse, M. D. Cyanoacrylate adhesives of potential medical use. In: *Adhesion 7* (Allen, K. W., ed.). Applied Science Publishers, London and New York, 109–133 (1983).
109. Vanholder, R., Misotten, A., Roels, H., Matton, G. Cyanoacrylate tissue adhesive for closing skin wounds: a double blind randomised comparison with sutures. *Biomaterials*, 14, 737–742 (1993).
110. Papatheofanis, F. J. Prothrombotic cytotoxicity of cyanoacrylate tissue adhesive. *J. Surg. Res.*, 47, 309–312 (1989).
111. Harper, M. C., Ralston, M. Isobutyl-2-cyanoacrylate as an osseous adhesive in the repair of osteochondral fractures. *J. Biomed. Mater. Res.*, 17, 167–177 (1983).
112. Bonutti, P. M., Weiker, G. G., Andrish, J. T. Isobutyl cyanoacrylate as a soft tissue adhesive. *Clin. Orthop. Rel. Res.*, 229, 241–248 (1988).
113. Tseng, Y., Hyon, S., Ikada, Y., Shimizu, Y., Tamura, K., Hitomi, S. In vivo evaluation of 2-cyanoacrylates as surgical adhesives. *J. Appl. Biomater.* 1, 111–119 (1990).
114. Lenaerts, V., Couvreur, P., Christiaens-Leigh, D., Jorris, E., Roland, M., Rollman, B., Speiser, P. Degradation of poly(isobutyl cyanoacrylate) nanoparticles. *Biomaterials*, 5, 65–68 (1984).
115. Diaz, F. G., Mastri, A. R., Chou, S. N. Neural and vascular tissue reaction to aneurysm-coating adhesive (ethyl-2-cyanoacrylate). *Neurosurg*, 3, 45–49 (1978).
116. Ekelund, A., Nilsson, O. S. Tissue adhesives inhibit experimental bone formation. *Int. Orthop.*, 15, 331–334 (1991).
117. Watson, D. P. Use of cyanoacrylate tissue adhesive for closing facial lacerations in children. *Brit. Med. J.*, 299, 1014 (1989).
118. Leonard, F., Kulkarni, R. K., Brandes, G., Nelson, J., Cameron, J. J. Synthesis and degradation of poly(alkyl α-cyanoacrylates). *J. Appl. Poly. Sci.*, 10, 259–272 (1966).
119. Vezin, W. R., Florence, A. T. In vitro heterogeneous degradation of poly(n-alkyl α-cyanoacrylates). *J. Biomed. Mater. Res.*, 14, 93–106 (1980).
120. Wade, C. W. R., Leonard, F. Degradation of poly(methyl 2-cyanoacrylates). *J. Biomed. Mater. Res.* 6, 215–220 (1972).
121. Formaldehyde and other aldehydes. (U.S.) National Research Council, Washington DC (1982).
122. Conolly, R. B., Anderson, M. E. An approach to mechanism-based cancer risk assessment for formaldehyde. *Environ. Health Perspect.*, 101, 169–176, (1993).
123. Wang, P. Y. Adhesion mechanism for polyurethane prepolymers bonding biological tissue. Ed., Gregor, H. P. In: *Biomedical Applications of Polymers, Polymer Science and Technology*. New York, Plenum Press, 7, 111 (1975).
124. Salvatore J. E., Mandarino, M. P. Polyurethane polymer, its use in osseous lesions: an experimental study. *Ann. Surg.*, 149, 107–109 (1959).
125. Salvatore, J. E., Gilmer, W. S., Kashgarian, M., Barbee, W. R. An experimental study of the influence of pore size of implanted polyurethane sponges upon subsequent tissue formation. *Surg. Gyn. Obst.*, 112, 463 (1961).
126. Kobayashi, H., Hyon, S.-H., Ikada, Y. Water-curable and biodegradable prepolymers. *J. Biomed. Mater. Res.*, 25, 1481–1494 (1991).
127. Tatooles, C. J., Braunwald, N. S. The use of crosslinked gelatin as a tissue adhesive to control hemorrhage from liver and kidney. *Surgery*, 60(4), 857–861 (1966).
128. Braunwald, N. S., Gay, W., Tatooles, C. J. Evaluation of crosslinked gelatin as a tissue adhesive and hemostatic agent: an experimental study. *Surgery*, 59(6), 1024–1030 (1966).
129. Cooper, C. W., Falb, R. D. Surgical adhesives. *Ann. NY Acad. Sci.*, 146(1), 214–224 (1968).
130. Ikada, Y. Tissue adhesives. In: *Wound Closure Biomater. Devices* (Chu, C., Von Fraunhofer, J. A., Greisler, H. P., eds.). CRC Press, 317–346 (1997).
131. Guilmet, D., Bachet, J., Goudot, B., Laurian, C., Gigou, F., Bical, O., Barbegelatta, M. Use of biological glue in acute aortic dissection. Preliminary clinical results with a new surgical technique. *J. Thorac. Cardiovasc. Surg.*, 77(4), 516–521 (1979).
132. Goldmacher, V. S., Thilly, W. G. Formaldehyde is mutagenic for cultured human cells. *Mutation Res.*, 116(3–4), 417–422 (1983).
133. Ennker, J., Ennker, I. C., Schoon, D., Schoon, H. A., Dörge, S., Meissler, M., Rimpler, M., Hetzer, R. The impact of gelatin-resorcinol glue on aortic tissue: a histomorphologic evaluation. *J. Vasc. Surg.*, 20, 34–43 (1994).
134. Ennker, I. C., Ennker, J., Schoon, D., Schoon, H. A., Rimpler, M., Hetzer, R. Formaldehyde-free collagen glue in experimental lung gluing. *Ann. Thorac. Surg.*, 57(6), 1622–1627 (1994).
135. Sawhney, A. S., Pathak, C. P., Hubbell, J. A. Bioerodible hydrogels based on photopolymerized poly(ethylene glycol)-*co*-poly(α-hydroxy acid) diacrylate macromers. *Macromolecules*, 26, 581–587 (1993).
136. Sawhney, A. S., Poff, B., Powell, M., Messier, K., Doherty, E., Yao, F., Enscore, D. J., Jarrett, P. K. Bioabsorbable synthetic hydrogel as a surgical lung sealant. *Proc. Am. Chem. Soc.*, 79, 256 (1998).
137. Grey, E. G. Fibrin as a haemostatic in cerebral surgery. *Surg. Gynecol. Obstet.* 21, 452–454 (1915).
138. Bergel, S. Ueber Wirkungen des Fibrins. *Dtschr. Med. Wochenschr.*, 35, 663–665 (1909).
139. Fricke, W. A. Fibrin sealant and the US Food and Drug Administration review process. In: *Surgical Adhesives and Sealants* (Sierra, D., Saltz, R., eds.). Technomic, 17, 163–172 (1996).
140. Radosevich, M., Goubran, H. A., Burnouf T. Fibrin sealant: scientific rationale, production methods, properties, and current clinical use. *Vox Sang*, 72(3), 133–143 (1997).
141. Causten, B. E. In: *Materials Science Technology*. (Wil-

liams, D. F., ed.). VCH, Weinheim, Germany, 14, 285–302 (1992).

142. Forseth, M., O'Grady, K., Toriumi, D. M. The current status of cyanoacrylate and fibrin tissue adhesives. *J. Long-Term Effects Medical Implants*, 2(4), 221–233 (1992).
143. Dresdale, A., Bowman, F. O., Malm, J. R., Reemtsma, K., Smith, C. R., Spotnitz, H. M., Rose, E. A. Hemostatic effectiveness of fibrin glue derived from single-donor fresh frozen plasma. *Ann. Thorac. Surg*., 40(4), 385–387 (1985).
144. Epstein, G. H., Weisman, R. A., Zwillenberg, S., Schreiber, A. D. A new autologous fibrinogen-based adhesive for otologic surgery. *Ann. Otol. Rhinol. Laryngol*., 95(1), 40–45 (1986).
145. Siedentop, K. H., Harris, D. M., Ham, K., Sanchez, B. Extended experimental and preliminary surgical findings with autologous fibrin tissue adhesive made from patient's own blood. *Laryngoscope*, 96(10), 1062–1064 (1986).
146. Siedentop, K. H., Harris, D. M., Sanchez, B. Autologous fibrin tissue adhesive: factors influencing bonding power. *Laryngoscope*, 98(7), 731–733 (1988).
147. Martinowitz, U., Saltz, R. Fibrin sealant. *Current Opinion in Hematology*, 3(5), 395–402 (1996).
148. Martinowitz, U., Spotnitz, W. D. Fibrin tissue adhesives. *Thrombosis Haemostasis*, 78(1), 661–666 (1997).
149. Sierra, D. H. Fibrin sealant adhesive systems: a review of their chemistry, material properties and clinical applications. *J. Biomater. Appl*., 7(4), 309–352 (1993).
150. Pipan, C. M., Glasheen, W. P., Matthew, T. L., Gonias, S. L., Hwang, L.-J., Jane, J. A., Spotnitz, W. D. Effects of antifibrinolytic agents on the life span of fibrin sealant. *J. Surgical Research*, 53, 402–407 (1992).
151. Martinowitz, U., Schulman, S., Horoszowski, H., Heim, M. Role of fibrin sealants in surgical procedures on patients with hemostatic disorders. *Clin. Orthopaedics Related Research*, 328, 65–75 (1996).
152. Gladner, J. A., Nossal, R. Effects of crosslinking on the rigidity and proteolytic susceptibility of human fibrin clots. *Thrombosis Research*, 30, 273–288 (1983).
153. Marx, G. Kinetic and mechanical parameters of fibrin glue. In: *Surgical Adhesives and Sealants* (Sierra, D., Saltz, R., eds.). Technomic, Ch. 6, 49–59 (1996).
154. Kjaergard, H. K., Weis-Fogh, U.S. Important factors influencing the strength of autologous fibrin glue; the fibrin concentration and reaction time—comparison of strength with commercial fibrin glue. *Eur. Surg. Res*., 26, 273–276 (1994).
155. Bass, L. S., Libutti, S. K., Knayton, M. L., Nowygrod, R., Treat, M. R. Soldering is a superior alternative to fibrin sealant. In: *Surgical Adhesives and Sealants* (Sierra, D., Saltz, R., eds.). Technomic, Ch. 5, 41–46 (1996).
156. Lontz, J. F., Verderamo, J. M., Camac, J., Arikan, I., Arikan, D., Lemole, G. M. Assessment of restored tissue elasticity in prolonged in vivo animal tissue healing: comparing fibrin sealant to suturing. In: *Surgical Adhesives and Sealants* (Sierra, D., Saltz, R., eds.). Technomic, Ch. 9, 79–90 (1996).
157. Jackson, M. R., MacPhee, M. J., Drohan, W. N., Alving, B. M. Fibrin sealant: current and potential clinical applications. *Blood Coagulation and Fibrinolysis*, 7(8), 737–746 (1996)
158. MacPhee, M. J., Singh, M. P., Brady, R., Akhyani, N., Liau, G., Lasa, C., Hue, C., Best, A., Drohan, W. Fibrin sealant: a versatile delivery vehicle for drugs and biologics. In: *Surgical Adhesives and Sealants* (Sierra, D., Saltz, R., eds.). Technomic, Ch. 12, 109–120 (1996).
159. van der Ham, A. C., Kort, W. J., Weijma, I. M., van den Ingh, H. F. G., Jeekel, H. Effect of antibiotics in fibrin sealant on healing colonic anastomoses in the rat. *Br. J. Surg*., 79, 525–528 (1992).
160. Thompson, D. F., Davis, T. W. The addition of antibiotics to fibrin glue. *Southern Medical J*., 90(7), 681–684 (1997).
161. Singh, M. P., Brady, R., Drohan, W., MacPhee, M. J. Sustained release of antibiotics from fibrin sealant. In: *Surgical Adhesives and Sealants* (Sierra, D., Saltz, R., eds.). Technomic, Ch. 13, 121–133 (1996).
162. MacPhee, M. J., Campagna, A., Best, A., Kidd, R., Drohan, W. Fibrin sealant as a delivery vehicle for sustained and controlled release of chemotherapy agents. In: *Surgical Adhesives and Sealants* (Sierra, D., Saltz, R., eds.). Technomic, Ch. 15, 145–154 (1996).
163. Lasa, C., Hollinger, J., Drohan, W., MacPhee, M. J. Bone induction by demineralized bone powder and partially purified osteogenin using a fibrin-sealant carrier. In: *Surgical Adhesives and Sealants*, (Sierra, D., Saltz, R., eds.). Technomic, Ch. 14, 135–144 (1996).
164. Marchac, D., Sandor, G. Face lifts and sprayed fibrin glue: an outcome analysis of 200 patients. *Br. J. Plastic Surgery*, 47, 306–309 (1994).
165. Gleich, L. L., Reeiz, E. E., Pankratov, M. M., Shapshay, S. M. Autologous fibrin tissue adhesive in endoscopic sinus surgery. *Otolaryngology—Head and Neck Surgery*, 112(2), 238–241 (1995).
166. Saltz, R., Dimick, A., Harris, C., Grotting, J. C., Psillakis, J., Vasconez, L. O. Application of autologous fibrin glue in burn wounds. *J. Burn Care Rehabilitation*, 10(6), 504–507 (1989).
167. Saltz, R. Clinical applications of tissue adhesives in aesthetic and reconstructive surgery. In: *Surgical Adhesives and Sealants* (Sierra, D., Saltz, R., eds.). Technomic, Ch. 20, 187–203 (1996).
168. Himel, H. N. Tissue adhesives in plastic and reconstructive surgery. In: *Surgical Adhesives and Sealants* (Sierra, D., Saltz, R., eds.). Technomic, Ch. 19, 177–185 (1996).
169. Brown, D. M., Barton, B. R., Young, V. L., Pruitt, B. A. Decreased wound contraction with fibrin glue-treated skin crafts. *Arch. Surg*., 127, 404–406 (1992).
170. Siedentop, K. H. Surgical adhesives in otolaryngology. In: *Surgical Adhesives and Sealants* (Sierra, D., Saltz, R., eds.). Technomic, Ch. 21, 205–219 (1996).
171. Terzis, J. K. A review of nonsuture peripheral nerve repair. In: *Surgical Adhesives and Sealants*, (Sierra, D., Saltz, R., eds.). Technomic, Ch. 22, 221–232 (1996).

172. Spotnitz, W. D. Clinical applications of fibrin sealant in thoracic and cardiovascular surgery. In: *Surgical Adhesives and Sealants* (Sierra, D. Saltz, R., eds.). Technomic, Ch. 24, 239–242 (1996).
173. Rakocz, M., Mazar, A., Varon, D., Spierer, S., Blinder, D., Martinowitz, U., Hashomer, T. Dental extractions in patients with bleeding disorders: the use of fibrin glue. *Oral Surgery Oral Medicine Oral Pathology*, 75(3), 280–282, (1993).
174. Waite, J. H. Adhesion in byssally attached bivalves. *Biol. Rev. Cambridge Philos. Soc.*, 58(2), 209–231 (1983).
175. Waite, J. H. Nature's underwater adhesive specialist. *Int. J. Adhesion Adhesives*, 7(1), 9–14 (1987).
176. Waite, J. H. Process for purifying and stabilizing catechol-containing proteins and materials obtained thereby. U.S. patent 4,496,397 (1985).
177. Waite, J. H. Decapeptides produced from bioadhesive polyphenolic proteins. U.S. patent 4,585,585 (1986).
178. Waite, J. H. Decapeptides produced from bioadhesive polyphenolic proteins. U.S. patent 4,678,740 (1987).
179. Waite, J. H. Decapeptides produced from bioadhesive polyphenolic proteins. U.S. patent 4,808,702 (1989).
180. Benedict, C. V., Picciano, P. T. Bioadhesives for cell and tissue adhesion. U.S. patent 5,108,923 (1992).
181. Strausberg, R. L., Link, R. Protein-based medical adhesives. *Trends Biotechnol.*, 8(2), 53–57 (1990).
182. Waite, J. H. Polypeptide monomers, linearly extended and/or crosslinked forms thereof, and applications thereof. U.S. patent 5,574,134 (1996).
183. Maugh, K. J., Anderson, D. M., Strausberg, R., Strausberg, S. L., McCandliss, R., Wei, T., Filpula, D. Bioadhesive coding sequences. U.S. patent 5,049,504 (1991).
184. Maugh, K. J., Anderson, D. M. *Escherichia coli* expression vector encoding bioadhesive precursor protein analogs comprising three to twenty repeats of the decapeptide (ALA-LYS-PRO-SER-TYR-PRO-PRO-THR-TYR-LYS). U.S. patent 5,149,657 (1992).
185. Benedict, C. V., Chaturvedi, N. Synthetic amino acid and/or peptide-containing graft copolymers. U.S. patent 4,908,404 (1990).
186. Chivers, R. A., Wolowacz, R. G. The strength of adhesive-bonded tissue points. *Int. J. Adhesion Adhesives*, 17(2), 127–132 (1997).
187. Strausberg, R. L., Strausberg, S. L. Composite yeast vectors. U.S. patent 5,013,652 (1991).
188. Deming, T. J., Yu, M. Synthetic polypeptide mimics of marine adhesives. *Macromolecules*, 31, 4739–4745 (1998).
189. Yamamoto, H. Synthesis and adhesive studies of marine polypeptides. *J. Chem. Soc., Perkin Trans.*, 1(3), 613–618 (1987).
190. Yu, M., Hwang, J., Deming, T. J. Role of L-3,4-dihydroxyphenylalanine in mussel adhesive proteins. *J. Am. Chem. Soc.*, 121, 5825–5826 (1999).
191. von Oppell, U. O., Zilla, P. Tissue adhesives in cardiovascular surgery. *J. Long Term Effects Medical Implants*, 8(2), 87–101 (1998).

27

Glucose-Sensitive Hydrogel Membranes

Jin Ho Lee, Jung Ju Kim, and Kinam Park
Purdue University, West Lafayette, Indiana

I. INTRODUCTION

A. Hydrogels

Hydrogels are three-dimensional networks of hydrophilic polymers that can absorb large amounts of water (usually more than 20% of the total weight) but remain insoluble (1–7). Three-dimensional networks are usually formed by chemical or physical cross-linking of hydrophilic polymer chains. In chemical gels, polymer chains are connected by covalent bonds, and thus it is difficult to change the shape of chemical gels. On the other hand, polymer chains of physical gels are connected through noncovalent bonds, such as van der Waals interactions, ionic interactions, hydrogen bonding, hydrophobic interactions, traces of crystallinity, and multiple helices. Since these bonds are reversible, physical gels possess sol–gel reversibility. The key characteristic property of hydrogels is the reversible swelling/deswelling in aqueous solution. The hydrogels swell in aqueous solution because hydrophilic polymer chains of the cross-linked network try to dissolve in water. Upon absorption of water (or swelling), the glassy polymer network becomes elastic. The extent of swelling is determined by the nature of polymer chains and cross-linking density. The extent of swelling is quantified by the ratio of the volume (or weight) of the swollen hydrogel (V_s) to the volume (or weight) of the dried hydrogel (V_d). Thus the swelling ratio (Q) is defined as V_s/V_d.

B. Environment-Sensitive Hydrogels

All hydrogels have the same fundamental property of swelling in the presence of water and deswelling (or shrinking) in the absence of water. Some hydrogels have additional properties, such as swelling or shrinking, in response to changes in environmental conditions. The presence of thermodynamically active functional groups on polymer chains makes the hydrogels sensitive to certain stimulants, or environmental factors (8). These hydrogels with additional functions are collectively called ''environment-sensitive'' hydrogels or ''stimuli-responsive'' hydrogels. Compared with ordinary hydrogels without such additional properties, environment-sensitive hydrogels are more advanced, and for this reason they are also called ''smart'' or ''intelligent'' hydrogels. One of the unique properties of these types of hydrogels is that they change their swelling ratio rather abruptly upon small changes in environmental factors. This abrupt volume change is known as volume transition. If the volume transition occurs to make the volume smaller, it is called ''volume collapse.'' The volume transition of smart hydrogels can occur by only a small change in environmental conditions, such as pH, temperature, ionic strength, electric field, magnetic field, ultrasound, solvents, electrolytes, external stress, light, pressure, specific molecules, or enzymes (8–12). It is this unique property of smart hydrogels that makes them useful in various areas, such as pharmaceutical, biomedical, and biotechnological applications.

1. pH-Sensitive Hydrogels

Of the many smart hydrogels, pH-sensitive hydrogels have been most frequently used in glucose-sensitive insulin release devices. pH-sensitive hydrogels are cross-linked polyelectrolytes (i.e., polymers with a large number of ionizable groups) that display big differences in swelling properties depending on the pH of the environment. When a polymer becomes charged, its hydrogel swells substantially more than at the neutral state. Figure 1 shows examples of pH-sensitive polymers. Hydrogels made of polyanions, such as poly(acrylic acid) (PAA), swell less as pH becomes lower due to the loss of charges. On the other hand, polycations, such as poly(N,N′-diethylaminoethyl methacrylate) (PDAEM) swell substantially more at lower pH due to the generation of charges on the polymer chains. Both types of polyelectrolyte hydrogels have been used for preparing glucose-sensitive hydrogels.

2. Glucose-Sensitive Hydrogels

Some of the environment-sensitive hydrogels are glucose-sensitive hydrogels that can be used for glucose sensing and modulated (or self-regulating) insulin delivery. The glucose-sensitive hydrogels undergo changes in the swelling ratio or changes in physical states (i.e., sol and gel) in response to small changes in the glucose concentration in the environment (13). Development of glucose-sensitive hydrogels is critical in the development of self-regulating

Low pH High pH

$$-\!\!\left(CH_2-CH(COOH)\right)\!\!- \;\underset{H^+}{\overset{OH^-}{\rightleftharpoons}}\; -\!\!\left(CH_2-CH(COO^-)\right)\!\!-$$

$$-\!\!\left(CH_2-CH(C(=O)-O-CH_2CH_2N^+H(CH_2CH_3)_2)\right)\!\!- \;\underset{H^+}{\overset{OH^-}{\rightleftharpoons}}\; -\!\!\left(CH_2-CH(C(=O)-O-CH_2CH_2N(CH_2CH_3)_2)\right)\!\!-$$

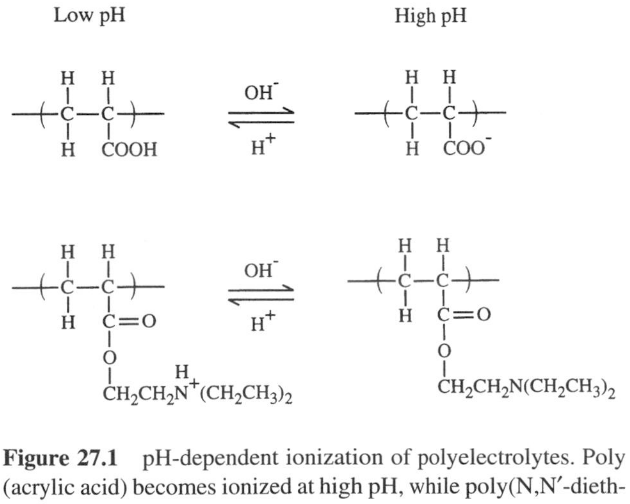

Figure 27.1 pH-dependent ionization of polyelectrolytes. Poly (acrylic acid) becomes ionized at high pH, while poly(N,N′-diethylaminoethyl methacrylate) becomes ionized at low pH. Ionized hydrogels swell more due to the presence of charges.

insulin delivery systems. Unlike other controlled release drug delivery systems, the insulin delivery systems require monitoring of glucose levels and the release of insulin just enough to reduce the elevated blood glucose level. Since the amount of insulin to be delivered will be different each time and depend on the glucose level, this type of self-regulating insulin delivery system presents the ultimate challenge in controlled drug delivery technologies. This chapter reviews various glucose-sensitive hydrogels that have been developed for glucose sensing and self-regulated insulin delivery.

II. GLUCOSE-SENSITIVE HYDROGELS FOR SELF-REGULATED INSULIN DELIVERY

Conventional methods of controlled drug delivery are mainly based on diffusion at a certain rate through polymer membranes or matrices. This mode of drug delivery results in relatively constant drug levels in blood, but it may not be suitable for insulin delivery. For management of diabetes, the level of insulin delivery has to be adjusted depending on the blood glucose level. Daily injections of insulin are known to be inadequate for maintaining normal blood glucose levels and preventing long-term complications of diabetes, and this increases demands for self-regulated insulin delivery systems. Since insulin release has to be in harmony with increase in the blood glucose level, the most desirable insulin delivery system requires a glucose-sensing ability and an ability to trigger release of the necessary amount of insulin (14). The methods that have been used for development of glucose-sensitive insulin delivery systems can be divided into biological and bioengineering approaches. In the biological approach, cultured, living pancreatic β-cells are used after encapsulation in polymeric membranes. Improvement has to be made for the long-term survival of encapsulated cells for this approach to be practical. In the bioengineering approach, mechanical glucose sensor and insulin delivery pumps are used. This approach has to overcome substantial hurdles in terms of reliability and convenience.

Recently, many researchers have focused on an alternative approach utilizing glucose-sensitive hydrogels. For the hydrogels to possess glucose sensitivity, they have to contain molecules that specifically interact with glucose molecules. Currently used glucose-sensitive molecules are concanavalin A (Con A), phenylboronic acid, glucose oxidase, and glucose dehydrogenase. These molecules have been incorporated into hydrogels for glucose sensing and modulated insulin delivery.

A. Con A as a Glucose Responsive Unit

1. Complexes of Con A and Glycosylated Insulin

a. *Initial Concept*

This approach, which was first introduced by Brownlee and Cerami (15,16), is based on competitive binding between glucose and glycosylated insulin for carbohydrate specific binding sites on Con A. Con A is a globular lectin that is composed of 237 amino acid residues and four polypeptide subunits. Each subunit has a molecular weight of 26,000 daltons and a dimension of 44 × 40 × 39 Å^3 (17,18). The glycosylated insulin molecules can be bound to each subunit of Con A and reversibly displaced from Con A by glucose in direct proportion to the concentration of external glucose. In this approach, the Con A/glycosylated insulin complex is enclosed in a polymer membrane that is permeable to glucose and glycosylated insulin, but not to Con A or antibodies. It is important to prevent release of Con A from the membrane of the device, because Con A is known to be immunotoxic (19–22). When the blood glucose level rises, glucose enters the reservoir and competitively displaces glycosylated insulin from Con A (23). The glycosylated insulin is then released by diffusion through the membrane into the body as shown in Fig. 2.

While the idea behind this approach is strikingly elegant and serves as a role model for other approaches, it has a few inherent limitations. First, each insulin molecule has to be modified with glucose, and this creates a new chemical entity. The glycosylated insulin is not the same as native insulin, and it has to go through the approval process as a new chemical entity. Second, Con A is a relatively large molecule but has only four glucose binding sites. This creates a problem of small reservoir capacity that is not suitable for long-term delivery. The binding constant of the glycosylated insulin to Con A needs to be higher than that of glucose to prevent easy displacement of glycosylated insulin with consequent danger of hypoglycemia.

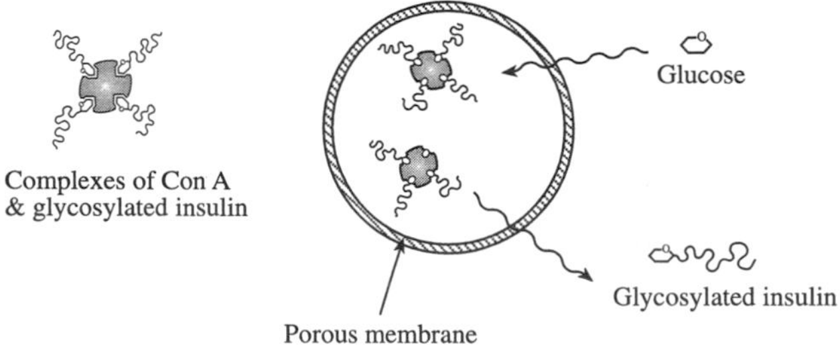

Figure 27.2 Schematic representation of self-regulated insulin delivery system.

b. *Various Insulin Formulations*

The glycosylated insulin approach has been further investigated by Kim and his coworkers (23–29). Various insulin derivatives having variable binding constants to Con A were synthesized and examined for their stability and biological activity. As shown in Fig. 2, the complexes of Con A and insulin derivatives are enclosed in a porous polymer membrane. Initially, poly(hydroxyl methacrylate) (polyHEMA) membrane was used, but it was replaced later with cellulose acetate tubing or thin nylon membrane due to its weak mechanical strength.

The main drawbacks of the encapsulation system were leakage of potentially immunogenic Con A through the polymer membrane and the slow onset of insulin release when the device was challenged by high glucose concentration. To avoid the leakage, Con A was immobilized to Sepharose beads, which were then enclosed in a macroporous membrane (30). This device still suffered in performance due to the settling of the beads by gravity and the bulk size of the beads. A cross-linked Con A gel was also synthesized to further prevent Con A leakage through the membrane (30). The long unwanted lag phase for the onset of glycosylated insulin release was partly the result of the limited solubility of Con A/glycosylated insulin complex (31). The limited solubility was due to the tetrameric nature of Con A and the formation of dimers or hexamers by glycosylated insulin derivatives (32,33). In an effort to prevent dimerization or hexamerization of insulin derivatives, insulin was derivatized with gylcosyl polyethylene glycol (PEG) for improved solubility and solution stability at physiological pH (34). To reduce the response time, the surface area was also increased by applying microcapsules and microspheres (30,35,36). A series of improvements made in this system is shown in Table 1.

2. Complexes of Con A and Glucose-Containing Polymers

A Con A-loaded hydrogel system was shown to undergo swelling and deswelling in response to different saccharides (38). The gel consisted of a covalently cross-linked network of poly(N-isopropyl acrylamide) and physically entrapped Con A associated with dextran sulfate. Since the hydrogel swelled due to ionic osmotic pressure induced by the anionic inclusion, it shrank when the dextran sulfate was displaced from the Con A binding site by uncharged saccharides.

Con A also reacts with glucose-containing polymers to form precipitates (39,40) or hydrogels (41). The hydrogels made of Con A and poly(2-glucosyloxyethyl methacrylate) swelled in the presence of glucose due to the dissociation

Table 27.1 Self-Regulated Insulin Delivery Systems Using Con A as a Glucose Sensing Unit

	Binding substrate	Insulin derivative	Membrane	Comments
I	Soluble Con A tetramer	SAPG-insulin SAPM-insulin	PolyHEMA, cellulose acetate.	Con A leakage. Decreased permeability of glucose and G-insulin resulting in long lag times.
II	Con A immobilized onto Sepharose beads	SAPG-insulin SAPM-insulin	Cellulose acetate, Necleopore membranes.	Con A leakage. Settling of Con A beads. Poor response by glucose increase.
III	Con A hydrogel	SAPG-insulin	Durapore membrane fabricated into a pouch	No Con A leakage. Increased permeability of glucose and G-insulin. Heat sealable membrane for easy fabrication.
IV	Con A microcapsules	SAPG-insulin	Durapore membrane fabricated into a pouch	Microcapsules (30–250 μm containing Con A/SAPG-insulin resulted in short response time.
V	Soluble Con A oligomer	SAPG-insulin	Hollow fiber recirculation system	Con A oligomers were not permeable to hollow fiber membrane. Short lag time. Refillable system.

G-insulin: glycosylated insulin.
SAPG-insulin: succinyl amidophenyl glucopyranosyl-insulin.
SAPM-insulin: succinyl amidophenyl mannopyranosyl-insulin.
polyHEMA: poly(hydroxyethyl methacrylate).
Source: Ref. 37.

of the Con A–polymer complex by competitive binding of free glucose. The hydrogels made of Con A and polysucrose complexes were used as rate-determining membranes for release of a solute in the reservoir as a function of the glucose concentration (42–44). The solute release through the membrane was controlled by the viscosity change resulting from the gel-to-sol transition of Con A/polysucrose complexes.

3. Sol–Gel Phase-Reversible Hydrogels

Con A and glucose-containing polymers were used to form stable hydrogels undergoing phase transition between sol and gel (14,45–48). To prepare phase-reversible glucose-sensitive hydrogels, Con A was used as a physical cross-linking agent for glucose-containing polymers, such as poly(allyl glucose-co-acrylamide) (poly(AG-co-AM)), poly(allyl glucose-co-vinylpyrrolidone) (poly(AG-co-VP)), and poly(allyl glucose-co-3-sulfopropylacrylate potassium) (poly(AG-co-SPAK). The gel–sol phase transition of this system is schematically described in Fig. 3 (46). In the absence of free glucose, Con A interacts with polymer-bound glucose to form a hydrogel. The increase in free glucose, however, results in detachment of polymer-bound glucose from Con A and thus transition to a sol. Upon removal of free glucose, the sol becomes a gel again, and this process can be repeated. In the sol state, the increased mobility of polymer chains allows solute molecules to diffuse more easily. Indeed, the release of insulin was shown to be faster in the sol state of the glucose-sensitive hydrogel membrane (14). The free glucose concentration necessary to induce phase transition of the hydrogels can

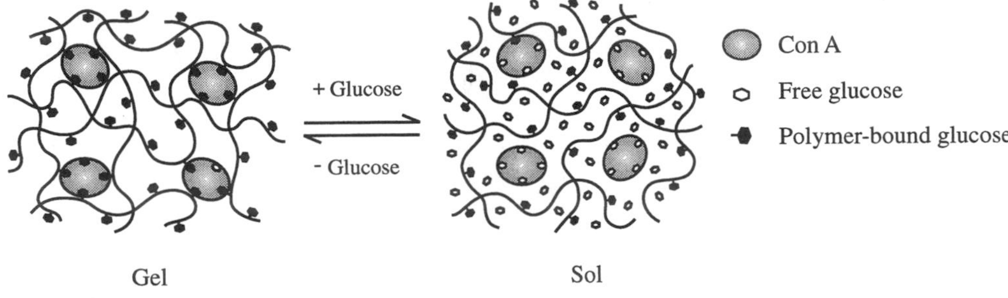

Figure 27.3 A gel–sol phase-reversible hydrogel system sensitive to glucose. (From Ref. 46.)

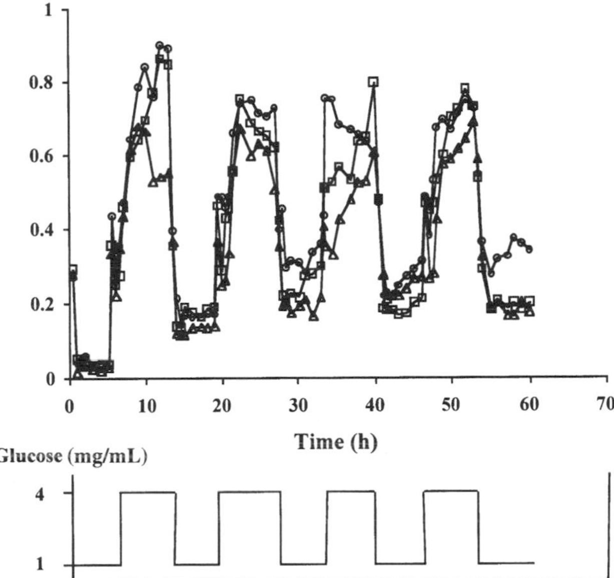

Figure 27.4 Insulin release profiles from glucose-sensitive hydrogel matrices. The matrices were prepared by MPEG5–Con A conjugate (14 w/v%) and different glucose-containing polymers. Δ, poly(AG-co-AM) (2 w/v%); □, poly(AG-co-VP2) (2 w/v%); ○, poly(AG-co-SPAK3) (3 w/v%). (From Ref. 49.)

be made within the physiological range (which is 1 ~ 4 mg/mL) by adjusting the molar ratio of Con A to glucose-containing polymer.

One of the difficulties in preparation of Con A–based glucose-sensitive hydrogels is the poor water solubility of Con A. To alleviate this problem, Con A was conjugated with monomethoxy poly(ethylene glycol) (MPEG) (49). The MPEG–Con A conjugate showed increased solubility and stability in aqueous solution. The hydrogels made of MPEG-Con A showed pulsatile release of insulin in response to the changes in glucose concentration between 1 mg/mL and 4 mg/mL as shown in Fig. 4 (49). The insulin release at the glucose concentration of 4 mg/mL, representing the hyperglycemic level, was several times higher than that at 1 mg/mL of glucose, representing the normal blood glucose level. Such pulsatile release of insulin can be continued as long as the membrane remains intact.

B. Phenylboronic Acid as a Glucose-Responsive Unit

In addition to Con A, phenylboronic acid (PBA) was used as a glucose-responsive unit. PBA is known to have a reversible binding property to dihydroxyl compounds when the two hydroxyl groups are in a coplanar configuration (50–53). Since most carbohydrates, including glucose, possess a cis-diol moiety in the structure, they can form a relatively strong complex with borate. A polymer having pendant PBA groups can form a gel with a polyol, such as polyvinyl alcohol (PVA), through the complex formation. When glucose is added to this complex gel, the gel swells and dissolves due to a decrease in the cross-linking density caused by the substitution reaction of glucose with the pendant hydroxyl groups of the polymer. This swelling results in increased release of loaded insulin. Reduction in free glucose concentration leads to reformation of borate–polyol cross-linking in the complex gel with subsequent reduction in insulin release. This concept was used to prepare glucose-sensitive insulin delivery formulations based on a reservoir system (54–57). An example of the reservoir system is schematically shown in Fig. 5. Poly(vinylpyrrolidone-co-*m*-acrylamidophenylboronic acid) is commonly used as a PBA-containing polymer. One of the problems of this system is that the complex gel is sensitive to glucose only under alkaline conditions (pH > ~9.0). This problem was partly overcome by incorporating amino groups into the PBA polymer (57).

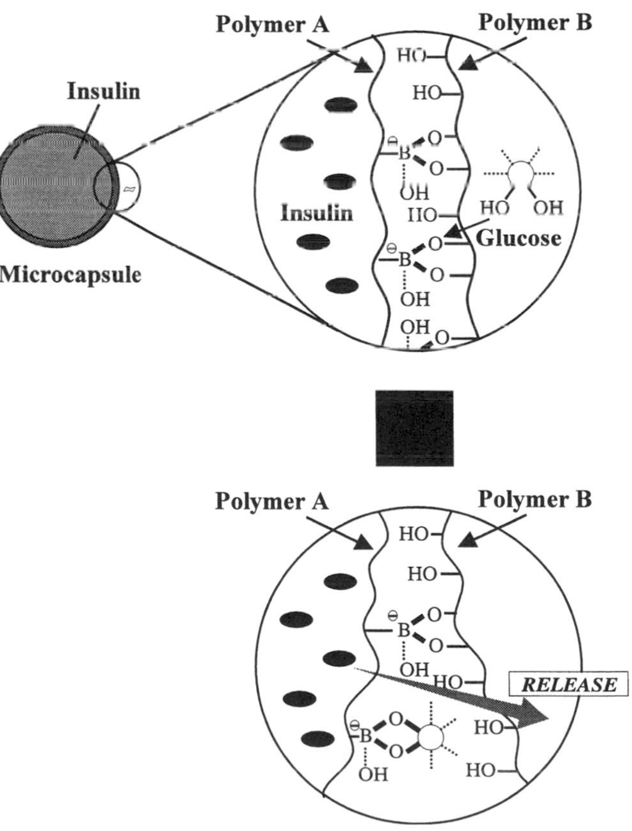

Figure 27.5 Concept of polymer capsule type glucose-sensitive insulin delivery system using PVA/poly(VP-co-PBA) complex. Polymer A, poly(VP-co-PBA); polymer B, PVA. (From Ref. 55.)

The borate–polyol cross-linked gel was also used as a glucose-sensitive erodible matrix system (58). In this erodible system, a gel is formed between *m*-aminophenylboronic acid-substituted polyacrylamide (poly(PBA-co-AM)) and diglucosyl hexanediamine (DGHDA). In the presence of free glucose, the gel matrix is disrupted by replacement of DGHDA with glucose through competitive binding to PBA, and this results in solubilization of the matrix and a significant increase in the insulin release rate. One of the advantages of using PBA is that it is a synthetic molecule that may not be immunotoxic as Con A is. For this system to be practical, however, the glucose-specificity of the PBA moiety has to be improved.

C. Glucose Oxidase as a Glucose Responsive Unit

pH-sensitive hydrogels (such as those shown in Fig. 1) can be used to prepare glucose-sensitive insulin delivery systems. In this approach, pH-sensitive hydrogels usually contain immobilized glucose oxidase (GOD), which converts glucose to gluconic acid in the presence of oxygen:

$$\text{Glucose} + O_2 + H_2O \xrightarrow{\text{GOD}} \text{Gluconate-} + H^+ + H_2O_2$$

The formation of acid results in lowering of the pH of the medium, and pH-sensitive hydrogels change their swelling ratio, i.e., the pore sizes inside the hydrogels change to affect the diffusion of solutes such as insulin molecules. Since the mechanism of insulin release depends on the type of polyelectrolyte used, this approach can be further classified based on the charge type used.

1. Hydrogels Made of Cationic Polyelectrolytes

Commonly used cationic polyelectrolytes are polymers made of cationic monomers, such as N,N-dimethylaminoethyl methacrylate (DMAEM) and N,N-diethylaminoethyl methacrylate (DEAEM) (59–66). Hydrogels made of these polymers undergo swelling as the pH is lowered as a result of formation of gluconic acid (Fig. 6). The increased swelling, in turn, results in more release of insulin. One of the problems of this system is the leveling off of the response as the glucose concentration is increased above 0.5 mg/mL, while the glucose concentration in the body can be as high as 10 mg/ml (65). This limited glucose sensitivity is due to the fact that GOD is a flavoprotein. GOD requires an electron acceptor, such as oxygen, to reoxidize the flavin adenine dinucleotide that is reduced as glucose is consumed (65). Thus the sensitivity of this system depends on the presence of oxygen in the medium. The same mechanism has been used to prepare variations of glucose-sensitive insulin delivery systems. Both matrix type (67) and reservoir type (68–71) devices have been prepared.

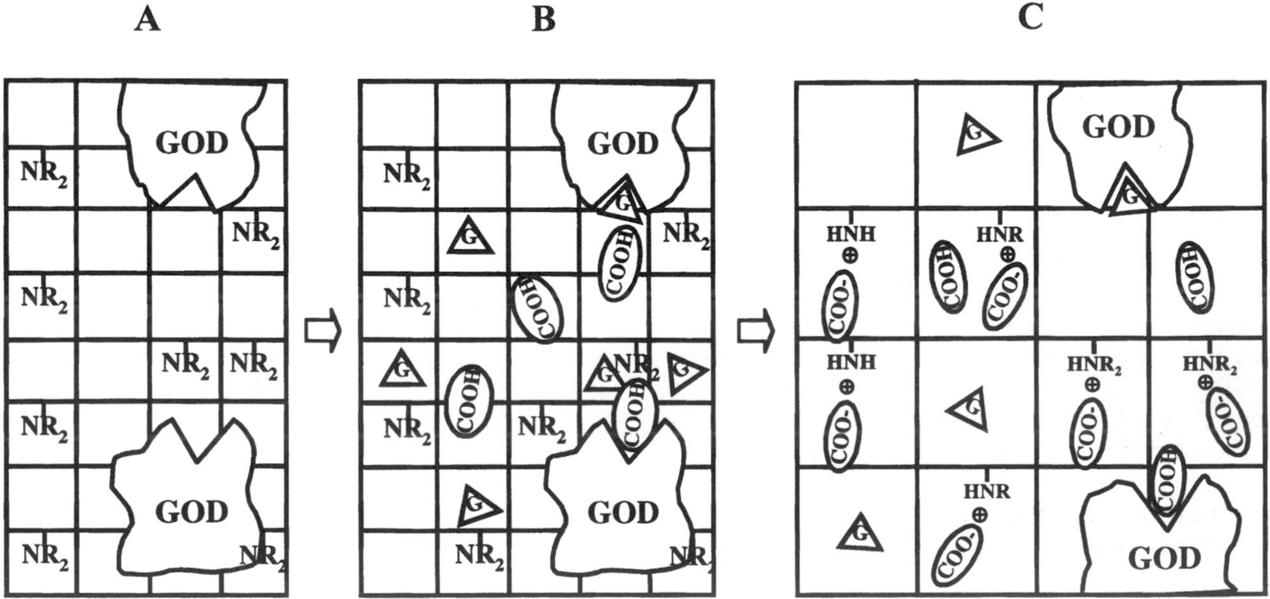

Figure 27.6 Schematic representation of mechanism of action of glucose-sensitive membrane. A, in the absence of glucose, at physiologic pH, few of the amine groups are protonated; B, in the presence of glucose (G), glucose oxidase (GOD) produces gluconic acid; C, gluconic acid can protonate the amine groups. The fixed positive charge on the polymeric network leads to electrostatic repulsion and membrane swelling. (From Ref. 59.)

2. Hydrogels Made of Anionic Polyelectrolytes

a. ''Squeezing'' Hydrogels.

Glucose-sensitive polyanionic hydrogels were prepared using poly(methacrylic acid-g-ethylene glycol) (poly(MAA-g-EG)) (10). Methacrylic acid and PEG monomethacrylate were copolymerized in the presence of GOD. At low pH, interpolymer complexes are formed in these hydrogels due to hydrogen bonding between the hydrogen atoms of the methacrylic acid units and the oxygen atoms on the ether groups of the PEG chains. Such hydrogen bonding results in the collapse of the gel due to increased hydrophobicity in the polymer network. As the carboxylic groups become ionized at high pH, the polymer chains in complexes become separated and the gel swells. If GOD is incorporated into the hydrogels at high pH, lowering of pH in the presence of free glucose results in ''squeezing'' of the hydrogels. if the glucose level decreases, less gluconic acid is produced, resulting in slow increase in pH of the environment. The squeezed hydrogel would then swell, and this cycle can be repeated.

b. ''Chemical Gate'' Hydrogel Layers

Poly(acrylic acid) and GOD were grafted to porous films, such as porous cellulose and poly(vinylidene fluoride) membranes (72,73). At neutral pH, negatively charged poly(acrylic acid) chains are fully extended to close up the pores of the membrane for poor diffusion of insulin. As pH decreases below 5 as a result of GOD action on glucose, poly(acrylic acid) chains become protonated and the chains collapse into coils on the pore walls, leading to the opening of the pores for the free diffusion of insulin (Fig. 7). The insulin diffusion through the open gate, however, was only a few times higher than the diffusion in the closed state. Poly(acrylic acid) was grafted onto the cellulose membrane by either plasma polymerization or ceric ion-induced radical polymerization, and GOD was immobilized by coupling with l-ethyl-3-(3-dimethylaminopropyl) carbodiimide.

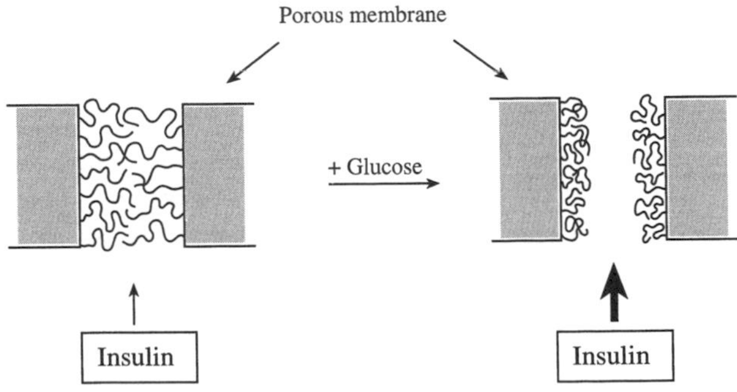

Figure 27.7 Principle of controlled release system of insulin. (Left) In the absence of glucose, the chains of poly(acrylic acid) grafts are rodlike, lowering the porosity of the membrane and suppressing insulin permeation. (Right) In the presence of glucose, gluconic acid produced by GOD protonates the poly(acrylic acid), making the graft chains coillike and opening the pores to enhance insulin permeation. (From Ref. 72.)

3. Hydrogels Made of pH/Temperature-Sensitive Polymers

pH/temperature-sensitive polymers were used to prepare glucose-sensitive insulin delivery systems (74,75). GOD was incorporated into the poly(N,N-dimethylaminoethyl methacrylate-co-ethylacrylamide) (poly(DMAEM-co-EAM)) matrix. Hydrogels made of poly(DMAEM-co-EAM) have different temperature-responsive swelling properties at different pH values, e.g., pH 4.0 and 7.4. The lower critical solution temperature (LCST) of the copolymer at pH 4.0 is higher than the LCST at pH 7.4 due to the increased electrostatic repulsion of DMAEM groups at pH 4.0. This means that at a given temperature the copolymer does not dissolve at pH 7.4. but become dissolved at pH 4.0. As shown in Fig. 8, the generation of gluconic acid from free glucose results in a lowering of the pH, and thus an increase in the LCST. The ultimate outcome is the dissolution of the copolymer and the subsequent release of the incorporated insulin.

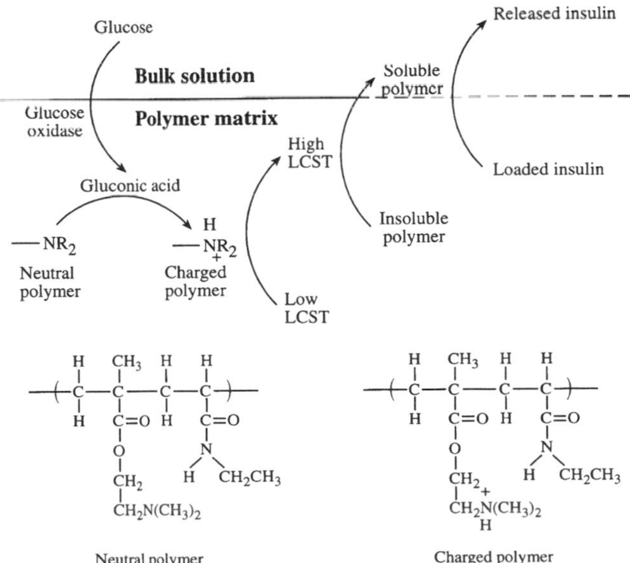

Figure 27.8 Schematic representation of glucose-responsive insulin delivery system using poly(DMA-co-EMA). (From Ref. 75.)

4. Erodible Polymers Surrounded by GOD-Containing Hydrogels

The pH-dependent erodible poly(ortho esters) were used to develop glucose-sensitive insulin delivery systems (76–80). The erosion rate of the poly(ortho esters) increases as the pH decreases. Insulin is physically dispersed in the poly(ortho esters) matrix, which is surrounded by a hydrogel containing immobilized GOD (80). When external glucose diffuses into the hydrogel to be oxidized by GOD to gluconic acid, the pH within the hydrogel drops and the erosion of the polymer matrix accelerates. When the concentration of external glucose drops, production of gluconic acid ceases, and the pH within the hydrogel rises again as gluconic acid diffuses from the hydrogel. This process leads to modulated insulin release from the device. One problem related with this device is that the amount of insulin released may not be proportional to the decrease in pH, and the on-and-off process of insulin release is not fast enough (80).

D. Glucose Dehydrogenase as a Glucose Responsive Unit

In addition to GOD, glucose dehydrogenase (GDH) was also used as a glucose-responsive unit (72,81). The reason for using GDH was to provide better glucose sensitivity than GOD does (81). For higher and faster glucose-sensitive insulin release, insulin molecules are grafted to the substrate surface through a disulfide bond. When glucose molecules are present in solution, GDH oxidizes the glucose molecules to generate electrons that reduce the disulfide bonds for release of the grafted insulin (Fig. 9). This process requires enzyme cofactors, nicotinamide adenine dinucleotide (NAD), and flavin adenine dinucleotide (FAD), which act as electron mediators. The glucose sensitivity was increased as NAD and FAD were immobilized to the surface and further enhanced by coimmobilization of GDH (Fig. 9). While this type of system can provide improved sensitivity to glucose and faster release of the immobilized insulin, the system has a major drawback in the limited amount of insulin that can be grafted onto the surface.

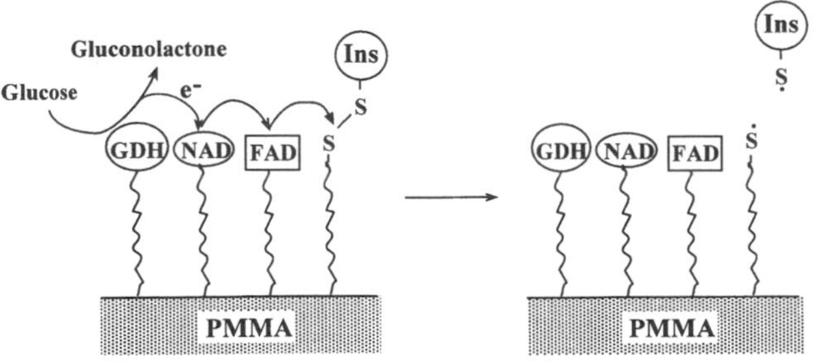

Figure 27.9 Design of glucose-sensitive insulin-releasing membrane systems. PMMA, poly(methyl methacrylate); Ins, insulin; S–S, disulfide bond; GDH, glucose dehydrogenase; NAD, nicotinamide adenine dinucleotide; FAD, flavin adenine dinucleotide. (From Ref. 81.)

III. GLUCOSE-SENSITIVE HYDROGELS AS GLUCOSE SENSORS

The detection and quantification of glucose in blood plays a vital role in the diagnosis and management of diabetes. Glucose levels are dynamically changing throughout the day depending on food intake, insulin availability, exercise, stress, and illness (82). Measurement of the accurate blood glucose levels and timely delivery of exact amounts of insulin are critical to the maintenance of normal life for diabetic patients. Currently, diabetic patients determine their blood glucose levels by obtaining a small amount of blood with a finger stick, followed by external analysis of the blood sample for the glucose content. This method, although accurate, is painful, inconvenient, and discontinuous, and it has a risk of infection (83). Devising a reliable noninvasive technique for continuous monitoring of glucose levels in the blood has important medical implications. The desired characteristics of in vivo glucose sensors are safety, clinical accuracy and reliability, feasibility of in vivo recalibration, long-term stability, small size, ease of insertion and removal, and a sufficiently fast response to allow timely intervention (84).

Several attempts have been made to prepare noninvasive glucose sensing devices. One of the most desirable noninvasive methods is near-infrared spectroscopy, which, unlike electrochemical sensors, responds directly to the glucose molecule (85–91). The main practical limitation of this system is that the glucose signal is very weak in the midst of much larger signals from proteins, lipids, and scattered light (92). Table 2 summarizes the blood glucose measurement techniques currently under evaluation (82). Regardless of the approaches, the lack of proper calibration in each system remains as the main limitation.

More realistic approaches in continuous glucose sensing may involve implanting glucose-sensitive devices. Two types of implantable glucose sensors have been reported, fully implantable and percutaneous (worn through the skin) (93). The fully implantable devices, which include electronics and a battery, are designed for longevity and intended to be implanted by a physician either subcutaneously or in a blood vessel. Glucose-sensitive hydrogels, e.g., sol–gel phase-reversible hydrogels, can be implanted,

Table 27.2 Techniques Used for Measuring Glucose Concentrations

1. Invasive techniques
   - Implanted electrochemical sensors
   - Suction blister extraction
   - Microdialysis extraction
   - Wick extraction
   - Competitive fluorescence implants
   - Needle puncture and extraction
2. Minimally invasive techniques
   - Iontophoretic extraction
   - Sontophoretic extraction
   - Chemically enhanced extraction
3. Noninvasive techniques
   - Infrared spectroscopy
   - Near-infrared spectroscopy
   - Raman spectroscopy
   - Photoacoustic spectroscopy
   - Scatter changes
   - Polarization changes

Source: Ref. 82.

and the glucose concentration–dependent changes of the hydrogels can be monitored using external spectroscopic devices (94). The percutaneous sensors are needlelike or inserted through a needle and are designed to operate for a few days and be replaced by the patient. Those implantable glucose sensors are mostly amperometric enzymatic sensors (95–98). GOD has been most widely used as a glucose sensing unit in amperometric glucose sensors because GOD transfers electrons to diffusing and nondiffusing mediators, withstands chemical immobilization techniques, and has a high turnover rate ($\sim 10^3\ s^{-1}$) at ambient temperatures (99). However, conventional GOD-based glucose sensors have some disadvantages. Variations in the level of dissolved oxygen may cause fluctuations in the electrode response, and the dynamic range of glucose detection can be limited by lack of dissolved oxygen (100,101). Diffusion mediators for electron transport such as quinones and ferrocenes have been used in enzyme electrodes to overcome oxygen dependence. However, in vivo leaching of the mediator has been a serious problem. To eliminate me-

Table 27.3 Glucose-Sensitive Hydrogels Used for Glucose Sensing

Glucose sensing unit	Hydrogel system	Detection method	References
GOD	Poly(HEMA)	H_2O_2	103, 104
	Poly(VP) or poly(VP-VI)/OS derivative/PEG redox hydrogels	Current	84, 99, 101, 105–109
	Poly(allylamine)/OS derivative	Current	110
	Heparin/PDMAA/PAzSt	Current	111
	Poly(DMAA/AzSt/VFe)	Current	112
	Poly(PEG diacrylate-VFe) redox hydrogel	Current	113
	± charged hydrogels/PTFE of PC membrane	Current	114, 115
	Polyacrylamide	Current	116
	APEG/SPA hydrogel	Current	102
	Gelatin	Photoacoustic wave	117
	Poly(carbamoyl sulphonate)	Chemiluminescence	118
	Polyacrylamide-based colloidal crystal hydrogel	Optical intensity	119, 120
Con A	FITC-dextran	Fluorescence	121
	Poly(GEMA)	Optical intensity	40
PBA	Poly(IPPm-co-PBA-co-DMA-PAA)	Optical intensity	122
GDH	Poly(ether amine quinone)	Current	123
	Poly(VI)/Os derivative/PEG redox hydrogel	Current	124

APEG: 8-armed, amine-terminated PEG. FITC-dextran: fluorescein-labeled dextran. Os: osmium. PAzSt: poly(m-azidostyrene). PC: polycarbonate. PDMAA: poly(dimethyl acrylamide-co-2-cinnamoylethyl methacrylate). Poly(DMAA/AzSt /VFe): poly(dimethylacrylamide-co-azidostyrene-co-vinylferrocene). Poly(GEMA): poly(glycosyl-ethyl methacrylate). Poly(HEMA): poly(hydroxyethyl methacrylate). Poly(IPPm-co-PBA-co-DMA-PAA): poly(N-isopropylacrylamide-co-3-acrylamido-phenylboronic acid-co-N-(3-dimethylaminopropyl)acrylamide. Poly(VP): poly(vinylpyrrolidone). Poly(VP-VI): poly(vinylpyrrolidone-co-1-vinylimidazole). PTFE: polytetrafluoroethylene. SPA: di-succinimidyl ester of PEG α, ω-dipropionic acid.

diator leaching, the mediator has been immobilized in different hydrogels. Sensors for in vivo monitoring of glucose also require biocompatible interfaces with the tissue in which they have been implanted (102). The interface material must be permeable to glucose and the product of its oxidation, gluconolactone. It also must not be fouled by proteins, cellular attachment, or fibrous encapsulation. Fast inactivation of the implanted sensors has been attributed to protein adsorption and cell adhesion. Hydrogels are known to have excellent biocompatibility due to large water contents and the ability to prevent protein adsorption and cell adhesion. Thus coating the surface of implantable glucose sensors with hydrogel layers would be a good approach for extending the lifetime of the sensors. Table 3 briefly summarizes the approaches used for glucose sensing using glucose-sensitive hydrogels. Glucose-sensitive hydrogel systems are those described in the previous section of this chapter.

IV. PROPERTIES TO IMPROVE FOR PRACTICAL APPLICATIONS

There are many hurdles to overcome to develop clinically useful self-regulated insulin delivery systems based on glucose-sensitive hydrogels. Those systems that work very well in the laboratory rarely work as well when applied in vivo. A number of improvements need to be made for practical applications. First, to be clinically useful, self-regulated insulin delivery devices should react fast to changes in the blood glucose concentration and release the necessary amount of insulin. The fast on-and-off function is critical for adequate control of the dynamically changing glucose levels in the blood. The response time of hydrogels depends on the time for diffusion of glucose molecules, and the hydrogel thickness has a big impact on the response time. Second, more biocompatible glucose-sensitive molecules need to be developed. Currently used glucose sensing moieties are proteins, such as Con A, GOD, and GDH. To minimize any side effects resulting from long-term use, glucose sensing moieties based on nonproteinaceous molecules are preferred. Implantable glucose sensing moieties should be biocompatible, nontoxic, cost-effective, and independent of environmental factors such as pH, ionic strength, or the presence of divalent cations (125). Third, the implantable devices, if used for the long term, tend to be isolated from the body by tissue remodeling around the implants. The formation of new tissues around the implant significantly retards the diffusion of glucose, which in turn delays the response time and sensitivity. This necessitates frequent recalibration of the implanted devices, since there is no clear understanding of the relationship between glucose levels in the blood and in the tissue (93). Continued research on these hurdles will undoubtedly present answers in the future. For now, however, biodegradable glucose-sensitive hydrogels can be developed so that they can be used to develop short-term glucose sensing as well as self-regulated insulin delivery.

REFERENCES

1. J. D. Andrade. *Hydrogels for Medical and Related Applications*. American Chemical Society, Washington, D.C., 1976.
2. D. DeRossi, K. Kajiwara, Y. Osada, A. Yamauchi. *Polymer Gels: Fundamentals and Biomedical Applications*. Plenum Press, New York, 1991.
3. V. Kudela. Hydrogels. *Encyclopedia of Polymer Science and Technology* (H. F. Mark, J. I. Kroschwitz, eds.) John Wiley, New York, 1985, pp. 783–807.
4. R. M. Ottenbrite, S. J. Huang, K. Park. *Hydrogels and Biodegradable Polymers for Bioapplications*. American Chemical Society, Washington, D.C., 1996.
5. N. A. Peppas. *Hydrogels in Medicine and Pharmacy*. CRC Press, Boca Raton, FL, 1986.
6. J. M. Guenet. *Thermoreversible Gelation of Polymers and Biopolymers*. Academic Press, New York, 1992.
7. K. Park, S. W. Shalaby, H. Park. *Biodegradable Hydrogels for Drug Delivery*. Technomic, Lancaster, PA, 1993, 1–12.
8. H. Park, K. Park. Hydrogels in bioapplications. *Hydrogels and Biodegradable Polymers for Bioapplications* (R. M. Ottenbrite, S. J. Huang, K. Park, eds.). American Chemical Society, Washington, D.C., 1996, pp. 2–10.
9. J. J. Kim, K. Park. Smart hydrogels for bioseparation. *Bioseparation 7*:177–184 (1999).
10. C. M. Hassan, F. J. Doyle, III, N. A. Peppas. Dynamic behavior of glucose-responsive poly(methacrylic acid-g-ethylene glycol) hydrogels. *Macromolecules 30*:6166–6173 (1997).
11. A. S. Hoffman. Intelligent polymers. *Controlled Drug Delivery* (K. Park, ed.). American Chemical Society, Washington, D.C., 1997, pp. 485–498.
12. Y. H. Bae. Stimuli-sensitive drug delivery. *Controlled Drug Delivery* (K. Park, ed.). American Chemical Society, Washington, D.C., 1997, pp. 147–162.
13. J. Heller. Feedback-controlled drug delivery. *Controlled Drug Delivery* (K. Park, ed.). American Chemical Society, Washington, D.C., 1997, pp. 127–146.
14. A. A. Obaidat, K. Park. Characterization of protein release through glucose-sensitive hydrogel membranes. *Biomaterials 18*:801–806 (1997).
15. M. Brownlee, A. Cerami. A glucose-controlled insulin-delivery system: semisynthetic insulin bound to lectin. *Science 206*:1190–1191 (1979).
16. M. Brownlee, A. Cerami. Glycosylated insulin complexed

to Concanavalin A. Biochemical basis for a closed-loop insulin delivery systems. *Diabetes 32*:499–504 (1983).
17. J. L. Wang, B. A. Cunningham, M. J. Waxdal, G. M. Edelman. The covalent and three-dimensional structure of concanavalin A. I. Amino acid sequence of cyanogen bromide fragments F_1 and F_2. *J. Biol. Chem. 250*:1490–1502 (1975).
18. B. A. Cunningham, J. L. Wang, M. J. Waxdal, G. M. Edelman. The covalent and three-dimensional structure of concanavalin A. II. Amino acid sequence of cyanogen bromide fragments F_3. *J. Biol. Chem. 250*:1503–1512 (1975).
19. W. H. Beckert, M. M. Sharkey. Mitogenic activity of the jack bean (Canavalia ensiformis) with rabbit peripheral blood lymphocytes. *Int. Arch. Allergy Appl. Immunol. 30*: 337–341 (1970).
20. A. E. Powell, M. A. Leon. Reversible interaction of human lymphocytes with the mitogen concanavalin A. *Exp. Cell Res. 62*:315–325 (1970).
21. R. D. Ekstedt. Enhancement of experimental tumors in mice by treatment with concanavalin A. *J. National Cancer Institute 63*:1065–1069 (1979).
22. E. V. Larsson, M. Gullberg, A. Coutinho. Heterogeneity of cells and factors participating in the concanavalin A–dependent activation of T lymphocytes with cytotoxic potential. *Immunobiology 161*:5–20 (1982).
23. S. Y. Jeong, S. W. Kim, D. L. Holmberg, J. C. McRea. Self-regulating insulin delivery systems. III. In vivo studies. *J. Controlled Rel. 2*:143–152 (1985).
24. S. W. Kim, S. Y. Jeong, S. Sato, J. C. McRea, J. Feijen. Self-regulating insulin delivery system—chemical approach. *Recent Advances in Drug Delivery Systems* (J. M. Anderson, S. W. Kim, eds.). Plenum Press, New York, 1984, pp. 123–136.
25. S. Y. Jeong, S. W. Kim, M. J. D. Eenink, J. Feijen. Self-regulating insulin delivery systems. I. Synthesis and characterization of glycosylated insulin. *J. Controlled Rel. 1*: 57–66 (1984).
26. S. Sato, S. Y. Jeong, J. C. McRea, S. W. Kim. Self-regulating insulin delivery systems. II. In vitro studies. *J. Controlled Rel. 1*:67–77 (1984).
27. S. Sato, S. Y. Jeong, J. C. McRea, S. W. Kim. Glucose stimulated insulin delivery system. *Pure Appl. Chem. 56*: 1323–1328 (1984).
28. L. A. Seminoff, G. B. Olsen, S. W. Kim. A self-regulating insulin delivery system I. Characterization of a synthetic glycosylated insulin derivative. *Int. J. Pharm. 54*:241–249 (1989).
29. L. A. Seminoff, J. M. Gleeson, J. Zheng, G. B. Olsen, D. Holberg, S. F. Mohammad, D. Wilson, S. W. Kim. A self-regulating insulin delivery system. II. In vivo characteristics of a synthetic glycosylated insulin. *Int. J. Pharm. 54*: 251–257 (1989).
30. S. W. Kim, C. M. Pai, K. Makino, L. A. Seminoff, D. L. Holmberg, J. M. Gleeson, D. E. Wilson, E. J. Mack. Self-regulated glycosylated insulin delivery. *J. Controlled Rel. 11*:193–201 (1990).
31. C. M. Pai, Y. H. Bae, E. J. Mack, D. E. Wilson, S. W. Kim. Concanavalin A microspheres for a self-regulating insulin delivery system. *J. Pharm. Sci. 81*:532–536 (1992).
32. M. Baudys, T. Uchio, L. Hovgaad, E. F. Zhu, T. Avramoglou, M. Jozefowicz, B. Rihova, J. Y. Park, H. K. Lee, S. W. Kim. Glycosylated insulins. *J. Controlled Rel. 36*: 151–157 (1995).
33. B. A. Cunningham, J. L. Wang, M. N. Pflumm, G. M. Edelman. Isolation and proteolytic cleavage of the intact subunit of concanavalin A. *Biochemistry 11*:3233–3239 (1972).
34. F. Liu, S. C. Song, D. Mix, M. Baudys, S. W. Kim. Glucose-induced release of glycosylpoly(ethylene glycol) insulin bound to a soluble conjugate of concanavalin A. *Bioconjugate Chem. 8*:664–672 (1997).
35. K. Makino, E. J. Mack, T. Okano, S. W. Kim. A microcapsule self-regulating delivery system for insulin. *J. Controlled Rel. 12*:235–239 (1990).
36. K. Makino, E. J. Mack, T. Okano, S. W. Kim. Self-regulated delivery of insulin from microcapsules. *Biomaterials. Artificial Cells, and Immobilization Biotechnology 19*: 219–228 (1991).
37. S. W. Kim, H. A. Jacobs. Self-regulated insulin delivery—artificial pancreas. *Drug Devel. Ind. Pharm. 20*: 575–580 (1994).
38. E. Kokufata, Y. Q. Zhang, T. Tanaka. Saccharide-sensitive phase transition of a lectin-loaded gel. *Nature 351*: 302–304 (1991).
39. J. E. Morris, A. S. Hoffman, R. R. Fisher. Affinity precipitation of proteins by polyligands. *Biotech. Bioeng. 41*: 991–997 (1993).
40. K. Nakamae, T. Miyata, A. Jikihara, A. S. Hoffman. Formation of poly(glycosyloxyethyl methacrylate)-concanavalin A complex and its glucose sensitivity. *J. Biomater. Sci., Polymer Edn. 6*:79–90 (1994).
41. T. Miyata, A. Jikihara, K. Nakamae, A. S. Hoffman. Preparation of poly(2-glucosyloxyethyl methacrylate)-concanavalin A complex hydrogel and its glucose-sensitivity. *Macromol. Chem. Phys. 197*:1135–1146 (1996).
42. M. J. Taylor, S. Tanna, P. M. Taylor, G. Adams. The delivery of insulin from aqueous and nonaqueous reservoirs governed by a glucose sensitive gel membrane. *J. Drug Target. 3*:209–216 (1995).
43. M. J. Taylor, S. Tanna, S. Cockshott, R. Vaitha. A self-regulated delivery system using unmodified solutes in glucose-sensitive gel membranes. *J. Pharm. Pharmac. 46* (suppl. 2):1051b (1994).
44. S. Tanna, M. J. Taylor. A self-regulating system using high-molecular weight solutes in glucose-sensitive gel membranes. *J. Pharm. Pharmac. 46*(suppl. 2):1051a (1994).
45. S. J. Lee, K. Park. Synthesis of sol–gel phase-reversible hydrogels sensitive to glucose. *Proc. Intern. Symp. Control. Rel. Bioact. Mater. 21*:93–94 (1994).
46. S. J. Lee, K. Park. Synthesis and characterization of sol–gel phase reversible hydrogels sensitive to glucose. *J. Mol. Recog. 9*:549–557 (1996).

47. S. J. Lee, K. Park. Glucose-sensitive phase-reversible hydrogels. *Hydrogels and Biodegradable Polymers for Bioapplications* (R. M. Ottenbrite, S. J. Huang, K. Park, eds.). American Chemical Society, Washington, D.C., 1996, pp. 11–16.
48. A. A. Obaidat, K. Park. Characterization of glucose dependent gel–sol phase transition of the polymeric glucose concanavalin A hydrogel system. *Pharm. Res. 13*:989–995 (1996).
49. J. J. Kim. Phase-reversible glucose-sensitive hydrogels for modulated insulin delivery. Purdue University, Doctor of Philosophy, West Lafayette, 1999.
50. J. Boeseken. The use of boric acid for the determination of the configuration of carbohydrates. *Advances in Carbohydrate Chemistry 47*:189–210 (1949).
51. A. B. Foster. Zone electrophoresis of carbohydrate. *Advances in Carbohydrate Chemistry 12*:81–115 (1957).
52. S. Aronoff, T. Chen, M. Cheveldayoff. Complexation of D-glucose with borate. *Carbohydrate Research 40*:299–309 (1975).
53. V. Bouriotis, I. J. Galpin, P. D. G. Dean. Applications of immobilised phenylboronic acids as supports for group-specific ligands in the affinity chromatography of enzymes. *J. Chromatography 210*:267–278 (1981).
54. S. Kitano, K. Kataoka, Y. Koyama, T. Okano, Y. Sakurai. Glucose-responsive complex formation between poly(vinyl alcohol) and poly(N-vinyl-2-pyrrolidone) with pendant phenylboronic acid moieties. *Makromolekulare Chemie, Rapid Communications 12*:227–233 (1991).
55. S. Kitano, Y. Koyama, K. Kataoka, T. Okano, Y. Sakurai. A novel drug delivery system utilizing a glucose responsive polymer complex between poly(vinyl alcohol) and poly(N-vinyl-2-pyrrolidone) with a phenyl boronic acid moiety. *J. Controlled Rel. 19*:162–170 (1992).
56. D. Shiino, Y. Murata, K. Kataoka, Y. Koyama, M. Yokoyama, T. Okano, Y. Sakurai. Preparation and characterization of a glucose-responsive insulin-releasing polymer device. *Biomaterials 15*:121–128 (1994).
57. I. Hisamitsu, K. Kataoka, T. Okano, Y. Sakurai. Glucose-responsive gel from phenylborate polymer and poly(vinyl alcohol): prompt response at physiological pH through the interaction of borate with amino group in the gel. *Pharm. Res. 14*:289–293 (1997).
58. Y. K. Choi, S. Y. Jeong, Y. H. Kim. A glucose-triggered solubilizable polymer gel matrix for an insulin delivery system. *Int. J. Pharm. 80*:9–16 (1992).
59. T. A. Horbett, J. Kost, B. D. Ratner. Swelling behavior of glucose sensitive membranes. *Polymers as Biomaterials*. (S. W. Shalaby, A. S. Hoffman, B. D. Ratner, T. A. Horbett, eds.). Plenum Press, New York, 1984, pp. 193–207.
60. T. A. Horbett, B. D. Ratner, J. Kost, M. Singh. A bioresponsive membrane for insulin delivery. *Recent Advances in Drug Delivery Systems* (J. M. Anderson, S. W. Kim, eds.). Plenum Press, New York, 1984, 209–220.
61. J. Kost, T. A. Horbett, B. D. Ratner, M. Singh. Glucose-sensitive membranes containing glucose oxidase: activity, swelling, and permeability studies. *J. Biomed. Mate. Res. 19*:1117–1133 (1985).
62. G. Albin, T. A. Horbett, B. D. Ratner. Glucose sensitive membranes for controlled delivery of insulin: insulin transport studies. *J. Controlled Rel. 2*:153–164 (1985).
63. G. Albin, T. A. Horbett, S. R. Miller, N. L. Ricker. Theoretical and experimental studies of glucose-sensitive membranes. *J. Controlled Rel. 6*:267–291 (1987).
64. G. Albin, T. A. Horbett, B. D. Ratner. Glucose-sensitive membranes for controlled release of insulin. *Pulsed and Self-regulated Drug Delivery*. (J. Kost, ed.). CRC Press, Boca Raton, FL, 1990, pp. 159–185.
65. L. A. Klumb, T. A. Horbett. Design of insulin delivery devices based on glucose-sensitive membranes. *J. Controlled Rel. 18*:59–80 (1992).
66. L. A. Klumb, T. A. Horbett. The effect of hydronium ion on the transient behavior of glucose-sensitive membranes. *J. Controlled Rel. 27*:95–114 (1993).
67. M. Goldraich, J. Kost. Glucose-sensitive polymeric matrices for controlled drug delivery. *Clinical Materials 13*: 135–142 (1993).
68. K. Ishihara, M. Kobayashi, I. Shionohara. Control of insulin permeation through a polymer membrane with responsive function for glucose. *Makromolekulare Chemie, Rapid Communication 4*:327–331 (1983).
69. K. Ishihara, M. Kobayashi, N. Ishimaru, I. Shinohara. Glucose induced permeation control of insulin through a complex membrane consisting of immobilized glucose oxidase and a poly(amine). *Polymer J. 16*:625–631 (1984).
70. K. Ishihara, M. Kobayashi, I. Shonohara. Insulin permeation through amphiphilic polymer membranes having 2-hydroxyethyl methacrylate moiety. *Polymer J. 16*:647–651 (1984).
71. K. Ishihara, K. Matsui. Glucose-responsive insulin release from a polymer capsule. *J. Polymer Sci., Polymer Letter Edn. 24*:413–417 (1986).
72. Y. Ito, M. Casolaro, K. Kono, Y. Imanishi. An insulin-releasing system that is responsive to glucose. *J. Controlled Rel. 10*:195–203 (1989).
73. H. Iwata, T. Matsuda. Preparation and properties of novel environment-sensitive membranes prepared by graft polymerization onto a porous membrane. *J. Memb. Sci. 38*: 185–199 (1988).
74. S. H. Cho, M. S. Jhon, S. H. Yuk, H. B. Lee. Temperature-induced phase transition of poly(N,N-dimethylaminoethyl methacrylate-co-acrylamide). *J. Polymer Sci., Polymer Phys. 35*:595 (1997).
75. S. H. Yuk, S. H. Cho, S. H. Lee. pH/temperature-responsive polymer composed of poly(N,N-dimethylamono) ethyl methacrylate-co-ethylacrylamide). *Macromolecules 30*:6856–6859 (1997).
76. J. Heller, D. W. H. Penhale, R. F. Helwing. Preparation of poly(ortho esters) by the reaction of ketane acetal and diol. *J. Polymer Sci., Polymer Lett. Edn 18*:611–624 (1980).
77. J. Heller, S. H. Pangburn, D. W. H. Penhale. Use of bioerodible polymers in self-regulated drug delivery systems.

Controlled Release Technology. (P. I. Lee, W. R. Good, eds.) American Chemical Society, Washington, D. C., 1987, pp. 172–187.
78. J. Heller. Chemically self-regulated drug delivery systems. *J. Controlled Rel. 8*:111–125 (1988).
79. J. Heller. Use of enzymes and bioerodible polymers in self-regulated and triggered drug delivery systems. *Modulated Controlled Release Systems*. (J. Kost, ed.). CRC Press, Boca Raton, FL, 1990, pp. 93–108.
80. J. Heller, A. C. Chang, G. Rodd, G. M. Grodsky. Release of insulin from pH-sensitive poly(orthoesters). *J. Controlled Rel. 13*:295–302 (1990).
81. D. J. Chung, Y. Ito, Y. Imanishi. An insulin-releasing membrane system on the basis of oxidation reaction of glucose. *J. Controlled Rel. 18*:45–54 (1992).
82. J. N. Roe, B. R. Smoller. Bloodless glucose measurements. *Crit. Rev. Therap. Drug Carrier Systems 15*:199–241 (1998).
83. R. T. Kurnik, B. Berner, J. Tamada, R. O. Potts. Design and simulation of a reverse iontophoretic glucose monitoring device. *J. Electrochem. Soc. 145*:4119–4125 (1998).
84. E. Csoregi, C. P. Quinn, D. W. Schmidtke, S. E. Lindquist, M. V. Pishko, L. Ye, I. Katakis, J. A. Hubbell, A. Heller. Design, characterization, and one-point in vivo calibration of a subcutaneously implanted glucose electrode. *Anal. Chem. 66*:3131–3138 (1994).
85. D. M. Back, D. F. Michalska, P. L. Polavaru. Fourier transform infrared spectroscopy as a powerful tool for the study of carbohydrates in aqueous solution. *Applied Spectroscopy 38*:173–180 (1984).
86. H. Zeller, P. Novak, R. Langer. Blood glucose measurement by IR spectroscopy. *Int. J. Pharm. 12*:129–134 (1989).
87. M. R. Robinson, R. P. Eaton, D. M. Haaland, G. W. Koepp, E. V. Thomas, B. R. Stallard, P. L. Robinson. Noninvasive glucose monitoring in diabetic patients: preliminary evaluation. *Clinical Materials 38*:1618–1622 (1992).
88. J. W. Hall. Near-infrared spectroscopy: a new dimension in clinical chemistry. *Clin. Chem. 38–39*:1623–1631 (1992).
89. L. A. Marquart, M. Arnold, G. W. Small. Near infrared spectroscopic measurement of glucose in a protein matrix. *Analytical Chemist 65*:3271–3278 (1993).
90. G. B. Christison, H. A. Mackenzie. Laser photoacoustic determination of physiological glucose concentrations in human whole blood. *Medical Biological Engineering Computing 31*:284–290 (1993).
91. H. M. Heise, R. Marbach, T. Koschinsky, F. A. Gries. Noninvasive blood glucose sensors based on near-infrared spectroscopy. *Artificial Organs 18*:439–447 (1994).
92. A. A. Sharkawy, M. R. Neuman, W. R. Reichert. Sensocompatibility: design consideration for biosensor-based drug delivery systems. *Controlled Drug Delivery: Challenges and Strategies*. (K. Park, ed.). American Chemical Society, Washington, D.C., 1997, pp. 163–181.
93. C. Henry. Getting under the skin: implantable glucose sensors. *Anal. Chem. 70*:594A–598A (1998).
94. H. Park, K. Park. Biocompatibility issues of implantable drug delivery systems. *Pharm. Res. 13*:1770–1776 (1996).
95. P. Vadgama, P. W. Crump. Biosensors—recent trends: a review. *Analyst 117*:1657–1670 (1992).
96. F. J. Schmidt, W. J. Sluiter, A. J. M. Schoonen. Glucose concentration in subcutaneous extracellular space. *Diabetes Care 16*:695–700 (1993).
97. J. C. Pickup, D. J. Claremont, G. W. Shaw. Responses and calibration of amperometric glucose sensors implanted in the subcutaneous tissue of man. *Acta Diabetologica 30*: 143–148 (1993).
98. D. Frazer. Biosensing in the body. *Medical Device Technology 5*:24–28 (1994).
99. M. V. Pishko, A. C. Michael, A. Heller. Amperometric glucose microelectrodes prepared through immobilization of glucose oxidase in redox hydrogels. *Anal. Chem. 63*: 2268–2272 (1991).
100. S. Dong, W. wBaoxing, L. Baifeng. Amperometric glucose sensor with ferrocene as an electron transfer mediator. *Biosensors Bioelectronics 7*:215–222 (1991).
101. B. Linke, W. Kerner, M. Kiwit, M. Pishko, A. Heller. Amperometric biosensor for in vivo glucose sensing based on glucose oxidase immobilized in a redox hydrogel. *Biosensors Bioelectronics 9*:151–158 (1994).
102. C. A. P. Quinn, R. E. Connor, A. Heller. Biocompatible glucose-permeable hydrogel for in situ coating of implantable biosensors. *Biomaterials 18*:1665–1670 (1997).
103. G. Urban, G. Jobst, F. Keplinger, E. Aschauer, O. Tilado, R. Fasching, F. Kohl. Miniaturized multi-enzyme biosensors integrated with pH sensors on flexible polymer carriers for in vivo applications. *Biosensors Bioelectronics 7*: 733–739 (1992).
104. G. Urban, G. Jobst, E. Aschauer, O. Tilado, P. Svasek, M. Varahram. Performance of integrated glucose and lactate thin-film microbiosensors for clinical analyzers. *Sensors Actuators B 18–19*:592–596 (1994).
105. B. A. Gregg, A. Heller. Redox polymer films containing enzymes. 2. Glucose oxidase-containing enzyme electrodes. *J. Phys. Chem. 95*:5976–5980 (1991).
106. T. J. Ohara, R. Rajagopalan, A. Heller. Glucose electrodes based on cross-linked [Os(bpy)2Cl]+/2+ complexed poly(1-vinylimidazole) films. *Anal. Chem. 65*:3512–3517 (1993).
107. T. J. Ohara, R. Rajagopalan, A. Heller. ''Wired'' enzyme electrodes for amperometric determination of glucose or lactate in the presence of interfering substances. *Anal. Chem. 66*:2451–2457 (1994).
108. T. Lumley-Woodyear, P. Rocca, J. Lindsay, Y. Dror, A. Freeman, A. Heller. Polyacrylamide-based redox polymer for connecting redox centers of enzymes to electrodes. *Anal. Chem. 67*:1332–1338 (1995).
109. D. W. Schmidtke, A. Heller. Accuracy of the one-point in vivo calibration of ''wired'' glucose oxidase electrodes implanted in jugular veins of rats in periods of rapid rise and decline of the glucose concentration. *Anal. Chem. 70*: 2149–2155 (1998).
110. C. Danilowicz, E. Corton, F. Battaglini, E. J. Calvo. An

Os(byp)2C1PyCH2NHPoly(allylamine) hydrogel mediator for enzyme wiring at electrodes. *Electrochimica Acta 43*:3525–3531 (1998).

111. Y. Nakayama, T. Matsuda. Surface fixation of hydrogels. Heparin and glucose oxidase hydrogelated surfaces. *ASAIO J. 38*:M421–M424 (1992).
112. Y. Nakayama, Q. Zheng, J. Nishimura, T. Matsuda. Design and properties of photocurable electroconductive polymers for use in biosensors. *ASAIO J. 41*:M418–M421 (1995).
113. K. Sirkar, M. V. Pishko. Amperometric biosensors based on oxidoreductases immobilized in photopolymerized poly(ethylen glycol) redox polymer hydrogels. *Anal. Chem. 70*:2888–2894 (1998).
114. R. Vaidya, E. Wilkins. Application of polytetrafluoroethylene (PTFE) membranes to control interference effects in a glucose biosensor. *Biomed. Instrument. Tech. 27*:486–494 (1993).
115. R. Vaidya, E. Wilkins. Use of charged membranes to control interference by body chemicals in a glucose biosensor. *Medical Engineering Physics 16*:416–421 (1994).
116. C. Jimenez, J. Bartrol, N. F. D. Rooij, M. Koudelka-Hep. Use of photopolymerizable membranes based on polyacrylamide hydrogels for enzymatic microsensor construction. *Analytica Chimica. Acta. 351*:169–176 (1997).
117. K. M. Quan, G. B. Christison, H. A. MacKenzie, P. Hodgson. Glucose determination by a pulsed photoacoustic technique: an experimental study using a gelatin-based tissue phantom. *Physics Medicine Biology 38*:1911–1922 (1993).
118. D. Janasek, U. Spohn. An enzyme-modified chemiluminescence detector for hydrogen peroxide and oxidase substrates. *Sensors Actuators B 38–39*: 291–294 (1997).
119. J. H. Holtz, S. A. Asher. Polymerized colloidal crystal hydrogel films as intelligent chemical sensing materials. *Nature 389*:829–832 (1997).
120. J. H. Holtz, J. S. W. Holtz, C. H. Munro, S. A. Asher. Intelligent polymerized crystalline colloidal arrays: novel chemical sensor materials. *Anal. Chem. 70*:780–791 (1998).
121. J. S. Schulz, S. Mansouri, I. J. Goldstein. Affinity sensor: a new technique for developing implantable sensors for glucose and other metabolites. *Diabetes Care 5*: 245–253 (1982).
122. T. Aoki, Y. Nagao, K. Sanui, N. Ogata, A. Kikuchi, Y. Sakurai, K. Kataoka, T. Okano. Glucose-sensitive lower critical solution temperature changes of copolymers composed of N-isopropylacrylamide and phenylboronic acid moieties. *Polymer J. 28*:371–374 (1996).
123. M. Tessema, T. Ruzgas, L. Gorton, T. Ikeda. Flow injection amperometric determination of glucose and some other low molecular weight saccharides based on oligosaccharide dehydrogenase mediated by benzoquinone systems. *Analytica. Chimica. Acta. 310*:161–171 (1995).
124. T. Ruzgas, E. Csoregi, I. Katakis, G. Kenausis, L. Gorton. Preliminary investigations of an amperometric oligosaccharide dehydrogenase-based electrode for the detection of glucose and some other low molecular weight saccharides. *J. Mol. Recog. 9*:480–484 (1996).
125. T. Li, H. B. Lee, K. Park. Comparative stereochemical analysis of glucose-binding proteins for rational design of glucose-specific agents. *J. Biomater. Sci., Polymer Edn. 9*:327–344 (1998).

28

Polymeric Micro- and Nanoparticles as Drug Carriers

G. Barratt, G. Couarraze, P. Couvreur, C. Dubernet, E. Fattal, R. Gref, D. Labarre, P. Legrand, G. Ponchel, and C. Vauthier
UMR CNRS, Chatenay-Malabry, France

I. INTRODUCTION

The fate of a drug after administration in vivo is determined by a combination of several processes: distribution and elimination when given intravenously; absorption, distribution, and elimination when an extravascular route is used. Regardless of the mechanisms involved, each of these processes depends mainly on the physicochemical properties of the drug and therefore, for the most part, on its chemical structure.

During the last few decades, research workers have been trying to develop delivery systems that would allow them to control the fate of drugs within the patient. Using dosage forms known as controlled release systems, pharmacists have succeeded, at least in part, in governing the first process, i.e., drug absorption. The aim of current research is to control the second process, drug distribution within the organism, by the use of carriers (Fig. 1).

As shown in Table 1, the carriers available at present can be divided into three main groups: first-, second-, and third-generation carriers.

The so-called first-generation carriers are systems capable of delivering the active substance specifically to the intended target but are not true carriers. Indeed, in order to do this, they have to be implanted as closely as possible to the site of action. Microcapsules and microspheres for chemoembolization belong to this group, as do similar systems used for the controlled release of proteins and peptides. Recently, much interest has been directed towards the use of small (10 μm or less) microparticles by the oral route, because of their selective uptake by Peyer's patches, which gives them great potential as carriers for oral vaccines.

In contrast, the carriers known as second-generation are true carriers (usually of colloidal size). Indeed they are capable not only of releasing an active product at the intended target but also of carrying it there after administration by a general route. This group includes so-called passive carriers such as liposomes, nanocapsules and nanospheres, and certain active carriers such as temperature-sensitive liposomes and magnetic nanospheres. It should be noted, however, that after intravenous administration, most colloidal carriers are rapidly removed from the circulation by phagocytic cells in the liver and spleen. This lim-

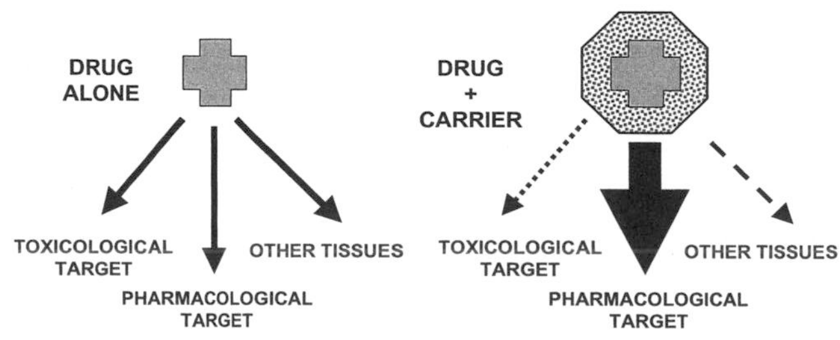

Figure 1 Illustration of the drug carrier concept.

Table 1 Definition of Different Carrier Systems

Generation	Size	Definition	Examples
First	>1 μm	Systems able to release a drug at the target site but necessitating a particular type of administration	Microspheres and microcapsules for chemoembolization and for controlled release of proteins and peptides
Second	<1 μm	Carriers that can be given by a general route able to transport a drug to the target site	*Passive carriers*: Erythrocytes, liposomes, nanoparticles, sterically stabilized liposomes, and nanoparticles *Active carriers*: Temperature or pH-sensitive liposomes, magnetic nanoparticles
Third	<1 μm	Carriers able to recognize the target specifically	Monoclonal antibodies, second-generation carriers, usually sterically stabilized, targeted with monoclonal antibodies or other ligands

Source: Adapted from Ref. 1.

its their targeting potential. Over the last ten years, system surface properties have been modified to avoid the deposition of plasma proteins and thus avoid recognition by phagocytes. These are known as sterically stabilized carriers and may remain in the blood compartment for a considerable time. Although such colloidal particles cannot cross normal continuous capillary endothelium, they have been shown to extravasate into sites where the endothelium is more permeable, such as solid tumors or regions of inflammation and infection.

The carriers referred to as third-generation are also true carriers and are capable of specific recognition of the target. For example, monoclonal antibodies belong to this group, as do certain second-generation carriers (liposomes, nanocapsules, nanospheres) piloted by monoclonal antibodies or other ligands. Of course, targeted colloidal carriers will be much more effective if they are also sterically stabilized.

This chapter, dealing with polymeric micro- and nanoparticles as drug carriers, is mainly based on the work carried out in our laboratory. Rather than giving a catalogue of our research we will concentrate on some of the most important areas. The chapter is divided into two main sections. The first deals with microparticles (microspheres and microcapsules) for the encapsulation of peptides and proteins, including vaccines. The second focuses on nanoparticles (nanospheres and nanocapsules). After a review of the general properties and therapeutic potential of these two different systems, we make special mention of nanoparticles with bioadhesive properties for drug delivery at mucosal surfaces and sterically stabilized nanoparticles that avoid phagocytosis, as described above.

As a preliminary remark, we would like to mention that the term ''nanoparticle'' is considered in this chapter as a submicron drug carrier system of polymeric nature. Thus this term is somewhat general, since it does not take into account the morphological and structural organization of the polymer. In this respect ''nanosphere'' is used to identify a nanoparticle system with a matrix character and constituted by a solid core with a dense polymeric network (Fig. 2). In contrast, ''nanocapsules'' are formed by a thin polymeric envelope surrounding an oil-filled cavity. Nanocapsules may thus be considered as a ''reservoir'' system (Fig. 2). Despite these definitions, in practice the term nanoparticles is also used (instead of nanospheres) to designate polymeric colloidal systems with a matrix structure. Most of the work described in this chapter is of a multidisciplinary nature and was carried out in collaboration with a large number of other groups both within and outside France, especially insofar as the biological evaluation of these systems is concerned. They are too many to mention individually but their names will be found in the reference list.

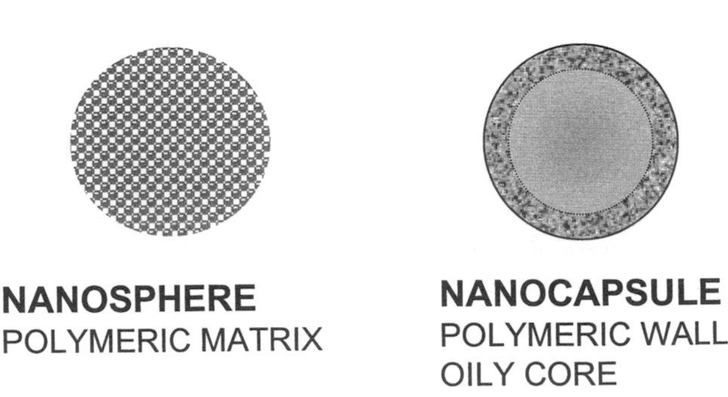

Figure 2 Schematic representation of nanospheres and nanocapsules.

For a more comprehensive description of micro and nanoparticles as drug carriers, readers are referred to several recent books and reviews (2–6).

II. CONTROLLED DELIVERY OF PEPTIDES AND VACCINES BY PLGA MICROSPHERES

During the last years, the design of preparation methods able to encapsulate hydrophilic drugs in biodegradable microparticles has allowed the development of several lines of research on the delivery of peptides and vaccines. This has led to the market several drugs mainly for the delivery of LHRH agonist peptides, and many research teams have approached the development of single-shot vaccines. Recent studies have focused on the delivery of neuropeptides and mucosal vaccines.

A. Delivery of Cholecystokinin (CCK) Analogs

One example of the delivery of neuropeptides is given by cholecystokinin (CCK) analogs. Peptide cholecystokinin, which is widely distributed in the central nervous system and in the gastrointestinal tract, plays a critical role as a neuromodulator in the central nervous system and seems to control emotional and adaptation processes through specific binding sites located in various brain regions. Cholecystokinin octapeptide (CCK-8), the predominant form in the brain, fulfills the criteria of a neurotransmitter or a neuromodulator (7). Numerous studies suggest that CCK is involved in satiety (8), motivated behavior (9), memory (10,11), nociception (12,13) and anxiety (14,15). Moreover, in humans, I.V. injection of CCK-4, the C-terminal fragment of CCK-8, was shown to induce panic attacks, which were suppressed by selective CCK-B antagonists (16). However, the site and the mechanism of action of CCK-4 remain unclear. To study the long-term effect of CCK agonist, a CCK-8 analog, pBC 264 (Boc-Tyr(SO_3H)-gNle-mGly-Trp-Nle-Asp-Phe-NH_2), which is a very potent and selective agonist for CCK-B receptors, was developed (17). This compound has been shown to give responses at low doses in behavioral and electrophysiological tests following local administration (18–20). However, chronic local stimulation of CCK-B receptors is hindered by the rapid elimination of pBC 264 from the brain after central administration (21). In order to study the long-term effects resulting from local stimulation of CCK-B receptors, the possibility of slowly delivering pBC 264 into a particular brain structure for a long period of time has been considered by its encapsulation into poly(lactide-co-glycolide) (PLGA) microspheres.

pBC 264 was encapsulated into microspheres sized between 1 and 10 μm. Using the basic multiple emulsion solvent evaporation procedure, the encapsulation efficiency of pBC 264 in microspheres was very low, However, when introducing in the inner aqueous phase a stabilizing agent such as ovalbumin (OVA) at a concentration of 2%, the encapsulation efficiency was significantly improved (22). An additional and dramatic improvement of the encapsulation rate was also observed when the pH of the internal aqueous phase was basic (pH 8 corresponding to the optimal solubility of pBC 264) and when the pH of the external aqueous phase was acid (pH 2.5 corresponding to a lower solubility of pBC 264) (23). In this case, the internal aqueous phase was playing the role of a real ''trap'' for the peptide. Therefore the simultaneous stabilizing effect of the OVA and the adjustment of the pH allowed an optimal entrapment to be obtained. OVA was believed to prevent the destruction of the internal globules during the process of preparation, therefore avoiding leakage of the peptide into the outer aqueous phase. The inner emulsion should also be stabilized due to the rigid structure formation (hydrophobic barrier) surrounding the drug cores following interaction between basic compounds and dissociated polymers. This improved stability resulted in a better entrapment of pBC 264.

Release kinetic experiments were performed in phosphate buffer solution (pH 7.4). pBC 264 was rapidly released from microspheres exhibiting a dramatic burst effect (22). Thus the microspheres obtained did not exhibit a typical controlled release profile. The presence of OVA as a stabilizer was shown to be responsible for the observed burst effect. Electron microscopy observations clearly showed that OVA induced the formation of pores at the surface of the microspheres (22). Actually, as OVA is totally insoluble in methylene chloride, it did not interact with the polymer as suggested by DSC measurements (22). The protein was assumed to be located at the interface between the first and second emulsion. Therefore OVA might have formed aggregates on the surface of the microspheres. The elution of the OVA aggregates from the surface during the microsphere washing process is probably responsible for the formation of pores. These pores could constitute a pathway for the observed fast release of pBC 264.

To reduce the porosity of the microspheres, other surfactants were used as stabilizers of the internal aqueous phase (PVA or Pluronic® F68) (23). The influence of the surfactant on the stability of the first emulsion as well as the encapsulation efficiency and the release of pBC 264 were examined. The encapsulation efficiency was very low for the formulation containing PVA. Pluronic® F68–containing microspheres displayed a lower encapsulation efficiency as compared to OVA-stabilized microspheres

(23). The most important factor for the successful encapsulation of hydrophilic compounds in microspheres prepared by the W/O/W solvent evaporation method is the stability of the first inner emulsion. From the stability studies, it was shown that when PVA was used as a stabilizer, the internal emulsion was unstable and the droplets coalesced. Thus it was assumed that in these conditions the peptide leaked out to the external aqueous phase. A similar phenomenon occurred with Pluronic® F68 but to a lesser exent, since the stability of the emulsion was improved. Pluronic® F68 and PVA are hydrophilic surfactants that are not suitable for stabilizing W/O emulsions. On the contrary, proteins have a tendency to localize at the interfaces, increasing the stability of emulsions, therefore reducing the leakage of the peptide to the external aqueous phase.

Release of pBC 264 from microspheres prepared with Pluronic® F68 was compared to the release obtained from OVA-stabilized microspheres. The burst observed was much smaller when the microspheres were prepared in the presence of Pluronic® F68 (23). The release profile and the extent of the initial peptide burst was significantly influenced by the structure of the microspheres. As mentioned above, the microspheres prepared with OVA had a porous structure and thus provide a fast release profile with a large peptide burst. In contrast, microspheres prepared in the presence of Pluronic® F68 did not have pores and were characterized by a smooth surface (23). With Pluronic® F68 it may be hypothesized that the inner drug cores were more finely dispersed throughout the polymer matrix without any aqueous channels. Therefore the differences in the initial burst from the tested formulations seemed to result from morphological differences.

For in vivo studies, microspheres were implanted into the rat brain in the anterior part of the Nucleus Accumbens (N. Acc.). After the administration of fluorescently labeled microspheres, observations of histological coronal sections confirmed that the microspheres were effectively localized in the anterior part of the N. Acc., and one hour after implantation the microspheres remained located around the injection site (24). The behavioral testing was carried out by placing rats in the open field (24). The behavioral effect was quantified by recording for 6 min: (a) the latency (or time)(s) to move out from the corner where the animal was placed and to cross two squares; (b) locomotion scores of the total number of squares crossed; (c) total number of rears; (d) number of defecation boli left on the field; and (e) number of grooming bouts. No behavioral change was observed 24 hours or 8 days after administration of unloaded microspheres as compared to the control group, whatever the parameters measured in the open field test were (24). Finally, implanted microspheres were well tolerated and minimally reactive, since only a localized inflammatory reaction was observed. This reaction was similar to the one induced by the administration of carboxymethylcellulose (24). Two different kinds of microspheres differing only by the stabilizing agent added to the inner aqueous phase (OVA or Pluronic® F68) were implanted into the rat brain. The in vivo studies showed that pBC 264 was almost completely released from the microspheres prepared with OVA within the first hour, whereas only 35% of peptide was released after 8 days when microspheres prepared with Pluronic® F68 as the stabilizing agent were administered (24). This was in agreement with the in vitro results (22,23). Intact pBC 218 was found to be released from PLGA microspheres containing Pluronic® F68 both in vitro and in vivo (23,24). In addition, the brain diffusion of pBC 218 was very limited, and its presence in the blood could not be detected. The peptide was essentially found concentrated near the site of administration in the N. Acc., showing that chronic behavioral studies using pBC 218 microspheres were made possible using the right formulation (24).

B. Delivery of Vaccines

One recent application of microspheres is the mucosal delivery of vaccines. For this application, the antigen delivery systems consisting of biodegradable microspheres are designed to increase antigen uptake by the M cells if the diameter of the particles is smaller than 10 μm (25), to enhance the mucosal immune response (26), and to induce long and slow antigen release (27). This approach was validated by showing that oral immunization of mice with phosphorylcholine (PC) could induce a mucosal secretion of specific anti-PC IgA. PC was chosen as an antigen because it is a ubiquitous hapten present on different pathogenic microorganisms that colonize or invade host mucosa at different epithelial sites (28–30). Since it has been demonstrated that an immune response disseminated to several effector mucosal sites can be obtained after stimulation of a unique mucosal inductor, one can now hope for unique mucosal vaccines with several potentialities. This is of interest because mucosal vaccines have a number of obvious advantages over invasive vaccines, particularly with respect to simplicity, safety, and cost effectiveness.

PC was conjugated to thyroglobulin (PC-Thyr) and entrapped in PLGA microspheres (31). The preparation used contained 97% of the microspheres with a diameter lower than 10 μm. The encapsulation efficiency of the conjugate was as high as 80% when the initial protein-to-polymer ratio was 1:8, the total protein load being approximately 9% w/w of the microspheres. It was also shown that only about 30% of the PC-Thyr entrapped in the microspheres was released within 30 minutes under acidic gastriclike conditions, and that the preparation was stable thereafer (31). Also, a major drawback of microencapsulation by the solvent evaporation method of antigens for vaccine development is the potential degradation of the antigen during the preparation process. However, this risk was lowered in the multiple emulsion solvent evaporation method used in the present study, because the contact between the compound and the organic solvent was limited. Indeed, the absorbance spectrum of PC-Thyr was not modified by the microencapsulation process, nor was its PC-immunologic reactivity, as judged in an ELISA using a monoclonal anti-phosphorylcholine antibody (31). This strongly suggested that only minor, if any, modifications of PC integrity occurred.

Oral administration of PC-Thyr, entrapped in PLGA microspheres, induced in mice a specific IgA response in intestinal, pulmonary, and vaginal secretions in addition to a strong specific systemic immune response. When mice were challenged with an oral infection induced by *Salmonella typhimurium*, only the animals that were vaccinated with microspheres containing the PC-Thyr conjugate were protected (32). However, only nasal administration, and not oral, was able to protect mice against the pulmonary infection by *Streptococcus pneumoniae* (33,34).

Very recently, a milk protein, β-lactoglobulin (BLG), was entrapped in PLGA microparticles sized around 7 μm (35,36). These particles were able to induce after oral administration an immunological oral tolerance against BLG. In addition, the amount of protein necessary to reduce the IgE response was 10,000 times lower when the BLG was entrapped into microspheres as compared with the native BLG (37). These results allow us to broaden the field of application of biodegradable microparticles to their use in the reduction of oral allergies.

III. POLYMERIC NANOPARTICLES AS DRUG CARRIERS

A. Nanospheres

1. Preparation and Characterization of Nanospheres

Submicronic macromolecular particles can be obtained following an emulsion or micellar polymerization of polymerizable monomers dispersed in an aqueous phase, or from preformed—natural and synthetic—macromolecules such as polymers, proteins, and polysaccharides. For drug-delivery purposes, the colloidal material needs to meet physicochemical and biological requirements adapted and optimized for these specific applications. Among these requirements, biocompatibility, safety, and biodegradability to nontoxic metabolites are of crucial importance, thus limiting the number of materials available. Methods for nanosphere production have been reviewed in Ref. 38.

a. Nanosphere Preparation by Polymerization of Monomers

In our laboratory, we have concentrated on poly(alkylcyanoacrylate) nanospheres. The most significant advantage of alkylcyanoacrylate is that, in contrast to other acrylic derivatives that require an energy input for the polymerization step, which could affect the stability of the adsorbed drug, alkylcyanoacrylates can be polymerized easily without such a contribution (39). These nanospheres are prepared by an emulsion polymerization process, in which droplets of water-insoluble monomers are emulsified in an aqueous phase (40). Anionic polymerization takes place in micelles after diffusion of monomer molecules through the water phase and is initiated by the water itself. The pH of the medium determines both the polymerization rate and the adsorption of the drug when the latter is ionizable (31). Drugs can be combined with nanospheres after dissolution in the polymerization medium either before the introduction of the monomer or after its polymerization. Since only nontoxic additives are used, no further purification is needed, and freshly prepared nanospheres may be freeze-dried with their drug content.

This preparation method is easily reproducible regarding size and drug adsorption rate, even after preparation at a semi-industrial level. Moreover, when prepared under sterile conditions in an aseptic room under laminar flow, poly(alkylcyanoacrylic) nanospheres met the usual requirements needed for intravenous administration such as sterility and lack of bacterial endotoxins (42).

Scanning electron microscopy of nanospheres generally shows spherical particles with a diameter of approximately 150 nm. Freeze-fracture studies revealed that the internal structure of the poly(alkylcyanoacrylic) nanospheres consisted of a matrix made up of a dense polymeric network (43). Molecular weight determinations made by gel permeation chromatography (GPC) suggested that nanospheres are built up from an entanglement of numerous small oligomeric subunits rather than from the rolling-up of one or a few long polymer chains (44). The low molecular weight of oligomeric subunits is consistent with observations concerning metabolism and elimination of nanospheres after the in vivo administration discussed below. The density of poly(isobutyl-cyanoacrylate) and poly (ethyl-butyl-cyanoacrylate) have been determined by isopycnic centrifugation in sucrose gradients and found to be 1.14 and 1.10 g/cm^3, respectively (45).

As far as drug association is concerned, it was found that nanospheres can entrap drug according to a Langmuir adsorption mechanism, due to their large specific surface area (46). The drug can either be incorporated into nanospheres during the polymerization process or be adsorbed onto the surface of preformed particles. In the former case, the drug–polymer interaction may result in the covalent linkage of the drug with the polymer. This was observed with vidarabine, an antiviral agent, whose nucleophilic N in position 3 and 7 may play the role of initiator for the anionic polymerization mechanism of the cyanoacrylic monomer (47). A similar interaction was observed, to a lower extent when a peptide, growth hormone releasing factor (GRF), was associated with nanospheres (48). In contrast, other drugs such as ampicillin are simply adsorbed. Based on the surface potential and surface pressure experiments, models of poly(isobutylcyanoacrylate)–ampicillin arrangements in the interfacial region have been proposed for low and high polymer surface densities (49).

Both the nature and the quantity of monomer influence the adsorption capacity of the carrier. Generally, the longer the alkyl chain length, the higher the affinity of the drug; the capacity for drug adsorption is related to the hydrophobicity of the polymer and to the specific surface area of the carrier. Moreover, the percentage of drug adsorption generally decreases with the quantity of drug dissolved in the polymerization medium, according to the Langmuir isotherm (50).

Poly(alkylcyanoacrylate) particles are degraded by surface erosion process (51), through enzymatic hydrolysis of the ester side chains of the polymer (52). Thus the polymeric backbone remains intact but gradually becomes more and more hydrophilic until it is water soluble. The rate of degradation is dependent on the length of their alkyl chain (53). Therefore it is possible to choose a monomer whose polymerized form has a biodegradability corresponding to the required profile for drug release (54). Indeed, by using a double radiolabel technique (^{14}C-labeled nanospheres loaded with ^{3}H-actinomycin), it was found that drug was released from nanospheres as a direct consequence of polymer bioerosion (52). This was confirmed using the peptide GRF (48).

Biodegradable nanospheres suitable for pharmaceutical applications can also be prepared from another monomer, poly(methylidene malonate 2.1.2), using a similar emulsion polymerization technique (55).

b. *Preparation of Nanospheres from Preformed Polymers*

In order to avoid some of the limitations associated with polymerization of a monomer, in particular the possibility of residual monomer and the risk of drug interaction and inactivation, methods of obtaining colloidal drug-delivery systems from preformed, well-defined macromolecular materials with known physicochemical and biological properties have been developed. These methods were initially adapted from industrial techniques available for the preparation of the artificial latexes used in surface-coating applications: paints, adhesives, textile sizing, paper coating. Original methods based on nanoprecipitation have been developed in our group, capable of yielding nanospheres from both preformed synthetic polymers and from natural macromolecules such as proteins or polysaccharides.

Among many other water-insoluble synthetic polymers, polyesters such as poly(lactide), poly(glycolide), and their copolymers meet the biological requirements of safety and biodegradability to nontoxic metabolites. The methods initially developed for the preparation of nanospheres from these polymers were derived from solution-emulsification techniques. Gurny et al. (56) and Krause et al. (57) prepared poly(lactide) nanospheres according to the solution-emulsification technique. A polyester is dissolved in an organic volatile solvent immiscible with water, such as chloroform, and the organic solution is dispersed in an aqueous phase to form an O/W emulsion. Continuous emulsification under mixing prevents coalescence of organic droplets and allows the spontaneous evaporation of the solvent at room temperature and the formation of the colloidal particles. Residual organic solvent is removed under reduced pressure. The emulsifiers can, if necessary, be removed by dialysis of the suspension, or by washing following the separation of nanospheres by ultracentrifugation.

One major limitation of the method is the size and the homogeneity of the droplets containing the macromolecular material. An ultrafine emulsification is necessary to obtain particles smaller than one micron, and it is very difficult to obtain a monodisperse population of nanospheres. More recently, microfluidization has been used in this step, for example, to prepare poly(lactic acid) nanoparticles coated with human serum albumin (58). As an alternative to the solution-emulsification technique, methods based on the desolvation of macromolecules have been developed.

Desolvation is a well-known method for isolating macromolecules from liquid media. Thus polymers can be precipitated from organic solutions following the addition of a third component, or of a nonsolvent miscible with the organic solvent. The same effect is obtained with an aqueous solution of proteins when neutral salts, i.e., sodium or ammonium sulfate, or alcohol are added. However, the process leads to a bulk precipitate of polymeric or protein material, but not to colloidal particles.

Fessi et al. (59) were the first to propose a new and simple method yielding polymeric nanospheres without the use of preliminary solution-emulsification or the use of autoemulsifying polymeric material. Briefly, an organic solution of the polymer is prepared and added to a nonsolvent. Both solvent and nonsolvent must have low viscosity and high mixing capacity in all proportions. As an example, acetone and water meet these conditions. The only complementary operation following the mixing of the two phases is thus to remove the volatile solvent by vaporization under reduced pressure. Further concentrations of the aqueous suspension can be carried out under the same conditions or by freeze-drying. This method has been successfully applied to various polymeric materials such as poly (lactic acid) and poly(lactic acid)-co(glycolic acid), poly (ε-caprolactone), ethylcellulose and cellulose acetophtalate, poly(alkylcyanoacrylate), poly(vinyl chloride-coacetate), poly(styrene), and poly(acrylic acid). The mean diameter of the particles obtained was about 200 nm with low polydispersity.

Small amounts of surfactants, either lipophilic or hydrophilic, can be added to the solvent and/or to the nonsolvent. The presence of a surfactant was not necessary to obtain the colloidal dispersion, which forms spontaneously, but only to stabilize it and to facilitate its redispersion following sedimentation. It was observed that the size of particles was mainly related to the polymer used and less to the experimental conditions. It was also noteworthy that the sense and the rate of mixing of the two phases had no influence; the organic solution of the polymer could be slowly or rapidly added to the nonsolvent or vice-versa, with or without the aid of additional stirring. In fact, the most critical conditions for obtaining the spontaneous formation of colloidal particles alone, avoiding any bulk precipitation of the material, were the concentration of the polymer in the organic phase and the ratio of solvent to nonsolvent. Depending on its physicochemical nature, the drug of interest can be added either to the solvent or to the nonsolvent. In the first case, it can be incorporated into the polymeric matrix, and its release will be determined by its rate of diffusion with this matrix and by the rate of degradation of the polymer. On the other hand, drug present in the nonsolvent will be adsorbed on the particle surface, and its release will be the result of partitioning. Zeta potential measurements can be a useful tool to determine the location of the drug in nanoparticulate systems (60).

Chang et al. (61) and Bodmeier et al. (62) have described a similar method to prepare nanospheres without any surfactant, using acrylic copolymers bearing quaternary ammonium groups. These autoemulsifying polymers are sensitive to the pH of the dispersing phase and may become soluble, thus limiting both the applications and the choice of the starting material. By using carboxylated polymers, i.e., acrylic Eudragit S, Fessi et al. (59) were able to prepare colloidal particles by simple addition of an aqueous alkaline solution of the macromolecular material to an acidic medium. However, subsequent neutralization of the suspension to physiological pH led to a clear colloidal solution.

Allemann et al. (63) have described a method of nanosphere preparation by reverse salting out. An organic solution of the polymer is included in a O/W emulsion by using an aqueous phase with a high salt concentration, with which the organic solvent is not miscible. Addition of pure water to the system reduces the ionic strength and leads to mixing of the two solvents and deposition of the polymer as nanoparticles. However, the resulting suspension is very dilute and requires subsequent concentration by removal of large amounts of water.

The method of Fessi et al. (59) can be adapted to prepare nanospheres from natural macromolecules. Thus Stainmesse et al. (64) developed a technique to prepare protein nanospheres in a single step by pouring an aqueous solution of the protein (e.g., albumin) into a nonsolvent. In particular, the nonsolvent can be boiling water, thus leading to heat-denaturation of albumin and formation of colloidal particles as small as 80 nm. Further strengthening of the albumin nanospheres can be obtained either by heat-sterilization (121°C) or by chemical cross-linking. It is noteworthy that no surfactant was necessary to obtain a stable dispersion of albumin nanospheres, due to the small and monodisperse size of the particles, even under the conditions of heat-sterilization. More recently, the nanoprecipitation technique has also been applied to amphiphilic cyclodextrins (65). An advantage of using these cyclic glucose oligomers would be that it provides a third possibility for drug association; as well as entrapment within the matrix and adsorption on the surface, suitable molecules could be included within the hydrophobic core of the cyclodextrin.

2. Therapeutic Applications of Nanospheres

Nanospheres have been designed mainly for administration by the intravenous route. In this case, particles whose surface has not been modified to avoid opsonisation will be limited in their applications by their distribution, which is mainly characterized by a rapid uptake into phagocytic cells of the liver, spleen, and bone marrow. These can be summarized as follows: concentrating drugs in accessible sites, rerouting drugs away from sites of toxicity, and increasing the circulation time of labile or rapidly eliminated drugs. Within these limits, nanospheres can not only change the distribution of the drug at the organ or tissue level, but also at the cellular level, by increasing uptake or modifying the intracellular localization of the drug. These properties will be illustrated by work done in our laboratory with three different classes of molecules: antibiotics, anticancer drugs (in particular doxorubicin), and oligo deoxynucleotides (ODN).

a. Application to the Treatment of Intracellular Infections

The need for intracellular chemotherapy has been recognized for many years, since a variety of bacterial diseases originate from so-called obligate or facultative intracellular parasites, which are often located within macrophages, often within the lysosomes (66). These infections are often resistant to commonly used antibiotics because of low intracellular uptake or reduced activity at the acidic pH of lysosomes. An alternative strategy to the development of new antibiotics would be the use of drug-delivery systems to obtain controlled release or targeting to specific sites, thereby improving the efficiency of existing drugs. In particular, the need for antibiotics with intracellular efficacy led to the association of antibiotics with colloidal systems: liposomes or nanoparticles, which can be endocytosed by the infected cells (reviewed in Ref. 67). In particular, our laboratory has developed antibiotics bound to poly(alkylcyanoacrylate) nanospheres.

It has been shown that linkage of ampicillin to poly(isohexylcyanoacrylate) (PIHCA) nanospheres increased its efficacy in treating *Salmonella typhimurium* infection in C57BL/6 mice by a factor of 120. A single injection of 0.8 mg of nanosphere-bound ampicillin was just as efficient as 3 doses of free drug in suppressing all mortality. The bacterial counts of the liver and spleen were 10 to 10,000 times lower in the mice given free ampicillin mixed with, or bound to, nanospheres than in control mice. None of the treatments tested was, however, able to sterilize the organs of the mice, since living bacteria were found in the livers and spleens of all mice still alive 60 days after bacterial inoculation, although the use of the nanosphere-bound form greatly increased the ampicillin concentration in these organs (68).

This high efficacy of nanosphere-bound ampicillin observed in the treatment of acute murine experimental salmonellosis is probably attributable to the combined effect of two types of targeting: first, the linkage of ampicillin to nanospheres led to the concentration of the drug in the liver and spleen; second, the uptake of ampicillin by infected macrophages was higher when the drug was bound to nanospheres than when it was in the free form (69). Transmission electron microscopy showed nanospheres colocalized with intracellular bacteria (70), and ultrastructural autoradiography showed that ampicillin itself could reach the bacterial cell walls, both by direct contact with nanospheres and by diffusion through the cell (71).

Ampicillin bound to PIHCA nanospheres was also effective against experimental listeriosis in athymic nude mice, a model involving a chronic infection of both liver and spleen macrophages (72). The nanosphere-bound form of ampicillin was capable of ensuring liver sterilization after two injections of 0.8 mg of nanosphere-bound drug, whereas no such sterilization was ever observed with any of the other regimens tested (free ampicillin or free ampicillin + unloaded nanospheres). On the other hand, spleen bacterial counts were less affected by the treatments (free or bound ampicillin). This was consistent with the in vivo distribution profile of nanospheres after intravenous administration. Autoradiographic studies performed with PIHCA nanospheres have shown that most of these particles were present in the liver 5 min after intravenous administration. Later, they also concentrated in the spleen, but to a lesser extent (73). The distribution profile of [^{14}C]-PIHCA nanospheres in mice showed that 60–80% of the polymer concentrated in the liver, whereas only 2–6% was in the spleen (74). This probably explains the fact that the spleen bacterial counts were less affected than the liver ones by ampicillin targeted by means of nanospheres.

In an attempt to kill both dividing and nondividing bacteria, a fluoroquinolone antibiotic, ciprofloxacin, has been associated with poly(isobutylcyanoacrylate) (PIBCA) and PIHCA nanospheres. In an animal model of persisting salmonella infection, although an effect on the early phase of the infection was observed, neither free nor nanoparticle-bound ciprofloxacin was able to eradicate truly persisting bacteria (75).

b. Application to the Treatment of Cancer

Intravenously administered anticancer drugs are distributed throughout the body as a function of the physicochemical properties of the molecule. A pharmacologically active concentration is reached in the tumor tissue at the expense of massive contamination of the rest of the body. For cytostatic compounds, this poor specificity raises a toxicological problem that presents a serious obstacle to effective therapy. The use of colloidal drug carriers could represent a more rational approach to specific cancer therapy. This section focus on the increase of the therapeutic index of doxorubicin when attached to poly(alkylcyanoacrylate) nanospheres. The possibility that multidrug resistance might be overcome by using doxorubicin-loaded nanospheres will also be considered.

The antitumor efficacy of doxorubicin-loaded nanospheres was first tested using the lymphoid leukemia L-

1210 as a tumor model. In this study, one intravenous injection of doxorubicin-loaded PIBCA nanospheres was found to be more effective against L1210 leukemia than when the drug was administered in its free form following the same dosing schedule (76). Although the increased life span (ILS%) of mice injected with doxorubicin-loaded PIBCA nanospheres was twice as high as the ILS% for free doxorubicin, there were no long-term survivors.

The effectiveness of doxorubicin-loaded poly(isohexylcyanoacrylate) (PIHCA) nanospheres against L1210 leukemia was even more pronounced than that of doxorubicin loaded onto PIBCA nanospheres. The drug toxicity was markedly decreased when it was bound to this sort of nanosphere, so that impressive results were obtained with this formulation at doses for which the therapeutic efficiency of free doxorubicin was completely masked by the overpowering toxicity of the drug (76). Furthermore, preliminary experiments suggested that one I.V. bolus injection of doxorubicin-loaded nanospheres was more active, in L1210-bearing mice, than perfusion of the free drug for 24 h.

The superiority of doxorubicin targeted with the aid of poly(alkylcyanoacrylate) nanospheres was later confirmed in a murine hepatic metastases model (M5076 reticulosarcoma) (77). Irrespective of the dose and the administration schedule, the reduction in the number of metastases was much greater with doxorubicin-loaded nanospheres than with free doxorubicin, particularly if treatment was given only when the metastases were well established. The improved efficacy of the targeted drug was clearly confirmed by histological examinations showing that both the number and the size of the tumor nodules were lower when doxorubicin was administered in its nanoparticulate form (77). Furthermore, liver biopsies of animals treated with the nanosphere-targeted drug showed a lower cancer cell density inside tumor tissue. Necrosis was often less widespread with the nanosphere-associated drug than in the control group and the group treated with free doxorubicin.

Studies performed on total homogenates of livers from both healthy and metastases-bearing mice showed extensive capture of nanoparticulate doxorubicin by the liver; no difference in hepatic concentrations was noted between healthy and tumor-bearing animals (77). In order to elucidate the mechanism behind the enhanced efficiency of doxorubicin-loaded nanospheres, doxorubicin measurements were made in both metastatic nodules and neighboring healthy hepatic tissue. This provided quantitative information concerning the drug distribution within these tissues (78). During the first 6 h after administration the exposure of the liver to doxorubicin was 18 times greater for nanosphere-associated doxorubicin. However, no special affinity for the tumor tissue was detected, and the nanospheres were seen by electron microscopy to be located within Küpffer cells (macrophages). However, at later time points, the amount of drug in the tumor tissue increased in nanosphere-treated animals to 2.5 times the level found in animals given free doxorubicin. Since uptake of nanospheres by neoplastic tissue is unlikely, this increase in the doxorubicin concentration in tumor tissue probably resulted from doxorubicin released from healthy tissue, in particular Küpffer cells. Hepatic tissue could play the role of drug reservoir from which prolonged diffusion of the free drug (from nanospheres entrapped in Küpffer cell lysosomes) toward the neighboring malignant cells occurs. Similar results were obtained with doxorubicin-containing liposomes (79).

This hypothesis raises the question of the long-term effect of an 18-fold increase of doxorubicin concentration in the liver. Although toxicological data have shown that doxorubicin-loaded nanospheres were not significantly or unexpectedly toxic to the liver in terms of survival rate at high doses, body weight loss, and histological appearance (80), this possibility should be borne in mind, especially since a temporary depletion in the number of Küpffer cells, and hence the ability to clear bacteria, was observed in rats treated with doxorubicin-loaded liposomes (81). A systematic study using unloaded poly(alkylcyanoacrylate) nanoparticles confirmed a reversible decline in the phagocytic capacity of the liver after repeated dosing, as well as a slight inflammatory response (82,83). Nanoparticle-associated doxorubicin also accumulated in bone marrow, leading to myelosuppression (84). However, this tropism of carriers might be useful to deliver myelostimulating compounds such as granulocyte colony stimulating factor to reverse the suppressive effects of intense chemotherapy (85). Nanospheres are also captured by splenic macrophages (86). In this study, the spleen architecture was shown to play a role in the localization of the nanospheres: in the mouse, uptake was mainly in metallophilic macrophages of the marginal zone, whereas in the rat, which has a sinusoidal spleen similar to that of the human, particles were found in the red pulp macrophages.

On the other hand, alteration of the drug distribution profile by linkage to nanospheres can in some cases considerably reduce the toxicity of a drug because of reduced accumulation in organs where the most acute toxic effects are exerted. This concept was indeed illustrated with doxorubicin, which displays severe acute and chronic cardiomyopathy. After intravenous administration to mice, plasma levels of doxorubicin were higher when the drug was adsorbed onto nanospheres, and at the same time the cardiac concentration of the drug was dramatically reduced (87). In accordance with the observed distribution profile, doxorubicin associated with nanospheres was found to be less toxic than free doxorubicin (80).

The ability of tumor cells to develop simultaneous resistance to multiple lipophilic compounds represents a major problem in cancer chemotherapy. Cellular resistance to anthracyclines has been attributed to an active drug efflux from resistant cells linked to the presence of transmembrane P-glycoprotein, which was not detectable in the parental drug-sensitive cell line. Drugs, such as doxorubicin, appear to enter the cell by passive diffusion through the lipid bilayer. Upon entering the cell, these drugs bind to P-glycoprotein, which forms transmembrane channels and uses energy from ATP hydrolysis to pump these compounds out of the cell (88). To solve this problem, many authors have proposed the use of competitive P-glycoprotein inhibitors, such as the calcium channel blocker verapamil, which are able to bind to P-glycoprotein and to overcome pleiotropic resistance. However, since the adverse effects of verapamil are serious, its clinical use to overcome multidrug resistance is limited.

During the past few years, many studies have been devoted to evaluating the antitumor potential of carrier–drug complexes (89). We have evaluated the effect of nanospheres loaded with doxorubicin, resistance to which is known to be related to the presence of P-glycoprotein. We compared the cytotoxicity of free-Dox, Dox-loaded PIHCA nanospheres (NP-Dox) (mean diameter 300 nm), and nanospheres without drug (NP) against sensitive (MCF7) and multidrug-resistant (Dox R MCF7) human breast cancer cell lines (90). MCF7 cells were more sensitive to free-Dox than Dox R MCF7 cells with a 150-fold difference in the IC_{50}. No significant difference was observed in the survival rate of MCF7 treated with free-Dox or NP-Dox. In contrast, for Dox R MCF7, the IC_{50} for Dox was 130-fold lower when NP-Dox was used instead of free-Dox (90). These results indicated that nanospheres provided an effective carrier for introducing a cytotoxic dose of doxorubicin into the pleiotropic resistant human cancer cell line Dox R MCF7.

Complementary experiments, conducted with other sensitive and resistant cell lines, have confirmed this efficacy of nanospheres (91,92). Doxorubicin resistance was circumvented in the majority of the cell lines tested, and some encouraging results were obtained in vivo in a P388 model growing as ascites (91). Further studies were undertaken to elucidate the mechanism of action of poly(alkylcyanoacrylate) nanospheres. The incubation time and number of particles per cell were important factors (93), and, when PIBCA nanospheres were used, doxorubicin accumulation within P388/ADR resistant leukemic cells was increased compared with free drug, although no endocytosis of nanospheres occurred (94). On the other hand, when the less rapidly degradable PIHCA nanospheres were used, reversion was observed in the absence of increased intracellular drug (95). The degradation products of poly(alkylcyanoacrylate) nanospheres [mainly poly(cyanoacrylic acid)] were also able to increase both accumulation and cytotoxicity of doxorubicin, although they were soluble in the culture medium. Hence the reversion of resistance seems to be due both to the adsorption of nanospheres on the cell surface and to the formation of a doxorubicin-poly(cyanoacrylic acid) ion pair that facilitates the transport of the drug across the cell membrane (95).

In the light of the results obtained with doxorubicin-loaded nanospheres in the liver metastases model described above (78), the role of macrophages as a reservoir for doxorubicin was tested in a two-compartment coculture system in vitro with both resistant and sensitive P388 cells (96). Even after prior uptake by macrophages, doxorubicin-loaded PIBCA nanospheres were able to overcome resistance. However, this reversion was only partial. It was decided to take advantage of the particulate drug carrier offers to associate an anticancer drug and a compound capable of inhibiting the P-gp. This approach was tested with doxorubicin and cyclosporin A bound to the same nanospheres and was found to be extremely effective in reversing P388 resistance (96). The association of cyclosporin A with nanospheres would ensure that it reaches the same sites as the anticancer drug at the same time and would also reduce its toxic side effects.

c. *Nanospheres for Oligonucleotide Delivery*

Oligodeoxynucleotides are potentially powerful new drugs because of their selectivity for particular gene products in both sense and antisense strategies. However, they pose a challenge to pharmaceutical technology because of their susceptibility to enzymatic degradation and their poor penetration across biological membranes. Liposomal systems have been shown to protect oligonucleotides from nucleases and to improve their efficacy in vitro (97,98). Nanoparticulate preparations might be an interesting alternative because of better stability in the presence of biological fluids. In the case of nanospheres, since oligonucleotides have no affinity for the polymeric matrix, association with nanoparticles has been achieved by ion pairing with a cationic surfactant, cetyl trimethyl ammonium bromide (CTAB) adsorbed onto the nanoparticle surface. Oligonucleotides bound to poly(alkylcyanoacrylate) nanospheres in this way were protected from nucleases in vitro (99), and their intracellular uptake was increased (100). Antisense oligonucleotides formulated in this way were able specifically to inhibit mutated Ha-*ras*-mediated cell proliferation and tumorigenicity in nude mice (101).

This approach has recently been applied to the association of a phosphodiester antisense oligonucleotide directed against the 3′ nontranslated region of the PKCα gene with nanoparticles prepared from poly(isobutylcyanoacrylate). These nanospheres were able to inhibit PKCα neoexpression in cultured Hep G6 cells (102).

Nanospheres containing oligonucleotides have also been formulated from a naturally occurring polysaccharide, alginate, which forms a gel in the presence of calcium ions. In this case, the oligonucleotides penetrate into the gel matrix by reptation, thus providing a high loading yield and good protection against nucleases (103).

B. Nanocapsules

1. Preparation and Characterization of Nanocapsules

As explained in the introduction, nanocapsules are characterized by an oily core surrounded by a thin polymeric wall (Fig. 2).

a. *Preparation of Nanocapsules*

A wide range of oils are suitable for the preparation of nanocapsules, including vegetal or mineral oils as well as pure compounds such as ethyl oleate and benzyl benzoate. The criteria of selection for oils are the absence of toxicity, the absence of risk of degradation of the polymer, and a high capacity to dissolve the drug in question.

Polymers, generally used at 0.2–2% (w/w), can be of different origins: natural (arabic gum, gelatine), hemisynthetic (ethylcellulose, hydroxypropylmethylcellulose phthalate, diacyl β cyclodextrins), or, more commonly, synthetic [poly(D,L-lactide), poly(ε caprolactone), poly (alkylcyanoacrylate)].

Both lipophilic and hydrophilic surfactants are used in the preparation of nanocapsules, usually at 0.2–2% (w/w). Generally the lipophilic surfactant is a natural lecithin of relatively low phosphatidylcholine content, while the hydrophilic one is synthetic: anionic (lauryl sulfate), cationic (quaternary ammonium), or, more commonly, nonionic (poly(oxyethylene)-poly(oxypropylene) glycol).

Two main techniques for the preparation of nanocapsules have been described: interfacial deposition processes and an emulsification diffusion technique.

The interfacial deposition processes may begin with either a preformed polymer (104,105) or a monomer (usually alkylcyanoacrylate) that polymerizes at the oil–water interface (106). In both cases, the procedure generally consists of mixing a water-miscible organic phase such as an alcohol or a ketone containing oil (with or without lipophilic surfactant) with an aqueous phase containing a hydrophilic surfactant. In the first case a preformed synthetic, hemisynthetic, or natural polymer is solubilized in the organic phase (or in a phase in which the polymer is soluble). After addition of the organic phase to the aqueous phase, the polymer diffuses with the organic solvent and is stranded at the interface between oil and water. In the second case, monomers of alkylcyanoacrylates are solubilized in ethanol and oil and then dispersed in water containing surfactants. Anionic polymerization of the cyanoacrylate in the oily phase is initiated at the interface by nucleophiles such as hydroxyl ions in the aqueous phase, leading to the formation of nanocapsules.

Gallardo et al. (107) reported that the most important factor that leads to the structure of nanocapsules is the diffusion of the organic solvent, with a complete miscibility of the organic phase in the aqueous phase and an insolubility of the polymer in both the oily phase and the aqueous phase. On the other hand, the stirring and the temperature of the aqueous phase do not seem to influence the characteristics of the nanoparticles formed.

Although the preparation of nanocapsules by interfacial deposition of preformed polymer is a simple, reproducible, easily scaled-up technique, applicable to many polymers, the interfacial polymerization process could also be interesting. In fact, in some cases (depending on the drug to be encapsulated) polymerization can lead to covalent reactions that could improve its encapsulation and retard its release (47).

The emulsification diffusion technique described by Quintanar et al. (108) is based on the initial formation of an O/W emulsion containing an oil, a polymer, and a drug in the organic solvent, in an aqueous solution of a stabilizing agent. Finally, the organic solvent is displaced into the external phase by addition of excess water. This alternative technique to prepare biodegradable nanocapsules, also starting from preformed polymers, is an interesting method of preparation that has several advantages: large choice and small quantities of solvents, simplicity, control of the size of the nanocapsules obtained (from 80 to 900 nm), control of the thickness of the polymeric wall by increased concentrations of the polymer (109), and the possibility of preparing nanocapsules with an inner aqueous core (110); however, a large amount of water has to be removed by evaporation if the organic solvent is highly miscible with water.

b. *Characterization of Nanocapsules*

The size of nanocapsules is usually found to be between 100 and 500 nm and depends on several factors: the nature and the concentration of the polymer and encapsulated drug, the amount of surfactants, the ratio of organic solvent to water, the concentration of oil in the organic solution

(111) or the speed of diffusion of the organic phase in the aqueous phase. This size evaluation is most frequently made by quasi elastic-light scattering. Scanning and transmission electron microscopy and scanning electron microscopy without or after freeze fracture have also been used to check their size and evaluate the structure of nanocapsules (105,108,112–114). Rollot et al. (112) estimated the thickness of polymeric wall of the particles prepared by interfacial deposition process as about 5 nm based on freeze-fracture microscopy. This agrees with theoretical calculations based on the relative volumes of oil and polymer in the formulations. Freeze-fracture microscopy also allowed visualization of different possible organizations of lipophilic surfactant, which can form vesicles, micelles, bilayers, or monolayers depending on the concentration (105). Quintanar et al. (108) have recently confirmed the spherical shape of these nanocapsules, which also appeared as smooth spheres by atomic force microscopy.

Measurement of the zeta potential is the technique most frequently used to characterize the surface of nanocapsules (60). The zeta potential reflects the electrical surface potential of particules, which is influenced by the charge of the different components of the nanocapsules located at the interface with the dispersing medium. High zeta potential values, above 30 mV (positive or negative value), lead to more stable nanocapsule suspensions, because repulsion between particles prevents their aggregation. Lecithins, poloxamer, and polymer are the major components that can affect this potential; while the polymers, especially the α-hydroxyacids such as poly(D,L-lactide) and lecithins impart a negative charge to the interface; poloxamer, a nonionic surfactant, tends to reduce the absolute value of the zeta potential (116). Chouinard et al. (117) showed that the molecular weight of poly(alkylcyanoacrylate) polymers influenced the surface charge of nanocapsules. Higher zeta potential values were observed with lower molecular weight polymers, presumably because the number of charged end groups was greater. As described by Benita and Levy (118) for emulsions, it is possible to modulate the negative charge by the choice of the lecithins and their purity, since the negative charge is due to phospholipids other than phosphatidylcholine in the lecithin. Monolayer studies carried out with mixed surfactant monolayers by Santos-Magalhaes et al. (119) have shown that the adsorption of the poloxamer surfactant is enhanced by the presence of phospholipids. Recently Calvo et al. (120) proposed alternative nanocapsules coated with positively charged hydrophilic polymer based on polysaccharide chitosan. Their surface charge depends mainly on the viscosity of the chitosan used.

Atomic force microscopy also allowed the surface of nanocapsules prepared by the emulsification diffusion technique to be observed (109). These surfaces are smooth with contrasting shiny regions, as compared with images of nanospheres, that could be due to free oil or to resonance frequencies during the scanning process.

Gallardo et al. (107) showed that ultracentrifugation of nanocapsules prepared from IBCA and Miglyol 812 yielded both a floating layer (nanocapsules) and a small pellet (nanospheres). This could be explained by the postulated mechanism of formation of nanoparticles at the interface of the rapidly diffusing organic solvent and water, which fragments to produce nanocapsules and nanospheres, depending on whether oil is entrapped or not.

Centrifugation in a density gradient is an accurate method to confirm the existence of nanocapsules by comparison with colloidal carriers prepared without polymer or without oil. Chouinard et al. (117) and Quintanar et al. (109) have reported isopycnic centrifugation in a density gradient of Percoll with respectively poly (alkylcyanoacrylate)/Miglyol and PLA/Miglyol nanocapsules. The density of nanocapsules was found to be intermediate between that of nanospheres and that of emulsions. They also demonstrated that the density of nanocapsules and the band thickness increased when the quantity of polymer increased.

One of the advantage of nanocapsules over nanospheres is their low polymer content and a high loading capacity for lipophilic drugs. The percentage of encapsulation is generally related to the solubility of the drug in the oily inner core, as demonstrated by Fresta et al. (114) with different antiepileptic drugs.

The conservation of nanocapsules can be also improved by freeze-drying (121,122). However, these carriers need large amounts of cryoprotectors such as trehalose to prevent the capsules from disintegrating and to avoid aggregation on their rehydration. Recently Calvo et al. (120) showed that chitosan coated nanocapsules can be freeze-dried with lower amounts of cryoprotective agents than uncoated negatively charged nanocapsules.

Release of encapsulated drugs from nanocapsules appears only to be by the partition coefficient of the drug between oily core and the aqueous external medium and the relative volumes of these two phases. The rate of diffusion of the drug through the thin polymeric barrier does not seem to be a limiting factor, nor does the nature of polymeric wall. On the other hand, the nature of the external aqueous phase is of prime importance. For example, Ammoury et al. (115,123) showed variations in the release profile of indomethacin as a function of the medium: drug release was faster and more complete in the presence of albumin, which acts as an acceptor in the aqueous phase. Similarly, release of halofantrine, a highly lipophilic antimalarial drug, was only observed in the presence of serum, since the drug has a high affinity for lipoproteins (124).

However, the nature of the nanocapsule surface can modulate the release rate in some cases. For example, the presence of excess lecithin at the interface could decrease the rate of release of indomethacin (123). Release of halofantrine from surface-modified nanocapsules bearing poly-(ethylene glycol) chains (see Section III. D) was reduced compared with conventional nanocapsules (124).

2. Therapeutic Applications of Nanocapsules

The potential of nanocapsules as carriers for liposoluble drugs will be discussed with respect to the route of administration: oral, parenteral, or ocular.

a. Oral Route

The oral route was the first route of administration of nanocapsules loaded with indomethacin and insulin in the mid-1980s with the aim of avoiding local side effects and/or improving the biological response. Several in vitro studies have evaluated the release rate of drugs encapsulated in nanocapsules prepared from polyester and poly(alkylcyanoacrylate) polymers in media mimicking pH of the gut (125–127). The results seemed to indicate that the kinetics were more sensitive to changes in drug partitioning related to the change of pH than the nature of polymers as suggested by Magalhaes et al. (113). Although acidification may catalyze hydrolysis of the ester linkages of the polymer, the release of drugs is not dependent on the destruction of the nanocapsules. In order better to understand the behavior of nanocapsules after oral administration, Marchais et al. (126), Kedzierewicz et al. (127), and Lowe and Temple (128) also analysed the interactions of nanocapsules with digestive enzymes. Drug release from nanocapsules was accelerated in the presence of proteases and esterases. In the latter case, this was correlated with a decrease in polymer molecular weight.

Damgé et al. (129) studied the fate of nanocapsules after oral administration in vivo using an iodized oil (Lipiodol), a contrast agent for tomography, as a marker. This showed that encapsulation increased the plasma concentration of the marker, which remained at a high level for a few hours. This improved absorption was attributed to the prolongation of the contact between the encapsulated iodized oil and mucosa, or to a transfer of the drug still associated with nanocapsules through the intestinal mucosa by a paracellular pathway. Evidence of the latter pathway was given by the observation of nanocapsules in the intestinal capillaries by electron microscopy coupled with atomic absorption spectroscopy, but this technique did not allow the extent of passage to the blood to be quantified.

Mucosal protection. Two major nonsteroidal antiinflammatory agents have been encapsulated in this carrier in order to reduce their side effects on the gastric mucosa. The pharmaceutical activity of diclofenac (130) and indomethacin (131) after oral administration does not seem to be influenced by the encapsulation of the drug in nanocapsules. In fact, nanocapsules containing these two drugs exhibited drug concentration–time profiles in the plasma of rats similar to those obtained with the corresponding aqueous solutions. These observations seemed to be independent of the nature of the polymer coating (132). In contrast, the side effects of both drugs were completely modified and reduced by encapsulation in nanocapsules. In fact, nanocapsules induced a marked protective effect on the gastrointestinal mucosa as compared with the ulcerative effect observed with the drug solutions. Fawaz et al. (133) also observed a reduced rectal irritability of indomethacin after administration of nanocapsules by the rectal route. These results suggested that mucosal side effects after oral administration were of local rather than systemic origin, since intravenous administration of nanocapsules did not reduce the gastrointestinal side effects. This protection could be attributed to a slow release of the drugs in the acidic gastric environment or to the reduced toxicity of the acidic form of diclofenac or indomethacin encapsulated in nanocapsules compared with the sodium salt in the aqueous solution (130).

Improvement of biological response. Poly(isobutylcyanoacrylate) nanocapsules were shown 10 years ago to be able to encapsulate insulin and to increase its activity as assessed by a reduction of glycemia (134). Several aspects of this phenomenon are surprising: encapsulation of a hydrophilic drug in the oily core of nanocapsules; reduction of glycemia was only obtained with diabetic animals; hypoglycemia appeared two days after administration and was maintained for up to 20 days depending on the insulin doses, although the amplitude of the pharmacological effect (minimum level of blood glucose) did not depend on the insulin dose. Damgé et al. (135) and Lowe and Temple (128) suggested that nanocapsules could protect insulin from proteolytic degradation in intestinal fluids, based on the protection of encapsulated insulin, observed in the presence of different enzymes in vitro. Later studies showed that insulin did not react with the alkylcyanoacrylate monomer during the formation of nanocapsules and was located within the oily core rather than adsorbed on their surface (136,137).

The capacity of insulin nanocapsules to reduce glycemia could be explained by their translocation through the intestinal barrier, as suggested by Damgé et al. (135), for example by paracellular pathway or via M cells in Peyer's patches (138). Recently, the use of Texas Red®–labelled insulin allowed this translocation to be visualized more readily (139). One hour after oral administration, nanocapsules reached the ileum. The presence of fluorescent areas within the mucosa and even in the lamina propria suggested that insulin-loaded nanocapsules could cross the intestinal epithelium. Although this passage is certainly an important factor, it does not explain the duration of the hypoglycemia. This prolonged action could be due to the retention of a part of the colloidal system in the gastrointestinal tract.

Interestingly, a prolonged hypoglycemic effect was also observed with insulin entrapped in poly(alkyl cyanoacrylate) nanospheres when these were dispersed in an oily phase containing surfactant (140). This suggests that some components of nanocapsules could act as promoters of absorption.

Recently, Damgé et al. (141) showed that the incorporation of octreotide, a somatostatin analogue, in poly(alkylcyanoacrylate) nanocapsules also improved and prolonged the therapeutic effect of this peptide, after administration by the oral route.

Another application by the oral route concerns anti-infectious agents such as atovaquone and rifabutine, two compounds active against *Toxoplasma gondii*, an opportunistic parasite whose antiparasitic activity is limited by their poor bioavailability due to their insolubility in water. Encapsulation of atovaquone resulted in an increase of both the survival of mice infected by the parasite and of its activity after an intragastric administration (142). This improvement of activity translated as a decrease in the brain parasitic burden, which was more pronounced in mice treated with loaded nanocapsules than in those treated with free drug.

b. Parenteral Route

As far as the parenteral route is concerned, nanocapsules are useful for formulating poorly soluble drugs for injection and for modulating the distribution of a drug after injection according to the properties of the carrier.

Encapsulation of poorly soluble drugs. Two poorly water-soluble nonsteroidal anti-inflammatory agents that showed different behavior after I.V. administration have been studied: indomethacin and diclofenac. Diclofenac in solution or in nanocapsules showed similar plasma concentration profiles, whereas encapsulated indomethacin showed lower plasma concentrations than free drug, due to enhanced hepatic uptake of loaded nanocapsules (143). In contrast to the oral route, the similar pharmacokinetic parameters obtained after I.V. administration were accompanied by similar deleterious effects at the level of the intestinal mucosa. Subcutaneous injection did not lead to a slow release of the drug either. One possible explanation about the lack of modification of the pharmacokinetics of anti-inflammatory agents could be the rapid rate of release of these drugs into the circulation in the presence of plasma proteins. Nevertheless, after I.M. administration, nanocapsules of diclofenac showed significantly reduced inflammation at the site of injection as compared with the free drug in solution (144). Similar observations were made by Hubert et al. (145) with nanocapsules of darodipine, a poorly soluble antihypertensive.

Passive targeting to macrophages. As early as 1986, Al Khouri et al. (106) observed that, like other colloidal carriers, nanocapsules, administered by the I.V. route in rabbits, were taken up rapidly by organs of the mononuclear phagocyte system.

As well as changing the distribution of a drug, nanocapsules may also affect its elimination. Since encapsulated drug is concentrated in the Küpffer cells of the liver, the proximity to the hepatic parenchymal cells may increase its secretion into bile. This was found to be the case in rabbits, when indomethacin-loaded PLA nanocapsules were administered intravenously (143). Both the biliary excretion and the enterohepatic circulation were increased with the nanocapule form.

One application that takes advantage of this uptake concerns nanocapsules of MTP-Chol (Muramyl tripeptide cholesterol) developed in our laboratory. This immunostimulating agent is able to activate macrophages and induce toxicity towards tumor cells and would therefore be a useful agent to treat metastatic cancer. The mechanisms by which activated macrophages arrest tumor proliferation include production of nitric oxide and TNF-α. We showed in in vitro models of rat alveolar macrophages and RAW 264.7 mouse monocyte macrophage line that nanocapsules based on poly(D,L-lactide) containing MTP-Chol are more efficient activators than the free drug (146,147). This action could be due to an intracellular delivery of the immunomodulator encapsulated in nanocapsules after phagocytosis and to an intermediate transfer of the drug to serum proteins (148). This system has also demonstrated its efficiency in vivo; in fact Barratt et al. (149) reported antimetastatic effects of nanocapsules containing MTP-Chol in a model of liver metastases. Some antimetastatic activity was also seen after oral administration.

c. Ocular Delivery

Tear turnover, lachrymal drainage, and the hydrophobic structure of the corneal epithelium combine to reduce greatly the ocular bioavailability of drugs formulated as eye drops. First studies carried out with nanocapsules, as ocular drug carriers, attempted to increase the penetration of lipophilic drugs into the eye by prolonging the precorneal residence time, as observed with other colloidal systems, liposomes, and nanospheres. These studies, which concerned antiglaucomatous agents such as betaxolol, carteolol, and metipranolol encapsulated in nanocapsules, only showed a reduction of the noncorneal absorption (systemic circulation) leading to reduced side effects as compared with the free drug (150–152). These systemic side effects are due to a poor ocular retention of drugs, which are directly absorbed into the systemic circulation by conjunctival and nasal blood vessels. In two cases (carteolol and betaxolol), encapsulation in nanocapsules produced a better pharmacological effect (reduction of intraocular pressure) than free drug and nanospheres (although the penetration of nanocapsules was not tested) and reduced cardiovascular systemic side effects (150,151). Metipranolol showed the same activity alone and associated with nanocapsules but, as in the case of carteolol and betaxolol, its side effects were reduced. When betaxolol was used, the nature of the polymer making up the nanocapsule wall was found to be important, and poly (ε-caprolactone) was more efficient than poly(isobutylcyanocrylate) or poly(lactic-co-glycolic acid) (151).

Calvo et al. (153) explored the mechanisms of interaction of nanocapsules with ocular tissues the better to understand the pharmacological responses obtained with antiglaucomatous agents. By confocal microscopy, they showed that poly(ε-caprolactone) nanocapsules could specifically penetrate the corneal epithelium by an endocytic process without causing any damage to the cells, in contrast with poly(isobutylcyanoacrylate) nanoparticles, the uptake of which was associated with cellular lysis (154). These results explained the improved therapeutic effect and the reduction of systemic side effects as a result of drug loss through the conjunctiva provided by poly(ε-caprolactone) nanocapsules by increasing corneal epithelium penetration of lipophilic drugs. Calvo et al. (155) also excluded the influence of the oily inner structure in the activity of the nanocapsules, in the light of the absence of differences in penetration between nanospheres and nanocapsules, in contrast with Marchal-Heussler et al. (150), who observed a better therapeutic effect with nanocapsules than with nanospheres. Moreover, Calvo et al. (155) demonstrated with indomethacin-loaded nanocapsules that the colloidal nature of the carrier was the main factor influencing its ocular bioavailability. The same authors were also interested in the influence of the nature and the charge of the surface of the nanocapsules on their physical stability and on their ocular bioavailability (156,157). They found that coating the negatively charged surface of poly ε-caprolactone nanocapsules with cationic polymers could prevent their degradation caused by the adsorption of lysozyme, a positively charged enzyme found in tear fluid (156). Moreover, they noticed that a cationic polymer, chitosan, adsorbed on the surface of nanocapsules, was able to provide the best corneal drug penetration without any local intolerance as compared to another positively charged polymer. This was achieved by a combination of effects: penetration of particles into the corneal epithelial cells, mucoadhesion of positively charged particles onto negatively charged membranes, and a specific effect on the tight junctions (157).

This effect of improvement of ocular absorption was also reported by Calvo et al. (158) with the immunosuppressive peptide cyclosporin A. The corneal level of the drug was increased fivefold as compared with an oily solution of the drug owing to a highly loaded nanocapsule preparation, also containing poly(ε-caprolactone). The efficacy of this topical formulation has also been observed on a penetrating keratoplasty rejection model in the rat (159). Le Bourlais et al. (160) also proposed an alternative preparation of cyclosporin nanocapsules based on poly(alkylcyanoacrylate) dispersed in poly(acrylic acid) gel able to reduce drastically the toxicity of poly(alkylcyanocrylates) on the cornea and to promote absorption of the drug.

C. Bioadhesive Interactions of Nanoparticles with Intestinal Mucosae

Controlled drug delivery not only consists in the release of drugs in a time-controlled manner but also often necessitates an adequate and finely tuned localization of the delivery system in the body. In this respect, nanoparticles are of considerable interest, since their physicochemical properties can be modulated for increasing their interactions with specific tissues or cells. In the last decade, the bioadhesive potential of nanoparticles in contact with intestinal mucosae has been thoroughly explored, for improving the delivery of poorly absorbed drugs.

Different targets can be considered in the gastrointestinal tract for immobilizing nano- or microparticles in close contact with the intestinal mucosa, including (1) epithelial cells, (2) specialized immunocompetent cells organized in Peyer's patches and (3) the mucus layer covering the mucosa with a continuous adherent blanket. Interactions can be mediated by nonspecific forces, which are driven by the surface properties of the particles and the intestine, or specific interactions can occur when a ligand attached to the particle is used for the recognition and attachment to a specific site at the mucosal surface.

1. Nonspecific Bioadhesive Interactions of Nanoparticles

a. *In Vitro Experiments*

Nonspecific adhesion of nanoparticles have been reported in vitro and in vivo. Adsorption experiments have been conducted under flow (161,162) or static conditions (163–166) by placing suspensions of nanospheres made of various polymers in contact with excised gut fragments. Interaction equilibria are generally reached rapidly [e.g., 10 min for 230 nm poly(styrene) nanospheres], and it has been shown that it was controlled by the kinetics of diffusion of the particles in the suspension medium (163).

Interactions of poly(styrene) latex (163–165) and poly (isobutyl cyanoacrylate) nanoparticle suspensions (167, 168) with rat intestinal mucosa have been described by using adsorption models. The shape of adsorption isotherms of poly(styrene) latexes were dependent on a particle size threshold (163,164). For nanoparticles up to 670 nm, the isotherm consisted in a linear increase in adsorbed amounts up to a plateau that was reached suddenly for particle concentrations in the bulk particle suspension of about 2.5 g/L, indicating a sudden saturation of the mucus layer by the particles. The saturation values, expressed as the mass of polymer per unit of apparent mucosal surface, ranged typically between 0.5 and 1.0 g/m^2 depending on the type of particles, which corresponded to multilayers of particles. In contrast, the adsorption of 2-μm polystyrene particles increased progressively, and saturation could be obtained for polymer concentrations in the bulk particle suspension higher than 10 g/L. The saturation values were in the range of 2 g/m^2 but, due to the larger size of the particles, it could be calculated that it corresponded approximately only to a single layer of particles on the mucosal surface. Adsorption isotherm of poly(isobutylcyanoacrylate) nanoparticles showed a linear increase in particle adsorption at least up to a bulk concentration of 20 g/L in the particle suspension (168). For poly(isobutylcyanoacrylate) nanoparticles, the slope of the linear segments of the isotherms is lower than the slope for poly(styrene) latex, suggesting a lower affinity of poly(isobutylcyanoacrylate) nanoparticles for the rat intestinal mucosa compared to poly(styrene) particles in the same size range (i.e., 200 nm).

The characteristics of adsorbing species (e.g., concentration, size, surface properties) and adsorbent (e.g., number of sites available for adsorption, geometry) determine the shape of adsorption isotherms. Giles et al. (169) has proposed a general classification of isotherms based on the shape of the isotherms. Nanoparticle adsorption isotherms have been analyzed in this framework (164). Isotherms of 200 nm poly(isobutylcyanoacrylate) nanoparticles and 230 and 670 nm poly(styrene) latexes had the characteristic isotherm shape of adsorbates that penetrate into a porous adsorbent. In this situation, the linear increase of the isotherm corresponds to the creation of new adsorption sites when the bulk particle concentration is increased. These sites are available for further adsorption up to the isotherm plateau, which corresponds to a saturation of the available sites. The porous nature of the hydrogel mucus layer for undermicron particles is likely. The possibility of a diffusion of particles into the mucus layer has been demonstrated by diffusion studies for polystyrene nanoparticles in gastrointestinal mucus (170). Confocal microscopy studies by Scherrer et al. (171) have shown that fluorescently labeled poly(isobutylcyanoacrylate) particles (211 nm in diameter) could penetrate at least 60 μm deep into the mucus layer of rat intestine mucosal fragments. Alternatively, scanning electron microscopic photographs made by Damgé et al. (135) showed 200-nm nanocapsules in close contact with the mucous network a short time after oral administration to rats.

In the case of larger particles, such as 2-μm poly(styrene) particles, Langmuir-type isotherms are observed. The adsorbate adsorbs in a monolayer on the adsorbent surface, which behaves like a smooth surface.

b. *In Vivo Considerations*

After oral administration, nanoparticles can transit directly and/or adhere to the mucosa before fecal elimination. However, oral absorption (translocation) can also occur (172–175), and nanoparticles are likely to cross the gastrointestinal barrier to deliver their drug content in the blood, lymph, or even target organ. A direct contact and/or adhesion of the particles to the mucosal surface is a prerequisite for the translocation process. Obviously, the capture of nanoparticles by M cells in the Peyer's patches and further translocation of the particles is rate limited and quantitatively secondary.

After peroral administration of radiolabeled poly(hexylcyanoacrylate) nanoparticles to mice, whole-body autoradiography showed that thirty minutes after administration the particles were exclusively localized in the stomach (174,176). After 4 h, a large quantity of radioactivity was found in the intestine in the form of clusters without macroradiographic evidence of accumulation at specific intestinal sites. On the contrary, a persistent film of nanoparticles adhering to the stomach wall was observed. In this study, very little of the radioactivity was found to be absorbed. In a similar study, microautoradiographs confirmed the presence of radioactivity throughout the whole gut (174,177). The amount of radioactivity dropped to 30–40% of the 90-min value within 4 to 8 h and to 5%, 24 h after dosing. Histological investigation showed radioactivity adjacent to the brush border, incorporated into the underlying cell layers, and in goblet cells up to 6 days after administration.

After intragastric administration of ^{14}C-labeled poly (lactide) micron-range microspheres a dynamic description of the intestinal transit of a suspension of colloidal particles has been proposed (178). Firstly, the suspension of particles immediately enters in contact with a portion of the oral mucosa (step 1). From this moment, the concentrated suspension acts as a reservoir of particles, and very rapidly an adsorption process takes place, leading to the adsorption of a fraction of the available particles (step 2). Adsorption occurs with the mucus layer and is an irreversible process. The luminal particle suspension transits through the intestine, sweeping progressively the whole mucosa. The simultaneous adsorption process results in a progressive covering of the intestinal mucosa by adherent particles (step 3). Finally, due to mucus turnover, detachment of the particles from the mucosa begins to occur in the proximal region and is progressively extended to the distal region (step 4). Nonadherent particles from the lumen pool and detached particles from the mucoadherent pool are finally eliminated in the feces. Quantitatively, in this description, particle translocation through the intestinal mucosa remains a marginal phenomenon.

2. Specific Bioadhesive Interactions of Nanoparticles

In order to increase the adhesivity of the nanoparticles as well as their site specificity, modifying nanoparticles for exploiting receptor-mediated interactions within the gastrointestinal tract has been suggested. For this purpose, nanoparticles have been conjugated to different ligands which show an affinity for a receptor located into the gastrointestinal cavity. The attachment is generally obtained by grafting onto the polymer making up the particles. Different targets within the gastrointestinal tract can be foreseen, including (1) mucus glycoproteins (mucins), (2) epithelial cell membranes, (3) M-cell membranes, Peyer's patches, or gut-associated lymphoid tissue (GALT), and (4) abnormal glycoproteins secreted by cancerous cells (local tumors). Different ligands have been suggested, including plant, viral, and bacterial lectins, invasins, antibodies (179–182), and vitamin B12 (183).

Lectins are proteins that bind with considerable specificity to carbohydrate moieties expressed by large molecules, such as glycoproteins. So far, the capacities and limits of tomato lectin (TL) conjugates have been extensively studied.

According to Kilpatrick and coworkers (184), the tomato lectin resists digestion and binds to rat intestinal mucosa without any obvious deleterious effects. Furthermore, it is recognized as a natural component of our diet. The potential of this lectin as an oral drug delivery agent was first shown by Naisbett and Woodley (185) who reported that this lectin bound avidly to the small intestine epithelial surface and that this interaction was mediated through N-acetylglucosamine-containing glycoconjugates. The use of fluorescently labeled polystyrene latex coated by adsorption with TL revealed that these conjugates avidly adhered to isolated fixed pig enterocytes in vitro, while the albumin coated control conjugates did not (186). Later, adhesion experiments were carried out by placing a suspension of conjugated microspheres in contact with isolated rat intestinal segments (187,188). Two sets of intestinal mucosa were assayed, with or without Peyer's patches (PP). For intestinal samples without PP, the extent of interaction of TL conjugates decreased from duodenum to ileum, which was attributed to a progressive decrease in the mucin concentration along the gastrointestinal tract, although interactions could also take place with enterocytes after diffusion of the particle conjugates through the mucus layer. About 1.5 g of TL conjugate per m^2 of apparent surface of intestine were adhered to the mucosa, which was significantly higher than the 0.3 g/m^2 observed for control BSA conjugates. Nevertheless, the interaction of TL conjugates with mucosa samples having PP was dramatically decreased. Fluorescence microscopy gave clear evidence of the absence of surface interactions between TL conjugates and PP regions (188).

Recently, some in vivo experiments have been performed in rats. On the one hand, fluorescently labeled TL–latex conjugates were intraduodenally administered to study the transit time of the conjugates. After killing the animals 4 h after a single dose administration, no significant difference was found in the distribution of TL conjugates and controls (189). On the other hand, after multiple dosing, TL–latex conjugates showed unusual absorption. Simple dosing produced no evidence of uptake, whereas 5 or 10 days daily dosing caused uptake of 18% of the dose, not due to tissue damage but perhaps to the induction of receptor expression by the lectin (172). Therefore TL conjugates induced a marked increase in systemic uptake (a 10-fold increase in absorption over plain latex after 5 days daily dosing), which could only be accounted for by induction of uptake through enterocytes as well as lymphoid tissue (190). Analysis of tissue samples revealed that significant amounts of TL conjugates remained bound to the mucosa for several hours in spite of peristalsism and mucus turnover (191). By microscopy, these authors have demonstrated the accumulation of conjugates within the mucus layer, their penetration through the gel layer, and their passage along the serosa layer indicative of the absorption. In contrast with controls, low uptake of TL conjugates by the PP was found (187). Further, TL conjugates showed high circulating levels in the blood, suggesting the low ability of the reticuloendothelial system to sequester them (192).

Finally, the specificity of asparagus pea (*Lotus tetragonologus*) lectin (AL) conjugates to various structures of the intestinal mucosa ex vivo was tested. It appears that these conjugates, which are specific for L-fucose residues, moderately interacted with regular intestinal segments (188). By using intestinal segments containing PP, a 25% increase in the interaction was observed when compared with segments without PP. This last result is in agreement with the reported presence of L-fucose residues on PP regions, especially in M cells (193) and may be of interest for oral vaccination.

3. Increased Oral Bioavailability after Drug Delivery by Nanoparticles

Immobilization of drug carrying particles at the mucosal surface would result in (1) a prolonged residence time at the site of drug action or absorption, (2) a localization of the delivery system at a given target site, (3) an increase in the drug concentration gradient due to intense contact of the particles with the mucosal surface, and (4) a direct contact with intestinal cells, which is the first step before particle absorption. Therefore association to nanoparticles may be of benefit for poorly absorbed or unstable drugs.

The pharmacokinetics of several drugs after oral administration have been improved by means of nanoparticles (128,129,131,135,141,194–197). Most of these studies were carried out with conventional formulations, which means that the carriers were generally not specifically designed for improving the bioadhesion performances of the particles. Some results obtained with nanocapsules by the oral route have been discussed in Section III.B.2.a; these concern mainly peptides such as insulin, calcitonin, and somatostatin. Cyclosporin loaded into poly(isohexylcyanoacrylate) nanocapsules had an increased bioavailability when compared to emulsions (194).

The bioavailability of vincamine was about 25% when administered as an aqueous solution to rabbits, whereas after oral administration adsorbed on poly(hexylcyanoacrylate) nanoparticles, its bioavailability reached 40%, probably due to a prolonged period of contact of the drug delivery system with the mucosa (195). The bioavailability of avarol was also improved by association with poly(alkylcyanoacrylate) nanoparticles (196). The anti-inflammatory effect of hydrocortisone was also increased when the drug was incorporated into microspheres (197), while less significative results were obtained for interferon (198). Interesting results have also been obtained for human growth hormone and heparin by using proteinoid microspheres in the micron range (199). Microparticles made of a copolymer of fumaric acid and sebacic acid (FA:SA) and specifically designed with bioadhesive properties were shown to be efficient in transfecting epithelial cells in vivo by a model plasmid DNA pCMV/βgal, coding for the β galactosidase (200).

These pharmacological observations suggest a high potential for nanoparticles as a peroral drug delivery system. However, most of those studies make no attempt to relate the pharmacological modifications to the fate of the particles in the GI tract, which is generally not characterized. In these examples, it can be estimated that bioadhesion is likely to occur to some extent (as demonstrated in vitro and in vivo for model particles), but the exact incidence of the extent and duration of the adhesion on drug absorption remains unknown. This point should be determined in detail in the future, since it will be crucial for evaluating the interest of the concept of bioadhesive particulate systems for oral delivery.

D. Nanoparticles Avoiding Uptake by Phagocytic Cells

Despite the promising results with some first-generation drug carrier systems, their usefulness is limited by their distribution and in particular by their recognition by the mononuclear phagocyte system. Recently, a great deal of work has been devoted to developing so-called Stealth™ particles, which are invisible to macrophages (Stealth™ is a registered trademark of Liposome Technology Inc.). A major breakthrough in the liposome field consisted in the use of phospholipids substituted with poly(ethylene glycol) (PEG) chains of molecular weight from 1 to 5 kDa (201). This provides a cloud of hydrophilic chains at the particle surface that repels plasma proteins, as discussed theoretically by Jeon et al. (202). These ''sterically stabilized'' liposomes have circulating half-lives of up to 45 h, as opposed to a few hours or even minutes for conventional liposomes. They have been shown to function as reservoir systems and can penetrate into sites such as solid tumors (203,204). A similar strategy has been applied to nanoparticles. PEG can be introduced at the surface in two ways, either by adsorption of surfactants or by the use of block or branched copolymers, usually with poly(lactide) (PLA).

As far as the adsorption of hydrophilic surfactants onto the particle surface is concerned, Illum et al. (205) studied the use of surfactants with polyoxyethylene blocks, such as the poloxamer and poloxamine series onto polystyrene latex surfaces. In particular, they found that coating with poloxamer 407 avoided uptake by Küppfer cells but promoted uptake by the bone marrow (206). Moghimi and coworkers (207) found that poloxamine 908 reduced liver uptake of polystyrene particles, which they interpreted as being due to reduced adsorption of opsonins and increased adsorption of dysopsonins. However, this surfactant, even when not associated with particles, also activated phagocytic cells so that a second dose some days later was cleared rapidly by the liver. The group of Müller (208,209) confirmed reduced plasma protein deposition and uptake by phagocytes of particles coated with these two surfactants, for polystyrene and for biodegradable poly(ester) particles. However, with the more hydrophilic biodegradable polymeric surfaces, reversible adsorption in vivo was observed (210). On the other hand, biodegradable nanospheres prepared from PLGA coated with PLA-PEG diblock copolymers showed a significant increase in blood circulation time and reduced liver uptake in a rat model, as compared to naked PLGA nanospheres (211).

Polysorbate 80 coated nanospheres (212) have been shown to accumulate in the brain, and to allow the delivery of analgesics that do not normally penetrate the blood–brain barrier, but this may be due to the toxicity of the surfactant.

The same approach has been applied to nanocapsules. Lenaerts et al. (213) encapsulated phthalocyanines, important agents in photodynamic tumor therapy, and modified the surface of nanocapsules by adsorption of different poloxamers with various lengths of their POE and POP domains. Their results indicated that some poloxamers, such as poloxamer 407, decreased the uptake of nanocapsules by organs rich in phagocytic cells and increased accumulation of phthalocyanines in primary tumors. Concentration of photosensitizers in the tumor were maximal 12 h after injection of loaded nanocapsules. These carriers allowed an accumulation 200-fold higher in the tumor compared with blood.

Despite these results, the use of copolymers in which PEG is covalently attached to a hydrophobic block would seem to be the better choice, since it avoids the possibility of desorption. Nanospheres prepared from copolymers of PLA, PLGA, or poly(ε-caprolactone) (PCL) with one or more PEG chains of molecular weight 2–20 kDa have been extensively studied in order to optimize their surface characteristics (length and density of PEG chains) for minimizing their interactions with plasma proteins and maximizing their circulating half-life (214–216). It was found that the average calculated distance between two terminally attached PEG 2 kDa chains had to be 2.2 nm or less for repulsion of complement proteins (217). Steric repulsion was also sufficient to avoid strong interactions with the plasma factors of the coagulation system and formation of aggregates between particles and calcium ions in the presence of sodium cholate (218). Recently, nanospheres have been prepared from a series of PEG–PLA copolymers in which the Mw of both the PLA and PEG blocks were varied (219,220). Moreover, the use of blends of PLA and PEG–PLA allowed nanospheres to be prepared coated with a PEG ''brush'' of varying thickness (related to the PEG Mw) and density (related to the content of PEG in the nanospheres). It was thus possible to study the effect of these coatings on competitive plasma protein adsorption, surface charge, and interaction with phagocytic cells.

Although PEG of 5 kDa was more effective than PEG of 2 kDa in terms of reduction of the total amount of plasma protein adsorption, increasing the chain length further, up to 20 kDa, gave no additional advantage (219). The major plasma proteins adsorbed onto these nanospheres were albumin, immunoglobulin G, immunoglobulin light chains, and the apolipoproteins apoA-I and apoE. It was shown further that at optimal PEG chain length (5 kDa) a PEG surface density corresponding to a distance of less than 1.5 nm between two terminally attached PEG chains is required to minimize the total amount of plasma protein adsorbed. A study of phagocytosis by polymorphonuclear (PMN) cells as measured by chemiluminescence and zeta potential carried on with these particles agreed well with these findings: the same PEG surface density threshold (1.5 nm) was found both to ensure efficient steric stabilization and to avoid the uptake by PMN cells. Finally, at optimal coating conditions, it was shown that the nature of the core material (PLA, PLGA, or PCL) played a major role in determining the nature and the amount of the plasma proteins adsorbed (220). On the other hand, a better avoidance of uptake by THP-1 monocytes was provided by 5 kDa PEG chains than by 2 kDa PEG chains at a distance of 1.2–1.4 nm (215).

Lidocaine was used as a model lipophilic drug in an attempt to establish the main factors that determine the encapsulation efficiencies and the release kinetics (221). Human serum albumin (HSA), as a model protein, was also encapsulated in PLA–PEG nanospheres (222,223). HSA loading as high as 7 wt% was thus obtained with 200 nm nanospheres, and controlled release over several days was observed. Surface analysis techniques established that HSA was mainly entrapped deep inside the cores and not adsorbed on their surface (222). As well as their potential as circulating microreservoirs, such systems might be useful for controlled release of antigens. Tetanus toxoid (TT) could also be entrapped in nanospheres of about 150 nm, with high efficiency. When gelatin was included as a stabilizer, active antigen could be released for at least 28 days (224). The transport of radiolabeled TT through the rat nasal mucosa was highly dependent on the surface properties of the nanospheres: the PEG coated ones led to a much higher penetration of TT into the blood circulation and the lymph nodes than the uncoated PLA ones.

PEG chains have also been attached covalently to poly (alkylcyanoacrylate) polymers by two different chemical strategies. In the first, PEG was added during the emulsion-polymerization of alkylcyanoacrylate monomers (225). When PEG blocked at one terminal (MePEG) was used, a brushlike configuration would be expected, whereas when unblocked PEG was added, both ends could be linked to the poly(alkylcyanoacrylate), yielding a loop structure. Studies of complement consumption suggested that this latter structure was the most effective in repelling proteins from the particle surface. The second strategy consisted in the synthesis of a cyanoacrylate monomer substituted with PEG and its copolymerization with hexadecylcyanoacrylate in a 1:4 ratio (226). Both types of particle have shown long-circulating properties in vivo (225,226).

Recently, we have prepared nanocapsules from PLA–PEG copolymers, with the aim of creating circulating reservoirs with a high capacity for lipophilic drugs. In this case, both PEG chain length and density were found to be important in reducing interactions with phagocytic cells in vitro and prolonging circulation time in vivo (227). We compared systems containing PLA–PEG with those stabilized by poloxamer F68. The latter showed some Stealth™ properties in vitro at low dilution, but these were lost at higher dilution and in vivo. The nanocapsules containing PLA–PEG of 5 kDa, with a spacing of 4.3 nm, avoided capture by the liver better and remained longer in the circulation than those containing PLA–PEG of 20 kDa (spacing 7.8 nm), and both were better than nanocapsules with adsorbed poloxamer F68. When a denser coverage of PLA–PEG 20 kDa was used, better Stealth™ properties were obtained than with PLA–PEG 5 kDa at the same spacing (228). These nanocapsules have been used to encapsulate an antimalarial drug, halofantrine (124), in order to prepare a well tolerated, injectable form for the treatment of severe disease.

Another strategy for preparing long-circulating colloidal systems can be considered as biomimetic in that it seeks to imitate cells or pathogens that avoid phagocytosis by reducing or inhibiting complement activation. One example is the development of liposomes with a membrane composition similar to that of erythrocytes, e.g., liposomes containing GM_1, (229) and those coated with poly(sialic acid) (230). These systems may show circulating half-lives as long as liposomes bearing PEG. Attempts to introduce sialic acid onto the surface of nanospheres, either by adsorption of glycoproteins (231) or by synthesis of a PLA-poly(sialic acid) copolymer (232), were not successful, in the first case because of desorption in the presence of plasma proteins and in the second because of either the water-solubility of the polymer or an inappropriate conformation of the polysaccharide.

Another biomimetic approach, which has yielded more promising results, is the use of heparin, the anionic polysaccharidic anticoagulant, also known to inhibit several steps of the complement cascade, as the hydrophilic part of amphiphilic diblock copolymers capable of forming nanospheres (233). The hydrophobic segment was poly (methylmethacrylate) (PMMA), a nonbiodegradable polymer, which was however suitable for validating the concept. These systems were indeed able to avoid complement activation, as shown by two-dimensional immunoelectrophoresis of C3 (234). Using a fluorescent marker randomly copolymerized in the PMMA segment, it was possible to demonstrate reduced uptake by macrophages in vitro (235), and an increased half-life in vivo (5 h or more depending on the injected dose, compared with a few minutes for PMMA nanospheres) for the heparin-bearing systems (236). It is interesting to note that nanospheres prepared from a diblock copolymer with dextran, intended as controls, had only low complement-activating potential, in contrast to cross-linked dextran, i.e. Sephadex®, and also showed long-circulating properties; thus a steric hindrance effect caused by a brushlike arrangement of end-attached polysaccharide chains may also contribute to the long-circulating properties of these nanospheres.

REFERENCES

1. F. Puisieux, G. Barratt, L. Roblot-Treupel, J. Delattre, P. Couvreur, J. P. Devissaguet (1990). Therapeutic aspects of liposomes. *Phospholipids: Biochemical, Pharmaceutical and Analytical Considerations* (I. Hanin and G. Pepeu, eds.). Plenum Press, New York, 1990, p. 133.
2. A. Rolland, ed. *Pharmaceutical Particulate Carriers. Therapeutic Applications*. Marcel Dekker, New York, 1993.
3. S. Benita, ed. *Microencapsulation. Methods and Industrial Applications*. Marcel Dekker, New York, 1996.
4. G. Gregoriadis, B. McCormack, eds. *Targeting of Drugs 6. Strategies for Stealth Therapeutic Systems*. Plenum Press, New York, 1998.
5. J. E. Diederichs, R. H. Müller, eds. *Future Strategies for Drug Delivery with Particulate Systems*. CRC Press, Boca Raton, FL, 1998.
6. R. H. Müller, S. Benita, B. H. L. Böhm, eds. *Emulsions and Nanosuspensions for the Formulation of Poorly Soluble Drugs*. Medpharm Scientific Publishers, Stuttgart, 1998.
7. J. N. Crawley. Neuronal cholecystokinin. *ISI Atlas Pharmac. 2*:84 (1988).
8. G. P. Smith, J. Gibbs. The satiety effect of cholecystokinin, recent progress and current problems. *Ann. N.Y. Acad. Sci. 448*:417 (1985).
9. V. Daugé, P. Steimes, M. Derrien, N. Bean, B. P. Roques, J. Féger. CCK_8 effects on motivational and emotional states of rats involve CCK_A receptors of the postero-median part of the nucleus accumbens. *Pharmacol. Biochem. Behav. 34*:1 (1989).
10. S. Itoh, H. Lal. Influences of cholecystokinin and analogues on memory processes. *Drug Dev. Res. 21*:257 (1990).

11. M. Lemaire, O. Piot, B. P. Roques, G. A. Böhme, J. C. Blanchard. Evidence for an endogenous cholecystokininergic balance in social memory. *NeuroReport 3*:929 (1992).
12. P. L. Faris, B. R. Komisaruk, L. R. Watkins, D. J. Mayer. Evidence for the neuropeptide cholecystokinin as an antagonist of opiate analgesia. *Science 219*:310 (1983).
13. R. Maldonado, M. Derrien, F. Noble, B. P. Roques. Association of the peptidase inhibitor RB 101 and a CCK-B antagonist strongly enhances antinociceptive responses. *NeuroReport 4*:93 (1993).
14. L. Singh, A. S. Lewis, M. J. Field, J. Hughes, G. N. Woodruff. Evidence for an involvement of the brain cholecystokinin B receptor in anxiety. *Proc. Natl. Acad. Sci. U.S.A. 88*:1130 (1991).
15. J. Harro, E. Vasar, J. Bradwejn. CCK in animal and human research on anxiety. *Trends Pharmacol. Sci. 14*:244 (1993).
16. J. Bradwejn, D. Koszycki, A. Couétoux du Tertre, H. van Megen, J. den Boer, H. Westenberg, L. Annable. The panicogenic effects of cholecystokinin-tetrapeptide are antagonized by L-365, 216, a central cholecystokinin receptor antagonist, in patients with panic disorder. *Arch. Gen. Psychiatry 51*:486 (1994).
17. B. Charpentier, C. Durieux, D. Pélaprat, A. Do, M. Reibaud, J. C. Blanchard, B. P. Roques. Enzyme-resistant CCK analogs with high affinities for central receptors. *Peptides 9*:835 (1988).
18. V. Daugé, G. A. Böhme, J. N. Crawley, C. Durieux, J. M. Stutzmann, J. Féger, J. C. Blanchard, B. P. Roques. Investigation of behavioral and electrophysiological responses induced by selective stimulation of CCK-B receptors by using a new highly potent CCK analog, BC 264. *Synapse 6*:73 (1990).
19. P. Branchereau, G. A. Böhme, J. Champagnat, M. P. Morin-Surun, C. Durieux, J. C. Blanchard, B. P. Roques, M. Denavit-Saubié. Cholecystokinin A and cholecystokinin B receptors in neurons of the brainstem solitary complex of the rat: pharmacological identification. *J. Pharmacol. Exp. Ther. 260*:1433 (1992).
20. L. Derrien, C. Durieux, B. P. Roques. Antidepressant-like effects of CCK-B antagonists in mice: antagonism by naltrindole. *Br. J. Pharmacol. 111*:956 (1994).
21. C. Durieux, M. Ruiz-Gayo, B. P. Roques. *In vivo* binding affinities of cholecystokinin agonists and antagonists determined using the selective CCK-B agonist [^{3}H]pBC 264. *Eur. J. Pharmacol. 209*:195 (1991).
22. M. J. Blanco-Prieto, E. Fattal, A. Gulik, J. C. Dedieu, B. P. Roques, P. Couvreur. Characterization and morphological analysis of a cholecystokinin derivative peptide-loaded poly(lactide-co-glycolide) microspheres prepared by a water-in-oil-in-water emulsion solvent evaporation method. *J. Control. Rel. 43*:81 (1997).
23. M. J. Blanco-Prieto, E. Leo, F. Delie, A. Gulik, P. Couvreur, E. Fattal. Study of the influence of several stabilizing agents on the entrapment and *in vitro* release of pBC 264 from poly(lactide-co-glycolide) microspheres prepared by a W/O/W solvent evaporation method. *Pharm. Res. 13*:1127 (1996).
24. M. J. Blanco-Prieto, C. Durieux, V. Daugé, E. Fattal, P. Couvreur, B. P. Roques. Slow delivery of the selective cholecystokinin agonist pBC 264 into the rat nucleus accumbens using microspheres. *J. Neurochem. 67*:2417 (1996).
25. J. H. Eldridge, C. J. Hammond, J. A. Meulbroek, J. K. Staats, R. M. Gilley, T. R. Tice. Controlled vaccine released in the gut-associated lymphoid tissues. I. Orally administered biodegradable microspheres target the Peyer's patches. *J. Control. Rel. 11*:205 (1990).
26. J. H. Eldridge, J. K. Staats, J. A. Eulbroek, T. R. Tice, R. M. Gilley. Biodegradable and biocompatible poly(D,L-lactide-co-glycolide) microspheres as an adjuvant for staphylococccal enterotoxin B toxoid which enhances the level of toxin-neutralizing antibodies. *Infect. Immun. 59*: 2978 (1991).
27. D. T. O'Hagan, H. Jeffery, M. J. J. Roberts, J. P. McGhee, S. S. Davis. Controlled released microparticles for vaccine development. *Vaccine 9*:768 (1991).
28. S. Pecquet, C. Ehrat, P. B. Ernst. Enhancement of mucosal antibody responses to Salmonella typhimurium and the microbial hapten phosphorylcholine in mice X-linked immunodeficiency by B-cell precursors from the peritoneal cavity. *Infect. Immun. 60*:503 (1992).
29. S. H. Gillepsie, S. Ainscough, A. Dickens, J. Lawin. Phosphorylcholine-containing antigens in bacteria from the mouth and respiratory tract. *J. Med. Microbiol. 44*:35 (1996).
30. J. Kolberg, E. A. Hoiby, E. Jantzen. Detection of the phosphorylcholine epitope in streptococci, Haemophilus and pathogenic Neisseria by immunoblotting. *Microb. Pathog. 22*:321 (1997).
31. K. Allaoui-Attarki, E. Fattal, S. Pecquet, S. Trollé, E. Chachaty, P. Couvreur, A. Andremont. Mucosal immunogenicity elicited in mice by oral vaccination with phosphorylcholine encapsulated in poly(D,L-lactide-co-glycolide) microspheres. *Vaccine 16*:685 (1998).
32. K. Allaoui-Attarki, S. Pecquet, E. Fattal, S. Trollé, E. Chachaty, P. Couvreur, A. Andremont. Protective immunity against Salmonella typhimurium elicited in mice by oral vaccination with phosphorylcholine encapsulated in poly(D,L-lactide-co-glycolide) microspheres. *Infect. Immun. 65*:853 (1997).
33. S. Trolle E. Caudron, E. Leo, P. Couvreur, A. Andremont, E. Fattal. *In vivo* fate and immune pulmonary response after nasal administration of microspheres loaded with phosphorylcholine-thyroglobulin. *Int. J. Pharm. 183*:73 (1999).
34. S. Trolle, E. Chachaty, N. Kassis-Chikhani, C. Wang, E. Fattal, P. Couvreur, B. Diamond, J. M. Alonso, A. Andremont. Intranasal immunization with protein-linked phosphorylcholine protects mice against a lethal intranasal challenge with Streptococcus pneumoniae. *Vaccine 18*: 2991 (2000).
35. J. Rojas, H. Alphandary, A. Gulik, P. Couvreur, S. Pec-

quet, E. Fattal. A polysorbate-based non-ionic surfactant can modulate loading and release of β lactoglobulin entrapped in multiphase poly(D,L-lactide-co-glycolide) microspheres. *Pharm. Res. 16*:255 (1999).

36. J. Rojas, H. Pinto-Alphandary, E. Leo, S. Pecquet, P. Couvreur, E. Fattal. Optimization of the encapsulation and release of β-lactoglobulin entrapped poly(D,L-lactide-co-glycolide) microspheres. *Int. J. Pharm. 183*:67 (1999).
37. S. Pecquet, E. Leo, R. Fritsché, A. Pfeifer, P. Couvreur, E. Fattal. Oral tolerance elicited in mice by β-lactoglobulin entrapped in biodegradable poly(lactide-co-glycolide) microspheres. *Vaccine 18*:1196 (2000).
38. P. Couvreur, G. Couarraze, J. P. Devissaguet, F. Puisieux. Nanoparticles: preparation and characterization. *Microencapsulation. Methods and Industrial Applications* (S. Benita, ed.). Marcel Dekker, New York, 1996, p. 183.
39. E. Fattal, M. T. Peracchia, and P. Couvreur. Poly(alkylcyanoacrylates). *Handbook of Polymers* (A. J. Domb, D. M. Kost, D. M. Wiseman, eds.). Harwood Academic Publishers, 1997, p. 183.
40. P. Couvreur, M. Roland, P. Speiser. Biodegradable submicroscopic particles containing a biologically active substance and compositions containing them. *U.S. patent* 4,329,332 (1982).
41. P. Couvreur, B. Kante, M. Roland, P. Guiot, P. Baudhuin, P. Speiser. Polycyanoacrylate nanocapsules as potential lysosomotropic carriers: preparation, morphological and sorptive properties. *J. Pharm. Pharmacol. 31*:331 (1979).
42. C. Verdun, P. Couvreur, H. Vranckx, V. Lenaerts. Development of nanoparticle controlled release formulation for human use. *J. Control. Rel. 3*:205 (1986).
43. J. M. Rollot, P. Couvreur, L. Roblot-Treupel, F. Puisieux. Physicochemical and morphological characterization of polysiobutylcyanoacrylate nanocapsules. *J. Pharm. Sci. 75*:361 (1986).
44. L. Vansnick, P. Couvreur, D. Christiaens-Leyh, M. Roland. Molecular weights of free and drug-loaded nanoparticles. *Pharm. Res. 1*:36 (1985).
45. C. Vauthier, C. Schmidt, P. Couvreur. Measurement of the density of polymeric nanoparticulate drug carriers by isopycnic centrifugation. *J. Nanoparticle Res. 1*:411 (1999).
46. L. Illum, M. A. Khan, E. Mak, S. S. Davis. Evaluation of carrier capacity and release characteristics for poly(butyl 2-cyanoacrylate) nanoparticles. *Int. J. Pharm. 30*:17 (1986).
47. V. Guise, J. Drouin, J. Benoît, J. Mahuteau, P. Dumont, P. Couvreur. Vidarabine loaded nanoparticles: a physicochemical study. *Pharm. Res. 7*:736 (1990).
48. J. L. Grangier, M. Puygrenier, J. C. Gauthier, P. Couvreur. Nanoparticles as carriers for growth hormone releasing factor (GRF). *J. Control. Rel. 15*:3 (1991).
49. A. Baszkin, M. Deyme, P. Couvreur, G. Albrecht. Surface pressure and surface potential studies of poly(isobutylcyanoacrylate)-ampicillin interactions at the water-air interface. *J. Bioact. and Comp. Poly. 4*:110 (1989).
50. S. Michelland, M. J. Alonso, A. Andremont, P. Maincent, J. Sauzières, P. Couvreur. Nanoparticles as carriers of antibiotics for intracellular antibiotherapy. *Int. J. Pharm. 35*: 121 (1987).
51. R. Müller, C. Lherm, J. Herbort, P. Couvreur. In vitro model for the degradation of alkylcyanoacrylate nanoparticles. *Biomaterials 11*:590 (1990).
52. V. Lenaerts, J. F. Nagelkerke, T. J. C. Van Berkel, P. Couvreur, L. Grislain, M. Roland, P. Speiser. *In vivo* uptake of polyisobutylcyanoacrylate nanoparticles by rat liver Kupffer, endothelial and parenchymal cells. *J. Pharm. Sci. 73*:980 (1984).
53. F. Leonard, R. K. Kulkarni, G. Brandes, J. Nelson, J. J. Mameron. Synthesis and degradation of poly(alkylcyanoacrylates). *J. Applied Polym. Sci. 10*:259 (1966).
54. P. Couvreur, B. Kante, M. Roland, P. Speiser. Adsorption of antineoplasic drugs to polyalkylcyanoacrylate nanoparticles and their release characteristics in a calf serum medium. *J. Pharm. Sci. 68*:1521 (1979).
55. F. Lescure, C. Seguin, P. Breton, P. Bourrinet, D. Roy, P. Couvreur. Preparation and characterization of novel poly (methylidene malonate 2.1.2)-made nanoparticles. *Pharm. Res. 10*:1270 (1994).
56. R. Gurny, N. Peppas, D. D. Harrington, G. S. Banks. Development of biodegradable and injectable latices for controlled release of potent drugs. *Drug Develop. Ind. Pharm. 7*:1 (1981).
57. H. J. Krause, A. Schwartz, P. Rohdewald. Polylactic acid nanoparticles, a colloidal drug delivery system for lipophilic drugs. *Int. J. Pharm. 27*:145 (1985).
58. T. Verracchia, G. Spenlehauer, D. V. Bazile, A. Murry-Brelier, Y. Archimbaud, M. Veillard. Non-stealth (poly (lactic acid/albumin)) and stealth (poly(lactic acid-polyethylene glycol)) nanoparticles as injectable drug carriers. *J. Control. Rel. 36*:49 (1995).
59. H. Fessi, J. P. Devissaguet, F. Puisieux, C. Thies. Procédé de préparation des systèmes colloïdaux dispersibles d'une substance sous forme de nanoparticules. French patent application 8618446 (1986).
60. G. Barratt. Characterization of colloidal drug carrier systems with zeta potential measurements. *Pharma. Tech. Eur. 11*:25 (1999).
61. R. K. Chang, J. C. Price, C. Hsiao. Preparation and preliminary evaluation of Eudragit RL and RS pseudolatices for controlled drug release. *Drug Develop. Ind. Pharm. 15*: 361 (1989).
62. R. Bodmeier, H. Chen, P. Tyle, P. Jarosz. Spontaneous formation of drug-containing acrylic nanoparticles. *J. Microencaps. 8*:161 (1991).
63. E. Allemann, R. Gurny, E. Doelker. Preparation of aqueous polymeric nanodispersions by a reversible salting-out process: influence of process parameters on particle size. *Int. J. Pharm. 87*:247 (1992).
64. S. Stainmesse, H. Fessi, J. P. Devissaguet, F. Puisieux. Procédé de préparation des systèmes colloïdaux dispersibles d'une protéine sous forme de nanoparticules. First addition to French patent application 8618446 (1988).
65. M. Skiba, D. Wouessidjewe, F. Puisieux, D. Duchêne, A.

Gulik. Characterization of amphiphilic β-cyclodextrin nanospheres. *Int. J. Pharm. 142*:121 (1996).
66. P. Tulkens. The design of antibiotics capable of an intracellular action. *Aims, Potentialities and Problems in Drug Targeting* (P. Buri, R. Gumma, eds.). Elsevier, Amsterdam, 1985, p. 179.
67. H. Pinto-Alphandary, A. Andremont, P. Couvreur. Targeted delivery of antibiotics using liposomes and nanoparticles: research and applications. *Int. J. Antimicrob. Agents 13*:155 (2000).
68. E. Fattal, M. Youssef, P. Couvreur, A. Andremont. Treatment of experimental salmonellosis in mice with ampicillin-bound nanoparticles. *Antimicrob. Agents Chemother. 33*:1540 (1989).
69. O. Balland, H. Pinto-Alphandary, S. Pecquet, A. Andremont, P. Couvreur. Intracellular distribution of ampicillin in murine macrophages infected with Salmonella typhimurium. *J. Antimicrob. Chemother. 33*:509 (1994).
70. H. Pinto-Alphandary, O. Balland, M. Laurent, A. Andremont, F. Puisieux, P. Couvreur. Intracellular visualization of ampicillin-loaded nanoparticles in peritoneal macrophages infected *in vitro* with Salmonella typhimurium. *Pharm. Res. 11*:38 (1994).
71. O. Balland, H. Pinto-Alphandary, H. Virion, E. Puvion, A. Andremont, P. Couvreur. Intracellular distribution of ampicillin in murine macrophages infected with Salmonella typhimurium and treated with (^{3}H) ampicillin-loaded nanoparticles. *J. Antimicrob. Chemother. 37*:105 (1996).
72. M. Youssef, E. Fattal, M. J. Alonso, L. Roblot-Treupel, J. Sauzières, C. Tancrede, A. Omnes, P. Couvreur, A. Andremont. Effectiveness of nanoparticle-bound ampicillin in the treatment of Listeria monocytogenes infections in nude mice. *Antimicrob. Agents Chemother. 32*:1204 (1988).
73. L. Grislain, P. Couvreur, V. Lenaerts, M. Roland, D. Deprez-Decampeneere, P. Speiser. Pharmacokinetics and distribution of a biodegradable drug-carrier. *Int. J. Pharm. 15*:335 (1983).
74. E. M. Gipps, R. Arshady, J. Kreuter, P. Groscurth, P. Speiser. Distribution of polyhexylcyanoacrylete nanoparticles in nude mice bearing human osteosarcoma. *J. Pharm. Sci. 25*:256 (1986).
75. M. E. Page-Clisson, H. Pinto-Alphandary, E. Chachaty, P. Couvreur and A. Andremont. Drug targeting by polyalkylcyanoacrylate is not efficient against persistent Salmonella. *Pharm. Res. 15*:542 (1998).
76. F. Brasseur, C. Verdun, P. Couvreur, C. Deckers, M. Roland. Evaluation expérimentale de l'efficacité thérapeutique de la doxorubicine associée aux nanoparticules de polyalkylcyanoacrylate. *Proceed. 4th Inter. Conf. Pharmaceut. Tech. 5*:177 (1986).
77. N. Chiannikulchai, Z. Driouich, J. P. Benoit, A. L. Parodi, P. Couvreur. Doxorubicin-loaded nanoparticles increased efficiency in murine hepatic metastases. *Select. Cancer Ther. 5*:1 (1989).
78. N. Chiannikulchai, N. Ammoury, B. Caillou, J. P. Devissaguet, P. Couvreur. Hepatic tissue distribution of doxorubicin-loaded nanoparticles after i.v. administration in M5076 metastasis-bearing mice. *Cancer Chemother. Pharmacol. 26*:122 (1990).
79. G. Storm, P. A. Steerenberg, F. Emmen, M. van Borssum Waalkes, D. J. A. Crommelin. Release of doxorubicin from peritoneal macrophages exposed in vivo to doxorubicin-containing liposomes. *Biochim. Biophys. Acta 965*: 136 (1988).
80. P. Couvreur, L. Grislain, V. Lenaerts, F. Brasseur, P. Guiot, A. Biornacki. Biodegradable polymeric nanoparticles as drug carrier for antitumor agents. *Polymeric Nanoparticles and Microspheres* (P. Guiot, P. Couvreur, eds.). CRC Press, Boca Raton, FL, 1986, p. 27.
81. T. Daeman, J. Regts, M. Meesters, M. T. Ten Kate, I. A. J. M. Bakker-Woudenberg, G. Scherphof. Toxicity of doxorubicin entrapped within long-circulating liposomes. *J. Control. Rel. 44*:1 (1997).
82. R. Fernandez-Urrusuno, E. Fattal, J. M. Rodrigues, Jr., J. Féger, P. Bedossa, P. Couvreur. Effect of polymeric nanoparticle administration on the clearance activity of the mononuclear phagocyte system in mice. *J. Biomed. Mater. Res. 31*:401 (1996).
83. R. Fernandez-Urrusuno, E. Fattal, D. Porquet, J. Feger, P. Couvreur. Evaluation of liver toxicological effects induced by polyalkylcyanoacrylate nanoparticles. *Toxicol. Appl. Pharmacol. 130*: 272 (1995).
84. S. Gibaud, J. P. Andreux, C. Weingarten, M. Renard, P. Couvreur. Increased bone marrow toxicity of doxorubicin bound to nanoparticles. *Eur. J. Cancer 30A*:820 (1994).
85. S. Gibaud, C. Rousseau, C. Weingarten, R. Favier, L. Douay, J. P. Andreux, P. Couvreur. Polyalkylcyanoacrylate nanoparticles as carriers for granulocyte colony stimulating factor (G-CSF). *J. Control. Rel. 52*:131 (1998).
86. M. Demoy, J. P. Andreux, C. Weingarten, B. Gouritin, V. Guilloux, P. Couvreur. Spleen capture of nanoparticles: influence of animal species and surface characteristics. *Pharm. Res. 16*:37 (1999).
87. C. Verdun, F. Brasseur, H. Vranckx, P. Couvreur, M. Roland. Tissue distribution of doxorubicin associated with polyisohexyl-cyanoacrylate nanoparticles. *Cancer Chemother. Pharmacol. 26*:13 (1990).
88. N. Kartner, D. Everden-Porelle, G. Bradley, V. Ling. Detection of P-glycoprotein in multidrug-resistant cell lines by monoclonal antibodies. *Nature 316*:820 (1985).
89. F. Brasseur, P. Couvreur, B. Kante, L. Deckers-Passau, M. Roland, C. Deckers, P. P. Speiser. Actinomycin D adsorbed on polymethylcyanoacrylate nanoparticles: increased efficiency against an experimental tumor. *Eur. J. Cancer 16*:1441 (1986).
90. L. Treupel, M. F. Poupon, P. Couvreur, F. Puisieux. Vectorisation de la doxorubicine dans des nanosphères et réversion de la résistance pléiotropique des cellules tumorales. *Comptes-Rendus de l'Académie de Médecine Série III*:313 (1991).
91. C. Cuvier, L. Roblot-Treupel, S. Chevillard, J. M. Millot,

M. Manfait, G. Bastein, P. Couvreur, M. F. Poupon. Doxorubicin-loaded nanoparticles bypass tumor multidrug resistance. *Biochem. Pharmacol. 44*:509 (1992).

92. S. Bennis, C. Chapey, P. Couvreur, J. Robert. Enhanced cytotoxicity of doxorubicin encapsulated in polyisohexylcyanoacrylate nanospheres against multidrugresistant tumor cells in culture. *Eur. J. Cancer 30A*:106 (1993).
93. F. Nemati, C. Dubernet, A. Colin de Verdière, M. F. Poupon, L. Treupel-Acar, F. Puisieux, P. Couvreur. Some parameters influencing cytotoxicity of free doxorubicin-loaded nanoparticles in sensitive and multidrug resistant leucemic murine cells: incubation time, number of nanoparticles per cell. *Int. J. Pharm. 102*:55 (1994).
94. A. Colin de Verdière, C. Dubernet, F. Nemati, M. F. Poupon, F. Puisieux, P. Couvreur. Uptake of doxorubicin from loaded nanoparticles in multidrug-resistant leukemic murine cells. *Cancer Chemother. Pharmacol. 33*:504 (1994).
95. A. Colin de Verdière, C. Dubernet, F. Nemati, E. Soma, M. Appel, J. Ferté, S. Bernard, F. Puisieux, P. Couvreur. Reversion of multidrug resistance with polyalkylcyanoacrylate nanoparticles: mechanism of action. *Brit. J. Cancer 76*:198 (1997).
96. C. E. Soma, C. Dubernet, G. Barratt, F. Nemati, M. Appel, S. Benita, P. Couvreur. Ability of doxorubicin-loaded nanoparticles to overcome multidrug resistance of tumor cells after their capture by macrophages. *Pharm. Res. 16*: 1710 (1999).
97. C. Ropert, C. Malvy, P. Couvreur. Inhibition of the Friend retrovirus by antisense oligonucleotides encapsulated in liposomes: mechanism of action. *Pharm. Res. 10*:1427 (1993).
98. M. C. De Oliveira, E. Fattal, P. Couvreur, P. Lesieur, C. Bourgaux, M. Ollivon, C. Dubernet. pH-sensitive liposomes as a carrier for oligonucleotides: a physico-chemical study of the interaction between DOPE and a 15-mer oligonucleotide in quasi-anhydrous samples. *Biochimica Biophysica Acta 1372*:301 (1998).
99. C. Chavany, T. Le Doan, P. Couvreur, F. Puisieux, C. Helene. Polyalkylcyanoacrylate nanoparticles as polymeric carriers for antisense oligonucleotides. *Pharm. Res. 9*:441 (1992).
100. C. Chavany, T. Saison-Behmoaras, T. Le Doan, F. Puisieux, P. Couvreur, C. Helene. Adsorption of oligonucleotides onto polyisohexylcyanoacrylate nanoparticles protects them against nucleases and increases their cellular uptake. *Pharm. Res. 11*:1370 (1994).
101. G. Schwab, C. Chavany, I. Duroux, G. Goubin, J. Lebeau, C. Helene, T. Saison-Behmoaras. Antisense oligonucleotides adsorbed to polyalkylcyanoacrylate nanoparticles specifically inhibit mutated Ha-ras-mediated cell proliferation and tumorigenicity in nude mice. *Proc. Natl. Acad. Sci. U.S.A. 91*:10460 (1994).
102. G. Lambert, E. Fattal, A. Brehier, J. Feger, P. Couvreur. Effect of polyisobutylcyanoacrylate nanoparticles and Lipofectin loaded with oligonucleotides on cell viability and PKC alpha neosynthesis in HepG2 cells. *Biochimie 80*: 969 (1998).
103. I. Aynie, C. Vauthier, H. Chacun, E. Fattal, P. Couvreur. Spongelike alginate nanoparticles as a new potential system for the delivery of antisense oligonucleotides. *Antisense Nucl. Acid Drug Devel. 9*:301 (1999).
104. H. Fessi, J. P. Devissaguet, F. Puisieux. Procédé de préparation des systèmes colloïdaux dispersibles d'une substance sous forme de nanocapsules. *French patent application* 8618444 (1986).
105. H. Fessi, F. Puisieux, J. P. Devissaguet, N. Ammoury, S. Benita. Nanocapsule formation by interfacial deposition following solvent displacement. *Int. J. Pharm. 55*:R1 (1989).
106. N. Al Khouri, H. Fessi, L. Roblot-Treupel, J. P. Devissaguet, F. Puisieux. An original procedure for preparing nanocapsules of polyalkyl-cyanoacrylates for interfacial polymerization. *Pharm. Acta. Helv. 61*:274 (1986).
107. M. M. Gallardo, G. Couarraze, B. Denizot, L. Treupel, P. Couvreur, F. Puisieux. Preparation and purification of isohexylcyanoacrylate nanocapsules. *Int. J. Pharm. 100*: 55 (1993).
108. D. Quintanar-Guerrero, E. Allemann, E. Doelker, H. Fessi. Preparation and characterization of nanocapsules from preformed polymers by a new process based on emulsification-diffusion technique. *Pharm. Res. 15*:1056 (1998).
109. D. Quintanar-Guerrero, H. Fessi, E. Doelker, E. Allemann. Procédé de préparation de nanocapsules de type vésiculaire utilisable notamment comme vecteurs colloïdaux de principes actifs pharmaceutiques ou autres. *French patent* 97.09.672 (1997).
110. H. Vranckx, M. De Moutier, M. De Leers. Pharmaceutical compositions containing nanocapsules. *U.S. patent* 5,500,224 (1996).
111. F. Chouinard, F. W. Kan, J. C. Leroux, C. Foucher, V. Lenaerts. Preparation and purification of polyisohexylcyanoacrylate nanocapsules. *Int. J. Pharm. 72*:211 (1991).
112. J. M. Rollot, P. Couvreur, L. Roblot-Treupel, F. Puisieux. Physicochemical and morphological characterization of polyisobutylcyanoacrylate nanocapsules. *J. Pharm. Sci. 75*:361 (1986).
113. N. S. Magalhaes, H. Fessi, F. Puisieux, S. Benita, M. Seiller. An *in-vitro* release kinetic examination and comparative evaluation between submicron emulsion and polylactic acid nanocapsules of clofibride. *J. Microencapsul. 12*:195 (1995).
114. M. Fresta, G. Cavallaro, G. Giammona, E. Wehrli, G. Puglisi. Preparation and characterization of polyethyl-2-cyanoacrylate nanocapsules containing antiepileptic drugs, *Biomaterials 17*:751 (1996).
115. N. Ammoury, H. Fessi, J. P. Devissaguet, F. Puisieux, S. Benita. Physicochemical characterization of polymeric nanocapsules and *in vitro* release evaluation of indomethacin as a drug model. *S. T. P. Pharma 5*:647 (1989).
116. V. C. F. Mosqueira, P. Legrand, H. Pinto-Alphandary, F. Puisieux, G. Barratt. Poly(D,L-lactide) nanocapsules prepared by a solvent displacement process: influence of the composition on physicochemical and structural properties. *J. Pharm. Sci. 89*:614 (2000).

117. F. Chouinard, S. Buczkowski, V. Lenaerts. Poly(alkylcyanoacrylate) nanocapsules: physicochemical characterization and mechanism of formation. *Pharm. Res. 11*:869 (1994).
118. S. Benita, M. Y. Levy. Submicron emulsions as colloidal drug carriers for intravenous administration: comprehensive physicochemical characterization. *J. Pharm. Sci. 82*: 1069 (1993).
119. N. Santos-Magalhaes, S. Benita, A. Baszkin. Penetration of poly(oxyethylene) poly(oxypropylene) block copolymer surfactant into soja phospholipid monolayers. *Colloids Surf. A 52*:195 (1991).
120. P. Calvo, C. Remunan-Lopcz, J. L. Vila-Jato, M. J. Alonso. Development of positively charged colloidal drug carriers: chitosan-coated polyester nanocapsules and submicron-emulsions. *Col. Polym. Sci. 275*:46 (1997).
121. M. Auvillain, G. Cave, H. Fessi, J. P. Devissaguet. Lyophilisation de vecteurs colloïdaux submicroniques. *S.T.P. Pharma. 5*:738 (1989).
122. S. De Chasteigner, H. Fessi, G. Cave, J. P. Devissaguet, F. Puisieux. Gastrointestinal tolerance study of a freeze-dried oral dosage form of indomethacin-loaded nanocapsules. *S.T.P. Pharma. Sci. 5*:242 (1995).
123. N. Ammoury, H. Fessi, J. P. Devissaguet, F. Puisieux, S. Benita. *In-vitro* release kinetic pattern of indomethacin from poly(D,L-lactide) nanocapsules. *J. Pharm. Sci. 79*: 763 (1990).
124. V. C. F. Mosqueira, P. Legrand, R. Gref, G. Barratt. *In-vitro* release kinetic studies of PEG modified nanocapsules and nanospheres loaded with a lipophilic drug: halofantrine base. *Proceed. Intern. Symp. Control. Rel. Bioact. Mater. 26*:1074 (1999).
125. L. Polato, L. M. Benedetti, L. Callegaro, P. Couvreur. *In-vitro* evaluation of nanoparticle formulations containing gangliosides. *J. Drug Target. 2*:53 (1994).
126. H. Marchais, S. Benali, J. M. Irache, C. Tharasse-Bloch, O. Lafont, A. M. Orecchioni. Entrapment efficiency and initial release of phenylbutazone from nanocapsules prepared from different polyesters. *Drug Dev. Ind. Pharm. 24*:883 (1998).
127. F. Kedzierewicz, P. Thouvenot, I. Monot, M. Hoffman, P. Maincent. Influence of different physicochemical conditions on the release of indium oxine from nanocapsules. *J. Biomed. Mater. Res. 39*:588 (1998).
128. P. J. Lowe, C. S. Temple. Calcitonin and insulin in isobutylcyanoacrylate nanocapsules: protection against proteases and effect on intestinal absorption in rats. *J. Pharm. Pharmacol. 46*:547 (1994).
129. C. Damgé, M. Aprahamian, G. Balboni, A. Hoetzel, V. Andrieu, J. P. Devissaguet. Polyalkylcyanoacrylate nanocapsules increase the intestinal absorption of a lipophilic drug. *Int. J. Pharm. 36*:121 (1987).
130. S. S. Guterres, H. Fessi, G. Barratt, F. Puisieux, J. P. Devissaguet. Poly (D,L-lactide) nanocapsules containing non-steroidal anti-inflammatory drugs: gastrointestinal tolerance following intravenous and oral administration. *Pharm. Res. 12*:1545 (1995).
131. N. Ammoury, H. Fessi, J. P. Devissaguet, M. Dubrasquet, S. Benita. Jejunal absorption, pharmacological activity, and pharmacokinetic evaluation of indomethacin-loaded poly(D,L-lactide) and poly(isobutyl-cyanoacrylate) nanocapsules in rats. *Pharm. Res. 8*:101 (1991).
132. V. Andrieu, H. Fessi, M. Dubrasquet, J. P. Devissaguet, F. Puisieux, S. Benita. Pharmacokinetic evaluation of indomethacin nanocapsules. *Drug Des. Deliv. 4*:295 (1989).
133. F. Fawaz, F. Bonini, M. Guyot, A. M. Lagueny, H. Fessi, J. P. Devissaguet. Disposition and protective effect against irritation after intravenous and rectal administration of indomethacin loaded nanocapsules to rabbit. *Int. J. Pharm. 133*:107 (1996).
134. C. Damge, C. Michel, M. Aprahamian, P. Couvreur. New approach for oral administration of insulin with polyalkylcyanoacrylate nanocapsules as drug carrier. *Diabetes 37*: 246 (1988).
135. C. Damgé, C. Michel, M. Aprahamian, P. Couvreur, J. P. Devissaguet. Nanocapsules as carriers for oral peptide delivery. *J. Control. Rel. 13*:233 (1990).
136. M. Aboubakar, F. Puisieux, P. Couvreur, C. Vauthier. Physico-chemical characterization of insulin-loaded poly-(isobutylcyanoacrylate) nanocapsules obtained by interfacial polymerization. *Int. J. Pharm. 183*:63 (1999).
137. M. Aboubakar, F. Puisieux, P. Couvreur M., Deyme, C. Vauthier. Study of the mechanism of insulin encapsulation in poly(isobutylcyanoacrylate) nanocapsules obtained by interfacial polymerization. *J. Biomed. Mater. Res. 47*:568 (1991).
138. C. Michel, M. Aprahamian, L. Defontaine, P. Couvreur, C. Damgé. The effect of site of administration in the gastrointestinal tract on the absorption of insulin from nanocapsules in diabetic rats. *J. Pharm. Pharmacol. 43*:1 (1991).
139. M. Aboubakar, P. Couvreur, H. Pinto-Alphandary, B. Gouritin, B. Lacour, R. Farinotti, F. Puisieux, C. Vauthier. Insulin-loaded nanocapsules for oral administration: in vitro and in vivo investigation. *Drug Devel. Res. 49*:109 (2000).
140. C. Damgé, H. Vranckx, P. Balschmidt, P. Couvreur. Poly-(alkylcyanoacrylate) nanospheres for oral administration of insulin. *J. Pharm. Sci. 86*:1403 (1997).
141. C. Damgé, J. Vonderscher, P. Marbach, M. Pinget. Poly (alkylcyanoacrylate) nanocapsules as a delivery system in the rat for octreotide, a long-acting somatostatin analogue. *J. Pharm. Pharmacol. 49*:949 (1997).
142. F. Dalencon, Y. Amjaud, C. Lafforgue, F. Derouin, H. Fessi. Atovaquone and rifabutine-loaded nanocapsules: formulation studies. *Int. J. Pharm. 153*:127 (1998).
143. F. Fawaz, F. Bonini, M. Guyot, A. M. Lagueny, H. Fessi, J. P. Devissaguet. Influence of poly(D,L-lactide) nanocapsules on the biliary clearance and enterohepatic circulation of indomethacin in the rabbit. *Pharm. Res. 10*:750 (1993).
144. S. S. Guterres. Etude pharmacotechnique, pharmacocinétique et de tolérances digestive et tissulaire des nanocapsules de diclofenac. Ph.D. thesis, University of Paris XI, France (1995).

145. B. Hubert, J. Atkinson, M. Guerret, M. Hoffman, J. P. Devissaguet, P. Maincent. The preparation and acute antihypertensive effects of a nanocapsular form of darodipine, a dihydropyridine calcium entry blocker. *Pharm. Res. 8*: 734 (1991).

146. C. Morin, G. Barratt, H. Fessi, J. P. Devissaguet, F. Puisieux. Improved intracellular delivery of a muramyl dipeptide analog by means of nanocapsules. *Int. J. Immunopharmacol. 16*:451 (1994).

147. I. Seyler, M. Appel, J. P. Devissaguet, P. Legrand, G. Barratt. Relationship between NO-synthase activity and TNF-alpha secretion in mouse macrophage lines stimulated by a muramyl peptide entrapped in nanocapsules. *Int. J. Immunopharmacol. 18*:385 (1996).

148. I. Seyler, M. Appel. J. P. Devissaguet, P. Legrand, G. Barratt. Macrophage activation by a lipophilic derivative of muramyldipeptide within nanocapsules: investigation of the mechanism of drug delivery. *J. Nanoparticle Res. 1*: 91 (1999).

149. G. Barratt, F. Puisieux, W. P. Yu, C. Foucher, H. Fessi, J. P. Devissaguet. Anti-metastatic activity of MDP-L-alanyl-cholesterol incorporated into various types of nanocapsules. *Int. J. Immunopharmacol. 16*:457–461 (1994).

150. L. Marchal-Heussler, H. Fessi, J.-P. Devissaguet, M. Hoffman, P. Maincent. Colloidal drug delivery systems for the eye. A comparison of the efficacy of three different polymers: polyisobutylcyanoacrylate, polylactic-co-glycolic acid, poly-epsilon-caprolactone. *S.T.P. Pharma 2*:98 (1992).

151. L. Marchal-Heussler, D. Sirbat, M. Hoffman, P. Maincent. Poly(epsilon-caprolactone) nanocapsules in carteolol ophthalmic delivery. *Pharm. Res. 10*:386 (1993).

152. C. Losa, L. Marchal-Heussler, F. Orallo, J. L. Vila Jato, M. J. Alonso. Design of new formulations for topical ocular administration: polymeric nanocapsules containing metipranolol. *Pharm. Res. 10*:80 (1993).

153. P. Calvo, C. Thomas, M. J. Alonso, J. L. Vila Jato, J. R. Robinson. Study of the mechanism of interaction of poly(ε-caprolactone) nanocapsules with the cornea by confocal laser scanning microscopy. *Int. J. Pharm. 103*:283 (1994).

154. A. Zimmer, J. Kreuter, J. R. Robinson. Studies on the transport pathway of PBCA nanoparticles in ocular tissues. *J. Microencapsulation* 8:497 (1991).

155. P. Calvo, M. J. Alonso, J. L. Vila-Jato, J. R. Robinson. Improved ocular bioavailability of indomethacin by novel ocular drug carriers. *J. Pharm. Pharmacol. 48*:1147 (1996).

156. P. Calvo, J. L. Vila-Jato, M. J. Alonso. Effect of lysozyme on the stability of polyester nanocapsules and nanoparticles: stabilization approaches. *Biomaterials 18*:1305 (1997).

157. P. Calvo, J. L. Vila-Jato, M. J. Alonso. Evaluation of cationic polymer-coated nanocapsules as ocular drug carriers. *Int. J. Pharm. 153*:41 (1997).

158. P. Calvo, A. Sanchez, J. Martinez, M. I. Lopez Calonge, J. C. Pastor, M. J. Alonso. Polyester nanocapsules as new topical ocular delivery systems for cyclosporin A. *Pharm Res. 13*:311 (1996).

159. J. J. Ramon, M. Calonge, S. Gomez, M. I. Lopez, P. Calvo, J. M. Herreras, M. J. Alonso. Efficacy of topical cyclosporine-loaded nanocapsules on keratoplasty rejection in the rat. *Curr. Eye Res. 17*:39 (1998).

160. C. A. Le Bourlais, F. Chevanne, B. Turlin, L. Acar, H. Zia, P. A. Sado. T. E. Needham, R. Leverage. Effect of cyclosporin A formulations on bovine corneal absorption: *ex-vivo* study. *J. Microencapsul. 14*:457 (1997).

161. C. L. C. Teng. Kinetics of adhesion of polymer-coated particles to intestinal mucous surface. Ph.D. thesis, University of Michigan, U.M.I. Dissertation Services, Order number 8712222, Ann Arbor (1987).

162. C. L. C. Teng, N. F. H. Ho. Mechanistic studies in the simultaneous flow and adsorption of polymer-coated latex particles on intestinal mucus: I. Methods and physical model development. *J. Control. Rel. 6*:133 (1987).

163. C. Durrer, J. M. Irache, F. Puisieux, D. Duchêne, G. Ponchel. Mucoadhesion of latexes: I. Analytical methods and kinetic studies. *Pharm. Res. 11*:674 (1994).

164. C. Durrer, J. M. Irache, F. Puisieux, D. Duchêne, G. Ponchel. Mucoadhesion of latexes: II. Adsorption isotherms and desorption studies. *Pharm. Res. 11*:680 (1994).

165. C. Durrer, J. M. Irache, D. Duchêne, G. Ponchel. Study of the interactions between nanoparticles and intestinal mucosa. *Prog. Colloid Polym. Sci. 97*:275 (1994).

166. C. Durrer, J. M. Irache, D. Duchêne, G. Ponchel. Mucin interactions with functionalized poly(styrene) latexes. *J. Colloid Interf. Sci. 170*:555 (1995).

167. A. Dembri, M.-J. Montisci, D. Duchêne, G. Ponchel. Degradation of poly(isobutylcyanoacrylate) nanoparticles in presence of gastrointestinal fluids and mucus. Proceed. XIème Journées Scientfiques du Groupe de Recherche Thématique sur les Vecteurs. Paris, 1996, p. 68.

168. A. Dembri, M.-J. Montisci, D. Duchêne, G. Ponchel. Mucoadhesion of poly isobutylcyanoacrylate nanoparticles on the intestinal mucosa. Proceed. Eur. Symp. Formulation of Poorly-available Drugs for Oral Administration. Paris, 1996, Editions de Santé, Paris, 1986, pp. 342–346.

169. C. H. Giles, D. Smith, A. Huitson. A general treatment and classification of the solute adsorption isotherm: I. Theoretical. *J. Colloid. Interface Sci.* 47:755 (1974).

170. D. A. Norris, P. J. Sinko. Effect of size, surface charge, and hydrophobicity on the translocation of polystyrene microspheres through gastrointestinal mucin. *J. Appl. Polym. Sci. b3*:1481 (1997).

171. D. Scherrer, F. C. Mooren, R. K. H. Kinne, J. Kreuter. *In vitro* permeability of PBCA nanoparticles through porcine small intestine. *J. Drug Targeting 1*:21 (1994).

172. A. T. Florence. The oral absorption of micro and nanoparticulates: neither exceptional nor unusual. *Pharm. Res. 14*: 259 (1997).

173. D. T. O'Hagan. Intestinal translocation of particulates—implications for drug and antigen delivery. *Adv. Drug Del. Rev. 5*:265 (1990).

174. J. Kreuter. Peroral administration of nanoparticles. *Adv. Drug Del. Rev. 7*:71 (1991).
175. J. P. Devissaguet, H. Fessi, N. Ammoury, G. Barratt. Colloidal drug delivery systems for gastrointestinal application. *Drug Targeting and Delivery—Concepts in Dosage Form Design* (H. E. Junginger, ed.). Ellis Horwood, New York, 1992, p. 71.
176. V. Lenaerts, P. Couvreur, L. Grislain, P. Maincent. Nanoparticles as a gastrointestinal drug delivery system. *Bioadhesive Drug Delivery Systems* (V. M. Lenaerts, and R. Gurny, eds.). CRC Press, Boca Raton, FL, 1990, p. 94.
177. J. Kreuter, U. Müller, K. Munz. Quantitative and microautoradiographic study on mouse intestinal distribution of polycyanoacrylate nanoparticles. *Int. J. Pharm. 55*:39 (1989).
178. G. Ponchel, M.-J. Montisci, A. Dembri, C. Durrer, D. Duchêne. Mucoadhesion of colloidal particulate systems in the gastro-intestinal tract. *Eur. J. Pharm. Biopharm. 44*: 25 (1997).
179. J. Pappo, T. H. Ermak, H. J. Steger. Monoclonal antibody–directed targeting of fluorescent polystyrene microspheres to Peyer's patch M cells. *Immunology 73*:277 (1991).
180. A. B. MacAdam, G. P. Martin, S. L. James, C. Marriott. Development of a novel delivery system for targeting to the gastrointestinal tract. *Proc. Int. Symp. Cont. Rel. Bioact. Mat. 19*, 1992, pp. 277–278.
181. A. B. MacAdam, G. P. Martin, S. L. James, C. Marriott. Preparation and characterisation of antiporcine gastric mucus antibodies. *J. Pharm. Pharmacol. 43*:62 (1991).
182. M. W. Smith, N. W. Thomas, P. G. Jenkins, N. G. A. Miller, D. Cremaschi, C. Porta. Selective transport of microparticles across Peyer's patch follicle-associated M cells from mice and rats. *Exp. Physiol. 80*:735 (1995).
183. G. J. Russell-Jones. Oral drug delivery via the vitamin B12 uptake system. *Pharmaceutical Manufacturing International*:81 (1994).
184. D. C. Kilpatrick, A. Pusztai, G. Grant, C. Graham, S. W. B. Ewen. Tomato lectin resists digestion in the mammalian alimentary canal and binds to intestinal villi without deleterious effects. *FEBS Lett. 185*:299 (1985).
185. B. Naisbett, J. F. Woodley. Binding of tomato lectin to the intestinal mucosa and its potential for oral drug delivery. *Biochem. Soc. Trans. 18*:879 (1990).
186. C. M. Lehr, J. A. Bouwstra, W. Kok, A. B. J. Noach, A. G. de Boer, H. E. Junginger. Bioadhesion by means of specific binding of tomato lectin. *Pharm. Res. 9*:547 (1992).
187. J. M. Irache, C. Durrer, D. Duchêne, G. Ponchel. *In vitro* study of lectin–latex conjugates for specific bioadhesion. *J. Control. Rel. 31*:181 (1994).
188. J. M. Irache, C. Durrer, D. Duchêne G. Ponchel. Bioadhesion of lectin-latex conjugates to rat intestinal mucosa. *Pharm. Res. 13*:1714 (1996).
189. C. M. Lehr, A. Puzstai. The potential of bioadhesive lectins for the delivery of peptide and protein drugs to the gastrointestinal tract. *Lectins: Biomedical Perspectives* (A. Puzstai, S. Bardocz, eds.). Taylor and Francis, London, 1995, p. 117.
190. A. T. Florence, A. M. Hillery, N. Hussain, P. U. Jani. Nanoparticles as carriers for oral peptide absorption: studies on particle uptake and fate. *J. Control. Rel. 36*:39 (1995).
191. A. T. Florence, A. M. Hillery, N. Hussain, P. U. Jani. Factors affecting the oral uptake and translocation of polystyrene nanoparticles: histological and analytical evidence. *J. Drug Target. 3*:65 (1995).
192. N. Hussain, P. H. Jani, A. T. Florence. Enhanced oral uptake of tomato lectin-conjugated nanoparticles in the rat. *Pharm. Res. 14*:613 (1997).
193. M. A. Clark, M. A. Jepson, N. L. Simmons, B. H. Hirst. Differential surface characteristics of M cells from mouse intestinal Peyer's patches. *Histochem J. 26*:271 (1994).
194. S. Bonduelle, M. Carrier, C. Pimienta, J. P. Benoit, V. Lenaerts. Tissue concentration of nanoencapsulated radiolabelled cyclosporin following peroral delivery in mice or ophthalmic applications in rabbits. *Eur. J. Pharm. Biopharm. 42*:313 (1996).
195. P. Maincent, R. Le Verge, P. Sado, P. Couvreur, J. P. Devissaguet. Deposition kinetics and oral bioavailability of vincamine-loaded polyalkyl cyanoacrylate nanoparticles. *J. Pharm. Sci. 75*:955 (1986).
196. P. H. Beck, J. Kreuter, W. E. G. Müller, W. Schatton. Improved peroral delivery of avarol with polyalkylcyanoacrylate nanoparticles. *Eur. J. Pharm. Biopharm. 40*:134 (1994).
197. H. O. Alpar, W. N. Field, K. Hayes, D. A. Lewis. A possible use of orally administered microspheres in the treatment of inflammation. *J. Pharm. Pharmacol. 41*(suppl):1 (1989).
198. J. E. Eyles, H. O. Alpar, B. R. Conway, M. Keswick. Oral delivery and fate of poly(lactic acid) microspheres-encapsulated interferon in rats. *J. Pharm. Pharmacol. 49*:669 (1997).
199. R. Ottenbrite, R. Zhao, S. Milstein. A new oral microsphere drug delivery system. *Macromol. Symp. 101*:379 (1996).
200. E. Mathiowitz, J. S. Jacob, Y. S. Jong, G. P. Carino, D. E. Chickering, P. Chaturvedi, C. A. Santos, K. Vijayaraghavan, S. Montgomery, M. Bassett, C. Morell. Biologically erodable microspheres as potential oral drug delivery systems. *Nature 386*:410 (1997).
201. M. C. Woodle, D. D. Lasic. Sterically stabilized liposomes. *Biochim. Biophys. Acta. 1113*:193 (1992).
202. S. I. Jeon, J. H. Lee, J. D. Andrade, P. G. De Gennes. Protein-surface interactions in the presence of polyethylene oxide. I Simplified theory. *J. Colloid Interf. Sci. 142*: 149 (1991).
203. Gabizon, R. Catane, B. Uziely, B. Kaufman, Y. Barenholz. Prolonged circulation time and enhanced accumulation in malignant exudates of doxorubicin encapsulated in polyethylene-glycol coated liposomes. *Cancer Res. 54*: 987 (1994).
204. T. M. Allen, D. Lopez de Menezes, C. B. Hansen, E. H. Moase. Stealth™ liposomes for the targeting of drugs in can-

cer therapy. *Targeting of Drugs 6: Strategies for Stealth Therapeutic Systems* (G. Gregoriadis, B. McCormack, eds.). Plenum Press, New York and London, 1998, p. 61.

205. L. Illum, L. O. Jacobsen, R. H. Müller, R. Mak, S. S. Davis. Surface characteristics and the interaction of colloidal particles with mouse peritoneal macrophages. *Biomaterials 8*:113 (1987).

206. M. E. Norman, P. Williams, L. Illum. Influence of block co-polymers on the adsorption of plasma proteins to microspheres. *Biomaterials 14*:193 (1993).

207. S. M. Moghimi, I. Muir, L. Illum, S. Davis, V. Kolb-Bachofen. Coating particles with a block copolymer (poloxamine 908) suppresses opsonisation but permits the activity of dysopsonins in the serum. *Biochim. Biophys. Acta 1179*:157 (1993).

208. S. Rudt, R. H. Muller. *In-vitro* phagocytosis assay of nano- and microparticles by chemiluminescence. II. Effect of surface coating of particles with poloxamer on the phagocytic uptake. *J. Control. Rel. 25*:51 (1993).

209. M. Luck, B. R. Paulke, W. Schroder, T. Blunk, R. H. Müller. Analysis of plasma protein adsorption on polymeric nanoparticles with different surface characteristics. *J. Biomed. Mater. Res. 5*:478 (1998).

210. S. E. Dunn, A. G. A. Coombes, M. C. Garnett, S. S. Davis, M. C. Davies, L. Illum. *In vitro* interaction and *in vivo* biodistribution of poly(lactide-co-glycolide) nanospheres surface modified by poloxamer and poloxamine copolymers. *J. Control. Release 44*:65 (1997).

211. S. Stolnik, S. E. Dunn, M. C. Garnett, M. C. Davies, A. G. A. Coombes, D. C. Taylor, M. P. Irving, S. C. Purkiss, T. F. Tadros, S. S. Davis, L. Illum. Surface modification of poly(lactide-co-glycolide) nanospheres by biodegradable poly(lactide)-poly(ethylene glycol) copolymers. *Pharm. Res. 11*:1800 (1994).

212. J. Kreuter, V. E. Petrov, D. A. Kharkevich, R. N. Alyautdin. Influence of the type of surfactant on the analgesic effects induced by the peptide dalargin after its delivery across the blood–brain barrier using surfactant-coated nanoparticles. *J. Control. Rel. 49*:81 (1997).

213. V. Lenaerts, A. Labib, F. Chouinard, J. Rousseau, H. Ali, J. van Lier. Nanocapsules with a reduced liver uptake: targeting of phthalocyanines to EMT-6 mouse mammary tumour *in vivo. Eur. J. Pharm. Biopharm. 41*:38 (1995).

214. R. Gref, Y. Minamitake, M. T. Peracchia, V. Trubetskoy, V. Torchilin, R. Langer. Biodegradable long-circulating nanospheres. *Science 263*:1600 (1994).

215. D. Bazile, C. Prud'Homme, M.-T. Bassoulet, M. Marlard, G. Spenlehauer, M. Veillard. Stealth Me. PEG-PLA nanoparticles avoid uptake by the mononuclear phagocytes system. *J. Pharm. Sci. 84*:493 (1995).

216. M. T. Peracchia, R. Gref, Y. Minamitake, A. Domb, N. Lotan, R. Langer. PEG-coated nanospheres from amphiphilic diblock and multiblock copolymers: investigation of their drug encapsulation and release characteristics. *J. Control. Rel. 46*:223 (1997).

217. M. Vittaz, D. Bazile, G. Spenlehauer, T. Verrecchia, M. Veillard, F. Puisieux, D. Labarre. Effect of PEO surface density on long-circulating PLA-PEO nanoparticles which are very low complement activators. *Biomaterials 17*:1575 (1996).

218. H. Sahli, J. Tapon-Bretaudière, A. M. Fischer, C. Sternberg, G. Spenlehauer, T. Verracchia, D. Labarre. Interactions of poly(lactic acid) and poly(lactic acid-co-ethylene oxide) nanoparticles with the plasma factors of the coagulation system. *Biomaterials 18*:281 (1997).

219. R. Gref, G. Miralles, E. Dellacherie. Polyoxyethylene-coated nanospheres: effect of coating on zeta potential and phagocytosis. *Polymer Int. 48*:251 (1999).

220. R. Gref, M. Lück, P. Quellec, M. Marchand, E. Dellacherie, S. Harnish, T. Blunk, R. H. Müller. "Stealth" core-corona nanoparticles surface modified by polyethylene glycol (PEG): influence of the corona (PEG chain length and surface density) and of the core composition on phagocytic uptake and plasma protein adsorption. *Colloids and Surfaces B: Biointerfaces 18*:301 (2000).

221. R. Gref, A. Domb, P. Quellec, T. Blunk, R. H. Müller, J. M. Verbavatz, R. Langer. The controlled intravenous delivery of drugs using PEG-coated sterically stabilized nanospheres. *Adv. Drug Deliv. Rev. 16*:215 (1995).

222. P. Quellec, R. Gref, L. Perrin, E. Dellacherie, F. Sommer, J. M. Verbavatz, M. J. Alonso. Protein encapsulation within polyethylene glycol-coated nanospheres I. Physicochemical characterization. *J. Biomed. Mater. Res. 42*:45 (1998).

223. P. Quellec, R. Gref, E. Dellacherie, F. Sommer, M. D. Tran, M. J. Alonso. Protein encapsulation within PEG-coated nanospheres. II. Controlled release properties. *J. Biomed. Mater. Res. 47*:388 (1999).

224. M. Tobio, R. Gref, A. Sanchez, R. Langer, M. J. Alonso. Stealth PLA-PEG nanoparticles as protein carriers for nasal administration. *Pharm. Res. 15*:274 (1998).

225. M. T. Peracchia, C. Vauthier, C. Passirani, P. Couvreur, D. Labarre. Complement consumption by poly(ethylene glycol) in different configurations chemically coupled to poly(isobutyl 2-cyanoacrylate) nanoparticles. *Life Sci. 61*: 749 (1997)

226. M. T. Peracchia, C. Vauthier, D. Desmaële, A. Gulik, J. C. Dedieu, M. Demoy, J. D'Angelo, P. Couvreur. Pegylated nanoparticles from a novel methoxypolyethyleneglycol cyanoacrylate-hexadecyl cyanoacrylate amphiphilic copolymer. *Pharm. Res. 15*:550 (1998).

227. V. C. F. Mosqueira, P. Legrand, R. Gref, B. Heurtault, M. Appel, G. Barratt. Interactions between a macrophage cell line (J774A1) and surface-modified poly(D,L-lactide) nanocapsules bearing poly(ethylene glycol). *J. Drug Target. 7*:65 (1999).

228. V. C. F. Mosqueira, P. Legrand, J.-L. Morgat, M. Vert, E. Mysiakine, R. Gref, J.-P. Devissaguet, G. Barratt. Biodistribution of novel long-circulating PEG-grafted nanocapsules in mice: effects of PEG chain length and density. Submitted to *Pharm. Res.*

229. T. M. Allen, A. Chonn. Large unilamellar liposomes with

low uptake into the reticuloendothelial system. *FEBS Lett.* *223*:42 (1987).
230. G. Gregoriadis, B. McCormack, Z. Wang, R. Lifely. Polysialic acids: potential in drug delivery. *FEBS Lett.* *315*: 271 (1993).
231. J. C. Olivier, C. Vauthier, M. Taverna, F. Puisieux, D. Ferrier, P. Couvreur. Stability of orosomucoid-coated polyisobutylcyanoacrylate nanoparticles in the presence of serum. *J. Control. Rel.* *40*:157 (1996).
232. P. Huve. Comprendre et éviter la capture des nanoparticules de poly(acide-lactique) par le système des phagocytes mononucléaires. Ph.D. thesis University of Paris-Sud. Sci. Pharm. (1994).
233. C. Passirani, L. Ferrarini, G. Barratt, J.-Ph. Devissaguet, D. Labarre. Preparation and characterization of nanoparticles bearing heparin or dextran covalently bound to poly(methyl methacrylate). *J. Biomat. Sci. Polymer Ed.* *10*:47 (1999).
234. C. Passirani, G. Barratt, J.-Ph. Devissaguet, D. Labarre. Interactions of nanoparticles bearing heparin or dextran covalently bound to poly(methyl methacrylate) with the complement system. *Life Sci.* *62*:775 (1998).
235. N. Jaulin, M. Appel, C. Passirani, G. Barratt, J.-Ph. Devissaguet, D. Labarre. Reduction of the uptake by a macrophagic cell line of nanoparticles bearing heparin or dextran covalently bound to poly(methyl methacrylate). *J. Drug Target.* *8*:165 (2000).
236. C. Passirani, G. Barratt, J.-Ph. Devissaguet, D. Labarre. Long-circulating nanoparticles bearing heparin or dextran covalently bound to poly(methyl methacrylate). *Pharm. Res.* *15*:1046 (1998).

29

Liposomes in Drug Delivery

Yuan-Peng Zhang
ALZA Corp., Mountain View, California
Boris Čeh
University of Ljubljana, Ljubljana, Slovenia
Danilo D. Lasic
Liposome Consultations, Newark, California

I. INTRODUCTION

Liposomes or (phospho)lipid vesicles are self-assembled colloidal particles that occur in nature or can be prepared artificially. In the beginning, liposomes were used to study biological membranes, while several practical applications, most notably in drug delivery, emerged in the 1970s. After three decades of research, several products are commercially available. They rely on liposome colloidal, microencapsulating, and surface properties. Products range from drug dosage forms (antifungals, anticancer agents, and vaccines) and cosmetic formulations (skin care products, shampoos) to diagnostics and food industry products. However, at present, it seems that drug delivery applications are now the most widely investigated area of their practical applications.

A. Properties of Amphiphiles, Bilayers, and Liposomes

Amphiphiles are a class of organic molecules that contain both polar (or in most cases charged) and nonpolar segments and are therefore simultaneously hydrophobic and hydrophilic. Typical examples of these molecules are detergents and phospholipids. Due to hydrophobic and hydrophilic interactions, these molecules orient and self-organize in polar and nonpolar solvents. Aqueous systems of amphiphiles exhibit a very rich phase behavior, and some of the structures formed are schematically shown in Fig. 1. The most common structures that most phospholipids adopt spontaneously are the concentric bilayer vesicles in excess water (see Fig. 2). In the bilayer structure the lipid hydrophilic head groups orient toward the aqueous phase, and the nonpolar hydrocarbons are associated at the center, away from water molecules. In comparison most detergents and lysophospholipids form micelles that vary in size and shape depending on the type of detergents. Certain amphiphiles, such as long-chain phosphatidylethanolamine (PE), do not form bilayers but adopt a structure known as the inverted hexagonal (H_{II}) phase (see Figs. 1 and 2). Various types of cubic phases have also been discovered (42). In drug delivery, the dispersed, self-closed lamellar phase—the liposomes—is most important and used for most applications. In other applications, such as gene transfer and regional drug delivery applications, other structures such as the dispersed H_{II} and cubic phases, and the micelles, may be also important.

The ability of lipids to adopt structures other than the bilayers upon hydration is known as lipid polymorphism. A generalized concept of geometric shape properties can be ascribed to various amphiphilic molecules that reflect the phase structure they prefer under given conditions, as illustrated in Fig. 3. Molecules, such as lipids, that prefer the bilayer structure, have cylindrical shapes, i.e., similar areas of polar heads and the cross sections of nonpolar

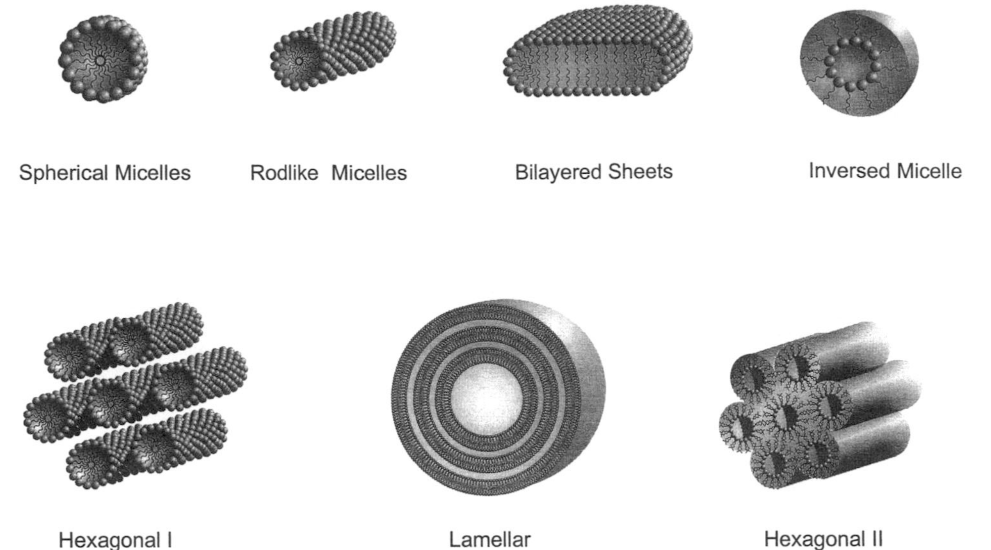

Figure 29.1 Common types of structures (phases) formed by amphiphiles in aqueous solution. Most detergents and lysophospholipids form various types of micelles. Most phospholipids form spontaneously into concentric bilayer vesicles. Certain amphiphiles, such as long-chain phosphatidylethanolamines (PE), can form inverted hexagonal (H_{II}) phase. Micellar and liquid-crystalline phases are stable thermodynamic phases, while lamellar colloidal suspension liposomes is a kinetically trapped state.

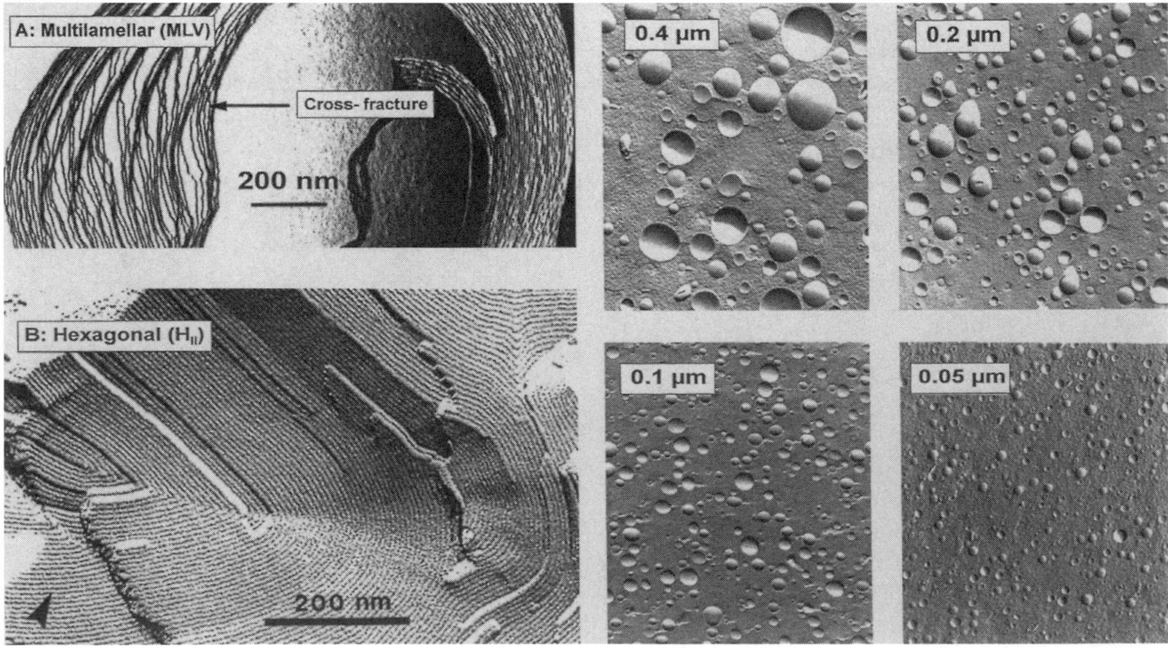

Figure 29.2 Freeze-fracture electron micrographs. (A) Multilamellar vesicles of EPC prepared by the freeze-thaw and hydration procedure; (B) hexagonal (H_{II}) phase of DOPE; LUVs and SUVs (right panels) prepared by extrusion of MLVs through filters of various pore sizes (0.4, 0.2, 0.1, and 0.05 μm). (Courtesy of M. Hope and P. Cullis).

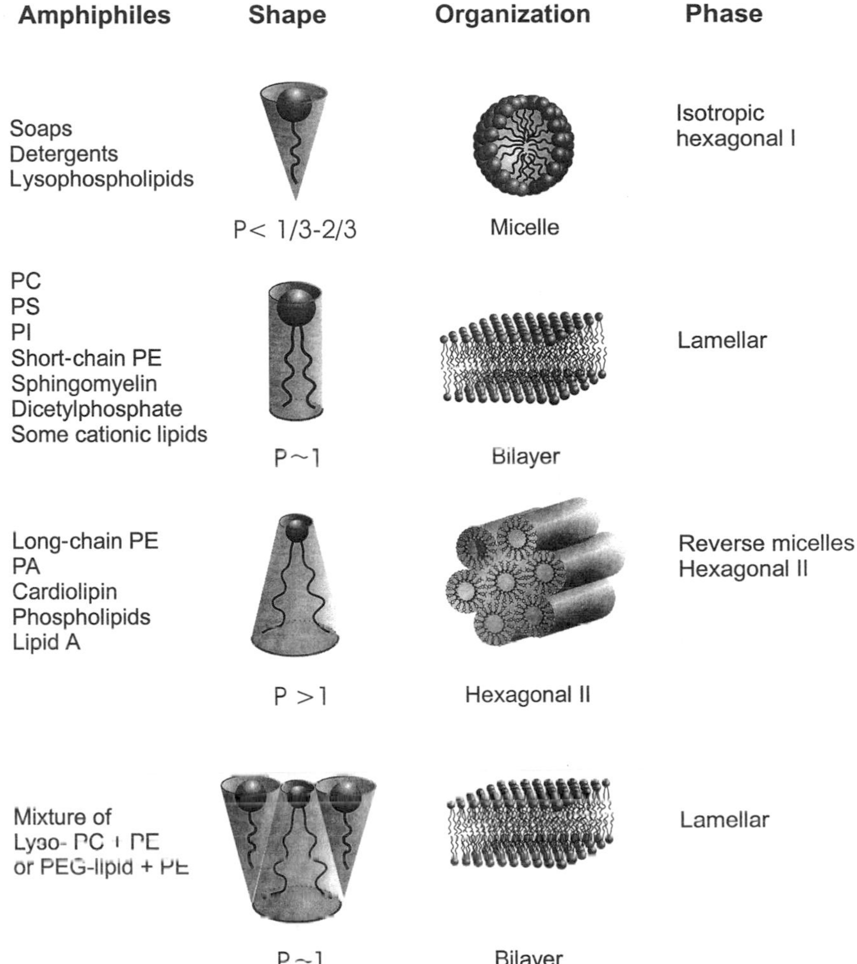

Figure 29.3 The concept of molecular geometry and structure of the amphiphilic phases as determined by packing of surfactant molecules (steric effects) (From Ref. 43.)

chains. In the case of a larger polar head with respect to a smaller nonpolar tail, the molecules tend to pack into structures with high radii of curvature, such as micelles. A structural parameter P is defined as

$$P = \frac{v}{al}$$

Where v is the volume of the amphiphile, a the cross-sectional area of the polar head, and l the length of the nonpolar tail. For values of P below 1/3, spherical micelles are favored structures; for values between 0.8 and 1.1, preferably around 1, bilayers are the most stable structure, while for values of $P > 1$ to 1.1, inverse structures form. In the case of ideal mixing within the bilayer, one can define an average P as the sum of the mole fractions of individual values of P of individual lipids:

$$\langle P \rangle = \Sigma_i x_i P_i$$

where fractions are taken for lipid i with structural parameter P_i. We shall use this concept later to explain liposomes with built-in instability.

Liposomes are simply lipid vesicles in which an aqueous volume is entirely enclosed by a membrane composed of lipid molecules. Liposomes were first discovered by Alec D. Bangham in the early 1960s when investigating the role of phospholipids in the clotting cascade (10). The bilayer structure of the liposomes is in principle identical to the lipid portion of natural cell membranes—the ''sea of phospholipids'' in the Singer and Nicholson mosaic model. Since then, liposomes have been used as models of biological membranes because liposomes can be prepared from natural constituents and membrane proteins can be reconstituted.

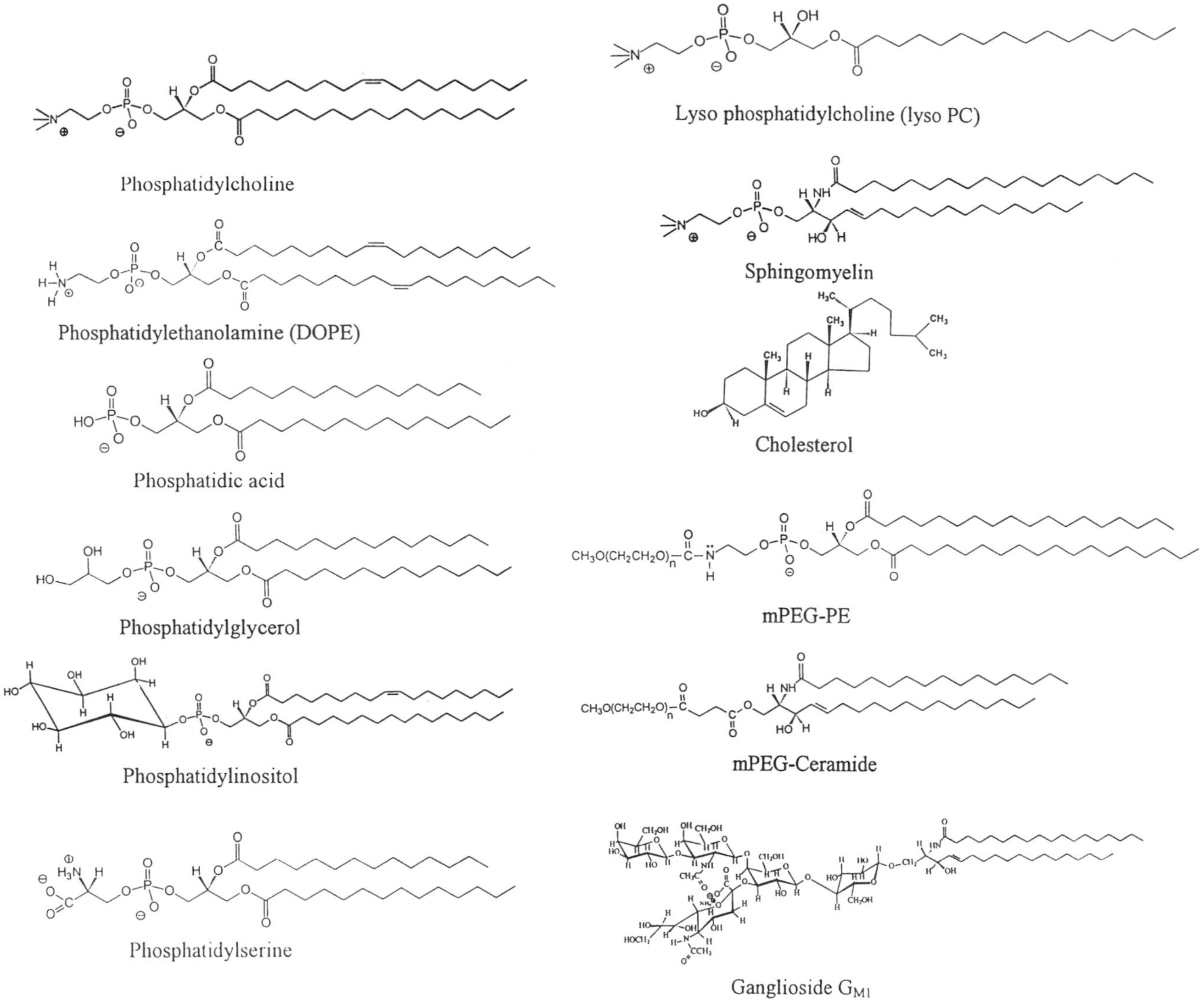

Figure 29.4 Chemical structures of neutral and anionic phospholipids commonly used in liposomes for drug delivery.

With respect to the size and number of membranes (lamellarity) we distinguish large multilamellar vesicles (LMV or MLV) and unilamellar vesicles (UV), which can be small (SUV), large (LUV), or giant (GUV), as shown in Fig. 2. MLVs are normally formed when dry lipid is hydrated in aqueous solution. Small liposomes are normally defined as the ones where curvature effects are important for their properties. This curvature depends on the lipid composition and can vary from 50 nm for soft bilayers to 80 to 100 nm for mechanically very cohesive bilayers. Giant vesicles are normally those with diameters above 1 μm. The thickness of the bilayer depends on the type of lipids, the hydrocarbon chain length, and the temperature, and is around 4 nm.

B. Chemical Structures of Lipids

1. Zwitterionic and Anionic Lipids

Lipids used for lipid vesicles are extracted from natural sources or chemically synthesized. The most common natural lipids are lecithins (phosphatidylcholines, PC), sphingomyelins, and phosphatidylethanolamines (PE), which are zwitterionic at physiological pH (see Fig. 4 for structures). The common sources for these lipids are egg yolks, soya beans, and (ox) brains.

Anionic natural lipids include phosphatidylserine (PS), phosphatidylglycerols (PG), and phosphatidylinositols (PI). One serious drawback of the natural lipids is that they contain a mixture of acyl chains. Natural lipids are less commonly used for liposome/vesicle preparations for several reasons. First, membranes prepared from natural lipids tend to be leakier to solutes, due to the low main phase transition temperatures (see below), and therefore they are not suitable for drug delivery when drug retention and carrier stability are emphasized. Second, for basic biophysics researches involving artificial lipid membranes, synthetic lipids are preferred, since in most cases one would like to be able to control the physical properties of the membranes and be able better to interpret the experimental results. The last problem for using natural lipids particularly as drug carriers for human use concerns the possible contamination from the source, especially if it is animal in origin.

The stability of natural lipids can be increased by hydrogenation (reduction of double bounds). The degree of hydrogenation is described by the iodine value (*I.V.*), where *I.V.* = 1 for fully saturated chains. A typical high *I.V.* number is 65 of EPC. Commercially, *I.V.* values of 1, 10, 20, 30, and 40 are normally available.

In contrast, chemical synthesis can produce lipids with well-defined acyl chains (number of carbons on the chains and degree of saturation) attached to the same polar heads. The most commonly used acyl chains are dimyristoyl (C14:0), dipalmitoyl (C16:0), distearoyl (C18:0), dioleoyl (C18:1), and palmitoyl-oleoyl (C16:0-C18:1). The structures of some common phospholipids are shown in Fig. 4. Cholesterol is a very important component in lipid vesicles for improving the mechanical stability of the lipid bilayers and therefore increasing the retention of the encapsulated agents.

2. Cationic Lipids

Cationic lipids, with the exception of sphingosine and some lipids in primary life forms, practically do not exist in nature. The first lipid that was used to complex DNA (80) and transfect DNA in cell culture (30) was DOTMA (dioleoxypropyltrimethylammonium chloride). Many more cationic lipids have been synthesized to improve the transfection capability and to decrease the toxicity (see Fig. 5 for structures). The best-known monocationic lipids include DOTAP, DMRIE, DODAP, DDAB, and DODAC, and all these are efficacious for DNA transfection. Natural cations, such as spermine^{4+} and spermidine^{3+}, can also be coupled to fatty acyl chains. In DC-CHOL the positive charge is associated on the sterol backbone (33). The chemical structures of the most frequently used cationic lipids for DNA delivery are given in Fig. 5. DOPC, DOPE, and cholesterol are the most often used neutral lipids and are often called helper lipids. mPEG-PE and mPEG-ceramides have often been used to increase circulation and to prevent aggregation. The majority of positive charges of cationic lipids are based on (poly)amines and quaternary ammonium salts. The pK_a values for most cationic lipids are currently unknown. Quaternary ammonium salts are very strong bases, and pK_a values are above 12. For simple alkyl amines, pK_a values are typically above 10 for primary and secondary amines and above 9 for tertiary.

C. Thermodynamic Properties of Lipids

1. Thermodynamic Properties of Pure Lipids

Liposome characteristics are determined by surface and membrane properties, including surface charge, steric interaction, and membrane rigidity. The most common lipids used for preparing liposomes for drug delivery are neutral or zwitterionic lipids, such as PCs and PEs. Therefore the characteristics of these types of liposomes are determined by the thermodynamic properties of these lipids. One of the most important thermodynamic parameters of chemically pure lipids is the phase transition temperature (T_m) and the associated enthalpy (ΔH). The low temperature phase for PC and PE is commonly referred to as the gel phase, where the acyl chains are packed tightly and lateral movement of lipid molecules is restricted (Fig. 6). For this reason, solute transport through lipid bilayers in the gel phase is limited. The high-temperature phase is called the liquid-crystalline, or fluid, phase, where the acyl chains become mobile and disordered, although the lipids are still arranged in a smooth bilayer structure. This phase transition between the ordered gel phase and the fluid phase is most commonly referred to as the main phase transition (T_m). Lipid phase transitions are cooperative events and can occur over a very narrow temperature range of less than one degree (58,59,61).

Factors affecting the phase transition properties are the type of polar head groups, the length of the acyl chains, and the degree of saturation of the acyl chains. For the same polar head groups T_m increases with increasing acyl chain length (58,61,100). Lipids with unsaturated acyl chains have much lower T_m values, due to the introduced disorder of the double bond. PEs usually have higher values of T_m than the corresponding PCs with identical acyl chains, because of the stronger interactions (intermolecular H-bonding) between PE head groups. The presence of charge on the polar heads (i.e., cationic and anionic lipids) tends to decrease the value of T_m, due to the repulsive interactions between molecules. It should be noted that some anionic lipids, such as phosphatidic acid (PA) and PS, are known to exhibit strong intermolecular hydrogen bonding and therefore may have higher values of T_m (15,16). There are relatively few studies on the phase properties of the cationic lipids commonly used for genosome formation.

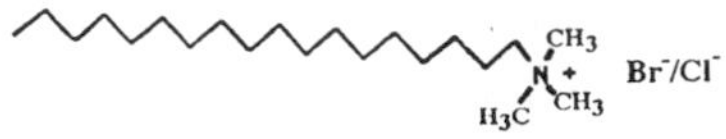

Br^-/Cl^- CTAB/C
Br^-/Cl^- DODAB/DODAC
DOTAP 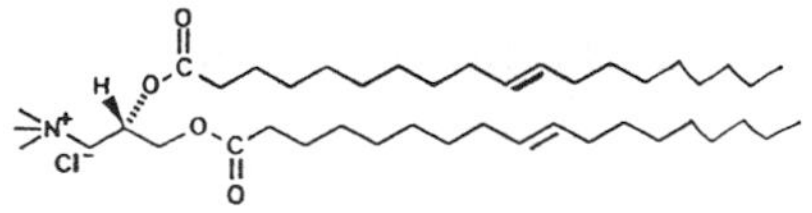
DODAP 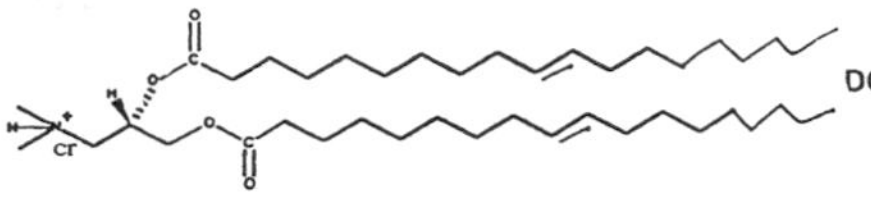
DOTMA 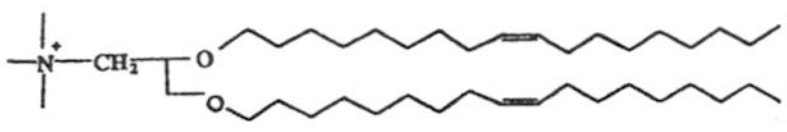
DC-Chol 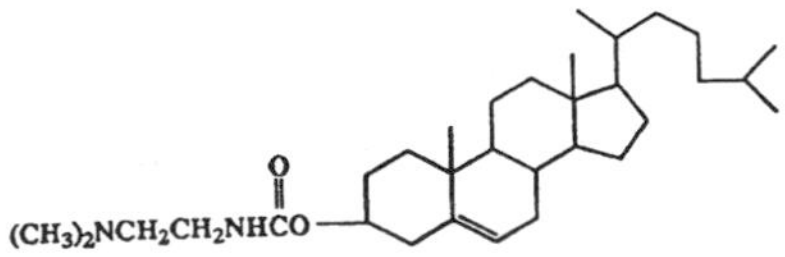 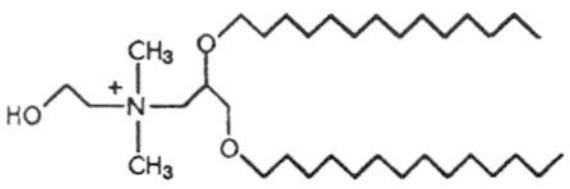
$(CH_3)_2NCH_2CH_2NHCO$
DMRIE
DOSPA 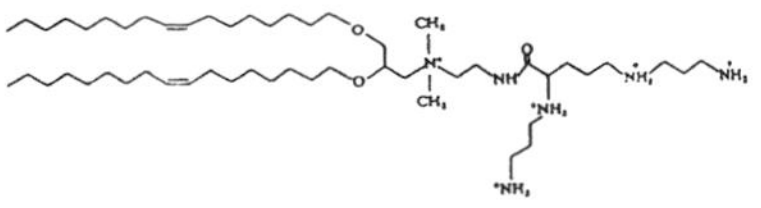
DOGS 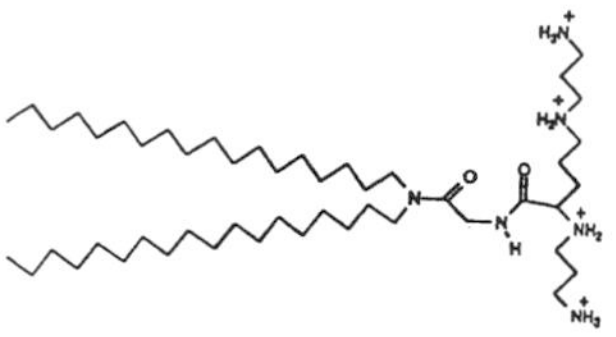
DPPES 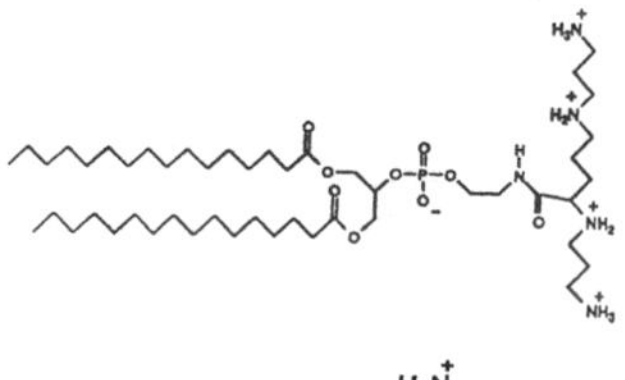
C_8Gly Sper $^{3+}$ 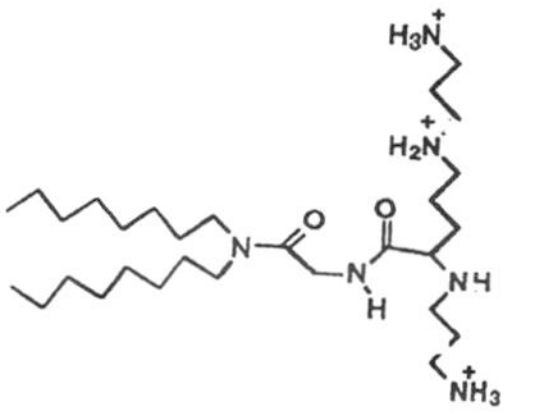
$(C_{18})_2$ Sper $^{3+}$ 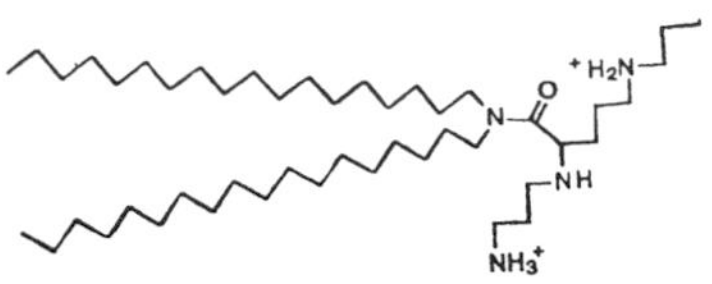
Sper-Chol 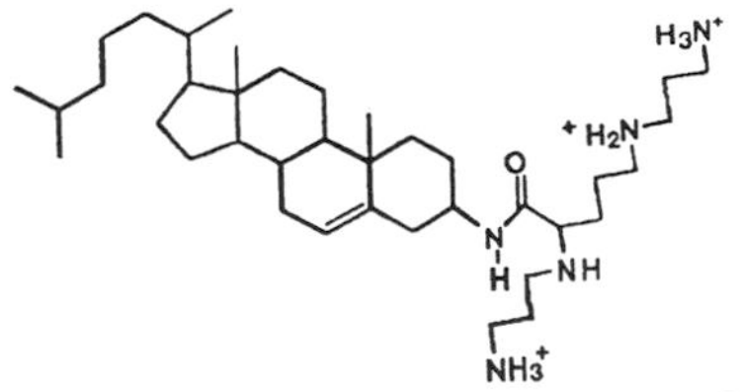
$C_{12}GluPhC_nN^+$ 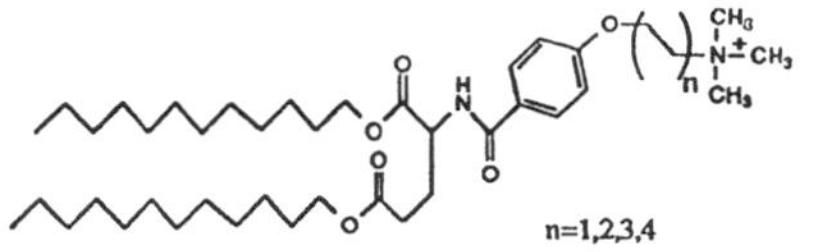 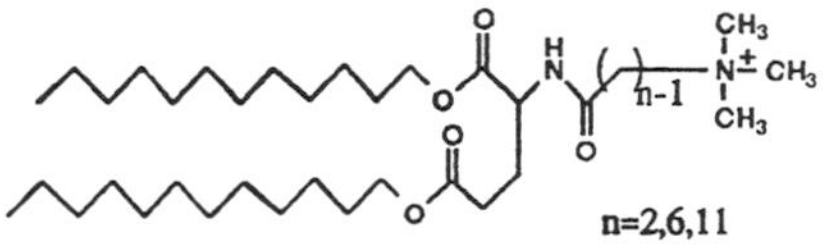
n=1,2,3,4
$C_{14}GluC_nN^+$ 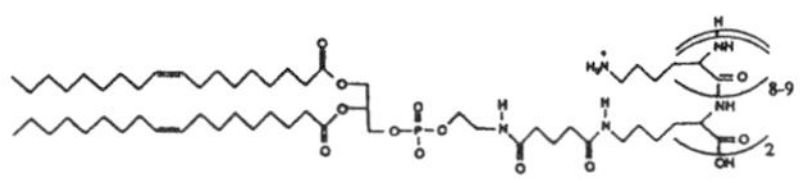
n=2,6,11
LPL
TMAG 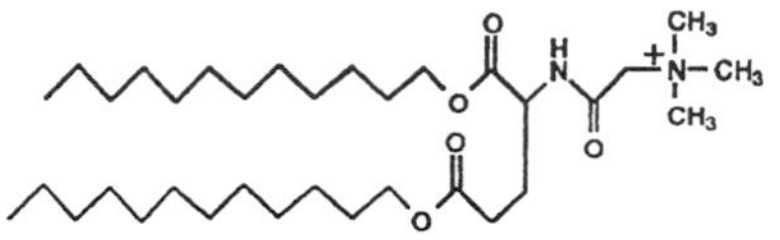
DOIC 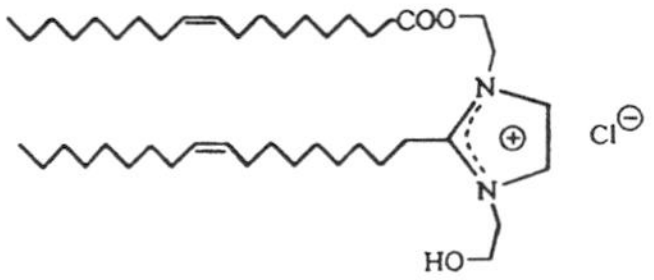
DMEPC 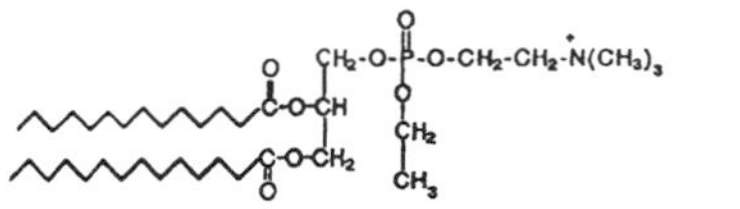
Lysinyl PE 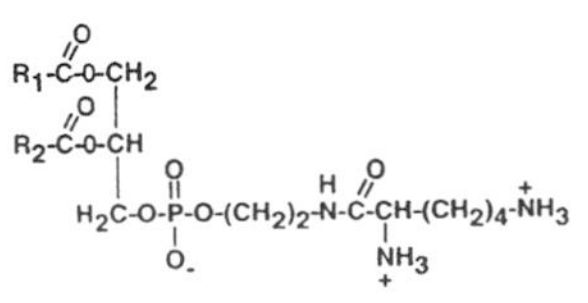
AE Chol 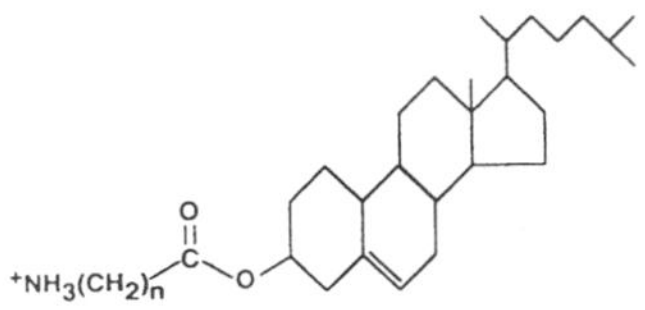
GENZYME #47 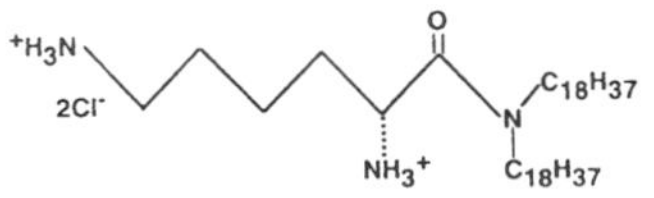
GENZYME

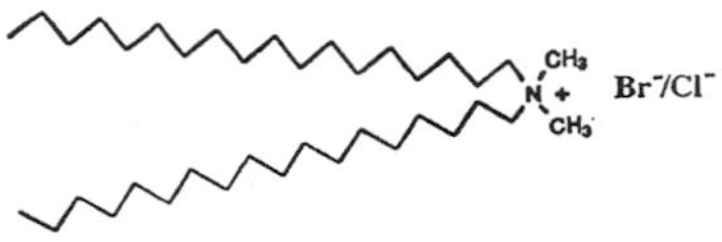

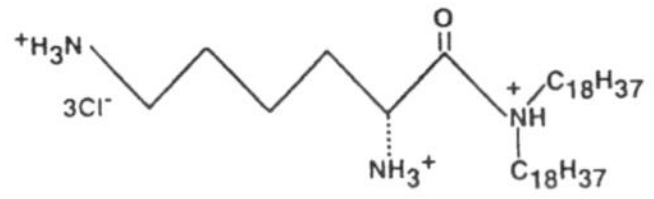

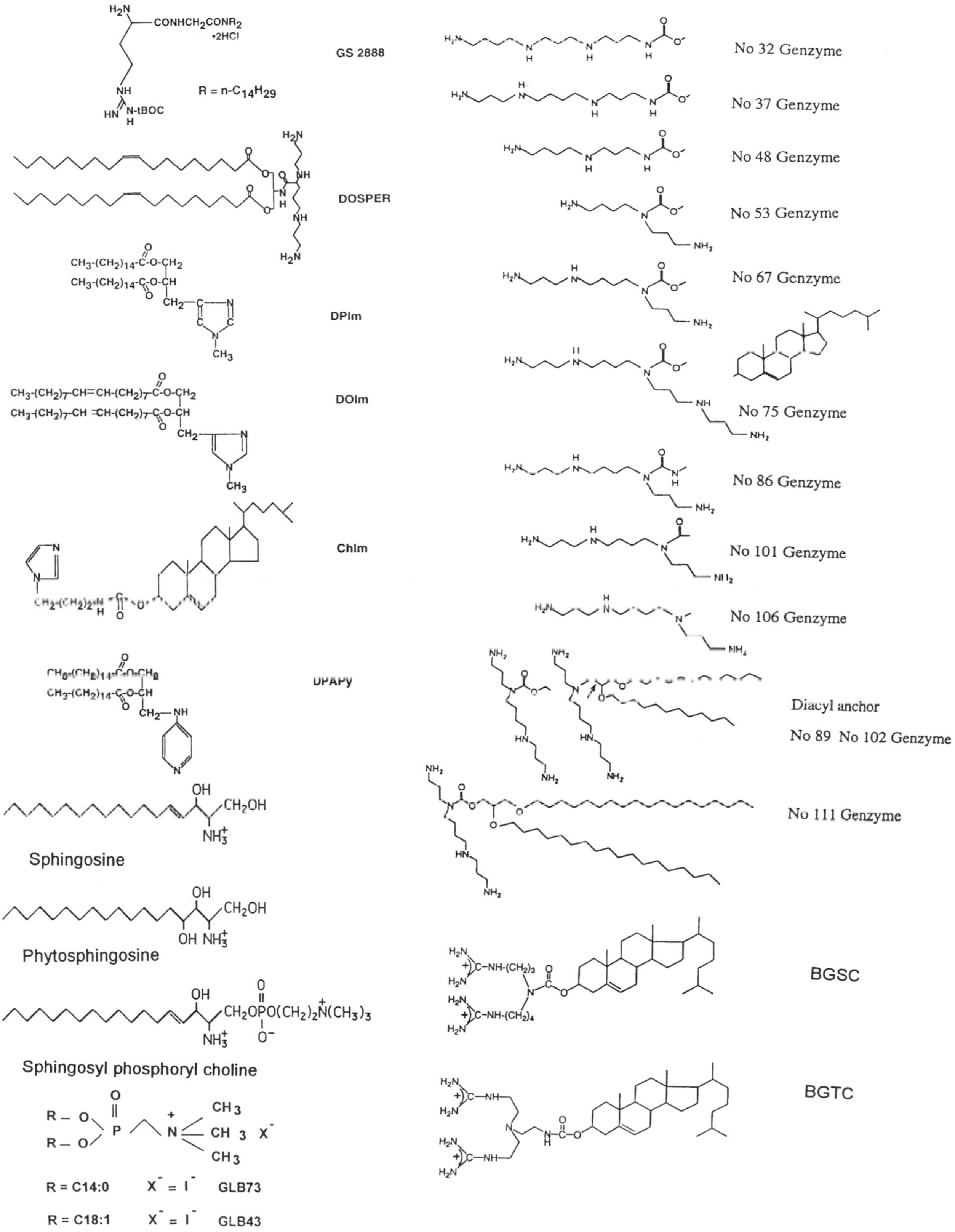

GS 2888

$R = n\text{-}C_{14}H_{29}$

DOSPER

DPIm

DOIm

ChIm

DPAPy

Sphingosine

Phytosphingosine

Sphingosyl phosphoryl choline

R = C14:0 $X^- = I^-$ GLB73

R = C18:1 $X^- = I^-$ GLB43

No 32 Genzyme

No 37 Genzyme

No 48 Genzyme

No 53 Genzyme

No 67 Genzyme

No 75 Genzyme

No 86 Genzyme

No 101 Genzyme

No 106 Genzyme

Diacyl anchor

No 89 No 102 Genzyme

No 111 Genzyme

BGSC

BGTC

Figure 29.5 Structures of cationic lipids used in gene transfection.

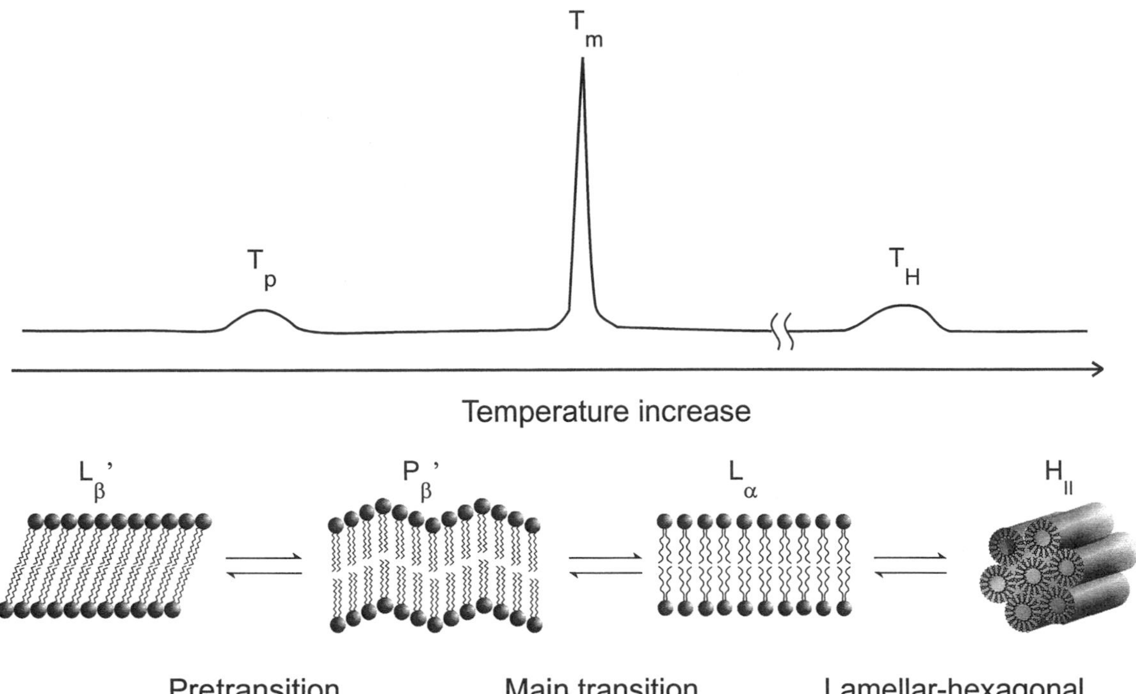

Figure 29.6 Illustration of phase transition for phospholipids in a lipid bilayer when fully hydrated. Crystalline phase (L_{β}') transforms into a gel phase (P_{β}') at temperature T_p. The P_{β}' transforms into liquid-crystalline phase (L_{α}) at T_m. In the L_{β}' phase, lipid molecules are packed tightly with hydrocarbon chains tilted to a certain degree. In the P_{β}' phase, the hydrocarbon chains also packed closely, but the bilayer surface becomes rippled. In the L_{α} phase, the lipid chains become mobile and disordered, and in addition lipid molecules can move freely within the bilayer. A transition from lamellar phase to reversed hexagonal phase (L_{α} to H_{II}) can occur for some lipids having small head groups and bulky hydrocarbon chains, such as DOPE. The phase temperatures and enthalpies can be conveniently monitored by differential scanning calorimetry (upper curve).

2. Thermodynamic Properties of Binary Lipid Mixtures

In most cases, liposomal formulations designed for drug delivery contain more than one type of lipid. An essential parameter to consider for formulation stability is the miscibility of the lipid matrix. Lipid mixtures that are poorly miscible will eventually fail in stability tests, due to lipid phase separation. For lipids with the same head groups, the degree of miscibility is mostly determined by the mismatch of the acyl chains as well as the degree of saturation of the acyl chain. PC, for instance, with an acyl chain length differing by two carbons, has high miscibility, while a four-carbon difference results in poor miscibility. In contrast, di-18:1Δ^9 PC has better miscibility with di-14:0 PC than with di-18:0PC (82). The binary phase behavior of PC with other types of lipids is rather complicated (60,61). PC and PE have medium or low miscibility unless the two have the same acyl chain length. It is interesting to note that PC and PG can be ideally miscible when the two have the same acyl chain length, because the thermodynamic properties of PG are very similar to their corresponding PCs.

Cholesterol as one of the most useful liposome components can be readily incorporated into PC bilayers to a maximum of 50 mol%. Cholesterol itself does not exhibit a phase transition, but it does broaden significantly the main phase transition of the lipid bilayer into which cholesterol is incorporated. The primary purpose of using cholesterol is to increase the membrane rigidity, and it therefore improves mechanical properties of liposomes, which leads to better drug retention and longer liposome circulation.

II. TECHNIQUES OF LIPOSOME PREPARATION

Unlike other polymer-based drug delivery systems, the emphasis in making liposomes is not towards assembling the membrane, since lipid bilayers form spontaneously upon hydration, but towards getting the swollen membranes to form vesicles/liposomes of the desired size and structure, and to achieve high solute trapping efficiency and good pharmacological properties of the liposomal delivery system.

The common procedures for generating various types of lipid vesicles are summarized in Table 1 and will be described in detail below. Two of these involve the initial solubilization of lipids in organic solvents or detergents that have to be removed subsequently. The third procedure does not require solubilizing agents. But first it is useful to introduce two primary parameters that have been used extensively by liposomologists to describe liposomes for their attributes as drug carriers. The first parameter is the trapping efficiency, which is defined as the percentage of aqueous solute that is entrapped within the liposomes. The second parameter is the trapped (or the captured) volume, which is defined as the internal aqueous volume enclosed by a given amount of lipids and is often expressed as L/mol lipid. We will describe first the generation of different types of vesicles (i.e., MLV, LUV, and SUV) and then the most common preparation procedures.

A. Mechanical Dispersion by Direct Hydration (MLVs)

The conventional procedure for generating MLVs is very simple and straightforward. Lipids are dissolved in organic solvent, such as chloroform, or a chloroform/methanol mixture (2/1 to 3/1 vol/vol) and then dried by rotary evaporation to form a thin film on the glass wall of a round-bottom flask. The residual amount of solvent is then removed by vacuum (< millitorr) for several hours. Aqueous medium is then added to hydrate the lipids at temperatures 5–10°C above the T_m for the lipid (or the T_m for the lipid with the highest T_m value, in case of mixed lipids). Often, agitation, such as shaking, vortexing, or even the use of glass beads, is required to remove all lipids from the glass wall and to generate homogeneous lipid suspensions. At this stage, vesicle formation lipids are said to ''swell'' and

Table 29.1 Common Methods for the Generation of Multilamellar and Unilamellar Vesicles

Technique	Type of vesicles	Trapped vol. (L/mol lipid)	Advantages	Disadvantages
Direct hydration (hand-shaking) (36,64)	MLV	0.5	Fast	Low trapped volume; low trapping efficiency; unequal distribution of solute
Direct hydration plus freeze-thaw (36,63)	MLV	5–10	Fast, high trapped volume; high trapping efficiency	Solute dependent
Extrusion of MLV (40)	LUV	1–2	Fast, high trapping efficiency, uniform size	Relatively low trapped volume
Sonication	SUV	0.2–0.5	Fast, small size, most suitable for small volume preparation	Low trapped volume, size less uniform than extruded LUVS, small scale
Ethanol injection (25)	LUV/SUV	5–10	Convenient	Limited lipid-solubility in ethanol, low trapping efficiency, low lipid concentration, ethanol needs to be removed
Ether injection (25)	MLV/LUV	5–10	High lipid concentration, high trapping efficiency	Small scale, heterogeneous size
Reverse-phase by sonication/rotary evaporation (84)	LUV/MLV	10	High trapping efficiency	Small scale, heterogeneous size, safety concern
Reverse-phase by sonication/N_2 evaporation (36)	MLV (SPLV)	10–20	Even distribution of solute	Low trapping efficiency (< 30%), heterogeneous size, safety concern
Detergent dialysis (27,44)	SUV/LUV	0.5–5	Reconstitution of proteins, high trapped volume	Slow process, detergents difficult to remove completely, size difficult to control, low trapping efficiency

Source: Ref. 23.

peel off the glass support in sheets and form vesicles. MLVs have been very useful as model membranes for fundamental research, such as studies on the thermodynamic and structural properties of lipids and lipid interactions with drugs, peptides, or proteins (72,85,96–98,100,101). They have very limited use as drug carriers, particularly for systemic delivery of water-soluble agents, because MLVs produced by the above method give rise to very low trapping volume (36,64). In addition, they have a very wide size distribution. However, lipid-soluble agents can be entrapped inside the lipid bilayers of MLVs to very high efficiency, provided that they are not presented in quantities oversaturating the structural capacity of the lipid bilayers. The trapping efficiencies and trapped volumes of MLVs produced by this direct hydration method can be dramatically increased by a procedure called freeze-and-thaw, in which the lipid suspension is subjected to cycles of freezing and thawing between subzero temperatures (i.e., freezing in liquid nitrogen or dry ice) and temperatures above the T_m of the lipid (see Fig. 2) (36,64).

For preparations that will be used in humans, the use of chloroform or methylene chloride is not recommended, although that it can be pumped off below 1 ppm relatively quickly. Thus an alternative method of generating homogeneous MLVs is to freeze-dry the lipids dissolved in a suitable organic solvent. Tertiary butanol is considered the most suitable solvent. Aqueous medium is added to the dry lipid, which is in an expanded foamlike form (cake), and MLVs are then produced by rapid mixing above the T_{m} of the lipids (3).

B. Extrusion/Homogenization (LUVs)

The majority of applications, however, require smaller and better-defined liposomes. For systemic delivery of drug, liposomes with sizes in the range of 100–150 nm are desired, while for genosomes, SUVs smaller than 50 nm are often preferred. LUVs can be conveniently produced by an extrusion technique, where MLVs produced as described above or by other methods are forced through polycarbonate filters at temperatures above the T_m (77). It was later improved by employing a purposely built high-pressure extruder that allowed rapid extrusion under high pressure up to 5.5 mpa from a nitrogen tank (40,63). LUVs with sizes ranging from 50 to 200 nm can be generated by the extrusion procedure using appropriate filters (see Fig. 2). The extruded liposomes have well-defined size distribution and thus have been widely used both in research laboratories and in industries. Further, high lipid concentrations (up to 400 mg/mL) can be employed, which is critical for achieving high trapping efficiencies for hydrophilic agents.

Homogenization and microfluidization (or microemulsification) are the easiest ways for the scale-up. In homogenization, MLV solution is forced at high pressure through a small hole; it collides into a wall, a small ceramic ball, or the tip of the pyramid in an interaction chamber. Microfluidization devices split the suspension of MLVs into two parts, which are collided head on at pressures up to 20,000 psi. This technique offers such advantages as simplicity, large capacity, and speed (depending on the size of the homogenizer one can prepare 10 mL to hundreds of liters in 1 hour). Some of the disadvantages are possible lipid degradation and wider size distributions compared with liposomes produced by the extrusion method.

Another device, called the French press, originally designed for disrupting plant and bacterial cells (SLM-Amino Inc. Urbana, Illinois), can also be used for producing homogeneous liposomes in the mid size range (30–80 nm) (11,51). In this procedure, the MLV suspension is enclosed in a pressure chamber (20,000–40,000 psi), and the suspension is forced through an outlet orifice. As in other homogenization processes, the vesicles are disrupted into lipid fragments by the large shear force generated by the high pressure, and small liposomes are thus formed. The size and the homogeneity of the liposomes for a given lipid are inversely proportional to the flow rate through the orifice. For different lipid matrixes the size of the liposomes produced by the French press procedure depends on variables like lipid composition, temperature, and pressure. Consequently, this method can only be used when high reproducibility is not necessary. In addition to the high initial cost of the device, the French press has not been as popular as some other methods described above.

C. Sonication (SUVs)

Sonication has long been used as a convenient method for producing SUVs with the smallest size possible in small scale (41). In recent years it has been the most often used method for generating cationic liposomes for gene delivery studies (see Fig. 9C) (28,29,35,51,73). Both bath and probe ultrasonic disintegrators can be used, and each has advantages and disadvantages. For larger volume and dilute samples, bath sonication is more suitable. Normally 5 to 10 min sonication above the T_m of lipids is sufficient for producing SUVs of 50 nm or even smaller. The advantages of bath sonication include a controllable operating temperature, freedom from contamination, and no loss of lipid material. Probe sonication, on the other hand, dissipates more energy into the sample and therefore is often used for achieving the smallest size possible and for concentrated samples or suspension in more viscous aqueous solutions. Different size probes are available for different volumes of samples. One problem with the probe sonication is the heating of the sample solution, which can cause lipid degradation. This situation can be overcome by placing the sample vessel in a circulating water bath and under a nitrogen blanket. Another problem with probe sonication is contamination from the titanium particles from the probe tip, which must be removed by centrifugation or filtering. The size of liposomes produced depends on lipid concentration and composition, temperature, sonication time and power, volume of sample, as well as tuning; i.e., by carefully adjusting the position of the tip, one can achieve better energy dissipation. Consequently this process is not very reproducible in some circumstances. One should also be aware that the vesicle size may start to increase after a minimum size has been achieved, due to vesicle fusion resulting from colliding of vesicles in the process.

D. Solvent Injection

Several methods for liposome preparation involve the solubilization of lipids in organic solvent, which is then brought into contact with the aqueous solution containing the materials to be entrapped within the liposomes. Liposomes are formed almost instantly when the solvent is diluted or evaporated. Ethanol and ether are the two most often used solvents. Because of the markedly different miscibility of the two solvents with water, different procedures, though based on similar concepts, have been developed.

The ethanol injection method has been used for nearly thirty years (13,20) and has the advantages of simplicity and being free of the risk of lipid oxidation. In this procedure lipid (or lipid mixture) is first dissolved in ethanol solution and then rapidly injected through a fine needle into an aqueous solution in excess volume at high temperature. Under optimum conditions, liposomes in the size range of 20–50 nm can be produced. The major drawbacks of this method are the limited lipid solubility in ethanol (40 mM for PC and much lower for cholesterol) and the maximum proportion of ethanol in the final solution ($< 10\%$). Consequently, the lipid concentration in the final product is rather low, and thus the trapping efficiency for water-soluble materials is very low. Moreover, complete removal of ethanol by dialysis or diafiltration is a lengthy task.

The ether injection procedure is different from ethanol injection in many ways, because ether is a water-immiscible solvent. The lipid-containing ether solution is injected into the aqueous medium rather slowly through a narrow-bore needle, at such a temperature that ether becomes vaporized, and thus liposomes are formed. The advantage of the ether injection method is the high lipid concentration achievable, since the solvent can be removed at the speed that it is introduced. In order to achieve the best results of liposome size and homogeneity, a mechanical infusion pump is required to control the injection speed. Large-scale preparation, however, is not recommended due to safety reasons.

E. Reverse-Phase Method

The method commonly called the reverse-phase method involves the preparation of water-in-oil emulsions, where the lipid is solubilized in the organic phase and the aqueous phase contains the solute to be entrapped (84). The marked differences in the reverse-phase procedure from the solvent injection method are the relative volume ratio of solvent to water and the way that solvent is removed. In the reverse-phase method, organic solvent is in great excess and is mixed with a small amount of aqueous solution (for instance 1/10 of the volume of the solvent). Many solvents have been used in this method, but ether is the most common one. The mixture is then subjected to sonication to produce water-in-oil emulsions, which contain monolayers of lipid surrounding the water droplets. The emulsion droplets are then collapsed and transform into LUVs or MLVs (often called MLV-REVs) after the removal of the solvent, depending on the way that solvent is removed, the amount of excess lipids, and the solvent-to-water volume ratio. A solute trapping efficiency of greater than 50% can be achieved for MLV-REVs under optimum conditions.

Another version of the reverse-phase method is to add an aqueous solution containing the solute to an ethanol solution containing the lipids; the mixture is then dried under vacuum or by rotary evaporation (36,39). This results in a dry lipid film within which the solute is entrapped. Subsequent hydration of the dry lipid film results in the formation of MLVs with efficient entrapment of the solute initially presented in the system. Other solvent systems, such as ethyl ether, ethyl acetate, and ethanol/ethyl acetate mixtures, have also been used (21). Alternatively, the solvent in the reverse-phase emulsions can also be removed by blowing of nitrogen gas while continuously sonicating the emulsions. In the presence of an excess of lipids, MLVs called stable plurilamellar vesicles (SPLVs) can be produced (36). The effect of sonication during solvent removal is to disrupt the lipid bilayers continuously and cause redistribution and equilibrium of solutes between adjacent bilayers within each SPLV. The trapping efficiency of SPLVs can be only around 30%.

It is important to understand the differences in the structure of MLVs and the nature of solute distribution within the vesicle for different type of MLVs prepared by the various methods that have been described so far. In MLV-REVs a single large aqueous core is surrounded by several or more bilayers, because the MLV formed is originated from a water droplet. In comparison, both types of MLVs (produced from direct hydration) and SPLVs lack such a large aqueous core. The solute distributions are also very different for these vesicles. Most of the solute entrapped in MLV-REVs is mainly enclosed in the center aqueous cores. SPLVs differ from MLVs in the way that the solute distributes evenly throughout each SPLV, because of the constant rupture to the bilayers by sonication, while in MLVs the solute might be more concentrated near the center of the MLV and depleted in the outer compartments. Certainly for membrane-permeable solutes, eventual equilibrium distribution within MLVs will be reached after a certain period of time.

F. Detergent Dialysis

Detergent dialysis involves the solubilization of lipids (and proteins) in nonionic or ionic detergents, such as OGP and bile salts (51,68,71,85). Liposomes form gradually during the process of detergent removal either by passing the sample through a gel filtration column or by dialysis in a large volume of solution. This method suffers the serious drawback that the size of the liposomes formed cannot be easily controlled and often depends on factors like lipid type and composition, the nature of the solute, and the type of detergent and the process for detergent removal. Detergent dialysis procedure is not very practical for trapping water-soluble agents, because that the process is cumbersome and time consuming, but it is the method of choice for the reconstitution of biological transmembrane proteins, because of its nondestructive conditions.

G. Mechanism of Lipid Vesicle Formation

Dilution and energy-accompanied dispersal of lyotropic liquid crystals can result in rather homogeneous suspensions of stable colloidal particles, which can retain the short-range order of the parent liquid crystalline phase. In the case of diacyl amphiphiles, such as phospholipids, sphingolipids, and many synthetic polar lipids, one can form either liposomes upon the dispersal of lamellar phases or hexasomes upon homogenization of hexagonal phase in excess water.

Although methods for the preparation of well-defined liposomes have existed for almost thirty years, the exact mechanisms are still not known. Because identical structures can be made by very different preparation methods, several different mechanisms operate, depending on the experimental conditions, as illustrated in Fig. 7 (49,50).

The most widely used methods are high-energy treatment of large multilamellar liposomes formed by hydration of lipids or their mixtures. In these aggregates, lipid molecules are already organized into bilayers, and high-energy treatment, such as sonication, homogenization, or extrusion, simply breaks down large bilayers into smaller bilayered fragments. Such fragments are unstable due to the exposure of the hydrocarbon chains to an aqueous environment and tend to eliminate this unfavorable interaction. This can be done by bending and closing. This process is opposed by bending elasticity of the bilayer. For large enough lipid fragments, however, the bending energy is smaller than the unfavorable edge interaction and they close into self-closed vesicles. This means that in the beginning small fragments fuse into larger ones, because at constant area this reduces the circumference. Because bending elasticity scales linearly with the number of molecules (N) in the fragment ($E_b \propto N$) till it reaches a constant value, and edge interaction scales parabolically ($E_e \propto N^{1/2}$), one can see that for sufficient size $E_b < E_e$ and the fragment self-closes.

This mechanism operates also in the case of detergent depletion. In this case, however, the edge is shielded by detergent molecules, and larger fragments are stable, resulting in the formation of larger liposomes.

Liposomes can also be made from preexisting bilayers by fission of bilayers. This, the so-called budding off mechanism, operates when self-closed multilamellar vesicles are subjected to asymmetric changes that cause the area change of the outer monolayer in the bilayer. For instance, change in the pH can result in ionization of the outer monolayer. The charges repel, and the area of the outer monolayer increases with respect to the inner one. This creates a bud, and when the mismatch reaches the critical area ratio the budding off occurs. This mechanism operates in the liposome preparation by the change of pH, addition of amphiphiles (lysolecithin, short chained lipids, which distribute in the outer monolayer), and exchange of counterions in some cationic lipids. The process is thoroughly studied in giant unilamellar vesicles, which can bud off daughter vesicles under temperature changes, as monitored in the optical microscope (51).

Reverse-phase methods, in which liposomes are formed from water-in-oil or oil-in-water-in-oil emulsions upon removal of the organic solvent, operate by the dispersal of bicontinuous phases into the aqueous phase. For instance, upon removal of organic phase, lipid gel is formed. It consists of connected bilayered structures (cubic phases, similarly to foam, etc.) that disintegrate into bilayered colloidal particles upon their dilution and breakage in the aqueous phase.

Many different liposome preparation methods exist, and all can be described by the mechanisms above or their

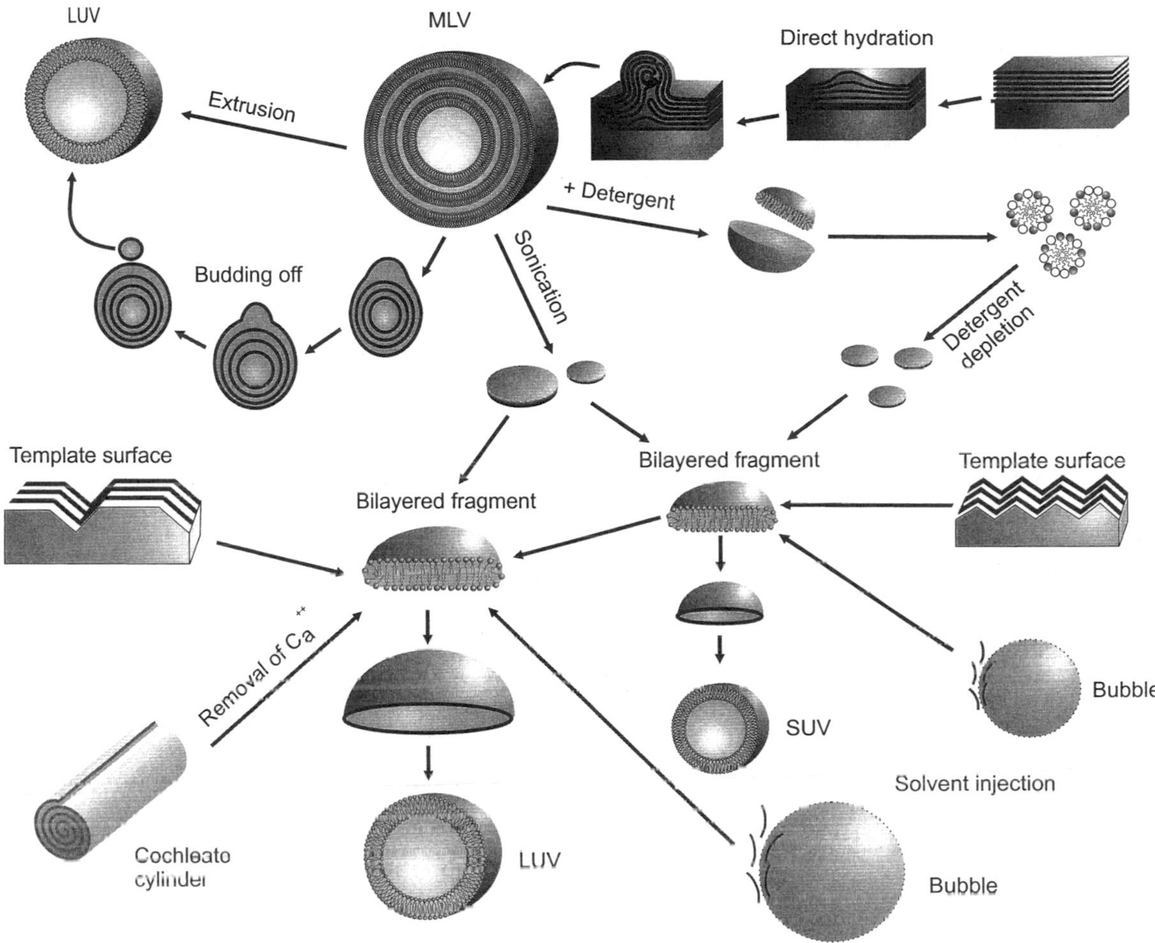

Figure 29.7 Mechanisms of liposome formation by various techniques.

combination(s). The exception is spontaneous vesicles, which are formed by simple mixing of appropriate amphiphiles. These liposomes, however, have very soft bilayers, and because they are very unstable upon application, and because they cannot stably encapsulate (hydrophilic) molecules, we will not describe them further.

III. METHODS FOR DRUG LOADING

All the techniques described above are useful for solute encapsulation. The primary considerations for choosing a suitable loading method are the physical and chemical properties of the solutes. For water-soluble drugs, active or remote loading procedures are preferred, due to the high trapping efficiency and high drug-to-lipid ratio achievable. However, for most water-soluble solutes that do not respond to the active loading process, passive loading procedures have to be used.

Hydrophobic solutes can be incorporated into the hydrocarbon chain region, either by direct hydration of the solute–lipid mixtures or by the reverse-phase procedure. Detergent dialysis is very useful, especially for the reconstitution of membrane proteins. The most troublesome solutes to load are these having very low water solubility and at the same time cannot be solubilized in organic solvent due either to low solubility or to chemical instability. Examples of this type of drug are cisplatin and amphotericin.

The primary variables to consider in drug loading are trapping (or encapsulation) efficiency, drug-to-lipid ratio, and drug retention (or release). The trapping efficiency is defined as the percentage of solute encapsulated, and the maximum trapping efficiency achievable depends on the loading technique. The drug-to-lipid ratio is determined after loading and is usually expressed as a weight ratio or a mole ratio.

Drug retention refers to the capability of liposomes to keep the drug from leaking out under defined conditions and is usually determined by incubation tests at 4°C and 20°C or by in vitro serum stability tests at 37°C. It should be noted that liposomes that provide satisfactory shelf stability do not necessarily always pass the serum stability test. This is because possible interactions of serum proteins with lipid components can significantly change the membrane integrity and therefore alter the drug leakage profile.

A. Passive Loading of Water-Soluble Solutes

In the process of passive loading, lipids and solute are codispersed in an aqueous medium, so solute entrapment is achieved while the liposomes are being formed. Typically lipid or lipid mixtures used in passive loading are in the form of thin films on glass walls, lyophilized powder, or simply dry cake in glass test tubes. The aqueous medium containing the water-soluble solute is then added to the lipid. Liposomes are generated by agitation at temperatures above the phase transition temperature. In order to achieve high trapping efficiency and drug-to-lipid ratio, a high concentration of solute is used whenever possible. Normally the trapping efficiency in a passive loading process is less than 10–20%. However, the trapping efficiency can be increased to the 30–50% range by the freeze-thaw (23,64) and the dehydration-rehydration (45) procedures. A high drug-to-lipid ratio is difficult to achieve unless the solute to be entrapped is extremely water soluble (e.g., >100 mg/mL). The trapping efficiencies can also be significantly improved using procedures involving organic solvents such as the stable plurilamellar vesicles (36) and the reverse-phase evaporation (84).

B. Passive Loading of Hydrophobic Molecules

Hydrophobic agents that are compatible with organic solvents can be passively loaded by the direct hydration procedure. Normally the solute and lipids are codissolved in organic solvents, such as chloroform or ether, and dried to form thin films or powdered cakes as described above. An aqueous solution is then added to hydrate the lipid mixture by agitation at temperatures above the main phase transition temperature. The loading efficiency and maximum drug-to-lipid ratio achievable depend on the structure of the solute. α-helical peptides, for instance, that match the lipid acyl chain length can be incorporated into lipid bilayers with near 100% efficiency (97,98). For solutes that are only soluble in aprotic solvents, a modified reverse-phase method can be used, where the solute and the lipids are dissolved in a mixture of ethanol and DMSO (or DMF) and then injected into the aqueous phase (83).

In case liposomes need to be downsized to LUVs, sonication and extrusion procedures can be applied. However, it is usually more difficult and sometimes troublesome to downsize solute-containing vesicles. This is because hydrophobic molecules are usually less flexible than the lipid hydrocarbon chains, and therefore the lipid bilayers are hardened and more difficult to break by the shear force applied by the sonication or extrusion procedures. Moreover, more solute may be spit out of lipid bilayers, since bilayers with high curvature can accommodate less solute.

C. Reconstitution of Proteins

The reason that detergent is used for protein separation and purification from biological membranes is to maintain the proteins in their native conformations. Similarly, for purposes of simulating the native membrane environment and maintaining protein conformation and biological activity, the characterization of membrane proteins can be carried out by solubilizing the proteins in detergent micelles. However, the ultimate ideal environment for protein characterization is provided by small phospholipid vesicles, and obviously the most suitable method for protein reconstitution is the detergent dialysis procedure (22,85). In this procedure, both protein and phospholipids are cosolubilized in appropriate detergent micelles so that the protein hydrophobic region (i.e., the part of the protein that is normally embedded in the membrane core) traverses the acyl chain region of the micelles. The detergent molecules are then slowly replaced with lipids by dialysis or column chromatography. The resulted lipid vesicles will contain the proteins embedded in the bilayer acyl chain region as illustrated. The choice of detergent mostly depends on the nature of the protein. When there is no specific head group requirement, nonionic detergents become the first choice because of the well-known tendency of long-chain ionic detergent to be protein denaturants. Polyoxyethylene derivatives are the most common type of nonionic detergents. It is essential to use detergents, such as $C_{12}E_8$ and Tweens (polyoxyethylene-sorbitan-alkyl esters), that form small micelles at high concentration. Another important consideration is the matching of the hydrophobic lengths of the protein and the detergent micelles to minimize the distortion of the proteins.

Some membrane proteins may require an anionic head group for maintaining their native conformation. In these cases, mixtures of anionic and neutral lipids can be used, since anionic lipids alone do not form stable vesicles, particularly when there are interactions with the protein. Similarly, anionic or mixtures of anionic and nonionic detergents can be used for the solubilization of proteins and lipids. It is worthwhile to mention that the commonly used ionic detergents sodium dodecyl sulfate (SDS) and alkyl trimethylammionuim salts will normally stabilize the conformation of the hydrophobic region of the protein, but at the same time they may have a tendency of denaturing the water-soluble portions of the protein. Caution should also be taken when deoxycholate and cholate are used, since the COO^- groups become protonated below pH 7.8 and the micelle size increases greatly.

D. Active Loading Procedures

In one major type of active loading procedure, the solute is loaded after the liposomes have been formed, and the accumulation of solutes inside liposomes is mediated by certain active transportation mechanisms. For this reason it is also often called remote loading. Lipophilic amino-containing solutes can be loaded with near 100% efficiency into liposomes exhibiting transmembrane pH gradients, specific chemical gradients, or membrane potential ($\Delta\Psi$).

In another type of active loading procedure, very different strategies are used for improving the trapping efficiency: special lipid components are deliberately incorporated to initiate the association of the solute with the carrier. Such a strategy has been extensively explored for the anticancer agent doxorubicin (Adriamycin), which is positively charged and has low water solubility. The trapping efficiency can be improved from less than 10% for neutral liposome systems by the passive loading procedure described above to 50–90% when anionic lipids are included in the lipid composition (23).

1. Active Loading by pH Gradient

Many lipophilic amino agents can be actively driven into the interior of liposomes in response to a pH gradient (inside acidic) (26,53,62,65,67) as illustrated in Fig. 8. LUVs are first prepared by extrusion in the presence of a low pH buffer (i.e., sodium citrate, pH 4.0). The transmembrane pH gradient is then established by adjusting the external pH value to three units higher than the internal pH. The loading of drug is simply achieved by introducing the drug into the liposome solution followed by incubation at temperatures above the T_m. Many lipophilic amino drugs have been successfully loaded by the pH gradient method. Doxorubicin, an antineoplastic agent, is the best characterized in the pH remote loading process, and greater than 95% trapping efficiency and a 0.29:1 drug-to-lipid ratio (wt/wt) have been achieved. The loading process is rather fast, and maximum drug loading is usually achieved in about

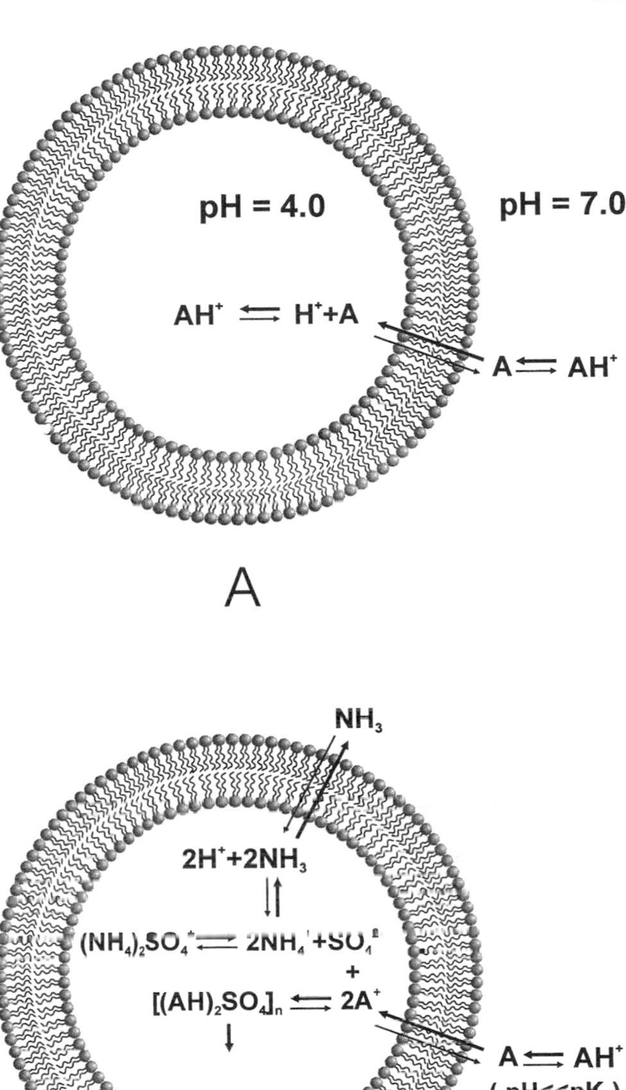

Figure 29.8 Mechanisms of active drug loading by pH gradient (A) and ammonium sulfate gradient (B). Lipophilic amino agents can pass through the liposome membrane in neutral form A. Once inside the liposome, the drug becomes positively charged, AH^+, due to the low internal pH. The drug will be retained inside the liposome provided sufficient ΔpH is maintained. When loaded with the ammonium sulfate gradient procedure, the positively charged drug AH^+ can form a complex with sulfate, $[(AH)_2SO_4]_n$, and precipitate inside the liposomes.

30–60 min incubation time (Fig. 8) and may vary from drug to drug. It should be noted that the loading time should be well controlled: a prolonged incubation at temperature above the T_m may cause drug leakage as well as lipid degradation. A potential drawback for some drugs in the pH remote loading procedure is the buffer/drug incompatibilities, which can lead to drug precipitation or degradation. Moreover, the addition of the alkalinizer for adjusting the external pH can cause a transmembrane osmotic gradient that can decrease the liposome stability. This problem can be overcome by exchanging the external buffer at constant osmolarity using column chromatography. However, such a procedure is cumbersome and has the limitations of scale-up and maintaining sample sterility.

2. Loading by Specific Chemical Gradient

Lipophilic amino drugs can also be efficiently loaded into preformed liposomes by the chemical gradient method (Fig. 8). In this procedure, an ammonium salt (i.e., ammonium sulfate, ammonium citrate, or ammonium phosphate) is enclosed in the hydration medium, and the ammonium gradient is established by the subsequent exchange of the external ammonium salt by means of dialysis/diafiltration or gel filtration column chromatography. Drug is then added to the liposome solution, and the accumulation of drugs inside the liposomes is achieved by incubation at a temperature above the T_m.

3. Ionophore-Mediated Loading by Membrane Potential

Another important type of active drug loading procedure involves the use of ionophores in the presence of a membrane potential (8,66). Ionophores are a class of organic molecules that vastly increase the permeability of membranes to particular ions and are mostly antibiotics of bacterial origin (88). The carrier type (the other type is the membrane channel formers, such as Gramicidin A) ionophores, such as K^+-valinomycin and Ca^{2+}-A23187, have been used for liposome drug loading. In the ionophore loading procedure, a membrane potential generated by an ion gradient needs first to be established, which can be easily achieved by passively trapping the ions, for instance, potassium glutamine, and subsequently exchanging the external K^+ with Na^+. Solute loading occurs in the presence of an appreciable concentration of ionophore molecules. Several lipophilic cationic molecules, such as safranine, methyltriphenylphosphonium, chlorpromazine, and vinblastine, have been successfully loaded by K^+-ionophore valinomycin. Ca^{2+} can also be successfully loaded into liposomes by the Ca^{2+}-A23187-mediated pH (i.e., H^+) gradient procedure (89). Similarly, ionophore A23187 can also be used for loading of Fe^{2+} and Ba^{2+}, because of the significant affinity of A23187 with the two ions (19). Fe^{2+}- and Ba^{2+}-loaded liposomes have some interesting properties, such as increased densities, which enable them to be isolated from serum and can be used for identifying serum proteins that associate strongly with liposomes. The Ba^{2+}-loaded liposomes have enhanced EM contrast, which could allow the direct detection of extracellular and intracellular liposomes and determine their integrity and metabolic fate. One serious problem with ionophore loading is that these ionophores are toxic, and complete removal of ionophores after loading can be difficult to accomplish.

The mechanism of carrier ionophores for transporting ions involves binding to the ion, diffusing through the membrane, and releasing the ion on the other side of the membrane. It should be noted that solute transport into liposomes also occurs in the absence an ionophore, though at a much slower rate. Studies using the membrane potential probes safranine and methyltriphenylphosphonium indicate that solute accumulation inside liposomes is an energy-dependent process in which solute is exchanged with K^+ on a 1:1 basis (8). In the case of ionophore-A23187-mediated loading of Ca^{2+} by pH gradient, exchange of Ca^{2+}-$2H^+$ occurs. It is conceivable that the role of ionophores in solute loading is simply to increase the efflux rate of ions across the membrane.

4. Theoretical Explanation of Remote Loading

The accumulation of drugs as well as the entrapment efficiency in the case of pH gradient loading can be simply explained and calculated by a Henderson–Hasselbalch equation (26). Unlike pH gradient, the ammonium sulfate (or exchange) gradient loading is based on an influx of the drug molecules from exterior into interior and a simultaneous efflux of appropriate weak base molecules into the opposite direction. Due to this, there is a constant reduction of concentrations of related neutral species on both sides of the membrane. At the same time, following the law of mass action, both processes are accompanied by a simultaneous redissociation of the H^+-charged drug and the corresponding inner charged exchange-related weak base molecules. During the loading, these processes determine the level of the protons as well as the amount of the corresponding influxed/effluxed molecules in either compartment: the dissociations are coupled and drive the molecules in opposite directions until the concentration equality of neutral species on both sides of the membrane is established.

For the theoretical approach (37), we shall assume that

the permeability of charged species is negligible, although we have shown that this requirement can be lifted, and by using permeability and diffusion coefficients the kinetic nature can be incorporated into the model. Below, we shall define the equilibria that dictate the drug encapsulation and allow one to calculate drug loading as a function of experimental parameters and pK value(s) of the drug molecules.

Assuming that each compartment consists of an aqueous solution, and taking into account that each of them contains an optional number of acid/base and inert species (strong ions), the time-dependent equilibria during the simultaneous uptake and efflux of different permeable molecules can be described by the following two electroneutrality equations:

$$\sum_{m,\mathrm{I}} c_{m\mathrm{I}}(1 - B_m\beta_m)\frac{g_{m\mathrm{I}}(x)}{f_{m\mathrm{I}}(x)} + K_v \sum_{k,\mathrm{O}} (A_k c_{k\mathrm{O}}\alpha_k)\frac{g_{k\mathrm{O}}(x)}{f_{k\mathrm{O}}(x)} + \upsilon_\mathrm{I}(x) = 0$$

for the inner compartment and

$$\sum_{k,\mathrm{O}} c_{k\mathrm{O}}(1 - A_k\alpha_k)\frac{g_{k\mathrm{O}}(y)}{f_{k\mathrm{O}}(y)} + \frac{1}{K_v}\sum_{m,\mathrm{I}} (B_m c_{m\mathrm{I}}\alpha_m)\frac{g_{m\mathrm{O}}(y)}{f_{m\mathrm{O}}(y)} + \upsilon_\mathrm{O}(y) = 0$$

for the outer solution. Here, I and O refer to the inner and outer compartment of the liposomes, m and k to the inner- and outer-compartment sister-species, respectively; x and y are the concentrations of the inner and outer H^+ ions, A_k and B_m are factors that allow the particular species to permeate the membrane (=1) or not (=0), $a_k(t)$ and $ß_m(t)$ represent the time-dependent percentages of the entrapped and released species, $c_{m\mathrm{I}}$ and $c_{k\mathrm{O}}$ are the starting concentration of the corresponding inner/outer species, K_v is an outer-to-inner liposome compartment volume ratio, and the functions $u_\mathrm{I}(x)$ and $u_\mathrm{O}(y)$, respectively, refer to H^+ and inert ions that do not cross the membrane. For details with regard to x- and y-dependent f and g functions the reader is referred to Ref. 18.

Before the chemical equilibration between the compartments is established, the concentrations of the particular uptaken and/or released molecules are time dependent. Assuming a homogenous liposome solution, the portion of an outer k-species influxed can be expressed by the differential equation

$$\frac{d\alpha_k}{dt} = \frac{S_\mathrm{O}P_k \prod_{j=0} K_{ij}}{K_v v_l}\left((1-\alpha_k)\frac{1}{y^{l_k-1}f_{k\mathrm{O}}(y)} - K_v\alpha_k\frac{1}{x^{l_k-1}f_{k\mathrm{I}}(x)}\right)$$

and a similar expression is valid also for inner m-molecules effluxed:

$$\frac{d\beta_m}{dt} = \frac{S_\mathrm{I}P_m \prod_{j=0} K_{ij}}{v_l}\left((1-\beta_m)\frac{1}{x^{l_m-1}f_{m\mathrm{I}}(x)} - \frac{1}{K_v}\beta_m\frac{1}{y^{l_m-1}f_{m\mathrm{O}}(y)}\right)$$

Here, $S_\mathrm{I}/S_\mathrm{O}$ refers to the inner/outer liposome bilayer surface, P_k/P_m is a permeability of k/m-species, v_l is the volume of the liposome vesicle, respectively, and ΠK_{ij} is a product of the corresponding acid–base dissociation constants. By introducing of an appropriate set of initial conditions (for $t = 0$, $a_k(0) = ß_m(0) = 0$), the above four equations can be solved using numerical methods.

The entirely equilibrated system (for $t \to \infty$) is described by two equations. The first one,

$$\sum_m c_{m\mathrm{I}}\frac{g_{m\mathrm{I}}(x)}{B_m K_v\left(\frac{y}{x}\right)^{l_m-1} f_{m\mathrm{O}}(y) + f_{m\mathrm{I}}(x)} + \sum_m A_k c_{k\mathrm{O}}\frac{g_{k\mathrm{I}}(x)}{\left(\frac{y}{x}\right)^{l_k-1} f_{k\mathrm{O}}(y) + k_v f_{k\mathrm{I}}(x)} + \upsilon_\mathrm{I}(x) = 0$$

is valid for the exterior solution, and the second one, for the interior solution.

$$\sum_k c_{k\mathrm{O}}\frac{g_{k\mathrm{O}}(y)}{A_k k_v\left(\frac{x}{y}\right)^{l_k-1} f_{k\mathrm{I}}(x) + f_{k\mathrm{O}}(y)} + \sum_m B_m c_{m\mathrm{I}}\frac{g_{m\mathrm{O}}(y)}{\left(\frac{x}{y}\right)^{l_m-1} f_{m\mathrm{I}}(x) + K_v f_{m\mathrm{O}}(y)} + \upsilon_\mathrm{O}(y) = 0$$

Now these general equations will be applied to a concrete example: uptake of the drug under simultaneous synergetic action of pH and bidirectional ammonium sulfate (exchange) gradient loading. The outer solution is buffered with a monovalent weak acid/salt of concentration c_{oa}/c_{os} and with a dissociation constant K_{oa}. It contains a drug of concentration c_d. In its molecular form the drug molecule is a weak base with the acid dissociation constant K_d. Similarly, the interior of the liposomes is filled with an aqueous solution containing also a monovalent weak acid/salt buffer (concentration c_{ia} for acid and c_{is} for salt with the acidity constant K_{ia} and a monovalent weak base of concentration c_i and constant K_i which is simultaneously allowed to release during the uptake of the drug molecules. For the situation in the exterior during the loading/release one obtains

$$c_d(1-\alpha)\frac{1}{1+K_d/y}+\frac{1}{K_v}c_i\beta\frac{1}{1+K_i/y}-(c_{oa}+c_{os})\frac{K_{oa}/y}{1+K_{oa}/y}+(c_{os}-c_d)+\left(1-\frac{K_w}{y}\right)=0$$

and the following is valid for the interior,

$$c_i(1-\beta)\frac{1}{1+K_i/x}+K_v c_d\alpha\frac{1}{1+K_d/x}-(c_{ia}+c_{is})\frac{K_{ia}/x}{1+K_{ia}/x}+(c_{ia}-c_i)+\left(1-\frac{K_w}{x}\right)=0$$

When time-dependent portions of the uptaken drug- and exchange-related molecules, α and β, in the above equations are set to 0, one can calculate the pH in both compartments before loading. Here, K_w is the autodissociation constant of water. The time dependency of the portions of the drug taken up and released molecules is given by

$$\frac{d\alpha}{dt}=\frac{3}{r}\left(1+\frac{2d}{r}+\frac{d^2}{r^2}\right)P_d K_d\left(\frac{1}{K_v}(1-\alpha)\frac{1}{K_d+y}-\alpha\frac{1}{K_d+x}\right)$$

and

$$\frac{d\beta}{dt}=\frac{3}{r}P_i K_i\left((1-\beta)\frac{1}{K_i+x}-\frac{1}{K_v}\beta\frac{1}{K_i+y}\right)$$

When loading is finished, one obtains for the exterior compartment,

$$c_d\frac{1}{k_v\left(\frac{x}{y}\right)(1+K_d/x)+(1+K_d/y)}+c_i\frac{1}{\left(\frac{x}{y}\right)(1+K_i/x)+K_v(1+K_i/y)}-(c_{oa}+c_{os})\frac{K_{oa}/y}{1+K_{oa}/y}+(c_{os}-c_d)+\left(y-\frac{K_w}{y}\right)=0$$

and similarly for the interior solution,

$$c_i\frac{1}{K_v\left(\frac{y}{x}\right)(1+K_i/y)+(1+K_i/x)}+c_d\frac{1}{\left(\frac{y}{x}\right)(1+K_d/y)+k_v(1+K_i/x)}-(c_{ia}+c_{is})\frac{K_{ia}/x}{1+K_{ia}/x}+(c_{is}-c_i)+\left(x-\frac{K_w}{x}\right)=0$$

E. Strategies for Improving Drug Retention

Perhaps the most challenging task in using liposomes as drug carriers is not to maximize the trapping efficiency and drug-to-lipid ratio, but to achieve acceptable drug retention. While loading technology can always be improved and new techniques may be developed, drug retention depends mostly on factors at the molecular level. Drug retention refers to the capability of liposomes for keeping the drug from leaking out and is usually determined by an incubation test or in in vitro serum stability tests. Drug retention is influenced by many factors including the membrane affinity of the drug, the water solubility, and the physical properties of the lipid matrix. With a given drug, the lipid composition has the most effect on drug retention. Lipids with long saturated acyl chains offer the possibility of good drug retention, because of their high phase transition temperatures. Cholesterol is often used for increasing drug retention and blood circulation because it increases the rigidity of bilayers. However, drugs with high membrane partitioning coefficients tend to leak out of liposomes very fast even with well-designed lipid compositions. Nevertheless, there are some strategies that can be used for improving drug retention. The best example is the commercial product DOXIL® (doxorubicin HCL liposome injection, ALZA Corp., Mountain View, CA), which is a liposomal doxorubicin loaded by the ammonium sulfate method. Excellent drug retention is achieved in part by the formation of crystals of the doxorubicin/sulfate complexes (see Fig. 9A) (54).

IV. CHARACTERIZATION OF LIPOSOMES

Liposome preparations need to be characterized, and in many cases the quality needs to be controlled, before their usage. Table 2 shows the most important parameters for liposome characterization and quality control. Some of the major characterization parameters include visual appearance, turbidity, size distribution, lamellarity, trapped volume,

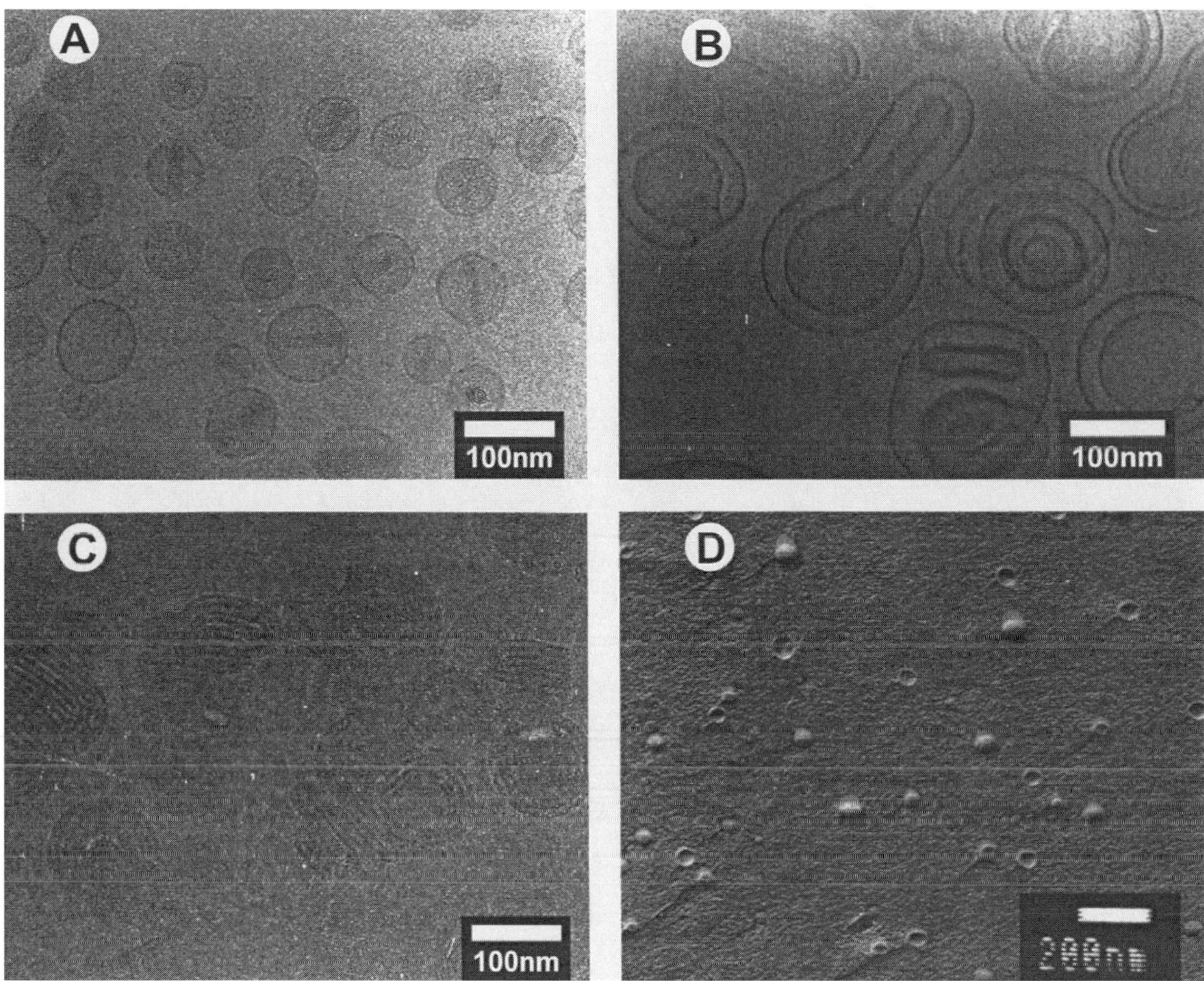

Figure 29.9 Cryoelectron micrographs. (A) Doxorubicin encapsulated in sterically stabilized liposomes. In the interior of the liposomes, fibers of $(DoxH^+)_2SO_4$ crystals can be observed that also change the shapes of the liposomes from spherical into oval. Higher magnification (not shown) shows that fibers in the bundles exhibit a helical arrangement. (B) DOTAP: Chol (50/50, mol%)–DNA (2/1, +/− charge) complex formed with extruded 100 nm liposomes. On the upper left, an unreacted liposome with a typical invaginated shape can be observed. A high degree of DNA encapsulation is characteristic for these complexes, which may originate in the invaginated liposome shape. C) DDAB:Chol (50:50, mol%)–DNA (2/1, +/− charge) complex prepared by mixing with sonicated liposomes. An intercalated lamellar phase, in which lipid bilayers in fluid state sandwich two-dimensionally condensed DNA, can be observed. (Courtesy of P. Fredrik.) (D) Freeze-fracture electron micrographs of stabilized plasmid–lipid particles (SPLP) prepared by the dialysis method (see Section VII.C.4) with lipids (DODAC/DOPE/PEG200-Cer–C20, 20:70:10 mol%) and CMVLuc plasmid. The photo shows the DNA containing the SPLP population isolated by the sucrose density gradient. The purified SPLPs have an average size of around 70 nm, while the empty vesicles are less than 30 nm (not shown).

composition, trapping efficiency, and stability. The extensiveness of physical characterization obviously depends on the purpose of the liposome application. For cationic liposomes prepared for DNA complex formation, for instance, a simple general characterization to ensure liposome homogeneity should be adequate, since liposomes will disintegrate after complexation with DNA. Also, liposomes for topical and oral applications may not have to undergo the stringent sterility requirement of parenteral samples.

A. Visual Examination

As in any formulation's specification data sheets, visual observation is important in liposomal preparations. While most of the stability parameters cannot be controlled, obser-

Table 29.2 Liposome Characterization and Quality Control Parameters

General characterization	*Technique*
pH, conductivity, and osmolarity	pH meter, conductivity meter, osmometer, respectively
Phospholipid concentration	Lipid phosphorous content (Bartlett method)
Phospholipid composition	TLC (combined with the Bartlett method), HPLC
Cholesterol concentration	Cholesterol oxidase assay, HPLC
Trapped volume	Measurement of liposomal internal aqueous phase by membrane-impermeable markers
Trapped solute concentration	Spectrophotometry, spectrofluorometry HPLC, GC,
Residual organic solvents and heavy metals	NMR, GC
Trapped solute/lipid ratio	Determination of solute and lipid concentrations
Liposomal internal pH	Fluorescent indicators, ESR indicators, ^{31}P-NMR, ^{19}F-NMR
Physical characterization and stability assays	
Vesicle size distribution—submicron range	Dynamic light scattering (DLS), static light scattering (SLC), gel exclusion chromatography, specific turbidity, electron microscopy
—micron range	Coutler counter, light microscopy, light diffusion, specific turbidity
Electrical surface potential and surface pH	Membrane-bound electrical field probes and pH-sensitive probes
Zeta (ζ) potential	Electrophoretic mobility
Lipid phase transition temperature, lipids miscibility/phase separation	DSC, x-ray diffraction, membrane fluorescent probes, Raman spectroscopy, NMR, ESR, FTIR
Trapping efficiency and percentage of free solute	Gel exclusion chromatography, ion exchange chromatography, dialysis, diafiltration, precipitation by polyelectrolyte
Membrane/aqueous phase partition coefficient (K_p)	Dilution effect (up to 10,000-fold) on liposomal solute/lipid ratio at equilibrium
Drug retention stability (K_{eff})	Monitoring change in solute/lipid ratio as a function of time
Chemical stability assay	
Phospholipid acyl chain autoxidation	Conjugated dienes, lipid peroxides, TBARS, and fatty acid composition (GC), UV/VIS
Phospholipid hydrolysis	TLC, HPLC, total PL and/or free acid concentration
Cholesterol autoxidation	TLC, HPLC
Antioxidant degradation	TLC, HPLC
Solute degradation	TLC, HPLC, spectrophotometry
Biological assay	
Sterility	Aerobic and anaerobic bottle cultures
Pyrogenicity and endotoxin	Rabbit and/or limulus amebocyte lysate (LAL) test
Medium-induced solute leakage	Incubation with buffer, serum and plasma and assay for solute leakage by chromatography
Toxicity	In vitro and/or in vivo tests
Pharmacokinetics and biodistribution	In vivo animal tests

Source: Ref. 12.

vations such as color change, presence of aggregates or sediment, or phase separation can definitively show that something went wrong with the formulation. Optical observations (phase contrast and polarization) are very useful for the size determination of large liposomes or the contamination of liposome dispersions with large particles or aggregates.

B. Size Determination of Liposomes

The sizes of liposomes can be measured directly by electron microscopy (EM) techniques or by dynamic light scattering. In negative staining EM, liposomes are deposited on a film of formvar, which is then stained by a hydrophilic electron dense solution (38). When the film is examined in an electron beam, the relatively electron-transparent liposomes will appear as bright areas against a dark background, hence the term negative stain. In addition to size distribution, it can also reveal information regarding whether the liposomes produced are multi- or unilameller. However, caution must be exercised due to possible artifacts of sample drying and coalescence.

Freeze-fracture (FF) EM procedures (Fig. 2), providing that the freezing was fast enough, can provide very useful data on sample heterogeneity and lamellarity of liposomes, especially on contamination with larger liposomes (23). Cryo-EM (Fig. 9) is sophisticated, but is the most artifact-free method for yielding the size distribution and the lamellarity of vesicles. It also allows the detection of entrapped material that is electron dense. It is based on the direct observation of rapidly frozen biological samples and therefore does not require any staining or preparation of replicas, as in FF-EM. Its limitation is the inability to observe particles larger than ~0.5 μm.

In contrast, light scattering techniques are usually used for determining liposome size due to their simplicity. Particle sizers from manufactures such as Coulter, Malvern, and NICOMP are based on quasi-elastic light scattering (QELS), where the time-dependent coherence of light scattered by particles is sensitive to their diffusion, which is dependent on the viscosity of the medium and particle size. Therefore QELS only measures the average size distribution of the bulk of the liposomes. For unimodal liposome systems the results are in good agreement with electron microscopic measurement. However, the size distribution profile can be severely skewed by the presence of a small number of extra-large liposomes or aggregates, and, on the other hand, the presence of very small liposomes, which may be significant in number, may be underweighed. A common and reliable approach is that the QELS size result is compared with EM examination first, and once a consistency has been achieved for an established liposome system, QELS can be reliably used as a day to day method.

C. Lamellarity of Liposomes

Electron microscopy as described above is the most convincing way that allows one directly to visualize the membrane lamellarity as well as the organization of the internal lamellae. However, negative staining is subject to artifacts and sometimes misleading indications that arise with the collapse of vesicles on top of each other. FF-EM can provide a unique view of internal lamellae when cross-fracturing occurs (Fig. 2). However, cross-fracturing is usually rare, particularly for small vesicles, and in such case thin-section EM (39) and especially cryo-EM are the methods of choice (Fig. 9). The measurement of trapped volume (see the following section) can be used as a good indicator for liposome lamellarity. If the measured value of trapped volume is substantially lower than the theoretical value for the LUV of the same size, it can be assumed that an appreciable proportion of the vesicles is multilamellar.

Alternatively, the liposome lamellarity can also be estimated by measuring the proportion of phospholipid exposed on the outside surface of the liposomes. For LUVs this will be exactly 50%, and for SUVs this will be higher than 50%. For MLVs the proportion will be much smaller than 50%. This proportion of outside lipid can be quantified spectroscopically using liposomes containing trace amounts of derivatizing PE with trinitrobenzene sulphonic acid (TNBS) (70). ^{31}P NMR spectroscopy can be used to provide an indication of the lamellarity of liposomes by measuring the signal of the PC head groups of liposomes before and after the addition of membrane-impermeable manganese ions (Mn^{2+}) to the external medium (40). Phospholipids facing the external medium interact with Mn^{2+} such that the resonance signal is broadened beyond detection. Thus direct comparison of the size of phosphorous signals before and after the manganese addition readily reveals the percentage of phospholipids on the outer leaflet.

D. Measurement of Trapped Volume, Trapping Efficiency, and Solute Release

In practice the trapped volume can be much lower than that expected for unilamellar vesicles with identical size distribution, due to the presence of multilamellar vesicles. The trapped volume of the liposome population can be determined by measuring the amount of the solute entrapped inside liposomes. The use of radioactive markers, though separation of free markers is necessary, offers advantages of being nondestructive and accurate. Optical, fluorescent, or HPLC measurements are also often used for trapped volume measurement (see following section).

It is clearly essential to determine the solute trapping efficiency before going on to investigate the physical and biological properties of the trapped material in a liposomal form, since the effects observed experimentally will usually be dose related. The trapping efficiency is measured as the percentage of trapped solute with respect of the total amount of solute added. This analysis therefore requires the separation of the liposome-encapsulated solute from the external medium, which can be carried out by several methods, including column chromatography, dialysis, and diafiltration. The gel filtration column, particularly the Sephadex G-50 spin column method (31), is rapid, requires small sample quantities, and can be performed on many different samples simultaneously. The ion exchange column can also be used for charged solutes. Dialysis offers the advantages of being able to retain 100% of the liposomes and being easy to scale up. However, this method is only applicable for solutes that have a leakage profile significantly longer than the time course of dialysis. Diafiltration is mostly used in pharmaceutical industries for removing unentrapped drug from large-scale liposome formulations. Therefore the drug trapping efficiency and the drug–lipid ratio of the final products can be evaluated. It should be noted that the measurement of trapping efficiency by the separation method is based on the assumption that after separation the remaining solute is 100% entrapped inside liposomes, which may not be true for many drugs.

There are fundamentally two different types of release studies emphasizing different aspects of liposomal systems, one being to characterize the ability of liposomes prepared with a particular lipid composition or a method to retain water-soluble markers and the other being to monitor the leakage rate of a primary drug that has been entrapped inside liposomes. The former is often used when release is induced deliberately as a result of a biochemical or biophysical change. In such case it is desirable to use special markers designed for the purpose, which normally do not leak out from intact membranes and do not associate with membranes, to simplify measurements and interpretation of these experiments. Various types of water-soluble markers have been used by researchers. Fluorescence markers such as calcein, carboxyfluorescein, and fluorescein dextrans, and radiolabeled markers such as glucose, sucrose, DTPA, and inulin, have been commonly used (76). On the other hand, when the objective of a particular liposome formulation is for delivery or targeting of the drug of interest, the marker for drug release and long-term stability is naturally the entrapped solute itself. Many cationic lipophilic drugs, such as doxorubicin and camptothecins, are strong chromophors and fluorophors, and therefore spectroscopic methods can be used for the detection of drug concentration.

V. TYPES OF LIPOSOMES ACCORDING TO FUNCTIONALITY

Several different classes of liposomes have been developed (Fig. 10). This development was based mostly on liposome reactivity with the environment. In order to explain that, we shall briefly introduce forces between liposomes

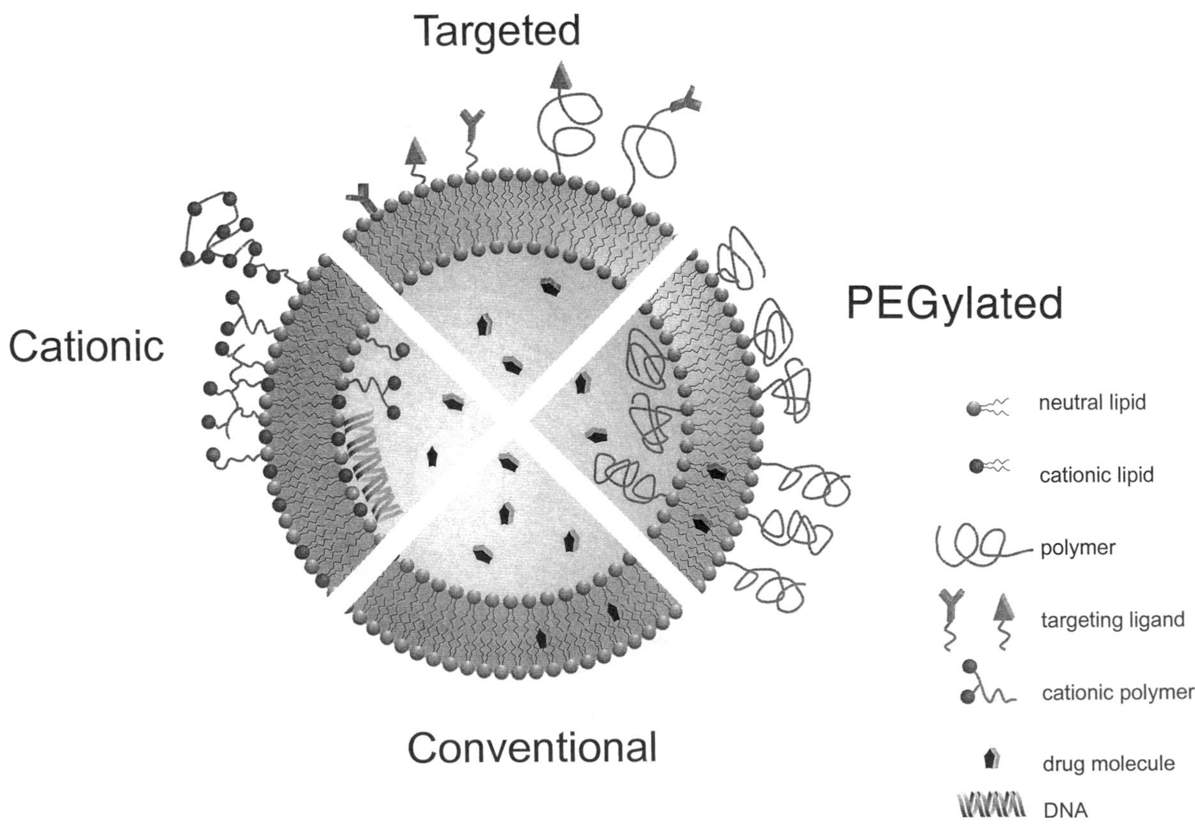

Figure 29.10 Four different classes of liposomes as defined according to their functionality. Conventional liposomes interact with the biological milieu nonspecifically. Sterically stabilized liposomes can avoid the uptake by the RES system due to surface coating with inert hydrophilic polymer. Liposomes with attached targeting ligands react specifically. Polymorphic liposomes, including cationic liposomes, pH-sensitive, temperature-sensitive liposomes and PEGylated liposomes with cleavable polymers, change their phase upon interaction with a specific agent or medium. Drugs can be entrapped inside liposomes or be incorporated into the hydrocarbon chain region, depending on their hydrophobicity.

themselves and other macromolecules. Attractive forces in these systems can be electrostatic in the case of oppositely charged colloids, van der Waals attraction and ion–ion correlation forces, as well as specific interactions in the case of the surface attached ligands. Weaker attractive forces are hydrogen bonds. Repulsive interactions are electrostatic repulsion between colloids of the same charge, steric and hydration repulsion and undulation forces of fluid bilayers.

In a biological milieu, liposomes interact by lipid exchange with surrounding membranes and lipoproteins or by adsorption of proteins and other macromolecules. In the presence of cells liposomes can adhere to cells or be either endocytosed by cells, or they fuse with cell membranes.

It seems that the main application of liposomes will be in systemic drug delivery. Therefore we shall briefly describe their interaction in the human body. Upon systemic (intravenous or intra-arterial) injection, liposomes are recognized as foreign particles and are quickly opsonized, i.e., tagged by proteins of the immune system for the subsequent uptake in the fixed macrophages in the liver, spleen, and bone marrow by adsorption of various proteins, such as immunolglobulins or proteins of the complement cascade.

For a particular application, liposomes can be optimized and programmed for optimal efficiency. This rational design is based on the knowledge of interactions upon their application. Along those lines, several different types of liposomes have been developed, as discussed below (Fig. 10).

A. Conventional Liposomes

Although relatively stable in the test tube, conventional liposomes are in general not stable upon application. It was realized early in the 1980s that the successful use of liposomes for drug delivery in most applications depends on two basic aspects of the carrier: quantitative retention of the entrapped drug and sufficiently long blood circulation time (7,46,56). Now it is well known that the in vivo pharmacokinetics of liposomes can be dramatically altered by the size of the liposomes as well as the lipid composition. Unilamellar vesicles of 100–200 nm size have much longer circulation time than MLVs (34). Drug retention and liposome circulation can be increased by the incorporation of cholesterol, which increases the packing of lipids in the bilayer. Another reason that cholesterol increases circulation is believed to be that the more rigid cholesterol-rich bilayer is more resistant to phospholipid removal by plasma high-density lipoproteins (HDL). With the same principle, the use of lipids with high phase transition temperatures, such as DSPC or sphingomyelin (SM), also produces liposomes with increased drug retention and circulation. In summary, for conventional liposomes, the blood circulation half-life decreases with increasing size, charge density, and fluidity. Thus the longest circulation achievable with conventional liposomes is with small unilamellar vesicles ($<$ 150 nm) prepared from DSPC (or sphingomyelin) with saturated cholesterol (45 mol%).

While mechanical stabilization of liposomes can decrease interactions with proteins and consequently increase their stability and blood circulation times, it cannot prevent long-range van der Waals forces and subsequent uptake by macrophages.

B. Sterically Stabilized Liposomes

Because conventional liposomes are characterized by relatively poor stability in the blood circulation, many researchers tried to prepare biologically stable liposomes. After numerous trial-and-error experiments, which concentrated on various liposome compositions, it was finally realized that coating of liposomes with a steric shield of inert polymers can reduce adsorption of various blood proteins and extend their circulation times. Nowadays, PEGylated liposomes, i.e., liposomes coated by (polyethylene) oxide chains, are almost exclusively used. PEGylated lipids with various PEG chains covalently attached to various hydrophobic anchors are commercially available. These liposomes are sometimes referred to as STEALTH® liposomes because of their ability to avoid the defense systems of the body. The general question is normally the amount of PEG-lipid in the formulation and the degree of polymerization, N, of the polymer chain. Typically the half-life in blood circulation is used to optimize the two parameters: typically formulations contain 1–5 mol% of PEG-lipids with N between 30 and 110 (molecular mass 2000 or 5000). The polymer is anchored on a hydrophobic moiety, such as diacyl phospho- or sphingo-lipid or cholesterol. Because PEG-lipids are more hydrophilic than other lipids in the membrane, they slowly dissociate from the membrane. Longer hydrocarbon chains result in longer residence time in the bilayer and better stability. In some cases prolonged liposome stability is not desired, and shorter chains are used, resulting in temporarily predictable liposome destabilization. Below we shall briefly discuss the theoretical basis of sterical stabilization (55).

In agreement with the hypothesis that repulsive steric pressure is responsible for the stabilization of liposomes, osmotic stress and surface force apparatus measurements have found increased repulsion between bilayers with grafted PEG chains (48). In the simplest picture one would like to understand the stability of sterically stabilized systems as a function of two controlling parameters: chain grafting density and chain length.

The corona of an attached polymer causes repulsion because the entropy of the chains is reduced upon approach (loss of configurational entropy) as well as excluded volume repulsion (osmotic repulsion). Depending on the size of the polymer (R) and the distance between attachment/grafting points (D), a polymer can have different conformations. At very low surface coverage ($D > R$), the polymer forms either a pancakelike structure or an inverse dropletlike structure, depending on whether it forms an adsorption or a depletion layer on the surface. At $D > R$ the so-called mushroom conformation is present, while for $D \approx R$ and $D < R$ the polymer chains starts to interact, forcing its extension into the so-called brush conformation (24). In the mushroom conformation, repulsive pressure (p) at a distance h is proportional to $(h_c/h)^{8/3}$, where h_c is the chain extension, while in the brush conformation,

$$p \approx \left[\left(\frac{h_c}{h}\right)^{9/4} - \left(\frac{h}{h_c}\right)^{3/4}\right]$$

and

$$h_c = N \cdot a \cdot \left(\frac{a}{D}\right)^{2/3}$$

where a is the size of the monomer.

The hypothesis of increased repulsive pressure above the membrane and reduced adsorption on the blood circulation lifetimes was tested by measuring the repulsive pressure between membranes with and without incorporated polymer-bearing lipid by using the osmotic stress technique, which can measure force (distance) profiles. Results showed that bilayers containing PEG-lipid show much larger interbilayer spacings, and even upon strong osmotic compression the bilayers are still 4 nm apart as compared to surface unmodified bilayers, which show practical collapse to the hard wall (Born) repulsion upon compression (74,75).

Theoretical analysis of polymers at interfaces defines two regimes: mushroom at low coverage and brush at higher. These simple calculations, however, do not take into account the solubility of PEG-lipid in the aqueous phase. In closed systems, however, this assumption is good because the solubility is below μM.

The interbilayer repulsion is calculated from the repulsive pressure of surface attached polymer in a mushroom configuration, i.e., from

$$P = \left(\frac{5}{2}\right)\frac{kTN}{aD^2}\left(\frac{a}{(h/2)}\right)^{8/3}$$

For $N = 44$, $a = 0.35$ nm, and $D = 3.57$ nm, this is in nice agreement with experimental data. For distances $>h_c$ the repulsive pressure is zero (57).

Similar results were obtained also by the surface force apparatus (47). These results also show increased repulsion with increasing amount of PEG polymer. Surface force measurements have found reversible repulsive force at all separations, and the thickness of the steric barrier was found to be controlled by N and D.

At low coverages (dilute mushroom), the Dolan and Edwards mean-field theory of steric forces and their scaling model described experimental data satisfactorily, while at higher coverages the Alexander–de Gennes theory based on scaling concepts described the data better. A more complex mean-field treatment did not bring any better fit. At low coverages, in the mushroom regime, the Dolan–Edwards expression for force between two curved cylindrical surfaces of radius R is

$$\frac{F_c(h)}{R} = 72\left(\frac{kT}{D^2}\right)\exp\left(\frac{-h}{R_g}\right)$$

where R_g is the radius of gyration of the polymer in a theta solvent and corresponds to the thickness of the extending polymer; it can be in good solvents substituted by R_F. Despite an underestimation of the polymer layer thickness in the low coverage regime, both models are able to fit the force–distance profiles rather well. The scaling analysis fit, which also describes the measured dependence rather satisfactorily, can be expressed as

$$\frac{F_c(h)}{R} = 1.6\left(\frac{2kT}{D^2}\right)\left[\left(\frac{R_F}{h}\right)^{5/3} - 1\right]$$

where the numerical prefactor is close to expected unity.

At higher grafting densities, i.e., in the brush regime, the force can be described by

$$\frac{F_c(h)}{R} = \frac{16kT\,\pi\,h_c}{35D^3}\left[7\left(\frac{2h_c}{h}\right)^{5/4} + 5\left(\frac{h}{2h_c}\right)^{7/4} - 12\right]$$

where

$$h_c = D\left(\frac{R_F}{D}\right)^{5/3}$$

The force between two cylindrical surfaces (F_c) and the repulsive pressure (P), as measured by the osmotic stress technique, can be calculated using then Derjaguin approximation,

$$\frac{F_c(h)}{R} = 2\pi\int P(h)\,dh$$

The agreement with theory was good up to the distance of two polymer layers. At larger separations, the theory, however, predicts a steep decrease that was not observed. Obviously, simply taking into account the polydispersity of the polymer as well as polymer configurations would alleviate this inconsistency. Good agreement with this simple scaling concept was observed also in the regime of the interacting mushrooms (47).

Experimental measurements using the osmotic stress technique, zeta-potential measurements, and surface force apparatus have shown that a polymer extends, in agreement with theoretical calculation, for a typical sterically stabilized liposome composition (5 mol% of $mPEG_{2000}$-lipid), about 5 nm above the bilayer.

Other attempts to prepare sterically stabilized liposomes include the use of single-chain detergents such as Tween 80 (a nonionic detergent with about 20 total ethylene glycol groups in four branches), which is somewhat effective (5). The use of PEG-SA was proven to be ineffective (92). The difference is due to higher dissociation rates of PEG-lipids containing a single acyl chain from the bilayers. However, PEG-lipid dissociation can be exploited for liposome destabilization, which is sometimes needed in drug delivery. Recently, mPEG conjugated to ceramides with acyl chain length of C14:0 and C20:0 (the other chain is C18:1) was shown to be another effective PEG-lipid, with stabilization related to the hydrophobicity of PEG the anchors. Recently, mPEG conjugated to ceramides with various acyl chain lengths was shown to be another effective PEG-lipid, with stabilization related to the hydrophobicity of PEG anchors (90). Longer acyl chains on the PEG-Cer were associated with longer circulation lifetimes of the liposomal carriers and, consequently, higher plasma drug concentration. This study also demonstrated the potential for developing controlled release formulations by manipulating the retention of the PEG-Cer conjugate in liposome bilayers.

Besides PEG, other hydrophilic polymers, particularly biodegradable and natural polymers, have been tested for liposome stabilization. The use of dextrans is only slightly effective, possibly due to the polymer being negatively charged. The conjugation of more hydrophilic carboxylic acid end group derivatives of mPEG-PE reduces liposome stability (14). Successful results have been achieved with polyvinyl alcohol (PVA), polylactic acid (PLA), and polyglycolic acid (PLG) (92). Other reports also indicated polyvinylpyrrolidone and polyacrylamide as effective (51,87). However, none of these new polymers tested could provide the same extent of liposome circulation as that did by PEG-lipids. An exception is polyaxozline, an entirely new class of hydrophilic polymer, which appears to be able to provide comparable circulation extension to mPEG-PE (93,94). This finding of polyaxozline is significant in a way that further confirms that the increase of circulation by hydrophilic polymers is indeed a steric effect rather than a specific chemical effect of PEG.

C. Targetable Liposomes

One of the major goals of liposome medical applications was the targeting of specific cells by specific ligands attached to the liposome surface. This is Ehrlich's idea, almost a century old, of the selective destruction of diseased cells. In the 1970s and 1980s researchers attached antibodies to liposome surfaces and achieved very nice liposome binding and uptake into target cells expressing targeted ligands. None of these experiments, however, was successful in vivo. Nonspecific interactions with the immune system resulted in very quick liposome clearance in the liver and spleen, regardless of the attached antibody or lectin. The development of PEGylated liposomes, however, renewed interest in specific targeting, because the polymer layer reduced nonspecific interactions. It was quickly discovered that binding ligands to the far end of surface attached PEG chains improves binding (4). In addition to antibodies and their fragments, single-chain antibodies, lectins, oligosaccharides, and polypeptides are used as targeting moieties, typically at a surface density between 20 to 100 ligands per 100 nm of liposome.

Liposomes with attached ligands can also be used in diagnostics to enhance the signal-to-noise ratio (51). Normally one ligand carries one or a few markers, while liposomes can encapsulate thousands of fluorescent or radioactive probes. Liposomes loaded with radioactive, radiopaque, sonoreflective, or paramagnetic substances can be delivered in ways similar to other types of drugs and can improve contrast.

D. Activosomes or Polymorphic Liposomes

These types of liposomes are designed for specific reactions or interactions. The goal is either to disintegrate the liposomes under particular conditions or enhance nonspecific, such as electrostatic, interactions with a target molecule.

Temperature sensitive liposomes are designed to release their content at a particular temperature. Liposomes become very leaky in the temperature region where the gel-to-fluid phase transition occurs. By mixing appropriate lipids, such as DPPC and DPHC (diacyl partial hydrogenated PC), one can vary the phase transition. One of the applications would be in the therapy of a particular organ which is heated to 41°C; circulating liposomes release their contents when they pass through the heated region. In addition to drug leakage, these liposomes can also be prepared to disintegrate into a nonbilayer structure and release their contents immediately. Similarly, laser sensitive liposomes (hydrolysis of light sensitive lipids) can be designed. Because the leakage is not instantaneous, however, it might happen that the encapsulated cargo is released too late.

pH sensitive liposomes are designed so that they release the encapsulated cargo upon an increase or decrease of pH. Some organs and organelles might be at lower pH and typically release of the liposome cargo in these conditions is desired. For such application lipids that are weak acids are used in mixtures with membrane destabilizing lipid, such as DOPE. For instance, oleic acid/DOPE mixtures form liposomes at pH $> pK_a$ of fatty acid because ionized fatty acid forms a conelike molecule ($P < 1$, see Section I.A) and neutralized the inverse cone shape of DOPE ($P > 1$) giving rise to $P \geq 1$. Upon protonation at lower pH, however, the area of the polar head of fatty acids is reduced (P increases) and it cannot compensate for the propensity of DOPE to form inverted phase and liposome disintegrates, because $\langle P \rangle$ increases and causes a lamellar–inverse hexagonal phase transition.

The opposite case, in which a weak basic lipid protonates at higher pH and disrupts the liposomes, is not generally used. Another strategy is possible by using PEG-lipids which have $P \ll 1$. Mixing them into bilayer and controlling either their dissociation from the bilayer or chemical degradation of the PEG attachment can again change P from < 1 to > 1 and cause liposome break up.

Another group of active liposomes are fusogenic liposomes, which are designed to fuse with cells. They can either carry fusogenic proteins or peptides, or can be the instability built into the bilayer by special lipid compositions, as discussed above. For instance, stable bilayers containing PEG-lipid and DOPE become fusogenic upon the loss of PEG or PEG-lipid from the bilayer.

E. Cationic Liposomes

Cationic liposomes are primarily used for preparing liposome/DNA complexes for gene delivery into cells (52). These liposomes are prepared in similar procedures (such as sonication and extrusion) as conventional or sterically stabilized liposomes. Normally they contain 50 mol% of neutral lipid, such as DOPE or cholesterol. The former proved to increase DNA delivery in vivo, while the latter was shown to be more efficient in in vivo DNA expression (52). In later sections we will discuss in detail the applications of cationic liposomes in gene delivery.

There are many other functionalized liposomes. We briefly mention virosomes, which are reconstituted vesicles containing viral lipids and proteins and are used in vaccination (17,69).

VI. LIPOSOMES AS DRUG DELIVERY VEHICLES

With respect to structure and composition, liposomes resemble cell membranes. They are typically made from natural, biodegradable, nontoxic and nonimmunogenic lipid molecules, and they can encapsulate or bind a variety of drug molecules into or onto their membranes. All these properties make them attractive candidates for applications as drug delivery vehicles.

Liposome applications in drug delivery depend, and are based on, physicochemical and colloidal characteristics such as composition, size, loading efficiency, and the stability of the carrier, as well as their biological interactions with the cells. There are four major interactions between liposomes and cells. Lipid exchange is a long-range interaction that involves the exchange of liposomal lipids for the lipids of various cell membranes; it depends on the mechanical stability of the bilayer and can be reduced by ''alloying'' the membrane with cholesterol (mixing it in the lipid bilayer, which gives rise to greatly improved mechanical properties, such as increased stretching elastic modulus, resulting in stronger membranes and reduced permeability). The second major interaction is adsorption onto cells, which occurs when the attractive forces (electrostatic, electrodynamic, van der Waals, hydrophobic insertion, hydrogen bonding, specific ''lock-and-key'', etc.) exceed the repulsive forces (electrostatic, steric, hydration, undulation, protrusion, etc.) and can be nonspecific or specific. Adsorption onto phagocytic cells is normally followed by endocytosis or, rarely, by fusion. Endocytosis delivers the liposome and its contents into the cytoplasm indirectly via a lysosomal vacuole in which low pH and enzymes may inactivate the solute. During fusion, however, the liposome contents are delivered directly into the cell, and the liposomal lipids merge into the plasma membrane. Therefore a substantial effort to utilize this mode of drug entry is being undertaken. Efforts range from the incorporation of fusogenic proteins into the bilayer to the preparation of metastable bilayers and pH sensitive polymer coatings (1).

For drug delivery, liposomes can be formulated in a suspension, as an aerosol, or in a (semi)solid form such as a gel, cream, or dry powder; in vivo, they can be administered topically or parenterally. After systemic (usually intravenous) administration, which seems to be the most promising route for this carrier system, liposomes are typically recognized as foreign particles and consequently endocytosed by cells of the mononuclear phagocytic system (MPS), mostly fixed Kupffer cells in the liver and spleen. This fate is very useful for delivering drugs to these cells but, in general, excludes other applications, including site-specific drug delivery by using ligands expressed on the liposome surface in order to bind to the receptors (over)expressed on the diseased cells. For this reason, a search for liposomes that could evade rapid uptake by the MPS started, and a few lipid compositions that prolonged liposome blood circulation times were discovered, culminating in the development of PEG-coated, sterically stabilized liposomes.

Based on the liposome properties introduced above, several modes of drug delivery can be envisaged; the major ones are enhanced drug solubilization (e.g., amphotericin B, minoxidil), protection of sensitive drug molecules (e.g., cytosine arabinose, DNA, RNA, antisense oligonucleotides, ribozymes), enhanced intracellular uptake (all agents, including antineoplastic agents, antibiotics, antivirals), and altered pharmacokinetics and biodistribution of the encapsulated drug. The latter accounts for the decreased toxicity of liposomal formulations, because liposome-associated drug molecules cannot normally spill to organs such as the heart, brain, and kidneys as well as to increased targeting of the encapsulated drug to certain cells and tissues. As discussed above, normal, or conventional, liposomes are taken up by macrophages and can therefore serve as excellent drug delivery vehicles to these cells. However, sterically stabilized liposomes, which are not avidly taken up by MPS cells, have different biodistribution properties and have shown enhanced accumulation in sites of trauma, such as tumors, infections, and inflammations, which are characterized by leaky capillaries. This accumulation is due simply to their prolonged liposome circulation and small size (<100 nm), which enable them to extravasate (51). Very small neutral and mechanically stabilized liposomes (by preparing a bilayer with high mechanic strength as can be measured via stretching elasticity of the membrane) also exhibit prolonged circulation times and may also accumulate in sites of trauma. In a first approximation, we can simply view this as a statistical problem: the longer a liposome can circulate (by avoiding uptake in the liver), the greater the possibility that it extravasates (leaves the vascular system) at the sites where the blood vessels are porous (51).

VII. MEDICAL APPLICATIONS OF LIPOSOMES

In many cases, effective chemotherapy is severely limited by the toxic side effects of the drugs. Liposome encapsulation can alter the spatial and temporal distribution of the encapsulated drug molecules in the body, which may significantly reduce unwanted toxic side effects and increase the efficacy of the treatment. Although many drugs have been studied in preclinical settings in numerous animal disease models, at present, in human therapy, liposomal therapeutics are used only in systemic fungal infections and cancer therapy. However, in preventative medicine, liposome-based vaccines show great promise. Table 3 lists liposomal products on the market and in advanced phases of clinical development. Before discussing their application in anticancer therapy, we shall present their use in parasitic infections.

A. Liposomes in Infectious Diseases

In infectious disease, liposomes were proven to be efficacious in the treatment of parasitic and fungal infections (2). Bacterial infectious antibiotics are in general effective and safe, while for viral disease there are not many potent drugs available. The main modes of liposome action are reduced toxicity and targeting of the drug to macrophages. Recently, small stable and long-circulating liposomes have been shown to be effective in targeting parasitic infections in the lung and in systemic treatment with the antibiotic amikacin. Tropical parasitic diseases like leishmaniasis and malaria are ideally suited for treatment with liposomal drugs. Indeed, early results have shown up to 700-fold improved therapeutic index in the case of leishmaniasis. Unfortunately, these formulations have never been fully developed, possibly due to a low marketing potential.

Targeting the drug to macrophages, where the infectious agents often reside, and the reduced toxicity of the formulation that results from the limited spillage of drug to other tissues, increases the therapeutic efficiency of the treatment. The drug of choice in the treatment of systemic fungal infections is amphotericin B, which is, owing to its aqueous insolubility, typically formulated into detergent micelles. However, micelles are unstable upon systemic administration, and severe neuro- and nephrotoxicity limit the dose that can be administered. If the drug is formulated in a stable colloidal particle, it is delivered much more efficiently to macrophages, and toxicity can be significantly reduced. Following this rationale, three lipid-based amphotericin B formulations are commercially available (Table 3): AmBisome® (Gilead, Foster City, CA) contains the drug formulated into small, negatively charged liposomes; Amphotec® (InterMune Inc., Burlingame, CA) is a stable mixed micelle of drug complexed with cholesterol sulfate; and Abelcet® (Elan Co., Dublin, Ireland) is a homogenized liquid crystalline suspension of drug and lipids.

Although these products are getting increased shares of the market, a number of improvements can be envisaged, including more potent antifungals (in a free form that is

Table 29.3 Liposomal Products on the Market or in Advanced Clinical Studies

Company	Product	Status
ALZA Corp.	DOXIL®—dox in STEALTH® liposomes for Kaposi's sarcoma for ovarian cancer	On the market (1995, USA and 1996, Europe) On the market (1999, USA)
	AMPHOCIL®—ampB in mixed micelles	On the market (1996, USA and 1993, Europe) (sold to InterMune 2/01)
Gilead Sciences	Ambisome®—ampB in liposomes	On the market (1997, USA and 1990, Europe)
	DaunoXome®—dauno in lip's.	On the market (1996, USA and Europe)
	MiKasome®—liposomal amikacin	Phase II
	NX 211—liposomal lurtotecan	Phase I (Canada)
Elan Co.	Myocet®—liposomal dox	Phase III not successful
	VENTUS®—liposomal PGE1	Phase III not successful
	ABELCET®—ampB in lipid dispersion	On the market (1995, USA and 1993, Europe)
	EVACET®—liposomal doxorubicin for breast cancer	Phase III
Asta Medic.	Topical anticancer cream	On German market
Aronex Pharm. Inc.	Nyotran®—liposomal nystatin	Phase III (USA and Europe)
	Liposomal annamycin for refractory breast cancer for leukemia	Phase II Phase I/II
	ATRAGEN®—liposomal retinoic acid (tretinoin) for	
	Acute promyelocytic leukemia (APL), non-Hodgkin's lymphoma and prostate cancer; and renal cell carcinoma and bladder cancer	Phase II Phase I/II
	Aroplatin®—liposomal platinum for lung cancer	Phase II
Inex Pharm. Corp.	ONCO TCS—liposomal vincristine for non-Hodgkin's lymphoma	Phase II/III
	INX-3280—liposomal oligonucleotide	Phase II/III
Swiss Serum Inst.	Epaxal®—hepatitis A vaccine	On Swiss market since 1994
	Trivalent influenza vaccine	Phase III
	Hepatitis A and B vaccine	Phase I
	Diphtheria, tetanus, hep A vaccine	Phase I
	Diph., tet., infl., hep. A vaccine	Phase I
Skyepharma PLC	DepoMorphine®, morphine sulfate for acute post-surgical pain	Phase II
	DepoCyt® for neoplastic meningitis	Phase III
	DepoAmikacin® for bacterial infections	Phase I (completed)
Biomira Inc.	Liposomal BLP25, a MUC1 based vaccine, for breast cancer	Phase I
Neopharm Inc.	Liposomal paclitaxel	Phase II
Vical Inc.	Leuvectin®—DNA/lipid complex of IL-2 for kidney and prostate cancer	Phase II
	Leuvectin-7®—DNA/lipid complex of HLA-B7 gene	Phase II/III
Valentis Inc.	Cationic lipid formulation of IL-2, IL-12, and IFN-α	Phase IIa (USA and Germany)
	Cationic lipid formulation of $VEGF_{165}$ for angiogenesis	Phase II

Abbreviations: dox-doxorubicin HCl, dauno—daunorubicin, ampB—amphotericin B.

too toxic for systemic treatment) and sterically stabilized liposomal antifungals and antibiotics for the targeting of targets away from the MPS system. Additionally, such liposomes can act as long-circulating platforms to bind and neutralize pathogens or reduce inflammation by blocking appropriate receptors overexpressed at the site of trauma. Liposome expression of oligosaccharides or antibodies increases their circulation times and, additionally, the multivalency of binding. For instance, sialyl Lewis x liposomes (liposomes with attached oligosaccharides to target inflammation receptors) were shown to be 750 times more effective in inhibition of selection-E mediated cellular adhesion than the free ligand, while antibodies expressed on stealth liposomes targeted to the same receptors on cultured endothelial cells have shown 275-fold higher selectivity and can be used to deliver drugs to these cells effectively with increased specificity (6).

In general, saturation of various receptors may be an important new modality in liposome applications, and such systems may become useful in treating inflammation, autoimmune diseases, and cardiovascular diseases. Owing to their systemic accessibility, these applications of targeted liposomes may be more important than in cancer therapy, for here most of the tumors are rarely directly accessible.

B. Liposomes in Anticancer Therapy

Most of the medical applications of liposomes that have reached the preclinical stage are in cancer treatment. Very early studies showed reduced toxicity of liposome-encapsulated drug, but in most of the cases the drug molecules were not bioavailable, resulting in reduced toxicity but also in severely compromised efficacy. Unfortunately, this was also found to be true for primary and secondary liver tumors. Although most of the drug ended in the liver, it did not diffuse into malignant cells (2,32).

Several clinical studies did not support the use of conventional liposomes in cancer treatment. Although it is still not clear whether such liposomes can be beneficial in the therapy of some cancers, it has been demonstrated that small, stable liposomes can passively target several different tumors because they can (owing to their biological stability) circulate for prolonged times and (owing to their small size [<50–150 nm]) extravasate in tissues with enhanced vascular permeability, which is often present in tumors.

Two liposomal formulations have been approved by the Food and Drug Administration (FDA, Washington DC, USA) and are commercially available in the USA, Europe, and Japan (Table 3). DOXIL® (ALZA) is a formulation of doxorubicin precipitated in sterically stabilized liposomes (see Fig. 9A), while DaunoXome® (Gilead) is daunorubicin encapsulated in small liposomes with very strong and cohesive bilayers, which can be referred to as mechanical stabilization (2).

DaunoXome is a small liposome (distearoyl phosphatidylcholine/cholesterol—DSPC/Chol, 2/1, ≈27 mM, size SUV) with daunorubicin (1 mg/mL) loaded by a pH gradient. These liposomes are relatively stable in the circulation because they are small, and their membrane is electrically neutral and mechanically very strong. This reduces the charge-induced and hydrophobic binding of plasma components but does not protect against van der Waals adsorption. Also, uncharged liposome formulations are colloidally less stable than charged ones.

DOXIL is a liquid suspension of 80–100 nm liposomes ($mPEG_{2000}$-DSPE/HSPC/Chol, 20 mM) with doxorubicin HCl at 2 mg/mL. The drug is encapsulated into preformed liposomes by an ammonium sulfate gradient technique and is, additionally, precipitated with encapsulated sulfate anions (see Fig. 9A). These liposomes circulate in patients for several days, which increases their chance to extravasate at those sites with a leaky vascular system. Their stability is due to their surface PEG coating, as well as to their mechanically very stable bilayers. DOXIL was the first liposomal drug approved by the FDA and has been on the market since 1995, while DaunoXome was approved approximately half a year later. Both formulations are used in the treatment of Kaposi's sarcoma. In 1999, Doxil was approved for the treatment of ovarian cancer refractory to paclitaxel- and platinum-based chemotherapy regimens.

DaunoXome was shown to be equally effective to conventional therapies with reduced drug toxicity and improved patient quality-of-life. The use of DOXIL in Kaposi's sarcoma has shown a high response rate in comparison with standard treatments: 58.7% response rate as compared to 23.3% for bleomycin–vincristine therapy and 46.2% vs. 23.3% for adriamycin–bleomycin–vincristine. Owing to a different biodistribution of DOXIL, the toxicity profile is quite different from that of the free drug or of conventional liposomes. While the dose-limiting toxicity is hand–foot syndrome and stomatitis (liposomes that are not taken by the MPS or do not extravasate eventually end up in the skin, especially in areas where the vascular system is constantly under pressure and slightly damaged), but nausea, vomiting, and alopecia, which are usual after conventional therapy, are notably mild after liposomal (DOXIL) treatment. DOXIL has also been shown to have a 4.5-fold lower medium pathology score for doxorubicin-induced cardiotoxicity than free drug, while neutropenia is similar to the free drug. Despite these convincing results, however, it seems that, in the long run, the potential of this formulation is greater for the treatment of solid tumors.

For instance, in salvage therapy of ovarian carcinoma in patients refractory to platinum and paclitaxel, a 13.8% overall response was recorded with DOXIL®.

The difference between the various formulations can be explained by the pharmacokinetic data and biodistribution reports, which have shown significant drug accumulation in tumors in the case of small and stable liposomes; free drug distributes to a large volume (causing systemic toxicity) and is quickly washed away, and because no specific tumor targeting occurs, the activity is relatively low. Conventional liposomes are distributed to a smaller volume (which depends on the drug leakage from liposomes, which is a function of lipid composition and the method of drug encapsulation, and in this case decreases in the following formulations: PC/PG > PC/Chol > PC/CL; CL is cardiolipin). Correspondingly, with improved liposome encapsulation, the volume of distribution decreases and the clearance rates are smaller. Because such liposomes typically do not target tumors (short blood circulation times and larger size) and end up in the MPS, the drug activity in most tumor models is compromised and generally, the reduction in drug toxicity does not justify the use of such formulations. When the drug is encapsulated in small and stable vesicles, the volume of distribution is very small (mostly blood, some in the liver and in tissues with leaky vasculature), reducing toxicity, and if accompanied by liposome accumulation in tumors, resulting in increased activity. Clearance rates are slow because the drug is stably encapsulated. The elimination times of the drug are not a good indicator of their activity, because typically they represent a second phase, in which smaller amounts of the drug are involved. The elimination times are calculated from the clearance rate and the volume of distribution and do not correlate to liposome stability. It is actually the blood circulation time that, in the case of small, stable liposomes, determines the accumulation in tumors. From this simple pharmacokinetic analysis, we can conclude that long clearance times, which are associated with low volumes of distribution and small changes in biodistribution toward tumors, characterize good liposomal anticancer formulations. Of course, for drug to remain active, it must not be encapsulated too stably. When liposomes accumulate and become trapped in the tissue, they start to disintegrate, and the drugs leak into the surroundings, as can be inferred from observed drug degradation products, fluorescent microscopy, as well as anticancer activity. The origin of liposome disintegration is one of the least understood processes, but it is very likely that various phospholipases degrade the lipids, and the softer bilayers start to disintegrate. PEG-lipids are lost slowly owing to their greater hydrophilicity, which allows the close approach of enzymes to liposomes and internalization by the cells. In parallel, gradients can dissipate, and the drugs can leak from topologically intact liposomes. Also, mechanical damage and lipid phase transitions in regions with reduced liquid concentration can occur. Although the real situation is far more complex, this simplified explanation of pharmacokinetic and biodistribution data may explain the differences between different formulations. Briefly, while the selectivity of free drug means only that it is more damaging to rapidly dividing cells (such as tumor, but also hair, blood, and nonsomatic cells), some drug delivery vehicles can deliver drug closer to the tumor and, in the process, spare some other tissues. Liposomes containing targeting ligands, however, can be internalized into the cell by the endocytic receptors.

Because of the complexity of systemic drug delivery, the potency of the immune system, and the anatomy of tumors, it is difficult to anticipate that formulations will be prepared that will be much more effective from the drug delivery perspective. The next developments may instead rely on combination therapy, which has been found to be effective in the treatment of AIDS. Consequently, liposomal vincristine (phase II) and annamycin (phase II) are being developed by Inex (Burnaby, BC, Canada) and Aronex (The Woodlands, TX, USA), respectively. A second liposomal agent in the combination therapy may be one of various cytokines, such as interferons, interleukins, or tumor necrosis factor (TNF), which are very active, but also very toxic, anticancer agents. Preliminary data on combination therapy of TNF in sterically stabilized liposomes and DOXIL® have shown arrested tumor growth, whereas either agent alone only decreased the growth rate.

Since the early days of liposome applications, scientists have been trying to develop liposomes that would target specific cells via surface attached ligands. Promising in vitro experiments however were not reproduced in vivo because of nonspecific liposome clearance by the MPS. PEGylated liposomes have revived the concept. The immunogenicity of current formulations can be reduced by using fragments of humanized antibodies, so the remaining problems are the accessibility of tumors and cellular uptake of the liposomes. Solid tumors are rather inaccessible and the therapeutic benefit can be expected mostly within the vascular endothelium or in other body fluids, especially in the case of targeting internalizing epitopes. If the therapy proves to be effective, we must also be aware that large-scale manufacturing of targeted liposomes is technically considerably more demanding, and the economics of this will be more critical than on nontargeted liposomal formulations.

Because of their accumulation in tumors and sites of infection and inflammation, sterically stabilized liposomes can be used to deliver other agents to these sites: antisense oligonucleotides and ribozymes are very potent agents, but the major problem is their delivery to and into the appropriate cells. It is very likely that encapsulation of these agents in PEGylated liposomes may not only dramatically improve the therapeutic index but also be the optimal way to deliver these molecules.

C. Liposomes in Gene Therapy

Recombinant DNA technology and studies of gene function and gene therapy all depend on the delivery of nucleic acids into cells in vitro and in vivo. Although in vitro techniques can rely on a number of physical and chemical methods, in vivo delivery is more demanding. DNA carrier systems include several colloidal particles; cationic liposomes have been shown to complex (negatively charged) DNA, and such complexes were able to transfect cells in vitro, resulting in the expression of the protein, encoded in the DNA plasmid, in the target cells. Obviously, for gene therapy (treatment of diseases on the molecular level by switching genes on or off), in vivo delivery is preferred. It was discovered that cationic lipid-based DNA complexes can transfect certain cells in vivo upon localized (mostly lung epithelial cells upon intratracheal instillation) and systemic (endothelial cells in the lung) administration (52). However, the first series of in vivo studies yielded rather low levels of gene expression. Improvements were sought both on the molecular level, where many novel lipids were synthesized and many different combinations of lipids tried, and on the colloidal level, where the DNA–lipid-complex structure–activity relationships were studied. Despite the fact that quantitative structure–activity relationship (QSAR) is still not understood, developments in DNA plasmid design, synthesis of novel cationic lipids, cholesterol stabilization of complexes, and improvements on the colloidal structures have resulted in more than a 1000-fold increased gene expression as compared to the first experiments. Although these developments are impressive, the absolute expression levels are still rather low and the duration of expression short; before initiating realistic clinical trials, more research on the delivery vehicles and the DNA vector will be required. For example, the latter may include DNA sequences that would give rise to nuclear localization. Lipid-based gene delivery is the focus of several specialized high-tech companies, among which Vical (San Diego, CA, USA), Genzyme (Framingham, MA, USA) and Valentis (Burlingame, CA, USA) have products in clinical trials.

Several types of lipid vectors for DNA delivery have been developed and, according to the way they are prepared and the structures of the final products, they fall into three major categories: (1) DNA complexed with preformed cationic liposomes, (2) DNA packed in a lipid-based carrier in particle form, and (3) DNA encapsulated in liposomes.

As the first-generation lipid-based DNA vector, DNA-cationic liposome complexes, have been playing a leading role in the area of gene vector development. A great deal has been learned about many respects of gene delivery, including the construction of the delivery vectors, vector–cell interactions, gene construction and expression, pharmacokinetics, and the in vivo aspects of gene delivery. The search for better-defined and commercially viable gene vectors has lead to the development of several novel and interesting lipid-based vectors that have show advancement in one or more aspects over the existing ones. However, being at their early development stage, these systems are yet to be more extensively investigated.

1. The Formation of Genosomes (Lipoplex)

The use of liposome–DNA complexes (genosomes or lipoplexes) for transfer of plasmids into cells was first described by Felgner et al. in 1987 (30,52), who observed gene expression in vitro. Genosomes are formed by mixing plasmid DNA with preformed liposomes containing around 50% of cationic lipid, the rest being the so-called helper lipids, which normally are neutral lipids such as DOPE and cholesterol. Genosome formation is a very complex process, which depends on thermodynamic as well as kinetic factors. Thermodynamic factors include charge ratio, concentration, ionic strength, pH, the presence of impurities, and temperature. Other important parameters are the type of cationic and to a lesser extent the neutral helper lipid, the degree of supercoiling, and the size of the DNA.

The procedure for genosome formation involves simple mixing of DNA and liposome solutions. Over the years, researchers have adopted various protocols. At low lipid concentrations, genosomes are often made by pouring lipid and DNA solutions together and then incubating for 5 to 15 min and applying onto the cell culture. Some papers also reported separate addition of plasmid DNA and liposomes to the cell culture. For in vivo applications, however, larger concentrations of colloidally suspended DNA are preferred, and different protocols are used. Complexes are normally prepared by rapid pipette injection of one solution into the other and quick aspiration–ejection cycles. In many cases the speed and the sequence of addition (DNA into liposomes or vice versa) critically influences the characteristics of the complexes formed. Normally a small-scale sample is used (50–200 μL), as scaling up to milliliter volumes often results in precipitation. There is also large variation between different operators. All these observations indicate that the interaction of DNA with liposomes is clearly a kinetic phenomenon. To increase consistency, automation is strongly recommended.

It is obvious that the charge ratio is the most important and influential thermodynamic parameter. Typically, precipitation occurs near charge neutralization ($\rho = 1$), and optimal charge ratios for transfection have been reported to be in the 1.2–1.5 range (cationic complexes) in most cases. In preparing anionic complexes, liposomes are injected into DNA and vice versa for the cationic complexes. Moreover, the process of complex formation is normally carried out in nonelectrolyte solutions such as water or 5 to 10% sucrose, because the presence of even a low concentration of electrolytes dramatically increases precipitation. The stability of the formed complexes depends on many factors, including lipid concentration, pH, ionic strength, temperature, and in particular charge ratio and can vary from minutes to up to several months (52).

2. Lipid–DNA Particles (LDPs)

It has been envisioned that the ideal carrier for gene delivery would be one that contains densely packed DNA, preferably the particle form, with a protective coating material, such as neutral and PEG lipids, on the outer surface. Such a carrier system would possess the basic characteristics of the long-circulating pegylated liposomes and other functional liposome systems. In recent years, several types of lipid–DNA particles have been developed.

The above-described preparation methods give rise to several different genosome topologies. Researchers have typically described samples that are heterogeneous with respect to particle size, shape, and density. EM and small-angle x-ray scattering studies have established mostly two types of complexes—random structures of aggregated spherical particles containing elongated fibrils (the so called ''spaghetti-meatball'' structures) and ordered intercalated lamellar phases, in which two-dimensionally condensed DNA is sandwiched by lipid bilayers (Fig. 9C) (78). Obviously, in the limit of lipid excess, the latter structures are prevalent, and the former ones at higher DNA–lipid ratios. However, we believe that the structure of these particles is kinetically controlled and a real sample may contain a large number of different structures ranging from the two extremes—lipid-coated DNA helices and intercalated lamellar phases.

A novel type of liposome–DNA complex has been reported by Templeton et al. (86). The invaginated liposomes used in the procedure are bilamellar (two concentric bilayers) and are produced by a stepped extrusion process using the Whatman Anotop filter, which has long cylindrical pores. The liposome–DNA complexes have a very different structure compared with the complexes prepared using the single-bilayered liposomes. In this case DNA molecules are entrapped inside the invaginated liposomes, and there is an intact outer bilayer around the complex (see Fig. 9B.). This type of genosome appears to be very stable upon storage and very efficient in transfecting cells.

3. Formation of Lipid–DNA Particles via Hydrophobic Intermediates

One of the significant advancements in gene delivery is the lipid–DNA particle (LDP) system (9,99). The procedure for LDP preparation is the direct outcome of the Bligh & Dyer extraction process, where polyanionic plasmid is extracted from the aqueous phase into the solvent phase by cationic lipids (79). The driving force for DNA into organic solvent ($CHCl_3$) is the electrostatic interaction of the cationic lipid with negatively charged DNA. The resultant stable DNA–lipid complex in organic solvent is thus hydrophobic. Such DNA–lipid complexes can serve as self-assembling intermediates for the formation of a well-defined LDP carrier.

It was first attempted to hydrate directly such lipid–DNA complexes after the removal of the solvent. This was not successful because the complexes stuck to the glass, even in the presence of excessive amounts of neutral and $mPEG_{2000}$-DSPE. Then an alternative route was taken, where plasmids were introduced to the cationic lipid presented in mixed micelles formed with nonionic detergent. This system of mixed micelles is in fact a microscopic version of the aqueous-solvent biphase system where the DNA–lipid interaction occurs at the extended aqueous-solvent interface (Fig. 11). In the micellar system, the hydrophilic–hydrophobic interfaces are provided by the surface of the micelles. Studies indicated that the instant and spontaneous interaction of DNA with cationic micelles upon the addition of DNA to the cationic micelles lead to the formation of well-defined small particles in the range of 50–70 nm in diameter. Detergent concentration appears to be a crucial parameter that determines the physical properties of the LDP formed. Under the optimum detergent concentration, which is near the critical micelle concentration (CMC = 20 mM for octyl glucoside), the LDP formed are small, uniform, and reproducible. The detergent in the system can be easily dialyzed away without affecting the size of the LDPs. Negative stain and freeze-fracture electron microscopic studies revealed that these particles are rather uniform in size, but not necessarily spherical in shape (99).

In contrast, at high detergent concentration (>50 mM), no indication of LDP was observed before dialysis, and large aggregates were seen 2–3 hours into dialysis. The

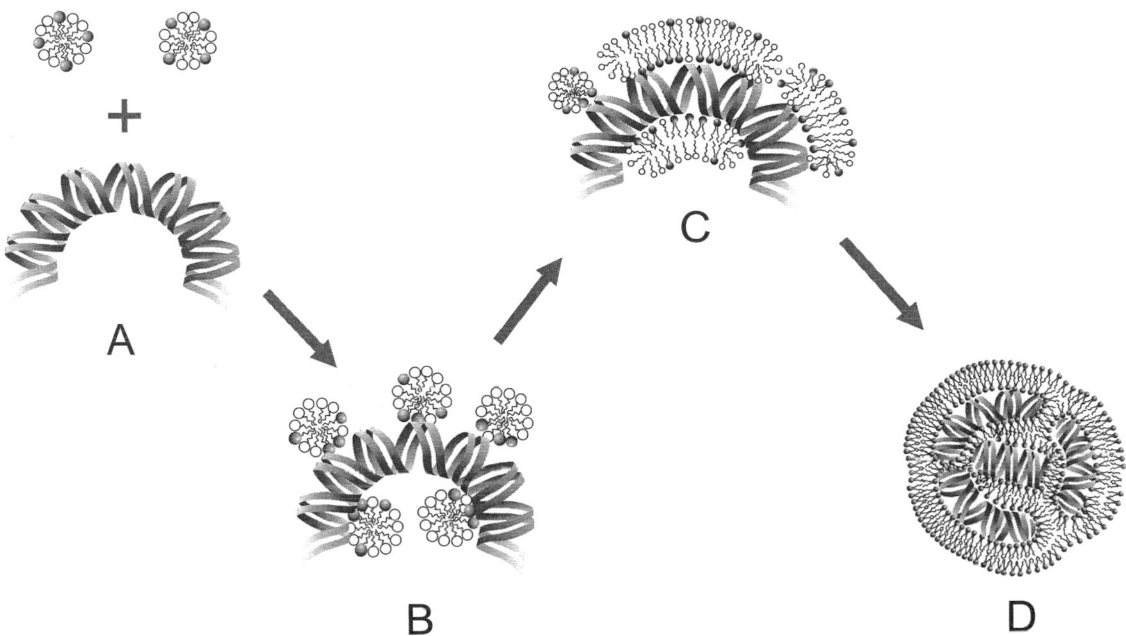

Figure 29.11 Model describing the intermediates that may be involved in the formation of cationic lipid/DNA particles (99). Small lipid/DNA particles (<100 nm) form spontaneously, driven by electrostatic interactions, when the detergent concentration is near the critical micelle concentration of OGP (approx. 20 mM). The formation of LDPs might involve the following steps. Mixed detergent micelles containing the cationic lipid and the helper lipid are mixed with DNA (step A). The binding of the mixed micelles with DNA is instant (step B), and such binding will likely cause asymmetric distribution of lipids on the micelles and a decrease in micelle stability. The next step (step C) is the fusion of micelles along the DNA strand into larger bilayered intermediates that serve as DNA condensation agents and lead to the formation of LDP (steps C to D). Detergent molecules are then removed by dialysis.

detergent concentration effect on the particle formation could be a kinetic effect resulting from the fast detergent removing process, and the result could be different if the rate of detergent removal is slower and controlled.

Another observation is that the LDPs formed by the above procedure are only stable in a nonelectrolyte solution in the absence of a stabilizing agent. Further investigation showed that stable LDPs can be produced using the same procedure in PBS buffer or saline, when a sufficient amount of mPEG$_{2000}$-Cer (5–10%) is incorporated in the system. Regarding the DNA loading capacity of this LDP system, the DNA-to-lipid ratio can be maximized by lowering the amount of cationic lipid to a level that is slightly greater than charge neutralization (1.2:1–1.5:1, +/−). Neutral lipids (SM or PC) can be incorporated with the aim of reducing the charge density on the outer surface.

It appears that efficient formation of lipid–DNA particles by the above procedure requires a significant proportion of the cationic lipid in the system, which clearly poses a major hurdle for such a carrier system to be used for systemic delivery. However, it might be possible to overcome this problem by the use of pH sensitive cationic lipids. In such a case LDPs can be formed at a low pH where the lipid is charged, and then the surface charge can be reduced by increasing the pH to a level where the lipid becomes neutral.

With regard to the mechanism of the formation of LDPs, a working model has been proposed as illustrated in Fig. 11 (99). The formation of LDP is spontaneous and nearly instantaneous. However, sequential steps might be involved during the process. The initial binding of the small cationic mixed micelles to DNA molecules is driven by electrostatic interactions. Such binding will likely cause asymmetric distribution of lipids, particularly the cationic lipid, on the micelles, being cationic-lipid enriched for the side binding to the DNA. This process will probably lead to the growth of larger bilayered intermediates due to the fusion of micelles along the DNA. These bilayered intermediates, at the same time, serve as DNA condensation agents and lead to the formation of LDP.

In vitro transfection studies in the CHO cell line indicated that LDPs (with β-galactosidase plasmid) prepared with SM/DODAC (1:1 to 8:1, +/−, SM/DODAC, 50:50 mol%) are effective in delivering DNA into cells. The transfection level increases with increasing charge ratio (maximum at 4:1 +/−), as is expected, as for other lipid-based DNA carriers. In vivo transfection of tissues and solid tumors in animals are currently under investigation (personal communication).

4. Stabilized Plasmid–Lipid Particles

Stabilized plasmid–lipid particles (SPLP) constitute a novel system in which DNA is truly encapsulated inside liposomes (81,91,95). The process for SPLP preparation appears to be similar to the conventional detergent dialysis method, but it differs significantly in the mechanisms of DNA entrapment. Consequently, the plasmid encapsulation efficiency achieved in this method is much higher than that normally achievable with conventional dialysis procedure, in which solute is passively entrapped. The plasmid trapping efficiency appears to be a sensitive function of the cationic lipid content (i.e., DODAC mol%) in the lipid composition, and 50–70% encapsulation can be achieved with only minimal amount of DODAC (6–8 mol%), which is insignificant enough so that the pharmacokinetic profiles of these liposomes are identical to conventional liposomes bearing no DODAC. The presence of a sufficient amount of mPEG-Cer, acting as a stabilizer, appears to be critical, and in its absence aggregation occurs during dialysis. Although mPEG-Cer is advantageous over mPEG-PE, because it is noncharged and will not interfere with the interaction between DNA and cationic lipid, mPEG-DSPE can be used as the stabilizer lipids.

An attempt to increase the content of cationic lipid in the system was made first by increasing the ionic strength of the dialysis medium. However, raising the NaCl concentration to 1 M was not effective in samples containing 10 mol% of DODAC. The use of polyvalent anionic counterions (i.e., citric acid and phosphoric acid), with an expected stronger shielding effect against the cationic lipid, can increase the amount of cationic lipid to as high as 45 mol% without significantly losing the encapsulation efficiency. It has been shown that citrate is more effective in the low range of DODOC content (7–30 mol%) (95) and phosphate is more effective in the high end of DODAC content (20–45 mol%) (81). There is basically a linear relationship between the percentage of cationic lipid in the lipid composition and the optimum citrate concentration. Above the optimum citrate concentration, empty liposomes are produced, and DNA encapsulation efficiency is low. In contrast, when the citrate concentration used is too low, the strong interaction of the cationic lipid with plasmid leads to aggregation.

SPLPs can be purified by removing the empty vesicles using procedures such as sucrose density gradient isolation, and a high DNA-to-lipid ratio (>60 μg/μmol) can be obtained. The formulation can be concentrated to at least 1 mg DNA/mL with no compromise in stability. The SPLPs exhibit shelf stability of at least 6 months upon storage at 4°C. Moreover, plasmids encapsulated inside the SPLPs are protected from degradation enzymes as tested in serum and DNase (91).

The purified SPLPs have a consistent size distribution of 80–100 nm between preparation as observed by QELS and confirmed by both freeze-fracture (see Fig. 9D) (95). In contrast, the empty vesicles are much smaller (<20–40 nm in diameter). Cryo-EM investigations (81) of the purified SPLPs revealed a dense internal structure that is present neither in extruded liposomes nor in the empty liposomes with the same lipid composition prepared by the same dialysis procedure, suggesting that the plasmids are entirely entrapped inside the liposomes.

With regard to the mechanism whereby SPLPs are formed as facilitated by the presence of polyvalent anionic counterions, a working model has been proposed as illustrated in Fig. 12 (91,95). This model is based on the proposed processes of micelle-to-vesicle transformation occurring during dialysis (see Section II.G.). Prior to detergent removal, plasmids probably do not directly associate with the mixed micelles, due to the extensive dilution of the positive charge density on the surface of the mixed micelles.

The next step of micelle–vesicle transition is believed to be the formation of various lipid-detergent intermediates, including cylindrical micelles, lamellar sheets, or open leaky vesicles. It is proposed that it is at one of these stages that DNA–micelle association starts to occur, as the size and the charge density of these intermediates grow to a significant level. The fact that only a minimal amount of cationic lipid is required in the formation of SPLPs indicates that the association of DNA with the intermediates is not a charge neutralization effect but rather a loose association facilitated by a minimal number of binding sites along the DNA molecule and the intermediates. At optimum cationic lipid concentration (i.e., 6–8% DODAC in the absence of counterions), DNA association possesses little interference with the natural growing process of the intermediates into open vesicles, which eventually close up as the detergent is depleted, resulting in the encapsulation of DNA. In the case of increased cationic lipid content, anionic counterions are needed to shield the surface charge density effectively on the intermediates to a level equivalent to the minimum cationic lipid situation (i.e., 6–8% of DODAC). In either case, low content of cationic lipid or higher concentration of the counterions, the association of DNA with these intermediates is not sufficient for efficient encapsulation. On the other hand, increased cationic lipid content or a too-low concentration of counterions will lead to the association of DNA with multiple lipid–detergent intermediates and/or the collapse of open vesicles, eventually resulting in lipid–DNA aggregation.

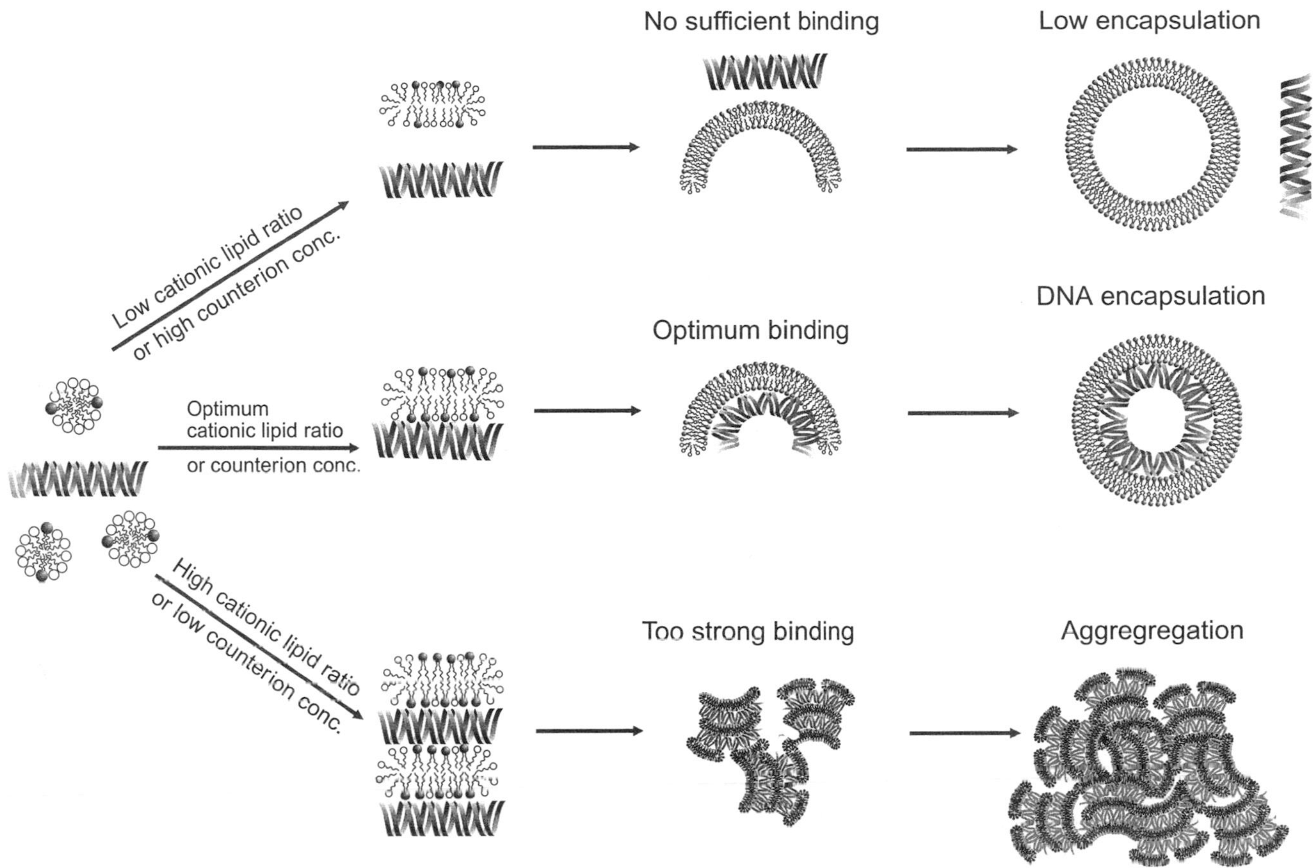

Figure 29.12 Mechanism of SPLP formation. Prior to detergent removal, DNA does not directly associate with the mixed micelles, due to the extensive dilution of the positive charge density on the surface of the mixed micelles. Plasmid DNA–micelle association starts to occur, as the size and the charge density of the bilayered intermediates grow to a significant level. At optimum cationic lipid concentration (i.e., 6–8% DODAC in the absence of counterions), DNA association possesses little interference with the natural growing process of the intermediates into open vesicles and eventually closes up as the detergent is being depleted, resulting in the encapsulation of DNA. In the case of increased cationic lipid content, anionic counterions are needed effectively to shield the surface charge density on the intermediates to a level equivalent to the minimum cationic lipid situation (i.e., 6–8% of DODAC). In either case, low content of cationic lipid or higher concentration of the counterions, the association of DNA with these intermediates will not be sufficient for efficient encapsulation. On the other hand, increased cationic lipid content or too low a concentration of counterions will lead to the association of DNA with multiple lipid–detergent intermediates and/or the collapse of open vesicles, eventually resulting in lipid–DNA aggregation.

4. Other Applications: Vaccination and Diagnostics

As a result of their special properties, liposomes are being studied for treating other diseases and for vaccinations (51). Examples are liposomal antibiotics, antivirals, prostaglandins, steroid and nonsteroid anti-inflammatory drugs and insulin. Administration routes being investigated include pulmonary, topical, and other parenteral administration routes. Although several approaches look promising, more basic research is necessary before meaningful clinical studies can be initiated. Some promising concepts are described below.

Infections and inflammations are characterized by a leaky vascular system. Therefore sterically stabilized liposomes may be an effective vehicle for the delivery of appropriate drugs in such circumstances. Many infectious agents reside in cells, such as deep tissue macrophages, which are practically inaccessible to chemotherapy. A large fraction of PEGylated liposomes ends up in the skin, where many are ultimately taken up by macrophages. It is also believed that these cells harbor many bacterial and viral infections and therefore the potential of long-circulating liposomes should be thoroughly exploited.

A liposomal vaccine against hepatitis A has been successfully launched by the Swiss Serum Institute (Bern). Purified hepatitis A virion capsule, viral phospholipids, and envelope glycoprotein from the influenza virus are mixed with phosphatidylcholine and phosphatidylethanolamine in the presence of excess detergent (octyl glucoside). Upon detergent removal, liposomes containing antigen as well as some viral lipids and proteins, including fusogenic glycoprotein (influenza hemagglutinin) are formed. The liposomes act as an adjuvant and carrier of coadjuvants (viral glycolipids and glycoproteins) and potentiate an immune response to the vaccine antigen. The same company is developing vaccines for influenza, hepatitis B, diphtheria, and tetanus. Other companies, such as Orasomal Technology (Chicago, IL, USA) are working on the development of other vaccines, including oral vaccination and drug delivery by polymerized liposomes.

A significant barrier to the development of liposomes formed from commonly available lipids for oral or mucosal delivery of macromolecules has been their susceptibility to enzymatic degradation, due to low pH in the stomach, and bile-salt dissolution in the intestine. Hence, common liposomes fail to protect macromolecules against enzymatic degradation in the gastrointestinal tract. Mechanically and sterically stabilized liposomes can survive these conditions but are consequently too stable in the intestine to deliver the encapsulated drug via the normal absorption process. A possible solution to this problem would be to design liposomes with time-dependent stability. Direct uptake by M cells in Peyer's patches, which sample the surroundings for antigens, is too low for a substantial absorption of drugs but may be effective for vaccination, where low concentrations of antigens may be adequate to stimulate an immune response. Recently, the potential of utilizing polymerized liposomes as oral drug and vaccine vehicles has been explored. Such liposomes survive intestinal transit intact and can be rapidly taken up by M cells and subsequently by macrophages. Providing M-cell targeting molecules by the covalent attachment of lectins and other ligands provides further increases in intestinal liposome uptake. Very stable liposomes and polymerized liposome vehicles may provide a route for the commercial development of liposome technology for oral vaccination, and eventually, for a wide variety of macromolecular drugs.

In conclusion, the synergistic input from colloid science, physics, chemistry, biology, pharmacology, and medicine has resulted in the successful development of liposomal drug delivery in less than thirty years, and a solid theoretical and experimental basis has developed promising novel products. Only time will tell which of the above applications and speculations will prove to be successful. However, based on the already available products, we can say that liposomes have definitely established their position in modern technology.

REFERENCES

1. Lasic, D. D., Barenholz, Y., eds. *Handbook of Nonmedical Applications of Liposomes: Theory and Basic Sciences*. CRC Press, Boca Raton, FL (1996).
2. Lasic, D. D., Papahadjopoulos, D., eds. *Medical Applications of Liposomes*. Elsevier Science, Amsterdam (1998).
3. Amselem, S., Gabizon, A., Barenholz, Y. Optimization and upscaling of doxorubicin-containing liposomes for clinical use. J. Pharm. Sci., 79 (1990), 1045–1052.
4. Allen, T. M., Hansen Christian, B., Zalipsky, S. Antibody-targeted stealth liposomes. In D. D. Lasic, F. Martin, eds. *Stealth Liposomes*. CRC Press, Boca Raton, FL (1995), 233–244.
5. Allen, T. M., Hansen, C., Martin, F., Redemann, C., Yau-Young, A. Liposomes containing synthetic lipid derivatives of poly(ethylene glycol) show prolonged circulation half-lives in vivo. Biochim. Biophys. Acta, 1066 (1991), 29–36.
6. Allen, T. M., Agrawal, A. K., Ahmad, I., Hansen Christian B., Zalipsky, S. Antibody-mediated targeting of long-circulating (stealth) liposomes. Journal of Liposome Research, 4 (1994), 1–25.
7. Allen, T. M., Hansen, C., Rutledge, J. Liposomes with prolonged circulation times: factors affecting uptake by reticuloendothelial and other tissues. Biochim. Biophys. Acta, 981 (1989), 27–35.
8. Bally, M. B., Hope, M. J., Echteld, C. J. A. V., Cullis, P. R. Uptake of safranine and other lipophilic cations into vesicles in response to a membrane potential. Biochim. Biophys. Acta, 812 (1985), 55–65.
9. Bally, M. B., Zhang, Y., Wong, F. M. P., Kong, S., Wasan, E., Reimer, D. L. DNA–lipid complexes as an intermediate in the preparation of particles for gene transfer: an alternative to cationic liposome–DNA aggregates. Adv. Drug Del. Rev. (1997), 275–290.
10. Bangham, A. D., Standish, M. M., Watkins, J. C. Diffusion of univalent ions across the lamellae of swollen phospholipids. J. Mol. Biol., 13 (1965), 238–252.
11. Barenholz, Y., Amselem, S., Lichtenberg, D. A new method for preparation of phospholipid vesicles. FEBS Lett., 99 (1979), 210–214.
12. Barenholz, Y., Lasic, D. D., eds. *Nonmedical Application of Liposomes*. CRC Press, Boca Raton, FL (1996), 23–41.
13. Batzri, S., Korn, E. D. Single bilayer liposomes prepared without sonication. Biochim. Biophys. Acta, 298 (1973), 1015–1019.
14. Blume, G., Cevc, G. Molecular mechanism of the lipid vesicle longevity in vivo. Biochim. Biophys. Acta, 1146 (1993), 157–168.

15. Boggs, J. M. Lipid intermoleculer hydrogen bonding: influence on structural organization and membrane function. Biochim. Biophys. Acta, 906 (1987), 353–404.
16. Boggs, J. M., Rangaraj, G., Koshy, K. M. Effect of hydrogen-bonding and non-hydrogen-bonding long chain compounds on the phase transition temperatures of phospholipids. Chem. Phys. Lipids, 40 (1986), 23–34.
17. Bron, R., Oratz, A., Wilschut, J. Cellular cytoplasmic delivery of polypeptide toxin by reconstituted influenza virus envelopes (virosomes). Biochemistry, 33 (1994), 9110–9117.
18. Čeh, B., Lasic, D. D. Kinetics of accumulation of molecule. J. Phys. Chem. B, (1998) 102 (16), 3036–3043.
19. Chakrabarti, A. C., Veiro, J. A., Wong, N. S., Wheeler, J. J., Cullis, P. R. Generation and characterization of iron- and barium-loaded liposomes. Biochim. Biophys. Acta, 1108 (1992), 233–239.
20. Chowhan, Z. U., Yotsuyanagi, T., Higuchi, W. I. Model transports studies utilizing lecithin spherules. 1. Critical evaluation of several physical models in the determination of the permeability coefficient for glucose. Biochim. Biophys. Acta, 266 (1972), 320–342.
21. Cortesi, R.,Esposito, E., Gambarin, S., Telloli, E., Menegatti, E., Nastruzzi, C. Preparation of liposomes by reverse-phase evaporation using alternative organic solvents. J. Microencapsulation, 16 (1999), 251–256.
22. Cullis, P. R., Fenske, D. B., Hope, M. J. Physical properties and functional roles of lipids in membranes. In D. E. Vance and J. E. Vance, eds. *Biochemistry of Lipids, Lipoproteins and Membranes*. Elsevier Science, New York (1996), 1–33.
23. Cullis, P. R., Mayer, L. D., Bally, M. B., Madden, T. D., Hope, M. J. Generating and loading of liposomal systems for drug-delivery applications. Adv. Drug Del. Rev., 3 (1989), 267–282.
24. de Gennes, P. G. Polymer at interfaces; a simplified view. Adv. Colloid Interface Sci., 27 (1987), 189–209.
25. Deamer, D. W. Preparation of solvent vaporization liposomes. In G. Gregoriadis, ed. *Liposome Technology*. CRC Press, Boca Raton, FL. (1984), 29–35.
26. Deamer, D. W., Prince, C. W., Croft, A. R.. The response of fluorescent amines to pH gradients across liposome membranes. Biochim. Biophys. Acta, 274 (1972), 323–335.
27. Enoch, H. G., Strittmatter, P. Formation and properties of 1000-A-diameter, single-bilayer phospholipid vesicles. Proc. Natl. Acad. Sci. USA, 76 (1979), 145–149.
28. Farhood, H., Serbina, N., Huang, L. The role of dioleoyl phosphatidylethanolamine in cationic liposome mediated gene transfer. Biochim. Biophys. Acta, 1235 (1995), 289–295.
29. Felgner, P. L. Particulate systems and polymers for in vitro and in vivo delivery of polynucleotides. Adv. Drug Del. Rev., 5 (1990), 163–187.
30. Felgner, P. L., Gadek, T. R., Holm, M., Roman, R., Chan, H. W., Wenz, M., Northrop, J. P., Ringold, G. M., Danielsen, M. Lipofection: a highly efficient, lipid-mediated DNA-transfection procedure. Proc. Natl. Acad. Sci. USA, 84 (1987), 7413–7417.
31. Fry, D., et al. Anal. Biochem., 90 (1980), 809.
32. Gabizon, A., Goren, D., Fuks, Z., Barenholz, Y., Dagan, A., Meshorer, A. Enhancement of adriamycin delivery to liver metastatic cells with increased tumoricidal effect using liposomes as drug carriers. Cancer Research, 43 (1983), 4730–4735.
33. Gao, X., Huang, L. A novel cationic liposome reagent for efficient transfection of mammalian cells. Biochim. Biophys. Res. Commun., 179 (1991), 280–285.
34. Gregoriadis, G. Fate of liposomes in vivo and its control: a historical perspective. In D. D. Lasic, F. Martin, eds. *Stealth Liposomes*. CRC Press, Boca Raton, FL (1995), 7–12.
35. Gregoriadis, G., Saffic, R., Hart, S. L. High yield incorporation of plasmid DNA within liposomes: effect on DNA integrity and transfection efficiency. Journal of Drug Targeting, 3 (1996), 469–475.
36. Gruner, S. M., Lenk, R. P., Janoff, A. S., Ostro, M. J. Novel multilayered lipid vesicles: comparison of physical characteristics of multilamellar liposomes and stable plurilamellar vesicles. Biochemistry, 24 (1985), 2833–2842.
37. Haran, G., Cohen, R., Bar, L. K., Barenholz, Y. Transmembrane ammonium sulfate gradients in liposomes produce efficient and stable entrapment of amphiphatic weak bas. Biochim. Biophys. Acta, 1151 (1993), 201–215.
38. Hayat, M. A., ed. *Principles and Techneques of Electron Microscopy Biological Applications*. Van Nostrand Reinhold, New York (1970), 323.
39. Hong, K., Friend, D. S., Glabe, C. G., Papahadjopoulos, D. Liposomes containing colloidal gold are a useful probe of liposome–cell interactions. Biochim. Biophys. Acta, 732 (1983), 320–323.
40. Hope, M. J., Bally, M. B., Webb, G., Cullis, P. R. Production of large unilamellar vesicles by a rapid extrusion procedure. Characterization of size distribution, trapped volume, and ability to maintain a membrane potential. Biochim. Biophys. Acta, 812 (1985), 55–65.
41. Huang, C. Studies on phosphatidylcholine vesicles. Formation and physical characteristics. Biochemistry, 8 (1969), 344–351.
42. Hui, S. W., Stewart, T. P., Boni, L. T. The nature of lipidic particles and their roles in polymorphic transitions. Chemistry and Physics of Lipids, 33 (1983), 113–126.
43. Israelachvili, J. N., Marcelja, S., and Horn, R. G. Physical principles of membrane organization. Q. Rev. Biophys., 13 (1980), 121–200
44. Kagawa, Y., Racker, E. Partial resolution of the enzymes catalyzing oxidative phospholylation, part 15, reconstitution of vesicles catalyzing phosphorus-32 ortho phosphate ATP exchange. J. Biol. Chem., 246 (1971), 5477–5487.
45. Kirby, C. J., Gregoriadis, G. A simple procedure for preparing liposomes capable of high encapsulation efficiency under mild conditions. In: G. Gregoriadis, ed. *Liposome Technology*. CRC Press, Boca Raton, FL (1984), 19–27.
46. Klibanov, A., Maruyama, K., Torchilin, V. P., Huang, L.

Amphipathic polyethyleneglycols effectively prolong the circulation time of liposomes. FEBS Lett., 268 (1990), 235–237.

47. Kuhl, T. L., Leckband, D. E., Lasic, D. D., Israelachvili, J. N. Modulation of interaction forces between lipid bilayers exposing short-chained ethylene oxide headgroups. Biophys. J., 66 (1994), 1479.
48. Kuhl, T. L., Leckband, D. E., Lasic, D. D., Israelachvili, J. N. Modulation and modeling of interaction forces between lipid bilayers exposing terminally grafted polymer chains. In: D. D. Lasic, F. Martin, eds. *Stealth Liposomes*. CRC Press, Boca Raton, FL (1995), 73–101.
49. Lasic, D. D. The mechanism of vesicle formation. Biochem J., 256 (1988), 1–11.
50. Lasic, D. D. Formation of liposomes. Nature, 351 (1991), 613.
51. Lasic, D. D., ed. *Liposomes, From Physics to Application*. Elsevier, Amsterdam (1993).
52. Lasic, D. D., ed. *Liposomes in Gene Delivery*. CRC Press, Boca Raton, FL (1997).
53. Lasic, D. D., Ceh, B., Stuart, M. C. A., Guo, L., Frederik, P. M., Barenholz, Y. Transmembrane gradient driven phase transitions within vesicles: lessons for drug delivery. Biochim. Biophys. Acta, 1239 (1995), 145–156.
54. Lasic, D. D., Frederik, P. M., Stuart, M. C. A., Barenholz, Y., McIntosh, T. J. Gelation of liposome interior: a novel method for drug encapsulation. FEBS Lett., 312 (1992), 255–258.
55. Lasic, D. D., Martin, F., eds. *Stealth Liposomes*. CRC Press, Boca Raton, FL (1995).
56. Lasic, D. D., Martin, F. J., Gabizon, A., Huang, S. K., Papahadjopoulos, D. Sterically stabilized liposomes: a hypothesis on the molecular origin of the extended circulation times. Biochim. Biophys. Acta, 1070 (1991), 187–192.
57. Lasic, D. D., Needham, D. The ''Stealth'' liposomes: a prototypical biomaterial. Chem. Rev., 95 (1995), 2601–2628.
58. Lewis, R. N. A. H., Mak, N., McElhaney, R. N. A differential scanning calorimetric study of the thermatropic phase behavior of model membranes composed of phosphatidylcholines containing linear saturated fatty acyl chains. Biochemistry (1987), 6118–6126.
59. Lewis, R. N. A. H., Sykes, B. D., McElhaney, R. N. Thermotropic phase behavior of model membranes composed of phosphatidylcholines containing cis-monounsaturated acyl chain homologues of oleic acid: differential scanning calorimetric and ^{31}P NMR spectroscopic studies. Biochemistry, 27 (1988), 880–887.
60. Marsh, D., ed. *Handbook of Lipid Bilayers*. CRC Press, Boca Raton, FL (1990).
61. Marsh, D. General features of phospholipid phase transitions. Chemistry and Physics of Lipids, 57 (1991), 109–120.
62. Mayer, L. D., Bally, M. B., Cullis, P. R. Uptake of adriamycin into large unilamellar vesicles in response to a pH gradient. Biochim. Biophys. Acta, 857 (1986), 123–126.
63. Mayer, L. D., Hope, M. J., Cullis, P. R. Vesicles of variable sizes produced by a rapid extrusion procedure. Biochim. Biophys. Acta, 858 (1986), 161–168.
64. Mayer, L. D., Hope, M. J., Cullis, P. R., Janoff, A. S. Solute distributions and trapping efficiencies observed in freeze-thawed multilamellar vesicles. Biochim. Biophys. Acta, 817 (1985), 193–196.
65. Mayer, L. D., Madden, T. D., Bally, M. B., Cullis, P. R. pH gradient-mediated drug entrapment in liposomes. In: G. Gregoriadis, ed. *Liposome Technology*. CRC Press, Boca Raton, FL (1993), 27–44.
66. Mayer, L. D., Bally, M. B., Hope, M. J., Cullis, P. R. Uptake of dibucaine into unilamellar vesicles in response to a membrane potential. Journal of Biological Chemistry, 260 (1985), 802–808.
67. Mayer, L. D., Tai, L. C. L., Bally, M. B., Mitilenes, G. N., Ginsberg, R. S., Cullis, P. R. Characterization of liposomal systems containing doxorubicin entrapped in response to pH gradients. Biochim. Biophys. Acta, 1025 (1990), 143–151.
68. Memoli, A., Palermiti, L. G., Travagli, V., Alhaique, F. Egg and soya phospholipids—sonication and dialysis: a study on liposome characterization. Int. J. Pharm., 117 (1995), 159–163.
69. Metsikk, K., van Meer, G., Simons, K. EMBO J., 5 (1986), 3429–3435.
70. Michaelson, D. M., Barkai, G., Barenholz, Y. Asymmetry of lipid organization in cholinergic synaptic vesicles membranes. Biochem., 211 (1983), 155–162.
71. Milsmann, M. H. W., Schwendener, R. A., Weder, H. The preparation of large single bilayer liposomes by a fast and controlled dialysis. Biochim. Biophys. Acta, 512 (1978), 147–155.
72. Mimms, L. T., Zampighi, G., Nozaki, Y., Tanford, C., Reynolds, J. A. Phospholipid vesicle formation and transmembrane proein incorporation using octyl glucoside. Biochemistry, 20 (1981), 833–840.
73. Nabel, G. J., Nabel, E. G., Yang, Z., Fox, B. A., Plautz, G. E., Gao, X., Huang, L., Shu, S., Gordon, D. Chang, A. E. Direct gene transfer with DNA–liposome complexes in melanoma: expression, biologic activity and lack of toxicity in humans. Proc. Natl. Acad. Sci. USA, 90 (1993), 11307–11311.
74. Needham, D., Hristova, K., McIntosh, T. J., Dewhirst, M., Wu, N., Lasic, D. D. Polymer-grafted liposomes: physical basis for the ''stealth'' property. J. of Liposome Research, 2 (1992), 411–430.
75. Needham, D., McIntosh, T. J., Lasic, D. D. Repulsive interactions and mechanical stability of polymer-grafted membranes. Biochim. Biophys. Acta, 1108 (1992), 40.
76. New, R. R. C. Characterization of liposomes. In R. R. C. New, ed. *Liposomes: A Practical Approach*. Oril Press at Oxford University Press, Oxford (1990), 105–162.
77. Olson, F., Hunt, C. A., Szoka, F. C., Vail, W. J., Papahadjopoulos, D. Preparation of liposomes of defined size distribution by extrusion through polycarbonate membranes. Biochim. Biophys. Acta, 557 (1979), 9–23.

78. Rädler, J. O., Koltover, L., Salditt, T., Safinya, C. R. Structure of DNA-cationic liposome complexes: DNA intercalation in multilamellar membranes in distinct interhelical packing regimes. Science, 275 (1997), 810–814.
79. Reimer, D. L., Zhang, Y., Kong, S., Wheeler, J. J., Graham, R. W., Bally, M. B. Formation of novel hydrophobic complexes between cationic lipids and plasmid DNA. Biochemistry, 34 (1995), 12877–12883.
80. Salazar, D., Cohen, S. A. Multiple tumoricidal effector mechanisms induced by adriamycin. Cancer Research, 44 (1984), 2561–2566.
81. Saravolic, E. G., Ludkovski, O., Skirrow, W., Ossanlou, M., Zhang, Y.-P., Giesbrecht, C., Thompson, J., Stark, H., Cullis, P. R., Scherrer, P. Encapsulation of plasmid DNA in stabilized plasmid-lipid particles composed of different cationic lipid concentration for optimal transfection activity. J. Drug Target., 7(6) (2000), 423–437.
82. Seddon, J. M., Cevc, G., Marsh, D. Calorimertric studies of the gel-fluid (L_β-L_α) and lamellar-inverted hexagonal (L_α-L_β) phase transitions in dialkyl- and diacylphosphatidylethanolamines. Biochemistry, 22 (1983), 1280–1289.
83. Szoka, F. C., Jr. Preparation of liposome and lipid complex compositions. U.S. patent 5,567,434, (1996).
84. Szoka, F. C., Jr., Papahadjopoulos, D. Procedure for preparation of liposomes with large internal aqueous space and high capture by reverse-phase evaporation. Proc. Natl. Acad. Sci. USA, 75 (1978), 4194–4198.
85. Tanford, C., Reynolds, J. A. Characterization of membrane proteins in detergent solutions. Biochim. Biophys. Acta, 457 (1976), 133–170.
86. Templeton, S. N., Lasic, D. D., Frederik, P. M., Strey, H. H., Roberts, D. D., Pavlakis, G. Improved DNA/liposome complexes for increased systemic delivery and gene expression. Nature Biotech, 15 (1997), 647–652.
87. Torchilin, V. P., Papisov, M. L. Hypothesis: why do polyethylene glycol-coated liposomes circulate so long? J. Liposome Res., 4 (1994), 725.
88. Voet, D., Voet, J., eds. *Biochemistry*. John Wiley (1995).
89. Veiro, J. A., Cullis, P. R. Biochim. Biophys. Acta, 1025 (1990), 109–115.
90. Webb, M. S., Saxon, D., Wong, F. M., Lim, H. J., Wang, Z., Bally, M. B., Choi, L. S., Cullis, P. R., Mayer, L. D. Comparison of different hydrophobic anchors conjugated to poly(ethylene glycol): effects on the pharmacokinetics of liposomal vincristine. Biochim. Biophys. Acta, 1372 (1998), 272–282.
91. Wheeler, J. J., Palmer, L., Ossanlou, M., MacLachlan, I., Graham, R. W., Zhang, Y., Hope, M. J., Scherrer, P., Cullis, P. R. Stabilized plasmid–lipid particles: construction and characterization. Gene Ther., 6 (1999), 271–281.
92. Woodle, M. C., Newman, M. S., Working, P. K. Biological properties of sterically stabilized liposomes. In: D. D. Lasic, F. Martin, eds. *Stealth Liposomes*. CRC Press, Boca Raton, FL (1995), 103–114.
93. Woodle, M. C., Engberg, C. M., Zalipsky, S. New amphipatic polymer–lipid conjugates forming long-circulating reticuloendothelia system–evading liposomes. Bioconjugate Chem., 5 (1994), 493–496.
94. Zalipsky, S., Brandeis, E., Newman, M. S., Woodle, M. C. Long circulating, cationic liposomes containing amino-PEG-phosphatidylethanolamine. FEBS Lett., 353 (1994), 71–74.
95. Zhang, Y. P., Sekirov, L., Saravolac, E. G., Wheeler, J. J., Tardi, P., Clow, K., Leng, E., Sun, R., Cullis, P. R., Scherrer, P. Stabilized plasmid–lipid particles for regional gene therapy: formulation and transfection properties. Gene Ther., 6 (1999), 1438–1447.
96. Zhang, Y.-P., Lewis, R. N. A. H., Henry, G. D., Sykes, B. D., Hodges, R. S., McElhaney, R. N. Peptide models of helical hydrophobic transmembrane segments of membrane proteins. 1. Studies of the conformation, intrabilayer orientation, and amide hydrogen exchangeability of Ac-K_2-$(LA)_{12}$-K_2 amide. Biochemistry, 34 (1995), 2348–2361.
97. Zhang, Y.-P., Lewis, R. N. A. H., Hodges, R. S., McElhaney, R. N. Peptide models of helical hydrophobic transmembrane segments of membrane proteins. 2. Differential scanning calorimetric and FTIR spectroscopic studies of the interaction of Ac K_2-$(LA)_{12}$-K_2-amide with phosphatidylcholine bilayers. Biochemistry, 34 (1995), 2362–2371.
98. Zhang, Y. P., Lewis, R. N. A. H., Hodges, R. S., McElhaney, R. N. Interaction of a peptide model of a hydrophobic transmembrane α-helical segment of a membrane protein with phosphatidylethnolamine bilayers: differential scanning calorimetric and Fourier transform infrared spectroscopic studies. Biophys. J., 68 (1995), 847–857.
99. Zhang, Y. P., Reimer, D. L., Zhang, G., Lee, P. H., Bally, M. B. Self-asembling DNA–lipid particles for gene transfer. Pharm. Res., 14 (1997), 190–195.
100. Zhang, Y.-P., Lewis, R. N. A. H., Hodges, R. S., McElhaney, R. N. Interaction of a peptide model of a hydrophobic transmembrane α-helical segment of a membrane protein with phosphatidylethanolamine bilayers: differential scanning calorimetric and Fourier transform infrared spectroscopic studies. Biophys. J., 68 (1995), 847–857.
101. Zhang, Y.-P., Lewis, R. N. A. H., McElhaney, R. N., Ryan, R. O. Calorimetric and spectroscopic studies of the interaction of Manduca sexta apolipophorin III with zwitterionic, anionic, and nonionic lipids. Biochemistry, 32 (1993), 3942–3952.

30

Liposomes for Cancer Therapy Applications

Lawrence D. Mayer, Rajesh Krishna,* and Marcel B. Bally
BC Cancer Agency, Vancouver, British Columbia, Canada and
University of British Columbia, Vancouver, British Columbia, Canada

I. INTRODUCTION

Drugs developed for diseases caused by pathogenic infections have the advantage that there are numerous physical and functional differences between the pathogens and the cells of their human host. As a result, most of these therapeutic agents can be administered chronically at relatively high doses with typically few side effects. This is not the case in diseases such as cancer, which arise from transformed or dysregulated growth/function of the body's own cells. Here, the differences between healthy and diseased tissues are often quite subtle and rarely are cellular features truly disease specific. It is not surprising then that for these diseases a very narrow window exists between drug doses that are effective and drug doses that are toxic. Although our increased understanding of the genetic alterations associated with diseases such as cancer will no doubt improve our ability to design better drugs, it is likely that we will continue to be faced with therapeutic targets that are differentially expressed in disease cells rather than uniquely expressed. This suggests that manipulating drug pharmacology/distribution to enhance tumor selectivity will continue to be an important feature in the development of new anticancer drugs.

In view of the above considerations, it is not surprising that the potential utility of liposomes for delivery of cancer drugs was identified very quickly after the first descriptions of ordered spherical membrane structures spontaneously forming upon hydration of purified lipids were made by Bangham in the 1960s (1,2). It was thought that encapsulating highly toxic anticancer agents could alleviate many of the life-threatening side effects associated with conventional chemotherapy. Since that time, our understanding of the chemical and biophysical properties of bilayer membrane vesicles has increased dramatically as has our knowledge of how the body handles systemically administered liposomes. Through this process, liposome researchers developed anticancer drug formulations based on first-generation, conventional phospholipid/cholesterol-based liposomes followed by second-generation, sterically stabilized systems and more recently third-generation, multifunctional liposomes (see Fig. 1).

Although significant advances in the generation of novel liposome-based delivery systems were demonstrated

* *Current affiliation*: Bristol-Meyers Squibb, Princeton, New Jersey

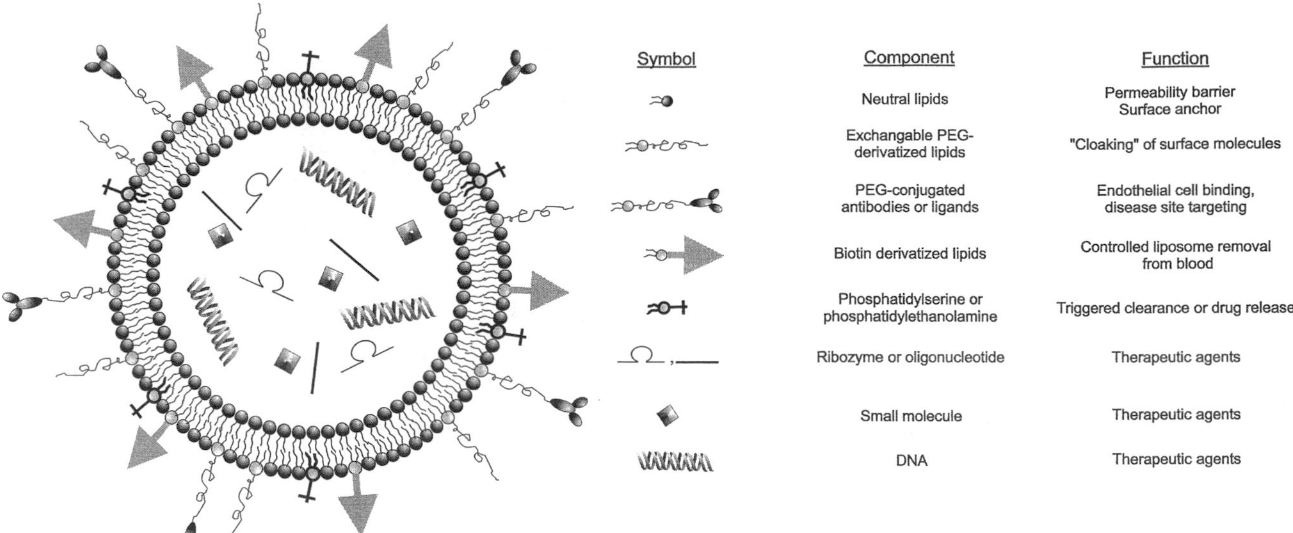

Symbol
Component
Function
Neutral lipids
Permeability barrier
Surface anchor
Exchangable PEG-derivatized lipids
"Cloaking" of surface molecules
PEG-conjugated antibodies or ligands
Endothelial cell binding, disease site targeting
Biotin derivatized lipids
Controlled liposome removal from blood
Phosphatidylserine or phosphatidylethanolamine
Triggered clearance or drug release
Ribozyme or oligonucleotide
Therapeutic agents
Small molecule
Therapeutic agents
DNA
Therapeutic agents

Figure 30.1 Illustration of a multifunctional liposome exhibiting the various therapeutic agents as well as structural and functional components that can be designed into current liposome formulations.

in laboratory studies, several hurdles regarding the production of homogeneous liposomes as well as effective entrapment and stability properties had to be overcome before liposomes could be considered as viable pharmaceutical delivery systems. These obstacles were overcome in the 1980s with the development of various liposome production procedures amenable to large-scale production including extrusion (3), dialysis (4), homogenization (5), and dehydration/rehydration techniques (6, see also 7 for a review). Also at this time efficient drug entrapment procedures were developed based on transmembrane ion gradients that provide nearly quantitative encapsulation efficiencies and extended drug retention stability (8,9). In addition, significant advances were made in the production of inexpensive, high-purity synthetic lipids that could support commercial-scale liposome production. Together, these accomplishments enabled liposomal cancer drugs to move not only from the bench to the bedside but also into the market place, where currently there are four approved liposome-based cytotoxic agents and one photodynamic agent (see Refs. 10 and 11 for reviews of clinical liposomal anticancer drug development). In addition, several liposomal anticancer drug formulations are in various stages of clinical testing for a range of tumor applications.

In the sections to follow, the different components that can be designed into liposomal anticancer drug formulations will be discussed in terms of the structural, physical, and functional properties that they impart to the carrier system. In addition, the biological processes and mechanisms whereby liposomes exert their pharmacological benefits will be reviewed.

II. COMPONENTS OF GENERATION-BASED LIPOSOMAL ANTICANCER DRUGS

A. Drug Permeability Barrier and Solubilization Properties of the Bilayer Membrane

The basic structure that is the foundation for the construction of virtually all liposomes is the bilayer membrane, which forms upon hydration of polar lipids such as phospholipids and sphingolipids in an aqueous medium. There are two key features associated with lipid bilayers that are particularly suited for drug delivery applications. The spherical lipid envelope formed by bilayer membranes constitutes a very effective permeability barrier that limits movement of highly polar or charged molecules between the internal and external aqueous compartments of liposomes. This is due to the presence of the hydrophobic bilayer core associated with the hydrocarbon chains of phospholipids and sphingolipids, which presents a very energetically unfavorable environment (dielectric constant of approximately 5) for polar compounds. Consequently, when very water soluble anticancer drugs are encapsulated in the internal aqueous compartment of liposomes, they remain there for extended times, the duration of which will be dictated by the polarity/charge of the drug and the composition of the bilayer membrane.

Since a number of commonly used chemotherapy agents are polar (amphipathic at a minimum) and water soluble, most early efforts in liposomal anticancer drug development focused on entrapping agents such as doxorubicin and cytosine arabinoside in an attempt to (a) prolong systemic exposure of drug (increase AUC) and (b) decrease drug accumulation in susceptible healthy tissues (e.g., cardiac tissue for doxorubicin). At that time it was unclear how tumor tissue would handle liposomes, and the working hypothesis was that liposomes would increase the therapeutic index of anticancer drugs by decreasing toxicity while maintaining or possibly increasing antitumor potency. These early investigations identified a number of key structure/biological function relationships for liposomal anticancer drug formulations that provided the basis for subsequent optimization studies. First, large (> 500 nm diameter) liposomes are cleared from the blood far more rapidly than small (100 nm diameter) liposomes (12). Second, liposomes containing < 30 mol% cholesterol are rapidly destabilized in the circulation, leading to significant drug release (13). Third, liposomes with charged surfaces (particularly negatively charged liposomes) are cleared from the circulation more rapidly than neutral liposomes (14).

Detailed investigations into the in vivo behavior of liposomes showed that many of the above relationships were the results of interactions of liposomes with blood components and cells of the reticuloendothelial system. When liposomes are injected intravenously they are immediately exposed to a plethora of circulating cells, lipoproteins, and soluble factors including proteins, carbohydrates, and small ions (Na^+, Cl^-, Ca^{2+}, Mg^{2+}, etc.). Assuming that liposomes contain sufficient amounts of cholesterol to avoid the bilayer destabilization effects of lipoproteins (15,16), the fate of liposomes in this compartment is dictated primarily by interactions between the liposome surface and serum protein components. Two deleterious responses can result when proteins adsorb to liposomes; (i) increased membrane permeability, which compromises drug retention in the liposomes, and (ii) recognition and subsequent clearance of liposomes by the RES.

The ability of adsorbed blood proteins to increase liposome permeability properties has been demonstrated by several laboratories (16–19). Such interactions can be simply modeled by determining the drug release kinetics for liposomes suspended in serum compared to protein-free buffer. An example of this is shown in Fig. 2. The leakage of vincristine from DSPC/Chol liposomes is approximately five fold faster in the presence of serum. Interestingly, comparison of these results with the release kinetics

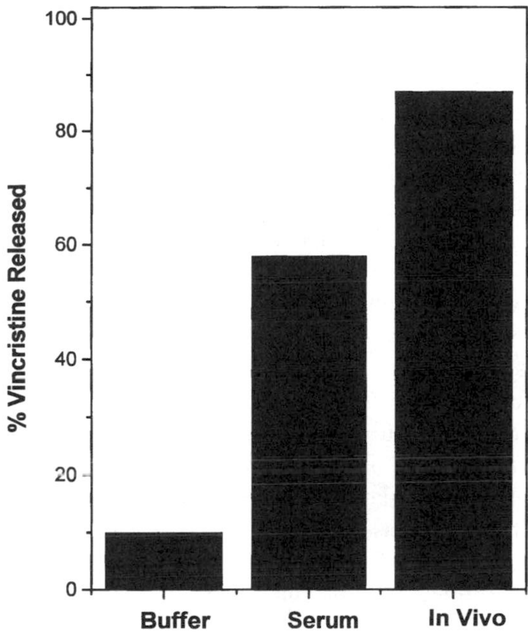

Figure 30.2 Release of vincristine from 100 nm DSPC/cholesterol (55:45 molar ratio) liposomes in buffer, in serum, and in the circulation after iv injection to BDF1 mice. Vincristine was encapsulated at a drug-to-lipid ratio of 0.05:1 (wt:wt) using the citrate-based pH gradient entrapment procedure, where trapping efficiencies were > 95%. The values represent the percent vincristine release after 24 h. In vitro experiments were conducted at 37°C.

of vincristine from DSPC/Chol liposomes after iv administration (as determined by monitoring changes in the circulating drug-to-lipid ratio) reveals that drug leakage is further increased in vivo (Fig. 2). These differences are not simply due to the presence of a ''tissue sink'' into which the released vincristine is absorbed, since increased dilutions or extended dialysis times in the presence of serum do not increase in vitro drug release rates (L. Mayer, unpublished observations). Consequently, we believe that in vivo drug retention properties as well as comparisons of drug release kinetics for different liposomes cannot always be predicted simply on the basis of in vitro data.

In addition to increasing the permeability of liposome bilayers in the blood, protein adsorption can also lead to increased susceptibility to transmembrane stresses caused by ion gradients or high levels of encapsulated drugs. The high concentrations of buffer components and/or drug entrapped in liposomes often result in significant osmotic gradients across the liposome membrane when exposed to physiological fluids. While most liposomes can withstand a significant transmembrane osmotic gradient in the absence of extraneous proteins, exposure of liposomes exhibiting large osmotic gradients to plasma or purified lipoprotein fractions results in a burst of leakage from the liposomes while osmotic balance is reestablished (20). This effect is more pronounced with less ordered membranes, where, for example, DSPC/Chol liposomes can withstand osmotic gradients of far greater magnitude than EPC/Chol liposomes in the presence of proteins (20). This may, in part, explain the differences observed between DSPC/Chol and EPC/Chol liposomal doxorubicin formulations in vivo where the circulating drug-to-lipid ratio (used to assess drug leakage) observed for EPC/Chol liposomes drops approximately 50% within 1 h of injection and subsequently decreases to a release rate comparable to that observed for DSPC/Chol (21).

The RES has long been recognized as the major site of liposome accumulation after systemic administration. The primary organs associated with the RES are the liver, spleen, and lung. The liver exhibits the largest capacity for liposome uptake, while the spleen can accumulate liposomes so that the tissue concentration (liposomal lipid/gm tissue) is 10-fold higher than that which can be achieved in other organs. Assuming that liposomes are designed to minimize protein binding and cell interactions, the extent of liposome accumulation in the lung is typically below 1% of the injected dose. Early studies demonstrated that large as well as charged liposomes (particularly those containing negatively charged lipids like PS, PA, or cardiolipin) were removed very rapidly by the liver and spleen with clearance half-lives of less than 1 hour (22,23). The rate of clearance from the circulation could be reduced to some extent by increasing the administered lipid dose, however, only when small (approx. 100 nm), neutral liposomes containing $\geq$ 30% cholesterol were utilized at doses of at least 10 mg/kg or more, could circulation lifetimes in the range of several hours be achieved (24). The removal of liposomes from the blood is attributed to phagocytic cells that reside in the RES and appears to be mediated through direct interactions between the phagocytic cell and the liposomes. In vitro studies have shown that liposome uptake into macrophages can occur in the absence of serum proteins, but recognition mediated by protein elements that associate with liposome surfaces likely plays a dominating role on interactions with the RES.

Once the processes dictating liposome drug retention and clearance in the central blood compartment were elucidated, several liposomal anticancer drug formulations using previously approved therapeutic agents were entered into formal developmental programs aimed at testing in human clinical trials. The liposomes in these formulations had the common features of being approximately 100 nm in size, being unilamellar, and containing 35–50 mol% cholesterol. In general, these preparations exhibited significantly reduced toxicity in preclinical models compared to the same drug administered in conventional, unencapsulated form (25–27). In the case of doxorubicin, dose-limiting cardiotoxicity was significantly reduced by liposome entrapment, and this correlated with decreased accumulation of doxorubicin in heart tissue when administered in liposome-encapsulated form (28,29). Similar observations were made with doxorubicin concerning other toxicity target organs such as the kidneys and GI tract (30,31) as well as for other drugs exhibiting significant nonhematological toxicities such as cisplatin (32). In addition to the benefits of reduced toxicity, early indications suggested that these liposomes also preferentially accumulated in sites of tumor growth. This was initially revealed by imaging and drug distribution studies that demonstrated increased liposome and entrapped anticancer drug uptake into solid tumors compared to other, healthy tissues (33,34). Subsequent studies investigated the mechanisms involved in this tumor selectivity in order to improve further the therapeutic potential of liposomal anticancer drugs (see Section III.B). These observations taken together with the toxicity buffering effects brought renewed enthusiasm into the development of liposome-based cancer therapeutics.

Although the majority of liposome technology applications for cancer treatment have been on the delivery of existing cytotoxics already approved for use in free (non-entrapped) form, interest is increasing on examining liposome formulations for investigational agents at earlier stages of development. An emerging trend in the development of new small-molecule anticancer agents is that the most potent and promising molecules are often very hydrophobic and insoluble in conventional aqueous pharmaceutical vehicles for systemic administration. Examples of such compounds include the taxanes, camptothecins, prophyrin-based photosensitizers, and podophyllotoxins. Given the adverse reactions associated with surfactant-based vehicles such as Cremophor EL utilized to solubilize paclitaxel, increasing attention has focused on formulating these types of compounds in liposomes. This is due to the hydrophobic core of the bilayer membrane, which can accommodate high levels of various hydrophobic drugs, depending on the membrane intercollation properties of the specific compound being formulated. As such, the solubilization capabilities of liposomes represents a second basic property of these carriers that can be applied beneficially for many anticancer drugs.

For hydrophobic drugs, liposomes can serve two purposes. First, the bilayer membrane can simply serve as a biocompatible solubilization vehicle where drugs can be incorporated and subsequently administered in an aqueous diluent such as saline or 5% dextrose in water (D5W). Second, the liposomes can act as true drug delivery carriers in vivo and dictate the pharmacodistribution properties of the drug if they are capable of retaining the agent for extended times (hours). Intuitively, one would predict that hydrophobic drugs would be ideally suited for liposomal delivery systems, perhaps even more so than hydrophilic drugs. Clearly, very hydrophobic drugs will be dependent on some form of solubilization system such as liposomes for iv use, and the DMPC/EPG formulation of the recently approved photosensitizer benzoporphyrin derivative (BPD) is one such example. This compound has water solubility below 1 μg/mL but can be constituted at several mg/mL when formulated in small DMPC/EPG liposomes (35). Interestingly, however, shortly after iv administration, BPD exchanges out of the liposome carrier and partitions into other hydrophobic domains, most notably lipoproteins, which may act as secondary carriers of the drug to disease sites (36).

The limitations that are often experienced with liposomal formulations of hydrophobic drugs can be grouped into two general categories. These are (1) low drug incorporation capacity and (2) poor drug retention after injection. Some agents, such as paclitaxel, although hydrophobic, are not readily incorporated into phospholipid-based liposome bilayers, and capacities on the order of 3 mol% of the total lipid are observed (37,38). Such formulations have been tested and shown to have extended drug circulation times and antitumor activity (39). However, the very high lipid dose required to deliver therapeutic amounts of paclitaxel may artificially maintain elevated plasma drug levels by virtue of the amount of liposomal membrane in the circulation and not be able to be utilized in a clinical setting. By analogy, cyclosporins have been incorporated into liposomes with similar incorporation capacities and have also been shown to be released rapidly from the lipid carrier in the circulation, resulting in drug pharmacokinetic properties that are comparable to unencapsulated drug (40,41). It should be noted that the rapid release of hydrophobic drugs often experienced in vivo is rarely observed in vitro. This is likely due to the presence of the large lipid pool existing in animal tissues that can act as a membrane ''sink'' into which hydrophobic drugs can partition, which is absent in in vitro test systems utilized to monitor drug release. Also, the permeability energy barrier provided by the liposomal bilayer for polar, water-soluble drugs is virtually nonexistent for hydrophobic agents, which can readily move across phospholipid membranes. Consequently, although aqueous concentrations of hydrophobic drugs are very low at equilibrium, their redistribution in the body is driven kinetically through the rapid exchange of drug between hydrophobic lipid pools.

In addition to being used for formulating drugs that are inherently hydrophobic, liposomes have also been used to deliver hydrophobic derivatives of otherwise water-soluble drugs. The rationale of this approach is based on speculations that decreased toxicity/increased antitumor potency may arise from favorable alterations of drug tissue distribution in favor of disease tissue, increased duration of exposure, or decreased metabolic degradation. Alkylated derivatives of Ara-C formulated in liposomes have been shown to be significantly more potent than free Ara-C against some tumor models (42,43). Methotrexate conjugated to phosphatidylethanolamine incorporated in liposomes was shown to be equally active against sensitive and MTX resistant tumor cells (44). Also, palmitoyl-asparaginase incorporated into liposomes exhibited significantly extended circulation longevity and reduced immunogenicity compared to native L-asparaginase. One of the most advanced lipophilic drug derivative formulations is liposomal NDDP, a diaminocyclohexane platinum analog (45,46). Perez-Solar and coworkers have studied this compound extensively and have demonstrated improved pharmacokinetic and therapeutic characteristics compared to conventional cisplatin (47). In addition, clinical trials with this formulation indicated that nephrotoxicity often observed with cisplatin was not dose limiting for liposomal NDDP (48). These examples demonstrate how the hydrophobic properties of liposomes can be utilized in various ways to enhance the therapeutic index of a wide variety of anticancer agents.

As a final comment on the in vivo drug retention and distribution properties of liposomes, it should be noted that, while liposomes clearly can provide sustained exposure of therapeutic agents in the blood compartment through controlled release kinetics of encapsulated drugs, currently it is difficult to justify development of liposomal drugs using a rationale that involves only sustained systemic exposure. This is largely due to significant advances made in the area of drug infusion technology. Compact and cost-effective infusion pumps are now widely used, and these can provide well-controlled systemic drug exposure over several days. We maintain that the most significant advantage for the use of liposome drug carriers arises as a consequence of disease-specific changes in vascular permeability that favor the accumulation of the intact liposome and associated drug into sites of tumor growth. We differentiate this property from the benefits of drug infusion technology, which are primarily concerned with the maintenance of circulating blood levels of free drug. However, this does not detract from the utility of liposomes to protect labile anticancer agents such as the camptothecins (49,50), nor does it discount the value of formulating drugs that would be otherwise incompatible for systemic administration regardless of the pharmacodistribution characteristics imparted by the liposomes.

B. Effects of Steric Stabilizing Lipids on PK and Therapeutic Properties of Liposomal Anticancer Agents

Incorporation of steric stabilizing lipids reflects the ''second generation'' of liposome development, where components were included in the basic phospholipid/cholesterol bilayer in an attempt to improve the tumor selective delivery of encapsulated agents. These developments arose as we began to understand how liposomes interacted not only with proteins and phagocytic cells of the RES but also with the vasculature of healthy tissues and solid tumors. While in the circulation, liposomes are continually exposed to cells lining the vasculature. The inner lining, or intima, of blood vessels is composed primarily of endothelial cells, which form a contiguous layer on the interior surface of all blood vessels. Underlying this layer is the basement membrane, and in larger (noncapillary) vessels the vasculature is supported by smooth muscle cells (51). The endothelial cells in most normal vasculature exhibit intact intercellular junctions, and only small molecules are able to permeate readily across capillaries of this type. However, this structure is significantly altered in certain normal tissues, most notably the liver and the spleen, as well as in disease sites such as infection and tumor growth. The latter are characterized by the presence of capillaries that exhibit fenestrae or larger intercellular openings and can be devoid of the basement membrane layer. The gaps in these endothelial layers can range in size from 30 nm for fenestrated capillaries to greater than 500 nm in liver and tumor vascular beds (52–54). In the liver, these openings provide access to sinusoids, wherein the phagocytic Kupffer cells lie. In solid tumor sites, the fenestrated/discontinuous capillary beds and postcapillary venules allow direct exposure of the underlying epithelial cells to the circulation. It is the unique nature of vascular structures, that particularly exist in solid tumors, that significantly impacts the pharmacological behavior of liposomal anticancer drug delivery systems.

The mediators that lead to increased permeability of the vascular barrier in solid tumors appears related to hypoxic environments arising from rapid tumor cell proliferation, which causes increased distances of cells from vascular sources of nutrients and oxygen. Such conditions appear to induce release vascular endothelial growth factor (VEGF) from tumor cells (55,56). VEGF is an endothelial cell-specific mitogen, and its release from tumor cells can lead to the angiogenic development of neovasculature by inducing proliferation and migration of existing vascular endothelial cells towards regions of tumor hypoxia. Interestingly, VEGF has proven to be identical to vascular permeability factor (57,58), a protein first identified as a factor capable of inducing defects in the permeability barrier of blood vessels. The end result of these conditions is the presence of blood vessels that are permeable to large molecules. This may be a consequence of fenestrae or larger ''gaps'' occurring between adjacent endothelial cells through which macromolecules can pass (59) or, alternatively, may involve increases in endothelial cell mediated transcytosis (60).

Increases in vascular permeability give rise to the selective accumulation of small liposomes in sites of tumor growth. However, this is not a selective process, and there is also a general increase in extravascular fluids in these regions. The hydrostatic pressure within these sites is elevated relative to the vascular pressure, resulting in a pressure gradient that impedes movement of molecules from the blood into the tissue interstitium (61,62). We must therefore assume that additional features lead to selective accumulation of macromolecules in the diseased extravascular space. Studies, for example, have demonstrated that the lack of a developed lymphatic system in conjunction with the large openings in the vascular endothelial cell lining may lead to an extravascular ''trapping'' phenomenon (62). In the absence of lymphatic drainage, interstitial diffusion of molecules leads to egress from the disease site, and this diffusion rate is dependent on molecule size, small molecules exiting more rapidly than large molecules.

Liposome extravasation and accumulation in solid tumors has been well studied, and there is a great deal of phenomenological evidence demonstrating that liposomes can enter an extravascular site in regions of tumor growth following iv administration. Although evidence for endothelial cell uptake of liposomes and transcytosis across endothelial cells has been documented, videomicroscopy investigations in solid tumor models indicate that the majority of liposome extravasation occurs directly through the openings present in tumor neovasculature (63,64). This extravasation process appears to be quite heterogeneous within the tumor and does not appear to be associated with any specific histological characteristics in the tumor mass. The net result of this phenomenon is that peak tumor drug concentrations achieved are greater, and drug exposure as measured by concentration vs. time AUCs is often dramatically increased when the drug is administered in liposomal compared to free form.

Since the selective uptake of liposomes in solid tumors appears related to a ''sieving'' phenomenon, the drug concentration vs. time curves in solid tumors are very different for liposomal anticancer agents compared to unencapsulated drugs. This, of course, assumes that the liposome serves as a pharmacologically active delivery vehicle and not simply as a formulation excipient. This results from the fact that peak plasma free drug concentrations are observed very shortly after injection, which subsequently fall rapidly due to tissue distribution and elimination processes. Small-molecule therapeutics readily access extravascular sites (including tumors) without significant impediment due to the vascular lining. Consequently, peak tumor drug levels are also observed soon after injection, and the retention (e.g., AUC) of the anticancer agent is dictated primarily by the drug binding avidity of the tumor cells/interstitium, since the egress of most small-molecule drugs from tumor sites is minimally hindered. In contrast, liposome extravasation in solid tumors is much slower, being limited by the abundance of leaky fenestrae in the tumor neovasculature, and demonstrates an ascending accumulation curve that may peak between 4 and 48 h, depending on the plasma pharmacokinetics of the liposomes and the permeability properties of the tumor vasculature. Therefore at early times (i.e., 1 h after drug administration), tumor-free drug levels may actually be elevated compared to liposomal drug, but this relationship soon reverses: later peak concentrations and AUCs for liposomal anticancer agents can be 15-fold and 100-fold greater, respectively, than for free drug (65–67).

Given the nature of liposome extravasation and retention within solid tumors, a great deal of effort has been placed on developing liposomes that circulate at elevated concentrations in the blood for extended times. The underlying hypothesis of this approach is that the extravasation process is driven by mass action diffusion of small liposomes across the leaky tumor neovasculature. Consequently, increasing the plasma concentration of liposomes should theoretically increase tumor uptake levels of liposomal anticancer drugs. This provided impetus for the development of liposomes with extended circulation times. Early observations identified certain gangliosides, particularly GM_1, that reduced the elimination rates of conventional PC/cholesterol liposomes (68,69). A significant breakthrough was achieved with the demonstration that liposomes containing surface grafted polyethylene glycol (PEG) were cleared 2- to 5-fold slower than conventional liposomes of similar size and bulk lipid composition (70–72). This appears to be due to the ability of the PEG polymer (typically 2000 MW) to form a ''steric stabilizing'' shield on the liposome surface that inhibits protein binding and cellular recognition (73–75). This effect is optimal at incorporation levels of approximately 5 mol%, which also coincides with the transition of the PEG polymer from a random coil, mushroomlike conformation to a more ordered brushlike conformation. The details of this phenomenon are presented by Dr. Lasic in this book. Using this technology, numerous anticancer drugs have been encapsulated in sterically stabilized liposomes to generate formulations that can provide significant circulating drug concentrations over several days. It is important to note that as the ability to extend the blood residence time increases, so does the need to retain encapsulated anticancer drugs over similar times. Therefore most sterically stabilized formulations employ more impermeable saturated phospholipid/cholesterol bilayer compositions that minimize drug leakage in the circulation (see Section III.A).

Although steric stabilization technology has clearly been able to provide increased circulation lifetimes of liposomal anticancer agents, the design of liposomes that will exhibit maximal extravasation in tumor sites is an area of some controversy. It has generally been assumed that increases in the concentration of liposomes in plasma over time will lead to increased accumulation of liposomes in the extravascular disease sites, and experimental evidence supporting this has been reported (76). Videomicroscopy has also suggested that the permeability coefficient of tumor vasculature is greater for PEG-PE containing liposomes than for conventional liposomes (77). In contrast, studies conducted in our laboratories as well as others have demonstrated that although plasma levels of PEG containing liposomes are several times higher than for comparable conventional liposomes, this often does not result in increased extravasation and accumulation in solid tumor tissue (66,78,79).

We have examined the tumor uptake properties for conventional and steric stabilized liposomal formulations of doxorubicin in a variety of solid tumor models. Three important observations can be made on the basis of the comparative biological properties of conventional and sterically stabilized liposomes. First, sterically stabilized liposomes uniformly display increased circulation longevity compared to conventional liposomes, regardless of the presence of encapsulated drug. Second, the rate and extent of liposome accumulation in tumor tissue are often comparable for both conventional and sterically stabilized liposomes, although the PEG containing systems typically exhibit a slight increase. Third, the tumor accumulation efficiency or TAE (defined as the AUC in the tumor divided by the AUC in plasma) is higher for conventional liposomes than for sterically stabilized systems. It is important to note that the relationship between tumor liposome uptake and plasma liposome AUC is linear for conventional and sterically stabilized liposomes, respectively (M. Bally, unpublished observation). This suggests that mass action does appear to drive the accumulation of specific types of nontargeted small liposomes into tumors. However, inclusion of lipids such as PEG-DSPE appears to decrease the efficiency of liposome extravasation from the blood into tumor tissue, or alternatively increase the rate of egress from the tumor as indicated by the decreased TAE values observed for sterically stabilized liposomes in several solid tumor models (Fig. 3).

The basis for discrepancies in tumor extravasation comparisons between conventional and sterically stabilized liposomes may be related to one of several potential explanations. The tumor models utilized may exhibit different vascular structures (80), and it is reasonable to assume that increases or decreases in conventional liposome extravasation in comparison to sterically stabilized liposomes may be tumor specific. However, many different tumor types have been evaluated, and it would appear that preferential accumulation efficiency of conventional liposomes is prevalent in most tumor types, regardless of differences in vascular structure. Our comparisons are typically based on extended AUC measurements of total tumor liposome uptake (following a nonexchangeable, nonmetabolizable lipid label and correcting for blood volume contributions), and we place great emphasis on measuring both liposomal lipid and drug over the specified time course. It should be noted that the lipid compositions utilized in our studies for conventional and sterically stabilized liposomes contain 45 mol% cholesterol, whereas studies by others often utilize liposomes with 33 mol% cholesterol. Reduced cholesterol content may result in increased drug permeability (see Section III.A). Several comparative studies demonstrating improved tumor accumulation for sterically stabilized liposomes have relied on the use of entrapped aqueous markers such as Ga^{67}, which are rapidly cleared when released from the liposomes. Consequently, it is difficult to determine if differences in tumor accumulation are due to altered elimination and/or extravasation properties affected by lipid composition or a result of lipid composition effects on drug retention. It is important to resolve these issues, and this will require a concerted effort to standardize the tumor models, liposome compositions, and methods/parameters for evaluating liposome extravasation into tumors.

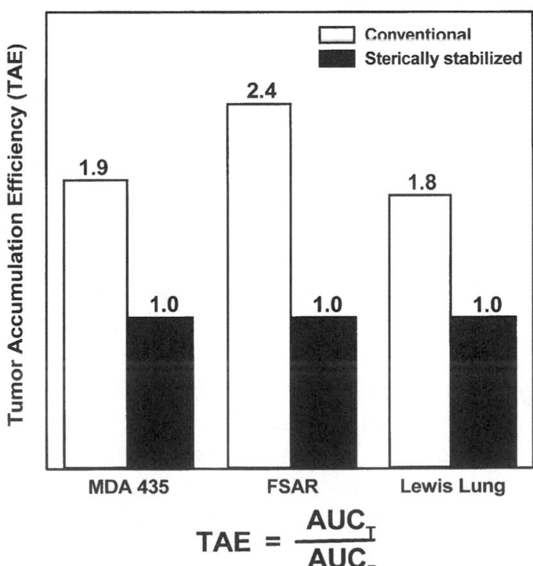

Figure 30.3 Relative tumor accumulation efficiency (TAE) of conventional (open bars) and sterically stabilized (closed bars) after iv injection to mice bearing the indicated solid tumor type. The tumor accumulation efficiency was determined for 100 nm liposomes by dividing the AUC in the tumor with the AUC in plasma for DSPC/cholesterol (55:45 molar ratio) and DSPC/PEG-DSPE/cholesterol (50:5:45 molar ratio) liposomes.

It should not be unexpected that conventional and sterically stabilized liposomes exhibit different efficiencies in extravasation. Videomicroscopy studies with steric stabilized liposomal doxorubicin systems have shown that some endothelial cells can take up liposomes (63,77). Endothelial cell interactions may contribute to the extravasation process either directly via transcytosis or indirectly by facilitating an increase in the local liposome concentration at the endothelial cell surface, thereby increasing access to openings in the vasculature. Given the effects of PEG on inhibiting liposome–cell interactions, this polymer may reduce endothelial cell interactions, and this in turn would reduce the rate of extravasation. In contrast, conventional liposome extravasation could be facilitated through increased interactions with the endothelial cell lining of the neovasculature in tumors. This is, of course, highly speculative but is consistent with the surface properties of conventional liposomes compared to steric stabilized liposomes. A logical extension of this argument, however, is that improved extravasation may be possible by designing liposomes that interact more extensively with vascular endothelium in tumors.

Although there are areas where additional information will be required in order to optimize the pharmacological properties of sterically stabilized liposomal anticancer agents, the utility of such systems has been clearly demonstrated for liposomal formulations of anthracyclines, vinca alkaloids, camptothecins, and platinum compounds. In all cases, significant increases in therapeutic index have been achieved through decreased toxicity and typically increased antitumor activity, which is associated with significantly increased circulating drug concentrations (81). The ability to control interactions with liposome surfaces using steric stabilizing lipids provides the added advantage of opportunities to engineer specific components into the liposomes to select for specific, desired interactions while avoiding nonspecific, and often undesirable, reactions with the biological milieu.

C. Multifunctional Liposomes; Targeted, Fusogenic, and Controlled Release Systems

As noted above, the distribution of liposomes that have extravasated into the tumor interstitium is heterogeneous. This is not unexpected, given the irregular and often redundant organization of tumor vasculature. Tumor vascular structure often engenders highly variable blood flow properties, and evaluation of histological sections from tumors reflects this heterogeneity. This would be more apparent for liposomes than for unencapsulated small molecules due to the decreased diffusion through the interstitial space for large macromolecules. Importantly, evaluations of drug accumulation properties suggest that drug release from the liposomes in the extravascular site results in greater drug penetration into the tissue and more rapid loss of the drug from the site when compared with the loss of liposomal lipid (66). The prolonged residence of liposomes in tumors significantly increases the duration of tumor drug exposure and AUC relative to free agents (66). In some tumor models, such properties have been shown to correlate with increased antitumor activity for liposomal formulations of drugs such as doxorubicin and daunorubicin.

The consensus emerging from studies in several laboratories on the mechanism of action of liposomal anticancer drug formulations is that liposomes exert their effect on therapeutic activity by providing an in situ drug infusion reservoir within the tumor. Once released, the anticancer drug can diffuse through the tumor and has direct access to tumor cells, where it can act in a manner that presumably is similar to drug in the absence of a liposomal carrier (Fig. 4). In vitro studies have demonstrated that macrophages can engulf doxorubicin loaded liposomes, process them, and rerelease doxorubicin extracelluarly in free form (82). In view of the high macrophage content residing in some tumors (83), such phenomena led to the proposal that liposomal anticancer drug release may involve macrophage processing after extravasation. However, we have shown that in solid tumors there are limited interactions between tumor associated macrophages and extravasated liposomes (78). Although macrophage enriched tumors do accumulate higher levels of liposomal doxorubicin, this effect appears more related to increased vascular permeability rather than to direct uptake and processing of the liposomes by the macrophages. This was further supported by the fact that both conventional and sterically stabilized liposomes displayed comparable distribution properties (as determined by fluorescence microscopy of tumor thin sections) after extravasation into the tumor.

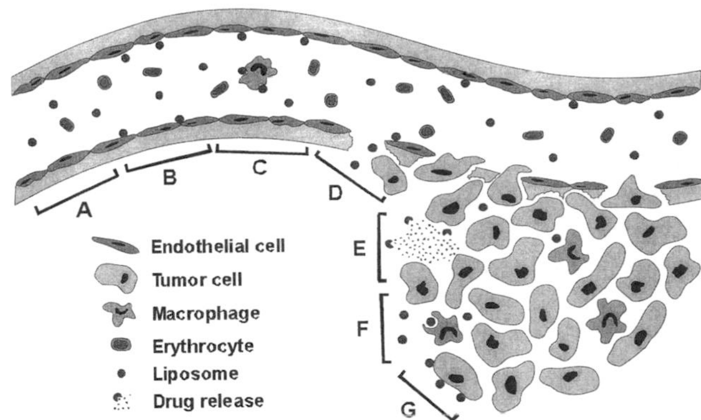

Figure 30.4 Model of tumor site infusion mechanism for improved therapeutic activity of conventional and sterically stabilized liposomal anticancer drug formulations. (A) Liposomes circulating freely in the blood. (B) Liposomes interacting with endothelial cells. (C) Liposomes interacting with circulating white blood cells. (D) Liposomes passively extravasating through the leaky vasculature of solid tumors. (E) Liposomes releasing entrapped drug in the tumor interstitium. (F) Liposomes being taken up by phagocytic cells in the solid tumor. (G) Liposomes associating with the surface of tumor cells.

In view of the heterogeneous tumor distribution of liposomes and the apparent in situ free drug infusion model of activity for most liposomal anticancer drug formulations, increasing interest and effort has been given to developing liposomes that can increase the intracellular delivery of anticancer agents to tumor cells. Two general approaches have been taken in this regard, namely (1) targeting to internalizable epitopes expressed preferentially on the surface of tumor cells so that the liposomes deliver their entrapped anticancer drugs directly to the tumor cell interior, and (2) increasing tumor site selective release of encapsulated drugs to increase bioavailable drug pools.

Targeting of liposomes to tumor cells has long been investigated as an avenue to improve the delivery of anticancer drugs to tumors. Unfortunately, many early attempts to employ targeted liposomes suffered from the increased clearance experienced by liposomes containing surface-derivatized proteins (e.g., antibodies). Consequently, accumulation into solid tumors for these liposomes was often less than obtained for conventional systems. This improved somewhat upon the implementation of steric stabilizing lipids, but it was observed that tumor uptake still was marginally improved compared to nontargeted sterically stabilized liposomes.

It is important to note that tumor cell directed targeting information does not inherently alter the extravasation events required for the liposomes to reach their cellular target. Consequently, the pharmacodistribution benefits provided by targeted liposomes should arise from a decreased rate of egress from the disease site (rather than increased influx) and cell specific binding. Given that nontargeted liposomes migrate slowly through interstitial spaces in solid tumors, it may not be surprising that pharmacological improvements are typically very modest. In addition, the avidity of liposome binding to target cells may actually inhibit liposome migration and subsequent drug exposure in areas more distant from blood vessels. This is anticipated on the basis of binding barrier effects that have been described by Saga et al. (84). Therefore targeting approaches may be most appropriate for small foci of disease where extensive interstitial diffusion is not required to expose all of the diseased cells to the therapeutic agent. This has been demonstrated with immunoliposomes targeted to lung cancer metastases growing in mice, where small tumors could be treated much more effectively with targeted liposomes than with conventional liposomes or free drug (85).

More recently, very encouraging signs of significant benefits associated with tumor targeted immunoliposomes containing doxorubicin have been obtained using a human breast cancer xenograft model. These studies conducted in the laboratories of the late Demitri Paphadjopoulos utilized an antibody fragment directed against the Her-2 receptor present on some breast cancer cells. Doxorubicin loaded sterically stabilized liposomes derivatized with this protein were internalized and endocytosed upon binding to tumor cells expressing this receptor (86). Preclinical xenograft solid tumor model studies showed that such formulations could provide significantly increased antitumor potency compared to free doxorubicin or doxorubicin entrapped in nontargeted sterically stabilized liposomes (87). These reports are some of the first to provide solid evidence that targeting of liposomes to tumors may lead to enhanced therapeutic activity. However, in view of the numerous other failed attempts to achieve this result, it would appear that successful application in this technology will require very careful consideration of the tumor antigen to be targeted, the specific nature of the targeting ligand, and the cellular processing that occurs subsequent to the liposomes binding to the tumor cells in order to achieve intracellular release of encapsulated drug.

In the case of targeted liposomal anticancer drugs directed at internalizable tumor surface proteins, additional lipid components can be included to make the liposomal carrier pH sensitive (88) or fusogenic (89). Release of entrapped contents or fusion with the endocytic/lysosomal membranes can be designed to occur when the liposomes are exposed to the low pH of the late endosome/lysozome. This has been shown to increase dramatically the potency of liposomal anticancer drugs in vitro (90). A second approach for intracellular delivery is based on the use of fusogenic lipids. In this case, introduction of liposome encapsulated agents into the cytoplasm of targeted cells is via membrane fusion of the liposome bilayer with the disease cell's plasma membrane. This first requires binding to the cell surface. Recent reports suggest that highly fusogenic lipid mixtures can be stabilized by incorporation of small amounts of exchangeable or cleavable PEG lipids (91,92). Loss of the PEG moiety leads to destabilization of the liposome membrane, which in turn will have the potential to fuse with nearby cell membranes. While these novel approaches for intracellular delivery of liposomal contents are providing exciting data in cell culture systems, their utility in vivo will depend on maintaining or increasing access and delivery of the liposomal carrier and encapsulated drug to the disease tissue. It should also be noted that the exchangeable/cleavable PEG containing liposomes can also be used to provide controlled release of the entrapped anticancer drug in the tumor interstitium after the liposomes have accumulated at the tumor site. This has been shown to be advantageous with a controlled release liposomal formulation of mitoxantrone (93). The key feature here is that the timing of increased drug release from the liposomes (dictated by PEG exchange rates) must coincide with the kinetics of liposome accumulation at the tumor site.

Another approach being investigated to achieve tumor site selective anticancer drug release from liposomes is the use of membrane systems whose permeability properties are sensitive to a physical/chemical property that can be manipulated in sites of tumor growth from extraneous sources. In particular, thermosensitive and photosensitive liposomes have been investigated in an attempt to achieve triggered drug release specifically at the tumor site (94,95). Liposomal doxorubicin preparations, for example, can be prepared such that there is an increase in drug release at 42°C, compared to 37°C. These liposomes are administered iv to tumor bearing mice, and the tumor site is then heated using a topical microwave heating device placed on the subcutaneous tumor. Application of a transient heating pulse after the liposomal doxorubicin had accumulated into the solid tumor resulted in a significant increase of therapeutic activity compared to free drug with hyperthermia and liposomal doxorubicin in the absence of heating. Very recently (95), lysolipid containing liposomal doxorubicin formulations have been developed that undergo a dramatic increase in drug release upon very mild hyperthermia (from 37°C to 39°C). These liposomes exhibit more extensive drug release than previous thermosensitive systems, and early results with the application of mild hyperthermia have provided solid tumor cure rates in excess of 75%, whereas free doxorubicin or doxorubicin entrapped in sterically stabilized or lysolipid-free thermosensitive liposomes yielded modest reductions in tumor growth rates and cure rates between 0 and 10% (95). These results highlight the importance of controlling the bioavailability of liposome entrapped anticancer drugs at the tumor site and emphasize the importance of this area for further improvements in the therapeutic activity of liposomal anticancer drugs.

III. THE APPLICATION PARADIGM FOR LIPOSOMAL ANTICANCER DRUG DEVELOPMENT

A. General Comments

Now that we have progressed from the conventional and sterically stabilized generations of liposomal anticancer drugs to third-generation multifunctional liposomes, we have at our disposal a wide range of properties that can be engineered into liposome formulations. We have the ability to trap drugs at nearly quantitative efficiencies, retain liposomes and their entrapped drugs in the circulation for days, effectively target liposomal drugs to tumor cells, and control the release of encapsulated cytotoxics for selective exposure. This sophistication provides numerous options for generating effective liposome-based anticancer agents, but it is unlikely that increased design complexity will be necessary in all applications, and in some instances it may be undesirable. Consequently, it would appear that increased attention to formulation design based on specific applications will be required in order to optimize the therapeutic potential of new liposomal anticancer drug formulations.

It is important at this point to reflect on the usefulness of preclinical model systems in predicting the behaviour of liposomal anticancer drug formulations in humans. Clearly, a focus on application-based development of therapeutic liposomes relies heavily on the dependability of in vitro and in vivo testing systems to show which formulation characteristics affect biological performance parameters important for human use. We and other laboratories have observed that most in vitro systems are of limited utility, because improvements provided by liposome encapsulation are typically related to alterations in drug pharmacodistribution, which is very difficult to predict with in vitro assays. In contrast, most in vivo models for determining toxicity and efficacy profiles of liposomal anticancer agents have been quite accurate predictors of activity in humans. For example, the liposomal doxorubicin formulation TLC D-99 exhibited significantly reduced cardiac and GI toxicity compared to conventional nonencapsulated doxorubicin, whereas hematological toxicities were modestly effected in preclinical studies (28,29). In addition, ascitic and solid tumor models demonstrated comparable antitumor potency between the two drug forms, and improved therapeutic activity was obtained at elevated doxorubicin doses allowed by the less toxic liposome formulation. These characteristics were consistent with the clinical performance of this liposomal formulation, where toxicity and therapeutic response comparisons with nonencapsulated doxorubicin mirrored the results from preclinical studies (10). Similar situations occurred with the liposomal daunorubicin formulation Daunoxome developed by Nexstar and the sterically stabilized liposomal doxorubicin formulation Caelyx, where clinical trial results were comparable to those from preclinical investigations (11). This strongly suggests that the therapeutic performance of liposomal anticancer drugs in humans can be predicted with reasonable accuracy using conventional preclinical animal toxicity and efficacy models. Consequently, emphasis on liposome design/biological performance optimization using animal models appears to be appropriate for identifying formulations that will likely provide significant therapeutic benefits in humans.

Whereas historically researchers have generated liposomal formulations of anticancer drugs and subsequently looked for tumor types where they may be most useful, an application-based development approach focuses on important problems existing in cancer chemotherapy and then identifies active agents and liposome properties that are best suited to meet these needs. That does not mean that attention to anticancer applications has been absent in generation-based liposome research programs, but rather that more emphasis was placed on developing formulation technology than on identifying biological problems or mechanisms for which liposomes could provide a unique solution. There will no doubt continue to be a need for additional liposome technology development in order to improve on existing design parameters and generate additional novel functional capabilities. However, it is suggested here that this development should be dictated by deficiencies that exist in current technology for specific anticancer applications of liposomes. Two examples of how liposome design attributes can be systematically matched to specific cancer treatment applications are presented in the following sections.

B. Development of Liposomal Vincristine

Vincristine is a potent cell cycle specific vinca alkaloid that is widely utilized in the treatment of many human malignancies such as lymphomas. It has a very steep dose–response curve, but its clinical use is limited by neurotoxicity that can be debilitating if the dose exceeds 2 mg in a single course of treatment. Its sensitivity to the tumor cell cycle is demonstrated by the fact that varying the duration of drug exposure to tumor cells in culture from 4 h to 72 h results in a 10^5-fold increase in cytotoxic potency, whereas other agents such as doxorubicin are affected by less than 50-fold (95). Given our in situ tumor drug infusion model of liposomal anticancer drug activity, we postulated that the cytotoxic potency of vincristine in vivo may be dramatically increased if we could increase the level and duration of drug exposure in tumors using small (100 nm) liposomes.

We found that increasing the retention of vincristine inside 100 nm liposomes by changing the phosphorylcholine containing lipid component from EPC to DSPC to sphingomyelin (while maintaining cholesterol content at 45 mol%) led to dramatic increases in antitumor activity, particularly when compared to the efficacy obtained with free vincristine (96). This is consistent with the steep dependence of vincristine antitumor potency on the duration of drug exposure as well as the fact that retention of vincristine in most tissues, including tumors, is rather poor (96). In this case it appeared that the ability to prolong the exposure of vincristine in vivo was more important than peak drug concentrations. Interestingly, although inclusion of PEG-PE in the liposomes increased the circulating liposomal lipid levels at extended time periods, this steric stabilizing lipid did not improve the vincristine pharmacokinetic or therapeutic properties over conventional DSPC/Chol or sphingomyelin/Chol systems (67). This was because PEG-PE increased the permeability of the lipid bilayer to vincristine, thus offsetting the potential benefits provided by increased longevity of the liposomal carrier. The reasons for this increased drug leakage are not well understood. It may be related to the fact that PEG-modified phosphatidylethanolamine is negatively charged, and this may alter drug partitioning properties at the inner monolayer membrane surface. In addition, it is not yet clear whether this phenomenon is specific for vincristine encapsulated via pH gradient techniques employing citrate buffers, compared to ammonium sulfate entrapment systems (97). Nonetheless, for vincristine, we determined that the optimized formulation for improving therapeutic activity consisted of a SM/cholesterol conventional liposome using pH gradient drug entrapment. This formulation is currently in Phase II clinical trials for non-Hodgkin's lymphoma and is exhibiting very promising therapeutic response rates in addition to evidence of toxicity buffering effects compared to conventional (nonencapsulated) vincristine.

C. Liposomes for Multidrug Resistance (MDR) Applications

Multidrug resistance is a significant obstacle to the successful treatment of many human malignancies with chemotherapy, affecting as high as 50% of all cancer patients (98). The phenomenon of MDR is the simultaneous resistance of tumor cells to a broad range of structurally and functionally unrelated drugs. Some tumor types such as colorectal, pancreatic, and renal are inherently resistant to chemotherapy, whereas other tumors initially respond only to relapse with tumors exhibiting MDR. Although it is clear that MDR is a multifactorial problem associated with various alterations in the biochemical makeup of resistant tumor cells, one very common and well-characterized mediator of MDR is the plasma membrane protein P-glycoprotein (PGP). This protein is often over-expressed in MDR tumor cells and acts by pumping anticancer drugs out of tumor cells in an energy dependent fashion (98). Consequently, numerous researchers have attempted to develop therapeutic treatment strategies whereby PGP could be either blocked or bypassed in order to circumvent MDR and improve the antitumor activity of drugs against such tumors.

Many properties of liposomal delivery systems described in the previous sections appear well suited to solving the problems associated with therapeutic strategies utilizing conventional anticancer drugs and MDR modulators. Liposomes have been applied to the treatment of MDR tumors in three basic approaches. These are (1) the use of liposomes alone as carriers for anticancer drugs to provide increased dose intensity and intracellular delivery, (2) inclusion of PGP modulators in the lipid bilayer or entrapped inside liposomes for delivery to MDR tumors, and (3) combining potent conventional MDR modulators with long circulating liposomal anticancer drugs.

1. Liposomal Anticancer Drugs Alone

One of the early indications that liposomes can be beneficial for MDR reversal was observed in a study that demonstrated enhanced activity of a long-circulating liposomal formulation incorporating either the ganglioside GM1 or PEG-DSPE for doxorubicin and epirubicin against the murine C26 colon carcinoma (99,100). These C26 colon carcinoma cells are known to express moderate levels of PGP, and doses of free doxorubicin up to 10 mg/kg are rendered ineffective (99) in vivo. Results indicated that mice receiving sterically stabilized formulations of doxorubicin or epirubicin resulted in tumor regression to nonmeasurable sizes. In comparison, free drugs did not have any effect on delaying tumor growth. The authors attributed this enhanced activity in a relatively resistant tumor model to be due to enhanced localization of the drug in tumor tissue, drug release from liposomes into the extravascular spaces, and uptake of the released contents by the tumor. Similar results were observed with free and sterically stabilized liposomal formulations of epirubicin (100).

This ability of the liposomes to inherently overcome a certain degree of MDR stems from the fact that liposomes can markedly enhance (between 3 and 10 times) the amount of drug that can be delivered to tumors compared to free drug. For example, if the tumor exhibits a resistance factor of 5-fold, this could theoretically be overcome by increasing anticancer drug dose intensity through liposomal delivery. The fact that stealth liposomes were able to provide tumor regression in C26 colon carcinoma cells by delivering higher amounts of the drug supports this argument. However, for tumors exhibiting higher resistance levels, liposomes by themselves may be unable to circumvent MDR significantly, as demonstrated in a rat glioblastoma tumor model (101). In further support of this argument, DSPC/Chol liposomal DOX alone at a dose of 10 mg/kg was unable to circumvent PGP mediated MDR in a murine P388/ADR solid tumor model, where resistance factors are in the order of 50- to 100-fold (102).

Antibody-coated liposomes have the ability to provide targeting and selective toxicity towards tumor cells, and this has been demonstrated in KLN-205 squamous cell carcinoma in vivo (85). By targeting tumor cells expressing a specific internalizable surface epitope, these immunoliposomes provide a unique approach of providing intracellular tumor delivery of the drug (103). This is exemplified by a study that reported that doxorubicin resistance was modulated by an immunoliposome targeting transferring receptor in MDR human leukemic cells (104). However, site-directed targeting may not result in enhanced antitumor activity unless targeting leads to endocytosis, since most tumor cells do not actively internalize nontargeted liposomes. In addition, given the penetration difficulties experienced for immunoliposomes in larger solid tumors, it is likely that this application to circumvent MDR may be limited to circulating hematopoietic malignancies, as has been suggested by Allen and coworkers (103,105).

2. Liposomal MDR Modulators

Studies with liposomes composed of certain acidic phospholipid–based systems such as phosphatidylserine (106,107) or cardiolipin liposomes (108) have shown that these lipids are able to increase the cytotoxicity of encapsulated or complexed anticancer drugs against MDR cells. MDR reversal using such phospholipids has been related to increased intracellular delivery of anticancer agents as well as a direct PGP blocking effect (109,110). While these results provided encouraging data implicating certain lipids as PGP modulating agents, acidic lipids such as cardiolipin and phosphatidylserine are readily recognized by phagocytic cells of the reticuloendothelial system, and liposomes containing these lipids are cleared within minutes from the circulation offering little systemic availability in vivo. Given the rapid clearance of negatively charged liposomes from the plasma, it is yet to be demonstrated that such liposome systems will be of use in treating solid tumors. Liposomes have also been employed to deliver antisense oligonucleotides and ribozymes against mdr-1 mRNA, and studies demonstrated increased therapy when the antisense was encapsulated in liposomes (111,112). However, effective reduction in mdr1 mRNA and protein levels is often only obtained when used in conjunction with cationic liposomes (113). Currently, the therapeutic utility of such cationic liposomes is limited to local delivery, since these systems are also cleared very rapidly after systemic administration and are unlikely to provide any systemic delivery.

We have also investigated the activity of conventional small-molecule PGP inhibitors encapsulated in liposomes as MDR modulators. However, to date such MDR modulators are either not amenable for liposomal encapsulation or exhibit high leakage rates after iv injection. We have demonstrated that modulators such as verapamil, prochlorperazine (67), and cyclosporin A (40,41) could be effectively entrapped within liposomes. However, their very high membrane permeabilities resulted in rapid leakage from the liposomes on iv administration (41, Mayer, unpublished observations), which compromises any potential advantages of such liposomes to provide improved selectivity of tumor delivery for these agents through the use of long circulating liposomes. It is speculated that stable liposomal modulator formulations will increase the selectivity of their reversal properties in solid tumors. However, in order for such liposomes to be formulated, the modulator must possess either higher affinity for the liposomal membranes or significantly reduced membrane permeability when entrapped in the liposome interior (see Section III.A).

3. Liposomal Anticancer Drugs Combined with Conventional PGP Inhibitors

Several compounds from various therapeutic classes already approved for human use were shown to effectively block PGP in tumor cells in vitro and restore their sensitivity to anticancer agents. However, virtually all of these first-generation modulators displayed significant toxicities at doses required to achieve PGP blocking concentrations in plasma as a result of their inherent pharmacological activity. Although second-generation agents such as the cyclosporin derivative PSC 833 have alleviated many of the problems experienced with first-generation MDR modulators, they often cause significant alterations in pharmacokinetics and biodistribution properties of the anticancer drugs when coadministered, which typically necessitates a significant (approx. 50%) reduction in the anticancer dose (reviewed in Ref. 114). Studies have suggested that this is due to PGP blockade in normal tissues, given the localization of PGP in such tissues as the liver and kidney, which have a secretory function. When anticancer drug doses are reduced to accommodate modulator-induced anticancer drug pharmacokinetic alterations and toxicity, it complicates interpreting whether these dose adjustments affect drug exposure to the tumor and antitumor activity compared to anticancer drug administration alone. Consequently, the disappointing results from a number of Phase II clinical trials in patients with colorectal, lung, ovarian, breast, renal cell, myeloma, and leukemic cancers cannot be used to resolve the importance of PGP in these tumor types.

Given the pharmacokinetic changes induced by PSC 833 on conventional (free) doxorubicin, we postulated that liposomes may limit these effects by virtue of their ability to reduce the exposure of encapsulated drug to the kidneys and alter clearance of doxorubicin in the liver. These tissues appear to be key factors involved in PSC 833-induced doxorubicin pharmacokinetic changes (115). This, combined with the ability of small liposomes to extravasate passively in tumors, opened the possibility that liposomal doxorubicin may avoid the adverse pharmacokinetic interactions with PSC 833, which may lead to increased selectivity of PGP modulation drug effects at the tumor site and result in improved therapy of MDR solid tumors.

Our initial studies with DSPC/cholesterol doxorubicin formulations demonstrated that liposome encapsulation using nonleaky liposomes could significantly ameliorate the adverse toxicity effects of PSC 833 on doxorubicin when compared to free drug form (102). This was accompanied by a lack of PSC 833–induced alterations in doxorubicin plasma pharmacokinetics for the liposomal formulation, whereas free doxorubicin displayed dramatic increases in terminal elimination phases and AUC when coadministered with the MDR modulator (102). Subsequent studies showed that protection from PSC 833–induced doxorubicin toxicity was related to the ability of liposomes to reduce the rates of biliary and urinary excretion (116). In this context, sterically stabilized liposome encapsulated doxorubicin displayed the least extent of urinary and biliary doxorubicin excretion compared to conventional PC/cholesterol systems, most likely reflecting the reduced uptake of such liposomes by the liver and minimal doxorubicin release in plasma compared to EPC/cholesterol liposomes. For sterically stabilized liposomal doxorubicin, drug excretion rates were well within the capacity of biliary and urinary doxorubicin clearance even under conditions of PSC 833 impairment.

In addition to the improved toxicity profile of sterically stabilized liposomal doxorubicin combined with PSC 833 compared to other liposomal systems, this formulation also exhibited superior antitumor activity compared to conventional liposomes and nonencapsulated doxorubicin. Solid MDR human breast cancer xenografts were virtually unaffected by combinations of PSC 833 and free doxorubicin. In contrast, the MDR tumors regressed and were kept in remission for weeks when combined with DSPC/cholesterol/PEG-DSPE (55:40:5 molar ratio) entrapped doxorubicin (117). Improved antitumor activity against the MDR tumors was also observed for EPC/cholesterol and DSPC/cholesterol formulations, but the degree of therapy was inferior to that obtained with the sterically stabilized liposomes. Confocal fluorescence microscopy of fresh solid tumor specimens after treatment with doxorubicin and PSC 833 revealed that increased intracellular accumulation of doxorubicin in the MDR tumor cells was observed with coadministration of PSC 833, and the extent of tumor cell drug uptake was significantly greater for sterically stabilized liposome formulations compared to either EPC/cholesterol entrapped or nonencapsulated doxorubicin. These results indicated that the enhanced therapeutic profile of liposomal doxorubicin combined with PSC 833 was related to the maintenance of elevated tumor doxorubicin exposure in conjunction with PGP inhibition by PSC 833. In addition, the ability to achieve increased antitumor activity of liposomal doxorubicin when combined with PSC 833 in the absence of altered total tumor drug accumulation levels strongly supported the direct role of PGP inhibition in sensitizing MDR tumors to doxorubicin. Taken together, the results observed for liposome applications to circumvent MDR suggest that optimal performance will be obtained using sterically stabilized liposomes.

IV. FUTURE DIRECTIONS

We now have numerous techniques available to optimize liposomal anticancer drug formulation characteristics such as entrapment efficiency, in vivo drug leakage, circulation longevity, protein binding, tumor accumulation, triggered drug release, and intracellular delivery. Since most production and characterization issues have been resolved with the first wave of commercial liposome formulations, it is likely that our success in developing effective liposomal anticancer agents will be limited primarily by matching the correct biological performance properties with the right active agent and applying it to the appropriate tumor population. For example, the exciting recent efficacy results obtained with thermosensitive liposome formulations of doxorubicin suggest that such an ability to control the release and tumor cell exposure of entrapped cytotoxics will dramatically improve the therapeutic activity of liposomal formulations. However, this application may not be straightforward for other cancer drug classes, particularly cell cycle specific drugs (e.g., vinca alkaloids and camptothecins), where duration of exposure may be more important in dictating antitumor activity than peak tumor drug levels. In this case, slower drug release rates or pulsed drug leakage may be required in order to optimize therapy. It remains to be seen whether thermosensitive liposomes can be readily manipulated to provide a range of drug release properties. Based on the results obtained to date, we predict that major improvements in antitumor activity of liposomal anticancer drugs will be obtained if drug bioavailability can be controlled at the tumor site in this fashion.

This review did not address the area of liposomal formulations of plasmids and oligonucleotides for cancer gene therapy. This topic has received tremendous interest in numerous laboratories over the past decade, and several applications have progressed to the stage of clinical trials. However, these trials have utilized formulations for direct tumor injection, s.c. injection, or ex vivo exposure due to the technical complexity of DNA–liposome formulations and their general incompatibility for intravenous administration. These formulations typically exist as large heterogeneous aggregates of DNA–cationic lipid complexes. When lipids are incorporated to generate small DNA–lipid complexes that do not aggregate, these systems tend to exhibit poor transfecting efficiencies. The numerous approaches and applications in this area warrant separate attention, and the reader is directed to informative reviews on this topic in Chapters 34 and 35 in this book. In addition, although some liposome formulations of antisense oligonucleotides (AS-ODN) have been described in the literature (118,119), it remains to be seen whether liposomes provide significant improvements over nonencapsulated AS-ODN other than reducing the dose required to achieve the desired pharmacological effect. Clearly, these two areas will continue to attract the attention of liposome researchers focused on gene modulation approaches for cancer treatment.

Finally, a significant challenge exists in the formulation of hydrophobic anticancer drugs in liposomes so that the benefits of selective tumor delivery can be realized. Achievements to date have focused primarily on utilizing liposomes as aqueous compatible excipients for such agents or as intermediaries for drug transfer to biological carriers like lipoproteins in vivo. Consequently, one may expect to see a fourth generation of liposome technology focused on such applications. Given the historical ability of liposome researchers to overcome significant technological problems to achieve the desired biological behavior, it is likely that these challenges will be successfully met in the near future, leading to further improvements in the therapeutic potential of liposomal anticancer drugs.

REFERENCES

1. A. D. Bangham, M. M. Standish, G. Weismann. Diffusion of univalent ions across the lamellae of swollen phospholipids. *J. Mol. Biol. 13*:238–252, 1965.
2. D. Paphadjopoulos, A. D. Bangham. Biophysical properties of phospholipids. II. Permeability of phosphatidylserine liquid crystals to univalent ions. *Biochim. Biophys. Acta 126*:185–188, 1966.
3. L. D. Mayer, M. J. Hope, P. R. Cullis. Vesicles of variable sizes produced by a rapid extrusion procedure. *Biochim. Biophys. Acta 858*:161–168, 1986.
4. H. G. Weder. Liposome production: the sizing up technology starting from mixed micelles. In: *Liposome Dermatics* (Braun-Falco, Korting, Maibach, eds.). Springer-Verlag, Berlin, 1992, pp. 101–109.
5. G. Mason. Advanced techniques for preparation and characterization of small unilamellar vesicles. *Food Microstructure 8*:11–14, 1989.
6. C. J. Kirby, G. Gregoriadis. A simple procedure for preparing liposomes capable of high encapsulation efficiency under mild conditions. In: *Liposome Technology*, Vol. I (G. Gregoriadis, ed.). CRC Press, Boca Raton, FL, 1984, pp. 19–27.
7. E. C. A. Van Winden, M. J. Zuidam, D. J. A. Crommelin. Strategies for large scale production and optimized stability of pharmaceutical liposomes developed for parenteral use. In: *Medical Applications of Liposomes* (D. Lasic, D. Papahadjopoulos, eds.). Elsevier, New York, 1998, pp. 567–604.
8. L. D. Mayer, M. B. Bally, T. D. Madden, P. R. Cullis. pH gradient mediated drug entrapment in liposomes. In: *Liposome Technology*, Vol. II (G. Gregoriadis, ed.). CRC Press, Boca Raton, FL, 1992, pp. 27–44.
9. E. M. Bolotin, R. Cohen, L. K. Bar, Y. Barenholz. Transmembrane ammonium sulfate gradients in liposomes produce efficient and stable entrapment of amphipathic weak bases. *Biochim. Biophys. Acta 4*:455–479, 1994.
10. C. E. Swenson, J. Freitag, A. S. Janoff. The Liposome Company: lipid-based pharmaceuticals in clinical development. In:*Medical Applications of Liposomes* (D. Lasic and D. Papahadjopoulos, eds.). Elsevier, New York, 1998, pp. 689–702.
11. F. Martin. Clinical pharmacology and antitumor efficacy of DOXIL (pegylated liposomal doxorubicin): Sequus Pharmaceuticals, Inc. In: *Medical Applications of Liposomes* (D. Lasic and D. Papahadjopoulos, eds.). Elsevier, New York, 1998, pp. 635–688.
12. R. M. Abra, C. A. Hunt. Liposome disposition in vivo: III Dose and vesicle size effect. *Biochim. Biophys. Acta 666*: 493–497, 1981.
13. G. L. Scherphof, B. van Leeuwen, J. Wilschut, J. Damen. Exchange of phosphatidylcholine between small unilamellar liposomes and human plasma high-density lipoprotein involves excusively the phospholipid in the outer monolayer of the liposome membrane. *Biochim. Bhiophys. Acta 732*:595–602, 1983.

14. J. Senior, G. Gregoriadis. Stability of small unilamellar liposomes in serum and clearance from the circulation: the effect of the phospholipid and cholesterol components. *Life Sci. 30*:2123–2129, 1982.
15. C. Kirby, J. Clark, G. Gregoriadis. Cholesterol content of small unilamellar liposomes controls phospholipid loss to high density lipoproteins. *FEBS Lett. 111*:324–328, 1980.
16. G. Scherphof, F. Roerdink, M. Waite, J. Parks. Disintegration of phosphatidylcholine liposomes in plasma as a result of interaction with high density lipoproteins. *Biochim. Biophys. Acta 842*:296–307, 1978.
17. D. C. Litzinger, A. M. Buiting, N. van Rooijen, L. Huang. Effect of liposome size on the circulation time and intraorgan distribution of amphipathic poly(ethylene glycol)-containing liposomes. *Biochim. Biophys. Acta 1190*:99–107, 1994.
18. S. C. Semple, A. Chonn, P. R. Cullis. Influence of cholesterol on the association of plasma proteins with liposomes. *Biochemistry 35*:2521–2525, 1996.
19. M. C. Woodle, M. S. Newman, J. A. Cohen. Sterically stabilized liposomes: physical and biological properties. *J. Drug Targeting 2*:397–403, 1994.
20. B. L. Mui, P. R. Cullis, P. H. Pritchard, T. D. Madden. Influence of plasma on the osmotic sensitivity of large unilamellar vesicles prepared by extrusion. *J. Biol. Chem. 269*:7364–7370, 1994.
21. M. B. Bally, H. Lim, P. R. Cullis, L. D. Mayer. Controlling the drug delivery attributes of lipid-based drug formulations. *J. Liposome Res. 8*:299–335, 1998.
22. K. J. Hwang. Liposome pharmacokinetics. In: *Liposomes: from Biophysics to Therapeutics* (M. J. Ostro, ed.). Marcel Dekker, New York, 1987, pp. 109–156.
23. G. Gregoriadis, J. Senior. The phospholipid component of small unilamellar liposomes controls the rate of clearance of entrapped solutes from the circulation. *FEBS Lett. 119*: 43–47, 1980.
24. J. Senior, J. C. W. Crawley, G. Gregoriadis. Tissue distribution of liposomes exhibiting long half lives in the circulation after intravenous injection. *Biochim. Biophys. Acta 839*:1–8, 1985.
25. G. Gregoriadis, E. D. Nerunjun. Treatment of tumor bearing mice with liposome-entrapped actinomycin D prolongs their survival. *Res. Commun. Chem. Pathol. Pharmacol. 10*:351–356, 1975.
26. A. Gabizon, A. Dagan, D. Goren, Y. Barenholz, Z. Fuks. Liposomes as in vivo carreirs of adriamycin: reduced cardiac uptake and preserved antitumor activity in mice. *Cancer Res. 42*:4734–4739, 1982.
27. P. K. Working. Pre-clinical studies of lipid-complexed and liposomal drugs: Amphotec, Doxil and SPI-77. In: *Medical Applications of Liposomes* (D. Lasic, D. Papahadjopoulos, eds.). Elsevier, New York, 1998, pp. 605–624.
28. E. H. Herman, A. Rahman, V. J. Ferrans, J. A. Vicks, P. S. Schein. Prevention of chronic doxorubicin cardiotoxicity in beagles by liposomal encapsulation. *Cancer Res. 43*: 5427–5432, 1983.
29. L. D. Mayer, L. C. L. Tai, D. S. C. Ko, D. Masin, R. S. Ginsberg, P. R. Cullis, M. B. Bally. Influence of vesicle size, lipid composition and drug-to-lipid ratio on the biological activity of liposomal doxorubicin. *Cancer Res. 49*: 5922–5930, 1989.
30. Q. G. C. M. Van Hossel, P. A. Steerenberg, D. J. A. Crommelin, A. van Dijk, W. van Oost, S. Klein, J. M. C. Douze, D. J. de Wildt, F. C. Hillen. Reduced cardiotoxicity and nephrotoxicity with preservation of antitumor activity of doxorubicin entrapped in stable liposomes in the LOU/M Wsl rat. *Cancer Res. 44*:3698–3705, 1984.
31. A. Rahman, G. White, N. Moore, P. S. Schein. Pharmacological, toxicological and therapeutic evaluation in mice of doxorubicin entrapped in liposomes. *Cancer Res. 45*: 796–803, 1985.
32. M. S. Newman, G. T. Colbern, P. K. Working, C. Engbers, M. A. Amantea. Comparative pharmacokinetics, tissue distribution, and therapeutic effectiveness of cisplatin encapsulated in long-circulating, pegylated liposomes (SPI-077) in tumor-bearing mice. *Cancer Chemother. Pharmacol. 43*:1–7, 1999.
33. C. A. Presant, A. F. Turner, R. T. Proffitt. Potential for improvement in clinical decision-making: tumor imaging with In-111 labeled liposomes. Results of a Phase II–III study. *J. Liposome Res. 4*:985–1008, 1994.
34. B. Briele, M. Graefen, A. Bockisch, J. P. Hartlapp, W. Roedel, A. Hotze, H. J. Biersack. Indium (111) labeled liposomes as a tumor imagin agent: first clinical results. *Eur. J. Nucl. Med. 16*:411–416, 1990.
35. M. Korbelik, G. Krosl. Accumulation of benzoporphyrin derivative in malignant and host cell populations of the murine RIF tumor. *Cancer Lett. 97*:249–254, 1995.
36. A. M. Richter, E. Waterfield, A. K. Jain, A. J. Canaan, B. A. Allison, J. G. Levy. Liposomal delivery of a photosensitizer, benzoporphyrin derivative monoacid ring A (BPD), to tumor tissue in a mouse tumor model. *Photochem. Photobiol. 57*:1000–1006, 1993.
37. A. Sharma, R. M. Straubinger. Novel taxol formulations: preparation and characterization of taxol-containing liposomes. *Pharmaceut. Res. 11*:889–896, 1994.
38. C. Bernsdorff, R. Reszka, R. Winter. Interaction of the anticancer agent Taxol (paclitaxel) with phospholipid bilayers. *J. Biomed. Mat. Res. 46*:141–149, 1999.
39. A. Sharma, R. M. Straubinger, I. Ojima, R. J. Bernacki. Antitumor efficacy of taxane liposomes on a human ovarian tumor xenograft in nude athymic mice. *J. Pharm. Sci. 84*:1400–404, 1995.
40. C. Ouyang, E. Choice, J. Holland, M. Meloche, T. D. Madden. Liposomal cyclosporine. Characterization of drug incorporation and interbilayer exchange. *Transplantation 60*:999–1005, 1995.
41. E. Choice, D. Masin, M. B. Bally, M. Meloche, T. D. Madden. Liposomal cyclosporine. Comparison of drug and lipid carrier pharmacokinetics and biodistribution. *Transplantation 60*:1006–1012, 1995.
42. W. Rubas, A. Supersaxo, H. G. Weder, H. R. Hartmann, H. Hengartner, H. Schott, R. Schwendener. Treatment of

murine L1210 lymphoid leukemia and melanoma B16 with lipophilic cytosine arabinoside prodrugs incorporated into unilamellar liposomes. *Int. J. Cancer 37*:149–154, 1986.
43. R. A. Schwendener, H. Schott, H. R. Hartmann, A. Supersaxo, W. Rubas, H. Hengartner. Liposomes as carriers of lipophilic cytosine arabinoside and fluorodeoxyuridine derivatives. Their cytostatic effect and possibilities of tumor cell specific therapy. *Onkologie 10*:232–239, 1987.
44. S. C. Kinsky, J. E. Loader. Circumvention of the methotrexate transport system by methotrexate-phosphatidylethanolamine derivatives: effect of fatty acid chain length. *Biochim. Biophys. Acta 921*:96–103, 1987.
45. R. Perez-Soler, A. R. Khokhar, G. Lopez-Berestein. Treatment and prophylaxis of experimental liver metastases of M5076 reticulosarcoma with cis-bis-neodecanoato-trans-R,R-1,2-diaminocyclohexane platinum (II) encapsulated in multilamellar vesicles. *Cancer Res. 47*:6462–6466, 1987.
46. R. Perez-Soler, A. R. Khokhar. Lipophilic cisplatin analogue entrapped in liposomes: role of intraliposomal drug activation in biological activity. *Cancer Res. 52*:6341–6347, 1992.
47. A. R. Khokhar, K. Wright, Z. H. Siddik, R. Perez-Soler. Organ distribution and tumor uptake of liposome entrapped cis-bis-neodecanoato-trans-R,R-1,2-diaminocyclohexane platinum (II) administered intravenously and into the proper hepatic artery. *Cancer Chemother. Pharmacol. 22*:223–227, 1988.
48. R. Perez-Soler, G. Lopez-Berestein, J. Lautersztain, S. Al-Baker, K. Francis, D. Macias-Loger, N. Raber-Martin, A. R. Khokhar. Phase I clinical and pharmacology study of liposome-entrapped cis-bis-neodecanoato-trans-R,R-1,2-diaminocyclohexane platinum (II). *Cancer Res. 50*:4254–4259, 1990.
49. G. T. Colbern, D. J. Dykes, C. Engbers, R. Musterer, A. Hiller, E. Pegg, R. Saville, S. Weng, M. Luzzio, P. Uster, M. Amantea, P. K. Working. Encapsulation of the topoisomerase I inhibitor GL147211C in pegylated (STEALTH) liposomes: pharmacokinetics and antitumor activity in HT29 colon tumor xenografts. *Clin. Cancer Res. 4*:3077–3082, 1998.
50. T. G. Burke, X. Gao. Stabilization of topotecan in low pH liposomes composed of distearoylphosphatidylcholine. *J. Pharm. Sci. 83*:967–969, 1994.
51. R. K. Jain. Transport of molecules across tumor vasculature. *Cancer and Metastasis Rev. 6*:559–593, 1987.
52. H. F. Dvorak, J. A. Nagy, J. T. Dvorak, A. M. Dvorak. Identification and characterization of the blood vessels of solid tumors that are leaky to circulating macromolecules. *Amer. J. Pathol. 133*:95–109, 1988.
53. G. Poste, R. Kirsh, T. Kuster. The challenge of liposome targeting in vivo. In: *Liposome Technology*, Vol. III (G. Gregoriadis, ed.). CRC Press, Boca Raton, FL, pp. 1–28, 1992.
54. F. Yuan, M. Dellian, D. Fukumura, M. Leunig, D. A. Berk, V. P. Torchilin, R. K. Jain. Vascular permeability in a human tumor xenograft: molecular size dependence and cutoff size. *Cancer Res. 55*:3752–3756, 1995.
55. D. Hanahan, J. Folkman. Patterns and emerging mechanisms of the angiogenic switch during tumorigenesis. *Cell 86*:353–364, 1996.
56. J. W. Rak, B. D. St. Croix, R. S. Kerbel. Consequences of angiogenesis for tumor progression, metastasis and cancer therapy. *Anticancer Drugs 6*:3–18, 1996.
57. M. Dellian, B. P. Witwer, H. A. Salehi, F. Yuan, R. K. Jain. Quantitation and physiological characterization of angiogenic vessels in mice: effect of basic fibroblast growth factor, vascular endothelial growth factor/vascular permeability factor, and host microenvironment. *Amer. J. Path. 149*:59–71, 1996.
58. H. F. Dvorak, M. Detmar, K. P. Claffey, J. A. Nagy, L. van de Water, D. R. Senger. Vascular permeability factor/vascular endothelial growth factor: an important mediator of angiogenesis in malignancy and inflammation. *Int. Arch. Allergy Immunol. 107*:233–235, 1995.
59. S. Kohn, J. A. Nagy, H. F. Dvorak, A. M. Dvorak. Pathways of macromolecular tracer transport across venules and small veins. *Lab. Invest. 67*:596–607, 1992.
60. S. K. Huang, F. J. Martin, G. Jay, J. Vogel, D. Paphadjopoulos, D. S. Friend. Extavasation and transcytosis of liposomes in Kaposi's sarcoma-like dermal lesions of transgenic mice bearing the HIV tat gene. *Am. J. Pathol. 143*:10–14, 1993.
61. R. K. Jain. Physiological resistance to treatment of solid tumors. In: *Drug Resistance in Oncology* (B. A. Teicher, ed.). Marcel Dekker, New York, 1993, pp. 87–105.
62. Y. Buocher, R. K. Jain. Microvascular pressure is the principal driving force for interstitial hypertension in solid tumors: implications for vascular collapse. *Cancer Res. 52*:5110–5114, 1992.
63. N. Z. Wu, T. L. Rudoll, D. Needham, A. R. Whorton, M. W. Dewhirst. Increased microvascular permeability contributes to preferential accumulation of stealth liposomes in tumor tissue. *Cancer Res. 53*:3765–3770, 1993.
64. F. Yuan, M. Leunig, S. K. Huang, D. A. Berk, D. Papahad jopoulos, R. K. Jain. Microvascular permeability and interstitial penetration of sterically stabilized (Stealth) liposomes in a human tumor xenograft. *Cancer Res. 54*:3352–3356, 1994.
65. L. D. Mayer, P. R. Cullis, M. B. Bally. The use of transmembrane pH gradient-drive drug encapsulation in the pharmacodynamic evaluation of liposomal doxorubicin. *J. Liposome Res. 4*:529–553, 1994.
66. M. J. Parr, D. Masin, P. R. Cullis, M. B. Bally. Accumulation of liposomal lipid and encapsulated doxorubicin in murine Lewis lung carcinoma: the lack of beneficial effects by coating liposomes with poly(ethylene glycol). *J. Pharmacol. Exp. Therapeut. 280*:1319–1324, 1997.
67. M. S. Webb, T. O. Harasym, D. Masin, M. B. Bally, L. D. Mayer. Sphingomyelin–cholesterol liposomes significantly enhance the pharmacokinetic and therapeutic properties of vincristine in murine and human tumor models. *Br. J. Cancer 72*:896–904, 1995.
68. T. M. Allen, J. L. Ryan, D. Paphadjopoulos. Gangliosides reduce leakage of aqueous-space markers from liposomes

in the presence of human plasma. *Biochim. Biophys. Acta 818*:205–212, 1985.
69. T. M. Allen, A. Chonn. Large unilamellar liposomes with low uptake into the reticuloendothelial system. *FEBS Lett. 223*:42–46, 1987.
70. M. C. Woodle, D. D. Lasic. Sterically stabilized liposomes. *Biochim. Biophys. Acta 1113*:171–199, 1992.
71. M. C. Woodle, K. K. Matthay, M. S. Newman, J. E. Hidayat, L. R. Collin, C. Redemann, F. J. Martin, D. Paphadjopoulos. Versatility in lipid compositions showing prolonged circulation with sterically stabilized liposomes. *Biochim. Biophys. Acta 1105*:193–200, 1992.
72. T. M. Allen. Long-circulating (sterically stabilized) liposomes for targeted drug delivery. *Trends Pharm. Sci. 15*: 215–220, 1994.
73. V. P. Torchilin, V. G. Omelyanenko, M. I. Papisov, A. A. Bogdanov, V. S. Trubetskoy, J. N. Herron, C. A. Gentry. Poly(ethylene glycol) on the liposome surface: on the mechanism of polymer-coated liposome longevity. *Biochim. Biophys. Acta 1195*:11–20, 1994.
74. V. P. Torchelin, V. S. Trubetskoy, K. R. Whiteman, P. Caliceti, P. Ferruti, F. M. Veronese. New synthetic amphiphilic polymers for steric protection of liposomes in vivo. *J. Pharm. Sci. 84*:1049–1053, 1995.
75. D. Needham, D. V. Zhelev, T. J. McIntosh. Surface chemistry of the sterically stabilized PEG-liposome. In: *Liposomes, Rational Design* (A. Janoff, ed.). Marcel Dekker, New York, 1999, pp. 13–62.
76. S. S. Williams, T. R. Alosco, E. Mayhew, D. D. Lasic, F. J. Martin, R. B. Bankert. Arrest of human lung tumor xenograt growth in severe combined immunodeficient mice using doxorubicin encapsulated in sterically stabilized liposomes. *Cancer Res. 53*:3964–3967, 1993.
77. F. Yuan, M. Leunig, S. K. Huang, D. A. Berk, D. Papahadjopoulos, R. K. Jain. Microvascular permeability and interstitial penetration of sterically stabilized (Stealth) liposomes in a human tumor xenograft. *Cancer Res. 54*:3352–3356, 1994.
78. L. D. Mayer, G. Dougherty, T. O. Harasym, M. B. Bally. The role of tumor-associated macrophages in the delivery of liposomal doxorubicin to solid murine fibrosarcoma tumors. *J. Pharmacol. Exp. Therapeut. 280*:1406–1414, 1997.
79. R. L. Hong, C. J. Huang, Y. L. Tseng, V. F. Pang, S. T. Chen, J. J. Liu, F. H. Chang. Direct comparison of liposomal doxorubicin with or without polyethylene glycol coating in C-26 tumor-bearing mice: is surface coating with polyethylene glycol beneficial? *Clin. Cancer Res. 5*:3645–3652, 1999.
80. R. K. Jain. Physiological barriers to delivery of monoclonal antibodies and other macromolecules in tumors. *Cancer Res., 50*:814s–819s, 1990.
81. D. J. A. Crommelin, G. Storm. Stealth™ therapeutic systems: rationale and strategies. In: *Targeting of Drugs 6: Strategies for Stealth Therapeutic Systems* (G. Gregoriadis, B. McCormack, eds.). Plenum Press, New York, 1998, pp. 25–34.
82. G. Storm, P. A. Steerenberg, F. Emmen, M. van Borssum Waalkes, D. J. A. Crommelin. Release of doxorubicin from peritoneal macrophages exposed in vivo to doxorubicin-containing liposomes. *Biochim. Biophys. Acta 965*: 136–145, 1988.
83. R. D. Leek, C. E. Lewis, R. Whitehouse, M. Greenall, J. Clarke, A. L. Harris. Association of macrophage infiltration with angiogenesis and prognosis in invasive breast carcinoma. *Cancer Res. 56*:4625–4629, 1996.
84. T. Saga, R. D. Neumann, T. Heya, J. Sato, S. Kinuya, N. Le, C. H. Paik, J. N. Weinstein. Targeting cancer micrometastases with monoclonal antibodies: a binding-site barrier. *Proc. Natl. Acad. Sci. USA 92*:8999–9003, 1995.
85. I. Ahmad, M. Longenecker, J. Samuel, T. M. Allen. Antibody-targeted delivery of doxorubicin entrapped in sterically stabilized liposomes can eradicate lung cancer in mice. *Cancer Res. 53*:1484–1488, 1993.
86. D. Kirpotin, J. W. Park, K. Hong, S. Zalipsky, W. L. Li, P. Carter, C. C. Benz, D. Paphadjopoulos. Sterically stabilized anti-HER2 immunoliposomes: design and targeting to human breast cancer cells in vitro. *Biochemistry 36*:66–75, 1997.
87. J. W. Park, K. Hong, D. B. Kirpotin, O. Meyer, D. Papahadjopoulos, C. Benz. Anti-HER2 immunoliposomes for targeted therapy of human tumors. *Cancer Lett. 118*:153–160, 1997.
88. J. Connor, L. Huang. pH-sensitive immunoliposomes as an efficient and target-specific carrier for antitumor drugs. *Cancer Res. 46*:3431–3435, 1986.
89. H. Mizuguchi, M. Nakanishi, T. Nakanishi, T. Nakagawa, S. Nakagawa, T. Mayumi. Application of fusogenic liposomes containing fragment A of diphtheria toxin to cancer therapy. *Br. J. Cancer 73*:472–476, 1996.
90. A. Archer, K. Miller, E. Reich, R. Hautmann. Photodynamic therapy of human bladder carcinoma cells in vitro with pH sensitive liposomes as carriers for 9-acetoxytetrapropylpophyrene. *Urological Res. 22*:25–32, 1994.
91. J. W. Holland, C. Hui, P. R. Cullis, T. D. Madden. Poly(ethylene glycol)-lipid conjugates regulate the calcium-induced fusion of liposomes composed of phosphatidylethanolamin and phosphatidylserine. *Biochemistry 35*: 2618–2624, 1996.
92. D. Kirpotin, K. Hong, N. Mullah, D. Papahadjopoulos, S. Zalipsky. Liposomes with detachable polymer coating: destabilization and fusion of dioleoylphosphatidylethanolamine vesicles triggered by cleavage of surface-grafyted poly(ethylene glycol). *FEBS Lett. 388*:115–118, 1996.
93. G. Adlakha-Hutcheon, M. B. Bally, C. R. Shew, T. D. Madden. Controlled destabilization of a liposomal drug delivery system enhances mitoxantrone antitumor activity. *Nature Biotech. 17*:775–779, 1999.
94. S. Unezaki, K. Maruyama, N. Takahashi, M. Koyama, T. Yuda, A. Suginaka, M. Iwatsuru. Enhanced delivery and antitumor activity of doxorubicin using long-circulating thermosensitive liposomes containing amphiphathic polyethylene glycol in combination with local hyperthermia. *Pharmaceutical Res. 11*:1180–1185, 1994.
95. D. Needham, G. Anyarambhatla, G. Kong, M. W. De-

whirst. A new temperature sensitive liposome for use with mild hyperthermic characterization and testing in a human tumor xenograft model. *Cancer Res*. in press, 2000.

96. L. D. Mayer, D. Masin, R. Nayar, N. L. Boman, M. B. Bally. Pharmacology of liposomal vincristine in mice bearing L1210 ascitic and B16/BL6 solid tumours. *Br. J. Cancer 71*:482–488, 1995.
97. T. M. Allen, M. S. Newman, M. C. Woodle, E. Mayhew, P. S. Uster. Pharmacokinetics and anti-tumor activity of vincristine encapsulated in sterically stabilized liposomes. *Int. J. Cancer 62*:199–204, 1995.
98. G. Bradley, V. Ling. P-glycoprotein, multidrug resistance and tumor progression. *Cancer Metastasis Rev. 13*:223, 1994.
99. S. K. Huang, E. Mayhew, S. Gilani, D. D. Lasic, F. J. Martin, D. Papahadjopoulos. Pharmacokinetics and therapeutics of sterically stabilized liposomes in mice bearing C-26 colon carcinoma. *Cancer Res. 52*:6744–6750, 1992.
100. E. Mayhew, D. D. Lasic, S. Babbar, F. J. Martin. Pharmacokinetics and antitumor activity of epirubicin encapsulated in long-circulating liposomes incorporating a polyethylene glycol-derivatized phospholipid. *Int. J. Cancer 51*:302–307, 1992.
101. Y. P. Hu, N. Henry-Toulme, J. Robert. Failure of liposome encapsulation of doxorubicin to circumvent multidrug resistance in an in vitro model of rat glioblastoma cells. *Eur. J. Cancer 31A*:389–393, 1995.
102. R. Krishna, L. D. Mayer. Liposomal doxorubicin circumvents PSC 833-free drug interactions, resulting in effective therapy of multidrug resistant solid tumors. *Cancer Res. 57*:5246, 1997.
103. J. Marjan, G. Charrios, D. Lopes de Menezes, T. M. Allen. Antibody-mediated targeting of liposomal doxorubicin to lymphoblastic cells can reverse multidrug resistance. *Proc. AACR 37*:A2103, 1996.
104. S. Suzuki, K. Inoue, A. Hongoh, Y. Hashimoto, Y. Yamazoe. Modulation of doxorubicin resistance in a doxorubicin-resistant human leukemia cell by an immunoliposome targeting transferring receptor. *Br. J. Cancer 76*: 83–88, 1997.
105. D. Lopes de Menezes, L. M. Pilarski, T. M. Allen. In vitro and in vivo targeting of immunoliposomal doxorubicin to human B-cell lymphoma. *Cancer Res. 58*:3320–3330, 1998.
106. C. A. Seid, I. J. Fidler, R. K. Clyne, L. E. Earnest, D. Fan. Overcoming murine tumor cell resistance to vinblastine by presentation of the drug in multiamellar liposomes consisting of phosphatidylcholine and phosphatidylserine. *Selective Cancer Ther. 7*:103–112, 1991.
107. D. Fan, P. J. Beltran, C. A. O'Brien. Reversal of multidrug resistance. In: *Reversal of Multidrug Resistance in Cancer* (J. A. Kellen, ed.). CRC Press, Boca Raton, FLS 1994, pp. 93–124.
108. A. Rahman, S. R. Husain, J. Siddiqui, M. Verma, M. Agresti, M. Center, A. R. Safa, R. I. Glazer. Liposome-mediated modulation of multidrug resistance in human HL-60 leukemia cells. *J. Natl. Cancer Inst. 84*:1909–1913, 1992.
109. A. R. Thierry, A. Dritschilo, A. Rahman. Effect of liposomes on P-glycoprotein function in multidrug resistant cells. *Biochem. Biophys. Res. Comm. 187*:1098–1105, 1992.
110. A. R. Thierry, A. Rahman, A. Dritschilo. A new procedure for the preparation of liposomal doxorubicin: biological activity in multidrug-resistant tumor cells. *Cancer Chemother. Pharmacol. 35*:84–93, 1994.
111. A. R. Thierry, A. Rahman, A. Dritschilo. Overcoming multidrug resistance in human tumor cells using free and liposomally encapsulated antisense oligonucleotides. *Biochem. Biophys. Res. Comm. 190*:952–957, 1993.
112. M. Kiehntopf, M. A. Brach, T. Licht, S. Petschauer, L. Karawajew, C. Krischning, F. Herrmann. Ribozyme-mediated cleavage of the MDR-1 transcript restores chemosensitivity in previously resistant cancer cells. *EMBO J. 13*:4645–4651, 1994.
113. S. K. Alahari, N. M. Dean, M. H. Fisher, R. Delong, M. Manoharan, K. L. Tivel, R. L. Juliano. Inhibition of expression of the multidrug resistance associated P-glycoprotein by phosphorothioate and 5′ cholesterol-conjugated phosphorothioate antisense oligonucleotides. *Mol. Pharmacol. 50*:808–815, 1996.
114. B. Lum, M. Gosland. MDR expression in normal tissues: pharmacologic implications for the clinical use of P-glycoprotein inhibitors. *Hematol. Oncol. Clin. North Am.* 9:319, 1995.
115. T. Colombo, O. G. Paz, M. D'Incalci. Distribution and activity of doxorubicin combined with SDZ PSC 833 in mice with P388 and P388/DOX leukemia. *Br. J. Cancer 73*:866–873, 1996.
116. R. Krishna, N. McIntosh, K. W. Riggs, L. D. Mayer. Doxorubicin encapsulated in sterically stabilized liposomes exhibits renal and biliary clearance properties that are independent of Valspodar (PSC 833) under conditions that significantly inhibit nonencapsulated drug excretion. *Clin. Cancer Res. 5*:2939–2947.
117. R. Krishna, M. St. Louis, L. D. Mayer. Increased intracellular drug accumulation and complete chemosensitization achieved in multidrug-resistant solid tumors by co-administering Valspodar (PSC 833) with sterically stabilized liposomal doxorubicin. *Int. J. Cancer 85*:131–141, 2000.
118. M. C. Woodle, L. Leserman. Liposomal antisense oligonucleotide therapeutics. In: *Medical Applications of Liposomes* (D. Lasic, D. Papahadjopoulos, eds.). Elsevier, New York, 1998, pp. 429–449.
119. M. C. Woodle, F. I. Raynaus, M. Dizik, O. Meyer, S. K. Huang, J. A. Jaeger, B. D. Brown, R. Orr, I. R. Judson, D. Paphadjopoulos. Oligonucleotide pharmacology and formulation: G3139 anti-Bcl-2 phosphorothioate in Stealth liposomes and gel implants. *Nucleosides Nucleotides 16*:1731–1734, 1997.

31

Systemic Cancer Therapy Using Polymer-Based Prodrugs and Progenes

Leonard W. Seymour
University of Birmingham, Birmingham, England

I. SUITABILITY OF POLYMERS FOR SYSTEMIC ADMINISTRATION

Many soluble polymers have been proposed for use in therapeutic systems for treatment of a range of diseases (1). Several have undergone exhaustive examination in animals, and a few have made it through to clinical evaluation. The most extensively developed of these is poly(ethylene glycol) (pEG), which is now being used clinically in treatment of several disorders using polymer–protein conjugates and also forms an important excipient in the formulation of many licensed drugs (2–8). There are no significant toxicities associated with pEG, although doses administered parenterally have usually been fairly small. Poly(vinylpyrrolidone) (pVP) has been evaluated for clinical use as a plasma expander and was administered to several hundreds of soldiers in high doses during the Second World War. pVP does appear to be associated with a storage disease termed ''la maladie vinylique,'' but the frequency of this is remarkably low and, considering the doses administered and the delay to onset, the overall toxicity profile is very mild (9–12). Copolymers based on poly [N-(2-hydroxypropyl)methacrylamide] (pHPMA) have also seen clinical application in recent years (see below) and there have been no reported instances of any polymer-related toxicities in over 50 cancer patients treated.

One factor common to many well-tolerated polymers in therapeutic use is their lack of biodegradability in the polymer backbone. Frequently the processes of biodegradation and immunogenicity seem to go hand in hand, probably reflecting either similar principles of macromolecular recognition involved in both processes or the requirement for intracellular degradation to permit cellular presentation of immunogenic epitopes (13). Although backbone-biodegradable polymers may appear to have advantages in terms of good distribution kinetics coupled with efficient renal excretion, they have not yet undergone thorough clinical evaluation. Poor recognition of polymers by enzymes and antibodies is likely to be an important factor contributing to the extended plasma circulation exhibited by large nondegradable polymers (such as pHPMA), and this plays a significant role in the successful application as carriers of drugs, as described in more detail below.

II. PLASMA CIRCULATION AND EXTRAVASATION OF POLYMERS AND OTHER MACROMOLECULES

Macromolecules showing significant binding to cells, membranes, or plasma proteins are generally cleared quickly from the circulation, often by phagocytes in the spleen or liver or by trapping in fine capillary beds. Conversely, macromolecules not binding biological components can show extended circulation in the bloodstream, being cleared in the fluid phase at rates and locations dictated largely by their physical properties (14,15). One significant clearance mechanism is renal excretion, where passage through the kidney glomerulus is restricted to materials of molecular mass of 50–60 kDa for proteins and 30–35 kDa for dextran and pHPMA (16). The reason for this difference in renal thresholds reflects the more extended structure of these hydrophilic polymers, compared with proteins which tend to form intramolecular bonds and adopt a more globular structure.

Extravasation from the bloodstream can also provide a significant pathway for clearance of macromolecules, with extravasated materials either becoming lodged within extracellular matrix or being recycled to the bloodstream via the lymphatics. The permeability of the blood vessels varies significantly between organs and tissues, although generally the wall of the arterioles are relatively impermeable, with capillaries generally thought to be more permissive to extravasation. Postcapillary venous endothelium shows the highest permeability within an individual tissue, although the low postcapillary blood pressure probably does not promote extravasation; in contrast, these are regions where osmotic force is collecting tissue fluid from the interstitium back into the circulation, and hence the net flow is usually away from the tissues into the blood.

Capillaries are generally described as tight, fenestrated, or sinusoidal. The tightest capillaries in the body are found in the brain, supported on a continuous basement membrane and with very low levels of fluid-phase extravasation. Fenestrated capillaries are found in many places, including the kidney and intestine, where limited extravasation is important. Generally speaking, however, size restrictions prevent extravasation of most proteins. This limited permeability is reflected in the protein content of interstitial fluid, which is about 10% that of whole serum (and underlies the 10% serum conditions employed for culture of most mammalian nonblood cells, where 100% serum is often toxic) and is also important to prevent uncontrolled loss of proteins in the urine. Sinusoidal endothelium is found in the bone marrow, liver, and spleen. In these organs the basement membrane is either highly discontinuous or absent, and the vessels are freely permeable. This complements the phagocytic activity of cells contained within these organs, providing access of blood-borne macromolecules and particles to cells designed to sequester them (17).

III. INFLUENCE OF PHYSICOCHEMICAL PROPERTIES ON PHARMACOKINETICS OF MACROMOLECULES

The pharmacokinetic of systemically administered macromolecules have undergone extensive study in animals in order to determine the most promising candidates for use as drug carriers (18,19). Two key factors emerge as determinants of distribution kinetics, molecular weight and molecular charge. Macromolecules smaller than the renal threshold are rapidly cleared from the circulation into the urine. Initially this would appear to indicate in favor of larger molecules, but when the macromolecule is nonbiodegradable, eventual excretion is useful to decrease unwanted toxicities. Hence some agents (such as the doxorubicin conjugates described below) are smaller than the renal threshold, and are therefore applied at high doses to achieve the required systemic exposure.

Macromolecules with significant positive charge show electrostatic interaction with cells, membranes, and connective tissues in vivo, undergoing unwanted rapid elimination from the bloodstream. Conversely, macromolecules with strong negative charge act as substrates for the polyanion scavenger receptors of Kupffer cells and other phagocytes and are also cleared quickly from the circulation. Optimum circulation is obtained using macromolecules with a hydrophilic structure, noncharged or slightly negatively charged under physiological conditions. Hence for targeted systemic delivery of therapeutic agents, neutral or slightly negative charged carriers are normally employed.

IV. TUMOR VASCULAR PERMEABILITY AND THE EPR EFFECT

Tumors are usually thought to arise from uncontrolled proliferation of a single malignant cell, produced by the acquisition of several genetic mutations. Some of the mutations are inherited, and others are caused by environmental factors during the lifetime of the individual. Consequently, the growing tumor is a relatively simple structure, composed of a small mass of proliferating cells. However, such a structure is unable to grow beyond a size of 2 mm approximately, due to the difficulty of obtaining a sufficient supply of oxygen and nutrition that must diffuse an increasing distance from the nearest blood vessel (20–23). Hence successful tumors must mutate to become angiogenic, in other words they must attract capillaries to provide them with blood and nutrients. In order to do this tumors produce various angiogenic factors, secreted from tumor cells and mediating mitogenic and chemoattractant effects through specific receptors expressed on the surface of endothelial cells (21–26). In this way capillary cells can be recruited and induced to differentiate to form capillary tubular sprouts and eventually fully functional capillary structures. This enables the tumor to obtain better nutrition and oxygenation, and hence a larger tumor mass can be viable.

In order to grow, tumors recruit macromolecules from the bloodstream and incorporate them into the extended tumor matrix by clotting them. This matrix extracelluar stroma provides a framework for the local migration and growth of new tumor cells and facilitates enlargement of the primary tumour and its invasion of nearby tissues. The tumor usually recruits its extracelluar matrix from the bloodstream, but the efficiency of extravasation of proteins such as albumin and fibrinogen is usually very low in normal vasculature. Tumors therefore secrete permeability-enhancing factors, designed to act on endothelial cells and increase the rate of extravasation in tumor vasculature. The best characterised tumor-secreted vascular permeabilizing factor is vascular endothelial growth factor (VEGF), originally also termed vascular permeability factor (VPF). The mechanism of action is not yet completely understood, although the molecule is known to bind specifically to two types of receptors, knows as flt-1 and KDR (VEGF-receptors 1 and 2), on the endothelial cell, exerting signaling cascades and mediating several biological responses in vitro. In vivo, VEGF is known to be a very potent inducer of extravasation, although whether fluid extravasates paracellularly (between the endothelial cells) or by transcytosis (by transcellular vesicular pathways, such as the recently reported vesiculovacuolar organelles (VVOs), is only presently being elucidated (27,28).

Another property of tumors that significantly affects their accessibility to macromolecules is their inadequate lymphatic drainage. The reason for this is unknown, but presumably it involves either a lack of lymphangiogenic agents produced by tumors, or possibly overgrowth of lymphatic vessels by the proliferating tumor cells (29–31). In either case there is deficiency of fluid drainage from many tumors, notably carcinomas, that leads to an interstitial blood pressure that is significantly elevated in tumors compared with the corresponding normal tissues. This in turn may contribute to collapse of capillaries, resulting in poor blood supply in the center of solid tumors, giving rise to hypoxic and necrotic regions. Production of permeability-enhancing VEGF is known to be induced by hypoxia, and hence the residual blood vessels serving the tumor tend to have regions of high vascular permeability (27,28,32).

This combination of enhanced vascular permeability and poor lymphatic drainage has been termed the enhanced permeability and retention (EPR) effect (19,33–36). One result of the EPR effect is that tumors capture substantial levels of circulating macromolecules, which extravasate and then are unable to rejoin the blood circulation in the normal way (35,37). Clearly one role for this capture is to provide the tumor with macromolecules for clotting and formation of new substratum. However, if the macromolecule is actually an anticancer drug, this provides a simple yet effective means of concentrating the therapeutic agent within the tumor and can mediate significant benefit (19).

V. PRINCIPLES OF TARGETING MACROMOLECULAR DRUGS

Most anticancer agents are low molecular weight drugs, able to enter cells readily either by simple diffusion or by receptor-mediated capture. Hence they tend to have a relatively high volume of distribution and short circulatory half life, mediating significant systemic toxicities by exerting their effects against any vulnerable cells they encounter. Helmut Ringsdorf proposed in 1975 that linkage of such drugs to a targeted macromolecular carrier, such as a synthetic polymer, could be used to deliver the agents selectively to their targets, simultaneously increasing their activity and decreasing unwanted side effects (38). In his original proposal Ringsdorf envisaged incorporation of drugs, targeting agents, and also solubilizing agents covalently onto the polymer carrier. In practice, incorporation of specific solubilizing agents has proved superfluous, and attempts to target polymers with agents such as antibodies is still under active development. The most advanced exemplification of the Ringsdorf approach is a system where copolymers based on poly[*N*-(2-hydroxypropyl)methacrylamide], derivatized with tetrapeptide spacers (Gly-Phe-Leu-Gly) (36,39–41), have been modified with the anticancer agent doxorubicin and the simple targeting agent galactosamine (42,43). These polymer conjugates, which are effectively macromolecular prodrugs, since they require activation by enzymic cleavage of the Gly-Phe-Leu-Gly spacer, show good anticancer activity in animals (40), and their pharmacokinetics and anticancer activity have recently been evaluated in humans (see below).

VI. SIGNIFICANCE OF THE DRUG–POLYMER LINKAGE

The linkage of drug to polymer is key to regulating the biological activity of the drug and to providing a trigger for activation of the macromolecular prodrug. Three main types of linkage are normally used, namely those susceptible to hydrolysis at neutral pH (such as ester linkages), those subject to acid-catalyzed cleavage (e.g., based on cis-aconityl groups) or those designed for cleavage by target-associated enzymes (such as the Gly-Phe-Leu-Gly linkage mentioned above, which is designed for cleavage by lysosomal cathepsins). Appropriate choice of the optimum linkage, with respect to its required site of activation and the speed of activation achieved, is fundamental to successful design of a macromolecular prodrug. The reader is referred to an excellent recent review of this subject (44).

VII. CLINICAL PHARMACOKINETICS OF POLYMER CONJUGATES

Clinical studies in the UK have recently been reported evaluating the use of doxorubicin–pHPMA conjugates, either targeted with galactosamine or without, for treatment of cancer (45,46). The incorporation of galactosamine mediates substantial hepatic targeting of the polymer conjugate, determined using an analogue bearing ^{123}I (45,47). At clinical doses (approximately 1 g polymer conjugate/m^2), hepatic targeting of 15–18% of the dose could be achieved, measured by gamma camera and single positron emission tomographic (SPECT) imaging. In contrast, nontargeted doxorubicin–pHPMA conjugate showed no measurable liver accumulation. Both agents showed rapid elimination from the bloodstream, with biphasic clearance kinetic and alpha half-life less than 30 min. Therapeutic activity of the agents has also been assessed and will shortly be disclosed.

VIII. USE OF POLYMERS FOR DELIVERY OF THERAPEUTIC PROTEINS

Targeting of polymer conjugates with antibodies is fraught with difficulties, as the polymer tends to prevent access of the antibody to its cell surface target, and modifications must be performed very carefully (48,49). However, this shielding effect can be turned to major advantage when the target is a low molecular weight agent, rather than a cell surface receptor. For example, conjugates of synthetic polymers with enzymes can stabilize the enzyme, prolonging blood circulation and preventing recognition by cells, without seriously impairing access of low molecular substrates to the active site of the enzyme. The most advanced example of this approach is the use of the conjugate pEG–asparaginase (3–5). Asparaginase has been evaluated and used for many years in the treatment of asparagine-dependent leukaemia, though asparaginase itself is normally immunogenic and displays a relatively short half-life, with complications of frequent dosing. pEG–asparaginase, in contrast, shows a prolonged half-life and can be administered repeatedly when required without significant toxicological complications. Given the success of pEG–asparaginase, other pEGylated enzymes are presently under development (8,50).

IX. GENE THERAPY: THE PROMISE AND THE HYPE

The concept of gene therapy for treatment of several diseases has attracted considerable attention in recent years, although the considerable promise of this approach is continually frustrated by inadequate systems for gene delivery (51–55). Systemic delivery of genes is a particular problem and in all of the clinical trials performed to data the gene vector has been administered locally, either by direct injection into the target tissue, into the artery supplying it, or into the body cavity housing it. Many vectors for gene therapy are based on viruses, usually attenuated to make them nonreplicative in order to improve their safety profile. These agents are often highly efficient following arrival at the target cell, but they usually have no intrinsic tropism for the diseased cell and tend to deliver their genes nonselectively. In addition, some virus vectors are attracting concerns over safety, and they are often neutralized by antibodies present in the bloodstream of patients. Consequently, in recent years, attention has largely shifted to the development of fully defined synthetic vectors.

X. SYNTHETIC VECTORS FOR SYSTEMIC GENE DELIVERY

There are two main types of synthetic vector for gene delivery—those based essentially on cationic lipids (51,52), and those based on cationic polymers (56). The former have the advantage of good ability to disrupt membranes and mediate relatively efficient delivery of DNA, while the latter are more selective in transfection and are more suitable for intravenous administration. Cationic lipid/DNA complexes are usually sequestered by the first capillary bed they encounter, normally the pulmonary bed following intravenous administration (57), and this severely hampers their ability to reach dispersed systemic cellular targets. Polyelectrolyte DNA complexes, on the other hand, show less tendency to occlude capillary beds following intravenous injection, often passing through the lungs into the oxygenated blood and then dispersing throughout the body in the arterial circulation. Factors affecting biodistribution will be assessed for both types of vector in turn.

A. Systemic Distribution of Cationic Lipid/DNA Complexes

When DNA binds to cationic liposomes, the structures reorientate and the DNA is thought to become coated in a double membrane of cationic lipid, with the outside surface presenting a combination of cationic charge and hydrophobic domains (58). This is thought to mediate the high membrane activity of the complexes, although it also mediates interaction of lipid complexes with proteins in the plasma, leading to the formation of large aggregates. This is likely to constitute a major contributory factor in the pulmonary first pass clearance, although the precise details of the solvent employed and the volume administered can significantly affect biodistribution, probably by regulating the speed at which the complexes encounter serum proteins, and to some extent affecting the solvent in which the mixing takes place (15,51,59). The organ immediately preceding and also following the pulmonary capillary bed is the heart, and cationic lipid formulations with significant membrane activity can often achieve measurable deposition and even transgene expression within the cardiac endothelium as well as the pulmonary endothelium.

B. Systemic Distribution of Cationic Polymer/DNA Complexes

Most simple cationic polymer/DNA complexes are cleared rapidly from the circulation following intravenous administration, usually becoming sequestered by cells in the liver (60). The precise reasons for this are presently under intensive investigation, although they are usually assumed to include binding of serum proteins to the complexes leading to their destabilization and rapid phagocytic scavenging by Kupffer cells (61). There is evidence that simple polyelectrolyte complexes that remain in the circulation are actually bound to the surface of erythrocytes (Ward and Seymour, unpublished observations), and this may represent one strategy for gaining systemic circulation of such gene delivery vectors; however, passage of complexes through the endothelial layer to gain access to extravascular target cells would present a significant problem.

XI. STERIC STABILIZATION OF POLYELECTROLYTE COMPLEXES

The concept of introducing a surface layer of hydrophilic polymers onto the surface of polyelectrolyte complexes in order to induce a level of stabilization, preventing their interaction with cells and proteins, has been under development for several years. Oriented self-assembly of DNA with cationic-hydrophilic block copolymers can lead to the formation of particles with a shielded surface charge (62,63), although improved pharmacokinetics of these materials have not been reported. Subsequently Ogris et al. (1999) were the first to examine the biodistribution of sterically stabilized particles following intravenous injection (64). These authors worked with pEGylated pEI/DNA complexes including also transferrin; compared with complexes lacking the pEG layer, pEGylated materials showed extended plasma circulation leading to detectable accumulation and gene expression in subcutaneous B16F10 melanoma tumors (64,65). This achievement paves the way for gaining extended circulation of sterically stabilized polyelectrolyte gene vectors that are further modified to carry target-specific ligands in order to mediate their association and uptake by preselected target cells and tissues (66). Probably the greatest challenge in this field is to design a sterically stabilized vector that is capable of extended plasma circulation, but which incorporates a powerful trigger mechanism capable of activating strong transfection activity following arrival at the vector at the desired target site. Such systems, analogous to the design of macromolecular prodrugs, could be regarded as inactive chemical precursors of transcriptionally active DNA, or ''progenes.'' Suitable triggers might involve selective processing by target-associated enzymes, or they could involve the falling endosomal pH after endocytic internalization of complexes into cells. Design of effective triggers in many ways reflects the development of effective drug–polymer spacers, described above, and it may be that the most effective systems are based ultimately on the same principles.

XII. SURFACE MODIFICATION AND STERIC STABILIZATION OF ADENOVIRUSES

Recent studies have shown that viruses, specifically adenoviruses, can be surface modified with hydrophilic polymers in order to decrease binding of antiadenovirus antibodies that often restrict clinical usefulness. Judicious application of the polymer can either maintain binding to the normal adenovirus receptor (CAR) or ablate it (67,68). Preventing CAR-mediated infection of cells essentially prevents normal infection, and the introduction of new targeting ligands onto the polymer-coated virus can then promote its infection of chosen cells by target-associated receptor-mediated endocytosis. These polymer-coated virus vectors offer considerable steps forward for gene therapy, combining the great infective power of adenoviruses with the versatile targeting and low antibody binding activity of nonviral vector systems.

XIII. CONCLUSIONS

It is enlightening that soluble polymers are finding useful applications through the fields of systemic therapies. At one level they can enhance solubility of hydrophobic drugs and change their pharmacokinetics, simultaneously endowing them with the possibility of targeting to desired sites of action. With larger molecules, such as proteins, they are capable of steric stabilization, preventing interaction with proteases and extending plasma circulation. In the context of polyelectrolyte complexes for gene delivery, hydrophillic polymers are capable of steric stabilization but also of cross-linking the surfaces of complexes, preventing their desegregation by components of the plasma. Finally, in new fields that are just being probed, hydrophillic polymers are clearly capable of enshrouding biological agents such as viruses, preventing antibody recognition and permitting the introduction of new targeting tropisms. Each of these approaches holds important clinical advances, often in different and sometimes complementary therapeutic fields, and each is presently under active development in widespread parts of the global research community.

ACKNOWLEDGMENTS

I am grateful to my colleagues for inspiration and insight, and to the Cancer Research Campaign for financial support.

REFERENCES

1. J. H. Guo, G. W. Skinner, W. W. Harcum, P. E. Barnum. Pharmaceutical applications of naturally occurring water-soluble polymers. *Pharmaceutical Science and Technology Today 1*:254 (1998).
2. O. Rosen, W. Langer, N. Gokbuget, R. Muck, N. Peter, D. Reichert, F. Rothmann, S. Schwartz, R. Arnold, J. Muller, J. Boos, D. Hoelzer. PEG-asparaginase in combination with high-dose methotrexate (HD-MTX): pharmacokinetics and toxicity in adult patients with all. *Blood 94*:4263 (1999).
3. A. Aguayo, J. Cortes, D. Thomas, S. Pierce, M. Keating, H. Kantarjian. Combination therapy with methotrexate, vincristine, polyethylene-glycol conjugated-asparaginase, and prednisone in the treatment of patients with refractory or recurrent acute lymphoblastic leukemia. *Cancer 86*:1203 (1999).
4. P. Bailon, W. Berthold. Polyethylene glycol-conjugated pharmaceutical proteins. *Pharmaceutical Science and Technology Today 1*:352 (1998).
5. J. Boos, H. J. Muller, L. Loning, A. Horn, D. Schwabe, M. Gunkel, V. von Schutz, G. Henze, J. C. daPalma, J. Ritter, H. Jurgens, M. Schrappe. PEG-asparaginase (ONCASPAR™) drug monitoring in children (ALL-BFM). *Blood 94*:1291 (1999).
6. M. L. Graham, B. L. Asselin, J. E. Herndon, J. R. Casey, S. Chaffee, G. H. Ciocci, C. W. Daeschner, A. R. Davis, S. Gold, E. C. Halperin, M. J. Laughlin, P. L. Martin, J. F. Olson, J. Kurtzberg. Toxicity, pharmacology and feasibility of administration of PEG-L-asparaginase as consolidation therapy in patients undergoing bone marrow transplantation for acute lymphoblastic leukemia. *Bone Marrow Transplantation 21*:879 (1998).
7. Y. Levy, C. Capitant, S. Houhou, I. Carriere, J. P. Viard, C. Goujard, J. A. Gastaut, E. Oksenhendler, L. Boumsell, E. Gomard, C. Rabian, L. Weiss, J. G. Guillet, J. F. Delfraissy, J. P. Aboulker, M. Seligmann. Comparison of subcutaneous and intravenous interleukin-2 in asymptomatic HIV-1 infection: a randomised controlled trial. *Lancet 353*: 1923 (1999).
8. Y. Tan, X. Sun, M. Xu, Z. An, X. Tan, Q. Han, D. A. Miljkovic, M. Yang, R. M. Hoffman. Polyethyleneglycol conjugation of recombinant methioninase for tumor-selective cancer therapy. *Annals of Oncology 9*:307 (1998).
9. F. Cabanne, R. Michiels, P. Dusserre, H. Bastien, E. Justrabo. La maladie polyvinylique. *Annales d'Anatomie pathologique 14*:419 (1969).
10. A. J. Balaton, P. Vaury, A. Bonan. Persistent perineal sinus following an abdominoperineal amputation: an unusual form of polyvinylpyrrolidone storage disease. *Pathology 20*:83 (1988).
11. E. Reske-Nielsen, M. Bosjen-Moller, M. Vetner, J. C. Hansen. Polyvinylpyrrolidone storage disease. *Acta Path. Microbiol. Scand. A 84*:397 (1976).
12. H. A. Ravin, A. M. Seligman, J. Fine. Polyvinylpyrrolidone as a plasma expander. *New England Journal of Medicine 247*:921 (1952).
13. B. Rihova. Biocompatibility of biomaterials: hemocompatibility, immunocompatibility and biocompatibility of solid polymeric materials and soluble targetable polymeric carriers. *Advanced Drug Delivery Reviews 21*:157 (1996).
14. M. Hashida, R. I. Mahato, K. Kawabata, T. Miyao, M. Nishikawa, Y. Takakura. Pharmacokinetics and targeted delivery of proteins and genes. *Journal of Controlled Release 41*:91 (1996).
15. O. Meyer, D. Kirpotin, K. L. Hong, B. Sternberg, J. W. Park, M. C. Woodle, D. Papahadjopoulos. Cationic liposomes coated with polyethylene glycol as carriers for oligonucleotides. *Journal of Biological Chemistry 273*:15621 (1998).
16. L. W. Seymour, R. Duncan, J. Strohalm, J. Kopecek. Effect of molecular-weight (mbarw) of n-(2-hydroxypropyl)methacrylamide copolymers on body distribution and rate of excretion after subcutaneous, intraperitoneal, and intravenous administration to rats. *Journal of Biomedical Materials Research 21*:1341 (1987).
17. V. P. Torchilin, V. S. Trubetskoy. Amphiphilic polyethyleneglycol derivatives: long-circulating micellar carriers for therapeutic and diagnostic agents. *Abstracts of Papers of the American Chemical Society 213*:59 (1997).
18. D. C. Drummond, O. Meyer, K. L. Hong, D. B. Kirpotin, D. Papahadjopoulos. Optimizing liposomes for delivery of chemotherapeutic agents to solid tumors. *Pharmacological Reviews 51*:691 (1999).
19. L. W. Seymour. Passive tumor targeting of soluble macromolecules and drug conjugates. *Critical Reviews in Therapeutic Drug Carrier Systems 9*:135 (1992).
20. M. W. Dewhirst, H. Kimura, S. W. E. Rehmus, R. D. Braun, D. Papahadjopoulos, K. Hong, T. W. Secomb. Microvascular studies on the origins of perfusion-limited hypoxia. *British Journal of Cancer 74*:S247 (1996).
21. R. K. Jain. Vascular and interstitial barriers to delivery of therapeutic agents in tumors. *Cancer and Metastasis Reviews 9*:253 (1990).

22. R. K. Jain. Barriers to drug-delivery in solid tumors. *Scientific American 271*:58 (1994).
23. R. K. Jain. 1995 Whitaker lecture: delivery of molecules, particles, and cells to solid tumors. *Annals of Biomedical Engineering 24*:457 (1996).
24. R. Folberg, M. J. C. Hendrix, A. J. Maniotis. Vasculogenic mimicry and tumor angiogenesis. *American Journal of Pathology 156*:361 (2000).
25. Z. C. Han, Y. Liu. Angiogenesis: state of the art. *International Journal of Hematology 70*:68 (1999).
26. R. K. Jain, Delivery of molecular and cellular medicine to solid tumors. *Microcirculation—London 4*:3 (1997).
27. D. Feng, J. A. Nagy, K. Pyne, I. Hammel, H. F. Dvorak, A. M. Dvorak. Pathways of macromolecular extravasation across microvascular endothelium in response to VPF VEGF and other vasoactive mediators. *Microcirculation 6*: 23 (1999).
28. D. Feng, J. A. Nagy, A. M. Dvorak, H. F. Dvorak. Different pathways of macromolecule extravasation from hyperpermeable tumor vessels. *Microvascular Research 59*:24 (2000).
29. H. C. Lichtenbeld, L. V. Leak, L. L. Munn, R. K. Jain. Lymphangiogenesis in tumors. *Faseb Journal 13*:A6 (1999).
30. Y. Ohta, V. Shridhar, R. K. Bright, G. P. Kalemkerian, W. Du, M. Carbone, Y. Watanabe, H. I. Pass. VEGF and VEGF type C play an important role in angiogenesis and lymphangiogenesis in human malignant mesothelioma tumours. *British Journal of Cancer 81*:54 (1999).
31. B. Olofsson, M. Jeltsch, U. Eriksson, K. Alitalo. Current biology of VEGF-B and VEGF-C. *Current Opinion in Biotechnology 10*:528 (1999).
32. H. F. Dvorak. VPF/VEGF and the angiogenic response. *Seminars in Perinatology 24*:75 (2000).
33. H. Maeda, Y. Matsumura. Tumoritropic and lymphotropic principles of macromolecular drugs. *Critical Reviews in Therapeutic Drug Carrier Systems 6*:193 (1989).
34. H. Maeda, L. W. Seymour, Y. Miyamoto. Conjugates of anticancer agents and polymers-Advantages of macromolecular therapeutics in vivo. *Bioconjugate Chemistry 3*:351 (1992).
35. Y. Matsumura, H. Maeda. A new concept for macromolecular therapeutics in cancer-chemotherapy-Mechanism of tumoritropic accumulation of proteins and the antitumor agent SMANCS. *Cancer Research 46*:6387 (1986).
36. L. W. Seymour, K. Ulbrich, P. S. Steyger, M. Brereton, V. Subr, J. Strohalm, R. Duncan. Tumor tropism and anticancer efficacy of polymer-based doxorubicin prodrugs in the treatment of subcutaneous murine B16F10 melanoma. *British Journal of Cancer 70*:636 (1994).
37. Y. Takakura, R. I. Mahato, M. Hashida. Extravasation of macromolecules. *Advanced Drug Delivery Reviews 34*:93 (1998).
38. H. Ringsdorf. Structure and properties of pharmacologically active polymers. *J. Polym. Sci. Polym. Symp. 51*:135 (1975).
39. R. Duncan, P. Kopeckovarejmanova, J. Strohalm, I. Hume, H. C. Cable, J. Pohl, J. B. Lloyd, J. Kopecek. Anticancer agents coupled to N-(2-hydroxypropyl)methacrylamide copolymers. 1. Evaluation of daunomycin and puromycin conjugates in vitro. *British Journal of Cancer 55*:165 (1987).
40. R. Duncan, L. W. Seymour, P. A. Flanagan, K. B. O'Hare, S. R. Wedge, K. Ulbrich, J. Strohalm, V. Subr, F. Spreadfico, M. Grandi, M. Ripamonti, M. Farao, A. Suarato. Preclinical evaluation of polymer-bound doxorubicin. *J. Controlled Release 19*:331 (1992).
41. J. Kopecek, P. Rejmanova, R. Duncan, J. B. Lloyd. Controlled release of drug model from N-(2-hydroxypropyl)-methacrylamide copolymers. *Annals of the New York Academy of Sciences 446*:93 (1985).
42. K. B. O'Hare, I. C. Hume, L. Scarlett, V. Chytry, P. Kopeckova, J. Kopecek, R. Duncan. Effect of galactose on interaction of N-(2-hydroxypropyl)methacrylamide copolymers with hepatoma-cells in culture—preliminary application to an anticancer agent, daunomycin. *Hepatology 10*:207 (1989).
43. L. W. Seymour, K. Ulbrich, S. R. Wedge, I. C. Hume, J. Strohalm, R. Duncan. N-(2-hydroxypropyl)methacrylamide copolymers targeted to the hepatocyte galactose-receptor—pharmacokinetics in DBA2 mice. *British Journal of Cancer 63*:859 (1991).
44. H. Soyez, E. Schacht, S. Vanderkerken. The crucial role of spacer groups in macromolecular prodrug design. *Advanced Drug Delivery Reviews 21*:81 (1996).
45. D. R. Ferry, L. W. Seymour, D. Anderson, S. Hesselwood, P. Julyan, C. Boivin, R. Poyner, P. Guest, J. Doran, D. J. Kerr. Phase I trial or liver targeted HPMA copolymer of doxorubicin PK2, pharmacokinetics, spect imaging of I-123-PK2 and activity in hepatoma. *British Journal of Cancer 80*:413 (1999).
46. P. A. Vasey, S. B. Kaye, R. Morrison, C. Twelves, P. Wilson, R. Duncan, A. H. Thomson, L. S. Murray, T. E. Hilditch, T. Murray, S. Burtles, D. Fraier, E. Frigerio, J. Cassidy. Phase I clinical and pharmacokinetic study of PK1 [N-(2-hydroxypropyl)methacrylamide copolymer doxorubicin]: first member of a new class of chemotherapeutic agents—Drug–polymer conjugates. *Clinical Cancer Research 5*:83 (1999).
47. P. J. Julyan, L. W. Seymour, D. R. Ferry, S. Daryani, C. M. Boivin, J. Doran, M. David, D. Anderson, C. Christodoulou, A. M. Young, S. Hesslewood, D. J. Kerr. Preliminary clinical study of the distribution of HPMA copolymers bearing doxorubicin and galactosamine. *Journal of Controlled Release 57*:281 (1999).
48. C. Delgado, R. B. Pedley, A. Herraez, R. Boden, J. A. Boden, P. A. Keep, K. A. Chester, D. Fisher, R. H. J. Begent, G. E. Francis. Enhanced tumour specificity of an anti-carcinoembrionic antigen Fab' fragment by poly(ethylene glycol) (PEG) modification. *British Journal of Cancer 73*: 175 (1996).
49. G. E. Francis, D. Fisher, C. Delgado, F. Malik, A. Gardiner, D. Neale. PEGylation of cytokines and other therapeutic proteins and peptides: the importance of biological optimisation of coupling techniques. *International Journal of Hematology 68*:1 (1998).

50. T. Sawa, J. Wu, T. Akaike, H. Maeda. Tumor-targeting chemotherapy by a xanthine oxidase-polymer conjugate that generates oxygen-free radicals in tumor tissue. *Cancer Research 60*:666 (2000).
51. F. Liu, H. Qi, L. Huang, D. Liu. Factors controlling the efficiency of cationic lipid-mediated transfection in vivo via intravenous administration. *Gene Therapy 4*:517 (1997).
52. S. L. Hart. Synthetic vectors for gene therapy. *Expert Opinion On Therapeutic Patents 10*:199 (2000).
53. R. Ruger. Gene therapy in oncology: current technical and clinical status of development. *Onkologie 22*:11 (1999).
54. E. Marshall. The trouble with vectors. *Science 269*:1052 (1995).
55. R. G. Vile, S. J. Russell, N. R. Lemoine. Cancer gene therapy: hard lessons and new courses. *Gene Therapy 7*:2 (2000).
56. M. C. Garnett. Gene-delivery systems using cationic polymers. *Critical Reviews in Therapeutic Drug Carrier Systems 16*:147 (1999).
57. S. Ferrari, E. Moro, A. Pettenazzo, J. P. Behr, F. Zacchello, M. Scarpa. ExGen 500 is an efficient vector for gene delivery to lung epithelial cells in vitro and in vivo. *Gene Therapy 4*:1100 (1997).
58. Y. P. Zhang, L. Sekirov, E. G. Saravolac, J. J. Wheeler, P. Tardi, K. Clow, E. Leng, R. Sun, P. R. Cullis, P. Scherrer. Stabilized plasmid–lipid particles for regional gene therapy: formulation and transfection properties. *Gene Therapy 6*:1438 (1999).
59. S. Li, M. A. Rizzo, S. Bhattacharya, L. Huang. Characterization of cationic lipid-protamine-DNA (LPD) complexes for intravenous gene delivery. *Gene Therapy 5*:930 (1998).
60. J. L. Coll, P. Chollet, E. Brambilla, D. Desplanques, J. P. Behr, M. Favrot. In vivo delivery to tumors of DNA complexed with linear polyethylenimine. *Human Gene Therapy 10*:1659 (1999).
61. P. R. Dash, M. L. Read, L. B. Barrett, M. Wolfert, L. W. Seymour. Factors affecting blood clearance and in vivo distribution of polyelectrolyte complexes for gene delivery. *Gene Therapy 6*:643 (1999).
62. V. Toncheva, M. A. Wolfert, P. R. Dash, D. Oupicky, K. Ulbrich, L. W. Seymour, E. H. Schacht. Novel vectors for gene delivery formed by self-assembly of DNA with poly(L-lysine) grafted with hydrophilic polymers. *Biochimica et Biophysica Acta—General Subjects 1380*:354 (1998).
63. M. A. Wolfert, E. H. Schacht, V. Toncheva, K. Ulbrich, O. Nazarova, L. W. Seymour. Characterization of vectors for gene therapy formed by self-assembly of DNA with synthetic block polymers. *Human Gene Therapy 7*:2123 (1996).
64. M. Ogris, S. Brunner, S. Schuller, R. Kircheis, E. Wagner. PEGylated DNA/transferrin–PEI complexes: reduced interaction with blood components, extended circulation in blood and potential for systemic gene delivery. *Gene Therapy 6*:595 (1999).
65. R. Kircheis, S. Schuller, S. Brunner, M. Ogris, K. H. Heider, W. Zauner, E. Wagner. Polycation-based DNA complexes for tumor-targeted gene delivery in vivo. *Journal of Gene Medicine 1*:111 (1999).
66. P. R. Dash, M. L. Read, K. D. Fisher, K. A. Howard, M. Wolfert, D. Oupicky, V. Subr, J. Strohalm, K. Ulbrich, L. W. Seymour. Decreased binding to proteins and cells of polymeric gene delivery vectors surface modified with a multivalent hydrophilic polymer and retargeting through attachment of transferrin. *Journal of Biological Chemistry 275*:3793 (2000).
67. K. Fisher, Y. Stallwood, V. Mautner, K. Ulbrich, D. Oupicky, L. W. Seymour. Protection and retargeting of adenovirus using a multifunctional hydrophilic polymer. *Proc. 4th Int. Symp. Polymer Therapeut. London*, 2000, p. 74.
68. C. R. O'Riordan, A. Lachapelle, C. Delgado, V. Parkes, S. C. Wadsworth, A. E. Smith, G. E. Francis. PEGylation of adenovirus with retention of infectivity and protection from neutralizing antibody in vitro and in vivo. *Human Gene Therapy 10*:1349 (1999).

32

Anticancer Drug Conjugates with Macromolecular Carriers

F. Kratz, A. Warnecke, K. Riebeseel, and P. C. A. Rodrigues
Tumor Biology Center, Freiburg, Germany

I. INTRODUCTION: DRUG TARGETING IN ANTICANCER CHEMOTHERAPY USING MACROMOLECULAR CARRIERS

The therapeutic concept of drug targeting is founded on Paul Ehrlich's vision of "the magic bullet," which he proclaimed at the beginning of the last century (1). Ehrlich's notion was based on a concept of affinity between a drug acting as a ligand and a molecular target, e.g., a receptor (see Fig. 1). Even if the precise molecular targets for all of today's drugs are not known, Ehrlich's concept appears to be valid in principle.

With the advances in protein crystallography and molecular modeling, the number of tailor-made drugs will certainly increase. However, even when the drug and its molecular target interact ideally according to a "key and lock principle," the issue of site-specific and optimal drug delivery, especially for drugs with a narrow therapeutic index, has not been satisfactorily solved for a great number of medical indications. Indeed, tailor-made drugs that fit neatly into their target cannot always be developed clinically due to poor water-solubility or rapid elimination from the body after systemic administration of the respective drug candidate. In this respect, fundamental goals of drug conjugation with suitable carriers are to circumvent such inherent disadvantages and to achieve a higher concentration of the drug at its site of action.

In the field of cancer research, coupling of anticancer drugs to macromolecular carriers received an important impetus from 1975 onwards with the development of monoclonal antibodies by Milstein and Köhler (2). In addition, research efforts have recently concentrated on conjugating anticancer agents with a wide spectrum of macromolecules including polysaccharides, lectins, serum proteins, peptides, growth factors, and synthetic polymers. Some typical examples are summarized in Table 1.

The rationale for simply using high molecular weight molecules as efficient carriers for the delivery of antitumor agents, even if they are not targeted towards an antigen or receptor on the surface of the tumor cell, has been strengthened by recent studies concerning the enhanced vascular permeability of circulating macromolecules for tumor tissue and their subsequent accumulation in solid tumors (3–9). This phenomenon has been termed "enhanced permeability and retention" in relation to passive tumor targeting (EPR effect) (3) and is depicted schematically in Fig. 2. Blood vessels in most normal tissues have an intact endothelial layer that allows the diffusion of small molecules but not the entry of macromolecules into the tissue. In contrast, the endothelial layer of blood vessels in tumor tissue is often leaky, so that small as well as large molecules have access to malignant tissue. Because tumor tissue does not generally have a lymphatic drainage system, macromolecules are thus retained and can accumulate in solid tumors.

The pathophysiology of tumor tissue, characterized by angiogenesis, hypervasculature, a defective vascular architecture, and an impaired lymphatic drainage, seems to be a universal feature of solid tumors that can be exploited

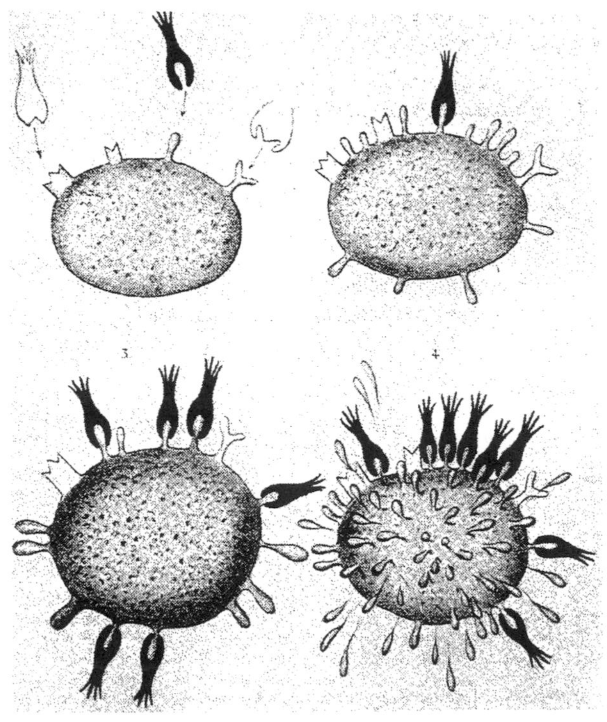

Figure 32.1 Hand drawing by Paul Ehrlich (1854–1915) illustrating his concept of receptor–ligand interaction. (We thank Professor Gerd Folkers of the ETH, Zürich, Switzerland, for supplying this figure.)

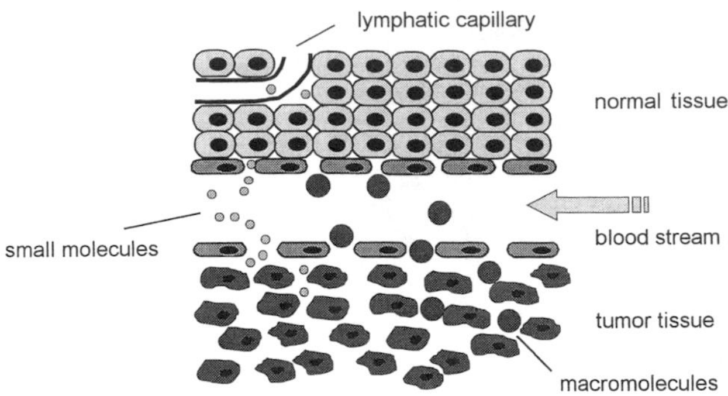

Figure 32.2 Schematic representation of the anatomical and physiological differences between normal and tumor tissue with respect to vascular permeability of small and large molecules.

for tumor-selective drug delivery. A detailed description of the EPR effect is given in the article by L. Seymour in this book.

This article focuses on the design of drug polymer conjugates, on drug conjugation techniques, and on drug polymer conjugates that have reached the clinical setting or an advanced stage of preclinical development. Monoclonal antibodies, serum proteins, growth factors, cytokines, and polysaccharides are discussed as macromolecular carriers. Drug conjugates with synthetic polymers are presented in detail in the contribution by L. Seymour.

Table 32.1 Examples of Macromolecular Carriers for Anticancer Agents

Macromolecular carrier	Examples
Biological macromolecules	
Proteins	Antibodies, serum proteins
Polysaccharides	Dextran, hyaluronic acid
Lectins	Concanavalin A
Peptides	Growth factors, cytokines
Nucleic acids	Deoxyribonucleic acid (DNA)
Synthetic macromolecules	
Poly(amino acids)	Polylysine, polyaspartic acid
Copolymers	Styrene–maleic acid anhydride copolymer (SMA)
	Divinylether–maleic acid anhydride copolymer (DIVEMA)
	N-(2-hydroxypropyl)methacrylamide copolymer (HPMA)
Polymers	Polyethylene glycol (PEG)
	Polyvinylalcohol (PVA)

Primarily, the review describes conjugates with organic anticancer agents of clinical relevance. Immunotoxins are only discussed with growth factors and cytokines because anticancer drugs have rarely been coupled to these carriers. It is beyond the scope of this article to review the vast amount of work that has been carried out with immunotoxins using antibodies, and the reader is referred to several review articles covering this topic (10–12).

In our discussion, we make no attempt to include every conjugate that has been developed and investigated in vitro; rather, examples are described that illustrate the characteristics of a particular drug delivery system. When covering the different macromolecular carriers that have been used for drug conjugation, we have taken care to refer to pertinent review articles that have appeared throughout the past 30 years.

II. DESIGN OF ANTICANCER DRUG CONJUGATES

A. Macromolecular Prodrugs with Anticancer Agents—Similarities and Differences in Their Design and Implications for Their Preparation

In 1975, Ringsdorf proposed a general model for a drug delivery system using synthetic polymers (see Fig. 3) (13). A definite number of drug molecules are bound to a macromolecule through a spacer molecule that incorporates a predetermined breaking point to ensure release of the drug after cellular uptake of the conjugate (the significance of incorporating predetermined breaking points will be discussed in Section D). The system can also contain targeting moieties, e.g., tumor-specific antibodies (14), antibody fragments (15), or saccharides (14), which render the conjugate biorecognizable and enhance cellular uptake by receptor-mediated endocytosis.

Macromolecules chosen for the preparation of drug conjugates should ideally be water soluble, nontoxic, and nonimmunogenic. Furthermore, after fulfilling their function, they should be degraded and/or eliminated from the organism (16). A final, and from the chemical point of view most important, feature of the carrier is the existence of drug attachment sites or at least the possibility of their creation through chemical modification of the polymer. The number of accessible attachment sites depends on the type of polymer. The majority of relevant synthetic polymers are homopolyfunctional: classical examples for mono- or bifunctional polymers are poly(ethylene glycols) (PEGs), which have the advantage that the drug loading rate is well-defined (17).

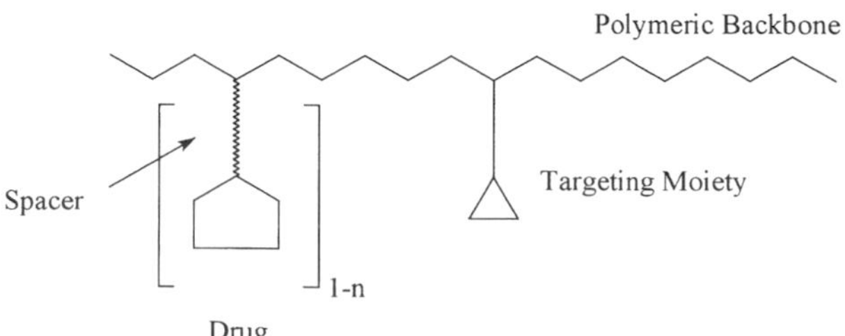

Figure 32.3 Ringsdorf's model for drug delivery systems based on synthetic polymers.

In contrast to synthetic polymers, biomolecules (e.g., serum proteins, antibodies, or polysaccharides) are classified as heteropolyfunctional due to the abundance of different functional groups that can compete with each other in coupling reactions. Treatment of biomolecules with unspecific coupling agents can yield mixtures, and conjugation methods that allow group-specific coupling of the drug to the biopolymer are therefore preferred. Furthermore, coupling of low molecular weight drugs to proteins or antibodies should yield stable products preserving their native state and activity (e.g., drug–antibody conjugates have to retain their affinity for corresponding antigens). Hence coupling reactions are carried out in aqueous solution under mild conditions (temperature range 5–37°C, pH range 6–8, low concentration of organic solvents) in order to prevent the biopolymer from denaturation and/or degradation.

In the following section, the description of coupling methods will be restricted to *bioconjugation* methods for developing drug–polymer conjugates. Examples including synthetic polymers as carriers will be additionally mentioned if the applied coupling method can be classified as a typical bioconjugation technique. This article will not cover special methods that are only applicable to the preparation of drug conjugates with synthetic polymers (e.g., copolymerization of drug derivatives with suitable monomers), and we refer the reader to excellent reviews covering this topic (16,18).

B. Bioconjugation Methods Used for the Preparation of Drug–Carrier Conjugates

Bioconjugate chemistry plays an important role in today's life sciences; applications range from basic protein chemistry to applied biotechnology and from immunology to medicine; relevant methods have been reviewed extensively (19–21). In the field of bioconjugation, the two terms cross-linking and conjugation are often used synonymously although there is a fine distinction in connotation between them. Cross-linking usually refers to the joining of two molecular species that have some sort of affinity between them. Covalent coupling of a ligand to its receptor is an example of cross-linking as well as the formation of chemical bonds between subunits of enzymes. Conjugation, however, refers to the coupling of two unrelated species. Thus coupling low molecular weight anticancer drugs to macromolecules belongs to this category. Nevertheless, coupling reagents for conjugation purposes are denoted as cross-linking agents, and we also use this expression in our article.

Cross-linking agents with two attachment sites are classified as homobifunctional and heterobifunctional, respectively. Glutaraldehyde is a popular member of the first group, forming Schiff bases with amines. However, homobifunctional cross-linkers are not always appropriate for the preparation of well-defined drug–polymer conjugates, and even from a theoretical point of view quantitative coupling is impossible (22). In contrast, heterobifunctional cross-linking agents have the advantage that different functional groups of the polymer as well as of the drug can be chemically exploited. Typically, the drug is first reacted with the heterobifunctional cross-linker and then coupled to another functional group of the polymer without prior purification of the derivatized drug. Although a number of heterobifunctional cross-linking agents are commercially available, most of them were designed for the conjugation of proteins. Several groups have therefore developed tailor-made heterobifunctional cross-linkers that allow an improved derivatization of the respective functional group of the drug. Furthermore, it has proved advantageous to isolate and characterize the polymer-binding drug derivative, which is subsequently coupled to the surface of the macromolecular carrier. In this way, the drug loading ratio, the purity of the conjugate, as well as the nature of the chemical bond between the drug and the carrier are better defined.

In the next two sections, the main functional targets of biopolymers and their reactivity in coupling reactions will be discussed (Section II.B.1), and a short overview of methods for the chemical modification of functional groups will be given (Section II.B.2). Section II.C deals with specific coupling reactions with respect to the functional groups discussed in Section II.B.1.

1. Functional Targets for Bioconjugation

a. Polypeptides (Proteins)

The twenty amino acids that constitute proteins can be divided into three classes according to the polarity of their side chains (Fig. 4).

The degree of hydrophilicity and hydrophobicity is one of the major determinants of the three-dimensional structure of proteins. In contrast to hydrophobic amino acids that are generally found in the interior of proteins, hydrophilic groups are located on the surface of the protein, where they can interact strongly with the aqueous environment. Eight of the hydrophilic side chains are chemically active. These are the guanidinyl group of arginine, the β- and γ-carboxyl groups of aspartic and glutamic acid, respectively, the sulfhydryl (thiol) group of cysteine, the imidazolyl group of histidine, the ε-amino group of lysine, the thioether moiety of methionine, and the phenolic hydroxyl group of tyrosine. However, the frequency with which they are incorporated in proteins differs substantially, and some of them are rarely found (Table 2) (23).

Important functional targets are the sulfhydryl group of cysteine, the amino group of lysine, and the carboxylic groups of aspartic and glutamic acid.

The nucleophilicity of functional groups as well as the electrophilic properties of the attacking polymer-binding group are fundamental parameters in coupling reactions. Nucleophilicity is influenced by various factors, such as charge (which again depends on the pH value), basicity, size, and polarizability. Furthermore, the influence of the solvent cannot be neglected, since the order of nucleophilicity can be reversed by changing from a protic solvent to an aprotic polar one (e.g., DMF or DMSO). Pearson and coworkers measured the rate constants of the reaction between methyliodide and various nucleophilic compounds in methanol to formulate the following order of nucleophilicity (24,25):

$$C_4H_9S^- > ArS^- > HS^- > (C_2H_5)_2NH > ArO^- > ArSH > (CH_3)_2S > \text{imidazole} > AcO^- > CH_3OH$$

The strongest nucleophile is the thiol group in its ionized, thiolate form, followed by the amino group. Thioethers and carboxylic acids are less potent nucleophiles. Applying this order of nucleophilicity to amino acid residues, we obtain the following general order:

$$\text{Cys} > \text{Lys} > \text{Tyr} > \text{Met} > \text{His} > \text{Glu, Asp} > \text{Ser, Thr}$$

The aliphatic hydroxyl groups of serine and threonine, having about the same nucleophilicity as water, are generally unreactive in aqueous solutions. A reactive reagent will preferably undergo hydrolysis before it reacts with amino acids containing hydroxyl group.

One of the most important considerations for planning coupling reactions is the choice of the pH value. For example, the protonated form of amino groups is unreactive towards electrophilic agents. According to the equation of Henderson and Hasselbalch (1), an ionizable group is 50% protonated at its pK_a value.

$$pH = pK_a + \log\left(\frac{[\text{base}]}{[\text{acid}]}\right) \tag{1}$$

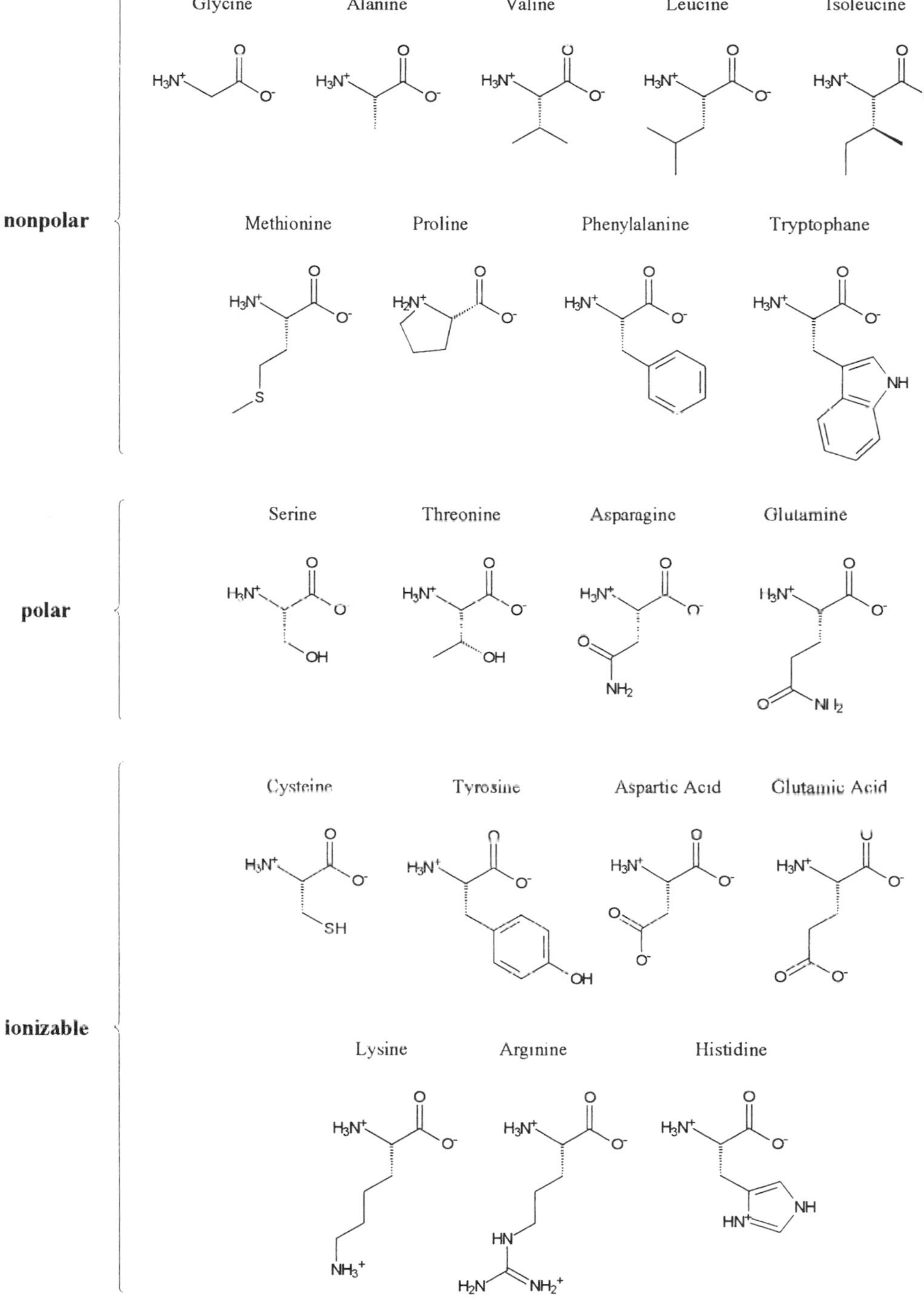

Figure 32.4 Structures of the twenty common amino acids.

Further implications of this equation are that at one or two pH units above the pK_a, an acidic group will be 91% and 99% deprotonated, respectively. pK_a ranges of the major functional targets of proteins are given in Table 3 (20). Notably, pK_a values can vary up to three pK_a units in extreme cases depending on the influence of the microenvironment.

Generally, coupling reactions are carried out by adjusting the pH value close to the pK_a value of the protonated form of the respective functional group.

Table 32.2 Average Frequency with which Amino Acids Are Incorporated in Proteins. Values were Determined by Comparing the Structures of 207 Nonrelated Proteins

Amino acid	Average frequency of incorporation (%)
Arginine	4.7
Aspartic acid	5.5
Glutamic acid	6.2
Histidine	2.1
Lysine	7.0
Tyrosine	3.5
Cysteine	2.8
Methionine	1.7

Source: Ref. 23.

Fortunately, the reaction profile of amino acid residues with coupling agents does not depend only on their nucleophilic properties (otherwise, it would be difficult to achieve group specific coupling) but also on the nature of the attacking electrophile. In some cases, the HSAB (hard and soft acids and bases) principle propagated by Pearson can serve as a tool for predicting preferences of a nucleophile towards an electrophilic group of a coupling agent (26,27).

Finally, it should be mentioned that, apart from nucleophilic reactions, group specific properties can be exploited for conjugation purpose. Cysteine, for example, is susceptible to oxidation forming disulfide bonds. Tyrosine residues represent activated aromatic systems that can undergo electrophilic reactions, e.g., addition of iodine or diazonium salts (labeling reactions), and Mannich condensation reactions. To our knowledge, the latter reaction has not been applied in the field of drug–carrier conjugation.

b. *Polysaccharides*

Polysaccharides are polymers in which monosaccharide units are covalently linked via acetale (ketale) bonds.

Table 32.3 pK_a Ranges of Amino Acid Side Chains in Proteins

Amino acid residue	pK_a range
Lysine	9.3–9.5
Histidine	6.7–7.1
Arginine	> 12
Tyrosine	9.7–10.1
Aspartic acid	3.7–4.0
Glutamic acid	4.2–4.5
Cysteine	8.8–9.1

Popular polysaccharide carriers for drugs are dextran, chitosan, and hyaluronic acid consisting of D-glucose, D-glucosamine, and *N*-acetyl-D-neuraminic acid, respectively. With the exception of hyaluronic acid and chitosan, polysaccharides merely contain hydroxyl groups as primary targeting moieties. Thus for performing coupling reactions in aqueous solutions, the biopolymer has to be modified by introducing functional groups of higher reactivity. However, coupling reactions using polysaccharides can also be carried out in organic solvents that allow the drug to be coupled to hydroxyl groups.

c. *Antibodies*

Most tumor-targeted monoclonal antibodies belong to the immunoglobulin-γ (IgG) class. IgGs are symmetric glycoproteins composed of heavy and light chains. At the ends of these two chains are hypervariable regions containing antigen-binding domains. A variable sized branched carbohydrate domain is attached to the complement-activating Fc region, and the so-called ''hinge region'' contains especially accessible interchain disulfides.

For attaching cytotoxic drugs to antibodies three different methods have been used (28):

1. Coupling to lysine moieties that are distributed over the entire antibody surface
2. Generation of aldehyde groups by oxidizing the carbohydrate region and subsequent reaction with amino-containing drugs or drug derivatives
3. Coupling to sulfhydryl groups that are obtained by selectively reducing the interchain disulfides at the ''hinge region''

Various problems are encountered with the first method. Binding to lysine residues results in a relatively nonselective distribution of drugs on the antibody surface, often including the antigen-binding domains. The latter will cause significant loss of antigen specificity (29). Furthermore, aggregation and precipitation of the conjugate can occur due to a reduced number of solubilizing amino groups (30).

Preparing conjugates using the second method should circumvent the first of the above mentioned problems, since the carbohydrate region is far away from the antigen-binding domain. However, the resulting immunoconjugates will have a distribution of drug loading ratios, and therefore the product will not be truly homogenous. A given glycosylation pattern may also be necessary for retaining the protein structure (31), and severe changes in the carbohydrate region can alter protein topology with consequences for antigen specificity, immunogenicity, and aggregation.

The great advantage of the third method is that four disulfides can be reduced to generate eight sulfhydryl groups selectively at the ''hinge region'' of an IgG under mild conditions and without significant perturbation of the protein structure (32). Hence, sulfhydryl groups obtained by reduction of interchain disulfides are the most important functional targets for conjugation of anticancer drugs to antibodies.

2. Methods for Modification of Functional Groups

In many cases, the native structure of a biopolymer is modified by introducing functional targets for conjugation because the carrier does not provide appropriate or sufficiently reactive groups. Chemical modifications are described in the following paragraphs including examples for applications in drug–carrier conjugation.

a. Introduction of Sulfhydryl Groups

Reduction of Disulfide Bonds with DTT (Cleland's Reagent). Dithiothreitol (DTT) is the *trans* isomer of 1,4-dimercapto-2,3-butanediol. The reducing potential of this versatile reagent was first described by Cleland in 1964. Due to its low redox potential, it cleaves all accessible biological disulfides and maintains free sulfhydryl groups, even in the presence of oxygen (Fig. 5). In contrast to other disulfide reductants, such as 2-mercaptoethanol, 2-mercaptoethylamine or thioglycolate, which are added in large excess, DTT forms an entropically favored cyclic product, and reduction succeeds with a slight excess of DTT.

For example, this method has been used for selectively reducing the four interchain disulfides of monoclonal antibodies (32,33).

Modification of Amines with 2-Iminothiolane (Traut's Reagent). 2-Iminothiolane (Traut's Reagent) is a water-soluble hydrochloride of a cyclic thioimidoester that reacts with amines in a ring-opening reaction forming stable amidine bonds while generating sulfhydryl groups (Fig. 6). This reaction offers the advantage of maintaining the net charge of the protein, since amidine bonds are positively charged under physiological conditions.

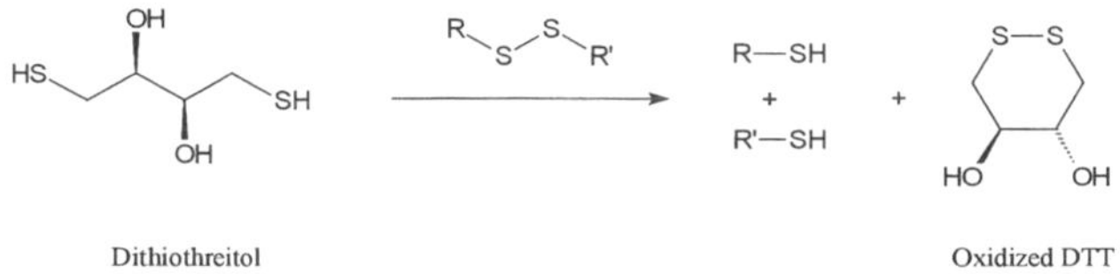

Dithiothreitol DTT

Oxidized DTT

Figure 32.5 Reduction of disulfides with dithiothreitol.

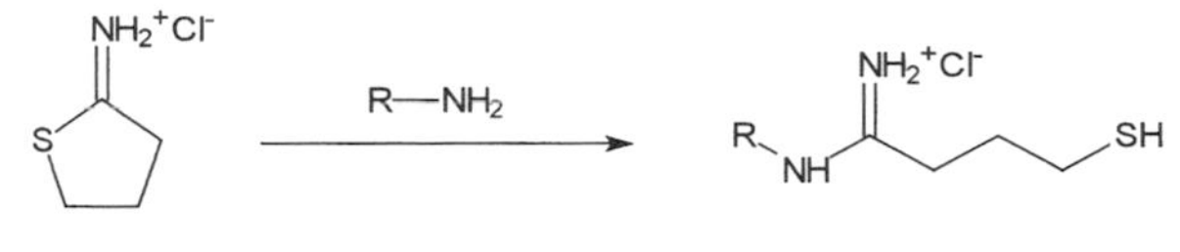

2-Iminothiolane

Modified Amine

Figure 32.6 Thiolation of amines with 2-iminothiolane.

This reaction has mainly been applied for modifying lysine residues of antibodies (32,34) and of serum proteins, such as human serum albumin (35) or transferrin (36).

Modification of Amines with N-Succinimidyl 3-(2-Pyridyldithio)propionate (SPDP). *N*-succinimidyl-3-(2-pyridyldithio)propionate (SPDP) is one of the most popular heterobifunctional cross-linking agents that forms a stable amide bond with amines of proteins. Free sulfhydryl groups are generated after straightforward DTT-reduction of the pyridyldithio moieties (Fig. 7) (37).

Using this technique, a number of drug conjugates have been prepared with antibodies (32,38,39) and with the iron transport protein transferrin (40). HPLC methods determining the degree of derivatization (ratio of pyridyl disulfide moieties to macromolecule) have recently been developed (41).

Furthermore, SPDP can be employed as a sulfhydryl-binding heterobifunctional cross-linking agent by first reacting the linker with an amino-containing drug and subsequently binding it to cysteine moieties (34).

b. Introduction of Aldehyde Groups

Carbohydrates and other biological molecules that contain polysaccharides, such as antibodies, can be specifically modified at their sugar residues to yield reactive aldehyde groups. The most commonly used method is the oxidation with sodium periodate that cleaves the carbon–carbon bond between vicinal hydroxyl groups (Fig. 8).

Although this method alters the polysaccharide structure substantially, it has been frequently used for the preparation of drug–polysaccharide conjugates with dextran (42–44) and antibodies (30).

c. Introduction of Carboxylic Groups

A commonly used method for introducing carboxylic groups is the reaction of amino groups with cyclic anhydrides, such as succinic, glutaric, or maleic anhydride (Fig. 9).

The sulfhydryl group of cysteine, the phenolate group of tyrosine, and the imidazole ring of histidine can also react, but the acylated products are sensitive to hydrolysis,

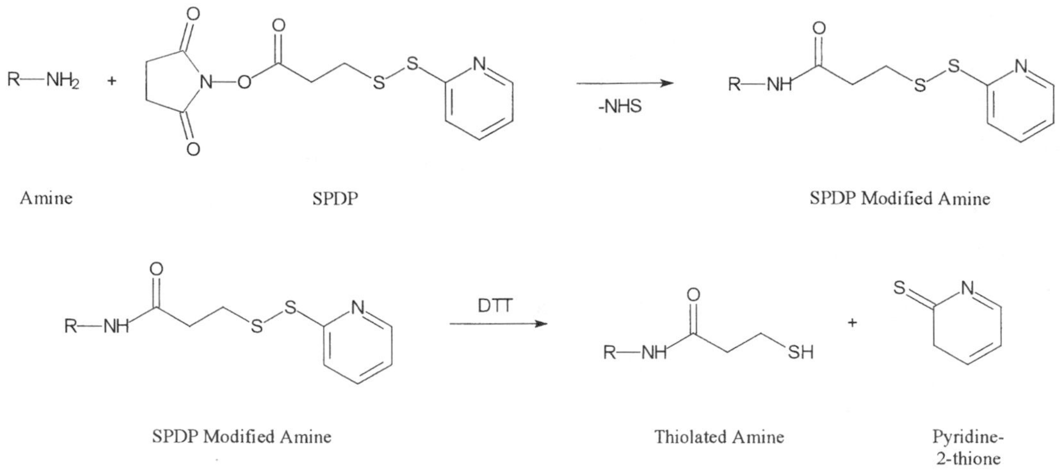

Figure 32.7 Reaction of SPDP with amino-containing polymers forming an amide bond and subsequent reduction with DTT yielding the thiolated macromolecule.

Figure 32.8 Periodate oxidation of vicinal dihydroxyl compounds.

thus regenerating the original group of the respective amino acid at physiological pH.

Only a few examples for preparing drug–carrier conjugates with succinylated macromolecules have been reported (45). For instance, bovine serum albumin was treated with succinic anhydride yielding 88–95% conversion of the lysine residues. Subsequently, peptide spacer containing daunorubicin derivatives were coupled to the introduced carboxylic groups (46).

It should be noted that converting positively charged amino groups into negatively charged carboxylic groups is often accompanied by a dramatic change in the three-dimensional structure of the protein (47) that results in an irreversible loss of biological activity (48).

C. Group-Specific Coupling Reactions

This chapter deals with the final step in the preparation of drug–carrier conjugates: the coupling of the polymer-

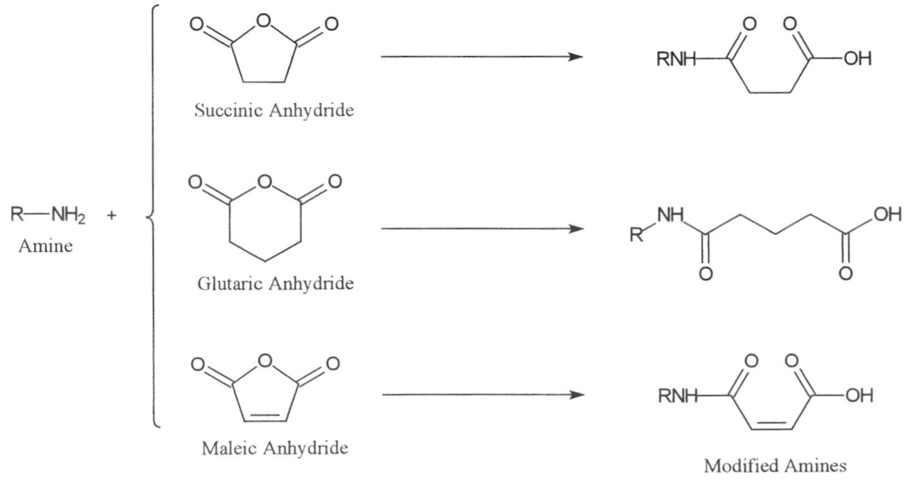

Figure 32.9 Introduction of carboxylic groups into amino-containing polymers.

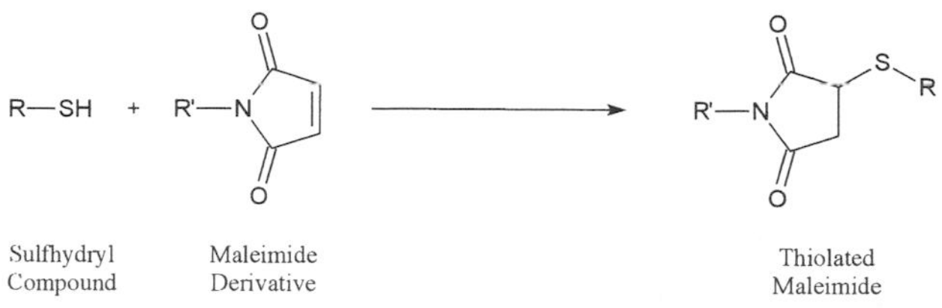

Sulfhydryl Compound   Maleimide Derivative   Thiolated Maleimide

Figure 32.10 Michael addition of a sulfhydryl group to the double bond of a maleimide group.

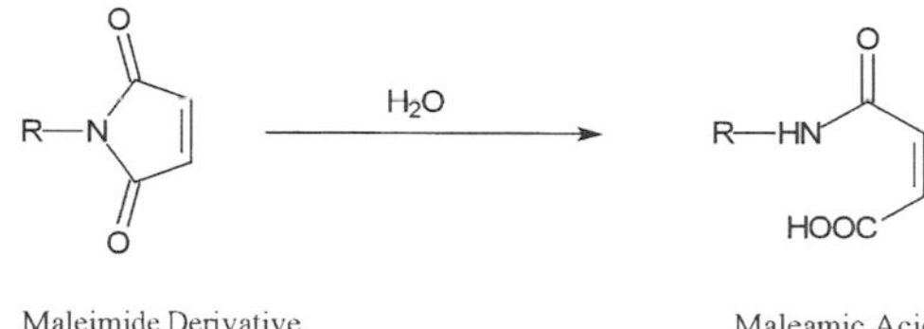

Figure 32.11 Hydrolysis of maleimide derivatives at alkaline pH.

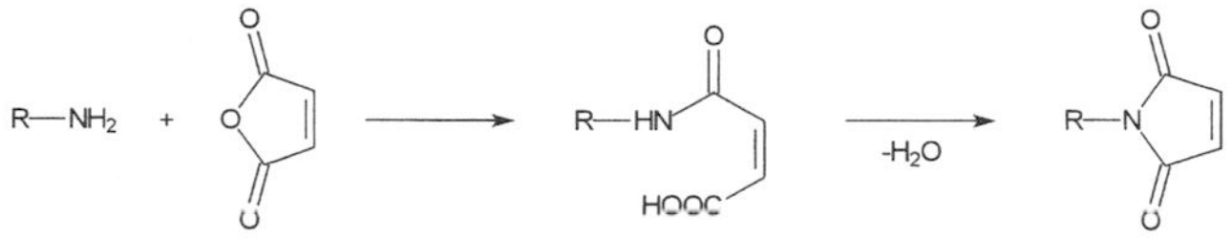

binding drug to the (modified) macromolecular carrier. Although a number of water-soluble spacer molecules have been developed (49), most polymer-binding drug derivatives have the disadvantage of being hydrophobic. In these cases, it has proved useful to add small amounts of organic solvents, such as dimethylformamide or dimethylsulfoxide, up to a concentration of approximately 10%. It is important that under these coupling conditions no denaturing and/or irreversible precipitation of the biopolymer takes place. Due to the poor water-solubility of many drugs, precipitation can also take place if the carrier is overloaded with hydrophobic drugs. Therefore the drug loading ratio needs to be carefully controlled (30).

In the following sections, the most commonly used group-specific coupling methods are described, including examples for their application in drug carrier conjugation.

1. Sulfhydryl-Reactive Methods

a. *Maleimides*

Maleimides are the most important sulfhydryl specific coupling agents. The sulfhydryl group adds to the double bond of the heterocycle's Michael system, forming a stable thioether bond (Fig. 10).

Maleimide reactions are specific for sulfhydryl groups when carried out in a pH range of 6.5–7.5 and can be followed spectrophotometrically by the decrease in absorbance at 300 nm as the double bond reacts and disappears. At pH 7, the reaction of the maleimide with sulfhydryl groups is approximately 1000 times faster than with amino groups. At higher pH values, cross-reactivity with amino groups can take place (20), whereas tyrosine, histidine, or methionine residues do not react (21). A possible side reaction of the maleimide itself is a hydrolytic ring-opening reaction forming maleamic acids (Fig. 11) that typically occurs at higher pH values.

For preparing tailor-made spacer molecules containing maleimide groups, three methods are commonly used:

1. Acylation of amines with maleic anhydride followed by dehydration of the intermediate maleamic acid (Fig. 12) (50).

Figure 32.12 Synthetic route for maleimide derivatives by treating primary amines with maleic anhydride in a first step and forming the maleimide ring by dehydratation in a second step (method 1).

Although this approach proved to be suitable for the preparation of a number of heterobifunctional cross-linking agents containing maleimide groups (51), it was not applicable to acid-sensitive compounds. A variation of this method carried out as a microwave-induced one-pot reaction has been reported recently (52).

2. Conversion of amino acids under mild conditions by treatment with *N*-alkoxycarbonylmaleimides in aqueous solution (Fig. 13) (53).

This method was chosen for synthesizing water-soluble oligooxyethylene-based heterobifunctional cross-linkers (49).

3. Alkylation of maleimide by alcohols using Mitsunobu reaction conditions (Fig. 14) (54,55).

Since alcohols as starting materials are generally easier to prepare than the analogous amino compounds, the importance of this reaction for the synthesis of heterobifunctional cross-linkers will probably increase in future.

Due to their excellent sulfhydryl binding properties, maleimide groups have been extensively used for coupling drug derivatives to thiol-group-containing macromolecules (28,32,33,56–63). For instance, the commercially

Figure 32.13 Synthetic route for maleimide derivatives by reacting amines with *N*-alkoxycarbonylmaleimides (method 2).

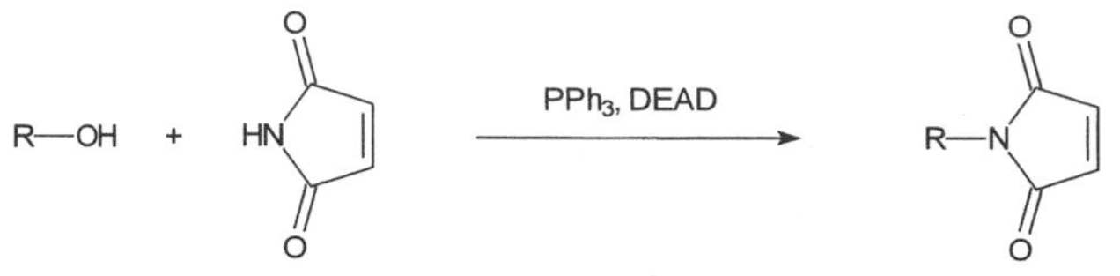

Figure 32.14 Mitsunobu aklylation of maleimide with alcohols (method 3).

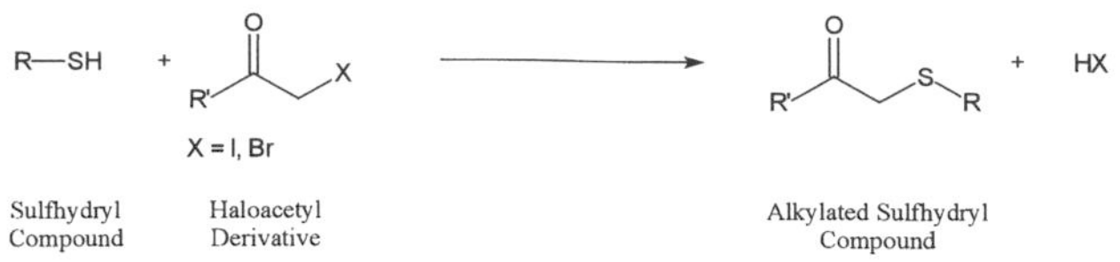

Figure 32.16 Alkylation of sulfhydryl groups by haloacetyl compounds forming a thioether bond.

available heterobifunctional cross-linker 6-maleimidocapric acid hydrazide has been employed in the preparation of the immunoconjugate BR96-Dox (32,33), which has undergone clinical trials (57). In more recent work, the synthesis of branched-maleimide-bearing linkers that are capable of binding two or more molecules of doxorubicin and attaching them to BR96 antibodies has been reported (58). Maleimide derivatives of doxorubicin were coupled to a number of thiolated poly(ethylene glycols) employing maleimide-containing heterobifunctional cross-linkers (59). Maleimide chemistry has also been used for coupling anticancer drugs to thiolated serum proteins (60,61). Recently, it could be shown that the selectivity of the maleimide group towards the cysteine 34 position of circulating albumin is sufficient for forming doxorubicin–albumin conjugates in vivo after intravenous application of the maleimide-containing drug derivative shown in Fig. 15 (62).

b. *Haloacetyl Derivatives*

Sulfhydryl groups can be alkylated by haloacetyl derivatives, e.g., bromoacetyl or iodoacetyl compounds, in a nucleophilic substitution reaction (Fig. 16).

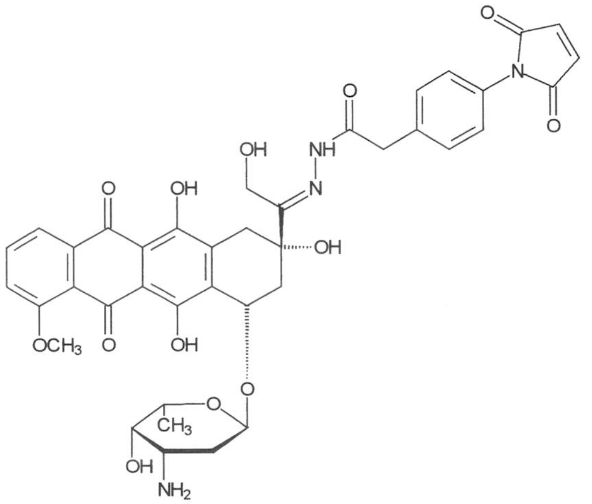

Figure 32.15 Albumin-binding doxorubicin prodrug incorporating an acid-cleavable hydrazone bond and a thiol-binding maleimide moiety.

In contrast to the reaction with maleimides, coupling of bromoacetamides to sulfhydryl groups takes place only at higher pH values (pH 9) with a desirable reaction rate (63) whereby surprisingly no cross-reactivity towards amino or imidazole groups could be observed in the investigated examples. According to Hermanson, the relative reactivity of haloacetates toward protein functionalities is sulfhydryl > imidazolyl > thioether > amine (32). Among halo derivatives, the relative reactivity is I > Br > Cl > F, with fluorine being almost unreactive. Haloacetyl drug derivatives can be conveniently prepared by esterification or by forming amides of a haloacetic acid with hydroxyl- or amino-containing drugs using carbodiimides for activation.

Although haloacetyl derivatives are generally appropriate for group-specific coupling, their use in the field of drug–carrier conjugation is limited due to the basic pH values needed for achieving effective coupling.

c. *Pyridyl Disulfides*

Pyridyl disulfides are perhaps the most popular type of sulfhydryl-disulfide-exchange reagents that ensure effective group specific coupling. Treatment of thiols with these compounds yields mixed disulfides and releases pyridine-2-thione as a nonreactive by-product (Fig. 17).

For the preparation of tailor-made heterobifunctional cross-linkers, pyridyl disulfides can be generated by activating thiols with methoxycarbonylsulfenylchloride and subsequent reaction with 2-mercaptopyridine (Fig. 18) (39).

Pyridyl disulfide containing drug derivatives have been conjugated with antibodies. For instance, doxorubicin was reacted with pyridyl disulfide containing carboxylic hydrazides to form respective hydrazone derivatives. The

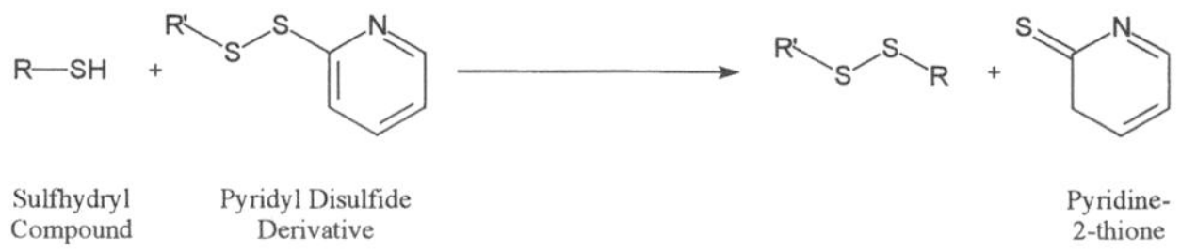

Figure 32.17 Disulfide exchange reaction of sulfhydryl compounds with pyridyl disulfide derivatives.

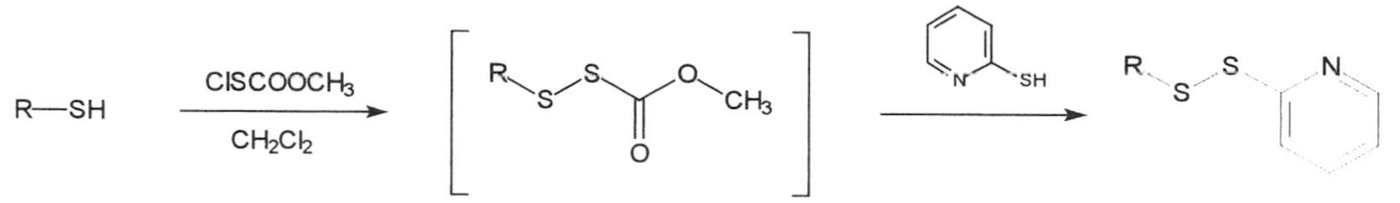

Figure 32.18 Pyridyl disulfides formed by reaction of sulfhydryl compounds with methoxycarbonylsulfenylchloride and 2-mercaptopyridine.

drug derivatives were then coupled to a thiolated monoclonal antibody (39). In another work, the hydrazone of doxorubicin and hydrazine was treated with SPDP to form a thiol-binding drug derivative, which was coupled to a SPDP-thiolated antibody (38).

2. Amino Groups

In most cases, drugs are coupled to amino-containing polymers by acylation. Other functional groups, e.g., sulfhydryl, imidazolyl, or phenolate groups, can also be acylated, but the products are sensitive to hydrolysis, thus regenerating the original group. Acylation can be achieved by in situ activation of carboxyl-group-containing drugs with carbodiimides (compare with Section II.C.3) or by the use of drug derivatives containing isothiocyanates, imidoesters, or active esters, especially (NHS) esters. NHS esters react readily with amines, forming a stable amide bond while releasing *N*-hydroxysuccinimide as a by-product (Fig. 19)..

NHS esters are usually prepared by esterification of the carboxyl-containing drug with *N*-hydroxysuccinimide and DCC (*N,N′*-dicyclohexyl carbodiimide) (Fig. 20). In situ activation using this method is also possible.

In some cases, the NHS ester derivatives exhibit poor water solubility. This problem can be circumvented by the use of sulfo-NHS (*N*-hydroxysulfosuccinimide), a highly hydrophilic NHS derivative. Sulfo-NHS-esters reacts with amines on target molecules with the same specificity and reactivity as NHS-esters (64).

The use of NHS- or sulfo-NHS esters in the field of drug–carrier conjugation is widespread. For example, the sulfo-NHS or NHS esters of a succinylated derivative of the anticancer drug paclitaxel was coupled to human serum albumin (65) and hyaluronic acid, respectively (66).

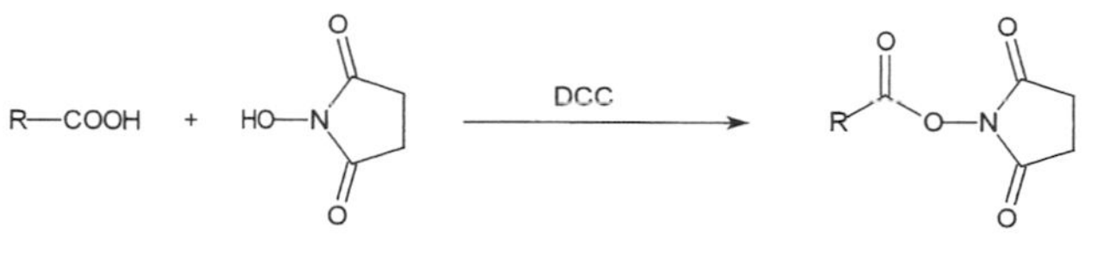

Figure 32.20 Synthesis of NHS esters using DCC activation.

3. Carboxylic Acid Groups

In contrast to other functional targets of polymers, carboxyl groups need to be activated by so-called *zero-length* cross-linking agents in order to react with amino-bearing drugs. Carbodiimides are popular members of this class of cross-linkers. In a first step, they react with carboxylic acids forming a reactive *O*-acylisourea intermediate that is capable of acylating amines (Fig. 21). In nonaqueous solutions, reaction with alcohols takes place, forming respective esters.

Due to the poor water-solubility of most carbodiimides, such as DCC or DIPC (*N,N′*-diisopropyl carbodiimide), water-soluble derivatives have been developed. EDC (1-ethyl-3-(3-dimethylaminopropyl) carbodiimide hydrochloride) (Fig. 22), a hydrophilic tertiary amine containing reagent, is the most popular zero-length cross-linker used for bioconjugation purposes. The urea by-product formed after reaction with the amine compound is also water-solu-

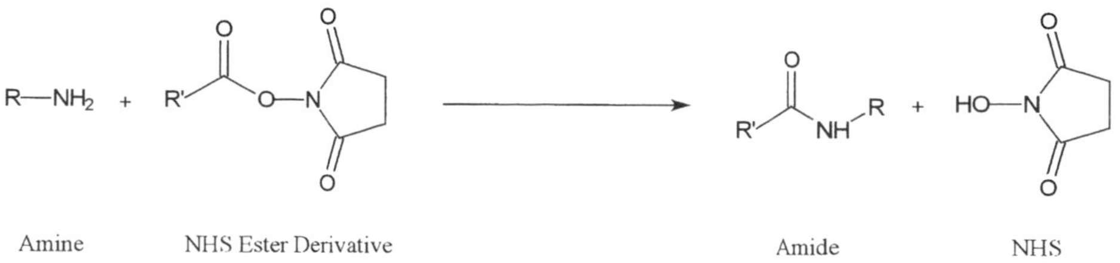

Figure 32.19 Reaction of amines with NHS esters forming stable amide bonds.

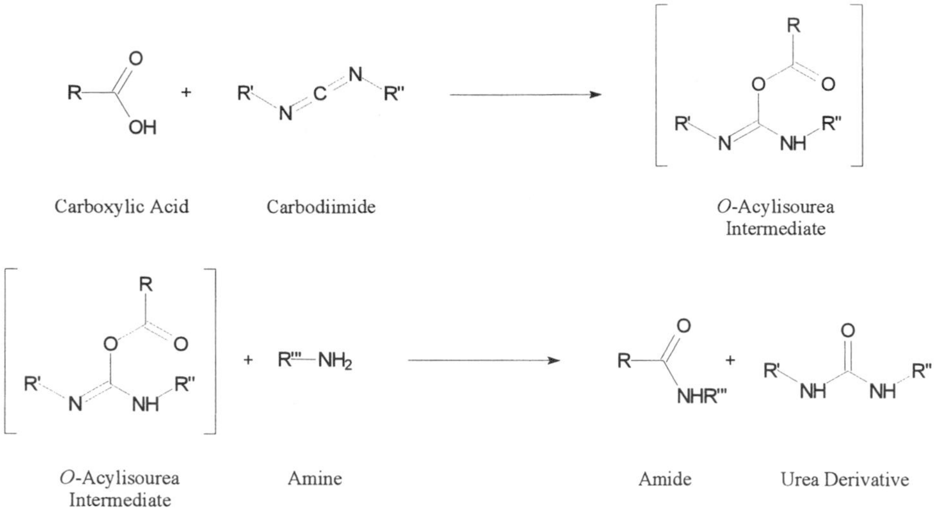

Figure 32.21 Activation of carboxylic acids with carbodiimides forms a reactive *O*-acylisourea intermediate that readily reacts with amines forming a stable amide bond.

ble, so it can be removed using standard methods such as dialysis or gel filtration.

A general problem that can occur when working with carboxyl group activating agents is cross-linking of the protein due to the presence of reactive groups on the protein surface, such as amino groups. This method has therefore been mainly used with homopolyfunctional polymers (67–69).

4. Aldehyde Groups

Amino-containing drugs can be coupled to aldehyde groups of the macromolecular carrier that are typically generated by periodate oxidation. The resulting imine bonds of the Schiff base are sensitive towards hydrolysis and are therefore converted into respective stable secondary amines by treatment with the mild reducing agent sodium cyanoborohydride (Fig. 23).

This method is named reductive amination, wherein the first step (the formation of the Schiff base) is pH dependent, and best results are obtained when it is carried out at a pH of ~ 9 (19). Instead of using sodium cyanoborohydride in the second step, sodium borohydride can be selected if other reducible groups are absent.

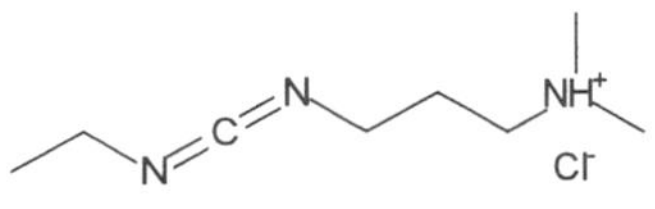

Figure 32.22 Structure of EDC (1-ethyl-3-(3-dimethylaminopropyl) carbodiimide hydrochloride).

Primarily, doxorubicin has been coupled to oxidized dextran (42–44) or to the carbohydrate region of antibodies using this method (30).

D. Incorporation of a Predetermined Breaking Point Between Drug and Carrier

One of the key issues that has been addressed in recent years regarding the pharmacological activity of polymer drug conjugates is the significance of the chemical bond between the drug and the polymer. For designing effective drug polymer conjugates, spacers can be incorporated between the anticancer agent and the macromolecule, which allow the drug to be released either extra- or intracellularly. Over the past ten years two types of bonds have been investigated in some detail: (a) acid-cleavable bonds [reviewed in (70)] and (b) peptide bonds, which can be cleaved by lysosomal enzymes [reviewed in (71)].

Essentially, both types of bonds exploit the cellular uptake mechanism for macromolecules, i.e., endocytosis, which allows the macromolecular bound drug to be released in intracellular compartments. In general, macromolecules are taken up by the cell through receptor-mediated endocytosis, adsorptive endocytosis, or fluid-phase endocytosis.

Endocytosis is a complex process that is shown as a simplified scheme in Fig. 24 [for a review see (72)].

At the cell surface, invaginations occur and endosomes are formed that migrate into the cytoplasm. Depending on the macromolecule and the kind of endocytosis process involved, a series of sorting steps takes place in which the

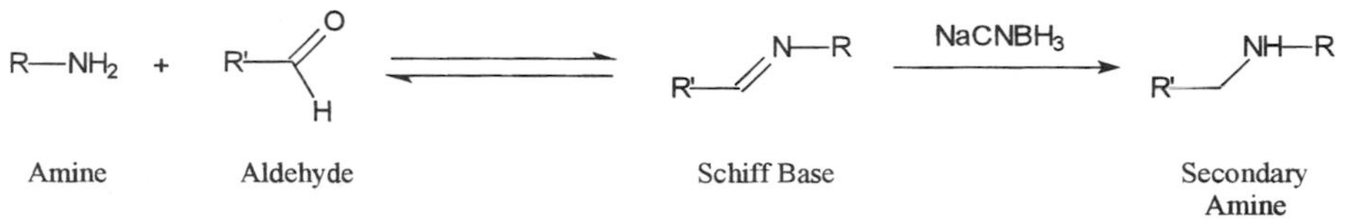

$$R{-}NH_2 + R'{-}CHO \rightleftharpoons R'{-}CH{=}N{-}R \xrightarrow{NaCNBH_3} R'{-}CH_2{-}NH{-}R$$

Amine Aldehyde Schiff Base Secondary Amine

Figure 32.23 Reductive amination of aldehydes.

endosome is transported to certain cell organelles (e.g., the Golgi apparatus), returns to the cell surface (recycling), or forms primary and secondary lysosomes, respectively.

During endocytosis a significant drop in the pH value takes place from the physiological pH (7.2–7.4) in the extracellular space to pH 6.5–5.0 in the endosomes and to around pH 4 in primary and secondary lysosomes. A great number of lysosomal enzymes become active in the acidic environment of these vesicles.

Both the low pH values in endosomes and lysosomes and the presence of lysosomal enzymes are therefore intracellular properties that can be exploited for developing drug polymer conjugates that release the polymer-bound drug specifically in tumor cells.

Classical examples for acid-cleavable bonds are carboxylic hydrazone and *cis*-aconityl bonds (70). Peptide spacers that are enzymatically degraded by lysosomal enzymes, such as cathepsin B, have been studied in great detail, and tailor-made sequences, e.g., Gly-Phe-Leu-Gly, have been identified (71).

III. DRUG CONJUGATES WITH MONOCLONAL ANTIBODIES

A. Introduction

Targeted tumor therapy is based on the presumption that characteristic differences exist between normal and cancer cells, which can be exploited for the selective delivery of antineoplastic agents to the tumor site (73–79). The initial discovery that tumor cells expressed certain determinants (tumor associated antigens) on their cell surface that were not found on normal cells and the subsequent advent of hybridoma technology (2,80) led to the development of a large number of monoclonal antibodies (mAbs), each with its own binding specificity for certain tumor-specific antigens. Monoclonal antibodies are very attractive as drug carriers due to the high binding affinity for their respective antigens. This should allow the accumulation of lethal drug concentrations in the tumor tissue. In addition, the long circulation time of antibodies in the blood stream

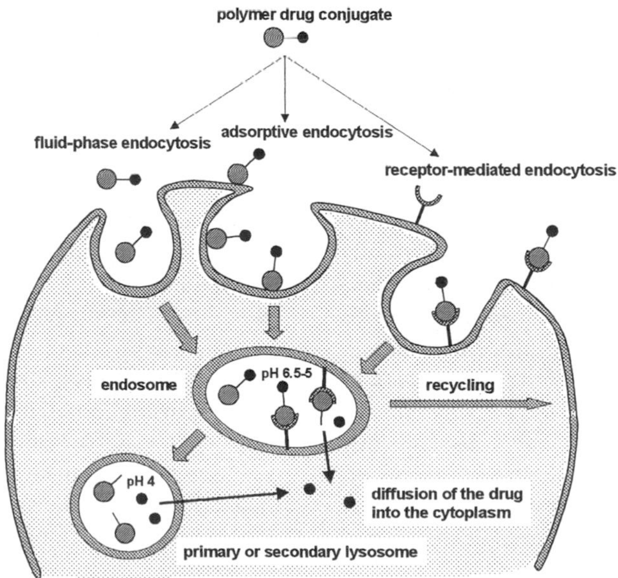

Figure 32.24 Schematic illustration of endocytosis of polymer drug conjugates.

Table 32.4 Examples of Cytotoxic Agents Linked to Monoclonal Antibodies

Agents	Examples	
Toxins	Pseudomonas exotoxin, ricin, saporin, diphtheria toxin	
Isotopes	^{131}Iodine, ^{90}yttrium, ^{213}bismuth, ^{211}astatine, ^{177}lutetium	
Drugs	Anthracyclines	Doxorubicin
		Daunorubicin
		Idarubicin
		Morpholinodoxorubicin
		Bleomycin
	Antimetabolites	Aminopterin
		Methotrexate
		5-Fluorouracil
		5-Fluorouridine
		5-Fluorodeoxyuridine
		Cystosine arabinoside
	Alkylating agents	Chlorambucil
		Cisplatin
		Melphalan
		Mitomycin C
		Trenimon
	Antimitotic drugs	Vinca alkaloids
		Desacetylvinblastine
		Podophyllotoxin
		Colchicine
	Enediynes	Neocarzinostatin
		Calicheamicin

provides a greater probability that the drug will reach the tumor site.

In the past twenty-five years, a great number of immunoconjugates with cytotoxic agents, such as bacterial or plant toxins, radioisotopes, and anticancer drugs, have been synthesized, and their biological activity has been investigated intensively (see Table 4) (28,82–84).

Radioimmunoconjugates are widely used in the diagnosis of malignant diseases. Additionally, they are currently being evaluated as therapeutic agents in radioimmunotherapy with encouraging results in the treatment of advanced hematological malignancies. The concept for radioimmunotherapy as well as preclinical and clinical investigations have been reviewed elsewhere (85–87). *Immunotoxins*, conjugates comprising toxins and antibodies, have recently been tested clinically and have demonstrated promising efficacy in some patients with malignant diseases refractory to surgery, radiation therapy, and conventional chemotherapy (10).

This chapter does not address the latter two topics but focuses on recent progress that has been made in the area of antibody-targeted delivery of conventional chemotherapeutic drugs, summarizing the results of clinical trials with these chemoimmunoconjugates.

B. Design of Chemoimmunoconjugates

As a result of their transformed state, cancer cells overexpress various membrane-associated proteins in comparison to normal tissue, a property that is exploited in targeted cancer chemotherapy.

To achieve optimal targeting, the antigen (Ag) on the tumor cell surface should (a) be a well-defined molecule expressed exclusively on tumor tissue, (b) not be expressed on normal tissues, (c) be bound by the targeting molecule with high affinity, (d) be expressed homogeneously on all the tumor cells, (e) be present on the tumors of all patients with the same type of cancer and (f) not be shed into the circulation.

1. Antigenic Targets

A variety of cell surface proteins have been exploited as targets for monoclonal antibodies (88,89). Notable classes of tumor-associated Ags include the carcinoembryonic antigen (CEA), α-fetoprotein, blood group carbohydrates such as Lewis y (Le^{y}) (recognized by the mAbs BR64 and BR96), gangliosides such as the L6 Ag, the adenocarcinoma-related KS1/4 antigen, mucins, selectins, glycosphingolipids, integrins, and other adhesion molecules. These antigens are generally overexpressed on the surface of tumor cells, although they are found to some extent on normal cells as well. The only truly tumor-specific antigens appear to be those found in hematopoietic tumors, such as idiotypes present on the surface of B-cell tumors (90).

2. Antibodies as Targeting Moieties

Monoclonal antibodies have been used for tumor targeting in their natural, fragmented, chemically modified, or recombinant forms in a variety of experimental settings.

The majority of employed natural antibodies were murine immunoglobulins of the IgG class. A murine immunoglobulin IgG molecule is a symmetric glycoprotein (MW ca. 150,000 Da) that has a Y-shaped structure and is composed of two heavy and two light chains that are linked through disulfide bonds (91). Each chain contains a constant and a variable region (see Fig. 25). The stem region is a constant domain that is responsible for triggering effector functions that eliminate the antigen-associated cells (80). It also contains a carbohydrate moiety.

Treatment with pepsin removes the constant domains that are not required for antigen binding, generating a bivalent antibody-binding fragment F(ab′)$_2$. In contrast, digestion of the antibody with papain produces two monovalent

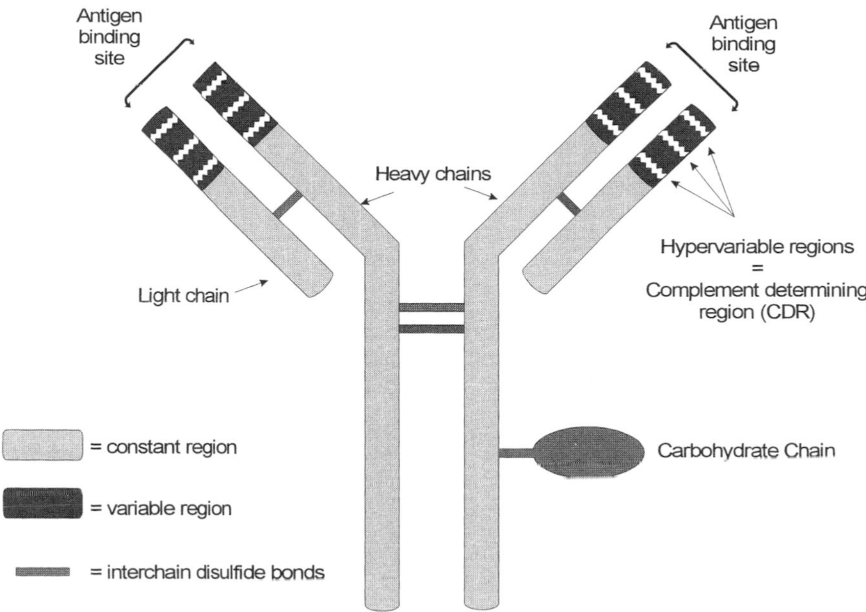

Figure 32.25 Schematic representation of an IgG.

antigen-binding fragments Fab, and one crystallizable fragment Fc (see Fig. 26). The smallest fragment that retains the full monovalent binding affinity of the intact antibody is the Fv molecule that comprises aligned and noncovalently associated V_H and V_L domains containing three hypervariable amino acid sequence regions that are responsible for antigen recognition, known as the complementary determining regions (CDR).

Most mAbs that have been evaluated in the clinical setting, were generated in mice. As a consequence, the use of these mAbs and respective conjugates is limited by the development of human antimouse antibodies (HAMA), which results in accelerated clearance of the administered immunoconjugates (92,93). The antigen-binding fragments ($F(ab')_2$, Fab, or Fv) are much less immunogenic than the intact antibody, because the cleavage of the Fc region generally removes the most immunogenic portion of the immunoglobulin molecule.

Recently, innovative recombinant techniques have opened the possibility of generating genetically engineered monoclonal antibodies (94,95). Chimeric antibodies are constructs containing a mouse Fab portion and a human Fc portion (see Fig. 27a). For preparing chimeric mAbs, the genes encoding the murine Fab variable regions are spliced to the genes encoding the human constant region (96,97). Humanized monoclonal antibodies contain mouse sequences in the binding site of a human mAb (see Fig. 27b). These antibodies are engineered by transferring the antigen-binding hypervariable sequences (CDR) of a murine antibody into human immunoglobulin framework regions (98,99). Since CDRs form the antigen binding site, a humanized antibody preserves the murine antigen specificity. Both constructs, chimeric and humanized antibodies, are predicted to minimize the immune response, because the most immunogenic domain of the murine antibody, the constant Fc region, is substituted by a human sequence.

3. Coupling of Drugs to the Monoclonal Antibody

The success of drug immunoconjugate therapy depends not only on highly tumor-specific mAbs but also on the drug coupling technique used (100,101). Conjugation of cytotoxic drugs with antibody molecules aims to introduce a maximum number of residues under conditions that ensure optimal retention of both drug activity and antibody reactivity. Only a limited number of drug molecules can be introduced by conjugation with the antibody without producing protein denaturation and loss of antibody reactivity. Individual antibodies vary in their ruggedness towards substitution. Some antibodies will lose antigen-binding activity even when using mild procedures, while others permit a relatively high substitution rate without losing binding affinity (102,103).

C. Clinical Trials with Chemoimmunoconjugates

Several chemoimmunoconjugates have been investigated in clinical trials. Although, in general, the observed thera-

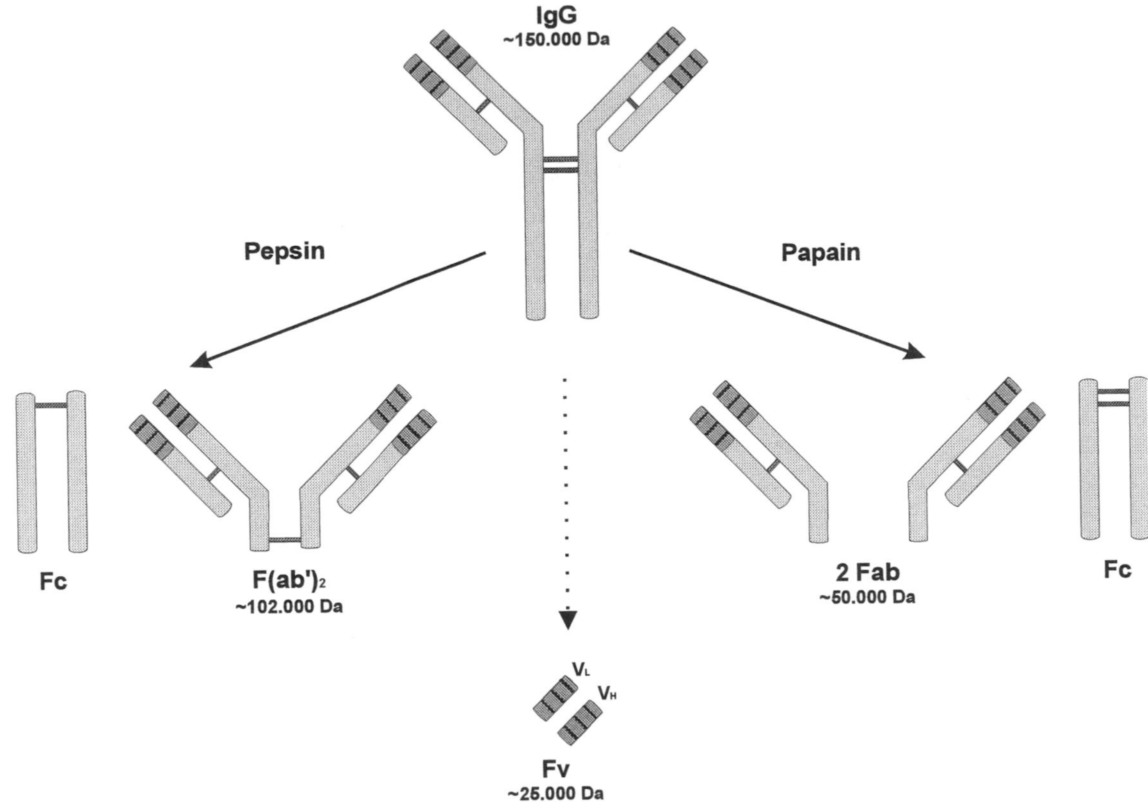

Figure 32.26 Cleavage of IgG into $F(ab')_2$ and Fab fragments by pepsin and papain, respectively.

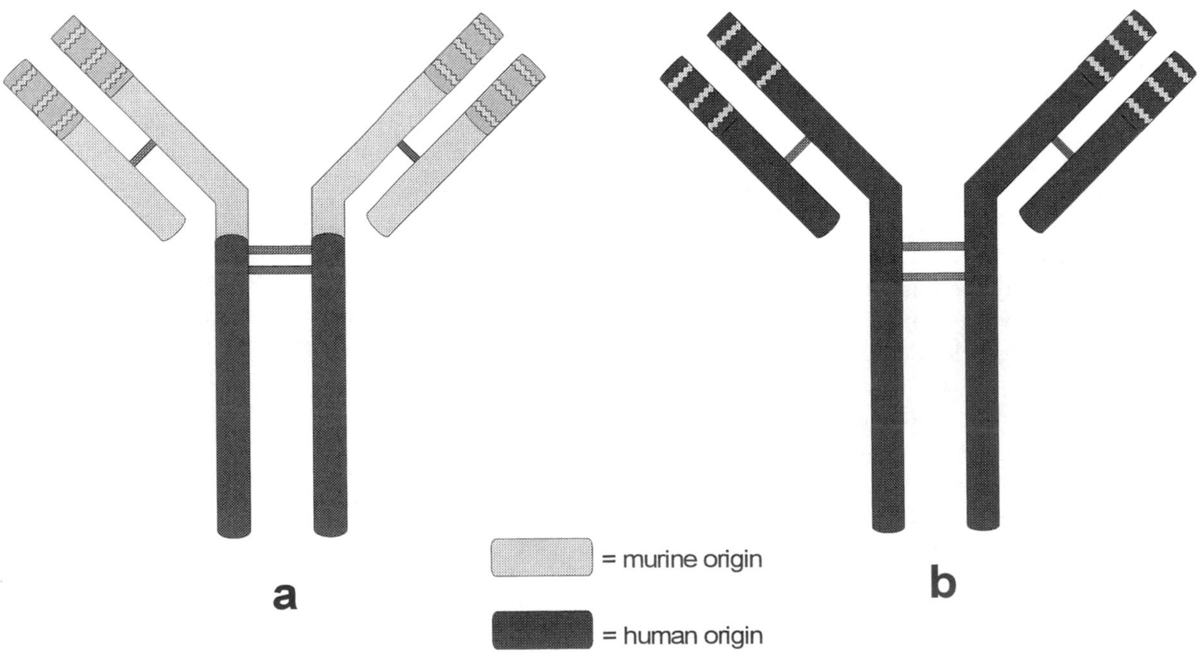

Figure 32.27 Schematic representation of (a) chimeric and (b) humanized antibody.

peutic efficacy must be considered disappointing, these studies provide valuable information on the fate of cytotoxic immunoconjugates in humans. The results of pertinent studies are summarized in Table 5. In the following, selected trials are described in more detail with respect to the exploited cytotoxic drug.

1. Doxorubicin-BR96-Immunoconjugate

One of the most promising chemoimmunoconjugates in preclinical studies has been the BR96-doxorubicin conjugate (BR96-DOX) (32,33,104–106). BR96 is a mouse/human chimeric monoclonal antibody (IgG1) that identifies a tumor-associated antigen related to Lewis Y (Le^y), which is overexpressed on the surface of human carcinoma cells of the breast, colon, lung, and ovary (> 200,000 molecules/cell); the antibody shows high tumor selectivity and is rapidly internalized after binding (107).

BR96-DOX is prepared in a multi step reaction. First, the four interchain disulfides of the mAb are reduced with dithiothreitol to create eight thiol groups. These are then reacted with the maleimidocaproyl hydrazone derivative of doxorubicin. The caproyl hydrazone bond has acid-sensitive properties that allow the drug to be released in the acid environment of lysosomes or endosomes (see Fig. 28). The antitumor activity of the BR96-DOX-immunoconjugate has been evaluated in several clinical trials.

In a first phase I study, the immunoconjugate was administered to 62 patients as an intravenous infusion every 21 days (108,109). Doses of BR96-DOX ranged from 66 to 875 mg/m^2, which is equivalent to 2 to 25 mg/m^2 of free doxorubicin. Two patients exhibited partial responses, one with breast and the other with gastric carcinoma. A stable disease was observed in a number of patients.

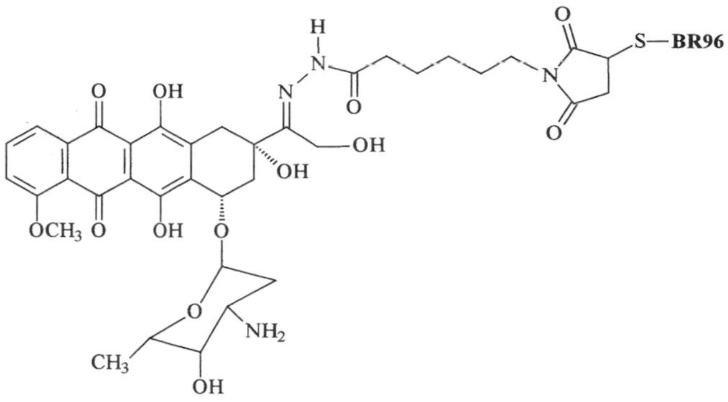

In a second phase I dose-escalation study, 34 patients with Le^y-expressing tumors were treated with BR96-DOX administered as a weekly infusion of 100–500 mg/m^2 of BR96-DOX (equivalent to 3–15 mg/m^2 doxorubicin) (110). Although antibody localization studies demon-

Figure 32.28 Structure of the BR96–doxorubicin immunoconjugate. (From Ref. 104.)

Table 32.5 Clinical Trails with Chemoimmunoconjugates

Drug	Tumor types	Antibody	MOLR	Patients	Response	Ref.
Doxorubicin	Various	BR96	8	62	2 PR	(41,42)
		BR96	8	34	None	(43)
	Metastatic breast cancer	BR96	8	14	1 PR	(44)
	Various	Various cocktails	*	24	2 MR	(59)
		Various cocktails	*	23	5 MR	(93)
Methotrexate	Non-small-cell lung cancer	KS1/4	6	5	None	(49)
		KS1/4	6	11	1 MR	(50)
Vinca alkaloid	Lung and colon carcinoma	KS1/4	4–6	22	None	(55,56)
		KS1/4	4–6	22	None	(54)
Daunomycin/Chlorambucil	Neuroblastoma	Human allogenic IgG	3	7	3 CR, 1 PR, 2 MR	(57)
Mitomycin C	Various	Various cocktails	2.5–15.8	19	1 SD	(60)
Neocarzinostatin	Colon carcinoma	A7 (α-colon Ca)	2	41	3 MR	(67)
N-acetylmelphalan	Colon carcinoma	α colon Ca mAb—cocktails	*	10	3MR	(62)
Calicheamicin	Acute myeloid leukemia	Humanized anti-CD33	2–3	40	3 CR, 5–8 PR	(74)

CR = complete response; MR = minor response; PR = partial response; SD = stable disease; MOLR = Drug/mAb mole ratio; * = value not given.

strated binding of the immunoconjugate at the tumor site, no objective responses were observed. In both studies, BR96-DOX showed dose-limiting gastrointestinal (GI) toxicity at the highest doses. Importantly, no cardiotoxicity was observed, and no HAMA response to the chimeric Ab was induced after repeated dosing.

Recently, a randomized phase II study was performed to evaluate the activity of BR96-DOX against metastatic breast cancer in patients with confirmed sensitivity to single-agent doxorubicin (57). Patients received either 700 mg/m^2 of BR96-DOX (equivalent to 20 mg/m^2 DOX) or 60 mg/m^2 doxorubicin every 3 weeks. There was one partial response in the 14 patients receiving BR96-DOX as well as one complete and three partial responses in the nine patients treated with doxorubicin alone. Interestingly, two of the four patients who crossed over to the BR96-DOX group of the trial after persistent stable disease during treatment with doxorubicin, achieved partial regression of hepatic metastases following immunoconjugate therapy (111). The cross-reactivity of BR96 with normal gastrointestinal tissue led to prominent toxicities and probably impaired the delivery of the immunoconjugate to the tumor sites.

Localization of the antibody and the drug was seen in tumor biopsies of patients receiving BR96-DOX, indicating that the antibody successfully delivered doxorubicin to tumors. Together with the low clinical response rates, the data suggest that the dose that could be safely administered every 3 weeks was insufficient for maintaining the intratumoral concentration of doxorubicin required to achieve tumor regression.

2. Methotrexate-KS1/4-Conjugate

A conjugate of methotrexate (MTX) and the monoclonal antibody KS1/4 was evaluated in two phase I studies in patients with non-small-cell lung cancer. KS1/4 is a murine monoclonal IgG2a antibody that recognizes a M_r 40000 antigen found on most adenocarcinomas (112,113). Its high density on the cell surface and homogeneous expression in non-small-cell lung carcinoma makes this antigen a suitable target for mAb-directed therapy. The conjugate was prepared by simple nonspecific carbodiimide coupling of MTX to the lysine moieties of the mAb (114).

In the first trial, five patients received the MTX–KS1/4 conjugate, while six other patients were treated with KS1/4 alone (115). The total dose of protein administered was 1661 mg, while the amount of methotrexate received with the preparation was 28 mg. In general, the infusions were well tolerated, and similar toxicities were observed in both groups. A majority of patients in both groups developed a HAMA response. Immunoperoxidase staining of carcinoma samples provided evidence of posttreatment localization of antibody and conjugate, respectively. However, no significant clinical response was observed in either group.

In a second clinical study, differing schedules at higher doses were administered to evaluate the therapeutic efficacy and toxicity of the MTX immunoconjugate in more detail (116).

Eleven patients were treated with the KS1/4–MTX conjugate. The resulting MTD of the conjugate was 1750 mg/m^2, equivalent to 40 mg/m^2 MTX.

Under therapy, two patients developed meningitis. One patient had an objective response in two lung nodules ($>$ 50% decrease) but developed immune-complex-mediated arthritis and serum sickness upon subsequent treatment.

3. Immunoconjugates with Desacetylvinblastine

In various preclinical studies, the tumor-associated antigen recognized by the KS1/4 mAb (112) has been investigated as a target for site-directed therapy strategies with chemoimmunoconjugates of vinca alkaloid derivatives (30,117,118). Two of these conjugates, KS1/4–desacetylvinblastine (KS1/4-DAVLB) and KS1/4–desacetylvinblastine-3-carboxyhydrazide (KS1/4-DAVLBHYD), have been examined in phase I clinical studies in patients with epithelial malignancies (119–121).

The first conjugate, KS1/4–DAVLB, contains four to six molecules of desacetylvinblastine covalently attached to the ϵ-amino groups of lysine residues of the KS1/4 mAb through hemisuccinate linkers (see Fig. 29a). Twenty-two patients with lung, colon, or rectum adenocarcinomas received KS1/4–DAVLB in phase I clinical trial (120,121). Single-infusion doses in 13 patients ranged from 40 to 250 mg/m^2 (equivalent to 1.4–14 mg DAVLB). Nine patients were treated with multiple doses of 1.5 mg/kg every 2–3 days. Antitumor activity was not observed in any of the patients. The development of gastrointestinal toxicity was predominant and limited the amount of KS1/4–DAVLB that could be administered. Results from preclinical studies suggested that substantially higher doses than those achieved in the clinical trial would be required to observe antitumor activity.

The second conjugate, KS1/4–DAVLBHYD, was synthesized by coupling 4-desactetyl-vinblastine-3-carbohydrazide to aldehyde moieties generated by mild oxidation of the immunoglobulin's carbohydrate region (30). The resulting conjugate contains 4–6 drug molecules linked to the antibody through an acid-cleavable hydrazone bond that allows release of the free drug in the acidic environment of tumor cells and tumor tissue (see Fig. 29b).

Fourteen patients were treated with a single dose of the

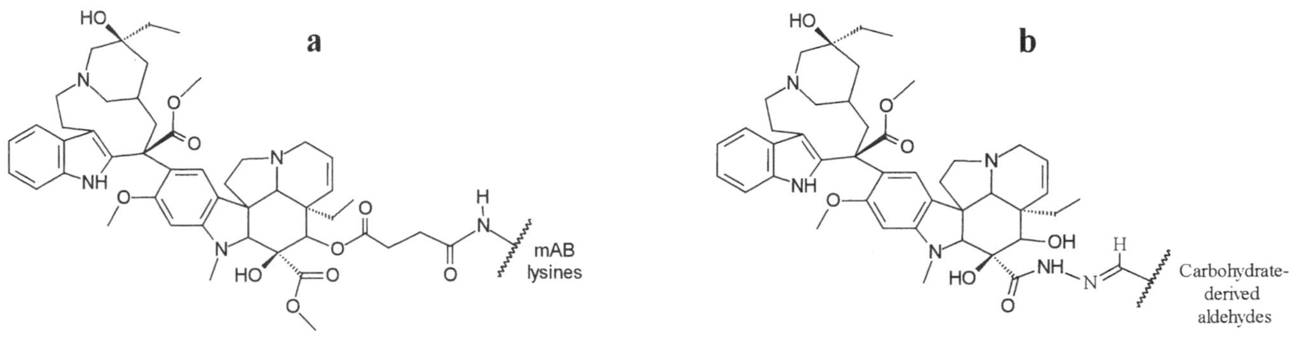

Figure 32.29 Structure of the KS1/4–DAVLB immunoconjugate (a) and the KS1/4–DAVLBHYD immunoconjugate (b).(From Ref. 30.)

conjugate, ranging from 5 to 178 mg KS1/4–DAVLB-HYD. Eight patients received multiple infusions with a total of 127–896 mg conjugate. 77% of the patients developed a HAMA response and 36% showed antivinca antibodies (119). No clinical responses were seen in these studies.

4. Immunoconjugates with Daunorubicin/Chlorambucil

Allogenic antibodies directed against neuroblastoma tumor cells were used as macromolecular carriers of daunorubicin and chlorambucil (122). Seven children bearing neuroblastoma have been treated with 1 mg daunorubicin/kg or 0.5 mg chlorambucil/kg per dose. Treatment was continued up to one year, achieving three cures and one long-term response. Two further patients showed minor responses.

5. Immunoconjugates with Mitomycin C

Mitomycin C immunoconjugates were prepared by coupling mitomycin C through an *N*-hydroxysuccinimide ester intermediate to amino groups of the antibody (123). In a phase I clinical trial, 19 patients with refractory solid malignancies received individualized combinations of mitomycin C murine monoclonal antibody conjugates selected by immunohistochemical and flow cytometric screening of tumor specimens. These ''cocktails'' were composed of up to six different MMC–antibody conjugates. Doses were escalated to 60 mg of MMC conjugated to 3–5 grams of mAbs over a period of 2–3 weeks. The most frequent toxicities were rash, fever, and chills. No clinical responses were seen with these immunoconjugates, although several patients had less tumor-induced pain after treatment (124,125).

6. N-Actelymelphalan-Immunoconjugates

N-Actelymelphalan, a derivative of the alkylating agent melphalan, has been coupled to murine anticolon cancer monoclonal antibodies (126). The selected antibodies demonstrated a high specificity for colon carcinomas as shown by immunoperoxidase staining.

In a phase I clinical trial with *N*-acetylmelphalan conjugates in nine patients with colorectal cancer, up to 1000 mg/m^2 of conjugate (equivalent to 20 mg/m^2 *N*-acetylmelphalan) could be safely administered via the hepatic artery without causing serious side effects (127).

Interestingly, the patients did not develop myelosuppression that typically occurs with the free drug at equivalent doses. HAMA responses were observed in all patients as well as serum sickness in one patient who received two courses of treatment. Minor objective responses of disseminated disease in the liver were noted in three of nine patients.

7. Neocarzinostatin-Immunoconjugate

The anticancer antibiotic neocarzinostatin (NCS) consists of a very reactive polycyclic chromophore that is noncovalently associated with an apoprotein that protects it from degradation (128,129). The proteinaceous drug (MW ≈ 11,700 Da) is obtained from the culture filtrate of *Streptomyces carzinostaticus* and inhibits DNA synthesis by direct DNA strand scission (130). In vivo, NCS is deactivated within minutes and quickly cleared from the circulation-causing renal toxicity.

A neocarzinostatin immunoconjugate containing the murine monoclonal antibody A7 (131) directed against human colorectal cancer has been evaluated in 41 patients (132). No adverse effects were reported, and immunostaining of biopsied cancerous tissue showed that the conjugate was localized in these cells.

Three of eight patients with postoperative liver metastases exhibited tumor reduction as shown by computer-aided tomography. However, patients with lung or peritoneal metastases had no benefit. A follow-up study (133) showed that, in general, the rate of 5-year survival was similar for the group treated with the conjugate in comparison with groups treated with traditional chemotherapy, while individual patients of the conjugate-treated group lived longer. In addition, patients who received higher doses of the A7–NCS conjugate had a higher survival rate. Even after 5 years, human antimouse responses against the conjugate were still detected.

8. Calicheamicin γ_1^I-Immunoconjugate

Calicheamicins are a family of enediyne antibiotics (129) that are probably the most toxic antitumor drugs described to date. Calicheamicin γ_1^I is the most prominent member of this class of compounds. It binds to the minor groove of DNA and causes double-strand DNA breaks that ultimately induce cell death (134–136). Antibody-directed delivery provides a potential means of exploiting the impressive potency of this compound while minimizing its systemic toxicity (137,138).

A phase I study of CMA-676, an immunoconjugate of calicheamicin γ^I_1 conjugated to a humanized anti-CD33 monoclonal antibody through an acid-labile hydrazone bond (see Fig. 30), has recently been completed (139). The CD33 target antigen is expressed on blast cells in most patients with acute myeloid leukemia (AML), but not on normal hematopoietic stem cells. This feature raises the possibility of selectively delivering the cytotoxic drug to malignant cells while sparing healthy stem cells.

Forty patients with refractory or elapsed AML were treated intravenously with 0.25–9.0 mg/m^2 of CMA-676. Besides fever and chills, toxicity was preliminary hematological but not considered to be a dose-limiting factor. Leukemic cells were eliminated from the blood and marrow in eight (20%) of the treated patients, while blood counts returned to normal in three (8%) of them.

A phase II study has been performed involving patients with AML in first relapse, who received 9 mg/m^2 CMA-676 intravenously every 14 days in three cycles. An interim analysis of 23 patients showed a response rate of 43% (86,140). In ten patients, treatment with CMA-676 resulted in elimination of blasts from peripheral blood and marrow; three patients achieved a complete response, with toxicity being lower than with standard chemotherapy.

D. Strategies for Enhanced Antibody Delivery to Tumors

Despite the encouraging preclinical results with distinct anticancer immunoconjugates, the antitumor efficacy of these conjugates in clinical trials has been modest. The shortcomings of the strategy may derive from different obstacles concerning the effective delivery of immunoconjugates to the tumor site (141–143). Some of these problems and their potential solutions are listed in Table 6.

The most disappointing aspect of the clinical studies using chemoimmunoconjugates is the poor mAb localization in tumors, resulting in minimal antitumor responses. Tumor imaging in patients has shown that the amount of mAb localizing to the tumors is sometimes less than 0.01% (81). The inability to translate the promising efficacy seen in animal models to patients can, in part, be rationalized

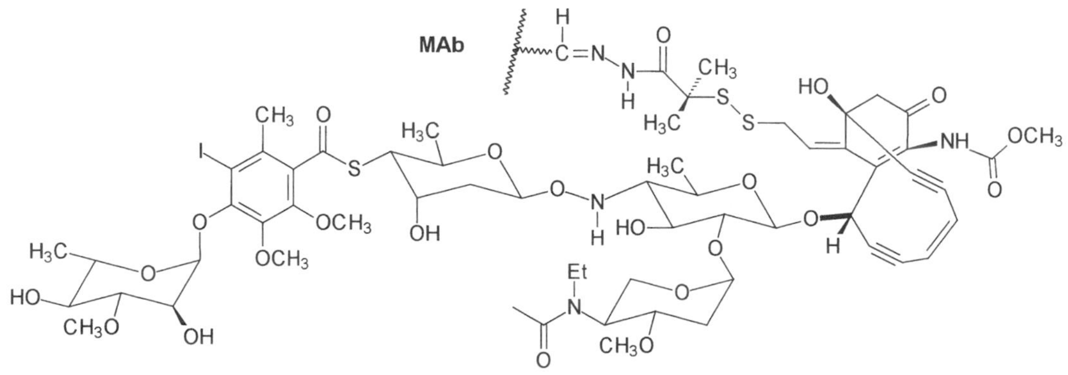

Figure 32.30 Structure of the CMA-676 calicheamicin immunoconjugate. (From Ref. 137.)

Table 32.6 Problems of Delivery of Monoclonal Antibodies and Some Potential Solutions

Phenomenon	Delivery problem/obstacle	Potential solutions
Low tumor uptake	Heterogeneous Ag expression	Cytokines
		Antibody cocktails
	Poor vascularization	Interleukin 1 + 2
		TNF-α
		Vasoactive agents
		Antibody fragments
	Small amounts of mAb binding to the tumor cell	More potent drugs
		Ab cocktails
		Carriers
		Higher doses
Systemic toxicity	Cross-reactivity of mAbs with normal tissue	More selective antibodies
		Interleukins to enhance Ag expression
Host immune response	Murine origin of antibodies	Antibody fragments
		Chimeric/humanized antibodies

by a dilution effect due to the larger volume of humans compared with mice. In addition, several other reasons that prevent effective delivery of the chemoimmunoconjugates to the tumor have been discussed:

1. Monoclonal antibodies are not absolutely tumor specific, and cross-reactivity of the conjugates with normal tissue causes systemic toxicity. For instance, severe gastrointestinal toxicity was observed in clinical trials with the doxorubicin conjugate BR96–DOX due to cross-reactivity of the antibody with these tissues (109–110).
2. The inner regions of solid tumors are often poorly vascularized and exhibit a relatively low blood flow (144–146). Both factors reduce the amount of chemoimmoncojugate reaching these parts of the tumor.
3. Heterogeneity of antigen expression by tumor cells restricts the amount of cells that can effectively be targeted by antibody conjugates. Tumor cells that express the associated antigen only at a low level or not at all will most likely escape therapy due to their failure to bind the mAbs.
4. To some extent, antigens are secreted into the circulation (147). The ''shedding'' of antigen from the surface of tumor cells limits the amount of immunoconjugate accumulating in the tumor because soluble antigens neutralize the chemoimmunoconjugate in the circulation.
5. Murine antibodies and respective immunoconjugates evoke a human antimouse antibody (HAMA) immune response when administered to humans (93), which leads to their rapid neutralization and clearance from the bloodstream (92).
6. Conventional anticancer drugs are only moderately cytotoxic, and a large number of drug molecules have to be internalized to cause cell death. This is difficult to achieve by antibody delivery, as only a limited number of drug molecules can be linked to the immunoglobulin without diminishing its immunoreactivity (103).

Consequently, a number of strategies for improving the delivery of chemoimmunoconjugates to the tumor have been considered.

1. Increasing the Administered Dose

The rate of diffusion of chemoimmunoconjugates from the blood to the tissues is dependent on the concentration gradient between the plasma and the extravascular space (144).

Thus increasing the administered dose of the conjugate generally results in higher drug levels at the tumor site, but it also increases the concentration of conjugate in the blood and normal tissues. Furthermore, high concentrations of immunoconjugate in the blood may lead to saturation of binding sites on the tumor, resulting in decreased tumor uptake (''binding-site barrier'') (148–150).

2. Regional Administration of Immunoconjugates

Administration of chemoimmunoconjugates via an intercavitary route provides the advantages of (1) increased concentration of the agent at the tumor site and (2) decreased systemic toxicity to normal organs, such as the bone marrow or the gastrointestinal tract.

The route of local arterial injection at the tumor site has been evaluated for the neocarzinostatin immunoconjugate

A7-NCS (151). In contrast to systemic injection, the regional administration of the A7-NCS immunoconjugate produced a significant tumor reduction and was effective in preventing tumor development with concomitant low toxicity.

3. Overcoming Antigenic Heterogeneity

The homogeneity of antigen expression is a critical determinant for the therapeutic efficacy of chemoimmunoconjugates (152). Several biological response modifies, such as interferon-alpha (IFN-α), interferon-gamma (IFN-γ) and tumor necrosis factor alpha (TNF-α), are capable of modulating the expression of certain tumor antigens, both in vitro and in vivo (153–155).

Interleukin-2 in combination with antibodies and an interleukin-2 immunotoxin was shown to overcome antigen heterogeneity by inducting a cellular immune response (156,157). Furthermore, a combination of antibodies may be useful for targeting all of the cancer cells. In clinical studies with chemoimmunoconjugates that were mentioned before, the use of immunoconjugate ''cocktails'' has been evaluated by different groups with good clinical results (124,158).

4. Engineered Antibodies

The large size of the IgG molecules impairs their diffusion into extravascular tumor regions (144). A variety of chemically digested or recombinantly prepared antibody fragments of lower molecular weight have therefore been developed and evaluated in preclinical models (99). Commonly used antibody-based structures include $F(ab')_2$, Fab, and Fv fragments (see Fig. 26). Immunoconjugates derived from these compounds usually exhibit a faster and deeper diffusion into the tumor tissue (160,161). However, renal clearance rates are also enhanced because the reduced molecular weight falls below the exclusion size of the kidney, which can lead to increased renal toxicity.

The HAMA response in humans was significantly reduced by the use of chimeric (96) and humanized (139) antibodies. However, it cannot be eliminated completely, since the mouse variable and hypervariable regions may still elicit an immune response.

5. Enhancement of Tumor Vascularization

Several cytokines, including interferons, tumor necrosis factor α, and interleukin-2 along with some novel drugs, can increase the vascular permeability and thus enhance the uptake of macromolecules in tumors (142). Several preclinical studies have demonstrated improved targeting of radiolabeled mAbs to tumor xenografts in mice that were treated with some of these compounds (82,156).

6. Increasing the Cytotoxicity of Chemoimmunoconjugates

In order to increase the drug loading ratio of chemoimmunoconjugates, intermediate carrier molecules, such as dextran (103,162) or human serum albumin (163–166), have been developed, but none of these constructs have reached an advanced stage in preclinical development.

Another possibility of enhancing the cytotoxicity of a chemoimmunoconjugate is to use drugs that are much more cytotoxic than conventional anticancer agents, such as calicheamicins. The feasibility of this approach has been demonstrated in recent clinical trials with a calicheamicin-immunoconjugate CMA-676 (see Section III.C)

IV. DRUG CONJUGATES WITH SERUM PROTEINS

During the past 40 years numerous studies have demonstrated that the blood proteins albumin and transferrin accumulate in tumor tissue (167–171). Tumor uptake has been demonstrated by labeling these proteins with ^{67}gallium, ^{68}gallium, ^{111}indium, ^{97}ruthenium. or ^{103}ruthenium, and ^{131}iodine, respectively, or with dyes such as evans blue or fluorescein compounds. Relevant diagnostic studies with albumin, transferrin and low-density lipoprotein (LDL) have been recently reviewed (171).

As an example of a fairly new technique to determine the uptake of labeled proteins, Fig. 31 shows the results of near-infrared imaging with an indotricarbocyanine albumin conjugate (structures are depicted in Fig. 32) in a tumor-bearing nude mouse model. In near-infrared imaging, suitable fluorescent dyes are administered intravenously to tumor-bearing mice, and these are then exposed to radiation in the near-infrared (> 740 nm). The resulting emission spectrum is recorded with a camera and analyzed with a software program, thus revealing the in vivo distribution pattern of the respective dye. As can be seen in Fig. 30, there is a high rate of tumor accumulation for the albumin conjugated form (MW ~ 67,500 g/mol).

Besides the diagnostic evidence for tumor uptake, albumin and transferrin are suitable as drug carriers for a number of reasons: (1) tumor cells express high amounts of specific transferrin receptors on their cell surface (172–177); (2) they accumulate in tumor tissue and are taken up by proliferating tumor cells through endocytosis, probably in order to cover nutritional needs, i.e., amino acids and iron (178–182); (3) they are biodegradable, nontoxic, and

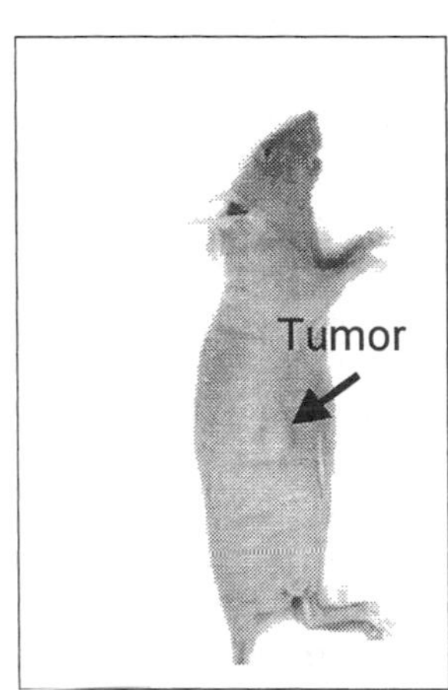

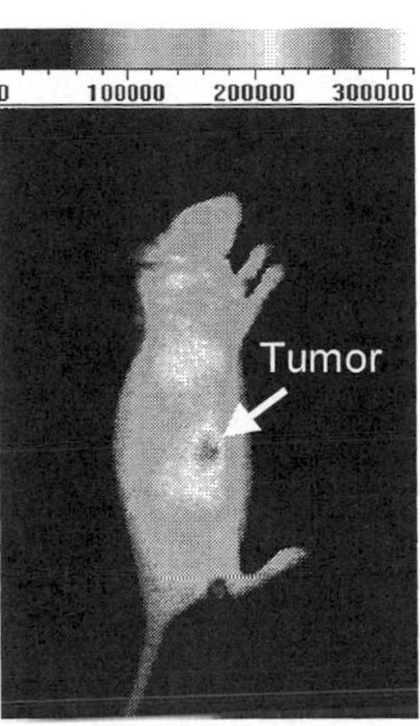

Figure 32.31 Near-infrared imaging in a nude mouse bearing a subcutaneously growing colon tumor 24 h after intravenous administration of an acid-sensitive indotricarbocyanine albumin conjugate (see Fig. 31). Tumor localization is indicated by an arrow. (From cooperation with the IDF of the University of Berlin, FRG.)

nonimmunogenic (4) they are readily available in a pure form exhibiting good biological stability. In the past, various transferrin and albumin conjugates have been developed with clinically established anticancer drugs and some representative examples are described in the following. Direct coupling methods employing glutaraldehyde for cross-linking doxorubicin and daunorubicin to either albumin or transferrin have resulted in active conjugates. Faulk et al. have shown that transferrin conjugates of doxorubicin show high in vitro activity (183). These transferrin doxorubicin conjugates have been used in the treatment of patients with acute leukemia, and the results have shown diminished numbers of leukemic cells in peripheral blood, no increase of leukemic cells in bone marrow biopsies, and no anaphylactic reaction in the patients (184). When doxorubicin was coupled to bovine serum albumin by the same glutaraldehyde method, Ohkawa et al. demonstrated that this albumin conjugate was able to overcome multidrug resistance in a rat hepatoma model (185).

The importance of realizing a point of cleavage in the spacer between the drug and the carrier protein was established through experiments by Trouet et al. in the development of peptide spacers (46). Daunorubicin was bound to succinylated albumin through a series of peptide spacer arms, and the resulting conjugates differed significantly in their antitumor activity against L1210 leukemia depending on the cleavability of the oligopeptide by lysosomal enzymes. Albumin daunorubicin conjugates containing distinct tri- and tetrapeptide spacers have shown high antitumor activity in in vivo models when administered i.p. (46).

We have recently developed a number of transferrin and albumin conjugates with the anthracyclines doxo- and daunorubicin and the alkylating agent chlorambucil (35,36,60,61). In these protein conjugates the drug was bound to the protein through an acid-sensitive carboxylic hydrazone bond or through an amide and ester bond, respectively.

The in vitro and in vivo activity of these conjugates has been studied intensively (35,36,60,61,186–188). A common theme that emerged from these studies was that the conjugates in which the anthracycline was linked to the protein through a stable amide bond showed no or only marginal inhibitory effects, whereas the acid-sensitive anthracycline protein conjugates demonstrated cytotoxic activity that was, on the whole, comparable to that of the free anthracycline. Interestingly, a comparison of analogous transferrin and albumin anthracycline conjugates showed a very similar picture of in vitro activity, i.e., inhibitory effects did not depend on the carrier protein but rather on the chemical link realized between the drug and the protein. This fact was confirmed with our work on chlorambucil protein conjugates (36,61).

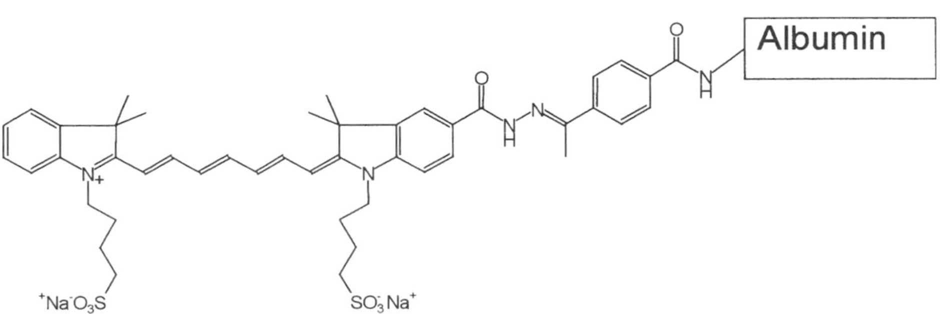

Figure 32.32 Structure of an acid-sensitive indotricarbocyanine albumin conjugate.

The antitumor activity of acid-sensitive doxorubicin transferrin and albumin conjugates has been evaluated in a number of xenograft mammalian carcinoma models in a direct comparison with free doxorubicin (187). The transferrin and albumin conjugates showed significantly reduced toxicity (reduced lethality and body weight loss) at low doses with a concomitantly stable and improved antitumor activity of the transferrin and albumin conjugates compared to free doxorubicin at equitoxic doses. In accordance with the in vitro data we did not observe a pronounced difference between the transferrin and albumin conjugates with regard to in vivo efficacy in the tested animal models. Additionally, an acid-sensitive doxorubicin albumin conjugate induced a complete remission of primary kidney tumors against murine renal cell carcinoma and prevented the formation of metastases in the lungs, in contrast to therapy with doxorubicin at its optimal dose (188).

The folate antagonist methotrexate (MTX) is a further anticancer drug that has been bound serum proteins such as HSA. Chu and Whiteley have shown that MTX–albumin conjugates, synthesized through direct coupling with carbodiimides, are more effective than the free drug in reducing the number of lung metastases in mice inoculated subcutaneously with Lewis lung carcinoma (189–191). However, this conjugate did not exhibit improved activity over free MTX in prolonging the life span of L1210-bearing mice (192). Research on MTX-HSA conjugates carried out a decade later focused on the required coupling techniques in more detail (193–195). Preparation methods using either the N-hydroxysuccinimide ester of MTX or the assistance of water-soluble carbodiimides were carried out. Both methods were found to have advantages and drawbacks in terms of purity, stability, efficiency, and simplicity. MTX–HSA conjugates, synthesized by direct coupling using carbodiimides, have shown a reduction in tumor size and pulmonary metastases in B16 melanoma-bearing mice compared to free MTX, but no difference in overall survival was observed (194). It should be noted that the MTX/HSA ratio in these conjugates was high, about 26 molecules of MTX being bound to one molecules of HSA.

Sinn and coworkers have very recently emphasized that the drug loading rate determines the tumor-targeting properties of MTX albumin conjugates in rats. In a systematic study, in which the tumor uptake of MTX–albumin conjugates loaded with 1, 5, 7, 10, or 20 molecules of MTX was compared in Walker-256 carcinoma-bearing rats, only loading rates of close to one equivalent of MTX per molecule of albumin offered optimal tumor-targeting properties comparable to unmodified albumin (196).

This MTX–HSA conjugate loaded with one equivalent of MTX by reacting an intermediate derivative of MTX, i.e., a reactive ester formed by reaction with N-hydroxysuccinimide and dicyclohexylcarbodiimide, with HSA, has shown high antitumor efficacy in xenograft nude mice models, and a clinical phase I study has been performed with 17 patients in Germany (197). An interesting finding of this study was that two patients with renal cell carcinoma and one patient with mesothelioma responded to MTX–HSA therapy (one partial response, two minor responses). Phase II studies are ongoing.

A further anticancer agent that has been bound to serum albumin as well as to transferrin is mitomycin C. Coupling was performed by binding an activated ester of glutarylated mitomycin C to bovine serum albumin or human serum transferrin (198,199). The resulting imide bond at the aziridine group of the drug is relatively unstable, and mitomycin C release appears to be through hydrolysis in the extracellular space of tumor tissue. The albumin mitomycin C conjugate was shown to accumulate in tumor tissue, to suppress tumor growth, and to increase the survival rate of tumor-bearing mice.

V. DRUG CONJUGATES WITH GROWTH FACTORS AND CYTOKINES

Another approach for delivering therapeutic agents to tumor cells that has been investigated in the past 20 years is the coupling of anticancer drugs or toxins to peptide growth factors and cytokines, which are often involved in the mitogenic signaling of tumors (200,201).

A. Drug Conjugates with Growth Factors

Peptide growth factors bind with high affinity to their receptors on the surface of malignant cells, resulting in intracellular responses that cause differentiation, growth, and survival of these cells (202,203). The specific interaction between ligand and growth factor receptor triggers a cascade of intracellular biochemical signals, resulting in the activation or suppression of various subsets of genes that are linked to cancer and that promote cell division and cell replication (200,204). Because large numbers of growth factors are found on the surfaces of various tumor cells, growth factors are potential carriers of cytotoxic agents for selective drug delivery.

In the past, different kinds of peptide growth factors have been studied with regard to their structure, to their origin, to their site of action, and to the characterization of the respective receptor. Some of these properties are listed in Table 7.

Coupling toxins to growth factors is a relatively new method for drug delivery due to their only recent characterization and restricted availability. Nevertheless, estab-

Table 32.7 Properties of a Selection of Polypeptide Growth Factors

Growth factor (GF)	Description	Target cells	Receptors
Epidermal GF (EGF) Transforming GF α (TGF-α)	Major forms are 6 kD with some larger species detected, EGF and TGF-α proteins are 40% identical; both are released by proteolysis of membrane-bound precursors	Epithelia, mesenchymal and glial cells	Protein tyrosine kinase (175 kD) Product of the c-ERB protooncogene; receptor for EGF, TGF-α and vaccina virus growth factor
Transforming GF (TGF): TGF-β1, TGF-β2, and TGF-β3	25 kD homodimers; secreted as latent complexes	Wide variety of cells	Type 1, 50 kD; type 2, 70 kD; type 3, 280–330 kD; each type binds TGF-β1, β2, and β3; type 1 may be main mediator of responses
Fibroblast GF (FGF): FGF1 and FGF2	Acidic and basic FGF, respectively; 16–17 kD; occasionally larger forms; 55% homologous; also related to other FGFs and interleukin-1 family	Variety of endothelial, epithelial, mesenchymal, and neuronal cell types	(150 kD) FGF1 and (130 kD) FGF2 receptors, both protein tyrosine kinases; high cross-reactivity of FGF1 and FGF2 with respect to receptor binding
Plateted-derived GF (PDGF): PDGF AA, AB, and BB	Dimers of A (17 kD) and B (16 kD) chains; B chain is product of c *SIS* proto-oncogene	Mesenchymal cells, glial cells, smooth muscle cells, placental trophoblasts	Two species of glycoproteins; both tyrosine kinases; type α (170 kD) binds all PDGF dimers; type β (180 kD) binds PDGF BB and AB weakly
Vascular endothelial-GF (VEGF)	Dimers composed of two subunits of identical weight (2 × 23 kD)	Wide variety of cells	Protein tyrosine kinase; *fms*-like, (160 kD)
Insulin-GF (IGF): IGF1 and IGF2	7 kD, both growth factors are related to each other and to proinsulin	Wide variety of cells	IGF1 receptor (130 kD + 90 kD), protein tyrosine kinase that binds IGF1 and IGF2; IGF2 receptor (250 kD) is identical to mannose-6-phosphate receptor and binds solely IGF2

Source: Ref. 205.

lished methods for providing recombinant peptides in large amounts, e.g., through gene splicing DNA techniques (206), has led to an increased interest in using peptide growth factors as carrier systems. In this chapter, several important peptide growth factors and their conjugates with anticancer agents and toxins are described.

1. Epidermal Growth Factor and Transforming Growth Factor

The epidermal growth factor family of proteins encompasses several growth factors that are structurally and functionally related proteins, such as the epidermal growth factor (EGF) and the transforming growth factor (TGF-α) (207). Both growth factors interact with the same cell surface receptor which is overexpressed in several tumor types.

a. Epidermal Growth Factor (EGF)

Among the growth factors that have been purified to date, EGF is one of the most biologically potent and best characterized with respect to its physical, chemical, and biological properties (208). EGF is a single-chain polypeptide composed of 53 amino acids with an isoelectric point at pH 4.6 and a molecular weight of 6045 Da (208). The biological effect of EGF, which is secreted by the salivary glands and by specialized enteroendocrine cells found in the small intestine (209,210), is to promote proliferation of the basal cell layer of various epithelia of ectodermal origin.

The epidermal growth factor receptor (EGF-R) is frequently implicated in human cancer and binds EGF as well as TGF-α. Enhanced expression of EGF-R has been implicated in human cancers of the brain, head, neck, lung, ovary, and breast (211). Furthermore, patients with tumors that overexpress EGF-R often have a poor prognosis. Considering the high affinity between EGF-R and its ligand ($K_d \cong 2–4 \times 10^{-10}$ M) and the fact that the receptor–ligand complex is internalized (208), exploitation of EGF as a vehicle for drug delivery is a potential strategy of achieving tumor targeting (212).

In the past years, EGF was coupled to different anticancer drugs, toxins, or radionuclides by several groups, and the antitumor activity of the respective conjugates was evaluated with respect to the amount of EGF-R on the cell membranes of different cell lines (213–219). A first example is described by Acostata and coworkers who prepared a conjugate of EGF with the anticancer drug doxorubicin using glutaraldehyde as the coupling reagent (213). Subsequent in vitro studies in tumor cell lines showed that the conjugate exhibited the same binding capacity to EGF-R as free EGF in breast cancer cells, and there was no loss of cytotoxicity of conjugated doxorubicin. Furthermore, the cytotoxicity of the conjugate was dependent on the amount of EGF-R on the cell membranes (214).

In another work, conjugation of EGF with ricin, a toxic glycoprotein from castor beans, is described (215). The EGF–ricin conjugate was cytotoxic at much lower concentrations than required for ricin itself.

Jinno et al. reported on human and murine EGF conjugates with mammalian pancreatic ribonuclease (RNase) (214). The EGF–RNase conjugate retained potent RNase activity and showed dose-dependent cytotoxicity against EGFR-overexpressing A431 human squamous carcinoma cells as well as against other squamous carcinoma and breast cancer cell lines. In all cases, the efficacy of the conjugates correlated positively with the level of expression of EGF-R in each cell line.

In vitro as well as in vivo experiments were performed in order to study the antitumor activity of EGF conjugates with phthalocyanines, i.e., disulfochloride cobalt phthalocyanine, against murine B16 melanoma (216). The conjugate possessed higher cytotoxic activity in vitro in comparison to the unbound agent. Subsequent in vivo results in mice treated intravenously with the conjugate showed tumor growth inhibition as well as increased mean life spans and survival of the experimental animals.

Further in vivo experiments were described using a conjugate of EGF and the isoflavone genistein (217). The in vivo toxicity profile, the pharmacokinetics, and the anticancer activity of the conjugate were studied. The conjugate significantly improved tumor-free survival in a severe combined immune deficiency (SCID) mouse model of subcutaneous human breast cancer at systemic exposure levels nontoxic to mice when administered intraperitoneally. Furthermore, plasma samples from EGF–genistein-treated monkeys exhibited potent and EGF-R-specific in vitro antitumor activity against EGF-R-positive human breast cancer cell lines (217).

Another example for using EGF as a carrier system is a targeted DNA delivery system developed for the possible treatment of lung cancer (218). EGF was chemically modified with poly-L-lysine (PLL), thus allowing the attachment of DNA to the conjugate. In vitro experiments showed that EGF/PLL complexed with DNA could efficiently deliver DNA into several lung cancer cells by receptor-mediated endocytosis.

Finally, Zhao et al. studied a mouse EGF-dextran-tyrosine conjugate in vitro. Murine EGF was coupled to dextran (70 kD) by reductive amination, and subsequently tyrosine residues were introduced onto the dextran surface to allow the attachment of radionuclides, such as ^{125}I (219). Exposure of the conjugate to dextranase showed that iodine was primarily attached to the dextran part of the conjugate. The conjugate was effectively internalized in human glioma cells, and radioactivity was retained inside the cells for up to 50 h.

b. *Transforming Growth Factor α (TGF-α)*

TGF-α is a 50 amino acid protein that is derived from a 170 amino acid precursor protein, pro-TGF-α, and that acts exclusively through the EGF receptor (220). TGF-α expression is prevalent in tumor-derived cell lines, e.g., in non-small-cell lung cancer cell lines, and in cells transformed by cellular oncogenes, retroviruses, and tumor promoters; its biological function is to induce cell proliferation (221).

Although it is more common to conjugate drugs or toxins with the epidermal growth factor, several studies have been performed in order to investigate the potential of TGF-α as a drug carrier. For this purpose, recombinant toxins containing TGF-α and Pseudomonas exotoxin 40 have been developed (222). This technique has the advantage of avoiding the difficulties of chemical conjugation methods that can produce heterogeneous products that affect receptor binding (223). The former chimeric protein exhibits the receptor-binding properties of TGF-α and the cell-killing activity of Pseudomonas exotoxin 40. Pseudomonas exotoxin is a single-chain protein made up of three major domains: domain Ia binds to the cell surface, domain II catalyzes the translocation of the toxin into the cytosol, and domain III contains the ADP-ribosylation activity that inactivates elongation factor 2 and leads to cell death (224). In vitro experiments showed that the conjugate is extremely cytotoxic against a variety of cancer cell lines including liver, ovarian, and colon cancer cell lines, indicating high levels of EGF receptor expression in these cells (225).

Similar results were obtained when using conjugates with derivatives of Pseudomonas exotoxin, which are slightly modified forms of the toxin protein (226). In vivo experiments with a modified conjugate were performed in tumor-bearing nude mice that either expressed the EGF receptor or not. The results showed that mice bearing tumors that express EGF receptors lived significantly longer when treated intraperitoneally with the conjugate compared to saline-treated controls (101 vs 47.5 days) (222). The conjugate did not prolong the survival of mice bearing tumor cells that lack EGF-R (15.5 vs. 19.5 days). When given i.p., the conjugate induced moderate hepatic necrosis.

In further studies, a recombinant toxin composed of human TGF-α fused to a fragment of Pseudomonas exotoxin (PE38), devoid of its cell-binding domain, was coupled to monomethoxy-polyethylene glycol (PEG) of different molecular weights (range 10–63 kDa), and the conjugates were tested for their in vivo activity (227). The experiments showed that PEGylated conjugates exhibited markedly enhanced antitumor activities when compared to the unmodified chimeric toxin and that they had extended lifetimes in the blood when administered intravenously to tumor-bearing mice.

2. Vascular Endothelial Growth Factor (VEGF)

The growth of solid tumors beyond a size of 2 mm in diameter is dependent on new blood vessel formation through angiogenesis (228). This complex process is mediated by various tumor-derived growth factors such as the vascular endothelial growth factor (VEGF) (203). VEGF was first purified by Gospodarowicz et al. and Ferrara and Henzel in the late 1980s from the conditioned medium of bovine pituitary folliculo stellate cells and shown to be an approximately 46-kDa protein that dissociates into two apparently identical 23-kDa subunits upon reduction (229–231). Binding of VEGF to its specific receptor causes a mitogenic signal resulting in cell proliferation and growth of capillary endothelial cells (232). VEGF is an angiogenic factor that simulates endothelial cell proliferation and migration and can induce local permeability changes in blood vessels that lead to capillary leak and fibrin deposition (233).

The use of VEGF as a carrier system in order to block tumor neovascularization was intensively studied by Ramakrishnan et al., who investigated whether VEGF could be used to deliver toxin polypeptides to vascular endothelial cells (234). A conjugate was synthesized linking a recombinant isoform of VEGF to a truncated form of the diphtheria toxin. This conjugate was investigated for its ability to inhibit vascular endothelial cell proliferation and angiogenesis in human umbilical vein endothelial cells and in human microvascular endothelial cells. Treatment of the cells with the conjugate resulted in a selective, dose-dependent inhibition of growth. In an in vivo model of angiogenesis, the conjugate showed marked inhibition of neovascularization. Further in vivo experiments were performed to investigate the antitumor activity of the chemically linked conjugate (235). Athymic nude mice bearing established subcutaneous tumors were treated i.p. with VEGF toxin, which evoked an antitumor response. Histological analyses showed tumor necrosis originating from vascular injury, and the conjugate had no effect on well-vascularized, normal tissues.

However, recent in vitro investigations with similar chimeric proteins containing isoforms of VEGF and diphtheria toxin were found to be highly toxic to proliferating endothelial cells but not to vascular smooth muscle cells (236). The fusion proteins were also active in in vivo experiments with mice bearing s.c. Kaposi's sarcoma, a tumor type that expresses high levels of VEGF receptor and that is commonly seen in patients with HIV-1 (237). The in vivo results of subcutaneously administered conjugate showed that the protein substantially retarded the growth of Kaposi's sarcoma (236).

3. Fibroblast Growth Factor (FGF)

Several factors inducing angiogenesis have been isolated and characterized, among them the acidic and basic form of the fibroblast growth factor (229). Acidic FGF (aFGF) is a 140 amino acid protein (~ 16 kDa) that was isolated from tissue of the bovine brain and the retina (238). Basic FGF (bFGF) is a 146 amino acid protein (~ 17 kDa) isolated from several tissues, such as bovine and human brain, retina, and kidney (239). The amino acid sequence of aFGF has a 53% homology to bFGF, and both growth factors interact with the same cell surface receptor. In fact, in normal adult tissue, the expression of FGF receptors is either absent or very low. However, in diseased tissues, such as cancer, FGF receptors are overexpressed (240). Both aFGF and bFGF are potent mitogens for vascular and capillary endothelial cell types and stimulate the formation of blood capillaries (angiogenesis) in vitro and in vivo (241,242).

Conjugation of toxins with FGF to achieve selective antitumor activity has been investigated in a few cases. In the early 1990s, several scientists investigated the in vitro and in vivo antitumor activity of a chemically linked conjugate of bFGF and the ribosome-inactivating protein saporin (243). In vitro experiments in several human cancer cell lines expressing FGF receptors showed high cytotoxicity of the conjugate. In addition, in in vivo experiments in nude mice bearing human tumor xenografts, the conjugate given intravenously demonstrated dramatic tumor growth inhibition and minimal toxicity (243,244). Later, a fusion protein containing the full-length sequences of bFGF and saporin was expressed in Escherichia coli that demonstrated high binding affinity to FGF receptors (245). In in vivo experiments against B16-F10 melanoma, the subcutaneously administered recombinant protein proved to have the same therapeutic profile as the chemically linked conjugate (246). Recently, in vitro studies with both chemically conjugated and fusion bFGF–saporin protein in human bladder cancer cell lines showed high antiproliferative activity (247).

In other studies, chimeric proteins composed of aFGF and several forms of Pseudomonas exotoxin were investigated for their cytotoxic activity in a variety of tumor cell lines including hepatocellular, prostatic, colon, and breast carcinomas (248). The cytotoxic effects were FGF-receptor-specific as demonstrated by competition with excess aFGF and by showing that aFGF toxin bound to the FGF receptor with the same affinity as aFGF. Furthermore, it could be shown that cell-killing was toxin mediated.

B. Immunotoxins with Interleukins

Among the cytokines, interleukins have been extensively studied as carriers of toxins, and we will restrict our discussion to respective immunotoxins with these cytokines.

The interactions between immune and inflammatory cells are mediated to a large part by proteins termed interleukins (IL) that are able to promote cell growth, differentiation, and functional activation. The name interleukin was chosen because it reflected the basic property of these mediators to serve as communication links between leukocytes (201). During the last two decades, the molecular structure of a number of interleukins (IL-1–IL-15) has been described in detail, and their genes have been cloned. Each interleukin has its unique biological activity, but some functions overlap. Physicochemical and biological properties of interleukin 1–8 are presented in Table 8.

The interest of synthesizing immunotoxins with interleukins is increasing due the promising results obtained in the past. Clinical phase 1 and 2 trials have been performed in some cases, e.g., with a chimeric conjugate of interleukin-2 and diphtheria toxin (250). In the following, immunotoxins with interleukins IL-2, IL-4, and IL-6 are presented as prominent examples.

1. Immunotoxins with Interleukin-2

Interleukin-2 (IL-2) is a 15.5 kDa glycoprotein with a slightly basic isoelectric point; its cDNA was first isolated and characterized by Taniguchi et al. (251). Among the various biological functions of IL-2, the most important are to promote the proliferation, differentiation, and recruitment of T and B cells, natural killer cells, and thymocytes (252–254). In addition, IL-2 causes cytolytic activity in a subset of lymphocytes (201). As a biological response modifier, IL-2 as well as several forms of recombinant IL-2 are in various stages of preclinical development and clinical investigation (255). Patients with renal cell cancer and melanoma, who responded poorly to conventional therapy, have responded to therapy with recombinant IL-2 (256).

The use of IL-2 as a drug carrier in drug delivery was studied mainly as a fusion protein with Pseudomonas exotoxin (PE) or diphtheria toxin. A chimeric immunotoxin was constructed genetically by fusing IL-2 to the amino terminus of a modified form of PE (257). Internalization of IL-2-PE40 via the p55 and p70 subunits of the IL-2 receptor, two individual polypeptide chains of the receptor complex, was studied in several mouse and human cell lines expressing either the p55 or the p70 or both IL-2 receptor subunits. The results demonstrated that internalization of IL-2-PE40 is mediated by either the p55 receptor subunit or by the p70 subunit but is much more efficient when high-affinity receptors composed of both subunits arc present.

The construction of a diphtheria-toxin-related fusion protein has been described in which the native receptor-binding domain of the toxin was genetically replaced with IL-2 (258). This conjugate, DAB389IL-2, was shown to bind to the IL-2 receptor and selectively to kill lymphocytes that express a high-affinity form of the IL-2 receptor. The cytotoxic action of the conjugate was mediated by receptor-mediated endocytosis with subsequent delivery of fragment-A-associated ADP-ribosyltransferase to the cytosol of targeted cells, inducing cell death (259). A multicenter phase I/II clinical trial with DAB389-IL-2 was performed primarily including a large number of lymphoma patients (250). The results of these clinical trials are promising with partial and complete responses being observed.

2. Immunotoxins with Interleukin-4

Interleukin-4 (IL-4) is a 20 kDa glycoprotein produced by activated T lymphocytes. Principally, IL-4 enhances the growth and/or differentiation of T cells (260), activates macrophages, and promotes the growth of granulocytes, mast cell, and erythrocyte colonies (201).

Immunotoxins with IL-4 have primarily been developed

Table 32.8 Properties of Interleukins

Interleukin	M_W[kDa]	Principal origin(s)	Principal site(s) of activity	Major effects
Interleukin-1	13–17	Monocytes, macrophages	T cells, B cells, endothelium, fibroblasts, chondrocytes, hepacytes, osteoclasts, synovial cells	Pyrogenesis, major proinflammation, cellular proliferation, lymphokine synthesis, lymphokine receptor synthesis
Interleukin-2	15.5	T cells	T cells, natural killer cells, lymphokine-activated killer cells, tumor-infiltrating lymphocytes	Cellular proliferation and differentiation
Interleukin-3	28	T cells	Hematopoietic precursors	Hematopoiesis, cellular differentiation
Interleukin-4	20	T cells	B cells	Cellular proliferation and differentiation
Interleukin-5	4	T cells	B cells, eosinophils	Cellular proliferation and differentiation
Interleukin-6	20–30	Monocytes, macrophages	B cells, T cells	Pyrogenesis, cellular proliferation and differentiation
Interleukin-7	25	Bone marrow stromal cells	B-cell precursors	Cellular proliferation
Interleukin-8	8–11	Monocytes	Neutrophils	Chemotaxis, respiratory bursting

Source: Ref. 249.

with Pseudomonas exotoxin (PE) as well as with several mutated forms of this toxin. Chimeric toxins have been constructed by fusing a gene encoding human interleukin-4 with a gene encoding PE40, a noncytotoxic mutant form of PE that retains ADP-ribosylating activity, as well as with genes encoding different mutant forms of PE40 (261). The toxin with PE40 was found to be highly toxic against a murine T cell line bearing high numbers of IL-4 receptors, but had no effect on human cell lines lacking receptors for murine IL-4. In addition, the chimeric protein composed of a mutant form of PE40 with very low ADP-ribosylating activity displayed mitogenic activity similar to that of IL-4 rather than cytotoxic activity. These results demonstrated that ADP-ribosylating activity was essential for cytotoxicity and that the mutant form of the conjugate bound specifically to the IL-4 receptor (261).

A few years later, another chimeric toxin was prepared using human IL-4 and the mutant form of PE (PE4E) in which glutamate residues were substituted for basic amino acids at various positions (262). This conjugate exhibited antitumor effects against cancer cells expressing human IL-4 receptors in vitro. A wide range of human cancer cells express IL-4 receptors. These include hematopoietic and nonhematopoietic malignancies. Among the latter group, distinct tumor cell lines, e.g., colon, breast, stomach, liver, skin, prostate, adrenals, and cervix cell lines, were highly sensitive to the cytotoxic action of the fusion protein.

Further investigations focused on targeting IL-4 receptors in human solid tumor xenografts with IL-4 toxins (263). In these studies, a mutant chimeric protein composed of human IL-4 and a mutant form of PE, named PE38QQR, was tested in vitro and in vivo together with the earlier developed conjugate IL-4-PE4E with respect to their receptor-dependent and dose-dependent antitumor activities. Neither of the chimeric toxins showed antitumor potency when the ADP-ribosylation activity of the toxin was inactivated by mutagenesis. Only the new conjugate caused a complete although transient regression of established solid tumors when administered intraperitoneally to mice (263).

Although the former studies demonstrated high antitumor activity in vivo, the fusion proteins bound to the IL-4 receptor with only ~ 1% of the affinity of native IL-4 (262). Several studies have indicated that the carboxyl terminus of IL-4 is important for receptor binding (264). Therefore, it is likely that the large toxin molecule attached to the carboxyl terminus impairs the binding of IL-4 to the IL-4 receptor. Thus a new strategy has been developed in which the toxin is fused to new carboxyl termini of circularly permuted forms of IL-4 (265). A circular permuted protein is a mutant protein in which the termini have been fused and new termini are created elsewhere in the molecule. This conjugate, IL-4(38-37)–PE38KDEL, manifested improved cytotoxicity against various tumor cell lines and bound to the IL-4 receptor with a 10-fold higher affinity than an IL-4 toxin in which the toxin was fused to the carboxyl terminus of IL-4 (265). In addition, the conjugate was 3–30 fold more cytotoxic in glioblastoma cell lines than IL-4–PE4E (266).

Further experiments were performed in order to assess the in vivo antitumor activity of the conjugate. To date, no curative therapy is available for treating malignant gliomas; recent studies have demonstrated a lack of IL-4 receptor expression on normal brain tissue compared to malignant human glioblastoma cells (267). Intratumoral administration of the IL-4 toxin into U251 glioblastoma flank tumors in nude mice produced a complete remission of small and large tumors in all animals without any evidence of toxicity (268).

Due to the positive results obtained in vivo, toxicologic experiments were performed in mice, rats, guinea pigs and monkeys. Intrathecal administration in monkeys produced high cerebrospinal fluid levels and low systemic toxicity (268). However, intravenous administration of IL-4(38-37)–PE38KDEL to monkeys resulted in reversible grade 3 or grade 4 elevations of hepatic enzymes in a dose-dependent manner. These results indicate that local administration can achieve nontoxic systemic levels of IL-4-toxin (267).

Based on these results, a phase I clinical trial evaluating the effects of an intratumoral administration of IL-4(38-37)–PE38KDEL against recurrent grade IV astrocytoma has been initiated (269).

3. Immunotoxins with interleukin-6

Interleukin-6 (IL-6) is a glycoprotein of 20–30 kDa that is produced by a wide variety of cells, including fibroblasts, monocytes/macrophages, endothelial cells, T and B cells, keratinocytes, and also by a number of tumor types (201). The heterogeneity in size is a function of differential glycosylation of the interleukin. IL-6 has a broad spectrum of cell targets and can thus influence an equally broad array of immune and inflammatory responses. Its varied designations—hybridoma growth factor, plasmacytoma growth factor, hepatocyte-stimulating factor, B-cell-stimulating factor 2, and β_2-interferon—attest to the wide range of activities associated with this interleukin (201,270,271). IL-6 receptors have been identified in large numbers on the surface of malignant bone marrow cells and on tumor cell lines derived from patients with multiple myeloma (272). IL-6 receptors have also been identified on the surface of hepatocellular carcinoma and prostate carcinoma cell lines.

In analogy to the development of immunotoxins with IL-2 and IL-4 (273,274), several fusion proteins containing different mutant forms of *Pseudomonas* exotoxin (PE40, $PE66^{4Glu}$) and IL-6 were prepared and tested for their cytotoxic activity (275). The conjugates were cytotoxic against IL-6 receptor expressing tumor cell types in culture, such as human myeloma or hepatoma cell lines. In vivo studies showed that two selected chimeric conjugates had antitumor activity against subcutaneously growing hepatocellular carcinoma (PLC/PRF/5). This tumor contained about 2300 IL-6 receptors per tumor cell (275).

VI. DRUG CONJUGATES WITH POLYSACCHARIDES

Besides the already-mentioned natural polymers, which have been discussed as carrier systems, e.g., growth factors and the serum proteins albumin and transferrin, several other naturally occurring macromolecules have been investigated as drug carriers of anticancer agents. These biopolymers serve as carriers of information (nucleic acids), catalysts (enzymes), structural elements (fiber-forming proteins, cellulose), food reservoirs (glycogen, polyesters) or transporting molecules (hemoglobin) (276). The most important classes of homologous biopolymers are nucleic acids, proteins, polysaccharides, polyprenes, lignins, and aliphatic polyesters.

In this chapter we focus on the polysaccharides dextran and hyaluronic acid as carriers of anticancer agents. Drug conjugates with other natural substances, such as lectins, nucleic acids, and the polysaccharide alginates, chitins, or chitosans (277–280), will not be addressed in this review.

Polysaccharides are homopolymers or copolymers of various sugars that are almost exclusively hexoses and pentoses. Polysaccharides are usually subdivided into fibrin-forming linear structural polysaccharides (e.g., cellulose, chitin), moderately to strongly branched reserve polysaccharides (e.g., amylose, pectin), and physically cross-linked, gel-forming polysaccharides (e.g., gums, mucopolysaccharides).

Several techniques for coupling anticancer agents to polysaccharides have been developed, the most frequent being periodate oxidation, cyanogen bromide activation, and diazotization (28).

A. Dextrans

Dextrans are natural macromolecules consisting of repeated linear units of covalently linked $(1 \rightarrow 6')$ glycopyranose which are branched at the α-$(1 \rightarrow 4')$ position. As a consequence, only one hydroxyl group of the respective dextran chain can be reduced yielding an aldehyde group at the C1 end (276).

At least three forms of dextran have been used in drug delivery: the native hydroxyl form (Dex-OH), the aldehyde form (Dex-CHO), obtained by mild periodate treatment, and the amino form (Dex-NH_2), derived from treatment of Dex-CHO with a diamine followed by mild borohydride reduction (28). Dextrans of different molecular weight (they range generally between 40 and 70 kDa) have been suggested as drug carrier because they have

Well-defined and repetitive chemical structure
Good water-solubility
High stability, because glucosidic bonds are hydrolyzed only under strong acidic or alkaline conditions
Presence of numerous reactive hydroxyl groups that allow derivatization
Availability of different molecular weights
Low pharmacological activity and toxicity
Protection of conjugated drugs from biodegradation (281)

In the past, various anticancer agents, such as the anthracyclines doxorubicin and daunorubicin, methotrexate, and mitomycin C or 5-fluorouracil, have been coupled to dextrans (282–284). The development of dextran conjugates with the anticancer agents daunorubicin, doxorubicin, and mitomycin C is described below.

1. Dextran Conjugates with Anthracyclines and Mitomycin C

In a series of investigations, daunorubicin and doxorubicin were coupled to dextrans. Dextran–daunorubicin conjugates of molecular weight 40 kDA and 70 kDa were prepared by Schiff base formation followed by reduction with sodium borohydride (285). In vitro studies with these conjugates revealed a cytotoxic activity similar to that of the free drug. Later, Levi-Schaffer et al. showed that the subacute toxicity of the dextran–daunorubicin conjugates was lower than that of free daunorubicin (286).

In related work, doxorubicin linked to oxidized dextran (70 kDa) via Schiff base formation without any further reduction of the conjugate was compared to free doxorubicin with regard to antitumor activity, acute toxicity, and plasma pharmacokinetics in rats following i.v. administration. Extensive in vivo studies in a number of animal tumor models demonstrated that the conjugate had both higher antitumor activity and less toxicity than free doxorubicin. In studies with tumor-bearing mice, growth inhibition was observed in various tumor models including Lewis lung and Walker carcinosarcoma 256 (44). In mouse leukemia models, treatment with the conjugate resulted in an increase of life span. The antitumor effect of the dextran–doxorubicin conjugate was also higher than that of doxorubicin in human tumor xenografts (gastric, colorectal, breast), including tumors resistant to doxorubicin.

Toxicological evaluations of the conjugate have been performed in mice, rats, and dogs (287) demonstrating that the LD_{10} was about five times higher for the conjugate than for doxorubicin. Furthermore, the biodistribution profile of the dextran–doxorubicin conjugate in animals differed from that of doxorubicin. In mice and rats, plasma and tumor levels of the conjugate were higher than those of the free drug (44,288,289).

Based on the preclinical activity of the dextran–doxorubicin conjugate, a clinical phase I trial was performed. The conjugate was administered i.v. as a single dose every 21–28 days. However, significant clinical toxicity (profound but reversible thrombocytopenia and hepatotoxicity) was noted presumably due to uptake of the conjugate by the reticuloendothelial system (287).

The antitumor antibiotic mitomycin C has been coupled to dextrans and the respective conjugates evaluated as chemotherapeutic agents (281). For synthesizing the conjugates, dextrans of various molecular weights (10, 70, 500 kDa) were activated with cyanogen bromide at pH 7.0, and ε-aminocaproic acid, acting as spacer, was subsequently coupled to the glucose moiety. Mitomycin C was attached to the spacer-introduced dextran by means of a carbodiimide-catalyzed reaction (290).

In a series of investigations, the physicochemical properties and the antitumor activity of dextran–mitomycin C conjugates were studied with regard to molecular weight, electric charge, and drug release rates (69). The conjugates liberated active mitomycin C in vitro by simple base-catalyzed hydrolysis, and the release rate was controlled by changing the length of the spacer (291). In addition, the conjugates showed significant antitumor activities in mice bearing P388 leukemia or B16 melanoma in an i.p.–i.p. system, and antitumor efficacy correlated positively with increasing molecular weights (69,291).

In further studies, the relationships between the physicochemical characteristics, the pharmacokinetic properties, and the therapeutic efficacy of the conjugates were investigated. Anionic dextran–mitomycin C conjugates were prepared and compared to the already presented conjugates, which are cationic conjugates due to the positive structure of the dextran chain. Anionic dextran conjugates were synthesized by derivatizing dextran with a spacer molecule, i.e., 6-bromohexanoic acid. Since not all of the spacer arms were conjugated with mitomycin C, the resulting dextran–mitomycin C conjugates are anionic. Only these conjugates demonstrated a prolonged retention in the blood circulation, marked accumulation in subcutaneously growing mouse tumors, and tumor growth inhibition (292,293).

In addition, the therapeutic effect of an intratumoral injection of cationic and anionic dextran–mitomycin C conjugates was evaluated in vivo (294). Usually, the clinical application of regional chemotherapy using antitumor drugs of low molecular weight is limited due to the rapid clearance of the drugs from the injection site while causing severe local damage to normal tissues. Macromolecular prodrugs may offer a therapeutic benefit in this respect (295,296).

Saikawa et al. established an experimental system for assessing drug disposition in the tumor after intratumoral injection using a tissue-isolated tumor preparation of Walker 256 carcinosarcoma (297). In contrast to free mitomycin C, cationic and anionic dextran–mitomycin C conjugates were retained in the tumor for longer periods, demonstrating that the intratumoral clearance of mitomycin C can be greatly retarded by dextran conjugation. The effect was more pronounced in the case of the cationic conjugate, and a certain level of active free drug was achieved in the tumor tissue.

This study was expanded by carrying out perfusion experiments using the same tumor model in order to determine the pharmacokinetics of mitomycin C and dextran–mitomycin C conjugates after intratumoral injection. While mitomycin C disappeared rapidly from the tumor preparation following direct intratumoral injection, cationic and anionic dextran–mitomycin C were retained in the tumor. Drug accumulation was more pronounced for the cationic conjugates of high molecular weight (500 kD). In addition, pharmacokinetic studies and whole autoradiography in rats bearing subcutaneous Walker 256 carcinosarcoma confirmed the significant retention of cationic dextran–mitomycin C in the tumor after intratumoral injection (298).

B. Hyaluronic Acid and Anticancer Drug Conjugates

Hyaluronic acid (HA) is a linear polysaccharide of alternating D-glucuronic and N-acetyl-D-glucosamine units and adopts a three-dimensional structure in solution showing extensive intramolecular hydrogen binding. This restricts the conformational flexibility of the polymer chains and induces distinctive secondary (helical) and tertiary (coiled coil) interactions. HA is one of several glycosaminoglycan components of the extracellular matrix and the synovial fluid of joints. The remarkable viscoelastic properties of HA and commercial cross-linked derivatives account for their usefulness in joint lubrification (299). Protein interactions with HA play crucial roles in cell adhesion, cell motility, inflammation, wound healing, and cancer metastasis (300,301). Due to its immunoneutrality, HA is an excellent building block for developing novel biocompatible and biodegradable biomaterials with potential application in tissue engineering and drug delivery systems (66).

HA is overexpressed at sites of tumor attachment to the mesentery and provides a matrix that facilitates invasion. HA is an important signal for activating kinase pathways (302) and regulating angiogenesis in tumors (303). Moreover, several types of cellular HA receptors respond to HA as a signal (304). Since HA receptors are overexpressed in transformed human breast epithelial cells and other cancers, HA drug conjugates might be able to deliver the drug selectively to these cells.

In a series of investigations, HA has been coupled to doxorubicin and daunorubicin and very recently to paclitaxel (34,305,66). Doxorubicin and daunorubicin were bound to HA ($\sim$ 140 kDa) via condensation or a cyanogen bromide technique (34). In vitro studies showed that the conjugates were less active than the free drug and that they were bound at the cell surface with only small amounts entering tumor cells (305).

The conjugation of the powerful antimitotic agent paclitaxel with HA has been described in a recent study (66) HA–paclitaxel conjugates were synthesized by coupling the 2′-OH paclitaxel-hydroxysuccinimide ester to adipic dihydrazide (ADH) modified HA of molecular weight 1500 kDa (HA–ADH). A fluorescent dye conjugate with HA was also synthesized in order to demonstrate cell targeting and uptake of chemically modified HA using confocal microscopy. HA–paclitaxel conjugates showed selective in vitro toxicity against human cancer cell lines (breast, colon, and ovarian) that are known to overexpress HA receptors, while no toxicity was observed in mouse fibroblast cell lines at comparable concentrations. The modified HA drug carrier was nontoxic. The selective cytotoxicity was consistent with results from confocal microscopy studies, which demonstrated that the fluorescent dye conjugate was taken up by cancer cells (66).

VII. SUMMARY

Coupling anticancer drugs to suitable macromolecular carriers is a promising approach of circumventing the toxic side effects of these agents to normal cells and of improving their efficacy towards malignant cells. In the 1980s, research efforts focused on the development of monoclonal antibodies directed against tumor-associated antigens or receptors as site-directed delivery systems in cancer therapy. Despite the encouraging preclinical results with distinct anticancer drug–antibody conjugates, the antitumor efficacy of these conjugates in clinical trials has been disappointing (for a discussion see Section III).

In the past ten years, tumor-targeting strategies have shifted from active targeting strategies to passive ones due to a more detailed understanding of the anatomy and physiology of solid tumors (see Introduction, EPR effect). Indeed, recent studies comparing the tumor uptake of a monoclonal antibody directed against a tumor-specific antigen and a long-circulating polymer in a human small lung carcinoma model suggests that delivery to the tumor can be equally effective (306).

Seen in this light, the decisive criteria for tumor accumulation in vivo would be the molecular size of the macromolecular carrier and not its specificity and selectivity for an antigen or a receptor on the surface of tumor cells. After tumor uptake, the release of a polymer-bound anticancer drug is then dictated by conditions prevailing in the extracellular space of tumor tissue or in the intracellular compartments after cellular uptake of the conjugate by receptor-mediated, fluid-phase, or adsorptive endocytosis.

A number of anticancer drug conjugates with synthetic polymers [N-(2-hydroxypropyl)methacrylamide (HPMA) doxorubicin conjugates PK1 and PK2; PEG-camptothecin conjugate—see the article by L. Seymour for details] and biopolymers such as albumin (methotrexate albumin conjugate—see Section IV) are currently undergoing phase I and II studies. Together with further preclinical assessment, these studies should lead to a better understanding of the therapeutic potential of competing drug carriers for the treatment of cancer.

ACKNOWLEDGMENTS

We would like to thank the Dr. Mildred-Scheel Stiftung der Deutschen Krebshilfe and the Deutsche Forschungsgemeinschaft for their support.

REFERENCES

1. P. Ehrlich. *Collected Studies on Immunity*. John Wiley, New York, 2, 1906, p. 442.
2. G. Köhler, C. Milstein. Continuous cultures of fused cells secreting antibody of predefined specificity. *Nature 256*: 495 (1975).
3. Y. Matsumura, H. Maeda. A new concept for macromolecular therapeutics in cancer chemotherapy: mechanism of tumoritropic accumulation of proteins of the antitumor agent smancs. *Cancer Res. 46*:6387 (1986).
4. R. K. Jain. Transport of molecules across tumor vasculature. *Cancer Metast. Rev. 6*:559 (1987).
5. H. Maeda,Y. Matsumura. Tumoritropic and lymphotropic principles of macromolecular prodrugs. *Crit. Rev. Ther. Drug Carrier Sys. 6*:193 (1989).
6. L. W. Seymour. Passive tumour targeting of soluble macromolecules and drug conjugates. *Crit. Rev. Ther. Drug Carrier Sys. 9*:135 (1992).
7. R. Duncan, F. Spreafico. Polymer conjugates. Pharmacokinetic considerations for design and development. *Clin. Pharmacokinet. 27*:290 (1994).
8. F. Yuan, M. Deilian, D. Fukumura, M. Leuning, D. A. Berk, V. P. Torchilin, R. K. Jain. Vascular permeability in a human tumor xenograft: molecular size dependence and cutoff size. *Cancer Res. 55*:3752 (1995).
9. H. Maeda, J. Wu, T. Sawa, Y. Matsumura, K. Hori. Tumor vascular permeability and the EPR effect in macromolecular therapeutics: a review. *J. Control Release 65*:271 (2000).
10. R. J. Kreitmann. Immunotoxins in cancer therapy. *Curr. Opin. Immunology 11*:570 (1999).
11. I. Pastan. Targeted therapy of cancer with recombinat immunotoxins. *Biochim. Biophys. Acta 1333*:C1 (1997).
12. G. R. Trush, L. R. Lark, B. C. Clinchy, E. S. Vitetta. Immunotoxins: an update. *Ann. Rev. Immunol. 14*:49 (1996).
13. H. Ringsdorf. Structure and properties of pharmacologically active polymers. *J. Polym. Sci. Polym. Symp. 51*:135 (1975).
14. V. Omelyanenko, P. Kopeckova, C. Gentry, J. Kopecek. Targetable HPMA copolymer–adriamycin conjugates. Recognition, internalization, and subcellular fate. *J. Controlled Release 53*:25 (1998).
15. Z.-R. Lu, P. Kopeckova, J. Kopecek. Polymerizable Fab′ antibody fragments for targeting of anticancer drugs. *Nat. Biotechnol. 17*:1101 (1999).
16. R. Duncan, J. Kopecek. Soluble synthetic polymers as potential drug carriers. *Adv. Polym. Sci. 57*:51 (1984).
17. S. Zalipsky. Chemistry of polyethylene glycol conjugates with biologically active molecules. *Adv. Drug Delivery Rev. 16*:157 (1995).
18. D. Putnam, J. Kopecek. Polymer conjugates with anticancer activity. *Adv. Polym. Sci. 122*:55 (1995).
19. S. S. Wong. *Chemistry of Protein Conjugation and Crosslinking*. CRC Press, Boca Raton, Florida, 1993.
20. G. T. Hermanson. *Bioconjugate techniques*. Academic Press, San Diego, 1996.
21. G. Mattson, E. Conclin, S. Desai, G. Nielander, M. D. Savage, S. Morgensen. A practical approach to crosslinking. *Mol. Bio. Rep. 17*:167 (1993).
22. D. M. Simons. Spacers, probability, and yields. *Bioconjugate Chem. 10*:3 (1999).
23. M. H. Klapper. The independent distribution of amino acid near neighbor pairs into polypeptides. *Biochem. Biophys. Res. Comm. 78*:1018 (1977).
24. J. O. Edwards, R. G. Pearson. The factors determining nucleophilic reactivities, *J. Am. Chem. Soc. 84*:16 (1962).
25. R. G. Pearson, H. Sobel, J. Songstad. Nucleophilic reactivity constants toward mehtyl iodide and trans-[Pt(py)$_2$Cl$_2$], *J. Am. Chem. Soc. 89*:319 (1967).
26. R. G. Pearson, J. Songstad. Application of the principle of hard and soft acids and bases to organic chemistry. *J. Am. Chem. Soc. 90*:1827 (1968).
27. T.-L. Ho. The hard soft acids bases (HSAB) principle and organic chemistry. *Chem. Rev. 75*:1 (1975).

28. G. M. Dubowchik, M. A. Walker. Receptor-mediated and enzyme-dependent targeting of cytotoxic anticancer drugs. *Pharmacol. Ther. 83*:67 (1999).
29. K. L. Law, K. E. Studtmann, R. E. Carlson, T. A. Swanson, A. W. Buirge, A. Ahmad. Tumor reactive cis-aconitylated monoclonal antibodies coupled to daunorubicin through a peptide spacer are unable to kill tumor cells. *Anticancer Res. 10*:845 (1990).
30. B. C. Laguzza, C. L. Nichols, S. L. Briggs, G. J. Cullinan, D. A. Johnson, J. J. Starling, A. L. Baker, T. F. Bumol, J. R. F. Corvalan. New antitumor monoclonal antibody–vinca conjugates LY203725 and related compounds: design, preparation, and representative in vivo activity, *J. Med. Chem. 32*:548 (1989).
31. S. E. Oconnor, B. Imperiali. Modulation of protein structure and function by asparagine-linked glycosylation, *Chem. Biol. 3*:803 (1996).
32. R. A. Firestone, D. Willner, S. J. Hofstead, H. D. King, T. Kaneko, G. R. Braslawsky, R. S. Greenfield, P. A. Trail, S. J. Lasch, A. J. Henderson, A. M. Casazza, I. Hellström, K. E. Hellström. Synthesis and antitumor activity of the immunoconjugate BR96-Dox. *J. Controlled Release 39*: 251 (1996).
33. D. Willner, P. A. Trail, S. J. Hofstead, H. D. King, S. J. Lasch, G. R. Braslawsky, R. S. Greenfield, T. Kaneko, R. A. Firestone. (6-Maleimidocaproyl)hydrazone of doxorubicin—a new derivative for the preparation of immunoconjugates of doxorubicin. *Bioconjugate Chem. 4*:521 (1993).
34. H. Jinno, M. Ueda, K. Enomoto, T. Ikeda, P. Kyriakos, M. Kitajima. Effectiveness of an adriamycin immunoconjugate that recognizes the C-erbB-2 product on breast cancer cell lines. *Surg. Today 26*:501 (1996).
35. F. Kratz, U. Beyer, P. Collery, F. Lechenault, A. Cazabat, P. Schumacher, U. Falken, C. Unger. Preparation, characterization and in vitro efficacy of albumin conjugates of doxorubicin. *Biol. Pharm. Bull. 21*:56 (1998).
36. U. Beyer, T. Roth, P. Schumacher, G. Maier, A. Unold, A. W. Frahm, H. H. Fiebig, C. Unger, F. Kratz. Synthesis and in vitro efficacy of transferrin conjugates of the anticancer drug chlorambucil. *J. Med. Chem. 41*:2701 (1998).
37. J. Carlsson, H. Drevin, R. Axen. Protein thiolation and reversible protein–protein conjugation. *Biochem. J. 173*: 723 (1978).
38. R. S. Greenfield, T. Kaneko, A. Daues, M. A. Edson, K. A. Fitzgerald, L. J. Olech, J. A. Grattan, G. L. Spitalny, G. R. Braslawsky. Evaluation in vitro of adriamycin immunoconjugates synthesized using an acid-sensitive hydrazone linker. *Cancer Res. 50*:6600 (1990).
39. T. Kaneko, D. Willner, I. Monkovic, J. O. Knipe, G. R. Braslawsky, R. S. Greenfield, D. M. Vyas. New hydrazone derivatives of adriamycin and their immunoconjugates—a correlation between acid stability and cytotoxicity. *Bioconjugate Chem. 2*:133 (1991).
40. Y. Kohgo, Y. Niitsu, T. Nishisato, Y. Urushizaki, H. Kondo, M. Fukushima, M. Tsushima, I. Urushizaki. Transferrin receptor of tumor cells—potential tools for diagnosis and treatment of malignancies. *Proteins of iron storage and transport* (G. Spik, J. Montreuil, R. R. Crichton, J. Mazurier, eds.), Elsevier Science Publishers, Amsterdam, 1985, p. 155.
41. D. H. Na, B. H. Woo, K. C. Lee. Quantitative analysis of derivatized proteins prepared with pyridyl disulfide-containing cross-linkers by high-performance liquid chromatography. *Bioconjugate Chem. 10*:306 (1999).
42. N. Bapat, M. Boroujerdi. Effect of colloidal carriers on the disposition and tissue uptake of doxorubicin: I. Conjugation with oxidized dextran particles. *Drug Dev. Ind. Pharm. 19*:2651 (1993).
43. M. Yang, H. L. Chan, W. Lam, W. F. Fong. Cytotoxicity and DNA binding characteristics of dextran-conjugated doxorubicins. *Biochim. Biophys. Acta 1380*:329 (1998).
44. Y. Ueda, K. Munechika, A. Kirukawa, Y. Kanoh, K. Yamanouchi, K. Yokoyama. Comparison of efficacy, toxicity and pharmacokinetics of free adriamycin and adriamycin linked to oxidized dextran in rats. *Chem. Pharm. Bull. 37*: 1639 (1989).
45. M. Sato, H. Onishi, M. Kitano, Y. Machida, T. Nagai. Preparation and drug release characteristics of the conjugates of mitomycin C with glycol-chitosan and N-succinyl-chitosan. *Biol. Pharm. Bull. 19*:241 (1996).
46. A. Trouet, M. Masquelier, R. Baurain, D. Deprez-De Campeneere. A covalent linkage between daunorubicin and proteins that is stable in serum and reversible by lysosomal hydrolases, as required for a lysosomotropic drug–carrier conjugate: in vitro and in vivo studies. *Proc. Natl. Acad. Sci. USA 79*:626 (1982).
47. I. M. Klotz, S. Keresztes-Nagy. Dissociation of proteins into su-units by succinylation: haemerythrin. *Nature 195*: 900 (1962).
48. J. F. Riordan, B. L. Vallee. Succinylcarboxypeptidase. *Biochem. 3*:1768 (1964).
49. B. Frisch, C. Boeckler, F. Schuber. Synthesis of short polyoxyethylene-based heterobifunctional cross-linking reagents. Application to the coupling of peptides to liposomes. *Bioconjugate Chem. 7*:180 (1996).
50. D. H. Rich, P. D. Gesellchen, A. Tong, A. Cheung, C. K. Buckner. Alkylating derivatives of amino acids and peptides. Synthesis of N-maleoylamino acids, [1-(N-maleoylglycyl)cysteinyl]oxytocin, and [1-(N-maleoyl-11-aminoundecanoyl)cysteinyl]oxytocin. Effects on vasopressin-stimulated water loss from isolated toad bladder. *J. Med. Chem. 18*:1004 (1975).
51. U. Beyer, M. Krueger, P. Schumacher, C. Unger, F. Kratz. Synthesis of new bifunctional maleimide compounds for the preparation of chemoimmunoconjugates. *Monatsh. Chem. 128*:91 (1997).
52. H. N. Borah, R. C. Boruah, J. S. Sandhu. Microwaved-induced one-pot synthesis of N-carboxyalkyl maleimides and phthalimides. *J. Chem. Res. 272* (1998).
53. O. Keller, J. Rudinger. Preparation and some properties of maleimido acids and maleoyl derivatives of peptides. *Helv. Chim. Acta 58*:531 (1975).

54. M. A. Walker. A high yielding synthesis of N-alkyl maleimides using a novel modification of the Mitsunobu reaction. *J. Org. Chem. 60*:5352 (1995).
55. M. A. Walker. The Mitsunobu reaction: a novel method for the synthesis of bifunctional maleimide linkers. *Tetrahedron Lett. 35*:665 (1994).
56. F. M. Veronese, M. Morpurgo. Bioconjugation in pharmaceutical chemistry. *Farmaco 54*:497 (1999).
57. A. W. Tolcher, S. Sugarman, K. A. Gelmon, R. Cohen, M. Saleh, C. Isaacs, L. Young, D. Healey, N. Onetto, W. Slichenmyer. Randomized phase II study of BR96–doxorubicin conjugate in patients with metastatic breast cancer. *J. Clin. Oncol. 17*:478 (1999).
58. H. D. King, D. Yurgaitis, D. Willner, R. A. Firestone, M. B. Yang, S. J. Lasch, K. E. Hellstroem, P. A. Trail. Monoclonal antibody conjugates of doxorubicin prepared with branched linkers: a novel method for increasing the potency of doxorubicin immunoconjugates. *Bioconjugate Chem. 10*:279 (1999).
59. P. C. A. Rodrigues, U. Beyer, P. Schumacher, T. Roth, H. H. Fiebig, C. Unger, L. Messori, P. Orioli, D. H. Paper, R. Muelhaupt, F. Kratz. Acid-sensitive polyethylene glycol conjugates of doxorubicin: preparation, in vitro efficacy and intracellular distribution. *Bioorg. Med. Chem. 7*:2517 (1999).
60. F. Kratz, U. Beyer, T. Roth, N. Tarasova, P. Collery, F. Lechenault, A. Cazabat, P. Schumacher, C. Unger, U. Falken. Transferrin conjugates of doxorubicin: synthesis, characterization, cellular uptake, and in vitro efficacy. *J. Pharm. Sci. 87*:338 (1998).
61. F. Kratz, U. Beyer, T. Roth, M. T. Schuette, A. Unold, H. H. Fiebig, C. Unger. Albumin conjugates of the anticancer drug chlorambucil: synthesis, characterization, and in vitro efficacy. *Arch. Pharm. Pharm. Med. Chem. 331*:47 (1998).
62. F. Kratz, R. Mueller-Driver, I. Hofmann, J. Drevs, C. Unger. A novel macromolecular prodrug concept exploiting endogenous serum albumin as a drug carrier for cancer chemotherapy. *J. Med. Chem. 43*:1253 (2000).
63. P. Schelte, C. Boeckler, B. Frisch, F. Schuber. Differential reactivity of maleimide and bromoacetyl functions with thiols: application to the preparation of liposomal diepitope constructs. *Bioconjugate Chem. 11*:118 (2000).
64. J. V. Staros. N-hydroxysulfosuccinimide active esters: bis(N-hydroxysulfosuccinimide) esters of two dicarboxylic acids are hydrophilic, membrane impermeant, protein cross-linkers. *Biochem. 21*:3950 (1982).
65. F. Dosio, P. Brusa, P. Crosasso, S. Arpicco, L. Cattel. Preparation, characterization and properties in vitro and in vivo of a paclitaxel–albumin conjugate. *J. Controlled Release 47*:293 (1997).
66. Y. Luo, G. D. Prestwich. Synthesis and selective cytotoxicity of a hyaluronic acid–antitumor bioconjugate. *Bioconjugate Chem. 10*:755 (1999).
67. G. Giammona, L. I. Giannola, B. Carlisi, M. L. Bajardi. Synthesis of macromolecular prodrugs of procaine, histamine and isoniazid. *Chem. Pharm. Bull. 37*:2245 (1989).
68. C. Cera, M. Terbojevich, A. Cosani, M. Palumbo. Anthracycline antibiotics supported on water-soluble polysaccharides: synthesis and physicochemical characterization. *Int. J. Biol. Macromol. 10*:66 (1988).
69. A. Kato, Y. Takakura, M. Hashida, T. Kimura, H. Sezaki. Physicochemical and antitumor characteristics of high molecular weight prodrugs of mitomycin C. *Chem. Pharm. Bull. 30*:2951 (1982).
70. F. Kratz, U. Beyer, M. T. Schütte. Polymer drug conjugates containing acid-cleavable bonds. *Crit. Rev. Ther. Drug Carrier Sys. 16*:245 (1999).
71. H. Soyez, E. Schacht, S. Vanderkerken. The crucial role of spacer groups in macromolecular design. *Adv. Drug Del. Rev. 21*:81 (1996).
72. S. Mukherjee, R. N. Grosh, F. R. Maxfield. Endocytosis. *Physiol. Rev. 77*:759 (1997).
73. K. Sikora, H. Smedley, P. Thorpe. Tumour imaging and drug targeting. *Brit. Med. Bull. 40*:233 (1984).
74. M. J. Poznansky, R. L. Juliano. Biological approaches to the controlled delivery of drugs: a critical review. *Pharmacol. Rev. 36*:277 (1984).
75. A. I. Freeman, E. Mathew. Targeted drug delivery. *Cancer (suppl.) 58*:573 (1986).
76. D. Parker. Tumour targeting. *Chem. Brit. 942* (1990).
77. F. Kratz. Drug Targeting in der antitumoralen Chemotherapie: Antigene und Rezeptoren auf der Tumorzelloberfläche als Angriffspunkte einer selektiven Chemotherapie. *Pharm. i. u. Z. 24*:14 (1995).
78. B. Rihova. Targeting of drugs to cell surface receptors. *Crit. Rev. Biotechn. 17*:149 (1997).
79. B. Rihova. Receptor-mediated targeted drug or toxin delivery. *Adv. Drug Delivery Rev. 29*:273 (1998).
80. C. Milstein. Monoclonal antibodies. *Scientific American 10*:56 (1980).
81. M. V. Pimm. Drug monoclonal antibody conjugates for cancer therapy: potentials and limitations. *CRC Crit. Rev. Ther. Drug Carrier Sys. 5*:189 (1988).
82. G. A. Pietersz, K. Krauer. Antibody-targeted drugs for the therapy of cancer, *J. Drug Targeting 2*:183 (1994).
83. C. Panousis, G. A. Pietersz. Monoclonal antibody-directed cytotoxic therapy, *Drugs & Aging 15*:1 (1999).
84. R. V. J. Chari. Targeted delivery of chemotherapeutics: tumor-activated prodrug therapy, *Adv. Drug Delivery Rev. 31*:89 (1998).
85. S. J. DeNardo, L. A. Kroger, G. L. DeNardo. A new era for radiolabeled antibodies in cancer?, *Curr. Opin. Immunol. 11*:563 (1999).
86. F. R. Appelbaum. Antibody-targeted therapy for myeloid leukemia, *Sem. Hematol. 38*:2 (1999).
87. K. Alpaugh, M. von Mehren. Monoclonal antibodies in cancer treatment, *BioDrugs 12*:209 (1999).
88. R. A. Reisfeld, D. A. Cheresch. Human tumour antigens, *Adv. Immunol. 40*:323 (1987).
89. T. I. Ghose, A. H. Blair. The design of cytotoxic agent-antibody conjugates, *CRC Crit. Rev. Ther. Drug Carrier Sys. 3*:263 (1986).
90. K. Thielemans, D. G. Maloney, T. Meeker, J. Fujimoto,

C. Doss, R. A. Warnke, J. Bindl, J. Gralow, R. A. Miller, R. Levi. Strategies for production of monoclonal anti-idiotype antibodies against human B-cell lymphomas, *J. Immunol. 133*:495 (1984).
91. L. J. Harris, E. Skaletsky, A. McPherson. Cystallographic structure of an intact IgG1 monoclonal antibody, *Journal of Molecular Biology 275*:861 (1998).
92. N. S. Courtenay-Luck, A. A. Epenetos, R. Moore, M. Larche, D. Pectasides, B. Dhorkia, R. A. Ritter. Development of primary and secondary immune responses to mouse monoclonal antibodies used in the diagnosis and therapy of malignant neoplasms, *Cancer Res. 46*:6489 (1986).
93. D. L. Shawler, R. M. Bartholomew, L. M. Smith, R. O. Dillman. Human immune response to multiple injections of murine monoclonal IgG. *J. Immunol. 153*:1530 (1985).
94. G. Winter, C. Milstein. Man-made antibodies. *Nature 349*: 293 (1991).
95. L. Riechmann, M. Clark, W. Herman, G. Winter. Reshaping human antibodies for therapy. *Nature 332*:323 (1988).
96. A. F. LoBuglio, R. H. Wheeler, J. Trang, A. Haynes, K. Rogers, E. B. Harvey, L. Sun, J. Ghrayeb, M. B. Khazaeli. Mouse/human chimeric monoclonal antibody in man: kinetics and immune response. *Proc. Natl. Acad. Sci. USA 86*:4220 (1989).
97. S. L. Morrison. Transfectomas provide novel chimeric antibodies. *Science 229*:379 (1986).
98. P. J. Hudson. Recombinant antibody constructs in cancer therapy. *Curr. Opin. Immunol. 11*:548 (1999).
99. R. A. Farah, B. Clinchy, L. Herrera, E. S. Vitetta. The development of monoclonal antibodies for the therapy of cancer. *Crit. Rev. Eukariotic Gene Expr. 8*:321 (1998).
100. A. H. Blair, T. I. Ghose. Linkage of cytotoxic agents to immunoglobulins. *J. Immunol. Methods. 59*:129 (1983).
101. G. A. Pietersz. The linkage of cytotoxic drugs to antibodies for the treatment of cancer. *Bioconjugate Chem. 1*:89 (1990).
102. G. F. Rowland, R. G. Simmonds, J. R. F. Corvalan, R. W. Baldwin, J. P. Brown, M. J. Embleton, C. H. J. Ford, K. E. Hellström, I. Hellström, J. T. Kemshead, C. E. Newman, C. S. Woodhouse. Monoclonal antibodies for targeted therapy with vindesine. *Protides Biol. Fluids 30*:375 (1982).
103. M. J. Embleton, M. C. Garnett. Antibody targeting of anticancer agents. *Monoclonal Antibodies for Cancer Detection and Therapy* (R. W. Baldwin, V. S. Byers, eds.). Academic Press, Orlando, FL, 1985, p. 317.
104. P. A. Trail. Cure of xenografted human carcinomas by BR96-doxorubicin immuno conjugates. *Science 261*:212 (1993).
105. C. R. Comerski, W. M. Peden, T. J. Davidson, G. L. Warner, R. S. Hirth, J. D. Frantz. BR96-doxorubicin conjugate (BMS-182248) versus doxorubicin: a comparative toxicity assessment in rats. *Tox. Path. 22*:473 (1994).
106. H. O. Sjögren, M. Isaksson, D. Willner, I. Hellström, K. E. Hellström, P. A. Trail. Antitumor activity of carcinoma-reactive BR96-doxorubicin conjugate against human carcinomas in athymic mice and rats and syngenic rat carcinomas in immunocompetent rats. *Cancer Res. 57*:4560 (1997).
107. I. Hellström, H. J. Garrigues, U. Garrigues, K. E. Hellström. Highly tumor-reactive, internalizing mouse monoclonal antibodies to Le(Y)-related cell surface antigens. *Cancer Res. 50*:2183 (1990).
108. S. Sugerman, J. L. Murray, M. Saleh, A. F. LoBuglio, D. Jones, C. Daniel, D. LeBherz, H. Brewer, D. Healeey, S. Kelly, K. E. Hellström, N. Onetto. A phase I study of BR96-doxorubicin (BR96-DOX) in patients with advanced carcinoma expressing the Lewis y-antigen. *Proc. Am. Soc. Clin. Oncol. 14*:A1532 (1995).
109. W. J. Slichenmyer, M. N. Saleh, M. A. Bookman, T. A. Gilewski, J. L. Murray, K. E. Hellström. Phase I studies of BR96 doxorubicin in patients with advanced solid tumors that express the Lewis Y antigen. *Sixth International Congress on Anti-Cancer Treatment*, Paris, 1996, p. 95.
110. B. J. Giantonio, T. A. Gilewski, M. A. Bookman, L. Norton, D. Kilpatrick, M. A. Dougan, W. J. Slichenmyer, N. M. Onetto, R. M. Canetta. A phase I study of weekly BR96-doxorubicin (BR96-DOX) in patients with advanced carcinoma expressing the Lewis Y (Le^Y) antigen. *Proc. Ann. Meet. Am. Soc. Clin. Oncol. 15*:A1380 (1996).
111. P. A. Trail, A. B. Bianchi. Monoclonal antibody drug conjugates in the treatment of cancer. *Curr. Opin. Immunol. 11*:584 (1999).
112. T. F. Bumol, P. Marder, S. V. DeHerdt, M. J. Borowitz, L. D. Apelgren. Characterization of human tumor and normal tissue reactivity of the KS1/4 monoclonal antibody. *Hybridoma 7*:407 (1988).
113. N. M. Varki, R. A. Reisfeld, L. E. Walker. Antigens associated with a human lung adenocarcinoma defined by monoclonal antibodies. *Cancer Res. 44*:681 (1984).
114. N. M. Varki, R. A. Reisfeld, L. E. Walker. Effects of monoclonal antibody–drug conjugates on the in vivo growth of human tumors established in nude mice. *Monoclonal Antibodies in Cancer Therapy* (R. A. Reisfeld, S. Sell, R. Alan, eds.). Alan R. Liss, New York, 1985, p. 207.
115. D. J. Elias, L. Hirshowitz, L. E. Kline, J. F. Kroener, R. O. Dillman, L. E. Walker, J. A. Robb, R. M. Timms. Phase I clinical comparative study of monoclonal antibody KS1/4 and KS1/4–methotrexate immunoconjugate in patients with non-small cell lung carcinoma. *Cancer Res. 50*: 4154 (1990).
116. D. J. Elias, L. E. Kline, B. A. Robbins, H. C. L. Johnson, K. Pekny, M. Benz, J. A. Robb, L. E. Walker, M. Kosty, R. O. Dillman. Monoclonal antibody KS1/4-methotrexate immunoconjugate studies in non-small cell lung carcinoma. *Am. J. Respir. Crit. Care Med. 150*:1114 (1994).
117. T. F. Bumol, A. L. Baker, E. L. Andrews, S. V. DeHerdt, S. L. Briggs, M. E. Spearman, L. D. Apelgren. KS1/4-DAVLB, a monoclonal antibody-vinca alkaloid conjugate for site-directed therapy of epithelial malignancies. *Target. Diagn. Ther. 1*:55 (1988).
118. M. E. Spearman, R. M. Goodwin, L. D. Apelgren, T. F. Bumol. Disposition of the monoclonal antibody–vinca alkaloid conjugate KS1/4-DAVLB (LY256787) and free 4-

desacetylvinblastine (DAVLB) in tumor bearing nude mice. *J. Pharm. Exp. Ther. 241*:695 (1987).

119. B. H. Petersen, S. V. DeHerdt, D. W. Schneck, T. F. Bumol. The human immune response to KS1/4-deacetylvinblastine (LY256787) and KS1/4-deacetylvinblastine hydrazide (LY203728) in single and multiple dose clinical studies. *Cancer Res. 51*:2286 (1991).
120. D. W. Schneck, F. Butler, W. Dungan, D. Littrel, S. Dorrbecker, B. Peterson, R. Bowsher, A. DeLong, J. Zimmermann. Phase I studies with a murine antibody vinca conjugate (KS1/4-DAVLB) in patients with adenocarcinoma. *Antibody Immunoconj. Radiopharm. 2*:93 (1989).
121. D. W. Schneck, F. Butler, W. Dungan, D. Littrell, B. Petersen, R. Bowsher, A. DeLong, S. Dorrbecker. Disposition of a murine monoclonal antibody vinca conjugate (KS1/4 DAVLB) in patients with adenocarcinomas. *Clin. Pharmacol. Ther. 47*:36 (1990).
122. G. Melino, J. R. Hobbs, M. Radford, K. B. Cooke, A. M. Evans, M. A. Castello, D. M. Forrest. Drug targeting for 7 neuroblastoma patients using human polyclonal antibodies. *Proteides Biol. Fluids 32*:413 (1984).
123. Y. Kato, Y. Tsukada, T. Hara, H. Hirai. Enhanced antitumor activity of mitomycin C conjugated with anti-α-fetoprotein antibody by a novel method of conjugation. *J. Appl. Biochem. 5*:313 (1983).
124. R. K. Oldham, M. Lewis, D. W. Orr, S.-K. Liao, J. R. Ogden, W. H. Hubbard, R. Birch. Individually specified drug immunoconjugates in cancer treatment. *Intern. J. Biol. Markers 4*:65 (1989).
125. D. W. Orr, R. K. Oldham, M. Lewis, J. R. Ogden, S.-K. Liao, K. Leung, S. Dupere, R. Birch, B. Avener. Phase I trial of mitomycin C immunoconjugate cocktails in human malignancies. *Mol. Biother. 1*:229 (1989).
126. M. J. Smyth, G. A. Pietersz, I. F. C. McKenzie. Selective enhancement of antitumor activity of N-acetyl melphalan upon conjugation to monoclonal antibodies. *Cancer Res. 47*:62 (1987).
127. J. J. Tjandra, G. A. Pietersz, J. G. Teh, A. M. Cuthbertson, J. R. Sullivan, C. Penfold, I. F. McKenzie, M. Smyth. Phase I clinical trial of drug-monoclonal antibody conjugates in patients with advanced colorectal carcinoma: a preliminary report. *Surgery 106*:533 (1989).
128. A. G. Myers, M. E. Kort, M. Hammond. A comparison of DNA cleavage by neocarzinostatin chromophore and its aglycon: evaluating the role of the carbohydrate residue. *J. Am. Chem. Soc. 119*:2965 (1997).
129. A. L. Smith, K. C. Nicolaou. The enediyne antibiotics. *J. Med. Chem. 39*:2103 (1996).
130. J. Takeshita, H. Maeda, K. Koike. Subcellular action of neocarzinostatin. Intracellular incorporation, DNA breakdown and cytotoxicity. *J. Biochem. 88*:1071 (1980).
131. K. Kitamura, T. Takahashi, T. Yamaguchi, T. Yokota, A. Noguchi, T. Amagai, J. Imanishi. Immunochemical characterization of the antigen recognized by the murine monoclonal antibody A7 against human colorectal cancer. *J. Exp. Med. 157*:83 (1988).
132. T. Takahashi, T. Yamaguchi, K. Kitamura, H. Suzuyama, M. Honda, Y. Yokota, Y. Kotanagi, M. Takahashi, Y. Hashimoto. Clinical application of monoclonal antibody-drug conjugates for immunotargeting chemotherapy of colorectal carcinoma. *Cancer 61*:881 (1988).
133. T. Takahashi, T. Yamaguchi, K. Kitamura, A. Noguchi, M. Honda. Follow-up study of patients treated with monoclonal antibody drug conjugate—report of 77 cases with colorectal cancer. *Jap. J. Cancer Res. 84*:976 (1993).
134. J. Drak, N. Iwasawa, S. Danielshewsky, D. M. Crothers. The carbohydrate domain of calicheamicin γ_1^I determines its sequence specificity for DNA cleavage. *Proc. Natl. Acad. Sci. USA 88*:7464 (1991).
135. S. Walker, R. Landovitz, W. D. Ding, G. A. Ellestad, D. Kahne. Cleavage behaviour of calicheamicin γ1 and calicheamicin T. *Proc. Natl. Acad. Sci. USA 89*:4608 (1992).
136. N. Zein, A. M. Sinha, W. J. McGahren, G. A. Ellestad. Calicheamicin γ_1^I: an antitumour antibiotic that cleaves double-stranded DNA site specifically. *Science 240*:1198 (1988).
137. L. M. Hinman, P. R. Hamann, R. E. Wallace, A. T. Menendez, F. E. Durr, J. Upselacis. Preparation and characterization of monoclonal antibody conjugates of the calicheamicins—a novel and potent family of antitumor antibiotics. *Cancer Res. 53*:3336 (1993).
138. L. M. Hinman, R. E. Wallace, P. R. Hamann, F. E. Durr, J. Upeslacis. Calicheamicin immunoconjugates: influence of analog and linker modification on activity in vivo. *Antibody Immunoconj. Radiopharm. 3*:59 (1990).
139. E. L. Sievers, F. R. Appelbaum, R. T. Spielberger, S. J. Forman, S. J. Flowers, F. O. Smith, K. Shannon-Dorcy, M. S. Berger, I. D. Bernstein. Selective ablation of acute myeloid leukemia using antibody-targeted chemotherapy: a phase I study of an anti-CD33 calicheamicin immunoconjugate. *Blood 93*:3678 (1999).
140. E. Sievers, R. A. Larson, E. Estey. Interim analysis of the efficacy and safety of CMA-676 in patients with AML in first relapse. *Blood (suppl.) 92*:613A (1998).
141. R. M. Reilly, J. Sandhu, T. M. Alvarez-Diez, S. Gallinger, J. Kirsh, H. Stern. Problems of delivery of monoclonal antibodies. *Clin. Pharmacokinet. 28*:126 (1995).
142. J. L. Murray. Factors for improving monoclonal antibody targeting. *Diagn. Oncol. 2*:234 (1992).
143. J. Schlom, P. H. Hand, J. W. Greiner, D. Colcher, S. Shrivastav, J. A. Carrasquillo, J. C. Reynolds, S. M. Larson, A. Raubitschek. Innovations that influence the pharmacology of monoclonal antibody guided tumor targeting. *Cancer Res. (suppl.) 50*:820S (1990).
144. R. K. Jain. Physiological barriers to delivery of monoclonal antibodies and other macromolecules in tumors. *Cancer Res. (suppl.) 50*:814S (1990).
145. H. F. Dvorak, J. A. Nagy, A. M. Dvorak. Structure of solid tumors and their vasculature implication for therapy with monoclonal antibodies. *Cancer Cells 3*:77 (1991).
146. P. W. Vaupel. The influence of tumor blood flow and microenvironmental factors on the efficacy of radiation, drugs and localized hyperthermia. *Klin. Päd. 209*:243 (1997).

147. L. L. Kiessling, E. L. Gordon. Transforming the cell surface through proteolysis. *Chem. Biol. 5*:R49 (1998).
148. M. Juweid, R. Neumann, C. Paik, M. J. Perez-Bacete, J. Sato, W. v. Osdol, J. N. Weinstein. Micropharmacology of monoclonal antibodies in solid tumors: direct experimental evidence for a binding site barrier. *Cancer Res. 52*: 5144 (1992).
149. W. v. Osdol, K. Fujimori, J. N. Weinstein. An analysis of monoclonal antibody distribution in microspheric tumor nodules: consequences of a "binding site barrier." *Cancer Res. 51*:4776 (1991).
150. T. Saga, R. D. Neumann, T. Heya, J. Sato, S. Kinuya, N. Le, C. H. Paik, J. N. Weinstein. Targeting cancer micrometastases with monoclonal antibodies: a binding-site barrier. *Proc. Natl. Acad. Sci. USA 92*:8999 (1995).
151. K. Kitamura, T. Takahashi, T. Kotani, T. Miyagaki, N. Yamaoka, H. Tsurumi, A. Noguchi, T. Yamaguchi. Local administration of monoclonal antibody–drug conjugate: a new strategy to reduce the local recurrence of colorectal cancer. *Cancer Res. 52*:6323 (1992).
152. R. Murano, L. Frati, R. Bei, F. Ficari, C. Valli, D. French, S. Mammarella, F. Caramia, G. Faziz, R. Mariani-Constantini. Regional heterogeneity and complementation of the expression of the tumor-associated glycoprotein 72 epitopes in colorectal cancer. *Cancer Res. 51*:5378 (1991).
153. R. J. Chang, S. H. Lee. Effects on interferon gamma and tumor necrosis factor-alpha on the expression of an 1a antigen on a murine macrophage cell line. *J. Immunol. 137*: 2853 (1986).
154. F. Guadagni, J. Schlom, S. Pothen, S. Pestka, J. W. Greiner. Parameters involved in the enhancement of monoclonal antibody targeting in vivo with recombinant interferon. *Cancer Immunol. Immunother. 26*:222 (1988).
155. F. Guadagni, M. Roselli, J. Schlom, J. W. Greiner. In vitro and in vivo regulation of human tumor antigen expression by human recombinant interferons: a review. *Intern. J. Biol. Markers 9*:53 (1994).
156. W. M. J. Vuist, F. v. Buitenen, A. Hekman, C. J. M. Melief. Two distinct mechanisms of antitumor activity mediated by the combination of interleukin 2 and monoclonal antibodies. *Cancer Res. 50*:5767 (1990).
157. J. C. Becker, N. Varki, S. D. Gilles, K. Furukawa, R. A. Reisfeld. An antibody–interleukin 2 fusion protein overcomes tumor heterogeneity by induction of a cellular immune response. *Proc. Natl. Acad. Sci. USA 93*:7826 (1996).
158. R. K. Oldham, M. Lewis, D. W. Orr, B. Avener, S.-K. Liao, J. R. Ogden, B. Avener, R. Birch. Adriamycin custom-tailored immunoconjugates in the treatment of human malignancies. *Mol. Biother. 1*:103 (1988).
159. R. K. Jain, L. T. Baxter. Mechanisms of heterogeneous distribution of monoclonal antibodies and other macromolecules in tumors: significance of elevated interstitial pressure. *Cancer Res. 48*:7022 (1988).
160. D. G. Covell, J. Barbet, O. D. Holton. Pharmacokinetics of monoclonal immunoglobulin G1, $F(ab')_2$ and Fab in mice. *Cancer Res. 46*:3969 (1986).
161. T. Yokota, D. E. Milenic, M. Whitlow, J. Schlom. Rapid tumor penetration of a single-chain Fv and comparison with other immunoglobulin forms. *Cancer Res. 52*:3402 (1992).
162. Y. Tsukada, K. Ohkawa, N. Hibi. Therapeutic effect of treatment with polyclonal or monoclonal antibodies to α-fetoprotein that have been conjugated to daunomycin via a dextran bridge: studies with a α-fetoprotein-product in rat hepatoma tumor model. *Cancer Res. 47*:4293 (1987).
163. M. C. Garnett, M. J. Embleton, E. Jakobs, R. W. Baldwin. Preparation and properties of a drug-carrier-antibody conjugate showing selective antibody-directed cytotoxicity. *Int. J. Cancer 31*:661 (1983).
164. M. C. Garnett, R. W. Baldwin. An improved synthesis of a methotrexate-albumin-791T/36 monoclonal antibody conjugate cytotoxic to human osteogenic sarcoma cell lines. *Cancer Res. 46*:2407 (1986).
165. J. J. Fitzpatrick, M. C. Garnett. Design, synthesis and in vitro testing of methotrexate carrier conjugate linked via oligopeptide linkers. *Anti-Cancer Drug Des. 10*:1 (1995).
166. J. J. Fitzpatrick, M. C. Garnett. Studies on the mechanism of action of an MTX-HSA-MoAb conjugate. *Anti-Cancer Drug Des. 10*:11 (1995).
167. U. Schilling, A. Friedrich, H. Sinn, H. H. Schrenk, J. H. Clorius, W. Maier-Borst. Design of compounds having enhanced tumour uptake, using serum albumin as a carrier—Part 2. *Nucl. Med. Biol. 19, Int. J. Rad. Appl. Instrum. B*: 685 (1990).
168. C. Andersson, B. M. Iresjo, K. Lundholm. Identification of tumor sites for increased albumin degradation in sarcoma-bearing mice. *J. Surg. Res. 50*:156 (1991).
169. E. Aulbert. *Transferrinmangelanämie bei malignen Tumorerkrankungen*. Georg Thieme Verlag, Stuttgart, 1986.
170. C. L. Edwards, R. L. Hayes. Tumor scanning with ^{68}Ga citrate. *J. Nucl. Med. 10*:103 (1969).
171. F. Kratz, U. Beyer. Serum proteins as drug carriers of anticancer agents, a review. *Drug Delivery 5*:1 (1998).
172. J. H. Brock. *Transferrins. Metal Proteins with Non-redox Roles*. Metalloproteins, Part 2 (P. M. Harrison, ed.). Verlag Chemie, Weinheim, 1985, p. 183.
173. L. Messori, F. Kratz. Transferrin—from inorganic biochemistry to medicine. *Metal-Based Drugs 1*:161 (1994).
174. U. Testa, E. Pelosi, C. Peschle. The transferrin receptor. *Crit. Rev. Oncogenesis 4*:241 (1993).
175. M. Cazzola, G. Bergamaschi, L. Dezza, P. Arosio. Manipulations of cellular iron metabolism for modulating normal and malignant cell proliferation: achievements and prospects. *Blood 75*:1903 (1990).
176. I. S. Trowbridge. Transferrin receptor as a potential therapeutic agent. *Prog. Allergy 45*:121 (1988).
177. E. Wagner, D. Curiel, M. Cotten. Delivery of drugs, proteins and genes into cells using transferrin as ligand for receptor-mediated endocytosis. *Adv. Drug Del. Rev. 14*: 113 (1994).
178. H. Sinn, H. H. Schrenk, A. Friedrich, U. Schilling, W. Maier-Borst. Design of compounds having an enhanced tumour uptake, using serum albumin as a carrier. *Part 1.*

Nucl. Med. Biol. 17, Int. J. Rad. Appl. Instrum. B:819 (1990).

179. A. Wunder, G. Stehle, H. Sinn, H., H. H. Schrenk, D. Hoff-Biederbeck, F. Bader, E. A. Friedrich, P. Peschke, W. Maier-Borst, D. L. Heene. Enhanced albumin uptake by rat tumors. *Int. J. Oncol. 11*:497 (1997).
180. G. Stehle, H. Sinn, A. Wunder, H. H. Schrenk, J. C. M. Stewart, G. Hartung, W. Maier-Borst, D. L. Heene. Plasma protein (albumin) catabolism by the tumor itself—implications for tumor metabolism and the genesis of cachexia. *Crit. Rev. Oncol. Hematol. 26*:77 (1997).
181. T. Tanaka, Y. Kaneo, S. Shiramoto, S. Iguchi. The disposition of serum proteins as drug carriers in mice bearing sarcoma 180. *Biol. Pharm. Bull. 16*:1270 (1993).
182. Y. Takakura, T. Fujita, T., M. Hashida, H. Sezaki. Disposition characteristics of macromolecules in tumor-bearing mice. *Pharm. Res. 7*:339 (1990).
183. C. J. Yeh, W. P. Faulk. Killing of human tumor cells in culture with adriamycin conjugates of human transferrin. *Clin. Immunol. Immunopath. 32*:1 (1984).
184. M. Fritzer, K. Barabas, V. Szüts, A. Berczi, T. Szekeres, W. P. Faulk, H. Goldenberg. Cytotoxicity of a transferrin-adriamycin conjugate to anthracycline-resistant cells. *Int. J. Cancer 52*:619 (1992).
185. K. Ohkawa, T. Hatano, K. Yamada, K. Joh, K. Takada, Y. Tsukada, M. Matsuda. Bovine serum albumin-doxorubicin conjugate overcomes multidrug resistance in a rat hepatoma. *Cancer Res. 53*:4238 (1993).
186. T. Roth, H. H. Fiebig, C. Eckert, B. Spitzmüller, U. Beyer, P. Schumacher, C. Unger, F. Kratz. In vitro antitumor efficacy of acid labile transferrin conjugates of doxorubicin in ten human tumor xenografts. *Proc. Amer. Assoc. Cancer Res. 38*:S433, 2898 (1997).
187. F. Kratz, I. Fichtner, U. Beyer, P. Schumacher, T. Roth, H. H. Fiebig, C. Unger. Antitumour activity of acid labile transferrin and albumin doxorubicin conjugates in in vitro and in vivo human tumour xenograft models. *Eur. J. Cancer 33*:S175, 784 (1997).
188. J. Drevs, I. Hofmann, D. Marmé, C. Unger, F. Kratz. In vivo and in vitro efficacy of an acid-sensitive albumin conjugate of adriamycin compared to the parent compound in murine renal cell carcinoma. *Drug Delivery 6*:1 (1999).
189. B. C. F. Chu, J. M. Whitely. High molecular weight derivatives of methotrexate as chemotherapeutic agent. *Mol. Pharmacol. 13*:80 (1977).
190. B. C. F. Chu, J. M. Whitely. Control of solid tumor metastasis with a high-molecular-weight derivative of methotrexate. *J. Natl. Cancer Inst. 62*:79 (1979).
191. B. C. F. Chu, J. M. Whitely. The interaction of carrier-bound methotrexate with L1210 cells. *Mol. Pharmacol. 17*:382 (1980).
192. B. C. F. Chu, S. P. Howell. Differential toxicity of carrier-bound methotrexate toward human lymphocytes, marrow and tumor cells. *Biochem. Pharmacol. 30*:2545 (1981).
193. L. Bures, J. Bostik, K. Motycka, M. Spundova, L. Rehak. The use of protein as a carrier of methotrexate for experimental cancer chemotherapy. III. Human serum albumin–methotrexate derivative, its preparation and basic testing. *Neoplasma 35*:329 (1988).
194. J. Bostik, L. Bures, M. Spundova, L. Rehak. The use of protein as a carrier of methotrexate for experimental cancer chemotherapy. IV. Therapy of murine melanoma B16 by human serum albumin–methotrexate. *Neoplasma 35*: 343 (1988).
195. L. Bures, A. Lichy, J. Bostik, M. Spundova. The use of protein as a carrier of methotrexate for experimental cancer chemotherapy. V. Alternative methods for preparation of serum albumin–methotrexate derivative. *Neoplasma 37*:225 (1990).
196. G. Stehle, H. Sinn, A. Wunder, H. H. Schrenk, S. Schütt, D. L. Heene, W. Maier-Borst. The loading rate determines tumor targeting of methotrexate–albumin conjugates in rats. *Anti-Cancer Drugs 8*:677 (1997).
197. G. Hartung, G. Stehle, H. Sinn, H. H. Schrenk, S. Heeger, D. Kränzle, H. H. Fiebig, W. Maier-Borst, D. L. Heene, W. Queißler. Phase I-trial of a methotrexate–albumin conjugate (MTX-HSA) in cancer patients, *Eur. J. Cancer S249*:1129 (1997).
198. T. Tanaka, Y. Kaneo, M. Miyashita. Synthesis of transferrin–mitomycin C conjugate as a receptor-mediated drug targeting system. *Biol. Pharm. Bull. 19*:774 (1996).
199. T. Tanaka, Y. Kaneo, S. Iguchi. Properties of mitomycin C–protein conjugates in vitro and in vivo. *Bioconjugate Chem. 2*:261 (1991).
200. S. A. Aaronson. Growth factors and cancer. *Science 254*: 1146 (1991).
201. S. B. Mitzel. The interleukins. *FASEB 3*:2379 (1989).
202. M. Mokotoff, J. Chen, J.-H. Zhou, E. D. Ball. Targeting growth factor receptors with bispecific molecules. *Curr. Med. Chem. 3*:87 (1996).
203. I. J. Fidler, L. M. Ellis. The implication of angiogenesis for the biology and therapy of cancer metastasis. *Cell 79*: 185 (1994).
204. P. Lupulescu. Hormones, vitamins, and growth factors in cancer treatment and prevention. *Cancer 78*:2264 (1996).
205. M. Cross, T. M. Dexter. Growth factors in development, transformation, and tumorigenesis. *Cell 64*:271 (1991).
206. I. Pastan, V. Chaudhary, D. J. FitzGerald. Recombinant toxins as novel therapeutic agents. *Annu. Rev. Biochem. 61*:331 (1992).
207. N. Normanno, A. De Luca, D. S. Salomon, F. Ciardiello. Epidermal growth factor-related peptides as targets for experimental therapy of human colon carcinoma. *Cancer Detection and Prevention 22*:62 (1998).
208. G. Carpenter, S. Cohen. Epidermal growth factor. *Annu. Rev. Biochem. 48*:193 (1979).
209. D. Gospodarowicz. Growth factors and their action in vivo and in vitro. *J. Pathol. 141*:201 (1983).
210. D. W. Wilmore. Metabolic support of the gastrointestinal tract. *Cancer 79*:1794 (1997).
211. W. J. Gullick. Prevalence of aberrant expression of the epidermal-growth-factor receptor in human cancers. *Brit. Med. Bull. 47*:87 (1991).
212. H. Modjtahedi, C. Dean. The receptor for EGF and its li-

gands: expression, prognostic value and target for therapy in cancer. *Int. J. Oncol. 4*:277 (1995).

213. C. Mateo de Acosta, A. Lage. Preparación y toxicidad in vitro de un conjugado entre el factor de crecimiento epidérmico y la adriamicina. *Interferón y Biotecnol. 6*:177 (1989).
214. H. Jinno, M. Ueda, S. Ozawa, K. Kikuchi, T. Ikeda, K. Enomoto, M. Kitajima. Epidermal growth factor receptor-dependant cytotoxic effect by an EGF-ribonuclease conjugate on human cancer cell lines. *Cancer Chemother. Pharmacol. 38*:303 (1996).
215. D. B. Cawley, H. R. Herschman, D. G. Gilliland, R. J. Collier. Epidermal growth factor-toxin A chain conjugates: EGF-ricin A is a potent toxin while EGF-diphtheria fragment A is nontoxic. *Cell 22*:563 (1980).
216. S. V. Lutsenko, N. B. Feldman, G. V. Finakova, G. A. Posypanova, S. E. Severin, K. G. Skryabin, M. P. Kirpichnikov, E. A. Lukyanets, G. N. Vorozhtsov. Targeting phthalocyanines to tumor cells using epidermal growth factor conjugates. *Tumor Biol. 20*:218 (1999).
217. F. M. Uckun, R. K. Narla, T. Zeren, Y. Yanishevski, D. E. Myers, B. Waurzyniak, O. Ek, E. Schneider, Y. Messinger, L. M. Chelstrom, R. Gunther, W. Evans. In vivo toxicity, pharmacokinetics, and anticancer activity of genistein linked to recombinant human epidermal growth factor. *Clin. Cancer Res. 4*:1125 (1998).
218. R. J. Cristiano, J. A. Roth. Epidermal growth factor mediated DNA delivery into lung cancer cells via the epidermal growth factor receptor. *Cancer Gene Therapy 3*:4 (1996).
219. Q. Zhao, E. Blomquist, H. Bolander, L. Gedda, P. Hartvig, J.-C. Janson, H. Lundqvist, H. Mellstedt, S. Nilsson, M. Nister, A. Sundin, V. Tolmachev, J.-E. Westlin, J. Carlsson. Conjugate chemistry, iodination and cellular binding of mEGF-dextran-tyrosine: preclinical test in preparation for clinical trials. *J. Mol. Med. 1*:693 (1998).
220. J. Massagué. Transforming growth factor-α. *J. Biol. Chem. 265*:21396 (1990).
221. G. J. Todaro, C. Fryling, J. E. De Larco. Transforming growth factors produced by certain human tumor cells: polypeptides that interact with epidermal growth factor receptors. *Proc. Natl. Acad. Sci. USA 77*:5258 (1980).
222. D. C. Heimbrook, S. M. Stirdivant, J. D. Ahern, N. L. Balishin, D. R. Patrick, G. M. Edwards, D. Defoe-Jones, D. J. FitzGerald, I. Pastan, A. Oliff. Transforming growth factor α-Pseudomonas exotoxin fusion protein prolongs survival of nude mice bearing tumor xenografts. *Proc. Natl. Acad. Sci. USA 87*:4697 (1990).
223. I. Pastan, D. FitzGerald. Recombinant toxins for cancer treatment. *Science 254*:1173 (1991).
224. V. Allured, R. J. Collier, S. F. Carroll, D. B. McKay. Structure of exotoxin A of Pseudomonas aeruginosa at 3.0 angstrom resolution. *Proc. Natl. Acad. Sci. USA 83*:1320 (1986).
225. C. B. Siegall, Y.-H. Xu, V. K. Chaudhary, S. Adhya, D. FitzGerald, I. Pastan. Cytotoxic activities of a fusion protein comprised of TGF-α and Pseudomonas exotoxin. *FASEB 3*:2647 (1989).
226. M. Draoui, C. B. Siegall, D. FitzGerald, I. Pastan, T. W. Moody. TGFα-PE40 inhibits non-small cell lung cancer growth. *Life Sciences 54*:445 (1994).
227. Q.-C. Wang, L. H. Pai, W. Debinski, D. FitzGerald, I. Pastan. Polyethylene glycol-modified chimeric toxin composed of transforming growth factor α and Pseudomonas exotoxin. *Cancer Res. 53*:4588 (1993).
228. L. A. Liotta, P. S. Steeg, W. G. Stetler-Stevenson. Cancer metastasis and angiogenesis: an imbalance of positive and negative regulation. *Cell 64*:327 (1991).
229. D. Gospodarowicz, J. A. Abraham, J. Schilling. Isolation and characterization of a vascular endothelial cell mitogen produced by pituitary-derived folliculo stellate cells. *Proc. Natl. Acad. Sci. USA 86*:7311 (1989).
230. N. Ferrara, W. J. Henzel. Pituitary follicular cells secrete a novel heparin-binding growth factor specific for vascular endothelial cells. *Biochem. Biophys. Res. Commun. 161*: 851 (1989).
231. E. Tischler, R. Mitchell, T. Hartman, M. Silva, D. Gospodarowicz, J. C. Fiddes, J. A. Abraham. The human gene for vascular endothelial growth factor. *J. Biol. Chem. 266*: 11947 (1991).
232. K. Kim, B. Li, J. Winer, M. Armanini, N. Gillett, H. S. Phillips, N. Ferrara. Inhibition of vascular endothelial growth factor-induced angiogenesis suppresses tumor growth in vivo. *Nature 362*:841 (1993).
233. R. A. Brekken, X. Huang, S. W. King, P. E. Thorpe. Vascular endothelial growth factor as a maker of tumor endothelium. *Cancer Res. 58*:1952 (1998).
234. S. Ramakrishnan, T. A. Olson, V. L. Bautch, D. Mohanraj. Vascular endothelial growth factor-toxin conjugate specifically inhibits KDR/flk-1-positive endothelial cell proliferation in vitro and angiogenesis in vivo. *Cancer Res. 56*:1324 (1996).
235. T. A. Olson, D. Mohanraj, S. Roy, S. Ramakrishnan. Targeting the tumor vasculature: inhibition of tumor growth by a vascular endothelial growth factor–toxin conjugate. *Int. J. Cancer 73*:865 (1997).
236. N. Arora, R. Masood, T. Zheng, J. Cai, D. L. Smith, P. S. Gill. Vascular endothelial growth factor chimeric toxin is highly active against endothelial cells. *Cancer Res. 59*: 183 (1999).
237. M. B. Rettig, H. J. Vescio, M. Põld, G. Schiller, D. Belson, A. Savage, C. Nishikubo, C. Wu, J. Fraser, J. W. Said, J. R. Berenson. Kaposi's sarcoma-associated herpes virus infection of bone marrow dendritic cells from multiple myeloma patients. *Science 276*:1851 (1997).
238. F. Esch, N. Ueno, A. Baird, F. Hill, L. Denoroy, N. Ling, D. Gospodarowicz, R. Guillemin. Primary structure of bovine brain acidic fibroblast growth factor (FGF). *Biochem. Biophys. Res. Commun. 133*:554 (1985).
239. F. Esch, A. Baird, N. Ling, N. Ueno, F. Hill, L. Denoroy, R. Klepper, D. Gospodarowicz, P. Böhlen, R. Guillemin. Primary structure of bovine pituitary basic fibroblast growth factor (FGF) and comparison with the amino-terminal sequence of bovine brain acidic FGF. *Proc. Natl. Acad. Sci. USA 82*:6507 (1985).

240. W. Casscells, D. A. Lappi, B. B. Olwin, C. Wai, M. Siegman, E. H. Speir, J. Sasse, A. Baird. Elimination of smooth muscle cells in experimental restenosis: targeting of fibroblast growth factor receptors. *Proc. Natl. Acad. Sci. USA 89*:7159 (1992).
241. K. A. Thomas, M. Rios-Candelore, G. Gimenez-Gallego, J. DiSalvo, C. Bennett, J. Rodkey, S. FitzPatrick. Pure brain-derived acidic fibroblast growth factor is a potent angiogenic vascular endothelial cell mitogen with sequence homology to interleukin l. *Proc. Natl. Acad. Sci. USA 82*:6409 (1985).
242. R. Montesano, J.-D. Vassalli, A. Baird, R. Guillemin, L. Orci. Basic fibroblast factor induces angiogenesis in vitro. *Proc. Natl. Acad. Sci. USA 83*:7297 (1986).
243. J. G. Beitz, P. Davol, J. W. Clark, J. Kato, M. Medina, A. R. Frackelton, Jr., D. A. Lappi, A. Baird, P. Calabresi. Antitumor activity of basic fibroblast growth factor-saporin mitotoxin in vitro and in vivo. *Cancer Res. 52*: 227 (1992).
244. D. A. Lappi, R. Matsunami, D. Martineau, A. Baird. Reducing the heterogeneity of chemically conjugated targeted toxins: homogeneous basic FGF-saporin. *Anal. Biochem. 212*:446 (1993).
245. D. A. Lappi, W. Ying, I. Barthelemy, D. Martineau, I. Prieto, L. Benatti, M. Soria, A. Baird. Expression and activities of a recombinant basic fibroblast growth factor-saporin fusion protein. *J. Biol. Chem. 269*:12552 (1994).
246. W. Ying, S. Martineau, J. Beitz, D. A. Lappi, A. Baird. Anti B16-F10 melanoma activity of a basic fibroblast growth-factor saporin mitotoxin. *Cancer 74*:848 (1994).
247. T. A. Tetzke, M. C. Caton, P. A. Maher, Z. Parandoosh. Effect of fibroblast growth factor saporin mitotoxins on human bladder cell lines. *Clin. Exp. Metastasis 15*:620 (1997).
248. C. B. Siegall, S. Epstein, E. Speir, T. HLA, R. Forough, T. Maciag, D. J. FitzGerald, I. Pastan. Cytotoxic activity of chimeric proteins composed of acidic fibroblast growth factor and Pseudomonas exotoxin on a variety of cell types. *FASEB J. 5*:2843 (1991).
249. P. E. Kintzel, K. A. Calis. Recombinant interleukin-2: a biological response modifier. *Clin. Pharm. 10*:110 (1991).
250. J. Nichols, F. Foss, T. M. Kuzel, C. F. LeMaistre, L. Platanias, M. J. Ratain, A. Rook, M. Saleh, G. Schwartz. Interleukin-2 fusion protein: an investigational therapy for interleukin-2 receptor expressing malignancies. *Eur. J. Cancer 33*:34 (1997).
251. T. Taniguchi, H. Matsui, T. Fujita, C. Takaoka, N. Kashima, R. Yoshimoto, J. Hamuro. Structure and expression of a cloned cDNA for human interleukin-2. *Nature 302*: 305 (1983).
252. R. J. Robb, R. M. Kutny, V. Chowdhry. Purification and partial sequence analysis of human T-cell growth factor. *Proc. Natl. Acad. Sci. USA 80*:5990 (1983).
253. R. Noelle, P. H. Krammer, J. Ohara, J. W. Uhr, E. S. Vitetta. Increased expression of 1a antigens on resting B cells: an additional role for B-cell growth factor. *Proc. Natl. Acad. Sci. USA 81*:6149 (1984).
254. K. A. Smith. Interleukin-2: inception, impact, and implications. *Science 240*:1169 (1988).
255. R. M. Bukowski. Natural history and therapy of mestastatic renal cell carcinoma: the role of interleukin-2. *Cancer 80*:1198 (1997).
256. U. Keilholz, G. Stoter, C. J. Punt, C. Scheibenbogen, F. Lejeune, A. M. Eggermont. Recombinant interleukin-2-based treatments for advanced melanoma: the experience of the European Organization for Research and Treatment of Cancer Melanoma Cooperative Group. Cancer *J. Scientific Am. 3* (suppl. 1):22 (1997).
257. H. Lorberboum-Galski, R. W. Kozak, T. A. Waldmann, P. Bailon, D. J. P. FitzGerald, I. Pastan. Interleukin 2 (IL2) PE40 is cytotoxic to cells displaying either the p55 or p70 subunit of the IL2 receptor. *J. Biol. Chem. 263*:18650 (1988).
258. D. P. Williams, K. Parker, P. Bacha, W. Bishai, M. Borowski, F. Genbauffe, T. B. Strom, J. R. Murphy. Diphtheria toxin receptor binding domain substitution with interleukin-2: genetic construction and properties of a diphtheria-toxin-related interleukin-2 fusion protein. *Protein Engineer. 1*(6):493 (1987).
259. D. P. Williams, C. E. Snider, T. B. Strom, J. R. Murphy. Structure/function analysis of interleukin-2-toxin (DAB_{486}-IL-2). *J. Biol. Chem. 265*:11885 (1990).
260. T. R. Mosmann, M. W. Bond, R. L. Coffman, J. Ohara, W. E. Paul. T-cells and mast cell lines respond to B-cell stimulatory factor 1. *Proc. Natl. Acad. Sci. USA 83*:5654 (1986).
261. M. Ogata, V. K. Chaudhary, D. J. FitzGerald, I. Pastan. Cytotoxic activity of a recombinant fusion protein between interleukin 4 and Pseudomonas exotoxin. *Proc. Natl. Acad. Sci. USA 86*:4215 (1989).
262. W. Debinski, R. K. Puri, R. J. Kreitman, I. Pastan. A wide range of human cancers express interleukin 4 (IL4) receptors that can be targeted with chimeric toxin composed of IL4 and Pseudomonas exotoxin. *J. Biol. Chem. 268*:14065 (1993).
263. W. Debinsky, R. K. Puri, I. Pastan. Interleukin-4 receptors expressed on tumor cells may serve as a target for anticancer therapy using chimeric Pseudomonas exotoxin. *Int. J. Cancer 58*:744 (1994).
264. L. Ramanathan, R. Ingram, L. Sullivan, R. Greenberg, R. Reim, P. P. Trotta, H. V. Le. Immunochemical mapping of domains in human interleukin 4 recognized by neutralizing monoclonal antibodies. *Biochem. 32*:3549 (1993).
265. R. J. Kreitman, R. J. Puri, I. Pastan. A circularly permuted recombinant interleukin 4 toxin with increased activity. *Proc. Natl. Acad. Sci. USA 91*:6889 (1994).
266. R. K. Puri, P. Leland, R. J. Kreitman, I. Pastan. I. Human neurological cancer cells express interleukin-4 (IL-4) receptors which are targets for the toxic effects of IL-4-Pseudomonas exotoxin chimeric protein. *Int. J. Cancer 58*:574 (1994).
267. R. K. Puri, D. S. Hoon, P. Leland, P. Snoy, R. W. Rand, I. Pastan, R. J. Kreitman. Preclinical development of a

recombinant toxin containing circularly permuted interleukin-4 and truncated Pseudomonas exotoxin for therapy of malignant astrocytoma. *Cancer Res. 56*:5631 (1996).

268. S. R. Husain, N. Behari, R. J. Kreitman, I. Pastan, R. K. Puri. Complete regression of established human glioblastoma tumor xenograft by interleukin-4 toxin therapy. *Cancer Res. 58*:3649 (1998).
269. R. K. Puri. Toxicologic pathology of cytokines, cytokine receptors, and other recombinant human proteins. *Toxicologic Pathol. 27*:53 (1999).
270. J. Gauldie, C. Richards, D. Harnish, P. Lansdorp, H. Baumann. Interferon β_2/B-cell stimulatory factor type shares identity with monocyte-derived hepatocyte-stimulating factor and regulates the major acute phase protein response in liver cells. *Proc. Natl. Acad. Aci. USA 84*:7251 (1987).
271. M. Varterasian. Advances in the biology and treatment of multiple myeloma. *Curr. Opinion Oncol. 11*:3 (1999).
272. M. Kawano, T. Hirano, T. Matsuda, T. Taga, Y. Horii, K. Iwato, H. Asaoku, B. Tang, O. Tanabe, H. Tanaka, A. Kuramoto, T. Kishimoto. Autocrine generation and requirement of BSF-2/IL-6 for human multiple myelomas. *Nature 332*:83 (1988).
273. C. B. Siegall, V. K. Chaudhary, D. J. FitzGerald, I. Pastan. Cytotoxic activity of an interleukin 6-Pseudomonas exotoxin fusion protein on human myeloma cells. *Proc. Natl. Acad. Sci. USA 85*:9738 (1988).
274. C. B. Siegall, D. J. FitzGerald, I. Pastan. Cytotoxicity of IL-6-PE40 and derivatives on tumor cells expressing a range of IL-6 receptor levels. *J. Biol. Chem. 265*:16318 (1990).
275. C. B. Siegall, R. J. Kreitman, D. J. FitzGerald, I. Pastan. Antitumor effects of interkeukin 6-Pseudomonas exotoxin chimeric molecules against the human hepatocellular carcinoma, PLC/PRF/5 in mice. *Cancer Res. 51*:2831 (1991).
276. H.-G. Elias. *An Introduction to Polymer Science*. VCH, Weinheim, 1997, p. 156.
277. M. Wirth, B. Schübl, F. Gabor. Doxorubicin-lectin prodrugs: influence of coupling rate and coupling agent on cell binding and cytostatic activity. *Sci. Pharm. 67*:277 (1999).
278. R. I. Mahato, Y. Takakura, M. Hashida. Development of targeted delivery systems for nucleic acid drugs. *J. Drug Targeting 4*:337 (1997).
279. H. Kitazawa, H. Sato, I. Adachi, Y. Masuko, I. Horikoshi. Microdialysis assessment of fibrin glue containing sodium alginate for local delivery of doxorubicin in tumor-bearing rats. *Biol. Pharm. Bull. 20*:278 (1997).
280. T. Ouchi, M. Tada, M. Matsumoto, Y. Ohya. Design of lysosomotropic macromolecular prodrug of doxorubicin using N-acetyl-α-1,4-polygalactosamine as a targeting carrier to hepatoma tissue. *J. Bioact. Compat. Polym. 13*: 257 (1998).
281. Y. Takakura, M. Hashida. Macromolecular drug carrier systems in cancer chemotherapy: macromolecular prodrugs. *Crit. Rev. Oncol. Hematol. 18*:207 (1995).
282. H. Iwata, E. P. Goldberg. Tissue-binding macromolecular antitumor drugs for localized therapy: mitomycin C-concanavalin A conjugates. *Polymer Prepr. 24*:74 (1983).
283. H. Onishi, T. Nagai. Preparation of dextran T70-methotrexate conjugate and dextran T70-mycophenolic acid conjugate, and in vitro effect of dextran T70-methotrexate on dihydrofolate reductase. *Chem. Pharm. Bull. 34*:2561 (1986).
284. M. Nichifor, E. H. Schacht, L. W. Seymour. Polymeric prodrugs of 5-fluorouracil. *J. Contr. Rel. 48*:165 (1997).
285. A. Bernstein, E. Hurwitz, R. Maron, R. Arnon, M. Sela, M. Wilchek. Higher antitumor efficacy of daunomycin when linked to dextran: in vivo and in vitro studies. *J. Natl. Cancer Inst. 60*:379 (1978).
286. F. Levi-Schaffer, A. Bernstein, A. Meshorer, R. Arnon. Reduced toxicity of daunorubicin by conjugation to dextran. *Cancer Treatment Reports 66*:107 (1982).
287. S. Danhauser-Riedl, E. Hausmann, H.-D. Schick, R. Bender, H. Dietzfelbinger, J. Rastetter, A.-R. Hanauske. Phase I clinical and pharmacokinetic trial of dextran conjugated doxorubicin (AD-70, DOX-OXD). *Investigat. New Drugs 11*:187 (1993).
288. A. Kikukawa, K. Munechika, Y. Ueda, K. Yamanouchi, K. Yokoyama, S. Tsukagoshi. Tissue concentration of adriamycin and adriamycin-oxidized dextran conjugate in tumor bearing rats and mice. *Drug Delivery Sys. 5*:255 (1990).
289. K. Munechika, Y. Sogame, N. Kishi, Y. Kawabata, Y. Ueda, K. Yamanouchi, K. Yokoyama. Tissue distribution of macromolecular conjugate, adriamycin linked to oxidized dextran in rat and mouse bearing tumor cells. *Biol. Pharm. Bull. 17*:1193 (1994).
290. T. Kojima, M. Hashida, S. Muranishi, H. Sezaki. Mitomycin C–dextran conjugate: a novel high molecular weight pro-drug of mitomycin C. *J. Pharm. Pharmacol. 32*:30 (1980).
291. M. Hashida, Y. Takakura, S. Matsumoto, H. Sasaki, A. Kato, T. Kojima, S. Muranishi, H. Sezaki. Regeneration characteristics of mitomycin C–dextran conjugate in relation to its activity. *Chem. Pharm. Bull. 31*:2055 (1983).
292. Y. Takakura, K. Mori, M. Hashida, H. Sezaki. Absorption characteristics of macromolecular prodrugs of mitomycin C following intramuscular administration. *Chem. Pharm. Bull. 34*:1775 (1986).
293. Y. Takakura, A. Takagi, M. Hashida, H. Sezaki. Disposition and tumor-localization of mitomycin C–dextran conjugates in mice. *Pharm. Res. 4*:293 (1987).
294. Y. Takakura, S. Matsumoto, M. Hashida, H. Sezaki. Enhanced lymphatic delivery of mitomycin C conjugated with dextran. *Cancer Res. 44*:2505 (1984).
295. Y. Takakura, M. Atsumi, M. Hashida, H. Sezaki. Development of a novel polymeric prodrug of mitomycin C–dextran conjugate with anionic charge, II. Disposition and pharmacokinetics following intravenous and intramuscular administration. *Int. J. Pharm. 37*:145 (1987).
296. S. Matsumoto, Y. Arase, Y. Takakura, M. Hashida, H. Sezaki. Plasma Disposition and in vivo and in vivo antitu-

mor activities of mitomycin C–dextran conjugate in relation to the mode of action. *Chem. Pharm. Bull. 33*:2941 (1985).

297. A. Saikawa, T. Nomura, F. Yamashita, Y. Takakura, H. Sezaki, M. Hashida. Pharmacokinetic analysis of drug disposition after intratumoural injection in a tissue-isolated tumor perfusion system. *Pharm. Res. 13*:1436 (1996).
298. T. Nomura, A. Saikawa, S. Morita, T. Sakaeda (né Kakutani), F. Yamashita, K. Honda, Y. Takakura, M. Hashida. Pharmacokinetic characteristics and therapeutic effects of mitomycin C–dextran conjugates after intratumoral injection. *J. Contr. Rel. 52*:239 (1998).
299. N. E. Larsen, E. Leshchiner, E. A. Balazs. Biocompatibility of hylan polymers in various tissue compartments. *Mater. Res. Soc. Symp. Proc. 394*:149 (1995).
300. C. B. Knudson, W. Knudson. Hyaluronan-binding proteins in development, tissue homeostasis, and disease. *FASEB J. 7*:1233 (1993).
301. J. Entwistle. C. L. Hall, E. A. Turley. Hyaluronan receptors: regulators of signaling to the cytoskeleton. *J. Cell Biochem. 61*:569 (1996).
302. C. L. Hall, B. H. Yang, X. W. Yang, S. W. Zhang, M. Turley, S. Samuel, L. A. Lange, C. Wang, G. D. Curpen, R. C. Savani, A. H. Greenberg, E. A. Turley. Overexpression of the hyaluronan receptor RHAMM is transforming and is also required for h-ras transformation. *Cell 82*:19 (1995).
303. P. Rooney, S. Kumar, J. Ponting, M. Wang. The role of hyaluronan in tumor neovascularization (review). *Int. J. Cancer 60*:632 (1995).
304. M. Culty, H. A. Nguyen, C. B. Underhill. The hyaluronan receptor (CD44) participates in the uptake and degradation of hyaluronan. *J. Cell Biol. 116*:1055 (1992).
305. C. Cera, M. Palumbo, S. Stefanelli, M. Rassu, G. Palù. Water-soluble polysaccharide–anthracycline conjugates: biological activity. *Anti-Cancer Drug Design 7*:143 (1992).
306. E. Marecos, R. Weissleder, A. Bogdanov, Jr. Antibody-mediated versus nontargeted delivery in a human small cell lung carcinoma model. *Bioconjugate. Chem. 9*:184, (1998).

33

Enzyme–Prodrug Therapies of Cancer

Richard J. Knox and Roger G. Melton
Gnact Pharma Plc, Salisbury, England
Ronit Satchi
Centre for Polymer Therapeutics, London, England

I. INTRODUCTION

Since the early twentieth century, many researchers have sought Ehrlich's elusive magic bullet for treating the ills of mankind. The principle of chemotherapy has depended on a compound exhibiting selective toxicity towards a foreign organism. With respect to cancer chemotherapy, it was envisaged that compounds would be identified or synthesized that would be able to distinguish between normal and neoplastic cells and selectively correct or eradicate the abnormal cells with minimal toxic side effects on normal cells and tissue. Cancer chemotherapy has rarely achieved this.

The first documented use of cancer chemotherapy was in 1942 when the alkylating agent nitrogen mustard was used to obtain brief clinical remission in a patient with lymphoma. Since then, a large number of clinical drugs have been tested for their potential antitumor activity. Of the 300,000 or so compounds that have been evaluated as possible anticancer agents, about forty are considered useful. Most anticancer drugs can be grouped into families based on their biochemical activity or on their origins. These major families are

1. The alkylating agents, which include nitrogen mustard, chlorambucil, melphalan, busulphan, and the nitrosoureas. All of these compounds react so that an alkyl or substituted alkyl group becomes covalently linked to cellular constituents. Although reactions with proteins and other cellular components occur, the important site of mustard action is almost certainly DNA (1).
2. The platinating agents, which include cisplatin and carboplatin. The important site of action of these compounds is also DNA (2,3).
3. The antimetabolites, which include methotrexate, 5-fluorouracil, cytosine arabinoside, and 6-thioguanine, are principally inhibitors of critical biochemical pathways usually leading to inhibition of DNA or RNA synthesis.
4. A number of natural anticancer compounds exist that are either plant or bacterial products and include mitotic inhibitors such as the vinca alkaloids, or the antibiotics such as mitomycin C or bleomycin.

Chemotherapy is used predominantly in the treatment of metastatic or disseminated malignant neoplastic disease. However, cytotoxic chemotherapy is an imperfect modality due to the systemic nature of its cytotoxicity. This is because the agents lack any intrinsic antitumor selectivity. They mostly act by an antiproliferative mechanism, and their action is on cells that are in cycle, or in some cases, on a specific phase of the cell cycle, rather than by a specific toxicity directed towards a particular type of cancer cell. Thus the limiting toxicity of the majority of anticancer agents is a result of a toxic effect on the normal host tissues that are the most rapidly dividing, such as bone marrow, gut mucosa, and the lymphatic system. Further, most human solid cancers do not have a high proportion of cells that are rapidly proliferating, and they are therefore not particularly sensitive to this class of agent. Thus, because of host toxicity, treatment has to be discontinued at dose levels that are well below the dose that would be required to kill all viable tumor stem cells. This intrinsic poor selectivity of anticancer agents has been recognized for a long time, and attempts to improve selectivity and allow greater doses to be administered have been numerous. For example, different classes of chemicals with antiproliferative activity (but that differed in their normal tissue toxicity) were put together in drug combinations, thus improving the total dose of cytotoxic agent administered (4). There have been clinical trials of chemicals that either sensitize the cancer to the administered chemotherapy or protect the sensitive host tissues (4). Local application of drugs, such as by injection into the hepatic artery for the treatment of liver metastases or limb perfusion for melanoma, also allows higher doses of cytotoxic drugs to be given. However, even using these techniques, the dose escalation obtained is still not great enough for the complete regression of resistant solid cancers.

A sufficiently high degree of selectivity might be obtained by the use of prodrugs. Prodrugs are chemicals that are toxicologically and pharmacodynamically inert but may be converted in vivo to active products. Conversion of the prodrug to the active form can take place by a number of mechanisms such as changes of pH, oxygen tension, temperature, or salt concentration, or by spontaneous decomposition of the drug, internal ring opening, or cyclization (5,6). However, the major approach in prodrug design for cancer chemotherapy is the synthesis of inert compounds that are converted to an active drug by enzyme action (5,6). Thus, in cancer chemotherapy, the prodrug would be inert but converted in vivo into a highly toxic metabolite by an enzyme present in the cancer cells but not in other cells (Table 1). A classic example of the high degree of selectivity that can be attained with prodrugs is the dinitro compound CB 1954 [5-(aziridin-1-yl)-2,4-dinitrobenzamide]. CB 1954 is only a weak monofunctional alkylating agent (by virtue of the aziridine function) and is not highly cytotoxic. However, in the presence of the enzyme DT diaphorase, the 4-nitro group is reduced to the hydroxylamino that is then rapidly acylated to a bifunctional alkylating agent (7–9). There is a least a 10,000-fold difference in cytotoxicity between the Walker cell line that has a high level of the activating enzyme and Chinese hamster V79 cells, which lack the enzyme (10,11), and CB 1954 can actually cure the Walker tumor when implanted into either rats or mice (12,13). The enzyme β-glucuronidase can also activate a relatively nontoxic alkylating agent to an extremely reactive and toxic metabolite that can cure cancers in experimental models (14). Other examples are given in Table 1. The appropriate use of prodrugs can deliver much higher doses of drug than normal and might therefore be effective against common solid cancers. However, clinical studies have been disappointing. For example, in the case of CB 1954 it has been shown that this prodrug is a much poorer substrate for the human form of the enzyme DT diaphorase than it is for the rat form (15). Thus it does not show the marked antitumor effect against human tumors that were demonstrated in rat tumors (11). An aniline mustard prodrug activated by β-glucuronidase was also subject to a preliminary clinical trial. Although a correlation was shown between glucuronidase activity and patient response, the occurrence of high β-glucuronidase activity was rare and not predictable by tumor type, and it was not considered a feasible therapy (16). However, comparison of biochemical, molecular biological, and chemosensitivity data obtained from screening a large number of cell lines (e.g., the NCI tumor cell line panel) may facilitate investigation of factors influencing drug antitumor activity (17). The knowledge gained may be of value in determining the distribution of prodrug activating enzymes, the development of new anticancer prodrugs, and the selection of patients to receive specific therapies. This approach may make prodrug therapy more feasible and has stimulated further work. Thus recently a new prodrug activated by β-glucuronidase has been proposed for prodrug therapy (18). A new prodrug-activating enzyme has also been described. This human enzyme called NQO2 can efficiently activate the prodrug CB 1954 (19). NQO2 activity was not previously seen because its activity is normally latent and it requires a nonbiogenic cosubstrate such as NRH [nicotinamide riboside (reduced)] for enzymatic activity (19). These developments suggest that prodrug therapy is still viable but is probably applicable only to specific cancers.

Another approach to overcome the intrinsic problems associated with the administration of cytotoxic drugs was drug targeting. This came with the advent of monoclonal antibody technology in the 1970s. It involves the conjugation of a cytotoxic agent to an antibody or other molecule that can specifically bind to antigen or receptor sites that are overexpressed on the target tumor cell. However, the goal has proved to be elusive, and a number of limitations are apparent. These include poor penetration of the tumor, lack of intertumor accumulation of the cytotoxic component, and heterogeneity of tumor-associated antigens.

Therefore there are major limitations to the use of both prodrugs and drug targeting for the treatment of human cancers. However, the disappointing clinical results might be overcome by combining these two approaches. Prodrug activating enzymes, not drugs, could be targeted to human tumors using tumor-associated monoclonal antibodies prior to administration of a prodrug (20,21). This approach

Table 33.1 Enzymes and Prodrugs That Have Been Proposed for Cancer Therapy. The Enzymes May Have Been Targets for Prodrug Therapy or Targeted Using ADEPT, MDEPT, GDEPT, or PDEPT

Enzyme	Prodrug	Drug	Application	Reference
Alkaline phosphatase	Phenol mustard phosphate	Phenol mustard	ADEPT	165
	Doxorubicin phosphate	Doxorubicin		166
	Mitomycin phosphate	Mitomycin alcohol		167, 168
	Etoposide phosphate	Etoposide		29, 167, 169
Aminopeptidase	2-L-Pyroglutamyl-methotrexate.	Methotrexate	Prodrug therapy ADEPT	170
Azoreductase	Azobenzene mustards	Phenylenediamine mustards (various)	Prodrug therapy	171, 172
Carboxylesterase	Irinotecan	SN-38	GDEPT	173
	CPT-11	Camptothecin		174
Carboxypeptidase A/B	Methotrexate-alanine	Methotrexate	ADEPT	153, 175–177
	α-linked derivatives of quinazoline antifolates	TS inhibitors	ADEPT	178
Carboxypeptidase A1 (mutant)	Methotrexate-α-3-cyclobutylphenylalanine	Methotrexate	ADEPT	46
	Methotrexate-α-3-cyclopentyltyrosine	Methotrexate	ADEPT GDEPT	46 179
Carboxypeptidase G2	Benzoic acid mustard glutamates	Benzene mustards (various)	ADEPT GDEPT	20, 38, 180–184 63, 185
	Self-immolative derivatives	Doxorubicin Daunorubicin	GDEPT	186
		Benzene mustards (various)	GDEPT	187
	Folinic acid	5-Formyltetrahydropteroic acid	AMIRACS	124, 142
Cathepsin B	PK1	Doxorubicin	PDEPT	133, 134
Cytochrome-P450	Cyclophosphamide Ifosfamide	Phosphamide mustard	GDEPT	188–193
	2-Aminoanthracene	Alkylating metabolites	GDEPT	194
	4-Ipomeanol	Alkylating metabolites	GDEPT	194
Cytosine deaminase	5-Fluorocytosine	5-Fluorouracil	GDEPT ADEPT	57, 85–88, 195 196, 197
Deoxycytidine kinase	Ara-C	Adenine arabinonucleoside triphosphate (araATP)	GDEPT	198
DT-diaphorase (NQO1)	CB 1954	5-(Aziridine-1-yl)-4-hydroxylamino-2-nitrobenzamide	Prodrug therapy	7, 8, 11, 12
	3-hydroxymethyl-5-aziridinyl-1-methyl-2-(1H indole-4,7-dione)prop-β-en-α-ol (EO9)	Reduced-EO9 products	Prodrug therapy GDEPT	199 200
	RH1	Reduced aziridinylbenzoquinone products	Prodrug therapy	201
α-Galactosidase	N-[4-(α-D-galactopyranosyl)-benyloxycarbonyl]-daunorubicin	Daunorubicin	ADEPT	202–204
Glucose oxidase	Glucose	Hydrogen peroxide	Radical generating system	28, 205–207
β-Glucosidase	Amygdalin	Cyanide	ADEPT	208, 209
β-Glucuronidase	Phenol mustard-glucuronide	Phenol mustard	Prodrug Therapy ADEPT	14, 210 211–216
	Epirubicin-glucuronide	Epirubicin	ADEPT	217
	Paclitaxel-glucuronide	Paclitaxel	ADEPT	218
	Daunorubicin-glucuronides	Daunorubicin	ADEPT	219–221
	Doxorubicin-glucuronides	Doxorubicin	ADEPT Prodrug Therapy	221, 222 221
	Camptothecin-glucuronides	Camptothecin	ADEPT Prodrug therapy	223 223
	Verapamil-glucuronide	Verapamil	ADEPT	253

Table 33.1 Continued

Enzyme	Prodrug	Drug	Application	Reference
γ-Glutamyl transferase	γ-Glutamyl p-phenylenediamine mustard	Phenylenediamine mustard	Prodrug therapy	224
β-Lactamase	Vinca-cephalosporin	4-Desacetylvinblastine-3-carboxyhydrazide	ADEPT	225–228
	Nitrogen-mustard-cephalosporin	Nitrogen mustards (various)	ADEPT	34, 229–231
	Doxorubicin-cephalosporin	Doxorubicin	ADEPT	232
Nitroreductase	5-(Aziridine-1-yl)-2,4-dinitrobenzamide (CB 1954)	5-(Aziridine-1-yl)-4-hydroxylamino-2-nitrobenzamide	ADEPT GDEPT	11, 90, 233 92, 93, 95, 97
	2,4-dinitrobenzamide mustards	Activated mustards	ADEPT & GDEPT	98, 234
	4-Nitrobenzyloxycarbonyl derivatives	e.g. Actinomycin D, mitomycin C, enediynes, tallimustine	ADEPT & GDEPT	99, 102, 235, 236
	2-Nitroimidazole carbamate derivative	Amino-seco-CBI TMI	ADEPT & GDEPT	237
NQO2	CB 1954	5-(Aziridine-1-yl)-4-hydroxylamino-2-nitrobenzamide	Prodrug therapy	19
Penicillin amidase	Palytoxin-4-hydroxyphenyl-acetamide	Palytoxin	ADEPT	238
	Doxorubicin-phenoxyacetamide	Doxorubicin	ADEPT	239, 240
	Melphalan-phenoxyacetamide	Melphalan	ADEPT	240
Plasmin	Peptidyl-p-phenylenediamine-mustard	Phenylenediamine mustard	Prodrug therapy	241
	Peptidyl-p-doxorubicin	Doxorubicin	Prodrug therapy	242
Purine nucleoside phosphorylase	6-Methylpurine-2′-deoxyriboside	6-Methylpurine	GDEPT	243–245
	Arabinofuranosyl-2-fluoroadenine monophosphate (F-araAMP),	F-araA	GDEPT	243
Ribonuclease A (mutant)	Ribotide prodrug	Aniline mustard	ADEPT	45
Thymidine kinase (viral)	Ganciclovir	Ganciclovir triphosphate	GDEPT	69–78, 246, 247
	Adenine arabinonucleoside (araM)	Adenine arabinonucleoside triphosphate (araATP)	GDEPT	248
Thymidine phosphorylase	5′-Deoxy-5-fluorouridine (5′-DFUR)	5-FU	GDEPT	249
	Thymidine	Thymine	AMIRACS	124, 142
Tyrosinase	[2′-(3″,4″-Dihydroxyphenyl)-ethyl] carbonic acid p-(bis-2-2-chloroethylamino)phenyl ester	Phenol mustard	Prodrug therapy	250
Xanthine guanine phosphoribosyltransferase	6-Thioxanthine	6-Thioxanthine monophosphate	GDEPT	251
Xanthine oxidase	Hypoxanthine	Superoxide, hydrogen peroxide	Radical generating system AMIRACS	147 124, 142

Source: Modified from Ref. 164.

has been termed ADEPT (antibody-directed enzyme–prodrug therapy). A related approach is gene-directed enzyme–prodrug therapy (GDEPT) or VDEPT (virus-directed enzyme–prodrug therapy) where the DNA encoding for a prodrug activating enzyme is selectively expressed and translated within a tumor cell. Biocompatible polymers can also localize tumors at a tumor site and has led to the development of PDEPT and MDEPT. As well as activating prodrugs, targeted enzymes can also be used to destroy a rescue agent at the tumor site. Thus, while normal tissues continue to be rescued, the tumor is subject to the effect of the therapy. This approach is particularly relevant to antimetabolites and has been termed AMIRACS.

II. ENZYME–PRODRUG THERAPIES OF CANCER

A. Antibody-Directed Enzyme–Prodrug Therapy (ADEPT)

In ADEPT, an enzyme of nonhuman or nonmammalian origin that can metabolize substrates not normally activated in humans, is linked to a tumor-associated antibody. This can be done either chemically or by using recombinant DNA techniques, and the conjugate is allowed to localize to the tumor (20–23). Thus ADEPT creates a tumor environment with a high concentration of an enzyme that would convert a normally inert substrate to a highly reactive metabolite (Fig. 1). The unlocalized conjugate is either given time to be eliminated from the body or is hastened on its way by a "clearance" antibody. A nontoxic prodrug is then administered and activated at the site of the tumor by the bound conjugate (Fig. 1). The active drug can be of low molecular weight and can therefore rapidly diffuse and reach tumor regions not accessible to the antibody–enzyme conjugate, producing a bystander effect (20–23) (Fig. 1). The targeted antibody–enzyme conjugate is designed to remain extracellular, as internalized conjugate would be expected to be rapidly degraded in the lysosomal compartment. Further, the prodrugs can exploit the external enzyme by being charged and thus excluded from the cell until activated. Secreted antigen, if it accumulates within the interstitial spaces of the tumor, can also be exploited to achieve higher levels of the conjugate at the tumor. A major advantage of the ADEPT approach (over, for example, antibody–drug conjugates or immunotoxins) is that it is catalytic, and a single enzyme molecule can, in theory, generate many hundreds of active molecules per second from the prodrug. Further, because the enzyme–conjugate is by itself inactive, it can be allowed to clear before administration of the prodrug. Not all tumor cells would be required to bind the antibody, since the drug released by the enzyme could diffuse to neighbouring cells not expressing the antigen.

It is fundamental to the concept of ADEPT that the prodrug cannot be activated by normal human enzymes. Therefore the enzyme must have little equivalent activity in humans, particularly in serum, as the prodrug may be designed to be excluded from entering cells. Further, the enzyme must be active under physiological conditions and remain active when conjugated to an antibody. A number of enzymes and prodrugs have been considered for ADEPT. They have been reviewed in detail elsewhere (24–27) and are summarized in Table 1. They will therefore not be discussed in full. The enzymes include carboxypeptidase G2 (Fig. 2), β-glucuronidase (Fig. 3), alkaline phosphatase, β-lactamase, penicillin amidase, and cytosine deaminase, which can activate a wide range of prodrugs (Table 1). Mention should also be made of glucose oxidase, which was the first example of an antibody-targeted enzyme (28) and was made before the advent of monoclonal antibody technology. The enzyme generates hydro-

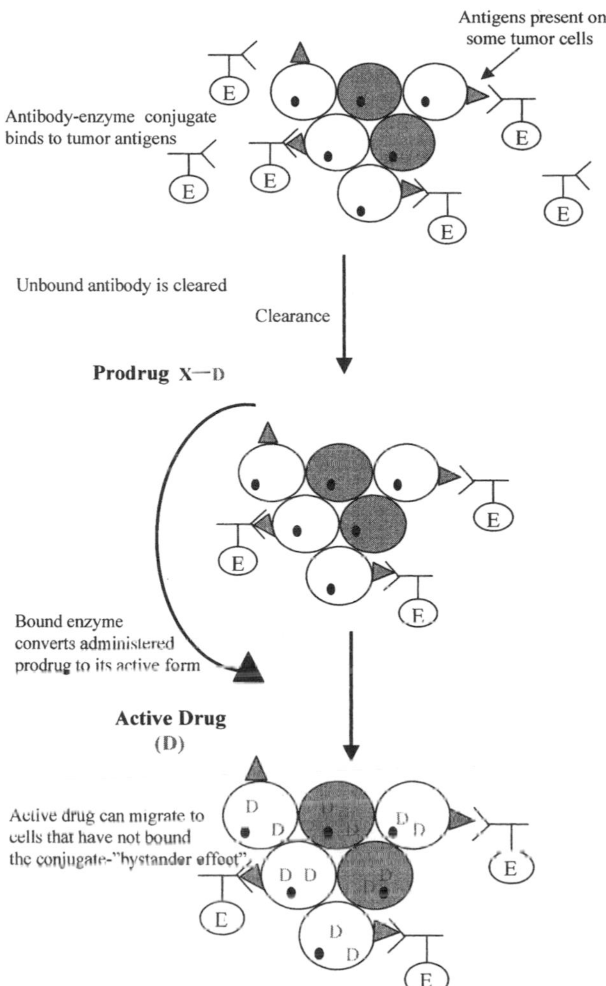

Figure 33.1 The generation of a cytotoxic drug by ADEPT. In the first phase, the antibody–enzyme conjugate is allowed to bind to the target cell population. After unbound conjugate is allowed to clear, a prodrug is administered, which is converted to an active drug (D) by the bound enzyme. Importantly the active drug can migrate and have cytotoxic effects on cells that have not bound the conjugate.

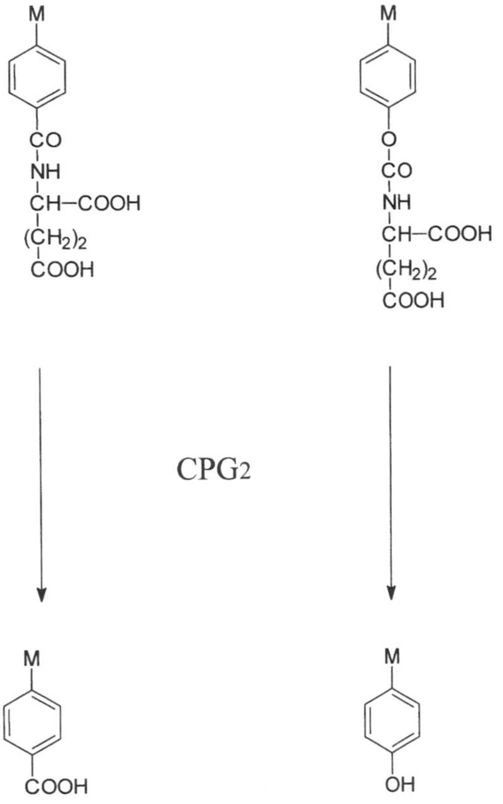

Figure 33.2 The activation of a glutamate prodrug to an active mustard by the enzyme carboxypeptidase G2 (CPG2). M = $N(CH_2CH_2Cl)_2$ (25).

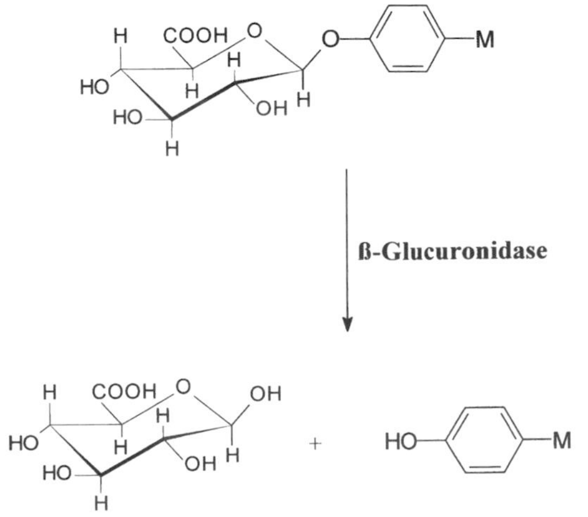

Figure 33.3 The activation of a glucuronide prodrug to an active mustard by the enzyme β-glucuronidase. (From Ref. 25.)

gen peroxide upon oxidation of glucose. However, this system was not very cytotoxic, and glucose is a poor choice of prodrug for ADEPT because it is normally present in serum.

Experimentally, ADEPT has been extraordinarily successful and has caused regression of many different types of human tumor xenografts, many of which are resistant to standard therapy (20,29–34). An example is shown in Fig. 4. These experimental studies have been followed by a limited clinical study of ADEPT in patients being treated for advanced metastatic colon or rectal cancer, who all had previously received extensive chemotherapy (35,36). The trial used a conjugate of an antibody directed to the human carcinoembryonic (CEA) antigen linked chemically to a bacterial enzyme, carboxypeptidase G2, which can hydrolyze folates and glutamate prodrugs. The prodrug used was 4-[N-(2-chloroethyl)-N-(2-mesyloxyethyl)amino]-benzoyl-L-glutamic acid, and this was well tolerated by patients (doses 200–2500 mg/m^2 given as either 6 or 12 doses divided over 3 days). Four patients received 20,000 units/m^2 of an anti-CEA $F(ab')_2$-carboxypeptidase G2 conjugate. After 2 days, a clearing antibody was administered and the prodrug given when serum carboxypeptidase G2 levels were <0.02 units/mL. Circulating levels of CEA and another tumor marker (19-9) fell in all four patients by 10–15 days after therapy. In two cases, there was a measurable decrease in the size of liver metastases, and in another case the patient became free of jaundice but with no decrease in the size of the metastases. There were also subjective responses, such as weight gain, lessening of pain, and generally improved health. However, adverse effects were also reported. After 10 to 12 days an antibody response was raised against both antibody and enzyme components of the conjugate (35,36). Thus the treatment was effectively limited to a single round of therapy, although it was proposed to prevent this antibody response by use of cyclosporin A to immunosuppress the patient. The two patients who received conjugate and the highest dose of prodrug developed myelosuppresion. Presumably, this was due to the active drug being released into the circulation and being able to migrate to the bone marrow. The active drug could be detected (35–37) in serum samples, and this suggests that the release of a more reactive drug would be beneficial. With the demonstration that carbamates were also reasonable substrates for CPG2, it was possible to make a series of derivatives that would be converted by the enzyme to highly toxic and reactive chemicals that would probably hydrolyze before reaching sensitive host tissues (38) (Fig. 2). It is planned that chemicals of this structure will be used in a Phase I clinical trial of ADEPT (39).

Problems associated with ADEPT include the immunogenicity of the conjugate, activation of the prodrug by un-

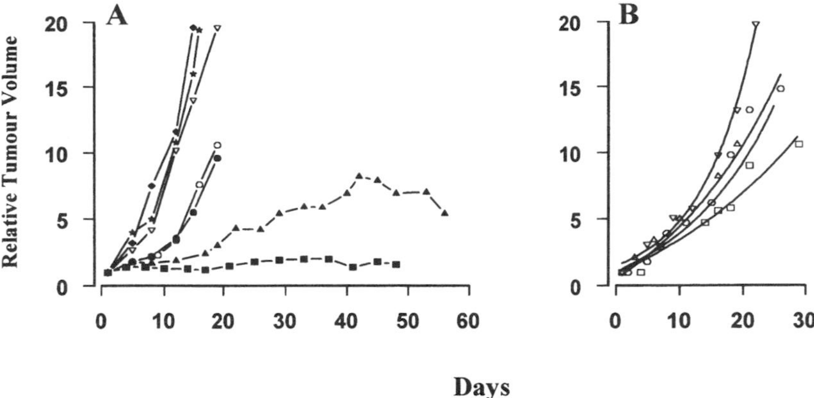

Figure 33.4 The effect of ADEPT therapy on the growth of a xenograft in vivo. Groups of mice bearing the CC3 choriocarcinoma xenograft were treated (A) intravenously with a W14 F(ab′)$_2$:carboxypeptidase G2 conjugate followed 56, 72 and 80 h later with either 5 mg/kg (▲) or 10 mg/kg (■) of p-N-bis(2-chloroethyl)aminobenzoyl glutamic acid prodrug. This can be compared with conventional therapy with either 5 mg/kg methotrexate (▽), 50 mg/kg hydroxyurea (◆), 7.5 μg/mL actinomycin D (●), 20 mg/kg cyclophosphamide (✱), or 20 mg/kg ara-C (○). (B) the mice were treated with saline (△), conjugate alone (○), prodrug alone (□), or active drug alone (▽) (both at their maximum tolerated doses). (Data from Refs. 20 and 180.)

bound circulating conjugate, conjugate heterogeneity, and optimization of the pharmacology of the active drug (24–26), and these were apparent in the initial ADEPT clinical trial. These issues need to be addressed. First is the need for the release of a more reactive drug as discussed above. Second, patients rapidly developed an antibody response against both enzyme and antibody components of the conjugate. These antibody responses developed within 10 to 12 days and effectively limited the treatment to a single cycle (35,36). These preliminary clinical experiments used a murine antibody. Techniques such as complementarity-determining region (CDR) grafting now exist to humanize such antibodies, reducing the likelihood of developing antibodies against them. The enzyme is more of a problem. It is possible to use immune-suppressive drugs such as cyclosporin (40–42). This may provide a partial remedy, but a less aggressive solution is desirable. It would also be possible to use a different activating enzyme for each cycle of treatment to circumvent the antibody response. However, this is complicated and would require a different clearance system for each cycle. Alternatively, a human enzyme could be employed. The use of a human enzyme has been best exemplified by β-glucuronidase, and this enzyme forms part of a fusion protein with a humanized antibody that has been tested in vivo (43,44) but only in nude mice. The fundamental problem of using a human enzyme is activation of the prodrug by endogenous enzyme. This can be overcome by design of the prodrug. For example, glucuronides are charged at physiological pH. Therefore they do not readily enter cells, and prodrugs exploiting this principle will not be activated by endogeneous enzymes that are intracellular. A more elegant approach is to modify a human enzyme to have new substrate specificity, but this approach carries the risk that the ''new'' enzyme will be recognized as foreign by the immune system. A mutated version of human ribonuclease has been reported (45). Human ribonuclease A is a low molecular weight (14 kDa) monomer with pH optimum of 6.5 that cleaves RNA into ribonucleotides. Ribonucleases are renowned for their robustness. In the mutated enzyme, the substrate specificity has been altered so that a prodrug is converted to an active drug by the mutant enzyme but is a relatively poor substrate for the native form. This was accomplished by mutating the positively charged lysine to a negatively charged glutamate at residue 66 of the enzyme. A basic substituent was incorporated into the prodrug so that it could interact with this mutated residue and hydrolyze to realize an active mustard that is 30-fold more cytotoxic. The mutant enzyme has both a greater affinity (K_m = 4.2 mM) and turns over the prodrug faster (k_{cat} = 15 s^{-1}) than the native enzyme (K_m = 17 mM, k_{cat} = 3.4 s^{-1}). A similar approach has been taken with a mutant of human carboxypeptidase A1 (46). Based upon a computer model of the human enzyme (built from the known crystal structure of bovine carboxypeptidase A), bulky phenylalanine- and tyrosine-based prodrugs of methotrexate that were metabolically stable in vivo and were not substrates for wild-type human carboxypeptidase A were designed and synthesized. In addition, the model was used to design a mutant of human carboxypeptidase A1, changed at position 268 from the wild-type threonine to a glycine. This novel enzyme is, unlike the wild type, capable of activating the synthesized prodrugs. The mutant enzyme could use these new prodrugs as efficiently as the wild type used its best substrates such as methotrexate-α-phenylalanine. The k_{cat}/K_m value for the wild-type enzyme with methotrexate-α-phenylalanine was 0.44 μM^{-1} s^{-1}, and k_{cat}/K_m values for mutant enzyme with methotrexate-α-3-cyclobutylphenylalanine and methotrexate-α-3-cyclopentyltyrosine were 1.8 and 0.16 μM^{-1} s^{-1}, respectively. In vitro, a conjugate of the mutant enzyme with ING-1 (an antibody that binds to the tumor antigen Ep-Cam) could kill human HT-29 colon adenocarcinoma cells (which express Ep-Cam) in the presence of either methotrexate-α-3-cyclobutylphenylalanine or methotrexate-α-3-cyclopentyltyrosine as efficiently as the active drug, methotrexate. This effect was not seen when the conjugate was made with an irrelevant antibody (46).

It is already possible to obtain highly specific tumor associated antibodies using filamentous phage display (47,48). Fusion proteins can now be constructed between enzymes and such antibodies (that have been engineered to be nonantigenic), allowing several cycles of ADEPT therapy to be given. The first fusion protein developed specifically for an ADEPT application consisted of human β-glucuronidase fused to the CH domain of BW431, a high-avidity anti-CEA antibody, humanized by complementarity determining region (CDR) grafting. The protein thus produced has antigen affinity very similar to that of the parent antibody and retains enzymatic activity (43,44). It is interesting that human β-glucuronidase is a tetramer but in this situation appears to be active as a monomer. In addition, a *B. cereus* β–lactamase II–single-chain antibody fusion protein has been constructed. The antibody used, L6, recognizes an antigen expressed on breast, colon, lung, and ovarian carcinomas. The fusion protein is produced by *E. coli* in an active form that does not require refolding and retains similar immunological and enzymatic properties to an equivalent chemically linked conjugate (49). The ultimate refinement, which is now possible by protein engineering techniques, is to make a bifunctional antibody in which one arm possesses an enzyme-mimicking catalytic function. Indeed, a catalytic antibody (termed an ''abzyme'') has been made which, in vitro, can produce cytotoxicity (50). The abzyme can hydrolyze the carbamate prodrug 4-[N, N-bis(2-chloroethyl)]aminophenyl-N-[(1S)-(1,3-dicarboxy)propyl]carbamate to generate the corresponding nitrogen mustard.

Such studies demonstrate the potential for making nonimmunogenic systems that can activate prodrugs that are not substrates for native enzymes. With so many prodrugs already shown to be effective, one might also envisage combination therapy using different enzyme, antibody, and prodrug combinations that might overcome any problems arising from tumor heterogeneity. Thus there is optimism that ADEPT, in some form or another, will eventually be used to treat the common solid cancers that are now refractory to chemotherapy.

B. Gene-Directed Enzyme–Prodrug Therapy (GDEPT)

A gene-therapy-based approach for targeting cancer cells and making them sensitive to prodrugs has been proposed for human gene therapy trials (51–54). Prodrug gene therapy, commonly referred to as GDEPT (gene-directed enzyme–prodrug therapy) or virus-directed enzyme–prodrug therapy (VDEPT) is based on the premise that a large therapeutic benefit can be gained by transferring into tumor cells a drug susceptibility gene. The gene encodes an enzyme that can catalyze the activation of a prodrug to its cytotoxic form (Fig. 5) and has also been termed a ''suicide gene'' (55–58). However, in contrast to cytotoxic gene therapy approaches that involve expression of a toxic

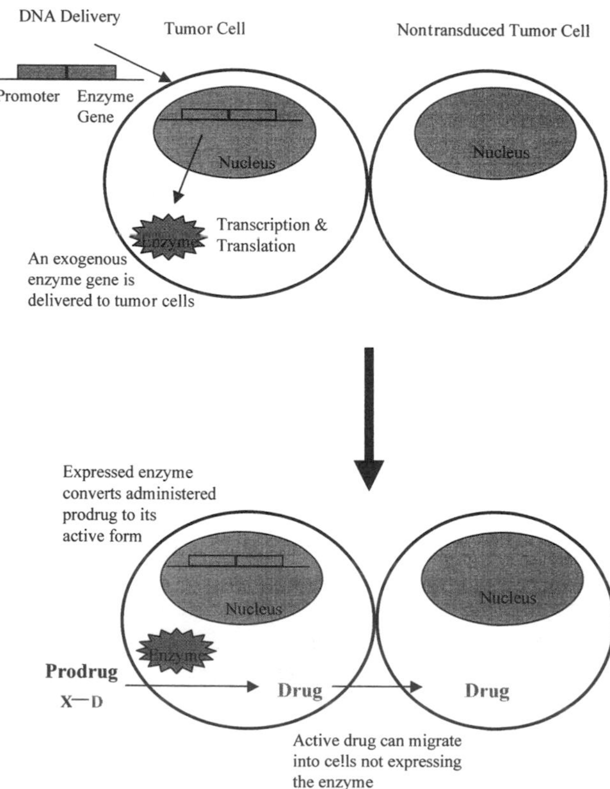

Figure 33.5 The generation of a cytotoxic drug by GDEPT. In the first phase, the cell is transduced with gene coding for a prodrug-activating enzyme. This enzyme is expressed, and a prodrug is administered that is converted to an active drug (D) by the bound enzyme. Importantly, the active drug can migrate and have cytotoxic effects on cells that have not been transduced.

product [for example, diphtheria toxin (59–62)], the enzyme itself is not toxic. Thus, in GDEPT, cytotoxicity only results after administration of the prodrug. There is also a bystander effect as the active drug can migrate into nontransduced cells.

GDEPT is obviously related to antibody-directed enzyme–prodrug therapy (ADEPT) in that both use a nonendogenous enzyme to activate a prodrug. However, there are fundamental differences between the two techniques, and this means that the enzyme/prodrug combinations used for ADEPT may not be suitable for GDEPT. In GDEPT the prodrug activating enzyme is expressed inside the cell, while in ADEPT it is extracellular. Thus for GDEPT the prodrug must be able to enter the cell. Therefore enzyme–prodrug systems developed for ADEPT that rely on a charged prodrug (such as a glucuronide) being excluded from the cell while the active drug is not are not directly applicable to GDEPT. However, it should be noted that a GDEPT system that allows for extracellular expression of the activating enzyme has now been described (63). As with ADEPT, a number of different enzyme/prodrug combinations have been proposed for GDEPT. Again, as in ADEPT, an enzyme is required for which there is no endogenous activity. Bacterial or viral enzymes are particularly suited to GDEPT (64). First, many of these enzymes have no corresponding activity in humans. Second, as expression is intracellular, there should be no problems with an immune response against a "foreign" protein. Third, bacterial enzymes tend to have few requirements for post-translational modification, and proteins that are more complex may not fold correctly in different species. Because activation takes place intracellularly, the prodrug should be freely diffusible and thus, ideally, a neutral species with an appropriate partition coefficient. As not all of the tumor cells will be transduced by the vector, a bystander effect is required. Thus the active drug should be able to diffuse away from the site of activation. However, its half-life needs to be such that it cannot migrate far enough to affect normal tissue. The active drug should also not be phase specific or proliferation dependent since many tumors have a low mitotic index (65), and all malignant cells need to be killed, not just those that are proliferating.

Success for this technique requires not only a choice of an enzyme/prodrug system but also a delivery system by which the gene encoding for the enzyme can be delivered efficiently and accurately to a human tumor. This requirement is common to all types of gene therapy. At present, targeting is achieved by altering the surface components of viruses and liposomes so as to achieve a level of target cell recognition while transcriptional elements can be incorporated so that the incorporated gene is expressed only in the target cells. Retroviral-based vectors can only infect replicating target cells; but integration occurs leading to stable expression of the transduced gene. Adenoviral vectors are more efficient at transduction and do not require cell division but do not integrate, so it is very difficult to regulate gene expression. However, for the long term success of GDEPT (and any gene therapy procedure) there needs to be advances in these systems particularly to improve their targeting. It has been proposed that future gene therapy vectors will be not based on any single virus but will be synthetic vehicles, custom designed to incorporate specific targeting features relevant to the target disease and tissue (Fig. 6) (66). Nonviral gene transfer into the central nervous system (CNS) of the mouse has been achieved using the cationic polymer polyethyleneimine (PEI) (67). Three different preparations of PEI (25, 50, and 800 kD) were compared for their transfection efficiencies in the brains of adult mice. The highest levels of transfection were obtained with the 25 kD polymer (67). Further work has shown this polymer to be a promising gene-delivery vehicle [reviewed in (68)]. Such compounds may form the basis of the future gene therapy vectors discussed above.

Why GDEPT? Cancer gene therapy strategies have been divided into two categories. One, correction of the genetic defect in tumor cells or, two, killing them. This cell killing can be mediated either by the host immune system or by direct cytotoxic effects. Each of these approaches imposes different requirements on the delivery

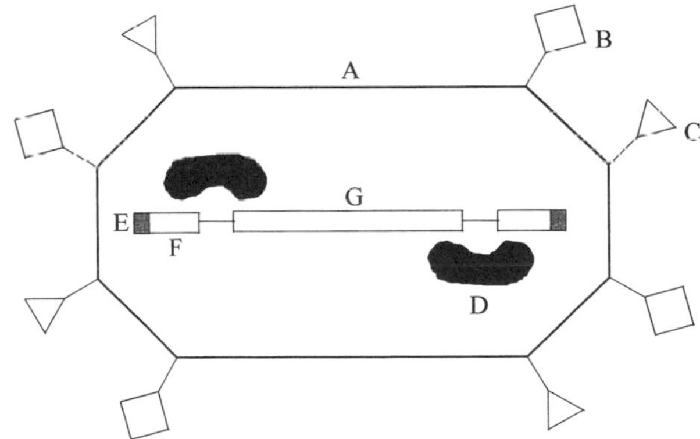

Figure 33.6 An ideal synthetic gene therapy vector as proposed by Miller and Vile (66). Its features include (A) a stable nonimmunogenic envelope, (B) ligands to confer a particular affinity on the vector, (C) moieties to encourage fusion with the target cell membrane, (D) site-specific recombinase for direct integration of the vector DNA, (E) sequences to enable homologous recombination of the vector DNA and a specific target loci of the genome, (F) tissue-specific promoter region to confer restricted expression on the therapeutic gene, and (G) the therapeutic DNA. Certain features of the vector such as homologous recombination are not required for GDEPT. (From Ref. 66.)

Table 33.2 The Requirements Needed by the Delivery System for the Various Types of Gene Therapy

	Gene correction	Cytotoxic	Immunotherapy	GDEPT
Accuracy	+	+++	+	++
Efficiency	+++	+++	+	++
Stability	+++	+	+	+

Source: Ref. 252.

system. These are summarized in Table 2. Corrective gene therapy may not be a feasible approach to curing cancer because most tumor cells contain multiple genetic mutations. However, if for example a mutant oncogene was a realistic target for corrective therapy, what is required from the delivery system? First, it will need to deliver a correct copy to every tumor cell. Therefore it must be very efficient. This is because gene correction will only work in cells that receive the gene and will have no effect on those that do not. Second, expression of the corrective gene will need to be stable, as the mutant phenotype will reappear once the corrective gene ceases to be expressed. However, third, delivery need not be accurate and specific to tumor cells. This is because, in theory, delivery of a therapeutic gene (such as a functional tumor suppresser gene) should not be deleterious to a normal cell. Opposite demands are made when a cytotoxic gene (coding for, e.g., diphtheria toxin) is delivered. Stable expression is not required, but delivery must be accurate, as expression would be deleterious to a normal cell. Immunotherapy puts the fewest constraints on the delivery system. In theory, immunotherapy attempts to activate an immune response by stimulating the immune system to recognise tumor-specific antigens as nonself by expression of immunomodulatory genes such as IL-2. These antigens are not present on normal cells. Thus expression of immunomodulatory genes should not be deleterious to them, so there is less of a requirement for accuracy. Expression and stability are also lesser requirements because, once activated, the immune system will amplify the antitumor response and kill any nontransduced tumor cells anywhere in the body. This represents an ideal situation, but in reality it may be limited autoimmunity. Further, human tumors have been extensively selected against immune recognition. It is possible to transduce genes coding for antigens, but in this case systemic protection is lost, and delivery would have to be accurate. However, the bystander effect mediated by the immune system can still be substantial.

GDEPT is a recognition of the limitations of the present delivery systems. In comparison to cytotoxic gene therapy, less accuracy is required, because the pharmacology of the prodrug and its route of administration can confer some tumor specificity. Efficiency is improved, because there is a bystander effect that can kill nontransduced tumor cells. Long-term stability is not required, as transduced cells will be killed after prodrug administration. Further, in contrast to all other types of gene therapy, GDEPT is controllable. The expressed enzyme is nontoxic, and cytotoxicity can only occur after administration of the prodrug. Therefore unexpected toxicity can be countered by lowering the dose of the prodrug.

GDEPT has been pioneered by the use of the enzymes viral thymidine kinase (tk) (Fig. 7) and cytosine deaminase, which can activate ganciclovir and 5-fluorocytidine, respectively. Herpes simplex virus (HSV)-tk/ganciclovir

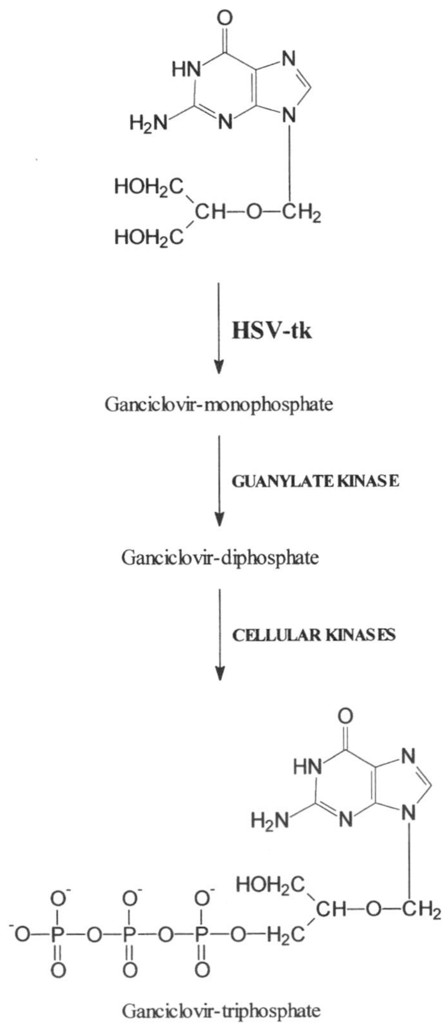

HSV-tk

Ganciclovir-monophosphate

GUANYLATE KINASE

Ganciclovir-diphosphate

CELLULAR KINASES

Ganciclovir-triphosphate

Figure 33.7 The activation of ganciclovir by herpes simplex virus thymidine kinase (HSV-tk). Cellular kinase can convert the monophosphate into the cytotoxic triphosphate.

is the most described GDEPT system, and delivery of the gene coding for HSV-tk to animal tumors in vivo has been achieved using retroviruses, adenoviruses, and naked DNA. Retroviral vectors are being used to deliver the gene to intracranial and leptomeningeal tumors in rats, and the system was proposed for the first clinical trials of GDEPT (51,52). Transfer of HSV-tk in tumor cells confers sensitivity to ganciclovir and its analogues both in vitro and in vivo (Fig. 8). In vitro, cells expressing HSV-tk are generally inhibited by 1–50 μM ganciclovir, and these levels can be achieved by patients treated with the drug (69–78). In the absence of prodrug, expression of HSV-tk is not detrimental, and expressing cells grow normally in vivo and in vitro. Recently a mutant of HSV-tk has been made with improved kinetics with respect to ganciclovir (79). In theory, this should allow effective therapy at lower prodrug concentrations. A bystander effect has been demonstrated for the HSV-tk system. As the active metabolite of ganciclovir, the triphosphate, is highly charged, it should not diffuse out of expressing cells, and thus it would not be predicted that the bystander effect is mediated by a diffusible metabolite. In vitro, a mixed population containing only 10% of HSV-tk expressing cells was totally killed by a dose of ganciclovir that is not cytotoxic to non-HSV-tk expressing cells (80). Metabolic cooperation is involved, and a ganciclovir metabolic product, presumably a phosphorylated form, can pass from HSV-tk expressing cells to nonexpressing cells and mediate cytotoxicity, but only as a consequence of direct contact (81,82). The uptake of apoptotic vesicles by the nonexpressing tumor cells has also been demonstrated (80). In vivo, there is also a cell-mediated immune component to the bystander effect (83).

Cytosine deaminase is found in many fungi and bacteria and catalyzes the deamination of cytosine to uracil. It is not found in mammalian cells. Cytosine deaminase can also convert the clinically used antifungal agent 5-fluorocytosine (5-FC) into the known antitumor drug 5-fluorouracil (5-FU). The yeast enzyme is superior to the bacterial enzyme in this respect (84). 5-FU is further metabolized to 5-fluorouridine-5′ triphosphate, 5-fluoro-2′-deoxyuridine-5′-triphosphate, and 5-fluoro-2′-deoxyuridine-5′-monophosphate. The first two compounds inhibit RNA and DNA synthesis, respectively, and the monophosphate is a potent inhibitor of thymidylate synthase, which is an important enzyme in DNA biosynthesis. Mouse 3T3 cells were shown to be sensitive to 5-FC when transfected with the gene encoding cytosine deaminase (85). When injected into syngeneic mice, the cytosine deaminase expressing tumors could be eliminated in vivo by systemic treatment with 5-fluorocytosine without significant toxicity to the host, although delaying the prodrug treatment reduced its effectiveness (Fig. 9). Interestingly, animals whose tumors

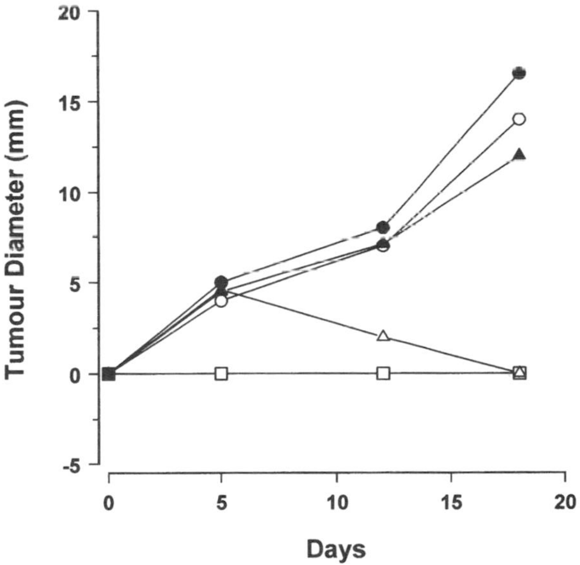

Figure 33.8 The antitumor effect of ganciclovir on HSV-tk expressing tumor cells. Mice were inoculated s.c. with KBALD tumor cells (2×10^5) either stably transduced with the HSV-tk gene (open symbols) or not (closed symbols). The mice were then treated with ganciclovir (150 mg/kg twice a day for 3.5 days) commencing on day 0 (□) or day 5 (△), or left untreated (○). (Data from Ref. 80.)

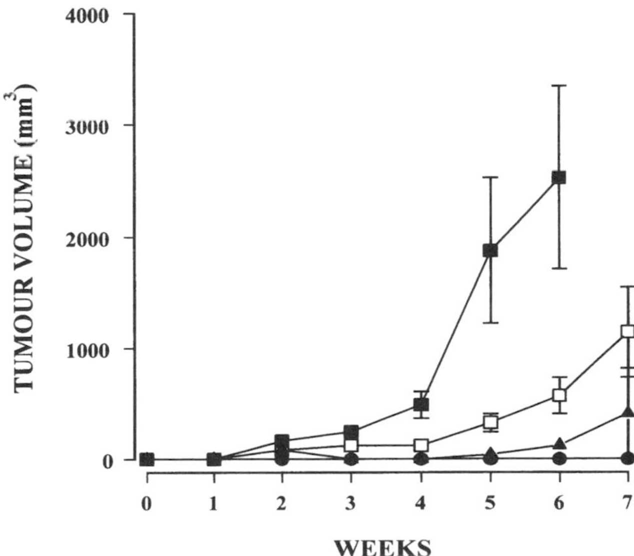

Figure 33.9 The effect of 5-fluorocytidine (5-FC) on the growth of a cytosine deaminase expressing mouse adenocarcinoma. Mice received 5-FC (37.5 mg) twice daily for 10 days 4 (●), 11 (▲), or 18 (□) days after tumor inoculation. Untreated controls (■). (Data from Ref. 57.)

NITROREDUCTASE
NAD(P)H
$NAD(P)^+$
DNA THIOESTERS
DNA ADDUCTS
+
DNA INTERSTRAND CROSS LINKS

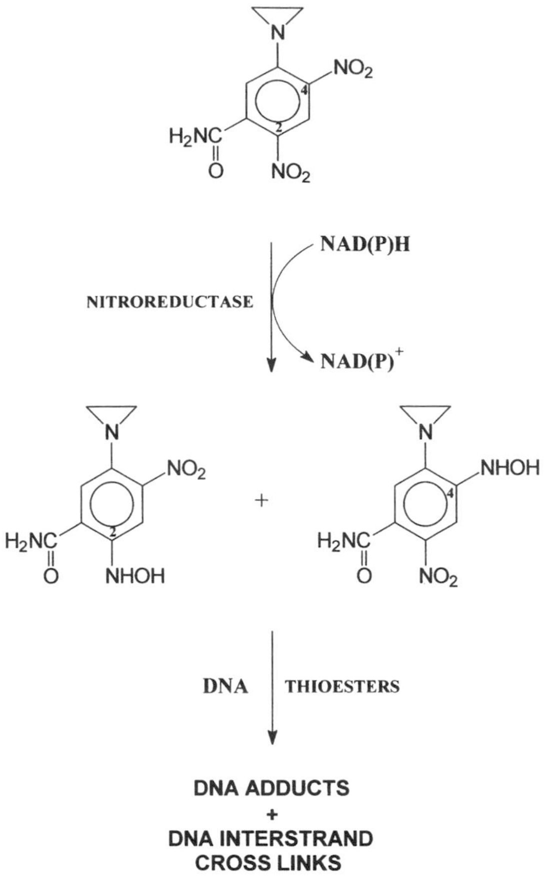

Figure 33.10 The activation of the prodrug CB 1954 (5-(aziridin-1-yl)-2,4-dinitrobenzamide) by the enzyme *E.* coli nitroreductase.

had been eliminated by prodrug treatment resisted subsequent rechallenge with unmodified wild type tumor. This posttreatment immunity appeared to be tumor specific (57). It was postulated that cell death led to a more effective antigen presentation or that the cytosine deaminase protein itself was immunogenic (57). Human colorectal tumors were also shown to be sensitive to 5-FC when they expressed cytosine deaminase (86). The sensitivity of these cells was retained in vivo when grown in nude mice (86). Importantly, a large bystander effect was observed both in vitro and in vivo (86). 5-FU was liberated into the surrounding environment when cytosine deaminase–expressing tumor cells were treated with 5FC. This liberated 5FU is able to kill neighboring, noncytosine-deaminase–expressing tumor cells in vitro and in vivo (87). When only 2% of the tumor mass contained cytosine deaminase–expressing cells (98% noncytosine-deaminase–expressing cells), significant regressions in tumors were observed when the host mouse was dosed with nontoxic levels of 5FC (87).

In a direct comparison of cytosine deaminase with HSV-tk, both transduced into human lung adenocarcinoma cell lines and driven by the CMV promoter, cytosine deaminase was superior in its ability to achieve high levels of specific enzyme activity, to induce growth inhibition, and to affect neighboring cell growth (88).

Other enzymes considered for GDEPT include thymidine phosphorylase, deoxycytidine kinase, carboxylesterase, cytochrome-P450, carboxypeptidase G2, and nitroreductase (Table 1). All demonstrate that selective killing of tumor cells can be achieved by GDEPT. Perhaps unfortunately, many of the active drugs formed are antimetabolites. These are only cytotoxic to dividing cells and tend to require long exposure times for optimum effect. Alkylating agents do not suffer from these disadvantages and are less prone to induce drug resistance (89). In this respect, the enzymes cytochrome-P450, carboxypeptidase G2, and nitroreductase are of particular interest because they can activate prodrugs of alkylating agents.

An example is the activation of the prodrug CB 1954 by *E. coli* nitroreductase (Fig. 10). This enzyme requires a cosubstrate to act as an electron donor. This is not a problem, because the enzyme is expressed intracellularly and thus can use the endogenous cofactors NADH or NADPH [of which nitroreductase can use either (90)]. In fact, the cofactor requirement may be an advantage, because any enzyme that escapes into the circulation (for example, from dying cells) will be incapable of activating circulating prodrug because of the lack of a cofactor. This is because NAD(P)H is very rapidly metabolized by serum components (91).

A recombinant retrovirus encoding NR was used to infect mammalian cells. NIH3T3 cells expressing NR were killed by CB1954. The bulk infected, unselected, cell population was about 100-fold more sensitive to CB 1954 than the parental cells (92). A selected clone was even more sensitive and, using a cell count assay, was over 1000-fold more sensitive to CB 1954 than parental NIH3T3 cells (Fig. 11) (92). Similar results were seen in human melanoma, ovarian carcinoma, and mesothelioma cells (93). The rapid action of CB 1954 (94) and the resulting need for a shorter exposure time may facilitate the use of this prodrug clinically. A significant bystander effect was observed, and admixed, unmodified NIH3T3 cells could also be killed by a normally nontoxic dose of prodrug (92). The bystander effect is mitigated by diffusible metabolites, and both the 2- and 4-hydroxylamino derivatives of CB 1954 are released into the medium of CB 1954 treated nitroreductase-expressing NIH3T3 cells (95). Importantly, and in contrast to the HSV-tk/ganciclovir enzyme/prodrug system, NR/CB1954 cell killing was cytotoxic towards non-

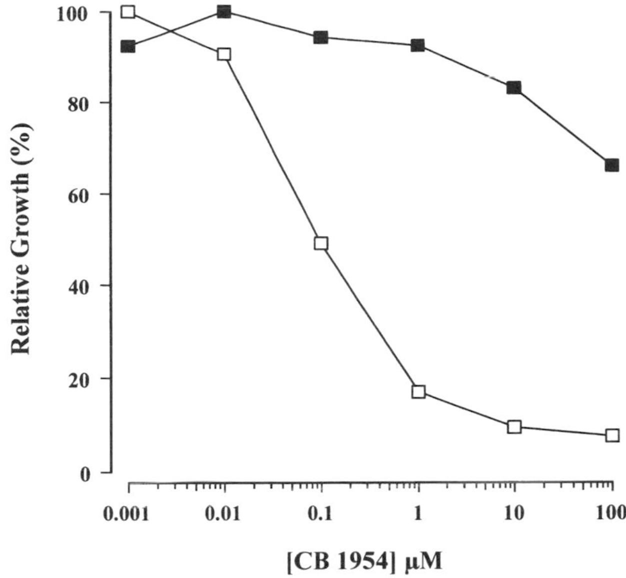

Figure 33.11 The cytotoxicity of CB 1954 against a nitroreductase (NR) expressing cell line. NIH3T3 cells were infected with a recombinant retrovirus containing NR and a cell clone (NIH3T3-NR) derived by limiting dilution. Parental NIH3T3 cells (■) or NIH3T3-NR cells (□) were treated with CB 1954 for 24 h prior to assay. (Data from Ref. 92.)

cycling cells (Fig. 12) (92). In an initial investigation of NTR/CB1954 for the treatment of tumors in vivo, a regression of tumors expressing NTR following administration of CB1954 resulted in significantly increased median survival (Fig. 13) (96).

In addition, other potential prodrugs for use with the nitroreductase have been proposed. Chinese hamster V79 cells transfected with a nitroreductase expression vector were 770-fold more sensitive to CB1954 than control nonexpressing cells. Other prodrugs, such as nitrofurazone (97-fold) and the nitroimidazole compounds, misonidazole (21-fold), and metronidazole (50-fold), also exhibited increased cytotoxicity against the nitroreductase-expressing cells and were found by HPLC to act as substrates for the purified NR enzyme (97). However, this correlation was not absolute. In particular, the quinone EO9 [3-hydroxymethyl-5-aziridinyl-1-methyl-2-(H-indole-4, 7-indione)-propenol] showed only a very small differential (< 3-fold). This is probably because this compound can undergo activation by endogenous enzymes such as NADPH: cytochrome P450 reductase (97). When misonidazole or metronidazole were activated by NR extracellularly there was little increase in cytotoxicity (98). This would suggest that the active species has a very short half-life and probably would not exhibit a large bystander effect. A series of 2,4-dinitrobenzamide mustard analogues of CB 1954 have also been evaluated as potential prodrugs (98). Other potential prodrugs that could be used in GDEPT are those activated by a self-immolative mechanism (99) that can potentially form active drugs such as mustards, actinomycins, mitomycin C (99), enediynes, (100), seco-CI alkylating agents (101), or tallimustine (102). The prodrugs are all 4-nitrobenzyloxycarbonyl derivatives of these drugs, which upon enzymatic reduction generated the drug through self-immolation of the 4-(hydroxylamino)benzyloxycarbonyl group.

The generation of an alkylating agent probably offers the most potent means of killing targeted cell types. However, given the different modes of action of, for example, nitroreductase/CB 1954 and HSV-tk/ganciclovir, a combination of these approaches offers a way of obtaining potentially synergistic effects. Cooperative killing was observed when cells expressing both NR and HSV-tk were treated with a combination of CB1954 and ganciclovir (92).

Thus CB 1954 is a good example of the requirements of an ideal prodrug for use in GDEPT when activated by nitroreductase, and this enzyme can also be used in combination with other prodrugs. However, systemic administration of the present generation of gene therapy vectors is not possible, and GDEPT, unlike ADEPT, is thus limited to isolable tumor deposits, such as intracerebral tumors (53) or prostate cancers (54), surrounded by largely nondividing normal tissue. In such cases, GDEPT is feasible using retroviral-based vectors. Improved vectors have been proposed and are certainly under development. On the other hand, it should be remembered that GDEPT is an answer to some of the limitations of the present gene therapy vectors, and a perfect tumor-specific gene delivery system would make GDEPT obsolete.

C. Targeting by Means of Polymers

The use of polymer drug conjugates as prodrugs is a relatively new concept for the treatment of cancer (103,104). N-(2-Hydroxypropyl)methacrylamide (HPMA) copolymer–drug conjugates containing either anthracycline antibiotics or alkylating agents bound to polymer through peptide linkers designed for cleavage by endogenous thiol dependent enzymes have been described (103,104). HPMA copolymer conjugates, like other molecules with prolonged plasma residence times, can accumulate preferentially in a tumor because of the phenomenon of the enhanced permeability and retention effect (EPR) (Fig. 14) (105). This effect occurs because the physiology of solid tumors differs from that of normal tissues in a number of important aspects, the majority of which stem from differences between the two types of vasculature. Compared

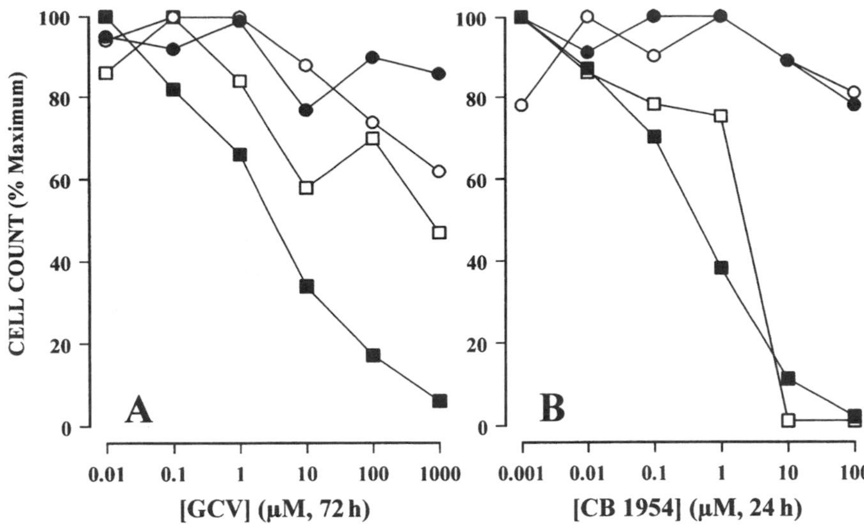

Figure 33.12 The effect of (A) ganciclovir (GCV) or (B) CB 1954 on NIH3T3 cells either expressing (■) or not expressing (●) either (A) herpes simplex virus thymidine kinase or (B) nitroreductase. Cells were treated during either normal growth (solid symbols) or following growth arrest in the G_0 phase of the cell cycle (open symbols). (Data from Ref. 92.)

with the regular, ordered vasculature of normal tissues, blood vessels in tumors are often highly abnormal, distended capillaries with leaky walls and sluggish flow (106). Tumor growth also requires continuous new vessel growth, or angiogenesis. The physiology of solid tumors at the microenvironmental level is thus sufficiently different from that of the normal tissues from which they arise to provide a unique and selective target for cancer treatment (107,108). The EPR effect results from enhanced permeability of macromolecules or small particles within the tumor neovasculature due to leakiness of its discontinuous endothelium, and this mechanism of tumor targeting of polymeric systems was first described by Matsumura and Maeda (105). In addition to the tumor angiogenesis (hypervasculature) and irregular and incompleteness of vascular networks, the attendant lack of lymphatic drainage promotes accumulation of macromolecules once they extravasate (105). This phenomenon of passive diffusion and localization in the tumor interstitium is observed in many solid tumors for macromolecular agents and lipids (109,110). Unless specifically addressed for tumor cell uptake by receptor-mediated endocytosis, macromolecules entering the intratumoral environment are taken up relatively slowly by fluid-phase pinocytosis and then transferred from endosomes to lysosomes (111). Modified pathways of fluid extravasation and tissue drainage in tumors are thought to be the physiological cause for the passive tumor tropism of macromolecules. Ineffective or absent pathways of lymphatic drainage results in poor convection and elevated interstitial hydrostatic pressures [reviewed by (112)]. This leads to poor oxygenation of the tumor mass and induces the release of angiogenic and capillary-permeabilizing factors such as vascular endothelial growth factor (VEGF) in order to improve the supply of oxygen

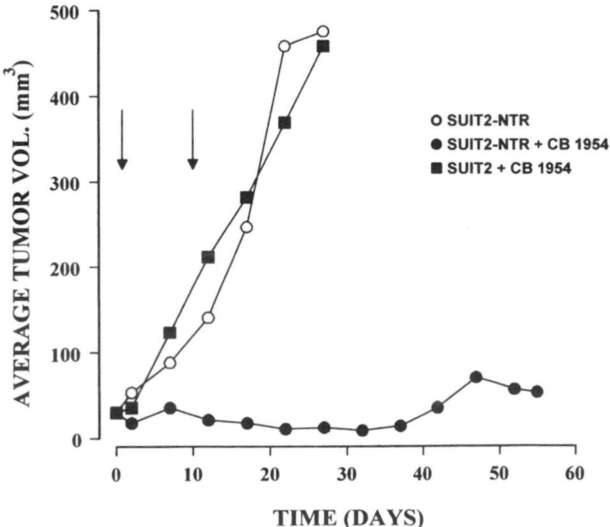

Figure 33.13 The therapy of a human, E. coli B nitroreductase-expressing, pancreatic xenograft (SUIT-2-NR) using CB 1954. This is compared to the treated parental line (SUIT-2). CB 1954 (80 mg/kg I.P.) was given on day 1 and 10 as indicated. (Data from Ref. 96.)

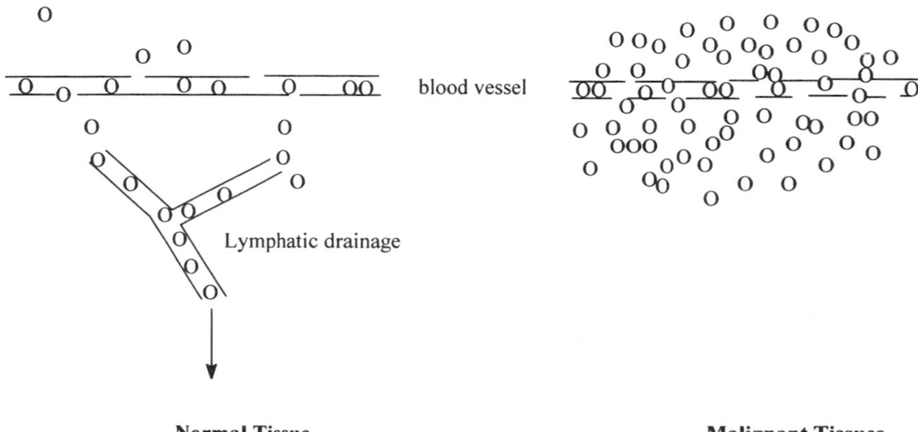

Normal Tissue

Blood vessels supplying normal organs have a tight vascular endothelium with only a few gaps through which chemicals of a high molecular weight can pass. Any molecules that do extravasate are rapidly removed by draining lymph vessels

Malignant Tissues

The newly formed blood vessels supplying cancers are less continuous and have many gaps through which chemicals of high molecular weight can pass. Because there is no lymphatic drainage there is an accumulation of chemicals of high molecular weight

Figure 33.14 Enhanced permeability and retention (EPR) of tumor cells.

and nutrients [reviewed by (113)]. The enhanced vascular permeability will support the great demand of nutrients and oxygen for the rapid growth of the tumor. To realize this demand, tumor tissues recruit blood vessels from the preexisting network as well as inducing extensive angiogenesis by releasing VEGF (114–116). Without angiogenesis, the tumors could not increase in size to more than a few millimetres (117–119). When using polymeric drugs, the EPR effect is a great advantage as a means for targeting. Low molecular weight analogues readily leave the bloodstream, as they quickly pass into the tissue or are excreted. In contrast, polymeric drugs can often achieve a prolonged half-life in the bloodstream because of decreased glomerular excretion. Extended circulation is a requisite for optimal targeting by the EPR effect. However, an optimum molecular weight exists in terms of achieving the most efficient accumulation by the EPR effect. For example, with N-(2-hydroxypropyl)methacrylamide (HPMA) copolymers (fractions of molecular weight 22–778 kDa), at 10 min after administration in vivo, all fractions were already detectable in the tumor, and those of molecular weight greater than the renal threshold showed progressive tumor accumulation up to 20% of dose administered per gram after 72 h in Sarcoma 180 model (120). HPMA copolymer fractions showed profiles of blood clearance that were strongly dependent on molecular size. For this reason, although the initial tumor levels of all HPMA copolymer fractions were virtually independent of substrate size, their progressive accumulation was different (120). Further, macromolecular carriers in the bloodstream may be recognized by scavenger cells, namely the RES (reticuloendothelial system). These cells are located within such organs as liver, spleen, and lung. RES recognition is particularly serious for colloidal and vesicular carrier systems, including microparticles and liposomes (121). Thus it is essential to develop appropriate carrier systems that achieve long circulation in the bloodstream, avoiding glomerular excretion and RES recognition [like HPMA copolymer (122)], allowing for a significant EPR effect. When using a polymer it is therefore important to choose an adequate carrier for both the drug and the enzyme, bearing in mind the possibility of immunogenicity in addition to the abovementioned RES capture.

A polymer-based system for targeting enzymes to tumors that is directly analogous to ADEPT is called MDEPT (macromolecular-directed enzyme–prodrug therapy). This utilizes EPR to localize an enzyme polymer conjugate but uses a nonpolymer prodrug and an inhibitor such as a clearing antibody to reduce conversion of the prodrug in the circulation (123). MDEPT would therefore appear to be the same as ADEPT except that the antibody-targeting component is replaced by a polymer and in theory could be used with any of the ADEPT enzyme/prodrug systems previously described. It was found that the enzyme CPG2 conjugated to methoxypolyethyleneglycol (MPEG) localized in nude mice bearing human xenografts of LS174T (colon), PC3 (prostate), COR L23 (lung), and MDA (breast). Different chain lengths of polyethylene glycol (PEG) and ratios of PEG to protein affect the level of uptake by tumors (124). In a therapy study using CPG2 conjugated to methoxypolyethyleneglycol and a phenol mustard prodrug, MDEPT was shown to be effective in treating the LS174T tumor (Fig. 15) (123).

Polymers are interesting also for their effect on the immunogenicity of foreign proteins (125,126). This is generally attributed to their ability to mask epitopes. There is also evidence that they can induce immune tolerance to challenge with bacterial enzymes. The substitution of a polymer for an antibody vector therefore reduces the immunogenicity associated with an antibody vector and has the potential to allow the repeated use of bacterial enzymes in man without the necessity for immunosuppression (124).

Recently, polymer-directed enzyme prodrug therapy (PDEPT) has been described. PDEPT is a novel two-step antitumor approach in which both the prodrug and the enzyme components are targeted (Fig. 16). Because of the relatively short plasma residence time of the polymeric prodrug, a dosing schedule can be used that avoids the need for a clearing antibody used in ADEPT, and the unwanted activation of the prodrug in the circulation can be controlled. As PDEPT relies on two components, a polymer–enzyme and a polymer–drug conjugate, the design of both components must be optimized. There are different requirements for the polymer–drug and for the polymer–enzyme conjugates, which can be summarised as follows.

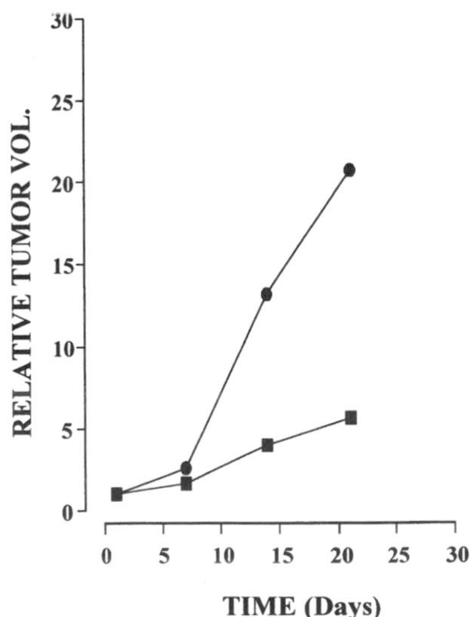

Figure 33.15 The effect of MDEPT therapy on the growth of a xenograft in vivo. (■) Nude mice bearing LS174T tumors were given 25 enzyme units of CPG2–PEG (22 PEG molecules, PEG 5000) followed at 20 h, 23 h, 26 h by galactosylated SB43. At 48 h post-CPG2–PEG they received 25 mg/kg of a phenol prodrug (4-[N,N- bis(2chloroethyl)amino]-phenoxycarbonyl-L-glutamic acid) intraperitoneally, repeated to a total of 3 doses at 1 h intervals. (●) Untreated controls. (Data from Ref. 123.)

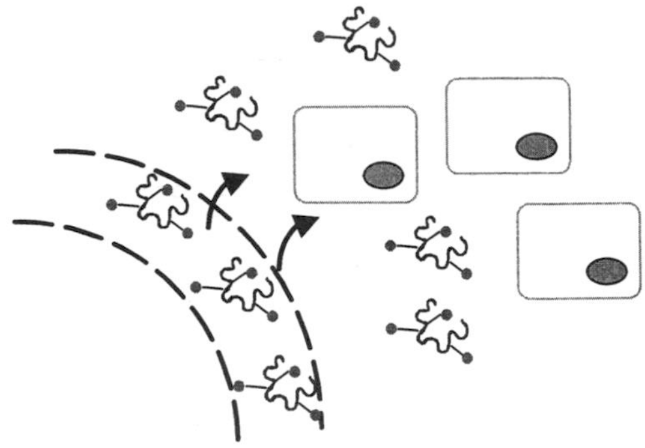

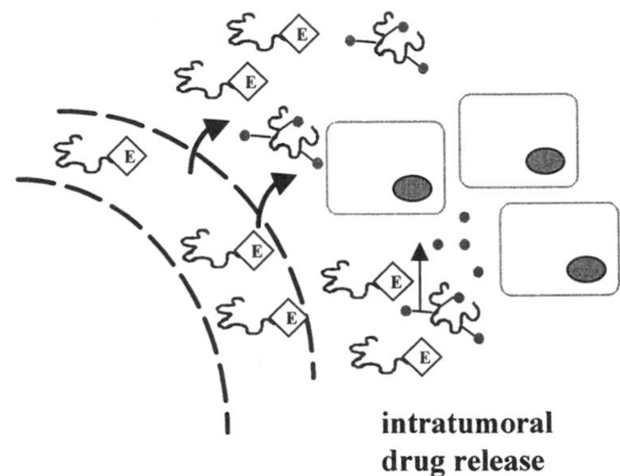

Figure 33.16 Release of anticancer drug from its polymeric conjugate by polymer-directed enzyme–prodrug therapy (PDEPT).

The polymer–drug conjugate should

1. Be inactive in the bloodstream
2. Accumulate selectively in the tumor by the EPR effect
3. Be excreted through the kidneys within 5 h after administration
4. Contain a linker that is degraded by the enzyme present in the polymer–enzyme conjugate with effective and favorable activation kinetics
5. Release an active drug that can induce a bystander effect
6. Ideally release a drug effective in both cycling and noncycling cells

The polymer–enzyme conjugate should

1. Display enzymatic activity in the polymer conjugate with a favorable K_m towards the polymer–drug conjugate as a substrate in vivo
2. Ideally contain a nondegradable linker in vivo
3. Accumulate selectively in the tumor by the EPR effect
4. Display a long circulation time in the body
5. Be nonimmunogenic

To test the feasibility of PDEPT, an HPMA copolymer–enzyme conjugate was synthesized containing cathepsin B as a model enzyme, and its in vitro and in vivo properties were evaluated. HPMA copolymer-Gly-Phe-Leu-Gly-doxorubicin (PK1) was used as a substrate for HPMA copolymer–cathepsin B to test the PDEPT combination in vitro and in vivo. PK1 has already shown promise in early clinical trials, and a phase II programme is currently ongoing (127). This compound was chosen following a systematic evolution of an optimum molecular weight of carrier to allow tumor-selective doxorubicin delivery, and optimization of linker design (Gly-Phe-Leu-Gly) to mediate controlled release of the drug intratumorally (intracellularly) by the lysosomal thiol-dependent proteases [reviewed in (128)]. PK1 has a molecular weight of 30 kDa and a doxorubicin content of ~8 wt% (~2 mol%). The conjugate displays a longer plasma half-life than free doxorubicin ($t_{1/2}$ approximately 1 h). Moreover, it shows significantly increased tumor deposition of doxorubicin (in comparison with free doxorubicin) due to the EPR effect (129). Covalent conjugation of doxorubicin via the peptidyl spacer ensures that no significant liberation of free drug in circulation takes place, thus leading to a marked increase in the therapeutic index compared with free doxorubicin. HPMA copolymer–protein conjugates showed little or no immunogenicity and are known to display selective accumulation in solid tumors (129,130). A recent Phase I clinical trial of PK1 showed that it is biocompatible in man and amenable to clinical development (127). PK1 is activated by endogenous proteases, mainly following internalization into the cell and then intralysosomal digestion. The compound is a suitable polymeric prodrug for PDEPT because the targeting of an additional external protease could significantly increase the rate of doxorubicin release.

The enzyme model selected to explore the PDEPT concept was cathepsin B, a mammalian lysosomal thiol-dependent enzyme that functions in the normal turnover of proteins in mammalian cells. Cathepsin B is a cysteine proteinase that plays an important role in lysosomal proteolysis (131). It is secreted as a preproenzyme that requires two proteolytic events for activation. Once activated, cathepsin B degrades type IV collagen, fibronectin, and lamin, at both acidic and neutral pH (132). This enzyme was a convenient choice, as it is the native activating enzyme for PK1. This gave the possibility to combine well with PK1 to study aspects of the PDEPT combination. The spacer chosen between the HPMA copolymer and the enzyme was a glycine–glycine linker, because it is nondegradable in the body.

Following polymer conjugation (yield of 30–35%), the enzyme retained 20–25% of its enzymatic activity. To investigate the pharmacokinetics in vivo, a ^{125}I-labelled HPMA copolymer–enzyme conjugate was administered intravenously (i.v.) to B16F10 tumor-bearing mice. Due to selective tumor tissue accumulation by the enhanced permeability and retention (EPR) effect, the HPMA copolymer–enzyme conjugate showed a 2–3-fold increase in tumor accumulation compared to the native enzyme, and also exhibited a longer plasma half-life.

The ability of HPMA copolymer–cathepsin B to access and degrade the prodrug in vivo was determined by HPLC evaluation of doxorubicin release. PK1 was injected i.v. to B16F10-bearing mice and after 5 h HPMA copolymer–cathepsin B was administered. This enzyme led to a rapid increase in the rate of doxorubicin release intratumorally (3.6-fold faster than seen for PK1 alone) (Fig. 17). Moreover, when the antitumor activity (doxorubicin-equivalent dose of 10 mg/kg) was measured using this tumor model, the combination PDEPT had the highest activity (T/C = 168%) compared to that seen for PK1 alone (T/C = 152%) or free doxorubicin (T/C = 144%). No animal weight loss

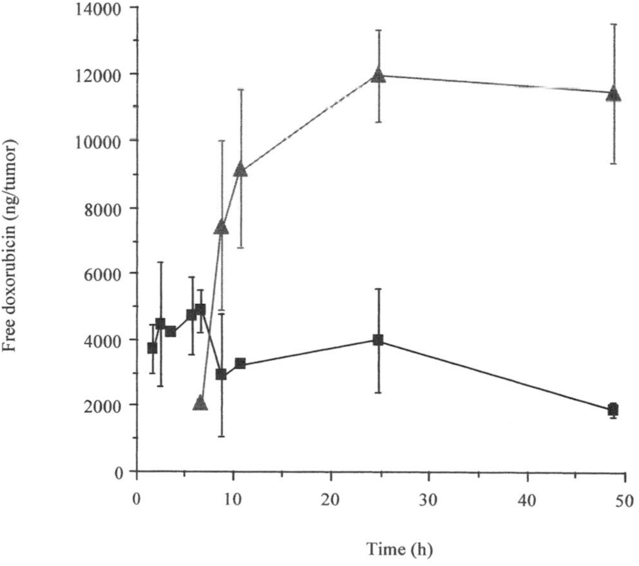

Figure 33.17 Free doxorubicin released in the tumors of C57 mice bearing B16F10 murine melanoma from PK1 in the absence (■) or presence (▲) of HPMA–cathepsin B.

or other toxicity was observed, indicating the possibility of dose escalation (133,134). These results demonstrate that the PDEPT concept is sound and could be applied to other enzyme polymeric prodrug systems. An HPMA conjugate of β-lactamase has been made and could activate polymeric prodrugs where the linker between the polymer and the drug is a cephalosporin linker, which is cleaved by β-lactamase (134).

D. Antimetabolite with Inactivation of Rescue Agent at Cancer Sites (AMIRACS)

Antimetabolite drugs have been in the forefront of anticancer therapies since the early 1950s and still remain an important and fruitful field for drug discovery. Such drugs act as competitive inhibitors for certain key enzymes involved in the process of cell growth and division, leading to arrest of the cell cycle and cell death. A number of such enzyme targets exist, particularly in the pathways associated with folic acid metabolism, as illustrated in Fig. 18. However, it has long been recognized that antimetabolite drugs do have some disadvantages. An important disadvantage is that they are not cancer specific in their activity. Although they tend to be selective for the rapidly dividing cells of tumors, they are not tumor specific because they also exert toxic effects on rapidly dividing normal cells in the body, for example the gut mucosa and hair follicles. The severity of these side effects is dose limiting, and the development of such dose-limiting toxicity necessitates cessation of treatment. A further disadvantage of antimetabolite drugs as currently used is that cancer cells can develop resistance to them, either by increased expression of the target enzyme, or by changes in the rate of drug transport into and out of the cell, or by a combination of these. The development of resistance necessitates escalating doses to retain efficacy, and the history of antimetabolite cancer therapy is one of a drive to find safe methods of administering ever greater doses of drug while circumventing the problems of dose-limiting toxicity. One approach has been to administer high doses of drug and then follow at an appropriate juncture with the appropriate metabolite, termed a ''rescue agent,'' which is capable of counteracting the effect of the antimetabolite. Such agents generally work by entering the metabolic pathway downstream of the antimetabolite-induced block. For example, folinic acid can rescue cells from methotrexate toxicity (Fig. 19). The use of methotrexate therapy in conjunction with folinic acid rescue is now well established in clinical practice (135,136), and the possible use of alternative rescue agents for methotrexate, such as 5-methyltetrahydrofolate and thymidine, has also been explored (137–140). For

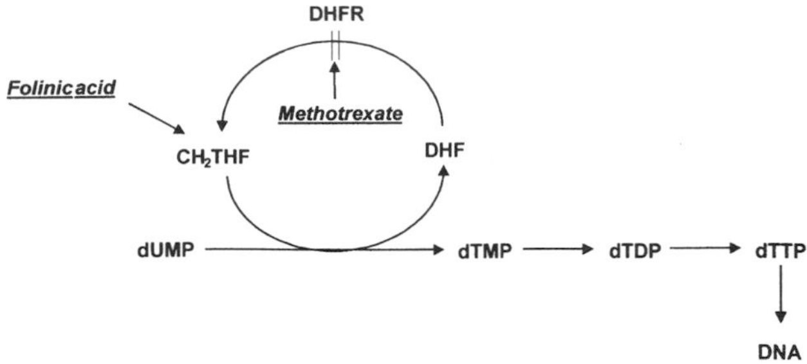

Figure 33.19 Folinic acid enters the folate cycle downstream of the methotrexate-induced blockade, allowing the cell cycle to restart.

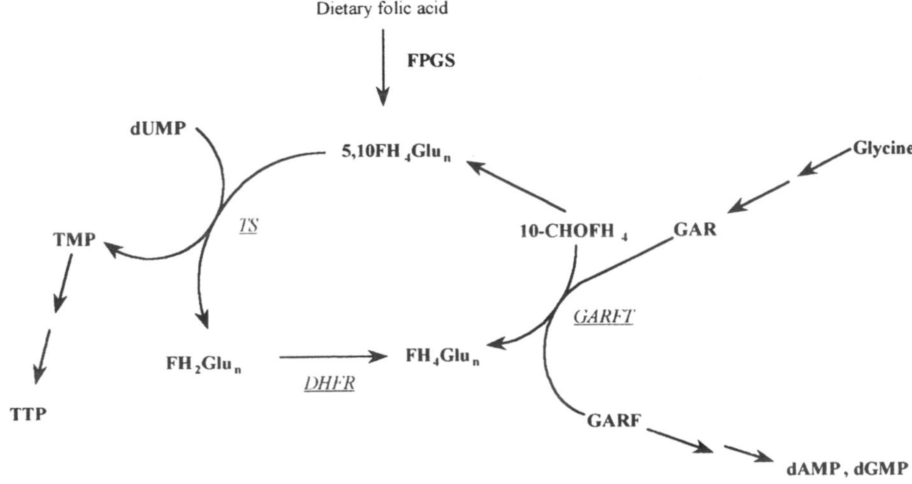

Figure 33.18 The pathways of folate metabolism. Key enzymes are underlined.

most cancers, the value of rescue agents is limited because the agent ''rescues'' the tumor as well as the normal tissue. However, the use of folinic acid to rescue the combination of the folate antimetabolite trimetrexate and the metabolite folinic acid is licensed for the treatment of Pneumocystis carini in HIV patients (141). In this instance the infective organism is, like its host, sensitive to the antimetabolite but, unlike the human cells, the infective organism cannot be rescued by folinic acid. The use of the trimetrexate/folinic acid combination allows much higher doses of drug to be administered without toxic side effects than would be the case with drug alone.

AMIRACS is a technology that uses targeted enzymes to break down rescue agents at tumor sites, preventing the tumor cells from being rescued while allowing the rescue agent to carry out its beneficial function in normal cells (Fig. 20) (124,142). Normal tissues are exposed to a non-toxic combination of antimetabolite drug plus rescue agent, whilst the tumor can be exposed to amplified doses of drug, because the rescue agent is destroyed by an enzyme that is specifically targeted to the tumor. Therefore AMIRACS offers the opportunity to make antimetabolite drugs tumor specific rather than tumor selective and enables high doses of drug to be used while confining toxic effects to the tumor and thereby increasing the therapeutic ratio of the antimetabolite. It can create in human tumors the system used successfully to eradicate Pneumocystis carini.

Two methods of targeting enzymes to tumors can be used—antibodies or polymers. The prototype system for demonstration of enzymatic inactivation of rescue agent uses the dihydrofolate antagonist trimetrexate as the antimetabolite component, with folinic acid as the rescue agent and CPG2 as the rescue agent. When using folinic acid as rescue agent in conjunction with CPG2 it is necessary to use a nonclassical antifolate drug such as trimetrexate. These lack the terminal glutamate of the classical folate analogues that are degraded by CPG2. CPG2 breaks down folinic acid by cleaving off the glutamate moiety from the pterin component of the molecule. The ability of tumor-targeted enzyme selectively to break down folinic acid at a tumor site in vivo is illustrated in Fig. 21. In this experiment, antibody–CPG2 conjugate was used, with second antibody clearance to give high tumor-to-blood ratios, but similar results would be anticipated from a polymer-based system. In fact, the lowest tumor folinic acid levels are obtained without the use of clearance (Fig. 21), but this is probably due to the depletion of serum folinic acid by the uncleared enzyme. In a clinical experiment, two patients were treated with antibody–CPG2 conjugate (20,000 U/m^2), followed by a mixture of trimetrexate and folinic acid (143). The protocol for this experiment is illustrated in Fig. 22. This experiment used a conjugate of the anti-CEA murine monoclonal antibody, A5B7, chemically linked to CPG2. Because both components of the conjugate are of xenogeneic origin and have been shown in earlier studies to be immunogenic (41,144), it was necessary to carry out the therapy under the cover of the immunosuppressive drug, cyclosporin A. Cyclosporin A was chosen because it had been shown to suppress the immune response to both antibody and enzyme for up to three weeks. This allows three cycles of therapy to be given (36,42). The dose of

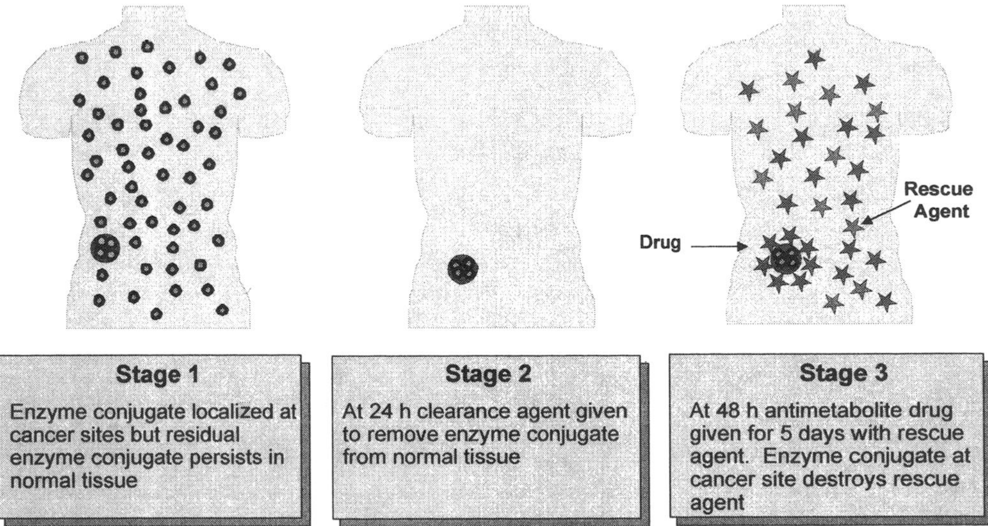

Rescue Agent
Drug
Stage 1
Enzyme conjugate localized at cancer sites but residual enzyme conjugate persists in normal tissue
Stage 2
At 24 h clearance agent given to remove enzyme conjugate from normal tissue
Stage 3
At 48 h antimetabolite drug given for 5 days with rescue agent. Enzyme conjugate at cancer site destroys rescue agent

Figure 33.20 Antimetabolite with inactivation of rescue agent at cancer sites (AMIRACS).

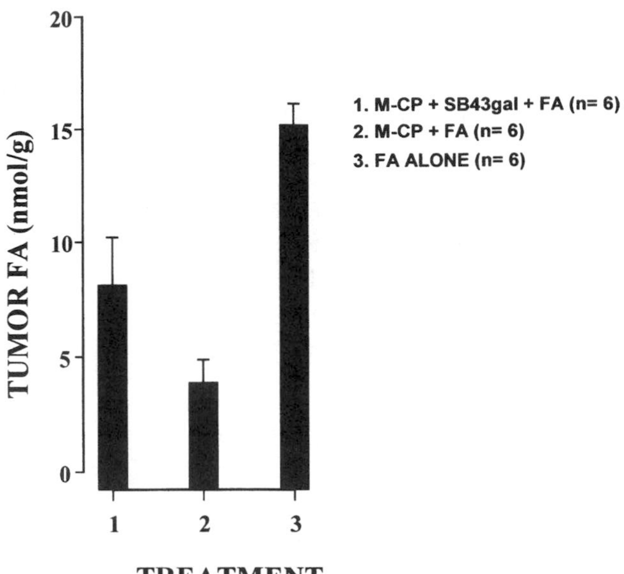

Figure 33.21 The ability of a tumor-targeted CPG2 conjugate selectively to break down folinic acid (FA) at a tumor site in vivo. In this experiment, an antibody–CPG_2 conjugate was used, with second antibody clearance to give high tumor-to-blood ratios.

folinic acid was 10–12-fold higher than the normal maximum tolerated dose in the absence of folinic acid. This regimen resulted in minimal toxicity towards the patient's platelet count; indeed the patient's platelet count was depressed at the commencement of therapy, due to prior chemotherapy, and rose during the course of therapy with high-dose trimetrexate plus folinic acid (143).

Although inhibition of dihydrofolate reductase is the longest established form of antimetabolite therapy, drugs have also been developed as inhibitors of the other key enzymes involved in folate metabolism, notably thymidylate synthase (TS) and glycinamide ribonucleotide formyl transferase (GARFT). AMIRACS can be adapted to work with both classes. Thymidine protects against the cytotoxic effects of TS inhibitors such as Tomudex (Raltitrexed) (Fig. 23a) (145). The addition of the enzyme thymidine phosphorylase abolishes the protective effect of thymidine (Fig. 23b) (145). Inhibitors of GARFT can be rescued by means of hypoxanthine, as shown in Fig. 24. This illustrates rescue of Lometrexol by hypoxanthine (146). Hypoxanthine can be destroyed by the enzyme xanthine oxidase (147).

The use of targeted enzymes in AMIRACS is of particular interest because the therapy can be applied to agents that are already in everyday clinical use. Unlike the other therapies using targeted enzymes, AMIRACS does not involve activation of a prodrug to a cytotoxic species. Thus the clearance of nontargeted enzyme may not be so important. Although the systemic concentration of the rescue agent may be reduced by circulating enzyme, this can be monitored and the dosage adjusted accordingly.

III. CONCLUSIONS

Since the concept was first described in 1987, the use of antibody–enzyme conjugates directed at tumor-associated antigens to achieve site-specific activation of prodrugs to potent cytotoxic species, termed antibody-directed

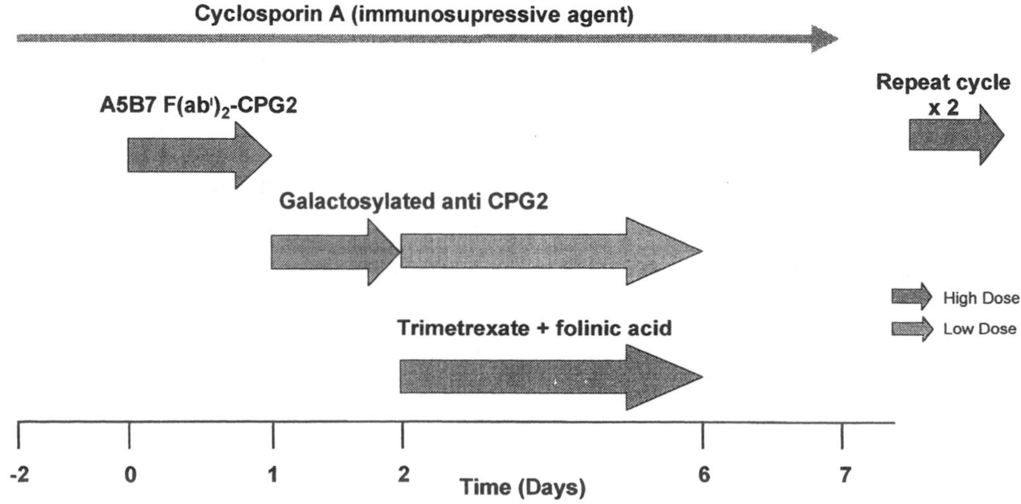

Figure 33.22 AMIRACS clinical experiment protocol.

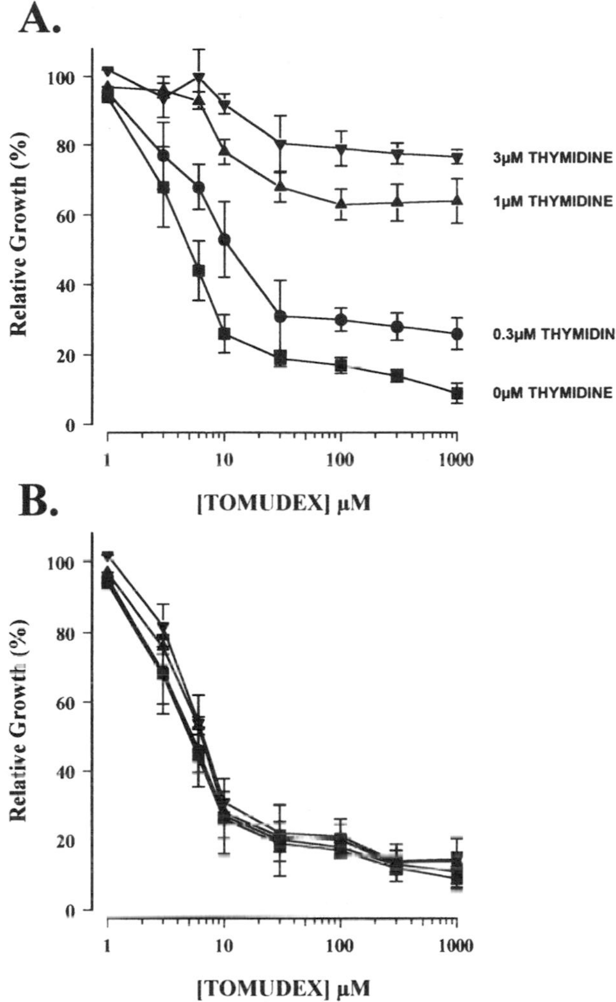

Figure 33.23 The effect of 0, 0.3, 1.0, or 3 μM thymidine on the cytotoxicity of Tomudex against human MCF-7 cancer cells in (A) the absence or (B) the presence of E. coli thymidine phosphorylase (3.5×10^{-4} U). (Data from Ref. 145.)

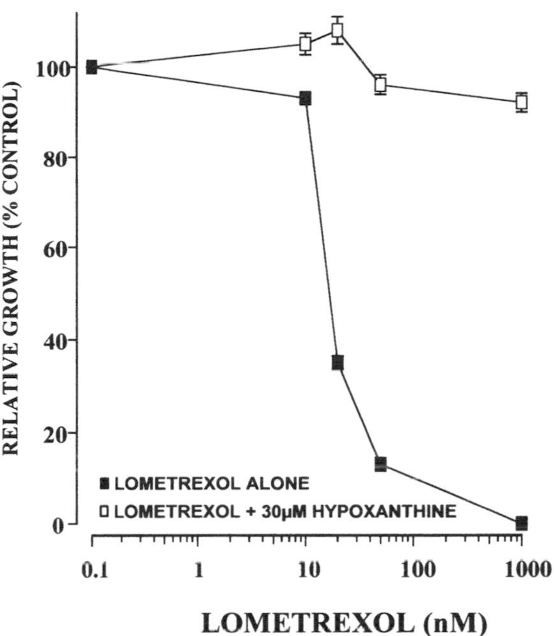

Figure 33.24 The effect of 30 μM hypoxanthine on the cytotoxicity of Lometrexol against human A549 cancer cells. (Data from Ref. 146.)

enzyme–prodrug therapy (ADEPT), has attracted considerable interest. ADEPT attempts to overcome a major problem associated with the administration of cytotoxic drugs for the treatment of cancer—the intrinsic lack of tumor selectivity. Thus cancer chemotherapeutic agents have the potency to kill tumor cells in vitro but, because of host toxicity, the treatment of human patients has to be discontinued at dose levels that are well below that expected to kill all their tumor cells. In ADEPT, a cytotoxic agent is generated selectively at the site of a tumor by an antibody-targeted enzyme. The antibody delivery combined with the amplification provided by the enzymatic activation of prodrugs enables adequate selection to be made between tumor and normal tissue.

For any form of antibody-mediated targeting to be successful, it is axiomatic that there must be selective expression of the target antigen by the tumor cells. However, the only well-characterized tumor-specific antigens described to date are the idiotypic determinants on the surface immunoglobulins of B-cell lymphomas (148). Many antigens and other potential targets such as growth factor receptors are present in elevated levels in tumor tissue, but most, if not all, appear to be found to a greater or lesser extent in other tissue. Furthermore, it is well documented that tumor cells exhibit considerable antigenic heterogeneity (149,150), with some devoid of antigen expression, so a single monoclonal antibody conjugate targeted at a tumor mass will not bind to all the cells present. In ADEPT, the active drug, being of low molecular weight, can diffuse to adjacent tissues, including cells that express the target antigen but that have not bound conjugate, cells that may express alternative tumor-associated antigens not recognized by the vector, or cells that are antigen-negative, providing a bystander effect. A further advantage of ADEPT is that a single enzyme molecule has the potential to cleave many prodrug molecules—up to 800 molecules per mole of enzyme per second in the case of the benzoic acid mustard substrates of carboxypeptidase G2 (151). This provides an amplification effect, giving high levels of drug localized at the tumor which may be an important advantage in the clinic in view of the typically low localization of immunoconjugate in humans (152). There is also some evidence that high levels of drug generated at the surface of tumor cells are more effective than equivalent concentrations of free drug (153). To make the use of prodrug-activating enzymes pan-tumor, polymers have been used to target the tumor. These can accumulate preferentially at a tumor because of the enhanced permeability and retention effect (EPR). Macromolecule-directed enzyme–prodrug therapy (MDEPT) is the direct equivalent of ADEPT, with the antibody-targeting component replaced by a polymer.

After administration of an enzyme conjugate, much can remain in the vascular compartment for several days. Although the concentration is much less than that bound to the tumor, the volume of plasma is much greater, and thus the actual enzyme activity may be greater in the plasma than in the tumor. In this case, administration of a prodrug may be no more selective than systemic administration of the active form. It is thus necessary to wait for plasma levels of the conjugate to fall to very low levels (during which time the enzyme activity at the tumor may also be decreasing). Alternatively, the plasma may be "cleared" of enzyme activity, and this can be achieved by use of a clearance antibody directed towards the enzyme (or antibody component in ADEPT). Many high tumor-to-blood levels of the conjugate have been reported (38 : 1 compared to 1.3 : 1 without clearance) using such an approach (154,155). Inhibition of the active site of the enzyme will enhance clearance of enzyme activity, but it is important that the clearance antibody have little effect on bound conjugate. This has been achieved by the rapid clearance of the second antibody through the carbohydrate receptors in the liver by the introduction, into the antibody, of galactose residues (154–156). Alternatively the conjugate itself can be galactosylated and the conjugate localized while the liver receptors are blocked by administration of a competitive binder such as asialobovine submaxillary gland mucin. A tumor-to-blood ratio of 45 : 1 for an anti-CEA-carboxypeptidase G_2 conjugate was achieved 24 h after injection of the conjugate using this type of approach; this increased to 100 : 1 at 72 h (155). Tumor-to-blood ratios of a conjugate (but not other tissue-to-blood ratios) could also be significantly increased by coadministration of tumor necrosis factor alpha, which enhances the tumor uptake of enzyme conjugates (157). Coadministration of 1.5 μg TNF with conjugate led to a twofold increase in tumor uptake with only transient increases in normal tissue localization. To avoid the need for a clearance step, polymer-directed enzyme–prodrug therapy (PDEPT) has been proposed. PDEPT uses a polymer–enzyme conjugate that accumulates by EPR to activate a polymeric prodrug that also accumulates. The polymeric prodrug is administered first, but unlocalized prodrug can be excreted through the kidneys within 5 h after administration. Thus when the enzyme conjugate is given there is no prodrug available to cause systemic activation.

An immune response was observed in patients treated with ADEPT therapy and developed against both the enzyme and antibody components of the conjugate (35, 40,144). The raising of an immune response would preclude extended therapy. Several strategies have been proposed to overcome the host immune response. The use of immunosuppressive agents such as cyclosporin can extend the time window available for therapy. This may enable multiple cycles to be given (35,40,144). Also, it may be possible to administer a second ADEPT conjugate consisting of a different antibody–enzyme combination. There is also the possibility of masking the immunogenicity of conjugates by modification with polyethylene glycol or other similar polymers (158). The immunogenicity of the antibody moiety can be reduced by using chimeric (e.g., rat antigen-binding domains coupled to human constant regions), humanized or fully human-derived antibody fragments (159). Efforts to reduce the immunogenicity of the enzyme can be attempted by reduction of the size of the active fragment; modification of the bacterial enzyme at key immunogenic residues by site-directed mutagenesis, engineering of the enzyme's active site into an antibody, or engineering a human enzyme such that its activity and specificity more closely match that of the bacterial enzyme. Model compounds consisting of a mouse Fab fused with nucleases or polymerases were reported some time ago (160). Such constructs could be produced in relatively high yield and with a greater homogeneity of conjugate by comparison with current chemical linking techniques, where yields are relatively low and the product ill defined in terms of its homogeneity (161,162). The first fusion protein developed specifically for an ADEPT application consisted of human β-glucuronidase fused to the CH domain of BW431, a high-avidity anti-CEA antibody, humanized by complementarity determining region (CDR) grafting. The protein thus produced has antigen affinity very similar to that of the parent antibody and retains enzymatic activity (163). A *B. cereus* β-lactamase II single-chain antibody fusion protein has also been constructed. The antibody used, L6, recognizes an antigen expressed on breast, colon, lung, and ovarian carcinomas. The fusion protein is produced by *E. coli* in an active form that does not require refolding and retains similar immunological and enzymatic properties to an equivalent chemically linked conjugate (49).

An important advantage of polymers is their effect on the immunogenicity of foreign proteins (125,126). This is generally attributed to their ability to mask epitopes. There is also evidence that they can induce immune tolerance to challenge with bacterial enzymes. The substitution of a polymer for an antibody vector should therefore reduce the immunogenicity associated with an antibody conjugate and has the potential to allow the repeated use of bacterial enzymes in humans without the necessity for immunosuppression or engineering. However, polymer conjugates are still produced chemically while an antibody–enzyme fusion protein can (in theory) be produced directly in bacteria. The fusion protein is a defined product of much greater homogeneity than can be obtained with a polymer–enzyme conjugate.

An alternative enzyme–prodrug system uses methods such as viral vectors to deliver genetic material encoding the activating enzyme to the target cells. This approach is called GDEPT. Unlike ADEPT, MDEPT, or PDEPT, there is no intrinsic tumor selectivity. At present, targeting is achieved by altering the surface components of viruses and liposomes to achieve a level of target cell recognition. Transcriptional elements can also be incorporated so that the incorporated gene is expressed only in the target cells. At present, there seem to be many problems that have yet to be overcome, in particular the issue of ensuring that only the target cells are transfected with the desired gene. Systemic administration of the present generation of gene therapy vectors is not possible. Thus GDEPT, unlike the other enzyme–prodrug therapies described, is limited to isolable tumor deposits, such as intracerebral tumors, surrounded by largely nondividing normal tissue.

There is a rich diversity of potential enzyme and prodrug combinations for use in targeted enzyme prodrug applications (Table 1). In selecting an enzyme one would look for activity under physiological conditions, low immunogenicity, and little or no equivalent endogenous enzyme in humans. The enzyme must localize efficiently at the tumor when coupled. The prodrug is an integral component of systems and requires careful design in its own right. An ideal prodrug would be one with a large differential in cytotoxicity between drug and prodrug, which is a good substrate for the enzyme under physiological conditions and for which there is no mammalian homologue capable of performing the same reaction. Equal cytotoxicity of the released active drug towards proliferating and quiescent cells is also desirable if residual deposits of viable but nonproliferating cells with the potential for outgrowth are to be eradicated. Once formed, it would be desirable for the drug to have a very short half-life, limiting the possibility of escape of active drug back into the circulation and access to healthy tissue. Antimetabolite with inactivation of rescue agent at cancer sites (AMIRACS) allows the use of targeted enzymes in combination with antimetabolite agents that are already in everyday clinical use. Unlike the other therapies using a targeted enzyme, AMIRACS does not involve the activation of a prodrug to a cytotoxic species but selectively destroys a rescue agent only at the tumor site.

In summary, the in vivo data available for the various targeted enzyme–prodrug therapies provide exciting grounds for encouragement. They offer new research ideas for the medicinal chemist, polymer chemist, enzymologist, molecular biologist, and protein engineer as well as the cancer chemotherapist. Although relatively complicated, targeted enzyme–prodrug therapies offer opportunities for the therapy of systemic cancer and may be a major advance for the treatment of solid tumors.

REFERENCES

1. T. A. Connors. Alkylating agents, nitrosoureas and dimethyltriazenes. *Cancer Chemotherapy*, Vol. 3 (H. M. Pinedo, ed.). Excerpta Medica, Amsterdam, 1981, pp. 32–74.
2. J. J. Roberts, R. J. Knox, F. Friedlos, D. A. Lydall. DNA as the target for the cytotoxic and antitumour action of platinum co-ordination complexes: comparative in vitro and in vivo studies of cisplatin and carboplatin. *Biochemical Mechanisms of Platinum Antitumour Drugs* (D. C. H. McBrien, T. F. Slater ed.). IRL Press, Oxford, 1986, pp. 29–64.
3. R. J. Knox, F. Friedlos, D. A. Lydall, J. J. Roberts. Mechanism of cytotoxicity of anticancer platinum drugs: evidence that cis-diamminedichloroplatinum(II) and cis-diammine-(1,1-cyclobutanedicarboxylato)platinum(II) differ only in the kinetics of their interaction with DNA. *Cancer Res 46*:1972–1979 (1986).
4. T. A. Connors. Has chemotherapy anywhere to go? *Cancer Surv 8*:693–705 (1989).
5. T. A. Connors, R. J. Knox. Prodrugs in cancer chemotherapy. *Stem Cells 13*:501–511 (1995).
6. T. A. Connors, R. J. Knox. Prodrugs in medicine. *Expert Opinion on Therapeutic Patents 5*:873–885 (1995).
7. R. J. Knox, M. P. Boland, F. Friedlos, B. Coles, C. Southan, J. J. Roberts. The nitroreductase enzyme in Walker cells that activates 5-(aziridin-1-yl)-2,4-dinitrobenzamide (CB 1954) to 5-(aziridin-1-yl)-4-hydroxylamino-2-nitrobenzamide is a form of NAD(P)H dehydrogenase (quinone) (EC 1.6.99.2). *Biochem Pharmacol 37*:4671–4677 (1988).
8. R. J. Knox, F. Friedlos, M. Jarman, J. J. Roberts. A new cytotoxic, DNA interstrand crosslinking agent, 5-(aziridin-1-yl)-4-hydroxylamino-2-nitrobenzamide, is formed from 5-(aziridin-1-yl)-2,4-dinitrobenzamide (CB 1954) by a nitroreductase enzyme in Walker carcinoma cells. *Biochem Pharmacol 37*:4661–4669 (1988).
9. R. J. Knox, F. Friedlos, T. Marchbank, J. J. Roberts. Bio-

activation of CB 1954: reaction of the active 4-hydroxylamino derivative with thioesters to form the ultimate DNA–DNA interstrand crosslinking species. *Biochem Pharmacol 42*:1691–1697 (1991).

10. J. J. Roberts, F. Friedlos, R. J. Knox. CB 1954 (2,4-dinitro-5-aziridinyl benzamide) becomes a DNA interstrand crosslinking agent in Walker tumor cells. *Biochem. Biophys Res Commun 140*:1073–1078 (1986).
11. R. J. Knox, F. Friedlos, M. P. Boland. The bioactivation of CB 1954 and its use as a prodrug in antibody-directed enzyme prodrug therapy (ADEPT). *Cancer Metastasis Rev 12*:195–212 (1993).
12. L. M. Cobb, T. A. Connors, L. A. Elson, A. H. Khan, B. C. Mitchley, W. C. Ross, M. E. Whisson. 2,4-dinitro-5-ethyleneiminobenzamide (CB 1954): a potent and selective inhibitor of the growth of the Walker carcinoma 256. *Biochem Pharmacol 18*:1519–1527 (1969).
13. T. A. Connors, D. H. Melzack. Studies on the mechanism of action of 5-aziridinyl-2,4-dinitrobenzamide (CB 1954), a selective inhibitor of the Walker tumor. *Int J Cancer 7*: 86–92 (1971).
14. T. A. Connors, M. E. Whisson. Cure of mice bearing advanced plasma cell tumors with aniline mustard: the relationship between glucuronidase activity and tumor sensitivity. *Nature 210*:866–867 (1966).
15. M. P. Boland, R. J. Knox, J. J. Roberts. The differences in kinetics of rat and human DT diaphorase result in a differential sensitivity of derived cell lines to CB 1954 (5-(aziridin-1-yl)-2,4-dinitrobenzamide). *Biochem Pharmacol 41*:867–875 (1991).
16. C. W. Young, A. Yagoda, E. S. Bittar, S. W. Smith, H. Grabstald, W. Whitmore. Therapeutic trial of aniline mustard in patients with advanced cancer. Comparison of therapeutic response with cytochemical assessment of tumor cell β-glucuronidase activity. *Cancer 38*:1887–1895 (1976).
17. S. A. Fitzsimmons, P. Workman, M. Grever, K. Paull, R. Camalier, A. D. Lewis. Reductase enzyme expression across the National Cancer Institute Tumor cell line panel: correlation with sensitivity to mitomycin C and EO9. *J Natl Cancer. Inst 88*:259–269 (1996).
18. K. Bosslet, R. Straub, M. Blumrich, J. Czech, M. Gerken, B. Sperker, H. K. Kroemer, J. P. Gesson, M. Koch, C. Monneret. Elucidation of the mechanism enabling tumor selective prodrug monotherapy. *Cancer Res 58*:1195–1201 (1998).
19. K. B. Wu, R. J. Knox, X. Z. Sun, P. Joseph, A. K. Jaiswal, D. Zhang, P. S. K. Deng, S. Chen. Catalytic properties of NAD(P)H-quinone oxidoreductase-2 (NQO2), a dihydronicotinamide riboside dependent oxidoreductase. *Archives of Biochemistry and Biophysics 347*:221–228 (1997).
20. K. D. Bagshawe, C. J. Springer, F. Searle, P. Antoniw, S. K. Sharma, R. G. Melton, R. F. Sherwood. A cytotoxic agent can be generated selectively at cancer sites. *Br J Cancer 58*:700–703 (1988).
21. K. D. Bagshawe. Antibody directed enzymes revive anti-cancer prodrugs concept. *Br J Cancer 56*:531–532 (1987).
22. K. D. Bagshawe. ADEPT and related concepts. *Cell Biophys 25*:83–91 (1994).
23. K. D. Bagshawe. Antibody-directed enzyme prodrug therapy—a review. *Drug Development Research 34*:220–230 (1995).
24. R. J. Knox, T. A. Connors. Antibody-directed enzyme prodrug therapy. *Clin Immunother 3*:136–153 (1995).
25. R. G. Melton, R. J. Knox, T. A. Connors. Antibody-directed enzyme prodrug therapy (ADEPT). *Drugs Fut 21*: 167–181 (1996).
26. R. G. Melton, R. F. Sherwood. Antibody-enzyme conjugates for cancer therapy. *J Natl Cancer Inst 88*:153–165 (1996).
27. R. J. Knox. Enzymes and prodrugs used for ADEPT. *Enzyme Prodrug Strategies for Cancer Therapy* (R. G. Melton, R. I. Knox, eds.). Kluwer Academic/Plenum Press, New York, 1999, pp. 97–131.
28. M. Stanislawski, V. Rousseau, M. Goavec, H. Ito. Immunotoxins containing glucose oxidase and lactoperoxidase with tumoricidal properties: in vitro killing effectiveness in a mouse plasmacytoma cell model. *Cancer Res 49*: 5497–5504 (1989).
29. P. D. Senter, M. G. Saulnier, G. J. Schreiber, D. L. Hirschberg, J. P. Brown, I. Hellstrom, K. E. Hellstrom. Anti-tumor effects of antibody–alkaline phosphatase conjugates in combination with etoposide phosphate. *Proc Natl Acad Sci USA 85*:4842–4846 (1988).
30. C. J. Springer, K. D. Bagshawe, S. K. Sharma, F. Searle, J. A. Boden, P. Antoniw, P. J. Burke, G. T. Rogers, R. F. Sherwood, R. G. Melton. Ablation of human choriocarcinoma xenografts in nude mice by antibody-directed enzyme prodrug therapy (ADEPT) with three novel compounds. *Eur J Cancer 27*:1361–1366 (1991).
31. D. C. Blakey, B. E. Valcaccia, S. East, A. F. Wright, F. T. Boyle, C. J. Springer, P. J. Burke, R. G. Melton, K. D. Bagshawe. Antitumor effects of an antibody–carboxypeptidase G2 conjugate in combination with a benzoic acid mustard prodrug. *Cell Biophys 22*:1–8 (1993).
32. S. A. Eccles, W. J. Court, G. A. Box, C. J. Dean, R. G. Melton, C. J. Springer. Regression of established breast carcinoma xenografts with antibody-directed enzyme prodrug therapy against c-erbB2 p185. *Cancer Res 54*:5171–5177 (1994).
33. D. C. Blakey, D. H. Davies, R. I. Dowell, S. J. East, P. J. Burke, S. K. Sharma, C. J. Springer, A. B. Mauger, R. G. Melton. Anti-tumor effects of an antibody–carboxypeptidase G2 conjugate in combination with phenol mustard prodrugs. *Br J Cancer 72*:1083–1088 (1995).
34. H. P. Svensson, I. S. Frank, K. K. Berry, P. D. Senter. Therapeutic effects of monoclonal antibody-β-lactamase conjugates in combination with a nitrogen mustard anticancer prodrug in models of human renal cell carcinoma. *J Med Chem 41*:1507–1512 (1998).
35. K. D. Bagshawe, S. K. Sharma, C. J. Springer, P. Antoniw, J. A. Boden, G. T. Rogers, P. J. Burke, R. G. Melton, R. F. Sherwood. Antibody directed enzyme prodrug therapy (ADEPT): clinical report. *Dis Markers 9*:233–238 (1991).

36. K. D. Bagshawe, S. K. Sharma, C. J. Springer, P. Antoniw. Antibody directed enzyme prodrug therapy: a pilot-scale clinical trial. *Tumor Targeting 1*:17–29 (1995).
37. C. J. Springer, G. K. Poon, S. K. Sharma, K. D. Bagshawe. Identification of prodrug, active drug, and metabolites in an ADEPT clinical study. *Cell Biophys 22*:9–26 (1993).
38. R. I. Dowell, C. J. Springer, D. H. Davies, E. M. Hadley, P. J. Burke, F. T. Boyle, R. G. Melton, T. A. Connors, D. C. Blakey, A. B. Mauger. New mustard prodrugs for antibody-directed enzyme prodrug therapy: alternatives to the amide link. *J Med Chem 39*:1100–1105 (1996).
39. D. C. Blakey, P. J. Burke, D. H. Davies, R. I. Dowell, S. J. East, K. P. Eckersley, J. E. Fitton, J. McDaid, R. G. Melton, I. A. Niculescu Duvaz, P. E. Pinder, S. K. Sharma, A. F. Wright, C. J. Springer. ZD2767, an improved system for antibody-directed enzyme prodrug therapy that results in tumor regressions in colorectal tumor xenografts. *Cancer Res 56*:3287–3292 (1996).
40. S. K. Sharma, K. D. Bagshawe, R. G. Melton, R. F. Sherwood. Human immune response to monoclonal antibody–enzyme conjugates in ADEPT pilot clinical trial. *Cell Biophys 21*:109–120 (1992).
41. S. K. Sharma, K. D. Bagshawe, R. G. Melton, R. H. Begent. Effect of cyclosporine on immunogenicity of a bacterial enzyme carboxypeptidase G2 in ADEPT. *Transplant Proc 28*:3154–3155 (1996).
42. K. D. Bagshawe, S. K. Sharma. Cyclosporine delays host immune response to antibody enzyme conjugate in ADEPT. *Transplant Proc 28*:3156–3158 (1996).
43. K. Bosslet, J. Czech, P. Lorenz, H. H. Sedlacek, M. Schuermann, G. Seemann. Molecular and functional characterisation of a fusion protein suited for tumor specific prodrug activation. *Br J Cancer 65*:234–238 (1992).
44. K. Bosslet, J. Czech, D. Hoffmann. Tumor-selective prodrug activation by fusion protein–mediated catalysis. *Cancer Res 54*:2151–2159 (1994).
45. D. C. Blakey, F. T. Boyle, D. H. Davies, R. I. Dowell, D. W. Heaton, M. S. Rose, A. M. Slater, H. J. Eggelte, A. Tarragona-Fiol, C. J. Taylorson. Mutant human ribonuclease for use in antibody-directed enzyme prodrug therapy (ADEPT). *Br J Cancer 75*(suppl 1):36 (1997).
46. G. K. Smith, S. Banks, T. A. Blumenkopf, M. Cory, J. Humphreys, R. M. Laethem, J. Miller, C. P. Moxham, R. Mullin, P. H. Ray, L. M. Walton, L. A. Wolfe. Toward antibody-directed enzyme prodrug therapy with the T268G mutant of human carboxypeptidase A 1 and novel in vivo stable prodrugs of methotrexate. *J Biol Chem 272*:15804–15816 (1997).
47. L. J. Partridge. Production of catalytic antibodies using combinatorial libraries. *Biochem Soc Trans 21*:1096–1098 (1993).
48. K. A. Chester, R. H. Begent, L. Robson, P. Keep, R. B. Pedley, J. A. Boden, G. Boxer, A. Green, G. Winter, O. Cochet, R. E. Hawkins. Phage libraries for generation of clinically useful antibodies. *Lancet 343*:455–456 (1994).
49. S. C. Goshorn, H. P. Svensson, D. E. Kerr, J. E. Somerville, P. D. Senter, H. P. Fell. Genetic construction, expression, and characterization of a single chain anticarcinoma antibody fused to β-lactamase. *Cancer Res 53*:2123–2127 (1993).
50. P. Wentworth, A. Datta, D. Blakey, T. Boyle, L. J. Partridge, G. M. Blackburn. Toward antibody-directed "abzyme" prodrug therapy, ADAPT: carbamate prodrug activation by a catalytic antibody and its in vitro application to human tumor cell killing. *Proc Natl Acad Sci USA 93*:799–803 (1996).
51. E. H. Oldfield, Z. Ram, K. W. Culver, R. M. Blaese, H. L. DeVroom, W. F. Anderson. Gene therapy for the treatment of brain tumors using intra-tumoral transduction with the thymidine kinase gene and intravenous ganciclovir. *Hum Gene Ther 4*:39–69 (1993).
52. I. R. Hart, R. G. Vile. Targeted gene therapy. *Br Med Bull 51*:647–655 (1995).
53. J. Fueyo, C. Gomez-Manzano, W. K. Yung, A. P. Kyritsis. Targeting in gene therapy for gliomas. *Arch Neurol 56*:445– 448 (1999).
54. J. R. Herman, H. L. Adler, E. Aguilar-Cordova, A. Rojas-Martinez, S. Woo, T. L. Timme, T. M. Wheeler, T. C. Thompson, P. T. Scardino. In situ gene therapy for adenocarcinoma of the prostate: a phase I clinical trial. *Hum Gene Ther 10*:1239–1249 (1999).
55. M. Consalvo, C. A. Mullen, A. Modesti, P. Musiani, A. Allione, F. Cavallo, M. Giovarelli, G. Forni. 5-Fluorocytosine-induced eradication of murine adenocarcinomas engineered to express the cytosine deaminase suicide gene requires host immune competence and leaves an efficient memory. *J Immunol 154*:5302–5312 (1995).
56. S. M. Freeman, K. A. Whartenby, J. L. Freeman, C. N. Abboud, A. J. Marrogi. In situ use of suicide genes for cancer therapy. *Semin Oncol 23*:31–45 (1996).
57. C. A. Mullen, M. M. Coale, R. Lowe, R. M. Blaese. Tumors expressing the cytosine deaminase suicide gene can be eliminated in vivo with 5-fluorocytosine and induce protective immunity to wild type tumor. *Cancer Res 54*:1503–1506 (1994).
58. C. A. Mullen. Metabolic suicide genes in gene therapy. *Pharmacol Ther 63*:199–207 (1994).
59. D. F. Robinson. and I. H. Maxwell. Suppression of single and double nonsense mutations introduced into the diphtheria toxin A-chain gene: a potential binary system for toxin gene therapy. *Hum Gene Ther 6*:137–143 (1995).
60. D. R. Cook, I. H. Maxwell, L. M. Glode, F. Maxwell, J. O. Stevens, M. B. Purner, E. Wagner, D. T. Curiel, T. J. Curiel. Gene therapy for B-cell lymphoma in a SCID mouse model using an immunoglobulin-regulated diphtheria toxin gene delivered by a novel adenovirus-polylysine conjugate. *Cancer Biother 9*:131–141 (1994).
61. I. H. Maxwell, L. M. Glode, F. Maxwell. Expression of diphtheria toxin A-chain in mature B-cells: a potential approach to therapy of B-lymphoid malignancy. *Leuk Lymphoma 7*:457–462 (1992).
62. I. H. Maxwell, L. M. Glode, F. Maxwell. Expression of the diphtheria toxin A-chain coding sequence under the control of promoters and enhancers from immunoglobulin

genes as a means of directing toxicity to B-lymphoid cells. *Cancer Res 51*:4299–4304 (1991).
63. R. Marais, R. A. Spooner, S. M. Stribbling, Y. Light, J. Martin, C. J. Springer. A cell surface tethered enzyme improves efficiency in gene-directed enzyme prodrug therapy. *Nature Biotechnology 15*:1373–1377 (1997).
64. T. A. Connors. The choice of prodrugs for gene directed enzyme prodrug therapy of cancer. *Gene Therapy 2*:1–9 (1995).
65. M. L. Mendelsohn. The growth fraction: a new concept applied to tumors. *Science 132*:1496 (1960).
66. N. Miller, R. Vile. Targeted vectors for gene therapy. *FASEB J. 9*:190–199 (1995).
67. B. Abdallah, A. Hassan, C. Benoist, D. Goula, J. P. Behr, B. A. Demeneix. A powerful nonviral vector for in vivo gene transfer into the adult mammalian brain:polyethylenimine. *Hum Gene Ther 7*:1947–1954 (1996).
68. W. T. Godbey, K. K. Wu, A. G. Mikos. Poly(ethylenimine) and its role in gene delivery. *J Controlled Release 60*:149–160 (1999).
69. E. Borrelli, R. Heyman, M. Hsi, R. M. Evans. Targeting of an inducible toxic phenotype in animal cells. *Proc Natl Acad Sci USA 85*:7572–7576 (1988).
70. F. S. Moolten, J. M. Wells, P. J. Mroz. Multiple transduction as a means of preserving ganciclovir chemosensitivity in sarcoma cells carrying retrovirally transduced herpes thymidine kinase genes. *Cancer Lett 64*:257–263 (1992).
71. A. Abe, T. Takeo, N. Emi, M. Tanimoto, R. Ueda, J. K. Yee, T. Friedmann, H. Saito. Transduction of a drug-sensitive toxic gene into human leukemia cell lines with a novel retroviral vector. *Proc Soc Exp Biol Med 203*:354–359 (1993).
72. D. Barba, J. Hardin, J. Ray, F. H. Gage. Thymidine kinase-mediated killing of rat brain tumors. *J Neurosurg 79*:729–735 (1993).
73. R. G. Vile, I. R. Hart. Use of tissue-specific expression of the herpes simplex virus thymidine kinase gene to inhibit growth of established murine melanomas following direct intratumoral injection of DNA. *Cancer Res 53*:3860–3864 (1993).
74. S. H. Chen, H. D. Shine, J. C. Goodman, R. G. Grossman, S. L. Woo. Gene therapy for brain tumors: regression of experimental gliomas by adenovirus-mediated gene transfer in vivo. *Proc Natl Acad Sci USA 91*:3054–3057 (1994).
75. W. R. Smythe, H. C. Hwang, K. M. Amin, S. L. Eck, B. L. Davidson, J. M. Wilson, L. R. Kaiser, S. M. Albelda. Use of recombinant adenovirus to transfer the herpes simplex virus thymidine kinase (HSVtk) gene to thoracic neoplasms: an effective in vitro drug sensitization system. *Cancer Res 54*:2055–2059 (1994).
76. X. W. Tong, A. Block, S. H. Chen, C. F. Contant, I. Agoulnik, K. Blankenburg, R. H. Kaufman, S. L. Woo, D. G. Kieback. In vivo gene therapy of ovarian cancer by adenovirus-mediated thymidine kinase gene transduction and ganciclovir administration. *Gynecol Oncol 61*:175–179 (1996).
77. X. W. Tong, A. Block, S. H. Chen, S. L. Woo, D. G. Kieback. Adenovirus-mediated thymidine kinase gene transduction in human epithelial ovarian cancer cell lines followed by exposure to ganciclovir. *Anticancer Res 16*: 1611–1617 (1996).
78. T. Tanaka, F. Kanai, S. Okabe, Y. Yoshida, H. Wakimoto, H. Hamada, Y. Shiratori, K. Lan, M. Ishitobi, M. Omata. Adenovirus-mediated prodrug gene therapy for carcinoembryonic antigen-producing human gastric carcinoma cells in vitro. *Cancer Res 56*:1341–1345 (1996).
79. M. Kokoris, P. Sabo, E. Adman, M. Black. Enhancement of tumor ablation by a selected HSV-1 thymidine kinase mutant. *Gene Ther 6*:1415–1426 (1999).
80. S. M. Freeman, C. N. Abboud, K. A. Whartenby, C. H. Packman, D. S. Koeplin, F. L. Moolten, G. N. Abraham. The ''bystander effect'': tumor regression when a fraction of the tumor mass is genetically modified. *Cancer Res 53*: 5274–5283 (1993).
81. W. L. Bi, L. M. Parysek, R. Warnick, P. J. Stambrook. In vitro evidence that metabolic cooperation is responsible for the bystander effect observed with HSV tk retroviral gene therapy. *Hum Gene Ther 4*:725–731 (1993).
82. J. Fick, F. G. Barker, 2nd, P. Dazin, E. M. Westphale, E. C. Beyer, M. A. Israel. The extent of heterocellular communication mediated by gap junctions is predictive of bystander tumor cytotoxicity in vitro. *Proc Natl Acad Sci USA 92*:11071–11075 (1995).
83. S. Gagandeep, R. Brew, B. Green, S. E. Christmas, D. Klatzmann, G. J. Poston, A. R. Kinsella. Prodrug-activated gene therapy: involvement of an immunological component in the ''bystander effect.'' *Cancer Gene Ther 3*:83–88 (1996).
84. E. Kievit, E. Bershad, E. Ng, P. Sethna, I. Dev, T. S. Lawrence, A. Rehemtulla. Superiority of yeast over bacterial cytosine deaminase for enzyme/prodrug gene therapy in colon cancer xenografts. *Cancer Res 59*:1417–1421 (1999).
85. C. A. Mullen, M. Kilstrup, R. M. Blaese. Transfer of the bacterial gene for cytosine deaminase to mammalian cells confers lethal sensitivity to 5-fluorocytosine: a negative selection system. *Proc Natl Acad Sci USA 89*:33–37 (1992).
86. B. E. Huber, E. A. Austin, S. S. Good, V. C. Knick, S. Tibbels, C. A. Richards. In vivo antitumor activity of 5-fluorocytosine on human colorectal carcinoma cells genetically modified to express cytosine deaminase. *Cancer Res 53*:4619–4626 (1993).
87. B. E. Huber, E. A. Austin, C. A. Richards, S. T. Davis, S. S. Good. Metabolism of 5-fluorocytosine to 5-fluorouracil in human colorectal tumor cells transduced with the cytosine deaminase gene:significant antitumor effects when only a small percentage of tumor cells express cytosine deaminase. *Proc Natl Acad Sci USA 91*:8302–8306 (1994).
88. D. K. Hoganson, R. K. Batra, J. C. Olsen, R. C. Boucher. Comparison of the effects of three different toxin genes and their levels of expression on cell growth and bystander

effect in lung adenocarcinoma. *Cancer Res 56*:1315–1323 (1996).
89. E. Frei, B. A. Teicher, S. A. Holden, K. N. Cathcart, Y. Y. Wang. Preclinical studies and clinical correlation of the effect of alkylating dose. *Cancer Res 48*:6417–6423 (1988).
90. G. M. Anlezark, R. G. Melton, R. F. Sherwood, B. Coles, F. Friedlos, R. J. Knox. The bioactivation of 5-(aziridin-1-yl)-2,4-dinitrobenzamide (CB1954)-1. Purification and properties of a nitroreductase enzyme from Escherichia coli—a potential enzyme for antibody-directed enzyme prodrug therapy (ADEPT). *Biochem Pharmacol 44*:2289–2295 (1992).
91. F. Friedlos, R. J. Knox. Metabolism of NAD(P)H by blood components. Relevance to bioreductively activated prodrugs in a targeted enzyme therapy system. *Biochem Pharmacol 44*:631–635 (1992).
92. J. A. Bridgewater, C. J. Springer, R. J. Knox, N. P. Minton, N. P. Michael, M. K. Collins. Expression of the bacterial introreductase enzyme in mammalian cells renders them selectively sensitive to killing by the prodrug CB1954. *Eur J Cancer 31a*:2362–2370 (1995).
93. N. K. Green, D. J. Youngs, J. P. Neoptolemos, F. Friedlos, R. J. Knox, C. J. Springer, G. M. Anlezark, N. P. Michael, R. G. Melton, M. J. Ford, L. S. Young, D. J. Kerr, P. F. Searle. Sensitization of colorectal and pancreatic cell lines to the prodrug 5-(aziridin-1-yl)-2,4-dinitrobenzamide (CB 1954) by retroviral transduction and expression of the E. coli nitroreductase gene. *Cancer Gene Ther 4*:229–238 (1997).
94. W. Cui, B. Gusterson, A. J. Clark. Nitroreductase-mediated cell ablation is very rapid and mediated by a p53-independent apoptotic pathway. *Gene Ther 6*:764–770 (1999).
95. J. A. Bridgewater, R. J. Knox, J. D. Pitts, M. K. Collins, C. J. Springer. The bystander effect of the nitroreductase CB 1954 enzyme prodrug system is due to a cell-permeable metabolite. *Hum Gene Ther 8*:709–717 (1997).
96. I. A. McNeish, N. K. Green, M. G. Gilligan, M. J. Ford, V. Mautner, L. S. Young, D. J. Kerr, P. F. Searle. Virus directed enzyme prodrug therapy for ovarian and pancreatic cancer using retrovirally delivered E-coli nitroreductase and CB1954. *Gene Ther 5*:1061–1069 (1998).
97. S. M. Bailey, R. J. Knox, S. M. Hobbs, T. C. Jenkins, A. B. Mauger, R. G. Melton, P. J. Burke, T. A. Connors, I. R. Hart. Investigation of alternative prodrugs for use with E. coli nitroreductase in 'suicide gene' approaches to cancer therapy. *Gene Ther 3*:1143–1150 (1996).
98. G. M. Anlezark, R. G. Melton, R. F. Sherwood, W. R. Wilson, W. A. Denny, B. D. Palmer, R. J. Knox, F. Friedlos, A. Williams. Bioactivation of dinitrobenzamide mustards by an E. coli B nitroreductase. *Biochem Pharmacol 50*:609–618 (1995).
99. A. B. Mauger, P. J. Burke, H. H. Somani, F. Friedlos, R. J. Knox. Self-immolative prodrugs: candidates for antibody-directed enzyme prodrug therapy in conjunction with a nitroreductase enzyme. *J Med Chem 37*:3452–3458 (1994).
100. K. Haack, U. Moebius, M. V. K. Doeberitz, C. Herfarth, H. K. Schackert, J. F. Gebert. Detection of cytosine deaminase in genetically modified tumor cells by specific antibodies. *Hum Gene Ther 8*:1395–1401 (1997).
101. M. Tada, M. Tada. Enzymatic activation of the carcinogen 4-hydroxylaminoquinoline-1-oxide and its interaction with cellular macromolecules. *Biochem Biophys Res Commun 46*:1025 (1972).
102. M. Lee, J. E. Simpson, S. Woo, C. Kaenzig, G. M. Anlezark, E. EnoAmooquaye, P. J. Burke. Synthesis of an aminopropyl, analog of the experimental anticancer drug tallimustine, and activation of its 4-nitrobenzylcarbamoyl prodrug by nitroreductase and NADH. *Bioorg Med Chem Lett 7*:1065–1070 (1997).
103. R. Duncan, F. Spreafico. Polymer conjugates. Pharmacokinetic considerations for design and development. *Clin Pharmacokinet 27*:290–306 (1994).
104. R. Duncan, S. Dimitrijevic, E. G. Evagorou. The role of polymer conjugates in the diagnosis and treatment of cancer. *STP Pharma Sciences 6*:237–263 (1996).
105. Y. Matsumura, H. Maeda. A new concept for macromolecular therapeutics in cancer chemotherapy: mechanism of tumoritropic accumulation of proteins and the antitumor agent SMANCS. *Cancer Res 46*:6387–6392 (1986).
106. J. M. Brown, A. J. Giaccia. The unique physiology of solid tumors: opportunities (and problems) for cancer therapy. *Cancer Res 58*:1408–1416 (1998).
107. L. J. Nugent, R. K. Jain. Extravascular diffusion in normal and neoplastic tissues. *Cancer Res 44*:238–244 (1984).
108. L. E. Gerlowski, R. K. Jain. Microvascular permeability of normal and neoplastic tissues. *Microvasc Res 31*:288–305 (1986).
109. H. Maeda, Y. Matsumura. Tumoritropic and lymphotropic principles of macromolecular drugs. *Crit Rev Ther Drug Carrier Syst 6*:193–210 (1989).
110. H. F. Dvorak, J. A. Nagy, J. T. Dvorak, A. M. Dvorak. Identification and characterization of the blood vessels of solid tumors that are leaky to circulating macromolecules. *Am J Pathol 133*:95–109 (1988).
111. R. Duncan, P. Kopeckova-Rejmanova, J. Strohalm, I. Hume, H. C. Cable, J. Pohl, J. B. Lloyd, J. Kopecek. Anticancer agents coupled to N-(2-hydroxypropyl)methacrylamide copolymers. I. Evaluation of daunomycin and puromycin conjugates in vitro. *Br J Cancer 55*:165–174 (1987).
112. R. K. Jain. Vascular and interstitial barriers to delivery of therapeutic agents in tumors. *Cancer Metastasis Rev 9*: 253–266 (1990).
113. N. Ferrara, K. Houck, L. Jakeman, D. W. Leung. Molecular and biological properties of the vascular endothelial growth factor family of proteins. *Endocr Rev 13*:18–32 (1992).
114. R. K. Jain. Transport of molecules across tumor vasculature. *Cancer Metastasis Rev 6*:559–593 (1987).
115. R. K. Jain. Transport of molecules in the tumor interstitium: a review. *Cancer Res 47*:3039–3051 (1987).
116. H. F. Dvorak, L. F. Brown, M. Detmar, A. M. Dvorak.

Vascular permeability factor/vascular endothelial growth factor, microvascular hyperpermeability, and angiogenesis. *Am J Pathol 146*:1029–1039 (1995).

117. J. Folkman, E. Merler, C. Abernathy, G. Williams. Isolation of a tumor factor responsible for angiogenesis. *J Exp Med 133*:275–288 (1971).
118. J. Folkman. Tumor angiogenesis. *Adv Cancer Res 43*: 175–203 (1985).
119. J. Folkman, M. Klagsbrun. Angiogenic factors. *Science 235*:442–447 (1987).
120. L. W. Seymour, Y. Miyamoto, H. Maeda, M. Brereton, J. Strohalm, K. Ulbrich, R. Duncan. Influence of molecular weight on passive tumor accumulation of a soluble macromolecular drug carrier. *Eur J Cancer 5*:766–770 (1995).
121. N. Oku, K. Doi, Y. Namba, S. Okada. Therapeutic effect of adriamycin encapsulated in long-circulating liposomes on Meth-A-sarcoma-bearing mice. *Int J Cancer 58*:415–419 (1994).
122. L. W. Seymour, R. Duncan, P. Kopeckova, J. Kopecek. Daunomycin- and adriamycin-N-(2-hydroxypropyl)methacrylamide copolymer conjugates; toxicity reduction by improved drug-delivery. *Cancer Treat Rev 14*:319–327 (1987).
123. K. D. Bagshawe. *Tumor Therapy WO 98/24478* (1998).
124. K. D. Bagshawe, S. K. Sharma, P. J. Burke, R. G. Melton, R. J. Knox. Developments with targeted enzymes in cancer therapy. *Curr Opin Immunol 11*:579–583 (1999).
125. A. Abuchowski, F. F. Davis, S. Davis. Immunosuppressive properties and circulating life of Achromobacter glutaminase-asparaginase covalently attached to polyethylene glycol in man. *Cancer Treat Rep 65*:1077–1081 (1981).
126. M. L. Nucci, J. Olejarczyk, A. Abuchowski. Immunogenicity of polyethylene glycol-modified superoxide dismutase and catalase. *J Free Radic Biol Med 2*:321–325 (1986).
127. P. A. Vasey, S. B. Kaye, R. Morrison, C. Twelves, P. Wilson, R. Duncan, A. H. Thomson, L. S. Murray, T. E. Hilditch, T. Murray, S. Burtles, D. Fraier, E. Frigerio, J. Cassidy. Phase I clinical and pharmacokinetic study of PK1 [N-(2-hydroxypropyl)methacrylamide copolymer doxorubicin]: first member of a new class of chemotherapeutic agents-drug-polymer conjugates. Cancer Research Campaign Phase I/II Committee. *Clin Cancer Res 5*:83–94 (1999).
128. R. Duncan. Drug–polymer conjugates: potential for improved chemotherapy. *Anticancer Drugs 3*:175–210 (1992).
129. L. W. Seymour, K. Ulbrich, P. S. Steyger, M. Brereton, V. Subr, J. Strohalm, R. Duncan. Tumor tropism and anticancer efficacy of polymer-based doxorubicin prodrugs in the treatment of subcutaneous murine B16F10 melanoma. *Br J Cancer 70*:636–641 (1994).
130. J. Cassidy, R. Duncan, G. J. Morrison, J. Strohalm, D. Plocova, J. Kopecek, S. B. Kaye. Activity of N-(2-hydroxypropyl)methacrylamide copolymers containing daunomycin against a rat tumor model. *Biochem Pharmacol 38*: 875–879 (1989).
131. S. B. Mordier, D. M. Bechet, M. P. Roux, A. Obled, M. J. Ferrara. The structure of the bovine cathepsin B gene. Genetic variability in the 3′ untranslated region. *Eur J Biochem 229*:35–44 (1995).
132. M. R. Buck, D. G. Karustis, N. A. Day, K. V. Honn, B. F. Sloane. Degradation of extracellular-matrix proteins by human cathepsin B from normal and tumor tissues. *Biochem J 282*:273–278 (1992).
133. R. Satchi, R. Duncan. PDEPT: polymer directed enzyme prodrug therapy—in vitro and in vivo characterisation. In: 3rd International Symposium on Polymer Therapeutics, London, Jan 7–11, 1998.
134. R. Satchi, R. Duncan. PDEPT: polymer directed enzyme prodrug therapy. In: 10th NCI-EORTC Symposium on New Drugs in Cancer Therapy, Amsterdam, June 16–19, 1998.
135. W. A. Bleyer. The clinical pharmacology of methotrexate. *Cancer 41*:36–51 (1978).
136. C. D. Flombaum, P. A. Meyers. High-dose leucovorin as sole therapy for methotrexate toxicity. *J Clin Oncol 17*: 1589–1594 (1999).
137. S. B. Howell, K. Herbst, G. R. Boss, E. Frei. Thymidine requirements for the rescue of patients treated with high-dose methotrexate. *Cancer Research 40*:1824–1829 (1980).
138. S. B. Howell, S. J. Mansfield, R. Taetle. Thymidine and hypoxanthine requirements of normal and malignant human cells for protection against methotrexate cytotoxicity. *Cancer Research 41*:945–950 (1981).
139. A. Reggev, I. Djerassi. Rescue from high-dose methotrexate with 5-methyltetrahydrofolate. *Cancer Treat Rep 70*: 251–253 (1986).
140. M. M. Simile, M. R. DeMiglio, A. Nufris, R. M. Pascale, M. R. Muroni, F. Feo. 1-5-formyltetrahydrofolate and 1-5-methyltetrahydrofolate rescue in L1210 leukemia treated with high methotrexate doses. *Res Commun Chem Pathol Pharmacol 81*:251–254 (1993).
141. B. Fulton, A. J. Wagstaff, D. McTavish. Trimetrexate. A review of its pharmacodynamic and pharmacokinetic properties and therapeutic potential in the treatment of Pneumocystis carinii pneumonia. *Drugs 49*:563–576 (1995).
142. K. D. Bagshawe, P. J. Burke, R. J. Knox, R. G. Melton, S. K. Sharma. Targeting enzymes to cancers—new developments. *Expert Opin Investigational Drugs 8*:161–172 (1999).
143. K. D. Bagshawe, M. Napier. Early clinical studies with ADEPT. *Enzyme Prodrug Strategies for Cancer Therapy* (R. G. Melton, R. J. Knox, eds.). Kluwer Academic/Plenum Press, New York, 1999, pp. 199–207.
144. S. K. Sharma, K. D. Bagshawe, R. G. Melton, R. F. Sherwood. Human immune response to monoclonal antibody-enzyme conjugates in ADEPT. *Cell Biophys 21*:109–120 (1993).
145. A. V. Patterson, D. C. Talbot, I. J. Stratford, A. L. Harris.

Thymidine phosphorylase moderates thymidine-dependent rescue after exposure to the thymidylate synthase inhibitor ZD1964 (Tomudex). *Cancer Research 58*:2737–2740 (1998).

146. R. N. Turner, G. W. Aherne, N. J. Curtin. Selective potentiation of lometrexol growth inhibition by dipyridamole through cell-specific inhibition of hypoxanthine salvage. *Br J Cancer 76*:1300–1307 (1997).

147. H. Ito, J. Morizet, L. Coulombel, M. Stanislawski. T cell depletion of human bone marrow using an oxidase-peroxidase enzyme immunotoxin. *Bone Marrow Transplant 6*: 395–398 (1990).

148. G. T. Stevenson, F. K. Stevenson. Antibody to a molecularly defined antigen confined to a tumor cell surface. *Nature 254*:714–716 (1975).

149. S. Fargion, D. Carney, J. Mulshine, S. Rosen, P. Bunn, P. Jewitt, F. Cuttitta, A. Gazdar, J. Minna. Heterogeneity of cell surface antigen expression of human small cell lung cancer detected by monoclonal antibodies. *Cancer Res 46*: 2633–2638 (1986).

150. P. G. Natali, R. Cavaliere, A. Bigotti, M. R. Nicotra, C. Russo, A. K. Ng, P. Giacomini, S. Ferrone. Antigenic heterogeneity of surgically removed primary and autologous metastatic human melanoma lesions. *J Immunol 130*: 1462–1466 (1983).

151. C. J. Springer, P. Antoniw, K. D. Bagshawe, D. E. Wilman. Comparison of half-lives and cytotoxicity of N-chloroethyl-4-amino and N-mesyloxyethyl-benzoyl compounds, products of prodrugs in antibody-directed enzyme prodrug therapy (ADEPT). *Anticancer Drug Des 6*:467–479 (1991).

152. R. H. J. Begent. Recent advances in tumor imaging: use of radiolabelled monoclonal antibodies. *Biochim Biophys Acta 780*:151–166 (1985).

153. E. Haenseler, A. Esswein, K. S. Vitols, Y. Montejano, B. M. Mueller, R. A. Reisfeld, F. M. Huennekens. Activation of methotrexate-α-alanine by carboxypeptidase A–monoclonal antibody conjugate. *Biochemistry 31*:891–897 (1992).

154. S. K. Sharma, K. D. Bagshawe, P. J. Burke, R. W. Boden, G. T. Rogers. Inactivation and clearance of an anti-CEA carboxypeptidase G2 conjugate in blood after localisation in a xenograft model. *Br J Cancer 61*:659–662 (1990).

155. S. K. Sharma, K. D. Bagshawe, P. J. Burke, J. A. Boden, G. T. Rogers, C. J. Springer, R. G. Melton, R. F. Sherwood. Galactosylated antibodies and antibody–enzyme conjugates in antibody-directed enzyme prodrug therapy. *Cancer 73*:1114–1120 (1994).

156. G. T. Rogers, P. J. Burke, S. K. Sharma, R. Koodie, J. A. Boden. Plasma clearance of an antibody–enzyme conjugate in ADEPT by monoclonal anti-enzyme: its effect on prodrug activation in vivo. *Br J Cancer 72*:1357–1363 (1995).

157. R. G. Melton, J. A. Rowland, G. A. Pietersz, R. F. Sherwood, I. F. McKenzie. Tumor necrosis factor increases tumor uptake of co-administered antibody-carboxypeptidase G2 conjugate. *Eur J Cancer 29a*:1177–1183 (1993).

158. A. H. Sehon. Suppression of antibody responses by chemically modified antigens. *Int Arch Allergy Appl Immunol 94*:11–20 (1991).

159. K. De-Sutter, W. Fiers. A bifunctional murine:human chimeric antibody with one antigen-binding arm replaced by bacterial β-lactamase. *Mol Immunol 31*:261–267 (1994).

160. M. S. Neuberger, G. T. Williams, R. O. Fox. Recombinant antibodies possessing novel effector functions. *Nature 312*:604–608 (1984).

161. R. G. Melton. Preparation and purification of antibody-enzyme conjugates for therapeutic applications. *Adv Drug Delivery Rev 22*:289–301 (1996).

162. R. G. Melton. Preparation and purification of antibody–enzyme conjugates for therapeutic applications. *Enzyme Prodrug. Strategies for Cancer Therapy* (R. G. Melton, R. J. Knox, ed.). Kluwer Academic/Plenum Press, New York, 1999, pp. 155–179.

163. K. Bosslet, J. Czech, G. Seemann, C. Monneret, D. Hoffmann. Fusion protein mediated prodrug activation (FMPA) in vivo. *Cell Biophys 24–25*:51–63 (1994).

164. R. G. Melton, R. J. Knox. Enzyme–prodrug strategies for cancer therapy. Kluwer Academic/Plenum Press, New York, 1999.

165. P. M. Wallace, P. D. Senter. In vitro and in vivo activities of monoclonal antibody–alkaline phosphatase conjugates in combination with phenol mustard phosphate. *Bioconjug Chem 2*:349–352 (1991).

166. P. D. Senter. Activation of prodrugs by antibody–enzyme conjugates: a new approach to cancer therapy. *Faseb J 4*: 188–193 (1990).

167. P. D. Senter, G. J. Schreiber, D. L. Hirschberg, S. A. Ashe, K. E. Hellstrom, I. Hellstrom. Enhancement of the in vitro and in vivo antitumor activities of phosphorylated mitomycin C and etoposide derivatives by monoclonal antibody–alkaline phosphatase conjugates. *Cancer Res 49*: 5789–5792 (1989).

168. U. Sahin, F. Hartmann, P. Senter, C. Pohl, A. Engert, V. Diehl, M. Pfreundschuh. Specific activation of the prodrug mitomycin phosphate by a bispecific anti-CD30/anti-alkaline phosphatase monoclonal antibody. *Cancer Res 50*:6944–6948 (1990).

169. H. J. Haisma, E. Boven, M. van Muijen, R. De Vries, H. M. Pinedo. Analysis of a conjugate between anti-carcinoembryonic antigen monoclonal antibody and alkaline phosphatase for specific activation of the prodrug etoposide phosphate. *Cancer Immunol Immunother 34*:343–348 (1992).

170. M. A. Smal, Z. Dong, H. T. Cheung, Y. Asano, L. Escoffier, M. Costello, M. H. Tattersall. Activation and cytotoxicity of 2-α-aminoacyl prodrugs of methotrexate. *Biochem Pharmacol 49*:567–574 (1995).

171. A. Bukhari, T. A. Connors, A. M. Gilsenan, W. C. Ross, M. J. Tisdale, G. P. Warwick, D. E. Wilman. Cytotoxic agents designed to be selective for liver cancer. *J Natl Cancer Inst 50*:243–247 (1973).

172. H. Autrup, G. P. Warwick. Some characteristics of two azoreductase systems in rat liver. Relevance to the activity

of 2-[4′-di(2″-bromopropyl)-aminophenylazo]benzoic acid (CB10-252), a compound possessing latent cytotoxic activity. *Chem Biol Interact 11*:329–342 (1975).
173. A. Kojima, N. R. Hackett, A. Ohwada, R. G. Crystal. In vivo human carboxylesterase cDNA gene transfer to activate the prodrug CPT-11 for local treatment of solid tumors. *J Clin Investigation 101*:1789–1796 (1998).
174. M. K. Danks, C. L. Morton, E. J. Krull, P. J. Cheshire, L. B. Richmond, C. W. Naeve, C. A. Pawlik, P. J. Houghton, P. M. Potter. Comparison of activation of CPT-11 by rabbit and human carboxylesterases for use in enzyme/prodrug therapy. *Clin Cancer Res 5*:917–924 (1999).
175. U. Kuefner, U. Lohrmann, Y. D. Montejano, K. S. Vitols, F. M. Huennekens. Carboxypeptidase-mediated release of methotrexate from methotrexate α-peptides. *Biochemistry 28*:2288–2297 (1989).
176. A. Esswein, E. Hanseler, Y. Montejano, K. S. Vitols, F. M. Huennekens. Construction and chemotherapeutic potential of carboxypeptidase-A/monoclonal antibody conjugate. *Adv Enzyme Regul 31*:3–12 (1991).
177. M. J. Perron, M. Page. Activation of methotrexate-phenylalanine by monoclonal antibody–carboxypeptidase A conjugate for the specific treatment of ovarian cancer in vitro. *Br J Cancer 73*:281–287 (1996).
178. C. J. Springer, V. Bavetsias, A. L. Jackman, F. T. Boyle, D. Marshall, R. B. Pedley, G. M. Bisset. Prodrugs of thymidylate synthase inhibitors: potential for antibody directed enzyme prodrug therapy (ADEPT). *Anticancer Drug Des 11*:625–636 (1996).
179. D. A. Hamstra, A. Rehemtulla. Toward an enzyme/prodrug strategy for cancer gene therapy: endogenous activation of carboxypeptidase A mutants by the PACE/Furin family of propeptidases. *Hum Gene Ther 10*:235–248 (1999).
180. K. D. Bagshawe. Towards generating cytotoxic agents at cancer sites. *Br J Cancer 60*:275–281 (1989).
181. C. J. Springer, P. Antoniw, K. D. Bagshawe, F. Searle, G. M. Bisset, M. Jarman. Novel prodrugs which are activated to cytotoxic alkylating agents by carboxypeptidase G2. *J Med Chem 33*:677–681 (1990).
182. C. J. Springer. CMDA 4-[2-chloroethyl)[2-(mesyloxy) ethyl]amino]benzoyl-L-glutamic acid. Antineoplastic prodrug. *Drugs Fut 18*:212–215 (1993).
183. C. J. Springer, I. Niculescu-Duvaz, R. B. Pedley. Novel prodrugs of alkylating agents derived from 2-fluoro- and 3-fluorobenzoic acids for antibody-directed enzyme prodrug therapy. *J Med Chem 37*:2361–2370 (1994).
184. C. J. Springer, R. Dowell, P. J. Burke, E. Hadley, D. H. Davis, D. C. Blakey, R. G. Melton, I. Niculescu Duvaz. Optimization of alkylating agent prodrugs derived from phenol and aniline mustards: a new clinical candidate prodrug (ZD2767) for antibody-directed enzyme prodrug therapy (ADEPT). *J Med Chem 38*:5051–5065 (1995).
185. R. Marais, R. A. Spooner, Y. Light, J. Martin, C. J. Springer, Gene-directed enzyme prodrug therapy with a mustard prodrug/carboxypeptidase G2 combination. *Cancer Res 56*:4735–4742 (1996).
186. I. Niculescu-Duvaz, D. Niculescu-Duvaz, F. Friedlos, R. Spooner, J. Martin, R. Marais, C. J. Springer. Self-immolative anthracycline prodrugs for suicide gene therapy. *J Med Chem 42*:2485–2489 (1999).
187. D. Niculescu-Duvaz, I. Niculescu-Duvaz, F. Friedlos, J. Martin, R. Spooner, L. Davies, R. Marais, C. J. Springer. Self-immolative nitrogen mustard prodrugs for suicide gene therapy. *J Med Chem 41*:5297–5309 (1998).
188. M. X. Wei, T. Tamiya, M. Chase, E. J. Boviatsis, T. K. Chang, N. W. Kowall, F. H. Hochberg, D. J. Waxman, X. O. Breakefield, E. A. Chiocca. Experimental tumor therapy in mice using the cyclophosphamide-activating cytochrome P450 2B1 gene. *Hum Gene Ther 5*:969–978 (1994).
189. L. Chen, D. J. Waxman. Intratumoral activation and enhanced chemotherapeutic effect of oxazaphosphorines following cytochrome P-450 gene transfer: development of a combined chemotherapy/cancer gene therapy strategy. *Cancer Res 55*:581–589 (1995).
190. M. X. Wei, T. Tamiya, R. J. Rhee, X. O. Breakefield, E. A. Chiocca. Diffusible cytotoxic metabolites contribute to the in vitro bystander effect associated with the cyclophosphamide/cytochrome P450 2B1 cancer gene therapy paradigm. *Clin Cancer Res 1*:1171–1177 (1995).
191. L. Chen, D. J. Waxman, D. Chen, D. W. Kufe. Sensitization of human breast cancer cells to cyclophosphamide and ifosfamide by transfer of a liver cytochrome P450 gene. *Cancer Res 56*:1331–1340 (1996).
192. Y. Manome, P. Y. Wen, L. Chen, T. Tanaka, Y. Dong, M. Yamazoe, A. Hirshowitz, D. W. Kufe, H. A. Fine. Gene therapy for malignant gliomas using replication incompetent retroviral and adenoviral vectors encoding the cytochrome P450 2B1 gene together with cyclophosphamide. *Gene Ther 3*:513–520 (1996).
193. M. Lohr, P. Muller, P. Karle, J. Stange, S. Mitzner, R. Jesnowski, H. Nizze, B. Nebe, S. Liebe, B. Salmons, W. H. Gunzburg. Targeted chemotherapy by intratumour injection of encapsulated cells engineered to produce CYP2B1, an ifosfamide activating cytochrome P450. *Gene Ther 5*:1070–1078 (1998).
194. N. G. Rainov, K. U. Dobberstein, M. Senaesteves, U. Herrlinger, C. M. Kramm, R. M. Philpot, J. Hilton, E. A. Chiocca, X. O. Breakefield. New prodrug activation gene therapy for cancer using cytochrome P450 4B1 and 2-aminoanthracene/4-ipomeanol. *Hum Gene Ther 9*:1261–1273 (1998).
195. D. A. Hamstra, D. J. Rice, S. Fahmy, B. D. Ross, A. Rehemtulla. Enzyme/prodrug therapy for head and neck cancer using a catalytically superior cytosine deaminase. *Hum Gene Ther 10*:1993–2003 (1999).
196. P. D. Senter, P. C. D. Su, T. Katsuragi, T. Sakai, W. L. Cosand, I. Hellstrom, K. E. Hellstrom. Generation of 5-fluorouracil from 5-fluorocytosine by monoclonal antibody–cytosine deaminase conjugates. *Bioconjugate Chem 2*:447–451 (1991).
197. P. M. Wallace, J. F. MacMaster, V. F. Smith, D. E. Kerr,

P. D. Senter, W. L. Cosand. Intratumoral generation of 5-fluorouracil mediated by an antibody–cytosine deaminase conjugate in combination with 5-fluorocytosine. *Cancer Res 54*:2719–2723 (1994).

198. Y. Manome, P. Y. Wen, Y. Dong, T. Tanaka, B. S. Mitchell, D. W. Kufe, H. A. Fine. Viral vector transduction of the human deoxycytidine kinase cDNA sensitizes glioma cells to the cytotoxic effects of cytosine arabinoside in vitro and in vivo. *Nat Med 2*:567–573 (1996).

199. N. Robertson, A. Haigh, G. E. Adams, I. J. Stratford. Factors affecting sensitivity to EO9 in rodent and human tumor cells in vitro: DT-diaphorase activity and hypoxia. *Eur J Cancer 7*:1013–1019 (1994).

200. K. H. Warrington, Jr., C. Teschendorf, L. Cao, N. Muzyczka, D. W. Siemann. Developing VDEPT for DT-diaphorase (NQO1) using an AAV vector plasmid. *Int J Radiat Oncol Biol Phys 42*:909–912 (1998).

201. S. L. Winski, R. H. Hargreaves, J. Butler, D. Ross. A new screening system for NAD(P)H:quinone oxidoreductase (NQO1)-directed antitumor quinones: identification of a new aziridinylbenzoquinone, RH1, as a NQO1-directed antitumor agent. *Clin Cancer Res 4*:3083–3088 (1998).

202. S. Andrianomenjanahary, X. Dong, J. C. Florent, G. Gaudel, J. P. Gesson, J. C. Jacquesy, M. Koch, S. Michel, M. Mondon. Synthesis of novel targeted pro-prodrugs of anthracyclines potentially activated by a monoclonal antibody galactosidase conjugate. *Bioorg Med Chem Lett 2*: 1093–1096 (1992).

203. J. P. Gesson, J. C. Jacquesy, M. Mondon, P. Petit, B. Renoux, S. Andrianomenjanahary, H. Dufat Trinh-Van, M. Koch, S. Michel, F. Tillequin. Prodrugs of anthracyclines for chemotherapy via enzyme-monoclonal antibody conjugates. *Anticancer Drug Des 9*:409–423 (1994).

204. M. Azoulay, J. C. Florent, C. Monneret, J. P. Gesson, J. C. Jacquesy, F. Tillequin, M. Koch, K. Bosslet, J. Czech, D. Hoffman. Prodrugs of anthracycline antibiotics suited for tumor-specific activation. *Anticancer Drug Des 10*: 441–450 (1995).

205. V. R. Muzykantov, D. V. Saharov, V. V. Sinitsyn, S. P. Domogatsky, N. V. Goncharov, S. M. Danilov. Specific killing of human endothelial cells by antibody-conjugated glucose oxidase. *Anal Chem 169*:383–389 (1988).

206. V. R. Muzykantov, O. V. Trubetskaya, E. A. Puchnina, D. V. Sakharov, S. P. Domogatsky. Cytotoxicity of glucose oxidase conjugated with antibodies to target cells: killing efficiency depends on conjugate internalization. *Biochim Biophys Acta 1053*:27–31 (1990).

207. H. Ito, J. Morizet, L. Coulombel, M. Goavec, V. Rousseau, A. Bernard, M. Stanislawski. An immunotoxin system intended for bone marrow purging composed of glucose oxidase and lactoperoxidase coupled to monoclonal antibody 097. *Bone Marrow Transplant 4*:519–527 (1989).

208. G. Rowlinson-Busza, A. Bamias, T. Krausz, A. A. Epenetos. Antibody-guided enzyme nitrile therapy (agent): in vitro cytotoxicity and in vivo tumor localisation. *Monoclonal Antibodies, Applications in Clinical Oncology* (A. A. Epenetos ed.). Chapman and Hall, London, 1992, pp. 111–118.

209. G. Rowlinson-Busza, A. A. Epenetos. Targeted delivery of biologic and other antineoplastic agents. *Curr Opin Oncol 4*:1142–1148 (1992).

210. M. E. Whisson, T. A. Connors, A. Jeney. Mechanism of cure of large plasma cell tumors. *Arch Immunol Ther Exp Warsz 14*:825–831 (1966).

211. S. R. Roffler, S. M. Wang, J. W. Chern, M. Y. Yeh, E. Tung. Anti-neoplastic glucuronide prodrug treatment of human tumor cells targeted with a monoclonal antibody–enzyme conjugate. *Biochem Pharmacol 42*:2062–2065 (1991).

212. S. M. Wang, J. W. Chern, M. Y. Yeh, J. C. Ng, E. Tung, S. R. Roffler. Specific activation of glucuronide prodrugs by antibody-targeted enzyme conjugates for cancer therapy. *Cancer Res 52*:4484–4491 (1992).

213. F. Schmidt, J. C. Florent, C. Monneret, R. Straub, J. Czech, M. Gerken, K. Bosslet. Glucuronide prodrugs of hydroxy compounds for antibody directed enzyme prodrug therapy (ADEPT): a phenol nitrogen mustard carbamate. *Bioorg Med Chem Lett 7*:1071–1076 (1997).

214. B. M. Chen, L. Y. Chan, S. M. Wang, M. F. Wu, J. W. Chern, S. R. Roffler. Cure of malignant ascites and generation of protective immunity by monoclonal antibody-targeted activation of a glucuronide prodrug in rats. *Int J Cancer 73*:392–402 (1997).

215. R. Lougerstay-Madec, J. C. Florent, C. Monneret, F. Nemati, M. F. Poupon. Synthesis of self-immolative glucuronide-based prodrugs of a phenol mustard. *Anticancer Drug Des 13*:995–1007 (1998).

216. T. L. Cheng, W. C. Chou, B. M. Chen, J. W. Chern, S. R. Roffler. Characterization of an antineoplastic glucuronide prodrug. *Biochem Pharmacol 58*:325–328 (1999).

217. H. J. Haisma, M. van-Muijen, H. M. Pinedo, E. Boven. Comparison of two anthracycline-based prodrugs for activation by a monoclonal antibody–β-glucuronidase conjugate in the specific treatment of cancer. *Cell Biophys 25*: 185–192 (1994).

218. D. B. de Bont, R. G. Leenders, H. J. Haisma, I. van der Meulen-Muileman, H. W. Scheeren. Synthesis and biological activity of β-glucuronyl carbamate-based prodrugs of paclitaxel as potential candidates for ADEPT. *Bioorg Med Chem Lett 5*:405–414 (1997).

219. P. H. Houba, R. G. Leenders, E. Boven, J. W. Scheeren, H. M. Pinedo, H. J. Haisma. Characterization of novel anthracycline prodrugs activated by human β-glucuronidase for use in antibody-directed enzyme prodrug therapy. *Biochem Pharmacol 52*:455–463 (1996).

220. E. Bakina, Z. Wu, M. Rosenblum, D. Farquhar. Intensely cytotoxic anthracycline prodrugs—glucuronides. *J Med Chem 40*:4013–4018 (1997).

221. J. C. Florent, X. Dong, G. Gaudel, S. Mitaku, C. Monneret, J. P. Gesson, J. C. Jacquesy, M. Mondon, B. Renoux, S. Andrianomenjanahary, S. Michel, M. Koch, F. Tillequin, M. Gerken, J. Czech, R. Straub, K. Bosslet. Prodrugs of anthracyclines for use in antibody-directed en-

zyme prodrug therapy. *J Med Chem 41*:3572–3581 (1998).

222. S. Desbene, H. D. Van, S. Michel, M. Koch, F. Tillequin, G. Fournier, N. Farjaudon, C. Monneret. Doxorubicin prodrugs with reduced cytotoxicity suited for tumor-specific activation. *Anticancer Drug Des 13*:955–968 (1998).
223. Y. L. Leu, S. R. Roffler, J. W. Chern. Design and synthesis of water-soluble glucuronide derivatives of camptothecin for cancer prodrug monotherapy and antibody-directed enzyme prodrug therapy (ADEPT). *J Med Chem 42*:3623–3628 (1999).
224. G. D. Smith, P. K. Chakravarty, T. A. Connors, T. J. Peters. Synthesis and preliminary characterization of a novel substrate for γ-glutamyl transferase. A potential anti-hepatoma drug. *Biochem Pharmacol 33*:527–529 (1984).
225. T. A. Shepherd, L. N. Jungheim, D. M. Meyer, J. J. Starling. A novel targeted delivery system utilizing a cephalosporin-oncolytic prodrug activated by an antibody β-lactamase conjugate for the treatment of cancer. *Bioorg Med Chem Lett 1*:21–26 (1991).
226. D. L. Meyer, K. L. Law, J. K. Payne, S. D. Mikolajczyk, H. Zarrinmayeh, L. N. Jungheim, J. K. Kling, T. A. Shepherd, J. J. Starling. Site-specific prodrug activation by antibody–β-lactamase conjugates: preclinical investigation of the efficacy and toxicity of doxorubicin delivered by antibody directed catalysis. *Bioconjug Chem 6*:440–446 (1995).
227. D. L. Meyer, L. N. Jungheim, K. L. Law, S. D. Mikolajczyk, T. A. Shepherd, D. G. Mackensen, S. L. Briggs, J. J. Starling. Site-specific prodrug activation by antibody–β-lactamase conjugates: regression and long-term growth inhibition of human colon carcinoma xenograft models. *Cancer Res 53*:3956–3963 (1993).
228. D. L. Meyer, L. N. Jungheim, S. D. Mikolajczyk, T. A. Shepherd, J. J. Starling, C. N. Ahlem. Preparation and characterization of a β-lactamase–Fab′ conjugate for the site-specific activation of oncolytic agents. *Bioconjug Chem 3*:42–48 (1992).
229. R. P. Alexander, N. R. A. Beeley, M. O'Driscoll, F. P. O'Neill, T. A. Millican, A. J. Pratt, F. W. Willenbrock. Cephalosporin nitrogen mustard carbamate prodrugs for ''ADEPT.'' *Tetrahedron Lett 32*:3269–3272 (1991).
230. H. P. Svensson, J. F. Kadow, V. M. Vrudhula, P. M. Wallace, P. D. Senter. Monoclonal antibody-β-lactamase conjugates for the activation of a cephalosporin mustard prodrug. *Bioconjug Chem 3*:176–181 (1992).
231. D. E. Kerr, Z. G. Li, N. O. Siemers, P. D. Senter, V. M. Vrudhula. Development and activities of a new melphalan prodrug designed for tumor-selective activation. *Bioconjugate Chem 9*:255–259 (1998).
232. V. M. Vrudhula, H. P. Svensson, P. D. Senter. Cephalosporin derivatives of doxorubicin as prodrugs for activation by monoclonal antibody–β-lactamase conjugates. *J Med Chem 38*:1380–1385 (1995).
233. R. J. Knox, F. Friedlos, R. F. Sherwood, R. G. Melton, G. M. Anlezark. The bioactivation of 5-(aziridin-1-yl)-2,4-dinitrobenzamide (CB1954)—II. A comparison of an Escherichia coli nitroreductase and Walker DT diaphorase. *Biochem Pharmacol 44*:2297–2301 (1992).
234. F. Friedlos, W. A. Denny, B. D. Palmer, C. J. Springer. Mustard prodrugs for activation by escherichia coli nitroreductase in gene-directed enzyme prodrug therapy. *J Med Chem 40*:1270–1275 (1997).
235. M. P. Hay, W. R. Wilson, W. A. Denny. A novel enediyne prodrug for antibody-directed enzyme prodrug therapy (ADEPT) using E. coli B nitroreductase. *Bioorg Med Chem Lett 5*:2829–2834 (1995).
236. M. Tercel, W. A. Denny, W. R. Wilson. A novel nitro-substituted seco-CI: application as a reductively activated ADEPT prodrug. *Bioorg Med Chem Lett 6*:2741–2744 (1996).
237. M. P. Hay, B. M. Sykes, W. A. Denny, W. R. Wilson. A 2-nitroimidazole carbamate prodrug of 5-amimo-1-(chloromethyl)-3-[(5,6,7-trimethoxyindol-2-yl)carbony l]-1,2-dihydro-3H-benz[E]indole (amino-seco-CBI-TMI) for use with ADEPT and GDEPT. *Bioorg Med Chem Lett 9*:2237–2242 (1999).
238. G. S. Bignami, P. D. Senter, P. G. Grothaus, K. J. Fischer, T. Humphreys, P. M. Wallace. N-(4′-hydroxyphenylacetyl)palytoxin: a palytoxin prodrug that can be activated by a monoclonal antibody-penicillin G amidase conjugate. *Cancer Res 52*:5759–5764 (1992).
239. D. E. Kerr, P. D. Senter, W. V. Burnett, D. L. Hirschberg, I. Hellstrom, K. E. Hellstrom. Antibody-penicillin-V-amidase conjugates kill antigen-positive tumor cells when combined with doxorubicin phenoxyacetamide. *Cancer Immunol. Immunother 31*:202–206 (1990).
240. V. M. Vrudhula, P. D. Senter, K. J. Fischer, P. M. Wallace. Prodrugs of doxorubicin and melphalan and their activation by a monoclonal antibody-penicillin-G amidase conjugate. *J Med Chem 36*:919–923 (1993).
241. P. K. Chakravarty, P. L. Carl, M. J. Weber, J. A. Katzenellenbogen. Plasmin-activated prodrugs for cancer chemotherapy. 1. Synthesis and biological activity of peptidylacivicin and peptidylphenylenediamine mustard. *J Med Chem 26*:633–638 (1983).
242. P. K. Chakravarty, P. L. Carl, M. J. Weber, J. A. Katzenellenbogen. Plasmin-activated prodrugs for cancer chemotherapy. 2. Synthesis and biological activity of peptidyl derivatives of doxorubicin. *J Med Chem 26*:638–644 (1983).
243. W. B. Parker, S. A. King, P. W. Allan, L. L. Bennett, J. A. Secrist, J. A. Montgomery, K. S. Gilberg, W. R. Waud, A. H. Wells, G. Y. Gillespie, E. J. Sorscher. In vivo gene therapy of cancer with E. coli purine nucleoside phosphorylase. *Hum Gene Ther 8*:1637–1644 (1997).
244. B. W. Hughes, S. A. King, P. W. Allan, W. B. Parker, E. J. Sorscher. Cell to cell contact is not required for bystander cell killing by Escherichia coli purine nucleoside phosphorylase. *J Biol Chem 273*:2322–2328 (1998).
245. J. A. Secrist, W. B. Parker, P. W. Allan, L. L. Bennett, Jr., W. R. Waud, J. W. Truss, A. T. Fowler, J. A. Montgomery, S. E. Ealick, A. H. Wells, G. Y. Gillespie, V. K. Gadi, E. J. Sorscher. Gene therapy of cancer: activation

of nucleoside prodrugs with *E. coli* purine nucleoside phosphorylase. *Nucleosides Nucleotides 18*:745–757 (1999).

246. K. W. Culver, Z. Ram, S. Wallbridge, H. Ishii, E. H. Oldfield, R. M. Blaese. In vivo gene transfer with retroviral vector-producer cells for treatment of experimental brain tumors. *Science 256*:1550–1552 (1992).

247. Z. Ram, K. W. Culver, S. Walbridge, J. A. Frank, R. M. Blaese, E. H. Oldfield. Toxicity studies of retroviral-mediated gene transfer for the treatment of brain tumors. *J Neurosurg 79*:400–407 (1993).

248. B. E. Huber, C. A. Richards, T. A. Krenitsky. Retroviral-mediated gene therapy for the treatment of hepatocellular carcinoma: an innovative approach for cancer therapy. *Proc Natl Acad Sci USA 88*:8039–8043 (1991).

249. A. V. Patterson, H. Zhang, A. Moghaddam, R. Bicknell, D. C. Talbot, I. J. Stratford, A. L. Harris. Increased sensitivity to the prodrug 5′-deoxy-5-fluorouridine and modulation of 5-fluoro-2′-deoxyuridine sensitivity in MCF-7 cells transfected with thymidine phosphorylase. *Br J Cancer 72*:669–675 (1995).

250. A. M. Jordan, T. H. Khan, H. M. Osborn, A. Photiou, P. A. Riley. Melanocyte-directed enzyme prodrug therapy (MDEPT): development of a targeted treatment for malignant melanoma. *Bioorg Med Chem 7*:1775–1780 (1999).

251. Y. Ono, K. Ikeda, M. X. Wei, G. R. Harsh, T. Tamiya, E. A. Chiocca. Regression of experimental brain tumors with 6-thioxanthine and Escherichia coli gpt gene therapy. *Hum Gene Ther 8*:2043–2055 (1997).

252. R. J. Knox. Gene-directed enzyme prodrug therapy (GDEPT) of cancer. *Enzyme Prodrug Strategies for Cancer Therapy* (R. G. Melton, R. J. Knox, eds.). Kluwer Academic/Plenum Press, New York, 1999, pp. 209–243.

253. S. Desbene, H. D. Van, S. Michel, F. Tillequin, M. Koch, F. Schmidt, J. C. Florent, C. Monneret, R. Straub, J. Czech, M. Gerken, K. Bosslet. Application of the ADEPT strategy to the MDR resistance in cancer chemotherapy. *Anticancer Drug Des 14*:93–106 (1999).

34

New Lipid/DNA Complexes for Gene Delivery

Kenneth W. Liang and Leaf Huang
University of Pittsburgh, Pittsburgh, Pennsylvania

ABBREVIATIONS

CHEMS	cholesteryl hemisuccinate
DC-Chol	3-β-[N-(N′,N′-dimethylaminoethane)carbamoyl]-cholesterol
DOPE	dioleoylphosphatidylethanolamine
DOTAP	1,2-dioleoyloxypropyl-3-N,N,N-trimethylammonium chloride
DOTMA	1,2-dioleyloxypropyl-3-N,N,N-trimethylammonium chloride
LPD	liposome-entrapped, polycation-condensed DNA
PNA	peptide nucleic acid
SAXS	small-angle x-ray scattering

I. INTRODUCTION

Recent advances in molecular biology and the human genome project have brought about improved understanding of the genetic basis for human disease. The wealth of knowledge about genetic causes of human disease provides compelling and substantive support for the development of gene therapy. Today, many incurable diseases such as inherited genetic diseases, cancer, and AIDS are under consideration as candidates for gene therapy. In some cases, gene therapy represents the only long-term hope of treatment for these diseases. After a decade of extensive research on gene therapy, it is now clear that the vehicle with which the therapeutic gene is introduced into the target cells of the body is the key to the success of an effective cure.

Gene delivery can be achieved with viral and nonviral vectors. Viral vectors are recombinant, replication-defective viruses with all or part of the viral coding sequences replaced by the therapeutic gene. They infect cells using the same mechanism as the parental virus and are thus highly efficient in transducing cells. For example, adenoviral vectors can efficiently transfer genes into quiescent, terminally differentiated cells such as hepatocytes and nerve cells. This is because these cells express the appropriate receptor for adenovirus, and the virus can efficiently move into the cellular nuclei by some active processes (1). Even though viral vectors are highly efficient in terms of gene transfer, they still suffer from many problems in clinical application. For example, gene expression of adenovirus vectors usually does not last for more than several weeks (2). However, high immunogenicity of the adenoviral vector makes repeat administration of the vector virtually impossible. Retroviral vectors are also very effective and are widely used for current clinical applications. However, there are serious concerns about the tumorigenic potential of retroviral vectors via the random disruption of an oncogene and/or the generation of infectious virons through recombination with latent viral sequences (3,4). Moreover, the commonly used ex vivo protocol of using retroviral vectors also limits the clinical application of these types of vector. Other problems associated with viral vectors are low yield in viral preparation (5,6), a limit in the size of a therapeutic gene that can be fitted into a viral vector, and difficulty in maintaining reproducible quality and quantity. All of these problems with viral vectors have made the development of nonviral vectors more desirable for gene therapy.

Various chemical and physical methods, such as electroporation, microinjection (7), and a complex of DNA with calcium phosphate (8) or DEAE-dextran have been used for gene transfer in cultured cells. Although most of these methods are still favored in laboratories, they obviously cannot be used to transfect cells for therapeutic purposes because of efficacy and safety concerns. On the other hand, gene transfer using a cationic liposome has been successfully applied in clinical trials.

Cationic lipids are amphiphilic molecules consisting of a hydrophobic anchor, usually made up of fatty acyl, alkyl, or alkoxy chain(s), a hydrophilic head group carrying a single or multiple positive charged groups, and a linker that connects the hydrophobic portion and the hydrophilic head together. In 1987, Felgner et al. reported the first successful gene transfer using a cationic lipid (9). They showed that N-[1-(2,3-dioleoyloxy)propyl]-N,N,N,-trimethyl ammonium chloride (DOTMA) and dioleyl phosphatidylethanolamine (DOPE) formed a cationic lipid/DNA complex with plasmid DNA that successfully transfected several cell lines. Since then, hundreds of new lipids and many new formulations have been developed. Currently, cationic lipid vectors generally contain a cationic lipid, a neutral lipid such as DOPE or cholesterol, and the plasmid DNA that encodes the transgene of interest. The assembly of a cationic lipidic vector is normally achieved by mixing the positively charged cationic lipid with the negatively charged plasmid DNA. Through electrostatic interaction between a lipid head group and a phosphate group on DNA, nanoparticles composed of lipid and DNA are formed. These particles are capable of entering cells efficiently and delivering a portion of the encapsulated DNA into the nucleus. Excess in positive charge is usually required for efficient gene transfer, because the resultant positively charged vectors can bind efficiently with the negatively charged cell membranes and induce endocytosis.

In order to achieve the ultimate goal in gene therapy, one needs to develop a vector that can be administered in vivo, and that will deliver therapeutic genes to the target cells, release the DNA from the endosome, and import the DNA into the nucleus. In addition to these features, the vector should also be safe, simple to use, and easy to produce on a large scale. Compared to viral vectors, cationic lipidic vectors are much less immunogenic and can be easily produced in large quantities with reproducible pharmaceutical quality. Furthermore, there is no structure and size limit on the therapeutic gene. For example, cationic lipids have been successfully used to deliver human artificial chromosomes of the order of 230 kb into cells (10). Besides that, the properties of cationic lipid vectors can easily be modified via chemical and pharmaceutical methods, such as incorporating targeting ligands or by manipulation of the formulation. All of these advantages make cationic lipids a good candidate for developing ideal synthetic vectors for gene therapy.

However, one major limitation of cationic lipid vectors is their transfection efficiency, which is much lower than that of viral vectors. When the efficiency is indicated by the number of cells transfected, the cationic lipid/DNA complex can only introduce nucleic acid into a portion of the cells, usually less than several percent of the total. Numerous attempts have been done to increase the transfection efficiency of cationic lipid vectors. These include the synthesis of new cationic lipids and the improvement of cationic lipid formulations. In this chapter, we will briefly review the current development of cationic lipids and their structure–activity relationships. Later, we will focus our discussion on several new developments in our laboratory.

II. CATIONIC LIPOSOMAL VECTORS

Since the first cationic lipid, DOTMA, was reported in 1987 (9), several hundreds of cationic lipids have been synthesized by different laboratories and companies. Although the detailed structures of the individual lipids vary considerably, all of them share a common feature in that they all consist of a cationic headgroup attached via a linker to a hydrophobic moiety. Cationic lipids can be classified into two different categories, double-chain amphiphiles and cholesterol derivatives, according to the difference in their hydrophobic anchor.

Double-chain amphiphiles are widely used in cationic liposome formulation. This type of lipid normally contains a headgroup of quaternary ammonium salt and double acyl or alkyl chains as the hydrophobic anchor. Two representative lipids in this category are DOTMA and DOTAP (Fig. 1). Lipids of cholesterol derivatives all have cholesterol as a hydrophobic anchor. A representative lipid in this category is 3β[N-(N′,N′-dimethylaminoethane)-carbamoyl]-cholesterol (DC-chol) (Fig. 1). DC-chol was first developed in our laboratory with the purpose of preparing a less toxic lipid. It contains a single tertiary amine group, which

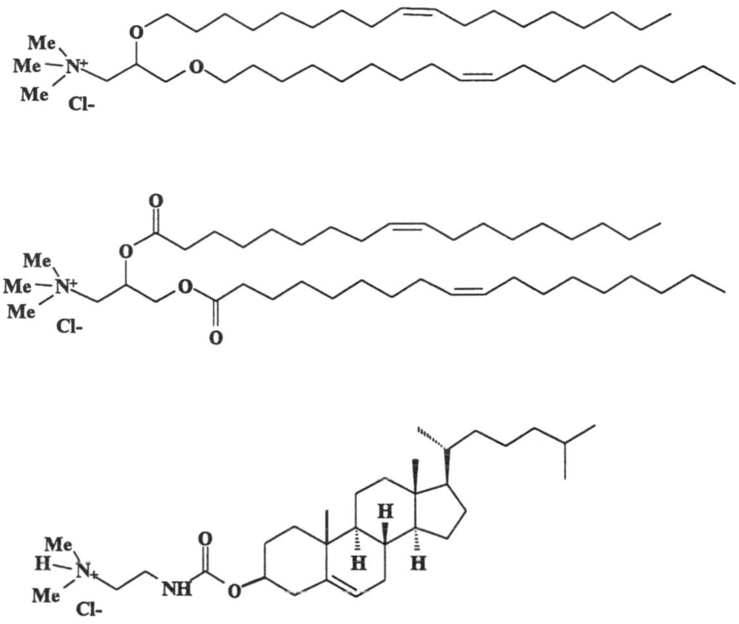

Figure 34.1 Structure of DOTMA, DOTAP, and DC-Chol.

is considered to be less toxic than quaternary ammonium groups. In addition, DC-chol is biodegradable because of the carbamoyl bond instead of a nondegradable ether bond to connect the headgroup and the cholesterol. Due to its low toxicity, DC-chol combined with DOPE was the first lipid-based gene transfer agent to be used in a human gene therapy trial (11).

The efficiency of gene transfection by a cationic lipid vector can be manipulated by changes either at the molecular level or at the formulation level. At the molecular level, it was observed that a small change in the structure of the cationic lipid could significantly affect its transfection activity. This observation has sparked an extensive interest in studying the structure–activity relationship of cationic lipids, with the goal of developing lipids with high transfection efficiency and low toxicity. To date, several general rules regarding the structure–activity relationships of cationic lipids have been established.

The three moieties of cationic lipids, the headgroup, the hydrophobic moiety, and the linker, all have an influence on the transfection activity of a cationic lipid. The headgroup has been found to play an important role, since the function of the headgroup in cationic lipids is to bind with and to condense the negatively charged plasmid DNA to form a complex, as well as to enhance the binding of the complex to the negatively charged cell surface. The headgroup can affect the activity of lipids in several ways. The type of amine (12), the number of amines (13,14), and the orientation of the headgroup (14,15) all can affect transfection activity. The hydrophobic anchor influences the transfection activity by affecting the phase transition of lipids between the gel and the liquid crystalline phases. For example, Akao and colleagues (16) found that lipids with phase transition temperatures (Tm) below 37°C were more efficient in gene delivery than those with a Tm above 37°C. The observation that lipids with lower Tm exhibit better transfection activity is supported by the results of other groups such as those of Solodin et al. (17), and Lee et al. (14). The linker is the moiety that connects the hydrophobic anchor and the polar headgroup. The linker can be an ether, amide, carbamate, ester, or phosphate bond. The linker group affects the biodegradability and consequently the stability of the cationic lipid. Lipids with an unhydrolyzable linker provide good chemical stability and generally exhibit higher transfection activities (9,18). However, the poor biodegradability of these lipids has generated some concerns about their safety in clinical usage. Easily degraded linkers, on the other hand, may make the lipid unstable in a biological environment (13). As a result, lipid/DNA complexes may dissociate before they reach target cells. In general, a linker is selected based on the balance between its biodegradability and its relative stability.

At the formulation level, many factors can affect the efficiency of gene expression. One important factor is the stoichiometry of lipids and DNA. The relative amounts of lipid and DNA directly affect the surface charge of a lipid/DNA complex, which subsequently affect the size and the stability of the complex, and its interaction with various components in vivo. Other factors, such as the method of mixing, the elapsed time of incubation (19), and helper lipid selection (20–22), can also affect the gene delivery activity.

One important characteristic of the cationic lipid/DNA complex that is affected at the formulation level is the structure of the lipid/DNA complex. In contrast to what Felgner and Ringold (23) originally proposed, that cationic liposomes remain the same size and shape after binding to DNA, experimental data now show that cationic liposome/DNA complexes are very heterogenous and dynamic, varying in size and shape depending on the molar ratio of cationic liposomes to DNA. Using ethidium bromide as a fluorescent probe for the exposed DNA, Gershon et al. showed that the most dramatic changes in structure appear to occur when the positive/negative charge ratio is around 1 (24). Based on these results, Gershon et al. proposed that the binding of cationic liposomes to DNA induced a cooperative collapse in the DNA structure and that the resulting condensed DNA is subsequently encapsulated by lipid (24). Using Kleinschmidt metal-shadowing electron microscopy, Gershon et al. showed that DNA was bound to liposome and formed a spherical structure at low lipid-to-DNA ratios. When the amount of lipid was increased, DNA became covered by liposomes, and a smooth rodlike structure was formed (24). A similar structure was later found by Sternberg et al., who examined the structure of DC-chol/DNA complex by freeze-fracture electron microscopy and revealed that some freshly prepared DC-chol/DNA complex contained so called ''spaghetti and meatball'' structures (25), indicating that in the complex the DNA molecules are not well condensed but exist in an extended conformation and covered by lipids. These observations have been supported by the recent results of cryo-EM (26) and small-angle x-ray scattering (SAXS). Gustafsson et al. have obtained cryotransmission electron microscopy images of DNA complexed with DOTAP, which revealed the presence of loosely bound DNA at a low positive/negative charge ratio and large lipid/DNA structures at a high charge ratio (26). The experimental results from SAXS further revealed the detail structure inside the DNA/lipid complex. It was found that the DNA/lipid complex formed a multilamellar structure with alternating lipid bilayers and DNA monolayers. A long periodicity of 6.5 nm was observed, which is consistent with a lipid bilayer thickness of 4 nm and a hydrated DNA helix of 2.5 nm (27). These studies further indicate that the transfection activity can be greatly influenced at the formulation level.

III. NEW LIPID/DNA COMPLEXES FOR GENE DELIVERY

A. Cationic Liposome-Entrapped, Polycation-Condensed DNA (LPD-I)

Improving the transfection activity is continuously a major effort in developing nonviral vectors. Although studies on the structure–activity relationship has established some correlations between the structure of a cationic lipid and its transfection activity, the search for new lipids still largely relies on a trial-and-error approach and has achieved only limited success. On the other hand, studies on the formulations of cationic lipid/DNA complex have revealed some drawbacks that require further correction. For example, the DC-chol/DNA complexes tend to form large aggregates with sizes well beyond 500 nm in diameter. The aggregation problem is particularly severe at high concentrations of lipid and DNA. This creates difficulties for preclinical and clinical studies in which complexes of high concentrations are often required (28). Since DC-chol does not condense DNA efficiently, additional condensing agents, such as polycations, have been studied extensively (24,29). For example, human chromatin can be reduced in length by a factor of 10^4or more after binding to histones. Cationic polymers such as polylysine can also condense DNA from an extended conformation to highly compact structures of about 30–100 nm in diameter. Based on all these results, Gao and Huang proposed that the introduction of polylysine into the DC-chol/DNA complexes might alter its overall physical state, thereby improving its transfection activity (30). This idea was tested by examining the transfection efficiency of DC-chol/DNA complexes alone or DC-chol/DNA complexes with the addition of different amounts of polylysine. Incorporation of polylysine into the DC-chol/DNA complexes was found to increase transfection activity of DC-chol/DNA from 2- to 28-fold, depending on the cell lines being transfected (30). This new and efficient formulation was named LPD-I. In the following we examine some characteristics of LPD-I and factors that affect its transfection activity.

1. Effect of Polymers

Different cationic polymers were tested for their ability to form active LPD formulation. High molecular weight poly-L-lysine, poly-D-lysine, poly-L-ornithine, polybrene, and protamine (free base) were among the most active polymers tested. Low molecular weight poly-L-lysine, histone, and oligocations such as spermine were much less active than poly-L-lysine (MW 200,000). The reason for this low activity may be that these polymers do not have enough cation density and thus cannot form stable complexes with DNA under physiological conditions. Polyamidoamine dendrimeric polymer was completely inactive. It is unclear why this type of polymer could not enhance transfection activity, since the polymer itself can complex with DNA and mediate DNA delivery (31). Sorgi et al. later substituted the condensing agent from polylysine to protamine sulfate and found that the transfection activity of LPD-I increased about 5- to 10-fold (32). The finding is significant because protamine sulfate is approved by the FDA as an intravenous antidote for a heparin overdose. Therefore the substitution made the formulation more suitable for clinical applications. The amount of cationic polymer also affects the transfection activity. Among the three lipids tested (DC-chol/DOPE, Lipofectin, and LipofectAMINE), gene expression reached a peak level when the polylysine-to-DNA ratio (W/W) reached 1:2, and the gene expression started to decrease when the ratio exceeded 1:1 (Fig. 2). In addition, the way that cationic polymer was added to form the complex was also important for gene transfection. An active formulation could only be done when DNA was incubated with polylysine first, followed by the addition of lipid, or when DNA was mixed with polylysine and lipid simultaneously. If DNA was mixed with the lipid first, followed by the addition of polylysine, the resulting complexes had very low activity. This further indicates the importance of the formation of the polylysine/DNA core.

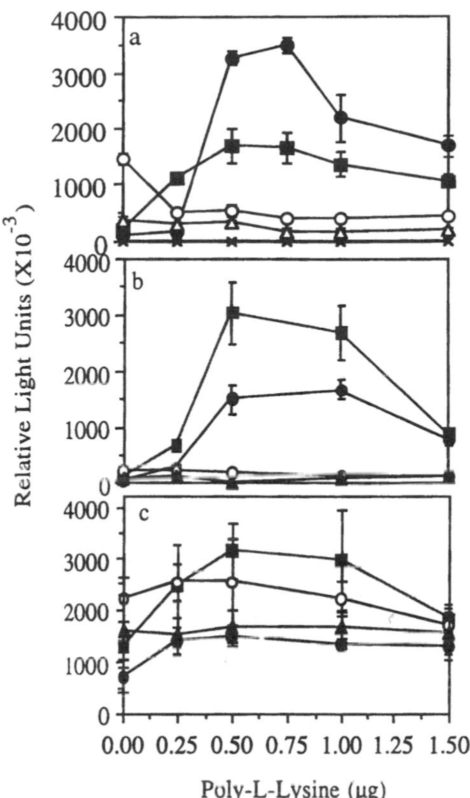

Figure 34.2 Potentiation of cationic liposome-mediated transfection by polylysine on CHO cells. CHO cells grown in a 48-well plate were lipofected for 4 h with 1 μg of pRSVL plasmid complexed with the indicated amount of DC-chol liposome (a), lipofectin (b), or lipofectamine (c) and with the indicated amount of polylysine. Thirty-six hours after the transfection, luciferase activity was measured over a period of 20 seconds using 2 μg of cellular proteins from each cell lysate. Symbols: (a) ● 3.3 nmol, ■ 6.6 nmol, ○ 10 nmol, △ 12.5 nmol, × no DC-chol/DOPE (4.5/5.5, mol/mol) liposome; (b) ● 4.5 nmol, ■ 6.0 nmol, ○ 8.5 nmol, ▲ 10 nmol Lipofectin; (c) ● 4 nmol, ■ 6 nmol, ○ 8.5 nmol, ▲ 10 nmol Lipofectamine. (From Ref. 30, used with permission.)

2. Effect of Lipid

The enhancing effect of polylysine on gene transfection with cationic lipids seemed to be a general phenomenon. Three different types of lipids were tested, DC-chol/DOPE, Lipofectin, and LipofectAmine. Compared to DNA/lipid complex without polylysine, DNA/lipid complex with the addition of polylysine in those three lipid formulations all resulted in different degrees of enhancement in gene transfection. The amount of lipid also affected the transfection activity of LPD formulations. When the lipid concentration was low, the enhancing effect by polylysine was not very significant. When the lipid concentration was high, the addition of polylysine actually decreased the transfection activity. The optimal amount of lipid seemed to be between 3 to 6 nmol per μg of DNA. Furthermore, the amount of lipid incorporated in the LPD ternary complex is very important for high activity. When LPD complex was subjected to a linear sucrose density gradient (5–30%) ultracentrifugation, a broad band was found in the middle of the gradient, due to the heterogeneity of LPD complex. Fractions located at the upper portion of the gradient were more enriched with liposome than those located at the lower portion of the gradient. LPD complex purified from the upper portion of the gradient had higher transfection activity than those purified from the lower portion of the gradient. Interestingly, the activity of LPD complex lacking a sufficient amount of lipid could be restored for high activity when additional DC-chol/DOPE liposomes were added. A possible explanation of this phenomenon is that the addition of lipid improves the stability of the polylysine/DNA complex. The attachment of lipid to the polylysine/DNA complex may provide a repulsive force and thus prevent the aggregation of polylysine/DNA complexes. In addition, excess lipid in the LPD complex may help the release of the complex from the endosome.

3. Structure of LPD-I

The structure of LPD-I was analyzed by negative stain electron microscopy. The LPD-I complex appeared to be small spherical particles consisting of a dense core surrounded by a lightly stained ring-shaped structure (Fig. 3) (30). The ring-shaped structure appeared to be lipid bilayers, because they had characteristics of a typical membrane staining pattern. From the structure, it is appeared that polylysine and DNA formed a very dense core and cationic lipid formed a shell covering the polylysine/DNA core. Though somewhat heterogeneous, most LPD-I particles were under 100 nm and very stable physically. No increase in size was observed even after 3 months of storage at 4°C. Compared with the structure of DC-chol/DNA complex (Fig. 4), the structure of LPD-I is very different. The more compact structure of LPD-I as compared to DC-chol/DNA complex may be due to the complete condensation of DNA by polylysine. Recent studies using cryoEM have also revealed a similar structure of LPD-I complex containing protamine sulfate instead of polylysine (data not shown).

Figure 34.4 Freeze-fracture electron micrograph of DC-Chol liposome/DNA complex. (From Ref. 25, used with permission.)

4. In Vivo gene transfer with LPD-I

Intraveneous injection of LPD-I resulted in gene expression in all major organs including the lung, liver, heart, speen, and kidney (Fig. 5). The lung and speen had the highest gene expression levels. The liver and heart showed modest gene expression, and the kidney had the lowest gene expression level (33). This pattern of gene expression is similar to that of cationic lipid/DNA complex. Gene expression in all organs was found to be DNA dose-dependent. However, a higher dose was always associated with higher toxicity. Gene expression was also transient. It normally lasted for 3 days to a week. The lipid-to:DNA charge ratio was important for efficient gene transfection in vivo. The activity of LPD-I increased as the charge ratio (+/−) increased. It was reported that a high charge ratio could overcome the inhibitory effect of serum (19). At low charge ratios, serum can neutralize the positive charges on LPD-I and thus block its entry into cells. As with in vitro results, LPD-I that is formed with different types of lipid resulted in different degrees of increase in activity. However, the type of lipid seems to have a more profound effect on in vivo gene transfer. For example, double-chain hydrocarbon-anchored lipids such as DOTAP were more active than cholesterol-anchored cationic lipids. Furthermore, the type of helper lipid also plays an important role in in vivo gene transfer. When DOPE was used as a helper lipid, the activity of LPD-I was significantly decreased. On the other hand, cholesterol can significantly increase the activity of LPD-I. Similar results were reported by other groups using

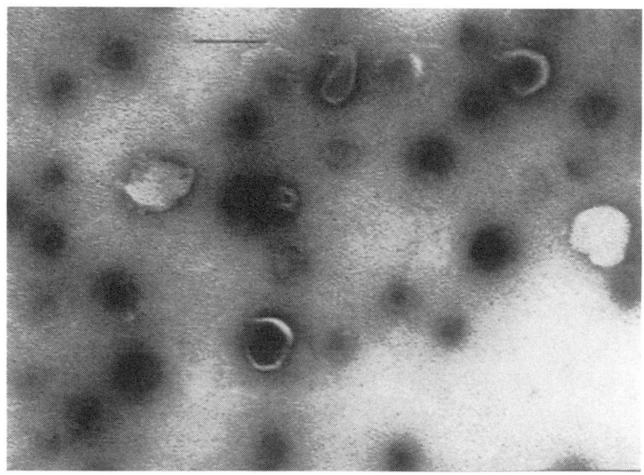

Figure 34.3 Electron micrographs of LPDI complexes. Purified LPDI was examined by negative stain electron micrography. The LPDI was prepared from 20 μg of DNA, 10 μg of polylysine (MW 26,500) and 800 nmol of DC-chol/DOPE liposomes (2/3 mol ratio) and purified with sucrose gradient ultracentrifugation. The samples were stained with 1% phosphotungstic acid. Bar = 100 nm. (From Ref. 30, used with permission.)

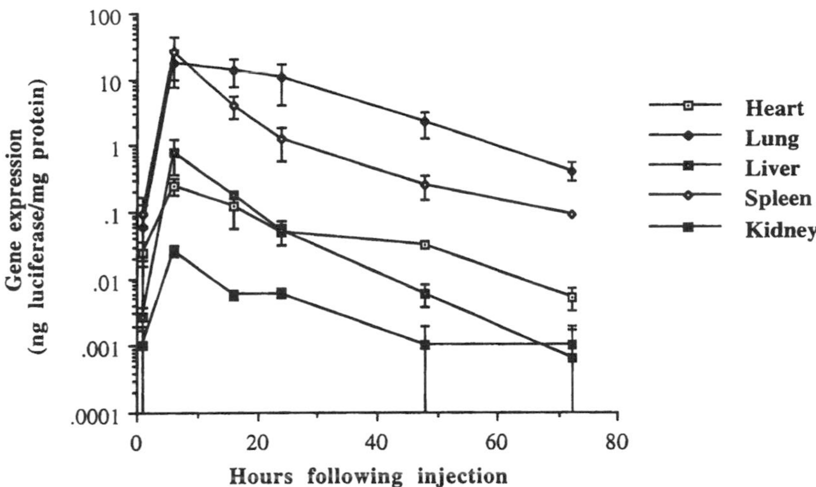

Figure 34.5 In vivo gene expression of LPD as a function of time. LPD was prepared as described in the legend to Fig. 2 and injected into the mice at 50 μg DNA per animal. At different times following the injection, mice were sacrificed and major organs were assayed for gene expression (n = 3). (From Ref. 33, used with permission.)

cationic lipid/DNA complex (34–36). It appeared that cholesterol could reduce the interaction of LPD-I with serum. Using SDS-PAGE, Li et al. found that the amount of serum associated with LPD-I formulated with cholesterol was much less than in LPD-I formulated with DOPE as a helper lipid (37). This may explain that LPD I formulated with cholesterol provides a higher activity. The amount of lipid is also very important for in vivo gene transfection. The activity of purified LPD-I by sucrose gradient centrifugation can be enhanced by adding free liposome. The excess of liposome is thought to prevent the release of DNA due to the interaction between LPD-I and serum. In addition, excess cationic liposomes also prolong the residence of LPD-I in the microvasculature of the lung.

B. Anionic Liposome-Entrapped, Polycation-Condensed DNA (LPD-II)

Although cationic liposomes are the most popular delivery vectors in nonviral gene therapy, there is always a strong desire to use anionic liposomes to deliver plasmid DNA. There are several reasons that anionic liposomes are more desirable for in vivo gene delivery. First, the cationic liposome/DNA complex is highly positively charged; it can interact nonspecifically with anionic proteins in vivo such as serum proteins. This kind of interaction normally causes a decrease in gene transfection activity and an increase in undesirable side effects. Second, cationic liposomes deliver DNA by nonspecific binding to the cell membrane. Therefore it is difficult to use cationic liposomes to target DNA to a specific cell type. In contrast, it is possible to conjugate a specific targeting ligand to the surface of an anionic liposome and deliver DNA to the target cells. Much effort has been devoted to developing an anionic liposome delivery system, but little success has been achieved. The major problem is the low encapsulation efficiency of DNA due to the large size and the high negative charge content of uncondensed DNA. Recently, in our laboratory, we have developed an anionic liposome formulation for the delivery of DNA without actually encapsulating DNA into the liposome (38). DNA was first condensed with an excessive amount of polylysine so that the resulting complex was slightly positively charged. Anionic liposomes were then added. Through electrostatic interaction, anionic liposomes could bind to the surface of the polylysine/DNA complex and form a shell on the polylysine/DNA core. The protocol and possible mechanism for the formation of this formulation is illustrated in Fig. 6. We call this type of formulation LPD-II.

To form LPD-II particles, a pH-sensitive, anionic liposome formulation composed of DOPE/cholesterol hemisuccinate (CHEMS)/folate-poly(ethylene)glycol-DOPE was added to the polylysine/DNA complexes. The resulting particles contained folate-targeting ligand. However, depending on the amount of anionic liposome used, the resulting LPD-II particles could either be positively or negatively charged. Negative-stain EM analysis revealed that the resulting LPD-II particles had a very similar morphology to the LPD-I particles. Both kinds of particle appeared to be spherical and consisted of a high-density core of polylysine/DNA complex coated with a low-density lipid layer. The LPD-II particles were used to transfect KB

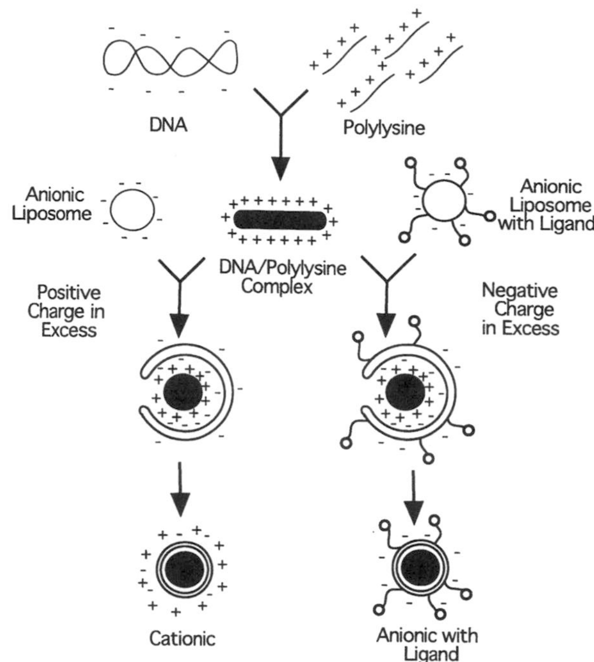

Figure 34.6 Possible mechanism for the formation of LPD-II. The targeting ligand in the current study is folate. (From Ref. 38, used with permission.)

cells, a cell line that over-expresses folate receptor. Both cationic and anionic LPDII particles could efficiently transfect KB cells. Both types of LPD-II were about 20- to 30-fold more efficient than the conventional DC-chol/DNA complex. However, anionic LPD-II particles containing folate as a targeting ligand were more efficient than the cationic LPD-II particles. Although both types of particle transfected cells efficiently, they apparently used different mechanisms. First, unlike the anionic LPD-II particle, the transfection by cationic LPD-II particles could not be blocked by excessive free folate. Second, another cell line, CHO, which does not express the folate receptor, could also be efficiently transfected by cationic LPD-II particles, but not the anionic LPD-II. This indicated that the transfection of cationic LPD-II was not through the receptor-mediated endocytosis but rather by nonspecific binding to the cell membrane. On the other hand, efficient transfection with anionic LPD-II particles depended on the presence of both the targeting ligand on the LPD-II and the receptor on the cell surface. All these results demonstrated that the anionic LPD-II particles could only transfect cells through the receptor-mediated endocytosis pathway.

C. Advantages and Limitations

Compared to the traditional lipid/DNA complex, the addition of a polycationic condensing agent to the cationic lipid/DNA complex offers several advantages. First, the LPD particles are much smaller than the simple lipid/DNA complex. LPD particles are more homogeneous, and most of them are about 100 nm in diameter, while the conventional lipid/DNA complexes are very heterogeneous and tend to form large aggregates around 1000 nm or larger. Since endocytosis is the major pathway responsible for the cellular uptake of DNA complexes, smaller particles such as LPD are favored for internalization. Second, LPD formulations are much more stable than cationic lipid/DNA complexes. As mentioned previously, LPD can be stored for more than three months at 4°C without any increase in size. Third, LPD formulations offer better protection to DNA from nuclease degradation. Even at the optimal ratio, DC-chol/DNA complex cannot completely protect DNA from degradation. In contrast, LPD formulations can completely protect DNA from enzymatic degradation. Thus the more functional plasmid DNA can be delivered into cells by LPD. Fourth, the addition of a condensing agent dramatically reduces the amount of cationic lipid needed for condensing, protecting, and delivering the plasmid DNA. The amount of cationic lipids needed to form the conventional lipid/DNA complex is usually too large for scaling up for clinical applications. Finally, highly positively charged polylysine and protamine may facilitate the transport of DNA from cytoplasm to nucleus, once the DNA is delivered into cells.

The advantages of LPD-II include that the preparation does not require purification and is a single-vial formulation. Unlike cationic delivery systems, LPD-II can be targeted to different organs simply by changing the ligand conjugated to the anionic lipid. For example, Transferrin (Tf) was conjugated to anionic liposomes, and LPD-II using transferrin as a targeting ligand has been developed. The LPD-II was able efficiently and selectively to deliver DNA into myoblasts and myotubes (39). Compared with the traditional anionic and neutral liposomal vectors, there is no problem with encapsulation. DNA is highly condensed and is quantitatively encapsulated without the use of excess amounts of lipid. Finally, since anionic lipids exist naturally, they are more biocompatible than cationic lipids. Taking all together, the anionic LPD-II may become the vector for tissue-specific gene delivery in vivo.

Although the LPD formulations increase gene expression both in vitro and in vivo, the level of gene expression is still lower than in viral vectors. Lack of targeting capacity is another limitation for LPD-I. And due to the cationic nature of LPD-I, targeting to a specific organ may not be efficient with a targeting ligand conjugated to these particles. LPD-II has the targeting capacity, but the instability of this formulation in serum greatly limits the application of the formulation in vivo. Improvement of the formulation is currently underway in our laboratory.

D. Toxicity of LPD

Recently, more and more reports suggest that cationic lipids are not so nontoxic or nonimmunogenic as previously thought (40,41). When introduced into the lung, cationic lipid/DNA complex usually causes an inflammatory response characterized by cellular infiltrates and elevated levels of several proinflammatory cytokines. In our studies, injection of a large dose of LPD-I can also cause severe toxicity that results in the deaths of animals. Injection of LPD was found to cause the secretion of several proinflammatory cytokines including TNFα, INFγ IL-6, and IL-12. However, injection of DNA, polylysine, or cationic lipid alone was not able to cause the production of these cytokines (42). Recent studies suggest that the presence of unmethylated CpG motif in the plasmid DNA is responsible for the production of the cytokines (43). Methylation of the CpG motif in the plasmid significantly reduced the cytokine response.

IV. CURRENT AND FUTURE DEVELOPMENT

A. Systemic delivery

Systemic delivery of therapeutic genes is the ultimate goal of gene delivery. Compared to in vitro gene transfer, in vivo gene delivery faces many more obstacles such as the size limitation of endothelial wall penetration, nonspecific interaction with components in the biological fluids and extracellular matrix, and binding to nontarget cell types. Although many studies have reported successful in vivo transgene expression in several organs, few studies have identified the cell type in the organ that has been transfected. Whether the transgene expression is in the parenchymal cells, which are normally the targets for gene transfer, remains unclear. Using fluorescence microscopy to track the fate of fluorescently labeled liposomes and DNA, McLean et al. found that most DNA uptake apparently was in endothelial cells of the microvasculature of the lung, lymph nodes, ovary, anterior pituitary, and adrenal medulla (44). Therefore transgene expressed mostly in endothelial cells and not in parenchymal cells. The result is not surprising, considering that most liposome/DNA complexes have diameters of about 100 to 500 nm. The endothelial barrier would normally prevent particles of this size from crossing, except in organs such as the spleen where the endothelium is discontinuous. Besides the endothelium barrier, the positive charged surface of the lipid/DNA complex also causes problems. Interaction of the lipid/DNA complex with plasma components such as serum proteins (37) and coagulation factors (45,46) may cause the growth of the particle and eventually obstruct the lung capillaries. Other problems associated with the positively charged lipid/DNA complex, such as erythrocyte aggregation, opsonization (47), and clearance by the reticuloendothelial system (RES), have also been reported (48). The results from in vivo gene transfer studies suggest that the ideal vectors for systemic application should be small enough to pass through physiological barriers and be specific for binding to the target cells but inert against both body fluids and unspecific interactions with tissues and cells, such as the RES.

B. Naked DNA as a Vector

Naked DNA appears to be a good vector for gene transfer, since it is small in size and polyanionic, which can diminish nonspecific interactions. However, due to its polyanionic charge and high molecular weight, naked DNA is not likely to be able to enter cells by itself. A condensing agent is thought to be required to condense DNA and make the particles positive in charge so that the DNA complex can be internalized by cells via absorptive endocytosis. However, many in vivo gene transfer studies have demonstrated that naked DNA can enter cells efficiently by some unknown mechanisms. Many different organs including the liver (49–51), lung (52–54), heart (55), kidney (50), thyroid gland (56), and tumor (57,58) have been successfully transfected by naked DNA. It is now clear that many cell types in vivo can be transfected by naked DNA, as long as a sufficient amount comes in contact with the cells for a prolonged period of time (52). Compared to lipid/DNA vectors, naked DNA is much smaller in size. It was found that supercoiled DNA in plectonomic form has superhelix dimensions of approximately 10 nm (59), which allows the naked DNA to cross microvascular walls and reach target cells. Systemic administration of the naked DNA is still limited due to the rapid degradation of naked DNA in the circulation. However, in some cases, systemic administration of naked DNA resulted in significant gene expression. For example, Liu et al. injected a large volume of DNA solution through the tail veils of mice and detected very high transgene expression in the liver (60). Budker et al. injected large volumes of DNA solution through the femoral artery while blocking the blood inflow and outflow and successfully transfected a large percentage of muscle fibers in the whole legs of rats (61).

C. Targeted Delivery of Naked DNA

Since many cells are capable of internalizing naked DNA, targeted delivery of naked DNA via receptor-mediated endocytosis may enhance the cellular uptake of DNA and therefore increase the gene expression. However, targeted delivery of naked DNA is hindered by the difficulty in con-

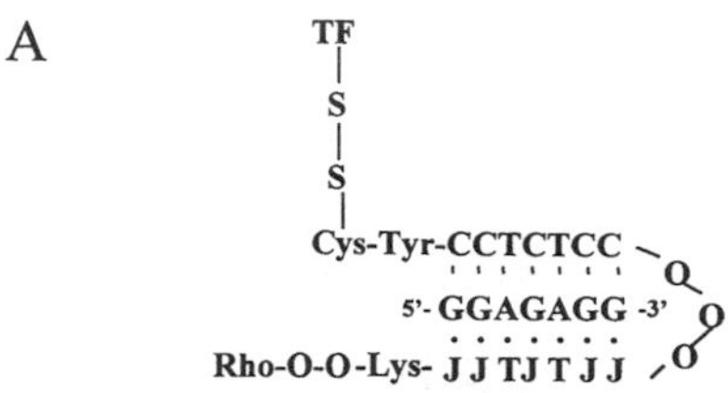

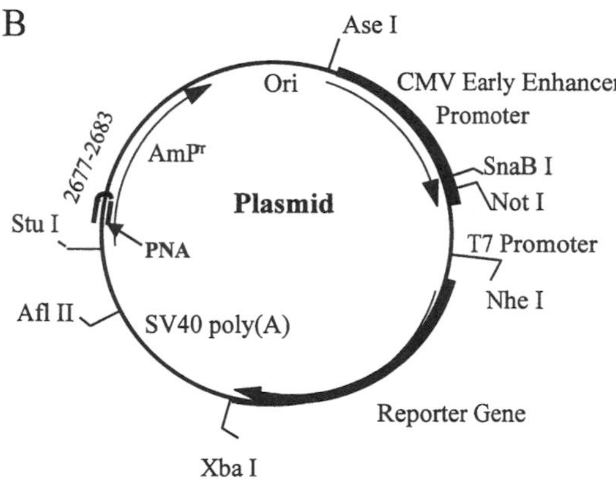

Figure 34.7 Schematic illustration of PNA/DNA/PNA triplex (A) and PNA binding to plasmid DNA (B). The PNA contains two stands that are linked through three linker units (8-amino-3, 6-dioxaoctanoic acid or O). One of the strands can hybridize with plasmid DNA in an antiparallel manner by Watson–Crick hydrogen bonding, while the other strand can bind in a parallel manner via Hoogsteen hydrogen bonding. Lysine and cystine are incorporated at the C and N terminals of PNA, respectively. Tf can be conjugated with PNA through a disulfide bond. (B) PNA can be designed to bind to a specific location in the plasmid, such as an antibiotic resistant gene, so that the binding of PNA does not affect gene activity. (From Ref. 63, used with permission.)

jugating a targeting ligand to naked DNA without affecting its activity. Current conjugation methods, whether they are chemical methods such as photolysis or enzymatic methods such as nick translation, attack DNA randomly and therefore may break the coding sequence on the plasmid. Recently, we have successfully used the peptide nucleic acid (PNA) as a linker to conjugate a targeting ligand to naked DNA without affecting the gene expression (63). PNA is an oligonucleotide analog containing normal DNA bases with a polyamide backbone (62). The bases on the PNA can bind to double-stranded DNA in a sequence-specific manner and form a stable triplex (Fig. 7). The polyamide backbone can be readily modified, which will allow conjugation of targeting ligands to PNA. We have designed a PNA sequence targeted to a unique sequence located at the antibiotic resistance region of a plasmid containing the gene luciferase, so that the binding of PNA to the plasmid will not affect the luciferase gene expression. A targeting ligand, transferrin, was conjugated to PNA via a disulfide bond. The binding of PNA to the naked DNA was highly sequence specific. Furthermore, the binding of PNA did not change the supercoiled conformation of DNA. This is important because DNA in the supercoiled form is much smaller than that in the relaxed form. The ability of Tf-PNA/DNA to mediate the cellular uptake of the naked DNA via receptor-mediated endocytosis was tested in myogenic cells in vitro. Compared to PNA/DNA complexes, Tf-PNA/DNA enhanced gene expression in both myoblasts and myotubes, and the enhancement effect was only seen when the whole complexes were negatively charged, although a polycationic polymer, polyethylenimine (PEI), was still required. Furthermore, the enhancement of gene expression could be inhibited by the excess amount of free transferrin, indicating that the enhancement was via Tf-mediated endocytosis.

V. SUMMARY AND CONCLUSION

Nonviral gene delivery systems have become more popular because of their favorable safety profiles, contrasting with those of viral vectors. Numerous nonviral vectors have been developed. However, cationic lipids remain the most widely used vectors in nonviral gene transfer. It is clear that the application of cationic lipid vectors for therapeutic use is limited by their low transfection efficiency. Several factors may be responsible for this drawback. First, the entry of the cationic lipid/DNA complex is inefficient. Second, the cationic lipid/DNA complex is not cell-type selective. Third, cationic lipids bind to DNA with relatively low efficiency, so that the amount of lipid required is often too large to be delivered for therapeutic purposes. Fourth, once internalized, the lipid/DNA complex remains trapped in the endosome compartment and is subsequently degraded. Substantial efforts have been devoted to overcome these hurdles by synthesizing new lipids and developing new formulations. The development of LPD represents one such effort. The LPD formulations not only increase the efficiency of gene transfer but also make the complex more stable, more homogenous, and smaller in size. LPD-II is the first formulation with an overall negative charge on the surface, while the DNA is efficiently protected. Although the development of this LPD formulation has overcome part of the problems associated with cationic lipid vectors, other problems such as the large size of the DNA/lipid complex and the lack of targeted delivery still remain. Continuous efforts are required for the development of new lipids and new formulations that may overcome these problems in the future.

The size limitation of endothelial wall penetration represents one of the major barriers for cationic lipid vectors. The small size of naked DNA offers an alternative to study gene transfection in those organs where cationic lipid vectors are normally unable to reach. While cationic lipids have been extensively studied as gene delivery vectors, studies on using naked DNA for gene transfection have lagged behind. The fact that naked DNA can enter many cell types efficiently has intrigued many investigators. Successful gene therapy with naked DNA will require specific delivery of the DNA to target organs. However, the property of naked DNA makes it difficult for the site-specific conjugation of the targeting ligand. PNA appears to be an ideal linker for conjugating a targeting ligand to naked DNA, since the binding of PNA to DNA is sequence specific and does not change the conformation of the naked DNA. However, many studies aiming at understanding the mechanism by which naked DNA enters into the cell and at overcoming the degradation of naked DNA must be carried out, before naked DNA can be widely used for in vivo gene transfer. On the other hand, further developments in cationic lipids and formulations may overcome the size limitation and degradation problems. For example, we have seen the development of new lipids and formulations capable of forming DNA/lipid complexes that contain only a single molecule of DNA. This type of formulation may become an efficient gene delivery vector in the future.

ACKNOWLEDGMENTS

The original work described in this chapter was supported by a grant from NIH CA 64654, CA 71731, DK 44935, PO1 AR 45925-01, a contract from Targeted Genetics Corporation, and a grant from the Muscular Dystrophy Association. We thank Dr. Michael Mokotoff for his critical reading of this paper.

REFERENCES

1. U. F. Greber, M. Willetts, P. Webster, A. Helenius. Stepwise dismantling of adenovirus 2 during entry into cells. *Cell 75*:477 (1993).
2. J. Herz, R. D. Gerard. Adenovirus-mediated transfer of low density lipoprotein receptor gene acutely accelerates cholesterol clearance in normal mice. *Proc. Natl. Acad. Sci. U.S.A. 90*:2812 (1993).
3. A. D. Miller. Retrovirus packaging cells. *Hum. Gene Ther. 1*:5 (1990).
4. P. K. Bandyopadhyay, S. Watanabe, H. M. Temin. Recombination of transfected DNAs in vertebrate cells in culture. *Proc. Natl. Acad. Sci. U.S.A. 81*:3476 (1984).
5. M. Ali, N. R. Lemoine, C. J. Ring. The use of DNA viruses as vectors for gene therapy. *Gene Ther. 1*:367 (1994).
6. T. R. Flotte, B. J. Carter. Adeno-associated virus vectors for gene therapy. *Gene Ther. 2*:357 (1995).
7. M. R. Capecchi. High efficiency transformation by direct microinjection of DNA into cultured mammalian cells. *Cell 22*:479 (1980).
8. F. L. Graham, A. J. v. d. Eb. A new technique for the assay of infectivity of human adenovirus 5 DNA. *Virology 52*: 456 (1973).
9. P. L. Felgner, T. R. Gadek, M. Holm, R. Roman, H. W. Chan, M. Wenz, J. P. Northrop, G. M. Ringold, M. Danielsen. Lipofection: a highly efficient, lipid-mediated DNA-transfection procedure. *Proc. Natl. Acad. Sci. U.S.A. 84*: 7413 (1987).
10. M. Chen, S. T. Compton, V. F. Coviello, E. D. Green, M. A. Ashlock. Transient gene expression from yeast artificial chromosome DNA in mammalian cells is enhanced by adenovirus. *Nucleic Acids Res. 25*:4416 (1997).
11. G. J. Nabel, E. G. Nabel, Z. Y. Yang, B. A. Fox, G. E. Plautz, X. Gao, L. Huang, S. Shu, D. Gordon, A. E. Chang. Direct gene transfer with DNA-liposome complexes in melanoma: expression, biologic activity, and lack of toxicity in humans. *Proc. Natl. Acad. Sci. U.S.A. 90*:11307 (1993).
12. H. Farhood, R. Bottega, R. M. Epand, L. Huang. Effect of cationic cholesterol derivatives on gene transfer and protein kinase C activity. *Biochim. Biophys. Acta 1111*:239 (1992).
13. X. Gao, L. Huang. Cationic liposome-mediated gene transfer. *Gene Ther. 2*:710 (1995).
14. E. R. Lee, J. Marshall, C. S. Siegel, C. Jiang, N. S. Yew, M. R. Nichols, J. B. Nietupski, R. J. Ziegler, M. B. Lane, K. X. Wang, N. C. Wan, R. K. Scheule, D. J. Harris, A. E. Smith, S. H. Cheng. Detailed analysis of structures and formulations of cationic lipids for efficient gene transfer to the lung. *Hum. Gene Ther. 7*:1701 (1996).
15. R. Bischoff, Y. Cordier, F. Perraud, C. Thioudellet, S. Braun, A. Pavirani. Transfection of myoblasts in primary culture with isomeric cationic cholesterol derivatives. *Anal. Biochem. 254*:69 (1997).
16. T. Akao, T. Osaki, J. Mitoma, A. Ito, T. Kunitake. Correlation between physicochemical characteristics of synthetic cationic amphiphiles and their DNA transfection ability. *Bull. Chem. Soc. Japan 64*:3677 (1991).
17. I. Solodin, C. S. Brown, M. S. Bruno, C. Y. Chow, E. H. Jang, R. J. Debs, T. D. Heath. A novel series of amphiphilic imidazolinium compounds for in vitro and in vivo gene delivery. *Biochemistry 34*:13537 (1995).
18. J. H. Felgner, R. Kumar, C. N. Sridhar, C. J. Wheeler, Y. J. Tsai, R. Border, P. Ramsey, M. Martin, P. L. Felgner. Enhanced gene delivery and mechanism studies with a novel series of cationic lipid formulations. *J. Biol. Chem. 269*:2550 (1994).
19. J. P. Yang, L. Huang. Overcoming the inhibitory effect of serum on lipofection by increasing the charge ratio of cationic liposome to DNA. *Gene Ther. 4*:950 (1997).
20. H. Farhood, N. Serbina, L. Huang. The role of dioleoyl phosphatidylethanolamine in cationic liposome mediated gene transfer. *Biochim. Biophys. Acta 1235*:289 (1995).
21. X. Zhou, L. Huang. DNA transfection mediated by cationic liposomes containing lipopolylysine: characterization and mechanism of action. *Biochim. Biophys. Acta 1189*:195 (1994).
22. A. Fasbender, J. Marshall, T. O. Moninger, T. Grunst, S. Cheng, M. J. Welsh. Effect of co-lipids in enhancing cat-

ionic lipid-mediated gene transfer in vitro and in vivo. *Gene Ther. 4*:716 (1997).
23. P. L. Felgner, G. M. Ringold. Cationic liposome-mediated transfection. *Nature 337*:387 (1989).
24. H. Gershon, R. Ghirlando, S. B. Guttman, A. Minsky. Mode of formation and structural features of DNA-cationic liposome complexes used for transfection. *Biochemistry 32*: 7143 (1993).
25. B. Sternberg, F. L. Sorgi, L. Huang. New structures in complex formation between DNA and cationic liposomes visualized by freeze-fracture electron microscopy. *FEBS Lett. 356*:361 (1994).
26. J. Gustafsson, G. Arvidson, G. Karlsson, M. Almgren. Complexes between cationic liposomes and DNA visualized by cryo-TEM. *Biochim. Biophys. Acta 1235*:305 (1995).
27. J. O. Radler, I. Koltover, T. Salditt, C. R. Safinya. Structure of DNA-cationic liposome complexes: DNA intercalation in multilamellar membranes in distinct interhelical packing regimes [see comments]. *Science 275*:810 (1997).
28. N. J. Caplen, E. W. Alton, P. G. Middleton, J. R. Dorin, B. J. Stevenson, X. Gao, S. R. Durham, P. K. Jeffery, M. E. Hodson, C. Coutelle. Liposome-mediated CFTR gene transfer to the nasal epithelium of patients with cystic fibrosis [see comments] [published erratum appears in *Nat. Med. 1*(3):272 (1995)]. *Nat. Med. 1*:39 (1995).
29. I. Baeza, M. Ibanez, C. Wong, P. Chavez, P. Gariglio, J. Oro. Possible prebiotic significance of polyamines in the condensation, protection, encapsulation, and biological properties of DNA. *Orig. Life Evol. Biosph. 21*:225 (1991).
30. X. Gao, L. Huang. Potentiation of cationic liposome-mediated gene delivery by polycations. *Biochemistry 35*: 1027 (1996).
31. J. Haensler, F. C. Szoka, Jr. Polyamidoamine cascade polymers mediate efficient transfection of cells in culture. *Bioconjug. Chem. 4*:372 (1993).
32. F. L. Sorgi, S. Bhattacharya, L. Huang. Protamine sulfate enhances lipid-mediated gene transfer. *Gene Ther. 4*:961 (1997).
33. S. Li, M. A. Rizzo, S. Bhattacharya, L. Huang. Characterization of cationic lipid-protamine-DNA (LPD) complexes for intravenous gene delivery. *Gene Ther. 5*:930 (1998).
34. Y. Liu, L. C. Mounkes, H. D. Liggitt, C. S. Brown, I. Solodin, T. D. Heath, R. J. Debs. Factors influencing the efficiency of cationic liposome-mediated intravenous gene delivery. *Nat. Biotechnol. 15*:167 (1997).
35. N. S. Templeton, D. D. Lasic, P. M. Frederik, H. H. Strey, D. D. Roberts, G. N. Pavlakis. Improved DNA: liposome complexes for increased systemic delivery and gene expression. *Nat. Biotechnol. 15*:647 (1997).
36. K. Crook, B. J. Stevenson, M. Dubouchet, D. J. Porteous. Inclusion of cholesterol in DOTAP transfection complexes increases the delivery of DNA to cells in vitro in the presence of serum. *Gene Ther. 5*:137 (1998).
37. S. Li, W. C. Tseng, D. B. Stolz, S. P. Wu, S. C. Watkins, L. Huang. Dynamic changes in the characteristics of cationic lipidic vectors after exposure to mouse serum: implications for intravenous lipofection. *Gene Ther. 6*:585 (1999).
38. R. J. Lee, L. Huang. Folate-targeted, anionic liposome-entrapped polylysine-condensed DNA for tumor cell-specific gene transfer. *J. Biol. Chem. 271*:8481 (1996).
39. W. G. Feero, S. Li, J. D. Rosenblatt, N. Sirianni, J. E. Morgan, T. A. Partridge, L. Huang, E. P. Hoffman. Selection and use of ligands for receptor-mediated gene delivery to myogenic cells. *Gene Ther. 4*:664 (1997).
40. R. K. Scheule, J. A. St George, R. G. Bagley, J. Marshall, J. M. Kaplan, G. Y. Akita, K. X. Wang, E. R. Lee, D. J. Harris, C. Jiang, N. S. Yew, A. E. Smith, S. H. Cheng. Basis of pulmonary toxicity associated with cationic lipid-mediated gene transfer to the mammalian lung. *Hum. Gene Ther. 8*:689 (1997).
41. J. J. Logan, Z. Bebok, L. C. Walker, S. Peng, P. L. Felgner, G. P. Siegal, R. A. Frizzell, J. Dong, M. Howard, Matalon. Cationic lipids for reporter gene and CFTR transfer to rat pulmonary epithelium. *Gene Ther. 2*:38 (1995).
42. S. Li, S. P. Wu, M. Whitmore, E. J. Loeffert, L. Wang, S. C. Watkins, B. R. Pitt, L. Huang. Effect of immune response on gene transfer to the lung via systemic administration of cationic lipidic vectors. *Am. J. Physiol 276*:L796 (1999).
43. Y. Tan, S. Li, B. R. Pitt, L. Huang. The inhibitory role of CpG immunostimulatory motifs in cationic lipid vector-mediated transgene expression in vivo [see comments]. *Hum. Gene Ther. 10*:2153 (1999).
44. J. W. McLean, E. A. Fox, P. Baluk, P. B. Bolton, A. Haskell, R. Pearlman, G. Thurston, E. Y. Umemoto, D. M. McDonald. Organ-specific endothelial cell uptake of cationic liposome–DNA complexes in mice. *Am. J. Physiol 273*: H387 (1997).
45. M. Ogris, S. Brunner, S. Schuller, R. Kircheis, E. Wagner. PEGylated DNA/transferrin-PEI complexes: reduced interaction with blood components, extended circulation in blood and potential for systemic gene delivery. *Gene Ther. 6*:595 (1999).
46. M. Ogris, P. Steinlein, M. Kursa, K. Mechtler, R. Kircheis, E. Wagner. The size of DNA/transferrin-PEI complexes is an important factor for gene expression in cultured cells. *Gene Ther. 5*:1425 (1998).
47. C. Plank, K. Mechtler, F. C. Szoka, Jr., E. Wagner. Activation of the complement system by synthetic DNA complexes: a potential barrier for intravenous gene delivery. *Hum. Gene Ther. 7*:1437 (1996).
48. R. I. Mahato, A. Rolland, E. Tomlinson. Cationic lipid-based gene delivery systems: pharmaceutical perspectives. *Pharm. Res. 14*:853 (1997).
49. G. Zhang, D. Vargo, V. Budker, N. Armstrong, S. Knechtle, J. A. Wolff. Expression of naked plasmid DNA injected into the afferent and efferent vessels of rodent and dog livers [see comments]. *Hum. Gene Ther. 8*:1763 (1997).
50. V. Budker, G. Zhang, S. Knechtle, J. A. Wolff. Naked DNA delivered intraportally expresses efficiently in hepatocytes. *Gene Ther. 3*:593 (1996).
51. M. A. Hickman, R. W. Malone, K. Lehmann-Bruinsma, T.

R. Sih, D. Knoell, F. C. Szoka, R. Walzem, D. M. Carlson, J. S. Powell. Gene expression following direct injection of DNA into liver. *Hum. Gene Ther. 5*:1477 (1994).

52. Y. K. Song, F. Liu, D. Liu. Enhanced gene expression in mouse lung by prolonging the retention time of intravenously injected plasmid DNA. *Gene Ther. 5*:1531 (1998).
53. K. B. Meyer, M. M. Thompson, M. Y. Levy, L. G. Barron, F. C. Szoka, Jr. Intratracheal gene delivery to the mouse airway: characterization of plasmid DNA expression and pharmacokinetics. *Gene Ther. 2*:450 (1995).
54. M. F. Tsan, J. E. White, B. Shepard. Lung-specific direct in vivo gene transfer with recombinant plasmid DNA. *Am. J. Physiol 268*:L1052 (1995).
55. H. Lin, M. S. Parmacek, G. Morle, S. Bolling, J. M. Leiden. Expression of recombinant genes in myocardium in vivo after direct injection of DNA. *Circulation 82*:2217 (1990).
56. M. L. Sikes, B. W. O'Malley, Jr., M. J. Finegold, F. D. Ledley. In vivo gene transfer into rabbit thyroid follicular cells by direct DNA injection. *Hum. Gene Ther. 5*:837 (1994).
57. J. P. Yang, L. Huang. Direct gene transfer to mouse melanoma by intratumor injection of free DNA. *Gene Ther. 3*: 542 (1996).
58. R. G. Vile, I. R. Hart. Use of tissue-specific expression of the herpes simplex virus thymidine kinase gene to inhibit growth of established murine melanomas following direct intratumoral injection of DNA. *Cancer Res. 53*:3860 (1993).
59. V. V. Rybenkov, A. V. Vologodskii, N. R. Cozzarelli. The effect of ionic conditions on the conformations of supercoiled DNA. I. Sedimentation analysis. *J. Mol. Biol. 267*:299 (1997).
60. F. Liu, Y. Song, D. Liu. Hydrodynamics-based transfection in animals by systemic administration of plasmid DNA. *Gene Ther. 6*:1258 (1999).
61. V. Budker, G. Zhang, I. Danko, P. Williams, J. Wolff. The efficient expression of intravascularly delivered DNA in rat muscle. *Gene Ther. 5*:272 (1998).
62. P. E. Nielsen, M. Egholm, R. H. Berg, O. Buchardt. Sequence-selective recognition of DNA by strand displacement with a thymine-substituted polyamide. *Science 254*: 1497 (1991).
63. K. Liang, E. P. Hoffman, L. Huang. Targeted delivery of plasmid DNA to myogenic cells via transferrin conjugated peptide nucleic acid (PNA). *Molecular Therapy*, in press.

35

Gene Delivery by Cationic Liposome–DNA Complexes

Nejat Düzgüneş
University of the Pacific, San Francisco, California
Sérgio Simões, Pedro Pires, and Maria C. Pedroso de Lima
University of Coimbra, Coimbra, Portugal

I. INTRODUCTION

During the last decade much interest has been generated in gene therapy approaches to the treatment of genetic metabolic disorders, cardiovascular diseases, cancer, immune system deficiencies, and infectious diseases (1–8). One of the major problems of gene therapy is the effective delivery and expression of therapeutic genes into target cells in vitro and in vivo (9–11). Viral vectors for gene delivery have certain advantages, including high levels of transduction in the case of adenoviruses, and efficient and stable integration of foreign DNA into a wide range of host genomes in the case of retroviruses. The usefulness of viral vectors is limited, however, by host immune and inflammatory reactions, toxicity, difficulty of large-scale production, size limit of the exogenous DNA (for adeno-associated viruses), random integration into the host genome (in the case of retroviruses), and the risk of inducing tumorigenic mutations and generating active viral particles through recombination (9,12–14). Therefore the use of non-viral gene delivery vehicles may become essential in many applications of gene therapy (15).

Cationic liposome–DNA complexes (''lipoplexes'') have been used extensively for in vitro and in vivo gene delivery and constitute a viable alternative to viral gene delivery vehicles (10,16–21). Reporter genes delivered via lipoplexes in vivo have been expressed in a number of tissues, and relatively long-term expression of several months has been obtained in some cases (22–29).

Lipoplexes present numerous advantages in gene delivery (9,13,20): (a) lipoplexes can protect DNA or RNA from inactivation or degradation in biological milieux; (b) lipopexes can carry large pieces of DNA, potentially up to chromosome size; (c) they are not immunogenic; (d) they are safe relative to viral vectors; (e) large-scale production of liposomes is easy compared to that of viruses; (f) they are highly versatile.

Lipoplexes currently have several limitations for gene therapy applications: (a) they utilize unnatural cationic lipids that can be toxic (16,30,31); (b) they incorporate suboptimal DNA concentrations; (c) their efficiency of gene delivery and expression is limited due to various cellular barriers (10,11,32); (d) many lipoplex formulations are heterogeneous and have a relatively large particle size; (e) they may have adverse interactions with negatively charged biological macromolecules in serum and on cell surfaces (33); (f) gene delivery by many types of lipoplexes is inhibited by serum (16,34–38).

II. GENE DELIVERY BY LIPOPLEXES IN CELL CULTURE

Cationic liposome–DNA complexes have been used for gene delivery in cell culture (in vitro) and in experimental animals (in vivo) since the landmark contribution of Felgner et al. (16) utilizing the cationic lipid N-[1-(2,3-dioleyl)-propyl]-N,N,N-trimethylammonium chloride (DOTMA) (Fig. 1). This study showed that small unilamellar liposomes composed of DOTMA and the zwitterionic lipid phosphatidylethanolamine (PE) (1 : 1) interact spontaneously with DNA to form lipid–DNA complexes, that the complexes mediate the expression of chloramphenicol acetyltransferase (CAT) from the pSV2cat plasmid in CV-1 or COS-7 kidney cell lines, that the level of gene expression depends on the amount of cationic liposome complexed with the DNA, and that transfection activity is much more effective than that achieved with calcium phosphate or DEAE–dextran. Felgner et al. (16) also reported that high concentrations of lipid are toxic depending on the cell type, time of incubation with the cells, and the cell density in the culture, and that the presence of serum during incubation of the complexes with cells is inhibitory to transfection. Based on the observation of diffuse fluorescence over the plasma membrane as well as intracellular membranes following incubation of rhodamine-labeled DOTMA/PE liposome–DNA complexes with cells, it was proposed that gene delivery occurs via fusion of the complexes with the plasma membrane. When phosphatidylcholine (PC) was substituted for PE in the cationic liposomes, only punctate fluorescence was observed; this was interpreted to reflect the inhibition of fusion by the presence of PC in the liposome membrane, consistent with the fusion-inhibitory properties of PC (39,40). DOTMA/dioleoyl-PE (DOPE) liposomes are commercially available as the reagent Lipofectin (Table 1).

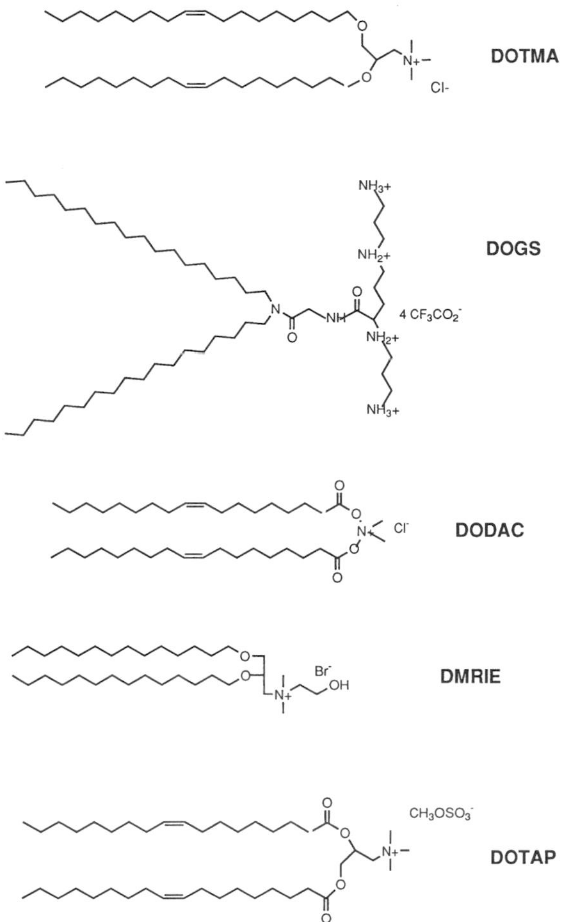

Figure 35.1 Structures of some cationic lipids used for transfection. DOTMA, N-[1-(2,3-dioleyl)propyl]-N,N,N-trimethylammonium chloride; DOGS, dioctadecylamidoglycyl-spermine; DODAC, dioctadecyldimethylammonium chloride; DMRIE, 1,2-dimyristyloxypropyl-3-dimethyl-hydoxyethylammonium bromide; DOTAP, 1,2-dioleoyloxy-3-(trimethylammonio)propane.

DNA complexed to lipopolyamine particles composed of spermine with covalently attached lipidic moieties [for example, dioctadecylamidoglycyl-spermine (DOGS)] (Fig. 1) could induce CAT activity in primary pituitary melanotrope cells (41). Efficient transfection required a net positive charge on the lipoplexes, but CAT activity could still be observed at theoretically net negative charges [e.g., lipid/DNA (+/−) charge ratio of 0.5]. In this system maximal transfection could be observed even after a 30 min incubation of the cells with complexes containing 5 μg DNA and a charge ratio (+/−) of 3. This transfection method was not cytotoxic, while complexes prepared with the cationic detergent dioctadecyldimethylammonium chloride (DODAC) (Fig. 1) at the same charge ratio were highly toxic. The DOGS lipoplexes were also reported to transfect various cell lines, including LMTK, Ras4, CHO, F9, BU4m S49, HeLa, and AtT-20 cells, as well as other primary chromaffin and neuronal cell cultures. In a later study it was shown that lipoplex formation by sequential addition of DOGS (''Transfectam'') to DNA, compared to immediate mixing of the two reagents, resulted in significant enhancement of transfection in 3T3, Hep-G2, K562, and COS-7 cells and primary fibroblasts (42). This effect was particularly pronounced when transfection was carried out in the presence of 10% serum. Preparation of the complexes in a small volume (10 μL vs. 100 μL) also resulted in significant enhancement of transfection in the presence of serum.

The cationic lipids 1,2-dioleoyloxy-3-(trimethylammonio)propane (DOTAP) (Fig. 1) and 1,2-dioleoyl-3-(4′-trimethylammonio)butanoyl-*sn*-glycerol (DOTB) formulated in liposomes with DOPE in a 1 : 3 ratio and complexed with a CAT expression plasmid (pSV2cat) could transfect CV-1 and 3T3 cells much more effectively than DEAE–dextran (43). The ability of these liposomes to undergo lipid mixing with negatively charged liposomes or

Table 35.1 Commercially Available Cationic Liposome Reagents

Commercial name	Supplier	Composition
Lipofectin	GIBCO BRL	DOTMA/DOPE (1:1)
Transfectam	Promega	DOGS
Tfx-50	Promega	Tfx-50/DOPE (1:1)
Lipofectamine	Gibco BRL	DOSPA/DOPE (1:0.65)
Lipofectamine PLUS	Gibco BRL	Lipofectamine + PLUS Reagent
Lipofectamine 2000	Gibco BRL	Proprietary
DMRIE-C	Gibco BRL	DMRIE/cholesterol (1:1)
Escort	Sigma	DOTAP/DOPE (1:1)
DC-Chol	Sigma	DC-Chol/DOPE (1:0.67)
Fugene 6	Roche Molecular Biochemicals	Proprietary
Vectamidine	BioTech Tools	diC14-amidine
LipofectACE	Gibco BRL	DDAB/DOPE (1:1)
LipoTAXI	Stratagene	Proprietary
Gene Porter 1 & 2	Gene Therapy Systems	Proprietary

cellular membranes was greatly inhibited by the presence of even small amounts of DNA (0.17–0.85 mol DNA phosphate per mol cationic lipid). This observation, coupled with the lack of correlation between the ability of different cationic lipid dispersions to intermix lipids with anionic liposomes and their ability to mediate transfection, suggested that DNA transfer to cells may not be mediated simply by the fusion of the cationic liposomes with cellular membranes.

Liposomes containing the cationic amphiphile, N-t-butyl-N′-tetradecyl-3-tetradecylaminopropionamidine (diC14-amidine, Vectamidine) complexed with a CAT-expressing plasmid transfected both adherent CHO cells and K562 cells in suspension (44). Maximal transfection activity was achieved within the first hour of incubation with cells.

Using the 2,3-dialkyloxypropyl quaternary ammonium backbone, Felgner et al. (45) developed a series of cationic lipid molecules to understand the relationship of structure to transfection activity. Liposomes composed of these compounds and DOPE (at a 1:1 mol ratio) were tested for their ability to transfect cultured COS-7 cells. The homologous series of dioleylalkyl (di-C18:1) lipids with hydroxyalkyl moieties with increasing chain lengths attached to the quaternary amine group showed transfection activity in the order ethyl > propyl > butyl > pentyl > DOTMA (no hydroxyalkyl group). Varying the alkyl chains attached to dimethyl hydroxyethyl ammonium bromide head groups resulted in the order of transfection activity dimyristyl (di-C14:0) > dioleyl (diC18:1) > dipalmityl (di-C16:0) > disteryl (di-C18:0). The lipid 1,2-dimyristyloxypropyl-3-dimethyl- hydoxyethylammonium bromide (DMRIE) (Fig. 1) in combination with DOPE (1:1) was more active than DOTMA/DOPE under a range of transfection conditions, and this was especially pronounced at lower lipid and DNA levels (under which conditions the potential for toxicity would also be reduced). The role of the colipid in transfection was also established in this study, the use of 50 mol% DOPE in conjunction with cationic lipids resulting in 2–5-fold higher transfection activity than formulations containing the same mol percentage of dioleoylphosphatidylcholine (DOPC) or those without any neutral lipid. The use of DOPE analogs with increasing acyl chain saturation resulted in progressively lower activity. These investigators found that lipoplexes prepared by the complexation of DNA with multilamellar liposomes of diameter 300–700 nm mediated about 2-fold higher levels of gene expression than those prepared with small unilamellar vesicles (50–100 nm).

The optimal charge ratio of cationic lipid to anionic DNA in lipoplexes prepared from DMRIE:DOPE was found to be 1.25 for the transfection of HeLa and primary airway epithelial cells (46), indicating that positively charged complexes interact with the negatively charged cell surface. The optimal charge ratio was found to be somewhat higher in studies with COS-7 cells (45).

Replacing the alcohol group on DMRIE with a primary amine produced a cationic lipid (β-aminoethyl-DMRIE) with transfection activity higher than that achieved with DMRIE over a broad range of cationic lipid and DNA concentrations (47). This lipid was maximally effective in the absence of the colipid DOPE.

The role of cationic lipid acyl chains in gene delivery was investigated using analogs of N-[1-(2,3-dioleoyloxy)-propyl]-N-[1-(hydroxy)ethyl]-N,N-dimethyl ammonium iodide (DORI) in liposomes containing 50 mol% DOPE (48). In NIH3T3 fibroblasts, Chinese hamster ovary (CHO) cells and human respiratory epithelial cells (16HBE14o-), optimal acyl chain length and symmetry varied with cell type. Cationic lipids with dissymmetric acyl chains (for example, 14:0, 12:0) mediated transfection activities equivalent or superior to the best symmetric acyl chain analogs. The phase transition temperature of the cationic lipid moiety did not appear to alter appreciably the transfection activity of the lipoplexes. For certain lipids and cells investigated in this study the use of multilamellar or small unilamellar liposomes provided an advantage over the other.

Utilizing the cationic lipid 3β[N-(n′, N′-dimethylaminoethane)-carbamoyl]cholesterol (DC-chol) (49) (Fig. 2) and DOPE, Caplen et al. (50) found that optimal weight ratios of DNA/cationic lipid (at fixed DNA concentrations, or at fixed liposome concentrations) were different among the various cell lines used in their study (monkey kidney SV40-transformed COS-7, cystic fibrosis nasal polyp CFNPE-9o-, and bronchial mucosal epithelial 16HBE14o-

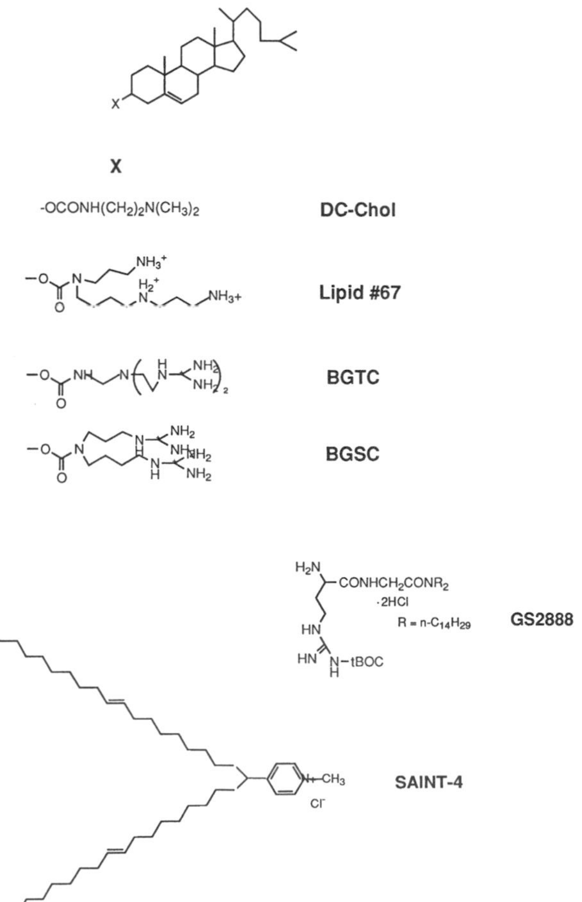

Figure 35.2 Structures of some cationic lipids used for transfection. DC-Chol, 3β[N-(n′,N′-dimethylaminoethane)-carbamoyl]cholesterol; lipid #67, 3-β-(sperminecarbamoyl) cholesterol; BGTC, bis-guanidium-tren-cholesterol; BGSC, bis-guanidium-spermidine-cholesterol; GS2888, Boc-arginine dimyristylamide; SAINT-4, dioleylmethylpyridinium chloride.

cells), but still within a fairly narrow range. Gene expression was enhanced when lipoplexes were formed at pH 9 in physiological solution, compared to pH 7 or plain water. Increased incubation time of the lipoplexes resulted in increased reporter gene expression, except at very high DNA and liposome concentrations where maximal activity was observed after 1 h.

Two guanidium derivatives of cholesterol, bis-guanidium-spermidine-cholesterol (BGSC) and bis-guanidium-tren-cholesterol (BGTC) (Fig. 2) were found to mediate transfection of a number of cell lines (human lung carcinoma A549, COS-7, canine kidney MDCK-1, rat pancreatic islet cell tumor RIB-m5F, rat osteosarcoma ROS, rat pheochromocytoma PC12, and mouse pituitary tumor AtT-20 cells), both when prepared in micellar form or formulated with DOPE as liposomes (51). The optimal charge ratio (+/−) of lipid/DNA phosphate was found to be between 6 and 8 for micellar BGTC, and about 3 for liposomal BGSC/DOPE.

The cationic lipid GS2888 (Boc-arginine dimyristylamide) (Fig. 2) was found to transfect COS-7 cells more effectively than Lipofectin, LipofectAce, Transfectam, or Lipofectamine (Table 1) in the presence of 10% serum at various charge ratios (52).

Pyridinium compounds with double hydrocarbon chains, formulated with equimolar DOPE, showed improved transfection activity and reduced toxicity over that of DOTMA/DOPE (53). The most active compound, SAINT-4, has two C18:1 chains with trans double bonds (Fig. 2). The percentage of cells transfected with liposomes composed of SAINT-2 (containing 85% cis double bonds) and DOPE was also higher than that obtained with DOTMA/DOPE.

Liposomes composed of the cationic phospholipid derivative ethylphosphatidylcholine mediated gene transfer to a number of cell lines, including BHK, CHO, and 3T3 cells (54). The advantages of this cationic lipid are that transfection is not inhibited by the presence of 10% serum during incubation with cells, that preincubation with 95% serum does not have any deleterious effect on the transfection capacity of the lipoplexes, and that the lipid is biodegradable.

Formation of lipoplexes in the presence of phosphate or calcium was found to affect their transfection activity. For example, preincubation of Lipofectin with phosphate buffer followed by complex formation with a reporter plasmid resulted in a 26-fold enhancement of luciferase gene expression (55). Optimal transfection was observed when complexes were prepared in the presence of 30–80 mM phosphate, and around pH 6. The presence of calcium in the concentration range 5–25 mM during complex preparation, using DODAC/DOPE liposomes, mediated up to 20-fold higher transfection levels in BHK cells compared

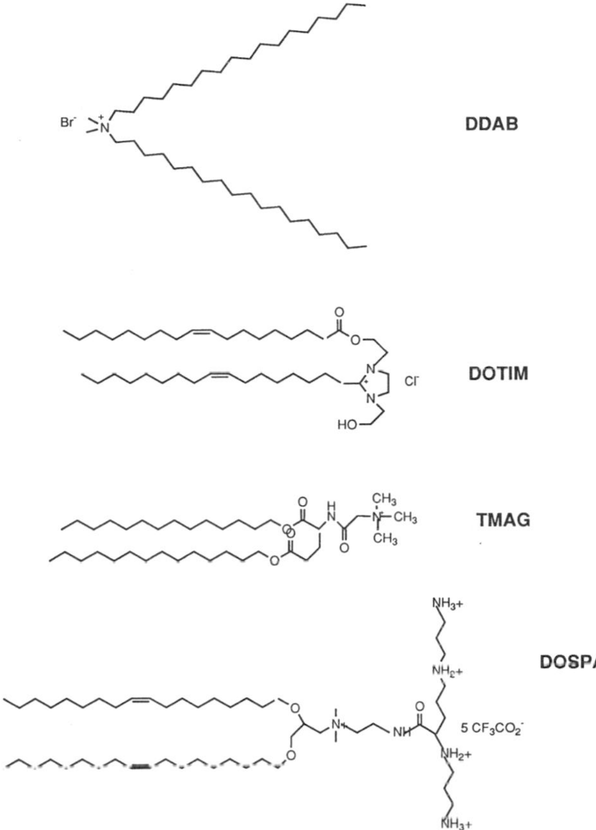

Figure 35.3 Structures of some cationic lipids used for transfection. DDAB, dimethyldioctadecylammonium bromide; DOTIM, 1-[2-(9(Z)-octadecenoyloxy)-ethyl]-2-(8(Z)-heptadecenyl)-3-(2-hydroxyethyl)-imidazolinium chloride; TMAG, didodecyl-N-(a-(trimethylammonio)acetyl)-D-glutamate chloride; DOSPA, 2,3 dioleyloxy-N-[2(spermine carboxaminino)ethyl]-N,N-dimethyl-1-propanaminium trifluroacetate.

to controls (56). This result was attributable in part to the enhancement of cellular uptake of lipoplexes and to a higher level of intact intracellular plasmid. Transfection of other cell lines such as CV-1 and SK-OV3 was also enhanced by this procedure. Calcium enhancement of transfection was not limited to DODAC/DOPE liposomes and was also observed with DOTMA/DOPE and dimethyldioctadecylammonium bromide (DDAB)/DOPE liposomes (Fig. 3).

III. DELIVERY OF PROTEINS BY CATIONIC LIPOSOMES

Cationic liposomes have also been used to deliver proteins to cells. A glucocorticoid receptor derivative, T7X556, a mammalian transcriptional regulator, was delivered to cultured cells by association with DOTMA liposomes, was localized rapidly in the nucleus, and mediated selective expression from glucocorticoid response element–linked promoters (57). Thus cationic liposome-mediated delivery may enable the functional analysis of proteins that have been generated by recombinant DNA techniques in bacteria or other mammalian cells.

A subsequent study showed the DOTMA/DOPE liposome-mediated delivery of prostatic acid phosphatase into human prostate carcinoma cells (58).

The human immunodeficiency virus type 1 (HIV-1) transactivator protein Tat was delivered into A431 cells expressing a reporter gene (CAT) under the control of the HIV-1 promoter, using DC-Chol or DOTMA-containing liposomes coincubated with the protein (59). The presence of DOPE in the liposome membrane was essential for protein delivery; liposomes containing phosphatidylcholine were ineffective. DOTMA/DOPE liposome-mediated delivery of Tat was confirmed in a system using a luciferase reporter gene in HeLa cells (60).

The delivery of the Rep68 protein of adenoassociated virus by complexation with the polycationic liposome Lipofectamine resulted in the rescue-replication of a codelivered inverted terminal repeat-flanked plasmid, as well its site-specific integration (61).

IV. GENE DELIVERY BY LIPOPLEXES IN VIVO

In the first demonstration of the in vivo activity of lipoplexes, intravenous or intratracheal injection of DOTMA/DOPE complexes with the plasmid pSV2CAT resulted in CAT expression in the lungs of mice (62). The authors speculated that the technique could be used to genetically engineer lung cells to produce intracellular or secreted proteins, with eventual ''therapeutic possibilities.''

Nabel et al. (22) injected lipoplexes containing a β-gal expression plasmid and DOTMA/DOPE liposomes into iliofemoral arteries of pigs via a catheter and found extensive gene expression after 4 days in the arterial tissue. Gene expression persisted for at least 6 weeks in this system. It was proposed that this method of gene delivery may be useful in the inhibition of smooth muscle proliferation or thrombus formation, which are complications of coronary angioplasty.

Transgene expression in the lungs was achieved by aerosolizing DOTMA/DOPE liposome complexes with the pCIS-CAT plasmid (23). The majority of airway epithelial and alveolar lining cells were transfected, and the gene product was detectable in the lung for at least 3 weeks. Intravenous injection of the same type of lipoplexes in mice resulted in transfection of lungs, spleen, liver, heart, kidney, lymph nodes, thymus, uterus, ovary, skeletal muscle, pancreas, bone marrow, stomach, small intestine, and colon (63). Metastatic lung tumors were also transfected by this procedure. The CAT gene was expressed in many cells in multiple tissues for at least 9 weeks after injection of the lipoplexes. No treatment-related toxicity was noted in injected animals. These authors suggested that specific tissues and cell types could be targeted by means of tissue-specific promoter-enhancer elements, by regional administration of lipoplexes, or by the use of targeting ligands.

Intravenous injection of DNA complexes with DC-Chol/DOPE liposomes into mice resulted in the localization of the DNA primarily in the lung and heart, as detected 2 to 3 weeks following injection (64). Complexes injected directly into experimental CT26 tumors caused DNA localization in the tumors. Lipoplex injection was found not to affect tissue-specific serum enzymes, tissue histology, or electrocardiography. When mice were injected with genes encoding murine MHC class I proteins, no significant immunopathology could be detected in various tissues, although an immune response was elicited (65). The introduction of various genes did not cause any gene localization in the testes or ovary, as detected by the polymerase chain reaction. Similar experiments were performed with DMRIE/DOPE–DNA complexes, which facilitated the injection of much higher doses of DNA without causing toxicity (66). The introduction of a foreign MHC gene (H-2Kd) into an MCA 205 tumor via DMRIE/DOPE liposomes resulted in a substantial antitumor response, while DC-chol/DOPE liposomes were ineffective.

Intraperitoneal injection of a CAT-expressing plasmid complexed to DOTMA/DOPE, DDAB/DOPE, or DOTAP/DOPE liposomes mediated gene expression in the spleen, lymph nodes, and liver (67). A large percentage of T lymphocytes from the paracortical region of lymph nodes was transfected. In addition, about 20% of bone marrow cells revealed intracellular CAT antigen.

Repeated intravenous injection of DDAB/DOPE–DNA complexes was shown to produce levels of transfection at least as high as that following a single injection, demonstrating that lipopexes are not neutralized by a host immune reaction (24). Among the different promoters tested (CMV immediate early promoter, SV40 early promoter, herpes simplex virus thymidine kinase promoter, and the adenovirus 2 major late promoter), the CMV promoter resulted in the expression of much higher levels of CAT in various tissues. The delivery of genes encoding murine granulocyte-macrophage colony stimulating factor and human granulocyte-colony stimulating factor resulted in significant levels of expression of these proteins in the circulation. The expression of β-gal under the CMV promoter, following aerosol delivery in DOTAP lipoplexes, was found to be more prolonged compared to that under the SV promoter (68). The CMV promoter was also found to be more effective in luciferase gene expression in vivo than the RSV promoter when complexed to DOGS/DOPE liposomes (28). The use of a human papovavirus (BKV)-derived episomal vector, however, resulted in prolonged expression in various tissues, up to 3 months (28). The episomal vector was shown to replicate extrachromosomally in the lungs 2 weeks following initial administration.

A comparison of the in vitro and in vivo transfection efficiencies of different cationic liposome reagents showed that the optimal lipid-to-DNA ratio for cell culture experiments did not translate to the highest transfection efficiencies (% of transfected cells) following intratumoral injection of lipoplexes containing the plasmid pCMVβIL-2 (69). DC-Chol was found to be the most efficacious cationic liposome in this study, which also reported that transfection efficiency was reduced as the size of 2E9 tumors increased. Interestingly, gene expression was inhibited by cationic liposomes upon injection into BL6 mouse melanoma tumors, while the size of the tumor did not affect gene expression by free DNA (70). Nevertheless, growth suppression of tumor xenografts was observed only upon injection of a therapeutic plasmid (expressing interleukin-2) complexed to DC-Chol liposomes, but not with naked DNA (71).

The role of linking polyamines to a cholesterol or other hydrophobic anchor in the transfection activity of lipoplexes in cystic fibrosis airway epithelial CFT1 cells and lungs was examined by Lee et al. (27). The orientation of the polyamine head group in relation to cholesterol was found to be highly significant, with lipids conjugated to spermine or spermidine in a T-shape configuration producing higher in vivo gene delivery than lipids coupled to the polyamines via a primary amine to generate a linear shape. Intranasal instillation of lipoplexes composed of 3-β-(sperminecarbamoyl) cholesterol (Lipid #67)/DOPE (Fig. 2) and a CAT plasmid produced lung levels of the enzyme 100-fold higher than that mediated by DMRIE/DOPE or DC-Chol/DOPE lipoplexes.

The inclusion of the nonionic surfactant Tween 80 in DC-chol-containing liposomes appeared to inhibit the formation of large DNA–lipid complexes and to resist the inhibitory effects of serum on transfection activity (72). Increasing the ratio of DOTMA to DNA, or DOTMA to Tween 80, enhanced gene expression in various organs, especially the lungs (73).

The use of DOTAP/cholesterol liposomes resulted in much higher levels of gene expression in the lungs after intravenous injection compared to DOTAP/DOPE, DDAB/cholesterol and DDAB/DOPE liposomes (29). Cryoelectron microscopy indicated that condensed DNA is sequestered inside invaginated DOTAP/cholesterol liposomes. Similarly, preparation of lipoplexes from multilamellar liposomes containing cholesterol and the cationic lipid 1-[2-(9(Z)-octadecenoyloxy)-ethyl]-2-(8(Z)-heptadecenyl)-3-(2-hydroxyethyl)-imidazolinium chloride (DOTIM) (Fig. 3) (74) resulted in much higher levels of gene expression in various tissues than that obtained with small unilamellar vesicles of the same composition, or with multilamellar liposomes containing DOPE (25). The uptake and retention of the delivered plasmid, as well as the circulation time, were much higher with the use of DOTIM/cholesterol multilamellar liposomes than that obtained with DOTIM/DOPE small unilamellar liposomes. The most commonly transfected cells were found to be vascular endothelial cells, monocytes, and macrophages, and the level of gene expression normalized to the amount of DNA taken up per tissue was three orders of magnitude higher in the lung than in the liver, suggesting that host tissue factors influence the efficiency of transgene expression.

V. GENE DELIVERY IN HUMANS

A gene encoding a foreign major histocompatibility complex protein (HLA-B7) was introduced into melanoma nodules in patients with advanced disease, using DC-Chol/DOPE liposomes (18). Immunohistochemistry showed that HLA-B7 was expressed in tumors. One patient even demonstrated regression of injected nodules This study indicated the feasibility, safety, and therapeutic potential of direct gene transfer in humans. A subsequent study employing DMRIE/DOPE lipoplexes demonstrated the presence of RNA encoding HLA-B7 or HLA-B7 protein in tumor nodules in nine of ten patients with stage IV melanoma (75). T cells migrated into treated tumors in six of seven patients, and tumor-infiltrating lymphocyte reactivity was enhanced in two of two patients analyzed. Local inhibition of tumor growth was noted in two patients.

DNA encoding the cystic fibrosis transmembrane conductance regulator (CFTR), complexed with DC-Chol/DOPE liposomes, was delivered to the nasal epithelium (76). Plasmid DNA and transgene-derived RNA were detected in the majority of treated patients. A modest restoration of function was achieved, with optimal activity being detected 3 days following transfection. Nevertheless, the function reverted to pretreatment levels by day 7. The authors suggested that transfection efficiency and the duration of expression need to be increased to achieve a therapeutic effect. Lipoplexes containing DOTAP were administered in a later clinical trial, and transgene DNA was detected by the polymerase chain reaction in 7 of 8 treated patients, while vector-derived CFTR mRNA was observed in 2 of the 7 patients (77). In 2 treated patients partial and sustained CFTR-related functional changes toward normal values were noted.

Delivery of the gene for α_1-antitrypsin to the respiratory epithelium of patients with α_1-antitrypsin deficiency resulted in the expression of the mRNA and the protein (78). The expression of the α_1-antitrypsin via this gene therapy approach, but not the delivery of the protein, had an anti-inflammatory effect. It is believed that this therapeutic effect is due to the localization of the transgene-generated α_1-antitrypsin within cells and intercellular spaces.

VI. MECHANISMS OF AND BARRIERS TO LIPOPLEX-MEDIATED GENE DELIVERY

A series of criteria have to be met and barriers to be overcome for lipoplexes to be able to deliver functional DNA to the nucleus (10,11). The cationic lipid component of lipoplexes must be able to condense DNA to enable its protection from nuclease degradation. Lipoplexes with any associated serum components must be able to bind to cell surfaces. Cell surface heparin/heparan sulfate-containing proteoglycans are thought to mediate the uptake of lipoplexes (79,80). In contrast, proteoglycan-deficient cells were found to internalize lipoplexes containing Lipofectamine as efficiently as wild type cells (81). At low +/− charge ratios, proteoglycan-deficient cells took up DNA and exhibited reporter gene expression to even higher levels than the wild-type cells. At higher charge ratios, however, gene expression in the mutant cells was inhibited, most likely because of the cytotoxic effect of the highly positively charged lipoplexes. Accordingly, addition of exogenous polyanionic glycosaminoglycans neutralized the cytotoxic effects of cationic liposomes in proteoglycan-deficient cells. Pretreatment of cells with enzymes that remove glycosaminoglycans or neuraminic acid did not affect the binding of liposomes containing the cationic lipid didodecyl-N-(a-(trimethylammonio)acetyl)-D-glutamate chloride (TMAG) (Fig. 3) to a macrophagelike cell line, while pretreatment of cells with trypsin reduced binding, indicating that trypsin-sensitive proteins are involved in the attachment of these liposomes to the cell surface (82).

Lipoplexes are thought to be internalized via endocytosis involving coated or uncoated endocytotic vesicles (32,83–87). Alternatively, the lipoplexes may undergo fusion with the plasma membrane with the appropriate topology to facilitate the intracytoplasmic delivery of DNA (10,16,87). Several studies have concluded that fusion (assessed by lipid mixing between lipoplexes and cellular membranes) is not correlated sufficiently with transfection (87,88). An alternative mechanism of DNA entry has been suggested to involve pore formation in the plasma membrane (89).

The endosome membrane has to be destabilized so that the lipoplex or DNA can gain access into the cytoplasm and avoid degradation in lysosomes. The destabilization of the endosomal membrane by the internalized lipoplexes has been proposed to induce the transbilayer movement of anionic lipids from the cytoplasmic monolayer to the lumenal monolayer (83). The interaction of these lipids with the cationic lipids in the lipoplex was proposed to result in the displacement of the DNA from its complex with the cationic lipid and its release into the cytoplasm. Fusion of DOPE-containing cationic lipid/oligonucleotide complexes with endosomal membranes under acidic conditions following endocytosis was observed by fluorescence microscopy (90), and this supports the hypothesis that DNA is released into the cytoplasm following endocytosis. Electron microscopic evidence for the destabilization of endosomes by various lipoplexes has also been presented (91). Pore formation at the endosomal membrane may be an alternative mechanism of escape of the complexes or of free DNA into the cytoplasm (10).

For successful transfection, the delivered DNA has to be transported to the nucleus and pass through the nuclear membrane. Since free DNA in the cytoplasm is degraded within a short period (92), it is essential for DNA to rapidly access the nucleus or continue to be protected from nucleases via complexation with the cationic lipid. The relative ease with which proliferating cell lines are transfected compared to nondividing primary cells can be attributed to enhanced nuclear permeability during mitosis (93). Whether DNA penetrates the nuclear membrane through pores via passive diffusion or active transport following nonspecific association with receptors for nuclear localization signal (NLS) peptides is not known. It is also not known if DNA in the cytoplasm is indeed plain or still associated with cationic lipid. Experiments utilizing lipoplexes or free plasmid microinjected into oocyte nuclei indicated that gene expression was inhibited completely when DNA was complexed with cationic lipids at a 5:1 lipid:DNA (w/w) ratio (approx. 1:1 +/− charge ratio), while at a lower ratio (1:1) (w/w) it was not (32). Clearly, most of the DNA has to be freed of its associated cationic lipids to be transcribed.

The presence of serum usually inhibits lipoplex-mediated gene delivery (16,34–38,94,95), although lipids can vary in their sensitivity to serum (96–99). The inhibition is most likely mediated by negatively charged components in serum (100). This inhibition is of major concern regarding the in vivo use of lipoplexes for gene delivery by intravenous injection. Lipoplexes interacting with mouse serum aggregated and disintegrated, accompanied by the release of DNA (101). Complexes containing DOPE disintegrated rapidly and were also ineffective in gene delivery in vivo.

Another physiologically relevant inhibitor of lipoplex-mediated transfection in the airways is pulmonary surfactant. Transfection of murine respiratory epithelial MLE-15 and human bronchial adenocarcinoma H441 cells by lipoplexes containing DMRIE/cholesterol or DOTAP/DOPE decreased as the concentration of natural pulmonary surfactant was increased (102). Incorporation of surfactant proteins SP-B or SP-C or the phospholipids dioleoylphosphatidylcholine and dioleoylphosphatidylglycerol in the cationic lipid formulation also inhibited transfection. These observations suggest that lung surfactant may be one of the barriers to gene transfer in the lung (102). Tsan et al. (103) found that transfection of rat fetal lung fibroblast RFL-6 and calf pulmonary artery endothelial CPAE cell lines, and rat primary type II alveolar epithelial cells by lipoplexes of DDAB, DOTAP, or DOTMA/DOPE, was inhibited by a naturally derived surfactant preparation (Survanta) and a synthetic surfactant (Exosurf). This finding was proposed as a partial explanation for the observation that the efficiency of gene delivery to the lungs by plain plasmid DNA was equivalent to that by lipoplexes (103).

Sulfated glycosaminoglycans, including heparan sulfate and chondroitin sulfate, completely inhibited transfection mediated by cationic liposomes such as DOTAP and DOTAP/cholesterol, while hyaluronic acid was not inhibitory (104). Nevertheless, it is interesting to note that DOPE-containing lipoplexes were not inhibited as much as those containing only cationic lipids by the sulfated glycosaminoglycans (104).

VII. STRATEGIES FOR THE ENHANCEMENT OF LIPOPLEX-MEDIATED GENE DELIVERY

Several strategies have been employed and are being developed to enhance the ability of lipoplexes to mediate transgene expression in mammalian cells. These include the attachment of a targeting ligand with the lipoplexes to promote cell binding and receptor-mediated endocytosis, the association of synthetic fusogenic peptides with the complexes to facilitate the destabilization of endosomes and release of the genetic material into the cytoplasm, the employment of cationic polypeptides to condense DNA, the association of albumin with cationic liposomes to confer serum resistance and enhanced gene transfer, and the use of nuclear localization signal (NLS) peptides to mediate the nuclear localization of the plasmid.

A. Lipoplex Targeting

Neutral plasmid–lipospermine complexes containing a trigalactolipid could transfect HepG2 hepatoma cells bearing the asialoglycoprotein receptor more efficiently than lipoplexes lacking the ligand (33). Galactosylated cholesterol derivatives also mediated higher levels of transfection of HepG2 cells when included in DC-Chol/DOPE liposomes (105). The association of monoclonal antibodies or ligands (asialofetuin) with lipoplexes increased both the selectivity and the efficiency of DNA delivery into cancer cells (106,107). Addition of succinylated asialofetuin to the surface of preformed DOTAP:Chol-DNA complexes resulted in high levels of transfection in the liver upon in vivo administration (29). Association of Lipofectin (DOTMA/DOPE) with complexes of integrin-targeting peptide and DNA resulted in a 100-fold enhancement of transfection activity and a substantial increase in transfection efficiency in a variety of cell lines (108). A further advantage of this type of complex was its tendency to form particles, unlike the fibrous network usually reported for Lipofectin/DNA complexes. The association of various lectins with Lipofectin followed by complexation with DNA resulted in the enhancement of transfection in a number of cell lines and could be inhibited by a competing sugar (109).

Association of transferrin with cationic liposomes followed by complexation with DNA resulted in ternary complexes with enhanced transfection in a variety of cells, including dividing and nondividing cells (110–113) (Fig. 4). Transfection was most effective with the use of optimized lipid/DNA (+/−) charge ratios at which the complexes presented a net negative charge. Studies on the mechanisms of gene delivery by such transferrin lipoplexes have suggested, however, that internalization of these ternary complexes is mediated by a nonspecific process (86) (Fig. 4). Transferrin was proposed also to facilitate endosome destabilization following acidification of the endosome lumen. Furthermore, triggering internalization of the lipoplexes through a nonspecific endocytotic process could be achieved by associating human serum albumin with cationic liposomes (38). It should be noted, however, that enhancement of the extent of binding and internalization of the lipoplexes did not necessarily result in the enhancement of transgene expression (11), indicating the importance of additional factors that facilitate the entry and expression of the transgene.

Initial complexation of DNA with poly(L-lysine) reduced the particle size of ternary complexes with DC-Chol/DOPE liposomes, conferred nuclease resistance to the DNA, and enhanced the transfection of different cell lines (114). The association of the polycationic peptide, protamine sulfate, with DNA followed by the addition of DC-Chol/DOPE, DOTAP/DOPE, or DOTMA/DOPE liposomes greatly increased gene expression in CHO cells compared to the plain lipoplexes (115). This effect was attributed to the ability of protamine to condense DNA efficiently, as well as to the presence of NLS peptides in its sequence. The cationic lipid–protamine–DNA formulation also mediated higher gene expression in mice than did DOTAP/DNA complexes (116). In an alternative protocol, the ternary complexes formed by the addition of protamine sulfate to DOTAP lipsomes followed by the addition of DNA were shown to be small, condensed particles (diameter ~ 135 nm), which exhibited even higher levels of in vivo gene expression (117). It was also shown that the use of cholesterol instead of DOPE as a helper lipid in cationic liposomes enhanced the in vivo transfection efficacy of cationic lipid–protamine–DNA complexes.

B. Use of Nuclear Localization Signal (NLS) Peptides

The studies of Zabner et al. (32) and Labat-Moleur et al. (118) have indicated that the nuclear membrane is a major barrier to lipoplex-mediated gene delivery to the nucleus. NLS peptides generally consist of a short stretch of basic amino acids in certain proteins and direct them to the nucleus through nuclear pore complexes. Several laboratories have investigated the potential of these peptides to facilitate DNA entry into the nucleus. DNA–NLS peptide complexes associated with DC-Chol/DOPE liposomes exhibited an eightfold enhancement at the optimal peptide concentration compared to plain lipoplexes (119). Aronsohn and Hughes (120) showed that the association of a NLS peptide with DOTAP/DOPE–DNA complexes enhanced transfection in SKnSH cells by about threefold. Zanta et al. (121) covalently coupled a single NLS sequence to capped DNA containing a hairpin oligonucleotide enriched with amino groups and showed a dramatic enhancement in the transfection of a variety of cell types, including nondividing cells, even with small amounts of DNA. This represents an important progress in the use of lipid-based vectors, taking into account the putative cytotoxic effects of cationic lipids at higher concentrations. An alternative technique to attach NLS peptides was described by Brandén et al. (122) and involved the use of a bifunctional peptide nucleic acid (PNA)–NLS peptide conjugate that could both hybridize with certain sites in the plasmid

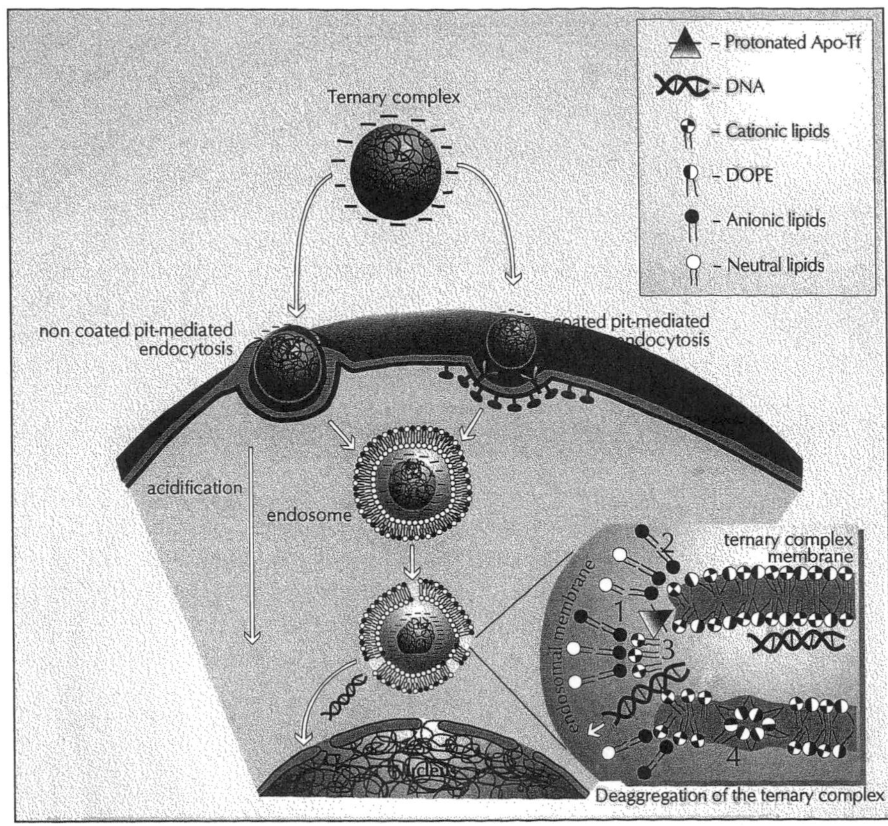

Figure 35.4 Proposed mechanism of cellular uptake and cytoplasmic delivery of DNA via transferrin lipoplexes. Lipoplexes are internalized either by receptor-mediated endocytosis via clathrin-coated pits or by phagocytosis via uncoated pits. Efficient delivery of DNA into the cytoplasm appears to be dependent on the acidification of the endosome. Acidification may trigger a series of reactions leading to the dissociation of DNA from the transferrin lipoplexes and to the destabilization of the endosome membrane. The protonation of apotransferrin causes a structural change that facilitates deaggregation of the complex. Protonated apotransferrin becomes membrane-active and destabilizes the endosomal membrane, thereby facilitating the transbilayer movement of anionic lipids from the cytoplasmic monolayer of the endosome membrane. Electrostatic interactions between these anionic lipids and cationic lipids induce the dissociation of the DNA from the lipoplexes and transformation of the DOPE into a nonbilayer phase, which further facilitates the intermixing of lipids between the lipoplex and endosome membranes. These events lead to the cytoplasmic delivery of dissociated or partially dissociated DNA, which then has to be transported into the nucleus for gene expression. (Adapted from Simões et al. (86).)

and transport DNA into the nucleus. Cell-specific nuclear translocation mechanisms were exploited by using plasmids that interact in the cytoplasm with particular transcription factors that facilitate nuclear import of the complex (123). Lipoplexes composed of one such plasmid (incorporating the SV40 enhancer/promoter region) and Lipofectin enhanced gene expression in nondividing CV1 cells 500-fold compared to lipoplexes containing only the CMV promoter. In another approach to nuclear targeting, the M9 sequence of heterogeneous nuclear ribonucleoprotein (hnRNP) A1 was conjugated to a cationic peptide scaffold derived from a scrambled sequence of the SV40 T-antigen consensus NLS to facilitate its binding to DNA. Lipoplexes prepared by complexing this construct with Lipofectamine enhanced transfection activity in bovine aortic endothelial cells 63-fold (124).

C. Facilitation of Gene Transfer by Membrane-Destabilizing or Fusogenic Peptides

Since lipoplexes are thought to enter cells primarily via endocytosis, it has been hypothesized that the use of peptides that can destabilize endosomes or facilitate the fusion of the lipoplexes with the endosomal membrane would enhance gene delivery. Incubation of COS-7 cells with a β-gal-expressing plasmid, and anionic or cationic derivatives of the N-terminal peptide of the HA2 subunit of the influenza virus fusion protein, hemagglutinin, in the presence of Lipofectin, resulted in the enhancement of transfection activity over that of Lipofectin alone by a factor of 2–7 (125). The levels of Transfectam (lipopolyamine)-mediated transfection of H225 human melanoma cells could be increased by up to 1000-fold [over that obtained with a suboptimal charge-equivalent (1.5) of Transfectam/DNA] by adding the hemagglutinin-derived peptide INF6 to the preformed lipoplexes (126). The association of the pH-sensitive peptide GALA (127,128) with DOTAP/DOPE (1:1) liposomes before complexation with plasmid DNA resulted in about a 10-fold enhancement in luciferase expression in COS-7 cells at the optimal charge ratio of 1/1 (+/−) (111). For net negatively charged complexes (1/2), however, the enhancement in transfection activity was several orders of magnitude, since gene transfer by plain lipoplexes at this charge ratio was very limited. GALA lipoplexes were considerably more effective in gene transfer to COS-7 cells than to HeLa cells, presumably because internalization of the complexes was not rate limiting in the former (111). The combined use of GALA and transferrin or albumin in conjunction with cationic liposomes resulted in considerably higher levels of gene expression in primary, nondividing human macrophages compared to plain lipoplexes or ternary lipoplexes containing transferrin or albumin (38,112).

D. Serum-Resistant Lipoplexes

Crook et al. (95) reported that the inclusion of cholesterol in DOTAP liposomes (at 2:1 to 1:2 weight ratios, DOTAP:cholesterol) mediated substantial transfection in COS-7 cells, while gene transfer by DOTAP lipoplexes was almost completely abolished in the presence of 40% serum. They also showed that more DOTAP/cholesterol than DOTAP lipoplexes were attached to and taken up by cells, although the differences did not appear to be sufficient to explain the drastic difference in gene expression mediated by the two complexes in 40 or 80% serum. Cholesterol-containing lipoplexes were shown to disintegrate much slower in serum than those containing DOPE (101).

Increasing the charge ratio of the cationic lipid to DNA was found to overcome the inactivation by 20% serum of gene delivery by DC-Chol and DOTAP/DOPE lipoplexes (100). Prolonging the incubation time of the cationic liposome–DNA complexes before addition to cultured cells, a process termed ''maturation,'' also rendered the lipoplexes resistant to the inhibitory effects of 20% serum (35). This maturation was observed for complexes containing monovalent cationic lipids such as DC-Chol, DOTAP, and DOTMA, but not for those containing the multivalent Lipofectamine. This process was accelerated by high charge ratios, high concentration of the lipoplexes, and elevated temperature. Lipoplexes containing DOGS/DOPE were found to be resistant to the effect of 10% serum, although complexes prepared with Lipofectamine were greatly inhibited in their transfection activity in HeLa cells (99).

Resistance to the inhibitory effects of serum was also conferred by the association of human transferrin or human serum albumin with DOTAP/DOPE or DOTAP/cholesterol liposomes before complexation with DNA (37,38,111). Increasing concentrations of serum actually enhanced the transfection of cultured cells by these ternary lipoplexes (37,38).

E. Steric Stabilization

Liposomes sterically stabilized by the incorporation of poly(ethylene glycol)-phosphatidylethanolamine (PEG-PE) are known to have prolonged circulation in vivo, most likely because they are not readily opsonized and are thus not taken up rapidly by the mononuclear phagocyte system (129–131). Lipoplexes prepared from cationic liposomes containing PEG-PE were stable upon prolonged storage and mediated gene delivery in vivo (132). However, in vitro studies performed in our laboratory have shown that lipoplexes prepared from cationic liposomes composed of DOTAP:DOPE:DSPE-PEG 2000 (1:1:0.02) were essentially inactive, while complexation with transferrin partially overcame this inhibitory effect (133). Inclusion of 5% PEG-PE in DODAC/DOPE liposomes completely inhibited transfection; this effect could be ascribed only partially to the inhibition of cellular uptake, suggesting that internalization of the complexes or subsequent steps are also affected by PEG-PE (11). A relatively short-chain PEG–ceramide conferred prolonged circulation to liposomes with a low cationic lipid content and entrapping DNA (employing a detergent dialysis procedure) (134–136). The PEG–ceramide could dissociate from the complexes inside endosomes, thus removing the inhibitory effect of the PEG from the liposomes.

CONCLUDING REMARKS

The studies outlined above indicate the progress made in gene delivery using cationic liposomes since the first report in this area in 1987 (16). One aspect of this nonviral method of gene delivery that has been rather puzzling is the difference in activity observed in vitro and in vivo. Nevertheless, it should be noted that experiments in vitro without even the use of physiologically relevant levels of serum should not be expected to reflect accurately the in vivo conditions, let alone the high complexity of biodistribution following intravenous delivery of the vectors. Although cell culture results obtained in serum-free conditions cannot be translated directly to in vivo gene delivery, they can at least indicate which lipoplexes have the potential to be useful for applications in animal models or humans. It is advisable to examine the transfection activity of such lipoplexes in culture in the presence of high concentrations of serum before embarking on expensive animal experiments (37,38,111).

Although an enormous amount of work has been done on developing new cationic lipids and their derivatives, as well as on the characteristics of their complexes with DNA, the mechanisms by which they facilitate the delivery of DNA and gene expression are not well understood. Clearly much basic research in this area is needed to understand how lipoplexes interact with biological milieux, how they destabilize endosomes to enter the cytoplasm, where and how the DNA dissociates from the cationic lipids, and how the lipoplexes or free DNA reach and enter the nucleus.

ACKNOWLEDGMENTS

The support of PRAXIS/PCNA/BIO/45/96, Portugal, Grant BIO4-CT97-2191 from the European Union, and a NATO Collaborative Linkage Grant CLG.976106 are gratefully acknowledged.

REFERENCES

1. W. F. Anderson. Human gene therapy. *Nature 392 (suppl.)*:25 (1998).
2. W. F. Anderson. The best of times, the worst of times. *Science 288*:627 (2000).
3. T. J. Davern, II, B. F. Scharschmidt. Gene therapy for liver disease. *Dig. Dis. 16*:23 (1998).
4. P. Demoly, M. Mathieu, D. T. Curiel, P. Godard, J. Bousquet, F. B. Michel. Gene therapy strategies for asthma. *Gene Therapy 4*:507 (1997).
5. P. Sinnaeve, O. Varenne, D. Collen, S. Janssens. Gene therapy in the cardiovascular system: an update. *Cardiovasc. Res. 44*:498 (1999).
6. I. M. Verma, N. Somia. Gene therapy—promises, problems and prospects. *Nature 389*:239 (1997).
7. B. A. Bunnell, R. A. Morgan. Gene therapy for infectious diseases. *Clin. Microbiol. Rev. 11*:42 (1998).
8. M. Cavazzana-Calvo, S. Hacein-Bey, G. de Saint Basile, F. Gross, E. Yvon, P. Nusbaum, F. Selz, C. Hue, S. Certain, J.-L. Casanova, P. Bousso, F. Le Deist, A. Fischer. Gene therapy of human severe combined immunodeficiency (SCID)-X1 disease. *Science 288*:669 (2000).
9. P. R. Clark, E. M. Hersh. Cationic lipid-mediated gene transfer: current concepts. *Curr. Opin. Mol. Therapeut. 1*: 158 (1999).
10. S. Simões, P. Pires, N. Düzgüneş, M. C. Pedroso de Lima. Cationic liposomes as gene transfer vectors: barriers to successful application in gene therapy. *Curr. Opin. Mol. Therapeut. 1*:147 (1999).
11. M. B. Bally, P. Harvie, F. M. P. Wong, S. Kong, E. K. Wasan, D. L. Reimer. Biological barriers to cellular delivery of lipid-based DNA carriers. *Adv. Drug Deliv. Rev. 38*:291 (1999).
12. B. C. Trapnell, M. Gorziglia. Gene therapy using adenoviral vectors. *Curr. Opin. Biotechnol. 5*:617 (1994).
13. A. Singhal, L. Huang. Gene transfer in mammalian cells using liposomes as carriers. *Gene Therapeutics: Methods and Applications of Direct Gene Transfer* (J. A. Wolf, ed.). Birkhäuser, Boston, 1994, p. 118.
14. N. S. Templeton, D. D. Lasic. New directions in liposome gene delivery. *Mol. Biotechnol. 11*:175 (1999).
15. D. A. Treco, R. F. Selden. Non-viral gene therapy. *Mol. Med. Today 1*:314 (1995).
16. P. L. Felgner, T. R. Gadek, M. Holm, R. Roman, H. W. Chan, M. Wenz, J. P. Northrop, G. M. Ringold, M. Danielsen. Lipofection: a highly efficient, lipid-mediated DNA-transfection procedure. *Proc. Natl. Acad. Sci. USA 84*: 7413 (1987).
17. N. Düzgüneş, P. L. Felgner. Intracellular delivery of nucleic acids and transcription factors by cationic liposomes. *Methods Enzymol. 221*:303 (1993).
18. G. J. Nabel, E. G. Nabel, Z. Yang, B. A. Fox, G. E. Plautz, X. Gao, L. Huang, S. Shu, D. Gordon, A. E. Chang. Direct gene transfer with DNA-liposome complexes in melanoma: expression, biologic activity, and lack of toxicity in humans. *Proc. Natl. Acad. Sci. USA 90*:11307 (1993).
19. X. Gao, L. Huang. Cationic liposome-mediated gene transfer. *Gene Therapy 2*:710 (1995).
20. D. D. Lasic, N. S. Templeton. Liposomes in gene therapy. *Adv. Drug Deliv. Rev. 20*:221 (1996).
21. J. Smith, Y. L. Zhang, R. Niven. Toward development of a non-viral gene therapeutic. *Adv. Drug Deliv. Rev. 26*: 135 (1997).
22. E. G. Nabel, G. Plautz, G. J. Nabel. Site-specific gene expression in vivo by direct gene transfer into the arterial wall. *Science 249*:1285 (1990).
23. R. Stribling, E. Brunette, D. Liggitt, K. Gaensler, R. Debs. Aerosol gene delivery in vivo. *Proc. Natl. Acad. Sci. USA 89*:11277 (1992).
24. Y. Liu, D. Liggitt, W. Zhong, G. Tu, K. Gaensler, R. Debs. Cationic liposome-mediated intravenous gene delivery. *J. Biol. Chem. 270*:24864 (1995).
25. Y. Liu, L. C. Mounkes, H. D. Liggitt, C. S. Brown, I. Solodin, T. D. Heath, R. J. Debs. Factors influencing the efficiency of cationic liposome-mediated intravenous gene delivery. *Nat. Biotechnol. 15*:167 (1997).
26. T. Takehara, N. Hayashi, Y. Miyamoto, M. Yamamoto, E. Mita, H. Fusamoto, T. Kamada. Expression of the hepa-

titis C virus genome in rat liver after cationic liposome-mediated in vivo gene transfer. *Hepatology 21*:746 (1995).
27. E. R. Lee, J. Marshall, C. S. Siegel, C. Jiang, N. S. Yew, M. R. Nichols, J. B. Nietupski, R. J. Ziegler, M. B. Lane, K. X. Wang, N. C. Wan, R. K. Scheule, D. J. Harris, A. E. Smith, S. H. Cheng. Detailed analysis of structures and formulations of cationic lipids for efficient gene transfer to the lung. *Hum. Gene Ther. 7*:1701 (1996).
28. A. R. Thierry, Y. Lunardi-Iskandar, J. L. Bryant, P. Rabinovich, R. C. Gallo, L. C. Mahan. Systemic gene therapy: biodistribution and long-term expression of a transgene in mice. *Proc. Natl. Acad. Sci. USA 92*:9742 (1995).
29. N. S. Templeton, D. D. Lasic, P. M. Frederik, H. H. Strey, D. D. Roberts, G. N. Pavlakis. Improved DNA: liposome complexes for increased systemic delivery and gene expression. *Nat. Biotechnol. 15*:647 (1997).
30. K. Konopka, E. Pretzer, P. L. Felgner, N. Düzgüneş. Human immunodeficiency virus type-1 (HIV-1) infection increases the sensitivity of macrophages and THP-1 cells to cytotoxicity by cationic liposomes. *Biochim. Biophys. Acta 1312*:186 (1996).
31. K. Konopka, G. S. Harrison, P. L. Felgner, N. Düzgüneş. Cationic liposome-mediated expression of HIV-regulated luciferase and diphtheria toxin A genes in HeLa cells infected with or expressing HIV. *Biochim. Biophys. Acta 1356*:185 (1997).
32. J. Zabner, A. J. Fasbender, T. Moninger, K. A. Poellinger, M. J. Welsh. Cellular and molecular barriers to gene transfer by a cationic lipid. *J. Biol. Chem. 270*:18997 (1995).
33. J.-S. Remy, A. Kichler, V. Mordvinov, F. Schuber, J.-P. Behr. Targeted gene transfer into hepatoma cells with lipopolyamine-condensed DNA particles presenting galactose ligands: a stage toward artificial viruses. *Proc. Natl. Acad. Sci. USA 92*:1744 (1995).
34. F. Van Bambeke, A. Kerkhofs, A. Schanck, C. Remacle, E. Sonveaux, P. M. Tulkens, M. P. Mingeot-Leclercq. Biophysical studies and intracellular destabilization of pH-sensitive liposomes. *Lipids 35*:213 (2000).
35. J.-P. Yang, L. Huang. Time-dependent maturation of cationic liposome–DNA complex for serum resistance. *Gene Therapy 5*:380 (1998).
36. O. Zelphati, L. S. Uyechi, L. G. Barron, F. C. Jr. Szoka. Effect of serum components on the physico-chemical properties of cationic lipid/oligonucleotide complexes and on their interactions with cells. *Biochim. Biophys. Acta 1390*:119 (1998).
37. C. Tros de Ilarduya, N. Düzgüneş. Efficient gene transfer by transferrin lipoplexes in the presence of serum. *Biochim. Biophys. Acta 1463*:333 (2000).
38. S. Simões, V. Slepushkin, P. Pires, R. Gaspar, M. C. Pedroso de Lima, and N. Düzgüneş. Human serum albumin enhances DNA transfection by lipoplexes and confers resistance to inhibition by serum. *Biochim. Biophys. Acta 1463*:459 (2000).
39. N. Düzgüneş, J. Wilschut, R. Fraley, D. Papahadjopoulos. Studies on the mechanism of membrane fusion: role of head-group composition in calcium- and magnesium-induced fusion of mixed phospholipid vesicles. *Biochim. Biophys. Acta 642*:182 (1981).
40. N. Düzgüneş. Molecular mechanisms of membrane fusion. *Trafficking of Intracellular Membranes. From Molecular Sorting to Membrane Fusion* (M. C. Pedroso de Lima, N. Düzgünes, D. Hoekstra, eds.). Springer-Verlag, Berlin, 1995, p. 97.
41. J.-P. Behr, B. Demeneix, J.-P. Loeffler, J. Perez-Mutul. Efficient gene transfer into mammalian pituitary cells with lipopolyamine-coated DNA. *Proc. Natl. Acad. Sci. USA 86*:6982 (1989).
42. O. Boussif, M. A. Zanta, J.-P. Behr. Optimized galenics improve in vitro gene transfer with cationic molecules up to 1000-fold. *Gene Therapy 3*:1074 (1996).
43. R. Leventis, J. R. Silvius. Interactions of mammalian cells with lipid dispersions containing novel metabolizable cationic amphiphiles. *Biochim. Biophys. Acta 1023*:124 (1990).
44. J.-M. Ruysschaert, A. El Ouahabi, V. Willeaume, G. Huez, R. Fuks, M. Vandenbranden, P. Di Stefano. A novel cationic amphiphile for transfection of mammalian cells. *Biochem. Biophys. Res. Commun. 203*:1622 (1994).
45. J. H. Felgner, R. Kumar, C. N. Sridhar, C. J. Wheeler, Y. J. Tsai, R. Border, P. Ramsey, M. Martin, P. L. Felgner. Enhanced gene delivery and mechanism studies with a novel series of cationic lipid formulations. *J. Biol. Chem. 269*:2550 (1994).
46. A. J. Fasbender, J. Zabner, M. J. Welsh. Optimization of cationic lipid-mediated gene transfer to airway epithelia. *Am. J. Physiol. Lung Cell. Mol. Physiol. 269*:L45 (1995).
47. C. J. Wheeler, L. Sukhu, G. L. Yang, Y. L. Tsai, C. Bustamente, P. Felgner, J. Norman, M. Manthorpe. Converting an alcohol to an amine in a cationic lipid dramatically alters the co-lipid requirement, cellular transfection activity and the ultrastructure of DNA cytofectin complexes. *Biochim. Biophys. Acta 1280*:1 (1996).
48. R. P. Balasubramaniam, M. J. Bennett, A. M. Aberle, J. G. Malone, M. H. Nantz, R. W. Malone. Structural and functional analysis of cationic transfection lipids: the hydrophobic domain. *Gene Therapy 3*:163 (1996).
49. X. Gao, L. Huang. A novel cationic liposome reagent for efficient transfection of mammalian cells. *Biochem. Biophys. Res. Commun. 179*:280 (1991).
50. N. J. Caplen, E. Kinrade, F. Sorgi, X. Gao, D. Geddes, C. Coutelle, L. Huang, E. W. F. W. Alton, R. Williamson. In vitro liposome-mediated DNA transfection of epithelial cell lines using the cationic liposome DC-Chol/DOPE. *Gene Therapy 2*:603 (1995).
51. J.-P. Vigneron, N. Oudrhiri, M. Fauquet, L. Vergely, J.-C. Bradley, M. Basseville, P. Lehn, J.-M. Lehn. Guanidinium-cholesterol cationic lipids: efficient vectors for the transfection of eukaryotic cells. *Proc. Natl. Acad. Sci. USA 93*:9682 (1996).
52. J. G. Lewis, K. Y. Lin, A. Kothavale, W. M. Flanagan, M. D. Matteucci, R. B. DePrince, R. A. Mook, Jr., R. W. Hendren, R. W. Wagner. A serum-resistant cytofectin for cellular delivery of antisense oligodeoxynucleotides and plasmid DNA. *Proc. Natl. Acad. Sci. USA 93*:3176 (1996).

53. I. Van der Woude, A. Wagenaar, A. A. P. Meekel, M. B. A. Ter Beest, M. H. J. Ruiters, J. B. F. N. Engberts, D. Hoekstra. Novel pyridinium surfactants for efficient, nontoxic in vitro gene delivery. *Proc. Natl. Acad. Sci. USA 94*:1160 (1997).
54. R. C. MacDonald, V. A. Rakhmanova, K. L. Choi, H. S. Rosenzweig, M. K. Lahiri. O-Ethylphosphatidylcholine: a metabolizable cationic phospholipid which is a serum-compatible DNA transfection agent. *J. Pharm. Sci. 88*:896 (1999).
55. K. Karikó, A. Kuo, E. S. Barnathan, D. J. Langer. Phosphate-enhanced transfection of cationic lipid-complexed mRNA and plasmid DNA. *Biochim. Biophys. Acta 1369*: 320 (1998).
56. A. M. I. Lam, P. R. Cullis. Calcium enhances the transfection potency of plasmid DNA–cationic liposome complexes. *Biochim. Biophys. Acta 1463*:279 (2000).
57. R. J. Debs, L. P. Freedman, S. Edmunds, K. L. Gaensler, N. Düzgüneş, K. R. Yamamoto. Regulation of gene expression in vivo by liposome-mediated delivery of a purified transcription factor. *J. Biol. Chem. 265*:10189 (1990).
58. M.-F. Lin, J. DaVolio, R. Garcia. Cationic liposome-mediated incorporation of prostatic acid phosphatase protein into human prostate carcinoma cells. *Biochem. Biophys. Res. Commun. 192*:413 (1993).
59. L. Huang, H. Farhood, N. Serbina, A. G. Teepe, J. Barsoum. Endosomolytic activity of cationic liposomes enhances the delivery of human immunodeficiency virus-1 trans-activator protein (tat) to mammalian cells. *Biochem. Biophys. Res. Commun. 217*:761 (1995).
60. S. E. Fong, P. Smanik, M. C. Smith, S. R. Jaskunas. Cationic liposome-mediated uptake of human immunodeficiency virus type 1 Tat protein into cells. *J. Virol. Methods 66*:149 (1997).
61. S. Lamartina, G. Roscilli, D. Rinaudo, P. Delmastro, C. Toniatti, Lipofection of purified adeno-associated virus rep68 protein: toward a chromosome-targeting nonviral particle. *J. Virol. 72*:7653 (1998).
62. K. L. Brigham, B. Meyrick, B. Christman, M. Magnuson, G. King, L. C. Berry, Jr. In vivo transfection of murine lungs with a functioning prokaryotic gene using a liposome vehicle. *Am. J. Med. Sci. 298*:278 (1989).
63. N. Zhu, D. Liggitt, Y. Liu, R. Debs. Systemic gene expression after intravenous DNA delivery into adult mice. *Science 261*:209 (1993).
64. M. J. Stewart, G. E. Plautz, L. D. Buono, Z. Y. Yang, L. Xu, X. Gao, L. Huang, E. G. Nabel, G. J. Nabel. Gene transfer in vivo with DNA–liposome complexes: safety and acute toxicity in mice. *Hum. Gene Ther. 3*:267 (1992).
65. E. G. Nabel, D. Gordon, Z.-Y. Yang, L. Xu, H. San, G. E. Plautz, B.-Y. Wu, X. Gao, L. Huang, and G. J. Nabel. Gene transfer in vivo with DNA–liposome complexes: lack of autoimmunity and gonadal localization. *Hum. Gene Ther. 3*:649 (1992).
66. H. San, Z.-Y. Yang, V. J. Pompili, M. L. Jaffe, G. E. Plautz, L. Xu, J. H. Felgner, C. J. Wheeler, P. L. Felgner, X. Gao, L. Huang, D. Gordon, G. J. Nabel, E. G. Nabel. Safety and short-term toxicity of a novel cationic lipid formulation for human gene therapy. *Hum. Gene Ther. 4*:781 (1993).
67. R. Philip, D. Liggitt, M. Philip, P. Dazin, R. Debs. In vivo gene delivery: efficient transfection of T lymphocytes in adult mice. *J. Biol. Chem. 268*:16087 (1993).
68. G. McLachlan, D. J. Davidson, B. J. Stevenson, P. Dickinson, H. Davidson-Smith, J. R. Dorin, D. J. Porteous. Evaluation in vitro and in vivo of cationic liposome-expression construct complexes for cystic fibrosis gene therapy. *Gene Therapy 2*:614 (1995).
69. N. K. Egilmez, Y. Iwanuma, R. B. Bankert. Evaluation and optimization of different cationic liposome formulations for in vivo gene transfer. *Biochem. Biophys. Res. Commun. 221*:169 (1996).
70. J.-P. Yang, L. Huang. Direct gene transfer to mouse melanoma by intratumor injection of free DNA. *Gene Therapy 3*:542 (1996).
71. N. K. Egilmez, R. Cuenca, S. J. Yokota, F. Sorgi, R. B. Bankert. In vivo cytokine gene therapy of human xenografts in SCID mice by liposome-mediated DNA delivery. *Gene Therapy 3*:607 (1996).
72. F. Liu, J. P. Yang, L. Huang, D. Liu. Effect of nonionic surfactants on the formation of DNA–emulsion complexes and emulsion-mediated gene transfer. *Pharm. Res. 13*:1856 (1996).
73. F. Liu, H. Qi, L. Huang, D. Liu. Factors controlling the efficiency of cationic lipid-mediated transfection in vivo via intravenous administration. *Gene Therapy 4*:517 (1997).
74. I. Solodin, C. Brown, M. Bruno, C. Chow, E.-H. Jang, R. Debs, T. Heath. High efficiency in vivo gene delivery with a novel series of amphiphilic imidazolinium compounds. *Biochemistry 34*:13537 (1995).
75. G. J. Nabel, D. Gordon, D. K. Bishop, B. J. Nickoloff, Z.-Y. Yang, A. Aruga, M. J. Cameron, E. G. Nabel, A. E. Chang. Immune response in human melanoma after transfer of an allogeneic class I major histocompatibility complex gene with DNA–liposome complexes. *Proc. Natl. Acad. Sci. USA 93*:15388 (1996).
76. N. J. Caplen, E. W. F. W. Alton, P. G. Middleton, J. R. Dorin, B. J. Stevenson, X. Gao, S. R. Durham, P. K. Jeffery, M. E. Hodson, C. Coutelle, L. Huang, D. J. Porteous, R. Williamson, D. M. Geddes. Liposome-mediated CFTR gene transfer to the nasal epithelium of patients with cystic fibrosis. *Nature Med. 1*:39 (1995).
77. D. J. Porteous, J. R. Dorin, G. McLachlan, H. Davidson-Smith, H. Davidson, B. J. Stevenson, A. D. Carothers, W. A. Wallace, S. Moralee, C. Hoenes, G. Kallmeyer, U. Michaelis, K. Naujoks, L. P. Ho, J. M. Samways, M. Imrie, A. P. Greening, J. A. Innes. Evidence for safety and efficacy of DOTAP cationic liposome–mediated CFTR gene transfer to the nasal epithelium of patients with cystic fibrosis. *Gene Therapy 4*:210 (1997).
78. K. L. Brigham, K. B. Lane, B. Meyrick, A. A. Stecenko, S. Strack, D. R. Cannon, M. Caudill, A. E. Canonico. Transfection of nasal mucosa with a normal $\alpha 1$-antitrypsin

gene in α1-antitrypsin-deficient subjects: comparison with protein therapy. *Hum. Gene Ther. 11*:1023 (2000).
79. K. A. Mislick, J. D. Baldeschwieler. Evidence for the role of proteoglycans in cation-mediated gene transfer. *Proc. Natl. Acad. Sci. USA 93*:12349 (1996).
80. L. C. Mounkes, W. Zhong, G. Cipres-Palacin, T. D. Heath, R. J. Debs. Proteoglycans mediate cationic liposome–DNA complex–based gene delivery in vitro and in vivo. *J. Biol. Chem. 273*:26164 (1998).
81. M. Belting, P. Petersson. Protective role for proteoglycans against cationic lipid cytotoxicity allowing optimal transfection efficiency in vitro. *Biochem. J. 342*:281 (1999).
82. H. Arima, Y. Aramaki, S. Tsuchiya. Contribution of trypsin-sensitive proteins to binding of cationic liposomes to the mouse macrophage–like cell line RAW264.7. *J. Pharm. Sci. 86*:786 (1997).
83. Y. H. Xu, F. C. Szoka, Jr. Mechanism of DNA release from cationic liposome/DNA complexes used in cell transfection. *Biochemistry 35*:5616 (1996).
84. D. S. Friend, D. Papahadjopoulos, R. J. Debs. Endocytosis and intracellular processing accompanying transfection mediated by cationic liposomes. *Biochim. Biophys. Acta 1278*:41 (1996).
85. X. Zhou, L. Huang. DNA transfection mediated by cationic liposomes containing lipopolylysine: characterization and mechanism of action. *Biochim. Biophys. Acta 1189*:195 (1994).
86. S. Simões, V. Slepushkin, P. Pires, R. Gaspar, M. C. Pedroso de Lima, N. Düzgüneş. Mechanisms of gene transfer mediated by lipoplexes associated with targeting ligands or pH-sensitive peptides. *Gene Therapy 6*:1798 (1999).
87. P. Pires, S. Simões, R. Gaspar, N. Düzgüneş, M. C. Pedroso de Lima. Interaction of cationic liposomes and their DNA complexes with monocytic leukemia cells. *Biochim. Biophys. Acta 1418*:71 (1999).
88. T. Stegmann, J.-Y. Legendre. Gene transfer mediated by cationic lipids: lack of correlation between lipid mixing and transfection. *Biochim. Biophys. Acta 1325*:71 (1997).
89. I. Van der Woude, H. W. Visser, M. B. A. Ter Beest, A. Wagenaar, M. H. J. Ruiters, J. B. F. N. Engberts, D. Hoekstra. Parameters influencing the introduction of plasmid DNA into cells by the use of synthetic amphiphiles as a carrier system. *Biochim. Biophys. Acta 1240*:34 (1995).
90. A. Noguchi, T. Furuno, C. Kawaura, M. Nakanishi. Membrane fusion plays an important role in gene transfection mediated by cationic liposomes. *FEBS Lett. 433*:169 (1998).
91. A. El Ouahabi, M. Thiry, V. Pector, R. Fuks, J. M. Ruysschaert, M. Vandenbranden. The role of endosome destabilizing activity in the gene transfer process mediated by cationic lipids. *FEBS Lett. 414*:187 (1997).
92. R. L. Page, S. P. Butler, A. Subramamian, F. C. Gwazdauskas, J. L. Johnson, W. H. Velander. Transgenesis in mice by cytoplasmic injection of polylysine/DNA mixtures. *Transgenic Res. 4*:353 (2000).
93. W. C. Tseng, F. R. Haselton, T. D. Giorgio. Mitosis enhances transgene expression of plasmid delivered by cationic liposomes. *Biochim. Biophys. Acta 1445*:53 (1999).
94. P. Hawley-Nelson, V. Ciccarone, G. Gebeyehu, J. Jessee. Lipofectamine reagent: a new, higher efficency polycationic liposome transfection reagent. *Focus 15*:73 (1993).
95. K. Crook, B. J. Stevenson, M. Dubouchet, D. J. Porteous. Inclusion of cholesterol in DOTAP transfection complexes increases the delivery of DNA to cells in vitro in the presence of serum. *Gene Therapy 5*:137 (1998).
96. E. Brunette, R. Stribling, R. Debs. Lipofection does not require the removal of serum. *Nucleic Acids Res. 20*:1151 (1992).
97. V. Ciccarone, P. Hawley-Nelson, J. Jessee. Cationic liposome-mediated transfection:effect of serum on expression and efficiency. *Focus 15*:80 (1993).
98. S. Li, L. Huang. Lipidic supramolecular assemblies for gene transfer. *J. Liposome Res. 6*:589 (1996).
99. A. R. Thierry, P. Robinovich, B. Peng, L. C. Mahan, J. L. Bryant, R. C. Gallo. Characterization of liposome-mediated gene delivery:expression, stability and pharmacokinetics of plasmid DNA. *Gene Therapy 4*:226 (1997).
100. J. P. Yang, L. Huang. Overcoming the inhibitory effect of serum on lipofection by increasing the charge ratio of cationic liposome to DNA. *Gene Therapy 4*:950 (1997).
101. S. Li, W.-C. Tseng, D. B. Stolz, S.-P. Wu, S. C. Watkins, L. Huang. Dynamic changes in the characteristics of cationic lipidic vectors after exposure to mouse serum: implications for intravenous lipofection. *Gene Therapy 6*:585 (1999).
102. J. E. Duncan, J. A. Whitsett, A. D. Horowitz. Pulmonary surfactant inhibits cationic liposome-mediated gene delivery to respiratory epithelial cells in vitro. *Hum. Gene Ther. 8*:431 (1997).
103. M. F. Tsan, G. L. Tsan, J. E. White. Surfactant inhibits cationic liposome-mediated gene transfer. *Hum. Gene Ther. 8*:817 (1997).
104. M. Ruponen, S. Yla-Herttuala, A. Urtti. Interactions of polymeric and liposomal gene delivery systems with extracellular glycosaminoglycans: physicochemical and transfection studies. *Biochim. Biophys. Acta 1415*:331 (1999).
105. S. Kawakami, F. Yamashita, M. Nishikawa, Y. Takakura, M. Hashida. Asiolglycoprotein receptor-mediated gene transfer using novel galactosylated cationic liposomes. *Biochem. Biophys. Res. Commun. 252*:78 (1998).
106. T. Hara, Y. Aramaki, S. Takada, K. Koike, S. Tsuchiya. Receptor-mediated transfer of pSV2CAT DNA to a human hepatoblastoma cell line HepG2 using asiolofetuin-labeled cationic liposomes. *Gene 159*:167 (1995).
107. G. Y. Kao, L. J. Change, T. M. Allen. Use of targeted cationic liposomes in enhanced DNA delivery to cancer cells. *Gene Therapy 3*:250 (1996).
108. S. L. Hart, C. Arancibia-Carcamo, M. A. Wolfert, C. Mailhos, N. J. O'Reilly, R. R. Ali, C. Coutelle, A. J. George, R. P. Arbottle, A. M. Knight, D. F. Larkin, R. J. Lewinsky, L. W. Seymour, A. J. Trasher, C. Kinnon.

Lipid-mediated enhancement of transfection by a nonviral integrin-targeting vector. *Gene Therapy 9*:575 (1998).

109. K. Yanagihara, P. W. Cheng. Lectin enhancement of the lipofection efficiency in human lung carcinoma cells. *Biochim. Biophys. Acta. 1472*:25 (1999).

110. P. W. Cheng. Receptor ligand-facilitated gene transfer: enhancement of liposome-mediated gene transfer and expression by transferrin. *Hum. Gene Ther. 7*:275 (1996).

111. S. Simões, V. Slepushkin, R. Gaspar, M. C. Pedroso de Lima, N. Düzgüneş, Gene delivery by negatively charged ternary complexes of DNA, cationic liposomes and transferrin or fusigenic peptides. *Gene Therapy 5*:955 (1998).

112. S. Simões, V. Slepushkin, E. Pretzer, P. Dazin, R. Gaspar, M. C. Pedroso de Lima, N. Düzgüneş, Transfection of human macrophages by lipoplexes via the combined use of targeting ligands and pH-sensitive peptides. *J. Leukocyte Biol. 65*:270 (1999).

113. S. Simões, V. Slepushkin, R. Gaspar, M. C. Pedroso de Lima, N. Düzgüneş. Successful transfection of lymphocytes by ternary lipoplexes. *Biosci. Rep. 19*:601 (1999).

114. X. Gao, L. Huang. Potentiation of cationic liposome-mediated gene delivery by polycations. *Biochemistry 35*:1027 (1996).

115. F. L. Sorgi, S. Bhattacharya, L. Huang. Protamine sulfate enhances lipid-mediated gene transfer. *Gene Therapy 4*: 961 (1997).

116. S. Li, L. Huang. In vivo gene transfer via intravenous administration of cationic lipid-protamine-DNA (LPD) complexes. *Gene Therapy 4*:891 (1997).

117. S. Li, M. A Rizzo, S. Bhattacharya, L. Huang. Characterization of cationic lipid-protamine-DNA (LPD) complexes for intravenous gene delivery. *Gene Therapy 5*:930 (1998).

118. F. Labat-Moleur, A.-M. Steffan, C. Brisson, H. Perron, O. Feugeas, P. Furstenberger, F. Oberling, E. Brambilla, J.-P. Behr. An electron microscopy study into the mechanism of gene transfer with lipopolyamines. *Gene Therapy 3*: 1010 (1996).

119. J. T. Conary, G. Erdos, M. McGuire, R. Faulks, L. Gagné, P. Price, B. Christman, K. Brigham, F. Ebner, H. Schreier. Cationic liposome plasmid DNA complexes: in vitro cell entry and transgene expression augmented by synthetic signal peptides. *Eur. J. Pharm. Biopharm. 42*:277 (1996).

120. A. I. Aronsohn, J. A. Hughes. Nuclear localization signal peptides enhance cationic liposome-mediated gene therapy. *J. Drug Targeting 5*:163 (1997).

121. M. A. Zanta, P. B. Valladier, J. P. Behr. Gene delivery: a single nuclear localization signal peptide is sufficient to carry DNA to the cell nucleus. *Proc. Natl. Acad. Sci. USA 96*:91 (1999).

122. L. J. Brandén, A. J. Mohamed, C. I. E. Smith. A peptide nucleic acid-nuclear localization signal fusion that mediates nuclear transport of DNA. *Nat. Biotechnol. 17*:784 (1999).

123. J. Vacik, B. S. Dean, W. E. Zimmer, D. A. Dean. Cell-specific nuclear import of plasmid DNA. *Gene Therapy 6*:1006 (1999).

124. A. Subramanian, P. Ranganathan, S. L. Diamond. Nuclear targeting peptide scaffolds for lipofection of nondividing mammalian cells. *Nature Biotechnol. 17*:873 (1999).

125. H. Kamata, H. Yagisawa, S. Takahashi, H. Hirata. Amphiphilic peptides enhance the eficiency of liposome-mediated DNA transfection. *Nucleic Acids Res. 22*: 536 (1994).

126. A. Kichler, K. Mechtler, J.-P. Behr, E. Wagner. Influence of membrane-active peptides on lipospermine/DNA complex mediated gene transfer. *Bioconjugate Chem. 8*:213 (1997).

127. N. K. Subbarao, R. A. Parente, F. C. Szoka, L. Nadasdi, K. Pongracz. pH-dependent bilayer destabilization by an amphipathic peptide. *Biochemistry 26*:2964 (1987).

128. R. A. Parente, S. Nir, F. C. Szoka, Jr. pH-dependent fusion of phosphatidylcholine small vesicles. *J. Biol. Chem. 263*: 4724 (1988).

129. D. Papahadjopoulos, T. M. Allen, A. Gabizon, E. Mayhew, K. Matthay, S. K. Huang, K.-D. Lee, M. C. Woodle, D. D. Lasic, C. Redemann, F. J. Martin. Sterically stabilized liposomes: improvements in pharmacokinetics and antitumor therapeutic efficacy. *Proc. Natl. Acad. Sci. USA 88*:11460 (1991).

130. T. M. Allen, C. Hansen, F. Martin, C. Redemann, A. Yau-Young. Liposomes containing synthetic lipid derivatives of poly(ethylene glycol) show prolonged circulation half-lives in vivo. *Biochim. Biophys. Acta 1066*:29 (1991).

131. V. A. Slepushkin, S. Simões, P. Dazin, M. S. Newman, L. S. Guo, M. C. Pedroso de Lima, N. Düzgüneş. Sterically stabilized pH-sensitive liposomes: intracellular delivery of aqueous contents and prolonged circulation in vivo. *J. Biol. Chem. 272*:2382 (1997).

132. K. L. Hong, W. W. Zheng, A. Baker, D. Papahadjopoulos. Stabilization of cationic liposome—plasmid DNA complexes by polyamines and poly(ethylene glycol)–phospholipid conjugates for efficient in vivo gene delivery. *FEBS Lett. 400*:233 (1997).

133. M. C. Pedroso de Lima, S. Simões, P. Pires, H. Faneca, N. Düzgüneş. Cationic lipid–DNA complexes in gene delivery: from biophysics to biological applications. *Adv. Drug Deliv. Rev. 47*:277 (2001).

134. N. Maurer, B. Mui, K. W. C. Mok, Q. F. Akhong, P. R. Cullis. Lipid-based systems for the intracellular delivery of genetic drugs. *Mol. Membrane Biol 16*:129 (1999).

135. Y. P. Zhang, L. Sekirov, E. G. Saravolac, J. J. Wheeler, P. Tardi, K. Clow, E. Leng, R. Sun, P. R. Cullis, P. Scherrer. Stabilized plasmid-lipid particles for regional gene therapy: formulation and transfection properties. *Gene Therapy 6*:1438 (1999).

136. J. J. Wheeler, L. Palmer, M. Ossanlou, I. MacLachlan, R. W. Graham, Y. P. Zhang, M. J. Hope, P. Scherrer, P. R. Cullis. Stabilized plasmid-lipid particles: construction and characterization. *Gene Therapy 6*:271 (1999).

36

Biological Stimulus-Responsive Hydrogels

Takashi Miyata and Tadashi Uragami
Kansai University, Osaka, Japan

I. INTRODUCTION

Hydrogels are attractive objects consisting of cross-linked networks of polymer swollen with aqueous solutions, and have both liquidlike and solidlike properties. As hydrogels have a variety of properties including their sorption capacities, swelling behavior, mechanical properties, permeabilities, surface properties, and optical properties, they are highly promising materials for wide applications in various fields such as medicine, agriculture, and biotechnology (1–5). Such fascinating properties of hydrogels are governed by the characteristics of polymer chains in aqueous solutions and the cross-linking structures. In general hydrogels, the cross-linked polymer networks are formed by covalent bonds, electrostatic interactions, hydrogen bonding, hydrophobic interactions, van der Waals forces, physical entanglements, crystallite, and so on. The polymer chains in hydrogels are generally hydrophilic and compatible with aqueous solutions to be swollen. Therefore such cross-linked structures and polymer chains must be designed for developing useful hydrogels for a given purpose.

Hydrogels have the unique property that they undergo abrupt changes in their volume from the collapsed and swollen states. Since the unique property, the volume phase transition, of hydrogels was discovered experimentally and theoretically (6–10), the stimulus-responsiveness of hydrogels has attracted considerable attention in the biological and biomedical fields. The hydrogels that undergo swelling changes in response to environmental stimuli are called stimulus-responsive, smart, or intelligent hydrogels. Stimulus-responsive hydrogels are considered for use in various devices such as switches, sensors, mechanochemical actuators, drug delivery devices, specialized separation systems, bioreactors, and, artificial muscles. In particular, the stimulus-responsive hydrogels that exhibit swelling changes in response to environmental changes such as pH (7), temperature (6,8,9,11), electric field (12,13), and other stimuli (14–17) can provide the tools for creating functional materials with a wide variety of uses because they can sense environmental changes and induce structural changes. The fascinating properties of such hydrogels suggest that they have many future opportunities as suitable materials for mimicking biomolecules and designing smart systems in the biochemical and biomedical fields.

An organ must respond to the presence of specific molecules as well as environmental changes like pH and temperature to maintain life. Stimulus-responsive hydrogels that can respond to specific molecules seem to be biomimetic materials because they have the functions of both sensing specific molecules and responding to them. Therefore, recently, the stimulus-responsive hydrogels that recognize a specific molecule and respond to it have become increasingly important, because of their potential application in the biomedical field and in the design of novel biomaterials. For example, stimulus-responsive hydrogels that respond to blood glucose are very useful in the treatment of diabetes in which a necessary amount of insulin should be administered in response to the blood glucose concentration. This chapter provides a short overview of current research in such biological stimulus-responsive gels and their applications.

II. PHYSICOCHEMICAL STIMULUS-RESPONSIVE HYDROGELS

The most important biosystems for maintaining life are closely associated with natural feedback system functions, so-called homeostasis. For example, the release of hormones from secretory cells is regulated by physiological cycles or by specific input signals. Mimicking such natural feedback systems enables us to develop novel intelligent systems having sensor, processor, and effector functions simultaneously. As temperature and pH are the most widely utilized triggering signals for modulated intelligent systems, studies on a variety of pH- and temperature-responsive hydrogels have been reported. The following subsections focus on pH- and temperature-responsive hydrogels as physicochemical stimulus-responsive hydrogels.

A. pH-Responsive Hydrogels

A change in pH is the most useful signal in the human body, because variation in pH occurs naturally in some areas of the body. For example, as the pH of the gastric condition is much lower than that of the enteric condition, a pH-responsive hydrogel is a candidate as an intelligent device to deliver drugs to the small intestines while avoiding release in the stomach. In general, such pH-responsive hydrogels can be prepared by using polymers with ionizable groups. Some fine reviews dealing with various aspects of pH-responsive hydrogels have appeared in the past decade (1,18).

The pH change as an external signal influences the degree of ionization of the polymer with ionizable groups to govern its solubility in water. Therefore when a hydrogel is prepared from a polymer with ionizable groups, the swelling behavior of the hydrogel is strongly dependent upon pH. Carboxyl, sulfonic, and amino groups are very popular as ionizable groups to prepare pH-responsive hydrogels. There have been many reports on pH-responsive hydrogels prepared from monomers with such ionizable groups. For example, Siegel et al. (18–20) prepared lightly cross-linked hydrophobic polyelectrolyte hydrogels by copolymerization of methyl methacrylate (MMA) and *N,N*-dimethylaminoethyl methacrylate (DMA) with a small amount of the cross-linking reagent divinyl benzene. The copolymer hydrogels with amino groups were collapsed above neutral pH but became swollen at acidic pH. Furthermore, caffeine release from the copolymer hydrogels with amino groups was investigated as a function of pH. Caffeine was not released at neutral pH, while it was released with near-zero-order kinetics at pH 3 and 5, the release rate being greater at pH 3. On the other hand, hydrogels with carboxyl groups were prepared by copolymerization of 2-hydroxyethyl methacrylate (HEMA) or poly(*N*-isopropylacrylamide) (PNIPAAm) with methacrylic acid or maleic anhydride and by the formation of interpenetrating polymer networks (IPNs) of poly(vinyl alcohol) (PVA) and poly(acrylic acid) (PAAc), and pH-responsive drug release systems were developed (21–23). The resultant hydrogels with carboxyl groups showed sharp swelling transition with small changes in pH, depending upon composition. Drug release systems controlled by pH-responsive swelling can be developed, which exhibit zero-order or near-zero-order release of the incorporated drug. Hoffman et al. (24) proposed a novel approach for preparation of pH-responsive hydrogels for enteric drug delivery. The pH-responsive hydrogels were composed of pH-responsive PAAc and temperature-responsive PNIPAAm. The special hydrogels were sensitive to both temperature and pH, and respond to the pH change to a much greater extent than normal hydrogels with carboxyl groups. Only a negligible amount of indomethacin as a model drug was released at pH 1.4, while more than 90% of the total drug in the gels was released at pH 7.4.

Very few studies on hydrogels with phosphate groups have been reported, although the phosphate group is a very important acidic group. Hydrogels with phosphate groups have been prepared by copolymerization of a monomer with a phosphate group and various comonomers (25,26). The swelling behavior of the hydrogels with the phosphate groups is strongly dependent upon the kind of comonomer. The swelling ratio of the hydrogels with the phosphate groups increased steeply at pH 5 and 10 because the phosphate group acts as an acidic charged divalent group. Furthermore, the hydrogels with the phosphate groups also showed swelling changes in response to temperature and solvent composition.

Immobilization of physiologically active compounds is very important in a variety of applications of the compounds in drug delivery systems, bioreactors, bioseparations, tissue engineering, and so on. Hydrogels with ionizable groups are promising supports to immobilize such physiologically active compounds based on their electrostatic interaction. For example, in order to immobilize a cationic protein, the hydrogels with phosphate groups were synthesized by copolymerizing the monomer with the phosphate group with *N*-isopropylacrylamide (NIPAAm) and *N,N′*-methylene-bis-acrylamide (MBAA) (27,28). Lysozyme, which is a cationic protein, could be immobilized

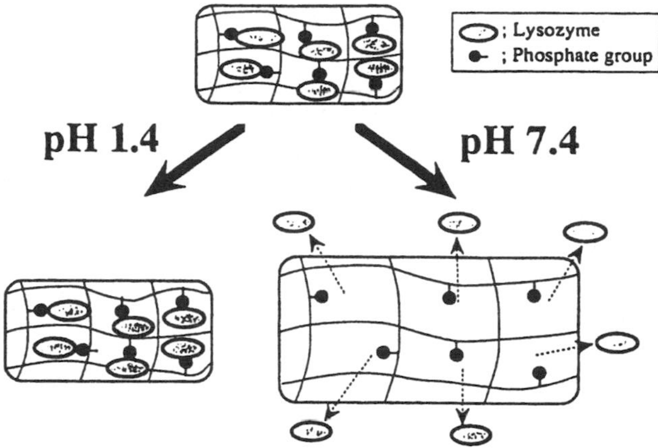

Figure 36.1 Schematic of pH-responsive release of lysozyme from the hydrogel with phosphate groups. (From Ref. 28.)

efficiently within the hydrogels with phosphate groups through polyion complexation. The hydrogels with phosphate groups had a higher binding capacity for cationic proteins than hydrogels with carboxyl groups. Lysozyme release from the hydrogels with phosphate groups was influenced by conditions such as ionic strength and pH, because electrostatic binding between the phosphate group and lysozyme was sensitive to environmental conditions. In particular, the pulsatile release pattern indicated that lysozyme is released at pH 7.4 (enteric conditions) and resists release at pH 1.4 (gastric conditions) (Fig. 1). These studies demonstrated that pH-responsive dissociation of ionic binding enables us to deliver drugs to the small intestines while avoiding release in the stomach.

B. Temperature-Responsive Hydrogels

Most temperature-responsive hydrogels have been prepared on the basis of poly(*N*-isopropylacrylamide) (PNIPAAm), which has a lower critical solution temperature (LCST) in its aqueous solution. PNIPAAm homopolymer is soluble in water at a temperature of less than LCST (about 32°C) but becomes insoluble at more than LCST due to strong polymer–polymer interaction (29). Therefore a cross-linked PNIPAAm hydrogel was swollen at a temperature of less than LCST and collapsed at more than LCST. Such a temperature-responsive swelling change of the cross-linked PNIPAAm hydrogel is attributable to a dramatic change in hydrophilicity of polymer chain in water. The temperature-responsive PNIPAAm hydrogel and its derivative hydrogels are currently of great interest in biochemical and biomedical fields. Some articles provide overviews of current research in temperature-responsive hydrogels prepared from PNIPAAm and their applications (30,31).

Hoffman et al. (32,33) used a cross-linked PNIPAAm hydrogel that exhibits LCST as a temperature-responsive drug carrier. They demonstrated that the temperature-responsive hydrogel is a candidate to absorb or release a variety of biologically and industrially important substances. After vitamin B_{12} and myoglobin as model drugs were loaded into the temperature-responsive hydrogels swollen below LCST, their release behavior from the hydrogels was investigated above LCST. Their model drugs were released rapidly from the hydrogels due to the squeeze effect by their collapse as soon as the hydrogels were transferred from an aqueous solution below LCST to the release media above LCST (Fig. 2).

On the other hand, Okano and Kim et al. (34–36) synthesized cross-linked copolymer hydrogels consisting of NIPAAm and butyl methacrylate (BMA) and achieved on–off regulation of drug release in response to stepwise temperature changes. As the temperature-responsive hydrogels were swollen at a low temperature, the drug could be released from the hydrogels. However, when the temperature increased beyond their LCST, dense skin formation during the deswelling process stopped the release of the drug. Furthermore, they have systematically studied the sharp swelling/deswelling changes of the temperature-responsive PNIPAAm copolymer hydrogels and the pulsatile drug release from the hydrogels in response to temperature changes (37–39). In addition, comb-type grafted PNIPAAm hydrogels were prepared to accelerate the deswelling rate, and the effect of hydrogel architecture on the

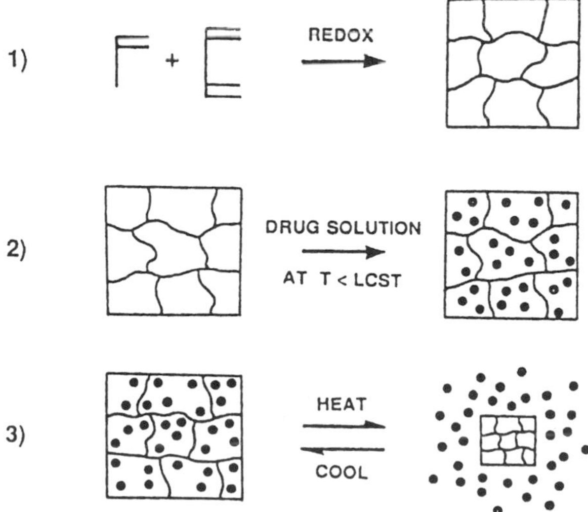

Figure 36.2 Schematic of delivery or removal of substances using temperature-responsive hydrogels. (From Ref. 32.)

deswelling kinetics was investigated (40–42). The comb-type grafted PNIPAAm hydrogels showed much rapider temperature responses during the deswelling beyond LCST than the conventional cross-linked PNIPAAm hydrogel. The rapid deswelling of the comb-type grafted PNIPAAm hydrogels was based on rapid hydrophobic aggregation of freely mobile PNIPAAm graft chains and the intrinsic elastic forces of the polymer network.

The opposite type of temperature-responsive hydrogels, which are swollen at high temperature and shrunken at low temperature, were also prepared by using hydrogen bonding between polymer chains (43–45). The temperature-responsive hydrogels were interpenetrating polymer networks (IPNs) composed of acrylamide (AAm) and acrylic acid (AAc). In the IPN hydrogels, AAm formed a complex with AAc by hydrogen bonding at low temperature, but the complex was dissociated at high temperature. As a result, the IPN hydrogels were swollen at high temperature due to the dissociation of the hydrogen bonding and shrunken gradually with decreasing temperature due to the formation of hydrogen bonding. (Fig. 3). Furthermore, they performed the release experiments of ketoprofen from the temperature-responsive IPN hydrogels in response to stepwise changes in temperature. In contrast to drug release profiles from the PNIPAAm hydrogels, the release rate of drug decreased at low temperature but increased at high temperature. The temperature-responsive IPN hydrogels showed that stimulus-responsive hydrogels can be prepared by using association and dissociation of a complex between polymer chains.

Some researchers reported that temperature-responsive PNIPAAm hydrogels are applicable as intelligent devices

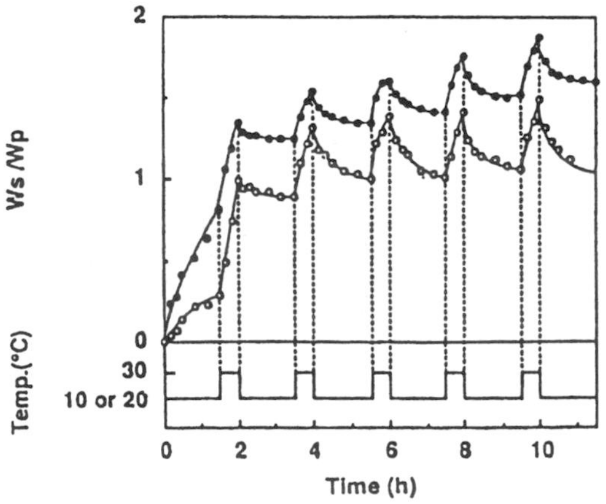

Figure 36.3 Swelling changes of temperature-responsive IPN hydrogel in response to stepwise temperature changes. (○) between 10°C and 30°C; (●) between 20°C and 30°C. (From Ref. 44.)

to separate molecules (46–48). Cussler (47) used a cross-linked PNIPAAm hydrogel as an extraction solvent for aqueous solutions. Low molecular weight solutes were adsorbed in the cross-linked PNIPAAm hydrogel during swelling, but proteins and other high molecular weight solutes were excluded. The adsorbed solutes were released when the cross-linked PNIPAAm hydrogel was collapsed at high temperature. Therefore the cross-linked PNIPAAm hydrogel can be easily regenerated by a slight increase in temperature. Kim et al. (48) showed that molecules of different sizes can be separated by a temperature-responsive hydrogel membrane consisting of NIPAAm and BMA. The swelling of the hydrogel membrane as a function of temperature strongly influenced the permeability of molecules of different sizes.

Temperature-responsive hydrogels are also novel intelligent supports to immobilize enzymes and proteins. For example, Hoffman et al. (49,50) immobilized asparaginase and β-galactosidase within temperature-responsive hydrogels consisting of NIPAAm and AAm. The enzyme activity was shut off when temperature increased above its LCST. When operational temperature was cycled between temperature below and above LCST, the activity of the immobilized enzymes almost fully recovered after each cycle.

III. GLUCOSE-RESPONSIVE HYDROGELS

Diabetes is caused by the inability of the pancreas to control blood glucose concentration. As insulin, which is a hormone secreted from the Wrangell Hans island of the pancreas, controls the glucose metabolizing, various medicines made of insulin are developed to control the blood sugar value effectively. In the treatment of diabetes by insulin injection, the blood glucose concentration is monitored and a necessary amount of insulin is administered. Recently, several investigations have been undertaken for the purpose of developing drug delivery systems that can release insulin in response to the blood glucose concentration. Glucose-responsive hydrogels are promising artificial pancreases that can administer a necessary amount of insulin in response to the blood glucose concentration. Some approaches to develop glucose-responsive hydrogels are described in the following subsections.

A. Glucose Oxidase–Containing Hydrogels

Some researchers designed glucose-responsive hydrogels by combining glucose oxidase to sense glucose with pH-responsive hydrogels to regulate the permeation rate of insulin. In their systems, gluconic acid is produced from glucose by glucose oxidase in the hydrogels. Since the produced gluconic acid induces a decrease in pH, the permeability of insulin through the pH-responsive hydrogels is enhanced due to their pH-responsive swelling. Glucose-responsive insulin permeation of their membrane can be explained by the tentative model in Fig. 4. Ishihara et al. (51,52) and Horbett et al. (53–55) prepared hydrogels with pendant amine groups as pH-responsive hydrogels, in which glucose oxidase was entrapped during the preparation. In their systems, when glucose diffuses into the hydrogels, glucose oxidase catalyzes its conversion to gluconic acid. Then the produced gluconic acid leads to lowering pH within the hydrogel microenvironment. The reduced pH results in increasing ionization of the pendant amine groups in the hydrogels. As hydrogels were swollen by the electrostatic repulsion between ionized amine groups, insulin can be permeated through the hydrogels. Ultimately, then, the insulin permeability is a function of glucose concentration external to the hydrogels, and insulin delivery is accelerated by an increase in glucose level. (Fig. 4)

Ishihara et al. (51) combined a copolymer membrane of *N,N*-diethylaminoethyl methacrylate (DEA) and 2-hydroxypropyl methacrylate (HPMA) with cross-linked poly(acrylamide) membrane in which glucose oxidase was immobilized. The insulin permeability through the complex membrane containing glucose oxidase was strongly independent of the glucose concentration. Furthermore, Ishihara and Matsui (52) prepared polymer capsules containing insulin and glucose oxidase by a conventional interfacial precipitation method and investigated insulin release from the capsules in response to the glucose concentration. Figure 5 shows the effect of the glucose concentration on the insulin release from the glucose-responsive polymer capsules. The insulin release rate was very low in the absence of glucose but was strongly enhanced by the presence of glucose. Glucose-responsive insulin release can be achieved by using pH-responsive co-

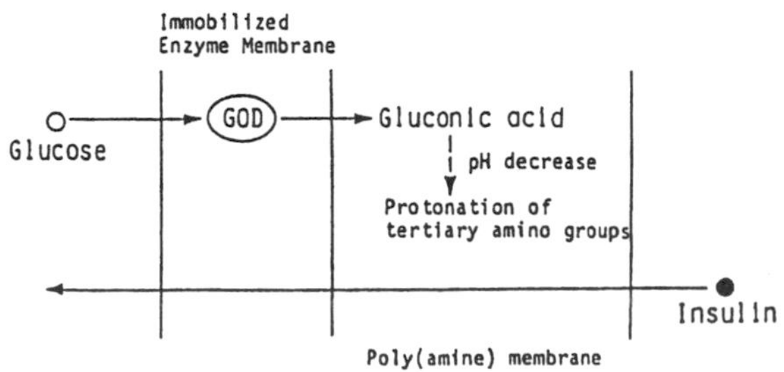

Figure 36.4 Schematic representation of the glucose-responsive hydrogel membrane consisting of a poly(amine) and glucose oxidase–loaded membranes. (From Ref. 51.)

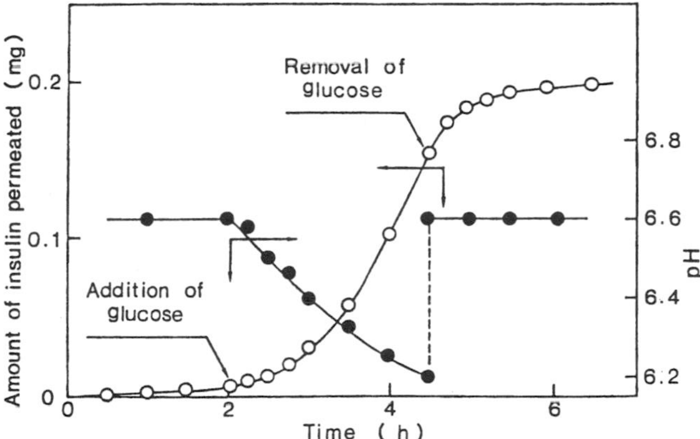

Figure 36.5 Effect of the addition and removal of glucose on insulin permeation through the copolymer membranes containing glucose oxidase and on the pH of the feed side solution. (From Ref. 52.)

polymer capsules containing insulin and glucose oxidase, based on the combination of pH response of the copolymer membrane and the enzymatic reaction between glucose and glucose oxidase.

On the other hand, Horbett et al. (53–55) reported that hydroxyethyl methacrylate-*N,N*-dimetylaminoethyl methacrylate copolymer hydrogel membranes containing entrapped glucose oxidase demonstrated promising responsiveness to glucose concentration. The presence of glucose resulted in swelling of their membranes and enhanced insulin delivery from a reservoir by diffusion through the swollen hydrogel membranes. The pH indicator dyes that can convert a pH decrease to a color change were introduced into the glucose-responsive hydrogel membranes to examine pH changes in the membranes. Based on the results of the experiments, they developed a mathematical model to describe the steady-state behavior of the glucose-responsive hydrogel membranes. Their theoretical and experimental studies suggested that a glucose-responsive hydrogel membrane prepared by using glucose oxidase would achieve a maximum response at subphysiological concentrations of glucose and not respond to higher glucose concentrations. Their model showed that loading of sufficiently low glucose oxidase enables a progressive response to glucose concentration in the physiological range.

Peppas et al. (56,57) synthesized glucose-responsive poly(methacrylic acid-*g*-ethylene glycol) (poly(MAAc-*g*-EG)) hydrogels by copolymerizing methacrylic acid and poly(ethylene glycol) monomethacrylate in the presence of activated glucose oxidase. At low pH, the protonation of the carboxyl groups caused complex formation between carboxyl groups and etheric groups, and this resulted in a collapse of the hydrogel due to the lowering of the hydrophobicity in the polymer network. At high pH, however, the complex was dissociated as the carboxyl groups became ionized. As a result, the hydrogels were collapsed at low pH due to complexation between carboxyl groups of MMAc and etheric groups of EG, but were swollen at high pH. When the glucose-responsive poly(MAAc-*g*-EG) hydrogels were swollen from an initially dry state, the hydrogels at the high glucose concentration of hyperglycemic condition (200–500 mg/dL) were swollen at a slower rate than those at the lower glucose concentrations of normal blood glucose (80 mg/dL). At a high glucose concentration, the glucose oxidase catalyzed glucose reaction with oxygen and then produced gluconic acid to result in a decrease in the environmental pH. Then the hydrogels were collapsed due to complex formation between carboxyl groups and etheric groups with lowering pH in the hydrogels. Peppas et al. expected that such a collapse can "squeeze out" any incorporated drugs including insulin when the hydrogels are collapsed abruptly with a decrease in pH. Alternatively, as the glucose concentration decreases by the action of the released insulin, less produced gluconic acid results in an increase in the environmental pH. The hydrogels are expected to be swollen with this change in pH.

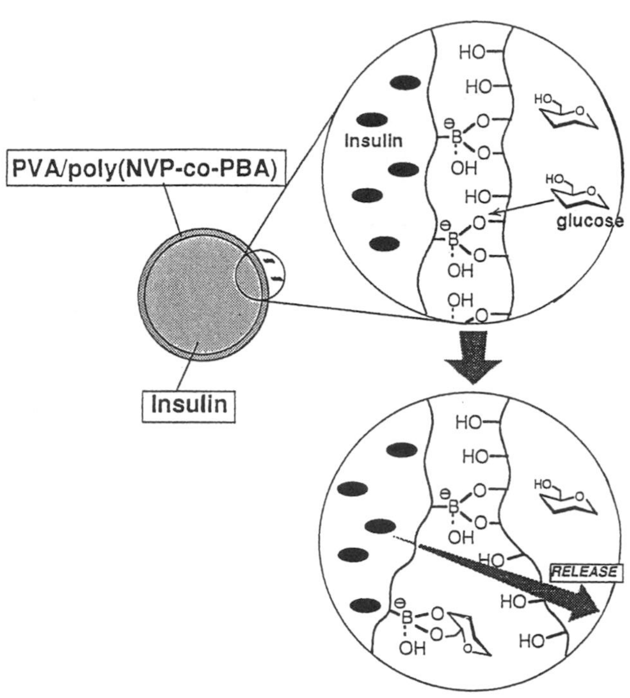

Figure 36.6 Concept of glucose-responsive insulin release system using PVA/poly(NVP-co-PBA) complex system (polymer capsule type). (From Ref. 59.)

B. Hydrogels with Phenylboronic Acid Moiety

Phenylboronic acid and its derivatives form complexes with polyol compounds such as glucose in aqueous solution. The complex between phenylboronic acid and a polyol compound can be exchanged by a competing polyol compound that can form complexes more strongly. The fascinating property of the complex between phenylboronic acid and a polyol compound suggests that the complex has many future opportunities as suitable glucose-responsive materials. Kitano et al. (58,59) prepared the copolymers with phenylboronic acid moieties (Poly(NVP-*co*-PBA)) by the copolymerization of *N*-vinyl-2-pyrrolidone (NVP) and 3-(acrylamido)phenylboronic acid (PBA), which formed its reversible complex with poly(vinyl alcohol) (PVA). Their strategy is to utilize the competitive binding of phenylboronic acid with glucose and PVA. As shown in Fig. 6, poly(NVP-*co*-PBA) formed a complex with PVA, and the resultant complex can be dissociated in the presence of free glucose. The formation and dissociation of the complex between poly(NVP-*co*-PBA) and PVA were estimated by the viscosity change of the system. These results suggested that the complex is glucose-responsive, and novel insulin release systems shown in Fig. 6 can be constructed (59). For example, a glucose-responsive electrode was prepared by coating with a polymer complex hydrogel consisting of the copolymer with phenylboronic acid and PVA (60). In the glucose-responsive electrode, as the presence of free glucose induced swelling of the coating polymer hydrogel, increasing diffusivity of ion species resulted in the current changes. The fact that the current changes were proportional to glucose concentration suggested that they are indicative of high selectivity for glucose. However, this hydrogel is difficult to use in controlled release systems because of its intrinsically unstable complex at physiological pH of 7.4.

The complex between the phenylboronic acid and glucose was stabilized by the introduction of amino groups into the polymer or in the vicinity of the phenylboronic acid moiety due to the charge transfer interaction of nitrogen atom to boron atom under physiological pH 7.4 (61). A glucose-responsive hydrogel containing amino groups was prepared for the development of an advanced insulin delivery system at physiological pH. Gluconated insulin was bound with phenylboronic acid groups. As the complex between the phenylboronic acid and gluconated insulin can be dissociated in the presence of glucose by complex exchange between gluconated insulin and glucose, it can control insulin release in response to glucose (62).

The phenylboronic acid compounds are in equilibrium between the undissociated (or uncharged) and the dissociated (or charged) form (Fig. 7). Only charged phenylboronic acid groups can form a stable complex with glucose, and the complex between the uncharged form and glucose

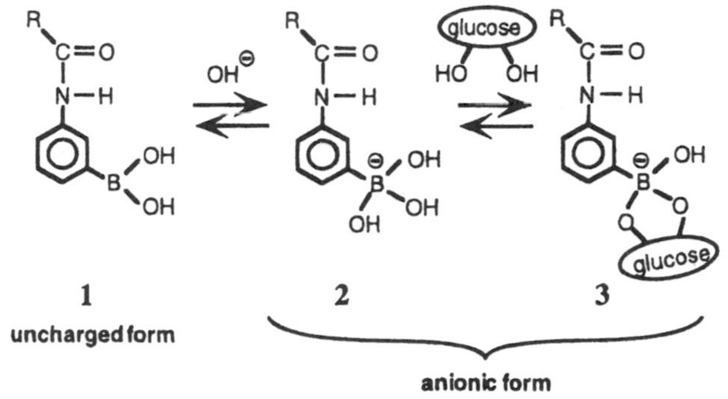

Figure 36.7 Equilibria of (alkylamido)phenyl boronic acid (1). (From Ref. 64.)

is unstable because of its high susceptibility to hydrolysis. When the charged phenylboronic acid groups form complexes with glucose, the equilibrium is shifted in the direction of increasing charged phenylboronic acid groups (2 + 3). As a result, the presence of glucose results in increasing total charged form (2 + 3) and in decreasing uncharged form. The ratio of the uncharged and charged forms strongly influences the solubility of the polymer in the aqueous solution, because the charged form is more hydrophilic than the uncharged form. To develop glucose-responsive devices, Kataoka et al. (61) used the shift in the equilibrium between the uncharged and charged forms of phenylboronic acid through its complex formation with glucose. They prepared a copolymer of *N,N*-dimethylacrylamide and PBA by radical copolymerization. The copolymer with phenylboronic acid groups was dissolved in a buffer solution of pH 7.4 at a temperature of less than 27°C but was insoluble at more than 27°C. This indicated that the copolymer with phenylboronic acid groups has LCST of about 27°C. The phenylboronic acid groups played an important role in the appearance of LCST, as the copolymer without the phenylboronic acid groups had no LCST. The addition of glucose to the aqueous solution of the copolymer with the phenylboronic acid groups led to a monotonic increase in LCST. The glucose-responsive LCST change is due to an increase in more hydrophilic charged phenylboronic acid groups by the complex formation between the phenylboronic acid groups and glucose.

Furthermore, the fascinating property of phenylboronic acid was combined with the temperature-responsive properties of PNIPAAm to improve the glucose-responsive LCST changes of the copolymers with the phenylboronic acid groups (63). A glucose-responsive ternary polymer was obtained by the copolymerization of NIPAAm, PBA, and *N*-(2-dimethylaminopropyl)acrylamide) (DMAPAA). The LCST of the ternary polymer was strongly dependent upon the glucose concentration (Fig. 8). This suggests that

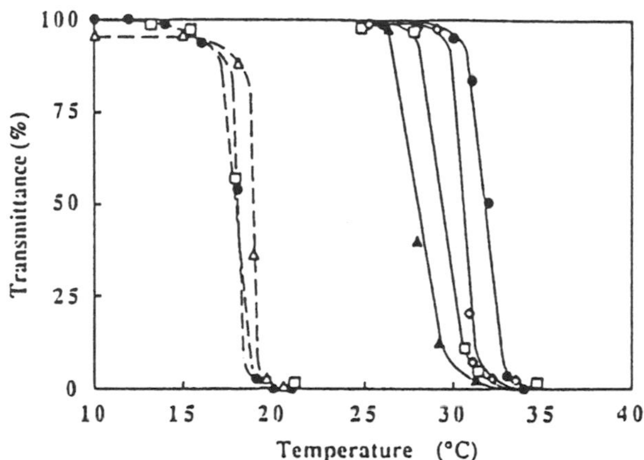

Figure 36.8 Temperature dependence of optical transmittance for IB (dotted line) and IAB (solid line) copolymers in phosphate buffer solutions (pH 7.4) in the absence and presence of glucose. Glucose concentration: 0 mg/mL,(●); 1 mg/mL, (◇); 2 mg/mL, (□);4 mg/mL, (▲); 10 mg/mL, (△). (From Ref. 63.)

the LCST of the ternary polymer can be controlled easily by a change in the external glucose concentration. Therefore a glucose-responsive hydrogel can be developed from the polymer containing both phenylboronic acid and NIPAAm, as the solubility of the polymer chain is strongly influenced by the glucose concentration at a constant temperature. Kataoka et al. (64) prepared totally synthetic polymer hydrogels composed of NIPAAm and phenylboronic acid, which undergo a sharp transition in the swelling ratio in response to glucose concentration. No insulin release from the hydrogel was observed in the buffer solution with a glucose concentration less than 1 g/L, but a remarkable release of insulin took place with a glucose concentration of 3 g/L. As shown in Fig. 9, furthermore, on-off regulation of insulin release was successfully repeated in response to stepwise changes in the glucose concentration.

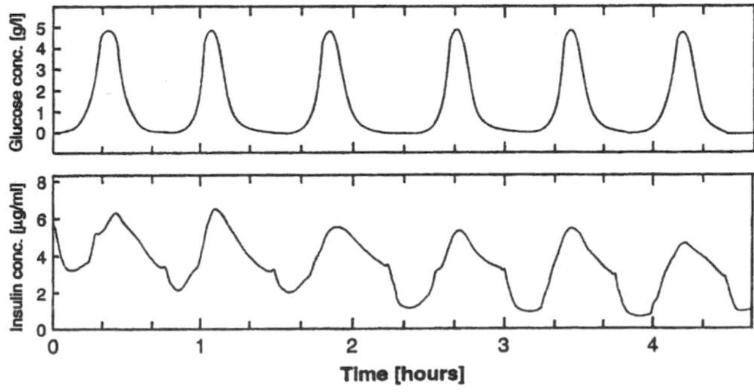

Figure 36.9 Repeated on–off release of FITC-insulin from the glucose-responsive hydrogel at 28°C, pH 9.0, in response to external glucose concentration. (From Ref. 64.)

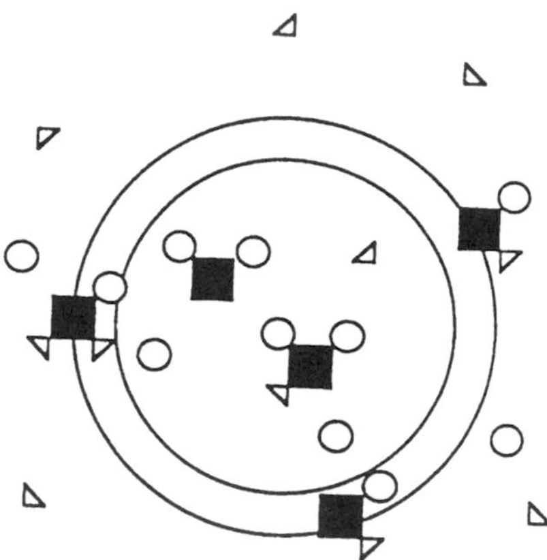

Figure 36.10 Schematic of Con A-glycosylated insulin derivative complex in a microcapsule. Con A (■), SAPG insulin (○), and glucose (△). (From Ref. 68.)

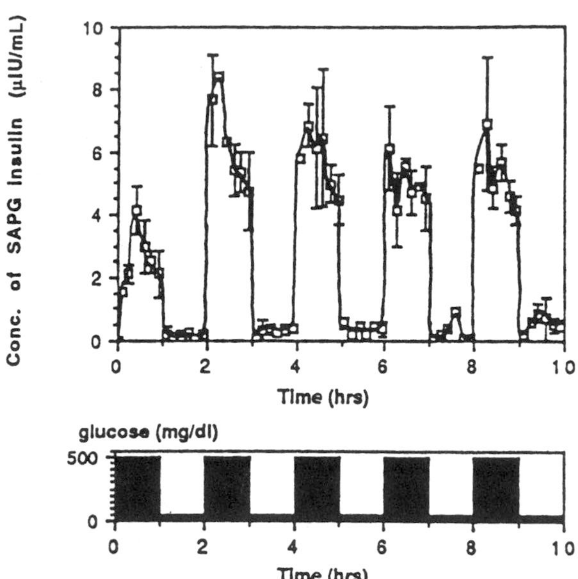

Figure 36.11 Release of SAPG insulin in response to stepwise glucose changes. (From Ref. 68.)

C. Lectin-Containing Hydrogels

Lectins are carbohydrate-binding proteins of nonimmune origin that do not exhibit enzymatic activity. They interact with glycoproteins and glycolipids on the cell surface and can induce a variety of effects such as cell agglutination, cell adhesion to surfaces, and hormonelike action. These biological effects are of considerable interest because of molecular recognition of hormones and toxins that bind to carbohydrate receptor sites on membranes.

Brownlee and Cerami (65) and Kim et al. (66,67) were pioneers in the development of glucose-responsive insulin release systems using lectin, and their work has been expanded further by various researchers. Their method of developing a physiological insulin delivery system is to synthesize a stable, biologically active glycosylated insulin derivative that can form a complex with lectin, concanavalin A (Con A). A glycosylated insulin derivative binding with lectin can be released as a function of the free glucose concentration. Furthermore, Kim et al. (68) enclosed the self-regulating insulin delivery systems in polymer membranes to soluble, bead-immobilized, or cross-linked Con A, based on the concept of competitive binding between succinyl-amidophenyl-glucopyranoside insulin (SAPG insulin) and glucose to Con A. SAPG insulin was released from polymer membranes or microcapsules containing Con A and SAPG insulin in a quick response to changes in glucose concentration (Fig. 10). Based on the concept of competitive and complementary binding properties of SAPG insulin and glucose to Con A, SAPG insulin was released through the membranes in response to the concentration of free glucose (Fig. 11). The polymer membranes and microcapsules containing Con A and SAPG insulin were good self-regulating insulin release devices in vitro and can be optimized for use in in vivo studies.

Kokufuta et al. (69) loaded Con A in the cross-linked PNIPAAm hydrogel, which undergoes a volume phase transition at 34°C. The Con A–loaded PNIPAAm hydrogel was abruptly swollen in the presence of the ionic saccharide dextran sulphate at temperatures close to the volume phase transition because of the ionic osmotic pressure exerted by ionized saccharide. However, replacing the ionic saccharide dextran sulphate with the nonionic saccharide induced the collapse of the hydrogel back to its native volume. The swelling and shrinking changes of Con A–loaded hydrogel were determined by a balance among the pressure due to repulsive electrostatic interactions of ionized dextran sulphate molecules, the osmotic pressure owing to counterions from saccharide dextran sulphate, and the attractive hydrophobic interaction among NIPAAm moieties.

Polymers with a pendant saccharide have attracted much attention because of their potential applications as suitable materials for biochemical and biomedical applications (70). In particular, a variety of synthetic polymers having well-defined saccharide residues can be used for the investigation of molecular recognition processes in the interactions between saccharides and proteins. The complexation between a polymer with pendant glucose groups, poly(2-glucosyloxyethyl methacrylate) (PGEMA), and Con A was investigated in the absence and presence of various free monosaccharides (71). PGEMA in water was flocculated by the addition of Con A, and the aqueous PGEMA solution became turbid owing to the complex formation between PGEMA and Con A. However, the addition of free glucose or mannose into the turbid PGEMA–Con A complex solution made the solution transparent, but the addition of free galactose did not. This is because the PGEMA–Con A complex is dissociated in the presence of free glucose or mannose by their exchange with the PGEMA, but the complex is not dissociated in the presence of free galactose. These results suggested that PGEMA can be recognized by Con A and that the affinity of Con A for PGEMA differs from the affinity for the free monosaccharides. Furthermore, the PGEMA–Con A complex can recognize the kind of free monosaccharides and is saccharide responsive. Therefore the PGEMA–Con A complex is a promising candidate for constructing a novel glucose sensor or an intelligent device for a glucose-responsive insulin release system. Similarly, Park et al. (72,73) investigated specific interaction between Con A and the copolymers containing allylglucose and revealed that hydrogels were formed immediately after mixing the copolymers and Con A. A sol–gel phase-reversible hydrogel responsive to glucose was prepared from the copolymers and Con A.

Recently, novel glucose-responsive hydrogels were prepared by application of the complex formation between Con A and pendant glucose groups in PGEMA to produce cross-linking points in hydrogels (74,75). The Con A–entrapped PGEMA hydrogels were obtained by copolymerizations of a monomer with a pendant glucose and a divinylmonomer in the presence of Con A. That the cross-linking density of the resultant hydrogels increased with increasing Con A suggested that Con A acts as a cross-linking point in the hydrogel due to complex formation between pendant glucose groups and Con A. When the Con A–entrapped PGEMA hydrogels were immersed in an aqueous glucose solution, their swelling ratio increased, depending on the glucose concentration. The glucose-responsive swelling of the Con A–entrapped PGEMA hydrogels was due to the dissociation of the complex between the pendant glucose groups and Con A by competitive exchange of pendant glucose groups with free glucose, as shown in Fig. 12. In addition, the Con A–entrapped PGEMA hydrogels were more strongly swollen in the presence of mannose than of glucose, but their swelling ratio did not change in the presence of galactose (Fig. 13). The presence of mannose can induce the dissociation of the complex between the pendant glucose groups and Con A more effectively than that of glucose, and the presence of galactose did not. This means that the Con A–entrapped PGEMA hydrogels can recognize a specific monosaccharide and respond to it. Therefore the Con A–entrapped PGEMA hydrogel has many applications in a novel glucose sensor and in a potential device for new closed-loop insulin-delivery systems. Obaidat and Park (76) also prepared glucose-responsive phase-reversible hydrogels based on the specific interaction between polymer-bound glucose and Con A. Release of lysozyme and insulin as model proteins through the glucose-responsive hydrogel membranes was examined using a diffusion cell consisting of two chambers. Poly(hydroxyethyl methacrylate) (PHEMA) membranes, which allowed diffusion of glucose and model proteins but prevented a glucose-containing polymer and Con A from being released, were used to sandwich the mixture of the glucose-containing polymer and Con A between the diffusion chambers. The release of model proteins through the glucose-responsive hydrogel membrane was strongly dependent on the concentration of

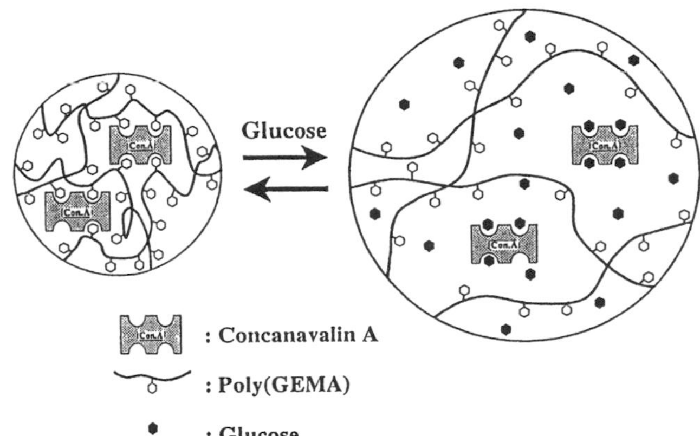

Figure 36.12 Schematic representation of glucose-responsive swelling changes of the poly(GEMA)–Con A hydogel. (From Ref. 74.)

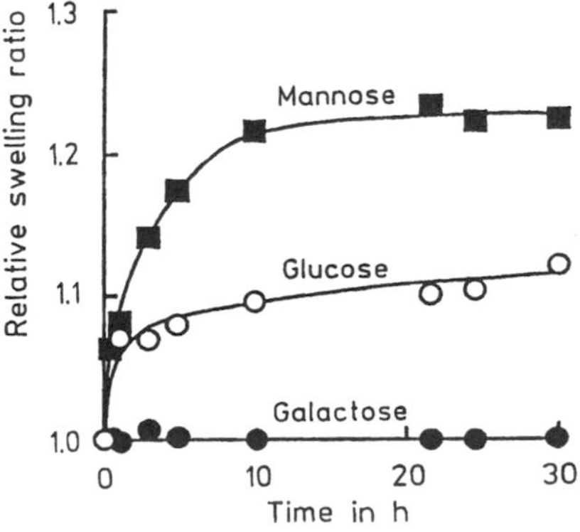

Figure 36.13 Swelling ratio changes of poly(GEMA)–Con A hydrogel as a function of time, when the hydrogels are immersed in the buffer solution (pH 7.5) containing 1 wt% of monosaccharide: (○) glucose; (■) mannose; (●) galactose. (From Ref. 74.)

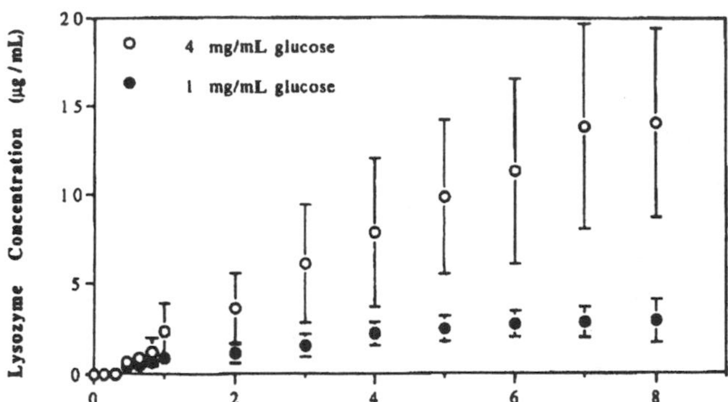

Figure 36.14 Release of lysozyme through a glucose-responsive hydrogel membrane at the initial glucose concentrations of 1 (●) and 4 (○) mg/mL. (From Ref. 76.)

free glucose (Fig. 14). Their studies demonstrated that the glucose-responsive phase-reversible hydrogels are applicable to regulate the insulin release in response to the free glucose concentration in the environment.

IV. ANTIGEN-RESPONSIVE HYDROGELS

Antibodies have unique recognition sites that bind antigen specifically, and their antigen recognition and binding are associated with the complex immune responses to protect the organism from infection. The antigen–antibody binding occurs through multiple noncovalent bonds such as electrostatic, hydrogen, hydrophobic, and van der Waals. Therefore antibodies are widely employed in a great number of immunological assays to detect and measure biological and nonbiological substances in many fields because of their exceptional specificity, stability, and versatility (77).

Recently, antigen–antibody bindings were used for preparing novel antigen-responsive hydrogels because the specific antigen recognition of an antibody can provide the basis for fabricating sensing devices with a wide variety of uses for immunoassay and antigen sensing. The novel antigen-responsive hydrogels were prepared by the application of the antigen–antibody binding at cross-linking points in the hydrogel (78,79). The fundamental strategy for the preparation of antigen-responsive hydrogels is shown in Fig. 15. Rabbit immunoglobulin G (IgG) as an antigen was chemically modified by coupling it with *N*-succinimidylacrylate (NSA) in phosphate buffer solution to introduce vinyl groups into the rabbit IgG. After the resultant vinyl rabbit IgG was purified by gel filtration, it was mixed with goat antirabbit IgG (GAR lgG) as an antibody to form an antigen–antibody binding. Then the vinyl-rabbit IgG was copolymerized with acrylamide (AAm) as

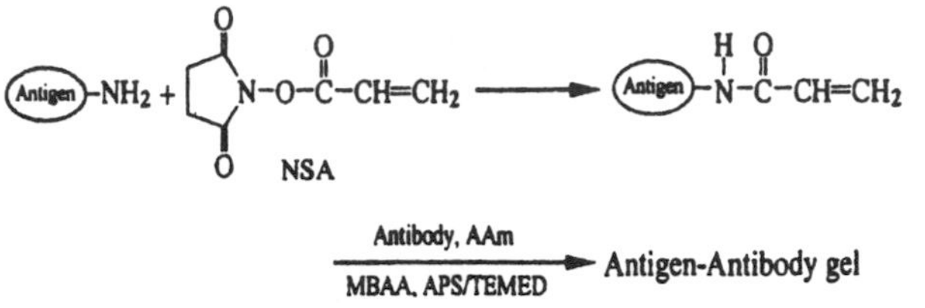

Antigen$-NH_2$ + NSA ($N-O-C(=O)-CH=CH_2$) ⟶ Antigen$-N(H)-C(=O)-CH=CH_2$

Antibody, AAm / MBAA, APS/TEMED ⟶ Antigen-Antibody gel

Figure 36.15 Synthesis of an antigen–antibody hydrogel. (From Ref. 78.)

a comonomer and *N,N′*-methylenebisacrylamide (MBAA) as a cross-linker in the presence of GAR IgG, to prepare a hydrogel having antigen–antibody bindings (antigen–antibody hydrogel). The antigen–antibody hydrogel was swollen by the addition of a free rabbit IgG in the buffer solution after its swelling attained equilibrium in the buffer solution. This means that the antigen–antibody hydrogel is rabbit-IgG responsive. Furthermore, the swelling changes to the antigen–antibody hydrogel caused by the presence of IgG from a different species were investigated in order to determine whether the effect is due to specific molecular recognition. The addition of free rabbit IgG resulted in a dramatic increase in the swelling ratio of the antigen–antibody hydrogel, but the addition of free goat IgG did not (Fig. 16). The antigen-responsive swelling of

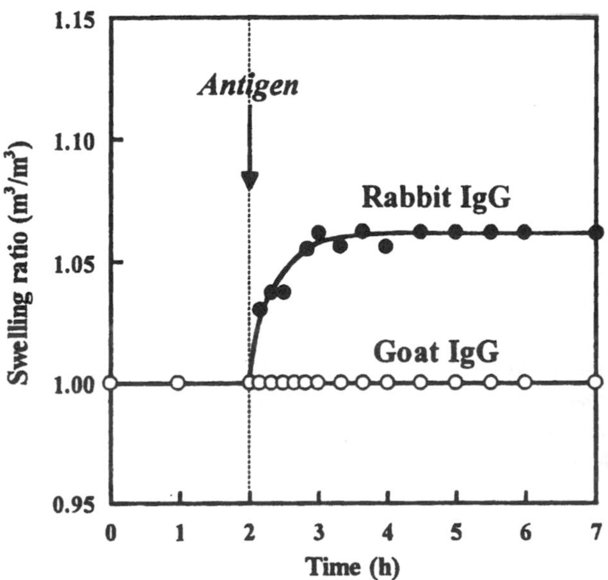

Figure 36.16 Changes in the swelling ratio of the antigen–antibody hydrogel by the addition of goat IgG (○) and rabbit IgG (●) after swelling attained equilibrium in the phosphate buffer solution. The concentration of the antigen in the phosphate buffer solution is 4 mg/mL. (From Ref. 78.)

the antigen–antibody hydrogel could be explained by the decrease in the cross-linking density in the hydrogel because of the dissociation of the antigen–antibody binding in the presence of a free antigen. These results suggest that the antigen–antibody hydrogel can recognize only rabbit IgG and exhibit structural changes. Other researches on molecule-responsive hydrogels reported previously focused on the hydrogels that undergo swelling changes in response to only low molecular weight compounds such as glucose described in Section III. Studies on the antigen-responsive hydrogels suggest a simple solution to the problem that is of relevance to many aspects of the development of specific macromolecule-responsive hydrogels that can respond to macromolecules such as protein and antigen.

Most potential applications of stimulus-responsive hydrogels require that their swelling behavior be reversible. However, the antibody was leaked out of the antigen–antibody hydrogel during the swelling in response to the antigen, as the antibody became free in the swollen hydrogel. The antigen–antibody entrapment hydrogel was not shrunken in the buffer solution without the antigen, and showed no reversibility in the antigen-responsive swelling changes. Therefore structures of the antigen–antibody hydrogel must be designed so that the antibody can form a binding with the antigen grafted with gel networks again in the buffer solution without the antigen. One approach is to introduce a semi-interpenetrating polymer network (semi-IPN) structure into the antigen–antibody hydrogel in order to immobilize the antibody in the gel network. A reversibly antigen-responsive hydrogel was prepared based on the strategy that the reversible binding between an antigen and an antibody is used as the cross-linking mechanism in a semi-IPN hydrogel (Fig. 17a) (79). The synthetic procedure of the antigen–antibody semi-IPN hydrogel is shown in Fig. 17b. As described previously, rabbit IgG as an antigen and GAR IgG as an antibody were chemically modified by coupling them with NSA to synthesize a vinyl antigen and a vinyl antibody. The resultant vinyl GAR IgG was copolymerized with AAm to synthesize the polymerized GAR IgG that acts as a linear chain in a semi-IPN hydrogel. Then the antigen–antibody semi-IPN hydrogel was prepared by the copolymerization of the vinyl rabbit IgG, AAm, and MBAA as a cross-linker in the presence of the polymerized GAR IgG. The linear polymerized antibody was interpenetrated through the network having the antigen in the resultant antigen–antibody semi-IPN hydrogel. The linear polymerized antibody was not leaked out of the semi-IPN hydrogel because it was entangled with the network. The antigen–antibody semi-IPN hydrogel was swollen immediately following the addition of free rabbit IgG into the buffer solution after its

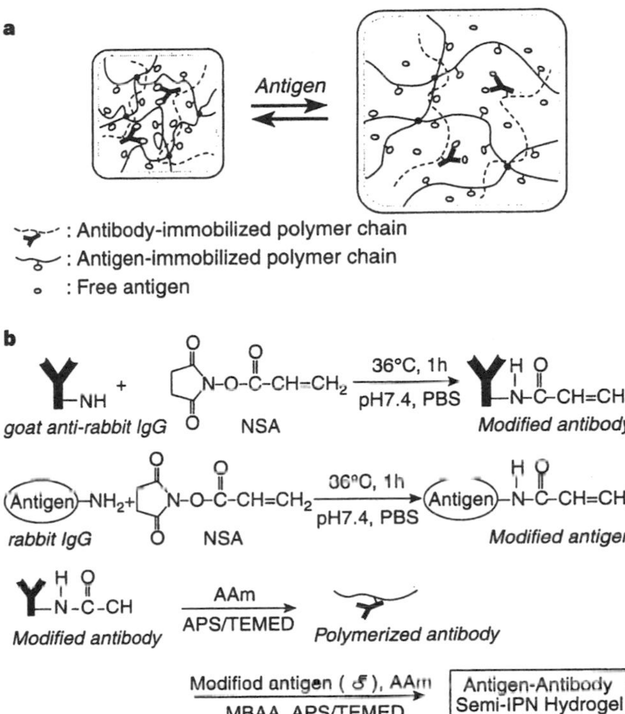

Figure 36.17 Strategy for the preparation of an antigen-responsive semi-IPN hydrogel. (From Ref. 79.)

swelling attained equilibrium in the buffer solution. In addition, the swelling ratio of the antigen–antibody semi-IPN hydrogel was strongly dependent upon the antigen concentration in the buffer solution. The antigen–antibody semi-IPN hydrogel did not recognize goat IgG but only rabbit IgG (Fig. 18). Because the presence of goat IgG did not result in the dissociation of the complex between the poly-

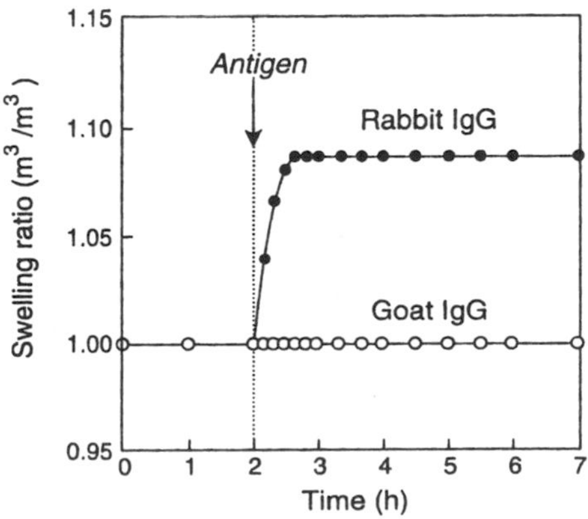

Figure 36.18 Antigen recognition by the antigen–antibody semi-IPN hydrogel. (From Ref. 79.)

merized rabbit IgG and polymerized GAR IgG due to the antigen recognition of the GAR IgG, the swelling ratio of the antigen–antibody semi-IPN hydrogel did not change by the addition of goat IgG. These mean that the antigen–antibody semi-IPN hydrogel can recognize a specific antigen and change its structure chemomechanically. Furthermore, the swelling changes in response to stepwise changes in the antigen concentration were examined for investigating the reversibility of the antigen-responsive swelling changes. Figure 19 shows the swelling ratio change of the antigen–antibody semi-IPN hydrogel as a function of time when it was alternately immersed in buffer solution with and without a free antigen. The antigen–antibody semi-IPN hydrogel was swollen immediately in the presence of the free antigen and shrunken gradually in its absence. In contrast, the antigen–antibody entrapment hydrogel without a semi-IPN structure described previously was also swollen in the buffer solution with the free antigen but was not shrunken following immersion in the buffer solution without the free antigen. Therefore only antigen–antibody semi-IPN hydrogel undergoes reversible changes in its swelling ratio in response to stepwise changes in the antigen concentration. This reveals that the semi-IPN structure plays an important role in the reversibility of the antigen-responsive swelling changes. In the semi-IPN hydrogel, the polymerized GAR IgG is trapped in a network containing grafted rabbit IgG, and so the hydrogel can shrink reversibly because the cross-links between polymerized GAR IgG and grafted rabbit IgG reform in buffer solution without the free antigen.

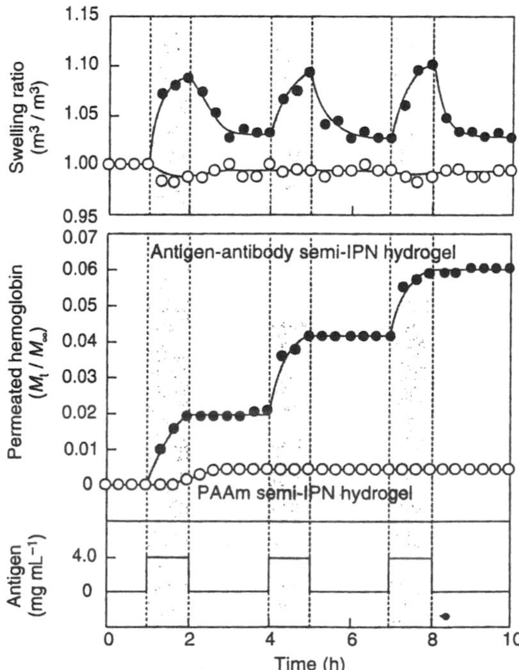

Figure 36.19 Reversible swelling changes and antigen-responsive permeation profiles of hemoglobin through the PAAm semi-IPN hydrogel (○) and the antigen–antibody semi-IPN hydrogel (●) in response to stepwise changes in the antigen concentration between 0 and 4 mg/mL. (From Ref. 79.)

Stimulus-responsive swelling changes of smart hydrogels can control the permeability of solutes such as drugs, and such stimulus-responsive hydrogels are promising candidates of intelligent devices for novel drug delivery systems. Therefore to investigate the possibility of the antigen-responsive hydrogel as the intelligent device, the pulsatile permeation of a model protein drug in response to stepwise changes in the antigen concentration was also investigated with a membrane of the antigen–antibody semi-IPN hydrogel (79). As a result, it became apparent from the permeation profile in Fig. 19 that a model drug is permeated through the antigen–antibody semi-IPN hydrogel membrane in the presence of free rabbit IgG, but the drug permeation is stopped in its absence. That is, the pulsatile permeation of a model drug in response to the specific antigen concentration can be achieved by using the antigen–antibody semi-IPN hydrogel. This means that the antigen–antibody semi-IPN hydrogel is useful as an intelligent device for a novel drug delivery system in which a drug can be released in the presence of a specific antigen and the drug release can be stopped in its absence. For example, drug targeting to a cancer with a specific antigen, in which an anticancer compound is released selectively to a cancer in response to its antigen, can be achieved by using such antigen-responsive hydrogels. Furthermore, some findings in these studies imply that a specific biochemical interaction, the binding between biomolecules, can be converted into a mechanical force. Developing specific molecule-responsive hydrogels using the method will contribute significantly to creating a novel intelligent system in which specific interactions can be converted into mechanical force.

V. OTHER STIMULUS-RESPONSIVE HYDROGELS

A. Molecular Imprint Hydrogels

Some proteins such as enzymes and antibodies have a molecular recognition ability that is based on fitting guest molecules into their molecular cavities. Molecular imprinting is an attractive technique to synthesize polymers having a molecular recognition function, and it has potential practical applications in separation science and technology (80,81). In most of the recent studies, after the functionalized monomer is prearranged around the print molecule by noncovalent interactions such as electrostatic, hydrophobic, and hydrogen bonding, it is polymerized by interacting with a print molecule. Then the print molecule is removed from the resultant polymer to form a molecular cavity. The cavity formed by the removal of the print molecule can recognize the guest molecule (print molecule) due to a combination of reversible binding and shape complementarity. Watanabe et al. (82) showed that the molecular imprinting is applicable to the preparation of the molecule-responsive hydrogels that undergo a large swelling change in accordance with the print molecules. They prepared temperature-responsive copolymer hydrogels consisting of NIPAAm and acrylic acid in the presence of guest molecules as print molecules. The hydrogels were swollen at a low temperature and collapsed at a high temperature. The swelling ratio of the hydrogel swollen at a low temperature was independent of the guest molecule concentration, but that of the hydrogel collapsed at a high temperature increased with increasing concentration. This suggests that the hydrogel in the collapsed state can respond to the guest molecule. It is noticeable that the guest molecule–responsiveness of the hydrogel prepared by the molecular imprinting technique was strongly affected by the preparation condition. The hydrogel prepared in 1,4-dioxane in the presence of the guest molecule exhibited the specific volume change, but that prepared in water did not. Figure 20 shows the swelling ratio changes at 50°C of the temperature-responsive hydrogels prepared in the presence of norephedrine and adrenaline as guest molecules. The hydrogel prepared in the presence of norephedrine was swollen with increasing norephedrine concentration but was not changed by increasing adrenaline, and vice versa for the hydrogel prepared in the presence of adrenaline. As the temperature-responsive hydrogels prepared in the presence of a guest molecule memorize the guest molecule, the hydrogels show specific volume change in response to the guest molecule owing to its specific adsorption into hydrogels.

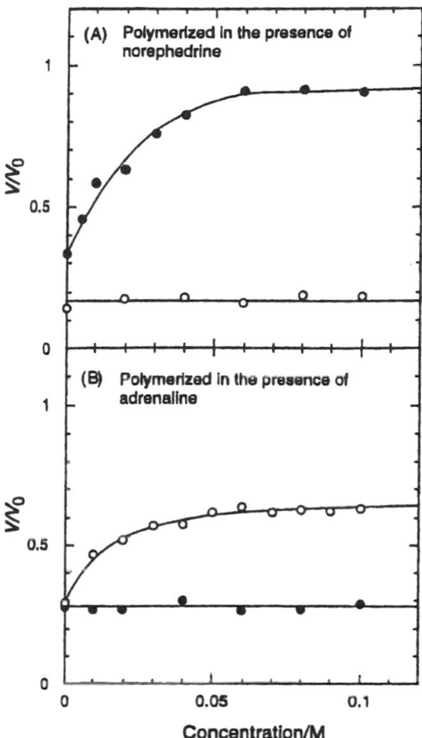

Figure 36.20 Equilibrium swelling ratios at 50°C as a function of concentration of either norephedrine (●) or adrenaline (○) for molecular recognition hydrogels prepared in the presence of norphedrine (A) and adrenaline (B). (From Ref. 82.)

B. Other Biomolecule-Responsive Hydrogels

Deoxyribonucleic acid (DNA) and ribonucleic acid (RNA) are composed of the nucleotides adenine, cytosine, guanine, thymine, and uracil and form double or triple strands with its complementary base pairs by hydrogen bonding and stacking of its bases. Aoki et al. (83) focused on uracil in the nucleic acid bases and synthesized poly(6-(acryloyloxymethyl)uracil) (PAU) having uracil moieties as side chains. PAU was insoluble in water due to the polymer complexes between uracil moieties at lower temperatures and became water soluble above a characteristic transition temperature, showing an upper critical solution temperature (UCST). UCST of PAU was shifted to a lower temperature by the presence of adenosine in aqueous solutions because the uracil moieties form complexes with complementary adenosine. Furthermore, additions of adenosin and guanosine had different effects on the phase transition changes due to the different interaction of uracil moieties with adenosin and guanosin. This means that PAU changes its phase transition temperature in response to species of the additive nucleic acid bases. These results imply that hydrogels prepared from PAU must recognize species of the nucleic acid bases, and that they may undergo swelling ratio changes in response to the concentration of the nucleic acid bases.

In addition, Aoki et al. (84) synthesized a temperature-responsive copolymer composed of *N*-(*S*)-*sec*-butylacrylamide ((*S*)-*sec*-BAAm) and NIPAAm that exhibits hydration change in response to foreign optically active compounds as external stimuli. The resultant poly((*S*)-*sec*-BAAm-*co*-NIPAAm) with a (*S*)-*sec*-BAAm content of 50 mol% has LCST at 23.1°C. The presence of L-tryptophan

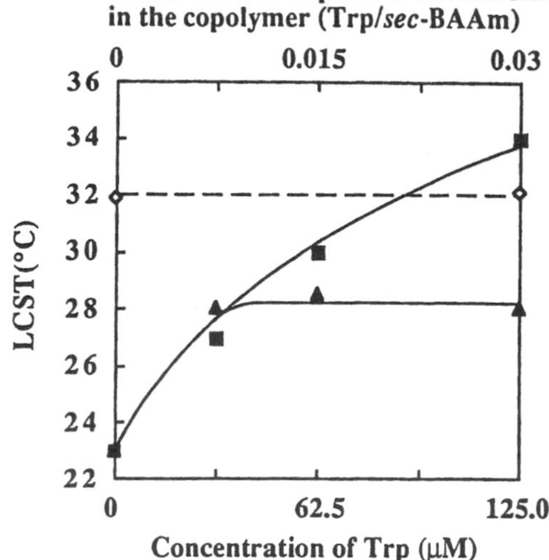

Figure 36.21 Change in LCST of (S) BP-50 in water as a function of Trp concentration. LCST is defined as the temperature at 50% optical transmittance. (■) L-Trp; (▲) D-Trp; (◇) PNIPAAm + L-Trp. (From Ref. 84.)

(L-Trp) resulted in a remarkable shift in LCST to 34.5°C, but that of D-Trp did so to 28.7°C (Fig. 21). On the other hand, a racemic (*R,S*)-*sec*-BAAm–containing copolymer showed the same phase transition temperatures in the presence of D- and L-Trp. Therefore, the remarkable shift in LCST of poly((*S*)-*sec*-BAAm-*co*-NIPAAm) is attributable to the stereospecifical interaction between the optically active (*S*)-*sec*-BAAm in the copolymer and L-Trp. Furthermore, the LCSTs of the copolymers were strongly dependent upon the L-Trp concentration. Therefore these results revealed that the temperature-responsive copolymer with optically active moieties can respond to the enantiomers. Based on its concept, Aoki et al. (85) prepared temperature-responsive poly((*S*)-*sec*-BAAm-*co*-NIPAAm) hydrogel and poly(*N*-(L)-(1-hydroxymethyl) propylmethacrylamide) (PHMPMA) hydrogel with optically active moieties that undergo phase transitions. The temperature dependence of the swelling ratio of their hydrogels was remarkably influenced by the addition of L-Trp. Therefore their hydrogels can recognize the difference between L- and D-Trp and can be called amino acid–responsive hydrogels.

REFERENCES

1. N. A. Peppas. *Hydrogels in Medicine and Pharmacy*. CRC Press, Boca Raton, FL, 1987.
2. D. DeRossi, K. Kajiwara, Y. Osada, A. Yamauchi. *Polymer Gels, Fundamentals and Biomedical Applications*. Plenum Press, New York, 1991.
3. K. Dusek. *Responsive Gels: Volume Transitions I*. Adv. Polym. Sci. 109. Springer-Verlag, Berlin, 1993.
4. K. Dusek. *Responsive Gels: Volume Transitions II*. Adv. Polym. Sci. 110. Springer-Verlag, Berlin, 1993.
5. T. Okano. *Biorelated Polymers and Gels*. Academic Press, Boston, 1998.
6. T. Tanaka. Collapse of gels and the critical endpoint. *Phys. Rev. Lett. 40*:820 (1978).
7. T. Tanaka, D. Fillmore, S.-T. Sun, I. Nishio, G. Swislow, A. Shah. Phase transitions in ionic gels. *Phys. Rev. Lett. 45*:1636 (1980).
8. Y. Hirokawa, T. Tanaka. Volume phase transition in a nonionic gel. *J. Chem. Phys. 81*:6379 (1984).
9. T. Amiya, Y. Hirokawa, Y. Hirose, Y. Li, T. Tanaka. Reentrant phase transition of N-isopropylacrylamide gels in mixed solvents. *J. Chem. Phys. 86*:2375 (1987).
10. M. Annaka, T. Tanaka. Multiple phases of polymer gels. *Nature 355*:430 (1992).
11. G. Chen, A. S. Hoffman. Graft copolymers that exhibit temperature-induced phase transitions over a wide range of pH. *Nature 373*:49 (1995).
12. T. Tanaka, I. Nishio, S.-T. Sun, S. Ueno-Nishio. Collapse of gels in an electric field. *Science 218*:467 (1982).
13. Y. Osada, H. Okuzaki, H. Hori. A polymer gel with electrically driven motility. *Nature 355*:242 (1992).
14. M. Irie. Stimuli-responsive poly(N-isopropylacrylamide). Photo- and chemical-induced phase transitions. *Adv. Polym. Sci. 110*:49 (1993).
15. A. Suzuki, T. Tanaka. Phase transition in polymer gels induced by visible light. *Nature 346*:345 (1990).
16. D. Szabo, G. Szeghy, M. Zrinyi. Shape transition of magnetic field sensitive polymer gels. *Macromoleules 31*:6541 (1998).
17. R. Yoshida, T. Takahashi, T. Yamaguchi, H. Ichijo. Self-oscillating gel. *J. Am. Chem. Soc. 118*:5134 (1996).
18. R. A. Siegel. Hydrophobic weak polyelectrolyte gels: studies of swelling equilibria and kinetics. *Adv. Polym. Sci. 109*: 233 (1993).
19. B. A. Firestone, R. A. Siegel. Dynamic pH-dependence swelling properties of a hydrophobic polyelectrolyte gel. *Polym. Commun. 29*:204 (1988).
20. R. A. Siegel, M. Falamarzian, B. A. Firestone, B. C. Moxley. *J. Controlled Release 8*:179 (1988).
21. L. B. Peppas, N. A. Peppas. Solute and penetrants diffusion in swellable polymers. IX. The mechanisms of drug release from pH-sensitive swelling controlled systems. *J. Controlled Release 8*:267 (1989).
22. L. F. Gudeman, N. A. Peppas. pH-sensitive membranes from poly(vinyl alcohol)/poly(acrylic acid) interpenetrating networks. *J. Membrane Sci. 107*:239 (1995).
23. C. S. Brazel, N. A. Peppas. Synthesis and characterization of thermo- and chemomechanically responsive poly(N-isopropylacrylamide-co-methacrylic acid) hydrogels. *Macromolecules 28*:8016 (1995).

24. L.-C. Dong, A. S. Hoffman. A novel approach for preparation of pH-sensitive hydrogels for enteric drug delivery. *J. Controlled Release 15*:141 (1991).
25. K. Nakamae, T. Miyata, A. S. Hoffman. Swelling behavior of hydrogels containing phosphate groups. *Makromol. Chem. 193*:983 (1992).
26. T. Miyata, K. Nakamae, A. S. Hoffman, Y. Kanzaki. Stimuli-sensitivities of hydrogels containing phosphate groups. *Macromol. Chem. Phys. 195*:1111 (1994).
27. K. Nakamae, T. Nizuka, T. Miyata, M. Furukawa, T. Nishino, K. Kato, T. Inoue, A. S. Hoffman, Y. Kanzaki. Lysozyme loading and release from hydrogels carrying pendant phosphate groups. *J. Biomater. Sci. Polym. Ed. 9*:43 (1997).
28. K. Nakamae, T. Nizuka, T. Miyata, T. Uragami, A. S. Hoffman, Y. Kanzaki. Stimuli-sensitive release of lysozyme from hydrogel containing phosphate groups. *Advanced Biomaterials in Biomedical Engineering and Drug Delivery Systems* (N. Ogata, S. W. Kim, J. Feijen, T. Okano, eds.). Springer-Verlag, Tokyo, 1996, p. 313.
29. S. Fujishige, K. Kubota, I. Ando. Phase transition of aqueous solutions of poly(N-isopropylacrylamide) and poly(N-isopropylmethacrylamide). *J. Phys. Chem. 93*:3311 (1989).
30. A. S. Hoffman. Application of thermally reversible polymers and hydrogels in therapeutics and diagnostics. *J. Controlled Release 6*:297 (1987).
31. T. Okano. Molecular design of temperature-responsive polymers as intelligent materials. *Adv. Polym. Sci. 110*:179 (1993).
32. A. S. Hoffman, A. Afrassiabi, L. C. Dong. Thermally reversible hydrogels: II. Delivery and selective removal of substances from aqueous solutions. *J. Controlled Release 4*:213 (1986).
33. L.-C. Dong, A. S. Hoffman. Synthesis and application of thermally reversible heterogels for drug delivery. *J. Controlled Release 13*:21 (1990).
34. Y. H. Bae, T. Okano, R. Hsu, S. W. Kim. Thermo-sensitive polymers as on-off switches for drug release. *Makromol. Chem., Rapid Commun. 8*:481 (1987).
35. Y. H. Bae, T. Okano, S. W. Kim. Temperature dependence of swelling of crosslinked poly(N, N′-alkyl substituted acrylamide) in water. *J. Polym. Sci., Polym. Phys. 28*:923 (1990).
36. T. Okano, Y. H. Bae, H. Jacobs, S. W. Kim. Thermally on-off switching polymers for drug permeation and release. *J. Controlled Release 11*:255 (1990).
37. T. Okano, R. Yoshida, K. Sakai, Y. Sakurai. Thermo-responsive polymeric hydrogels and their application to pulsatile drug delivery. In: *Polymer Gels* (D. DeRossi, ed.). Plenum Press, New York, 1991, pp. 299–308.
38. T. Okano, R. Yoshida. Polymer for pharmaceutical and biomolecular engineering. *Biomedical Applications of Polymeric Materials* (T. Tsuruta, T. Hayashi, K. Kataoka, K. Ishihara, Y. Kimura, eds.). CRC Press, Boca Raton, FL, 1993, p. 407.
39. R. Yoshida, Y. Okuyama, K. Sakai, T. Okano, Y. Sakurai. Sigmoidal swelling profiles for temperature-responsive poly(N-isopropylacrylamide-co-butyl methacrylate) hydrogel. *J. Membrane Sci. 89*:267 (1994).
40. R. Yoshida, K. Uchida, T. Kaneko, K. Sakai, A. Kikuchi, Y. Sakurai, T. Okano. Comb-type grafted hydrogels with rapid de-swelling response to temperature changes. *Nature 374*:240 (1995).
41. Y. Kaneko, K. Sakai, A. Kikuchi, R. Yoshida, Y. Sakurai, T. Okano. Influence of freely mobile grafted chain length on dynamic properties of comb-type grafted poly(N-isopropylacrylamide) hydrogels. *Macromolecules 28*:7717 (1995).
42. Y. Kaneko, K. Sakai, A. Kikuchi, Y. Sakurai, T. Okano. Fast swelling/deswelling kinetics of comb-type grafted poly(N-isopropylacrylamide) hydrogels. *Macromol. Symp. 109*:41 (1996).
43. H. Katono, A. Maruyama, K. Sanui, N. Ogata, T. Okano, Y. Sakurai. Thermo-responsive swelling and drug release switching of interpenetrating polymer networks composed of poly(acrylamide-co-butyl methacrylate) and poly(acrylic acid). *J. Controlled Release 16*:215 (1991).
44. H. Katono, K. Sanui, N. Ogata, T. Okano, Y. Sakurai. Drug release OFF behavior and deswelling kinetics of thermo-responsive IPNs composed of poly(acrylamide-co-butyl methacrylate) and poly(acrylic acid). *Polym. J. 23*:1179 (1991).
45. T. Aoki, M. Kawashima, H. Katono, K. Sanui, N. Ogata, T. Okano, Y. Sakurai. Temperature-responsive interpenetrating polymer networks constructed with poly(acrylic acid) and poly(N,N-dimethylacrylamide). *Macromolecules 27*:947 (1994).
46. K. L. Wang, J. H. Burban, E. L. Cussler. Hydrogels as separation agents. *Adv. Polym. Sci. 110*:67 (1993).
47. R. S. Freitas, E. L. Cussler. Temperature sensitive gels as size selective adsorbants. *Sep. Sci. Technol. 22*:911 (1987).
48. H. Feil, Y. H. Bae, J. Feijen, S. W. Kim. Molecular separation by thermosensitive hydrogel membranes. *J. Membrane Sci. 64*:283 (1991).
49. L. C. Dong, A. S. Hoffman. Thermally reversible hydrogels; III. Immobilization of enzymes for feedback reaction control. *J. Controlled Release 4*:223 (1986).
50. T. G. Park, A. S. Hoffman. Immobilization and characterization of β-galactosidase in thermally reversible hydrogel beads. *J. Biomed. Mater. Res. 24*:21 (1990).
51. K. Ishihara, M. Kobayashi, N. Ishimaru, I. Shinohara. Glucose induced permeation control of insulin through a complex membrane consisting of immobilized glucose oxidase and a poly(amine). *Polym. J. 16*:625 (1984).
52. K. Ishihara, K. Matsui. Glucose-responsive insulin release from polymer capsule. *J. Polym. Sci.: Polym. Lett. Ed. 24*: 413 (1986).
53. G. Albin, T. A. Horbett, B. D. Ratner. Glucose sensitive membranes for controlled delivery of insulin: insulin transport studies. *J. Controlled Release 2*:153 (1985).
54. G. W. Albin, T. A. Horbett, S. R. Miller, N. L. Ricker. Theoretical and experimental studies of glucose sensitive membranes. *J. Controlled Release 6*:267 (1987).
55. S. Cartier, T. A. Horbett, B. D. Ratner. Glucose-sensitive

membrane coated porous filters for control of hydraulic permeability and insulin delivery from a pressurized reservoir. *J. Membrane Sci. 106*:17 (1995).

56. C. M. Hassan, F. J. Doyle III, N. A. Peppas. Dynamic behavior of glucose-responsive poly(methacrylic acid-g-ethylene glycol) hydrogels. *Macromolecules 30*:6166 (1997).
57. R. S. Parker, F. J. Doyle, III, N. A. Peppas. A model-based algorithm for blood glucose control in type I diabetic patients. *IEEE Trans. Biomed. Eng. 46*:148 (1999).
58. S. Kitano, K. Kataoka, Y. Koyama, T. Okano, Y. Sakurai. Glucose-responsive complex formation between poly(vinyl alcohol) and poly(N-vinyl-2-pyrrolidone) with pendent phenylboronic acid moieties. *Makromol. Chem., Rapid Commun. 12*:227 (1991).
59. S. Kitano, Y. Koyama, K. Kataoka, T. Okano, Y. Sakurai. A novel drug delivery system utilizing a glucose responsive polymer complex between poly(vinyl alcohol) and poly(N-vinyl-2-pyrrolidone) with a phenylboronic acid moiety. *J. Control. Release 19*:162 (1992).
60. A. Kikuchi, K. Suzuki, O. Okabayashi, H. Hoshino, K. Kataoka, Y. Sakurai, T. Okano. Glucose-sensing electrode coated with polymer complex gel containing phenylboronic acid. *Anal. Chem. 68*:823 (1996).
61. K. Kataoka, H. Miyazaki, T. Okano, Y. Sakurai. Sensitive glucose-induced change of the lower critical solution temperature of poly[N,N-dimethylacrylamide-co-3-(acrylamido)phenylboronic acid] in physiological saline. *Macromolecules 27*:1061 (1994).
62. D. Shiino, Y. Murata, A. Kubo, Y. J. Kim, K. Kataoka, Y. Koyama, A. Kikuchi, M. Yokoyama, Y. Sakurai, T. Okano. Amine containing phenylboronic acid gel for glucose-responsive insulin release under physiological pH. *J. Control. Release 37*:269 (1995).
63. T. Aoki, Y. Nagao, K. Sanui, N. Ogata, A. Kikuchi, Y. Sakurai, K. Kataoka, T. Okano. Glucose-sensitive lower critical solution temperature changes of copolymers composed of N-isopropylacrylamide and phenylboronic acid moieties. *Polym. J. 28*:371 (1996).
64. K. Kataoka, H. Miyazaki, M. Bunya, T. Okano, Y. Sakurai. Totally synthetic polymer gels responding to external glucose concentration: their preparation and application to on-off regulation of insulin release. *J. Am. Chem. Soc. 120*: 12694 (1998).
65. M. Brownlee, A. Cerami. A glucose-controlled insulin delivery system: semisynthetic insulin bound to lectin. *Science 206*:1190 (1979).
66. L. A. Seminoff, G. B. Olsen, S. W. Kim. A self-regulating insulin delivery system. I. Characterization of a synthetic glycosylated insulin derivative. *Int. J. Pharm. 54*:241 (1989).
67. S. W. Kim, C. M. Pai, K. Makino, L. A. Seminoff, D. L. Holmberg, J. M. Gleeson, D. E. Wilson, E. J. Mack. Self-regulated glycosylated insulin delivery. *J. Control. Release 11*:193 (1990).
68. K. Makino, E. J. Mack, T. Okano, S. W. Kim. A microcapsule self-regulating delivery system for insulin. *J. Control. Release 12*:235 (1990).
69. E. Kokufuta, Y.-Q. Zhang, T. Tanaka. Saccharide-sensitive phase transition of a lectin-gel. *Nature 351*:302 (1991).
70. T. Miyata, K. Nakamae. Polymers with pendant saccharides—'Glycopolymers.' *Trend. Polym. Sci. 5*:198 (1997).
71. K. Nakamae, T. Miyata, A. Jikihara, A. S. Hoffman. Formation of poly(glucosyloxyethyl methacrylate)–concanavalin A complex and its glucose-sensitivity. *J. Biomater. Sci., Polym. Ed. 6*:79 (1994).
72. S. J. Lee, K. Park. Synthesis and characterization of sol–gel phase-reversible hydrogels sensitive to glucose. *J. Molecular Recognition 9*:549 (1996).
73. A. A. Obaidat, K. Park. Characterization of glucose dependent gel–sol phase transition of the polymeric glucose–concanavalin A hydrogel system. *Pharm. Res. 13*:989 (1996).
74. T. Miyata, A. Jikihara, K. Nakamae, A. S. Hoffman. Preparation of poly(2-glucosyloxyethyl methacrylate)–concanavalin A complex hydrogel and its glucose-sensitivity. *Macromol. Chem. Phys. 197*:1135 (1996).
75. T. Miyata, A. Jikihara, K. Nakamae, T. Uragami, A. S. Hoffman, K. Kinomura, M. Okumura. Preparation of glucose-sensitive hydrogels by entrapment or copolymerization of concanavalin A in a glucosyloxyethyl methacrylate hydrogel. *Advanced Biomaterials in Biomedical Engineering and Drug Delivery Systems* (N. Ogata, S. W. Kim, J. Feijen, T. Okano, eds.). Springer: Verlag, Tokyo, 1996, p. 237.
76. A. A. Obaidat, K. Park. Characterization of protein release through glucose-sensitive hydrogel membranes. *Biomaterials 18*:801 (1997).
77. E. P. Diamandis, T. K. Christopoulos. *Immunoassay*. Academic Press, New York, 1996.
78. T. Miyata, N. Asami, T. Uragami. Preparation of an antigen-sensitive hydrogel using antigen–antibody bindings. *Macromolecules 32*:2082 (1999).
79. T. Miyata, N. Asami, T. Uragami. A reversibly antigen-responsive hydrogel. *Nature 399*:766 (1999).
80. G. Wulff, A. Sarhan, K. Zabrocki. Enzyme-analogue built polymers and their use for the resolution of racemates. *Tetrahedron Lett. 44*:4329 (1973).
81. B. Sellergren, M. Lepisto, K. Mosbach. Highly enantioselective and substrate-selective polymers obtained by molecular imprinting utilizing noncovalent interactions. NMR and chromatographic studies on the nature of recognition. *J. Am. Chem. Soc. 110*:5853 (1988).
82. M. Watanabe, T. Akahoshi, Y. Tabata, D. Nakayama. Molecular specific swelling change of hydrogels in accordance with the concentration of guest molecules. *J. Am. Chem. Soc. 120*:5577 (1998).
83. T. Aoki, K. Nakamura, K. Sanui, A. Kikuchi, T. Okano, Y. Sakurai, N. Ogata. Adenosine-induced changes of the phase transition of poly(6-(acryloyloxymethyl)uracil) aqueous solution. *Polym. J. 31*:1185 (1999).
84. T. Aoki, T. Nishimura, K. Sanui, N. Ogata. Phase-transition changes of poy(N-(S)-sec-buthylacrylamide-co-N-isopropylacrylamide) in response to amino acids and its chiral recognition. *React. Functional Polymers 37*:299 (1998).
85. T. Aoki, M. Muramatsu, K. Sanui. Volume-phase transition of an optically active hydrogel in response to external environmental changes. *Polym. Prep., Jpn. 48*:3124 (1999).

37

Biocompatible Polymers in Liver-Targeted Gene Delivery Systems

Edwin C. Ouyang, George Y. Wu, and Catherine H. Wu
University of Connecticut Health Center, Farmington, Connecticut

I. POLYCATIONS AS DNA BINDING COMPONENTS

A. Polylysine

Poly-L lysine is a polymer of varying numbers of lysine residues. Poly-L-lysine with its positive surface charge has been utilized to improve cell adhesion to solid substrates and can also be used for nucleic acid delivery to cells by noncovalently linking to negatively charged DNA phosphate backbones.

B. Polyethylenimine

Polyethylenimine (PEI) is another useful polycation applied in DNA-binding delivery systems. It also has positive charges on the surface with sizes from 6,000 to 60,000 daltons.

Poly-L-lysine can compact the size of DNA complexes to make it more accessible for internalization and provide protective effects to the complexes against degradation by serum nucleases. Without coupling to polymer–ligand, DNA alone is usually degraded completely within 15 minutes in serum, or 60 minutes in culture medium plus 10% fetal bovine serum, whereas 90% of DNA complexed with poly-L-lysine–ligand remained full length after 1.5 h or 3 h in serum or medium with 10% serum, respectively (1).

The stabilization of the DNA complexes is dependent on the length of poly-L-lysine. At least twenty residues of poly-L-lysine is required to maintain the proper complexation with DNA fragments. Poly-L-lysine–DNA complexes were found to be about 25–80 nm in diameter and increased to 50–90 nm when conjugated with ligand. Attaching of asialoglycoprotein (ASGP) was found to decrease the aggregation of poly-L-lysine–DNA complexes in saline (2).

The efficiency of delivery is also dependent on the length of the poly-L-lysine and the ratio of galactose residues conjugated to the polymer (3). Complexes that have longer polymers (MW 13,000–29,000) with more galactose residues (13 to 26) are more efficient in gene transfer than shorter polymers (MW 1,800) with fewer galactose residues (only 5).

Galactosylated poly-L-lysine (MW 13,000) was complexed with plasmid DNA encoding a CAT gene at weight ratio of 1:0.6. The size of the complex was measured to be 180 nm with a weakly negative charge (zeta potential of −20 mV) (4). Radioisotope-labeled DNA/galactosylated poly-L-lysine was rapidly eliminated from serum and taken up by hepatocytes, which were abolished by an excess of free galactosylated bovine serum albumin.

Chemically modified poly-L-lysine can be used to achieve liver-selective delivery. Galactose-modified poly-L-lysine can be delivered specifically into mouse hepatocytes via tail vein injection (5). The complex was shown to accumulate in the liver, and the uptake was inhibited by treating with an excess of galactose-modified bovine serum albumin. Artificial galactose-modified ligand has been applied in DNA–poly-L-lysine complex to target to hepatocytes (6). A ligand carrying tetra-antennary branches with terminal galactose residues was conjugated to poly-L-lysine. The complex was then coupled with a plasmid DNA encoding a reporter gene (luciferase) to target human and murine hepatocyte cell lines, HepG2, BNL CL2, respectively.

II. LIVER TARGETING BY RECEPTOR-MEDIATED DNA DELIVERY SYSTEMS

Liver-directed gene therapy can theoretically be applied to many liver diseases (congenital deficiencies of hepatic enzymes, viral infections, liver tumors, etc.) to deliver genes of interest into hepatocytes directly. It has the potential to result in efficient and persistent expression with fewer toxic side effects than with conventional medical therapy. One of the best-studied liver-selective delivery systems utilizes polymers conjugated to asialoglycoproteins to target genes of interest to ASGP receptors located selectively on the surface of mammalian hepatocytes.

A. ASGP-Mediated Liver-Specific Delivery Systems

1. The ASGP Receptor

a. Expression and Distribution of ASGP Receptor.

ASGP receptors are located selectively on the plasma membrane of hepatocytes, mostly on the basolateral side (sinusoidal side). They bind specifically to desialylated serum proteins containing galactose (Gal) or N-acetyl-galactosamine (GalNAc) at the nonreducing termini of oligosaccharide chains (7). Detailed aspects of ASGP receptor functions have been reviewed elsewhere (7–11). The human ASGP receptor is a 46 kD transmembrane protein consisting of approximately 300 amino acids, including an N-terminal 40 amino acid on the cytoplasmic side, a 20 amino acid hydrophobic transmembrane domain, and a long 220 amino acid C-terminal extracellular domain. A second ASGP receptor has been identified with a molecular weight of 50 kD as the minor form with very low expression in human liver (12,13). High homology has been identified among ASGP receptors across different species. For example, rat and human ASGP receptors show 70 to 80% homology in essential amino acid sequences. The ASGP receptor itself is a glycoprotein with the C-terminal domain containing three glycosylation sites. However, carbohydrate side chains do not seem to play an important role in ligand binding ability (14–16).

b. Physiological Function of the ASGP receptor.

The expression of the ASGP receptor may change at different stages of differentiation and in the presence of certain diseases. For example, after partial hepatectomy, the number of ASGP receptors decreases (8,17). A decreased number and binding ability of ASGP receptors can also be detected in various carcinogen-induced rat hepatocellular carcinoma models (18,19). In a diabetes mellitus rat model, the binding ability for ASGP receptors was found to be decreased during early stages and dramatically reduced in later or severe stages (20,21).

2. The Ligand for the ASGP Receptor

The ASGP receptor may recognize and bind to natural proteins with terminal Gal/GalNAc residues, including asialoorosomucoid, asialofetuin, etc., as well as synthetic galactose-modified materials.

a. Asialoorosomucoid (ASOR).

A commonly used ASGP in liver-targeted gene therapy is the desialylated form of orosomucoid (OR, alphal-acid glycoprotein). OR consists of 180 amino acids with a molecular weight of 46 kD (22). Glycosylation of OR decreases during liver diseases (23), while it increases in the presence of certain cancers (24).

b. Other Natural ASGPs.

Many serum glycoproteins can be desialylated to expose terminal Gal/GalNAc residues and can be internalized into hepatocytes through the binding with ASGP receptors. The binding affinities of these ASGPs differ greatly, largely depending on the density of terminal Gal residues on saccharide chains. Among these, ASOR with 15 exposed clusters of Gal residues has the strongest binding affinity, while asialotransferrin (ASTF) with only 4 Gal residues, has the weakest (25).

c. Galactose-Modified Materials.

Synthetic di-, tri- or tetra-branched oligosaccharides with Gal/GalNAc can be coupled to nongalactose bearing proteins enabling binding to ASGP receptor on hepatocytes, although the binding affinities are much lower than those of natural desialylated glycoproteins. Clustering of Gal residues can enhance the binding affinity. The correct orientation of these Gal residues is critical (25).

3. Endocytosis and Recycling Pathway of Ligand–Receptor Complexes

Natural desialylated glycoproteins, such as asialoorosomucoid (ASOR), bind to the ASGP receptor through galactose residues on termini of ASOR. Then, the ligand–receptor complex is internalized by hepatocytes in clathrin-coated pits (26–28). Next, coated pits are pinched off from the plasma membrane and form coated vesicles in cytoplasm, which is further decoated to become smooth vesicles. Then the ligand–receptor complex is delivered into CURL (compartment of uncoupling receptor and ligand) in which a low pH environment dissociates the complex. Ligands are transferred to endosome-lysosome compartments and degraded. Meanwhile, unoccupied ASGP receptors are recycled back to plasma membrane through an alternative pathway (10,29). The entire endocytosis-recycling process happens rapidly, being completed within eight minutes, and is ATP- and Ca^{2+}-dependent (30).

B. Low-Density Lipoprotein Receptor (LDL-R)–mediated Liver Delivery Systems

Another liver-targeting system utilizes LDL receptors to deliver genes of interest into hepatocytes. LDL receptors are expressed on hepatocytes as well as other cells such as spleen, adrenal, and certain tumor cells. An apoE liposome carrying a lipophilic derivative of daunorubicin (LAD) was administered to B16 melanoma bearing mice (31,32). The binding affinity of apoE liposome to LDL receptor was 15 to 50-fold stronger than that of LDL ligand alone or liposome alone, respectively. LAD were detected mainly in LDL-receptor expressing tissues such as liver or spleen, with the highest uptake in tumor cells.

C. Other Targeting Systems Using Polymers

Transferrin receptor has been used as a ligand linked to poly-L-lysine for liver-directed delivery systems. A complex consisting of maleimidotransferrin coupled with poly-L-lysine was constructed and had significantly enhanced transfection efficiency compared with conventional transfection methods including electroporation, lipofection, and $CaPO_4$ precipitation (33). The efficiency was correlated to the length of the lysine polymer, the concentration of the transferrin–poly-L-lysine complexes, the ratio of poly-L-lysine to DNA, and the treatment of target cells with chloroquine. This method was efficient and had low cytotoxicity.

III. APPLICATIONS OF POLY-L-LYSINE–ASGP CONJUGATES

To deliver genes to hepatocytes, a DNA–poly-L-lysine–ASGP carrier was constructed by conjugation of ASGP to polylysine, subsequently forming DNA complexes through noncovalent interaction between negatively charged DNA strands and positively charged polylysine. Orosomucoid from human donor blood was desialylated by neuraminidase. ASOR was chemically coupled to poly-L-lysine and then bound to DNA to form a complex. The DNA–poly-L-lysine–ASOR complex provided specific targeting to hepatocytes through the binding between ASOR on the complex and ASGP receptor on the surface of parenchymal liver cells.

Studies conducted in cell lines and animal models showed that ASGP–poly-L-lysine–DNA system can specifically deliver genes of interest to hepatocytes. For example, 10 minutes after intravenous administration of ^{32}P-labeled ASOR–poly-L-lysine–DNA complexes into rats, 85% of the radioactivity was detected in the liver and only 15% remained in serum, whereas in control groups with ^{32}P-labeled DNA alone, only 17% of the radioactivity was measured in the liver and 55% still remained in serum. Microscopy studies demonstrated that most of the targeted cells were parenchymal cells (hepatocytes) with few Kupffer cells involved (34). The uptake was shown to be exclusively through ASGP receptors on hepatocytes (35).

A. Delivery of Medications

Side effects are frequent in the medical treatment of diseases. Specific delivery of medication, for example, to hepatocytes could significantly decrease the concentration of the drug in extrahepatic tissues, reducing the cytotoxic effects while increasing the concentration of the drug in liver cells to enhance the pharmaceutical effects. The ASGP–poly-L-lysine system has been utilized specifically to deliver certain drugs, such as cytotoxic agents or prodrugs, into hepatocytes.

Antiviral drugs have been delivered into hepatocytes of HBV-infected patients by receptor-mediated systems (36). Adenine arabinoside monophosphate (ara-AMP) modified with lactosaminated human albumin was administered into chronic HBV patients for 28 days, exerting similar antiviral efficiency to free drug groups without any clinical side effects. Recent attempts have been made to deliver drugs by conjugating the drug with lactosaminated poly-L-lysine. This modified poly-L-lysine can be easily prepared by chemical synthesis, permitting high drug loading, and can be administered via intramuscular injection (37).

B. Delivery of Genes of Interest

ASGP–poly-L-lysine as a DNA delivery system for therapeutic purposes was first tested by directing a normal albumin gene into analbuminemic rats (38). Nagase analbuminemic rats underwent partial hepatectomy, and ASOR–poly-L-lysine complexed with a plasmid carrying a normal human albumin gene was administered subsequently by tail vein injection. The plasmid was taken up by rat hepatocytes through ASGP receptor. Two weeks after treatment, the expression of human albumin in serum was detected as high as 34 μg/mL and was stable four weeks thereafter. Other studies also showed that an ASGP delivery system could successfully transfer LDL receptor gene into Watanabe rabbits via intravenous injection (39) and a bilirubin-uridine diphosphoglucuronate (UDP)-glucuronosyltransferase-1 gene into Gunn rats (40). The gene of interest was expressed in sufficient amounts to correct the endogenous deficiency in these animal models.

C. Delivery of Antisense Oligonucleotides

Antisense oligodeoxynucleotides (ODN) are attractive nucleic acids to deliver to cells because they can match corresponding RNA sequences and block translation or induce enzymatic digestion of the RNA–DNA hybrid by RNase H. For example, the ASGP-receptor-mediated system has been used to deliver antisense oligonucleotides into hepatocytes to block viral expression (41). A 21-mer antisense ODN against the HBV polyA signal sequence (nt 1903 to 1923) was synthesized and complexed with ASOR–poly-L-lysine carrier. The complex was administered to HepG2 2.2.15 cell lines possessing ASGP receptors and stable HBV viral expression and infectious particle formation. Twenty four h later, HBV DNA replication and HBsAg expression was found to be suppressed by 80%, and the effect persisted for up to 6 days thereafter. To investigate further whether treating cells with antisense ODN may actually protect cells from HBV infection, a phosphorothioate-modified 21-mer antisense ODN complementary to the HBV polyA signal and 5′-upstream sequence was prepared (42). The 21-mer was delivered to Huh7 cells [ASGP receptor (+) and HBV viral expression (−)] after forming a complex with ASGP–poly-L-lysine carrier. Subsequent transfection of these cells with HBV-encoding plasmids (6.5×10^5 copies per cell) showed strong suppression of HBV replicaton and gene expression. Compared with control groups treated with antisense ODN alone, in cells pretreated with antisense ODN delivered by the ASOR–poly-L-lysine system, HBV viral DNA was undetectable (less than 0.1 pg as determined by quantitative PCR), HBV viral transcription was suppressed by 60%, and HBsAg expression was decreased by 97%. Similar results were also demonstrated by Madon and Blum (43). In this case, an antisense ODN complementary to the HBV encapsidation signal was synthesized and coupled to N-acetyl-glucosaminated albumin–poly-L-lysine conjugates, which was cotransfected with an HBV-encoding plasmid into a chicken hepatoma cell line LMH. Viral replication was decreased significantly.

To study antiviral effects in animals, Bartholomew et al. administered antisense ODN (complementary to the polyA signal and adjacent upstream sequences of woodchuck hepatitis virus, WHV) complexed with ASOR–poly-L-lysine conjugates intravenously to infected woodchucks for five consecutive days. This significantly suppressed WHV viremia, detected as a 10-fold decrease in circulating viral DNA on day 25, and was persistent for at least 2 weeks (44).

IV. RECENT IMPROVEMENTS IN POLY-L-LYSINE CONJUGATED RECEPTOR-MEDIATED SYSTEMS

A. Coupling of Fusogenic Proteins

A major obstacle in applying ASGP receptor and other receptor-mediated systems has been the degradation of internalized DNA in lysosomes. Several attempts have been made to overcome this problem by adding viral fusogenic proteins or coinfecting with viruses that can fuse and disrupt endosome-lysosome vesicles. Influenza hemagglutinin (HA) protein (45,46) and vesicular stomatitis virus (VSV) G protein (47) were used to improve the gene delivery in targeting cells. Bongartz *et al.* (45) chemically conjugated HA protein to phosphorothioate-modified antisense oligodeoxynucleotides (anti-TAT) and cultured the DNA–protein complex with HIV-infected lymphocytes. The anti-HIV activity of the antisense ODN was enhanced up to 10-fold. Coupling of HA2 protein to poly-L-lysine–transferrin–DNA complex increased expression of reporter genes in targeted cells. Also, replication-defective adenoviral particles were conjugated with poly-L-lysine–ASOR–DNA complexes to deliver a canine factor IX gene into murine primary hepatocytes, resulting in significant expression of canine factor IX in culture medium (48). Wild-type adenovirus (type 5) or replication-defective adenoviral particles were conjugated to the ASOR–poly-L-lysine–DNA system. The number of infected cells increased by 200-fold, and expression of the targeting gene, HBsAg, was enhanced dramatically 13- to 30-fold.

B. Modification of Polymers

Novel vectors have been prepared by covering DNA–poly-L-lysine complexes with protective hydrophilic polymers, such as polyethylene glycol (PEG), dextran or poly[N-(2-hydroxypropyl)methacrylamide] (pHPMA). These modified polymers resulted in self-assembly of DNA, shielded surface charge, increased solubility, and enhanced transfection efficiency (49,50).

PEG also has protective effects to DNA–poly-L-lysine complexes. Results showed that coating poly-L-lysine–DNA complex with PEG significantly enhanced the resistance to nucleases in vitro. The shielding effect was correlated to the length of the polymer (51).

C. Application of Other Polymers

Polyethyleneimine (PEI) has been applied in receptor mediated systems recently. PEI was conjugated with 5% galactose and coupled with plasmid DNA encoding a reporter gene (luciferase or beta galactosidase). This complex was then used to transfect a human hepatoblastoma cell line HepG2 with 50% efficiency. The transfection was galactose-residue-dependent and ASGP-receptor-dependent, as it was abolished in the presence of an excess of asialofetuin (52). In addition, galactocerebroside-incorporated liposomes were used to transfer PEI-coupled DNA complexes to the human hepatoma cell line Huh7 and to isolated rat hepatocytes (53). Recently, Bandyopadhyay et al. applied galactocerebroside-modified liposomes to transfer a lactosylated PEI-complexed chimeric RNA/DNA oligonucleotide carrying an A/C conversion at Ser365 of rat factor IX gene into isolated rat hepatocytes (54). Transfection efficiency was 100%, and the conversion rate was 19–24%.

REFERENCES

1. Chiou, H., Tangco, M. V., Levine, S. M., Robertson, D., G., Kormis, K., Wu, C. H., Wu, G. Y. Enhanced resistance to nuclease degradation of nucleic acids complexed to asialoglycoprotein-polylysine carriers. *Nucl. Acids. Res.*, 1994, *22*(24):5439–5446.
2. Kwoh, D., Coffin, C. C., Lollo, C. P., Jovenal, J., Banaszczyk, M. G., Mullen, P., Phillips, A., Amini, A., Fabrycki, J., Bartholomew, R. M., Brostoff, S. W., Carlo, D. J. Stabilization of poly-L-lysine/DNA polyplexes for in vivo gene delivery to the liver. *Biochim. Biophys. Acta*, 1999, *1444*(2):171–190.
3. Nishikawa, M., Takemura, S., Takakura, Y., Hashida, M. Targeted delivery of plasmid DNA to hepatocytes in vivo: optimization of the pharmacokinetics of plasmid DNA/galactosylated poly(L-lysine) complexes by controlling their physicochemical properties. *J. Pharmacol. Exp. Ther.*, 1998, *287*(1):408–415.
4. Hashida, M., Takemura, S., Nishikawa, M., Takakura, Y. Targeted delivery of plasmid DNA complexed with galactosylated poly(L-lysine). *J. Controlled Release*, 1998, *53*(1–3):301–310.
5. Akamatsu, K., Imai, M., Yamasaki, Y., Nishikawa, M., Takakura, Y., Hashida, M. Disposition characteristics of glycosylated poly(amino acids) as liver cell-specific drug carrier. *J. Drug Target*, 1998, *6*(3):229–239.
6. Plank, C., Zatloukal, K., Cotten, M., Mechtler, K., Wagner, E. Gene transfer into hepatocytes using asialoglycoprotein receptor mediated endocytosis of DNA complexed with an artificial tetra-antennary galactose ligand. *Bioconjug. Chem.* 1992, *3*(6):533–539.
7. Schwartz, A. The hepatic asialoglycoprotein receptor. *CRC Crit. Rev. Biochem.*, 1984, *16*(3):207–233.
8. Stockert, R., Morell, A. G., Ashwell, G. Structural characteristics and regulation of the asialoglycoprotein receptor. *Targeted Diagn. Ther.*, 1991, *4*:41–64.
9. Ashwell, G., Harford, J. Carbohydrate-specific receptors of the liver. *Annu. Rev. Biochem.*, 1982, *51*:531–554.
10. Breitfeld, P., Simmons, C. F., Jr., Strous, G. J., Geuze, H. J., Schwartz, A. L. Cell biology of the asialoglycoprotein receptor system: a model of receptor-mediated endocytosis. *Int. Rev. Cytol.*, 1985, *97*:47–95.
11. Drickamer, K. Two distinct classes of carbohydrate-recognition domains in animal lectins. *J. Biol. Chem.*, 1988, *263*(20):9557–9560.
12. Spiess, M., Schwartz, A. L., Lodish, H. F. Sequence of human asialoglycoprotein receptor cDNA. An internal signal sequence for membrane insertion. *J. Biol. Chem.*, 1985, *260*(4):1979–1982.
13. Spiess, M., Lodish, H. F. Sequence of a second human asialoglycoprotein receptor: conservation of two receptor genes during evolution. *Proc. Natl. Acad. Sci. USA*, 1985, *82*(19):6465–6469.
14. Halberg, D., Wager, R. E., Farrell, D. C., Hildreth, J., 4th, Quesenberry, M. S., Loeb, J. A., Holland, E. C., Drickamer, K. Major and minor forms of the rat liver asialoglycoprotein receptor are independent galactose-binding proteins. Primary structure and glycosylation heterogeneity of minor receptor forms. *J. Biol. Chem.*, 1987, *262*(20):9828–9838.
15. Bischoff, J., Lodish, H. F. Two asialoglycoprotein receptor polypeptides in human hepatoma cells. *J. Biol. Chem.*, 1987, *262*(24):11825–1132.
16. Hsueh, E., Holland, E. C., Carrera, G. M., Jr., Drickamer, K. The rat liver asialoglycoprotein receptor polypeptide must be inserted into a microsome to achieve its active conformation. *J. Biol. Chem.*, 1986, *261*(11):4940–4947.
17. Gartner, U., Stockert, R. J., Morell, A. G., Wolkoff, A. W. Modulation of the transport of bilirubin and asialoorosomucoid during liver regeneration. *Hepatology*, 1981, *1*(2):99–106.
18. Stockert, R., Becker, F. F. Diminished hepatic binding protein for desialylated glycoproteins during chemical hepatocarcinogenesis. *Cancer Res.*, 1980, *40*(10):3632–3634.

19. Schwarze, E., Tolleshaug, H., Seglen, O. Uptake and degradation of asialoorosomucoid in hepatocytes from carcinogen-treated rats. *Carcinogenesis*, 1985, *6*(5):777–782.
20. Dodeur, M., Durand, D., Dumont, J., Durand, G., Feger, J., Agneray, J. Effects of streptozotocin-induced diabetes mellitus on the binding and uptake of asialoorosomucoid by isolated hepatocytes from rats. *Eur. J. Biochem.*, 1982, *123*(2):383–387.
21. Scarmato, P., Feger, J., Dodeur, M., Durand, G., Agneray, J. Kinetic evidence of a surface membranous step during the endocytosic process of 3H-labelled asialoorosomucoid and its alteration in diabetic mellitus rats. *Biochim. Biophys. Acta*, 1985, *843*(1–2):8–14.
22. Schmid, K., Kaufmann, H., Isemura, S., Bauer, F., Emura, J., Motoyama, T., Ishiguro, M., Nanno, S. Structure of alphal-acid glycoprotein. The complete amino acid sequence, multiple amino acid substitutions, and homology with the immunoglobulins. *Biochemistry*, 1973, *12*(14): 2711–2724.
23. van Dijk, W., Havenaar, E. C., Brinkman-van der Linden, E. C. Alpha 1-acid glycoprotein (orosomucoid): pathophysiological changes in glycosylation in relation to its function. *Glycoconj. J.*, 1995, *12*(3):227–233.
24. Matei, L. Plasma proteins glycosylation and its alteration in disease. *Rom. J. Intern. Med.*, 1997, *35*(1–4):3–11.
25. Lee, R. Ligand structural requirements for recognition and binding by the hepatic asialoglycoprotein receptor. *Targeted Diagn. Ther.*, 1991, *4*:65–86.
26. Geuze, H., Slot, J. W., Strous, G. J., Lodish, H. F., Schwartz, A. L. Intracellular site of asialoglycoprotein receptor-ligand uncoupling: double-label immunoelectron microscopy during receptor-mediated endocytosis. *Cell*, 1983, *32*(1):277–287.
27. Hubbard, A., Stukenbrok, H. An electron microscope autoradiographic study of the carbohydrate recognition systems in rat liver. II. Intracellular fates of the 125I-ligands. *J. Cell Biol.*, 1979, *83*(1):65–81.
28. Wall, D., Wilson, G., Hubbard, A. L. The galactose-specific recognition system of mammalian liver: the route of ligand internalization in rat hepatocytes. *Cell*, 1980, *21*(1):79–93.
29. Ciechanover, A., Schwartz, A. L., Lodish, H. F. Sorting and recycling of cell surface receptors and endocytosed ligands: the asialoglycoprotein and transferrin receptors. *J. Cell Biochem.*, 1983, *23*(1–4):107–130.
30. Schwartz, A., Fridovich, S. E., Lodish, H. F. Kinetics of internalization and recycling of the asialoglycoprotein receptor in a hepatoma cell line. *J. Biol. Chem.*, 1982, *257*(8): 4230–4237.
31. Versluis, A., Rensen, P. C., Rumn, E. T., Van Berkel, T. J., Bijsterbosch, M. K. Low-density lipoprotein receptor-mediated delivery of a lipophilic daunorubicin derivative to B16 tumors in mice using apolipoprotein E-enriched liposomes. *Br. J. Cancer*, 1998, *78*(12):1607–1614.
32. Versluis, A., Rump, E. T., Rensen, P. C., Van Berkel, T. J., Bijsterbosch, M. K. Synthesis of a lipophilic daunorubicin derivative and its incorporation into lipidic carriers developed for LDL receptor-mediated tumor therapy. *Pharm. Res.*, 1998, *15*(4):531–537.
33. Taxman, D., Lee, E. S., Wojchowski, D. M. Receptor-targeted transfection using stable maleimido-transferrin/thio-poly-L-lysine conjugates. *Anal. Biochem.*, 1993, *213*(1):97–103.
34. Wu, G., Wu, C. H. Receptor-mediated gene delivery and expression in vivo. *J. Biol. Chem.*, 1988, *263*(29):14621–14624.
35. Wu, G., Wu, C. H. Evidence for targeted gene delivery to Hep G2 hepatoma cells in vitro. *Biochemistry*, 1988, *27*(3): 887–892.
36. Fiume, L., Verme, G. Liver targeting of nucleoside analogues coupled to galactosyl terminating macromolecules: a new approach to the treatment of a chronic viral hepatitis. *Ital. J. Gastroenterol. Hepatol.*, 1997, *29*(3):275–280.
37. Fiume, L., Di Stefano, G., Busi, C., Mattioli, A., Bonino, F., Torrani-Cerenzia, M., Verme, G., Rapicetta, M., Bertini, M., Gervasi, G. B. Liver targeting of antiviral nucleoside analogues through the asialoglycoprotein receptor. *J. Viral. Hepat.*, 1997, *4*(6):363–370.
38. Wu, G., Wilson, J. M., Shalaby, F., Grossman, M., Shafritz, D. A., Wu, C. H. Receptor-mediated gene delivery in vivo. Partial correction of genetic analbuminemia in Nagase rats. *J. Biol. Chem.*, 1991, *266*(22):14338–14342.
39. Wilson, J., Grossman, M., Wu, C. H., Chowdhury, N. R., Wu, G. Y., Chowdhury, J. R. Hepatocyte-directed gene transfer in vivo leads to transient improvement of hypercholesterolemia in low density lipoprotein receptor-deficient rabbits. *J. Biol. Chem.*, 1992, *267*(2):963–967.
40. Chowdhury, N., Hays, R. M., Bommineni, V. R., Franki, N., Chowdhury, J. R., Wu, C. H., Wu, G. Y. Microtubular disruption prolongs the expression of human bilirubin-uridinediphosphoglucuronate-glucuronosyltransferase-1 gene transferred into Gunn rat livers. *J. Biol. Chem.*, 1996, *271*(4): 2341–2346.
41. Wu, G., Wu, C. H., Specific inhibition of hepatitis B viral gene expression in vitro by targeted antisense oligonucleotides. *J. Biol. Chem.*, 1992, *267*(18):12436–12439.
42. Nakazono, K., Ito, Y., Wu, C. H., Wu, G. Y. Inhibition of hepatitis B virus replication by targeted pretreatment of complexed antisense DNA in vitro. *Hepatology*, 1996, *23*(6).
43. Madon, J., Blum, H. E. Receptor-mediated delivery of hepatitis B virus DNA and antisense oligodeoxynucleotides to avian liver cells. *Hepatology*, 1996, *24*(3):474–481.
44. Bartholomew, R., Carmichael, E. P., Findeis, M. A., Wu, C. H., Wu, G. Y. Targeted delivery of antisense DNA in woodchuck hepatitis virus-infected woodchucks. *J. Viral. Hepat.*, 1995, *2*(6):273–278.
45. Bongartz, J., Aubertin, A. M., Milhaud, P. G., Lebleu, B. Improved biological activity of antisense oligonucleotides conjugated to a fusogenic peptide. *Nucl. Acids. Res.*, 1994, *22*(22):4681–4688.
46. Wagner, E., Plank, C., Zatloukal, K., Cotten, M., Birnstiel,

M. L. Influenza virus hemagglutinin HA-2 N-terminal fusogenic peptides augment gene transfer by transferrin-polylysine-DNA complexes: toward a synthetic virus-like gene-transfer vehicle. *Proc. Natl. Acad. Sci. USA*, 1992, *89*(17): 7934–7938.

47. Douar, A., Themis, M., Sandig, V., Friedmann, T., Coutelle, C. Effect of amniotic fluid on cationic lipid mediated transfection and retroviral infection. *Gene Ther.*, 1996, *3*(9):789–796.
48. Cristiano, R., Smith, L. C., Kay, M. A., Brinkley, B. R., Woo, S. L. Hepatic gene therapy: efficient gene delivery and expression in primary hepatocytes utilizing a conjugated adenovirus-DNA complex. *Proc. Natl. Acad. Sci. USA*, 1993, *90*(24):11548–11552.
49. Toncheva, V., Wolfert, M. A., Dash, P. R., Oupicky, D., Ulbrich, K., Seymour, L. W., Schacht, E. H. Novel vectors for gene delivery formed by self-assembly of DNA with poly(L-lysine) grafted with hydrophilic polymers. *Biochim. Biophys. Acta*, 1998, *1380*(3):354–368.
50. Choi, Y., Liu, F., Park, J. S., Kim, S. W. Lactose-poly(ethylene glycol)-grafted poly-L-lysine as hepatoma cell-targeted gene carrier. *Bioconjug. Chem.*, 1998, *9*(6):708–718.
51. Katayose, S., Kataoka, K. Remarkable increase in nuclease resistance of plasmid DNA through supramolecular assembly with poly(ethylene glycol)-poly(L-lysine) block copolymer. *J. Pharm. Sci.*, 1998, *87*(2):160–163.
52. Zanta, M., Boussif, O., Adib, A., Behr, J. P. In vitro gene delivery to hepatocytes with galactosylated polyethylenimine. *Bioconjug. Chem.*, 1997, *8*(6):839–844.
53. Bandyopadhyay, P., Kren, B. T., Ma, X., Steer, C. J. Enhanced gene transfer into HuH7 cells and primary rat hepatocytes using targeted liposomes and polyethylenimine. *Biotechniques.*, 1998, *25*(2):282–284, 286–292.
54. Bandyopadhyay, P., Ma, X., Linehan-Stieers, C., Kren, B. T., Steer, C. J. Nucleotide exchange in genomic DNA of rat hepatocytes using RNA/DNA oligonucleotides. Targeted delivery of liposomes and polyethyleneimine to the asialoglycoprotein receptor. *J. Biol. Chem.*, 1999, *274*(15): 10163–10172.

38

Bioartificial Pancreas

Riccardo Calafiore
University of Perugia School of Medicine, Perugia, Italy

I. FOREWORD

The biohybrid artificial pancreas (BAP) represents one of the most challenging, complex, and promising applications of a relatively new science: tissue engineering. Tissue engineering has been defined as ''an interdisciplinary field which applies the principles of engineering and life sciences towards the development of biological substitutes that aim to maintain, restore or improve tissue function'' (1). Sometimes, even excellent artificial devices must comply with restrictions and specific requirements associated with the biological counterpart, in order to secure optimal functional performance.

In general, every organ that can be broken apart into single cells or cell clusters with no disruption of the original function is usually suitable for generating a bioartificial organ. However, it may be hard to preserve the functional identity of cellular units placed in environmental conditions that usually are quite different from their native site. Furthermore, only selected materials enable creation of a tissue/material interface that is harmless to both the embodied cells and the host's tissue. Specific issues associated with natural or engineered tissues that are entrusted with prevalently mechanical tasks, such as skin, bone or cartilage, significantly differ from those involving sophisticated cell systems that govern complex metabolic functions, like endocrine and liver cells. Nevertheless, common pathways should be explored regardless of the specific feature that is incorporated by a certain cellular system. In particular, special care must be taken not only to assess the material's physicochemical properties and biocompatibility but also to select the tissue's sources, in compliance with physiological competence and safety principles. Finally, the overall suitability of the newly developed biohybrid devices for clinical application must be also carefully assessed.

A. Field of Application of the BAP

The BAP, composed of intact and viable insulin-producing cells that are protected from the host's immune reaction by highly biocompatible, selective permeable and chemically stable artificial membranes, would ideally apply to the potential cure of insulin-dependent (type I or type II) diabetes mellitus (DM) by transplantation. In fact, a BAP that contains insulin-producing cells would provide for continuous insulin delivery, under strict regulation by the extracellular glucose levels. The goal of such a ''totally automatic treatment'' (Table 1) for DM could be also theoretically accomplished by creating an ''artificial pancreas.'' In this instance, a glucose sensor (1) implanted subcutaneously would continuously record extracellular glucose levels and convert the chemical information into an electronic signal by a minicomputerized system. This would regulate in turn the release of prestored insulin by a minipump in order to

Table 38.1 Automatic Treatment of IDDM

1. Transplant of the Endocrine Pancreas
   - Whole organs
   - Isolated islets of Langerhans
2. Artificial Glucose Sensor

maintain normoglycemia. This machinery, while able to reverse hyperglycemia, has been preliminarily proven to function, so far, for only very limited periods of time, due to still pending technical problems.

II. DM: EPIDEMIOLOGY, PATHOGENESIS, AND CLINICAL FEATURES

DM is the most common endocrine disorder, hallmarked by metabolic abnormalities such as elevated blood glucose levels and their biochemical consequences, that may provoke acute illness or result, during the course of the disease, in secondary chronic complications (2). These mainly affect eyes, kidney, nerves and blood vessels. It is not an immediately life-threatening disease anymore, in the majority of instances, provided that the patients are promptly and adequately treated, but DM still represents a potentially lethal and certainly highly disabling disease. It is estimated that over 100 million people in the world actually suffer from either type I or type II DM, the difference between these two forms depending upon ethiopathogenesis (Table 2). The prevalence of cumulative type I and type II DM in 1995 in Europe and the Americas was 32.8 and 30.7/1000 population, respectively, but these figures are expected to rise to 47.5 and 63.5/1000, respectively, in 2025 (3). At the clinical onset of type I insulin-dependent DM that affects mainly, but not exclusively, adolescent/young individuals, the majority of islet β-cells have been completely destroyed by autoimmunity. On the basis of a quite clear genetic susceptibility that can be identified in patients at risk of developing type I DM in the form of defined gene haplotypes, the disease process is triggered by only partially known environmental, infectious (i.e., viruses), or chemical toxic agents. Environmental agents indeed play a major role in triggering the onset of overt DM, which is confirmed by observations in haploidentical twins. In fact, if type I DM were a pure genetic disorder, it would affect not 50%, as it actually does, but rather 100% of identical twins. Upon intervention of these agents, the autoimmune attack will begin, although this process is clinically silent, while the pancreatic islets are progressively being infiltrated with mononuclear cells (i.e., macrophages, monocytes) and activated cytotoxic T lymphocytes (''insulitis''). At this time, in the serum of the affected individuals, several autoantibody fractions, specifically targeting selected β-cell antigens, are detectable (i.e., anti-insulin, antityrosin phosphatase, antiglutammic-decarboxylase acid, etc.). The condition where autoimminity is active, but the patients still look healthy, is defined the ''prediabetes.'' This stage of the disease may last for variable periods of time, even years, but it abruptly ends when over 90% of the original β-cell mass has been destroyed. At this point, the residual endogenous insulin production is no more able to counteract elevations in blood glucose levels. The patients will develop hyperglycemia, polyuria, polydipsia, ketonuria, and body weight loss and will undergo severe derangements of fluid and electrolyte balance, culminating with ketoacidotic coma. If not promptly treated this condition may lead to ketotic coma and death.

Since type I DM patients will require lifelong, multiple daily insulin injections, they unfortunately represent a heavy burden to society. In fact, if these individuals have seen their life expectancy gradually improve from the introduction of insulin therapy, their increased longevity has not resulted in elimination of the risk of developing chronic, often very serious complications of the disease, such as premature blindness, terminal renal insufficiency, vascular disease, and disabling neuropathy. These complications may occur in spite of the selected insulin therapy regimens because it is very difficult to mimic performance of the glucose sensing apparatus incorporated in normal β-cells. Consequently, the physiological stimulus-coupled

Table 38.2 Classification of Diabetes Mellitus

Primary	Secondary
Autoimmune (Type I)	Pancreatic disorders
Insulin dependent	Endocrine abnormalities
Non-insulin-dependent	Toxic agents, drugs
	Insulin receptor abnormalities
Nonautoimmune	Genetic syndromes
Insulin-dependent	Miscellaneous agents
Non-insulin-dependent	
''Maturity onset'' (MODY)	

insulin secretory response is far from being reproduced by subcutaneous exogenous insulin injections.

III. DM: INSULIN THERAPY REGIMENS

The most actual and effective way to prevent the onset of secondary complications in patients with type I, insulin-dependent DM (T1DM) is based on strict blood glucose control, especially in coincidence with meals, so as to avoid hyper-/hypoglycemia. This goal may be achieved by injecting either short- or long-acting insulin molecules, as frequently as four times a day, according to the indications of the Diabetes Control Complications Trial (DCCT) Study Group (4,5). An important mirror of the ability to keep T1DM patients under good blood glucose control is represented by concentration of the glycosylated hemoglobin (HbA1c): the higher this parameter (reflecting % of glucose-bound hemoglobin), the worse the degree of metabolic control. Insulin administration, according to intensive, multiple daily-dose regimen protocols, as suggested by DCCT, actually represents the most effective tool to treat T1DM. Unfortunately, this therapeutic option is associated with at least two major pitfalls: (1) patients are exposed to a high risk of developing severe hypoglycemia, and (2) the risk of developing secondary complications of the disease is attenuated, but not eliminated. Finally, patients' compliance with intensive insulin therapy may be quite poor, because of both strict blood glucose monitoring that involves multiple finger pricks, and insulin injections. The imperfect blood glucose control achieved by subcutaneously injected exogenous insulin is a fundamental pitfall consisting of the physical distance between the insulin injection site and the liver, that represents the first physiological site of insulin action.

IV. T1DM: "AUTOMATIC TREATMENT"

In light of the above-reviewed concerns, the "T1DM problem" could be solved only if the original physiological model, namely viable and functional insulin producing β-cells, was fully reconstituted. Such a goal might be accomplished by transplanting the endocrine pancreas as a whole organ or as isolated human/nonhuman, primary or engineered islet cells (6) (Table 3). BAP could offer the opportunity to graft the islet cells within artificial membranes so as to prevent immune destruction of the transplanted tissue with no recipient's general immunosuppression.

A. Whole Pancreas Transplantation

Whole pancreatic transplantation would apparently look closer than islets to short-term or mid-term clinical applicability. Combined (simultaneous or temporally deferred) pancreas + kidney (PK) transplant in T1DM patients that are also affected by terminal renal insufficiency, and are in a waiting list for renal transplant, represents a relatively safe and viable procedure, as compared to kidney transplant alone, in terms of both patient and graft survival (7). However, this approach, in the best conditions, could only apply to patients with T1DM who require a simultaneous kidney graft, thereby inevitably cutting off younger patients, who are specifically eligible for pancreas transplant alone (PTA). Furthermore, PTA has been associated with significantly lower clinical success and higher morbidity in comparison with PK. The invariable need for life-long pharmacologic immunosuppression, based on still poorly selective agents, makes this approach ethically acceptable only to those patients with T1DM that either need PK or alternatively, suffer from brittle diabetes, and are not adequately responsive to conventional insulin therapy. These concepts have been included in the recommendations of the American Diabetes Association (8). It is worth mentioning that although much improved, the morbidity and mortality associated with whole pancreatic graft continue to be relevant. Consequently, whether to apply the whole-pancreas transplant to selected T1DM patient cohorts remains difficult. Hopefully, the introduction of more specific immunosuppressive agents will broaden the applicability spectrum of this therapeutic option in the not too distant future.

Table 38.3 Potential Alternate Tissue as a Resource for Donor Islets in Humans

1. Nonhuman, nonengineered islets
Porcine (adult, neonatal)
Bovine
2. Human/nonhuman engineered islets
Transfected cell lines
β-cell origin
non-β-cell origin

B. Islet Cell Transplantation

Isolated islet transplantation would be incomparably less invasive, in comparison with a whole pancreatic graft, since the islets, which represent only 1–2% of the total pancreatic mass, are usually injected through percutaneous cannulation of the porta vein, directly into the liver, under local anesthesia. However, in spite of the potential advantages associated with islets over whole organs (Table 4), the immune problems involved are similar, while clinical results, as reported by the last International Islet Transplant Registry Newsletter (9) look even poorer. According to the

Table 38.4 Islets over Whole Pancreatic Grafts—Potential Advantages

1. Lower graft mass (1–2% of the whole pancreas)
2. Minor surgery
3. In vitro immunomanipulation
4. Cryogenic storage
5. Immunoisolation in artificial membranes

Islet Registry, as of June 1999 only about 8% of the patients with T1DM, receiving islet allografts (islets separated and purified from cadaveric human donor pancreata) have achieved full insulin-independence at 1 year posttransplant. The vast majority of these patients were affected by terminal renal insufficiency, and they either required a simultaneous or already carried a stable kidney graft. On the contrary, about 40% of totally pancreatectomized patients (usually undergoing upper abdominal exenteration for malignancy), receiving intrahepatic islet allografts, under general immunosuppression, have been reported to enjoy insulin independence at the same clinical follow-up time period (PIDM-Allo). The 1-yr remission percentage was raised to 50 when autologous islets were injected into patients that had undergone total pancreatectomy subsequent to chronic pancreatitis (PIDM-Auto) (Fig. 1). A major variable associated with these clinical trials was the host's immune competence, which obviously was more stringent for patients with T1DM than for those with pancreatectomy-induced diabetes, where no autoimmunity would ever occur. Autoimmunity has been clearly reported to recur in whole pancreatic grafts, and it has been deemed to play a major role in islet graft-directed immune destruction as well (10).

In summary, both immune rejection and recurrent autoimmunity have severely hampered the progress of islet transplantation into clinical trials, also in terms of the total number of patients (close to 500) that have undergone, so far, this procedure, according to the Registry. The reasons for this limitation are multiple, beginning with restrictions for recruiting the potential candidates to graft. Only a major or life-saving organ graft could, in fact, justify general immunosuppression with its life-threatening side effects. Therefore only a few patients with T1DM have been thus far enrolled into clinical trials. Furthermore, the islets are not a solid organ, and while problems of islet yield and purity are gradually being overcome, in many instances it may be still difficult to separate an islet mass, from a single pancreas, that is adequate for grafting a single T1DM recipient. As far as immunity is concerned, very recent attempts to perform human islet allografts alone, using putatively safer and more selective immunosuppressants (Table 5) are in progress. It is too early to anticipate the outcome of these trials. However, even in the best case, where the stable reversal of hyperglycemia and the absence of severe immunosuppression-related side effects were achieved, a major issue would still be the restricted availability of cadaveric human donor organs. Finally, prevention of allograft-directed immune destruction, by use of old or newly developed immunosuppressive agents, may not equally succeed in obviating the autoimmune recurrence of T1DM.

A potential solution could be to use alternate islet sources to both avoid tissue shortage, while preventing recurrent autoimmunity, and in parallel develop alternate strategies to protect islet grafts from the immune destruction, with no recipient's general immunosuppression. It is current opinion that either adult (11) or neonatal (12,13) porcine islets might successfully replace human donor tissue, provided that they were properly immunoprotected. On one hand, pork insulin would be acceptable to humans, and on the other hand, it would be possible to create specific pathogen free (SPF) pig breeding stocks that comply with standard safety requirements.

With regard to alternatives to general immunosuppression, two main strategies seem to be feasible at this juncture (Table 6). Creation of properly assembled physical immunobarriers will be the subject of this review on the actual perspectives associated with the BAP. The possibility of inducing a state of donor's immune unresponsiveness continues to be appealing and controversial at the same time. Recent reports seem to speak in favor of using novel approaches to modulate the host's immune response. These would consist of employing new agents in conjunction with the induction of microchimerism, the latter being obtained by grafting homologous bone marrow cells, before islets. While this approach preliminarily looks encouraging (14), caution should be used in assessing whether such manipulation of the host's immune system may satisfactorily fulfill safety criteria that would apply even to pilot clinical trials.

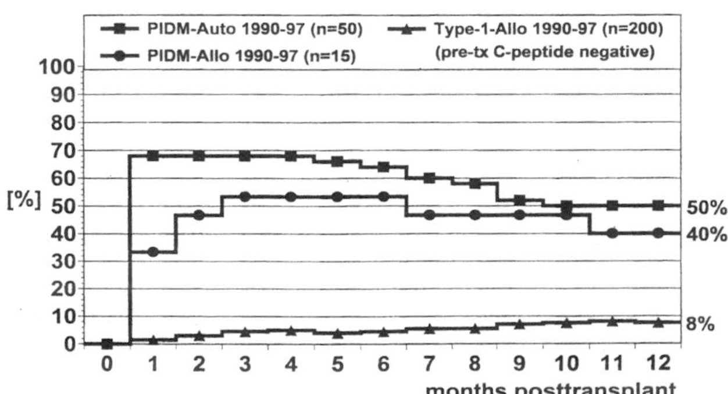

Figure 38.1 Islet transplant registry: outcome of human islet transplantation, August 1999. (Only well-documented patients.)

Table 38.5 General Immunosuppression

Old agents	New agents
Cyclosporin	Tacrolimus
Anti-lymphocyte globulin (ALG)	Sirolimus
Azathioprine	Anti-Tc monoclonal antibodies
Corticosteroids	

V. THE BAP: GENERAL FEATURES

The basic principles that development of a safe and reliable prototype of a BAP should conform to are the following: (1) unlimited access to islet tissue sources, whether they are primary or engineered human/nonhuman islet cells, in strict compliance with recommendations of regulatory agencies and ethical committees; (2) availability of artificial materials that are properly engineered, so as to avoid any harmful effects on both the transplanted tissue and the host; (3) elimination of any kind of recipient's general immunosuppression. Simply, a BAP is a hybrid device that embodies viable and functional islet cells within selective permeable and highly biocompatible artificial membranes.

The biological component, the islets, should be able to survive in a possibly hostile and certainly new microenvironment, while retaining the ability to secrete insulin, according to physiological algorithms, in order to avoid delays of insulin action.

The artificial component should fulfill specific criteria, such as membrane physicochemical properties that are compatible with a membrane's selective porosity, filtration, and permeability. In fact, it is mandatory that the embodied islets receive adequate nutrient supply, in order to permit biochemical exchange, with special regard to oxygen and macro- and microsolutes, glucose, and insulin included. Nonetheless, on the other hand, the membrane's selective cutoff properties should strictly interdict access to cellular as well as humoral mediators of the islet graft-directed immune destruction.

The ambitious goal of generating a biohybrid device that is fully able to substitute the human endocrine pancreas has been relentlessly pursued for decades (15). No clear and ultimate success has been so far achieved, although there are initial promising results and novel research hints. What had initially appeared as quite a feasible task has been proven to represent one of the most difficult challenges in medicine. In fact, the islets cells constitute a highly differentiated cellular system, equipped with a sophisticated glucose sensing apparatus that is very difficult to reproduce. Moreover, adult islet β-cells do not replicate (mitotic rate 1–2%) (16), and there is consequently no hope that any eventual loss in the transplanted islet mass may be replaced by islet cell proliferation. Although some mitotic figures have been detected within microencapsulated islets (17), these are totally insufficient to secure sufficient islet β-cell turnover. As a further complication, the islets contained in any type of BAP, differently from naked islets, cannot be revascularized or reinnervated upon transplant, and their nourishment will solely depend upon gas/metabolite passive diffusion. With regard to this matter, a pivotal role seems to be played by the graft site, in strict relation to the device's geometry. Some sites, such as the subcutaneous tissue, would provide O_2 gradients that seemingly are insufficient for the islet cell metabolic requirements. Progressively relevant, over time, also is the issue of device morphology and configuration that is strictly associated with both immunoisolation capacity and provision for adequate nutrient supply. The window between the membrane's molecular weight cutoff that permits ingress of indispensable molecules, while providing for immunoisolation, is very narrow, and it continues to represent the Holy Grail for the entire field of BAP clinical application to the therapy of T1DM.

Table 38.6 Alternatives to General Immunosuppression

1. Induction of immune tolerance	
a. In vitro	b. In vivo
Culture conditions	Microchimerism
Irradiation	Pharmacologic agents (i.e., tacrolimus, etc.)
Cyogenic storage	Antibodies (i.e., anti-CD40L, etc.)
2. Immunoisolation in artificial membranes	

VI. BAP: PHYSICAL CONFIGURATION

Different prototypes of a BAP have been designed and developed over the past few decades (Fig. 2) (18). In general, the classification is still valid that keeps macrodevices and microcapsules as distinct entities, although many of the old notions in the field have become obsolete, in light of the latest accomplishments.

The greater attention devoted to microcapsules than to macrodevices in this chapter does not necessarily reflect the formers' better performance but rather outlines the wider body of information that has been achieved by microencapsulation. Nevertheless, the potential impact of newer macrodevice strategies on BAP development is well acknowledged.

A. Macrodevices

A wide range of biomaterials have been considered suitable for the fabrication of devices that would allow for islet graft immunoisolation. The most common of them (Table 7) are still in use, with special regard to PAN-PVC, although many once innovative concepts are progressively fading away. In particular, the macrodevice geometry (Fig. 2) includes a few types, some of which hold mere historical value, at this juncture, although they have been associated with remarkable results in experimental islet transplantation trials in the past. In particular, hollow fibers have been extensively employed for subcutaneous (19) or intraperitoneal (20) islet grafts. The fibers were made of a selective permeable macromembrane (molecular weight cutoff—MWCO—50–70 kD) that while impeding ingress to immune cells and antibodies was supposed to be highly biocompatible and allow for biochemical exchange. While successful in a few rodent trials, these devices have shown to correct diabetes in higher mammalians only sporadically. Major limitations derive from the low islet survival rate inside the chambers, possibly due to restricted nutrient

Table 38.7 Biomaterials for Fabrication of Macrodevices

1. Polyacrylonitrile-polyvinylchloride (PAN-PVC)
2. Acrylates
3. Cellulose acetate
4. Cellulose mixed esters
5. Polytetrafluoroethylene (PTFE)
6. Alginates
7. Polyethylene glycol (PEG)

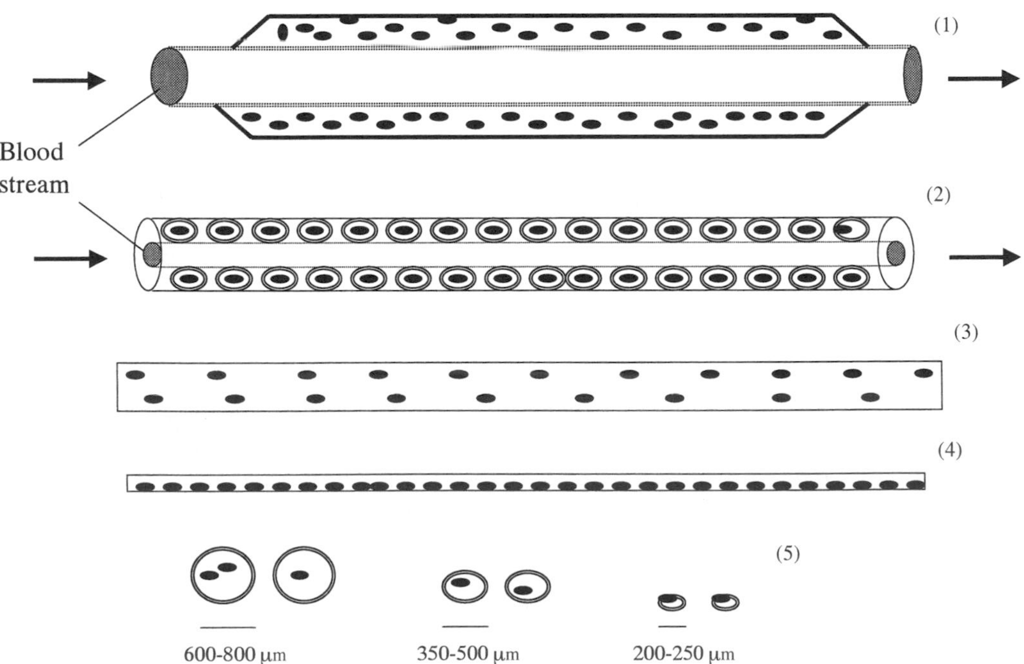

Figure 38.2 Types of immunoisolation devices for islet transplant: (1) vascular chamber containing matrix-embedded islets (23); (2) vascular prosthesis containing AG/PLO encapsulated islets (24); (3) hollow fiber containing matrix embedded islets (19, 27); (4) Laminar AG thin sheet containing monolayered islets (Hanuman Medical, personal communication); (5) AG-based microcapsules of different size (CSM, MSM, CM) (25, 69, 40, 33, 57, 64, 61).

supply. The latter likely depended on the graft site but also on the physicochemical properties of the employed materials. A major, often erratically solved problem, was fibroblast overgrowth of the chambers that resulted in membranes' pore clogging, thereby further limiting O_2/nutrient supply.

It was commonly observed that when loaded ''plain'' in the device, the islets were associated with quite a limited survival rate. This specific problem was overcome using a gel matrix to embed the islets prior to the fiber's loading process. The matrix, usually an alginate gel, greatly improved the islet viability. Unfortunately, the trade-off was that the matrix-embedded islet loading capacity was very low, thereby imposing the use of larger-size devices to obtain sufficient metabolic results in the grafted diabetic animal models (21). It is likely that if this order of magnitude were to be applied to a high mammal transplant setting, every patient would require several ''meters'' of fibers filled with matrix-embedded islets.

As far as the fibrotic overgrowth, and in particular the fiber environmental conditions for islet bedding, are concerned, significant progress has been made in an attempt to provide the embodied islets with sufficient O_2 nutrient supply: the idea of prevascularizing the device, prior to the islet entrapment, has resulted in interesting preliminary results in rodents (22). Whether these findings will foster renewed efforts in this specific research field will be the subject of future assessment.

Another possible strategy to use macrodevices for islet transplantation has consisted of seeding the islets in a special compartment chamber, directly anastomosed to blood vessels, usually as arterio-vein shunts. In this way, the islets were continuously perfused by blood ultrafiltrate, which seemingly facilitated biochemical exchange. The membrane, at contact with the bloodstream, was associated with an appropriate MWCO (commonly below 100 kD) so as to prevent immune cells or antibodies from crossing the islet containing chamber's wall. The vascular approach, while brilliant in principle, and preliminarily associated with results that were convincingly positive in diabetic dogs (23), although not in humans, where only partial and transient remission of hyperglycemia was achieved (24) (in this setting the islets had been microencapsulated prior to loading in the vascular prosthesis), has left open many issues. In particular, the risk for the devices' clotting/breakage, with subsequent need for intense anticoagulation of the recipients, has never permitted thorough assessment of the procedure's safety.

Recently, ultrathin immunoisolatory sheets, made of highly biocompatible alginate, where the islets are homogeneously monolayered, seem to be particularly appealing. The alginate sheet represents a very versatile macrodevice that couples safety requirements such as (1) retrievability; (2) biocompatibility (one of the most studied biomaterials, Na alginate, is employed); (3) physical configuration (O_2/nutrient supply is facilitated by the minimal diffusion distance) to reduce trauma. The sheet may be virtually implanted anywhere in the mesenteric area. Preliminary data with this new generation of macrodevices in pilot diabetic canine graft trials look encouraging, although it is premature to make any forecast on the future impact of this technology on islet transplantation (personal communication, Hanuman Medical, San Francisco, CA, USA, March 2000). In summary, while the past generation of macrodevices was associated with a number of disadvantages, the latest developed prototypes hold promises for broader application in diabetic higher mammalians.

B. Microcapsules

1. Technical Procedures

Microcapsules fabricated with alginic acid derivatives, chemically linked to polyaminoacids, have represented the most widely known and studied kind of microimmunobarrier for islet cell transplantation. According to the original Lim and Sun recipe (25), microcapsules were fabricated by suspending the islets in sodium alginate (AG), a polysaccharide seaweed derivative, and subsequently by extruding the alginate–islet mixture through a microdroplet generator (Fig. 3). The islet containing microdroplets were collected on a calcium chloride bath, which immediately turned them into gel microspheres. The calcium alginate gel microbeads were then subsequently coated with aminoacidic polycations, such as poly-L-lysine (PLL), or uniquely in our laboratory, poly-L-ornithine (PLO). The outer coat was comprised of highly purified sodium alginate that provided the multilayered selective permeable membrane with additional biocompatibility (Figs. 4 and 5). These ''conventional-size microcapsules'' (CSM), measuring an average of 700–800 μm in equatorial diameter, have

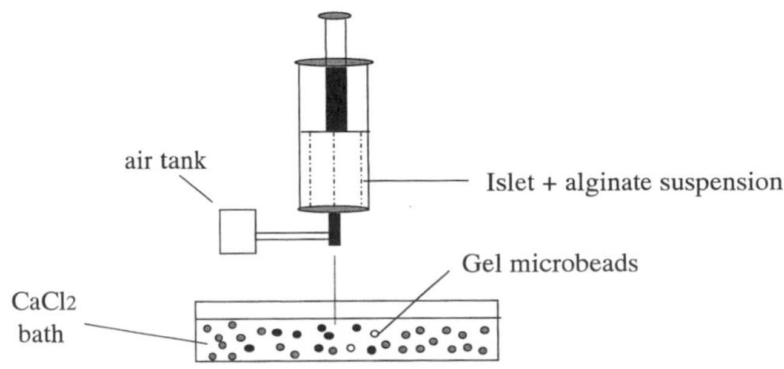

Figure 38.3 Microdroplet generator for microcapsule fabrication. (From Univ. of Perugia.)

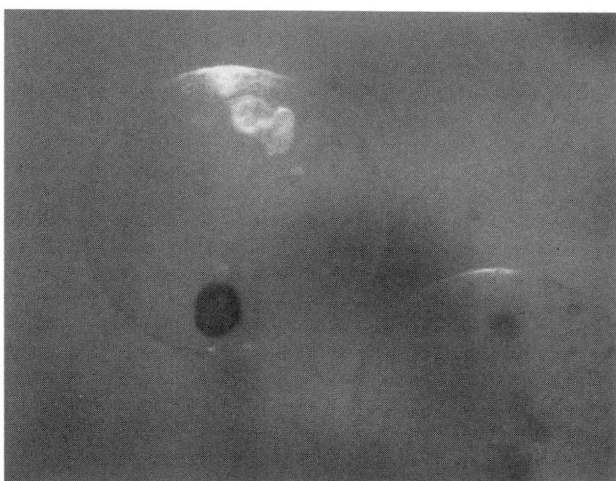

Figure 38.4 Freshly prepared AG/PLO microcapsules containing human islets (staining with Dithizone, 25×); morphological integrity is evident. (From Univ. of Perugia.)

undergone several adjustments, including scale-up protocols, in our and a few other laboratories (26,27).

Other procedures, still involving the use of alginate, although by different chemistries or consisting of chemically different reagents, will be discussed later. In general, regardless of more or less pronounced differences in the fabricative procedure, a limited number of experimental trials has been reported on the use of these alginate alternatives for microencapsulation in diabetic animal graft trials.

2. Achieved Results

Most either allogeneic or xenogeneic islets containing AG-PLL/PLO microcapsules have been usually implanted in the peritoneal cavity, into diabetic hosts that were not generally immunosuppressed. Unfortunately, the lack of standardized materials and procedures has hampered impartial assessment of the results communicated by different laboratories. In general, rodent studies were associated with greater and more consistent success, in terms of full remission of hyperglycemia, following microencapsulated islet TX, as compared to higher mammal trials (28). Nevertheless, the transplanted rodents carrying spontaneous autoimmune diabetes (NOD mice, BB rats) have shown lower remission rates than mice and rats, where diabetes had been artificially induced by streptozotocin (29,30). Briefly, the most experienced laboratories provided convincing evidence that the microencapsulated islet system may work satisfactorily in small-size animals. On the contrary, very few reports documented the reversal of hyperglycemia with the discontinuation of exogenous insulin injections in dogs with spontaneous or pancreatectomy-induced diabetes, receiving intraperitoneal microencapsulated canine or porcine islet grafts (Table 8). In this high mammal animal model, full reversal of hyperglycemia was partially accomplished when the encapsulated islets had been embodied in a special vascular chamber, directly anastomosed to blood vessels (31). The canine background permitted us to initiate pilot human clinical trials in our laboratory, that were associated with only partial and transient metabolic results. Preliminary success has been reported in dogs (32) and T1DM patients (33) receiving intraperitoneal canine or human islet allografts, respectively, but only when the recipients underwent a course of general immunosuppression with cyclosporin. An important exception to this uncertain trend was represented by a successful trial of AG-PLL encapsulated porcine islet xenograft into spontaneously diabetic, nonimmunosuppressed monkeys. In these animals, full remission of hyperglycemia and exogenous insulin withdrawal were achieved in all recipients, and sustained in some of them for extraordinary long periods of time, throughout 3 years of post-TX follow-up (34). This striking, so far unique result, indeed proves the principle that under the best conditions, microcapsules may constitute an effective, biocompatible, and immunoselective physical barrier that enables immunoprotection of islet xenografts with no general immunosuppression of the host.

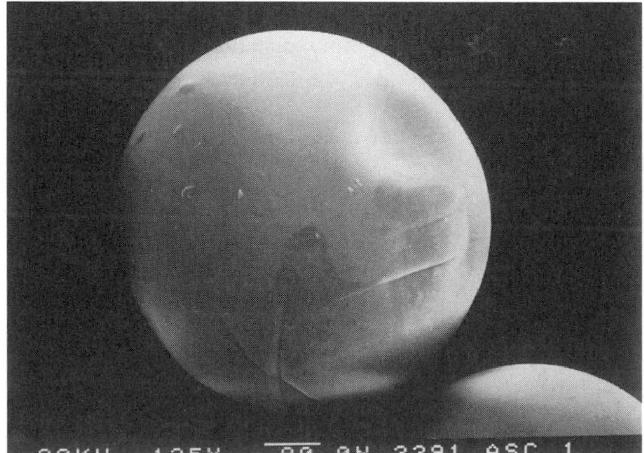

Figure 38.5 Scanning electron microscopy examination of AG/PLO microcapsules: intactness and smoothness of the coating membranes are evident (20 kV). (From Univ. of Perugia.)

3. Pitfalls

The trials of encapsulated islet cell allo- or xenografts in nonimmunosuppressed diabetic recipients have shown, with anecdotal exceptions, that the encapsulation system still needs to surmount unsolved problems. In particular, it has not yet determined whether AG-PPL/PLO represents the best chemical formulation to prepare highly biocompatible and selective permeable microcapsules. For this

Table 38.8 TX of Allo/xenogeneic Islet Containing STD Microcapsules TX in Diabetic High Mammalians

RX	TX site	Success	Reference
Primates:	peritoneum	Substantial	(34)
Canines:	peritoneum	*Partial	(32)
	vascular prosth.	Partial	(31)
Humans:	peritoneum	*Partial	(33)
	vascular prosth.	Partial	(24)

* Recipients underwent general immunosuppression.

purpose, studies focusing on generation of new biopolymers for microcapsulation should be highly pursued. It remains unclear whether the actual microcapsule design should be modified or implemented, in order both to improve the functional, possibly long-term performance of these membranes and better to regulate the enveloped islet mass (35). Moreover, the potential role of immune cells that may be part of the islet cell populations, within microcapsules, should also be studied (36).

VII. GENERAL ISSUES RELATED TO IMMUNOISOLATED ISLET GRAFTS

There are issues, in the development of BAP that require special attention.

A. Mechanisms of Islet Graft Destruction

1. Immunomediated

a. Allograft

Naked islet transplantation. Allogeneic and discordant/nondiscordant xenogeneic islet transplants have been studied in either patients or animal models of diabetes. From extensive observations, even if many aspects of the islet graft-directed immune destruction remain unknown, there are few doubts that two conditions need to be examined, Immune rejection, and the Recurrence of autoimmune diabetes.

Role of cellular immunity. According to Gill (37), T lymphocytes (Tc) detect foreign antigens as peptides that are presented in association with major histocompatibility complex (MHC) molecules. In particular, CD8 Tc recognize antigens presented in association with class I MHC molecules, whereas CD4 Tc generally recognize antigens presented with class II MHC. However, two signals are required for Tc activation: signal 1 is provided by Tc receptor/antigen binding; signal 2, ''costimulator,'' is provided by non-antigen-specific signals from an antigen presenting cell (APC) (ie, macrophage, dendritic cell, etc.). The process defined as costimulation comprises a complex cell surface receptor–ligand interactions network, which has been partly clarified in recent years (Fig. 6). Typical participants in such interactions are CD80/CD86 molecules, expressed by APC which bind CD28 coreceptor molecules on Tc; interactions between CD40 on APC with CD40L on Tc also result in important costimulatory signals. The role played by costimulatory molecules in inducing islet allograft rejection is shown by observations where blocking CD40/CD40L interactions in vivo could result in indefinite islet allograft survival (38). Gill (39) has elegantly explained that two antigen presentation forms can fulfill this two-signal requirement for Tc activation:

Direct (or donor MCH-restricted) antigen presentation, in which Tc engage native MHC molecules directly on the surface of donor-derived APC.

Indirect (or host MHC-restricted) antigen presentation, in which donor-derived antigens are captured by recipient APC, processed, and re-presented in association with recipient MHC molecules.

Exogenous antigens are usually processed and presented by class II MHC molecules and so are associated with predominant CD4 response.

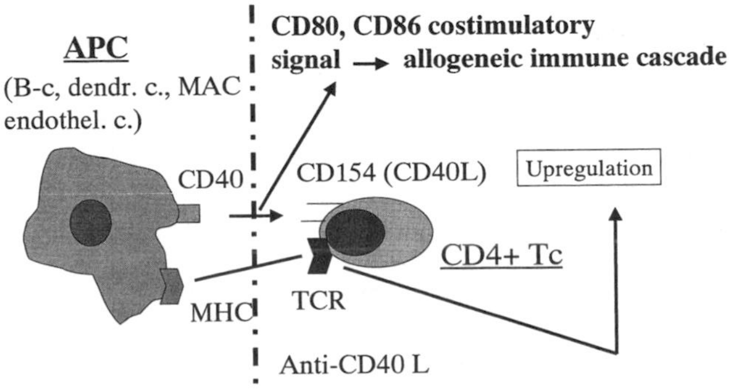

Figure 38.6 Pathways of β-cell-directed cellular immune attack.

Autoimmunity. A relevant issue that needs to be properly examined in islet cell transplantation in type I DM regards autoimmune recurrence of the disease on the transplanted tissue. Recurring disease has been shown to destroy transplanted islet tissue in either animal models of autoimmune diabetes or in man. Evidence can be now marshaled about the fact that Tc-dependent autoimmune islet injuries behave similarly to the xenograft-mediated response, in that islet damage can be triggered by autoreactive CD4 Tc, which are specific for islet-associated antigens, processed and presented by APC, similar to the indirect pathway immune response process. Although CD8 Tc may contribute to developing islet graft-directed autoimmune destruction, the latter clearly appears to be a CD4 Tc-mediated event.

Overall strategies so far employed to prevent islet allograft-directed immunity have included both general immunosuppressive agents, as mentioned elsewhere, and molecular mechanisms to block the costimulator pathway, such as blocking CD40-CD40L and B7/B28 interactions. The former have been usually applied to human islet allografts, by using different pharmacological agent cocktails. The latter have been preliminarily studied in lower and higher mammal animal models of diabetes (40).

b. *Xenograft*

When dealing with xenogeneic islet transplantation, both cellular and humoral immunity play a major role in inducing rapid destruction of the foreign tissue.

Role of humoral immunity. Hyperacute rejection will rapidly destroy xenogeneic discordant organ/tissue vascularized transplants by intervention of naturally occurring xenoreactive antibodies present in the host. These antibodies bind to antigens located on the surface of endothelial cells within the xenograft, an event that in turn activates complement with subsequent graft destruction (41). The most important target of these antibodies is represented by the terminal carbohydrate galactose $\alpha(1,3)$-galactose. The presence of α-Gal in high levels, in all adult porcine endothelial cells, accounts for the immediate activation of a heterologous host's immune system. α-Gal expression seems to be lower in neonatal pig islets, which should confer them a relatively higher extent of protection when implanted in xenogeneic hosts (41). Among several possible strategies that now are available for immunoprotection of xenogeneic discordant grafts, with special regard to potential use of pig tissue in humans, a recent one includes generation of transgenic pigs that express human complement-regulatory proteins (41). These proteins, CD 55 (DAF, decay accelerating factor), CD 46 (MCP-1, membrane cofactor protein), and CD 59 (protectin), do not impede anti-αGal antibody binding to endothelial cells, and thereby complement activation. These factors, however, inhibit subsequent steps of the complement cascade and finally prevent cell lysis. Results so far achieved with transgenically modified organs, and particularly with islet cells separated from the pancreata obtained from these animals, are still premature to interpret and seemingly warrant further investigation.

Role of cellular immunity. Islet xenograft rejection is less dependent on donor APC than allograft. In fact, depletion of passenger leukocytes is not associated with the beneficial effects described for allografts. Xenograft rejection may occur in the absence of CD 8 Tc, while CD 4 Tc are predominantly involved with the immune destruction process, which reflects the steps already described for the indirect pathway (Gill).

In conclusion, islet xenograft rejection is an extremely rapid and devastating process, that involves both the humoral and cellular loops of the host's immune response system and is very difficult to contrast with conventionally available strategies. Potential progress in this field could be represented by the availability of transgenically modified pigs, that should turn a ''xenogeneic discordant'' into an ''allogeneic'' islet graft system. Use of biohybrid device strategies could dramatically impact this particular islet graft setting.

c. *Immunoisolated Islet Transplantation*

Allograft immunity. Since the above mentioned immune mechanisms mainly consist of direct engagement of Tc subpopulations (ie, CD 8 Tc) with donor cells, physical isolation of the islet graft within a BAP device should result, by preventing direct cell-to-cell contact between graft and the host's immune system, in much easier biologic acceptance.

Xenograft immunity. If BAP may effectively prevent ingress of both various antibody (including xenogeneic naturally occurring antibodies) and complement fractions, owing to its membrane's cutoff exclusion properties, it cannot fully eliminate the risk for activation of CD 4–driven cellular reactivity. Small-size molecules (''shed antigens'' such as proteins, enzymes, etc.) might cross the BAP membranes due to their small size, ranging on a few kD. These molecules could be incorporated, processed, and presented on a host's APC, thereby triggering CD 4–driven significant inflammatory reaction. Either small cytokines, or nitric oxide, or free oxygen radicals, released by the inflammatory cells, individually or collectively, could easily result in the final destruction of the immunoisolated islets. For this purpose, Gill has proposed that ''immunoisolating'' membranes should be simply defined as ''isolating'' (39).

Autoimmune recurrence of disease. As said above this

type of immune reactivity may be assimilated to xenograft immunity as far as the involved cellular mechanisms are concerned. It may eventually affect the extent of cellular inflammatory response that is elicited in the immediate vicinity of BAP subsequent to graft into a diabetic recipient (Gill).

However, these putative immune mechanisms of immunoisolated islet graft destruction are in contrast with several reports documenting long-term protection of the encapsulated islets from the host's immune response in either low or high diabetic mammalians both in vivo and in vitro (42). For instance, we had made an original observation, in our laboratory, where in a static incubation system, either free or AG-PLO encapsulated rat islets had been exposed to 17 kD, β-recombinant human interleukin 1 (IL-1). Theoretically, such a small molecule should have easily crossed the capsular barrier, and provoked effects similar to those observed in unencapsulated islets that, upon IL-1 exposure, underwent loss of both cellular architecture (Figs. 7ab and 8ab) and functional competence (Fig. 9). On the contrary, the encapsulated islets were intact and retained their insulin secretion capacity. It probably was just a matter of physical distance, since in that experiment we had employed CSM, although a stereochemical molecule binding process might not be excluded. Therefore further study is required of the limits and pitfalls of immunoisolation strategies to prevent the islet graft-directed immune phenomena, as specifically reported below.

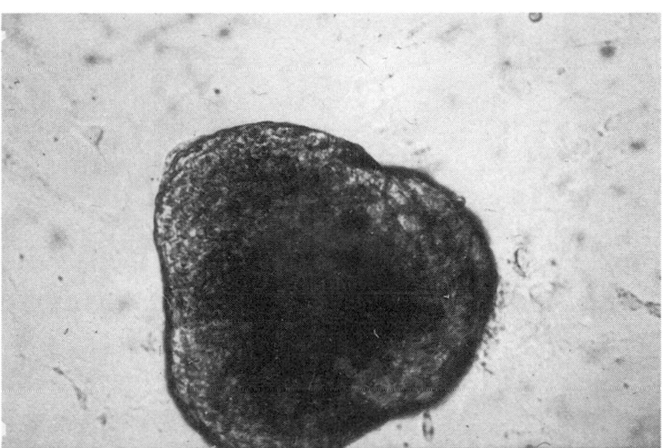

(a)

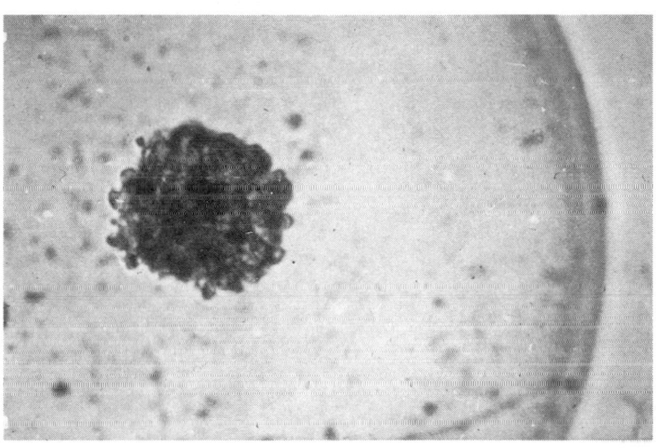

(b)

Figure 38.7 In vitro (a) free and (b) AG/PLO microencapsulated rat islets prior to incubation with IL-1 (25×). (From Univ. of Perugia.)

2. Not Immunomediated

Many potential pathways that do not include immune events may theoretically lead to restricted survival and functional performance of immunoisolated islet grafts. The BAP's (whether it be composed of macrodevices or microcapsules) overgrowth with fibroblasts is a common cause for graft failure. As said elsewhere, this phenomenon may depend on the material biocompatibility but also on the islet cell capacity to adapt to new environmental conditions. These obviously are quite distant from those of the native endocrine pancreas. For this purpose, a major problem, strictly correlated with the implant site, is represented by O_2 supply (43,44). In fact, necrosis of the central islet cell core has been extensively documented in support of this assumption. Preliminary documented attempts have been made, as elsewhere mentioned, to connect somehow the naked or encapsulated islet containing immunoisolatory devices directly to the blood stream, or prevascularize them so as to overcome this specific problem, although other reports focusing on the same topic remain questionable (45). In general, while selected studies have proven to be on the right path, the overall results so far obtained are still too limited and poorly quantifiable to allow for systematic evaluation.

Another major problem lies on the adult islet β-cell low mitotic rate (1–2%) (16). Although the islet cells have been shown to be able to replicate (46), even within artificial membranes, the islet death rate overtook the islet cell proliferation rate. Therefore any islet mass loss occurring in the posttransplant period, as a consequence of either immune- or not immunomediated events, virtually is not replaceable by the newly generated β-cells. Islet cell regeneration within the neoimplanted tissue would be greatly facilitated by physical communication with the native pancreatic matrix. Increasing evidence, for this purpose, seems to demonstrate that the pancreatic ductal network may play an important role in inducing islet β-cell replication by virtue of the release of selective growth factors (ie, IGF II) (47). This being the case, reconstruction of the original

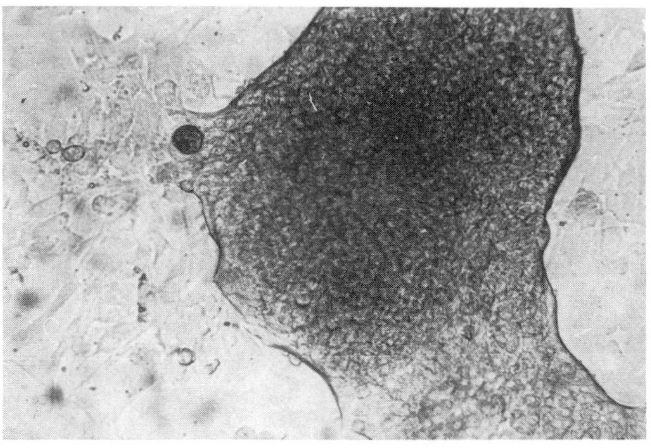

(a)

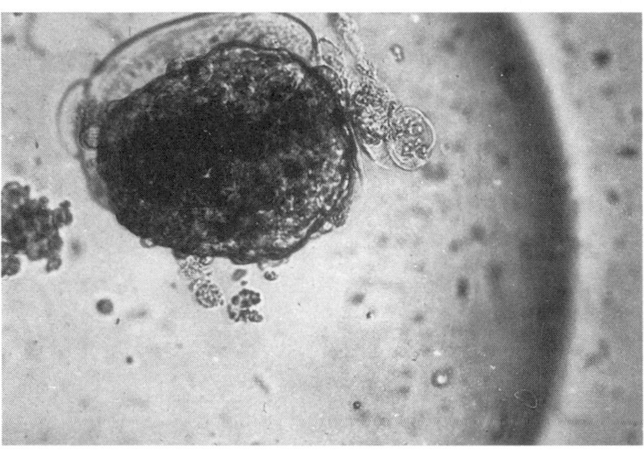

(b)

Figure 38.8 In vitro exposure of (a) free and (b) AG/PLO microencapsulated rat islets to 17 kD β-recombinant human IL-1 (100 mU/mL); profound derangements (a) or normality (b) of the islet architecture are observed (25×). (From Univ. of Perugia.)

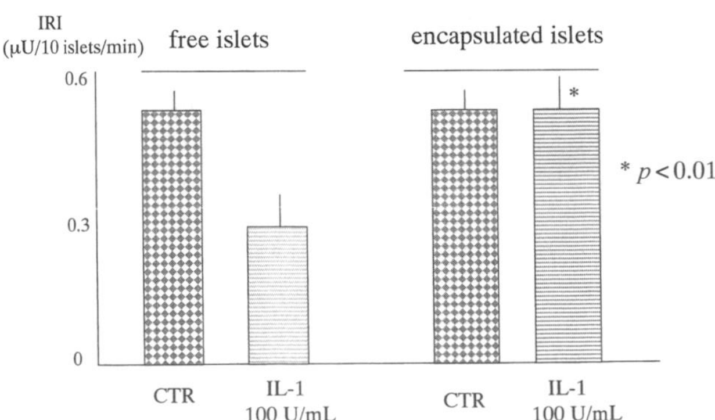

Figure 38.9 In vitro exposure of free and AG/PLO microencapsulated rat islets to 17 kD β-recombinant human IL-1 (100 mU/mL); insulin secretory patterns are compromised for free but not for encapsulated islets ($p < 0.01$). (From Univ. of Perugia).

pancreatic milieu inside a BAP might greatly help to prolong survival of the enveloped islets.

It is finally worth mentioning the possible occurrence of apoptotic phenomena. These could account for loss of the originally implanted islet cells mass, not otherwise explainable by the occurrence of other β-cell targeting noxious events. However, the impact of apoptosis inducing decline in the original transplanted islet cell mass will require further study (48). Finally, a problem that seems to pertain closely to macrodevices, rather than microcapsules, is the fibroblasts' growth within the islet graft. Obviously, these opportunistic cells will thrive at the expense of the islet tissue, with additional subtraction of O_2 nutrient supply.

B. Biocompatibility and Specific Aspects of Membrane Configuration

Any artificial material that is introduced into the body is ''foreign'' and may therefore elicit some kind of response. Whether such a response is of a ''benign'' nature, self-limiting and composed of humoral/cellular elements that are relatively harmless to both the recipient and the implanted tissue, will depend upon either the nature of the constituent biomaterials or that of the graft site or finally the transplant procedure. On both the macrodevice's and the microcapsule's side, and increasing number of biopolymers, associated with more or less specified physicochemical properties, have been introduced in the past few years. Not always, however, have the improved membranes' technical features coincided with improvements of their biological acceptance. Neither has compliance with safety requirements. The interactions between the membrane and the embodied cells on the one hand, and the membrane and the host's tissue on the other hand, both the graft functional performance, and the host's reactivity being equally critical, represent the most important issues, regardless of the device's size and physical configuration. The ideal endpoint would be that the biohybrid organ is not simply tolerated by the host but rather becomes part of it. Following this principle, the artificial membrane, separating the grafted tissue from the host, would not represent something foreign that is simply in charge of regulating transmembrane nutrient/metabolite fluxes, while interdicting access to mediators of the immune system. Rather, the membrane should be fully integrated in a hybrid entity that incorporates the implanted cells and the host's tissue, within a system that finely adjusts its functional and immunological properties to physiological requirements of the hosting counterpart. If such an integrated system were in effect, many of the still open issues could possibly be overcome. There is no question that the device's size and geometry, other than its bare chemical composition, deeply influence either the extent of the device-directed host's reactivity or its survival or finally the functional performance of the device-embodied live cells.

Size is very important. In fact, the bigger the grafted tissue volume, the more frequent are the unfavorable side effects observed in the recipient. Newly developed prototypes of the BAP aim to decrease the overall graft size. However, problems differ between macrodevices and microcapsules. In the former, the main issue is to couple an ideal tissue cell distribution, in compliance with the optimal biochemical/metabolic exchange, to the smallest possible appropriate size. In this regard, a possible innovation, as already mentioned, is the thin double-layer alginate sheet. The particular flat, and minimally invasive configuration of this macrodevice type seems to be suitable for islet grafting purposes.

As far as microcapsules are concerned, special care is now being taken in optimizing already established polymers, such as the alginates, by either improving the basic chemical formulations or creating new chemical blends to coat alginate gel microbeads. In particular, alginate-based microcapsules, associated with the possibility of dynamically altering the outer coating thickness so as to regulate the membrane's permeability/porosity, are in progress (49,50).

Practically, only a few attempts to use alternate polymers for islet microencapsulation in high mammalians have been made. Many different polymers and materials belonging to different chemical categories have been proposed along the past 4 or 5 years, but none has proven able, at this juncture, to replace alginates effectively (see below). Whether it reflects insufficient examination of new chemistries, or the achievement of no tangible improvements, remains unclear. Microcapsules, because of their typical spherical shape, hold the important advantage of coupling ideal cell distribution to favorable diffusion properties. However, major efforts have been made, in few world laboratories, including ours, to reduce the original CSM size (600–800 μm) (51). CSM were, in fact, originally tailored for rodent graft trials, where they have indeed performed satisfactorily, and proved to reverse hyperglycemia in mice and rats with either spontaneous autoimmune or artificially induced diabetes. Unfortunately, in larger animals, the islet mass putatively able to reverse hyperglycemia coincided with a final encapsulated islet volume that was overwhelmingly big (ie, 80/100 mL per recipient dog), thereby resulting in mechanical and inflammatory reaction at the level of the peritoneal cavity. Obviously, the latter represented the only possible graft site for these capsules. To try to address this issue, attention was turned to develop smaller capsule prototypes, measuring 300–400 μm, all the way through "conformal" coatings that enveloped each individual islet within a skin-like hydrogel cast. This advanced capsule prototype virtually coincided with the islet's size itself. However, it is not demonstrated that conformal coatings, ideally fulfilling volume requirements, are also advantageous in terms of either metabolic performance of, or immunobarrier competence for, the embodied islets. We now believe that medium-size microcapsules, measuring an average 300 to 400 μm in diameter, represent, as mentioned elsewhere, a good compromise between volume and functional/immunoisolation performance.

C. Specific Issues and Pitfalls Associated with the BAP

1. Macrodevices vs. Microcapsules

a. *Pros and Cons*

The search for an optimal device that couples the least invasive transplant procedure to the best functional performance is not over yet; it rather continues relentlessly. On the basis of the features that are associated with macrodevices and microcapsules, the advantages and disadvantages of each immunoisolation system can initially be recognized (Table 9).

As progress in validating refinements of previously developed devices or in generating newly configured devices, whether they are macromembranes or microcapsules, is gradually being established, some of the pro's and con's associated with either membrane's type likely require regular updates. Furthermore, new device engineering may introduce substantial changes in some principles previously indicated as sine qua non conditions. Some macromembranes are now able to fulfill the need for size restrictions that make them suitable for intraperitoneal or subcutaneous islet grafting in high mammalians with no significant adverse side effects. Likewise, a wide spectrum of variably sized microcapsules is gradually being made available in order to fulfill multiple tasks (Fig. 10). One of these relates to the implant site that should not necessarily be restricted to the large peritoneal cavity, as it was for CSM. Furthermore, the smallest microcapsules could also eventually be deposited in bigger devices, so as to generate a multicompartment entity that incorporates cells and other agents within a common immunoisolatory shield. This could represent an advanced version of the BAP.

2. Chemistries

While, as elsewhere mentioned, AG-PLL/PLO continues to represent the most widely studied chemistry for the fabrication of islet containing microcapsules, alternate chemi-

Table 38.9 Microcapsules vs. Macrodevices

	Pros	Cons
Microcapsules	Easier fabrication	Suitable chemistry
	Favorable size/distribution	Different retrievability
	Individual islet immunoprotection	Quality control of each capsule
	Minimal fibroblast growth	Assessment of long-term durability
	Easier in vitro culture	Tissue loading restrictions
Macrodevices	Uniform intact endocrine tissue distribution	Size inadequacy for TX
		Fibroblast growth
	Easy retrievability	Traumatic TX procedure
		Thrombogenicity (vasc. dev.)

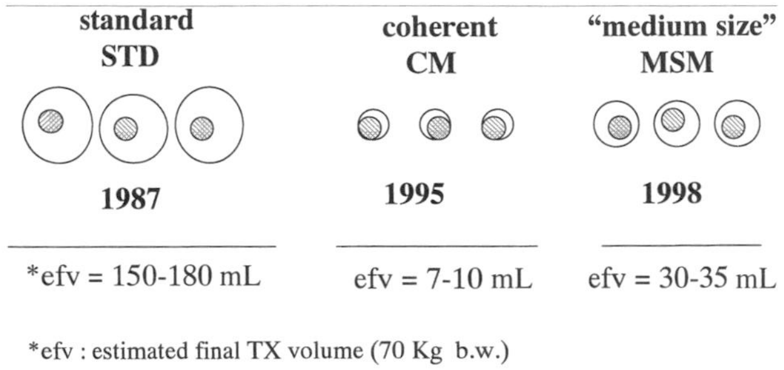

standard STD — 1987 — *efv = 150-180 mL

coherent CM — 1995 — efv = 7-10 mL

"medium size" MSM — 1998 — efv = 30-35 mL

*efv : estimated final TX volume (70 Kg b.w.)

Figure 38.10 Schematic representation of variably sized microcapsules with respective final graft volumes. (From Univ. of Perugia).

cal formulations have been described that may enable the formation of microcapsules.

a. Uncoated Ca-Alginate Microcapsules

Because of the very well known ability of the alginates to form highly biocompatible hydrogels, whether these are prevalently composed of guluronic or mannuronic acid derivatives, some groups have prepared alginate beads that were uncoated with poly-L-lysine or other polyaminoacids (52). Usually big Ca-AG microcapsules, measuring up to 3 mm in diameter, were fabricated and employed successfully to immunoprotect either allo- (53) or xenogeneic islet grafts in diabetic dogs or rodents, respectively (54). Although the capsules were proven to protect the islet allografts from acute rejection, a treatment course with general immunosuppressants was scheduled for the recipients in order to achieve sustained remission of hyperglycemia. The uncoated AG beads have posed a special problem that still remains open. In fact, MWCO associated with uncoated AG capsules should amount to several hundred kD, thereby virtually allowing for either immunoglobulins or large-size complement fractions to cross the membrane. Why such events did not result in acute destruction of the capsules–embodied tissue remains unclear. The authors postulated AG negative surface charges that would prevent intracapsular access to similarly charged molecules. Moreover, prevention of cell-to-cell contact may play a role in circumventing the xenograft-directed immune destruction. While the immunoprotection capacity of these microcapsules warrants further investigation, their large size constitutes a problem per se, for the reasons reviewed above. Moreover, disposal of such a great amount of Ca-AG could be a problem per se, since biodegradation of this polymer is only partially known. Finally, need for using immunosuppressive agents in conjunction with the AG microcapsules would limit the impact of this microcapsule's type on islet graft immunoisolation technology.

b. Barium Alginate

Based on the principle that AG can be complexed with divalent cations to form hydrogels, some authors have employed Ba ions to generate Ba-AG microbeads. This is a one-step procedure, because Ba covalently binds to AG and it does not require additional coating (55). Although the simplicity of the method is attractive, and further refinements have permitted the authors to initiate the study of smaller microcapsules (56), there are two major issues with this microencapsulation procedure: (1) toxicity associated with Ba ions (Ba is toxic to live cells), since the Ba ion complexation rate cannot be precisely quantified; (2) questionable ''immunoisolation'' properties of the microcapsules that have been only partially elucidated. In particular, Ba-AG microcapsules do not seem able to provide the islet xenografts with sufficient immunoprotection (57).

c. Agarose

Agarose, like alginate, is a gelling agent extracted from seaweeds, and it is composed of repeating units of alternating β-D-galactopyranosyl and 3,6-anhydro-α-L-galactopyranosyl. Depending upon temperature conditions, agarose forms thermally reversible gels, with gelling occurring at temperatures that are far below the gel fusion point (58). Iwata and Ikada have developed several technical procedures to form agarose microcapsules to immunoprotect transplanted islets from the host's immune attack. Typically these authors have employed 5% agarose gel solutions to formulate their capsules, with no major problems adversely affecting the enveloped islet cell viability as assessed by in vitro studies. However, a potential technical problem occurred with the capsules' polydispersity, with variable amounts of empty microspheres burdening the overall graft volume. In terms of immunobarrier competence, agarose capsules have performed satisfactorily, within an islet allograft system (typically mouse islets into allogeneic mice with streptozotocin-induced diabetes) upon intraperitoneal transplantation (59). Noticeably, extent of immunoprotection afforded by these microcapsules seemed to decline when the recipients were NOD mice with spontaneous, autoimmune diabetes, thus raising an issue on the membrane's immunoselectivity (59). Further advances to cope with immune problems, specially associated with islet xenografts, have consisted of coating the agarose microcapsule's gel core with outer layers. These were composed of sulfated polyanions, complement activators (ie, poly-styrene sulfonic acid–polybrene polyion), and carboxymethylcellulose, the latter being added to improve the capsule's biocompatibility (59). Although these specially engineered agarose microcapsules have apparently prolonged the islet xenograft survival (hamster islets into C57BL/6 diabetic mice), there is no compelling evidence that blocking complement factors would per se provide the final solution to the many problems posed by encapsulated islet xenografts in nonimmunosuppressed diabetic recipients. Furthermore, no experience whatsoever has been achieved in diabetic higher mammalians.

d. Polyethylene Glycol

The most appealing property of polyethylene glycol (PEG) for fabrication of immunoprotective microcapsules is its capability of forming a protein-repellent surface. In fact, protein adhesion has been indicated to initiate the cell aggregation process, on the capsule's surface, which could impair the membrane's diffusion properties (60). PEG has been used either to coat alginate-poly-L-lysine gel microbeads (61) or to create conformal coatings (62) that tightly envelop each individual islet. The latter process specifically addresses the capsular volume issue, already indicated as a potential cause for encapsulated islet graft failure. However, PEG has so far gained only limited confidence as a basic material for microcapsule formulation. Aside in vitro and in vivo biocompatibility studies, conducted with empty PEG capsules, few data on grafts of islet containing PEG microcapsules into diabetic recipients have been so far reported. Furthermore, while chemically elegant and sharp, the fabrication procedure of PEG microcapsules requires a laser-induced photopolymerization process that is initiated by eosin. No sufficient data on process safety have been provided to endorse this approach as a potential substitute for alginates, agarose, and other biomaterials. This being the state of the art with PEG microcapsules, in vivo studies, targeting both biocompatibility and immunobarrier competence of these membranes in transplants in lower and higher animal models of diabetes, are warranted.

e. Other Polymers

Chitosan has been proposed as a biomaterial that may be suitable for the immobilization of cells that retain viability and function, using a semiautomated process involving the use of alginates within the microcapsule's chemical formulation (63). The multiple functional and therapeutic properties (ie, wound healing, etc.) of chitosan require further investigation to assess whether the systematic use of this polymer, either alone or in combination with other molecules, would result in improving the microcapsules' physicochemical structure. This being the case, the environmental conditions to which the islets are exposed would consistently improve.

Hydroxyethyl-methacrylate-methyl-methacrylate (HEMA-MMA) is a polyacrylate copolymer (64) prepared by solution polymerization after careful monomer purification. Although the polymer is to some extent hydrophilic (25–30% water uptake), it also associates with mechanical strength, elasticity, and durability over time (Sefton). The polymer has proven suitable to fabricate small capsules that might represent the ultimate answer to the final graft volume–related encapsulated islet problems. A possible concern eventually lies in the use of potentially harmful technical steps, during the capsule fabrication process, such as exposure to shear forces and organic solvents that might damage live cells. However, several cell types, such as tumoral human and animal cell lines, and primary rat hepatocytes and islets, have been successfully encapsulated in HEMA-MMA, showing viability retention over long periods of time. Engineered cell lines that produce several factors have been also successfully encapsulated in HEMA-MMA (65). While HEMA-MMA might provide the enveloped cells with acceptable microenvironmental conditions, owing to its suitability for incorporating additional factors, the specific usefulness of these membranes to encapsulate islet cells for transplantion in diabetic recipients remains undetermined, and it certainly will require further experimental study.

f. Miscellaneous Strategies

Many technical skills have been introduced by several authors in an attempt either to upscale the microcapsule fabrication process, and/or the individual microcapsule physicochemical characteristics, or to implement the immunoprotection shield provided by these membranes. The latter have been mainly addressed to treat diabetic NOD mice, which indeed represent a precious animal model of human type I diabetes, with molecules such as CTLAIg that interfere with costimulatory immune pathways resulting in the Tc system activation (Weber). These studies (66) have been of great interest because they have induced a significant prolongation of xenogeneic neonatal pig and rabbit encapsulated islet graft survival in diabetic NOD mice. This approach clearly addresses and aims to solve the problem posed by the complex immune interactions between an immunologically stringent host (NOD) and the reactivity associated with immunologically distant islet cells, as it happens in a xenogeneic islet transplant setting. It is likely, as mentioned elsewhere, that the simple mechanical barrier provided by microcapsules, based on an even refined polymer formulation, does not suffice for adequate tissue protection, and could, on the contrary, require the addition of other factors. Whether these should consist of agents that induce either substantial general changes of the host's immune system or modifications of the membrane formulations, including embodiment of molecules that prevent and/or attenuate activation of the host's immune system, requires additional study.

g. Conformal Microcapsules (CM)

Several methods have been described to envelop the isolated islets within conformal coatings. The rationale behind this procedure is to create immunoisolatory membranes that tightly envelop each individual islet, thereby eliminating any idle dead space between the islet and the outer capsule's membrane. These volume-sparing microcapsules could still provide the grafted islets with immunoprotection, although they would additionally be suitable for transplant in sites that are alternative to the peritoneal cavity. Among a few reported procedures to fabricate CM, our own consists of suspending the islets in a two-phase aqueous system comprising AG, dextran, and PEG that are emulsified extemporaneously. The microemulsion droplets engulf each islet, during brief incubation on a rocking plate, prior to immediate gelling upon reaction on Ca-

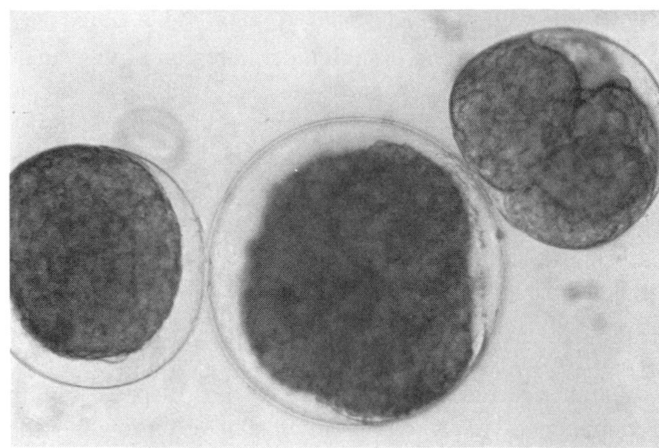

Figure 38.11 Freshly prepared AG/PLO coherent microcapsules (25×). (From Univ. of Perugia.)

chloride. The gel microbeads are sequentially coated with PLO, at particular molar ratios, in order to avoid any membrane's shrinkage, and finally with diluted AG (67) (Figs. 11 and 12). Alternatively to alginates, PEG can be used to fabricate photopolymerized conformal coatings, as mentioned above.

h. MSM

In order to reduce the capsule's size from 700–800 μm to 350–400 μm and thereby obtain smaller but not conformal microcapsules, we have modified physicochemical variables of our original method for fabrication of CSM (67). In particular, we have adjusted the following parameters:

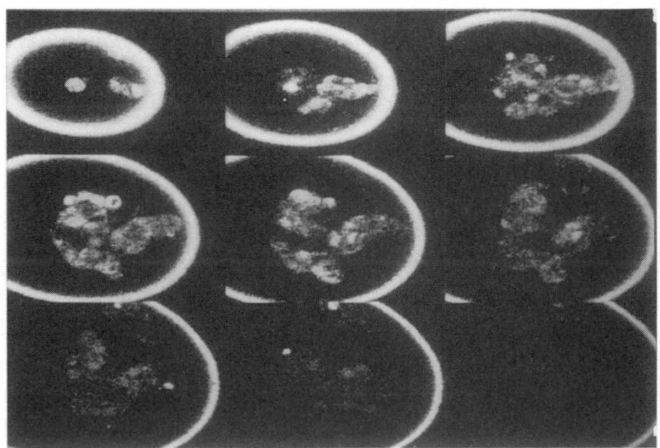

Figure 38.12 Laser confocal microscopy micrograph of AG/PLO microcapsules containing canine islets: membrane integrity and viability of the islets are evident. (From Univ. of Perugia.)

AG viscosity (increased)
AG/islet suspension extrusion flow rate (increased)
Air flow (increased)
Temperature of the suspension (decreased)

By making these technical modifications, we have been able to obtain microcapsules that measured approximately half the size of the original CSM (Fig. 13). The question whether MSM might prevail over CM in immunoprotection of the enveloped islets while retaining a smaller size, is still the subject of intense study in our laboratory. However, canine islet allografts, immunoprotected in MSM, in two spontaneously diabetic, insulin-dependent dogs, was associated with full reversal of hyperglycemia and discontinuation of exogenous insulin that lasted in one animal for as long as 600 days of transplantation (Fig. 14) in our laboratory.

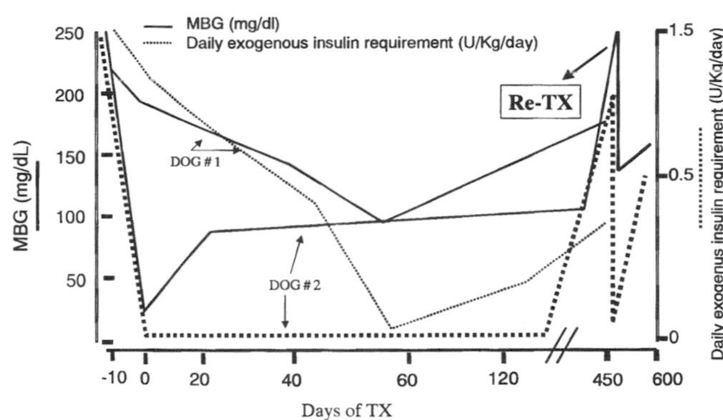

Figure 38.14 Long-term remission of hyperglycemia and discontinuation of exogenous insulin administration in two dogs with spontaneous insulin-dependent diabetes. (From Univ. of Perugia.)

3. Microcapsules: Summary of Current Issues and Outlook

Overall, the following issues related to islet microencapsulation are now under scrutiny: (1) microcapsules' size, configuration, biomaterials, and membrane engineering; (2) TX sites; (3) the use of xenogeneic islets; (4) protection from both immune and not immune specific mediators of the host's reaction; (5) islets life span and functional survival within the special microenvironment of the capsules.

Some of these issues are closely related to each other. (1) From what was said above TX of islet containing conventional-size microcapsules (CSM) has consistently functioned only in small diabetic recipients, like rodents, thereby implying that new sizes and/or configurations need to be explored, in order to put the system to work in higher mammalians. In our laboratory and a very few others, research efforts for reducing the capsule size went even further and finally resulted in the development of new conformal microcapsules. In particular, a major advantage was that, because of their minimal size, CM could be grafted virtually everywhere, including in parenchymatous organs, otherwise interdicted to bigger capsules. A general major advantage of this encapsulation system has been that we have employed the same chemical polymer formulation (AG/PLO), although adjustments of the reagents' concentrations, to fabricate our differently sized microcapsule prototypes, were from time to time required. CM that would occupy a final volume of a few mL/patient have been shown to form highly biocompatible and immunoslective biomembranes in our laboratory (68). We have preliminarily documented that these membranes retain immunobarrier competence and protect allo- or xenogeneic islets, at subtherapeutic doses, upon TX into adult pigs (Figs. 15, 16, and 17) (69). However, the afforded immunoprotection lasted shorter for xenografts (ie, canine islets in CM) than allografts (ie, porcine islets in CM) implanted in the adult pig liver (unpublished data). In summary, CM and other conformal microcapsules, while favorably impacting the final TX volume, which would permit access to alternate TX sites, could be more vulnerable, because of the extra-thin nature of the coating membrane, in comparison with CSM or MSM, to noxious environmental either immune or not immune factors, that are part of the hosts' reactivity to the encapsulated islet TX. Just very recently we have determined the extent of the immunoprotection afforded to islets by differently sized microcapsules, namely CM vs. MSM vs. CSM, with the latter prototype still showing

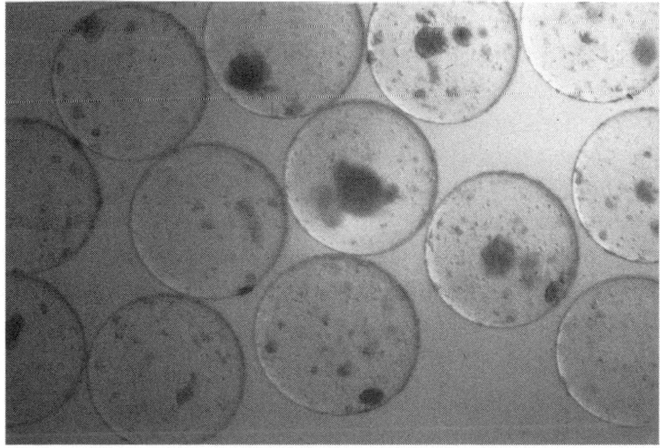

Figure 38.13 AG/PLO "medium size" (400 μm in diameter) microcapsules containing adult pig islets (25×). (From Univ. of Perugia.)

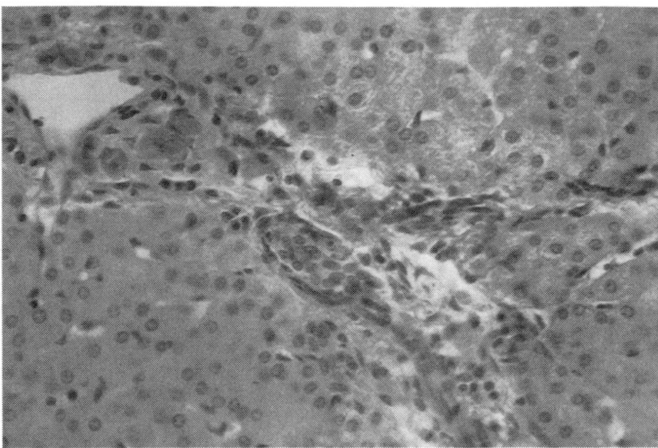

Figure 38.15 Porcine islets contained in AG/PLO CM, implanted into the liver of adult normal pigs at 30 days post-TX: full retention of islet cell viability and CM intactness (H&E 25×). (From Univ. of Perugia.)

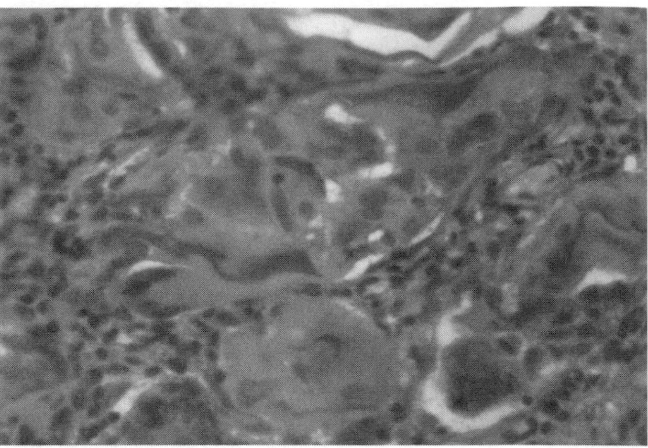

Figure 38.17 Canine islets contained in AG/PLO CM, implanted into the liver of adult normal pigs at 15 days post-TX: full retention of islet cell viability and CM intactness (H&E 25×). (From Univ. of Perugia.)

better immunoisolation properties. In light of these findings, microcapsule's geometry certainly requires further investigation. This looks like a priority together with further implementation of the membrane's functional competence in terms of improved immunoisolation capacity. Assessment of permselective properties associated with the currently available AG/PLO/PLL microcapsules is in progress and will likely be implemented in the near future. (2) This issue is related to the former, since the TX site has been virtually imposed so far by the final encapsulated islet volume and therefore the individual microcapsules' size. Obviously, CSM or MSM, because of their size, may

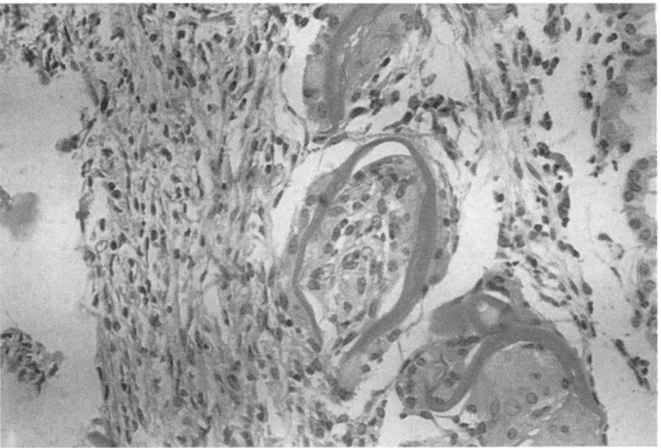

Figure 38.16 Porcine islets contained in AG/PLO CM, implanted into the omentum of adult normal pigs at 30 days post-TX: full retention of islet cell viability and CM intactness (H& E 25×). (From Univ. of Perugia.)

be grafted only intraperitoneally, whereas CM or other conformal capsules, as mentioned above, could be transplanted elsewhere, including, for instance, the mesenteric leaflets or some parenchymatous organs. However, CM application to a higher mammal TX system may be more difficult than expected, as documented by early trials conducted in large animals in our laboratory. Also, it is premature to assess which TX site for encapsulated islets could better suit metabolic requirements of diabetic higher mammals, although it is quite plausible that the closer the site to the portal system, the better the TX functional performance. Moreover, although microencapsulated islets are not comparable to free islets, the lessons we have learned from the latter, with special regard to certain TX settings, as for instance, immunoprivileged conditions, like those provided by testis-derived Sertoli's cells (70), should be highly regarded. Consequently, attempts to transfer the immunoprivilege associated with given organs/tissues into the special capsules' microenvironment should be closely pursued. (3) Immunoisolation in microcapsules, differently from any other islet TX immunoprotection strategies, at this juncture, may offer the unique opportunity to employ nonhuman tissue as a resource for donor islets. In the specific instance of T1DM patients, access to xenogeneic porcine islets could successfully overcome the problems of the restricted procurement of human donor islets. Pig islets would in fact secrete an insulin molecule that differs negligibly from the human. Furthermore, pigs are omnivores and their glucose metabolism is very close to human. Finally, pigs are attractive because they can be genetically engineered, thus potentially providing for organs/tissues that are closer to human. Methods for bulk retrieval of highly purified and intact islets, from selected donor pig breeding stocks, have been made available by our laboratory and a few others upon generation of the many technical skills that were necessary to fulfill all the specifications that are indispensable for the retrieval of viable and functionally competent pig islet tissue (71). Since it is well established that adult pig islets can easily lose their viability upon culture maintenance and in vivo transplant, some groups are now focusing their attention on neonatal pig islets. These islets might survive and eventually expand and mature, so as to acquire progressively their functional competence (72). Finally, even if pig islets would be the best candidates for xenogeneic functional islet tissue for humans, within immunoisolatory microcapsules, still some issues require additional and specific attention, prior to even thinking to embark upon clinical TX. In particular, the potential transmission of infectious agents, and in particular retroviruses, from pigs to humans has raised more than scientific discussions (73). However, there are strong reasons to believe that cellular transplants of such islet cells are different from solid organ grafts, since the former may enjoy immunoisolation in microcapsules or other artificial membranes, which makes a big difference, also under regulatory restrictions. This point has been recently recognized by international regulatory agencies and by groups, including our own, that are working in this field (74). (4) This issue is particularly difficult, because the immunoisolation membrane, depending on its MWCO, can be crossed by a very wide range of humoral substances and cell derivatives, many of which may trigger an intense inflammatory cell reaction. Although the microcapsule membrane's MWCO may be empirically tuned to the desired level, there are several technical limits. It is difficult, in fact, to determine the membrane's porosity as well as filtration/permeability which collectively represent a dynamic rather than static property. It also is implicit that not only immune specific mediators but also general cytotoxic agents such as free oxygen radicals, nitric oxide, etc. may be harmful to the encapsulated islet TX survival, and obviously there are no physical barriers that could interdict access to gases or very small molecules (75). An increasingly important islet graft destruction mechanism is represented by β-cell induced apoptosis. Cytokine-mediated expression of Fas receptor on β-cells may represent a target for FasL expressing CD4 and CD8, not to mention that an additional β-cell destruction mechanism could be represented by FasL expressed by β-cells themselves (48). To try to obviate these insidious mechanisms, major interventions are required on the capsules' outer membrane formulation, in an attempt to prevent the immediate impact of these mediators on the encapsulated islet grafts. Moreover, a simple three-dimensional monobarrier, like a sphere, may not suffice for immunoprotection of highly immunogenic tissue such as xenogeneic islets, and an additional intervention on the membrane's engineering may be necessary. Finally, as mentioned below, an integrated physicochemical and biological immunodefense barrier is increasingly required to obviate the islet TX directed, specific immune destruction. Issue (5) is of critical importance for the encapsulated islet TX longevity, which should be at least sufficient to provide for substantial long-term reversal of hyperglycemia in the transplanted recipients. An important point is the islet mass that needs to be loaded into a single microcapsule or macrodevice, since only balanced islet cell populations that are sufficiently supplied with O_2 and nutrients can result in effective functional performance. Also, the islet life span may be limited upon retrieval from the native pancreas by a number of variables, including hostile environmental conditions associated with the TX site. If neonatal islets were employed, they would even more require factors that implement their time-related maturation and acquisition of functional competence. Some groups have initially reported on the use of Sertoli's cells, originally situated in the testis, as a biological and immunological support to islets in transplant trials into diabetic rodents (76). These or other ''ancillary'' cells could greatly help the rearrangement of the special microenvironment that may mimic, although under completely different conditions, the biological hallmarks of the original pancreatic matrix. It is possible that the addition of growth factors may play a key role in enhancing the survival potential not only of fetal/neonatal (77) but also of adult islets (78). Comicroencapsulation of islets and purified Sertoli's cells has been preliminarily shown to improve islet cell viability and function (Fig. 18). The role of cotransplantating islets with other cells producing substances, like hormones, that

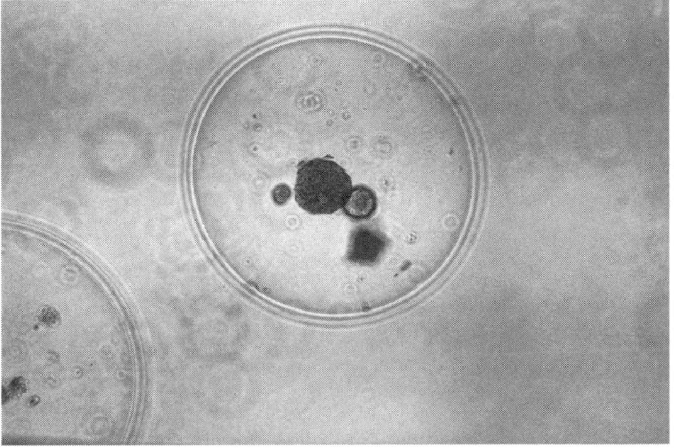

Figure 38.18 AG/PLO microcapsules containing rat islets + prepubertal Sertoli's cells (12×). (From Univ. of Perugia.)

may consistently improve islet viability and functional performance is being increasingly recognized (79). Finally, as mentioned, it is necessary to counteract apoptosis mechanisms that may increase β-cell destruction. With respect to this specific matter, some cotransplanted cells, such as Sertoli's cells, may express FasL or release FasL as a soluble product that might help divert CD4-CD8 islet cell–directed immune attacks. As an actual ancillary strategy, which may be proved or disproved in the not too far future, artificial β-cell lines, compatible with higher mammal metabolic requirements, could be developed so as to provide for a virtually-inexhaustible islet tissue source (80). This could be another application for microcapsules, since the artificial β-cells could also be of nonhuman origin.

VIII. FUTURE TRENDS

The BAP seems to offer the practical opportunity to transplant either human or nonhuman islets without general immunosuppression in the recipient. While the former is clearly limited by the restricted availability of cadaveric human donor pancreata, the latter could rely on the improved performance of methods for separation and purification of adult or neonatal islets from selected SPF pig breeding stocks. Should the problem of potential transmission of adventitious microbial agents be overcome, pig islets could represent an ideal tissue resource for human transplants, for all the above-mentioned reasons. Two alternatives are possibly ready to take over the pig islet source in the more or less far future: (1) engineered insulin producing cells; (2) in vitro regenerated islet cells. (1) could rely on methods that are more and more refined for generating cellular systems that are genetically engineered not only to secrete insulin but lately also to reproduce the glucose sensing apparatus that is uniquely associated with β-cells (81). Regulation of insulin secretion appears to be an indispensable feature of any insulin-producing tissue. Caution is still required prior to endorsing the use of this artificial tissue that is derived from tumor cells lines, for transplanting diabetes patients. Much additional study in this area is warranted for the years to come, although this cell source seems to be the most appealing solution to fulfill not only islet mass problems but also immune identity problems. Immune determinants of the artificial β-cells could be, in fact, engineered to that the cells would not be recognized by the immune system (81). Recently, methods to engineer endogenous cells that would be ''educated'' to synthesize and release insulin upon pharmacologic stimulation have been described (82). If successful, this approach could even obviate the need for transplantation. This ingenious strategy has been validated, at the moment and very preliminarily, only in rodents and requires further safety and efficacy studies, but it remains a very interesting potential strategy for the final cure of T1DM. (2) is closely linked to (1). Recently attempts to regenerate islets from pancreatic duct stem cells has preliminarily suceeded (83). This procedure could make a virtually unlimited beta-cell mass available in vitro, to fulfill the demand for cell transplants in diabetes. The system's performance, which seems to require long-term in-vitro islet cell growing protocols, as well as safety restrictions, needs to be further elucidated in the near future. An important common feature of these new engineered cell strategies is that with a few exceptions, all of them will require an immunoprotection, and most likely an immunoisolation membrane. Therefore the BAP likely represents the best approach to put the engineered or otherwise artificially generated insulin-producing cells to work.

IX. CONCLUSION

Multiple-competence strategies, incorporating diversified research fields, are now in progress to develop a cure for T1DM. Cell/tissue engineering gene therapy, or artificial device engineering might individually or synergistically fulfill the task of definitely restoring normal blood glucose levels in diabetic patients. Achievement of this goal would mean prevention of both acute and chronic complications of the disease. Among the available strategies, the BAP seems to be closest to human application. In fact, variably sized microcapsules, associated with increasingly refined immunoisolation and biocompatibility properties, are now available for islet graft immunoprotection with no need to treat the host with general immunosuppressants. Moreover, as problems of neonatal or adult porcine islet mass and purity are gradually being surmounted, the opportunity to employ nonhuman tissue as a resource for donor islets has come into sharper focus. Access to both immunoselective membranes and functionally competent cell tissue has greatly enhanced the BAP's potential. Toward this goal, artificial cells that are genetically engineered to synthesize and secrete insulin, contingent upon extracellular glucose levels, will also be available in the not too far future for application to in vivo trials. Conceivably, artificial cells, entrusted with the highly specialized task of replacing the normal islet β-cell function, are likely to take over human and animal tissue sources for transplant of patients with T1DM. Obviously, safety issues will require thorough scrutiny and specific solutions prior to enrolling the artificial cells into clinical use.

The ideal microcapsule should be versatile, in terms of long-term material stability, size, and immutable membrane selective properties (filtration/permeability/porosity), in order to access multiple and diversified implant sites. Suitability for replenishment would represent an additional useful feature. The ideal insulin producing cell/

tissue should be able to survive and thrive under likely difficult environmental conditions, regardless of the immune or not immune nature of adverse TX setting-related factors. To accomplish this objective, additional tissue and membrane engineering may be necessary. Multicompartment microcapsules, where the primary or engineered insulin-producing cells require the support of ''nursing'' cell systems, or rather pharmacological agents that may substantially improve their longevity and functional performance, could represent the final version of the BAP, suitable for second-level clinical trials.

Advances of the above reviewed research lines, in compliance with standard international safety and efficacy criteria, will unfold which prototype of BAP is readily available to begin human studies within the next few years to come.

REFERENCES

1. M. Machluf, A. Atala. Tissue engineering: emerging concepts. *Graft 1*,1:31 (1998).
2. P. Felig, M. Bergman. The endocrine pancreas: diabetes mellitus. *Endocrinology and Metabolism* (P. Felig, J. D. Baxter, L. A. Frohman, eds.). McGraw-Hill, New York, 1995, p. 1107.
3. R. Williams. The burden of diabetes in the next millennium. *Diabetes Reviews International 7*(3):21 (1998).
4. Diabetes Control and Complications Trial Research Group. The effect of intensive treatment of diabetes on the development of long-term complications in insulin dependent diabetes mellitus. *N. Engl. J. Med. 329*:976 (1993).
5. Diabetes Control and Complications Trial Research Group. The absence of a glycemic threshold for the development of long-term complications: the perspective of the DCCT. *Diabetes 45*:1289 (1996).
6. R. Calafiore. Perspectives in pancreatic and islet cell transplantation for the therapy of T1DM. *Diabetes Care 20*,5: 889 (1997).
7. D. E. R. Sutherland. The case for pancreas transplantation. *Diabete Metabolisme 22*:132 (1996).
8. American Diabetes Association. Pancreas transplantation for patients with diabetes mellitus. *Diabetes Care 21*,1:S79 (1998).
9. International Islet Transplant Registry. *Newsletter #8*,1 (1999).
10. G. Tyden, F. P. Reinholt, G. Sundkvist, J. Bolinder. Recurrence of autoimmune diabetes mellitus in recipients of cadaveric pancreatic grafts. *N. Engl. J. Med. 335*:888 (1996).
11. H. Brandhorst, D. Brandhorst, B. J. Hering, R. G. Bretzel. Significant progress in porcine islet mass isolation utilizing Liberase HI for enzymatic low temperature pancreas digestion. *Transplantation 68*:355 (1999).
12. G. S. Korbutt, J. F. Elliott, Z. Ao, D. K. Smith, G. L. Warnock, R. V. Rajotte. Large-scale isolation, growth and function of neonatal porcine islets. *J. Clin. Invest. 97*:2119 (1996).
13. K. H. Yoon, R. R. Quickel, K. Tatarkiewicz, T. R. Ulrich, J. Hollister-Lock, N. Trivedi, S. Bonner-Weir, G. Weir. Differentiation and expansion of Beta-cell mass in porcine neonatal pancreatic cell clusters transplanted into nude mice. *Cell Trans. 8*:673 (1999).
14. N. S. Kenyon, L. A. Fernandez, R. Lehmann, M. Masetti, A. Ranuncoli, M. Chatzipetrou, G. Iaria, D. Han, J. L. Wagner, P. Ruiz, M. Berho, L. Inverardi, R. Alejandro, A. D. Kirk, D. M. Harlan, L. D. Burkly, C. Ricordi. Long-term survival and function of intrahepatic islet allografts in baboons treated with humanized anti-CD154. *Diabetes 48*: 1473 (1999).
15. C. K. Colton. Implantable biohybrid artificial organs. *Cell Trans. 4*:415 (1995).
16. I. Swenne. Effect of aging on the regenerative capacity of the pancreatic β-cell of the rat. *Diabetes 32*:14 (1983).
17. P. De Vos, J. F. M. Van Stratten, A. G. Nieuwenhuizen, M. de Groot, R. J. Ploeg, B. J. De Haan, R. Van Schilfgaarde. Why do microencapsulated islet grafts fail in the absence of fibrotic overgrowth? *Diabetes 48*:1381 (1999).
18. R. P. Lanza, S. J. Sullivan, W. L. Chick. Islet transplantation with immunoisolation. *Diabetes 41*:1503 (1990).
19. P. E. Lacy, O. H. Hegre, A. Gerasimide-Vazeou, F. Gentile, K. E. Dionne. Subcutaneous xenografts of rat islets in acrylic copolymer capsules maintain normoglycemia in diabetic mice. *Science 254*:1782 (1991).
20. R. P. Lanza, A. M. Beyer, J. E. Staruk, W. L. Chick. Biohybrid artificial pancreas: long-term function of discordant islet xenografts in streptozotocin-diabetic rats. *Transplantation 56*:1067 (1993).
21. R. P. Lanza, K. M. Borland, J. E. Staruk, M. C. Appel, B. A. Solomon, W. L. Chick. Transplantation of encapsulated canine islets into spontaneously diabetic BB/Wor rats without immunosuppression. *Endocrinology 131*:637 (1992).
22. P. De Vos, J. L. Hillebrands, B. J. De Haan, J. H. Strubbe, R. Van Schilfgaarde. Efficacy of a prevascularized expanded polytetrafluoroethylene solid support system as a transplantation site for pancreatic islets. *Transplantation 63*(6):824 (1997).
23. T. Maki, I. Otsu, J. J. O'Neill, K. Dunleavy, C. J. P. Mullon, B. A. Solomon, A. P. Monaco. Treatment of diabetes by xenogeneic islets without immunosuppression: use of a vascularized bioartificial pancreas. *Diabetes 45*:342 (1996).
24. R. Calafiore. Transplantation of microencapsulated pancreatic human islets for the therapy of diabetes mellitus: a preliminary report. *ASAIO Journal 38*:34 (1992).
25. F. Lim, A. M. Sun. Microencapsulated islets as bioartificial endocrine pancreas. *Science 210*:908 (1980).
26. R. Calafiore, G. Basta, A. Falorni, F. Calcinaro, M. Pietropaolo, P. Brunetti. A method for the large-scale production of microencapsulated islets: in vitro and in vivo results. *Diab. Nutr. Metab. 5*:23 (1992).
27. R. P. Lanza, W. M. Kuhtreiber, W. L. Chick. Encapsulation technologies. *Tissue Engineer 1*:181 (1995).
28. R. P. Lanza, D. M. Ecker, W. M. Kuhtreiber, J. P. Marsh, J. Ringeling, W. L. Chick. Transplantation of islets using microencapsulation: studies in diabetic rodents and dogs. *J. Molecular Medicine 77*:206 (1999).

29. M. Y. Fan, Z. P. Lum, X. W. Fu, A. M. Sun. Reversal of diabetes in BB rats by transplantation of encapsulated pancreatic islets. *Diabetes 39*:519 (1990).
30. Z. P. Lum, I. T. Tai, M. Krestow, J. Norton, I. Vacek, A. M. Sun. Prolonged reversal of diabetic state in NOD mice by xenografts of microencapsulated rat islets. *Diabetes 40*: 511 (1991).
31. P. Brunetti, G. Basta, A. Falorni, F. Calcinaro, M. Pietropaolo, R. Calafiore. Immunoprotection of pancreatic islet grafts within artificial microcapsules. *Int. J. Artif. Organs 14*:789 (1991).
32. P. Soon-Shiong, E. Feldman, R. Nelson, J. Komtedebbe, O. Smidsrod, G. Skjak-Braek, T. Espevik, R. Heintz, M. Lee. Successful reversal of spontaneous diabetes in dogs by intraperitoneal microencapsulated islets. *Transplantation 54*:769 (1992).
33. P. Soon-Shiong, R. Heintz, N. Merideth, Q. X. Yao, Z. Yao, T. Zheng, M. Murphy, M. K. Moloney, M. Schmehl, M. Harris, R. Mendez, R. Mendez, P. Sanford. Insulin-independence in a type I diabetic patient after encapsulated islet transplantation. *Lancet 343*:950 (1994).
34. Y. Sun, X. Ma, D. Zhou, I. Vacek, A. M. Sun. Normalization of diabetes in spontaneously diabetic cynomologous monkeys by xenografts of microencapsulated porcine islets without immunosuppression. *J. Clin. Invest. 98*:1417 (1996).
35. K. Suzuki, S. Bonner-Weir, J. Hollister-Lock, C. K. Colton, G. C. Weir. Number and volume of islets transplanted in immunobarrier devices. *Cell Trans. 7*(1):47 (1998).
36. R. B. Fraser, M. A. MacAulay, J. R. Wright, A. M. Sun, G. Rowden. Migration of macrophage-like cells within encapsulated islets of Langerhans maintained in tissue culture. *Cell Trans. 5*(5):529 (1995).
37. R. G. Gill, L. Wolf. Immunobiology of cellular transplantation. *Cell Trans. 4*(4):361 (1995).
38. I. S. Grewall, R. A. Flavell. CD40 and CD154 in cell-mediated immunity. *Ann. Rev. Immunol. 16*:111 (1997).
39. R. Gill. Antigen presentation pathways for immunity to islet transplants: relevance to immunoisolation. *Bioartificial Organs II, Technology, Medicine and Materials* (D. Hunkeler, A. Prokop, A. Cherrington, R. Rajotte, M. Sefton, eds.). *Annals of the New York Acad. Sciences 875*, 1999, p. 255.
40. C. J. Weber, S. Safley, M. Hagler, J. Kapp. Evaluation of graft–host response for various tissue resources and animal models. *Bioartificial Organs II, Technology, Medicine and Materials* (D. Hunkeler, A. Prokop, A. Cherrington, R. Rajotte, M. Sefton, eds.). *Annals of the New York Acad. Sciences 875*, 1999, p. 233;
41. G. R. Rayat, R. V. Rajotte, G. S. Korbutt. Potential application of neonatal porcine islets as treatment for type 1 diabetes: a review. *Bioartificial Organs II, Technology, Medicine and Materials* (D. Hunkeler, A. Prokop, A. Cherrington, R. Rajotte, M. Sefton, eds.). *Annals of the New York Acad. Sciences 875*, 1999, p. 175.
42. B. Kulseng, B. Thu, T. Espevik, G. Skjak-Braek. Alginate polylysine microcapsules as immune barrier: permeability of cytokines and immunoglobulins over the capsule membrane. *Cell Trans. 6*,4:387 (1997).
43. H. Wu, E. S. Avgustiniatos, L. Swette, S. Bonner-Weir, G. C. Weir, C. K. Colton. In situ electrochemical oxygen generation with an immunoisolation device. *Bioartificial Organs II, Technology, Medicine and Materials* (D. Hunkeler, A. Prokop, A. Cherrington, R. Rajotte, M. Sefton, eds.). *Annals of the New York Acad. Sciences 875*, p. 105.
44. P. De Vos, B. J. De Haan, R. Van Schilfgaarde. Upscaling the production of microencapsulated pancreatic islets. *Biomaterials 18*:1085 (1997).
45. T. Loudovaris, S. Jacobs, S. Young, D. Maryanov, J. Brauker, R. C. Johnson. Correction of diabetic NOD mice with insulinomas implanted within Baxter immunoisolation devices. *J. Molecolar. Med. 77*(1):219 (1999).
46. M. Polak, L. Bouchareb-Banaei, R. Scharfman, P. Czernichow. Early pattern of differentiation in the human pancreas. *Diabetes 49*:225 (2000).
47. A. Ilieva, S. Yuan, R. N. Wang, D. Agapitos, D. J. Hill, L. Rosenberg. Pancreatic islet cell survival following islet isolation: the role of cellular interactions in the pancreas. *J. Endocrinology 161*:357 (1999).
48. W. Suarez-Pinzon, O. Sorensen, R. C. Bleackley, J. F. Elliott, R. V. Rajotte, A. Rabinovitch. Beta-cell destruction in NOD mice correlates with Fas (CD95) expression on Beta-cells and proinflammatory cytokine expression in islets. *Diabetes 48*:21 (1999).
49. D. Hunkeler. Polymers for bioartificial organs. *Trends Polymer Sci. 5*(9):286 (1997).
50. D. Hunkeler, A. Prokop, A. Powers, M. Haralson, S. DiMari, T. Wang. A screening of polymers as biomaterials for cell encapsulation. *Polymer News 22*:232 (1997).
51. G. H. Wolters GH, W. M. Fritschy, D. Gerrits, R. Van Schilfgaarde. A versatile alginate droplet generator applicable for microencapsulation of pancreatic islets. *J. Appl. Biomater. 3*(4):281 (1991).
52. R. P. Lanza, R. Jackson, A. Sullivan, J. Ringeling, C. MacGrath, W. Kuhtreiber, W. L. Chick. Xenotransplantation of cells using biodegradable microcapsules. *Transplantation 67*(8):1105 (1999).
53. R. P. Lanza, W. L. Chick. Transplantation of encapsulated cells and tissues. *Surgery 121*(1):1 (1997).
54. R. P. Lanza, D. Ecker, W. M. Kuhtreiber, J. E. Staruk, J. Marsh, W. L. Chick. A simple method for transplanting discordant islets into rats using alginate gel spheres. *Transplantation 59*(10):1482 (1995).
55. T. Zekorn, A. Horcher, U. Siebers, R. Schnetter, B. Hering, U. Zimmermann, R. G. Bretzel, K. Federlin. Barium cross-linked alginate beads: a simple one-step method for successful immunoisolated transplantation of islets of Langerhans. *Acta Diabetologica 29*:99 (1992).
56. T. Zekorn, A. Horcher, U. Siebers, R. Schnetter, B. Hering, U. Zimmermann, R. G. Bretzel, K. Federlin. Alginate coating of islets of islets of Langerhans: in vitro studies on a new method of immunoisolated transplantation. *Acta Diabetologica 29*:41 (1992).
57. T. Zekorn, R. G. Bretzel. Immunoprotection of islets of

Langerhans by microencapsulation in barium alginate beads. *Cell Encapsulation Technology and Therapeutics* (W. M. Kuhtreiber, R. P. Lanza, W. L. Chick, eds). Birkhäuser, Boston, 1999, p. 90.

58. S. Dumitriu, P. F. Vidal, E. Chornet. Hydrogels based on polysaccharides. *Polysaccharides in Medical Applications* (S. Dumitriu, ed.). Marcel Dekker, New York, 1996, p. 125.
59. H. Iwata, Y Ikada. Agarose. *Cell Encapsulation Technology and Therapeutics* (W. M. Kuhtreiber, R. P. Lanza, W. L. Chick, eds.). Birkhäuser, Boston, 1999, p. 97.
60. A. S. Sawhney. Poly(ethylene glycol). *Cell Encapsulation Technology and Therapeutics* (W. M. Kuhtreiber, R. P. Lanza, W. L. Chick, eds.). Birkhäuser, Boston, 1999, p. 108.
61. A. Sawhney, J. A. Hubbell. Poly(ethylene oxide)-graft-poly(l-lysine) copolymers to enhance the biocompatibility of poly(l-lysine)-alginate microcapsule membrane. *Biomaterials 13*:863 (1992).
62. A. Sawhney, C. P. Pathak, J. A. Hubbell. Modification of Langerhans islet surfaces with immunoprotective poly(ethylene glycol) coatings. *Bioech. Bioeng. 44*:383 (1994).
63. S. K. Kim, J. Choi, E. A. Balmaceda, C. Rha, Chitosan. *Cell Encapsulation Technology and Therapeutics* (W. M. Kuhtreiber, R. P. Lanza, W. L. Chick, eds.). Birkhäuser, Boston, 1999, p. 151.
64. M. V. Sefton, W. T. K. Stevenson. Microencapsulation of live animal cells using polyacrylates. *Adv. Polymer Sci. 107*:145 (1993).
65. T. Roberts, U. De Boni, M. V. Sefton. Dopamine secretion by PC12 cells microencapsulated in a hydroxyethyl methacrylate–methyl methacrylate copolymer. *Biomaterials 17*: 267 (1996).
66. C. J. Weber, J. A. Kapp, M. K. Hagler, S. Safley, J. T. Chryssochoos, E. L. Chaikof. Long-term survival of poly-L-lysine-alginate microencapsulated islet xenografts in spontaneously diabetic NOD mice. *Cell Encapsulation Technology and Therapeuticsc* (W. M. Kuhtreiber, R. P. Lanza, W. L. Chick, eds.). Birkhäuser, Boston, 1999, p. 117.
67. R. Calafiore, G. Basta. Alginate/poly-L-ornithine microcapsules for pancreatic islet cell immunoprotection. *Cell Encapsulation Technology and Therapeutics* (W. M. Kuhtreiber, R. P. Lanza, W. L. Chick, eds.). Birkhäuser, Boston, 1999, p. 138.
68. R. Calafiore, G. Basta, G. Luca, C. Boselli, A. Bufalari, G. M. Giustozzi, L. Moggi, P. Brunetti. Alginate/polyaminoacidic coherent microcapsules for pancreatic islet graft immunoisolation in diabetic recipients. *Bioartificial Organs I, Technology, Medicine and Materials* (D. Hunkeler, A. Prokop, A. Cherrington, R. Rajotte, M. Sefton, eds.). *Annals of the New York Acad. Sciences 831*, 1997, p. 313.
69. R. Calafiore, G. Basta, G. Luca, C. Boselli, A. Bufalari, G. M. Giustozzi, R. Gialletti, F. Moriconi, P. Brunetti. Transplantation of allogeneic/xenogeneic pancreatic islet containing coherent microcapsules in adult pigs. *Trans. Proc. 30*;482 (1998).
70. D. Bellgrau, D. Gold, H. Selawry, J. Moore, A. Franzusoff, R. C. Duke. A role for CD95 ligand in preventing graft rejection. *Nature 377*(6550):630 (1995).
71. G. Basta, A. Falorni, L. Osticioli, P. Brunetti, R. Calafiore. Method for mass retrieval, morphologic and functional characterization of porcine islets of Langerhans: a potential xenogeneic tissue resource for transplantation in T1DM. *J. Invest. Med. 43*,6:555 (1995).
72. G. S. Korbutt, Z. Ao, G. L. Warnock, R. V. Rajotte. Large-scale isolation of viable porcine neonatal islet cell (NIC) aggregates. *Trans. Proc. 27*:3267 (1995).
73. C. Patience, Y. Takeuchi, R. A. Weiss. Infection of human cells by an endogenous retrovirus of pigs. *Nature Medicine 3*(3):282 (1997).
74. D. Hunkeler, A. M. Sun, R. Rajotte, R. Gill, R. Calafiore, P. Morel. Risks involved in the xenotransplantation of immunoisolated tissue. *Nature Biotechnology 17*:1045 (1999).
75. U. Zumsteg, S. Frigerio, G. A. Hollande. Nitric oxide production and Fas surface expression mediate two independent pathways of cytokine-induced murine β-cell damage. *Diabetes 49*:39 (2000).
76. G. Luca, M. Calvitti, E. Becchetti, G. Basta, G. Angeletti, F. Santeusanio, P. Brunetti, R. Calafiore. Method for separation morphological and functional characterization of Sertoli's cells from the rat pre-pubertal testis: a potential nursing cell system for pancreatic islets. *Diab. Nutr. Metabol. 11*:307 (1998).
77. G. M. Beattie, J. S. Rubin, M. I. Mally, T. Otolonski, A. Hayek. Regulation of proliferation and differentiation of human fetal pancreatic islets by extracellular matrix, hepatocyte growth factor and cell to cell contact. *Diabetes 45*: 1223 (1996).
78. A. Hayek, G. M. Beattie, V. Cirulli, A. D. Lopez, C. Ricordi, J. S. Rubin. Growth factor/matrix-induced proliferation of human adult β-cells. *Diabetes 44*:1458 (1996).
79. G. S. Korbutt, J. F. Elliott, R. V. Rajotte. Co-transplantation of allogeneic islets with allogeneic testicular cell aggregates allows long-term survival without systemic immunosuppression. *Diabetes 46*:317 (1997).
80. V. Poitut, L. E. Stout, M. B. Armstrong, T. F. Walseth, R. L. Sorenson, R. P. Robertson. Morphological and functional characterization of β-TC-6 cells: an insulin-secreting cell line derived from transgenic mice. *Diabetes 44*:306 (1995).
81. S. Efrat. Genetically engineered pancreatic β-cell lines for cell therapy of diabetes. *Bioartificial Organs II. Technology, Medicine and Materials* (D. Hunkeler, A. Prokop, A. Cherrington, R. Rajotte, M. Sefton, eds.). *Annals of the New York Acad. Sciences 875*, 1999, p. 286.
82. V. M. Rivera, X. Wang, S. Wardwell, N. L. Courage, A. Volchuk, T. Keenan, D. A. Holt, M. Gilman, L. Orci, F. Cerasoli, J. E. Rothman, T. Clackson. Regulation of protein secretion in the endoplasmic reticulum. *Science 287*:826 (2000).
83. V. K. Ramiya, M. Maraist, K. E. Arfors, D. A. Schatz, A. B. Peck, J. G. Cornelius. Reversal of insulin-dependent diabetes using islets generated in vitro from pancreatic stem cells. *Nature Medicine 6*(3):278 (2000).

39

Transdermal Delivery of Drugs

B. B. Michniak
University of Medicine and Dentistry of New Jersey, Newark, New Jersey

A. El-Kattan
Pfizer Inc., Ann Arbor, Michigan

I. SKIN ANATOMY

Skin, the largest body organ, covers a surface area of approximately $2m^2$ and receives about one-third of the blood circulating through the body (1). It maintains the integrity and function of the internal organs by restricting the passage of chemicals into and out of the body and stabilizes body temperature and blood pressure.

Histologically, the skin can be divided into two layers, the outer epidermis and the inner dermis with a total thickness of 2–3 mm, as illustrated in Fig. 1 (2–3). The epidermis is approximately 100 μm thick and can be further subdivided into several layers of cells (4). These cells originate in the stratum germinativum (the basal layer of the epidermis) and undergo differentiation as they move up through the stratum spinosum, granulosum, and lucidum. The uppermost layer, the stratum corneum, is considered to be the rate-limiting barrier to the permeation of drugs (5–7). It consists of corneocytes, which are nonnucleated, flattened cells approximately 40 μm long and 0.5 μm in diameter, surrounded by an intercellular lipid-rich matrix. Elias described the stratum corneum in terms of a two-compartment model containing bricks (corneocytes) and mortar (intercellular lipid) (8,9).

The corneocytes have lost their cellular organelles and cytoplasm during the process of cornification. They are composed of keratin filaments that are stabilized by the formation of disulfide bridges and surrounded by cellular envelopes that consist of cross-linked proteins and covalently bound lipid. The corneocytes are also interconnected by polar structures termed the corneodesmosomes (10,11).

The intercellular spaces of the stratum corneum appear to be completely filled with intercellular lipids that are generated from the exocytosis of lamellar bodies during the differentiation of keratinocytes and to a lesser extent from sebaceous secretions (12–14). One of the most notable differences between the stratum corneum intercellular lipids and all other biological membranes is that these lipids are almost devoid of phospholipids. Instead, they are selectively enriched in ceramides, fatty acids, sterols, lesser quantities of glycolipids, sterol esters, triglycerides, cholesterol sulfate, and hydrocarbons (15,16). The lipid compositions of pig and human stratum corneum are shown in Table 1. The layer between the epidermis and the inner dermis is referred to as the epidermal–dermal junction, also known as the basement membrane (17). Beneath the basement membrane lies the dermis, which consists essentially of a matrix of fibroelastic connective tissue woven

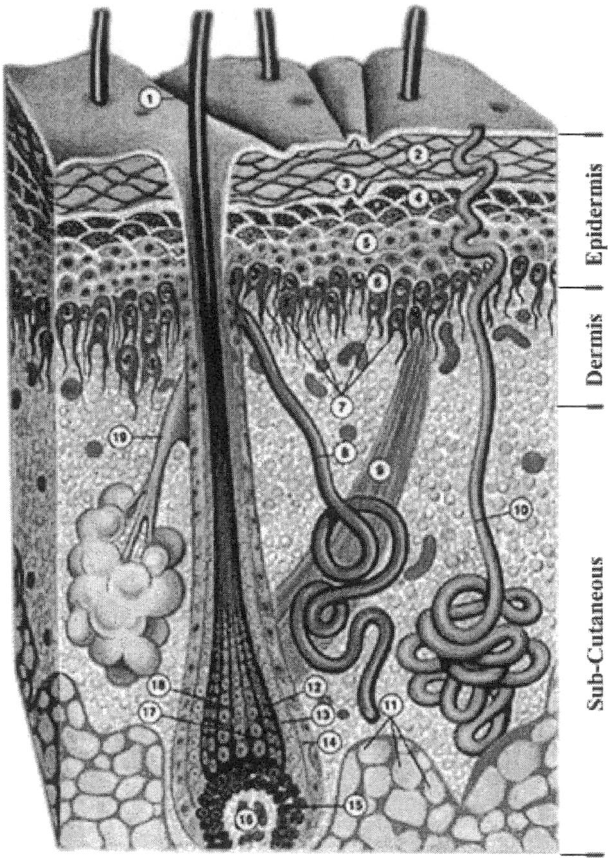

Figure 39.1 A cross-sectional illustration of human skin structure. **1.** Hair. **2.** Stratum corneum. **3.** Stratum lucidum. **4.** Stratum granulosum. **5.** Stratum spinosum. **6.** Stratum germinativum. **7.** Papillary dermis. **8.** Apocrine sweat gland. **9.** Arrector pili muscle. **10.** Eccrine sweat gland. **11.** Adipocyte. **12.** Hair cuticle. **13.** Inner root sheath. **14.** Outer root sheath. **15.** Hair bulb. **16.** Follicular papilla. **17.** Cortex. **18.** Medulla. **19.** Sebaceous gland.

from collagen fibers, elastic fibers, and an interfibrillar gel of glycosaminoglycans (18). The dermis can be further divided into two main layers: the papillary dermis and the reticular dermis (19). The papillary dermis consists of relatively loose connective tissue that contains smaller collagen and elastic fibers. It is also interspersed by the circulatory, lymphatic, and nervous system of the skin. In addition, it contains glycosaminoglycans and adhesive proteins known as fibronectins that attach fibroblasts to the collagen. The reticular dermis contains dense collagenous and elastic connective tissue and unlike the papillary dermis does not enclose many cells and blood vessels or glycosaminoglycans. It spreads over the subcutaneous fat, which functions as a heat insulator and reservoir for extra caloric intake.

The skin is breached by appendages such as sweat glands, hair follicles, and associated sebaceous glands. These appendages extend through the epidermis and dermis to the subcutaneous tissue but occupy only about 0.1% of the total human skin surface (20).

II. PERCUTANEOUS ABSORPTION PATHWAYS

Chemicals can permeate the skin through one of three possible routes: appendageal, transcellular, and intercellular (Fig. 2). The overall flux across the skin is the sum of the individual fluxes through each pathway that itself depends upon their number per surface area and path length as well as the compound diffusivity and physicochemical properties such as compound solubility. These

Table 39.1 Composition of Human and Pig Stratum Corneum Lipids (wt%)

	Species (site)			
Constituent	Human (abdominal)	Human (planar)	Human (?)	Pig (?)
Ceramides	18	35	41	39
Ceramide 1	14		3.2	6
Ceramide 2	4.3		8.9	13
Ceramide 3			4.9	5.2
Ceramide 4			6.1	4.1
Ceramide 5			5.7	3.5
Ceramide 6			12	7.2
Glycosylceramides	Trace		0	0.7
Fatty acids	19	19	9.1	11
Stearic acid	1.9			
Palmitic acid	7			
Myristic acid	0.7			
Oleic acid	6.3			
Linoleic acid	2.4			
Palmitoleic acid	0.7			
Others	<0.1			
Cholesterol	14	29	27	28
Cholesterol Sulfate	1.5	1.8	1.9	2
Sterol/wax esters	5.4	6.5	10	15
Di- and triglycerides	25	3.5	0	1.3
Squalene	4.8	0.2		
n-Alkanes	6.1	1.7		
Phospholipids	4.9	3.2		0

Source: Refs. 188–193.

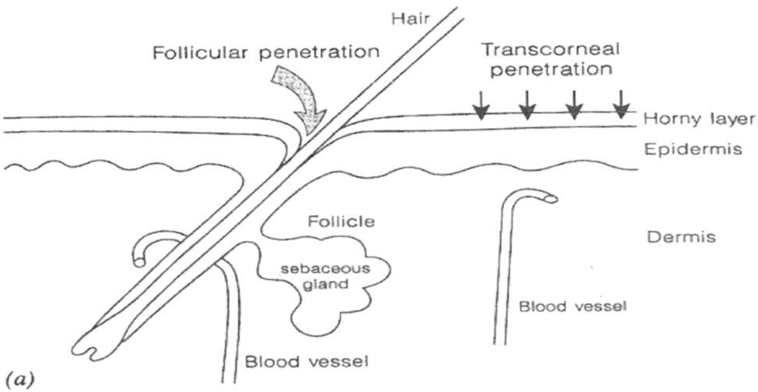

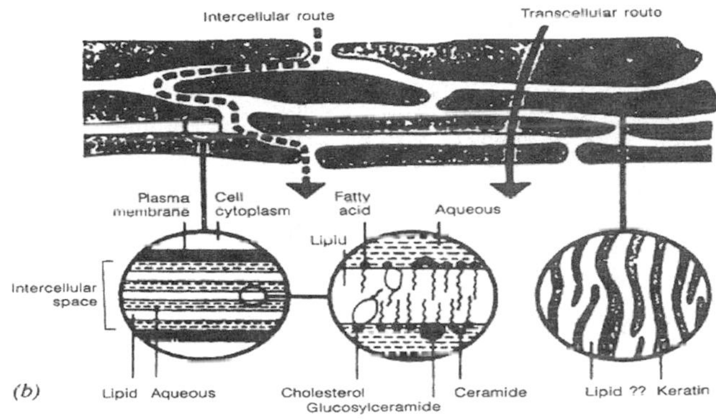

Figure 39.2 Model of transcellular and intracellular pathways. (a) Penetration occurs via appendages that exhibit a reduced barrier to diffusion but occupy a relatively small surface area. (b) Permeation through the stratum corneum (transcorneal permeation) may be considered to occur through the intercellular lipid domain or through the corneocytes (transcellular route). (Reproduced with permission from Schaefer et al. (187).)

pathways should not be treated as mutually exclusive (21).

A. The Appendageal Pathways

The lower ducts of hair follicles and sweat glands are sites of physical discontinuity in the stratum corneum and as such serve as penetration pathways. The available diffusional area of the appendageal route is approximately 0.1% of the total skin area and only 0.01–0.1% of the total skin volume (22–24). The possibility that appendages form a pathway for percutaneous absorption has been demonstrated using fluorescence microscopy (25). These findings were consistent with previous investigations using autoradiography, which showed penetration of hair follicles by several substances such as adapalene, pyridostigmine, and 8-arginine vasopressin (25–27). Scheuplein et al. evaluated the percutaneous permeation of small nonelectrolytes through the appendageal and intercellular pathways. They found that appendageal pathways provide a poor barrier to the diffusion of nonelectrolytes relative to the intercellular pathways. However, the relative surface area of the follicular pathways is relatively small (24). In view of these facts, the follicular pathway is predicted to be quantitatively more important at the transition (non-steady-state) phase of percutaneous absorption (22–24,28,29). On the contrary, the diffusivity through the intercellular lipid domain is relatively low, and as a result there is a corresponding longer lag time. The comparatively large surface area of this pathway contributes to a maximum flux for compounds with the appropriate physicochemical characteristics.

B. Transcellular vs. Intercellular Pathways

The transcellular and intercellular pathways are considered the two major pathways for chemical penetration in the stratum corneum. The transcellular pathway involves the penetration of the substances through the inner lumen of the stratum corneum corneocytes. The intercellular pathway involves the diffusion of the penetrants via the tortuous but continuous intercellular lipid pathway (30). The pathway through which the penetration occurs is largely determined by the physicochemical characteristics of the penetrants. It has been shown that a very important relationship exists between the hydrophobicity of the compounds and their permeability coefficients (31). This may suggest that the rate-limiting step for penetration through the skin involves a hydrophobic barrier, i.e., intercellular lipid. However, small polar compounds such as urea produced higher permeability coefficients than expected on the basis of their octanol/water partition coefficient. This may suggest that compounds penetrating the skin must pass through the intercellular lipid. However, that does not exclude that compounds can also pass through the inner lumen of the stratum corneum corneocytes.

III. PRINCIPLES OF DRUG DIFFUSION ACROSS SKIN

Percutaneous absorption of most drugs takes place by a passive diffusion process that can be described by Fick's first law of diffusion (32). This principle is illustrated mathematically in Eq. (1), which shows that the amount of drug transported through a unit area of skin per unit time (flux, J) is the product of the diffusion coefficient of the drug in the skin (D), the concentration of the diffusing drug (C), and the space coordinate measured normal to the skin, which is actually the path length of diffusion from the skin surface (x). The negative sign in the equation indicates that the flux occurs in the direction of lowest concentration.

$$J = -\frac{D\delta C}{\delta x} \tag{1}$$

Fick's first law is compounded by certain factors when used to describe the drug permeation across the skin. It is generally applied to steady-state flux from an infinite dose along a unidirectional path through a homogenous membrane. Furthermore, Fick's first law assumes that only the drug molecules diffuse through the stratum corneum, the vehicle neither diffuses nor evaporates significantly, the components from the skin do not alter the properties of the donor solution, and the drug concentration remains constant in the vehicle during the experiment. The stratum corneum is not homogeneous by nature but it is composed of lipid and protein domains. Therefore diffusion is not constant over all dimensions, as demonstrated by the first law. In addition, changes in the composition of the vehicle can occur during the permeation experiments, affecting the solubility of the drug and hence the calculated flux. To examine the rate of change of drug concentration at a point in the system, Fick's second law is used. Prior to steady state, the second law relates the rate of change of concentration at a point to the spatial variation of the concentration at that point, and this can be described as

$$\delta C = \frac{D\delta^2 C}{\delta^2 x} \tag{2}$$

In in vitro diffusion studies, the concentration differential is present during the experiment when sink conditions are maintained. This can be achieved by using saturated drug suspensions in the donor compartment and the maintaince of low receptor concentrations (sink conditions). It is expressed mathematically by Eq. 3:

$$M = \frac{ADKC_O}{x(t - x^2/6D)} \tag{3}$$

M is the cumulative amount of drug diffusing through the membrane of area A, of thickness x, in time t. C_0 is the concentration at time $= 0$ and is assumed to be constant. D is the diffusion coefficient. K is the partition coefficient. This equation describes the diffusion process at steady state. Differentiation of this equation with respect to time gives the steady-state permeation rate as

$$\frac{dM}{dt} = \frac{ADKC_O}{x} \tag{4}$$

or in terms of steady-state flux (J) as

$$J = \frac{DKC_O}{h} \tag{5}$$

Flux is the slope of the graph of M vs. t after steady state is reached. If the steady state line is extrapolated back to the time axis, the intercept gives the lag time. Lag time (t_{lag}) is used to determine the time that is needed to establish steady state across the membrane. This value can also be used to estimate the diffusion coefficient, provided that the thickness of the stratum corneum is known (x) as shown in Eq. (6):

$$D = \frac{X^2}{6t_{\text{lag}}} \tag{6}$$

IV. TRANSDERMAL THERAPEUTIC SYSTEMS

The transdermal route offers distinct advantages over other routes for the systemic delivery of drugs. For example, drug administration through the skin eliminates variables that influence gut absorption, such as changes in pH along the gastrointestinal tract, food and fluid intake, stomach emptying time, intestinal motility, and transit time. Once absorbed, the first pass effect of the liver is eliminated, thus avoiding another profound site of potential metabolism and improving drug bioavailability (33). In addition, the transdermal route reduces the dose frequency by providing a continuous controlled zero-order absorption through the stratum corneum into the systemic circulation, which is somewhat similar to that provided by an I.V. infusion. However, unlike, an I.V. infusion, delivery is noninvasive, and no hospitalization is required. The uniform drug plasma levels reduce the side effects associated with oral dosage forms, permit the use of drugs with short half-lives, and improve patient compliance. Furthermore, the transdermal route can use drugs with a low therapeutic index.

The systemic delivery of drugs through the skin, i.e., transdermal delivery, was first introduced in 1981 when Transderm V® (now called Transderm Scop®) was marketed for motion sickness. Since then more than 25 different transdermal drug delivery systems (TDS) have been commercialized for the systemic treatment of a variety of diseases such as scopolamine (34), nitroglycerin, isosorbide dinitrite, clonidine, fentanyl (35), estradiol, (36,37), testosterone (38,39), timolol (40–42), triprolidine (43), and nicotine (44–46).

A. Drug Candidate Selection

The selection of a potential candidate for the development of a TDS has been extensively reviewed (34,47). Table 2 summarizes the major physicochemical properties and pharmacokinetic characteristics of drugs incorporated in TDSs. It can be speculated that drug candidates with low molecular weights permeate the skin at greater rates than compounds with higher molecular weights. Generally, they should have appropriate lipid solubility, i.e., log [octanol/water partition coefficient] between −1 and 2, since compound with log[octanol/water partition coefficient] lower than −1 will have difficulty in distributing from the TDS into the stratum corneum. Moreover, compounds with log[octanol/water partition coefficient] of less than 2 may have problems in achieving steady plasma concentrations in reasonable time spans because the drug is delayed in the stratum corneum where a reservoir may be established (47). Drug crystallinity or melting point influences drug skin permeability. In general, the lower the drug melting point, the higher its permeability through the skin. In addition, drug candidates should have short half-lies, and this was suggested by the observation that a longer biological half-life increases the steady state concentration of the compound. Simultaneously, the time needed to reach steady state is significantly increased under these conditions (it takes five half-lives to reach steady state), which may be overly long as compared to the duration of patch use. The drug candidate should not exhibit any skin irritation, toxicity, or allergy. The drug should also be potent, and the therapeutic response should be obtained by the systemic delivery of no more than 25 mg/day. The equation used to determine the flux required to maintain a therapeutic concentration of a drug in the plasma is

$$J_{skin} = CL_{SS} \times CL_T \tag{7}$$

where J_{skin} is the flux (μg/cm²/h) required to obtain a therapeutic drug concentration, CL_{ss} is the unchanged drug

Table 39.2 Physicochemical Properties and Pharmacokinetics of Drugs Transdermal Therapeutic System Products

Drug	M.W. (Daltons)	p*Ka*	M.P. (°C)	log P (o/w)	Perm. coeff. (cm/h·10³)	Cl (L/h)	V$_d$ (L)	$t_{1/2}$ (h)	Max. 24 h dose (mg)	Oral bioavail. (%)	Effective Plasma level (ng/mL)
Clonidine	230	8.2	140	0.83	35	13	147	6–20	0.3	95	0.2–2
Cytisine	190	—	155	−0.94	12.9	—	—	—	12	—	—
Estradiol	272	—	176	2.49	5.2	615–790	4.8	0.05	0.1	—	0.04–6
Fentanyl	337	84	83	2.93	10	25–75	280	3–12	2.4	—	1.0
Isosorbide dinitrate	236	—	70	1.55	—	204	100–300	0.5–1.0	20	20	2–4
Nicotine	162	6.2/11	≤80	—	3	78	182	2	22	30	10–30
Nitroglycerin	227	—	13.5	2.05	20	966	231	0.04	15	<1	1.2–11
Norethindrone acetate	340	—	161	—	—	—	—	—	0.25	—	0.5–1.0
Scopolamine	303	7.8	59	1.24	0.5	67.2	98	2.9	0.17	27	0.04
Testosterone	288	—	153	3.31	400	—	—	—	6.0	<1	10–100

Source: Refs. 194–196.

plasma concentration (ng/mL), and CL_T is the drug clearance rate (L/h).

B. Design of Transdermal Delivery Systems

Transdermal drug delivery systems are pharmaceutical dosage forms that can deliver therapeutically safe and effective concentrations of drug molecules from polymeric patches applied on the human skin over extended periods of time. These systems can be classified into four main categories as shown in Fig. 3.

1. Membrane-Moderated Transdermal Drug Delivery Systems (MMT)

It is generally recognized that the membrane-moderated transdermal delivery systems (MMT) consist of a drug reservoir compartment that lies between an uppermost impermeable flexible backing layer that prevents the drug permeation from the top and a semipermeable rate-controlling membrane with an adhesive layer that resides below the drug reservoir (2). The lowermost layer is a release linear that protects the system during storage.

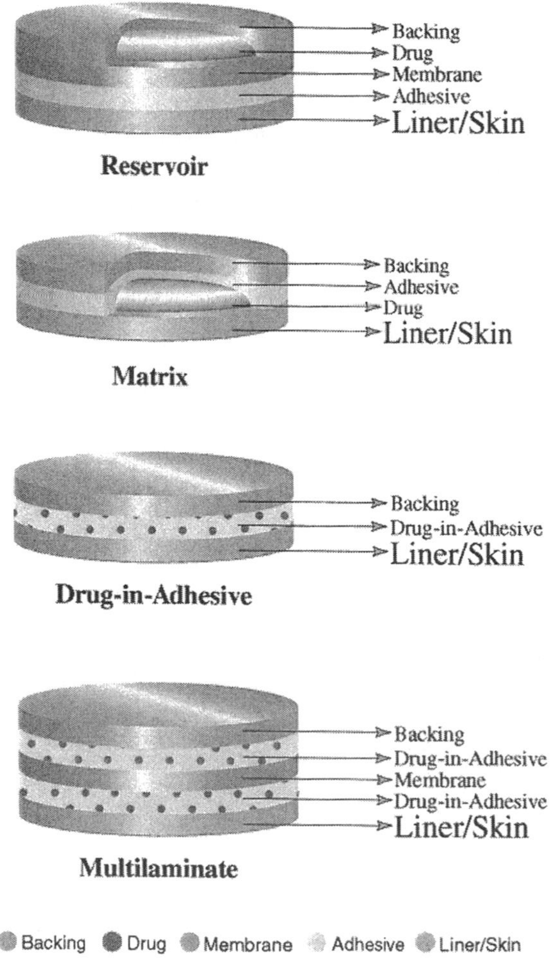

Figure 39.3 Cross-sectional illustration of the four main marketed transdermal system designs. (Reproduced with permission from 3M.)

In the drug reservoir, the drug is either completely solubilized or suspended in a liquid phase (e.g., Carbomer gel) (48). The drug diffuses passively through the rate-controlling membrane toward the skin. In a suspension formulation, the drug release follows zero-order kinetics. These release kinetics are considered the one of the primary advantages of this delivery system (40,48–50).

The semipermeable rate-controlling membrane must demonstrate preferential permeability of the drug over the reservoir media. The membrane can be either microporous or a nonporous continuous film. The microporous membranes contain interconnected pores that are made of polyethylene or polypropylene. These pores are filled with liquid such as mineral oil or ethanol. The drug is transported through the interconnected pores by diffusion through the liquid phase. The nonporous continuous membranes are made of polyethylene, polydimethylsiloxane, polypropylene, or ethylene vinyl acetate copolymer. The drug transport mechanism involves partitioning of the drug in the upper side of the membrane and then diffusion through the polymer film (51).

A thin layer of drug–skin compatible adhesive lines the external surface of the rate-controlling membrane. These adhesives are composed of polymers that can adhere to the skin when slight pressure is applied. Adhesives are a critical component in MMT. In addition to their usual requirements of functional adhesive properties, they must demonstrate good biocompatibility with the skin, chemical compatibility with the drug, various components of the formulation, and provide controlled and consistent transdermal delivery of the drug with several years of shelf stability. Furthermore, they should be easy to remove when desired, without causing trauma to the skin. Adhesives should maintain their functional properties of tack, release, adhesion, and cohesive strength in the presence of drugs and excipients. Appropriate selection of the adhesive is required for a consistent and predictable delivery of drug to systemic circulation. Most of the adhesives utilized are within one of the following general classes: silicone, polyisobutylenes, or acrylate (52–55). Silicone polymers are inert and compatible with many drugs and excipients. However, they have a high tendency to react with basic amine functional groups in compounds that may compromise their shelf stability by reducing their tack and skin adhesion properties or increasing the force required to remove the release liner. To increase their compatibility, the

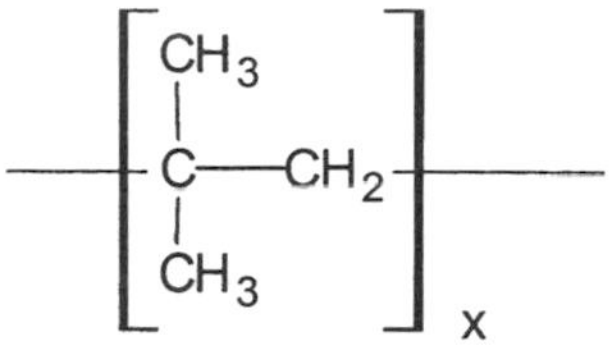

$$\left[-C(CH_3)_2-CH_2- \right]_x$$

Figure 39.4 The chemical structure of polyisobutylene.

reactive hydroxy-functional groups can be eliminated by an end-capping process. Polyisobutylenes are homopolymers of isobutylene and are composed of a long straight hydrocarbon backbone with an unsaturated terminal (Fig. 4).

Polyisobutylenes are relatively nontoxic, odorless and inert and are used when prolonged skin contact is desired, since they are less likely to cause skin irritation than other adhesives (56). However, Fang et al. evaluated skin adhesion properties of polyisobutylenes and compared them with these provided by silicone adhesive. They found that the adhesion properties of silicone adhesive were higher than the polyisobutylenes adhesive. To improve polyisobutylene adhesion properties, tackifiers such as polyisoprene and secondary tackifiers (terpene polymer; Px 1150) were incorporated into the formulation (57). Acrylate-based adhesives are extensively used for a variety of medical applications, since they are highly resistant to oxidation and do not require the addition of stabilizers. In addition, they are less irritant than silicone adhesives and are available in porous grades that are air and water vapor permeable. They are produced by copolymerizing acrylic esters with other comonomers (Fig. 5).

The flow chart in Fig. 6 outlines the manufacturing process and equipment for the production of MMT patches.

2. Matrix Diffusion-Controlled Transdermal Drug Delivery Systems (MDC)

The matrix diffusion-controlled transdermal delivery system (MDC) has the simplest design of the four TDS categories. Its components are similar to those found in MMT (Fig. 3). Unlike MMT, the drug reservoir contains a homogeneously dispersed drug in a semisolid hydrophilic or lipophilic matrix. The adhesive layer is spread on the circumference of the drug reservoir disc. This may eliminate all possible drug/excipients adhesive interactions. Furthermore, the drug reservoir is not lined by a rate controlling membrane layer. As a result, the rate of drug delivery is regulated by the stratum corneum (56). Consequently, the rate of drug delivery across the skin can be modified by the incorporation of excipients that may alter the stratum corneum barrier properties such as transdermal enhancers (see below).

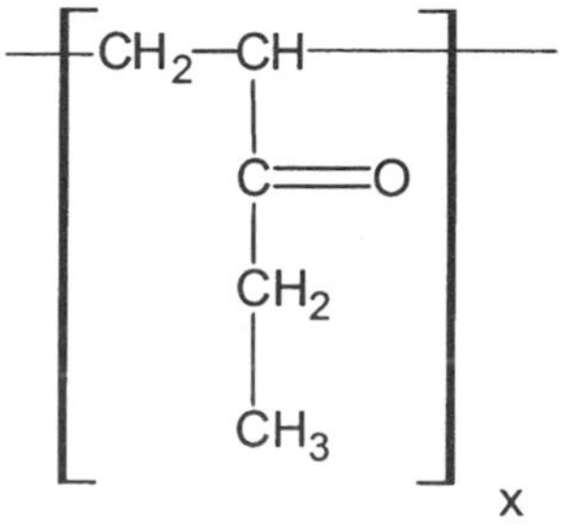

$$\left[-CH_2-CH(-C(=O)-CH_2-CH_3)- \right]_x$$

Figure 39.5 The chemical structure of polyacrylate.

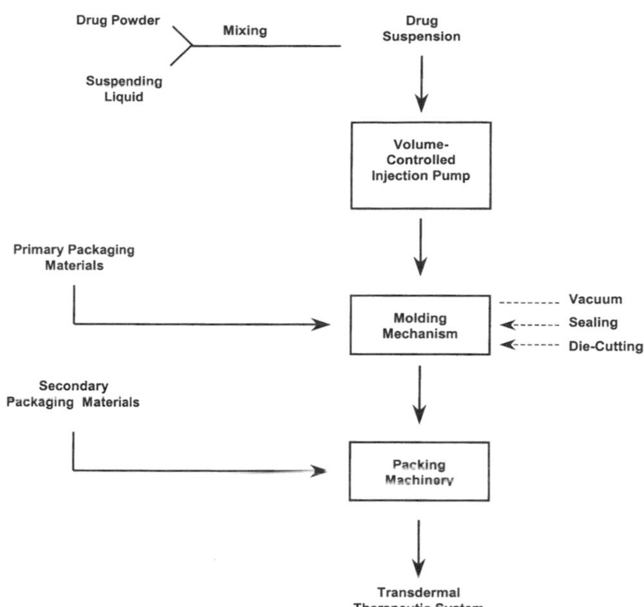

Figure 39.6 Flow chart of the process and equipment involved in the production of the Transderm-Nitro® system, a membrane moderated transdermal therapeutic system. (Reproduced with permission from Chien (2).)

The flow chart in Fig. 7 shows the process and equipment involved in the production of MDC patches.

3. Adhesive Dispersion-Type Transdermal Drug Delivery Systems (ADT)

Adhesive dispersion-type transdermal drug delivery systems (ADT) technology is based on suspending the drug and excipients in the skin adhesive polymer (e.g., polyisobutylenes or polyacrylate). The adhesive layer must provide skin contact for a specified period of time. Simultaneously, the drug must be released from the adhesive layer in a consistent and reproducible manner. So as to fulfill these two requirements, ADT formulation is considered to

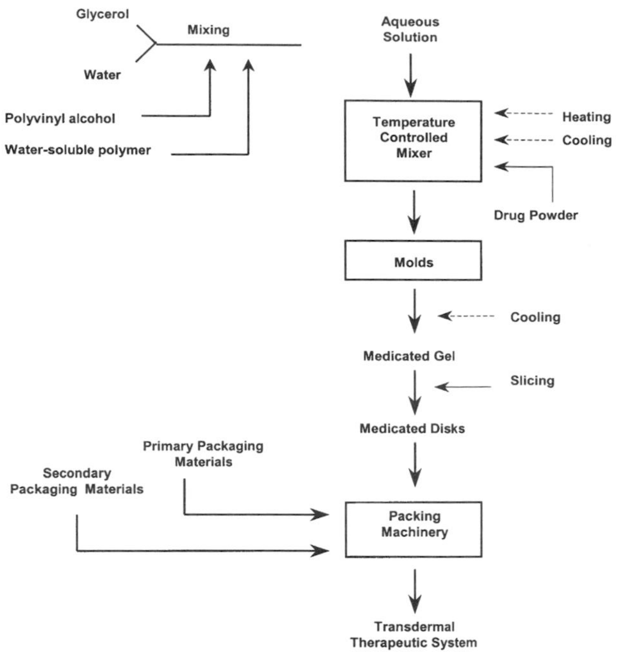

Figure 39.7 Flow chart of the process and equipment involved in the production of the Nitro-Dur® system, a matrix-diffusion-controlled type transdermal drug delivery system. (Reproduced with permission from Chien (2).)

be the most complex to formulate within the four basic TDS.

Unlike MMT, the drug release rate from ADT patches follows first-order kinetics. The drug release rate will decline as the drug concentration in the adhesive layer decreases due to drug release. To overcome first-order kinetics, a reservoir of suspended drug particles may be incorporated in the adhesive formulation, or a multilaminate concept may be used.

The flow chart in Fig. 8 shows the process and equipment involved in the production of ADT patches.

4. Microreservoir Dissolution-Controlled Transdermal Drug Delivery Systems (MDCT)

Microsealed transdermal drug delivery systems combine the properties of the membrane moderated and matrix diffusion-controlled transdermal delivery systems (Fig. 3) (58,59). The system is made of two basic components, a biologically compatible polymer container and an inner biologically compatible silicone polymer matrix with microsealed compartments. The microsealed compartment contains the drug such as nitroglycerin in a hydrophilic solvent system, in which the ratio of the partition coefficient of the drug between the hydrophilic solvent system and the inner biologically compatible silicone polymer matrix is between 1 and $10^{\times 4}$ mL/µm. The biologically compatible polymer container should be biologically and pharmaceutically acceptable. They must also be nonallergenic and nonirritating. Furthermore, the polymer should not be soluble, because both dissolution and/or erosion of the system may affect the release rate of the drug. The polymer container should have a glass transition temperature below room temperature. Examples of the polymer container components include polyethylene, polypropylene, ethylene/propylene copolymer, ethylene/ethylacrylate copolymers, ethylene/vinyl acetate copolymers, silicone rubbers, especially medical grade polydimethyl siloxane, neoprene rubber, vinyl chloride copolymer with vinyl acetate, polymethacrylate polymer, vinylidene chloride, ethylene and propylene, polyethylene terephthalate, butyl rubber, epichlorohydrin rubbers, ethylene/vinyl alcohol copolymer, ethylene/vinyloxyethanol copolymer, and the like. The pharmaceutical delivery device is in an aqueous environment that is composed of a hydrophilic solvent system

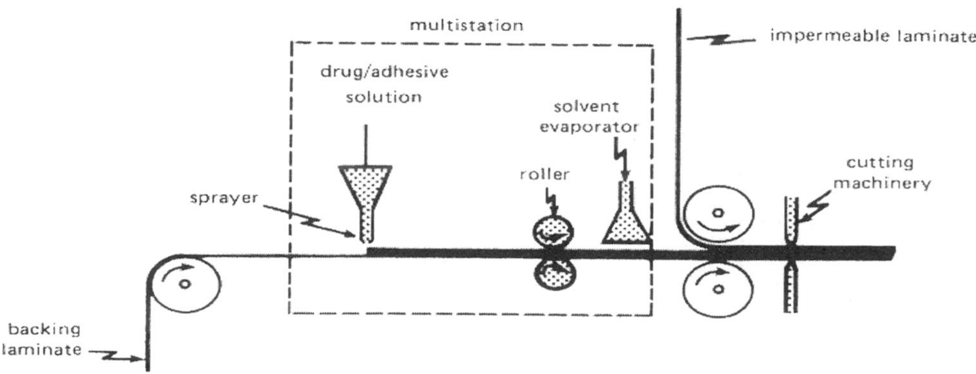

Figure 39.8 Flow chart of the process and equipment involved in the production of an adhesive dispersion-type transdermal drug delivery system. [Reproduced with permission from Chien (Ref. 2).]

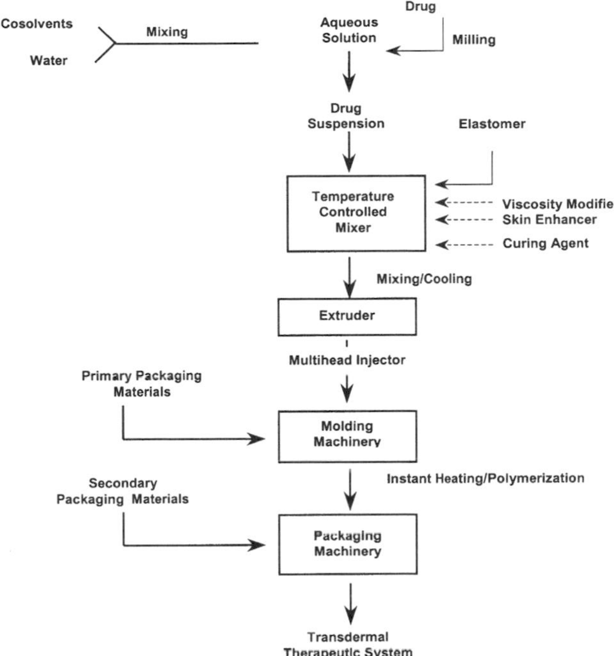

Figure 39.9 Flow chart of the process and equipment involved in the production of the Nitrodisc® system, a microsealed transdermal drug delivery system. (Reproduced with permission from Chien (2).)

that is inserted with a hydrophobic solvent system into the silicone matrix. The hydrophilic solvent is mostly composed of 10–30% polyethylene glycol 400 in distilled water. The hydrophobic solvent system is made of about 5 to 15% w/w of a product chosen from isopropyl palmitate, mineral oil, cholesterol, or a triglyceride of a saturated coconut oil, or a mixture thereof. The flow chart in Fig. 9 shows the process and equipment involved in the production of MDCT patches.

V. ENHANCEMENT OF SKIN PERMEATION

The transdermal route offers several advantages over other routes for the delivery of drugs with systemic activity. These advantages are well documented and are discussed in Section IV. However, the use of this route is often hindered by the poor permeability of the skin, primarily attributed to the outermost layer, the stratum corneum (SC). A number of approaches, chemical and physical, have been used that reversibly alter the barrier layer of the skin and allow the delivery of drugs into the systemic circulation.

This section focuses on the various approaches, that have been developed to reduce the skin barrier properties and promote drug permeation.

A. Chemical Penetration Enhancers

Chemical penetration enhancement using accelerants or sorption promoters is the most widely utilized technique to increase drug permeation across the skin (60). These compounds alter the permeability characteristics of the stratum corneum, allowing an increased rate of percutaneous absorption of coadministered drugs. Ideally, an enhancer should have the following properties (34,61,62):

1. Be nonirritant, nonallergenic, nonphototoxic, and noncomedogenic.
2. Be colorless, tasteless, odorless,and cosmetically acceptable, with a suitable skin "feel."
3. Be compatible with all formulation components.
4. Be chemically and physically stable and pharmacologically inert.
5. Be potent, with an immediate, predictable, and reversible effect on the skin barrier properties.
6. Have a solubility parameter close to that of skin (i.e., 10 $(cal/cm^3)^{1/2}$ (63).

Williams and Barry have published extensively on the mode of action of chemical enhancers on the skin barrier properties (64). They proposed that enhancers may act by one or a combination of the following mechanisms: interaction with the highly ordered intercellular lipids to impart disorganization, thus increasing the paracellular diffusivity through the membrane; interaction with the intracellular proteins to promote the transcellular permeation through the corneocytes; and/or increasing the partitioning of the drug into the SC.

1. Water

Water alters the barrier properties of the SC and is a safe and natural penetration enhancer. Normally, the SC is partially hydrated, containing 5–15% water, and when fully hydrated, the SC can hold as much as 50% water. Barry et al. reported a 10-fold increase in the rate of permeation of the fully hydrated SC relative to its value when it is perfectly dry (65).

Hydration increases the permeation of both polar and nonpolar substances, and Barry attributed this to the water effect on the lipid fluidity and keratin of the SC (66). Water increases the lipid fluidity by reducing the intermolecular forces via hydrogen bonding with lipid polar head groups. This loosens the lipid packing and increases the permeation of drugs paracellularly. The dry SC is a dense structure with a significant constraint on the rearrangement and folding of the intercellular protein molecules. As skin water content increases, restrictions on protein rearrangements lessen and the intermolecular hydrogen bonding between protein molecules is reduced. This is thought to increase drug permeation transcellularly.

2. Sulfoxides

Dimethylsulfoxide (DMSO) was one of the first compounds to be tested for its penetration enhancing activity (Fig. 10). There have been several publications in which the enhancing activity of DMSO was evaluated using a variety of drugs including steroids (67), flufenamic acid (68), acyclovir (69,70), vidarabine (71), and salicylates (72). Several theories were advanced regarding the mechanism of enhancement of DMSO. Among these are the displacement of water from the lipid head groups and loosening of the dense polymeric structure of the protein molecules within the corneocytes (73). The second theory suggests that DMSO extracts stratum corneum lipids, lipoproteins, and nucleoproteins, altering the barrier function and increasing drug permeability (74,75). The third theory proposes that DMSO may promote the drug permeation by osmotically delaminating the SC, thus disrupting the structure of the SC and increasing drug permeability (76).

DMSO use as a penetration enhancer was precluded by primary disadvantages. DMSO penetration enhancing activity is concentration dependent where high concentrations, greater than 60%, are required to produce enhancement activity (71). Unfortunately, at these high concentrations, DMSO was found to produce irreversible skin damage (77) and to cause side effects such as erythema (78,79) and wheals (80). Furthermore, unpleasant taste and mouth odor were reported to follow the topical application of DMSO due to dimethylsulfide, a metabolite of DMSO (81).

Sekura and Scala investigated several alkyl homologues of methyl sulfoxide (Fig. 10) with respect to their absorption enhancement and transdermal permeation of sodium nicotinate and thiourea across guinea pig skin in vitro (82). In the series (C_2-C_{14}), the best enhancement activity for the permeation of both sodium nicotinate and thiourea across guinea pig skin was provided by C_{10} analog, decyl methyl sulfoxide (DCMS). DMCS provided 4.6 times greater penetration than DMSO. Later, the effect of DCMS on the skin permeation of a number of drug molecules was extensively studied. DCMS enhanced the permeation of progesterone (82,83), 5-fluorouracil (84), phenyl propanolamine, propranolol, acyclovir, and hydrocortisone (74,85). The experimental findings suggested that DCMS is an effective enhancer for hydrophilic compounds rather than lipophilic molecules. The major advantages of DCMS over DMSO are that DCMS was reported to be a more potent enhancer, and it can be used at concentrations as low as 0.1%. Unlike DMSO, it has a reversible action on the skin (86).

3. Amides

N, N-dimethylacetamide (DMAC) and *N, N*-dimethylformamide (DMF) are short chain amides (Fig. 11). Like DMSO, they are aprotic solvents that were reported to enhance the permeation of lidocaine (87), griseofulvin (88), caffeine (89), and hydrocortisone (90) in vitro. The mode of action of DMAC and DMF is similar to that of DMSO (73). However, they are less effective than DMSO (91).

Michniak et al. prepared several clofibric acid amide analogues and showed their effectiveness as penetration enhancers using male athymic nude mouse skin as a model membrane (92). Using hydrocortisone 21-acetate and betamethasone 17-valerate as permeants, they found that clofibric acid octyl amide demonstrated the highest enhancement activity; coadministration increased the cumulative amount of hydrocortisone 21-acetate after 24 h 51.6-fold and that of betamethasone 17-valerate 10.3-fold. In another study, Michniak et al. screened a series of acyclic alkyl amide using hairless mouse skin in vitro and hydrocortisone 21-acetate as a model drug. These authors demonstrated that some compounds provided higher enhancing activity than Azone® (increasing the flux up to 35.2-fold over control with no enhancer) (93).

Ogiso and coworkers studied the effect of *N, N*-diethyl-

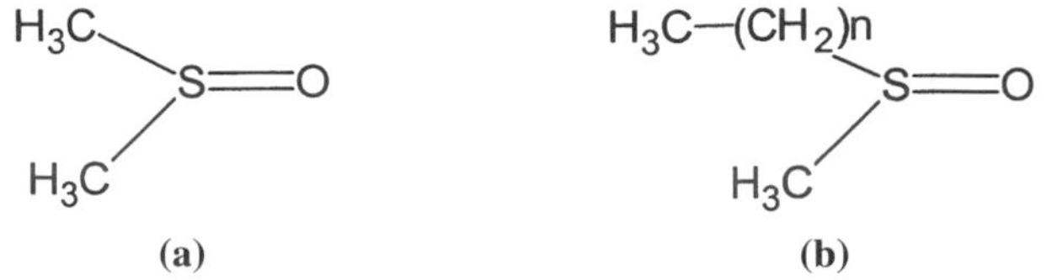

H3C
S=O
H3C
(a)

H3C—(CH2)n
S=O
H3C
(b)

Figure 39.10 The chemical structures of (a) DMSO and (b) alkyl-DMSO.

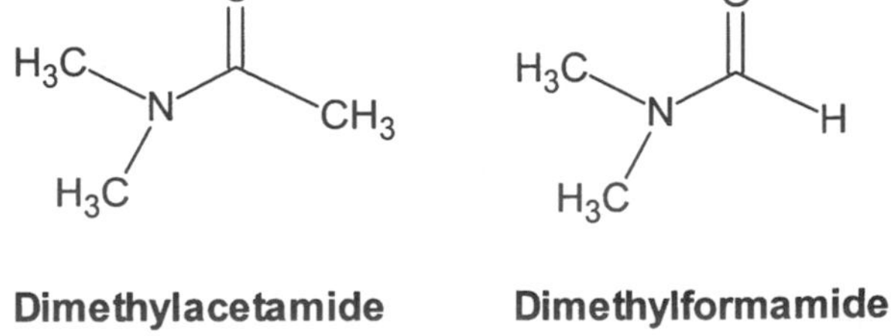

O
H3C
N
CH3
H3C
Dimethylacetamide

O
H3C
N
H
H3C
Dimethylformamide

Figure 39.11 Chemical structures of DMAC and DMF.

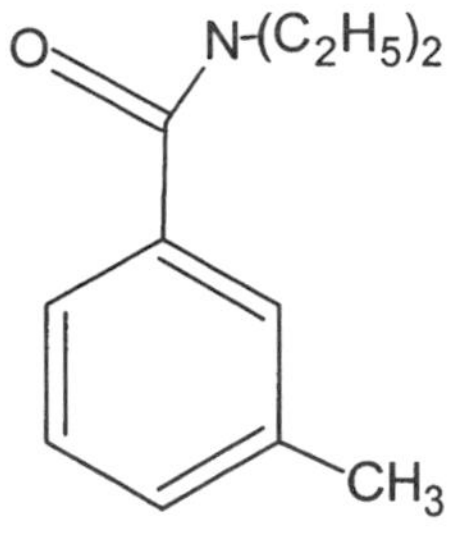

Figure 39.12 Chemical structure of *N,N*-diethyl-*m*-toluamide.

m-toluamide (DEET) (Fig. 12) and other transdermal enhancers such as *N*-octyl-β-D-thioglucoside, DMSO, *N*-methyl-2-pyrrolidone, *n*-octanol, Azone®, isopropyl myristate, sodium oleate, oleic acid, monoolein, oleyl oleate, cineole, and limonene on the permeation of indomethacin and urea using male hairless rat skin in vitro (94). Aqueous ethanol solution (40% ethanol) was used as control. They found that DEET enhanced the permeation of indomethacin and urea relative to the control. In addition, DEET increased the partitioning of indomethacin in the skin and enhanced the thermodynamic activity of drug in the vehicle. In previous research reported by Windheuser et al., DEET was evaluated against a variety of drugs including hydrocortisone, hydrocortisone acetate, hydrocortisone 17-butyrate, hydrocortisone 17-valerate, dibucaine, benzocaine, indomethacin, erythromycin, tetracycline hydrochloride, grisefulvin, myco-21-phenolic acid, and methylsalicylate using hairless mouse skin in vitro (95). DEET provided significant increase in the permeation of all drugs tested. The investigators attributed this to the DEET solvent effect on the skin.

In summary, several amides show promise as potential penetration enhancers. However, further studies need to be performed to evaluate their toxicities, appropriate enhancer concentrations, and model drugs that can be used with these enhancers.

4. Amines

Several amines, namely diphenhydramine base, chlorpheniramine, and nicotine-were evaluated for their enhancing activities using procaterol base and hairless mouse skin in vitro. The mode of action of these compounds is still nuclear. The highest flux and permeability coefficient was provided by nicotine > chlorpheniramine > diphenhydramine base (96).

In a study on the enhancing activity of 12 long chain amine enhancers with secondary and tertiary structures using hairless mouse skin in vitro and hydrocortisone as a model drug, Michniak et al. found that the cyclic tertiary and secondary acyclic amines are less active for increasing hydrocortisone flux than tertiary acyclic amines (97). Furthermore, alkyl amines were poorer flux enhancers than their corresponding amides. In general, both cyclic and acyclic amines were weak transdermal penetration enhancers.

5. Urea and Analogs

Urea (NH_2CONH_2) is a natural substance that is commonly used in the treatment of scaling conditions such as ichthyosis and psoriasis (98,99). Several investigators evaluated urea as a potential enhancer of drug permeation. Feldman and Maibach were the first to investigate the enhancing effect of urea on the permeation of hydrocortisone acetate from a cream formulation across the skin of volunteers in vivo (100). They found a twofold increase in the permeation of the drug when 10% urea in a cream formulation was applied. A similar study by other groups concluded that 10% urea can enhance the efficacy of the coadministered steroid (101,102).

It is worth noting the the activity of urea is largely dependent on the vehicle as well as the concentration of urea. Kim et al. evaluated the efect of vehicle composition on the urea enhancing activity of ketoprofen permeation using rat skin in vitro (103). Ketoprofen permeability was not enhanced when urea in PG or PG/ethanol/water mixtures was used. However, its permeation was slightly enhanced when skin was pretreated with a 20% aqueous urea solution. Naito and Tsai evaluated the effect of urea concentrations in an ointment base on the percutaneous absorption of indomethacin using rabbit skin in vivo. A marked increase in the drug absorption was provided with 0.5% urea in the formulation. However, a decrease in the drug absorption was reported when urea concentration was increased to 2.5 or 5%. This was attributed to the high pH of the aqueous phase of the ointment, which may have reduced the partition of the drug into the skin (104).

Using a series of urea analogues, Williams and Barry evaluated the transdermal permeation of a model drug 5-fluorouracil in human epidermal membranes in vitro (105). They also correlated the effect of vehicle composition with the activity of the tested urea analogues. The series include 1-dodecylurea, 1,3-didodecylurea, and 1,3-diphenyl urea. The vehicles tested were light liquid paraffin, dimethylisosorbide, and propylene glycol (PG). The chemical structures of the urea analogues are illustrated in Figure 13.

When light liquid paraffin was used as a vehicle, urea analogues did not have significant permeation enhancement activity relative to the liquid paraffin. However, when other vehicles were used, a pronounced effect on the enhancing activity of the urea analogues was noticed. The

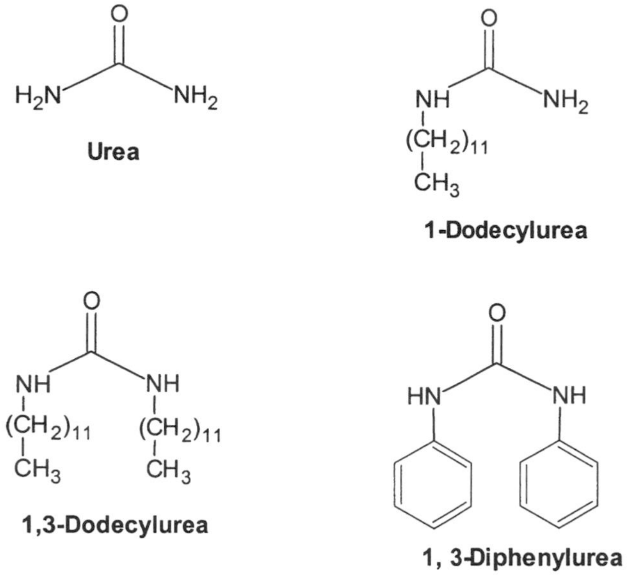

Urea

1-Dodecylurea

1,3-Dodecylurea

1, 3-Diphenylurea

Figure 39.13 The chemical structures of the urea analogues.

permeation of 5-fluorouracil from saturated solutions from PG was higher than drug saturated solutions in dimethylisosorbide and light liquid paraffin. In addition, no correlation was established between the log partition coefficient (octanol/water) and the enhancement ratios of the urea analogues. Wong et al. prepared a series of cyclic, unsaturated, biodegradable analogs of urea penetration enhancers. Their investigations were designed to produce penetration enhancers that maintained their excellent penetration effect and possessed low toxicity. The enhancer consisted of two parts: a highly polar parent moiety that is an unsaturated cyclic urea (1-alkyl-4-imidazolin-2-one) and a long chain alkyl ester group (106). They evaluated the penetration enhancing activity of these enhancers on the transport of indomethacin in petrolatum ointment through shed skin of the black rat snake and reported a penetration enhancement of indomethacin with three of these analogues to be comparable to or better than that of Azone® (107).

6. Pyrrolidone Derivatives

N-methyl-2-pyrrolidone (NMP) is a hygroscopic and aprotic solvent used for pesticides, chemicals, and polymers. Recently, pyrrolidone and its derivatives were intensively evaluated for their potential penetration enhancing activity.

Sasaki et al. prepared a series of alkyl-substituted pyrrolidones and evaluated their effects on the transdermal permeation and skin accumulation of phenosulfonphthalein (phenol red) using rat skin in vivo and in vitro (108). A relationship was established between the partition coefficients of the pyrrolidone derivatives and the flux of phenol red. This suggested that high enhancer lipophilicity is crucial for enhancing the penetration of the hydrophilic compound. The skin accumulation of the enhancers and phenol red were reported to be concentration dependent (109). In a later study, NMP was reported to promote the permeation and to develop a reservoir of drug in SC with 5-fluorouracil, triamcinolone acetonide, indomethacin, flurbiprofen, aminopyrine, tetracycline, betamethasone 17-benzoate, ibuprofen, and sulfaguanidine (108–114.)

Yano et al. demonstrated that *N*-[2-(decylthio)ethyl]-2-pyrrolidone (HPE-101) showed similar enhancing activity relative to Azone® (115,116). In another study, the same group of investigators reported a synergistic effect for the enhancing activity of prostaglandin E_1 percutaneous permeation by NMP when glycols, glycerin, water, or triethanolamine were used as vehicles. However, NMP enhancement activity was lost when isopropylmyristate, alcohols, peppermint oil, or hexylene glycol were used (117,118). Like DMSO, the widespread clinical use of pyrrolidone derivatives was limited because of their damaging effects on the skin especially at high concentrations (81).

7. Azone and Its Analogues

Azone® (1-dodecylazacycloheptan-2-one or laurocapram) is the most extensively studied penetration enhancer since being patented in 1976 (119). It is a chemical combination of cyclic amide (pyrrolidone) and alkyl sulfoxide (DCMS). Azone® is a highly lipophilic structure with a log[octanol/water partition coefficient] of 6.4. The chemical structure of Azone® is thought to relate directly to its mode of action and imparts both hydrophilic and lipophilic characteristics to the molecule. Molecular modeling studies suggested that Azone® assumes a spoon-shape configuration which allows it to insert into the lipids of the stratum corneum and interact with the hydrophilic domains, thereby disrupting their packing and increasing drug penetration (120). When compared with other transdermal enhancers, Azone® provides many advantages. It is not irritating or allergenic when applied to human skin, it is absorbed dermally, but minimally absorbed into the systemic circulation. In addition, it has been incorporated into various topical formulations such as creams, ointments, suspensions, and patches and has promoted drug permeation at low concentrations (1–5%). The enhancing activity of Azone® was investigated using drugs with different lipophilicities including 9-desglycinamide-8-arginine vasopressin, heparin, indomethacin, nitroglycerin, propranolol, timolol, and acyclovir (26,42,121–124). Barry and Bennett speculated that Azone® may be more effective in promoting the permeation of hydrophilic compounds than for enhancing the permeation of hydrophobic compounds (125).

Due to the permeation enhancing activities of Azone®, several investigators developed new Azone® analogs and tested their penetration enhancing activities. Mirejovsky and Takruri developed a series of hexamethylene lauramide (pyrrolidine, piperidine, and hexahydro-1*H*-azepine) with alkyl chain (seven, eleven, and fifteen carbons) derivatives using hairless mouse skin in vitro and in vivo (126). Three model drugs were investigated: hydrocortisone, griseofulvin, and erythromycin, and the effect of pretreatment time with enhancer on the permeation of hydrocortisone was also evaluated. The results suggested that hexamethylene lauramide had a broad spectrum of enhancer activity. Furthermore, the longer the pretreatment, the lower the amount of hydrocortisone permeated. The in vitro results also suggested that the permeation of hydrocortisone through hairless mouse skin in vitro was slower than the penetration through living skin in vivo.

Okamoto et al. evaluated the enhancing effect of a series of azacycloalkanone analogues using guinea pig skin in vitro and 6-mercaptopurine and 5-fluorouracil as model drugs (127). Two parameters were determined, drug diffusivity and partitioning into the skin using aqueous and ethanolic systems. In the aqueous system, the partitioning of the drugs into the skin was enhanced by pretreatment with enhancers, which also increased drug accumulation in the skin. However, drug diffusivities were little affected. In the ethanolic system, the enhancement was less than that provided by the aqueous system. In another study by the same group, Okamoto et al. showed that the enhancement effect of various azacycloalkan-2-one derivatives was related parabolically to model drug lipophilicity (128).

Michniak et al. evaluated a series of Azone® analogues for their enhancing activities by substituting various sized rings, new heteroatoms such as sulfur or nitrogen, and incorporating changes in the chain structure of the alkyl residue using hairless mouse skin in vitro and hydrocortisone as a model drug. They reported that most of the derivatives were not as effective as Azone® in delivering the hydrocortisone across the hairless mouse skin. In addition, thioamide enhancers were less effective than their oxygen-containing counterparts (129,130).

In summary, Azone® has been shown to be an effective penetration enhancer for a variety of drugs and is often used as a standard for comparison when newer agents are being evaluated.

8. Iminosulfuranes

Kim et al. and Strekowski et al. evaluated the penetration enhancing activity of three novel series of iminosulfuranes using hydrocortisone as a model drug and hairless mouse skin in vitro (131,132). All enhancers tested were applied to the skin as saturated suspensions in propylene glycol to ensure their maximum thermodynamic activity. From the evaluated series, *S,S*-dimethyl-*N*-(4-bromobenzoyl)iminosulfurane, *S,S*-dimethyl-*N*-(5-nitro-2-pyridyl)iminosulfurane, and *S,S*-dimethyl-*N*-(4-phenylazaphenyl) iminosulfurane provided significantly higher flux over 24 h relative to the control (propylene glycol without enhancer). However, *S,S*-dimethyl-*N*-(benzenesulfonyl)iminosulfurane, *S,S*-dimethyl-*N*-(2-methoxycarbonyl benzenesulfonyl)iminosulfurane, and *S,S*-dimethyl-*N*-(4-chlorobenzenesulfonyl)imino sulfurane all decreased the permeation of hydrocortisone significantly relative to the control. The authors suggested that these three compounds could function as skin drug permeation retardants.

9. Surfactants

Surfactants are extensively used in topical and dermatological formulations to increase product stability and to promote drug permeation through the skin. Surfactants can be generally classified based on the charge of their hydrophilic head at physiological pH into cationic, anionic, nonionic, and zwitterionic. It is generally accepted that the use of nonionic surfactants results in less skin irritation compared to other surfactant groups, yet they provide lower transdermal enhancing activity. Ashton et al. investigated the transdermal enhancement activity of a cationic (dodecyltrimethylammonium bromide ''DTAB''), anionic (sodium dodecyl sulfate ''SDS''), and nonionic (Brij 36T™) surfactants (133). By evaluating the flux of the model drug methyl nicotinamide using human cadaver skin in vitro they reported that Brij 36T™ showed the lowest enhancing activity compared to the other two surfactants. Furthermore, Brij 36T™ provided the shortest lag time, which suggested that nonionic surfactant penetrates the membrane more rapidly but binds less with the skin components. Similar findings were demonstrated by Gershbein, who evaluated the skin irritation induced by the application of cetyltrimethyl ammonium bromide and sodium lauryl sulfate (SLS) and correlated it with their transdermal promoting activity (134). He reported that the two enhancers damaged the skin,leading to a significant increase in transdermal drug permeation.

In view of the damaging effects of cationic and anionic surfactants, Rhein et al. investigated the effect of amphoteric surfactants on the swelling of isolated human stratum corneum secondary to sodium lauryl sulfate (SLS) (135). They found that the swelling was significantly reduced when 1% alkyl ethoxy sulfate was coadministered with 1% SLS. They attributed the lowering of stratum corneum swelling to either a competition between the two surfactants for binding sites in the stratum corneum or mixed micelle formation that may reduce the number of SLS monomers available to interact with the stratum corneum. Borras-Blasco et al. investigated the enhancing activity of SLS on compounds with different lipophilicity using rat skin in vitro (136). They concluded that the enhancing activity of SLS was more significant on compounds with lower lipophilicity, i.e., log[octanol/water partition coefficient] < 3.

Kushla evaluated the effect of alkyl chain length of three classes of cationic surfactants: alkyl dimethylbenzyl ammonium halides, alkyl trimethyl ammonium halides, and alkyl pyridinium halides, on the percutaneous permeation of lidocaine and water through human epidermis in vitro (137). Lidocaine was applied to the skin in the form of a suspension in propylene glycol : water mixtures containing cationic surfactants of varying alkyl chain lengths from the three classes given above. Increasing the concentration of surfactant was associated with higher enhancement ratios of both penetrants relative to control. In summary, alkyl chain lengths of 12 or 14 carbons provided the highest enhancement activity.

10. Alcohols

Small alcohols such as methanol and ethanol have been commonly used in topical formulations as transdermal penetration enhancers and solvents. Bommannan et al. evaluated the mechanism of action of these compounds (138). Alcohols exert their action by extracting the stratum corneum polar lipids, hence the increase in drug permeation is lower than that observed with Azone®. Ethanol was shown to be an effective enhancer for a variety of drugs including estradiol (139), nitroglycerin (140,141), buprenorphine (142), diclofenac (143,144), alclofenac, and ibuprofen (145). It is important to note that the enhancing activity of ethanol is concentration dependent. For example, at an ethanol volume fraction of 0.7 or below, nitroglycerin flux was improved, whereas at higher alcohol concentrations the drug permeation was hindered (141).

The enhancement activity of alcohols is a function of their alkyl chain length. Friend et al. evaluated the effect of alkyl chain length using levonorgestrel as a model drug (146). The enhancement activity of alkanols was reduced as the alkyl chain length was increased beyond 4 carbons.

11. Cyclodextrins

Cyclodextrins (CDs) are naturally occurring biocompatible substances that are made of six, seven, or eight glucopyranose units: α, β, and γ CDs, respectively with a hydrophilic outer surface and a lipophilic cavity in the center. Depending on the drug and treatment protocol, these agents can act as enhancer, or mild dermal retardants. They are known to increase lipophilic drug solubility by forming inclusion complexes. Williams et al. evaluated the permeation enhancement activity of β and 2-hydroxypropyl-β-CDs alone and complexed with other penetration enhancers for the model drugs 5-fluorouracil and estradiol using human cadaver skin in vitro (147). They demonstrated that CDs did not enhance the permeation of either drug. Furthermore, complexation of lipophilic terpene with the CDs reduced enhancer efficacy. They concluded that CDs do not act as penetration enhancers. In addition, some CDs can be incorporated into topical formulations to retard the absorption of toxic materials. These findings conflict with data published by Vollmer et al. (148). They evaluated the enhancement activity of different concentrations of 2-hydroxypropyl-β-CDs and 2,6-dimethyl-β-cyclodextrin using liarozole (cytochrome P450 inhibitor) and a rat model in vivo. At 20% aqueous solution of 2-hydroxypropyl-β-CDs, the absorption of liarozole was increased threefold. However, a 20% aqueous solution of 2,6-dimethyl-β-cyclodextrin retarded the percutaneous absorption of liarozole in blood by a factor of 0.6. It is interesting to note that pretreatment with 2,6-dimethyl-β-cyclodextrin (20%, 4 h) promoted the transdermal absorption 9.4-fold. Differential scanning calorimetery (DSC) studies suggested that 2,6-dimethyl-β-cyclodextrin enhanced the permeation of liarozole by interacting with protein and lipid fraction, whereas 2-hydroxypropyl-β-CDs enhanced the drug permeation by affecting the drug partitioning into the skin.

12. Terpenes

Terpenes are naturally occurring hydrocarbon constituents of essential oils that are not aromatic in character. They have been found to be promising penetration enhancers with low cutaneous irritancy and good toxicological profiles at low concentrations (1–5%) (149). Terpene enhancers were reported to increase the percutaneous absorption of lipophilic as well as hydrophilic compounds (150–153). In their study on the effect of terpene enhancers on the percutaneous absorption of indomethacin in rats in vivo, Nagai et al. found that limonene was as effective a penetration enhancer as Azone at a concentration of 1% in an indomethacin gel ointment (154). In addition, limonene was reported not to affect the partitioning of indomethacin into the skin, and it was suggested that it enhanced indomethacin permeation mainly by affecting the highly ordered structure of the stratum corneum lipid. Oxygen-containing terpenes in this study did not increase the permeability of indomethacin, a nonpolar drug. However, the oxygen-containing terpenes carvone, α-terpineol, and 1,8-cineole did enhance the transdermal delivery of the polar drug 5-fluorouracil across human cadaver skin in vitro (155). Williams and Barry conducted differential scanning calorimetry and Fourier transform infrared studies to evaluate the mechanism of action of terpene enhancers (156). These studies supported the theory that terpenes increase the drug percutaneous permeation mainly by interrupting the highly ordered intercellular packing of the SC lipids. The same authors evaluated the reversibility of terpene enhancing effects using 5-fluorouracil as a model drug and human epidermal sheets in vitro. They demonstrated that 5- fluorouracil flux in terpene-treated skins decreased with time after washing out of the terpenes from the skin surface (64). Unlike Azone, terpene enhancers were reported to produce lower skin irritation (149).

13. Fatty Acids and Fatty Alcohols

A parabolic relationship has been established between enhancement of drug permeation across the skin and the chain lengths of saturated fatty acids and fatty alcohols. Aungst evaluated the enhancement activity of naloxone penetration across human skin using saturated fatty acids (C_7-C_{18}) and fatty alcohols (C_8-C_{18}) (157). From the two series the C_{12} alkyl chain lengths demonstrated the highest flux of naloxone. In addition, they reported that C_{18} unsaturated acids and alcohols provided more enhancement for naloxone permeation relative to the saturated compounds. Ogiso and Shintani reported in another study the parabolic relationship between the chain length of the saturated fatty acids (C_6-C_{18}) and the flux of propranolol across rabbit skin (158). Since the C_{10} and C_{12} alkyl chain lengths are similar to the alkyl chain length of cholesterol, it has been proposed that these fatty acids and fatty alcohols disrupt ceramide–cholestrol and/or cholesterol–cholesterol interactions in the SC. Aungst evaluated the effect of branched and substituted fatty acids using human cadaver skin in vitro and naloxone (157). He did not report any difference between branched and linear fatty acids of the various chain lengths studied except for the C_{18} compounds. Isostearic acid provided more penetration enhancement for naloxone than stearic acid.

14. Polymeric Enhancers

Aoyagi et al. investigated the use of polymers made of benzalkonium chloride surfactants as percutaneous permeation enhancers using 5-fluorouracil as a model drug and rabbit skin in vitro (159). In this study, *p*-vinylbenzyldimethylalkylammonium chloride having a long alkyl chain was polymerized to variable average molecular weights ranging from 3,000 to 32,000. This enhancer increased the permeability of 5-fluorouracil significantly relative to the control with no enhancer. In addition, an inverse relationship was established between the molecular weight of the polymer and the permeation enhancement. Furthermore, these polymers demonstrated lower skin irritation, which was attributed to their low skin permeability due to their large molecular weight, as suggested by DSC studies. The polymeric enhancer approach offers many significant advantages such as low skin permeability of enhancers and the chance to modify the structure of these compounds to generate molecules of different lipophilicity and different charge distribution characteristics (which have a major impact on their interaction with the skin, donor vehicle, and drug molecules).

Polydimethylsiloxane (PDMS) has been used extensively in the medical and pharmaceutical fields. It has many favorable physicochemical properties such as high flexibility, hydrophobicity, good biocompatability, and physiologic inertness (160). Recent studies on polymeric enhancers have focused on using PDMS-based polymers. In general, most of the evaluated polymeric enhancers possess polar and nonpolar groups. Such a combination provides the polymeric enhancers with different physicochemical properties that enable them to interact with both the skin and the drug. In another study by the same group, Aoyagi et al. prepared PDMS enhancers with different polar substituents at one chain end, such as *N*-methylpyridinio, quaternary amino, phosphoramid, carboxyl, and 1-alkyl-2-pyrrolidon-3-yl. These polymeric enhancers with polar end groups were effective surface active agents and penetration enhancers. They were also nonirritating due to their low skin penetration and the physiologic inertness of the PDMS backbone (161). These early PDMS polymers were shown to be effective only for lipophilic drugs. This was attributed to the mechanism of action of these polymers, primarily interacting with the surface of the skin, resulting in a change of the partitioning of the drug into the skin. As a consequence, the same research group developed a new class of polymeric enhancers containing a hydrophilic segment such as polyethyleneglycol (PEG). These copolymers containing a quaternary ammonium moiety at one end of the chain were shown to be effective in enhancing the permeability of both hydrophilic and lipophilic drugs (146). The effect of polymeric enhancer molecular weights on the percutaneous parameters of evaluated drugs depended on the type of polymeric enhancer used. Studies with diethyl phosphoramidate or carboxy terminated dimethylsiloxane and benzalkonium chloride–based polymers demonstrated a decrease in the activity of the polymer with an increase in the degree of polymerization. On the other hand, ammonio or pyridinio terminated polydimethylsiloxane showed an increase in the enhancing activity with an increase in molecular weight (159–162). The end chain group had a major effect on the penetration enhancing activity of the polymeric enhancers. Cationic groups such as the ammonio or pyridinio groups precluded the permeation of coadministered anionic drugs (indomethacin). However, pretreating the skin with cationic enhancer resulted in an increase in the amount of the anionic drug permeated through the skin. Several mechanisms were advanced to explain to explain this behavior. DSC studies and curve fitting analysis revealed that these polymers did not have an effect on the stratum corneum lipids, as shown by the absence of changes in the transition temperatures obtained. However, these polymers are known to increase the partition coefficient of the drug into the skin. These data support the idea that these polymers are adsorbed onto the surface of the skin and consequently create a new layer into which the drug can partition (163).

B. Physical Approaches

Although the chemical penetration enhancer method has been widely studied, not many enhancers have been approved for use in marketed products. This is mainly due to the unresolved question concerning the reversibility of their actions as well as their safety. Recently, physical approaches that include phonophoresis, iontophoresis, and electroporation have been extensively evaluated for their ability to enhance drug permeation across the skin.

1. Phonophoresis

Phonophoresis, also known as sonophoresis or ultrasonophoresis, is defined as permeation of drug molecules with a coupling/contact agent across the skin under the influence of ultrasonic perturbation (164–170). Phonophoresis has been used in medicine for many years, and its use as an aid for delivering medication transdermally is gaining increasing recognition. The mechanism of ultrasound induced skin permeation enhancement is not completely understood, but a combination of thermal, mechanical, or cavitational effects are involved in this process (165).

Levy et al. evaluated the effect of ultrasound on skin permeation of D-mannitol, insulin, and physostigmine in rats and guinea pigs (164). They found that ultrasound altered the permeability of skin temporarily and reversibly and it nearly eliminated the long lag time usually associated with the delivery of drugs across the skin. Furthermore, the transdermal permeation of insulin and mannitol in rats was increased by 5–20-fold relative to control within 1–2 h after ultrasound treatment. Ultrasound also substantially promoted the inhibition of cholinesterase after the first hour following its application in both physostigmine-treated rats and guinea pigs ($p < 0.05$). The in vivo ultrasound-treated guinea pig group had a $15 \pm 5\%$ decline in the blood cholinesterase activity following 1 h ultrasound application. However, in the control guinea pigs, no significant inhibition of cholinesterase was recorded after the 2 h application of physostigmine. In rats, cholinesterase level inhibition following 1 h ultrasound application was $53 \pm 5\%$ in the ultrasound-treated group and $35 \pm 5\%$ in the control group. The authors suggested that mixing and cavitation effects could be involved in the diffusion phenomenon through the membranes both in vitro and in vivo.

Mitragotri et al. evaluated the role of various ultrasound-related phenomena, including cavitation, thermal effects, generation of convective velocities, and mechanical effects, in the ultrasonic enhancement of drug delivery across the skin (165). The results suggested that cavitation appeared to play a profound role in the enhancement of drug permeation across the skin following ultrasonic application. Furthermore, confocal microscopy results indicated that cavitation occurs in the keratinocytes of the stratum corneum upon ultrasound exposure. The authors also hypothesized that oscillations of the cavitation bubbles induce disorder in the stratum corneum lipid bilayers, leading to enhancement in drug permeation across the skin.

There are a number of factors that must be considered prior to the use of phonophoresis, including the frequency and intensity of applied ultrasonic energy, the media between the ultrasonic applicator and the skin, and the length of exposure time.

Frequencies ranging from 20 kHz to 10 MHz are commonly used in phonophoresis. Mitrogotri et al. used low frequency ultrasound to enhance insulin permeation across skin. This study was based that on the hypothesis that if cavitation plays the major role in ultrasound enhanced skin permeation and since the cavitation threshold increases rapidly with an increase of ultrasound frequency, lower frequency may lead to more cavitation, and as a result higher drug delivery across the skin may be expected

(167). They found that the increase in the ultrasound intensity was associated with an exponential increase in the insulin permeation. This may suggest that there is a nonlinear dependence of cavitation on ultrasound intensity. They also investigated the efficacy of ultrasound in promoting the insulin flux in an in vivo hairless rat model. An intensity dependent decrease in the blood glucose concentration was observed upon the application of ultrasound. They concluded that low-frequency ultrasound can be an effective tool in delivering insulin across hairless rat skin in an intensity-dependent model. On the contrary, Bommannan et al. used high ultrasound frequency to increase drug permeation across the skin. They referred this to the hypothesis that higher ultrasound frequency would localize the acoustic energy more within the stratum corneum and this may significantly alter the skin barrier layer and promote drug permeation (166). They evaluated this hypothesis using salicylic acid and hairless rat skin in vivo. Salicylic acid permeation across hairless rat skin in vivo was more significant when 10 MHz ultrasound was used compared to that provided by 16 MHz. However, 16 MHz was more effective in enhancing salicylic acid relative to 2 MHz.

The intensity used for sonophoresis typically ranges from 0.1 to 3 W/cm^2. Several research groups have independently evaluated the effect of ultrasound intensity on phonophoresis enhancement activity (164,166,168,169). Miyazaki et al. investigated the relation between the plasma concentration of indomethacin permeated across hairless mouse skin and ultrasound intensity (171). The plasma level of indomethacin increased threefold relative to control after 3 h of phonophoresis (0.25 W/cm^2). Further increase in the intensity to 0.75 W/cm^2 increased the indomethacin level 33-fold. The contact agents that are widely used in phonophoresis are ultrasonic gels and propylene glycol–water mixtures (170), and contact times of 5 min are usually reported for most phonophoresis studies (166,170).

2. Iontophoresis

Iontophoresis can be defined as the enhanced transport of charged molecules across the skin under an externally applied constant small amount of physiologically acceptable electric current (< 1.0 mA) (172–174). This technique has been extensively used for decades to deliver drugs topically. Lately, the use of iontophoresis for the delivery of drugs across skin has gained much attention, since this method of delivery is particularly promising for the delivery of large macromolecules such as peptide and protein drugs across the skin (175).

During passive application, the nonionized species penetrates the skin much more readily than the ionized counterparts. However, following iontophoresis, a large concentration of ionized species is found at the regions of lower diffusional resistance, including sweat glands and hair follicles. This data was supported by studies performed by different research groups (172,173,176–178). It should be emphasized that the pore pathways do not just include appendages; the ultimate pathway is still the intercellular space between the hair follicles and epidermal cells (179).

Different parameters affect the systemic iontophoretic delivery of drugs across the skin including formulation pH (180) and ionic strength, applied current strength (181), drug concentration (182), and molecular size (183).

It is postulated that the morphology and permeability of the skin are only temporarily changed during iontophoresis (184). This change could be due to a localized heating within the pores as the current passes through the skin.

3. Electroporation

Electroporation involves the application of high-intensity electric field pulses (equivalent to several hundreds of volts for micro or milliseconds) to the cell membrane so as to create transient aqueous pathways in the cell membrane lipid bilayer. As a result, the permeation of the exogenous materials from the surrounding media to the inside of the cells will be significantly increased. Recently, this technique has been recognized as a powerful one for delivering macromolecules such as oligonucleotides and proteins or even smaller molecules across the skin for systemic activity (172–174,185).

The mechanism of drug permeation enhancement across lipid membranes using electroporation has been extensively investigated. Briefly, when an electric field is applied on a lipid membrane, it forces water to enter the lipid structure. Newly created aqueous pores will be formed when the water molecules from both sides of the lipid membrane meet. The pore size may increase as long as the electric field is applied. However, pores will collapse once the electric field is stopped.

Prausnitz et al. evaluated the reversibility of the action of electric voltage range on the flux of calcein across human epidermal membrane in vitro. The electric voltage was applied every 5 seconds for 1 h and ranged from 0 to 500 volts. The authors found that full reversibility of calcein flux was seen over the 0–100 volt range. Above 100 volts the reversibility of calcein flux was significant yet not full, suggesting permanent changes in the skin structure had occured.

Vanbever et al. compared the effects of short (approximately 1 ms, high-voltage (approximately 100 volt) and long (approximately 100 ms), medium-voltage (>30 volt) pulses on the skin electrical properties and the transdermal permeation of sulforhodamine in vitro. They reported that long pulses of medium-range voltage were more efficient in enhancing the permeation of sulforhodamine across skin. However, skin resistance was reduced threefold (with short pulses) and twofold (with long pulses) followed by incomplete recovery in both situations (186).

In another study by Prausnitz et al., the effect of high-voltage pulses on the permeation of a highly charged macromolecule (heparin) was investigated using human skin in vitro. The flux of heparin occurred at therapeutic rates (100–500 mg/cm^{2}h) that were reported to induce systemic anticoagulation. It is interesting to note that the heparin that was permeated across the human skin was biologically active, but it retained one eighth the anticoagulant activity of heparin in the donor compartment. This was attributed to the preferential transport of small (less active) heparin molecules (185).

REFERENCES

1. S. Jacob, C. Francone. *Structure and Function of Man*. Philadelphia, Saunders, 1970.
2. Y. Chien. *Transdermal Controlled Systemic Medications*. New York, Marcel Dekker, 1987.
3. A. Breathnach. Embryology of human skin. *J. Invest. Dermatol. 57*:133 (1971).
4. D. P. Smack, B. P. Korge, W. D. James. Keratin and keratinization. *J. Am. Acad. Dermatol. 30*:85 (1994).
5. P. Minghetti, A. Casiraghi, F. Cilurzo, L. Montanari, M. Marazzi, L. Falcone, V. Donati. Comparison of different membranes with cultures of keratinocytes from man for percutaneous absorption of nitroglycerine. *J. Pharm. Pharmacol. 51*:673 (1999).
6. R. Pouliot, L. Ermain, F. A. Auger, N. Tremblay, J. Juhasz. Physical characterization of the stratum corneum of an in vitro human skin equivalent produced by tissue engineering and its comparison with normal human skin by ATR-FTIR spectroscopy and thermal analysis (DSC). *Biochim. Biophys. Acta 1439*:341 (1999).
7. M. Fartasch, M. L. Williams, P. M. Elias. Altered lamellar body secretion and stratum corneum membrane structure in Netherton syndrome: differentiation from other infantile erythrodermas and pathogenic implications. *Arch. Dermatol. 135*:823 (1999).
8. R. Ghadially, B. E. Brown, K. Hanley, J. T. Reed, K. R. Feingold, P. M. Elias. Decreased epidermal lipid synthesis accounts for altered barrier function in aged mice. *J. Invest. Dermatol. 106*:1064 (1996).
9. E. Proksch, W. M. Holleran, G. K. Menon, P. M. Elias, K. R. Feingold. Barrier function regulates epidermal lipid and DNA synthesis. *Br. J. Dermatol. 128*:473 (1993).
10. P. W. Wertz, D. C. Swartzendruber, D. J. Kitko, K. C. Madison, D. T. Downing. The role of the corneocyte lipid envelopes in cohesion of the stratum corneum. *J. Invest. Dermatol. 93*:169 (1989).
11. P. W. Wertz, K. C. Madison, D. T. Downing. Covalently bound lipids of human stratum corneum. *J. Invest. Dermatol. 92*:109 (1989).
12. R. B. Phillips, R. D. Mootz, J. Nyiendo, R. Cooperstein, J. Konsler, M. Mennon. The descriptive profile of low back pain patients of field practicing chiropractors contrasted with those treated in the clinics of west coast chiropractic colleges. *J. Manipulative Physiol. Ther. 15*:512 (1992).
13. L. Landmann. Epidermal permeability barrier: transformation of lamellar granule-disks into intercellular sheets by a membrane-fusion process, a freeze-fracture study. *J. Invest. Dermatol. 87*:202 (1986).
14. S. Grayson, A. G. Johnson-Winegar, B. U. Wintroub, R. R. Isseroff, E. H. Epstein, Jr., P. M. Elias. Lamellar body-enriched fractions from neonatal mice: preparative techniques and partial characterization. *J. Invest. Dermatol. 85*:289 (1985).
15. P. M. Elias, J. Goerke, D. S. Friend. Mammalian epidermal barrier layer lipids: composition and influence on structure. *J. Invest. Dermatol. 69*:535 (1977).
16. P. M. Elias, N. S. McNutt, D. S. Friend. Membrane alterations during cornification of mammalian squamous epithelia: a freeze-fracture, tracer, and thin-section study. *Anat. Rec. 189*:577 (1977).
17. R. A. Briggaman. Biochemical composition of the epidermal-dermal junction and other basement membrane. *J. Invest. Dermatol. 78*:1 (1982).
18. A. Jarret. *The Physiology and Pathophysiology of the Skin*. London, Academic Press, 1974.
19. L. Smith, K. Holbrook, P. Byers. Structure of the dermal matrix during development and in the adult. *J. Invest. Dermatol. 79 (suppl 1)*:93s (1982).
20. H. Schaefer, F. Watts, J. Brod, B. Illel. Follicular penetration. *Prediction of Percutaneous Penetration: Methods, Measurements, and Modelling* (R. Scott, R. Guy, J. Hadgraft, eds.) IBC Technical Services, London, 1990, p. 163.
21. B. Barry. *Dermatological Formulations: Percutaneous Absorption*. New York, Marcel Dekker, 1983.
22. R. J. Scheuplein, I. H. Blank. Permeability of the skin. *Physiol. Rev. 51*:702 (1971).
23. R. Scheuplein. Percutaneous absorption. *Biochemistry and Physiology of the Skin* (L. Goldsmith, ed.). Oxford University Press, Oxford, 1983, p. 1255.
24. R. J. Scheuplein. Mechanism of percutaneous absorption. II. Transient diffusion and the relative importance of various routes of skin penetration. *J. Invest. Dermatol. 48*:79 (1967).
25. A. Rolland, N. Wagner, A. Chatelus, B. Shroot, H. Schaefer. Site-specific drug delivery to pilosebaceous structures using polymeric microspheres. *Pharm. Res. 10*: 1738 (1993).

26. W. H. Craane-van Hinsberg, J. C. Verhoef, L. J. Bax, H. E. Junginger, H. E. Bodde. Role of appendages in skin resistance and iontophoretic peptide flux: human versus snake skin. *Pharm. Res. 12*:1506 (1995).
27. F. L. Bamba, J. Wepierre. Role of the appendageal pathway in the percutaneous absorption of pyridostigmine bromide in various vehicles. *Eur. J. Drug Metab. Pharmacokinet. 18*:339 (1993).
28. B. D. Anderson, W. I. Higuchi, P. V. Raykar. Heterogeneity effects on permeability-partition coefficient relationships in human stratum corneum. *Pharm. Res. 5*:566 (1988).
29. G. Kastings, P. Robinson. Can we assign an upper limit to skin permeability. *Pharm. Res. 10*:930 (1993).
30. G. Flynn, S. H. Yalkowsky. Correlation and prediction of mass transport across membranes. I. Influence of alkyl chainlength on flux determination properties and diffusant. *J. Pharm. Sci. 61*:838 (1972).
31. G. Flynn. Physicochemical determinants of skin absorption. *Principles of Route to Route Extrapolation for Risk Assessment* (T. Gerrity, C. Henry, eds.). Elsevier, New York, 1990, p. 93.
32. J. Crank. *The Mathematics of Diffusion*. London, Oxford University Press, 1975.
33. B. Berner, V. A. John. Pharmacokinetic characterization of transdermal delivery systems. *Clin. Pharmacokinet. 26*: 121 (1994).
34. W. R. Pfister, D. S. Hsieh. Permeation enhancers compatible with transdermal drug delivery systems. Part I: selection and formulation considerations. *Med. Device Technol. 1*:48 (1990).
35. J. Varvel, S. Shafer, S. Hwang, P. Coen, D. Stanski. Absorption characteristics of transdermally administered fentanyl. *Anesthesiology 70*:928 (1989).
36. W. Utian, K. Burry, D. Archer, J. Gallagher, R. Boyett, M. Guy, G. Tachon, H. Chadha-Boreham, A. Bouvet. Efficacy and safety of low, standard, and high dosages of an estradiol transdermal system (Esclim) compared with placebo on vasomotor symptoms in highly symptomatic menopausal patients. The Esclim Study Group. *Am. J. Obstet. Gynecol. 181*:71 (1999).
37. N. Variankaval, K. Jacob, S. Dinh. Crystallization of beta-estradiol in an acrylic transdermal drug delivery system. *J. Biomed. Mater. Res. 44*:397 (1999).
38. V. De Sanctis, C. Vullo, L. Urso, F. Rigolin, A. Cavallini, K. Caramelli, C. Daugherty, and N. Mazer. Clinical experience using the Androderm testosterone transdermal system in hypogonadal adolescents and young men with beta-thalassemia major. *J. Pediatr. Endocrinol. Metab. 11*:891 (1998).
39. D. Buchter, S. Von Eckardstein, A. Von Eckardstein, A. Kamischke, M. Simoni, H. M. Behre, E. Nieschlag. Clinical trial of transdermal testosterone and oral levonorgestrel for male contraception. *J. Clin. Endocrinol. Metab. 84*: 1244 (1999).
40. X. Ji, Q. Jing, G. Liu, S. Yu. The bioavailability of transdermal therapeutic system of timolol. *Yao Hsueh Hsueh Pao 28*:609 (1993).
41. J. Kemken, A. Ziegler, B. Muller. Pharmacodynamic effects of transdermal bupranolol and timolol in vivo: comparison of microemulsions and matrix patches as vehicle. *Methods Find Exp. Clin. Pharmacol. 13*:361 (1991).
42. Y. Xu, H. Xu. Transdermal properties of timolol and factors affecting its permeability in vitro. *Yao Hsueh Hsueh Pao 27*:467 (1992).
43. M. Robinson, K. Parsell, D. Breneman, C. Cruze. Evaluation of the primary skin irritation and allergic contact sensitization potential of transdermal triprolidine. *Fundam. Appl. Toxicol. 17*:103 (1991).
44. P. Gariti, A. Alterman, W. Barber, N. Bedi, G. Luck, A. Cnaan. Cotinine replacement levels for a 21 mg/day transdermal nicotine patch in an outpatient treatment setting. *Drug Alcohol Depend. 54*:111 (1999).
45. J. Y. Fang, S. S. Chen, Y. B. Huang, P. C. Wu, Y. H. Tsai. In vitro study of transdermal nicotine delivery: influence of rate-controlling membranes and adhesives. *Drug Dev. Ind. Pharm. 25*:789 (1999).
46. J. Stapleton, A. Lowin, M. Russell. Prescription of transdermal nicotine patches for smoking cessation in general practice: evaluation of cost-effectiveness. *Lancet 354*:210 (1999).
47. R. Guy, J. Hadgraft. Selection of candidates for transdermal drug delivery. *Transdermal Drug Delivery* (J. Hadgraft, R. Guy, eds.). Marcel Dekker, New York, 1989, p. 59.
48. F. Ocak, I. Agabeyoglu. Development of a membrane-controlled transdermal therapeutic system containing isosorbide dinitrate. *Int. J. Pharm. 180*:177 (1999).
49. S. Chen, R. Lostritto. Maintaining a near zero-order drug delivery from minidose reservoirs: simultaneous drug diffusion and binary vehicle evaporation. *J. Pharm. Sci. 86*: 739 (1997).
50. A. Ji, X. Jiang, X. Wang, H. Zou. Studies on transdermal delivery system of isosorbide dinitrate. *Yao Hsueh Hsueh Pao 27*:858 (1992).
51. T. Peterson, S. Burton, R. Ferber. In vitro permeability of poly (ethylen-vinyl acetate) and microporous polyethylene membranes. *Proceed. Inter. Symp. Control. Rel. Bioact. Mater. 17*:411 (1990).
52. K. A. Wick, S. M. Wick, R. W. Hawkinson, J. L. Holtzman. Adhesion-to-skin performance of a new transdermal nitroglycerin adhesive patch. *Clin. Ther. 11*:417 (1989).
53. S. Venkatraman, R. Gale. Skin adhesives and skin adhesion. 1. Transdermal drug delivery systems. *Biomaterials 19*:1119 (1998).
54. S. D. Roy, M. Gutierrez, G. L. Flynn, G. W. Cleary. Controlled transdermal delivery of fentanyl: characterizations of pressure-sensitive adhesives for matrix patch design. *J. Pharm. Sci. 85*:491 (1996).
55. T. Kokubo, K. Sugibayashi, Y. Morimoto. Interaction between drugs and pressure-sensitive adhesives in transdermal therapeutic systems. *Pharm. Res. 11*:104 (1994).
56. T. Peterson, S. Wick, C. Ko. Design, development, manufacturing, and testing of transdermal drug delivery systems. *Transdermal and Topical Drug Delivery Systems*

(T. Ghosh, W. Pfister, S. Yum, eds.). Interpharm Press, Buffalo Grove, 1999, p. 249.

57. J. Fang, S. Chen, Y. Huang, P. Wu, Y. Tsai. In vitro study of transdermal nicotine delivery: influence of rate-controlling membranes and adhesives. *Drug Dev. Ind. Pharm. 25*: 789 (1999).
58. D. Sanvordeker, E. Grove, J. G. Cooney, D. Plaines, R. Wester. Transdermal nitroglycerin pad. US Patent 4,336,243 (1982).
59. Y. Chien, W., and D. Lambert. Microsealed pharmaceutical delivery device. US Patent 3,946,106 (1976).
60. N. Buyuktimkin, S. Buyuktimkin, J. Howard Rytting. Chemical means of transdermal drug permeation enhancement. *Transdermal and Topical Drug Delivery Systems* (T. Ghosh, W. Pfister, S. Yum, eds.). Interpharm, Buffalo Grove, 1999, p. 357.
61. W. R. Pfister, D. S. Hsieh. Permeation enhancers compatible with transdermal drug delivery systems: part II: system design considerations. *Med. Device Technol. 1*:28 (1990).
62. R. C. Wester, H. I. Maibach, L. Sedik, J. Melendres, S. DiZio, M. Wade. In vitro percutaneous absorption of cadmium from water and soil into human skin. *Fundam. Appl. Toxicol. 19*:1 (1992).
63. K. B. Sloan, S. A. Koch, K. G. Siver, F. P. Flowers. Use of solubility parameters of drug and vehicle to predict flux through skin. *J. Invest. Dermatol. 87*:244 (1986).
64. A. C. Williams, B. W. Barry. Terpenes and the lipid-protein-partitioning theory of skin penetration enhancement. *Pharm. Res. 8*:17 (1991).
65. B. W. Barry, D. Southwell, R. Woodford. Optimization of bioavailability of topical steroids: penetration enhancers under occlusion. *J. Invest. Dermatol. 82*:49 (1984).
66. B. Barry. Mode of action of penetration enhancers in human skin. *J. Contr. Rel. 6*:85 (1987).
67. H. Maibach, R. Feldman. The effect of DMSO on the percutaneous penetration of hydrocortisone and testosterone in man. *Ann. N. Y. Acad. Sci. 141*:423 (1967).
68. C. Hwang, A. Danti. Percutaneous absorption of flufenamic acid in rabbits: effect of dimethyl sulfoxide and various nonionic surface-active agents. *J. Pharm. Sci. 72*:857 (1986).
69. S. Spruance, M. McKeough, J. Cardinal. Dimethyl sulfoxide as a vehicle for topical antiviral chemotherapy. *Ann. N.Y. Acad. Sci. 411*:28 (1983).
70. S. Spruance, M. McKeough, J. Cardinal. Penetration of guinea pig skin by acyclovir in different vehicles and correlation with the efficacy of topical therapy of experimental cutaneous herpes simplex virus infection. *Antimicrob. Agents Chemother. 25*:10 (1984).
71. T. Kurihara Bergstrom, G. Flynn, W. Higuchi. Physicochemical study of percutaneous absorption enhancement by dimethyl sulfoxide: dimethyl sulfoxide mediation of vidarabine (ara-A) permeation of hairless mouse skin. *J. Invest. Dermatol. 89*:274 (1987).
72. W. W. Shen, A. G. Danti, F. N. Bruscato. Effect of nonionic surfactants on percutaneous absorption of salicylic acid and sodium salicylate in the presence of dimethyl sulfoxide. *J. Pharm. Sci. 65*:1780 (1976).
73. H. H. Sharata, R. R. Burnette. Effect of dipolar aprotic permeability enhancers on the basal stratum corneum. *J. Pharm. Sci. 77*:27 (1988).
74. G. Embery, P. H. Dugard. The isolation of dimethyl sulfoxide soluble components from human epidermal preparations: a possible mechanism of action of dimethyl sulfoxide in effecting percutaneous migration phenomena. *J. Invest. Dermatol. 57*:308 (1971).
75. N. H. Creasey, A. C. Allenby, C. Schock. Mechanism of action of accelerants. The effect of cutaneously-applied penetration accelerants on the skin circulation of the rat. *Br. J. Dermatol. 85*:368 (1971).
76. S. Chandrasekaram, P. Campbell, A. Michaels. Effect of dimethyl sulfoxide on drug permeation through human skin. *AIChE. J. 23*:810 (1977).
77. S. M. Al-Saidan, A. B. Selkirk, A. J. Winfield. Effect of dimethylsulfoxide concentration on the permeability of neonatal rat stratum corneum to alkanols. *J. Invest. Dermatol. 89*:426 (1987).
78. M. B. Sulzberger, T. A. Cortese, Jr., L. Fishman, H. S. Wiley, P. S. Peyakovich. Some effects of DMSO on human skin in vivo. *Ann. N.Y. Acad. Sci. 141*:437 (1967).
79. S. I. Subotnick. Dimethyl sulfoxide dehydration of the foot. A case history. *J. Am. Podiatr. Med. Assoc. 75*:263 (1985).
80. G. Volden, H. F. Haugen, S. Skrede. Reversible cellular damage by dimethyl sulfoxide reflected by release of marker enzymes for intracellular fractions. *Arch. Dermatol. Res. 269*:147 (1980).
81. A. C. Williams, B. W. Barry. Skin absorption enhancers. *Crit. Rev. Ther. Drug. Carrier Syst. 9*:305 (1992).
82. D. Sekura, J. Scala. The percutaneous absorption of alkyl methyl sulfoxides. *Adv. Biol. Skin 12*:257 (1972).
83. R. Woodford, B. W. Barry. Optimization of bioavailability of topical steroids: thermodynamic control. *J. Invest. Dermatol. 79*:388 (1982).
84. M. Goodman, B. W. Barry. Action of penetration enhancers on human skin as assessed by the permeation of model drugs 5-fluorouracil and estradiol. I. Infinite dose technique. *J. Invest. Dermatol. 91*:323 (1988).
85. H. K. Choi, G. L. Amidon, G. L. Flynn. Some general influences of n-decylmethyl sulfoxide on the permeation of drugs across hairless mouse skin. *J. Invest. Dermatol. 96*:822 (1991).
86. E. Touitou, L. Abed. Effect of propylene glycol, Azone® and n-decylmethyl sulfoxide on skin on skin permeation kinetics of 5-fluorouracil. *Int. J. Pharm. 27*:89 (1985).
87. B. Akerman, G. Haegerstam, B. G. Pring, R. Sandberg. Penetration enhancers and other factors governing percutaneous local anaesthesia with lidocaine. *Acta Pharmacol. Toxicol. 45*:58 (1979).
88. D. Munro, R. Stoughton. Dimethylacetamide (DMAC) and dimethylformamide (DMF) effect on percutaneous absorption. *Arch. Dermatol. 92*:585 (1965).
89. D. Southwell, B. Barry. Penetration enhancers for human

skin: mode of action of 2-pyrrolidone and dimethylformamide on partition and diffusion of model compounds water, alcohols, and caffeine. *J Invest. Dermatol. 66*:243 (1983).

90. B. W. Barry, R. Woodford. Optimization of bioavailability of topical steroids: non-occluded penetration enhancers under thermodynamic control. *J. Invest. Dermatol. 79*:388 (1982).
91. H. Baker. The effects of dimethylsulfoxide dimethylformamide and dimethylacetamide on the cutaneous barrier to water in human skin. *J. Invest. Dermatol. 50*:283 (1968).
92. B. B. Michniak, J. M. Chapman, K. L. Seyda. Facilitated transport of two model steroids by esters and amides of clofibric acid. *J. Pharm. Sci. 82*:214 (1993).
93. B. Michniak, M. Player, L. Fuhrman, C. Christensen, J. Chapman, J. Sowell. In vitro evaluation of a series of azone analogs as dermal penetration enhancers: III. Acyclic amides. *Int. J. Pharm. 110*:231 (1994).
94. T. Ogiso, M. Iwaki, T. Paku. Effect of various enhancers on transdermal penetration of indomethacin and urea, and relationship between penetration parameters and enhancement factors. *J. Pharm. Sci. 84*:482 (1995).
95. J. J. Windheuser, J. L. Haslam, L. Caldwell, R. D. Shaffer. The use of N,N-diethyl-m-toluamide to enhance dermal and transdermal delivery of drugs. *J. Pharm. Sci. 71*:1211 (1982).
96. B. Mauser, M. Mahjour, M. Fawzi. Cholesterol solubility and skin penetration enhancement. *Proc. Int. Symp. Cont. Rel. Bioact. Mater. 16*:302 (1989).
97. B. Michniak, M. Player, D. Godwin, C. Phillips, J. Sowell. In vitro evaluation of a series of azone analogs as dermal penetration enhancers: IV. Amines. *Int. J. Pharm. 116*:201 (1995).
98. M. Rosten. The treatment of ichthyosis and hyperkeratotic conditions with urea. *Australas. J. Dermatol. 11*:142 (1970).
99. I. Hagemann, E. Proksch. Topical treatment by urea reduces epidermal hyperproliferation and induces differentiation in psoriasis. *Acta. Derm. Venereol. 76*:353 (1996).
100. R. J. Feldmann, H. I. Maibach. Percutaneous penetration of hydrocortisone with urea. *Arch. Dermatol. 109*:58 (1974).
101. G. Swanbeck. A new treatment of ichthyosis and other hyperkeratotic conditions. *Acta. Derm. Venereol. 48*:123 (1968).
102. T. C. Hindson. Urea in the topical treatment of atopic eczema. *Arch. Dermatol. 104*:284 (1971).
103. C. Kim, J. Kim., S. Chi, C. Shim. Effect of fatty acids and urea on the penetration of ketoprofen through rat skin. *Int. J. Pharm. 99*:109 (1993).
104. S. I. Naito, Y. H. Tsai. Percutaneous absorption of indomethacin from ointment bases in rabbits. *Int. J. Pharm. 8*: 263 (1981).
105. A. Williams, B. Barry. Urea analogues in propylene glycol as penetration enhancers in human skin. *Int. J. Pharm. 36*: 43 (1989).
106. O. Wong, J. Huntington, R. Konishi, J. H. Rytting, T. Higuchi. Unsaturated cyclic ureas as new nontoxic biodegradable transdermal penetration enhancers I: synthesis. *J. Pharm. Sci. 77*:967 (1988).
107. O. Wong, N. Tsuzuki, B. Nghiem, J. Kuehnhoff, T. Itoh, K. Masaki, J. Huntington, R. Konishi, J. Rytting, T. Higuchi. Unsaturated cyclic urea as new nontoxic biodegradable transdermal penetration enhancers. II. Evaluation study. *Int. J. Pharm. 52*:191 (1989).
108. H. Sasaki, M. Kojima, Y. Mori, J. Nakamura, J. Shibasaki. Enhancing effect of pyrrolidone derivatives on transdermal drug delivery. I. *Int. J. Pharm. 44*:15 (1988).
109. H. Sasaki, M. Kojima, Y. Mori, J. Nakamura, J. Shibasaki. Enhancing effect of pyrrolidone derivatives on transdermal drug delivery II. Effect of application concentration and pre-treatment of enhancer. *Int. J. Pharm. 60*:177 (1990)
110. H. Sasaki, M. Kojima, J. Nakamura, J. Shibasaki. Enhancing effect of pyrrolidone derivatives on transdermal penetration of phenolsulfonphthalein and indomethacin from aqueous vehicle. *Chem. Pharm. Bull. (Tokyo) 38*:797 (1990).
111. H. Sasaki, M. Kojima, J. Nakamura, J. Shibasaki. Enhancing effect of pyrrolidone derivatives on the transdermal penetration of sulfaguanidine, aminopyrine and Sudan III. *J. Pharmacobiodyn. 13*:200 (1990).
112. H. Sasaki, M. Kojima, Y. Mori, J. Nakamura, J. Shibasaki. Enhancing effect of pyrrolidone derivatives on transdermal penetration of 5-fluorouracil, triamcinolone acetonide, indomethacin, and flurbiprofen. *J. Pharm. Sci. 80*:533 (1991).
113. S. A. Akhter, B. W. Barry. Absorption through human skin of ibuprofen and flurbiprofen; effect of dose variation, deposited drug films, occlusion and the penetration enhancer N-methyl-2-pyrrolidone. *J. Pharm. Pharmacol. 37*:27 (1985).
114. S. L. Bennett, B. W. Barry, R. Woodford. Optimization of bioavailability of topical steroids: non-occluded penetration enhancers under thermodynamic control. *J. Pharm. Pharmacol. 37*:298 (1985).
115. T. Yano, N. Higo, K. Fukuda, M. Tsuji, K. Noda, M. Otagiri. Further evaluation of a new penetration enhancer, HPE-101. *J. Pharm. Pharmacol. 45*:775 (1993).
116. T. Yano, N. Higo, K. Furukawa, M. Tsuji, K. Noda, M. Otagiri. Evaluation of a new penetration enhancer 1-[2-(decylthio)ethyl]azacyclopentan-2-one (HPE-101). *J. Pharmacobiodyn. 15*:527 (1992).
117. K. Uekama, H. Adachi, T. Irie, T. Yano, M. Saita, K. Noda. Improved transdermal delivery of prostaglandin E1 through hairless mouse skin: combined use of carboxymethyl-ethyl-beta-cyclodextrin and penetration enhancers. *J. Pharm. Pharmacol. 44*:119 (1992).
118. H. Adachi, T. Irie, K. Uekama, T. Manako, T. Yano, M. Saita. Inhibitory effect of prostaglandin E1 on laurate-induced peripheral vascular occlusive sequelae in rabbits: optimized topical formulation with beta-cyclodextrin derivative and penetration enhancer HPE-101. *Acta. Derm. Venereol. 44*:1033 (1992).

119. V. Rajadhyaksha, M. Viejo, J. V. Peck. Method of synthesis of 1-dodecylazacycloheptan-2-one. *US Patent 4, 422, 970* (1978).
120. K. Brian, K. Walters. Molecular modeling of skin permeation enhancement by chemical agents. *Pharmaceutical Skin Penetration Enhancement* (K. Walter, J. Hadgraft, ed.). Marcel Dekker, New York, 1993, p. 389.
121. F. Bonina, L. Montenegro. Penetration enhancer effects on in vitro percutaneous absorption of heparin sodium salt. *Int. J. Pharm. 82*:171 (1994).
122. S. Buyuktimkin, N. Buyuktimkin, J. H. Rytting. Synthesis and enhancing effect of dodecyl 2-(N,N-dimethylamino)-propionate on the transepidermal delivery of indomethacin, clonidine, and hydrocortisone. *Pharm. Res. 10*:1632 (1993).
123. N. Higo, R. S. Hinz, D. T. Lau, L. Z. Benet, R. H. Guy. Cutaneous metabolism of nitroglycerin in vitro. II. Effects of skin condition and penetration enhancement. *Pharm. Res. 9*:303 (1992).
124. M. I. Afouna, S. C. Mehta, A. H. Ghanem, W. I. Higuchi, E. R. Kern, E. De Clercq, H. H. El-Shattawy. Assessment of correlation between skin target site free drug concentration and the in vivo topical antiviral efficacy in hairless mice for (E)-5-(2-bromovinyl)-2′-deoxyuridine and acyclovir formulations. *J. Pharm. Sci. 87*:917 (1998).
125. B. W. Barry, S. L. Bennett. Effect of penetration enhancers on the permeation of mannitol, hydrocortisone and progesterone through human skin. *J. Pharm. Pharmacol. 39*:535 (1987).
126. D. Mirejovsky, H. Takruri. Dermal penetration enhancement profile of hexamethylenelauramide and its homologues: in vitro versus in vivo behavior of enhancers in the penetration of hydrocortisone. *J. Pharm. Sci. 75*:1089 (1986).
127. H. Okamoto, M. Hashida, H. Sezaki. Structure-activity relationship of 1-alkyl- or 1-alkenylazacycloalkanone derivatives as percutaneous penetration enhancers. *J. Pharm. Sci. 77*:418 (1988).
128. H. Okamoto, M. Hashida, H. Sezaki. Effect of 1-alkyl- or 1-alkenylazacycloalkanone derivatives on the penetration of drugs with different lipophilicities through guinea pig skin. *J. Pharm. Sci. 80*:39 (1991).
129. B. Michniak, M. Player, L Fuhrman. C. Christensen, J. Chapman, J. Sowell. In vitro evaluation of a series of azone analogs as dermal penetration enhancers. II. (Thio) amides. *Int. J. Pharm. 94*:203 (1993).
130. B. Michniak, M. Player, J. Chapman, J. Sowell. In vitro evaluation of a series of azone analogs as dermal penetration enhancers. *Int. J. Pharm. 91*:85 (1993).
131. N. Kim, M. El-Khalili, M. M. Henary, L. Strekowski, B. B. Michniak. Percutaneous penetration enhancement activity of aromatic S, S-dimethyliminosulfuranes. *Int. J. Pharm. 187*:219 (1999).
132. L. Strekowski, M. Henary, N. Kim, B. B. Michniak. N-(4-bromobenzoyl)-S,S-dimethyliminosulfurane, a potent dermal penetration enhancer. *Bioorg. Med. Chem. Lett. 9*: 1033 (1999).
133. P. Ashton, J. Hadgraft, K. A. Walters. Effects of surfactants in percutaneous absorption. *Pharm. Acta. Helv. 61*: 228 (1986).
134. L. Gershbein. Percutaneous toxicity of thioglycolate mixtures in rabbits. *J. Pharm. Sci. 68*:1230 (1979).
135. L. D. Rhein, C. R. Robbins, K. Fernee, R. Cantore. Surfactant structure effects on swelling of isolated human stratum corneum. *J. Soc. Cosmet. Chem. 37*:125 (1986).
136. J. Borras-Blasco, A. Lopez, M. Morant, O. Diez-Sales, M. Herraez-Dominguez. Influence of sodium lauryl sulfphate on in vitro percutaneous absorption of compounds with different lipophilicity. *Eur. J. Pharm. Sci. 5*:15 (1997).
137. G. P. Kushla, J. L. Zatz. Correlation of water and lidocaine flux enhancement by cationic surfactants in vitro. *J. Pharm. Sci. 80*:1079 (1991).
138. D. Bommannan, R. O. Potts, R. Guy. Examination of the effects of ethanol on human stratum corneum in vivo using infrared spectroscopy. *J. Control. Rel. 16*:299 (1991).
139. P. Liu, T. Kurihara-Bergstrom, W. R. Good. Cotransport of estradiol and ethanol through human skin in vitro: understanding the permeant/enhancer flux relationship. *Pharm. Res. 8*:938 (1991).
140. B. Berner, J. H. Otte, G. C. Mazzenga, R. J. Steffens, C. D. Ebert. Ethanol: water mutually enhanced transdermal therapeutic system. I: Nitroglycerin solution properties and membrane transport. *J. Pharm. Sci. 78*:314 (1989).
141. B. Berner, G. C. Mazzenga, J. H. Otte, R. J. Steffens, R. H. Juang, C. D. Ebert. Ethanol: water mutually enhanced transdermal therapeutic system II: skin permeation of ethanol and nitroglycerin. *J. Pharm. Sci. 78*:402 (1989).
142. S. D. Roy, E. Roos, K. Sharma. Transdermal delivery of buprenorphine through cadaver skin. *J. Pharm. Sci. 83*: 126 (1994).
143. K. Takahashi, S. Tamagawa, T. Katagi, H. Yoshitomi, A. Kamada, J. H. Rytting, T. Nishihata, N. Mizuno. In vitro transport of sodium diclofenac across rat abdominal skin: effect of selection of oleaginous component and the addition of alcohols to the vehicle. *Chem. Pharm. Bull.* (Tokyo) *39*:154 (1991).
144. H. O. Ho, F. C. Huang, T. D. Sokoloski, M. T. Sheu. The influence of cosolvents on the in-vitro percutaneous penetration of diclofenac sodium from a gel system. *J. Pharm. Pharmacol. 46*:636 (1994).
145. C. K. Lee, T. Uchida, K. Kitagawa, A. Yagi, N. S. Kim, S. Goto. Effect of hydrophilic and lipophilic vehicles on skin permeation of tegafur, alclofenac and ibuprofen with or without permeation enhancers. *Biol. Pharm. Bull. 16*: 1264 (1993).
146. D. Friend, P. Catz. J. Heller, J. Reid, R. Baker. Transdermal delivery of levonorgestrel. I. Alkanols as permeation enhancers in vitro. *J. Control. Rel. 7*:243 (1988).
147. A. C. Williams, S. R. Shatri, B. W. Barry. Transdermal permeation modulation by cyclodextrins: a mechanistic study. *Pharm. Dev. Technol. 3*:283 (1998).
148. U. Vollmer, B. W. Muller, J. Peeters, J. Mesens, B. Wilffert, T. Peters. A study of the percutaneous absorption-

enhancing effects of cyclodextrin derivatives in rats. *J. Pharm. Pharmacol. 46*:19 (1994).
149. H. Okabe, Y. Obata, K. Takayama, T. Nagai. Percutaneous absorption enhancing effect and skin irritation of monocyclic monoterpenes. *Drug Design Del. 6*:229 (1990).
150. D. A. Godwin, B. B. Michniak. Influence of drug lipophilicity on terpenes as transdermal penetration enhancers. *Drug Dev. Ind. Pharm. 25*:905 (1999).
151. H. R. Moghimi, A. C. Williams, B. W. Barry. Enhancement by terpenes of 5-fluorouracil permeation through the stratum corneum: model solvent approach. *J. Pharm. Pharmacol. 50*:955 (1998).
152. S. Kitagawa, H. Li, S. Sato. Skin permeation of parabens in excised guinea pig dorsal skin, its modification by penetration enhancers and their relationship with n-octanol/water partition coefficients. *Chem. Pharm. Bull.* (Tokyo) *45*:1354 (1997).
153. M. Hori, S. Satoh, H. I. Maibach, R. H. Guy. Enhancement of popranolol hydrochloride and diazepam skin absorption in vitro: effect of enhancer lipophilicity. *J. Pharm. Sci. 80*: 32 (1991).
154. T. Nagai, H. Okabe, A. Ogura, K. Takayama. Effect of limonene and related compounds on the percutaneous absorption of indomethacin. *Proc. Int. Symp. Controlled Release Bioactive Mater. 16*:181 (1989).
155. P. Cornwell, B. Barry. Sesquiterpene components of volatile oils as skin penetration enhancers for the hydrophilic permeant 5-flurouracil. *J. Pharm. Pharmacol. 46*:261 (1991).
156. A. C. Williams, B. W. Barry. Permeation, FTIR and DSC investigation of terpene penetration enhancers in human skin. *J. Pharm. Pharmacol. 41* (Suppl.):12P (1989).
157. B. J. Aungst. Structure/effect studies of fatty acid isomers as skin penetration enhancers and skin irritants. *J. Pharm. Sci. 6*:244 (1989).
158. T. Ogiso, M. Shintani. Mechanism for the enhancement effect of fatty acids on the percutaneous absorption of popranolol. *J. Pharm. Sci. 79*:1065 (1990).
159. T. Aoyagi, O. Terashima, N. Suzuki, K. Matui, Y. Nagase. Polymerization of benzalkonium chloride-type monomer and application to percutaneous drug absorption enhancer. *J. Cont. Rel. 13*:63 (1990).
160. T. Aoyagi, T. Nakamura, Y. Yabuchi, Y. Nagase. Novel silicones for transdermal therapeutic system III. Preparation of pyridinio or ammonio terminated polydimethylsiloxanes and the evaluation as transdermal penetration enhancers. *Polymer J. 24*:545 (1992).
161. T. Aoyagi, T. Nakamura, Y. Nagase. Novel silicones for transdermal therapeutic system II. Preparation of oligodimethylsiloxanes containing diethylphosphoramidate group at the chain end and the evaluation as transdermal penetration enhancers. *Polymer J. 24*:375 (1992).
162. T. Aoyagi, T. Nakamura, Y. Yabuchi, Y. Nagas. A novel method of preparation of carboxy-terminated polydimethylsiloxane and its enhancing effect on transdermal drug penetration. *Macromolcule. Chemistry. Rapid Communications 5*:441 (1992).
163. T. Aoyagi, Y. Nagase. Silicone based polymers. *Percutaneous Penetration Enhancers* (E. Smith, H. Maibach, eds.). CRC Press, New York, 1995, p. 267.
164. D. Levy, J. Kost, Y. Meshulam, R. Langer. Effect of ultrasound on transdermal drug delivery to rats and guinea pigs. *J. Clin. Invest. 83*:2074 (1989).
165. S. Mitragotri, D. A. Edwards, D. Blankschtein, R. Langer. A mechanistic study of ultrasonically-enhanced transdermal drug delivery. *J. Pharm. Sci. 84*:697 (1995).
166. D. Bommannan, H. Okuyama, P. Stauffer, R. H. Guy. Sonophoresis. I. The use of high-frequency ultrasound to enhance transdermal drug delivery. *Pharm. Res. 9*:559 (1992).
167. S. Mitragotri, D. Blankschtein, R. Langer. Ultrasound-mediated transdermal protein delivery. *Science 269*:850 (1995).
168. R. Brucks, M. Nanavaty, D. Jung, F. Siegel. The effect of ultrasound on the in vitro penetration of ibuprofen through human epidermis. *Pharm. Res. 6*:697 (1989).
169. H. A. Benson, J. C. McElnay, R. Harland, J. Hadgraft. Influence of ultrasound on the percutaneous absorption of nicotinate esters. *Pharm. Res. 8*:204 (1991).
170. P. Tyle, P. Agrawala. Drug delivery by phonophoresis. *Pharm. Res. 6*:355 (1989).
171. S. Miyazaki, O. Mizuoka, M. Takada. External control of drug release and penetration: enhancement of the transdermal absorption of indomethacin by ultrasound irradiation. *J. Pharm. Pharmacol. 43*:115 (1990).
172. A. K. Banga, S. Bose, T. K. Ghosh. Iontophoresis and electroporation: comparisons and contrasts. *Int. J. Pharm. 179*:1 (1999).
173. M. R. Prausnitz. Reversible skin permeabilization for transdermal delivery of macromolecules. *Crit. Rev. Ther. Drug Carrier Syst. 14*:455 (1997).
174. M. V. Préat. Augmentation of transdermal drug penetration. Methods Find Exp. Clin. Pharmacol. *Bull. Mem. Acad. R. Med. Belg. 153*:295 (1998).
175. A. K. Banga. Transdermal and topical delivery of therapeutic peptides and proteins. *Therapeutic Peptides and Proteins* (A. K. Banga, ed.). Technomic, Lancaster, 1995, p. 245.
176. H. A. Abramson, M. H. Gorin. Skin reactions. IX. The electrophoretic demonstration of the patent pores of the living human skin: its relation to the charge of the skin. *J. Phys. Chem. 44*:1094 (1940).
177. R. R. Burnette, B. Ongpipattanakul. Characterization of pore transport properties and tissue alteration of excised human skin during iontophoresis. *J. Pharm. Sci. 77*:132 (1988).
178. S. Grimmnes. Pathways of ionic flow through human skin in vivo. *Acta Dermato-Venereol.* (Stockh) *198*:93 (1984).
179. N. A. Monteiro-Riviere, A. O. Inman, J. E. Riviere. Identification of the pathway of iontophoretic drug delivery: light and ultrastructural studies using mercuric chloride in pigs. *Pharm. Res. 11*:251 (1994).
180. O. Siddiqui, M. S. Roberts, A. E. Polack. Iontophoretic transport of weak electrolytes through excised human stratum corneum. *J. Pharm. Pharmacol. 41*:430 (1989).

181. P. Singh, S. Boniello, P. Liu, S. Dinh. Iontophoretic transdermal delivery of methylphenidate hydrochloride. *Pharm. Res. 14*(Suppl):309 (1997).
182. T. Koizumi, M. Kakemi, K. Katayama, H. Inada, K. Sudeji, M. Kwasaki. Transfer of diclofenac sodium across excised guinea pig skin on high-frequency pulse iontophoresis. *Chem. Pharm. Bull. 38*:1022 (1990).
183. N. H. Yoshida, M. S. Roberts. Structure-transport relationships in transdermal iontophoresis. *Adv. Drug Del. Rev. 9*:239 (1992).
184. S. C. Jacobsen, R. D. Luntz. Ion mobility limiting iontophoretic bioelectrode. US Patent 4,416,274 (1983).
185. M. R. Prausnitz, E. R. Edelman, J. A. Gimm, R. Langer, J. C. Weaver. Transdermal delivery of heparin by skin electroporation. *Biotechnology 13*:1205 (1995).
186. R. Vanbever, U. F. Pliquett, V. Préat, J. C. Weaver. Comparison of the effects of short, high-voltage and long, medium-voltage pulses on skin electrical and transport properties. *J. Controlled Release 60*:35 (1999).
187. H. Shaefer, T. E. Redelmeier. *Skin Barrier Principles of Percutaneous Absorption*. Basel, Karger, 1996.
188. R. L. Bronaugh, H. I. Maibach. *Percutaneous Absorption Drugs-Cosmetics-Mechanisms-Methodology*. New York, Marcel Dekker, 1999.
189. M. A. Lampe, A. L. Burlingame, J. Whitney, M. L. Williams, B. E. Brown, E. Rotiman, P. M. Elias. Human stratum corneum lipids: characterization and regional variations. *J. Lipid Res. 24*:120 (1983).
190. P. W. Wertz, M. Kremer, C. A. Squier. Comparison of lipids from epidermal and palatal stratum corneum. *J. Invest. Dermatol. 98*:375 (1992).
191. M. A. Lampe, M. L. Williams, P. M. Elias. Human epidermal lipids: characterization and modulations during differentiation. *J. Lipid Res. 24*:131 (1983).
192. B. C. Melnik, J. Hollmann, E. Erler, B. Verhoeven, G. Plewig. Microanalytical screening of all major stratum corneum lipids by sequential high performance thin-layer chromatography. *J. Invest. Dermatol. 92*:231 (1989).
193. P. W. Wertz, D. T. Downing. Stratum corneum: biological and biochemical considerations. *Transdermal Drug Delivery* (J. Hadgraft, R. Guy, eds.). Marcel Dekker, New York, 1989, p. 1.
194. W. Pfister. Transdermal and dermal therapeutic systems: current status. *Transdermal and Topical Drug Delivery Systems* (T. Ghosh, W. Pfister, S. Yum, eds.). Interpharm Press, Buffalo Grove, 1999, p. 47.
195. G. W. Cleary. The first two decades of transdermal drug delivery systems and a peek into the 1990s. *Dermal and Transdermal Drug Delivery* (R. Gurney, A. Teubner, eds.). Wissenschaftliche Verlagsgesellschaft, Stuttgart, 1993, p. 13.
196. G. W. Cleary. Transdermal Drug Delivery. *Cosmetics and Toiletries 106*:97 (1991).

40

Drug Delivery via Mucosal Routes

Nimit Worakul
University of Wisconsin, Madison, Wisconsin and Prince of Songkla University, Haad Yai, Thailand

Joseph R. Robinson
University of Wisconsin, Madison, Wisconsin

I. INTRODUCTION

The route of drug delivery for a particular drug is of great importance for its onset, intensity, and duration of pharmacological action. Although direct placement into the bloodstream by intravenous injection can give the fastest onset, and control over intensity and duration, it is clearly not convenient nor usually desirable from a patient point of view; thus alternative routes of drug delivery become attractive. Historically, the most common alternative route of drug delivery is the oral route, presumably because of its convenience to the patient. However, this is a relatively hostile portal of entry for a drug. In addition, collecting intestinal blood vessels go directly to the liver and not to the general circulation, and thus orally absorbed drugs are subject to first-pass metabolism. For these and other reasons there has been increasing interest in other noninjectable routes of drug delivery including nasal, pulmonary, buccal, sublingual, transdermal, rectal, and vaginal. All of these portals of entry into the body, with the exception of transdermal and vaginal, are referred to as mucosal routes of delivery and are characterized by tissue with high blood flow and possessing numerous barriers to drug absorption (Fig. 1).

To design a drug delivery system properly for these mucosal routes of drug delivery, it is necessary to:

1. Have an understanding of the anatomy and physiology of these routes
2. Have a good knowledge of the barriers to drug absorption and strategies to deal with these barriers
3. Be able to integrate a drug into a delivery system that is user-friendly while permitting an optimum performance of the drug

This chapter will first describe the characteristics and permeability properties of selected mucous membranes and then review the relevant anatomy and physiology of certain routes of drug delivery. This will be followed by examples of drug delivery systems for these mucosal routes.

II. CHARACTERISTICS OF MUCOUS MEMBRANE (1,2)

Surface epithelium and its underlying connective tissue are regarded as a functional unit called a ''membrane.'' There are two kinds of epithelial membranes, serous and mucous, present in the body. These membranes are kept moist by secretions from specialized epithelial cells and fluid extruded from underlying capillaries. Mucous membranes, also called *mucosae*, line the cavities of most organs that open to the outside of the body, namely, the alimentary canal, the respiratory tract, and the genitourinary tract. The multiple functions of mucosa are absorption, secretion, lubrication, entrapment, and antibacterial. This tissue can be highly specialized, i.e., the absorptive structures, lining the

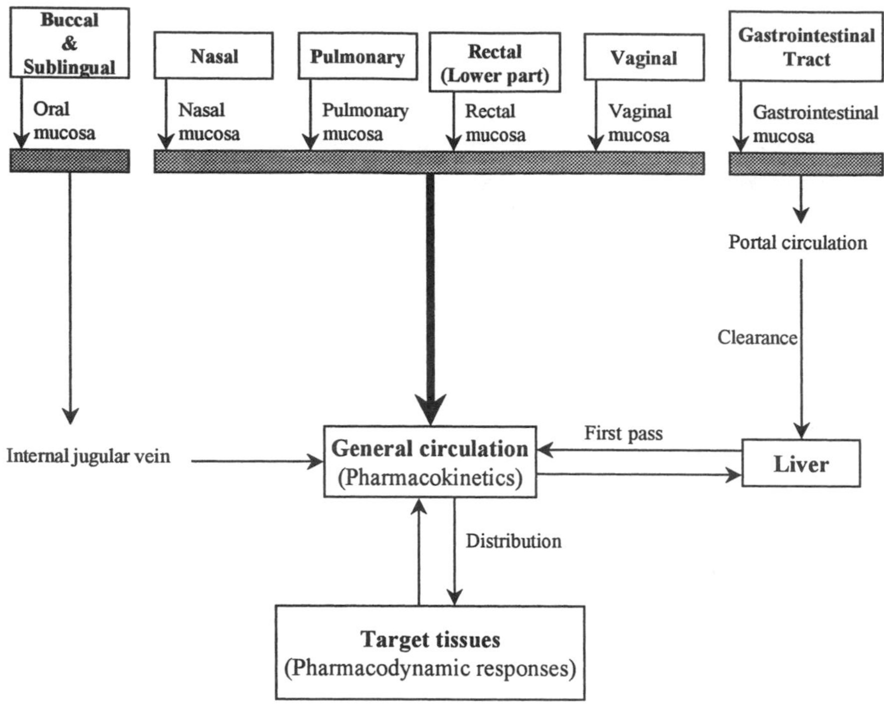

Figure 40.1 Drug delivery via mucosal routes.

digestive tract, or secretory. The secretions produced may be of a rather watery consistency, with an abundance of active enzymes, as in serous cells, or more viscous, where lubrication is the principal function, as in mucous cells. In the respiratory system, mucus traps foreign particles from inspired air and may also have bacteriocidal properties. In the stomach and intestines, the enzyme-rich digestive juices of the mucosae aid in the breakdown of food into simple nutrients available to the cells. A typical mucosa consists of four layers, namely, surface epithelium, basement membrane, lamina propria, and muscularis mucosa.

A. Surface Epithelium

The surface epithelium is a moist inner layer, which may be formed by many types of epithelium. These cells vary in morphology and function depending on their location. For example, the mucosa of the respiratory system has many ciliated cells but sparse microvilli. In contrast, interspersed among the simple columnar cells lining the digestive tract are many globlet cells whose mucoid secretion protects the intestinal lining cells and aids in movement of the intestinal contents. Also the apical ends of the columnar cells are completely covered by closely packed microvilli that greatly increase the absorptive area of the gut.

Classification of the various epithelia of the body is based on the arrangement and shape of the cells. Thus the terminology is related to structure and not function. Epithelium can be described as *simple*, consisting of one cell layer, or *stratified*, consisting of two or more cell layers. The individual cell may be described as squamous, where the width and depth of the cell are greater than its height; cuboidal, where the width, depth, and height are approximately the same; or columnar, where the height of the cell appreciably exceeds the width and depth. Two special categories are typically included in this classification, i.e., pseudostratified and transitional epithelium. A pseudostratified epithelium is actually a simple epithelium because all of the cells rest on the basement membrane. However, some of the cells do not reach the free surface at the lumen site, so the appearance of this epithelium looks similar to stratified. Transition epithelium is a special name applied to the epithelium lining the pelvis of the kidney, ureters, urinary bladder, and part of the urethra. The morphology of the epithelium and its resident cells can also be related to its functional activity, and thus epithelia involved in secretion or absorption are typically simple. Examples of epithelial cell types in different areas are (3)

1. *The simple squamous epithelium* forms a thin layer and lines most blood vessels.
2. *The simple columnar epithelium* is a single layer of columnar cells, which is found in areas such as the stomach and small intestine.
3. *The stratified epithelium* is several cells thick and is found in areas that have to withstand large amounts of wear and tear, i.e., the inside of the mouth and esophagus.
4. *The pseudostratified columnar epithelium* typically lines the respiratory passageways.

B. Basement Membrane

The basement membrane is composed of two layers. The inner layer located next to the base of the epithelial cells is called the ''basal lamina,'' a homogenous, electron-dense, 50–100 nm thick layer. The outer layer is the ''reticular lamina,'' a tangle of fine reticular fibers interposed between the basal lamina and the underlying loose connective tissue layer, the lamina propria. The basement membrane acts as a two-way filter or diffusion barrier to protect the body from the entrance of harmful macromolecular substances into the bloodstream. It also provides an elastic support for protection against trauma from hard and rough materials as they pass through the lumen of an organ.

C. Lamina Propria

The lamina propria is the thickest layer, consisting of loose connective tissue in which lymphoid nodules are often located. It has an abundant supply of blood and lymph capillaries, which lie close to the basal surface of epithelial cells. Therefore the exchange of nutrients and gases across the capillary wall and the cell membrane is greatly facilitated by their close proximity. It also has a large population of lymphocytes and plasma cells that are important in the body's defense against pathogenic bacteria, viruses, and other foreign substances. This layer can provide an effective shock absorber for the passage of bulky materials, especially through the gastrointestinal tract, because of its loose and spongy structure.

D. Muscularis Mucosa

The muscularis mucosa is the outermost component of the mucosa. It is a layer of muscle surrounding the mucosa and usually consists of a few smooth muscle fibers, which may be oriented in the circular or longitudinal directions. It modifies the shape of the mucosa into folds or rugae, to increase surface area, and may also contribute slightly to the peristaltic action of the gut.

III. PERMEABILITY OF THE MUCOUS MEMBRANE

A. Mechanisms

The epithelia of different mucosae are well characterized. Each mucosa varies in cell type, thickness, and function. However the basic drug transport mechanisms are the same (4–8). The principal routes of a substrate flux from the mucosal site into the bloodstream or lymph are shown in Fig. 2. The substrate can transport through the mucosa by passive diffusion, a process that begins with drug diffusion through mucus or unstirred water layer, followed by two major routes, transcellular (intracellular) and paracellular (intercellular) pathways. After that, the substrate is transported into the lamina propria, which is richly endowed

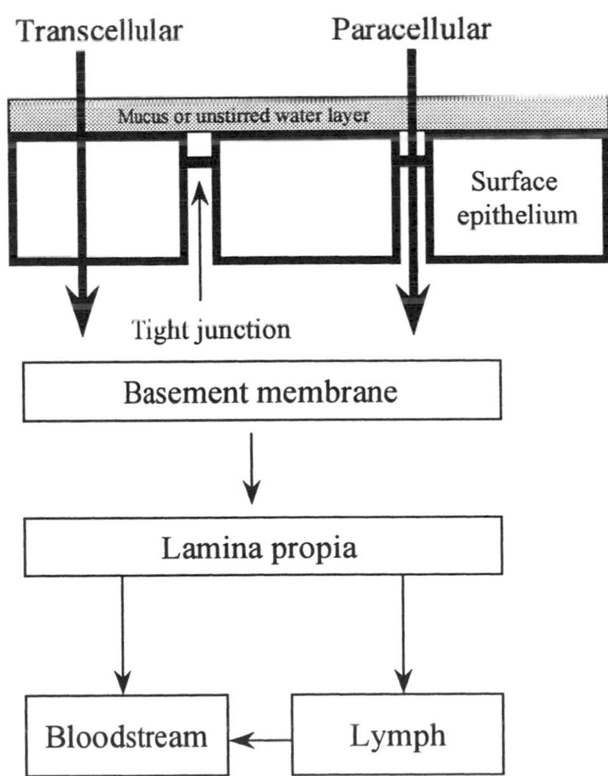

Figure 40.2 The principal pathways of substrate flux from the mucosal site into bloodstream or lymph.

with capillaries, and where the substrate can readily be taken. Substrates which cannot pass the capillary wall end up in the lymph and are carried through the mesenteric lymph vessels into the thoracic duct, which empties into the bloodstream (9). Therefore the two major pathways that could be the rate-limiting steps in transmucosal permeation of substances across the mucosa, namely the transcellular and paracellular routes, will be discussed.

1. Transcellular Pathway

The transcellular pathway involves permeation across the apical cell membrane, the intracellular space, and the basolateral membrane by passive transport (partition and diffusion) or by active transport (carrier-mediated diffusion, endocytosis, and transcytosis) (10–11). In general, passive transport is the movement of a solute along its concentration gradient, while active transport is mediated by membrane transport proteins and energy sources. Therefore, in active transport, the solutes can be transported against a concentration gradient. This pathway is important for the absorption of lipophilic molecules (passive transport) or molecules capable of specific recognition of a membrane site (active transport).

2. Paracellular Pathway

The paracellular pathway involves passage between the cells through the intercellular lipid material of the intercellular space. The two components of the paracellular pathway are the tight junction (zonula occludens) and the underlying intracellular space (10,12). Actually, the tight junctions are dynamic structures that consist of plasma membranes brought into extremely close apposition, but not fused, so as to occlude the extracellular space. They can assemble or disassemble in response to various physiological stimuli. For instance, lowering the extracellular Ca^{2+} concentration induces the opening of tight junctions (13–16). This pathway is important for the absorption of hydrophilic compounds, which cannot transport by an active transport process.

B. Barriers

A problem that commonly limits the development of suitable drug delivery systems for mucosal routes is poor drug bioavailability. This occurs because of the presence of barriers. Two main barrier classifications have been made in drug transport across mucosa, namely physicochemical and metabolic (17). An understanding of these barriers is essential to formulating ways to overcome these problems.

1. Physicochemical Barrier

The transcellular route consists primarily of two main barriers, the apical and basolateral cell membranes. These are barriers for hydrophilic drugs, which cannot easily partition into the cell membrane. The paracellular pathway can be considered as either a single barrier, the tight junction, or as two barriers, the tight junction and the intercellular space (7). Each of these epithelial barriers has two properties that can be measured experimentally. First is the general permeability or magnitude of the barrier, which is largely controlled by the tight junction and may be simply quantified by electrical resistance. Epithelia can be classified as leaky or tight by the magnitude of these values. Leaky epithelia are those with a tissue resistance less than 1,000 ohm cm^2 (7). The second property is the permselectivity of the barrier, which is a quantitative measure of the ability of the epithelia to discriminate or show preference for the transport of molecules of different charge type. The physicochemical properties of the different mucosae are compared in Table 1. Note the wide variation in magnitude of barrier permeability among different epithelia. The rank order of the intrinsic membrane permeability is intestinal $\approx$ nasal $\approx$ bronchial > vaginal > rectal > buccal. With respect to the permselectivity barrier, all epithelia are selective to positive charged solutes, the magnitude of selectivity being comparable in most epithelia (18).

Several studies suggest that the mucus layer can also be a rate-limiting barrier to drug absorption (19–21). For example, results reported by Wikman and coworkers (20), who used the mucus layer producing human intestinal goblet cell line (HT29-H), indicated that the mucus layer is a significant barrier to the absorption of testosterone. Nimmerfall and Rosenthaler (19) showed that the absorption of a series of alkaloids correlated with their diffusion coefficients in a mucus layer. Therefore, the reduction in permeability coefficients through mucosa due to mucus needs to be considered for all compounds that must traverse any mucus layer prior to absorption or action.

2. Metabolic Barrier

The metabolism of drugs and other foreign substances can be considered to be part of the body's defense mechanisms. Drug metabolism takes place in many parts of the body including the liver, bloodstream, spleen, kidney, brain, lumen of the GI tract (bacteria), wall of the GI tract, and many other tissues. However, the major organ in which drug metabolism takes place is the liver (22). During the

Table 40.1 Permeability and Permselectivity of Various Mucosae in Rabbit

Mucosa	R(ohm · cm^2)[a]	pd(-mV)[b]	DP(-mV)[c]
Nasal	261 (55)[d]	13.8 (4.4)	10.22 (2.43)
Pulmonary (bronchial)	266 (97)	12.4 (3.7)	10.25 (1.96)
Buccal	1,803 (175)	28.4 (12.1)	15.90 (3.21)
Duodenal	211 (91)	3.9 (1.8)	9.08 (1.36)
Jejunal	224 (104)	3.8 (2.2)	9.10 (1.71)
Ileal	266 (95)	4.4 (1.5)	10.20 (1.76)
Colon	288 (72)	5.3 (2.6)	10.27 (2.12)
Rectal	406 (70)	18.1 (7.6)	7.95 (1.06)
Vaginal	372 (85)	16.5 (6.8)	11.39 (2.44)

a = Steady-state electrical resistance.
b = Steady-state transepithelial potential difference (epithelial with respect to serosal).
c = Steady-state diffusion potential.
d = Standard deviation.
Source: Ref. 18.

transmucosal permeation process, drugs may be subjected to metabolism by several enzymes in the lumen of the route of administration, the mucosal surface environment, or in the mucosal membrane. The presence of peptide and protein drugs makes the metabolic barrier more important, because this barrier plays a critical role in the delivery of these drugs via the mucosal routes. In this case, the barrier is composed of exopeptidases and endopeptidases, which are able to cleave peptides and proteins at their N and C termini, and at an internal peptide bond, respectively (23). The exopeptidases are grouped into amino and carboxy peptidases according to their substrate specificity. The endopeptidases are classified into serine, cysteine, and aspartic proteinases according to mechanism rather than substrate specificity. These enzymes can inactivate peptide and protein molecules during absorption, presenting a presystemic first pass effect (17). It should be also noted that peptides and proteins are susceptible to degradation not only at the site of application but also in the blood, liver, and endothelial membrane (24,25). The role of the proteolytic barrier in limiting peptide and protein absorption appears to be dependent on the specific mucosa and the nature of the peptide and protein drugs. For example, comparative studies on peptidase activity against enkephalins (26), insulin, and proinsulin (27) in the absorptive mucosae (nasal, buccal, vaginal, and rectal) of the rabbit demonstrated that the vaginal mucosa is consistently low and the rectal mucosa is consistently high in proteolytic activity against all three peptides, but the rank order of proteolytic activity between nasal and buccal mucosae varies with the peptides.

C. Penetration Enhancers

Penetration enhancers are compounds that can increase the absorption of coadministered drugs by increasing mucosal permeability and hence absorption of poorly permeable drugs (28). Examples of penetration enhancers and their possible mechanisms are shown in Table 2. The mechanism of penetration enhancers can be divided (8,28,58) as follows:

1. Altering the Permeability of the Membranes

a. Action on the Mucus Layer

The mucus layer covering the surface of the mucosal membrane can be seen as an unstirred layer, acting as a barrier to the diffusion of drug molecules. Ionic surfactants such as sodium deoxycholate, sodium taurodeoxycholate, and sodium glycocholate are able to reduce mucus viscosity and elasticity by disrupting the associated structure of the glycoproteins (40). Consequently, the barrier function of the mucus layer is reduced, and the permeability of drugs through this layer is increased.

b. Action on Tight Junction (Enhanced Paracellular Transport)

At a tight junction, the interacting membranes are so closely apposed that there is no intercellular space and the membranes are within 2 Å of each other. Therefore a tight junction is one of the major barriers to paracellular transport of macromolecules and polar compounds. These junc-

Table 40.2 Classification of Penetration Enhancers and Their Mechanisms

Penetration enhancers	Mechanisms	Ref.
Surfactants	Extraction of membrane proteins or lipids	29–32
ionic	Membrane fluidization	
Sodium lauryl sulfate		
Sodium laurate		
nonionic		
Tween 80		
Polysorbates		
Bile salts and derivatives	Extraction of membrane proteins or lipids	8, 29, 31–41
Sodium glycocholate	Membrane fluidization	
Sodium deoxycholate	Reverse micellization in membrane creating aqueous channels	
Sodium taurocholate		
Sodium taurodihydrofusidate	Disrupting the associated structure of glycoproteins in mucus layer	
Sodium glycodihydrofusidate		
Fatty acids and derivatives	Disrupting lipid packaging	8, 29, 32, 35, 37, 41–46
Oleic acid	Decreases nonprotein sulfhydryls in membrane	
Mono(di)glycerides		
Sodium caprate	Alteration of the tight junctions	
Lauric acid		
Acylcarnitines		
Chelating and complexing agents	Chelates calcium at tight junctions, increasing paracellular transport	17, 29, 31–32, 35–37, 47–53
EDTA		
Citric acid		
Cyclodextrins		
Salicylate and analogues	Decreases nonprotein sulfhydryls in membrane	8, 29, 32, 35–37, 54–55
Sodium salicylate	Interacts with membrane proteins	
3-methoxysalicylate	Calcium chelating	
5-methoxysalicylate		
Others		
Glycyrrhetinic acid	Inhibits leucine aminopeptidase	57
Azone (1-dodecylazacycloheptan-2-one)	Increases the fluidity of lipid layers and lower lipid transition temperature	8, 44, 56

tions can be disrupted by regulating the controlling physiological factors including concentration of AMP (59), intercellular calcium concentrations (14), and transient mucosal osmotic pressure (60). Hence, treatment by chelator (EDTA), cytochalasin, or increasing glucose or amino acids can cause a temporary change in the integrity of the tight junction and increase paracellular transport. The increasing water influx through the intercellular space, which is Na dependent, is another factor that involves enhancement in paracellular absorption (61). An increase in water influx may affect drug absorption by one of the following mechanisms: (1) increase in concentration gradient for penetration, (2) increase in solvent drag, and (3) increase in blood flow at the absorption site (62).

c. Action on the Membrane Structure (Enhanced Transcellular Transport)

Major components of the biological cell membrane include lipids (cholesterol, phospholipid, and sphingolipid), proteins, and carbohydrates attached to the proteins or lipids as glycoproteins or glycolipids. Enhancers that increase transcellular permeability to drugs probably do so by affecting membrane lipids and proteins. Some fatty acids and their derivatives have been found to act primarily on the phospholipid component of membranes, thereby creating disorder and resulting in increasing permeability (63). However, some fatty acids act on the protein component in membranes (64).

2. Altering the Physicochemical Properties of the Drugs

Solubility and dissolution rate of a drug are very important properties related to absorption. A penetration enhancer can promote absorption by increasing drug solubility and/or dissolution rate (65). Penetration enhancers such as sodium salicylate and bile salts are thought to increase absorption by this effect (66).

3. Inhibition of Enzymatic Activity

Protein and peptide drugs are subject to degradation by proteolytic enzymes during passage through mucosal membrane (25). A significant inhibition of this degradation can promote permeability by increasing the effective concentration of protein and peptide drugs at the absorption site. This may result in an increase in the absorption and bioavaibility of the drugs. LueBen et al. (67) found that two bioadhesive poly(acrylates), polycarbophil and carbomer, showed an inhibitory effect on the proteolytic activity of the luminal enzyme trypsin. This can be explained by the pronounced Ca^{2+} binding affinity of the polymers. Therefore Ca^{2+} is depleted out of the trypsin structure and enzyme activity is reduced.

IV. DRUG DELIVERY SYSTEMS

A. Nasal

1. Relevant Anatomy and Physiology

The nasal passage, which measures from the nasal vestibule to the nasopharynx, has a length of approximately 12 to 14 cm (5,68) with a surface area of about 150 cm^2 and a total volume of about 15 mL (69). This upper respiratory tract provides a modification of the inspired air by filtration, controlled temperature, and humidity. It also provides a highly efficient defense system that protects the lungs against inhaled particles and toxicants. The main regions of the nasal cavity are the nasal vestibule, the respiratory region, and the olfactory region. First, the vestibular region serves as a baffle system, and its surface is covered with a keratinized stratified squamous epithelium where the long hairs may provide the function of filtering inhaled particles (10). Second, the respiratory region comprises the largest part of the nasal cavity. This area is covered with pseudostratified columnar cells containing mucus and numerous cilia. The respiratory epithelium is highly vascularized, supplying the lamina propria of the epithelium with a dense network of fenestrated capillaries. Thus the respiratory region is assumed to be the major site for drug absorption. Finally, the olfactory region is generally free of inspiration airflow. It is covered with pseudo-stratified columnar cells, specialized olfactory cells, supporting cells, and serous and mucous glands (70). Absorption may also occur in this region through the olfactory nerve ending into the brain (71,72).

2. Major Biological Barriers and Appropriate Strategies

The nasal mucosa is considered to be permeable to small molecules, while penetration enhancers can be used to increase the absorption of larger molecules. Several enzymes also exist in nasal secretions and have been reportedly involved in the metabolism of different kinds of drugs (73–78). For example, several researchers observed that the respiratory mucosa contains detectable amounts of cytochrome P-450 and monooxygenase activity; however, the levels are generally lower than those found in the liver. The level of these enzymes is also related to species differences and variation in the nasal mucosa regions (74,79). The proteolytic activity in the nasal region is also comparatively low compared with other mucosae in the body (79,80).

One of the most important defense mechanisms of the nose is mucociliary clearance. This is a function of the physical properties of the mucus coupled with appropriately functioning cilia (81,82). The mucus layer on the nasal mucosa consists of two layers. The outer layer is relatively viscous (gel) and rests on a thin watery (sol) layer that tends to facilitate the action of the underlying cilia and is probably the most significant physical factor to affect nasal drug absorption (83). The speed of the mucus flow is about 5 mm/min with a range from 0 to more than 20 mm/min, while the mucus will tend to reconstitute itself every 10 to 15 minutes (70,84). Temperature, relative humidity, airway secretion, and specific drug formulations can affect the nasal mucociliary function. For example, drying of the nasal mucosa will cause cessation of ciliary activity; however, moistening will promptly restore its normal activity (17,70). A study by Hardy et al. (85) showed that approximately 40% of the dose administered via the intranasal route cleared rapidly with average half-times ranging from 6 to 9 min. Following this rapid phase, clearance of a spray was much slower than drops. Thus development of nasally administered drugs must take into account this clearance mechanism.

3. Examples

The efficacy of drugs delivered intranasally, whether for local or for systemic therapy, will depend upon the initial pattern of deposition and the subsequent rate of clearance.

It therefore depends on dosage forms and delivery devices. Several dosage forms can be used in nasal drug delivery, such as solutions, suspensions, emulsions, powders, microspheres, and liposomes, while delivery devices, such as dropper, spray pump, metered dose inhaler, dry powder inhaler, and nebulizer, have been used.

Solution dosage forms, applied as drops, showed a short residence time in the nasal cavity (85–87) because the nasal drops dispersed more extensively into the posterior part of the nose, where they were cleared more rapidly by mucociliary clearance. This can be improved by using spray systems (85–87), adding bioadhesive (88–90), or using a powder form (87,88,90–92). Harris et al. (86) showed that desmopressin solutions containing ^{99m}Tc-labeled human serum albumin administered intranasally as a spray were deposited anteriorly, and small portions were cleared slowly into the nasal pharynx. In addition, the plasma level of desmopressin and the biological response were clearly enhanced after administration of the spray. Studies have shown that spray preparations containing 0.25% methylcellulose decreased mucociliary clearance (93), resulting in a prolonged absorption of nasally administered desmopressin (94). However, highly viscous solutions have the disadvantage of being difficult to administer. Therefore solutions of thermogelling systems, which have low viscosity at room temperature and form a gel at body temperature, have been introduced. These systems are easy to apply to the nasal cavity at room temperature, whereas the gel state at body temperature adheres to the mucosa. Examples of these systems are Poloxamer 407 solutions (95) and combinations of ethyl (hydroxyethyl) cellulose and a small amount of sodium dodecyl sulfate in water (96,97). Gellan gum (Gelrite®) can also be used as a phase-change polymer system, because the gellan gum solution can form a gel if a physiological level of cation, such as Ca^{2+}, is added. Therefore a promising way of lengthening the residence time of drugs with gellan gum could be forming a gel through the contact of a solution with the nasal mucosa (98).

A powder dosage form of insulin was evaluated by Nagai et al. (91). Hydroxypropyl cellulose and Carbopol 934 were the bioadhesives, which they used as excipients in a powder form. They found that a powder dosage form showed higher bioavalability of insulin than a liquid dosage form. Nakamura et al. (88) studied the adhesion of powder bioadhesives to the nasal tissue. They suggested that at the initial stage, bioadhesive powders absorb water from mucus on the nasal mucosa and gradually gel. Therefore these systems can prolong residence time in the nasal cavity. Xanthan gum showed the longest residence time, followed by tamarind gum, hydroxypropyl cellulose, and polyvinyl alcohol. Moreover, they found that mixing two or more bioadhesives with different properties could modify the bioadhesive property of a system. Microspheres also show an increased residence time in the nasal cavity (99–103). Bioadhesive microspheres show the same mechanism as the bioadhesive powder, as they absorb water and form a gellike structure after swelling. Illum et al. (102) used albumin, starch, and DEAE-dextran microspheres to increase residence time in the nasal cavity and thus increase bioavailability of the drugs. The results indicated that these microspheres exhibited a half-life period of about 4 hours, which was much longer than the about 15 minutes observed for solutions and powder. The enhancing effect of microspheres on nasal drug absorption is caused not only by an increase in residence time but also by a temporary widening of the tight junctions, as shown in a Caco-2 cell monolayer in the presence of bioadhesive microspheres (99,100). Hence when the microspheres absorb water from mucus and swell, the nasal mucosa is dehydrated, which causes the tight junctions to separate. Recently, Witschi and Mrsny (103) evaluated microparticles of starch, alginate, chitosan, and Carbopol 971 as nasal platforms for protein delivery. They found that protein transport across polarized Calu-3 cells, a well-characterized human respiratory tract epithelial cell line, was enhanced using Carbopol and chitosan microparticles.

Liposomes are lipid vesicles composed of layers of membranes enclosing an aqueous phase. The use of liposomes as a nasal delivery dosage form has been considered because they may either prolong residence time in the nasal cavity or sustain the duration of action. Vyas et al (104) prepared multilamellar liposomes containing nifedipine using a conventional cast film method. An in vivo experiment in rabbits showed that the nasal administration of liposomes could maintain an effective drug concentration for prolonged periods of time with improved bioavailability. Liposomes can also be used as vaccine adjuvants. An influenza vaccine composed of liposome-entrapped glycoproteins from the envelope of influenza virus was administered intranasally. It induced a strong systemic immunity and a local IgA response that protected against virus infection. Alpar et al. (105) incorporated tetanus toxoid into liposomes composed of equimolar concentration of distearoyl phosphadylcholine and cholesterol. They found that liposome formulations delivered via the nasal cavity significantly improved the immune response when compared to free antigen. Note that this work was in guinea pigs, and a concentration of tetanus toxoid used in nasal delivery was ten times higher than intramuscular delivery. In spite of the positive response to liposomes, the stability of liposomes should be an important concern because of the presence of an aqueous phase in the system. Therefore proliposomes, a free-flowing form that creates a liposomal dispersion on adding water, were introduced (106). Ahn et al. (107) studied the possibility of using proliposome as an intranasal dosage form for the sustained delivery of propranolol. The results showed that nasal administration of proliposomes to the rats could sustain the plasma concentration profile of propranolol better than the solution, proliposome-loaded sorbitol, and hydrated proliposomes. Moreover, the bioavailability of the nasal proliposomes (~97.5%) was much higher than that of an oral solution (~14.2%).

Emulsions were tested as a nasal delivery formulation for lipid-soluble drugs. Kararli et al. (108) used oleic acid/monoolein emulsion formulations to delivery a renin inhibitor. Using the rat as a model animal, the percent absolute bioavailability of a renin inhibitor was enhanced from 3–6% in control polyethylene glycol 400 solution to 15–27% when the emulsion formulations were used. Ko et al. (109) applied three different charged testosterone submicron-size emulsion formulations intranasally in rabbits, and compared the bioavailability to intravenous injection. The results suggested that emulsions might have considerable potential for nasal delivery, especially for positively and negatively charged emulsions, because these formulations showed bioavailability of 55 and 51%, respectively, as compared to only 37% for neutral charged emulsion.

Delivery devices are also an important factor to improve accuracy, reproducibility, and patient acceptability of drug delivery to the nasal cavity. For a better understanding of delivery devices, the reader is directed to other reviews (110–112).

B. Pulmonary

1. Relevant Anatomy and Physiology

The respiratory tract consists of all structures that conduct air between the atmosphere and the alveoli. The parts of the respiratory tract that lie outside the lung are the nose, pharynx, larynx, trachea, and main bronchi. Within the lung, each main bronchus branches extensively to a smaller diameter and a higher number, as shown in Table 3. The exchange zones of the lung with the blood stream are the alveolar ducts and alveolar sacs, which consist of numerous spherical alveoli. The surface area for gas exchange, as well as drug absorption, of the alveoli is about 60 to 80 m^2. Dense vascular beds of pulmonary capillaries also match this large surface area of the alveoli. The bronchial epithelium consists of pseudo stratified ciliated columnar cells and mucus-secreting goblet cells (115). The alveoli are covered by a very thin epithelium (0.1–0.2 μm) without cilia and a mucus blanket (116). This epithelium consists of granular pneumocytes and membranous pneumocytes, which are responsible for secretion of surfactant phospholipids and are involved in gas exchange, respectively. The pore size of this sievelike epithelium is about 0.5 to 0.6 nm (115).

2. Major Biological Barriers and Appropriate Strategies

The primary barrier mechanisms for pulmonary drug delivery are the geometry of the airways, mucociliary clearance, macrophage activity, and pulmonary metabolism. The branching airways act like an impaction filter for inhaled particles. There are three basic mechanisms, which effect the deposition of particles in airways, i.e., inertial impaction, sedimentation, and Brownian motion (117). Impaction occurs when particles have inertia such that they are unable to travel with the airstream when they change direction as an airway branches. Thus they impact on the mucus surface of ciliated epithelium. This mucus is continuously moving out of the lung into the throat, where it is swallowed or coughed up. Impaction is important for

Table 40.3 Subdivision of the Respiratory Tree

Name	Diameter	Length	Total cross-section area (cm^2)	Number
Trachea	1.8 cm	12.0 cm	2.54	1
Main bronchi	1.2 cm	4.8 cm	2.33	2
Lobar bronchi	0.8 cm	1.9 cm	2.13	3 on right, 2 on left
Segmental bronchi	0.6 cm	0.8 cm	2.00	10 on right, 8 on left
Terminal bronchioles	1.0 mm	2.0 mm	180.00	about 48,000
Respiratory bronchioles	0.5 mm	1.0 mm	about 1,000	about 300,000
Alveolar ducts	400 μm	0.5 mm	about 10,000	about 9×10^6
Alveoli	250 μm	—	about 8×10^5	about 3×10^9

Source: Refs. 113, 114.

particles above 5 μm in size, which tend to deposit in the upper respiratory tract. Smaller particles (0.6–5 μm) are able to move with the airstream; however, sedimentation can occur in the lower respiratory region, which has a lower air velocity. Finally, particles below 0.6 μm can deposit by Brownian motion. Therefore the efficiency of drug delivery via this route is often not high enough because of the difficulty in targeting particles to the sites of maximum absorption (alveoli). The degree to which inhaled particles penetrate into airway spaces is controlled by a number of factors, i.e., particle size and size distribution, shape, and density (118). The particle size is usually referred to as an aerodynamic diameter, defined as the geometric diameter of a sphere with unit mass density, and has the same velocity as the test particle (118,119). Small particles (mean geometric diameter <5 μm) with high density tend to aggregate and can be rapidly phagocytosed in the deep lungs by alveolar macrophages, while larger particles tend to deposit in the upper respiratory tract. A new type of inhalation aerosol has recently been introduced and identified that may help to solve this problem (120). This aerosol is formed by considering the aerodynamic diameter window of 1 to 3 μm as a lower limit, because it was demonstrated that particles of mean aerodynamic diameter of 1 to 3 μm deposit maximally in the deep-lung region, whereas particles with an aerodynamic diameter smaller than 1 μm are mostly exhaled, and particles larger than 10 μm deposit maximally in the mouth and throat. Particles with low mass density (porous) and larger geometric diameter (about 10–15 μm) have been used and gave excellent results. The advantage of this new aerosol is a decrease in the tendency to aggregate while minimizing clearance by alveolar macrophages. Biodegradable polymers or lipophilic carriers can also be used to protect drug from metabolism. Therefore these systems are potentially good for long-acting therapeutic delivery following inhalation (120,121).

There are no cilia or mucus blankets in the alveoli, but insoluble substances will be engulfed by alveolar macrophages, which are then carried to the ciliated epithelium and cleared through the mucociliary clearance pathway (122). Systemic absorption of inhaled substances may also be inhibited by pulmonary metabolism. The lungs contain a variety of drug-metabolizing enzymes, including microsomal mixed function, amine oxidase, monoamine oxidase, reductases, esterases, a variety of conjugates, and proteolytic enzymes (123,124).

3. Examples

Drug administration to the lungs has been employed for many years for local treatment of pulmonary diseases (110,125). For example, bronchodilators, corticosteroids, anticholinergics, and antiallergic drugs are administered by means of oral inhalation (110). The advantages are a decrease in systemic side effects and rapid onset of action. In recent years, the systemic effect of the drugs delivered via this route has been studied. Patton et al. (126) reviewed the potential of using the pulmonary area as a route for the delivery of insulin. By comparing the data from several studies, they concluded that aerosol insulin has a potential advantage over subcutaneous injection because it is more rapidly absorbed (serum peak at 5–60 min) and cleared than a subcutaneous injection (serum peak at 60–150 min). However, aerosol delivery involves some losses, including loss from the device, loss from exhaling or deposits in the mouth and throat, and loss from lung metabolism. These problems can be largely overcome by proper dosage form design, which accurately delivers drug to a specific site, the lungs, and prevents lung metabolism. Moreover, delivery devices are also a crucial factor. It is interesting to note that the same delivery devices as nasal delivery are used for pulmonary delivery; however, the aims of pulmonary delivery are achieving a high deposition in the lower airways and a minimum deposition in the upper airways. Most of the drugs are administered to the lungs as aerosols, and their delivery is to a great extent dependent on particle size distribution. Different aerosol size distributions may be expected to deposit preferentially in different regions of the respiratory tract, and aerosols inhaled intranasally also have different deposition patterns from those inhaled orally (127). For further information about delivery devices a number of interesting reviews are recommended (110–112,128–130); this report will concentrate on the development of formulations.

Sanders et al. (131) compared the pulmonary deposition of aerosols generated from two formulations in healthy volunteers and asthmatic subjects. The two formulations were a teflon particulate suspension with a mean particle diameter of 5.76 μm, and a solution that generated aerosol particles of 2.13 μm mean diameter. Both formulations were radiolabeled with ^{99m}Tc and delivered by standard metered dose inhaler. The results showed that the aerosol generated from the solution produced a significantly greater total lung deposition than the suspension aerosol. Adjai and Garren (132) studied the effect of particle size on bioavailability of leuprolide acetate from a solution and suspension formulation. By using impaction and light scattering methods, they demonstrated that the solution aerosol consisted largely of coarse particles, and they lost most of the drug at the actuator, while the suspension aerosol had greater fractions of particles within the respiratory range (<5 μm) and delivered about 40–50% to the lungs. The bioavailability of each aerosol formulation based on intravenous control was compared. The bioavailability results from the suspension aerosol were significantly more than for the solution aerosol, which correlated with the particle size distribution results. However, when the correct bioavailabilities were used to compare the effect of the type of formulation to the absorption of drug from the lungs, there were no differences between these two formulations. Note that the correct bioavailability was calculated from the effective dose, which equals the amount of the drug reaching the lungs.

Emulsion is another liquid formulation that could be used as a pulmonary delivery system, for example, water in fluorocarbon emulsions (133). These emulsions are stabilized by perfluoroalkylated surfactants with a dimorpholinophosphate head group. Fluorocarbons, such as perfluorooctyl bromide and perfluorooctyl ethane, were used as the external phase of reverse emulsions. Significantly slower release of 5,6-carboxyfluororescein entrapped in the water droplets was found compared to using hydrocarbon oils as the external phase. This can be explained by the extremely hydrophobic character of fluorocarbons, which act as a physical barrier to diffusion of encapsulated hydrophilic compounds.

The ban of chlorofluorocarbon propellants has spurred the development of alternative pulmonary drug delivery systems for metered dose inhalers. One possibility is the use of dry powders in the inhalation device. Many trials to develop an ideal dry powder formulation that can deliver the drug accurately to the lungs have been conducted by designing powders and constructing inhalers (134–139). Kawashima et al. (135) presented a new powder design method to improve inhalation efficacy of hydrophobic drug powders such as Pranlukast hydrate. They modified the surface of powders by using hydroxypropylmethylcellulose phthalate, a pH-dependent cellulose derivative (soluble at pH > 5.5). Dramatically improved inhalation properties of the surface modified powders were found in vitro compared with the original unmodified powder. The improvements were attributed to the increased surface roughness, the hydrophilicity of the surface-modified powders (at pH > 5.5), and the resultant increased dispersibility in air and emission from a device. As described before, a large and porous particle with biodegradable polymers or lipophilic carriers can be aerosolized from a dry powder inhaler more efficiently than smaller nonporous particles, resulting in higher respiratory fractions of inhaled drug (120,121). In addition, these formulations can be particularly useful for controlled release purposes. For example, insulin was encapsulated into porous and nonporous particles of poly(lactic acid-co-glycolic acid). Inhalation of large porous insulin particles resulted in suppressed systemic glucose levels in rats for 96 hours, whereas only 4 hours was the result from small nonporous insulin particles. The bioavailabilities relative to subcutaneous injection were 87.5% and 12% for large porous and small nonporous particles, respectively. High systemic bioavailability of testosterone was also achieved by inhalation of large porous particles. Note that both porous and nonporous particles have similar aerodynamic diameter. Ben-Jebria et al (121) used dipalmitoyl phosphatidylcholine as a lipophilic carrier for sustaining the action of the bronchodilator, albuterol sulfate, in the lungs. They found that inhalation of large porous particles produced a significant inhibition of carbochol-induced bronchoconstriction for at least 16 hours, while inhalation of small nonporous particles protected from carbochol-induced bronchoconstriction for up to 5 hours.

Drug entrapment in liposomes has been most extensively investigated, and evidence has been presented both in animals and in humans that liposomes may enhance drug efficacy by greater intrapulmonary retention (125, 140–146). Taylor et al. (140) studied the pharmacokinetics of sodium cromoglycolate in healthy volunteers. It was inhaled as a solution and encapsulated in dipalmitoylphosphatidylcholine/cholesterol (1:1) liposomes. The data indicated that encapsulation of this drug can prolong drug retention within the lungs and alter its pharmacokinetics as compared to a solution formulation. The antitumor drug cytosine arabinoside was administered to the lungs of rats via intratracheal instillation as free or liposome encapsulated form (141). It was demonstrated that prolonged action of this drug could be confined to the lungs through the use of a liposomal formulation. This formulation also showed a reduction in extrapulmonary side effects. A similar clearance study was done by Lambros et al. (144). They evaluated the clearance kinetics and organ distribution of inhaled liposomal Amphotericin B in mice. Amphotericin B was not detected in the serum and other organs, such as kidney, liver, and brain. Surface charge of the liposome also affects lung clearance kinetics, where the positively charged and neutral liposomes exhibited a longer half-life than the negatively charged liposome. Saari et al. (147) used gamma scintigraphy to study the pulmonary distribution and clearance of ^{99m}Tc-labeled beclomethasone diproprionate liposome formulations in healthy volunteers. The liposome formulations contained dilauroylphosphatidylcholine or dipalmitoylphosphatidylcholine. By using an Aerotech jet nebulizer, both liposome aerosols had a suitable droplet size (mass median aerodynamic diameter about 1.3 μm). The remaining liposomes in the lung after 24 hours were 79% and 83% for dilauroylphosphatidylcholine and dipalmitoylphosphatidylcholine liposomes, respectively. These results indicate that liposomal formulations delivered to the lungs have the potential to sustain drug release, prolong drug action, and minimize side effects.

C. Oral Cavity

1. Relevant Anatomy and Physiology

The oral cavity is lined by a relatively thick, dense, and multilayered highly vascularized mucosa. The total surface area is about 100–160 cm^2 (148,149) and can be divided into three general areas according to functional demands. First, the masticatory mucosa consists of stratified squamous epithelium, which is generally keratinized, and presents on the gingival and hard palate. Second, a specialized mucosa is present on the borders of the lips and the dorsal surface of the tongue with its highly selective keratinization. Third, the lining mucosa is found on the lips, cheeks, floor of the mouth, ventral surface of the tongue, alveolar, and soft palate. This mucosa consists of nonkeratinized stratified squamous epithelium (148,150). Transport of drugs via the oral mucosa most likely occurs through the nonkeratinized areas (151), so we will discuss this pathway by emphasizing sublingual and buccal delivery. Sublingual delivery is drug administration through the mucosa of the floor of the mouth and the ventral surface of the tongue, while buccal delivery is drug administration through the mucosa lining the inside of the cheeks. The thickness of the oral epithelium varies depending on location and species. In humans, dogs, and rabbits, the thickness of the buccal epithelium is about 500 to 800 μm, whereas the sublingual area measures about 100 to 200 μm. This is because the buccal epithelium is composed of about 40 to 50 cell layers, while the sublingual epithelium contains considerably fewer cell layers (152,153). Separately, the buccal and sublingual mucosae of the human have a surface area about 50 and 26.5 cm^2, respectively (149). The surface of the mucosa is continuously washed by a stream of about 0.5 to 2 liters of saliva daily and is cleared by swallowing. Saliva also functions as a moistening liquid for the mucosa with a pH between 5.8 to 7.2. The thickness of the coating of saliva on the mucosa throughout the mouth is about 70 μm.

2. Major Biological Barriers and Appropriate Strategies

The permeability of the oral mucosa in general is probably intermediate between that of the skin and the intestinal mucosa. The sublingual mucosa is relatively permeable, giving relatively rapid absorption and acceptable bioavailability for many drugs, while the buccal mucosa is considerably less permeable and is generally not able to provide rapid absorption (154). The permeability barrier of the oral mucosa is thought to reside within the superficial layers, constituting about one-fourth to one-third, of the epithelium (155), as shown by using horseradish peroxidase (156) and lanthanum salts (157), which are both water-soluble markers. In these studies, they found that both markers did not penetrate beyond the top three cellular layers when applied topically. When applied subepithelially, markers can penetrate through the lower area of the mucosa but they did not penetrate outward to the upper epithelium. Dowty et al. (158) showed that the upper 50 μm of the epithelial tissue was a barrier to transport of thyrotropin-releasing hormone in rabbit buccal mucosa in vitro. A possible explanation of this behavior is the presence of membrane coating granules or Odland bodies, which are spherical or oval organelles, 100 to 300 nm in diameter (152–154). This barrier is apparently an accumulation of neutral lipids and glycolipids in the intercellular spaces of the superficial cell layers and is derived from the extrusion of the contents of the membrane coating granules as the epithelial cells move superficially (155).

Besides membrane coating granules as a barrier for permeation, one must also consider enzymatic activity in the oral mucosa. The presence of hydrolytic enzymes, esterases, and peptidases has been reported in the buccal mucosa. The small available surface area and short residence time are also factors to be considered for designing an appropriate dosage form.

3. Examples

Oral cavity drug delivery systems have been employed to obtain local or systemic effects (159,160). Despite some limitations in the oral cavity, such as low permeability, short retention time, and small surface area, the delivery systems can be optimized to provide desired drug concentration in the oral cavity (local) or in the blood (systemic). Several dosage forms have been used in the oral cavity including solutions, tablets, patches, films, and semisolid systems.

Buccal tablets, incorporating active ingredients and bioadhesives, have been studies extensively. The purpose of using a bioadhesive is to increase contact time and sustain drug release. Several studies showed promising results for the delivery of local-acting drugs. Nagai (161) used two-layered tablets to deliver triamcinolone acetonide, which proved to be effective in the treatment of apthous stomatitis, an infection of the mouth. The tablet consisted of a bioadhesive and a non-adhesive layer. The bioadhesive layer contained the drug Carbopol 934 and hydroxypropylcellulose, while the nonadhesive layer contained lactose. Bouckaert et al. (162,163) prepared a buccal bioadhesive miconazole slow-release tablet by mixing miconazole, an antifungal drug, with thermally modified starch and 5% polyacrylic acid (Carbopol 934, 910, or 907). The in vivo data of release characteristics and adhesion time showed that no significant difference could be seen among the formulations (163). For the Carbopol 934 formulation, mean adhesion time was 586 ± 133 minutes (range 360–870 minutes), and the salivary miconazole concentration remained constant at an effective level for more than 10 hours (162). Recently, buccal bioadhesive tablets of metronidazole were prepared by using Carbopol 934P, hydroxypropylmethylcellulose, and sodium carboxymethylcellulose as bioadhesives (164). In situ release studies using the bovine cheek pouch membrane in a flow through cell demonstrated that the concentration of metronidazole was maintained at an effective level for a period of 6 hours, the time for which it was held in the buccal cavity.

Buccal tablets are also used as a drug delivery system for systemic delivery. Ishida et at. (165) examined the possibility of buccal delivery of insulin using a bioadhesive dosage form. Insulin was incorporated into core materials of cocao butter and sodium glycocholate, and then a mixture of hydroxypropylcellulose and Carbopol 934 (1:2) was compressed around the core. Subsequent studies in beagles revealed that the tablet adhered for up to 6 hours; however, the percentage of insulin absorbed was about 0.5% compared to an intramuscular injection. There has been interest in buccal administration of morphine; however, studies of buccal tablets have indicated poor bioavailability and clinical efficacy (166,167). Bioadhesive tablets seem to be the answer to these problems. Anlar et al. (168) prepared a bioadhesive tablet of morphine sulfate by compression with bioadhesives, hydroxypropylmethylcellulose, and Carbopol 910, in various amounts. This tablet showed significant bioadhesive characteristics in contact with bovine sublingual mucus. The release behavior was found to be non-Fickian, suggesting that the release of morphine sulfate was controlled by a combination of diffusion of morphine sulfate in the matrix and swelling of the matrix. The average percentage of morphine sulfate absorbed in healthy volunteers was 30 ± 5% of the drug load within 8 hours. The current comparison study between three formulations, a 30 mg oral controlled-release tablet, a 20 mg aqueous solution retained in the mouth for 10 minutes, and a 60 mg bioadhesive buccal tablet, showed that morphine absorption from the solution was very low, while a sustained-release pattern was found in the case of other formulations (169). However, information about excipients in the formulations was not available. Yukimatsu et al (170) developed a bioadhesive tablet consisting of two layers, a fast-release layer and a sustained-release layer. It was designed for gingival application. The fast-release layer contained 20% isosorbide dinitrate dispersed in the solid matrix of D-mannitol and low molecular weight polyvinylpyrrolidone, while the sustained-release layer consisted of 80% isosorbide dinitrate in the polymer matrix of high molecular weight polyvinylpyrrolidone and polyacrylic acid. The release profile of drug from this tablet indicated that the fast-release fraction was released within 15 minutes, contributing to a rapid absorption, and the sustained-release fraction was released gradually over a period of 12 hours, contributing to a prolonged duration of drug plasma concentration. This tablet of isosorbide dinitrate was confirmed to achieve a desired pharmacokinetics profile, such as a rapid onset and long duration of therapeutic effect in both beagles and humans. Moreover, it showed a better bioavailability than a marketed oral sustained-release tablet and the same bioavailability as a sublingual tablet. Han et al. (171) used a manual single-punch tableting machine to prepare a disk with diameter and thickness of 8 mm and 1 mm, respectively. Various amounts of Carbopol 934, hydroxypropylcellulose, and drug were mixed to prepare different formulations of buccal disks. The impermeable backing layer of ethylcellulose was cast on one side of the disks. By varying the drug hydrophilicity and the amount of solubility enhancer (β-cyclodextrin) inside the disks, the results demonstrated that drug solubility was an important factor in controlling drug release. Moreover, in vitro studies showed that the disks can attach to porcine buccal tissues for at least 48 hours.

Oral mucosal patches for drug delivery systems have been developed extensively (172,173). These patches may show unidirectional (releasing the drug only into the mucosa) or bidirectional (releasing the drug into the mouth as well) drug release. Depending on its size and shape, the patch may be applied at different sites, including the buccal, sublingual, and gingival mucosa. For optimum patient compliance, high flexibility and appropriate size of the patch need to be considered. Buccal patches may have a size of up to 12–15 cm^2, while sublingual or gingival patches should be in the range of 1–3 cm^2 (160,174). Veillard et al (175) developed a three-layer buccal patch consisting of a bioadhesive basement membrane, a rate-limiting center membrane, and an impermeable facing membrane, which permitted unidirectional drug release into the mucosa. Polycarbophil was used as a bioadhesive in the bottom layer. This patch can remain in place for approximately 17 hours regardless of eating or drinking. However, the study did not incorporate drugs into the patch. Buccal adhesive patches, consisting of two-ply laminates of an impermeable backing layer and a bioadhesive layer containing drugs, were developed (174). The polymers used were hydroxyethylcellulose, hydroxypropylcellulose, polyvinylpyrrolidone, and polyvinylalcohol. It was found that the duration of mucosal adhesion in vivo was affected by the type of polymer, their viscosity grades, polymer content, and the drying procedure for preparation. Biolaminated films, consisting of a drug-containing bioadhesive layer and a drug-free backing layer, were produced by a casting/solvent evaporation technique (176). The bioadhesive layer consisted of a mixture of drug and chitosan, with or without an anionic polymer such as polycarbophil, sodium alginate, and gellan gum. The backing layer was made of ethylcellulose to avoid loss of drug due to washout with saliva. This patch was also expected to provide drug release in a unidirectional fashion to the mucosa.

Several ointment bases for oral mucosa administration have been developed. Ishida et al. (177) mixed Carbopol 934 with three kinds of ointment bases including white petrolatum, hydrophilic petrolatum, and absorptive ointment. They found that Carbopol 934 could improve the physical properties of the ointments and the release profiles of the model drug, prednisolone. Bremecker et al. (178) explored the possibility of using a neutralized polymethacrylic acid methyl ester as an adhesive ointment base. Sveinsson and Holbrook (179) used a copolymer of methacrylic acid and methacrylic acid methyl ester to produce an oral bioadhesive ointment containing liposomes with triamcinolone acetonide. Gelatin was also added to the formulation to improve its adhesive property. Recently, Petelin et al. (180) studied the properties of bioadhesive ointment for delivery of liposomes containing drugs. Polymethyl methacrylate, Carbopol 934P, and Orabase were used as the bioadhesive ointment bases. The results showed that polymethyl methacrylate was the most appropriate bioadhesive ointment base for local application in the oral cavity because the liposomes were most stable in this base. Sublingual gel formulation of leuprolide was developed by using hydroxypropylcellulose as a bioadhesive (181). A prolonged absorption of up to 6 hours after sublingual dosing in the human was observed. This observation may be attributed to an increase in residence time at the site of absorption.

Recently, high-velocity powder injection, namely PowderJect, has been introduced as a new drug delivery technology to overcome the barrier limitation of skin and mucosal sites (182). This new technology was developed based on the direct injection of solid powder through the stratum corneum or into mucosal membranes without using a needle. The oral PowderJect delivery system used a gas-pressure pulse to accelerate and deliver fine particles 20 to 100 μm diameter into oral mucosal tissue. This device proved useful for delivery of testosterone and lidocaine hydrochloride through the oral mucosal, showing promising pharmacokinetic results.

D. Gastrointestinal Tract

1. Relevant Anatomy and Physiology

The GI tract comprises the stomach, small intestine, and large intestine. It is responsible for the major digestion of food, absorption of the end products of digestion, absorption of ingested fluids and reabsorption of secreted fluids, and excretion of some digestive end products. Table 4 lists the specific functions of different regions of the GI tract.

The human stomach is usually described as a J-shaped sac with a maximum length of 25 cm and a width of 15 cm. The average adult stomach has a capacity of about 1.5 L. It is divided into the cardia, the fundus, the body, the antrum, and the pylorus, and the function of each region is described in Table 4. The surface of the gastric mucosa is covered by simple columnar epithelium. Its surface is increased by downgrowths, which are variously called foveolae, pits, or crypts. The major function of the gastric epithelium is the secretion of acid and digestive enzymes. It also secretes mucus to lubricate ingested food and to protect itself from autodigestion by acid and pepsin. The glands in the area of the cardia and the pylorus only produce mucus, while the glands in the body and the fundus produce mucus, hydrochloric acid, and proteolytic enzymes (184). The stomach is emptied by vigorous peristaltic contractions that begin in the antral region, or perhaps in the esophagus, with simultaneous opening of the pylorus. The pylorus is immediately closed after chyme (the mixture of food with gastric secretions) has left the stomach. Emptying is most intensive immediately after a meal and then becomes weaker. Gastric emptying time depends on the amount, the composition, and the particle size and shape of the food. A carbohydrate-rich food is emptied after a relatively short time, while the poorly chewed or fat-rich foods can prolong the emptying time up to 5 hours. In the fasting state, the stomach displays a minimum secretion (5–15 mL/h). This secretion is free of hydrochloric acid and pepsin and therefore shows a neutral to slightly alkaline pH. After ingestion of a meal, the rate of gastric secretion increases rapidly. The three phases of gastric secretion in response to food are the cephalic phase, the gastric phase, and the intestinal phase (185). The cephalic phase takes place under the influence of nervous impulses from the central nervous system such as the sight, smell, and taste of food. The acid secretion rate during this phase can be as much as 40% of the maximum rate. The gastric phase is initiated when food reaches the stomach. The principal stimuli include distension of the stomach and the re-

Table 40.4 Regions of the Gastrointestinal Tract and Their Primary Functions

Organ	Region	Function
Stomach	Fundus	Initial storage of food
	Body	Storage/secretion
	Antrum	Vigorous mixing of food with secretions to form a semisolid chyme
	Pylorus	Separate the stomach from the duodenum
Small intestine	Duodenum	Segment receiving liver and pancreatic secretions
		Important site for the regulation and overall coordination of GI function
	Jejunum	Absorption of the majority of the end products of digestion
	Ileum	Fluid reabsorption; junction with the large intestine is termed the ileocecal sphincter
Large intestine	Ascending colon	Fluid reabsorption
	Transverse colon	Fluid reabsorption
	Descending colon	Fluid reabsorption
	Sigmoid colon	Storage of feces
	Rectum	Storage and elimination of feces

Source: Ref. 183.

lease of gastrin. The intestinal phase is the final phase of secretion, which is initiated by the passage of chyme into the duodenum. The presence of chyme in the duodenum results in both stimulation and inhibition of acid secretion in the stomach. The stimulatory influence dominates when the pH of gastric chyme is above 3. This is probably based on the release of gastrin. However, when the pH falls below 3, the inhibitory influence dominates by the release of secretin, which inhibits acid secretion.

The small intestine is the major site for both digestion and absorption of food (Table 4). It is 6 to 7 meters long with an average diameter of 4 cm. It is divided into the duodenum, the jejunum, and the ileum (182,185). The duodenum is a 20 to 30 cm long segment connected to the pylorus. This region is the place where secretions of the liver and pancreas enter the small intestine. The jejunum is 250 to 300 cm long, while the length of the ileum is about 300 to 400 cm. The small intestinal mucosa is covered by a simple columnar epithelium and is characterized by a large surface area. The first stage of the surface enlargement results from circular mucosal folds called the plicae circularis or Kerckring's folds. In this fold, the epithelium consists of an enormous number of small fingerlike villi, roughly 1 mm high. These two factors increase the luminal area 30-fold over that of a smooth cylinder of the same external diameter. A further increase in the surface area can be produced by the presence of microvilli or brush border which are about 1 μm long and 0.1 μm in width. The microvilli can enlarge the surface area of the lumen by a factor of 600. This makes a total absorptive surface area of the small intestine up to 200 m^2. The transport of chyme through the small intestine varies but can be estimated as 5 minutes in the duodenum, 2 hours in the jejunum, and 3 to 6 hours in the ileum (186).

The large intestine is the terminal part of the intestinal tract and is composed of the cecum, the colon, and the rectum (182,185). It is about 150 cm long, and connects to the small intestine at the ileoceal valve, where the small intestine contents are released into the large intestine. The colon has a total length of about 130 cm and a lumen diameter of 6 to 8 cm. It is divided into an ascending, a transverse, a descending, and a sigmoid colon. The function of each region is shown in Table 4. Even though the large intestine has a larger diameter and length than the small intestine, it has a smaller surface area than the small intestine because of the lack of villi in its entire region. The surface area of the large intestine can be increased by the presence of crypts. The epithelium of the crypts consists of goblet cells, which produce mucus. The peristaltic movement starting from the caecum and moving to the sigmoid colon is superimposed two to three times a day. Therefore a colonic transit time of 35 hours or longer is typical (187).

The rectum is the dilated distal portion of the alimentary canal. Its upper part is distinguished from the rest of the colon by the presence of transverse folds (188). The rectal epithelium is the same as other regions of the large intestine and is covered by simple columnar cells, consisting of numerous goblet cells. There are no villi or microvilli on the rectal mucosa, and only a limited surface area of 200 to 400 cm^2 is available for drug absorption. The human rectum is approximately 15 to 19 cm long and 1.5 to 3.5 cm in diameter. The rectum contains 1 to 3 mL of mucus fluid with a pH of 7 to 8 (189).

The GI tract is richly supplied by a blood-capillary network that is associated with the splanchnic circulation. All venous drainage from the GI tract is via the mesenteric veins, which empty into the portal vein, and passes through the liver before entering the systemic circulation. In addition to the blood supply, the lymphatic vessels that drain to the thoracic duct extensively supply the GI tract. As shown in Fig. 2, absorption of drug by the lymphatic system will then enter directly into the venous system of the systemic circulation without passing through the liver.

Both blood and lymphatic vessels are also abundant in the submucosal region of the rectal mucosa. Blood supply is delivered by the superior rectal artery and is drained by a superior, a middle, and an inferior hemorrhoidal vein. The superior veins drain into the portal circulation, while the middle and inferior veins drain into the inferior vena cava. However, there are extensive anastomoses among these veins (189,190).

2. Major Biological Barriers and Appropriate Strategies

The oral route is the most popular and convenient route of drug administration; however, several factors can influence absorption of drugs through the GI tract. These include the pH of the luminal content, the presence of enzymes, motility and transit time, the complexity of the absorbing surface, blood and lymph flow, and colonic microflora.

Variation of pH in different regions was found throughout the entire GI tract, as shown in Table 5. The small intestine has the largest absorptive surface area in the GI tract (Table 5) and therefore is the primary region responsible for drug absorption. However, the stomach controls the rate at which oral drug delivery systems reach the small intestine. In addition, the acidic environment and presence of pepsin in the stomach can influence drug stability and solubility. A number of factors influence the rate of gastric emptying, including stomach content volume, pH, caloric content, osmolarity, viscosity, and calcium sequestrant capacity (194). Davis et al. (195) studied the transit of pharmaceutical dosage forms through the small intestine in normal subjects using gamma scintigraphy. They found that solutions and pellets (<2 mm in size) were emptied rapidly (<1 h) even when the stomach was in the digestive mode, while matrix tablets and osmotic pumps were retained for longer periods of time, depending on the size of the meal. Small intestinal transit times were found constant (3 to 4 h) for all dosage forms and in different states. They concluded that the nature of the dosage form and the presence of food in the stomach affected gastric emptying, while small intestinal transit times were independent of the dosage form and fed state. Since small intestinal transit time does not vary, the arrival of an oral dosage form to the colon is determined mostly by the rate of gastric emptying. When the dosage form reaches the colon, the transit time depends on size, not density, especially in the proximal area of the colon. Small units pass through the colon slower than large units. Segmental transit measured by radioopaque markers gave values of 35 hours for the whole colon. Men had slightly shorter transit times than women, and this was significantly apparent in the proximal area of the colon (187). There is no experimental evidence for the existence of carrier or receptor mediated drug transport through the colon mucosa. Since the intercellular spaces are considerably smaller in the colon than in the duodenum and ileum, the principal route for colonic absorption appears to be transcellular rather than paracellular.

Table 40.5 pH and Absorbing Surface Area in the Gastrointestinal Tract

Region	pH	Absorbing surface area (m^2)
Stomach	1–3.5	~0.11
Duodenum	5–7	~0.09
Jejunum	6.1–7.1	~60
Ileum	7–8	~60
Colon	5.8–7.7	~0.25
Rectum	7–8	~0.03

Source: Refs. 191–193.

Exterior to the small intestinal lining, the microvilli are covered by a mucus layer about 300 μm thick. This layer is bound to a 0.1 μm thick glycocalyx, which is covalently linked to the microvilli. The negatively charged glycoprotein of the glycocalyx is balanced by positively charged counterions resulting in an acidic microclimate at the surface of the microvilli. It also has an unstirred water layer present at the surface. This layer can be the rate-limiting step in absorption of lipophilic drugs (196). However, studies in dogs and humans showed that the unstirred water layer has a maximal apparent thickness of only about 40 μm (197). With this thin layer, the barrier property of this layer is still questionable.

A recently recognized factor, which may play an important role in drug absorption through the small intestine, is the presence of P-glycoprotein, which is a membrane-associated transport protein. P-glycoproteins that are located in the apical membrane of enterocytes can actively transport a broad range of drugs out of the cell into the intestinal lumen. This results in a low total absorption of drugs. Examples of drugs that are P-glycoprotein substrates are anthracyclines, digoxin, verapamil, phenytoin, erythromycin, cyclosporin A, quinine, and dexamethasone (198).

The GI tract presents the largest metabolic barrier to the absorption of drugs, especially proteins and peptides, because the GI tract contains several enzymes, i.e., trypsin, chymotrypsin, elastase, carboxypeptidase, and aminopeptidase (80,199,200). The metabolism of many drugs occurs either in the lumen or in the wall of the GI tract, which can reduce the overall bioavailability and hence the pharmacological activity of a wide range of drugs. When such metabolism occurs within the lumen of the intestine, it simply reduces the amount of drug available for absorption. In the GI wall, metabolism can occur during the absorptive process, so the amount of active drug available within the body can be further decreased (201). Note that huge interindividual and interspecies differences in these processes have been recorded (202,203). Since venous drainage from the GI tract enters the portal circulation, first-pass metabolism in the liver will be a problem for many drugs. This problem can be avoided if a drug molecule is absorbed through the lymphatic circulation. However, the flow rate of lymphatic fluid (1 to 2 mL/min) compared with blood (2 L/min) is significantly lower. This may be one of the main reasons for the lower extent of absorption through the lymphatics (204). The activities of various proteases in the small intestine were generally higher than those in the large intestine. This tendency was more remarkable in the case of endopeptidases such as trypsin, chymotrypsin, and elastases (205). Once a drug is released into the colonic lumen, it is subjected to possible metabolism by host enzymes in the enterocytes and by microbial enzymes from the colon flora. A large number of aerobic and anaerobic bacteria are present throughout the entire length of the human GI tract, especially in the colon. The concentration of bacteria in the human colon is about 10^{11} to 10^{12} colony-forming units/mL and is dominated by anaerobic bacteria. Hence microbial metabolism proceeds primarily by reduction and hydrolysis without oxidation (206,207).

As in the colon, drug absorption from rectal administration may occur via the transcellular and/or paracellular pathways. Hence passive transport seems to be the main mechanism for rectal drug absorption (208). The fate of a drug absorbed from the rectum depends on its position in the rectum. The chance of a drug avoiding hepatic first-pass metabolism increases as the absorption site distance from the colon increases (189). Ritschel et al. (209) studied the first-pass elimination of lidocaine in the rabbit after peroral and rectal administration. The result indicated that about 30 percent of the rectally administered dose bypassed the liver after rectal absorption. This indicates that absorption from the rectum is partly via the portal route and partly via direct systemic delivery. This is similar to the finding in man, which was studied by de Boer et al. (210). However, the presence of extensive anastomoses may attenuate this effect. Absorption from the rectum is also accomplished by the lymphatics, which are located all along the gastrointestinal tract. This pathway can circumvent hepatic first-pass metabolism; however, it should be noted that the total amount of drug absorbed through the lymphatic appears to be quite small compared with the blood circulation (189,211).

According to metabolic studies, the rectal mucosa had consistently high activities of aminopeptidase and esterase (26,27,212). This coupled with the other disadvantages, i.e., interruption of absorption by defecation, lack of acceptability by patients, limited area, and local irritation, make this route less desirable than other routes. However, this route can be useful for the delivery of drugs to patients with difficulty using the oral or other routes, such as children (213–215) and psychiatric patients (216).

3. Examples

Despite several barriers to drug delivery that exist within the GI tract, drug administration through the GI tract continues to be the most popular route. To be able to maintain the successful therapeutic effect of a drug, the correct amount should be released and maintained during a specific time. Moreover, the drug should be delivered or targeted to a specific region in the GI tract. There are a number of developments relative to the GI tract that have been reported. This report will review some of these, including bioadhesive systems, enteric-coated systems, and rectal delivery systems.

The original idea of using bioadhesives as platforms for oral controlled drug delivery was to use these polymers to control and prolong the gastrointestinal transit of the systems (217–219). However, in vivo results obtained in humans did not show that bioadhesive systems were able to control or slow down significantly the gastrointestinal transit time of dosage forms (220). Possible explanations of this failure are overhydration of the bioadhesive, inactivation of the bioadhesive by soluble mucin, and fast turnover rate of mucus (221). Microspheres prepared by using a combination of polyglycerol esters of fatty acid (PGEF) and Carbopol 934P have been demonstrated to be good candidates for increasing the gastrointestinal residence time, especially in the stomach and upper small intestine (222). Akiyama et al. (222) prepared three types of microsphere based on this combination including Carbopol-coated microspheres, PGEF-based microspheres, and Carbopol dispersion in PGEF microspheres. Only the dispersion type microspheres, referred as an Adhesive Micromatrix System, showed a promising result. They claimed that this system prolonged gastrointestinal transit time by adhering to the gastric or intestinal mucosae in the rat.

Studies in man, using furosemide and riboflavin as markers, were done to comfirm this conclusion (223). Since furosemide and riboflavin are compounds that are absorbed in the upper small intestine, drug absorption after administration of bioadhesive microspheres should be higher than that of nonadhesive microspheres, if the gastrointestinal transit time of bioadhesive microspheres is prolonged. Results from humans confirm the former conclusion. Therefore this system may be useful for drugs with low bioavailability and drugs expecting a local action. In fact, Nagahara et al. (224) tried to use this system for increasing the anti-*Helicobacter pylori* effect of amoxicillin. This idea was done following the publication by Kimura et al. (225), who demonstrated that 1-hour topical treatment with the infused solution of high concentration of antimicrobial agents provided more complete eradication of *H. pylori* than conventional therapy due to the extended gastric residence times of the drugs. The percentage of amoxicillin remaining in the stomach after oral administration of the bioadhesive microspheres to Mongolian gerbils under fed conditions was about three times higher than that of a 0.5% methylcellulose suspension. Moreover, the required dose of amoxicillin to eradicate *H. pylori* was effectively reduced by a factor of ten when the bioadhesive microspheres were used (224). Chickering et al. (226) demonstrated that polymeric microspheres made of poly(fumaric-co-sebacic), a bioerodible polyanhydride, displayed good adhesion. In vivo gastrointestinal studies using barium as a marker showed that these microspheres had a residence time of 24 to 36 hours. Optical microscopy was used to explore the absorption efficacy of these microspheres. After 24 hours, large amount of microspheres were observed in the enterocytes, the Payer's patches, and also in the spleen and liver tissues (227). Further studies were performed to determine whether these microspheres could be used as oral drug carriers. Microspheres composed of poly(fumaric-co-sebacic) in a 20:80 molar ratio were used to encapsulate dicumarol and plasmid DNA. Oral administration of these drug-containing microspheres to rats showed an improvement in the oral bioavailability of these drugs. In the same study, insulin was encapsulated in a blend of poly(fumaric anhydride) and poly(lactide-co-glycolide), and the microspheres were administered to the group of fasted rats that were injected subcutaneously with an initial glucose load. The results showed that this preparation was better able to regulate the glucose load than the controls, soluble insulin in saline (227). Coating liposomes with bioadhesive also showed a good relationship with prolonged the residence time. Takeuchi et al. (228) evaluated the adhesive property of the bioadhesive-coated liposomes in vitro using rat intestine. Multilamellar liposomes of dipalmitoyl phosphatidylcholine and dicetyl phosphate in a molar ratio of 8:2 were coated with three different types of bioadhesives including chitosan, polyvinyl alcohol, and poly(acrylic acid). They found that chitosan-coated liposomes had the highest adhesive property. An in vivo study was carried out using insulin as a model drug. A marked reduction in basal blood glucose level was observed within 30 minutes after administration of the chitosan-coated liposomes containing insulin and was prolonged for up to 12 hours. This prolonging effect could be attributed to the bioadhesive property. It is interesting to note that the relative effectiveness of this system was only 10% compared to the case of a 5 International Unit (IU) subcutaneous injection. An oil-in-water bioadhesive submicron emulsion was used as a carrier for desmopressin acetate (229). Oral administration of these formulations using Carbopol 940 as a bioadhesive showed substantial enhancement, up to 12-fold, of the rat oral bioavailability of desmopressin acetate with regard to a simple saline solution of the drug.

These examples are based on the use of a nonspecific bioadhesive interaction as a tool for localizing drug delivery systems to the surface of the GI tract and hence increasing bioavailability of the drugs. However, targeting to a specific area of the mucosa is still difficult. To deal with this problem, specific bioadhesion using ligand–receptor interactions was investigated. Lectin is an interesting ligand that can bind specifically to biological surfaces bearing sugar residues located at the surface of epithelial cells (221,230–233). It was demonstrated that tomato lectin will bind to the small intestinal epithelium surface in vitro (232,234). This binding was shown to be specific and mediated through N-acetylglucosamine-containing glycoconjugates. The uptake in vitro study of tomato lectin was shown to be taken up by enterocytes via specific adsorptive endocytosis (235). Note that an in vivo study did not show a significant difference in the transit time of tomato lectin and polyvinylpyrrolidone (236). Specific bioadhesion particulate systems for oral delivery purposes can be designed by grafting lectin onto the surface of particulate carriers such as microspheres, nanoparticles, and liposomes. Irache et al. (237) proved this idea through lectin–latex conjugates by covalent coupling of different lectins (tomato, asparagus, pea, and mycoplasma) to polystyrene latexes. In vitro interaction between lectin–latex conjugates and pig gastric mucin was tested. The results showed that the lectin–latex conjugates still had the properties of free lectin such as high binding affinity and high specificity for carbohydrate moieties of mucus glycoproteins. These results support the former publication, in which Lehr et al. (232) found that tomato lectin–coated polystyrene microspheres showed specific binding to enterocytes in vitro.

Hussain et al. (238) investigated the oral uptake of tomato lectin–conjugated nanoparticles in rat intestine. They found the enhancement of intestinal transcytosis at the site where these nanoparticles were bound. Another example is lectin-bearing polymerized liposomes. Lectins can be incorporated into a liposome membrane by modifying them with a hydrophobic anchor, N-glutaryl-phosphatidylethanolamine. Chen et al. (239) incorporated modified lectins into liposome bilayers, and then the liposomes were stabilized via polymerization. Studies in mice established that these systems had the potential to be used as vehicles for an oral vaccine because they can promote binding to Peyer's patches.

Enterio-coated systems are those that remain intact in the stomach but will dissolve and release drugs once they reach the intestinal tract (240,241). The traditional purposes of using these systems are preventing destruction of the drug by gastric enzymes or by the acidity of the gastric fluid, preventing nausea and vomiting caused by the drug's irritation, and delivering the drug to the primary absorption site. Recently, such coatings have been used as a controlled-release dosage form. In principle, the action of enteric coating results from a difference in pH and enzymatic properties in the stomach and intestinal tract. pH-sensitive polymers used as enteric coating are undissociated in the low-pH environment of the stomach and dissolve when the pH rises to appropriate values (Table 6). For example, CAP will dissolve when the pH of the environment increases to 6, while CAT will dissolve at a lower pH (4.8). Since the variation of pH in the GI tract is obvious and is changed by foods or drugs, enteric coated systems for the distal area of the GI tract are still unpredictable (243).

Coatings subject to enzymatic breakdown are considered as potential colonic drug delivery systems because of the presence of bacteria in the colon. Saffran et al. (244) developed a copolymer of styrene and 2-hydroxyethyl methacrylate cross-linked with divinylazobenzene to coat oral dosage forms. They showed the success of delivering insulin to rats and dogs via oral administration. The coating polymer is degraded by bacterial azoreductase, and the drug is subsequently released. Later, Van den Mooter et al. (245,246) studied the influence of hydrophilicity on the degradation of azopolymers. They found that degradation of the azo polymers was strongly affected by hydrophilicity of the polymers. The hydrophilic moiety ensures a good availability of the azo group for bacterial reduction, while the hydrophobic moiety provides resistance to degradation by the gastric and small intestinal fluids. Therefore, a balance between hydrophilic and hydrophobic moieties in the polymer must be considered. Kopecek et al. (247) introduced two new ideas to develop a polymeric carrier for site-specific drug delivery to the colon. The first system is based on the concept of binding polymeric carriers to the colonic mucosa followed by the degrading action of microbial enzymes. One example of these polymers is N-(2-hydroxypropyl)methacrylamide copolymers containing high amounts of both bioadhesive moiety (fucosylamine) and 5-aminosalicylic acid (a drug for the treatment of ulcerative colitis). Fucosylamine will bind to the colonic mucosa, and the system will release drug by the degrading action of azoreductase (248). The second system, a biodegradable pH-sensitive hydrogel, is based on the swelling properties of the polymer at different pH and an enzymatically degradable azoaromatic cross-link. This polymer has a low degree of swelling in the low-pH environment of the stomach, and thus the drug is protected against digestion

Table 40.6 Commonly Used Enteric Coating Materials

Polymer	Dissolved pH
Cellulose acetate trimellitate (CAT)	4.8
Polyvinyl acetate phthalate (PVAP)	5.0
Carboxymethylethylcellulose (CMEC)	5.0
HPMCP 50	5.2
HPMCP 55, HPMCP 55S	5.4
Hydroxypropylmethylcellulose acetate succinate (HPMCAS)	5.0, 5.5, 7.0 (three grades)
Eudragit L30D	5.6
Cellulose acetate phthalate (CAP)	6.0
Eudragit L	6.0
Eudragit S	6.8

HPMCP = Hydroxypropylmethylcellulose phthalate
Eudragit = Tradename of methalic acid copolymers
Source: Refs. 241, 242.

by enzymes. The degree of swelling increases as the polymer passes down the GI tract according to an increase in pH. In the colon, the polymer has a high degree of swelling that makes the cross-links accessible to azoreductase. When the polymer is degraded, drug is released (249,250).

Rectally administered drugs can be given by several dosage forms such as suppositories, enemas, and foams. The spread of the rectal formulations within the distal colon is dependent on both type and volume of the formulations. Studies using scintigraphy have shown that the spread of both enemas and foams was more extensive than that of suppositories (251–253). In fact, the spread of suppositories was limited to the rectal vault (252,253), while enemas and foams can spread as far as the transverse colon (253,254). Therefore suppository dosage forms should be more appropriate for systemic absorption of drugs to minimize first-pass metabolism. Enemas and foams, however, are appropriate for local activity in the colon.

Ulcerative colitis is a primary disease of nonsmokers, for which nicotine is a therapeutic agent. However, long-term transdermal nicotine treatment is likely to be limited by side effects (255,256). Pharmacokinetic studies in humans demonstrated that administration of nicotine as a liquid enema resulted in less systemic absorption and thus decreased side effects (257,258). Moreover, Sandborn et al. (256) observed the clinical improvement in patients with mild to moderately active left-sided ulcerative colitis as nicotine tartrate liquid enema was administered at a dose of 3 mg nicotine base per day for 1 week and then 6 mg per day for 3 weeks. Otten et al. (259) compared two products containing 5-aminosalicylic acid, which is widely used as a topical treatment in distal inflamatory bowel disease. These products had different properties such as viscosity and particle size. They found that the product with a larger volume and smaller particle size reached a substantially larger proportion of the colon. Furthermore, with higher viscosity, this product also produced a significantly higher retention time in the proximal parts of the large intestine.

In an attempt to restrict drug absorption from suppositories to only the lower rectum, several approaches have been explored. These included the use of hydrogel, bioadhesive, or thermoreversible polymer systems as vehicles. The formulations with sustained-release properties have been designed using the hydrogel as a vehicle. Polyvinyl alcohol hydrogel suppositories containing indomethacin were prepared and tested in rats and dogs (260). The plasma concentration of indomethacin showed a sustained-release profile, and formulations with higher pH showed higher plasma concentration. Propranolol HCl was also used as a model drug in this vehicle. By adding phospholipid into the formulations, Morimoto et al. (261) found better bioavailability than that with a conventional suppository (Witepsol H15). Cole et al. (262) used a cross-linked polyethylene glycol 4000 to prepare a sustained-release formulation of morphine sulfate that was administered to human volunteers. The plasma morphine concentrations in these subjects increased to approximately steady values within 2 to 4 hours after administration and were maintained for at least 12 hours. Bioadhesives have been incorporated into suppository formulations to improve bioavailability of the drugs. Recently, Hosny (263) prepared an insulin suppository by adding deoxycholic acid, sodium taurocholate, and/or polycarbophil to the formulations. Using the rabbit as an animal model, the greatest relative hypoglycemia was obtained from the suppository formulation containing all three materials. This formulation showed a relative bioavailability of 56.4% compared with subcutaneous injection. Another example was by Yahagi et al. (264). They prepared bioadhesive lidocaine suppositories using Witepsol H15 as a base, and Carbopol 934P and white beeswax as additives. Carbopol has a bioadhesive property, while white beeswax was added to increase the melting point of the suppositories. Suppositories containing 10% Carbopol and 20 to 30% white beeswax were retained in the lower rectum for at least 2 hours, but adding white beeswax reduced drug release. Therefore double-phased suppositories consisting of a bioadhesive front layer containing 10% Carbopol and 20% white beeswax, and a second layer containing Carbopol and lidocaine, were designed. The double-phased formulation, with the addition of 5% Carbopol in the second layer, exhibited the lowest metabolites/lidocaine area under the curve ratio. This can be explained in that the bioadhesive front layer containing white beeswax prevented movement of the second layer toward the upper rectum and limited the absorption region to the lower rectum, and hence reduced first-pass metabolism. A new approach employed a bioadhesive liquid suppository composed of a thermoreversible polymer and bioadhesives. These systems exist as liquids at room temperature and form gels at body temperature, so they can be easily applied into the rectum as liquids and remain in the rectum without leakage as gels. Choi et al. (265) developed acetaminophen liquid suppositories composed of Poloxamer 188, 407, and polycarbophil as a vehicle. Pharmacokinetic studies in rats showed good bioavailability and did not cause damage to the rectal tissues. Insulin was also incorporated into these formulations (266). The pharmacodynamic studies and quantitative histological assessment of rectal mucosa of rats were carried out. The optimum result was found for the formulation of insulin: P407: P188: polycarbophil: sodium salicylate (100 IU/g: 15: 20: 0.5: 10%). From these studies, one can conclude that thermoreversible liquid suppositories may be more convenient, safe, and effective rectal delivery systems.

E. Vaginal

1. Relevant Anatomy and Physiology

The vagina is a fibromuscular tube that extends to the lower part of the uterine cervix (267–269). In adult premenopausal women, the vagina is approximately 7 to 8 cm in length and 2 cm wide, and shrinks to approximately 4.5 to 6 cm in length and 1 to 1.5 cm in width in postmenopausal women (270). The vaginal wall consists of three layers: an epithelial layer, the muscular coat, and the tunica adventitia. The epithelium is a nonkeratinized stratified squamous type with thickness about 200 to 300 μm. The thickness of the vaginal epithelium varies with age and menstrual cycle. In postmenopausal women and ovariectomized animals, the vaginal epithelium becomes extremely thin and porous, leading to a substantial increase in permeability of this tissue. The menstrual cycle is the human female reproductive cycle. In most women, this cycle lasts about 28 days (21 to 35 days). The first menstrual day is numbered as day 1. At day 14, ovulation occurs, and splits the ovarian cycle into the follicular and luteal phase. These phases represent different hormone levels as described in Table 7.

The vaginal epithelium consists of five different cell layers: basal, parabasal, intermediate, transitional, and superficial layers. The cyclical variations of the vaginal epithelium generally involve proliferation, differentiation, and desquamation. The intermediate, transitional, and superficial layers are strongly affected by the cycle and become thickest at ovulation. During the luteal phase, desquamation occurs on the superficial epithelial layer extending as far as the intermediate cells. Loosening of intercellular grooves and a porelike widening of intercellular crevices following ovulation precedes this cyclic desquamation. The intercellular channels are narrow during the early follicular phase but become wide at ovulation and during the luteal phase. This can be explained by the change of estrogen and progesterone hormone level during the menstrual cycle. Estrogen causes an increase in mucosal thickness and the amount of endocervical fluid, while progesterone causes an increase in fluid secretion from the epithelial layers and an increase in glycogen, which is metabolized to decrease pH. Estrogen also decreases fluid viscosity, but progesterone increases viscosity. A related physiological event associated with a decline in estrogen levels is a substantial reduction in vaginal blood flow with concomitant drying of vaginal tissue (270). The surface area of the vagina is increased by numerous foldings and by microridges, which run either longitudinally or in circles.

Although the vaginal epithelium lacks specific moisturizing glands, the surface is usually covered with a film of moisture exuded from the vascular bed of the lamina propria and the cervix. The origin and composition of vaginal fluid is complex and its volume varies with age, menstrual phase, and degree of sexual excitement (272). pH of the

Table 40.7 Phases of the Human Menstrual Cycle

Days	Phase	Hormone	Numbers of cell layers
1 to 4	Follicular phase begins	Estradiol, progesterone, FSH[B] and LH[C]: low	18
5 to 13	Follicular phase continues	Estradiol: rising Progesterone, FSH and LH: low	22
14	Ovalatory phase	Estradiol: falls after follicle ovulation Progesterone: rising FSH and LH: sharply increase (peak)	45
15 to 28	Luteal phase	Estradiol and progesterone: high FSH and LH: low	23 to 33

A = Day 1 is the first menstrual day.
B = Follicle stimulating hormone.
C = Luteinizing hormone.
Source: Refs. 269, 271.

vagina varies between 4 to 5 in adult premenopausal women depending on the menstrual cycle and location, whereas in the postmenopausal women neutral or alkaline values are typical (273). Blood supply to the vagina is via the uterine and pudendal arteries, which arise from the internal iliac arteries. The vagina is drained by a rich venous plexus, which empties into the internal iliac veins.

2. Major Biological Barriers and Appropriate Strategies

The transcellular and paracellular pathways for drug absorption through the vaginal mucosa are apparently the same as those found in other mucosae (269,274). It was proposed that the absorption barriers of the vagina consisted of two barriers in series, which are an effective aqueous barrier and a membrane barrier (275). The membrane barrier consists of a lipoidal pathway (transcellular) and aqueous pore pathway (paracellular). In addition, Hwang et al. (276) found that the permeability coefficient of the aqueous pore for 1-alkanoic acids was significantly smaller than that for alcohols. Therefore they concluded that the aqueous pore might be semipolar in character or negatively charged, so that aqueous anions had more difficulty being transported via this pathway.

Cyclic changes in the thickness and porosity of the vaginal epithelium can also affect absorption; however, different results have been reported. For example, the effect of the estrus cycle on the permeability of Vidarabine in mice vaginal mucosa has been demonstrated. The results revealed that the permeability coefficients during the diestrus phase were 10 to 100 times higher than those obtained in the estrus phase (277). Okada et al. (278) showed the effect of estrus cyclic changes on vaginal absorption of insulin, phenolsulfonphthalein, and salicylic acid in rats. They concluded that the permeability for hydrophobic compounds, which are transported mainly through the transcellular pathway, is less affected by the estrus cycle; whereas the permeability for hydrophobic compounds, which are transported mainly through intercellular channels, is highly dependent on the estrus cycle. Note that the estrus cycle of the rat is completed in 4 to 5 days, and during this cycle, changes in the vaginal mucosal membrane, the ovaries, and the uterus occur. These changes are similar to the vaginal mucosa changes in women during the menstrual cycle. In contrast to these results, the vaginal absorption of methanol, octanol, progesterone, and estrone in monkeys showed no relationship between the menstrual cycle and the vaginal membrane permeability. The result is also the same as the case of prostaglandin E_1 in humans (268).

Regarding the metabolic barrier, several enzymes have been found in the vaginal mucosa. The outer cell layers of the vagina contain varying amounts of β-glucuronidase, acid phosphatase, phosphoamidase, succinic dehydrogenase, and diphosphopyridine nucleotide diaphorase (279). The protease enzyme activities could also potentially be an important barrier in the delivery of protein and peptide drugs through the vaginal route (24,269). The enzymatic activity can be changed with hormonal fluctuation in premenopausal and postmenopausal women (279–281). Acarturk and Robinson (281) studied the vaginal permeability and enzymatic activity in intact and ovariectomized rabbits. The ovariectomized rabbits showed a thinner mucosal epithelium and a higher vaginal permeability to enkephalin. In addition, the activity of aminopeptidase N was significantly decreased in these rabbits, while the activity of leucine aminopeptidase, aminopeptidase A, and aminopeptidase B remained unchanged in both cases. They suggested that the ovariectomized rabbit might be useful as an animal model for postmenopausal vaginal studies. Since the activity of several vaginal enzymes was found to vary with the menstrual cycle and postmenopausally, variation in enzyme activity with hormonal changes should be added to the consideration of drug delivery design.

3. Examples

Typical vaginal delivery systems include solutions, gels, foams, and suppositories. Each of these systems suffers from a short residence time in the vaginal region. Therefore formulations that can provide a prolonged contact time will be discussed.

Robinson and Bologna (270) described a vaginal delivery system using the bioadhesive polycarbophil. This system can remain on vaginal tissues for 3 to 4 days. Moreover, intrinsic properties of polycarbophil can alter the local pH and hydration level of the tissue. A gel product was an emulsion gel of approximately 60,000 to 80,000 cps, containing 1 to 3% polycarbophil plus humectants and lipid lubricants in an aqueous base. In postmenopausal women, this gel showed a reduction in pH from about 6 to 4 and maintenance in this pH for about 3 to 4 days after the last dose. It also created a significant increase in blood flow, which then provided hydration of the vaginal tissues. Therefore a low viscosity gel of this polymer can be used as an effective treating agent for dry vagina in postmenopausal women and those female cancer patients on estrogen blockers. In contrast to this report, Brown et al. (282) used the γ-scintigraphy method to study the spreading and retention of this formulation in postmenopausal women. They found a lack of significant retention time and considerable intersubject variability.

Bioadhesive gel formulation also has a capacity to serve as a platform for drug delivery (270). For example, vaginally delivered progesterone containing polycarbophil allowed a prolonged delivery of progesterone to achieve clinical effect with low side effects. In another case, a combination of this polymer with nonoxynol 9, a spermicidal/antiviral agent, offered suitable potential benefit in the area of contraception and prophylaxis against sexually transmitted diseases. Lee and Chien (283) used Carbopol 934P as a gel base for a combination of nonoxynol 9 and chelating agent (EDTA). This system showed longer contact time and greater surface contact with the vaginal mucosa, so the effective drug concentration could be maintained for a prolonged period of time. These results confirmed that a bioadhesive gel formulation is suitable for controlled release of nonoxynol 9 and enhances fertility control by providing prolonged spermicidal action.

Another formulation for nonoxynol 9 was prepared by coprecipitation with polyvinylpyrrolidone with or without iodine to produce powders, and then incorporated into bilayer tablets or hard gelatin capsules (284). Bilayer tablets consisted of fast and slow releasing compartments, while hard gelatin capsules were composed of fast, intermediate, and slow releasing compartments. After intravaginal administration in rabbits, these systems released their nonoxynol 9 and iodine content rapidly, reaching spermicidal levels within 3 minutes and that lasted for at least 4 hours.

Richardson et al. (285) prepared vaginal delivery systems for calcitonin using the microspheres of hyaluronane esters. The vaginal absorption of calcitonin was compared after administration of these microspheres and calcitonin solution to rats. The results from the microspheres showed more rapid pronounced hypocalcaemic effects than a simple calcitonin solution. This enhancement of absorption may be due to intimate contact between the microspheres and the vaginal mucosa, as shown by microscopic examination. In addition, the distribution and retention time of these formulations were evaluated in sheep by using γ-scintigraphy (286). They found that about 68 to 80% of the radioactive remained within the vagina, and there was no indication of movement of the microspheres to the upper levels of the genital tract. Note that the microspheres in dry powder form showed slightly longer retention time than the vaginal pessary form. These results demonstrated the potential of using the microspheres as a sustained-release intravaginal delivery system.

Liposomes containing progesterone were developed and evaluated for use as an intravaginal contraceptive system (287). These liposomes were incorporated into polyacrylamide gel and progestational activity determined. It was observed that the formation of functioning corpora lutea was decreased in treated animals. Moreover, the effect of liposomal preparation was found to be greater and more prolonged as compared to a control gel.

V. CONCLUSION

In this chapter, several areas have been reviewed including relevant anatomy and physiology of mucosae in different areas, major biological barriers, and some examples of drug delivery systems through these routes. However, our knowledge of these areas is still insufficient to warrant optimizm about the routine delivery of drugs.

Drug delivery via the mucosal routes offers many advantages, which support these routes as alternatives to the injection route. However, the presence of several barriers creates some difficulties in the design of promising drug delivery systems. Each of these barriers can be overcome by one or more possible solutions. At present, there is no universal solution to all problems. Therefore it is reasonable to conclude that there remains a need for properly designed drug delivery systems, and careful consideration of the relevant anatomy and physiology of mucosae will help pharmaceutical scientists in the development of more efficient and safe drug delivery systems.

REFERENCES

1. I. R. Telford, C. F. Bridgman. Specializations of epithelia-glands, serous and mucous membrane (Chapter 5). In: *Introduction to Functional Histology*. 2[d] ed. Harper Collins College, New York, 1995, pp. 69–77.
2. M. H. Ross, L. J. Romrell, G. I. Kaye, (eds. Epithelial tissue Chapter 4). In: *Histology a Text and Atlas*. 3[d] ed. Williams and Wilkins, Baltimore, 1995, pp. 58–93.
3. C. G. Wilson, C. Washington, N. Washington. Overview of epithelial barriers and drug transport (Chapter 1). In: *Physiological Pharmaceutics: Biological Barriers to Drug Absorption* (C. G. Wilson, N. Washington, eds.). Ellis Harwood, Chichester, U.K., 1989, pp. 11–20.
4. K. L. Audus, R. L. Bartel, I. J. Hidalgo, R. T. Borchardt. The use of cultured epithelial and endothelial cells for drug transport and metabolism studies. *Pharm. Res. 7*: 435–451 (1990).
5. Y. W. Chien. Biopharmaceutics basis for transmucosal delivery. *S.T.P. Pharm. Sci. 5*(4):257–275 (1995).
6. L. Narawane, V. H. L. Lee. Absorption barriers. In: *Drug Absorption Enhancement: Concepts, Possibilities, Limitations and Trends* (A. G. de Boer, ed.). Harwood Academic Publishers, 1994, pp. 1–66.
7. D. W. Powell. Barrier function of epithelia. *Am. J. Physiol. 241*:G275–G288 (1981).

8. S. Muranishi. Absorption enhancers. *Crit. Rev. Ther. Drug. Carrier. Syst.* *7*(1):1–33 (1990).
9. T. Z. Csaky. Intestinal permeation and permeability: an overview (Chapter 2). In: *Handbook of Experimental Pharmacology, Vol. 70/I* (T. Z. Csaky, ed.). Springer-Verlag, New York, 1984, pp. 51–59.
10. A.-L. Cornaz, P. Buri. Nasal mucosa as an absorption barrier. *Eur. J. Pharm. Biopharm.* *40*(5):261–270 (1994).
11. D. Cremaschi, C. Rossetti, M. T. Draghetti, C. Manzoni, V. Aliverti. Active transport of polypeptides in rabbit respiratory nasal mucosa. *J. Cont. Rel.* *13*:319–320 (1990).
12. S. Travis, I. Menzies. Intestinal permeability: functional assessment and significance. *Clin. Sci.* *82*:471–488 (1992).
13. V. H. L. Lee. A. Yamamoto, U. V. Kompella. Mucosal penetration enhancers for facilitation of peptide and protein drug absorption. *Crit. Rev. Ther. Drug. Carrier. Syst.* *8*:91–92 (1991).
14. J. L. Madara. Loosening tight junctions: lessons from the intestine. *J. Clin. Invest.* *83*(4):1089–1094 (1989).
15. N. J. Lane, T. S. Reese, B. Kachar. Structural domains of the tight junctional intramembrane fibrils. *Tissue and Cell* *24*(2):291–300 (1992).
16. J. Hochman, P. Artursson. Mechanisms of absorption enhancement and tight junction regulation. *J. Cont. Rel.* *29*: 253–267 (1994).
17. A. P. Sayani, Y. W. Chien. Systemic delivery of peptides and proteins across absorptive mucosae. *Crit. Rev. Ther. Drug. Carrier. Syst.* *13*(1&2):85–184 (1996).
18. Y. Rojanasakul, L.-Y. Wang, M. Bhat, D. D. Glover, C. J. Malanga, J. K. H. Ma. The transport barrier of epithelia: a comparative study on membrane permeability and charge selectivity in the rabbit. *Pharm. Res.* *9*(8):1029–1034 (1992).
19. F. Nimmerfall, J. Rosenthaler. Significance of the goblet-cell mucin layer, the outermost luminal barrier to passage through the gut wall. *Biochem. Biophys. Res. Comm.* *94*: 960–966 (1980).
20. A. Wikman, J. Karlsson, I. Carlstedt, P. Artorsson. A drug absorption model based on the mucus layer produing human intestinal goblet cell line HT29-H. *Pharm. Res.* *10*(6): 843–852 (1993).
21. P. G. Bhat, D. R. Flanagan, M. D. Donovan. The limiting role of mucus in drug absorption: drug permeation through mucus solution. *Int. J. Pharm.* *126*:179–187 (1995).
22. P. G. Welling, ed. Drug metabolism (Chapter 9). In: *Pharmacokinetics: Processes, Methamatics, and Applications.* 2[d] ed. American Chemical Society, Washington D.C., 1997, pp. 145–162.
23. J. S. Bond, R. J. Beynon. Proteolysis and physiological regulation. *Molec. Aspects Med.* *9*:173–287 (1987).
24. V. H. L. Lee. Enzymatic barriers to peptide and protein absorption. *Crit. Rev. Ther. Drug. Carrier. Syst.* *5*:69–98 (1988).
25. V. H. L. Lee. A. Yamamoto. Penetration and enzymatic barriers to peptide and protein absorption. *Adv. Drug. Del. Rev.* *4*:171–207 (1990).
26. S. D. Kashi, V. H. L. Lee. Enkephalin hydrolysis in homogenates of various absorptive mucosae of the albino rabbit: similarities in rates and involvement of aminopeptidases. *Life Sci.* *38*:2019–2028 (1986).
27. A. Yamamoto, E. Hayakawa, V. H. L. Lee. Insulin and proinsulin proteolysis in mucosal homogenates of the albino rabbit: implications in peptide delivery from normal routes. *Life Sci.* *47*:2465–2474 (1990).
28. W.-C. Shen, Y.-J. Lin. Basic mechanisms in transepithelium transport enhancement. In: *Drug Permeation Enhancement: Theory and Application* (D. S. Hsieh, ed.). Marcel Dekker, New York, 1994, pp. 25–40.
29. E. S. Swenson, W. J. Curatolo. Intestinal permeability enhancement for proteins, peptides and other polar drugs: mechanisms and potential toxicity. *Adv. Drug. Del. Rev.* *8*: 39–92 (1992).
30. E. S. Swenson, W. B. Milisen, W. Curatolo. Intestinal permeability enhancement: efficacy, acute local toxicity, and reversibility. *Pharm. Res.* *11*(8):1132–1142 (1994).
31. D. J. Freeman, R. W. Niven. The influence of sodium glycocholate and other additives on the in vivo transfection of plasmid DNA in the lungs. *Pharm. Res.* *13*(2):202–209 (1996).
32. B. J. Aungst, H. Saitoh, D. L. Burcham, S.-M. Huang, S. A. Mousa, M. A. Hussain. Enhancement of the intestinal absorption of peptides and nonpeptides. *J. Cont. Rel.* *41*: 19–31 (1996).
33. M. J. Sequra-Bono, T. M. Garrigues, V. Merino, M. V. Bermejo. Compared effects of synthetic and natural bile acid surfactants on xenobiotic absorption. III. Studies with mixed micelles. *Int. J. Pharm.* *107*:159–166 (1994).
34. S. Senel, Y. Capan, M. F. Sargon, G. Ikinci, D. Solpan, O. Guven, H. E. Bodde, A. A. Hincal. Enhancement of tranbuccal permeation of morphine sulphate by sodium glycodeoxycholate in vitro. *J. Cont. Rel.* *45*:153–162 (1997).
35. Y.-S. Quan, K. Hattori, E. Lundborg, T. Fujita, M. Murakami, S. Muranishi, A. Yamamoto. Effectiveness and toxicity screening of various absorption enhances using Caco-2 cell monolayers. *Biol. Pharm. Bull.* *21*(6):615–620 (1998).
36. S. Okumura, Y. Fukuda, K. Takahashi, T. Fujita, A. Yamamoto, S. Muranishi. Transport of drugs across the Xenopus pulmonary membrane and their absorption enhancement by various absorption enhancers. *Pharm. Res.* *13*(8): 1247–1251 (1996).
37. T. Sugiyama, A. Yamamoto, Y. Kawabe, T. Uchiyama, Y.-S. Quan, and S. Muranishi. Effects of various absorption enhancements on the intestinal absorption of water soluble drugs by in vitro Ussing chamber method: correlation with and in situ absorption experiment. *Biol. Pharm. Bull.* *20*(7):812–814 (1997).
38. A. Fasano, G. Budillon, S. Guandalini, R. Cuomo, G. Parrilli, A. M. Cangiotti, M. Morroni, A. Rubino. Bile acids reversible effects on small intestinal permeability: an in vitro study in the rabbit. *Dig. Dis. Sci.* *35*(7):801–808 (1990).

39. S. Feldman, M. Reinhard, C. Willson. Effect of sodium taurodeoxycholate on biological membranes: release of phosphorus, phospholipid, and protein from everted rat small intestine. *J. Pharm. Sci. 62*(12):1961–1964 (1973).
40. G. P. Martin, C. Marriot, I. W. Kellaway. Direct effect of bile salt and phospholipids on the physical properties of mucus. *Gut 19*(2):103–107 (1978).
41. H. Yoshitomi, T. Nishihata, G. Frederick, M. Dillsaver, T. Higuchi. Effect of triglyceride on small intestinal absorption of cefoxitin in rats. *J. Pharm. Pharmacol. 39*: 887–891 (1987).
42. T. Lindmark, J. D. Soderholm, G. Olaison, G. Alvan, G. Ocklind, P. Artursson. Mechanism of absorption enhancement in humans after rectal administration of Ampicillin in suppositories containing sodium caprate. *Pharm. Res. 14*(7):930–935 (1997).
43. G. S. M. J. E. Duchateau, J. Zuidema, F. W. H. M. Merkus. Bioavailability of propanolol after oral, sublingual, and intranasal administration. *Pharm. Res. 3*(2):108–111 (1986).
44. H. Fukui, M. Murakami, K. Takada, S. Muranishi. Combinative promotion effect of Azone and fusogenic fatty acid on the large intestinal absorption in rat. *Int. J. Pharm. 31*: 239–246 (1986).
45. J. A. Fix, K. Eagle, P. A. Porter, P. S. Leppert, S. J. Selk, C. R. Gardner, J. Alexander. Acylcarnitines: drug absorption-enhancing agents in the gastrointestinal tract. *Am. J. Physiol. 251*:G332–G340 (1986).
46. E. K. Anderberg, T. Lindmark, P. Artursson. Sodium caprate elicits dilations in human intestinal tight junctions and enhances drug absorption by the paracellular route. *Pharm. Res. 10*(6):857–864 (1993).
47. J. C. Verhoef, N. G. M. Schipper, S. G. Romeijn, F. W. H. M. Merkus. The potential of cyclodextrins as absorption enhancers in nasal delivery of peptide drugs. *J. Cont. Rel. 29*:351–360 (1994).
48. Z. Shao, R. Krishnamoorthy, A. K. Mitra. Cyclodextrins as nasal absorption promoters of insulin: mechanistic evaluations. *Pharm. Res. 9*:1157–1163 (1992).
49. Y. Watanabe, Y. Matsumoto, M. Seki, M. Takase, M. Matsumoto. Absorption enhancement of polypeptide drugs by cyclodextrins: I. Enhanced rectal absorption of insulin from hollow type suppositories containing insulin and cyclodextrins in rabbits. *Chem. Pharm. Bull. 40*: 3042–3047 (1992).
50. Y. Watanabe, Y. Matsumoto, K. Kawamoto, S. Yazawa, M. Matsumoto. Enhancing effect of cyclodextrins on nasal absorption of insulin and its duration in rabbits. *Chem. Pharm. Bull. 40*:3100–3104 (1992).
51. T. Irie, K. Wakamatsu, H. Arima, K. Uekama. Enhancing effects of cyclodextrins on nasal absorption of insulin in rats. *Int. J. Pharm. 84*:129–139 (1992).
52. E. Windsor, G. E. Cronheim. Gastro-intestinal absorption of Heparin and synthetic Heparinoids. *Nature 190*(4772): 263–264 (1961).
53. M. M. Cassidy, C. S. Tidball. Cellular mechanism of intestinal permeability alterations produced by chelation depletion. *J. Cell. Biol 32*:685–698 (1967).
54. T. Nishihata, J. H. Rytting, T. Higuchi. Enhanced rectal absorption of theophylline, lidocaine, cefmetazole, and levodopa by several adjuvants. *J. Pharm. Sci. 71*(8):865–868 (1982).
55. T. Nishihata, J. H. Rytting, T. Higuchi. Effect of salicylate on the rectal absorption of lidocaine, levodopa, and cefmetazole in rats. *J. Pharm. Sci. 71*(8):869–872 (1982).
56. M. Murakami, K. Takada, S. Muranishi. Promoting effect of Azone on intestinal absorption of poorly absorbable drugs in rats. *Int. J. Pharm. 31*:231–238 (1986).
57. M. Mishima, S. Okada, Y. Wakita, M. Nakano. Promotion of nasal absorption of insulin by glycyrrhetinic acid derivative. I. *J. Pharmacobio-Dyn. 12*:31–36 (1989).
58. J. R. Robinson, X. Yang. Absorption enhancers. In: *Encyclopedia of Pharmaceutical Technology, Vol. 18 (suppl. 1)* (J. Swarbrick, J. C. Boylan, eds.). Marcel Dekker, New York, 1999, pp. 1–27.
59. M. E. Duffey, B. Hainau, S. Ho, C. J. Bentzel. Regulation of epithelial tight junction permeability by cyclic AMP. *Nature 294*:451–453 (1981).
60. J. L. Madara. Increases in guinea pig small intestinal transepithelial resistance induced by osmotic loads are accompanied by rapid alterations in absorptive-cell tight-junction structure. *J. Cell. Biol. 97*:125–136 (1983).
61. R. W. Freel, M. Hatch, D. L. Earnest, A. M. Goldner. Role of tight junctional pathway in bile salt-induced increase in colonic permeability *Am. J. Physiol. 245*:G816–G823 (1983).
62. M. Shiga, M. Hayashi, T. Horie, S. Awazu. Promotion of drug rectal absorption related to water absorption. *Chem. Pharm. Bull. 34*:2254–2256 (1986).
63. Y. Watanabe, E. J. V. Hoogdalem, A. G. de Boer, D. D. Breimer. Absorption enhancement of rectally infused cefoxitin by medium chain monoglycerides in concious rats. *J. Pharm. Sci. 77*(10):847–849 (1988).
64. H. Kajii, T. Horie, M. Hayashi, S. Awazu. Fluorescence study of the membrane-perturbating action of sodium caprylate as related to promotion of drug absorption. *J. Pharm. Sci. 77*:390–392 (1988).
65. S. Muranishi, A. Yamamoto. Mechanisms of absorption enhancement through gastrointestinal epithelium. In: *Drug Absorption Enhancement: Concepts, Possibilities, Limitations and Trends* (A. G. de Boer, ed.). Harwood Academic Publishers, Philadelphia, 1994, pp. 67–100.
66. E. Touitou, F. Alhaique, P. Fisher, A. Memoli, F. M. Riccieri, E. Santucci. Prevention of molecular self-association by sodium salicylate: effect on insulin and 6-carboxyfluorescein. *J. Pharm. Sci. 76*:791–793 (1987).
67. H. L. LueBen, J. C. Verhoef, G. Borchard, C.-M. Lehr, A. G. de Boer, H. E. Junginger. Bioadhesive polymers in peroral peptide drug delivery II: carbomer and polycarbophil are potent inhibitors of the intestinal proteolytic enzyme trypsin. *Pharm. Res. 12*(9):1293–1298 (1995).
68. P. C. Graziadeip. The mucus membrane of the nose. *Am. Otol. Rhinol. Laryngol. 79*:433–442 (1970).

69. N. Mygind, R. Dahl. Anatomy, physiology and function of the nasal cavities in health and disease. *Adv. Drug Del. Rev. 29*:3–12 (1998).
70. N. Geurkink. Nasal anatomy, physiology, and function. *J. Allergy Clin. Immunol 72*:123–128 (1983).
71. H. Char, S. Kumar, S. Patel, D. Piemontese, K. Iqbal, A. W. Malick, R. A. Salvador, C. R. Behl. Nasal delivery of C^{14}-dextromethorphan hydrochloride in rats—levels in plasma and brain. *J. Pharm. Sci. 81*:750–752 (1992).
72. P. G. Gopinath, G. Gopinath, T. C. AnadKumar. Target site of intranasally-sprayed substances and their transport across the nasal mucosa: a new insight into the intranasal route of drug delivery. *Curr. Ther. Res. 23*:596–607 (1978).
73. E. B. Brittebo. Metabolism of progesterone in the nasal mucosa in mice and rats. *Acta. Pharmacol. Toxicol. 51*: 441–445 (1982).
74. W. M. Hadley, A. R. Dahl. Cytochrome P-450 dependent monooxygenase activity in nasal membranes of six species. *Drug Met. Disp. 11*(3):275–276 (1983).
75. J. A. Bond, J. R. Harkema, V. I. Russell. Regional distribution of xenobiotic metabolizing enzymes in respiratory airways of dogs. *Drug Met. Disp. 16*(1):116–124 (1988).
76. J. Jenner, G. H. Dodd. Xenobiotic-metabolism in the nasal epithelia. *Drug Met. Drug Interact. 6*(2):123–148 (1988).
77. P. G. Gervasi, V. Longo, F. Naldi, G. Panattoni, F. Ursino. Xenobiotic-metabolizing enzymes in human respiratory nasal mucosa. *Biochem. Pharmacol. 41*:177–184 (1991).
78. A. R. Dahl, W. M. Hadley. Nasal cavity enzymes involved in xenobiotic metabolism: effects on the toxicity of inhalants. *Crit. Rev. Toxicol. 21*(5):345–372 (1991).
79. M. A. Sarkar. Drug metabolism in the nasal mucosa. *Pharm. Res. 9*(1):1–9 (1992).
80. X. H. Zhou. Overcoming enzymatic and absorption barriers to non-parenterally administered protein and peptide drugs. *J. Cont. Rel. 29*:239–252 (1994).
81. Y. Sakakura, K. Ukai, Y. Majima, S. Murai, T. Harada, Y. Miyoshi. Nasal mucociliary clearance under various conditions. *Acta. Otolaryngol. 96*:167–173 (1983).
82. A. B. Lansley. Mucociliary clearance and drug delivery via the respiratory tract. *Adv. Drug Del. Rev. 11*:299–327 (1993).
83. S. Gizurarson. The relevance of nasal physiology to the design of drug absorption studies. *Adv. Drug. Del. Rev. 11*(3):329–347 (1993).
84. D. F. Proctor, I. Anderson, G. Lundqvist. Clearance of inhaled particles from the human nose. *Arch. Intern. Med. 131*:132–139 (1973).
85. J. G. Hardy, S. W. Lee, C. G. Wilson. Intranasal drug delivery by spray and drops. *J. Pharm. Pharmacol. 37*:294–297 (1985).
86. A. S. Harris, I. M. Nilsson, Z. G. Wagner, U. Alkner. Intranasal administration of peptides: nasal deposition, biological response, and absorption of desmopressin. *J. Pharm. Sci. 75*(11):1085–1088 (1986).
87. E. Marttin, S. G. Romeijin, J. C. Verhoef, F. W. H. M. Merkus. Nasal absorption of dihydroergotamine from liquid and powder formulations in rabbits. *J. Pharm. Sci. 86*(7):802–807 (1997).
88. F. Nakamura, R. Ohta, Y. Machida, T. Nagai. In vitro and in vivo nasal mucoadhesion of some water-soluble polymers. *Int. J. Pharm. 134*:173–181 (1996).
89. P. Dondeti, H. Zia, T. E. Needham. Bioadhesive and formulation parameters affecting nasal absorption. *Int. J. Pharm. 127*:115–133 (1996).
90. M. I. Ugwoke, E. Sam, G. Van den Mooter, N. Verbeke, R. Kinget. Nasal mucoadhesive delivery systems of the anti-parkinsonian drug, apomorphine: influence of drug-loading on in vitro and in vivo release in rabbits. *Int. J. Pharm. 181*(1):125–138 (1999).
91. T. Nagai, Y. Nishimoto, N. Nambu, Y. Sozuki, K. Sekine. Powder dosage form of insulin for nasal administration. *J. Cont. Rel. 1*:15–22 (1984).
92. N. G. M. Schipper, S. G. Romeign, J. C. Verhoef, F. W. H. M. Merkus. Nasal insulin delivery with dimethyl-beta-cyclodextrin as an absorption enhancer in rabbits: powder more effective than liquid formulations. *Pharm. Res. 10*(5):682–686 (1993).
93. A. S. Harris, E. Svensson, Z. G. Wagner, S. Lethagen, I. M. Nilsson. Effect of viscosity on particle size, deposition, and clearance of nasal delivery systems containing desmopressin. *J. Pharm. Sci. 77*:405–408 (1988).
94. A. S. Harris, M. Olin, E. Svensson, S. Lethagen, I. M. Nilsson. Effect of viscosity on the pharmacokinetics and biological response to intranasal desmopressin. *J. Pharm. Sci. 78*:470–471 (1989).
95. J. Juhasz, V. Lenaerts, H. Ong. Poloxamer gels for peptide delivery: in vitro studies. *J. Cont. Rel. 13*:321 (1990).
96. L. Pereswetoff-Morath, P. Edman. Influence of osmolarity on nasal absorption of insulin from the thermogelling polymer ethyl(hydroxyethyl) cellulose. *Int. J. Pharm. 125*: 205–213 (1995).
97. L. Pereswetoff-Morath, S. Bjurstrom, R. Khan, M. Dahlin, P. Edman. Toxicological aspects of the use of dextran microspheres and thermogelling ethyl(hydroxyethyl) cellulose (EHEC) as nasal drug delivery systems. *Int. J. Pharm. 128*:9–21 (1996).
98. H. Kublik, B. W. Muller. Rheological properties of polymer-solutions as carriers for nasal drug delivery systems. *Eur. J. Pharm. Biopharm. 39*(5):192–196 (1993).
99. L. Pereswetoff-Morath. Microsphere as nasal drug delivery systems. *Adv. Drug. Del. Rev. 29*:185–194 (1998).
100. P. Edman, E. Bjork, L. Ryden. Microsphere as a nasal delivery system for peptides. *J. Cont. Rel. 21*:165–172 (1992).
101. N. F. Farraj, B. R. Johansen, S. S. Davis, L. Illum. Nasal administration of insulin using bioadhesive microsphere as a delivery system. *J. Cont. Rel. 13*:253–261 (1990).
102. L. Illum, H. Jorgensen, H. Bisgaard, O. Krogsgaard, N. Rossing. Bioadhesive microspheres as a potential nasal drug delivery system. *Int. J. Pharm. 39*:189–199 (1987).
103. C. Witschi, R. J. Mrsny. In vitro evaluation of microparticles and polymer gels for use as nasal platforms for protein delivery. *Pharm. Res. 16*(3):382–390 (1999).

104. S. P. Vyas, S. K. Goswami, R. Singh. Liposomes based nasal delivery system of nifedipine: development and characterization. *Int. J. Pharm. 118*:23–30 (1995).
105. H. O. Alpar, J. C. Bowen, M. R. W. Brown. Effectiveness of liposomes as adjuvants of orally and nasally administered tetanus toxoid. *Int. J. Pharm. 88*:335–344 (1992).
106. N. I. Payne, P. Timmis, C. V. Ambrose, M. D. Warel, F. Ridgway. Proliposomes: a novel solution to an old problem. *J. Pharm. Sci. 75*:325–329 (1986).
107. B.-N. Ahn, S.-K. Kim, C.-K. Shim. Proliposomes as an intranasal dosage form for the sustained delivery of propanolol. *J. Cont. Rel. 34*:203–210 (1995).
108. T. T. Kararli, T. E. Needham, G. Schoenhard, D. A. Baron, R. E. Schmidt, B. Katz, B. Belonio. Enhancement of nasal delivery of a renin inhibitor in the rat using emulsion formulations. *Pharm. Res. 9*(8):1024–1028 (1992).
109. K.-T. Ko, T. E. Needham, H. Zia. Emulsion formulations of testosterone for nasal administration. *J. Microencapsulation 15*(2):197–205 (1998).
110. F. Moren. Aerosol dosage forms and formulations (Chapter 13). In: *Aerosols in Medicine: Principles, Diagnosis, and Therapy,* 2[d] ed. (F. Moren, M. B. Dolovich, M. T. Newhouse, S. P. Newman, (eds.). Elsevier, Amsterdam, 1993, pp. 321–350.
111. P. J. Atkins, N. P. Barker, D. Mathisen. The design and development of inhalation drug delivery systems (Chapter 6). In: *Pharmaceutical Inhalation Aerosol Technology* (A. J. Hickey, ed.). Marcel Dekker, New York, 1992, pp. 155–185.
112. J. L. Rau, Jr., ed. Administration of aerosolized agents. In: *Respiratory care pharmacology,* 4th ed. Mosby-Year Book, 1994, pp. 33–73.
113. C. L. Schauf, D. F. Moffett, S. B. Moffett, eds. Respiratory structure and dynamics (Chapter 16). In: *Human Physiology: Foundation and Frontiers*. Times Mirror/Mosby College Publishing, 1990, pp. 400–425.
114. J. S. Patton. Mechanisms of macromolecule absorption by the lungs. *Adv. Drug Del. Rev. 19*:3–36 (1996).
115. T. R. Gerrity. Pathophysiological and disease constrains on aerosol delivery (Chapter 1). In: *Respiratory Drug Delivery* (P. R. Byron, ed.). CRC Press, Boca Raton, 1990, pp. 1–38.
116. J. S. Patton, R. M. Platz. Pulmonary delivery of peptides and proteins for systemic action. *Adv. Drug Del. Rev. 8*: 179–196 (1992).
117. J. M. Padfield. Principles of drug administration to the respiratory tract (Chapter 8). In: *Drug Delivery to the Respiratory Tract* (D. Ganderton, T. Jones, eds.). Ellis Harwood, Chichester, U.K., 1987, pp. 75–86.
118. P. K. Gupta, A. J. Hickey. Contemporary approaches in aerosolized drug delivery to the lung. *J. Cont. Rel. 17*: 129–148 (1991).
119. D. A. Edwards, A. Ben-Jebria, R. Langer. Recent advances in pulmonary drug delivery using large, porous inhaled particles. *J. Appl. Physiol. 84*(2):379–385 (1998).
120. D. A. Edwards, J. Hanes, G. Caponetti, J. Hrkach, A. Ben-Jebria, M. L. Eskew, J. Mintzes, D. Deaver, N. Lotan, R. Langer. Large porous particles for pulmonary drug delivery. *Science 276*:1868–1871 (1997).
121. A. Ben-Jebria, D. Chen, M. L. Eskew, R. Vanbever, R. Longer, D. A. Edwards. Large porous particles for sustained protection from Carbachol-induced bronchoconstriction in guinea pigs. *Pharm. Res. 16*(4):555–561 (1999).
122. A. T. Florence, D. Attwood, eds. Drug absorption and routes of administration (Chapter 9). In: *Physicochemical Principles of Pharmacy,* 3[d] ed. Creative Print and Design (Wales), Ebbw Vale, U.K. 1998, pp. 372–447.
123. J. Baron, J. M. Voigt. Localization, distribution, and induction of xenobiotic-metabolizing enzymes and aryl hydrocarbon hydroxylase activity within lung. *Pharmac. Ther. 47*:419–445 (1990).
124. G. M. Cohen. Pulmonary metabolism of foreign compounds: its role in metabolic activation. *Env. Health Perspect. 85*:31–41 (1990).
125. R. J. Gonzalezrothi, H. Schreier. Pulmonary delivery of liposome encapsulated drugs in asthma therapy. *Clin. Immuno. 4*(5):331–337 (1995).
126. J. S. Patton, J. Bukar, S. Nagarajan. Inhaled insulin. *Adv. Drug Del. Rev. 35*:235–247 (1999).
127. I. Gonda. A semi-empirical model of aerosol deposition in the human respiratory tract for mouth inhalation. *J. Pharm. Pharmacol. 33*:692–696 (1981).
128. M. L. Levy. Metered dose inhalers: current and future uses. *Br. J. Clin. Practice 89*(suppl.):16–21 (1997).
129. O. Selroos, A. Pietinalho, H. Riska. Delivery devices for inhaled asthma medication—clinical implications of differences in effectiveness. *Clin. Immuno. 6*(4):273–299 (1996).
130. M. Keller. Innovations and perspectives of metered dose inhalers in pulmonary drug delivery. *Int. J. Pharm. 186*(1): 81–90 (1999).
131. P. Sanders, N. Washington, M. Frier, C. G. Wilson, L. C. Feely, C. Washington. The deposition of solution-based and suspension-based aerosols from metered dose inhalers in healthy subjects and asthmatic patients. *STP Pharm. Sci. 7*(4):300–306 (1997).
132. A. Adjai, J. Garren. Pulmonary delivery of peptide drugs: effect of particle size on bioavailability of leuprolide acetate in healthy male volunteers. *Pharm. Res. 7*(6):565–569 (1990).
133. V. M. Sadtler, M. P. Krafft, J. G. Riess. Reverse water-in-fluorocarbon emulsions as a drug delivery system: an in vitro study. *Colloids and Surfaces A: Physico. Engineer. Aspects 147*:309–315 (1999).
134. M. P. Timsina, G. P. Martin, C. Marriott, D. Ganderton, M. Yianneskis. Drug delivery to the respiratory tract using dry powder inhalers. *Int. J. Pharm. 101*:1–13 (1994).
135. Y. Kawashima, T. Serigano, T. Hino, H. Yamamoto, H. Takeuchi. A new powder design method to improve inhalation efficiency of Pranlukast hydrate dry powder aerosols by surface modification with hydroxypropylmethylcellulose phthalate nanospheres. *Pharm. Res. 15*(11): 1748–1752 (1998).

136. C. LiCalsi, T. Christensen, J. V. Bennett, E. Phillips, C. Witham. Dry powder inhalation as a potential delivery method for vaccines. *Vaccine 17*(13–14, sp.iss.):1796–1803 (1999).
137. R. N. Jashnani, P. R. Byron, R. N. Dalby. Testing of dry powder aerosol formulations in different environmental conditions. *Int. J. Pharm. 113*(1):123–130 (1995).
138. H. Steckel, J. Thies, B. W. Muller. Micronizing of steroids for pulmonary delivery by supercritical carbon dioxide. *Int. J. Pharm. 152*(1):99–110 (1997).
139. N. M. Concessio, M. M. VanOort, M. R. Knowles, A. J. Hickey. Pharmaceutical dry powder aerosols: correlation of powder properties with dose delivery and implications for pharmacodynamic effect. *Pharm. Res. 16*(6):828–834 (1999).
140. K. M. G. Taylor, G. Taylor, I. W. Kellaway, J. Stevens. The influence of liposomal encapsulation on sodium cromoglycate pharmacokinetics in man. *Pharm. Res. 6*(7): 633–636 (1989).
141. R. L. Juliano, H. N. McCullough. Controlled delivery of an antitumor drug: localized action of liposome encapsulated cytosine arabinoside administered via the respiratory system. *J. Pharmacol. Exp. Ther. 214*:381–387 (1980).
142. P. N. Shek, Z. E. Suntres, J. I. Brooks. Liposomes in pulmonary applications: physicochemical considerations, pulmonary distribution and antioxidant delivery. *J. Drug Target. 2*:431–442 (1994).
143. H. Schreier, R. J. Gonzalez-Rothi, A. A. Stecenko. Pulmonary delivery of liposomes, *J. Cont. Rel. 24*:209–223 (1993).
144. M. P. Lambros, D. W. A. Bourne, S. A. Abbas, D. L. Johnson. Disposition of aerosolized liposomal amphotericin B. *J. Pharm. Sci. 86*(9):1066–1069 (1997).
145. Z. E. Suntres, P. N. Shek. Liposomes promote pulmonary glucocorticoid delivery. *J. Drug Target. 6*(3):175–182 (1998).
146. X. M. Zeng, G. P. Martin, C. Marriott. The controlled delivery of drugs to the lung. *Int. J. Pharm. 124*:149–164 (1995).
147. M. Saari, M. T. Vidgren, M. O. Koskinen, V. M. H. Turjanmaa, M. M. Nieminen. Pulmonary distribution and clearance of two beclomethasone liposome formulations in healthy volunteers. *Int. J. Pharm. 181*(1):1–9 (1999).
148. H. P. Merkle, R. Anders, A. Wermerskirchem. Mucoadhesive buccal patches for peptide delivery (Chapter 6). In: *Bioadhesive Drug Delivery Systems* (V. Lenaerts, R. Gurny, (eds.). CRC Press, Boca Raton, FL, 1990, pp. 105–136.
149. L. M. C. Collins, C. Dawes. The surface area of the adult human mouth and thickness of the salivary film covering the teeth and oral mucosa. *J. Dent. Res. 66*(8):1300–1302 (1987).
150. C. A. Squier, N. W. Johnson, M. Hackemann. Structure and function of normal human oral mucosa (Chapter 1). In: *Oral Mucosa in Health and Disease* (A. E. Dolby, ed.). Blackwell Scientific, London, 1975, pp. 1–112.
151. M. J. Rathbone, J. Hadgraft. Absorption of drugs from the human oral cavity. *Int. J. Pharm. 74*:9–24 (1991).
152. D. Harris, J. R. Robinson. Drug delivery via the mucous membranes of the oral cavity. *J. Pharm. Sci. 81*(1):1–10 (1992).
153. R. B. Gandhi, J. R. Robinson. Oral cavity as a site for bioadhesive drug delivery. *Adv. Drug Del. Rev. 13*(1&2): 43–74 (1994).
154. C. A. Squier, N. W. Johnson. Permeability of oral mucosa. *Br. Med. Bull. 31*(2):169–175 (1975).
155. P. W. Wertz, D. C. Swartzendruber, C. A. Squier. Regional variation in the structure and permeability of oral mucosa and skin. *Adv. Drug Del. Rev. 12*:1–12 (1993).
156. C. A. Squier. The permeability of keratinized and nonkeratinized oral epithelium to horseradish peroxidase. *J. Ultrastruct. Res. 43*:160–177 (1973).
157. C. A. Squier, L. Rooney. The permeability of keratinized and non-keratinized oral epithelium to lanthanum in vivo. *J. Ultrastruct. Res. 54*:286–295 (1976).
158. M. E. Dowty, K. E. Knuth, B. K. Irons, J. R. Robinson. Transport of thyrotropin-releasing hormone in rabbit buccal mucosa in vitro. *Pharm. Res. 9*:1113–1122 (1992).
159. T. Nagai, Y. Machida. Buccal delivery systems using hydrogels. *Adv. Drug Del. Rel. 11*:179–191 (1993).
160. M. J. Rathbone, G. Ponchel, F. A. Ghazali. Systemic oral mucosal drug delivery and delivery systems (Chapter 11). In: *Oral Mucosal Drug Delivery* (M. J. Rathbone, ed.). Marcel Dekker, New York, 1996, pp. 241–284.
161. T. Nagai. Adhesive topical drug delivery system. *J. Cont. Rel. 2*:121–134 (1985).
162. S. Bouckaert, H. Schautteet, R. A. Lefebvre, J. P. Remon, R. van Clooster. Comparison of salivary miconazole concentrations after administration of a bioadhesive slow-release buccal tablet and an oral gel. *Eur. J. Clin. Pharmacol. 43*:137–140 (1992).
163. S. Bouckaert, R. A. Lefbvre, J. P. Remon. In vitri/in vivo correlation of the bioadhesive properties of a buccal bioadhesive miconazole slow-release tablet. *Pharm. Res. 10*(6): 853–856 (1993).
164. A. Ahuja, R. K. Khar, R. Chaudhry. Evaluation of buccoadhesive metronidazole tablets: microbiological response. *Pharmazie 53*(4):264–267 (1998).
165. M. Ishida, Y. Machida, N. Nambu, T. Nagai. New mucosal dosage form of insulin. *Chem. Pharm. Bull. 29*:810–816 (1981).
166. K. H. Simpson, I. C. Tring, F. R. Ellis. An investigation of premedication with morphine given by the buccal or intramuscular route. *Br. J. Clin. Pharmac. 27*:377–380 (1989).
167. P. J. Hoskin, G. W. Hanks, G. W. Aherne, D. Chapman, P. Littleton, J. Filshie. The bioavailability and pharmacokinetics of morphine after intravenous, oral, and buccal administration in healthy volunteers. *Br. J. Clin. Pharmac. 27*:499–505 (1989).
168. S. Anlar, Y. Capan, O. Guven, A. Gogus, T. Dalkara, A. A. Hincal. Formulation and in vitro–in vivo evaluation of

buccoadhesive morphine sulfate tablets. *Pharm. Res.* *11*(2):231–236 (1994).

169. E. Beyssac, F. Touaref, M. Meyer,L. Jacob, P. Sandouk, J.-M. Aiache. Bioavailability of morphine after administration of a new bioadhesive buccal tablet. *Biopharm. Drug Disp. 19*:401–405 (1998).
170. K. Yukimatsu, Y. Nozaki, M. Kakumoto, M. Ohta, Development of a tran-mucosal controlled-release device for systemic delivery of antianginal drugs: pharmacokinetics and pharmacodynamics. *Drug Del. Ind. Pharm. 20*(4): 503–534 (1994).
171. R.-Y. Han, J.-Y. Fang, K. C. Sung, and O. Y. P. Hu. Mucoadhesive buccal disks for novel nalbuphine prodrug controlled delivery: effect of formulation variables on drug release and mucoadhesive performance. *Int. J. Pharm. 177*:201–209 (1999).
172. H. P. Merkle, G. J. M. Wolany. Muco-adhesive patches for buccal peptide administration. In: *Buccal and Nasal Administration as an Alternative to Parenteral Administration* (D. Duchene, (ed.). Paris, 1992, pp. 110–124.
173. G. DeGrande, L. Benes, F. Horriere, H. Karsenty, C. Lacoste, R. McQuinn, J.-W. Guo, R. Scherrer. Specialized oral mucosal drug delivery systems: patches (Chapter 12). In: *Oral Mucosal Drug Delivery* (M. J. Rathbone, ed.). Marcel Dekker, New York, 1996, pp. 285–317.
174. R. Anders, H. P. Merkle. Evaluation of laminated mucoadhesive patches for buccal drug delivery. *Int. J. Pharm. 49*:231–240 (1989).
175. M. M. Veillard, M. A. Longer, T. W. Martens, J. R. Robinson. Preliminary studies of oral mucosal delivery of peptide drugs. *J. Cont. Rel. 6*:123–131 (1987).
176. C. Remunan-Lopez, A. Portero, J. L. Vila-Jato, and M. J. Alonso. Design and evaluation of chitosan/ethylcellulose mucoadhesive bilayered devices for buccal drug delivery. *J. Cont. Rel. 55*:143–152 (1998).
177. M. Ishida, N. Nambu, T. Nagai. Ointment-type oral mucosal dosage form of Carbopol containing prednisolone for treatment of aphtha. *Chem. Pharm. Bull. 31*(3):1010–1014 (1983).
178. K.-D. Bremecker, H. Strempel, G. Klein. Novel concept for a mucosal adhesive ointment. *J. Pharm. Sci. 73*(4): 548–552 (1984).
179. S. J. Sveinsson, W. P. Holbrook. Oral mucosal adhesive ointment containing liposomal corticosteroid. *Int. J. Pharm. 95*:105–109 (1993).
180. M. Petelin, M. Sentjure, Z. Stolic, U. Skaleric. EPR study of mucoadhesive ointments for delivery of liposomes into the oral mucosa. *Int. J. Pharm. 173*:193–202 (1998).
181. Y. Qiu, H. W. Johnson, T. L. Reiland, M.-Y. F. Lu. Sublingual absorption of leuprolide: comparison between human and animal models. *Int. J. Pharm. 179*:27–36 (1999).
182. T. L. Burkoth; B. J. Bellhouse, G. Hewson, D. J. Longridge, A. G. Muddle, D. F. Sarphie. Transdermal and transmucosal powdered drug delivery. *Crit. Rev. Ther. Drug. Carrier. Syst. 16*(4):331–384 (1999).
183. C. L. Schauf, D. F. Moffett, S. B. Moffett, (eds.) Gastrointestinal organization, motility, and secretion (Chapter 21). In: *Human Physiology: Foundation and Frontiers*. Times Mirror/Mosby, 1990, pp. 532–561.
184. A. Stevens, J. S. Lowe, (eds.) Alimentary tract (Chapter 11). In: *Human Histology*, 2d ed. Mosby, New York, 1997, pp. 177–214.
185. G. Thew, E. Mutschler, P. Vaupel, (eds.) Gastrointestinal tract and digestion (Chapter 11). In: *Human Anatomy, Physiology, and Pathophysiology*. Elsevier Science, Amsterdam, The Netherlands, 1985, pp. 349–412.
186. P. G. Welling, (ed.) Enteral routes of drug administration (Chapter 4). In: *Pharmacokinetics: Processes, Methamatics, and Applications*, 2d ed. American Chemical Society, Washington DC, 1997, pp. 43–62.
187. A. M. Metcalf, S. F. Phillips, A. R. Zinsmeister, R. L. MacCarty, R. W. Beart, B. G. Wolff. Simplified assessment of segmental colonic transit. *Gastroenterology 92*: 40–47 (1987).
188. M. H. Ross, L. J. Romrell, G. I. Kaye, (eds.). Digestive system II: Esophagus and gastrointestinal tract (Chapter 16). In: *Histology, a Text and Atlas* 3d ed. Williams and Wilkins, Baltimore, 1995, pp. 440–495.
189. E. J. van Hoogdalem, A. G. de Boer, D. D. Breimer. Pharmacokinetics of rectal drug administration, Part I: general considerations and clinical applications of centrally acting drugs. *Clin. Pharmacokinet. 21*(1):11–26 (1991).
190. S. Muranishi, A. Yamamoto, H. Okada. Rectal and vaginal absorption of peptides and proteins (Chapter 9). In: *Biological Barriers to Proteins Delivery* (K. L. Audus, T. J. Raub, eds.). Plenum Press, New York, 1993, pp. 199–227.
191. P. Gruber, M. A. Longer, J. R. Robinson. Some biological issues in oral controlled drug delivery. *Adv. Drug Del. Rev. 1*:1–18 (1987).
192. D. F. Evan, G. Pye, R. Bramley, A. G. Clark, T. J. Dyson, J. D. Hardcastle. Measurement of gastrointestinal pH profiles in normal ambulant human subjects. *Gut 29*:1035–1041 (1988).
193. W. A. Ritschel. Targeting in the gastrointestinal tract: new approaches. *Meth. Find. Exp. Clin. Pharmacol. 13*(5): 313–336 (1991).
194. D. Fleisher, C. Li, Y. Zhou, L.-H. Pao, A. Karim. Drug, meal and formulation interactions influencing drug absorption after oral administration: clinical implication. *Clin. Pharmacolkinet. 36*(3):233–254 (1999).
195. S. S. Davis, J. G. Hardy, J. W. Fara. Transit of pharmaceutical dosage forms through the small intestine. *Gut 27*(8): 886–892 (1986).
196. A. MacAdam. The effect of gastrointestinal mucus on drug absorption. *Adv. Drug Del. Rev. 11*:201–220 (1993).
197. M. D. Levitt, J. K. Furne, A. Strocchi, B. W. Anderson, D. G. Levitt. Physiological measurements of luminal stirring in the dog and human small bowel. *J. Clin. Invest. 86*:1540–1547 (1990).
198. J. Hunter, B. H. Hirst. Intestinal secretion of drugs: the role of P-glycoprotein and related drug efflux systems in limiting oral drug absorption. *Adv. Drug Del. Rev. 25*: 129–157 (1997).
199. J. C. Verhoef, H. E. Bodde, A. G. de Boer, J. A. Bouwstra,

H. E. Junginer, F. W. H. M. Merkus, D. D. Breimer. Transport of peptide and protein drugs across biological membranes. *Eur. J. Drug Met. and Pharmacokin.* *15*(2):83–93 (1990).

200. W. Wang. Oral protein drug delivery. *J. Drug Target.* *4*(4):195–232 (1996).
201. K. F. Ilett, L. B. G. Tee, P. T. Reeves, R. F. Minchin. Metabolism of drugs and other xenobiotics in the gut lumen and wall. *Pharmac. Ther.* *46*:67–93 (1990).
202. Y. K. Tam. Individual variation in first-pass metabolism. *Clin. Pharmacokinet.* *25*(4):300–328 (1993).
203. T. T. Karaali. Comparison of the gastrointestinal anatomy, physiology, and biochemistry of humans and commonly used laboratory animal. *Biopharm. Drug Disp.* *16*:351–380 (1995).
204. T. T. Kararli. Gastrointestinal absorption of drugs. *Crit. Rev. Ther. Drug. Carrier. Syst.* *6*(1):39–86 (1989).
205. A. Yamamoto, S. Muranishi. Rectal drug delivery systems: improvement of rectal peptide absorption by absorption enhancers, protease inhibitors and chemical modification. *Adv. Drug Del. Rev.* *28*:275–299 (1997).
206. G. Van den Mooter, R. Kinget. Oral colon specific drug delivery: a review. *Drug Del.* *2*:81–93 (1995).
207. J. W. Faigle, Drug metabolism in the colon wall and lumen. In: *Colonic Drug Absorption and Metabolism* P. R. Bieck, ed. Marcel Dekker, New York, 1993, pp. 29–54.
208. S. Muranishi. Characteristics of drug absorption via the rectal route. *Meth. Find. Exp. Clin. Pharmacol.* *6*:763–772 (1984).
209. W. A. Ritschel, H. Elconin, G. J. Alcorn, D. D. Denson. First-pass elimination of lidocaine in the rabbit after peroral and rectal route of administration. *Biopharm. Drug Disp.* *6*:281–290 (1985).
210. A. G. de Boer, D. D. Breimer, H. Mattie, J. Pronk, J. M. Gubbens-Stibbe. Rectal bioavailability of lidocaine in man: partial avoidance of first-pass metabolism. *Clin. Pharmacol. Ther.* *26*:701–709 (1979).
211. A. G. de Boer, D. D. Breimer. Hepatic first-pass effect and controlled drug delivery following rectal administration. *Adv. Drug Del. Rev.* *28*:229–237 (1997).
212. X. H. Zhou, A. L. W. Po. Comparison of enzyme activities of tissues lining portals of absorption of drugs: species differences. *Int. J. Pharm.* *70*:271–283 (1991).
213. S. Lundeberg, O. Beck, G. L. Olsson, L. O. Boreus. Rectal administration of morphine in children: pharmacokinetic evaluation after a single dose. *Acta. Anaes. Scan.* *40*(4):445–451 (1996).
214. P. K. Birmingham, M. J. Tobin, T. K. Henthorn, D. M. Fisher, M. C. Berkelhamer, F. A. Smith, K. B. Fanta, C. J. Cote. Twenty four hour pharmacokinetics of rectal acetaminophen in children: an old drug with new recommendations. *Anesthesiology* *87*(2):244–252 (1997).
215. E. Fabre, S. Chevret, J. F. Piechaud, E. Rey, F. Vauzelle-Kervroedan, P. D'Athis, G. Olive, G. Pons. An approach for dose finding of drugs in infants: sedation by midazolam studies using the continual reassessment method. *Brit. J. Clin. Pharmacol.* *46*(4):395–401 (1998).
216. D. Thompson, A. Dimartini. Nonenteral routes of administration for psychiatric medications: a literature review. *Psychosomatics* *40*(3):185–192 (1999).
217. K. Park, J. R. Robinson. Bioadhesive polymers as platforms for oral controlled drug delivery I: method to study bioadhesion. *Int. J. Pharm.* *19*:107–127 (1984).
218. H. S. Ch'ng, H. Park, P. Kelly, J. R. Robinson. Bioadhesive polymers as platforms for oral controlled drug delivery II: synthesis and evaluation of some swelling, water insoluble polymers. *J. Pharm. Sci.* *74*:399–405 (1985).
219. M. A. Longer, H. S. Ch'ng, J. R. Robinson. Bioadhesive polymers as platforms for oral controlled drug delivery III: oral delivery of chlorothiazide using bioadhesive polymers. *J. Pharm. Sci.* *74*:406–411 (1985).
220. D. Harris, J. T. Fell, H. L. Sharma, D. C. Taylor. GI transit of potential bioadhesive formulations in man: a scintigraphic study. *J. Cont. Rel.* *12*:45–53 (1990).
221. C.-M. Leh. From sticky stuff to sweet receptors: achievements, limits, and novel approaches to bioadhesion. *Eur. J. Drug Metab. Pharmacikinet.* *21*:139–148 (1996).
222. Y. Akiyama, N. Nagahara, T. Kashihara, S. Hirai, H. Toguchi. In vitro and in vivo evaluation of mucoadhesive microspheres prepared for the gastrointestinal tract using polyglycerol esters of fatty acids and a poly(acrylic acid) derivative. *Pharm. Res.* *12*(3):397–405 (1995).
223. Y. Akiyama, N. Nagahara, E. Nara, M. Kitano, S. Iwasa, I. Yamamoto, J. Azuma, Y. Ogawa. Evaluation of oral mucoadhesive microsphere in man on the basis of the pharmacokinetics of furosemide and riboflavin, compounds with limited gastrointestinal absorption sites. *J. Pharm. Pharmacol.* *50*:159–166 (1998).
224. N. Nagahara, Y. Akiyama, M. Nakao, M. Tada, M. Kitano, Y. Ogawa. Mucoadhesive microspheres containing amoxicillin for clearance of *Helicobacter pylori. Antimicrob. Agents Chemother.* *42*(10):2492–2494 (1998).
225. K. Kimura, K. Ido, K. Saifuku, Y. Tanigushi, K. Kihira, K. Satoh, T. Takimoto, Y. Yoshida. A 1-h topical therapy for the treatment of *Helicobacter pylori* infection. *Am. J. Gastroenterol.* *90*(1):60–63 (1995).
226. D. Chickering, J. Jacob, E. Mathiowitz. Poly(fumaric-co-sebacic) microspheres as oral drug delivery systems. *Biotech. Bioeng.* *52*:96–101 (1996).
227. E. Mathiowitz, J. S. Jacob, Y. S. Jong, G. P. Carino, D. E. Checkering, P. Chaturvedi, C. A. Santos, K. Vijayaraghavan, S. Montgomery, M. Bassett, C. Morrell. Biologically erodable microspheres as potential oral drug delivery systems. *Nature* *386*(27 March):410–414 (1997).
228. H. Takeuchi, H. Yamamoto, T. Niwa, T. Hino, Y. Kawashima. Enteral absorption of insulin in rats from mucoadhesive chitosan-coated liposomes. *Pharm. Res.* *13*(6):896–901 (1996).
229. E. Ilan, S. Amselem, M. Weisspapir, J. Schwarz, A. Yogev, E. Zawoznik, D. Friedman. Improved oral delivery of desmopressin via a novel vehicle: mucoadhesive submicron emulsion. *Pharm. Res.* *13*(7):1083–1087 (1996).
230. X. Yang, J. R. Robinson. Bioadhesion in mucosal drug

delivery (Chapter 5). In: *Biorelated Polymers and Gels: Controlled Release and Applications in Biomedical Engineering* (T. Okano, ed.). Academic Press, New York, 1998, pp. 135–192.

231. G. Ponchel, J.-M. Irache. Specific and non-specific bioadhesive particulate systems for oral delivery to the gastrointestinal tract. *Adv. Drug Del. Rev. 34*:191–219 (1998).
232. C.-M. Lehr, J. A. Bouwstra, W. Kok, A. B. J. Noach, A. G. de Boer, H. E. Junginger. Bioadhesion by means of specific binding of tomato lectin. *Pharm. Res. 9*(4):547–553 (1992).
233. A. Pusztai, S. Bardocz, S. W. B. Ewen. Plant lectins for oral drug delivery to different parts of the gastrointestinal tract. In: *Bioadhesive Drug Delivery Systems: Fundamentals, Novel Approaches, and Development* (E. Mathiowitz, D. E. Chickering, C.-M. Lehr, eds.). Marcel Dekker, New York, 1999, pp. 387–407.
234. B. Naisbett, J. Woodley. The potential use of tomato lectin for oral drug delivery 1: lectin binding to rat small intestine in vitro. *Int. J. Pharm. 107*(3):223–230 (1994).
235. B. Naisbett, J. Woodley. The potential use of tomato lectin for oral drug delivery 2: mechanism of uptake in vitro. *Int. J. Pharm. 110*(2):127–136 (1994).
236. B. Naisbett, J. Woodley. The potential use of tomato lectin for oral drug delivery 3: bioadhesion in vivo. *Int. J. Pharm. 114*:227–236 (1995).
237. J. M. Irache, C. Durrer, D. Duchene, G. Ponchel. In vitro study of lectin–latex conjugates for specific bioadhesion. *J. Cont. Rel. 31*:181–188 (1994).
238. N. Hussain, P. U. Jani, A. T. Florence. Enhanced oral uptake of tomato lectin conjugated nanoparticles in the rat. *Pharm. Res. 14*(5):613–618 (1997).
239. H. Chen, V. Torchilin, R. Langer. Lectin-bearing polymerized liposomes as potential oral vaccine carriers. *Pharm. Res. 13*(9):1378–1383 (1996).
240. S. C. Porter. Coating of pharmaceutical dosage forms (Chapter 93). In: *Remington: The Science and Practice of Pharmacy*, 19th ed. (A. R. Gennaro, ed.). Mack, Easton, PA, 1995, pp. 1650–1659.
241. J. N. C. Healey. Enteric coatings and delayed release (Chapter 7). In: *Drug Delivery to the Gastrointestinal Tract*. (J. G. Hardy, S. S. Davis, C. G. Wilson, eds.). Ellis Horwood, Chichester, U.K. 1989, pp. 83–96.
242. K. H. Bauer, K. Lehmann, H. P. Osterwald, G. Rothgang, (eds. Film coating Chapter 4). In: *Coated Pharmaceutical Dosage Forms: Fundamentals, Manufacturing Techniques, Biopharmaceutical Aspects, Test Methods, and Raw Materials*. Medpharm, Stuttgart, Germany, 1998, pp. 63–120.
243. C. J. Kenyon, E. T. Cole, I. R. Wilding. The effect of food on the in vivo behaviour of enteric coated starch capsules. *Int. J. Pharm. 112*:207–213 (1994).
244. M. Saffran, G. S. Kumar, C. Savariar, J. C. Burnham, F. Williams, D. C. Necker. A new approach to the oral administration of insulin and other peptide drugs. *Science 233*:1081–1084 (1986).
245. G. Van den Mooter, C. Samyn, R. Kinget. Azo polymers for colon-specific drug delivery. *Int. J. Pharm. 87*:37–46 (1992).
246. G. Van den Mooter, C. Samyn, R. Kinget. Azo polymers for colon-specific drug delivery II: influence of the type of azo polymer on the degradation by intestinal microflora. *Int. J. Pharm. 97*:133–139 (1993).
247. J. Kopecek, P. Kopeckova, H. Brondsted, R. Rathi, B. Rihova, P.-Y. Yeh, K. Ikesue. Polymers for colon-specific drug delivery. *J. Cont. Rel. 19*:121–130 (1992).
248. P. Kopeckova, R. Rathi, S. Takada, B. Rihova, M. M. Berenson, J. Kopecek. Bioadhesive N-(2-hydroxypropyl) methacrylamide copolymers for colon-specific drug delivery. *J. Cont. Rel. 28*:211–222 (1994).
249. E. O. Akala, P. Kopeckova, J. Kopecek. Novel pH-sensitive hydrogels with adjustable swelling kinetics. *Biomaterials, 19*:1037–1047 (1998).
250. H. Ghandehari, P. Kopeckova, J. Kopecek. In vitro degradation of pH-sensitive hydrogels containing aromatic azo bonds. *Biomaterials 18*:861–872 (1997).
251. E. Wood, C. G. Wilson, J. G. Hardy. The spreading of foam and solution enemas. *Int. J. Pharm. 25*:191–197 (1985).
252. J. G. Hardy, L. C. Freely, E. Wood, S. S. Davis. The application of γ-scintigraphy for the evaluation of the relative spreading of suppository bases in rectal hard gelatin capsules. *Int. J. Pharm. 38*:103–108 (1987).
253. J. Brown, S. Haines, I. R. Wilding. Colonic spread of three rectally administered mesalazine (Pentasa) dosage forms in healthy volunteers as assessed by gamma scintigraphy. *Aliment. Pharmacol. Ther. 11*:685–691 (1997).
254. M. Campieri, C. Corbelli, P. Giochetti, C. Brignola, A. Belluzzi, G. Di Febo, P. Zagni, G. Brunetti, M. Miglioli, L. Barbara. Spread and distribution of 5-ASA colonic foam and 5-ASA enema in patients with ulcerative colitis. *Dig. Dis. Sci. 37*(12):1890–1897 (1992).
255. G. A. Thomas, J. Rhodes, V. Mani, G. T. Williams, R. G. Newcombe, M. A. Russell, C. Feyerabend. Transdermal nicotine as maintenance therapy for ulcerative colitis. *New Eng. J. Med. 332*(15):988–992 (1995).
256. W. J. Sandborn, W. J. Tremaine, J. A. Leighton, G. M. Lawson, B. J. Zins, R. F. Compton, D. C. Mays, J. J. Lipsky, K. P. Batts, K. P. Offord, R. D. Hurt, J. Green. Nicotine tartrate liquid enemas for mildly to moderately active left-sided ulcerative colitis unresponsive to first-line therapy: a pilot study. *Aliment. Pharmacol. Ther. 11*:663–671 (1997).
257. B. J. Zins, W. J. Sandborn, D. C. Mays, G. M. Lawson, J. A. McKinney, W. J. Tremaine, D. W. Mahoney, A. R. Zinsmeister, R. D. Hurt, K. P. Offord, J. J. Lipsky. Pharmacokinetics of nicotine tartrate after single-dose liquid enema, oral, and intravenous administration. *J. Clin. Pharmacol. 37*(5):426–36 (1997).
258. J. T. Green, G. A. Thomas, J. Rhodes, B. K. Evans, M. A. Russell, C. Feyerabend, G. S. Fuller, R. G. Newcombe, W. J. Sandborn. Pharmacokinetics of nicotine carbomer enemas: a new treatment modality for ulcerative colitis. *Clin. Pharmacol. Ther. 61*(3):340–348 (1997).

259. M. H. Otten, G. De Haas, R. Van Den Ende. Colonic spread of 5-ASA enemas in healthy individuals, with a comparison of their physical and chemical characteristics. *Aliment. Pharmacol. Ther. 11*:693–697 (1997).
260. K. Morimoto, A. Nagayasu, S. Fukanogi, K. Morisaka, S.-H. Hyon, Y. Ikada. Evaluation of polyvinyl alcohol hydrogel as a sustained-release vehicle for rectal administration of indomethacin. *Pharm. Res. 6*(4):338–341 (1989).
261. K. Morimoto, S. Fukanoki, Y. Hatakeyama, A. Nagayasu, K. Morisaka, S.-H. Hyon, Y. Ikada. Design of a polyvinyl alcohol hydrogel containing phospholipid as controlled-release vehicle for rectal administration of (±)-propranolol HCl. *J. Pharm. Pharmacol. 42*:720–722 (1990).
262. L. Cole, C. D. Hanning, S. Robertson, K. Quinn. Further development of a morphine hydrogel suppository. *Br. J. Clin. Pharmac. 30*:781–786 (1990).
263. E. A. Hosny. Relative hypoglycemia of rectal insulin suppositories containing deoxycholic acid, sodium taurocholate, polycarbophil, and their combinations in diabetic rabbits. *Drug Del. Ind. Pharm. 25*(6):745–752 (1999).
264. R. Yahagi, H. Onishi, Y. Machida. Preparation and evaluation of double-phased mucoadhesive suppositories of lidocaine utilizing Carbopol and white beeswax. *J. Cont. Rel. 61*:1–8 (1999).
265. H.-G. Choi, Y.-K. Oh, C.-K. Kim. In situ gelling and mucoadhesive liquid suppository containing acetaminophen: enhanced bioavailability. *Int. J. Pharm. 165*:23–32 (1998).
266. M.-O. Yun, H.-G. Choi, J.-H. Jung, and C.-K. Kim. Development of a thermo-reversible insulin liquid suppository with bioavailability enhancement. *Int. J. Pharm. 189*:137–145 (1999).
267. W. S. Platzer, S. Poisel, E. S. E. Hafez. Functional anatomy of the human vagina. In: *Human Reproductive Medicine: The Human Vagina* (E. S. E. Hafez, T. N. Evans, eds.). Vol. 2. North-Holland, New York, 1978, pp. 39–54.
268. D. P. Benziger, J. Edelson. Absorption from the vagina. *Drug Met. Rev. 14*(2):137–168 (1983).
269. J. L. Richardson, L. Illum. Routes of delivery: case studies, the vaginal route of peptide and protein drug delivery. *Adv. Drug Del. Rev. 8*:341–366 (1992).
270. J. R. Robinson, W. J. Bologna. Vaginal and reproductive system treatments using a bioadhesive polymer. *J. Cont. Rel. 28*:87–94 (1994).
271. C. L. Schauf, D. F. Moffett, S. B. Moffett, eds. Reproduction and its endocrine control (Chapter 25). In: *Human Physiology: Foundation and Frontiers*. Times Mirror/Mosby, New York, 1990, pp. 642–677.
272. G. Wagner, R. J. Levin. Vaginal fluid. In: *Human Reproductive Medicine: The Human Vagina* (E. S. E. Hafez, T. N. Evans, eds.). Vol. 2. North-Holland, New York, 1978, pp. 121–137.
273. R. W. Kristner. Physiology of the vagina. In: *Human Reproductive Medicine: The Human Vagina* (E. S. E. Hafez, T. N. Evans, eds.). Vol. 2. North-Holland, New York, 1978, pp. 109–120.
274. K. Knuth, M. Amiji, J. R. Robinson. Hydrogel delivery systems for vaginal and oral applications: formulation and biological considerations. *Adv. Drug Del. Rev. 11*:137–167 (1993).
275. S. Hwang, E. Owada, T. Yotsuyanagi, L. Suhardja, N. F. H. Ho, G. L. Flynn, W. I. Higuchi. Systems approach to vaginal delivery of drugs II: in situ vaginal absorption of unbranced aliphatic alcohols. *J. Pharm. Sci. 65*:1574–1578 (1976).
276. S. Hwang, E. Owada, L. Suhardja, N. F. H. Ho, G. L. Flynn, W. I. Higuchi. Systems approach to vaginal delivery of drugs V: in situ vaginal absorption of 1-alkanoic acids. *J. Pharm. Sci. 66*(6):781–784 (1977).
277. C. C. Hsu, J. Y. Park, N. F. H. Ho, W. I. Higuchi, J. L. Fox. Topical vaginal drug delivery I: effect of the estrus cycle on vaginal membrane permeability and diffusivity of Vidarabine in mice. *J. Pharm. Sci. 72*(6):674–680 (1983).
278. H. Okada, T. Yashiki, H. Mima. Vaginal absorption of a potent luteinizing hormone-releasing hormone analogue (luprolide) in rats III: effect of estrous cycle on vaginal absorption of hydrophilic model compounds. *J. Pharm. Sci. 72*:173–176 (1983).
279. E. H. Schmidt, F. K. Beller. Bichemistry of the vagina. In: *Human Reproductive Medicine: The Human Vagina* (E. S. E. Hafez, T. N. Evans, eds.). Vol. 2. North-Holland, New York, 1978, pp. 139–149.
280. W. H. Fishman, G. W. Mitchell. Studies on vaginal enzymology. *Ann. N. Y. Acad. Sci. 83*:105–121 (1959).
281. F. Acarturk, J. R. Robinson. Vaginal permeability and enzymatic activity studies in normal and ovariectomized rabbits. *Pharm. Res. 13*(5):779–783 (1996).
282. J. Brown, G. Hooper, C. J. Kenyon, S. Haines, J. Burt, J. M. Humphries, S. P. Newman, S. S. Davis, R. A. Sparrow, I. R. Wilding. Spreading and retention of vaginal formulations in postmenopausal women as assessed by gamma scintigraphy. *Pharm. Res. 14*(8):1073–1078 (1997).
283. C.-H. Lee, Y. W. Chien. Development and evaluation of a mucoadhesive drug delivery system for dual-controlled delivery of Nonoxynol-9. *J. Cont. Rel. 39*:93–103 (1996).
284. G. A. Digenis, D. Nosek, F. Mohammadi, N. B. Darwazeh, H. S. Anwar, P. M. Zavos. Novel vaginal controlled-delivery systems incorporating coprecipitates of nonoxynol-9. *Pharm. Del. Tech. 4*(3):421–430 (1999).
285. J. L. Richardson, P. A. Ramires, M. R. Miglietta, M. Rochira, L. Bacelle, L. Callegaro, L. Benedetti. Novel vaginal delivery systems for calcitonin: I. Evaluation of HYAFF/calcitonin microspheres in rats. *Int. J. Pharm. 115*:9–15 (1995).
286. J. L. Richardson, J. Whetstine, A. N. Fisher, P. Watts, N. F. Farraj, M. Hinchcliffe, L. Benedetti, L. Illum. Gamma-scintigraphy as a novel method to study the distribution and retention of a bioadhesive vaginal delivery system in sheep. *J. Cont. Rel. 42*:133–142 (1996).
287. S. K. Jain, R. Singh, B. Sahu. Development of a liposome based contraceptive system for intravaginal administration of progesterone. *Drug Del. Ind. Pharm. 23*(8):827–830 (1997).

41

Bioadhesive Drug Delivery Systems

A. David Woolfson, R. Karl Malcolm, Paul A. McCarron, and David S. Jones
The Queen's University of Belfast, Belfast, Northern Ireland

I. BIOADHESION AND MUCOADHESION

Bioadhesion, the adherence of two phases, at least one of which is biological in nature, is a topic of increasing interest to the pharmaceutical scientist. Bioadhesion can be regarded as a useful phenomenon, facilitating the design of novel polymeric drug delivery systems. However, it can also be a clinically unhelpful or potentially dangerous problem, responsible, in part, for the formation of microbial biofilm on medical devices, a primary cause of clinical infection in, for example, intensive care situations.

Bioadhesive drug delivery systems may be prepared as flexible films, solid compacts, gels, and viscoelastic semisolids. Applied advanced rheological methods, notably texture profile analysis and oscillatory rheometry, may be employed in the study of bioadhesive systems in order to obtain a fundamental understanding of their adhesive, cohesive, and flow properties. Sites for application of bioadhesive drug delivery systems are primarily topical and include ocular, nasal buccal, urogenital, and anal applications. A number of excellent reviews have been published on the general area of bioadhesion (1–10).

Bioadhesion, in the context of drug delivery, is the molecular force that resists separation across the interface between a biological surface and a carrier, usually polymeric in nature. The primary aim of bioadhesive drug delivery systems is to provide prolonged residence time, improved contact, and hence improved absorption at a given application site. Most frequently, bioadhesive drug delivery systems operate in an environment in which a mucus gel layer acts as an interface between adherent and epithelial surfaces. This three-phase system is therefore a special case of the phenomenon of bioadhesion.

Mucosal epithelial sites are rich in a viscous secretion, mucus, which adheres to the adjacent epithelium, protecting it from mechanical, bacterial, viral, and chemical attack. Mucus consists of water (up to 95% by weight), glycoproteins (0.5 to 5%), low proportions of lipids, mineral salts (1%) and 0.5 to 1% of free proteins (11). The rheological properties of mucus, together with its adhesive and cohesive properties, are primarily attributable to the glycoprotein fraction.

II. NONSPECIFIC AND SPECIFIC BIOADHESION

Bioadhesive drug delivery systems may offer certain advantages over more traditional dosage forms, including:

- Enhanced absorption rates by minimizing diffusion barriers
- Increased residence time leading to enhanced adsorption
- Improved intimacy of contact with various biological membranes

Improved bioavailability through the protection of bioactive molecules from physical and chemical degradation

The process of bioadhesion can result either from nonspecific interactions with the mucosal tissue, which are governed by the physicochemical properties of the bioadhesive material and the surface, or from specific interactions when a ligand attached to a specific site is used for the recognition and attachment to a specific site at the mucosal surface.

The first generation of bioadhesive drug delivery systems were based on mucoadhesive polymers, e.g., natural or synthetic macromolecules, often already well accepted and used as pharmaceutical excipients for other purposes. These agents had the interesting ability to adhere to humid or wet mucosal tissue surfaces. Certain mucoadhesive polymers were also found to have other, possibly more significant, biological activities, namely, to inhibit proteolytic enzymes and/or to modulate the permeability of tight epithelial tissue barriers. Such features may be particularly useful in the context of peptide and protein drug delivery.

As an alternative to the nonspecific binding to tissue achieved with mucoadhesive polymers, interest has increasingly turned to molecules with specific cell-binding affinities. In particular, specific bioadhesives such as plant or bacterial lectins, adhesion molecules that specifically bind to sugar moieties of the epithelial cell membrane, have been extensively studied, notably for drug targeting applications (12,13). These second-generation bioadhesives provide not only for cellular binding but also for subsequent endo- and transcytosis. However, lectins have yet to deliver on their initial promise as specific bioadhesive drug delivery systems, as the literature to date indicate the general variability in drug absorption associated with their use (14).

III. MECHANISMS OF BIOADHESION

There is no unified theory to explain the process of bioadhesion. The mechanism responsible for the process depends on the tissue substrate, the environment, and, critically, the design of the dosage form. The mechanical or wetting theory is perhaps the oldest available adhesion theory and explains adhesion as an embedding process. Adhesive molecules penetrate into surface irregularities and ultimately harden, producing numerous anchors (15). For the adhesive to move freely into such pores, it must overcome any surface tension effects present at the interface. The wetting theory is most applicable to liquid bioadhesives. It describes the interactions between the bioadhesive polymer, and its angle of contact with the substrate, to thermodynamic work of adhesion. Such liquid bioadhesives are, in practice, only weakly adherent.

The formation of an electrical double layer is the basis of the electrostatic theory (16). This theory, applicable to mucoadhesion, assumes that, upon contact, the mucin glycoproteins and polymer material transfer electrons across the interface. This movement of coulombic charge sets up an electrical double layer and, subsequently, a system of attractive forces that maintains the contact.

The interpenetration or diffusion theory is perhaps the most widely accepted physical theory and the most relevant to bioadhesive hydrogel systems. In the mucoadhesive context, polymeric chains, from both bioadhesive and mucus, intermingle and reach a sufficient depth within the opposite matrix to allow formation of a semipermanent bond. The process can be visualized from the point of initial contact. The bioadhesive concentration gradient will drive polymeric chains into the mucous network until an equilibrium penetration depth is achieved. Once intimate contact occurs, the mucin and polymer chains move along their concentration gradients into the opposing phase. The depth of penetration into this phase depends on their respective diffusion coefficients. Reinhart and Peppas found that the diffusion coefficient depended on the molecular weight of the polymer strand and that it decreased with increasing cross-linking density (17). This process of simple physical entanglement has been described as analogous to forcing two pieces of steel wool to intermingle upon contact (18). In the context of bioadhesive drug delivery systems, the formation of permanent ionic or covalent bonds, by contrast, is not required. Bioadhesive delivery systems require a defined residence time at the target site, after which they are removed or rendered inactive. Adhesives that form permanent covalent bonds with the tissue substrate, such as those used in dentistry and orthopaedics, are too strong for this purpose (19,20).

The adsorption theory of adhesion involves surface forces (21,22). Adherence to tissue is due to the establishment of weak attractive forces between substrate and polymer, primarily via van der Waal's forces and hydrogen bonding (21,22). Once intimate contact has been established between polymer and substrate, these may be considered to be the strongest contributors to bioadhesion (23).

A study of the formation of fractures or cracks at the polymer mucin interface (in the case of mucoadhesion) leads to the fracture theory of adhesion (24). The mechanical deformation during the separation process between bioadhesive polymer system and tissue substrate dissipates heat at, or into, the immediate area of the interface, thereby raising the temperature at the fracture area. In an interface, microcracks will always exist. The temperature rise, especially in the region of the crack tip, will increase the plasticity of polymer chains and accelerate the crack propagation. Thus the process of interface fracture is determined by crack formation and crack propagation. This theory is dependent upon fracture occurring exactly at the polymer–mucin interface. Experience, however, has shown that this cannot always be guaranteed (25).

The surface microstructure of any tissue is an important parameter for considering those bioadhesives that achieve their bonding mostly through physical attachment. Such surfaces often contain undulations and crevices that provide both attachment points and a key for the deposition and inclusion of the polymeric adhesive. Mucoadhesion, involving polymeric carrier adherence to a mucin-coated epithelium, can be considered a ''wet-stick'' adhesion process, in which polymer chains are released from restraining dry lattice forces by hydration. These chains then move and entangle into the matrix of the substance. Once entangled, opposing chemical groupings become aligned, increasing the probability of van der Waals attraction or hydrogen bonding. As this liberation of polymer chains is hydration dependent, the amount of water present in the bond site is decisive in determining its strength. Overhydration will form a wet slippery mucilage with no adhesive strength.

Overall, bioadhesion can be understood as a two-step process, in which the first adsorptive contact is governed by surface energy effects and a spreading process. In the latter phase, the diffusion of polymer chains across the interface may enhance the final bond (26). Water plays an important role in mucoadhesion, the invading water molecules liberating polymer chains from their twisted and entangled state and thus exposing reactive sites that can bond to tissue macromolecules. The adhesion of dried hydrogels to moist tissue can be quite substantial, with water uptake from the tissue surface facilitating surface dehydration and exposing surface depressions that may act as anchoring locations (27). Thus the use of strongly mucoadhesive polymeric drug delivery systems in the oral cavity can, for example, produce a profound and, for some individuals, unpleasant dryness at the site of attachment.

IV. DETERMINATION OF BIOADHESIVE FORCE OF ATTACHMENT

The evaluation of bioadhesive properties is fundamental to the development of novel bioadhesive delivery systems. Measurement of the mechanical properties of a bioadhesive material after interaction with a substrate is one of the most direct ways to quantify the bioadhesive performance. To measure the force of bioadhesive attachment to tissue involves the application of a stress to the bonding interface. Numerous designs of apparatus have been proposed for this purpose (11,28,29). Since no standard apparatus is available for testing bioadhesive strength, an inevitable lack of uniformity between test methods has arisen. Nevertheless, three main testing modes are recognized: tensile, shear, and peel tests.

When the force of separation is applied perpendicularly to the tissue/adhesive interface, a state of tensile stress is set up. This is perhaps the most common configuration used in bioadhesive testing. During shear stress the direction of the forces is reoriented so that it acts along the joint interface. In both tensile and shear modes, an equal pressure is distributed over the contact area (29).

The peel test is more applicable to systems involving adhesive tape where removal of the device is an important parameter. By pulling the two interfaces apart at an acute angle, the force is focused along a single line of contact. This effectively concentrates the applied force at the point of separation and removal of the tape can be achieved easily. The peel test is of limited use in bioadhesive systems, being better suited to transdermal patch technology or the evaluation of adhesive dressing systems.

Once formed, the interface of the bioadhesive bond is likely to exist in a microenvironment that will inevitably influence its further performance. Factors such as pH, temperature, ionic strength, and water content will all affect bond durability. Experimental rigs have been devised to investigate these important considerations. Although the force used to break the bond can be applied in one of three fashions, as described, the majority of researchers prefer to use the tensile stress method. Irrespective of which configuration is used to apply the force, the measurement of the performance of the bioadhesive bond is generally performed in one of three environments:

Most in-vitro methods involve measurement of shear or tensile stress. Smart et al. described a method to study bioadhesion whereby a polymer-coated glass slide was withdrawn from a mucus solution (30). The force required to accomplish this was equated to mucoadhesion. A fluorescence probe technique was developed by Park and Robinson to determine bonding between epithelial cells and a test polymer (31). Other methods have involved measurement of the force required to detach polymeric materials from excised rabbit corneal endothelium. Mikos and Peppas described a method whereby a polymer particle was blown across a mucus filled channel (25). The motion, recorded photographically, gave details regarding the adhesion process. Woolfson et al. described the in vitro measurement of bioadhesive strength via a linear variable differential transformer (32). The apparatus (Fig. 1), constructed from Perspex, consists of a movable platform

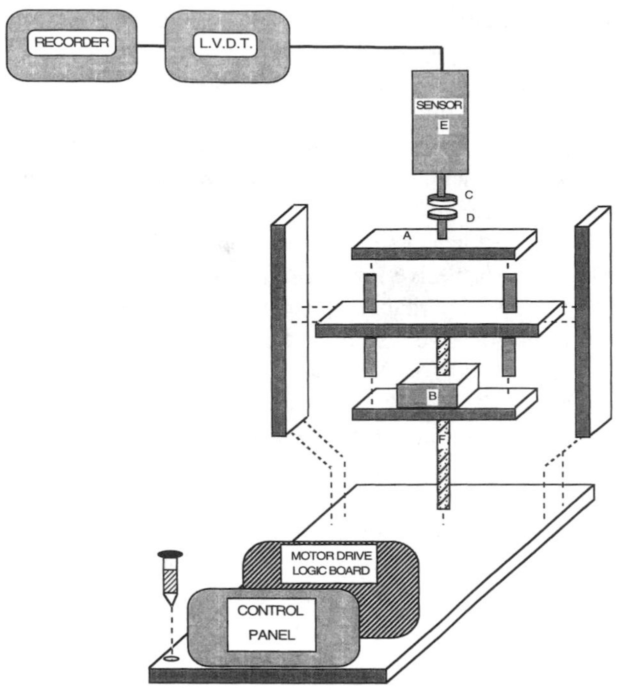

Figure 41.1 Design of a tensile test device for bioadhesion using a linear variable displacement transformer. (A) Platform. (B) Stepping motor. (C) Upper pedestal (tissue mount). (D) Lower pedestal (bioadhesive mount). (E) Sensor housing. (F) Screw-threaded shaft.

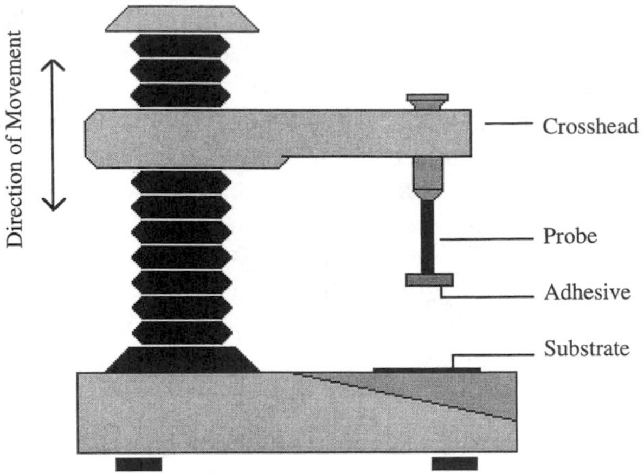

Figure 41.2 Texture profile analyzer in bioadhesion test mode.

with a pedestal to which the bioadhesive formulation is attached. Tissue was secured to a second, upper pedestal linked to the LVDT via a sensor. During the test procedure the platform was moved downwards at a predetermined rate controlled by the motor drive logic board. As movement occurs to the upper pedestal this was detected by the output from the LVDT, which is recorded potentiometrically. The force generated was measured by a previously calibrated spring contained within the sensor housing. Different calibrated springs were used to increase or decrease the sensitivity of the apparatus. Initial evaluation of the testing device was performed on a range of candidate bioadhesive film formulations. Jones et al. have reported on the use of a commercial apparatus, in the form of a Texture Profile Analyser (Fig. 2) operating in bioadhesive test mode, to perform similar functions (33–35).

The ultimate destination for successful bioadhesive devices is a tissue surface such as the gastrointestinal tract or the buccal cavity. Some investigators have used previous results obtained from their own in vitro work to predict and subsequently test for in vivo performance. Examples of in vivo studies in the literature are not as plentiful as in vitro work because of the amount and expense of animal trials and the difficulty in maintaining experimental consistency. Commonly measured variables include gastrointestinal transit times of bioadhesive-coated particles and drug release from in situ bioadhesive devices.

Ch'ng et al. studied the in vivo transit time for bioadhesive beads in the rat (36). A radiolabeled bioadhesive (using ^{51}Cr) was carefully introduced into the stomach. At selected time intervals, the gastrointestinal tracts were removed and cut into 20 equal segments, and the radioactivity in each was measured.

Reich et al. described an instrument for measuring the force of adhesion between intraocular lens materials and the endothelium of excised rabbit corneas (37). It used a metal or glass fiber deflection technique for measuring pressures that were usually lower than 1 g cm^{-2}. The contacting surfaces were submerged in saline to eliminate surface tension effects. In contrast to many bioadhesive testing situations, the authors were attempting to find a polymer with a lower adhesion than poly(methylmethacrylate) (PMMA), the material most commonly used to make intraocular lenses. A direct relationship had been shown between the adhesive forces of a polymer to the endothelium and the extent of cell damage. Poly(hydroxyethylmethacrylate) and Duragel®, both hydrophilic materials used to make soft lenses, gave less adhesion and subsequently less cell damage than the hydrophobic PMMA.

Davis (1985) described a method to study gastric-controlled release systems (38). It can be advantageous in certain diseased states for a single dosage unit to be retained in the stomach. This enables released drug to empty from the stomach and have the length of the small intestine available for absorption. Mucoadhesives can help achieve this objective. Therefore, as an alternative to using invasive in vivo techniques, a formulation was used containing a gamma emitting radionuclide. The release characteristics and the position of the device could be monitored using gamma scintigraphy. This technique could measure two different radionuclides simultaneously, so that a subject could be given two different formulations together, with their release and transit times then being studied in a single-occasion cross-over study.

V. CHARACTERIZATION METHODS FOR BIOADHESIVE DRUG DELIVERY SYSTEMS

A range of techniques are available for the study of bioadhesive delivery systems; the choice is influenced by the site of attachment and the substrate tissue, together with the design of the dosage form. Most viable bioadhesive drug delivery applications involve mucoadhesion to accessible (topical) epithelia. Certain tests will be required for regulatory purposes, while others, though optional, often constitute an important part of the development of a pharmaceutics package for a novel bioadhesive carrier. Bioadhesive (mucoadhesive) dosage forms vary in bioadhesive performance, water content being a primary determinant of bioadhesion. Delivery platforms include liquids and gels (low adhesion), viscoelastic semisolids (moderate adhesion), flexible hydrogel films, particulates, and compacts (strong adhesion). Applicable characterization methodologies include

- Drug release studies
- Drug diffusion (membrane penetration) studies
- Examination of mechanical and textural properties
- Examination of continuous shear properties
- Examination of structural (viscoelastic) properties at defined temperatures
- Evaluation of adhesion to model substrates
- Examination of product/packaging interactions

Texture profile analysis can be used to determine the following characteristics of semisolid bioadhesive delivery systems (39).

- Hardness/compressibility—a measure of the resistance of the formulation to probe depression, which can be used to characterize product spreadability
- Cohesiveness—a measure of the effects of successive deformations on the structural properties of a product
- Adhesiveness—a measure of the work required to remove the probe from the sample, a property related to bioadhesion

Flow rheometry can be used to obtain the following information on bioadhesive semisolids.

- Effects of successive shearing stresses on the rate of deformation of a product
- Information concerning product viscosity, which in turn affects drug release and ease of application at the site
- Information concerning the rate of structural recovery following deformation (thixotropy), which in turn affects product retention at the application site
- Drug release
- Manufacture

Oscillatory rheometry, a nondestructive test (unlike flow rheometry) can be used to obtain the information on bioadhesive semisolids used at sites subjected to variable stresses, for example, in the oral cavity, where chewing, swallowing, and talking all affect the structural rheology of bioadhesive systems (35,40). Oscillatory rheometry determines strain in the system, resolved into two components: the storage modulus (G') is the part of the strain that is in phase with the stress and is a measure of the solid character, and the loss modulus (G'') is the part of the strain that lags the stress by 90° and is a measure of the liquid character.

Dielectric spectroscopy is an analytical technique that involves the application of an oscillating electric field to a sample and the measurement of the corresponding response over a range of frequencies, from which information on sample structure and behavior can be extrapolated. Craig and Tamburic described studies on sodium alginate gels whereby a model was proposed in order to relate the low-frequency response to the gel structure (41). They also reported on the application of dielectric spectroscopy to the study of cross-linked polyacrylic acids with respect to the effects of additives such as propylene glycol and chlorhexidine gluconate, the influence of the choice of neutralizing agent, and the effects of aging on the gel structure.

VI. BIOADHESIVE (MUCOADHESIVE) POLYMERS FOR PHARMACEUTICAL APPLICATIONS

Mucoadhesives may be defined as natural or synthetic materials that adhere to mucosal surfaces for prolonged periods of time. Mucoadhesion is the specific form of bioadhesion most relevant to drug delivery applications. Mucoadhesive character is almost exclusively displayed by hydrophilic polymers whose large molecular size and ability to form intermolecular bonds provides them with the necessary interpenetrative and retentive properties required to adhere to the mucus layer. The chemical structures of some of the more common mucoadhesive polymers are presented in Fig. 3. Others include gelatin, pectin, agarose, polystyrene graft copolymers, and glyceryl oleates.

Mucoadhesive polymers can be distinguished into two main categories: polysaccharides and vinyl polymers. The polysaccharides, exemplified by the cellulose derivatives, chitosan, starch, and alginic acid, are simple sugars joined by glycosidic linkages and are of natural origin, although chemical modification of the side chains allows polymers having different physicochemical properties to be produced. For example, chitosan is prepared by the deacetylation of chitin, the principal component of the exoskeleton of crustaceans. The vinyl polymers are of synthetic origin, commonly formed by the free-radical polymerization of the corresponding vinyl monomer.

In order to function as mucoadhesives, polymer molecules must achieve intimate and prolonged contact with the mucosal substrate. This may be facilitated by more than one adhesion mechanism. Certain structural features and physicochemical properties are known to contribute to or improve bio(muco)adhesion. Thus by careful selection and modification of the appropriate polymer, bioadhesive systems can be tailored for a wide range of drug delivery applications.

A. Hydrophilic Functional Groups

The presence of hydrophilic functionalities, typically ether, carboxyl, hydroxyl, amine, and sulphate groups, are a prerequisite for good mucoadhesion. These groups allow interaction with and attachment to the mucosal tissue, predominantly via hydrogen and dipole–dipole bonding. Where the polymers have been rendered water-insoluble by cross-linking, the hydrophilic moieties also take up water, causing the networks to swell. The swelling increases the conformational mobility of the polymer chains and enhances mucoadhesive strength due to a concomitant increase in chain diffusivity and degree of entanglement. Swelling also exposes a greater number of hydrophilic functionalities to the mucosal surface.

Additional hydrophilic groups can be incorporated into polymers to enhance their bioadhesion. For example, Matsuda et al. have reported that the in vitro bonding strength of gelatin films is enhanced up to sixfold by grafting free-dangling aldehyde groups (42). When the aldehyde groups were reacted with either glycine or sodium borohydride to form the corresponding imine and alcohol, respectively, the bonding strengths subsequently decreased, suggesting that the Schiff bases formed between the aldehyde groups in the gelatin film and the amino groups of the porcine skin test material strongly contributed to enhanced bioadhesion.

If the addition of hydrophilic moieties increases the bioadhesive strength of polymers, the converse should also hold true. Pritchard et al. have demonstrated that esterification of the carboxylic acid groups of hyaluronic acid caused a decrease in the mean detachment force from rat epithelia (43). However, additional factors, including the degree of hydrophilicity and swelling of the esterified polymers, were also considered to be factors.

Bioerodible copolymers of fumaric and sebacic acid have been reported whose bioadhesion is attributed to hydrogen bonding between the copolymer and mucus glycoproteins (44). Interestingly, the continuous degradation of these materials enhanced their bioadhesive properties by changing surface energy, increasing surface roughness, and increasing the number of carboxylic acid moieties available for binding.

Some proteins have been shown to possess bioadhesive character. The strong mucoadhesive properties of the mussel adhesive protein produced by the blue mussel, *Mytilus edulis*, have been attributed to hydroxylated amino acid residues and the presence of 10–20% dopa residues (45). However, the presence of hydrophilic functions does not necessarily ensure good bioadhesion. In certain bioadhesive formulations the ability of the hydrophilic groups to bind with mucin may be significantly reduced by interactions or complex formation with other excipients. Also, certain coiled-chain configurations can hide the functionalities within the coil structure, thereby decreasing the possibility of interactions with a substrate.

B. Cross-linking

Linear bioadhesive polymers, such as those in Fig. 3, may be cross-linked according to their type. For vinyl polymers, this is most readily achieved by including a divinyl molecule into the polymerization reaction, typically at concentrations of between 0.1 and 1.0%. Both polysaccharides and vinyl polymers can be cross-linked postpolymerization by the addition of a difunctional molecule capable of reacting with functional groups in the polymer. The addition of divalent cations, such as Ca^{2+}, to poly(acrylic acid) will also cause cross-linking. While the linear polymers are generally soluble in water, cross-linking causes them to become water insoluble, although they will swell in aqueous environments. Although cross-linking tends to improve the retentive character of the bioadhesive at a biological site, the restricted chain mobility and loss of

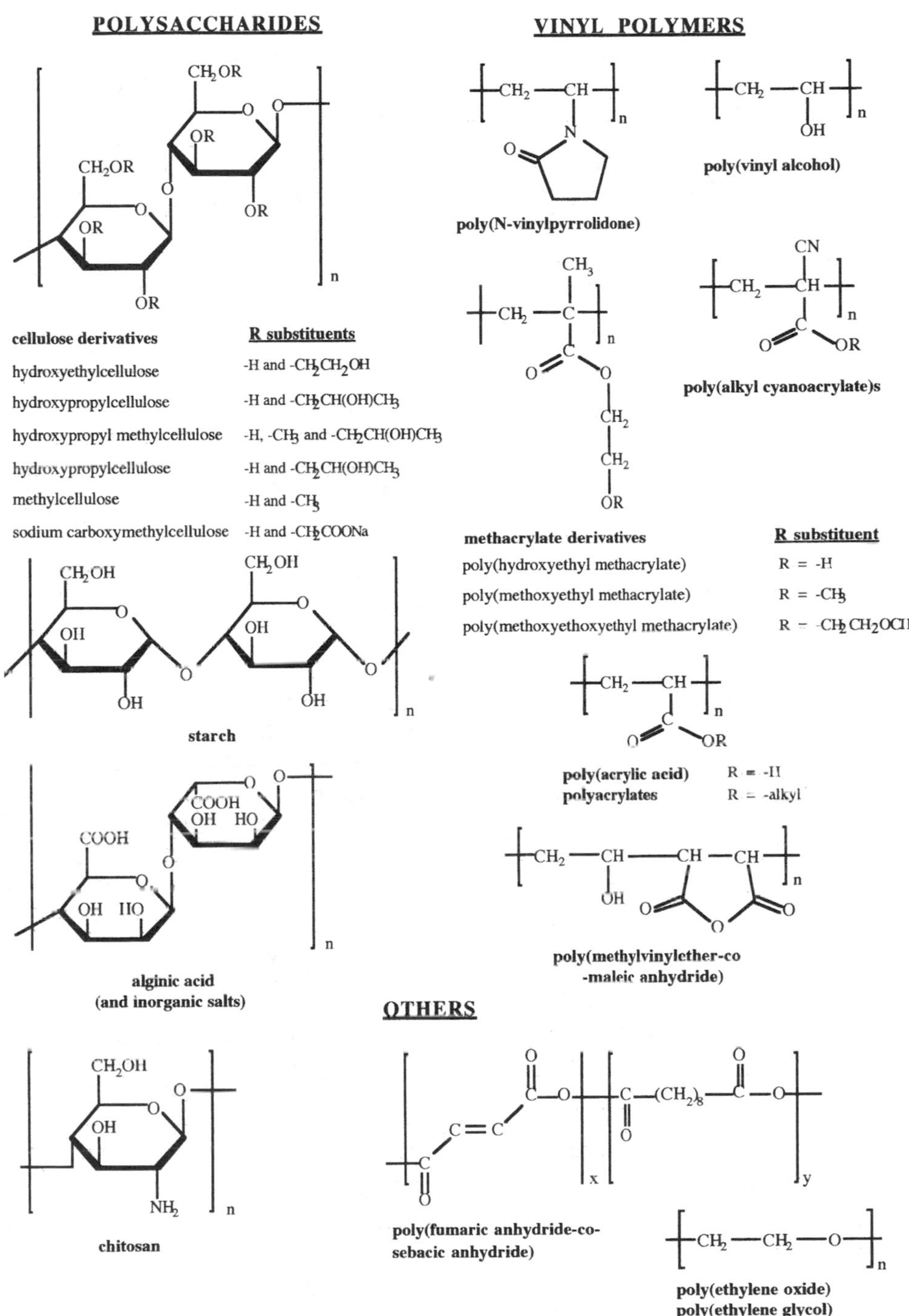

Figure 41.3 Chemical structures of some bioadhesive polymers.

functionality may to some degree counteract this advantage. Optimal bioadhesive properties can be achieved by controlling the degree of cross-linking.

Much of the work on cross-linked mucoadhesive polymers has focused on poly(acrylic acid) derivatives. Several studies have shown that the degree of bioadhesion observed decreases with increasing cross-linking concentration (46–48). Cross-linking, however, need not be limited to modifying the mechanical and dissolution properties of bioadhesive systems. Kakoulides et al. reported on the performance of poly(acrylic acid) networks where the azo groups of the divinylazobenzene cross-links are designed to be cleaved in the reductive environment of the colon, causing biodegradation of the network and the simultaneous formation of bioadhesive amine groups (49).

C. Degree of Hydration/Polymer Concentration

Mucoadhesive polymers display different adhesive properties at different hydration levels, but maximum adhesion will exist at an optimum hydration level. If excess water is present, adhesiveness may be lost due to the formation of a slippery mucilage. The amount of solvent present during the initial phase of bioadhesion is also vital. For interpenetration of the polymer chains, they must first be mobilized with sufficient solvent to allow exposure of the functional groups on the chains. Many studies have shown that the bioadhesive character of mucoadhesive drug delivery formulations decreases with increasing formulation water content (47,50,51). In some cases, the greatest bioadhesion on wet mucosal tissue was observed for hydrophobic formulations, i.e., where water is absent (50). Needleman et al. reported that the differences in bioadhesive strength and degree of retention for chitosan, xanthan gum, and poly(ethylene oxide) aqueous formulations could be attributed to their different rates of hydration (52).

D. Polymer Molecular Weight

The relationship between polymer molecular weight and bioadhesive properties is not as simple as might first be imagined, as evidenced by many conflicting results in the literature. The problem is exacerbated in that there is no standard test of bioadhesive strength (force of detachment from substrate). Differences in testing procedures and biological substrates further complicate intersystem comparisons. A standardized and more comprehensive evaluation of the influence of molecular weight is required.

Early studies reported that adhesive strength increased as the molecular weight of the adhesive polymer increased up to approximately 100,000, after which there was little effect (22). Although a critical molecular weight may be necessary to produce the interpenetration and entanglement of adhesive and substrate, others factors must also be considered. For example, the linear chain configurations in poly(ethylene oxide)s and the subsequent increase in penetration length are reported to be responsible for increasing adhesive strength up to a molecular weight of 4×10^6 (20). Polymers with coiled configurations, such as dextrans, displayed similar bioadhesive strengths at both low and high molecular weights, owing to ''shielding'' of the hydrophilic functional groups and intra- rather than intermolecular hydrogen bonding.

Pritchard et al. found that low molecular weight hyaluronic acids performed better than higher molecular weights in detachment tests, although their effect on mucociliary transport was similar (43). This contrasts with the work of Saettone et al. (53) and Durrani et al. (54) who reported opposite findings. The differences may be attributed to differences in hydration rates and/or testing methodologies. The bioadhesive properties of N-(2-hydroxypropyl)methacrylamide copolymers with different amounts of pendant saccharide moieties were also found to decrease with increasing molecular weight (55).

Poly(acrylic acid)s have also been extensively studied with regard to the effect of molecular weight on bioadhesive strength. In general, higher molecular weights were reported to produce better bioadhesion (56–58), although the opposite trend has also been reported (47).

E. Segment Mobility/Flexibility

For mucin and polymer chains to interpenetrate, they must be able to diffuse. Interdiffusion will increase with increasing chain segment mobility in the bioadhesive, which is intrinsically related to the degree of hydration and the cross-linking density. Gu et al. have shown the importance of the expanded nature of both mucin and the bioadhesive polymer on the subsequent strength of mucoadhesion (59). A series of 0.2% cross-linked copolymers were synthesized from varying ratios of acrylic acid and methyl methacrylate monomers. The amount of water a polymer absorbed was determined as the difference between the dried weight of polymer particles and the equilibrium weight after hydration in a buffer solution. Dried, unswollen particles were shown to possess no bioadhesion, but when swollen, adhesion was significant. As the percentage of acrylic acid in the copolymer increased, the number of available charged carboxyl groups also increased and bioadhesion likewise increased. Increased charge density drives water molecules into the polymer network and expands its structure. Solvent uptake, polymer network openness, and polymer bioadhesion are all interdependent properties. The addition of a plasticizer to poly(methyl vinyl ether-co-maleic anhydride) bioadhesive formulations increased mechanical flexibility but did not significantly effect in vitro bioadhesion (60).

F. Charge/pH

Polyelectrolytes, i.e., polymers having ionizable groups along the chain, are often excellent mucoadhesives (30,61). Examples include poly(acrylic acid) and chitosan, which possess a carboxylic acid and an amine function, respectively. The zeta potential and extent of hydration of 0.3% cross-linked poly(acrylic acid)s were examined in the presence of isotonic saline, simulated gastric fluid, and simulated intestinal fluid (62). The relationship between the zeta potential and the bioadhesive strength showed that bioadhesion was linked to the extent of ionization that occurred during the interaction of the polymer with the mucin/epithelial cell surface. A suitable extent of ionization was required for the initial adhesion of polymer to the biological tissue. However, extensive ionization and swelling could lead to the formation of very loose particles or dissolution and therefore to difficulty in interacting with the animal tissue. Thus the degree of neutralization of poly(acrylic acid) influences its mucoadhesions to pig gastric tissue (48).

Ch'ng et al. demonstrated that a neutral polymer, such as poly(2-hydroxyethyl methacrylate), required a much lower force for detachment than charged polymers based on acrylic acid (36). The poly(acrylic acid) derivative polycarbophil was tested in buffers of increasing pH values; bioadhesion to rabbit stomach increased correspondingly until, at pH 7, bioadhesion unexpectedly and suddenly dropped. The authors explained this phenomenon as an imbalance between repulsive and swelling forces. Pendant groups on the mucus backbone are fully ionized at pH 2.6. The pKa of polycarbophil is 4.75, and above this pH value, electrostatic repulsion between the two negatively charged polymer and mucin chains will increase. However, above pH 4, the polycarbophil network becomes highly swollen, which favors the interpenetration of polymeric chains across the interface. At pH 7, the authors believed that the repulsion was so great that bioadhesion was no longer feasible.

Changes in pH of the substrate can also affect bioadhesion. Anionic polymers were found to strengthen the mucus gel more than a neutral or cationic polymer, with a consequent potential increase in chain entanglement with mucoadhesive polymers (57). Positively charged lysine residues are implicated in the bioadhesive character of mussel adhesive protein (45).

G. Drug/Excipient Concentration

BlancoFuente et al. showed that the addition of propranolol hydrochloride to Carbopol (a lightly cross-linked poly(acrylic acid) polymer) hydrogels increased adhesion when water was limited in the system, due to an increase in the elasticity caused by complex formation between drug and polymer (47). However, when large quantities of water were present, the complex precipitated out, leading to a slight decrease in adhesive character. By contrast, miconazole nitrate did not influence bioadhesion up to a concentration of 30% in poly(acrylic acid)-based tablets (56). Voorspoels et al. demonstrated that increasing concentrations of testosterone and its esters produced a decrease in adhesive characteristics of buccal tablets (63).

For ionic bioadhesive polymers, dissolved salts can have a significant effect on bioadhesion by a variety of mechanisms, including salting out of polymeric components and charge neutralization, preventing expansion of coiled polymer chains and thus entanglement with epithelial mucin. Woolfson et al. added sodium chloride to polymer blends containing Gantrez and found no effect on bioadhesive character, indicating that the sensitivity of bioadhesive polymers to the presence of dissolved salts varies considerably, depending on structure (51). However sodium chloride can affect bioadhesion through a direct action on the substrate. Thus sodium chloride, when added to several mucoadhesive polymers, was observed to reduce the stiffness of mucus gel (57).

H. Polymer Combinations

Although individual polymers may display mucoadhesive properties, combinations of mucoadhesives and nonmucoadhesive polymers have been used advantageously to tailor the physicochemical properties (56,64–68).

When several polymers are included in mucoadhesive drug delivery devices, interactions between the polymers can cause either a decrease or an increase in the degree of mucoadhesion observed. Hydrogen bond-accepting polymers, such as poly(vinyl pyrrolidone) and poly(ethylene oxide), can form interpolymer complexes with hydrogen bond donating polymers, such as poly(acrylic acid), thereby reducing the strength of the bond between the mucoadhesive and mucin (69). In the same study, the incorporation of starch or poly(vinyl alcohol), which cannot form strong intermolecular forces with poly(acrylic acid), did not significantly reduce mucoadhesion.

I. Comparison of Bioadhesive Strength of Mucoadhesive Polymers

Despite the difficulties in comparing bioadhesive strengths of individual polymers, polymer combinations, and formulated bioadhesive or mucoadhesive delivery systems, a number of ''ranking orders'' have been published. Wong et al. used an in vitro chicken pouch model to simulate conditions for buccal drug delivery and reported decreasing bioadhesive strength in the order Carbopols > gelatin > NaCMC > HPMC > alginic acid, Eudragits, chitosan (65). Needleman et al., using an in vivo periodontal model, produced the ranking xanthan gum > poly(ethylene oxide) > chitosan (70). An in vitro modified intestinal perfusion technique reported by Cvetkovic et al. ranked poly(acrylic acid) > sodium carboxymethylcellulose > hydroxypropylcellulose (71).

The variability among the literature reports strongly suggests a cautious interpretation of comparative bioadhesion among polymers. Of more profound significance is likely to be the delivery platform itself into which the polymeric components are formulated, together with the extent of water in the delivery system, the amount of water (or biological fluid) at the site, and the capacity of the delivery system to swell in the presence of water and to retain water without forming a slippery mucilage. In general, polycations and polyanions provide the strongest bioadhesion if correctly formulated. Arguably, highly hydrated systems such as liquids and light gels are at best capable of tack but not significant bioadhesion/mucoadhesion. Solid systems with low water contents, or semisolids formulated with a low water content/high polymer content, are most likely to optimize the bioadhesion or mucoadhesion of a given polymer or polymeric combination.

VII. BIOADHESIVE AND MUCOADHESIVE DRUG DELIVERY APPLICATIONS

A. Bioadhesive Devices for the Oral Cavity

The oral cavity is a convenient and accessible area, ideally suited to bioadhesive drug delivery. The most commonly used areas are the buccal and sublingual areas. The nonkeratinized regions in the oral cavity, such as the soft palate, the mouth floor, the ventral side of the tongue, and the buccal mucosa also offer least resistance to drug absorption (72). In many instances, these areas are most suitable for locating bioadhesives. Molecular transport across the barrier membrane is achieved by simple diffusion along a concentration gradient from carrier to tissue, possibly aided by some form of penetration enhancement. Larger, hydrophilic molecules encounter greater resistance to diffusion and are believed to cross the oral epithelium by intercellular pathways. The advantages of drug delivery through the oral mucosa have been detailed by Veillard et al. (73) and include bypassing hepatic first-pass metabolism, excellent accessibility, unidirectional drug flux, and improved barrier permeability compared, for example, to intact skin.

Many drugs, such as glyceryl trinitrate, testosterone, and buprenorphine (74), have been delivered via the buccal route. Absorption is rapid and drains into the reticulated vein. This avoids hepatic first-pass metabolism and intestinal enzymatic attack on the drug species. Because of its accessibility, the buccal area offers excellent patient compliance in comparison to epithelial routes that involve insertion of oleaginous devices, as sometimes found in rectal and vaginal delivery.

The process of drug absorption from conventional oral delivery systems normally occurs from saliva after dissolution from the formulation. Bioadhesive devices differ in that drug diffuses directly through the swollen polymer and then into the membrane. The intimacy of contact concentrates the drug on the epithelial surface. At other mucosal surfaces, the activity of goblet cells builds up a diffusion-limiting mucus layer that can impede drug absorption. This potential problem does not manifest itself at the oral mucosa, which contains no goblet cells, the mouth receiving its mucus primarily from the parotid, submaxillary, and sublingual salivary glands (73).

Ponchel et al. described the delivery of metronidazole from a bioadhesive tablet containing hydropropylmethylcellulose and poly(acrylic acid) (75). The tablets contained a compressed mixture of 50% drug and 50% of various blends of the two polymers. Using a method described by Ponchel et al., the work of adhesion required to remove the tablets from bovine sublingual mucus was measured and found to plateau to a maximum force (76). This maximum adhesion was produced from a tablet containing 25% poly(acrylic acid), and further increases in poly(acrylic acid) content did not increase bioadhesion. Drug release was found to be non-Fickian, controlled by a combination of diffusion and polymer chain relaxation.

The buccal route may be an attractive site for peptide delivery because it is deficient in enzymatic degradation pathways. The cavity is lined with a relatively thick mucous membrane that is highly vascularized and approximately 100 cm^2 in area (72). Prompted by studies showing rectal absorption of insulin in oleaginous vehicles, various workers tried to achieve systemic blood levels of insulin using a bioadhesive buccal patch (74,77–80). Insulin was mixed with sodium glycocholate (used to enhance penetration) in an oil-based core and encapsulated with a bioadhesive cap. Once located, the overlapping outer layer stuck to the membrane and swelled into a gelled mass that kept the molten inner core intact. Unfortunately, concentrations of peptide in the blood only achieved 0.5% bioavailability compared with an intramuscular administration of insulin. The poor bioavailability of insulin from this device was explained as a combination of several factors. The center core was made from cocoa butter, which is poorly soluble in saliva, and the area of the core in contact with the mucosa was quite small.

Oral mucosal ulceration is a common condition: up to 50% of healthy adults suffer from recurrent minor aphthous ulcers. A bioadhesive compound, Zilactin®, has provided a novel way to alleviate symptoms of this oral mucosal ulceration. It is based on hydroxypropylcellulose complexed with tannic, boric, and salicylic acids. Rodu et al. applied portions of gel to oral mucosa in vivo and found that its adhesion was not affected by food or drink intake (81).

Eversole has compared the mucosal binding properties of Orabase® and Zilactin® and evaluated their temporary pain relief and protective properties (82). Samples of ointment were placed on the labial mucosa and allowed to dry. Zilactin® formed a resilient film that did not adhere to the opposite gingiva, whereas Orabase® formed a sticky, inconvenient mass that readily adhered to teeth and gingiva. Zilactin® was shown to give significant symptomatic relief of pain and protected the ulcerated area from irritants present in the oral cavity. On the principle that low-water-content solid systems offer better mucoadhesion than semisolids, a novel bioadhesive patch for the treatment of recurrent apthous ulceration has been reported (83). The patch provided passive protection to the site, allowing healing and resisting in vivo challenge by acidic citrus juice.

The formulation and in vitro/in vivo evaluation of buccal bioadhesive captopril tablets was reported by Iscan et al. (66). The objectives were to elucidate factors affecting the bioadhesion property of compressed tablet consisting of hydroxypropylmethylcellulose and carbomer and to evaluate the bioavailability and pharmacokinetics of the formulated buccal adhesive captopril tablets. Buccal adhesive controlled-release systems for the delivery of captopril were prepared by compression of hydroxypropylmethylcellulose with carbomer, which served as the bioactive adhesive compound. The mean pharmacokinetic parameters after use of the buccal adhesive tablet were C-max, 310.7 ng/mL; t(max), 1.2 h; AUC(0–8), 890.1 ng h/mL. This study demonstrated that the formulated buccoadhesive therapeutic system was suitable for buccal administration of captopril.

Wong et al. reported the formulation and evaluation of controlled release Eudragit buccal patches (65). Controlled release buccal patches were fabricated using Eudragit NE40D. Various bioadhesive polymers, namely hydroxypropylmethylcellulose, sodium carboxymethylcellulose, and Carbopol of different grades, were incorporated into the patches, using metoprolol tartrate as the model drug. The in vitro drug release was determined using the USP 23 dissolution test apparatus 5 with slight modification, while the bioadhesive properties were evaluated using texture analyser equipment with chicken pouch as the model tissue. The incorporation of hydrophilic polymers was found to affect the drug release as well as enhance the bioadhesiveness. Although high-viscosity polymers can enhance the bioadhesiveness of the patches, they also tend to cause nonhomogeneous distribution of the polymers and drug, resulting in nonpredictable drug-release rates. Of the various bioadhesive polymers studied, Cekol 700 appeared to be the most satisfactory in terms of modifying the drug release and enhancement of the bioadhesive properties.

The design and evaluation of chitosan/ethylcellulose mucoadhesive bilayered devices for buccal drug delivery has recently been described (84). Buccal bilayered devices comprise a drug-containing mucoadhesive layer and a drug-free backing layer. Bilaminated films were produced by a casting/solvent evaporation technique, and bilayered tablets were obtained by direct compression. The mucoadhesive layer was composed of a mixture of drug and chitosan, with or without an anionic cross-linking polymer (polycarbophil, sodium alginate, gellan gum), and the backing layer was made of ethylcellulose. The double-layered structure design was expected to provide drug delivery in a unidirectional fashion to the mucosa and avoid loss of drug due to washout with saliva. Using nifedipine and propranolol hydrochloride as slightly and highly water-soluble model drugs, respectively, it was demonstrated that the devices show promising potential for use in controlled delivery of drugs to the oral cavity. The uncross-linked chitosan-containing devices absorbed a large amount of water, gelled, and then eroded, allowing drug release. Bilaminated films showed a sustained drug release in a phosphate buffer (pH 6.4).

Diseases of the oral cavity can be broadly differentiated into two categories, namely inflammatory and infective conditions. In many cases, the demarcation between these two categories is unclear, as some inflammatory diseases may be microbiological in origin (85). Most frequently, therapeutic agents are delivered into the oral cavity in the form of solutions and gels, as this ensures direct access of the specific agent to the required site in concentrations that vastly exceed those that can be achieved using systemic administration (34,85). However, the retention of such formulations within the oral cavity is poor, due primarily to their inability to interact with the hard and soft tissues and additionally to overcome the flushing actions of saliva. Therefore the use of bioadhesive formulations has been promoted by several authors to overcome these problems and hence improve the clinical resolution of superficial diseases of the oral cavity (33,86). In particular, within the oral cavity, bioadhesive formulations have been reported for the treatment of periodontal diseases and superficial oral infection, in which specific interactions between the bioadhesive formulations and the oral mucosa may be utilized to ''anchor'' the formulations to the site of application.

Periodontitis is an inflammatory disease of the oral cavity that results in the destruction of the supporting structures of the teeth (87). It is characterized by the formation of pockets between the soft tissue of the gingiva and the tooth; if untreated it may result in tooth loss (33,87). Drug delivery problems associated with the periodontal pocket may be overcome by the use of novel, bioadhesive, syringeable semisolid systems (33,34). Such systems can be formulated to exhibit requisitory flow properties (and hence may be easily administered into the periodontal pocket using a syringe), mucoadhesive properties (ensuring prolonged retention within the pocket), and sustained release of therapeutic agent within this environment. In a series of papers, Jones et al. (28,33,34,40) described the formulation and physicochemical characterization of semisolid bioadhesive networks (containing tetracycline or metronidazole), in which the physical state of the bioadhesive component, polycarbophil, i.e., neutralized (swollen) or unneutralized (particulate), was controlled by the amount of available water in the formulations. Increasing the concentrations of the other hydrophilic polymeric components reduced the amount of available water for the swelling of polycarbophil following its incorporation into the gel matrix, which, in turn, decreased the swelling of this component. Upon contact with mucus, water diffuses into the formulation and controls swelling of the bioadhesive polymer, which in turn allows interpenetration of the fluidized polymer chains with mucus and hence ensures physical and chemical adhesion. The authors concluded that, when used in combination with mechanical treatments, tetracycline-containing bioadhesive semisolid systems described in the study would augment periodontal therapy by improving the removal of pathogens and hence improving periodontal health.

B. Bioadhesive Devices for the Gastrointestinal Tract

Bioadhesive polymers may provide useful delivery systems for drugs that have limited bioavailability from more conventional dosage forms. To attain a once-daily dosing strategy using peroral bioadhesion, it is desirable for a dosage form to attach itself to the mucosa of the gastrointestinal tract (GIT). If attachment is successful in the gastric region, then a steady supply of drug is available to the intestinal tract for absorption. However, gastric motility and muscular contractions will tend to dislodge any such device. Motility in the GIT during the fasted state is known as the interdigestive migrating motor complex (IMMC) (72). During certain phases of the IMMC, a ''housekeeper wave'' migrates from the foregut to the terminal ileum and is intended to clear all nondigestible items from the gut. The strength of this ''housekeeper wave'' is such that poorly adhered dosage forms are easily removed, and only strong bioadhesives will be of any practical use.

In addition to physical abrasion and erosion exerted on a bioadhesive dosage form, the mucin to which it is attached turns over quickly, especially in the gastric region. Any device that has lodged to surface mucus will be dislodged as newly synthesized mucus displaces the older surface layers. The acid environment will also affect bioadhesion, especially with polyacid polymers, such as poly(acrylic acid), where the mechanism of bond formation is thought to occur chiefly through hydrogen bonding and electrostatic interactions.

Longer et al. described the formulation of a bioadhesive sustained release system containing polycarbophil and drug incorporated in albumin beads, all enclosed in a hard gelatin capsule (88). Once swallowed, the capsule wall disintegrated and the exposed polymer swelled to bind the beads together. The drug studied was chlorthiazide, which showed saturable, or site-specific, absorption from the human GIT. High doses (500 mg) were 16% bioavailable, whereas lower doses (50 mg) were 56% bioavailable. In vitro release studies showed that the albumin beads controlled drug release, and the presence of polycarbophil did not affect this release significantly. However, in the rat intestine, after 6 hours, 90% of the beads remained in the stomach encapsulated in a gelled polymer mass. In contrast, polycarbophil bound the beads together, the majority located halfway along the gut; only a few beads were detected in the stomach.

Any swallowed dosage form will be subjected to shear forces as the intestinal contents move. This motility will either prevent the attachment of the bioadhesive or attempt to remove it if bonding to the gut wall has occurred. Previous attempts to reduce the GIT transit time of normal dosage forms have included devices that swelled and floated on the stomach contents, their increasing size preventing them from passing through the pylorus. Harris et al. studied the GIT transit time of formulations containing potential bioadhesive agents (89). Of the agents investigated in animal studies, the acrylic acid polymers were most likely to delay GIT transit time. The study was extended to human subjects, and the transit times of test devices were monitored using gamma scintigraphy. The materials under study were of four types, large and small resin particles, both coated and uncoated with bioadhesive polymers. Bioadhesives were unable to slow significantly the GIT transit of radiolabeled large resin beads (size range 750–1000 μm) when compared to uncoated controls. However, when the smaller beads were used (size range 5–50 μm), bioadhesives increased the orocaecal transit times by 30% when compared to uncoated controls. It was further observed that the transit speed for the smaller nonbioadhesive particulate material was significantly quicker than for larger nonbioadhesive particles. Thus the action of a bioadhesive coating was to extend the transit time of the smaller particles to levels comparable with larger bioadhesive particles. Images from the gamma camera revealed that uncoated controls dispersed readily, releasing the beads, whereas the bioadhesives transformed the beads into a gelled cluster in the stomach. Overall, it was concluded that the results did not necessarily confirm the potential of bioadhesive materials for either delaying GIT transit or improving bioavailability.

The in vitro controlled release and bioadhesive properties of furosemide formulations were evaluated with standard dissolution tests and with a specially developed model using rabbit intestine (90). The results showed that the controlled release properties were not affected by the application of the bioadhesive polymer but that the bioadhesive properties were substantially different. In order to assess the gastrointestinal transit time in vivo, a gamma-scintigraphy study was performed in six volunteers testing the same controlled-release formulation with and without bioadhesive polymer. Plasma levels of furosemide, evaluation of urinary flux, and measures of urinary excretion of furosemide in the six volunteers allowed correlations to be made between gastrointestinal transit and furosemide absorption.

The oral route constitutes the preferred route for drug delivery. However, numerous drugs remain poorly available when administered orally. Drugs associated with bioadhesive polymeric nanoparticulates or small particles in the micrometer size range may be advantageous in this respect due to their ability to interact with the mucosal surface. Targeting applications are possible by this method if there are specific interactions occurring when a ligand attached to the particle is used for the recognition and attachment to a specific site at the mucosal surface (91).

Drug delivery particulates of alginate, polylysine, and pectin have been reported (92). Theophylline, chlorothiazide, and indomethacin were used as the model drugs for in vitro assessments, and mannitol was the model for assessing paracellular drug absorption across Caco-2 cell monolayers. Alginate and pectin served as the core polymers, and polylysine helped to strengthen the particulates. Use of pectin specially helped in forming a more robust particulate that was more resistant in acidic pH and modulated the release profiles of the encapsulated model drugs in alkaline pH. Alginate and pectin were also found to enhance the paracellular absorption of mannitol across Caco-2 cell monolayers by about three times. The release rate could be described as a first-order or square-root time process depending on the drug load. Use of alginate–polylysine–pectin particulates is expected to combine the advantages of bioadhesion, absorption enhancement, and sustained release. This particulate system may have use as a carrier for drugs that are poorly absorbed after oral administration.

Alginate–chitosan microcapsules have been investigated as a bioadhesive drug delivery system (93). Calcium alginate gel beads uncoated and coated with chitosan were tested for adhesive properties, using novel techniques, with negatively charged chromatography particles and in vitro with pig oesophagus and stomach mucosa. The addition of a chitosan coating increased the adhesive properties significantly. The adherence of both coated and uncoated beads was much greater to the stomach mucosa than to oesophageal mucosa. The difference in adhesive properties between the coated and uncoated microcapsules was also found considerably larger for the stomach mucosa.

In general, oral applications of bioadhesive technology have been less successful than where the delivery system can be applied directly to an accessible site. Issues such as overhydration, attachment to gastrointestinal mucus and subsequent shedding, and poor resistance to mechanical forces in the gastrointestinal tract have tended to limit the utility of bioadhesive gastric-retentive systems to date.

C. Bioadhesive Devices for Rectal Drug Delivery

The function of the rectum is mostly concerned with removing water. It is only 10 cm in length, with no villi, giving it a relatively small surface area for drug absorption (72). Drug permeability differs from that found in both the oral cavity and the intestinal regions. Most rectal absorption of drugs is achieved by a simple diffusion process through the lipid membrane. However, the rectal route is readily accessible, penetration enhancers can be used, and there is access to the lymphatic system. In contrast, drugs absorbed via the small intestine are transported mostly to the blood, and only a small proportion enters the lymphatic system.

Drugs that are liable to extensive first-pass metabolism can benefit greatly if delivered to the rectal area, especially it they are targeted to areas close to the anus. This is because the blood from the lower rectum drains directly into the systemic circulation, whereas blood from the upper regions drains into the portal systems via the superior hemorrhoidal vein and the inferior mesenteric vein. Drug absorbed from this upper site is subjected to liver metabolism. A bioadhesive suppository will attach to the lower rectal area and once inserted will reduce the tendency for migration upwards to the upper rectum; this migration can occur with conventional suppositories.

Morimoto et al. recognized that high molecular weight hormones, such as insulin, were not readily absorbed through rectal mucosa (94). Emulsions containing surfactants have been employed to promote the absorption of this peptide, but their long-term administration was associated with irritation and toxicity. To overcome these problems, insulin was formulated in a poly(acrylic acid) gel comprising long-chain fatty acids as absorption enhancers. The absorption rate of insulin was encouraging, and consequently a simple and painless dosage form in the long-term therapy of diabetics may be a possibility in the future. However, bioadhesives have not been extensively employed for rectal delivery.

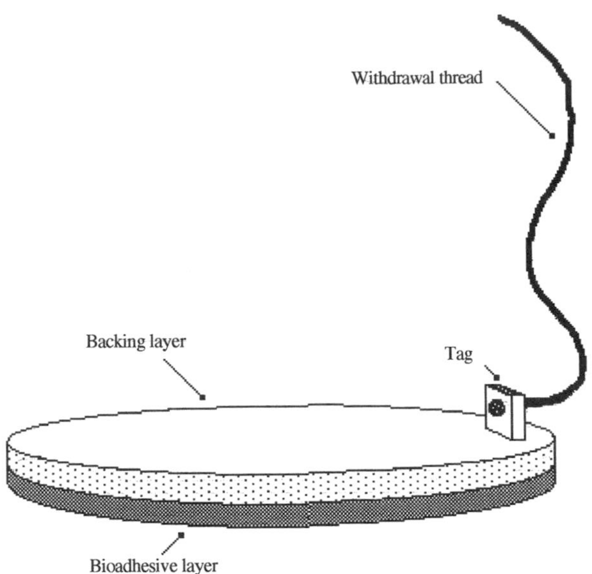

Figure 41.4 Bioadhesive cervical drug delivery system.

D. Bioadhesive Devices for Cervical Drug Delivery

Nagai has investigated various bioadhesive forms to treat uterine and cervical cancers (78). Uterine cancers comprise about 25% of all malignant tumors in Japan, among which *carcinoma colli* accounts for 95% of this figure (95). The target cells remain at, or near, the cervical epithelium and can be readily targeted with an appropriate dosage form. Pessaries have been of limited use because drug release is rapid and leakage into the surrounding tissue causes inflammation of vaginal mucosa. In order to meet three important criteria, drug release, swelling of the preparation, and adhesion to the diseased tissue, a bioadhesive disc was prepared. Bleomycin was incorporated into a hydroxypropylcellulose/Carbopol® mix and molded into a suitable shape. This could be placed on or into the cervical canal, where it adhered and released the cytotoxic drug. After treatment and following colposcopic examination, areas of necrosis on the lesion were observed, with surrounding normal mucosal cells unaffected. With pessaries, however, this was not the case. In approximately 33% of cases, cancerous foci had completely disappeared.

Woolfson et al. described a novel bioadhesive cervical patch containing 5-fluorouracil for the treatment of cervical intraepithelial neoplasia (CIN) (96). The patch was of bilaminar design (Fig. 4), with a drug-loaded bioadhesive film cast from a gel containing 2% w/w Carbopol® 981 plasticized with 1% w/w glycerin. The casting solvent was ethanol:water 30:70, chosen to give a nonfissuring film with an even particle size distribution. The film, which was mechanically stable on storage under ambient conditions, was bonded directly to a backing layer formed from thermally cured poly(vinyl chloride) emulsion. Bioadhesive strength was independent of drug loading in the bioadhesive matrix over the range investigated but was influenced by both the plasticizer concentration in the casting gel and the thickness of the final film. Release of 5-fluorouracil from the bioadhesive layer into an aqueous sink was rapid but was controlled down to an undetectable level through the backing layer. The latter characteristic was desirable to prevent drug spill from the device onto vaginal epithelium in vivo. Despite the relatively hydrophilic nature of 5-fluorouracil, substantial drug release through human cervical tissue samples was observed over approximately 20 hours. Drug release, which was clearly tissue dependent rather than device dependent, may have been aided by a shunt diffusion route through aqueous pores in the tissue. The bioadhesive and drug release characteristics of the 5-fluorouracil cervical patch indicated that it would be suitable for further clinical investigation as a drug treatment for CIN (97).

E. Bioadhesive Devices for Vaginal Drug Delivery

Bioadhesives can control the rate of drug release from, and extend the residence time of, vaginal formulations. These formulations may contain drug or quite simply act in conjunction with moisturizing agents as a control for vaginal dryness. Gursoy and Bayhan described the formulation of a vaginal tablet comprising sodium carboxymethylcellulose and poly(acrylic acid) with zinc sulphate as a model drug (98). Gursoy et al. recognized that no in vitro dissolution apparatus was available to quantify drug release from such formulations (99). A mechanical device was designed and built and compared to the dissolution apparatus described in the BP and USP. The in situ drug released was also determined using freshly obtained bovine vaginal tissue. Tablets were placed into the tissue and removed at specified time intervals. The drug remaining in the tablet following removal was then quantified. The release behavior in the new apparatus correlated well to in situ data, whereas the BP and USP grossly exaggerated the release profiles.

Spreading and retention (and consequent ''messiness'') of semisolid vaginal bioadhesive systems, such as gels, are major issues in intravaginal drug delivery. Brown et al. reported a scintigraphic evaluation of vaginal dosage forms in postmenopausal women (100). The vaginal spreading and clearance of a radiolabeled pessary formulation and a commercial polycarbophil gel was assessed in six healthy postmenopausal female volunteers over a 6 hour period by gamma scintigraphy. In five out of the six subjects studied, clearance of the two formulations exhibited very little intrasubject variation. However, there was considerable intersubject variability in clearance. Importantly, there was no evidence to suggest that either of the formulations dispersed material beyond the cervix, into the uterus, in any of the subjects studied. The authors concluded that the lack of significant retention of these products in most of the volunteers had obvious implications for the delivery of therapeutic agents.

F. Bioadhesive Devices for Nasal Drug Delivery

The area of the normal human nasal mucosa is approximately 150 cm^2. The nasal mucosa and submucosa are liberally populated with goblet cells along with numerous mucous and serous glands. These keep the nasal mucosal surfaces moist. Drug administration to this region is normally reserved for local treatment, such as nasal allergy or inflammation. The nasal mucosa is thin and incorporates a dense vascular network, indicating that drug absorption may be good from this site (101). Absorbed drugs avoid first-pass metabolism and lumenal degradation associated with the oral route (72). The nasal mucosa itself is sensitive to drug molecules and to surfactants, which are often used to enhance drug absorption. Moreover, the mucociliary escalator travels at 5 mm min^{-1} as it drags mucous fluid backwards toward the throat. Therefore if a drug is applied in either a simple powder or a liquid formulation it will be quickly cleared from this site of absorption. Thus any useful dosage form must be nonirritant to the nasal mucosa and be retained for extended periods of time.

Nagai et al. reported the low bioavailability of insulin across the oral epithelia and attempted to enhance its absorption using the nasal route (102). He prepared a powder containing freeze-dried bovine insulin and powdered bioadhesive excipients. The powder was puffed into the nostrils of beagles, and blood-glucose levels were then determined. The bioavailability of insulin was enhanced when given in a bioadhesive powder. Indeed, a powder of freeze dried Carbopol® and insulin produced a hypoglycemia to an extent of one-third that achieved by an equivalent dose given intravenously.

Nasal delivery of protein therapeutics can be compromised by the brief residence time at this mucosal surface. Some bioadhesive polymers have been suggested to extend residence time and improve protein uptake across the nasal mucosa. Witschi and Mrsny examined several potential polymer platforms, in the form of starch, alginate, chitosan, or Carbopol microparticles, for their in vitro protein (bovine serum albumin) release, relative bioadhesive properties, and induction of cytokine release from respiratory epithelium (103). Carbopol gels and chitosan microparticles provided the most desirable characteristics for protein therapeutic and protein antigen delivery, respectively, of the formulations examined.

An erythrocyte-based bioadhesive system has been reported for controlled systemic delivery of propranolol hydrochloride through the nasal route (104). Rat erythrocytes were loaded with propranolol HCl by a method based on hypotonic swelling, isotonic resealing, and reannealing. The loaded erythrocytes were cross-linked by treating with glutaraldehyde and were characterized in vitro for drug payload efficiency, propranolol release, drug diffusion through rat intestine, bioadhesion, and morphological characteristics. Loaded erythrocytes were found to release propranolol HCl slowly, the release being dependent on the degree of cross-linking. The system was found to possess good bioadhesive properties. In vivo studies conducted on rats revealed that the developed system maintained constant plasma levels of propranolol for up to 10 hours.

Overall, the nasal route remains highly promising, particularly for macromelecular absorption, but there are problems regarding effects of bioadhesive delivery systems on nasal cilial beat.

G. Bioadhesive Devices for Ocular Drug Delivery

Extended drug delivery to the eye is difficult for several reasons. Systemically administered drugs must cross the blood–aqueous humor barrier, and to achieve local therapeutic concentrations of drug, high levels of drug in the blood are needed. In addition, lacrimation, blinking, and tear turnover will all reduce the bioavailability of topically administered drug to approximately 1–10% (18). Conventional delivery methods are not ideal. Solutions and suspensions are readily washed from the cornea, and ointments alter the tear refractive index and blur vision.

Hui and Robinson have described the synthesis of cross-linked acrylic acid and evaluated their potential for ocular drug delivery of progesterone (105). The aqueous humor levels in the rabbit obtained after dosing with 50 μL of 0.3% progesterone suspension and 50 μL of the 0.3% progesterone suspension in the bioadhesive polymer showed a 4.2 times increase in bioavailability from the bioadhesive suspension. Further, drug levels in the aqueous humor declined at a slower rate when given in the latter suspension.

Saettone and Chiellini investigated the release of pilocarpine from poly(acrylic acid) and hydroxypropylcellulose discs and found that the solubility of the discs could be controlled by the addition of high molecular weight poly(vinyl alcohol) (PVA) (106). Once added, PVA improved in vivo bioavailability when compared to simple nonbioadhesive pilocarpine drops, as demonstrated by the miotic activity in rabbits. Ünlü et al. described the acrylic polymers and their hydrogels as good adhering agents for ocular drug therapy (107). They were nonirritant, well tolerated, and, according to the polymer type, presented no interference to visibility when in situ. The viscosity of Carbopol vehicles significantly decreased when subjected to a simulated lacrimal fluid, but the mucoadhesive properties of the polymer were still retained. More recently, Lebourlais et al. have reviewed the applications of bioadhesion in the development of novel ocular drug delivery systems (9).

VIII. CONCLUSIONS

Bioadhesive drug delivery systems have been widely studied in the past decade, and a number of interesting, novel drug delivery systems incorporating the use of bioadhesive polymers have been described. Primarily, the application of bioadhesion to drug delivery involves the process of mucoadhesion, a ''wet-stick'' adhesion requiring, primarily, spreading of the mucoadhesive on a mucin-coated epithelium, followed by interpenetration of polymer chains between the hydrated delivery system and mucin. An alternative mechanism can involve the use of specific bioadhesive molecules, notably lectins, primarily for oral drug delivery applications. Results, however, remain variable by this specific method, as with the use of bioadhesives, generally, for oral systemic drug delivery. The most notable applications of mucoadhesive systems to date remain those involving accessible epithelia, such as in ocular, nasal, buccal, rectal, or intravaginal drug delivery systems. The ancillary effects of certain mucoadhesive polymers, such as polycations, on promoting drug penetration of epithelial tissues, may be increasingly important in the future, particularly for the transmucosal delivery of therapeutic peptides and other biomolecules.

REFERENCES

1. A. G. Mikos, N. A. Peppas. Systems for controlled release of drugs, Part 5. Bioadhesive systems. *S.T.P. Pharma 2*: 705–716 (1986).
2. S. H. S. Leung, J. R. Robinson. Bioadhesive drug delivery. *ACS Symp. Ser. 467*:350–366 (1991).
3. J. L. Greaves, C. G. Wilson. Treatment of diseases of the eye with mucoadhesive delivery systems. *Adv. Drug. Del. Rev. 11*:349–383 (1993).
4. M. Helliwell. The use of bioadhesives in targeted delivery within the gastrointestinal tract. *Adv. Drug. Del. Rev. 11*: 221–251 (1993).
5. A. J. Moes. Gastroretentive dosage forms. *Crit. Rev. Ther. Drug. Carrier Sys. 10*:143–195 (1993).
6. C. M. Lehr. Bioadhesion technologies for the delivery of peptide and protein drugs to the gastrointestinal tract. *Crit. Rev. Ther. Drug Carrier Sys. 11*:119–160 (1994).
7. C. M. Lehr. From sticky stuff to sweet receptors—achievements, limits and novel approaches to bioadhesion. *Eur. J. Drug Met. Pharm. 21*:139–148 (1996).
8. G. Ponchel. Formulation of oral mucosal drug-delivery systems for the systemic delivery of bioactive materials. *Adv. Drug. Del. Rev. 13*:75–87 (1994).
9. C. A. Leboulais, L. Treupelacar, C. T. Rhodes, P. A. Sado, R. Leverge. New ophthalmic drug-delivery systems. *Drug. Dev. Ind. Pharm. 21*:19–59 (1995).
10. N. A. Peppas, J. J. Sahlin. Hydrogels as mucoadhesive and bioadhesive materials: a review. *Biomaterials 17*:1553–1561 (1996).
11. D. Duchêne, F. Touchard, N. A. Peppas. Pharmaceutical and medical aspects of bioadhesive systems for drug administration. *Drug Dev. Ind. Pharm. 14*:283–318 (1988).
12. B. Naisbett, J. Woodley. The potential use of tomato lectin for oral-drug delivery. *Int. J. Pharm. 107*:223–230 (1994).
13. T. J. Nicholls, K. L. Green, D. J. Rogers, J. D. Cook, S. Wolowacz, J. D. Smart. Lectins in ocular drug delivery: an investigation of lectin binding sites on the corneal and conjunctival surfaces. *Int. J. Pharm. 138*:175–183 (1996).
14. X. Yang, J. R. Robinson. Bioadhesion in mucosal drug delivery. *Biorelated Polymers and Gels* (T. Okana, ed.). Academic Press, San Diego, CA, pp. 149–150 (1998).
15. J. W. McBain, D. G. Hopkins. On adhesives and adhesive action. *J. Phys. Chem. 29*:188–204 (1925).
16. B. V. Derjaguin, V. P. Smilga. Adhesion: Fundamentals and Practice. McLaren, London, 1969.

17. C. P. Reinhart, N. A. Peppas. Solute diffusion in swollen membranes. II. Influence of crosslinking on diffusion properties. *J. Membrane Sci. 18*:227–239 (1984).
18. J. R. Robinson. Ocular drug delivery mechanism(s) of corneal drug transport and mucoadhesive delivery systems. S.T.P. *Pharma* 5:839–846 (1989).
19. M. Mutimer, C. Riffkin, J. Hill, M. E. Glickman, G. N. Cyr. *J. Amer. Pharm. Assoc. Sci. Ed. 45*:212 (1956).
20. J. L. Chen, G. N. Cyr. Compositions producing hydration. Adhesion in Biological Systems (R. S. Manley, ed.). Academic Press, New York, 1970.
21. A. J. Kinloch. The science of adhesion: I. Surface and interfacial aspects. *J. Mater. Sci. 15*:2141–2141 (1980).
22. J. R. Huntsberger. Mechanisms of adhesion. *J. Paint Techn. 39*:199–211 (1967).
23. M. A. Longer, J. R. Robinson. Fundamental aspects of bioadhesion. *Pharm. Int.* 7:114–117 (1986).
24. N. A. Peppas, G. Ponchel, D. Duchêne. Bioadhesive analysis of controlled-release systems. II. Time-dependent bioadhesive stress in poly(acrylic acid)-containing systems. *J. Controlled Release* 5:143–149 (1987).
25. A. G. Mikos, N. A. Peppas. Comparison of experimental techniques for the measurement of the bioadhesive forces of polymeric materials with soft tissues. *Proc. Int. Symp. Controlled Release Bioactive Mater. 13*:97 (1986).
26. C. M. Lehr, J. A. Bouwstra, H. E. Boddé, H. E. Junginger. A surface energy analysis of mucoadhesion. contact angle measurements on polycarbophil and pig intestinal mucosa in physiologically relevant fluids. *Pharm. Res.* 9:70–75 (1992).
27. A. J. Coury, P. T. Cahalan, A. H. Jevne, J. J. Perrault, M. J. Kallok. Recent developments in hydrophilic polymers. *Med. Dev. Dia. Ind.* 6:28 (1984).
28. D. S. Jones, A. D. Woolfson, A. F. Brown, M. J. O'Neill. Mucoadhesive, syringeable drug delivery systems for controlled application of metronidazole to the periodontal pocket: in vitro release kinetics, syringeability, mechanical and mucoadhesive properties. *J. Cont. Rel. 49*:71–79 (1997).
29. K. Park, H. Park. Test methods of bioadhesion. Bioadhesive Drug Delivery Systems (V. Lenaerts, R. Gurney, eds.). CRC Press, Boca Raton, FL, 1990.
30. J. D. Smart, I. W. Kellaway, H. E. C. Worthington. An in vitro investigation of mucosa—adhesive materials for use in controlled drug delivery. *J. Pharm. Pharmacol. 36*: 295–299 (1984).
31. H. Park, J. R. Robinson. Physicochemical properties of water soluble polymers important to mucin/epithelium adhesion. *J. Cont. Rel.* 2:47–57 (1985).
32. A. D. Woolfson, D. F. McCafferty, S. P. Gorman, P. A. McCarron, J. H. Price. Design of an apparatus incorporating a linear variable differential transformer for the measurement of type III bioadhesion to cervical tissue. *Int. J. Pharm. 84*:69–76 (1992).
33. D. S. Jones, A. D. Woolfson, J. Djokic, W. A. Coulter. Development and physical characterisation of bioadhesive semi-solid, polymeric systems containing tetracycline for the treatment of periodontal diseases. *Pharm. Res. 13*: 1732–1736 (1996).
34. D. S. Jones, A. D. Woolfson, A. F. Brown. Textural, viscoelastic and mucoadhesive properties of pharmaceutical gels and polymers. *Int. J. Pharm. 151*:223–233 (1997).
35. D. S. Jones, A. D. Woolfson, A. F. Brown. Viscoelastic properties of bioadhesive, chlorhexidine-containing semisolids for topical application to the oropharynx. *Pharm. Res. 15*:1131–1136 (1998).
36. H. S. Ch'ng, H. Park, P. Kelly, J. R. Robinson. Bioadhesive polymers as platforms for oral controlled drug delivery. II Synthesis and evaluation of some swelling, water-insoluble bioadhesive polymers. *J. Pharm. Sci. 74*:399–405 (1985).
37. S. Reich, M. Levy, A. Meshorer, M. Blumental, M. Yalon, J. W. Sheets, E. P. Goldberg. Intraocular-lens endothelial interface—adhesive force measurements. *J. Biom. Mat. Res. 18*:737–744 (1984).
38. S. S. Davis. The design and evaluation of controlled release systems for the gastrointestinal tract. *J. Cont. Rel.* 2:27–38 (1985).
39. D. S. Jones, A. D. Woolfson. Measuring sensory properties of semi-solid products using texture profile analysis. *Pharmaceutical Manufacturing Review* 9:S3–S6 (1997).
40. D. S. Jones, C. R. Irwin, A. F. Brown, A. D. Woolfson, W. A. Coulter, C. McClelland. Design, characterisation and preliminary clinical evaluation of a syringeable, mucoadhesive topical formulation containing tetracycline for the treatment of periodontal disease. *J. Cont. Rel.* (in press).
41. D. Q. M. Craig, S. Tamburic. Dielectric analysis of bioadhesive gel systems. *Eur. J. Poharm. Biopharm. 44*:61–70 (1997).
42. S. Matsuda, H. S. N. Iwata, Y. Ikada. Bioadhesion of gelatin films crosslinked with glutaraldehyde. *J. Biomed. Mat. Res. 45*:20–27 (1999).
43. K. Pritchard, A. B. Lansley, G. P. Martin, M. Helliwell, C. Marriott, L. M. Benedetti. Evaluation of the bioadhesive properties of hyaluronan derivatives: detachment weight and mucociliary transport rate studies. *Int. J. Pharm. 129*: 137–145 (1996).
44. D. E. Chickering, J. S. Jacob, T. A. Desai, M. Harrison, W. P. Harris, C. N. Morrell, P. Chaturvedi, E. Mathiowitz. Bioadhesive microspheres. 3. An in vivo transit and bioavailability study of drug-loaded alginate and poly(fumaric-co-sebacic anhydride) microspheres. *J. Cont. Rel. 48*:35–46 (1997).
45. J. Schnurrer, C. M. Lehr. Mucoadhesive properties of the mussel adhesive protein. *Int. J. Pharm. 141*:251–256 (1996).
46. S. Anlar, Y. Capan, A. A. Hincal. Physicochemical and bioadhesive properties of polyacrylic-acid polymers. *Pharmazie 48*:285–287 (1993).
47. H. BlancoFuente, S. AnguianoIgea, F. J. OteroEspinar, J. BlancoMendez. In-vitro bioadhesion of carbopol hydrogels. *Int. J. Pharm. 142*:169–174 (1996).
48. M. J. Tobyn, J. R. Johnson, P. W. Dettmar. Factors affect-

ing in vitro gastric mucoadhesion. 2. Physical properties of polymers. *Eur. J. Pharm. Biopharm. 42*:56–61 (1996).
49. E. P. Kakoulides, J. D. Smart, J. Tsibouklis. Azocross-linked poly(acrylic acid) for colonic delivery and adhesion specificity: in vitro degradation and preliminary ex vivo bioadhesion studies. *J. Cont. Rel. 54*:95–109 (1998).
50. T. NguyenXuan, R. Towart, A. Terras, Y. Jacques, P. Buri, R. Gurny. Mucoadhesive semi-solid formulations for intraoral use containing sucralfate. *Eur. J. Pharm. Biopharm. 42*:289–295 (1996).
51. A. D. Woolfson, D. F. McCafferty, C. R. McCallion, E. T. McAdams, J. Anderson. Moisture-activated, electrically-conducting bioadhesive hydrogels as interfaces for bio-electrodes: effect of film hydration on cutaneous adherence in wet environments. *J. Appl. Polymer Science 58*: 1291–1296 (1995).
52. I. G. Needleman, G. P. Martin, F. C. Smales. Characterisation of bioadhesives for periodontal and oral mucosal drug delivery. *J. Clin. Periodont. 25*:74–82 (1998).
53. M. F. Saettone, P. Chetoni, M. T. Torracca, S. Burgalassi, B. Giannaccini. Evaluation of mucoadhesive properties and in vivo activity of ophthalmic vehicles based on hyaluronic acid. *Int. J. Pharm. 51*:203–212 (1989).
54. A. M. Durrani, S. J. Farr, I. W. Kellaway. Influence of molecular-weight and formulation pH on the precorneal clearance rate of hyaluronic-acid in the rabbit eye. *Int. J. Pharm. 118*:243–250 (1995).
55. B. Rihova, R. Rathi, P. Kopeckova, J. Kopecek. In vitro bioadhesion of carbohydrate-containing N-(2-hydroxy-propyl) methacrylamide copolymers to the GI tract of guinea-pigs. *Int. J. Pharm. 87*:105–116 (1992).
56. S. Bouckaert, J. P. Remon. In-vitro bioadhesion of a buccal, miconazole slow-release tablet. *J. Pharm. Pharmacol. 45*:504–507 (1993).
57. S. A. Mortazavi, J. D. Smart. Factors influencing gel-strengthening at the mucoadhesive–mucus interface. *J. Pharm. Pharmacology 46*:86–90 (1994).
58. M. J. Tobyn, J. R. Johnson, P. W. Dettmar. Factors affecting in vitro gastric mucoadhesion. 3. Influence of polymer addition on the observed mucoadhesion of some materials. *Eur. J. Pharmaceut. Biopharm. 42*:331–335 (1996).
59. J. M. Gu, J. R. Robinson, S. H. S. Leing. Binding of acrylic polymers to mucin epithelial surfaces—structure–property relationships. *CRC Crit. Rev. Ther. Drug Carrier Sys. 5*:21–67 (1988).
60. A. D. Woolfson, D. F. McCafferty, C. R. McCallion, E. T. McAdams, J. Anderson. Moisture-activated, electrically-conducting bioadhesive hydrogels as interfaces for bio-electrodes: effect of formulation factors on cutaneous adherence in wet environments. *J. Appl. Polymer Science 56*:1151–1160 (1995).
61. J. D. Smart, I. W. Kellaway. In vitro techniques for measuring mucoadhesion. *J. Pharm. Pharmacol. 34*:70P (1982).
62. K. M. Tur, H. S. Chng. Evaluation of possible mechanism(s) of bioadhesion. *Int. J. Pharm. 160*:61–74 (1998).
63. J. Voorspoels, J. P. Remon, W. Eechaute, W. DeSy. Buccal absorption of testosterone and its esters using a bioadhesive tablet in dogs. *Pharm. Res. 13*:1228–1332 (1996).
64. B. Parodi, E. Russo, P. Gatti, S. Cafaggi, G. Bignardi. Development and in vitro evaluation of buccoadhesive tablets using a new model substrate for bioadhesion measures: the eggshell. *Drug Dev. Ind. Pharm. 25*:289–295 (1999).
65. C. F. Wong, K. H. Yuen, K. K. Peh. Formulation and evaluation of controlled release Eudragit buccal patches. *Int. J. Pharm. 178*:11–22 (1999).
66. Y. Y. Iscan, Y. Capan, S. Senel, M. F. Sahin, S. Kes, D. Duchane, A. A. Hincal. Formulation and in vitro in vivo evaluation of buccal bioadhesive captopril tablets. *STP Pharma Sci. 8*:357–363 (1998).
67. M. Efentakis, E. C. Hatzi, A. G. Andreopoulos. Development, evaluation and release characteristics in vitro of oral mucosal bioadhesive matrices containing tetracycline for application to the oral cavity. *STP Pharma Sci. 8*:227–232 (1998).
68. J. Y. Fang, Y. B. Huang, H. H. Lin, Y. H. Tsai. Transdermal iontophoresis of sodium nonivamide acetate. IV. Effect of polymer formulations. *Int. J. Pharm. 173*:127–140 (1998).
69. M. L. Cole, T. L. Whateley. Interaction of nonionic block copolymeric (poloxamer) surfactants with poly (acrylic acid), studied by photon correlation spectroscopy. *J. Colloid Interface Sci. 180*:421–427 (1996).
70. I. G. Needleman, F. C. Smales, G. P. Martin. An investigation of bioadhesion for periodontal and oral mucosal drug delivery. *J. Clin. Periodont. 24*:394–400 (1997).
71. N. Cvetkovic, M. Nesic, V. Moracic, M. Rosic. Design of a method for in vitro studies of polymer adhesion. *Pharmazie 52*:536–537 (1997).
72. S. H. S. Leung, J. A. Robinson. Polyanionic polymers in bioadhesive and mucoadhesive drug delivery. *ACS Symp. Ser. 480*:269–284 (1992).
73. M. M. Veillard, M. A. Longer, T. W. Martens, J. R. Robinson. Preliminary studies of oral mucosal delivery of peptide drugs. *J. Controlled Release 6*:123–131 (1987).
74. T. Nagai, Y. Machida. Advances in drug delivery—mucosal adhesive dosage forms. *Pharm. Int. 6*:196–200 (1985).
75. G. Ponchel, F. Touchard, D. Wouessidjewe, D. Duchêne, N. A. Peppas. Bioadhesive analysis of controlled-release systems. III Bioadhesive and release behaviour of metronidazole-containing poly(acrylic acid)-hydroyxpropylmethylcellulose systems. *Int. J. Pharm. 38*:65–70 (1987).
76. G. Ponchel, F. Touchard, D. Duchêne, N. A. Peppas. Bioadhesive analysis of controlled release systems. I. Fracture and interpretation analysis in poly(acrylic acid)-containing systems. *J. Cont. Rel. 5*:129–141 (1987).
77. T. Nagai. Topical mucosal adhesive dosage forms. *Med. Res. Rev. 6*:227–242 (1986).
78. T. Nagai. 46th Int. Congress of Pharmaceutical Science of FIP. Helsinki, 1–5 September (1986).
79. T. Nagai. Adhesive topical drug delivery systems. *J. Cont. Rel. 2*:121–134 (1985).
80. M. Ishida, Y. Machida, N. Nambu, T. Nagai. New muco-

sal dosage forms of insulin. *Chem. Pharm. Bull. 29*:810–816 (1981).
81. B. Rodu, C. M. Russell, A. J. Desmarais. Clinical and chemical properties of a novel mucosal bioadhesive agent. *J. Oral Pathol. 17*:564–567 (1988).
82. L. R. Eversole. Performance of a hydroxypropylcellulose film former in normal and ulcerated oral mucosa. *Oral Surg. 65*:699–703 (1988).
83. A. B. Mahdi, W. A. Coulter, A. D. Woolfson, P. J. Lamey. Efficacy of bioadhesive patches in the treatment of recurrent apthous stomatitis. *J. Oral. Path. Med. 25*:420–423 (1996).
84. C. RemunanLopez, A. Portero, J. L. VilaJato, M. J. Alonso. Design and evaluation of chitosan/ethylcellulose mucoadhesive bilayered devices for buccal drug delivery. *J. Cont. Rel. 55*:143–152 (1998).
85. M. Addy. Local delivery of antimicrobial agents to the oral cavity. *Adv. Drug Deliv. Rev. 13*:123–134 (1994).
86. R. B. Gandhi, J. R. Robinson. The oral cavity as a site for bioadhesive drug delivery. *Adv. Drug Deliv. Rev. 13*:43 74 (1994).
87. N. J. Medlicott, M. J. Rathbone, I. J. Tucker, D. W. Holborow. Delivery systems for the administration of drugs to the periodontal pocket. *Adv. Drug Del. Rev. 13*:181–203 (1994).
88. M. A. Longer, H. S. Ch'ng, J. R. Robinson. Bioadhesive polymers as platforms for oral controlled drug delivery III: oral delivery of chlorthiazide using a bioadhesive polymer. *J. Pharm. Sci. 74*:406–411 (1985).
89. D. Harris, J. T. Fell, H. Sharma, D. C. Taylor, J. Linch. Studies on potential bioadhesive systems for oral drug delivery. *S. T. P. Pharma 5*:852–856 (1989).
90. G. Santus, C. Lazzarini, G. Bottoni, E. P. Sandefer, R. C. Page, W. J. Doll, U. Y. Ryo, G. A. Digenis. An in vitro in vivo investigation of oral bioadhesive controlled release furosemide formulations. *Eur. J. Pharm. Biopharm. 44*: 39–52 (1997).
91. G. Ponchel, J. M. Irache. Specific and non-specific bioadhesive particulate systems for oral delivery to the gastrointestinal tract. *Adv. Drug Del. Rev. 34*:191–219 (1998).
92. P. Liu, T. R. Krishnan. Alginate-pectin-poly-L-lysine particulate as a potential controlled release formulation. *J. Pharm. Pharmacol. 51*:141–149 (1999).
93. O. Gaserod, I. G. Jolliffe, F. C. Hampson, P. W. Dettmar, G. SkjakBraek. The enhancement of the bioadhesive properties of calcium alginate gel beads by coating with chitosan. *Int. J. Pharm. 175*:237–246 (1998).
94. K. Morimoto, E. Kamiya, T. Takeeda, Y. Nakamoto, K. Morisaka. Enhancement of rectal absorption of insulin in polyacrylic acid aqueous gel bases containing long chain fatty acids in rats. *Int. J. Pharm. 14*:149–157 (1983).
95. Y. Machida, H. Masuda, N. Fujiyama, S. Ito, M. Iwata, T. Nagai. *Chem. Pharm. Bull. 27*:93 (1979).
96. A. D. Woolfson, D. F. McCafferty, P. A. McCarron, and J. H. Price. A bioadhesive patch cervical drug delivery system for the administration of 5-fluorouracil to cervical tissue. *J. Cont. Rel. 35*:49–58 (1995).
97. H. Sidhu, J. H. Price, P. A. McCarron, D. F. McCafferty, A. D. Woolfson, D. Biggart, W. Thompson. A randomised control trial evaluating a novel cytotoxic drug delivery system for the treatment of cervical intraepithelial neoplasia. *Brit. J. Obstet. Gynaecol. 104*:145–149 (1997).
98. A. Gursoy, A. Bayhan. Testing of drug release from bioadhesive vaginal tablets. *Drug Dev. Ind. Pharm. 18*:203–221 (1992).
99. A. Gursoy, I. Sohtorik, N. Uyanik, N. A. Peppas. Bioadhesive controlled release systems for vaginal delivery. *S. T. P. Pharma 5*:886–892 (1989).
100. J. Brown, G. Hooper, C. J. Kenyon, S. Haines, J. Burt, J. M. Humphries, S. P. Newman, S. S. Davis, R. A. Sparrow, I. R. Wilding. Spreading and retention of vaginal formulations in post-menopausal women as assessed by gamma scintigraphy. *Pharm. Res. 14*:1073–1078 (1997).
101. T. Nagai, Y. Machida. Bioadhesive dosage forms for nasal administration. *Bioadhesive Drug Delivery Systems* (V. Lenaerts, R. Gurney, eds.). CRC Press, Boca Raton, FL, 1990.
102. T. Nagai, Y. Nishimoto, N. Nambu, Y. Suzuki, K. Sekine. Powder dosage form of insulin for nasal administration. *J. Cont. Rel. 1*:15–22 (1984).
103. C. Witschi, R. J. Mrsny. In vitro evaluation of microparticles and polymer gels for use nasal platforms for protein delivery. *Pharm. Res. 16*:382–290 (1999).
104. S. P. Vyas, N. Talwar, J. S. Karajgi, N. K. Jain. An erythrocyte based bioadhesive system for nasal delivery of propranolol. *J. Cont. Rel. 23*:231–237 (1993).
105. H.-W. Hui, J. R. Robinson. Ocular delivery of progesterone using a bioadhesive polymer. *Int. J. Pharm. 26*:203–213 (1985).
106. M. F. Saettone, E. Chiellini. Ophthalmic inserts. *Proc. Int. Symp. Control. Rel. Bioact. Mater. 12*:302–303 (1988).
107. N. Ünlü, A. Ludwig, M. Vanooteghem, A. A. Hincal. Formulation of Carbopol-940 ophthalmic vehicles, and in vitro evaluation of the influence of simulated lacrimal fluid on their physicochemical properties. *Pharmazie 46*:784–788 (1991).

42

Recent Developments in Drug Delivery to the Nervous System

Dusica Maysinger, Radoslav Savic, Joseph Tam, Christine Allen, and Adi Eisenberg
McGill University, Montreal, Quebec, Canada

I. NEUROACTIVE AGENTS AND THEIR DELIVERY TO THE CNS AND PNS

In the past decade, the contribution of the material sciences to drug delivery to the brain was realized mainly through the use of nonbiodegradable cylinders for intracerebral implantation of genetically engineered cells, or through the use of polymeric matrices that contained drugs. More recently, further progress has been made in the arena of prodrugs or conjugates that can exploit existing transport systems. An understanding of the basic mechanisms of the blood–brain barrier (BBB) transport biology provides a broad platform for current and future nervous system drug targeting strategies. In general, current approaches are either invasive (e.g., neurosurgical), pharmacological (e.g., by applying lipid carriers, liposomes, or different kinds of nanoparticles), or physiological (e.g., by taking advantage of normal endogenous pathways of carrier-mediated transport or receptor-mediated transport). Lipid-soluble molecules that have molecular mass under 500 daltons access the brain via lipid-mediated transport, but hydrophilic molecules such as peptides are mainly transported via receptor-mediated endocytosis. The main concepts of and underlying strategies for the administration of clinically relevant growth factors to the PNS, (1), and to overcome the BBB in the CNS, are summarized in several reviews (2–6).

A. Problems with Hydrophilic Agents

Although the surface area of the BBB in the human brain is large [approximately 20 m^2 (7)], small hydrophilic molecules cannot access the brain in pharmacologically adequate amounts when administered systemically or orally. This applies also to small peptidomimetic agents such as nerve growth factor (NGF)-mimetics (8) or neurotensine mimetics (9); hence effective delivery of these agents will require a drug delivery and targeting vehicle, or they should be conjugated to a BBB-targeting system. Development of novel drug delivery strategies requires adequate biological models to test their suitability. In vitro models include (i) primary cultures, (ii) immortalized neuronal, glial, and cerebromicrovascular endothelial cultures, (iii) hippocampal immortalized neuronal cultures (10,11), (iv) human cerebromicrovascular endothelial cell lines as a model of the BBB (12), and (v) more complex cocultures of neuronal and glial cells (13). In addition to these in vitro models, a number of in vivo model systems have been employed for testing neuroactive agents and their delivery systems. Rodent models, although indispensable and most commonly used to investigate neurological diseases, have limitations: (i) In general, they show some, but rarely all, of the pathological features of human neurological diseases; (ii) the time course of the progression of the disease is limited due to the difference in life span between two species; and (iii) tests for verbal communication skills cannot be applied. A number of neurological disorders are associated with either a lack of neuroactive peptides (e.g., growth factors, neurotrophins) or malfunctioning of their receptors (defective binding between receptor and ligand, impaired internalization and transport of the receptor–ligand complex, or impaired signaling pathways downstream from the receptor site) (14–20). For example, abnormal growth factor levels in the CNS and/or PNS have been associated with Alzheimer's disease, Parkinson's disease, Huntington's disease, and diabetic neuropathy (21–24). Results from preclinical studies employing both in vitro and in vivo models discussed above suggest that individual growth factors (as representatives of hydrophilic molecules) can indeed correct, prevent, or delay some of the pathological features characteristic of diabetic neuropathy, Alzheimer's, Parkinson's, and Huntington's diseases. However, due to the complexities involved in these pathologies, a simple replacement therapy employing drug delivery systems containing individual hydrophilic neurotherapeutics will most likely be used in conjunction with gene therapy and/or stem-cell therapies.

B. Problems with Lipophilic Agents

In general, lipophilic agents have little difficulty in penetrating cell membranes, including those of the BBB. The more lipophilic a drug is, the more readily it will cross the BBB and reach the brain. Thus the main problem with these agents lies not in their permeability but rather with aspects of (i) specificity and selectivity of action in the brain, (ii) neurotoxicity, and (iii) poor solubility and unfavorable pharmacokinetic properties. Some of these problems can be solved, at least partially, by incorporating the drug into a carrier polymer so that the release is slower and the toxicity is reduced. An attempt to increase the specificity and selectivity of neuroactive lipophilic drugs has been made by conjugation of the drug either with a specific ligand or with an antibody toward a protein specifically expressed at the cell surface (3). More recently, a class of lipophilic compounds, neurosteroids, i.e., steroids known to be particularly effective in the nervous system, were found to influence the brain's functions significantly, memory in particular. These agents do not have a problem in crossing the BBB or in specifically binding to their receptors. Studies by Toran-Allerand and colleagues showed that estrogen receptors are localized in central cholinergic neurons, and that signaling pathways activated by growth factors can be also activated by estrogens (25,26). Neurosteroids have been tested in several models (27), and numerous studies are currently underway to provide a proof of concept for neurosteroids as potential therapeutics in neurodegenerative diseases (27).

II. NONVIRAL DELIVERY SYSTEMS TO DELIVER AGENTS TO THE NERVOUS SYSTEM

Although the expression of specific proteins by transfection with viral vectors has been a commonly used technique, this method of drug delivery has certain disadvantages (28–30). A number of nonviral approaches to drug delivery to the nervous system have been developed, including (i) intraventricular infusion of neuroactive agents, (ii) injection or implantation of polymeric systems, (iii) implantation of genetically engineered cells or stem cells, and (iv) use of liposomes. These approaches are summarized in the following sections.

A. Intraventricular Infusion of Neuroactive Agents

Poorly soluble agents and unstable peptides are often administered into the lateral ventricles either as single injections or via permanently installed cannulae (31,32). The advantages of these approaches are that the dosage and rate of drug administration can be controlled, and the results resemble a slow intravenous infusion if the drug is readily distributed into the peripheral bloodstream. However, intracerebroventricular (ICV) injection of drug results in distribution to the ependymal surface of only the ipsilateral brain because of the unidirectional flow of cerebrospinal fluid within the brain. The major disadvantage of ICV drug administration is its invasiveness and the possibility of infection at the site of penetration of the BBB.

B. Injectable and Implantable Polymeric Systems as Drug Carriers

1. Drug–Polymer Conjugates

Synthetic polymer materials have been used as drug carriers in several modalities (Fig. 1). Injectable drug–polymer conjugates are produced by covalent binding of water-soluble polymers to a drug. The nature of the covalent bond between the drug and the polymer should be such that the bond is strong enough to be stable in the bloodstream but easily cleaved once the conjugate has reached the target site. This is often difficult to achieve. Moreover, only a relatively small number of biologically active molecules can be attached to the polymer molecule, thus requiring relatively large amounts of drug–polymer conjugate to be injected at the site of action. Approaches overcoming some of these problems are discussed in the following sections.

2. Implants

Simple replacement therapy with polymeric implants of nerve growth factor have been implemented in animal

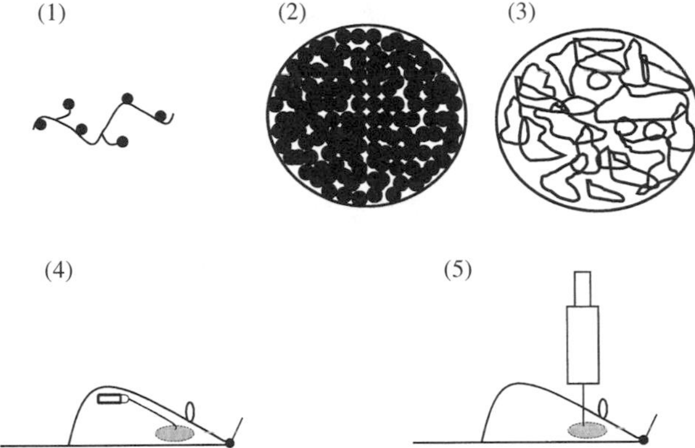

Figure 42.1 Some common approaches to administer neuroactive drugs. 1. Drug covalently bound to the polymer. 2. Microspheres (made of biodegradable polymers) containing neuroactive agents can be injected either systemically, into the lateral ventricle, or into the selected brain structure. 3. Microsponges can be impregnated with neuroactive agent and administered locally. 4. Osmotic pumps allow for steady release of neuroactive agent for a prolonged time period (1–2 weeks). 5. Injections of neuroactive agents directly into the lateral ventricle or parenchyma.

models of central cholinergic deficiencies (33,34) and of peripheral nerve impairment in diabetes (24), in both humans and several animal species (35,36). Recently nerve growth factor (NGF) was delivered locally by implantation of a small polymer pellet providing slow release at a controlled distance from the target site (37). The implants placed 1–2 mm away from the target cholinergic site were effective, whereas the same implants placed 3 mm away from the target site had no detectable effect. These findings strengthen the notion that NGF delivery within a spatially restricted area should be considered a desirable feature if the drug is to be effective. Due to the larger size of the target areas in the human brain than in rodent animal models, the concept of pharmacotechtonics has been tested. This strategy involves the creation of an array of local drug-releasing loci to create large but spatially restricted and anatomically defined fields of biological activity. Drug distribution can be more controlled, and moreover this approach lends itself to comparison with mathematical models (38). The geometry and sites of implantations can be determined by noninvasive diagnostic procedures, such as MRI, prior to the surgical procedure. Local delivery, in conjunction with pharmacokinetic modeling (39) and stereotaxic atlases linked to MRI scanners, will eventually allow for customized drug therapy for individual patients. Replacement of other factors such as ciliary neurotrophic factor (CNTF), lymphocyte inhibitory factor (LIF), and brain derived neurotrophic factor (BDNF) has also been achieved in different ways. However, all of these simple replacement approaches have three major limitations: (i) site-specific delivery, (ii) the amount of drug that can be administered by single administration, and (iii) susceptibility of full-length peptides to enzymatic cleavage due to the presence of various peptidases in the tissue. To solve some of the stability problems, drugs can be incorporated into biodegradable polymers, and an overview of these polymers is given in Chapter 5.

3. Osmotic Pumps

Osmotic pumps are also often used in experimental animals. For instance, implantable pumps have been used in primates to deliver dopamine or dopamine agonists (40,41). The pump reservoir is typically installed subcutaneously, and a catheter links a cannula with the pump. There are different sizes of pumps, suitable for small rodents or larger animals (commercially available "Alzet" minipumps); the pumps are refillable, and newer models allow for the adjustment of the delivery rates. The major limitation of pumps is the possibility of a local immune reaction at the site of delivery. In addition, due to the limited diffusion of most peptidergic neurotrophic agents, the majority of the agent is degraded before reaching the intended site of action.

4. Micro- and Nanoparticles

These delivery systems were reviewed previously (4,42–48). Controlled release polymer systems not only improve drug safety and efficacy but may also lead to new therapeutics. Some of the frequently used polymers are poly(sebacic acid-*co*-1,3-bis(p-carboxyphenoxy)propane), poly(b-hydroxybutyrate-hydroxyvalerate), poly(lactide-*co*-glycolide), poly(methyl methacrylate-acrylic acid), poly(acrylamide-*co*-acrylic acid), and poly(fumaric-*co*-sebacic anhydride). Numerous micro- and nanoparticles (some examples of which are shown in Fig. 2) have been designed and tested in vitro and in vivo to demonstrate superior effectiveness with concomitant reduction in neurotoxicity. Conventional oral or transdermal delivery is inadequate for the delivery of macromolecules such as proteins. Due to the short half-life of macromolecules such as growth factors, micro- and nanocontainers made of different polymers have been investigated as a means of their controlled and prolonged release. Johnson et al. (49) developed biodegradable microspheres composed of lactic coglycolic acid polymer in which lyophilized macromolecules (human growth hormone) were complexed with zinc to solve the problem of moisture-induced protein aggregation. The system was tested in vitro and in a primate in vivo model. A release of the protein for one month was demonstrated, suggesting the possibility that such a system

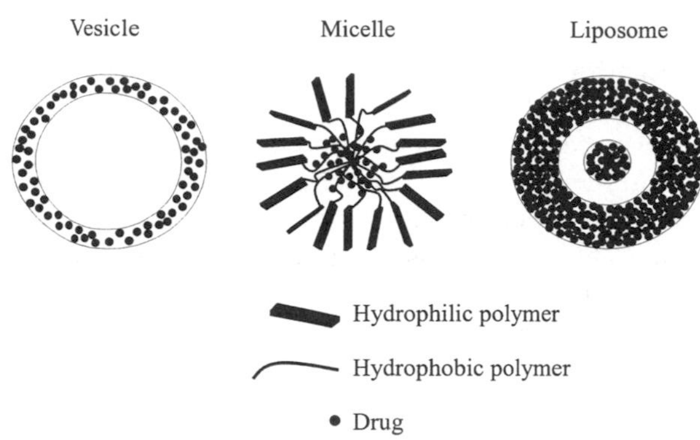

Figure 42.2 Some examples of injectable nanoparticles as carriers for neuroactive agents.

may be considered for chronic clinical use. Numerous other nano- and microparticulate biodegradable and biocompatible delivery systems have been developed in the last several years. For instance, rhodium (II) citrate, a recent member of promising antitumor agents, was complexed and encapsulated into poly (D,L-lactic-*co*-glycolic) acid (PLGA) and poly(anhydride) microspheres (50). Complexation in this case significantly increased the encapsulation efficiency and duration of release in both polymer systems (50). However, problems that need to be dealt with include the limited supply of the neuroactive agent, the invasive aspects of micro- and nanoparticle administration, and release kinetics that were not amenable to regulation by physiological changes at the site or elsewhere. Two interesting and novel approaches have been recently considered and tested: controlled release microchips and neurospheres. Briefly, in contrast to previous methods of controlling drug release from polymeric devices such as pulsatile stimuli by an electric or magnetic field, exposure to ultrasound, light, enzymes, changes in pH or in temperature, new biotechnological approaches have led to the development of a solid-state silicon microchip that can provide controlled release of a single or multiple agents on demand (51). Although it is too early to evaluate its usefulness for the delivery of neuroactive substances, it certainly seems promising. Neurospheres of multipotent and restricted precursors may provide solutions for a longer lasting and more physiological supply of biologically active compounds, either singly or in combination (52–54).

5. Liposomes

Cationic liposomes may have a significant potential for clinical applications in gene therapy for the disordered central nervous system (CNS) (55). Recently it has been reported that intracerebroventricular or intrathecal injection of cationic liposome–DNA complexes can produce significant levels of expression of biologically and therapeutically relevant genes within the CNS such as nerve growth factor (NGF), granulocyte colony-stimulating factor (G-CSF), and choline acetyltransferase (ChAT) (56). Technical aspects to achieve maximal gene transfer into brain cells using a plasmid DNA–cationic liposome complex have been discussed by Imaoka et al. (57). These authors have administered plasmid DNA–cationic liposome (lipopolyamine of dioctadecylamidoglycyl spermine) complex to 3–6 months old male rats using an osmotic pump. They report an increase of approximately up to two orders of magnitude in transfection efficiency compared to one obtained by a single injection. The authors propose that the continuous injection approach may be safe and effective in increasing the transfection efficiency. Another group led by Yokota (58) examined the effects of a calcium-dependent cysteine protease (calpain) inhibitor entrapped in liposomes in delaying neuronal death in gerbil hippocampal CA1 neurons following a transient forebrain ischemia. Selective neuronal damage induced by forebrain ischemia in the CA1 region of the hippocampus, and calpain-induced proteolysis of neuronal cytoskeleton, were prevented by administration of the inhibitor in a dose-dependent manner (58). Evaluation of transfection efficacy of a plasmid vector complexed with three different cationic liposomes into two experimental rodent and human malignant glioma cell lines and the mouse 3T3 fibroblast were studied by Bell et al. (59). The transfection efficacy and cytotoxicity of the liposomes were reported to vary quantitatively and qualitatively between cell lines. These authors suggest that their results support a potential application of cationic liposomes in both experimental and human malignant glioma gene transduction. Further studies on liposomal transfection of normal and neoplastic cells derived from the CNS will likely be very useful in helping to ascertain the particular merits of liposome-mediated gene transfection (59). Although the emphasis has been on utilizing liposomes in gene delivery to the CNS, this by no means limits their use to gene transfection (60–63).

C. Therapeutic Approaches Employing Cells

1. Genetically Engineered Cells

In order to provide longer term neurotrophin delivery without the need to refill the containers or reduce the frequency of reimplantation of delivery devices, several groups (5,64–66) have developed implantable polymeric devices containing genetically engineered cells that can produce, for example, a missing trophic factor (Fig. 3). This strategy has been tested in animal models, including primates (67). Either primary cultures or genetically engineered cells producing a missing factor can serve as ''long term effective mini-factories,'' and various cell types used for these pur-

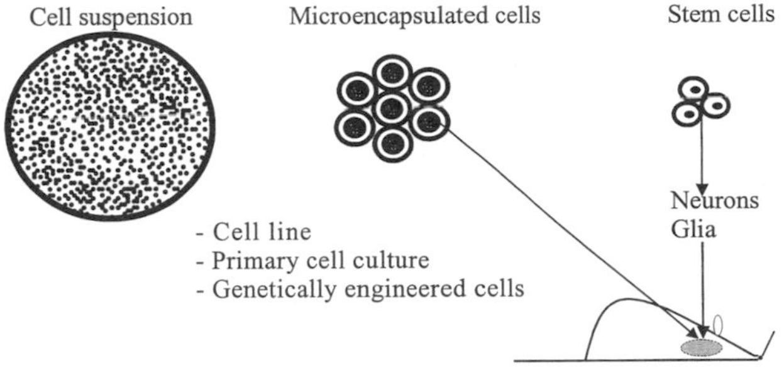

Figure 42.3 Genetically engineered cells and stem cells. Different cell lines, primary cell cultures, and genetically engineered cells producing a neuroactive agent can be directly injected into the brain as a cell suspension, or prior to administration cells can be microencapsulated in biocompatible polymers. Neural stem cells with the capacity to renew and produce the major cell types of the brain can be used for cell replacement therapy in neurological disorders. (See Section C.3, Stem Cells.)

poses have been reported and reviewed, including pheochromocytoma cells (PC12) (68), fibroblasts, and NIH 3T3 cells genetically altered to produce growth factors (66,69,70). Although fully mature primary cultures or genetically engineered proliferating cells of nonneuronal origin can replace missing peptides, they are either (i) deliberately physically separated from the environment at the implantation site to prevent tumor formation (e.g., by encapsulation or by placement of cells within a retractable implantation device) or (ii) in contact with the immediate microenvironment, their phenotype not allowing them to integrate and make functional connections (e.g., PC12 cells, fibroblasts).

3. Stem Cells

Replacement strategies using stem cells have recently become an attractive way to overcome the problems of cell integration and of acquisition of normal brain functions (71). Adult CNS stem cells can replace neurons and glia in the adult brain and spinal cord (72) and can also give rise to other cell types such as skin melanocytes and a range of mesenchymal cells in the head and neck (73). Stem cells may integrate appropriately into both the developing and the degenerating central nervous system and may be uniquely responsive to some types of neurodegenerative conditions (74). Neural-derived stem cells are self-renewing under the influence of mitotic agents such as fibroblast growth factor (75), epidermal growth factor (76,77), BDNF (78), and other factors (71,79–82). These cells can differentiate into either neuronal or glial cells and therefore can be used to replace neurons that are damaged or destroyed in defined neuronal structures, such as dopaminergic nigral neurons in Parkinson's disease, or hippocampal neurons (70,76,83–87). Neural stem cells cultured from human embryos can be grown for extended periods of time while retaining the capacity for neuronal and glial differentiation. The ability to generate human neural tissue in vitro allows for screening of neuroactive compounds and provides a source of tissue for testing cellular and genetic therapies for CNS disorders (88). Neurospheres of multipotent and restricted precursors may provide solutions for a longer lasting and more physiological supply of missing biologically active compounds, single or multiple, (52,54,89). Most importantly, stem cells have the advantage of establishing functional connections within the nervous system, a property that cannot be achieved with any polymeric drug delivery system, at least not in cases when a large proportion of neurons is lost. Accounts of the current status of stem cells and their biology and potential in treating neurological disorders are available in recent reviews (70,87,90). Obviously, ethical issues are of importance in implementing stem-cell strategies (84,86).

III. BLOCK COPOLYMER MICELLES AND VESICLES

Block copolymer micelles (Fig. 4) have a great potential as delivery systems for the administration of neuroactive agents. Previous work (91,92) provided some seminal information in this regard, but much fundamental work relating to physical, morphological, and biological (pharmacological) properties must be done before block copolymer micelles and vesicles can be used either as diagnostics or as therapeutics. Thus far, only spherical micelles have been studied from the physicochemical and biological aspect. Other morphologies were only recently produced and identified using EM (93–95) and some sporadic in vivo studies have been reported (cylinder shapes delivered to lungs) without physical-chemical characterization. Our group has been involved in fundamental studies addressing the questions of interrelationships between morphological features (shape, size) and physiochemical properties of tailor-made micelles containing either fluorescent labels (96,97), highly lipophilic radiolabeled agents such as benzopyrene, or poorly soluble bioactive agents such as dihydrotestosterone and FK506 (96,98,99). A recent overview of the physical properties of block copolymer micelles used in vitro and in vivo in studies by several groups, including ours, is available in (92) and (98).

A. Biodegradable Block Copolymers for Development of Micellar Delivery Systems

One of the promising biodegradable and biocompatible polymers for micellar delivery systems is polycaprolac-

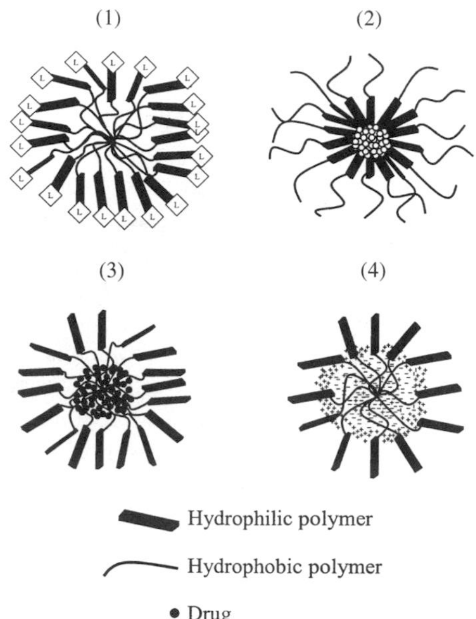

Figure 42.4 Representative types of block copolymer micellar delivery systems. 1. Corona forming block has attached ligand to provide site-specific delivery of neuroactive agents. 2. Inverse micelle for the delivery of hydrophilic neuroactive agents. 3. Block copolymer micelle for the delivery of lipophilic neuroactive agents. 4. Block ionomer micelle suitable for the delivery of antisense oligonucleotides and DNA.

tone-*b*-poly(ethylene oxide) (98). The individual components, polycaprolactone and polyethylene oxide, were explored previously for a variety of biomedical applications (110,111). A list of some core forming polymers is given in Table 1. Polycaprolactone, the hydrophobic core–forming block of the micelles, is a biodegradable polymer used as (i) a structural material in the production of medical devices such as implants, sutures, stents, and prosthetics, (ii) a carrier for a variety of drugs (112), (iii) in paste form for drugs (113,114), and (iv) a nanoparticulate ocular delivery system (115). Polyethylene oxide, the hydrophilic shell–forming block of the micelles, imparts blood compatibility to material surfaces (116) and is commonly used in micellar drug delivery systems (91). This polymer lends itself to chemical modifications that can enhance site-directed delivery. Micellar delivery systems for neuroactive agents are described in the following paragraph.

B. Micelles for the Delivery of Neuroactive Agents

Several types of micelles formed from different polymers and copolymers have been developed (Table 2), some of which could be useful for the delivery of neuroactive agents. Several considerations arise when considering polymer micelles as drug carriers for CNS-based therapeutics. For instance, a high partitioning of the drug into the micelle is required because otherwise one would have to administer large amounts of micelles. Similarly, the polymer must be biodegradable, biocompatible, and with sufficiently low critical micellar concentration (CMC) to achieve a longer length of time in the bloodstream to allow the drug to reach its site of action. Modifications to optimize polymer–drug interactions and high stability of micelles in the blood have been recently reviewed (127). In order to enhance endocytosis and ultimately transcytosis of micelles containing neuroleptic agents across the BBB, ligands were attached to the polymer micelles. An improved internalization of derivatized pluronic micelles has been demonstrated using primary cultures of brain microvessel endothelial cells (BBMEC) (128–130). Nevertheless, micelle carriers for drug delivery to the brain still have some limitations:

1. Brain endothelial cells take up a relatively small amount of micelles designed so far. Consequently, the amount of drug delivered is relatively small. However, if the drug has a high potency, it should be still possible to obtain the desired biological effects.
2. Drugs to be incorporated into the micelles should have very low capacity to cross the BBB. This is not the case with highly lipophilic agents, but many peptides and oligonucleotides are good candidates for micellar delivery systems to the brain.

In addition to the micelle-based strategies outlined here, there is another strategy that has been explored to a limited extent, namely the exploitation of specific interactions of the polymers themselves with the membrane and membrane transporter proteins found in the brain microvessel endothelial cells that form the BBB. In this case it is not necessary to have micelles, since the monomer itself can modify the BBB permeability. Two proteins within the BBB are targets of such an approach: P-glycoprotein (P-gp) and multidrug resistance–associated protein (MRP) (128,129). Modifications of P-gp and MRP and possibly other transporters in the BBB with polymeric formulations may considerably facilitate the accumulation of neuroactive agents within the brain. Among the polymers tested so far, pluronic polymers seem to be particularly effective in inhibiting P-gp drug efflux in the brain microvessel endothelial cells (106). Some drugs can also change the permeability of the BBB, such as cyclosporin A. Unfortunately, cyclosporin A in equivalent concentrations to Pluronic P85 seriously disrupts the integrity of the brain microvessel endothelial cells (127). Some examples of drugs with effects on the nervous system and of their delivery vehicles are given in Table 3. Polymers as trans-

Table 42.1 Core-Forming Polymers

Polymer	Application	Reference
Polyamino acid cores		
Poly(aspartic acid)	Poly(ethylene glycol)–poly(aspartic acid) block copolymer micelles conjugated with adriamycin were prepared, characterized, and tested for anticancer activity. Result in vivo: high anticancer activity of adriamycin micelles in P 388 mouse leukemia and smaller weight loss with adriamycin micelles than with free adriamycin.	100
Poly(gamma-benzyl-L-glutamate)	Clonazepam containing block copolymer nanoparticles were prepared and tested in vitro. Result: release of clonazepam is dependent on the drug loading contents and PBLG chain length.	101
Polyamines		
Poly(L-lysine)	Supramolecular associates of DNA with PEG-PLL block copolymers increase nuclease resistance of plasmid DNA.	102
	Turbidometric analysis of polyelectrolyte complexes formed between PLL and DNA.	103
	Examination of the use of oligopeptides containing the RGD sequence in promoting the uptake of PLL/DNA complexes in vitro.	104
Polyspermine	Discussion of physicochemical aspects of formation and behavior of interpolyelectrolyte and block ionomer complexes for gene delivery.	105
Others		
Polypropylene oxide	Interactions of pluronic block copolymers with brain microvessel endothelial cells.	106
	Haloperidol's neuroleptic activity increases after its solubilization in surfactant micelles.	107
	Establishment of the structure–property relationship for E_mP_n (E, oxyethylene; P, oxypropylene) diblock copolymers–micellization and gelation of E_mP_n diblock copolymers in aqueous solution.	108
Polyethylene	New class of synthetic thin-shelled capsules. Polymersomes: tough vesicles made from diblock copolymers.	109

port-modifying agents have both advantages and limitations:

1. Polymers exhibit relative selectivity. However, since P-gp and MRP are also present in liver, the drug could accumulate in this organ, affecting normal functions, in particular metabolic processes.
2. Less damage occurs to the microvessel endothelial cell integrity.
3. Some polymers are effective below the CMC.

C. Micelles for the Delivery of Antisense Oligonucleotides

1. Problems in the Delivery of Antisense Oligonucleotides

The use of antisense oligonucleotides to control protein expression has received considerable attention due to their relative ease of synthesis and specific use. Commonly, short (between 15 and 20 bases) chemically modified nucleic acids are employed. They hybridize to complementary nucleotide sequences in accessible regions of mRNA molecules, thus blocking expression of the encoded proteins. Two major problems are associated with the use of antisense oligonucleotides: (i) the identification of accessible regions in mRNA to which the oligonucleotides can hybridize, and (ii) the delivery of the oligonucleotides into the cell and to their target (136). The accessibility of an mRNA sequence is determined by the primary nucleotide sequence, its three-dimensional structure, and the presence of associated proteins. This combination of factors has made the design of effective antisense oligonucleotides a difficult task, and experiments involving the use of poorly designed antisense oligonucleotides often produce misleading results. However, once a candidate oligonucleotide has been synthesized, several challenges arise in delivering it effectively to its target. A first consideration is the presence of various endo- and exonucleases in serum that can degrade the oligonucleotides and destroy their biological activity (137). Also, some types of antisense oligonucleotides (e.g., phosphothiorate) are highly charged

Table 42.2 Types of Micelles

Micelles	Utilization	Reference
Pluronic based micelles	Examination of the solubilization of hydrocarbons by aggregates of Pluronic® block copolymers in water [Pluronic®; symmetric copolymers with poly(ethylene oxide), PEO, as the hydrophilic end blocks and poly(propylene oxide), PPO, as the hydrophobic middle block (PEO-PPO-PEO)].	117
	Stabilization and activation of Pluronic micelles for tumor-targeted drug delivery.	118
	Pluronic based (Pluronic L61 and Pluronic F127) formulation of doxorubicin.	119
Block ionomer complexes	Potential application of polymer–surfactant complexes between block ionomers and oppositely charged surfactants as a drug delivery system.	120
Polyion micelles	Surface charge modulation of poly(ethylene glycol)-poly(D,L-lactide) block copolymer micelles: conjugation of charged peptides.	121
Thermoresponsive micelles	Control of adriamycin cytotoxic activity using thermally responsive polymeric micelles composed of poly(N-isopropylacrylamide-*co*-N,N,-dimethylacrylamide)-*b*-poly(D,L-lactide).	122
	Gelation behavior of thermosensitive poly(ethylene glycol-*b*-(D,L-lactic acid-*co*-glycolic acid)-b-ethylene glycol) triblock copolymers.	123
Polycaprolactone-b-polyethylene oxide micelles	PCL-b-PEO copolymer micelles as a delivery vehicle for lipophilic drugs.	96, 124
Shell-cross linked polymer micelles	Formation of cross-links throughout the shell of polymer micelles reinforces the weak interactions that facilitate micelle existence.	125
Polycation comb-type copolymer	Copolymer-mediated stabilization of DNA duplexes and triplexes.	126

polyanions and bind extensively to serum proteins (138–140). Such oligonucleotides have a very limited ability to cross cellular membranes, and this reduces the amount of oligonucleotides that can reach their nuclear target (141, 142). Furthermore, the binding of serum proteins by oligonucleotides may also alter the protein activity, which may be misinterpreted as an effect of the oligonucleotides on nuclear targets. A second consideration is that the amount of oligonucleotides taken up by different tissues and cells varies considerably. Studies in rodents have shown that the majority of intravenously injected oligonucleotides distribute to the kidney and liver, where they may be degraded and lose biological activity (143–144). Also, these studies showed that the biodistribution of oligonucleotides is generally much greater in the intestine, bone marrow, skeletal muscle, and skin than in the brain. This presents additional concerns for the delivery of oligonucleotides to targets in the central nervous system.

Table 42.3 Examples of Drugs with Effects on the Nervous System and Their Delivery Vehicles

Drug	Delivery vehicle	Reference
FK 506	Polycaprolactone-*b*-poly(ethylene oxide) block copolymer micelles; a novel drug delivery vehicle for FK506 and L-685,818.	96
Paclitaxel	Poly(D,L-lactide)-*b*-methoxy polyethylene glycol micelles; human plasma distribution of free and micelle incorporated paclitaxel.	131
Neurosteroids	Polycaprolactone-*b*-poly(ethylene oxide) copolymer micelles; delivery vehicle for dihydrotestosterone.	124
	Poly (D,L-lactic acid) microspheres; evaluation of microspheres' physical properties and their in vivo characteristics for long-acting drug delivery (testosterone).	132
	Caprolactone/lactide block copolymer; delivery kinetics of levonorgestrel and estradiol.	133
	Tricalcium phosphate lysine (TCLP) drug delivery system; the synergistic effect of sustained delivery of steroid hormones on the ventral prostate of adult male rodents.	134
Cyclosporine	Microemulsion formulation; comparison of the steady-state pharmacokinetics and tolerability of the cyclosporine microemulsion formulation with the cyclosporine commercial formulation in clinically stable renal allograft recipients.	135
Clonazepam	Core-shell type nanoparticles; in vitro clonazepam release is dependent on the drug loading contents and poly(gamma-benzyl-L-glutamate) chain length.	101

The actual mechanisms by which antisense oligonucleotides enter into cells are currently under debate, but they are thought to involve fluid-phase pinocytosis and/or receptor-mediated endocytosis (145). A third concern is that once the oligonucleotides have successfully entered into a cell, they are subject to degradation. Confocal and electron microscopy studies have shown that the majority of internalized antisense oligonucleotides enter into the cellular endosomal and lysosomal systems (146). These compartments may have acidic pH and contain enzymes that degrade the oligonucleotides and destroy their biological activity. It is clear, however, that a portion of the administered oligonucleotides either escapes from the endosomal/lysosomal systems or bypasses them altogether and enters the cytoplasm to diffuse into the nucleus, by poorly understood mechanisms (141). The cytosolic environment also contains a variety of exo- and endonucleases and proteins, which presents problems similar to those encountered in the serum. An additional concern arising from microscopy studies is the existence of an efflux of oligonucleotides from the nucleus, likely through a passive diffusion mechanism via nuclear pores (146), and from the cell (147–149). These phenomena must also be considered when targeting antisense oligonucleotides to the nucleus.

2. Early Approaches in Antisense Oligonucleotide Delivery Systems

Short-term antisense therapy is often marked by the development of two common toxicities, namely activation of the complement system and increase in blood clotting time (150,151). However, clinical studies have shown that these effects are minor due to the relatively short half-life (30–60 minutes) of oligonucleotides in the serum, which may be due to binding of serum proteins to oligonucleotides (145). As well, chronic administration of antisense oligonucleotides in rodent models leads to the induction of immune responses, characterized by lymphoid hyperplasia, splenomegaly, and monocyte recruitment to a number of tissues (151). In consequence, there has been considerable interest in developing delivery methods that can minimize these degradative and immune drawbacks to the use of oligonucleotides. Early approaches focused on improving the cellular uptake of antisense oligonucleotides by coupling them to polycations such as polylysine (152) and polyethylenime (153), or, more successfully, to polycationic lipids/lipid formulations (154,155). Other approaches involved targeting antisense oligonucleotides to cell surface receptors such as the folate receptor (156), transferrin receptor (157), and asialoglycoprotein receptor (158), to stimulate receptor-mediated endocytosis. Another strategy involved the coupling of oligonucleotides to typical membrane entities such as cholesterol (159) or to fusogenic peptides (160). These in vitro uptake enhancers have been successful in tissue culture systems, with significant increases in the uptake and nuclear localization of the oligonucleotides. However, it is noteworthy that the internalization profiles of oligonucleotides in vitro often differ considerably from their behavior in vivo (145). Furthermore, these early approaches did not focus upon minimizing the immunogenicity and instability of the oligonucleotides, which has been addressed by the micellar delivery approach described below.

3. Delivery of Antisense Oligonucleotides Employing Copolymers

Although some oligonucleotides are relatively stable following their administration by a variety of routes (intravenous, intraperitoneal, subcutaneous, intracerebroventricular) (161), delivery of antisense oligonucleotides by delivery systems has been proposed in order to reduce the immunogenicity and protect them from the physiological degradative processes referred to above. The major work in this area has been carried out by Kabanov's group, using, first, hydrophilic polymeric vesicles (162), and subsequently, polyion complex micelles with a protein modified corona (163). A hydrophilic polymer, Nanogel, was used to create vesicles with an average particle size of 120 nanometers (162). This polymer is formed from cross-linked polyethyleneimine (PEI) and carbonyldiimidazole-activated poly(ethylene glycol) (PEG), thus building on the PEI-coupling approach of Boussif (153), which focused on enhancement of cellular uptake. Nanogel particles have been employed for the delivery of antisense phosphothiorate oligonucleotides (SODN), targeting the mRNA of the human multidrug resistance gene, mdr1 (162). These studies have shown that the Nanogel vesicular approach attains a greater cellular localization of SODN in a human carcinoma cell line in comparison to the administration of free SODN. The effectiveness of the Nanogel-incorporated SODN was demonstrated by its ability to inhibit the expression of P-glycoprotein, a major cellular pump involved in the efflux of cytosolic drugs.

A significant reduction in delivery vehicle size was achieved by Kabanov's group with the synthesis of polyion complex micelles incorporating SODN (163). These micelles contain poly(ethylene glycol)–PEI graft copolymer complexes that self-assemble with SODN to form 40 nanometer particles, each consisting of a PEI/SODN neutralized core surrounded by a poly(ethylene glycol) corona. Transferrin molecules attached to the poly(ethylene glycol) corona create polyion complex micelles (75 to 103 nanometers) facilitating the internalization of SODN into cells. These studies (163) have indicated that SODN incorporated in polyion micelles with a protein modified corona have a significantly greater ability to inhibit the expression of the P-glycoprotein drug efflux transporter in several cancer cell lines.

4. Micelles for the Delivery of DNA

Some of the most frequently applied techniques to transfect cells are (i) precipitation with calcium phosphate (164–170), (ii) polybrene (171), (iii) electroporation (172), (iv) microinjection (173), (v) modified viral vectors (174,175), (vi) microspheres (47), (vii) liposomes (176–178), and (viii) polycation delivery systems (121,179–181). There are several commonly used protocols for cell transfections, the most frequently used being precipitation with calcium phosphate (164,182).

Kabanov's group has contributed greatly to the delivery systems of genetic materials. The methodology is based on formation of soluble interpolyelectrolyte complexes (IPECs). The term IPEC in polymer science relates to the products of reaction of oppositely charged polyions, and in the case referred to here polycation–DNA complexes represent a special kind of IPECs relevant to biological issues. A critical assessment of transfection approaches in mammalian cells using DNA–IPECs and more common methods such as calcium phosphate precipitation and lipofectin is given in several reviews, e.g., Ref. 179. Polycations as building blocks for DNA complexes can be relatively easily conjugated with ligands and undergo receptor-mediated endocytosis (179). Among the most frequently attached ligands are asialoglycoprotein, insulin, and transferrin. Asialoglycoprotein receptors play a critical role in hepatocytes and allow for targeted delivery of DNA to the liver (183,184). Insulin and transferrin receptors are present in many cell types and provide an endocytotic internalization of the delivery system (insulin receptor) or vesicular transport followed by the return to the cell surface (185). Antibody molecules have also been linked to IPECs to achieve cell and tissue specificity (186). Since neural cells express a wide variety of receptors, some of them specific for (sub)classes of neurons (e.g., cholinergic, dopaminergic, gabaergic), IPECs linked to ligands recognizing these receptors offer an attractive approach in drug and gene delivery to the nervous system.

IV. FUTURE TRENDS

If a disease is diagnosed early enough, and if significant improvements can result simply from a slowing down of pathological changes, drug administration using suitable polymeric devices is a viable approach. Moreover, in conjunction with stem cells, drug delivery systems could improve the viability of implanted stem cells, enhance the rate of functional reconnectivity, and promote neuronal differentiation. Regardless of particular specific needs, localized (site-specific) physiologically controlled drug release is always desirable. These goals are being gradually achieved by the increasing availability of new functionalized biopolymers, and ''smart polymers.''

Smart polymers are hydrogels that undergo fast, reversible changes in microstructure from a hydrophilic to a hydrophobic state (187). Triggers that can produce these changes include neutralization of charged groups by a shift in pH, the addition of an oppositely charged polymer, a change in temperature or ionic strength, or the formation of interpenetrating polymer networks (187). Stimulus-responsive or smart polymers have been used mainly for bioseparations but also for the development of drug delivery systems. One of the models based on smart polymers is a glucose-responsive insulin loaded polymer matrix (188). The number of drug delivery systems utilizing smart polymers is very limited, and much work remains to be done in the area of their synthesis and structure–property relationship studies before they can be considered for clinical applications.

In addition to drug release from polymeric devices such as pulsatile stimuli by an electric or magnetic field, exposure to ultrasound, light, enzyme, or change in pH or temperature, new biotechnological approaches have led to the development of solid-state silicon microchips that can provide controlled release of single or multiple agents on demand (51). Although it is too early to say if microchips will supplement or replace classical delivery systems, they represent a significant step forward in biotechnological approaches to administer neuroactive agents in a more controlled manner.

REFERENCES

1. G. Terenghi. Peripheral nerve regeneration and neurotrophic factors. *Journal of Anatomy 194*:1 (1999).
2. W. M. Pardridge. CNS drug design based on principles of blood–brain barrier transport. *Journal of Neurochemistry 70*:1781 (1998).
3. L. Prokai. Peptide drug delivery into the central nervous system. *Progress in Drug Research 51*:95 (1998).
4. D. Maysinger, A. Morinville. Drug delivery to the nervous system. *Trends in Biotechnology 15*:410 (1997).
5. C. W. Pouton, L. W. Seymour. Key issues in non-viral gene delivery. *Advanced Drug Delivery Reviews 46*:187 (2001).
6. M. Mak, L. Fung, J. F. Strasser, W. M. Saltzman. Distribu-

tion of drugs following controlled delivery to the brain interstitium. *Journal of Neurooncology 26*:91 (1995).

7. W. M. Pardridge. *Peptide Drug Delivery to the Brain* (W. M. Pardridge, ed.). Raven Press, New York, 1991, pp. 1–357.
8. S. Maliartchouk, Y. Feng, L. Ivanisevic, T. Debeier, A. C. Cuello, K. Burgess, H. U. Saragovi. A designed peptidomimetic agonistic ligand of trkA nerve growth factor receptors. *Molecular Pharmacology 57*:385 (2000).
9. B. Cusack, D. J. McCormick, Y. P. Pang, T. Souder, R. Garcia, A. Fauq, E. Richelson. Pharmacological and biochemical profiles of unique neurotensin 8–13 analogs exhibiting species selectivity, stereoselectivity, and superagonism. *Journal of Biological Chemistry 270*:18359 (1995).
10. S. Gokhan, Q. Song, M. F. Mehler. Generation and regulation of developing immortalized neural cell lines. *Methods 16*:345 (1998).
11. C. Thuerl, U. Otten, R. Knoth, R. P. Meyer, B. Volk. Possible role of cytochrome P450 in inactivation of testosterone in immortalized hippocampal neurons. *Brain Research 762*:47 (1997).
12. A. Muruganandam, L. M. Herx, R. Monette, J. P. Durkin, D. B. Stanimirovic. Development of immortalized human cerebromicrovascular endothelial cell line as an in vitro model of the human blood-brain barrier. *FASEB Journal 11*:1187 (1997).
13. S. Duport, F. Robert, D. Muller, G. Grau, L. Parisi, L. Stoppini. An in vitro blood–brain barrier model: cocultures between endothelial cells and organotypic brain slice cultures. *Proceedings of the National Academy of Sciences of the United States of America 95*:1840 (1998).
14. M. P. Mattson, W. Duan. ''Apoptotic'' biochemical cascades in synaptic compartments: roles in adaptive plasticity and neurodegenerative disorders. *Journal of Neuroscience Research 58*:152 (1999).
15. M. P. Mattson, K. Furukawa. Signaling events regulating the neurodevelopmental triad. Glutamate and secreted forms of beta amyloid precursor protein as examples. *Perspectives in Developmental Neurobiology 5*:337 (1998).
16. I. Semkova, J. Krieglstein. Neuroprotection mediated via neurotrophic factors and induction of neurotrophic factors. *Brain Research—Brain Research Reviews 30*:176 (1999).
17. C. F. Ibanez. Emerging themes in structural biology of neurotrophic factors. *Trends in Neurosciences 21*:438 (1998).
18. W. P. Chen, Y. C. Chang, S. T. Hsieh. Trophic interactions between sensory nerves and their targets, *Journal of Biomedical Science 6*:79 (1999).
19. S. D. Skaper, F. S. Walsh. Neurotrophic molecules: strategies for designing effective therapeutic molecules in neurodegeneration. *Molecular and Cellular Neurosciences 12*:179 (1998).
20. B. Connor, M. Dragunow. The role of neuronal growth factors in neurodegenerative disorders of the human brain. *Brain Research—Brain Research Reviews 27*:1 (1998).
21. R. Kramer, Y. Zhang, J. Gehrmann, R. Gold, H. Thoenen, H. Wekerle. Gene transfer through the blood–nerve barrier: NGF-engineered neuritogenic T lymphocytes attenuate experimental autoimmune neuritis. *Nature Medicine 1*: 1162 (1995).
22. F. H. Gage. Intracerebral grafting of genetically modified cells acting as biological pumps. *Trends in Pharmacological Sciences 11*:437 (1990).
23. F. H. Gage. Cell therapy. *Nature 392*:18 (1998).
24. S. C. Apfel, J. A. Kessler. Neurotrophic factors in the treatment of peripheral neuropathy. *Ciba Foundation Symposium 196*:98 (1996).
25. R. C. Miranda, F. Sohrabji, C. D. Toran-Allerand. Neuronal colocalization of mRNAs for neurotrophins and their receptors in the developing central nervous system suggests a potential for autocrine interactions. *Proceedings of the National Academy of Sciences of the United States of America 90*:6439 (1993).
26. M. Singh, G. J. Setalo, X. Guan, M. Warren, C. D. Toran-Allerand. Estrogen-induced activation of mitogen-activated protein kinase in cerebral cortical explants: convergence of estrogen and neurotrophin signaling pathways. *Journal of Neuroscience 19*:1179 (1999).
27. M. Gasior, R. B. Carter, J. M. Witkin. Neuroactive steroids: potential therapeutic use in neurological and psychiatric disorders. *Trends in Pharmacological Sciences 20*: 107 (1999).
28. R. J. Cristiano, B. Xu, D. Nguyen, G. Schumacher, M. Kataoka, F. R. Spitz, J. A. Roth. Viral and nonviral gene delivery vectors for cancer gene therapy. *Cancer Detection and Prevention 22*:445 (1998).
29. O. Isacson, X. O. Breakefield. Benefits and risks of hosting animal cells in the human brain. *Nature Medicine 3*:964 (1997).
30. J. M. Schumacher, O. Isacson. Neuronal xenotransplantation in Parkinson's disease. *Nature Medicine 3*:474 (1997).
31. A. Bjorklund. Neural transplantation—an experimental tool with clinical possibilities. *Trends in Neurosciences 14*:319 (1991).
32. T. Kushikata, J. Fang, J. M. Krueger. Brain-derived neurotrophic factor enhances spontaneous sleep in rats and rabbits. *American Journal of Physiology 276*:R1334 (1999).
33. J. Winkler, L. J. Thal, F. H. Gage, L. J. Fisher. Cholinergic strategies for Alzheimer's disease. *Journal of Molecular Medicine 76*:555 (1998).
34. H. K. Raymon, S. Thode, F. H. Gage. Application of ex vivo gene therapy in the treatment of Parkinson's disease. *Experimental Neurology 144*:82 (1997).
35. P. Anand. Neurotrophins and peripheral neuropathy. *Philosophical Transactions of the Royal Society London—Series B: Biological Sciences 351*:449 (1996).
36. P. Anand, G. Terenghi, G. Warner, P. Kopelman, R. E. Williams-Chestnut, D. V. Sinicropi. The role of endogenous nerve growth factor in human diabetic neuropathy. *Nature Medicine 2*:703 (1996).
37. M. J. Mahoney, W. M. Saltzman. Millimeter-scale positioning of a nerve-growth-factor source and biological ac-

tivity in the brain. *Proceedings of the National Academy of Sciences of the United States of America 96*:4536 (1999).
38. M. J. Mahoney, W. M. Saltzman. Controlled release of proteins to tissue transplants for the treatment of neurodegenerative disorders. *Journal of Pharmaceutical Sciences 85*:1276 (1996).
39. P. G. Welling. *Pharmocokinetics: Processes, Mathematics, and Applications* (P. G. Welling, ed.). American Chemical Society, Washington, DC., 1997, pp. 1–393.
40. J. G. De Yebenes, S. Fahn, V. Jackson-Lewis, P. Jorge, M. A. Mena, J. Reiriz. Continuous intracerebroventricular infusion of dopamine and dopamine agonists through a totally implanted drug delivery system in animal models of Parkinson's disease. *Journal of Neural Transmission Supplementum 27*:141 (1988).
41. R. Hargraves, W. J. Freed. Chronic intrastriatal dopamine infusions in rats with unilateral lesions of the substantia nigra. *Life Sciences 40*:959 (1987).
42. C. Allen, A. Eisenberg, D. Maysinger. Copolymer drug carriers: conjugates, micelles and microspheres. *Scientific and Technical Pharmacy—Pharmaceutical Sciences 9*: 139 (1999).
43. J. Filipovic-Grcic, D. Maysinger, I. Jalsenjak. Microparticles with neuroactive agents. *Journal of Microencapsulation 12*:343 (1995).
44. R. Langer. Controlled release of a therapeutic protein. *Nature Medicine 2*:742 (1996).
45. W. M. Saltzman, M. W. Mak, M. J. Mahoney, E. T. Duenas, J. L. Cleland. Intracranial delivery of recombinant nerve growth factor: release kinetics and protein distribution for three delivery systems. *Pharmaceutical Research 16*:232 (1999).
46. J. Kreuter. Nanoparticulate systems for brain delivery of drugs. *Advanced Drug Delivery Reviews 47*:65 (2001).
47. S. Ando, D. Putnam, D. W. Pack, R. Langer. PLGA microspheres containing plasmid DNA: preservation of supercoiled DNA via cryopreparation and carbohydrate stabilization. *Journal of Pharmaceutical Sciences 88*:126 (1999).
48. J. P. Benoit, N. Faisant, M. C. Venier-Julienne, P. Menei. Development of microspheres for neurological disorders: from basics to clinical applications. *Advanced Drug Delivery Reviews 46*:187 (2001).
49. O. L. Johnson, J. L. Cleland, H. J. Lee, M. Charnis, E. Duenas, W. Jaworowicz, D. Shepard, A. Shahzamani, A. J. Jones, S. D. Putney. A mouth-long effect from a single injection of microencapsulated human growth hormone. *Nature Medicine 2*:795 (1996).
50. R. D. Sinisterra, V. P. Shastri, R. Najjar, R. Langer. Encapsulation and release of rhodium(II) citrate and its association complex with hydroxypropyl-beta-cyclodextrin from biodegradable polymer microspheres. *Journal of Pharmaceutical Sciences 88*:574 (1999).
51. J. T. J. Santini, M. J. Cima, R. Langer. A controlled-release microchip. *Nature 397*:335 (1999).
52. M. S. Rao. Multipotent and restricted precursors in the central nervous system. *Anatomical Record 257*:137 (1999).
53. E. D. Laywell, V. G. Kukekov, D. A. Steindler. Multipotent neurospheres can be derived from forebrain subependymal zone and spinal cord of adult mice after protracted postmortem intervals. *Experimental Neurology 156*:430 (1999).
54. B. J. Chiasson, V. Tropepe, C. M. Morshead, D. van der Kooy. Adult mammalian forebrain ependymal and subependymal cells demonstrate proliferative potential, but only subependymal cells have neural stem cell characteristics. *Journal of Neuroscience 19*:4462 (1999).
55. K. Yang, G. L. Clifton, R. L. Hayes. Gene therapy for central nervous system injury: the use of cationic liposomes: an invited review. *Journal of Neurotrauma 14*:281 (1997).
56. C. Meuli-Simmen, Y. Liu, T. T. Yeo, D. Liggitt, G. Tu, T. Yang, M. Meuli, S. Knauer, T. D. Heath, F. M. Longo, R. J. Debs. Gene expression along the cerebral-spinal axis after regional gene delivery. *Human Gene Therapy 10*: 2689 (1999).
57. T. Imaoka, I. Date, T. Ohmoto, T. Yasuda, M. Tsuda. In vivo gene transfer into the adult mammalian central nervous system by continuous injection of plasmid DNA–cationic liposome complex. *Brain Research 780*:119 (1998).
58. M. Yokota, E. Tani, S. Tsubuki, I. Yamaura, I. Nakagaki, S. Hori, Saido, TC. Calpain inhibitor entrapped in liposome rescues ischemic neuronal damage. *Brain Research 819*:8 (1999).
59. H. Bell, W. L. Kimber, M. Li, I. R. Whittle. Liposomal transfection efficiency and toxicity on glioma cell lines: in vitro and in vivo studies. *Neuroreport 9*:793 (1998).
60. M. Zucchetti, A. Boiardi, A. Silvani, I. Parisi, S. Piccolrovazzi, M. D'Incalci. Distribution of daunorubicin and daunorubicinol in human glioma tumors after administration of liposomal daunorubicin. *Cancer Chemotherapy and Pharmacology 44*:173 (1999).
61. R. Krishna, L. D. Mayer. Liposomal doxorubicin circumvents PSC 833-free drug interactions, resulting in effective therapy of multidrug-resistant solid tumors. *Cancer Research 57*:5246 (1997).
62. N. K. Jain, A. C. Rana, S. K. Jain. Brain drug delivery system bearing dopamine hydrochloride for effective management of parkinsonism. *Drug Development & Industrial Pharmacy 24*:671 (1998).
63. J. Huwyler, D. Wu, W. M. Pardridge. Brain drug delivery of small molecules using immunoliposomes. *Proceedings of the National Academy of Sciences of the United States of America 93*:14164 (1996).
64. P. L. Chang, J. M. Van Raamsdonk, G. Hortelano, S. C. Barsoum, N. C. MacDonald, T. L. Stockley. The in vivo delivery of heterologous proteins by microencapsulated recombinant cells. *Trends in Biotechnology 17*:78 (1999).
65. D. Hoffman, X. O. Breakefield, M. P. Short, P. Aebischer. Transplantation of a polymer-encapsulated cell line genetically engineered to release NGF. *Experimental Neurology 122*:100 (1993).
66. D. Maysinger, P. Piccardo, A. C. Cuello. Microencap-

sulation and the grafting of genetically transformed cells as therapeutic strategies to rescue degenerating neurons of the CNS. *Reviews in the Neurosciences 6*:15 (1995).

67. J. H. Kordower, S. R. Winn, Y. T. Liu, E. J. Mufson, J. R. J. Sladek, J. P. Hammang, E. E. Baetge, D. F. Emerich. The aged monkey basal forebrain: rescue and sprouting of axotomized basal forebrain neurons after grafts of encapsulated cells secreting human nerve growth factor. *Proceedings of the National Academy of Sciences of the United States of America 91*:10898 (1994).
68. F. Hefti, J. Hartikka, M. Schlumpf. Implantation of PC12 cells into the corpus striatum of rats with lesions of the dopaminergic nigrostriatal neurons. *Brain Research 348*: 283 (1985).
69. D. F. Emerich, J. P. Hammang, E. E. Baetge, S. R. Winn. Implantation of polymer-encapsulated human nerve growth factor–secreting fibroblasts attenuates the behavioral and neuropathological consequences of quinolinic acid injections into rodent striatum. *Experimental Neurology 130*:141 (1994).
70. F. H. Gage. Mammalian neural stem cells. *Science 287*: 1433 (2000).
71. H. A. Cameron, R. McKay. Stem cells and neurogenesis in the adult brain. *Current Opinion in Neurobiology 8*:677 (1998).
72. S. Temple, A. Alvarez-Buylla. Stem cells in the adult mammalian central nervous system. *Current Opinion in Neurobiology 9*:135 (1999).
73. M. Murphy, K. Reid, R. Dutton, G. Brooker, P. F. Bartlett. Neural stem cells. *Journal of Investigative Dermatology Symposium Proceedings 2*:8 (1997).
74. K. I. Park, S. Liu, J. D. Flax, S. Nissim, P. E. Stieg, E. Y. Snyder. Transplantation of neural progenitor and stem cells: developmental insights may suggest new therapies for spinal cord and other CNS dysfunction. *Journal of Neurotrauma 16*:675 (1999).
75. F. H. Gage, P. W. Coates, T. D. Palmer, H. G. Kuhn, L. J. Fisher, J. O. Suhonen, D. A. Peterson, S. T. Suhr, J. Ray. Survival and differentiation of adult neuronal progenitor cells transplanted to the adult brain. *Proceedings of the National Academy of Sciences of the United States of America 92*:11879 (1995).
76. B. A. Reynolds, S. Weiss. Generation of neurons and astrocytes from isolated cells of the adult mammalian central nervous system. *Science 255*:1707 (1992).
77. S. Weiss, B. A. Reynolds, A. L. Vescovi, C. Morshead, C. G. Craig, K. van der. Is there a neural stem cell in the mammalian forebrain? *Trends in Neurosciences 19*:387 (1996).
78. Y. Arsenijevic S. Weiss. Insulin-like growth factor-I is a differentiation factor for postmitotic CNS stem cell–derived neuronal precursors: distinct actions from those of brain-derived neurotrophic factor. *Journal of Neuroscience 18*:2118 (1998).
79. C. N. Svendsen, M. A. Caldwell, T. Ostenfeld. Human neural stem cells: isolation, expansion and transplantation. *Brain Pathology 9*:499 (1999).
80. D. Panchision, T. Hazel, R. McKay. Plasticity and stem cells in the vertebrate nervous system. *Current Opinion in Cell Biology 10*:727 (1998).
81. H. A. Cameron, R. D. McKay. Restoring production of hippocampal neurons in old age. *Nature Neuroscience 2*: 894 (1999).
82. O. Brustle, K. N. Jones, R. D. Learish, K. Karram, K. Choudhary, O. D. Wiestler, I. D. Duncan, R. D. McKay. Embryonic stem cell–derived glial precursors: a source of myelinating transplants. *Science 285*:754 (1999).
83. M. Barinaga. Fetal neuron grafts pave the way for stem cell therapies. *Science 287*:1421 (2000).
84. M. S. Frankel. In search of stem cell policy. *Science 287*: 1397 (2000).
85. E. Marshall. The business of stem cells. *Science 287*:1419 (2000).
86. N. Lenoir. Europe confronts the embryonic stem cell research challenge. *Science 287*:1425 (2000).
87. I. L. Weissman. Translating stem and progenitor cell biology to the clinic: barriers and opportunities. *Science 287*: 1442 (2000).
88. C. N. Svendsen, A. G. Smith. New prospects for human stem-cell therapy in the nervous system. *Trends in Neurosciences 22*:357 (1999).
89. G.-W. Lu, S.-S. Jiao, G.-F. Zhang. Morphological evidence for newly discovered double projection spinal neurons. *Neuroscience Letters 93*:181 (1988).
90. D. van der Kooy, S. Weiss. Why stem cells? *Science 287*: 1439 (2000).
91. V. Y. Alakhov, E. Y. Moskaleva, E. V. Batrakova, A. V. Kabanov. Hypersensitization of multidrug resistant human ovarian carcinoma cells by pluronic P85 block copolymer. *Bioconjugate Chemistry 7*:209 (1996).
92. K. Kataoka, A. Harada, Y. Nagasaki. Block copolymer micelles for drug delivery: design, characterization and biological significance. *Advanced Drug Delivery Reviews 47*:113 (2001).
93. L. Zhang, C. Bartels, Y. Yu, H. Shen, A. Eisenberg. Mesosized crystal-like structure of hexagonally packed hollow hoops by solution self assembly of diblock copolymers. *Physical Review Letters 79*:5034 (1997).
94. Y. Yu, A. Eisenberg. Control of morphology through polymer–solvent interactions in crew-cut aggregates of amphiphilic block copolymers. *Journal of the American Chemical Society 119*:8383 (1997).
95. L. Zhang, A. Eisenberg. Multiple morphologies of ‘‘crew-cut’’ aggregates of polystyrene-b-poly(acrylic acid) block copolymers. *Science 118*:3168 (1995).
96. C. Allen, Y. Yu, D. Maysinger, A. Eisenberg. Polycaprolactone-b-poly(ethylene oxide) block copolymer micelles as a novel drug delivery vehicle for neurotrophic agents FK506 and L-685,818. *Bioconjugate Chemistry 9*:564 (1998).
97. J. Zhao, C. Allen, A. Eisenberg. Partitioning of pyrene between ‘‘crew-cut’’ block copolymer micelles and H_2O/DMF solvent mixtures. *Macromolecules 30*:7143 (1997).
98. C. Allen, D. Maysinger, A. Eisenberg. Nano-engineering

block copolymer aggregates for drug delivery. *Colloids and Surfaces B: Biointerfaces 16*:3 (1999).

99. C. Allen, J. N. Han, Y. S. Yu, D. Maysinger, A. Eisenberg. d. Drug, Dihydrotestosterone, c. m. Block, and Polycaprolactone, Polycaprolactone-b-poly(ethylene oxide) copolymer micelles as a delivery vehicle for dihydrotestosterone. *Journal of Controlled Release 63*:275 (2000).
100. M. Yokoyama, M. Miyauchi, N. Yamada, T. Okano, Y. Sakurai, K. Kataoka, S. Inoue. Characterization and anticancer activity of the micelle-forming polymeric anticancer drug adriamycin-conjugated poly(ethylene glycol)-poly(aspartic acid) block copolymer. *Cancer Research 50*: 1693 (1990).
101. Y. I. Jeong, J. B. Cheon, S. H. Kim, J. W. Nah, Y. M. Lee, Y. K. Sung, T. Akaike, C. S. Cho. Clonazepam release from core-shell type nanoparticles in vitro. *Journal of Controlled Release 51*:169 (1998).
102. S. Katayose, K. Kataoka. Remarkable increase in nuclease resistance of plasmid DNA through supramolecular assembly with poly(ethylene glycol) poly(L-lysine) block copolymer. *Journal of Pharmaceutical Sciences 87*:160 (1998).
103. C. M. Ward, K. D. Fisher, L. W. Seymour. Turbidometric analysis of polyelectrolyte complexes formed between poly(L-lysine) and DNA. *Colloids and Surfaces B: Biointerfaces 16*:253 (1999).
104. R. C. Carlisle, M. L. Read, M. A. Wolfert, L. W. Seymour. Self-assembling poly(L-lysine)/DNA complexes capable of integrin-mediated cellular uptake and gene expression. *Colloids and Surfaces B: Biointerfaces 16*:261 (1999).
105. A. V. Kabanov, V. A. Kabanov. Interpolyelectrolyte and block complexes for gene delivery—physicochemical aspects. *Advanced Drug Delivery Reviews 30*:49 (1998).
106. D. W. Miller, E. V. Batrakova, T. O. Waltner, V. Y. Alakhov, A. V. Kabanov. Interactions of pluronic block copolymers with brain microvessel endothelial cells—evidence of two potential pathways for drug absorption. *Bioconjugate Chemistry 8*:649 (1997).
107. A. V. Kabanov, V. P. Chekhonin, V. Y. Alakhov, E. V. Batrakova, A. S. Lebedev, N. S. Melik-Nubarov, S. A. Arzhakov, A. V. Levashov, G. V. Morozov, E. S. Severin. The neuroleptic activity of haloperidol increases after its solubilization in surfactant micelles. Micelles as microcontainers for drug targeting. *FEBS Letters 258*:343 (1989).
108. H. Altinok, S. K. Nixon, P. A. Gorry, D. Attwood, C. Booth, A. Kelarkis, V. Havredaki. Micellisation and gelation of diblock copolymers of ethylene oxide and propylene oxide in aqueous solution, the effect of P-block length. *Colloids and Surfaces B: Biointerfaces 16*:73 (1999).
109. B. M. Discher, Y. Y. Won, D. S. Ege, J. C. Lee, F. S. Bates, D. E. Discher, D. A. Hammer. Polymersomes: tough vesicles made from diblock copolymers. *Science 284*:1143 (1999).
110. W. J. van der Giessen, A. M. Lincoff, R. S. Schwartz, H. M. van Beusekom, P. W. Serruys, D. R. J. Holmes, S. G. Ellis, E. J. Topol. Marked inflammatory sequelae to implantation of biodegradable and nonbiodegradable polymers in porcine coronary arteries. *Circulation 94*: 1690 (1996).
111. J. Lee, H. Lee, J. D. Andrade. Blood compatibility of polyethylene oxide surfaces. *Progress in Polymer Science 20*: 1043 (1995).
112. S. C. Woodward, P. S. Brewer, F. Moatamed, A. Schindler, C. G. Pitt. The intracellular degradation of poly(epsilon-caprolactone). *Journal of Biomedical Materials Research 19*:437 (1985).
113. C. I. Winternitz, J. K. Jackson, A. M. Oktaba, H. M. Burt. Development of a polymeric surgical paste formulation for taxol. *Pharmaceutical Research 13*:368 (1996).
114. J. K. Jackson, W. Min, T. F. Cruz, S. Cindric, L. Arsenault, D. D. Von Hoff, D. Degan, W. L. Hunter, H. M. Burt. A polymer-based drug delivery system for the antineoplastic agent bis(maltolato)oxovanadium in mice. *British Journal of Cancer 75*:1014 (1997).
115. P. Calvo, M. J. Alonso, J. L. Vila-Jato, J. R. Robinson. Improved ocular bioavailability of indomethacin by novel ocular drug carriers. *Journal of Pharmacy and Pharmacology 48*:1147 (1996).
116. D. L. Elbert, J. A. Hubbell. Surface treatments of polymers for biocompatibility. *Annual Review of Materials Science 26*:365 (1996).
117. R. Nagarajan. Solubilization of hydrocarbons and resulting aggregate shape transitions in aqueous solutions of Pluronic (PEO-PPO-PEO) block copolymers. *Colloids and Surfaces B: Biointerfaces 16*:55 (1999).
118. N. Rapoport. Stabilization and activation of Pluronic micelles for tumor-targeted drug delivery. *Colloids and Surfaces B:Biointerfaces 16*:93 (1999).
119. V. Alakhov, E. Klinski, S. Li, G. Pietrzynski, A. Venne, E. Batrakova, T. Bronitch, A. Kabanov. Block copolymer-based formulation of doxorubicin. From cell screen to clinical trials. *Colloids and Surfaces B: Biointerfaces 16*: 113 (1999).
120. T. K. Bronich, A. Nehls, A. Eisenberg, V. A. Kabanov, A. V. Kabanov. Novel drug delivery systems based on the complexes of block ionomers and surfactants of opposite charge. *Colloids and Surfaces B: Biointerfaces 16*:243 (1999).
121. Y. Yamamoto, Y. Nagasaki, M. Kato, K. Kataoka. Surface charge modulation of poly(ethylene glycol)-poly(D,L-lactide) block copolymer micelles: conjugation of charged particles. *Colloids and Surfaces B: Biointerfaces 16*:135 (1999).
122. F. Kohori, K. Sakai, T. Aoyagi, M. Yokoyama, M. Yamato, Y. Sakurai, T. Okano. Control of adriamycin cytotoxic activity using thermally responsive polymeric micelles composed of poly(N-isopropylacrylamide-co-N,N-dimethylacrylamide)-b-poly(D,L-lactide). *Colloids and Surfaces B: Biointerfaces 16*:195 (1999).
123. B. Jeong, Y. H. Bae, S. W. Kim. Biodegradable thermosensitive micelles of PEG-PLGA-PEG triblock copolymers. *Colloids and Surfaces B: Biointerfaces 16*:185 (1999).
124. C. Allen, J. Han, Y. Yu, D. Maysinger, A. Eisenberg. Po-

lycaprolactone-b-poly(ethylene oxide) copolymer micelles as a deliver vehicle for dihydrotestosterone. *Journal of Controlled Release 63*:275 (2000).

125. K. B. Thurmond II, H. Huang, C. G. Clark, Jr., T. Kowalewski, K. L. Wooley. Shell cross-linked polymer micelles: stabilized assemblies with great versatility and potential. *Colloids and Surfaces B: Biointerfaces 16*:45 (1999).
126. A. Maruyama, Y. Ohnishi, H. Watanabe, H. Torigoe, A. Ferdous, T. Akaike. Polycation comb-type copolymer reduces counterion condensation effect to stabilize DNA duplex and triplex formation. *Colloids and Surfaces B: Biointerfaces 16*:273 (1999).
127. D. W. Miller, A. V. Kabanov. Potential applications of polymers in the delivery of drugs to the central nervous system. *Colloids and Surfaces B: Biointerfaces 16*:321 (1999).
128. H. Huai-Yun, D. T. Secrest, K. S. Mark, D. Carney, C. Brandquist, W. F. Elmquist, D. W. Miller. Expression of multidrug resistance–associated protein (MRP) in brain microvessel endothelial cells. *Biochemical and Biophysical Research Communications 243*:816 (1998).
129. P. Lemieux, S. V. Vinogradov, C. L. Gebhart, N. Guerin, G. Paradis, H. K. Nguyen, B. Ochietti, Y. G. Suzdaltseva, E. V. Bartakova, T. K. Bronich, Y. St Pierre, V. Yu. Alakhov, A. V. Kabanov. Block and graft copolymers and Nanogel™ Copolymer networks for DNA delivery into cell. *Drug Targeting 8*:91 (2000).
130. E. V. Batrakova, H. Y. Han, V. Y. Alakhov, D. W. Miller, A. V. Kabanov. Effects of pluronic block copolymers on drug absorption in Caco-2 cell monolayers. *Pharmaceutical Research 15*:850 (1998).
131. M. Ramaswamy, X. Zhang, H. M. Burt, K. M. Wasan. Human plasma distribution of free paclitaxel and paclitaxel associated with diblock copolymers. *Journal of Pharmaceutical Sciences 86*:460 (1997).
132. D. Kobayashi, S. Tsubuku, H. Yamanaka, M. Asano, M. Miyajima, M. Yoshida. In vivo characteristics of injectable poly(D,L-lactic acid) microspheres for long-acting drug delivery. *Drug Development and Industrial Pharmacy 24*:819 (1998).
133. W. P. Ye, Y. W. Chien. Dual-controlled drug delivery across biodegradable copolymer. I. Delivery kinetics of levonorgestrel and estradiol through (caprolactone/lactide) block copolymer. *Pharmaceutical Development and Technology 1*:1 (1996).
134. W. Cavett, Z. Cason, M. Tucci, A. Puckett, H. Benghuzzi. The synergistic effect of sustained delivery of DHT, DHEA, and E on the ventral prostate of adult male rodents. *Biomedical Sciences Instrumentation 34*:30 (1997).
135. J. M. Kovarik, E. A. Mueller, J. B. van Bree, S. S. Fluckiger, H. Lange, Schmidt, W. H. Boesken, A. E. Lison, K. Kutz. Cyclosporine pharmacokinetics and variability from a microemulsion formulation—a multicenter investigation in kidney transplant patients. *Transplantation 58*:658 (1994).
136. A. M. Gewirtz, C. A. Stein, P. M. Glazer. Facilitating oligonucleotide delivery: helping antisense deliver on its promise. *Proceedings of the National Academy of Sciences of the United States of America 93*:3161 (1996).
137. P. S. Eder, R. J. DeVine, J. M. Dagle, J. A. Walder. Substrate specificity and kinetics of degradation of antisense oligonucleotides by a 3′ exonuclease in plasma. *Antisense Research and Development 1*:141 (1991).
138. L. Yakubov, Z. Khaled, L. M. Zhang, A. Truneh, V. Vlassov, C. A. Stein. Oligodeoxynucleotides interact with recombinant CD4 at multiple sites. *Journal of Biological Chemistry 268*:18818 (1993).
139. J. R. Perez, Y. Li, C. A. Stein, S. Majumder, A. van Oorschot, R. Narayanan. Sequence-independent induction of Sp1 transcription factor activity by phosphorothioate oligodeoxynucleotides. *Proceedings of the National Academy of Sciences of the United States of America 91*:5957 (1994).
140. M. A. Guvakova, L. A. Yakubov, I. Vlodavsky, J. L. Tonkinson, C. A. Stein. Phosphorothioate oligodeoxynucleotides bind to basic fibroblast growth factor, inhibit its binding to cell surface receptors, and remove it from low affinity binding sites on extracellular matrix. *Journal of Biological Chemistry 270*:2620 (1995).
141. C. A. Stein, Y. C. Cheng. Antisense oligonucleotides as therapeutic agents—is the bullet really magical? *Science 261*:1004 (1993).
142. W. C. Broaddus, S. S. Pabhu, S. Wu Pong, G. T. Gillies, H. Fillmore. Strategies for the design and delivery of antisense oligonucleotides in central nervous system. *Methods in Enzymology 314*:121 (2000).
143. S. Agrawal, J. Temsamani, J. Y. Tang. Pharmacokinetics, biodistribution, and stability of oligodeoxynucleotide phosphorothioates in mice. *Proceedings of the National Academy of Sciences of the United States of America 88*:7595 (1991).
144. H. Sands, L. J. Gorey-Feret, A. J. Cocuzza, F. W. Hobbs, D. Chidester, G. L. Trainor. Biodistribution and metabolism of internally 3H-labeled oligonucleotides. I. Comparison of a phosphodiester and a phosphorothioate. *Molecular Pharmacology 45*:932 (1994).
145. K. J. Myers, N. M. Dean. Sensible use of antisense: how to use oligonucleotides as research tools. *Trends in Pharmacological Sciences 21*:19 (2000).
146. C. Beltinger, H. U. Saragovi, R. M. Smith, L. LeSauteur, N. Shah, DeDionisio, L. Christensen, A. Raible, L. Jarett, A. M. Gewirtz. Binding, uptake, and intracellular trafficking of phosphorothioate-modified oligodeoxynucleotides. *Journal of Clinical Investigation 95*:1814 (1995).
147. L. A. Yakubov, E. A. Deeva, V. F. Zarytova, E. M. Ivanova, A. S. Ryte, L. V. Yurchenko, V. V. Vlassov. Mechanism of oligonucleotide uptake by cells: involvement of specific receptors? *Proceedings of the National Academy of Sciences of the United States of America 86*:6454 (1989).
148. G. Marti, W. Egan, P. Noguchi, G. Zon, M. Matsukura, S. Broder. Oligodeoxyribonucleotide phosphorothioate fluxes and localization in hematopoietic cells. *Antisense Research and Development 2*:27 (1992).

149. R. M. Crooke. In vitro toxicology and pharmacokinetics of antisense oligonucleotides. *Anti-Cancer Drug Design 6*:609 (1991).
150. W. M. Galbraith, W. C. Hobson, P. C. Giclas, P. J. Schechter, S. Agrawal. Complement activation and hemodynamic changes following intravenous administration of phosphorothioate oligonucleotides in the monkey. *Antisense Research and Development 4*:201 (1994).
151. S. P. Henry, J. Taylor, L. Midgley, A. A. Levin, D. J. Kornbrust. Evaluation of the toxicity of ISIS 2302, a phosphorothioate oligonucleotide, in a 4-week study in CD-1 mice. *Antisense and Nucleic Acid Drug Development 7*:473 (1997).
152. J. P. Clarenc, G. Degols, J. P. Leonetti, P. Milhaud, B. Lebleu. Delivery of antisense oligonucleotides by poly(L-lysine) conjugation and liposome encapsulation. *Anti-Cancer Drug Design 8*:81 (1993).
153. O. Boussif, F. Lezoualc'h, M. A. Zanta, M. D. Mergny, D. Scherman, B. Demeneix, J. P. Behr. A versatile vector for gene and oligonucleotide transfer into cells in culture and in vivo: polyethylenimine. *Proceedings of the National Academy of Sciences of the United States of America 92*:7297 (1995).
154. C. F. Bennett, M. Y. Chiang, H. Chan, J. E. Shoemaker, C. K. Mirabelli. Cationic lipids enhance cellular uptake and activity of phosphorothioate antisense oligonucleotides. *Molecular Pharmacology 41*:1023 (1992).
155. J. G. Lewis, K. Y. Lin, A. Kothavale, W. M. Flanagan, M. D. Matteucci, R. B. DePrince, R. A. J. Mook, R. W. Hendren, R. W. Wagner. A serum-resistant cytofectin for cellular delivery of antisense oligodeoxynucleotides and plasmid DNA. *Proceedings of the National Academy of Sciences of the United States of America 93*:3176 (1996).
156. S. Wang, R. J. Lee, G. Cauchon, D. G. Gorenstein, P. S. Low. Delivery of antisense oligodeoxyribonucleotides against the human epidermal growth factor receptor into cultured KB cells with liposomes conjugated to folate via polyethylene glycol. *Proceedings of the National Academy of Sciences of the United States of America 92*:3318 (1995).
157. G. Citro, D. Perrotti, C. Cucco, I. D'Agnano, A. Sacchi, G. Zupi, B. Calabretta. Inhibition of leukemia cell proliferation by receptor-mediated uptake of c-myb antisense oligodeoxynucleotides. *Proceedings of the National Academy of Sciences of the United States of America 89*:7031 (1992).
158. G. Y. Wu, C. H. Wu. Specific inhibition of hepatitis B viral gene expression in vitro by targeted antisense oligonucleotides. *Journal of Biological Chemistry 267*:12436 (1992).
159. N. H. Ing, J. M. Beekman, D. J. Kessler, M. Murphy, K. Jayaraman, J. G. Zendegui, M. E. Hogan, B. W. O'Malley, M. J. Tsai. In vivo transcription of a progesterone-responsive gene is specifically inhibited by a triplex-forming oligonucleotide. *Nucleic Acids Research 21*:2789 (1993).
160. J. P. Bongartz, A. M. Aubertin, P. G. Milhaud, B. Lebleu. Improved biological activity of antisense oligonucleotides conjugated to a fusogenic peptide. *Nucleic Acids Research 22*:4681 (1994).
161. S. Agrawal. Antisense oligonucleotides: towards clinical trials. *Trends in Biotechnology 14*:376 (1996).
162. S. Vinogradov, E. Batrakova, A. Kabanov. Poly(ethylene glycol)-polyethyleneimine NanoGel™ particles: novel drug delivery systems for antisense oligonucleotides. *Colloids and Surfaces B: Biointerfaces 16*:291 (1999).
163. S. Vinogradov, E. Batrakova, S. Li, A. Kabanov. Polyion complex micelles with protein-modified corona for receptor-mediated delivery of oligonucleotides into cells. *Bioconjugate Chemistry 10*:851 (1999).
164. F. L. Graham, A. J. Eb. A new technique for the assay of infectivity of human adenovirus 5 DNA. *Virology 52*:456 (1973).
165. C. D. Thompson, M. R. Frazier-Jessen, R. Rawat, R. P. Nordan, R. T. Brown. Evaluation of methods for transient transfection of a murine macrophage cell line, RAW 264.7. *BioTechniques 27*:824 (1999).
166. J. H. Lee, M. J. Welsh. Enhancement of calcium phosphate-mediated transfection by inclusion of adenovirus in coprecipitates. *Gene Therapy 6*:676 (1999).
167. A. Haberland, T. Knaus, S. V. Zaitsev, R. Stahn, A. R. Mistry, C. Coutelle, H. Haller, M. Bottger. Calcium ions as efficient cofactor of polycation-mediated gene transfer. *Biochimica et Biophysica Acta 1445*:21 (1999).
168. S. Y. Watanabe, A. M. Albsoul-Younes, T. Kawano, H. Itoh, Y. Kaziro, S. Nakajima, Y. Nakajima. Calcium phosphate–mediated transfection of primary cultured brain neurons using GFP expression as a marker: application for single neuron electrophysiology. *Neuroscience Research 33*:71 (1999).
169. J. A. Nickoloff, L. N. Spirio, R. J. Reynolds. A comparison of calcium phosphate coprecipitation and electroporation. Implications for studies on the genetic effects of DNA damage. *Molecular Biotechnology 10*:93 (1998).
170. A. Watson, D. Latchman. Gene delivery into neuronal cells by calcium phosphatase–mediated transfection. *Methods: A Companion to Methods in Enzymology 10*:289 (1996).
171. R. A. Aubin, M. Weinfeld, R. Mirzayans, M. C. Paterson. Polybrene/DMSO-assisted gene transfer. Generating stable transfectants with nanogram amounts of DNA. *Molecular Biotechnology 1*:29 (1994).
172. H. Potter, L. Weir, P. Leder. Enhancer-dependent expression of human kappa immunoglobulin genes introduced into mouse pre-B lymphocytes by electroporation. *Proceedings of the National Academy of Sciences of the United States of America 81*:7161 (1984).
173. M. R. Capecchi. High efficiency transformation by direct microinjection of DNA into cultured mammalian cells. *Cell 22*:479 (1980).
174. A. G. Gitman, A. Graessmann, A. Loyter. Targeting of loaded Sendai virus envelopes by covalently attached insulin molecules to virus receptor-depleted cells: fusion-mediated microinjection of ricin A and simian virus 40

DNA. *Proceedings of the National Academy of Sciences of the United States of America 82*:7309 (1985).

175. E. Gilboa, M. A. Eglitis, P. W. Kantoff, W. French Anderson. Transfer and expression of cloned genes using retroviral vectors. *BioTechniques 4*:504 (1986).
176. K. M. L. Gaensler, G. Tu, S. Bruch, D. Liggitt, G. S. Lipshutz, A. Metkus, M. Harrison, T. D. Heath, R. J. Debs. Fetal gene transfer by transuterine injection of cationic liposome–DNA complexes. *Nature Biotechnology 17*:1188 (1999).
177. R. Fraley, R. M. Straubinger, G. Rule, E. L. Springer, D. Papahadjopoulos. Liposome-mediated delivery of deoxyribonucleic acid to cells: enhanced efficiency of delivery related to lipid composition and incubation conditions. *Biochemistry 20*:6978 (1981).
178. J. Y. Legendre, F. C. Szoka, Jr. Delivery of plasmid DNA into mammalian cell lines using pH-sensitive liposomes: comparison with cationic liposomes. *Pharmaceutical Research 9*:1235 (1992).
179. A. V. Kabanov, V. A. Kabanov. DNA complexes with polycations for the delivery of genetic material into cells. *Bioconjugate Chemistry 6*:7 (1995).
180. A. A. Yaroslavov, E. G. Yaroslavova, A. A. Rakhnyanskaya, F. M. Menger, V. A. Kabanov. Modulation of interaction of polycations with negative unilamellar lipid vesicles. *Colloids and Surfaces B: Biointerfaces 16*:29 (1999).
181. K. B. Thurmond II, H. Y. Huang, C. G. Clark, Jr., T. Kowalewski, K. L. Wooley. Shell cross-linked polymer micelles: stabilized assemblies with great versatility and potential. *Colloids and Surfaces B: Biointerfaces 16*:45 (1999).
182. C. Chen, H. Okayama. High-efficiency transformation of mammalian cells by plasmid DNA. *Molecular and Cellular Biology 7*:2745 (1987).
183. C. H. Wu, J. M. Wilson, G. Y. Wu. Targeting genes: delivery and persistent expression of a foreign gene driven by mammalian regulatory elements in vivo. *Journal of Biological Chemistry 264*:16985 (1989).
184. G. Y. Wu, J. M. Wilson, F. Shalaby, M. Grossman, D. A. Shafritz, C. H. Wu. Receptor-mediated gene delivery in vivo. Partial correction of genetic analbuminemia in Nagase rats. *Journal of Biological Chemistry 266*:14338 (1991).
185. L. F. Cotlin, M. A. Siddiqui, F. Simpson, J. F. Collawn. Casein kinase II activity is required for transferrin receptor endocytosis. *Journal of Biological Chemistry 274*:30550 (1999).
186. V. S. Trubetskoy, V. P. Torchilin, S. Kennel, L. Huang. Cationic liposomes enhance targeted delivery and expression of exogenous DNA mediated by N-terminal modified poly(L-lysine)-antibody conjugate in mouse lung endothelial cells. *Biochimica et Biophysica Acta 1131*:311 (1992).
187. I. Y. Galaev, B. Mattiasson. 'Smart' polymers and what they could do in biotechnology and medicine. *Trends in Biotechnology 17*:335 (1999).
188. S. J. Lee, K. Park. *Hydrogels and Biodegradable Polymers for Bioapplications* (R. M. Ottenbrite, S. J. Huang, K. Park, eds.), American Chemical Society, Washington, D.C., 1996, pp. 1–268.

43

Glucose-Mediated Insulin Delivery from Implantable Polymers

Larry R. Brown
Harvard–Massachusetts Institute of Technology, Cambridge, Massachusetts

I. INTRODUCTION

In 1976, Langer and Folkman pioneered the sustained release of protein molecules from biocompatible hydrophobic polymers (1). Further work by Rhine et al. refined the techniques for fabricating polymer matrices for the sustained release of very water soluble proteins such as albumin (2). These polymeric delivery systems were ultimately used to deliver therapeutic proteins such as insulin (3–6). However, insulin represented a somewhat different macromolecule from those that had been studied before. Insulin's aqueous solubility was much lower than that of albumin, and its release kinetics were slower than what had been observed for more water soluble proteins like albumin. Thus the research efforts in developing a sustained release insulin delivery system became a study in the solubility properties of insulin.

Ultimately, it became apparent that altering insulin's solubility within the polymer matrix could be used to alter the insulin release in response to glucose concentration changes. It was determined that changes in the solubility of insulin as a function of pH could be used to control and alter the delivery of insulin. The pH dependent solubility properties of insulin were then used to develop a polymeric glucose mediated insulin delivery system. An understanding of the interrelationship between insulin and its solubility properties is necessary to appreciate fully the development of a pH controlled insulin delivery system. Therefore the topics covered in this article include

1. Insulin's role in physiology
2. Insulin's structural and physical-chemical elements
3. Sustained insulin release from ethylene vinyl acetate copolymer matrices
4. A glucose mediated insulin delivery implant

II. INSULIN'S ROLE IN PHYSIOLOGY

Insulin is a hormone produced by the beta cells of the endocrine pancreas. The hormone plays a major role in the metabolism and absorption of carbohydrate, protein, and fat. It is especially important in enabling the transport of d-glucose by cells for energy and the transport of amino acids for protein synthesis. It also increases glycogen synthase activity, inhibits lipolysis, and inhibits proteolysis. Insulin dependent diabetes mellitus occurs when the beta cells of the pancreas do not make sufficient insulin to meet the needs of the organism. Given these diverse insulin functions, it is not surprising to discover that the secretion of insulin from the beta cell islets varies in response to many metabolic events.

When insulin was first discovered by Frederick Banting and Charles Best in 1921 it was not fully appreciated that insulin needed to be delivered in pulses in response to carbohydrate laden meals, and in an extended release form in order to mimic basal insulin delivery (7). Thus one sees an increase in plasma insulin concentration at meal times, which returns to a lower basal concentration in between

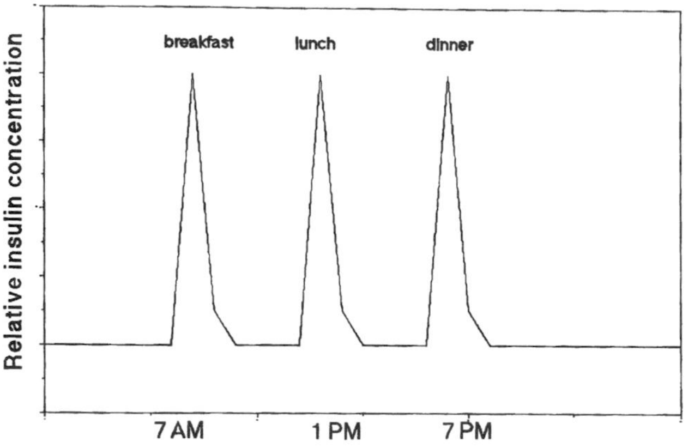

Figure 43.1 An idealized representation of the relative concentration of circulating insulin is depicted over the course of a day. Pulses of insulin are observed in response to meals, which is superimposed on basal insulin secretion.

meals. The focus in insulin delivery research has been to develop the systems and means that match the physiologic plasma insulin concentration profiles seen in Fig. 1.

III. INSULIN'S STRUCTURAL AND PHYSICAL-CHEMICAL ELEMENTS

A brief review of the structural, physical, and chemical properties of insulin is presented as a background for the solubility dependent release properties of the molecule from polymeric matrix implants.

A. Insulin Structure

Insulin is a polypeptide hormone produced by cells in the islets of Langerhans in the pancreas. It is synthesized as proinsulin shown in Fig. 2. Proinsulin is composed of 84 amino acids, of which 33 are cleaved as a posttranslational modification of the synthesis of insulin (8–13).

This 33 amino acid peptide that is cleaved is known as the connecting peptide or C-peptide (14). The C-peptide is excreted into the urine, and its presence in the urine can be used to determine if the organism is synthesizing any endogenous insulin.

Two amino acid peptide chains remain after the cleavage of the C-peptide chain shown in Fig. 2. Chain A consists of 21 amino acids, and Chain B consists of 30 amino acids. Fig. 3 shows the primary amino acid structure of human insulin. Insulin is a polypeptide hormone with a molecular weight of approximately 5800 daltons in its monomeric form (17). However, it is often found in its dimeric or hexameric forms. There is remarkable homology between insulin of different species. Historically, insulin for human use was extracted from beef and pork pancreases. Today, recombinant human insulin has largely replaced animal source insulin in the United States and Europe. Table 1 compares the differences between human,

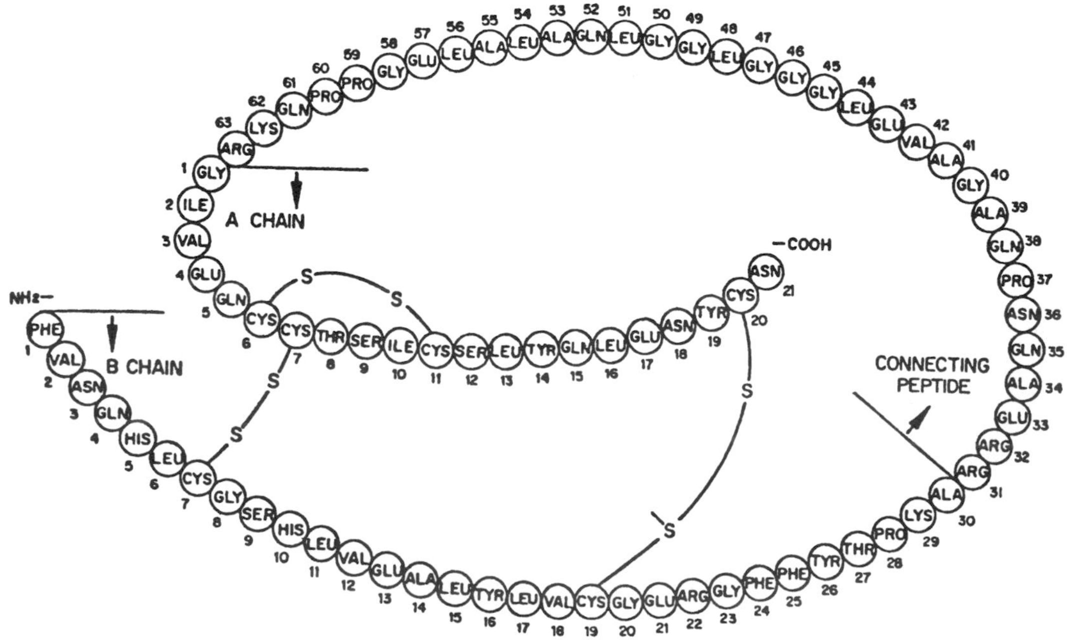

Figure 43.2 The complete pork proinsulin structure shows the A, B, and C-peptide chains of insulin. There is minimal biological activity associated with proinsulin compared to that of insulin. (From Refs. 15 and 16.) (Reprinted with permission from Lippincott Williams and Wilkins.)

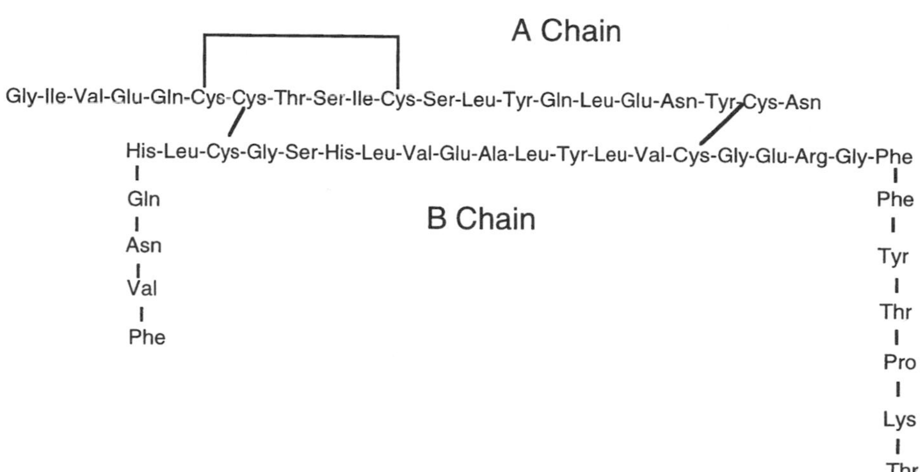

Figure 43.3 The primary amino acid structure of A and B chains of human insulin is shown along with the two disulfide linkages connecting the A-Cys(7) to B-Cys(7) and A-Cys(20) to B-Cys(19). A third intrachain disulfide linkage is found between A-Cys(6) and A-Cys(11).

beef, and pork insulin. Pork insulin differs from human insulin by one amino acid and from beef insulin by two amino acids. As a result of this homology, there is no appreciable difference in biological activity between insulin forms derived from different species.

B. Fast Acting Insulin Forms

Recent developments have led to subtle changes in insulin's primary structure which yield insulin forms with unique pharmacokinetic properties that do not alter the biological activity of these new insulin forms. For example, on the B-chain of insulin, the number 28 and 29 amino acids are Proline(28) and Lysine(29). Insulin forms have been shown to have greater aqueous solubility when the order of Lysine and Proline in the B-chain is reversed. Insulins have been engineered so that now Lysine is the number 28 amino acid and Proline is the number 29 amino acid. The increase in Lys(28)Pro(29)-insulin's aqueous solubility has been attributed to a 300-fold inhibition of hexameric insulin unit formation compared to regular insulin. Thus Lys(28)Pro(29)-insulin tends to exist in the monomer form, and this increase in insulin solubility enables fast absorption from an injection site. This results in a faster onset of insulin action compared to regular insulin (18). Lys(28)Pro(29) insulin is commercially known as Humalog®. There are other insulin analogues that have been developed to exhibit a faster onset of action through the same basic mechanism. These analogues include Asp (B10)Lys(B28)Pro(B29)-human insulin and Insulin Aspart [Asp(B28)]-insulin (19). Furthermore, the decreased tendency of these insulin analogues to associate into dimers and hexamers has been attributed to its decreased tendency to be coordinated with divalent zinc ions.

Table 43.1 Primary Structure Amino Acid Variations Between Insulin of Different Species

Insulin source	Chain A—position 8	Chain B—position 30
Human	Thr	Thr
Pork	Thr	Ala
Beef	Ala	Ala

C. Physical-Chemical Elements

The solubility of insulin is strongly influenced by its interaction with zinc and the pH of the aqueous solutions in which it is dissolved. These factors and others have been used to develop new insulin forms that either are faster acting or possess extended release properties.

1. The Role of Zinc in Insulin Solubility

As previously described, the A and B chains of insulin are organized in a hexameric structure when in the presence of divalent zinc anions (Zn^{++}) and in a pH neutral solution. In this hexameric structure, three insulin dimers are held together by two Zn^{++} ions. Figure 4 shows a three-dimensional depiction of the zinc anions in association with an insulin dimer.

The presence of these zinc atoms profoundly affects the pH solubility profile of insulin. Removal of the zinc from the insulin molecule by chelating it with EDTA yields an

Figure 43.4 Three-dimensional depiction of an insulin dimer in the presence of zinc. The darker gray chain is the A chain of insulin and the helical chain is the B chain. The A and B chains are chemically bonded with two disulfide linkages. Solutions of bovine insulin have been shown to contain two molecules of zinc (Zn^{++}) ions per each six base-molecule of insulin. (From Refs. 23 and 24.)

insulin whose water solubility is approximately 100-fold greater than zinc containing insulin (4).

2. pH and Insulin Delivery

There have been numerous efforts to correlate exogenous insulin action with physiologic action and delivery in order to achieve the goal of normalized blood glucose levels. Initially, investigators attempted to prolong the 2–4 hour regular crystalline insulin action so that a 24 hour period could be covered by a single insulin injection. Table 2 outlines several commercial insulin types and their pharmacodynamics in terms of peak action and duration of action.

Regular crystalline insulin has an isoelectric point of approximately 5.3. That is its point of electroneutrality and of lowest solubility in aqueous solution occurs at about pH 5.3. At the physiologic pH of 7.4, insulin's aqueous solubility is high enough to result in a relatively short duration of action due to its complete dissolution and absorption from the site of injection.

Table 43.2 Insulin Type and Pharmacodynamics

Insulin type	Onset (hours)	Peak action (hours)	Duration (hours)
Humalog®	0.25	0.5–1.5	3–5
Regular	0.5	2–4	6–8
Lantus®	2–4	4–24	24
NPH®	2–4	6–8	12–15
Lente®	1–3	6–12	18–26
UltraLente®	4–8	12–18	22–30

3. Early Efforts to Prolong Insulin Action

Soon after the discovery of insulin, researchers tried to increase its half-life by decreasing insulin's solubility in the body. Lewis (23) reported unsuccessful attempts to sustain insulin release by injecting the hormone into oily suspensions, acacia solutions, and lecithin solutions. Other investigators attempted to administer insulin in the presence of epinephrine, posterior pituitary extracts, and astringent metals. These attempts also proved to be unsuccessful. Hagedorn (24) showed that insulin combined with protamine would allow continuous action for 3–12 hours. Protamine insulin actually has an isoelectric point that is near physiologic pH (pI $\sim$ 7.3), and therefore it dissolves slowly following subcutaneous injection. This was a significant advance toward a more continuous insulin release. However, protamine insulin proved to be unstable. Scott and Fischer (25) therefore developed protamine zinc insulin, which overcame the instability problem of protamine insulin. It provided a daily requirement of insulin in a single injection. The preparation involved combining 0.20 mg zinc and 1.5 mg protamine per 100 units of insulin. The onset of activity of protamine zinc insulin is 6–8 hours after injection, and it lasts for 24 hours. When used in conjunction with rapidly acting insulin preparations, this provided a more effective method of continuous insulin availability throughout the day for the diabetic (26).

Other insulin preparations include globin zinc insulin, which is an intermediate acting insulin that can normalize glucose levels within 2 hours of injection and maintain them for 16 hours. Isophane zinc insulin is modified protamine zinc insulin consisting of 0.05 mg protamine per 100 units of insulin. Its action is similar to globin insulin with an onset of action within 2 hours and lasting for 20 hours (27).

Lente® insulin was developed in 1952 by Hallas Moller (28). This insulin begins to take advantage of the interaction of zinc and insulin. Lente® insulin is a combination of zinc and insulin in an acetate buffer. The duration of Lente® insulin action can be varied by simply modifying particle size. SemiLente® insulin consists of amorphous insulin particles and has 18 hours activity. UltraLente® is made of large particles with a 30 hour duration of activity. Lente® insulin, which contains 3 parts SemiLente® insulin and 7 parts UltraLente® insulin, lasts 24 hours. Its longevity is due to the large insulin crystals and its neutrality of suspension (pH 7.2), which retard its absorption.

Although these insulin preparations provide an "insulin presence" in the blood of a diabetic for a 24 hour period, their action is neither physiologic nor at a constant basal level when not operating at peak activity. The final 4–6 hours of these 24 hour preparations are often marked by insufficient insulin concentrations in the blood to maintain normal blood glucose levels throughout the night. Elevated fasting blood glucose levels are often seen in diabetics maintained by this regimen. Complete reduction in plasma glucose levels by a single subcutaneous injection of rapid or intermediate acting insulin is virtually never achieved (29,30), and the long-term complications of the disease remain a serious threat to the normal life of the diabetic.

The Diabetes Control and Complications Trial (DCCT) was a multicenter, randomized, clinical study designed to determine whether an intensive treatment regimen directed at maintaining blood glucose concentrations as close to normal as possible will affect the appearance or progression of early vascular complications in patients with insulin dependent diabetes. As a result of the DCCT study, it was clearly shown that good metabolic control of the insulin dependent diabetic resulted in a significant reduction of the secondary complications of diabetes (31). Improved metabolic control by continuous insulin delivery has been achieved by the use of externally worn miniature infusion pumps (32) and with new genetically engineered insulin forms.

4. A New Insulin Analogue for Longer Term Delivery

Recently a new insulin analogue was approved for use in the United States (33). Lantus® is a sterile solution of insulin glargine for use as an injection. Insulin glargine is a 24 hour long acting recombinant human insulin analogue. Insulin glargine differs from human insulin in that the amino acid asparagine at position A21 is replaced by glycine and two arginines are added to the C-terminus of the B-chain. It has a molecular weight of 6063. Insulin glargine is designed to have a low aqueous solubility at neutral pH. It is injected in a pH 4 injection solution, where it is completely soluble. The acidic solution is neutralized after injection into the subcutaneous tissue, which leads to the formation of microprecipitates from which small amounts of insulin glargine are slowly released. This results in a relatively constant concentration–time profile over 24 hours with no pronounced insulin peak. This profile allows once daily dosing as a patient's basal insulin.

IV. SUSTAINED INSULIN RELEASE FROM ETHYLENE VINYL ACETATE COPOLYMER MATRICES

The efforts to develop a feedback controlled, sustained release insulin delivery system began with the observations that proteins could be continually released from polymer matrices (1). However, initial attempts to develop a long-term insulin delivery matrix implant using ethylene vinyl acetate were impeded by two factors. One was the low solubility of insulin which resulted in slow release kinetics (3–5). The second factor was that the polymer matrix geometry employed up to this time resulted in release kinetics that decreased over time (2). Therefore studies were undertaken to address both the marginal solubility of insulin and the nonoptimal insulin release profile from the matrices.

A. Methods and Materials

The following section describes the experimental techniques used to fabricate insulin releasing ethylene vinyl acetate copolymer matrices.

1. Insulin

Dry powdered zinc insulin from a beef source was the starting material used in the fabrication of the delivery devices. Each mg of insulin contained approximately 26 international units of activity.

2. Conversion of Zinc Insulin to Zinc Free Sodium Insulin

Zinc free insulin is much more soluble than crystalline zinc insulin. Therefore zinc insulin was converted to sodium insulin using the following method. Zinc insulin crystals were dissolved at a concentration of 20 mg/mL in distilled water that was adjusted to pH 3.0 with 0.1 M HCl. Sodium EDTA was added to the insulin solution to achieve a final concentration of 0.02 M, and NaCl was also added to yield a final concentration of 3 M NaCl. This insulin solution was adjusted to pH 8.2 with NaOH. NaCl was then added until the solution turned hazy. Liquefied phenol was added to a final concentration of 0.3%. This suspension was stirred for 3 hours. The suspension was centrifuged, and the supernatant was discarded. The resulting pellet was resuspended in a 2% NaCl solution. The aqueous solubility of this zinc free insulin preparation was determined to be 120 mg/mL (4). The insulin was then lyophilized to a dry powdered form.

3. Insulin Polymer Matrix Preparation

Zinc insulin powder or sodium insulin powder was sieved to various particle size ranges through U.S. Standards Sieves with mesh sizes of 75, 250, 300, and 425 μm. Flat-slab insulin containing matrices were prepared by a solvent-casting procedure (2). Washed ethylene vinyl acetate copolymer was dissolved in methylene chloride to form a 10% (wt/vol) solution. Various weighed amounts of insulin powder were added to the ethylene vinyl acetate copolymer solution. The mixture was vortexed for 15 seconds to give a homogeneous suspension of powdered insulin particles in the polymer solution. The suspension was quickly poured from corner to corner onto a leveled square glass mold (5 × 5 × 1.5 cm) that had been precooled on a slab of dry ice for 10 minutes. The frozen slab remained in the mold for 10 minutes. Then the slab was pried loose with a spatula precooled on dry ice, transferred to a nickle-coated wire screen, placed into a −20°C freezer, and allowed to dry for 2 days. The methylene chloride evaporated without significant migration of the insulin particles within the polymer matrix. The flat slab was dried for an additional 2 days at room temperature under mild vacuum (600 millitorr). Individual disks, 5 mm in diameter and 0.8 mm thick, were excised from the resulting slab with a #3 size cork borer.

4. Release Studies

The insulin polymer slabs were placed in release media that generally consisted of 50 mM Tris-HCl buffer adjusted to pH 7.4. The polymer matrices were placed in a 22 mL glass vial containing the release medium. The matrices were transferred to fresh release medium at appropriate time points. The old release medium was then assayed for insulin by ultraviolet absorption at 220 nm (34).

B. Results

The following experimental results show the effect of insulin solubility on release and a discussion of the mechanism of insulin release from the polymeric release devices. A near-zero-order releasing polymer matrix implant was investigated, and results both in vitro and in vivo are reviewed.

1. Effect of Insulin Solubility on Release

The release kinetics of insulin from polymer matrices containing either zinc insulin or sodium insulin are compared in Fig. 5. The weight percent loading of insulin in the matrices was 50% (wt/wt). The insulin particle size was selected to be between 250 and 425 μm. The solubility of the zinc insulin was measured to be 0.33 mg/mL, and the solubility of the sodium insulin was measured to be 120 mg/mL in 50 mM Tris-HCl buffer.

The sodium insulin matrices released 75% of the originally incorporated insulin, whereas the zinc insulin matrices released only 36% over a 36 day period. The initial burst effect of insulin release lasted approximately 100 hours ($t^{0.5}$, 10 hours$^{0.5}$) for the zinc insulin matrices and 50 hours ($t^{0.5}$, 7.1 hours$^{0.5}$) for the sodium insulin matrices. The second phase of release showed a lower release rate than the burst effect and a linear dependence of cumulative percent release on the square root of time. During this second phase of release, an additional 5.5% of the zinc insulin and 41.4% of the sodium insulin was released over the 17 days after the burst effect. These results supported the use of sodium insulin in order to obtain greater cumulative release from these sustained release polymer matrices.

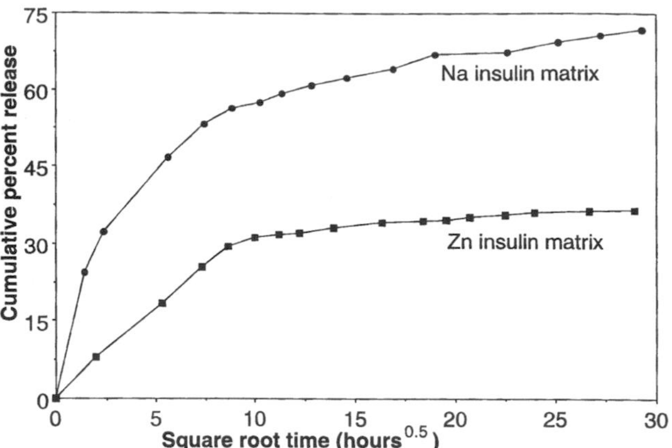

Figure 43.5 The cumulative percentage release of insulin versus the square root of time graphed for matrices made with zinc insulin and with sodium insulin. The matrices were released into 50 mM Tris buffer at pH 7.4 and 23°C. The sodium insulin matrices showed greater cumulative percent release than the zinc insulin matrices. Each point represents the mean of seven matrix samples. (Reproduced with permission from John Wiley & Sons.)

2. Mechanism of Release

Bawa et al. previously showed that the release mechanism of macromolecular drugs from these matrices involved the influx of release medium into the matrix through pores where the drug was dispersed (35). The dissolution of the drug was followed by the slow diffusion of the drug through the complex network of pores to the exterior of the matrix. Figure 5 showed that insulin release from flat-slab solvent cast matrices was divided into two phases. The first phase was an initial burst of insulin release that can be explained as the dissolution of particulate insulin at the surface of the matrix. The second phase of release was the further dissolution and diffusion of insulin through the tortuous porous network. In the second release phase, the cumulative percent release of incorporated insulin was observed to be linear with the square root of time for the flat-slab geometry. This is consistent with the diffusion of molecules through a porous planar matrix as the rate limiting step. An alternate model proposed by Higuchi supports a similar diffusion mechanism through a porous network for drugs like insulin, which possess low solubility

(36). A cross-sectional representation of a typical slab release device is shown in Fig. 6.

A simplified model of release from a matrix slab can be represented by the equation

$$\frac{dQ}{dt} = \frac{C_s D a_i^2}{R}$$

where dQ/dt = release rate in mg/s, C_s = drug solubility in mg/mL, D = diffusion coefficient in cm²/s, a_i = length of the side of the matrix slab in cm, and R = distance to the releasing surface between the dissolved and dispersed drug in cm.

This equation predicts that release rates will decrease with time as the distance R increases.

An approach to achieving near constant release rates was then hypothesized by altering the matrix geometry that restricted the releasing surface to a small aperture in the matrix (37). Figure 7 shows a cross section of a hemisphere-shaped matrix that is coated with an impermeable layer of polymer except for a small aperture on the top surface of the matrix.

This geometry restricts release to a small aperture on the surface of the polymer matrix. The increase in R, which in the slab matrix resulted in decreased release rates over time, now is compensated by an increase in the available drug for release. This geometry predicts that release rates will be near constant when R is much greater than a_i. The release rate for the hemisphere-shaped geometry is shown by

$$\frac{dQ}{dt} = 2\pi C_s D_{a_i}\left(\frac{R}{R - a_i}\right)$$

where dQ/dt = release rate in mg/s, C_s = drug solubility in mg/mL, D = diffusion coefficient in cm²/s, a_i = inner aperture radius in cm, and R = distance of aperture surface to dispersed drug in cm.

Figure 43.7 Cross section of hemisphere-shaped matrix coated with an impermeable layer of polymer that restricts the release of drug to the aperture. Release rates are predicted to be constant when $R \gg a_i$ hr. (Reprinted with permission from Harcourt Academic Press.)

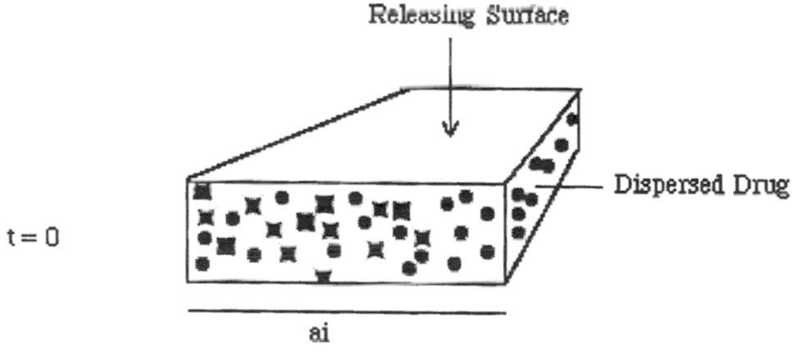

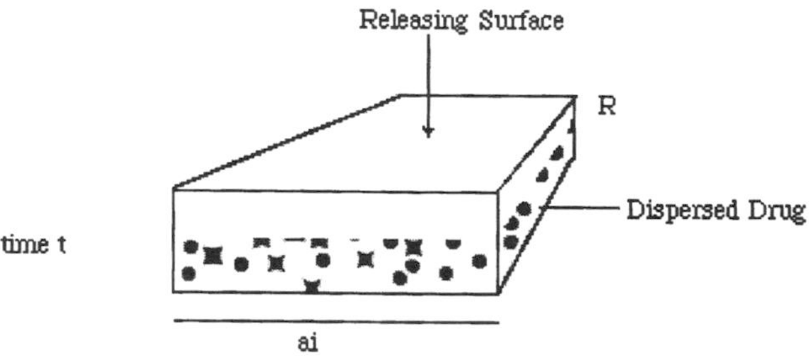

Figure 43.6 Cross section of a one-sided releasing flat-slab matrix. A $t = 0$ drug is evenly dispersed throughout the matrix. At some later time t the distance R to the releasing surface between the dissolved drug and the dispersed drug increases. Thus R increases with time and release rates decrease.

3. Insulin Release from Coated Hemisphere Matrices

In vitro and in vivo studies demonstrated that indeed near-zero-order release rates are observed when a matrix is formed utilizing the hemisphere geometry (3,4). Matrices designed to release insulin at constant rates were fabricated by coating them with an impermeable layer of ethylene vinyl acetate copolymer and then opening an aperture on the top surface. A 30-gauge needle was inserted into the flat face of the matrix. The frozen matrix was then dipped for 1 second into a solution of pure ethylene vinyl acetate copolymer in methylene chloride. An aperture was drilled in the same location after removal of the needle. The in vitro release kinetics seen in Fig. 8 demonstrated that the cumulative percentage release of insulin from these hemispheric releasing matrices was directly proportional to time. Thus constant release rates in vitro were observed.

4. In Vivo Insulin Delivery

The same sodium insulin hemisphere matrices were implanted into streptozotocin induced diabetic rats (38). Figure 9 shows that plasma glucose concentrations among these animals were maintained in the normal range for greater than 3 months. Thus a constant release of insulin was observed both in vitro and in vivo (4,6).

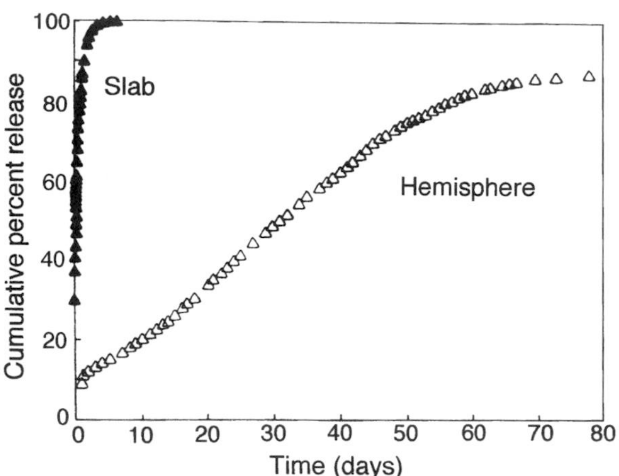

Figure 43.8 Cumulative percentage release of sodium insulin from an aperture restricted hemisphere matrix design compared to insulin release from an uncoated slab. The hemisphere matrix sustained insulin release at near constant rates of approximately 5 units per day. In contrast, the uncoated slabs released greater than 99% of the incorporated insulin in less than 4 days.

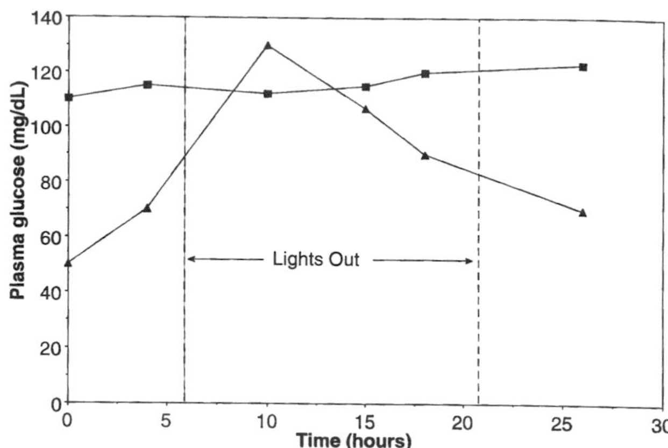

Figure 43.10 Diurnal variation of plasma glucose is graphed over a 26 hour period for normal rats (■) and insulin hemisphere matrix treated diabetic rats (▲). The lights were off for the time period shown in the graph. The rats' feeding period occurred during this nocturnal time frame.

5. Diurnal Glucose Variations

Figure 10 shows the plasma glucose concentrations in normal rats and insulin matrix treated rats over a 26 hour period. It is evident that the control animals remained normoglycemic during the nocturnal time period when the rats are awake and feeding. In contrast, the insulin matrix treated rats showed significant plasma glucose elevations during this time period. In this group, plasma glucose concentrations started out below the normal glucose concentrations. Then when the lights were turned off, and the feeding period began, plasma glucose levels rose and exceeded the concentrations of the normal control rats. This data demonstrated that the sustained release of a basal level of insulin from the polymer matrices was incapable of controlling plasma glucose in a completely physiologic manner. This example demonstrated the need to alter insulin delivery rates in response to metabolic changes.

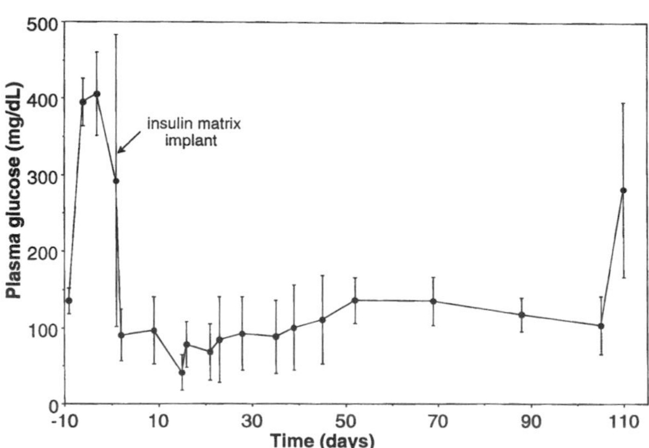

Figure 43.9 Ethylene vinyl acetate copolymer matrix implants using a hemisphere geometry were implanted into 8 diabetic rats. Diabetes was induced by an intravenous injection of streptozotocin on day −9. The insulin matrices were surgically implanted on day 0. The data shows normalization of insulin release for over 100 days by a single implant. (From Refs. 4 and 6.) (Reproduced with permission from the American Diabetes Association.)

Thus implantable insulin polymer matrices could be effective in delivering consistent basal insulin levels to diabetic rats. However, these implants could not increase insulin release rates in response to feeding periods. Therefore the demonstrated solubility dependent nature of insulin release from the ethylene vinyl acetate copolymer matrices encouraged further studies toward the development of a glucose responsive system.

6. Effect of pH on Insulin Release

The next series of experiments focused on studying the effect of pH on the release and solubility properties of insulin from the insulin polymer matrices. The following experiment was conducted in order to test the hypothesis that the release of insulin could be controlled by the pH of the release medium (4). In this experiment, crystalline zinc insulin loaded matrices were fabricated according to the procedures described previously. These matrices were 50%

by weight insulin sieved to a particle size less than 75 μm. Nine matrices were released in pH 5.9 release medium containing 0.9% NaCl. The matrices released into this medium for 9 days. The release medium buffer was changed to one buffered at pH 7.4. The results of this experiment are shown in Fig. 11.

Figure 11 shows that during the first 9 days of release at pH 5.9, release rates were less than 20 μg/h. However, upon switching the release media to pH 7.4, an immediate burst of insulin was observed. Release rates increased from 10 μg/h to 90 μg/h. During this burst, 13% of the originally incorporated insulin was suddenly released in a 6 hour period, whereas only 5.8% had been released over the previous 9 days. After this burst, release rates fell to 45 μg/h over the next 3 days. This observation was further supported by the pH and solubility profile of crystalline zinc insulin.

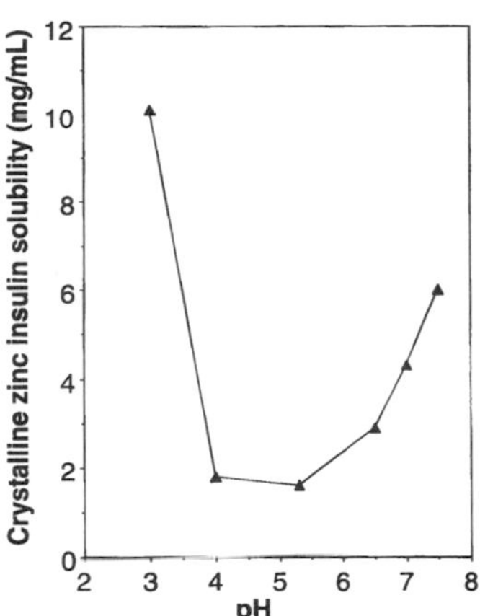

Figure 43.12 The solubility dependence of crystalline zinc insulin on pH. The graph shows that insulin solubility increases markedly at pH's above and below its isoelectric point of 5.3.

7. Solubility Profile of Crystalline Zinc Insulin

Figure 12 shows that the aqueous solubility of zinc crystalline insulin is strongly affected by pH. The isoelectric point of this insulin form is approximately 5.3. At pH's above and below the isoelectric point of the protein, its solubility increases significantly. The observations in Fig. 11 show that increasing the pH of the release medium from 5.9 to 7.4 appeared to be a result of the increased dissolution of the insulin incorporated within the polymer matrix. This enabled the additional dissolved insulin to be released at much higher release rates.

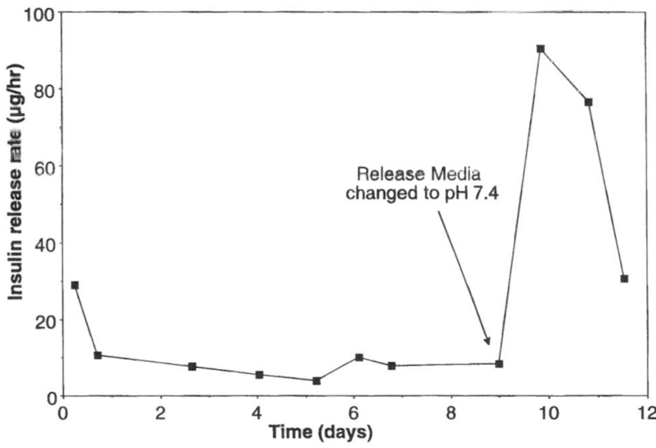

Figure 43.11 The graph shows the release rate of zinc insulin from ethylene vinyl acetate copolymer matrices over an 11.5 day period. The insulin loading was 50% by weight and the particle size of the insulin was less than 75 μm. The first 9 days of release were carried out in pH 5.9 buffer. This buffer was replaced with a pH 7.4 buffered medium at day 9. At this time, a large increase in insulin release rate was observed.

V. A GLUCOSE MEDIATED INSULIN DELIVERY IMPLANT

The observations that insulin release could be extended for months with the use of an implantable matrix, and that the release of insulin could be increased by changes in pH, gave rise to the possibility that a feedback controlled insulin delivery system could be developed. The key was to find a way to alter the microenvironmental pH of the polymer matrix in a manner that would produce a burst of insulin release such as the one seen in Fig. 10 in response to changes in glucose concentration. The approach used in these studies was to take advantage of the enzyme glucose oxidase that converts glucose to gluconic acid:

$$\text{glucose} + O_2 + H_2O \xrightarrow{\textit{Glucose Oxidase}} \text{gluconic acid} + H_2O_2$$

The glucose responsive system was realized by immobilizing glucose oxidase enzyme to sepharose beads, which were incorporated along with zinc insulin into the polymer matrix. When glucose in solution entered the insulin delivery system, gluconic acid was produced, causing a drop in the microenvironmental pH of the matrix. This fall in pH resulted in a rise in insulin solubility and consequently a rise in the insulin release rate from the matrix (39–41).

The following studies review how the observations of the sustained release of insulin from the polymer matrices and the pH dependent solubility properties of insulin were combined to demonstrate a glucose responsive insulin delivery system.

A. Methods

The methods and procedures for incorporating the immobilized glucose oxidase enzyme into the ethylene vinyl acetate sustained release matrix are outlined.

1. Immobilization of Glucose Oxidase

Glucose oxidase was immobilized to cross-linked sepharose CL-6B or 4B beads using the method of Kohn and Wilchek (42). The sepharose beads were suspended in deionized water, frozen at −20°C, and lyophilized to yield a fluffy, pale-yellow-colored powder.

2. Polymer Matrix Preparation

Ethylene vinyl acetate copolymer matrices for the in vitro and the in vivo studies containing immobilized glucose oxidase and/or insulin were prepared according to published methods described previously. For these experiments the sepharose beads containing the immobilized glucose oxidase were coincorporated into the polymer matrix with either regular crystalline zinc insulin or trilysine insulin. Trilysine insulin has an isoelectric point of 7.4 (43).

3. In Vitro Release Studies

A flow-through spectrophotometer system was used to examine continuously the feasibility of increasing the insulin release kinetics in response to a glucose stimulus in real time. An insulin loaded matrix with immobilized glucose oxidase was placed inside a filter holder. Distilled water, adjusted to pH 5.3 and containing a glucose concentration of 0 or 10 mg/mL, was delivered with a peristaltic pump past the samples into a 65 μL flow cell of a spectrophotometer. The in vitro experiments were conducted at pH 5.3 as regular crystalline insulin was used in these in vitro experiments.

B. Results

In vitro and in vivo studies confirm that insulin release can be increased in response to a glucose stimulus.

1. Solubility and pH Dependence of Trilysine Insulin

Figure 13 shows the pH solubility profile of trilysine insulin. For the proposed system to work in vivo, an insulin form with a pI at physiologic pH was necessary. In this way, production of gluconic acid by the glucose oxidase

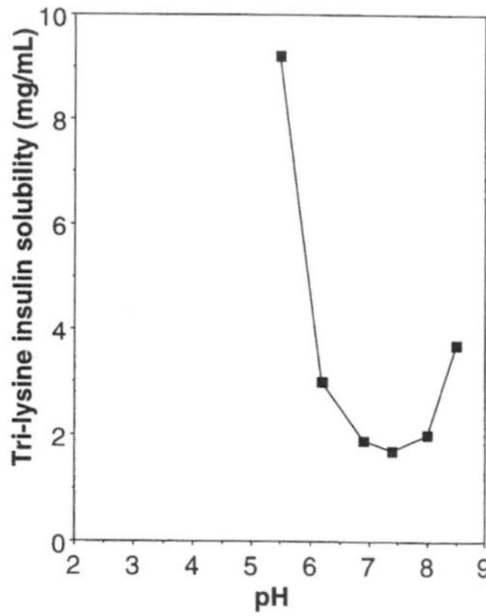

Figure 43.13 The solubility dependence of trilysine insulin on pH. The graph shows that insulin solubility increases markedly at pH's above and below its isoelectric point of 7.4. Trilysine insulin is appropriate for systems to be tested at physiologic pH.

enzyme would result in an increase in trilysine insulin's solubility. If regular crystalline insulin with a pI of 5.3 were used, one would predict that a decrease in insulin solubility would result, and release rates from the polymer matrix would decrease (see Fig. 12).

2. In Vitro Release of Insulin in Response to a Glucose Stimulus

The in vitro results in Fig. 14 demonstrate the kinetic response of the immobilized glucose oxidase–regular crystalline insulin–ethylene vinyl acetate copolymer matrices

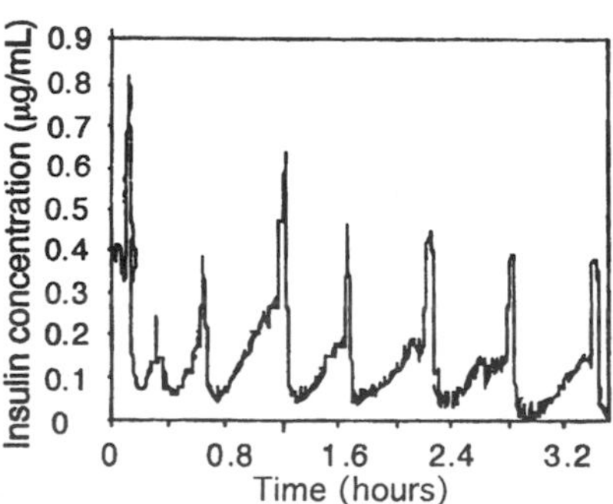

Figure 43.14 Glucose oxidase was immobilized to sepharose beads. The beads were then incorporated with insulin into an ethylene vinyl acetate copolymer matrix. This matrix was tested in a flow-through system. Each insulin peak was preceded by a 35 minute continuous glucose infusion at a concentration of 500 mg/dL through the flow-through apparatus, which monitored the insulin concentration spectrophotometically. The influx of glucose resulted in a significant increase in insulin release from the polymer matrix.

to a change in glucose concentration. Flow-through release medium containing no glucose was pumped through the flow cell at a rate of 10 mL/hour to establish the baseline insulin release from the system. After the baseline insulin release rate was established, a 500-mg/dL solution of glucose was infused through the system. Peak insulin release from the matrix reached a maximum 35 minutes before the release medium was changed to one containing no glucose. Then the insulin concentrations decreased. Glucose infusion was alternated with nonglucose containing medium eight different times over a 3 hour period. Turning the glucose on and off was shown to stimulate repeatedly the release of insulin from the polymer matrix.

3. Refractory Time Interval

Figure 14 showed that each glucose infusion was characterized by a discrete burst of insulin release. The following experiments were conducted to test whether the glucose responsive matrix required a rest or refractory period for maximal response to repeated glucose challenges. The duration between successive glucose stimuli was varied, and the magnitude of the response was measured to determine the amplitude of insulin response. Figure 15 showed that increasing the time between glucose stimuli from 5 to 60 minutes resulted in increasingly greater insulin release from the matrix. Increasing the rest period in between glucose stimuli from 60 to 130 minutes did not result in increased insulin release from the matrix.

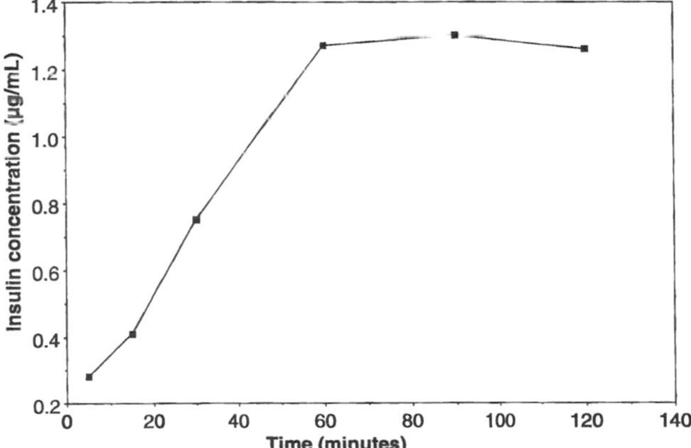

Figure 43.15 The maximum response of the glucose oxidase insulin release is graphed as the concentration of insulin observed in the release media as a function of the time intervals between repeated glucose challenges applied to the insulin–glucose oxidase matrix. The data shows that increasing the time interval between glucose stimuli resulted in greater insulin release until a plateau was reached after a 60 minute interval. (Reproduced by permission from John Wiley & Sons.)

Table 43.3 Weight Percent Composition of Insulin–Glucose Oxidase Matrices

Matrix No.	Sepharose-GO (wt%)	Sepharose (wt%)	Ethylene vinyl acetate copolymer (wt%)	Insulin (wt%)
1	0	29	48	23
2	9	20	48	23
3	20	0	48	23
4	29	0	48	23

Sepharose-GO matrices contained beads with immobilized glucose oxidase enzyme. Sepharose matrices contained no immobilized enzyme.

4. Effect of Glucose Oxidase Concentration

In the following experiment, the effect of increasing the glucose oxidase content on the concentration of insulin released was determined. Four different matrices were tested using increasing percentages of glucose oxidase immobilized sepharose beads in the matrix, as shown in Table 3.

Figure 16 shows that increasing the glucose oxidase concentration resulted in increasing insulin release from the matrix up to a matrix containing 20% by weight of glucose oxidase immobilized to sepharose. Twenty percent glucose oxidase–sepharose appeared to be the optimal content for maximal response to a glucose stimulus.

C. In Vivo Study Methods

Sixteen female Sprague-Dawley rats weighing 210 ± 10 g were used in this experiment. Diabetes was induced by

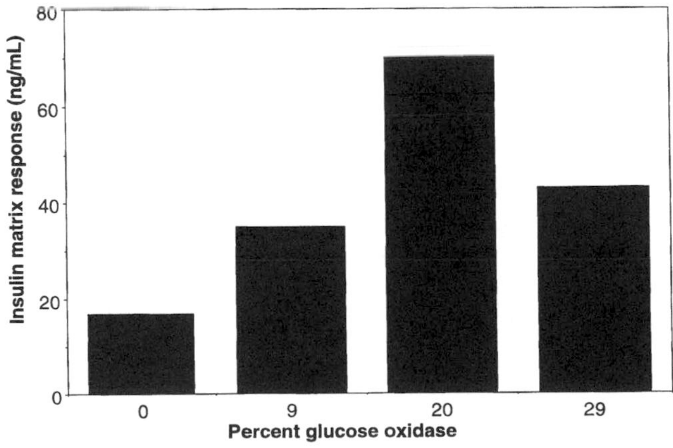

Figure 43.16 The glucose oxidase–insulin response increases with increased glucose oxidase content in the matrix to a maximal response when the polymer matrix contained 20% by weight glucose oxidase. (Reproduced by permission from John Wiley & Sons.)

tail vein injection of streptozotocin (38). A glucose solution at a concentration of 2 M was infused through a catheter to these animals for 30 minutes. Blood samples for glucose and insulin analyses were collected from the tail vein. The glucose oxidase insulin matrices were implanted subcutaneously in each rat. Seven rats received matrices containing trilysine insulin and glucose oxidase–immobilized sepharose beads. Three rats received matrices containing regular insulin and glucose oxidase–immobilized sepharose beads. Three rats received matrices containing regular insulin alone. Three rats were used as diabetic controls with no implants.

1. In Vivo Results

The in vivo effect on serum insulin concentrations with this glucose responsive insulin–glucose oxidase matrix is shown in Fig. 17, in which the serum insulin concentrations (ng/mL) are plotted over time for the four groups of rats containing different insulin matrix implants into diabetic rats. The 2 M glucose infusion was begun 15 minutes into the experiment. The rats that received trilysine insulin–glucose oxidase matrices showed a 180% rise in serum insulin concentration that peaked 45 minutes into the experiment. There was a subsequent decrease in plasma insulin concentration 30 minutes after the initiation of the glucose infusion. Those rats that received regular insulin–glucose oxidase matrices exhibited a decrease in serum insulin concentration. Those rats that received matrices containing regular insulin but no glucose oxidase enzyme or no insulin showed no change in serum insulin concentration. The diabetic control rats showed no change in serum insulin concentration. There was no effect observed in the plasma glucose concentrations in these rats (data not shown). All four groups of rats showed an increase of 500 to 600 mg/dL in plasma glucose in response to the 2 M glucose infusion. The diabetic controls began the experiment with a glucose concentration of 450 mg/dL and at 105 minutes had reached a glucose concentration of 1050 mg/dL. In contrast, the three groups of rats that had received insulin implants exhibited initial plasma glucose concentrations of 50 mg/dL. Their plasma glucose levels rose to 600 to 700 mg/dL in response to the glucose infusion. Normal controls also displayed similar increases in plasma glucose.

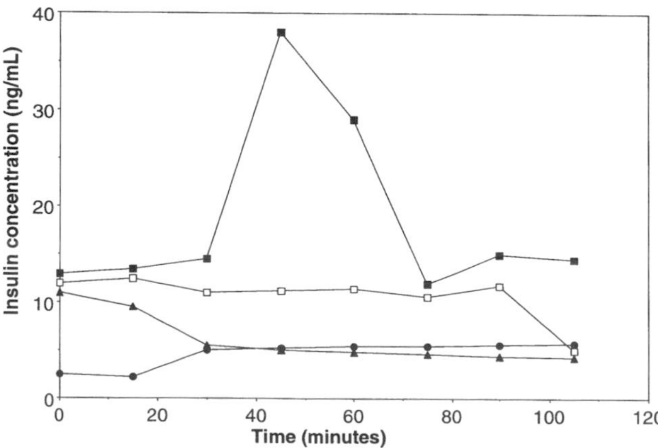

Figure 43.17 Serum insulin concentration is graphed as a function of time among the four treatment groups. 2 M glucose was infused at 15 minutes and was continued for the duration of the experiment. Key (■) trilysine insulin and immobilized glucose oxidase ($n = 7$ rats); (▲) regular insulin and immobilized glucose oxidase ($n = 3$ rats); (□) regular insulin and no enzyme ($n = 3$ rats); (●) diabetic controls, no treatment ($n = 3$ rats). (Reproduced by permission from John Wiley & Sons.)

VI. DISCUSSION

The experimental results described helped characterize the release of insulin from ethylene vinyl acetate copolymer matrices in response to glucose (40,41). Glucose oxidase was immobilized on sepharose beads that were incorporated along with one of two different insulin preparations into an ethylene vinyl acetate copolymer matrix. These two insulin preparations had different isoelectric points. The immobilized beads were fixed in the matrix and could not diffuse out of the matrix. Only the insulin could diffuse out of the matrix. The effect of an increase in glucose concentration within the matrix on the immobilized glucose oxidase was to lower the pH in the microenvironment of the matrix by the production of gluconic acid according to the following enzyme catalyzed reaction (44):

$$\text{glucose} + O_2 + H_2O \xrightarrow{\text{Glucose Oxidase}} \text{gluconic acid} + H_2O_2$$

The production of gluconic acid caused a decrease in the microenvironmental pH of the matrix and a subsequent increase in the insulin release rates from the ethylene vinyl acetate copolymer matrix. The increased insulin release was a result of the accelerated dissolution and diffusion of the dry powdered insulin from the delivery system (4).

The solubility and dissolution of polypeptides or proteins like insulin is dependent on the pH of the aqueous environment. Proteins and peptides exhibit their lowest solubility at the pI where there is no net charge on the peptide (17). Thus the drug solubility is one important parameter that can be used to alter release rates of drugs from diffusion controlled (36) and dissolution controlled (45) release matrix systems. This system responded in vitro repeatedly, consistently, and in real time to challenges with glucose solutions (Fig. 14).

The insulin used in the in vitro experiments had a pI of

5.3. Peak insulin levels occurred 35 minutes after a 500 mg/mL glucose solution was introduced into the flow-through system. Repeated glucose stimuli were shown to induce a degree of refractory behavior (Fig. 15). It was determined that at least 60 minutes were required between glucose stimuli to observe a maximal response. The rise to a peak in insulin concentration and the ensuing fall to baseline levels following a glucose stimulus (Fig. 14) could be attributed to several factors: (1) the inability of the matrix to secrete insulin because soluble and undissolved insulin might be depleted in the vicinity of the immobilized glucose oxidase; (2) a reequilibration of the pH in the microenvironment of the matrix due to the buffering capacity of dissolved insulin in the pores of the matrix; (3) a decrease in oxygen tension that is necessary to maintain the enzymatic reaction; or (4) the inability of the glucose oxidase enzyme to retain a pH gradient. The maximal system response observed after a 60 minute rest period also suggests that there is only a finite quantity of insulin available for release after any glucose stimulus. This quantity of insulin could be localized near or within the sepharose beads or precipitated along the empty pores of the matrix.

The refractory period may represent the time necessary for the local areas of insulin depletion to be replenished before the next glucose stimulus, which is presumably related to the dissolution rate of insulin, the diffusion coefficient of insulin, and the porosity or the tortuosity of the matrix. A similar finite insulin release observed in vivo is shown in Fig. 16. These experiments also supported previous observations that the bioactivity of insulin is preserved in the ethylene vinyl acetate copolymer matrix (5,6). The absence of any effect on the plasma glucose concentrations in the diabetic test animals can be attributed to the large excess of glucose infused into the animals. The high concentration of glucose in these initial studies was used to ensure that the subcutaneously implanted insulin delivery device would be exposed to a sufficient glucose concentration to obtain an insulin response. One can calculate that the quantity of glucose infused in the experiments described here would result in an instantaneous plasma glucose concentration of 1800 mg/dL when distributed in the extracellular fluid compartment of a 200 gram rat if there is no intracellular glucose uptake. Thus it is not surprising that little or no effect on plasma glucose was observed in any experimental animal group. Furthermore, similar large increases in plasma glucose concentrations were also observed in normal control rats. Thus even a physiologically intact pancreas could not respond to the excess glucose load presented to the rats during the time period described in this study.

A maximal insulin response was observed when 20% (w/w) of glucose oxidase immobilized on sepharose beads was incorporated into the matrix (Fig. 16). The observation that there is an apparent optimal immobilized glucose oxidase concentration may be due to stearic hindrance of glucose diffusion into the sepharose beads. A threefold increase in serum insulin concentrations was observed for those rats that received trilysine insulin–glucose oxidase implants. The dependence of regular and trilysine insulin solubility on pH is shown in Fig. 13. These results are consistent with a fall in the microenvironmental pH of the matrix and a rise in trilysine solubility and dissolution rate within the matrix. Trilysine insulin has a pI of 7.4. Thus when the glucose oxidase matrix system operates at physiologic pH, the decrease in microenvironmental pH of the matrix resulted in an increase in insulin solubility and consequently in an increase in insulin release. Regular insulin has a pI of 5.3. Thus the lowering of the pH below pH 5.3 resulted in increased release rates of insulin from the matrix; the matrix was in a release media at pH 5.3. At physiologic pH, those rats treated with regular insulin–glucose oxidase implants experienced a decrease in serum insulin concentrations in response to glucose infusion (Fig. 17). Rats with polymers containing regular insulin but no enzyme, and with no insulin or no enzyme, did not show a serum insulin response to glucose infusion.

VII. CONCLUSIONS

Indeed, other approaches have been used to develop glucose responsive insulin delivery systems, including microencapsulated islets (46), a lectin bound insulin in which glucose displaces a carbohydrate, such as maltose, to release the insulin (47), and membranes that alter their permeabilities in response to glucose (48). In another system, insulin was esterified with methanol and connected to glucose oxidase with a disulfide compound, 5,5′-dithiobis(2-nitrobenzoic acid). Then adding glucose to an aqueous solution containing the hybrid enzyme resulted in the release of modified insulin (49).

This review has summarized the scientific observations and research efforts that have enabled the development of an alternate approach to achieving a glucose responsive polymeric insulin delivery system. The physiological role of insulin in the treatment of diabetes mellitus was discussed. Scientific and historical efforts resulted in the development of modified insulin forms that are either fast acting or possess extended biological half-lives. The role of zinc in determining insulin's solubility properties proved to be a critical parameter in allowing the long-term release of insulin from ethylene vinyl acetate copolymer matrices. Finally, the solubility properties of insulin were exploited in order to demonstrate a glucose responsive feedback controlled insulin delivery system. Additional studies are necessary to determine if an optimized system could control the diurnal fluctuations in plasma glucose concentration in diabetic animals. In addition, the in vivo operating duration may be limited by the half-life of the glucose oxidase and the degree of encapsulation of the polymer implant over time (50). Others have observed several months' glucose oxidase activity in implantable glucose biosensors (51). The ability of this system to demonstrate a blood-glucose controlling effect in future experiments might be observed by simply increasing the surface area of the matrix so that more insulin is released in response to a change in glucose concentration. Immobilizing the enzyme on the polymer backbone instead of onto the sepharose beads might also increase the sensitivity of the system and might allow an increased payload of insulin in the matrix with the elimination of the sepharose beads. Other pH-sensitive insulins with greater solubilities might be another possible approach.

The mechanism of an increase in dissolution rate of protein or polypeptide drugs in the microenvironment of a polymer matrix may be a model for use with other pH-sensitive proteins having an appropriate triggering enzyme and substrate. The further development of this feedback controlled release system may find application with the variety of new genetically engineered proteins that are already approved for use or among those that are being researched.

ACKNOWLEDGMENTS

This work was supported by grants from the NIH (GM26698), the American Diabetes Association, and the Juvenile Diabetes Foundation. L. R. Brown was a fellow of the Juvenile Diabetes Foundation. The author is indebted to Susan Garramone, Barbara Lapidas-Brown, and Karen Kuzmich for their assistance in the preparation of this manuscript.

REFERENCES

1. R. S. Langer, J. Folkman. Polymers for the sustained release of macromolecules. *Nature 263*:797 (1976).
2. W. D. Rhine, D. S. Hsieh, R. S. Langer. Polymers for sustained macromolecule release. Procedure to fabricate reproducible delivery systems and control release kinetics. *J. Pharm. Sci. 69*:265 (1980).
3. L. Brown, L. Siemer, C. Munoz, R. Langer. Controlled release of insulin from polymer matrices: in vitro kinetics. *Diabetes 35*:684–691 (1986).
4. L. Brown. Controlled release polymers: in vivo studies with insulin and other macromolecules. Doctoral thesis Massachusetts Institute of Technology, 1983.
5. H. M. Creque, R. Langer, J. Folkman. One month sustained release of insulin from a polymer implant. *Diabetes 29*:37–40 (1980).
6. L. Brown, C. Munoz, L. Siemer, C. Munoz, E. Edelman, R. Langer. Controlled release of insulin from polymer matrices: control of diabetes in rats. *Diabetes 35*:692–697 (1986).
7. F. G. Banting, C. H. Best. The internal secretion of the pancreas. *J. Lab. Clin. Med. 7*:251 (1922).
8. D. F. Steiner, C. Patzelt, S. J. Chan, P. S. Quinn, H. S. Tager, D. Nielsen, A. Lernmark, B. E. Noyes, K. L. Agarwal, K. H. Gabbay, A. H. Rubenstein. Formation of biologically active peptides. *Proc. R Soc. Lond. B. Biol. Sci. 210*: 45–59 (1980).
9. Y. Zambre, Z. Ling, X. Hou, A. Foriers, B. Van Den Bogaert, C. Van Schravendijk, D. Pipeleers. Effect of glucose on production and release of proinsulin conversion products by cultured human islets. *J. Clin. Endocrinol. Metab. 83*: 1234–1238 (1998).
10. E. F. Usac, B. Nadal, R. Gasa, J. Fernandez-Alvarez, R. Gomis. Effect of high glucose concentration on proinsulin biosynthesis and conversion by human islets. *Biochem. Biophys. Res. Commun. 248*:186–189 (1998).
11. P. T. Lomedico, S. J. Chan, D. F. Steiner, G. F. Saunders. Immunological and chemical characterization of bovine preproinsulin. *J. Biol. Chem. 252*:7971–7978 (1977).
12. D. F. Steiner. On the role of the proinsulin C-peptide. *Diabetes 27*(suppl. 1):145–148 (1978).
13. T. Tanese et al. Synthesis and release of proinsulin and insulin by isolated rat islets of Langerhans. *J. Clin. Invest. 7*: 1394–404 (1970).
14. A. E. Kitabchi. Proinsulin and C-peptide: a review. *Metabolism 5*:547–587 (1977).
15. R. E. Chance, R. M. Ellis, W. W. Bromer. Porcine proinsulin: characterization and amino acid sequence. *Science 161*: 165–167 (1968).
16. R. E. Chance. Amino acid sequences of proinsulins and intermediates. *Diabetes 21*:461–467 (1972).
17. G. M. Grodsky. The chemistry and function of hormones. *Review of Physiological Chemistry* (H. A. Harper, V. W. Rodwell, P. A. Mayes, eds.) 16th ed. Lange Medical, Los Altos, CA, 1975, pp. 19, 471.
18. M. R. DeFelippis, D. L. Bakaysa, M. A. Bell, M. A. Heady, S. Li, S. Pye, K. M. Youngman, J. Radziuk, B. H. Frank. Preparation and characterization of cocrystalline suspension of [LysB28,ProB29]-human insulin analogue. *J. Pharm. Sci. 87*:170–176 (1998).
19. P. D. Home, L. Barriocanal, A. Lindholm. Comparative pharmacokinetics and pharmacodynamics of the novel rapid acting insulin analogue, insulin aspart, in healthy volunteers. *Eur. J. Clin. Pharmacol. 55*:199–203 (1999).
20. B. K. Milthorpe, L. W. Nichol, P. D. Jeffrey. The polymerization pattern of zinc(II)-insulin at pH 7.0. *Biochim. Biophys. Acta. 2*:195–202 (1977).
21. D. C. Hodgkin. The structure of insulin. *Diabetes 21*:1131–1150 (1972).
22. Images provided by K. Harper, T. Richmond, and Maestro-X Technologies, Claremont, CA, copyright 1998, 1999.

23. J. Lewis. Insulin administration problem. *Physiol. Rev. 29*: 75 (1949).
24. H. C. Hagedorn. Protamine insulinate. *J.A.M.A. 106*:177 (1936).
25. D. Scott, A. J. Fisher. Prolongation of insulin action by protamine and zinc. *J. Biol. Chem. Sc. Proc. 114*:88 (1936).
26. A. Marble. *Chemistry and Chemotherapy of Diabetes Mellitus*. Charles C. Thomas, Springfield, IL, 1962, pp. 83, 100.
27. N. Kirkpatrick, R. Wilder. Experience with a new insulin. *Proc. Staff Meet. Mayo Clinic 24*:365 (1949).
28. K. Hallas Moller, M. Jersild, K. Peterson. Insulin preparations for single daily injection. Clinical studies of new preparations with prolonged action. *J.A.M.A. 150*:1667 (1952).
29. F. J. Service, G. D. Molnar, J. W. Rosevear. Mean amplitude of glycemic excursion, a measure of diabetic instability. *Diabetes 19*:644 (1970).
30. M. D. Siperstein, D. W. Foster, H. C. Knoles. Control of blood glucose and diabetic vascular disease. *N. Engl. J. Med. 296*:1060 (1977).
31. Diabetes Control and Complications Trial Research Group. The effect of intensive treatment of diabetes on the development and progression of long-term complications in insulin-dependent diabetes mellitus. *N. Engl. J. Med. 329*:977–986 (1993).
32. F. R. Kaufman, M. Halvorson, L. Fisher, P. Pitukcheewanont. Insulin pump therapy in type 1 pediatric patients. *J. Pediatr. Endocrinol. Metab. 12* (suppl 3):759–764 (1999).
33. P. S. Gillies, D. P. Figgitt, H. M. Lamb. Insulin glargine. *Drugs 59*:253–260 (2000).
34. W. B. Grutzer. Spectrophotometric determination of protein concentration in the short wavelength ultraviolet. *Handbook of Chemistry and Molecular Biology. Physical and Chemical Data, Vol. II*. CRC Press, Cleveland, OH, 1976, p. 197.
35. R. Bawa, R. Siegel, B. Marasca, M. Karel, R. Langer. An explanation for the controlled release of macromolecules from polymers. *J. Controlled Release 1*:259–267 (1985).
36. T. Higuchi. Mechanism of sustained action medications: theoretical analysis of rate of rate of release of solid drugs in dispersed in solid matrices. *J. Pharm. Sci. 52*:1145–1149 (1963).
37. W. D. Rhine, V. Sukhatme, D. Hsieh, R. Langer. A new approach to achieve zero order release kinetics from diffusion controlled polymer systems. *Controlled Release of Bioactive Materials* (R. Baker, ed.). Academic Press, New York, 1980, pp. 177–187.
38. A. Junod, A. Lambert, L. Orli, R. Pictet, A. Gonet, A. Renot. Studies on the diabetogenic action of streptozotocin. *P.S.E.B.M. 126*:201–205 (1967).
39. L. Brown, F. Fischel-Ghodsian, R. S. Langer. Feedback controlled-release implant for delivery of protein drugs. U.S. Patent 4,952,406, August 8, 1990.
40. F. Fischel-Ghodsian, L. Brown, E. Mathiowitz, D. Brandenburg, R. Langer. Enzymatically controlled drug delivery. *P.N.A.S. USA 85*:2403–2406 (1988).
41. L. Brown, E. R. Edelman, F. Fischel-Ghodsian, R. Langer. Characterization of glucose-mediated insulin release from implantable polymers. *J. Pharm. Sci. 85*:1341–1345 (1996).
42. J. Kohn, M. Wilchek. A new approach (cyano-transfer) for cyanogen bromide activation of Sepharose at neutral pH, which yields activated resins, free of interfering nitrogen derivatives. *Biochem. Biophys. Res. Commun. 107*:878–884 (1982).
43. D. Levy, F. H. Carpenter. The synthesis of triaminoacyl insulins and the use of the t-butyloxycarbonyl group for the reversible blocking of the amino groups of insulin. *Biochemistry 6*:3559–3568 (1967).
44. F. W. Scheller, R. Renneberg, F. Schubert. Coupled enzyme reactions in enzyme electrodes using sequence, amplification, competition, and anti-interference principles. *Methods Enzymol. 137*:29–43 (1988).
45. S. K. Chandrasekaran, D. R. Paul. Dissolution controlled transport from dispersed matrixes. *J. Pharm. Sci. 71*:1399–402 (1982).
46. G. M. Cruise, O. D. Hegre, F. V. Lamberti, S. R. Hager, R. Hill, D. S. Scharp, J. A. Hubbell. In vitro and in vivo performance of porcine islets encapsulated in interfacially photopolymerized poly(ethylene glycol) diacrylate membranes. *Cell Transplant 8*:293–306 (1999).
47. M. Brownlee, A. Cerami. Glycosylated insulin complexed to Concanavalin A. Biochemical basis for a closed-loop insulin delivery system. *Diabetes 32*:499–504 (1983).
48. J. Kost, T. A. Horbett, B. D. Ratner, M. Singh. Glucose-sensitive membranes containing glucose oxidase: activity, swelling, and permeability studies. *J. Biomed. Mater. Res. 19*:1117–1133 (1985).
49. Y. Ito, D. J. Chung, Y. Imanishi. Design and synthesis of a protein device that releases insulin in response to glucose concentration. *Bioconjug. Chem. 5*:84–87 (1994).
50. S. J. Updike, M. C. Shults, R. K. Rhodes, B. J. Gilligan, J. O. Luebow, D. von Heimburg. Enzymatic glucose sensors. Improved long-term performance in vitro and in vivo. *ASAIO J. 40*:157–163 (1994).
51. D. A. Gough, J. C. Armour. Development of the implantable glucose sensor. What are the prospects and why is it taking so long? *Diabetes 44*:1005–1009 (1995).

44

Drug Targeting to the Kidney: The Low-Molecular-Weight Protein Approach

R. F. G. Haverdings, R. J. Kok, M. Haas, F. Moolenaar, D. de Zeeuw, and D. K. F. Meijer
University of Groningen, Groningen, The Netherlands

I. INTRODUCTION

Drug therapy in general has improved greatly during the last century. However, although it is successful in many cases, drug treatment is often hampered by serious side effects. This can be explained in that many drugs do not exert their action at the desired site of action exclusively, but either can perturb physiological processes in other tissues/organs or can produce severe toxicity as well. The occurrence of drug adverse effects can greatly affect the patient's therapeutic compliance.

To solve the problems related to side effects and create more selectively acting drugs, it was already stated in 1906 by Ehrlich that specific targeting of drugs may result in an increased drug accumulation in the cells to be aimed at, relative to nontarget cells. This can lead to a reduction of the required drug dosage and less side effects or toxicity (1). A second application of drug targeting technology lies in the development of pharmacological tools to obtain more insight in the cellular aspects of pathophysiological processes.

The kidneys constitute an important target, since pharmacotherapy of renal disorders is frequently hampered by side effects, e.g., cardiovascular [''Angiotensin I Converting Enzyme'' (ACE) inhibitors] or gastrointestinal (nonsteroidal anti-inflammatory drugs, NSAIDs) adverse effects. Although the kidneys receive approximately 25 percent of the cardiac output, and although many drugs are more or less concentrated in the kidney, systemic concentrations are usually so high that extrarenal side effects cannot be prevented.

In this chapter an introduction to renal anatomy and physiology will be given in order to provide a rationale for renal drug targeting. Subsequently, we will focus on the renal targeting strategies applied in our laboratories, considering potentials and limitations of renal targeting with low-molecular-weight proteins (LMWPs) and, in particular, with lysozyme as the drug carrier. The chemical features of LMWP conjugates will be described using ACE inhibitors as model drugs. Furthermore, a number of targetable drugs that can affect kidney functions will be discussed. Finally, we will present data that demonstrate the effectiveness of the drug–lysozyme conjugates. For an extensive overview of various aspects of renal targeting strategies, we recommend other comprehensive reviews (2,3).

II. THE KIDNEY AS THE TARGET ORGAN

A. Anatomy and Functions of the Kidney

The kidneys play a major part in the regulation of the volume and composition of the body fluids through the excretion and retention of a large variety of solutes. It is also an eliminating organ for the excretion of several endogenous and exogenous organic substances. To entertain this function, the kidney exhibits a specific anatomical structure. Each kidney contains several millions of nephrons, the functional urine-producing units in this organ. The nephron consists of glomerular and tubular parts. The latter can be subdivided in the proximal convoluted tubule, the proximal straight tubule, Henle's loop, the distal convoluted tubule, and the collecting duct.

In every nephron blood is supplied to the glomerulus via the afferent arteriole. In the glomerulus, a portion of the blood (approximately 20%) is filtered through the negatively charged glomerular basement membrane. The remaining blood flows through the efferent arterioles and extends in a close network of arterial and venous microcapillaries, the so-called peritubular capillaries, providing a close connection between the blood circulation and the tubular cells.

Macroscopically, the kidney can be divided into three major parts. The outer shell, the cortex, consists mainly of glomeruli and the proximal and distal tubuli; it receives more than 80% of the arterial blood delivered to the kidney. The inner part, the medulla, contains predominantly Henle's loops and collecting ducts of several nephrons. Finally, in the papilla, all of the collecting ducts come together and urine is excreted into the urether. A schematic overview of the anatomical structure of the kidney is depicted in Fig. 1.

In the context of drug targeting, the kidney offers several favorable characteristics: a relatively high blood flow, an efficient filtration system, a spectrum of tubular transport systems for inorganic and organic compounds, as well as the presence of a versatile metabolic system.

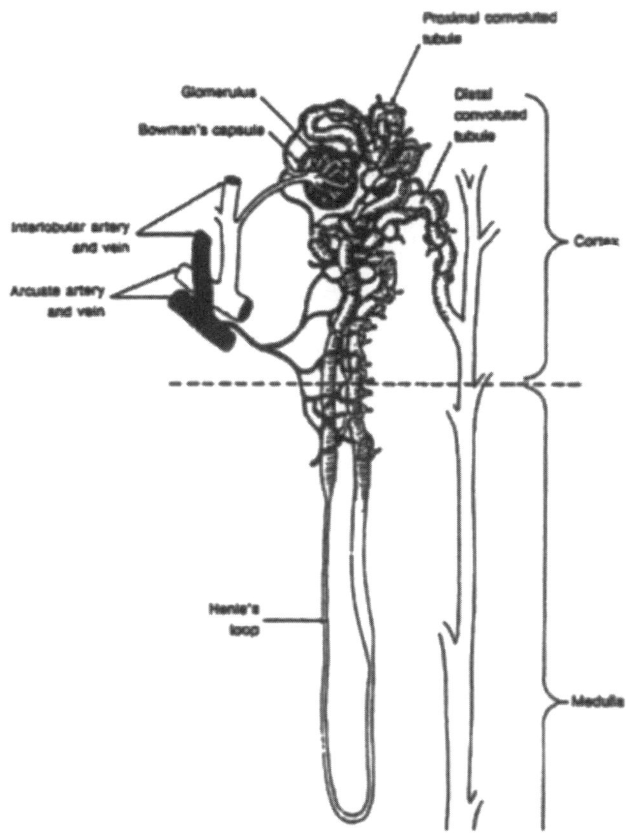

Figure 44.1 The functional nephron with representative blood supply in the cortical region.

B. Blood Flow and Filtration: The Glomerulus

In the glomerulus, the connecting tissue of Bowman's capsule that consists predominantly of endothelial cells, glomerular epithelial cells, and mesangial cells surrounds a capillary network of afferent and efferent arterioles. In the glomerulus, blood components with a size up to approximately 67 kDa can be filtered through the glomerular basement membrane. The latter contains predominantly negatively charged glycoproteins. Consequently, under normal physiological conditions, small molecules will be filtered, whereas large molecules (blood cells, immunoglobulins, and relatively large plasma proteins) will remain in the circulation. Net charge, size, and rigidity of the molecular structure of proteins are important factors in determining the sieving process (4).

In the glomerulus, the mesangial cells represent a suitable target for drug delivery. Not only do these cells play an essential role in the development and progression of both acute and chronic inflammatory processes, they can also be reached quite easily since a continuous flow of blood plasma flows along the mesangial cells through mesangial fenestrations (5).

C. Transport Systems: The Tubules

After filtration in the glomerulus, the preurine flows along the proximal tubular cells, a cell group specializing in the transport of sodium, water, glucose, and a variety of other molecules. In the tubule, a monolayer of epithelial cells separates the tubular lumen from the blood. In the tubular cells, three different transport mechanisms can be distinguished, both located at the basolateral membrane and the apical brush border membrane, facing the blood site and the tubular lumen, respectively. Apart from simple passive diffusion across the tubular membrane, the active transport of compounds into and out of the tubular cells is mediated by carrier proteins. In addition, endocytosis and transcytosis occurs at both poles of these cells (6,7). At least three categories of saturable and energy dependent carrier systems exist in the tubular membranes: for organic cations, for organic anions, and for uncharged compounds (8). In fact, multiple carrier systems contribute to each of these categories. Other basolateral secretory transport systems include the γ-glutamyl transport system and those for glycoproteins (9,10). Finally, certain proteins (insulin, EGF) seem to be transcytosed across the tubular cells via receptor-mediated uptake (11).

On the other hand, valuable endogenous compounds, though freely filtered in the glomeruli, only appear in the urine to a small extent. These compounds are ''rescued'' during passage through the tubuli. These retention mechanisms consist of a variety of, mostly, carrier-mediated processes at the luminal site of the tubular cell. Substances transported by reabsorptive systems include sugars (12), amino acids (13), dipeptides (14), urate (15), folate (16), nucleosides (17), and certain proteins (18). The tubular reabsorption of low-molecular-weight proteins (LMWPs) as a crucial mechanism in relation to renal drug delivery will be discussed separately (see Section IV.A).

The proximal tubular cell also plays a central role in the regulation of inflammatory processes in the kidney. A wide variety of noxious triggers have been identified, including cytokines, growth factors, hypoxia, ischemia, nephron damage, and luminal obstruction. Finally, tubular protein overload as a result of glomerular proteinuria and high tubular delivery of glucose in the diabetic state is considered to be an important factor causing tubular activation (19,20). The proximal tubular cell therefore is a central target for drug delivery (21).

D. The Metabolic Function of the Kidneys

The kidneys are involved in the metabolism of many endogenous and exogenous substances. These compounds are concentrated in the kidney before being eliminated via the urine. Therefore the driving force for metabolic conversion can be high. Within the kidney, various enzymes are involved in the biotransformation of many different compounds. For metabolism of exogenous compounds such as drugs, but also for some endogenous substrates, enzymes involved in both phase I and phase II metabolic routes are present in the kidney, e.g., cytochrome P450, cytochrome b5, glucoronyl transferase, and sulfotransferase (22–26). In addition, renal tubular cells contain various proteases for the degradation of proteins and oligopeptides. These enzymes are either located at the brush border membrane or can be found in the smooth endoplasmic reticulum and the lysosomes of these cells (27). Degradative enzymes include a variety of endopeptidases, exopeptidases, and esterases (28–32). Due to this versatile metabolic apparatus, the kidneys are also able to degrade macromolecular drug carriers and can regenerate covalently coupled drugs within intracellular compartments, like the lysosomes and the cytoplasm.

III. LOW-MOLECULAR-WEIGHT PROTEINS AS CARRIERS FOR RENAL DRUG TARGETING

A. Renal Handling of Low-Molecular-Weight Proteins

Low-molecular-weight proteins (LMWPs) are small proteins with molecular weights up to approximately 25,000 Da. Their application for renal drug targeting was considered since it was demonstrated that radiolabeled LMWPs accumulated specifically in the kidney (33). Autoradiography of the kidney after administration of labeled probes revealed a specific accumulation within the renal proximal tubular cells. Besides in the kidney, LMWPs do not seem to accumulate to a large extent in tissues elsewhere in the body (Fig. 2). Comparison of the kinetic features of different LMWPs revealed that all of the injected proteins are rapidly cleared from the circulation by the kidney leading to temporary accumulation in this organ. The fraction of the LMWPs that is taken up by the kidney in such studies ranges from 40 to 80% of the injected dose (34–36). In some studies these values are inherently underestimations, since part of the endocytosed protein was degraded during the study, followed by transport of radiolabel out of the kidney.

Within the kidney, the LMWPs are freely filtered in the glomerulus and subsequently reabsorbed by the proximal tubular cells through receptor-mediated endocytosis (37). Christensen et al. have identified the megalin/gp 330 receptor to be responsible for this tubular uptake (38). After endocytosis, an endocytic vesicle (endosome) is formed that is subsequently transported to the lysosomes. The

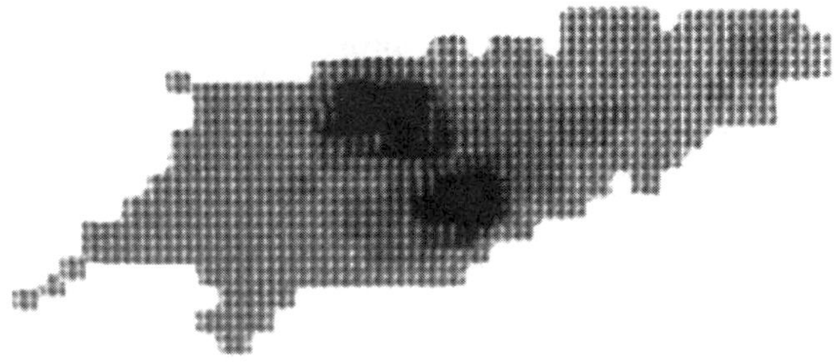

Figure 44.2 Gamma-camera imaging after an intravenous injection of a ^{123}I-radiolabeled LMWP in the rat, showing the predominant uptake of the LMWP by the kidneys.

LMWP is degraded intralysosomally into small peptides and free amino acids by a variety of proteolytic enzymes, such as proteases and hydrolases (39–41).

B. Low-Molecular-Weight Proteins as Renal Specific Drug Carriers

We considered the possibility that through coupling a drug to an LMWP, an inactive ''prodrug'' could be formed that could potentially deliver coupled drugs to the kidney. The tissue distribution of the drug could thereby be manipulated in favor of the kidney. This in fact results from the overruling of the physicochemical properties of the linked drug by those of the LMWP. In the kidney, the drug–LMWP conjugate follows the same pathway as endogenous LMWPs and ends up in the lysosomes of the proximal tubular cells. In the lysosomes, the drug is then released from the protein, either by proteolytic or hydrolytic actions, and is excreted from the lysosomes into the cytoplasm of the cells. Subsequently, the released drug can act either within the tubular cell or further downstream after excretion into the urine or back into the renal venous circulation (42). Figure 3 gives a schematic representation of the renal handling of an LMWP–drug conjugate.

A number of essential variables in this concept require closer examination. First, the LMWP carrier must contain functional groups allowing drug attachment. The chemistry of drug–LMWP conjugates will be described in more detail in Section IV.C, using ACE inhibitors as the model drugs. Second, the physicochemical properties of the LMWP should largely remain intact to ensure tubular uptake and prevent loss of intact conjugate into the urine. Third, the drug conjugate complex must be stable within the circulation to prevent premature systemic side effects of the free drug. Finally, after successful tubular uptake, the drug must be released in the catabolically active lysosomes. For the individual drug targeting constructs, these aspects will be addressed in more detail in Section V.

In addition, the ultimate effectiveness of the targeted drug is also determined by the local redistribution of the drug after endorenal release from the carrier. Transport systems or simple passive reabsorption, described in detail in Section II.B, can relocate the drug into either the primary urine or the systemic circulation. Reabsorption processes therefore may impose limits on the cellular selectivity of drugs released endorenally.

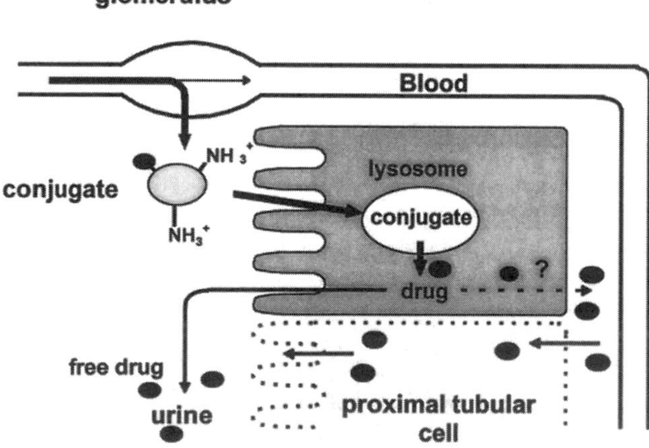

Figure 44.3 Schematic representation of the concept of renal drug targeting with LMWPs. After filtration in the glomerulus, the conjugate is reabsorbed by the proximal tubular cell and degraded in the lysosomes. The free drug diffuses out of the lysosomes and can diffuse out of the tubular cell at the luminal side into the urine, or at the basolateral side into the renal interstitium. From the interstitium the drug can diffuse into the blood vessel or back into the proximal tubule. For some drugs the transport out of the proximal tubular cell might also be mediated by carrier proteins.

C. Potentials and Limitations of Low-Molecular-Weight Proteins as Renal Drug Carriers

The processes responsible for the selective tubular reabsorption of LMWP–drug conjugates have been summarized in the previous section. However, to be able to determine the in vivo effectiveness of such constructs, the LMWP conjugates have to be administered for a longer period of time in therapeutically appropriate dosages. The required therapeutic drug dosage is governed by the pharmacokinetic properties of the drug targeting preparations as well as the pharmacokinetic and pharmacodynamic features of the targeted drug.

Pharmacokinetic aspects of the free drug are of importance, in particular after the release from the protein carrier in the desired target cell. First, an effective concentration of the drug will only be reached if the drug is retained at the target site after release at least to some extent. This retention is dominated by the physicochemical properties of the drug (diffusion out of the target cell) and its tendency to act as a substrate for excreting carrier proteins (see Sections II.C and III.B). Since the kidney is an organ with a wide variety of carrier systems specialized to transport compounds both into the cell and back into the systemic circulation, it is obvious that these processes will influence renal drug retention. Second, the targeted drug may be inactivated due to metabolizing enzymes, thus influencing the cellular availability and duration of action of the drug. The metabolizing capacity of the kidney for the targeted drug has been described in Section II.D.

The therapeutic response is, apart from delivery to the desired cell type, determined by the intrinsic activity of the drug, often expressed as the concentration in which 50% of the maximal effect is reached (EC_{50}). For a drug with a low EC_{50} relatively long-lasting therapeutic levels have to be attained at the target site. Furthermore, this intrinsic potency of the drug, together with the number of available coupling sites in the carrier molecule, determines and therefore also influences the necessary amount of carrier protein. Thus both pharmacokinetic and pharmacodynamic parameters have to be taken into account when designing an LMWP–drug conjugate.

For the attachment of drugs, the amino groups of 7 lysine moieties are available in the lysozyme molecule for drug coupling. However, these amino acids do also contribute to the solubility of the LMWP and may influence the interaction with the megalin receptor that mediates tubular reabsorption (39,43,44). Therefore only a limited number of amino groups can be substituted to maintain most of the physicochemical properties of the lysozyme molecule. Kok et al. studied the effect of charge of the carrier protein on the renal accumulation of a lysozyme conjugate using fluorescein isothiocyanate (FITC) as a model drug (45). They showed that 45 $\pm$ 4% of a single dose of 0.33 mg $\cdot$ kg^{-1} negatively charged FITC–lysozyme was excreted intactly in the urine, versus 29 $\pm$ 4% of nonderivatized FITC–lysozyme ($P < 0.05$). After introduction of extra positive charge in the FITC–lysozyme preparation, the conjugate seemed to be completely reabsorbed, since it was not excreted into the urine intactly. However, an organ distribution study with the latter ^{125}I-radioiodinated conjugate showed that only 12.0 $\pm$ 0.1% of the dose accumulated in the kidney. Apparently, the distribution was no longer selective for the kidney and indeed significant amounts were found in the liver (53 $\pm$ 1%) and the spleen (5 $\pm$ 2%). Consequently, a proper balance of positive and negative charges has to be found for optimal renal delivery.

Drug therapy with LMWP conjugates may require the administration of relatively large amounts of lysozyme for a longer period of time. It should be noted, however, that Cojocel et al. showed that infusion of relatively high dosages of lysozyme might lead to a dose dependent and time dependent decrease of tubular reabsorption, as well as to adverse effects on systemic blood pressure and renal function (46,47). Therefore we have recently investigated the maximal amount of lysozyme that can be administered with an efficient tubular reabsorption and minimal effect on blood pressure and renal function. We compared continuous infusion of different dosages of lysozyme with single-dose injections in unrestrained rats. These studies showed that continuous low-dose infusion of lysozyme is preferable over single high-dose injections, since infusion did not affect systemic blood pressure in dosages up to 1000 mg $\cdot$ kg^{-1} $\cdot$ $6h^{-1}$. Additionally, total urinary lysozyme excretion was sixfold lower after infusion of lysozyme. However, infusion of the LMWP could not entirely prevent the loss of intact lysozyme into the urine. About 8% of the dose was renally excreted after 100 mg lysozyme $\cdot$ kg^{-1} $\cdot$ $6h^{-1}$, while up to 33% of the dose was lost in the urine following 1000 mg lysozyme $\cdot$ kg^{-1} $\cdot$ $6h^{-1}$. Dosages exceeding 100 mg $\cdot$ kg^{-1} also produced a decrease of renal blood flow and glomerular filtration rate. Therefore the amount of lysozyme that can be safely administered is limited. Using drugs with a high intrinsic activity (low EC_{50}) may therefore limit the required dose of lysozyme and may prevent the occurrence of unwanted side effects on renal hemodynamics (R. F. G. Haverdings et al., manuscript in preparation).

Obvious disadvantages of the LMWP strategy for the treatment of chronic renal disease are the requirement of parenteral administration and the possible immunogenicity of the drug conjugate. With respect to alternative administration routes, it is encouraging that technologies for parenteral administration are being developed. This includes pulmonary spray technology and programmable wearable infusion pumps. Also it should be realized that parenteral dosing of proteins is not uncommon, since insulin dependent diabetic patients perform daily administration via subcutaneous administration for extensive periods. If immunogenicity is a serious limitation, in particular for chronic treatment, alternatively a synthetic polymer may be used as a ''reabsorptive'' carrier. For short-term clinical interventions, aiming to protect the kidney during acute reperfusion or preventing allograft rejection after transplantation, the prerequisite of parenteral administration does not constitute a serious limitation anyway.

D. How Chemically to Link an ACE Inhibitor to a Low-Molecular-Weight Protein

The characteristics of the bond between drug and protein are very important for the stability of the conjugate, but also for a proper release of the parent drug after the conjugate has accumulated in the target organ. Ideally, the bond between drug and protein should be stable while the conjugate distributes throughout the body, but at the same time it should be easily degradable after the drug has reached its target site. Drug–protein conjugates, like those that have been designed for renal drug delivery, are accumulated in the target organ by adsorptive endocytosis and followed by transport to the lysosomes (48). This cellular compartment contains a variety of proteases and esterases that can be instrumental in the enzymatic degradation of drug–protein conjugates. The regeneration of the parent drug can be ensured by the insertion of a spacer molecule between drug and protein (Fig. 4). Such a spacer moiety can provide a linkage that can be degraded either enzymatically or chemically under, for instance, influence of the acidic pH of the lysosomes (49).

In the past years, many structurally different ACE inhibitors have been developed. Figure 5 gives an overview of the chemical structures of currently available ACE inhibitors (50). Since these drugs have a similar pharmacological action, the most suitable drug for a particular conjugation strategy can be selected. Table 1 summarizes the coupling strategies that are discussed in this chapter. This table also gives examples of ACE inhibitors that can be coupled via the various strategies, and the type of bond that is formed between the drug and the linker. The strategies are based on well-known synthetic procedures that have been used for the preparation of drug–protein conjugates. They will be discussed in relation to the specific functional groups that are used for the conjugation process. Some functional groups are present in the structures of all the listed ACE inhibitors, such as the carboxylic acid group; other functional groups are unique for specific agents in this therapeutic class.

1. Conjugation via the Carboxylic Acid Group

The first ACE inhibitors, captopril, enalapril, and lisinopril, were designed as analogues of peptides found in the venom of a Brazilian viper (51). These drugs all contain a proline carboxylic acid group. In the more recently developed ACE inhibitors, like quinapril and cilazapril, the proline carboxyl group was replaced by conformationally restricted analogues (52). Since lysosomal proteases should

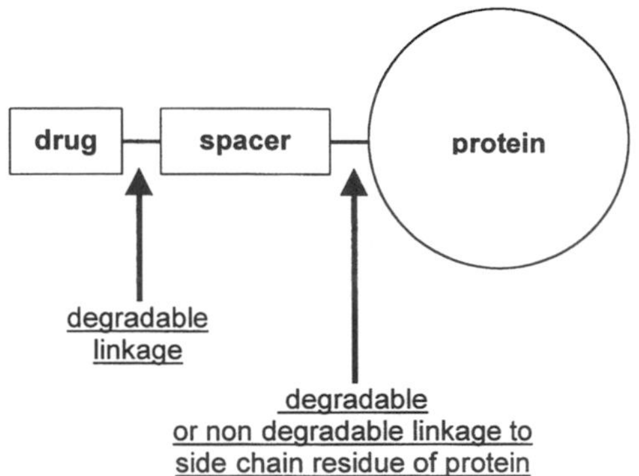

Figure 44.4 Schematic representation of a drug–spacer–protein conjugate. The bond between drug and spacer must be biodegradable. The bond between spacer and protein might be either nondegradable or degradable.

be able to degrade bonds with a peptidelike character, the ACE inhibitors with a proline carboxylic acid group can be coupled bioreversibly via a peptide spacer. We propose two types of peptide spacers that can be tailored to different classes of peptidases (Fig. 6). In both spacer concepts, the proline carboxylic acid group is coupled via a natural peptide bond to the α-amino group of the spacer, thus forming a bond that should be cleavable by peptidases.

The first spacer consists of a specific sequence of amino acids that is a substrate for a lysosomal endopeptidase, for instance one of the cathepsins. The drug is coupled to the N-terminal amino group of the spacer, while the spacer itself is conjugated at its C-terminal carboxyl group to the carrier protein. The release of the parent drug can be the result from a single-step degradation of the prolyl–spacer bond (Fig. 7A) or via the formation of a drug–amino acid intermediate product, which is further degraded to the parent drug by a carboxypeptidase (Fig. 7B). Several peptide sequences have been reported as endopeptidase sensitive spacers (Table 2). Of these sequences, the spacers that have been used for the cytostatic drug methotrexate (MTX), which also has peptidic carboxyl groups, are attractive candidates for the conjugation of an ACE inhibitor to a LMWP carrier. The other peptide spacers in Table 2 have been optimized for the coupling of drug molecules that contain a primary amino group, for instance the cytostatic drug doxorubicin. Lysosomal degradation studies of those spacers with isolated lysosomal endopeptidases, like cathepsin B, D, and L, showed that these enzymes cleave the peptide spacer gly-phe-leu-gly either next to the drug moiety or between phe and leu, followed by removal of the remaining peptide fragment by an aminopeptidase (53,54). If the gly-phe-leu-gly spacer is used for the proline carboxyl drugs, the release of the parent drug is likely to result from the two-step degradation route as depicted in Fig. 7B.

The other type of peptide spacers that we propose for conjugation of an ACE inhibitor via the proline carboxyl group are the exopeptidase sensitive spacers. Figure 8 shows how the drug is coupled to the free α-amino group of the spacer, which is subsequently linked to the LMWP carrier via its side chain functional group. Since the spacer has a free α-carboxyl group, it should be degradable by lysosomal carboxypeptidases (Fig. 8A). We tested the carboxypeptidase sensitive spacer concept by incubating the model compound di(Z-pro)-lysine with lysosomal enzymes. This conjugate was degraded to Z-proline and Z-pro-ε-lysine, which proves that carboxypeptidases are indeed capable of digesting peptides in which the side chain group is modified. Interesting candidates for this spacer concept are amino acids with a reactive side chain, for instance the thiol-containing amino acid cysteine. After attachment of the drug to the amino group of this spacer, the

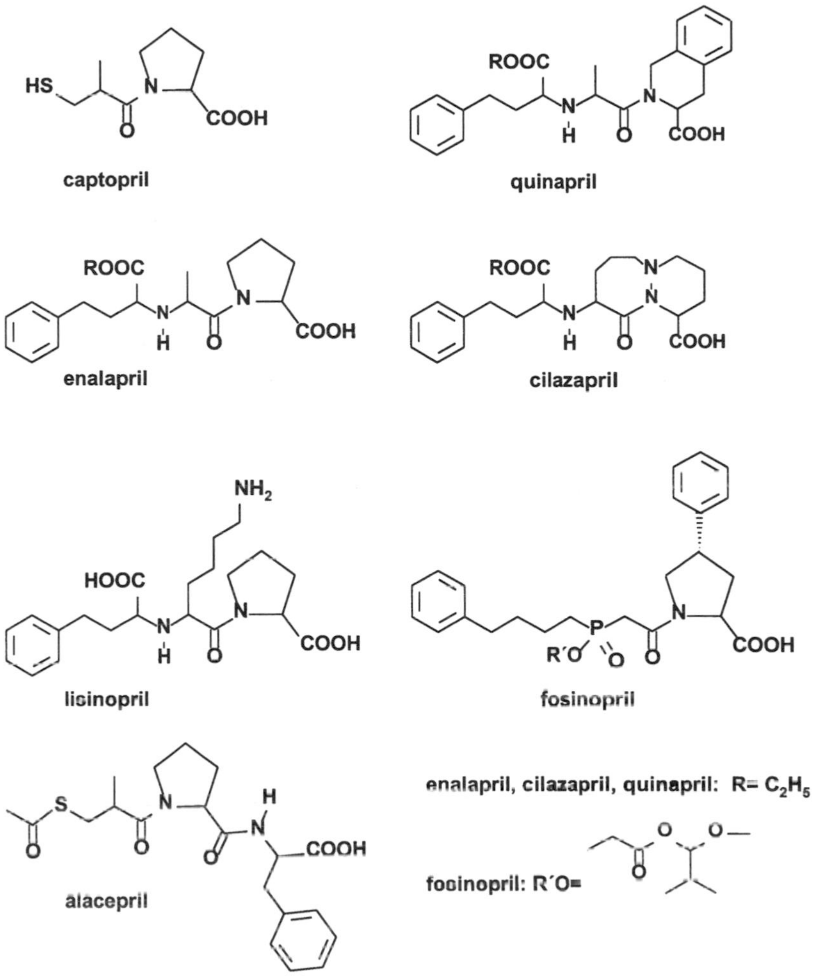

Figure 44.5 Chemical structures of ACE inhibitors.

Table 44.1 Linkages That Can Be Used for the Conjugation of an ACE Inhibitor to an LMWP

Functional group	Drug	Linkage
Proline carboxylic acid	Captopril, enalapril, lisinopril, alacepril	Amide bond Ester bond
Other carboxylic acid	Other ACE inhibitors	Ester bond
Primary amine	Lisinopril	Acid sensitive amide bond
Thiol	Captopril	Disulfide bond
Phosphinic acid	Fosinopril	Phosphoramide bond

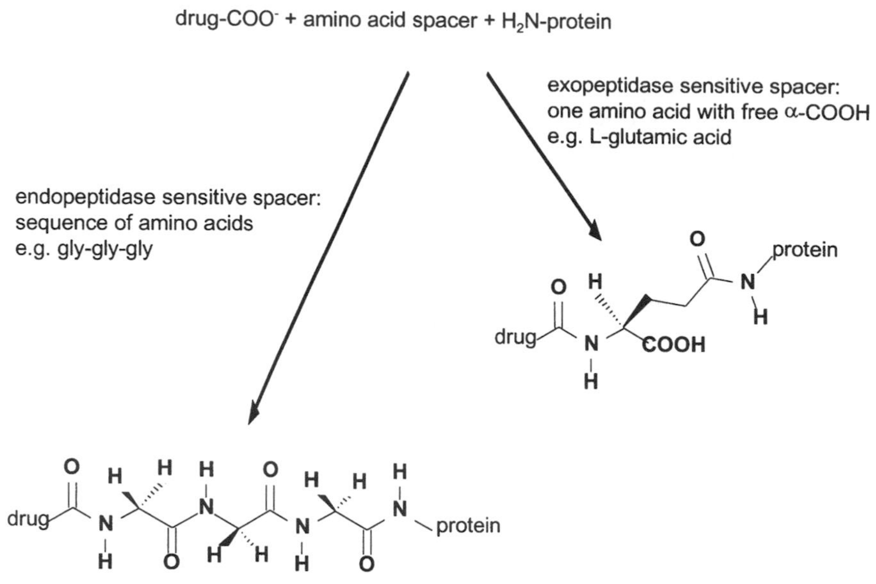

drug-COO⁻ + amino acid spacer + H_2N-protein

exopeptidase sensitive spacer:
one amino acid with free α-COOH
e.g. L-glutamic acid

endopeptidase sensitive spacer:
sequence of amino acids
e.g. gly-gly-gly

Figure 44.6 Concepts for peptidase sensitive spacers. The release of the parent drug is the result of degradation of the bond between drug and spacer mediated by either an endopeptidase or an exopeptidase.

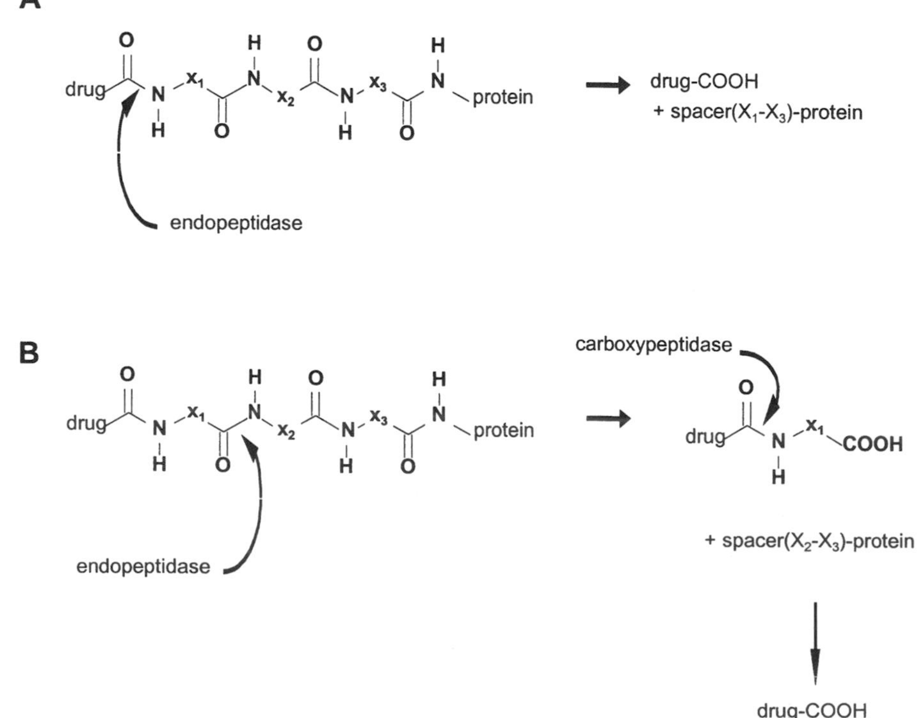

Figure 44.7 Endopeptidase sensitive spacers: routes of drug release. (A) Direct attack of endopeptidase at bond between drug and spacer. (B) Attack of endopeptidase at bond in spacer, followed by exopeptidase-mediated degradation of bond between drug and remaining spacer amino acid.

Table 44.2 Peptide Spacers That Have Been Used as Spacers in Lysosomotropic Drug Delivery

Drug	Spacer	Reference
Conjugation at carboxylic acid group of the drug		
Methotrexate	ala-leu-ala-leu	138
	gly-gly-gly	
	gly-gly-phe	
	gly-phe-ala	
Conjugation at amino group of the drug		49
Doxorubicin, daunomycin	ala-leu-ala-leu	
	gly-gly-gly-leu	
	gly-gly-phe-gly	
	gly-phe-leu-gly	
5-Fluorouracil	gly-phe-leu-gly-leu-gly	
Mitomycin C	ala-leu-ala-leu	
	gly-phe-ala-leu	
	gly-phe-leu-gly	
Primaquine	ala-leu-ala-leu	

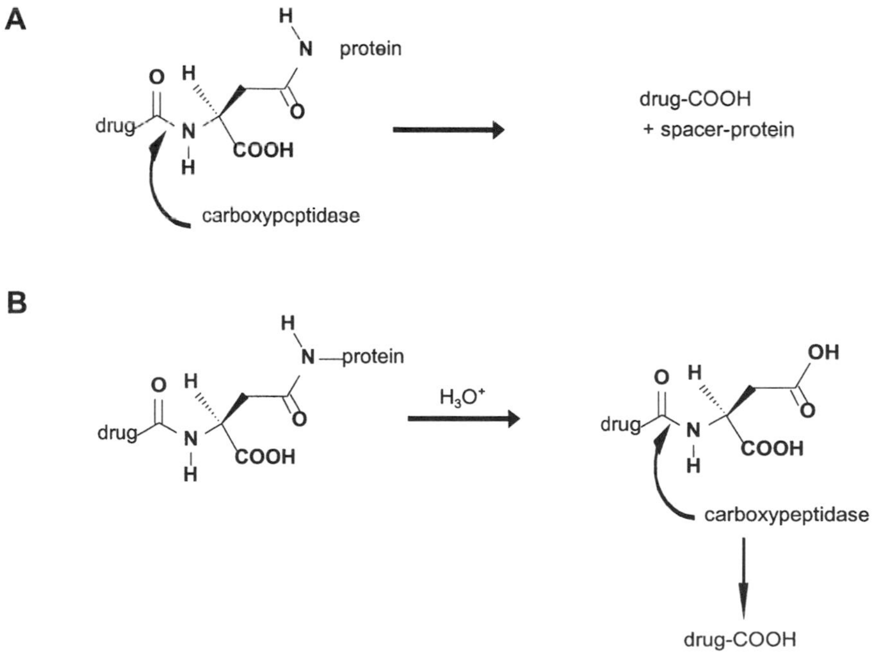

Figure 44.8 Exopeptidase sensitive spacers: routes of drug release. (A) Direct attack of endopeptidase at bond between drug and spacer. (B) Aspartic acid spacer. The bond between spacer and protein can undergo acid-catalyzed degradation. In a second degradation step, the drug is released from the spacer by exopeptidase-mediated degradation.

conjugation of the spacer to the LMWP can be performed under mild conditions by reacting the thiol group of cysteine with a LMWP that has been thiolated with N-succinimidyl 3-(2-pyryldithio) propionate (SPDP) or modified with maleimido groups with m-maleidobenzoyl-N-hydroxysuccinimide ester (MBS). A similar conjugation procedure, using cystamine as the linker, has been described for the coupling of MTX and several peptides to macromolecules (55–57).

Another amino acid that can be used as an amino acid spacer is aspartic acid, which has been reported as an acid sensitive spacer for the conjugation of doxorubicin (58). This spacer might undergo enzymatic degradation, as well as acid-catalyzed degradation of the spacer–LMWP bond followed by enzymatic degradation of the drug–spacer bond (Fig. 8B).

A major drawback of the conjugation of a proline-containing ACE inhibitor via its carboxylic acid group is the formation of undesired condensation products. We considered enalapril as the representative example of this class of ACE inhibitors, since this drug does not have functional groups that have to be protected during the conjugation reaction. However, the activated proline carboxylic acid group of enalapril can react with the secondary amino group of the ACE inhibitor structure itself, resulting in the formation of a piperazinedione internal condensation product (R. J. Kok, unpublished data) (Fig. 9). This type of condensation products has also been described as one of the few metabolic products of enalapril (59). The cyclization reaction can also occur with lisinopril, but not with captopril. However, the free thiol group of captopril can interfere with the conjugation reactions at the carboxyl group and therefore it needs to be protected during the synthesis of the conjugate. Since a conjugate with a free thiol group can induce an immunogenic response, thiol group protection is also preferable in vivo (60). This could be achieved by protecting the thiol group as a thioester, since such a group is metabolized in vivo to the free thiol group, as was shown for the ACE inhibitor prodrug alacepril (Fig. 5) (61). The preparation of a captopril-protein conjugate could be started with alacepril, in which case the prodrug moiety at the carboxylic acid group can be regarded as the first amino acid of the peptide spacer.

In order to test whether the amino acid spacer could also be useful for drugs with carboxyl groups that have nonpeptide characteristics, we synthesized the model compound cilazapril–phenylalanine. Degradation experiments with lysosomal proteases indicated that this nonpeptide conjugate could not be cleaved by proteases (R. J. Kok et al., unpublished data). Similar lysosomal degradation experiments were performed with amino acid conjugates of naproxen, an anti-inflammatory drug with a nonpeptide carboxylic acid group (62). All of the model compounds proved to be resistant to lysosomal degradation. The use of an α-hydroxy acid, instead of an α-amino acid, as spacer proved to be a feasible solution for obtaining a biodegradable conjugate of naproxen (62,63). Such a spacer, of which L-lactic acid is a typical example, contains a hydroxyl group that can be used for the conjugation of the drug via an ester bond, and a carboxyl group that is used for the conjugation of the spacer to the protein (Fig. 10). The naproxen-L-lactic conjugate was sufficiently stable for renal drug delivery. Lysosomal degradation experiments with this conjugate showed that the lysosomal degradation of the ester bond was even slower than the degradation of the carrier backbone itself (63).

Although ester bonds have been used frequently for the preparation of small prodrugs, they have not been frequently employed as biodegradable spacers in macromolecular drug conjugates (64). A disadvantage of the use of an ester bond is its relative instability in the circulation due to degradation by plasma esterases (65).

2. Conjugation via the Amino Group: Lisinopril

The ACE inhibitor lisinopril contains a free primary amino group that has similar characteristics to the lysine side chain. Lysosomal degradation experiments have demonstrated the inability of lysosomal enzymes to degrade amide bonds that are formed with such an amino group.

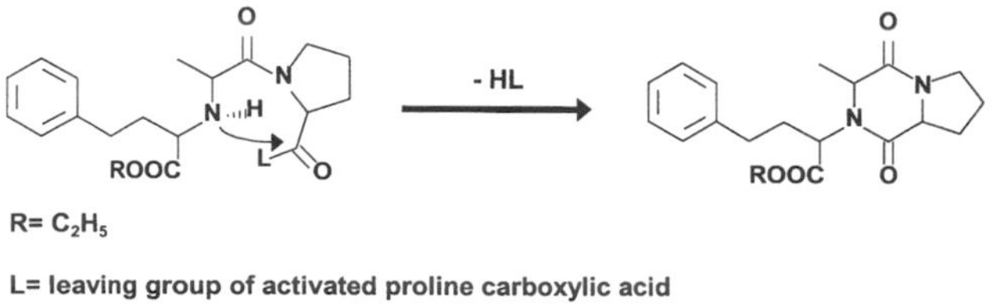

R= C_2H_5

L= leaving group of activated proline carboxylic acid

Figure 44.9 Internal condensation reaction of enalapril. The activated carboxylic acid group is attacked by the secondary amino group of the ACE inhibitor.

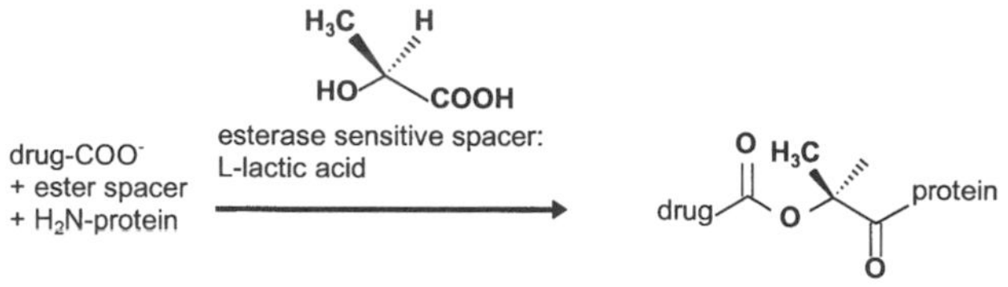

Figure 44.10 Concept of an esterase sensitive spacer: L-lactic acid spacer. After conjugation of the drug to the hydroxyl group of L-lactic acid, the spacer is coupled to a lysine side chain by activation of the carboxylic acid group.

Therefore we propose the use of an acid sensitive spacer instead of aiming at enzymatic degradation for conjugated lisinopril via its amino group. Acid sensitive spacers are stable at neutral pH but undergo autocatalytic degradation at the acidic pH of the lysosomal compartment. Acid sensitive spacers that have been used for drug delivery are depicted in Fig. 11. Of these spacers, the *cis*-aconitic anhydride (*cis*-ACO) spacer has been used frequently for the conjugation of doxorubicin to different macromolecular carriers (66–70). This spacer is commonly reacted with the amino group of the drug, followed by an 1-ethyl-3-(3-dimethylaminopropyl) carbodiimide (EDCI) catalyzed conjugation of one of the *cis*-carboxyl groups with an amino group of the carrier. If this *cis*-ACO spacer is applied for the conjugation of lisinopril, the carboxyl groups of the drug itself should be protected beforehand, for instance by esterification with ethanol. The above-described conjugation route is depicted in Fig. 12 and has recently been carried out (R. J. Kok et al., manuscript in preparation).

The extra protection of the carboxylic acid groups of lisinopril is not necessary when one of the other *cis*-diacid spacers is used, since the conjugation reaction of those spacers to the carrier protein is aimed at functional groups that are not present in lisinopril. Another disadvantage of the *cis*-ACO linker is that the conjugation of the spacer to the LMWP will not take place exclusively at the γ-carboxyl group but also at one of the *cis*-carboxyl groups (66). This results in the abolishment of acid sensitivity in some of the conjugated drug molecules. Therefore we also aim at

HOOC COOH COOH P D 1
COOH COOH N H P D 2
H COOH H2N COOH P D 3
HO COOH COOH HO OH D P 4
R COOH N COOH O P D 5

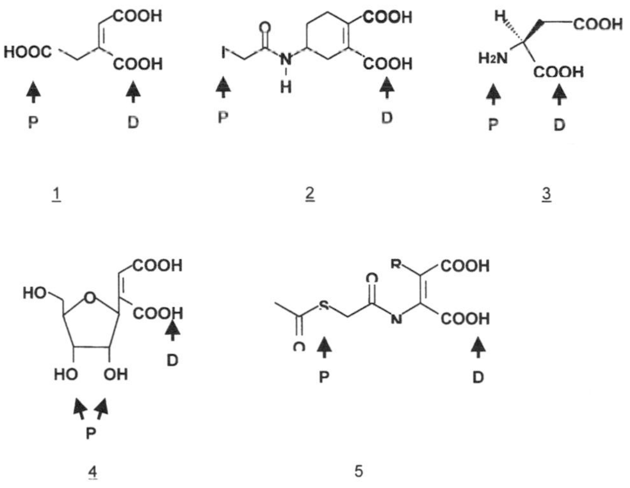

Figure 44.11 Acid sensitive spacers. P denotes the attachment site for the protein, D denotes the drug attachment site. (1) *cis*-ACO (66); (2) 4-(iodoacetamide)-1-cyclohexene-1,2-dicarboxylic acid (139); (3) L-aspartic acid (58); (4) *cis*-ribofuranomaleic acid (140); (5) 3-substituted 2-N-(S-acetyl-thioacetyl)amino maleic anhydride (141).

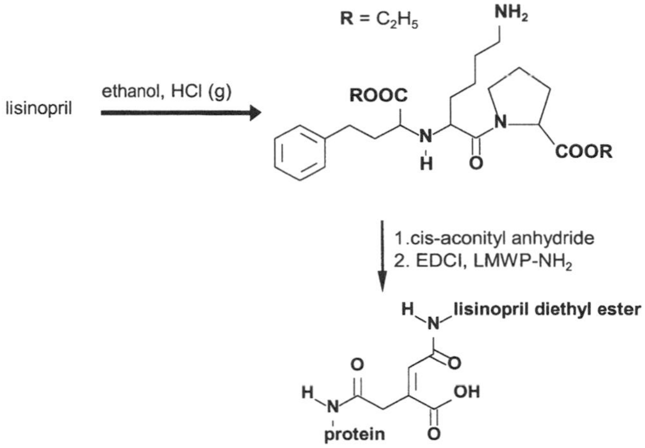

lisinopril → ethanol, HCl (g) → R = C_2H_5, NH_2, ROOC, N H, O, N, COOR
↓ 1.cis-aconityl anhydride 2. EDCI, LMWP-NH_2
H N lisinopril diethyl ester, O, O, H N protein, OH, O

Figure 44.12 Conjugation of lisinopril via the *cis*-ACO acid sensitive spacer. In the first reaction step, the carboxylic acid groups of lisinopril are esterified with ethanol. Following conjugation of *cis*-ACO to the primary amino group of lisinopril, the spacer is activated with EDCI and coupled to an amino group of the LMWP.

preparing a lisinopril conjugate using the acid sensitive spacer 4-(iodoacetamide)-1-cyclohexene -1,2-dicarboxylic acid (R. J. Kok et al., manuscript in preparation).

3. Conjugation via the Thiol Group: Captopril

The ACE inhibitor captopril contains a free sulfhydryl group that offers several possibilities for the conjugation of the drug to an LMWP. Since thiol groups are the most reactive nucleophilic groups in proteins, these groups can be reacted selectively in the presence of other amino acid side chains at neutral pH (71). Many protein cross-linking procedures that have been developed are based on the specific reactivity of thiol groups (72). These reagents form either a disulfide bond, which can be cleaved by reducing agents, or a thioether bond (73). Since the latter thioether bonds were not degradable when applied in drug targeting conjugates like immunotoxins, this type of conjugation seems not suitable for the preparation of ACE inhibitor protein conjugates (74). In contrast, disulfide bond cross-linked immunotoxins proved effective in vivo. Several disulfide cross-linking reagents have been applied for the preparation of immunotoxins (72). Of these reagents, the heterobifunctional cross-linker SPDP was used most frequently (75). Other reagents that were developed to introduce free thiol groups into macromolecules are 2-iminothiolane (2-IT) and S-acetylmercaptosuccinic anhydride (SAMSA) (76,77). Since the unhindered disulfide bonds that are formed with these reagents can undergo premature reduction in the circulation, other cross-linkers have been developed that contain sterically hindered disulfide bonds (74,78–84). Figure 13 shows representative examples of disulfide cross-linking reagents. The stability of the disulfide bond can be increased by the introduction of bulky substituents next to the disulfide moiety. The steric effect of the phenyl group in the ethyl S-acetyl 3-mercaptopropionthioimidate (AMPT) spacer is greater than that of the methyl group, but the reduced hydrophilicity of the phe-AMPT decreases the reactivity of the spacer (82). Although the introduction of substituents results in an increased stability of immunotoxins both in vitro and in vivo, no major differences were found in cytotoxicity, a process for which intracellular disulfide bond degradation is a prerequisite (74,81).

Some of the disulfide spacers have a 2-pyridyl-sulfide protecting group at their thiol group. This protecting group reacts readily with free thiols to yield a disulfide bond. The other reagents contain a thioacetyl group that can be deprotected with hydroxylamine and subsequently activated with 2-2′ dipyridyl disulfide (2-DP) (72). The second coupling site of these bifunctional linkers is directed to a primary amino group of the protein. Figure 14 shows the conjugation routes of captopril with three different spacers, resulting in conjugates with a differently charged linker. Since the net charge of a protein might influence its pharmacokinetic behavior, it is interesting to compare the influence of these linkers on the drug targeting properties of these conjugates.

4. Conjugation via the Phosphinate Group: Fosinopril

The ACE inhibitor fosinopril belongs to a class of ACE inhibitors that contains a phosphinate group as the Zn binding ligand (50). The prodrug fosinopril is hydrolyzed in the gastrointestinal mucosa and liver to the active metabolite

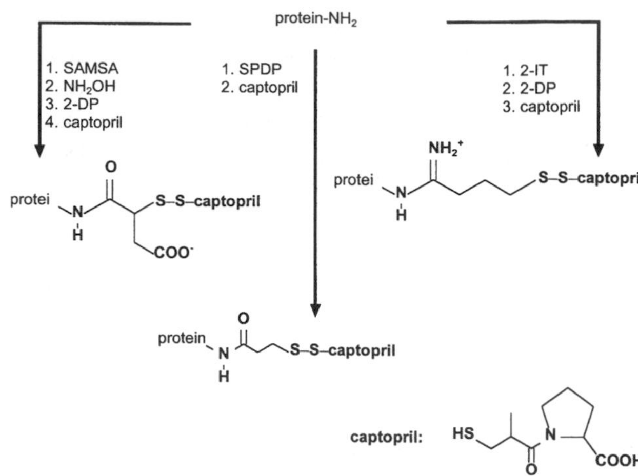

Figure 44.14 Conjugation of captopril via a disulfide linkage. The three proposed conjugation routes yield conjugates with a differently charged linkage. The conjugation of captopril to the SPDP spacer can be performed directly after the reaction of SPDP with the protein. The SAMSA spacer has to be deprotected with hydroxylamine (NH_2OH), and subsequently activated with 2,2-dipyridyl disulfide (2-DP). Conjugation of captopril to the 2-IT spacer also proceeds via 2-DP activation of the spacer.

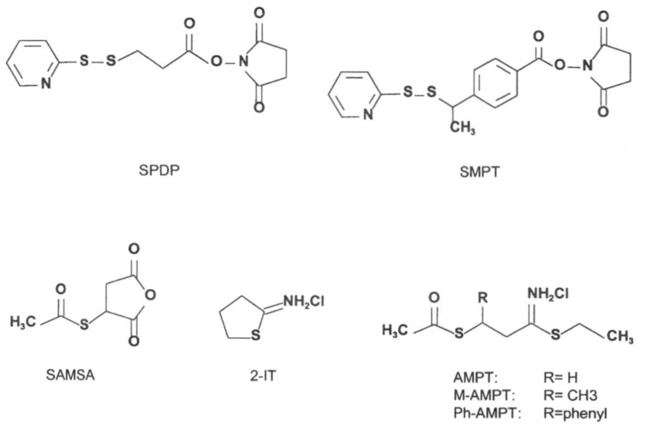

Figure 44.13 Disulfide cross-linking reagents (72).

fosinoprilate (85). In this section we propose procedures for the conjugation of fosinoprilate via its phosphinate group, thus replacing the prodrug group of fosinopril by a linkage to a proteinaceous carrier. These conjugation strategies are based on the structures of prodrugs that have been developed for nucleotides that also contain phosphate or phosphonate groups (86,87). Due to the polarity of the negatively charged phosphate group, drugs with this group exert a low bioavailability and a low diffusion rate across lipid bilayers (86). Various substituents have been attached to the phosphate group to increase the lipophilicity of these drugs. Of the linkages used in these prodrugs, the phosphoramide would be the most attractive for the preparation of a fosinoprilate–protein conjugate, since this linkage is acid sensitive (87). The phosphoramide linkage has been applied for the lysosomotropic drug delivery of nucleotides to the liver and macrophages using neoglycoproteins or lactosaminated polylysine as drug carriers (88–90). Figure 15 shows the conjugation strategy that can be followed to prepare a fosinoprilate conjugate with a phosphoramide linkage. To prevent the concomitant reaction of the proline-carboxyl group, this group should be esterified in a procedure similar to the one applied to lisinopril in the conjugation strategy described in Section III.D.2.

5. Discussion and Conclusions

In this review multiple conjugation routes that can be applied for the conjugation of an ACE inhibitor to a protein

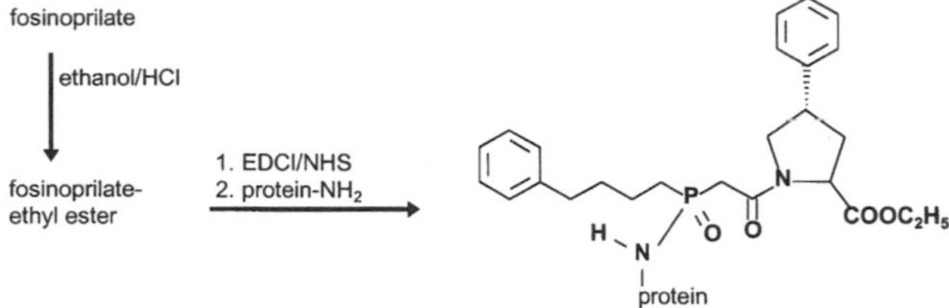

Figure 44.15 Conjugation of fosinoprilate via a phosphoramide linkage. After protection of the carboxylic acid group of fosinopril by esterification, the phosphonate group can be activated and reacted to a primary amino group of the LMWP.

are listed. The availability of structurally different but pharmacologically equivalent ACE inhibitors offers the possibility to choose between various biodegradable linkages between the drug and the carrier molecule. Although the presented strategies were designed for a specific class of drugs, the particular linkages can also be used to prepare drug–protein conjugates for other types of drugs, that is, if they contain a free carboxylic acid, amino, sulfhydryl, or phosphate group.

Of the above presented conjugates, the ones in which the proline carboxylic acid group of the drug is conjugated via a peptide spacer will be the most stable in the bloodstream. Premature liberation of the drug from the conjugate, i.e., before the conjugate is accumulated in the lysosomal compartment of the target cells, is least likely to happen with this type of linkage. However, the stability of the amide bond also limits the applicability of peptide linkages to a small subset of drugs. Another disadvantage of this conjugation route is that the first step in the synthesis route will suffer a serious loss of starting compound when the synthesis is performed with enalapril. We suggest the use of alacepril in order to avoid this problem.

For the ACE inhibitor captopril, we have suggested several strategies that are aimed at the free thiol group. The formation of such a linkage can be accomplished in a straightforward reaction with a commercially available linker. Although disulfide bonds can undergo premature degradation in the circulation, this aspect is of limited importance if the conjugate is accumulated rapidly in the target organ. Disulfide bonds as drug carrier linkers have been incorporated into many conjugates and exhibit only a moderate immunogenic response (91,92). The availability of a spectrum of coupling reagents with differently charged linkers could be instrumental in the fine-tuning of the physicochemical properties of the conjugate.

Alternative conjugation strategies are those in which either lisinopril or fosinoprilate are conjugated via an acid sensitive linkage. The degradation of this type of linkages is a nonenzymatic process and is therefore less dependent on the physicochemical properties of the drug than enzymatic degradation. With the conjugation of lisinopril, the selective conjugation of the spacer to the protein is the critical step. Recently, we prepared such lisinopril-ACO derivatives and linked them to the LMWP lysozyme. These conjugates might show an improved stability compared to disulfide bound captopril–lysozyme conjugates.

IV. DRUGS FOR RENAL DRUG TARGETING AND THEIR EFFECTIVENESS AFTER CONJUGATION TO LYSOZYME

A. Introduction

Currently, several kinds of drugs can be used for the treatment of renal disorders. At present, angiotensin-converting enzyme (ACE) inhibitors are the first-choice drugs for the treatment of chronic kidney diseases that are characterized by loss of proteinuria and renal function. In practice, these drugs exhibit only moderate side effects. However, renal targeting of an ACE inhibitor may improve the therapy in certain cases. For example, when proteinuria is accompanied by normal blood pressure, hypotension due to ACE inhibition limits the amount of drug that can be given.

Renal inflammations like glomerulonephritis and tubulointerstitial inflammation may be treated with corticosteroids or NSAIDs. These drugs, however, can have serious side effects, and renal targeting of these drugs may allow a more aggressive treatment of the inflammation. Also, a local suppression of the immune system may be useful to prevent transplant rejection. However, it is as yet unknown whether suppression of the local immune system is sufficient or whether the systemic system should also be suppressed to prevent rejection.

Renal tumors are characterized by insensitivity to the common antitumor drugs. This is probably due to an unfavorable kinetic profile of these drugs. By renal targeting, an antitumor drug may reach the renal tumor in higher concentrations while the extrarenal side effects may be reduced.

B. Nonsteroidal Anti-Inflammatory Drugs

Nonsteroidal anti-inflammatory drugs (NSAIDs) are among the most commonly used drugs in the world. Their application ranges from the treatment of headaches to that of rheumatoid arthritis. The inhibition of cyclo-oxygenase, responsible for the synthesis of prostaglandins, causes the well-known analgesic, anti-inflammatory, and antipyretic effects of the NSAIDs. However, the inhibited prostaglandin synthesis is also responsible for frequently observed adverse effects such as gastrointestinal ulcerations and bleedings, impairment of renal perfusion, and induction of acute airway obstruction responses. The number and severity of the adverse reactions increases with dose and time. With respect to the kidney, the inhibition of prostaglandin synthesis can have disastrous consequences, in particular when prostaglandin synthesis is required to maintain renal perfusion. This is crucial, for example, during hypovolemia or sodium restriction or in diabetic and cirrhotic patients (93–95). A significant number of people having used NSAIDs chronically now require hemodialysis.

However, the application of NSAIDs in the case of several renal diseases is feasible. Before the development of ACE inhibitors, treatment with NSAIDs formed the common therapy to lower proteinuria in diabetic patients (96). Their application is still of importance when an ACE inhibitor is poorly effective or side effects hamper further treatment. In addition, the treatment of glomerulonephritis may benefit from a short-term application of NSAIDs. Finally, NSAIDs may be used for the treatment of tubular defects seen in Fanconi and Bartter's syndromes. Therefore the development of renal-selective NSAID conjugates with minimum effect on renal hemodynamics forms a challenge.

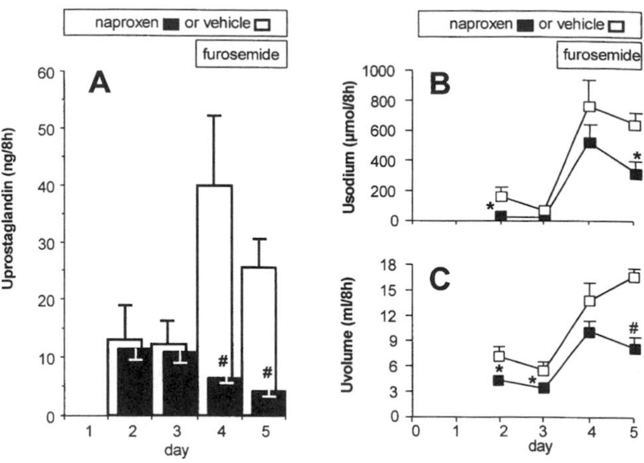

Figure 44.17 Urinary excretion of PGE_2 (panel A), sodium (panel B), and water (panel C) after twice daily intravenous administration of vehicle (white bars, $n = 6$) or naproxen ($10\ mg \cdot kg^{-1} \cdot 24h^{-1}$, black bars, $n = 6$) treated rats. Furosemide $10\ mg \cdot kg^{-1}$ was given subcutaneously on days 4 and 5 additionally to all groups of rats. Urine parameters were determined in urine collected during the daytime. Data are expressed as mean ± SEM. Difference of naproxen versus vehicle treated rats: *, $p < 0.05$; #, $p < 0.005$.

1. Naproxen–lysozyme

Although a conjugate with an ester spacer is favorable compared with a conjugate with a direct peptide linkage, we continued our research using a conjugate with naproxen coupled directly to lysozyme, since the synthesis of the conjugate with an ester spacer (naproxen-L-lactic acid-lysozyme) is cumbersome (Fig. 16). Fortunately, the catabolite of the conjugate with direct peptide linkage (naproxen–lysine) appeared to have an equipotent inhibitory effect on prostaglandin synthesis in vitro as the parent drug (63,97).

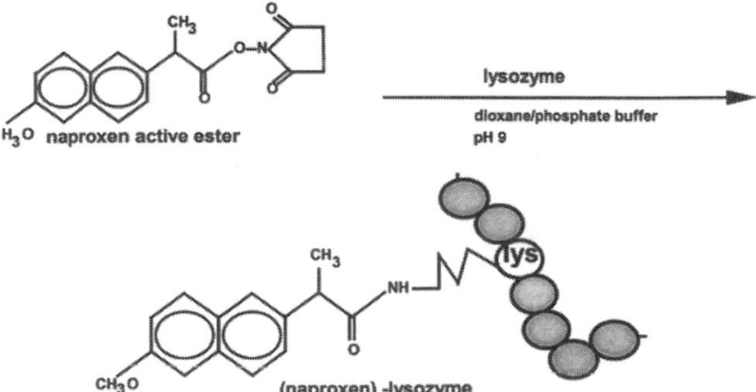

Figure 44.16 Chemical structure of naproxen–lysozyme. Naproxen-N-hydroxysuccinimide was reacted with lysozyme, resulting in the naproxen–lysozyme conjugate with a peptide linkage between the free carboxylic group of naproxen and one of the free amino groups of lysozyme.

The coupling of 2 moles of naproxen to 1 mole of lysozyme did not affect the renal uptake or catabolism of the lysozyme carrier in the rat. The conjugate, like native lysozyme, rapidly accumulated in the kidney. Focusing on the drug moiety of the conjugate, it was shown that the conjugation of naproxen to lysozyme distinctly altered the kinetics of the drug. Conjugation to lysozyme resulted in a 70-fold increase in naproxen concentrations in the kidney (Fig. 17). After delivery to the kidney, naproxen was gradually released from the conjugate in the form of naproxen–lysine. This catabolite was subsequently eliminated from the kidney, and after a single injection, drug levels in the renal tissue gradually decreased with a half-life of 160 min. Since no detectable amounts of naproxen or its lysine conjugates are found in the plasma after administration of conjugate, it can be inferred that excretion into the urine is the crucial process determining the elimination half-life. The lack of diffusion to the bloodstream is favorable in relation to unwanted extrarenal effects (98).

Having obtained this promising kinetic profile, it was investigated whether naproxen–lysozyme exhibits renal effects in the rat. Naproxen, as an inhibitor of cyclo-oxygenase, blocks prostaglandin synthesis. Among others, naproxen reduces furosemide-stimulated urinary excretion of prostaglandin E_2 (PGE_2) as well as the natriuretic and diuretic effects of furosemide. Studies with the conjugate

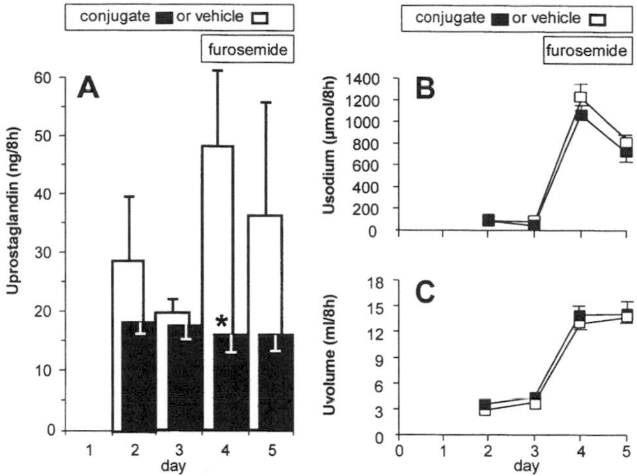

Figure 44.18 Urinary excretion of PGE_2 (panel A), sodium (panel B), and water (panel C) after twice daily intravenous administration of vehicle (white bars, $n = 6$) or naproxen–lysozyme ($10\ mg \cdot kg^{-1} \cdot 24h^{-1}$, black bars, $n = 6$) treated rats. Furosemide $10\ mg \cdot kg^{-1}$ was given subcutaneously on days 4 and 5 additionally to all groups of rats. Urine parameters were determined in urine collected during the daytime. Data are expressed as mean ± SEM. Difference of naproxen versus vehicle treated rats: *, $p < 0.05$; #, $p < 0.005$.

showed that naproxen–lysozyme treatment clearly prevents furosemide-induced excretion of PGE_2. This occurred in a dose of naproxen that was not effective in the unconjugated form (Fig. 17A and 18A) (99).

Surprisingly, this effect occurred in the absence of a change in natriuretic and diuretic response to furosemide (Figs. 17B and 18B). In this respect, the pharmacological effect differed from treatment with a high dose of free naproxen. An explanation for these differences remains to be found. One possibility is that there is a difference in intrarenal kinetics of the NSAID conjugate compared with the parent drug. Free naproxen is extensively reabsorbed in the distal tubule of the kidney, via which route it may effectively inhibit prostaglandin synthesis in the medullary interstitial cells. On the other hand, naproxen–lysine is more hydrophilic and may be unable to reach these sites of prostaglandin synthesis involved in the furosemide-induced excretion of sodium and water. These data show that renal drug targeting preparations can also be used as tools to unravel the mechanisms of renal therapeutic effect.

C. Angiotensin Converting Enzyme Inhibitors

1. The Renin Angiotensin System

Angiotensin converting enzyme (ACE) inhibitors form a relatively new class of antihypertensive drugs acting on the renin angiotensin system (RAS). The RAS is an enzymatic system that results in the formation of angiotensin II, via two consecutive steps. In the first step, angiotensinogen, synthesized in the liver, is cleaved by the soluble protease renin into the decapeptide angiotensin I. In the second step angiotensin I is converted to the bioactive octapeptide angiotensin II by the membrane bound protease ACE, localized at the luminal side of epithelial cells in the gastrointestinal tract, the renal proximal tubule, and at the endothelium of vascular cells (100). Finally, degradation of angiotensin II into angiotensin 1–7 and angiotensin 2–8 by ACE or other endopeptidases occurs rapidly (Fig. 19).

In the classical view, the RAS serves as a ''circulating endocrine system.'' The presence of many components of the RAS in other tissues has led to the concept of a local acting tissue RAS (101–105). Particularly, the detection of much higher concentrations of angiotensin II in the renal tubular fluid than in the plasma has focused attention on a possible physiological role for the RAS in the kidneys (106,107). However, the exact contribution of this local renal RAS has not been elucidated so far (see Section IV.C.2). In providing a renal specific ACE inhibitor, drug targeting constructs may offer a pharmacological approach to determine the effects of renal ACE inhibition.

2. Effects of Angiotensin II and ACE Inhibitors

Angiotensin II exerts numerous actions both systemically and intrarenally. The regulation of systemic blood pressure is among the most important roles of angiotensin II. A number of different mechanisms contribute to the blood-pressure-stimulating activity. First, angiotensin II produces sodium (and water) retention through a direct effect on the proximal tubular cells as well as the release of aldosterone from the adrenal gland (108,109). Additionally, angiotensin II regulates systemic blood pressure and renal hemodynamics resulting from its strong vasoconstrictive properties. Intrarenally, this leads to an increased glomerular capillary pressure and a reduction of the renal blood flow (105,109). Several studies have investigated the contribution of intrarenal angiotensin II production on systemic blood pressure (110), renal blood flow, and urinary sodium excretion (111–114) using intrarenal or intra-arterial administration of low dosages of ACE inhibitor to obtain a renal specific ACE inhibition. These studies made clear that local ACE inhibition is likely to have profound effects on renal physiology and pathology as well as on systemically related parameters. However, the techniques used in these investigations all required special surgical procedures. Moreover, the contribution of extrarenal ACE inhibition cannot be fully excluded in these investigations due to the administration of active ACE inhibitor. A systemically inactive renal specific ACE inhibitor–LMWP conjugate may provide a pharmacological tool to solve the problems of surgery and extrarenal effects.

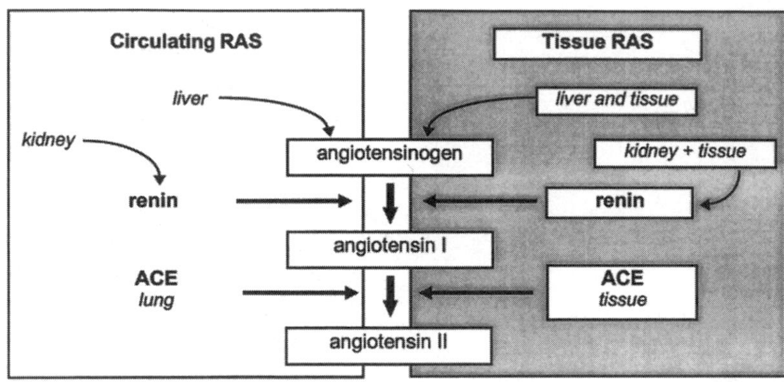

Figure 44.19 The renin angiotensin system (RAS). In the circulating RAS, liver-derived angiotensinogen is converted by kidney-derived renin into angiotensin I in the circulation. Angiotensin I is converted into angiotensin II by angiotensin converting enzyme (ACE) in the lungs during the passage of the blood through the pulmonary circulation. ACE is a membrane bound protein, oriented with its active site at the outside of the cell. In the tissue RAS, the soluble components (renin, angiotensinogen, or angiotensin I) of the RAS are either synthesized locally or acquired from the circulation. Angiotensin II can be generated locally when both ACE and angiotensin I are present.

Besides the hemodynamic actions, angiotensin II also has trophic effects on vascular smooth muscle cells and mesangial cells (108,115). Angiotensin II has direct effects on cell growth, as well as indirect effects by stimulating the expression of other growth factors and an activation of the immune system (21,115,116). Finally, angiotensin II is probably an important mediator of cytokine release and progressive inflammation processes causing renal function loss during chronic renal disease (21).

The development of ACE inhibitors caused a breakthrough in the drug treatment of renal failure in particular. These drugs were shown to reduce proteinuria and retard the chronic progression of renal function loss in both diabetic and nondiabetic renal disease (117–120). The mechanisms by which the ACE inhibitors could show their renoprotective effect are summarized in Table 3. An important renoprotective mechanism of ACE inhibitors is their influence on the glomerular capillary pressure, caused by vasodilation of the arterioles in the glomerulus (105,121,122). This pressure lowering results in an improved glomerular permselectivity and a marked decrease of proteinuria (120,123,124). Since proteinuria is now considered as an independent risk factor for renal function loss (125), the reduction of proteinuria may be related to the long-term renoprotective effect (120). However, ACE inhibition using standard drug dosages does not always guarantee an adequate reduction of proteinuria. Further increment of the dose is hampered by the concomitant fall of systemic blood pressure. Since intrarenal production of angiotensin II is believed to play an important role in the development and progression of renal disorders, renal specific ACE inhibitor–lysozyme conjugates may provide a way to improve renal drug therapy.

Table 44.3 Postulated Renoprotective Mechanisms of ACE Inhibitors

Normalization of systemic blood pressure
Control of glomerular capillary pressure
Decreased mesangial macromolecular deposition
Decreased mesangial proliferation
Improved glomerular permselectivity
Decreased proteinuria
Decreased tubulointerstitial injury
Decreased procollagen formation
Improved serum–lipid profiles

3. Captopril–Lysozyme

We have coupled the ACE inhibitor captopril to the LMWP lysozyme via a disulfide bond (see Section III.D.3, Fig. 20). Conjugation of captopril to lysozyme resulted in a sixfold increase of captopril accumulation in the kidney. This modest enrichment, as compared to naproxen–lysozyme, is due to the fact that free captopril is highly cleared by the kidney itself. After renal uptake, captopril was rapidly released from the conjugate as indicated by the rapid decrease of renal captopril levels in time. The difference in renal half-life of naproxen and captopril after delivery is likely to be due to an unequal rate of release from the lysozyme conjugates (126). Whereas naproxen–lysozyme requires a peptidase for cleavage, captopril is released from the conjugate enzymatically by β-lyase and/or nonenzy-

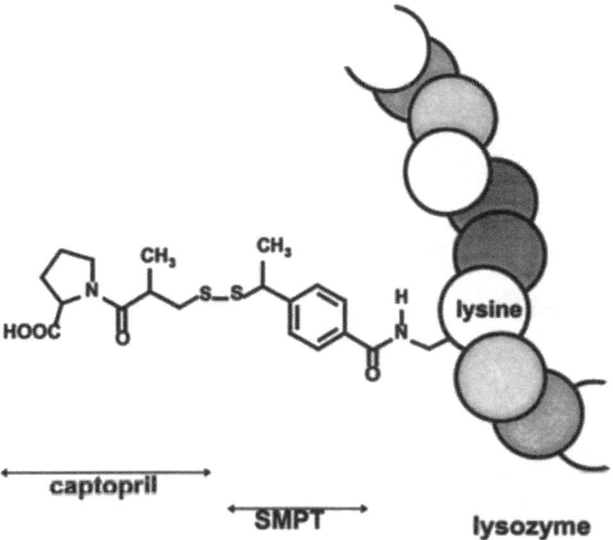

Figure 44.20 Chemical structure of the captopril–lysozyme conjugate. Captopril is coupled via a disulfide bond to the thiol group of the SMPT spacer. The amine-reactive end of the spacer is linked to a lysine residue of lysozyme.

matically by thiol-disulfide exchange with endogenous thiols. To reduce the rate of captopril–lysozyme breakdown, two different cross-linking reagents, SPDP and SMPT, were tested. Although an SMPT link between two proteins is in principle less susceptible to disulfide reduction, no difference in degradation rate was found between the SPDP and SMPT captopril–lysozyme conjugates (127).

In an ex vivo experiment in the rat, we measured the extent of ACE inhibition in plasma and the kidney after a single intravenous injection of the conjugate and an equimolar dosage of free captopril. Captopril–lysozyme caused a similar but more sustained renal ACE inhibition (Fig. 21B). The inhibition of plasma ACE activity was also markedly reduced but not entirely prevented (Fig. 21A). Possibly, the disulfide bond is partly degraded within the circulation. However, transportation of free captopril back into the circulation after renal degradation of the conjugate provides an alternative explanation for the observed plasma ACE inhibition. The rapid intracellular release may provide a sufficient driving force for transport across the basolateral membranes (R. F. G. Haverdings et al., manuscript in preparation).

Importantly, captopril–lysozyme did not significantly affect systemic blood pressure, whereas an equimolar dose of captopril itself largely decreased blood pressure (Fig. 22). While free captopril (5 mg · kg^{-1} · 6h^{-1}) completely prevented the angiotensin-I-induced pressor response, this response was more maintained after infusion of equimolar amounts of captopril–lysozyme. However, in line with the direct ACE activity measurements in renal tissue and plasma, the angiotensin-I-induced pressor response was still present in captopril–lysozyme treated rats, but also less than in untreated rats, suggesting that systemic activity was not fully prevented (R. F. G. Haverdings et al., manuscript in preparation).

Both free and conjugated captopril did not affect glomerular filtration. Renal plasma flow increased to the same extent during treatment of free and conjugated captopril (1 mg · kg^{-1} · 6h^{-1}). Although a complete dose–effect relation was not performed, we can conclude that captopril still provides clear effects on the renal plasma flow after conjugation to lysozyme. Whether this effect is determined by intrarenal or systemic ACE inhibition remains to be studied.

At the moment, the synthesis of lysozyme conjugates with ACE inhibitors other than captopril is studied. Some of these ACE inhibitors may be advantageous for renal delivery; an example is lisinopril. The amount of conjugate required for therapy can be significantly reduced when using lisinopril, which has a higher affinity for ACE. Furthermore, the plasma stability of the conjugate may be increased by using an ACE inhibitor that is conjugated to lysozyme via a linkage that is more stable in plasma (e.g., lisinopril can in principle be coupled via an acid sensitive spacer).

D. Doxorubicin

The antineoplastic agent doxorubicin (adriamycin) is a standard drug in the chemotherapy of a variety of cancer diseases, including bladder carcinoma. However, although effective, the repeated administration of these drugs retrogradically via the urethra is a rather patient-unfriendly method (128,129). Unfortunately, the intravenous application is restricted due to the low urinary excretion of doxorubicin and the occurrence of cardiac toxicity, which is related to the total cumulative doxorubicin dose. Drug targeting may be able to prevent the occurrence of these side effects after intravenous administration and may form an alternative approach to bladder instillation.

During in vivo testing of doxorubicin conjugates it is important to realize, however, that the doxorubicin toxicity in rodents is shifted toward the kidney. The degeneration of the negatively charged glycoproteins in the glomerular basement membrane is followed by progressive loss of glomerular permselectivity. As a consequence, serum proteins are excreted in the urine (proteinuria) once the tubular protein reabsorption transport is overloaded. Moreover, the excessive amounts of reabsorbed protein congest the lysosomal compartment, causing tubular protein cast formation

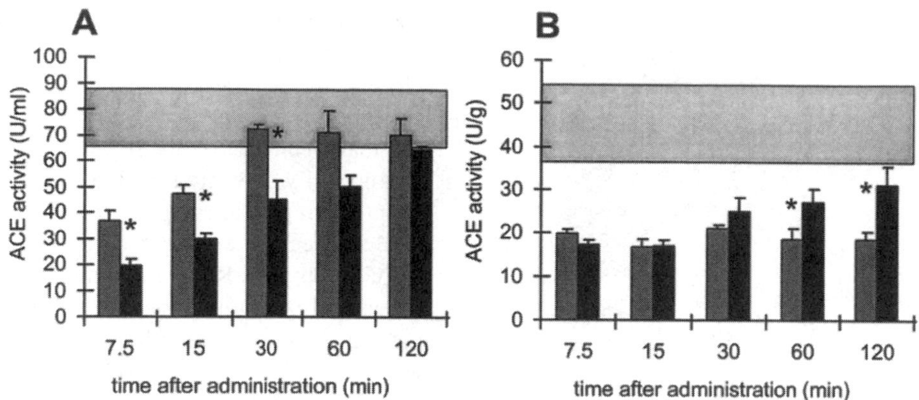

Figure 44.21 ACE activity in plasma (panel A) and kidneys (panel B) after single intravenous administration of captopril–lysozyme (lysozyme (0.2 mg · kg^{-1} captopril, corresponding to 33 mg · kg^{-1} lysozyme, striped bars, $n = 6$) and unbound captopril–lysozyme (0.2 mg · kg^{-1} captopril plus 33 mg · kg^{-1} lysozyme, black bars, $n = 6$). The shaded area indicates the 95% confidence interval of the ACE activity of control rats. Data are expressed as mean ± SEM. Difference of conjugate versus captopril treated rats: *, $p < 0.05$.

and rupture of lysosomes. Due to the release of aggressive enzymes from the lysosomes in the cytosol, cell death occurs. This vicious circle leads to a progressive loss of nephrons and a fall of renal function with time (130–134).

Since the doxorubicin-induced proteinuria in rodents resembles the nephrotic syndrome in man, this animal model is generally used to study the effectiveness of antiproteinuric drugs (135,136) such as our newly developed renal specific antiproteinuric drug conjugates.

1. Doxorubicin–Lysozyme

In previous sections the selective accumulation of LMWPs and drug–lysozyme conjugates in the renal proximal tubular cell has been discussed. Furthermore, in Section III.B we stressed the importance of positive charges in the lysozyme molecule for this tubular uptake. For tubular reuptake only a limited number of amino groups can be substituted to prevent loss of conjugate via the urine, and excessive drug loading may alter the favorable tissue distribution (45).

Note that tubular reabsorption should be prevented when using LMWPs for drug delivery to the distal urinary tract. We have considered the possibility of reacting all available free amino groups of the LMWP with drug molecules to prevent the tubular uptake process. Doxorubicin was coupled to lysozyme via the *cis*-aconityl acid-labile spacer, with an average coupling degree of 6.3 mol doxorubicin to 1 mol of lysozyme (Fig. 23). Since lysozyme contains seven free amino groups, an almost complete derivatization was obtained. In vitro studies showed a time dependent release of doxorubicin up to 45% after 6 hours at pH 4.5 and 37°C (137).

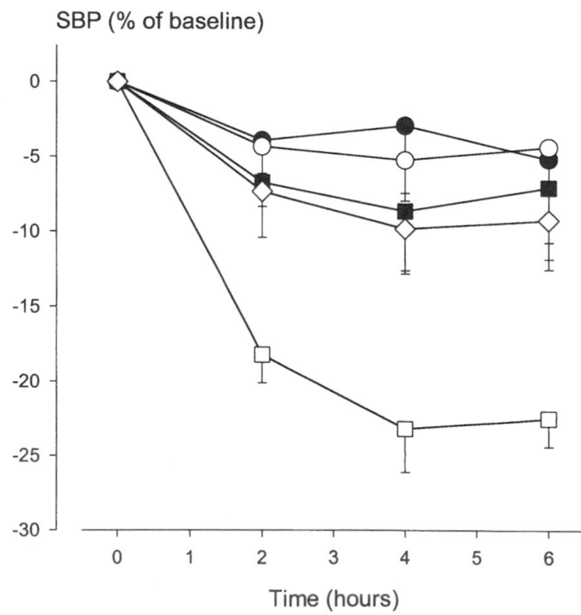

Figure 44.22 Systemic blood pressure during 6 hour infusion of vehicle (open circles, $n = 6$), lysozyme (100 mg · kg^{-1} · 6h^{-1}, closed circles, $n = 6$), captopril (1 mg · kg^{-1} · 6h^{-1}, open diamonds), captopril, (5 mg · kg^{-1} · 6h^{-1}, open squares, $n = 6$) or captopril–lysozyme (5 mg · kg^{-1} · 6h^{-1}, corresponding to 100 mg · kg^{-1} · 6h^{-1} lysozyme, closed squares, $n = 6$). Values given at $t = 0$ h represent mean baseline values of 2 hour vehicle infusion in all groups. Data are expressed as mean ± SEM.

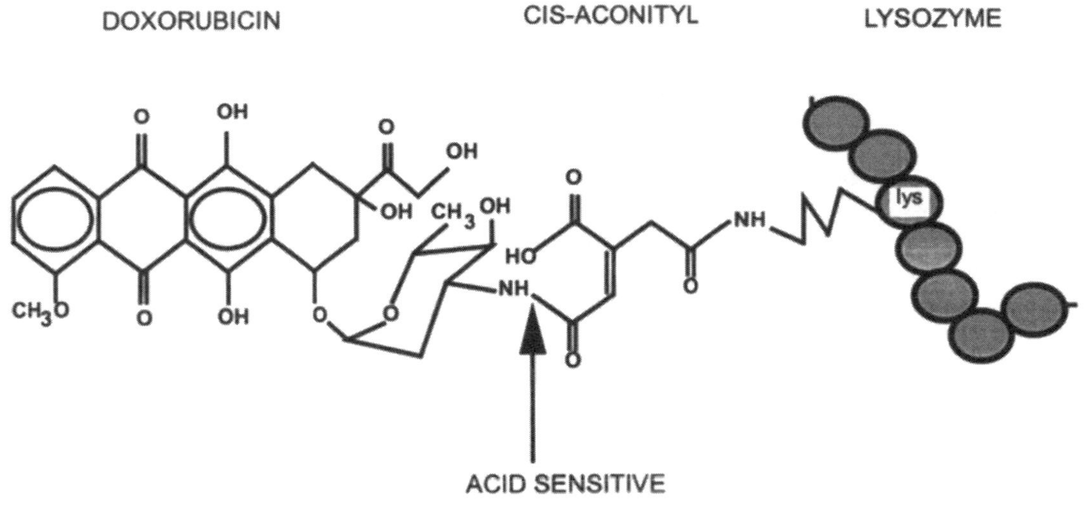

Figure 44.23 Chemical structure of the doxorubicin–aconityl–lysozyme conjugate.

The urinary kinetics of doxorubicin–lysozyme and free doxorubicin were studied in healthy rats. Following a single intravenous injection of 2 mg · kg^{-1} free doxorubicin, only 4% of the dose was recovered in the urine 24 hours after administration. In contrast, doxorubicin–lysozyme resulted in a much higher urinary excretion of doxorubicin: 27% and 39% of the dose after injection of 2 and 10 mg · kg^{-1}, respectively. In the same study, the development of proteinuria was followed. After a single injection of 2 mg · kg^{-1} free doxorubicin, all rats developed a marked proteinuria and glomerulosclerosis with a maximum of approximately 450 mg protein per day. Rats treated with either 2 or 10 mg · kg^{-1} doxorubicin–lysozyme did not develop proteinuria or glomerulosclerosis (Fig. 24) (137). Studies on the efficacy of this conjugate for the treatment of bladder cancer are now in progress. Rapid release of sufficient doxorubicin from the conjugate present in the bladder lumen can in principle be produced by acidification of the urine. This may provide sufficiently high concentrations in the bladder, enabling diffusion into the tumor.

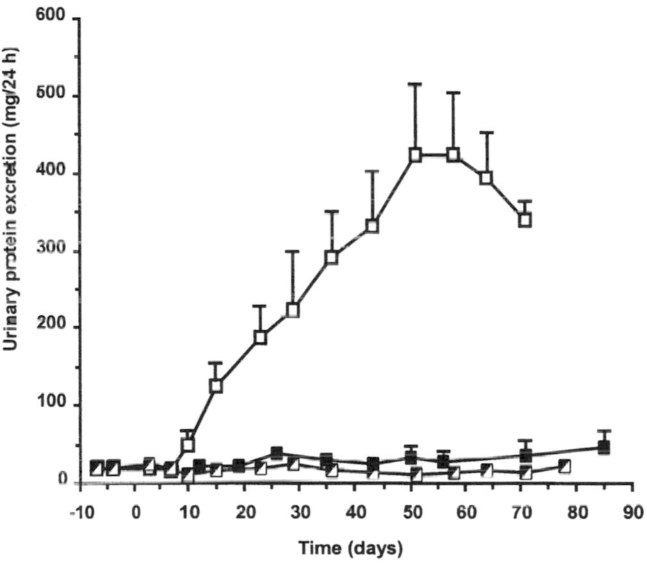

Figure 44.24 Time course of urinary protein excretion after single intravenous administration. Open squares: 2 mg · kg^{-1} unbound doxorubicin (n = 5); black/white squares: 2 mg · kg^{-1} doxorubicin–aconityl–lysozyme (n = 6); closed squares: 10 mg · kg^{-1} doxorubicin–aconityl–lysozyme (n = 5). Data are expressed as mean ± SEM.

V. CONCLUDING REMARKS

In this chapter we have pointed out that the kidney is a feasible target organ for drug delivery. The kidney offers a number of characteristics including a high blood perfusion and efficient uptake mechanisms that may retain the tubular delivered drug. Since a number of renal diseases can still not be effectively treated, or serious adverse effects hamper drug treatment, selective drug delivery to the kidney, leading to local accumulation of drugs, could help to overcome some of these problems. In this chapter we have described the low-molecular-weight protein approach to target the proximal tubular cell in the kidney. Since this cell type is essential in the regulation of sodium and fluid homeostasis as well as in the induction of renal inflammatory processes, it appears a suitable target for drug delivery. At present, we have investigated the effectiveness of ACE inhibitors and NSAIDs, drugs that are presently used in the drug treatment of renal disorders. Coupling of these drugs to the renal carrier resulted in a marked increase in renal selectivity as well as a change in their pharmacological profile. Additionally, we showed that charge modifications in the lysozyme molecule can be applied in the selective delivery of doxorubicin to the urinary bladder.

In future, these developments may not only improve the safety and efficacy of drug treatment, but therapy may also benefit from new pathophysiological insights provided by organ selective pharmacological tools.

REFERENCES

1. Ehrlich, P. 1906. *Collected Studies on Immunity*. John Wiley, New York.
2. Haas, M., Meijer, D. K. F., Moolenaar, F., De Jong, P. E., De Zeeuw, D. 1996. Renal drug targeting: optimalisation of renal pharmacotherapeutics. In: *International Yearbook of Nephrology 1996* (Andreucci, V. E., Fine, L. G., eds.). Oxford University Press, pp. 3–11.
3. Haas, M., Kato, Y., Haverdings, R. F. G., Moolenaar, F., Suzuki, K., De Zeeuw, D., Sugiyama, Y., Meijer, D. K. F. 2001. Delivery of drugs and antisense oligonucleotides to the proximal tubular cell of the kidney using macromolecular and prodrug approaches In: *Organ-specific Drug Targeting Strategies* (Molema, G., Meijer, D. K. F., eds.). Wiley-vch, Weinheim. (In press.)
4. Tencer, J., Frick, I. M., Oquist, B. W., Alm, P., Rippe, B. 1998. Size-selectivity of the glomerular barrier to high molecular weight proteins: upper size limitations of shunt pathways. *Kidney Int. 53*:709–715.
5. Kitamura, M., Fine, L. G. 1999. The concept of glomerular self-defence. *Kidney Int. 55*:1639–1671.
6. Ito, S. 1999. Drug secretion systems in renal tubular cells: functional models and molecular identity. *Pediatr. Nephrol. 13*:980–988.
7. Pritchard, J. B., Miller, D. S. 1991. Comparative insights into the mechanisms of renal organic anion and cation secretion. *Am. J. Physiol. 261*:R1329–R1340.
8. Somogyi, A. 1996. Renal transport of drugs: specificity and molecular mechanisms. *Clin. Exp. Pharmacol. Physiol. 23*:986–989.
9. Welbourne, T. C., Matthews, J. C. 1999. Glutamate transport and renal function. *Am. J. Physiol. 277*:F501–F505.
10. Suzuki, K., Susaki, H., Okuno, S., Yamada, H., Watanabe, H. K., Sugiyama, Y. 1999. Specific renal delivery of sugar-modified low-molecular-weight peptides. *J. Pharmacol. Exp. Ther. 288*:888–897.
11. Orlando, R. A., Rader, K., Authier, F., Yamazaki, H., Posner, B. I., Bergeron, J. J., Farquhar, M. G. 1998. Megalin is an endocytic receptor for insulin. *J. Am. Soc. Nephrol. 9*:1759–1766.
12. Hoffman, B. B., Ziyadeh, F. N. 1995. Facilitative glucose transport proteins and sodium-glucose co-transporters in the kidney. *Curr. Opin. Nephrol. Hypertens. 4*:406–411.
13. Murer, H. 1982. Renal transport of amino acids: membrane mechanism. *Contrib. Nephrol. 33*:14–28.
14. Adibi, S. A. 1997. Renal assimilation of oligopeptides: physiological mechanisms and metabolic importance. *Am. J. Physiol. 272*:E723–E736.
15. Maesaka, J. K., Fishbane, S. 1998. Regulation of renal urate excretion: a critical review. *Am. J. Kidney Dis. 32*: 917–933.
16. Christensen, E. I., Birn, H., Verroust, P., Moestrup, S. K. 1998. Membrane receptors for endocytosis in the renal proximal tubule. *Int. Rev. Cytol. 180*:237–284.
17. Leung, S., Bendayan, R. 1999. Role of P-glycoprotein in the renal transport of dideoxynucleoside analog drugs. *Can. J. Physiol. Pharmacol. 77*:625–630.
18. Birn, H., Vorum, H., Verroust, P. J., Moestrup, S. K., Christensen, E. I. 2000. Receptor-associated protein is important for normal processing of megalin in kidney proximal tubular cells. *J. Am. Soc. Nephrol. 11*:191–202.
19. Vleming, L. J., Bruijn, J. A., Van Es, L. A. 1999. The pathogenesis of progressive renal failure. *Neth. J. Med. 54*:114–128.
20. Johnson, R. J. 1997. Cytokines, growth factors and renal injury: where do we go now? *Kidney Int. 63*(suppl.):52–56.
21. Palmer, B. F. 1997. The renal tubule in the progression of chronic renal failure. *J. Invest. Med. 45*:346–361.
22. Pacifici, G. M., Viani, A., Franchi, M., Gervasi, P. G., Longo, V., Simplicio, P. Di., Temellini, A., Romiti, P., Santerini, S., Vanucci, L. et al. 1989. Profile of drug-metabolizing enzymes in the cortex and medulla of the human kidney. *Pharmacol. 39*:299–308.
23. Moolenaar, F., Cancrinus, S., Visser, J., De Zeeuw, D., Meijer, D. K. F. 1992. Clearance of indomethacin occurs predominantly by renal glucuronidation. *Pharm. Weekbl. [Sci.] 14*:191–195.
24. Powel, G. M., Roy, A. B. 1980. Sulphate conjugation. In: *Extrahepatic Metabolism of Drugs and Other Foreign Compounds* (Gram, T. E., ed.). MTP Press, Lancaster, pp. 389–425.
25. Mulder, G. J. 1984. Sulfation—metabolic aspects. In: *Progress in Drug Metabolism, Vol. 8* (Bridges, J. W., Chasseaud, L. F., eds.). Taylor and Tracies, London pp. 35–100.
26. Pacifici, G. M., Franchi, M., Collizi, C. 1988. Sulphotransferase in humans: development and tissue distribution. *Pharmacol. 36*:411–419.
27. Minard, F. N., Grant, D. S., Cain, J. C., Jones, P. H., Kyncl, J. 1980. Metabolism of τ-glutamyl dopamide and its carboxylic acid esters. *Biochem. Pharmacol. 29*:69–75.
28. Haga, H. J. 1989. Kidney lysosomes. *Int. J. Biochem. 21*: 343–345.

29. Barrett, A. J., McDonald, J. K. 1980. *Mammalian Proteases: A Glossary and Bibliography*, Vol. 1. *Endopeptidases*. Academic Press, London.
30. McDonald, J. K., Barrett, A. J. 1986. *Mammalian Proteases*. Vol. 2. *Exopeptidases*. Academic Press, London.
31. Barrett, A. J., Heath, M. F. 1977. *Lysosomes: A Laboratory Handbook*. Elsevier, Amsterdam.
32. Segal, H. L. 1975. Lysosomes and intracellular protein turnover. In: *Lysosomes in Biology and Pathology* (Dingle, J. T., ed.). North Holland, Amsterdam, pp. 295–302.
33. Haas, M., De Zeeuw, D., Van Zanten, A., Meijer, D. K. 1993. Quantification of renal low-molecular-weight protein handling in thc intact rat. *Kidney Int. 43*:949–954.
34. Bianchi, C., Donadio, C., Tramonti, G., Auner, I., Lorusso, P., Deleide, G., Lunghi, F., Salvadori, P. 1988. Renal handling of cationic and anionic small proteins: experiments in intact rats. In: *Contributions to Nephrology* (Bianchi, C., Bocci, V., Carone, F. A., Rabkin, R., eds.). Karger, Basel, pp. 37–44.
35. Hysing, J., Tolleshaug, H., Curthoys, N. P. 1990. Reabsorption and intracellular transport of cytochrome c and lysozyme in rat kidney. *Acta Physiol. Scand. 140*:419–427.
36. Hysing, J., Tolleshaug, H. 1986. Quantitative aspects of the uptake and degradation of lysozyme in the rat kidney in vivo. *Biochim. Biophys. Acta 887*:42–50.
37. Maack, T., Park, C. H., Camargo, M. J. F. 1985. Renal filtration, transport and metabolism in the kidney. In: *The Kidney: Physiology and Pathophysiology* (Seldin, D. W., Giebisch, G., eds.). Raven Press, New York, pp. 1773–1803.
38. Cui, S. Y., Verroust, P. J., Moestrup, S. K., Christensen, E. I. 1996. Megalin/gp330 mediates uptake of albumin in renal proximal tubule. *Amer. J. Physiol.-Renal. Fl. Elect. 40*:F900–F907.
39. Carone, F. A., Peterson, D. R., Oparil, S., Pullman, T. N. 1980. Renal tubular transport and catabolism of small peptides. In: *Functional Ultrastructure of the Kidney* (Maunsbach, A. B., Olsen, T. S., Christensen, E. I., eds.). Academic Press, London, pp. 327–340.
40. Christensen, E. I., Maunsbach, A. B. 1974. Intralysosomal digestion of lysozyme in renal proximal tubule cells. *Kidney Int. 6*:396–407.
41. Maack, T., Johnson, V., Kau, S. T., Figueiredo, J., Sigulem, D. 1979. Renal filtration, transport, and metabolism of low-molecular-weight proteins: a review. *Kidney Int. 16*:251–270.
42. Kok, R. J., Haas, M., Moolenaar, F., De Zeeuw, D., Meijer, D. K. F. 1998. Drug delivery to the kidneys and the bladder with the low molecular weight protein lysozyme. *Renal Failure 20*:211–217.
43. Christensen, E. I., Carone, F. A., Rennke, H. G. 1981. Effect of molecular charge on endocytic uptake of ferritin in renal proximal tubule cells. *Lab. Invest. 44*:351–358.
44. Christensen, E. I., Rennke, H. G., Carone, F. A. 1983. Renal tubular uptake of protein: effect of molecular charge. *Am. J. Physiol. 244*:F436–F441.
45. Kok, R. J., Grijpstra, F., Nederhoed, K. H., Moolenaar, F., De Zeeuw, D., Meijer, D. K. F. 1999. Renal drug delivery with low-molecular-weight proteins: the effect of charge modifications on the body distribution of drug–lysozyme conjugates. *Drug Del. 6*:1–8.
46. Cojocel, C., Docui, N., Baumann, K. 1982. Early nephrotoxicity at high plasma concentrations in the rat. *Lab. Invest. 46*:149–157.
47. Cojocel, C., Franzen-Sieveking, M., Baumann, K. 1984. Dependence of renal protein reabsorption on glomerular filtration rate and infusion time. *Pflügers Arch.-Eur. J. Physiol. 402*:34–38.
48. McGraw, T. E., Maxfield, F. R. 1991. Internalization and sorting of macromolecules: endocytosis. In: *Targeted Drug Delivery* (Juliano, R. L., ed.). Springer-Verlag, Berlin, pp. 11–41.
49. Soyez, H., Schacht, E., Van Der Kerken, S. 1996. The crucial role of spacer groups in macromolecular prodrug design. *Adv. Drug. Deliv. Rev. 21*:81–106.
50. Salvetti, A. 1990. Newer ACE inhibitors. A look at the future. *Drugs 40*:800–828.
51. Ondetti, M. A. 1988. Structural relationships of angiotensin converting-enzyme inhibitors to pharmacologic activity. *Circulation 77*:I74–8.
52. Lawton, G., Paciorek, P. M., Waterfall, J. F. 1992. The design and biological profile of ACE inhibitors. *Adv. Drug Res. 23*:161–220.
53. Rejmanova, P., Kopacek, J., Pohl, J., Baudys, M., Kostka, V. 1983. Polymers containing enzymatically degradable bonds, 8. Degradation of oligopeptide sequences in N-(2-hydroxypropyl)methacrylamide copolymers by bovine spleen cathepsin B. *Makromol. Chem. 184*:2009–2020.
54. Duncan, R., Cable, H. C., Lloyd, J. B., Rejmanová, P., Kopacek, J. 1983. Polymers containing enzymatically degradable bonds, 7. Design of oligopeptide side-chains in poly[N-(2-hydroxypropyl)methacrylamide] copolymers to promote efficient degradation by lysosomal enzymes. *Makromol. Chem. 184*:1997–2008.
55. Shen, W. C., Ryser, H. J. P., La Manna, L. 1985. Disulfide spacer between methotrexate and poly-(D-lysine). A probe for exploring the reductive process in endocytosis. *J. Biol. Chem. 260*:10905–10908.
56. Fattom, A., Shiloach, J., Bryla, D., Fitzgerald, D., Pastan, I., Karakawa, W. W., Robbins, J. B., Schneerson, R. 1992. Comparative immunogenicity of conjugates composed of the Staphylococcus aureus type 8 capsular polysaccharide bound to carrier proteins by adipic acid dihydrazide or N-succinimidyl-3-(2-pyridyldithio)propionate. *Infect. Immun. 60*:584–589.
57. Trimble, S. P., Marquardt, D., Anderson, D. C. 1997. Use of designed peptide linkers and recombinant hemoglobin mutants for drug delivery: in vitro release of an angiotensin II analog and kinetic modeling of delivery. *Bioconjug. Chem. 8*:416–423.

58. Daussin, F., Boschetti, E., Delmotte, F., Monsigny, M. 1988. p-Benzylthiocarbamoyl-aspartyl-daunorubicin-substituted polytrisacryl. A new drug acid-labile arm-carrier conjugate. *Eur. J. Biochem. 176*:625–628.
59. Drummer, O. H., Kourtis, S., Iakovidis, D. 1988. Biotransformation studies of di-acid angiotensin coverting enzyme inhibitors. *Arzneimittelforschung 38*:647–650.
60. Yeung, J. H., Coleman, J. W., Park, B. K. 1985. Drug-protein conjugates—IX. Immunogenicity of captopril-protein conjugates. *Biochem. Pharmacol. 34*:4005–4012.
61. Matsumoto, K., Miyazaki, H., Fujii, T., Yoshida, K., Amejima, H., Hashimoto, M. 1986. Disposition and metabolism of the novel antihypertensive agent alacepril in rats. *Arzneimittelforschung 1*:40–46.
62. Franssen, E. J. F., Koiter, J., Kuipers, C. A. M., Bruins, A. P., Moolenaar, F., De Zeeuw, D., Kruizinga, W. H., Kellogg, R. M., Meijer, D. K. F. 1992. Low molecular weight proteins as carriers for renal drug targeting. Preparation of drug–protein conjugates and drug–spacer derivatives and their catabolism in renal cortex homogenates and lysosomal lysates. *J. Med. Chem. 35*:1246–1259.
63. Franssen, E. J. F., Moolenaar, F., De Zeeuw, D., Meijer, D. K. F. 1993. Low molecular weight proteins as carriers for renal drug targeting: naproxen coupled to lysozyme via the spacer L-lactic acid. *Pharm. Res. 10*:963–969.
64. Bundgaard, H. 1991. Novel chemical approaches in prodrug design. *Drugs of the Future 16*:443–458.
65. Arano, Y., Wakisaka, K., Ohmono, Y., Uezono, T., Akizawa, H., Nakayama, M., Sakahara, H., Tanaka, C., Konishi, J., Yokoyama, A. 1996. Assessment of radiochemical design of antibodies using an ester bond as the metabolizable linkage: evaluation of maleimidoethyl 3-(tri-n-butylstannyl)hippurate as a radioiodination reagent of antibodies for diagnostic and therapeutic applications. *Bioconjug. Chem. 7*:628–637.
66. Shen, W.-C., Ryser, J.-P. 1981. Cis-aconityl spacer between daunomycin and macromolecular carriers: a model of pH-sensitive linkage releasing drug from a lysosomotropic conjugate. *Biochem. Biophys. Res. Commun. 102*: 1048–1054.
67. Ogden, J. R., Leung, K., Kunda, S. A., Telander, M. W., Avner, B. P., Liao, S. K., Thurman, G. B., Oldham, R. K. 1989. Immunoconjugates of doxorubicin and murine antihuman breast carcinoma monoclonal antibodies prepared via an N-hydroxysuccinimide active ester intermediate of cis-aconityl-doxorubicin: preparation and in vitro cytotoxicity. *Mol. Biother. 1*:170–174.
68. Hudecz, F., Clegg, J. A., Kajtar, J., Embleton, M. J., Szekerke, M., Baldwin, R. W. 1992. Synthesis, conformation, biodistribution, and in vitro cytotoxicity of daunomycin-branched polypeptide conjugates. *Bioconjug. Chem. 3*: 49–57.
69. Gallego, J., Price, M. R., Baldwin, R. W. 1984. Preparation of four daunomycin-monoclonal antibody 791T/36 conjugates with anti-tumour activity. *Int. J. Cancer 33*: 737–744.
70. Franssen, E. J. F., Moolenaar, F., De Zeeuw, D., Meijer, D. K. F. 1992. Renal specific delivery of sulfamethoxazole in the rat by coupling to the low molecular weight protein lysozyme via an acid-sensitive linker. *Int. J. Pharm. 80*: R15–R19.
71. Brinkley, M. 1992. A brief survey of methods for preparing protein conjugates with dyes, haptens, and cross-linking reagents. *Bioconjug. Chem. 3*:2–13.
72. Wong, S. S. 1993. Heterobifunctional cross-linkers. In: *Chemistry of Protein Conjugation and Cross-Linking*. CRC Press, Boca Raton, FL, pp. 147–194.
73. Melton, R. G. 1996. Preparation and purification of antibody–enzyme conjugates for therapeutic applications. *Adv. Drug Deliv. Rev. 22*:289–301.
74. Cumber, A. J., Westwood, J. H., Henry, R. V., Parnell, G. D., Coles, B. F., Wawrzynczak, E. J. 1992. Structural features of the antibody-A chain linkage that influence the activity and stability of ricin A chain immunotoxins. *Bioconjug. Chem. 3*:397–401.
75. Carlsson, J., Drevin, H., Axén, R. 1978. Protein thiolation and reversible protein–protein conjugation. N- Succinimidyl 3-(2-pyridyldithio)propionate, a new heterobifunctional reagent. *Biochem. J. 173*:723–737.
76. Jue, R., Lambert, J. M., Pierce, L. R., Traut, R. R. 1978. Addition of sulfhydryl groups of Escherichia coli ribosomes by protein modification with 2-iminothiolane (methyl 4-mercaptobutyrimidate). *Biochemistry 17*:5399–5406.
77. Klotz, I. M., Heiney, R. E. 1962. Introduction of sulfhydryl groups into proteins using acetylmercaptosuccinic anhydride. *Arch. Biochem. Biophys. 96*:605–612.
78. Thorpe, P. E., Wallace, P. M., Knowles, P. P., Relf, M. G., Brown, A. N., Watson, G. J., Knyba, R. E., Wawrzynczak, E. J., Blakey, D. C. 1987. New coupling agents for the synthesis of immunotoxins containing a hindered disulfide bond with improved stability in vivo. *Cancer Res. 47*: 5924–5931.
79. Goff, D. A., Carroll, S. F. 1990. Substituted 2-iminothiolanes: reagents for the preparation of disulfide cross-linked conjugates with increased stability. *Bioconjug. Chem. 1*: 381–386.
80. Greenfield, L., Bloch, W., Moreland, M. 1990. Thiol-containing cross-linking agent with enhanced steric hindrance. *Bioconjug. Chem. 1*:400–410.
81. Carroll, S. F., Bernhard, S. L., Goff, D. A., Bauer, R. J., Leach, W., Kung, A. H. 1994. Enhanced stability in vitro and in vivo of immunoconjugates prepared with 5-methyl-2-iminothiolane. *Bioconjug. Chem. 5*:248–256.
82. Arpicco, S., Dosio, F., Brusa, P., Crosasso, P., Cattel, L. 1997. New coupling reagents for the preparation of disulfide cross-linked conjugates with increased stability. *Bioconjug. Chem. 8*:327–337.
83. Delprino, L., Giacomotti, M., Dosio, F., Brusa, P., Ceruti, M., Grosa, G., Cattel, L. 1993. Toxin-targeted design for anticancer therapy. II: Preparation and biological comparison of different chemically linked gelonin–antibody conjugates. *J. Pharm. Sci. 82*:699–704.
84. Delprino, L., Giacomotti, M., Dosio, F., Brusa, P., Ceruti,

M., Grosa, G., Cattel, L. 1993. Toxin-targeted design for anticancer therapy. I: Synthesis and biological evaluation of new thioimidate heterobifunctional reagents. *J. Pharm. Sci. 82*:506–512.

85. Shionoiri, H., Naruse, M., Minamisawa, K., Ueda, S., Himeno, H., Hiroto, S., Takasaki, I. 1997. Fosinopril. Clinical pharmacokinetics and clinical potential. *Clin. Pharmacokinet. 32*:460–480.
86. Krise, J. P., Stella, V. J. 1996. Prodrugs of phosphates, phosphonates, and phosphinates. *Adv. Drug Deliv. Rev.* 287–310.
87. Jones, R. J., Bischofberger, N. 1995. Minireview: Nucleotide prodrugs. *Antiviral Res. 27*:1–17.
88. Fiume, L., Busi, C., Di Stefano, G., Mattioli, A. 1994. Targeting of antiviral drugs to the liver using glycoprotein carriers. *Adv. Drug Deliv. Rev. 14*:51–65.
89. Midoux, P., Negre, E., Roche, A.-C., Mayer, R., Monsigny, M., Balzarini, J., De Clercq, E., Mayer, E., Ghaffar, A., and Gangemi, J. D. 1990. Drug targeting: anti-HSV-1 activity of mannosylated polymer-bound 9-(2-phosphonylmethoxyethyl)adenine. *Biochem. Biophys. Res. Commun. 167*:1044–1049.
90. Kuipers, M. E., Swart, P. J., Hendriks, M. M. W. B., Meijer, D. K. F. 1995. Optimization of the reaction conditions for the synthesis of neoglycoprotein-AZT-monophosphate conjugates. *J. Med. Chem. 38*:883–889.
91. Peeters, J. M., Hazendonk, T. G., Beuvery, E. C., Tesser, G. I. 1989. Comparison of four bifunctional reagents for coupling peptides to proteins and the effect of the three moieties on the immunogenicity of the conjugates. *J. Immunol. Methods 120*:133–143.
92. Uckun, F. M., Myers, D. E., Irvin, J. D., Kuebelbeck, V. M., Finnegan, D., Chelstrom, L. M., Houston, L. L. 1993. Effects of the intermolecular toxin-monoclonal antibody linkage on the in vivo stability, immunogenicity and anti-leukemic activity of B43 (anti-CD19) pokeweed antiviral protein immunotoxin. *Leuk. Lymphoma 9*:459–476.
93. Dunn, M. J., Zambraski, E. J. 1980. Renal effects of drugs that inhibit prostaglandin synthesis. *Kidney Int. 18*:609–622.
94. Schlondorff, D. 1986. Renal prostaglandin synthesis. Sites of production and specific actions of prostaglandins. *Am. J. Med. 81*:1–11.
95. Patrono, C., Dunn, M. J. 1987. The clinical significance of inhibition of renal prostaglandin synthesis. *Kidney Int. 32*:1–12.
96. Vriesendorp, R., Donker, A. J. M., De Zeeuw, D., De Jong, P. E., Van der Hem, G. K., Brentjes, J. R. M. 1986. Effects of nonsteroidal anti-inflammatory drugs in proteinuria. *Am. J. Med. 81*:84–94.
97. Franssen, E. J. F., Van Amsterdam, R. G. M., Visser, J., Moolenaar, F., De Zeeuw, D., Meijer, D. K. F. 1991. Low molecular weight proteins as carriers for renal drug targeting: naproxen-lysozyme. *Pharm. Res. 8*:1223–1230.
98. Haas, M., Kluppel, A. C. A., Wartna, E. S., Moolenaar, F., Meijer, D. K. F., De Jong, P. E., De Zeeuw, D. 1997. Drug targeting to the kidney: renal delivery and degradation of a naproxen-lysozyme conjugate in vivo. *Kidney Int. 52*:1693–1699.
99. Haas, M., Moolenaar, F., Meijer, D. K. F., De Jong, P. E., De Zeeuw, D. 1998. Renal targeting of an NSAID: effects on renal prostaglandins. *Clin. Sci. 95*:603–609.
100. Sibony, M., Gasc, J. M., Soubrier, F., Alhenc Gelas, F., Corvol, P. 1993. Gene expression and tissue localization of the two isoforms of angiotensin I converting enzyme. *Hypertension 21*:827–835.
101. Dzau, V. J., Burt, D. W., Pratt, R. E. 1988. Molecular biology of the renin–angiotensin system. *Am. J. Physiol. 255*: F563–73.
102. Dzau, V. J. 1988. Circulating versus local renin–angiotensin system in cardiovascular homeostasis. *Circulation 77*: I4–13.
103. Campbell, D. J. 1987. Tissue renin–angiotensin system: sites of angiotensin formation. *J. Cardiovasc. Pharmacol. 10*(suppl. 7):S1–8.
104. Zimmerman, B. G., Dunham, E. W. 1997. Tissue renin–angiotensin system: a site of drug action? *Annu. Rev. Pharmacol. Toxicol. 37*:53–69.
105. Navar, L. G., Rosivall, L. 1984. Contribution of the renin–angiotensin system to the control of intrarenal hemodynamics. *Kidney Int. 25*:857–868.
106. Seikaly, M. G., Arant, B. S., Jr., Seney, F. D., Jr. 1990. Endogenous angiotensin concentrations in specific intrarenal fluid compartments of the rat. *J. Clin. Invest. 86*:1352–1357.
107. Braam, B., Mitchell, K. D., Fox, J., Navar, L. G. 1993. Proximal tubular secretion of angiotensin II in rats. *Am. J. Physiol. 264*:F891–F898.
108. Wong, P. C., Hajj-ali, A. F. 1994. Functional correlates of angiotensin receptor stimulation. In: *Medicinal Chemistry of the Renin–Angiotensin system.* (Timmermans, P. B. M. W. M., Wexler, R. R., eds.). Elsevier Science BV, Amsterdam, pp. 313–346.
109. Mitchell, K. D., Navar, L. G. 1991. Influence of intrarenally generated angiotensin II on renal hemodynamics and tubular reabsorption. *Renal Physiol. Biochem. 14*:155–163.
110. Lu, S., Mattson, D. L., Cowley, A. W., Jr. 1994. Renal medullary captopril delivery lowers blood pressure in spontaneously hypertensive rats. *Hypertension 23*:337–345.
111. Rassier, M. E., Li, T., Zimmerman, B. G. 1986. Analysis of influence of extra- and intrarenally formed angiotensin II on renal blood flow. *J. Cardiovasc. Pharmacol. 8*(suppl. 10):S106–S110.
112. Li, T., Zimmerman, B. G. 1987. Intrarenal converting enzyme inhibition in rabbit. *J. Cardiovasc. Pharmacol. 10*(suppl. 7):S133–S136.
113. Siragy, H. M., Howell, N. L., Peach, M. J., Carey, R. M. 1990. Combined intrarenal blockade of the renin–angiotensin system in the conscious dog. *Am. J. Physiol. 258*: F522–F529.
114. Zhang, X., Dunham, E. W., Zimmerman, B. G. 1995. Threshold sodium excretory and renal blood flow effects

of angiotensin converting enzyme inhibition. *J. Hypertens. 13*:1413–1419.

115. Norman, J. T. 1991. The role of angiotensin II in renal growth. *Renal Physiol. Biochem. 14*:175–185.
116. Hilgers, K. F., Mann, J. F. 1996. Role of angiotensin II in glomerular injury: lessons from experimental and clinical studies. *Kidney Blood Press. Res. 19*:254–262.
117. Savage, S., Schrier, R. W. 1992. Progressive renal insufficiency: the role of angiotensin converting enzyme inhibitors. *Adv. Intern. Med. 37*:85–101.
118. Baba, T., Ishizaki, T. 1992. Recent advances in pharmacological management of hypertension in diabetic patients with nephropathy. Effects of antihypertensive drugs on kidney function and insulin sensitivity. *Drugs 43*:464–489.
119. Orth, S., Nowicki, M., Wiecek, A., Ritz, E. 1993. Nephroprotective effect of ACE inhibitors. Drugs *46* (suppl. 2): 189–195.
120. Navis, G., Faber, H. J., De Zeeuw, D., De Jong, P. E. 1996. ACE inhibitors and the kidney. A risk–benefit assessment. *Drug Saf. 15*:200–211.
121. Anderson, S., Rennke, H. G., Brenner, B. M. 1986. Therapeutic advantage of converting enzyme inhibitors in arresting progressive renal disease associated with systemic hypertension in the rat. *J. Clin. Invest. 77*:1993–2000.
122. Scicli, A. G., Carretero, O. A. 1986. Renal kallikrein–kinin system. *Kidney Int. 29*:120–130.
123. Hostetter, T. H., Rosenberg, M. E. 1990. Hemodynamic effects of glomerular permselectivity. *Am. J Nephrol. 10*: 24–27.
124. Anderson, S., Hershberger, R. E. 1994. Angiotensin converting enzyme inhibitors: new clinical applications. In: *Medicinal Chemistry of the Renin–Angiotensin System* (Timmermans, P. B. M. W. M., Wexler, R. R., eds.). Elsevier Science B. V., Amsterdam, pp. 47–63.
125. Gansevoort, R. T., Navis, G. J., Wapstra, F. H., De Jong, P. E., De Zeeuw, D. 1997. Proteinuria and progression of renal disease: therapeutic implications. *Curr. Opin. Nephrol. Hypertens. 6*:133–140.
126. Kok, R. J., Grijpstra, F., Walthuis, R. B., Moolenaar, F., De Zeeuw, D., Meijer, D. K. F. 1999. Specific delivery of captopril to the kidneys with the prodrug captopril-lysozyme. *J. Pharmacol. Exp. Ther. 288*:281–285.
127. Kok, R. J., Grijpstra, F., Walthuis, R. B., Heersema, J., Visser, J., Somsen, G. W., Moolenaar, F., De Zeeuw, D., Meijer, D. K. F. 2000. Synthesis and characterization of the kidney-selective prodrug captopril-lysozyme. *Bioconjug. Chem.*, accepted.
128. Soloway, M. S. 1988. Introduction and overview of intravesical therapy for superficial bladder cancer. *Urology 31*:5–16.
129. Whitmore, W. F., Jr., Yagoda, A. 1989. Chemotherapy in the management of bladder tumours. *Drugs 38*:301–312.
130. Weening, J. J., Rennke, H. G. 1983. Glomerular permeability and polyanion in adriamycin nephrosis in the rat. *Kidney Int. 24*:152–159.
131. Bertani, T., Cutillo, F., Zoja, C., Broggini, M., Remuzzi, G. 1986. Tubulo-interstitial lesions mediate renal damage in adriamycin glomerulopathy. *Kidney Int. 30*:488–496.
132. Young, D. M. 1975. Pathologic effects of adriamycin (NSC-123127) in experimental systems. *Cancer Chemotherapy Reports Part 3 6*:159–175.
133. Grond, J., Weening, J. J., Elema, J. D. 1984. Glomerular sclerosis in nephrotic rats. Comparison of the long-term effects of adriamycin and aminonucleoside. *Lab. Invest. 51*:277–285.
134. Guezmes, A., Fernandez, F., Garijo, F., Val Bernal, F. 1992. Correlation between tubulointerstitial nephropathy and glomerular lesions induced by adriamycin. *Nephron 62*:198–202.
135. Bertani, T., Poggi, A., Pozzoni, R., Delaini, F., Sacchi, G., Thoua, Y., Mecca, G., Remuzzi, G., Donati, M. B. 1982. Adriamycin-induced nephrotic syndrome in rats: sequence of pathologic events. *Lab. Invest. 46*:16–23.
136. Wapstra, F. H., Van Goor, H., Navis, G., De Jong, P. E., De Zeeuw, D. 1996. Antiproteinuric effect predicts renal protection by angiotensin-converting enzyme inhibition in rats with established adriamycin nephrosis. *Clin. Sci. 90*: 393–401.
137. Haas, M., Moolenaar, F., Elsinga, A., Kluppel, C. A., Tiebosch, T., De Jong, P. E., Meijer, D. K. F., De Zeeuw, D. 1998. Drug targeting to the urinary bladder: an increased urinary excretion and reduced renal toxicity of doxorubicin in the rat. *Cancer Res.*, submitted.
138. Fitzpatrick, J. J., Garnett, M. C. 1995. Design, synthesis and in vitro testing of methotrexate carrier conjugates linked via oligopeptide spacers. *Anti-Cancer Drug Design. 10*:1–9.
139. McIntyre, G. D., Scott, C. F., Jr., Ritz, J., Blattler, W. A., Lambert, J. M. 1994. Preparation and characterization of interleukin-2-gelonin conjugates made using different cross-linking reagents. *Bioconjug. Chem. 5*:88–97.
140. Monsigny, M., Roche, A. C., Midoux, P. 1988. Endogenous lectins and drug targeting. *Ann. N. Y. Acad. Sci. 551*: 399–414.
141. Boon, P. J., Kaspersen, F. M., Bos, E. S., Hermans, F. G. M. Acid-labile linker molecules. 91203439.4(495265 Al), 1–34. 1991. EUROPE: Patent.

Index